D1219436

Teacher Wraparound Edition

BIOLOGY
THE DYNAMICS OF LIFE

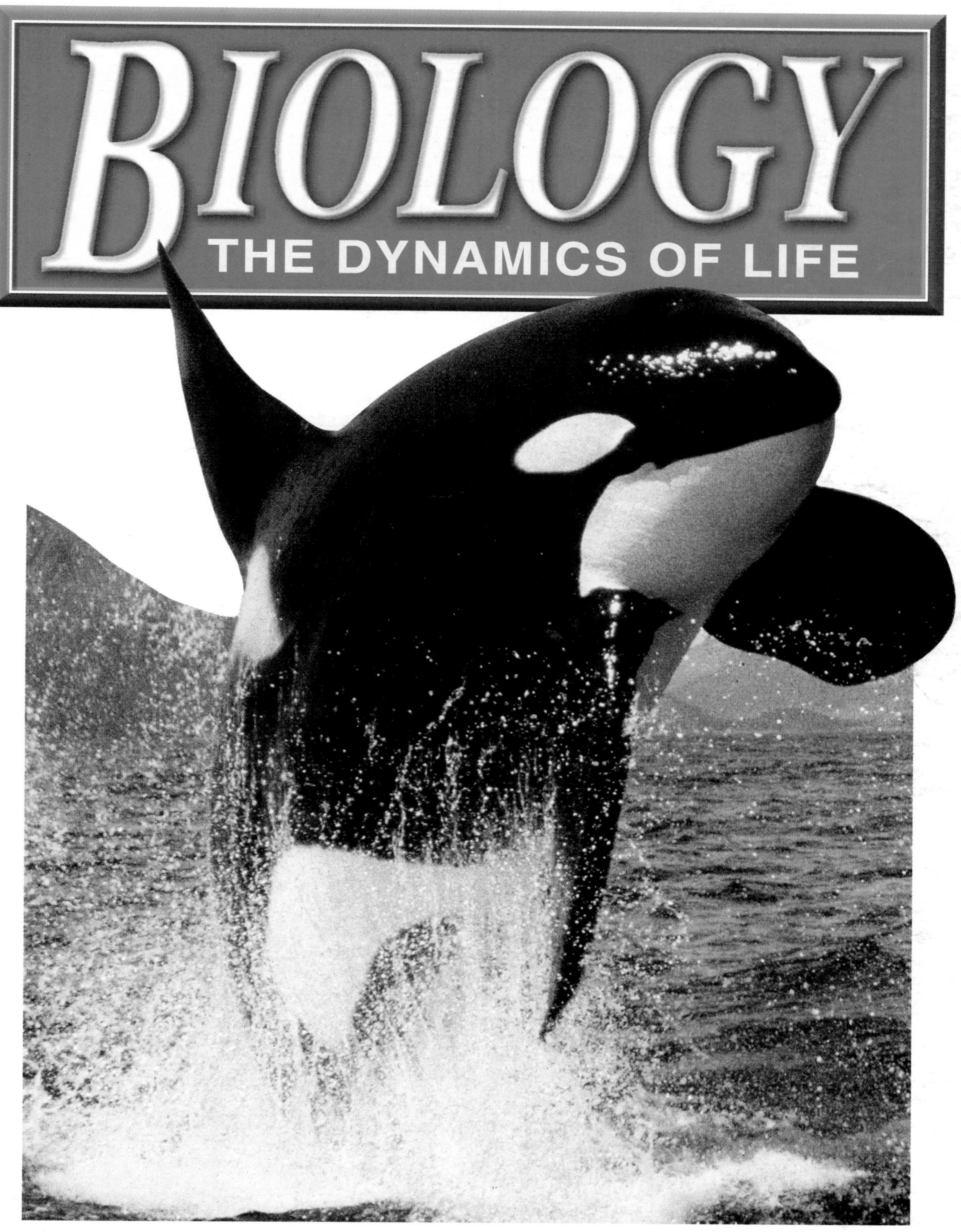

GLENCOE

McGraw-Hill

New York, New York Columbus, Ohio Woodland Hills, California Peoria, Illinois

A GLENCOE PROGRAM

Biology: The Dynamics of Life

Student Edition
Teacher Wraparound Edition
Reinforcement and Study Guide, SE and TE
Biolab and Minilab Worksheets
Laboratory Manual, SE and TE
Videodisc Correlations
Biology Projects
Chapter Assessment
Science and Technology Videodisc Series, Teacher Guide
Section Focus Masters
Basic Concepts Transparencies and Masters
Basic Skills Transparencies and Masters
Reteaching Transparencies and Masters
Concept Mapping
Tech Prep Applications
Critical Thinking/Problem Solving
Spanish Resources
Great Developments in Biology
Lesson Plans
CD-ROM Multimedia System
Videodisc Program
MindJogger Videoquizzes
Computer Test Bank IBM/Apple/Macintosh
English/Spanish Audiocassettes
Glencoe Science Professional Series:
- **Exploring Environmental Issues**
- **Performance Assessment in the Biology Classroom**
- **Alternate Assessment in the Science Classroom**
- **Cooperative Learning in the Science Classroom**

Authors

Alton Biggs • Allen High School • Allen, TX
Dan Blaustein • Science Writer • Evanston, IL
Chris Kapicka • Northwest Nazarene College • Boise, ID
Albert Kaskel • Evanston Township High School • Evanston, IL
Linda Lundgren • Bear Creek High School • Lakewood, CO

Glencoe/McGraw-Hill
A Division of The ***McGraw-Hill*** *Companies*

Send all inquiries to:
Glencoe/McGraw-Hill • 936 Eastwind Drive • Westerville, OH 43081

ISBN 0-02-825432-5
Printed in the United States of America.
2 3 4 5 6 7 8 9 10 11 12 071/046 04 03 02 01 00 99 98

Teacher Guide

Contents in Brief

Teacher Wraparound Edition

Student Edition

Biology: The Dynamics of Life
addresses the challenge of achieving scientific literacy

The *National Science Education Standards*, published by the National Research Council and representing the contribution of thousands of educators and scientists, offer a comprehensive vision of a scientifically literate society. The standards describe not only what students should know but also offer guidelines for biology teaching and assessment. If you are using, or plan to use, the standards to guide changes in your biology curriculum, you can be assured that ***Biology: The Dynamics of Life*** aligns with the *National Science Education Standards.*

Biology: The Dynamics of Life is an example of how Glencoe's commitment to effective science education is changing the materials used in biology classrooms today. More than just a collection of facts in a textbook, ***Biology: The Dynamics of Life*** is a program that provides numerous opportunities for students, teachers, and school districts to meet the *National Science Education Standards.*

Content Standards

Correlations on each Chapter Organizer show the close alignment between ***Biology: The Dynamics of Life*** and the content standards. The approach of ***Biology: The Dynamics of Life*** allows students to discover concepts within each of the content standards, giving them opportunities to make connections between biology concepts and biology and the real world. Hands-on labs and inquiry-based lessons reinforce the science processes emphasized in the standards.

Teaching Standards

Alignment with the *National Science Education Standards* requires much more than alignment with the outcomes in the content standards. The way in which concepts are presented is critical to effective learning. The teaching standards within the *National Science Education Standards* recommend an inquiry-based program facilitated and guided by teachers. ***Biology: The Dynamics of Life*** provides such opportunities through activities and discussions that allow students to discover critical concepts by inquiry and apply the knowledge they've constructed to their own lives. Throughout the program, students are building critical skills that will be available to them for lifelong learning. The ***Teacher Wraparound Edition*** helps you make the most of every instructional moment. It offers an abundance of effective strategies and suggestions for guiding students as they explore biology.

Assessment Standards

The assessment standards are supported by many of the components that make up the ***Biology: The Dynamics of Life*** program. The ***Teacher Wraparound Edition*** and ***Teacher Classroom Resources*** provide multiple chances to assess students' understanding of important concepts as well as their abilities to perform a wide range of skills. Ideas for portfolios, performance assessment, written reports, and other assessment activities accompany every lesson. For more suggestions about assessment ideas and resources, see pages 26T-27T.

Tech Prep Education

Tech Prep is a rigorous and focused program of study that aims to create a workforce in the United States that is technically literate. It is designed to prepare students enrolled in a general curriculum for the demands of further education or for employment by providing them with essential academic and technical foundations, along with problem-solving, group-process, and lifelong-learning skills. These goals can be achieved by integrating vocational study with higher-level academic study.

What Are the Characteristics of the Tech Prep Curriculum?

In 1990, Congress passed the Carl D. Perkins Vocational and Applied Technology Act to set aside funds for the development and administration of Tech Prep programs. The criteria outlined in Title III of the Perkins Act specify that Tech Prep programs take place during the last two years of high school, followed by two years of postsecondary occupational education, and that this education culminate in a certificate or associate degree. The Secretary's Commission on Achieving Necessary Skills (SCANS), an arm of the U.S. Department of Labor, published a report in June 1991 that outlined several competencies that characterize successful workers. The Tech Prep curriculum seeks to address these competencies, which include:

- the ability to use resources productively.
- the ability to use interpersonal skills effectively, including fostering teamwork, teaching others, serving customers, leading, negotiating, and working well with individuals from culturally diverse backgrounds.
- the ability to acquire, evaluate, interpret, and communicate data and information.
- the ability to understand social, organizational, and technological systems.
- the ability to apply technology to specific tasks.

How does *Biology: The Dynamics of Life* address Tech Prep issues?

Biology: The Dynamics of Life helps you develop scientific and technological literacy in your students through a variety of performance-based activities that emphasize problem solving, critical thinking skills, and teamwork. In the ***Student Edition***, many applications of biology are used throughout the text and in the features to illustrate the relevance of biology to everyday life and the world of work. *BioLabs* and *MiniLabs* provide opportunities for practical applications of biology concepts. *Assessing Knowledge and Skills* in the Chapter Review provides additional problem-solving activities that connect to the real world. *People in Biology; Biology and Society;* and connections to *Math*, *Physics*, *Health*, and other disciplines are some of the features of the ***Student Edition*** that offer Tech Prep connections. The ***Teacher Classroom Resources*** includes a ***Tech Prep Applications*** booklet that gives additional activities and career ideas designed to fulfill the goals of the tech prep education.

Designed to help your students achieve scientific literacy.

Philosophy and Themes

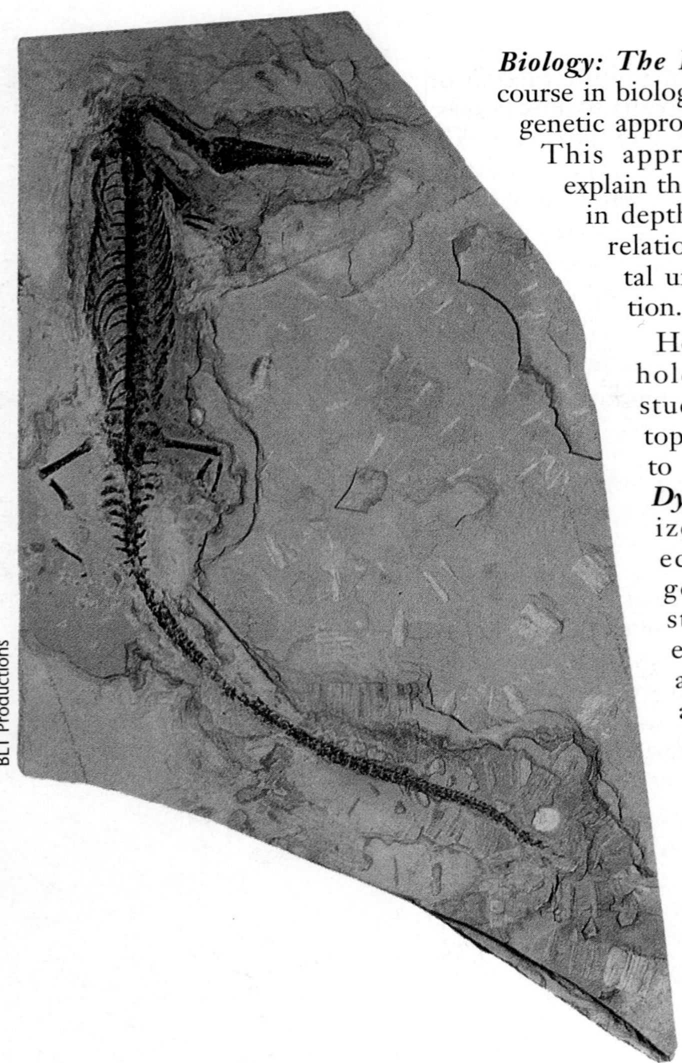

BLT Productions

Biology: The Dynamics of Life is a course in biology that follows a phylogenetic approach in its organization. This approach allows you to explain the diversity of life-forms in depth while revealing their relationships and fundamental unity in form and function.

How do you attract and hold the attention of students unless you find topics that are of interest to students? ***Biology: The Dynamics of Life*** capitalizes on topics, such as ecology and human genetics, that interest students. Ecology, for example, is of popular and political interest and often the lead story of local and national news broadcasts. Rather than present ecology at the end of the text, the authors have chosen to move it to the front, where it can be emphasized. Heredity and human genetics, too, have emerged as biological topics with meaningful and far-ranging consequences for society. Students will find it easy to see how and why biological science is an important component of their studies and their futures.

In ***Biology: The Dynamics of Life***, there is an emphasis on six themes, as shown in the chart below. Emphasis on themes contributes to the big picture by focusing learning on connections among major ideas and concepts. The thematic approach contributes to students' comprehension of fundamental life processes, understanding of interactions among organisms, and appreciation of how scientists work.

Biology: The Dynamics of Life emphasizes the following themes.

Unity Within Diversity

Life on Earth is represented by millions of species. In order to survive, all species must possess the same basic functions. General patterns of carrying out life functions are easily recognized, and these patterns form a common bond that unites all forms of life. This theme is evident in discussions of topics ranging from cell structure and function, to the genetic code, to the six kingdoms of life.

THEME	CHAPTER																					
	1	2	3	4	5	6	7	8	9	10	11	12	13	14	15	16	17	18	19	20	21	22
Evolution	■				■			■			■	■	■	■	■	■	■	■	■	■	■	
Unity Within Diversity	■	■	■	■			■	■		■	■	■	■					■	■		■	■
Energy	■	■	■	■		■	■	■	■	■	■						■				■	
Homeostasis	■		■	■		■		■	■	■	■	■									■	■
Systems and Interactions	■		■	■	■	■	■		■	■			■	■	■	■		■	■	■		■
The Nature of Science	■	■		■	■		■	■			■	■	■	■	■	■	■		■	■	■	

Evolution

Similarities and differences among species indicate evolutionary relationships. All organisms are related in that they are descendants of the first forms of life on Earth. Evolution is the process by which organisms are adapted to changing environments and by which one species gives rise to another. The diversity of life on Earth is a product of evolutionary change. This theme is developed not only where the process of evolution is discussed, but also in discussions of related groups of organisms, the six kingdoms of life, and comparisons of adaptations among species.

Energy

In order to carry out basic functions, organisms require a continual source of energy and a means of extracting and utilizing that energy. Energy flow pervades every level of biological organization from a cell to the biosphere. This theme is explored in numerous contexts: photosynthesis and respiration, ATP and its uses, and energy flow through ecosystems.

Homeostasis

In the strictest sense, homeostasis refers to the tendency to maintain stability in the internal environment of an organism. However, the theme of homeostasis is extended here to the balance of chemistry at the cellular level and maintenace of stability within an ecosystem. The theme has been applied to the balance within organisms and to that which exists among organisms and between organisms and their environments.

Systems and Interactions

No single part at any level of organization operates independently. Proper functioning depends on coordinated relationships. Parts of organisms interact with one another, organisms are interdependent, and populations interact with their environments. This theme is stressed in the study of the interaction of organelles within cells, cells within tissues, organs within systems, and systems within organisms. It is also apparent in such topics as feeding relationships, symbiotic associations, recycling of nutrients, life cycles, patterns of population growth, and ecological succession.

The Nature of Science

Students learn about the fundamental methods common to scientific inquiry and see evidence of those methods as classic experiments leading to the development of major principles are discussed and analyzed. More significantly, students are afforded the opportunity to practice the methods of science themselves throughout the course. Far from the "cookbook" labs that require students to confirm what they have already learned, many of the lab activities require students to formulate hypotheses and then design their own controlled experiments.

Emphasis on themes contributes to the big picture.

THEME	CHAPTER																				
	23	24	25	26	27	28	29	30	31	32	33	34	35	36	37	38	39	40	41	42	43
Evolution		■	■	■	■	■	■	■	■	■	■	■	■	■	■	■	■		■		■
Unity Within Diversity									■	■			■								
Energy	■											■					■	■			■
Homeostasis	■				■		■	■	■	■	■	■	■		■	■		■	■	■	■
Systems and Interactions		■	■	■	■	■	■	■	■	■	■	■		■	■	■	■	■	■	■	■
The Nature of Science														■							

Program Resources
help you customize your biology program to meet your needs

The *Teacher Classroom Resources* are designed to enhance the ***Biology: The Dynamics of Life*** program by providing exciting options for every teacher's needs. Use the components of the ***Teacher Classroom Resources*** as time-saving teaching aids in the planning, instruction, evaluation, and enrichment of your biology classes. Each component of the ***Teacher Classroom Resources*** is described below.

Review and Reinforcement

- The ***Reinforcement and Study Guide*** contains masters that provide students with guided reading activities for each numbered section of the text. It helps students focus on the main ideas. The ***Reinforcement and Study Guide*** is also available in a consumable student edition.
- ***Concept Mapping*** reinforces connections within and among concepts and processes. Each master challenges students to construct visual representations of relationships among particular chapter concepts.
- ***Critical Thinking/Problem Solving*** provides masters that require students to apply knowledge and develop higher-level thinking skills to analyze new situations and solve problems.

Hands-on Learning

- A ***Laboratory Manual*** offers more than 70 additional laboratory explorations and investigations, all different from those in the student text. Available in blackline and consumable editions, it provides students with a variety of laboratory experiences that reinforce the biology principles encountered in ***Biology: The Dynamics of Life***. The lab manual provides students with numerous opportunities for the manipulation of apparatus, observation, data gathering and processing, and interpretation to form conclusions.
- ***Biolab and Minilab Worksheets*** include blackline master versions of all the Biolabs and Minilabs in the student text. Complete instructions and extra space for answers and data tables have been provided to ensure you won't have to worry about students taking their student texts into the laboratory.
- ***Biology Projects*** provide suggestions of long-term group or individual activities that expand on a theme or major topic of each unit. These projects are designed to involve and excite students about biology in their lives. They require students to be creative, do research, design and conduct experiments, and report on their results using a variety of methods.

Application and Enrichment

- ***Tech Prep Applications*** encourage students to become familiar with everyday applications of biology. These worksheets show students how the main ideas of their biology course are essential to the understanding of modern developments in technology and medicine. Many worksheets directly involve the students with hands-on activities involving modeling or easily obtained materials.
- ***Exploring Environmental Issues*** incorporate many of the environmental issues covered daily by the media into open-ended projects. These activities will reveal how these issues can affect students' lives.
- ***Great Developments in Biology*** is made up of posters and an accompanying workbook. The large, full-color posters tell the story of important events in biology and the people that made them happen. The student workbook provides background readings and activities for added enrichment.

Teacher Classroom Resources

Teaching Aids

- ***Lesson Plans*** offer you a complete planning resource by correlating objectives, activities, and program resources for every student text lesson. Each plan is geared to the teaching cycle employed in the ***Teacher Wraparound Edition*** and contains a complete list of all the resources available for that numbered section. Block Scheduling strategies are also included.
- Three sets of transparencies—***Basic Concepts, Basic Skills, and Reteaching***—include more than 150 full-color transparencies, many with label overlays, to help you present, reinforce, or review all key concepts developed in the text. Many overheads are designed to reinforce scientific methods and processes. Each set of transparencies has an accompanying workbook that provides additional opportunities for review and reinforcement.
- ***Section Focus*** masters provide a way for you to begin each lesson by capturing the attention of your students with simple activities. Make your own overheads or copy the blackline masters for each student to study as class begins. There is one master for each lesson in the student text.
- ***Spanish Resources*** help your Spanish-speaking students get more out of your biology lessons. In addition to a complete English-Spanish glossary, the book contains translations of all chapter and lesson objectives, key terms and definitions, and summary statements for each chapter of the student text.
- ***English/Spanish Audiocassettes*** give students who are acquiring English, have reading difficulties, or those who need additional reinforcement, the opportunity to listen to a summarized recording of the student text in either English or Spanish.
- ***Cooperative Learning in the Science Classroom*** provides strategies for implementing cooperative learning techniques that put the effectiveness of group learning to work in your biology classroom.

Assessment

- ***Chapter Assessment*** includes a five-part assessment tool for every chapter that helps you assess process as well as content objectives. Unique simulations require students to analyze experimental designs and interpret data.
- ***Computer Test Banks***, available in DOS and Macintosh versions, provide the ultimate flexibility in designing and creating your own test instruments. Program features include selection of sections to be tested, number of questions for each section, objectives to be tested, total number of questions on a test, inclusion or exclusion of specific questions, and multiple test forms. There is also flexibility built in to edit or add your own questions.
- ***Alternate Assessment in the Science Classroom*** provides practical strategies for using alternate forms of assessment such as the use of performance-based methods and portfolios. In addition, there are report forms and scoring rubrics.
- ***Performance Assessment in the Biology Classroom*** offers a unique variety of performance-based task and skill assessments especially designed for the biology curriculum.

Includes easy-to-use ways for you to present biology and innovative ways for your students to apply it.

Technology Resources

From software to CD-ROM, ***Biology: The Dynamics of Life*** offers a wide range of innovative technology resources that will enrich your biology course.

Videodisc Programs

Biology: The Dynamics of Life Videodisc Program Provide your students with visual explanations of major biology concepts by showing them a wide variety of videos and animations. Each video clip and animation is correlated to a chapter in your text.

Science and Technology Videodisc Series Use this laserdisc series to enhance your biology lessons. Use the 280 video reports in the 7-disc series to let your students observe real scientists conducting actual research looking for technological solutions to real problems. The *Teacher Guide* includes specific strategies for integrating the video reports with lessons from ***Biology: The Dynamics of Life***.

The Infinite Voyage, the award-winning PBS series, provides 20 hours of some of the best live-action and animation sequences available to enrich your biology program. They are available in VHS and videodisc formats.

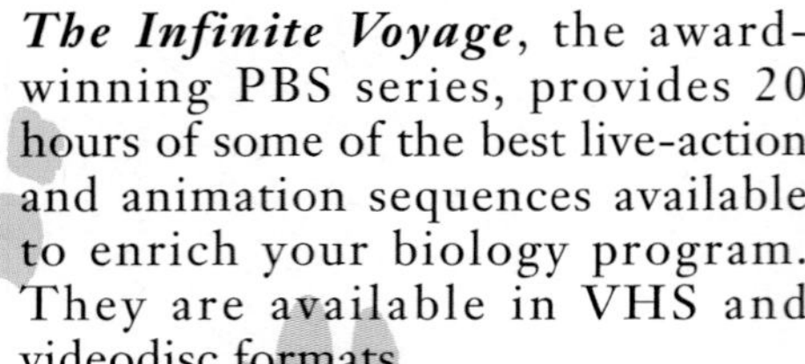

Teacher Classroom Resources

CD-ROM Multimedia System

A creative combination of sound, video, photographs, animation, and text make the Glencoe CD-ROM multimedia system an exciting addition to your biology program.

Biology: The Dynamics of Life's own CD-ROM Multimedia System offers an enormous array of interactive simulations that engage and help students to learn. Full-motion videos, animations, explorations, and lab simulations let students explore biological concepts and practice critical-thinking skills, all at the computer.

MindJogger Videoquizzes

The interactive quiz-show format of ***Biology: The Dynamics of Life MindJogger Videoquizzes*** provides fun for your students while reviewing core concepts for every chapter.

The Internet

The Internet can allow you access to the most up-to-date information on many topics covered in ***Biology: The Dynamics of Life***. Glencoe has provided you with a convenient web address that leads to fascinating sites correlated to each chapter of the textbook. In addition, you can access Glencoe's own web site for more information about all our products at **www.glencoe.com/sec/science.**

and More

Videodisc Correlations give you instant access by bar codes to all the images in Videodiscovery's *BioSciII* and Optical Data's *The Living Textbook* life sciences videodiscs that correlate to ***Biology: The Dynamics of Life***.

The Secret of Life videotapes and videodisc produced by WGBH, Boston, can demonstrate the myriad ways in which our expanding knowledge of DNA and heredity have affected all areas of biology and created many ethical dilemmas.

Technology-based components help your students explore new ways to learn biology.

Developing and Applying Thinking Processes

Applying Scientific Methods and Thinking Processes

Biology: The Dynamics of Life encourages the understanding and use of scientific methods. It is well known that hands-on activities are a way of providing a bridge between science content and student comprehension. This text encourages the interaction between content and critical-thinking processes and experiences with scientific methods by offering hundreds of hands-on activities that are easy to set up and do. In the Student Edition, the *Thinking Labs* and *Minilabs* require students to make observations, and collect a variety of data. *Biolabs* require students to design, conduct, and evaluate their own experiments and investigations. *Skill Review* questions at the end of each section provide students with another opportunity to practice the thinking processes relevant to the material they are studying. The *Skill Handbook* provides examples of all the processes that students may need to practice during these activities.

At the end of each chapter, students use critical-thinking skills as they complete *Understanding Concepts*, *Applying Concepts*, *Thinking Critically*, and *Assessing Knowledge and Skills* questions. *Biology and Society* and cross-curricular Connections connect the science content to other disciplines. In each case, students apply critical-thinking processes as they answer the questions at the end of each feature.

The important thinking process skills are described below.

Observing

How do thinking processes begin? The most basic skill that leads to critical thinking is observing. Through observation—seeing, hearing, touching, smelling, tasting—the student begins to acquire information about an object or event. Observation allows a student to gather information regarding size, shape, texture, or quantity of an object or event. Observing is a part of every hands-on experience in the text and in many of the ancillaries.

Organizing Information

The process of organizing information encompasses ordering, organizing, and comparing. How the objects or events are ordered, categorized, or compared is determined by the purpose for doing so. When ordering information, events are placed in a sequence that tells a logical story. Students will gain this experience through the study of many life cycles and science processes throughout the course. To classify or categorize information, objects or ideas are compared in order to identify common features. The similarities and differences among organisms is stressed throughout the program.

Communicating

Communicating information is an important part of science. Once all the information is gathered, it is necessary to organize the observations so that the findings can be considered and

Concept Mapping

Concept mapping helps the student make abstract information more concrete and useful by visually representing relationships among concepts. A concept map can show the interaction of a series of events, describe the stages of a process, or present a hierarchy of procedures. Students are given an opportunity to create concept maps for all major concepts in the ***Concept Mapping*** booklet of the Teacher Classroom Resources and in every *Chapter Review*.

Critical Thinking Processes in the Teacher Wraparound Edition

Thinking processes are also featured throughout the Teacher Wraparound Edition. In the margins are suggestions for students to write in a Student Journal. Keeping a journal encourages students to communicate their ideas, a key process in science. The journal may be used to record findings, explore a related topic, or make inferences based on an activity they have just completed. Project ideas require the use of thinking process skills as students research and complete each activity.

Skills Map

On pages 14T and 15T, you will find the Skills Map, which indicates how frequently thinking process skills are introduced and developed in ***Biology: The Dynamics of Life***.

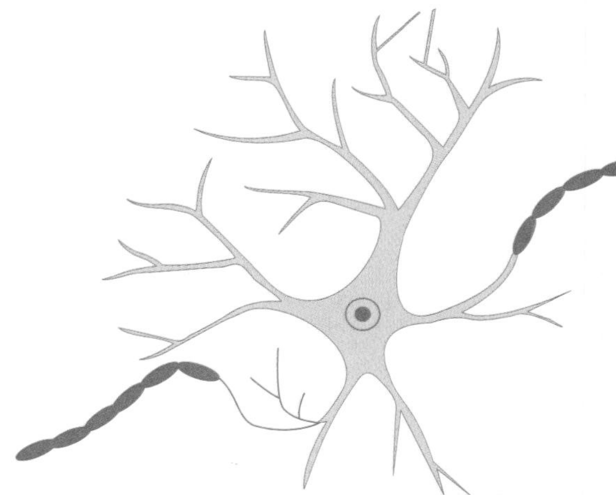

Using scientific methods and applying higher-order thinking processes become second nature.

shared by others. Information can be presented in tables, charts, a variety of graphs, or models that make it easier to consider the facts.

Inferring

Inferences are logical conclusions based on observations and are made after careful evaluation of all the available facts or data. Inferences are a means to explain or interpret observations. They are a prediction or hypothesis that can be tested and evaluated.

Relating Cause and Effect

Another process skill essential to critical thinking is recognizing cause and effect. This process focuses on how events or objects interact with one another. It also involves examining dependencies and relationships between objects and events. Students will be practicing the process of relating cause and effect while carrying out experiments throughout the course. While testing hypotheses, students will be relating cause and effect as they interpret their results.

Applying

Finally, the findings can be applied. Applying is a process that puts scientific information to use. Sometimes the findings can be applied in a practical sense or they can be used to tie together complex data.

Development of Thinking Processes and Skills

Chapters

SKILL	1	2	3	4	5	6	7	8	9	10	11	12	13	14	15	16	17	18	19	20	21	22
THINKING CRITICALLY																						
Observing and Inferring	SR	TL, ML BL	ML	ML BL	ML BL	TL, ML BL, SR	TL, ML BL	TL, ML BL, SR	TL, ML BL, SR	TL, ML BL, SR	TL, ML BL, SR	TL, ML BL, SR	TL, ML BL	TL, ML BL	TL, ML BL, SR	TL, ML BL, SR	TL, ML BL	TL, ML BL	ML BL	TL, ML BL	ML BL	TL, ML BL, SR
Comparing and Contrasting	BL	ML	TL BL	ML BL	BL	ML BL	TL, ML BL	BL	ML BL	ML BL	TL, ML BL	TL ML	TL, ML BL	ML BL	TL, ML BL	TL, ML BL	ML BL	TL, ML BL	ML BL	TL, ML BL	ML BL	ML BL
Recognizing Cause and Effect		TL	BL	ML	ML BL	TL ML	TL, ML BL		TL SR	ML BL	ML	ML	TL, ML BL		TL ML	BL		TL, ML BL	TL, ML BL		TL, ML BL	
Interpreting Scientific Illustrations		TL			BL		SR	TL,ML BL, CR		ML BL	TL, ML CR	SR CR	BL, SR CR	TL SR	TL BL	TL	TL		BL SR	TL BL	ML, BL CR	TL
PRACTICING SCIENTIFIC METHODS																						
Care and Use of a Microscope			ML BL	BL	BL	ML		ML SR	ML		BL	ML			ML		BL				ML	ML BL
Making a Wet Mount			ML BL	BL	BL			ML	ML								BL					ML BL
Measuring in SI	BL	BL	ML	BL	ML BL		ML	BL		ML BL	ML				ML		ML BL	ML	BL		BL	
Forming a Hypothesis		ML BL	TL, BL SR	BL	TL		BL		BL	BL		BL	BL SR	BL SR							BL	BL
Designing an Experiment	BL	BL	BL SR	TL BL	TL	SR	BL		BL	BL SR		BL	ML BL	ML BL				SR			BL	BL
Separating and Controlling Variables		ML BL	BL	TL BL	TL		BL		BL	BL		BL	BL	BL								BL
ORGANIZING INFORMATION																						
Classifying		ML			ML BL		ML							ML					SR	ML, BL SR		
Sequencing	CR												SR			SR	SR	SR				TL SR
Concept Mapping	BL	CR	CR	CR	CR	CR	CR	CR	CR	CR	CR	CR	CR	CR	CR	CR	CR	CR	CR	CR	CR	CR
Making and Using Tables	CR	BL CR	BL	TL, BL SR	BL	BL CR	BL SR	TL BL	TL	BL SR	BL SR	TL BL		BL	BL, SR CR		ML, BL SR	TL BL	TL, ML BL, CR	SR	TL, BL SR	BL, SR CR
Making and Using Graphs		ML, SR CR	BL CR	SR CR	BL, SR CR	TL CR	SR CR	CR	CR	TL CR	BL			TL, ML CR	ML CR	CR	CR	ML, BL CR		TL CR	SR	CR

Key: TL = Thinking Lab, ML = Minilab, BL = Biolab, SR = Skill Review, CR = Chapter Review

Thinking Processes

Chapters

SKILL	23	24	25	26	27	28	29	30	31	32	33	34	35	36	37	38	39	40	41	42	43
THINKING CRITICALLY																					
Observing and Inferring	ML, BL SR	TL, ML BL	ML BL	ML, BL SR	ML BL	ML BL	TL, ML BL, SR	ML BL	TL, ML BL	ML BL	TL, ML BL	TL, ML BL	TL, ML BL, SR	ML, BL SR	TL ML	TL, ML BL	TL, ML BL	TL, ML BL	TL, ML BL	ML BL	
Comparing and Contrasting	TL, ML BL	ML, BL SR	TL, ML BL, SR	TL, ML BL	ML, BL SR	ML BL	TL BL	ML BL	TL, ML BL, SR	ML BL	ML BL	ML BL	TL, ML BL	ML BL	TL ML	TL, ML BL, SR	ML BL	TL, ML BL, SR	TL, ML BL	ML BL	
Recognizing Cause and Effect	TL		TL ML	BL			ML BL	BL	BL SR		ML	ML	ML BL		ML BL	TL	ML		TL, ML SR	ML	
Interpreting Scientific Illustrations	CR	TL		BL CR		CR	TL	ML, SR CR	TL CR	CR		CR	CR	BL	ML SR	ML			SR	TL	
PRACTICING SCIENTIFIC METHODS																					
Care and Use of a Microscope	ML	ML BL	ML	ML	ML BL	ML BL	ML			ML BL	BL	ML	ML		ML			BL	ML	ML	
Making a Wet Mount	ML	ML BL	ML	ML	BL		ML												ML		
Measuring in SI	BL SR	BL				ML		BL	ML		ML BL	ML BL	BL		BL		ML BL	ML			
Forming a Hypothesis	BL	TL	BL		TL	TL BL	BL	TL BL	BL		ML BL	BL	BL	BL	BL	BL		BL			
Designing an Experiment	BL		BL			BL		TL BL	BL	TL SR		BL	BL	TL, BL SR	BL			BL		TL SR	
Separating and Controlling Variables	BL							BL	BL			BL	BL					BL			
ORGANIZING INFORMATION																					
Classifying			BL					ML SR		SR		SR				SR					
Sequencing		SR	SR			TL SR					SR				SR		SR	SR	SR	SR	
Concept Mapping	CR	CR	CR	CR	CR	CR	CR	CR	CR	CR	CR	CR	CR	CR	CR	CR	CR	CR	CR	CR	
Making and Using Tables	BL CR	ML	ML CR	ML		BL SR	BL SR	BL SR	BL		ML, BL SR	BL SR	BL		BL	ML, BL CR	BL CR	BL CR	ML, BL CR	ML	
Making and Using Graphs		CR	CR	BL CR	CR			CR	CR		CR	CR	TL, BL CR	TL CR	BL CR	SR CR	TL, SR CR		TL, ML BL, CR	BL CR	

Planning Your Course

Biology: The Dynamics of Life provides flexibility in the selection of topics and content that allows teachers to adapt the text to the needs of individual students and classes. In this regard, the teacher is in the best position to decide what topics are to be presented, the pace at which the content is covered, and what material should be given the most emphasis. To assist the teacher in planning the course, the following planning guides have been provided.

Biology: The Dynamics of Life may be used in a full-year course of two semesters covering the text activities and chapter-end materials. It is assumed that a year-long course in biology will have 180 periods of approximately 45 minutes each. This type of scheduling is represented in the table under the heading of Single Class Sessions. As an alternative, the table also outlines a plan for block scheduling, which is described in the next section.

William Weber

Block Scheduling

To build flexibility into the curriculum, many schools are introducing a block scheduling approach. This type of approach allows curriculum supervisors and teachers to tailor the curriculum to meet students' needs while achieving local and/or state curriculum goals. Long, concentrated periods of study can facilitate the learning of complex material. Furthermore, students may be able to take a wider variety of course work under a block scheduling plan than under a traditional full-year plan, thus enriching their high school experience and giving them a broader foundation for college-level work.

If you follow a block schedule, you may want to consider either combining lessons or eliminating certain topics and spending more time on the topics you do cover. ***Biology: The Dynamics of Life*** provides the flexibility that allows you to tailor the program to your needs. ***Biology: The Dynamics of Life*** also provides a wide variety of support materials in the ***Teacher Classroom Resources*** that will help you and your students whether you follow a block schedule or single-class schedule.

In the table shown here, it is assumed that for block scheduling, the course will be taught for one semester and include 90 periods of approximately 90 minutes each.

Please remember that planning guides are provided as aids in planning the best course for your students. Use the planning guide that relates to your curriculum and the ability levels of the classes you teach, the materials available for activities, and the time allotted for teaching. The planning guide will assist you in developing and following a schedule that will enable you to complete your goals for the school year or semester.

Planning Guide for Biology: The Dynamics of Life

Unit	Chapter	Single-Class Sessions	Block Sessions	Core	Enrichment
1	1	2	1	1.1, 1.2	
	2	2	$1\frac{1}{2}$	2.1, 2.2	
2	3	5	$2\frac{1}{2}$	3.1, 3.2 (except energy and trophic levels: ecological pyramids)	Energy and trophic levels: ecological pyramids
	4	5	$2\frac{1}{2}$	4.1 (except Succession: A Change in Communities over Time), 4.2	Succession: A Change in Communities over Time
	5	5	2	5.1, 5.2	
	6	6	2	6.1, 6.2 (except Acid precipitation, Ozone depletion, greenhouse effect)	Acid precipitation, Ozone depletion, The greenhouse effect
3	7	5	$2\frac{1}{2}$	7.1 (except Isotopes of an Element)	Isotopes of an Element
				7.2 (except Acids and bases), 7.3	Acids and bases
	8	4	2	8.1, 8.2	
	9	5	3	9.1, 9.2	
	10	6	3	10.1 (except Forming and Breaking Down ATP), 10.2 (except Light Reactions, The Calvin Cycle)	Forming and Breaking Down ATP, Light Reactions, The Calvin Cycle, 10.3
	11	6	$2\frac{1}{2}$	11.1	11.2
4	12	6	4	12.1, 12.2	
	13	6	4	13.1, 13.2, 13.3	
	14	5	3	14.1 (except Environmental Influences), 14.2	Environmental Influences
	15	4	2	15.1 15.2	
	16	4		16.1 (except Applications of DNA Technology)	Applications of DNA Technology, 16.2
5	17	4	2	17.1, 17.2	
	18	5	3	18.1, 18.2 (except Patterns of Evolution)	Patterns of Evolution
	19	3		19.1, 19.2	
6	20	4	2	20.1, 20.2	
	21	5	2	21.1, 21.2	
	22	5	$2\frac{1}{2}$	22.1, 22.2, 22.3	
	23	4	$1\frac{1}{2}$	23.1, 23.2	
7	24	3	$1\frac{1}{2}$	24.1, 24.2	
	25	3	$1\frac{1}{2}$	25.1, 25.2	
	26	4	$1\frac{1}{2}$	26.1, 26.2	
	27	3	$1\frac{1}{2}$	27.1 (except Photoperiodism), 27.2	Photoperiodism
	Biodigest		2		
8	28	3	2	28.1, 28.2	
	29	4		29.1, 29.2, 29.3, 29.4	
	30	3	$1\frac{1}{2}$	30.1, 30.2	
	31	3	$1\frac{1}{2}$	31.1, 31.2	
	32	3		32.1, 32.2	
	Biodigest		3		
9	33	4	$1\frac{1}{2}$	33.1, 33.2	
	34	4	$1\frac{1}{2}$	34.1, 34.2	
	35	3	$1\frac{1}{2}$	35.1, 35.2	
	36	3		36.1, 36.2	
	Biodigest		3		
10	37	5	3	37.1 (except Skin Injury and Restoration of Homeostasis), 37.2 (except Homeostasis, Aging, and the Skeletal System), 37.3	Skin Injury and Restoration of Homeostasis, Homeostasis, Aging, and the Skeletal System
	38	4	$1\frac{1}{2}$	38.1, 38.2 (except Endocrine Control of Homeostasis and Metabolism), 38.3	Endocrine Control of Homeostasis and Metabolism
	39	6	3	39.1, 39.2 (except ABO Blood Types), 39.3 (except The Urinary System and Homeostasis)	ABO Blood Types, The Urinary System and Homeostasis
	40	5	$1\frac{1}{2}$	40.1, 40.2, 40.3	
	41	6	3	41.1, 41.2 (except Genetic Counseling), 41.3	Genetic Counseling
	42	4		42.1 (except Patterns of Diseases), 42.2	Patterns of Diseases
	Biodigest		4		
Epilogue	43	1			

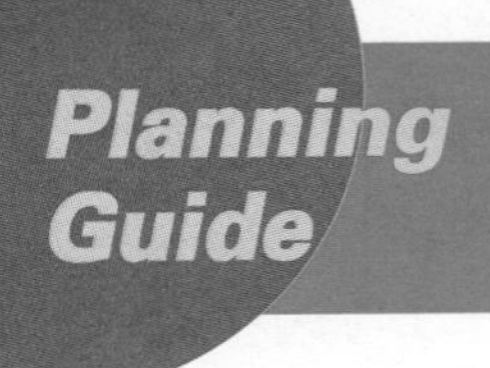

National Science Content Standards

The Need for New Directions in Science Education

Educators, public policy makers, corporate America, and parents have recognized the need for reform in science education. These groups have united in a call to action to solve this national problem. *The National Science Education Standards*, published by the National Research Council and representing the contribution of thousands of educators and scientists, offers a comprehensive vision of a scientifically literate society. The standards not only describe what students should know, but they also offer guidelines for science teaching. If you are using, or plan to use, the standards to guide changes in your science curriculum, you can be assured that *Biology: The Dynamics of Life* aligns exceedingly well with *The National Science Education Standards*.

National Science Content Standards

Unifying Concepts and Processes

UCP.1	Systems, order, and organization
UCP.2	Evidence, models, and explanation
UCP.3	Change, constancy, and measurement
UCP.4	Evolution and equilibrium
UCP.5	Form and function

Science as Inquiry

A.1	Abilities necessary to do scientific inquiry
A.2	Understandings about scientific inquiry

Physical Science

B.1	Structure of atoms
B.2	Structure and properties of matter
B.3	Chemical reactions
B.4	Motions and forces
B.5	Conservation of energy and increase in disorder
B.6	Interactions of energy and matter

Life Science

C.1	The cell
C.2	Molecular basis of heredity
C.3	Biological evolution
C.4	Interdependence of organisms
C.5	Matter, energy, and organization in living systems
C.6	Behavior of organisms

Earth and Space Science

D.1	Energy in the earth system
D.2	Geochemical cycles
D.3	Origin and evolution of the earth system
D.4	Origin and evolution of the universe

Science and Technology

E.1	Abilities of technological design
E.2	Understandings about science and technology

Science in Personal and Social Perspectives

F.1	Personal and community health
F.2	Population growth
F.3	Natural resources
F.4	Environmental quality
F.5	Natural and human-induced hazards
F.6	Science and technology in local, national, and global challenges

History and Nature of Science

G.1	Science as a human endeavor
G.2	Nature of scientific knowledge
G.3	Historical perspectives

In cooperative learning, students work together in small groups to learn academic material and interpersonal skills. Group members learn to "sink or swim together" in that they are responsible for the group accomplishing an assigned task as well as for learning the material themselves. When compared to competitive or individual accountability, in which students work against each other or alone, cooperative learning fosters academic, personal, and social success for all of your students.

Establishing Cooperative Groups

Cooperative groups in high school usually contain from two to five students. Generally, it is best to assign students to heterogeneous groups that contain a mixture of abilities, genders, and ethnicities. The use of heterogeneous groups exposes students to ideas different from their own.

Students may also be randomly assigned to groups, or they can be allowed to select their own groups. Consider using random grouping to let students get acquainted at the beginning of the year or for students to learn to work with all students in the class.

Initially, cooperative learning groups should work together for only a day or two. After the students are more experienced, they can do group work for longer periods of time. Some teachers change groups every week, while others keep groups together during the study of a chapter or unit.

Regardless of the duration you choose, it is important to keep groups together long enough for each group to experience success and to change groups often enough that students have the opportunity to work with others.

Preparing Students for Cooperative Learning

Before beginning, specify which interpersonal behaviors are necessary for people to work together. Discuss the basic rules for effective cooperative learning:

(1) Listen while others are speaking.

(2) Respect other people and their ideas.

(3) Stay on task.

(4) Be responsible for your own actions.

Students can learn and practice other interpersonal skills such as speaking quietly, encouraging other group members to participate, checking for understanding, disagreeing constructively, reaching a group consensus, and criticizing ideas rather than people.

Cooperative Learning in *Biology: The Dynamics of Life*

The ***Teacher Wraparound Edition*** codes the activities and teaching suggestions for which cooperative learning strategies are appropriate. The code **COOP LEARN** appears at the end of such suggestions. For additional help, as you prepare to teach the lessons cooperatively, refer again to these pages of background information.

The ***Cooperative Learning in the Science Classroom*** booklet provides valuable background and much practical help for selecting cooperative learning strategies. The booklet suggests methods for teachers to use to facilitate the cooperative groups as well as methods for troubleshooting and evaluation.

Includes cooperative learning skills in every lesson

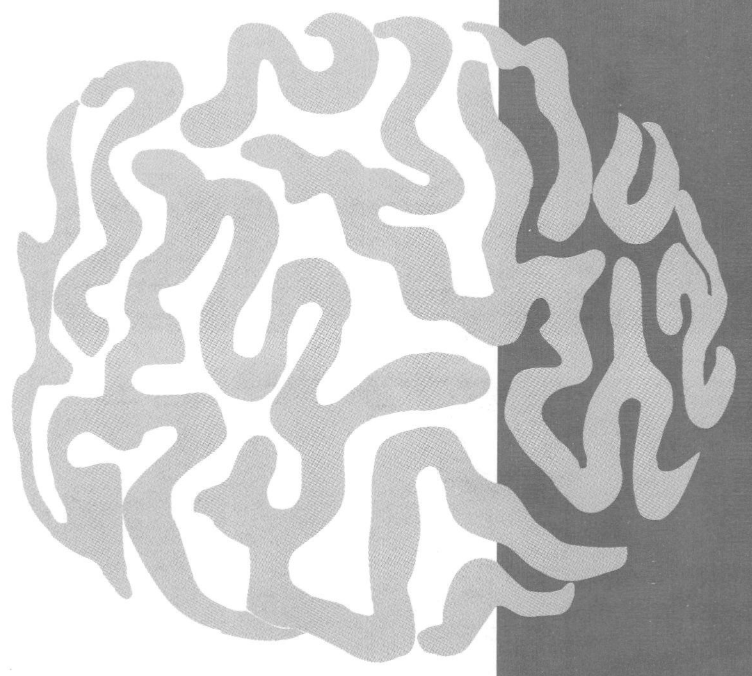

Preparing Students for Open-ended Lab Experiences

To prepare students for *Biolabs*, follow the guidelines in the ***Teacher Wraparound Edition***, especially in the sections titled *Teaching Strategies* and *Possible Procedures*. In these sections, you will be given information about what demonstrations to do and what questions to ask students so that they will be able to design their own experiments. Different groups of students will develop alternative hypotheses and alternative procedures. Check each group of students' procedures before they begin. In contrast to some "cookbook" labs, there will not be just one "right" answer.

Starting with Paper and Pencil

If you feel that your students are not prepared to begin designing experiments, you may want to practice some paper and pencil lab designs using the following format:

(a) Provide a background reading, and ask students to design an experiment that will test a hypothesis they make from the problem presented.

(b) Give them a set of questions for an experimental plan that will lead them to an appropriate experimental design.

An example of a paper-and-pencil design is shown on the next page.

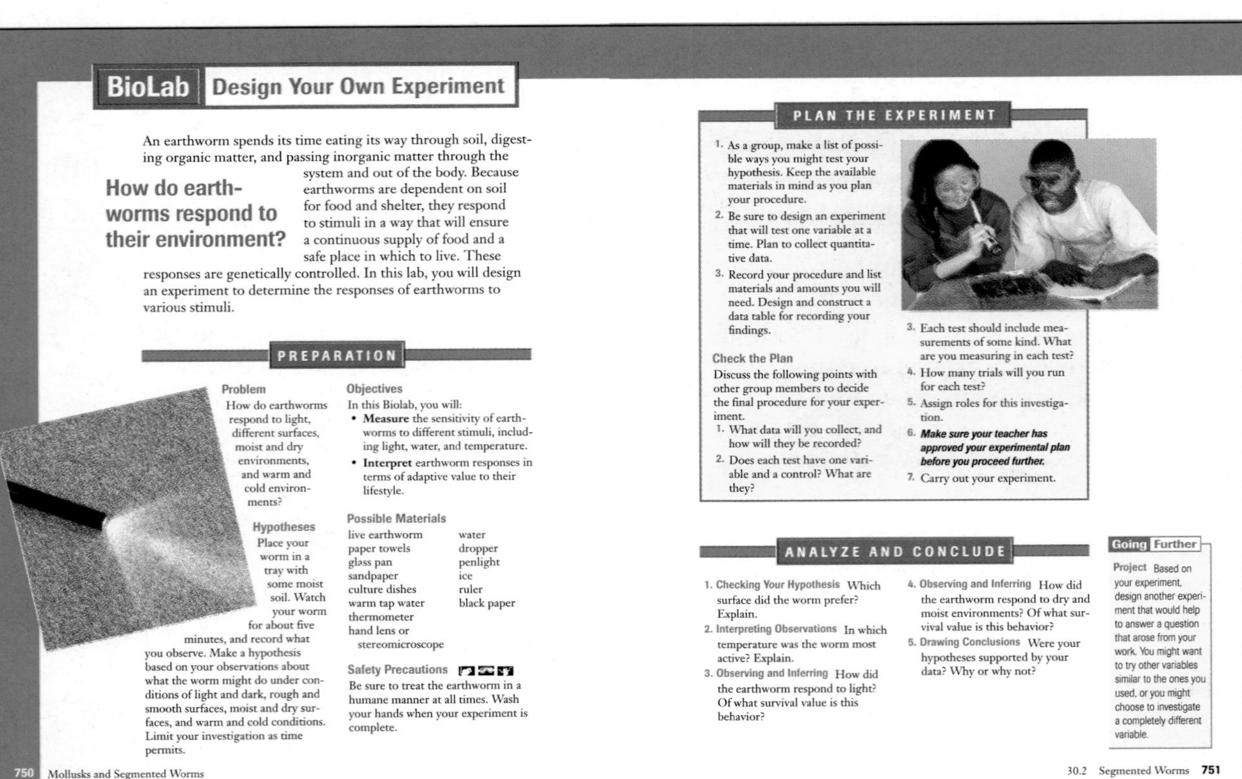

BioLab Design Your Own Experiment

How do earthworms respond to their environment?

An earthworm spends its time eating its way through soil, digesting organic matter, and passing inorganic matter through the system and out of the body. Because earthworms are dependent on soil for food and shelter, they respond to stimuli in a way that will ensure a continuous supply of food and a safe place in which to live. These responses are genetically controlled. In this lab, you will design an experiment to determine the responses of earthworms to various stimuli.

PREPARATION

Problem

How do earthworms respond to light, different surfaces, moist and dry environments, and warm and cold environments?

Hypotheses

Place your worm in a tray with some moist soil. Watch your worm for about five minutes, and record what you observe. Make a hypothesis based on your observations about what the worm might do under conditions of light and dark, rough and smooth surfaces, moist and dry surfaces, and warm and cold conditions. Limit your investigation as time permits.

Objectives

In this Biolab, you will:
- **Measure** the sensitivity of earthworms to different stimuli, including light, water, and temperature.
- **Interpret** earthworm responses in terms of adaptive value to their lifestyle.

Possible Materials

live earthworm
paper towels
glass pan
sandpaper
culture dishes
warm tap water
thermometer
hand lens or stereomicroscope
water
dropper
penlight
ice
ruler
black paper

Safety Precautions

Be sure to treat the earthworm in a humane manner at all times. Wash your hands when your experiment is complete.

750 Mollusks and Segmented Worms

PLAN THE EXPERIMENT

1. As a group, make a list of possible ways you might test your hypothesis. Keep the available materials in mind as you plan your procedure.
2. Be sure to design an experiment that will test one variable at a time. Plan to collect quantitative data.
3. Record your procedure and list materials and amounts you will need. Design and construct a data table for recording your findings.

Check the Plan

Discuss the following points with other group members to decide the final procedure for your experiment.

1. What data will you collect, and how will they be recorded?
2. Does each test have one variable and a control? What are they?
3. Each test should include measurements of some kind. What are you measuring in each test?
4. How many trials will you run for each test?
5. Assign roles for this investigation.
6. ***Make sure your teacher has approved your experimental plan before you proceed further.***
7. Carry out your experiment.

ANALYZE AND CONCLUDE

1. **Checking Your Hypothesis** Which surface did the worm prefer? Explain.
2. **Interpreting Observations** In which temperature was the worm most active? Explain.
3. **Observing and Inferring** How did the earthworm respond to light? Of what survival value is this behavior?
4. **Observing and Inferring** How did the earthworm respond to dry and moist environments? Of what survival value is this behavior?
5. **Drawing Conclusions** Were your hypotheses supported by your data? Why or why not?

Going Further

Project Based on your experiment, design another experiment that would help to answer a question that arose from your work. You might want to try other variables similar to the ones you used, or you might choose to investigate a completely different variable.

30.2 Segmented Worms 751

Practice Lab

Background

Stingrays, both male and female, are equipped with organs that detect the minute electrical fields given off by living organisms. Scientists speculate that during mating season, male stingrays find female stingrays by locating the electrical fields given off by the females.

Experimental Plan

1. State the problem posed by the reading.
2. To what other factors (excluding electricity) might male stingrays be responding in their search for females?
3. State a hypothesis about how male stingrays find female stingrays.
4. In a cooperative group, brainstorm a list of ways to test the stated hypothesis.
5. What variables are important?
6. From the list generated in Step 4, design an experiment to test one variable. List materials needed and a step-by-step procedure in a numbered list similar to a recipe.
7. What variables will be measured?
8. How many stingrays will be tested?
9. How many trials will be run?
10. How will controls be incorporated?
11. How can the results be graphed?

Possible Experimental Designs

1. Prepare two models of female stingrays. Incorporate electricity into one and not the other. Introduce both of them into the environment of a male stingray. Measure the amount of time he spends with each model.
2. Cover the electrical organs of female stingrays to prevent electrical fields from escaping. Compare the amount of time males spend with normal females and those that have their electrical organs covered.

Open-ended labs help students discover the biologists in themselves.

More Help, More Labs

If students need further help, you may wish to guide them step-by-step through the *Biolabs* in Chapters 1 and 2. These *Biolabs* are designed to be more structured than *Design Your Own Experiment Biolabs* in other chapters. They offer students additional help in designing and setting up experimental procedures. For additional labs in which students are asked to design their own experiments, use the *Investigations* in the accompanying ***Laboratory Manual***. For students who are still uncomfortable with developing their own experiments, ***Biology: The Dynamics of Life*** also offers structured *Biolabs* in the student text, in the ***Teacher Wraparound Edition***, and in the ***Laboratory Manual***.

Specific strategies that help every student succeed

Each student brings his or her own unique set of abilities, perceptions, and needs into the classroom. Teachers want to make the classroom environment as receptive to these differences as possible and to ensure a good learning environment for all students.

Individual learning styles are different, and learning style often does not reflect a student's ability level. While some students learn primarily through visual or auditory senses, others are kinesthetic learners and do best if they have hands-on exploratory interaction with materials. Some students work best alone and others learn best in a group environment.

In an effort to provide all students with a positive science experience, ***Biology: The Dynamics of Life*** offers a variety of ways for students to interact with materials so that they utilize their preferred method of learning concepts. A variety of approaches allows students to become familiar with other learning approaches as well.

Ability Levels

The activities are broken down into three levels to accommodate all student ability levels. ***Biology: The Dynamics of Life Teacher Wraparound Edition*** designates the activities as follows:

L1 activities are basic activities designed to be within the ability range of all students including those with learning difficulties. These activities reinforce the concepts presented.

L2 activities are application activities designed for students who have mastered the concepts presented. These activities give students an opportunity for practical application of the concepts presented.

L3 activities are challenging activities designed for the students who are able to go beyond the basic concepts presented. These activities allow students to expand their perspectives on the basic concepts presented.

Students Acquiring English

In providing for the student with limited English proficiency, the focus is on overcoming a language barrier. It is important not to confuse ability in speaking/reading English with academic ability. The following ideas may help you to structure learning for the **SAE** student:

1. Visual and experimental teaching is important. Use demonstrations and models to clarify ideas.
2. Make use of small group activities. Small group work allows for more participation and discussion by **SAE** students.
3. Present concepts in a variety of ways. Try to relate the concepts to situations of relevance to the student. Provide examples and model the concept whenever possible.
4. Check for understanding frequently. Provide opportunities for students to do sample evaluation items similar to those on tests.

The ***Teacher Classroom Resources*** offer a variety of ways to reach students of different ability levels. The ***Reinforcement and Study Guide*** contains exercises that help students of all ability levels master important concepts. The *Spanish Glossary* at the back of this booklet will help your Spanish-speaking students with unfamiliar terminology. ***Section Focus Masters***, ***Concept Mapping,*** and three sets of transparencies—***Basic Concepts, Basic Skills,*** and ***Reteaching***—are particularly helpful with students who have learning difficulties. Average and above-average students will especially benefit from activities contained in the ***Laboratory Manual***, ***Tech Prep Applications***, ***Biology Projects***, and ***Exploring Environmental Issues***. The ***Critical Thinking/Problem Solving*** booklet and *A Broader View* in the ***Student Edition*** will both appeal to above-average students. In the extended margins of the ***Teacher Wraparound Edition***, each chapter has several *Meeting Individual Needs* strategies that help you address all ability levels.

Biology: The Dynamics of Life offers a number of teaching resources that are especially helpful for students acquiring English. In addition to a complete English-Spanish glossary, the ***Spanish Resources*** booklet contains translations of all chapter and lesson objectives, key terms and definitions, and summary statements for each chapter of the student text. ***English/Spanish Audiocassettes*** give students an opportunity to listen to a summarized recording of the student text in either English or Spanish. The ***CD-ROM Multimedia System*** contains the entire student text translated into Spanish.

Learning Styles

We at Glencoe believe it is our responsibility to provide you with a program that allows you to apply diverse instructional strategies to a population of students with diverse learning styles. Several learning styles are emphasized in the ***Teacher Wraparound Edition***: Kinesthetic, Visual-Spatial, Logical-Mathematical, Interpersonal, Intrapersonal, Linguistic, and Auditory-Musical.

A student with a kinesthetic style learns from touch, movement, and manipulating objects. Visual-spatial learners respond to images and illustrations. Using numbers and reasoning are characteristics of the logical-mathematical learner. A student with an interpersonal style has confidence in social settings, whereas a student with an intrapersonal style may prefer to learn on his or her own. Linguistic learning involves the use and understanding of words. Finally, auditory-musical learning involves learning through the spoken word and responding to tones and rhythms.

Any student may display any or all of these styles depending on the learning situation. The ***Student Edition*** and ***Teacher Wraparound Edition*** provide a number of strategies for encouraging students with diverse learning styles. These include *Math Connections*, *Biolabs*, *Minilabs*, and *Thinking Labs*. The ***Teacher Wraparound Edition*** contains *Visual Learning*, *Projects*, and *Student Journal* strategies, plus a variety of other activities. Look for this logo LS that identifies strategies for different learning styles.

In general, the best method for dealing with SAE students, as well as variations in learning styles and ability levels, is to provide all students with a variety of ways to learn, apply, and be assessed on the concepts. The table on pages 24T-25T gives additional tips you may find useful in structuring the learning environment in your classroom to meet students' special needs.

A variety of approaches for today's diverse student populations

Structuring the Learning Environment

	DESCRIPTION	SOURCES OF HELP/INFORMATION
Learning Disabled	All learning disabled students have an academic problem in one or more areas, such as academic learning, language, perception, social-emotional adjustment, memory, or attention.	*Journal of Learning Disabilities* *Learning Disability Quarterly*
Behaviorally Disordered	Children with behavior disorders deviate from standards or expectations of behavior and impair the functioning of others and themselves. These children may also be gifted or learning disabled.	*Exceptional Children* *Journal of Special Education*
Physically Challenged	Children who are physically disabled fall into two categories—those with orthopedic impairments and those with other health impairments. Orthopedically impaired children have the use of one or more limbs severely restricted, so the use of wheelchairs, crutches, or braces may be necessary. Children with other health impairments may require the use of respirators or other medical equipment.	Batshaw, M.L. and M.Y. Perset. *Children with Handicaps: A Medical Primer.* Baltimore: Paul H. Brooks, 1981. Hale, G. (Ed.). *The Source Book for the Disabled.* New York: Holt, Rinehart & Winston, 1982. *Teaching Exceptional Children*
Visually Impaired	Children who are visually disabled have partial or total loss of sight. Individuals with visual impairments are not significantly different from their sighted peers in ability range or personality. However, blindness may affect cognitive, motor, and social development, especially if early intervention is lacking.	*Journal of Visual Impairment and Blindness* *Education of Visually Handicapped* *American Foundation for the Blind*
Hearing Impaired	Children who are hearing impaired have partial or total loss of hearing. Individuals with hearing impairments are not significantly different from their hearing peers in ability range or personality. However, the chronic condition of deafness may affect cognitive, motor, and social development if early intervention is lacking. Speech development also is often affected.	*American Annals of the Deaf* *Journal of Speech and Hearing Research* *Sign Language Studies*
Students Acquiring English	Multicultural and/or bilingual children often speak English as a second language or not at all. The customs and behavior of people in the majority culture may be confusing for some of these students. Cultural values may inhibit some of these students from full participation.	*Teaching English as a Second Language Reporter* R.L. Jones (Ed.). *Mainstreaming and the Minority Child.* Reston, VA: Council for Exceptional Children, 1976.
Gifted	Although no formal definition exists, these students can be described as having above-average ability, task commitment, and creativity. Gifted students rank in the top 5% of their class. They usually finish work more quickly than other students, and are capable of divergent thinking.	*Journal for the Education of the Gifted* *Gifted Child Quarterly* *Gifted Creative/Talented*

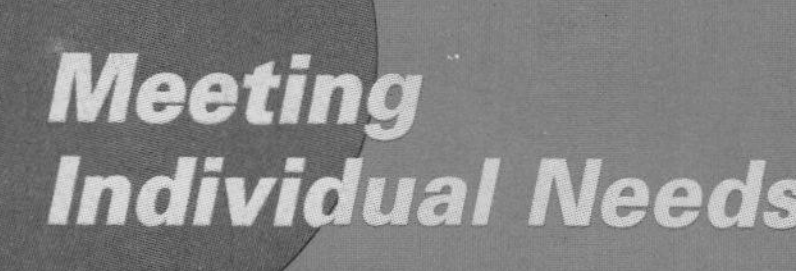

TIPS FOR INSTRUCTION

1. Provide support and structure; clearly specify rules, assignments, and duties.
2. Practice skills frequently. Use games and drills to help maintain student interest.
3. Allow students to record answers on tape and allow extra time to complete tests and assignments.
4. Provide outlines or tape lecture material.
5. Pair students with peer helpers, and provide class time for pair interaction.

---

1. Provide a clearly structured environment with regard to scheduling, rules, room arrangement, and safety.
2. Clearly outline objectives and how you will help students obtain objectives. Seek input from them about their strengths, weaknesses, and goals.
3. Reinforce appropriate behavior and model it for students.
4. Do not expect immediate success. Instead, work for long-term improvement.
5. Balance individual needs with group requirements.

---

1. Openly discuss with the student any uncertainties you have about when to offer aid.
2. Ask parents or therapists and students what special devices or procedures are needed, and whether any special safety precautions need to be taken.
3. Allow physically disabled students to do everything their peers do, including participating in field trips, special events, and projects.
4. Help nondisabled students and adults understand physically disabled students.

---

1. As with all students, help the student become independent. Some assignments may need to be modified.
2. Teach classmates how to serve as guides.
3. Limit unnecessary noise in the classroom.
4. Encourage students to use their sense of touch. Provide tactile models whenever possible.
5. Describe people and events as they occur in the classroom.
6. Provide taped lectures and reading assignments.
7. Team the student with a sighted peer for laboratory work.

---

1. Seat students where they can see your lip movements easily, and avoid visual distractions.
2. Avoid standing with your back to the window or light source.
3. Using an overhead projector allows you to maintain eye contact while writing.
4. Seat students where they can see speakers.
5. Write all assignments on the board, or hand out written instructions.
6. If the student has a manual interpreter, allow both student and interpreter to select the most favorable seating arrangements.

---

1. Remember that students' ability to speak English does not reflect their academic ability.
2. Try to incorporate the student's cultural experience into your instruction. The help of a bilingual aide may be effective.
3. Include information about different cultures in your curriculum to help build students' self-image. Avoid cultural stereotypes.
4. Encourage students to share their cultures in the classroom.

---

1. Make arrangements for students to take selected subjects early and to work on independent projects.
2. Let students express themselves in art forms such as drawing, creative writing, or acting.
3. Make public services available through a catalog of resources, such as agencies providing free and inexpensive materials, community services and programs, and people in the community with specific expertise.
4. Ask "what if" questions to develop high-level thinking skills. Establish an environment safe for risk taking.
5. Emphasize concepts, theories, ideas, relationships, and generalizations.

Performance Assessment: Practical Strategies and Tools

The ***Biology: The Dynamics of Life*** program has been designed to provide you with a variety of assessment tools, both formal and informal, to help you develop a clearer picture of your students' progress.

Performance Assessment

Various methods of assessing individual student performance are becoming more common in today's schools. These performance assessments differ in formality and complexity, but in most cases, the teacher observes a student or group of students involved in an activity and rates the performance and/or the products that result from the activity. Background information and specific examples of performance assessment are included in Glencoe's ***Alternate Assessment in the Science Classroom***.

Biology: The Dynamics of Life provides numerous opportunities to observe student behavior both in informal and formal settings. Each *Biolab*, *Thinking Lab*, and *Minilab* contains suggestions that will enable you to assess students' understanding of both concepts and process skills.

Another approach for assessing student mastery of concepts and skills in the laboratory is provided in Glencoe's ***Performance Assessment in the Biology Classroom***. It features 30 laboratory exercises that enable you to evaluate students' skills in handling laboratory equipment and students' knowledge of laboratory processes.

Group Performance Assessment

Recent research has shown that cooperative learning structures produce improved student learning outcomes for students of all ability levels. ***Biology: The Dynamics of Life*** provides many opportunities for cooperative learning and, as a result, many opportunities to observe group work processes and products. Glencoe's ***Cooperative Learning in the Science Classroom*** provides strategies and resources for implementing and evaluating group activities.

In cooperative group assessment, all members of the group contribute to the work process and its products. For example, if a mixed-ability, four-member laboratory work group conducts an activity, you can use a rating scale or checklist to assess the quality of both group interaction and work skills. All four members of the group are expected to review and agree on the data sheet produced by the group. You can require each member to certify the group's results by signing the data sheet or lab report. In this approach, all members of the group receive the same grade on the work product. Research shows that cooperative group assessment is as valid as individual assessment. Additionally, it reduces the marking and grading workload of the teacher.

Portfolios: Putting It All Together

The purpose of a student or cooperative group portfolio is to present examples of the individual's or group's work in a nontesting environment. A portfolio is simply a method for assembling and presenting selected examples of work products. The process of assembling the portfolio should be both integrative (of process and content) and reflective. The performance portfolio is not a complete collection of all worksheets and other assignments for a grading period.

At its best, the portfolio should include integrated performance products that show growth in concept attainment and skill development. You can structure the portfolio development process by establishing categories and other limiting specifications. An essential component in portfolio development is the composition of a submission letter or reflective paper that lists the contents of the portfolio and discusses growth in knowledge, attitudes, and skills.

Biology: The Dynamics of Life presents a wealth of opportunities for performance portfolio development. Each chapter contains projects; enrichment activities; laboratory investigations; skill reviews; suggestions for library research; features with critical-thinking opportunities; and connections to life, social studies, and the arts. Each of these student activities results in a product. A mixture of these products can be used to document student growth during the grading period.

In addition, ***Biology: The Dynamics of Life*** strongly suggests the use of student journals. Students are encouraged to write observations, descriptions, and reflections in their journals. They are also encouraged to include diagrams and drawings. Excerpts from the student journal can be included in an individual or group portfolio. Additionally, as many writers have discovered, the journal will be an excellent resource for developing the reflective submission letter or paper.

Content Assessment

The ***Biology: The Dynamics of Life*** program contains numerous strategies and an assortment of traditional aids for evaluating student progress toward mastery of science concepts. Throughout the chapters in the ***Student Edition***, *Section Review* questions and application tasks are presented. This spaced review process helps build learning bridges that allow all students to confidently progress from one lesson to the next.

After instruction for the chapter is completed, a summation of the major concepts is presented. Small groups of students can research the major concepts in the chapter and present restatements of their meaning in writing and as oral reports to the class.

After the main idea presentations, the formal review process for the written content assessment can begin. ***Biology: The Dynamics of Life*** presents a three-page *Chapter Review* at the end of each chapter. By evaluating the student responses to this extensive review, you can determine whether any substantial reteaching is needed.

For the formal content assessment, a six-page test is provided in the ***Chapter Assessment*** booklet for each chapter. If your individual assessment plan requires a test that differs from this test in the resource package, customized tests can be easily produced using the ***Computer Test Bank***, available in DOS and Macintosh formats.

New tools for new approaches to assessing student performance

Cultural Diversity

"Multicultural education is an idea stating that all students, regardless of the groups to which they belong, such as those related to gender, ethnicity, race, culture, social class, religion, or exceptionality, should experience education equality in the schools."—James Banks

Diverse Cultural Heritages

American classrooms reflect the rich and diverse cultural heritages of the American people. Students come from different ethnic backgrounds and different cultural experiences into a common classroom that must assist all of them to learn. The diversity itself is an important focus of the learning experience. Diversity can be repressed, creating a hostile environment; ignored, creating an indifferent environment; or appreciated, creating a receptive and productive environment.

Cultural Diversity in *Biology: The Dynamics of Life*

Responding to diversity and approaching it as a part of every curriculum may be challenging to a teacher, experienced or not. The goal of science is understanding. The goal of multicultural education is to promote an understanding of how people from different cultures approach and solve the basic problems all humans have in living and learning. ***Biology: The Dynamics of Life*** addresses this issue by including numerous *Cultural Diversity* connections in the ***Teacher Wraparound Edition***, by showing successful scientists from diverse ethnic backgrounds in *People in Biology*, and by the addition of cross-curricular connections relating biology to art and literature.

These features offer opportunities to integrate multicultural materials that relate to the topics presented in each chapter into the curriculum. By providing these opportunities, ***Biology: The Dynamics of Life*** is helping to meet the five major goals of multicultural education:

1. promoting the strength and value of cultural diversity
2. promoting the human rights of and respect for those who are different from oneself
3. promoting alternative life choices for people
4. promoting social justice and equal opportunity for all people
5. promoting equity in the distribution of power among groups

Readings

One book that provides additional information on multicultural education is:

Banks, James A. *Multicultural Education: Issues and Perspectives.* Boston: Allyn and Bacon, 1989.

Managing Activities in the Biology Classroom

The activities in ***Biology: The Dynamics of Life*** are designed to minimize dangers in the laboratory. Careful planning and preparation as well as being aware of hazards can keep accidents to a minimum. Numerous books and pamphlets are available on laboratory safety with detailed instructions on preventing accidents. Know the rules and be familiar with the common violations that occur. Know the Safety Symbols used in this book (see p. 30T). Know where emergency equipment is stored and how to use it. Practice good laboratory housekeeping and management by observing these guidelines:

Classroom/Laboratory

1. Store chemicals properly. (For details see p. 31T.)
2. Store equipment properly.
   a. Clean and dry all equipment before storing.
   b. Protect electronic equipment and microscopes from dust, humidity, and extreme temperatures.
   c. Label and organize equipment so that it is accessible.
3. Provide adequate work space.
4. Provide adequate room ventilation.
5. Post safety and evacuation guidelines.
6. Be sure safety equipment is accessible and works.
7. Provide containers for disposing of chemicals, waste products, and biological specimens. Disposal methods must meet local guidelines.
8. Use hot plates whenever possible as a heat source. If burners are used, a central shutoff valve for the gas supply should be available to the teacher. Never use open flames when a flammable solvent is in the same room.

First Day of Class (with students)

1. Distribute and discuss safety rules, safety symbols, and first aid guidelines. Have students refer to Appendix C on pages A8-A9 to review safety symbols and guidelines.
2. Review safe use of equipment and chemicals.
3. Review use and location of safety equipment.
4. Discuss safe disposal of materials and laboratory cleanup policy.
5. Discuss proper laboratory attitude and conduct.
6. Document students' understanding of the preceding points. Have students sign a safety contract and return it.

Before Each Biolab

1. Perform each investigation yourself before assigning it.
2. Arrange the lab in such a way that equipment and supplies are clearly labeled and easily accessible.
3. Have available only equipment and supplies needed to complete the assigned investigation.
4. Review the procedure with students, emphasizing any caution statements or safety symbols that appear.
5. Be sure all students know the proper procedures to follow if an accident should occur.

During the Biolab

1. Make sure the lab is clean and free of clutter.
2. Insist that students wear goggles and aprons.
3. Never allow a student to work alone in the lab.
4. Never allow students to use a cutting device with more than one edge.
5. Students should not point the open end of a heated test tube toward anyone.
6. Remove broken glassware or frayed cords from use. Also clean up any spills immediately. Dilute solutions with water before removing.
7. Be sure all glassware that is to be heated is of a heat-treated type that will not shatter.
8. Remind students that hot glassware looks cool.
9. Prohibit eating and drinking in the lab.

After the Biolab

1. Be sure that the lab is clean.
2. Be certain that students have returned all equipment and disposed of broken glassware and chemicals properly.
3. Be sure that all hot plates and electrical connections are off.
4. Insist that each student wash his or her hands when lab work is completed.

Safety Symbols

The ***Biology: The Dynamics of Life*** program uses safety symbols to alert you and your students to possible laboratory dangers. These symbols are provided in the student text on pages A8–A9 and are explained below. Be sure your students understand each symbol before they begin an activity that displays a symbol.

 **DISPOSAL ALERT** This symbol appears when care must be taken to dispose of materials properly.	 **ANIMAL SAFETY** This symbol appears whenever live animals are studied and the safety of the animals and the students must be ensured.
 **BIOLOGICAL HAZARD** This symbol appears when there is danger involving bacteria, fungi, or protists.	 **RADIOACTIVE SAFETY** This symbol appears when radioactive materials are used.
 **OPEN FLAME ALERT** This symbol appears when use of an open flame could cause a fire or an explosion.	 **CLOTHING PROTECTION SAFETY** This symbol appears when substances used could stain or burn clothing.
 **THERMAL SAFETY** This symbol appears as a reminder to use caution when handling hot objects.	 **FIRE SAFETY** This symbol appears when care should be taken around open flames.
 **SHARP OBJECT SAFETY** This symbol appears when a danger of cuts or punctures caused by the use of sharp objects exists.	 **EXPLOSION SAFETY** This symbol appears when the misuse of chemicals could cause an explosion.
 **FUME SAFETY** This symbol appears when chemicals or chemical reactions could cause dangerous fumes.	 **EYE SAFETY** This symbol appears when a danger to the eyes exists. Safety goggles should be worn when this symbol appears.
 **ELECTRICAL SAFETY** This symbol appears when care should be taken when using electrical equipment.	 **POISON SAFETY** This symbol appears when poisonous substances are used.
 **PLANT SAFETY** This symbol appears when poisonous plants or plants with thorns are handled.	 **CHEMICAL SAFETY** This symbol appears when chemicals used can cause burns or are poisonous if absorbed through the skin.

Chemical Storage and Disposal

General Guidelines

Be sure to store all chemicals properly. The following are guidelines commonly used. Your school, city, county, or state may have additional requirements for handling chemicals. It is the responsibility of each teacher to become informed as to what rules or guidelines are in effect in his or her area.

1. Separate chemicals by reaction type. Strong acids should be stored together. Likewise, strong bases should be stored together and should be separated from acids. Oxidants should be stored away from easily oxidized materials, and so on.
2. Be sure all chemicals are stored in labeled containers indicating contents, concentration, source, date purchased (or prepared), any precautions for handling and storage, and expiration date.
3. Dispose of any outdated or waste chemicals properly according to accepted disposal procedures.
4. Do not store chemicals above eye level.
5. Wood shelving is preferable to metal. All shelving should be firmly attached to the wall and should have anti-roll edges.
6. Store only those chemicals that you plan to use.
7. Hazardous chemicals require special storage containers and conditions. Be sure to know what those chemicals are and the accepted practices for your area. Some substances must even be stored outside the building.
8. When working with chemicals or preparing solutions, observe the same general safety precautions that you would expect from students. These include wearing an apron and goggles. Wear gloves and use the fume hood when necessary. Students will want to do as you do whether they admit it or not.
9. If you are a new teacher in a particular laboratory, it is your responsibility to survey the chemicals stored there and to be sure they are stored properly or disposed of. Consult the rules and laws in your area concerning what chemicals can be kept in your classroom. For disposal, consult up-to-date disposal information from the state and federal governments.

Disposal of Chemicals

Local, state, and federal laws regulate the proper disposal of chemicals. These laws should be consulted before chemical disposal is attempted. Although most substances encountered in high school biology can be flushed down the drain with plenty of water, it is not safe to assume that this is always true. It is recommended that teachers who use chemicals consult the following books from the National Research Council:

Prudent Practices for Handling Hazardous Chemicals in Laboratories, Washington, DC: National Academy Press, 1981.

Prudent Practices for Disposal of Chemicals from Laboratories, Washington, DC: National Academy Press, 1983.

These books are useful and still in print, although they are several years old. Current laws in your area would, of course, supersede the information in these books.

DISCLAIMER

Glencoe Publishing Company makes no claims to the completeness of this discussion of laboratory safety and chemical storage. The material presented is not all-inclusive, nor does it address all of the hazards associated with handling, storage, and disposal of chemicals, or with laboratory management.

Be prepared, and minimize safety hazards in the biology classroom.

Preparing solutions for all lab activities

It is most important to use safe laboratory techniques when handling all chemicals. Many substances may appear harmless but are, in fact, toxic, corrosive, or very reactive. Always check with the supplier. Chemicals should never be ingested. Be sure to use proper techniques to smell solutions or other agents. Always wear safety goggles and an apron. The following general cautions should be used.

1. Poisonous/corrosive liquid and/or vapor. Use in the fume hood. Examples: acetic acid, hydrochloric acid, ammonia hydroxide, nitric acid.
2. Poisonous and corrosive to eyes, lungs, and skin. Examples: acids, limewater, iron(III) chloride, bases, silver nitrate, iodine, potassium permanganate.
3. Poisonous if swallowed, inhaled, or absorbed through the skin. Examples: glacial acetic acid, copper compounds, barium chloride, lead compounds, chromium compounds, lithium compounds, cobalt(II) chloride, silver compounds.
4. Always add acids to water, never the reverse.
5. When sulfuric acid and sodium hydroxide are added to water, a large amount of thermal energy is released. Sodium metal reacts violently with water. Use extra care if handling any of these substances.

Aceto-orcein stain (11): Dissolve 1 g orcein in 45 mL hot glacial acetic acid in a large beaker. Foaming may result when the orcein is added. Cool the solution and add 55 mL distilled water. Store in a capped bottle. Filter before use.

Benedict's solution (38): Dissolve 173 g sodium citrate and 100 g sodium carbonate in 700 mL water over a hot plate. Filter. Dissolve 17.3 g copper sulfate in 100 mL water. Slowly add to the first solution. Add water to a total volume of 1 L.

Bromothymol blue (1, 10, 23) Add 0.5 g bromothymol blue powder to 500 mL distilled water to make a BTB stock solution. Dilute 10 mL of BTB stock solution to 500 mL distilled water. Use the diluted solution for the procedure. (23): Add 0.1 g bromothymol blue powder to 2000 mL water. Solution should be bright blue. If not, add one drop of HCl at a time, swirling to mix. Check color.

Carnoy's fluid (11): Mix 6 parts absolute alcohol, 3 parts chloroform, and 1 part glacial acetic acid.

Chalkey's solution (3): Dissolve 1 g sodium chloride, 0.04 g potassium chloride, and 0.06 g calcium chloride in 1 L of distilled water. Dilute the prepared solution by adding 100 mL of solution to 900 mL distilled water.

Cola, dilute solution (40): Add 1 part cola to 1 part distilled water.

Congo red (22): Add 0.1 g Congo red powder to 50 mL distilled water.

Cough medicine, dilute (40): Add 2 mL of cough medicine (syrup) to 98 mL distilled water. Stir before use.

Ethyl alcohol, dilute (40): Add 2 mL ethyl alcohol to 98 mL distilled water. Stir.

Fertilizer solution (4): To make a 1% fertilizer solution, mix 1 g 5-10-5 fertilizer with 99 mL water. For a 0.1 % serial dilution, mix 1 mL 1% solution with 9 mL water. For a 0.01% serial dilution, mix 1 mL 0.1% solution with 9 mL water.

Frog testes solution (13): Slowly add 0.5 mL (10 drops) sodium acetate to 0.5 mL frog testes solution.

Glucose solution (38): For 1% glucose solution, dissolve 1 g of glucose in 99 mL water.

Hydrochloric acid-alcohol solution (11): Mix 1 part 0.1*M* hydrochloric acid and 1 part absolute alcohol.

Iodine solution (8, 9, 10, 38): Dissolve 5 g of potassium iodide in 50 mL of water. Add 1.5 g of iodine crystals. Dilute to 500 mL.

Preparation of Solutions

Iodine-potassium iodide solution (26): Add 3 g potassium iodide and 6 g of iodine to 2000 mL of water. Stir the mixture until the substances dissolve in the water.

Methylene blue stain (8, 29): Dissolve 1.5 g methylene blue in 100 mL ethyl alcohol. Dilute by adding 10 mL of solution to 90 mL water.

Methyl cellulose (5, 9, 22): Methyl cellulose may be purchased commercially. To prepare, mix 20 g methyl cellulose with 10 mL boiling distilled water. Let stand for 30 minutes, then mix in 80 mL cool distilled water.

Nitric acid solution (25): add 2 mL nitric acid to 98 mL water, then add 5 g potassium dichromate.

Pancreatic solution (38): Blend a pig or sheep pancreas with 150 mL 30% ethyl alcohol. Allow the solution to stand for 14 hours, shaking occasionally. Strain the solution through cheesecloth and then filter. Neutralize with KOH until you get near the end point, then use 0.5% sodium carbonate.

Potassium permanganate (11): For a 0.02*M* solution of potassium permanganate, dissolve 0.3 g $KMnO_4$ in 100 mL water.

Salmon testes solution (13): Add 0.5 mL sodium acetate to 0.5 mL salmon testes solution.

Salt solution (26): For a 35% salt solution that simulates the concentration of ocean water, dissolve 35 g salt in 65 mL water. (9): For a 3% salt solution, dissolve 3 g of salt in 97 mL of water. (17): For a 1% solution, dissolve 1 g salt in 99 mL water. (24): For a 6% solution, dissolve 6 g salt in 94 mL water. (26): For a 5% solution, dissolve 5 g salt in 95 mL water.

Sodium bicarbonate solution (10): Mix 0.5 g sodium hydrogen carbonate with 99.5 mL of water.

Starch solution (9, 38): Make a liter of 1% starch solution by stirring a slurry of cornstarch and cold water into boiling water. Cool thoroughly before using.

Sterile pond water (3): Filter pond water and place it in flat pans. Boil for 15 minutes. Allow to cool before using.

Sucrose solution (27): For a 1% sucrose solution, dissolve 1 g sucrose in 99 mL water. For a 2% sucrose solution, mix 2 g sucrose in 98 mL water. For a 5% sucrose solution, mix 5 g sucrose with 95 mL water. For a 10% sucrose solution, dissolve 10 g of sucrose in 90 mL water. For a 20% sucrose solution, dissolve 20 g of sucrose in 80 mL of water. For a 30% sucrose solution, mix 30 g of sucrose in 70 mL of water. For a 40% sucrose solution, mix 40 g of sucrose in 60 mL of water.

Sugar solution (10): Add 1 tablespoon of sugar to 1 cup of warm water in a deep jar or flask. Stir to mix.

Tobacco solution (40, 21): Grind tobacco from one cigarette into a fine powder. Mix the powder with 100 mL of a 1% glucose solution.

Yeast culture (23): Add 1/5 package dry baker's yeast to 200 mL distilled water.

Equipment List

Nonconsumable

ITEM	BIOLAB	MINILABS	ALTERNATE LAB
apron	All biolabs and minilabs as appropriate		
aquarium			880
aquarium with pump	696	549, 819, 827	820
baby food jars	116		30
balance	32		100, 228, 712
bar codes			382
beads, various colors	912		
beakers	18, 66, 150, 182, 218, 246, 416, 572, 878	173, 253, 255, 263, 693, 742, 761	100, 126, 228, 252, 712
bird guide books		861	
bolts		506	
books		788	
bottles			492
bottle cork		191	
brush		287	
bucket		141, 827	
Bunsen burner			436
calculator	208	973	
cellophane, assorted colors	246		
clay			100, 840
clock or watch with second hand	182, 246, 416, 878, 948, 984	263, 788, 911, 993	
clothespins, spring-type	912		
cloth squares			100
cloth toweling			762
compass	247		
coverslips	66, 88, 116, 416, 588, 660	141, 191, 227, 229, 538, 549, 564, 595, 610, 642, 718	198, 268, 358, 414, 586, 638
crucibles			492
depression glass slide	696		
dichotomous keys	496	495	
dichotomous keys transparency		743	
dissecting needles		704	
dropper	66, 88, 116, 588, 598, 660, 726, 750, 974, 1028	42, 103, 168, 227, 229, 253, 742, 748, 873, 1051	198, 492, 962
dropper bottle			16, 962
Erlenmeyer flask	246, 416		492
fabric, red, white	912		
feathers, contour and down		855	
field guides		495, 875	
finger bowl			252
fish net			820
flashlight	828	773	
flasks, 250- or 500-mL			492
flowerpots	292		
forceps	340, 578, 598	229, 263, 672	198, 268, 638, 712
funnel			492
glass hockey sticks			1080
glass pan	750		
glass tubing	572		
goggles	All biolabs and minilabs as appropriate		
Gooey Grabber toys		693	
graduated cylinder	36, 88, 116, 218, 416		30, 58, 228, 492, 962
hand lens	292, 660, 750	141, 364, 488, 588, 672, 704, 731, 855, 937	638
hot plate	182, 246, 416, 572, 726		252
incubator	518		
inoculation loop			436

Consumable (Cont.)

ITEM	BIOLAB	MINILABS	ALTERNATE LAB
fish food		819	
flower extract-treated filter paper	764		
food samples	974		
frozen juice cans			840
fruit samples		672	
gelatin			180
glass marking pencil	572		1080
glucose test paper			962
glue	912		
grapes			180
graph paper, metric	948, 1060, 1088	264, 365, 426, 442	1046
grass seed			142
hay	88		
human hairs		761	382
ice	18, 182, 246, 572, 726, 750	693	
ink pad		937	
index cards	588		
Instant Sealife toy capsules	18		
juice, red			382
kidney beans, canned			672
kidney beans, dried			672
labels	340	634	30, 58, 142, 228, 252, 336, 638, 672
lettuce			586
lightbulbs	828		
liquid soap		168, 173	
liver, raw	182		
mammal hairs		873	
markers, black	912		
marking pens	518		
masking tape	218	86	
millet		861	
milk			962
milk cartons		400	
molasses	572		
mouthwash, 3 types			1088
mushrooms		569	
mustard seeds			142, 638
oatmeal cereal			552
oats, hulled		861	
orange slices			180
overhead transparencies		875	
paintbrushes	340		
paper clips			252, 300, 1016
paper, black	726	429	638, 740, 820, 840
paper, brown	974		
paper, construction (assorted colors, 6)	320		
paper, notebook	366	149	
paper, white	380	326, 378, 429, 457, 485, 518, 569, 970	16, 840
paper fastener	858		
paper packing material			150
paper towels	598, 750	42, 86, 168, 229, 674, 718	58, 142, 268, 336
paste		495	
pencil with eraser	858	610, 827	
petroleum jelly		141, 400	
pH paper		173	414
pinecones		616	
pinto beans		442	58

Equipment List

Consumable (Cont.)

ITEM	BIOLAB	MINILABS	ALTERNATE LAB
pine pollen		616	
pipe cleaners		506	
pineapple, canned, chunk			180
pineapple, fresh			180
plaster of paris		400	
plastic wrap			228
pond water	88		
pond water, sterile	66		
popcorn	150		
poster board			300
potato	182	263	228
potting trays	292, 340		
PTC paper	366		
radish seeds			142
rice	88, 116		
rubber bands	878		
salmon testes solution		311	
sand		103	880
sandpaper	750		740
sawdust			880
sea urchin eggs		1051	
sea urchin sperm		1051	
seeds	32	488	
shampoo		173	
shortening	878		
skull diagrams	466		
soapy water	974		
socks, long gray	912		
soda bottle, 2-L		86	
soil, garden		731	
soil, potting	32, 150, 292, 340	86, 103	100
soil samples			514
spiderwebs		761	
spoons, plastic	876		
straws, flexible	572, 984		16
string	246, 984	57, 121	300
tagboard	858, 912	911	
tape, clear	320, 380, 912	264, 457	300, 762, 840
tape, masking			820, 1080
tea	1028		
thistle		861	
thread, heavy			762
thread, nylon	912		
tobacco	1028		
tobacco seeds	292		336
tongue depressors			358
toothpicks, flat	588, 726	168, 549, 731	198, 300, 962, 1080
twigs		761	
twist ties	218		100
wax paper	182	168, 263	180, 672
waxed marking pencil			126, 436, 712, 740
wheat		861	
wheat seed particles		538	
wool			300
yarn			318
yeast	572		

Equipment List

Living Organisms

ITEM	BIOLAB	MINILABS	ALTERNATE LAB
algae		595	
Bacillus subtilis culture			436
bacteria cultures	518		
bess beetles			762
clams		742	
coleus			252
conifer twigs	622		
conifer branches	622		
daphnia culture	1028		
didinium culture	66		
earthworms	696, 750	704	
elodea	246	229	16, 198
euglena culture	544		
fern frond with sporangia		616	
fern prothallus		610	
fishes		819	820
flowers	660	30, 287, 672	
frogs, *Rana pipiens*	696	827	
goldfish	696		
hydra culture		718	
insects	764	30, 86	
land snails			740
leaves		30, 426, 495, 694	
lichen		69	
moss plants	598	595	
mouse	696		880
mussels, freshwater			16
nematodes, microscopic		704	
onion roots			268
paramecium culture	66, 116, 544	227, 538	
Physarum polycephalum			552
planarian culture	726		
plants, assorted		588	
plants, dicot		634	
plants, monocot		634	
plant seedlings	32		
potted plant		57, 86	
snails		86, 549	

Chemicals

ITEM	BIOLAB	MINILABS	ALTERNATE LAB
acetone		937	
aceto-orcein stain			268
agar, sterile nutrient plates	518	552, 1080	
agar, sterile nutrient slant tubes			436
aged tap water	1028		
alcohol 95%			252
alcohol, rubbing		1074	1080
ammonia		173	
aquarium water		253	
aspartic acid	416		
baking soda solution	246		514
biuret solution	974		
bleach	32		
bromothymol blue (BTB)		253	16
carmine powder suspension		742	
Carnoy's fluid			268

Equipment List

Chemicals (Cont.)

ITEM	BIOLAB	MINILABS	ALTERNATE LAB
distilled water		229	16, 514, 552, 672
ethanol		311	
ethyl alcohol	1028		
fertilizer solution	32, 88		
glucose solution			962
glutamic acid	416		
glycerine		873	
glycine	416		
gum arabic solution			414
HCl			414
HCl-alcohol solution			268
hydrogen peroxide	182	103	
iodine	218, 974		252
Lactaid			962
lemon juice		173	
methyl cellulose	116, 588	227	
methylene blue stain		718	198, 358
molar NaOH	416		
nail polish remover		937	
potassium permanganate		263	
Ringer's solution	828		
salt, table			228
salt solution		229, 642	
Schultz liquid plant food			514
starch solution	218		
streptomycin agar			436
2, 3, 5-triphery/tetrazolium chloride			672
vinegar		173, 718	142, 514

Preserved Specimens

ITEM	BIOLAB	MINILABS	ALTERNATE LAB
amphioxis development slides	794	796	
bacteria (prepared slides)		514	
blood cells, human		1080	
cheek cells, prepared stained slides (female)			382
compact bone slides		944	
dicot stems, cross-section slides		634	
earthworm, cross-section slides			702
frog legs, *Rana pipens* or *Xenopus laevis*	828		
gymnosperm leaf cross-section, prepared slides			618
human skin slides			936
hydra cross-section slides		702	
monocot stem cross-section prepared slides		634	
nematode, cross-section prepared slides		702	
onion root tip slides	274		
planarians, cross-section prepared slides		702	
sea sponge specimens			
grass			712
hard head			712
sheep's wool			712
yellow			712
sea urchin development slides	794		
sheep's kidneys		999	
stem cross-section, prepared slides	644		

Suppliers

Equipment Suppliers

Carolina Biological Supply Co.
2700 York Road
Burlington, NC 27215

Edmund Scientific Company
101 E. Gloucester Pike
Barrington, NJ 08007

Fisher Scientific Company
4901 W. LeMoyne St.
Chicago, IL 60651

Nasco
901 Janesville Avenue
Fort Atkinson, WI 53538

Nebraska Scientific
3823 Leavenworth Street
Omaha, NE 68105-1180

Sargent-Welch Scientific Co.
911 Commerce Ct.
Buffalo Grove, IL 60089

Ward's Natural Science Establishment, Inc.
5100 W. Henrietta Road
Rochester, NY 14692

Audiovisual Distributors

Ambrose Video Publisher
381 Park Avenue South
New York, NY 10016

Apple Computer, Inc.
20525 Mariana Avenue
Cupertino, CA 95014

BFA Educational Media
468 Park Avenue S.
New York, NY 10016

BBC Education and Training Sales
Division of BBC Enterprises
Suite 111
65 Hewarg Avenue
Toronto, Ontario
CANADA M4M 2T5

Biology Media
918 Parker St.
Berkeley, CA 94710

Bullfrog Films
Oley, PA 19547

Charles Clark Co., Inc.
170 Keyland Court
Bohemia, NY 11716

Coronet/MTI Film and Video Distributors of LCA
108 Wilmot Road
Deerfield, IL 60015

The Discovery Channel
770 Wisconsin Avenue
Bethesda, MD 20814

EDUCORP Computer Services
7434 Trade Street
San Diego, CA 92121-2410

Educational Dimensions
408 Hightstown
Hightstown, NJ 08526

Encyclopaedia Britannica
Educational Corp. (EBEC)
310 S. Michigan Avenue
Chicago, IL 60604

Films for the Humanities and Sciences, Inc.
12 Perrine Rd.
Monmouth Junction
S. Brunswick, NJ 08852

Frey Scientific Co.
905 Hickory Lane
Mansfield, OH 44905

Indiana University Audio-Visual Center
Bloomington, IN 47405-4203

Insight Media, Inc.
2162 Broadway
New York, NY 10024

International Film Bureau, Inc.
332 S. Michigan Avenue
Chicago, IL 60604

Media Design Associates, Inc.
Box 3189
Boulder, CO 80307

National Geographic Society
Educational Services
1145 17th Street, NW
Washington, DC 20036

NET Film Service, Audio-Visual Center
350 Fifth Avenue
New York, NY 10016

Optical Data Corporation
P.O. Box 4919
30 Technology Drive
Warren, NJ 07059

Scholastic, Inc.
555 Broadway
New York, NY 10012

Time-Life Video
Time and Life Building
1271 Avenue of the Americas
New York, NY 10020

TVOntario
Ontario Educational Communications Authority
Box 200, Sta Q
Toronto, Ontario
CANADA M4T 2T1

VideoDiscovery
1700 Westlake N.
Seattle, WA 98109

Software Distributors

Applied Optical Media Corporation
1450 Boot Road, Bldg 400
West Chester, PA 19380

Aquarius Instructional
P.O. Box 128
Indian Rocks Beach, FL 34635

Bio-Soft, Inc.
P.O. Box 7294
Winter Haven, FL 33880

Conduit
The University of Iowa
Oakdale Campus
Iowa City, IA 52242

Cross Educational Software, Inc.
504 E. Kentucky Ave.
P.O. Box 1536
Ruston, LA 71270

Educational Activities, Inc.
P.O. Box 392
Freeport, NY 11520

Educational Dimensions Group
P.O. Box 126
Stanford, CT 06904

Educational Materials and Equipment Company (EME)
P.O. Box 2805
Danbury, CT 06813-2805

Focus Media, Inc.
Suite 12
485 S. Broadway
Hicksville, NY 11801

Human Relations Media (HRM)
175 Tompkins Avenue
Pleasantville, NY 10570

IBM Educational Systems
Department PC
4111 Northside Parkway
Atlanta, GA 30327

Intelligent Software
562 Boston Avenue
Bridgeport, CT 06610

Intellimation
130 Cremora Drive
P.O. Box 1922
Santa Barbara, CA 93116-1922

J&S Audio Visual Communications Co., Inc.
Box 815249
Dallas, TX 75381-5249

Mayfield Publishing Company
1240 Villa Street
Mountain View, CA 94041

MacIntosh Educational Software
Apple Computer, Inc.
20525 Mariana Avenue
Cupertino, CA 95014

Micro-ED, Inca
P.O. Box 24750
Edina, MN 55424

Nutridata Software Corp.
1215 Rte.9, Suite F
Wappingers Falls, NY 12590

Queue, Inc.
338 Commerce Drive
Warren, NJ 07059

Scholastic Software
730 Broadway
New York, NY 10012

Sunburst Communications Inc.
39 Washington Avenue
Pleasantville, NY 10570

Wings for Learning
1600 Green Hilll Rd.
P.O. Box 66002
Scotts Valley, CA 95067

J. Weston Walch
P.O. Box 658
Portland, ME 04104

Readings

Bibliography

Student Readings

Bidwell, R., *Hydroponic Gardening*, Santa Barbara, Woodbridge Press, 1989.

Caldwell, Mark, "How Does a Single Cell Become a Whole Body?" *Discover*, November 1992.

Cohen, Joel E., "How Many People Can Earth Hold?" *Discover*, November 1992.

Crews, David, "Animal Sexuality," *Scientific American*, January 1994.

Duellman, William E., "Reproductive Strategies of Frogs," *Scientific American*, July 1992.

Freedman, David H., "The Aggressive Egg," *Discover*, June 1992.

Gould, Stephen J., "The Reversal of Hallucinogenia," *Natural History*, January 1992.

Hanson, Betsy, "Message in a Barrel," *Discover*, June 1992.

Jaroff, Leon, "Happy Birthday Double Helix," *Time*, March 15 1993.

Martin, Glen, "Spring Fever," *Discover*, October 1992.

Oliwenstein, Lori, "Drugs by Design," *Discover*, November 1991.

Ravven, Wallace, "In the Beginning," *Discover*, October 1992.

Robbins, Jim, "The Real Jurassic Park," *Discover*, March 1991.

Shimeck, Ronald L., "Sex Among the Sessile," *Natural History*, March 1987.

"Special Issue 1995 The Year in Science," *Discover*, January 1996.

Vogel, Shawna, "Cold Storage," *Discover*, February 1988.

Vickers-Rich, P., and Hewilt-Rich, T., "Australia's Polar Dinosaurs," *Scientific American*, July 1993.

Watson, James D., *The Double Helix*, New York: Antheneum, 1968.

Wilson, Edward O., "The Diversity of Life," *Discover*, September 1992.

Zimmer, Carl, "Masters of an Ancient Sky," *Discover*, February 1994.

Teacher Readings

Allard, David W., and Royce L. Granberry, "Osmosis Revisited," *The American Biology Teacher*, November/December 1992.

Beardsley, Tim, "Teaching Real Science," *Scientific American*, October 1992.

Brouse, Deborah E., "Population Education, " *The Science Teacher*, December 1990.

Carson, Rachel, *Silent Spring*, New York: Houghton Mifflin Company, 1962.

Chiras, Daniel D., "Teaching Critical Skills in the Biology and Environmental Science Classrooms," *The American Biology Teacher*, November/December 1992.

DiSpezio, Michael A., "Retroviruses—Gaining an Understanding," *The Science Teacher*, October 1990.

Eigen, M., "Viral Quasispecies," *Scientific American*, July 1993.

Gould, Stephen J., and Niles Eldredge, "Punctuated Equilibria: An Alternative to Phyletic Gradualism," San Francisco: Freeman, Cooper, and Co., 1972.

Johanson, D., and Blake Edgar, *From Lucy to Language*, Simon & Schuster Editions, 1996.

Levinton, Jeffrey S., "The Big Bang of Animal Evolution," *Scientific American*, November 1992.

Margulis, L., and K.V. Schwartz, *Five Kingdoms: An Illustrated Guide to the Phyla of Life on Earth*, New York: W. H. Freeman and Co., 1988.

Okebukola, Peter A., "Concept Mapping with a Cooperative Learning Flavor," *The American Biology Teacher*, April 1992.

Roach, Mary, "Secrets of the Shamans," *Discover*, November 1993.

Stossel, Thomas P., "How Cells Crawl," *American Scientist*, September/October 1990.

Wilson, E.O., *The Diversity of Life*, Cambridge: Belknap Press, 1992.

BIOLOGY
THE DYNAMICS OF LIFE

GLENCOE

McGraw-Hill

New York, New York Columbus, Ohio Woodland Hills, California Peoria, Illinois

A GLENCOE PROGRAM

BIOLOGY: THE DYNAMICS OF LIFE

Student Edition
Teacher Wraparound Edition
Laboratory Manual, SE and TE
Reinforcement and Study Guide, SE and TE
Section Focus Masters
Chapter Assessment
Lesson Plans
Videodisc Correlations
Science and Technology Videodisc Series, Teacher Guide
Basic Skills Transparencies and Masters
Basic Concepts Transparencies and Masters
Reteaching Transparencies and Masters
Critical Thinking/Problem Solving
Spanish Resources
Concept Mapping
Biolab and Minilab Worksheets
Tech Prep Applications
Great Developments in Biology
Biology Projects
Computer Test Bank
IBM/APPLE/MACINTOSH
English/Spanish Audiocassettes
CD-ROM Multimedia System
Videodisc Program
MindJogger Videoquizzes

Glencoe Science Professional Series:
Exploring Environmental Issues
Performance Assessment in the Biology Classroom
Alternate Assessment in the Science Classroom
Cooperative Learning in the Science Classroom

Glencoe/McGraw-Hill
A Division of The **McGraw-Hill** *Companies*

Send all inquiries to:
Glencoe/McGraw-Hill
936 Eastwind Drive
Westerville, OH 43081

ISBN 0-02-825431-7
Printed in the United States of America.

3 4 5 6 7 8 9 10 11 12   071/046   07 06 05 04 03 02 01 00 99 98

Authors

Alton Biggs teaches biology at Allen High School, Allen, Texas, where he also served as Science Department Chairperson for 16 years. He has a B.S. in Natural Sciences and an M.S. in Biology from Texas A & M University-Commerce. Mr. Biggs received NABT's Outstanding Biology Teacher Award for Texas in 1982 and 1995, he was president of the Texas Association of Biology Teachers in 1985, and in 1992 was the president of the National Association of Biology Teachers. Mr. Biggs has led several excursions to the Galápagos Islands, the Amazon River basin, and Australia. He is also coauthor of *Glencoe Life Science.*

Chris Kapicka is a biology professor at Northwest Nazarene College, Nampa, Idaho, and does collaborative heart research with the Veteran's Administration Hospital in Boise, Idaho. Previously, she was a biology teacher at Boise High School, Boise, Idaho. She has a B.S. in Biology from Boise State University, an M.S. in Microbiology from Washington State University, and a Ph.D. in Cell Physiology and Pharmacology from the University of Nevada-Reno. Dr. Kapicka received the Presidential Award for Excellence in Science Teaching in 1986, NABT's Outstanding Biology Teacher Award in 1987, and the Sigma Xi Distinguished Science Teaching Award in 1987.

Linda Lundgren teaches biology at Bear Creek High School, Lakewood, Colorado. She has a B.A. in Journalism and Zoology from the University of Massachussets and an M.S. in Zoology from Ohio State University. In 1988, Mrs. Lundgren was awarded a research fellowship as a visiting scientist at the National Renewable Energy Laboratory, Golden, Colorado. In 1991, she was named Colorado Science Teacher of the Year by the Colorado Association of Science Teachers. In 1996, Mrs. Lundgren was a Research Associate in the Mathematics, Science, and Technology Program at the University of Colorado at Denver.

Contributing Authors

Daniel Blaustein
Science Writer
Evanston, IL

Rebecca Johnson
Science Writer
Sioux Falls, SD

Devi Mathieu
Science Writer
Sebastopol, CA

Content Specialists

John Ash
Department of Biochemistry
The Ohio State University
Columbus, OH

Timothy J. Cain
Department of Cell Biology, Neurobiology, and Anatomy
The Ohio State University
Columbus, OH

Joe W. Crim, Ph.D.
Zoology Department
University of Georgia
Athens, GA

Mary Downey Coyne, Ph.D.
Department of Biological Sciences
Wellesley College
Wellesley, MA

R.E. Duhrkopf, Ph.D.
Department of Biology
Baylor University
Waco, TX

David G. Futch, Ph.D.
Department of Biology
San Diego State University
San Diego, CA

Mary Garvin, Ph.D.
Department of Biological Sciences
University of Notre Dame
South Bend, IN

Christopher W. Gregory
Department of Cell Biology, Neurobiology, and Anatomy
The Ohio State University
Columbus, OH

Leland G. Johnson, Ph.D.
Biology Department
Augustana College
Sioux Falls, SD

John J. Just, Ph.D.
School of Biology
University of Kentucky
Lexington, KY

Robert A. Klips, Ph.D.
Department of Plant Biology
The Ohio State University
Marion, OH

Randy Kraus, Ph.D.
City Lab
Boston University
School of Medicine
Boston, MA

Katharine B. Payne
Sapsucker Woods Ornithological Garden
Ithaca, NY

Peter N. Petersen, Ph.D.
Adrian C. Wilcox High School
Santa Clara, CA

Fred D. Sack, Ph.D.
Department of Plant Biology
The Ohio State University
Columbus, OH

Malcolm D. Schug, M.S.
Department of Zoology
The Ohio State University
Columbus, OH

E. Peter Volpe, Ph.D.
Department of Basic Medical Sciences
Mercer University School of Medicine
Macon, GA

Sandra S. West, Ph.D.
Department of Biology
Southwest Texas State University
San Marcos, TX

Ronald W. Wilson, Ph.D.
Botany and Plant Pathology Department
Michigan State University
East Lansing, MI

Leslie Wolf, Ph.D.
Genzyme Tissue Repair
Cambridge, MA

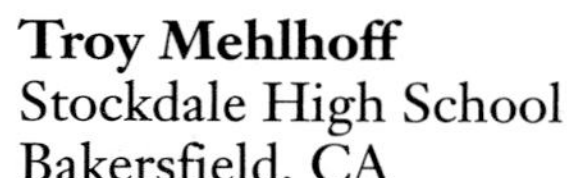

Teacher Reviewers

L. Anthony Black, M.S.
Westbrook High School
Westbrook, ME

Martha B. Buchanan
Freedom High School
Morgantown, NC

Thomas Custer, Ph.D.
Science Coordinator
Anne Arundel Public Schools
Annapolis, MD

Mary Davidson, Ph.D.
Mississippi School
for Mathematics and Science
Columbus, MS

Steven B. Henderson
S.H. Rider High School
Wichita Falls, TX

Peggy R. Henry-Watkins
Providence Senior High School
Charlotte, NC

Fleurette Hershman, M.S.
Montclair College Prep. School
Van Nuys, CA

Karen Hewitt
Coldspring/Jones High School
Coldspring, TX

Barbara B. Hopson
Menchville High School
Newport News, VA

Rose Anne Mannina
Carlynton High School
Carnegie, PA

Troy Mehlhoff
Stockdale High School
Bakersfield, CA

Ivey Gayle Nichols
New Hanover High School
Wilmington, NC

Patsye D. Peebles
LSU Laboratory School
Louisiana State University
Baton Rouge, LA

William F. Rigney
Marlborough High School
Marlborough, MA

G. Stuart Rogers
Tioga Central High School
Tioga Center, NY

Demby Sejd, M.S.E.D.
McLean High School
McLean, VA

William V. Soranno, M.A.
Union High School
Union, NJ

H. Samuel Stewart
Abington Senior High School
Abington, PA

Thomas Phil Talbot
Skyline High School
Salt Lake City, UT

Marilyn Thibodeaux, M.Ed.
Hallettsville High School
Hallettsville, TX

Cynthia Varnado
John Marshall Metro High School
Chicago, IL

Rod Waltermyer
William Penn Senior High School
York, PA

W. Fred H. Winey, M.A.
Dartmouth High School
North Dartmouth, MA

Nancy B. Wynne, M.A.
J.H. Rose High School
Greenville, NC

Contents in Brief

FOUNDATION

ENRICHMENT

Contents

Contents

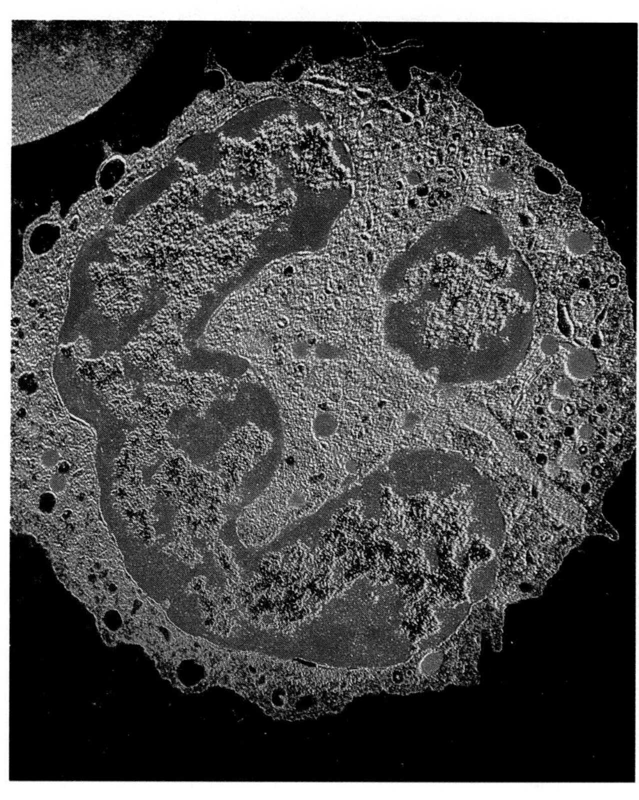

Contents

Contents

Contents

Contents

Contents

Contents

Appendices

Skill Handbook

BioLabs

Working in the lab is often the most enjoyable part of biology. BIOLABS *give you an opportunity to act like a biologist and develop your own plans for studying a question or problem. Whether you're designing experiments or following well-tested procedures, you'll have fun doing these lab activities.*

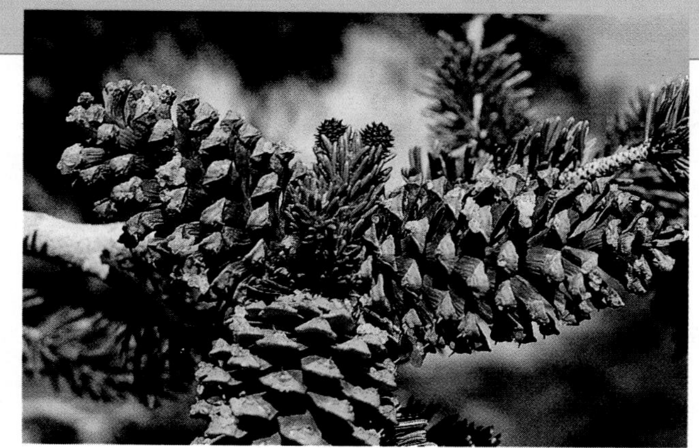

Chapter

MiniLabs

Do you often ask how, what, or why about the living world around you? Sometimes it takes just a little time to find out the answers for yourself. These short activities can be tried on your own at home or with help from a teacher at school. When you're feeling inquisitive, try a MINILAB.

ThinkingLabs

Sharpen up your pencil and your wit because you'll need them. THINKING LABS *offer a unique opportunity to evaluate another scientist's experiments and data without lab bench mess.*

Interdisciplinary Connections

It may not have occurred to you that biology is connected to all your courses. Learn in these features how biology is connected to art, literature, and other subjects that interest you.

MATH

PHYSICS

HEALTH

CHEMISTRY

EARTH SCIENCE

people in biology

What are biologists like? What excites them about biology, and how did they ever get interested in biology in the first place? Read these interviews with* PEOPLE IN BIOLOGY, *and find out what makes a biologist "tick."

Chapter

Focus On

Some topics of biology deserve more attention than others because they're unusual, informative, or just plain interesting. Here are several features that* FOCUS ON *topics from dinosaurs to your brain.

Chapter

A BROADER VIEW

The seemingly small events, anecdotes, and details found in the following articles will give you A BROADER VIEW *and better understanding of the big picture.*

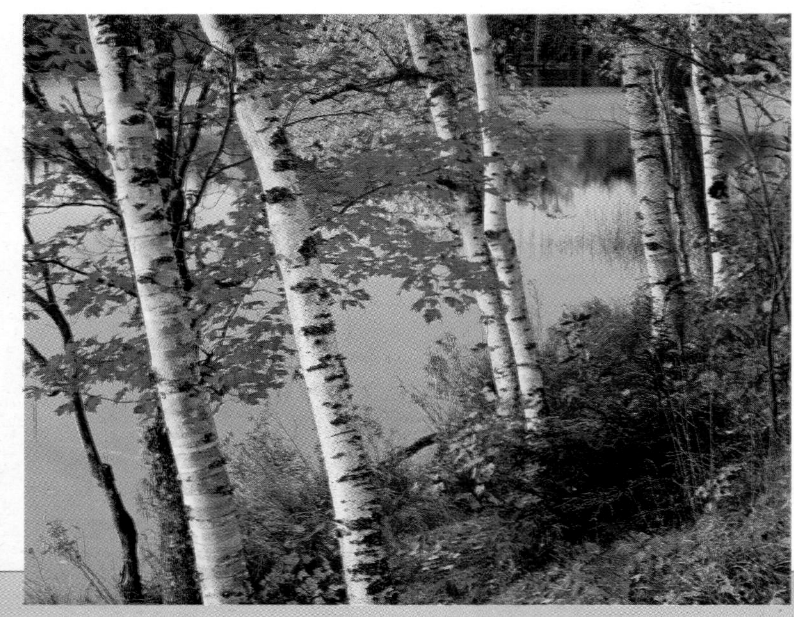

BIOLOGY & SOCIETY

How does biology impact you and society? Consider some of the issues that have been covered by the media. Take this opportunity to understand the many different sides of issues, learn how technology may change your life, and develop your own viewpoints with BIOLOGY & SOCIETY.

ISSUES

BIOLOGY & SOCIETY

ETHICS

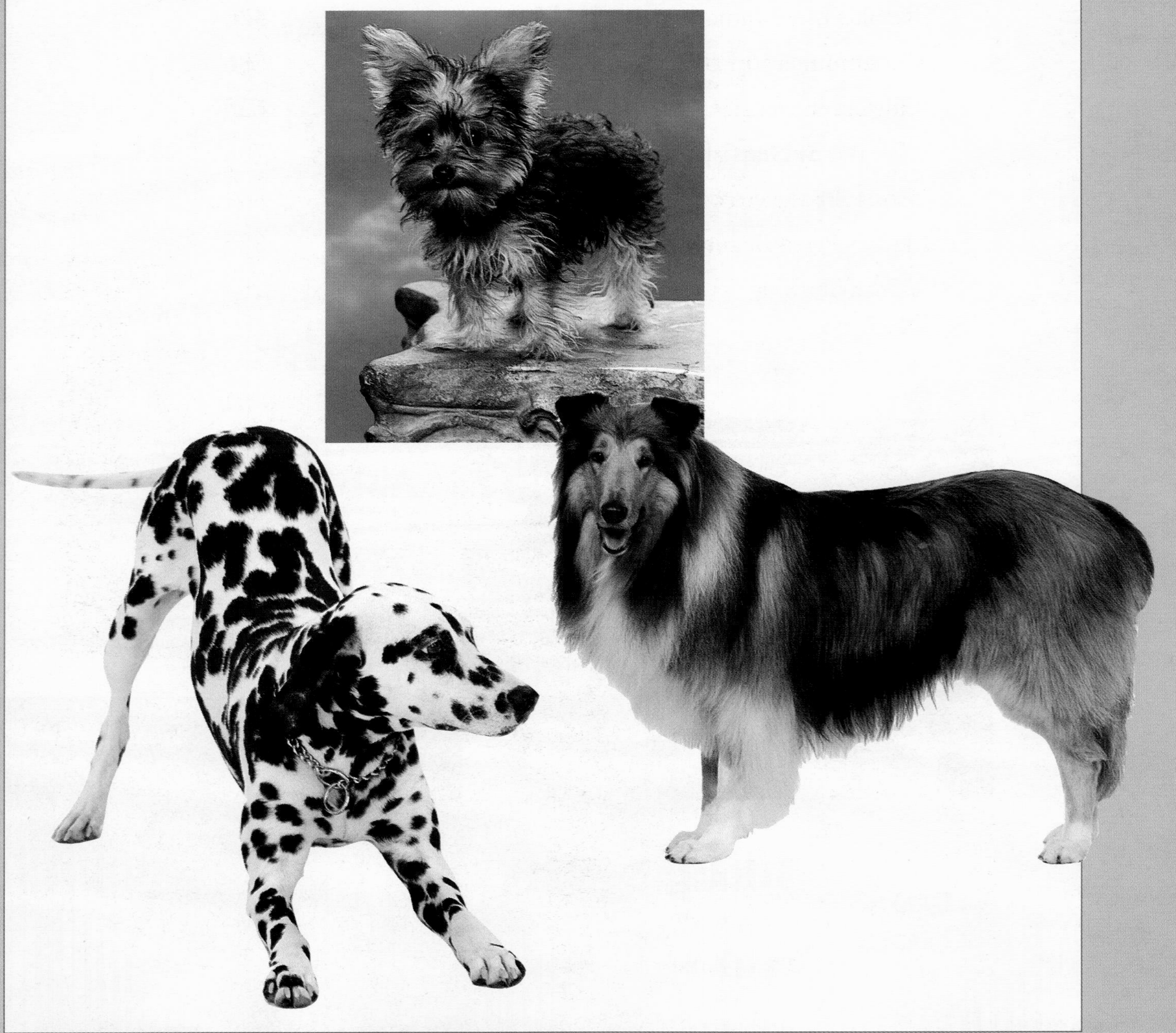

TECHNOLOGY

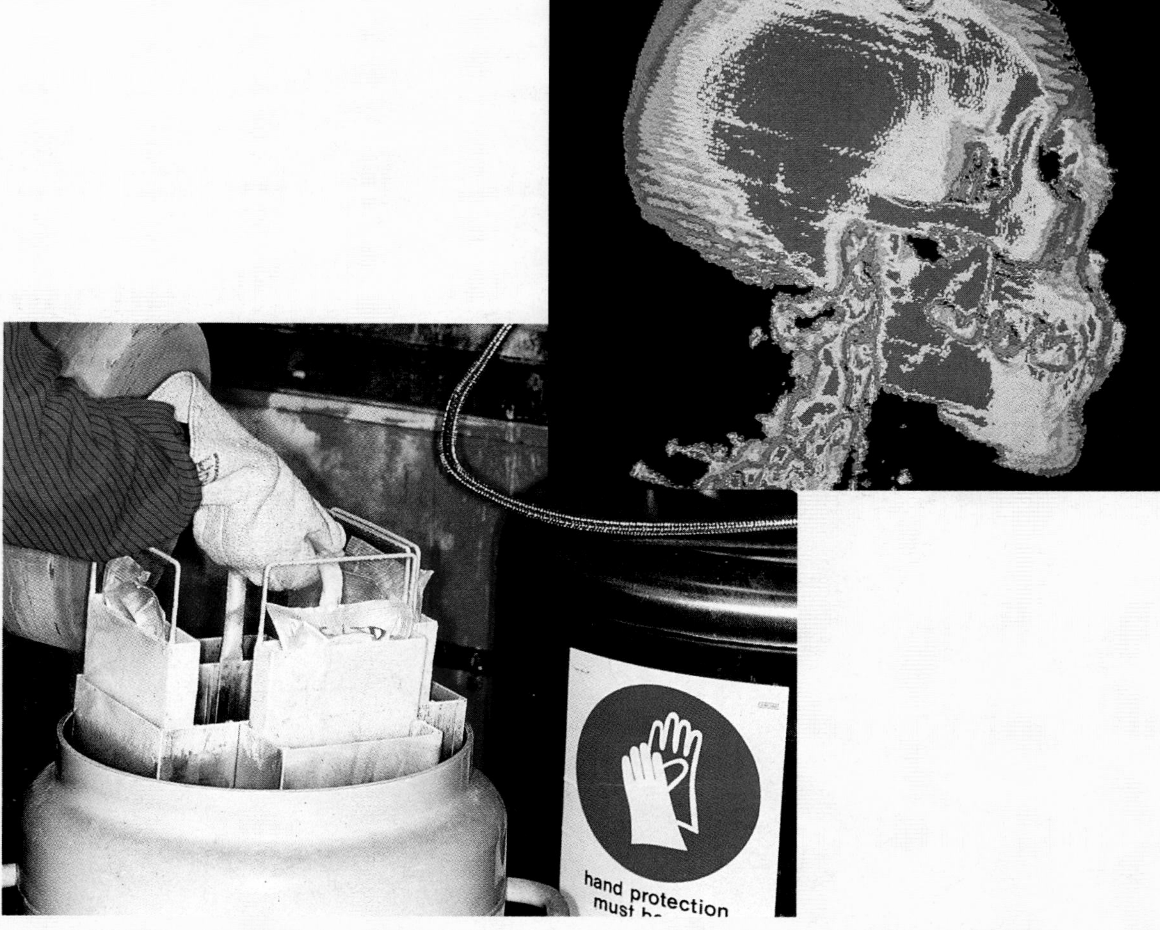

Dear student biologist:

One thing all humans share is a curiosity about the living world. Questions puzzle us continually, like why does a dog turn in circles and scratch the carpet before it lays down, or do horses dream? You may even ponder questions such as why the ozone is so important, why AIDS is such a devastating disease, and even how life on Earth began. Questions like these have been considered by biologists for centuries. As in all other sciences, the answer to one question often leads to another. Biology at work is the probing and searching for answers to question after question.

As you watch television or read the newspaper, you will hear about new discoveries in biology—such as how cells work, new possible cures or treatments for AIDS, and new genetic treatments for diseases. Throughout this year, we challenge you to discover biology in action around you. As you ride in a car, watch the landscape go by and think about how living things are related to each other—again, biology at work. Perhaps you will watch a hummingbird drinking the nectar from a petunia flower, or a squirrel picking up an acorn and scurrying up the trunk of a large oak tree. As you watch a basketball game, notice biology at work as the basketball players rush down the gym floor while excited fans cheer them on.

Our goal is that you experience the joy and awe of learning about the living world. Take time to marvel at the complexity of living things, as well as the beauty in biological forms. As you consider the tiniest of bacteria to the largest dinosaur, remember that all life is connected.

We hope that you do not get caught up in just learning the terms and facts of biology, but can appreciate the beauty of the processes and orderliness of living things. Continue to ask "Why...?", "How...?", and "What if...?". Seeking answers to questions about the living world will help us understand how we are connected to the world around us. We wish you a successful year in your course of biology. As you continue the study of life, we challenge you to keep asking questions and searching for answers. That is what the science of biology is all about.

Sincerely,

Alton L. Biggs   Chris L. Kapicka   Linda Lundgren

Authors of *Biology: The Dynamics of Life*

UNIT 1 What Is Biology?

Unit Overview

This unit includes two chapters that introduce students to the nature, excitement, and methods of biology. In Chapter 1, students are introduced to a newly discovered form of hydroid and how biologists determined the living nature of these organisms. In Chapter 2, interest in how science attempts to answer questions and solve problems concerning the world of life is sparked as students trace the progress of a study that deals with elephant communication.

Theme Development

All the themes students will encounter during their study of biology are introduced in this unit. In Chapter 1, students begin to explore the characteristics biologists use to identify something as living, and become acquainted with the themes of *systems and interactions*, *homeostasis*, *unity within diversity*, *evolution*, *energy*, and *the nature of science*. The nature of science theme is expanded upon in Chapter 2 as students examine the methods of inquiry used by a scientist studying elephant communication.

UNIT 1 What Is Biology?

The eagle's sharp eyes probe the lake, searching for signs of food. Without warning, the great bird of prey hurtles toward a movement in the water. Seconds later, the eagle flies back to its nest, a small fish gripped tightly in its powerful talons. What factors are responsible for this single dramatic moment in the life of an eagle?

Answering questions about the natural world involves methods of investigation and discovery that are at the core of biology. As biologists go about their work, they use scientific processes, special tools and techniques, and their knowledge of the basic principles of biology and other sciences.

An eagle's great strength comes from the energy in its food—a steady supply of fish and other small prey. In what other ways are eagles and other living things dependent on their environment?

2

Advance Planning

Chapter 1

- Purchase Instant Sea Life animals and gather stirring rods, beakers, and thermometers for the Biolab.

Chapter 2

- Purchase, or grow from seed, plant seedlings for the Biolab. Gather plastic trays, foam cups, balances, metric rulers, and graduated cylinders.
- Order flower, leaf, or insect specimens (living or preserved) for the Minilab on page 30.
- Purchase four different brands of paper towels for the Minilab on page 42.

The ability to single out a movement in the water from a great height is just one adaptation that has evolved among all birds of prey. How do biologists find out which adaptations are important to a species's survival?

The idea for the invention of flying machines almost certainly came from observations of feathers and birds in flight. What are some other applications of biology in our daily lives?

Unit Contents

UNIT 1

Introducing the Unit

Ask students to examine the position of the eyes on the eagle's head. Point out that like most predatory birds an eagle's eyes are located on the front of its head, rather than on the sides. Ask students to speculate why having eyes at the front of the head is important to birds of prey. Next ask students to examine the shape of the eagle's talons. Ask students to describe how the talons of the eagle are suited to capturing food such as fishes and mice.

Motivational Activity

Discussion: Explain that biology is a part of students' everyday lives. To help emphasize this point, have students work in cooperative groups to make lists that show how birds are part of their lives. Encourage students to include birds' uses as food, their value for recreation, ways people use feathers, and how birds are represented in literature, cartoons, as symbols, in common sayings, in advertising, and in art, song, dance, carvings, stamps, and coins. L1 COOP LEARN

Unit Project

Community Involvement: Ask students to find out about birds that may be declining in number in your area. You might have them contact a local chapter of the Audubon Society, the nearest United States Fish and Wildlife Service office, or your state wildlife agency. Ask them to determine if any local birds are declining due to destruction of their habitat such as that resulting from the draining of wetlands, clearing of forests or meadows, or pollution problems such as oil spills and chemical dumping. When they determine what the local problems are, ask students to design a program that will help prevent a further decline in the species. Suggest that they include in their program a letter, written to elected officials, which explains the problem and their recommendations for how it might be solved. L1 COOP LEARN

Chapter Organizer

SECTION	OBJECTIVES	ACTIVITIES/FEATURES
1.1 What Is Biology? National Science Standards: UCP.2; A.1, A.2; E.1, E.2; F.3, F.4; F.6; G.1	1. **Identify** the topics studied in biology. 2. **Recognize** some possible benefits from studying biology.	**Art Connection:** *Nesting Geese,* p. 8 **Focus On** Biologists in Action, p. 10
1.2 What Is Life? National Science Standards: UCP.1-5; A.1, A.2; C.4-6; G.1-3	3. **Summarize** the characteristics of living things. 4. **Relate** the characteristics of life to specific examples in organisms. 5. **Recognize** the major themes of biology.	**Biolab:** Design Your Own Experiment–How does temperature affect a living thing?, p. 18

ACTIVITY MATERIALS

BIOLAB	ALTERNATE LAB
page 18 plastic cups or beakers ice water/warm water thermometer stirring rods or coffee stirrers Instant Sealife toy capsules	**page 16** 2 mussels 2 *Elodea* stems 4 test tubes test-tube holder straw bromothymol blue solution dropper bottle white paper distilled water

Chapter 1 Biology: The Science of Life

TEACHER CLASSROOM RESOURCES

Reproducible Masters	Transparencies
Section Focus Master 1: Biology in Everyday Life L1 SAE **Reinforcement and Study Guide,** p. 1 L1 **Laboratory Manual:** Preparing an Insect Collection, pp. 1-4 L2 **Content Mastery,** pp. 1-4 L1	
Section Focus Master 2: Characteristics of Life L1 SAE **Reinforcement and Study Guide,** pp. 2-4 L1 **Biolab and Minilab Worksheets,** pp. 1-2 L1 **Concept Mapping:** Characteristics of Living Things, p. 1 L1 **Critical Thinking/Problem Solving:** Looking for Life on Mars, p. 1 L3	**Reteaching Transparency #1:** Characteristics of Life L1 SAE

ASSESSMENT MATERIALS

Chapter Assessment, pp. 1-6
Performance Assessment in the Biology Classroom, p. 1
MindJogger Videoquiz
Alternate Assessment in the Science Classroom
Computer Test Bank

Spanish Resources SAE
English/Spanish Audiocassettes SAE
Cooperative Learning in the Science Classroom COOP LEARN
Lesson Plans
Great Developments in Biology: Biological Tools and Technologies L1 SAE

KEY TO TEACHING STRATEGIES

- L1 Level 1 activities should be within the ability range of all students including those with learning difficulties.
- L2 Level 2 activities are within the ability range of average to above-average students.
- L3 Level 3 activities are designed for the ability range of above-average students.
- SAE SAE activities should be within the ability range of Students Acquiring English.
- COOP LEARN Cooperative Learning activities are designed for small group work.
- P These strategies represent student products that can be placed into a best-work portfolio.
- These strategies are useful in a block scheduling format.

GLENCOE TECHNOLOGY

The following multimedia resources are available from Glencoe.

Biology: The Dynamics of Life
CD-ROM SAE
Videodisc Program

The Infinite Voyage Series
Miracles by Design

The Secret of Life Series
What's in Stetter's Pond?

Science and Technology Videodisc Series (STVS)
Human Biology
Patient Simulator
Animals
Blood Fluke Life Cycle
Plants & Simple Organisms
Simple Forms of Life in the Antarctic

CHAPTER

1 Biology: The Science of Life

Chapter Overview

Chapter 1 provides students with an overview of the concerns of biology. Students learn that biology is the organized study of living things and their interactions with the environment. They learn that biology can help them discover more about themselves, other living things, and the importance of the environment.

Students study the features that characterize all living things: organization, reproduction, growth and development, and the ability to adjust to the environment. Students are also introduced to the six general themes that are woven into the study of biology: energy, homeostasis, unity within diversity, systems and interactions, evolution, and the nature of science.

Key Terms

adaptation
biology
development
energy
environment
evolution
growth
homeostasis
organism
organization
reproduction
response
species
stimulus

4

Learning Styles

Look for the following logo for strategies that emphasize different learning modalities. LS

Kinesthetic	Activity, p. 7; Meeting Individual Needs, p. 8
Visual-Spatial	Portfolio, pp. 7, 15; Activity, pp. 12, 17; Meeting Individual Needs, p. 13; Microscope Activity, p. 14; Display, p. 15
Interpersonal	Cultural Diversity, p. 14
Intrapersonal	Student Journal, p. 6
Linguistic	Student Journal, p. 20

LS

CHAPTER

1 Biology: The Science of Life

The world's oceans teem with many unusual animals, but perhaps none are as unusual as the sea slug. Often brilliantly colored, sea slugs appear as floating ribbons to the observer, but these slow-moving, delicate-looking relatives of snails and clams are anything but helpless. All known species of sea slugs are carnivores and prey on animals such as sea anemones, corals, and sponges. A remarkable characteristic of some species of sea slugs is the ability to incorporate the stinging cells of their prey into their own bodies. They then use these cells for defense.

You won't always find sea slugs living alone. Several species interact in a curious relationship with another organism. Living within the skin of some species of sea slugs are microscopic algae—green, single-celled organisms that produce some of the sugars and other nutrients sea slugs need for their activities.

In some ways, the world of the sea slug mirrors what you can expect as you study biology—a world of fantastic living things, unusual behaviors, and unexpected relationships. Welcome to the world of biology!

Chapter Preview

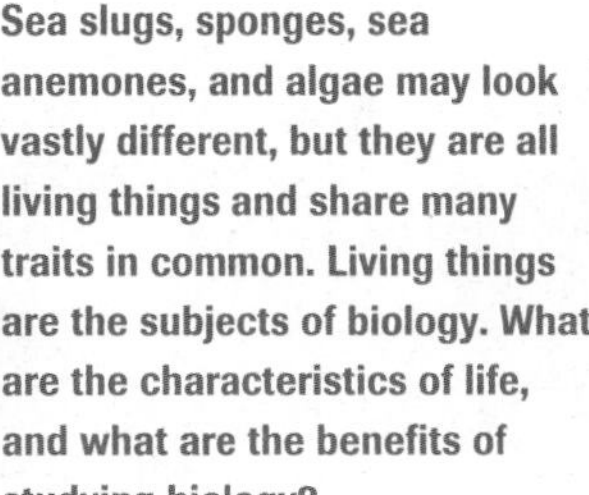

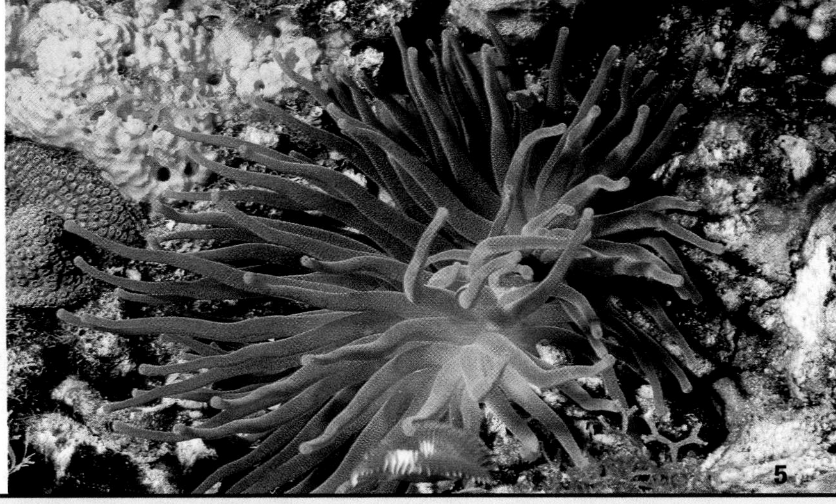

Sea slugs, sponges, sea anemones, and algae may look vastly different, but they are all living things and share many traits in common. Living things are the subjects of biology. What are the characteristics of life, and what are the benefits of studying biology?

5

CHAPTER

1

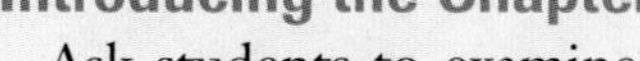

Introducing the Chapter

Ask students to examine the sea slug. Ask them how they can tell the sea slug is alive. Draw attention to the sponges and sea anemone. Ask students if they know what kind of living things these are and where they would find them. *Students are likely to suggest they are ocean plants.* Explain that, like the sea slug, the sponges and sea anemone are animals.

Theme Development

Students are introduced to six major themes of biology: *systems and interactions, homeostasis, the nature of science, unity within diversity, evolution, and energy.*

Concept Check

Students will learn how biology is relevant to their lives. They will see that many questions they have about living things and the environment can be answered through the study of biology.

interNET CONNECTION

Follow the link for this chapter on the Glencoe Homepage at **www.glencoe.com/sec/science** to find out more about the science of biology.

Assessment Planner

Choose assessment strategies from the following pages to evaluate the progress of your students.

Assess, pp. 8, 21
Alternate Lab, pp. 16-17
Portfolios, pp. 7, 15
Biolab, pp. 18-19
Chapter Review, pp. 22-23

SECTION
1.1

Prepare

Key Concepts

In this section, biology is presented as an organized way of studying living things. Students will recognize the benefits of studying biology.

Block Scheduling

Look for this symbol for strategies that are useful in a block scheduling format. For more information on block scheduling, refer to Section 1.1 in the **Lesson Plans** booklet.

Materials

- Obtain squash seeds, bubble wands and bubble liquid, pieces of fresh fruit, test tubes, cut flowers, warming candles, and jar lids for the Focus Activity.

1 Focus

Bellringer

Before presenting the lesson, display **Section Focus Master 1** on the overhead projector and have students answer the accompanying questions. L1 SAE

SECTION
1.1 What Is Biology?

Section Preview

Objectives

Identify the topics studied in biology.

Recognize some possible benefits from studying biology.

Key Term

biology

Why don't people get goose bumps on their faces? How do plants that make seedless fruits reproduce? Why are there so many different kinds of insects? For humans who share this planet with an amazing diversity of living things, the natural world often poses questions that arouse our curiosity. More often than not, such questions have simple explanations, but sometimes nature defies common sense. Whether nature's puzzles are simple or complex, many may be explained with the concepts and principles of biological science.

The Science of Biology

People have always been curious about living things—how many different species there are, where they live, what they are like, how they relate to each other, and how they behave. These and many other questions about life are answerable, and the concepts, principles, and theories that allow people to understand the natural environment form the core of **biology,** the study of life. What will you, as an amateur biologist, learn about in your study of biology?

A key aspect of biology is simply learning about the different types of living things around us. With all the

Figure 1.1

None of the creatures that you read about in works of fantasy and science fiction are as unusual as some of the kinds of living things that actually live on Earth.

Mudskippers are a type of fish that crawl over land on strong, muscular fins.

Program Resources

Section Focus Master 1 L1 SAE
Reinforcement and Study Guide, p. 1 L1
Laboratory Manual, pp. 1-4 L2

STUDENT JOURNAL

Investigating New Life Forms Ask students to find out about some of the most recent discoveries of new species. Ask them to report in their journals about how and where these new life-forms were discovered and who discovered them. Ask them to discuss the significance of each discovery. L2 LS

Figure 1.2

Questions about the features and behaviors of living things can sometimes be answered only by finding out about their interactions with their surroundings.

The banded pipefish can hide horizontally in its environment of seaweeds. Its stripes blend in with the stems of the seaweeds.

This Australian frog burrows underground and encases itself in a waterproof envelope to prevent water loss during dry weather.

This Scamman's spring beauty is able to survive at high altitudes, in inhospitable-looking environments.

facts in biology textbooks, you might think that biologists have answered almost all the questions about life. Of course, this is not true. Millions of life forms haven't even been named yet, let alone studied. The ones that are studied are often distinctive looking organisms that have unusual behaviors, *Figure 1.1.*

When studying the different types of living things, you'll ask what, why, and how questions about life. That is, you may ask, "What are some of the features of this living thing? Why does this living thing possess such features? How do these structures work?" By asking such questions, you will develop general principles and rules, which indicate that, as strange as some forms of life appear to be, there is order in the natural world.

Biologists study the interactions of life

One of the most general concepts in biology is that living things do not exist in isolation; they are all functioning parts in the delicate balance of nature. As you can see in *Figure 1.2*, living things depend upon other living and nonliving things in a variety of ways and for a variety of reasons.

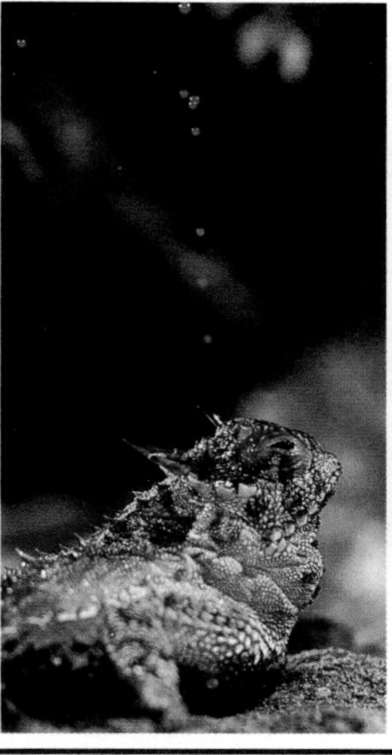

The meat-eating sundew plant lures insects with a powerful fragrance and then traps them on hundreds of sticky hairs.

When threatened, the Texas horned lizard raises its blood pressure and squirts blood out of the corners of its eyes.

Activity

LS **Kinesthetic** Give students in their groups the following materials: a squash seed, a bottle of bubbles and a bubble wand, a fresh fruit such as a tomato, a cut flower in a test tube filled with water, and a warming candle in a jar lid.

Light the candles for each group and instruct students to observe the flame. Next, instruct them to blow a few bubbles. Ask them to observe each of the other objects. Have students decide whether each item is alive and have them explain their reasons. Discuss responses as a class. *Students will probably know that the bubbles and warming candles are not alive, but because they both move, students may say they are alive.*

You may find it necessary to distinguish between nonliving things and once-living or dead things as part of the discussion.

2 Teach

GLENCOE TECHNOLOGY

Videodisc

The Infinite Voyage: Miracles by Design

Artificial Body Parts and Durability: The Zina Bethune Story (Ch. 3)

Burn Patients and Artificial Skin (Ch. 5)

PORTFOLIO

Identifying Habitats Provide students with pictures of unusual organisms they may not have seen. Ask them to glue each picture to a sheet of paper and speculate on the type of habitat in which each organism might live. Have them explain their choices and place their pictures and descriptions in their portfolios. L1 LS P

Art

Nesting Geese

Purpose

Students will examine the integration of environmental concerns and art.

Teaching Strategies

- Initiate a class discussion to find out what students know about environmental problems. Address the topic of habitat loss. COOP LEARN L1
- Elicit what might happen if the geese in the painting did not have an appropriate habitat for nesting. *They would likely not lay eggs and produce new young.*
- Ask students to research other Bateman paintings. Have students report on the environmental implications of the works.

Possible Answer

People enjoy such paintings, and they may raise people's awareness of the need to preserve habitats and animal populations.

3 Assess

Check for Understanding

Take students around the school grounds. Ask them to list in their notebooks all the different kinds of organisms they observe. They should indicate the characteristics they used to categorize each organism as a living thing.

Art

Nesting Geese

by Robert Bateman (1930–)

The best things in life—clean air, clean water, and the song of a bird—are not free anymore. They will be beyond price in the next century if we don't start rethinking our priorities now.

Internationally known painter Robert Bateman is an artist with an environmental message. Through his dramatic and powerful portrayals of wildlife in their natural settings, he raises our awareness of a vanishing world in conflict with the 20th century. Toronto-born Bateman travels the world to exhibit and lecture on the plight of our endangered planet, its species, and their habitats.

A vanishing world Critics note that many of Bateman's wildlife paintings have a disturbing sense of immediacy; the viewer often feels as if she or he has stumbled into the presence of some wild animal or bird and been given a rare glimpse of a fleeting moment in its world.

The painting "Nesting Geese" clearly shows the fierce territoriality of nesting Canada geese. Bateman's attention to the minute details of the feathers, as well as to the accuracy of the bird's behavior, gives the painting realism.

CONNECTION TO Biology

Bateman has donated many works to the conservationist cause, helping to raise millions of dollars for the preservation of endangered animals and their habitats. What other ways do the paintings of naturalists such as Bateman contribute to the preservation of our wildlife and their surroundings?

Why Study Biology?

Many people study biology simply for the pleasure of learning about the world of living things. As you've seen, the natural world is filled with examples of living things that can be amusing, or amazing, or with other examples that challenge one's thinking. Through your study of biology, you will come to appreciate the great diversity of species on Earth and the way each species fits into the dynamic pattern of life on our planet.

Biology and the future

Human existence depends on the existence of other living things on Earth. Living things are our supply of food and raw materials, such as wood, cotton, and oil. Plants replenish the essential oxygen in the air, ***Figure 1.3.*** Only with a thorough understanding of living things and the intricate web of nature can humans expect to understand the future health of our planet.

Figure 1.3

By understanding the interactions of living things, we are better able to impact our planet in a positive way.

Meeting Individual Needs

Learning Disabled Have students make a web from string to illustrate the intricate web of nature. On the web, have students place index cards and pictures that illustrate the organisms common to your area and their needs for life. L1 LS

The future of the human species holds many promises, but problems will also arise, *Figure 1.4*. For instance, scientists may one day be able to produce complex living things in their labs, but many species that already exist will become extinct. New agricultural techniques may help farmers see insects, droughts, and floods become problems of the past, but the number of people to be fed is expected to rise sharply in the near future. With a basic knowledge of biology, you'll be able to make critical choices relevant to our future on Earth.

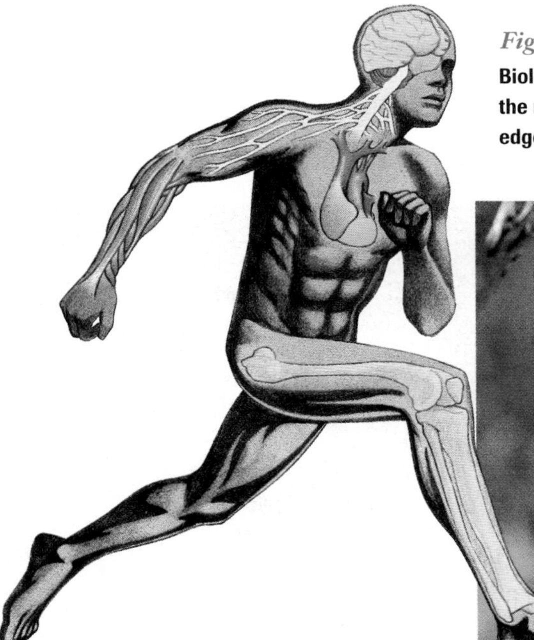

Figure 1.4

Biology will teach you about how humans function and fit in with the rest of the natural world. It will also equip you with the knowledge needed to handle any future biological problems of Earth.

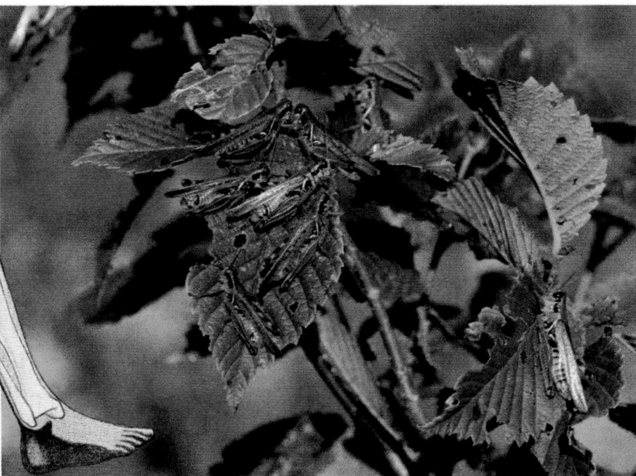

Section Review

Understanding Concepts

1. What kinds of questions are answerable using the science of biology?
2. Identify an important reason for studying biology.
3. Identify at least two ways in which other living things affect you every day.

Thinking Critically

4. Describe a situation you have been in where knowledge of human biology was important.

Skill Review

5. **Observing and Inferring** Choose a type of living thing that you either live with or have seen on television or in a magazine, and identify three questions that you would like answered about this living thing. For more help, refer to Thinking Critically in the *Skill Handbook*.

SECTION 1.1

Reteach

Have students look through newspapers and magazines to find examples of articles, advertising, and comics that are related to biology. Have students make a small scrapbook of the materials they discover and write a caption beside each that indicates its relevance to biology. L1

Extension

Explain that life is found in some unexpected places. Ask students to bring in dustballs from their homes. Have them place each sample in a small plastic bag that is labelled with the location from which it was taken. Students should then examine the dustballs under the microscope for dust mites. Ask them to compare the numbers of mites found in the samples from various locations. L2

Assessment

Portfolio: Give students a copy of a current article about an issue in medicine, the environment, or new agricultural techniques. Ask them to write in their portfolios how the article is related to biology. L1 P

4 Close

Discussion

Ask students in their groups to brainstorm a list of how biology is important to their lives. COOP LEARN

Answers to Section Review

1. Questions about life and the natural world.
2. For the pleasure of learning about the world of living things, or to appreciate the diversity of species on Earth and the way each species fits into the dynamic pattern of life on our planet.
3. Enjoyment of pets, food from plants and animals, and clothing from plants and animals.

Thinking Critically

4. Students may describe a situation requiring medical care, knowledge of human behavior and the nervous system, or knowledge of human nutrition.

Skill Review

5. As an example, students may want to know more about sharks: whether they are all aggressive predators, whether they can be tamed and taught like whales, and whether they sleep.

Focus On

Biologists in Action

Purpose

Students will learn how biologists work and will be introduced to the variety of strategies and equipment they use and the places they go to conduct their research.

Teaching Strategies

- Before students read this feature, ask them to generate a list of careers and occupations that are related to biology. Write their responses on the chalkboard.

Visual Learning

Figure Caption Question Why do you think experienced Antarctic researchers always bring along several sets of spare parts for all of their instruments? *If a piece of equipment breaks down, or is lost, research can continue using the replacement equipment.*

Figure Caption Question What advantage can you see in being able to analyze data at the site where you carried out an experiment or made your observations, before returning to your office or laboratory? *If data need to be reevaluated due to inconsistencies, this can be done on-site. Such on-site analysis can save time and money.*

Focus On

Biologists in Action

Biologists who do research follow the general methods followed by all scientists. But the ways in which they gather, record, and analyze data range from simple to high-tech. The environments in which biologists carry out their investigations are as varied and diverse as Earth itself.

Biologists work in a variety of environments Some biologists work in clean, air-conditioned laboratories. But others carry out their research in the field, which includes places ranging from mountaintops to the ocean floor—and everything in between. Field work often involves doing experiments and recording observations in remote and inhospitable places. In some cases, special equipment is essential in order to venture into environments where human beings would not normally be able to survive.

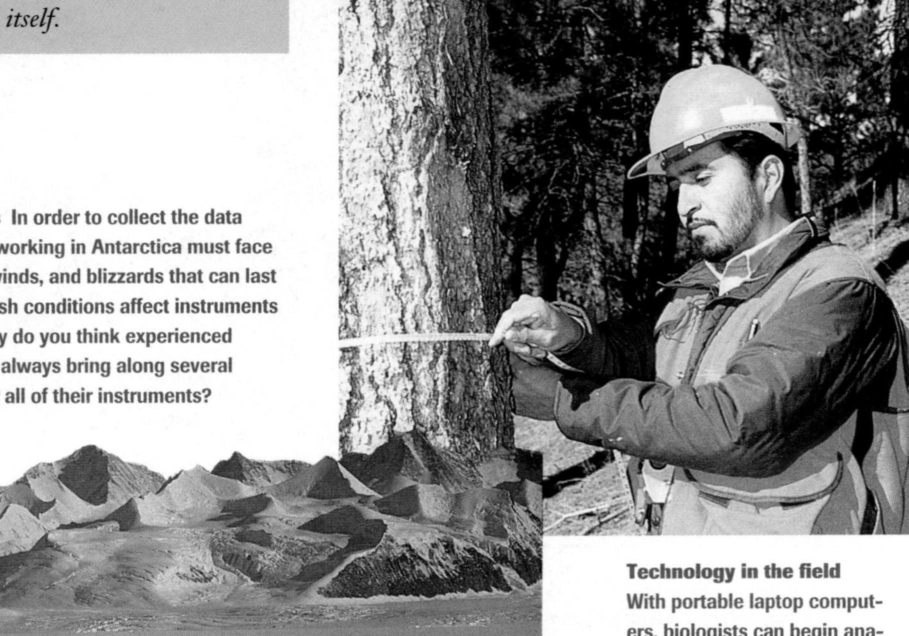

Arctic environments In order to collect the data they need, biologists working in Antarctica must face extreme cold, fierce winds, and blizzards that can last many days. These harsh conditions affect instruments as well as people. Why do you think experienced antarctic researchers always bring along several sets of spare parts for all of their instruments?

Technology in the field With portable laptop computers, biologists can begin analyzing data while still in the field. What advantage can you see in being able to analyze data at the site where you carried out an experiment or made your observations, before returning to your office or laboratory?

GLENCOE TECHNOLOGY

Videodisc

Biology: The Dynamics of Life
Disc 1, Side 1
Biologists at Work (Ch. 2)

Radio tracking How far does a wolf travel in a day, a month, or a year? By using instruments like this radio-transmitter collar, biologists can track animals from a great distance over long periods of time. The data collected in this type of experiment can be used to produce maps that show daily movements and seasonal migration patterns. Why do you think it would be important to use long-life batteries in such a radio-transmitter collar?

Research in the laboratory A microbiologist counts bacterial colonies growing in a culture dish in which antibiotics are present. Different antibiotics inhibit the growth of different kinds of bacteria. By counting the number of colonies in the dish, this researcher can tell how effective this antibiotic is against the bacteria. What would it mean if no colonies of bacteria were growing on this petri dish?

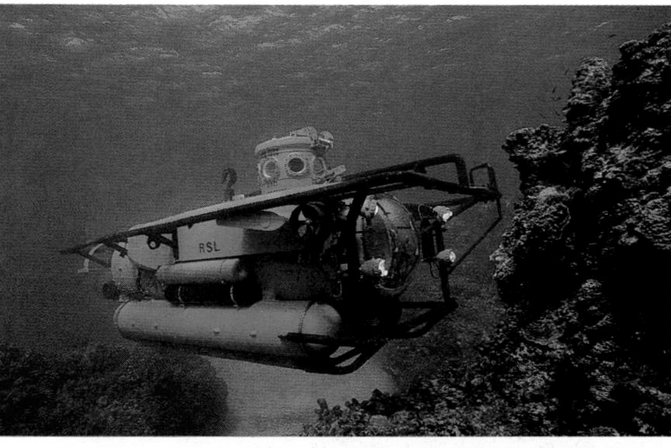

Deep-sea studies To study the deep sea, biologists rely on complex equipment like this deep-diving research submersible. These small subs can descend hundreds of meters below the surface. They are equipped with mechanical arms, claws, and suction tubes that are used to collect organisms that live in the darkness at these great depths. How might collecting organisms with a submersible be better than using nets lowered from ships at the surface?

EXPANDING YOUR VIEW

1. **Writing About Biology** In a short paragraph, summarize methods of research used by biologists.
2. **Going Further** Conduct library research on current studies of marine environments. What risks might be associated with some of these studies?

Focus On

Visual Learning

Figure Caption Question Why do you think it would be important to use long-life batteries in such a radio-transmitter collar? *It is important for the batteries to last the duration of the experiments. It also allows for more long-term data.*

Figure Caption Question What would it mean if no colonies of bacteria were growing on this petri dish? *The antibiotic would be effective in inhibiting growth of that type of bacteria.*

Figure Caption Question How might collecting organisms with a submersible be better than using nets lowered from ships at the surface? *The submersible can be focused in a specific area more easily than can a net.*

Answers to Expanding Your View

1. **Writing About Biology** Responses will vary but may include work in the field, work in science laboratories, or work in private industry. Tools may involve instruments used for measuring, observing, and recording data. Likely responses will summarize the information in this feature.
2. **Going Further** Students may cite risks of attacks by certain fishes such as sharks, or of stings of animals such as jellyfishes and sea anemones. There is also a risk of drowning and of having medical problems such as "the bends."

SECTION

1.2

Prepare

Key Concepts

This section presents the characteristics that living things have in common and the major themes of biology.

Block Scheduling

Look for this symbol for strategies that are useful in a block scheduling format. For more information on block scheduling, refer to Section 1.2 in the **Lesson Plans** booklet.

Materials

- For the Biolab, gather plastic cups or beakers, hot water, ice water, stirring rods or coffee stirrers, thermometers, and Instant Sea Life animals.

1 Focus

Bellringer

Before presenting the lesson, display **Section Focus Master 2** on the overhead projector and have students answer the accompanying questions. L1 SAE

Activity

LS **Visual-Spatial** Set up stations around your room with living and nonliving items: a rock with lichen, a bird's nest, a skeleton, a potted plant, a piece of uncooked steak, a few strands of hair, a drop of pond water with protozoans or algae focused on a microscope slide, a potato, a peanut, a stalk of celery, seeds, a goldfish in a bowl, and other interesting material you may have in your area. Ask students to go from one station to the next and decide if the specimens are or ever were alive. Have students justify their answers.

SECTION

1.2 What Is Life?

Section Preview

Objectives

Summarize the characteristics of living things.

Relate the characteristics of life to specific examples in organisms.

Recognize the major themes of biology.

Key Terms

organism
organization
reproduction
species
growth
development
environment
stimulus
response
adaptation
homeostasis
energy
evolution

It was just another hot summer day along Mexico's Sonoron coast when marine biologists Katrina Mangin and Pete Raimondi took time out from their research to view something curious. Here and there, amid the dense growth of acorn barnacles that normally covers the coastal rocks, were bare areas totally devoid of barnacles. These bare areas seemed to occur in a regular pattern. Upon closer examination, the scientists discovered a tiny, hard, coal-black spot in the center of each of the bare areas. Could these black spots be living things? Could they be causing the formation of the bare areas?

Characteristics of Living Things

Most people feel confident that they could identify a living thing from a nonliving thing, but sometimes it's not so easy. In identifying life, you might ask, "Does it move? Does it grow? Does it reproduce?" These are all excellent questions, but consider a flame. A flame can move, it can grow, and it can produce more flames. So are flames alive?

Biologists have formulated a list of characteristics by which we recognize living things. Sometimes, nonliving things have one or more of life's characteristics. But only when something has all of them can it be considered alive. Anything that possesses all of the characteristics of life is known as an **organism**, *Figure 1.5*. What are the characteristics of living things?

Figure 1.5
These plants are called *Lithops* from the Greek *lithos,* meaning "stone." Although they don't appear to be so, *Lithops* are just as alive as elephants. They both possess all of the characteristics of life.

Program Resources

Section Focus Master 2 L1 SAE
Reinforcement and Study Guide, pp. 2-4 L1
Biolab and Minilab Worksheets, pp. 1-2 L1
Concept Mapping, p. 1 L1
Critical Thinking/Problem Solving, p. 1 L3
Performance Assessment in the Biology Classroom, p. 1 L1 P
Reteaching Transparency 1 and **Master** L1 SAE

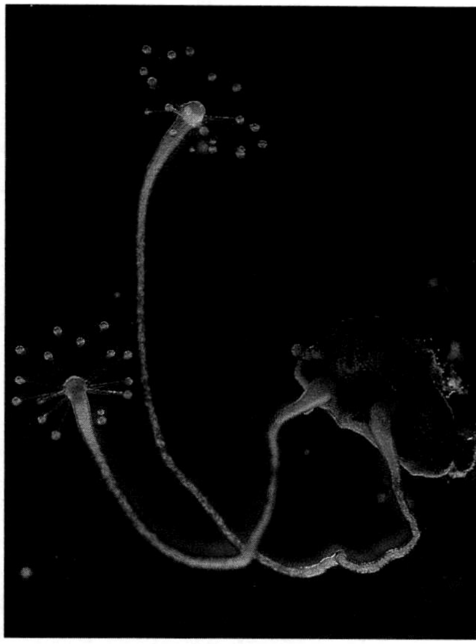

Figure 1.6

Like all organisms, hydroids show organization—they possess structures for every function. Each function that a hydroid performs, such as feeding or digestion, is vital to its existence, but these functions don't occur independently. In all organisms, body functions interact with one another to create a single, orderly, living system.

Living things are organized

When biologists Mangin and Raimondi were searching for signs of life in the little black spots they collected, one of the first things they did was put them under a microscope to observe their structure. That's because they knew that all living things show an orderly structure, or **organization.**

After careful scrutiny under the microscope, the black spots were identified as tiny relatives of jellyfishes, known as hydroids, *Figure 1.6.* As Mangin and Raimondi suspected, they found organization within these little animals. Like other organisms, hydroids have specialized parts that perform particular functions.

Although the living world is filled with many examples of organisms, life may be defined on the basis of several characteristics shared by all organisms. One of these characteristics is organization. Whether an organism is unicellular or multicellular, all structures and functions of the organism together form an orderly living system.

Living things make more living things

Perhaps the most obvious of all the characteristics of life is **reproduction,** the production of offspring, *Figure 1.7.* Organisms don't live forever. For life to continue, organisms must replace themselves. Biologists Mangin and Raimondi realized that their little black spots reproduced because they found many bare areas within the patches of barnacles. The tiny hydroids had produced more hydroids.

Figure 1.7

Living things reproduce to make more of their own kind. Organisms have evolved a variety of mechanisms for reproducing and ensuring the continuation of their own species. Some organisms, such as rabbits, tend to produce many offspring in one lifetime.

2 Teach

Different Viewpoints in Biology

Old-Growth Forests

The old-growth forests of the Northwest have an organization based on tiny rodents, a fungus, and giant trees.

For years, foresters removed rotting logs and fungus from the forest floor in an effort to clean up the forest. Recently, biologists have discovered that Douglas firs and western hemlocks depend on a fungus that lives on the logs. The fungus, in turn, depends upon the trees.

The fungus obtains sugar from the photosynthesis carried out by the trees. The fungus also provides a shield around the trees' roots and secretes antibiotics into the soil near the roots, thus preventing infection. Small voles eat the fungus and disperse the spores of the fungus throughout the forest. When an area of the forest is clear-cut and all old logs removed, the fungus dies and no new trees of this type can begin to grow. Some biologists say that loggers should remove only some of the mature trees and leave many live trees and rotting logs, so that the fungus and the voles that spread the spores can survive to renew the forest.

Meeting Individual Needs

Students Acquiring English/Learning Disabled Show students slides of organisms. Have students identify the different levels of organization represented by the organisms and environments shown in each slide.

L1 SAE LS

GLENCOE TECHNOLOGY

Videotape

The Secret of Life

Use the videotape *What's in Stetter's Pond: The Basics of Life* to introduce students to the main characteristics that define life.

Microscope Activity

Visual-Spatial Have students refer to the Skill Handbook to review the proper procedures for caring for and using a microscope. Then ask students to examine prepared slides of fertilization and development in a variety of organisms. You might have them examine the development of a chick, sea urchin, frog, and mammal embryo. Make sure that students understand that all these organisms developed from the union of sperm and egg. L1

Science, Technology, and Society

Cell Division

When cells reproduce, the hereditary material in the cell must be duplicated precisely. If a mistake is made, a checking protein scans the DNA and initiates repairs. This process might be compared to the spell-checker function of a computer.

If the scanning protein is defective, mistakes begin to occur and the cell's directions for control of cell division may no longer work. Such an occurrence may allow cells to divide uncontrollably, resulting in cancer.

In 1993, the gene that regulates development of a hereditary form of colon cancer was found. Researchers immediately went to work to develop a screening test to determine who has the defective gene. This test could save the lives of as many as one million people in the United States each year.

Figure 1.8

All life begins as a single cell. As cells multiply, organisms grow and develop and begin to take on the characteristics that identify them as members of a particular species of organisms.

desiccate:
desiccatus (L) to dry up
Desiccate means to dry up.

Reproduction is not essential for the survival of an individual organism. However, it is essential for the continuation of an organism's **species,** a group of similar-looking organisms that can interbreed and produce fertile offspring. If individuals in a species never reproduced, it would mean an end to that species's existence on Earth.

Living things change during their lives

The hydroids studied by Mangin and Raimondi were adults, but they didn't always look like this. A hydroid's life begins as a single cell, as do the lives of all organisms, and over time, it grows and takes on the characteristics of its species. **Growth** results in an increase in the amount of living material and the formation of new structures.

All organisms, such as the young rabbit in *Figure 1.8,* grow, and different parts of organisms may grow at different rates. Organisms made up of only one cell may change little during their lives, but they do grow.

On the other hand, organisms made up of numerous cells go through many changes during their lifetimes, *Figure 1.9.* Think about some of the structural changes your body has already undergone in your short life. All of the changes that take place during the life of an organism are known as its **development.**

Living things adjust to their surroundings

One of the things that really perplexed scientists Mangin and Raimondi about the new hydroids was the fact that these animals were able to survive full exposure to the sun for up to six hours at temperatures reaching up to 100°C. Until the discovery of the black spots, no known hydroid was able to withstand these dry, hot conditions without desiccating. How did they do that?

Figure 1.9

During an organism's first few months of life, development occurs rapidly. After maturity, little change occurs for many years.

CULTURAL DIVERSITY

The History of Biological Discovery

Interpersonal Have students work in groups of three or four and do basic library research to prepare a presentation about the history of biological science in a particular country. Student presentations can include a chronology of important biological science discoveries in the country, or they can be more specific accounts of a particular area of research. Presentations should include names of important scientists, details about the area of research, and how the research may have benefited society. Encourage students to incorporate posters, models, photographs, or videos in their presentations. L1 COOP LEARN

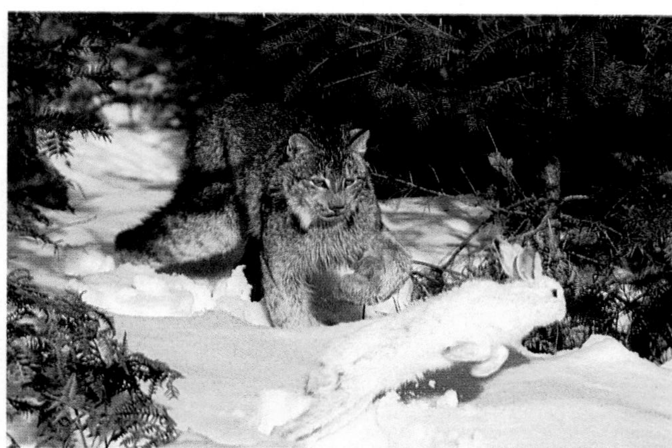

Figure 1.10

Living things are able to respond to stimuli and make adjustments to environmental conditions. Adaptations such as long hind legs enable rabbits to quickly avoid predators in their environment. Fur is an adaptation that allows rabbits and other mammals to regulate their body temperatures.

Living things live in a constant interface with their surroundings, or **environment,** which includes the air, water, weather, temperature, any organisms in the area, and many other factors, as shown in *Figure 1.10.* But anyone who has ever lived through the freezing temperatures of winter or has had to cross a busy street knows that it's not enough to just exist in an environment; sometimes you must adjust to the environment, *Figure 1.11.* Any condition in the environment that requires an organism to adjust is known as a **stimulus.** A reaction to a stimulus is called a **response.**

The ability to respond to stimuli in the environment is an important characteristic of living things and it's one of the more obvious ones, as well. That's because many of the structures and behaviors that you see in organisms enable them to make adjustments to stimuli in the environment.

Consider the hydroid animals, for example. The delicate, flower-like forms that they display occur only when the tide is up and they are underwater. When the tide is out and conditions are hotter and drier, hydroids withdraw their bodies into shell-like structures that effectively seal them off from the external environment—the black spots!

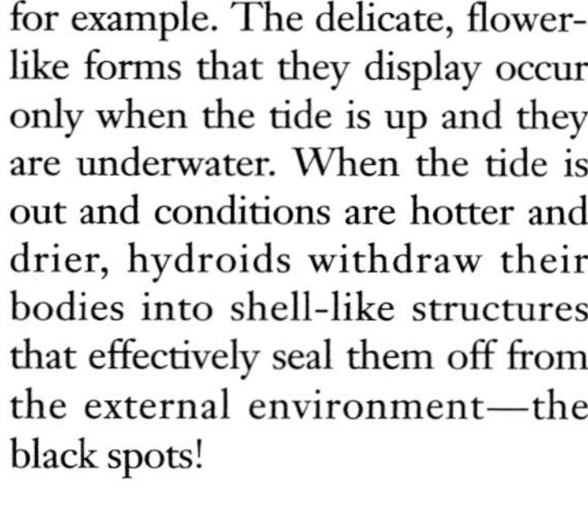

Figure 1.11

Humans adjust to weather changes for survival and for comfort.

Display

LS **Visual-Spatial** Make a bulletin board display that shows unusual structural and behavioral adaptations of plants and animals.

Bioethics

Life Under the Law

Doctors and hospital employees often must determine when a patient is no longer alive. Patients with little or no brain activity can be kept on life support systems that keep their hearts beating and provide oxygen and nutrients to their bodies.

Laws vary from one state to another about the nature of life support that must be provided, under what circumstances, and who decides when life support can be turned off. Ask students to research the laws of your state regarding this issue and ask for their opinions.

PORTFOLIO

Recognizing Stimuli and Responses Have students cut five pictures from magazines that show organisms responding to stimuli. Ask students to mount each picture on a clean sheet of paper and label the stimulus and response. Have students place their labeled illustrations in their portfolios. L1 LS P

Brainstorming

Have students choose some familiar living things, such as cats, goldfish, birds, and pine trees. Ask them to list the ways in which each of these organisms maintain homeostasis. Encourage students to first think of environmental factors that may change in the plant or animal's environment. Next, they should speculate about how each living thing adjusts to these changes. L1

Discussion

Discuss with students ways in which the human body maintains homeostasis. Sweating and shivering are two homeostatic mechanisms for maintaining body temperature. L1

Assessment

Performance Assessment in the Biology Classroom: p. 1, Brine Shrimp Life Functions. Have students carry out this activity to observe brine shrimp carrying out some of their life functions. L1

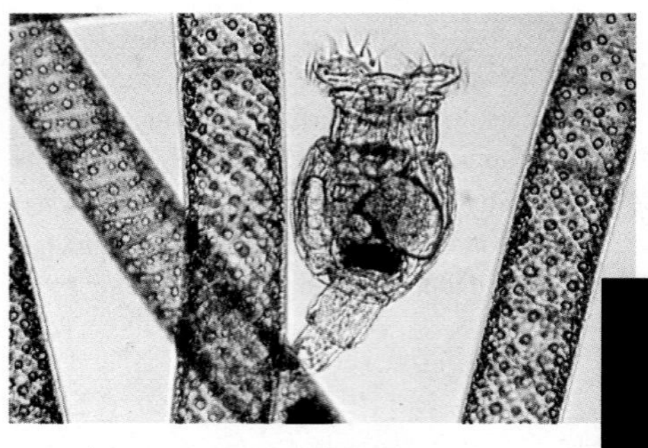

Microscopic organisms known as rotifers create a water current with their wheels of cilia. They feed on microscopic food particles from the water.

Many nocturnal animals, such as this tarsier, possess large eyes for efficient vision at night.

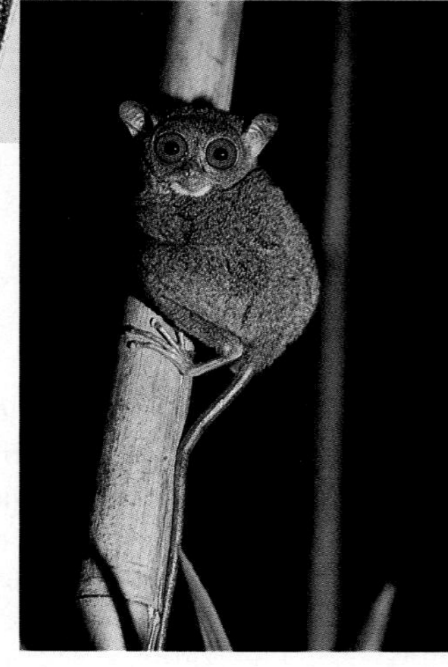

The sharp spines of the cactus are reduced leaves. This adaptation enables cacti to conserve water in their desert environment.

Figure 1.12
Living things adapt to their environments in diverse ways.

Any structure, behavior, or internal process that enables an organism to respond to stimuli and better survive in an environment is known as an **adaptation.** *Figure 1.12* shows some other adaptations in organisms.

An organism must respond to stimuli from its internal environment, as well. Factors such as external temperature or infection from bacteria can cause changes in body temperature; the quantities of water, nutrients, and minerals inside the body; or other internal changes. Such changes can disrupt proper functioning. Adjustments to internal stimuli help organisms maintain a steady internal environment.

An example of this kind of adjustment is human sweating. When it's hot outside or after strenuous physical activity, your body temperature increases slightly. In response to this internal change, you begin to sweat and your face gets flushed as many tiny blood vessels fill with blood. Both of these responses have the effect of cooling the body, and these adaptations help your body maintain its proper internal temperature necessary for metabolism. The regulation of an organism's internal environment to maintain conditions suitable for life is called **homeostasis.**

As you learn more about Earth's organisms in the chapters of this book, always reflect on these general characteristics of life, and ask yourself questions about how each organism meets the requirements for life. By asking questions such as, "How does this organism reproduce?" "What is the organization of this organism?" and "What are the adaptations of this organism that enable it to survive in its environment?" you'll learn that there are more similarities in the natural world than differences.

Alternate Lab: Mussel Environments

Purpose

Students will analyze the effect mussels have on their environment.

Materials

2 mussels, 2 *Elodea* stems, 4 test tubes, test-tube holder, clean straw, bromothymol blue solution in a dropper bottle, sheet of white paper, distilled water

Procedure

Give students the following directions.

1. Fill a test tube with distilled water. Add bromothymol blue (BTB) until you get a light-blue solution. **CAUTION: Bromothymol blue stains clothing and skin.**
2. Blow through a clean straw into the BTB solution until a color change appears. **CAUTION: Do not inhale through the straw.**
3. Add distilled water and BTB to four test tubes. Make sure all the tubes are the same color by holding them against a piece of white paper.
4. To one tube, add a mussel, to another add an *Elodea* sprig, to another add a mussel and a sprig of *Elodea*. Add

The Themes of Biology

In this course, you will be presented with many facts about organisms. Such facts are useful for gaining a good, working vocabulary of biology. However, biology, like other sciences, isn't merely a collection of isolated facts. Several major themes in biology serve to unify it as a science by linking isolated facts and ideas.

Energy

Energy is a central concept of the physical sciences, but it also pervades other sciences, including biology. Defined in physical terms, **energy** is the ability to do work or the ability to make things move.

Energy is important because it powers life processes. It provides organisms with the ability to maintain homeostasis, grow, reproduce, move, and carry out other life functions. Organisms obtain energy from the foods they eat or, in the case of plants and several other types of organisms, the foods that they produce.

As you'll learn, energy doesn't just flow through individual organisms; it also flows through communities of organisms, or ecosystems, and determines how organisms interact with each other and the environment. *Figure 1.13* shows one example.

Systems and interactions

You're probably familiar with a variety of systems in your life: the telephone system, a stereo system, the public-transportation system. As you know, each of these systems is made up of separate parts interacting to form a functioning whole.

In biology, you'll come across the idea of systems frequently. Organisms themselves may be thought of as systems. You're probably aware that your body contains several systems, including a nervous system, digestive system, and circulatory system. Each of these systems, as you'll learn, does not function independently; they interact in some rather complex ways to help perform the functions of life.

Unity within diversity

As you study the various types of ecosystems in the natural world, you'll be introduced to the different kinds of organisms that live there and how they interact to form a stable system. The theme of unity within diversity reflects the idea that, although ecosystems contain countless numbers of species—each with its own structural and behavioral specializations—all life is unified by the general characteristics you learned about earlier.

Figure 1.13

In coastal communities, such as the one where hydroid animals live, plants and algae convert the sun's energy into energy that can be used by other organisms in the community.

Physics Connection

The Law of Conservation of Energy

In 1845, H. Helmholtz and J.R. Mayer formulated the law of conservation of energy. This law states that energy can be neither created nor destroyed. The importance of the interrelationships of living and nonliving systems has been enhanced by an understanding of the law of conservation of energy.

Activity

LS **Visual-Spatial** The flow of energy and materials can be demonstrated by having students sow radish seeds and then place the plants under a grow-light or by a sunny window. As the seeds begin to sprout, explain the energy relationships that are at work. The plant uses the energy in sunlight to change carbon dioxide in the air and water in the soil to sugar and plant tissue. In this process, oxygen is given off. This oxygen may be used to meet the energy demands of other living things.

nothing to the last tube.

5. Make a prediction about any color change you might expect in each test tube over the next two days.
6. Keep the tubes in a warm, lighted area. Hold the white paper behind the tubes to check the color on each of the next two days. Record any changes in color you observe.

Analysis

1. Explain what happened in each tube. *The mussel and the Elodea produced CO_2 and caused the BTB to change to yellow. The test tube with both plant and mussel is yellowish blue because the plant uses some of the CO_2 that the mussel produces.*
2. Compare your outcome with your prediction. *If students predicted that the solutions with the plant, mussel, and plant and mussel would turn yellow, their prediction was supported.*
3. How might the test tube be similar to a pond environment? *Mussels produce carbon dioxide, which plants use.*

Assessment

Performance: Have students plan and carry out another test that would determine the relationship of snails and plants in a pond. L1

SECTION 1.2

BioLab Design Your Own Experiment

How does temperature affect a living thing?

Time Allotment

One class period

Objectives

Review objectives with students before they begin the Biolab.

Process Skills

design an experiment, compare and contrast, recognize cause and effect, form a hypothesis, interpret data

PREPARATION

- Order Instant Sea Life animals from Instant Products, Inc., 4804 Strawberry Lane, Louisville, Kentucky 40214 or telephone them at (800) 862-6688. Make sure you order the instant animals called "Sea Life" as they make a variety of kinds. You may also find this product in toy and novelty shops.

Possible Hypotheses

Students will not be using the term *hypothesis* in this lab but will call it a prediction.

- Students may predict that warm water will cause the animal to come out of its capsule sooner than colder water will.

BioLab Design Your Own Experiment

How does temperature affect a living thing?

Have you ever wondered what happens to animals in a pond when it dries up, when it gets cold, or when some other unfavorable condition occurs? Some animals are capable of burrowing into the mud; others can form a capsule around themselves when conditions are unfavorable. Assume that the capsule you have been given is an animal that has formed a protective cover under cold conditions in the pond in which it lives. It will come out of its capsule when conditions are favorable again.

PREPARATION

Problem

Under what conditions will an "animal" in a protective capsule emerge?

Objectives

In this Biolab, you will:

- **Determine** the temperature of the water that causes the animal to come out of its protective capsule.
- **Compare** the time it takes for the animals to come out of their capsules under different conditions.

Possible Materials

plastic cups or beakers of warm water and ice water
stirring rods or coffee stirrers
thermometers
Instant Sea Life toy capsules

Safety Precautions

Always wear goggles and an apron when carrying out an experiment.

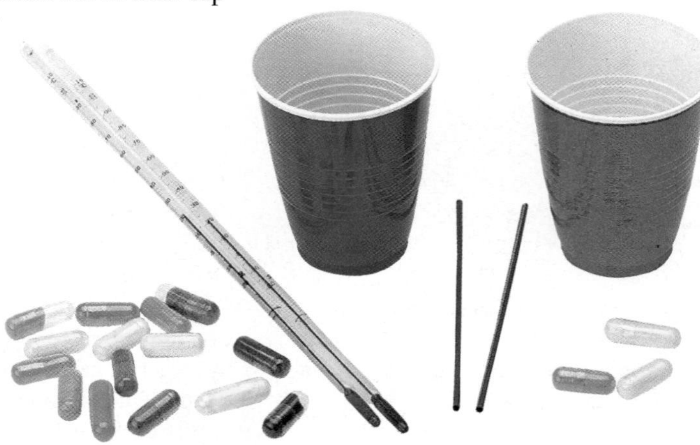

PLAN THE EXPERIMENT

Teaching Strategies

- If this is your first lab of the year, make sure students formulate their plans and get them approved before you give them the capsules. You may want to display the capsules first.
- Make sure students' plans include collection of quantitative data.
- You may want to review the relevant sections of the Skill Handbook with students before they begin to carry out the Biolab.

PLAN THE EXPERIMENT

1. In your group, make a prediction about how temperature might affect your capsule animals.
2. Design a way, based on your prediction, to test your animals with the materials provided by your teacher.
3. Make a numbered list of directions.
4. Make a list of the materials you will use.
5. Design and construct a table in which to record what happens during your experiment. In your table, you may want to record how long it takes for your animals to come out of their capsules, the temperature of the water, and details of the appearance of the animals.

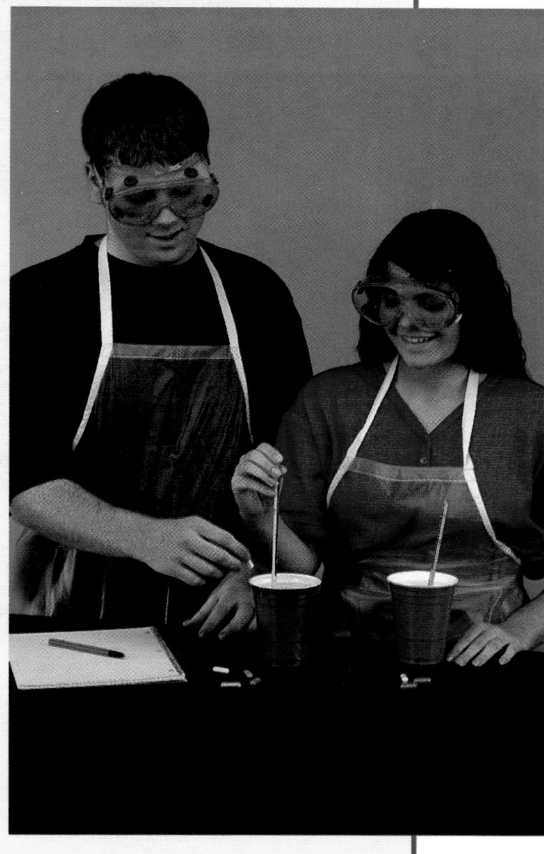

Check the Plan

1. ***Make sure your teacher has approved your experimental plan before you proceed further.***
2. Carry out your experiment.

ANALYZE AND CONCLUDE

1. **Calculating Results** How long did it take for your animals to come out of their capsules?
2. **Analyzing Data** What was the relationship between water temperature and the length of time it took for your animals to come out of their capsules?
3. **Relating Concepts** How might your results compare with those of real animals in a pond that gets cold during the winter?
4. **Making Inferences** Why would it be important for an animal in a protective capsule to stay in the capsule until water temperatures become warm?

Going Further

Project Plan a study in which you could test the effects of warm and cold temperatures on a variety of insects.

ANALYZE AND CONCLUDE

1. 20 minutes—ice water; 5–10 minutes—warm water
2. Warmer water was more effective in getting animals out of capsules than ice water.
3. Real animals might not come out of their protective coverings until the water is warmer.
4. There may be inadequate food. The water may be too cold for the animal to be able to move effectively to escape a predator.

Assessment

Performance: Ask students to design an experiment in which they determine the best temperature for enabling an animal to come out of its capsule. Have students place their experimental designs in their portfolios. L1 P

Going Further

Have students design an experiment to measure how far beetles crawl in a certain amount of time after one group has been kept at cool temperatures, another group at room temperature, and a third group under a lightbulb for similar lengths of time. L1

Possible Procedures

- Students may decide to put one capsule in warm tap water and another in ice water at the same time. They may then time how long it takes for the animals to come out.
- Students may decide to measure the temperature of the water in each beaker. Accept other experimental plans also as long as they make sense.

Data and Observations

Small sponge sea animals in a variety of colors come out of the capsules. It takes about 20 minutes for the ones in ice water to come out and about 5–10 minutes for the ones in warm water to come out.

Concept Development

Use the feeding process in the black spot hydroids as an example of systems and interactions. Have students brainstorm what different systems may have to be used in this process.

This animal's feeding involves several body systems. The hydroid's nervous system allows it to respond to the presence of a prey animal. Stinging cells on the tentacles discharge a chemical that paralyzes the prey. The hydroid's digestive system breaks down food into molecules that can be used by its cells for energy.

Misconception

Students often think that all plants found in an area are native to that area. Point out that many plants are not native to the region in which they are grown but were imported from other areas. For example, oats and rye were imported into California from Europe and have become established in many areas at the expense of native grasses.

Figure 1.14

Organisms have a variety of adaptations, both behavioral and structural, that help them maintain homeostasis.

Many animals, such as this box turtle, dig shallow burrows to escape heat and maintain optimum body temperatures.

The normally jet-black darkling beetle produces a yellowish, waxy covering over its body that acts as a natural sunblock when temperatures rise.

The camel's fatty hump acts as a storage compartment for water. Water is released from the fat through metabolism.

Homeostasis

You've learned that homeostasis is the regulation and maintenance of the internal environment of organisms. The concept of energy is involved here as well. Without energy, hydroid animals wouldn't be able to withdraw into their tiny, black shells to maintain homeostasis. Without energy, you could not perspire or shiver to maintain homeostasis in your body. *Figure 1.14* gives several examples of these mechanisms.

Figure 1.15

The paws of cats, the feet of frogs, and the hands of people, although appearing different on the outside, contain similar sets of bones. This suggests that these animals all share a common ancestry.

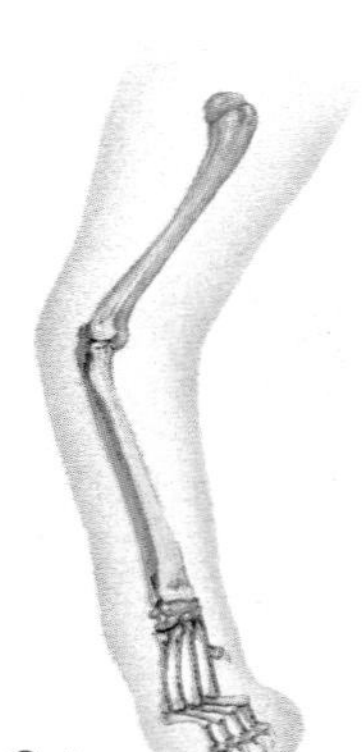

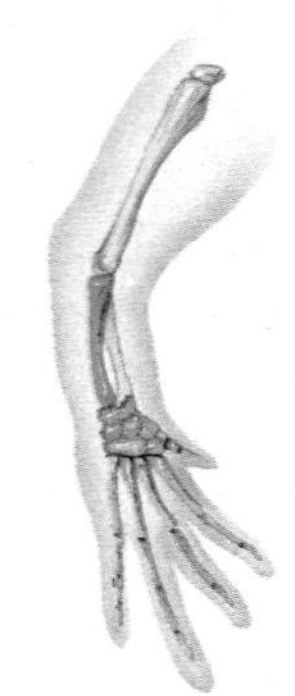

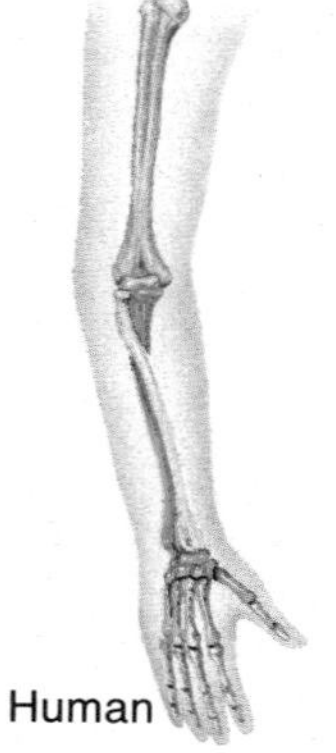

Evolution

Have you ever noticed that the paws of cats, the feet of frogs, and the hands of people, although different, are similar in overall structure, *Figure 1.15*? Clues to the diversity of life on Earth may be understood through the study of **evolution,** the gradual change in the characteristics of species over time.

Evolution is another important theme in biology and is perhaps the major unifying one. This is because all of the structures, behaviors, interactions, and internal processes observed in the millions of species of organisms on Earth are the result of the process of evolution.

In the science of biology, organisms are the principle objects of study, just as elements are in the science of chemistry and numbers are in the science of mathematics. But biologists have seen only the tip of the iceberg in terms of what the forces of evolution can produce. Tens of millions of organisms await discovery

STUDENT JOURNAL

No More Light Ask students to write a science fiction story about what might happen if Earth stopped getting energy from sunlight. Remind them that temperatures would decrease and photosynthesis would stop. L1 LS

and will only add to the great diversity that we already know is present on Earth, *Figure 1.16.*

The nature of science

Biology, like all sciences, is a continuous process that seeks to discover facts about the natural world. The discovery of a new species of hydroid with a new set of behaviors and characteristics may have invalidated earlier ideas about hydroids, but the scientists who developed such ideas were not wrong. Scientific facts can be determined only by making careful observations of present phenomena and by building on previous knowledge. The modification of ideas, rather than their outright rejection, is the norm in science. As new species are discovered and studied, more change is inevitable.

Figure 1.16
By most estimates, biologists have identified and studied about 1.5 million different types of organisms; however, the number of undiscovered species is estimated by some scientists to be in the tens of millions.

Connecting Ideas

Our world abounds with a great variety of life. Giant tube worms thrive at the bottom of our oceans near bubbling underwater volcanoes, colonies of green algae populate the tips of polar bear fur, and microscopic bacteria make their home in the pores of your skin. These and the millions of other organisms that inhabit the natural environment form the core of the science of biology.

However, biology is not just a body of knowledge. Biology is a process. It is a way of knowing. In the next chapter, you'll see that biologists, like other scientists, have an organized way of finding out about the natural world.

Section Review

Understanding Concepts

1. What is homeostasis, and how is it maintained in hydroid animals?
2. In what ways does the theme *systems and interactions* apply to both the living and the nonliving world?
3. What is meant by *unity within diversity?*

Thinking Critically

4. Why is evolution considered to be a unifying theme in biology? Explain your answer.

Skill Review

5. **Observing and Inferring** Suppose you discover an unidentified object on your way home from school one day. What steps would you take to determine whether the object is a living or nonliving thing? For more help, refer to Thinking Critically in the *Skill Handbook.*

SECTION 1.2

3 Assess

Check for Understanding

Have students explain how a home heating system models a homeostatic mechanism. L1

Reteach

Have students draw a diagram to illustrate how the pupil of the eye responds to bright light. Students can observe this homeostatic mechanism in the classroom, using flashlights or lamps. L1

Extension

Ask students working in groups to investigate the latest developments in robotics. Have them list the characteristics of robots and compare and contrast these characteristics to those of living organisms. Have them point out the key characteristics of robots that distinguish the robot as nonliving. L2 COOP LEARN

Assessment

Oral: Ask students to explain how they would test to see which color of light would cause a bean plant to grow fastest. L1

4 Close

Demonstration

Show slides and photographs that illustrate the six major themes of biology. Discuss each theme as a class.

Answers to Section Review

1. Homeostasis is the regulation and maintenance of the internal environment of organisms. Hydroids withdraw into a tiny shell to keep from drying out when the tide is low.
2. Organisms can be thought of as systems that interact with other living systems and the nonliving environment around them.
3. The theme of unity within diversity means that all life is unified by certain characteristics that all organisms share.

Thinking Critically

4. Evolution is considered a unifying theme because all of the structures, behaviors, interactions, and internal processes observed in the millions of species on Earth are the result of the process of evolution.

Skill Review

5. Examine its organization to see if it is cellular or not; see if it moves, grows, or develops; see if it responds to stimuli, or adapts to a change in its environment; see if it maintains homeostasis.

Chapter 1 Review

Reviewing Main Ideas

Summary statements can be used by students to review the major concepts of the chapter.

Key Terms

Answers should go beyond defining the terms. Accept any answer that uses the term correctly and in the proper context.

Understanding Concepts

1. a
2. c
3. b
4. d
5. c
6. a
7. d
8. c
9. a
10. d

Applying Concepts

11. They are composed of cells, which are organized into organs, which are organized into systems.
12. seashore, meadow, forest; Each community has living things that are carrying on life functions, maintaining homeostasis, and interacting in complex ways with each other and their environment. The seashore has animals and plants adapted to the rise and fall of the tide and the sandy shore. Forest plants and animals are adapted to the shady forest interior. Meadow plants and animals are adapted to the sunny dry conditions of the meadow.
13. Removing weeds will allow garden plants to grow better due to less competition for resources. Other kinds of organisms that may depend on the roots or leaves of the weeds for food or habitat may be adversely affected by weed removal.

Chapter 1 Review

REVIEWING MAIN IDEAS

1.1 What is biology?

- Biology is the organized study of living things and their interactions with their natural and physical environments.
- Biology teaches you about yourself, preserving the environment, and developing useful medical and agricultural techniques.

1.2 What is life?

- All living things have four characteristics in common: organization, reproduction, growth and development, and the ability to adjust to the environment.
- Six basic themes are woven into all biology topics: energy, homeostasis, unity within diversity, systems and interactions, evolution, and the nature of science.

Key Terms

Write a sentence that shows your understanding of each of the following terms.

adaptation	homeostasis
biology	organism
development	organization
energy	reproduction
environment	response
evolution	species
growth	stimulus

Understanding Concepts

1. Reproduction is an important life characteristic because ____________.
   a. for life to continue, organisms must replace themselves
   b. all living things show orderly structure
   c. all living things grow
   d. all living things adjust to their surroundings

2. Your heart beats more quickly and you breathe more rapidly after exercising. This characteristic of life is ____________.
   a. reproduction
   b. growth and development
   c. maintenance of homeostasis
   d. response to a stimulus

3. Which of the following is not alive?
   a. hydroids **c.** rabbits
   b. flames **d.** cactus

4. Energy is important to organisms because ____________.
   a. it enables them to regulate their internal environment
   b. living things adapt to their environments
   c. it provides the heat they need to stay warm
   d. it powers life processes

5. A caterpillar eventually turns into a butterfly. This is an example of ____________.
   a. homeostasis
   b. reproduction
   c. development
   d. response to a stimulus

6. The environment includes ____________.
   a. air, water, weather, temperature, and organisms
   b. responses to a stimulus
   c. adjustments of living things to their surroundings
   d. adaptations and evolution

7. A dog barks at a mail carrier. This is an example of ____________.
   a. homeostasis
   b. evolution
   c. an adaptation
   d. a response to a stimulus

8. The camel's fatty hump acts as a storage compartment for water. The hump is an example of ____________.
   a. homeostasis
   b. unity within diversity
   c. an adaptation
   d. a response to a stimulus

9. ____________ is the study of the gradual change of the characteristics of species over time.
   a. Evolution c. Homeostasis
   b. Adaptation d. Biology
10. A group of similar-looking organisms that can interbreed and produce fertile offspring is called ____________.
   a. a living system c. organization
   b. an adaptation d. a species

Applying Concepts

11. Describe how humans show the life characteristic of organization.
12. Identify several different types of natural communities. What do they have in common? How do they differ?
13. What effects, if any, might the removal of weeds from a garden have on other kinds of organisms?

Thinking Critically

14. ***Concept Mapping*** Make a concept map that relates the following words. Supply the appropriate linking words for your map.
    organism, species, evolution, reproduction, homeostasis, organization
15. ***Comparing and Contrasting*** Examine the following items: a flame, bubbles being blown from a bubble wand, and a balloon released into the air. List characteristics of each that might indicate life. List the characteristics that indicate they are not alive.
16. ***Recognizing Cause and Effect*** Hydroids have been shown to be the cause of bare areas within patches of acorn barnacles. If this trend continues, how might it affect the evolution of acorn barnacles?
17. ***Forming Hypotheses*** Sea slugs receive food from the algae that live within their skin. What benefits, if any, do you think algae receive from sea slugs?

ASSESSING KNOWLEDGE & SKILLS

The new species of hydroid discovered along the Sonoran coast was tested to see how long a dry period it could survive.

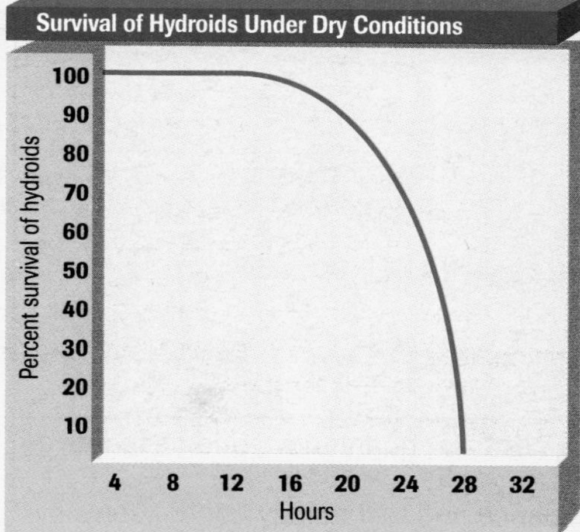

Using a Graph Study the graph and answer the following questions.

1. How many hydroids survive under dry conditions after four hours?
   a. 100 percent c. 28 percent
   b. 70 percent d. 10 percent
2. What is the maximum length of time these organisms survived in dry conditions?
   a. 100 hours c. 10 hours
   b. 16 hours d. 28 hours
3. How many hours does it take before hydroids begin to die?
   a. 28 hours c. 13 hours
   b. 32 hours d. 100 hours
4. ***Making a Graph*** Draw a line graph that shows that a population of seashore animals does not begin to die off until 50 hours after exposure to dryness and that there are no survivors after 75 hours.

Chapter 1 Review

Thinking Critically

14. Evaluate students' concept maps. The maps should show an understanding of the relationships among the concepts listed.
15. A flame has energy and may appear to grow and reproduce. Bubbles blown from a wand move and may grow. A balloon released into air moves. These objects cannot adapt to changes in the environment or maintain homeostasis.
16. Acorn barnacles may develop immunity to the venom of the hydroid.
17. Algae may receive water and minerals and protection from the sea slug.

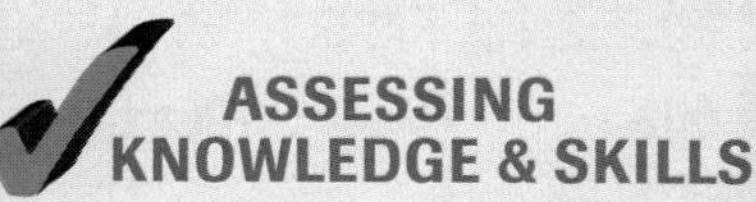

1. a
2. d
3. c
4. Effect of Dry Period on Survival

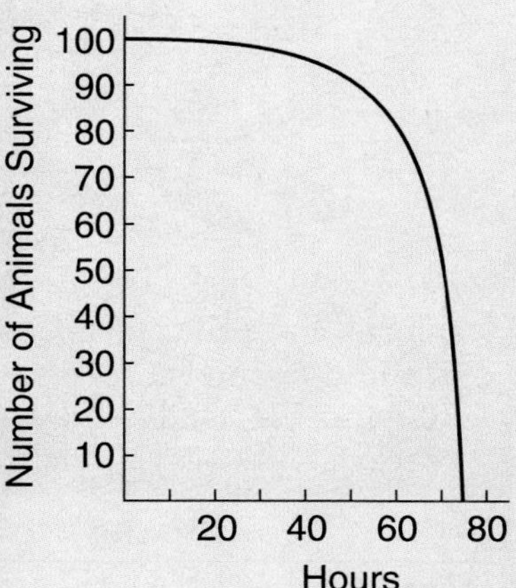

Program Resources

Chapter Assessment, pp. 1-6 L1
Alternate Assessment in the Science Classroom
Computer Test Bank L1
Content Mastery, pp. 1-4 L1

Chapter Organizer

SECTION	OBJECTIVES	ACTIVITIES/FEATURES
2.1 Problem-Solving Methods in Biology National Science Standards: UCP.2; A.1, A.2; B.2; C.6; F.4; G.1, G.2	1. **Compare** different scientific methods. 2. **Differentiate** among hypothesis, theory, and principle.	**Physics Connection:** The Sounds of Life, p. 28 **Minilab:** How can you use inductive reasoning?, p. 30 **Biolab:** Design Your Own Experiment—How does fertilizer affect early plant development?, p. 32 **Thinking Lab:** Does a weed killer perform as advertised?, p. 34 **Focus On:** What is a scientific theory?, p. 36
2.2 The Nature of Biology National Science Standards: UCP.2; A.1-2; E.1, E.2; F.3, F.4; G.1, G.2	3. **Compare and contrast** quantitative and descriptive research. 4. **Explain** why science and technology cannot solve all problems.	**Minilab:** How do you decide which paper towel to buy?, p. 42 **A Broader View:** Life as a Field Biologist, p. 44

ACTIVITY MATERIALS

BIOLAB	MINILABS	ALTERNATE LAB
page 32 seeds plant seedlings plastic trays sand or potting soil foam cups fertilizer balance metric ruler graduated cylinder plastic zipper bags 1% bleach solution	**page 30** flowers, leaves, or small insects hand lens **page 42** 1 piece from four brands of paper towels	**page 30** 2 baby food jars graduated cylinder labels coffee, caffeinated coffee, decaffeinated

Chapter 2 Scientific Methods in Biology

TEACHER CLASSROOM RESOURCES

Reproducible Masters	Transparencies
Section Focus Master 3: Designing an Experiment L1 SAE **Reinforcement and Study Guide,** pp. 5-6 L1 **Laboratory Manual:** Can Scientific Methods Be Used to Solve a Problem?, pp. 5-8 L2 **Biolab and Minilab Worksheets,** pp. 3, 5-6 L1 **Concept Mapping:** Applying Scientific Methods, p. 2 L1 **Tech Prep Applications:** Applying Scientific Methods to Everyday Applications of Biology, pp. 1-2 L2 **Critical Thinking/Problem Solving:** Using Scientific Methods, p. 2 L3	**Reteaching Transparency #2:** Scientific Methods L1 SAE
Section Focus Master 4: Systems of Measurement L1 SAE **Reinforcement and Study Guide,** pp. 7-8 L1 **Biolab and Minilab Worksheets,** p. 4 L1 **Laboratory Manual:** Using SI Units, pp. 9-12 L2 **Content Mastery,** pp. 5-8 L1	

ASSESSMENT MATERIALS

Chapter Assessment, pp. 7-12
Performance Assessment in the Biology Classroom, p. 3
MindJogger Videoquiz
Alternate Assessment in the Science Classroom
Computer Test Bank

Spanish Resources SAE
English/Spanish Audiocassettes SAE
Cooperative Learning in the Science Classroom COOP LEARN
Lesson Plans
Biology Projects: Investigating Insect Behavior, pp. 1-4 L2

KEY TO TEACHING STRATEGIES

- L1 Level 1 activities should be within the ability range of all students including those with learning difficulties.
- L2 Level 2 activities are within the ability range of average to above-average students.
- L3 Level 3 activities are designed for the ability range of above-average students.
- SAE SAE activities should be within the ability range of Students Acquiring English.
- COOP LEARN Cooperative Learning activities are designed for small group work.
- P These strategies represent student products that can be placed into a best-work portfolio.
- These strategies are useful in a block scheduling format.

GLENCOE TECHNOLOGY

The following multimedia resources are available from Glencoe.

Biology: The Dynamics of Life
CD-ROM SAE
Videodisc Program

The Infinite Voyage Series
Unseen Worlds

The Secret of Life Series
On the Brink: Portraits of Modern Science

Science and Technology Videodisc Series (STVS)
Ecology
Modeling Pollutants
Research in the Pinelands
Animals
Tagging Ants

CHAPTER

2 Scientific Methods in Biology

Chapter Overview

The first section of this chapter describes specific research on elephant communication to illustrate the problem-solving methods used by scientists. Students learn that, starting with what are often simple observations, scientists form hypotheses using both inductive and deductive reasoning. The steps in an investigation are described, as are the roles of independent variables, dependent variables, and controls.

The use of tools, gathering of data, and analysis of results are emphasized as part of the problem-solving process. The section ends with a discussion of how theories evolve from within the scientific process.

Scientists use both quantitative and qualitative data including descriptive research. The International System of measurement (SI) and its role in science is presented. The chapter concludes with a discussion of the importance of science and technology to society and presents the need for giving ethical considerations to scientific discoveries.

Key Terms

- control
- data
- deductive reasoning
- dependent variable
- ethics
- experiment
- hypothesis
- independent variable
- inductive reasoning
- safety symbol
- scientific methods
- technology
- theory

24

Learning Styles LS

Look for the following logo for strategies that emphasize different learning modalities. LS

Kinesthetic	**Meeting Individual Needs, p. 27**
Interpersonal	**Meeting Individual Needs, pp. 29, 34; Project, p. 45**
Intrapersonal	**Student Journal, p. 29**
Linguistic	**Portfolio, p. 35; Student Journal, pp. 35, 38**
Logical-Mathematical	**Demonstration, p. 27; Alternate Lab, p. 30; Minilab, pp. 30, 42; Thinking Lab, p. 34; Activity, p. 41; Portfolio, p. 43; Meeting Individual Needs, p. 44**
Auditory-Musical	**Demonstration, p. 26; Project, p. 28**

CHAPTER

2 Scientific Methods in Biology

Lions in the wild are free to wander the landscape, chase prey, eat, breed, take care of cubs, and sleep, which is one of their major occupations. The behavior of a caged lion is much different from the behavior of lions in their natural environment. Most modern zoos provide open spaces for large animals, but sometimes they must be confined to cages while cleaning and maintenance are performed. Have you ever seen a lion or tiger pacing back and forth in a cage? Did you wonder why they pace in their cages when they don't behave this way in nature? Biologists have found that pacing is a reaction many animals have to restricted space.

How do biologists know these things? They've acquired this knowledge using well-established methods of study. In this chapter, you'll learn more about experimenting as well as the other methods that scientists use to study the natural world.

Concept Check

You may wish to review the following concepts before studying this chapter.

- Chapter 1: biology, themes of biology

Chapter Preview

Laboratory Activities

Biolab: Design Your Own Experiment

- How does fertilizer affect early plant development?

Minilabs

- How can you use inductive reasoning?
- How do you decide which paper towel to buy?

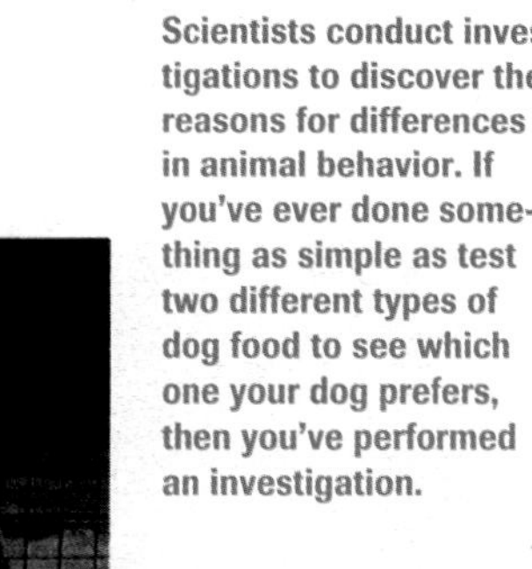

Scientists conduct investigations to discover the reasons for differences in animal behavior. If you've ever done something as simple as test two different types of dog food to see which one your dog prefers, then you've performed an investigation.

Assessment Planner

Choose assessment strategies from the following pages to evaluate the progress of your students.

Assess, pp. 39, 45
Alternate Lab, pp. 30-31
Minilabs, pp. 30, 42
Portfolio, pp. 35, 43
Thinking Lab, p. 34
Biolab, pp. 32-33
Chapter Review, pp. 47-49

CHAPTER 2

Introducing the Chapter

After students read the introduction, direct their attention to the photographs. Emphasize that many methods used by scientists conducting research are the same as those used by nonscientists in daily life. As an example, provide students with the following problem. You enter a dark room, turn on the wall switch, but no lamp goes on.

Ask students the questions that follow and discuss the responses as a class. (a) How might you explain the problem? (b) Is the problem more likely with the lamp or the switch? How do you know? (c) How will you test your thoughts about the problem? (d) How is the problem usually resolved? (e) Does the solution give you insights about how to solve the problem if it occurs again? L1 COOP LEARN

Theme Development

The *nature of science* is the most prominent theme of the chapter. The role of science and its importance to society are illustrated throughout the chapter. *Unity within diversity* is stressed as the contributions made by different branches of science (diversity) to solving problems (unity) are discussed.

Concept Check

Students will recognize that many everyday problems have solutions resulting from scientific methods.

SECTION

2.1

Prepare

Key Concepts

Students will study the steps used in scientific methods. They will trace methods used in scientific problem-solving by reading about how a specific scientist works to answer a question about elephant communication.

Block Scheduling

Look for this symbol for strategies that are useful in a block scheduling format. For more information on block scheduling, refer to Section 2.1 in the **Lesson Plans** booklet.

Materials

- Gather specimens of flowers, leaves, or small insects for the Minilab.
- Purchase or grow plants for the Biolab. Purchase fertilizer, seeds, sand or potting soil, and Styrofoam cups. Gather plastic trays, rulers, graduated cylinders, and balances.

1 Focus

Bellringer

Before presenting the lesson, display **Section Focus Master 3** on the overhead projector and have students answer the accompanying questions. L1 SAE

Demonstration

LS **Auditory-Musical** Have students plug their ears with their fingertips. Ask them to hum with a high pitch and then with a low pitch. Have them explain any differences they detect and relate their observations to the elephant's ability to produce vibrations.

SECTION 2.1

Problem-Solving Methods in Biology

Section Preview

Objectives

Compare different scientific methods.

Differentiate among hypothesis, theory, and principle.

Key Terms

scientific methods
hypothesis
inductive reasoning
deductive reasoning
experiment
control
independent variable
dependent variable
safety symbol
data
theory

Have you ever watched elephants in a zoo? Maybe you've heard them trumpet and snort or watched them swing their heads, examine things with their sensitive trunks, or flare their large ears. Have you ever seen them do things that made you think they were communicating with each other? In this section, you will see how one scientist applied scientific methods to learn how elephants communicate.

Observing and Hypothesizing

Why are biologists interested in answering a question such as, "How do elephants communicate?" For a scientist, the simple reason is curiosity about how and why things happen in nature. In addition, the answer to the question will lead to a better understanding of elephant behavior. This knowledge, in turn, may enable zookeepers and wildlife conservationists to better care for elephants.

The methods that biologists use

To solve problems, different biologists may take different approaches, yet there are some steps that are common to these different approaches. The common steps that biologists and other scientists use to gather information to solve problems are **scientific methods.**

Scientists often discover problems to solve—that is, questions to ask and answer—simply by observing the world around them. For example, a scientist may be working on the reproduction of mosses in the laboratory and come up with another question about their development. Another scientist may ask a question about the feeding habits of prairie dogs after observing a pattern of behavior in prairie dogs in the field.

The question of elephant communication

For many years, observers had noticed that elephant herds appear to move, turn, or stop suddenly and all together without any apparent audible or visual signal. For example, if one elephant noticed a lion, all the elephants in the herd seemed to become alarmed. Some might even charge, as shown in *Figure 2.1.*

Program Resources

Section Focus Master 3 L1 SAE
Reinforcement and Study Guide, pp. 5-6 L1
Biolab and Minilab Worksheets, pp. 3, 5-6 L1
Concept Mapping, p. 2 L2
Critical Thinking/Problem Solving, p. 2 L3
Tech Prep Applications, pp. 1-2 L2
Laboratory Manual, pp. 5-8 L2
Reteaching Transparency 2 and **Master** L1 SAE

Figure 2.1

Your first thought about elephant communication might be of the trumpeting sound they make when alarmed. A biologist might ask, "Is this the only sound elephants make to communicate?"

One biologist, Katharine Payne of Cornell University, made an important observation about elephant communication at the Metro Washington Park Zoo in Portland, Oregon. While visiting the zoo, Payne felt the air throbbing around her. It reminded her of the rumbling of thunder, a sound more felt than heard. Payne had spent 17 years studying the calls of whales and knew that some whales make sounds that are too low-pitched for people to hear. When she felt the vibrations, she also noticed that the skin on an elephant's forehead seemed to be fluttering. Look at *Figure 2.2*. She suspected that the vibrations were generated by the elephants and that they might be using the sounds to communicate. In other words, Payne hypothesized that elephants communicate by means of low-frequency sounds. A **hypothesis** is a testable explanation for a question or problem.

As you can see from Katharine Payne's example, a scientist's hypothesis is usually not just a random guess. More likely, before a scientist makes a hypothesis, he or she has some idea what the answer to a question might be because of experience, extensive reading, and previous experiments. Applied to all this knowledge is the scientist's reasoning powers.

Using inductive reasoning

Stop to think for a moment about how you solve problems in your everyday life. For example, suppose you can't find your house key. The last time you had it, you were wearing your gray jacket; and on two earlier occasions, coins and a pen had slipped through a hole in the pocket of that jacket into its lining. So you hypothesize that that's where the key is. You've used inductive reasoning. **Inductive reasoning** is reasoning from a particular set of facts to a general rule. Payne used inductive reasoning when she hypothesized about how elephants communicate.

Figure 2.2

Payne hypothesized that elephants communicate by means of low-pitched sounds when she saw the skin on the elephant's forehead fluttering. These sounds are too low-pitched to be heard by humans.

Demonstration

LS Logical-Mathematical Try to set a sugar cube on fire using a lighted match. (It does not burn.) Rub the edge of the cube in ashes and attempt to light the cube. (It burns.) Ask students to offer explanations of why the cube burns after it is rubbed in ashes. *Responses may include that the ashes—not the sugar cube—burn. Other suggestions might be that the ashes served as kindling, allowing the fire to become hot enough to ignite the sugar cube.* Ask students how they might test their explanations. *Responses may include trying to set either the ashes or the sugar on fire separately or exposing the sugar to a hotter flame for a longer time.*

2 Teach

Physics Connection

Observing Sound Energy Obtain a tuning fork and a soft mallet from the physics laboratory. Fill a large beaker with water and have students gather close to the demonstration. Show students the tuning fork and explain that when struck with an object, the tuning fork vibrates, producing sound.

Strike the tuning fork with the mallet and place the prong ends of the fork to the water's surface. Ask students to describe their observations. *The tuning fork produced a sound. When touched to the water, the water began to splatter from the rapid vibrations of the fork.*

Have students explain what the demonstration shows. *Sound is a form of energy.*

Meeting Individual Needs

Hearing Impaired To illustrate the vibrations that were sensed by Ms. Payne, have students place their hands onto the skin of a drum while the drum is hit with a stick. Ask students to correlate what they felt with the sensation that Ms. Payne felt when near the elephants. L1 LS

STUDENT JOURNAL

Feeling Sound A fog horn blast or a very low note on an organ can be felt as well as heard. Have students record in their journals what these sounds feel like. Ask them to provide a scientific explanation for their observations. L1 LS

Physics

The Sounds of Life

Purpose

Students will learn about the nature of sound and will understand how differences in pitch and loudness are created.

Teaching Strategies

- Ask a physics teacher to carry out a series of demonstrations to show students how sound waves behave and how wave characteristics relate to pitch and volume.
- Have above-grade level students explain some of the concepts presented to below-grade level students. COOP LEARN

Possible Answers

Bats and dolphins use sonar to locate objects. This principle is used to make depth-finders for ships.

Reinforcement

Examples of inductive and deductive reasoning may help students see how these processes differ. Provide these examples to students. A person knows the poisonous coral snake has brightly colored bands of red, yellow, and black. *Inductive reasoning:* The person assumes that all brightly banded snakes are poisonous. *Deductive reasoning:* The person sees a frog with a banded coloring similar to the coral snake and concludes (by assumption) that this frog is poisonous and should be avoided.

Physics

The Sounds of Life

Chuff-chuff-chuff-chuff, chuff, chuff! Resounding through the thick northern woods, these sounds signal the presence of a male ruffed grouse claiming his territory. To other males, it is a warning and a challenge; to females, it is an invitation.

Waves of energy The ruffed grouse makes sounds by spreading its wings and rapidly moving them down and in. This compresses the air trapped between its wings and body. Each beat of the wings produces a pulse of high pressure followed by a period when the air springs back to normal pressure. Each compression is a sound wave. But there is more to sound than just compressions. Sounds can be high or low, loud or soft. What causes these differences?

High or low? The pitch of a sound depends on its frequency—how many compressions are produced each second. The ruffed grouse's sound is low pitched because its wings beat only a few times each second. The rapidly beating wings of a mosquito produce many more air compressions per second; thus, a high-pitched sound results.

Loud or soft? The more energy a sound wave has, the more it compresses the air and the louder it sounds. In fact, some sounds such as over-amplified music or sharp explosions such as those from a gun have enough energy to severely damage your hearing.

The amount of energy in a sound wave also determines its range. Sound waves spread out in a spherical pattern as they travel. Therefore, the energy of the original sound becomes diluted in a larger space. A sound that has more energy will be heard at a greater distance. Thus, the powerful bellow produced by a 400-kg moose will be heard farther away than a chirp produced by a 250-g chipmunk.

CONNECTION TO Biology

Bats and dolphins make sounds and locate objects by the reflection of those sounds. In what ways do humans use this or similar techniques for practical purposes?

Using deductive reasoning

Sometimes a general rule is known before a particular case is apparent. For example, you know that dogs pant when they are hot and thirsty. One day, you see your dog panting heavily and think, "If the dog is panting, she must be hot and need water." So you check her water bowl and, sure enough, it's dry. You've used deductive reasoning. **Deductive reasoning** involves suggesting that something may be true about a specific case from known general rules. This kind of reasoning is often expressed as an "If . . . then . . ." statement. Suppose you live in an area that has a history of flash floods. Then you would use deductive reasoning to say, "If it rains another two inches in the next hour, then we'll have a flood."

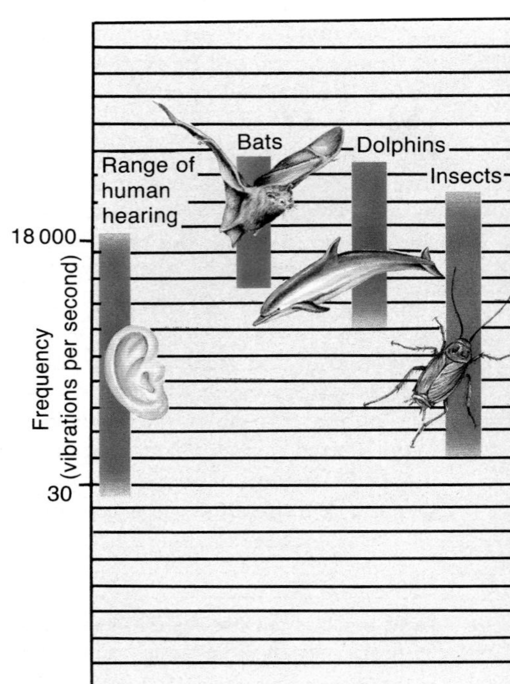

PROJECT

Make a Musical Instrument

Auditory-Musical Crease the ends of two straws so they are flattened. Snip off the corners at the tops of the flattened ends to form a point as shown in the diagram. Cut off the bottom of one straw so that the straw is 5-cm long. Predict which straw will have a high pitch and which will have a low pitch. Blow into the straws as a saxophone player would through a reed mouthpiece. Explain in technical terms why the two straw sounds differ in pitch. L1

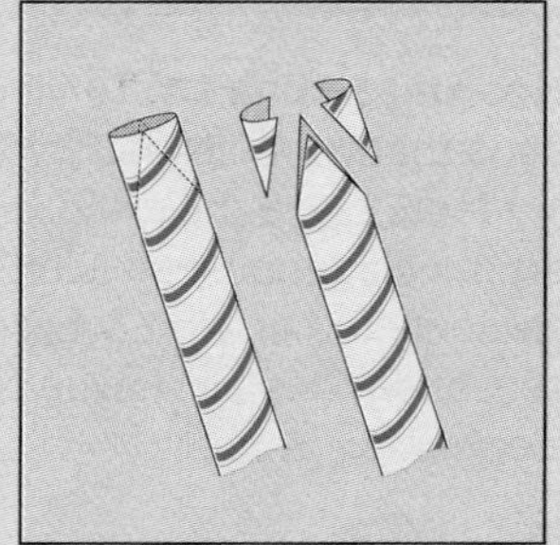

After stating a hypothesis, scientists use deductive reasoning. In Payne's case, she applied some general rules from physics to the hypothesis she had made about the elephant sounds. She already knew that humans can hear sounds beginning around 20 to 30 vibrations per second, but that lower-pitched sounds can be produced by animals. Look at *Figure 2.3* for a comparison. She also knew that lower-pitched sounds aren't easily absorbed by objects they strike, and therefore can travel farther than higher sounds before they become too faint to hear. Thus, Payne suggested that *if* elephants produce low-pitched sounds to communicate with other elephants, *then* there should be some evidence of elephants' reacting to the low-pitched sounds made by elephants that are far away. Payne's next step was to design a controlled experiment to test her hypothesis.

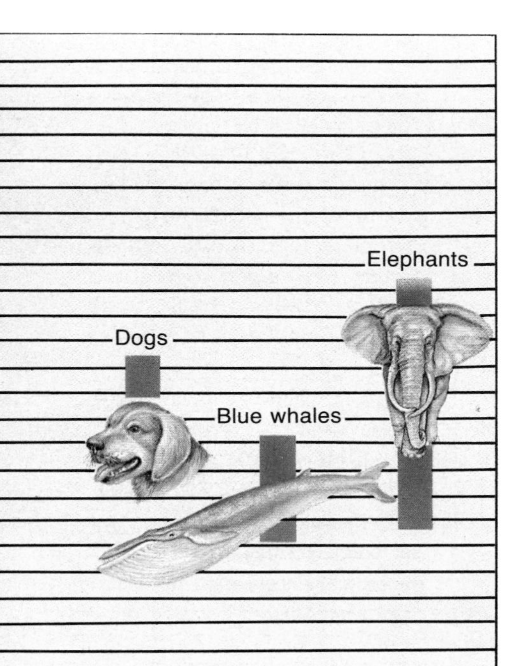

Experimenting

People do not always use the word *experiment* in their daily lives in the same way as scientists do in their work. As an example, you may have heard someone say that he or she was going to experiment with a cookie recipe. The person is planning to substitute raisins for chocolate chips, use margarine instead of butter, add cocoa powder, reduce the amount of sugar, and bake the cookies for a longer time. This is not an experiment in the scientific sense because there is no way to know what effect any one of the changes alone has on the resulting cookies. To a scientist, an **experiment** is a procedure that tests a hypothesis by the process of collecting information under controlled conditions.

What is a controlled experiment?

Some experiments involve two groups: the control group and the experimental group. The **control** is the standard, in which all conditions are kept the same. The experimental group is the test group, in which all conditions are kept the same except for the single condition being tested.

Figure 2.3

The normal range of human hearing is from about 20 to 30 vibrations per second up to about 18 000 to 20 000 vibrations per second. Some animals, such as bats and dolphins, can make sounds that are higher-pitched than humans can hear. Elephants and whales, on the other hand, can make audible sounds as well as sounds that are too low-pitched for humans to hear.

SECTION 2.1

Tying to Previous Knowledge

Using terminology and concepts from *The Sounds of Life* feature on the previous page, have students answer the following. (a) What causes sound? *Vibrations* (b) Explain the difference between a soprano and a baritone voice using the term *frequency* in your answer. *The soprano has a higher pitch because sound waves produced by the soprano have a greater frequency than those produced by the baritone.* (c) Explain the difference in loudness between the tap of a pencil on a desk and a book that falls on the floor. Use the terms *energy* and *compression* in your answer. *The pencil makes a softer sound because it has less energy and displaces less air (compression) than the falling book.*

Reinforcement

Ask students to note the sentence toward the end of the first column of text that begins "Thus, Payne suggested that . . ." Have them decide whether this sentence illustrates inductive or deductive reasoning, and explain. *Deductive; the information is based on known information.*

GLENCOE TECHNOLOGY

Videodisc

Biology: The Dynamics of Life
Disc 1, Side 1
Elephant Behavior (Ch. 3)

Meeting Individual Needs

Learning Disabled Have groups of students working below grade level work with students who are at or above grade level. Have the groups use scientific methods to alter a cookie recipe. Ask students to name the control to be used, list the independent variable, and name the dependent variable.

L1 SAE COOP LEARN LS

STUDENT JOURNAL

Working as a Food Chemist Tell students to imagine they are the chief food chemist working in a candy factory. They are trying to improve on the taste of a "Smacking Good Bar." Have students describe some things they would suggest to their staff to attempt, in a controlled manner, to find a better candy bar recipe. L2 LS

MiniLab

Purpose

IS **Logical-Mathematical** Students will use inductive reasoning. COOP LEARN

Process Skills

observe and infer, form a hypothesis, interpret data

Teaching Strategies

- Vary the types of leaves, flowers, and insects being used. Consider the use of hand lenses to assist in making observations.
- Have students work in small cooperative groups.

Expected Results

Sample observations: leaf is green; one side is shiny, the other is dull; has toothed edges. Possible questions: What insects feed on this leaf? Why does each side of the leaf have a different texture?

Analysis

1. Observations were based on first-hand accounts and included descriptive phrases.
2. Questions were based on data from observations.
3. Inductive reasoning leads scientists to form questions.

Assessment

Performance: Describe an experimental procedure that tests this hypothesis: "If leaves make food for plants, then removing leaves should result in poor growth." L1

MiniLab

How can you use inductive reasoning?

Scientists must be excellent observers. They need both their observation skills and their reasoning power to form hypotheses. This procedure will give you a chance to use your own.

Procedure

1. Obtain a single specimen of a flower, leaf, or small insect.
2. With the specimen in front of you, observe as many characteristics as you can about its shape, size, color, texture, or any other features that you can observe.
3. Write as many observations as you can in five minutes.
4. Now write two or three questions you have about your specimen.
5. After you have made and recorded your observations and questions, use inductive reasoning to form one hypothesis that you could test.

Analysis

1. How do you think your observations were similar to those a scientist might make?
2. Explain how your observations led you to form a question.
3. Why do scientists use inductive reasoning?

Suppose you wanted to learn how salt water affects a new variety of bluegrass. The control group would consist of several grass plants watered with plain water. The experimental group might consist of groups of plants watered with several different strengths of salt water. The condition being tested is the concentration of salt in the water. All other conditions would be held constant for both groups. For example, in *Figure 2.4* you can see control and experimental groups of plants for an experiment on growing soybeans.

Designing an experiment

Katharine Payne used insight and imagination in developing her hypothesis. Most scientists would agree that both of those qualities are also needed to design an experiment to test a hypothesis.

In a controlled experiment, only one condition is changed at a time. The condition in an experiment that is changed is the **independent variable.** While changing the independent variable, the scientist observes or measures a second condition that results from the change. This condition is the **dependent variable.** In the experiment to test the effect of salt water on bluegrass, the concentration of salt is the independent variable. The resulting growth rate of the grass is the dependent variable.

Figure 2.4

Shown here are the results of experiments to test the effect of various soil bacteria on the growth of different varieties of soybeans. For each experiment there are three rows of plants. The center rows of each 3-row plot are the experimental plants. The left and right pairs of rows are controls. All other conditions in the field—soil, light, water, and fertilizer—are the same.

Alternate Lab: Conducting an Experiment

Purpose

IS **Logical-Mathematical** Students will follow the steps of scientific methods to solve a problem. The question to be answered is, Will the caffeine present in coffee prevent mold growth?

Materials

2 small jars (baby food jars), graduated cylinder, labels, caffeinated coffee (freshly brewed), decaffeinated coffee (freshly brewed)

Procedure

Give the following directions to students.

1. Form and record a hypothesis.
2. Write your name and the date on the labels. Write #1 on one label and #2 on the other. Place the labels on the jar.
3. Add the following to each jar. Jar 1: 30 mL caffeinated coffee; Jar 2: 30 mL decaffeinated coffee.
4. Place both jars in the same location within your classroom.

In the experiment designed by Payne, the production of low-frequency sound was the independent variable, and observable change in the behavior of the elephants was the dependent variable. To perform her experiment, Payne went back to the zoo with a tape recorder and microphone. She recorded hours of what seemed to her to be silence among the elephants. At the same time, Payne was making careful notes about the behavior of the elephants in their compound.

Just as problems may be arrived at differently, the approaches taken to solve a particular problem can vary widely. The experimental design that a scientist selects depends on what other experimenters have done and what information the scientist hopes to gain. Sometimes, a scientist will design a second experiment even while a first one is being conducted if the scientist thinks the new experiment will help answer the question.

Using tools

In order to carry out her experiment, Payne required tools that would enable her to record the sounds of the elephant. For instance, her tape recorder and microphone had to be able to respond to low-pitched sounds.

Biologists use a wide variety of tools to obtain information in an experiment. Some of the common tools are beakers, test tubes, hot plates, petri dishes, balances, thermometers, metric rulers, and graduated cylinders. More complex tools include specialized microscopes, centrifuges, radiation detectors, spectrophotometers, DNA analyzers, and gas chromatographs. *Figure 2.5* shows some of these more complex tools.

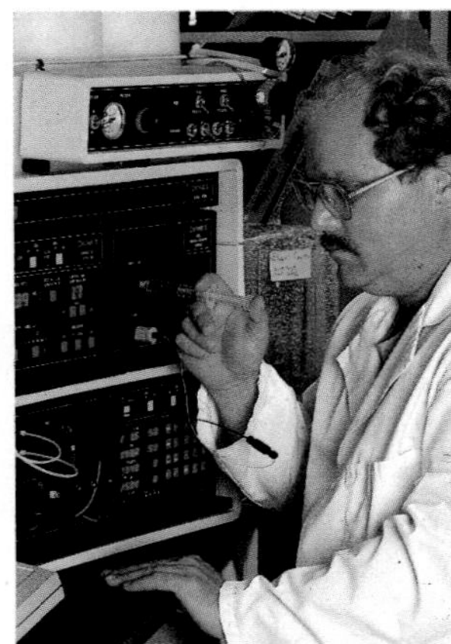

Figure 2.5

The microscope (below left), the gas chromatograph (right), and gel electrophoresis (below right) are three of the many tools that biologists can use in their studies.

The gas chromatograph can be used to detect and measure pesticide residues in plants or fish.

The microscope magnifies organisms or parts of organisms, making small details visible.

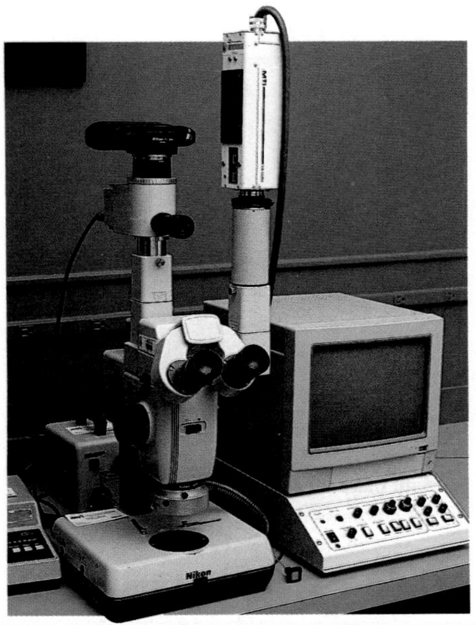

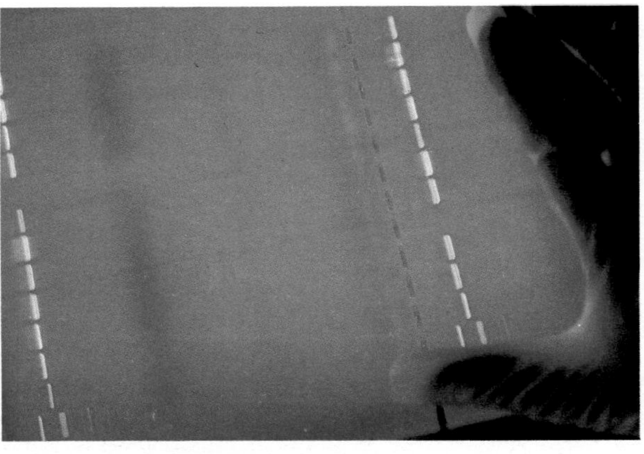

Gel electrophoresis can be used to analyze DNA, producing a DNA print as shown. Comparing DNA can tell a biologist how closely related two organisms are.

SECTION 2.1

Tying to Previous Knowledge

Present the following hypothesis to your students: "If yeast is living, then it will be composed of cells." Ask students to: (a) Describe the experimental procedure that could be used to solve the problem. *Observe yeast under the microscope.* (b) Describe any inductive reasoning that was used to form the hypothesis. *Yeast require food and give off carbon dioxide gas.* (c) Describe any deductive reasoning that was used to form the hypothesis. *All living matter is composed of cells.*

GLENCOE TECHNOLOGY

Videodisc

The Infinite Voyage: Unseen Worlds

Technology Reconstructs Egyptian Mummies (Ch. 1)

Digital X Rays, 3-D X Rays: Detection Made Easy (Ch. 8)

5. Make daily observations of your jars for one week. Check for the presence of mold.
6. Record your observations in a suitable data table by making diagrams of the coffee liquid surfaces.

Analysis

Ask students to answer the following questions.

1. Which type of coffee allows mold to grow? *Both*
2. Was your hypothesis supported by your data? *Will depend on hypothesis; probably not*
3. What was the control, independent variable, and dependent variable? *Decaffeinated coffee, caffeinated coffee, mold growth*

Assessment

Portfolio: Have students write a report of their experimental findings. Ask students to record any other questions that arose during this experiment and explain how they might be answered. L1 P

SECTION 2.1

BioLab Design Your Own Experiment

How does fertilizer affect early plant development?

Time Allotment

Initial session: one period; followup sessions: ten minutes daily for one week

Objectives

Review objectives with students before they begin the Biolab.

Process Skills

form a hypothesis, design an experiment, separate and control variables, observe and infer, interpret data

Safety Precautions

If spillage of bleach occurs, immediately wash hands or rinse clothing with water. If bleach is spilled on a table or floor, the area should be quickly washed with water.

PREPARATION

Alternate Materials

- Seeds can be purchased in a grocery store. Use kidney beans, pinto beans, or popcorn (found in plastic bags).
- Vermiculite can serve as a sand/soil substitute.
- Young plants purchased in garden shops are ideal for plant growth experiments.
- Presoak seeds in a 1% solution of bleach to prevent mold contamination. Soak seeds for two minutes, remove and blot dry with paper toweling.

BioLab Design Your Own Experiment

How does fertilizer affect early plant development?

Why do people fertilize plants? You may have observed someone fertilize a houseplant and wondered why he or she went to the trouble. Does fertilizer really make any difference in the growth of a plant? Does it make flowers bloom more rapidly or vegetable plants grow larger and produce more vegetables? These questions are all within the realm of science because they are testable by controlled experiments, observation, and data gathering.

PREPARATION

Problem

Do seeds germinate faster if fertilizer is applied? Do more seeds germinate when fertilizer is applied? Do different strengths of fertilizer cause different rates of growth in plants? Do roots, stems, or leaves grow faster or bigger if fertilizer has been applied to the plant?

Hypotheses

Make a group decision about which of these questions you will test, or make up your own questions. Finally, form testable hypotheses about the questions.

Objectives

In this Biolab, you will:

- **Carry out** a controlled experiment.
- **Observe** the effect of fertilizer on plant growth or seed germination.

Possible Materials

seeds
plant seedlings
water
plastic trays
sand or potting soil
foam cups
fertilizer
balance
metric ruler
graduated cylinder

Safety Precautions

Do not eat seeds or taste chemicals.

PLAN THE EXPERIMENT

Teaching Strategies

- Advise students about what to look for as evidence of germination.
- Review techniques needed for measuring volume, using a balance, reading metric units, and calculating surface area.
- Most commercial fertilizers will provide a dilution reference guide on the container, thus eliminating the need for balances.

Possible Procedures

- For seed germination studies, students will soak seeds in small cups for 12-24 hours using different concentrations of fertilizers. Remove the seeds after soaking, wrap them in paper towels, and slip them into plastic bags. Towels should be moistened with the same liquid in which the seeds were soaked. Control seeds will be soaked

PLAN THE EXPERIMENT

1. Write your experimental plan in the form of a numbered list similar to a recipe. First, list the materials you will need. Then, give details of steps you will take to collect your data.
2. Work in a group. Each group member should have one or more clearly defined tasks such as collecting materials, making observations, recording observations, or cleaning up.
3. Identify the conditions you will hold constant and a single independent variable. The independent variable could be how much fertilizer you apply or the strength of the fertilizer. Decide which dependent variable you will measure and how you will know if your data support your hypothesis.

Check the Plan

1. Review the summary of scientific methods in *Figure 2.11* to see if you have included most of the scientific methods.
2. Does your plan test only one variable, such as the amount of fertilizer added?
3. Have you determined how many seeds or plants you will use in each group and what dependent variable you will measure? Also, have you decided how often you will make measurements?
4. Did you make a data table that compares the observations you will make on the control and experimental groups?
5. ***Make sure your teacher has approved your experimental plan before you proceed further.***
6. Carry out the experiment.

ANALYZE AND CONCLUDE

1. **Identifying Variables** What factors did you hold constant during this experiment?
2. **Checking Your Hypothesis** What conclusions can you draw by using the data you collected? Does your conclusion support your hypothesis?
3. **Interpreting Data** Make a statement explaining how your data did or did not support your hypothesis.

Going Further

Changing Variables Design another experiment to test the effects of fertilizer on seeds or plants of different species. With your teacher's permission, carry out the experiment. Can you conclude that fertilizer is always beneficial to plants?

in water. Daily germination counts can then be made.

- For plant growth and development studies, students will measure concentrations of the fertilizer in the water used on the plants. Daily measurements of leaf area, leaf number, plant height, or stem length will be made and compared to a control.

Data and Observations

Fertilizers, in general, have no effect on germination rates. Fertilizer will stimulate growth and development of plants already germinated. However, too high a concentration of fertilizer can harm plants and result in death (from plasmolysis).

PREPARATION

- Punch small holes in cup bottoms for drainage.
- Start young seedlings 10-12 days prior to the lab. Soak seeds for 24 hours and then place them in small pots or plastic cups for germination and growth. Add sand (or vermiculite) to pots or cups. Keep young plants in light until ready to use.
- Use distilled water on control plants.

Possible Hypotheses

- If fertilizer speeds seed germination, then soaking seeds in fertilizer solution will cause seeds to germinate sooner.
- If fertilizer causes an increase in growth, then plants receiving fertilizer should grow taller and develop larger leaves than those not receiving fertilizer.

ANALYZE AND CONCLUDE

1. Temperature, amount of water or water and fertilizer added to each group of seeds, amount of water or water and fertilizer added to each plant, amount of light each plant received.
2. Data do not support the hypothesis that fertilizer will speed seed germination.
3. As many seeds germinated under control conditions as those soaked in fertilizer.

Assessment

Portfolio: Ask students to write a report describing this experiment. Have them include a graph of their data. L1

Going Further

Have students carry out an experiment to determine if seeds purchased in cans will germinate at the same rate as those purchased in plastic bags. L2

SECTION 2.1

ThinkingLab Draw a conclusion

Purpose

Logical-Mathematical Use scientific data to draw a conclusion about the effectiveness of a weed killer.

Process Skills

interpret scientific illustrations, interpret data

Teaching Strategies

- Review the meanings of the terms *independent variable*, *dependent variable*, and *control.* Provide students with an example of each factor for the experiment the gardener performed.

Thinking Critically

Responses should indicate that the Kilimal appears to be effective in killing dandelions, although one dandelion remains. No evidence shows the Kilimal to be effective at killing other weeds.

Assessment

Skill: Have students develop an experimental procedure for the problem posed in the lab. Students should identify independent and dependent variables as well as the control. L1

Visual Learning

Figure 2.6 Which symbol warns about dangers from radiation? *The second symbol from the top.* Which tells you to wear eye protection? *The symbol showing goggles.* Which alerts you to hazards from acids? *The last symbol; the chemical safety symbol.*

ThinkingLab Draw a Conclusion

Does a weed killer perform as advertised?

Scientific methods are useful for solving problems and answering questions that arise in everyday life. One method all scientists use is observation. A gardener made an observation that dandelions grow as unwanted weeds in many lawns. On television, the gardener heard about a new chemical, Kilimal, that is supposed to be useful for ridding lawns of weeds. The gardener decided to do an experiment to find out if the chemical was really effective on dandelions.

Analysis

The series of diagrams below represent a controlled experiment to examine the effects of Kilimal on dandelions.

1. Observation: A manufacturer claims that its herbicide will get rid of dandelions.
2. Hypothesis: Kilimal will kill dandelions while leaving grass healthy.
3. Experiment: Two identical sections of a lawn overgrown with dandelions are selected. Killimal is applied to the experimental section; the other section is left untreated as a control.
4. Results (data) are obtained.

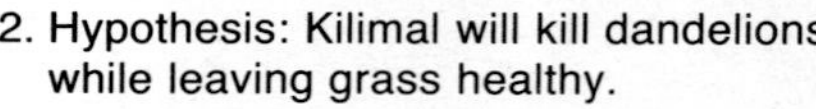

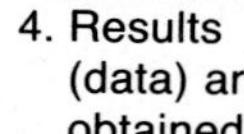
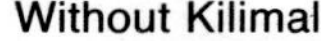

Thinking Critically

According to your interpretation of the illustration, what is your conclusion about the effect of the weed killer on dandelions? What can you say about its effects on other weeds?

SHARP OBJECT SAFETY: This symbol appears when a danger of cuts or punctures caused by the use of sharp objects exists.

RADIOACTIVE SAFETY: This symbol appears when radioactive materials are used.

CLOTHING PROTECTION SAFETY: This symbol appears when substances used could stain or burn clothing.

EYE SAFETY: This symbol appears when a danger to the eyes exists. Safety goggles should be worn when this symbol appears.

CHEMICAL SAFETY: This symbol appears when chemicals used can cause burns or are poisonous if absorbed through the skin.

Maintaining safety

Safety is another important factor that scientists consider when carrying out experiments. Biologists try to minimize hazards to themselves, the people working around them, and the organisms they are studying.

In the experiments in this textbook, you will be alerted to possible safety hazards by safety symbols like those shown in *Figure 2.6.* A **safety symbol** is a symbol that warns you about a danger that may exist from chemicals, electricity, heat, or procedures you will use. Refer to the safety symbols in the *Skill Handbook* at the back of this book before beginning any lab activity in this text. It is your responsibility to maintain the highest safety standards to protect yourself as well as your classmates.

Data gathering

To answer their questions about scientific problems, scientists seek information from their experiments. This information is called **data.** Sometimes, data from experiments are referred to as experimental results.

Often, data are in numerical form, such as the number of elephant calls per minute or the length that corn plants grow per day. Numerical data may be measurements of time, temperature, length, mass, area, volume, or other factors. Numerical data may also be counts, such as the number of bees that visit a flower per day or the number of wheat seeds that germinate.

Figure 2.6

Which symbol warns you about dangers from radiation? Which tells you to wear eye protection? Which alerts you to hazards from acids? You can review all the safety symbols used in this textbook in Appendix C.

Meeting Individual Needs

Students Acquiring English Supply students with the following data showing how many seeds were germinated by three laboratory groups over a three-day period. Group A had two seeds germinate on day 1, four seeds on day 2, and four seeds on day 3. Group B had one seed germinate on day 1, six on day 2, and five on day 3. Group C had two seeds germinate on day 1, four on day 2, and three on day 3. Have students work in groups to prepare a class histogram of the number of seeds germinating each day for all three groups. L1

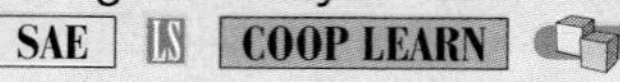

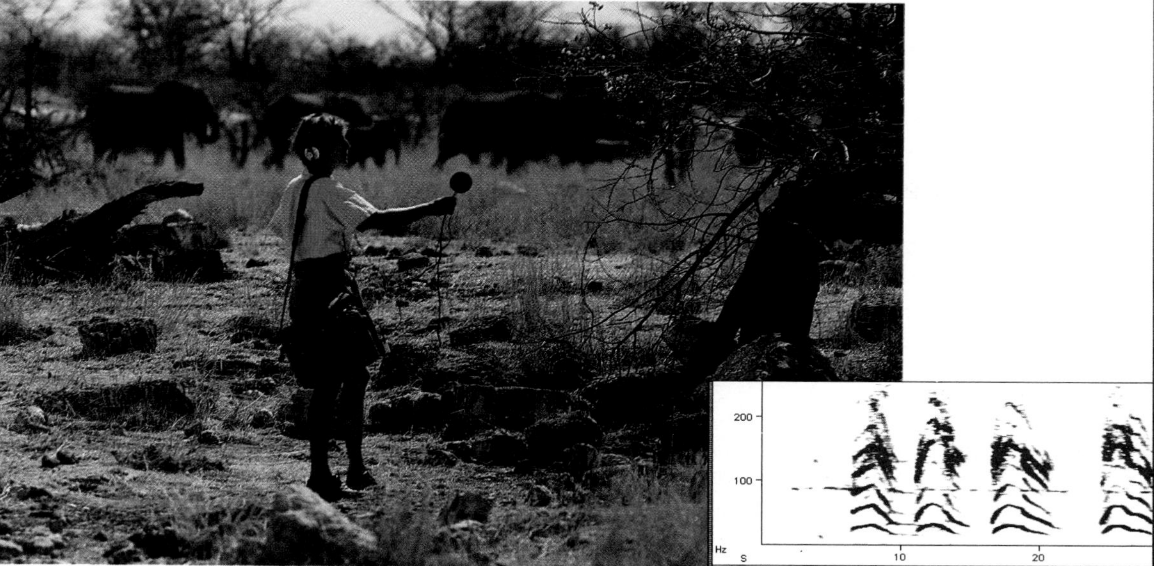

Figure 2.7

In the field, Payne recorded elephant sounds and then played them into a machine that produced a sonogram like the one shown here. A sonogram is a picture of a sound. You may have seen sonograms of bird songs in some field guides. With the sonogram, Payne was able to see when the elephants made sounds and how their pitch varied.

Sometimes data are expressed in verbal form, using words to describe observations during an experiment. The data that Payne gathered were the tape recordings and observations about elephant behavior. When she first played the tape recordings she had made of the elephants, Payne heard only background noise. This result was not unexpected, for she hadn't heard any other sound when she made the recordings. However, when the tapes were played faster to increase the frequency of the sound waves, Payne discovered that she had recorded hundreds of elephant calls. Some of these calls are represented by the sonogram in *Figure 2.7.*

Thinking about what happened

Having the data from the experiment did not end the scientific process for Katharine Payne. Often, the thought process that goes into analyzing experimental data takes the greatest amount of a scientist's time. After careful review of the results, the scientist must come to a conclusion: Was the hypothesis supported by the data? Or was it not supported? Or are more data needed? Data from an experiment may be considered confirmed only if repeating that experiment several times yields similar results.

Payne wanted to learn more than she could from her experiment at the zoo. After analyzing her data, she, like most scientists, had more questions than she had before her experiment. In order to compare her results and conclusions with studies by other scientists in the same field, she researched the published literature for more information on elephant communication. She also began to think of experiments she might carry out.

Different Viewpoints in Biology

Time Will Tell

Do scientists always arrive at the correct conclusion after an analysis of their data? This story will help answer the question.

In 1960, studies showed that a large number of convicted male criminals had an extra Y chromosome. It was also noted that these men tended to be taller than the general population. Through an analysis of data, scientists concluded that the extra Y chromosome predisposed a person to a life of crime.

Later, scientists compared prisoners with the extra Y chromosome to a control group of prisoners who had only one Y chromosome. Analysis of these data led scientists to reject some of their earlier conclusions. These scientists concluded that the extra Y chromosome predisposes males to extra height, but not a life of criminal behavior.

GLENCOE TECHNOLOGY

Videotape

The Secret of Life
Use the videotape *On the Brink: Portraits of Modern Science* to show the contributions of scientists working on the cutting edge.

PORTFOLIO

Communication in Science Locate a short article from a scientific journal that is not overly technical (*Science News* or *Discover*). Make a copy of the article for each student. Ask them to read the article and then write an abstract or summary of the article. L2 LS

STUDENT JOURNAL

Animal Communication Ask students to write a short story or paragraph that relates a personal experience that they have had with animal communication. Their paragraph should attempt to explain the "message" being conveyed by the animal to them. L1 LS

Focus On

What is a scientific theory?

Purpose

Students will learn from examples how theories are formed from repeated observations, comparative studies, and discussions among scientists over many decades.

Teaching Strategies

- Ask students to discuss what they know about dinosaurs. Ask them how they have learned about these amazing and extinct animals.
- Have students put forward their own ideas on how dinosaurs lived from day to day. Ask them if their ideas are hypotheses, speculations, beliefs, or theories. Have them begin to distinguish among these different ideas by asking them to state in a complete sentence what they know about another topic, such as computers.
- If possible, visit a dinosaur display in a local museum, and have students make notes in their journals about the different theories about dinosaur behavior explained on the exhibits.

Visual Learning

Figure 2.6 Have students study the skeleton of *Iguanodon*, art of *Anatosaurus*, and art of the hadrosaur foot. Have them list the structures that support the hypotheses that hadrosaurs lived in water and on land. L1

Focus On

What is a scientific theory?

Have you ever had a conversation in which you mentioned that you had a theory about something? In casual usage, the word theory *means an unproven assumption about a set of facts. A scientific theory, however, is an explanation of a natural phenomenon that is supported by a large body of scientific evidence obtained from many different investigations and observations.*

The scientific process begins with observations of the natural world. These observations lead to hypotheses, data collection, and experimentation. If weaknesses are observed, the hypotheses will be rejected or modified. When a hypothesis has been tested over and over, and there is little evidence to cause it to be rejected, eventually that hypothesis may become a theory.

In this feature, you can read about the process that led some scientists to develop new theories about the dinosaurs.

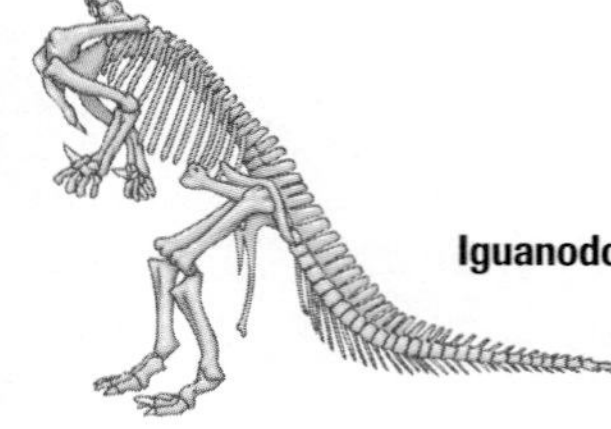

Iguanodon

Observing

Imagine finding a leg bone taller than you are! Fossils, such as this leg bone of a sauropod dinosaur, have been discovered by people for hundreds of years. The first person to reconstruct a dinosaur skeleton called it *Iguanodon,* meaning "iguana tooth," because it had bones and teeth like an iguana. By 1842 these extinct animals were named *dinosaurs,* meaning "terrible lizards."

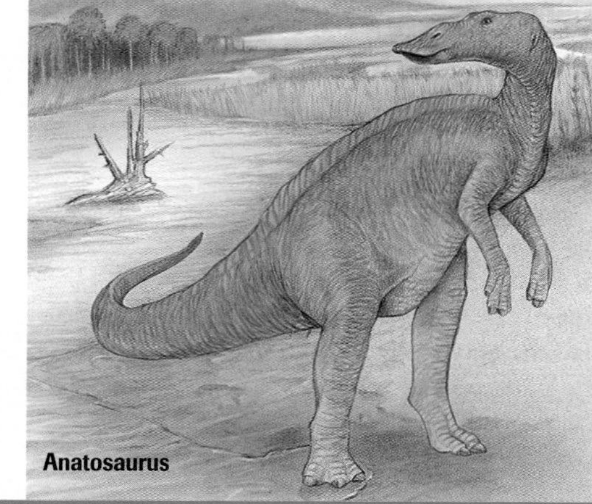

Anatosaurus

Making Hypotheses

Reptiles are ectotherms—animals with a body temperature that depends upon the temperature of their surroundings. Most ectotherms are slow-growing, slow-moving animals. Because dinosaur skeletons looked like those of reptiles, scientists argued that they, too, had been slow-growing, slow-moving animals. Because the most complete dinosaur fossils occurred in rocks formed at the bottom of bodies of water, scientists hypothesized that dinosaurs lived in water, perhaps because water helped support their great weight. This hypothesis gained support when skeletons of the duck-billed dinosaurs, the hadrosaurs, were discovered. Hadrosaurs had broad, flat, ducklike bills, which scientists suggested were used to collect and eat plants in ponds or lakes, like ducks.

STUDENT JOURNAL

A Dino-bird? A new dinosaur fossil discovered in China appears to have had downy feathers along its neck and backbone. Ask students to make a hypothesis about how and why feathers first evolved. Have them write down supporting evidence for their hypotheses. L2

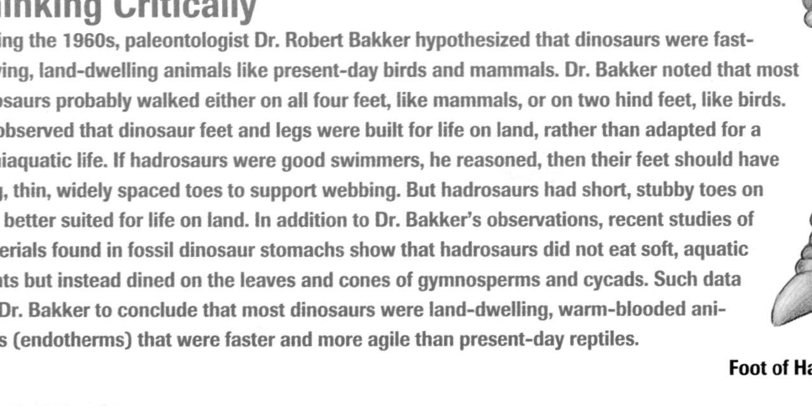

Thinking Critically

During the 1960s, paleontologist Dr. Robert Bakker hypothesized that dinosaurs were fast-moving, land-dwelling animals like present-day birds and mammals. Dr. Bakker noted that most dinosaurs probably walked either on all four feet, like mammals, or on two hind feet, like birds. He observed that dinosaur feet and legs were built for life on land, rather than adapted for a semiaquatic life. If hadrosaurs were good swimmers, he reasoned, then their feet should have long, thin, widely spaced toes to support webbing. But hadrosaurs had short, stubby toes on feet better suited for life on land. In addition to Dr. Bakker's observations, recent studies of materials found in fossil dinosaur stomachs show that hadrosaurs did not eat soft, aquatic plants but instead dined on the leaves and cones of gymnosperms and cycads. Such data led Dr. Bakker to conclude that most dinosaurs were land-dwelling, warm-blooded animals (endotherms) that were faster and more agile than present-day reptiles.

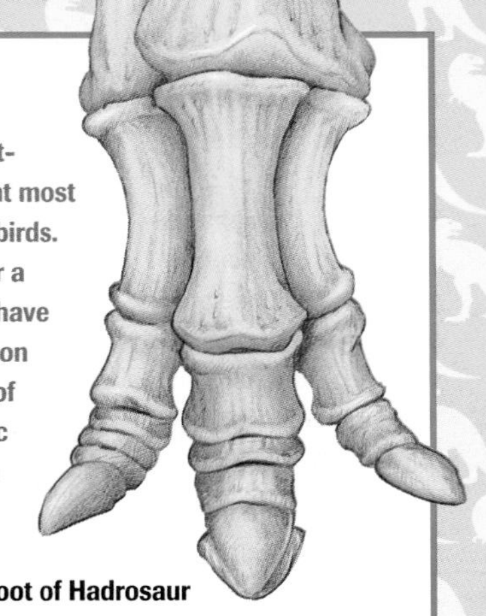

Foot of Hadrosaur

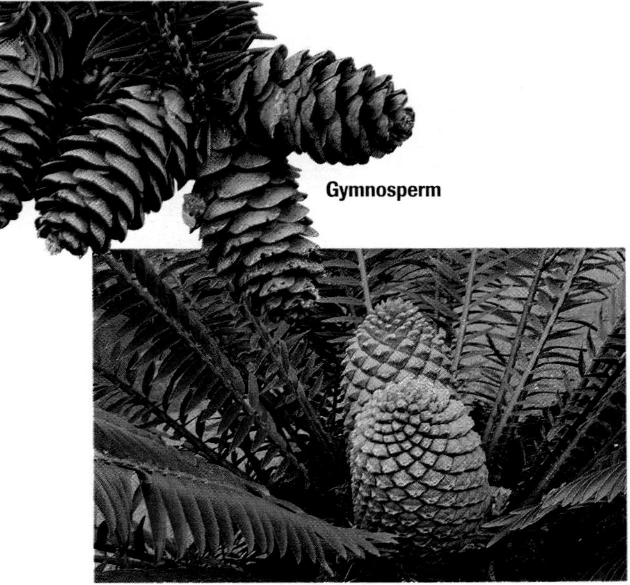

Gymnosperm

Cycad

Collecting Data

To test his hypothesis that dinosaurs were land-dwelling endotherms, Dr. Bakker collected data on dinosaur skeletons and bones. He discovered reports published in the1950s by two scientists who had compared thousands of cross sections of dinosaur bones with those of reptiles, birds, and mammals. They found that slow-growing bones of ectotherms are dense, with few channels for blood vessels. Fast-growing bones of endotherms are less dense, with many channels for blood vessels. These scientists concluded that dinosaur bones looked most like mammal bones. Dr. Bakker confirmed these findings by collecting similar data from other sources.

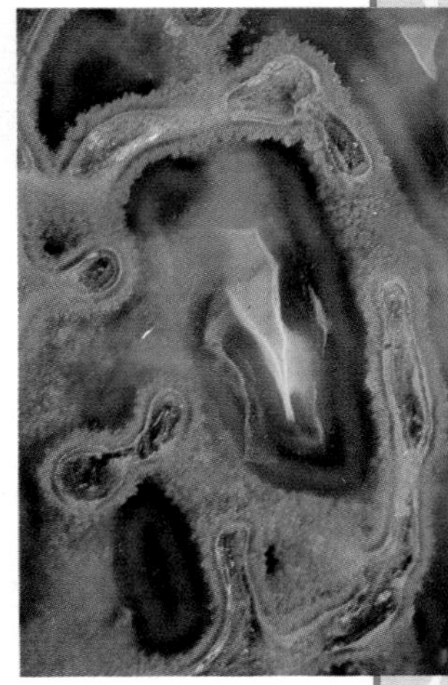

Forming Theories

Data collected by many other scientists about dinosaur bones, growth patterns, and behavior have changed our theories on dinosaurs. Were some dinosaurs endotherms and others ectotherms? Did dinosaurs have their own unique physiology that was neither reptile-like nor mammal-like? Scientific theories about the lives of the dinosaurs continue to be proposed as new fossils are found and new tools to study fossils are developed.

EXPANDING YOUR VIEW

Applying Concepts Dr. Bakker's research led to a different theory regarding the body temperature of dinosaurs. As new fossils are found and new tools to study fossils are developed, paleontologists will continue to replace existing theories with newer ones. What are some causes for a scientific theory to be changed?

STUDENT JOURNAL

New Research on Dinosaurs Although Dr. Bakker's bone evidence shows a link between dinosaurs and endotherms, other researchers dispute this and point to factors, such as nasal cavities, that show dinosaurs were more like ectotherms. Have students research the newest hypotheses about dinosaurs by reading, "A Cold Hard Look at Dinosaurs" by Virginia Morell, in *Discover,* December 1996. Ask them to write a summary of the article and list the factors that support both ectothermy and endothermy as the homeostatic mechanisms for dinosaur metabolism. L3

Focus On

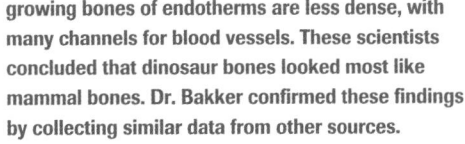

GLENCOE TECHNOLOGY

Videodisc

STVS: Animals
Disc 5, Side 1
Tagging Ants (Ch. 16)

STVS: Ecology
Disc 6, Side 1
Research in the Pinelands (Ch. 5)

The Infinite Voyage: The Great Dinosaur Hunt
Where the Great Hunt Began (Ch. 1)

Answers to Expanding Your View

1. **Applying Concepts** Students may list a number of reasons why scientific theories change. New data, more up-to-date observations, improved testing methods, and new discoveries all contribute to changes made to scientific theories.

SECTION 2.1

Demonstration

Bring copies of scientific journals to class and allow students to examine them. Ask students to speculate as to which branch of science each journal addresses.

Bioethics

Drug Testing Procedures A control group is an important element in the evaluation of any experiment. But, is a control group ethical when human lives are at stake? This problem was faced by researchers who discovered a drug called azidothymidine (AZT).

In early experimental studies, AZT was noted to extend the lives of AIDS patients. However, the drug had not been tested on a large group of AIDS patients. To correctly test AZT, a control group was needed. The control group would have consisted of 2000 AIDS patients who would receive a placebo, or sugar pill, instead of AZT. The experimental group would consist of 2000 AIDS patients who would receive AZT.

From an ethical viewpoint, a question arises as to whether a researcher should give an AIDS patient a sugar pill knowing that this placebo will do nothing to help the patient. The Food and Drug Administration agrees that this is ethically wrong. In the case of AZT, no control group was used and all 4000 AIDS patients were given the experimental drug.

Figure 2.8

This elephant herd in Africa is making its daily trip to the water hole. If these elephants in nature make sounds similar to the ones produced by elephants in zoos, do you think those sounds might have similar meanings? How could you find out?

It is clear that the elephants in the zoo make sounds that people cannot hear, but do elephants in nature, like those shown in *Figure 2.8*, make similar sounds? If the sounds are communication, how does the behavior of the elephants change as a result of the sound? To conduct field experiments to answer these questions, Payne and several colleagues traveled to Namibia in southwest Africa, where elephant herds roam.

Reporting results

Results and conclusions of experiments are reported in scientific journals, where they become open to examination by other scientists. Hundreds of scientific journals, such as those shown in *Figure 2.9*, are published weekly or monthly. In fact, scientists usually spend a large part of their time just reading journal articles trying to keep up with new information being reported.

Figure 2.9

The amount of information published every day in scientific journals is more than any single scientist could read. Fortunately, scientists also have access to computer databases that contain summaries of scientific articles, both old and new. It is important for a scientist to report data in the most easily understood way. Therefore, data are usually presented in tables, charts, and graphs such as those shown here.

STUDENT JOURNAL

Field Research Have students look in *Discover, National Geographic,* or similar publications to find an article that describes field research. Ask them to write a report on the article, contrasting the scientists' work in the field with the work they had to continue when back at the laboratory. L1 LS

Verifying results

Data and conclusions are shared with other scientists for an important reason: after results of an investigation have been published, other scientists may try to verify the results by repeating the experiments. If the results from the original experiment occur again, the later experiments provide support for the hypothesis. When a hypothesis is supported by additional data from the same or other scientists, the hypothesis is considered valid and is generally accepted by the scientific community.

To verify her original data from the Portland Zoo, Payne and her research team in Namibia set up equipment similar to that used at the zoo. As shown in ***Figure 2.10***, the researchers placed their equipment at distances of more than a mile apart. In that way, they could observe whether widely separated elephants heard and responded to the sounds made by other elephants. The team made many trials before reporting their results.

When scientists publish the results of their investigations, other scientists, such as Katharine Payne, can relate their work to the published data. For example, other biologists studying elephants in Africa had published observations of their behaviors. Thus, Payne knew what behaviors might be the result of sound communication. Her team found that female elephants emit certain sounds in order to attract mates. Other sounds are produced by bull elephants to warn other males away from receptive females.

Theories and laws

People use the word *theory* in everyday life much differently from the way scientists use this word. You may have heard someone say that he or she has a theory that a particular football team will win the Super Bowl this year. What the person really means is that he or she believes one team will play better for some reason.

Figure 2.10

Payne's research team observed groups of elephants from a tower. Microphones were set up at various locations in the field. Elephant sounds were recorded and activity was videotaped at the same time to see if there was a relationship between elephant sounds in one location and elephant behavior in another location.

SECTION 2.1

Brainstorming

A student repeats an experiment several times and each time records different data. Have students offer possible reasons why an experiment might yield different data for different trials. *Reasons include failure to keep all factors but one the same, errors in the recording of the data, or errors in mathematical treatment of data.*

Science, Technology, and Society

Technology—Friend or Foe?

Technology in the health field has contributed to a tenfold increase in the human population over the past three centuries. However, this trend may not always be a blessing. Consider the impact of increased population on the environment. Increased populations require more food, more living space, more energy, and more manufactured products. How might future technology help resolve some of the problems to which earlier technology has contributed?

3 Assess

Check for Understanding

Provide students with the steps of scientific method in a scrambled order. Ask students to sequence the steps in the correct order. L1

Reteach

Ask students to outline the steps of the scientific method. At each level on the outline, have them provide an example taken from Dr. Payne's studies of elephant communication. L1 SAE

PROJECT

Conducting Library Research in Science

Bring a variety of periodicals to class. They could be checked out of your school library. Ask students to record in their journals which magazines might aid Dr. Payne in her research with elephant communication. Ask them to explain how each magazine might specifically help her. L1

SECTION 2.1

Extension

Have students look up cell theory in this text. Ask them to speculate about the hypotheses that may have been made by scientists who first discovered cells. L1

Assessment

Skill: Provide each student with a piece of laboratory equipment and have them list five observations about the equipment and use inductive reasoning to suggest how the equipment might be used. L1 SAE

4 Close

Demonstration

Display pieces of laboratory equipment (compound microscope, graduated cylinder, stereomicroscope) and safety equipment or clothing (fire extinguisher, safety goggles, laboratory apron, etc.) that students will use throughout their study of biology. For each piece of equipment, identify its function and proper methods of use. As a followup activity, you may wish to set up lab stations at which students are required to demonstrate their knowledge of each piece of equipment.

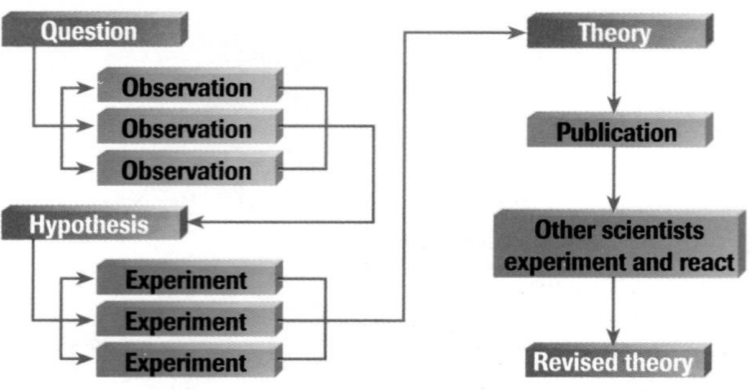

Figure 2.11

This chart presents an idealized version of the scientific method. Not all scientists carry out every step to answer a question. Not all investigations lead to published theories. Theories are often revised as new information is gathered.

Of course, much more evidence is needed to support a theory in science.

In science, a hypothesis that is supported by many separate observations and experiments, usually over a long period of time, becomes a theory. A **theory** is an explanation of a natural phenomenon that is supported by a large body of scientific evidence obtained from many different investigations and observations. A theory results from continual verification and refinement of a hypothesis.

For example, Katharine Payne made a hypothesis that elephants communicate by sounds that are below the hearing capabilities of humans. She collected data while observing elephants in zoos. These data supported her hypothesis. She then went with a research team to Africa to confirm her results with elephants in the wild. Because these new data and those from other groups of researchers supported Payne's hypothesis, the hypothesis became a theory about how elephants communicate.

Scientists also recognize certain facts of nature called laws or principles. The fact that a dropped apple falls to Earth is an illustration of the law of gravity. *Figure 2.11* presents a schematic diagram of scientific processes.

Section Review

Understanding Concepts

1. Suppose you have planted a row of peppers in your vegetable garden. The plants in part of the row are tall and green but produce only a small number of fruits. The remainder of the plants are smaller and slightly yellowish but produce many more peppers. You recall that when you were preparing the soil, you ran out of one kind of fertilizer and finished up with a second type. You decide that this second type caused the plants to grow tall but bear few peppers. Is this conclusion an example of inductive reasoning or deductive reasoning? Explain your choice.
2. Describe a controlled experiment you could perform to determine whether ants are more attracted to butter or to honey.
3. What is the difference between a theory and a hypothesis?

Thinking Critically

4. Read the paragraph under *Experimenting* on page 29 about making cookies. Describe a way that the baker might do a scientific experiment with the cookie recipe.

Skill Review

5. **Interpreting Scientific Illustrations** Use *Figure 2.11.* What happens when a hypothesis is not confirmed? What is the function of other scientists in the scientific process? What does the position of the word *theory* in the diagram indicate about the strength of a scientific theory? For more help, refer to Thinking Critically in the *Skill Handbook.*

Answers to Section Review

1. Inductive reasoning, because several separate facts were considered to develop a single testable conclusion.
2. Set up an experimental chamber. Within a specific amount of time, count and record how many ants move to the butter when placed a specific distance from the ants. Repeat several times. Perform the same experiment using honey in place of the butter.
3. A hypothesis may be a single testable explanation for a question while a theory is a refined explanation supported by many different experiments.

Thinking Critically

4. The baker should vary only one factor at a time to determine the effects of that factor alone.

Skill Review

5. A new, revised hypothesis is tested. Other scientists confirm and build upon the work. The theory is an explanation that has been strongly tested by experimentation.

SECTION

2.2 The Nature of Biology

Part of science is learning many of the known facts about the world around us. More important, though, is that scientists use these known facts to discover new problems, make hypotheses, design experiments, interpret data, and draw conclusions. In short, science is not just a body of facts and ideas but also a process of study by which we come to understand the natural world.

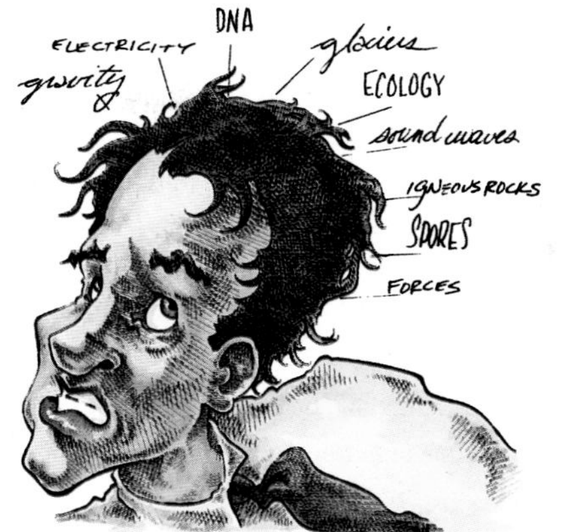

Section Preview

Objectives

Compare and contrast quantitative and descriptive research.

Explain why science and technology cannot solve all problems.

Key Terms

technology
ethics

Kinds of Research

You learned in the first part of this chapter that scientists use a variety of methods to test their hypotheses about the natural world. Scientific research can usually be classified into one of two main types, quantitative or descriptive.

Quantitative research

Most biologists conduct controlled experiments that result in counts or measurements—that is, numerical data. These kinds of experiments occur in quantitative research.

Data obtained in quantitative research may be used to make a graph or table. Suppose, for example, that a biologist is conducting research to count the number of microscopic organisms called *Paramecium* that survive at a given temperature. The study is an example of quantitative research. *Figure 2.12* shows the data as a graph.

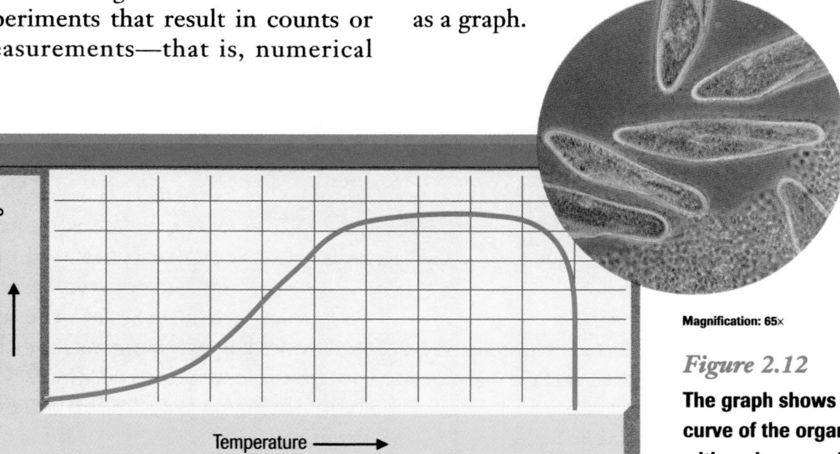

Figure 2.12
The graph shows the survival curve of the organism *Paramecium* with an increase in temperature.

Program Resources

Section Focus Master 4 L1 SAE
Reinforcement and Study Guide, pp. 7-8 L1
Biolab and Minilab Worksheets, p. 4 L1
Laboratory Manual, pp. 9-12 L2

SECTION 2.2

Prepare

Key Concepts

Students are exposed to types of research that may be carried out by scientists. Differences between quantitative and descriptive research are explained and illustrated. The role of research and its application and use by society as technology are then examined.

Block Scheduling

Look for this symbol for strategies that are useful in a block scheduling format. For more information on block scheduling, refer to Section 2.2 in the **Lesson Plans** booklet.

Materials

- Obtain metric rulers and pine needles for the Focus activity.
- Purchase four different brands of paper toweling and gather medicine droppers for the Minilab.

1 Focus

Bellringer

Before presenting the lesson, display **Section Focus Master 4** on the overhead projector and have students answer the accompanying questions. L1 SAE

Activity

LS **Logical-Mathematical** Provide students with a metric ruler and a clump of pine needles. Ask them to make two lists of their observations: one which describes the leaves with words and another that uses measurements. Have them compare their lists with classmates.

SECTION 2.2

2 Teach

MiniLab

Purpose

Logical-Mathematical Students will use scientific methods to evaluate products.

Process Skills

measure in SI, interpret data

Teaching Strategies

- Prior to the experiment, ask students to rank the towel brands from least to most absorbent. Ask them why guessing is not as reliable as experimentation.
- Tell students to add drops to the center of the square.

Expected Results

Data will vary. Typically, it takes 12 to 18 drops to reach total absorption.

Analysis

1. no
2. No, the towel may be poor at absorbing oil or too expensive.
3. experimentation using quantitative data

Assessment

Portfolio: Have students rank the towels from least to most absorbent and from least to most expensive. Have them summarize how absorbency and cost are related. L1 P

MiniLab

How do you decide which paper towel to buy?

Suppose that you are the new manager of a restaurant. The kitchen staff has complained that the paper towels they're given to clean up spills are not absorbent. The procedures manual for your restaurant chain says that you are allowed to choose from among the four lowest-priced brands available in your area. You decide to conduct an experiment to decide which brand of paper towel you should purchase.

Procedure

1. Cut a 5 cm × 5 cm piece from each of four brands of paper towel. Lay each piece on a smooth, level, waterproof surface.
2. Add one drop of water to each square.
3. Continue to add drops of water until the square can absorb no more.
4. Record your observations and make a graph of your results.

Analysis

1. Did all the squares absorb equal amounts of water?
2. If one brand of paper towel absorbs more water than the others, can you conclude that it is the brand you should buy? Explain.
3. Which scientific methods did you use to answer the question of which brand of paper towel is most absorbent?

Figure 2.13

Each of these animals can be seen in zoos. Do you think these animals behave in the same way in zoos as in nature?

Toucans live in the rain forests of South America.

Giant tortoises live on several of the Galapagos Islands in the Pacific Ocean. The bird is a Galapagos hawk.

Measuring in the International System

It is important that scientific research be understandable to scientists around the world. For example, what if scientists in the United States reported quantitative data in inches, feet, yards, ounces, pounds, pints, quarts, and gallons? People in many other countries would have trouble understanding these data because they are unfamiliar with the English system of measurement. Instead, scientists always report measurements in a modern form of the metric system called the International System of Measurement, abbreviated SI.

One advantage of SI is that there are only a few basic units, and nearly all measurements can be expressed in these units or combinations of them. The greatest advantage, though, is that SI is a decimal system. Measurements can be expressed in multiples of tens or tenths of a basic unit by applying a standard set of prefixes to the unit. In biology, the metric units you will encounter most often are meter (length), gram (mass), liter (volume), second (time), and Celsius degree (temperature). For a thorough review of measurement in SI, see the Skill Handbook.

Descriptive research

Do you think the behavior of the animals shown in *Figure 2.13* would be easier to explain with numbers or with written descriptions of what the animals did? Observational data—that is, written descriptions of what scientists observe—are often just as important in the solution of a scientific problem as numerical data.

CULTURAL DIVERSITY

Units and Standards from a Cross-cultural Perspective

The SI system is now used in 95 percent of the countries in the world. The modern system of SI measurement provides a standardized system of measurement that makes scientific communication easier. Many early systems of measurement were not standardized. For example, the ancient Egyptians used a unit of measurement called the cubit, which was based on the length of the arm from the elbow to the fingertips. Because sizes of individuals varied, the size of the unit also varied. In England, the foot was equal to the length of the foot of the King of England. When a new king came to power, the length changed according to the size of his foot. Encourage students to conduct research on other measuring systems used around the world. Have students create a visual display of their findings that can be used as part of a class bulletin board. L1 P

Figure 2.14

Gorillas in nature need large areas in which to roam. They also form family groups. This grouping makes foraging for food more efficient and helps protect them from enemies. Describe another biological study that could better be conducted in the field than in the laboratory.

When scientists use purely observational data, they are carrying out descriptive research. Descriptive research is useful because some phenomena aren't appropriate for quantitative research. For example, how a particular wild animal reacts to events in its environment cannot easily be illustrated with numbers.

An example of descriptive research in the field would be a study of how gorilla families behave. Look at *Figure 2.14* and you will quickly realize that it would be very difficult to duplicate the natural environment of the gorillas in a laboratory.

Science and Society

The road to scientific discovery includes making observations, formulating hypotheses, doing experiments, collecting and analyzing data, drawing conclusions, and reporting those conclusions in scientific journals. No matter what methods scientists choose, their research often provides society with important information that can be put to practical use.

Maybe you have heard people blame scientists for the existence of nuclear bombs or controversial drugs. To comprehend the nature of science in general and biology in particular, people must understand that knowledge gained through scientific research is never inherently good or bad. Notions of good and bad arise in the context of human social, ethical, and moral concerns. Scientists might not consider all the possible applications for the products of their research when planning their investigations. Society as a whole must take responsibility for the use of discoveries.

Can science answer all questions?

Some questions are simply not in the realm of science. Many of these involve questions of good versus evil, ugly versus beautiful, or similar judgments. If a question is not testable using scientific methods, the question is not science. However, this does not mean that the question is unimportant.

Enrichment

Have students each list quantitative and descriptive data about five common animals. Have them each present their data to the class and have the class use the data to identify the animals being described. Ask which type of data was most useful in attempting to identify the animals and have students explain why. *Quantitative data such as number of appendages and size may be too similar for many of the animals, whereas descriptive data such as presence of fur or feathers, common colors, and descriptions of behavioral traits may be more specific.*

interNET CONNECTION

Follow the link for this chapter on the Glencoe Homepage at **www.glencoe.com/sec/science** to find out more about scientific methods in biology.

PORTFOLIO

Making Predictions Ask students to carry out the following activity. Have them predict the chance that a coin, when flipped, should come up heads. *50%* How many heads should appear if a coin is flipped 10 times? *1/2 × 10 or 5* How many heads will appear if a coin is flipped 100 times? *1/2 × 100 or 50* Ask them to carry out the coin tosses and record their results. Have them use the activity to explain in their portfolios if scientists can predict the results of an experiment with 100% certainty. Have them explain what the advantage is to using a large sample or many trials in an experiment. *Large samples increase the likelihood that the sample is representative.* L1 LS COOP LEARN

SECTION 2.2

A BROADER VIEW

Life as a Field Biologist

Purpose

To illustrate how a scientist gathers data and how the data can be put to practical use in attempting to save a species from extinction.

Teaching Strategies

- Provide students with background as to the characteristics of clams and other mollusks.
- Point out the natural behavior of these animals as illustrated in the feature. Ask how the behavior of shell closing is a protective adaptation.
- Elicit from students what they believe is the value of attempting to prevent a giant clam from becoming extinct. Have them explain how human intervention may be accelerating this species' extinction.

Thinking Critically

The research team plans to capture and remove certain clams that show promise of breeding. Properly recording data as to location of a specimen will ensure that they are removing the correct animal. Other scientists may have done similar research on other animals that can provide answers to questions that may arise, thus saving the research team time and energy.

A BROADER VIEW

Life as a Field Biologist

Because many species of plants and animals, such as the giant panda, are endangered, biologists sometimes attempt to save endangered animals by breeding them in captivity.

Giant clams of the Pacific With a splash, Dr. Suzanne Williams rolls backward over the side of the boat and begins her descent to the coral reef below. Dr. Williams is a marine biologist who studies giant clams, the largest living mollusks. Some species grow to five feet across and weigh 1000 pounds. Several species of giant clams have nearly disappeared because of over-fishing for their meat and shells. Dr. Williams is working to restore giant clam populations by breeding them in captivity.

Working underwater Because a giant clam closes its shell rapidly when disturbed, Dr. Williams and the divers working with her approach each clam slowly. They cautiously move to within inches of it before quickly inserting a metal wedge between the two halves of the shell. Now, try as it might, the clam cannot close its shell all the way. Dr. Williams uses scissors to snip away a small piece of tissue. She places the sample in a self-sealing bag while another diver measures the clam and records data about its size, species, and location on an underwater slate.

In the ship's laboratory In the ship's laboratory, Dr. Williams carefully lowers the samples into a container of liquid nitrogen, where they freeze instantly. By the time she has finished preserving the samples, the scuba tanks have been refilled and everyone is ready for another dive.

Back on land After ten rigorous days at sea, Dr. Williams returns to her laboratory on land. There she runs a variety of genetics tests on all of the frozen samples. This information will help biologists select the best clams to use in captive-breeding programs.

Thinking Critically

Thousands of coral reefs are scattered throughout the Pacific Ocean. Why is it important for marine researchers like Dr. Suzanne Williams to accurately record information about where and how samples were collected?

Consider a particular question that is not testable. Some people assert that if a black cat crosses your path, you will have bad luck as shown in *Figure 2.15*. On the surface, that hypothesis appears to be one that you could test. But what is bad luck, and how long would you have to wait for the bad luck to occur? How would you distinguish between bad luck caused by the black cat and bad luck that occurs at random? Once you examine the question, you can see there is no way to test it scientifically because you cannot devise a controlled experiment that would yield valid data.

Can technology solve all problems?

Science attempts to explain how and why things happen. Scientific study that is carried out mainly for the sake of knowledge—with no immediate interest in applying the results to daily living—is called pure science.

However, much of pure science eventually does have an impact on people's lives. Have you ever thought about what it was like to live in the world before refrigerators, electric lights, and modern lifesaving medical equipment existed? These and other inventions are indirect results of research done by scientists in many different fields.

Meeting Individual Needs

Gifted Have students list the heights of their classmates in centimeters. Ask them to use their list to calculate average height, describe the range of heights, determine the most common height, and explain whether the original data and computational results are descriptive or quantitative. L2 LS

Figure 2.15

In cartoons, black cats always cause bad luck when they cross someone's path. If bad luck due to black cats occurred as reliably and as swiftly in real life as it does in cartoons, it really would be scientifically testable.

Other scientists work in research that has obvious and immediate applications. **Technology** is the application of scientific research to society's needs and problems. It is concerned with making improvements in human life and the world around us. Technology has helped increase the production of food, reduced the amount of manual labor needed to make products and raise crops, and aided in the reduction of wastes and environmental pollution.

The advance of technology has benefited humans in numerous ways, but it has also resulted in some serious problems. For example, suppose irrigation technology is used to boost the production of food crops in one area. If the irrigation is used for too long a time, the soil may become depleted of minerals or the evaporation of the irrigation water may leave deposits of mineral salts in the soil, making it useless for growing crops, as illustrated in *Figure 2.16.*

Considering ethics in science

Most scientists would state that scientific research and its results are neither right nor wrong, neither good nor bad. **Ethics** is a study of the standards of what is right and wrong.

Figure 2.16

One example of a possible harmful side effect of technology is the deterioration of soil due to a buildup of salts to such high levels that crops cannot grow. In the field shown here, the technology of irrigation initially appeared to solve one problem—that of low crop yield—but actually later caused a different problem of too many mineral salts.

Misconception

Many people equate science and technology, believing that all science should lead to a product or service. Explain that science does not have as its goal the production of some sort of technology. Rather, science occurs and the technology follows as an aside.

GLENCOE TECHNOLOGY

Videodisc

The Infinite Voyage: Unseen Worlds

Supercomputer Models: Photosynthesis, Space Exploration, Weather (Ch. 2)

3 Assess

Check for Understanding

Have students provide an example of: quantitative research, descriptive research, a contribution of technology, and an ethical issue in science. L1

PROJECT

Recognizing Technology

Interpersonal Ask students to work in cooperative groups to prepare a time line which shows the technological advances in one area of science. As an alternative, have the group prepare a report on one technological advance that has had a direct impact on the life of one of the members of the group. L1 COOP LEARN

SECTION 2.2

Reteach

Ask students to prepare an outline of the major concepts of this section. L1 SAE

Extension

Have students research the idea of "being able to beat cancer with a strong positive mental attitude." Have them explain why it may be difficult to scientifically evaluate how a positive mental attitude contributes to recovery from disease. L2

Assessment

Skill: Have students measure their arm span and palm width in centimeters. Convert these measurements to millimeters and meters. L1 SAE

4 Close

Discussion

Describe one possible benefit or spinoff that might come from the study of: how birds find their way during migration; longer lasting batteries; or bat echolocation.

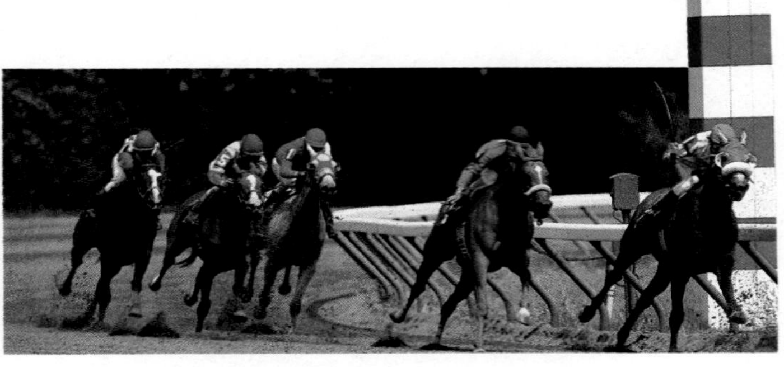

Figure 2.17

Steroids and tranquilizers are drugs that have been developed by scientists to treat certain disorders. Sometimes these drugs are used illegally and unethically to artificially enhance the performance of racehorses. This treatment may produce short-term benefits at the expense of the long-term health of the animal.

Frequently, ethics issues move from the scientific arena to the political, moral, and social arenas and back, as is the case with bioengineering technology. In the 1980s, the United States approved the use of recombinant human growth hormone (hGH) for children showing dwarfism. This synthetic hormone significantly increases the height of these children deficient in natural growth hormone. A concern that some people have is that parents might seek the drug for their shorter-than-average children just to give them the perceived advantage of being taller. *Figure 2.17* illustrates a different ethical problem.

Today, more scientists feel that they should become involved in the decisions about the consequences of their work and how that work is applied. Even so, once a scientific discovery is made, it is the people of a society who make those decisions.

Connecting Ideas

Science is not only a body of facts, but also a process. The methods that scientists employ to help them explore and understand the physical world include observing, forming hypotheses, experimenting, and publishing results and conclusions so that they may (or may not) be verified. What kinds of hypotheses do biologists form about the living world? How do they use scientific methods to learn about the interactions among organisms and between organisms and their environment? Answers to these questions will be revealed as you study and understand basic ecology.

Section Review

Understanding Concepts

1. Why is it important that scientists repeat their experiments whenever possible?
2. Compare and contrast quantitative and descriptive research.
3. Why is science considered to be a combination of fact and process?

Thinking Critically

4. Biomedical research has led to the development of technology that can keep elderly, very ill patients alive. How does the statement "The results of research aren't good or bad; they just are" apply to such research?

Skill Review

5. **Making and Using Graphs** Look again at *Figure 2.12.* Why do you think the high-temperature side of the graph drops off more sharply than the low-temperature side? For more help, refer to Organizing Information in the *Skill Handbook.*

Answers to Section Review

1. in order to see if the experiment's results are repeatable, thus confirming their authenticity
2. Quantitative research reports data in numerical values based on measuring while descriptive research reports data in written descriptions based on observations.
3. It's important that a scientist have a background of factual knowledge in his or her field. The process serves to increase that knowledge.

Thinking Critically

4. The biomedical researchers did not seek to deal with the possible applications of that research. They sought to increase knowledge. The application of the resulting technology is a question society must answer.

Skill Review

5. Paramecia die at temperatures above a certain point. This results in a rapid drop in numbers once this temperature is reached. They are better able to survive as low temperatures rise, thus the graph reflects this increased survival.

Chapter 2 Review

REVIEWING MAIN IDEAS

2.1 Problem-Solving Methods in Biology

- Based on observations of the natural world, scientists use inductive reasoning to form hypotheses. They then use deductive reasoning to help them develop their hypotheses and design investigations to test them.
- Biologists use controlled experiments to obtain data that either do or do not support a hypothesis. By publishing the results and conclusions of an experiment, a scientist allows others to try to verify the results. Repeated verification over time leads to the development of a theory.

2.2 The Nature of Biology

- Biologists do their work in laboratories and in the field. They collect both quantitative and descriptive data from their experiments and investigations.
- Scientists conduct investigations in order to increase knowledge about the natural world. Questions that are not testable are not in the realm of science. Scientific results may help solve some problems, but not all, and ethical issues must be decided by all of society, not just scientists.

Key Terms

Write a sentence that shows your understanding of each of the following terms.

control
data
deductive reasoning
dependent variable
ethics
experiment
hypothesis
independent variable
inductive reasoning
safety symbol
scientific methods
technology
theory

Understanding Concepts

1. Which of the following is not an appropriate question for science to consider?
   - a. How many seals can a killer whale consume in a day?
   - b. Which type of orchid flower is most beautiful?
   - c. What birds prefer seeds to worms as a food source?
   - d. When do hoofed mammals in Africa migrate northward?
2. If data from repeated experiments support a scientist's hypothesis, what is his or her next step?
   - a. Conclusions are drawn and results are published.
   - b. Another experiment is performed.
   - c. A theory is formed.
   - d. The hypothesis is revised.
3. If data from repeated experiments do not support the hypothesis, what is the scientist's next step?
   - a. The results are published and the scientist gives up.
   - b. The hypothesis is revised.
   - c. The experiment is repeated.
   - d. A theory is overturned.
4. The single factor that is altered in an experiment is the ____________.
   - a. control
   - b. dependent variable
   - c. hypothesis
   - d. independent variable
5. Science attempts to ____________.
   - a. decide right from wrong
   - b. explain how and why things happen
   - c. find out the facts
   - d. distinguish facts from theories

Chapter 2 Review

Reviewing Main Ideas

Summary statements can be used by students to review the major concepts of the chapter.

Key Terms

Answers should go beyond defining the terms. Accept any answer that uses the term correctly and in the proper context.

Understanding Concepts

1. b
2. a
3. b
4. d
5. b

Chapter 2 Review

6. b
7. d
8. b
9. b
10. a
11. d
12. a
13. c
14. c
15. c
16. a
17. b
18. c
19. a
20. b

Applying Concepts

21. predicting a person's luck, deciding whether a person should be kept alive if he or she is near death; accept all reasonable answers
22. finding one's way home if lost, locating a friend in school, retrieving a book left in class; accept all reasonable answers
23. This statement is typically a belief rather than a theory. The friend would need strong evidence to support her statement if it were a theory.
24. descriptive research; words rather than measurements were used

6. Most scientists report results of their research using the ____________ system of measurement.
- **a.** English
- **b.** SI
- **c.** Kelvin
- **d.** Celsius

7. Reasoning from a general set of facts to a specific one ____________.
- **a.** is the purpose of science
- **b.** is deductive reasoning
- **c.** cannot yield results
- **d.** is inductive reasoning

8. A procedure that tests a hypothesis is a(n) ____________.
- **a.** theory
- **b.** experiment
- **c.** variable
- **d.** control

9. Katharine Payne's ____________ was that elephants are able to make very low-pitched sounds.
- **a.** theory
- **b.** hypothesis
- **c.** inductive reasoning
- **d.** control

10. For the results of experiments to be considered valid, the data must be _________.
- **a.** verified
- **b.** inductive
- **c.** published
- **d.** exactly the same every time

11. ____________ is the study of the standards of what is right and wrong.
- **a.** Science
- **b.** Religion
- **c.** Technology
- **d.** Ethics

12. The information gained from an experiment is called ____________.
- **a.** data
- **b.** theory
- **c.** a control
- **d.** the publication

13. The application of scientific research to society's needs is ____________.
- **a.** work
- **b.** an experiment
- **c.** technology
- **d.** ethics

14. Theories are ____________.
- **a.** fact
- **b.** scientific laws
- **c.** supported by a large body of evidence
- **d.** often unsupported

15. Science is ____________.
- **a.** a body of facts
- **b.** a process
- **c.** both a and b
- **d.** neither a nor b

16. The group that is not altered in an experiment is the ____________.
- **a.** control
- **b.** experimental group
- **c.** dependent variable
- **d.** independent variable

17. Reasoning from a specific set of facts to a general rule ____________.
- **a.** is the purpose of science
- **b.** is deductive reasoning
- **c.** cannot yield results
- **d.** is inductive reasoning

18. Processes that scientists use to solve a problem are called ____________.
- **a.** inductive reasoning
- **b.** deductive reasoning
- **c.** scientific methods
- **d.** ethics

19. Which of the following is an appropriate question for scientists to consider?
- **a.** How do paramecia behave when a pond begins to dry up?
- **b.** Which perfume smells the best?
- **c.** Which religion is most sound?
- **d.** Are llamas less valuable than camels?

20. An explanation of a natural phenomenon with a high degree of confidence is a(n) ____________.
- **a.** hypothesis
- **b.** theory
- **c.** experiment
- **d.** reasonable explanation

Applying Concepts

21. Give an example of a problem that technology may not solve.

22. Give an example of how you have used inductive reasoning in your everyday life.

23. If your friend said that she had a theory about who would win the Academy Award for best actor, how would you know that she wasn't speaking scientifically?

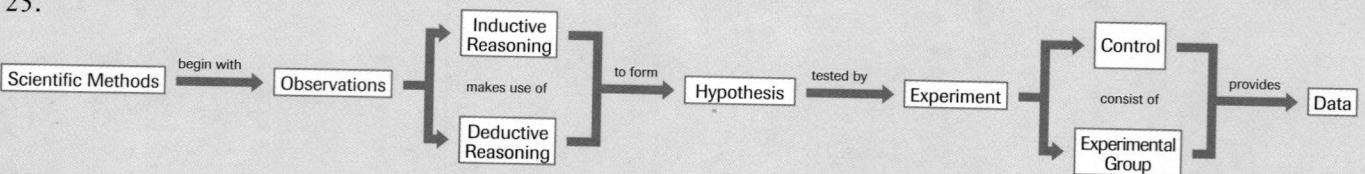

Thinking Critically

25.

Scientific Methods —begin with→ Observations —makes use of→ Inductive Reasoning / Deductive Reasoning —to form→ Hypothesis —tested by→ Experiment —consist of→ Control / Experimental Group —provides→ Data

24. Every day, Carmina observed the sparrows that lived in her neighborhood and wrote her observations down in her journal. What kind of research was Carmina doing? Explain your answer.

Thinking Critically

25. ***Concept Mapping*** Make a concept map that relates the following terms and phrases. Supply the appropriate words for your map.
    inductive reasoning, observation, deductive reasoning, scientific methods, control, experiments, hypothesis, data, experimental group
26. ***Observing and Inferring*** How are scientific methods useful in everyday life?
27. ***Formulating Hypotheses*** Form a hypothesis to explain why elephants sometimes trumpet loudly.
28. ***Interpreting Data*** Look at Figure 2.12 and explain why the survival curve drops suddenly at one point.
29. ***Comparing and Contrasting*** What is the major difference between a hypothesis and a theory?
30. ***Interpreting Data*** Suppose your data from the Minilab "How do you decide which paper towel to buy?" looked like that in the bar graph below.

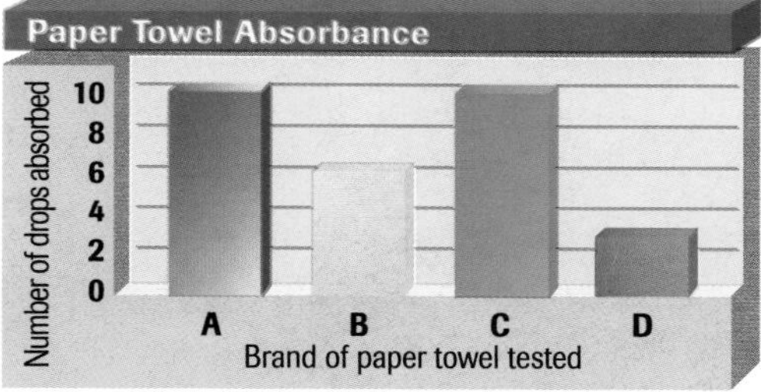

a. Which brand of paper towel were the employees probably complaining about?
b. Which brand would you switch to? Explain your answer.

ASSESSING KNOWLEDGE & SKILLS

A team of students measured the number of seeds that germinated over ten days in a control group and in an experimental group that contained added fertilizer. They graphed their data as shown below.

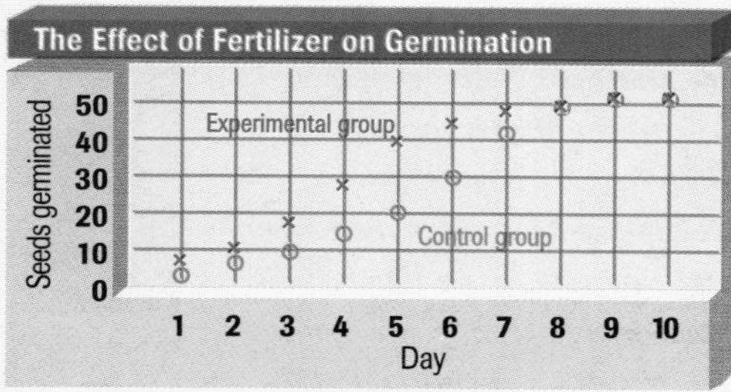

Interpreting Data Look at the graph and answer questions 1 to 4.

1. Which of the following would represent the hypothesis tested?
   a. Black beans are best.
   b. Fertilized beans germinate faster.
   c. Fertilization of beans requires water.
   d. Beans make their own fertilizer.
2. When was the first appropriate day to end the experiment?
   a. Day 3
   b. Day 6
   c. Day 7
   d. Day 9
3. Which of the following was the independent variable?
   a. kind of seeds
   b. number of seeds germinating
   c. fertilizer
   d. time
4. Which of the following was the dependent variable?
   a. kind of seeds
   b. number of seeds germinating
   c. fertilizer
   d. time
5. ***Interpreting Data*** Use the data in the graph above to describe the germination rate between Days 3 and 5 in the control group.

Chapter 2 Review

26. Certain situations or everyday problems may be solved through some of the same processes used in a scientific approach.
27. Student hypotheses will vary, but one example might be that elephants trumpet to warn off rivals.
28. The survival curve drops suddenly at the point where the temperature increases beyond the ability of the paramecia to survive.
29. A hypothesis is an untested explanation for a problem, but a theory has been tested repeatedly and is assumed to be true because it has never been found to be false.
30. (a) Brand D (b) Brands A and C were equally absorbent, so you would choose the cheaper of the two.

ASSESSING KNOWLEDGE & SKILLS

1. b
2. d
3. c
4. b
5. Between days 3 and 5, approximately 5 seeds germinated per day.

Program Resources

Chapter Assessment, pp. 7-12 L1
Performance Assessment in the Biology Classroom, p. 3 L1 P
Alternate Assessment in the Science Classroom
Computer Test Bank
Content Mastery, pp. 5-8 L1

UNIT

2 Ecology

Unit Overview

This unit focuses on the relationships and interactions that exist among organisms and their environments. In Chapter 3, students are introduced to the field of ecology and the interactions that exist among the biotic and abiotic factors in an ecosystem.

Chapter 4 centers on homeostasis within a community through a discussion of limiting factors and how such factors are involved in the development of biomes. In Chapter 5, environmental factors that limit population growth are presented and the study of demographics is undertaken.

In Chapter 6, the unit is brought to a close via a survey of ecological problems related to human activities. The discussion includes an examination of methods that may be implemented to slow or correct these problems.

Theme Development

The theme of *energy* is emphasized through the discussions of photosynthesis, food chains and food webs. The theme of *systems and interactions* are developed in the discussions of ecological systems. The theme of *homeostasis* is demonstrated in the sections dealing with succession.

UNIT

2 Ecology

A tropical rain forest—wet, warm, and, above all, green. At first glance, you may equate this vast expanse of green-ness with sameness. Nothing could be further from the truth. In fact, rain forests are home to more species per square meter than any other environment on Earth, although most of them are hidden from view. One scientist counted more than a hundred species of beetles living among the leaves of just one species of tree.

Even more hidden than the diversity of life in the rain forest is the complex web of interactions that exists among all these different living things and their environment. It is this web of interactions that defines and sustains a rain forest ecosystem and all other ecosystems. If enough of these interactions are disrupted, the whole ecosystem could change drastically.

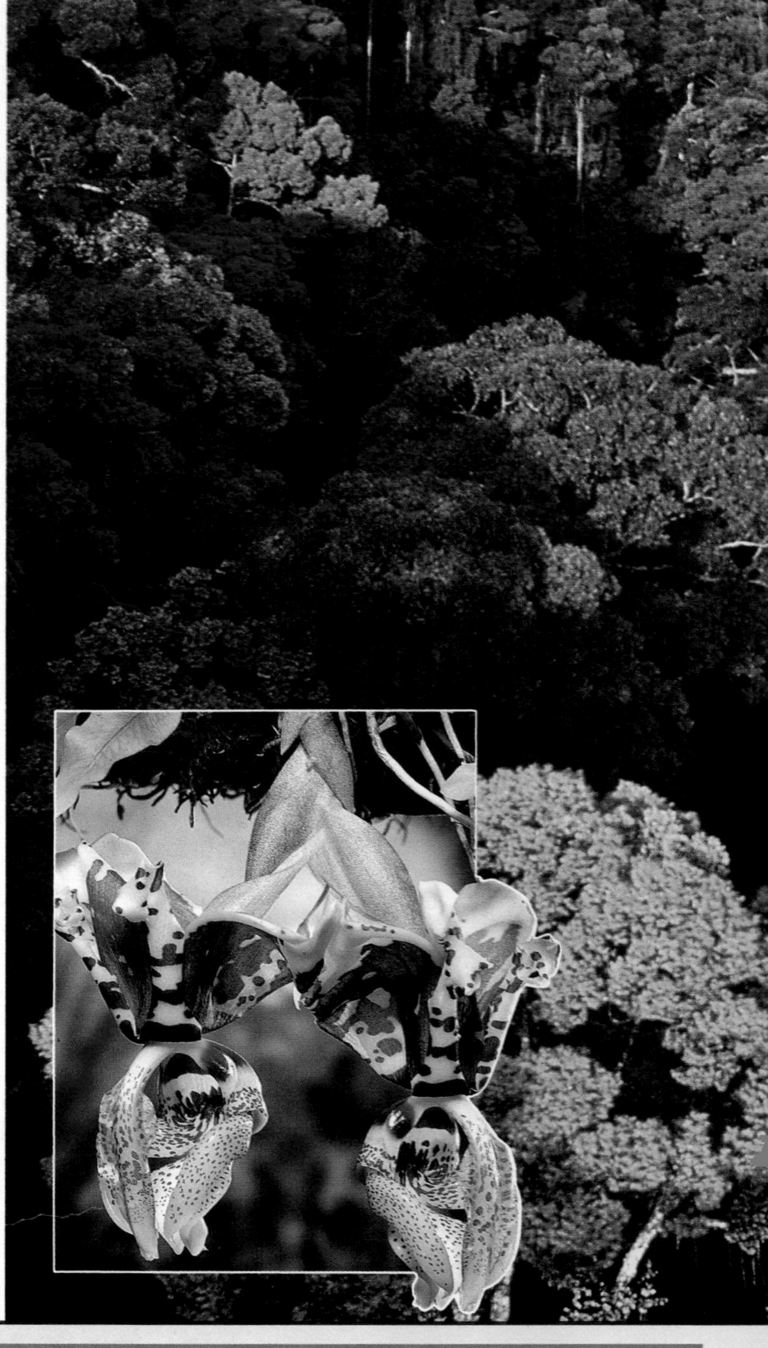

The unique shape and color of this orchid attracts just one kind of wasp, which visits the flower for its nectar. What are other ways that plants and animals depend on each other?

50

Advance Planning

Chapter 3

- Buy small plants or grow bean plants from seed for the Minilab on page 57.
- Order lichens for the Minilab on page 69.
- Order cultures of *Paramecium* and *Didinium* for use in the Biolab.

Chapter 4

- Collect 2-L plastic beverage bottles and purchase insects, snails, and plants for the Minilab on page 86.
- Prepare a 2% hydrogen peroxide solution for the Minilab on page 103.
- Purchase fertilizer for the Biolab and order or obtain pond water.

Chapter 5

- Order cultures of *Paramecium* for the Biolab.
- Purchase bananas for the Minilab on page 118.

Chapter 6

- Collect packaging materials (Styrofoam peanuts, cornstarch packing material, popcorn, paper packaging material, paper bags, and plastic bags) for the Biolab.
- Prepare microscope slides of automobile exhaust for the Minilab on page 141.

This frog's bright coloration warns predators of poisonous chemicals in its body. What other adaptations do animals and plants have to avoid being eaten?

Well over half of the world's species of plants and animals live in tropical rain forests. What are the effects of deforestation on these organisms? What can be done to save them?

Unit Contents

UNIT 2

Introducing the Unit

Use a world map to point out where the world's tropical rain forests are located. Explain that in many parts of the world, rain forests are being cut down for use of their lumber or cleared for use as farmland or for construction of homes. Ask students to explain how the clearing of rain forest land damages the environment, the species that live in rain forests, and people who live near rain forests.

Explain that poisonous frogs, such as the one shown, that live in tropical rain forests are often very colorful. Ask students how coloration may help poisonous frogs survive.

Motivational Activity

Activity: Advise students as to the meaning of the phrase "green products." *Products that are environmentally safe or made from recycled materials.* Ask students to survey their homes to determine the names of all products their family purchases that can be classified as green products. Ask students how the purchase of green products benefits the environment. L1

Unit Project

Group Project: Have students work in cooperative groups to build their own mini-compost chambers. Have students fill their chambers with biodegradable materials, observe it weekly, and record their observations for the duration of this unit (or longer, if possible).

Suggest that students use empty 3-L soda pop bottles as their chambers. Grass clippings, dead leaves, fruit peels, potato skins, and some soil might be suggested as materials to add to the compost. Water will have to be added to the chamber to keep it moist.

Have students design a cover for their chambers to help prevent the materials in the chamber from drying out. Students should expect to observe mold growth and be prepared for some unpleasant odors, which indicate that decomposition is occurring. Ask students to hypothesize as to what events are occurring within the chamber and why. Elicit what is accomplished through the act of composting. L1 COOP LEARN

Chapter Organizer

SECTION	OBJECTIVES	ACTIVITIES/FEATURES
3.1 Organisms and Their Environments National Science Standards: UCP.1; A.1-2, B.2, C.4-6, F.3	1. **Distinguish** between biotic and abiotic factors in the environment. 2. **Compare** the different levels of biological organization used in ecology. 3. **Explain** the difference between a niche and a habitat.	**Literature Connection:** *A Sand County Almanac,* p. 55 **Minilab:** How can you measure water loss by plants?, p. 57
3.2 How Organisms Interact National Science Standards: UCP.1-2; A.1-2, B.3, B.6, C.4-5, D.1-2	4. **Compare and contrast** the different types of symbiotic relationships. 5. **Explain** the matter and energy relationships shown by ecological pyramids. 6. **Compare and contrast** the ways nitrogen, carbon, and water cycle through biotic and abiotic parts of the biosphere.	**Biolab:** Design Your Own Experiment—How can one population affect another?, p. 66 **Minilab:** What type of symbiosis is found in a lichen?, p. 69 **Biology & Society Issues:** Saving the Everglades, p. 74 **Thinking Lab:** Rates of Nutrient Cycling, p. 77

ACTIVITY MATERIALS

BIOLAB	MINILABS	ALTERNATE LAB
page 66 microscope slides coverslips culture of *Didinium* culture of *Paramecium* beakers or jars eyedropper sterile pond water	**page 57** potted plant measuring cup plastic bag string cups, paper or plastic **page 69** lichen sample straight pins wet mount slide microscope colored pencils	**page 58** seeds, corn and pinto 6 paper cups 6 plastic sandwich bags paper towels graduated cylinder labels

Chapter 3 Principles of Ecology

TEACHER CLASSROOM RESOURCES

Reproducible Masters	Transparencies
Section Focus Master 5: Interactions within Ecosystems L1 SAE **Reinforcement and Study Guide,** pp. 9-10 L1 **Biolab and Minilab Worksheets,** p. 7 L1 **Content Mastery,** pp. 9-12 L1	
Section Focus Master 6: Energy Pathways L1 SAE **Reinforcement and Study Guide,** pp. 11-12 L1 **Biolab and Minilab Worksheets,** pp. 8-10 L1 **Concept Mapping:** Food Needs in a Community, p. 3 L1 **Critical Thinking/Problem Solving:** How the Parasitic Cowbird Affects Songbird Populations, p. 3 L3 **Laboratory Manual:** Physical Factors of Soil, pp. 13-14; The Lesson of the Kaibab, pp. 15-18 L2	**Basic Concepts Transparency #1:** A Food Web L1 SAE **Basic Concepts Transparency #2:** Biological Pyramids L1 SAE **Reteaching Transparency #3:** Trophic Levels L1 SAE **Basic Skills Transparency #1:** The Carbon Cycle L1 SAE **Basic Skills Transparency #2:** The Nitrogen Cycle L1 SAE

ASSESSMENT MATERIALS

Chapter Assessment, pp. 13-18
Performance Assessment in the Biology Classroom, p. 59
MindJogger Videoquiz
Alternate Assessment in the Science Classroom
Computer Test Bank

Spanish Resources SAE
English/Spanish Audiocassettes SAE
Cooperative Learning in the Science Classroom COOP LEARN
Lesson Plans
Exploring Environmental Issues: pp. 1, 13, 15, 37, 43 L2

KEY TO TEACHING STRATEGIES

- L1 Level 1 activities should be within the ability range of all students including those with learning difficulties.
- L2 Level 2 activities are within the ability range of average to above-average students.
- L3 Level 3 activities are designed for the ability range of above-average students.
- SAE SAE activities should be within the ability range of Students Acquiring English.
- COOP LEARN Cooperative Learning activities are designed for small group work.
- P These strategies represent student products that can be placed into a best-work portfolio.
- These strategies are useful in a block scheduling format.

GLENCOE TECHNOLOGY

The following multimedia resources are available from Glencoe.

Biology: The Dynamics of Life
CD-ROM SAE
Videodisc Program

The Infinite Voyage Series
Secrets from a Frozen World
Life in the Balance
To the Edge of the Earth

National Geographic Series
GTV: Planetary Manager
STV: Water

Science and Technology Videodisc Series (STVS)
Ecology
Fish Versus Mosquitoes
Preserving Duck Habitats

CHAPTER

3 Principles of Ecology

Chapter Overview

Students begin their study of ecology by discovering the historical link between ecology and natural history, and then develop a formal definition of ecology. They explore and then study examples of both biotic and abiotic factors and the importance of an organism's ability to adapt to these factors in the environment.

In the second part of the chapter, students explore the hierarchy of life, beginning with interactions among organisms. This leads them to a discussion of populations, communities, and ecosystems, including habitats and niches. Students then explore feeding and energy relationships and the symbiotic relationships within ecosystems.

Students complete their study of the principles of ecology by using the carbon, nitrogen, and water cycles to illustrate homeostasis within ecosystems.

Key Terms

abiotic factor
autotroph
biosphere
biotic factor
commensalism
community
decomposer
ecology
ecosystem
food chain
food web
habitat
heterotroph
mutualism
niche
parasitism
population
scavenger
symbiosis
trophic level

Look for the following logo for strategies that emphasize different learning modalities. 

Kinesthetic	**Project, p. 56; Meeting Individual Needs, p. 74**
Visual-Spatial	**Enrichment, p. 55; Meeting Individual Needs, pp. 56, 77; Visual Learning, pp. 59, 60, 61, 68; Microscope Activity, p. 61; Project, p. 68; Student Journal, p. 70; Portfolio, p. 72; Reinforcement, p. 75**
Linguistic	**Meeting Individual Needs, p. 60; Student Journal, pp. 65, 76**
Logical-Mathematical	**Concept Development, p. 56; Portfolio, p. 57; Using an Analogy, p. 59; Reinforcement, p. 65; Student Journal, p. 71; Enrichment, p. 75; Meeting Individual Needs, p. 75; Project, p. 77**

LS

CHAPTER

3 Principles of Ecology

Are mosquitoes of any use to anyone or anything? If all mosquitoes were killed, would any negative effects result? You might say they're just pests, but if you could ask the opinion of a sunfish, a tadpole, a dragonfly, or a swallow, you'd get a different answer. For these animals and others, mosquitoes and their larvae are a major food source.

Every organism is connected in some way to many other organisms. In nature, living organisms interact in a variety of ways. Some relationships are complex such as the pollination of a flower by a hummingbird, while others are quite simple such as the dispersal of seeds from a plant by a passing animal. In this chapter, you'll learn how groups of organisms interact and depend on each other for survival. You'll also consider how organisms are affected by physical factors of the environment.

Concept Check

You may wish to review the following concepts before studying this chapter.

- Chapter 1: adaptation, homeostasis, species
- Chapter 2: scientific methods

Chapter Preview

3.1 Organisms and Their Environments
The Beginnings of Ecology
Levels of Organization: The Hierarchy of Life

3.2 How Organisms Interact
Species Relationships
Matter and Energy in Ecosystems

Laboratory Activities

Biolab: Design Your Own Experiment

- How can one population affect another?

Minilabs

- How can you measure water loss by plants?
- What type of symbiosis is found in a lichen?

These planes are spraying chemicals that kill mosquitoes. Besides being an annoyance to people, some mosquitoes carry diseases that can affect both humans and other animals. Although this is true, mosquitoes and their larvae provide food for some animals, including this fish. Do you suppose that killing mosquitoes can have a harmful effect on the environment?

Assessment Planner

Choose assessment strategies from the following pages to evaluate the progress of your students.

Assess, pp. 62, 78
Portfolio, pp. 57, 72
Alternate Lab, pp. 58-59
Thinking Lab, p. 77
Minilabs, pp. 57, 69
Biolab, pp. 66-67
Chapter Review, pp. 79-81

CHAPTER 3

Introducing the Chapter

Direct students' attention to the photographs and caption and pose the following question: Are humans able to survive independently of all other organisms? *Lead students to realize that like all organisms, humans are unable to exist without some dependence on other forms of life.* Expand the discussion by asking: Do humans depend on nonliving matter? *Again, lead students to realize that humans are dependent on nonliving matter, such as air and water, for survival.* Use this exercise to establish that all organisms interact with living and nonliving parts of the environment.

Theme Development

Energy is a major theme of this chapter, specifically as it relates to photosynthesis and the subsequent passage of energy through each trophic level of the food chain. The theme of *systems and interactions* is developed as students relate the meaning of the term *ecosystem* (systems) to the symbiotic relationships and interactions that exist within populations, communities, ecosystems, and the biosphere.

Concept Check

This chapter will expand students' understanding of ecological relationships. They will discover how the science of ecology evolved due to the studies of many scientists using scientific methods, as described in Chapter 2.

SECTION

3.1

Prepare

Key Concepts

The basic meaning of the term *ecology* as well as the roles of biotic and abiotic factors within ecosystems are presented, and the interactions that occur at different levels of the environment are developed.

Block Scheduling

Look for this symbol for strategies that are useful in a block scheduling format. For more information on block scheduling, refer to Section 3.1 in the **Lesson Plans** booklet.

Materials

- Gather small potted plants, measuring cups (or beakers), string, and plastic bags to fit over the plants to be used with the Minilab.
- An established aquarium, microscope slides, and compound microscopes are needed for the microscope activity.

1 Focus

Bellringer

Before presenting the lesson, display **Section Focus Master 5** on the overhead projector and have students answer the accompanying questions. L1 SAE

SECTION 3.1

Organisms and Their Environments

Section Preview

Objectives

Distinguish between biotic and abiotic factors in the environment.

Compare the different levels of biological organization used in ecology.

Explain the difference between a niche and a habitat.

Key Terms

ecology
biosphere
biotic factor
abiotic factor
population
community
ecosystem
niche
habitat

When was the last time you heard or read something about the environment? Almost every evening newscast, daily newspaper, and news magazine presents you with stories about recycling efforts, campaigns to save tropical forests, concerns about water or air pollution, and many other environmental issues. Learning more about basic ecological principles that explain how organisms interact with each other and the environment can help you understand these issues and form your own opinions about them. Ecology is both an old and a new science. In this chapter, you will see how the observations of early and present ecologists, and the use of technology and new research tools, help us understand the interactions in nature.

The Beginnings of Ecology

Many people make a hobby out of observing and studying plants and animals in their natural habitats. Plant enthusiasts explore fields, forests, city parks, and even vacant lots to learn about the different trees and wildflowers that grow there, such as the goldenrods in *Figure 3.1.* Plant identification guides help these people identify different species of plants and learn about a plant's habitat requirements and flowering times.

Casual backyard bird observers may notice a nest of robins and the behavior patterns of the male and female. While avid bird-watchers would certainly be interested in a nest of robins, their hobby is to identify and observe as many bird species as possible. In order to find new birds, bird-watchers listen for bird songs and investigate all types of environments from woodlands to fields to wetlands. The pursuits of bird watching and plant identification are known as nature study or natural history.

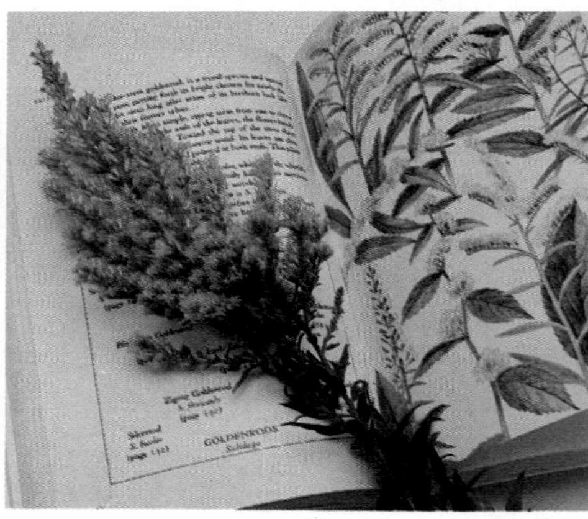

Figure 3.1
Goldenrod plants flower at different times. Depending on rainfall and temperature, goldenrods may flower as early as August or as late as November in different parts of the United States.

Program Resources

Section Focus Master 5 L1 SAE
Reinforcement and Study Guide, pp. 9-10 L1
Exploring Environmental Issues, p. 1 L2
Biolab and Minilab Worksheets, p. 7 L1

Natural history led to ecology

The goal of natural history is to find out as much as possible about living things. Many early natural historians spent their lives studying how various organisms live out their life cycles and how they depend on one another and on their environment. Natural historians then began to report their findings in a systematic way.

The branch of biology that developed from natural history is called ecology. **Ecology** is the scientific study of interactions between organisms and their environments. Ecological study reveals interrelationships between living and nonliving parts of the world. Ecology combines information from many scientific fields including chemistry, physics, geology, and other branches of biology.

You may remember from Chapter 2 that scientific methods include both descriptive and quantitative research. Most ecologists use both types of research. They obtain descriptive information by observing organisms in the field and laboratory. They obtain quantitative data by making measurements and carrying out carefully controlled experiments in both field and lab. Using these methods, ecologists can learn a great deal about relationships, such as how pesticides seep through soil from a farm to a nearby stream, how day length influences the behavior of migrating birds, how tiny shrimp help rid ocean fish of parasites, or how acid rain threatens some of Earth's forests.

Literature

A Sand County Almanac

by Aldo Leopold

The idea of unity of organisms and environment is one of long standing. However, the concept that humans are not separate from nature was not seriously considered by biologists until the American naturalist Aldo Leopold wrote *A Sand County Almanac* over a period of years. It was published in 1949, a year after his death. Leopold described himself as one who could not live without wild things.

On a farm in Wisconsin The book describes the annual events of nature Leopold observed on an abandoned farm in Wisconsin that he purchased in 1935 and visited during weekends over a ten-year period. Like many naturalists, Leopold describes nature as a continuum of events from the past to the future, with every organism relying on its environment. He not only describes the appearance of plants and animals, but also gives them life with needs as relevant as our own. When Leopold describes a rough-legged hawk sailing over the meadow in January in search of a mouse for lunch, he points out that the hawk has no opinion about why grass grows. The hawk, like the one shown here, only knows that snow melts to reveal his prey. It migrated south from the Arctic in hope of such a thaw. For the hawk, a thaw means freedom from want and fear.

A plea for the community Leopold wrote in his journal about the effects that improved mechanization would have on the environment. Leopold was concerned about human abuse of the land because he realized land is too often considered a commodity. He pleaded for an understanding that land is not just something to be bought and sold. He wanted people to understand that the land is essential to our psychological and spiritual well-being, as well as to our physical survival, and that we must take good care of it.

CONNECTION TO Biology

How might observations of naturalists such as Aldo Leopold lead to the conservation of lands destined for development?

2 Teach

Literature

A Sand County Almanac

Purpose

Students will get an overview of how human actions impact an ecosystem. Students may also be inspired to read Leopold's, *A Sand County Almanac.*

Teaching Strategies

- Ask students to name specific examples of how humans have interfered with the ecology of an area. *Responses may include pollution of air, land, and water; allowing soil to erode from land; or destruction of rain forests and other habitats.*

Possible Answers

The public may realize that land is an important resource that should serve the needs of plants and animals as well as humans.

Enrichment

Have students observe organisms, such as insects in a field or animals in a forest or park, in their natural environments. Ask students to list two or three examples of how an organism they observed interacts with other organisms and the nonliving parts of its environment. L1 IS

Audiovisual

Show the filmstrip *Introduction to Ecology* (REX, Carolina Biological Supply Company). L1 SAE

GLENCOE TECHNOLOGY

Videodisc

Biology: the Dynamics of Life
Disc 1, Side 1
How Organisms Interact (Ch. 3)

STUDENT JOURNAL

Connecting the Disciplines Have students list in their journals specific areas of chemistry, physics, and geology that might be studied as part of ecology. Ask them to describe the types of quantitative data that would be used in the specific areas that they have listed. L2

SECTION 3.1

Using Science Terms

Write the terms *biosphere* and *biotic* on the chalkboard and write the meaning of each word part beneath the terms. For example, write the word *life* below *bio*, *area* below *sphere*, and the phrase *relating to* below *tic*. Have students use this information to define each term. SAE

Concept Development

LS Logical-Mathematical Ask students working in cooperative groups to make lists of other animals that live in the ocean, underground, or high in the mountains. Have them compare the different adaptations of each of the three groups. Have them discuss how similar adaptations are shared by members of these three environmental groups. SAE COOP LEARN

Brainstorming

Ask students to explain what the survival advantage might be to newborn mountain goats that are able to feed, run, and climb within hours of being born. Have students explain the disadvantage that newborns face when they are unable to feed themselves or run about shortly after birth.

GLENCOE TECHNOLOGY

Videodisc

The Infinite Voyage: To the Edge of the Earth
Exploring the Galápagos Islands (Ch. 4)

The Tropical Rain Forest (Ch. 5)

ecology:
oikos, eco (GK) household, habitat
Ecology is the study of organisms and their environments.

The living environment: Biotic factors

As far as we know, life exists only on Earth. Living things can be found in the air, on land, and in both fresh water and salt water. The **biosphere** is the portion of Earth that supports life. It extends from high in the atmosphere to the bottom of the oceans. This life-supporting layer may seem extensive to us, but if you could shrink Earth down to the size of an apple, the biosphere would be thinner than the apple's peel.

Many different environments, both aquatic and terrestrial, exist in the various regions of the biosphere. Each environment includes both living and nonliving factors that affect the organisms living there. All the living organisms that inhabit an environment are called **biotic factors.**

Different organisms are adapted for life in different parts of the biosphere. For example, the mountain goats shown in *Figure 3.2* are adapted to climb on steep mountainsides and withstand freezing temperatures and strong winds. Animals like the mole live in, rather than on, the soil. Fishes that live in the darkest depths of the ocean have adaptations for finding food and mates in darkness. They also have adaptations to withstand cold temperatures and tremendous water pressure.

Figure 3.2
All organisms are adapted for life in a particular environment. They have adaptations for obtaining food, for protecting themselves, and for reproduction.

Moles are small mammals that dig tunnels in the soil. Their digging provides the soil with aeration, which is beneficial for soil organisms and plant roots. However, moles also eat roots as they dig, and can damage large populations of plants such as grasses.

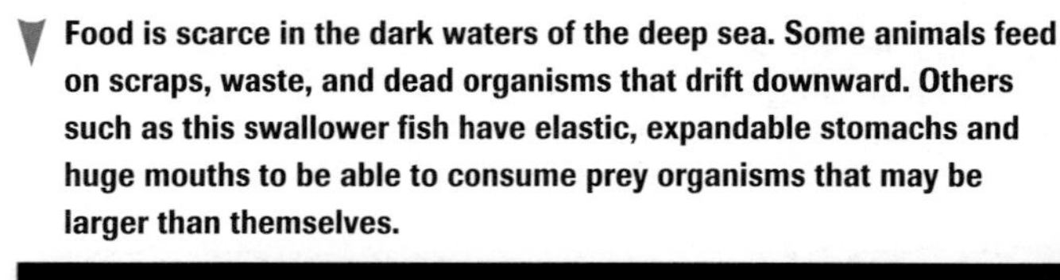

Food is scarce in the dark waters of the deep sea. Some animals feed on scraps, waste, and dead organisms that drift downward. Others such as this swallower fish have elastic, expandable stomachs and huge mouths to be able to consume prey organisms that may be larger than themselves.

Mountain goats live in the mountains above the timberline. They prefer this seemingly uninhabitable environment. These goats, not true goats but members of the antelope family, have a dense, woolly underfur covered by an overcoat of long, white hair to protect them from the wind and cold of the mountaintops. Females bear one or two young in May or June. The babies stand and nurse within minutes of birth, and run and climb hours later.

PROJECT

Modeling the Biosphere

Have students prepare a model or diagram of Earth that shows the total depth encompassed by the biosphere. Models should extend from the ocean floor to the height of the atmosphere. Have students do library research to determine and record specific quantitative data for these depths and heights. L2 LS

Meeting Individual Needs

Learning Disabled Provide students who may require additional reinforcement of the concepts of biotic and abiotic factors with photographs taken from old copies of nature magazines. Ask students to identify all the biotic factors appearing in each photograph. Ask them to explain why they identified these factors as biotic. SAE LS

The nonliving environment: Abiotic factors

An ecological study of moles wouldn't be complete without an examination of the types of soil in which these animals dig their tunnels. Similarly, research on the life cycle of trout would need to include where these fish lay their eggs—on rocky or sandy stream bottoms. Ecology includes the study of features of the environment that are not living. Ecologists study how the nonliving factors in the environment affect living things. **Abiotic factors** are the nonliving parts of the environment. Examples of abiotic factors include air currents, temperature, moisture, light, and soil.

Abiotic factors can have obvious effects on living things and often determine which species can survive in a particular environment. For example, lack of rainfall can cause drought in a grassland, as shown in *Figure 3.3.* Can you think of changes in a grassland that might result from a drought? Grasses would grow less quickly, wildflowers would not produce as many seeds, and the animals that depend on plants for food would find it harder to survive.

MiniLab

How can you measure water loss by plants?

All organisms lose water by respiration. Plants also lose water by a process called transpiration. Transpiration occurs when water travels to a plant's leaves and escapes through pores as water vapor.

Procedure

1. Allow the soil of a potted plant to dry out so that the plant just begins to wilt.
2. Add water in measured amounts to the pot until it just begins to drip from the bottom.
3. Allow the water to drip back into the measuring cup.
4. When the water stops dripping, record the amount of water that remained in the pot by subtracting the amount of water that dripped back into the measuring cup from the original amount.
5. Place the potted plant in a plastic bag and tie the top with string.
6. Put the plant in sunlight.
7. When water has collected on the bag, estimate the amount of water that has transpired from the plant.

Analysis

1. If the potted plant had not been placed in the bag, where would the water have gone?
2. Why do you not have to water plants in a terrarium as often as plants in a pot?
3. How might an ecologist alter the setup you used to measure transpiration of plants in an ecosystem?

Figure 3.3

As shown here, droughts are not uncommon in grasslands. As the grasses dry out, they turn yellow and appear to be dead, but new blades grow from the low-lying growing points when it rains again. Some animal species are adapted to living in grasslands by their ability to burrow underground and sleep through the dry period.

MiniLab

Purpose

Students will determine the amount of water lost by a plant during transpiration.

Process Skills

observe and infer, measure in SI

Teaching Strategies

- Plants may be grown from seed several weeks before needed.
- **Troubleshooting:** Remember to punch small holes in the bottoms of the containers. Use a heated nail for plastic.

Expected Results

Volume of water transpired will depend on the leaf surface size and the length of time the plants remain in the bags. If water amounts are too small to measure after 24 hours, keep plants in bags for two days.

Analysis

1. Into the air
2. Water remains in the terrarium chamber instead of being lost to the air.
3. By measuring the size of the ecosystem, running trials on many different plants, or using isotopes to trace water loss

Assessment

Performance: Have students design an experiment to determine the water-holding capacity for different soil types. L1

PORTFOLIO

Studying the Local Environment Ask students to observe the natural environment of the area in which they live on several different days. Tell students to prepare a table in which they list the biotic and abiotic factors of the environment. Have students summarize the importance of the abiotic factors listed in their tables. L1 SAE LS P

Different Viewpoints in Biology

The Biosphere II Project

The Biosphere II project was an ecology experiment on a grand scale. Biosphere II is a huge building consisting of regions representing six biomes in which 4000 different species, and eight humans, were sealed for two years.

During the Biosphere II experiment, the people were to depend totally on the other species for all their needs. Similarly, all species were to depend on each other as well as on the soil, water, and the original air sealed in the building for their needs. The experiment was not completely successful because it did not remain isolated from the outside environment.

Some scientists believed that Biosphere II would be a testing ground for life in space. Some people thought that the place was nothing more than a tourist trap whose main function is to provide income to its sponsors. Hotels have been built nearby to accommodate tourists. Biosphere II is now being converted into a major research unit in the desert.

Audiovisual

Show the film or video *An Ecosystem: A Struggle for Survival,* National Geographic.

Figure 3.4

These marsh marigolds represent a population of organisms. What characteristics are shared by this group of flowers that make them a population?

Levels of Organization: The Hierarchy of Life

The study of an individual organism, such as a male deer, or buck, might reveal what food items he prefers, how often he eats, and how far he roams to search for food or shelter. However, even though he spends a large part of his time alone, he does interact with other individuals of his species. For example, he periodically goes in search of a mate, which may require battling with other bucks.

All organisms depend on others for food, shelter, reproduction, or protection. So you can see that the study of an individual would provide only part of the story of its life cycle. To get a more complete picture requires studying its relationships with other organisms.

Ecologists study interactions among organisms at several different levels, as shown in The Inside Story. They study individual organisms, interactions between organisms of the same species, and interactions between organisms of different species. Ecologists also study how abiotic factors affect groups of interacting species.

These levels of organization provide ecologists with a tool to use in planning their research. For example, one ecologist might focus on the interactions among the deer within a single herd, while another might want to study how the herd interacts with other animals living in the same area. Yet another ecologist might investigate how the deer are affected by severe winter storms or by extremely hot, dry summer weather.

Interactions within populations

The marsh marigolds in *Figure 3.4* form a population. A **population** is a group of organisms of one species that interbreed and live in the same place at the same time.

Members of the same population compete with one another for food, water, mates, and other resources. The ways in which organisms in a population share the resources of their environment determine how far apart populations are and how large each population can become.

Alternate Lab: Moisture and Seed Germination

Purpose

Students will determine if the amount of time seeds are permitted to soak in water influences future germination.

Materials

120 seeds (corn, pinto), six paper cups, six plastic sandwich bags, water, paper toweling, graduated cylinder, labels

Procedure

Give students the following directions.

1. Label each of six cups with your name, the date, and one of the following times: 1, 6, 12, 24, 48 hours.
2. Fill each of the cups with 10 mL of water. Add 20 seeds to each cup.
3. Remove the seeds from each cup when the time marked is reached. Wrap the seeds in a layer of paper toweling and place each batch of seeds into a separate labeled plastic bag.
4. Seal the bags. Prepare a control bag with 20 unsoaked seeds.
5. Starting the next day, and continuing for five days, observe each group of seeds. Look for signs of germination

The Organization of Life

All living matter can be organized into levels starting with the smallest subatomic particles and going all the way to Earth as a whole. Ecology deals with the levels of organisms, populations, communities, ecosystems, biomes, and the biosphere.

White-tailed deer

Ecologists may study the behavior of an individual organism. They may study its daily movements, feeding, or breeding behavior.

Ecologists may study the effects of populations of organisms on the environment. They may also study growth rates of populations and predict the future of certain populations.

All organisms in a community depend in some way on the other organisms living there. Ecologists are concerned with studying the effects on the community when a new species is added or one is removed.

All the biotic and abiotic factors in an area form an ecosystem. Ecologists are concerned with ecosystem stability and knowing what keeps ecosystems stable.

The biosphere is the highest level of organization. It is made up of the entire planet and all its living and nonliving parts. Ecologists are concerned with all interactions within the biosphere.

THE INSIDE STORY

The Organization of Life

Purpose

Students will see the organization of levels used in studies pertaining to ecology and will better understand how organisms of one species affect each other as well as organisms from other species.

Visual Learning

LS **Visual-Spatial** Ask students to choose one diagram from *The Inside Story* and describe the biotic and abiotic factors present. Have students compare their lists and account for differences.

Using an Analogy

LS **Logical-Mathematical** Ask students to arrange the following from smallest to largest in terms of numbers of members: city, state, neighborhood, street block, house or apartment, county, country. Ask students to apply this grouping method to the levels of organization in ecology. Have them explain how the two groupings are alike and how they differ. Ask them to suggest reasons why humans categorize information into such groups. *Students should recognize that classifying things makes them easier to study.* SAE

(small roots growing from seed) and record the number of seeds from each container that are germinating each day.

Analysis

Ask students to answer the following questions.

1. What abiotic factor is being investigated? *Effect of water on seed germination*
2. Which seems to be the optimum (best) time for seed germination? *The time will depend on the seed type used.*
3. Will the seeds all germinate at the same time? Explain. *No, each seed type will have its own optimum soaking time.*

Assessment

Performance Have students design and then perform an experiment to determine how the abiotic factor light affects seed germination. L1 COOP LEARN

Visual Learning

Figure 3.5 Ask students to use the photograph to describe the adaptations of the adult frog that enable it to (a) swim well *strong hind legs, webbed feet* (b) hide or escape from its predators. *body coloration, strong hind legs* What adaptations seen in the tadpole enable it to swim well? *large tail fin*

Figure 3.6 Ask students to explain what evidence in the photograph supports the following statements:

(a) Beech and maple trees are not part of the same population. *Because they are not of the same species, beech and maple trees cannot belong to the same population.* (b) Beech and maple trees may compete with each other for resources. *The trees may compete for light, soil, nutrients, and water.* LS

Figure 3.5

Adult frogs and their young have different food requirements. This limits competition for food resources for the species.

▼ Eggs that adult frogs lay in the water hatch into tadpoles. Tadpoles have gills, live in water, and eat algae and small aquatic creatures.

▲ Adult frogs live both on land and in the water. They breathe air and eat insects such as dragonflies, grasshoppers, and beetles.

Some species have adaptations that reduce competition within a population. One example can be seen in the life cycle of a frog, shown in *Figure 3.5.* The juvenile stage of the frog is the tadpole, which not only looks very different from the adult but also has completely different food requirements. Many species of insects, including dragonflies and moths, also have juveniles that differ from the adult in body form and food requirements.

Individuals interact within communities

No population of organisms of one species lives independently of other species. Just as a population is made up of individuals, a community is made up of several populations. A **community** is a collection of interacting populations. An example of a community is shown in *Figure 3.6.*

A change in one population in a community will cause changes in the other populations. Some of these changes can be subtle, as seen when a small increase in one population causes a small decrease in another.

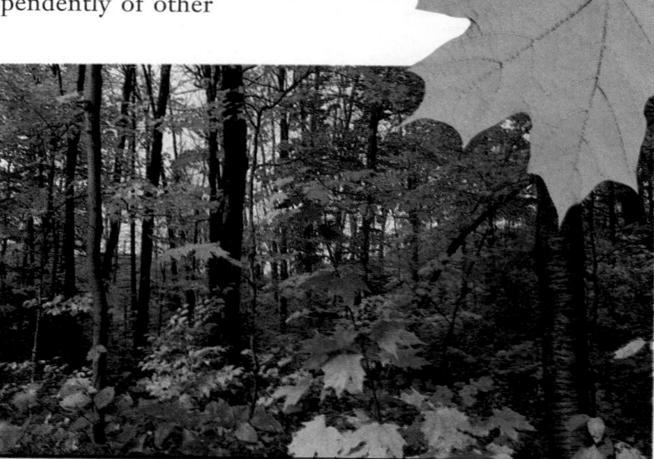

Figure 3.6

Beech and maple trees dominate this forest community; therefore, it is called a beech-maple forest. Beech-maple forests are found in the eastern United States, Europe, and northeast China. Many other populations are also found in this community. These populations interact with one another and with the abiotic factors of this environment.

Meeting Individual Needs

Hearing Impaired Provide students with photographs of an insect larval stage (caterpillar) and adult stage (moth or butterfly) taken from old nature magazines. Instruct them to prepare a chart that includes the following information: a description of the differences observed in the juvenile and adult stage, a comparison of the food requirements of each stage, and a description of the type of economic losses or benefits to trees, shrubs, or crops that may be caused by each stage. Have students prepare a written summary that explains how differences in the developmental stages serve as an advantage to the organism. L1 LS

For example, if the population of mouse-eating hawks increases slightly, the population of mice will, as a consequence, decrease slightly. Other changes might be more dramatic, as when the size of one population grows so large it begins affecting the food supply for another species in the community.

Interactions among living things and abiotic factors form ecosystems

In a healthy forest community, interactions between populations include birds eating insects, squirrels eating nuts from trees, mushrooms growing from decaying leaves or bark, and raccoons fishing in a stream. In addition to population interactions, ecologists also study interactions between populations and their physical surroundings. An **ecosystem** is made up of the interactions among the populations in a community and the community's physical surroundings, or abiotic factors.

Terrestrial ecosystems are those located on land. Examples include forests, meadows, and desert scrub.

Not all ecosystems occur on land. Aquatic ecosystems may occur either in fresh water or salt water. Freshwater ecosystems include ponds, lakes, and streams.

Saltwater ecosystems, also called marine ecosystems, occupy approximately 75 percent of Earth's surface. *Figure 3.7* gives an example of both a marine and a freshwater ecosystem.

Figure 3.7

There may be hundreds of populations interacting in a pond or tide pool. How do you think the abiotic factors in these environments affect the biotic factors?

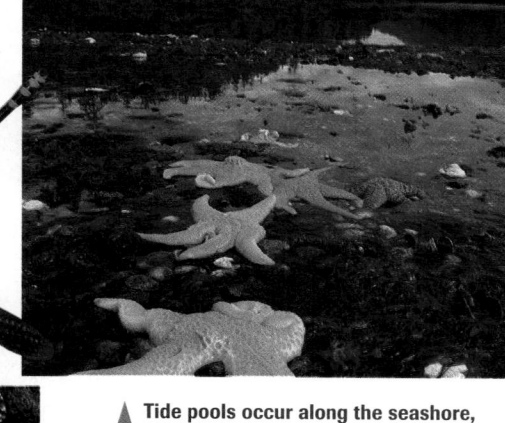

Dragonflies live near moist meadows and ponds. They feed on small insects they catch while flying. Dragonflies lay their eggs in the pond or on pond plants.

Tide pools occur along the seashore, often in rocky areas. Organisms living in tide pools must survive dramatic changes in abiotic factors. When the tide is high, ocean waves replenish the water in the pool. When the tide is low, water in the pool evaporates. Oxygen levels decrease, temperatures increase, and some organisms may even be exposed to the air.

SECTION 3.1

Microscope Activity

LS **Visual-Spatial** Have students prepare wet mounts of scrapings from an aquarium wall or from debris on the bottom. Ask them to observe the slide, noting the number of different organisms present. Have them sketch what they see. Ask students to explain whether their sample represents a community or an ecosystem. *Community* Challenge them to describe some of the interactions that may be occurring.

Visual Learning

Figure 3.7 Ask students to describe adaptations animals living within tide pools may have for survival when the tide is low for several hours. *Responses may include shells that help the animal retain moisture or features for burrowing in moist sand.*

Figure 3.7 How do you think the abiotic factors in these environments affect the biotic factors? *Responses may include that sunlight affects water temperature as well as whether or not photosynthesis will occur. Air dissolved in the water affects the respiratory and photosynthetic processes of the pond organisms, which, in turn, affect the flow of energy.* LS

Chemistry Connection

Salinity

What happens to the salinity (salt concentration) in a tide pool as water evaporates? Have students either research the answer or design and perform an experiment to determine the changes. L3

PROJECT

A Miniature Ecosystem

Have students work in cooperative groups. Have each group design a classroom aquarium or a terrarium created in a large, wide-mouthed jar such as a peanut butter jar. For the aquarium, students will need to purchase materials from a local pet store. For the terrarium, have students obtain soil, plants, and animals from the local environment. Instruct each team to keep a log that records the observed interactions between populations and the abiotic factors of their environment. L2 COOP LEARN

SECTION 3.1

Using an Analogy

Explain that the definition of the term *niche* includes the role of an organism in its environment. With this idea in mind, have students describe one niche of lions, insectivorous plants, sharks, earthworms, and dandelions. Provide students with information they need about each organism's habitat and role in the environment. SAE

Software

Ecology, SEI Editor (Queue, Inc.)

3 Assess

Check for Understanding

Have students explain the relationship between the words in each of the following groups:

(a) ecology – homeostasis
(b) population – ecosystem
(c) habitat – niche L1

Reteach

Ask students to work in groups to provide examples of biotic and abiotic factors within their classroom. SAE COOP LEARN

Extension

Ask students to research how abiotic factors limit life in the Arctic tundra or in a desert environment. Have students include a written summary of the information they gather in their portfolios. L2 P

Figure 3.8

This series of photographs shows how a habitat can be seen as a collection of several niches. As you can see, each species uses the available resources in a different way. Can you think of other ways the resources could be shared? Further division of resources would provide additional niches.

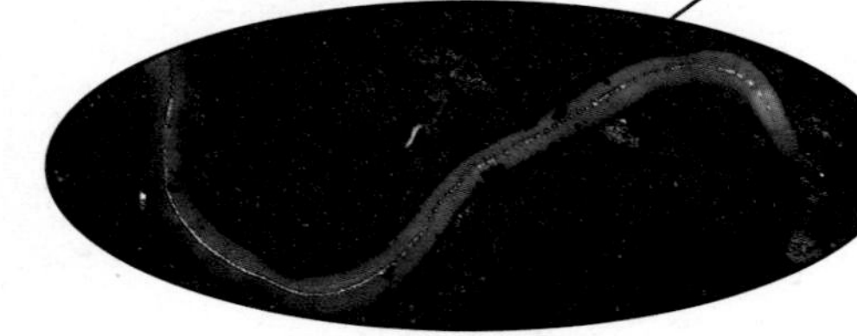

A worm obtains nourishment from the organic material it eats as it burrows through the soil.

A centipede is a predator that captures and eats beetles and other animals.

Where and how organisms live

Every species has a particular function in its community. In a grassland community, for example, the role of fungi is the breakdown of organic matter contained in the bodies of dead organisms. This process recycles nutrients and materials, releasing them back into the environment where they can be used by other organisms. What do you suppose is the role that coyotes play in a grassland community? They keep down populations of rodents by eating them on a regular basis. Both fungi and coyotes help maintain homeostasis in their communities. A **niche** is the role a species plays in a community. The space, food, and other conditions an organism needs to survive and reproduce are part of its niche. A species niche includes more than just feeding relationships. It also includes how a species uses and affects its environment.

A prairie dog living in a grassland makes its home in burrows it digs underground. Many birds make their homes in the trees of a beech-maple forest. These are examples of habitats. A **habitat** is the place where an organism lives out its life.

Meeting Individual Needs

Students Acquiring English Have students develop a table with the following columns: *Organism, Habitat, Niche.* Have students list the following names beneath *Organism:* bee, eagle, seaweed, hyena, and mushroom. Have students use the table as a review vehicle by having them complete the table as a homework assignment. SAE

1. Likely responses may include the following: Biotic—tree, grass, human, dog, ant. Abiotic—daylight hours, amount of rainfall, humidity, air, soil.
2. Populations consist of a single species of organisms that can interbreed that are present in the same place at the same time. A community is made up of several populations that interact with one another.

A millipede eats decaying leaves near the log.

Although several species may share a habitat, the food, shelter, and other resources of that habitat are divided into separate niches. For example, if you turn over a log like the one shown in *Figure 3.8*, you will find millipedes, centipedes, insects, and worms living there. At first, it looks as though all these animals are competing for food because they live in the same habitat. But close inspection reveals that each feeds in different ways, on different materials, and at different times. These differences lead to distinct behaviors and reduced competition.

Section Review

Understanding Concepts

1. List several different biotic and abiotic factors in an ecosystem.
2. Compare and contrast populations and communities.
3. Give examples that would demonstrate the differences between the terms *niche* and *habitat*.

Thinking Critically

4. On a visit to a desert, you observe a coyote, three snakes, several small types of cactus, and hundreds of creosote bushes. What name would an ecologist give to such an area?

Skill Review

5. **Forming a Hypothesis** Cape May warblers, bay-breasted warblers, and yellow-rumped warblers all feed in the same kind of spruce tree. Hypothesize how these birds reduce competition in the spruce tree habitat. For more help, refer to Practicing Scientific Methods in the *Skill Handbook*.

SECTION 3.1

Assessment

Performance: Have students obtain pictures from magazines of organisms that are likely to share the same ecosystem. Ask them to use the pictures to create a possible food web for the ecosystem. Instruct students to include a caption for the diagram they create that identifies the ecosystem, explains what is shown, and identifies abiotic factors that each organism included in the diagram interacts with.

Activity

Exploring Environmental Issues: p. 1, Water, Water Everywhere? Have students complete the activity to illustrate their knowledge of the importance of water. L1 P

4 Close

Discussion

Have students work in cooperative groups to explain how the following illustrate the theme of systems and interactions.

(a) forest ecosystem
(b) coral reef ecosystem
(c) classroom aquarium
(d) classroom terrarium

COOP LEARN

Answers to Section Review

3. Likely responses may include the following: Squirrel—habitat: forest, city park with trees; niche: to distribute seeds. Mushroom—habitat: moist, forest soil; niche: to break down and recycle organic matter. Bat—habitat: cave and air; niche: to pollinate flowers, control insects.

Thinking Critically

4. The area may be considered a community (a number of interacting populations) of an ecosystem *(desert* implies different populations interacting with abiotic and biotic factors).

Skill Review

5. Responses may suggest that the three types of warblers feed on different food sources at different times of day, or at different levels of the tree, such as on the trunk, lower branches, or upper branches.

SECTION

3.2

Prepare

Key Concepts

Energy is needed for the survival of all living things. The ways in which organisms obtain energy and how it is passed from one organism to another via feeding relationships (food chains and food webs) is presented along with a discussion of symbiotic relationships. The discussion of the relationship between food and energy is then expanded to include ecological pyramids that address trophic levels, biomass, and energy. The section concludes with a discussion of the nitrogen, carbon, and water cycles.

Block Scheduling

Look for this symbol for strategies that are useful in a block scheduling format. For more information on block scheduling, refer to Section 3.2 in the **Lesson Plans** booklet.

Materials

- Gather or purchase lichens for the Minilab.
- Order or prepare cultures of *Paramecium* and *Didinium* for the Biolab. Prepare sterile pond water or follow the directions in the Biolab for pond water substitutions that may be made.

SECTION

3.2 How Organisms Interact

Section Preview

Objectives

Compare and contrast the different types of symbiotic relationships.

Explain the matter and energy relationships shown by ecological pyramids.

Compare and contrast the ways nitrogen, carbon, and water cycle through biotic and abiotic parts of the biosphere.

Key Terms

autotroph
heterotroph
scavenger
decomposer
symbiosis
commensalism
mutualism
parasitism
food chain
trophic level
food web

Wile E. Coyote and Roadrunner aren't just characters in a Saturday morning cartoon. In the deserts of the American southwest, roadrunners are one source of food for coyotes. Coyotes will usually eat any animal they can catch, which is good for Wile E. Coyote because he never seems to have much luck catching the roadrunner! The connection between a coyote and its prey is only one example of a feeding relationship in a desert.

Species Relationships

Coyotes, roadrunners, and other living things interact in a variety of ways. These interactions allow organisms to obtain energy and materials necessary for life processes. These interactions also maintain homeostasis within populations, communities, and ecosystems.

herbivore:
herba (L) grass
vorare (L) to devour
Herbivores feed on plants.

Feeding relationships: How organisms obtain energy

You have learned that the role an organism plays in its community is its niche. Feeding relationships between organisms reflect these niches, or roles in the community. *Figure 3.9* shows a grassland community in Africa. The plants in the photograph are autotrophs. **Autotrophs** are organisms that use energy from the sun or energy stored in chemical compounds to manufacture their food. Autotrophs are also called producers. Plants are the most common terrestrial autotrophs, but chlorophyll-containing, single-cell organisms also produce their own food. All other organisms depend on autotrophs for nutrients and energy.

Organisms that depend on autotrophs as their source of nutrients and energy are **heterotrophs.** Some heterotrophs, such as grazing, seed-eating, and algae-eating animals, feed directly on autotrophs. The wildebeests depend on plants for their food. Heterotrophs are also known as consumers because they can't make their food; they must consume it. A consumer that feeds only on plants is called an herbivore. Herbivores include grazing animals such as rabbits, cows, and grasshoppers, as well as rodents such as beavers, mice, and squirrels.

1 Focus

Bellringer

Before presenting the lesson, display **Section Focus Master 6** on the overhead projector and have students answer the accompanying questions. L1 SAE

Program Resources

Section Focus Master 6 L1 SAE
Reinforcement and Study Guide, pp. 11-12 L1
Performance Assessment in the Biology Classroom, p. 59 L1 P SAE
Biolab and Minilab Worksheets, pp. 8-10 L1
Critical Thinking/Problem Solving, p. 3 L3
Laboratory Manual, pp. 13-14, 15-18 L2
Concept Mapping, p. 3 L1
Basic Skills Transparencies 1 and **2** and **Masters** L1 SAE
Basic Concepts Transparencies 1 and **2** and **Masters** L1 SAE

Carnivores and scavengers

Some heterotrophs eat other heterotrophs. Animals such as lions that kill and eat only other animals are carnivores. Some animals do not kill for food; instead, they eat animals that have already died. **Scavengers** such as black vultures are animals that feed on carrion, refuse, and similar dead organisms. Scavengers play a beneficial role in the ecosystem. Imagine for a moment what the environment would be like if there were no vultures to devour animals killed on the African plains, buzzards to clean up dead animals along roads, or ants and beetles to remove dead insects and small animals from sidewalks and basements.

Omnivores and decomposers

Humans are an example of yet another type of consumer. The teenagers in *Figure 3.9* are eating a variety of foods that include both animal and plant materials. They are omnivores. Raccoons, coyotes, and bears are other examples of omnivores.

A fungus is another type of consumer. Organisms that break down and absorb nutrients from dead organisms are called **decomposers.** Decomposers break down the complex compounds of dead and decaying plants and animals into simpler molecules that can be absorbed. Many bacteria, some protozoans, and most fungi carry out this essential process of decomposition.

carnivore:
caro (L) flesh
vorare (L) to devour
Carnivores feed on other animals.

omnivore:
omnis (L) all
vorare (L) to devour
Omnivores eat both plants and animals.

Figure 3.9

Feeding relationships determine how organisms interact and govern how energy flows through the ecosystem.

People are omnivores because we eat both producers and consumers.

Without fungi and other decomposers, dead organic material would pile up in all ecosystems.

The grass plants in this African savanna are producers. They use energy from sunlight to convert inorganic molecules into the nutrients necessary for their own survival. They also provide energy and nutrients for all heterotrophs. Wildebeests are herbivores that eat the grass and make its energy available for carnivores, such as this lion. The vultures are scavengers, consuming organisms that have already died. They will eat anything left by the carnivores.

2 Teach

Using Science Terms

Call students' attention to the derivations of the terms *herbivore, carnivore,* and *omnivore,* presented in the margins of these pages. Ask students to explain the appropriateness of each term to the feeding habits of the organisms it describes. *Herbivores are organisms that feed upon plants; carnivores feed on the flesh of animals; omnivores feed on both plant and animal products.* Ask students to think of other words that begin with the prefixes *omni-, herb-,* and *caro-.*

Reinforcement

LS **Logical-Mathematical** Have students list common pets. Ask students to identify the foods that are typically provided to each pet on their list. Challenge students to classify each pet as an omnivore, carnivore, or herbivore based upon the foods it eats. L1 SAE

interNET CONNECTION

Follow the link for this chapter on the Glencoe Homepage at **www.glencoe.com/sec/science** to find out more about the principles of ecology.

Meeting Individual Needs

Learning Disabled Have students who understand the concept of producer and consumer work with students who are having difficulty with the terms. Group the terms *producer, plant,* and *autotroph* and the terms *consumer, heterotroph,* and *animal.* Have students analyze the meanings of the terms. SAE COOP LEARN

STUDENT JOURNAL

Living as a Decomposer Have students assume that they are decomposers. Ask them to write a short paragraph that describes: a) what they look like, b) where they live, c) what their surroundings are like, d) what they are going to have for breakfast, lunch, and dinner, e) what their job is now and what it will be when they grow up. L1 SAE LS

BioLab Design Your Own Experiment

How can one population affect another?

Time Allotment
Initial preparation: one class session; ten-minute sessions for one to two weeks every other day.

Objectives
Review objectives with students before they begin the Biolab.

Process Skills
observe, record and analyze data, design an experiment, separate and control variables

Safety Precautions
- Use oven mitts when handling hot, sterile pond water.

PREPARATION

Alternate Materials
- Artificial pond water, called Chalkey's solution, may be prepared as follows. Dissolve 1 g sodium chloride, with 0.04 g potassium chloride, and 0.06 g calcium chloride in 1 L of distilled water. Dilute the prepared solution by adding 100 mL of solution to 900 mL of distilled water.
- To prepare sterile pond water, filter the water and place it in flat pans. Boil for 15 minutes. Allow to cool before using.

Possible Hypotheses
- If a predator population is added to a prey population, the size of the predator population will increase, while the prey population decreases.
- If a small predator population is added to a large prey population, no observable difference will occur in the sizes of the populations.

BioLab Design Your Own Experiment

How can one population affect another?

Why don't prey populations disappear when predators are present? Prey organisms have evolved a variety of defenses to avoid being eaten. For example, some caterpillars are distasteful to birds, and some fish confuse predators by appearing to have eyes at both ends of their bodies. Just as prey have evolved defenses to avoid predators, predators have evolved mechanisms to overcome those defenses.

Even single-celled protists such as *Paramecium* have predators. *Didinium* is another unicellular protist that attacks and devours a *Paramecium* larger than itself. Do populations of *Paramecium* change when a population of *Didinium* is present? In this investigation, you will use various methods to determine how both of these species interact.

Didinium

PREPARATION

Problem
How does a population of *Paramecium* react to a population of *Didinium?*

Hypotheses
Have your group agree on a hypothesis to be tested. Record your hypothesis.

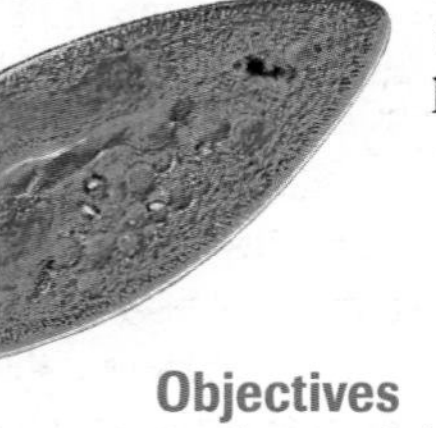
Paramecium

Objectives
In this Biolab, you will:
- **Design** an experiment to establish the relationships between *Paramecium* and *Didinium.*
- **Use** appropriate variables, constants, and controls in experimental design.

Possible Materials
microscope
microscope slides
coverslips
culture of *Didinium*
culture of *Paramecium*
beakers or jars
eyedroppers
sterile pond water

Safety Precautions
Take care when using electrical equipment. Review the Care and Use of a Microscope section in the *Skill Handbook.*

PLAN THE EXPERIMENT

Teaching Strategies
- The amount of initial culture of the two species can be quantified by premeasuring the volume in the pipets.
- A lower magnification provides a wider field of view, making counting easier.
- Methyl cellulose, available from supply houses, may be used to slow the protozoans.
- **Troubleshooting** Have students examine unmixed cultures first so they can later distinguish between *Paramecium* and *Didinium.*
- Suggest that several low-power field counts be made and an average of these counts be used in the data tables.
- Cover or stack culture dishes to prevent drying out.

PLAN THE EXPERIMENT

1. Review the discussion of feeding relationships in this chapter.
2. Decide which materials you will use in your investigation. Record your list.
3. Be sure that your experimental plan contains a control, tests a single variable such as population size, and allows for the collection of quantitative data.
4. Prepare a list of numbered directions. Explain how you will use each of your materials.

Check the Plan

Discuss the following points with other group members to decide final procedures. Make any needed changes to your plan.

1. What will you measure to determine the effect of the *Didinium* on *Paramecium*?
2. What single factor will you vary? For example, will you put no *Didinium* in one culture of *Paramecium* and 5 mL of *Didinium* culture in one culture of *Paramecium?*
3. How long will you observe the populations?
4. How will you estimate the changes in the populations of *Paramecium* and *Didinium* during the experiment?
5. ***Make sure your teacher has approved your experimental plan before you proceed further.***
6. Carry out your experiment.
7. Make a data table that has Date, Number of *Paramecium*, and Number of *Didinium* across the top. Place the data obtained for each culture in rows. Consider making a different table for each of your cultures. Complete your data table. Design and complete a graph of your data.

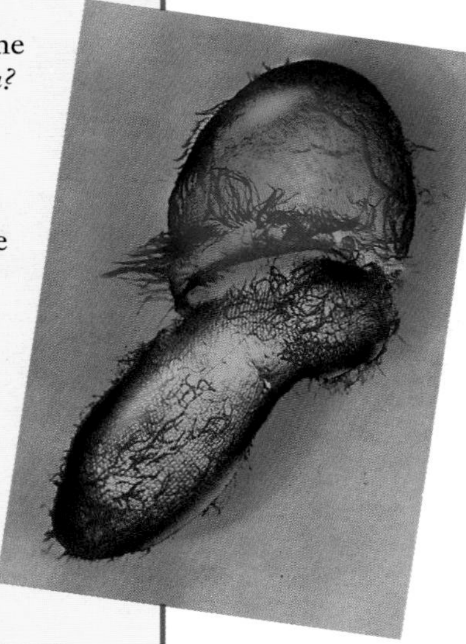

A *Didinium* captures a *Paramecium*.

ANALYZE AND CONCLUDE

1. **Analyzing Data** What differences did you observe among the experimental groups? Were these differences due to the presence of *Didinium?* Explain.
2. **Drawing Conclusions** Did the *Paramecium* die out in any culture? If they did, why do you think this happened?
3. **Checking Your Hypothesis** Was your hypothesis supported by your data? If not, suggest a new hypothesis that is supported by your data.
4. **Thinking Critically** List several ways that your methods may have affected the outcome of the experiment. Suggest ways that you might improve your methods. You may suggest other equipment or materials.
5. **Drawing Conclusions** Write a brief conclusion to your experiment.

Going Further

Application Based on this lab experience, design another experiment that would help you answer any questions that arose from your work. What factors might you allow to vary if you kept the number of *Didinium* constant?

ANALYZE AND CONCLUDE

1. Only cultures containing both *Didinium* and *Paramecium* showed a decline in numbers after a period of time.
2. *Paramecium* died out in the mixed culture. They were preyed upon by *Didinium.*
3. In most cases, hypotheses will be supported.
4. The list may include: counting errors, too few samples, or cultures becoming contaminated or being affected by temperature.
5. *Didinium* are predators of *Paramecium.* In time, *Didinium* will consume all *Paramecium* and then will themselves die off.

Assessment

Skill: Ask students to prepare a summary of this experiment in their journals. L1

Going Further

Have students alter the type of protozoans used or change the initial volume of predator culture. L1

Possible Procedures

- Controls will consist of *Paramecium* cultures with no predators added. Food will have to be added to these cultures. Use one alfalfa pill (available from pharmacies) per liter of water. Note: if one culture receives food, all cultures must receive food to maintain control conditions.

Data and Observations

Depending on the experiment, data and observations will vary. Typically, when they are mixed together, both populations will initially increase in number. A decrease in prey will then be detected as the predator population feeds upon them, with a final drop in the population of predators as their food supply runs out.

SECTION 3.2

Using an Analogy

Provide students with the following example of a symbiotic relationship illustrating commensalism. *You ask a friend who is going to buy tickets for a rock concert to buy tickets for you, because your friend is already making a trip to the box office.* Ask students to give other examples of analogous symbiotic relationships from their daily life that illustrate the idea of commensalism.

Visual Learning

Figure 3.11 Have students speculate about how Spanish moss benefits from the relationship described. *Spanish moss may receive better light intensity from its position or may receive more moisture in the form of dew or fog.* LS

Reinforcement

Provide students with the following situations. A bird builds its nest in the crook of a tree branch. Algae grow on the shell of a marine turtle. Ask students to explain why each situation illustrates commensalism. *Nests at tree height are protected from some predators and the tree is neither helped nor harmed. Algae benefit by receiving light as the turtle swims near the water's surface. The turtle is not harmed nor helped. Both examples show relationships between different species.*

Figure 3.10

Red-breasted geese (left) and peregrine falcons (right) both nest in the Siberian arctic in the spring. They share a symbiotic relationship. The falcon is a predatory bird that includes geese among its prey. The falcon also fiercely defends the area around its nest from predators. Since the falcon hunts away from its nest, geese nesting nearby are not attacked by the falcon. The falcon's behavior also protects the geese from other predators.

Close relationships for survival

Biologists once assumed that all organisms living in the same environment are in a continuous battle for survival. Some interactions are harmful to one species while beneficial to another. Predators such as lions and insect-eating birds kill and eat other animals—prey. Predator-prey relationships such as the one between lions and wildebeests do involve a fight for survival. But there are also other relationships among organisms that help maintain survival in many species. The relationship in which there is a close and permanent association between organisms of different species is **symbiosis.** *Symbiosis* means "living together."

There are several kinds of symbiosis. A symbiotic relationship between the peregrine falcon and red-breasted goose, shown in *Figure 3.10*, has evolved in the cold arctic region of Siberia in Russia. During the nesting season, the falcon's behavior protects nearby geese from predators. The geese benefit from the relationship, while the falcon is neither benefited nor harmed. This relationship is called a commensal relationship. **Commensalism** is a symbiotic relationship in which one species benefits and the other species is neither harmed nor benefited.

Commensal relationships also occur among plant species. The Spanish moss in *Figure 3.11* is a kind of flowering plant growing on the branch of a tree. Orchids, ferns, mosses, and other plants sometimes grow on the branches of larger plants. The larger plants are not harmed, but the smaller plants benefit from the additional habitat.

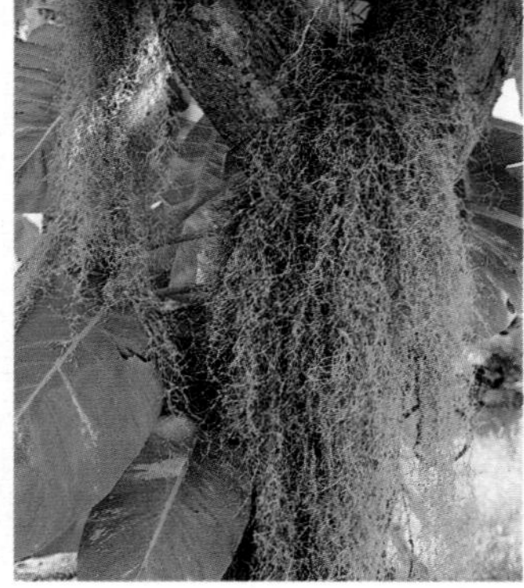

Figure 3.11

Spanish moss grows on the limbs of trees but does not obtain any nutrients or cause any harm to the trees unless it is in large enough numbers. This commensal relationship, like all symbiotic interactions, occurs between two organisms of different species.

PROJECT: Living With a Legume

A symbiotic relationship exists between clover root nodules and bacteria. It is a mutualistic symbiotic relationship commonly found in leguminous plants. Prepared slides are available of root nodules that show this relationship when examined under the microscope. Have students observe prepared slides of nodule bacteria and make diagrams of their observations. Ask them to research the name of the bacterium that lives within the nodules. Students should write a report of their observations and explain how the relationship benefits both plant and bacterium and why it illustrates symbiosis. L2 LS

Sometimes, two species of organisms benefit from living in close association. A symbiotic relationship in which both species benefit is called **mutualism.** *Figure 3.12* illustrates mutualism between a type of ant and a species of acacia tree living in the subtropical regions of the world. The ant protects the tree by attacking any herbivore that tries to feed on it. The ants also clear vegetation away from the trunk of the plant and kill any plant that begins to grow too close to the acacia. The tree provides nectar and a home for the ants.

Sometimes, one organism harms another. Have you ever owned a dog or cat that had been attacked by ticks or fleas? Ticks and fleas are examples of parasites. A symbiotic relationship in which one organism derives benefit at the expense of the other is called **parasitism.** Parasites have evolved in such a way that they harm, but usually do not kill, the host. Why would it be a disadvantage for a parasite to kill its host? If the host dies, the parasite also will die unless it can quickly find another host. Some parasites live inside other organisms. Examples of parasites that live inside the bodies of their hosts are tapeworms and roundworms, which live in the intestines of dogs, cats, and other vertebrates.

MiniLab

What type of symbiosis is found in a lichen?

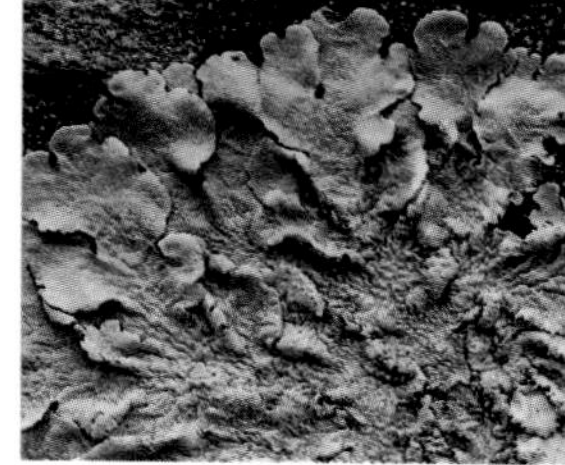

You may have seen lichens growing on bare rock, open soil, or tree bark. A lichen looks like a gray-green, orange, or yellow patchwork of spots on rocks. However, a lichen is a type of organism. In fact, it is more than one! It is a combination of a fungus and either an alga or cyanobacterium. Algae and cyanobacteria are producers that usually exist alone. Fungi, such as mushrooms, also usually exist alone.

Procedure

1. Take a bit of lichen that you have found on tree bark or a rock, or one provided by your teacher, and tease it apart with straight pins.
2. Prepare a wet mount slide of the lichen pieces.
3. Observe the lichen under the microscope.

Analysis

1. Describe the appearances of the two kinds of organisms that make up the lichen.
2. How could the alga or cyanobacterium benefit the fungus, and how could the fungus benefit the alga or cyanobacterium?
3. What kind of symbiosis exists between the fungus and the alga or cyanobacterium?

Figure 3.12

These ants and acacia trees both benefit from living in close association. In an experiment, ecologists removed the ants from some acacia trees. Results showed that the trees with ants grew faster and survived longer than trees without ants. This mutualistic relationship is so strong that in nature, the trees and ants are never found apart.

MiniLab

Purpose

Students will observe a mutualistic relationship.

Process Skills

observe and infer, form a hypothesis

Teaching Strategies

- Take some time to acquaint students with the traits of both algae and fungi.
- Remind students to first use low-power magnification. Final observations should be made under high power.
- Ask students to make colored diagrams of their observations.

Expected Results

Round green or blue-green single or colonial cells that represent the alga or cyanobacterium part of the lichen will be observed, and long colorless strands of the fungus will also be seen.

Analysis

1. Algal and cyanobacterial cells are green or blue-green, and round. Algae have a nucleus; cyanobacteria lack a nucleus. Fungi appear as long, colorless threads with many nuclei.
2. Algae and cyanobacteria make food for the fungus. The fungus provides moisture for the algae or cyanobacteria.
3. Mutualism

Assessment

Performance: Give students samples of the fern *Azolla.* Ask them to observe the fern using the same techniques as in the Minilab, to determine if this organism is involved in a symbiotic relationship. Azolla *contain colonies of* Anabaena *(a cyanobacterium) within its leaves.* L2

GLENCOE TECHNOLOGY

Videodisc

Biology: The Dynamics of Life
Disc 1, Side 1
Symbiosis (Ch. 4)

STUDENT JOURNAL

Learning to Cooperate Have students record events that they have participated in during school or after school hours that illustrate examples of mutualistic and parasitic relationships. For example, a classmate borrows a sheet of paper from another classmate but that classmate in turn borrows a pencil. SAE

SECTION 3.2

Discussion Questions

Ask students to explain what the arrow direction in all food chains represents. *The arrow shows in which direction matter and energy are moving through the food chain.* Why must all second-level organisms be consumers? *By definition, these organisms feed on or consume other organisms.* Why must all third-level organisms be carnivores and not herbivores? *By definition, these organisms feed on other animals and are therefore meat or flesh eaters.*

Bioethics

Saving the Whales
Baleen whales that live in Antarctic waters have been the prey of human hunters for use as food and as a source of fuel for 150 years. As a result, their numbers have been reduced to the point where they border on extinction.

The whaling industry has begun regulating itself to limit the numbers and types of whales that can be caught. Marine ecologists are monitoring the whale's progress. However, the numbers of whales may have been reduced to a point where the whale may not be able to reproduce in large enough numbers to prevent extinction. Should humans have anticipated this problem and reacted sooner to the plight of the whale?

Matter and Energy in Ecosystems

When you pick an apple from a tree and eat it, you are consuming carbon, nitrogen, and other elements the tree has used to produce the fruit. That apple also contains energy from the sunlight trapped by the tree's leaves while the apple was growing and ripening.

Matter and energy are constantly cycling through stable ecosystems. You have already learned that feeding relationships and symbiotic relationships describe ways in which organisms interact. Ecologists study these interactions to make models that trace the flow of matter and energy through ecosystems.

Food chains: Pathways for matter and energy

The wetlands community pictured in *Figure 3.13* illustrates an example of a food chain. A **food chain** is a simple model that scientists use to show how matter and energy move through an ecosystem. Nutrients and energy proceed, from autotroph to heterotroph and, eventually, to decomposers.

A food chain is typically drawn using arrows to indicate the direction in which energy is transferred from one organism to the next. One food chain in *Figure 3.13* could be shown as

algae → fish → heron

Food chains can consist of three links, but most have no more than

Figure 3.13
A community is able to function because each organism within the ecosystem depends on other organisms. In order for a wetland ecosystem to function, its organisms must depend on each other for a supply of energy. Energy must first pass from autotrophs to heterotrophs in a food chain. Follow the steps in the wetland food chain shown here.

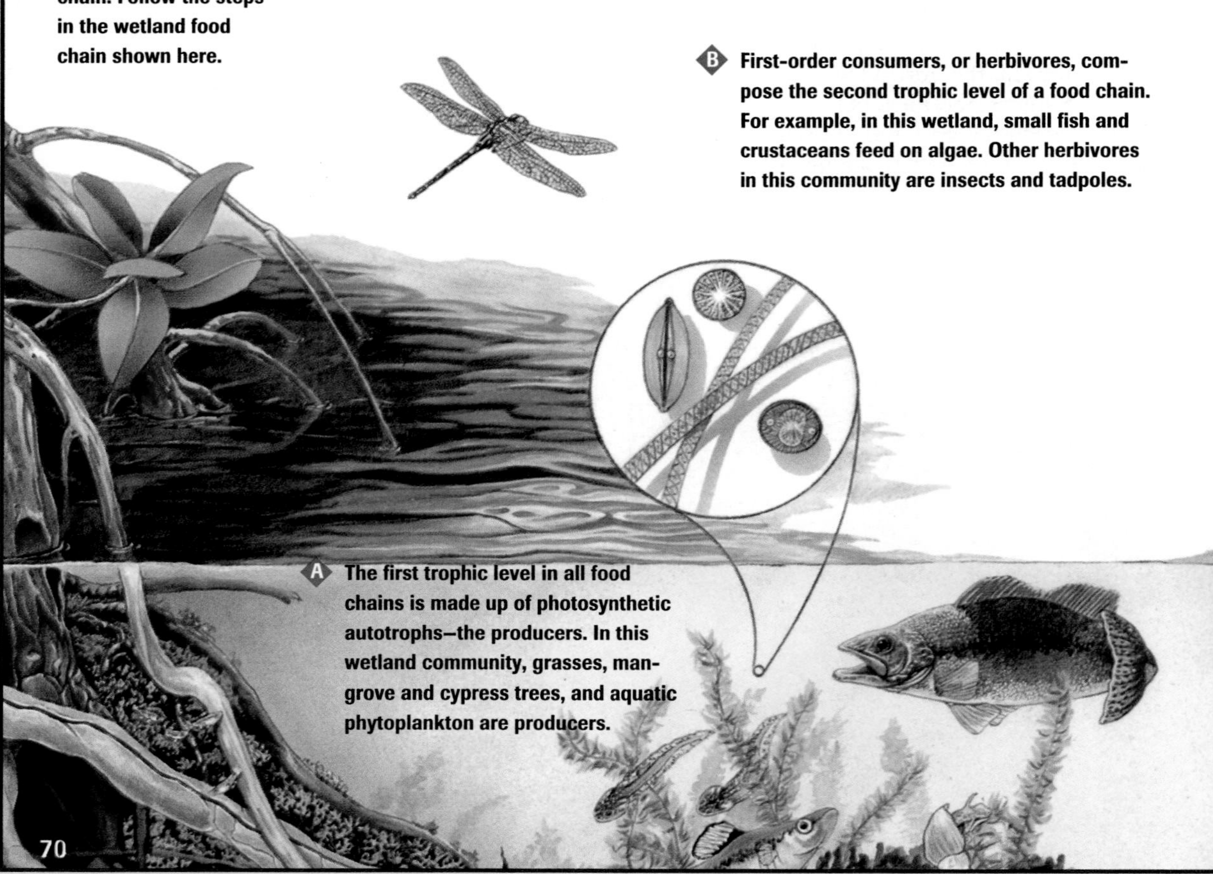

B First-order consumers, or herbivores, compose the second trophic level of a food chain. For example, in this wetland, small fish and crustaceans feed on algae. Other herbivores in this community are insects and tadpoles.

A The first trophic level in all food chains is made up of photosynthetic autotrophs—the producers. In this wetland community, grasses, mangrove and cypress trees, and aquatic phytoplankton are producers.

70

STUDENT JOURNAL

From Niche to Ecosystem Have students prepare a concept map that illustrates the levels comprising the hierarchy of life. As a suggestion, advise students that an inverted triangle may serve as a model structure with two or three populations at the top. L1 SAE

GLENCOE TECHNOLOGY

Videodisc

Biology: The Dynamics of Life
Disc 1, Side 1
Wetlands (Ch. 5)

five links. This is because the amount of energy left by the fifth link is only a small portion of what was available at the first link. A portion of the energy is lost as heat at each link. It makes sense, then, that typical food chains are three or four links long.

Trophic levels represent links in the chain

Each organism in a food chain represents a feeding step, or **trophic level,** in the passage of energy and materials. A food chain represents only one possible route for the transfer of matter and energy in an ecosystem. Many other routes exist. As *Figure 3.13* indicates, many different species occupy each trophic level in a wetlands ecosystem. In addition, many different kinds of organisms eat a variety of foods, so a single species may feed at several trophic levels. For example, the great blue heron eats largemouth black bass, but it also eats minnows, bluegills, and frogs. The alligator may feed on the heron, fish, or even a deer that comes too close. Can you think of other possible food chains?

C Second-order consumers, which are carnivores, make up the third trophic level. They feed on first-order consumers; they are meat eaters. The heron is a carnivorous bird that feeds on fish, frogs, and other small animals of the wetland habitat. Can you think of other animals that might occupy the third trophic level?

D Third-order consumers, carnivores that feed on second-order consumers, make up the fourth trophic level. An alligator eating a shorebird is one example of a third-order consumer. Bacteria and fungi decompose all the links of the food chain when organisms die and can function when necessary at any point in a food chain.

SECTION 3.2

Text Question

Can you think of other possible food chains? *Students are likely to substitute different organisms in place of those mentioned in the text. Accept all logical responses.*

Visual Learning

Figure 3.13 Can you think of other animals that might occupy the third trophic level? *Responses may include other birds of prey such as owls or eagles or land animals such as lions, bears, or cats. Accept all logical responses.*

Science, Technology, and Society

Stable Isotope Tracing

How do ecologists determine what each organism within a food chain or web eats? Ecologists no longer have to directly observe an organism in its food chain position. Thanks to Stable Isotope Tracing (SIT), they can now construct food chains and webs in the laboratory.

SIT is based on the observation that each primary consumer has a characteristic ratio of carbon-13 to carbon-12. Through mass spectrometer analysis, the relative amounts of C-13 to C-12 can be calculated in any sample taken from a food chain or web. By analyzing the amounts of each isotope present in organisms, ecologists can reconstruct the food chain or web.

One result of SIT is the discovery that kelp, and not phytoplankton, is the primary producer in some marine ecosystems. SIT technology is also being used to construct food chains and webs from fossil samples.

Meeting Individual Needs

Visually Impaired Have visually impaired students work with sighted students. Ask the visually impaired students to name some of the foods they consumed in the past 24 hours. Then have the student pair work together to determine if they were a first-order, second-order, or third-order consumer of each food. L1 COOP LEARN

STUDENT JOURNAL

Human Diets Have students write out a restaurant menu that incorporates several food items that would illustrate humans acting as first-order, second-order, and third-order consumers. L1 SAE

Chalkboard Example

Write the word *human* at the top of the chalkboard. Ask students to complete a food web that includes the two trophic levels below humans, using as many different organisms for each level as possible. *Examples of first-order consumers may include chickens, cows, sheep, pigs; examples of producers may include grass, shrubs, lettuce, pears, corn.* Ask volunteers to draw in organisms and arrows in the proper direction on the food web. Ask students to explain why this represents a food web rather than a food chain. *A food chain involves only one organism for each trophic level.*

Tying to Previous Knowledge

Have students review the meanings of the terms *scavenger* and *decomposer.* Ask them to describe the role of each of the types of organisms in a food chain or food web and to illustrate where the organisms should be placed in relation to the various trophic levels. SAE

Visual Learning

Figure 3.14 How many food chains can you find in this food web? *Students should see about eight different food chains.*

Food webs

Simple food chains are easy to study, but they cannot indicate the complex relationships that exist among organisms that feed on more than one species. Notice how the food web of the forest ecosystem in *Figure 3.14* represents a network of interconnected food chains. In an actual ecosystem, many more plants and animals would be involved in the food web. Ecologists who are particularly interested in energy flow in an ecosystem set up experiments with as many organisms in the community as they can. The model they create, a **food web,** expresses all the possible feeding relationships at each trophic level in a community. A food web is a more natural model than a food chain since most organisms depend on more than one other species for food.

Figure 3.14

A forest community food web includes many organisms at each trophic level. Arrows indicate the flow of materials and energy. How many food chains can you find in this food web?

72 Principles of Ecology

PORTFOLIO

Making a Food Web Have students design a food web using the following organisms: wheat, rat, fox, human, cow, corn, rabbit, hawk, grass. Ask them to use a colored pencil or marker to outline one single food chain. Ask them to indicate trophic levels as well as omnivores, herbivores, and carnivores. L1 LS

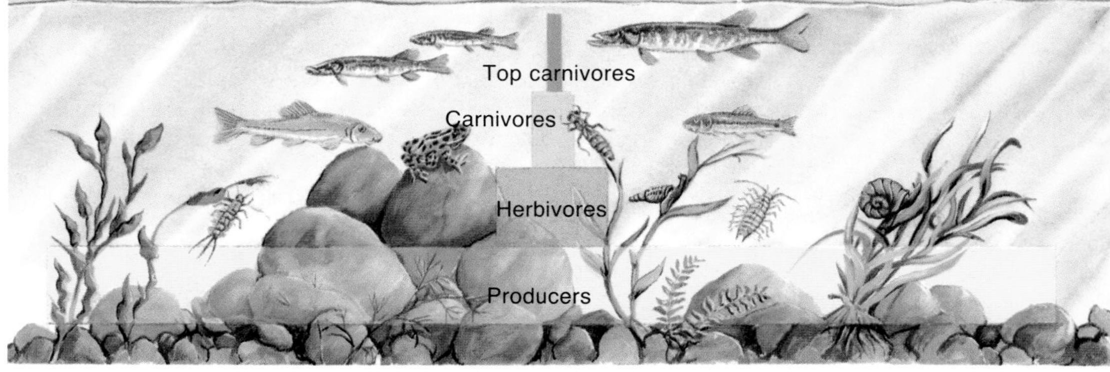

Figure 3.15

Pyramid of energy A pyramid of energy illustrates that energy decreases at each succeeding trophic level. The total energy transfer from one trophic level to the next is only about ten percent. What happens to the other 90 percent of the energy? Organisms fail to capture and eat all the food available at the trophic level below them. Not all the food that is captured and eaten gets digested. And some of the digested food is used by the organism as a source of energy. When the organism is eaten, that energy is used by the organism that consumed it. The energy lost at each successive trophic level enters the environment as heat.

Energy and trophic levels: Ecological pyramids

How can you show how energy is used in an ecosystem? Ecologists use food chains and food webs to model the distribution of matter and energy within an ecosystem. They also use another kind of model, called an ecological pyramid, to depict energy conversions in an ecosystem. The base of the ecological pyramid represents the producers, or first trophic level. Higher trophic levels are layered on top of one another. Examine each type of ecological pyramid in *Figures 3.15, 3.16 and 3.17.* Each pyramid gives different information about an ecosystem. Observe that each summarizes interactions of matter and energy at each trophic level. Notice that the source of energy for these ecological pyramids is energy from the sun.

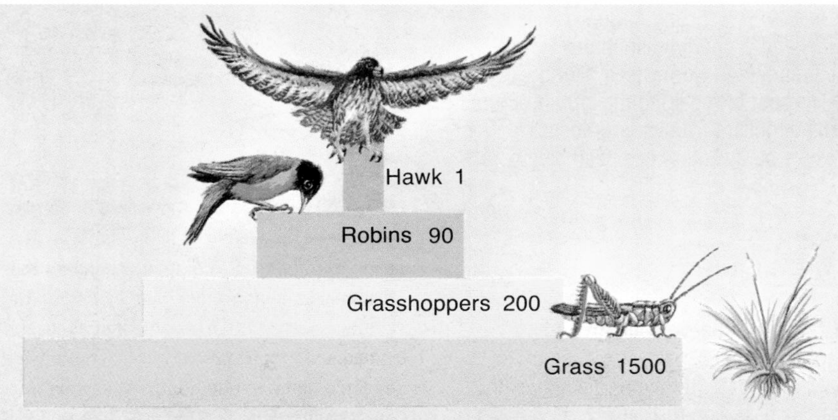

Figure 3.16

Pyramid of numbers Ecologists construct a pyramid of numbers based on the population sizes of organisms in each trophic level. This pyramid of numbers shows that population sizes decrease at each higher trophic level. This is not always true. For example, one tree can be food for 50 000 insects. In this case, the pyramid would be inverted. Can you think of another situation where the pyramid would be inverted?

Concept Development

Ask students to reread the following sentence in the first paragraph of this page: "Ecologists use food chains and food webs to model the distribution of matter and energy within an ecosystem." Elicit examples from students about the types of matter that might be distributed through a food chain or web. *Food, minerals, water.* Then ask where the original source of all energy being distributed comes from, and why food chains must begin with producers. *The sun; only producers can trap the energy of sunlight during photosynthesis.*

Discussion Question

Have students analyze the error in logic for the following scenario. Humans are not at the mercy of producers for their food because they eat animals as a food source. Thus, people would not suffer if all autotrophs were suddenly to die out. *Students should recognize that those animals upon which humans feed are themselves dependent on producers for their food. Thus, if there is no photosynthesis, there will be no possibility for any food chain or web to be completed.*

GLENCOE TECHNOLOGY

Videodisc

The Infinite Voyage: Life in the Balance

Rondonia: Home of a Dying Rain Forest (Ch. 3)

BioLat II: Taking Inventory of the Threatened Peruvian Rain Forest (Ch. 4)

CULTURAL DIVERSITY

Cultural Adaptations to the Environment

Humans occupy almost all types of habitats. People adapt to Earth's varying environments in many ways. For example, humans are able to meet their nutritional needs in many ways. People have designed clothing suited to virtually all climate conditions—from heavy rainfall to subzero temperatures. Architectural designs that make use of available materials help people create shelters adapted to a variety of environments. For example, the Innuit of North America created housing using their most available resources: snow and ice. Peoples of the southwestern United States often built their houses into mountainsides using a mud-clay mixture called adobe.

Issues
BIOLOGY & SOCIETY

Saving the Everglades

Purpose

Students should recognize that humans have altered a major ecosystem and understand some of the steps being taken to correct these alterations.

Background

Lake Okeechobee, to the north of the Everglades, provides the headwaters for this wetland. Because of water diversion, the Everglades now relies on local rainfall for its water. As a result, the vegetation is changing from native sawgrass and wet prairie vegetation to cattails. This is the first step toward eutrophication of the ecosystem, caused mostly by fertilizer runoff.

Teaching Strategies

- Discuss the need to conserve the nation's natural treasures.

INVESTIGATING the Issue

Writing About Biology The questionnaire may ask such questions as: Do you support efforts to restore ecological areas even if this means that jobs will be lost due to a freeze on building in those areas? Who should pay for restoration efforts?

Going Further

Have students contact the Superintendent of Everglades National Park and determine what efforts are being made to save the Everglades. Have students discover whether there are plans to build more park facilities such as raised trails and observation towers. Are there plans to restrict the number of water buggies or hydroplanes that use the park's waterways?

BIOLOGY & SOCIETY

Saving the Everglades

Nearly a million people each year visit Everglades National Park in southern Florida. There are vast colonies of wading birds, alligators, manatees, fishes, and other species. Together with insects and plants, they make this massive ecosystem—covering 4 million acres—one of America's greatest.

Endangered ecosystem

Today, the entire ecology of the region has been altered by the damming, draining, and polluting of its precious water supply for agriculture and urban development. Acres of natural sea grass have been overrun by exotic vegetation, which thrives on the fertilizers that drain into the water system from surrounding farms. The number of wading birds has dropped from a million to fewer than 4000 as their nesting areas have been drained. Thirteen native species are classified as threatened or endangered.

Different Viewpoints

Everglades restoration In 1978, a coalition of state and national organizations began a multimillion-dollar project to restore Everglades National Park. Their aim was to restore natural water flow to the park south of Lake Okeechobee. Besides money, this project requires the support of farmers and city dwellers in South Florida, who must conserve water and limit pollution.

Distribution of funds In 1992, Hurricane Andrew caused extensive damage to both parks and property in southern Florida. Everglades National Park headquarters in Homestead, Florida, was one of the hardest hit areas. The cost of repairing the infrastructure of the park is diverting millions of dollars from environmental research. Repairing structures for park visitors has been given higher priority.

INVESTIGATING the Issue

Writing About Biology Write a questionnaire to find out people's opinions about urban development at the expense of the environment. Collect and analyze the data from your questionnaire. Then, draw your own conclusions based on the data.

Second order consumers

Herbivores

Producers

Figure 3.17

Pyramid of biomass A pyramid of biomass expresses the weight of living material at each trophic level. Ecologists calculate the biomass at each trophic level by finding the average weight of an organism of each species at that trophic level and multiplying by the estimated number of organisms in each population.

STUDENT JOURNAL

Follow the Light Energy Have students assume they are a "packet" of light energy from the sun with a "value" of 100 energy units. Have them trace their path through a simple food chain and indicate their value at each level. Remind them that energy is also lost to the environment at each level. L1 SAE

Meeting Individual Needs

Visually Impaired Obtain or make a pyramid-shaped model. Allow visually impaired students to handle the model to identify its shape. Explain that this model depicts the relative amount of energy change, biomass, and population numbers as one progresses up through a food chain. L1 LS

Cycling maintains homeostasis

Food chains, food webs, and ecological pyramids all show how energy moves in only one direction through the trophic levels of an ecosystem, and how energy is lost at each transition from one trophic level to the next. This energy is lost to the environment in the form of heat generated by the body processes of organisms. Keep in mind that sunlight is the source of all this energy, so energy is always being replenished.

Matter, in the form of nutrients, also moves through the organisms at each trophic level of an ecosystem. But matter cannot be replenished like the energy from sunlight. The atoms of carbon, nitrogen, and other elements that make up the bodies of organisms alive today are the same atoms that have been on Earth since life began. Matter is constantly recycled. Because of this recycling, some of the atoms in your body right now could once have been part of a dinosaur! *Figures 3.18, 3.19, and 3.20* show how three important materials—water, carbon, and nitrogen—cycle through ecosystems.

The water cycle

Water (H_2O) occurs on Earth as a liquid or a solid and in the atmosphere as a gas. Water begins its cycle through an ecosystem when plants absorb it through their roots. Animals drink water or get it indirectly with the food they consume. As water moves through the ecosystem, plants and animals lose it back to the atmosphere through respiration. Organisms also lose water through excretion. After an organism dies, decomposition releases water back into the environment.

Figure 3.18

The Water Cycle

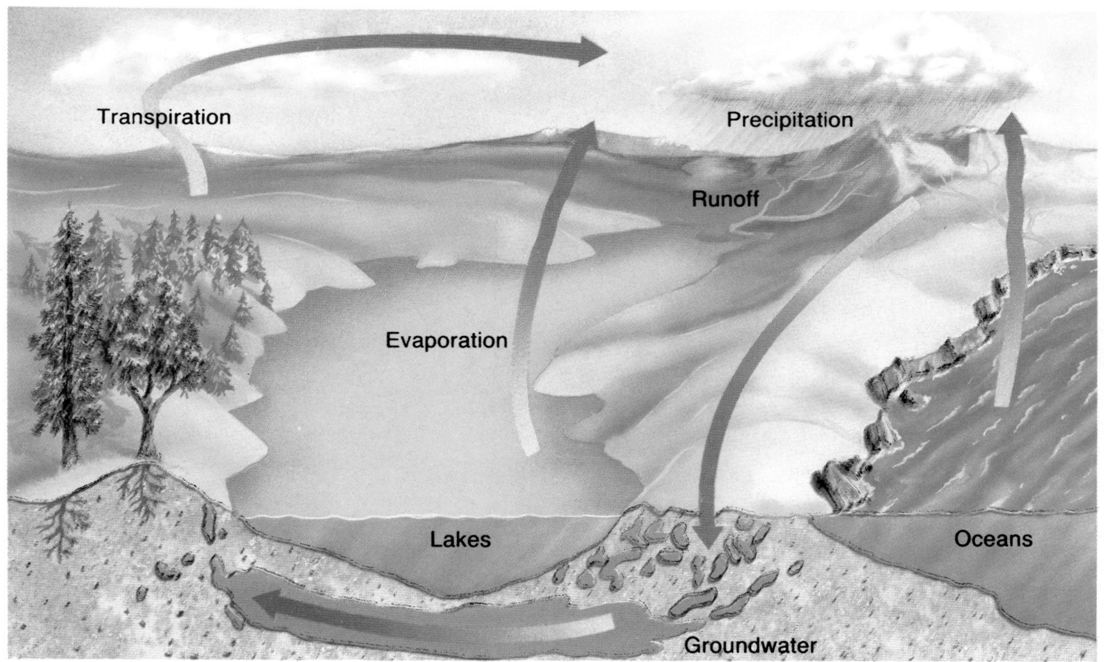

Meeting Individual Needs

Students Acquiring English Have students work in pairs to count the number of grass plants, ants, earthworms, or beetles in a one-decimeter-square area on the school grounds. Students should then use this small sample to estimate the biomass for the whole school yard. SAE COOP LEARN LS

SECTION 3.2

Reinforcement

LS **Visual Spatial** Have students diagram the water cycle for inclusion in their portfolios. Encourage students to include labels on their diagrams identifying the major processes involved in the water cycle such as Evaporation, Condensation, Precipitation, Respiration, Transpiration, Excretion, and Decomposition. Students may use library reference materials for help if necessary. L1 SAE P

Enrichment

LS **Logical-Mathematical** Challenge students to design an experiment that shows the water cycle at work. After reviewing the experimental designs, have students carry out their experiments as a classroom demonstration. Have students create a diagram that shows their experimental setup and a description of what the experiment shows. L3

GLENCOE TECHNOLOGY

Videodisc

The Infinite Voyage: Crisis in the Atmosphere

Studying Man-Made Carbon Dioxide (Ch. 2)

Our Future Climate (Ch. 3)

The Greenhouse Effect: Future and Past (Ch. 4)

SECTION 3.2

Tying to Previous Knowledge

Decomposition releases water back to the environment. Ask students what organisms carry out the decomposition process. *Decomposers such as bacteria and fungi.* How do scavengers recycle water through an ecosystem? *Scavengers feed on dead or decaying matter, thus retaining the water present in these foods. Scavengers recycle this water as they carry on respiration and excretion.*

Chemistry Connection

The Properties of Water
Show students a beaker with a dry outer surface. Add water and ice cubes to the beaker and have students note the condensation of water on the outside of the beaker. Ask students where the water comes from. *The atmosphere.* Ask them to describe several properties of water in the atmosphere. *It is invisible and colorless, and can be changed to a liquid when cooled.* Ask students to explain the steps shown by the demonstration of the water cycle. *Water vapor is present in the atmosphere and can be converted to a liquid (as precipitation) when its temperature is lowered.*

Assessment

Performance Assessment in the Biology Classroom: p. 59, First-Level Biological Magnification. Have students complete this activity to expand upon their knowledge of environmental problems that can harm organisms. L1 P

The carbon cycle

Carbon is found in the environment as carbon dioxide gas (CO_2) in the atmosphere and ocean. From the atmosphere, carbon dioxide moves to aquatic and terrestrial producers. Producers use carbon dioxide in photosynthesis, a process that chemically combines carbon dioxide with water to make sugar. Photosynthesis changes these molecules from low- to high-energy forms. Energy from the sun joins carbon dioxide, oxygen, and hydrogen from water into energy-rich sugars. Organisms obtain carbon when they consume producers or other consumers. Respiration and decay are two processes that usually return carbon to the atmosphere in the gas CO_2. Decay sometimes occurs in swamps, bogs, or other areas that have low amounts of oxygen. If decay occurs without oxygen, the carbon can be bound up in a fossil fuel formed over time by geological processes. Carbon also returns to the atmosphere in large amounts as carbon dioxide when fossil fuels are burned.

Figure 3.19
The Carbon Cycle

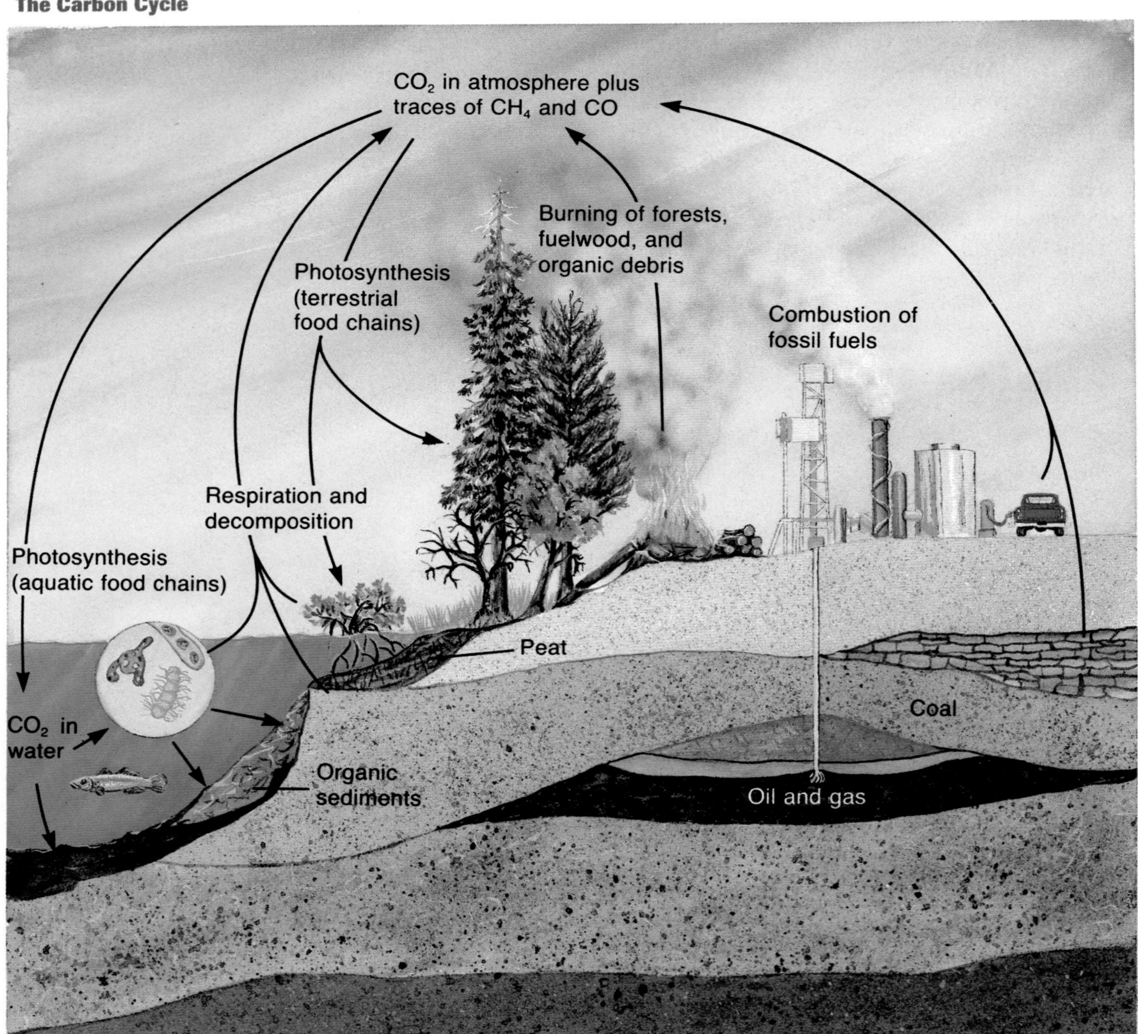

STUDENT JOURNAL

A Wet Life Cycle Have students write a description of what it would be like to be a molecule of water that is recycled through an ecosystem. Where would they spend most of their time, who would they be with, what sites would they see and visit, and how many changes in phase (gas, liquid, solid) might they experience? L1 SAE LS

The nitrogen cycle

Atmospheric nitrogen makes up nearly 78 percent of air. However, living things cannot use nitrogen in the atmospheric form. Notice that lightning and some bacteria convert atmospheric nitrogen into usable nitrogen-containing compounds that can be used by living things. Plants take up nitrates made by bacteria and lightning. Plants convert the nitrates into molecules that contain nitrogen. Herbivores eat plants and convert nitrogen-containing plant proteins into nitrogen-containing animal proteins. During digestion, you convert plant proteins and animal proteins to forms that combine to make human proteins. Organisms return nitrogen to the atmosphere when they die and decay. Can you trace the roles of producers, consumers, and decomposers in the cycling of nitrogen?

ThinkingLab Make a Hypothesis

Rates of Nutrient Cycling

Chemical cycling rates differ among ecosystems. Once decomposition releases chemicals into the environment, they can be taken up again by living organisms. The amount of nitrates in soils, for example, is generally reduced if the amount of decomposition in an area decreases. The amount of biomass, any form of living material, in an ecosystem also determines the amount of soil nitrates. If there is a lot of biomass, then there are fewer nitrates in the soil.

Analysis

The table below presents nitrate data from some hypothetical communities. Compare the amounts of nitrates in the soil and biomass of each community.

Nitrate Content in Three Communities

Community	Soil	Biomass	Atmosphere
	Kilograms/Hectare		Percent
#1	25	85	0%
#2	10	163	0%
#3	1	275	0%

Thinking Critically

Hypothesize which community has the fastest cycling rate for nitrogen. Which of the numbered ecosystems is a tropical rain forest, a desert, and a grassland?

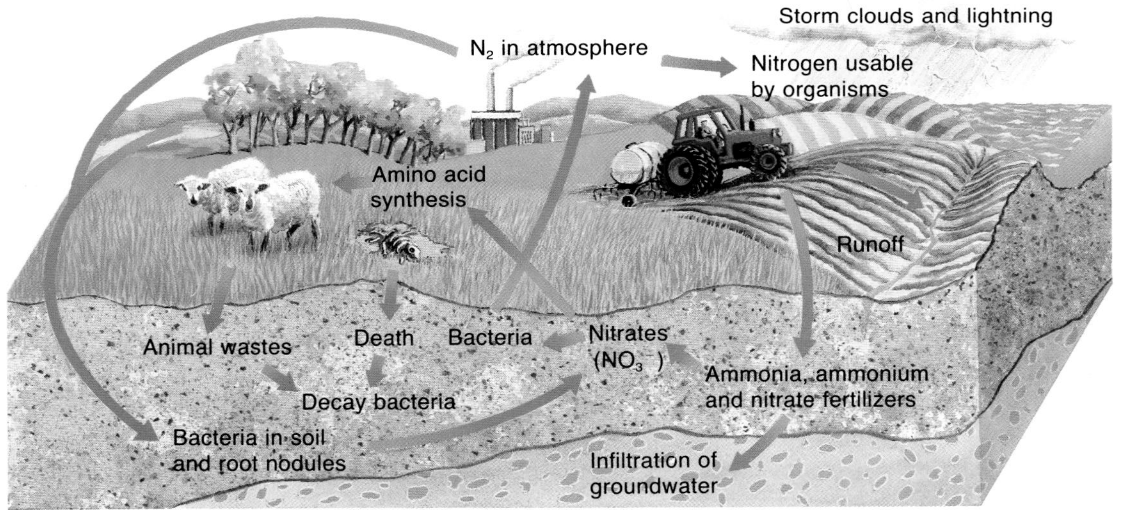

Figure 3.20
The Nitrogen Cycle

Text Question

Can you trace the roles of producers, consumers, and decomposers in the cycling of nitrogen? *Producers take in nitrogen compounds in soil and pass these compounds to consumers that eat the producers. Decomposers break down nitrogen compounds and release nitrogen gas to the air.*

ThinkingLab Make a Hypothesis

Purpose

This lab shows the connection between recycling of nutrients and the availability of a specific nutrient such as nitrate.

Process Skills

observe and infer, form a hypothesis, interpret data

Reinforcement

Ask students to prepare a chart that shows the correlation between increased and decreased decomposition and its effect on nitrate amounts. Have students use their chart as a means for interpreting their data. L1

Thinking Critically

Ecosystem 3 has the highest rate of nitrate cycling because it shows the greatest amount of biomass. Nitrates are recycled from soil to biomass and vice versa. Ecosystem 1 is a desert, 2 is a grassland, and 3 is a tropical rain forest.

Assessment

Portfolio: Ask students to write summaries of their findings for this activity and place them in their portfolios. L1 P

Meeting Individual Needs

Students Acquiring English Have students working in groups prepare a concept map that describes the carbon cycle. Included in this map should be such terms as: *consumers, photosynthesis, respiration, decay,* and *producers.* Groups should consist of students proficient in English working with those who are not. SAE COOP LEARN IN

PROJECT Plant Growth

Using a plant such as duckweed, have students test the idea that nitrates supplied to a plant will increase growth. Nitrates in the form of commercial fertilizers from plant stores can be added to the water in which the plant grows. Advise students to prepare controls. L2 IN

SECTION 3.2

3 Assess

Check for Understanding

Have students explain the relationship between words in the following pairs:

(a) autotroph – producer
(b) heterotroph – consumer
(c) symbiosis – mutualism
(d) recycling – carbon L1

Reteach

Have students identify the trophic level of each organism in a food chain and indicate the direction of energy flow. L1

Extension

Have students speculate as to the consequences to a food web if an organism at the second trophic level were to be eliminated. *This would eliminate a food source for some organisms at the third trophic level.* L1

Assessment

Performance: Have students diagram, label, and explain one of the following: a food web, water cycle, energy or biomass pyramid, or carbon cycle. L1

4 Close

Activity

List on the chalkboard 20 different organisms. Have students use these organisms to create a food web.

Figure 3.21

Phosphorus is an element essential to all organisms, such as this moose and the plants it feeds on, because it is part of all their cell membranes. Phosphorus is important in energy transport in a cell.

The cycling of phosphorus

Materials other than water, carbon, and nitrogen cycle through ecosystems also. Substances such as sulfur, calcium, and phosphorus, as well as others, must also cycle through an ecosystem. One essential element, phosphorus, cycles in two ways. Plants use phosphorus from the soil in their body tissues. Animals get phosphorus by eating plants, *Figure 3.21.* When these animals die, they decompose and the phosphorus is returned to the soil to be used again. This is the short-term phosphorus cycle. Phosphorus can also have a long-term cycle. Phosphates washed into the sea become incorporated into rock as insoluble compounds. Millions of years later, as the environment changes, the rock containing phosphorus is exposed and the phosphorus can again be made a part of the local ecological system.

Connecting Ideas

Organisms in different ecosystems are affected by changes in the abiotic as well as the biotic factors of the environment. The interactions among organisms and the physical factors in the environment result in a balance in each ecosystem.

In this chapter, you found out how ecologists model the balanced movement of matter and energy in ecosystems. Homeostasis in ecosystems is essential to the health of all species. What might happen when ecosystems are disrupted? Communities of organisms react in different ways to changes in their environments. Populations of different species in communities have evolved a variety of adaptations that enable them to survive these changing conditions.

Section Review

Understanding Concepts

1. What is the difference between a parasitic relationship and a commensal relationship?
2. Why do producers always occupy the lowest layer of ecological pyramids?
3. Give two examples of how nitrogen cycles from the abiotic portion of the environment into living things and back.

Thinking Critically

4. Clownfishes are small, tropical marine fishes usually found swimming among the stinging tentacles of sea anemones. What type of symbiotic relationship do these animals have if the clownfishes are protected by the sea anemone, but the anemone does not benefit from the clownfishes?

Skill Review

5. **Designing an Experiment** Design an experiment to test whether or not clownfish are immune from the stinging cells of the sea anemone. For more help, refer to Practicing Scientific Methods in the *Skill Handbook.*

Answers to Section Review

1. Parasitic relationship—one organism obtains a benefit at the expense of the other. Commensal relationship—one organism benefits, the other is neither harmed nor helped.
2. Producers are the only organisms capable of capturing light energy and thus forming their own food.
3. Nitrogen in air passes to bacteria, which, in turn, form chemical compounds used by plants. Death and decay of plants allow nitrogen to return to air. Nitrogen, from the air, passes to bacteria and again moves to plants. Plants are eaten by consumers, thereby passing along nitrogen. Death and decay of these consumers return nitrogen to the air.

Thinking Critically

4. Commensal

Skill Review

5. Place clownfishes in contact with tentacles and stinging cells of anemones and observe and record their reactions. Repeat several times. As a control, observe and record the effects of stinging cells on a different species of fish.

Chapter 3 Review

REVIEWING MAIN IDEAS

3.1 Organisms and Their Environments

- Natural history, the observation of how organisms live out their lives in nature, led to the development of the science of ecology—the study of the interactions of organisms and the environment.
- Ecologists classify and study the biological levels of organization from the individual up to the ecosystem. The role of an organism in its habitat is its niche.

3.2 How Organisms Interact

- Species that live in the same community have evolved feeding and symbiotic relationships. Energy and matter are transferred through ecosystems via these relationships.
- Ecologists use models to illustrate the flow of matter and energy from one trophic level to the next. Food chains are simple models that show one way that materials move from producers to consumers and eventually to decomposers.
- Food webs illustrate all of the possible ways materials are transferred within an ecosystem. Ecological pyramids model the transfer of energy from trophic level to trophic level. Homeostasis requires a constant energy input and is maintained in ecosystems by the cycling of matter through biotic and abiotic portions of the ecosystem.

Key Terms

Write a sentence that shows your understanding of each of the following terms.

abiotic factor	food web
autotroph	habitat
biosphere	heterotroph
biotic factor	mutualism
commensalism	niche
community	parasitism
decomposer	population
ecology	scavenger
ecosystem	symbiosis
food chain	trophic level

UNDERSTANDING CONCEPTS

1. The scientific study of interactions between organisms and their environments is ____________.
a. natural history **c.** ecology
b. a niche **d.** an ecosystem

2. Nonliving factors in the environment are ____________.
a. abiotic
b. homeostatic
c. always easily measurable
d. the same as dead things

3. A group of organisms of a species that can be found in one area at a given time is a(n) ____________.
a. community **c.** population
b. ecosystem **d.** trophic level

4. Biotic factors in a wetland community might include ____________.
a. water **c.** temperature
b. crayfishes **d.** all of these

5. Food webs are more realistic models than food chains because food webs describe ____________.
a. an ecosystem **c.** abiotic factors
b. more interactions **d.** none of these

6. How many species may occupy a niche without competing?
a. one **c.** more than ten
b. no more than three **d.** zero

7. As energy flows through an ecosystem, at each level it ____________.
a. remains the same **c.** fluctuates
b. increases **d.** decreases

Chapter 3 Review

Reviewing Main Ideas

Summary statements can be used by students to review the major concepts of the chapter.

Key Terms

Answers should go beyond defining the terms. Accept any answer that uses the term correctly and in the proper context.

Understanding Concepts

1. c
2. a
3. c
4. b
5. b
6. a
7. d

Chapter 3 Review

Understanding Concepts (contd.)

8. d
9. a
10. b
11. a
12. b
13. c
14. b
15. c
16. c
17. a
18. c
19. a
20. d

Applying Concepts

21. Energy available beyond the fourth trophic level is too small to support life.
22. The pesticide may be carried in the water to other bodies of water or may be carried from one organism to another.
23. Algae → Caterpillar; Algae → Moth → Bird. The symbiotic relationship between the algae and the sloth is commensal because the algae have a place to grow undisturbed, while the sloth is neither helped nor harmed.
24. No, because decomposers are responsible for conversion of biomass to chemicals.
25. Nitrates are compounds necessary for the growth of plants. Plants cannot remove nitrogen from the air directly and must rely on another source for this element.

Thinking Critically

26. Evaluate students' concept maps. The maps should show an understanding of the relationships among the concepts listed.
27. Ten kilocalories will be available to the bluegill. Ten percent of what was originally taken in as available energy is passed to each trophic level (0.10×1000 kilocalories = 100; 0.10×100 kilocalories = 10).

Chapter 3 Review

8. Which chemical cycle includes lightning and bacteria?
   a. carbon c. water
   b. phosphate d. nitrogen
9. Each organism in a food chain represents a ___________.
   a. trophic level c. pyramid
   b. consumer d. chemical cycle
10. The essential element that has two different cycles is ___________.
    a. carbon c. water
    b. phosphorus d. nitrogen
11. Two organisms that benefit from close association are ___________.
    a. mutualistic c. parasitic
    b. commensalistic d. herbivores
12. Organisms that consume only plants are ___________.
    a. omnivores c. carnivores
    b. herbivores d. predators
13. The place where an organism lives is its ___________.
    a. niche c. habitat
    b. community d. trophic level
14. The highest level of biological organization is the ___________.
    a. individual c. ecosystem
    b. biosphere d. community
15. *Vorare* is a Latin word that means ___________.
    a. flesh c. to devour
    b. all d. grass
16. The role of a species is its ___________.
    a. community c. niche
    b. habitat d. trophic level
17. Ecology is the branch of biology that developed from ___________.
    a. natural history
    b. philosophy
    c. scientific methodology
    d. superstition
18. A symbiotic relationship in which one organism harms the other is ___________.
    a. commensalism c. parasitism
    b. habitat d. mutualism
19. In an interaction called ___________, one species benefits and the other species is neither harmed nor benefited.
    a. commensalism c. parasitism
    b. symbiosis d. mutualism
20. A close and permanent interaction between two organisms is called ___________.
    a. ecology c. a trophic level
    b. niche d. symbiosis

APPLYING CONCEPTS

21. Why don't food chains have 20 trophic levels?
22. Explain how pesticides sprayed on the water in a wetland ecosystem could affect a different ecosystem.
23. Sloths are slow-moving herbivores that have algae growing in their fur. Caterpillars of some kinds of moths eat the algae. Birds eat the moths. Using this example, draw two food chains and explain one symbiotic relationship.
24. Deserts contain high levels of chemicals in soil, but low levels of biomass. Would you expect deserts to have large numbers of decomposers? Why or why not?
25. Why do fertilizers contain nitrates?

THINKING CRITICALLY

26. ***Concept Mapping*** Make a concept map that relates the following terms and phrases. Supply the appropriate linking words for your map.
    mutualism, predator, vulture, symbiosis, parasitism, lichen, scavenger, tapeworm, Spanish moss, commensalism, relationships, tiger
27. ***Measuring in SI*** Assume that 1000 kilocalories of solar energy are converted to biomass by algae in a pond. Assume that a minnow consumes the algae. How many kilocalories would be available to a bluegill that consumes the minnow?

Program Resources

Chapter Assessment, pp. 13-18 L1
Alternate Assessment in the Science Classroom
Computer Test Bank L1
Content Mastery, pp. 9-12 L1

28. *Observing and Inferring* If parasites typically do not kill their hosts, then why is it important to keep your pets free from parasites?
29. *Sequencing* Place each of the following organisms into its correct trophic level in a wetland ecosystem: alligator, heron, frog, turtle, phytoplankton, water lily, crayfish, dragonfly, algae, duckweed, minnow, bass, snail.
30. *Sequencing* Place the following organisms in correct order in a food chain: mouse, hawk, wheat, snake.
31. *Observing and Inferring* What do you predict will happen in the community illustrated below?

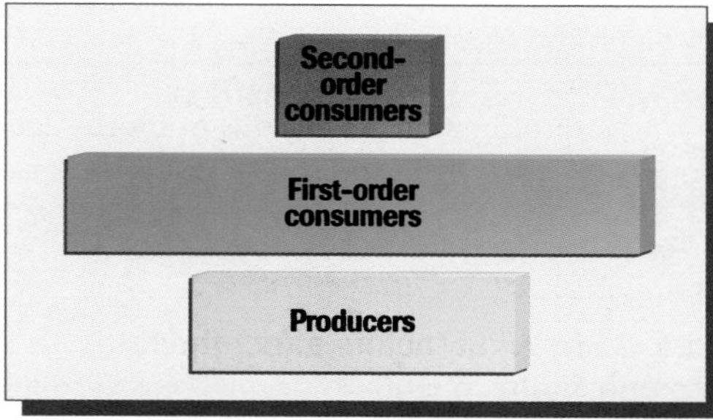

32. *Recognizing Cause and Effect* Predict what would happen if you tried to grow a lichen's fungus separately from its unicellular producer.
33. *Observing and Inferring* You have decided to become a vegetarian. When you mention this decision to your biology teacher, she says you will be eating "low on the food chain." Explain what your teacher means by this statement.
34. *Sequencing* Describe three examples of feeding relationships that cycle matter through an ecosystem.

ASSESSING KNOWLEDGE & SKILLS

Assume that two hypothetical organisms are found in close association. An experiment is performed to see how the populations of the organisms will grow when together and when apart. The graph below represents the results.

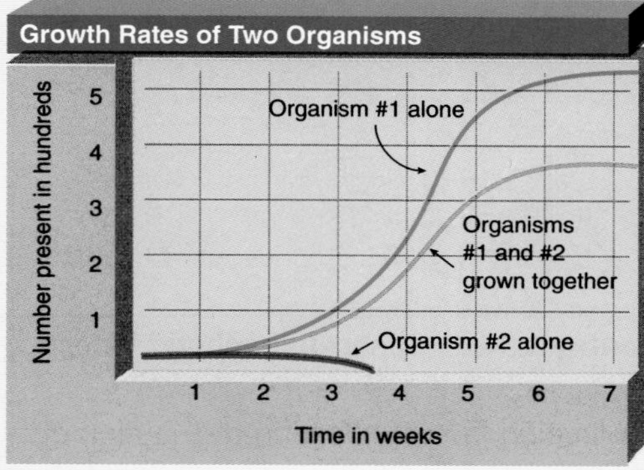

Interpreting Data Answer the following questions.

1. When grown separately, approximately how long after the extinction of Organism #2 did it take the population of Organism #1 to reach its highest point?
   a. 3 days
   b. 1 week
   c. 3 weeks
   d. 5 weeks
2. When the organisms were grown together, what was the approximate rate of growth between weeks 2 and 6?
   a. 75 per day
   b. 100 per month
   c. 50 per week
   d. 25 per day
3. *Observing and Inferring* From the data, it is clear that the association between the organisms is ____________.
   a. commensalism
   b. parasitism
   c. mutualism
   d. a new type of symbiosis

Chapter 3 Review

28. Parasites will ultimately weaken an animal by drawing nutrients away or by introducing disease.
29. First trophic level: phytoplankton, water lily, algae, duckweed; second trophic level: snail, minnow; third trophic level: dragonfly, bass, crayfish, frog; fourth trophic level: alligator, turtle, heron.
30. wheat → mouse → snake → hawk
31. If this diagram represents a pyramid of numbers, the community is not stable and the populations of first-order and second-order consumers would decline. If this diagram represents a pyramid of biomass, it could show a stable aquatic ecosystem in which whales feed exclusively on microscopic organisms.
32. Both depend upon one another. Therefore, they may not survive if separated from one another.
33. She means that you will be eating as a first-order consumer and will be eating only producers such as beans, corn, rice, and so on.
34. Plants provide nitrogen to animals when herbivores feed on plant producers. Bacteria and molds release nitrogen and carbon dioxide when decomposing dead plants and animals. Animals obtain carbon when they consume either producers or consumers.

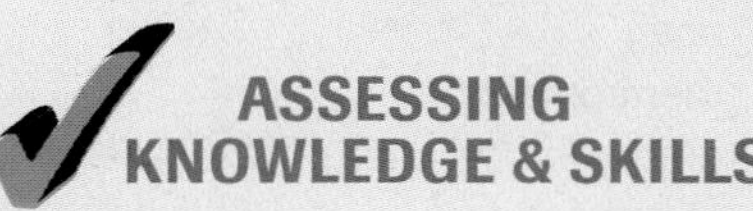

1. c
2. c
3. b

Chapter Organizer

SECTION	OBJECTIVES	ACTIVITIES/FEATURES
4.1 Homeostasis in Communities National Science Standards: UCP.1-2, A.1-2, C.4-6, F.2-3, F.5	1. **Explain** how limiting factors and ranges of tolerance affect distribution of organisms. 2. **Sequence** the stages of succession in different communities.	**Minilab:** How can you create a closed ecosystem?, p. 86 **History Connection:** Surtsey, the Newest Place on Earth, p. 87 **Biolab:** Design Your Own Experiment—Do abiotic factors affect succession in a puddle community?, p. 88
4.2 Biomes National Science Standards: UCP.3, A.1-2, B.3-4, C.4-6, F.1, F.3-4, F.6, G.1	3. **Compare and contrast** the euphotic and aphotic zones of ocean biomes. 4. **Identify** the major limiting factors affecting distribution of terrestrial biomes. 5. **Distinguish** among the terrestrial biomes.	**People in Biology:** Sharyn Richardson, p. 94 **Focus On** World Biomes, p. 98 **Minilab:** How much humus is in your soil?, p. 103 **Thinking Lab:** What factors affect the rate of decomposition?, p. 106 **Biology & Society Technology:** The View from on High, p. 107

ACTIVITY MATERIALS

BIOLAB	MINILABS	ALTERNATE LAB
page 88 plastic cups pond or puddle water fertilizer hay rice thermometer graduated cylinder microscope coverslips slides eyedropper	**page 86** 2-L plastic soda pop bottle potting soil grass or small plant insects, snails, etc. scissors masking tape paper towels **page 103** plastic cup sand 2% H_2O_2 solution eyedropper potting soil	**page 100** 3 cloth squares large beaker sand clay potting soil samples balance twist ties

Chapter 4 Community Distribution

TEACHER CLASSROOM RESOURCES

Reproducible Masters	Transparencies
Section Focus Master 7: Surviving Environmental Changes L1 SAE **Reinforcement and Study Guide,** pp. 13-14 L1 **Biolab and Minilab Worksheets,** pp. 11, 13-14 L1 **Concept Mapping:** Natural Changes in Communities, p. 4 L1 **Tech Prep Applications:** Ecosystem in a Jar, pp. 3-4 L2 **Laboratory Manual:** What Organisms Make Up a Microcommunity?, pp. 19-22 L2	**Basic Concepts Transparency #3:** Primary Succession L1 SAE **Basic Skills Transparency #3:** Abiotic Factors L1 SAE **Reteaching Transparency #4:** Primary and Secondary Succession L1 SAE
Section Focus Master 8: Comparing Biomes L1 SAE **Reinforcement and Study Guide,** pp. 15-16 L1 **Biolab and Minilab Worksheets,** p. 12 L1 **Critical Thinking/Problem Solving:** Where Am I?, p. 4 L3 **Content Mastery,** pp. 13-16 L1	**Basic Skills Transparency #4:** Terrestrial Biomes L1 SAE

ASSESSMENT MATERIALS

Chapter Assessment, pp. 19-24
Performance Assessment in the Biology Classroom, p. 57
MindJogger Videoquiz
Alternate Assessment in the Science Classroom
Computer Test Bank

Spanish Resources SAE
English/Spanish Audiocassettes SAE
Cooperative Learning in the Science Classroom COOP LEARN
Lesson Plans
Biology Projects: Improving Local Habitats, pp. 5-8 L2
Exploring Environmental Issues: pp. 39, 41 L2

KEY TO TEACHING STRATEGIES

- L1 Level 1 activities should be within the ability range of all students including those with learning difficulties.
- L2 Level 2 activities are within the ability range of average to above-average students.
- L3 Level 3 activities are designed for the ability range of above-average students.
- SAE SAE activities should be within the ability range of Students Acquiring English.
- COOP LEARN Cooperative Learning activities are designed for small group work.
- P These strategies represent student products that can be placed into a best-work portfolio.
- These strategies are useful in a block scheduling format.

GLENCOE TECHNOLOGY

The following multimedia resources are available from Glencoe.

Biology: The Dynamics of Life
CD-ROM SAE
Videodisc Program

The Infinite Voyage Series
To the Edge of the Earth
Life in the Balance

National Geographic Society Series
GTV: Rain Forest

Science and Technology Videodisc Series (STVS)
Ecology
Unusual Estuary
Lake Erie Recovery
Plants & Simple Organisms
Managing a Forest
Earth and Space
Flying Observatory

CHAPTER

4 Community Distribution

Chapter Overview

Students start their exploration of community distribution with a study of homeostasis within a community. They read about examples of limiting factors and their relationship to ranges of tolerance. Then they learn about succession, exploring both primary and secondary succession and the factors that lead to the establishment of a climax community.

In the second part of the chapter, students survey the world's biomes. As students study each biome, they are introduced to the biotic and abiotic factors that define the biome. First they learn about aquatic biomes, including marine and freshwater systems, as well as estuaries. Students then consider terrestrial biomes with a focus on the six main biomes: tundras, taigas, deserts, grasslands, temperate forests, and tropical rain forests.

Key Terms

aphotic zone
biome
climax community
desert
estuary
grassland
intertidal zone
limiting factor
permafrost
photic zone
plankton
primary succession
succession
taiga
temperate forest
tropical rain forest
tundra

82

Look for the following logo for strategies that emphasize different learning modalities. LS

Kinesthetic	**Meeting Individual Needs, pp. 98, 103**
Visual-Spatial	**Microscope Activity, p. 93; Tying to Previous Knowledge, p. 93; Student Journal, p. 95; Enrichment, p. 97**
Interpersonal	**Meeting Individual Needs, p. 97; Enrichment, p. 101; Project, p. 102**
Linguistic	**Portfolio, pp. 86, 95; Using Science Terms, p. 87; People in Biology, p. 94; Student Journal, p. 106**
Logical-Mathematical	**Meeting Individual Needs, p. 92; Project, p. 96; Portfolio, pp. 97, 105; Student Journal, pp. 99, 104**

CHAPTER 4 Community Distribution

Have you ever wondered why plants, animals, and other organisms live where they do? Why do lichens grow on bare rock or wooden fences but not in rich soil? Why do polar bears and walruses live only in cold, snowy polar regions? How do catfish manage to survive in waters that are too warm for trout to survive? Some of the most fundamental questions ecologists try to answer have to do with why populations of organisms live where they do, and what kinds of biotic and abiotic factors are important to their survival. Answers to questions like these form the basis for decisions humans must make about our future and the future of our planet.

In this chapter, you will learn how ecological communities are formed, and how and why they change. You will find out why some communities remain stable for long periods of time, while others change from year to year. You will learn how to recognize some of Earth's major aquatic and terrestrial ecosystems.

Concept Check

You may wish to review the following concepts before studying this chapter.

- Chapter 1: homeostasis
- Chapter 3: abiotic factors, biotic factors, ecosystems

Chapter Preview

Laboratory Activities

Biolab: Design Your Own Experiment

- Do abiotic factors affect succession in a puddle community?

Minilabs

- How can you create a closed ecosystem?
- How much humus is in your soil?

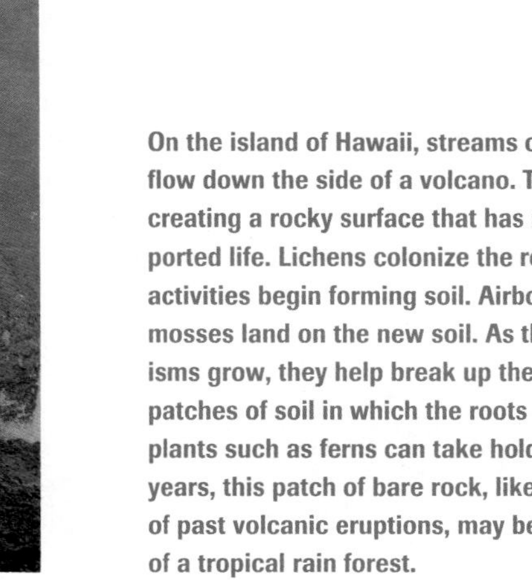

On the island of Hawaii, streams of molten lava flow down the side of a volcano. The lava cools, creating a rocky surface that has never supported life. Lichens colonize the rock, and their activities begin forming soil. Airborne spores of mosses land on the new soil. As these organisms grow, they help break up the rock into patches of soil in which the roots of larger plants such as ferns can take hold. After many years, this patch of bare rock, like many others of past volcanic eruptions, may become the site of a tropical rain forest.

Assessment Planner

Choose assessment strategies from the following pages to evaluate the progress of your students.

Assess, pp. 90, 107
Alternate Lab, pp. 100-101
Minilabs, pp. 86, 103
Portfolio, pp. 86, 95, 97, 105, 106
Thinking Lab, p. 106
Biolab, pp. 88-89
Chapter Review, pp. 109-111

Introducing the Chapter

Ask students to describe the types of plant and animal life they would expect to find along an ocean shore. *Responses may include crabs, clams, insects, shore birds, and algae.* Then have them identify plants and animals they would expect to find living in a forest in New England. *Responses may include trees such as maples, oaks, and hemlock, wildflowers, mosses, mushrooms, snakes, squirrels, and deer.* Challenge students to explain why the types of organisms present in each environment differ. *Students may mention that abiotic factors, such as temperature and moisture, differ in the areas.* Explain to students that in this chapter they will discover how and why different kinds of organisms are suited to different kinds of environments.

Theme Development

The theme of *systems and interactions* is illustrated as students learn about the changes involved in succession. Students will discover that succession results from interactions between biotic and abiotic factors within an ecosystem. *Homeostasis* is illustrated through the study of biomes, each of which represents a climax stage where biotic and abiotic factors interact to keep the system stable.

Concept Check

Have students describe differences in types of vegetation they may have noted while traveling through different parts of their state or the United States. Ask students to form a hypothesis that explains why differences in vegetation occur in different areas.

SECTION

4.1

Prepare

Key Concepts

In this section, the concept of limiting factors, the biotic and abiotic factors that restrict and control life activities of organisms, is introduced. The orderly successions that occur in ecosystems are also discussed, with an emphasis on differences between primary and secondary succession.

Block Scheduling

Look for this symbol for strategies that are useful in a block scheduling format. For more information on block scheduling, refer to Section 4.1 in the **Lesson Plans** booklet.

Materials

- Gather small potted plants, tape, soil, and insects or snails for the Minilab. Have students bring in 2-L plastic beverage bottles.
- Gather hay, rice, thermometers, graduated cylinders, and eyedroppers for the Biolab. Purchase plastic cups and fertilizer. Obtain a large container of pond water.

1 Focus

Bellringer

Before presenting the lesson, display **Section Focus Master 7** on the overhead projector and have students answer the accompanying questions. L1 SAE

SECTION

4.1 Homeostasis in Communities

Section Preview

Objectives

Explain how limiting factors and ranges of tolerance affect distribution of organisms.

Sequence the stages of succession in different communities.

Key Terms

limiting factor
succession
primary succession
climax community

A cactus can live in the driest desert, but it still needs water to survive. Its cells and tissues can absorb and store large amounts of water. Chipmunks can survive cold winters in the forest by going into hibernation. Most organisms are adapted to maintain homeostasis within the environments in which they live.

But what if the ecosystem changes? What happens when a flash flood sends torrents of water through the desert, uprooting the cacti that have lived there for a hundred years? What happens if a new group of chipmunks moves into the forest and begins competing with the other chipmunks for food? Even more dramatically, what happens when a forest fire destroys hundreds of acres of trees?

Changing with the Environment

Ecosystems are always changing. Sometimes they change quickly and dramatically, as with fire or flood. They also can change slowly. As young saplings grow into mature trees that shade the ground below them, grasses are slowly replaced by shade-loving plants. Changing conditions within an ecosystem affect the communities of organisms that live there. As ecologists study these changes, they discover patterns that help explain how the ecosystem has developed. These patterns can be used not only to deepen our understanding of the interactions that take place within an ecosystem, but also to predict what might happen if an ecosystem is disturbed.

Limiting factors

Why do more people live in middle latitudes than near the north pole? You would be correct if you pointed out that there isn't much to eat and it's too cold there! Obtaining food and warmth in the frozen reaches of the Arctic is difficult for humans, but the polar bears in *Figure 4.1* thrive in this icy environment. Environmental factors, such as food availability and temperature, that affect an organism's ability to survive in its environment are limiting factors. A **limiting factor** is any biotic or abiotic factor that restricts the existence, numbers, reproduction, or distribution of organisms. The timberline in *Figure 4.1* illustrates how limiting factors affect the plant life of an ecosystem.

Program Resources

Section Focus Master 7 L1 SAE
Reinforcement and Study Guide, pp. 13-14 L1
Biolab and Minilab Worksheets, pp. 11, 13-14 L1
Concept Mapping, p. 4 L1
Tech Prep Applications, pp. 3-4 L2
Laboratory Manual, pp. 19-22 L2
Basic Concepts Transparency 3 and **Master** L1 SAE
Reteaching Transparency 4 and **Master** L1 SAE
Basic Skills Transparency 3 and **Master** L1 SAE

Figure 4.1

Polar bears, among the largest carnivores on Earth, survive in the frozen lands and waters near the north pole (top). Their white fur offers camouflage against the ice and snow, enabling them to stalk the seals and walruses that serve as their primary food.

The forest stops at the timberline on this mountainside (bottom). At high elevations, temperatures are too low, winds too strong, and the soil too thin to support the growth of large trees. Vegetation is limited to small, shallow-rooted plants, mosses, ferns, and lichens.

Factors that limit one population in a community may also have an indirect effect on another population. For example, a lack of water could limit the growth of grass in a grassland, reducing the number of seeds produced. The population of mice dependent on those seeds for food will also be reduced. What about hawks that feed on mice? Their numbers may be reduced too as a result of a decrease in their food supply.

Ranges of tolerance

Corn plants need two to three months of sunny weather and a steady supply of water to produce a good yield. Corn grown in the shade or during a long dry period may survive, but probably won't produce much of a crop. The ability to withstand fluctuations in biotic and abiotic environmental factors is known as tolerance. *Figure 4.2* illustrates how the size of a population varies according to its tolerance for environmental change.

Some species can tolerate conditions that another species cannot. For example, catfish can live in warm water with low amounts of dissolved oxygen, which other fish species, such as bass or trout, could not tolerate. The bass or trout would have to swim to cooler water with more dissolved oxygen to avoid exceeding their range of tolerance.

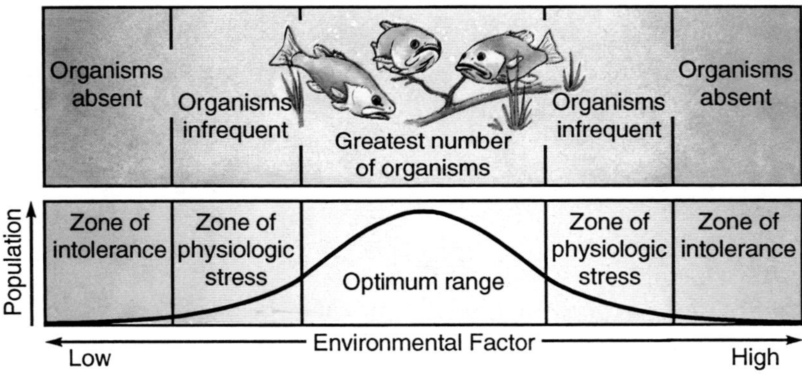

Figure 4.2

Limits of tolerance are reached when an organism receives too much or too little of some environmental factor. Notice that organisms become fewer and fewer as conditions move from optimal toward either extreme of the range of tolerance.

2 Teach

Tying to Previous Knowledge

Review the meanings of the terms *biotic factors* and *abiotic factors*. Ask students to list some abiotic factors that contrast polar regions with regions close to the equator. *Responses will include differences in temperature and precipitation.*

Visual Learning

Direct students' attention to Figure 4.1. Ask them to describe adaptations of the polar bear that allow it to survive in its cold climate. *Thick fur, much body fat, large paws for walking on snow and ice.* Have students infer the climate of the North Pole region. *Cold temperatures, with ice and snow rather than rain.*

Using Science Terms

Have students find the meaning of the term *timberline. Approximate point in elevation where trees cease to grow.* Explain that a timberline is created by limiting factors that prevent soil from supporting tree growth.

Chalkboard Example

Ask a volunteer to construct a food chain on the chalkboard using hawks, mice, and grass. *grass → mice → hawks* Have another volunteer identify the producers *(grass)* and consumers *(mice and hawks)*. Elicit from students how an increase in the hawk population may affect the mouse population. *The size of the mouse population will probably decrease in size as more mice are eaten by more hawks.*

STUDENT JOURNAL

Life as a Cactus Tell students to imagine they are cacti growing in the desert of the southwestern United States. Tell them to describe what their range of tolerance might be to water availability, humidity in the air, and day and nighttime temperatures. L1

Meeting Individual Needs

Learning Disabled Ask students working below grade level to prepare a personal list of the limits that they would place on themselves for tolerance to outside temperature, amount of humidity, amount of shade, and amount of bright light. Have students compare their lists and discuss why they are very similar. COOP LEARN

SECTION 4.1

MiniLab

Purpose

Students will observe and identify limiting factors.

Process Skills

observe and infer, compare and contrast, interpret data

Teaching Strategies

- The black base can be removed more easily by filling the bottle with hot water.
- Add paper toweling to the base before adding soil.

Safety Precautions

Instruct students to be careful when using scissors.

Expected Results

Chambers may remain in balance for several weeks. The system may fail due to too much water, mold growth, or insufficient lighting.

Analysis

1. Decomposition
2. Ecosystems will remain in balance for varying time periods. Limiting factors such as temperature, amount of moisture, soil type, and amount of light may disrupt the balance.
3. In nature, plants and animals are adapted to limiting factors in their area. In the bottle ecosystem, limiting factors outside an organism's range of tolerance may have been introduced.

Assessment

Knowledge: Ask students to trace the pathway of carbon dioxide within their system.

Audiovisual

Show the 35-mm transparency set *Primary Ecological Succession*, Carolina. L1

MiniLab

How can you create a closed ecosystem?

Earth can be thought of as a giant, closed system in which all living things interact with the environment within their ranges of tolerance. It seems like an easy matter to create a similar closed system in a bottle, but is it?

Procedure

1. Remove the black plastic base from an empty, 2-L plastic soda pop bottle and add some potting soil to it.
2. Choose small plants such as clover and plant them in the soil.
3. Determine the number of insects, snails, or other small animals you will include in your closed ecosystem. Put them in the base.
4. Cut the bottle about four inches from the spout. Place it over the plastic bottom.
5. Tape the ecosystem closed, and observe how long it appears to remain in balance.

Analysis

1. What process releases the nutrients of organisms that die?
2. Did your ecosystem remain in balance? For how long? What happened to the plants and animals?
3. How is your closed ecosystem different from Earth?

Succession: A Change in Communities over Time

Changes in an ecosystem also take place as populations grow and diminish in size or move in and out of the area. Ecologists refer to orderly, natural changes that take place in the communities of an ecosystem as **succession.** Succession is often difficult to observe. It can take decades, or even centuries, for one type of community to completely succeed another.

Primary succession

Lava flowing from the mouth of a volcano is so hot it destroys everything in its path, but when it cools it forms new land. Streams gradually deposit silt

Figure 4.3

The first organisms to colonize a new, rocky site are hearty pioneer species such as lichens. Larger plants such as mosses, ferns, and shrubs replace the pioneer species.

PORTFOLIO

Human Succession Have students write a report that chronicles the succession of their personal life over the past 14–16 years. The report should focus on physical, mental, and developmental changes. L1 LS P

along their banks, creating new soil in which plants can take root. The colonization of new sites by communities of organisms is called **primary succession.**

After some time, primary succession slows down, and the community becomes fairly stable. A stable, mature community that undergoes little or no succession is called a **climax community.** *Figure 4.3* illustrates primary succession of bare rock into a climax community.

As pioneer organisms die, their decaying bodies add to the bits of rock accumulating in cracks and crevices, initiating the first patches of soil. The presence of soil makes it possible for weedy plants, small ferns, and insects to become established. The soil builds up, and seeds borne by the wind blow into these larger patches of soil and begin to grow.

Over time, as the community of organisms changes and develops, additional habitats emerge and new species move in. Eventually, the area becomes a forest of vines, trees, and shrubs inhabited by birds and other forest-dwelling animals.

Secondary succession

What happens when a natural disaster such as a forest fire or hurricane destroys a community? What happens when farmers abandon a field or when a building is demolished in a city and nothing is built on the site? Secondary succession is the sequence of community changes that takes place when a community is disrupted by natural disasters or human actions.

History

Surtsey, the Newest Place on Earth

On a cold November morning in 1963, a sailor stood on the deck of a fishing boat bobbing in the sea off the southern coast of Iceland. A strange smell—a little like rotten eggs—was in the air. Rubbing his eyes in disbelief, the sailor witnessed land emerging from the sea. He saw the birth of a new volcanic island, named Surtsey after a giant in an Icelandic myth.

Lava flows; An island grows For nearly two years, lava flowed and volcanic rocks the size of automobiles shot into the sky. When the lava stopped, the barren island measured almost one square mile.

Because the island presented a rare chance for scientists to study how life colonizes a new area, Iceland designated it a nature preserve. Access was strictly controlled, and scientists were forbidden to take away even a pebble.

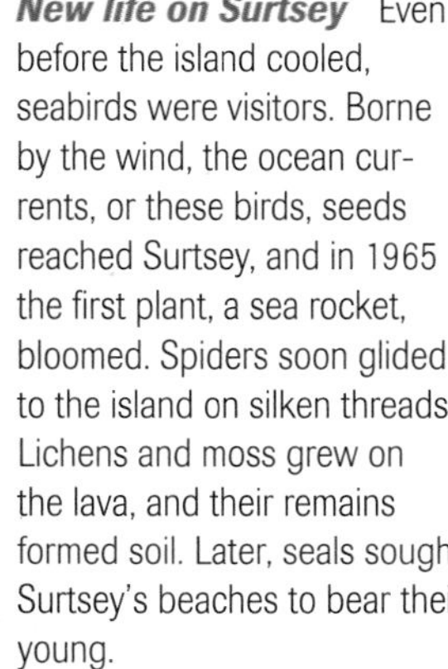

New life on Surtsey Even before the island cooled, seabirds were visitors. Borne by the wind, the ocean currents, or these birds, seeds reached Surtsey, and in 1965 the first plant, a sea rocket, bloomed. Spiders soon glided to the island on silken threads. Lichens and moss grew on the lava, and their remains formed soil. Later, seals sought Surtsey's beaches to bear their young.

A natural laboratory Today, eroding winds and waves are nibbling away at Surtsey. One-fourth of the island's area has disappeared. One day, Surtsey may vanish. Until then, biologist and ecologist Sturla Fridriksson says that scientists "will be able to tell how a society is built up—and how it is destroyed."

CONNECTION TO Biology

In what other places might conditions similar to those on Surtsey provide a chance to study succession in a natural environment?

History

Surtsey, the Newest Place on Earth

Purpose

The essay on Surtsey illustrates the changes that occur during succession.

Teaching Strategies

- Have students locate Surtsey on a map.
- Ask students to name other islands that formed from volcanoes. *The Hawaiian, Galapagos, and Canary Islands are just three volcanic archipelagoes.*

Possible Answer

Responses may suggest an area devastated by warfare, fire, or a mudslide.

Using Science Terms

LS **Linguistic** Have students find the meaning of the word *primary. First or foremost.* Have students explain how this meaning relates to primary succession. *Primary succession includes the first changes occurring in colonization.* Have students repeat this process to explain how the term *climax* relates to a climax community. *Climax means "final." Succession stops when the climax, or final, community is established.* SAE

Enrichment

Ask students to research the adaptation of the lodgepole pine tree that requires a periodic forest fire for survival. Have students report their findings to the class. L3

Meeting Individual Needs

Students Acquiring English Have students use a dictionary to explain why the term *pioneer organism* is well suited to the role such organisms perform. Pioneers are the first to colonize or settle an area. In succession, pioneer organisms are the first to inhabit a region. L1 SAE

BioLab Design Your Own Experiment

Do abiotic factors affect succession in a puddle community?

Time Allotment

Initial session: one class period; followup sessions: 15 minutes once per week

Objectives

After completing the Biolab, students will be able to:

- understand the dynamics associated with succession.
- form a hypothesis about how an abiotic factor affects the rate of succession.
- use data to support or reject a hypothesis.

Process Skills

form a hypothesis, design an experiment, observe and infer, use the microscope, separate and control variables, interpret data

PREPARATION

Alternate Materials

- Refrigerator (use the school cafeteria refrigerator if possible) or incubators
- Water samples from a local pond, or bottom debris from an aquarium if pond water is not available. Purchase protozoan cultures if necessary.
- Purchase alfalfa pellets as food for the microorganisms.
- Baby food jars with covers make excellent chambers for the experiment.

Possible Hypotheses

- If the events of succession are temperature dependent, then lowering temperatures will slow succession.
- If succession is dependent on food supply, then supplying more food should speed succession.

BioLab Design Your Own Experiment

Do abiotic factors affect succession in a puddle community?

Succession is the replacement of one community by another as environmental conditions change. You have read about the processes of primary succession and secondary succession in terrestrial communities. For most communities, succession takes a long time, but succession occurs much more rapidly in microscopic populations of small aquatic communities. In this Biolab, you'll investigate succession in a puddle community.

PREPARATION

Problem

How can you determine the effects of abiotic factors on succession? Among your group, determine one abiotic factor that you can measure in a controlled experiment.

Hypotheses

Decide on one hypothesis that you will test. Your hypothesis might be that added fertilizer increases the speed of succession, or that a higher or lower temperature affects the speed of succession.

Objectives

In this Biolab, you will:

- **Analyze** how abiotic factors affect the rate of succession in a small aquatic community.
- **Observe** how populations change during succession.

Possible Materials

plastic cups
pond or puddle water
fertilizer
hay
rice
thermometer
graduated cylinder
microscope
coverslips
microscope slides
eyedropper

Safety Precautions

Always wash your hands after handling materials that contain live bacteria. Wear eye protection to keep pond water or fertilizer from splashing into your eyes.

PLAN THE EXPERIMENT

Teaching Strategies

- Advise students that their sampling techniques must be consistent. Samples should always be removed from the same part of the container (bottom, middle, or surface) and records should indicate from where the samples were taken.
- Containers originally must have some bottom debris, such as mud, from the original puddle or pond sample. Have students keep the amount of debris/dead plant material/mud consistent in both the experimental and control containers unless the debris is the independent variable.

Possible Procedures

- Students must choose one variable to test and keep all others constant.

PLAN THE EXPERIMENT

1. Decide on a way to test your group's hypothesis. Keep the available materials in mind as you plan your procedure. Be sure to include a control. For example, you might use two cups—one with untreated water and one with added rice, hay, or fertilizer.
2. Decide how long you will observe your ecosystem. Prepare a table that includes a place to record data from your observations, such as population size, and an area to draw each of the kinds of organisms you observe. You will want to record the date, number of drops of water observed from each cup, and numbers of each kind of organism you observe. It is not necessary to know the names of the organisms if your drawings are clearly different.

Check the Plan

Discuss the following points with other group members to decide the final procedure for your experiment.

1. What is your one independent variable?
2. What control will be used?
3. How much water will be put in each cup? How many drops of water will you observe each day?
4. What data will you collect, and how will they be recorded?
5. ***Make sure your teacher has approved your experimental plan before you proceed further.***
6. Carry out your experiment.

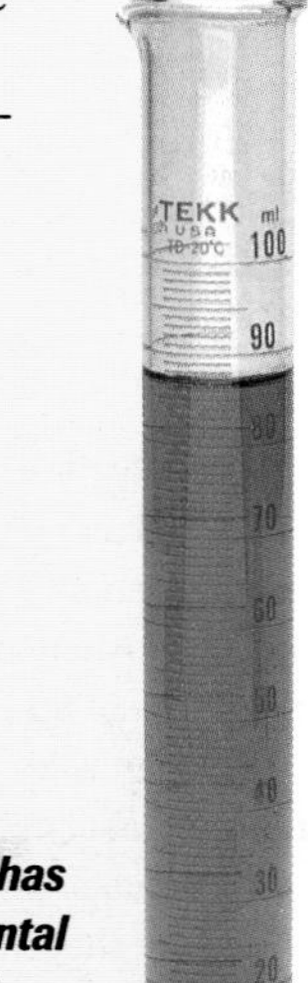

ANALYZE AND CONCLUDE

1. **Checking Your Hypothesis** Was your hypothesis supported by your data? Explain using specific experimental data to confirm your hypothesis about how the abiotic factor you chose affected succession in the puddle.
2. **Interpreting Observations** How many different species did you observe?
3. **Analyzing Data** Were the numbers and kinds of species the same in both cups? If not, how did they differ?
4. **Thinking Critically** Do you think other abiotic factors would have provided the same results? Explain.

Going Further

Project Design an experiment that you could perform to learn whether succession could be accelerated in a similar community of soil microorganisms. If you have all of the materials you will need, you may want to carry out this experiment.

ANALYZE AND CONCLUDE

Student answers may vary.

1. Verify that students use their data to accept or reject their hypotheses.
2. Species types will vary depending on the abiotic factor being tested.
3. Answers will depend on the hypothesis and the procedure.
4. No; each abiotic factor could have a different effect on the rate of succession.

Assessment

Portfolio: Ask students to provide a written evaluation of their experiment. Have them describe any procedural changes they might make if they were to repeat this experiment and explain why they would make the change. L1 P

Going Further

Have students use library resources to identify the organisms they observed. Encourage students to include labeled sketches of the organisms in their portfolios. L1 P

- Keep liquid levels the same and supply the same materials to both containers unless the supplied material is the variable being tested.
- A time frame needs to be established for periodic observations. It is suggested that students observe their communities for 15 minutes, one day each week.
- Observations can best be made using the lowest possible magnification. Most observations will be descriptive, recording only new species that are observed. Some students may wish to count the number of organisms per low-power field. A drop of methyl cellulose may be added to wet mounts to slow the organisms for ease in counting.

Data and Observations

Data will vary with the type of experiment being conducted. A normal progression of organisms: bacteria, paramecia, rotifers. Paramecia may appear as soon as a week into the experiment.

3 Assess

Check for Understanding

Have students explain how the terms in the following pairs are related.

(a) limiting factors - range of tolerance

(b) primary succession - secondary succession

(c) pioneer community - climax community L1

Reteach

Have students prepare a chart showing similarities and differences between primary and secondary succession. L1

Extension

Have students prepare a flow chart showing the sequence of changes that occur during succession of a pond into a hardwood forest. Students will need to use library references. L2

Assessment

Performance: Ask students to recall the opening discussion regarding what a vacant lot might look like over a 20-year period. Ask them to rethink the changes they described and state whether they would predict the same changes now. Ask them to name this process of change. *Succession* L1

4 Close

Discussion

Have students explain how human activities may disrupt or contribute to succession. Then have them list examples of natural events that bring about or hasten succession.

During secondary succession, as in primary succession, the community of organisms inhabiting an area gradually changes. But the species involved in secondary succession are different. Pioneer species, for example, are not the same because the area has not reverted back to the original conditions. Secondary succession also takes less time to reach a climax community than primary succession.

In 1988, a forest fire burned out of control for two months in Yellowstone National Park. Thousands of acres of trees, shrubs, and grasses were burned. As you can see in *Figure 4.4,* the fire has given biologists an excellent opportunity to study secondary succession in a community. Biologists have been observing succession from the time the ground was still smoking. They have been able to observe and compare secondary succession in areas that suffered damage of different levels of severity.

Figure 4.4

The first plants to grow in the newly bare soil following the great fire of 1988 in Yellowstone National Park were annual wildflowers that need plenty of sunshine and were not able to grow in the shade of the forest. Within three years, perennial wildflowers, grasses, ferns, and pine seedlings began coming up through the annuals. Once the pine seedlings grow above the shade cast by the grasses and these herbaceous perennials, the trees will grow more quickly, and eventually a mature forest will once again develop.

Section Review

Understanding Concepts

1. Compare the limiting factors and ranges of tolerance of a lichen and a pine tree.
2. How can a pond community be altered by a dry season?
3. Compare and contrast primary and secondary succession.

Thinking Critically

4. Explain how the success of one species can bring about the disappearance of other species during succession.

Skill Review

5. **Making and Using Graphs** Using the following data, graph the limits of tolerance for temperature for carp. The first number in each pair is temperature in degrees C; the second number is the number of carp surviving at that temperature: 0, 0; 10, 5; 20, 25; 30, 34; 40, 27; 50, 2; 60, 0. For more help, refer to Organizing Information in the *Skill Handbook*.

Answers to Section Review

1. Water: lichens need very little, pine trees require much. Soil: lichens grow on poor soil or rocks, pine trees need nutrient-rich and deep soil.
2. Lack of water can reduce the number and type of plant species in or along pond shores. Animal life within the pond may die from suffocation.
3. Both end in climax communities. Primary succession takes longer and begins with no soil being present, while secondary succession begins with soil present and therefore takes less time to reach climax. Plant types are different during the pioneer stages.

SECTION

4.2 Biomes

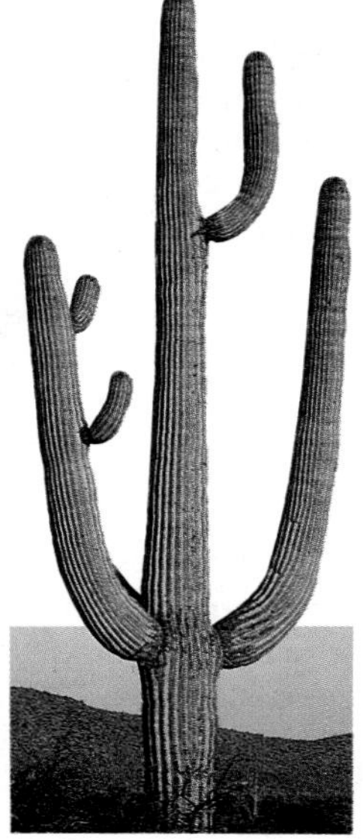

Climate—a combination of temperature, sunlight, prevailing winds, and precipitation—is an important factor in determining what climax community will develop in any one ecosystem. Soil type is also important. Many regions of the world share similar soil and climate characteristics and, as a result, also share similar types of climax communities. Although the species of organisms living in each desert ecosystem may vary, all are adapted for life in an environment with dry weather, poor soil, and hot daytime temperatures.

Aquatic Biomes: Life in the Water

Ecosystems that have similar kinds of climax communities can be grouped into a broader category of organization called a biome. A **biome** is a large group of ecosystems that share the same type of climax community. Biomes located on land are called terrestrial biomes; those located in oceans, lakes, streams, ponds, or other bodies of water are called aquatic biomes.

As a human who lives on land, you may tend to think of Earth as primarily a terrestrial planet. But one look at a globe, a world map, or a photograph of Earth taken from space tells you there is an aquatic world too; approximately 75 percent of Earth's surface is covered with water. Most of that water is salty. Oceans, seas, and even some inland lakes contain salt water. Fresh water is confined to rivers, streams, ponds, and most lakes. Saltwater and freshwater environments have similarities, but they also have important differences. As a result, aquatic biomes are separated into marine biomes and freshwater biomes.

Marine biomes

If you've watched TV programs about ocean life, you may have gotten the impression that the oceans are mostly full of great white sharks, whales, and other large animals. The oceans contain the largest amount of biomass, or living material, of any biome on Earth, but most of this biomass is made up of extremely small, often microscopic organisms that humans usually don't see.

One of the ways ecologists study marine biomes is to separate them into shallow, sunlit zones and deeper, unlighted zones. The portion of the marine biome that is shallow enough for sunlight to penetrate is called the **photic zone.** Deeper water that never receives sunlight makes up the **aphotic zone.** Different parts of the ocean differ in physical factors and in the organisms found there.

Section Preview

Objectives

Compare and contrast the euphotic and aphotic zones of ocean biomes.

Identify the major limiting factors affecting distribution of terrestrial biomes.

Distinguish among the terrestrial biomes.

Key Terms

biome
photic zone
aphotic zone
estuary
intertidal zone
plankton
tundra
permafrost
taiga
desert
grassland
temperate forest
tropical rain forest

aphotic:
a (GK) without
phos (GK) light
photic:
phos (GK) light
Aphotic means without light, photic means with light.

Thinking Critically

4. One species can crowd, block sun, and absorb nutrients and water needed by other species.

Skill Review

5. Check student graphs for logic and accuracy.

SECTION 4.2

Prepare

Key concepts

Students are introduced to world biomes—both aquatic and terrestrial. Limiting factors such as annual rainfall, temperature range, and sunlight availability are discussed in terms of how they result in the establishment of life zones throughout the world.

Block Scheduling

Look for this symbol for strategies that are useful in a block scheduling format. For more information on block scheduling, refer to Section 4.2 in the **Lesson Plans** booklet.

Materials

- Gather sand and a local soil sample for the Minilab. Wash the sand to prepare it for use. Purchase hydrogen peroxide and small plastic cups.

1 Focus

Bellringer

Before presenting the lesson, display **Section Focus Master 8** on the overhead projector and have students answer the accompanying questions. L1 SAE

Discussion

Elicit from students how climate affects their lives. *Responses are likely to indicate that climate determines how they dress, how much energy they use to heat or cool their homes, and what activities they carry out, such as sports.* Explain that climate is important to all organisms and is partly responsible for the distribution of organisms around the world.

2 Teach

Chemistry Connection

Salts in Ocean Water

The salinity of ocean water averages 3.47 percent. The most abundant dissolved salts in ocean water are composed of the following ions: chloride, sodium, sulfate, magnesium, calcium, and potassium. Chloride and sodium ions make up 85 percent of these dissolved solids. Because of the mixing of freshwater and salt water that occurs in estuaries, it is not possible to assign a salinity percentage to these regions. However, estuaries that show low mixing usually have a layer of freshwater atop the salt water. This occurs because freshwater, with fewer dissolved salts, has a lower density than salt water.

Visual Learning

Figure 4.5. Ask students to explain what types of organisms are likely to benefit most from the availability of light and nutrients in estuary waters. *Producers are most likely to benefit because they require the light and nutrients for growth and development.*

Activity

Exploring Environmental Issues: p. 41, Why are Wetlands Important? Have students use this activity to learn how abiotic factors affect wetlands. L2

Figure 4.5

Because estuaries provide an abundant supply of food and shelter, many fishes, clams, and other commercially important organisms live there while young. They venture out of the estuary and into the ocean once they reach adulthood.

Shallow marine environments exist along the coastlines of all landmasses on Earth. These coastal ecosystems include rocky shores, sandy beaches, and mudflats, and all are part of the photic zone.

A mixing of waters

If you were to follow the course of any river, you eventually would reach a sea or ocean. Wherever rivers join oceans, fresh water mixes with salt water. In many such places, an estuary is formed. An **estuary** is a coastal body of water, partially surrounded by land, in which fresh water and salt water mix. Salinity ranges between that of seawater and that of fresh water, and depends on how much fresh water the river brings into the estuary. Salinity in the estuary also changes with the tide. Estuaries contain salt marsh ecosystems, which are dominated by grasses, as illustrated in *Figure 4.5.* These plants grow so thick that their stems and roots form a tangled mat that traps food material and provides additional habitat for small organisms. Decay of dead organisms proceeds quickly, recycling nutrients through the food web.

Twice a day, the gravitational pull of the sun and moon causes the rise and fall of ocean tides. Tides vary in height, depending on the season and the slope of the land. The portion of the shoreline that lies between high and low tide lines is called the **intertidal zone.** *Figure 4.6* discusses organisms living in the intertidal zone of a sandy beach and a rocky shoreline.

Figure 4.6

Intertidal ecosystems have high levels of sunlight, nutrients, and oxygen, but productivity is limited by waves crashing against the shore.

If the shore is rocky, wave action constantly threatens to wash organisms into deeper water. Snails, starfishes, and other intertidal animals of the rocky shore have body parts that act as suction cups for holding onto the wave-beaten rocks. Other animals, such as mussels and barnacles (inset), secrete a strong glue that helps them remain anchored.

Meeting Individual Needs

Learning Disabled Ask students to prepare a pie graph that depicts the composition of Earth's surface. Provide students with the data for the graph as follows: salt water 73.5%, freshwater 2.5%, land 25.0%. LS SAE

STUDENT JOURNAL

Comparing Estuaries and Oceans Have students compare estuaries and oceans. As part of this comparison, ask students to focus on both biotic and abiotic factors associated with each ecosystem. L1

In the light

As you move away from the intertidal zone and into deeper water, the ocean bottom is no longer affected by waves or tides. Many organisms live in this shallow-water region that surrounds most continents and islands. Nutrients washed from the land by rainfall contribute to the abundant life and high productivity of this region of the photic zone.

The photic zone of the marine biome includes the vast expanse of open ocean that covers most of Earth's surface. Most of the organisms that live in the marine biome are plankton. **Plankton,** shown in *Figure 4.7,* are microscopic organisms that float in the waters of the photic zone. Producers of plankton include types of unicellular algae, such as diatoms. Consumers include tiny shrimplike creatures, jellyfishes, worms, and the juvenile stages of animals such as crabs, snails, jellyfishes, and marine worms.

In the dark

What is the ocean like in the dark depths of the aphotic zone? Imagine a darkness blacker than night and pressure so intense it exerts hundreds of pounds of weight on every square centimeter of your body's surface. Does this sound like a hospitable place to live? Almost 90 percent of the ocean is more than a mile deep. In some places, the ocean may extend miles below the sunlit surface. Even though the animals living here are very far below the photic zone where plankton abound, they still depend on plankton for food, either directly or indirectly by eating organisms that feed directly on plankton. Fish living in the deep areas of the ocean are adapted to a life of darkness and a scarcity of food. What adaptations might help these organisms survive in this environment?

Magnification: 10×

Figure 4.7

Plankton form the base of most marine food chains, but not all organisms that eat plankton are small. Baleen whales and whale sharks, some of the largest organisms that have ever lived, consume vast amounts of plankton.

If the shore is sandy, wave action keeps the bottom in constant motion. Most of the clams, worms (inset), snails, crabs, and other organisms that live along sandy shores survive by burrowing into the sand.

SECTION 4.2

Audiovisual

Show the video *4000 Meters Under the Sea,* Films for the Humanities and Sciences.

Show the video *Face of the Deep,* Oxford Scientific Films, Ltd.

Microscope Activity

LS Visual-Spatial Diatoms can be found in scrapings taken from inner aquarium walls or trapped within the filter. They can also be purchased in preserved form from biological supply houses. Have students observe diatoms under the microscope. Ask students to sketch their observations and place their sketches in their portfolios. Have them write a paragraph that describes the diatoms and explains, using evidence they observe, why these organisms make up the first link of a marine food chain. *Diatoms are producers.* L1

Tying to Previous Knowledge

LS Visual-Spatial Have students diagram a food chain that illustrates the relationships among photic zone organisms. Ask students to identify each organism in their food chain as a producer, first-order consumer, second-order consumer, or third-order consumer.

Text Question

What adaptations might help these organisms survive in this environment? *Responses may include a keen sense of smell or use of echolocation or similar adaptive behavior for locating food.*

Program Resources

Section Focus Master 8 L1 SAE
Performance Assessment in the Biology Classroom, p. 57 L1 P
Reinforcement and Study Guide, pp. 15-16 L1
Biolab and Minilab Worksheets, p. 12 L1
Critical Thinking/Problem Solving, p. 4 L3
Exploring Environmental Issues, pp. 39, 41 L2
Basic Skills Transparency 4 and **Master** L1 SAE

People in Biology

Meet Dr. Sharyn Richardson, College Professor

Teaching Strategies

- Use a map to show the location of the Florida Everglades. Explain why reduced water flow from the north has impacted the area.
- Ask students to describe how information from chemistry, physics, biology, and geology is helpful when studying an ecosystem. It may be necessary to first explain the major emphasis of some of these disciplines.
- Explore the meaning and significance of habitat loss.

LS **Linguistic** Read the sentence that begins, "The Florida panther represents what can happen to all kinds of final feeders . . ." Have students analyze the sentence to determine the meaning of the term *final feeders*. Elicit why humans are considered final feeders. *The animal at the end of the food chain is a final feeder. Humans are included here because they are not usually preyed upon by other organisms.*

people in biology

Meet Dr. Sharyn Richardson, College Professor

A typically busy afternoon finds Dr. Sharyn Richardson conducting a tour through the Florida Everglades for two Chinese engineers who are planning mass-transit systems in their country. She is pleased that they want to see the difference between old-style, dredge-and-fill operations and the newer, environmentally sensitive methods that allow a more natural flow of water. Environmental interest is growing all over the world, thanks in part to teachers like Dr. Richardson.

In the following interview, Dr. Richardson talks about her work and interest in the environmental sciences.

On the Job

Q Dr. Richardson, would you tell us about your career?

A I teach environmental science at Barry University in Miami. A favorite course is "Ocean World," which focuses on the interrelationships among the four disciplines of biology, chemistry, physics, and geology.

Q You also have a related outside interest, don't you?

A Yes. I'm the national secretary of Friends of the Everglades, founded by environmental activist Marjory Stoneman Douglas. The group promotes the protection and restoration of the Florida Everglades. One current issue is preservation of the panther. Schoolchildren lobbied the legislature to make the panther the state animal, but it may be the first state animal to become extinct! However, the real issue isn't any one individual species; it's the entire habitat. The Florida panther represents what can happen to all kinds of final feeders at the top of their food pyramids, from predator animals in the wild, and particulary to human beings if we fail to understand the intricate workings of an ecosystem. One theory suggests that if we select stock out of the wild and breed it in captivity, we would have animals to put back. But because the cause of the dwindling numbers of many animals is habitat loss, restocking the wild doesn't address this fundamental problem.

Early Influences

Q Could you tell us how you became interested in the field of biology?

A When I was a high-school sophomore, I did a project for the school's science fair. My kitchen chemistry project dealt with the differences between the ways proteins survive cooking in hard water and in soft water. Although I won the competition, the runner-up was sent to the county fair because she was a senior. I didn't give up,

94

CULTURAL DIVERSITY

Saving the World's Biomes

The treaties negotiated at the Earth Summit held in Rio de Janeiro in 1992 represented a major step in protecting the world's biomes. A global warming treaty included firm guidelines to reduce gas emissions that contribute to the greenhouse effect. A biodiversity treaty called for protection of the millions of species that inhabit Earth, half of which live in rain forest regions. Many developing nations resent having to suppress their economies to protect the environment when most environmental problems were created by industrialized nations. Initiate a debate on this subject and discuss the importance of international cooperation in the effort to preserve Earth and its resources.

L1 COOP LEARN

even though I was hurt. The next year I worked hard to improve my project, and this time I was invited to the science fair at the state level. I finally won fourth place in the international division. Life throws curves like that at all of us, and it's how we deal with the problem that counts.

Later, when I moved to Florida and began teaching in a middle school, I saw changes happening in the Everglades and wanted to know why. I went back to school to find the answers through environmental science courses.

Q What people have influenced your life?

A Marjory Stoneman Douglas certainly has. Even before I met her, I stood in awe of the beautiful prose in her book *The Everglades: River of Grass.* She has devoted so much of her life to saving the Everglades. One lesson I've learned from Marjory, who is now 103, is to be an optimist—always.

Marjory was involved in one of the singular pleasures of my life. She mentioned that she was disappointed at never having heard the alligators roar. I arranged to transport her in her wheelchair to the Everglades on a beautiful, moonlit night. I got a huge, electronic device to amplify the sounds for Marjory, who is nearly deaf. Being blind, she couldn't see the alligators, but she could hear their roar and smell the aroma of the night-blooming moon vine. It was a wonderful moment for both of us.

Personal Insights

Q What makes you tick, both as a person and as a teacher?

A I've never stopped asking questions. The more changes I see in the world, the more I want to know what caused them. To me, the world is a great puzzle, and science is basically the asking of questions. There are very few careers in which one can be paid to continue asking questions and guiding other people in finding answers to their own questions.

Q Do you have any advice for students who are beginning to think about choosing a career?

A I always advise young people to find work they can play at, because we spend more of our lives doing that than almost anything. For those who choose a career in sciences, I encourage them to find a discipline in science that encourages them to think and grow in an interdisciplinary fashion. The glasses I put on as an environmentalist require me to understand Earth's interrelations and interactions. That's where our environment will succeed or fail: on the interrelations, not on any single part.

CAREER CONNECTION

An interest in educating people about the environment could lead to a career in one of these fields.

College Professor *Master's degree or doctoral degree in Environmental Sciences or Biology*

High-school or Grade-school Teacher *Bachelor's degree*

Teacher's Aide *High-school diploma; on-the-job training*

CAREER CONNECTION

- **Career Path** Students interested in becoming environmentalists should explore the following course pathways: high school courses in Earth science, biology, chemistry, and physics, and college courses in biology geology, environmental sciences, chemistry, geology, and ecology.

For More Information

- Read the following articles in *National Geographic:* "Age-old Challenge: Water and the West," B. Babbitt, June 1991, and "South Florida Water: Paying the Price," N. Duplaix, July 1990.
- Contact the following organizations for more information regarding a career in this field:

  Conservation Foundation, 1250 24th Street, N.W., Washington, D.C. 20037; Environmental Action, 1525 New Hampshire Avenue, N.W., Washington, D.C. 20036; U.S. Environmental Protection Agency, 401 M Street, S.W., Washington, D.C. 20460; Worldwatch Institute, 1776 Massachusetts Avenue, N.W., Washington, D.C. 20036.

GLENCOE TECHNOLOGY

 **Videodisc**

The Infinite Voyage: Life in the Balance

Southern California's Chaparral: Extinction by Development (Ch. 6)

Biodiversity: Battling Extinction in the Costa Rican Dry Forest (Ch. 7)

PORTFOLIO

A River of Grass Have interested students read *The Everglades: River of Grass,* by Marjory Stoneman Douglas. Have students prepare a report on their reading or record how their personal view of the Everglades may have changed based on what they now know. L2 LS P

STUDENT JOURNAL

Diagraming Food Chains Ask students to diagram a food chain that would be typical of the life-forms present in the Everglades. Their chain must end with the panther as the animal at the top of the chain. Have them explain why there are no trophic levels above the panther. L1 LS

Science, Technology, and Society

Controlling Thermal Pollution

Power plants may release large volumes of heated water into nearby lakes or estuaries. Adding heated water to a lake reduces the amount of oxygen available to organisms living in the lake. While slight increases in water temperature may benefit some species, sharp rises in temperature are harmful to organisms.

To help resolve problems created by heated water, two methods are being used to reduce the discharge of heated water into lakes and estuaries. In one method, hot water is released into large human-made outdoor ponds, where the water can slowly cool. In the second method, hot water is passed over baffles in tall cooling towers to promote cooling. The use of such cooling methods is helping to preserve biodiversity in lakes and estuaries by reducing deaths caused by heated, oxygen-deficient water.

Discussion Question

Elicit from students why organisms that live on lake bottoms are often scavengers. *Dead organisms that drift to the bottom of the lake provide an ample food source for these organisms. Also, plant life is scarce in this part of a lake, so organisms that feed on plants cannot survive here.*

Figure 4.8

Cattails and sedges are the plants most commonly found growing around the edges of lakes. As you move from the margins of a lake or pond toward the center, you find concentric bands of different species of plants. The shallow water in which these plants grow serves as home for tadpoles, aquatic insects, turtles that bask on a partially submerged rock, and worms and crayfishes that burrow into the muddy bottom. Insect larvae, whirligig beetles, dragonflies, and fishes such as minnows, bluegill, and carp also live here.

Freshwater biomes

Have you ever gone swimming or boating in a lake or pond? If so, you may have noticed concentric rings of different kinds of plants growing around the shoreline and even into the water, as shown in *Figure 4.8.* The shallow waters in which these plants grow are highly productive and include fishes, algae, protists, mosquito larvae, tadpoles, and crayfishes.

If you ever jumped into a deep lake on a warm summer day, you probably got a cold surprise the instant you entered the water. Though the summer sun heats the lake's surface water, a foot or so below the surface remains cold. If you were to dive all the way to the bottom of the lake, you would discover more layers of increasingly cold water as you descended. These temperature variations within a lake are an abiotic factor that limits the kinds of organisms that can survive in deep lakes.

Another abiotic factor that limits life in deep lakes is light. Not enough sunlight penetrates to the bottom to support photosynthesis, so few aquatic plants or algae grow. As a result, the density of populations is lower in deeper waters. Decay takes place at the bottom of a lake. As dead organisms drift to the bottom, bacteria break them down and recycle the nutrients they contain.

PROJECT — Measuring Oxygen

Purchase kits from a biological supply house that measure the dissolved oxygen in water. Have students use these kits to test water samples from different local bodies of water. Ask students to record from where the samples came. L3 LS

Terrestrial Biomes

If you set off on an expedition beginning at the north pole and travel south to the equator, what kinds of environmental changes will you notice? The weather gets warmer, of course. You also see a gradual change in the kinds of plants that cover the ground. At the snow- and ice-covered polar cap, temperatures are always freezing and no plants exist. A little farther south, where temperatures sometimes rise above freezing but the soil never thaws, you might see soggy ground with just a few small cushions of low-growing lichens and plants.

As you continue on your journey, temperatures rise a little higher and you enter forests of coniferous trees. Farther south are grasslands and deserts with scorching summertime temperatures and little rain. Finally, as you approach the equator, you find yourself surrounded by the lush growth of a tropical forest, where it rains almost every day.

As you move south from the north pole, you find yourself traveling through one climax community after another. The graph in *Figure 4.9* shows how two abiotic factors—temperature and precipitation—influence the kind of climax community that develops in a particular part of the world. These climax communities can be used to group terrestrial ecosystems into six major biomes, as you will see in the Focus On World Biomes on the next page.

Figure 4.9

If you know the average annual temperature and rate of precipitation of a particular area, you should be able to determine the climax community that will develop. Predict which climax community would result from an area that has an annual precipitation of 150 cm and an annual average temperature of 15°C.

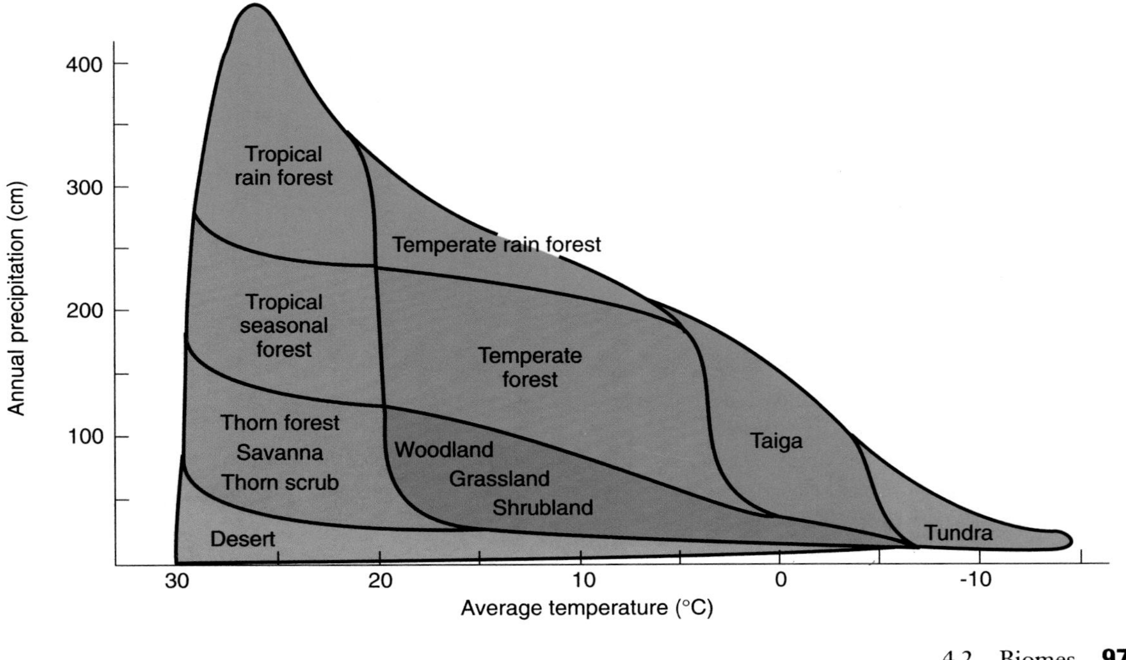

SECTION 4.2

Audiovisual

Show the filmstrip *Terrestrial Biomes of the World*, Educational Dimensions. L1

Visual Learning

Figure 4.9 Predict which climax community would result from an area that has an annual precipitation of 150 cm and an average temperature of 15°C. *A temperate forest.*

Discussion

You may wish to discuss with students what effect increased altitude has on temperature. Ask students to use this information to explain why mountaintops in tropical areas may be covered by snow.

Enrichment

LS **Visual-Spatial** Have interested students research the rain shadow effect that produces desert conditions in some areas. Ask students to prepare a diagram for their portfolios that explains this process. L2 P

PORTFOLIO

Recognizing Biome Characteristics Have students do library research to determine the average annual temperature and precipitation for the area in which they live. Have students determine if the values correlate to those given in the text for the biome present in their area. Have students explain any discrepancies. L2 LS

Meeting Individual Needs

Behaviorally Disordered Group students with different ability levels. Ask them to prepare a concept map showing the six terrestrial biomes. The biomes are to be correlated with their general location on the globe starting with the North Pole and moving south toward the equator. L1 LS COOP LEARN

Focus On
World Biomes

Purpose
Students will begin to associate terrestrial biomes with their geographic locations.

Visual Learning

- Have students identify the biome that makes up most of the area directly above the equator. *Desert*
- Have students use the map to name the three biomes that occupy the greatest area on Earth. *Taiga, temperate forest, and grassland* Then ask them to identify the biome that occupies the smallest percentage of Earth's surface. *Tropical rain forest*
- Use the map to point out that from largest to smallest the biomes that make up the United States are: grassland, temperate forest, desert, taiga, and coniferous forest. Elicit from students what this biome distribution suggests about the ability of the United States to provide food for its people. *Much of the United States is covered by the fertile soil of grasslands, allowing for the growth of a variety of crops.*

Background
Famine
Areas in which severe famine has occurred are in the northern parts of Africa. Ask students to correlate the famine problems of northern Africa to its predominant biome. *The area is mostly desert, which is not a productive biome for growing crops.*

Focus On
World Biomes

When you think of a biome, your first images may be of a land with one or two prominent animal species—elephants or lions on an African plain, musk oxen or reindeer on the tundra, or monkeys in a rain forest. Actually, it is the climax community of plants that ecologists use to characterize a biome. Plants don't migrate and are a better indicator of the long-term characteristics of the biome.

Earth is marvelously diverse. Millions of species are spread over its surface. But this distribution is not random. As shown on the map of the world, there is a pattern to the biomes of Earth.

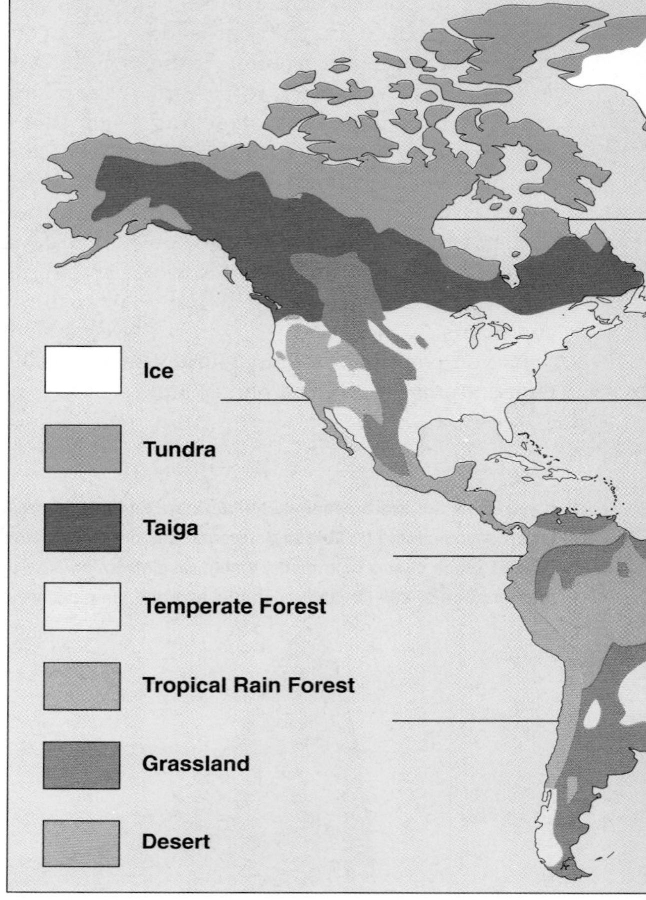

Terrestrial biomes

What defines a biome? In general, three factors determine which biome will be dominant in a terrestrial location—latitude, altitude, and precipitation. A low-lying area near the equator that gets lots of rain will have a tropical rain forest as climax vegetation. Just a few kilometers away on the side of a mountain, you may find plants typical of a biome thousands of kilometers to the north or south. Similarly, latitudes that produce rain forests in the western hemisphere can produce deserts or tropical savannas and grasslands in Africa, where rainfall patterns are different.

Meeting Individual Needs

Students Acquiring English Provide a globe that can be handled by students. Use the globe to review the locations of Earth's equator, the Tropics of Cancer and Capricorn, and the Arctic and Antarctic Circles. Review the orientation of latitude lines on the globe and correlate these with those on the map.
SAE LS

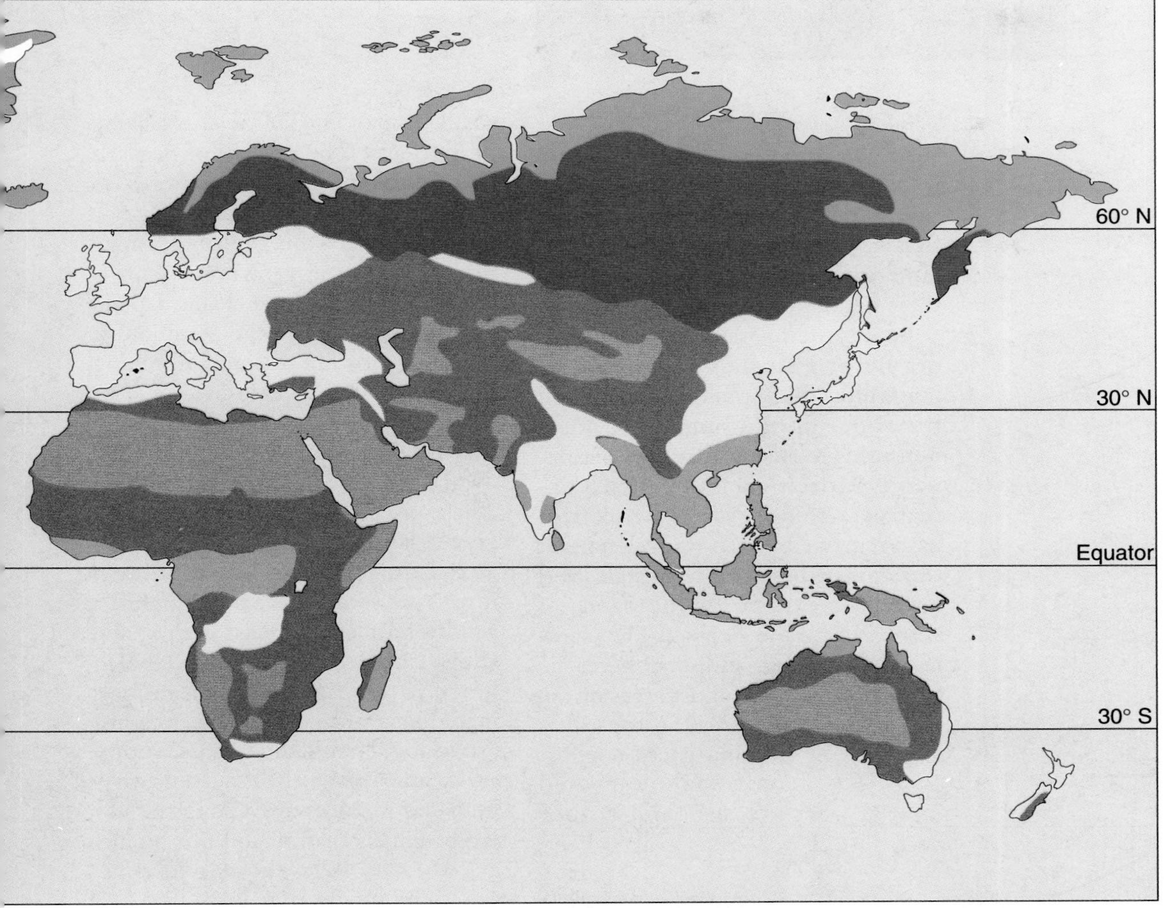

EXPANDING YOUR VIEW

1. **Thinking Critically** Which biome would recover the most slowly from destruction due to natural or human-caused events? Explain.
2. **Applying Concepts** Think about the general pattern of biome types that exists from the equator to the poles. Do you think you would find a similar pattern if you were to climb from the base to the top of a very high mountain? Explain your answer.

Focus On

Biome Distribution

Biome distribution is determined by air currents, Earth's rotation, prevailing winds, and proximity to the ocean. Discuss these four factors. Cooler, more dense air sinks and pushes warmer, less dense air upward. Heated air usually contains more moisture than cooler air. As warm, moist air becomes cooled, it drops its moisture as precipitation.

Earth's rotation affects the movement of warm and cool air to create convection cells: three north and three south of the equator. At the equator, moisture-laden air that is cooled as it is pushed upward results in great amounts of precipitation. Tropical rain forests are generally located in this region.

Prevailing winds result from convection cells and Earth's rotation. In addition, ocean currents also play a role in regulating climate and in determining biome locations.

Answers to Expanding Your View

1. **Thinking Critically** A tropical rain forest. Since this biome contains the greatest species diversity, destruction of habitat can result in extinction of large numbers of species.
2. **Applying Concepts** Yes. Temperature decreases with altitude increase just as it does with latitude increase.

STUDENT JOURNAL

Biome Distribution Ask students to describe in their journals how terrestrial and marine biomes are distributed over the globe in relation to the equator. That is, what general percentage of land and water appear above and below the equator? L1 IS

Tying to Previous Knowledge

Have students name examples of survival adaptations used by organisms in the tundra. *Likely responses will mention color camouflage, migration, heavy fur coat, hibernation, and flat leaves on plants that reduce water loss from wind.*

Misconception

A popular belief is that lemmings, small mammals common in the tundra, periodically march into the ocean in mass suicides intended to cull out or reduce the population. Actually, lemmings are migratory in nature and, after severely depleting an area of food, they move in large numbers to other areas in search of food. Sometimes, the animals stumble into the ocean during these mass migrations. However, the occurrence is not a programmed effort to reduce their population.

GLENCOE TECHNOLOGY

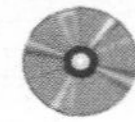

Videodisc

Biology: The Dynamics of Life
Disc 1, Side 1
Tundra (Ch. 6)

Disc 1, Side 1
Taiga (Ch. 7)

Figure 4.10

Grasses, grasslike sedges, small annuals, and reindeer moss, a type of lichen on which reindeer feed, are the most numerous producers of the tundra. The short growing season may last less than 60 days.

Life on the tundra

As you begin traveling south from the north pole, you reach the first of two biomes that circle the pole. This first area is the **tundra,** a treeless land with long summer days and short periods of winter sunlight.

Because temperatures in the tundra never rise above freezing for long, only the topmost layer of soil thaws during the summer. Underneath this topsoil is a layer of permanently frozen ground called **permafrost.** Some areas of permafrost have remained frozen for so long that the frozen bodies of animals that have been extinct for thousands of years, such as the elephant-like mammoth, are sometimes found there.

In most areas of the tundra, the topsoil is so thin it can support only shallow-rooted grasses and other small plants. The soil is also lacking in nutrients. The process of decay is slow due to the cold temperatures, so nutrients are not recycled quickly.

Summer days on the tundra may be long, but the growing season is short. Because all food chains depend on the producers of the community, the short growing season is a limiting factor for life in this biome. For example, typical flowering tundra plants, *Figure 4.10,* are grasses, dwarf shrubs, and cushion plants. These organisms live a long time and are resistant to drought and cold.

Mosquitoes and other biting insects are some of the most common tundra animals, at least during the short summer. The tundra is also home to a variety of small animals including ratlike lemmings, weasels, arctic foxes, snowshoe hares, snowy owls, and hawks. Musk-oxen, caribou, and reindeer are among the few large animals that inhabit this biome. *Figure 4.11* shows two common tundra animals.

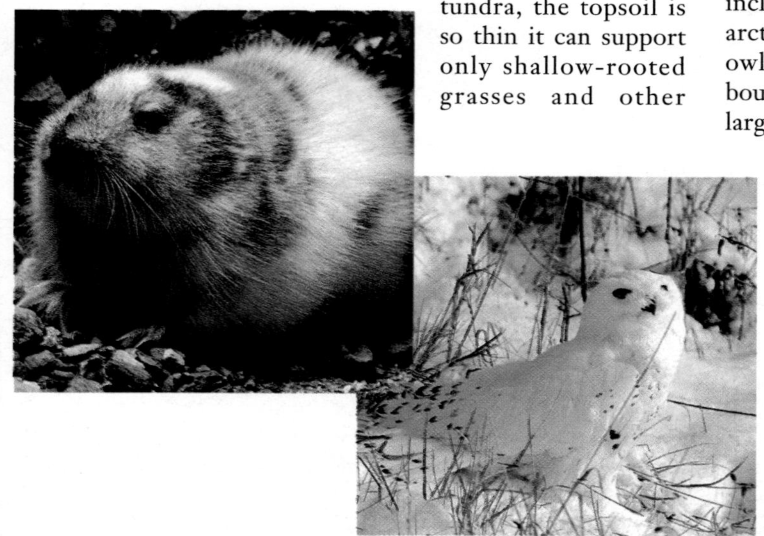

Figure 4.11

Lemmings are the most numerous mammals living in tundra communities (left). Populations of these small, furry, ratlike animals sometimes rise to exceedingly high numbers. At those times, populations of lemming predators, such as snowy owls and arctic foxes, also increase (right).

Alternate Lab: Water-Holding Capacity of Soils

Purpose

Students are introduced to an abiotic factor that affects plant life in different biomes.

Materials

3 cloth squares (30 cm per side), large beaker, water, sand, clay, potting soil samples, balance, twist ties

Procedure

Give the following directions to students.

1. Wrap a sample of sand into a cloth square. Fold the ends to form a bag and secure it with a twist tie.
2. Determine and record the mass of the sand bag.
3. Place the sand bag into a large beaker filled with water and allow it to soak for 5 minutes.
4. Remove the bag and allow it to drain for 1 minute. Determine and record the mass of the sand bag. Calculate the gain in mass of the wet sand.
5. Calculate and record the water-holding capacity (WHC) of sand using the formula:

Life on the taiga

Just south of the tundra lies another biome that circles the north pole. The **taiga,** also called the northern coniferous forest, is a land of mixed pine, fir, hemlock, and spruce trees, as shown in *Figure 4.12.* How can you tell when you leave the tundra and enter the taiga? The line between any two biomes is indistinct, and patches of one blend almost imperceptibly into the other. For example, if the soil in the taiga is waterlogged, a peat swamp habitat develops that looks much like tundra. Taiga communities are usually somewhat warmer and wetter than tundra, but the prevailing climatic conditions are still harsh, with long, severe winters and short, mild summers.

In the taiga, which stretches across much of Canada, Northern Europe, and Asia, permafrost is usually absent. The topsoil, which develops from decaying coniferous needles, is acidic and poor in minerals. When the taiga community is disrupted by fire or logging, the first trees to recolonize the land may be birch, aspen, or other deciduous species. The abundance of trees in the taiga provides more food and shelter for animals than the tundra. More large species of animals are found in the taiga as compared with the tundra. *Figure 4.13* shows some animals of the taiga.

Figure 4.12

The dominant climax plants of the taiga are primarily fir and spruce trees. The needlelike leaves of these conifers are drought resistant.

Figure 4.13

The taiga stretches across most of Canada, Northern Europe, and Asia.

The lynx is a predator that depends on the snowshoe hare as its primary source of food.

The snowshoe hare, one of the taiga's smaller herbivores, also lives in this biome all year long. During the winter, it grows a thick, white winter coat that includes extra hair on its feet for warmth.

Large, herbivorous mammals of the taiga include caribou, which migrate down from the tundra during the winter, and moose (right), which may be found in the taiga during most of the year.

Enrichment

LS **Interpersonal** Have students test the pH of a variety of soil samples. Instruct students on the use of and significance of the test. Have them familiarize themselves with pH paper by testing samples of vinegar (a known acid) and wet hand soap (a known base). Make sure that soil samples are moist and include a sample collected from under a coniferous tree. L1 COOP LEARN

Brainstorming

Ask students to supply common names for coniferous trees. *Pines, firs, spruces, evergreens* Ask students to compare and contrast conifers and deciduous trees in terms of appearance, adaptations, and where they grow.

Time Line

Have students prepare a time line that shows the progression of plant types in the taiga after a fire or logging has occurred. Have students indicate if the succession is primary or secondary. *Succession will progress as follows: small plants such as grasses, shrubs, and birch, aspen, fir, or spruce. The succession is a secondary succession.* L2

WHC = mass gain × 100/dry mass

6. Repeat steps 1-5 using the clay and then the potting soil sample.

Analysis

Ask students to answer the following questions.

1. How might water-holding capacity be important to plants? *Plants may be restricted in growth to certain soil types.*
2. Which soil samples had the highest and the lowest percentage of water-holding capacity? *Highest was potting soil; lowest was sand.*
3. How do your data support the fact that cacti had to evolve a water storage system to survive in a desert? *Sand does not hold much water for plant use.*

Assessment

Performance: Ask students what change they might suggest for trying to improve on the procedure for this experiment. Why? *A likely response might suggest deducting the mass of wet and dry cloth from soil mass data.* L1

Demonstration

Bring a cactus to class along with a small mesophytic plant such as *Coleus* or *Geranium*. Mesophytic plants are those plants that are adapted to growth in soils rich in water and mineral salts. Ask students to describe the typical care needed to maintain each plant. *Both need sunlight, warm temperatures, water, and an occasional fertilizing.* Explain that the cactus can tolerate higher temperature and much less watering and fertilizing than the mesophyte.

Using an Analogy

Tell students that the waxy material that coats creosote leaves is called cutin. Compare the action of cutin to a sheet of wax paper by doing the following. Moisten two paper towels and flatten them out on a table. Cover one towel with wax paper on both sides. Note the time needed for the uncovered towel to dry out and the time needed for the towel covered with wax paper. Ask students to interpret the analogy and results.

GLENCOE TECHNOLOGY

Videodisc

Biology: The Dynamics of Life
Disc 1, Side 1
Desert (Ch. 8)

The Infinite Voyage: To the Edge of the Earth
The Tropical Rain Forest
(Ch. 5)

Figure 4.14

Creosote bushes cover many square miles of desert in the southwestern United States. These plants, shown above in bloom with yellow flowers, have small leaves coated with a waxy resin that helps reduce water loss. The roots give off a growth-inhibiting chemical that prevents other plants from growing too close, reducing competition for scarce water resources.

Life in the desert

The driest of the biomes south of the taiga is the desert biome. A **desert** is an arid region with sparse to almost nonexistent plant life. Deserts usually get less than 25 cm of precipitation annually. One desert, the Atacama Desert in Chile, is the world's driest place. This desert receives an annual rainfall of zero.

Vegetation in deserts varies greatly, depending on precipitation levels. Areas that receive more rainfall produce a shrub community that may include drought-resistant trees such as mesquite. Less rainfall supports scattered plant life and produces an environment with large areas of bare ground. The driest deserts are drifting sand dunes with virtually no life at all.

Plants have evolved various adaptations for living in arid areas, as shown in *Figure 4.14*. Many desert plants are annuals that germinate from seed and grow to maturity quickly after sporadic rainfall. Cacti have leaves reduced to spines, photosynthetic stems, and thick waxy coatings that reduce water loss. The leaves of some desert plants curl up, or even drop off altogether, to reduce water loss during extremely dry spells. Desert plants sometimes have spines, thorns, or poisons that act to discourage herbivores.

Most desert mammals are small herbivores that remain under cover during the heat of the day, emerging at night to forage on plants. The kangaroo rat is a desert herbivore that does not have to drink water. These rodents obtain all the water they need to live from the water content in their food. A few larger herbivores such as pronghorn antelopes may be found in American deserts. Foxes, coyotes, hawks, owls, and roadrunners are carnivores that feed on the snakes, lizards, and small mammals of the desert. Scorpions are an example of a desert carnivore that uses venom to capture prey. *Figure 4.15* shows two of the many reptiles that make the desert their home.

Figure 4.15

Lizards, tortoises, and snakes are numerous in desert communities. Desert tortoises (left) feed on insects and plants. Venomous snakes such as the diamondback rattlesnake (right) are major predators of small rodents.

PROJECT

Comparing Biomes

LS **Interpersonal** Have students construct a chart that includes the following information for each of the six terrestrial biomes: typical plant species, typical animal species, important facts, and interesting facts. Have them add to the chart as each biome is introduced. L1 COOP LEARN

Life in the grassland

If an area receives between 25 and 75 cm of precipitation annually, a grassland usually forms. **Grasslands** are large communities covered with grasses and similar small plants. Grasslands, such as the ones shown in *Figure 4.16*, principally occur in climates that experience a dry season, where insufficient water exists to support forests. This biome occupies more area than any other terrestrial biome, and it has a higher biodiversity than deserts, often with more than 100 species per acre.

The soils of grasslands have considerable humus content because many grasses die off each winter, leaving decay products to build up in the soil. Grass roots survive through the winter, enlarging every year to form a continuous underground mat called sod.

Because they are ideal for growing cereal grains such as oats, rye, and wheat, which are different species of grasses, grasslands have become known as the breadbaskets of the world. Some species of grasses grow as high as eight feet or more. Many other plant species live in this environment, including drought-resistant and late-summer-flowering species of wildflowers such as blazing stars and sunflowers.

MiniLab

How much humus is in your soil?

Decayed materials release nutrients that combine with topsoil to form humus. The amount of humus in soil depends on the rate of decay and the rate of absorption of nutrients by the plants in an ecosystem. Tundra, taiga, deserts, and tropical rain forests usually have small amounts of humus in the soil. Grasslands and temperate forests usually have larger amounts of humus.

Procedure

1. Place a teaspoon of washed sand in a plastic cup.
2. Add 2 percent hydrogen peroxide solution to the sand drop by drop. Bubbles will form as the hydrogen peroxide breaks down any organic matter.
3. Record the number of drops you add until the bubbles stop forming.
4. Repeat the procedure for the soil sample from your area.

Analysis

1. Did your soil have more or less humus than the washed sand?
2. Why did you use washed sand as a comparison?
3. Compared to your soil sample, would you have to add more or less hydrogen peroxide to a similar soil sample from an African savanna?

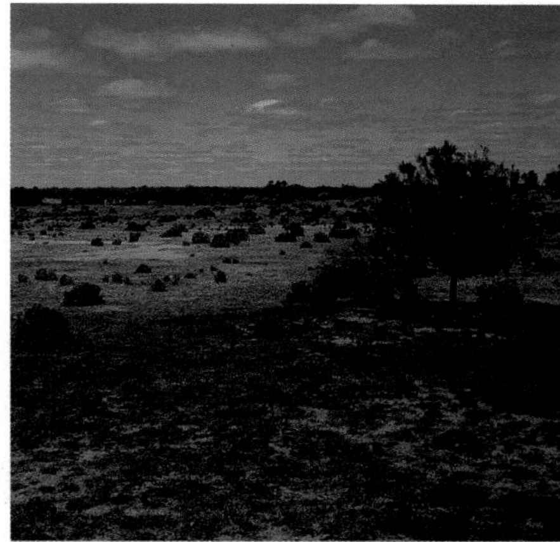

Figure 4.16

Summers are hot, winters are cold, and rainfall is often uncertain in a grassland. Called *prairies* in Australia (left), Canada, and the United States (right), these communities are called *steppes* in Russia, *savanna* in the Serengeti of Africa, and *pampas* in Argentina. Grasslands contain fewer than ten to 15 trees per hectare, though larger numbers of trees are found near streams and other water sources.

MiniLab

Purpose

To have students compare the amount of humus present in soil samples.

Process Skills

observe and infer, measure in SI, compare and contrast

Teaching Strategies

- Wash sand by placing it in a bucket and rinsing several times with tap water.
- Use fresh hydrogen peroxide.
- Use potting soil as the other soil sample.

Safety Precautions

Have students wear goggles and an apron.

Expected Results

Sand will require few drops of hydrogen peroxide. Other samples may require 10-30 drops of hydrogen peroxide.

Analysis

1. more
2. All humus was removed from the sand, which was used as a control.
3. Answers will vary depending on soil type used. A savanna is a grassland biome and typically contains much humus.

Assessment

Performance: Have students write a brief report of this experiment for their portfolios. Have them include their data as a graph. L1

Meeting Individual Needs

Visually Impaired Bring samples of sandy, clay, and loam soils to class. Allow visually impaired students to feel the soils to establish their textures. Have the visually impaired students work with sighted students to rank the samples in terms of decayed material present and explain how such material contributes to the richness and fertility of each soil. Ask students to correlate this information to the amount of vegetation typically found in each soil. L1 LS COOP LEARN

Different Viewpoints in Biology

Is a Pretty Lawn Environmentally Sound?

Well-manicured, weedless lawns are common in the United States. However, the strain such lawns place on natural resources is causing some ecologists to look for alternatives.

Chemical fertilizers, pesticides, and herbicides often pollute water systems. Lawn irrigation wastes a great deal of water. Some ecologists have suggested replacing lawns with native plants. In the Midwest, for example, prairie grasses and flowers such as bluets and dogtooth violets are native to the region. The growth of such plants would also reduce the amount of fossil fuel needed to run gasoline-powered mowers.

GLENCOE TECHNOLOGY

Videodisc

Biology: The Dynamics of Life
Disc 1, Side 1
Temperate Grassland (Ch. 9)

Disc 1, Side 1
Temperate Forest (Ch. 10)

Disc 1, Side 1
Rain Forests (Ch. 11)

Figure 4.17
The prairies of America support bison as well as many species of birds and insects.

Most grasslands are populated by large herds of grazing animals. *Figure 4.17* shows bison, a species of mammal that roams the American prairies. Millions of bison, commonly known as buffalo, once ranged over the American prairie, where they were preyed upon by wolves, coyotes, and humans. Other important prairie animals include prairie dogs, which are seed-eating rodents whose underground "towns" have been known to stretch across mile after mile of grassland, and the foxes and ferrets that prey on them. Many species of insects, birds, and reptiles, including tortoises, lizards, and snakes, also make their homes in grasslands.

Life in the temperate forest

When precipitation ranges from about 70 to 150 cm annually in the temperate zone, temperate deciduous forests develop. **Temperate forests** are dominated by broad-leaved hardwood trees that lose their foliage annually, *Figure 4.18*.

When European settlers first arrived on the east coast of North America, they cleared away large tracts of temperate forest for farmland. The thin soil of the mountainous regions was soon depleted by crops, and farmers abandoned their land. Since then, secondary succession has restored much of the original forest.

The soil of temperate forests usually consists of a top layer that is rich in humus and a deeper layer of clay. If mineral nutrients released by the decay of the humus are not immediately absorbed by the roots of the living trees, they may be washed into the clay and lost from the food web for many years.

Figure 4.18
Temperate forests occur where an even amount of precipitation falls during each of the four seasons. There are many types of temperate forests, each described by the two or three dominant species of trees. Typical trees of the temperate forest include birch, hickory, oak, beech, and maple.

STUDENT JOURNAL

Building Vocabulary Review the meaning of the term *limiting factors.* Then, have students describe those limiting factors that make the grassland and temperate forest biomes different. Have students prepare a concept map of these ideas. They should also include the location of these biomes in North America as part of their concept maps. L1 SAE

Meeting Individual Needs

Hearing Impaired Bring examples of tree leaves to class. Include samples of both evergreen and deciduous leaves. Have students describe similarities and differences for these two leaf types. Then have them group the following descriptions into their proper categories: evergreen, hardwood, coniferous, deciduous, broad-leaved, needlelike leaves. L1

Audiovisual

Show the video *Rain Forest*, National Geographic.

Bioethics

Rain Forest Pencils

Some pencils are made from a wood called jelutong, which comes from the tropical rain forests of Indonesia. Today, 85 percent of all pencil wood comes from cedar grown in Oregon and California of the United States. The remaining 15 percent is jelutong. Supporters of this use of jelutong maintain that the wood is being harvested from plantations rather than from clear-cutting of the rain forest. Ecologists maintain that even so, harvesting of jelutong still involves removal of rain forest foliage. How can you determine if a pencil is made from jelutong? The easiest way is to smell it after sharpening. Cedar pencils from U.S. forests smell like cedar; jelutong has no odor. Cedar pencils also show a grain, which is missing in jelutong. Reading package labels is also helpful. If the package says "Made from U.S.-grown wood" it is probably cedar.

Activity

Exploring Environmental Issues: p. 39, The Rain Forest. Have students use this activity to determine which organisms live in tropical rain forests. L2

Figure 4.19

Black bears and deer have always been residents of temperate forests in the United States. The numbers of these large animals are growing today because their habitat is protected in many places. Other abundant animals in temperate forests are squirrels and salamanders.

The animals that live in the temperate deciduous forest, as shown in *Figure 4.19,* include familiar squirrels, mice, rabbits, deer, and bears. Many birds live in the forest all year long, while others migrate south to tropical regions during the winter.

Life in tropical rain forests

The most biologically diverse of the terrestrial biomes, the **tropical rain forest** is a region of uniformly warm, wet weather dominated by lush plant growth. These forests occupy the equatorial regions around the world. The rain forest receives at least 200 cm of rain annually, while some rain forests can receive as much as 400 cm. Temperatures remain warm throughout the year, about 25°C.

Rain forests form jungles of dense, tangled vegetation only near stream banks or in areas that have been disturbed. In mature, undisturbed rain forest communities like the one in *Figure 4.20,* the leafy tops of tall trees shade the forest floor so completely that few plants grow there. Much of the precipitation the rain forest receives is retained and recycled by the heavy canopy of leaves. When trees are removed over a large area, this type of precipitation ceases, and the area may receive too little rain for regrowth.

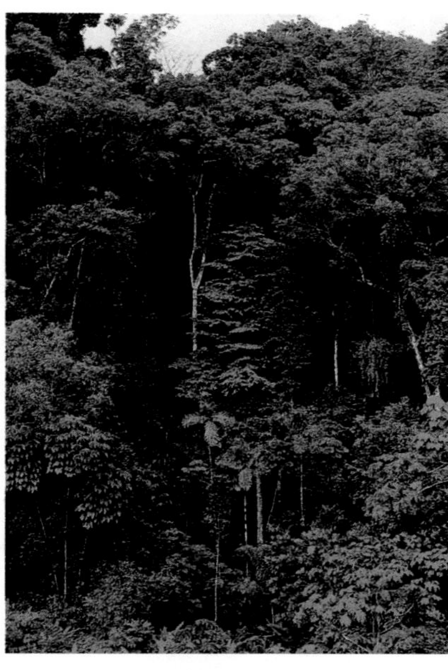

Figure 4.20

Warm temperatures, high humidity, and abundant rainfall allow the lush growth and great species diversity found in tropical rain forests.

PORTFOLIO

Locating Tropical Rain Forests Provide students with an outline map of North, South, and Central America. Ask them to color in regions that have tropical rain forests. Also have students devise a method for determining the approximate percentage of land covered by this biome in the Western Hemisphere. L1 LS P

interNET CONNECTION

Follow the link for this chapter on the Glencoe Homepage at **www.glencoe.com/sec/science** to find out more about biomes.

ThinkingLab Design an Experiment

Purpose

Design an experiment that determines what factors influence decomposition.

Process Skills

observe and infer, interpret data, recognize cause and effect

Teaching Strategies

- Show students a soil testing kit. Explain who uses them and why they are used.
- Ask students if it is possible to measure the exact amount of a nutrient in soil that results from decomposition. Responses will likely include that it is possible to quantitatively identify the nutrients, but difficult to attribute them to decomposition.

Thinking Critically

Experimental designs may suggest a method similar to the following. Select a nutrient for study. Test a soil sample for the amount of the nutrient in the soil. Place part of the sample in a high-temperature environment, and leave the other part at room temperature. Retest both samples for the nutrient after one week, and again after two weeks. Compare the data.

Assessment

Skill: Ask students if their experimental procedure would yield descriptive or quantitative data. *Quantitative* Have them identify the independent and dependent variables. *Responses will depend upon individual experimental designs.* L1

ThinkingLab Design an Experiment

What factors affect the rate of decomposition?

Decomposition occurs when any organism dies; it is the result of both abiotic and biotic factors. Most decomposition is the result of the feeding of bacteria and fungi. Nutrients released by decomposition either build up in soil, are washed away, or are absorbed by producers for recycling through the food chain. The rate of decomposition varies with several factors, including temperature and amount of water.

Analysis

Study the table below. The table shows the approximate rates of decomposition in each of the terrestrial ecosystems.

Relative Rates of Decomposition

Ecosystem	Rate
Tundra	– –
Taiga	–
Desert	–
Temperate Forest	+
Tropical Rain Forest	+ + +
Grassland	+ +

Thinking Critically

Design an experiment to determine what factor or factors have the most effect on the rate of decomposition for a specific terrestrial ecosystem.

Tree roots are often shallow, forming a thick mat on the surface of the soil. Roots that support tall trees are sometimes greatly enlarged or may form buttresses, which resemble scaffolding, *Figure 4.21.*

Most of the nutrients of the rain forest are tied up in living material. The hot, humid climate enables ants, termites, fungi, and other decomposers to break down leaves and dead animals quickly, before they can become humus. As a result, in some areas of the rain forest, the ground is completely bare.

Many animals live in dense vegetation of tropical rain forests. *Figure 4.21* shows some common rain forest animals.

Figure 4.21

Tropical rain forests are among Earth's richest ecosystems (far right). Sloths (left), monkeys, and other mammals, as well as a multitude of bird species such as this black-headed cacique (top), live in the rain forest canopy. Insects such as this Hercules beetle are also numerous in the rain forest understory (bottom).

STUDENT JOURNAL

Living in a Rain Forest Based on all the written descriptions and photographs referencing the tropical rain forest, have students describe what living in a tropical rain forest must be like as seen through the eyes of a sloth. Be sure students include what they might see on the ground as well as in the trees. L1 LS

PORTFOLIO

A Rain Forest Vacation Have students prepare travel brochures for the tropical rain forest for their portfolios. As a guide to what appears in such brochures, bring samples to class for reference. However, don't use brochures that might, by coincidence, discuss visiting the rain forests. L1

Technology

BIOLOGY & SOCIETY

The View from on High

In 1972, NASA launched a satellite called *Landsat* that would scan Earth from a distance of 914 km. Many people wondered what a satellite in space could reveal.

The big picture *Landsat* sent back detailed pictures of huge areas of Earth, and the relationships of distant landforms were seen for the first time. *Landsat* used a television camera called a Return Beam Vidicon (RBV) and a multispectral scanner (MSS) that recorded the reflectivity of objects on Earth. The multispectral scanner, far more sensitive to color than the human eye, was invaluable in studying crops and locating minerals and ancient topographical features.

Tracking pollution Environmentalists have used *Landsat* data to evaluate the effectiveness of their work in cleaning up polluted waterways. Lake Chicot in Arkansas, once a fishing paradise, had become polluted with sediment. Changes in drainage patterns and runoff from farm fields were found to be responsible. In 1985, the U.S. Army Corps of Engineers began a cleanup of the lake using *Landsat* data.

Applications for the Future

Landsat data could be used to check for features that indicate the presence of precious metals such as copper, gold, and molybdenum.

Landsat can be used to identify inaccuracies in map locations of lakes and rivers. Developing countries where resources for costly land surveys are not available could use *Landsat* to amend their maps.

INVESTIGATING the Technology

Making Inferences Explain how data from *Landsat* could help locate the source of pollution in a river in the northeastern United States.

Meeting Individual Needs

Gifted Have students prepare an illustrated chart that describes the major features of the three zones of the tropical rain forest: the canopy, understory, and forest floor. Have students include typical plant and animal life as well as the limiting factors that produce the zonation effect. L3

Technology

BIOLOGY & SOCIETY

The View from on High

Purpose

Students will see how technology is able to monitor changes in the world's biomes.

Teaching Strategies

- Review with students the electromagnetic spectrum so that they have a general understanding of the visible versus the infrared spectrum.
- Bring to class examples of photographs that have been taken through Landsat. Ask students to explain how these photographs may have been used in scientific studies.
- Ask students to describe the type of data that might be helpful to an agronomist who is trying to determine the extent of damage caused by a crop fungus. How might this type of data be used by an ecologist studying the wetland area available for migrating birds?

INVESTIGATING the Technology

Making Inferences Photographs would indicate the area where pollutants are entering the river.

Going Further

Have students describe how Landsat could be used to monitor problems associated with the shrinking rain forest. What kind of quantitative data could be gathered about rain forests?

3 Assess

Check for Understanding

Have students explain the following word relationships.

(a) photic zone - aphotic zone
(b) intertidal zone - estuary
(c) river - estuary - ocean
(d) food chain - plankton
(e) biome - limiting factor L1

SECTION 4.2

Reteach

Ask students to pick a biome and describe its location, climatic characteristics, animal and plant examples typical of the biome, and special or unusual traits. L1

Extension

Have students explain the relationship among the following: greenhouse effect, rain forest destruction, and the carbon cycle. L2

Assessment

Portfolio: Have students describe how the destruction of a biome such as the ocean would affect them directly. Have them suggest one or two ways that they themselves can make a difference in the saving of a specific biome. L1

Performance Assessment in the Biology Classroom: p. 57, Investigating Salinity and Marine Algae. Have students carry out this activity to determine the effect of salinity on marine algae. L1 P

4 Close

Activity

Advise students that they are to lead a group of tourists to a biome of their choice. Have them: (a) pick a specific biome as their travel destination, (b) prepare a guide for the travelers as to what to pack for clothing, (c) explain what they should be prepared to see and experience.

Figure 4.22
This blue morpho butterfly lives in the canopy of a tropical rain forest.

More species of reptiles, amphibians, and birds are found in the tropical rain forest than in any other terrestrial biome. Although large animals like gorillas and cougars are ground dwellers, most rain forest organisms live in the trees rather than on the forest floor. Butterflies such as the one in *Figure 4.22* and other insects are by far the most numerous animals in the rain forest. Biologists estimate that there may be as many as 3 million species of insects in the tropical rain forest.

Connecting Ideas

A population's range of tolerance determines whether or not it will exist in a particular community. As limiting factors change, and populations of different species of organisms succeed one another, a climax community eventually results. Climax communities are in homeostatic balance. Large climax communities, called biomes, remain stable over long periods of time. However, changes occur within populations of species. How do ecologists determine how interactions among populations can change the structure of a community?

Section Review

Understanding Concepts

1. Explain why the photic and aphotic zones of marine biomes are interdependent.
2. What is the most important abiotic factor that limits distribution of the tundra biome?
3. Describe some common plants and animals from a tropical rain forest and a grassland terrestrial biome.

Thinking Critically

4. Shaneka and her family were planning a trip to a foreign country. In her reading, Shaneka found that the winter was cold, the summer was hot, and most of the land was planted in fields of wheat. Infer which biome Shaneka's family would visit. Explain your answer.

Skill Review

5. **Making and Using Tables** Make a table to show the climate, plant types, plant adaptations, animal types, and animal adaptations for the terrestrial biomes. For more help, refer to Organizing Information in the *Skill Handbook.*

Answers to Section Review

1. The photic zone provides food for the scavengers and decomposers in the aphotic zone. The decomposers of the aphotic zone return nutrients to water that can be used by plants in the photic zone.
2. Temperature
3. Tropical rain forest: evergreen trees, vines, ferns, insects, monkeys. Grasslands: grass, windflowers, bison, prairie dogs, foxes.

Thinking Critically

4. Grasslands; most of the grassland biome has been replaced with grasslike crop plants such as wheat. The conditions of this biome support the cultivation of commercial crops.

Skill Review

5. Students' tables should resemble the following example. Taiga biome: long, harsh winters and short, cool summers; coniferous trees with needlelike leaves that resist drought; snowshoe hares grow white fur in winter and dark fur in summer; moose have heavy fur for warmth.

REVIEWING MAIN IDEAS

4.1 Homeostasis in Communities

- Communities, populations, and individual organisms occur in areas where biotic or abiotic factors fall within their range of tolerance. Abiotic or biotic factors that define whether or not an organism can survive are limiting factors.
- Primary succession is the development of living communities from bare rock. Secondary succession occurs when communities are disrupted. Left undisturbed, both primary succession and secondary succession will eventually result in a climax community.

4.2 Biomes

- Biomes are large areas that have characteristic climax communities. Aquatic biomes may be marine or fresh water. Estuaries occur at the boundaries of marine and freshwater biomes. Approximately three-quarters of Earth's surface is covered by aquatic biomes, and the vast majority of these are marine communities.
- Terrestrial biomes include tundra, taiga, desert, grassland, deciduous forest, and tropical rain forest. Climate is the major limiting factor for the formation of terrestrial biomes.

Key Terms

Write a sentence that shows your understanding of each of the following terms.

aphotic zone	photic zone
biome	plankton
climax community	primary succession
desert	succession
estuary	taiga
grassland	temperate forest
intertidal zone	tropical rain forest
limiting factor	tundra
permafrost	

UNDERSTANDING CONCEPTS

1. The zone of light penetration in aquatic ecosystems is ____________.
   a. unlimited
   b. the aphotic zone
   c. the photic zone
   d. unpopulated by autotrophs
2. A primary abiotic limiting factor of the tundra biome is ____________.
   a. temperature
   b. lack of plant life
   c. too much water
   d. altitude
3. The orderly changes that take place in communities are ____________.
   a. called succession
   b. a climax
   c. a range of tolerance
   d. limiting factors
4. A species that first occupies and lives in an area is a ____________.
   a. limiting species
   b. pioneer species
   c. consumer
   d. predator
5. The zone in which most organisms of a population congregate is the ____________.
   a. aphotic zone
   b. photic zone
   c. optimum range
   d. zone of intolerance
6. The major difference between climates of grassland and forest is that grasslands are ____________.
   a. cooler
   b. wetter
   c. drier
   d. brighter

Chapter 4 Review

Reviewing Main Ideas

Summary statements can be used by students to review the major concepts of the chapter.

Key Terms

Answers should go beyond defining the terms. Accept any answer that uses the term correctly and in the proper context.

Understanding Concepts

1. c
2. a
3. a
4. b
5. c
6. c

Program Resources

Chapter Assessment, pp. 19-24 L1

Alternate Assessment in the Science Classroom

Computer Test Bank L1

Content Mastery, pp. 13-16 L1

Chapter 4 Review

7. c
8. a
9. a
10. d
11. b
12. d
13. d
14. b
15. b
16. a
17. c
18. b
19. b
20. d

Applying Concepts

21. cooler temperatures, less chance for water loss
22. After crops are harvested, the bare soil may be subject to erosion by wind and water.
23. Annual fires would prevent the growth of the climax species of deciduous trees.
24. Biodiversity is much higher in a tropical forest biome than in a temperate forest biome.
25. The younger an area, the lower its biodiversity. One would therefore expect low biodiversity on Surtsey.

Chapter 4 Review

7. Stable community populations result in a(n) ____________.
   a. country
   b. desert
   c. climax community
   d. ecosystem
8. Coastal bodies of water, partially surrounded by land, in which fresh water and salt water mix are ____________.
   a. estuaries
   b. marine
   c. planktonic
   d. usually deep
9. Organisms inhabiting desert biomes might include ____________.
   a. cacti and reptiles
   b. trees and large mammals
   c. toucans and boas
   d. ferns and frogs
10. Primary succession occurs ____________.
    a. in old farm ponds
    b. in cleared fields
    c. in burned forests
    d. on fresh lava rock
11. A factor that determines the survival of an organism is a ____________.
    a. range of tolerance
    b. limiting factor
    c. biotic factor
    d. photic zone
12. Which of the following has lush plant growth but poor soil?
    a. temperate forest
    b. grassland
    c. tundra
    d. tropical rain forest
13. The biome that may be at greatest risk of destruction by humans is the ____________.
    a. desert
    b. taiga
    c. grassland
    d. tropical rain forest
14. Locations of biomes are usually determined by ____________.
    a. altitude and temperature
    b. temperature and precipitation
    c. precipitation and altitude
    d. soil type and temperature
15. Secondary succession occurs on ____________.
    a. glacial till
    b. abandoned fields
    c. the ocean floor
    d. fresh lava rocks
16. The principal difference between the aphotic and photic zones is the ____________.
    a. amount of light
    b. difference in chemicals
    c. construction of food webs
    d. effect of water pressure
17. The greatest diversity is found in the ____________ and ____________ biomes.
    a. freshwater, tropical forest
    b. tropical forest, grassland
    c. tropical forest, marine
    d. deciduous forest, desert
18. What is another name for the taiga?
    a. tundra
    b. northern coniferous forest
    c. deciduous forest
    d. second polar circle forest
19. The layer of frozen soil found in the tundra is the ____________.
    a. permanent layer
    b. permafrost
    c. taiga
    d. Arctic Circle
20. The basis of most marine food chains is ____________.
    a. fish
    b. whales
    c. krill
    d. plankton

APPLYING CONCEPTS

21. If you were lost in a desert, list at least two reasons why you should travel only at night.
22. How may agriculture lead to soil erosion in a grassland biome?
23. How might annual fires affect the succession of a temperate deciduous forest?
24. Compare the biodiversity of the temperate forest biome with that of a tropical rain forest biome.
25. Surtsey, a volcanic island, is very young. How might its age limit the biological diversity of the island?

THINKING CRITICALLY

26. ***Concept Mapping*** Make a concept map that relates the following terms and phrases. Supply the appropriate linking words for your map.
marine biome, lake, river, pond, freshwater biome, sandy shore, estuary, rocky shore, euphotic zone, plankton, aphotic zone

27. ***Comparing and Contrasting*** In what ways are limits of tolerance and succession similar?

28. ***Forming a Hypothesis*** Make a hypothesis about why tropical rain forests are the most energy-efficient biomes.

29. ***Interpreting Scientific Illustrations*** From the graph below, determine the upper and lower temperature tolerance for a species of minnow.

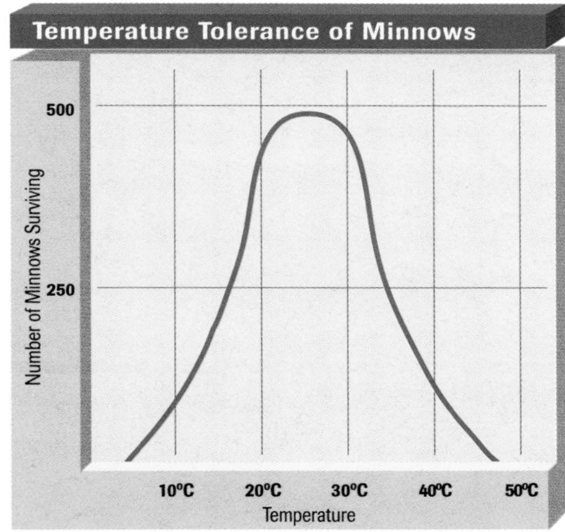

30. ***Sequencing*** Use *Figure 4.9* to sequence the types of biomes that would occur at an average annual temperature of 10°C as annual precipitation increases from 100 cm to 300 cm.

ASSESSING KNOWLEDGE & SKILLS

The figure below represents a pyramid of energy for an ecosystem. In this system, trophic levels contain large catfishes, humans, minnows, algae, and mosquito larvae.

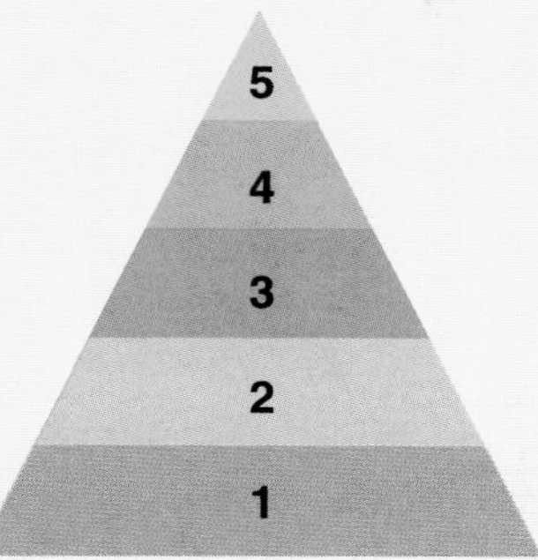

Interpreting Scientific Illustrations Use the illustration to answer the following questions.

1. Which trophic level would be represented by the minnows?
a. 1 **c.** 3
b. 2 **d.** 4

2. Which level would contain the mosquito larvae?
a. 1 **c.** 3
b. 2 **d.** 4

3. Which of the following species relationships is not included?
a. autotrophs **c.** omnivores
b. carnivores **d.** decomposers

4. ***Interpreting Data*** Which of the following would be an accurate food chain that matches the energy flow shown in the illustration?
a. humans → catfishes → minnows → mosquito larvae → algae
b. algae → mosquito larvae → minnows → catfishes → humans
c. mosquito larvae → algae → minnows → catfishes → humans
d. algae → minnows → mosquito larvae → catfishes → humans

Chapter 4 Review

Thinking Critically

26. See below.
27. Limits of tolerance control what type of organisms can survive in an area. Succession also limits life-forms to a specific, orderly pattern of change.
28. Student answers will vary, but they might indicate that tropical rain forests are energy efficient because of the numbers of producers that can trap energy or because more nutrients are tied up in living matter.
29. lower tolerance about 20°C; upper tolerance about 30°C
30. woodland (borderline), temperate forest, temperate rain forest

1. c
2. b
3. d
4. b

26.

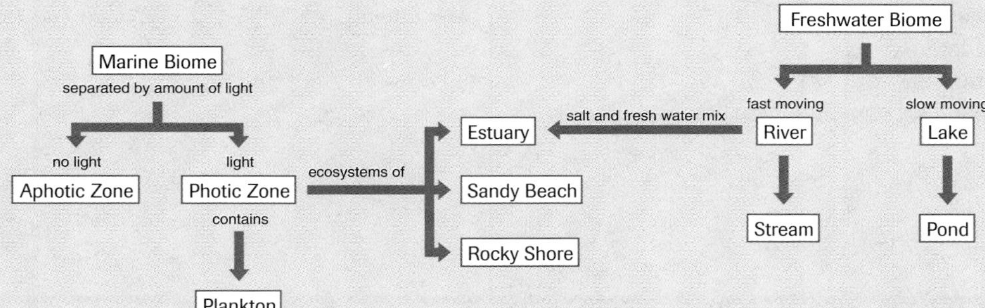

Chapter Organizer

SECTION	OBJECTIVES	ACTIVITIES/FEATURES
5.1 Population Dynamics National Science Standards: UCP.1-3; A.1, A.2; C.4, C.5; F.2-6; G.1, G.2	1. **Compare and contrast** exponential and linear population growth. 2. **Relate** the reproductive strategies of different populations of organisms to models of population growth. 3. **Predict** effects of environmental factors on population growth.	**Biolab:** Population Growth in *Paramecium,* p. 116 **Minilab:** How fast do populations use resources?, p. 118 **Thinking Lab:** How can you control an organism's life-history pattern?, p. 119 **A Broader View:** Mussel Beach, p. 120 **Minilab:** How does the population density of weeds in an area affect other plants?, p. 121
5.2 Human Population Growth National Science Standards: UCP.1-3; F.1, F.2, F.6; G.1	4. **Relate** characteristics of populations to population growth rates. 5. **Compare** the age structure of rapidly growing, slow-growing, and no-growth countries. 6. **Hypothesize** about problems that can be caused by immigration and emigration.	**Math Connection:** How many are there?, p. 126

ACTIVITY MATERIALS

BIOLAB	MINILABS	ALTERNATE LAB
page 116 microscope slides coverslips eyedropper 10-mL graduated cylinder plastic cups or baby food jars rice grains *Paramecium* culture methyl cellulose	**page 118** jar 1/2 banana **page 121** string small sticks meterstick or tape measure	**page 126** lima beans paper bag 250-mL beaker wax pencil

Chapter 5 Population Biology

TEACHER CLASSROOM RESOURCES

Reproducible Masters	Transparencies
Section Focus Master 9: Predator-Prey Relationships L1 SAE **Reinforcement and Study Guide,** pp. 17-18 L1 **Biolab and Minilab Worksheets,** pp. 15-18 L1 **Concept Mapping:** Population Control, p. 5 L1 **Critical Thinking/Problem Solving:** The Effect of Predators on Prey Populations, p. 5 L3 **Laboratory Manual:** How Does the Environment Affect an Eagle Population?, pp. 23-26 L2	**Reteaching Transparency #5:** Limits to Population Growth L1 SAE **Basic Skills Transparency #5:** Linear versus Exponential Growth L1 SAE
Section Focus Master 10: Birthrates and Death Rates L1 SAE **Reinforcement and Study Guide,** pp. 19-20 L1 **Content Mastery,** pp. 17-20 L1	

ASSESSMENT MATERIALS

Chapter Assessment, pp. 25-30
Performance Assessment in the Biology Classroom, p. 53
MindJogger Videoquiz
Alternate Assessment in the Science Classroom
Computer Test Bank

Spanish Resources SAE
English/Spanish Audiocassettes SAE
Cooperative Learning in the Science Classroom COOP LEARN
Lesson Plans
Exploring Environmental Issues: p. 23 L2

KEY TO TEACHING STRATEGIES

L1 Level 1 activities should be within the ability range of all students including those with learning difficulties.
L2 Level 2 activities are within the ability range of average to above-average students.
L3 Level 3 activities are designed for the ability range of above-average students.
SAE SAE activities should be within the ability range of Students Acquiring English.
COOP LEARN Cooperative Learning activities are designed for small group work.
P These strategies represent student products that can be placed into a best-work portfolio.
These strategies are useful in a block scheduling format.

GLENCOE TECHNOLOGY

The following multimedia resources are available from Glencoe.

Biology: The Dynamics of Life
CD-ROM SAE
Videodisc Program

The Infinite Voyage Series
Crisis in the Atmosphere

National Geographic Society Series
Newton's Apple: Life Sciences

Science and Technology Videodisc Series (STVS)
Ecology
Controlling Fruit Flies
Fish Survey
Wasp Biological Control
Garbage Science
Animals
Migration of Killer Bees

CHAPTER

5 Population Biology

Chapter Overview

Students begin their study of population biology by exploring the factors that define population growth. Examples are used to illustrate the differences between population growth that is linear and growth that is exponential.

The concept of carrying capacity is introduced and developed as it relates to controlling population size. Limits on population growth are then presented through a comparison of the reproductive patterns of different organisms. Density-dependent limiting factors and density-independent limiting factors are also presented, along with organism interactions that influence population size.

In the second part of the chapter, demographics are discussed as they relate to human populations. The influence of birth and death rates on population size are illustrated, and age structure diagrams are used to illustrate the effects of decreased death rates on population growth. Finally, immigration and emigration are described as processes that can change population sizes.

Key Terms

- age structure
- carrying capacity
- demography
- density-dependent factor
- density-independent factor
- emigration
- exponential growth
- immigration

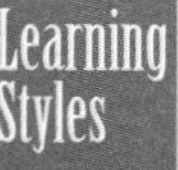

Learning Styles

Look for the following logo for strategies that emphasize different learning modalities. LS

Kinesthetic	**Meeting Individual Needs, p. 121**
Visual-Spatial	**Meeting Individual Needs, p. 122**
Interpersonal	**Project, p. 118; Thinking Lab, p. 119**
Intrapersonal	**Student Journal, pp. 115, 119**
Linguistic	**Portfolio, p. 129**
Logical-Mathematical	**Meeting Individual Needs, p. 115; Minilab, pp. 118, 121; Portfolio, pp. 118, 125; Activity, p. 124; Math Connection, p. 125; Visual Learning, p. 125; Alternate Lab, p. 126**

LS

CHAPTER

5 Population Biology

Consider large cities of the world such as New York City, Los Angeles, and Tokyo—cities populated by millions of people. Teeming streams of people bustle along the crowded streets, and it seems as if every last inch of space is occupied by stores, offices, and apartments. Now contrast one of these cities with a rural area populated by only a few people. Living in a rural area, you wouldn't experience traffic jams, crowded streets, or polluted air from automobile fumes. On the other hand, you probably wouldn't have your choice of concerts, movies, or sporting events to attend either. Cities can be exciting places, and according to the most recent census, the population of many cities is increasing.

Why do populations grow, and how can we predict future population trends? Can we learn how to deal with the problems of larger populations? Many characteristics of population growth are the same for other organisms as they are for humans. By studying the principles of population growth in other species, you may be able to suggest possible answers to these questions.

Concept Check

You may wish to review the following concepts before studying this chapter.

- Chapter 3: population
- Chapter 4: limiting factor, range of tolerance

Chapter Preview

What additional problems might be created if the human population of New York City doubled? How would increased population impact trash removal or electrical supply? Would clean water and food be as easy to obtain?

113

Introducing the Chapter

Describe the following scenario. Twenty years ago the Florida alligator was in danger of becoming extinct. Strict laws were passed to keep people from killing these reptiles. Today, the alligators have reproduced to the point where they are no longer endangered, but are now considered pests. Ask students to use this example to explain how removal of a predator can affect a population. *The population may be permitted to grow unchecked. If too much growth occurs, the population may deplete the resources it needs for survival.*

Theme Development

The theme of *systems and interactions* is illustrated as changes in populations (the system) result from interactions occurring within the population. *Evolution* is illustrated through the constant changes within populations.

Concept Check

Review with students the concept of population from Chapter 3. Students will need to recall the concepts of limiting factors and ranges of tolerance from Chapter 4.

Assessment Planner

Choose assessment strategies from the following pages to evaluate the progress of your students.

Assess, pp. 123, 128
Alternate Lab, pp. 126-127
Minilabs, pp. 118, 121
Portfolio, pp. 118, 125, 129
Thinking Lab, p. 119
Biolab, pp. 116-117
Chapter Review, pp. 129-131

SECTION

5.1

Prepare

Key Concepts

Population growth is the change in population size over time. Students learn that population growth, while exponential at times, is controlled by limiting factors that determine the carrying capacity of the environment. Such limits to population growth may result from predator-prey interactions or overcrowding.

Block Scheduling

Look for this symbol for strategies that are useful in a block scheduling format. For more information on block scheduling, refer to Section 5.1 in the **Lesson Plans** booklet.

Materials

- Purchase a *Paramecium* culture and rice for the Biolab. Gather eyedroppers, plastic cups, and small graduated cylinders.
- Gather small jars (e.g. baby food jars) for the Minilab.

1 Focus

Bellringer

Before presenting the lesson, display **Section Focus Master 9** on the overhead projector and have students answer the accompanying questions. L1 SAE

Discussion Questions

Ask students to speculate why human populations gather in cities. *Responses may indicate that people are generally social organisms and that resources are more readily available.* What are some of the drawbacks to this lifestyle? *Living space is at a higher premium.*

SECTION

5.1 Population Dynamics

Section Preview

Objectives

Compare and contrast exponential and linear population growth.

Relate the reproductive strategies of different populations of organisms to models of population growth.

Predict effects of environmental factors on population growth.

Key Terms

exponential growth
carrying capacity
density-dependent factor
density-independent factor

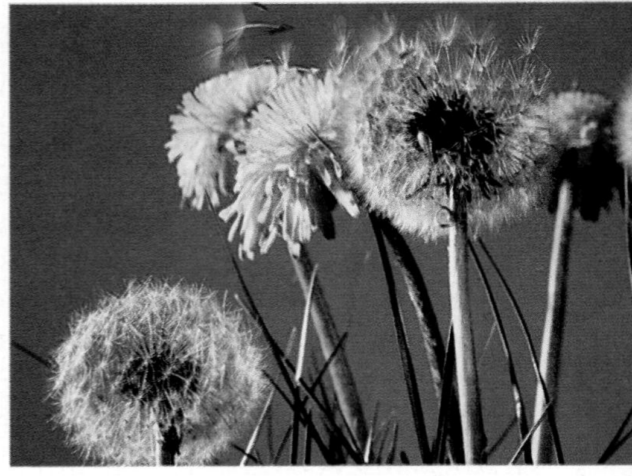

Weeds! Those pesky little plants are a real problem to gardeners. You've probably observed a scene like this before. What was recently a clear, grass-filled field or lawn is now crowded with hundreds, perhaps thousands, of bright yellow dandelions. Why do these plants appear so quickly and in such large numbers?

Dandelions are a problem to gardeners, but to population biologists, they are a useful species because they're easy to see and study. By studying other organisms such as dandelions, scientists can infer much about human population growth. Do similarities exist between human population growth and population growth in other species? In many ways, the changes that occur in human populations are similar to those of other populations. All populations—whether human, plant, or animal—are in a state of constant change.

Principles of Population Growth

How and why do populations grow? Population growth is defined as the change in the size of a population with time. Scientists use a variety of methods to investigate population growth in organisms, *Figure 5.1.* One method involves placing a microorganism, such as a bacterium or yeast cell, into a tube or bottle of nutrient solution, and observing how rapidly the population grows. Another interesting method involves introducing a plant or animal species into a new environment that contains abundant resources, and then observing the population growth of that species. Through studies such as these, scientists have identified clear patterns in how and why populations grow.

Figure 5.1

Ecologists can study population growth by inoculating a Petri dish containing nutrient medium with a few organisms and watching their growth.

Program Resources

Section Focus Master 9 L1 SAE
Reinforcement and Study Guide, pp. 17-18 L1
Performance Assessment in the Biology Classroom, p. 53 L1 P
Biolab and Minilab Worksheets, pp. 15-18 L1
Critical Thinking/Problem Solving, p. 5 L3
Concept Mapping, p. 5 L1
Laboratory Manual, pp. 23-26 L2
Basic Skills Transparency 5 and **Master** L1 SAE
Reteaching Transparency 5 and **Master** L1 SAE

How fast do populations grow?

What's interesting about the growth of a population of living organisms is that it is unlike the growth of some other familiar things. Consider, for example, the growth of a weekly paycheck for an after-school job. Suppose that you are working for a company that pays you $5 per hour. You know that if you work for two hours, you will be paid $10; if you work for four hours, you will be paid $20; if you work for eight hours, you will be paid $40; and so on. If you were to plot this rate of increase on a graph, as shown in *Figure 5.2*, you can see that the result is a steady, linear increase; that is, growth occurs in a straight line.

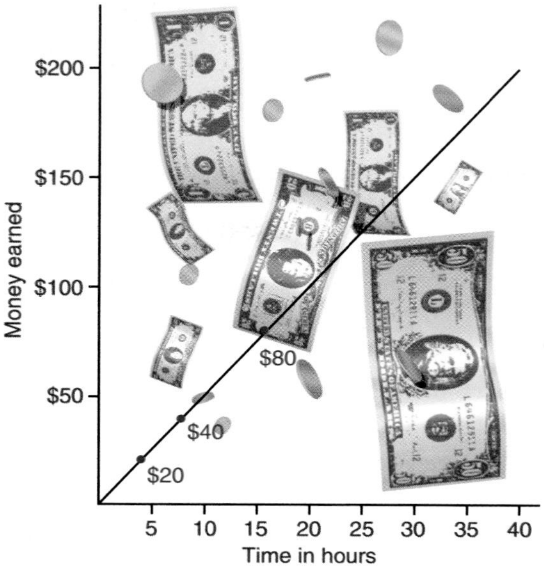

Figure 5.2

This graph shows that the way you earn money at an hourly rate is a straight line. Other examples might include the growth of your weekly allowance or the number of cars produced by an assembly line each month.

Populations of organisms do not experience this linear growth. Rather, the resulting graph of a growing population first resembles a J-shaped curve. Whether the population is one of weeds in a field, of frogs in a pond, or of humans in a city, the initial increase in the number of organisms is slow because the number of reproducing organisms is small. Soon, however, the rate of population growth increases rapidly because the total number of potentially reproducing organisms increases. This pattern illustrates the exponential nature of population growth. **Exponential growth** of a population of organisms occurs when the number of organisms rises at an ever-increasing rate. Exponential growth, *Figure 5.3*, results in a population explosion.

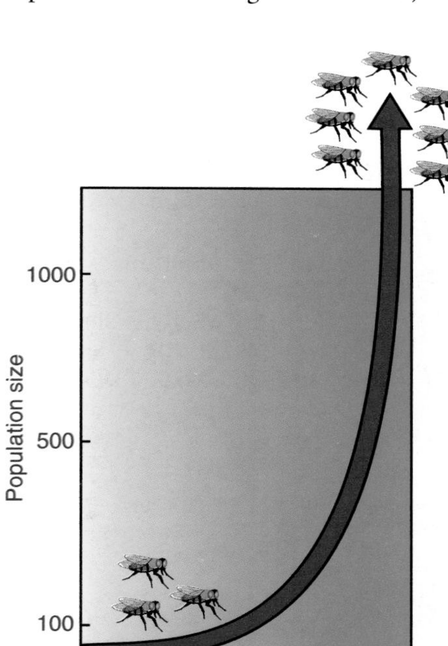

Figure 5.3

Because they grow exponentially, populations such as houseflies have the potential for explosive growth. One estimate is that a single pair of houseflies could have more than 6 trillion descendants in one summer if all of their eggs hatched and survived to reproduce. This type of population growth results in a J-shaped curve, showing no limit to a population's size.

2 Teach

Reinforcement

Explain that while linear growth is shown by height increases in people, it does not reflect the growth pattern of most populations.

Chalkboard Example

Graph two lines onto the same axis using different colors of chalk. Have one line show linear growth, while the other shows exponential growth. Ask students to describe how the patterns differ. Point out that when the curves begin, differences are smaller than at the end points.

Enrichment

Have students make the calculations needed to decide if they would rather be paid a linear salary of 5 dollars per hour for a 40-hour week or an exponential salary that starts at 1 cent the first hour and doubles each hour up to 40 hours. *The exponential salary will exceed the linear salary many times over.* L3

Visual Learning

Figures 5.2 and 5.3 Ask students to compare linear growth with exponential growth. *Populations showing exponential growth increase in size at a much faster rate than those showing linear growth.*

Meeting Individual Needs

Learning Disabled Review the proper construction of a graph with students and/or refer them to the Skill Handbook. Assign cooperative groups of mixed ability levels. Have students prepare graphs from data that you supply to illustrate linear and exponential changes. LS COOP LEARN

STUDENT JOURNAL

Evaluating Population Growth Have students write a scenario that depicts what life in the United States might be like if the population doubled its current status. Have them think in terms of available recreational space, demands made on natural resources, housing needs, and need for food. L1 LS

SECTION 5.1

BioLab

Population Growth in *Paramecium*

Time Allotment
Initial session: one class period; followup sessions of ten minutes daily for three weeks

Objectives
Review objectives with students before they begin the Biolab.

Process Skills
observe and infer, gather data, interpret data

Safety Precautions
Students should wash their hands in warm, soapy water after handling the cultures.

PREPARATION

Alternate Materials
- Cultures of *Paramecium* are available from biological supply houses. Baby food jars may be substituted for the plastic cups.
- Pre-boil rice grains prior to use. Substitute alfalfa tablets sold in drug or health food stores as a food source.
- Purchase methyl cellulose from a biological supply house.

GLENCOE TECHNOLOGY

Videodisc

Biology: The Dynamics of Life
Disc 1, Side 1
Carrying Capacity (Ch. 13)

BioLab

Population Growth in *Paramecium*

Paramecium is a group of unicellular protists that lives in freshwater environments. They consume bacteria that live on decaying material in the water. Although a paramecium can reproduce sexually, its usual method of reproduction is by simple cell division from one to two cells. Division occurs when there is a favorable environment with a good supply of food. When conditions such as crowding and decreasing food availability occur, *Paramecium* often begins to reproduce sexually. In this activity, you will determine the population of these protozoans and compare their growth with models of population growth.

PREPARATION

Objectives
In this Biolab, you will:
- **Calculate** the population size of *Paramecium.*
- **Compare** growth rate of a *Paramecium* population to models of population growth.
- **Relate** the reproductive strategy of *Paramecium* to environmental factors.

Materials
microscope
microscope slides
coverslips
eyedropper
10-mL graduated cylinder
plastic cup
rice grains
Paramecium culture

Safety Precautions
Use care when handling the microscope.

116

PROCEDURE

Teaching Strategies
- Cooperative groups consisting of two students are ideal. Make sure students working cooperatively alternate responsibilities on subsequent viewing days.
- Diagrams, photographs, or slides may be used to familiarize students with the general appearance of *Paramecium.* Include photographs that show *Paramecium* reproducing both sexually and asexually. Display visuals before students make their first microscopic observations or refer them to Chapter 22, which describes protozoans, as a reference.

Troubleshooting
- Methyl cellulose can be used to slow the movement of protozoans.

PROCEDURE

1. Measure 10 mL of *Paramecium* culture, and pour into a plastic cup.
2. Add two grains of rice to the cup. Rice grains serve as a food source for the bacteria, which are a food for *Paramecium.*
3. Make a data table like the one shown.
4. Make three wet mount slides using one drop of culture in each. Using the microscope under low power, count the number of *Paramecium* on each slide, and record the number and average of the three slides in your data table.
5. Twenty drops of culture equals 1 mL of liquid, so multiply the average by 20 to find the approximate number of *Paramecium* in 1 mL. Now multiply the number per mL by 10 to estimate the number of *Paramecium* in the culture.

Magnification: 85×

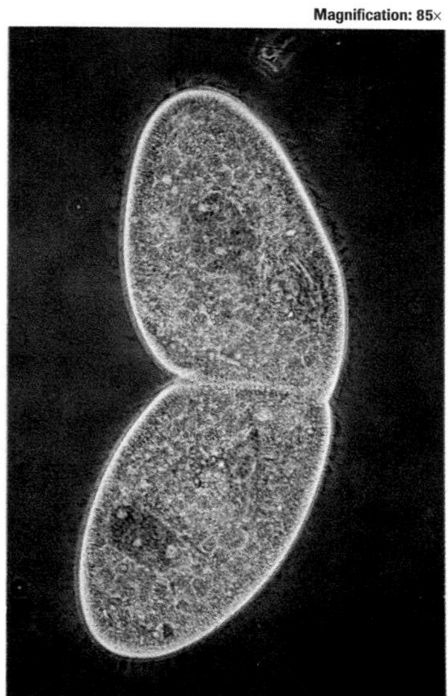

Number of Paramecium in Culture

	Slide 1	Slide 2	Slide 3	Average	Total in Culture
Mon.					
Tues.					
Wed.					
Thurs.					
Fri.					

6. Repeat steps 4 and 5 every school day for two weeks.
7. Use the data in your table to graph the population growth of *Paramecium* in your culture.

Going Further

Application What do you think would happen if you added more rice or hay to the water when the population began to decrease? Do you think dividing the culture and adding more water and rice every few weeks would make a difference? How could you find out?

ANALYZE AND CONCLUDE

1. **Drawing Conclusions** From your data, what reproductive strategy would you say *Paramecium* was using?
2. **Analyzing Data** What was the shape of your graph?
3. **Thinking Critically** When did the population appear to grow most quickly? Can you suggest reasons why the population may have grown most rapidly then?
4. **Interpreting Observations** If you observed *Paramecium* reproducing either sexually or asexually, make a drawing of your observation.

ANALYZE AND CONCLUDE

1. Asexual reproduction correlates with rapid increases in numbers.
2. J-shaped curve
3. After the first week; the lag was due to time needed for sufficient increase in numbers and in food supply.
4. Asexual reproduction will be the most common process viewed. Check students' drawings for logic and accuracy.

Assessment

Portfolio: Have students write a report of their results. P

Going Further

Populations of *Paramecium* will increase again as new food sources are introduced. Dividing cultures every few weeks will maintain the organisms at a maximum level of numbers. Experimentation would confirm this.

- Loss of liquid from the cups may be controlled by covering them with plastic wrap or by adding bottled spring water available from a grocery store.
- Students may need help calculating the total number of organisms in the cup. Review this process using actual numbers as an example.
- Use a 4×, rather than 10×, objective on the microscope if available.
- Demonstrate the experiment using a video camera-television setup for students who are physically or visually challenged.

Data and Observations

Paramecium populations should increase in size for two to three weeks and then decline in numbers as food runs out. The population decline may not be observed if the experiment is not continued over a long enough time period.

MiniLab

Purpose

Logical-Mathematical Students will learn that food is a factor in population growth.

Process Skills

observe and infer, interpret data

Teaching Strategies

- Jars may be kept inside the classroom with the possibility of attracting fruit flies or ants.
- Encourage students to record their observations in a data table.

Expected Results

Insects such as fruit flies, house flies, or ants may find the food. Fruit flies and house flies will lay their eggs on the food, and students may observe larval stages on the food.

Analysis

Responses may vary.

1. Answers may include ants, bees, fruit flies, house flies.
2. Insects should appear soon after the jars are placed outdoors, depending on outdoor temperature.
3. They are small bodied, mature rapidly, reproduce early.

Assessment

Skill: Have students describe how they could determine if food type affects the type of animal attracted.

MiniLab

How fast do populations use resources?

Fruit flies and similar insects have rapid life-history patterns. Insects are ideal organisms for population studies because they reproduce quickly and are easy to count. In this activity, you will observe the growth of insect populations as they exploit a food supply.

Materials

jar
half of a banana

Procedure

1. Place half of a banana in a jar and allow it to sit outside in a warm, shaded area.
2. Observe the jar each day for insects that feed off the banana.
3. Record the type of insect and the numbers of each type that appear in the jar. Be sure to record this information each day. If you do not know the name of the insect, draw it so that someone else will be able to recognize it.

Analysis

1. What types of insects were attracted to your jar?
2. How soon did insects appear, and how long were insects present?
3. Why are insects considered to have a rapid life-history pattern?

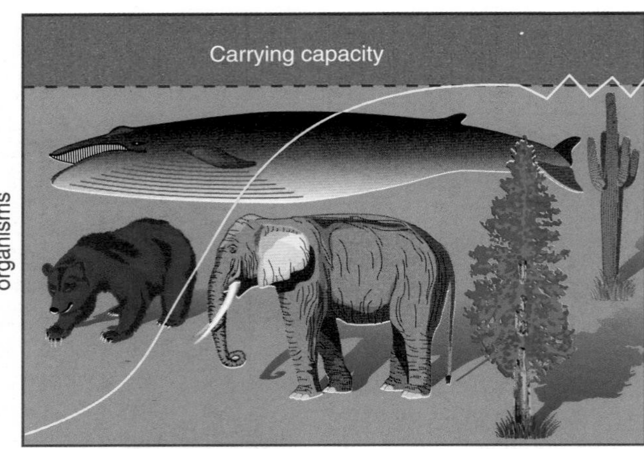

Figure 5.4

Population growth may follow an S-shaped curve. There is an initial slow-growth phase, a period of exponential growth, and a plateau where the number of organisms the environment can support is reached.

Limits of the environment

Can a population of organisms grow indefinitely? What prevents the world from being overrun with all kinds of living things? Through population experiments, scientists have found that, fortunately, population size does have a limit. Eventually, populations do have limiting factors in their environment, such as availability of food and space. This leveling off of population size results in an S-shaped growth curve. The number of organisms of a population that a particular environment can support over an indefinite period of time is known as its **carrying capacity.** Carrying capacity is often represented by the letter *K.* When populations are under the carrying capacity of a particular environment, births will exceed deaths until the carrying capacity is reached. If the population temporarily overshoots the carrying capacity, deaths will exceed births until population levels are once again at carrying capacity. *Figure 5.4* shows the S-shaped curve characteristic of many populations.

Patterns of population growth

In nature, many animal and plant populations remain in equilibrium. However, this is not true for all organisms. Why, for example, does it seem like mosquitoes are more numerous at some times of the year than others? Why don't all populations of organisms reach the carrying capacity and become stable? To answer these questions, population biologists study the most important factor that determines population growth—an organism's reproductive pattern.

Populations of organisms usually follow one of two growth patterns. Some species, such as mosquitoes, reproduce very rapidly and produce many offspring in a short period of

Brainstorming

Ask students to identify limiting factors for human population growth. *Food supply, living space, space for crop growth, pollution of water and air, disease.* Have them identify limiting factors of a dandelion population. *Lack of soil nutrients or water, crowding, herbicide use.*

PORTFOLIO

Interpreting Graphs Ask students to diagram an S-shaped curve. Have students identify the following areas on the graph: slow growth phase, exponential growth phase, plateau, point where carrying capacity (K) is reached. Ask how this graph might change if there were a sudden increase or decrease in food supply. L2 LS P

PROJECT

Fruit Fly Demographics

Interpersonal Fruit flies are easy to maintain in captivity. Allow students to carry out an experiment that will compare a population of fruit flies kept in balance with their environment with one not kept in balance. Students will first have to determine how this balance will be maintained. L3

time. Other species such as elephants have a slow rate of reproduction and produce relatively few young over their lifetimes. What causes species to have either one of these life-history strategies?

The kind of strategy a species has depends mainly on environmental conditions. For example, species such as mosquitoes are successful in environments that are unpredictable and change rapidly. This life-history pattern is found in organisms with similar adaptations. Typically, these organisms have a small body size, mature rapidly, reproduce early, and have a short life span. Populations of these organisms increase rapidly and then decline rapidly as environmental conditions suddenly change and become unsuitable. The small surviving population will begin reproducing exponentially when conditions are again favorable.

Species of organisms such as elephants that live in more stable environmental conditions have a different life-history pattern. Elephants, bears, whales, and long-lived plants such as cacti and California redwoods are large in size, and reproduce and mature slowly. These organisms maintain population sizes near the carrying capacity of the environment.

Although populations could follow either a rapid or slow life-history pattern, under uncrowded conditions, such as the pioneer stage in succession, rapid population growth seems to be most common. The organisms shown in *Figure 5.5* represent both types of life-history patterns. Which of these organisms would be most successful in a rapidly changing environment?

ThinkingLab Design an Experiment

How can you control an organism's life-history pattern?

Populations with a rapid life-history pattern, such as minnows, tend to grow exponentially until they reach the limits of the environment. This results in most or all of the population dying. But if organisms are given unlimited resources and environmental factors are controlled, populations can reach a maximum reproduction rate. Minnows give no parental care to their offspring, and female minnows may bear as many as 100 offspring in their lifetimes.

Analysis

Minnows will grow and reproduce in an aquarium or other large container of water. If you feed them commercial fish food, bits of crusty bread, or small insects, most minnows will live for several months and begin reproducing when they are only a month old.

Thinking Critically

Hypothesize how you can get minnows to achieve a maximum reproductive rate. Consider the environmental factors you will need to control in your aquarium.

Figure 5.5

Wild mustard plants taking over an abandoned field represent a species with a rapid life-history pattern. These plants also take advantage of any waste area, such as a vacant lot or a roadside. Organisms that have a slow life-history pattern, such as these Canada geese, provide much parental care for their young. The adults stay with their young from the time they hatch through their migration to wintering grounds.

ThinkingLab Design an Experiment

Purpose

IS **Interpersonal** Design an experiment to create an environment in which minnows achieve a maximum reproductive rate.

Process Skills

hypothesize, design an experiment, separate and control variables, interpret data

Teaching Strategies

- Allow students to work in cooperative groups of mixed ability levels.
- Review the basis for a properly executed experiment (Chapter 2).

Thinking Critically

Plans should include using the same ratio of males to females in experimental and control groups, and controlling temperature, amount and type of food, and the filtering mechanism. The independent variables may be temperature of water, original population density, or food type used.

Assessment

Knowledge: Have students list reasons why this exercise would be easier to conduct with minnows than with *Paramecium. Minnows are larger and more easily observed.* Why might it be more difficult? *The experimental design requires more space and the life history pattern is slower.*

STUDENT JOURNAL

Life as a Mosquito Have students imagine they are breeding mosquitoes. Tell them to describe the conditions that make their environment unpredictable and subject to rapid change. Students will have to research life stages and breeding habits of mosquitoes to complete this task. L2 IS

Text Question

Which of these organisms would be most successful in a rapidly changing environment? *Likely responses will suggest the wild mustard plants because they produce large numbers of offspring and therefore their ability to adapt is much greater.*

SECTION 5.1

A BROADER VIEW

Mussel Beach

Purpose

To alert students to problems associated with both population growth and population control using a real-life situation rather than a simulated population growth model.

Teaching Strategies

- Obtain preserved specimens of zebra mussels, *Dreissena polymorphia,* or the shells of this mussel to indicate what this animal looks like.
- Define the terms *mollusk* and *phytoplankton* for students.
- Keeping in mind that these organisms are not a food source for humans, ask students to suggest possible uses for mussels harvested from inlet water pipes. L1
- Ask students to speculate on whether the zebra mussels have reached their carrying capacity. L1

Thinking Critically

The odor may cause mussels to spawn. There will be insufficient food available for the larvae to survive. Eventually, the zebra mussel population will decline. However, the chemical may have harmful effects on other organisms.

Visual Learning

Figure 5.7 What do you think will happen to populations of mice and rabbits in the foxes' area? *As the fox population declines, the rabbit and mice populations will grow.* L1

Audiovisual

Show the filmstrip *Population Ecology*, Biology Media. L1

A BROADER VIEW

Mussel Beach

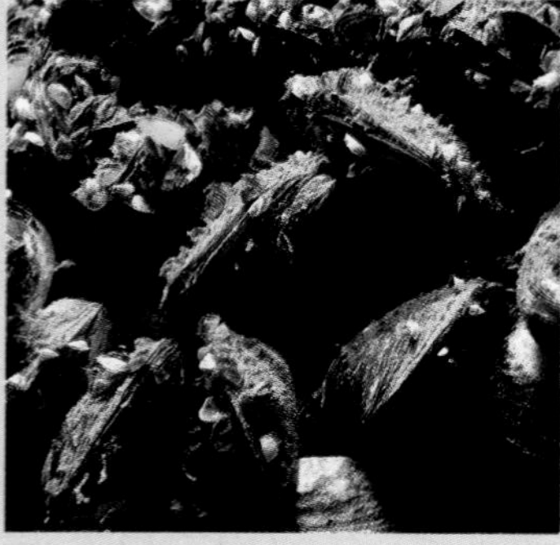

In 1989, the residents of Monroe, Michigan, made a startling discovery—zebras were living in the city's water pipes! Not just a few zebras, either, but tens of thousands—enough zebras, in fact, to shut down the city's water supply! Needless to say, normal activity in Monroe came to a screeching halt.

Bothersome bivalves The zebras that shut down schools, businesses, and industries in Monroe were not the striped mammals that roam the African plains. They were mussels, little bivalve mollusks 5 cm and smaller, that take their name from the distinctive dark-brown stripes that band their light-tan shells. Marine biologists speculate that the first zebra mussels arrived in the Great Lakes as stowaways on ships from Europe navigating the St. Lawrence Seaway.

Unlike their edible cousins, zebra mussels are not considered a viable food source for humans. Their chief occupation seems to be clogging water inlet pipes in the rivers and lakes.

The mussels attach themselves to pipes and other surfaces by means of sticky, elastic strands called byssal fibers. Once attached, the mussels take advantage of strong currents to filter passing phytoplankton through their ciliated gills. The Great Lakes are now home to billions of zebras, and the mussels' voracious appetites leave very little food for other plankton-reliant species.

A population explosion An adult female can produce up to 50 000 eggs each year. In the summer of 1989, officials inspecting the pipes and water-treatment machinery in Monroe counted about 1350 mussels per square yard. Before the year was over, the count per square yard had reached 600 000!

Controlling the zebra mussel population is proving no easy task. Toxic chemicals such as chlorine will kill both adults and larvae, but are equally fatal to other harmless, lake-dwelling species. Removing the mussels manually is a slow and largely ineffective task.

Perhaps the most promising solution is biological deception. Zebra mussels spawn when they detect odors coming from phytoplankton. The larvae need this rich source of food to survive to adulthood.

Thinking Critically

If biologists can successfully simulate that same odor, and release it when phytoplankton are actually scarce, what is likely to happen to the zebra mussel population? Might this kind of biological interference be dangerous?

Environmental limits to population growth

Limiting factors, you remember from Chapter 4, are biotic or abiotic factors that determine whether or not an organism can live in a particular environment. Limiting factors also regulate the size of a population. Limited food supply, extreme temperatures, and even storms can affect population size. Ecologists have identified two kinds of limiting factors: density-dependent and density-independent factors.

Density-dependent factors include disease, competition, and parasites, which have an increasing effect as the population increases. Disease, for example, spreads more quickly in a population whose members live close together, see *Figure 5.6*, than in smaller populations whose members live farther from each other. In very dense populations, disease may wipe out an entire

Figure 5.6

Corn smut is a fungus that produces large, deformed growths on the ears of corn. To prevent it from spreading through a cornfield, affected plants must be burned or buried before the fungus reproduces.

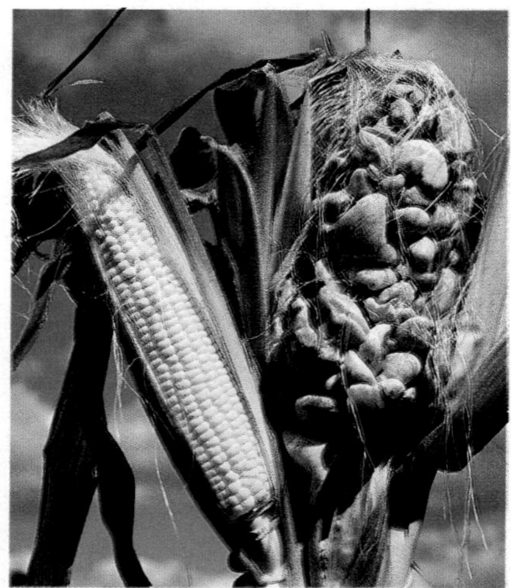

CULTURAL DIVERSITY

Protecting Endangered Species

The illegal trafficking of primates is currently a multimillion-dollar-a-year business. Since its founding in 1973, the International Primate Protection League has successfully lobbied many governments to enact laws banning the trade of primates destined for wildlife parks, exotic pet stores, and research labs. Initiate a class discussion about efforts to protect the world's endangered species. Discuss methods organizations use to uphold animal import and export laws.

entire population. In crops such as corn or soybeans, in which large numbers of the same plant are grown together, a disease will spread rapidly throughout the whole crop. In less-dense populations, fewer individuals may be affected.

Density-independent factors affect all populations, regardless of their density. Most density-independent factors are abiotic factors such as temperature, storms, floods, drought, and habitat disruption, *Figure 5.7.* No matter how many earthworms live in a field, they will all drown in a flood.

Figure 5.7

Populations are affected by both density-dependent and density-independent factors.

Hurricane Andrew, which hit south Florida in 1992, did extensive damage to both heavily populated areas and less populated ones.

As a population of foxes increases, competition for the available food, such as mice and rabbits, also increases. When food, a density-dependent factor, is in short supply, young foxes begin to die, and population growth slows. Problems for one population of organisms also can affect populations of other organisms. What do you think will happen to populations of mice and rabbits in the foxes' area?

MiniLab

How does the population density of weeds in an area affect other plants?

An ecological study of your front yard, the school grounds, or a similar area can reveal the density of organisms in the area. Population density is the measure of the number of individuals in a given area. High population densities of weeds tend to affect populations of other plants. In this lab, you will calculate the average density of weeds in an area, and observe how this affects other plants.

Materials

string
small sticks
meterstick or sewing tape measure

Procedure

1. Identify a weed plant in your study area.
2. Randomly choose three plots. For each plot, place four sticks in the ground so that they form one square meter.
3. Mark off your plots with string.
4. Count and record the number of weed plants in each square.
5. Calculate the average number of weed plants in your three plots.

Analysis

1. What was the average density of weeds in the area you measured?
2. Did the density of weed plants in your plots affect the populations of other plants in the same plot? Explain your answer.
3. Why do you suppose a high density of weeds can affect the way other plants grow?

Meeting Individual Needs

Learning Disabled/Visually Impaired
Have students make two squares measuring 10 cm per side. Ask them to place pennies in one square to illustrate a dense population. Have them illustrate a less dense population in the second square. Ask them to predict which population is more vulnerable to a disease. *The dense population.* L1 LS

MiniLab

Purpose

LS Logical-Mathematical
Students will determine that as population density increases in an area, other populations in the same area may decrease in size.

Process Skills

observe and infer, interpret data, relate cause and effect

Teaching Strategies

- Review the conversion of feet to meters if measuring tapes are in English units. Review methods for averaging.
- Advise students not to touch or remove any plants from the area.
- Identify plants that are considered weeds in the area where students are performing the experiment. This might best be done using a sample plot.

Expected Results

Number of weed plants may vary from plot to plot. As weed population numbers increase, other plant populations decrease.

Analysis

1. 5-20 plants
2. Yes; the greater the number of weed plants per plot, the lower the number of other plant types.
3. Competition for light, water, nutrients, and space

Assessment

Knowledge: Ask students to predict how their data might vary if they visited their plots during different weeks of the growing season.

SECTION 5.1

Tying to Previous Knowledge

Explain how the decrease in available energy at the top trophic level of a food chain serves as a limiting factor on predator population size.

Bioethics

Butterflies Wanted: Dead or Alive!
The Karner blue butterfly faces extinction. Nevertheless, three collectors of rare butterflies spotted, captured, and promptly killed five specimens of the Karner blue butterfly. The butterflies were killed so identification could be verified. Some scientists question why one should be allowed to collect and kill an endangered species. Collectors say this is the only way to verify the identity of species and to see if the insects are evolving new characteristics. Others claim that when adults are killed in the name of science, added stress is placed on the endangered population.

GLENCOE TECHNOLOGY

Videodisc

STVS: Ecology
Disc 6, Side 1
Wasp Biological Control
(Ch. 21)

Interactions Among Organisms that Limit Population Size

Population sizes are not limited only by environmental factors, but are also controlled by various interactions among organisms that share a community.

Predation affects population size

Predation of one organism by another is important for the proper functioning of a community. Recall that energy is moved through a community by way of food chains and food webs. Predation ensures the continuation of the flow of energy, but it may also be a limiting factor of population size.

Populations of predators and prey experience changes in their numbers over a period of years. Many predator-prey relationships show a cycle of population increases and decreases over time. The rise and fall of populations of snowshoe hare and lynx over a 90-year period are shown in *Figure 5.8*. The populations of both animals rise and fall almost together. Does this mean that the lynx overhunted their prey, losing their food supply? Or, did the food supply of the snowshoe hare become less abundant, causing their numbers to decline? While the graph can't indicate what factors caused the population fluctuations, it is clear that most populations are controlled in some way by predators.

Prey-predator relationships are important for the health of natural populations. Usually, in prey populations, the young, old, or injured members are caught. This predation keeps the population size within the limits of the available resources.

Figure 5.8

The data in this graph came from the number of hare and lynx pelts sold to the Hudson Bay Company in northern Canada. Notice that as the number of hares increased, so did the number of lynx. The number of hares is regulated by their food supply. The number of lynx is controlled by their food supply also—the number of hares.

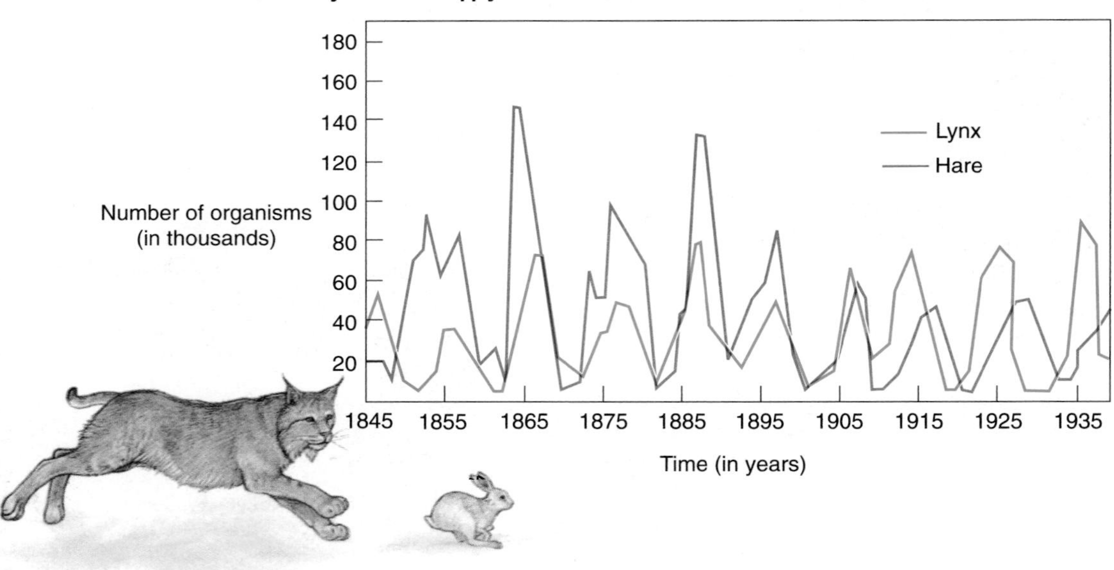

Meeting Individual Needs

Students Acquiring English Using Figure 5.8, ask students to identify on the graph where increasing numbers of snowshoe hare and lynx occur. Have them indicate where decreasing numbers occur. Use the graph to point out the cyclic nature of these events. Elicit from students why these events are cyclic. SAE LS

1. Linear growth graphs form a straight line, and exponential graphs form a curved line described as being J-shaped.
2. Rapid life-history organisms show a rapid increase and rapid decline caused by environments that are unpredictable. Slow life-history organisms show a slow population change within stable environments and usually maintain sizes near

Figure 5.9

Most organisms have needs for a certain amount of space for nesting, hunting, and defending. Stress caused by overcrowding in a rat population can limit population size. When overcrowded, animals fight and kill each other, they reproduce less, and they stop caring for offspring.

The effects of competition

Organisms within a population constantly compete for resources. When population numbers are low, resources are plentiful. However, as population size increases, competition for resources such as food, water, and territory can become fierce. Competition is a density-dependent factor. When only a few individuals compete for resources, no problem arises. When population size increases to the point at which demand for these resources exceeds the supply, the population size decreases.

The effects of crowding and stress

When populations of organisms become crowded, individuals may exhibit stress. The factors that create stress are not well understood, but the effects have been documented experimentally in populations of rats and mice, *Figure 5.9.* As populations increase in size, individual animals begin to exhibit a variety of symptoms, including aggression, decrease in parental care, decreased fertility, and decreased resistance to disease. All of these symptoms can lead to a decrease in population size.

Section Review

Understanding Concepts

1. How are the graphs of exponential growth and linear growth different?
2. Explain how rapid and slow life-history patterns differ.
3. Describe how density-dependent and density-independent factors regulate population growth.

Thinking Critically

4. Discuss how an environmental factor such as a hurricane can affect a population of organisms.

Skill Review

5. **Making and Using Graphs** Graph the following population growth for the unknown organism, and state whether the organism has a rapid or slow life-history pattern. For more help, refer to Organizing Information in the *Skill Handbook.*

Population Size of Unknown Organism

Year	Spring	Summer	Autumn	Winter
1995	564	14 598	25 762	127
1996	750	16 422	42 511	102
1997	365	14 106	36 562	136

3 Assess

Check for Understanding

Ask students to explain the difference between the words in each of the following pairs.

(a) linear growth-exponential growth
(b) carrying capacity-population size
(c) density-dependent factors-density-independent factors.
(d) predation-competition L1

Reteach

Have students provide an example that illustrates an understanding of the meanings of each word in the above section. L1 SAE

Extension

Have students research the meanings of *r-strategists* and *K-strategists.* Have them provide characteristics and examples for both groups and correlate this information with what has been studied in this chapter. L3

Assessment

Portfolio: Have students outline the concepts in this section. L1 SAE P

Performance Assessment in the Biology Classroom: p. 53, *Estimating Populations.* Have students carry out the activity to show their knowledge of how populaton size is determined. L2 P

Answers to Section Review

the carrying capacity of the environment.

3. Density-dependent factors have an increasing effect on populations as numbers of individuals increase. Density-independent factors do not depend on the numbers of individuals in a population.

Thinking Critically

4. Hurricanes can destroy or disrupt habitats and eliminate food sources.

Skill Review

5. The graphs should show that the organism has a rapid life-history pattern. It reproduces rapidly and then declines rapidly and exceeds or falls below its environmental carrying capacity.

4 Close

Activity

Ask students to count the number of potential offspring (the number of seeds) in one green pepper. Have them explain why the world is not covered with green pepper plants. SAE

SECTION

5.2

Prepare

Key Concepts

Population growth characteristics, called demography, are studied using growth rates, age structure, and geographic distribution. The effects of immigration and emigration on population growth are also considered.

Block Scheduling

Look for this symbol for strategies that are useful in a block scheduling format. For more information on block scheduling, refer to Section 5.2 in the **Lesson Plans** booklet.

Materials

- Purchase small lima bean seeds and gather 250-mL beakers for the Alternate Lab.

1 Focus

Bellringer

Before presenting the lesson, display **Section Focus Master 10** on the overhead projector and have students answer the accompanying questions. L1 SAE

Activity

LS **Logical-Mathematical** Have students estimate the populations of Mexico City and Calcutta in 1985 compared to their expected populations in 2000. Provide them with the correct data and ask them to compute the percent change expected in this 15-year period.

Mexico City = 17.3 million, 25.8 million *33% change.* Calcutta = 11 million, 16.5 million *33 % change.*

SECTION

5.2 Human Population Growth

Section Preview

Objectives

Relate characteristics of populations to population growth rates.

Compare the age structure of rapidly growing, slow-growing, and no-growth countries.

Hypothesize about problems that can be caused by immigration and emigration.

Key Terms

demography
age structure
immigration
emigration

Does Earth have a carrying capacity for the human population? How many people can live on Earth? No one knows how many people Earth can support, and it is presently impossible to tell when the human population will stop growing. However, an increasing number of scientists suggest that food production will not always keep pace with the population increase.

Demographic Trends

A good way to predict the future of the human population is to look at past population trends. For example, are there observable patterns in the growth of populations? That is, are there any similarities among the population growths of different countries—similarities that might help scientists predict, and therefore control, future population catastrophes? As you have seen, populations tend to increase until the environment cannot support any additional growth. The study of population growth characteristics is the subject of **demography.** Demographers study such population characteristics as growth rate, age structure, and geographic distribution.

What is the history of population growth for humans? Although local human populations often show fluctuations, the worldwide human population has increased exponentially over the past several hundred years, as shown in ***Figure 5.10.*** Unlike other organisms, humans are able to reduce environmental effects by eliminating competing organisms, increasing food production, and controlling disease organisms.

Effects of birthrates and death rates

How can you tell if a population is growing? A population's growth rate is the difference between the birthrate and the death rate. In many industrialized countries, such as the United States, declining death rates have a greater effect on total population

Program Resources

Section Focus Master 10 L1 SAE
Reinforcement and Study Guide, pp. 19-20 L1
Exploring Environmental Issues, p. 23 L2

interNET CONNECTION

Follow the link for this chapter on the Glencoe Homepage at **www.glencoe.com/sec/science** to find out more about population biology.

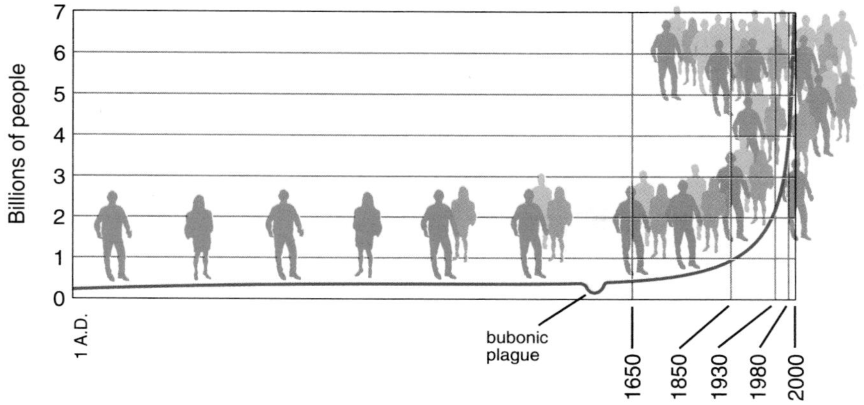

Figure 5.10

Ten thousand years ago, approximately 10 million people inhabited Earth. Today, there are more than 5 billion, and scientists estimate that by the year 2050, there will be more than 10 billion people on Earth.

growth than increasing birthrates. For example, in the United States, life expectancy increases almost every year. This means that you probably will live slightly longer than students who are presently in college. Interestingly, although people in the United States are living longer, the fertility rate is decreasing. This is because people are waiting until their thirties and forties to have children. Childbirth in women in their twenties was more common just a few decades ago. Today's families also have fewer children than they did in previous decades.

What are the birthrates and death rates of some countries around the world? Look at *Table 5.1* to see the birthrate, death rate, and fertility rate of some rapidly growing and slower-growing countries.

Table 5.1 Birthrates and Death Rates Around the World

	Birthrate (per 1000)	Death Rate (per 1000)	Fertility (per woman)	Population Increase (percent)
Rapidly Growing Countries				
Gaza	50	7	7.0	4.3
Iraq	46	7	7.3	3.9
Kenya	46	7	6.7	3.8
Rwanda	51	16	8.5	3.7
Uganda	52	17	7.4	3.6
Slowly Growing Countries				
Hungary	12	12	1.8	-0.2
Germany	12	12	1.5	0.0
Italy	10	9	1.3	0.1
Sweden	14	11	2.0	0.2

Source: 1990 World Population Data Sheet (Washington D.C.: Population Reference Bureau, Inc., April, 1990).

2 Teach

Math Connection

Calculating Annual Rate of Population Change

The annual rate (AR) of population change can be expressed as a percentage. It is calculated as follows, where birthrate is BR and death rate is DR.

$AR\% = BR - DR/1000 \times 100$

Have students use the data provided to calculate the AR percent for the following regions of the world.

Region	BR	DR	AR
	(per thousand)		
Africa	48	18	*3%*
Asia	29	12	*2%*
N. America	15	8	*0.7%*
Latin America	3	8	*-0.5%*
Europe	14	10	*0.4%*
World	27	9	*1.8%*

Ask students how different regions' growth rates compare with the world growth rate. *Africa, Asia, and Latin America show growth patterns larger than the world average. North America and Europe show growth patterns smaller than the world average.* L2 LS P

Visual Learning

Using the Table Ask students to use Table 5.1 to identify the factors that contribute to a high population growth rate. *Responses may include high birthrates, low death rates, and high fertility rates.* LS

GLENCOE TECHNOLOGY

Videodisc

The Infinite Voyage: Crisis in the Atmosphere

Our Future Climate (Ch. 3)

PORTFOLIO

Graphing Population Growth Have students make a bar graph that shows the population growth of each country listed in Table 5.1 in order from highest to lowest percentage. Have students find the population growth rate of the United States and add this percentage to the appropriate location in their graph. L1 LS P

Math

How many are there?

Purpose

Students will explore the practical side of demography and see how techniques used for counting the population of the United States compare with those that might be used to count wildlife populations.

Teaching Strategies

- Have students perform the Alternate Lab to experience the sampling technique used in estimating an animal population.
- Have students debate the issue regarding deer populations raised in the Connection to Biology.

Possible Answers

A population will be maintained naturally when it reaches its carrying capacity. Providing food during the winter disrupts this natural balance.

Math

How many are there?

Every ten years, the U.S. Government undertakes one of the most complicated counting jobs in the world. Every decade, as called for in the Constitution, a census is carried out. The goal is to count every individual person residing in the country and gather other information that measures progress or helps anticipate future needs.

The people who study populations and their characteristics and needs are mathematicians called demographers. Demographers can study any type of population—plant, animal, or human.

Using demographics One group of scientists that uses demographics as a foundation for their work is wildlife managers. They keep track of populations on a regular basis and use the information to make decisions in managing the size of those populations. One of the most important functions of wildlife managers is to determine the carrying capacity of the lands under their care. Another responsibility may be to protect populations from predators, people, or other factors that are stressing them. Wildlife managers gather facts about weather conditions, water availability, habitat, and many other conditions in areas in their care. Then they use these facts to balance one with another to produce the healthiest conditions possible for all populations.

They also use these facts to see that the needs of all are being met. For example, orchard owners want to control populations of deer feeding on their trees. Hunters want more rabbits. It is the goal of the wildlife manager to work out a population size to benefit all. Wildlife managers try to balance and control wild populations of organisms by various means.

CONNECTION TO Biology

In many areas of the United States, deer populations have increased to the point at which many starve each winter due to a lack of food. It has been suggested that surplus farm products such as corn or other grains could be used to save these animals. Discuss whether this is or is not a good solution to the problem.

Fertility is the number of offspring a female produces during her reproductive years. When fertility rates are high, populations grow more rapidly unless the death rate is also high. Some countries such as Uganda and Rwanda have high death rates among children because of disease and malnutrition. However, both countries have extremely high birthrates, and they are both growing rapidly. Some other countries such as Sweden and Italy have low death rates, but their birthrates are also low, so these countries' populations grow slowly, if at all.

The birthrate, death rate, and fertility rate of a country provide clues to that country's rate of population growth. As you can see, different combinations of birthrates and death rates have different effects on populations.

Does age affect population growth?

Imagine a country filled mostly with teenagers! Will it make a difference to population growth if the largest proportion of the population is in one age group? In order to make predictions about populations of the future, demographers must know the age structure of a population. **Age structure** refers to the proportions of a population that are either in their pre-reproductive years, their reproductive years, or their post-reproductive years. By knowing the age structure, you can tell if a population is growing rapidly, growing slowly, or not growing at all, as shown in *Figure 5.11.* What would happen to the population growth rate

Alternate Lab: Determining a Population Size

Purpose

LS **Logical-Mathematical** To simulate a sampling technique used to determine a population's size.

Materials

small lima beans, paper bag, 250-mL beaker, wax pencil

Procedure

Give the following directions to students.

1. A beaker represents a community, beans within the beaker represent animals in the community. One can determine a population's size without actually having to count all animals.
2. Fill the beaker with lima beans.
3. Use a wax pencil to mark 20 beans in the beaker with a large red dot. Then, place all beans into a paper bag and shake. Note: red beans are animals that were caught, marked, and then returned to the population.
4. Without looking in the bag, pick out 20 beans. Count the number of marked and unmarked beans in this sample.

if you and other students your age from around the country lived to be 100? You can see from studying the age structure distribution that if most of the population is living longer, then the overall population will be growing slowly, if at all, depending on the proportion of individuals in their pre-reproductive and reproductive years.

Mobility has an effect on population size

Unlike most organisms, humans can move in and out of different communities. The effects of human migrations can make it difficult for a demographer to make predictions, but patterns do exist. Movement of individuals into a population is **immigration.**

Figure 5.11

Notice that in a rapidly developing country such as Mexico, the large number of individuals in their pre-reproductive years will add significantly to the population when they reach reproductive age. Populations that are not growing, such as Germany's, have an almost even distribution of ages among the population.

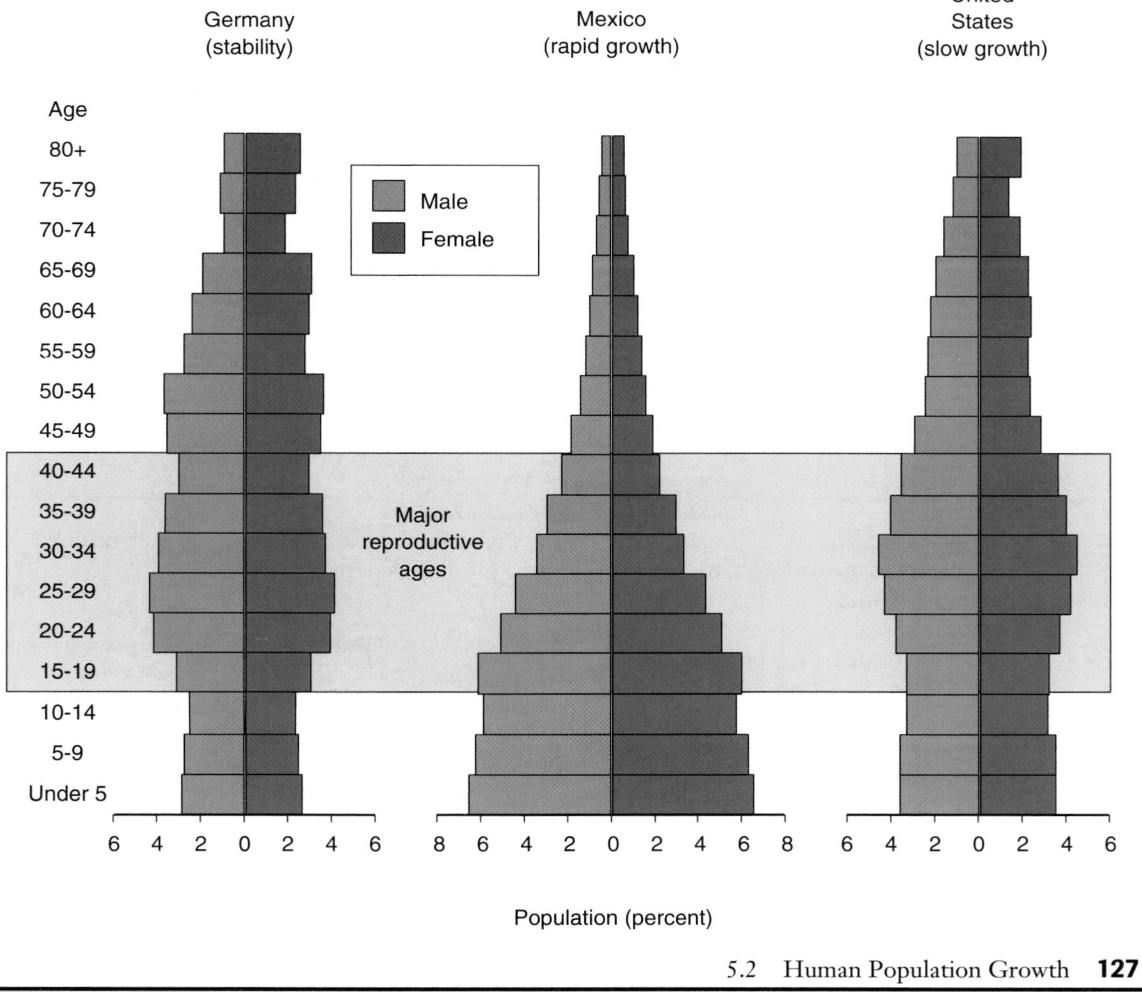

SECTION 5.2

Different Viewpoints in Biology

How large should the human population get?
Some people suggest that people are the main resource for finding solutions to world problems, thus more people mean more solutions. Others claim that population limits are a violation of personal rights.

People who favor setting limits on the population often claim that one out of five people today do not receive adequate basic necessities. At the same time, populations of the less developed countries increase, while more and more resources are used by developed countries. Thus, there is not a balance between world population and distribution of resources.

Science, Technology, and Society

Increasing Crop Productivity
Food is important for the continued growth of the human population. Efforts are being made to improve varieties of crops using a technique called hybrid protoplast fusion. Rice, the world's most important cereal crop, is being altered to make it more disease-resistant and salt tolerant. This is achieved by fusion of protoplasts of wild species and cultivated species of rice.

5. To calculate the animal population size, multiply the total number of marked beans (20) by the total number of the second sample (20), and divide by the number of marked beans counted in the second sample.

Analysis

Ask students the following questions.

1. Why might a sampling technique be more practical than having to count all animals in a population? *Saves time, equipment, and manpower; is impossible in certain cases*
2. Why might it be difficult to get a correct population count with animals? *Movement in/out of community, difficult to locate*

Assessment

Knowledge: Ask: How could you verify if the sampling technique was accurate or not? *Once you decide, try it. Count entire bean population and compare with data obtained through sampling.*

SECTION 5.2

Text Question

What other problems result from suburban growth? *More garbage is produced, more pollution occurs, and housing becomes more costly as availability decreases.*

3 Assess

Check for Understanding

Have students explain the relationship between the words in each of the following pairs.

(a) population growth - birth and death rate

(b) age of population - population growth L1 SAE

Reteach

Ask students to explain how a high fertility rate with low death rate will influence a population's size. *Population will increase.* L1 SAE

Extension

Have students research the pros and cons of allowing large numbers of immigrants to enter the U.S. L1

Assessment

Journal: Have students write a paragraph that expresses their opinion as to whether the human population should or should not be controlled in terms of future size. L1

4 Close

Activity

Have students list changes that can increase and decrease a population's size.

Figure 5.12

More people have immigrated to the United States during the past 165 years than to any other country in the world. These immigrants came mostly from European countries. How do you think these countries were affected by this emigration of their people?

Movement from a population is **emigration.** Obviously, movement of people between countries has no effect on total world population, but it does affect national population growth rates, *Figure 5.12.* Local populations can also feel the effects of a moving population. Many suburbs of large cities are expanding rapidly. This places stress on schools, roads, and police and fire services. What other problems result from suburban growth?

Connecting Ideas

In this chapter, you investigated some characteristics of populations and learned how environmental factors place limits on population growth. Unlike other organisms, humans can manipulate and regulate some of these biotic and abiotic factors, and this has resulted in a large increase in the world's human population.

Many ecologists suggest that the pollution observed today is directly related to the world's increasing human population. If human population growth continues at the present pace, additional environmental problems will result from such things as a diminishing energy supply and increasing pollution.

Section Review

Understanding Concepts

1. What characteristics of populations do demographers study?
2. What clues can an age structure graph provide about a country's population growth?
3. Discuss some possible problems for local populations caused by immigration and emigration of people.

Thinking Critically

4. Hungary has the world's slowest-growing human population. Which age groups do you think make up the largest portions of Hungary's population?

Skill Review

5. **Making and Using Graphs** Construct a bar graph showing the age structure of Kenya using the following data: pre-reproductive years (0-14)—42 percent; reproductive years (15-44)—39 percent; post-reproductive years (45-85+)—19 percent. For more help, refer to Organizing Information in the *Skill Handbook.*

Answers to Section Review

1. Demographers study population characteristics such as growth rate, age structure, and geographic distribution.
2. It can determine if a population is growing rapidly, growing slowly, or not growing at all.
3. Answers will vary. Immigration can increase needs for schools, medical facilities, public transportation. Emigration will leave many unused facilities and vacant homes.

Thinking Critically

4. pre- and post-reproductive

Skill Review

5. Evaluate students' graphs for logic and accuracy.

Chapter 5 Review

REVIEWING MAIN IDEAS

5.1 Population Dynamics

- Populations grow exponentially until they reach the carrying capacity of the environment. Populations either exhibit slow growth that tends to approach the carrying capacity with minor fluctuations, or rapid growth that tends to expand exponentially and then experiences massive diebacks.
- Density-dependent factors such as disease and food supply, and density-independent factors such as weather, have effects on population size. Interactions among organisms such as predation, competition, stress, and crowding also limit population size.

5.2 Human Population Growth

- Demography is the study of population characteristics such as growth rate, age structure, and movement of individuals. Birthrate, death rate, and fertility differ considerably among different countries, resulting in uneven population growth patterns across the world.

Key Terms

Write a sentence that shows your understanding of each of the following terms.

age structure
carrying capacity
demography
density-dependent factor
density-independent factor
emigration
exponential growth
immigration

UNDERSTANDING CONCEPTS

1. The population growth of flies is similar to the growth of ____________.
   a. money from an hourly job
   b. money in a savings account
   c. money placed under a mattress
   d. money saved in a piggy bank
2. Populations that grow at an ever-increasing pace experience ____________.
   a. experimental growth
   b. exponential growth
   c. straight-line growth
   d. unlimited growth
3. The highest level at which a population can be sustained is its ____________.
   a. carrying capacity
   b. straight-line growth
   c. high-point maximum
   d. mid-range level
4. Resource use by most organisms increases as populations ____________.
   a. remain steady
   b. decrease in size
   c. increase exponentially
   d. migrate
5. Storms, temperature, and water are all ____________.
   a. density dependent
   b. biotic factors
   c. exponential
   d. density independent
6. Which of the following is a density-dependent factor?
   a. drought
   b. flood
   c. food
   d. wind speed
7. Populations limited by having members eaten experience ____________.
   a. predation
   b. competition
   c. carrying capacity
   d. exponential growth
8. When populations increase, resource depletion may cause ____________.
   a. exponential growth
   b. straight-line growth
   c. competition
   d. predators to increase

Chapter 5 Review

Reviewing Main Ideas

The summary can be used by students to review the major concepts of the chapter.

Key Terms

Answers should go beyond defining the terms. Accept any answer that uses the term correctly and in the proper context.

Understanding Concepts

1. b
2. b
3. a
4. c
5. d
6. c
7. a
8. c

PORTFOLIO

Increasing Longevity Have students prepare a "public service radio commercial" that encourages the elderly to participate in exercise programs as well as to lose weight. As an alternative, students may wish to design a poster that promotes the same goals among the elderly. L1 IN P

Chapter 5 Review

9. b
10. b
11. c
12. a
13. d
14. b
15. c
16. d
17. d
18. b
19. d
20. c

Applying Concepts

21. Future transportation needs, medical facilities, and school facilities can be planned ahead of time.
22. Density-dependent factors. Populations in developing countries are most likely more crowded and thus density is higher, making them more susceptible to diseases.
23. Humans may disrupt the habitats of other species. Other species may decline as a result of habitat loss. They may not have enough space for mating, rearing of offspring, or finding shelter and enough food.
24. Those in transportation, education, health care, recreational programs, and law enforcement would all be interested. They would be able to anticipate the need for increasing staff and facilities.

Chapter 5 Review

9. Which of these countries has the fastest population growth rate?
   a. Hungary
   b. Uganda
   c. Iraq
   d. Italy
10. What term refers to the proportions of a population in their pre-reproductive, reproductive, and post-reproductive years?
    a. old growth
    b. age structure
    c. demographics
    d. population trend
11. What can be said about the growth of a country whose age structure graph approximates a rectangle?
    a. It is decreasing.
    b. It is increasing slowly.
    c. It is stable.
    d. It is increasing rapidly.
12. The study of population growth characteristics is ___________.
    a. demography
    b. ecology
    c. growth strategy
    d. age determination
13. Movement out of a country is termed ___________.
    a. population degeneration
    b. immigration
    c. population impact
    d. emigration
14. The life-history pattern that tends to approach carrying capacity with minor fluctuations is ___________.
    a. population dependent
    b. slow growth
    c. fast growth
    d. exhibited by most insects
15. One of the major stressors of a population is usually ___________.
    a. reaching carrying capacity
    b. when stability is reached
    c. overcrowding
    d. caused by size decreases
16. According to *Figure 5.8*, when did the largest population of lynx occur in northern Canada?
    a. 1865
    b. 1880
    c. 1914
    d. 1935
17. Between A.D.1 and A.D.1650, the world's population had a major dip because of ___________.
    a. fertility
    b. decreased death rate
    c. density-independent factors
    d. bubonic plague
18. Organisms that are small and have extreme population swings usually have a reproductive pattern exhibited by ___________.
    a. slow growth
    b. fast growth
    c. stability
    d. immigration
19. Which word best describes the population growth of humans during the past 50 years?
    a. straight line
    b. stable
    c. slowing
    d. exponential
20. Which of the following environments would likely be more advantageous to an organism that exhibits fast growth?
    a. desert
    b. large, deep lake
    c. intermittent stream
    d. tropical rain forest

APPLYING CONCEPTS

21. Sweden and Norway have populations that are in equilibrium. Discuss some advantages of a stable population size.
22. Which environmental factors would most affect the populations of developing countries?
23. As human populations grow, what might happen to the populations of other species? Discuss some possible effects.
24. The population of a small town is beginning to grow rapidly due to a new highway that provides access to a larger city. Who in this community would be most

Program Resources

Chapter Assessment, pp. 25-30 L1
Alternate Assessment in the Science Classroom
Computer Test Bank L1
Content Mastery, pp. 17-20 L1

interested in the changing demographics of the area? How would they use information supplied to them by demographers to provide services for the growing population?

25. How are the energy resources of fossil fuels, such as petroleum and coal, affected by the increasing world population?

THINKING CRITICALLY

26. ***Concept Mapping*** Make a concept map that relates the following terms and phrases. Supply the appropriate linking words for your map.
    environmental factors, demography, age structure, emigration, carrying capacity, population growth, exponential growth, predation, crowding, immigration

27. ***Observing and Inferring*** Why are rapid life-history species, such as mosquitoes and some weeds, successful organisms even though they often experience massive population declines?

28. ***Comparing and Contrasting*** Compare and contrast species having slow life histories with those that have rapid life histories.

29. ***Recognizing Cause and Effect*** Predict what might happen to the human population if the supply of food fails to keep up with the population growth.

30. ***Recognizing Cause and Effect*** Predict what might happen to a *Paramecium* population if a predator is added to the culture.

ASSESSING KNOWLEDGE & SKILLS

The following bar graph indicates the number of dandelions counted by five groups of students on school grounds. Assume that the school grounds have 21 950 square meters planted in grass.

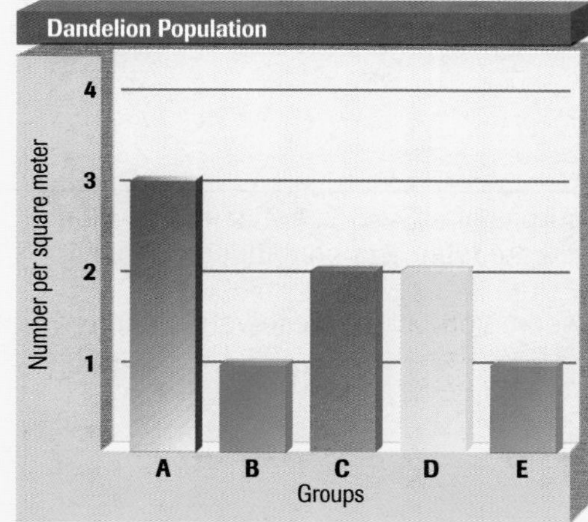

Interpreting Data Use the graph to answer the following questions.

1. How many dandelions per square meter did group A count?
   a. one c. three
   b. two d. four
2. On average, how many dandelions did the students count per square meter?
   a. one c. three
   b. two d. four
3. Using the average number of dandelions per square meter, what is the estimated size of the dandelion population on the school's grounds?
   a. 21 950 c. 50 485
   b. 43 900 d. 65 850
4. ***Interpreting Data*** In another section of the school grounds there is an average of five dandelions per square meter. Now, what is the estimated size of the population?

Chapter 5 Review

25. Fossil fuel demands are increasing, causing these resources to be extracted more rapidly. As they become more scarce, the price will also increase.

Thinking Critically

26. See below.
27. Rapid life-history species provide enough potential offspring in the form of eggs and/or larvae for the next generation even when there are massive die-offs.
28. Species with slow patterns of life history usually have few young that take a long time to mature, while those with rapid life-history patterns have many offspring that mature quickly. Students should also explain differences in the environments and population growth trends of these organisms.
29. Student answers will vary. They may hypothesize that human populations may experience famine, war caused by famine, or other population decreases because of stress or other factors.
30. Student answers will vary. They may hypothesize that the *Paramecium* population will stabilize, decline, or even go extinct.

ASSESSING KNOWLEDGE & SKILLS

1. c
2. b
3. b
4. The population size increased because the average number of dandelions per square meter increased. The total population is now estimated at 50 485.

26.

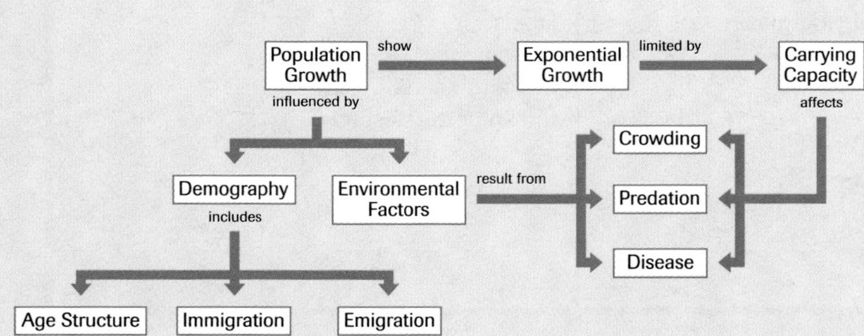

Chapter Organizer

SECTION	OBJECTIVES	ACTIVITIES/FEATURES
6.1 Effects of Human Activities on Our Resources National Science Standards: UCP.1, UCP.3; F.2-6; G.3	1. **Compare and contrast** renewable and nonrenewable resources. 2. **Determine** the effects of increasing demand and decreasing supply on natural resources.	**Biology & Society Technology:** What do we do with millions of tires?, p. 136
6.2 Maintaining the Natural Balance National Science Standards: UCP.1-3; A.1, A.2; B.2, B.3, B.6; D.1; E.1; F.3-6; G.1, G.2	3. **Identify** major sources of air, water, and land pollution. 4. **Differentiate** biodegradable and nonbiodegradable pollutants. 5. **Distinguish** between methods of preservation and conservation.	**Minilab:** Can you observe particulates from car exhaust?, p.141 **Thinking Lab:** How many pesticides are used on our food?, p. 145 **Chemistry Connection:** Separating Plastics, p. 147 **Minilab:** What pollutants do you have at home?, p. 149 **Biolab:** Degrading Time of Packing Materials, p. 150

ACTIVITY MATERIALS

BIOLAB	MINILABS	ALTERNATE LAB
page 150 soil Styrofoam peanuts cornstarch packing material popcorn paper packing material paper garbage bags plastic garbage bags 4 large beakers or clay pots	**page 141** 2 microscope slides petroleum jelly box or bucket microscope or magnifying glass coverslips **page 149** pencil paper	**page 142** plastic bags paper towels small cups labels vinegar mustard seeds radish or grass seed

Chapter 6 Wise Use of Our Resources

TEACHER CLASSROOM RESOURCES

Reproducible Masters	Transparencies
Section Focus Master 11: Earth's Resources L1 SAE **Reinforcement and Study Guide,** pp. 21-22 L1 **Concept Mapping:** Natural Resources, p. 6 L1 **Tech Prep Applications:** Building a Compost Pile, pp. 5-6 L2 **Content Mastery,** pp. 21-24 L1	
Section Focus Master 12: Insecticides: Helpful or Harmful? L1 SAE **Reinforcement and Study Guide,** pp. 23-24 L1 **Biolab and Minilab Worksheets,** pp. 19-22 L1 **Critical Thinking/Problem Solving:** Solid Waste Recycling, p. 6 L3 **Laboratory Manual:** Recycling Garbage, pp. 27-30; How Does Detergent Affect Seed Germination?, pp. 31-34; How Does Ionizing Radiation Affect Plant Growth?, pp. 35-38 L2	**Reteaching Transparency #6:** Groundwater Pollution L1 SAE **Basic Skills Transparency #6a, b, c:** Acid Rain L1 SAE

ASSESSMENT MATERIALS

Chapter Assessment, pp. 31-36
Performance Assessment in the Biology Classroom, p. 55
MindJogger Videoquiz
Alternate Assessment in the Science Classroom
Computer Test Bank

Spanish Resources SAE
English/Spanish Audiocassettes SAE
Cooperative Learning in the Science Classroom COOP LEARN
Lesson Plans
Exploring Environmental Issues: pp. 3, 11, 25, 27, 31, 33, 47 L2

KEY TO TEACHING STRATEGIES

L1 Level 1 activities should be within the ability range of all students including those with learning difficulties.
L2 Level 2 activities are within the ability range of average to above-average students.
L3 Level 3 activities are designed for the ability range of above-average students.
SAE SAE activities should be within the ability range of Students Acquiring English.
COOP LEARN Cooperative Learning activities are designed for small group work.
P These strategies represent student products that can be placed into a best-work portfolio.
These strategies are useful in a block scheduling format.

GLENCOE TECHNOLOGY

The following multimedia resources are available from Glencoe.

Biology: The Dynamics of Life
CD-ROM SAE
Videodisc Program

The Infinite Voyage Series
Life in the Balance
Miracles by Design

National Geographic Society Series
GTV: Planetary Manager
Greenhouse of Eden
Tidy World

Science and Technology Videodisc Series (STVS)
Chemistry
Wind Power
Ecology
Fish and Acid Rain
Hole in the Ozone
Greenhouse Effect

CHAPTER

6 Wise Use of Our Resources

Chapter Overview

Students are introduced to the ways humans use natural resources. Resources are described as being either renewable or nonrenewable. Fossil fuels are an example of nonrenewable resources, and populations are renewable resources.

Problems associated with use of worldwide resources as Earth attempts to maintain an ecological balance are the focus of the second part of the chapter. Air pollution, water pollution, and land use problems are discussed. The chapter suggests how individuals can make better use of resources and preserve Earth and its biodiversity.

Key Terms

acid precipitation
biodegradable
conservation
endangered species
extinction
fossil fuel
greenhouse effect
groundwater
natural resource
nonbiodegradable
nonrenewable resource
ozone layer
particulate
pollution
preservation
renewable resource
smog
threatened species

132

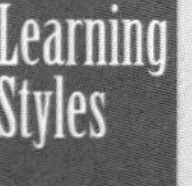

Look for the following logo for strategies that emphasize different learning modalities. LS

Kinesthetic	Meeting Individual Needs, p. 144; Cultural Diversity, p. 149
Visual-Spatial	Meeting Individual Needs, pp. 138, 146; Minilab, p. 141; Enrichment, pp. 142, 148; Portfolio, p. 144; Demonstration, p. 146
Interpersonal	Visual Learning, p. 138; Student Journal, p. 146; Project, p. 148
Intrapersonal	Minilab, p. 149
Logical-Mathematical	Project, pp. 135, 145; Portfolio, p. 137; Math Connection, p. 138; Alternate Lab, p. 142; Thinking Lab, p. 145; Reinforcement, p. 147

LS

CHAPTER

6 Wise Use of Our Resources

All living things have a basic set of needs that must be met. If you were to make a list of basic human needs, what would be on your list? You need food, water, clothing, shelter, and oxygen. How do human needs and demands affect the environment? We burn oil, gas, and coal to fuel power plants and cars and to provide electricity. We dam rivers for irrigation water. We dig deep into the ground to mine metals and minerals. We clear forests to build shopping centers and housing developments.

Does our use of Earth's natural resources always have to be destructive? In this chapter, you'll look more closely at the natural resources needed to support modern human life. You will investigate how humans have influenced Earth's ecosystems, and explore some of the methods we already use to maintain a healthy environment.

Concept Check

You may wish to review the following concepts before studying this chapter.

- Chapter 3: chemical cycling, food web
- Chapter 4: biomes
- Chapter 5: population growth

Chapter Preview

Laboratory Activities

Biolab

- Degrading Time of Packing Materials

Minilabs

- Can you observe particulates from car exhaust?
- What pollutants do you have at home?

Clearing forests for new buildings and mining the earth for minerals, coal, and oil are not the only human activities that have an effect on our natural resources. Many commonly used products are unnecessarily overpackaged. Single-serving packages, layers of plastic wrap, and nonrecyclable materials contribute to an overuse of natural resources. Products packaged in recyclable materials or biodegradable materials place less stress on resources and the environment.

133

Assessment Planner

Choose assessment strategies from the following pages to evaluate the progress of your students.

Assess, pp. 139, 151
Alternate Lab, pp. 142-143
Minilabs, pp. 141, 149
Portfolio, pp. 137, 144, 147
Thinking Lab, p. 145
Biolab, pp. 150-151
Chapter Review, pp. 153-155

CHAPTER 6

Introducing the Chapter

Elicit from students whether they think humans are Earth's most important organisms. Ask students to justify their responses. If students cite intelligence as part of their response, ask if intelligence should allow humans to destroy other organisms or if intelligence should be used to improve quality of life for all organisms. Use responses as a basis for a discussion about ways in which human activities affect the environment.

Theme Development

Energy is a dominant theme of the chapter, particularly as it relates to pollution resulting from fossil fuels. *Homeostasis* becomes a major theme as a correlation is made between increased human population and ecological problems. The point is made that increased resource demands and uses are disrupting the ecological balance, which tends to maintain homeostasis in the biosphere.

Concept Check

In Chapter 4, students learned that biomes are ecological areas that have reached a balance between abiotic and biotic factors. Now students will discover that ecological balance is currently being upset, largely through human activities. To fully understand and appreciate the impact of human activities on ecosystem balance, students will need to recall what they learned about chemical cycles and food webs in Chapter 3 and the factors related to population growth which were discussed in Chapter 5.

SECTION 6.1

Prepare

Key Concepts

Students will begin to explore the importance of the environment in supplying the resources needed for the survival of all organisms. Water, wildlife, and crops are identified as renewable resources. Fossil fuels and minerals that are not replaced or recycled by nature are identified as nonrenewable resources. Students will discover that although organisms are considered natural resources, their numbers may decline until species become extinct.

Block Scheduling

Look for this symbol for strategies that are useful in a block scheduling format. For more information on block scheduling, refer to Section 6.1 in the **Lesson Plans** booklet.

Materials

- none required

1 Focus

Bellringer

Before presenting the lesson, display **Section Focus Master 11** on the overhead projector and have students answer the accompanying questions. L1 SAE

Activity

Exploring Environmental Issues: p. 31, Going, Going, Gone: Endangerment and Extinction; p. 33, Sharing the World with Wildlife. Have students carry out these activities to develop their awareness of problems related to loss of biodiversity. L2

SECTION 6.1

Effects of Human Activities on Our Resources

Section Preview

Objectives

Compare and contrast renewable and nonrenewable resources.

Determine the effects of increasing demand and decreasing supply on natural resources.

Key Terms

natural resource
renewable resource
nonrenewable resource
fossil fuel
extinction
threatened species
endangered species

The sight of a quiet, misty lake, a snow-capped mountain, the geraniums growing in a flower box on your windowsill, or a squirrel chewing on a pinecone in the park usually make people stop to admire the natural resources of Earth. Our survival and the survival of all living things on Earth is dependent on our wise use of these resources.

Earth's Resources

Whenever you ride a bus, turn on a light, read a book, or eat lunch, you are using a **natural resource.** A natural resource is any part of the natural environment used by humans for their benefit. Natural resources include soil, water, crops, wildlife, oil, gas, and minerals, as shown in *Figure 6.1.* The gasoline that powers a bus or automobile, and the plastics and metals the vehicle is made of, are examples of natural resources. So is the wood pulp used to make the paper for this book, and the coal or natural gas that supplied the electricity for the printing press. Wilderness and recreation areas such as beaches, mountains, forests, and lakes are also natural resources.

Renewable resources

Why don't we run out of oxygen to breathe or water to drink? Many organisms produce oxygen during a process called photosynthesis, which constantly replenishes the oxygen consumed by all aerobic organisms. The physical materials on which life depends are limited to what is presently on Earth. You learned in Chapter 3 that water is naturally recycled from the atmosphere to the surface of Earth, through food webs and back to the atmosphere. Nitrogen, carbon, and other essential substances are cycled in a similar manner. A natural resource that is replaced or recycled by natural processes is called a **renewable resource.** Other examples of renewable resources include plants, animals, food crops, sunlight, and soil.

Program Resources

Section Focus Master 11 L1 SAE
Reinforcement and Study Guide, pp. 21-22 L1
Exploring Environmental Issues, pp. 31, 33 L2
Tech Prep Applications, pp. 5-6 L2
Concept Mapping, p. 6 L1

Meeting Individual Needs

Behaviorally Disordered Have students work in groups to categorize a list of resources you provide as renewable resources or nonrenewable resources. Include resources not given in the text such as wind, flowing water, sunlight, grasses, and forests *(all renewable),* along with uranium, clay, and glass *(all nonrenewable).* L1 COOP LEARN

Figure 6.1

Different regions are rich in different types of natural resources.

Corn you eat for dinner may have been grown on a farm in Kansas.

Wood used to build a house in Iowa may come from the forests of the Pacific Northwest.

If you live in Los Angeles or some other part of the arid Southwest, the water that runs from your kitchen faucet may come from a river hundreds of miles away.

Many of the resources you use every day are imported from outside the United States. For example, the aluminum in your soft drink can may have come from a mine on the island of Jamaica.

Nonrenewable resources

What does the aluminum in a soda can have in common with the plastic used to make grocery bags or the gasoline that fuels a car? These are examples of nonrenewable resources. A **nonrenewable resource** is available only in limited amounts and is not replaced or recycled by natural processes. Metals—including aluminum, tin, iron, silver, uranium, gold, and even the copper used to make wires and coins—are nonrenewable. Some minerals such as phosphorus, which is essential for plant growth, are recycled so slowly in the natural environment that they are considered nonrenewable. It is estimated to take between 500 and 1000 years for a 2.5-cm-deep layer of topsoil to develop from decomposed plant material. Because it would take several generations to replace, topsoil is also a nonrenewable resource.

Fossil fuels and petroleum products

You are probably using a nonrenewable resource whenever you watch TV or use some other electrical appliance. Electricity is produced in power plants, many of which are fueled by coal, oil, or natural gas.

Activity

Have students list all the items they used or relied upon today. Emphasize that the key term they should be thinking about is *relied upon.* Suggest that students divide the items they list into categories such as food, transportation, clothing, school items, chemicals, and household items.

After students read the paragraph about renewable resources, ask them to place a check mark beside items on their list they believe are renewable resources. Discuss the responses to provide concrete examples of the types of materials that are classified as renewable resources. L1 SAE COOP LEARN

2 Teach

Reinforcement

Briefly review the processes that contribute to the recycling of nitrogen, carbon, and oxygen. Ask students to: (a) Name the types of organisms responsible for decomposition. *Fungi and bacteria* (b) Explain why decomposers break down matter. *They obtain their energy from the chemicals released from organic matter.*

Enrichment

Advise students about the role of phosphorus in DNA, RNA, and cell membranes. Explain the roles of DNA and RNA and identify where these molecules are located in cells. Explain that plants obtain phosphorus from phosphate released by weathering rocks.

Brainstorming

Elicit from students how electricity can be generated from renewable resources. *Solar, gravitational-tidal power, and wind.* Ask students to explain why nuclear power is nonrenewable. *It relies on uranium which is a nonrenewable resource.*

PROJECT

Plant Growth Nutrients

Have students experiment to determine the importance of phosphorus as a plant growth nutrient. Students may carry out their experiments on a macro-level using a young plant such as *Coleus* or beans grown from seed, or they may work at a micro-level using *Euglena* cultures in test tubes. L2 LS

Technology
BIOLOGY & SOCIETY

What do we do with millions of tires?

Purpose

Students will discover how an item commonly discarded as garbage can be used as an energy source.

Teaching Strategies

- Have students research the source of rubber. Ask students to prepare reports that answer the following questions: (a) What process originally trapped energy in rubber plants? *Photosynthesis* (b) What process originally trapped energy in fossil fuels? *Photosynthesis* Have students relay to the class similarities among rubber, coal, and oil. L2

INVESTIGATING the Technology

Applying Concepts

Responses may include use in artificial reefs and as barriers against ocean erosion. Remove the chemicals that contribute most to pollution prior to burning.

Going Further

Have students research the chemical composition of tires and identify which chemicals are synthetically and naturally derived.

BIOLOGY & SOCIETY **Technology**

What do we do with millions of tires?

They don't decay, are practically indestructible, are piled up in heaps six stories high, and can be found submerged in ponds. Of the 240 million tires discarded in the U.S. every year, about 11 percent are burned as fuel and 15 percent are retreaded. The remainder are simply dumped. Each tire has the energy potential of 2.5 gallons of oil. This potential became obvious when a tire dump in Virginia burned for 275 days. The 690 000 gallons of oil that oozed out of the burning tires were collected and sold.

Tires as fuel In 1982, a tire-collection company realized the energy potential in that 2.5 gallons of oil and began selling tire chips to paper mills to be burned as fuel. Even though tires produce about the same polluting emissions as coal, they contain more energy than the same weight of bituminous coal. Whole or shredded tires are ignited at high temperatures, and the heat from their combustion is used to produce steam that powers turbines. The process is controlled by computers, which also monitor emissions. The ash components are used as gravel and to enrich soil.

Applications for the Future

Tires are also used as a source of direct heat in the manufacture of cement. Burning tires circulate with the rock while supplying heat for the roasting process. While only a few cement plants currently use tires, the process has high economic potential because the cement-producing industry could use 3 billion tires per year in this way, with no significant change in emissions.

INVESTIGATING the Technology

Applying Concepts What other uses might there be for waste tires besides using them for fuel? What steps might be taken to make using tires for fuel more economically and environmentally acceptable?

Figure 6.2

Fossil fuel deposits are always being formed, but humans are consuming them far faster than they can be replaced.

These **fossil fuels** are substances made up of the remains of organisms that have been buried underground for millions of years. Fossil fuels are nonrenewable resources because they form only over long spans of time. *Figure 6.2* shows the mining of coal.

Extinct organisms

A living population of organisms is a renewable resource; new individuals are born as older ones die. But what happens when all members of a species die? No individuals are left to reproduce, and the species is lost forever. Have you ever seen a flock of passenger pigeons? How about a woodland caribou, relic leopard frog, or Louisiana prairie vole? Unless you've seen a photograph or a stuffed museum specimen, your answer will be no. These animals are extinct. **Extinction** is the disappearance of a species when the last of its members dies. In the last 20 years, almost 30 species of plants and animals living in the United States have become extinct.

PROJECT Strip-Mining

Have students contact a coal mining company that practices strip-mining. Have students inquire about how this type of mining differs from the more traditional deep-tunnel mining and find out what the company does to restore the mined land to a usable natural setting. L2

Although extinction can occur as a result of natural processes, humans have been responsible for the extinction of many species.

Most extinctions come about because of destruction of the natural habitat of a species. Because the environment is always changing, there will always be species in danger of extinction. However, when you look at the increasing number of species that are becoming extinct every year, an unpleasant picture emerges. It has been estimated that, during the extinction of the dinosaurs, about one species was lost every year. Some scientists hypothesize that today, species are being lost at the rate of one per day. Human activities—including hunting, the building of cities and housing developments, and the destruction of forests to create farmland—are primarily responsible for this high extinction rate.

When the population of a species begins declining rapidly, the species is said to be a **threatened species.** African elephants, for example, are listed as a threatened species. In the early 1970s, the wild elephant population numbered about 3 million. Twenty years later, it numbered only about 700 000. Elephants have traditionally been hunted for their ivory tusks, which are used to make jewelry and ornamental carvings. Many countries have banned the importation and sale of ivory, and this has helped slow the decline of the elephant population. *Figure 6.3* shows some threatened animal species under protection in the United States.

Figure 6.3

Several plant and animal species in the United States are threatened.

The bald eagle was endangered in 44 states. As a result of conservation efforts, its status has been improved to a threatened species in all states except Alaska.

Wildlife experts classify loggerhead turtles as well as more than 40 other turtle species as threatened.

Sea otters have been hunted for centuries for their fur.

Grizzly bears are threatened in every state in the United States except Alaska.

Bioethics

Preserving Desert Habitats

The United States government is considering the purchase of 7 million hectares of desert land located near the Joshua Tree National Monument of southern California as part of the California Desert Protection Act. The purchase would protect the habitats of plants and animals in the region that are threatened with extinction due to habitat destruction. The attitude of some politicians toward this purchase seems to be, "Why preserve the desert habitat of rattlesnakes, tarantulas, and insects?" The issue becomes more heated when it is recognized that the purchase would eliminate use of the site for mining of a rare metal ore and ranching. The price tag for this purchase is definitely high. The question is: Is the price worth it?

Science, Technology, and Society

Technology and Biodiversity

Societal growth and advances in technology have taken their toll on biodiversity. To help solve this problem, the Endangered Species Act of 1973 empowered the United States Fish and Wildlife Service (FWS) to identify and list all endangered and threatened land and freshwater species of the United States.

Between 1970 and 1990, 500 new species were listed. Many more species are being considered for addition to the list. Unfortunately, in the time the FWS has been at work, more than 34 species have already disappeared.

GLENCOE TECHNOLOGY

Videotape

The Secret of Life

Use the videotape *Gone Before You Know It: The Biodiversity Crisis* to illustrate the effects human activities are having on biodiversity.

PORTFOLIO

Observing Habitat Destruction Have students record direct evidence they observe in their area that serves as an example of habitat destruction brought on by humans. Have students describe how the conditions they observe might be contributing to the reduction in numbers of some native plant or animal species. L1 SAE LS P

Math Connection

Doubling Time

Logical-Mathematical Doubling time is the time needed in years for a population to double in size. Doubling time is calculated using the following formula:

70 ÷ growth rate (GR) % = doubling time (DT) (in years)

Using the formula above, the world's doubling time, with the growth rate at 1.8%, is about 39 years. Have students use the formula to calculate the doubling time for the following regions.

Region	%	DT
Africa	2.9	*24*
Latin America	2.1	*33*
Asia	1.8	*39*
United States	0.7	*100*
Europe	0.3	*233*

Visual Learning

Figure 6.4 Ask students to prepare a table that describes the following information for each organism shown: biome of organism, area of United States where organism lives, and possible cause for the organism's endangered status.

Figure 6.4

A species becomes endangered when its numbers are so low that it is in danger of extinction. In the United States, scientists have developed programs designed to save endangered species.

A species is considered to be an **endangered species** when its numbers become so low that extinction is possible. In Africa, the black rhinoceros has become an endangered species. Like the African elephant, the black rhino is protected in national parks and preserves in several African countries. However, poachers continue to hunt and kill these animals for their horns. Rhinoceros horns are composed of matted hair rather than bone or ivory. In the Middle East, the horns are carved into handles for ceremonial daggers. In parts of Asia, they are used to make traditional medicines. *Figure 6.4* shows several endangered species found in the United States. In the United States, endangered species receive protection under the law. They cannot be hunted or trapped, and their habitats are protected.

Manatees, sometimes called "sea cows," are endangered due to loss of habitat and injury from barges and motorboats.

In 1850, approximately 20 million bison roamed the western plains. Bison were nearly exterminated by hunting, but today, approximately 80 000 of these animals live on fenced game preserves.

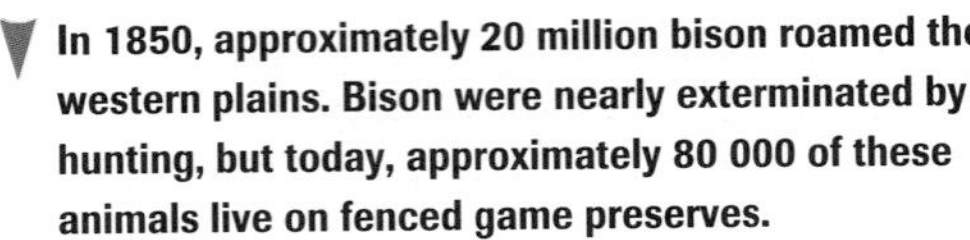

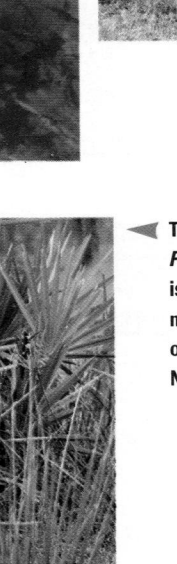

The Florida panther, *Felix concolor coryi*, is a subspecies of the mountain lion, which once roamed most of North America.

Urban growth has destroyed much of the California condor's habitat. To protect the species, all California condors were captured.

Meeting Individual Needs

Hearing Impaired Using the analogy of a teeter-totter, illustrate how the balance between human population and resources can change. Use diagrams that show the human population "down" with natural resources "up." Use additional drawings to indicate what happens as population increases and natural resources decrease. SAE

1. A natural resource that is replaced or recycled is renewable; a nonrenewable resource is available in limited amounts and is not replaced or recycled.
2. Recyclable, nonrenewable resources are capable of being reused and are thus like renewable resources.

When Demand Exceeds Supply

In Chapter 4, you learned that the human population is growing at an increasingly rapid pace. The more people that live on Earth, the more demand there is for food, water, living space, clothing, transportation, and other essentials. How long will Earth's resources last? At some point, will we start running out of the things we need to live?

When demand for a resource exceeds the available supply, competition for that resource increases. As a result, the cost of that resource goes up. For example, the price of housing in prosperous metropolitan areas has increased drastically in the last 30 years. A house in Levittown, New York, a suburb of New York City, that cost $8000 when it was built in 1950 is now valued at $200 000. A similar home in a small, midwestern city might cost about half of that. Increasing demand for resources means not only higher prices; it may also mean that people simply have to do without, as shown in *Figure 6.5*.

Figure 6.5

The demand for resources in overpopulated areas can have drastic effects on people's lives.

The need for living space so far exceeds supply in some parts of the world that portions of the population live in makeshift homes of cardboard, plastic, or other discarded materials. Some people have no homes at all.

In parts of Africa, drought and political strife have created food shortages so severe that thousands of people have starved.

Section Review

Understanding Concepts

1. What is the difference between a renewable and a nonrenewable resource?
2. How are renewable resources similar to recyclable, nonrenewable resources?
3. Explain some possible effects of increasing demand and decreasing supply of a nonrenewable resource.

Thinking Critically

4. Many people value elephant tusks as decorative objects in their homes. How would this affect the growth rate of the endangered elephant population in Africa?

Skill Review

5. **Observing and Inferring** Infer what might happen to the price of an electrical component if a new metal that is abundant is substituted for a rare metal that has been used exclusively in the manufacture of the electrical component. For more help, refer to Thinking Critically in the *Skill Handbook*.

SECTION 6.1

3 Assess

Check for Understanding

Ask students to evaluate the impact the following activities have on natural resources:

(a) increasing demand for electricity and fuel,

(b) increasing demand for housing and business space. L1 SAE

Reteach

Have students explain the relationship between the words in each of the following pairs.

(a) renewable resources—nonrenewable resources

(b) threatened species—endangered species

(c) fossil fuels—nonrenewable resources L1 SAE

Extension

Have students explore how fossil fuels such as coal and oil form. Ask students to create a visual device or written report that explains how fossil fuels form. Make sure students include information about the role and importance of plants, sunlight, and photosynthesis in their reports. L2

Assessment

Knowledge: Ask students to construct a concept map that includes these terms: *natural resource, renewable resource, nonrenewable resource, fossil fuel, food crops, oxygen, aluminum, coal, oil.* L1 SAE

4 Close

Discussion

Hold a class discussion about methods by which individuals can work to conserve natural resources. Discuss programs and projects that can be undertaken by groups and communities to meet the same goal.

Answers to Section Review

3. Increasing demand may result in an increase in cost, a shortage, or an eventual disappearance of the nonrenewable resource.

Thinking Critically

4. This could reduce the growth rate even more due to an increased demand for the ivory.

Skill Review

5. The price of the electrical component should drop. As supply increases, cost should decrease.

SECTION

6.2

Prepare

Key Concepts

Students will study ways in which humans pollute the environment and the global implications of problems resulting from pollution. Emphasis is given to air pollution problems and how these problems affect populations. Programs aimed at conserving natural areas and threatened and endangered species are also discussed.

Block Scheduling

Look for this symbol for strategies that are useful in a block scheduling format. For more information on block scheduling, refer to Section 6.2 in the **Lesson Plans** booklet.

Materials

- Obtain microscope slides, coverslips, and petroleum jelly for the Minilab.
- Gather soil, Styrofoam peanuts, cornstarch packing material, popcorn, paper and plastic garbage bags, newspaper, and large beakers or clay pots for the Biolab.

1 Focus

Bellringer

Before presenting the lesson, display **Section Focus Master 12** on the overhead projector and have students answer the accompanying questions. L1 SAE

Discussion

Ask students to list examples of how they have personally contributed to the pollution of the environment. Collect the lists and read some of the responses. As a class, discuss ways in which some of these means of pollution might be eliminated or reduced. COOP LEARN

SECTION 6.2

Maintaining the Natural Balance

Section Preview

Objectives

Identify major sources of air, water, and land pollution.

Differentiate between biodegradable and nonbiodegradable pollutants.

Distinguish between methods of preservation and conservation.

Key Terms

pollution
particulate
smog
acid precipitation
ozone layer
greenhouse effect
groundwater
biodegradable
nonbiodegradable
preservation
conservation

All living things produce waste, but humans produce a huge variety of wastes from many different activities. The wastes produced are solids, liquids, and gases and come from homes, factories, automobiles, and numerous other sources. The average American produces so much solid waste each year that we are running out of places to put it.

Effects of Pollution on the Biosphere

All living organisms produce waste products, which are usually recycled by natural processes. What happens when so much waste is produced that natural recycling processes are overloaded? Wastes build up faster than they can be broken down. **Pollution** is the contamination of any part of the environment—air, water, or land—by an excess of waste materials. A pollutant is a waste product that causes pollution. *Figure 6.6* describes how nitrogen, a nutrient essential to all living organisms, can become a pollutant.

Figure 6.6

Cattle manure contains nitrogen and other nutrients that make it valuable as a plant fertilizer. But too much of a good thing can cause pollution problems.

The cattle on this crowded feed lot produce more waste than the decomposers in the soil can handle. As a result, much of the waste is washed away by rainfall.

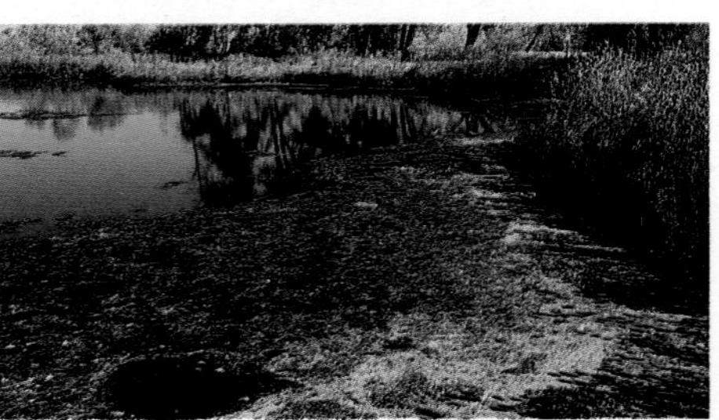

The large amounts of nitrogen in runoff from the feed lot stimulate the rapid growth of algae in waterways downstream. This lush growth of algae consumes all or most of the oxygen in the water, making it impossible for insects, fish, and other animals to live there.

Program Resources

Section Focus Master 12 L1 SAE
Reinforcement and Study Guide, pp. 23-24 L1
Performance Assessment in the Biology Classroom, p. 55 L1 P
Exploring Environmental Issues, pp. 3, 11, 25, 27 L2
Biolab and Minilab Worksheets, pp. 19-22 L1
Critical Thinking/Problem Solving, p. 6 L3
Laboratory Manual, pp. 27-30, 31-34, 35-38 L2
Basic Skills Transparencies 6a, 6b, 6c and **Masters** L1 SAE
Reteaching Transparency 6 and **Master** L1 SAE

Air pollution

Although pollutants can enter the atmosphere in many ways—including volcanic eruptions, forest fires, and the evaporation of volatile chemicals—the burning of fossil fuels is by far the greatest source of air pollution problems.

Figure 6.7

***Smog* is a term first used in the early 1900s to describe the mixture of smoke and fog that often covered industrial England.**

Take a moment to think about some of the reasons why humans burn fuel: to heat homes and businesses; to run plane, car, bus, and train engines; to produce electricity in power plants; and to drive manufacturing and industrial processes of all kinds. The smoke released by burning fuels contains gases and **particulates,** which are solid particles of soot that can harm living organisms directly, or change the environment in ways that are later harmful to life. Some major air pollutants released into the air by fossil fuel combustion are carbon monoxide, carbon dioxide, nitrogen oxides, sulfur oxides, and hydrocarbons. Some of these chemicals interact to form smog. **Smog** is a form of air pollution that hangs over many of the world's large cities, *Figure 6.7.* It consists mostly of particulates, sulfur dioxide, and other chemicals.

MiniLab

Can you observe particulates from car exhaust?

Particulates reduce visibility, and so are the most apparent form of air pollution. You can measure and compare the amounts of particulates in air and a car's exhaust without much trouble. For instance, you may find dust, ash, soot, lint, smoke, pollen, and other particulates suspended in air. Human activity produces approximately 100 million tons of particulates per year.

Procedure

1. Obtain two microscope slides, and coat one side of each slide lightly with petroleum jelly.
2. Place one of the slides with the petroleum side up on a table or window ledge for several hours.
3. Have an adult start a car engine in an open area, not a garage. While the engine is idling, place the other microscope slide about six inches from the exhaust stream for 30 s. Set the slide on a bucket or box near the tailpipe. Move away while the slide is exposed. Wait until the engine is turned off before retrieving your slide.

CAUTION: *Car exhaust is hot and poisonous. Do not set the slide so close that you burn yourself.*

4. Cover both slides, and keep them covered until you observe them with a microscope or magnifying glass.

Analysis

1. What was the difference between what you observed on the two microscope slides?
2. Suggest what the particulates on the slide might be.
3. How can air pollutants from cars be reduced?

GLENCOE TECHNOLOGY

Videodisc

The Infinite Voyage: Life in the Balance

Southern California's Chaparral: Extinction by Development (Ch. 6)

Biodiversity: Battling Extinction in the Costa Rican Dry Forest (Ch. 7)

2 Teach

MiniLab

Purpose

LS **Visual-Spatial** Students will observe particulate matter released from the burning of fuel in an automobile.

Process Skills

observe and infer, interpret data

Safety Precautions

Automobile exhaust contains carbon monoxide gas and other harmful substances. As a safety precaution, all slides requiring auto exhaust should be prepared by the instructor.

Teaching Strategies

- The entire activity can be demonstrated using a video camera/microscope assembly.

Expected Results

The slide from the windowsill may contain small particulates such as dust and pollen. The auto exhaust slide will contain large particles of carbon.

Analysis

1. The slide receiving exhaust had large black particles. The other slide had small particles of dust.
2. Particulate from exhaust contains carbon.
3. Reduce automobile use, improve the efficiency of auto engines, provide a filtering system for exhaust.

Assessment

Portfolio: Have students write a report of this Minilab for their portfolios. P

Enrichment

LS Visual-Spatial Teach students how to use pH paper and then ask them to determine the pH of several common substances. Suggested substances include lemon juice, tap water, distilled water, water plus detergent, vinegar, tomato juice, and window cleaner. L1

Demonstration

Determine the pH of distilled water using pH paper. Add an antacid tablet to a test tube of water. Place a stopper that has a plastic tube extending from it into the test tube with the antacid. Allow the gas generated in the tube (carbon dioxide) to bubble through a tube of distilled water. Check the pH of the tube that received the gas bubbles. Explain that this demonstration shows what happens when water and carbon dioxide join to form carbonic acid.

Visual Learning

Figure 6.8 Which is more acidic: lemon juice or saliva? *Lemon juice* Which solutions are bases? *Ammonia, milk of magnesia, seawater*

Audiovisual

Show the video *Acid Rain*, Scott Resources.

Show the two-part video *After the Warming: Episode 1 and Episode 2*, Full Motion.

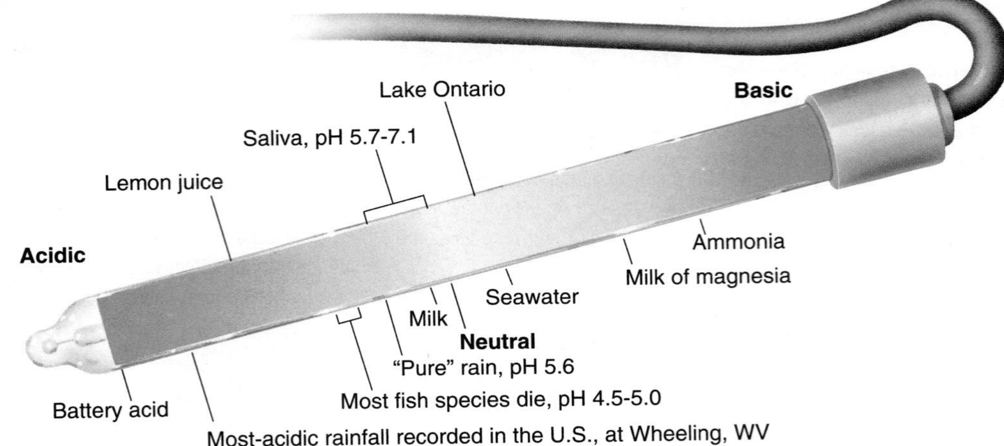

Figure 6.8

The pH scale indicates whether a solution is an acid or a base. Which is more acidic: lemon juice or rainwater? Which solutions are bases?

Acid precipitation

The atmosphere contains moisture, in the form of water vapor, which condenses and returns to Earth's surface as rain, snow, and other forms of precipitation. While in the atmosphere, water molecules can come into contact with air pollutants.

Carbon dioxide gas dissolves into water droplets as they form into rain, producing weak carbonic acid. As a result, rain normally is slightly acidic. But air pollutants, particularly sulfur dioxide and nitrogen oxides, increase the acidity of rain. In the presence of sunlight, these pollutants react with water and oxygen in the air to form sulfuric acid and nitric acid. The **acid precipitation** that results is rain or snow that is more acidic than unpolluted rainwater. Sulfur is released primarily by coal-burning factories and power plants. The major source of nitrogen oxides is automobile exhaust.

The acidity of normal rain, acid rain, and some common substances is compared in *Figure 6.8*. In some cities and heavily industrialized areas, the amount of these pollutants released into the air is so great that the rain or snow may become as acidic as vinegar. Even fog and dew can become acidic as a result of air pollution. Acidity is measured in units called pH. The pH scale ranges from 0 to 14; the lower the pH number, the more acid the substance. Nonpolluted rainfall has a pH of about 5.6 to 5.7. Precipitation with a pH below 5.6 is considered acidic. Most of the rain that now falls in the northeastern portion of the United States has a pH between 4.0 and 4.5. Rain with a pH as low as 1.9 has even been detected.

Effects of acid precipitation

Acid precipitation leaches calcium, potassium, and other valuable nutrients from the soil. As these substances are washed away, the soil becomes less fertile. This loss of nutrients can lead to the death of trees, especially conifers. Acid rain also damages plant tissues and interferes with their growth and nitrogen fixation. Many trees in the forests of the northeastern United States, as well as in many European countries where the concentration of industry is high, are dying as a result of acid rain.

Acid precipitation also has severe effects on lake ecosystems. Acid rain falling into a lake, or entering it as

Alternate Lab: Acids and Seed Germination

Purpose

LS Logical-Mathematical To illustrate the effect of acid precipitation on seed germination.

Materials

plastic bags, paper toweling, small cups, labels, water, vinegar, small seeds

Procedure

Give the following directions to students.

1. Place 20 seeds in a small labeled cup of water and 20 seeds in a small labeled cup of vinegar. **Caution:** *Wear goggles while handling and pouring vinegar.*
2. Allow the seeds to remain in the liquids overnight.
3. The next day, remove seeds from the water. Wrap the seeds in a paper towel. Moisten the towel with water and slide the toweling into a self-sealing plastic bag. Label the bag with your name, date, and the word *water.*
4. Repeat step 3 for the seeds in vinegar. Moisten the towel with vinegar and label the bag with your name, date, and the word *vinegar.*
5. Prepare a data table to record your observations. Note and record the

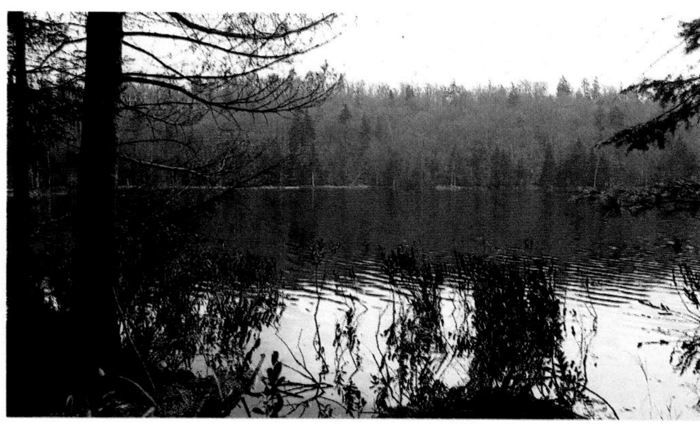

Figure 6.9

Acid precipitation has damaging effects on trees, organisms in affected lakes, and even on stone buildings and statues. This lake in Adirondack State Park shows the effects of acid precipitation.

runoff from streams or neighboring slopes, causes the pH of the lake water to fall below normal. The excess acidity can severely disrupt the lake ecosystem. *Figure 6.9* shows damage due to acid precipitation.

Ozone depletion

The atmosphere contains a sort of sunscreen—known as the **ozone layer**—that prevents living organisms on Earth's surface from receiving lethal doses of ultraviolet radiation. Ozone is a molecule made up of three oxygen atoms. Too near to Earth's surface, it is an air pollutant that causes lung damage and contributes to smog formation. But a thin layer of ozone located high in the stratosphere layer of the atmosphere, shown in *Figure 6.10,* absorbs most of the sun's harmful radiation.

Figure 6.10

The atmosphere extends several miles above Earth's surface and is made up of four layers. The layer closest to the surface, the troposphere, contains the air we breathe. The next layer is the stratosphere. At the top of the stratosphere lies the ozone layer. CFC molecules gradually rise up into the stratosphere, where they are exposed to ultraviolet (UV) light. The absorption of UV light releases chlorine atoms, which react with and destroy ozone molecules.

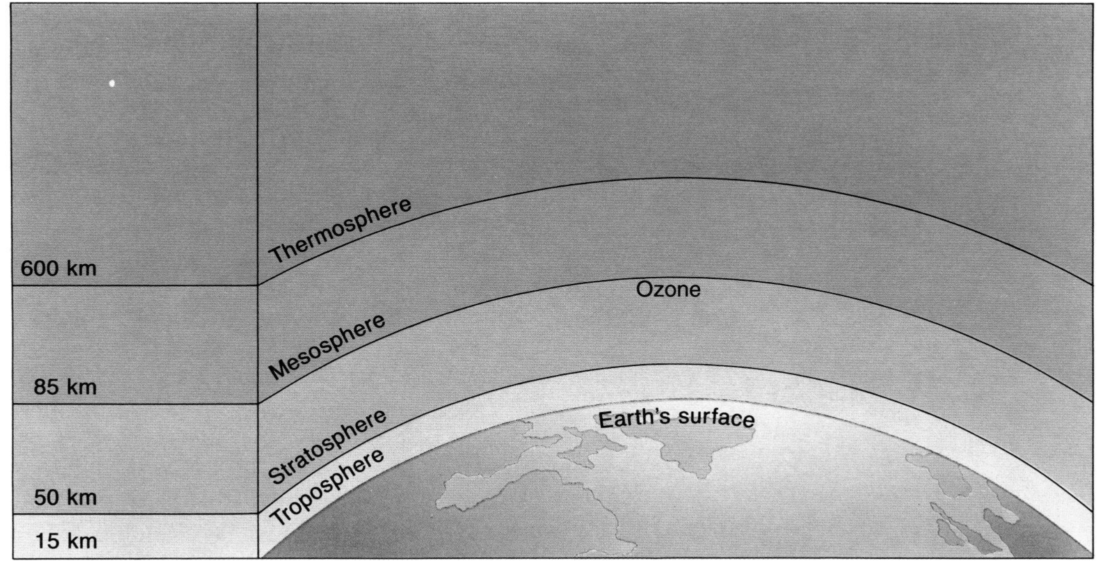

SECTION 6.2

Misconception

Students may think that tap water has a pH of 7, but tap water is usually slightly acidic. Illustrate this point by using pH paper to test the tap water in your area. Explain that the pH of tap water is affected by chlorine or other additives used in water purification and by contaminants located in the pipes.

Chalkboard Example

Show students the structural formula for an oxygen atom (O), an oxygen gas molecule (O-O), and an ozone molecule (O-O-O). Ask students to predict whether the chemical properties of all three examples are the same. *They are not.*

GLENCOE TECHNOLOGY

Videodisc

Biology: The Dynamics of Life
Disc 1, Side 1
The Atmosphere (Ch. 14)

STVS: Ecology
Disc 6, Side 2
Fish and Acid Rain (Ch. 11)

number of seeds that germinate in the next four days. Germinating seeds show a root growing from the seed.

Analysis

Ask students the following questions.

1. Compare the number of germinated seeds after 4 days. How do the water-soaked and vinegar-soaked seeds compare? *Less germination in vinegar-soaked seeds*
2. Vinegar is an acid. Explain how acid precipitation might affect seed germination. *Acid precipitation could delay or prevent seed germination.*

Assessment

Skill: Design an experiment that determines the tolerance limit for mustard seed germination in vinegar. Conduct the experiment if time permits.

Different Viewpoints in Biology

Ozone and CFCs

The link between ozone depletion and CFC use has been established. The guilty atom within the CFC molecule is chlorine. It seems obvious that an alternative chemical to CFCs should not contain chlorine. However, the elimination of chlorine from CFC replacement molecules has not yet occurred.

The chemicals that now replace CFCs are hydrochlorofluorocarbons, or HCFCs. Although HCFCs contain chlorine, they are less ozone depleting than CFCs because they break down in the lower atmosphere. HCFCs do, however, contribute to some ozone depletion.

Hydrofluorocarbons, HFCs, are another alternative to CFCs. HFCs do not contribute to ozone depletion because they do not contain chlorine. Although HFCs may reduce the ozone depletion problem, both HFCs and HCFCs contribute to possible global warming by trapping infrared radiation reflected from Earth's surface. Thus, one problem is partly solved at the expense of another.

Discussion Question

Ask students to (a) identify a major source of carbon dioxide in the air. *Burning of fossil fuels* (b) Explain why the increased concentration of this gas is implicated in raising global temperatures. *It is a greenhouse gas, which means that the more of it present, the higher the air temperatures will go.*

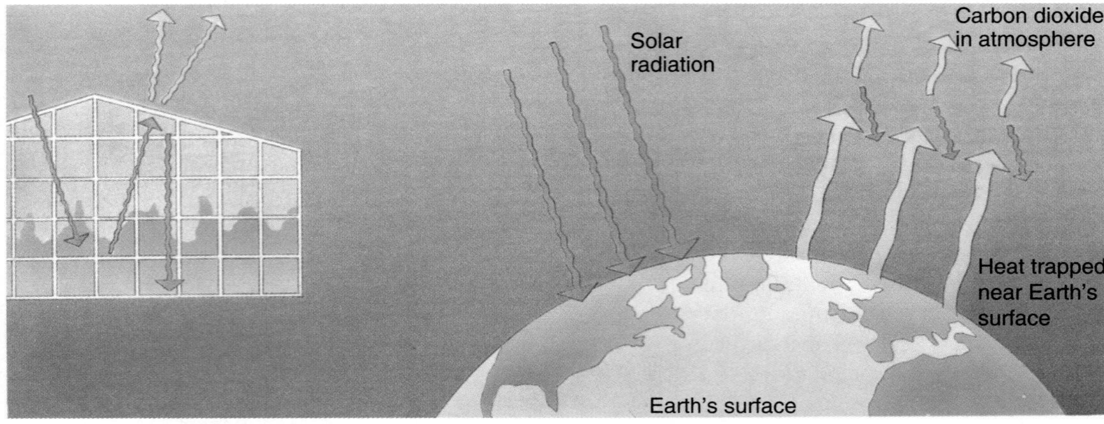

Figure 6.11

The greenhouse effect is a normal phenomenon and a major reason why life has been able to survive on Earth. Carbon dioxide and other atmospheric gases, including CFCs and nitrous oxide, prevent heat from escaping into space. Otherwise, Earth would probably be a cold, lifeless planet.

Scientists have recently discovered that this protective layer of the atmosphere is becoming thinner. This depletion of the ozone layer is caused by a number of air pollutants that are not necessarily associated with fossil fuel combustion. The primary culprits are chlorofluorocarbons, or CFCs. These chemicals do not occur naturally on Earth, but are manufactured for a variety of purposes. They are used as coolants in refrigerators and air conditioners, and in the production of Styrofoam.

The ozone layer is extremely important. Without it, large amounts of ultraviolet radiation could reach the surface of Earth and cause significant damage to living organisms. As a result, governments around the world have agreed to eliminate the manufacture of CFCs by the year 2000. Efforts are underway to regulate the recycling of existing CFCs rather than releasing them into the atmosphere.

The greenhouse effect

What happens to the air inside a closed car that's been parked in the sun for a few hours? The radiant energy of sunlight heats the air inside the car, making it much warmer than the air outside. The glass of the car windows, like the glass walls of a greenhouse, keeps much of that heat trapped inside.

Similarly, gases in the atmosphere trap much of the radiant energy from the sun that reaches the surface of Earth. The land, water, and all the things on Earth absorb the sun's energy. These warmed objects radiate heat energy back into space. However, the atmosphere prevents a good part of this heat from escaping back into space. This process of heat retention by atmospheric gases is called the **greenhouse effect.** Without the greenhouse effect, all of the sun's energy would be radiated back into space, and Earth would be too cold for living things to survive and evolve. Gases that contribute to the greenhouse effect are called greenhouse gases and include carbon dioxide, as shown in *Figure 6.11.*

Water pollution

Water covers more than two-thirds of Earth's surface. Only about three percent of it is fresh water, and most of that is frozen in the polar ice caps. The fresh water of our lakes, rivers, and underground wells represents only about 0.1 percent of all the

Meeting Individual Needs

Visually Impaired Provide students who are visually impaired with "marshmallow molecules" to simulate the differences between an oxygen atom, an oxygen molecule, and an ozone molecule. Join marshmallows with short lengths of toothpicks. Allow students to feel the differences among these three structures. SAE IN

PORTFOLIO

The Life of a Chlorine Atom Have students diagram the various steps that occur as light brings about the release of chlorine from CFCs and the subsequent chain reaction that occurs in the destruction of ozone molecules. Have students use library references or chemistry textbooks to find any information they may need. L2 IN P

water on Earth. This is the water supply that plants, animals, humans, and other organisms depend on. It's also the water supply humans use to carry away many of our wastes. Without adequate care, those wastes can make the water unsuitable for other uses, such as drinking.

Have you ever heard travelers use the phrase, "Don't drink the water"? In many parts of the world, pure drinking water is not available. The same rivers and streams in which sewage and industrial wastes are discharged also serve as the source for drinking and washing. Waterborne diseases such as cholera, dysentery, and hepatitis are common in these areas. Other major pollutants of water include inorganic substances such as fertilizers that provide excessive amounts of nutrients, sediments including silt and small particles from runoff, and heat from warm or hot water that has been used to cool machinery in power plants and factories. *Figure 6.12* shows the causes of two forms of water pollution.

ThinkingLab Draw a Conclusion

How many pesticides are used on our food?

In July of 1993, reports were filed by the National Research Council (NRC) and the Environmental Working Group citing problems that even minute amounts of pesticides pose for young children. The NRC study found that it is difficult to set safety standards without valid information on what infants and children actually eat or what constitutes safe limits for the chemicals. The Environmental Protection Agency sets limits of pesticide residues that may be found on fruits and vegetables, but the NRC suggests that the limits should be redefined at lower levels.

Analysis

The chart shows ten types of produce samples tested for pesticides. For each fruit or vegetable, the percent that contained pesticides is shown. The chart also shows the number of types of pesticides detected on the ten kinds of produce. Compare the numbers of pesticides detected.

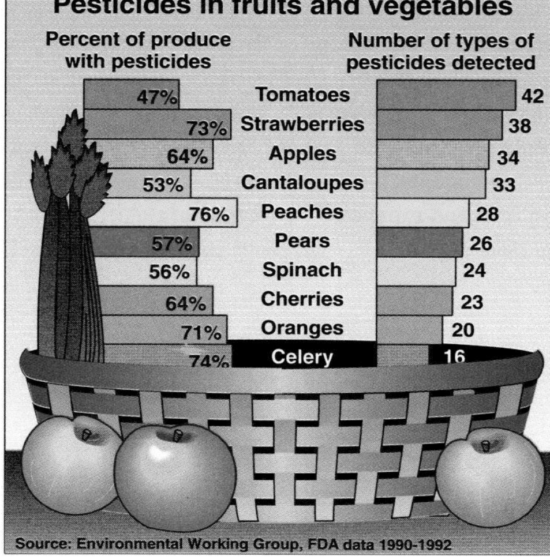

Thinking Critically

Conclude from the chart which kinds of produce are at greatest risk of having pesticide residues, and suggest what might be done to protect young children and infants from exposure.

Figure 6.12

Many power plants draw water from a river or lake to cool their machinery (left). The discharged water is much warmer and can alter or destroy the habitat for many organisms in the body of water. Sediments washed away from farm fields can enter a river or lake. Water thick with sediments can clog fishes' gills and kill them (right).

ThinkingLab Draw a Conclusion

Purpose

Logical-Mathematical Students will use provided data to determine the extent of pesticide use on foods.

Process Skills

interpret data, draw conclusions

Teaching Strategies

- Have students explain why the problem of pesticide use is difficult to analyze from a consumer's point of view.
- Discuss why farmers use pesticides. Have students speculate what produce might look like without the use of these chemicals.

Thinking Critically

Produce having the greatest risk of pesticide residue includes celery, strawberries, oranges, and peaches. Protection may include stronger laws regulating pesticide use, more research on the effects of pesticides, washing and peeling food before it is eaten, and labeling foods on which pesticides are used.

Assessment

Performance Ask students to investigate pesticides sold in supermarkets. Ask them to provide specific examples of which pesticides are used to control specific kinds of pests.

GLENCOE TECHNOLOGY

Videodisc

Biology: The Dynamics of Life
Disc 1, Side 1
The Greenhouse Effect (Ch. 15)

PROJECT Salinity and Germination

Suggest to students that they experimentally determine the effects of salt water on seed germination or on established plants. Remind students that salt concentration should be about 3.5% (ocean water salinity). Other concentrations of salt water might be used on bean seeds or bean plants.

Demonstration

Visual-Spatial Show that soil may not filter out pollutants from groundwater. Use water containing several drops of food coloring to simulate pollution. Pour this contaminated water through a funnel containing soil. Have students note the color of the water leaving the funnel. Students will observe that the food coloring (pollutant) has not been removed.

Tying to Previous Knowledge

Have students categorize groundwater as a renewable or nonrenewable resource. *Most groundwater supplies are considered renewable. The Ogallala Aquifer, because of its fossil origin and slow rate of recovery, is considered nonrenewable.*

Assessment

Performance Assessment in the Biology Classroom: p. 55, Finding Out Why a Pond is Dying. Have students carry out this activity to determine how pollutants affect pond ecosystems.

Activity

Exploring Environmental Issues: p. 3, Sources of Water Pollution; p. 11, Toxic Waste—Where Does It Go? Have students carry out these activities to determine how they contribute to pollution problems.

Groundwater pollution

Water in lakes and streams is known as surface water. Fresh water that is found underground is called **groundwater.** Do you know whether your community water supply comes from surface water or groundwater? About half the population of the United States depends on groundwater for drinking water. Until the 1970s, it was assumed that soil would filter out any pollutants before they reached underground reservoirs, so groundwater could not become polluted. *Figure 6.13* illustrates how pollutants can enter underground reservoirs.

Other problems are also associated with the use of groundwater. The Ogallala Aquifer is an underground reservoir that lies beneath much of the farmland in the midwestern United States. So much water is being removed from this aquifer for irrigating farmland and supplying water to cities and towns that the underground water level is going down faster than it can be recharged by rainfall. The total amount of water removed from this aquifer every year is estimated to equal the annual flow of the Colorado River. As the water level goes down, the soil above it may collapse, creating a giant sinkhole. Water from the Ogallala Aquifer and other underground reservoirs contains large amounts of dissolved salts. As this water is used to irrigate farmland, these salts gradually accumulate in the soil, eventually making it unsuitable for plant growth.

Figure 6.13

It takes hundreds, even thousands, of years for water to accumulate in underground reservoirs. Once groundwater is contaminated with pollutants, it is difficult and expensive to clean.

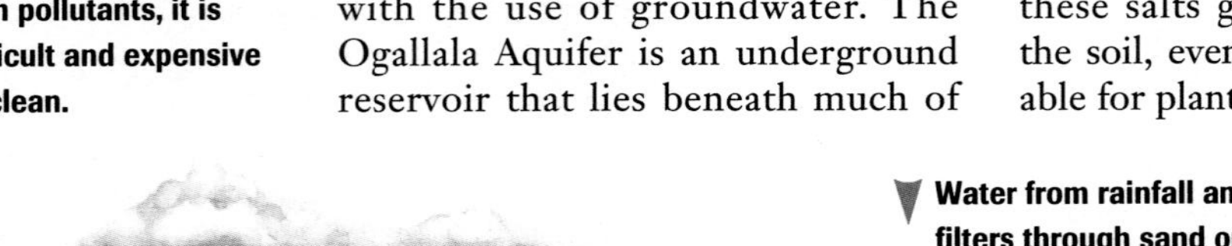

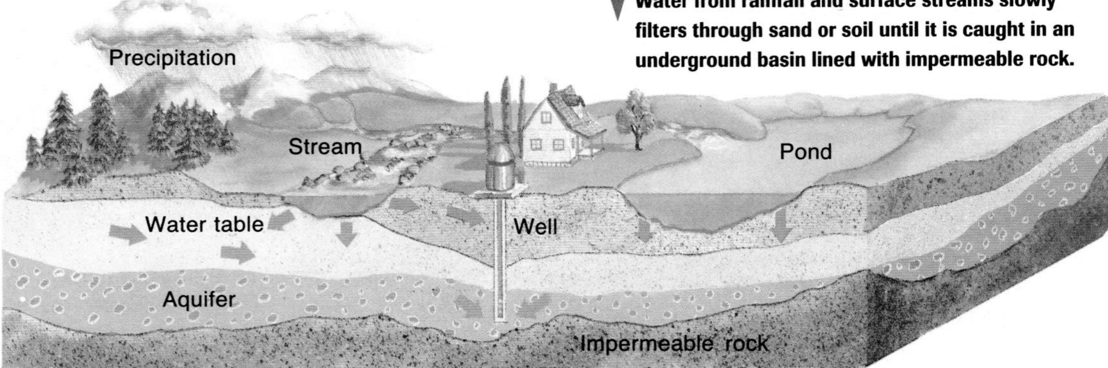

Water from rainfall and surface streams slowly filters through sand or soil until it is caught in an underground basin lined with impermeable rock.

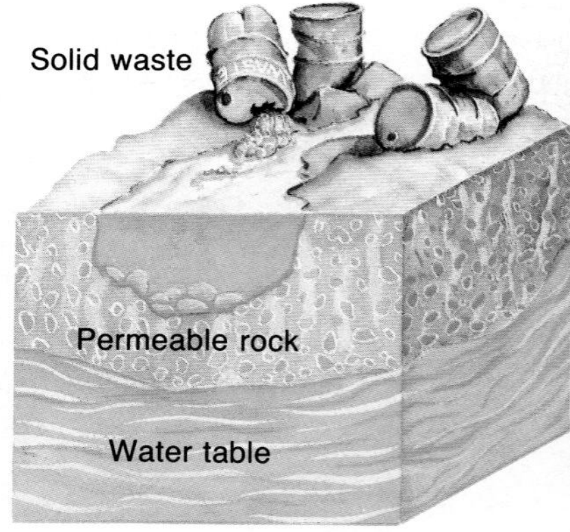

Pollutants from leaking storage containers or agricultural runoff also travel through the soil, eventually reaching and contaminating nearby groundwater (left). It is estimated that between 20 and 45 percent of the water wells in some areas of the United States may be polluted with agricultural or industrial chemicals. Efforts are being made to develop safer methods for storing hazardous materials to prevent them from contaminating groundwater supplies (right).

STUDENT JOURNAL

Problems with Lead Have students research problems associated with lead in drinking water and paint. Have them write a "public service commercial" that calls attention to the problem and describes the measures people can take to protect themselves. Have students work in cooperative groups to videotape some of their commercials. COOP LEARN

Meeting Individual Needs

Hearing Impaired Prepare a column of soil packed into a glass tube with a large diameter. Add water containing food coloring to the column and have students observe the slow percolation of the water through the soil. This visually reinforces how underground reservoirs obtain their water supply.

Figure 6.14

Many landfills are being filled to capacity and closed.

Land pollution

How much garbage do you and your family produce every day? Remember to include the notebook paper you use for schoolwork, and the plastic bags or other containers you discard after you've eaten your lunch. Do you put your aluminum soda cans in the trash or in the recycling bin?

Trash, or solid waste, is made up of the cans, bottles, paper, plastic, metals, dirt, and spoiled food that people throw away every day. The average American produces about 1.8 kg of solid waste daily. That's a total of about 657 kg of solid waste per person per year. Where do we put all this garbage? Does it ever decompose? Your trash becomes part of the billions of tons of solid waste that are burned or buried in landfills, *Figure 6.14,* all over the world every year.

Chemistry

Separating Plastics

It is predicted that 75 billion tons of plastic will be discarded by the year 2000. Currently, only about three percent of this plastic gets recycled. Unlike aluminum, which is easily separated from trash and reclaimed, the problem in recycling plastic is that the plastics have to be sorted by type, by hand, which is an expensive process. Different types of plastic polymers cannot be mixed together to form a strong, stable plastic.

Chemical sorting Researchers have devised a method by which chemical solvents sort the six plastics found in typical household trash. In a pilot program, researchers placed shredded, mixed plastics in a vat with a common solvent called xylene. Immediately, one type of plastic polymer, polystyrene, the type of polymer that is used to make Styrofoam, was separated from the other polymers.

When the temperature in the vat reached 170°C, xylene dissolved another type of plastic polymer, polyethylene. Polyethylene is used in plastic grocery bags and plastic wrap. At 220°C, another form of polyethylene used in milk jugs and detergent bottles was dissolved. Polypropylene, the polymer that is used in labels and some types of bottles, was dissolved at 250°C. The last two polymers, PET (polyethylene terephthalate) and PVC (polyvinyl chloride), required a different solvent because xylene is not effective on them. With the different solvent, they too were separated into their pure forms and were ready to be used again.

An efficient process Researchers have estimated that this process could yield nearly four tons of plastic polymer per hour. This process could also cost almost 30 percent less than producing the polymers from new raw materials.

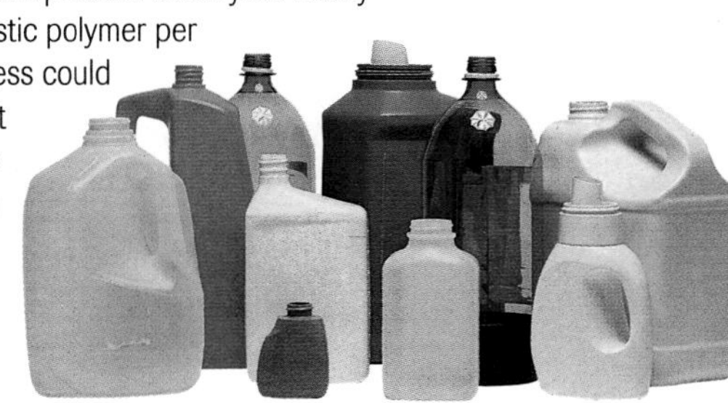

CONNECTION TO Biology

How could chemical research save other kinds of Earth's resources and limit our waste production?

SECTION 6.2

Reinforcement

LS Logical-Mathematical Have students prepare a pie graph showing the percentage by weight of urban solid waste. Data: paper/cardboard 41%; yard waste 18%; food waste 9%; metal 9%; glass 8%; plastic 7%; wood 3%; rubber/leather 3%; cloth 1%; and other 1%. L1 SAE

Chemistry

Separating Plastics

Purpose

LS Logical-Mathematical To explain how plastics are separated for recycling.

Teaching Strategies

- Have students research the chemical components common to all plastics and link this composition to oil.
- Have students research the chemical and physical properties of the six plastic types described in this feature. L2
- Ask students to create a display that shows the types of plastics along with their category names. L2

Possible Answer

Chemical researchers could design non-ozone-damaging propellants to replace CFCs, provide a means for removing pollutants from public water systems, and perfect desalination techniques to produce more fresh water.

PORTFOLIO

Measuring Solid Waste Generation Have students prepare a table to illustrate the amount of solid waste generated by their family per day, week, month, year, and decade. If necessary, have students use the value of 1.8 kg of solid waste per person for each day. *Day = 1.8 kg, week = 12.6 kg, month = 54 kg, year = 657 kg, decade = 6570 kg.* 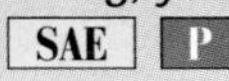

interNET CONNECTION

Follow the link for this chapter on the Glencoe Homepage at **www.glencoe.com/sec/science** to find out more about use of natural resources.

GLENCOE TECHNOLOGY

 **Videodisc**

Biology: The Dynamics of Life
Disc 1, Side 1
Recycling (Ch. 16)

Activity

Exploring Environmental Issues: p. 25, Using Land Wisely, p. 27; Sorting It Out with Recycling; p. 47, Be an Earthwise Consumer. Have students carry out these activities to examine the importance of recycling. L2

Concept Development

Review the process of aerobic respiration using a simple word equation. Point out that oxygen is needed as one of the reactants. Relate the inability of an organism to remain alive without carrying on respiration. Relate this need for oxygen to the slow decomposition of landfill matter.

Enrichment

LS **Visual-Spatial** Provide students with sterile petri dishes containing agar. Have students place soil samples onto the agar surface and incubate the dishes upside-down for 24 - 48 hours at 37°C. Control dishes may be prepared using sterile soil. Have students observe the petri dishes and note the amount of bacterial growth that appears. L2

Software

Race to Save the Planet (MM), Scholastic.

Figure 6.15
Scientists who dug up garbage from old landfills were surprised to find newspapers from the 1950s that were still readable. They also discovered that food thrown away along with those newspapers hadn't decomposed at all. Neither had grass clippings from lawns mowed 40 years ago.

You might expect that the trash that goes into landfills is decomposed by bacteria. But scientists who have dug up and sifted through the garbage in old landfill sites as shown in *Figure 6.15* have discovered that much of this material is not broken down by natural processes. Because oxygen is sealed out of landfills when they are covered with soil, bacteria and other decay-promoting organisms cannot survive there. As a result, materials that could be recycled back into the environment are trapped for decades.

Some types of solid waste—such as wood products, food, animal wastes, dead leaves, and other yard wastes—are called **biodegradable** wastes because they can be broken down by natural processes. Other types of waste—such as pesticides, toxic metals, and radioactive residues from nuclear power plants—are not easily broken down. These materials are considered **nonbiodegradable** because they can persist in the environment for hundreds or even thousands of years. There is currently much debate over where and how to store toxic, nonbiodegradable wastes. One method that is receiving a great deal of attention involves burial of these wastes in geologically stable areas.

What can we do to help?

How can we continue to meet the human population's increasing demand for resources? Modern technology offers some solutions. For example, selective breeding of plant crops—a topic you'll learn more about in Chapter 14—has resulted in the development of strains of wheat that produce more grain per wheat plant. Such developments not only increase food production, but also reduce the amount of land needed for growing crops. Recycling, as

PROJECT Recycling

Ask students to survey their homes to determine the types and number of plastic and paper products that are marked with a recycling logo. Have students bring their list to class and, working in cooperative groups, prepare a class list of all recyclable items.
L1 SAE COOP LEARN LS

GLENCOE TECHNOLOGY

Videodisc

The Infinite Voyage: Miracles by Design
Biodegradable Plastic: The Miracle Material?
(Ch. 4)

Figure 6.16

Recycling metals, plastics, paper, and other materials can help conserve important resources. If all Americans recycled just one-tenth of their newspapers, we could avoid harvesting about 25 million trees every year. We also could save energy. The production of recycled paper requires only about half the energy needed to make paper from wood pulp.

MiniLab

What pollutants do you have at home?

Landfills are not supposed to contain dangerous materials that may leak into the groundwater. However, tons of spray paint cans, pesticides, batteries, cleaning solvents, insect repellents, and similar toxic compounds are thrown out into the trash every day. One way to slow the disposal of such wastes is to become aware of how prevalent they are.

Procedure

1. Take a tour of your house, and write down the type and amount of any pollutant you find.
2. Divide your list into organic substances and inorganic substances. If you do not know if a particular compound is organic or inorganic, make your best guess.
3. Be sure to look in your garage, under the kitchen counter, and in the bathroom for substances that might be pollutants.

Analysis

1. Where were most of the pollutants in your house stored?
2. Were most of your pollutants organic or inorganic?
3. Suggest a way that you might be able to protect the landfill in your area from pollutants.

described in *Figure 6.16,* also helps meet demands without using up our nonrenewable resources.

Preservation of natural areas

How can human demands for additional living space and other resources be met without loss of habitat? **Preservation** is the act of keeping an area or organism from harm or destruction. One way of preserving wildlife habitats is by setting aside parks, refuges, and other areas that can be regulated through conservation. **Conservation** is the planned management of a natural area to prevent exploitation or destruction. Yellowstone National Park, the world's first national park, was founded in 1872 when Congress set aside 2 million acres to protect natural features and meet public recreational needs. Today, almost 800 million acres of U.S. land have been set aside, *Figure 6.17.*

About three percent of Earth's land has been set aside by world governments as parks or preserves.

Figure 6.17

Millions of tourists visit the national parks in the United States each year.

CULTURAL DIVERSITY

George Washington Carver

African American agriculturalist George Washington Carver (1865-1943) is best known for his work in establishing crops, such as cotton, peanuts, and sweet potatoes, in the southern United States. Carver's research was directed toward the poor farmer and showed how reliance on a single cash crop ultimately depleted the soil of nutrients, leaving it open to destructive erosion. Initiate a discussion about the agricultural methods used by farmers in different societies and how such methods are related to protecting soil against erosion. Have students research farming methods and irrigation practices and make models that illustrate the methods used by people of different cultures. IS

MiniLab

Purpose

IS **Intrapersonal** Students will become familiar with household products that are potential pollutants.

Process Skills

observe and infer, recognize cause and effect, classify

Teaching Strategies

- Prepare a transparency of labels from a variety of products. Review the chemical compositions of the products to indicate how one can determine if they are organic or inorganic.
- Have students work in groups to decide if the product is organic or inorganic. Have all groups pool their lists. COOP LEARN

Expected Results

Students will discover that the list of potential pollutants within their homes is quite lengthy. Examples of such products include: motor oil, rubbing alcohol, paint, lye, hydrogen peroxide, window cleaner, furniture polish, antifreeze, and detergents.

Analysis

1. Answers may include garage, kitchen, and bathroom.
2. Most answers should be organic.
3. Set up programs for the collection and proper disposal of all potential pollutants.

Assessment

Performance: Organize an "anti-pollution day." Design and distribute posters alerting the community to the potential pollution hazards present in their homes. L1 COOP LEARN

BioLab

Degrading Time of Packing Materials

Time Allotment
Initial session: one class period; followup: 15 minutes once a week for four weeks

Objectives
Review objectives with students before they begin the Biolab.

Process Skills
observe and infer, recognize cause and effect, classify, record data, interpret data

Safety Precautions
Have students wash their hands after each followup examination. Cornstarch packing is not toxic, but students should be reminded not to taste materials.

PREPARATION

Alternate Materials
- Plastic containers may be used in place of beakers or clay pots. Large-sized empty cottage cheese containers or 2-L plastic soda pop bottles with their top-third sections removed may be used. Avoid the use of cardboard containers as they are biodegradable.
- Many of the packing materials may be collected from students. Stores that specialize in mailing packages may supply some of the packing materials needed.

BioLab

Degrading Time of Packing Materials

In the United States, approximately one ton of municipal waste is dumped into landfills each year for every man, woman, and child. A significant portion of this waste is in the form of packing materials for food, clothing, and other household items. As suitable places for waste disposal become increasingly scarce, the waste stream must be slowed or changed. One way to do this is to be sure that packing materials can be decomposed or recycled.

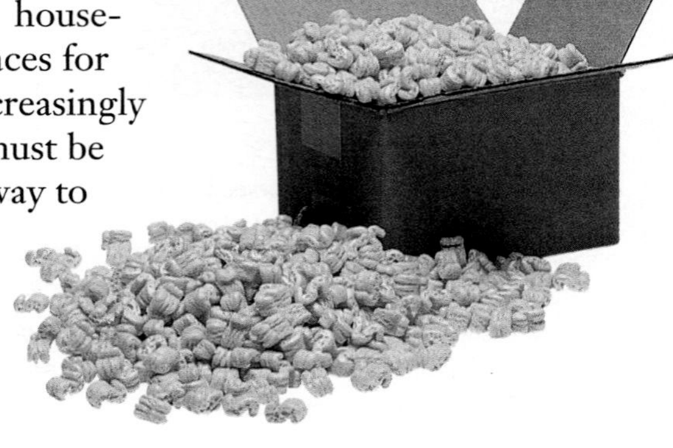

PREPARATION

Problem
How long does it take for packing materials to degrade?

Objective
In this Biolab, you will:
- **Determine** the conditions for rapid degradation of certain packing materials.

Materials
soil
Styrofoam peanuts
cornstarch packing material
popcorn
paper packing material
paper garbage bags
plastic garbage bags
4 large beakers or clay pots

Safety Precautions
Be sure to wash your hands after doing this Biolab.

PROCEDURE

Teaching Strategies
- Ask students to collect and bring in packing materials from their homes several days before the investigation begins.
- Prior to the experiment, have students list materials they believe will degrade. Have them compare this list with their final observations.
- Cooperative groups will reduce needed storage space and materials.
- Have students prepare statements prior to the experiment as to the possible roles of soil and water in the degrading process. Compare the statements with their final observations.
- Purchase popped popcorn or bring or have students bring the materials for it to class.

PROCEDURE

1. Collect four samples of each type of packing material.
2. Make four piles of packing material, mixing each kind of material in each pile.
3. Place each pile of material in a different beaker or clay pot.
4. Add soil to two of the pots so that all the samples are well covered and mixed with the soil.
5. Place a label that says DRY on one of the pots with soil and one of the pots without soil.
6. Place a label that says MOIST on the remaining two pots.
7. Keep the second two pots moist at all times.
8. Make a data table like the one shown.
9. Observe each pot once a week for four weeks, and record in your data table any evidence of decomposition.

Pot #	Observations
#1 Dry with soil	
#2 Dry without soil	
#3 Moist with soil	
#4 Moist without soil	

ANALYZE AND CONCLUDE

1. **Interpreting Observations** Describe your observations of decomposition of any of the packing materials.
2. **Analyzing Data** Did any of the packing materials appear to be nonbiodegradable? Explain.
3. **Thinking Critically** Suggest ways that commercial producers could help to reduce the waste stream.
4. **Drawing Conclusions** Which type of packing material appears to decompose fastest, and which type decomposes most slowly?

Going Further

Application Find out how your community disposes of its wastes and how much disposal costs. Do the disposal costs reflect stresses placed on the environment's food webs or only the monetary cost of disposal?

ANALYZE AND CONCLUDE

1. Cornstarch and popcorn showed evidence of degrading in the moist container and the moist soil. Mold growth and a foul odor was evident.
2. Paper, plastic, and Styrofoam appear to be nonbiodegradable, showing no signs of change.
3. Use packing materials that are rapidly degraded.
4. Popcorn and starch (foods) appear to decompose fastest, while paper and plastic decompose slowly, if at all, in the time allowed.

Assessment

Portfolio: Have students write a report of this experiment. Have them make a statement regarding the roles of both soil and water in the decomposition process and identify organisms that participate in the process of degradation. L1 P

Going Further

Have students determine the extent of degradation of popcorn if sterile water, sterile popcorn, and a closed container are used in the experiment. L1

3 Assess

Check for Understanding

Have students explain how the section title, "Maintaining the Natural Balance," relates to the contents of the section. L1

Possible Procedures

- Students should use similar amounts by volume of packing materials and soil for each container.
- It may be necessary for students to dig out some of the packing materials from the soil as future observations are made to check on the changes that may be occurring.

Data and Observations

Containers of dry packing materials with or without soil will show no signs of decomposition. Materials such as popcorn and cornstarch will show degradation with and without soil in the moist containers. Degradation should be more advanced in moist soil for cornstarch and popcorn.

SECTION 6.2

Reteach

Have students prepare an outline of the major topics in this section. L1

Extension

Have students explore and write a report on the problem of smog formation. Ask students to determine the chemical changes responsible for the formation of smog as well as the geographic features that must exist for smog to form. L2 P

Assessment

Portfolio: Have students describe in their portfolios the relationship between the words in each of the following pairs: (a) biodegradable - nonbiodegradable (b) pollution - particulate (c) ozone layer - acid rain (d) preservation - conservation. L1 P

4 Close

Activity

Have students provide examples of conditions or practices that lead to, result in, or increase problems associated with pollution and endangered species.

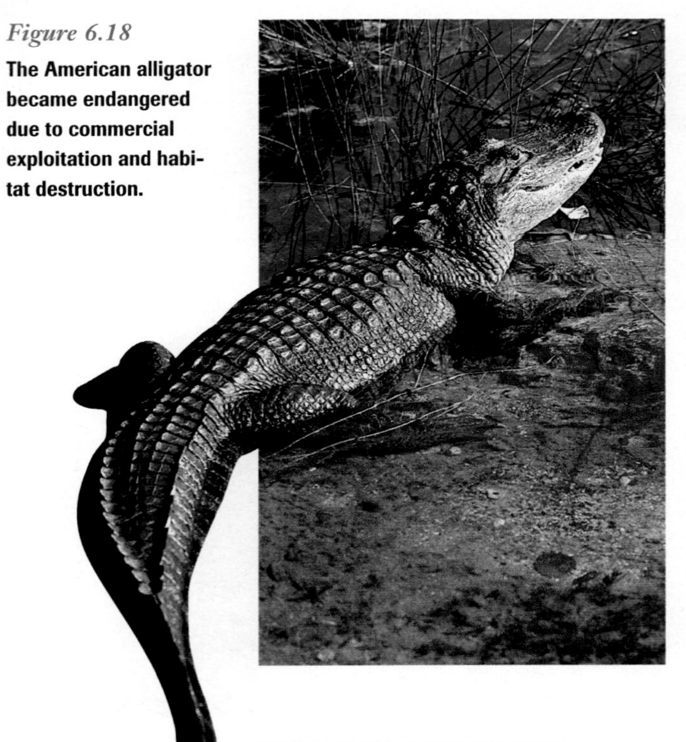

Figure 6.18
The American alligator became endangered due to commercial exploitation and habitat destruction.

Preserving threatened and endangered species

Recovery programs have been developed to keep some endangered species from becoming extinct. One of the greatest success stories is that of the American alligator. Once endangered throughout the entire United States, the species, shown in *Figure 6.18*, has made a comeback and is no longer listed as endangered or threatened in some parts of its range.

At present, far more species are becoming extinct than we can possibly save. Which species should be rescued? How do we decide whether one species is more important than another? These are not questions that ecologists can answer easily because they realize that every organism is important to the fabric that makes up the tapestry of life.

Connecting Ideas

The human population is increasing at a staggering rate. Because of the increased demands this places on the biosphere, natural resources once considered inexhaustible are now in short supply. Human activities have caused pollution of air, water, and land.

How can these trends be reversed? What can humans do to lessen stress on the environment while maintaining our standard of living? Finding the answers will require learning as much as possible about our environment and the organisms it supports. One area of study is the homeostatic balance in an ecosystem, which depends on a fundamental balance of chemicals and chemical reactions within each community of organisms.

Section Review

Understanding Concepts

1. What are two major sources of air pollution?
2. How are biodegradable and nonbiodegradable pollutants different?
3. How can the application of fertilizer and pesticides to a farm field affect drinking water?

Thinking Critically

4. Are all recyclable substances biodegradable? Explain.

Skill Review

5. **Designing an Experiment** Design an experiment, using a terrarium, to test the effects of different concentrations of acid precipitation. For more help, refer to Practicing Scientific Methods in the *Skill Handbook*.

Answers to Section Review

1. The burning of fossil fuels and natural events, such as volcanic eruptions and forest fires, are the main sources of air pollution.
2. Biodegradable pollutants can be broken down by natural processes. Nonbiodegradable pollutants cannot be broken down.
3. Fertilizers and pesticides may eventually be washed away and enter groundwater that is used for drinking water.

Thinking Critically

4. No; items such as plastic and metals are not biodegradable. They can, however, be recycled.

Skill Review

5. Experimental designs must contain a control and an experimental setup. Tests likely will suggest adding different concentrations of an acid solution to a terrarium at regular intervals, while observing changes that occur in plant growth and the ability of other organisms to survive.

Chapter 6 Review

REVIEWING MAIN IDEAS

6.1 Effects of Human Activities on Our Resources

- Resources can be classified as renewable or nonrenewable. Renewable resources may be replaced if managed properly. Nonrenewable resources cannot be replaced by natural means, but many of them can be recycled.
- Trends in resource use suggest that as demand increases, supply tends to diminish. Substitution, recycling, and technology can reduce demand for some resources.

6.2 Maintaining the Natural Balance

- The use of resources faster than the environment's ability to recycle them has resulted in problems of air, water, and land pollution. Air pollution is caused by particulates and gases such as carbon dioxide and nitrogen oxides. Water pollution is caused by organic and inorganic substances, sediments, heat, and radioactive substances. Land pollution occurs when solid wastes are dumped and not decomposed by natural systems.
- Human activities have increased the rate of habitat destruction and caused the extinction of many species of wildlife. The creation of national parks, forests, and similar areas helps reduce the number of species lost to extinction.

Key Terms

Write a sentence that shows your understanding of each of the following terms.

acid precipitation
biodegradable
conservation
endangered species
extinction
fossil fuel
greenhouse effect
groundwater
natural resource
nonbiodegradable
nonrenewable resource
ozone layer
particulate
pollution
preservation
renewable resource
smog
threatened species

UNDERSTANDING CONCEPTS

1. Which of the following is an endangered species?
   a. American alligator
   b. bald eagle
   c. manatee
   d. black bear
2. Forests, herds of cattle, and fields of wheat are all ____________.
   a. renewable resources
   b. nonrenewable resources
   c. nonbiodegradable resources
   d. limitless resources
3. Which of the following is a kind of air pollution?
   a. fog
   b. smog
   c. rain of pH 7
   d. snow
4. Which of the following is NOT biodegradable?
   a. plastic
   b. paper
   c. garbage
   d. grass clippings
5. Natural resources that can be replaced or recycled by natural processes are ____________.
   a. nonbiodegradable
   b. biodegradable
   c. nonrecyclable
   d. limited
6. Ultraviolet radiation reacting with nitrogen oxides and hydrocarbons from automobile exhausts results in ____________.
   a. smog
   b. a rusty haze
   c. rain
   d. increased sunshine

Chapter 6 Review

Reviewing Main Ideas

Summary statements can be used by students to review the major concepts of the chapter.

Key Terms

Answers should go beyond defining the terms. Accept any answer that uses the term correctly and in the proper context.

Understanding Concepts

1. c
2. a
3. b
4. a
5. b
6. a

PROJECT Testing for Salt

Advise students on the use of silver nitrate as a means for detecting the presence of salt: Three drops of a 1% solution of AgNO causes a white precipitate to form in the presence of salt. Have students test a sample of salt water to detect salt. **Caution:** *Silver nitrate solution is caustic to the skin. Students should wear gloves, aprons, and safety goggles when handling this chemical.* Distill the salt water using a distillation apparatus; see your chemistry department for the equipment. Retest the distilled water for salt. It should test negative for salt. Relate this project to the expense of desalination efforts as well as the high energy demand needed to purify the salt water. L3

Chapter 6 Review

7. d
8. d
9. b
10. c
11. b
12. b
13. a
14. c
15. b
16. a
17. d
18. c
19. a
20. d

Chapter 6 Review

7. When large amounts of oxygen are transformed into molecules composed of three oxygen atoms, ____________ is the result.
   a. carbon monoxide
   b. carbon dioxide
   c. nitrogen dioxide
   d. ozone
8. Nonrenewable resources can be made to last longer by ____________.
   a. increasing the use rate
   b. increasing the availability
   c. encouraging their use
   d. increasing the price
9. Which of these is an effect of acid precipitation on sensitive ecosystems?
   a. increased fish production
   b. decreasing pH of lakes
   c. leaching of chemicals
   d. increased growth of plants
10. Which biome is experiencing the most rapid destruction and increased extinction rate?
    a. grassland
    b. taiga
    c. tropical rain forest
    d. temperate deciduous forest
11. How may the thinning of the ozone layer in the Antarctic affect humans in South America?
    a. decreased tanning
    b. increased cancer rate
    c. decreased fertility
    d. decreased crop production
12. What would be the result of increasing participation in efforts to recycle metals, plastics, and paper?
    a. prices would decrease
    b. resources would be conserved
    c. use would decrease
    d. increased biodegradability
13. How does agricultural runoff cause pollution?
    a. contaminates water supply
    b. kills crop plants
    c. increases pest insects
    d. decreases wildlife
14. Which of these is an effect of using aquifers for irrigation?
    a. Salt levels in soil decrease.
    b. Crop production can be maintained indefinitely.
    c. Land may become unsuitable for agriculture.
    d. Sinkholes fill in, leveling the land.
15. What may be an indication that pollution is causing a system to be out of balance?
    a. stable numbers of animals
    b. increased extinctions
    c. decreased extinctions
    d. stable numbers of plants
16. How do present extinction rates differ from past extinctions?
    a. They are much higher.
    b. They are about the same.
    c. They are much lower.
    d. They are somewhat lower.
17. What causes acid rain?
    a. increased ozone levels
    b. agricultural runoff
    c. living in areas with fog
    d. burning high-sulfur coal
18. Contamination of any part of the biosphere is called ____________.
    a. acid rain
    b. particulate generation
    c. pollution
    d. pollination
19. Which nutrient derived from cattle manure is useful as an organic fertilizer?
    a. nitrogen
    b. methane
    c. water
    d. particulates
20. The ozone layer is found in the ____________.
    a. thermosphere
    b. troposphere
    c. mesosphere
    d. stratosphere

Program Resources

Chapter Assessment, pp. 31-36 L1
Alternate Assessment in the Science Classroom
Computer Test Bank L1
Content Mastery, pp. 21-24 L1

APPLYING CONCEPTS

21. Why do those who fish for sport NOT consider temperature pollution a problem in some lakes?
22. How would putting milk into glass bottles instead of plastic bottles be a possible benefit to the environment?
23. Why should you be concerned about the kinds of material placed in the sanitary landfill near where you live?
24. Many plastics can be broken down and formed into plastic products over and over again. What effect would this have on landfills?
25. How does recycling aluminum cans and other nonrenewable minerals save energy?

THINKING CRITICALLY

26. ***Concept Mapping*** Make a concept map that relates the following terms and phrases. Supply the appropriate linking words for your map.
    endangered species, smog, threatened species, fossil fuel, extinction, acid precipitation, air pollution, pollution, particulates, ozone layer, greenhouse effect, natural resources
27. ***Observing and Inferring*** Some kinds of packing material degrade when moisture is added. Why is such material considered environmentally friendly?
28. ***Observing and Inferring*** What kind of interactions are out of balance when biodegradable materials build up in landfills?
29. ***Comparing and Contrasting*** Compare and contrast the types and amounts of pollution produced by cars that are well tuned with those that are poorly tuned.
30. ***Recognizing Cause and Effect*** Predict the effect on an aquifer if water is pumped at increasing rates each year for 50 more years.

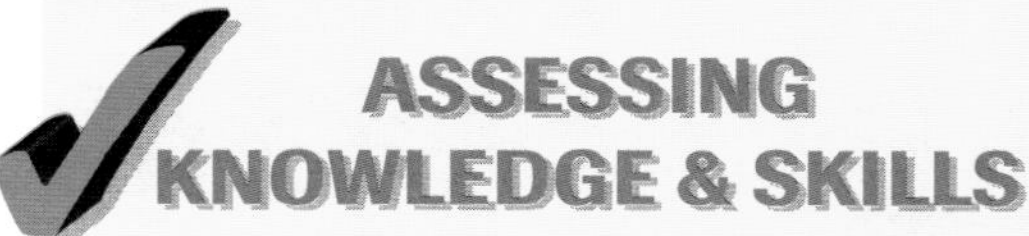

ASSESSING KNOWLEDGE & SKILLS

The following bar graph indicates the sources of sulfur, nitrogen, and carbon dioxide in the air.

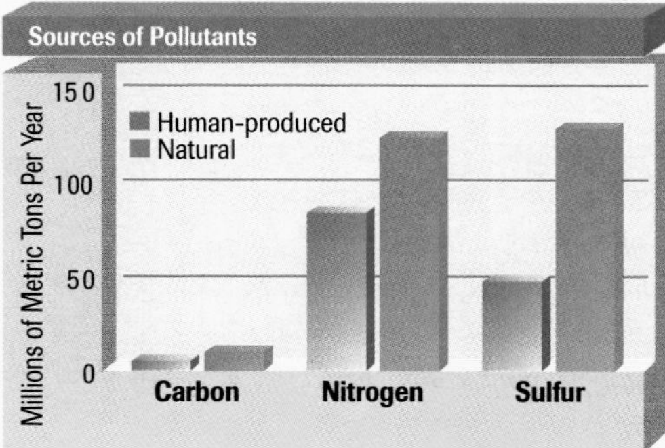

Interpreting Data Use the graph to answer the questions below.

1. From the data, one could infer that the levels of these pollutants have ____________.
   a. no effect on air quality
   b. been stable recently
   c. been decreasing
   d. increased due to human activity
2. Which of the following sciences probably provided the data in the graph for use by biologists?
   a. physics c. Earth science
   b. chemistry d. hydrology
3. What is the total amount of nitrogen released into the air each year (in millions of metric tons)?
   a. 80 c. 190
   b. 130 d. 210
4. ***Observing and Inferring*** Suppose power plants switched from burning coal to using hydroelectric power. Which of these pollutants would be affected?
   a. carbon c. sulfur
   b. nitrogen d. none of these

Chapter 6 Review

Applying Concepts

21. Higher temperatures may increase growth and reproduction of fish or provide more food.
22. Glass bottles would be recycled. Plastic bottles are usually not recycled and thus clog landfills with nonbiodegradable material. Plastics are made from fossil fuels. Reducing their use would reduce the demand for fossil fuels.
23. Many of the materials are poisonous and can, in time, filter into the groundwater system that we use as a water supply.
24. Landfills will no longer be supplied with plastics that are nonbiodegradable. This will provide space for more biodegradable materials.
25. The process of extracting aluminum from its ore requires heat energy. This heat typically requires the burning of coal or oil (fossil fuels). Recycling reduces the need for this energy.

Thinking Critically

26. Evaluate students' concept maps. The maps should show an understanding of the relationships among the concepts listed.
27. They are considered friendly because they will biodegrade.
28. Decomposers are unable to perform their tasks of breaking down matter.
29. Well-tuned cars produce fewer particulates and noxious gases, but more carbon dioxide and water, than less well-tuned cars.
30. If the rate of pumping increases on an aquifer, the production will eventually decrease.

ASSESSING KNOWLEDGE & SKILLS

1. d
2. b
3. d
4. c

UNIT 3 The Life of a Cell

Unit Overview

Unit 3 introduces students to basic chemistry, the structure and function of cells, and cell energetics. In Chapter 7, students learn the basic concepts of chemistry important in biology. Chapter 8 introduces the structure and function of cell organelles. This discussion is expanded upon in Chapter 9 through an in-depth view of the plasma membrane.

Chapter 10 acquaints students with the details of energy flow that result from respiration and photosynthesis. Chapter 11 focuses on cell reproduction.

Theme Development

The theme of *homeostasis* pervades this unit and is evident in the discussions of membrane function and the concept of cell size limitations. *Energy* is the main focus of Chapter 10. The theme of *unity within diversity* is evident in the discussion of cell types as well as the differences between animal and plant cells. The *nature of science* is stressed through the discussion of how the development of the microscope led to a better understanding of cells.

UNIT 3 The Life of a Cell

Greenish, rectangular shapes cram into every available space, filling your view. Within each rectangle, microscopic structures of varying shape and color float like tiny islands in a flowing sea of colorless liquid. At first glance, this scene may look like an alien landscape, but what you are actually looking at are the cells of a common aquarium plant.

Every living thing is made up of one or more cells. A cell is the smallest unit that can carry on all the processes of life. Living things, like nonliving things, are composed of chemicals such as carbon, hydrogen, and oxygen, but it is the organization of these elements into cells that distinguishes living things from all other matter.

Magnification: 14 500×

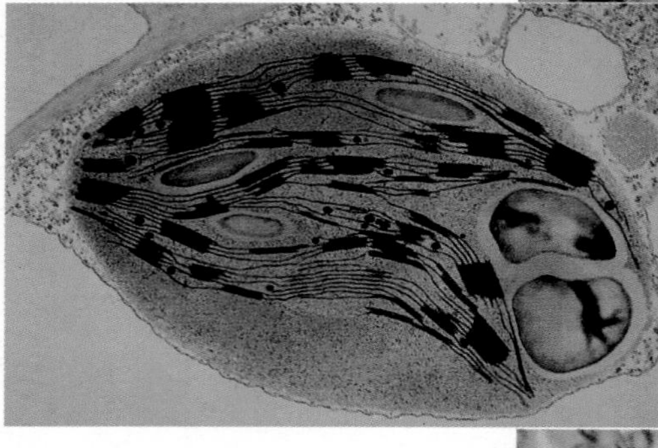

All organisms need energy to carry out life processes. How is this chloroplast important in providing the energy needed for the lives of all cells?

156

Advance Planning

Chapter 7
- Purchase wax paper and liquid detergent for the Minilab on page 168.
- Purchase lemon juice, ammonia, liquid detergent, shampoo, vinegar, and pH paper for the Minilab on page 173.
- Prepare potato slices for the Biolab.

Chapter 8
- Purchase *Elodea* for the Alternate Lab.

Chapter 9
- Order *Paramecium* culture for the Minilab on page 227, *Elodea* for the Minilab on page 229, and *Euglena* for the Microscope Activity.
- Purchase starch and iodine solutions for the Biolab.

Chapter 10
- Order bromthymol blue (BTB) solution for the Minilab on page 253.
- Purchase *Elodea* for the Microscope Activity and the Biolab.
- Purchase *Coleus* plants for the Alternate Lab.

Chapter 11
- Purchase potatoes and prepare potassium permanganate solution for the Minilab on page 263.
- Order prepared onion root tip slides for the Biolab.

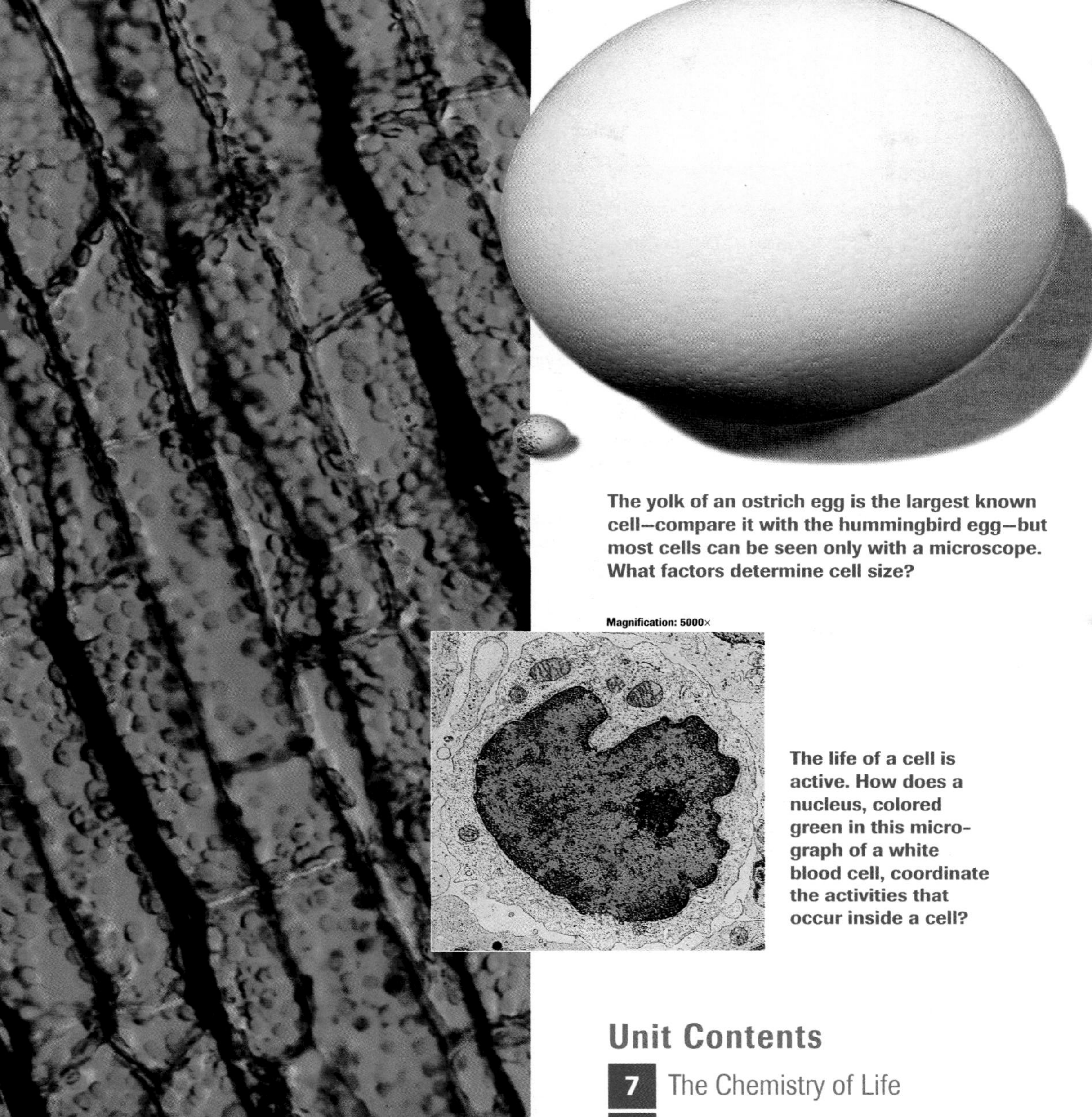

The yolk of an ostrich egg is the largest known cell—compare it with the hummingbird egg—but most cells can be seen only with a microscope. What factors determine cell size?

Magnification: 5000×

The life of a cell is active. How does a nucleus, colored green in this micrograph of a white blood cell, coordinate the activities that occur inside a cell?

Unit Contents

UNIT 3

Introducing the Unit

Explain that like *Elodea* all living things are made up of one or more cells. Point out that some organisms, like the nitrogen-fixing bacteria students studied in Unit 2, are made up of only one cell whereas others, such as humans, may be made up of trillions of cells.

Elicit from students what processes living things must carry out to survive. Explain that cells, because they are living, must carry out these same functions. Direct students' attention to the cell nucleus. Explain that this tiny structure controls most of the life activities within cells.

Motivational Activity

Demonstration: Without students seeing, half fill a Petri dish bottom with water. Place one drop of Duco cement (cement used must contain toluol) in the water. Place the dish in view of students on an overhead projector. The drop of Duco cement will appear to swim around in the water. Ask students if the "thing" is alive. Ask them how they would know. Use the demonstration as an opportunity to review the characteristics of living things. Remind students that all living things are made of cells. L1

Unit Project

Microscope Project Students will be introduced to cells, microscopes, and energy in this unit. Provide students with samples of pond water. Over the next two to three weeks, have students view pond water samples under the microscope. Have students draw ten different organisms viewed under the microscope. Ask them to classify each as multicellular or unicellular. Challenge them to also classify each organism as a producer or consumer. Ask students to write a report of their findings and include their reports and drawings in their portfolios. L1 P

Chapter Organizer

SECTION	OBJECTIVES	ACTIVITIES/FEATURES
7.1 Elements and Atoms National Science Standards: UCP.1, UCP.2; B.1, B.2; C.5;	1. **Relate** the particle structure of an atom to the identity of elements. 2. **Explain** how isotopes differ.	
7.2 Interactions of Matter National Science Standards: UCP.2, UCP.3; A.1; B.1-3; D.2; G.1, G.2	3. **Relate** the formation of covalent and ionic chemical bonds to the stability of atoms. 4. **Interpret** the formulas of chemical compounds and the meaning of chemical equations. 5. **Relate** water's polarity to its ability to dissolve substances and to the formation of acids and bases.	**Minilab:** How does liquid soap affect the surface tension of water?, p. 168 **Focus On** The Uniqueness of Water, p. 170 **Earth Science Connection:** Exploding Sodas, Potholes, and Soil, p. 172 **Minilab:** How can you determine the pH of common household items?, p. 173 **People in Biology:** Baldomero Olivera, p. 174
7.3 Life Substances National Science Standards: UCP.1, UCP.2; A.1, A.2; B.2, B.3; C.5; E.1, E.2; G.1-3	6. **Explain** why there is a large variety of organic compounds. 7. **Explain** how polymers are formed and broken down in organisms. 8. **Compare** the chemical structures of carbohydrates, lipids, proteins, and nucleic acids, and explain the importance of these substances in living things.	**A Broader View:** Was there life on Mars?, p. 178 **Thinking Lab:** How does dilution affect enzyme action?, p. 181 **Biolab:** Design Your Own Experiment—Does temperature affect an enzyme reaction?, p. 182

ACTIVITY MATERIALS

BIOLAB	MINILABS	ALTERNATE LAB
page 182 thermometer ice 400-mL beaker hot plate potato slices or raw liver kitchen knife waxed paper 3% hydrogen peroxide clock or timer	**page 168** waxed paper eyedropper toothpick liquid soap paper towels **page 173** small beakers lemon juice household ammonia solution liquid detergent shampoo vinegar pH paper pH color chart	**page 180** paper cups (4) gelatin, 6-oz. box 1 pineapple, fresh knife waxed paper chunk pineapple, canned grapes or orange sections refrigerator

Chapter 7 The Chemistry of Life

TEACHER CLASSROOM RESOURCES

Reproducible Masters	Transparencies
Section Focus Master 13: Elements L1 SAE **Reinforcement and Study Guide,** p. 25 L1 **Content Mastery,** pp. 25-28 L1	**Basic Concepts Transparency #4:** Atomic Structure L1 SAE
Section Focus Master 14: Forming Sodium Chloride L1 SAE **Reinforcement and Study Guide,** p. 26-27 L1 **Biolab and Minilab Worksheets,** pp. 23-24 L1 **Concept Mapping:** Properties of Water Important to Living Systems, p. 7 L1	**Basic Concepts Transparencies #5a, 5b:** Covalent Bonding; Ionic Bonding L1 SAE
Section Focus Master 15: Elements in Different Combinations L1 SAE **Reinforcement and Study Guide,** p. 28 L1 **Biolab and Minilab Worksheets,** pp. 25-26 L1 **Tech Prep Applications:** How Lean Is Lean Ground Beef?, pp. 7-8 L2 **Critical Thinking/Problem Solving:** Sugars and Isomers, p. 7 L3 **Laboratory Manual:** Tests for Organic Compounds, pp. 39-42; What Is the Action of Diastase?, pp. 43-46	**Reteaching Transparency #7:** Life Molecules L1 SAE

ASSESSMENT MATERIALS

Chapter Assessment, pp. 37-42
Performance Assessment in the Biology Classroom, p. 5
MindJogger Videoquiz
Alternate Assessment in the Science Classroom
Computer Test Bank

Spanish Resources SAE
English/Spanish Audiocassettes SAE
Cooperative Learning in the Science Classroom COOP LEARN
Lesson Plans

KEY TO TEACHING STRATEGIES

- L1 Level 1 activities should be within the ability range of all students including those with learning difficulties.
- L2 Level 2 activities are within the ability range of average to above-average students.
- L3 Level 3 activities are designed for the ability range of above-average students.
- SAE SAE activities should be within the ability range of Students Acquiring English.
- COOP LEARN Cooperative Learning activities are designed for small group work.
- P These strategies represent student products that can be placed into a best-work portfolio.
- These strategies are useful in a block scheduling format.

GLENCOE TECHNOLOGY

The following multimedia resources are available from Glencoe.

Biology: The Dynamics of Life
CD-ROM SAE
Videodisc Program

The Infinite Voyage Series
Unseen Worlds
The Geometry of Life
Fires of the Mind

Science and Technology Videodisc Series (STVS)
Chemistry
Images of Atoms
Ecology
Treating Acid Lakes
Human Biology
PET Scanner

CHAPTER

7 The Chemistry of Life

Chapter Overview

Students begin their study of the chemistry of living things by reviewing atoms and elements. Students examine some basic chemistry principles and learn to interpret chemical formulas and equations. Students also learn to distinguish between acids and bases.

In the last part of the chapter, students examine the role of carbon in living things. They learn about the structure of carbohydrates, lipids, proteins, and nucleic acids.

Key Terms

acid
amino acid
atom
base
carbohydrate
cellulose
compound
covalent bond
disaccharide
DNA
enzyme
glycogen
hydrogen bond
ion
ionic bond
isomer
isotope
lipid
metabolism
mixture
molecule
monosaccharide
nucleic acid
nucleotide
nucleus
peptide bond
pH
polar molecule
polymer
polysaccharide
protein
RNA
solution
starch

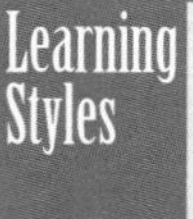

Learning Styles

Look for the following logo for strategies that emphasize different learning modalities. LS

Kinesthetic	**Meeting Individual Needs, p. 161; Building a Model, p. 165**
Visual-Spatial	**Demonstration, pp. 160, 165, 169, 177; Display, p. 161; Visual Learning, pp. 161, 176; Portfolio, p. 162; Enrichment, p. 166; Minilab, pp. 168, 173; Student Journal, p. 170; Activity, p. 177**
Interpersonal	**Meeting Individual Needs, p. 178; Project, p. 179**
Linguistic	**Student Journal, pp. 162, 178; Portfolio, p. 168; Using Science Terms, p. 177**
Logical-Mathematical	**Portfolio, p. 166; Meeting Individual Needs, p. 167; Alternate Lab, p. 180; Thinking Lab, p. 181**

CHAPTER

7 The Chemistry of Life

What makes a living thing unique? Suppose you held a rock in one hand and a bouquet of flowers in the other. How would you explain that the flowers are alive, and the rocks are part of the nonliving world? Are living and nonliving things composed of entirely different substances?

Early biologists struggled with the same questions. Two hundred years ago, many biologists believed that some mysterious life force guided all life processes. They also thought that this life force somehow made the matter of living things special so that it could never be prepared in the laboratory. If you've seen a Frankenstein movie, you know that Dr. Frankenstein was searching for this life force so that he could restore life to dead tissue. Scientists now know that there is no life force, and that living things actually have a great deal in common with rocks, cassette tapes, computer chips, and other nonliving objects.

Concept Check

You may wish to review the following concepts before studying this chapter.

- Chapter 1: homeostasis, energy, organism

Chapter Preview

Laboratory Activities

Biolab: Design Your Own Experiment

- Does temperature affect an enzyme reaction?

Minilabs

- How does liquid soap affect the surface tension of water?
- How can you determine the pH of common household items?

Living and nonliving things are composed of the same basic materials, and the substances found in horses, grass, and trees obey the same laws of chemistry and physics as those found in rocks. Why, then, do you suppose that living and nonliving things seem so different?

159

CHAPTER 7

Introducing the Chapter

Direct students' attention to the photographs. Ask students to list the features of living things. Lead the discussion to a comparison of the composition of living things and nonliving things. Ask students to speculate about why early scientists thought a mysterious force guided chemical changes in organisms.

Theme Development

Unity within diversity is stressed through the discussion that although living things and nonliving things differ, they are alike in that all are made up of the same elements. *Energy* should be discussed along with chemical reactions. Be sure to stress that as elements unite with other matter or break apart during chemical reactions, energy is either given off or used.

Concept Check

Review the concepts of homeostasis and energy from Unit 2 and the concept of organisms from Unit 1. Lead the discussion from the idea of homeostasis within an ecosystem to the idea of homeostasis and conservation within the cells of organisms.

Assessment Planner

Choose assessment strategies from the following pages to evaluate the progress of your students.

Assess, pp. 162, 172, 183
Alternate Lab, pp. 180-181
Minilabs, pp. 168, 173
Portfolio, pp. 162, 166, 168
Thinking Lab, p. 181
Biolab, pp. 182-183
Chapter Review, pp. 185-187

SECTION

7.1

Prepare

Key Concepts

Students are introduced to the subatomic particles that make up the atoms of elements. Special emphasis is placed on the elements that compose organisms. In addition, students become acquainted with the uses of isotopes in biology.

Block Scheduling

Look for this symbol for strategies that are useful in a block scheduling format. For more information on block scheduling, refer to Section 7.1 in the **Lesson Plans** booklet.

Materials

- Gather iron nails, copper pipe, a piece of aluminum foil, a piece of coal or charcoal, and a mercury thermometer for the demonstration.
- Purchase navy beans, pinto beans, and kidney beans for the closing activity.

1 Focus

Bellringer

Before presenting the lesson, display **Section Focus Master 13** on the overhead projector and have students answer the accompanying questions. L1 SAE

Demonstration

LS **Visual-Spatial** Display various elements using common objects such as iron nails, a piece of copper pipe, a piece of aluminum foil, a piece of coal or charcoal, and a mercury thermometer. Tell students that each material is composed of a different kind of element. Direct students' attention to the appropriate object when discussing each element.

SECTION

7.1 Elements and Atoms

Section Preview

Objectives

Relate the particle structure of an atom to the identity of elements.

Explain how isotopes differ.

Key Terms

atom
nucleus
isotope

Industrial robots are often used to do work that humans could do if it weren't for danger, hazardous materials, or the need for a strictly clean environment. While the behavior of industrial robots may seem human in many ways, they are still nonliving things. Even so, they are composed of the same basic building blocks as humans and other living organisms—elements.

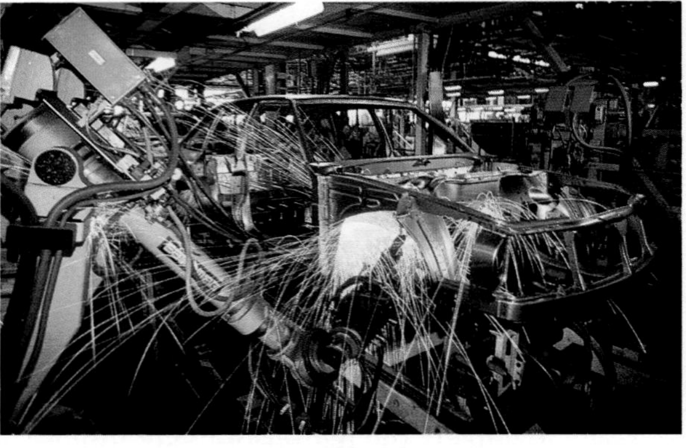

Elements

Everything—whether it is a rock, robot, or flower—is made of substances called elements. Suppose you found a nugget of pure gold. You could grind it into a billion bits of powder and every particle would still be gold. You could treat the gold with every known chemical, but you could never break it down into two simpler substances. That's because gold is an element, one of the simplest chemical substances. An element is a substance that can't be broken down into simpler substances.

On Earth, 90 elements occur naturally. The periodic table, shown in Appendix D, lists all currently known elements. With the exception of elements 43 and 61, elements numbered 1 through 92 are found in nature. Elements numbered 93 and above are synthetic elements.

Natural elements in living things

Of the 90 naturally occurring elements, only about 25 are essential to living organisms. *Table 7.1* lists several elements found in the human body. Notice that only four elements—carbon, hydrogen, oxygen, and nitrogen—make up more than 96 percent of the mass of a human. Each element is abbreviated by a one- or two-letter symbol. For example, the symbol C represents the element carbon, Ca represents calcium, and Cl represents chlorine.

Trace elements

Notice that some of the elements listed in *Table 7.1,* such as iron and magnesium, are present in living things in very small amounts. Such elements are known as trace elements. They play a vital role in maintaining healthy cells in all organisms, as shown by the examples in *Figure 7.1.*

Program Resources

Section Focus Master 13 L1 SAE
Reinforcement and Study Guide, p. 25 L1
Basic Concepts Transparency 4 and **Master** L1 SAE

Table 7.1 Elements that Make Up Living Things					
Element	**Symbol**	**Percent by Mass in Human Body**	**Element**	**Symbol**	**Percent by Mass in Human Body**
Oxygen	O	65.0	Iron	Fe	trace
Carbon	C	18.5	Iodine	I	trace
Hydrogen	H	9.5	Copper	Cu	trace
Nitrogen	N	3.3	Manganese	Mn	trace
Calcium	Ca	1.5	Molybdenum	Mo	trace
Phosphorus	P	1.0	Cobalt	Co	trace
Potassium	K	0.4	Boron	B	trace
Sulfur	S	0.3	Zinc	Zn	trace
Sodium	Na	0.2	Fluorine	F	trace
Chlorine	Cl	0.2	Selenium	Se	trace
Magnesium	Mg	0.1	Chromium	Cr	trace

Atoms—The Building Blocks of Elements

Elements, whether they are found in living things or not, are made up of atoms. An **atom** is the smallest particle of an element that has the characteristics of that element.

The structure of the atom

Each element has distinct characteristics because of the structure of the atoms of which it is composed. For example, iron differs from aluminum because the structure of iron atoms differs from that of aluminum atoms. Still, all atoms have the same general arrangement. The center of an atom is called the **nucleus** (plural, nuclei). It is made of positively charged particles called protons (p^+) as well as particles called neutrons (n^0), which have no charge. All nuclei are positively charged because of the presence of protons.

Forming a cloud around the nucleus are even smaller particles called electrons (e^-), which have negative charges. If you've ever looked at a spinning fan, you've probably noticed that as the fan blades turn, they appear to form a blurry disk that occupies a space around the center of the fan. An electron cloud is simply the space around the atom's nucleus that is occupied by these fast-moving electrons.

Figure 7.1

Plants obtain trace elements by absorbing them through their roots. Animals get these important elements in the foods they eat.

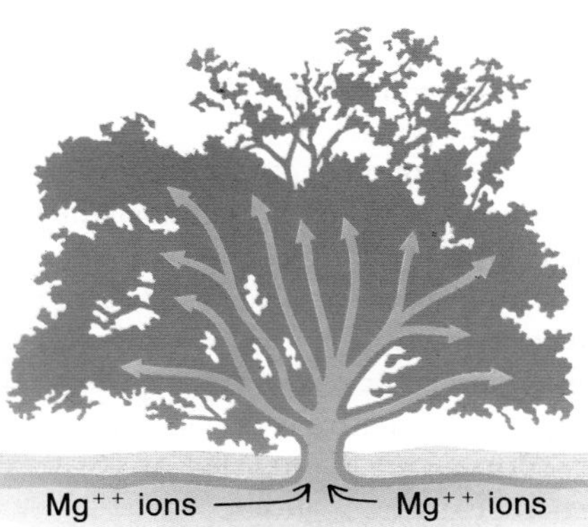

Plants must have magnesium (Mg) in order to form the green pigment chlorophyll that captures light energy for the production of sugars.

Iodine (I) is used in mammals to produce a substance that affects the rates of growth, development, and chemical activities in the body.

SECTION 7.1

2 Teach

Audiovisual

Show the filmstrip *An Introduction to Chemistry*, National Geographic Education Society.

Concept Development

Stress that the elements carbon, hydrogen, nitrogen, oxygen, phosphorus, and sulfur are the main components of living matter. All these elements form molecules through covalent bonding.

Tying to Previous Knowledge

Relate the discussion of the chemistry of living things to the characteristics of living things discussed in Chapter 1. Emphasize that the presence of carbon is a characteristic shared by all organisms.

Display

LS **Visual-Spatial** Make a bulletin board display that models the structure of an atom. Label the nucleus, protons, neutrons, and electrons of the model. Refer to the display when discussing atomic structure.

Visual Learning

Using the Table Discuss possible dietary sources of each element listed in Table 7.1. Have students redesign the table to include pictures that show two sources of each element listed. Have students conduct additional research if necessary. L1 LS

Meeting Individual Needs

Visually Impaired To help students who are visually impaired understand the structure of an atom, make atomic models by gluing marbles, jelly beans, and yarn to a piece of cardboard. Use the yarn to outline the nucleus and the energy levels. Use marbles for the neutrons and jelly beans for the electrons and protons. Allow visually impaired students to manipulate the model. Have sighted students work with visually impaired students to assist in the identification of the parts of the model. L1 COOP LEARN LS

Chalkboard Activity

On the chalkboard, prepare a table with the heads: Particle, Location, Charge, and Symbol. Beneath the *Particle* head, list: Electron, Neutron, and Proton. Have volunteers come to the chalkboard to complete the table. L1 SAE

Science, Technology, and Society

You *Can* See Atoms

The scanning tunneling microscope (STM) provides a three-dimensional map of a molecule's surface by measuring the flow of electrons across a small vacuum gap. A tungsten tip similar to a phonograph needle is positioned a few angstroms from the substance being studied. When a small voltage is applied, some electrons jump across the gap between the tip and the surface. As the tip moves across the surface, the current varies according to the contours of the atoms present. This varying current is recorded, stored, computer-enhanced, and then plotted.

3 Assess

Check for Understanding

Refer students to the periodic table. Provide students with a list of several common elements. For each element, ask students to give the correct symbol and to diagram the atom's structure to show its protons, neutrons, and electron energy levels. L2

Reteach

On the chalkboard, show students how to construct energy level diagrams of atoms. Emphasize how the electrons of an atom are distributed using several different models. L1 SAE

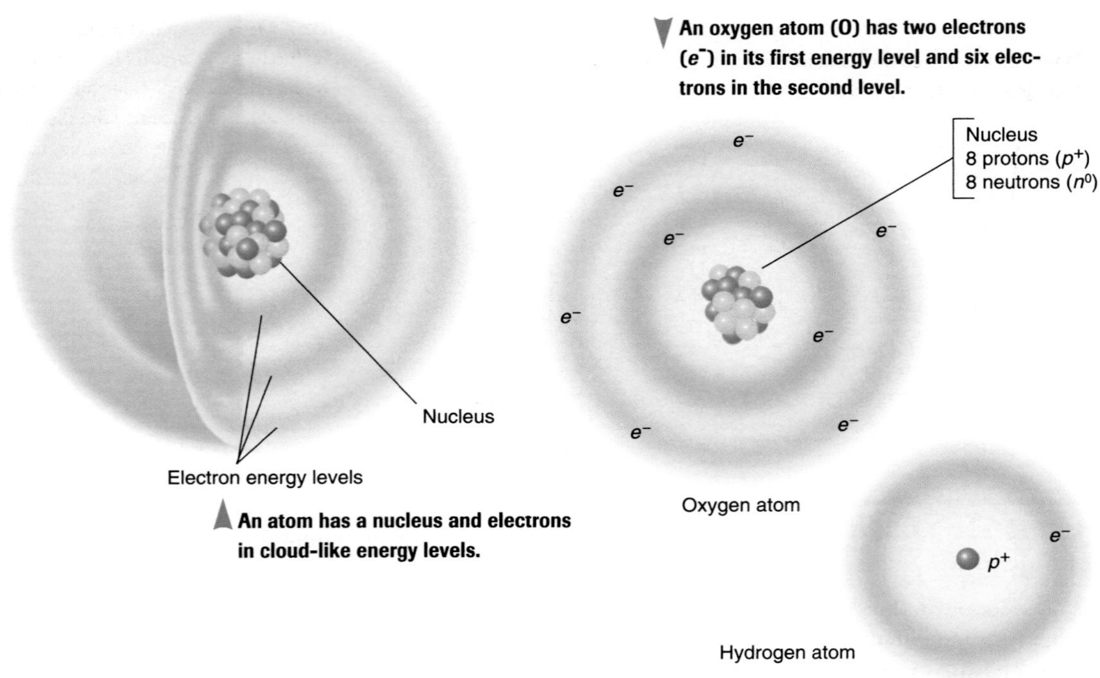

An atom has a nucleus and electrons in cloud-like energy levels.

Figure 7.2

Electrons move rapidly around atoms, forming electron clouds. Electron clouds are actually composed of several energy levels, which are regions in which electrons travel.

Hydrogen (H), the simplest atom, has just one electron in its first energy level and one proton in its nucleus.

Electron energy levels

Electrons travel around the nucleus in certain regions known as energy levels, *Figure 7.2.* Each energy level has a limited capacity for electrons.

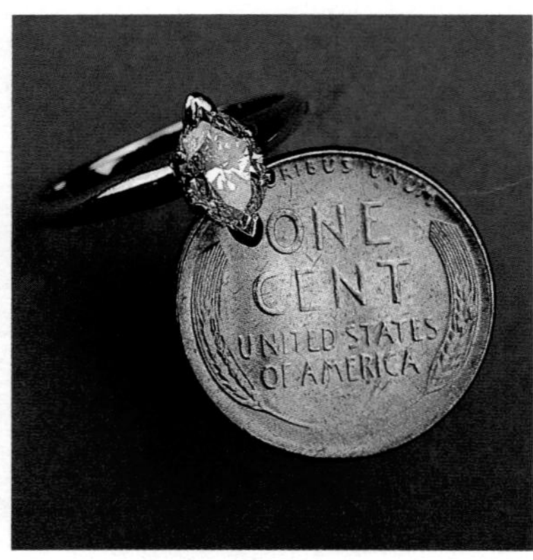

Figure 7.3

The properties of an element are determined by the structure of its atoms. Copper atoms in the coin have 29 protons and 29 electrons; gold atoms in the ring have 79 protons and 79 electrons. The diamond is nearly pure carbon, with atoms that contain six protons and six electrons. As you can see, these three elements have very different properties.

Because the first energy level is the smallest, it can hold a maximum of two electrons. The next level is larger and can hold a maximum of eight electrons. The third level is larger yet and can hold 18. For example, the oxygen atom in *Figure 7.2* has a total of eight electrons. Two electrons fill the first energy level. The remaining six electrons must, therefore, occupy the second level.

Atoms contain equal numbers of electrons and protons; therefore, they have no net charge. The hydrogen (H) atom in *Figure 7.2* has just one electron and one proton. Oxygen (O) has eight electrons and eight protons. *Figure 7.3* compares the atomic structure and properties of three other elements.

STUDENT JOURNAL

Life as an Atom Ask students to pretend that they are a particular type of atom. Ask them to write a paragraph or two to describe their basic structure. Encourage the students to use their imaginations, while retaining scientific accuracy. L1 LS

PORTFOLIO

Diagramming Atomic Structure Make a detailed drawing of an aluminum atom, showing the contents of the nucleus of the atom and the energy levels of the electrons. L1 LS SAE P

Figure 7.4

Radioactive isotopes are used in medicine to diagnose and/or treat some diseases.

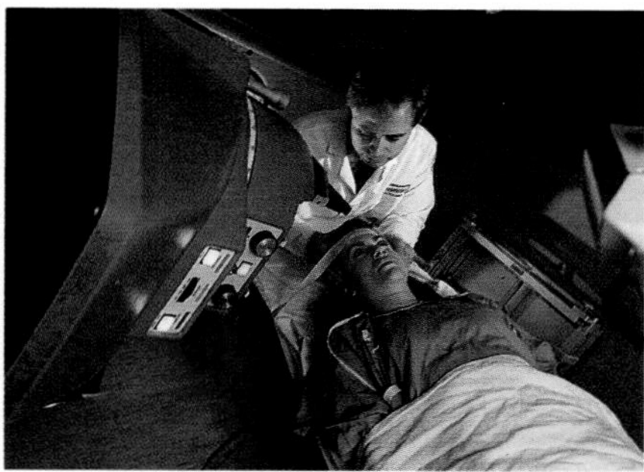

Neutrons given off when radioactive isotopes break apart are deadly to rapidly growing cancer cells. The patient is being treated with radiation from a radioactive isotope of cobalt (Co).

The thyroid gland in mammals uses iodine to make thyroid hormones. Radioactive iodine (I) introduced into the body is absorbed by the thyroid gland. By detecting the radioactive iodine taken up, the function of the thyroid gland can be measured.

Isotopes of an Element

Atoms of an element sometimes contain different numbers of neutrons in the nucleus. Atoms of the same element that have different numbers of neutrons are called **isotopes** of that element. For example, most carbon nuclei contain six neutrons. However, some have seven or eight neutrons. Each of these is an isotope of the element carbon. Scientists refer to isotopes by giving the combined total of protons and neutrons in the nucleus. Thus, the most common carbon atom is referred to as carbon-12 because it has six protons and six neutrons. Other isotopes of carbon include carbon-13 and carbon-14.

Isotopes are often useful to scientists. The nuclei of some isotopes are unstable and tend to break apart. For instance, carbon-14 is unstable. As nuclei break, they give off radiation. These isotopes are said to be radioactive. Because radiation is detectable and can damage or kill cells, scientists have developed some useful applications for radioactive isotopes, as you can see in *Figure 7.4.*

Section Review

Understanding Concepts

1. A nitrogen atom contains seven protons, seven neutrons, and seven electrons. Describe in detail the structure of a nitrogen atom. Use a labeled drawing if you wish.
2. An atom of aluminum has 13 protons in its nucleus. How many electrons does the atom have? Explain how you know.
3. How do two isotopes of the same element differ?

Thinking Critically

4. A fluorine atom has nine electrons. Make an energy level diagram similar to the one in *Figure 7.2.* How many electrons would be needed to fill its outer level?

Skill Review

5. **Making and Using Graphs** Make a pie graph of the elements listed in *Table 7.1.* Group all the elements below phosphorus in one segment. For more help, refer to Organizing Information in the *Skill Handbook.*

Answers to Section Review

1. The nucleus of the N atom contains seven protons and seven neutrons. The first energy level contains two electrons. The next energy level contains five electrons.
2. 13 electrons; the number of protons is equal to the number of electrons.
3. Isotopes of an element differ in the number of neutrons.

Thinking Critically

4. The diagram should show two energy levels containing two and seven electrons, respectively. One electron is needed to fill the outer level.

Skill Review

5. Elements below phosphorus should make up about 1.2% of the graph.

Bioethics

Use of Radioisotopes

Radiation can penetrate and disrupt the functions of living cells. However, certain radioisotopes have been shown to have practical uses in medicine as diagnostic tools. For example, radioactive iodine is used to identify problems with the thyroid gland. Discuss with students how the benefits of such uses of radiation compare with the risks of exposure.

Extension

Have students who have mastered the concepts in this section make drawings or three-dimensional models of some common elements and their isotopes. Ask students to display and explain their models to the class to provide reinforcement of the concept to other students. L3

Assessment

Performance: Ask students to come to the chalkboard to draw atoms with a specific number of electrons. Have other students use the periodic table in Appendix D to identify each element that is drawn. L1

4 Close

Activity

Have students build a model of an element, such as oxygen, at their desks using navy beans for electrons, pinto beans for neutrons, and kidney beans for protons. L1

SECTION

7.2

Prepare

Key Concepts

Students will study compounds and ionic, covalent, and hydrogen bonds. Students develop an understanding of the properties of water that make it an excellent solvent and a necessary component of organisms. Students also compare the properties of acids and bases.

Block Scheduling

Look for this symbol for strategies that are useful in a block scheduling format. For more information on block scheduling, refer to Section 7.2 in the **Lesson Plans** booklet.

Materials

- Purchase sand, salt, and gelatin for the Enrichment.
- Purchase a piece of silk cloth, a 9-volt battery, and two pieces of wire for the demonstrations.
- Purchase waxed paper for the Minilab on page 168.
- Purchase lemon juice, household ammonia, liquid detergent, shampoo, and vinegar for the Minilab on page 173.
- Purchase toothpicks and gumdrops or marshmallows for the activities.

1 Focus

Bellringer

Before presenting the lesson, display **Section Focus Master 14** on the overhead projector and have students answer the accompanying questions. L1 SAE

SECTION

7.2 Interactions of Matter

Section Preview

Objectives

Relate the formation of covalent and ionic chemical bonds to the stability of atoms.

Interpret the formulas of chemical compounds and the meaning of chemical equations.

Relate water's polarity to its ability to dissolve substances and to the formation of acids and bases.

Key Terms

compound
covalent bond
molecule
ion
ionic bond
mixture
solution
polar molecule
hydrogen bond
metabolism
pH
acid
base

For many people, nothing beats the delight of biting into a fresh, hot pizza cooked to perfection and built according to your instructions. Pizza is fun to eat because it can be made in so many different ways. Each of the individual ingredients in pizza may be boring to eat alone, but when assembled in the right combination and amounts, they become a personalized meal. In many ways, elements are similar to the ingredients of pizza. Standing alone, most elements are unremarkable. But when combined with one or more other elements, they form a large variety of useful substances.

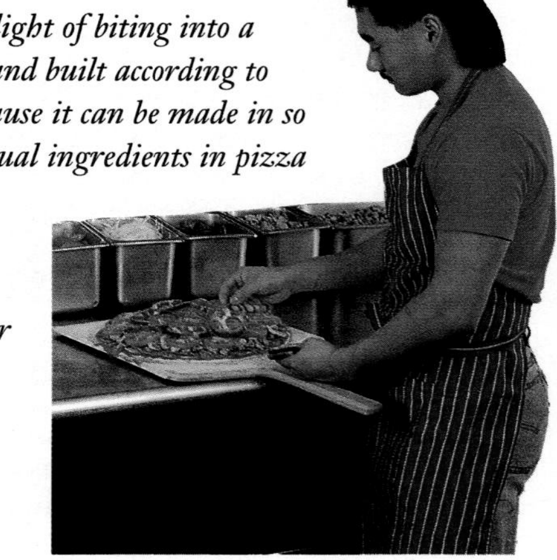

Compounds and Bonding

Water is a substance everyone is familiar with. Is it an element? If you pass an electric current through it, water breaks down into hydrogen and oxygen. Neither hydrogen nor oxygen can be broken down further, so they must be elements. From this description, you can infer that water is not an element. Rather, it is a type of substance called a compound. A **compound** is a substance that is composed of atoms of different elements chemically combined. Water (H_2O) is a compound composed of the elements hydrogen and oxygen. Just as the combined ingredients in a pizza result in a tasty meal, you can see in *Figure 7.5* that the properties of a compound are different from those of its individual elements.

Figure 7.5
Table salt is made from the elements sodium (Na) and chlorine (Cl). The flask contains the poisonous, yellow-green gas chlorine. The lump of silver-white metal is the element sodium. When sodium and chlorine are placed together, they combine. One atom of sodium combines with one atom of chlorine to form sodium chloride (NaCl), commonly known as table salt. Table salt consists of white crystals that no longer resemble either sodium or chlorine.

CULTURAL DIVERSITY

Kenichi Fukui and Chemical Reactions

In the 1950s, Japanese chemist Kenrich Fukui developed the idea that chemical reactions occur as a result of interactions of the outer-level electrons of one atom or molecule with the outer-level electrons of another atom or molecule. In 1981, Fukui received the Nobel Prize for Chemistry for his investigations of the mechanisms of chemical reactions. Discuss with students the work of Kenichi Fukui toward the understanding of chemical interactions.

How covalent bonds form

Most matter is in the form of compounds, but how and why do atoms combine? Atoms combine with other atoms only when conditions are right, and they do so because they become more stable by combining.

Remember electron energy levels? For most elements, an atom becomes stable when its outer energy level has eight electrons. An exception is hydrogen, which becomes stable with two electrons in the first energy level. One way that atoms can become stable is by sharing electrons with other atoms.

For example, two hydrogen atoms can combine with each other by sharing their electrons, as shown in *Figure 7.6.* As you know, individual atoms of hydrogen contain only one electron. Each atom becomes stable by sharing its electron with the other atom. The two shared electrons move about the first energy level of both atoms. The attraction of the positively charged nuclei for the shared, negatively charged electrons holds the atoms together. When two atoms share electrons, the force that holds them together is called a **covalent bond.**

Most compounds in organisms have covalent bonds. Examples of these compounds include sugars, fats, proteins, and water. A **molecule** is a group of atoms held together by covalent bonds and having no overall charge. A molecule of water is represented by the chemical formula H_2O. The subscript *2* represents two atoms of hydrogen (H) combined with one atom of oxygen (O). As you will see, many compounds in living things have more complex formulas.

How ionic bonds form

Not all atoms bond with each other by sharing electrons. Sometimes, atoms combine with each other by gaining or losing electrons in their outer energy levels. An atom (or group of atoms) that gains or loses electrons has an electrical charge and is called an **ion.**

Figure 7.6

Sometimes atoms combine by sharing electrons to form covalent bonds.

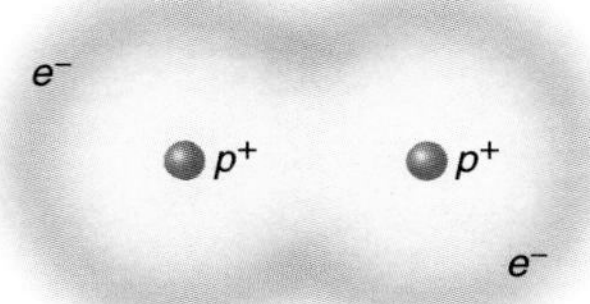

▲Hydrogen gas exists in nature as two hydrogen atoms sharing electrons with each other. The electrons move around the nuclei of both atoms. Because of this bonding, hydrogen gas is written as H_2 rather than just H.

▼When two hydrogens share electrons with oxygen, they form covalent bonds to produce a molecule of water. Each hydrogen atom shares one electron with the single oxygen atom.

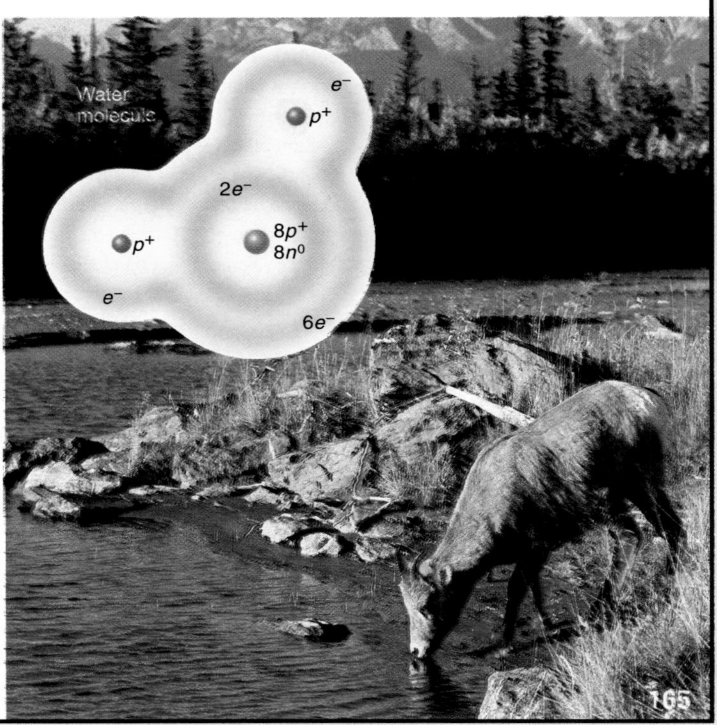

SECTION 7.2

Demonstration

LS Visual-Spatial Place a beaker of lightly salted water on a table where it can be seen by students. Connect one end of a piece of wire to the positive terminal of a 9-volt battery. Connect a second piece of wire to the negative terminal of the battery. Place the other ends of the wires into the water; do not allow wires to touch. Instruct students to observe the ends of the wires for the appearance of bubbles. Explain that passing an electric current through water breaks the water apart, resulting in the elements oxygen and hydrogen.

2 Teach

Reinforcement

Give students several chemical formulas related to living things, such as CO_2 and $C_6H_{12}O_6$. Have them practice identifying the elements in the compounds and the numbers of atoms of each type of element shown in each formula. L1 SAE

Building a Model

LS Kinesthetic Have students build models of water molecules. Students may use toothpicks and gumdrops or colored marshmallows to represent the atoms in the molecule. L1 COOP LEARN

Program Resources

Section Focus Master 14 L1 SAE
Reinforcement and Study Guide, pp. 26-27 L1
Biolab and Minilab Worksheets, pp. 23-24 L1
Performance Assessment in the Biology Classroom, p. 5 L1 P
Concept Mapping, p. 7 L1
Basic Concepts Transparencies 5a, 5b and **Master** L1 SAE

GLENCOE TECHNOLOGY

Videodisc

Biology: The Dynamics of Life
Disc 1, Side 1
Covalent Bonding (Ch. 17)

Enrichment

Visual-Spatial Prepare mixtures of sand and water (suspension), salt and water (solution), and water and gelatin (colloid). Explain that the contents of all three containers represent mixtures. Shake each container and ask students to describe any changes they observe in the appearance of the mixtures.

Use the observable traits of the mixtures to explain that there are different types of mixtures. Explain that the sand and water represent a suspension: a heterogeneous mixture consisting of finely divided particles of a solid temporarily suspended in a liquid. The salt and water are a solution—a mixture in which one or more substances are evenly distributed in another substance. The gelatin mixture is an example of a colloid—a mixture in which the particles are larger than those in a solution but smaller than those in a suspension.

As in a solution, the particles in a colloid do not settle out. Many biological liquids contain suspended proteins and are classified as colloids. Colloids may be distinguished from solutions by the fact that colloids scatter a beam of light passed through them, whereas solutions do not. This scattering is called the Tyndall effect, and it may be demonstrated by students comparing gelatin with plain water. L2

A different type of chemical bond holds ions together. The bond formed between a sodium atom (Na) and a chlorine atom (Cl) provides a good example of this. Sodium becomes stable by losing the one electron in its outer energy level, and chlorine becomes stable by accepting this lost electron. Because sodium has lost one negatively charged electron, it now has one more proton than it has electrons. Thus, it is a positively charged ion. Chlorine has gained one electron, and it now has one more electron than it has protons. Thus, it is a negatively charged ion. Because opposite charges attract, there is an attractive force between two ions of opposite charge known as an **ionic bond.** The compound formed when sodium and chlorine react to form an ionic bond is known as sodium chloride, as shown in *Figure* 7.7. Sodium chloride is represented by the chemical formula NaCl.

Ionic compounds are less abundant in living things than are covalent molecules, but ions are important in biological processes. For example, sodium and potassium ions are required for transmission of nerve impulses. Calcium ions are necessary for muscles to contract. Plant roots absorb needed minerals as ions.

Mixtures and Solutions

When elements combine to form a compound, they no longer have their original properties. What happens if substances are just mixed together and do not combine chemically? A **mixture** is a combination of substances in which individual substances retain their own properties. For example, if you stirred sand and sugar together, they would neither change nor combine chemically.

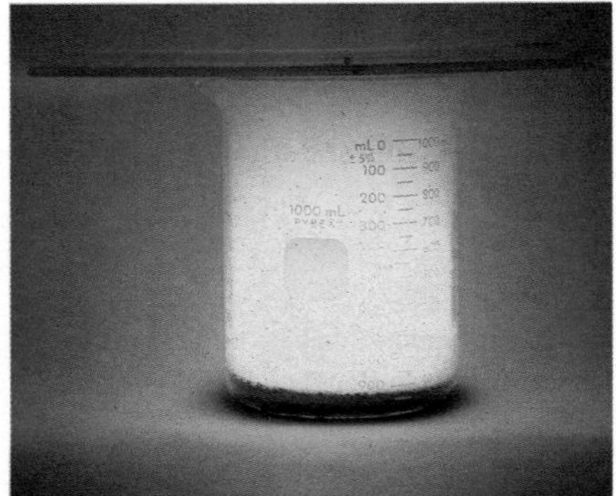

Figure 7.7

A sodium atom (Na) contains 11 electrons—two in the first energy level, eight in the second, and one in the third. A chlorine atom (Cl) has 17 electrons, with the outer level holding seven electrons (below). When sodium and chlorine combine, the sodium atom loses one electron, and the chlorine atom gains it. Thus, with eight electrons in its outer level, the chlorine ion formed is stable and has a negative charge. Sodium has lost the one electron that was in its third energy level. Thus, the sodium ion has a positive charge and is stable. The actual reaction is shown at right.

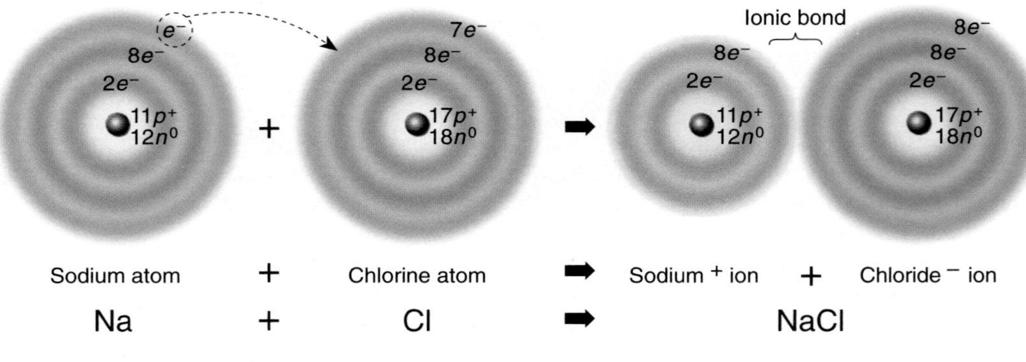

GLENCOE TECHNOLOGY

Videodisc

Biology: The Dynamics of Life
Disc 1, Side 1
Ionic Bonding (Ch. 18)

PORTFOLIO

Combinations of Elements Have students create a concept map that shows the relationships among elements, mixtures, compounds, and solutions. Have students include the following terms in their concept maps: *atoms, elements, molecules, compounds, mixtures, solutions, solvent.* Students may add other terms and should supply their own connecting words.

L1 P

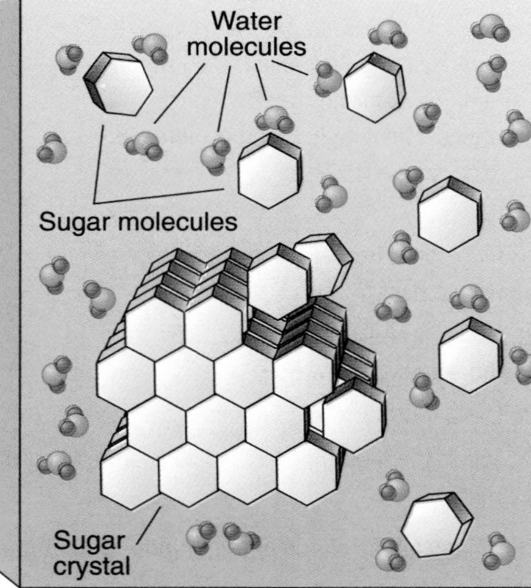

Figure 7.8
The sugar molecules in the Kool-Aid mix (left) spread evenly throughout the water, making a kind of mixture called a solution. The substance being dissolved, sugar, is called the solute; water is called the solvent. Neither the sugar nor the water changes chemically. Water molecules simply attract the sugar molecules and pull them into solution (below). The sugar can be recovered easily by evaporating the water.

A solution is a kind of mixture that is very important in living things. A **solution** is a mixture in which one or more substances are distributed evenly in another substance. You may remember making Kool-Aid when you were younger. The sugar molecules in Kool-Aid will dissolve easily in water to form a solution, as shown in *Figure 7.8.*

In organisms, many important substances, such as sugars and mineral ions, are dissolved in water. The more solute that is dissolved in a given amount of solvent, the greater is the solution's concentration (strength). The concentration of a solute is important to organisms. Organisms can't live unless the concentration of dissolved substances stays within a narrow range. Organisms have many mechanisms to keep the concentrations of molecules and ions within this range. For example, the amount of sugar dissolved in your bloodstream must stay within a critical range. Substances produced in the pancreas and other organs cause sugar to be stored or released in order to keep blood sugar levels in this range. This regulation is another example of homeostasis.

Water and Its Importance

Water is perhaps the most important compound in living organisms. Most life processes can occur only when molecules and ions are free to move and collide with one another. This condition exists when they are dissolved in water. Water serves as a means of transport of materials in organisms. For example, both blood and plant sap are mostly water. In fact, water makes up from 70 to 95 percent of most organisms. The water molecule, as you have learned in *Figure 7.6,* is composed of two atoms of hydrogen linked by covalent bonds to one atom of oxygen.

Using Science Terms

Tell students to imagine they are making a glass of iced tea from a mix. Ask: Which part of the resulting solution is the solute? Which is the solvent? *The tea mix is the solute because it is the material being dissolved. The water in which the mix is dissolved is the solvent.* L1 SAE

Audiovisual

Show the film *Chemistry in Everyday Life,* EBEC. 

Assessment

Performance Assessment in the Biology Classroom: p. 5, Just the Right Solution. Have students carry out this activity to determine how different activities affect rates of dissolving. L1 P

GLENCOE TECHNOLOGY

Videodisc

The Infinite Voyage: Unseen Worlds
Studying the Basic Building Blocks: The Atom (Ch. 9)

The Scanning Tunneling Microscope: Observing Atomic Particles (Ch. 10)

Meeting Individual Needs

Learning Disabled Have students construct a table to compare and contrast mixtures and compounds. Encourage students to reread pages 164-167 to find the information needed to complete their tables. L1 LS SAE

MiniLab

Purpose

Visual-Spatial Students will observe and explain the natural shape of water droplets and explain the effect soap has on the surface tension of water.

Process Skills

observe and infer, recognize cause and effect

Teaching Strategies

- As an extension to the activity, have students place water droplets on materials other than waxed paper to observe whether the shape of the droplet changes. Other materials might include notebook paper or paper toweling.

Expected Results

Students will observe the round shape of the water drop placed on the waxed paper. When liquid soap is added to the water, students will observe that the water droplet flattens out and changes shape.

Analysis

1. round or spherical
2. The droplet flattened.
3. Soap bonds to the water, decreasing the attraction of water to itself.
4. Soap disrupts the surface tension of water.

Assessment

Portfolio: Have students write an evaluation of the lab to place in their portfolio. Students can also research why soap, when added to water, helps remove oils and grease from dishes, clothing, or their bodies. L1 P

MiniLab

How does liquid soap affect the surface tension of water?

Surface tension of water is a force exerted by the surface of water on the particles below. It results from the attraction of water molecules to other water molecules. The force of surface tension tends to pull drops of water into spherical (ball) shapes. When drops of falling water are photographed with a high-speed camera, their spherical shape is visible.

Procedure

1. Place one drop of water on a piece of waxed paper.
2. Look carefully at the water drop from the side and draw its shape.
3. Dip a toothpick into liquid soap.
4. Touch the toothpick to the water drop while viewing from the side.

Analysis

1. What was the shape of the original water drop?
2. How did the shape change when the soap was added?
3. What do you think caused the drop to change?
4. What can you conclude about the effect of soap on the surface tension of water?

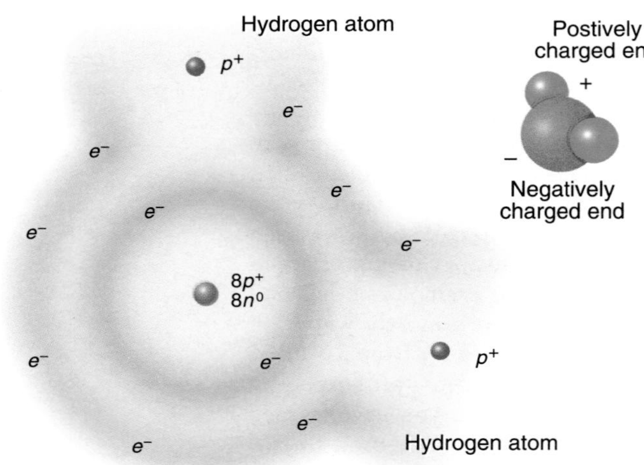

Water is polar

Sometimes, when atoms form covalent bonds, they do not share the electrons equally. In the water molecule, shown in *Figure 7.9,* you can see that the oxygen atom attracts the shared electrons more strongly than the hydrogen atoms do. As a result, the electrons spend more time near the oxygen atom than they do near the hydrogen atoms.

Water is an example of a polar molecule. A **polar molecule** is a molecule with an unequal distribution of charge; that is, each molecule has a positive end and a negative end. Polar water molecules attract each other as well as ions and other polar molecules. Because of this attraction, water has the ability to dissolve many ionic compounds, such as salt. Water also dissolves many polar molecular substances, such as sugar.

Water molecules also attract each other. The positively charged hydrogen atoms of one water molecule attract the negatively charged oxygen atoms of another water molecule. This attraction of opposite charges between hydrogen and oxygen forms a weak bond called a **hydrogen bond.** Hydrogen bonds are important to organisms because they help hold many large molecules, such as proteins, together.

Figure 7.9

In water molecules, the electron-rich oxygen atom has a slight negative charge. Because they lack electrons, the hydrogen atoms have a slight positive charge. Because of water's bent shape, the protruding oxygen end of the molecule has a slight negative charge, while the end with protruding hydrogen atoms has a slight positive charge.

PORTFOLIO

The Importance of Water Have students write an essay that explains why water is important to them. Ask them to think about their family's daily usage of water. Challenge them to think about what usages are especially important in view of the fact that they might someday be asked to cut their water usage in half. L1 P LS

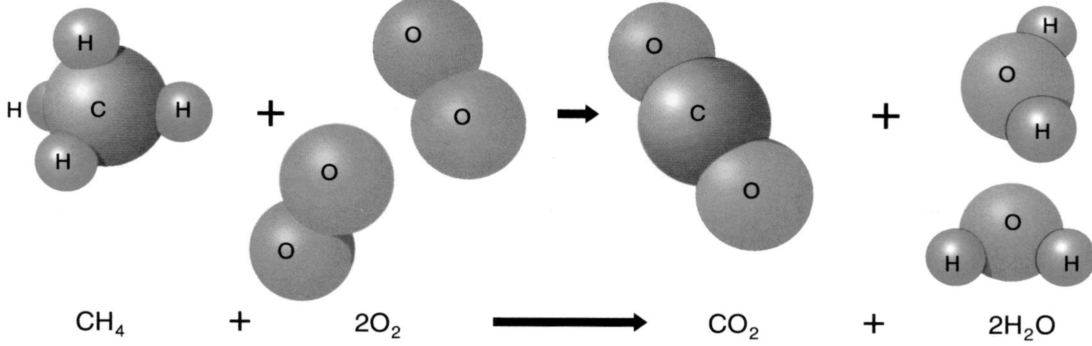

Figure 7.10

When methane burns in air, it reacts with two oxygen molecules. Notice how all the atoms that were present in the reactants (methane and oxygen) have been rearranged to produce the products (carbon dioxide and water). No atoms were lost, gained, or changed to different atoms. The equation tells you that one molecule of methane reacts with two molecules of oxygen to yield (produce) one molecule of carbon dioxide and two molecules of water.

Chemical Reactions

When chemical reactions occur, bonds between atoms are formed or broken, and substances change into different substances. In organisms, chemical reactions occur over and over inside cells. All the chemical reactions that occur within an organism are referred to as that organism's **metabolism.** These reactions break down and build molecules that are important for the functioning of organisms. Scientists represent a chemical reaction by writing a chemical equation. Chemical equations use chemical symbols and formulas to represent each element or substance.

Writing chemical equations

The events that take place when methane burns in air are pictured in *Figure 7.10.* Substances that undergo chemical reactions, such as the methane and oxygen in *Figure 7.10,* are called reactants. Substances formed by chemical reactions, such as the carbon dioxide and water in the figure, are called products.

It's easy to tell how many molecules are involved in the reaction because the number before each chemical formula indicates the number of molecules of each substance. The subscript numbers in a formula indicate the number of atoms of each element in a molecule of the substance. A molecule of table sugar can be represented by the formula $C_{12}H_{22}O_{11}$. The lack of a number before a formula or under a symbol indicates that only one atom or molecule is present.

Looking at the equation in *Figure 7.10,* you can see that methane is composed of one atom of carbon and four atoms of hydrogen. Oxygen is composed of two-atom molecules. Perhaps the easiest way to understand chemical equations is to know that atoms are never created or destroyed in chemical reactions. They are simply rearranged. Therefore, an equation is written so that the same numbers of atoms of each element appear on both sides of the arrow. In other words, equations must always be written so that they are balanced.

metabolism:
metabole (GK) change
Metabolism involves many chemical changes.

SECTION 7.2

Demonstration

LS **Visual-Spatial** Run a thin stream of water from a faucet into a sink. Rub a glass rod with a silk cloth and place it next to the stream of water. Be careful to keep the rod and silk dry. Explain that the rod has a positive charge, and point out that opposite charges attract, so the positive charge on the rod attracts the negatively charged oxygen in the water molecule, causing the water stream to move toward the rod.

Chalkboard Example

Write a few chemical equations on the chalkboard and work with students to balance the equations. Once the equations are balanced, have students confirm the balance by counting the number of atoms of each kind on each side of the equation. Reinforce the idea that atoms are never created or destroyed in ordinary reactions. Some possible equations are:

$Mg + 2HCl \rightarrow MgCl_2 + H_2$
$2C_2H_2 + 5O_2 \rightarrow 2H_2O + 4CO_2$
$2H_2S + 3O_2 \rightarrow 2H_2O + 2SO_2$

L1 SAE

interNET CONNECTION

Follow the link for this chapter on the Glencoe Homepage at **www.glencoe.com/sec/science** to find out more about the chemistry of life.

Meeting Individual Needs

Learning Disabled Provide students with a list of several simple chemical formulas, including some with coefficients, such as NO_2, $3H_2S$, $2H_2O$, and C_3H_8. Have students identify the number of atoms of each type of element represented in each formula. L1 SAE

Focus On
The Uniqueness of Water

Purpose

Students will become familiar with some of the unique properties of water and the biological applications of water that result from these properties.

Teaching Strategies

- Elicit from students whether they have observed the movement of water through a piece of paper toweling. Explain that this common occurrence results from capillary action.
- Explain that an example of the resistance of water to temperature change is observable in the difference in the rates at which the metal of an empty pot gets hot compared to that of a similar pot filled with water. The empty pot becomes hot faster than the pot filled with water.
- The solubility of oxygen in water increases as water temperature decreases. Ask students to explain why having water at a temperature of 2°C under a sheet of ice would be important for living organisms in terms of oxygen solubility. *The water at 2°C will dissolve more oxygen than the warmer layers of water beneath this layer. As the surface of the lake is sealed by ice, the temperature of the water becomes important to the ability of the water to provide enough oxygen to the aquatic organisms during winter.* L2

Focus On
The Uniqueness of Water

Water is the most common liquid on Earth. It covers three-fourths of Earth's surface and makes up from 70 to 95 percent of the weight of living things. Water has some extraordinary properties. In particular, a water molecule's ability to attract and bond to other water molecules makes it especially important to living organisms. You already know that water is an excellent solvent. Here are a few more examples of how water is important to living systems.

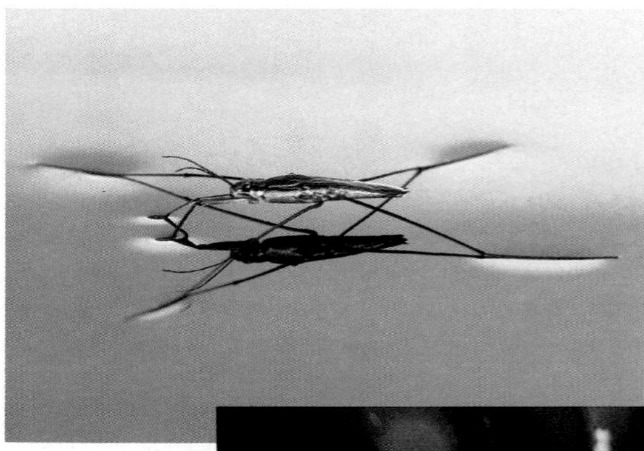

Water has a high surface tension

Because water is a polar molecule, it easily attracts other water molecules. This attraction causes the surface layer of water molecules to act like a stretched film over the surface of the water. This property of water is called surface tension. Thus, the water strider can stand on top of the water, with surface tension supporting its body weight. The nonpolar wax molecules on the surface of a leaf have little attraction for water, so the water beads up because of its surface tension.

Water creeps up in thin tubes

Plants also take advantage of the great attraction water has for itself and other molecules. In fact, it is this property of water that allows plants to get water from the ground. Water creeps up the thin tubes in plant roots and stems. This property is called capillary action and plays a major role in getting water from the soil to the tops of even the tallest trees.

STUDENT JOURNAL

"Water, water everywhere. . . " Have students cut out pictures from magazines that illustrate uses of water. Ask students to prepare a display similar to *Focus on the Uniqueness of Water.* Students may want to include industrial uses of water, uses of water by organisms, or environmental uses of water. L1 LS SAE

Water has a high heat of vaporization

Because water strongly attracts other water molecules, it takes a lot of energy to make water evaporate. As water heats, its molecules vibrate faster and faster. As more heat is applied, surface molecules begin to separate from the rest and evaporate into the air. This high heat of vaporization helps cool the human body. When sweat evaporates from human skin, it draws away the heat of vaporization from the body itself, thus helping keep the body cool.

Water resists temperature change

Water requires more heat to increase its temperature than do most other common substances. Likewise, water must lose a lot of heat when it cools. This property of water is extremely important because temperature variations have a strong effect on the kinds of organisms that can live in a given region. Because they are near large bodies of water, coastal areas vary less in temperature than inland areas at the same latitude. Thus, organisms such as wine grapes that cannot tolerate temperature extremes can live in these areas.

Water expands when it freezes

Water is one of the few substances that expands when it freezes. Because of this property, ice is less dense than liquid water and floats as it forms in a pond. Water expands as it freezes inside the cracks of rocks. As it expands, it often breaks apart the rocks. Over long time periods, this process will form soil.

EXPANDING YOUR VIEW

1. **Inferring** If you observe plants after a rain or in the early morning dew, you'll see water beaded up on the leaves. What does this tell you about the surface of plant leaves?
2. **Applying Concepts** What characteristic of water molecules helps explain all the properties discussed here?

Focus On

Background
Water and Life

Life on Earth could not have evolved without water. Wherever water is found, life is found. Some unicellular organisms can live on the amount of water that clings to a grain of sand. Life is found in water at all temperatures. For example, bacteria can live on the underside of snow and in the near-boiling water of hot springs.

The Polarity of Water

Water is a polar molecule because the arrangement of the polar bonds is not symmetrical. The bond angle of water is 104.5°.

Answers to Expanding Your View

1. **Inferring** The surfaces of leaves are covered by substances such as oils or waxes that are composed of nonpolar molecules.
2. **Applying Concepts** the polar nature of the water molecule

GLENCOE TECHNOLOGY

Videodisc

Biology: The Dynamics of Life
Disc 1, Side 1
Properties of Water (Ch. 19)

Earth Science

Exploding Sodas, Potholes, and Soil

Purpose

Students will learn how water contributes to the formation of soil and potholes.

Teaching Strategies

- Explain that ice cube trays are often made from plastic because plastic can change shape as water expands.
- Have students explain why a road in Arizona would have fewer potholes than a road in Ohio. *A road in Arizona would have fewer potholes than one in Ohio because Arizona does not have the extreme freezing and thawing temperatures common in the winters of the more northerly states.* L1

Possible Answer

When water freezes, it crystallizes, forming jagged edges that puncture the cell membranes and walls. When the fruit thaws, the cell contents leak out.

3 Assess

Check for Understanding

Ask students to construct models of compounds such as water and sodium chloride. L1

Reteach

Using gumdrop and toothpick molecules, demonstrate a chemical reaction such as $CH_4 + 2O_2 \rightarrow CO_2 + 2H_2O$. Stress the conservation of matter as you tear the original molecules apart to build new molecules. L1 SAE

Extension

Ask students to make a display to show the chemical formulas of ten common substances. L2

Earth Science

Exploding Sodas, Potholes, and Soil

Have you ever put a warm can of soda in the freezer to cool and then forgotten it? If you have, you know that water expands as it freezes. The can may only bulge, but it may explode, spraying soda all over the inside of the freezer. What's happening? How is it related to potholes and the formation of soil?

The pothole effect Most substances contract, taking up less space as they freeze. But, unlike other substances, water expands as ice crystals form. Potholes in streets and highways are created by the same action that caused the can of soda to explode. Water seeps into cracks in the road surface. When the water freezes, it expands, exerting a great deal of force, which makes the crack wider and deeper. With enough freezing and thawing, an entire chunk of roadway breaks loose. Also because of freezing and thawing—plus traffic—the chunk becomes pulverized, leaving a pothole in the road.

Making soil In the same way, exposed rocks are cracked and pulverized, eventually weathering into small particles of soil. The soil in an area can be less than a centimeter to more than 5 m thick. Soil holds moisture, provides minerals and nutrients for plant growth, and acts as a filter for water flowing through it to underground reservoirs.

CONNECTION TO Biology

Cells can be thought of as sacks containing mostly water. Why do strawberries and some other fruits become soft and mushy when frozen and then thawed?

Figure 7.11

The pH scale has a range from 0 to 14. Pure water has a pH of 7. A solution with a pH of 7 is neutral because there are equal numbers of hydrogen and hydroxide ions. As the pH of a solution becomes lower, the concentration of hydrogen ions becomes greater. A solution with a pH below 7 is acidic. As the pH of a solution becomes higher, the concentration of hydroxide ions becomes greater. A solution with a pH above 7 is basic.

Acids and bases

Chemical reactions can occur only when conditions are right. For example, a reaction might depend on temperature, the availability of energy, or a certain concentration of a substance dissolved in solution. Chemical reactions in organisms also depend on the pH of the environment. The **pH** is a measure of how acidic or basic a solution is. A scale with values ranging from 0 to 14 is used to measure pH. *Figure 7.11* depicts a typical dispenser of pH paper.

In Chapter 6, you learned how acids and bases are important in the environment. For example, some plants grow well only in acidic soil, while others require soil that is basic. An **acid** is any substance that forms hydrogen ions (H^+) in water. When

1. Atoms combine because their energy levels become stable.
2. Ionic bonds form as one atom gains electrons from another or gives up electrons to another. Covalent bond formation involves sharing of electrons.
3. There are more hydrogen than hydroxide ions in an acidic solution of pH 2.

the compound hydrogen chloride (HCl) is added to water, hydrogen ions (H^+) and chloride ions (Cl^-) are formed. Thus, hydrogen chloride in solution is called hydrochloric acid. This solution contains an abundance of H^+ ions and has a pH below 7.

A **base** is any substance that forms hydroxide ions (OH^-) in water. For example, if sodium hydroxide (NaOH) is dissolved in water, it forms sodium ions (Na^+) and hydroxide ions (OH^-). This solution contains an abundance of OH^- ions and has a pH above 7.

MiniLab

How can you determine the pH of common household items?

The pH of a solution is a measurement of how acidic or basic that solution is. An easy way to measure the pH of a solution is to use pH paper. This paper has been treated with chemical indicators whose color varies according to pH.

Procedure

1. Pour a small amount (about 5 mL) of each of the following into separate clean, small beakers or other small glass containers: lemon juice, household ammonia solution, liquid detergent, shampoo, and vinegar.
2. Dip a fresh strip of the pH paper into each solution briefly and remove.
3. Compare the color of the wet paper with the pH color chart, and record the pH of each material.

Analysis

1. Which solutions were acids?
2. Which solutions were bases?
3. What ions in the solution were causing the pH paper to change? Which solution contained the highest concentration of hydroxide ions? How do you know?

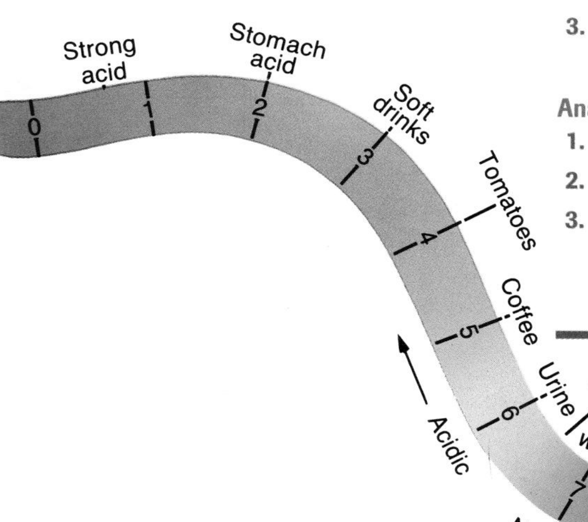

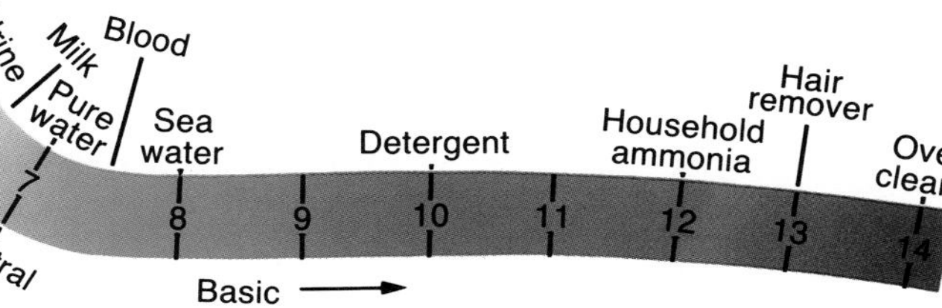

Section Review

Understanding Concepts

1. Why do atoms combine chemically with other atoms?
2. How does the formation of an ionic bond differ from the formation of a covalent bond?
3. What can you say about the proportion of hydrogen ions and hydroxide ions in a solution that has a pH of 2?

Thinking Critically

4. Explain why water dissolves so many different substances.

Skill Review

5. **Interpreting Scientific Illustrations** Study the diagram in *Figure 7.8,* which shows the process of a polar compound dissolving in water. In your own words, describe the process step-by-step. Tell what the water molecules are doing and why. Describe what is happening to the sugar molecules and why. Describe the nature of the mixture after the compound dissolves. For more help, refer to Thinking Critically in the *Skill Handbook.*

Assessment

Skill: Give students the following equations to balance.

$N_2O_4 \rightarrow NO_2$

$C_3H_8 + O_2 \rightarrow CO_2 + H_2O$

MiniLab

Purpose

LS **Visual-Spatial** Students will determine the pH of common solutions.

Process Skills

observe and infer, interpret data

Safety Precautions

Have students wear aprons and safety goggles. **CAUTION:** *Both high and low pH solutions can injure the skin and eyes.*

Teaching Strategies

- Use a pH paper that measures from pH 0 to 14.

Expected Results

The approximate pH of the solutions are: lemon juice, pH 3; household ammonia, pH 11; liquid detergent, pH 10; shampoo, pH 7; and vinegar, pH 3.

Analysis

1. lemon juice and vinegar
2. household ammonia and liquid detergent
3. H^+ ions and OH^- ions. Household ammonia contains the most OH^- ions; it had the highest pH.

Assessment

Portfolio: Have students make a pH scale in their portfolios and show where each solution falls on the scale. L1 P

4 Close

Activity

Have students create a pH scale poster using pictures of common foods and materials. L1

Answers to Section Review

Thinking Critically

4. Because water molecules are polar and attract other charged particles, water easily dissolves many substances.

Skill Review

5. Polar water molecules are attracted to the charges on the surface of the polar compound. The water molecules gradually surround the sugar molecules. The attraction between the water-surrounded molecules and the rest of the molecules in the crystal is overcome and they are pulled away from the crystal. This process is repeated until all the molecules in the crystal are in solution. A solution of a molecular substance consists of water molecules and solute molecules.

People in Biology

Meet Dr. Baldomero Olivera, Biochemist

Teaching Strategies

- Have students refer back to this feature during their study of the human nervous system.
- Have students explain what is meant by the phrase "pure research." *Pure research is research with the primary function of learning new information about objects and phenomena in nature.* Ask students how the goal of "pure research" differs from the role of technology. *Technology is applied science. The goal of technology is to use information provided by science in a practical way.* L1
- Elicit how emergency medical technicians might use biochemistry in their work. *EMTs must understand how certain drugs and chemicals might affect the body so that they can determine whether people are suffering from adverse drug reactions, drug overdoses, insulin shock, or some form of chemical poisoning.* L2

people in biology

Meet Dr. Baldomero Olivera, Biochemist

Have you ever had a chance to look at the circuit board of a computer? All that complicated circuitry is simple when compared to the working of your nervous system! Dr. Baldomero Olivera is conducting research that may someday help unravel the puzzle of how the nervous system works.

In the following interview, Dr. Olivera describes his research and tells how his early childhood interests pointed the way to his career as a biochemist.

On the Job

Q Dr. Olivera, could you tell us about your current work?

A I'm a biochemist, a chemist who deals with chemical processes that take place in living things. In our lab, we research the biochemistry of the nervous system. We're interested, ultimately, in how the nervous system works. We'd like to understand how the components are put together to enable nerve cells to signal each other. A problem in studying a whole system is that we can't break it apart to take a close look because then, of course, it is no longer a system. So we've had to find a way of studying the system while keeping it intact.

Q What approach is your research taking?

A It's actually somewhat unusual. Years ago, we began studying the effects of venoms of marine snails on nerve cells. Initially, we were just trying to figure out why these venoms could kill people. We soon discovered that the venoms are loaded with components that very specifically interfere with the function of particular molecules in nerve cells. That led to an entirely new study, the one we're concerned with now, which is tracking the detailed functions of all the component molecules of the nervous system. Basically, what we do is use molecules from venoms as a way to attach to the molecules of the nerve cells. We mark molecules from the venom with radioactive tags, and then trace the radioactivity to the specific targets in the nervous system. The process is rather like a tracking device to pinpoint areas that are crucial for a particular function in the nervous system.

Q What kinds of practical applications might come from your research?

A It's pure research, so our main goal is, first and foremost, to understand how nervous systems work. Of course, that kind of understanding has a lot of potential applications because knowing how the brain works could lead to figuring out what goes wrong to cause mental illness. Then there would be a chance of coming up with ways to alleviate the symptoms of these illnesses.

Q Do you think your current research will be wrapped up in a few years?

A No, certainly not within my lifetime. To discover how the brain works is a really big job—one we're just starting.

Early Influences

Q Can you remember how you first got interested in science?

A Even though it was a long time ago, I still remember very clearly an experiment that my second-grade teacher demonstrated to our class. She brought in a bunch of substances—ground-up chalk, sugar, flour, and a few more. As she put each powderlike substance into a tube with water, she said that some things dissolve in water and some don't. She suggested that we could figure out for ourselves which ones do and which ones don't. I still remember the excitement of discovery of that day.

Q Did you have any interests when you were growing up that directly relate to your present research?

A Growing up in the Philippines, I used to collect snails. That's how I knew about venomous sea snails. If I hadn't had that hobby, the present research using snail venom might never have begun.

Personal Insights

Q Do you have hobbies today that relate to other fields of science?

A My family and I got interested in fossil hunting when we were in Germany several years ago. And then we discovered that here in Utah, there are a lot of interesting opportunities for amateur paleontologists. Many biochemists are excited about trying to extract DNA from old bones. It's a situation in which different disciplines of science are interfacing.

CAREER CONNECTION

These careers are all essential ones in the health-care field.

Research Biochemist *Bachelor's degree in chemistry; PhD in biochemistry*

Medical Technologist *Bachelor's degree through a college or hospital program*

Emergency Medical Technician *High-school diploma and EMT program at college, hospital, or fire department*

Q What advice would you give to students who are interested in a career in biochemistry?

A I think research is a lot of fun. I like discovering something new. If kids have a strong streak of curiosity, they might enjoy the field. As preparation for any kind of career, I would advise kids to spend time developing hobbies and doing the things that interest them, because sometimes hobbies can lead to a profession.

CAREER CONNECTION

- **Career Path** A career in biochemistry would require these high school courses: biology, Earth science, chemistry, and physics. College courses would include biology, general chemistry, microbiology, organic chemistry, inorganic chemistry, physics, and biochemistry. Graduate courses would include: additional organic chemistry, microbiology, and perhaps pharmacology.

For More Information

Contact the chemistry department of a local college or university to find out what the requirements for a bachelor's degree or an advanced degree in biochemistry are and what types of career opportunities are available to a person who holds a degree in this field.

GLENCOE TECHNOLOGY

 Videodisc

Biology: The Dynamics of Life
Disc 1, Side 1
Pet Scanner (Ch. 20)

SECTION

7.3

Prepare

Key Concepts

Students will examine the classes of carbon compounds present in organisms. The structural and functional aspects of carbohydrates, lipids, proteins, and nucleic acids will be studied.

Block Scheduling

Look for this symbol for strategies that are useful in a block scheduling format. For more information on block scheduling, refer to Section 7.3 in the **Lesson Plans** booklet.

Materials

- Obtain ball-and-stick models from a science supply house for the demonstration.
- Purchase potatoes for the Biolab.
- Purchase gelatin, waxed paper, fresh pineapple, canned chunk pineapple, grapes, oranges, and paper cups for the Alternate Lab.
- Make flash cards for the Check for Understanding.

1 Focus

Bellringer

Before presenting the lesson, display **Section Focus Master 15** on the overhead projector and have students answer the accompanying questions. L1 SAE

Visual Learning

Figure 7.13 Direct students' attention to the illustration. Ask them what compound in addition to sucrose is formed when glucose and fructose combine. *Water* LS

SECTION

7.3 Life Substances

Section Preview

Objectives

Explain why there is a large variety of organic compounds.

Explain how polymers are formed and broken down in organisms.

Compare the chemical structures of carbohydrates, lipids, proteins, and nucleic acids, and explain the importance of these substances in living things.

Key Terms

isomer
polymer
carbohydrate
monosaccharide
disaccharide
polysaccharide
starch
glycogen
cellulose
lipid
protein
amino acid
peptide bond
enzyme
nucleic acid
nucleotide
DNA
RNA

Did you ever hear the saying, "You are what you eat"? It's at least partially true because the compounds that form the cells and tissues of the body are produced from similar compounds in the foods you eat. Common to most of these foods and to most substances in organisms is the element carbon. The first carbon compounds that scientists studied came from organisms. Because of this, they were called organic compounds.

Role of Carbon in Organisms

A carbon atom has four electrons available for bonding in its outer energy level. In order to become stable, carbon atoms form four covalent bonds. Follow the illustration of carbon atoms and bonding in *Figure 7.12.* Carbon can bond with other carbon atoms, as well as with many other elements. When carbon atoms bond to each other, they can form straight chains, branched chains, or rings. In addition, these chains and rings can have almost any number of carbon atoms and can include atoms of other elements as well. This property makes a huge number of carbon structures possible.

A great variety is possible in organic molecules. In fact, compounds with the same simple formula often differ in structure. Compounds that have the same simple formula but different three-dimensional structures are called **isomers.** Both

Single bond	Double bond	Triple bond
e^- e^-	e^- e^- e^- e^-	e^- e^- e^- e^- e^- e^-
C—C	C=C	—C≡C—

Figure 7.12

When two carbon atoms bond, they share one electron each and form a covalent bond. They can also share two or three electrons. When each atom shares two electrons, a double bond is formed. A double bond is represented by two bars between carbon atoms. When each shares three electrons, a triple bond is formed. Triple bonds are represented by three bars drawn between carbon atoms.

Program Resources

Section Focus Master 15 L1 SAE
Reinforcement and Study Guide, p. 28 L1
Biolab and Minilab Worksheets, pp. 25-26 L1
Laboratory Manual, pp. 39-42 and 43-46 L2
Critical Thinking/Problem Solving, p. 7 L3
Tech Prep Applications, pp. 7-8 L2
Reteaching Transparency 7 and **Master** L1 SAE

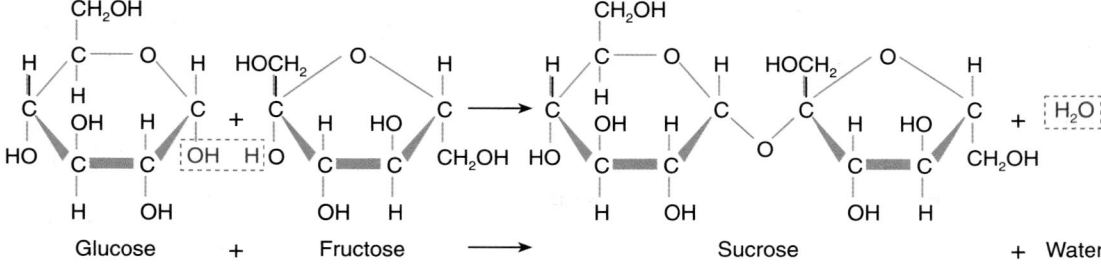

Glucose + Fructose ⟶ Sucrose + Water

Figure 7.13

The different arrangement of hydrogen and oxygen atoms around each carbon atom gives glucose and fructose molecules different chemical properties. When glucose and fructose combine, they form the disaccharide, sucrose, also known as table sugar.

glucose and fructose, shown in *Figure 7.13,* have the same simple formula, $C_6H_{12}O_6$, but different structures.

Molecular chains

Carbon compounds also vary greatly in size. Some compounds contain one or two carbon atoms, while others contain tens, hundreds, or even thousands of carbon atoms. These large molecules are called macromolecules. Proteins are examples of macromolecules in organisms. Cells build macromolecules by bonding small molecules together to form chains called polymers, as shown in *Figure 7.14.* A **polymer** is a large molecule formed when many smaller molecules bond together, usually in long chains.

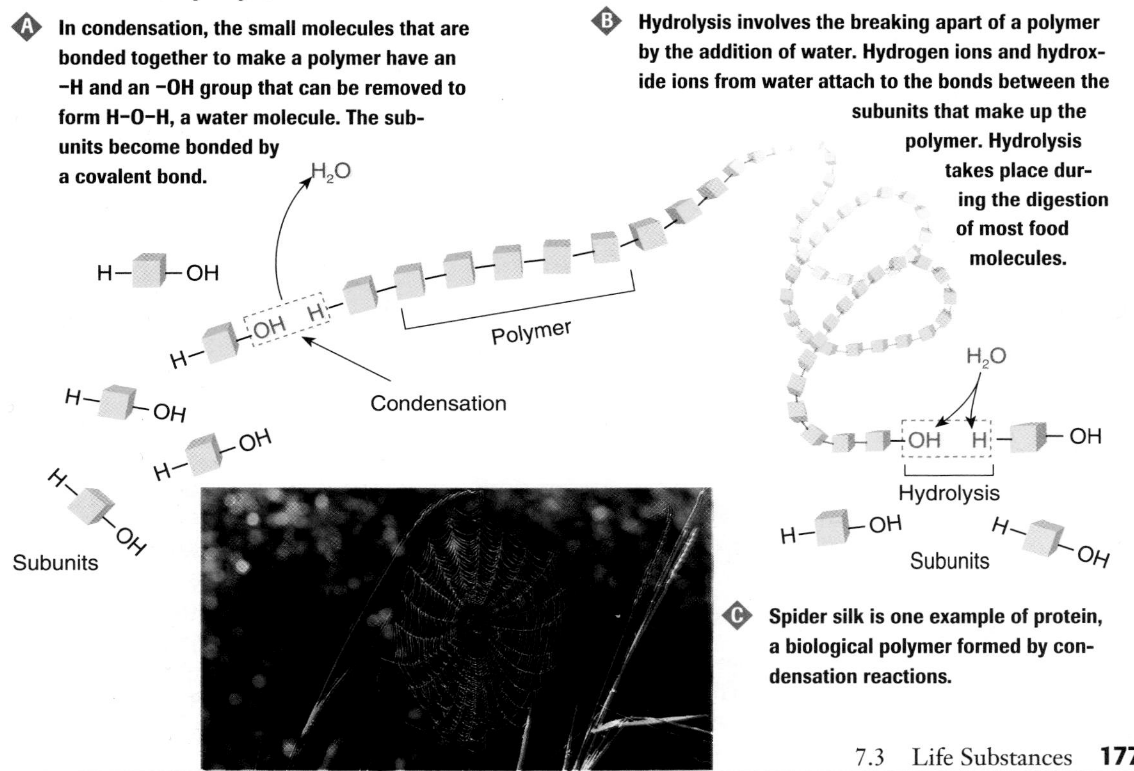

Figure 7.14

Condensation and Hydrolysis

A **In condensation, the small molecules that are bonded together to make a polymer have an –H and an –OH group that can be removed to form H–O–H, a water molecule. The subunits become bonded by a covalent bond.**

B **Hydrolysis involves the breaking apart of a polymer by the addition of water. Hydrogen ions and hydroxide ions from water attach to the bonds between the subunits that make up the polymer. Hydrolysis takes place during the digestion of most food molecules.**

C **Spider silk is one example of protein, a biological polymer formed by condensation reactions.**

SECTION 7.3

Activity

LS **Visual-Spatial** Have volunteers draw a model of a carbon atom on the chalkboard. Have one student draw the protons, a second student draw the neutrons, and a third student add the electrons in their correct energy levels. Ask students to use the model to explain how many electrons are needed to make the carbon atom stable. *Four* Have students predict how many hydrogen atoms could form covalent bonds with the carbon atom. *Four* Ask what the formula for this molecule would be. *CH_4* L1

2 Teach

Demonstration

LS **Visual-Spatial** Use a ball-and-stick model of a methane molecule (CH_4) to show students the regular tetrahedron arrangement formed by the bonds of carbon atoms. Contrast the arrangement of the atoms in the methane molecule to those in a molecule of water.

Using Science Terms

LS **Linguistic** Have students look up the meanings of the prefix *hydro-* and the suffix *-lysis* in a dictionary. *Hydro- refers to "water," and lysis means "to break down."* Ask students to relate the meanings of these word parts to the word *hydrolysis.* Remind students that during hydrolysis polymers are broken down by the addition of water. L1 SAE

A BROADER VIEW

Was there life on Mars?

Purpose

Students will use their knowledge of organic molecules to examine how scientists searched for signs of life on Mars.

Teaching Strategies

- Ask students to identify the types of carbon compounds that would suggest the presence of organisms on Mars. *Likely responses will include proteins, carbohydrates, lipids, and nucleic acids.*
- Show students pictures and video images from the Viking missions to Mars. Contact the nearest NASA Teacher Resource Center for additional materials. Viking images are also available on commercially produced videotapes and laserdiscs.
- Have students use the library to research articles on meteors from Mars.

Thinking Critically

The features on Mars look exactly like dry riverbeds on Earth, suggesting that Mars once had a warmer temperature and water flowing in rivers.

Audiovisual

Show students the film *The Chemistry of Life*, Biology Media.

A BROADER VIEW

Was there life on Mars?

In the summer of 1975, two *Viking* spacecraft, each consisting of an orbiter and an attached lander, were launched to Mars from Cape Canaveral. One of their missions was to look for signs of life.

Signs of life Something that has the capacity to replicate or reproduce, repair, evolve, and adapt is considered to be alive. These properties appear to originate in large, organic (carbon-based) molecules. The Martian atmosphere consists of 95% carbon dioxide, 2.5% nitrogen, and 1.5% argon, with traces of oxygen, carbon monoxide, and inert gases. The elements necessary for life are available on the planet, but there is virtually no water vapor, a substance essential to life as we know it. The only water on Mars is locked up in the form of ice in the polar caps. However, Mars has channels resembling dry riverbeds that cut through its desert terrain, suggesting that the atmosphere of Mars was once much different from what it is today. The instruments that the landers carried were designed to look for evidence of life: complex, carbon-based molecules and by-products of metabolic activity in the soil.

Looking for life processes The robot arm of the lander scraped up soil. The soil was placed in a container, where it was mixed with nutrients and incubated under conditions that the scientists believed would encourage growth of microorganisms. Seven months later, no evidence of living things had been found.

In 1996, scientists examined a meteor believed to have come from Mars. They discovered what they thought to be evidence of ancient bacterial life. However, not all scientists are convinced by the markings in the rock. They would like to examine more rocks from Mars' polar ice caps.

Thinking Critically

Today, Mars is a cold, dry, and windy planet. Why would the appearance of dry riverbeds and features caused by erosion suggest that the climate of Mars was much different in the past?

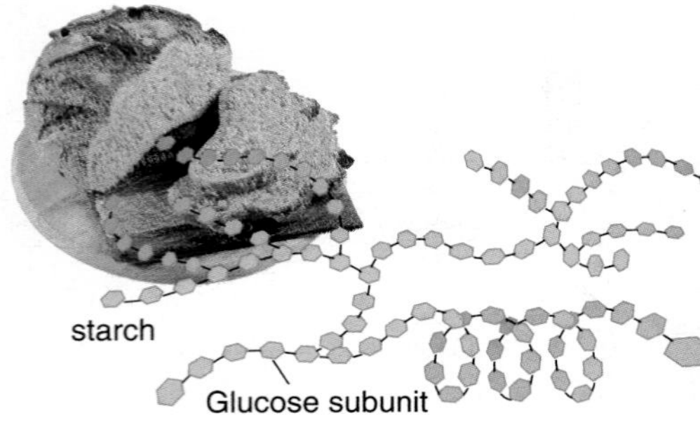

The structure of carbohydrates

You may have heard of runners eating large quantities of spaghetti or other starchy food the day before a race. This practice is called carbohydrate loading and works because carbohydrates are used by cells to store and release energy. A **carbohydrate** is an organic compound composed of carbon, hydrogen, and oxygen with a ratio of about two hydrogen atoms and one oxygen atom for every carbon atom.

The simplest type of carbohydrate is a simple sugar called a **monosaccharide.** Common examples are the isomers glucose and fructose, which were shown in *Figure 7.13.* Two monosaccharide molecules can link together to form a **disaccharide**, a two-sugar carbohydrate. When glucose and fructose combine by a condensation reaction, a molecule of sucrose is formed, as shown in *Figure 7.13.* Sucrose is more commonly known as table sugar.

The largest carbohydrate molecules are **polysaccharides**, polymers composed of many monosaccharide subunits. Starch, cellulose, and glycogen are all examples of polysaccharides, as shown in *Figure 7.15.* **Starch** consists of highly branched chains of glucose units and is used as food storage by plants. Animals store food in the form of **glycogen,** another glucose polymer similar to starch but more highly branched. **Cellulose** is another glucose poly-

Meeting Individual Needs

Hearing Impaired Cut out pictures of foods that represent carbohydrates, lipids, and proteins and mount them on 3" × 5" cards. Place a question on each card that asks what organic compound is predominant in this food or what monomers make up these compounds. Have students work in groups to answer the questions. LS

STUDENT JOURNAL

Evaluating Space Environments Have students write a short story about what it would be like to travel on a space ship to Mars. They should include facts about the search for life on Mars or the view of the terrain on Mars as it indicates a change in climate from the past. L1 LS

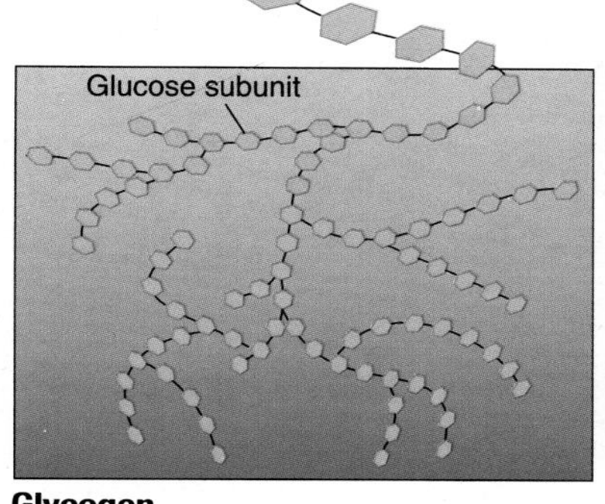

Glycogen

Cellulose

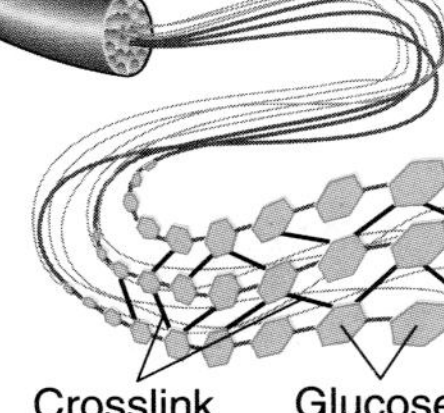

Figure 7.15

Notice the structural differences among the polysaccharides starch, glycogen, and cellulose. Notice that all three are polymers of glucose.

mer that forms the cell walls of plants and gives plants structural support. Cellulose is also made of glucose units hooked together somewhat like a chain-link fence.

The structure of lipids

If you've ever tried to lose weight, you may have wished that lipids (fats) never existed. Actually, they are extremely important for the proper functioning of organisms. **Lipids** are organic compounds that have a large proportion of C–H bonds and less oxygen than carbohydrates. For example, a common lipid found in beef fat has the formula $C_{57}H_{110}O_6$.

Lipids are commonly called fats and oils. They are insoluble in water because their molecules are nonpolar and, therefore, are not attracted by water molecules. Cells use lipids for long-term energy storage, insulation, and protective coatings. In fact, lipids are the major components of the membranes that surround all living cells. The most common type of lipid, shown in *Figure 7.16*, consists of three fatty acids bonded to a molecule of glycerol.

Lipids that contain fatty acid chains of carbon with only single bonds are referred to as saturated fats. Examples include the fats in butter and steak. These fats are generally solid at room temperature.

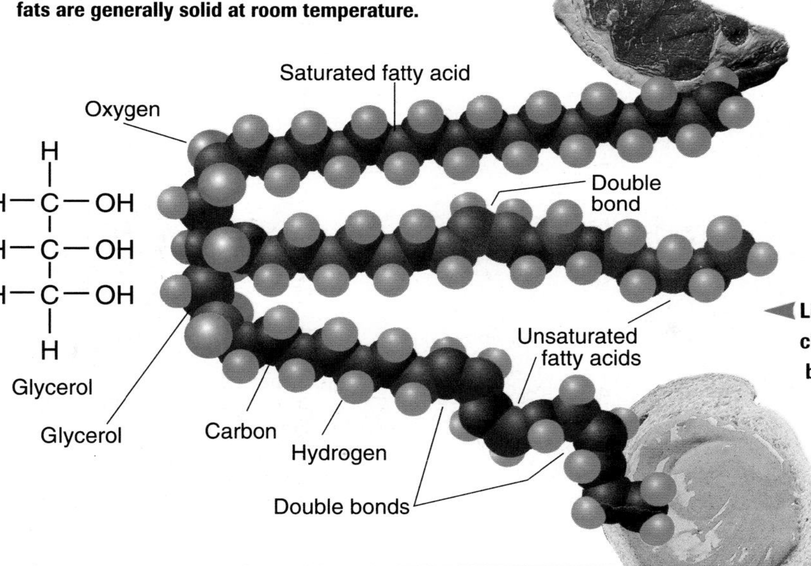

Lipids that contain fatty acid chains of carbon with double bonds are called unsaturated fats. Unsaturated fats are usually liquids at room temperature. Examples are peanut, corn, and olive oils.

Figure 7.16

Glycerol is a 3-carbon molecule that serves as a backbone for the lipid molecule. Attached to the glycerol are three fatty acids.

Misconception

Students often think molecules are flat, two-dimensional structures because the formulas in books are flat. Show students structural models of various molecules so they can observe the three-dimensional appearance of molecules.

Enrichment

Have students research the differences in the structures of starch, glycogen, and cellulose to determine how there can be more than one polymer of glucose. L2

Software

Cell Chemistry, Queue.

Visual Learning

Figure 7.16 Ask students to use Figure 7.16 to answer the following questions. (a) What compound serves as the backbone for lipid molecules? *Glycerol* (b) How do the bonds in saturated fats differ from those in unsaturated fats? *Saturated fats have single bonds; unsaturated fats have double bonds.*

PROJECT — Exploring Nutrients

Divide students into groups. Have each group research and prepare a presentation on one of the following: sugars and other nutritive sweeteners in processed foods, cholesterol in the diet, saturated and unsaturated fats in the diet, or the functions of proteins such as keratin, actin, myosin, insulin, and collagen.

L1 COOP LEARN LS

SECTION 7.3

Using an Analogy

To increase understanding of how only 20 amino acids can create such a variety of proteins, relate the 20 naturally occurring amino acids to the letters of the alphabet. Elicit from students why the letters of the alphabet can be used to create such a large number of words. Relate this phenomenon to how a similar number of amino acids can create so many different proteins. L1

GLENCOE TECHNOLOGY

Videodisc

The Infinite Voyage: The Geometry of Life
DNA and the Double Helix
(Ch. 2)

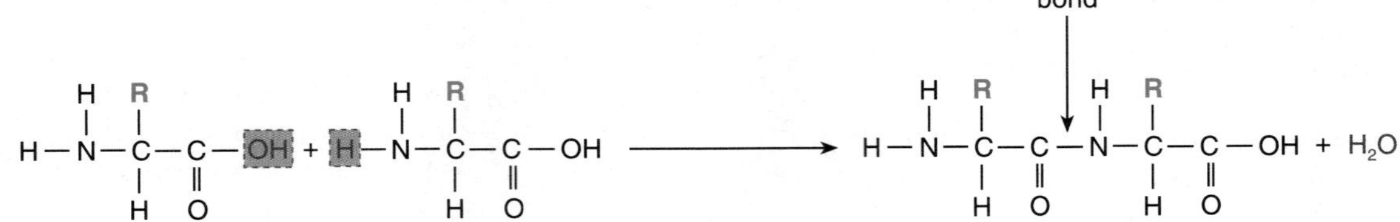

Figure 7.17

Each amino acid contains a central carbon atom, to which is attached a carboxyl group (–COOH), a hydrogen atom, and an amino group (–NH_2). Also attached to the central carbon atom is a group (–R) that makes each amino acid different. To form protein chains, the hydrogen from an amino group and a hydroxyl group from a carboxyl group are removed to bond the amino acids together.

hydrolysis:
hydor (GK) water
lysis (GK) to split, loosen
In hydrolysis, molecules are split by water.

polymer:
poly (GK) many
meros (GK) part
A polymer has many bonded subunits (parts).

The structure of proteins

Proteins are essential to all life. They build structure and carry out cell metabolism. A **protein** is a large, complex polymer composed of carbon, hydrogen, oxygen, nitrogen, and sometimes sulfur. The basic building blocks of proteins are called **amino acids,** as shown in *Figure 7.17.* There are 20 common amino acids. Because there are 20 types of building blocks, proteins come in a large variety of shapes and sizes. In fact, proteins vary more in structure than do the other classes of biological organic molecules.

Amino acids are linked together by condensation, the removal of an –H and –OH group to form a water molecule. The covalent bond formed between amino acids is called a **peptide bond,** as shown in *Figure 7.17.* Protein chains contain hundreds of amino acids, and it is the order of amino acids that determines the kind of protein. Many proteins consist of two or more amino acid chains that are held together by hydrogen bonds.

Proteins are the building blocks of many structural components of organisms, as shown in *Figure 7.18.* Proteins are also important in muscle contraction, transporting oxygen in the bloodstream, providing immunity, and carrying out chemical reactions.

Figure 7.18

Besides lipids, proteins are also a major component of the membrane coverings of all cells. Proteins make up much of the structure of organisms, such as hair, horns, hoofs, and nails.

Alternate Lab: What fruits contain enzymes that act on protein?

Purpose 

LS **Logical-Mathematical** Students will investigate the effect of the enzyme bromelin on gelatin.

Preparation

One pineapple should be enough for 20 students working in groups of four. A 6 oz. box of gelatin makes enough for two groups. Make gelatin and pour 100 mL into each paper cup. Use hot/cold cups large enough to hold 200 mL.

Materials

paper cups (4) with gelatin, fresh pineapple, knife, waxed paper, canned chunk pineapple, refrigerator, grapes or orange sections

Procedure

Give students the following directions.

1. Number the cups 1 to 4.
2. Select 3 chunks of canned pineapple. Cut 3 chunks of fresh pineapple the same size as the canned chunks. Cut 3 grapes in half, or choose three orange slices. Set these aside on waxed paper.

Enzymes are important proteins found in living things. An **enzyme** is a protein that speeds up a chemical reaction. In some cases, enzymes increase the speed of reactions that would otherwise occur so slowly you could say they didn't occur at all. Some of these reactions *would* take place at high temperatures or in strongly acidic or basic solutions. Enzymes enable these reactions to occur under the conditions present in living cells. Enzymes are involved in nearly all metabolic processes. They speed the reactions in digestion of food, synthesis of molecules, and storage and release of energy. *Figure 7.19* shows how enzymes function.

The structure of nucleic acids

Nucleic acids are another important type of organic compound that is necessary for life. A **nucleic acid** is a complex macromolecule that stores information in cells in the form of a code. Nucleic acids are polymers made of smaller subunits called **nucleotides.**

ThinkingLab — Draw a Conclusion

How does dilution affect enzyme action?

You have read that starch digestion begins in the mouth with enzymes that are secreted in saliva. You decide to see what effect diluting the saliva will have on the digestion of soda crackers, which are mostly baked flour.

Analysis

To verify that crackers contain starch, you grind up a cracker and add 5 mL of plain water. You add a drop of iodine solution, and it produces a blue-black color, indicating the presence of starch. Next, you repeat the test but use 5 mL of saliva instead of water. This time, the iodine produces a yellow-brown color instead of the blue color of a positive starch. Apparently, your saliva enzymes had broken down all the starch in the time it took to make the mixture and test it.

Now you decide to see what effect dilution of saliva will have on the digestion of the starch. You place one drop of saliva into 5 mL of water in a test tube, and add the same amount of ground cracker as before. You test the tube at various time intervals and obtain the data shown in the following table.

Time	Iodine Test for Starch
20 Seconds	blue-black
40 Seconds	blue-black
60 Seconds	blue-black
80 Seconds	blue-black
100 Seconds	yellow-brown
120 Seconds	yellow

Thinking Critically

Draw a conclusion about the effect of dilution of the enzyme found in saliva on the digestion of starch, and explain the results.

Figure 7.19

Enzymes enable molecules, called substrates, to undergo a chemical change to form new substances, called products.

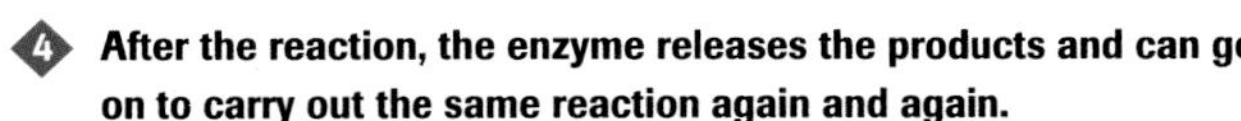

1 **Each enzyme acts on a specific molecule or set of molecules called substrates.**

4 **After the reaction, the enzyme releases the products and can go on to carry out the same reaction again and again.**

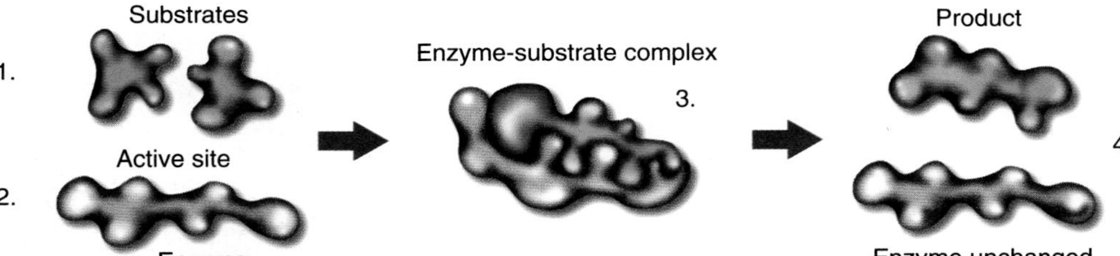

2 **Each substrate fits into an area of the enzyme called the active site. This fitting together is often compared to a lock-and-key mechanism. However, the enzyme changes shape a little to fit with the substrate.**

3 **In the enzyme-substrate complex, the enzyme holds the substrate or substrates in a position where a reaction can occur easily.**

ThinkingLab — Draw a Conclusion

Purpose

LS Logical-Mathematical Students will examine how dilution affects the rate of an enzyme reaction.

Process Skills

interpret data, analyze, draw a conclusion

Background

- Salivary amylase, also called ptyalin, breaks down starch into the disaccharide maltose. Its optimum pH is 6.7.

Teaching Strategies

- Use Figure 7.19 to point out that enzyme molecules can carry out a reaction many times per second because the enzyme is not changed in the reaction.

Thinking Critically

The rate of the enzyme reaction depends on the concentration of the enzyme. Diluting the enzyme lowers its concentration and slows the reaction rate.

Assessment

Portfolio: Have students write a summary of what they learned about the effect dilution has on the rates of reactions. L1 P

3. Add the following to each cup. Make sure the fruits are submerged.
   Cup 1—nothing
   Cup 2—canned pineapple chunks
   Cup 3—fresh pineapple chunks
   Cup 4—grape halves or orange slices
4. Set cups in the refrigerator. Check the cups at the end of the period.

Expected Results

Students should observe that the gelatin to which the fresh pineapple was added remained liquefied.

Analysis

Ask students to answer the following questions.

1. What was the purpose of cup 1? *Cup 1 was the control.*
2. Gelatin is a protein. Bromelin is a protein-digesting enzyme. What happened to the bromelin in the canned pineapple? *It was destroyed by heat during the canning process.*

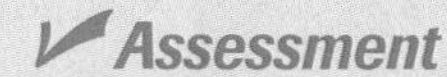

Assessment

Discussion: Ask students this question: Based on this activity, what can you conclude about which fruits have enzymes that act on protein? L1

BioLab Design Your Own Experiment

Does temperature affect an enzyme reaction?

Time Allotment 
One class period

Objectives
Review objectives with students before they begin the Biolab.

Process Skills
form a hypothesis, design an experiment, interpret data, recognize cause and effect

Safety Precautions
Students should wear aprons and safety goggles.

PREPARATION

Alternate Materials
Pieces of raw liver can be used instead of potato.

Possible Hypotheses

- If temperatures are very high or very low, the enzymes will be deactivated.
- If the temperature is raised, the speed at which the enzyme will work will increase.

GLENCOE TECHNOLOGY

Videodisc

Biology: The Dynamics of Life
Disc 1, Side 1
Enzyme Action (Ch. 2)

BioLab Design Your Own Experiment

Does temperature affect an enzyme reaction?

The compound hydrogen peroxide, H_2O_2, is a by-product of metabolic reactions in most living things. However, hydrogen peroxide is damaging to delicate molecules inside cells. As a result, nearly all organisms contain the enzyme peroxidase, which breaks down H_2O_2 as it is formed. Potatoes are one source of peroxidase. Peroxidase speeds up the breakdown of hydrogen peroxide into water and gaseous oxygen. This reaction can be detected by observing the oxygen bubbles generated.

PREPARATION

Problem
Does the enzyme peroxidase work in cold temperatures? Does peroxidase work better at higher temperatures? Does peroxidase work after being frozen or boiled?

Hypotheses
Make a hypothesis regarding how you think temperature will affect the rate at which the enzyme peroxidase breaks down hydrogen peroxide. Consider both low and high temperatures.

Objectives
In this Biolab, you will:

- **Observe** the activity of an enzyme.
- **Compare** the activity of the enzyme at various temperatures.

Possible Materials
thermometer
ice
beaker, 400-mL
hot plate
potato slices, 5-mm thick
kitchen knife
waxed paper
3% hydrogen peroxide
clock or timer

Safety Precautions
Be sure to wash your hands before and after handling the lab materials.

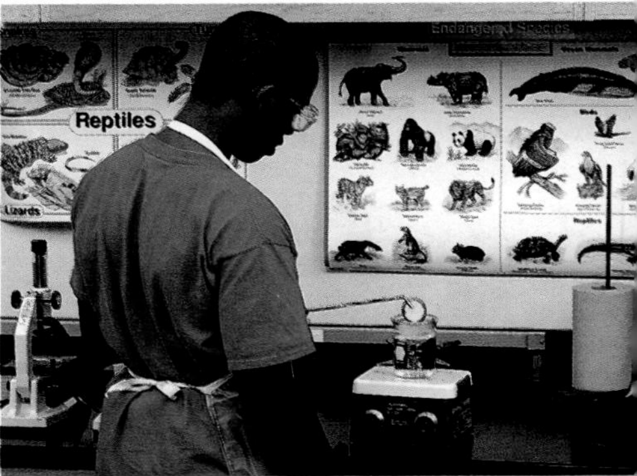

PLAN THE EXPERIMENT

Teaching Strategies

- Introduce the lab with a discussion of what factors might affect the rate of a reaction controlled by an enzyme.
- Students who cool the potato to low temperatures should be sure to run the test while the potato is still cool.
- Allow groups to discuss how their results differed when different experimental procedures were used.

PLAN THE EXPERIMENT

1. Decide on a way to test your group's hypothesis. Keep the available materials in mind.
2. When testing the activity of the enzyme at a certain temperature, consider the length of time it will take for the potato to reach that temperature, and how the temperature will be measured.
3. To test for peroxidase activity, add 1 drop of hydrogen peroxide to the potato slice and observe what happens.
4. When heating the potato slice, first place it in a small amount of water in a beaker. Then heat the beaker slowly so that the temperature of the water and the temperature of the slice are always the same. Try to make observations at several temperatures between 10°C and 100°C.

Check the Plan

Discuss the following points with other groups to decide on the final procedure for your experiment.

1. What data will you collect, and how will you record them?
2. What factors should be controlled?
3. What temperatures will you test?

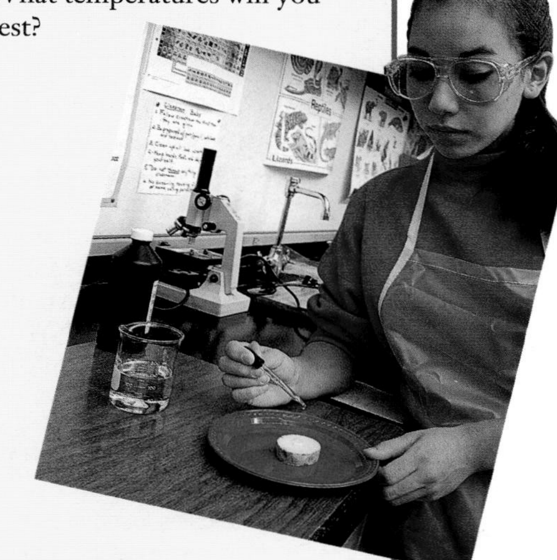

4. How will you achieve those temperatures?
5. ***Make sure your teacher has approved your experimental plan before you proceed further.***
6. Carry out your experiment.

ANALYZE AND CONCLUDE

1. **Checking Your Hypothesis** Do your data support or reject your hypothesis? Explain your results.
2. **Analyzing Data** At what temperature did peroxidase work best?
3. **Identifying Variables** What factors did you need to control in your tests?
4. **Recognizing Cause and Effect** If you've ever used hydrogen peroxide as an antiseptic to treat a cut or scrape, you know that it foams as soon as it touches an open wound. How can you account for this observation?

Going Further

Changing Variables
To carry this experiment further, you may wish to use hydrogen peroxide to test for the presence of peroxidase in other materials such as cut pieces of different vegetables. Also, test raw beef and diced bits of raw liver.

Possible Procedures

- Students may place a piece of potato on ice for 5 minutes, boil a second potato for 5 minutes, and allow a third piece of potato to sit at room temperature for 5 minutes. Each potato will then be tested for enzyme activity.

Data and Observations

Cooling will not deactivate the enzymes, but can slow the overall reaction. Potato slices heated over 70°C will not generate oxygen bubbles.

ANALYZE AND CONCLUDE

1. Students should explain whether their data support or reject their hypotheses.
2. Between 20°C–50°C
3. Answers may include the amount of time each potato was exposed to the temperature, the sizes of the potato slices, and the amount of peroxide added.
4. Human tissue contains peroxidase, so the hydrogen peroxide is broken down and releases oxygen.

Assessment

Oral: Have students discuss their results, especially the cold treatment. Discuss how results differed between cool pieces tested immediately and those allowed to warm to room temperature. L1

Going Further

Ask students why vegetables are boiled for a short time before freezing. *One reason is that boiling inactivates enzymes that begin to break down the other molecules in the vegetables.*

3 Assess

Check for Understanding

Use flash cards containing the names of monomers on one side and the corresponding polymer on the other side. First, show students the polymer name (e.g., protein) and then have them respond with the appropriate monomer name (amino acid), and vice versa. L1 SAE

Reteach

The concept of large polymers being composed of repeating units of monomers can be reinforced by having students list items that are composed of smaller units, such as beads making up a necklace, chain links, jigsaw puzzle pieces, or letters making up words. L1 SAE

Extension

Encourage above-level students to read *The Double Helix* by James Watson (Atheneum, 1968) which tells the story of the discovery of the DNA structure. L3

Assessment

Knowledge: Prepare a handout showing structural formulas for lipids, proteins, carbohydrates, and nucleic acids. Ask students to identify the type of organic compound shown in each diagram. L1

4 Close

Discussion Question

Ask students to explain the differences between saturated and unsaturated fats. *Saturated fats are composed of lipids containing fatty acids with only single bonds. Unsaturated fats are composed of fatty acid chains of carbon with double bonds.*

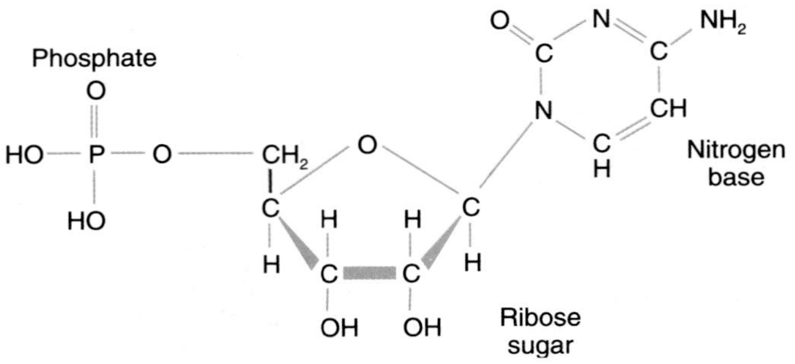

Figure 7.20

Each nucleic acid is built of subunits called nucleotides, which are formed from a sugar molecule bonded to a phosphate group and a nitrogen base.

Nucleotides consist of carbon, hydrogen, oxygen, nitrogen, and phosphorus atoms arranged in three groups—a base, a simple sugar, and a phosphate group, as shown in *Figure 7.20.* You have probably heard of the nucleic acid DNA, which stands for deoxyribonucleic acid. **DNA** is the master copy of an organism's information code. The information coded in DNA contains instructions used to form nearly all of an organism's enzymes and structural proteins. Thus, DNA contributes to how an organism looks and acts. DNA's instructions are passed on every time a cell divides and from one generation of an organism to the next. Thus, we say that DNA forms the genetic code.

Another important nucleic acid is RNA, which stands for ribonucleic acid. **RNA** is a nucleic acid that forms a copy of DNA for use in protein synthesis. The chemical differences between RNA and DNA are minor but important. Chapter 13 discusses how DNA and RNA work together to produce proteins.

Connecting Ideas

Organic molecules found in cells are used to build and repair cell parts and to carry out the life processes in cells. Organic compounds in living things behave according to the same laws of chemistry and physics as compounds found in nonliving objects. How do you think the properties of organic compounds relate to their functions in cells? In the next three chapters, you will learn more about cell structures, their chemical makeup, and the reactions that store and release energy for cell metabolism.

Section Review

Understanding Concepts

1. List three important functions of lipids in living organisms.
2. Describe the process by which polymers in living things are formed from smaller molecules.
3. If there are only 20 different common amino acids, why is it possible to have thousands of different proteins?

Thinking Critically

4. As polymers containing carbon atoms pass through the carbon cycle, what happens to the covalent bonds between carbon atoms?

Skill Review

5. **Making and Using Tables** Make a table comparing polysaccharides, lipids, proteins, and nucleic acids. List these four types of biological substances in the first column. In the next two columns, list the subunits that make up each substance and the functions of each in organisms. In the last column, provide some examples of each from the chapter. For more help, refer to Organizing Information in the *Skill Handbook.*

Answers to Section Review

1. long-term energy storage, insulation, protective coatings
2. Polymers form when one monomer loses an H^+ ion and another loses an OH^- to form water. A covalent bond forms between the monomers.
3. Amino acids link together in many different sequences to form proteins.

Thinking Critically

4. Bonds are constantly broken and re-formed as carbons pass around the cycle.

Skill Review

5. Subunits and functions: Polysaccharides, monosaccharides—for energy storage and structural components; Lipids, glycerol and fatty acids—for long-term energy storage;

Chapter 7 Review

REVIEWING MAIN IDEAS

7.1 Elements and Atoms

- Atoms are the basic building blocks of all matter.
- Atoms consist of a nucleus containing protons and neutrons. The positively charged nucleus is surrounded by a cloud of rapidly moving, negatively charged electrons.
- Isotopes, atoms of the same element that differ in the number of neutrons in their nuclei, are useful in diagnosing and treating some diseases.

7.2 Interactions of Matter

- Atoms become stable by bonding to other atoms through covalent or ionic bonds.
- Water is the most abundant compound in living things. Water is an excellent solvent due to the polar property of its molecules.
- Chemical reactions are expressed by a balanced chemical equation.

7.3 Life Substances

- All organic compounds contain carbon atoms. There are four principal types of organic compounds that make up living things: carbohydrates, lipids, proteins, and nucleic acids.

Key Terms

Write a sentence that shows your understanding of each of the following terms.

acid	lipid
amino acid	metabolism
atom	mixture
base	molecule
carbohydrate	monosaccharide
cellulose	nucleic acid
compound	nucleotide
covalent bond	nucleus
disaccharide	peptide bond
DNA	pH
enzyme	polar molecule
glycogen	polymer
hydrogen bond	polysaccharide
ion	protein
ionic bond	RNA
isomer	solution
isotope	starch

Understanding Concepts

1. What are the basic building blocks of all matter?
   a. DNA molecules c. atoms
   b. proteins d. carbon
2. Three amino acids linked by peptide bonds form how many water molecules?
   a. one c. three
   b. two d. four
3. A calcium atom has 20 protons. How many electrons does this atom have?
   a. 10 c. 30
   b. 20 d. 40
4. All of the chemical reactions that occur within an organism are referred to as that organism's ____________.
   a. metabolism c. surface tension
   b. polarity d. polymers
5. A covalent bond involves sharing of ____________.
   a. protons c. neutrons
   b. elements d. electrons
6. How many electrons are held in the first energy level of atoms?
   a. one c. eight
   b. two d. eighteen
7. Which of the following is an isotope of an oxygen atom that has 8 electrons, 8 protons, and 8 neutrons?
   a. 8 electrons, 8 protons, and 9 neutrons
   b. 7 electrons, 8 protons, and 8 neutrons
   c. 8 electrons, 7 protons, and 8 neutrons
   d. 7 electrons, 7 protons, and 8 neutrons
8. Which of these is NOT an element?
   a. sodium c. chlorine
   b. hydrogen d. water

Chapter 7 Review

Reviewing Main Ideas

Summary statements can be used by students to review the major concepts of the chapter.

Key Terms

Answers should go beyond defining the terms. Accept any answer that uses the term correctly and in the proper context.

Understanding Concepts

1. c
2. b
3. b
4. a
5. d
6. b
7. a
8. d

Proteins, amino acids—structure and enzymes; Nucleic acids, nucleotides—store information in cells. Examples: Polysaccharides— starch, glycogen and cellulose; Lipids—animal fats and vegetable oils; Proteins—muscle proteins, immunity proteins, enzymes; Nucleic acids—DNA and RNA.

Chapter 7 Review

9. c
10. d
11. c
12. d
13. d
14. c
15. b
16. d
17. b
18. b
19. d
20. a

Applying Concepts

21. The underlying energy level is a filled level.
22. Water is a polar molecule; it will not attract the nonpolar grease.
23. The starch serves as an energy source for the developing seedling.
24. The substance was a compound because two new substances were formed by the chemical reaction.
25. Milk has a pH of nearly 7; being close to neutral, milk can reduce the acidity or alkalinity of the ingested substance.

Chapter 7 Review

9. Which of the following will form a solution?
   a. sand and water
   b. oil and water
   c. salt and water
   d. salt and sand
10. Which of the following describes a water molecule?
    a. Water is a nonpolar molecule.
    b. The atoms in water are bonded by ionic bonds.
    c. The bond between two water molecules is a covalent bond.
    d. Water molecules have negatively charged ends.
11. Which feature of water explains why water has high surface tension?
    a. water's high heat of vaporization
    b. water's resistance to temperature changes
    c. Water is a polar molecule.
    d. Water expands when it freezes.
12. Which of the following carbohydrates is a polysaccharide?
    a. glucose
    b. fructose
    c. sucrose
    d. starch
13. Which of the following pairs is unrelated?
    a. sugar—carbohydrate
    b. fat—lipid
    c. amino acid—protein
    d. starch—nucleic acid
14. An acid is any substance that forms ____________ in water.
    a. hydroxide ions
    b. oxygen ions
    c. hydrogen ions
    d. sodium ions
15. Which of these cannot be considered a lipid?
    a. olive oil
    b. sugar
    c. butterfat
    d. corn oil
16. Which of these is NOT made up of proteins?
    a. hair
    b. enzymes
    c. fingernails
    d. cellulose
17. In a water molecule, each ____________ atom shares one electron with the single ____________ atom.
    a. oxygen, hydrogen
    b. hydrogen, oxygen
    c. hydrogen, carbon
    d. oxygen, carbon
18. Proteins are long chains or polymers of ____________.
    a. monosaccharides
    b. amino acids
    c. fatty acids
    d. nucleotides
19. Which of the following is NOT a smaller subunit of a nucleotide?
    a. phosphate
    b. nitrogen base
    c. ribose sugar
    d. glycerol
20. A substrate fits into an area of an enzyme called ____________.
    a. the active site
    b. the enzyme-substrate complex
    c. the product
    d. the peptide bond

Applying Concepts

21. A magnesium atom has 12 electrons. When it reacts, it usually loses two electrons. How does this loss make magnesium more stable?
22. Explain why water and a sponge would not be effective in cleaning up a grease spill.
23. What is the purpose of the large amount of starch found in the seeds of many plants?
24. Heating a white substance produces a vapor and black material. Was the substance an element or a compound? Explain your answer.
25. Sometimes when a child accidentally drinks a household substance, the local Poison Control Center advises the parents to have the child drink a glass of milk. Why?

Program Resources

Chapter Assessment, pp. 37-42 L1
Alternate Assessment in the Science Classroom
Computer Test Bank L1
Content Mastery, pp. 25-28 L1

Thinking Critically

26. ***Concept Mapping*** Make a concept map that relates the following terms and phrases. Supply the appropriate linking words for your map.
    atom, element, compound, amino acid, peptide bond, protein, enzyme
27. ***Interpreting Data*** The following graph compares the abundance of four elements in living things to their abundance in Earth's crust, oceans, and atmosphere. Which element is the most abundant in organisms? What can you say about the general composition of living things compared to nonliving matter near Earth's surface?

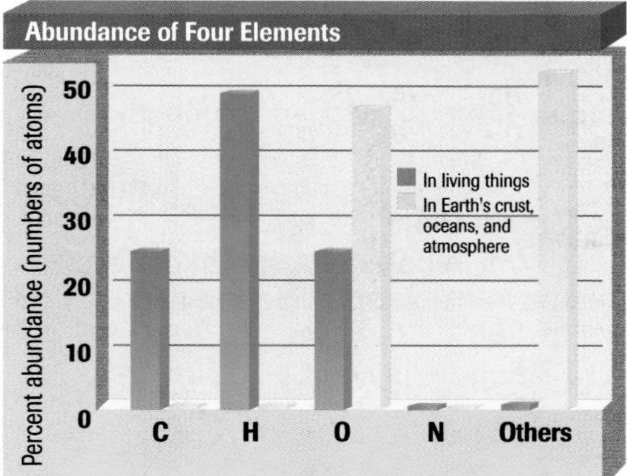

28. ***Designing an Experiment*** The enzyme peroxidase catalyzes the breakdown of hydrogen peroxide to form water and oxygen gas. Devise a way to perform a quantitative experiment that measures the rate of the reaction.
29. ***Making Inferences*** Chemical indicators usually change color sharply at a certain pH. pH papers usually contain a mixture of several indicators. Why do you think several indicators are necessary in pH paper?

ASSESSING KNOWLEDGE & SKILLS

Two students were studying the effect of temperature on two naturally occurring enzymes. The graph below summarizes their data.

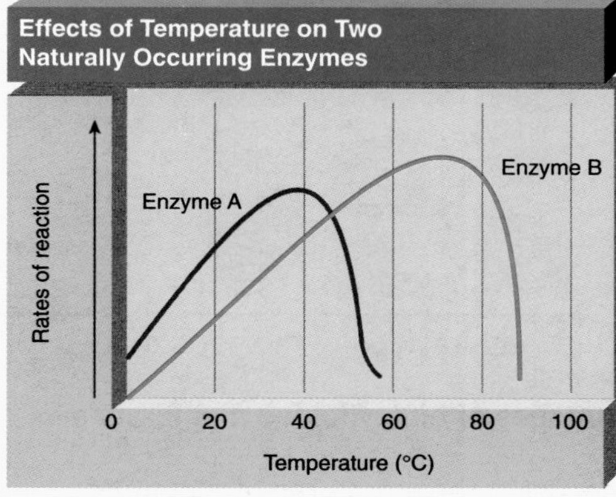

Using a Graph Use the information on the graph to answer the following questions.

1. At what temperature does the maximum activity of enzyme B occur?
   a. 0° c. 60°
   b. 35° d. 70°
2. At what temperature do both enzymes have an equal rate of reaction?
   a. 10° c. 45°
   b. 20° d. 60°
3. Which of the following descriptions best explains the patterns of temperature effects shown on this graph?
   a. Each enzyme has its own optimal temperature range.
   b. All enzymes have the same optimal temperature ranges.
   c. Each enzyme will function at room temperature.
   d. All enzymes are inactivated by freezing temperatures.
4. ***Designing an Experiment*** Design an experiment to test the optimal pH of enzyme B.

Chapter 7 Review

Thinking Critically

26. See below.
27. Hydrogen is the most abundant element in organisms. Living things contain much more hydrogen and carbon, about half the oxygen, and similar amounts of nitrogen when compared with nonliving substances.
28. Possible answers might include counting the bubbles given off or collecting and measuring the volume of oxygen given off.
29. Each indicator has a fairly narrow range, so by including more than one indicator, the range of pH change that can be measured is greater.

ASSESSING KNOWLEDGE & SKILLS

1. d
2. c
3. a
4. Place an equal amount of enzyme and substrate at different pH levels and assess the rate of the reaction at each pH.

26.

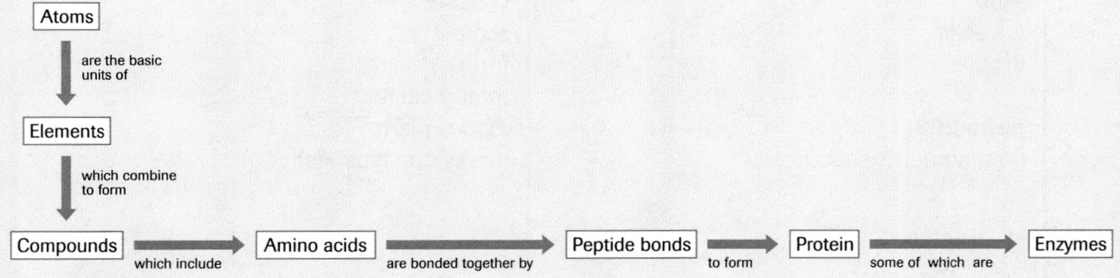

Chapter Organizer

SECTION	OBJECTIVES	ACTIVITIES/FEATURES
8.1 The Discovery of Cells National Science Standards: UCP.1; B.2; C.1, C.5; G.1-3	1. **Relate** advances in microscope technology to discoveries about cells and cell structure. 2. **Compare** the operation of a compound light microscope with that of an electron microscope. 3. **Identify** the main ideas of the cell theory.	**Minilab:** What do cork cells look like?, p. 191 **Focus On** The History of Microscopes, p. 192 **Physics Connection:** Resolution of Microscopes, p. 194
8.2 Eukaryotic Cell Structure National Science Standards: UCP.1-3, UCP.5; A.1, A.2; C.1, C.5; E.1, E.2; G.1-3	4. **Relate** the structure and function of the parts of a typical eukaryotic cell. 5. **Explain** the advantages of highly folded membranes in cells. 6. **Compare and contrast** the structures of plant and animal cells.	**Thinking Lab:** What organelle directs cell activity?, p. 198 **People in Biology:** Marian Diamond, p. 202 **Biology & Society Technology:** Natural Fiber—Natural Color, p. 205 **Minilab:** Which parts of a sperm cell are visible in a transmission electron micrograph?, p. 206 **Biolab:** Sizing Cells and Cell Structures, p. 208

ACTIVITY MATERIALS

BIOLAB	MINILABS	ALTERNATE LAB
page 208 small metric ruler calculator	**page 191** bottle cork razor blade microscope slide coverslip dropper **page 206** no materials needed	**page 198** microscope slides coverslips dropper scalpel forceps toothpick, flat *Elodea* plant methylene blue stain

Chapter 8 A View of the Cell

TEACHER CLASSROOM RESOURCES

Reproducible Masters	Transparencies
Section Focus Master 16: Gathering Information with Scientific Tools L1 SAE **Reinforcement and Study Guide,** p. 29 L1 **Biolab and Minilab Worksheets,** p. 27 L1 **Critical Thinking/Problem Solving:** Cell Organelles and Their Functions, p. 8 L3 **Laboratory Manual:** Use of the Light Microscope, pp. 47-50; How Can a Microscope Be Used in the Laboratory?, pp. 51-54 L2	**Basic Skills Transparency #7:** The Optical Microscope L1 SAE
Section Focus Master 17: Surface Area L1 SAE **Reinforcement and Study Guide,** pp. 30-32 L1 **Biolab and Minilab Worksheets,** pp. 28-30 L1 **Concept Mapping:** Recycling in the Cell, p. 8 L1 **Content Mastery,** pp. 29-32 L1	**Basic Concepts Transparency #6:** The Cell L1 SAE **Reteaching Transparency #8:** Eukaryotic Cell Structure and Organelles L1 SAE

ASSESSMENT MATERIALS

Chapter Assessment, pp. 43-48
MindJogger Videoquiz
Alternate Assessment in the Science Classroom
Computer Test Bank

Spanish Resources SAE
English/Spanish Audiocassettes SAE
Cooperative Learning in the Science Classroom COOP LEARN
Lesson Plans
Biology Projects: A Close Look at Cells, pp. 9-12 L2

KEY TO TEACHING STRATEGIES

- L1 Level 1 activities should be within the ability range of all students including those with learning difficulties.
- L2 Level 2 activities are within the ability range of average to above-average students.
- L3 Level 3 activities are designed for the ability range of above-average students.
- SAE SAE activities should be within the ability range of Students Acquiring English.
- COOP LEARN Cooperative Learning activities are designed for small group work.
- P These strategies represent student products that can be placed into a best-work portfolio.
- These strategies are useful in a block scheduling format.

GLENCOE TECHNOLOGY

The following multimedia resources are available from Glencoe.

Biology: The Dynamics of Life
CD-ROM SAE
Videodisc Program

National Geographic Society Series
STV: The Cell
Viewing the Cell
Parts of the Cell

Science and Technology Videodisc Series (STVS)
Human Biology
Diagnosing Disease with Glowing Cells

CHAPTER

8 A View of the Cell

Chapter Overview

Students study the history of cell theory as it developed along with the microscope. Students become familiar with the compound light microscope and how it differs from electron microscopes.

In the second part of the chapter, students are introduced to cell structure and function. The organelles found in typical animal and plant cells are introduced. The concept of the interdependence of cell parts is presented. Students also learn to calculate cell size to develop a concept of the sizes of cells and organelles.

Key Terms

cell
cell theory
cell wall
chlorophyll
chloroplast
chromatin
cilia
compound light microscope
cytoplasm
cytoskeleton
electron microscope
endoplasmic reticulum
eukaryote
flagella
Golgi apparatus
lysosome
microfilament
microtubule
mitochondria
multicellular
nucleolus
nucleus
organ
organ system
organelle
plasma membrane
plastid
prokaryote
ribosome
tissue
unicellular
vacuole

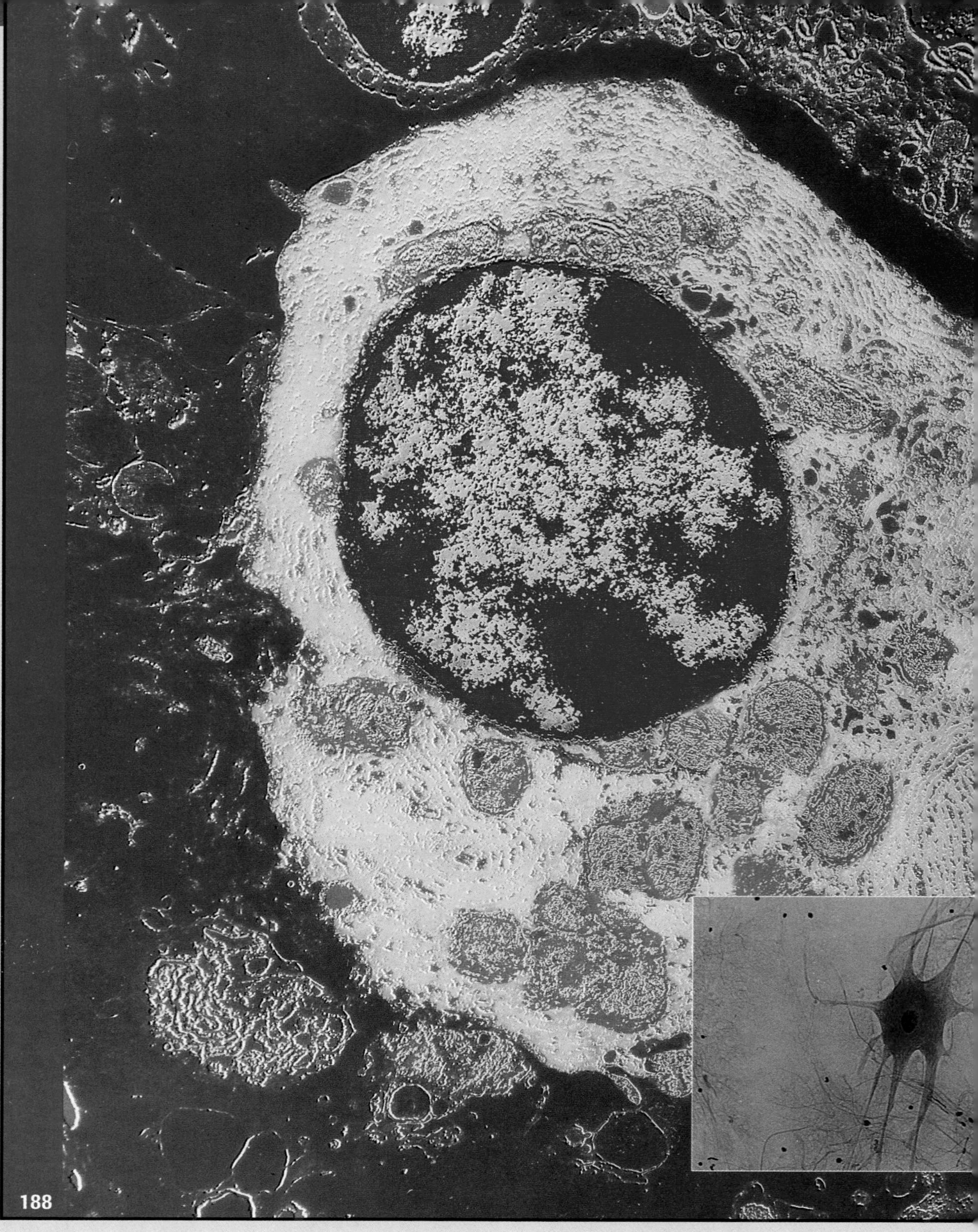

Look for the following logo for strategies that emphasize different learning modalities. LS

Kinesthetic	Meeting Individual Needs, p. 207
Visual-Spatial	Minilab, pp. 191, 206; Project, p. 192; Portfolio, pp. 193, 200; Meeting Individual Needs, p. 197; Alternate Lab, p. 198; Reinforcement, p. 200; Demonstration, p. 204; Student Journal, p. 204
Interpersonal	Enrichment, p. 200
Intrapersonal	Student Journal, p. 206
Linguistic	Student Journal, p. 194; Using Science Terms, p. 204
Logical-Mathematical	Time Line, p. 194; Thinking Lab, p. 198; Math Connection, p. 199; Project, p. 205

CHAPTER

8 A View of the Cell

What comes to mind when you think of visiting a busy pizza parlor during a fall evening? Can you smell the aroma of cheese, onions, and pepperoni or picture the employees busy making pizzas? Can you hear music playing in the background as well as the sounds from people laughing and video games being played? As the front door opens, cold, crisp air rushes in. People come in and go out, taking boxes of pizza-to-go. What a flurry of activity is going on!

Cells, the basic units of life, can be likened to a pizza parlor. They, too, are in a constant flurry of activity. These cells, so small that at least 50 000 of them would fit into this letter O, are busy building and breaking down macromolecules. They are at work releasing energy from foods, and then using that energy to make needed cell parts. Together, your cells function to make your body operate like a well-run pizza business.

Concept Check

You may wish to review the following concepts before studying this chapter.

- Chapter 1: homeostasis
- Chapter 7: proteins, enzymes

Chapter Preview

Laboratory Activities

Biolab

- Sizing Cells and Cell Structures

Minilabs

- What do cork cells look like?
- Which parts of a sperm cell are visible in a transmission electron micrograph?

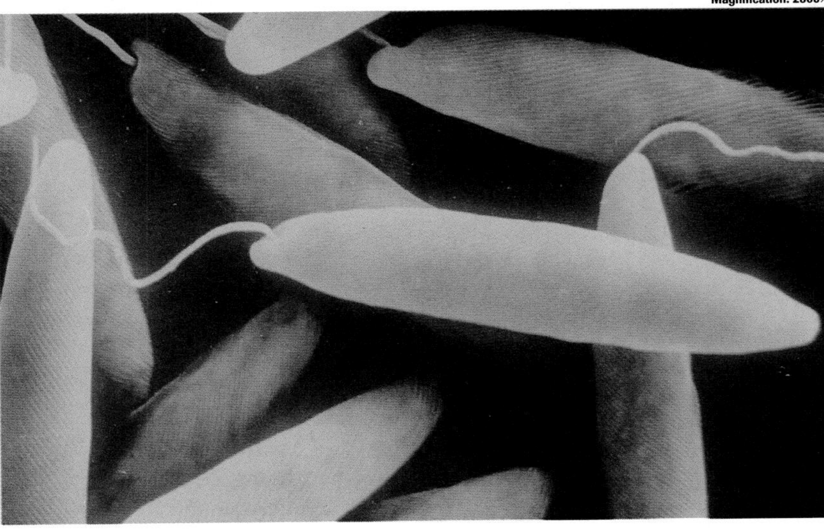

This large, white blood cell, magnified 225 000 times, is one of many millions and trillions of cells that make up a mature human. The nerve cell in the inset photo (enlarged 150 times) has a shape that is adapted for transmitting electrical impulses to the brain. The euglenas shown on this page are unicellular organisms. The single cell performs all the functions necessary for the organism to survive.

CHAPTER 8

Introducing the Chapter

Ask students to examine the photograph of the white blood cell and describe any cell structures they observe. Encourage students to speculate about the function of each cell structure based on its appearance. Remind students that cells are living, and therefore, must carry on life functions.

Theme Development

The first section of the chapter stresses *the nature of science* through a discussion of the development of the microscope and the cell theory. *Evolution* is apparent in the discussion of the number, variety, and complexity of living organisms, particularly as these factors relate to cell structure and function.

Concept Check

The concept of homeostasis is expanded upon as students learn that homeostasis in an organism, whether unicellular or multicellular, depends on the functions of the cell. The functions of proteins and enzymes, discussed in Chapter 7, are related to the functions of various cell organelles.

Assessment Planner

Choose assessment strategies from the following pages to evaluate the progress of your students.

Assess, pp. 195, 207
Alternate Lab, pp. 198-199
Minilabs, pp. 191, 206
Portfolio, pp. 193, 200
Thinking Lab, p. 198
Biolab, pp. 208-209
Chapter Review, pp. 211-213

SECTION

8.1

Prepare

Key Concepts

Students are provided with an overview of the historical development of the microscope and the events that led to the formation of the cell theory.

Block Scheduling

Look for this symbol for strategies that are useful in a block scheduling format. For more information on block scheduling, refer to Section 8.1 in the **Lesson Plans** booklet.

Materials

- Collect pictures of microscopes for the Display.
- Gather pictures of electron micrographs for the Demonstration.
- Purchase materials needed to carry out the microscopy project.

1 Focus

Bellringer

Before presenting the lesson, display **Section Focus Master 16** on the overhead projector and have students answer the accompanying questions. L1 SAE

2 Teach

Audiovisual

Show the video *Using a Compound Microscope* or *Preparing & Using Microscope Slides*, Media Design Associates, Inc.

SECTION

8.1 The Discovery of Cells

Section Preview

Objectives

Relate advances in microscope technology to discoveries about cells and cell structure.

Compare the operation of a compound light microscope with that of an electron microscope.

Identify the main ideas of the cell theory.

Key Terms

cell
compound light microscope
cell theory
electron microscope
prokaryote
eukaryote
organelle
nucleus

From the smallest bacterium to the largest redwood tree, all living things are made of cells. Each cell carries out life functions. However, microscopes had to be invented before scientists could study—or even discover—cells, because most cells are so tiny that they cannot be seen without the aid of a microscope. The unaided human eye can see something as small as 0.01 cm, or about the size of a tiny pencil dot on a piece of paper. An average-sized cell, however, has a diameter of about 0.002 cm. No wonder most cells can't be seen.

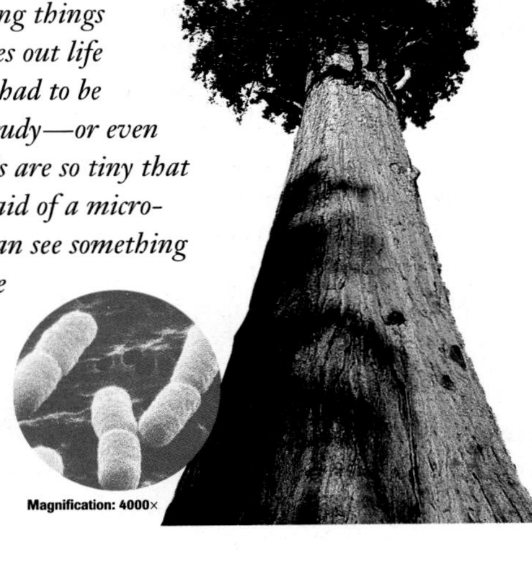

microscope:
mikros (GK) small
skopein (GK) to look
A microscope is used to examine small objects.

The History of the Cell Theory

Before microscopes were invented, people believed that diseases were caused by curses and supernatural spirits. They had no idea that organisms such as bacteria existed. The invention and development of the microscope enabled scientists to discover and study **cells,** the basic units of living organisms.

Development of light microscopes

The first microscopes used for viewing cells were simple microscopes—basically a single magnifying lens. In the 1600s, Anton van Leeuwenhoek described living cells as seen through a simple microscope.

During the next 200 years, microscopes were greatly improved. Biologists began to use compound light microscopes. A **compound light microscope** has a series of lenses that magnify an object in steps. Visible light is passed through the object and then through the lenses. The object may be a bacterium or a cell part as small as 0.00002 cm (0.2 µm) in diameter. The compound light microscope allowed biologists to achieve greater magnification and to see more detail inside cells.

The cell theory is developed

In 1665, Robert Hooke, an English scientist, used a crude compound microscope to examine thin slices of cork from the bark of an oak

Program Resources

Section Focus Master 16 L1 SAE
Reinforcement and Study Guide, p. 29 L1
Biolab and Minilab Worksheets, p. 27 L1
Laboratory Manual, pp. 47-50; 51-54 L2
Critical Thinking/Problem Solving, p. 8 L3
Basic Skills Transparency 7 and **Master** L1 SAE

tree, as shown in *Figure 8.1*. He observed that the cork was composed of tiny, hollow boxes similar to a honeycomb. Hooke called the boxes *cells* because they reminded him of the small, boxy living quarters of monks. Their rooms were, in fact, called cells. Today, we know that the structures Hooke observed were the walls of dead plant cells. Hooke published his drawings and descriptions, which led other scientists of that time to look for evidence of cells.

In the 1830s, two German scientists, Matthias Schleiden and Thomas Schwann, were able to view many different organisms with microscopes and draw some important conclusions. Schleiden used a microscope to observe plants and concluded that all plants are composed of cells. Schwann made similar observations about animal cells.

The observations and conclusions of scientists from the late 17th century to the time of Schwann and Schleiden are summarized into the cell theory, one of the most basic ideas of modern biology.

Figure 8.1

Robert Hooke observed cork cells such as these using a crude compound microscope that magnified only 30 times.

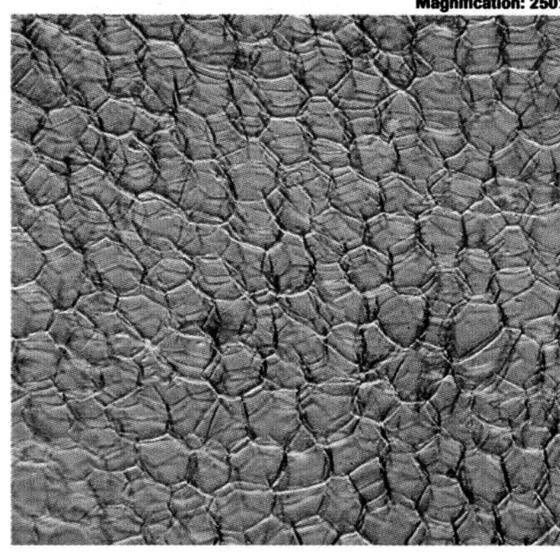

MiniLab

What do cork cells look like?

Robert Hooke discovered and named cells when he looked at cork with a crude microscope. Cork is a dead, protective tissue that is abundant on the trunk of some trees such as cork oak. It is periodically stripped off for making commercial products, such as bottle corks.

Procedure

1. Carefully shave a paper-thin piece of cork off a bottle cork with a razor blade.
2. Place the piece of cork in a drop of water on a microscope slide. Add a coverslip.
3. Observe the cells at the edge of the piece of cork on low and high power. Make a drawing of the cells at each power.

Analysis

1. What shape are the cork cells?
2. What is inside the cork cells?
3. What function might the cork provide for the tree?

The **cell theory** is made up of three main ideas.

1. *All organisms are composed of one or more cells.* An organism may be a single cell, such as an amoeba or a bacterium. Larger organisms, such as humans, are composed of many cells.
2. *The cell is the basic unit of organization of organisms.* In the same way that the basic unit of matter is the atom, the basic unit of life is the cell.
3. *All cells come from preexisting cells.* Cells come from the reproduction of previously existing cells, a process in which cells reproduce exact copies of themselves.

Development of electron microscopes

During the 1940s, more powerful microscopes were developed. Rather than using light, they used electrons that passed over or through an object.

Meeting Individual Needs

Visually Impaired Pair visually impaired students with sighted students to do microscope labs. Ask the sighted students to describe what they see in the microscope to the visually impaired students. Slides can also be prepared for a projection microscope and projected onto a classroom screen to help visually impaired students. L1

MiniLab

Purpose

LS **Visual-Spatial** Students will view cork cells using the methods used by Robert Hooke.

Process Skills

observe and infer, collect and organize data

Safety Precautions

Caution students to be careful while using the razor blade.

Teaching Strategies

- Prior to beginning the lab, direct students to the Skill Handbook so they can review the parts and functions of the compound microscope.
- Show students how to shave the top of the cork. If sections are not thin enough, students will be unable to see individual cells.
- Provide students with instruction about the proper procedures to be used when focusing the microscope on both low and high power. For use and care of the microscope, refer students to the Skill Handbook.

Expected Results

Students should observe empty box-shaped cells using both low and high power.

Analysis

1. box-shaped
2. nothing but air
3. protection from water loss

Assessment

Performance: Have each student demonstrate a focused section of cork on high power. L1

Focus On

The History of Microscopes

Purpose

Students will be provided with a historical perspective of the events that led to the development of the compound microscope, the transmission electron microscope (TEM), and the scanning electron microscope (SEM) and the uses of each microscope type.

Teaching Strategies

- **Discussion Question** Ask students whether they think development of a successful microscope was delayed due to the problem of creating lenses or due to a lack of knowledge about how to prepare specimens for study. *Problems resulted more from a lack of knowledge about how to prepare specimens since lens construction dates back to the time of Galileo.*
- Discuss the techniques involved in preparing specimens for electron microscopy studies. The use of metals for shadowing, freeze-fracture techniques, and ultramicrotomes are all good topics for discussion.
- **Enrichment** Have students research the backgrounds and training of the various individuals involved in the development of the microscope or who contributed to the development of the cell theory. L2

GLENCOE TECHNOLOGY

Videodisc

Biology: The Dynamics of Life
Disc 1, Side 1
The Light Microscope (Ch. 22)

Focus On

The History of Microscopes

1600 — 1665 — 170

The invention and development of the light microscope about 300 years ago enabled scientists to discover and study cells. Improved versions of the microscope have allowed scientists to increase the range of visibility enormously, and they can now see molecules.

Hooke's microscope Robert Hooke designed this microscope and drew the cork cells he observed. Although it has three lenses and is beautifully finished in gold-embossed leather, the lenses are of poor quality, and Hooke could actually see little detail.

TEM A transmission electron microscope (TEM) aims a beam of electrons through an object in a vacuum. The more dense portions allow fewer electrons to pass through. These denser areas appear darker on the screen. Transmission electron microscope images are two dimensional. The TEM is used to study the details of cell parts. It can magnify objects hundreds of thousands of times.

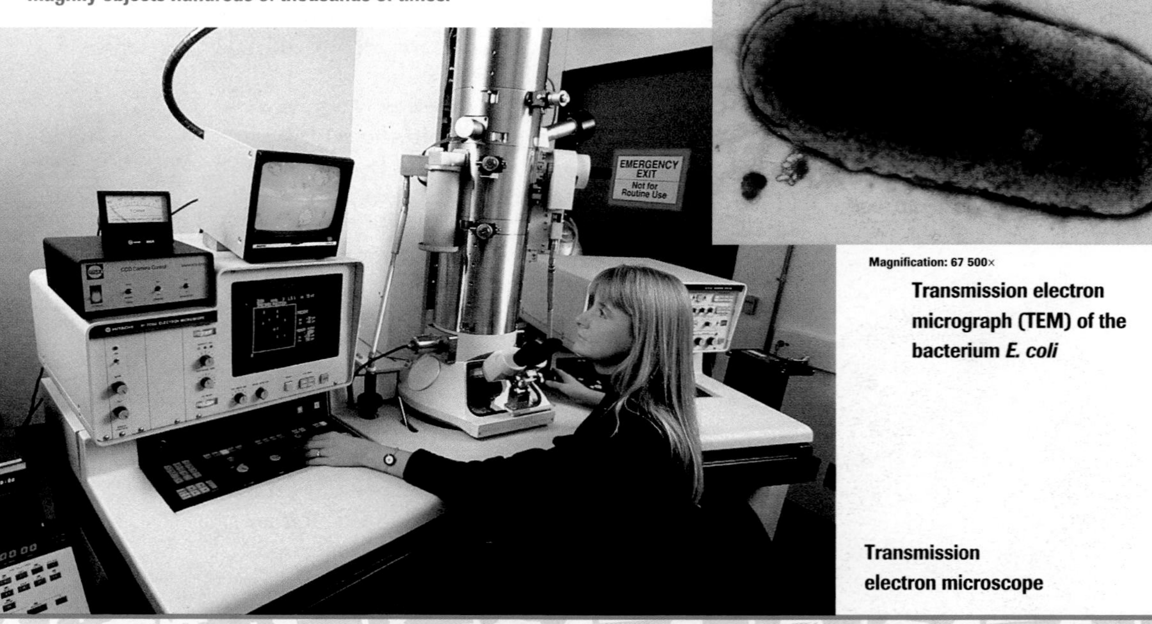

Magnification: 67 500×

Transmission electron micrograph (TEM) of the bacterium *E. coli*

Transmission electron microscope

PROJECT

Developing Skills In Microscopy

Have students examine various types of cells to develop their microscope skills. Set out iodine solution and methylene blue to allow students to experiment with the stains. Possible cells to examine might include: onion, *Elodea* leaf, cheek, potato, *Lactobacillus* in cultured yogurt, teeth bacteria, *Paramecium,* amoeba, *Volvox,* tomato skin, tomato pulp, and pear pulp. You may wish to have students work in their cooperative groups to prepare their stained slides. Encourage students to review the proper procedures for using the microscope in the Skill Handbook. L1 LS COOP LEARN

1800 — 1900 — 1940

Electron Microscope

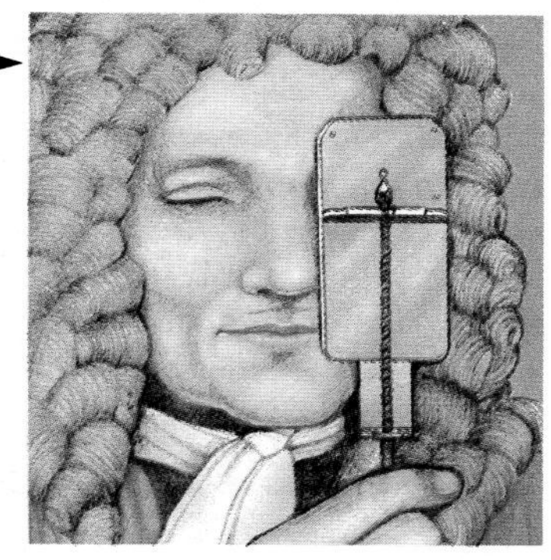

Better lenses Anton van Leeuwenhoek improved upon the simple microscope. He built more than 240 microscopes, grinding the lenses himself. Although his microscopes had only a single lens, the quality was far better than that of Hooke's compound microscope. Leeuwenhoek was the first to discover and describe red blood cells and bacteria taken from scrapings off his teeth.

Modern light microscopes A compound light microscope uses two or more glass lenses to magnify objects. Light microscopes can be used to look at living cells and small organisms, as well as at preserved cells. These microscopes can magnify up to about 1500 times.

Magnification: 63×

Light micrograph of *Paramecium*

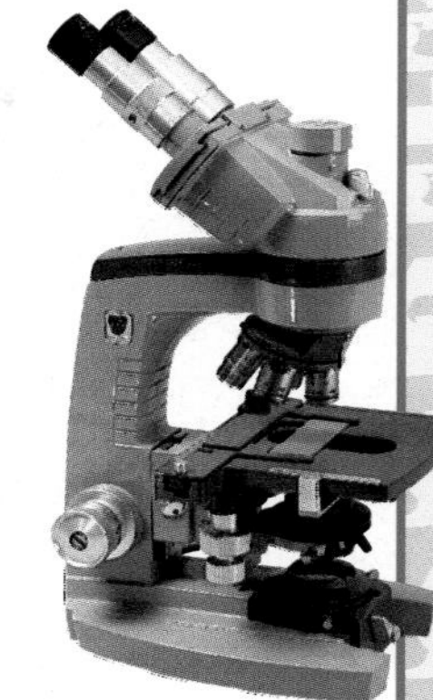

STM The scanning tunneling microscope (STM) can show the arrangement of atoms on the surface of a molecule, such as this DNA molecule. A metal probe is brought near the object's surface. Electrons flow from the surface to the probe. In this way, the hills and valleys of the surface can be mapped.

Magnification: 1 000 000×

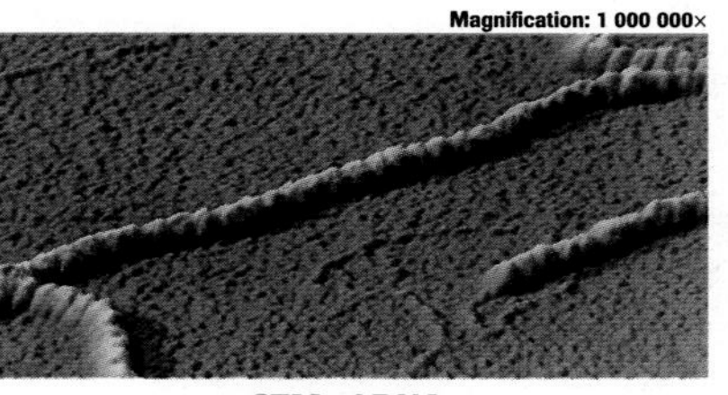

STM of DNA

SEM A scanning electron microscope (SEM) sweeps a beam of electrons over the surface of a specimen. This causes electrons to be emitted from the specimen. Scanning electron microscopes produce a realistic, three-dimensional picture of the object, but only the surfaces of specimens can be observed. The SEM can magnify only about 60 000 times without losing clarity.

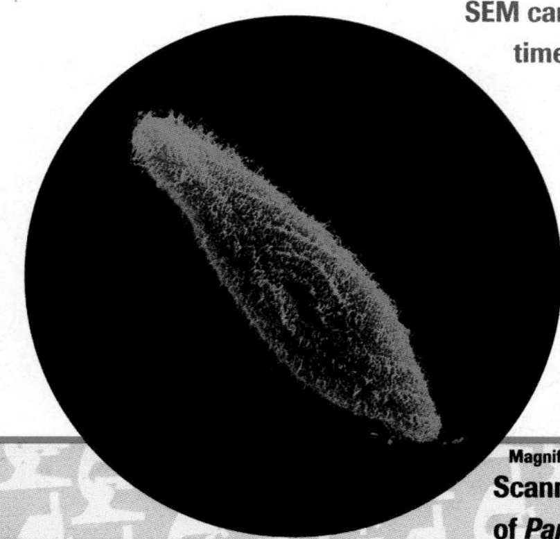

Magnification: 150×

Scanning electron micrograph (SEM) of *Paramecium*

EXPANDING YOUR VIEW

1. **Comparing and Contrasting** Compare the images seen with an SEM and a TEM.
2. **Thinking Critically** Can live specimens be examined with an electron microscope? Explain. Consider how the specimen must be prepared for viewing.

PORTFOLIO

The Compound Microscope Have students draw and label the parts of their microscope. Under each label have them describe the function of the part. For the objective lenses, have students calculate and identify the power of magnification for an object viewed with each lens. L1 LS P

Focus On

Visual Learning

- **Demonstration** Show students a series of photomicrographs of the same organelle as it appears with a light and an electron microscope. Also show them artists' sketches of cells and their organelles to aid students in recognizing two-dimensional and three-dimensional representations. Photomicrographs and drawings may be obtained from college level texts.
- **Display** Create a display consisting of pictures of the many types of early microscopes. Such pictures may be obtained from a museum or history of science text. Large displays of various types of modern microscopes are available from scientific supply houses. Use the pictures to contrast the location of the specimen, the methods of illumination, the types of lenses used, and how light passes through each (using light-ray optical diagrams). If possible, also show examples of specimens as they appear under each microscope.

Answers to Expanding Your View

1. **Comparing and Contrasting** Images in the SEM are three-dimensional, while TEM images are two-dimensional. Images in the TEM can be magnified greater than those in the SEM.
2. **Thinking Critically** Live specimens cannot be viewed with an electron microscope because specimens are dehydrated and placed in a vacuum.

Physics

Resolution of Microscopes

Purpose
Students will understand that resolving power and not magnification determines the quality of a microscope.

Teaching Strategies
- Emphasize that resolving power is the ability to form distinct images of two very close objects.
- Ask students to explain why there are limits to the magnification of a light microscope and what determines these limits. *Limits of resolution are in part determined by the limitations of the human eye to separate two points of reference in the field of view. The quality of the glass used in a lens also affects resolution.*

Possible Answer
The electron microscope allowed scientists to see much greater detail inside cells. The actual structure of cell organelles could be seen.

Time Line

LS **Logical-Mathematical** Have students reread this section and create a time line listing the significant events in the history of cell study. Remind students to record events associated with microscope development in their time lines. L1

Physics

Resolution of Microscopes

The quality of a microscope and its ability to magnify depend on its resolving power. Resolution is the minimal distance that two points can be separated and still be seen as two separate and distinct points. If two details inside an organism are closer than the resolution of the microscope, they will be seen as one.

The cells of the retina in the eye only react if light strikes them. As a result, the image produced in the retina is a pattern of light and dark spots. These spots resemble the appearance of a magnified newspaper photograph, such as the one shown above.

Light from two points separated by a distance of less than 0.01 cm activates the same retinal cell. The two points then are perceived as only a single point. It is only when light from two points activates two retinal cells separated by at least one unactivated cell that the two points can actually be seen as two distinct points.

Resolution and the compound light microscope One way to increase resolution is to add a lens between your eye and the object. If small details can be made to appear larger, the distance between points will be increased. The theoretical limit of the compound light microscope depends on the wavelength of light. Light must be able to go between two points in order for them to be seen distinctly. One can calculate that the maximum magnification available for a compound light microscope is about 1500x. If objects are magnified further, the details look bigger but not clearer. Finer detail cannot be resolved.

Resolution and the electron microscope The electron microscope allows for greater resolution because it uses a beam of electrons, which have a much shorter wavelength than that of light. The theoretical resolving power of an electron microscope is about 1000 times greater than that of the light microscope.

CONNECTION TO Biology

Why was the development of the electron microscope with its greater resolving power so important to the advancement of cell biology?

These powerful microscopes are called electron microscopes. An **electron microscope** aims a beam of electrons through a magnetic field to focus them, then through or over the surface of a specimen in a vacuum, and finally onto a fluorescent screen where it forms an image. The image can be photographed and developed into an electron micrograph, SEM or TEM. Because the specimen must be in a vacuum, only dead cells or organisms can be viewed. There are several types of electron microscopes. Electron micrographs have provided much detail about the structure of cell parts and have allowed scientists to view objects even as small as molecules.

Two Basic Cell Types

As biologists studied cells, they found two basic kinds—prokaryotic and eukaryotic, depending on their internal organization. Organisms that

Figure 8.2

Prokaryotic cells do not have a true nucleus or other internal organelles surrounded by a membrane. Their DNA is concentrated in a region called the nucleoid, but it is not separated from the rest of the cell by a membrane. Most of a prokaryote's metabolic functions take place in the cytoplasm.

Magnification: 30 500×

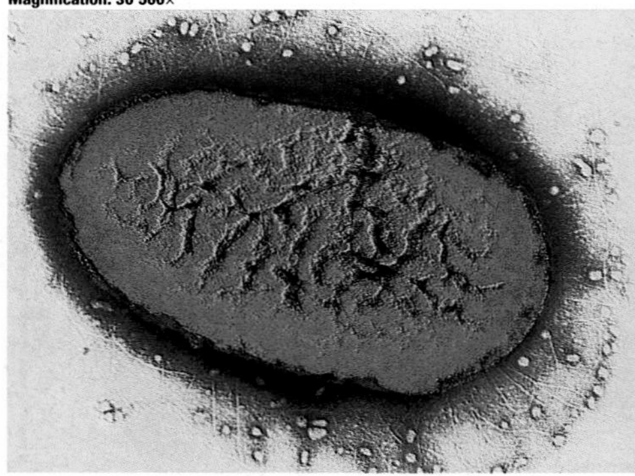

STUDENT JOURNAL

Building Vocabulary Have students write the definitions for the following Greek and Latin prefixes and suffixes: *micro, scop, eu, karyon, pro, plasma, chrom, endo, lysis, mitos, chondros, chloros,* and *plastos.* Students should identify a word from the chapter that uses each root and explain how the meaning of the root relates to the meaning of the term. L1 LS

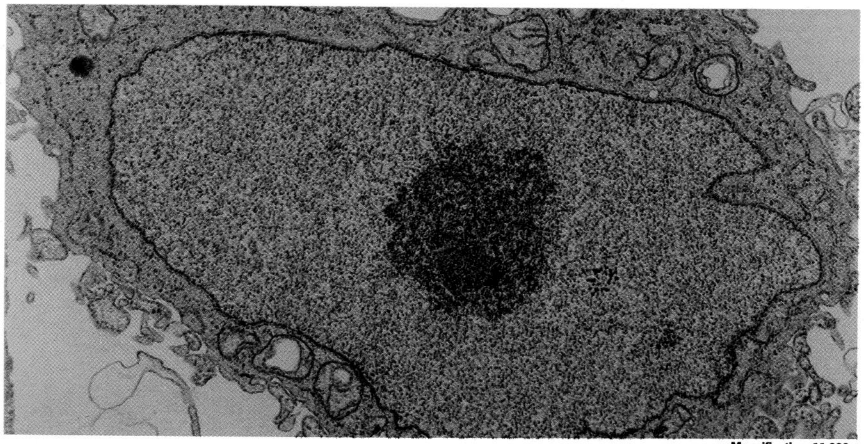

Figure 8.3

Eukaryotic cells have a true nucleus and discrete membrane-bound organelles, which allow different parts of the cell to specialize in different functions. The majority of cells in the living world are eukaryotic.

Magnification: 11 000×

are made of these two kinds of cells are called prokaryotes and eukaryotes, respectively. A **prokaryote**, shown in *Figure 8.2*, is an organism with a cell that lacks internal structures surrounded by membranes. Most prokaryotes are single-celled organisms. A **eukaryote**, shown in *Figure 8.3*, is an organism that has cells containing internal, membrane-bound structures. These structures are called **organelles.** Each organelle has a membrane surrounding it, isolating it from the rest of the cell. The largest organelle is a membrane-bound **nucleus,** which contains the cell's DNA and manages the cell's functions. Eukaryotes are either single celled or made of many cells.

The evolution of organelles in eukaryotes provided a way for the cell to be divided into compartments. Most of the internal organelles are surrounded by membranes, making them separate "rooms" and isolating them from the rest of the cell's contents. Thus, the activity of an individual compartment can become specialized. Chemical reactions that could not occur in the same area can be carried out at the same time in different compartments.

organelle:
organon (GK) tool, implement
ella (GK) small
Organelles are small, membrane-bound structures in cells.

Section Review

Understanding Concepts

1. Why was the development of microscopes necessary for the study of cells?
2. Compare the way images are formed in light microscopes and electron microscopes.
3. How are prokaryotic and eukaryotic cells different?

Thinking Critically

4. Suppose you were to discover a new microorganism in some ocean water. Applying the cell theory, what can you say for certain about this organism?

Skill Review

5. **Care and Use of a Microscope** A compound light microscope has three objective lenses that magnify 6, 40, and 95 times. What magnifications are available if an eyepiece that magnifies 15 times is used? For more help, refer to Practicing Scientific Methods in the *Skill Handbook*.

3 Assess

Check for Understanding

Quiz students orally about the importance of the cell theory and its acceptance by the scientific community. L1

Reteach

Review the parts of the cell theory and the significance of each statement making up the theory as a reinforcement exercise. L1

Extension

Have interested students research new types of microscopes such as the atomic force microscope. L3

Assessment

Knowledge: Quiz students on the names and functions of the parts of the compound microscope. L1

4 Close

Activity

In a class discussion, summarize the contributions of each scientist mentioned in this section. Have students make a table that lists the name of each scientist and his contributions. Encourage students to retain their tables in their journals and to add information about other scientists throughout the remainder of the course. L1

Answers to Section Review

1. The development of the microscope allowed scientists to observe cells, because most cells are not visible with the unaided eye.
2. In light microscopes, images are formed by passing visible light through an object. In electron microscopes, a beam of electrons is passed through or over an object.
3. Eukaryotic cells have a nucleus surrounded by a double bilayer and have membrane-bound organelles. Prokaryotic cells have neither of these features.

Thinking Critically

4. The microorganism is made of one or more cells, which are the basic units of organization of organisms. The microorganism's cells came from preexisting cells.

Skill Review

5. 90×, 600×, 1425×

SECTION

8.2

Prepare

Key Concepts

The section focuses on the structure and function of eukaryotic cells. The complexity of eukaryotic cells is emphasized through the use of electron micrographs and text descriptions of organelles and their functions.

Block Scheduling

Look for this symbol for strategies that are useful in a block scheduling format. For more information on block scheduling, refer to Section 8.2 in the **Lesson Plans** booklet.

Materials

- Purchase gelatin and fruit for the Demonstration.
- Purchase toothpicks and *Elodea* for the Alternate Lab.
- Gather materials for construction of the classroom "cell" as suggested in the Project strategy.

1 Focus

Bellringer

Before presenting the lesson, display **Section Focus Master 17** on the overhead projector and have students answer the accompanying questions. L1 SAE

Brainstorming

Ask students to list the body systems they need to live. Write their responses on the chalkboard. Use this list to introduce cell organelles as the structures that perform or help to perform the life functions of cells. L1

SECTION

8.2 Eukaryotic Cell Structure

Section Preview

Objectives

Relate the structure and function of the parts of a typical eukaryotic cell.

Explain the advantages of highly folded membranes in cells.

Compare and contrast the structures of plant and animal cells.

Key Terms

plasma membrane
cell wall
chromatin
nucleolus
ribosome
cytoplasm
endoplasmic reticulum
Golgi apparatus
vacuole
lysosome
mitochondria
chloroplast
chlorophyll
plastid
cytoskeleton
microtubule
microfilament
cilia
flagella
unicellular
multicellular
tissue
organ
organ system

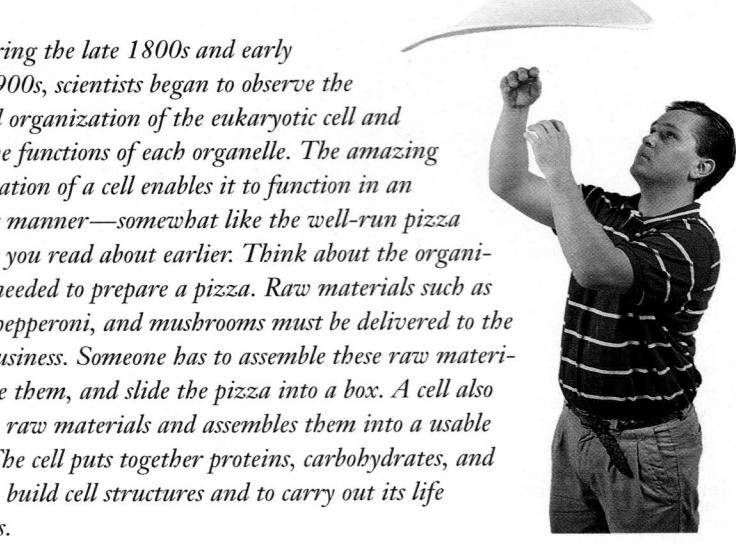

During the late 1800s and early 1900s, scientists began to observe the internal organization of the eukaryotic cell and learn the functions of each organelle. The amazing organization of a cell enables it to function in an efficient manner—somewhat like the well-run pizza business you read about earlier. Think about the organization needed to prepare a pizza. Raw materials such as cheese, pepperoni, and mushrooms must be delivered to the pizza business. Someone has to assemble these raw materials, bake them, and slide the pizza into a box. A cell also takes in raw materials and assembles them into a usable form. The cell puts together proteins, carbohydrates, and lipids to build cell structures and to carry out its life processes.

Boundaries and Control

When you enter a pizza parlor, you go through a door in the wall—the boundary of the business. Cells also have boundaries with passageways for materials to enter or exit.

Cells must have boundaries

A cell has a **plasma membrane** that serves as a boundary between the cell and its external environment. Unlike the walls of the pizza business, though, the plasma membrane is quite flexible and allows the cell to vary its shape if needed. The plasma membrane, shown in *Figure 8.4,* controls the movement of materials that enter and exit the cell. It allows useful materials such as oxygen and nutrients to enter, and waste products such as excess water to leave. Some materials enter and leave through protein passageways. Other materials pass directly through the membrane. The plasma membrane helps maintain a chemical balance within the cell.

Some cells have another external boundary outside their plasma membrane. This additional boundary, the **cell wall,** shown in *Figure 8.5,* is a relatively inflexible structure that surrounds the plasma membrane. The cell wall is much thicker than the plasma membrane and is made of different substances in different organisms. The cells of plants, fungi,

Program Resources

Section Focus Master 17 L1 SAE
Reinforcement and Study Guide, pp. 30-32 L1
Biolab and Minilab Worksheets, pp. 28-30 L1
Concept Mapping, p. 8 L1
Basic Concepts Transparency 6 and **Master** L1 SAE
Reteaching Transparency 8 and **Master** L1 SAE

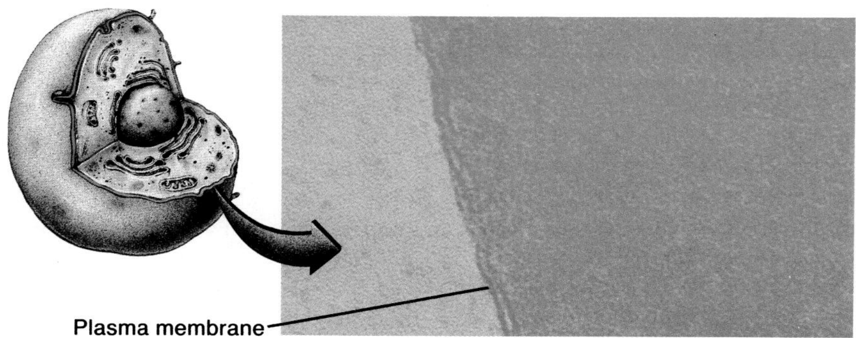

Figure 8.4

The plasma membrane that surrounds a cell is made of two layers of lipid and protein, which you can distinguish in the photomicrograph. Two-layered membranes similar to the plasma membrane also enclose all of the membrane-bound organelles inside a cell. Is the photo a TEM or an SEM? How do you know?

almost all bacteria, and some protists have cell walls. Animal cells have no cell walls. Plant cell walls contain cellulose molecules, which form fibers. The fibers are interwoven to produce a strong network that protects the cell and gives the plant support. It is this fibrous cellulose of plants that provides the bulk of the fiber in our diets. Chitin, a nitrogen-containing polysaccharide, makes up the cell walls of fungi.

Organelles that control cell functions

Just like the pizza business, the cell needs a manager within its boundaries. The manager directs the affairs of the pizza business. The nucleus is the organelle that manages cell functions in a eukaryotic cell. The nucleus is surrounded by a nuclear envelope, which is a double membrane, and each membrane is made up of two layers. Thus, the nuclear envelope is four layers thick.

Figure 8.5

The cell wall is a firm structure that protects the cell and gives the cell its shape. Plant cell walls are made mainly of multiple layers of cellulose, top right, but also contain pectin, the substance that causes jams and jellies to thicken. The fibrous nature of cellulose, magnified 120 000×, can be seen in the photomicrograph, bottom right.

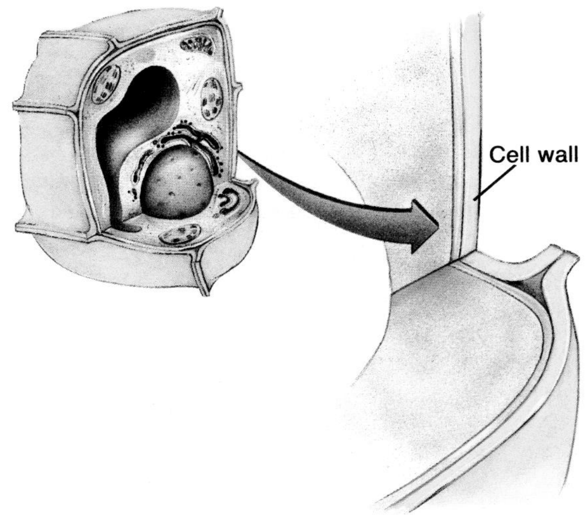

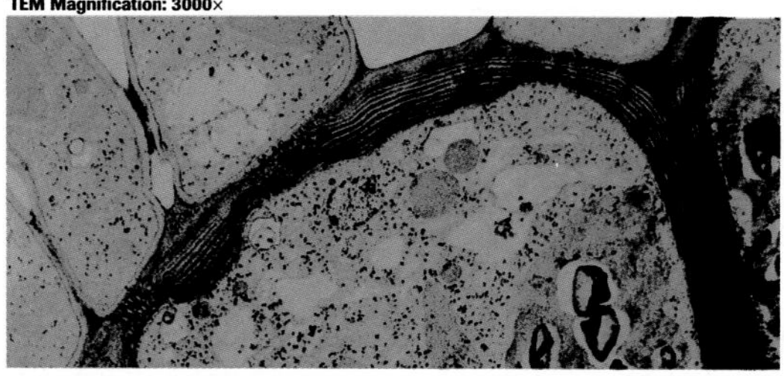

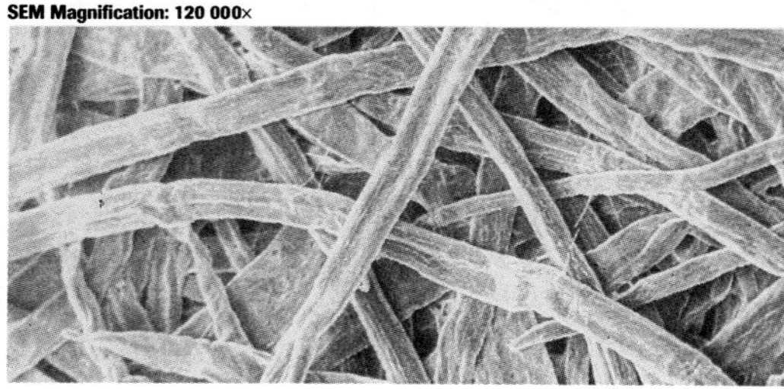

2 Teach

Audiovisual

Show the video *Structure of the Cell Video*, Frey Scientific.

Misconception

Students often think of a cell as being solid and impenetrable. Point out that although a cell contains some solids, most of a cell (almost 80%) is liquid water. The water, along with the solids and some dissolved gases, is enclosed within a membrane that permits certain materials to enter and exit the cell.

Software

Cell, Queue, Apple II, Mac, IBM

Teacher FYI

The cell walls of fungi contain chitin and those of diatoms contain silica. Bacteria have single-layered walls composed of a complex material that includes sugars, lipids, and amino acids rather than cellulose.

Visual Learning

Figure 8.4 Is the photo a TEM or an SEM? How do you know? *The photo is a TEM. It shows a section through a cell rather than a view of the cell surface.*

interNET CONNECTION

Follow the link for this chapter on the Glencoe Homepage at **www.glencoe.com/sec/science** to find out more about cells.

Meeting Individual Needs

Learning Disabled/Hearing Impaired
Hearing impaired students and those who have difficulty learning cell parts and their functions can review cell parts using the tutorials provided by *The Cell: Examination, Structure and Function,* Queue, Apple II, IBM, Mac. Students will be able to see cell parts as they are quizzed. L1 LS

SECTION 8.2

ThinkingLab Interpret the Data

Purpose

LS **Logical-Mathematical** Students will become familiar with the role of the nucleus in directing cell activities.

Process Skills

relate cause and effect, interpret data

Teaching Strategies

- You may wish to present this Thinking Lab as a chalkboard example or as an overhead transparency. Draw the diagrams on the chalkboard or transparency and review the results as a class. Allow students to discuss why the final cap ended up as it did.

Thinking Critically

The nucleus produces substances that control the type of cap the cell has. The first cap was of intermediate form because the cytoplasm contained substances from both the previous nucleus and the present nucleus. The cytoplasm of the second cap had only substances from the present nucleus.

Assessment

Portfolio: Ask students to write a summary of this Thinking Lab. Encourage them to include diagrams with their summaries. L1 P

ThinkingLab Interpret the Data

What organelle directs cell activity?

Acetabularia, a type of marine alga, grows as single, large cells 2 to 5 cm in height. The nuclei of these cells are in the "feet." Different species of these algae have different kinds of caps, some petal-like and others that look like umbrellas. If a cap is removed, it quickly grows back. If both cap and foot are removed from the cell of one species of alga and a foot from another species is attached, a new cap will grow. This new cap will have a structure with characteristics of both species. Then, if this new cap is removed, the cap that grows back will be like the cell that donated the nucleus.

The scientist who discovered these properties was Joachim Hämmerling. He wondered why the first cap that grew had characteristics of both species, yet the second cap was clearly like that of the cell that donated the nucleus.

Analysis

Look at the diagram below and interpret the data to explain the results.

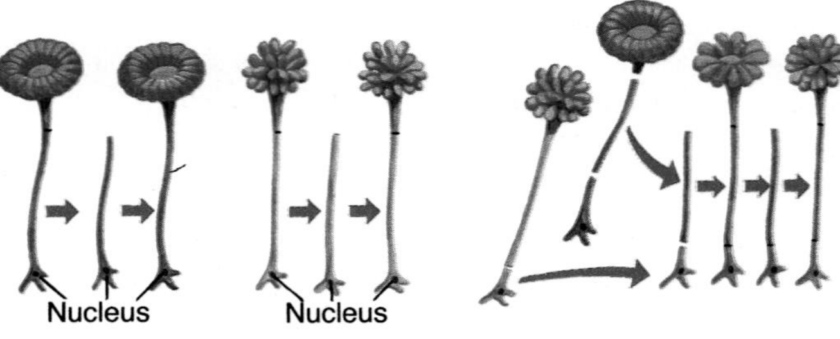

Thinking Critically

Why is the final cap like that of the cell from which the nucleus was taken? (HINT: Recall the function of the nucleus.)

The nuclear envelope has large pores so materials can pass back and forth between the nucleus and the rest of the cell. The nucleus contains DNA, the master instructions for building proteins. DNA forms tangles of long strands called **chromatin,** which is packed into identifiable chromosomes when the cells are ready to reproduce. Also within the nucleus, *Figure 8.6,* is the **nucleolus,** a region that produces tiny cell particles that are involved in protein synthesis. These particles, called **ribosomes,** are the sites where the cell assembles enzymes and other proteins according to the directions of the DNA. Although ribosomes are considered cell organelles, they are not bounded by a membrane.

Assembly, Transport, and Storage

A major function of most cells is to make proteins and other materials. Many of the cell organelles are involved in protein synthesis or storage of materials. Much of this assembly and storage takes place in the fluid inside the cell—the cytoplasm.

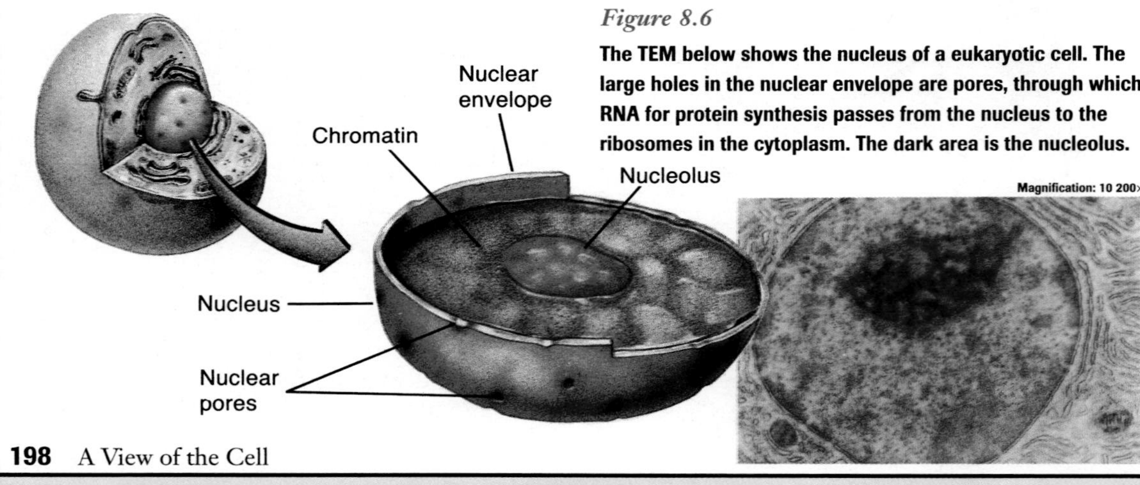

Figure 8.6
The TEM below shows the nucleus of a eukaryotic cell. The large holes in the nuclear envelope are pores, through which RNA for protein synthesis passes from the nucleus to the ribosomes in the cytoplasm. The dark area is the nucleolus.

Alternate Lab: Differences Between Plant and Animal Cells

Purpose

LS **Visual-Spatial** This lab will allow students to compare plant and animal cells.

Materials

microscope, glass slides and coverslips, droppers, scalpel, forceps, flat toothpick, *Elodea* plant, methylene blue stain

Procedure

Give students the following directions.

1. Use a dropper to place a drop of water in the center of a slide. Use the forceps to remove a leaf from the tip of an *Elodea* sprig and place it in the water on your slide. Add a coverslip.
2. Under low power, look for a thin area of the leaf where you can see the cells most clearly. Change to high power and locate a single cell. Observe carefully for a minute.
3. Draw the cell and label the structures you see.
4. Use a dropper to place a drop of methylene blue stain on a slide. *Gently* scrape the inside lining of your cheek

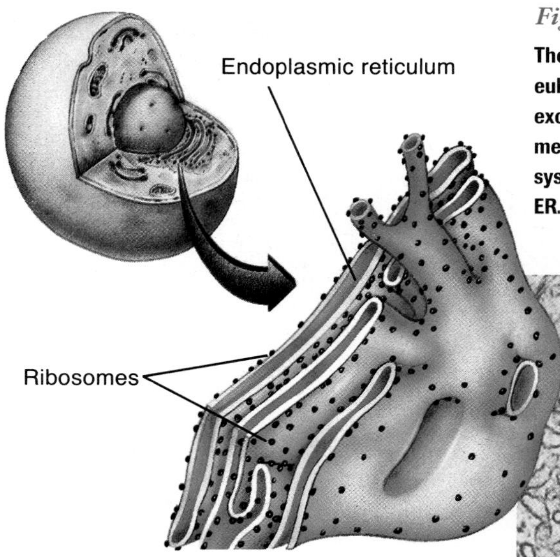

Figure 8.7

The ER is a complex system of membranes in the cytoplasm of eukaryotic cells. These membranes are too thin to be visible except by electron microscopy. The ER is attached to the outer membrane of the nuclear envelope and forms a kind of transport system within the cytoplasm. Note the many ribosomes on rough ER. Ribosomes are also found free in the cytoplasm.

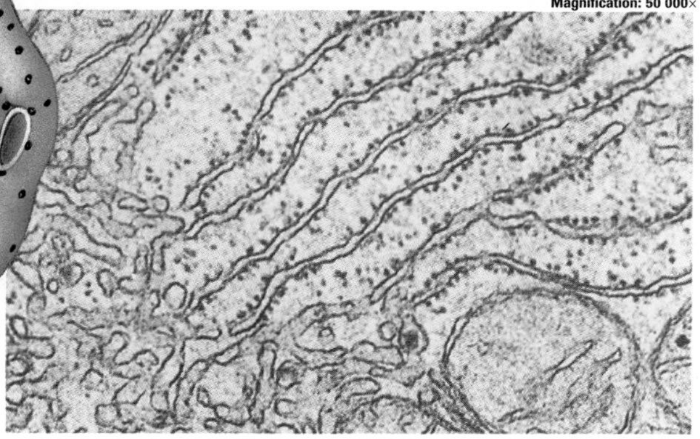

Endoplasmic reticulum

Ribosomes

Magnification: 50 000×

Structures for assembly and transport of proteins

The material that lies outside the nucleus and surrounds the organelles is the **cytoplasm,** a clear fluid that is a bit thinner than toothpaste gel. It usually constitutes a little more than half the volume of a typical animal cell. Many important chemical reactions, such as protein assembly, take place in the cytoplasm.

Much of the cytoplasm is occupied by a folded system of membranes. The **endoplasmic reticulum** (ER), shown in *Figure 8.7,* is a folded membrane that forms a network of interconnected compartments inside the cell. Suppose you took all of the tissues out of a box of facial tissues and spread them out on a table. They would cover a large surface area, probably several square meters. Yet, all of that surface area was packed into a small box. A large surface area can be packed into a small area by folding the surfaces. Like facial tissues that are folded into a box, the ER is folded into the cell with cytoplasm surrounding it. This system of membranes provides a large surface area on which chemical reactions can take place. The ER membranes also contain the enzymes for almost all of the cell's lipid synthesis, so they serve as the site of lipid synthesis in the cell.

Some of the ER is coated with ribosomes. The parts of the ER that are studded with ribosomes have a bumpy appearance when viewed with an electron microscope, and are referred to as rough ER. In areas without ribosomes, the ER is called smooth ER. The ER functions as the cell's delivery system, much like the trucks that deliver the raw products such as cheese and mushrooms to the pizza parlor. To make the pizza, these raw products must be assembled on a counter or workbench in the pizza parlor. In the cell, the sites of protein assembly are the ribosomes.

Math Connection

Calculating Surface Area

Have students calculate and compare the total surface area of the tissues in a tissue box to the surface area of the box itself. The total surface area is equal to the surface area of one unfolded tissue times the total number of tissues in the box.

Assume the box is 24 cm × 12 cm × 10 cm. Each tissue is 21 cm × 23 cm × 2 sides. The box contains 280 tissues. Contrast the large total surface area of all the tissues (270 480 cm^2) to the surface area of the box (1296 cm^2). Point out that folding the tissues, like the inner membrane of the mitochondrion, greatly increases the total surface area. L1 IS

Visual Learning

Figure 8.7 Discuss with students how showing cell organelles in both photographs and art is useful. Point out that illustrations can often accentuate structures to make them clearer than they appear in photographs.

with the flat edge of a toothpick. Mix the material from the toothpick into the drop of stain and add a coverslip to the slide. **CAUTION:** *Do not reuse toothpicks. Immediately dispose of the toothpick in the wastebasket.*

5. View under low power, moving the slide to center a single cell. Change to high power and observe the cell. Draw the cell and label the structures you see.

Analysis

Ask students the following questions.

1. What structures did you see in each cell? What are their functions? *Likely responses may include cell walls, cell membranes, vacuoles, and nuclei. Accept all logical descriptions of the functions of each cell part named.*
2. Methylene blue is one of many stains used when observing cells. What was the function of this stain? *The stain makes it possible to see some parts of the cell that would otherwise not be visible.*

Assessment

Journal: Have students prepare a lab report describing what they have seen. They should include drawings of both the *Elodea* and cheek cell and the answers to the analysis questions. L1

Bioethics

Money for Your Cells

A man went into the hospital to have his enlarged spleen removed. When the doctors cultured the man's spleen cells, they found that the cells secreted large amounts of a protein that stimulates the growth of blood cells. A drug company has cultured the man's cells and carried out years of research on this protein.

The man whose spleen was used as a source of the cultured cells believes he should receive money. His spleen was used without his permission for this research. The doctor had signed a form stating that the diseased spleen would be destroyed.

Ask students whether they believe the man should receive money and whether researchers should be able to use body tissues without permission from patients.

Reinforcement

LS Visual-Spatial Use models of plant cells and animal cells (available from biological supply houses) to review and reinforce knowledge of cell parts.

Enrichment

Divide students into groups of three or four. Assign each group a cell part and have students develop a commercial to "sell" their cell part. L1 COOP LEARN LS

Figure 8.8

The Golgi apparatus, as seen by electron microscopy (right), looks like a side view of a stack of pancakes. Associated with the Golgi apparatus are small, membrane-bound vesicles that are involved in protein packaging. The various proteins in the vesicles are sorted and sent to their final destination, like mail moving through a post office.

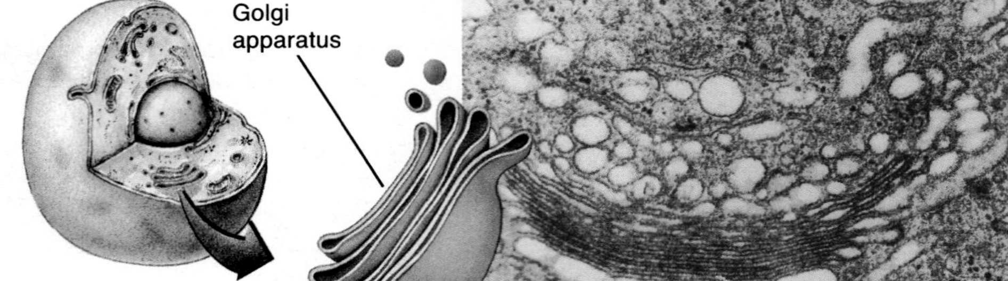

Structures for protein storage

Businesses usually have a storage room for keeping extra materials. Cells, too, have storage areas. Organelles that store materials include the Golgi apparatus, vacuoles, and lysosomes. The **Golgi apparatus** is a series of closely stacked, flattened membrane sacs that receives newly synthesized proteins and lipids from the ER and distributes them to the plasma membrane and other cell organelles. Proteins are transferred from the ER to the Golgi apparatus in small, membrane-bound transport packages. These packages, called vesicles, have pinched off from the membrane of the ER and contain proteins, *Figure 8.8*. The Golgi apparatus modifies the proteins chemically, then repackages them in new vesicles for their final destination in the cell. They may be incorporated into cell structures, expelled, or remain stored for later usage.

Vacuoles and storage

In the pizza parlor kitchen, the cheese and other ingredients are stored in bins. Cells have spaces, called vacuoles, for temporary storage of materials. A **vacuole,** like those in *Figure 8.9*, is a sac of fluid surrounded by a membrane. Vacuoles often store food, enzymes, and other materials needed by a cell, and some vacuoles store waste products. In some single-celled organisms, a specialized vacuole collects excess water and pumps it out of the cell. A plant cell has a single, large vacuole that stores water and other substances.

Figure 8.9

The vacuole is a membrane-bound, fluid-filled space within the cytoplasm, like a microscopic water balloon. Plant cells (left) usually have one large vacuole; animal cells (right) may contain many smaller vacuoles.

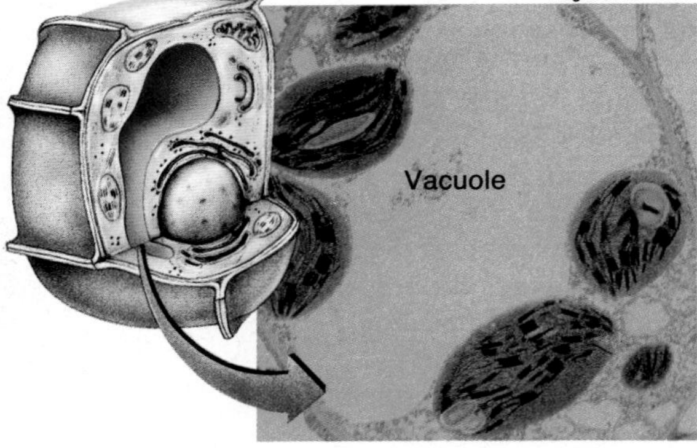

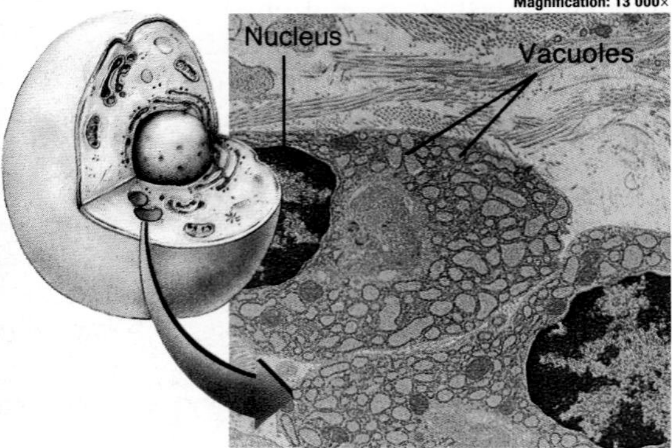

PORTFOLIO

Observing Plant and Animal Cells Have students observe several kinds of plant cells and animal cells with a microscope. Ask students to make labeled diagrams of each cell they observe. Beside each label, have them describe the function of each organelle. Have students use their observations to create a table in which they compare the organelles of plant and animal cells. Students' tables should include all organelles shown in their diagrams and identify whether the organelle is common to both plant and animal cells or unique to only one kind of cell. L1 P LS

Lysosomes and recycling

In addition to assembling and storing macromolecules, cells also can disassemble things. **Lysosomes,** organelles that contain digestive enzymes, digest excess or worn out cell parts, food particles, and invading viruses or bacteria. The membrane surrounding a lysosome prevents the digestive enzymes inside from destroying the cell's proteins. Lysosomes can fuse with vacuoles and dispense their enzymes into the vacuole, digesting its contents. For example, when an amoeba engulfs a food morsel and encloses it in a vacuole, a lysosome fuses with the vacuole, releases its enzymes, and digests the hapless prey. Sometimes, lysosomes digest the cells that contain them. When a tadpole develops into a frog, lysosomes within the cells of the tadpole's tail cause its digestion. The molecules thus released are used to build different cells, perhaps in the newly formed legs of the adult frog.

Energy Transformers

When you entered the pizza parlor, lights were on inside. The ovens in the kitchen were filled with baking pizzas. All of this equipment requires energy with which to run. The cell also requires energy to carry out its many functions.

Mitochondria and energy

Eukaryotic cells have membrane-bound organelles that transform energy for the cell. **Mitochondria,** shown in *Figure 8.10,* are organelles in which food molecules are broken down to release energy. This energy is then stored in other molecules that can power cell reactions easily.

A mitochondrion has an outer membrane and a highly folded inner membrane. As with the ER, the folds of the inner membrane provide a large surface area in a small space. Here on the inner folds, energy-storing molecules are produced.

Figure 8.10

Mitochondria are found in every cell except prokaryotes, in varying numbers (up to 2500 per cell in liver cells). They are granular and rod or thread shaped, with an inner membrane that forms long, narrow folds called cristae. The cristae are covered with enzyme systems involved in producing energy molecules for many cell functions. The TEM (right) shows a cross section of a mitochondrion.

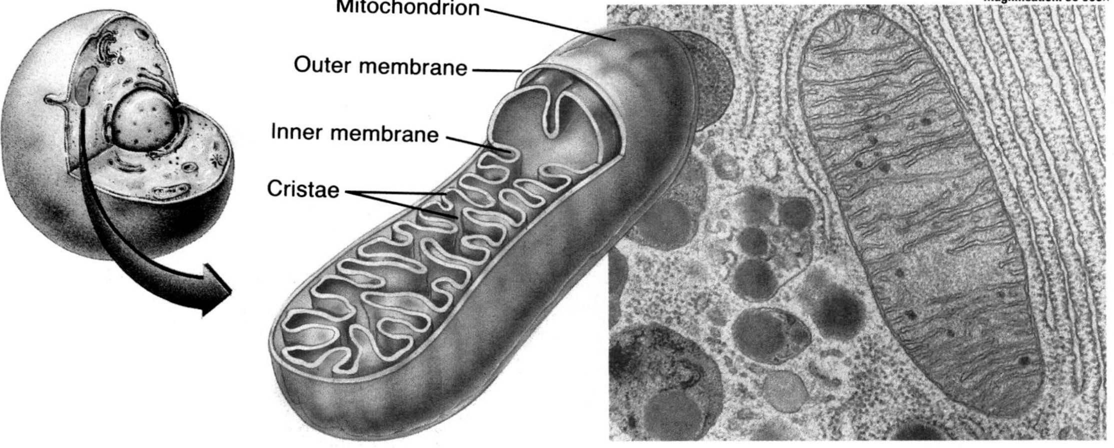

SECTION 8.2

Different Viewpoints in Biology

Aging and Energy

Many people believe energy loss is a natural part of aging. However, molecular biologist Anthony Linnane of Monash University believes defects in mitochondria may cause people to slow as they age.

Mitochondria contain their own DNA, which produces enzymes critical to cellular respiration. Linnane hypothesizes that over the course of an organism's life, errors collect in the mitochondria, lessening the ability of the cells to respirate efficiently. If experimental evidence confirms this hypothesis, treatment for this slowing process may include the use of substances that replenish depleted enzymes to restore strength.

GLENCOE TECHNOLOGY

Videodisc

Biology: The Dynamics of Life
Disc 1, Side 1
The SEM (Ch. 23)

STVS: Human Biology
Disc 7, Side 1
Diagnosing Disease with Glowing Cells (Ch. 13)

CULTURAL DIVERSITY

Santiago Ramón y Cajal

Introduce Spanish cell biologist Santiago Ramón y Cajal and his work on nerve cells in the early 1900s. Cajal, together with Italian biologist Camillo Golgi, received the Nobel Prize for Physiology in 1906 for establishing that neurons are the basic units of the nervous system. This research was important to understanding the transmission of nerve impulses. Ramón y Cajal was also responsible for developing cell staining techniques that are still used in today's laboratories.

People in Biology

Meet Dr. Marian Diamond, Neuroscientist

Teaching Strategies

- Have students refer back to this feature during their study of the nervous system.
- **Display** Obtain a photograph or drawing that shows the structure of a neuron. Point out the dendrites and synapses, which are referred to in the article.
- **Enrichment** Have students research the concept of fish as "brain food." Have these students explain why fish is considered by some people to be brain food and whether there are scientific data to support this belief. L2

Background

Dr. Marian Diamond has devoted much of her life to science education. The primary function of the Lawrence Hall of Science is to act as a research and development center for science and math curricula for grades K-12. All employees of the center work to design curricula for teachers worldwide.

Two books that Dr. Diamond has written or helped develop are *Enriching Heredity*, which focuses on the impact of the environment on the anatomy of the brain, and *The Physical Science of Living California*.

people in biology

Meet Dr. Marian Diamond, Neuroscientist

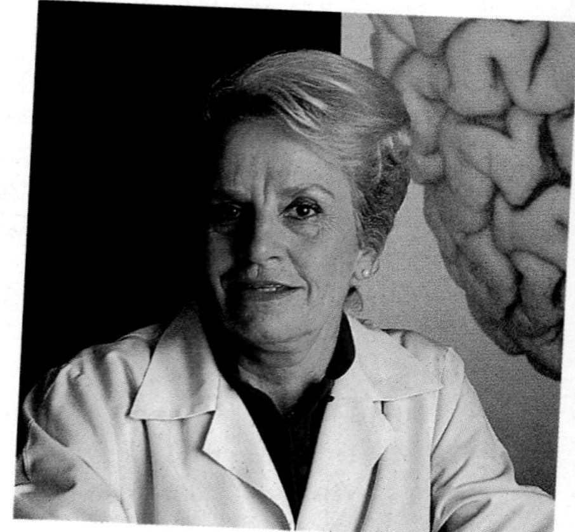

Brains and basketballs both have places in Dr. Marian Diamond's office. In a bucket of formaldehyde beside her desk is a human brain, and in the corner is a basketball, ready for her weekly game with students. Dr. Diamond is a professor of anatomy at the University of California in Berkeley, as well as executive director of the Lawrence Hall of Science, a unique, hands-on center for learning.

In the following interview, Dr. Diamond talks about her research on aging brain cells, as well as her enthusiasm for the museum she directs.

On the Job

Q Dr. Diamond, what is your specialty in the field of biology?

A I study the effects of the environment on the anatomy of the brain as it develops from before birth through old age. I've discovered that brain mass can increase with learning, and this can happen at any age. The more the brain is stimulated, the more the neurons increase in size. Within the brain, the dendrites become thicker, synapses expand, and the basic chemistry changes.

Q How can we keep our brains functioning at full capacity?

A The view that there is a loss of brain cells in advanced old age is a myth. From all my research on the development of rats' brains, I would recommend keeping your brain challenged with new experiences for a lifetime—not just in school but in every situation. If you sit back and let life pass you by, it's a strong probability that your brain may even shrink. You can avoid the possibility of a shrinking brain by simply using it and challenging yourself daily. Most people seem unaware of the importance of what they carry in the top of their heads. I like to show people the brain because I want them to see it as the place where education takes place.

Q What about treatment of damaged brains in the future?

A Doctors now can transplant pieces of brains, such as using embryonic brain tissue to treat Parkinson's disease. I think Alzheimer's may be treated when this complicated disease is better understood. Perhaps research into what triggers the expression of particular genes will be the key.

Q How did you become the director of the Lawrence Hall of Science?

A I guess I was asked because of my experience with both science and education. I've always had my own programs, using my university students to teach in grade schools and high schools and help other students enjoy the human body and brain. At

202

this museum, I can put together my theories of education with my teaching of brain functions and brain research.

Q Do you find it difficult to get kids interested in science?

A Not at all. Every child is naturally a scientist. At the Lawrence Hall of Science, children can find whatever fascinates them and play with an exhibit all day if they want. They discover the joy of doing something themselves. That's one of life's greatest pleasures.

Early Influences

Q Could you talk about how you became interested in brains?

A I saw my first human brain when I was in my teens, and I fell in love with brains because this mass represented cells that could think. I wanted to find out how cells could think and what was going on behind people's eyes. I've been studying human brains for 40 years. I even keep one beside my desk.

Q Do you remember any one person who may have introduced you to a love of biology?

A I especially remember a junior-high school teacher who taught me about the intricate structure of a fly's wing. Until then, I just thought a fly was for swatting. I never thought of a fly wing being built for special aerodynamics until she helped me see it differently.

Personal Insights

Q Do you have any advice for students who might be interested in a career in biology?

A You're surrounded by biology for a lifetime. In fact, you *are* biology. Everything around you is biology. It's an exciting thing to find out that there are chemical similarities in, say, a leaf, an ant, a monkey, and a person. Understanding that makes you look at the world differently. I would advise kids to follow their own hearts. Just as no two brains are alike, no two loves are the same. If you have a passion for something, follow that passion. You'll probably take a lot of courses that might not seem relevant at the time, but later on you'll understand.

Q How do you personally keep a balance in your life?

A I eat right and do the things I love to do. That includes playing a mean game of basketball with my students every week, even though they complain about my sharp elbows!

CAREER CONNECTION

To follow in Dr. Diamond's footsteps, you would need a Bachelor of Science degree in biology and a PhD in anatomy. Science museums like the one Dr. Diamond directs hire artists and writers to work on their displays and prepare written material, such as brochures and study guides.

Science Writer *College degree in English; courses in general science*

Neuroscientist *A doctoral degree in anatomy and physiology; courses in zoology*

Science museum volunteer *An interest in science and volunteering*

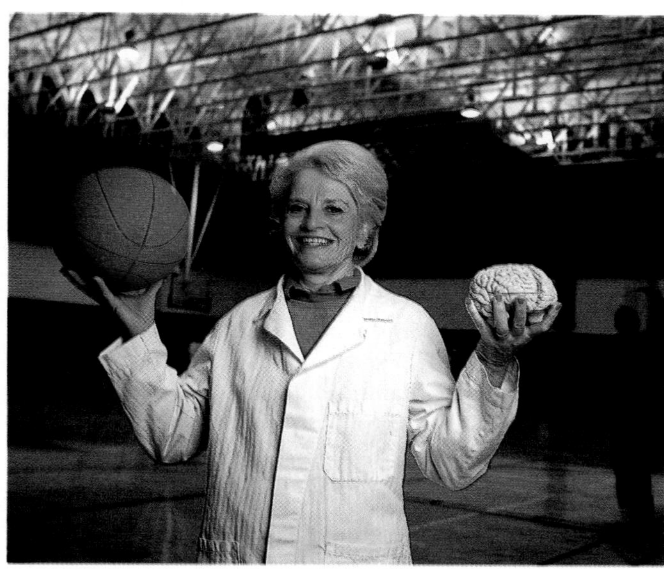

CAREER CONNECTION

- **Career Path** courses in high school: biology, chemistry, physics, and higher math

  College: bachelor's degree in biology or related field

  Graduate school: Ph.D. in anatomy and physiology or related field
- **Career Issue** Some people believe that it is improper to exhibit stuffed animals in museums, and some animals have been killed for the sole purpose of being exhibited. Discuss with students whether this practice should be continued. Ask them, If an animal dies of natural causes, could it then be used as a museum exhibit? Have students consider the benefits of museum exhibits. L1

For More Information

- **Neuroscientist** For more information on a degree in biology, students can write to the biology department of your local college or university.
- **Science Museum Volunteer** Have students visit or write to your local museum for more information on working at a museum.

Misconception

Students often believe that because plants carry on photosynthesis, they do not respire. Point out that like animals, the cells of plants contain mitochondria. Mitochondria present in all eukaryotic cells produce the ATP that provides the energy for the cells. Thus, plants must carry on cell respiration as well as photosynthesis.

Enrichment

Have students research methods of cell fractionation. Encourage students to present their findings to the class. L3

Demonstration

LS **Visual-Spatial** To aid students in understanding the orientation of organelles in a cell, create a cell model of gelatin containing whole grapes and chunks of different fruits. Have students compare cell parts suspended in cytoplasm to the fruit suspended in the gelatin. Have students speculate as to how the angle on which the cell is cut relates to the shapes organelles may take in an electron micrograph. L1

Using Science Terms

LS **Linguistic** Explain that the base word *plasto* means "formed body" and the prefix *chloro* means "green." Relate the meanings of these word parts to the appearance of chloroplasts.

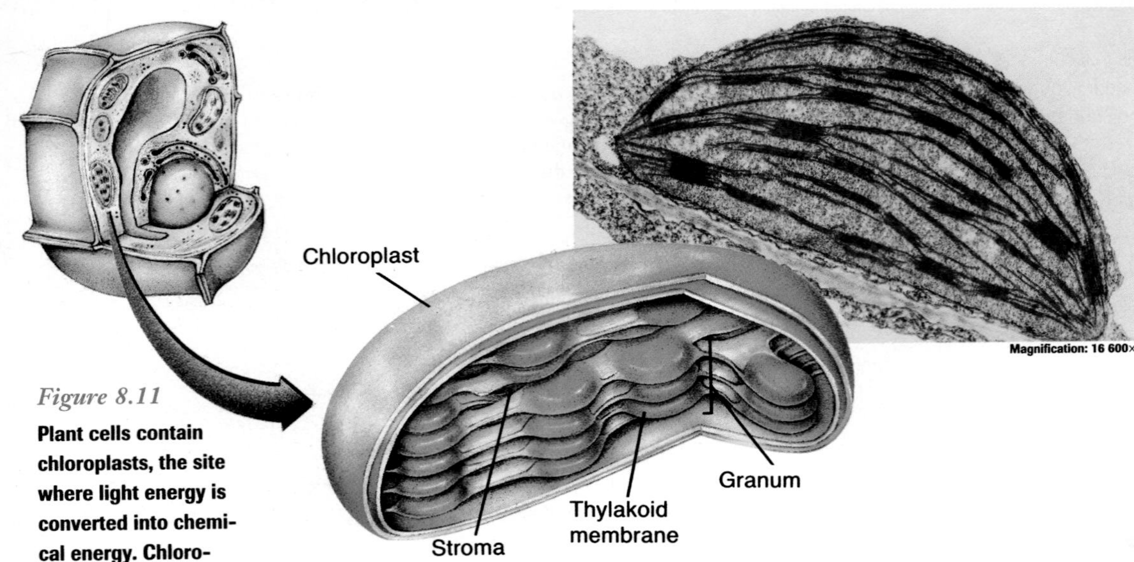

Figure 8.11
Plant cells contain chloroplasts, the site where light energy is converted into chemical energy. Chloroplasts are usually disc shaped but have the ability to change shape and position in the cell as light intensity changes. The pigment chlorophyll is embedded in the inner series of membranes, called thylakoid membranes, where light energy is trapped.

Chloroplasts and energy

Cells of green plants and some protists have organelles, called **chloroplasts,** that transform light energy directly into usable chemical energy and store that energy in food molecules. These foods include sugars and starches. Chloroplasts contain molecules of **chlorophyll,** a green pigment that traps the energy from sunlight and gives plants their green color.

A diagram and a TEM of a chloroplast, which has a double outer membrane and a folded inner membrane system, are shown in *Figure 8.11.* It is within the inner membranes that the energy from sunlight is trapped. These inner membranes are arranged in stacks of membranous sacs called grana, which resemble stacks of coins. The fluid that surrounds the grana membranes is called stroma.

The chloroplast belongs to a group of plant organelles called **plastids,** which are used for storage. Some plastids store starches or lipids, whereas others contain pigments, molecules that give color. Plastids are named according to their color or the pigment they contain. The chloroplast contains the green pigment chlorophyll, which gives leaves and stems their green color. Other plastids have different pigments and give flowers and fruits their beautiful colors of red, purple, blue, and yellow.

chloroplast:
chloros (GK) green
platos (GK) formed object
A chloroplast is a green structure found in plant cells.

Structures for Support and Locomotion

Less than 50 years ago, scientists thought plastids and other organelles just floated in a sea of cytoplasm. In the last 20 years, however, scientists have discovered that cells have a support structure called the cytoskeleton within the cytoplasm. The cytoskeleton is composed of tiny rods and filaments that form a framework for the cell, like a bony skeleton that forms the framework for your body. However, unlike your skeleton, the cytoskeleton is a dynamic and constantly changing structure.

204 A View of the Cell

STUDENT JOURNAL

Modeling Organelles Have students make a large drawing of a mitochondrion or a chloroplast. Drawings should illustrate the internal and external shapes of the organelles. L1
LS

Cytoplasm support

The **cytoskeleton,** shown in *Figure 8.12,* is a network of thin, fibrous elements that act as a sort of scaffold to provide support for organelles. It also helps maintain cell shape in a similar way that poles maintain the shape of a tent. The cytoskeleton is mainly composed of microtubules and microfilaments. They both assist organelles to move from place to place within the cell. **Microtubules** are thin, hollow cylinders made of protein. **Microfilaments** are thin, solid protein fibers.

Cilia and flagella

Some cells have cilia and flagella, which are structures adapted for locomotion. Cilia and flagella can be distinguished by their structure and by the nature of their action.

Figure 8.12

This photograph of a eukaryotic cell has been treated with fluorescent dye to show the cytoskeleton. The microtubules are stained yellow and the microfilaments are red. The cytoskeleton is constantly being formed and taken apart by the cell, depending on its needs.

Magnification: 5000×

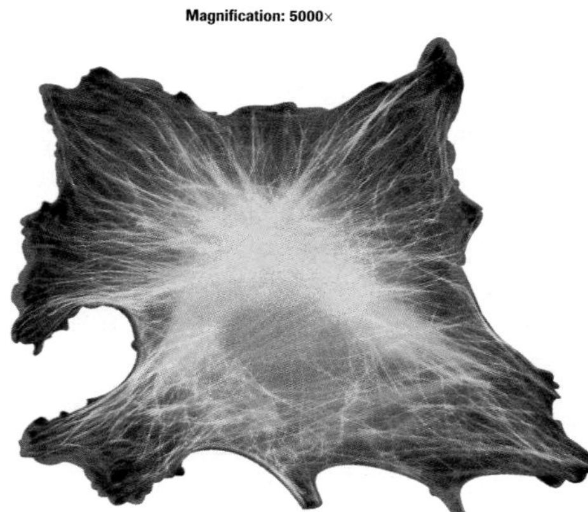

Technology BIOLOGY & SOCIETY

Natural Fiber—Natural Color

Cotton is one of the most widely produced fabrics in the world. Before the Civil War, cotton production was labor intensive because the cotton bolls—the white, fluffy seedpods—had to be handpicked from the plant and the seeds removed by hand.

Production of cotton By the late 18th century, technological developments began to change the way cotton was grown, harvested, and treated. In 1793, Eli Whitney invented the cotton gin, which automated the removal of seeds from the cotton fibers and made the production of cotton far less costly. Today, cotton gins employ the same principles as Whitney's original machine.

Structure of cotton fibers When cotton seeds mature, the seedpod bursts open, displaying a mass of soft, white fluff. Cotton fibers are unicellular hairs that project from the outer coat of cotton seeds. The walls of the hairs are almost pure cellulose, which is arranged in interwoven layers and gives cotton thread its strength. High-quality, strong thread is made from the longer fibers.

Cotton cloth After the cotton fibers are spun into thread, the thread is woven into cloth. The cloth is then bleached, dyed, or printed. The dyeing process is important in producing the colors and patterns that make cotton fabric so desirable.

Applications for the Future

An Arizona cotton breeder, Sally Fox, is using genetics and technology to breed cotton plants that grow colored rather than white fibers. She has already developed plants with green and brown fibers. Why does Fox want to produce plants that grow colored fibers? Chemicals that cause pollution are used to dye fabrics. So using naturally colored fibers decreases pollution.

INVESTIGATING the Technology

Applying Concepts What properties of cotton fibers other than color might be developed through breeding of cotton plants?

PROJECT

Making a Macroscopic Cell

Turn part of your classroom into a giant animal cell. From the ceiling, hang four strings to define the size of your cell. For ease in calculation, make the cell a cube. Have students calculate the magnification factor by dividing the length of a side in μm by 20 μm (the length of a liver cell). Have students research the sizes of organelles and calculate how large they need to make each organelle using the magnification factor (cell organelle size × magnification factor). Have students use paper and other materials to build cell organelles and then hang them inside the classroom cell. L1 COOP LEARN LS

Technology BIOLOGY & SOCIETY

Natural Fiber—Natural Color

Purpose

To show students how technology can be used to improve qualities of natural materials.

Background

A drawback in cotton harvesting involves the leaves of the plant. Cotton leaves are full of a resin that stains the cotton fibers. To destroy the leaves, cotton farmers often use a defoliant before they harvest the cotton bolls. Sally Fox is working to develop an organically grown white cotton that sheds its leaves just before harvest. This cotton produces a whiter, cleaner cotton.

Teaching Strategies

- Bring in products made from cotton to display.
- Elicit from students how naturally colored fibers help the environment. *The naturally colored fibers reduce water use and decrease pollution caused by chemicals released into the environment.*

INVESTIGATING the Technology

Applying Concepts Students may mention properties such as elasticity, wrinkle-resistance, and durability. Developing varieties with longer fibers would contribute to smoother fabrics that require little ironing.

Going Further

Have students research other plants that are valued for the products they yield. Have students working in cooperative groups create a visual display of plants that are used other than as food. Encourage students to include photographs or drawings of their plants and the products each yields.

SECTION 8.2

MiniLab

Purpose

IS **Visual-Spatial** Students will identify cell structures and infer their functions.

Process Skills

observe and infer, interpret scientific illustrations

Teaching Strategies

- Divide students into cooperative groups to allow sharing of observations.

Expected Results

Students should recognize mitochondria along the flagellum of the sperm and should recognize the flagellum itself. They may need help recognizing the lysosome as separate from the nucleus in the head of the sperm.

Analysis

1. mitochondria; produce ATP
2. sperm flagellum
3. Enzymes released from the lysosome help sperm fertilize the egg by digesting the egg's membrane.

Assessment

Oral: Large TEM photos of sperm cells are available from biological supply houses. Such photos can be used for a group discussion of this lab. L1

MiniLab

Which parts of a sperm cell are visible in a transmission electron micrograph?

Sperm cells are small, motile cells that combine with female reproductive cells. Careful inspection of longitudinal and cross sections of a sperm will allow identification of some of the cell structures.

Procedure

1. Look at the longitudinal section of a sperm flagellum shown here and the cross section shown below in *Figure 8.13.*
2. Identify the cell structures that are visible.

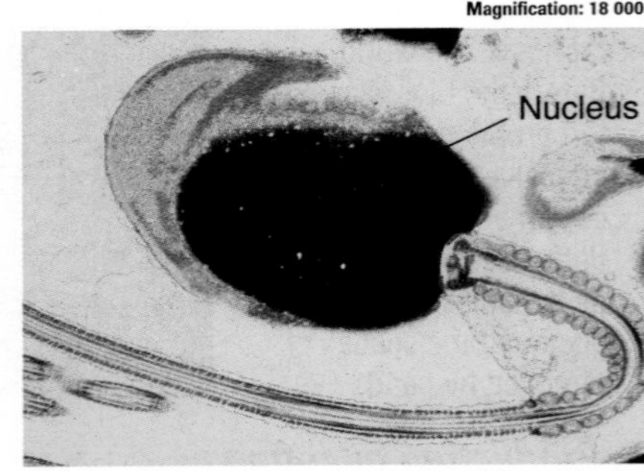

Analysis

1. Along the tail of the sperm are many oval-shaped structures. What are they? What is their cellular function?
2. The oval structures are used to power what cellular structure?
3. On the head of the sperm is a large lysosome. What do you think the function of this structure might be?

Cilia are short, numerous, hairlike projections from the plasma membrane. As shown in *Figure 8.13,* cilia tend to occur in large numbers on a cell's surface, and their beating activity is usually coordinated, much like the movement in a stadium "wave." **Flagella,** also shown in *Figure 8.13,* are longer projections that move with a whiplike motion. Cells that have flagella have only one or two per cell.

In single-celled organisms, cilia and flagella are the major means of locomotion. Sperm cells of animals and some plants move by means of flagella. Organisms that contain many cells, including humans, have cilia that move fluids over a cell's surface, rather than moving the cell itself.

Both plant and animal cells are adapted to carry out the functions of the organisms that they make up. Their structures and functions are similar, yet different. Many of the cell structures discussed in this chapter are found in both plant and animal cells, but each cell type also has some unique structures. Compare and contrast the animal cell and plant cell in *Figure 8.14.*

Figure 8.13

Even though cilia and flagella project out of the plasma membrane, they still are covered by the membrane.

In eukaryotic cells, both cilia and flagella are composed of microtubules arranged in a ring of nine pairs surrounding two single microtubules, as seen in this cross section of a sperm flagellum.

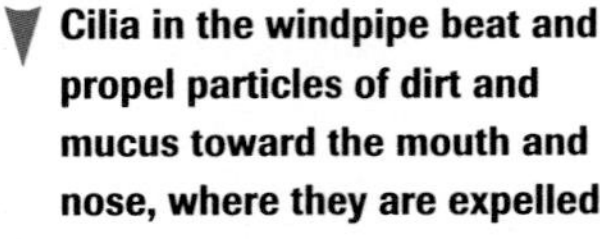

Cilia in the windpipe beat and propel particles of dirt and mucus toward the mouth and nose, where they are expelled.

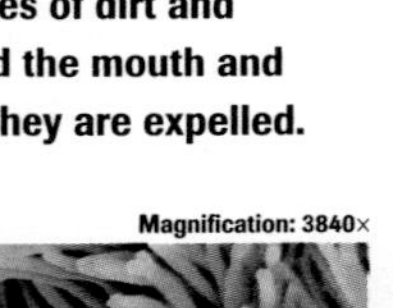

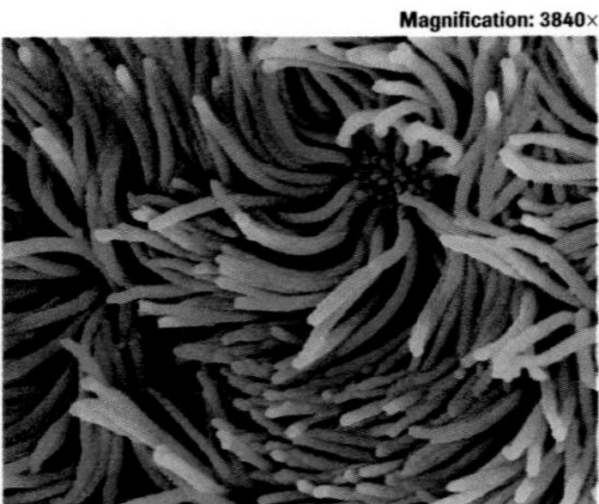

The flagella of these sperm cells move the cells forward with their whiplike action.

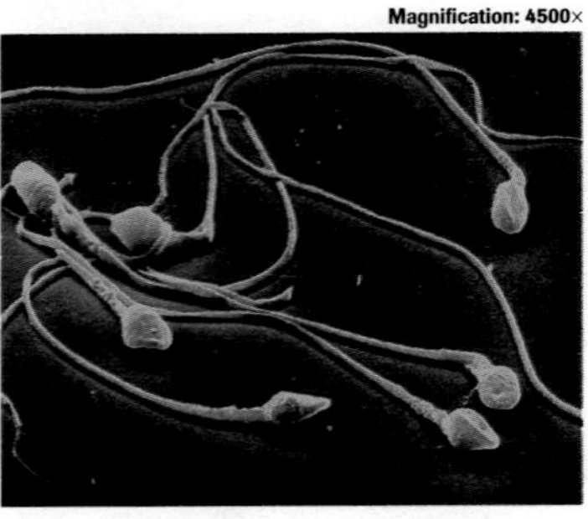

STUDENT JOURNAL

Microbodies Have interested students research and write a report on various types of microbodies. Reports should include the functions and locations of microbodies in the cells. Students will need to conduct library research to write their reports. L3 IS

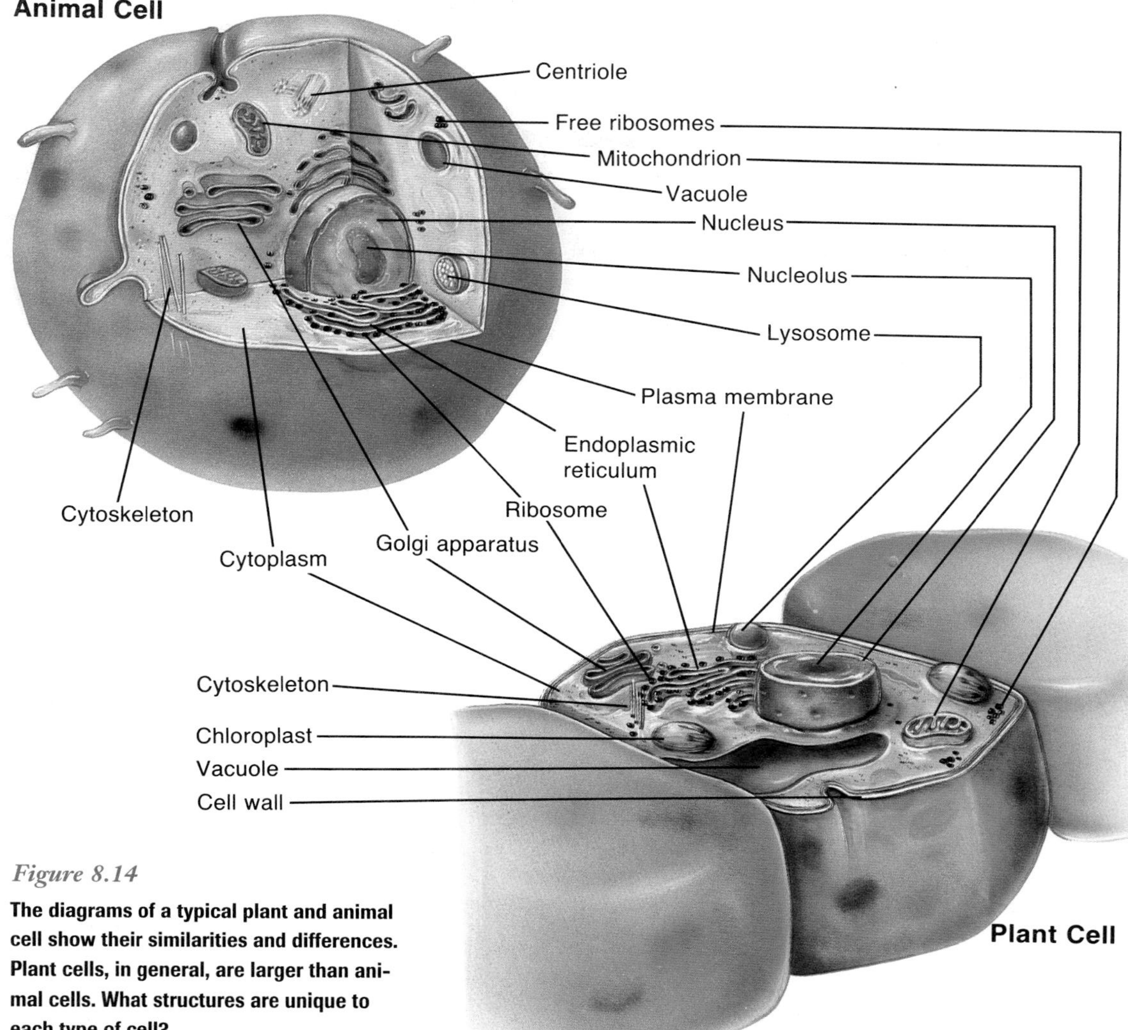

Figure 8.14

The diagrams of a typical plant and animal cell show their similarities and differences. Plant cells, in general, are larger than animal cells. What structures are unique to each type of cell?

Cellular Organization

Some cells exist as single-celled organisms. Such single-celled organisms are called **unicellular.** Other organisms are made up of many cells, each of which is specialized to perform a distinct metabolic function. One cell within an organism may be adapted for movement, while another cell carries out digestion. The individual cells do not carry out all life functions, but rather depend on each other. Many-celled organisms are called **multicellular.**

When a group of cells functions together to perform an activity, they form a **tissue.** The cells of your body are organized into tissues such as muscle and nerve tissues. Plant tissues include those of the stem and root. Many cells in tissues are directly linked to one another at contact sites called cell junctions.

uni-, multi-:
uni (L) one
multi (L) many
Unicellular organisms have one cell; multicellular organisms have many cells.

Visual Learning

Figure 8.14 What structures are unique to each type of cell? *Plant cell: cell wall, chloroplasts, few and large vacuoles. Animal cell: many small vacuoles.*

3 Assess

Check for Understanding

Ask students to hypothesize what would happen to a cell if it had a decreased number of mitochondria. *The cell would probably not have enough energy to carry out all life functions.* L1

Reteach

Ask students working in groups to role-play the organelles of a typical animal or plant cell. Students should take roles as particular organelles and show how these cell parts work together to keep the entire cell functioning. L1 COOP LEARN

Extension

Have students play vocabulary football with cell terms. Divide class into two teams. Ask a question to one team. If they answer correctly, advance the ball 10 yards on a football field drawn on the chalkboard. If they miss the question, the question should be given to the other team. The team that scores the most touchdowns wins. L1

Assessment

Portfolio: Have students construct a table to summarize cell structures and their functions. Working in pairs, students can quiz each other on information in their tables. L1 P

Meeting Individual Needs

Learning Disabled/Gifted Students who have difficulty with mathematical concepts will need extra help with the SI calculations and measurements in this chapter. Hands-on manipulations with metric rulers can increase their understanding of the mathematical concepts involved. Have students who are having difficulty with mathematics work cooperatively with students possessing strong math skills to make a variety of calculations related to microscopy. Calculations should include determining magnification, calculating field of view, and using the microscope as a measuring tool. L2 IN

BioLab

Sizing Cells and Cell Structures

Time Allotment
One class period

Objectives
Review objectives with students before they begin the Biolab.

Process Skills
observe and infer, measure in SI, interpret scientific illustrations

PREPARATION

- Collect electron micrographs from journals and college biology textbooks as additional material for students to measure and calculate real size. Try to find pictures with scale bars, with given magnifications, and with an object of known size included.

BioLab

Sizing Cells and Cell Structures

All eukaryotic cells have membrane-bound organelles. Textbooks such as this one often have many electron micrographs of cells and cell structures so that the reader can see their shapes and structures. It is often helpful for the reader to know the actual size of an object being studied. How are the sizes of these cells and cell structures determined? By using the information given in a scale bar or the magnification of the photo, you can calculate the real size of an object.

PREPARATION

Problem
How can you determine the size of cell structures?

Objectives
In this Biolab, you will:
- **Measure** cells and organelles.
- **Calculate** the real size of objects.

Materials
small metric ruler
calculator

PROCEDURE

1. Copy the data table.
2. Measure to the nearest 0.1 cm the length and width of the "graphic square" labeled A, and record the measurements in the data table.
3. Magnification can be represented by a scale bar. Measure the length of the scale bar beneath the graphic sausage.
4. Calculate the real size of the graphic square by taking each of the dimensions and dividing them by the length of the scale bar. Multiply the answer by the scale (in this case 1 μm). Keep track of the units of measurement. Record your answer in the data table.

A. Graphic square

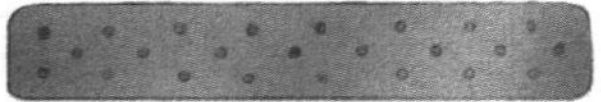
B. Graphic sausage

1 μm

PROCEDURE

Teaching Strategies
- Review with students how to measure to the nearest 0.1 cm using a metric ruler.
- Team students who have difficulty in math with students who can help them in their calculations.

Data and Observations
See the following table.

Object	Length	Width
Graphic square measurement	2.7 cm	2.7 cm
Graphic square scale bar	3.8 cm	–
Graphic square real size	0.71 μm	0.71 μm
Graphic sausage measurement	4.6 cm	0.8 cm
Graphic sausage real size	1.5 μm	0.3 μm
Chloroplast measurement	4.8 cm	2.5 cm
Chloroplast real size	5.6 μm	2.9 μm
Red blood cell measurement	5.5 cm	4.1 cm
Red blood cell real size	7.5 μm	5.6 μm
Granum measurement	0.2 cm	0.3 cm
Granum real size	0.2 μm	0.4 μm

5. Measure to the nearest 0.1 cm the length and width of the "graphic sausage" labeled B, and record the measurements in the data table.
6. A second way that magnification can be represented is by the number of times the object has been magnified. The graphic sausage has been magnified 30 000×. Calculate the real size by dividing the length and width measurements by 30 000. Keep track of the units of measurement. Record your answer in the data table.
7. Calculate the real sizes of the chloroplast (C) and the red blood cell (D). Record your measurements in the data table.
8. Calculate the real length and height of a granum in the chloroplast. Record your measurements in the data table.

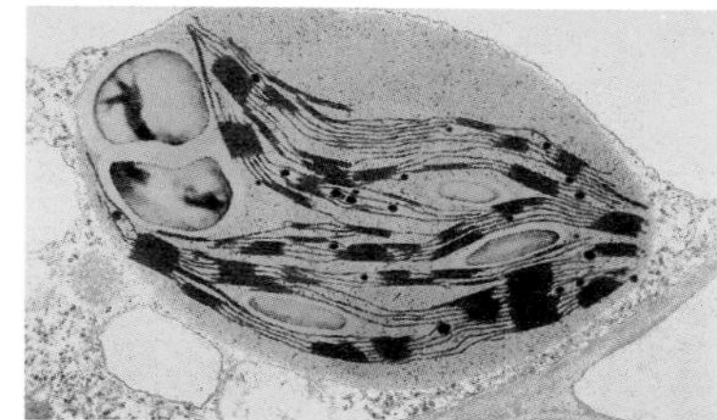

C. Chloroplast

Magnification: 7300×

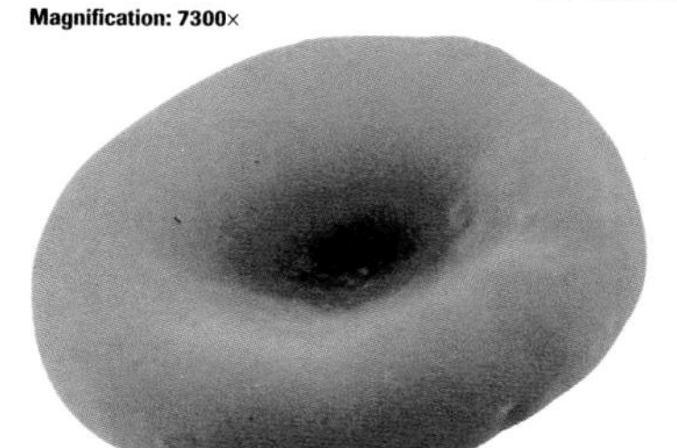
D. Red blood cell

Object	Length	Width
Graphic square measurement	____	____
Graphic square scale bar	____	____
Graphic square real size	____	____
Graphic sausage measurement	____	____
Graphic sausage real size	____	____
Chloroplast measurement	____	____
Chloroplast real size	____	____
Red blood cell measurement	____	____
Red blood cell real size	____	____
Granum measurement	____	____
Granum real size	____	____

ANALYZE AND CONCLUDE

1. **Observing** How many granum packages can you count in the chloroplast? What is their function?
2. **Calculating Results** Look at page 198 and calculate the real diameter of a cell nucleus. How does its size compare with the size of the chloroplast?
3. **Calculating Results** Sometimes it is difficult to realize how greatly cell parts are magnified in electron micrographs. If your real height were magnified 30 000×, how tall would you appear in kilometers? (1 km = 1000 m)

Going Further

Application Make a table listing all of the cell parts from the chapter. Use the electron micrographs in this chapter to help you calculate the real size of each cell part. Place the sizes in your table.

ANALYZE AND CONCLUDE

1. Approximately 30. They carry out photosynthesis.
2. 5.3 µm. The nucleus is usually the same size or larger than a chloroplast.
3. Most students will fall between 39 and 55 km tall.

Assessment

Portfolio: Have students write an evaluation of the lab. They should include the table from Going Further to demonstrate that they understand the procedure in the lab. L1 P

Going Further

Students can use SEM and TEM photos from college textbooks or from pictures obtained from a biological supply house to calculate the sizes of various cell organelles. L2

SECTION 8.2

Visual Learning

Figure 8.15 Why can't a cell from an organ carry out all life processes? *A division of labor exists among the cells of an organ. The cells are specialized to perform specific functions, while other cells carry out other functions. The cells work together, rather than individually, to carry out the life functions.*

4 Close

Discussion

Discuss the analogy of comparing a cell to a factory. Have students analyze the analogy to explain where it is accurate and where it fails. L1

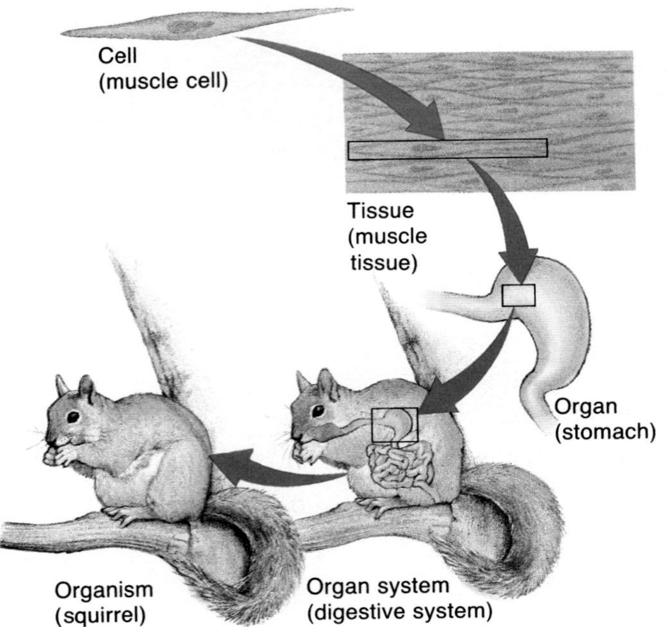

Figure 8.15

Multicellular organisms are highly organized into tissues, organs, and organ systems. Why can't a cell from an organ carry out all life processes?

Cell junctions help maintain differences in the internal environment between adjacent cells, help anchor cells together, and allow cells to communicate with one another by passing small molecules from one cell to another.

Groups of two or more tissues that function together make up **organs.** Your stomach and the leaf of a plant are organs. Cooperation among organs makes life functions within an organism efficient. An **organ system** is a group of organs that work together to carry out major life functions. Your nervous system and the flower of a plant are examples of organ systems. *Figure 8.15* presents the organization of cells to form a multicellular organism.

Connecting Ideas

Cells are the basic building blocks of both unicellular and multicellular organisms. Cells share basic "business" tasks that keep them alive. Each cell must interact with its environment directly through the plasma membrane by transporting needed materials into the cell and getting rid of harmful waste products. How does the cell monitor what comes in and what goes out? Understanding how the plasma membrane interacts with its external environment will help you understand how cells maintain homeostasis.

Section Review

Understanding Concepts

1. What is the advantage of highly folded membranes in a cell?
2. What organelles would be especially numerous in a cell that produces large amounts of a protein product?
3. Why are digestive enzymes in a cell enclosed in a membrane-bound organelle?

Thinking Critically

4. Compare the functions of mitochondria and chloroplasts. Why are they referred to in the text as energy transformers rather than as energy producers or energy generators?

Skill Review

5. **Observing and Inferring** Some cells have large numbers of mitochondria with many internal folds. Other cells have few mitochondria with few internal folds. What can you conclude about the functions of these two types of cells? For more help, refer to Thinking Critically in the *Skill Handbook*.

Answers to Section Review

1. The folding increases the surface area available on which chemical reactions occur.
2. Rough endoplasmic reticulum and Golgi apparatus
3. If these enzymes were free in the cytoplasm, they could damage the cell.

Thinking Critically

4. The chloroplasts store sunlight energy in energy-rich food molecules. The mitochondria break down food molecules to release energy that can be used by the cell to perform work. They only convert energy from one form into another. The energy must be present to begin with.

Skill Review

5. The cells with many mitochondria produce more energy and perform more work than the cells with fewer mitochondria.

REVIEWING MAIN IDEAS

8.1 The Discovery of Cells

- Microscopes enabled biologists to develop the cell theory.
- The cell theory states that the cell is the basic unit of organization, all cells come from preexisting cells, and all organisms are made up of one or more cells.
- Using electron microscopes, scientists can study detailed cell structures.
- Cells are classified as prokaryotic or eukaryotic, based on whether or not they have membrane-bound organelles.

8.2 Eukaryotic Cell Structure

- Eukaryotic cells have a nucleus and organelles, are enclosed by a plasma membrane, and some have cell walls that provide support and protection.
- Cells make proteins on ribosomes, which are often attached to the highly folded endoplasmic reticulum. They store materials in the Golgi apparatus, vacuoles, and lysosomes.
- Mitochondria break down food molecules to release energy, and chloroplasts convert light energy into chemical energy.
- The cytoskeleton helps maintain cell shape, and cilia and flagella help cells move.
- The cells of most multicellular organisms are organized into tissues, organs, and organ systems.

Key Terms

Write a sentence that shows your understanding of each of the following terms.

cell
cell theory
cell wall
chlorophyll
chloroplast
chromatin
cilia
compound light microscope
cytoplasm
cytoskeleton
electron microscope
endoplasmic reticulum
eukaryote
flagella
Golgi apparatus
lysosome
microfilament
microtubule
mitochondria
multicellular
nucleolus
nucleus
organ
organ system
organelle
plasma membrane
plastid
prokaryote
ribosome
tissue
unicellular
vacuole

Understanding Concepts

1. What type of cell would you examine to find a chloroplast?
   a. prokaryote
   b. animal
   c. plant
   d. fungus
2. An organism that exists as a single cell is called ___________.
   a. an organ
   b. a tissue
   c. unicellular
   d. multicellular
3. Which of the following structures utilizes the sun's energy to make carbohydrates?
   a. vacuole
   b. chloroplast
   c. cilia
   d. mitochondria
4. Membrane-bound structures found inside eukaryotic cells are called ___________.
   a. prokaryotes
   b. organelles
   c. nuclei
   d. cell walls
5. Long, whiplike projections called ___________ help cells move from place to place.
   a. cilia
   b. flagella
   c. tentacles
   d. plastids
6. The function of the ribosomes is to synthesize ___________.
   a. glucose
   b. lipids
   c. amino acids
   d. proteins

Chapter 8 Review

Reviewing Main Ideas

Summary statements can be used by students to review the major concepts of the chapter.

Key Terms

Answers should go beyond defining the terms. Accept any answer that uses the term correctly and in the proper context.

Understanding Concepts

1. c
2. c
3. b
4. b
5. b
6. d

Chapter 8 Review

7. a
8. d
9. c
10. b
11. d
12. d
13. c
14. a
15. a
16. c
17. b
18. b
19. d
20. c

Applying Concepts

21. All cells could not be seen until the development of more powerful microscopes.
22. If a particular protein is not needed immediately, its building blocks can be recycled into other proteins.
23. Schleiden and Schwann observed that all the living organisms they viewed were composed of cells.
24. Mitochondria are needed to release energy for cell activities from food molecules. Chloroplasts are needed to make those food molecules.
25. Lysosomal enzymes were released to digest the tissue between the fingers.

Chapter 8 Review

7. Who gave cells their name?
   a. Hooke c. Schleiden
   b. Schwann d. Leeuwenhoek
8. Cell functions are managed by the ______.
   a. mitochondria c. lysosomes
   b. ribosomes d. nucleus
9. Which of the following pairs of terms is not related?
   a. nucleus—DNA
   b. chloroplasts—chlorophyll
   c. flagella—chromatin
   d. cell wall—cellulose
10. A group of cells that work together to perform an activity is called ___________.
   a. unicellular c. an organ
   b. a tissue d. an organ system
11. Magnifications greater than 10 000× can be obtained when using ___________.
   a. light microscopes
   b. metric rulers
   c. well-ground lenses
   d. electron microscopes
12. What structure covers the outside of a fungal cell?
   a. plasma membrane c. cilia
   b. mitochondria d. cell wall
13. Which of the following structures is composed of DNA?
   a. ribosomes c. chromatin
   b. Golgi apparatus d. vacuole
14. Which of the following structures is NOT found in both plant and animal cells?
   a. chloroplast c. ribosomes
   b. cytoskeleton d. mitochondria
15. The ___________ forms the outer boundary of an animal cell.
   a. plasma membrane
   b. cell wall
   c. nuclear membrane
   d. cytoskeleton
16. In an amoeba, which structure digests food in a vacuole by releasing enzymes?
   a. Golgi apparatus
   b. endoplasmic reticulum
   c. lysosome
   d. mitochondria
17. A(n) ___________ is a type of microscope that sweeps a beam of electrons over the surface of a specimen.
   a. simple microscope
   b. SEM
   c. compound microscope
   d. TEM
18. A ___________ is an organism with a cell that lacks membrane-bound organelles.
   a. eukaryote c. nucleolus
   b. prokaryote d. chromatin
19. Protein assembly in a cell takes place in the ___________.
   a. nucleus c. vacuoles
   b. mitochondria d. cytoplasm
20. Which of the following is NOT stored in plastids?
   a. lipids c. amino acids
   b. pigments d. starches

Applying Concepts

21. Why did it take almost 200 years after Hooke discovered cells for the cell theory to be developed?
22. Sometimes packets of proteins collected by the Golgi apparatus merge with a lysosome. Suggest reasons for this activity.
23. Why did Schleiden and Schwann conclude that cells are the basic units of life?
24. Why must plant cells have both mitochondria and chloroplasts?
25. During human development, the embryo's hands develop from solid, paddle-shaped parts. How do you think fingers might be formed from these parts?

212 A View of the Cell

Program Resources

Chapter Assessment, pp. 43-48 L1
Alternate Assessment in the Science Classroom
Computer Test Bank L1
Content Mastery, pp. 29-32 L1

Thinking Critically

26. ***Concept Mapping*** Make a concept map that relates the following terms and phrases. Supply the appropriate linking words for your map.
    nucleus, nucleolus, chromatin, ribosome, endoplasmic reticulum, Golgi apparatus
27. ***Interpreting Data*** Sometimes the size of an object in a photograph is indicated by a known object, such as a dime, being placed beside the object to be photographed. How does this procedure allow you to calculate the real size of the object being photographed?
28. ***Interpreting Scientific Illustrations*** The inner membrane of a mitochondrion is highly folded. Estimate the length of the folded membrane in illustration A, and estimate how much longer it is compared with the membrane in illustration B.

A.

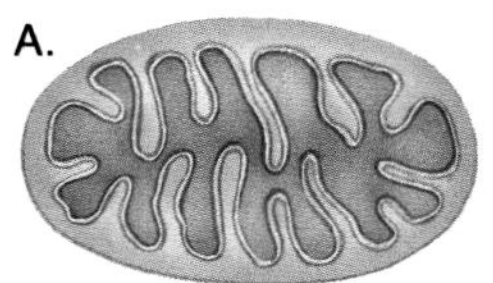

B.

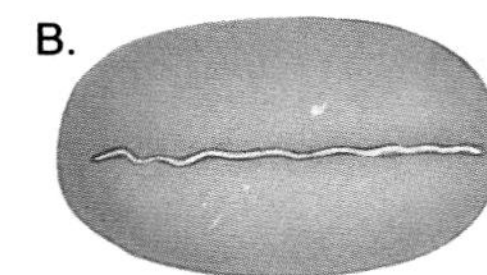

29. ***Comparing and Contrasting*** Compare and contrast prokaryotic and eukaryotic cells.
30. ***Making Predictions*** Predict whether you would expect muscle cells or fat cells to contain more mitochondria and explain why.
31. ***Recognizing Cause and Effect*** Cigarette smoke damages the cilia in the air passages between the throat and lungs. How does this damage help cause "smoker's cough"?

ASSESSING KNOWLEDGE & SKILLS

Identify the structures shown in the scientific diagram below.

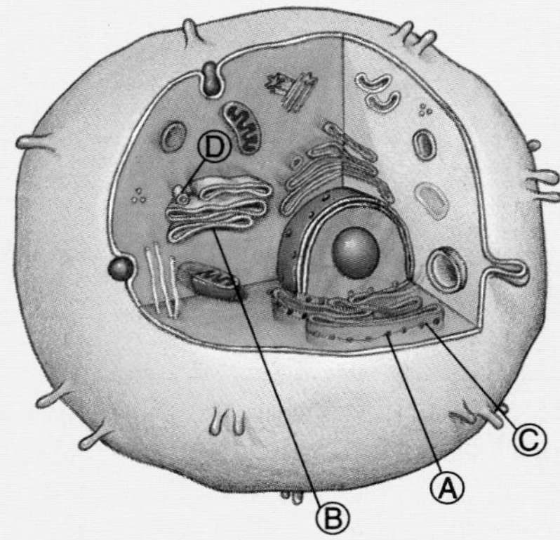

Interpreting Scientific Illustrations Use the diagram to answer the following questions.

1. The membrane labeled *C* in the diagram of a cell represents the __________.
   a. plasma membrane
   b. nuclear membrane
   c. endoplasmic reticulum
   d. nucleolus
2. The function of the circular structures on membrane *C* is to __________.
   a. synthesize cellulose
   b. transform energy
   c. synthesize proteins
   d. capture the sun's energy
3. The structure labeled *B* in the diagram represents the __________.
   a. lysosome c. nucleus
   b. Golgi apparatus d. vacuole
4. The type of cell shown is a __________ cell.
   a. plant c. animal
   b. fungal d. prokaryotic
5. ***Sequencing*** Structures A, B, C, and D in the diagram are involved in making a product to be released to the outside of the cell. What is the sequence of the production of this product?

26.

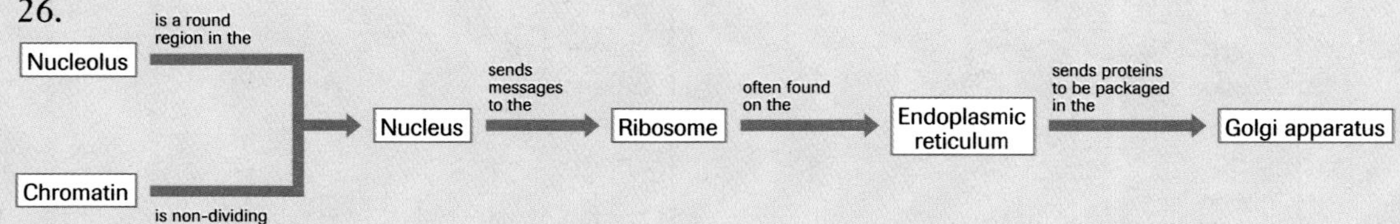

Chapter 8 Review

Thinking Critically

26. See below.
27. You can measure a dime and find out how magnification changes its size in the picture. This information can be used to calculate the actual size of the object.
28. The length of the folded membrane is about 25 to 30 cm. This is approximately ten times the length of the unfolded membrane.
29. Prokaryotic cells are cells that lack internal structures surrounded by membranes. Most prokaryotic cells are unicellular. Eukaryotic cells contain internal organelles. Eukaryotic cells are either unicellular or multicellular.
30. Muscle cells are very active cells, utilizing a lot of energy, whereas fat cells are utilized mainly for storage of fat. Mitochondria are the cell organelles that transform energy for the cell. Therefore, muscle cells will contain more mitochondria.
31. Cilia move mucus up from the air passages to the throat. When the cilia become damaged by smoking, the smoker must cough up the mucus.

ASSESSING KNOWLEDGE & SKILLS

1. c
2. c
3. b
4. c
5. The ribosomes (A) synthesize proteins, which are transported by the endoplasmic reticulum (C) to the Golgi apparatus (B), which packages the proteins into vesicles (D), which are then transported to cell membranes for release to the outside of the cell.

Chapter Organizer

SECTION	OBJECTIVES	ACTIVITIES/FEATURES
9.1 The Plasma Membrane National Science Standards: UCP.1-3, UCP.5; A.1, A.2; B.2; C.1, C.5; G.1	1. **Explain** how a cell's plasma membrane functions. 2. **Relate** the function of the plasma membrane to the fluid mosaic model.	**Biolab:** Design Your Own Experiment—Are plastic bags selectively permeable?, p. 218 **Literature Connection:** *Lives of a Cell,* p. 221
9.2 Cellular Transport National Science Standards: UCP.1-3, UCP.5; C.1, C.5; E.1, E.2;	3. **Explain** how the processes of diffusion, passive transport, and active transport occur and why they are important to cells. 4. **Predict** the direction of diffusion of a dissolved substance.	**Biology & Society Technology:** Frozen in Time, p. 225 **Minilab:** How many contractile vacuoles does a paramecium contain?, p. 227 **Minilab:** What happens to plant cells in a hypertonic solution?, p. 229 **Thinking Lab:** How does fertilizer affect earthworms in soil?, p. 231

ACTIVITY MATERIALS

BIOLAB	MINILABS	ALTERNATE LAB
page 218 400-mL beakers small plastic bags twist ties graduated cylinder starch solution iodine solution masking tape	**page 227** methyl cellulose slide coverslip microscope eyedropper *Paramecium* culture **page 229** *Elodea* sprig forceps slide coverslip microscope 3% salt solution eyedropper paper towel distilled water	**page 228** potato 100-mL beakers (2) or paper cups measuring spoon salt graduated cylinder label pen stirring rod balance plastic wrap or aluminum foil knife

Chapter 9 Homeostasis and the Plasma Membrane

TEACHER CLASSROOM RESOURCES

Reproducible Masters	Transparencies
Section Focus Master 18: Movement of Materials L1 SAE ☞ **Reinforcement and Study Guide,** p. 33 L1 ☞ **Biolab and Minilab Worksheets,** pp. 33-34 L1 ☞ **Content Mastery,** pp. 33-36 L1	**Basic Concepts Transparency #7:** Plasma Membrane L1 SAE ☞
Section Focus Master 19: Water in the Cell L1 SAE ☞ **Reinforcement and Study Guide,** pp. 34-36 L1 ☞ **Biolab and Minilab Worksheets,** pp. 31-32 L1 ☞ **Concept Mapping:** Transport Through Membranes, p. 9 L1 **Tech Prep Applications:** Osmosis and the Case of the Sad Salad Greens, pp. 9-10; Inside the Artificial Kidney Machine, pp. 11-12 L2 **Critical Thinking/Problem Solving:** Separating Sea Urchin Cells, p. 9 L3 **Laboratory Manual:** Normal and Plasmolyzed Cells, pp. 55-56 L2	**Basic Concepts Transparency #8:** Osmosis L1 SAE ☞ **Basic Concepts Transparency #9:** Active Transport L1 SAE ☞ **Basic Skills Transparency #8:** Active versus Passive Transport L1 SAE ☞ **Reteaching Transparency #9:** Osmosis in Hypotonic, Hypertonic, and Isotonic Solutions L1 SAE ☞

ASSESSMENT MATERIALS

Chapter Assessment, pp. 49-54 ☞
Performance Assessment in the Biology Classroom, p. 7
MindJogger Videoquiz ☞
Alternate Assessment in the Science Classroom
Computer Test Bank

Spanish Resources SAE
English/Spanish Audiocassettes SAE
Cooperative Learning in the Science Classroom COOP LEARN
Lesson Plans ☞

KEY TO TEACHING STRATEGIES

- L1 Level 1 activities should be within the ability range of all students including those with learning difficulties.
- L2 Level 2 activities are within the ability range of average to above-average students.
- L3 Level 3 activities are designed for the ability range of above-average students.
- SAE SAE activities should be within the ability range of Students Acquiring English.
- COOP LEARN Cooperative Learning activities are designed for small group work.
- P These strategies represent student products that can be placed into a best-work portfolio.
- ☞ These strategies are useful in a block scheduling format.

GLENCOE TECHNOLOGY

The following multimedia resources are available from Glencoe.

Biology: The Dynamics of Life
CD-ROM SAE
Videodisc Program ☞

National Geographic Society Series
STV: The Cell
Parts of the Cell
Active Transport
Passive Transport
Cell Membrane

Science and Technology Videodisc Series (STVS)
Human Biology
Natural Time-Release Capsules

CHAPTER

9 Homeostasis and the Plasma Membrane

Chapter Overview

Students begin their study of the plasma membrane by learning about the fluid mosaic model. As the model is discussed, students become familiar with the function of each type of molecule composing the plasma membrane.

In the second section of the chapter, the structure of the plasma membrane is related to methods of cellular transport: diffusion, osmosis, passive transport, and active transport. Students also discover why these cell transport processes are important to cell function.

Key Terms

- active transport
- contractile vacuole
- diffusion
- dynamic equilibrium
- endocytosis
- exocytosis
- facilitated diffusion
- fluid mosaic model
- hypertonic solution
- hypotonic solution
- isotonic solution
- osmosis
- passive transport
- phospholipid
- plasmolysis
- selective permeability
- transport protein
- turgor pressure

214

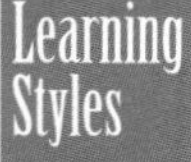

Learning Styles

Look for the following logo for strategies that emphasize different learning modalities. 

Kinesthetic	**Project, p. 220**
Visual-Spatial	**Demonstration, pp. 217, 224, 226; Meeting Individual Needs, p. 226; Minilab, pp. 227, 229**
Interpersonal	**Meeting Individual Needs, pp. 217, 227; Portfolio, p. 224**
Intrapersonal	**Meeting Individual Needs, p. 227**
Logical-Mathematical	**Alternate Lab, p. 228; Student Journal, p. 230; Thinking Lab, p. 231**

CHAPTER

9 Homeostasis and the Plasma Membrane

"The sun sparkled on the waves as they swept against the boat. The shark cage stood in position above the deep blue water. There were no sharks in sight, but I knew they were down there, and now it was time to step into the cage and find out firsthand. My stomach had butterflies as I stepped off the stern of the boat and swam into the strong, steel cage. Inside, the bars gave me assurance of protection while I filmed the great white sharks as they swam by.

"As the day went on, I became less tense as I filmed the details of the shark—its eyes, its sleek body, and those wicked white teeth. I sensed the strength and power of this animal—and its instinct to find food. Several times, a shark struck the cage, causing it to jerk wildly. I might have become the shark's food if I had not been inside a cage that let me look out but did not let the sharks inside."

Concept Check

You may wish to review the following concepts before studying this chapter.

- Chapter 1: homeostasis
- Chapter 7: polar molecules, protein, lipid

Chapter Preview

Laboratory Activities

Biolab: Design Your Own Experiment

- Are plastic bags selectively permeable?

Minilabs

- How many contractile vacuoles does a paramecium contain?
- What happens to plant cells in a hypertonic solution?

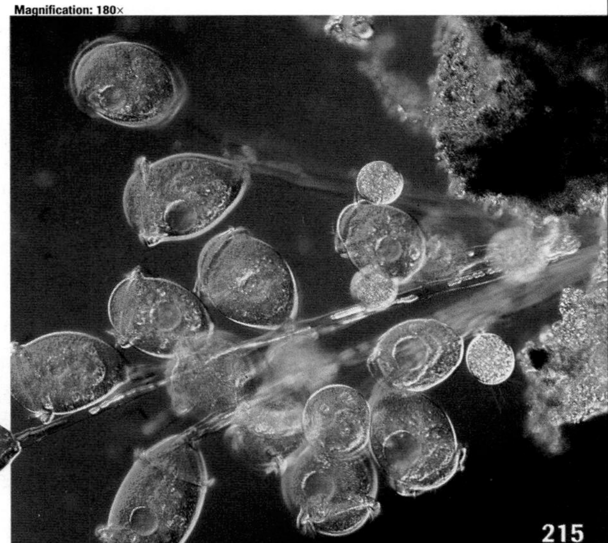

A steel cage allows a diver to observe sharks in their own environment because the cage will not allow the sharks to enter. In a similar manner, the plasma membrane of a cell enables it to exist in a hostile environment. The plasma membranes of these unicellular organisms called *Vorticella* protect the contents of cells and regulate what can come into and out of each cell. How is the structure and function of the plasma membrane similar to that of the shark cage?

215

CHAPTER 9

Introducing the Chapter

Direct students' attention to the photograph of the diver in the shark cage. Point out how the cage separates the diver from the sharks, allowing the diver to survive in this hostile environment. Elicit how the structure and function of plasma membranes are similar to those of the shark cage. *The plasma membrane separates the inside contents from the outside environment and controls what goes in and out.*

Theme Development

As suggested by the chapter title, *homeostasis* is the major theme of this chapter. The constancy maintained within a cell as a result of the selective permeability of the plasma membrane is the major focus of this chapter. Emphasis is placed on linking molecular structures, polarity, solubility, and function to the role of the plasma membrane in maintaining homeostasis.

Concept Check

Review the cell, its organelles and their functions, and the major macromolecules that make up cell components. Remind students that the main role of the plasma membrane is to maintain a stable internal environment that allows for proper functioning of the cell.

Assessment Planner

Choose assessment strategies from the following pages to evaluate the progress of your students.

Assess, pp. 221, 232
Alternate Lab, pp. 228-229
Minilabs, pp. 227, 229
Portfolio, p. 224
Thinking Lab, p. 231
Biolab, pp. 218-219
Chapter Review, pp. 233-235

SECTION

9.1

Prepare

Key Concepts

Students will examine the fluid mosaic model of the plasma membrane and discover how this model explains the functions of the membrane.

Block Scheduling

Look for this symbol for strategies that are useful in a block scheduling format. For more information on block scheduling, refer to Section 9.1 in the **Lesson Plans** booklet.

Materials

- Purchase plastic bags for the Biolab.
- Obtain glass jars, water, alcohol, food coloring, and salad oil for the Demonstration.
- Purchase Ping-Pong balls or apples for the Reteach.

1 Focus

Bellringer

Before presenting the lesson, display **Section Focus Master 18** on the overhead projector and have students answer the accompanying questions. L1 SAE

Discussion

Discuss the primary functions of the plasma membrane. Elicit the contribution of the plasma membrane to maintaining homeostasis in the cell. *Responses should reflect the idea that the plasma membrane controls the passage of materials into and out of the cell.*

SECTION

9.1 The Plasma Membrane

Section Preview

Objectives

Explain how a cell's plasma membrane functions.

Relate the function of the plasma membrane to the fluid mosaic model.

Key Terms

selective permeability
phospholipid
fluid mosaic model

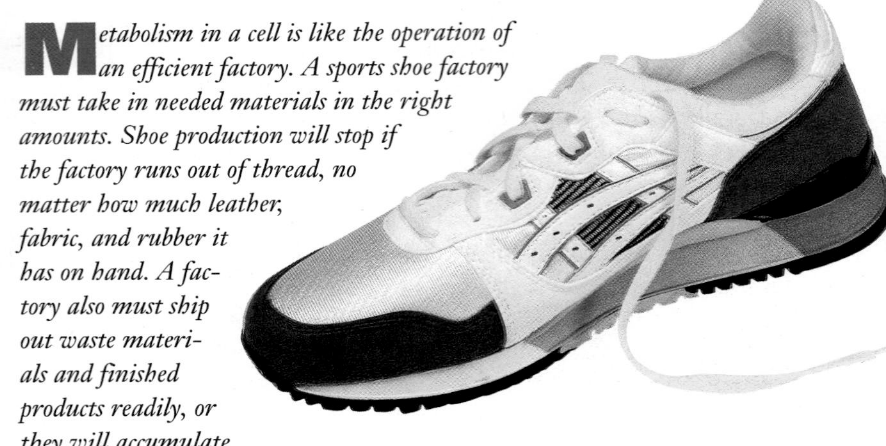

Metabolism in a cell is like the operation of an efficient factory. A sports shoe factory must take in needed materials in the right amounts. Shoe production will stop if the factory runs out of thread, no matter how much leather, fabric, and rubber it has on hand. A factory also must ship out waste materials and finished products readily, or they will accumulate and force the factory to stop production. At the same time, though, the factory must keep its raw materials, plans, tools, and machinery inside. Therefore, a factory, like a cell, must control what leaves, as well as what enters.

Maintaining a Balance

All organisms are subject to constant changes in their environment. Aquatic organisms are subjected to changes in water temperature and the chemicals dissolved in the water. Terrestrial organisms can be subjected to changes in air temperature and amount of sunlight. Organisms must adjust to changes in their environment; failure of living things to adjust means death.

Why cells must control materials

Living cells maintain a balance by controlling materials that enter and leave. Without this ability, the cell cannot maintain homeostasis and will die.

It's important for a cell to keep its internal concentrations of substances, such as water, glucose, and other nutrients, while eliminating wastes as they are produced. Concentrations of these materials in a cell's external environment often change. It is the plasma membrane that maintains the proper concentrations of materials inside a cell by controlling the passage of materials into and out of the cell. The job of the plasma membrane is illustrated in ***Figure 9.1.*** Thus, homeostasis in a cell is maintained by the plasma membrane, which allows only certain particles to pass through.

The property of a membrane that allows some materials to pass

Program Resources

Section Focus Master 18 L1 SAE
Reinforcement and Study Guide, p. 33 L1
Biolab and Minilab Worksheets, pp. 33-34 L1
Basic Concepts Transparency 7 and **Master** L1 SAE

Figure 9.1

Without the property of selective permeability, all substances could pass in and out of the cell freely, and its contents would always be the same as its surroundings.

▲ **Like a plasma membrane, the strainer is selectively permeable because it allows water to run off while holding back the cooked spaghetti.**

Oxygen

Sugars

Amino acids

Water

Plasma membrane

Wastes

Carbon dioxide

Wastes

▲ **A cell must remove wastes and keep harmful substances from entering. Likewise, it must take in and keep needed materials, such as glucose, while maintaining a balance with its surroundings.**

through while keeping others out is known as **selective permeability.** *Figure 9.1* also shows another familiar example of selective permeability.

Selective permeability also allows different cells to carry on different activities within the same organism. For example, only nerve cells in your body may respond to a certain chemical, even though the chemical is present in the bloodstream and all cells in the body are exposed to it. The membranes of the nerve cells admit the chemical, but the membranes of other cells do not.

Structure of the Plasma Membrane

The structure and properties of a plasma membrane show how it can be selective and maintain cell homeostasis. The plasma membrane is a bilayer, meaning a structure made up of two layers. Powerful electron microscopes have revealed many of the details of this two-layer structure, shown in *Figure 9.2.* Each layer is made up of a sheet of lipid molecules. Protein molecules are embedded in the lipid bilayers like raisins in a slice of raisin bread.

permeable:
per (L) through
meare (L) to glide
Materials move easily (glide) through permeable membranes.

Magnification: 100 000×

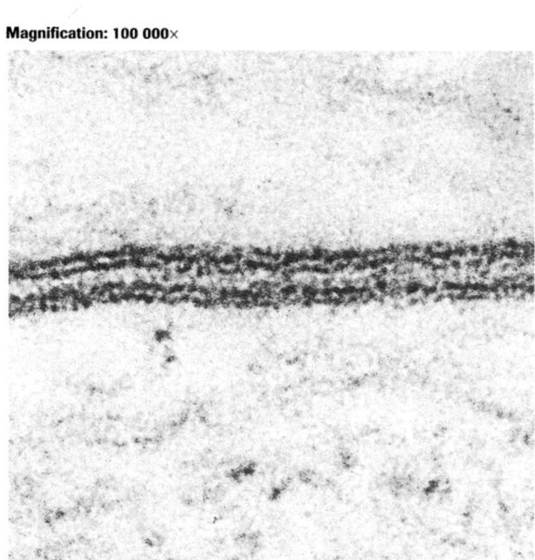

Figure 9.2

This transmission electron micrograph shows two cells side by side, with a space between the two cells. Each cell is clearly bounded by a plasma membrane made of two layers.

SECTION 9.1

2 Teach

Demonstration

LS **Visual-Spatial** To help students understand the nature of polar and nonpolar substances, mix together polar substances such as colored water and alcohol to show that the substances dissolve in each other. Demonstrate the interaction between polar and nonpolar substances by adding a small amount of salad oil to a container of water. Point out that the oil rests on top of the water. Stir the mixture vigorously. Explain that the oil forms spheres because the oil molecules have an affinity for each other, but not for the water molecules.

Misconception

Students frequently view the plasma membrane as merely a barrier surrounding a cell. Emphasize that the plasma membrane is in no way a solid barrier but has many functions, such as selectivity, molecular recognition, the export of wastes and cell products, the import of nutrients, and the ability to change in response to its environment.

Meeting Individual Needs

Hearing Impaired Pair hearing-impaired students with non-hearing-impaired students to work on the activities and questions in this chapter. Non-hearing-impaired students can assist hearing-impaired students by writing a summary of all oral directions given. L1 COOP LEARN LS

SECTION 9.1

BioLab Design Your Own Experiment

Are plastic bags selectively permeable?

Time Allotment

One class period

Objectives

Review objectives with students before they begin the Biolab.

Process Skills

observe and infer, formulate a hypothesis, interpret data, experiment

Safety Precautions

Caution: Iodine is poisonous and should be handled very carefully. If students get iodine on their hands, they should immediately wash their hands thoroughly with warm soapy water before proceeding with the experiment.

PREPARATION

When purchasing plastic bags, buy an inexpensive brand. Inexpensive bags tend to work better for this type of experiment than do more expensive plastic bags.

Possible Hypotheses

- If iodine and starch molecules are small enough, they will pass through the plastic bag. Movement is from a higher concentration to a lower concentration.

BioLab Design Your Own Experiment

Are plastic bags selectively permeable?

Plastic bags, such as sandwich bags, are made of thin, plastic membranes. If the plastic is selectively permeable, it will allow certain ions and molecules to diffuse across it but will hold back others. If particles diffuse through, they will have to pass between the molecules of polyethylene that make up the membrane. Molecules that are small enough to pass between the polyethylene molecules and cross the membrane should diffuse, but molecules that are too large will be held back.

PREPARATION

Problem

Will the polyethylene membrane allow iodine to cross the membrane? Will the polyethylene membrane allow starch, which is a larger molecule than iodine, to cross the membrane?

Hypotheses

Formulate a hypothesis that predicts the movement of starch and iodine across the plastic membrane. Consider all possible movements, as well as what factors are important in diffusion across a selectively permeable membrane.

Objectives

In this Biolab, you will:

- **Experiment** to determine whether diffusion occurs across a plastic membrane.
- **Interpret** the results to determine whether the plastic membrane was selectively permeable.

Possible Materials

400-mL beakers
small plastic bags
twist ties
graduated cylinder
starch solution
iodine solution
masking tape

Safety Precautions

Be sure to wash your hands thoroughly after you have started your experiment.

PLAN THE EXPERIMENT

Teaching Strategies

- Direct students to carefully plan how they will evaluate whether starch and/or iodine crosses the membrane.
- If students have trouble sealing their bags with twist ties, have them use rubber bands or masking tape.

Possible Procedures

- Students can set up the experiment with the starch in the bag and the iodine in the beaker or vice versa. Be sure students wash the outside of the bag thoroughly after it is filled and sealed.

PLAN THE EXPERIMENT

1. When you plan your experiment, be sure to consider how you will know whether starch and/or iodine crosses the membrane.
2. After you put solution in the bag, carefully seal the bag so that the solution does not leak out the top of the bag. Wash the bag under running water to clean off any solution that may be on the outside.

Check the Plan

1. Remember that when iodine combines with starch, the color of the starch solution will change to bluish-black.
2. Allow the bag to remain immersed in the solution overnight.
3. ***Make sure your teacher has approved your experimental plan before you proceed further.***
4. Carry out your experiment.

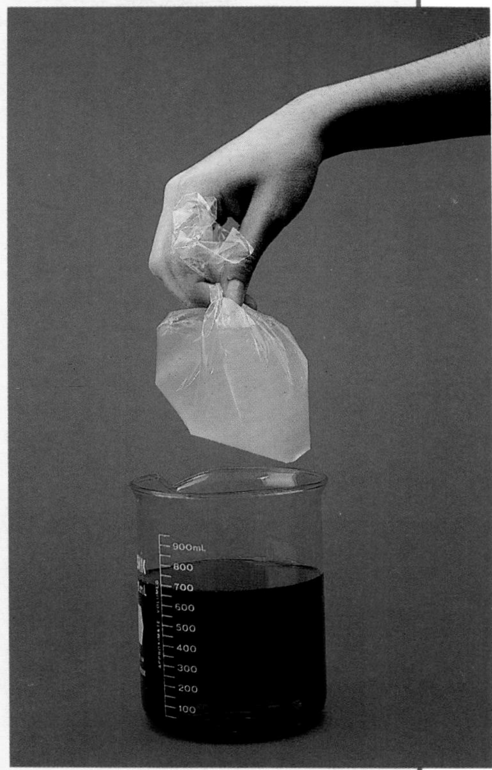

ANALYZE AND CONCLUDE

1. **Analyzing Data** Did iodine molecules move through the membrane? Explain how you know.
2. **Analyzing Data** Did starch molecules pass through the membrane? Explain how you know.
3. **Drawing Conclusions** What can you infer from this experiment about movement of large and small molecules through a thin polyethylene membrane?
4. **Checking Your Hypothesis** Do your data support your hypothesis? Explain why or why not.

Going Further

Changing Variables Will glucose pass through the polyethylene membrane? Set up an experiment that will answer this question. Use glucose test strips to detect the presence of glucose.

ANALYZE AND CONCLUDE

1. Iodine moved through the membrane and combined with the starch, changing the color of the starch solution to a blue-black.
2. Because the iodine remained yellow-brown in color, starch molecules evidently did not move into the compartment with the iodine.
3. The polyethylene membrane is selectively permeable. It allows the movement of small molecules such as iodine through the membrane, but restricts the movement of large molecules.
4. Students should explain whether the results of the experiment supported or rejected their hypotheses.

Assessment

Knowledge: Have students write a summary that explains their observations for this Biolab. Summaries should indicate an understanding of the movement of particles by diffusion. L1

Going Further

Ask students to predict what happened to the masses of the solutions in the various compartments in this experiment before and after diffusion occurred. Students can design an experiment to measure the mass changes. L2

Data and Observations

If students place the starch in the bag and the iodine in the beaker, they should observe that iodine moves into the bag. As the iodine combines with the starch, the color changes to a blue-black. The solution outside the beaker will be yellow-brown, indicating that starch did not move out of the bag. The opposite will occur if the students placed iodine in the bag and starch in the beaker.

interNET CONNECTION

Follow the link for this chapter on the Glencoe Homepage at **www.glencoe.com/sec/science** to find out more about the plasma membrane.

SECTION 9.1

Different Viewpoints in Biology

Membrane Models

Several anomalies are not explained by the fluid mosaic model. One involves a portion of osmotically inactive water found in cells. This portion may be between 10 and 60 percent of the total cell volume. Another problem involves the ability of cells with experimentally poisoned "pumps" to continue to regulate cell volume. Also some cells that have cut or removed membranes show an ability to maintain ion distributions.

Some scientists describe the cell as a drop of a polyphasic colloid. This view of the cell best explains the property of cytoplasmic streaming and the simultaneous occurrence of many biochemical reactions in different cells.

Visual Learning

Figure 9.3 Review the structure of the plasma membrane with students. Be sure students recognize that the phospholipid molecules are free to move sideways within the layer.

Figure 9.3

The structure of a plasma membrane (below) is a bilayer of phospholipids with proteins inserted into either side of or completely penetrating the membrane. Notice also that the phospholipid molecules (right) are not chemically bonded to each other; they are free to move sideways within the layer. The proteins within the membrane poke out like icebergs in a sea of phospholipids.

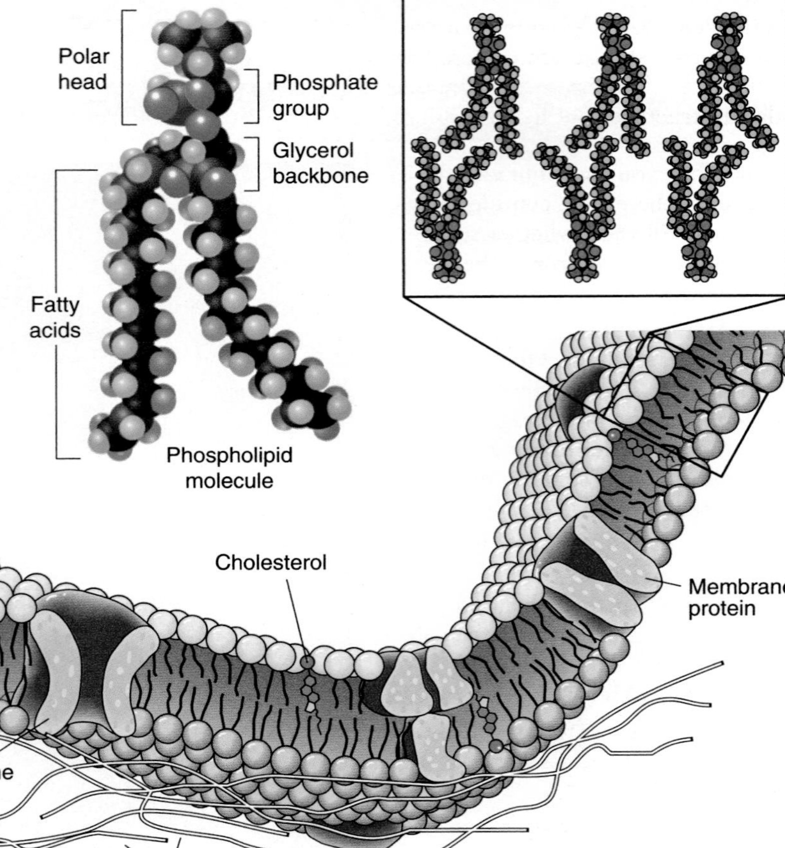

Makeup of the lipid bilayer

Unlike typical triglyceride lipids, most of the lipids that make up the two layers in the plasma membrane have two fatty acids attached to glycerol instead of three. In place of a third fatty acid, a membrane lipid has a small organic section attached to a phosphate group and thus is called a **phospholipid.** As you can see in *Figure 9.3,* phospholipids have polar, water-soluble heads attached to long, nonpolar, insoluble tails. The phosphate group is soluble in water because it is polar, and therefore is attracted to water molecules. The fatty acid chains are not soluble in water because they are nonpolar.

Most cells have a watery environment both on the inside and outside. Because water attracts the phosphate ends, the phospholipids align to form a double layer with the water-soluble phosphate ends toward the outside of each layer, as in *Figure 9.3.* The nonpolar tails lie inside the bilayer.

Within each layer, phospholipid molecules can move sideways through their layer. These properties make the bilayer behave like a fluid, a material that flows. This description of a plasma membrane as a structure made up of many similar molecules that are free to move sideways within the membrane is called the **fluid mosaic model.** Some cell organelles, such as

Meeting Individual Needs

Students Acquiring English Have students use a dictionary to find the meaning of the term *mosaic.* Ask students to describe the application of this term to a piece of art as well as to the plasma membrane. Have students describe the fluid nature of the membrane and explain why fluidity of the membrane is important to living cells. L1 SAE

PROJECT

Building a Model

LS **Kinesthetic** Have students in groups build a model of a plasma membrane using materials such as Styrofoam "peanuts," yarn, pipe cleaners, and popsicle sticks. Encourage students to be creative in putting their plasma membranes together. L1 COOP LEARN

the nucleus, vacuoles, mitochondria, and chloroplast, also are enclosed by membranes that have a fluid mosaic bilayer structure. Because the kinds and arrangements of proteins and lipids vary from one membrane to another, each type of membrane has its own permeability properties.

Saturated versus unsaturated fatty acids

The fatty acids that make up the phospholipids of the plasma membrane can be saturated or unsaturated. When a fatty acid tail contains unsaturated fatty acids, the fatty acid chain bends at the double bond, like a knee or elbow joint. As described in *Figure 9.4*, the more unsaturated fatty acids a membrane has, the more fluid it is.

Figure 9.4

In a caribou, an arctic animal, the membranes of cells near the animal's hooves have phospholipids with many unsaturated fatty acids. Unsaturated fatty acids remain liquid at low temperatures. The unsaturated lipids allow the caribou's feet and legs to drop to almost 0°C in the arctic winters and still maintain plasma membrane function. The cell membranes of the rest of the caribou's body have more saturated lipids.

Literature

Lives of a Cell

by Lewis Thomas

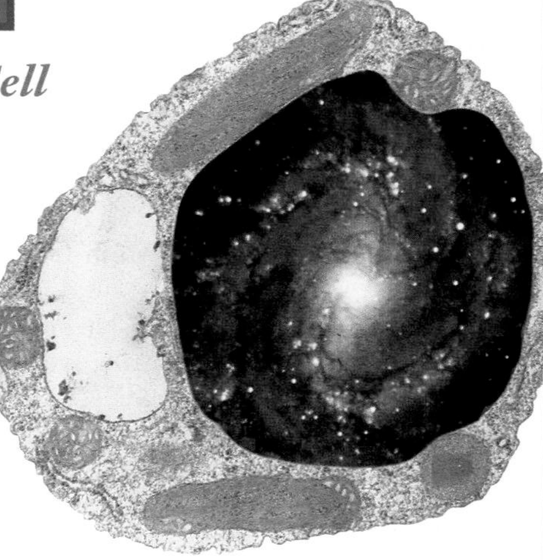

Does the following paragraph make you want to try an experiment?

"We leave traces of ourselves wherever we go, on whatever we touch. One of the odd discoveries made by small boys is that when two pebbles are struck sharply against each other they emit, briefly, a curious smoky odor. The phenomenon fades when the stones are immaculately cleaned, vanishes when they are heated to furnace temperature, and reappears when they are simply touched by the hand again before being struck."

Words are like experiments Despite all his technical knowledge, Dr. Thomas, a physician and medical researcher, writes simply and engagingly about everything from the tiny universe inside a single cell to the possibility of visitors from a distant planet. About his writing style, Dr. Thomas says, "Although I usually think I know what I'm going to be writing about . . . most of the time it doesn't happen that way at all. . . . I get surprised by an idea that I hadn't anticipated getting, which is a little like being in a laboratory."

Medicine, a young science Dr. Thomas grew up with the practice of medicine. As a boy, he accompanied his father, a family physician, on house calls to patients. Years later, Dr. Thomas described those days in his autobiography, *The Youngest Science.* The title reflects his belief that the practice of medicine is "still very early on" and that some basic problems of disease are just now yielding to exploration.

CONNECTION TO Biology

After you have studied Section 9.2, write a paragraph using Dr. Thomas's style to describe how osmosis restores freshness and beauty to a wilted flower.

Literature

Lives of a Cell

Purpose

Students will be introduced to the thoughts and scientific writings of Lewis Thomas.

Teaching Strategies

- Lewis Thomas said that his writings are "Notes of a Biology Watcher." Have students discuss what this phrase means and explain how all people are "Biology Watchers." L1
- Have students read all of Lewis Thomas's *Lives of a Cell,* or his book *Medusa and the Snail.* Ask students to prepare a written report on the book they read. L2

Possible Answer

Responses should use imagery to indicate that osmosis, by providing water to a wilted flower, increases turgor pressure within the flower, making the flower appear stronger and healthier.

Reinforcement

Display a tub of margarine and a bottle of corn oil. Ask students to use their knowledge of saturated and unsaturated fatty acids to identify which substance probably has more unsaturated fatty acids. *The oil; it is more fluid.* L1

3 Assess

Check for Understanding

Ask students to write a short description of the fluid mosaic model. Have students include in their descriptions the following terms: *plasma membrane, phospholipid, bilayer, polar, nonpolar,* and *proteins.* Ask volunteers to present their summaries to the class. L1

CULTURAL DIVERSITY

Ernest Everett Just

Discuss with students the history of the study of the plasma membrane and the important contributions of African American embryologist, Ernest Everett Just (1883-1941). Just is best known for his experimental studies on fertilization in sea urchins.

Just was interested in the role of the plasma membrane in fertilization, and was among the first researchers to realize that the plasma membrane played an active role in cell physiology. Read passages from Just's 1939 book, *The Biology of the Cell Surface,* and compare Just's work with current models of the plasma membrane.

SECTION 9.1

Reteach

Model the action of a fluid mosaic by half-filling a plastic tub with water. Add just enough Ping-Pong balls to the tub to completely cover the water's surface. Move one ball across the surface and have students note how the other balls jostle each other to make way for the moving ball. A similar demonstration could be done with a tub of red apples and one yellow apple.

Extension

Have students research and report on the development of the fluid mosaic model. Ask students to include the work of Gorter and Grendel, Danielli and Davison, and Singer and Nicholson in their reports. L3

Assessment

Portfolio: Ask students to draw and label the structure of the plasma membrane. L1 P

4 Close

Using an Analogy

Explain that the cell membrane can be compared to a wall that has doors and windows. Ask students to evaluate the strengths and weaknesses of this analogy. L1

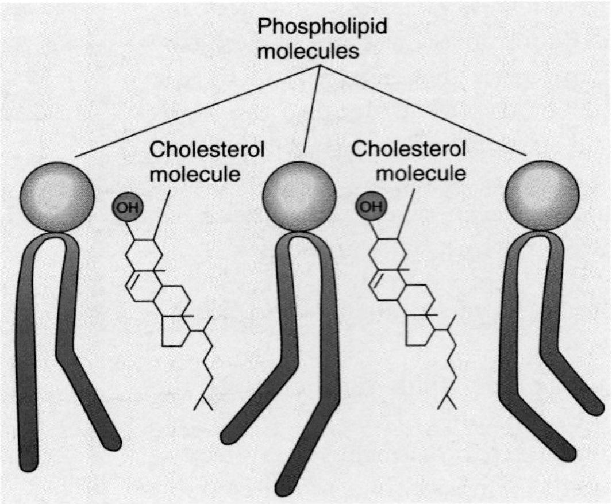

Figure 9.5

Eukaryotic plasma membranes can contain large amounts of cholesterol—up to one molecule for every phospholipid molecule. Although cholesterol tends to make lipid bilayers less fluid, it prevents the fatty acid chains from sticking together and makes the bilayer more stable. Cholesterol also decreases the permeability of lipid bilayers to water-soluble substances.

Phospholipid molecules

Cholesterol molecule

Cholesterol molecule

OH

OH

Membranes that are very fluid are not always an advantage. In many eukaryotes, especially animals, cholesterol is an important part of the plasma membrane, as shown in *Figure 9.5*. Fatty acid chains are flexible, but the cholesterol molecule is rigid. Therefore, the presence of cholesterol molecules strengthens the fluid mosaic and makes it more stable. In addition, the cholesterol helps keep the fatty acid tails of the phospholipids separated.

The function of membrane proteins

Although the basic structure of a plasma membrane is a lipid bilayer, most of the functions of the membrane are carried out by proteins. Some of the proteins extend through the bilayer and on both sides of the membrane, as you saw in *Figure 9.3*, while others do not extend into the interior of the lipids.

Many of the proteins determine which particles can pass across the membrane. Some proteins serve as enzymes. Others act as markers that are recognized by chemicals from both inside and outside the cell. Some of these markers are involved in fighting disease. For example, your immune system can distinguish your own cells from cells of a bacterium by certain membrane proteins. Thus, in many respects, the plasma membrane can be thought of as a communication center between a cell and its environment.

Section Review

Understanding Concepts

1. Why is the structure of the plasma membrane referred to as a bilayer and as a fluid mosaic structure?
2. Explain why selective permeability is necessary for homeostasis within the cell.
3. What is the function of cholesterol in the plasma membrane?

Thinking Critically

4. Suggest what might happen if cells grow and reproduce in an environment where no cholesterol is available.

Skill Review

5. **Recognizing Cause and Effect** Consider that plasma membranes allow materials to pass through. Explain how this property contributes to homeostasis. For more help, refer to Thinking Critically in the *Skill Handbook*.

Answers to Section Review

1. The plasma membrane is referred to as a bilayer because the phospholipids forming the membrane consist of a double layer. Because the lipids and proteins in the membrane are free to change position, they are said to be fluid. The pattern of the molecules is like a mosaic.
2. Cells must get rid of wastes while keeping other molecules in. Likewise, they must let some things in while keeping others out.
3. Cholesterol makes the fluid mosaic more stable and rigid.

Thinking Critically

4. The membranes would be very fragile and would not hold together.

Skill Review

5. Plasma membranes allow useful materials such as oxygen, sugars, amino acids to enter cells. These materials are needed for cell metabolism and survival.

SECTION 9.2 Cellular Transport

From your study of simple chemistry in Chapter 7, you learned that the electrons of atoms are constantly moving about the nucleus. In fact, all particles of matter are in constant motion. The atoms, ions, and molecules that make up all materials are moving. It makes no difference whether a material is solid, liquid, or gas; much like bees in a swarm, its particles move constantly in a totally random fashion. This random movement helps explain how materials enter or leave cells.

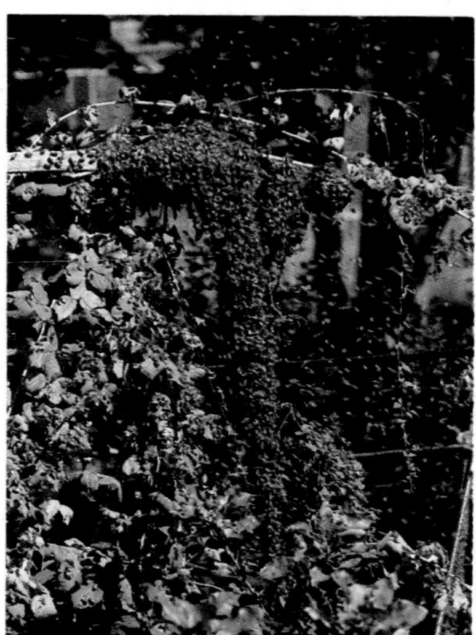

Section Preview

Objectives

Explain how the processes of diffusion, passive transport, and active transport occur and why they are important to cells.

Predict the direction of diffusion of a dissolved substance.

Key Terms

diffusion
dynamic equilibrium
osmosis
isotonic solution
hypotonic solution
turgor pressure
contractile vacuole
hypertonic solution
plasmolysis
passive transport
transport protein
facilitated diffusion
active transport
endocytosis
exocytosis

Diffusion

All objects in motion, like the fish in *Figure 9.6,* have energy of motion called kinetic energy. A moving particle of matter moves in a straight line until it collides with another particle, much like balls on a pool table. After the collision, both particles rebound. Imagine a room full of Ping-Pong balls, all in constant motion, colliding with no loss in energy. Particles of matter move in the same way.

The discovery of Brownian motion

In 1827, Robert Brown, a Scottish scientist, used a microscope to observe pollen grains suspended in water. He noticed that the grains moved constantly in little jerks, as if being struck by invisible objects. This motion, he thought, was the result of the life hidden within the pollen grains. However, when he repeated his experiment using dye particles, which are nonliving, he saw the same erratic motion. This motion has been called Brownian motion ever since. Brown had no explanation for the motion he saw, but today we know that Brown was observing evidence of the random motion of molecules. These were the invisible objects that were moving the tiny visible particles.

Figure 9.6
These leaping salmon have kinetic energy, the energy of motion. Like atoms and molecules, all moving objects have kinetic energy.

Program Resources

Section Focus Master 19 L1 SAE
Reinforcement and Study Guide, pp. 34-36 L1
Biolab and Minilab Worksheets, pp. 31-32 L1
Laboratory Manual, pp. 55-56 L2
Basic Skills Transparency 8 and **Master** L1 SAE
Concept Mapping, p. 9 L1
Critical Thinking/Problem Solving, p. 9 L3
Tech Prep Applications, pp. 9-12 L2
Basic Concepts Transparencies 8 and **9** and **Master** L1 SAE
Reteaching Transparency 9 and **Master** L1 SAE

SECTION 9.2

Prepare

Key Concepts

Students will recognize how the structure of the plasma membrane permits diffusion, passive transport, and active transport. They will develop an understanding of the importance of these processes in maintaining homeostasis and proper cell function.

Block Scheduling

Look for this symbol for strategies that are useful in a block scheduling format. For more information on block scheduling, refer to Section 9.2 in the **Lesson Plans** booklet.

Materials

- Order *Paramecium* for the Minilab and *Euglena* for the Microscope Activity.
- Obtain India ink, eggs, hydrochloric acid, and corn syrup for the Demonstrations, celery sticks for the Reteach, raisins for the Portfolio activity, and potato and measuring spoons for the Alternate Lab.
- Gather the materials needed for Building a Model and the Active Transport Project.

1 Focus

Bellringer

Before presenting the lesson, display **Section Focus Master 19** on the overhead projector and have students answer the accompanying questions. L1 SAE

SECTION 9.2

Demonstration

Visual-Spatial Brownian movement may be demonstrated using a microprojector and a very dilute solution of India ink. The colloidal carbon particles will collide with water molecules. Even though water molecules cannot be seen, the effects of these collisions (jiggling) can be seen. Place the wet mount over a light bulb, and quickly refocus the slide. Students should be able to see increased kinetic energy with increased temperature. Relate this to diffusion, caused by the random movement of molecules.

2 Teach

Tying to Previous Knowledge

The kinetic theory of matter serves as the fundamental explanation for the transport of molecules from one place to another. Elicit what the major differences, at the molecular level, are among a solid, a liquid, and a gas. Students should recognize that the freedom of random particle movement is the only difference among the three states of matter.

Software

Photosynthesis, Biology Explore Series, Wings for Learning.

Figure 9.7

A few crystals of copper(II) sulfate were dropped into a beaker.

B After two days, the colored ions have diffused upward slightly (center). As you can see, diffusion of solutes through water is a very slow process.

A In a few minutes, the blue color formed by the ions of the dissolving compound has begun to diffuse.

C After a much longer time—sometimes months or even years—the ions will have diffused completely throughout the beaker if it is left undisturbed and covered.

kinetic:
kinein (GK) to move
Moving objects have kinetic energy.

The process of diffusion

Most substances in and around a cell are in water solution, a mixture in which the ions or molecules of a solute are distributed evenly among water molecules. All of these particles, both water and solute, move randomly, colliding with each other. When a soluble substance is placed in water, these random collisions tend to scatter particles of solute and water until they are evenly mixed. One way of observing this effect is to drop a few crystals of a colored substance into a beaker of water, as shown in *Figure 9.7*.

The movement of individual particles is random. The overall movement of the dissolved ions in the beaker in *Figure 9.7* is from a region of high concentration (near the crystals) to a region of lower concentration (throughout the water). **Diffusion** is the net movement of particles from an area of higher concentration to an area of lower concentration. Diffusion results because of the random movement of particles.

The results of diffusion

Eventually, the colored ions will become evenly distributed throughout the molecules of water in the beaker. After this point, the ions continue to move randomly and collide with one another. However, no further change will occur in concentration throughout the beaker. This condition, in which there is continuous movement but no overall change, is called **dynamic equilibrium** as illustrated in *Figure 9.8*. The word *dynamic* refers to movement or change, while *equilibrium* refers to balance. The preservation of a dynamic equilibrium is one of the characteristics of homeostasis.

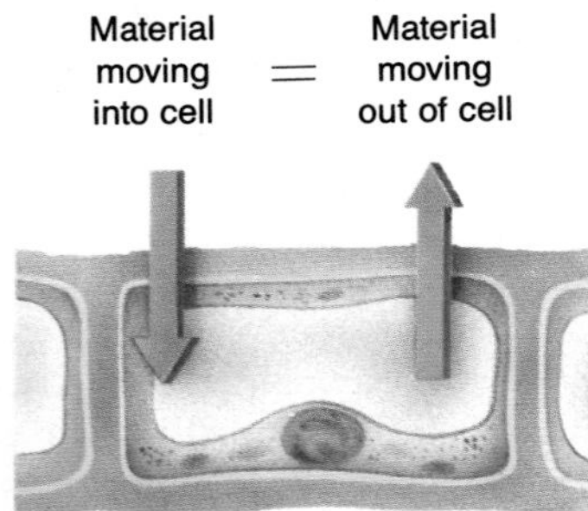

Figure 9.8

When a cell is in dynamic equilibrium with its environment, materials move into and out of the cell at equal rates. As a result, there is no net change in concentration inside the cell.

PORTFOLIO

Observing Osmotic Changes Have students design and conduct an experiment to show the effect of osmosis on a raisin. *Students should place raisins in warm water for several minutes. After removing the raisins from the water, they should explain any changes they observe in the appearance of the raisins in terms of osmosis.* L1 P LS

Diffusion depends on concentration gradients

Diffusion cannot occur unless a substance is in higher concentration in one region than it is in another. For example, oxygen diffuses into the capillaries of the lungs because there is a greater concentration of oxygen in the air sacs of the lungs than in the capillaries. The difference in concentration of a substance across space is called a concentration gradient. Because ions and molecules diffuse from an area of higher concentration to an area of lower concentration, they are said to move *with* a gradient. If no other processes interfere, diffusion will continue until there is no concentration gradient. At this point, dynamic equilibrium occurs.

The selectivity of membranes

You can compare this process to the deflation of a helium-filled balloon. The balloon slowly loses helium because the helium atoms are tiny and diffuse out of the balloon by passing between the molecules that make up its membrane.

A plasma membrane is selectively permeable. The lipid bilayer makes it difficult for charged ions or polar molecules to pass through by diffusion because they are not attracted to the nonpolar structure of the fatty acid tails. Only molecules of water, oxygen, nitrogen, carbon dioxide, and a few other small, nonpolar molecules can diffuse directly across the lipid bilayer. As you will see, cells have specific ways to allow needed molecules, such as amino acids, to enter.

Technology — BIOLOGY & SOCIETY

Frozen in Time

Frostbite is a condition in which body tissues, usually fingers and toes, freeze and die. How can a process that naturally kills healthy tissue be used to preserve living cells and tissues such as blood, skin, and sperm for later use?

Ordinary freezer burn Your freezer at home cools materials slowly, allowing the water in cells to form large ice crystals that may puncture the cells' plasma membranes. Also, if foods are not properly protected or are frozen for an extended period, the ice crystals evaporate, and the cells dry up due to freezer burn.

Methods of preserving tissue To avoid the problems of large ice crystals, biological specimens are frozen rapidly. An antifreeze such as glycerol is added to lower the temperature at which freezing occurs. The ice crystals formed by this method are much smaller and less damaging. The antifreeze and rapid temperature drop solidify the materials in the cells and the surrounding fluids into a glasslike rather than crystalline state. These procedures, called *cryopreservation,* produce cells that return to normal when thawed, even after long periods of storage.

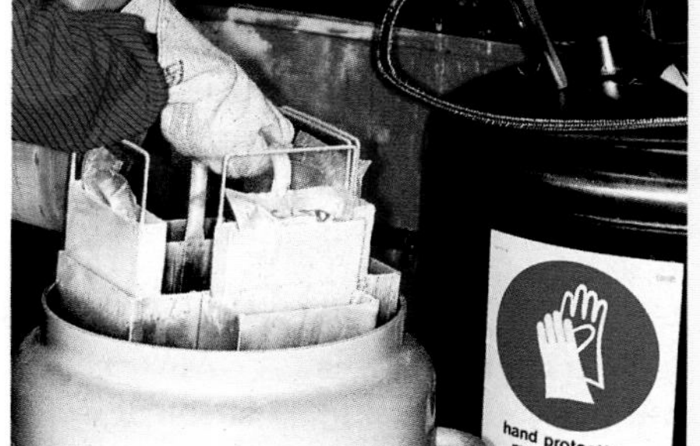

Applications for the Future

The freezing of organs is not yet possible because freezing does not occur uniformly in a large object. Still, cryopreservation has allowed blood banks to maintain a supply of the less common blood types.

Another application of this technology is cryosurgery, a technique that kills cells by rapid freezing. Using liquid nitrogen as a cooling agent, surgeons can destroy a specific area by placing a needlelike freezing apparatus in contact with the tissue. This method is precise and results in little or no bleeding.

INVESTIGATING the Technology

Debating the Issue Frozen fertilized human eggs may be preserved for many years and may sometimes be forgotten. Should there be a time limit on storage of frozen eggs? Prepare arguments on both sides to debate this issue.

Technology — BIOLOGY & SOCIETY

Frozen in Time

Purpose

Students will recognize the potential applications of cryopreservation as well as the problems associated with the process.

Teaching Strategies

- Ask students to consider the following situation. A couple has had embryos frozen. The couple is then killed in an automobile accident. Ask students if the embryos should have inheritance rights. (This occurred in 1984 when a multimillionaire couple was killed in a plane crash. They had two embryos frozen in a clinic in Australia. The Australian court ruled that embryos had no life or rights at that stage.)

INVESTIGATING the Technology

Debating the Issue Students should support their viewpoints logically. Frozen eggs may lose their viability over time, or perhaps damage to chromosomes will occur from prolonged freezing. If a couple deposits frozen embryos in hopes of implanting them at a future date, there may be a point at which the mother becomes too old to consider a full-term pregnancy. Students may suggest a time limit to force couples to make up their minds. Others may feel that this is "playing God" and object to the discussion.

Going Further

Have students contact their local blood bank, hospital, or Red Cross to learn how blood is fractionated and stored. Have them find out what techniques are used to maintain the integrity of blood cells.

Demonstration

LS Visual-Spatial Dissolve a shell from a hard-boiled egg by placing the egg in dilute hydrochloric acid. Change the acid several times until the hard shell is completely dissolved. Half-fill a 400-mL beaker with distilled water (a hypotonic solution). Half-fill a second 400-mL beaker with corn syrup (a hypertonic solution). Have students note the size of the egg as you carefully place the shell-less egg in the distilled water. Explain that osmosis causes the egg to swell. Have students note the size of the egg as you move it from the water to the corn syrup. In the corn syrup, the egg loses water by osmosis and shrinks. Allow the egg to shrink to a size smaller than normal. You may want to dissolve the shells of two or three eggs. Place one in an isotonic solution to demonstrate that no overall change occurs.

History Connection

Agricultural Lifestyles

People in early agricultural societies observed that dried or salted meats resisted decay. Drying food or adding salt or sugar lowers the available moisture, preventing spoilage caused by bacterial growth. Salting is still used to preserve fish, corned beef, and green olives. High sugar content inhibits growth in sugar-cured hams, jams, and jellies. The initial heating of jams and jellies also reduces the number of microbes. Although adding salt and sugar to foods lowers the likelihood of spoilage, both dry salt- and sugar-cured products will spoil if allowed to stand in humid atmospheres.

Figure 9.9

During osmosis, water diffuses across a selectively permeable membrane when one side has a higher concentration of a dissolved material that cannot pass through the membrane. Notice that the number of sugar molecules (orange dots) did not change on each side of the membrane, but the number of water molecules (blue dots) did change.

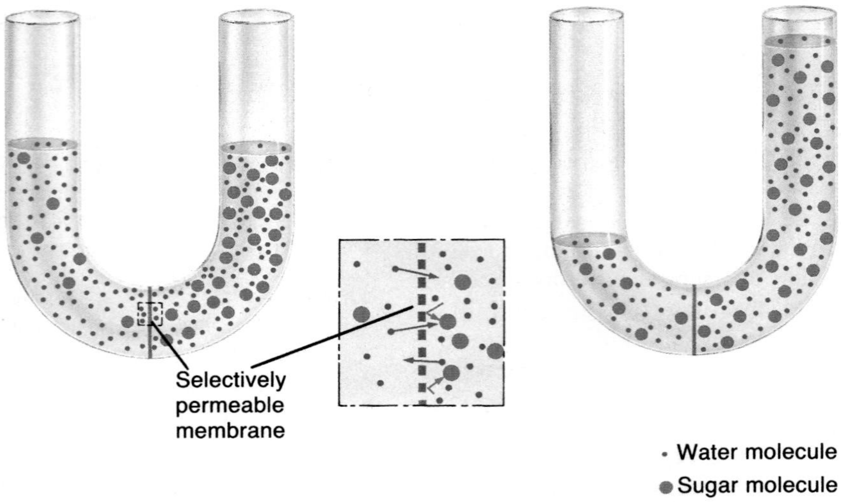

osmosis:
osmos (GK) pushing
Osmosis can push out a cell's plasma membrane.

iso-, hypo-, hyper-:
isos (GK) equal
hypo (GK) under
hyper (GK) over

Osmosis—Diffusion of Water

The diffusion of water molecules into and out of cells is so common that the process is given its own name, osmosis. **Osmosis** is the diffusion of water molecules through a selectively permeable membrane from an area of higher water concentration to an area of lower water concentration.

How osmosis occurs

A strong sugar solution has a lower concentration of water than a weak sugar solution. If these two solutions are placed in direct contact, water molecules diffuse in one direction and sugar molecules diffuse in the other direction. Now, suppose you separate the two solutions with a selectively permeable membrane that will allow water molecules to cross but will not allow sugar molecules to cross. What would happen to the solutions? Study *Figure 9.9.* Osmosis would occur as water molecules diffused across the membrane toward the more concentrated sugar solution on the right. The result would be a buildup of water on the right side of the membrane.

A plasma membrane has properties similar to those of the membrane shown in *Figure 9.9.* As you can see, a cell will lose water by osmosis if it is placed in an environment in which the water concentration is lower than that of the cell contents. Likewise, it will gain water if the water concentration is greater than that of the cell contents.

Osmosis in an isotonic solution

Most cells, whether in multicellular or unicellular organisms, are subject to osmosis because they are surrounded by water solutions. An **isotonic solution** is a solution in which the concentration of dissolved substances is the same as the concentration inside the cell. Likewise, the concentration of water is the same as inside the cell. Do you think osmosis will occur in a cell placed in an isotonic solution?

Meeting Individual Needs

Learning Disabled Using colored chalk, draw a U-tube on the chalkboard similar to the "Before Osmosis" diagram in Figure 9.9. Draw molecules in at least two colors, showing one that cannot cross the selectively permeable membrane. Have students make a similar diagram using colored pencils. Challenge students to diagram the "After Osmosis" stage. Walk around the room, checking to see if students understand the concept of osmosis as they draw. Help students as needed. L1 LS

If a cell is placed in an isotonic solution, water molecules still move into and out of the cell at random, but there is no net movement of water. Therefore, no osmosis occurs. The cell is in dynamic equilibrium with the surrounding liquid because there is movement of molecules, but no overall change is taking place. Cells in isotonic solution have their normal shape as shown in *Figure 9.10.* Most solutions injected into the body are isotonic, so that cells are not damaged by the loss or gain of water.

Cells in hypotonic solution

A **hypotonic solution** is a solution in which the concentration of dissolved substances is lower than the concentration inside the cell. If a cell is placed in a hypotonic solution, osmosis will cause water to move through the plasma membrane into the cell. As water diffuses into the cell, the cell swells and its internal pressure increases. The pressure that exists in a cell is called **turgor pressure.** This pressure increases in a hypotonic solution.

MiniLab

How many contractile vacuoles does a paramecium contain?

A *Paramecium* is an organism that has contractile vacuoles. Each vacuole fills with water, contracts, and expels its contents to the outside of the paramecium. When the vacuole contracts, it disappears from view but then refills with water and repeats the process.

Procedure

1. Make a ring of methyl cellulose on a slide. Add a drop of *Paramecium* culture in the center of the ring. Add a coverslip.
2. Find a paramecium, observe it under high power, and locate a contractile vacuole. You may need to adjust your light source or microscope mirror for best results. A contractile vacuole looks like a clear, star-shaped structure that will appear and disappear.

Analysis

1. How many contractile vacuoles does a paramecium have?
2. Do the contractile vacuoles move throughout the cell?
3. How do contractile vacuoles help a paramecium survive in fresh water?

Figure 9.10

Cells in Isotonic Solution

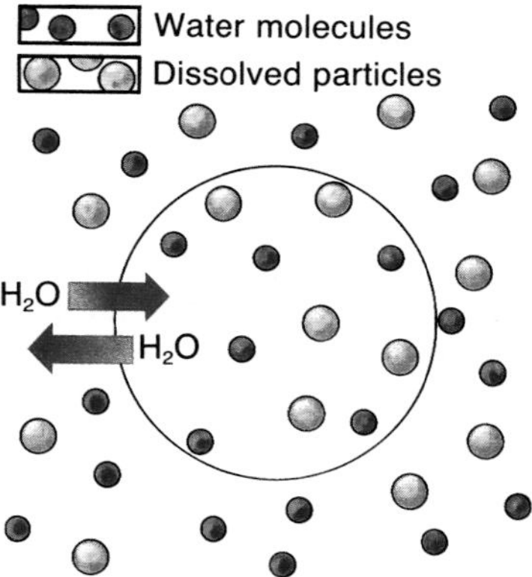

In an isotonic solution, water molecules move into and out of the cell at the same rate.

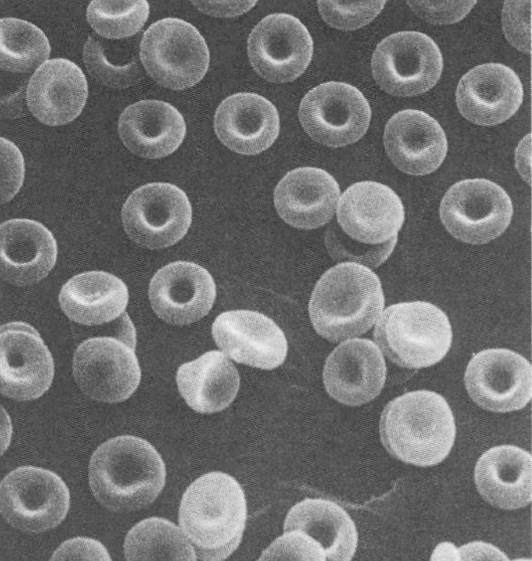

Cells have their normal shape in isotonic solution. Notice the concave disc shape of a red blood cell.

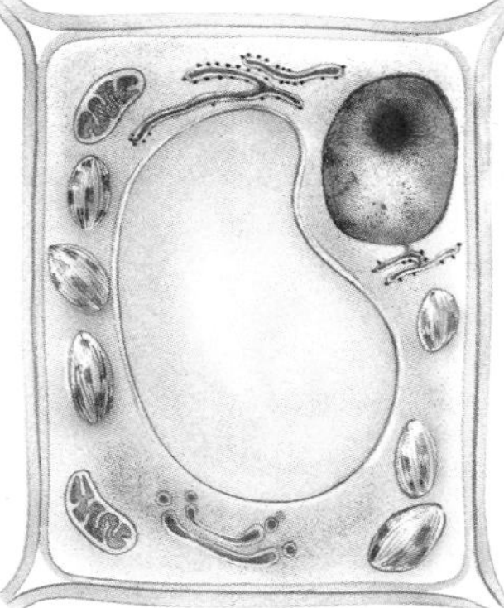

A plant cell has its normal shape and turgor pressure in isotonic solution.

MiniLab

Purpose

Visual-Spatial Students will observe an aquatic organism's response to a hypotonic environment.

Process Skills

observe and infer

Teaching Strategies

- Explain to students that they will need to adjust the light source to cut down the amount of light focused on the slide in order to see the contractile vacuoles.

Expected Results

Students should see two contractile vacuoles, one at each end of the cell.

Analysis

1. Two
2. No
3. Because of the hypotonic environment *Paramecium* lives in, water constantly diffuses into the cell. The contractile vacuole pumps the water out so that the paramecium will not burst.

Assessment

Journal: Have students draw a paramecium and explain the function of the contractile vacuole in terms of osmosis. L1

Meeting Individual Needs

Gifted Have advanced students research how ion channels are involved in a loss of turgor pressure in the leaf of *Mimosa pudica* when it is touched. Ask students to write a report of their findings. L3 LS

Meeting Individual Needs

Visually Impaired Pair visually impaired students with sighted students for the Minilab. Have the sighted students describe the appearance of *Paramecium* and its contractile vacuoles to the visually impaired students. Students who are not visually impaired should be required to draw the contractile vacuoles in *Paramecium.* L1 LS

Health Connection

Cystic Fibrosis
Cystic fibrosis is the most common lethal genetic disease among Caucasians. Researchers have discovered that a faulty gene for a chloride ion channel protein leads to the mucus-clogged organs characteristic of cystic fibrosis. The flawed protein fails to open and close properly, resulting in a buildup of chloride that impairs sodium ion flow. Together, the sodium and chloride ion imbalances destroy the pancreas and thicken the body's mucus. Scientists have successfully inserted corrected copies of the CF gene into the lungs of animals and used experimental enzyme treatments to alleviate patients' symptoms.

Software

Regulation and Homeostasis: Systems in Balance, IBM

Concept Development

Explain to students that *net movement* is the overall movement of a material without regard to the movements of each particle.

Figure 9.11
Cells in Hypotonic Solution

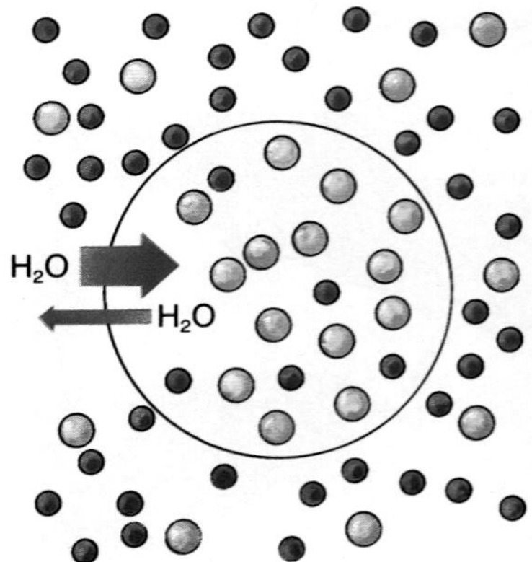

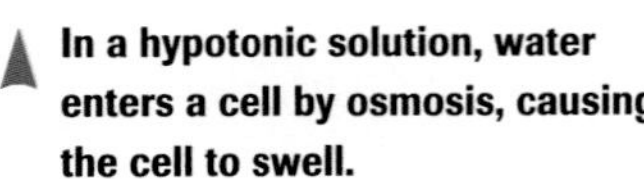

In a hypotonic solution, water enters a cell by osmosis, causing the cell to swell.

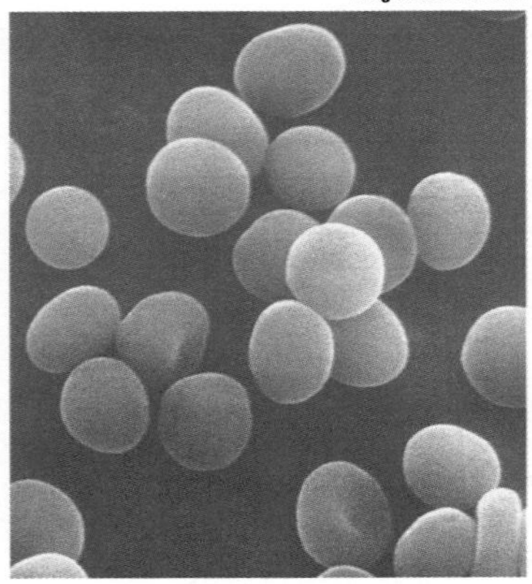

Animal cells, like these red blood cells, may continue to swell until they burst.

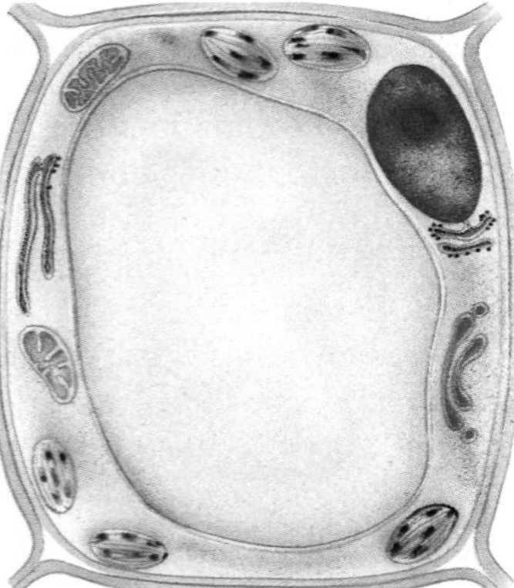

Plant cells swell beyond their normal size as turgor pressure increases.

You will recall from Chapter 8 that plant cells have strong cell walls. The rigid walls can resist the turgor pressure caused by osmosis. Turgor pressure causes the cytoplasm and plasma membrane to press outward against the cell wall, making the cell rigid. This action gives shape and support to plants. The rigidity produced by turgor pressure in cells gives shape and support to plants that are not woody, such as tulips and tomatoes. Plants wilt when they are deprived of water. Wilting occurs because the plant cells lose turgor pressure. When a wilting plant is supplied with water, its cells regain their turgor. Because plain water is hypotonic in comparison to cell contents, turgor pressure can increase to greater than normal, causing the cell to swell as shown in *Figure 9.11*. Grocers take advantage of this process by spraying their produce with water so that the vegetables stay crisp and appealing to customers.

Figure 9.12
***Paramecium* contains contractile vacuoles, the star-shaped structures in the photo. Water is collected in ducts (the rays of the star) and passed to the center vacuole. When the vacuole contracts, the water is expelled through a pore in the plasma membrane.**

Magnification: 150×

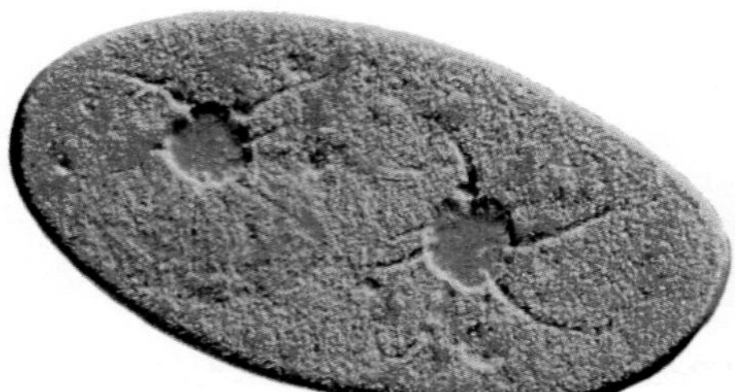

Animal cells do not have cell walls. Therefore, an animal cell in a hypotonic solution will swell as shown in *Figure 9.11*, and may even burst. Even so, many organisms that lack cell walls are adapted to live in hypotonic solutions such as fresh water. These organisms have adaptations that keep their cells from bursting. One mechanism is the continual excretion of excess water from cells. Some protists contain organelles, called **contractile vacuoles**, that work like medicine droppers. They collect excess water from the cell and then contract, squeezing the water out of the cell through the outer membrane, *Figure 9.12*.

Alternate Lab: Observing Osmosis

Purpose

LS Logical-Mathematical Students will observe and measure the effect of osmosis on a potato.

Materials

potato, water, two 100-mL beakers or paper cups, measuring spoon, salt, graduated cylinder, label, pen, stirring rod, balance, plastic wrap or aluminum foil, knife

Procedure

Give the following directions to students.

1. Label one beaker "Water" and the other "Salt." Place 100 mL of water into each beaker.
2. Place 3 tablespoons of salt into the "salt" beaker and stir until the salt is dissolved.
3. With a knife, cut two cubes of potato (without skin) that measure 2 cm on each side.
4. Using a balance, measure and record the mass of each potato piece. Then place one piece in the "Water" beaker and the other in the "Salt" beaker. Record the texture of the potato cubes

Osmosis in a hypertonic solution

Organisms that live in salty water have the opposite problem. A **hypertonic solution** is a solution in which the concentration of dissolved substances is higher than the concentration inside the cell. If a cell is placed in a hypertonic solution, osmosis will cause water to leave the cell.

Animal cells placed in a hypertonic solution will shrivel because of decreased pressure in the cells, as shown in *Figure 9.13*. Cookbooks suggest that you not salt meat before cooking. The salt forms a hypertonic solution on the meat's surface, and the water inside the meat's cells diffuses out. The result is cooked meat that is dry and tough.

If a plant cell is placed in a hypertonic environment, it will lose water, mainly from its central vacuole. The plasma membrane and cytoplasm will shrink away from the cell wall, as shown in *Figure 9.13*. Loss of water from a cell resulting in a drop in turgor pressure is called **plasmolysis.** This process causes a plant to wilt.

MiniLab

What happens to plant cells in a hypertonic solution?

Plant tissue is normally rigid or crisp because of the turgor pressure within the cells. When plant cells lose their turgor pressure, the plant wilts. What happens on the cellular level when a plant cell is exposed to a hypertonic solution?

Procedure

1. Use forceps to take a leaf from the tip of an *Elodea* sprig, and place it on a slide with a drop of water. Apply a coverslip.

2. Observe it first with low power and then with high power. Draw a cell and its contents.
3. Place a drop of 3% salt solution on the slide near one side of the coverslip. Place a piece of paper towel on the opposite side of the coverslip to draw out the plain water and pull the salt water underneath the coverslip.
4. Observe the cell for a few minutes and then sketch it.

Analysis

1. What happened to the cells when they were exposed to the 3% salt solution?
2. Based on what you have learned about diffusion and osmosis, explain what you observed.

Figure 9.13

Cells in Hypertonic Solution

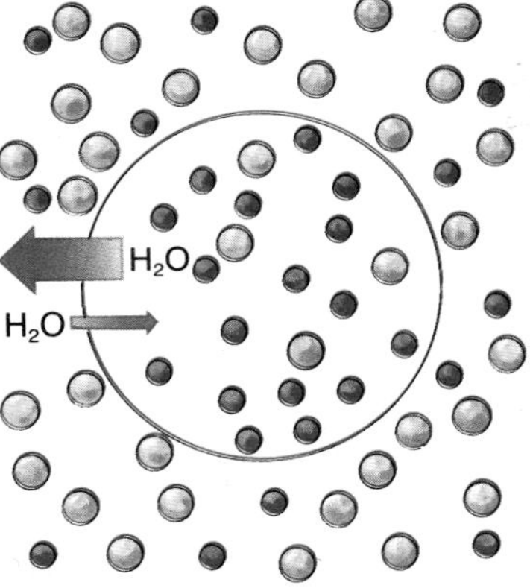

In a hypertonic solution, water leaves a cell by osmosis, causing the cell to shrink.

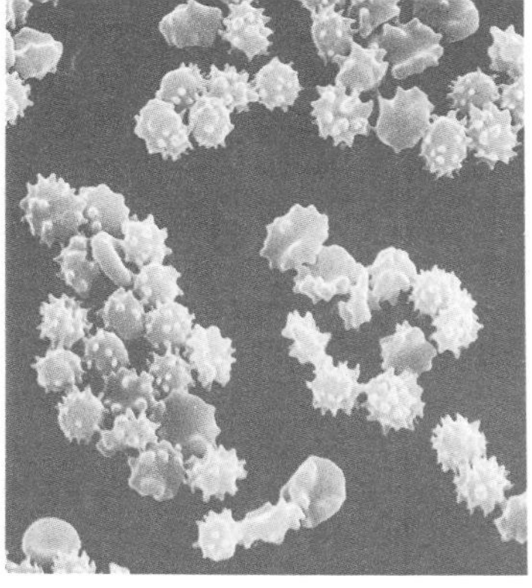

Animal cells like these red blood cells shrivel up as they lose water.

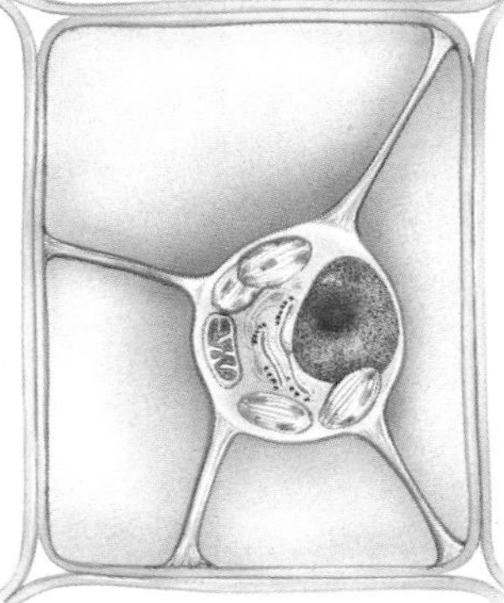

Plant cells lose turgor pressure as the plasma membrane shrinks away from the cell wall.

MiniLab

Purpose

LS **Visual-Spatial** Students will observe osmosis in an isotonic plant cell placed in a hypertonic solution.

Process Skills

observe and infer, experiment, analyze

Teaching Strategies

- Ask students to form a hypothesis as to what will happen when plant cells are placed in a hypertonic solution.

Expected Results

Students should observe plasmolysis of the *Elodea* cell.

Analysis

1. The cell membrane shrinks away from the cell wall.
2. Water left the cell by osmosis. As a result, the cell membrane shrank away from the cell wall.

Assessment

Performance: Have students demonstrate their focused plasmolyzed cell to you. Students should also design an experiment to return the cell to its original turgid stage (by adding distilled water to the edge of the slide, drawing out the salt water). If time permits, have them carry out this experiment. L1

before soaking (hard or soft).

5. Cover the beakers with plastic wrap or aluminum foil and allow them to sit undisturbed for two days.
6. On the second day, carefully remove the potato cubes one at a time and blot them dry on the outside. Weigh the pieces and record their masses. Observe any changes in the texture of each cube.

Analysis

Ask students to answer the following questions.

1. Describe what happened to the mass of each cube after soaking. *The mass of the potato placed in salt water decreased, while the one in plain water increased.*
2. Describe what happened to the texture of each cube after soaking. *The potato in the salt water became softer than that in the plain water.*
3. Explain the changes you observed in terms of osmosis. *Water in the potato placed in salt water left the potato by osmosis because of the high salt content of the water outside the potato.*

Assessment

Journal: Have students prepare a written laboratory report containing a data table and the answers to the analysis. L1

SECTION 9.2

Science, Technology, and Society

Ion Channels

Channel proteins form water-filled pores across the plasma membranes of cells. These channel proteins are very small and highly selective. Almost all of them are concerned with transport of specific ions and so are referred to as ion channels. More than 10^6 ions pass through such a channel each second, a rate more than 100 times faster than that for any known carrier protein.

Ion channels are not continuously open. Instead, they have "gates" that open briefly and then close again. Ion channels open under conditions such as a change in charge across the membrane, during mechanical stimulation, such as the stretching of the cells, or during the binding of a signaling molecule, such as a secretion from a nerve. Each channel may respond to one or more of these stimuli.

About 50 types of ion channels have been described and more are being discovered. The most common ion channels are those permeable mainly to potassium ions. Ion channels are responsible for the excitability of nerve and muscle cells and mediate most forms of electrical signaling in the nervous system. A single nerve cell, for example, typically contains more than five kinds of ion channels. Ion channels also propagate the leaf closing response of *Mimosa* and allow *Paramecium* to reverse direction after a collision.

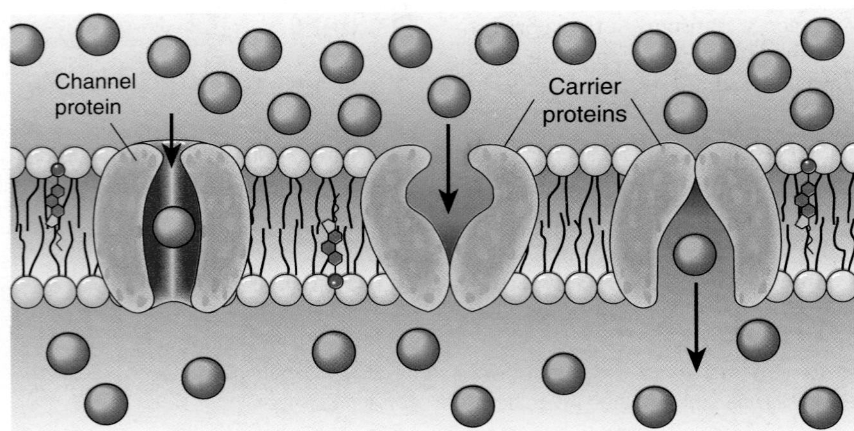

Figure 9.14
There are two main types of transport proteins: channel proteins and carrier proteins. Channel proteins (left) are tubelike and provide openings through which small, dissolved particles, especially ions, can diffuse. In general, each type of carrier protein (right) has a shape that fits a specific molecule or ion. When the proper molecule binds with the protein, it changes shape and moves the molecule across the membrane.

Passive Transport

Water, lipids, and lipid-soluble substances can pass through plasma membranes by diffusion. The cell does no work in moving the particles. This movement of particles across membranes by diffusion is called **passive transport** because the cell uses no energy to move the particles.

However, only a few substances are able to pass directly through a phospholipid bilayer. Particles of many other substances are not attracted to the lipid bilayer or are too large to pass through. Passive transport of other substances occurs in different ways.

Passive transport by proteins

Many kinds of proteins are embedded in the lipid bilayer of the plasma membrane like icebergs in the ocean. Some of these are **transport proteins,** which allow needed substances or waste materials to move through the plasma membrane. These proteins function in a variety of ways to transport molecules and ions across the membrane.

The passive transport of materials across the plasma membrane by means of transport proteins is called **facilitated diffusion.** As illustrated in *Figure 9.14,* the transport proteins provide convenient openings for particles to pass through. The process of facilitated diffusion is common in the movement of sugars and amino acids across membranes. Movement of materials by facilitated diffusion is the same as in any other diffusion because random motion of particles brings them into the transport proteins.

Active Transport

Can a cell ever move particles from a region of lesser concentration to a region of greater concentration? Yes, but to do so, the cell must expend energy to counteract the random motion of the particles that tends to make them diffuse in the opposite direction. For example, cells require nutrients, such as minerals, that are scarce in the environment.

Cells are adapted to move these nutrients from areas of lower concentration (the environment) to areas of higher concentration (inside the cell). Transport of materials against a concentration gradient requires energy and is called **active transport.**

STUDENT JOURNAL

Comparing Modes of Transport Have students create a table that lists and compares the modes of passive transport with those of active transport. Students should identify the kinds of materials transported by each mode. *For passive transport, tables should include diffusion through a bilayer (osmosis and diffusion of small molecules) and facilitated diffusion (channel transport proteins and carrier proteins). Active transport modes should include transport proteins, endocytosis, and exocytosis.* L1 LS

How active transport occurs

In active transport, a transport protein first binds with a particle of the substance to be transported. Chemical energy from the cell is then used to change the shape of the proteins so that the particle to be moved is released on the other side of the membrane. Once the particle is released, the protein's original shape is restored, as shown in *Figure 9.15*. Thus, a molecule or an ion can be moved into or out of a cell against a gradient.

Transport of large particles

Some cells can take in large molecules, groups of molecules, or even whole cells. **Endocytosis** is a process in which a cell surrounds and takes in material from its environment. This material does not pass directly through the membrane. Instead, it is engulfed and enclosed by a portion of the cell's plasma membrane. That portion of the membrane then breaks away, and the resulting vacuole with its contents moves to the inside of the cell.

ThinkingLab Draw a Conclusion

How does fertilizer affect earthworms in soil?

Earthworms play an important role in soil ecosystems. As they burrow through the soil, their tunnels allow air to enter the soil. At the same time, earthworms must respond to changes in the soil environment.

Analysis

A group of students placed 50 earthworms on a balance and recorded their mass. They then calculated the average mass of one worm. The students placed 25 of the earthworms into ordinary soil and 25 into soil that had been heavily fertilized. After one day, they removed the earthworms, and again determined the average mass of a worm. The data they obtained are shown in the following table.

Soil Conditions	Average Mass	
	Start	**End**
Unfertilized	2.4 g	2.5 g
Fertilized	2.4 g	1.8 g

Thinking Critically

Interpret the results shown in the table using what you have learned about osmosis.

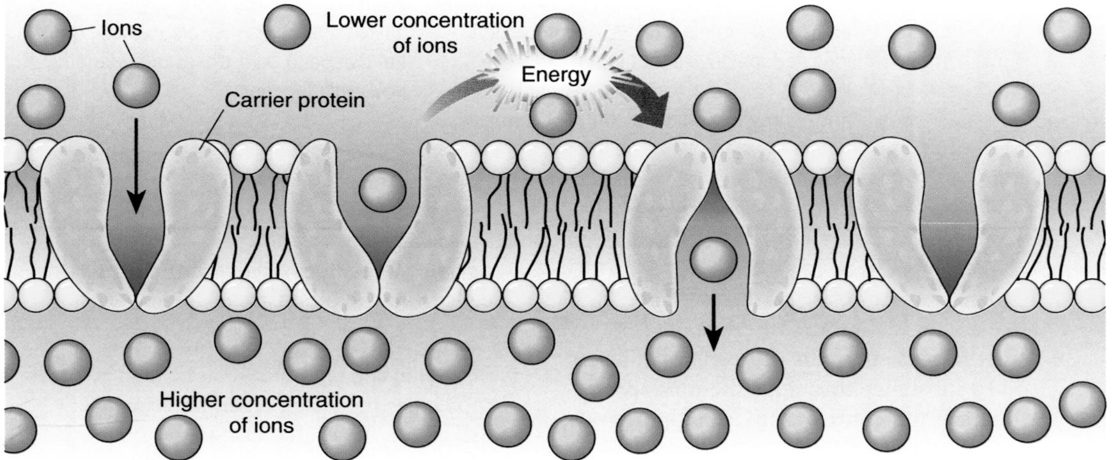

Figure 9.15

Some carrier protein molecules are used in active transport. They pick up ions or molecules from near the cell membrane, carry them across the membrane, and release them on the other side. Note that energy is needed for this process.

GLENCOE TECHNOLOGY

Videodisc

Biology: The Dynamics of Life
Disc 1, Side 1
Passive Transport (Ch. 24)

Disc 1, Side 1
Active Transport (Ch. 25)

ThinkingLab Draw a Conclusion

Purpose

LS **Logical-Mathematical** Students will exercise their knowledge about the movement of water through selectively permeable membranes.

Process Skills

recognize cause and effect, interpret data, analyze data

Teaching Strategies

- Point out that earthworms require moisture in order to acquire oxygen through their skin by diffusion.
- After students consider the effects of a hypertonic situation on earthworms, discuss the hypotonic situation earthworms encounter after a heavy rain.

Thinking Critically

The fertilized soil is hypertonic to the cells of the earthworm, so water leaves the cells.

Assessment

Knowledge: Make a worksheet showing cells in solutions with varying concentrations of solute. Label the percent of solute inside and outside cells. Have students use arrows to show the movement of water and label each drawing as hypotonic, hypertonic, or isotonic. L1

Assessment

Performance Assessment in the Biology Classroom: p. 7, Tasty Spuds. Have students carry out the activity to show their understanding of osmosis and diffusion. L1

3 Assess

Check for Understanding

Evaluate students' understanding of the differences between the terms in the following groups: *passive transport* versus *active transport; diffusion* versus *facilitated diffusion;* and among the terms *isotonic, hypotonic,* and *hypertonic.* Have students apply their knowledge of these terms by predicting the direction of movement between cells and solutions. L1

Reteach

Place several celery sticks in salt water and several in tap water. Ask students to describe any changes they observe in the celery. L1

Extension

Have students research intercellular structures such as desmosomes, gap junctions, tight junctions, and plasmodesmata. L3

Assessment

Journal: Place a recipe for making pickles on an overhead. Ask students to copy the recipe and explain where in the recipe osmosis occurs. L1

4 Close

Discussion

Explain to students that when cut carrot sticks begin to wilt they can be made firm again by placing them in water. Ask students to explain why.

Figure 9.16

Phagocytosis is a type of endocytosis in which large particles move into a cell. Some unicellular organisms, such as amoebas, ingest food by phagocytosis. Exocytosis can involve the release of wastes or cell products from a vacuole. The vacuole fuses with the cell membrane, and the contents are released to the outside.

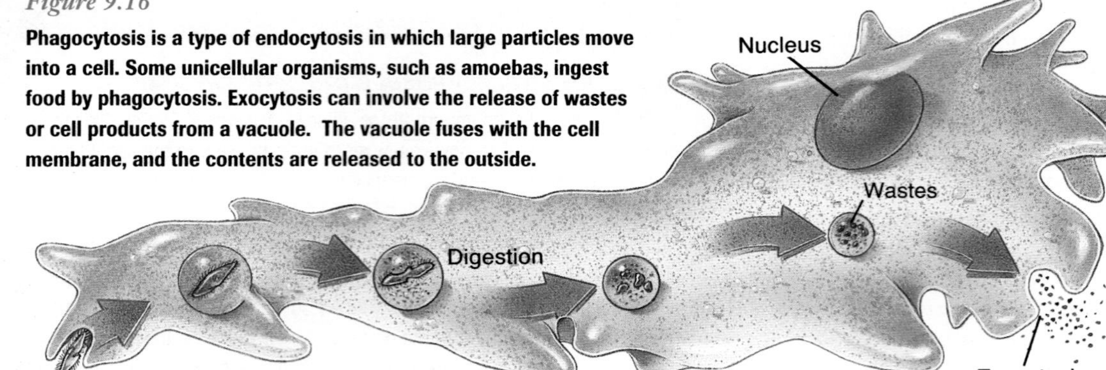

endo-, exo-:
endon (GK) within
exo (GK) out
Endocytosis moves materials into the cell; exocytosis moves materials out of the cell.

The reverse process of endocytosis is exocytosis, as shown in *Figure 9.16.* Cells use **exocytosis** to expel wastes, such as indigestible particles, from the interior to the exterior environment. They also use this method to secrete substances produced by the cell, such as hormones. Because endocytosis and exocytosis both move masses of material, they both require energy and are therefore classified as forms of active transport.

Connecting Ideas

The plasma membrane plays an important role in maintaining cell homeostasis. Its bilayer structure of phospholipids and embedded proteins allows it to regulate ions and molecules that enter and leave the cell. Passive transport does not require any energy from the cell, but active transport does. How does the cell get energy for this process and all other life processes? What is the source of the energy needed by all living things? Answers to these questions depend on further understanding of how cells function.

Section Review

Understanding Concepts

1. Explain how osmosis and diffusion are alike and how they are different.
2. What factors affect the diffusion of a dissolved substance through a membrane?
3. Compare and contrast active transport and facilitated diffusion.

Thinking Critically

4. Cells in the human body constantly use oxygen during metabolism. Blood contains oxygen. Why does oxygen diffuse into the cells from the bloodstream?

Skill Review

5. **Observing and Inferring** Kinetic energy of molecules increases as temperature increases. What effect do you think an increase in temperature has on diffusion? For more help, refer to Thinking Critically in the *Skill Handbook.*

Answers to Section Review

1. Both diffusion and osmosis involve the movement of materials from areas of higher to lower concentration. Osmosis is the diffusion of water molecules through a membrane.
2. The concentration of a substance on either side of the membrane; the size of the particles; polarity or ionic charge
3. Facilitated diffusion does not require an energy source; active transport depends on an energy source. Both facilitated diffusion and active transport use carrier proteins.

Thinking Critically

4. The oxygen content of the cell becomes depleted. Oxygen diffuses from a higher concentration (in the bloodstream) to a region of lower concentration (in the cell).

Chapter 9 Review

REVIEWING MAIN IDEAS

9.1 The Plasma Membrane

- The plasma membrane controls what enters and leaves a cell.
- The plasma membrane is a phospholipid bilayer with embedded proteins.

9.2 Cellular Transport

- Particles of matter are in constant motion and diffuse from areas of higher concentrations to areas of lower concentrations.
- Osmosis is the diffusion of water through a membrane.
- Facilitated diffusion is accomplished by transport proteins in the plasma membrane.
- Passive transport requires no energy from the cell.
- Active transport moves materials against a concentration gradient.
- Large particles may enter a cell by endocytosis and leave by exocytosis.

Key Terms

Write a sentence that shows your understanding of each of the following terms.

active transport	isotonic solution
contractile vacuole	osmosis
diffusion	passive transport
dynamic equilibrium	phospholipid
endocytosis	plasmolysis
exocytosis	selective permeability
facilitated diffusion	transport protein
fluid mosaic model	turgor pressure
hypertonic solution	
hypotonic solution	

Understanding Concepts

1. The ____________ is made up of a double layer of lipid molecules.
   a. nuclear wall
   b. plasma membrane
   c. cell wall
   d. cytoskeleton
2. Which of the following is NOT a mechanism of passive transport?
   a. direct diffusion
   b. diffusion through protein channels
   c. active transport
   d. transport by carrier proteins
3. Of these, which does NOT have a membrane with a fluid mosaic bilayer structure?
   a. nucleus   c. chloroplast
   b. mitochondria   d. cytoskeleton
4. The main structural component(s) of the plasma membrane is (are) ____________.
   a. carbohydrates   c. phospholipids
   b. cholesterol   d. proteins
5. What kinds of particles cross the lipid bilayer easily?
   a. large carbohydrates
   b. small sugars
   c. charged molecules
   d. small, uncharged molecules
6. ____________ is the diffusion of water molecules through a selectively permeable membrane.
   a. Passive transport
   b. Selective permeability
   c. Active transport
   d. Osmosis
7. When the concentration of solutes inside the cell is equal to the concentration outside the cell, the solution is said to be ____________.
   a. hypertonic   c. isotonic
   b. hypotonic   d. contractile
8. An amoeba ingests large food particles by ____________.
   a. osmosis   c. endocytosis
   b. diffusion   d. exocytosis

Chapter 9 Review

Reviewing Main Ideas

Summary statements can be used by students to review the major concepts of the chapter.

Key Terms

Answers should go beyond defining the terms. Accept any answer that uses the term correctly and in the proper context.

Understanding Concepts

1. b
2. c
3. d
4. c
5. d
6. d
7. c
8. c

Skill Review

5. An increase in temperature causes an increase in kinetic energy and in the rate of diffusion.

Chapter 9 Review

9. c
10. c
11. d
12. b
13. b
14. a
15. c
16. b
17. a
18. d
19. b
20. b

Applying Concepts

21. Cells that carry on a great deal of active transport would have more mitochondria to supply the necessary amounts of energy.
22. In order to excrete the excess salt, the body excretes more water than it takes in.
23. Some protein transporters carry materials out of the cell and some carry materials into the cell. The lipid bilayer itself allows diffusion of small molecules in either direction.
24. The amoeba would have to eliminate the excess water that enters by osmosis.
25. The celery loses turgor pressure as water moves from the cells into the salt water.

Chapter 9 Review

9. The type of diffusion that occurs through transport proteins is called ___________.
  a. dynamic movement
  b. osmotic movement
  c. facilitated diffusion
  d. active transport

10. ___________ is the net movement of particles from an area of higher concentration to an area of lower concentration.
  a. Active transport **c.** Diffusion
  b. Plasmolysis **d.** Exocytosis

11. In the plasma membrane, cholesterol ___________.
  a. allows insertion of proteins
  b. allows diffusion of nonpolar substances
  c. binds to carrier proteins
  d. makes the membrane more stable and less fluid

12. A ___________ membrane allows some materials to pass through while keeping others out.
  a. saturated
  b. selectively permeable
  c. trilayer
  d. homeostatic

13. A red blood cell placed into a 3 percent salt solution will ___________.
  a. swell, then burst
  b. shrink
  c. maintain its normal shape
  d. shrink, then swell back to normal

14. A membrane that is more fluid would contain more ___________.
  a. unsaturated fatty acids
  b. saturated fatty acids
  c. cholesterol
  d. protein

15. What is responsible for Brownian motion?
  a. fluid movement of the phospholipid molecules
  b. the plasma membrane surrounds a food particle
  c. the random movement of molecules
  d. movement of molecules against the concentration gradient

16. Which of the following is NOT a function of membrane proteins?
  a. act as membrane carrier molecules
  b. energy storage for the cell
  c. act as markers for recognition
  d. form channel molecules

17. What is the eventual result of diffusion?
  a. dynamic equilibrium
  b. the cell gains energy
  c. a gradient
  d. osmosis of water

18. The pressure that exists inside a cell when it swells is ___________ pressure.
  a. osmotic **c.** diffusion
  b. isotonic **d.** turgor

19. Loss of water from a cell that results in a drop in turgor pressure is ___________.
  a. exocytosis
  b. plasmolysis
  c. active transport
  d. passive transport

20. Osmosis does not occur when a cell is placed into a(n) ___________ solution.
  a. hypotonic **c.** hypertonic
  b. isotonic **d.** facilitated

Applying Concepts

21. How would you expect the number of mitochondria in a cell to be related to the amount of active transport it carries out?

22. Explain why drinking ocean water to quench thirst would be dangerous to humans. (Hint: The body must excrete salt as a water solution.)

23. How does the structure of the plasma membrane allow materials to move across it in both directions?

24. When an amoeba from the ocean is placed in fresh water, it forms a contractile vacuole. Explain how this ability is an adaptation for survival.

25. Describe the process by which a stalk of celery becomes limp and rubbery when placed in salt water.

Thinking Critically

26. ***Concept Mapping*** Make a concept map that relates the following terms and phrases. Supply the appropriate linking words for your map.
    diffusion, active transport, passive transport, facilitated diffusion, water, dissolved ions and polar molecules, osmosis, transport protein

27. ***Comparing and Contrasting*** Study the graph and describe how the rate of diffusion increases as the concentration of the diffusing substance increases for both simple diffusion and facilitated diffusion. Explain why the two are different.

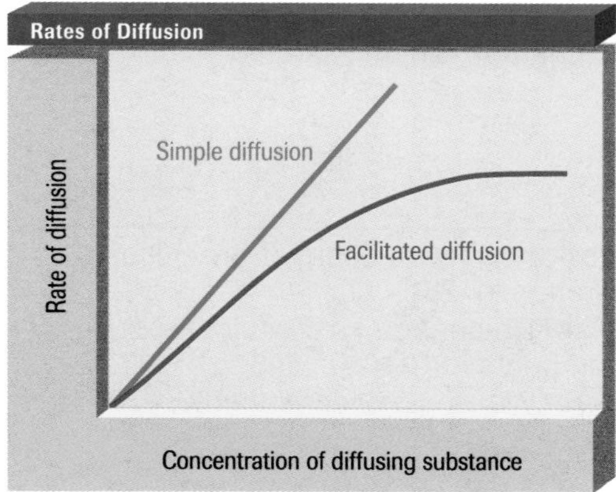

28. ***Formulating Models*** When phospholipid molecules are placed in water, they form bilayer membranes. Describe what you think happens to cause the formation of bilayer membranes.

29. ***Recognizing Cause and Effect*** You place a bag of starch into a beaker of iodine, and the bag is permeable to both iodine and starch. Describe what would happen after 24 hours.

30. ***Making Predictions*** When a freshwater paramecium is placed in salt water, what do you think will happen to the rate at which the contractile vacuole pumps?

ASSESSING KNOWLEDGE & SKILLS

Five potato cubes of equal size and weight were placed in five different concentrations of salt water. The potato cubes were weighed again after they sat in salt solutions for 30 minutes, and the following graph was made.

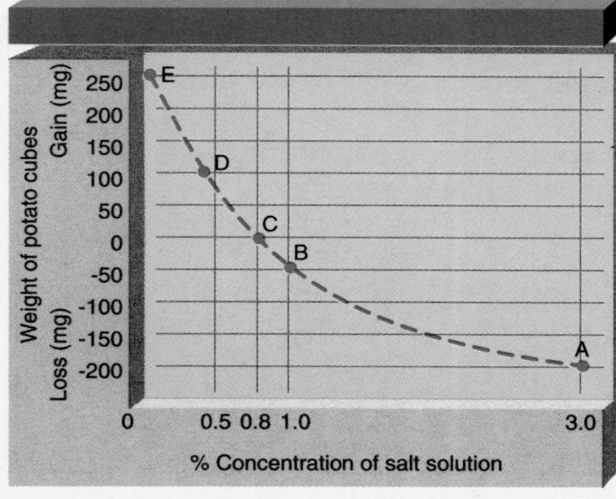

Interpreting Data Use the graph to answer the following questions.

1. Which of the variables is the independent variable?
   a. weight of the potatoes
   b. size of the potatoes
   c. concentrations of the salt water
   d. weight loss of each of the potatoes
2. Potato cubes A and B were below the original weight because ____________.
   a. there was a net flow of water from the potatoes
   b. salt left the potatoes
   c. cubes A and B weighed more than cubes C and D to begin with
   d. there was a net flow in both directions, resulting in the differences in weight
3. ***Observing and Inferring*** A produce manager constantly sprays the vegetables in the produce section. Why?

Program Resources

Chapter Assessment, pp. 49-54 L1
Alternate Assessment in the Science Classroom
Computer Test Bank L1
Content Mastery, pp. 33-36 L1

Chapter 9 Review

Thinking Critically

26. Evaluate students' concept maps. The maps should show an understanding of the relationships among the concepts listed.
27. The rate of simple diffusion increases steadily as the concentration of diffusing molecules increases. The rate of facilitated diffusion depends on the availability of additional transport molecules specific for the diffusing substance. When all the protein molecules are in use, the rate levels off and does not increase further.
28. Because the phospholipids have a polar head and a nonpolar tail, the nonpolar tails orient themselves toward each other and away from water, forming a bilayer.
29. Because both iodine and starch could move, the solution inside and outside the bag would be black due to the combination of iodine with starch.
30. Because water would now leave the paramecium cell, the contractile vacuole pump rate would decrease greatly and the pump would probably stop.

ASSESSING KNOWLEDGE & SKILLS

1. c
2. a
3. By spraying them with water, the vegetables are put in a hypotonic solution. Water will continuously move into the vegetables, giving them turgor pressure (so they remain crispy).

Chapter Organizer

SECTION	OBJECTIVES	ACTIVITIES/FEATURES
10.1 ATP: Energy in a Molecule National Science Standards: UCP.2, UCP.3; B.3, B.6; C.1, C.5	1. **Explain** why organisms need a supply of energy. 2. **Explain** how energy is stored in ATP and released from ATP.	
10.2 Photosynthesis: Trapping Energy National Science Standards: UCP.2, UCP.3; A.1, A.2; B.3, B.6; C.1, C.5;	3. **Relate** chlorophyll to the process of photosynthesis. 4. **Explain** how the light reactions and Calvin cycle are related.	**Thinking Lab:** How does photosynthesis vary with light intensity?, p. 244 **Biolab:** Design Your Own Experiment—What factors influence photosynthesis?, p. 246 **Biology & Society Technology:** Collecting the Sun, p. 248
10.3 Getting Energy to Make ATP National Science Standards: UCP.1 -3; B.3, B.6; C.1, C.5; E.1, E.2; F.6; G.1	5. **Compare and contrast** aerobic and anaerobic processes. 6. **Explain** how cells obtain energy from respiration.	**Math Connection:** Energy from Cellular Respiration, p. 251 **Minilab:** Do aquariums contain measurable amounts of CO_2?, p. 253 **Minilab:** Will apple juice ferment?, p. 255

ACTIVITY MATERIALS		
BIOLAB	**MINILABS**	**ALTERNATE LAB**
page 246 250-mL Erlenmeyer flask 3 sprigs of *Elodea* string washers 1000-mL beaker hot plate ice oven mitts thermometer colored cellophane, assorted colors lamp with reflector & 150-watt bulb 0.25% baking soda solution ring stand with clamp watch with second hand metric ruler	**page 253** aquarium water 100-mL beaker eyedropper bromothymol blue **page 255** test tube beaker plastic pipet apple juice metal washers	**page 252** black paper labels iodine solution *Coleus* plants paper clip hot plate beaker 95% alcohol finger bowl

Chapter 10 Energy in a Cell

TEACHER CLASSROOM RESOURCES

Reproducible Masters	Transparencies
Section Focus Master 20: Using Energy L1 SAE **Reinforcement and Study Guide,** p. 37 L1 **Content Mastery,** pp. 37-40 L1	**Basic Concepts Transparency #10:** ATP-ADP Cycle L1 SAE **Reteaching Transparency #10:** Electron Transport Chain L1 SAE
Section Focus Master 21: Photosynthesis L1 SAE **Reinforcement and Study Guide,** pp. 38-39 L1 **Biolab and Minilab Worksheets,** pp. 37-38 L1 **Critical Thinking/Problem Solving:** Two Factors Affecting Photosynthesis, p. 10 L3 **Laboratory Manual:** Chloroplast Pigment Analysis, pp. 57-60 L2 **Concept Mapping:** Photosynthesis: Trapping Energy, p. 10 L1	**Basic Concepts Transparency #11:** Photosynthesis L1 SAE **Basic Skills Transparency #9:** Light and Plant Growth L1 SAE
Section Focus Master 22: Respiration L1 SAE **Reinforcement and Study Guide,** p. 40 L1 **Biolab and Minilab Worksheets,** pp. 35-36 L1 **Laboratory Manual:** How Does Concentration of Sugar Affect Fermentation?, pp. 61-64 L2 **Tech Prep Applications:** Bioluminescence and Behavior, pp. 13-14 L2	**Basic Concepts Transparency #12:** Cellular Respiration L1 SAE **Basic Skills Transparency #10:** Photosynthesis and Respiration L1 SAE

ASSESSMENT MATERIALS

Chapter Assessment, pp. 55-60
Performance Assessment in the Biology Classroom, p. 11
MindJogger Videoquiz
Alternate Assessment in the Science Classroom
Computer Test Bank

Spanish Resources SAE
English/Spanish Audiocassettes SAE
Cooperative Learning in the Science Classroom COOP LEARN
Lesson Plans

KEY TO TEACHING STRATEGIES

- L1 Level 1 activities should be within the ability range of all students including those with learning difficulties.
- L2 Level 2 activities are within the ability range of average to above-average students.
- L3 Level 3 activities are designed for the ability range of above-average students.
- SAE SAE activities should be within the ability range of Students Acquiring English.
- COOP LEARN Cooperative Learning activities are designed for small group work.
- P These strategies represent student products that can be placed into a best-work portfolio.
- These strategies are useful in a block scheduling format.

GLENCOE TECHNOLOGY

The following multimedia resources are available from Glencoe.

Biology: The Dynamics of Life
CD-ROM SAE
Videodisc Program

The Infinite Voyage Series
The Champion Within

National Geographic Society Series
STV: Human Body Volume 1
Plants
Newton's Apple: Life Sciences

Science and Technology Videodisc Series (STVS)
Plants & Simple Organisms
Farming Indoors
Chemistry
Solar House
Spiral Solar Concentrator

CHAPTER

10 Energy in a Cell

Chapter Overview

As students start their study of energy in cells, they learn about the structure of ATP, its formation, and its breakdown into ADP and a phosphate group as energy is released. The concept of ATP as energy currency is presented. Students also survey how cells use energy.

In the next section of the chapter, students become familiar with the reactions of photosynthesis. These reactions include the light reactions and the Calvin cycle. Students then consider one process of chemosynthesis carried out by autotrophs.

The final section of the chapter presents the reactions of respiration, including glycolysis, aerobic respiration, alcoholic fermentation, and lactic acid fermentation. Students compare aerobic respiration to photosynthesis. They conclude their study with an examination of the ways in which aerobic respiration and photosynthesis are similar as well as the ways in which they are opposites.

Key Terms

ADP
aerobic process
alcoholic fermentation
anaerobic process
ATP
Calvin cycle
chemosynthesis
citric acid cycle
electron transport chain
glycolysis
lactic acid fermentation
light reactions
photolysis
photosynthesis
respiration

236

Learning Styles

LS

Look for the following logo for strategies that emphasize different learning modalities. LS

Kinesthetic	**Demonstration, p. 239; Portfolio, p. 239**
Visual-Spatial	**Demonstration, pp. 238, 242, 254; Visual Learning, p. 240; MiniLab, p. 255**
Interpersonal	**Portfolio, p. 244; Project, p. 248**
Linguistic	**Meeting Individual Needs, p. 243; Student Journal, p. 254; Portfolio, p. 255**
Logical-Mathematical	**Student Journal, p. 240; Thinking Lab, p. 244; Project, p. 245; Alternate Lab, p. 252; MiniLab, p. 253**

CHAPTER

10 Energy in a Cell

Marathon runners participate in a strenuous race of 26.2 miles, pounding their feet on the pavement continuously for more than two hours. As some runners cross the finish line, they collapse onto the ground, gasping for breath. Where do these runners get energy for their muscles to finish such a race? Perhaps you have heard of carbohydrate loading in preparation for a marathon. Why would a runner wish to "carbo-load" by eating platefuls of carbohydrates such as pasta before a race? Why do carbohydrates represent energy to the marathoner?

Life on Earth—whether it is a marathon runner, a flashing firefly, or a growing tomato plant—depends on a flow of energy. This flow begins some 93 million miles away with the sun. Each day, the sun delivers to Earth an amount of energy that is 1.5 billion times the amount of electrical energy generated in the United States each year. A small fraction of this solar energy—less than one percent—powers the activities performed by the cells of Earth's living things.

Concept Check

You may wish to review the following concepts before studying this chapter.

- Chapter 3: autotroph, heterotroph
- Chapter 7: carbohydrate
- Chapter 8: mitochondria, chloroplast

Chapter Preview

Laboratory Activities

Biolab: Design Your Own Experiment
- What factors influence photosynthesis?

Minilabs
- Do aquariums contain measurable amounts of CO_2?
- Will apple juice ferment?

This triathlete required an incredible amount of energy to finish the race. The muscle cells of the runner obtain much of their energy from the carbohydrates in the food eaten earlier. Considering that the sun is the ultimate source of energy on Earth, how did its energy become stored in carbohydrates?

Assessment Planner

Choose assessment strategies from the following pages to evaluate the progress of your students.

Assess, pp. 241, 249, 255
Alternate Lab, pp. 252-253
Minilabs, pp. 253, 255
Portfolio, pp. 239, 244, 255
Thinking Lab, p. 244
Biolab, pp. 246-247
Chapter Review, pp. 257-259

Introducing the Chapter

Ask students to look at the photo of the runner and identify the changes felt in the body as a person begins to run. *The runner begins to breathe harder and faster, and the heart beats faster.* Discuss how, as a result of these changes, the body takes in oxygen at a faster rate. Relate this faster heart rate to an increase in the rate at which oxygen is delivered to the muscles of the body.

Have students suggest answers to the questions at the end of the opening paragraph. *Those who are runners may be able to explain how loading up on complex carbohydrates provides a source of energy that can be drawn upon during the race.*

Theme Development

The main theme of this chapter is *energy.* ATP is presented as the energy storage molecule and the most prevalent source of energy for cells. Both the synthesis and breakdown of ATP are covered. An underlying theme is that there is cellular *unity within diversity.* All cells carry out respiration and many cells also carry out photosynthesis.

Concept Check

Students will build on their knowledge, from Chapter 3, of the ability of autotrophs to make their own food, and they will consider the processes of photosynthesis and chemosynthesis. They will also connect the material in this chapter with their study of carbohydrates, mitochondria, and chloroplasts in Chapters 7 and 8.

Prepare

Key Concepts

Students will examine the source of cellular energy—the ATP molecule. They will also learn about the processes in which cells use the energy stored in ATP.

Block Scheduling

Look for this symbol for strategies that are useful in a block scheduling format. For more information on block scheduling, refer to Section 10.1 in the **Lesson Plans** booklet.

Materials

- Obtain a luciferase kit, a potato, peanuts, a dissection needle, a cork, and a Bunsen burner for the Demonstrations.
- Gather construction paper to make models for Reteach.

1 Focus

Bellringer

Before presenting the lesson, display **Section Focus Master 20** on the overhead projector and have students answer the accompanying questions. L1 SAE

Demonstration

LS **Visual-Spatial** Using a kit containing luciferase and ATP, show that ATP may be used to convert chemical energy to light energy. Explain to students that this reaction is similar to that which causes the characteristic light of glowworms and fireflies.

SECTION

10.1 ATP: Energy in a Molecule

Section Preview

Objectives

Explain why organisms need a supply of energy.

Explain how energy is stored in ATP and released from ATP.

Key Terms

ATP
ADP

You wouldn't think of trying to make a car run by dropping a lighted match into the gasoline tank. So much energy would be released at once that the car would explode. Instead, gas must be delivered to the engine in small, measured amounts so that the energy is released from the fuel in a useful, controlled way. Likewise, living organisms have mechanisms that release energy from food molecules in a controlled way.

Cell Energy

Work is done whenever anything is moved. In animals, muscle cells contract in order to move the body and pump blood. In all cells, life processes constantly move and rearrange atoms, ions, and molecules. All this biological work demands energy. Energy is the ability to do work, so organisms and their cells need a steady supply of energy in order to function. Just consider the amount of energy the leafcutter ant in *Figure 10.1* needs.

Energy for biological work

Your muscle cells need energy to contract as you walk, bend, and exercise. Energy is used to move ions and molecules across membranes in your nerves and kidneys. Your cells use energy even when you sleep. Your heart muscle continues to pump blood. Your brain is actively processing the events of your day. Where do your cells get the energy they need to do this biological work? They get it from the food you eat.

Figure 10.1

Energy can take different forms, such as chemical, mechanical, or radiant energy. This leafcutter ant has transformed chemical energy from food into mechanical energy by moving the large piece of leaf.

Program Resources

Section Focus Master 20 L1 SAE
Reinforcement and Study Guide p. 37 L1
Basic Concepts Transparency 10 and **Master** L1 SAE
Reteaching Transparency 10 and **Master** L1 SAE

Releasing food energy

Although your diet may vary from spaghetti to spinach, almost all the food you eat is broken down by your body into smaller and smaller bits. By the time food reaches your bloodstream, it has been broken down into nutrient molecules that can enter your cells. Cell reactions then break down the food molecules, releasing energy for the biological work your cells need to perform.

ATP as energy currency

A cell can't use all the energy available in food molecules at once. Consider the following analogy. You go to the bank to withdraw $100, and the teller gives you a $100 bill. You stop on the way home to buy a drink from a vending machine, but you can't stuff the $100 bill in the slot. Smaller bills and coins are much more convenient for such purchases.

In the same way, cell processes need "smaller change" for their energy expenses. A mechanism has evolved that balances energy supply and demand. The cell "makes change" by breaking down food molecules bit by bit and distributing the energy to other molecules, as shown in *Figure 10.2.* These energy-storing molecules are adenosine triphosphate, or **ATP** for short. ATP can be called the cell's energy currency because it is from ATP molecules that the cell gets energy for its work. In fact, without a constant and abundant supply of ATP, a cell will die.

Forming and Breaking Down ATP

Cells store energy by bonding a third phosphate group to adenosine diphosphate, **ADP,** to form ATP. When the phosphate-phosphate bonds form, energy is stored. ADP contains adenine, a sugar called ribose, and two phosphate groups. ATP molecules are made of adenine, ribose, and three phosphate groups bonded in a row. When the third phosphate group breaks off an ATP

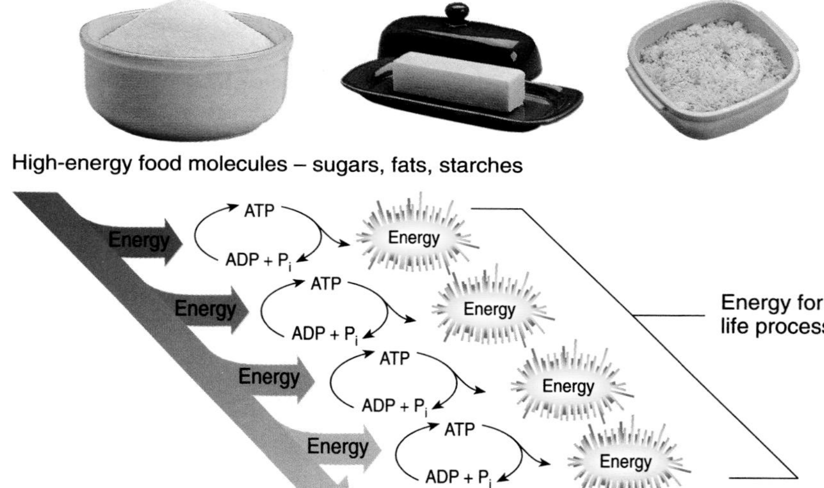

Figure 10.2

Food molecules are broken down in several steps. At each step, the energy released is stored in ATP, the cell's energy currency. This energy can be released from ATP for use in cell work.

SECTION 10.1

2 Teach

Enrichment

Provide students with additional information about the amount of energy that can be gained from each bond in ATP. Explain that the two phosphate-phosphate bonds release about 7.3 kcal each. The final phosphate-ribose bond releases much less.

Misconception

Many students believe that energy is derived directly from food. Explain that the energy produced by chemical reactions is stored in the bonds of ATP, which is then broken down to release energy. Many students think that energy is created or "produced." Introduce the law of conservation of energy, which states that energy is neither gained nor lost from a system but is converted from one form to another.

Demonstration

LS **Kinesthetic** To demonstrate the transfer of energy in small amounts as shown in Figure 10.2, heat a potato in boiling water or in a microwave oven in the school cafeteria. Have students pass the heated potato to one another. Ask students who first handle the potato to describe its temperature using the terms *hot, warm, cool,* or *cold.* Have a student near the end make the same observation. *At the end, the potato is cooler, while a lot of hands are warmer.* L1

Using an Analogy

Develop the analogy of cellular "currency." Emphasize that even plants cannot use the energy of the sun directly but must first use it to make ATP.

Students will understand the analogy of ATP to using money for exchange of goods and services. ATP is the energy exchange medium for cells.

PORTFOLIO

Modeling Chemical Compounds Provide students with toothpicks and Styrofoam balls of different sizes and colors. Have students use the materials to make models of ATP and ADP. Have students include in their portfolios a description of how the molecules differ from each other and how these differences relate to energy. L2 P LS

Meeting Individual Needs

Hearing Impaired Give hearing impaired students a photograph of an organism from a magazine or similar source and ask them to list the activities of the organism that will require energy from ATP. L1

Visual Learning

Figure 10.3 Direct students' attention to the diagram of the ATP-ADP cycle. Ask students where the energy comes from as ATP breaks down to ADP. *The energy released was stored in the chemical bond that is broken down to form ADP and a phosphate group.* LS

Different Viewpoints in Biology

Energy from the Deep
For many years most scientists thought that the source of energy for all communities was the sun, whereas some scientists thought other energy sources were possible. In 1986, a team of oceanographers from the Marine Science Institute at the University of California at Santa Barbara confirmed that thriving sea vent communities rely on energy sources other than the sun. These organisms obtain nutrients from bacteria that use the energy from chemicals in the hot, mineral-rich water that rises from cracks in the ocean floor.

GLENCOE TECHNOLOGY

Videodisc

STVS: Chemistry
Disc 2, Side 2
Solar House (Ch. 10)

Spiral Solar Concentrator (Ch. 12)

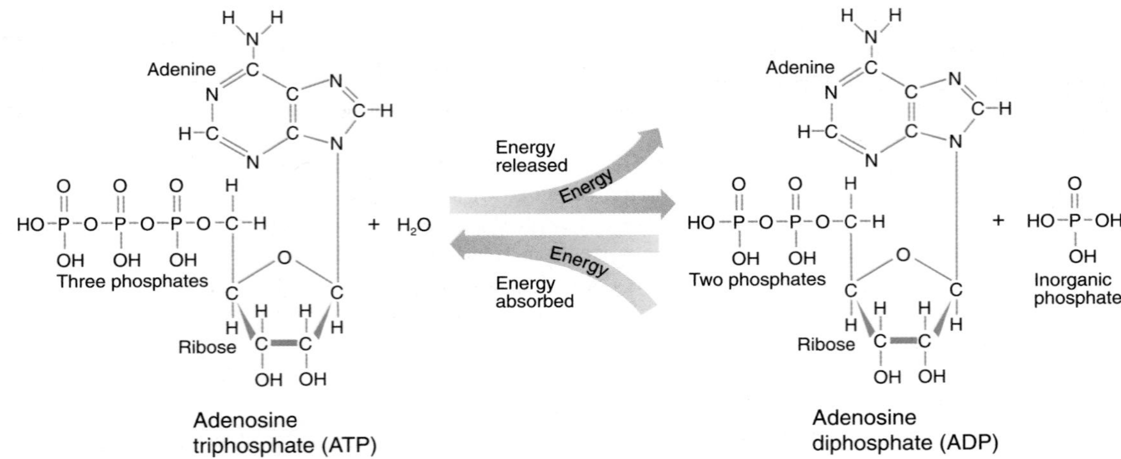

Figure 10.3
Energy is stored when ATP is made from ADP and phosphate. Energy is released when ATP breaks down to ADP and phosphate. It is estimated that in a resting adult human, this ATP ⇆ ADP cycle processes more than 40 kg of ATP each day.

molecule, the result is ADP, a phosphate group, and the release of energy, *Figure 10.3.* ADP is available again to store energy by forming ATP. Within a cell, formation of ATP from ADP and phosphate occurs over and over, storing energy each time. As the cell uses energy, ATP breaks down repeatedly to release energy, ADP, and phosphate. The conversion reactions of ADP and ATP are shown by the following equation and in *Figure 10.3.* The symbol P_i represents an inorganic phosphate group. As you can see, the formation and breakdown form a cycle.

$$\underset{\text{Lower energy}}{ADP + P_i} + \text{energy} \leftrightharpoons \underset{\text{Higher energy}}{ATP}$$

ATP links energy use and energy release

Biological work, as you know, takes energy. Energy that is available to do work is called free energy. Energy that is locked up in molecules is unavailable for work; it is potential energy. Unlocking this potential energy to produce free energy is vital to a cell.

Consider the reaction shown in *Figure 10.4,* in which a plant combines glucose and fructose to make the disaccharide sucrose. Making sucrose requires energy to form the bond between glucose and fructose. As a result, the product, sucrose, has more potential energy than the reactants. This energy comes from the breakdown of some of the plant's supply of ATP.

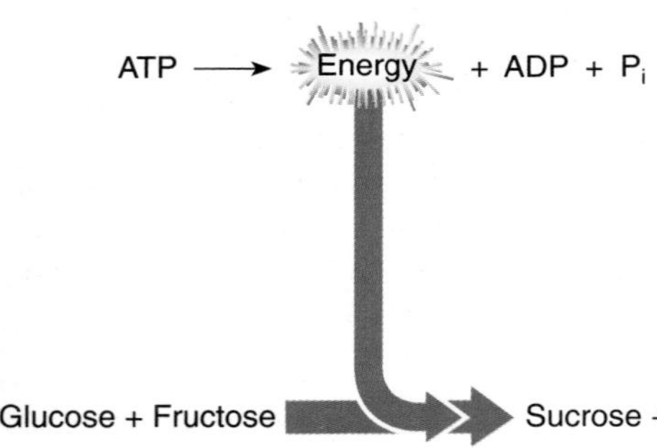

Figure 10.4
As the cells of a plant such as this holly tree make sucrose from glucose and fructose, energy must be released from the breakdown of ATP to ADP and inorganic phosphate.

STUDENT JOURNAL

Surviving Without Bread or Water Have students conduct research to find out how the amount of time a person can survive without food compares to the amount of time a person can survive without water. Have students report on their findings and evaluate why people cannot survive without water as long as they can without food. L1 LS

ATP seems to have evolved as the main energy link between energy-using and energy-releasing reactions. That's because the amount of free energy released when it breaks down is suitable for use in most cellular reactions.

The Uses of Cell Energy

Making sucrose is an example of one way in which cells use energy—making new molecules. Some of these molecules are enzymes, which carry out cell reactions. Others build membranes and cell organelles. In what other ways do cells use energy that is available from the breakdown of ATP to ADP?

Cells use energy to maintain homeostasis. Your kidneys—and those of other animals—use energy to move molecules and ions in order to eliminate waste substances while keeping needed substances in the bloodstream. *Figure 10.5* shows other ways that cells use energy.

Figure 10.5

Some Ways that Cells Use Energy

Warm-blooded animals make use of the heat given off in cell reactions to maintain a warm body temperature. With mechanisms of homeostasis to control body temperature, warm-blooded animals such as this polar bear can be active in many different climates.

Nerve cells are able to transmit impulses by using ATP to power the active transport of certain ions.

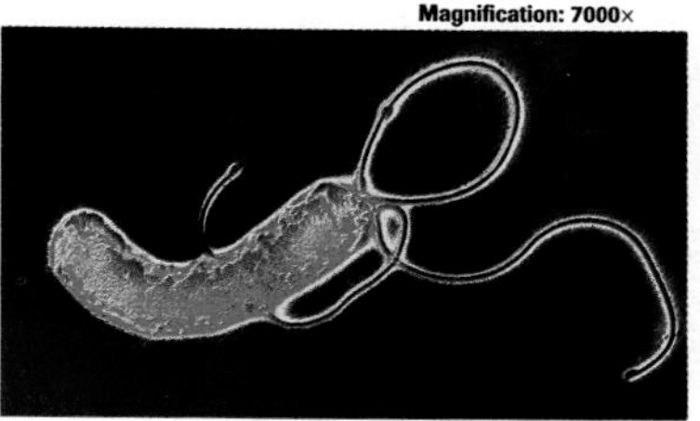

Some cells such as muscle cells and cells with cilia or flagella (right) use energy from ATP in order to move.

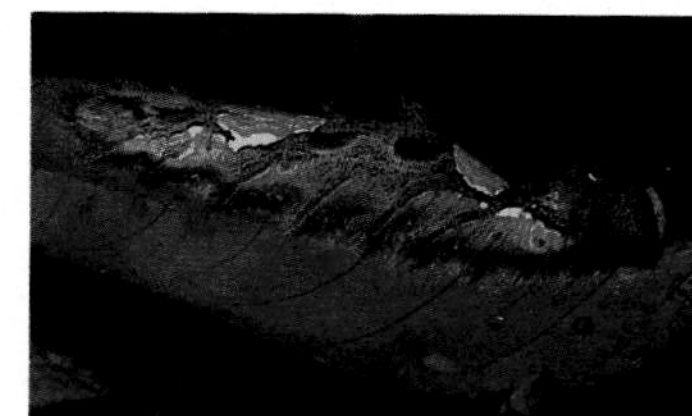

Fireflies, many deep-sea animals, and some caterpillars such as the one shown here produce light by a process called bioluminescence. The light results from a chemical reaction that is powered by the breakdown of ATP.

Section Review

Understanding Concepts

1. How does ATP store energy?
2. Why must there be a mechanism for storing energy that is given off during the breakdown of food molecules?
3. Suppose a cell combines molecules A and B to produce molecule C in an energy-requiring reaction. Explain how ATP may be used in this type of reaction.

Thinking Critically

4. Make a list of possible ways that an eagle might use energy from the breakdown of ATP.

Skill Review

5. **Observing and Inferring** When animals shiver in the cold, muscles move almost uncontrollably. Suggest how shivering helps an animal survive in the cold. For more help, refer to Thinking Critically in the *Skill Handbook*.

Answers to Section Review

1. ATP stores energy in its phosphate-phosphate bonds.
2. Energy must be stored so it can be released slowly to the cell, rather than in large amounts at one time.
3. ATP breaks down to ADP + P_i + energy to form the bond between A and B.

Thinking Critically

4. Active transport, muscle movement, cell movement, digestion, heat generation, nerve transmission, and synthesis are all cellular processes that require the energy of ATP.

Skill Review

5. When muscles move during shivering, heat is generated. This heat helps to warm the animal.

3 Assess

Check for Understanding

To evaluate whether students understand why energy must be stored in small amounts, ask them why it is easier to pay for a small purchase with a $1 bill than with a $50 bill. *The value of the $1 bill is near the value of the purchase and therefore more convenient.* L1

Reteach

Have students build a model of ATP using construction paper. Have them tape the adenosine molecule to a sheet of paper and add phosphates to make adenosine monophosphate, ADP, and finally ATP. Students should see that by removing a phosphate and breaking a phosphate-phosphate bond, energy becomes available for use. L1

Extension

Have interested students find further information about bioluminescence. Ask students to write a report about a bioluminescent organism. L1

Assessment

Knowledge: Ask students to summarize how each organism or cell part in Figure 10.5 uses energy. L1

4 Close

Demonstration

Burn a peanut to demonstrate that energy is stored in food. Impale a peanut on the end of a dissection needle. Stick the other end of the needle into a cork. Ignite the peanut with a Bunsen burner. Elicit from students an explanation of the form of energy released during the burning. *Energy that was stored in the peanut is released as light and heat when it burns.*

SECTION

10.2

Prepare

Key Concepts

Students will relate the structure of the chloroplast to the process of photosynthesis. They will also explore the overall reactions that occur in the light reactions and the Calvin cycle.

Block Scheduling

Look for this symbol for strategies that are useful in a block scheduling format. For more information on block scheduling, refer to Section 10.2 in the **Lesson Plans** booklet.

Materials

- Acquire a prism to use in the Demonstration.
- Acquire colored lights for the plant growth project.
- Purchase *Elodea* for the Biolab.
- Collect leaves for the leaf pigment project.

1 Focus

Bellringer

Before presenting the lesson, display **Section Focus Master 21** on the overhead projector and have students answer the accompanying questions. L1 SAE

Demonstration

LS **Visual-Spatial** Use a prism to show how visible light can be separated into a spectrum. Students may recall from their study of physical science that the different colors of the visible spectrum represent different wavelengths of light.

SECTION 10.2

Photosynthesis: Trapping Energy

Section Preview

Objectives

Relate chlorophyll to the process of photosynthesis.

Explain how the light reactions and Calvin cycle are related.

Key Terms

photosynthesis
light reactions
Calvin cycle
electron transport chain
photolysis
chemosynthesis

You say you're hungry? Well, how about a nice hot, steaming plate of—carbohydrates? Mmm. That doesn't sound appetizing? Well think again. You may be turning down a deep-dish pizza, a seafood pasta plate, or a Danish pastry. All those foods contain carbohydrates. Not only can carbohydrates taste delicious, but they are also vital to your health.

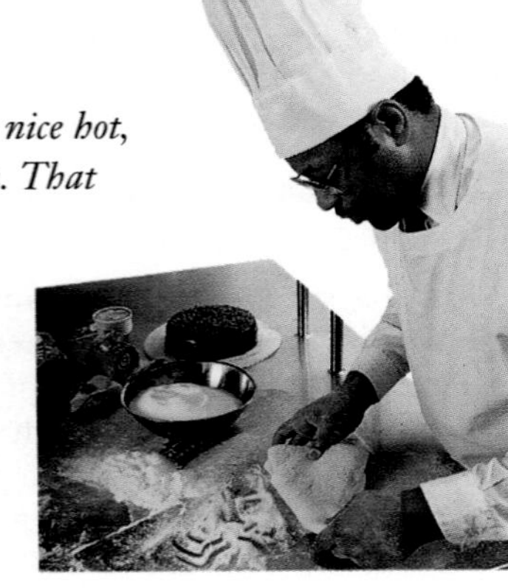

Autotrophs and Sunlight

Your cells break down carbohydrates to release energy and form ATP. Of course, a chef doesn't make these carbohydrates, so where do they come from? Think about autotrophs, organisms that make their own food. Autotrophs, including green plants, algae, and some bacteria, are found in nearly every ecosystem. These organisms, too, must break down carbohydrates to form ATP. These carbohydrates are usually in the form of simple sugars, especially glucose. However, autotrophs don't take in these sugars as food; they make the sugars themselves.

photosynthesis: *photos* (GK) light *syntithenai* (GK) to put together Photosynthesis puts together sugar molecules using energy from light.

Sunlight and chlorophyll

How do autotrophs make sugars? A major ingredient in their recipe is light. Autotrophs trap energy from sunlight and use this trapped energy to build carbohydrates in a process called **photosynthesis.** Sunlight, then, is the natural energy source for photosynthesis.

You may already know that white light from the sun consists of a mixture of colors ranging from red through orange, yellow, green, blue, and indigo to violet. Objects that appear colored, *Figure 10.6,* do so because they reflect some of the colors of white while absorbing other colors.

The green energy trap

What color are almost all the plants you see? Green, of course. Most plants are green because they contain chlorophyll. Chlorophyll reflects green and some yellow light and absorbs the energy of other colors of sunlight. It is this energy that is stored in the carbohydrates produced by photosynthesis.

Chlorophyll is found in the chloroplasts of green plants and algae and on membranes in the cytoplasm of photosynthetic bacteria. A chloroplast is surrounded by two lipid bilayer membranes. Internally, the chloroplast has a series of membranes called thylakoid membranes. It is within these membranes that the energy from sunlight is trapped by chlorophyll.

Program Resources

Section Focus Master 21 L1 SAE
Reinforcement and Study Guide, pp. 38-39 L1
Biolab and Minilab Worksheets, pp. 37-38 L1
Laboratory Manual, pp. 57-60 L2
Concept Mapping, p. 10 L1
Critical Thinking/Problem Solving, p. 10 L3
Basic Concepts Transparency 11 and **Master** L1 SAE
Basic Skills Transparency 9 and **Master** L1 SAE

Figure 10.6

As white light from the sun passes through a prism, it is separated into the colors of the visible spectrum. As white light strikes a green leaf, most of the colors except green and yellow are absorbed. Green and yellow are reflected by the leaf and are seen by the viewer.

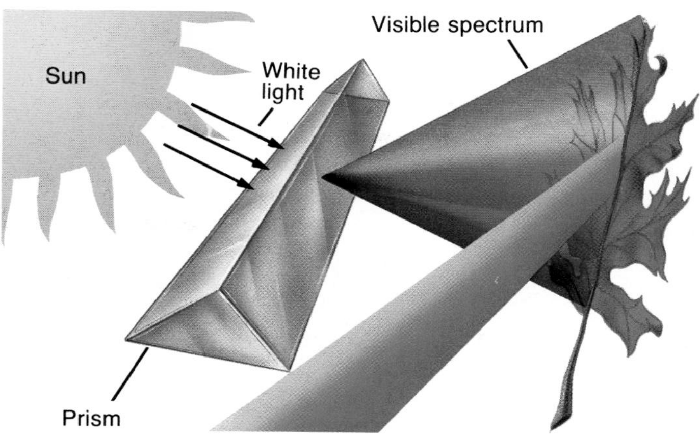

Other pigments may be found in the same membrane layers that hold chlorophyll. These pigments trap energy from colors of light that chlorophyll does not absorb well. They give cells the colors red and yellow that you see in carrots, tomatoes, some fruits, and fall leaves, *Figure 10.7.*

What happens during photosynthesis?

Trapping the energy of the sun in the chloroplasts is the first step of photosynthesis. This light energy is converted to chemical energy and then stored. Water and carbon dioxide are the other two ingredients needed for photosynthesis to occur.

The general equation for photosynthesis follows.

$$6CO_2 + 6H_2O + \text{light energy} \xrightarrow{\text{chlorophyll}} C_6H_{12}O_6 + 6O_2$$

You can see that the overall reaction of photosynthesis is simple. However, it represents the sum of many separate chemical reactions that take place when simple sugars are formed.

chlorophyll:
chloros (GK) pale (yellowish) green
phyllon (GK) leaf
Chlorophyll is a green substance found in leaves.

The reactions of photosynthesis

The process of photosynthesis includes two main groups of reactions. These are the light reactions and the Calvin cycle. **Light reactions** are the reactions in which light energy is converted to chemical energy. The light reactions are the *photo* part of photosynthesis. These reactions split water molecules, providing hydrogen and an energy source for the Calvin cycle. Oxygen from water is given off. The **Calvin cycle** is the series of reactions that forms simple sugars using carbon dioxide and hydrogen from water. The Calvin cycle is the *synthesis* part of photosynthesis.

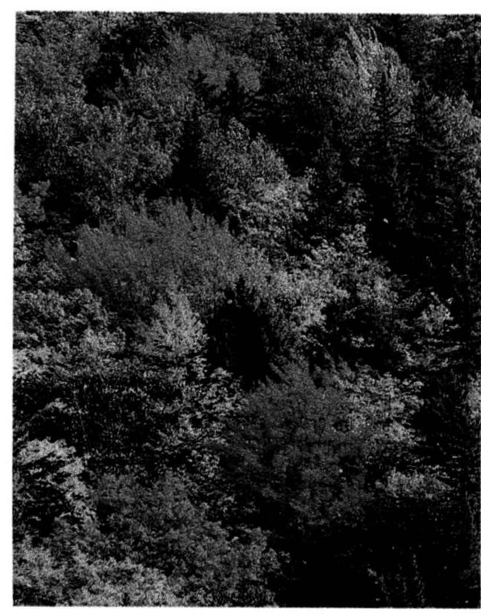

Figure 10.7

The red, yellow, and purple pigments in plants are usually masked by the green color of chlorophyll. However, they are visible in the autumn when the leaves of trees lose their chlorophyll.

2 Teach

Chalkboard Example

Write the equation that summarizes photosynthesis on the chalkboard. Have students analyze the equation. Ask which are the raw materials (reactants) in the process and which are the products. *Carbon dioxide and water are the raw materials; simple sugars and oxygen are the products.* Guide students to an understanding of what the equation means in terms of energy capture and conversion.

Science, Technology, and Society

Getting Work from Light

The exact process by which oxygen is produced in photosynthesis is still something of a mystery. The energy available from one unit of light, a photon, is not sufficient to split a water molecule; four photons are required to split two water molecules, thereby releasing four electrons and four protons. This creates a difficulty because the light reaction system can only handle one electron at a time.

In the 1970s scientists proposed what is called the water-oxidizing clock, a biochemical ratcheting mechanism for stabilizing intermediate stages of the water-splitting reaction. The water-oxidizing clock is a cyclic mechanism that supplies electrons to the chlorophyll, one electron in each of four steps. When the clock reaches the final step, it spontaneously releases an oxygen molecule and reverts to the first step of the cycle.

Meeting Individual Needs

Gifted The citric acid cycle is sometimes called the Krebs cycle. Have students research the work of Melvin Calvin or Hans Krebs and write a report. L3 LS

SECTION 10.2

ThinkingLab — Draw a Conclusion

Purpose

LS Logical-Mathematical Students will examine the effect of increasing light intensity on the rate of photosynthesis.

Process Skills

recognize cause and effect, interpret data, analyze data

Teaching Strategies

- Ask students how horticulturists increase light intensity, especially during winter. *Use grow lights and sun reflectors.* Ask students why keeping temperature constant during the experiment is important. *To avoid introducing another variable.*

Thinking Critically

The rate of photosynthesis increases with increasing light intensity until another factor limits the rate. Factors include the availability of water, carbon dioxide, phosphate, ADP, and enzymes.

Assessment

Performance: Have students working in groups design and carry out an experiment that uses grow lights to test whether increasing light intensity increases the rate of photosynthesis. L1 COOP LEARN

GLENCOE TECHNOLOGY

Videodisc

Biology: The Dynamics of Life

Disc 1, Side 1

Light Reactions (Ch. 26)

Deep Sea Rifts (Ch. 27)

ThinkingLab — Draw a Conclusion

How does photosynthesis vary with light intensity?

Photosynthesis is the process in which autotrophs synthesize organic compounds from water and carbon dioxide using energy absorbed by chlorophyll from sunlight. This process is basic for the existence of nearly all forms of life.

Analysis

Green plants were exposed to increasing light intensity, as measured in candelas, and the rate of photosynthesis was measured. The temperature of the plants was kept constant during the experiment. The following graph depicts the data obtained.

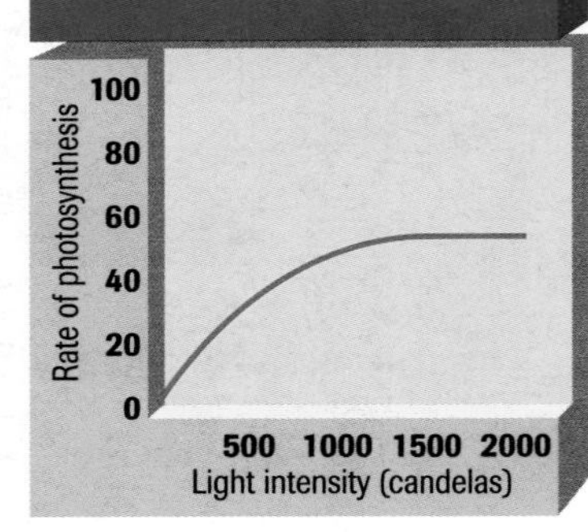

Thinking Critically

Considering the overall equation for photosynthesis, make a statement summarizing what the graph shows. Under normal field conditions, what factors may limit the relative rate of photosynthesis when the light intensity is increased and temperature remains constant?

Light Reactions

The light reactions of photosynthesis are quite complex, and scientists are still studying the process. Many reactions and enzymes are involved. You can follow along with the important steps of the light reactions in *Figure 10.8.*

Electrons in chlorophyll absorb light energy

When light strikes chlorophyll, electrons within the chlorophyll molecule absorb energy. When electrons absorb enough energy, they leave the chlorophyll molecule and are passed along a series of molecules in the thylakoids, releasing energy as they go. This series of molecules is known as the **electron transport chain.** As the electrons pass down the chain, the extra energy they received from light is stored in the bonds of ATP. In other words, energy from light becomes available to do biological work. The ATP produced will be used in the second stage of photosynthesis, the Calvin cycle.

Figure 10.8

In the light reactions of photosynthesis, there are two events involving chlorophyll and light energy. Between these reactions is another reaction involving water.

The Light Reactions

A **Light striking chlorophyll causes electrons to gain energy and leave the chlorophyll molecule. As these electrons pass down the electron transport chain, they lose energy, which is used to make ATP.**

e^- Energy ATP ADP + P_i To Calvin cycle

Cluster of chlorophyll and carotenoid molecules

Sun $2e^-$ H_2O $O + 2H^+$

B **Electrons from water replace electrons lost by chlorophyll. Water breaks up into hydrogen and oxygen.**

$e^- + H^+ + NADP^+ \longrightarrow NADPH + H^+$

C **Light causes electrons to gain energy and leave another chlorophyll molecule. Electrons along with hydrogen ions from water are added to $NADP^+$ to produce $NADPH + H^+$.**

Sun

Cluster of chlorophyll and carotenoid molecules

PORTFOLIO

Communicating About Science Divide the class into three groups. Ask each group to discuss chloroplasts, the light reactions, or the Calvin cycle. Provide time for each group to prepare a short report to be presented by one member from the group. Individual students should place copies of their group report in their portfolios. L1 P

Water molecules are split

Notice in *Figure 10.8* that water is one of the reactants in photosynthesis. Chlorophyll that has lost electrons now picks up electrons from water, splitting the water into hydrogen ions and oxygen. The splitting of water during photosynthesis is called **photolysis.** A freed hydrogen ion and an electron are picked up by the ion carrier $NADP^+$ (nicotinamide dinucleotide phosphate) to form $NADPH + H^+$, which carries the ion and electron to the Calvin cycle.

NADP is one of a class of organic molecules called coenzymes. Coenzymes act as carriers in many biological processes. The oxygen freed during photolysis is given off as a by-product. This process is the source of nearly all the oxygen in Earth's atmosphere. Note that the light reactions trap energy but do *not* involve CO_2, and no sugars are produced.

The Calvin Cycle

Take a deep breath of air. That oxygen you're inhaling probably was released during the light reactions of photosynthesis. This oxygen is vital to much of life on Earth, for a reason that you will soon learn. During the light reactions, a plant produces ATP and $NADPH + H^+$. It is now able to use CO_2 molecules to produce food molecules in the second stage of photosynthesis—the Calvin cycle.

Reactions of the Calvin cycle

In the Calvin cycle, *Figure 10.9*, an enzyme adds the carbon atom of carbon dioxide to a 5-carbon molecule. Because the carbon is now fixed in place in an organic molecule, the process is called carbon fixation. Carbon fixation is vital because it takes carbon dioxide from the air and converts it to a form that is usable by living things.

photolysis:
photos (GK) light
lyein (GK) to split
Water molecules are split in photolysis.

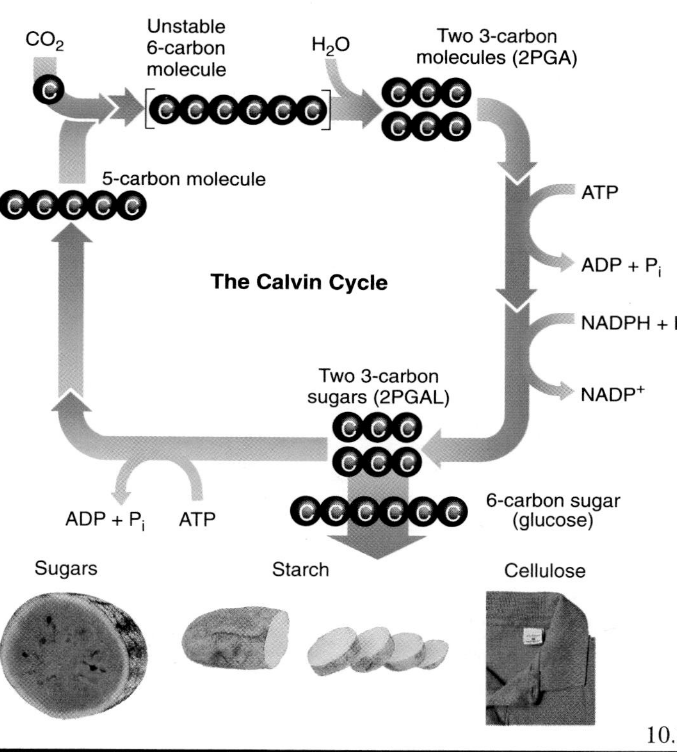

Figure 10.9

When carbon dioxide combines with the 5-carbon molecule, a 6-carbon molecule forms and splits immediately into two 3-carbon molecules. The two 3-carbon molecules formed are phosphoglyceric acid (PGA) molecules. These molecules are converted into two 3-carbon sugars, phosphoglyceraldehyde (PGAL), using the hydrogens of NADPH + H^+ and energy from ATP. Some of these sugars leave the cycle and are used to form glucose, fructose, starch, cellulose, or other complex carbohydrates.

History Connection

Early Plant Nutrition Experiment

Jean Baptist van Helmont (1577-1644) performed one of the first experiments in plant nutrition. He dried 200 pounds of soil in an oven and placed the soil in an earthen pot. He planted a willow shoot that weighed 5 pounds in the soil. At the end of 5 years the tree weighed about 169 pounds, but the soil in the pot weighed only 2 ounces less than the original 200 pounds.

Since he had added only water to the pot, van Helmont concluded that 164 pounds of wood, bark, and roots had come from water alone. Although this experiment is notable because it was carefully conducted, van Helmont reached the wrong conclusion. Discuss what we know now, that van Helmont did not know, about the building materials other than water that plants use.

GLENCOE TECHNOLOGY

Videodisc

STVS: Plants and Simple Organisms
Disc 4, Side 2
Farming Indoors (Ch. 7)

PROJECT

Growing Plants in Colored Light

Have students conduct the following experiment. Place a plant in each of several closed boxes containing a colored lightbulb. Tape the opening through which the electric cord passes so no outside light can enter. Provide the plants in each box with equal amounts of water. Have students make a hypothesis as to how the plants in each box will grow compared to a control plant.

Ask students to write a summary of the experiment. Students should include an evaluation of the hypothesis. L2 LS

Tying to Previous Knowledge

Review the structure and function of carbohydrates from Chapter 7.

Software

The Plant—Nature's Food Factory, EPIE (Apple II).

BioLab Design Your Own Experiment

What factors influence photosynthesis?

Time Allotment
One class period

Objectives
Review objectives with students before they begin the Biolab.

Process Skills
form a hypothesis, use variables and controls, collect and organize data, interpret data, design an experiment

Safety Precautions
Warn students that the hot plate and lamp can become very hot.

PREPARATION

Prepare sodium hydrogen carbonate solution by mixing 2.5 g sodium hydrogen carbonate with 1000 mL of water.

Alternate Materials
- Students can use a large classroom clock if the second hand is visible.

Possible Hypotheses
- If plants are exposed to colored light instead of white light, the rate at which photosynthesis occurs will change.
- If the temperature of the environment in which a plant grows is changed, the rate at which the plant carries out photosynthesis will change.
- The rate of photosynthesis increases with increasing light intensity.

BioLab Design Your Own Experiment

What factors influence photosynthesis?

Water molecules act as a source of electrons in the light reactions. The resulting products are hydrogen ions and oxygen gas. Oxygen is one of the end products of photosynthesis. Because oxygen is only slightly soluble in water, visible bubbles of oxygen are formed by aquatic plants, such as *Elodea*, as they carry out photosynthesis. By measuring the rate at which bubbles form, you can measure the photosynthetic rate.

PREPARATION

Problem
What colors of light are most effective in photosynthesis? What environmental factors, such as temperature and the availability of CO_2, influence the rate of photosynthesis in aquatic plants?

Hypotheses
Choose one factor that may influence the rate of photosynthesis. Hypothesize as to how it will influence that rate. Be sure to consider factors that you can easily change using the aquatic aquarium plant *Elodea.*

Objectives
In this Biolab, you will:
- **Identify** a factor that influences photosynthesis.
- **Predict** how the factor will influence the rate of photosynthesis.
- **Measure** the rate of photosynthesis.

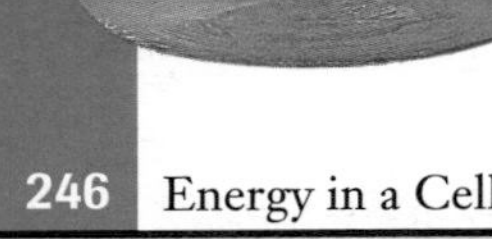

Possible Materials
250-mL Erlenmeyer flask
3 sprigs of *Elodea*
string
washers
1000-mL beaker
hot plate
ice
water
oven mitts
thermometer
colored cellophane, assorted colors
lamp with reflector and 150-watt bulb
0.25 percent sodium hydrogen carbonate (baking soda) solution
ring stand with clamp
watch with second hand
metric ruler

Safety Precautions
Exercise caution when using hot plates.

PLAN THE EXPERIMENT

Teaching Strategies
- **Troubleshooting:** Place the *Elodea* sprigs in a large bowl and place them under a lamp for about 10 minutes before students begin the lab.

Possible Procedures
- Students should set up the experiment as shown in the photo. They can cover the large beaker with one of the various colors of cellophane. Students should not put cellophane over the lamp, as it will scorch.
- If students are not looking at the effect of temperature, they should monitor the temperature of the solution in the beaker to make sure the lamp is not heating up the solution. The temperature should remain relatively constant.

PLAN THE EXPERIMENT

1. Study the diagram for the basic setup to measure photosynthesis from *Elodea*, and decide how your group will modify this setup to measure the factor you have chosen.
2. The *Elodea* needs to be lightly tied together with a weight and placed in sodium hydrogen carbonate solution inside the flask so that the plants are underneath the solution. Sodium hydrogen carbonate solution provides CO_2 for the aquarium plants.

Check the Plan

1. Consider all factors that you believe may influence the rate of photosynthesis. All of these factors, except the one you are testing, need to be kept constant throughout your experiment.
2. Consider how you will set up a control for the factor that you are testing. Be sure to run a control setup.
3. Design and construct a data table for recording your observations.
4. Plan to measure the number of bubbles of oxygen generated under both control conditions and experimental conditions at various time periods for a total of at least five minutes.
5. ***Make sure your teacher has approved your experimental plan before you proceed further.***
6. Carry out your experiment.

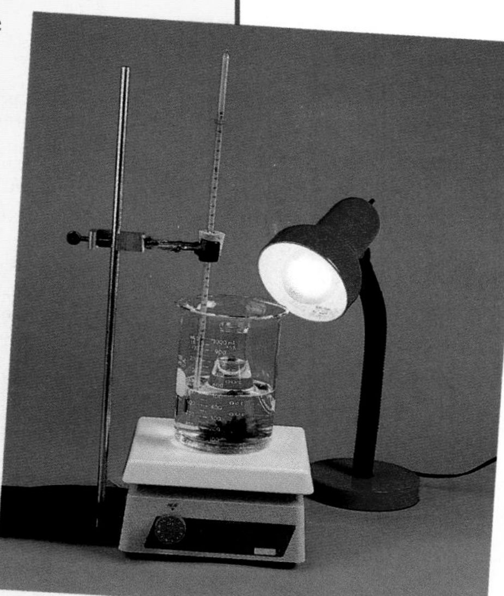

ANALYZE AND CONCLUDE

1. **Identifying Variables** What conditions did you keep constant in the experiment? What one condition did you vary?
2. **Interpreting Observations** From where did the bubbles of oxygen emerge?
3. **Making Inferences** Explain how counting bubbles measures the rate of photosynthesis.
4. **Analyzing Data** Make a graph of your data with the photosynthetic rate per minute plotted against the factor tested for both the control and experimental setups. Write a sentence or two explaining the graph.
5. **Checking Your Hypothesis** Did your results support your hypothesis? If not, suggest a new hypothesis that is supported by your data.

Going Further

Application Design a method to measure the rate of photosynthesis of a land plant, such as a potted geranium.

- If students are looking at the effect of temperature, they should use ice or warm water in the beaker to cool down or warm up the plant. They should also carefully monitor the temperature to see that it remains constant during the time of each temperature measurement.

Data and Observations

Students should record data under control conditions and then under experimental conditions for the same amount of time—at least 5 minutes.

ANALYZE AND CONCLUDE

1. If the student varies the color of the light, then temperature and the distance of the light from the plants should remain constant. If the student varies the temperature, then the color of the light and the distance of the light from the plants should remain constant. If the student varies the light intensity, then the color and temperature should remain constant.
2. The bubbles emerged from the end of the stem of the *Elodea* plant.
3. Oxygen is an end product of photosynthesis. As the rate of photosynthesis changes, so will the rate at which oxygen is produced.
4. The rates should be fairly constant for the control and experimental setup (relatively flat lines). The experimental setup rate will be greater than or lower than the control, so its line should correspondingly be above or below the control line.
5. Students should check their results against their original hypotheses.

Assessment

Portfolio: Have students write an evaluation of the lab. Their evaluations should include the hypothesis, analysis, and conclusion answers. L1 P

Going Further

Have students examine how the concentration of sodium hydrogen carbonate solution influences the rate of photosynthesis in an aquatic plant. L2

Technology
BIOLOGY & SOCIETY

Collecting the Sun

Purpose

Students will examine the uses and possible applications of sun collectors other than chlorophyll.

Teaching Strategies

- Ask students to brainstorm and list on the chalkboard all the things for which they use energy. Use the list to emphasize how heavily people depend on energy.

Different Viewpoints in Biology

Exploring the Calvin Cycle

For many years, the reactions of the Calvin cycle were called "dark reactions" of photosynthesis. This term has been dropped by many scientists because it implies that the reactions occur only at night. In fact, some of the enzymes of the Calvin cycle are regulated by light, and the materials used in the Calvin cycle are produced only in the light. Take a few moments to explain the role of light in the Calvin cycle to students.

INVESTIGATING the Technology

Applying Concepts Plants store energy captured during daylight hours in ATP and then release it in the amounts needed at any time.

Going Further

Ask students to research and find out what some of the problems have been with electric cars that run on solar batteries. Have students write reports of their findings for inclusion in their portfolios. L2 P

Technology
BIOLOGY & SOCIETY

Collecting the Sun

Sunlight provides energy for nearly all life on Earth, so why can't we use sunlight for all our energy needs? The sources of energy that we use most—coal, oil, and natural gas—will eventually be used up. If energy could be obtained directly from the sun instead, these fuels would be conserved and pollution would be reduced.

Solar collectors Gathering solar energy requires the use of collectors with large surface areas exposed to the sun. The most popular type of solar collector is the *flat-plate collector.* You may see this type on the roof of a house. It consists of a flat, black surface that absorbs sunlight and changes it to heat. Air or water flows behind the plate, absorbing the heat. This type of collector is used to heat homes and provide hot water.

Another kind of collector is a *photovoltaic cell,* which converts light into electricity. Arrays of these cells provide electric power for satellites and highway signs.

Applications for the Future

Scientists are designing a satellite system to collect solar energy in space without atmospheric interference. The system will then convert the solar energy to microwave radiation and transmit it to Earth, where a receiver will convert the radiation to electricity.

Engineers are also testing electric cars and buses. These vehicles run on batteries charged by large, stationary photovoltaic arrays. They may prove to be a good solution for commuters.

INVESTIGATING the Technology

Applying Concepts Solar energy technology must include storing the energy for times when sunlight is not available. How do green plants do this?

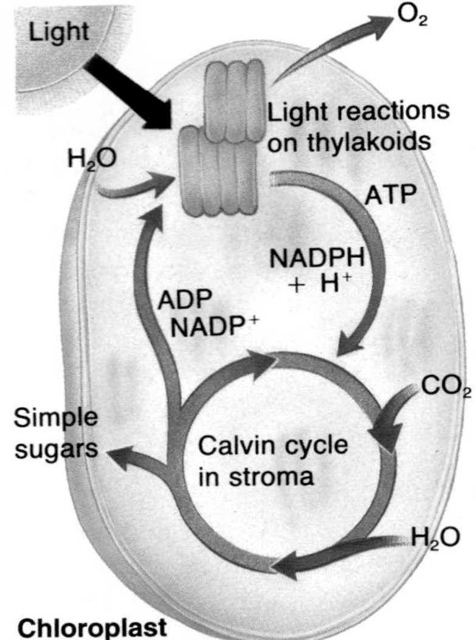

Figure 10.10

This diagram of a chloroplast summarizes the process of photosynthesis. In photosynthesis, the light reactions furnish energy and some raw materials to the remainder of the process, the Calvin cycle.

The Calvin cycle reactions take place in the stroma of the chloroplasts. The overall effect of the Calvin cycle is that carbon dioxide combines with hydrogen to form simple sugars that are used to make other carbohydrates such as more complex sugars, starch, and cellulose.

Carbon dioxide enters the leaves and stems of green plants through hundreds of small pores. Besides carbon dioxide, ATP and hydrogen from the light reactions are needed for the Calvin cycle. The Calvin cycle reactions, shown in ***Figure 10.9,*** continue as long as the materials are available.

You can get a good overall view of the relationship between the light reactions and the Calvin cycle, as well as the reactants and end products of the whole process of photosynthesis, in ***Figure 10.10.***

PROJECT: Separating Leaf Pigments

It is possible to remove the pigments from a leaf and separate them by paper chromatography. Less soluble compounds remain near the bottom of the paper and more soluble compounds are carried higher.

Have students conduct the following experiment. Soak ground fresh leaves in warm alcohol and apply the extract to a 12-cm-long strip of filter paper in a narrow band about 2 cm from the bottom of the strip. Place the filter paper into a test tube containing a small amount of acetone (available as clear fingernail polish remover–check the label). The test tube should be tightly sealed after placing the paper in it. Just the bottom edge of the paper should dip into the solvent. L2 LS

During the Calvin cycle, a plant produces carbohydrate molecules—mainly sugars, starch, and cellulose. Some of these molecules are now available as sources of energy to the plant and to organisms that eat the plant. In the cells of producers and consumers, carbohydrates are broken down to release the energy that originally radiated from the sun and make it available for life processes on Earth.

Figure 10.11

Most of these sea-vent communities rely on energy from reactions of inorganic sulfur compounds that come from openings in the ocean floor. The sulfur compounds are metabolized by bacteria that live symbiotically in the gills of clams, mussels, and tube worms.

Life Without Light

Some bacteria are able to capture energy from sources other than the sun. Instead of being photosynthetic, these bacteria are called chemosynthetic. **Chemosynthesis** is a process by which an autotroph obtains energy from inorganic compounds instead of from light.

One group of chemosynthetic prokaryotes is methane-producing bacteria. These organisms are adapted to live under conditions where there is no oxygen. In fact, they are poisoned by oxygen. They are found in marshes, lake sediments, and the digestive tracts of ruminant mammals, such as cows. These bacteria are the final participants in the decomposition process. They react CO_2 and H_2 to produce methane (CH_4). They are also important in the breakdown of sewage at sewage-disposal plants.

Scientists have discovered thriving animal communities, as shown in *Figure 10.11,* near cracks deep on the floor of the Pacific Ocean. In fact, these communities are a part of a whole ecosystem that doesn't need the sun to survive.

Section Review

Understanding Concepts

1. Why do you see green when you look at a leaf on a tree?
2. How do the light reactions of photosynthesis relate to the Calvin cycle?
3. What is the function of water in photosynthesis?

Thinking Critically

4. In the general equation for photosynthesis, why is chlorophyll written over the arrow rather than on the left side of the equation?

Skill Review

5. **Designing an Experiment** Design an experiment that would compare photosynthesis in red, green, and blue light. What would you use as a control? For more help, refer to Practicing Scientific Methods in the *Skill Handbook.*

SECTION 10.2

3 Assess

Check for Understanding

Make sure students understand that the Calvin cycle depends on products from the light reactions. Repeat the general equation for photosynthesis. Ask one student to give the products, and ask another student to give the reactants. L1

Reteach

Reinforce the idea that the light reactions of photosynthesis furnish energy and some raw materials to the Calvin cycle. Have students look back to Figure 10.10 on page 248. Explain to students that in the chloroplasts, the Calvin cycle unites carbon dioxide and hydrogen to form the simple sugars needed to make more complex carbohydrates.

Extension

Have students research the difference between the chlorophyll of plants and the light-capturing pigments of photosynthetic bacteria. L3

Assessment

Journal: Ask students to write a paragraph summarizing the process of chemosynthesis in their journals. L1

4 Close

Discussion

Ask students where the energy in a consumer's food comes from. *Students should realize that energy for the majority of living things initially comes from the sun.* Work through some simple food chains as part of the discussion. L1

Answers to Section Review

1. The chlorophyll in the leaf reflects green and yellow while absorbing other colors.
2. The light reactions form ATP and produce hydrogen ions and oxygen from the splitting of water. The ATP and hydrogen ions are used in the Calvin cycle.
3. Water is split, providing hydrogen ions for the Calvin cycle.

Thinking Critically

4. The chlorophyll is not a reactant or a product.

Skill Review

5. The experiment would be similar to the Biolab. The control would be plants under white light and in the dark.

SECTION

10.3

Prepare

Key Concepts

Students will learn the similarities and differences between aerobic respiration and anaerobic processes that release energy. They will also learn how these reactions are related to photosynthesis.

Block Scheduling

Look for this symbol for strategies that are useful in a block scheduling format. For more information on block scheduling, refer to Section 10.3 in the **Lesson Plans** booklet.

Materials

- Obtain baker's yeast for the Demonstration and the second Minilab.
- Purchase *Coleus* plants and black paper for the Alternate Lab.
- Collect materials for the solar food-cooker project.

1 Focus

Bellringer

Before presenting the lesson, display **Section Focus Master 22** on the overhead projector and have students answer the accompanying questions. L1 SAE

Discussion

Read a recipe for making bread. Explain to students that the yeast used in the recipe consists of microorganisms classified as fungi. Explain to students that unlike most organisms, yeast carry out processes that release energy in the absence of oxygen. Explain that bakers use yeast in their recipes for almost all breads because as yeast grow anaerobically, they release CO_2, which causes the bread to rise.

SECTION

10.3 Getting Energy to Make ATP

Section Preview

Objectives

Compare and contrast aerobic and anaerobic processes.

Explain how cells obtain energy from respiration.

Key Terms

respiration
aerobic process
anaerobic process
glycolysis
citric acid cycle
lactic acid fermentation
alcoholic fermentation

Food molecules produced by photosynthesis contain energy from the sun converted to chemical energy. But like the person who wants to buy only a can of soda but is stuck with a $100 bill, a cell can't use all the energy in a food molecule at once. As you will see, organisms have processes to "make change"—to break down food molecules to yield usable amounts of energy.

anaerobic:
an (GK) without
aeros (GK) air
Anaerobic organisms can live without air, i.e., oxygen.

The First Steps of Respiration

When scientists talk about food molecules in cells, they don't mean hamburgers or asparagus. Molecules of glucose and other 6-carbon sugars are the major source of energy for most organisms, although fatty acids and amino acids can also be used as energy sources. Cells break down glucose through a series of chemical reactions. The stored energy of glucose is released bit by bit and used to attach phosphate groups to ADP molecules to form ATP molecules, the cell's energy currency.

Figure 10.12

Notice that the first step of glycolysis uses the energy from two molecules of ATP. However, four ATP molecules are formed using the energy released by the second step of glycolysis. Thus, in glycolysis, the energy payoff for the breakdown of one molecule of glucose to two molecules of pyruvic acid is only two molecules of ATP. Notice that the intermediate product, PGAL, is the same 3-carbon sugar that was formed in photosynthesis.

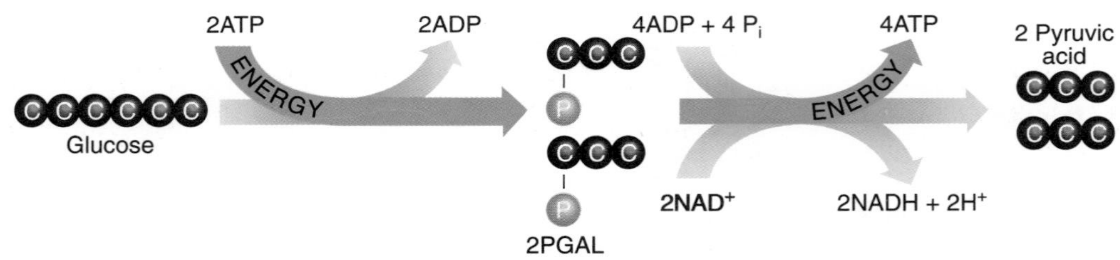

Program Resources

Section Focus Master 22 L1 SAE
Reinforcement and Study Guide, p. 40 L1
Biolab and Minilab Worksheets, pp. 35-36 L1
Performance Assessment in the Biology Classroom, p. 11 L1 P
Laboratory Manual, pp. 61-64 L2
Basic Concepts Transparency 12 and **Master** L1 SAE
Tech Prep Applications, pp. 13-14 L2
Basic Skills Transparency 10 and **Master** L1 SAE

The process by which food molecules are broken down to release energy is called **respiration.** In many cells, both aerobic and anaerobic processes take place during respiration. **Aerobic processes** require oxygen in order to take place. **Anaerobic processes** do not require oxygen. Anaerobic processes are simple and yield energy quickly. However, as you will see, the energy payoff is much greater when molecules are broken down aerobically.

The first step—Glycolysis

When a 6-carbon glucose molecule breaks down, it is changed into pyruvic acid molecules, which are 3-carbon molecules. This breakdown occurs without oxygen. The anaerobic process of splitting glucose and forming two molecules of pyruvic acid is called **glycolysis,** which you can see in *Figure 10.12.*

Glycolysis also produces hydrogen ions and electrons. These combine with organic ion carriers called NAD^+ (nicotinamide dinucleotide) to form NADH + H^+. The compound NAD, like NADP used in photosynthesis, is a coenzyme. Glycolysis occurs in the cytoplasm of cells and does not require oxygen. However, if oxygen is present, energy-yielding aerobic processes may follow.

Releasing Energy with Oxygen

The process of breaking down glucose begins with glycolysis, yielding two molecules of ATP for each molecule of glucose broken down. When

Math

Energy from Cellular Respiration

It's the first day of soccer practice. You need a lot of energy to follow that ball up and down the soccer field. How will the food you eat turn into the energy you need to keep running? Because much of the food you eat is broken down into glucose molecules, it will help to find out how much energy glucose provides.

If you burn glucose in a calorimeter, the following equation describes what happens:

$$C_6H_{12}O_6 + 6O_2 \rightarrow 6CO_2 + 6H_2O$$

In this combustion of glucose, all of the energy that is in the structure of the glucose is unlocked, carbon dioxide and water are formed, and 686 Calories of heat are measured in the calorimeter.

Compare that release of energy with the energy released by respiration in a cell in your body:

$$C_6H_{12}O_6 + 6O_2 + 38P + 38ADP \rightarrow 6CO_2 + 6H_2O + 38ATP$$

In cellular respiration, only 409 Calories are released as heat from glucose, as compared to 686 Calories in the complete combustion of glucose. How many Calories of heat have not been accounted for in cellular respiration?

The energy that was not released as heat is stored in ATP to be used when needed by the cell. As you can see, for every unit of glucose burned, 38 units of ATP are formed. How many Calories are there in 38 units of ATP? How many Calories are there in one unit of ATP?

CONNECTION TO Biology

When gasoline is burned in a car engine, it is only about 25 percent efficient in running the car. Seventy-five percent of the gasoline is not completely burned. How efficient is the formation of ATP from glucose during cell respiration? (HINT: Divide the number of Calories in 38 units of ATP by the number of Calories in a unit of glucose.)

2 Teach

Math

Energy from Cellular Respiration

Purpose

Students will compare the number of Calories produced by the complete combustion of glucose in a calorimeter with the Calories produced by glucose during cellular respiration, in which part of the energy produced is channeled into chemical energy as ATP.

Teaching Strategies

Ask: Which is more important in organisms: to produce more Calories in the form of heat or to produce sufficient heat plus ATP? They should realize that organisms must have sufficient ATP to carry out their vital functions and various activities.

Possible Answers

1. How many Calories of heat have not been accounted for in cellular respiration? *277 Calories*
2. How many Calories are there in 38 units of ATP? *277 Calories*
3. How many Calories are there in one unit of ATP? *277 Calories ÷ 38 units of ATP = 7.3 Calories/unit*
4. How efficient is the formation of ATP from glucose during cell respiration? *277 Calories ÷ 686 Calories × 100 = 40 percent*

CULTURAL DIVERSITY

Severo Ochoa

Have students research the efforts of Spanish-American biochemist Severo Ochoa (1905-) toward the modern understanding of the citric acid cycle and photosynthesis. Ochoa showed how the oxidation of one glucose molecule could yield 38 ATP molecules. He also elucidated the mechanisms of the citric acid cycle and photosynthesis by identifying the function of key enzymes.

Ochoa's research in cellular respiration in the 1930s and 1940s ultimately resulted in the discovery of the mechanisms of RNA and DNA synthesis, for which Ochoa and colleague Arthur Kornberg received a Nobel prize in 1959.

Bioethics

Deforestation of the Tropics

Many forests, especially tropical rain forests, are in danger today. The areas are being leveled to provide fuel and uncover farmland.

Many scientists are considering how the massive clearance of forested lands will affect the balance between photosynthesis and respiration in the biosphere. As a result, some organizations have begun buying up land in the tropics to try to prevent deforestation.

Follow the link for this chapter on the Glencoe Homepage at **www.glencoe.com/sec/science** to find out more about energy in the cell.

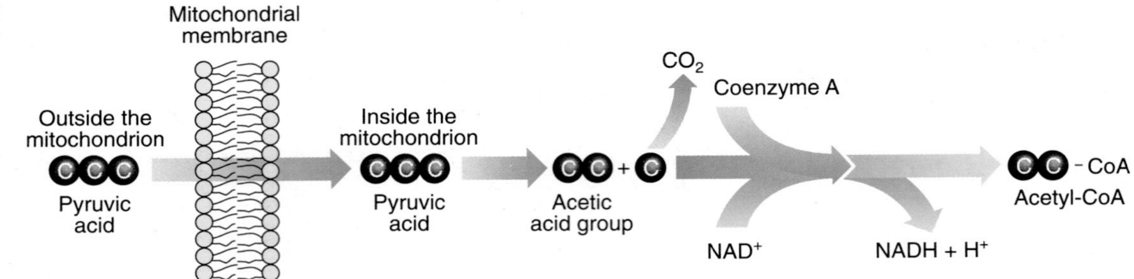

Figure 10.13

Pyruvic acid enters a mitochondrion. Here in the mitochondrion, pyruvic acid changes to acetic acid by losing a CO_2 molecule. It then forms acetyl-CoA before entering the third step of aerobic respiration.

oxygen is present, aerobic respiration, which uses oxygen, takes place in a mitochondrion, the powerhouse of the cell.

Breakdown of pyruvic acid

Aerobic respiration involves two more steps after glycolysis. The first step includes reactions that change pyruvic acid, as shown in *Figure 10.13.* Pyruvic acid, a 3-carbon compound, is changed to acetic acid, a 2-carbon compound. The third carbon atom forms carbon dioxide, CO_2. The acetic acid from the reaction is combined with a substance called coenzyme A to form a compound called acetyl-CoA.

The citric acid cycle

The second step of aerobic respiration is the **citric acid cycle,** shown in *Figure 10.14.* This cycle of chemical reactions produces more ATP and releases additional electrons. The electrons are picked up by NAD^+ and FAD (flavin adenine dinucleotide, an ion carrier similar to NAD^+).

In the citric acid cycle, acetyl-CoA combines with a 4-carbon molecule to form a 6-carbon molecule, citric acid. Locate acetyl-CoA and citric acid in *Figure 10.14* and follow each of the reactions. Notice that citric acid is broken down first to a 5-carbon molecule and then to a 4-carbon molecule, releasing CO_2 at each step.

Figure 10.14

Citric acid is changed to a 5-carbon molecule and then to a 4-carbon molecule. Notice that a carbon dioxide molecule is given off in each of these reactions. This loss of two CO_2 molecules produces a 4-carbon molecule that reacts to form the original 4-carbon substance again. As a result, a new 4-carbon molecule is available to combine with each new acetyl-CoA that enters the cycle. Follow the sequence of reactions in the cycle and locate where NAD^+ and FAD pick up electrons.

Alternate Lab: Carbon Fixation

Purpose 

LS **Logical-Mathematical** Students will demonstrate that without light, carbon fixation slows or stops.

Materials

black paper, labels, iodine solution, *Coleus* plants, paper clip, hot plate, beaker, 95% alcohol, small bowl

Procedure

Give students the following directions.

1. Cut out two identical pieces of black paper in the shape of a small square.
2. Stick a label on one piece and write your initials and the date on it.
3. Use a paper clip to fasten the black shapes to the top and bottom surfaces of the leaf of a *Coleus* so that they are matched up exactly.
4. Leave the plant in sunlight for 7 days. Then remove the leaf that was partially covered and take off the paper clip and papers.
5. Place the leaf into a beaker of boiling 95% alcohol enclosed in a fume hood and boil it until it turns white. **CAUTION:** *Do not use a heat source that has an open flame or a hot plate with an unsealed element.*

The electron transport chain

In both glycolysis and the citric acid cycle, some energy is trapped in the ATP that is formed. However, important sets of reactions also release electrons to NAD^+ and FAD. The fourth part of aerobic respiration is an electron transport chain.

Remember that an electron transport chain is a series of molecules along which electrons are transferred, releasing energy. Carrier molecules bring electrons from the reactions of both glycolysis and the citric acid cycle to the electron transport chain, as shown in *Figure 10.15*. The molecules of the electron transport chain are located on the inner membranes of mitochondria.

This process is aerobic because the final electron acceptor in the transport chain is oxygen. When oxygen accepts electrons, it combines with two hydrogen ions to form a molecule of water, H_2O. Much of the water vapor and carbon dioxide that you exhale is the result of aerobic respiration in your cells. You can see in *Figure 10.16* another example of the aerobic breakdown of sugar.

What happens if no oxygen is present? If the final electron acceptor, oxygen, is used up, the chain becomes jammed. It is somewhat like a

MiniLab

Do aquariums contain measurable amounts of CO_2?

Bromothymol blue solution is an acid-base indicator solution. When CO_2 dissolves in H_2O, they combine to form a weak acid, carbonic acid. Bromothymol blue changes to a yellow color in the presence of carbonic acid.

Procedure

1. Place 50 mL of aquarium water from a fish tank into a 100-mL beaker.
2. Add six to ten drops of bromothymol blue.

Analysis

1. Were there detectable amounts of CO_2 in the aquarium water? How do you know?
2. Did the aquarium contain numerous living plants? Would you expect these plants to affect the amount of CO_2 in the water? Explain.
3. Why would the number of fish in the aquarium affect the amount of CO_2 in the water?

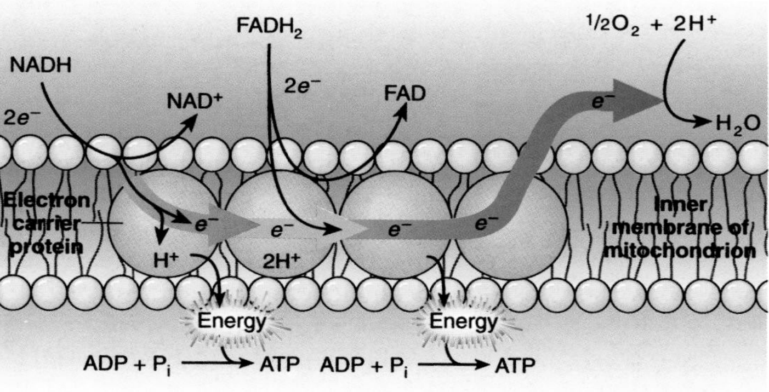

Figure 10.15

At the electron transport chain, the carrier molecules NADH and $FADH_2$ give up electrons that pass through a series of reactions. At least 15 carrier proteins are involved in the electron transport chain. At the top of the chain, the electrons have high energy. As the electrons pass down the chain, the energy given off is captured in molecules of ATP.

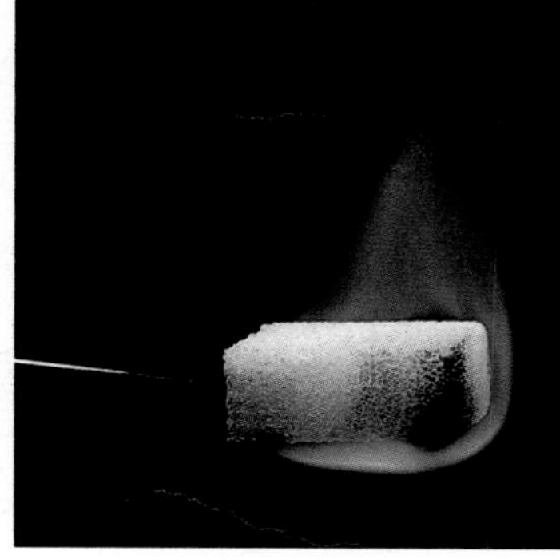

Figure 10.16

When sugar burns in air, the reaction releases CO_2, H_2O, and a lot of energy all at once. In aerobic respiration, the same overall process occurs, but it occurs step-by-step. The energy available from the breakdown of each glucose molecule can be used to make as many as 38 ATPs.

MiniLab

Purpose

LS Logical-Mathematical Students will measure the presence of CO_2 in aquarium water.

Process Skills

measure in SI, interpret data, experiment, analyze

Teaching Strategies

- Discuss the possible sources of the CO_2 present in aquarium water. These sources include the fish and other organisms, such as snails and bacteria, that utilize oxygen.

Expected Results

Depending on the number of fish, snails, and other oxygen users, the bromothymol blue will change to yellow in the water containing large amounts of dissolved CO_2.

Analysis

1. Yes. The bromothymol blue changed color.
2. Answers will vary according to the aquarium from which students got the water. Plants will use the dissolved CO_2, reducing the amount of CO_2 in the water.
3. The more fish in the water, the more CO_2 that will be given off via respiration.

Assessment

Journal: Have students write a summary of the Minilab. L1

6. Remove the leaf and place it in a small bowl. Pour iodine solution on the leaf and let it absorb the iodine for a few minutes.
7. Rinse the leaf with tap water and observe.

Analysis

Ask students to answer the following questions.

1. What happens when the leaf is boiled in alcohol? *Its chlorophyll dissolves in the alcohol.*
2. What part of the leaf stained dark? *The part that contained starch stained dark.*
3. In what part of the leaf did carbon fixation slow or stop? *In the part that was covered so that it received no light and therefore could not carry on photosynthesis, which is a carbon fixation process.*

Assessment

Journal: Have students write a summary of the lab, including answers to the *Analysis* questions, in their journals. L1

Demonstration

Visual-Spatial Prepare a sugar solution by mixing a tablespoon of sugar with a cup of warm water in a deep jar or flask. Add some baker's yeast a few hours before class. Have students note the odor of alcohol and the bubbles of carbon dioxide. Point out that these products result as the yeast carry out alcoholic fermentation.

Assessment

Performance Assessment in the Biology Classroom: p. 11, Vibration and Fermentation. Have students carry out this activity to determine whether a relationship exists between motion and fermentation rate. L1 P

Reinforcement

Make sure students understand that pyruvic acid is the intermediate product for both types of fermentation. Challenge students to explain how pyruvic acid changes in each type of fermentation. L1

GLENCOE TECHNOLOGY

Videodisc

The Infinite Voyage: The Champion Within

Physiology of Consistent Performance (Ch. 5)

parking lot with a blocked exit. If the first car can't exit, the cars behind can't get out either. Also, just as additional cars can't enter, neither can electrons. As you can see, the reactions of the electron transport chain cannot take place in the absence of oxygen.

Releasing Energy Without Oxygen

Remember that at the end of glycolysis, each glucose molecule has been broken down to two pyruvic acid molecules. If no oxygen is available to carry on aerobic respiration, the pyruvic acid molecules can be changed further without oxygen in an anaerobic process called fermentation. No additional ATP is formed during the process of fermentation. However, fermentation uses the electrons carried by NADH + H^+ and thus regenerates NAD^+ so that glycolysis can continue.

There are two major types of fermentation. In one, pyruvic acid changes to lactic acid. In the other, it changes to ethyl alcohol and carbon dioxide.

Lactic acid fermentation

When lactic acid is formed as an end product of fermentation, the process is called **lactic acid fermentation.** *Figure 10.17* shows what happens during lactic acid fermentation.

Do your muscles become fatigued after you've run for a while or lifted heavy objects? Anytime your muscle cells require energy at a faster rate than it can be supplied by aerobic respiration, they begin to carry out lactic acid fermentation. The resulting buildup of lactic acid causes muscle fatigue.

Alcoholic fermentation

When ethyl alcohol and carbon dioxide are the end products of fermentation, the process is called **alcoholic fermentation.** Many bacteria and fungi such as yeast carry out alcoholic fermentation. *Figure 10.18* shows the process. Notice that NAD^+ is produced from NADH + H^+ in this reaction.

Figure 10.17
Lactic Acid Fermentation

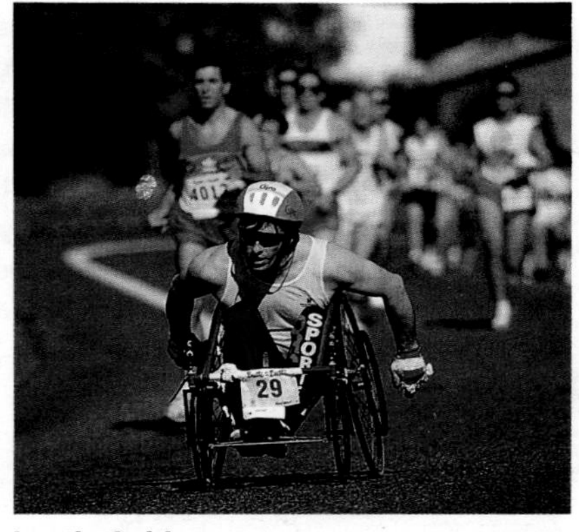

Strenuous exercise has caused lactic acid to build up inside this racer's muscle cells, resulting in muscle fatigue. When oxygen is available, the body will break down the lactic acid. For this reason, we say that he has built up an oxygen debt.

Lactic Acid Fermentation

NAD^+ | NADH + H^+ | NADH + H^+ | NAD^+

Glucose → Glycolysis → 2 Pyruvic acid → Lactic acid fermentation → 2 Lactic acid

2ADP + $2P_i$ | 2ATP

Lactic acid fermentation occurs in some bacteria, in plants, and in most animals, including humans.

Discussion

Discuss lactic acid fermentation and question the students about the problems of overexercise and improper cool down. Relate their answers to oxygen debt.

Meeting Individual Needs

Learning Disabled For students having difficulty with the concepts of this chapter, review the roles of a producer and consumer. Relate the processes in this chapter to the concepts of Unit 2, Ecology. L1

STUDENT JOURNAL

Life as a Carbon Atom Have students write a story from the viewpoint of a carbon atom. The story should detail the travels of a carbon atom involved in photosynthesis and then through cellular respiration. L1 LS

Comparing Photosynthesis and Aerobic Respiration

Both photosynthesis and aerobic respiration are complex sets of reactions that are alike in several ways. Both involve energy, require enzymes, occur in specific organelles, and involve the movement of electrons in transport chains.

In many ways, photosynthesis is the opposite of aerobic respiration. The end products of photosynthesis are the starting materials of respiration. In photosynthesis, starch or sugars are eventually formed, and oxygen is released as a waste product. In aerobic respiration, oxygen is used to break down sugars. Carbon dioxide and water are given off as wastes. So, the end products of aerobic respiration are the starting materials of photosynthesis. Compare the two reactions in the summary diagram, *Figure 10.19*, on the next page.

MiniLab

Will apple juice ferment?

Organisms such as yeast have the ability to break down food molecules and synthesize ATP when no oxygen is available. When the appropriate food is available, yeast can carry out alcoholic fermentation, producing CO_2. Thus, the production of CO_2 can be used to judge whether alcoholic fermentation is taking place.

Procedure

1. Carefully study the diagram and set up the experiment as shown.
2. Hold the test tube in a beaker of warm (not hot) water and observe.

Analysis

1. What were the gas bubbles that came from the plastic pipet?
2. Predict what would happen to the rate of bubbles given off if more yeast were present in the mixture.
3. Why was the test tube placed in warm water?

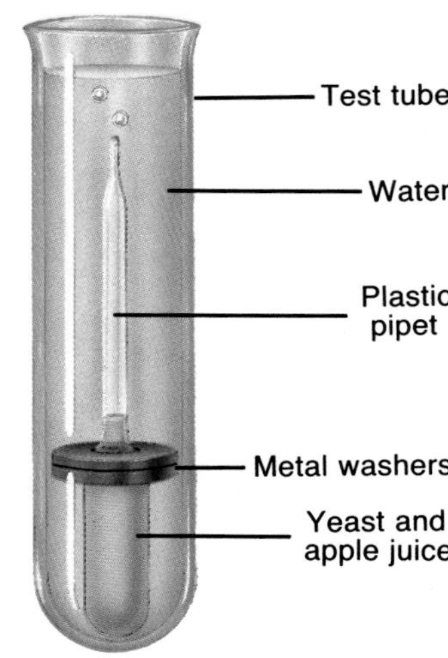

Figure 10.18

Alcoholic Fermentation

Yeast is used in fermentation to produce alcohol. Alcoholic fermentation also produces CO_2, which can be used to raise bread dough.

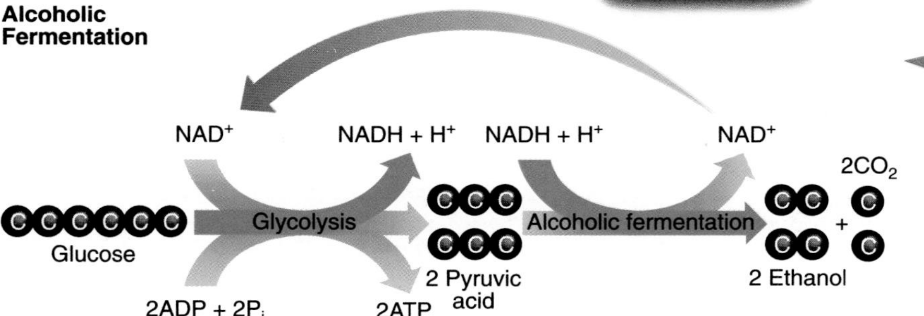

The ethyl alcohol produced by alcoholic fermentation is the alcohol in beer, wine, and other alcoholic beverages.

MiniLab

Purpose

Visual-Spatial Students will observe the production of CO_2 resulting from yeast fermenting sugar.

Process Skills

observe and infer, interpret data, experiment, analyze data

Teaching Strategies

- Ask students to brainstorm and list the uses of industrial alcoholic fermentation. Ask them which products (alcohol or CO_2) of the process are important.

Expected Results

Students will see gas bubbles coming from the yeast/apple juice mixture.

Analysis

1. CO_2
2. The rate would increase.
3. Warm water increases the metabolic rate of the yeast.

Assessment

Journal: Ask students to prepare a summary of the Minilab in their journals, including answers to the *Analysis* questions. L1

3 Assess

Check for Understanding

Provide students with a list of the major substances discussed and the names of the processes glycolysis, respiration, alcoholic fermentation, lactic acid fermentation, and electron transport chain. Have students construct a table that identifies whether each substance is a product or reactant of each process. L1

PORTFOLIO

Comparing Photosynthesis and Respiration Have students write a summary of Figure 10.18 on this page and Figure 10.19 on page 256. Have students explain the relationship of photosynthesis to the processes of alcoholic fermentation and respiration. Students should include their summaries in their portfolios. L1 LS P

SECTION 10.3

Reteach

Provide students with unlabeled diagrams of glycolysis, alcoholic fermentation, and lactic acid pathways. Have them fill in the labels as you discuss the processes. L1

Extension

Have students find out and report on how cyanide affects respiration. L2

Assessment

Portfolio: Ask students to write summaries of glycolysis, citric acid cycle, electron transport chain, and anaerobic respiration. L1 P

4 Close

Discussion

Discuss the importance of lactic acid fermentation in the food industry. Students can be asked to research the production of particular foods prior to the discussion. L1

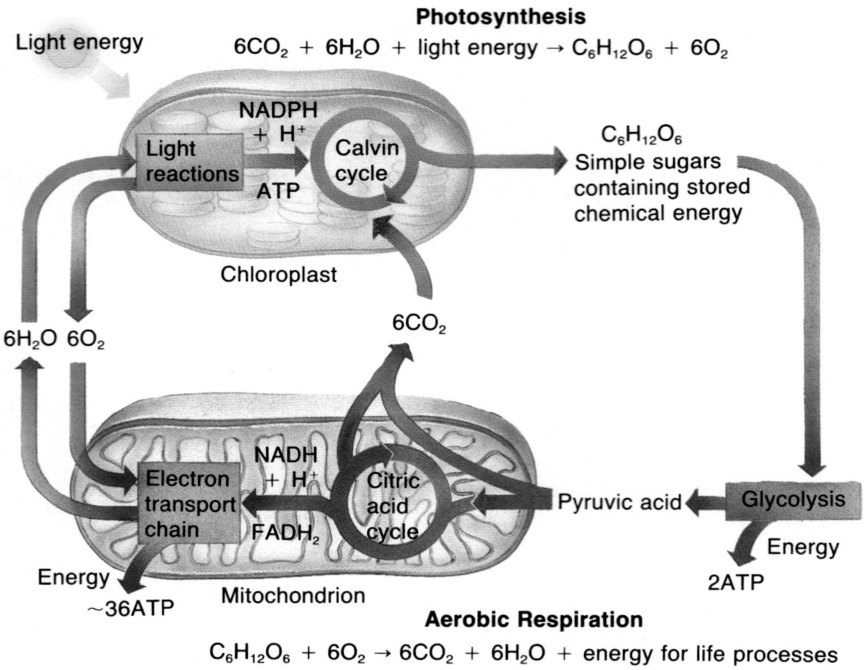

Figure 10.19

Photosynthesis in eukaryotic cells occurs in the chloroplast, whereas aerobic respiration occurs in the mitochondrion. In photosynthesis, energy from sunlight is converted to chemical energy in the bonds of carbohydrate molecules. In respiration, chemical energy from the bonds of carbohydrate molecules is released and used to make ATP.

Connecting Ideas

Nearly all of the millions of present-day species break down organic nutrients for energy, but only about half a million species are able to build nutrients through photosynthesis. If photosynthesis stopped today, both heterotrophs and autotrophs alike would die from lack of energy from food. Life on Earth depends largely on light energy channeled through photosynthesis.

One of the uses of this flow of energy by living cells is in the progress of cell reproduction. You will see that cell structures, chemistry, and cell energy also play a role in cell reproduction.

Section Review

Understanding Concepts

1. Compare the ATP yields of glycolysis and aerobic respiration.
2. How do alcoholic and lactic acid fermentation differ?
3. How is most of the ATP from aerobic respiration produced?

Thinking Critically

4. Compare the energy-producing processes in a jogger's leg muscles to those of a sprinter's leg muscles. Which is likely to build up more lactic acid? Which runner is more likely to be out of breath after running? Explain.

Skill Review

5. **Making and Using Tables** Use the section called Comparing Photosynthesis and Aerobic Respiration to make a table summarizing the similarities and differences between the two processes. For more help, refer to Organizing Information in the *Skill Handbook*.

Answers to Section Review

1. Glycolysis produces two ATP molecules; aerobic respiration, as many as 38 molecules.
2. Alcoholic fermentation produces alcohol and carbon dioxide from glucose. Lactic acid fermentation produces lactic acid from glucose.
3. Most of the ATP formed is produced by the electron transport chain.

Thinking Critically

4. Aerobic respiration occurs in the leg muscles of both runners. They may have insufficient oxygen, and lactic acid fermentation will take place. The sprinter will more likely be out of breath because of an oxygen debt.

Skill Review

5. *Differences:* Photosynthesis forms sugars and oxygen, uses water and CO_2. Respiration uses oxygen to break down sugar, forms water and CO_2. Photosynthesis uses light to form chemical bond energy. Respiration uses chemical bond energy to form ATP. *Similarities:* Both are complex

Chapter 10 Review

REVIEWING MAIN IDEAS

10.1 ATP: Energy in a Molecule

- ATP is the molecule that receives energy released in cell reactions.
- ATP is formed when a phosphate group is added to ADP. When ATP is broken down, ADP and phosphate are formed, and energy is released.
- ATP is the main link between energy-releasing and energy-using reactions.

10.2 Photosynthesis: Trapping Energy

- Photosynthesis is the process by which most producers use energy from light to make their own food. Green chlorophyll in the chloroplasts of a plant's cells traps light energy needed for photosynthesis.
- The light reactions of photosynthesis use light to produce ATP and result in the splitting of water molecules.
- The reactions of the Calvin cycle make carbohydrates using CO_2 along with ATP and hydrogen from the light reactions.
- Certain autotrophs can obtain energy from inorganic compounds by chemosynthesis.

10.3 Getting Energy to Make ATP

- Respiration is the process in which cells break down molecules to release energy.
- Aerobic respiration takes place in mitochondria, uses oxygen, and yields many more ATPs than do anaerobic processes.
- Energy can be released anaerobically by glycolysis followed by fermentation.
- The end products of photosynthesis are the starting products for respiration.

Key Terms

Write a sentence that shows your understanding of each of the following terms.

ADP
aerobic process
alcoholic fermentation
anaerobic process
ATP
Calvin cycle
chemosynthesis
citric acid cycle
electron transport chain
glycolysis
lactic acid fermentation
light reactions
photolysis
photosynthesis
respiration

Understanding Concepts

1. Which of the following is NOT one of the substances needed for the Calvin cycle?
   a. ATP
   b. water
   c. carbon dioxide
   d. NADPH + H^+
2. What are the products of photosynthesis?
   a. glucose and oxygen
   b. carbon dioxide and water
   c. glucose and carbon dioxide
   d. water and oxygen
3. Plants must have a constant supply of ____________ because it is a raw material needed for photosynthesis.
   a. oxygen
   b. sugar
   c. water
   d. ATP
4. ____________ processes require oxygen, whereas ____________ processes do not.
   a. Anaerobic, aerobic
   b. Aerobic, anaerobic
   c. Photolysis, aerobic
   d. Aerobic, respiration
5. During all energy conversions, some of the energy is converted to ____________.
   a. carbon dioxide
   b. water
   c. heat
   d. sunlight
6. Four ATPs are produced by glycolysis, but the net yield is ____________ ATPs.
   a. one
   b. two
   c. three
   d. four
7. What is the final electron acceptor in the electron transport chain?
   a. water
   b. hydrogen
   c. oxygen
   d. carbon dioxide

Chapter 10 Review

Reviewing Main Ideas

Summary statements can be used by students to review the major concepts of the chapter.

Key Terms

Answers should go beyond defining the terms. Accept any answer that uses the term correctly and in the proper context.

Understanding Concepts

1. b
2. a
3. c
4. b
5. c
6. b
7. c

reactions, require enzymes, occur in specific organelles, and involve movement of electrons in transport chains.

Chapter 10 Review

8. d
9. b
10. d
11. a
12. a
13. a
14. c
15. d
16. b
17. c
18. a
19. b
20. d

Applying Concepts

21. During physical activity, muscle cells must release energy at a higher rate than skin cells.
22. Photosynthesis stores energy, whereas respiration releases energy, and the products of one process are the reactants of the other process.
23. Light energy is used to produce the food that plants break down to supply chemical energy and raw materials.
24. Other pigments in the plant absorb some of the light of other wavelengths and pass the energy to chlorophyll for use in photosynthesis. The remaining light that is not trapped is reflected or absorbed as heat.
25. carbon dioxide and water
26. Possible answer: The amount of oxygen in the atmosphere would start to decrease, because it would be used up in respiration and not replaced.

Chapter 10 Review

8. What is the molecule that is directly used by a cell for energy?
   a. sugar c. starch
   b. water d. ATP
9. Which of the following is NOT an energy-utilizing reaction?
   a. glycolysis
   b. light reaction
   c. Calvin reactions
   d. muscle contraction
10. What compounds act as carriers in many biological reactions?
    a. lactic acids c. citric acids
    b. alcohols d. coenzymes
11. When yeast ferments the sugar in a bread mixture, what is produced that causes the bread dough to rise?
    a. carbon dioxide c. ethyl alcohol
    b. water d. oxygen
12. ___________ is the process by which autotrophs trap energy from sunlight to build carbohydrates.
    a. Photosynthesis
    b. Glycolysis
    c. The Krebs cycle
    d. Anaerobic respiration
13. What is the function of chlorophyll in the light reaction?
    a. capture light energy
    b. bind CO_2 to H_2O
    c. split to produce O_2
    d. act as a source of CO_2
14. What cell organelle is involved mainly in photosynthesis?
    a. nucleus c. chloroplast
    b. ribosome d. mitochondrion
15. Photolysis in photosynthesis is the splitting of ___________.
    a. carbon dioxide c. sunlight
    b. sugar d. water
16. Some bacteria are capable of capturing energy from a source other than the sun. These bacteria are called ___________.
    a. photosynthetic c. glycolytic
    b. chemosynthetic d. photolytic
17. Where does glycolysis occur in the cell?
    a. inside the endoplasmic reticulum
    b. inside the mitochondria
    c. in the cytoplasm
    d. outside of the cell
18. In fermentation, ___________ changes either to lactic acid or to ethyl alcohol and carbon dioxide.
    a. pyruvic acid c. oxygen
    b. citric acid d. water
19. Plants and animals are similar in that they both utilize ___________ to release chemical bond energy from carbohydrates.
    a. carbon dioxide c. methane
    b. oxygen d. ADP
20. Which of the following does NOT require ATP?
    a. bioluminescence
    b. transmitting nerve impulses
    c. movement of flagella
    d. diffusion

Applying Concepts

21. Why would human muscle cells contain many more mitochondria than skin cells?
22. Why are aerobic respiration and photosynthesis opposite processes?
23. A person tells you that plants "live off light energy instead of food." What could you say to help this person understand what photosynthesis produces?
24. What happens to sunlight that strikes a leaf but is not trapped by photosynthesis?
25. If you were planning on studying the compounds that could possibly be the source of the oxygen released during photosynthesis, which compounds would you need to consider?
26. What might happen to Earth's atmosphere if photosynthesis suddenly stopped?

Program Resources

Chapter Assessment, pp. 55-60 L1
Alternate Assessment in the Science Classroom
Computer Test Bank L1
Content Mastery, pp. 37-40 L1

Thinking Critically

27. ***Concept Mapping*** Make a concept map that relates the following terms and phrases. Supply the appropriate linking words for your map.

ATP, alcoholic fermentation, anaerobic processes, citric acid cycle, electron transport chain, glycolysis, lactic acid fermentation, respiration

28. ***Formulating Hypotheses*** *Elodea* sprigs were placed under a white light, and the rate of photosynthesis was measured by counting the number of oxygen bubbles per minute for ten minutes as shown in the graph. Predict what would happen to the rate of photosynthesis if a piece of red cellophane were placed over the white light.

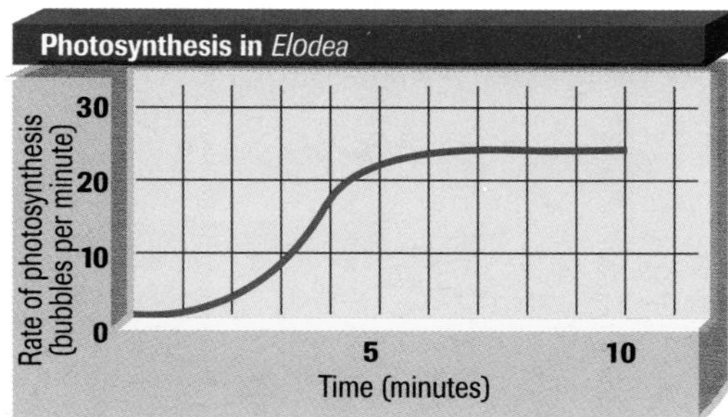

29. ***Observing and Inferring*** Yeast cells must be forced to ferment by placing them in an environment without any oxygen. Why would the yeast cells carry out aerobic respiration rather than fermentation when oxygen is present?

30. ***Recognizing Cause and Effect*** A Window plant native to the desert of South Africa has leaves that grow almost entirely underground with only the transparent tip of the leaf protruding above the soil surface. Suggest how this adaptation aids the survival of these plants.

ASSESSING KNOWLEDGE & SKILLS

Yeast cells and sucrose were placed into a test tube, then the tube was plugged. The yeast-sucrose mixture incubated for 24 hours. Gas bubbles began to rise to the top of the tube. After 24 hours, no sucrose was left in the solution.

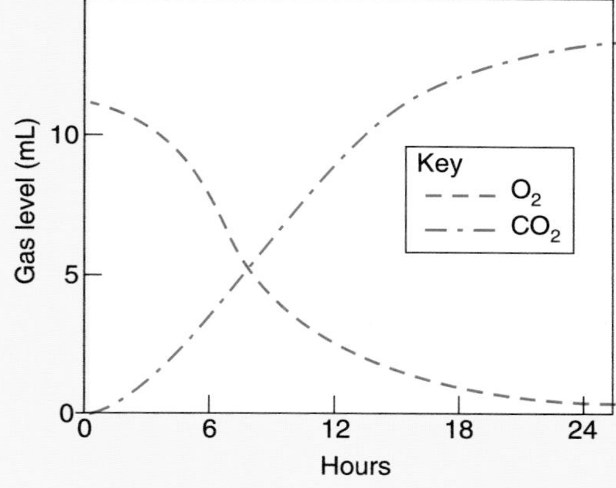

Interpreting Data Use the graph to answer the questions below.

1. What process was the yeast using to digest the sucrose at the beginning of the experiment?
   a. photosynthesis
   b. anaerobic respiration
   c. aerobic respiration
   d. light reactions
2. Which of the following would be left in the solution after 24 hours?
   a. sucrose
   b. lactic acid
   c. oxygen
   d. ethyl alcohol
3. What gas would be found in the top of the tube after the incubation period?
   a. carbon dioxide
   b. oxygen
   c. hydrogen
   d. nitrogen
4. ***Making a Table*** Construct a table from the graph of the oxygen levels and carbon dioxide levels every 4 hours.

Chapter 10 Review

Thinking Critically

27. See below.
28. Photosynthetic rate would decrease. Although the cellophane transmits red light, it would cut down the light in the blue end of the spectrum that is also used in photosynthesis.
29. Aerobic respiration is a much more efficient process and produces many more ATPs per sugar molecule.
30. This adaptation conserves water while allowing the plant to get light for photosynthesis, thereby solving the main problem of desert plants.

ASSESSING KNOWLEDGE & SKILLS

1. c
2. d
3. a
4. Student tables should be consistent with the information in the graph.

27.

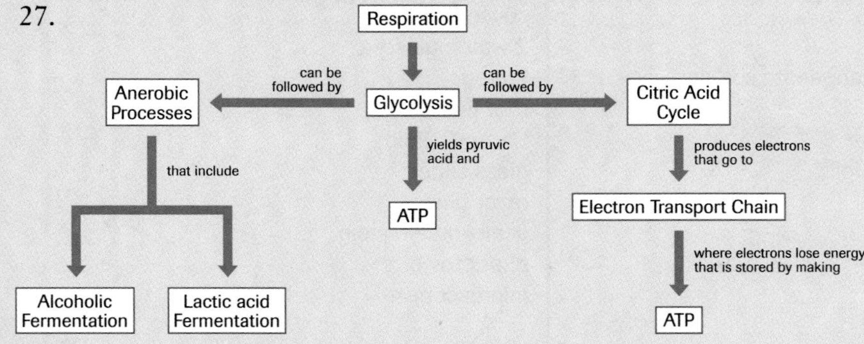

Chapter Organizer

SECTION	OBJECTIVES	ACTIVITIES/FEATURES
11.1 Cell Growth and Reproduction National Science Standards: UCP.1-3; A.1, A.2, C.1	1. **Analyze** the reasons why cells are small. 2. **Sequence** the events of the cell cycle.	**Minilab:** How quickly does diffusion occur?, p. 263 **Minilab:** What happens to the surface area of a cube as the volume increases?, p. 264 **Thinking Lab:** How does the length of the cell cycle vary?, p. 268 **Biology & Society Technology:** Replacement Skin, p. 272 **Biolab:** The Time for the Cell Cycle, p. 274
11.2 Control of the Cell Cycle National Science Standards: UCP.1, UCP.2; C.1; E.1, E.2; F.1, F.5, F.6; G.1, G.2	3. **Describe** the role of enzymes in the regulation of the cell cycle. 4. **Distinguish** between events of a normal cell cycle and abnormal events that result in cancer.	**Health Connection:** Skin Cancer, p. 277

ACTIVITY MATERIALS

BIOLAB	MINILABS	ALTERNATE LAB
page 274 compound microscope onion root tip slides compass protractor	**page 263** potato razor blade cup or beaker potassium permanganate solution forceps clock waxed paper or foil **page 264** graph paper tape scissors	**page 268** HCl-alcohol solution coverslips 2 watch glasses forceps root tips Carnoy's fluid glass slide razor blade aceto-orcein stain paper towel microscope

Chapter 11 Cell Reproduction

TEACHER CLASSROOM RESOURCES

Reproducible Masters	Transparencies
Section Focus Master 23: Diffusion and Cell Size L1 SAE **Reinforcement and Study Guide,** pp. 41-43 L1 **Biolab and Minilab Worksheets,** pp. 39-42 L1 **Concept Mapping:** Cell Cycle, p. 11 L1 **Laboratory Manual:** Why Don't Cells Grow Indefinitely?, pp. 65-68; How Does the Environment Affect Mitosis?, pp. 69-72 L2	**Basic Concepts Transparency #13:** Mitosis L1 SAE **Reteaching Transparency #11:** Cell Cycle L1 SAE
Section Focus Master 24: Uncontrolled Cell Division L1 SAE **Reinforcement and Study Guide,** p. 44 L1 **Critical Thinking/Problem Solving:** Linking a Tumor Suppressor Gene to the Cell Cycle, p. 11 L3 **Content Mastery,** pp. 41-44 L1	

ASSESSMENT MATERIALS

Chapter Assessment, pp. 61-66
Performance Assessment in the Biology Classroom, p. 13
MindJogger Videoquiz
Alternate Assessment in the Science Classroom
Computer Test Bank

Spanish Resources SAE
English/Spanish Audiocassettes SAE
Cooperative Learning in the Science Classroom COOP LEARN
Lesson Plans

KEY TO TEACHING STRATEGIES

- L1 Level 1 activities should be within the ability range of all students including those with learning difficulties.
- L2 Level 2 activities are within the ability range of average to above-average students.
- L3 Level 3 activities are designed for the ability range of above-average students.
- SAE SAE activities should be within the ability range of Students Acquiring English.
- COOP LEARN Cooperative Learning activities are designed for small group work.
- P These strategies represent student products that can be placed into a best-work portfolio.
- These strategies are useful in a block scheduling format.

GLENCOE TECHNOLOGY

The following multimedia resources are available from Glencoe.

Biology: The Dynamics of Life
CD-ROM SAE
Videodisc Program

The Infinite Voyage Series
Miracles by Design

The Secret of Life Series
On the Brink: Portraits of Modern Science

National Geographic Society Series
STV: The Cell
How Cells Reproduce

Science and Technology Videodisc Series (STVS)
Plants & Simple Organisms
Bitter Melon Cancer Treatment

CHAPTER

11 Cell Reproduction

Chapter Overview

Students start the chapter with a probe into the limits of cell size. They explore how diffusion of materials in a cell, the amount of DNA in a cell nucleus, and surface area-to-volume ratio are responsible for the small sizes of cells. The cell cycle, consisting of interphase and mitosis, is described. Students learn of the activities cells carry out during interphase. The prophase, metaphase, anaphase, and telophase stages of mitosis are then discussed, and students examine them in the Biolab.

Next, students examine the events that influence the amount of time spent by various cells in each part of the cell cycle. The roles of enzymes and contact inhibition are explored. Students then examine why cancer is considered as a mistake in the cell cycle. The chapter concludes with a discussion of the causes of cancer.

Key Terms

anaphase
cancer
cell cycle
centriole
centromere
chromosome
gene
interphase
metaphase
mitosis
prophase
sister chromatid
spindle
telophase

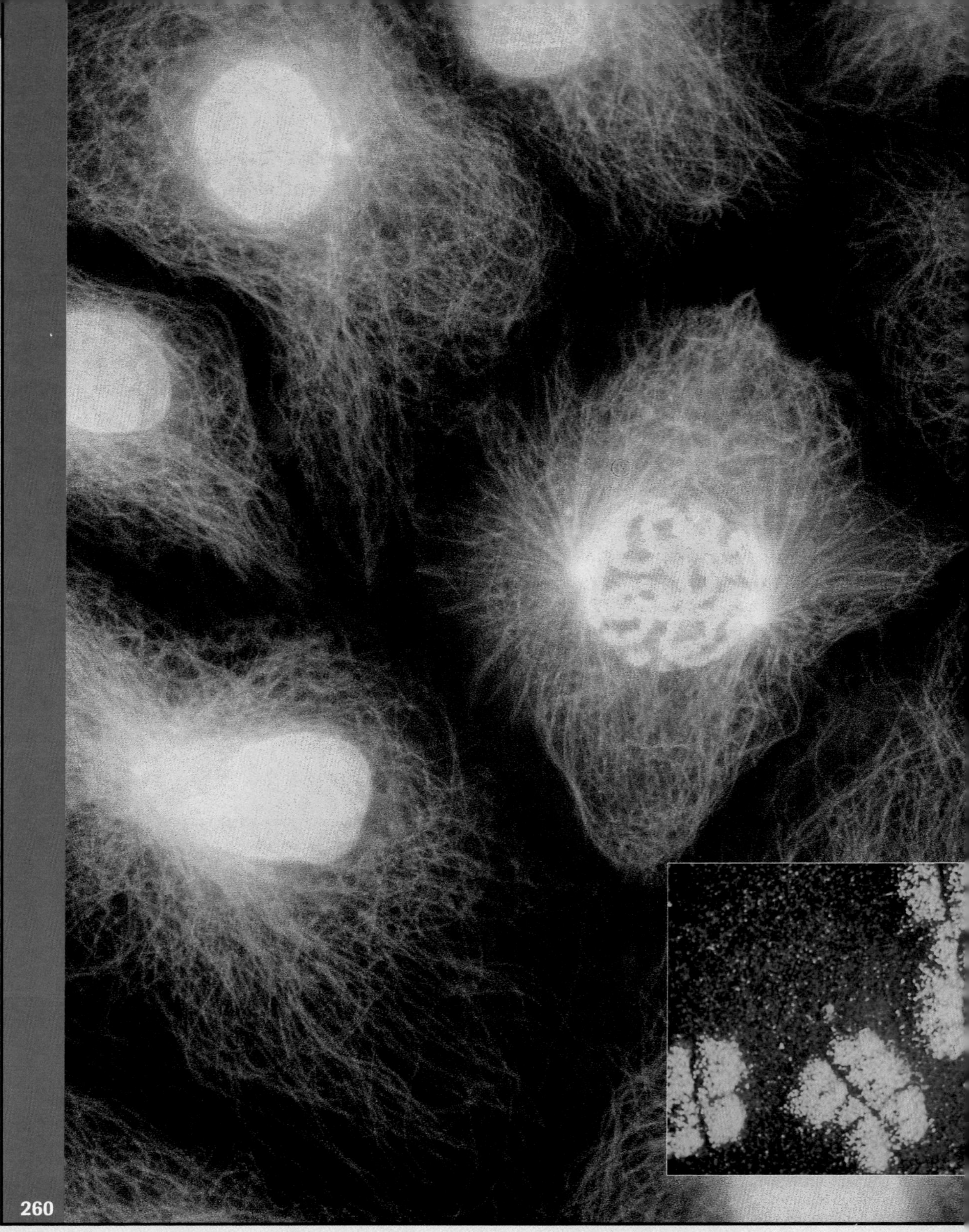

260

Look for the following logo for strategies that emphasize different learning modalities. LS

Kinesthetic	**Meeting Individual Needs, pp. 266, 270; Project, p. 267; Reinforcement, p. 271**
Visual-Spatial	**Making a Model, p. 265; Alternate Lab, p. 268; Visual Learning, p. 270**
Linguistic	**Student Journal, p. 264; Using Science Terms, p. 266; Portfolio, p. 277**
Logical-Mathematical	**MiniLab, pp. 263, 264; Math Connection, p. 265; Student Journal, p. 265; Thinking Lab, p. 268; Portfolio, p. 272; Discussion, p. 276**

LS

CHAPTER

11 Cell Reproduction

When you see a newborn baby, is it hard for you to imagine that just nine months ago, that baby was a single cell? And isn't it hard to imagine that a small acorn will eventually grow into a towering oak tree? When you mow the lawn every week, do you sometimes wonder how the grass grows so fast? Multicellular organisms such as humans, oaks, and grass contain billions and trillions of cells. Where do they all come from?

Living organisms are always making new cells. A healthy human adult produces 25 million new cells each second. Every time an organism grows in size or repairs worn out or damaged tissues, new cells are made. Regardless of whether new cells are being made in plants or in animals, the method by which they are produced is remarkably similar.

Concept Check

You may wish to review the following concepts before studying this chapter.

- Chapter 8: cell theory, mitochondria, nucleus
- Chapter 9: diffusion

Chapter Preview

Laboratory Activities

Biolab

- The Time for the Cell Cycle

Minilabs

- How quickly does diffusion occur?
- What happens to the surface area of a cube as the volume increases?

If you were to observe cells of a growing organism, you would see that they are reproducing in the same manner as the cell in the center of the large photo. The inset photo on the left shows several chromosomes—key players in the drama of cell reproduction.

Magnification: (large) 300×, (inset) 6000×

(Large photomicrograph courtesy of J.M. Murray, Department of Cell and Developmental Biology, University of Pennsylvania)

Assessment Planner

Choose assessment strategies from the following pages to evaluate the progress of your students.

Assess, pp. 272, 278
Alternate Lab, pp. 268-269
Minilabs, pp. 263, 264
Portfolio, pp. 272, 277
Thinking Lab, p. 268
Biolab, pp. 274-275
Chapter Review, pp. 279-281

CHAPTER 11

Introducing the Chapter

Review the cell theory, emphasizing that all cells come from preexisting cells. Ask students how a healthy human adult can produce 25 million new cells every second and not grow larger. *These cells replace dead cells.* You may want to explain that most of these dead cells are sloughed off the skin and the lining of the digestive tract.

Direct students' attention to the dividing cell in the center of the photograph (in prophase) and ask them to point out how it looks different, structurally, from the nondividing cells that they have viewed previously. *Instead of having a nucleus that appears to be mainly filled with granules, the center of the dividing cell seems to be filled with thick threadlike objects. The nucleolus has disappeared in the dividing cell.*

Theme Development

A major theme of the chapter is *homeostasis,* as it relates to the need for cells to survive size limitations. Cells divide in order to maintain homeostasis. Another theme of the chapter is *unity within diversity.* This theme is evident as the striking similarities of the process of mitosis in cells of both plants and animals are presented.

Concept Check

Students will review the structure of an interphase cell from Chapter 8 and contrast it to the structure of a dividing cell.

SECTION

11.1

Prepare

Key Concepts

Students will learn that there are limits to cell size. In particular, these limits are diffusion, DNA content, and surface area-to-volume ratio. Students will look at what cells do when they reach maximum size—they divide. The events of the cell cycle will then be considered, including the stages of mitosis.

Block Scheduling

Look for this symbol for strategies that are useful in a block scheduling format. For more information on block scheduling, refer to Section 11.1 in the **Lesson Plans** booklet.

Materials

- Purchase food coloring for the Demonstration, potato for the Minilab, pipe cleaners or yarn for the Reinforcement and Meeting Individual Needs.
- Collect old insulated electrical cords for the Demonstration.
- Start onion roots for the Alternate Lab.

1 Focus

Bellringer

Before presenting the lesson, display **Section Focus Master 23** on the overhead projector and have students answer the accompanying questions. L1 SAE

SECTION

11.1 Cell Growth and Reproduction

Section Preview

Objectives

Analyze the reasons why cells are small.

Sequence the events of the cell cycle.

Key Terms

chromosome
cell cycle
interphase
mitosis
prophase
sister chromatid
centromere
centriole
spindle
metaphase
anaphase
telophase

If you could examine the cells of this lion and her cub, how do you think they would compare in size? Would you be surprised to learn that the cells of both mother and cub are, on average, exactly the same size? The cells are too small to be seen without the aid of a microscope. Why do organisms have such small cells?

Cell Size Limitations

Cells come in a wide variety of sizes. Some, such as red blood cells, measure only 8 µm in diameter. Others, such as nerve cells in large animals, can reach lengths of up to 1 m. The largest known cell is the yolk of an ostrich egg—measuring a whopping 8 cm in diameter! Most living cells, however, are between 2 and 200 µm in diameter. The body of an adult lion contains trillions of these small cells. Why can't organisms be just one giant cell?

Why diffusion limits cell size

To understand why cell size is limited, think of the materials that cells need to stay alive and do their jobs. You know that cells require a constant supply of glucose and oxygen to carry out cellular respiration and produce large amounts of ATP. These substances, and waste products such as carbon dioxide, move through the cytoplasm by diffusion. It follows, then, that cells can metabolize only as quickly as they receive raw materials by diffusion.

It has been estimated that it takes a molecule of oxygen only a small fraction of a second to diffuse through the cytoplasm from the plasma membrane to the center of a typical cell. Therefore, the innermost mitochondrion in the center of an average-sized cell with a diameter of 20 µm will receive supplies of oxygen and other important molecules a fraction of a second after they pass through the plasma membrane.

What would happen if the cell got bigger? Although diffusion is a fast and efficient process over short distances, it becomes slow and inefficient as the distances become larger. A mitochondrion at the center of a hypothetical cell with a diameter of 20 cm would have to wait months before receiving molecules that enter the cell. Because of this time restric-

Program Resources

Section Focus Master 23 L1 SAE

Reinforcement and Study Guide, pp. 41-43 L1

Biolab and Minilab Worksheets, pp. 39-42 L1

Performance Assessment in the Biology Classroom, p. 13 L1 P

Laboratory Manual, pp. 65-68, 69-72 L2

Concept Mapping, p. 11 L1

Basic Concepts Transparency 13 and **Master** L1 SAE

Reteaching Transparency 11 and **Master** L1 SAE

tion, organisms can't be just one giant-sized cell. They would die long before the oxygen ever got to the mitochondria.

Why a cell's DNA content limits size

A second reason why cells are small is because most cells contain only one nucleus. Molecules of RNA that are involved in protein production are made by the DNA in the nucleus. The RNA then leaves the nucleus and moves through the cytoplasm to the ribosomes, where it directs the production of enzymes and other proteins. If a cell doesn't have enough DNA to program its metabolism, it cannot live.

What happens in larger cells where an increased amount of cytoplasm requires increased supplies of enzymes? In many large cells, such as the giant amoeba *Pelomyxa* shown in *Figure 11.1,* more than one nucleus has evolved. Large amounts of DNA in many nuclei ensure that cell activities will be carried out quickly and efficiently.

Surface area-to-volume ratio

The third reason why cells are limited in size is that as a cell's size increases, its volume increases much faster than its surface area.

MiniLab

How quickly does diffusion occur?

In this lab, you will place a small potato cube in a solution of potassium permanganate and observe how far the dark purple color diffuses into the potato after a given length of time. The potato cube represents a cell, and the purple solution represents nutrients that must reach the center of the cell quickly.

Procedure

1. Using a single-edge razor blade, cut a cube 1 cm on each side from a raw, peeled potato. **CAUTION:** *Be careful with sharp objects.*

2. Place the cube in a cup or beaker containing the purple solution. The solution should cover the cube. Note and record the time. Let the cube and solution stand for between 10 and 30 minutes.
3. Using forceps, remove the cube from the solution and note the time. Cut the cube in half.
4. Measure, in millimeters, how far the purple solution has diffused, and divide this number by the time you allowed your potato to remain in the solution. This is the diffusion rate.

Analysis

1. How far did the purple solution diffuse?
2. What was the rate of diffusion per minute?
3. Why are nearly all cells much smaller than 1 cm in diameter?

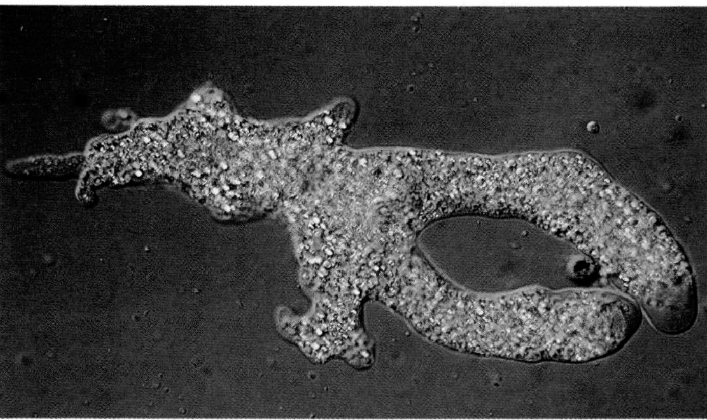

Magnification: 100×

Figure 11.1
This giant amoeba, *Pelomyxa,* is several millimeters in diameter. It can have up to 1000 nuclei.

2 Teach

MiniLab

Purpose

LS **Logical-Mathematical** Students will determine the rate of diffusion of a solution.

Process Skills

measure in SI, collect and organize data, interpret data, experiment, and analyze

Teaching Strategies

- Caution students that the solution can be caustic. If they get some on their hands, they should wash immediately. To keep the solution off their hands, students should set the cube on waxed paper or foil when they remove it from the solution and hold the cube with the forceps as it is cut.

Expected Results

The color will diffuse only a few millimeters into the cube, the exact distance depending upon the amount of time it is left in the solution.

Analysis

1. Answers will depend on the amount of time the cube is left in the solution.
2. The rate will be in tenths to hundredths of millimeters per minute.
3. The cell must be small so that nutrients can reach all parts of the cell and waste products can leave all parts of the cell fast enough for survival.

Assessment

Portfolio: Have students write a report of the Minilab for their portfolios. L1 P

MiniLab

Purpose

LS Logical-Mathematical Students will compare the increase in volume of an object with the increase in its surface area.

Process Skills

measure in SI, use numbers, recognize cause and effect, interpret data, analyze

Teaching Strategies

- Ask students if they have ever wondered why cells can't continue to grow larger and larger to become giant cells. Then ask them to consider the fact that most cells, whether from an elephant or an earthworm, are microscopic in size. L1

Expected Results

Students will find that the ratio of surface area to volume decreases as cell size increases.

Analysis

1. 8
2. The eight 1-cm cubes have a combined surface area of 48 square cm, and the 2-cm cube has a surface area of 24 square cm. The assembled surface area is twice that of the single cube.
3. Volume. It will become more difficult for the cell to survive as cell size increases. It will take more time for substances to get into the interior of the cell and for wastes to get from the interior of the cell to the outside.

Assessment

Journal: Students should write a summary of the Minilab, including the Analysis questions, for their journals. L1

MiniLab

What happens to the surface area of a cube as the volume increases?

One reason cells are small is that, as they grow, their volume increases faster than does their surface area. In this lab, you will measure how much greater the increase in volume is than the increase in surface area of an object.

Procedure

1. Using graph paper and the diagram at the right, cut out two cubes. The first cube should be 1 cm on each side. The second cube should be 2 cm on each side.
2. Put together each cube using tape.
3. Calculate the surface area of each cube by multiplying the area of one face by six. Express your answers in square centimeters.
4. Working with other groups, assemble your 1-cm cubes to form a cube as large as your 2-cm cube. Observe how many small cubes are needed to obtain a volume equal to the larger cube.

Analysis

1. How many small cubes were required to equal the volume of the large cube?
2. How much greater is the total surface area of the assembled cubes than the surface area of the single 2-cm cube?
3. As the assembled cube became larger, which increased more quickly: the surface area or the volume? What is the significance of this for cell size?

Picture a cube-shaped cell like those shown in *Figure 11.2*. The smallest cell has 1-mm sides, a surface area of 6 mm^2, and a volume of 1 mm^3. If the side of the cell is doubled to 2 mm, the surface area will increase fourfold to $6 \times 2 \times 2 = 24$ mm^2. Observe what happens to the volume; it increases eightfold to 8 mm^3.

Is big better?

What does this mean for cells? How does it affect cell function? If cell size doubled, the cell would require eight times more nutrients and would have eight times more waste to excrete. The surface area, however, would increase by a factor of only four. Thus, the plasma membrane would not have enough surface area through which oxygen, nutrients, and wastes could diffuse. The cell would either starve to death or be poisoned from the buildup of waste products. In fact, though, cells divide before they reach this point.

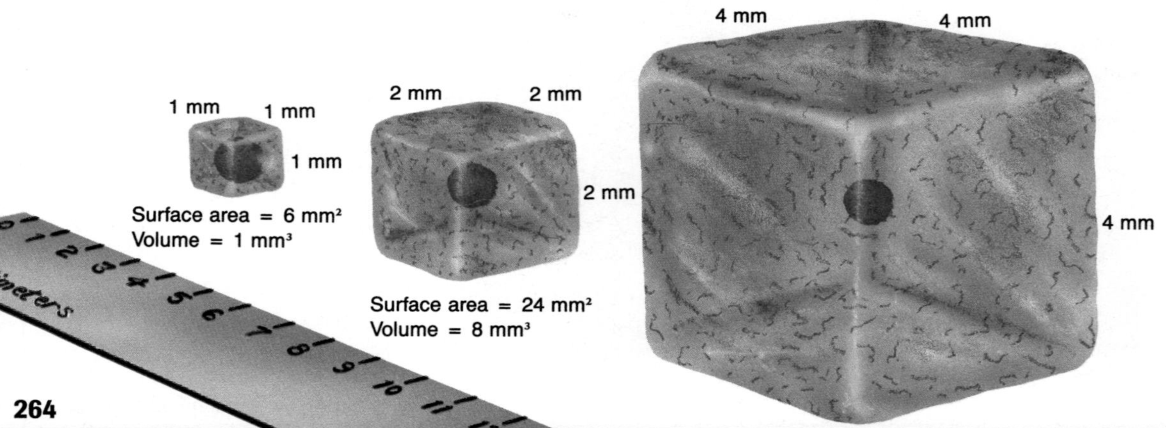

Figure 11.2
Surface area-to-volume ratio is one of the major factors limiting cell size. Note how the surface area and the volume change as the sides of a cell double in length from 1 mm to 2 mm. Calculate the change in surface area and volume as the cell doubles in size again to 4 mm on a side.

264

STUDENT JOURNAL

The Limits to Cell Size Ask students to explain why the invasion of a one-celled giant creature such as the one in the movie, "The Blob," would be impossible, using what they know about the relationship of surface area to volume of an object. L1 LS

Figure 11.3

Cells reproduce at different times in the life of an organism, depending on their function in the organism.

Cells reproduce when a leaf bud grows to be a leaf.

Cells reproduce when an insect egg grows first to be a caterpillar and then to be an adult butterfly.

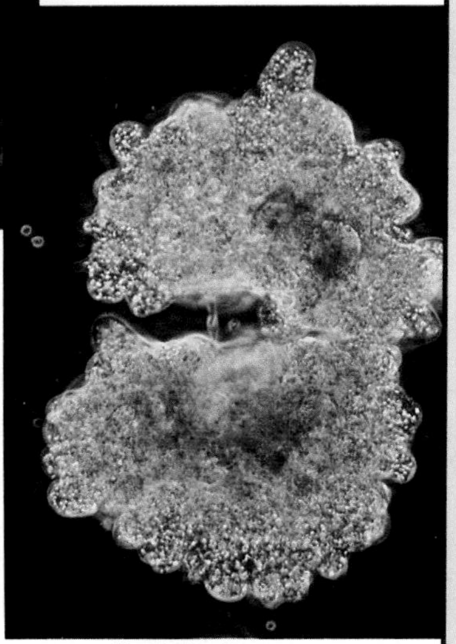

Cells reproduce when one amoeba divides to become two amoebas.

Magnification: 150×

When Do Cells Divide?

Right now, as you are reading this page, many of the cells in your body are growing, dividing, and dying. Old cells on the soles of your feet and on the palms of your hands are being shed and replaced, cuts and bruises are healing, and your intestines are producing millions of new cells each second.

Other organisms go through similar changes. New cells are produced as tadpoles become frogs, and as a tomato plant grows and wraps around a garden stake. All organisms grow and change; worn out tissues are repaired or are replaced by newly produced cells. *Figure 11.3* shows other instances when new cells are made.

Cell Reproduction—The Role of Chromosomes

By the middle of the 19th century, scientists knew that cells divide to form more cells. As microscopes improved, the details of cell structure were discovered, allowing scientists to interpret and better understand the process of cell division.

Math Connection

Volume and Surface Area Bring in a few small boxes and one large box that is approximately the same size as the small boxes put together. Stack the small boxes together beside the large box. Ask students which set will need more paper if they were to wrap each box separately. Some students may guess that it will take about the same amount of paper.

Have students wrap all the boxes, then unwrap them to demonstrate the difference in the amounts of paper needed. Students should see how the volumes of the large box and the set of small boxes are approximately equal, but the total surface area is much larger for the set of small boxes than for the large box. L1 SAE COOP LEARN LS

Making a Model

LS **Visual-Spatial** Hold two textbooks together with the covers touching. Let this represent a large cell. Separate the two books to divide the cell. Hold the books up so students can see the two surfaces (the covers) that were previously not visible. Elicit from students how the surface area of the two books separated differs from that of the two books together. *The surface area of the separated books is greater.* L1

STUDENT JOURNAL

Cell Cycle The average human cell that is capable of dividing has a cell cycle of 20 hours. Have students calculate how many cells there would be in one week, if they start with one cell at time zero. Students should show their work. L1 LS

Science, Technology, and Society

The Timing of the Cell Cycle

Both the S phase of interphase (in which the DNA is replicated) and mitosis seem to be triggered by the action of an enzyme called cdc2 kinase. Another protein called cyclin is also involved in cell division control. In a new cell, cyclin is synthesized continuously and once it builds up to a critical level, it links up with cdc2 kinase protein to form a complex. The cdc2 is activated and, in turn, adds a phosphate group to the cyclin.

Several other biochemical steps then serve to activate the molecular complex (cdc2 and cyclin) transforming it into maturation-promoting factor (MPF). Mitosis then takes place. After the cell completes mitosis, the level of MPF in the cytoplasm drops abruptly. The level of cdc2 remains constant, but the cyclin level plunges.

What process degrades cyclin? What sets the cell cycle clock into motion? What modifies the clock to run fast or slow? The full picture of how the timing of the cycle is regulated and how its signal is transmitted remains obscure but is the target of active research.

Using Science Terms

LS Linguistic Direct students' attention to the derivation of the term *chromosome* presented in the margin. Elicit from students how the literal meaning, "colored body," applies to chromosomes. *Chromosomes are dark-staining structures that contain genetic material.* L1

Figure 11.4

This bacterial cell was ruptured osmotically, and its DNA molecule leaked out of the cell. Although prokaryotic cells such as this one contain only one-thousandth as much DNA as a typical eukaryotic cell, you can still see that uncoiled DNA molecules are very long structures.

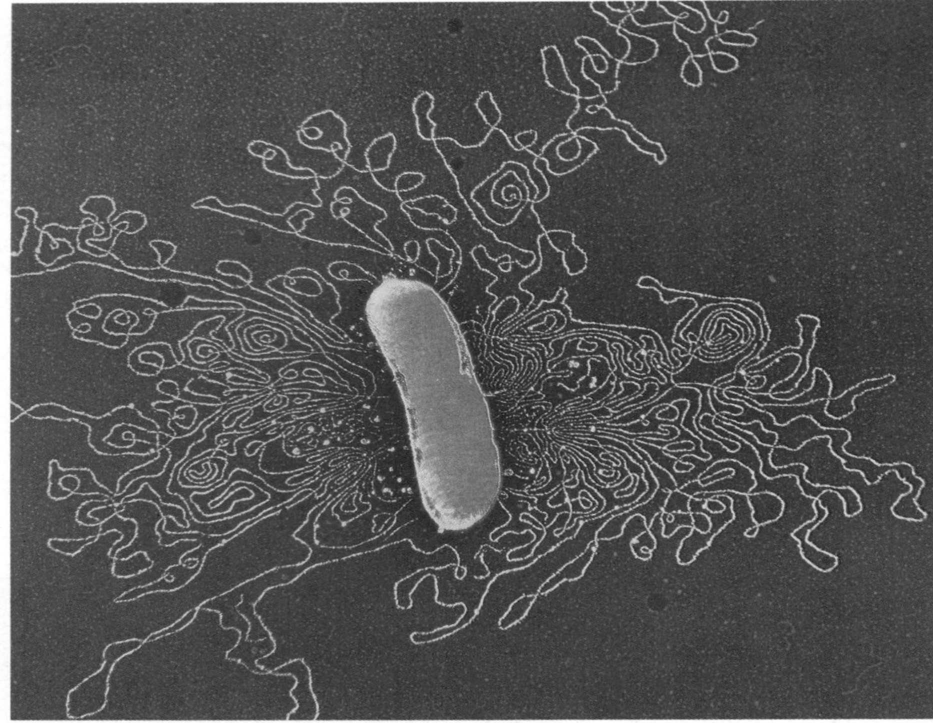

Magnification: 12 500×

chromosome:
chroma (GK) colored
soma (GK) body
Chromosomes are dark-staining structures that contain genetic material.

The discovery of chromosomes

Most interesting to the early biologists was their observation that just before cell division, several short, stringy structures suddenly appeared in the nucleus. What was strange was that these structures seemed to vanish, as mysteriously as they appeared, soon after division of a cell. These structures, which contain DNA and become darkly colored when stained, are now called **chromosomes.**

With the introduction of more advanced microscopes, scientists eventually came to understand that chromosomes are the carriers of genetic material, which is copied and passed from generation to generation of cells. Scientists also learned that chromosomes are always present in cells, but most of the time they exist in a form too thin to be seen with simple microscopes.

The structure of eukaryotic chromosomes

During most of a cell's life cycle, chromosomes exist as chromatin, strands of DNA wrapped around protein molecules. Under an electron microscope, chromatin resembles a plate of tangled-up spaghetti. How can such long structures function without becoming tangled? Before a cell begins to divide, the threads of chromatin begin to coil. The coils shorten and thicken, forming distinct chromosomes that are tightly packed with chromatin.

The electron micrograph in *Figure 11.4* shows a single, uncoiled DNA molecule from a bacterium. Why is it necessary for DNA to coil and uncoil during the life cycle of a cell? When DNA is uncoiled, it can be actively involved in making RNA and in replicating itself.

Meeting Individual Needs

Visually Impaired To demonstrate chromosome coiling and thickening to students who are visually impaired, remove a telephone cord from a telephone. Have students stretch out the cord. Explain that the stretched cord represents an interphase chromosome. Have students allow the cord to return to its normal shape. Explain that this coiling is what happens to chromosomes during prophase, when the chromatin condenses. L1 LS SAE

The Cell Cycle

Fall follows summer, night follows day, and low tide follows high tide. Many events in nature follow a recurring cyclical pattern. Living organisms are no exception. One cycle common to most living things is the cycle of the cell, *Figure 11.5.* The **cell cycle** is the sequence of growth and division of a cell.

As a cell proceeds through its cycle, it goes through two general periods: a period of growth and a period of division. The growth period of the cell cycle is known as **interphase.** Most of a cell's life is spent carrying on the activities of interphase. During interphase, a cell grows in size and carries on metabolism. Also during this period, chromosomes are duplicated in preparation for the period of division.

As you recall, chromosomes are made of DNA. Because DNA contains the master instructions for the cell, it is important that new cells have complete copies of DNA when parent cells divide. Accurate replication of the parent cell's chromosomes, and thus its DNA molecules, accomplishes this.

Following interphase, a cell enters its period of division, in which its nucleus and then its cytoplasm divide to form two daughter cells, each containing a complete set of chromosomes. The process of nuclear division followed by division of the cytoplasm, when chromosomes are distributed equally to daughter cells, is known as **mitosis.** Interphase and mitosis together make up the cell cycle.

mitosis:
mitos (GK) thread
Mitosis is a process that divides threadlike nuclear material equally between two daughter cells.

Interphase—A Busy Time

Interphase is the busiest part of the cell cycle. Cells are making ATP, repairing themselves, and excreting their wastes. They are making proteins, producing new organelles such

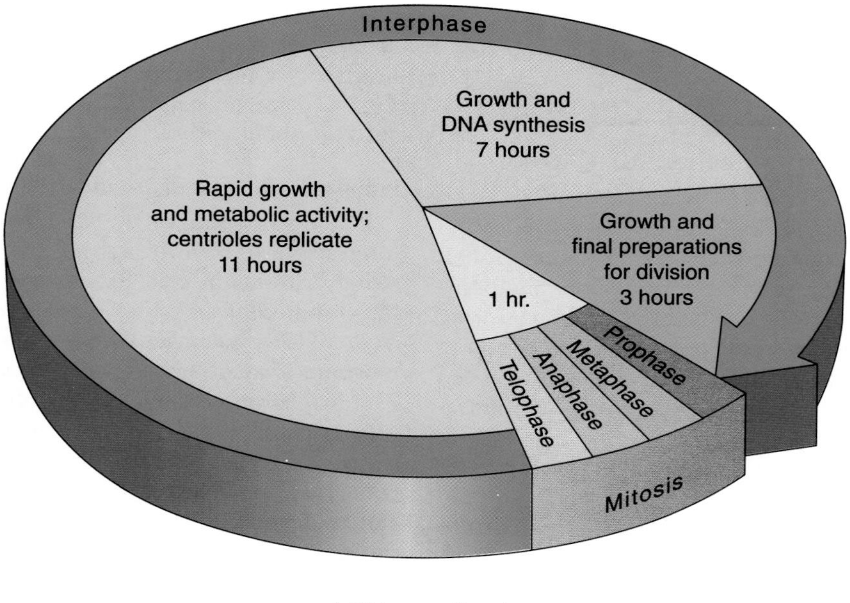

Interphase
Growth and DNA synthesis 7 hours
Rapid growth and metabolic activity; centrioles replicate 11 hours
Growth and final preparations for division 3 hours
1 hr.
Prophase
Metaphase
Anaphase
Telophase
Mitosis

A 22-hour cell cycle

Figure 11.5
The cell cycle is divided into phases representing the important events in the life of a cell. As soon as the cell completes one phase of the cycle, it proceeds to the next phase. During which phase does a cell spend most of its time?

SECTION 11.1

Misconception

Students often think that mitosis occurs in all cells of an organism throughout its life. Explain that in some tissues, once the cells are formed no mitosis occurs. For example, once formed, nerve cells function throughout the life of the organism. In plants, mitosis occurs only in the meristems and not throughout the entire plant.

Concept Development

A model using hand motions to represent the stages of mitosis can be found in the article "A Handy Model for Mitosis," *The American Biology Teacher,* March, 1988, pp. 170-172.

Audiovisual

Show the video *Mitosis,* EBEC or *Mitosis and Meiosis: How Cells Divide,* Charles Clark Co., Inc.

Visual Learning

Figure 11.5 During which phase does a cell spend most of its time? *Interphase*

interNET CONNECTION

Follow the link for this chapter on the Glencoe Homepage at **www.glencoe.com/sec/science** to find out more about cell reproduction.

PROJECT Models of Mitosis

Have students work in groups to design models of the four stages of mitosis. Students might use colored macaroni or licorice for chromosomes. Have students include their completed models in their portfolios. L1
SAE P COOP LEARN IN 

ThinkingLab Interpret the Data

Purpose

LS Logical-Mathematical Students will compare cell cycles of two different types of cells.

Process Skills

compare and contrast, interpret data, analyze

Teaching Strategies

- Relate this Thinking Lab to the uncontrolled growth of cancer cells, which spend a very short time in interphase.

Thinking Critically

The first part of interphase, in which the cell is growing, is the most variable in length. The cell with the longer period of growth would carry on more metabolic activities than the more rapidly dividing cell. Students should justify their answers with statements such as that certain types of cells are always being damaged and need to be replaced.

Assessment

Journal: Have students write a report of this Thinking Lab in their journals. L1

ThinkingLab Interpret the Data

How does the length of the cell cycle vary?

The cell cycle varies greatly in length from one kind of cell to another. Some kinds of cells divide rapidly, while others divide more slowly.

Analysis

Examine the cell cycle diagrams, shown below, of two different types of cells. Observe the total length of each cell cycle, as well as the length of time each cell spends in each phase of the cell cycle.

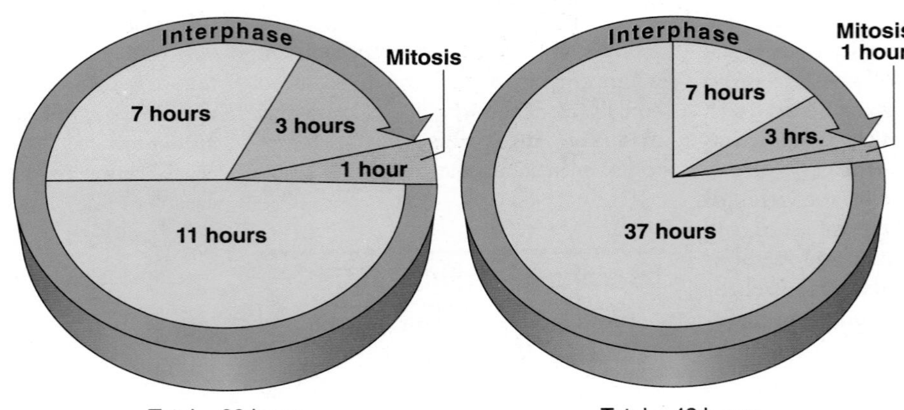

Thinking Critically

Which part of the cell cycle is most variable in length? What inferences can you make concerning the functions of these two types of cells? Why do you think the cycle of some types of cells is faster than in others? Explain your answer.

as ribosomes and mitochondria, and copying their chromosomes. A cell from a plant leaf might be photosynthesizing, a liver cell might be storing extra sugar from lunch in the form of glycogen, and an onion root tip cell might be taking in water and minerals from the soil.

If you look back at the cell cycle diagrammed in *Figure 11.5,* you can see that interphase is divided into three parts. During the first part, the cell grows in size and protein production is high.

In the next part of interphase, the cell copies its chromosomes. DNA synthesis does not occur all through interphase but is confined to this discrete time.

After the chromosomes have been duplicated, the cell enters another shorter growth period in which mitochondria and other organelles are manufactured and cell parts needed for cell division are assembled. Following this activity, interphase ends and mitosis begins.

The Phases of Mitosis

Although cell division is a continuous process, biologists recognize four distinct phases—each phase merging into the next. The four phases of mitosis are prophase, metaphase, anaphase, and telophase. Refer to the photos and diagrams in *Figure 11.8* on pages 270–271 as you read about mitosis.

Prophase—The first phase of mitosis

During **prophase,** the first and longest phase of mitosis, the long, stringy chromatin coils up into visible chromosomes. As you can see in *Figure 11.6,* each duplicated chromosome is made up of two halves. The two halves of the doubled structure are called **sister chromatids.** Sister chromatids and the DNA they contain are exact copies of each other and are formed when DNA is copied during interphase. Notice also that sister chromatids are held together by a structure called a **centromere.**

Alternate Lab: Onion Root Tip Mitosis

Purpose

LS Visual-Spatial Students will prepare slides of onion root tips, showing mitosis.

Materials

HCl-alcohol solution, coverslips, 2 watch glasses, forceps, root tips, Carnoy's fluid, glass slide, razor blade, aceto-orcein stain, paper towel, microscope

Preparation

Place an onion over a glass of water to stimulate roots to grow. Cut root tips and fix for 24 hours in a 3:1 solution of absolute alcohol to glacial acetic acid. Store in 70% alcohol.

Procedure

Give students the following directions.

1. Pour HCl-alcohol solution into one watch glass and Carnoy's fluid into the second watch glass. **Caution:** *Wear goggles and avoid getting the solutions on hands.*
2. Use forceps to transfer three root tips to the first watch glass. Allow to remain 5 minutes.
3. Transfer the root tips to Carnoy's fluid.

As prophase continues, the nucleus begins to disappear as the nuclear envelope and the nucleolus disintegrate. By late prophase, these structures are completely absent. Two important pairs of structures, the centrioles, begin to migrate to opposite ends of the cell. **Centrioles** are small, dark, cylindrical structures that are made of microtubules and are located just outside the nucleus, *Figure 11.7*. Only animal cells contain centrioles; plant cells do not.

As the pairs of centrioles move to opposite ends of the cell, another important structure, called the spindle, begins to form between them. The **spindle** is a football-shaped, cagelike structure consisting of thin fibers made of microtubules. The spindle fibers play a vital role in the separation of sister chromatids during mitosis.

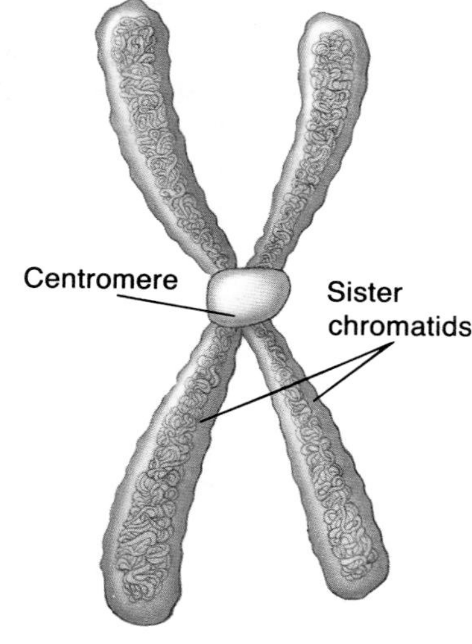

Figure 11.6

The fully coiled chromosome in this electron micrograph appears "hairy" because it consists of a single, long chromatin fiber that is coiled up. The two sister chromatids are held together by a centromere. Each sister chromatid is a complete copy of the original parent chromosome, and the two chromatids are, therefore, identical.

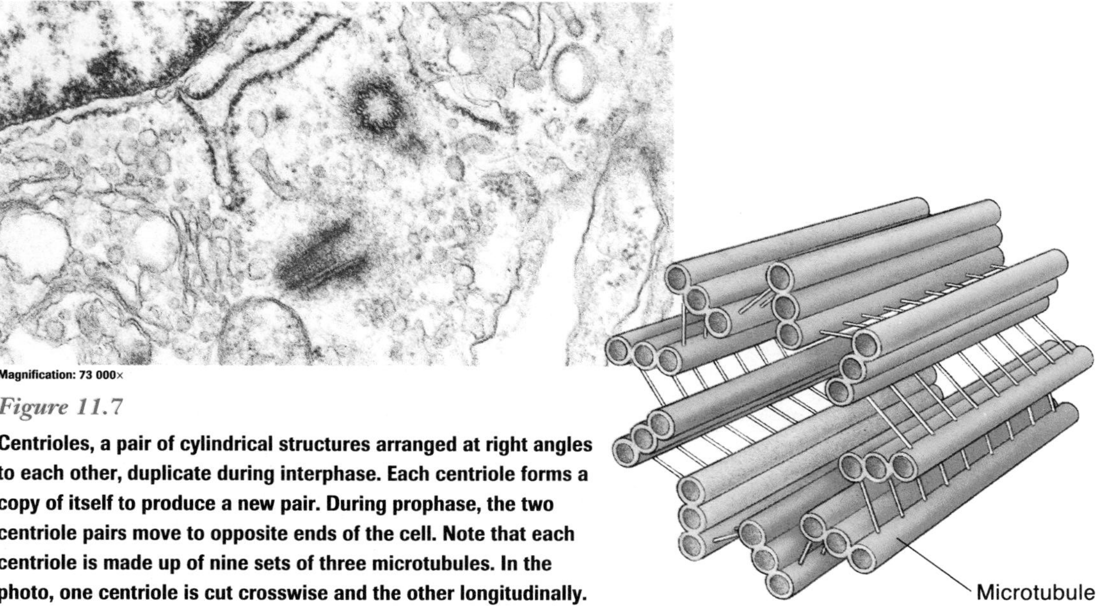

Figure 11.7

Centrioles, a pair of cylindrical structures arranged at right angles to each other, duplicate during interphase. Each centriole forms a copy of itself to produce a new pair. During prophase, the two centriole pairs move to opposite ends of the cell. Note that each centriole is made up of nine sets of three microtubules. In the photo, one centriole is cut crosswise and the other longitudinally.

SECTION 11.1

Enrichment

Have students research recent discoveries on the structure of spindle fibers and on how movement occurs along these fibers. Have students report on their findings. L3 P

Different Viewpoints in Biology

The Origin of Microtubules

The kinetochore is a structure on the centromere that links each chromatid to the microtubules of the spindle. The mechanics of the interactions between these two structures are still being worked out.

Where do the microtubules originate? Do they originate from the kinetochore or from the centrosome, the cloud of amorphous material surrounding the pair of centrioles? With light microscopy, individual microtubules are hard to see because of their high density in forming the spindle and the complexity of their arrangement.

Using video contrast enhancement with light microscopy, researchers have recorded the growth of microtubules out from the centrosome. As the tubules move past a kinetochore, they latch onto it, and the chromosome is dragged toward the centrosome.

4. After 3 minutes, use forceps to transfer the roots to a microscope slide.
5. Use a razor blade to cut off the last 2 mm of the tips and discard the rest.
6. Add 1-2 drops of aceto-orcein stain to each tip. Leave the tips in the solution for 5 minutes. Do not let the solution dry up.
7. Place a coverslip over the root tip. Fold a piece of paper towel several times and insert the slide inside the final fold. Press down with your thumb on the towel to spread the cells.
8. View the squashed tips under low and high power.

Analysis

Have students answer the following questions.

1. The HCl-alcohol treatment breaks down the materials that hold the cells together. Why is this necessary? *To separate cells so it is possible to view a thickness of only one cell*
2. What structures did the aceto-orcein stain? *Chromosomes*

Assessment

Knowledge: Have students draw and demonstrate the mitotic stages. L1

SECTION 11.1

Demonstration

Show how chromatids separate during anaphase. Separate slightly the middle portion between the two covered wires of an old piece of insulated electrical cord. Tie a piece of string to each piece of the separated cord. Slip a rubber band around the cord and through the strings. Pull the strings slowly apart until the wire splits in two, similar to the way chromatids are pulled along the spindle in a dividing cell.

Visual Learning

- **Figure 11.8** What two differences do you notice? *Plant cells do not have centrioles and plant cells divide their cytoplasm with a cell plate.*
- As a class, discuss the changes that occur during the cell cycle. Have students read the captions aloud.
- Encourage students to make a flow chart of the series of events that occur during the cell cycle. Have students include their diagrams in their portfolios. L1 P LS

Software

Mitosis and Meiosis, Intellimation.

GLENCOE TECHNOLOGY

Videodisc

Biology: The Dynamics of Life
Disc 1, Side 1
The Cell Cycle (Ch. 28)

The Cell Cycle

Figure 11.8

The cell cycle is a complex series of continuous events consisting of interphase and mitosis. Identify the characteristics of each phase in the cell cycle, and note the differences between mitosis in animal cells (diagrams) and in plant cells (photos). What two differences do you notice?

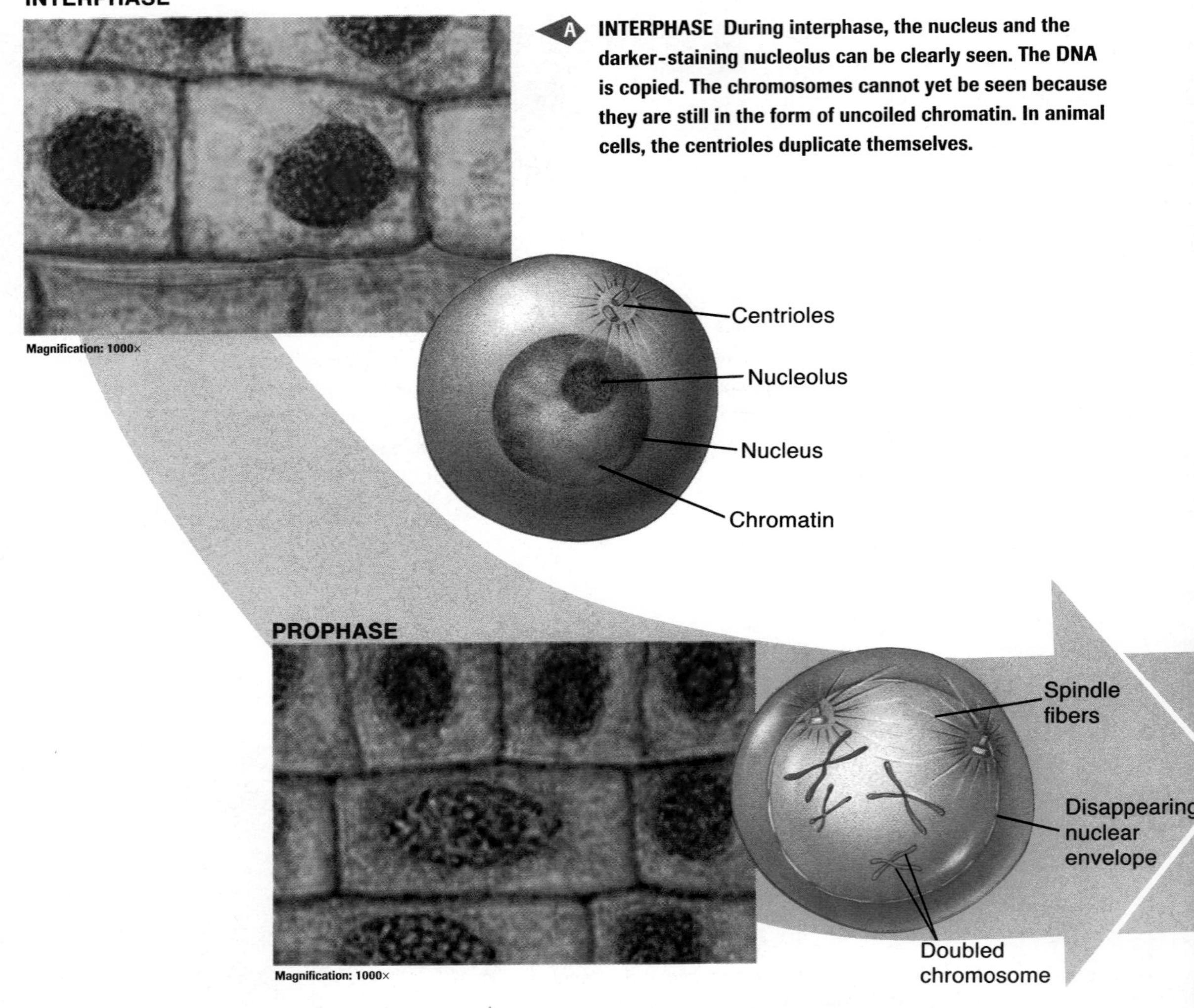

A INTERPHASE During interphase, the nucleus and the darker-staining nucleolus can be clearly seen. The DNA is copied. The chromosomes cannot yet be seen because they are still in the form of uncoiled chromatin. In animal cells, the centrioles duplicate themselves.

B PROPHASE In prophase, the chromatin coils to form visible chromosomes, the nuclear envelope and nucleolus disappear, and a spindle forms between the pairs of centrioles, which have moved to opposite ends of the cell.

Meeting Individual Needs

Visually Impaired Pair students who are visually impaired with other students. Have the pairs practice making the various stages of mitosis using pipe cleaners for chromosomes. L1 COOP LEARN LS

STUDENT JOURNAL

Structure of a Chromatid Have students look back at Figure 11.6 and write a description of the chromatid. Their description should include the terms *chromosome* and *centromere.* L1

TELOPHASE In this final phase of mitosis, two daughter cells are formed. With a complete set of chromosomes at each end of the cell, the cytoplasm divides, the nucleolus and nuclear envelope reappear, and the chromosomes begin to uncoil. When the new cells are separated, they enter interphase and a new cycle begins.

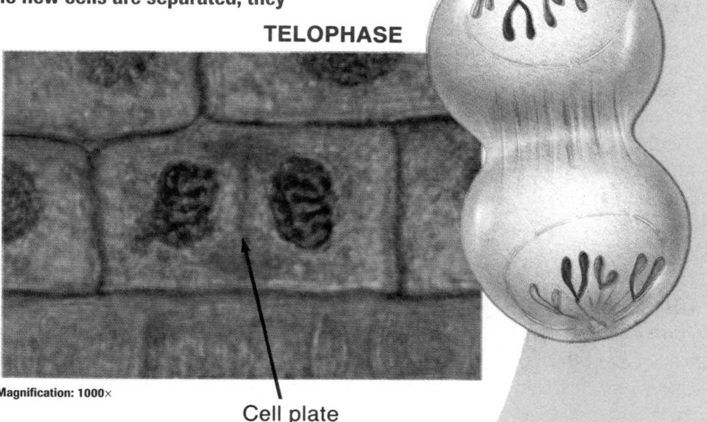

ANAPHASE During anaphase, the centromeres split and the sister chromatids are pulled apart to opposite poles of the cell. Each chromatid is now a separate chromosome.

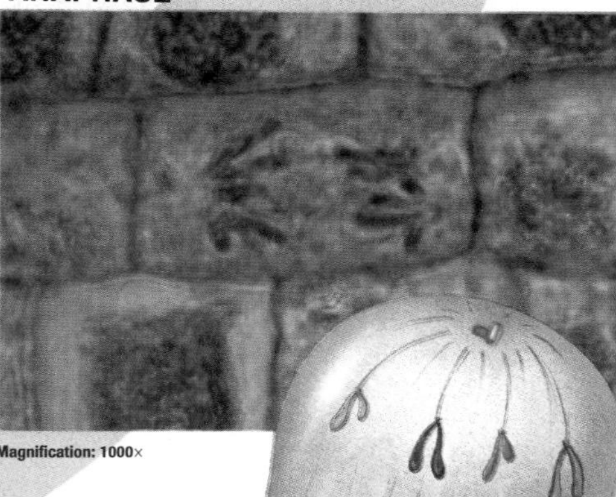

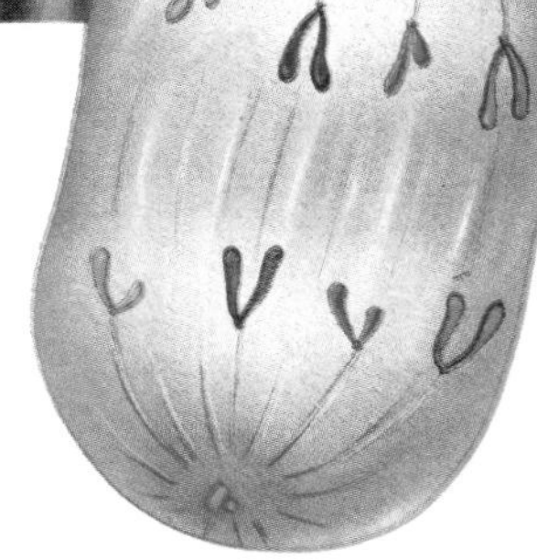

METAPHASE In metaphase, the chromosomes move to the equator of the spindle, with each chromatid attached to a separate spindle fiber by its centromere.

Bioethics

Cloning Human Embryos

The knowledge that the cell cytoplasm contains proteins responsible for regulating cell division has been used in the cloning of animals. In the cloning process, the cells of a young embryo are separated. An unfertilized egg is obtained and its nucleus is replaced by a nucleus from one of the separated cells of the embryo. The embryo then develops into an exact copy, or clone, of the embryo from which the cell was separated.

Recently, the cloning process was reported to be carried out on human embryos. Start a discussion on the ethical implications of this process. For instance, parents might be able to save identical copies of embryos. Then, if their child needed an organ transplant, the mother could give birth to the child's identical twin, a perfect match for organ donation. Parents could keep a frozen embryo as a backup in case their child died. Have students debate whether further research into cloning of human embryos should be controlled by law.

Reinforcement

Have groups of students make models of cells at various stages of cell division. Pipe cleaners or yarn pieces can represent chromosomes. L1 COOP LEARN

CULTURAL DIVERSITY

Jane Cooke Wright and the Development of Chemotherapy

Discuss with students the role of African American scientist Jane Cooke Wright in the development of chemotherapy techniques to treat cancer. Wright's work in the 1950s and 1960s involved testing various anticancer drugs on people with different kinds of cancer. In the 1970s, Wright found that examinations of cancer cells grown in tissue culture could help to predict which drugs would be most effective against that type of cancer. Since that time, Wright has been an active publisher of her work in the field, and in 1975 she was honored by the American Association for Cancer Research for her contributions.

Technology

BIOLOGY & SOCIETY

Replacement Skin

Purpose

Students examine a current use of cell-division technology to grow skin cells for burn victims.

Teaching Strategies

- Ask students to summarize the role of the skin and the problems of a person with severe burns.

INVESTIGATING the Technology

Writing About Biology These proteins and nutrients stimulate the skin cells to grow and divide quickly.

Going Further

Have students find out how skin cultures are grown and what types of media are used.

Software

Cell Functions: Growth and Mitosis, Biology Series from IBM.

3 Assess

Check for Understanding

Test students for their ability to recognize the various phases of mitosis. Place photomicrographs on an overhead projector and ask the class to identify each stage. L1 SAE

Reteach

Review the phases of mitosis, emphasizing that the process is continuous and that one phase blends into the next. Place photomicrographs of the stages of mitosis on the bulletin board directly above diagrams of each stage. Have students compare the photomicrographs and diagrams to correctly identify the phases and learn the terms associated with structures in mitosis. L1 SAE

BIOLOGY & SOCIETY

Technology

Replacement Skin

Skin is the body's first defense against injury and infection. When both the outer and inner layers of the skin are burned away, as they are in third-degree burns, bacteria and other harmful organisms have access to the body's systems. Burn victims must receive fluids continuously to stay alive.

The need for new skin

Until recently, skin to cover burns came from other areas of the patient's body. However, transferring skin from unburned areas isn't possible for patients who have lost most of their skin. The cells that usually produce new skin cells by mitosis have been destroyed in third-degree burn patients.

Cloned skin Now, burn specialists can make as much skin as they need in just 17 to 24 days. They cut a postage stamp-size piece of unburned skin into tiny pieces, then suspend it in a nutrient solution. These skin cells grow into colonies by repeated mitosis in the protein-rich environment. The colonies continue growing together until a large but thin sheet is formed that can be used for grafting.

Although cloned skin is a perfect match for the recipient, there are problems. Because this skin is grown only from the epithelial cells of the dermis, it develops without sweat glands and pores. Normally, sweat rises through pores to the skin's surface, where it evaporates and cools the body. Cloned-skin patients lack this cooling system.

Applications for the Future

Artificial skin, a flexible gelatin made up of collagen composites, cannot replace human skin but is useful for covering open wounds temporarily. It is permeable to body fluids and releases medications to help the body heal.

INVESTIGATING the Technology

Writing About Biology Find out why cultured skin is grown in a medium of proteins and nutrients similar to blood. Write a journal entry summarizing your research.

Metaphase—The second stage of mitosis

During **metaphase,** the short second phase of mitosis, the doubled chromosomes become attached to the spindle fibers by their centromeres. The chromosomes are pushed and pulled by the spindle fibers and begin to line up on the midline, or equator, of the spindle. As you saw in *Figure 11.8*, each sister chromatid is attached to its own spindle fiber. One sister chromatid's spindle fiber extends to one pole, and the other extends to the opposite pole. This arrangement is an important one because it ensures that each new cell will get an identical and complete set of chromosomes.

Anaphase—The third phase of mitosis

The separation of sister chromatids marks the beginning of **anaphase,** the third phase of mitosis. During anaphase, the centromeres split apart, and chromatid pairs from each chromosome separate from each other. How the chromatids move is not fully understood, but most scientists believe that spindle fibers shorten when the microtubules forming the spindle fibers begin to break down. The chromatids are pulled apart by the force of shortening.

Telophase—The fourth phase of mitosis

The final phase of mitosis is **telophase.** Telophase begins as the chromatids reach the opposite poles of the cell. During telophase, many of the changes that occurred during

PORTFOLIO

Nuclear Envelope During Mitosis Have the students make hypotheses as to what might be happening to the nuclear envelope during mitosis. Have students design an experiment that could test their hypotheses. L2 P IN

Recent Research on Mitosis Have students research the newest findings on what triggers the onset of mitosis and how microtubules are assembled and disassembled during spindle formation. Students could also research the formation of cell walls after cell division in plants. L3 P

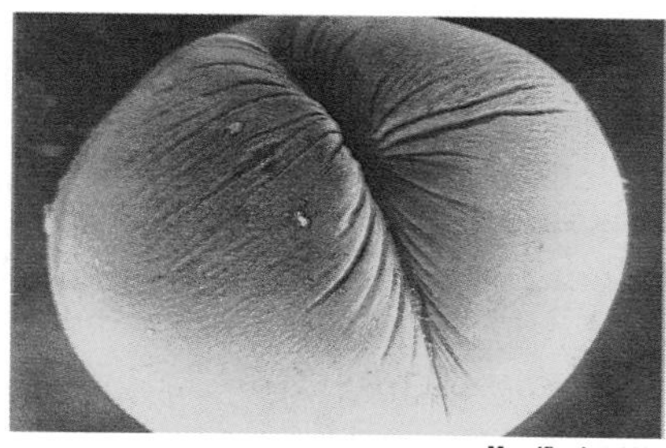

Figure 11.9

At the end of telophase in animal cells such as this frog egg, contractile proteins positioned just under the plasma membrane at the equator of the cell slide past each other to cause a deep furrow. The furrow deepens until the cell is pinched in two.

prophase are reversed as the new cells prepare for their own independent existence. The chromosomes, which had been tightly coiled since the end of prophase, now unwind so they can begin to direct the metabolic activities of the new cells. The spindle begins to break down, the nucleolus reappears, and a new nuclear envelope forms around each set of chromosomes. Finally, the plasma membrane begins to separate the two new nuclei.

Division of the cytoplasm

The final event to occur in mitosis, the division of the cytoplasm, differs between plants and animals. Toward the end of telophase in animal cells, the plasma membrane pinches in along the equator, as shown in *Figure 11.9.* Two new cells are formed, each identical to the original.

The division of the cytoplasm is different in plants. Because plant cells have a rigid cell wall, the plasma membrane does not pinch in. Rather, a structure known as the cell plate forms across the cell's equator, as shown in *Figure 11.10.*

Results of mitosis

Mitosis is a process that guarantees genetic continuity, resulting in the production of two new cells with chromosome sets that are identical to those of the parent cell. These new daughter cells will carry out the same cellular processes and functions as those of the parent cell and will grow until the limitations of cell size force them to divide.

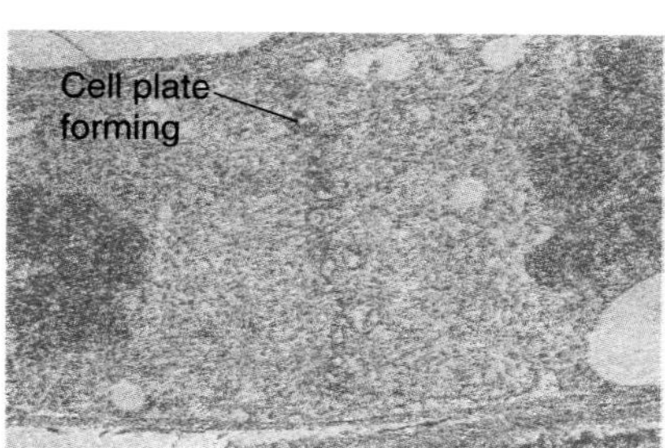

Figure 11.10

In plant cells, the cytoplasm is divided in half by formation of a cell plate. Vesicles from the Golgi apparatus fuse to form the cell plate, which grows outward and joins the old cell wall. New cell wall material is secreted on each side of the cell plate until separation is complete.

Section Review

Understanding Concepts

1. Why is cell size limited?
2. Why is it necessary for a cell's chromosomes to be distributed to its daughter cells in such a precise manner?
3. How is the division of the cytoplasm different in plants and in animals?

Thinking Critically

4. At one time, interphase was referred to as the resting phase of the cell cycle. Why do you think this description is no longer used?

Skill Review

5. **Making and Using Tables** Make a table showing the phases of the cell cycle. Mention one important event that occurs at each phase. For more help, refer to Organizing Information in the *Skill Handbook.*

Extension

Encourage students to research the stages of the cell cycle. They may find information on how long each stage lasts for various species and what events occur at each stage. L2

Assessment

Performance: Call out various stages of mitosis and have students demonstrate that stage to their partners under the microscope using onion root slides. Walk around the room to check their results. L1

4 Close

Activity

Prepare a worksheet with drawings of various stages of mitosis. Ask students to draw the next stage of mitosis that follows a certain drawing. L1

Answers to Section Review

1. Nutrients must rapidly reach the innermost parts of the cell, the plasma membrane must provide sufficient surface area for entering and exiting molecules, and the nucleus must be able to control the cell's activities.
2. Each daughter must get an identical copy of the set of chromosomes.
3. A cell plate forms when a plant cell divides, whereas the plasma membrane of an animal cell pinches in to divide the cytoplasm.

Thinking Critically

4. The cell is not resting, but growing, producing organelles, and replicating its chromosomes in preparation for division.

Making and Using Tables

5. Sample Table
Interphase: chromosomes are copied; Prophase: nuclear envelope disappears; Metaphase: chromosomes lined up; Anaphase: chromatids separate; Telophase: chromosomes at the poles of the cell

BioLab

The Time for the Cell Cycle

Time Allotment

One class period

Objectives

Review objectives with students before they begin the Biolab.

Process Skills

observe and infer, compare and contrast, collect and organize data, analyze

PREPARATION

Alternate Materials

- Whitefish blastula slides can be used in place of onion root tip slides.
- Have students review the appearance of cells during various phases of the cell cycle using the photographs on pages 270-271 before beginning this lab.

BioLab

The Time for the Cell Cycle

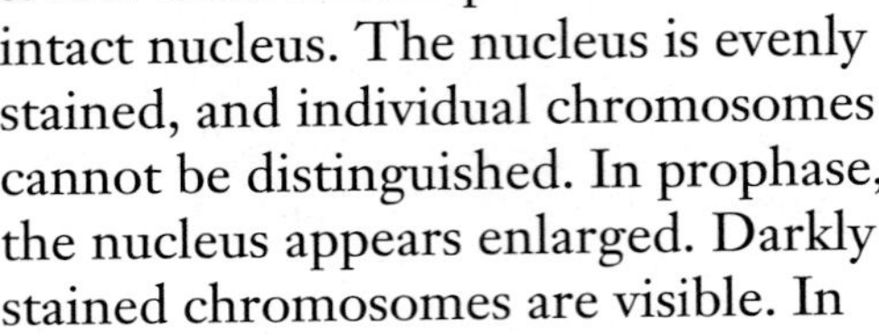

Onion root tips are regions of rapid growth, so cells in all phases of the cell cycle can be observed. Cells in interphase have an intact nucleus. The nucleus is evenly stained, and individual chromosomes cannot be distinguished. In prophase, the nucleus appears enlarged. Darkly stained chromosomes are visible. In metaphase, the chromosomes are lined up at the equator of the spindle. During anaphase, the chromosomes are found at both sides of the equator, the centromeres leading and the arms of the chromosomes trailing behind. In telophase, two dark, compact nuclei are visible, one at each pole of the cell. A cell plate may be visible between the nuclei.

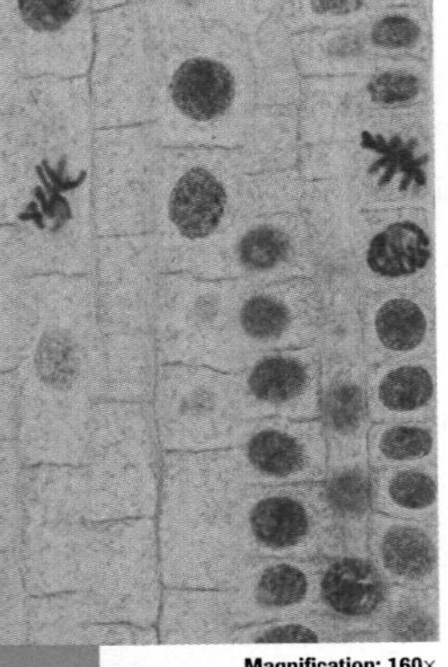

Magnification: 160×

Onion root tip cells

PREPARATION

Problem

How long do onion root tip cells spend in each phase of the cell cycle?

Objectives

In this Biolab, you will:

- **Identify** the phases of the cell cycle.
- **Determine** the time a cell spends in each part of the cell cycle.

Materials

compound microscope
prepared onion root tip slides
compass
protractor

Safety Precautions

Always use care when handling microscopes and glass slides.

Germinating onion seeds

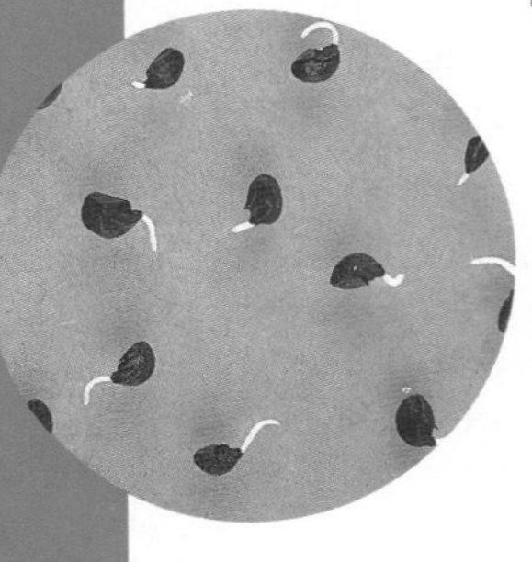

PROCEDURE

1. Examine a prepared onion root tip slide. Using the low power of your microscope, find an onion root tip. Then find the area just behind the tip of the root. This is where the cells are dividing most rapidly.
2. Now, switch your microscope to high power, and start looking at individual cells. You'll learn to recognize each phase of the cell cycle.
3. Copy the data table. Working with a partner, examine each cell in your field of view, and identify the phase of the cell cycle that each cell is in. Have your partner

274 Cell Reproduction

PROCEDURE

Teaching Strategies

- Students must be able to recognize the stages of mitosis before starting this Biolab.
- Students should not count cells that are damaged or unrecognizable.
- Counting will be easier if students count cells as they look across each row, then move down a row so they don't count the same cells twice.

Assessment

Performance Assessment in the Biology Classroom: p. 13, *Investigating Mitosis and Meiosis.* Have students carry out this activity to determine which of a set of unlabeled slides show mitosis and which show meiosis, and also which were made from animal cells and which from plant cells. L1 P record in the data table the phase of each cell you examine.

4. Now switch with your partner. Have your partner examine a different field of view while you record the phase of each cell.
5. Continue in this manner until you have counted at least 100 cells.
6. Add the number of cells in each phase, and record these numbers in the row marked *Total.* Total the number of cells you counted.
7. Calculate the fraction of cells in each phase by dividing the number of cells in that phase by the total number of cells you counted. Express your answer as a decimal, then as a percentage.
8. Calculate the number of degrees in a circle graph for each phase by multiplying the fraction of time the cell spends in each phase by 360°.
9. Make a circle graph showing the percentage of time a cell spends in each phase of the cell cycle. Label each portion of the circle with the phase it represents.

	Interphase	Prophase	Metaphase	Anaphase	Telophase
Number of cells					
First field					
Second field					
Total					
Fraction of cells in this phase					
Percent of time cell spends in this phase					
Number of degrees in a circle graph for this phase					

ANALYZE AND CONCLUDE

1. **Drawing Conclusions** In which phase of the cell cycle does a cell spend most of its time? The least time?
2. **Making Inferences** Explain why a slide of stained cells can be used to estimate how much time a living cell spends in each phase of the cell cycle.
3. **Comparing and Contrasting** Describe the appearance of the cell plate in the telophase cells you observed. How would the division of the cytoplasm appear in an animal cell?

Going Further

Application
Examine a prepared slide of a whitefish blastula to observe mitosis in animal cells. How are dividing animal cells different from dividing plant cells?

Data and Observations

	Interphase	Prophase	Metaphase	Anaphase	Telophase
Number of cells					
First field	89	15	4	1	3
Second field	94	18	6	1	4
Total	183	33	10	2	7
Fraction of cells in this phase	0.78	0.14	0.04	0.01	0.03
Percent of time cell spends in this phase	78	14	4	1	3
Number of degrees in a circle graph for this phase	281	50	14	4	11

ANALYZE AND CONCLUDE

1. A cell spends most of its time in interphase and the least time in anaphase.
2. The slide of onion cells represents the percentage of time spent in each phase of cell activity. The more time that cells spend in a phase, the greater the number of cells in that phase that will show up in the slide.
3. The cell plate is barely visible as a faint line across the middle of the cell. In an animal cell, the plasma membrane would pinch inward.

Assessment

Portfolio: Have students make an analogy about how this Biolab is similar to having someone take pictures of a student once every hour during a week and using the pictures to determine how much time the student spends in various activities. L1 P

Going Further

An onion cell cycle requires 12 hours for completion (from interphase to interphase). Using percentages, have students calculate the time required for each phase. Ask students to show their data in a pie graph. L2

SECTION

11.2

Prepare

Key Concepts

Students will learn about the events that regulate the cell cycle and compare these normal events with the abnormal events that result in cancer.

Block Scheduling

Look for this symbol for strategies that are useful in a block scheduling format. For more information on block scheduling, refer to Section 11.2 in the **Lesson Plans** booklet.

Materials

- Obtain bean seeds for the activity in Meeting Individual Needs.

1 Focus

Bellringer

Before presenting the lesson, display **Section Focus Master 24** on the overhead projector and have students answer the accompanying questions. L1 SAE

Discussion

LS **Logical-Mathematical** Write the following lists of times on the board. Normal chicken stomach cells, in minutes: Interphase 120, Prophase 60, Metaphase 10, Anaphase 3, Telophase 12. Cancerous chicken stomach cells in minutes: Interphase 16, Prophase 15, Metaphase 2, Anaphase 1, Telophase 3. Ask students to compare the lists and suggest possible reasons why they are different. L1

SECTION

11.2 Control of the Cell Cycle

Section Preview

Objectives

Describe the role of enzymes in the regulation of the cell cycle.

Distinguish between events of a normal cell cycle and abnormal events that result in cancer.

Key Terms

cancer
gene

How fast do cells divide? This depends on the particular type of cell involved. Cells that line the intestine will typically complete a cell cycle in 24 to 48 hours, whereas cells of frog embryos reproduce in less than an hour. Cells that divide slowly, such as those in an adult liver, reproduce only once a year. Other cells, such as nerve cells, do not divide at all once they mature. Despite this diversity, the factors that control the cell cycle are generally similar among all organisms.

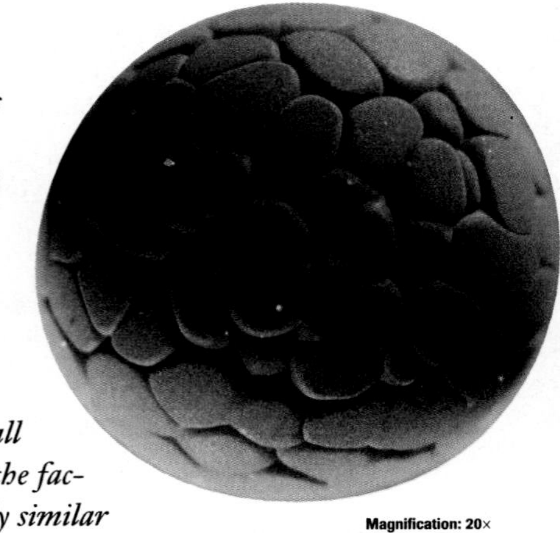

Magnification: 20×

Dividing frog embryo

Normal Control of the Cell Cycle

For the past 25 years, intensive research has been conducted into what causes a cell to divide. Despite the research, the full story still is not known. However, scientists do have some clues.

Enzymes control the cell cycle

Most biologists agree that it is the production of a series of enzymes that monitors a cell's progress from phase to phase during the cell cycle. Some enzymes are necessary to trigger the progression of the cell cycle, whereas other enzymes function to inhibit progression. Cell death or an uncontrolled dividing of cells, known as **cancer,** can result from the failure to produce certain enzymes, or from the overproduction or production at the wrong time of other enzymes. Enzyme production is directed by genes located on the chromosomes. A **gene** is a segment of DNA that controls the production of a protein.

Many studies point to the portion of interphase just before DNA replication as being a key control period in the cell cycle. Scientists have identified several enzymes that act to trigger DNA replication.

Contact inhibition

Another factor involved in control of the cell cycle is the phenomenon of contact inhibition, as shown in *Figure 11.11.* Many studies have shown that when normal cells are allowed to grow in a glass culture dish, they will stop dividing when they cover the bottom of the dish and come in contact with each other.

Contact inhibition is a type of cell-to-cell communication. Cells communicate with each other by producing and secreting chemical signals. When changes occur in the genetic control of these chemicals, cancer can result.

Program Resources

Section Focus Master 24 L1 SAE
Reinforcement and Study Guide, p. 44 L1
Critical Thinking/Problem Solving, p. 11 L3

Figure 11.11

In experiments demonstrating contact inhibition, cells are allowed to grow in a dish containing nutrients. Normal cells divide until they form a continuous layer on the bottom of the dish (a). When several rows of cells are removed, the cells at the edge of the gap begin to divide (b). They stop dividing when the bottom of the dish is covered once again (c). Cancer cells continue to divide, piling on top of each other (d).

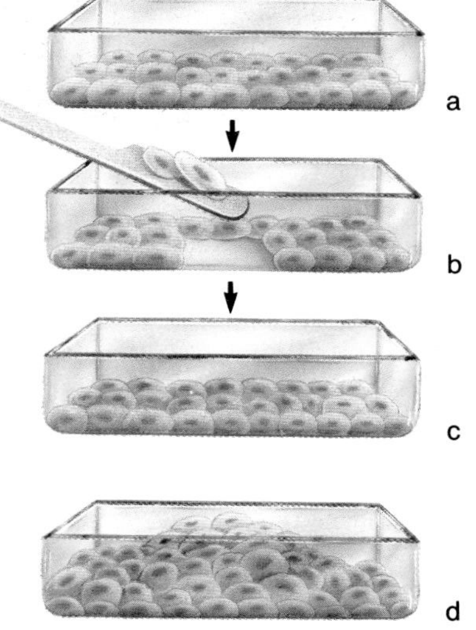

Cancer—A Mistake in the Cell Cycle

The current view of cancer is that it is caused by changes in one or more of the genes controlling production of enzymes involved in the cell cycle. These changes are expressed as cancer when environmental factors trigger the damaged genes into action. Cancerous cells affect normal cells, forming masses of tissue called tumors that deprive normal cells of nutrients. In the final stages, cancer cells enter the circulatory system and spread throughout the body, forming new tumors that disrupt the functioning of organs.

Health

Skin Cancer

Skin cancer makes up one-third of all malignancies diagnosed in the United States, and its incidence is increasing. Most cases are caused by exposure to harmful ultraviolet rays emitted by the sun, so skin cancer most often develops on the exposed face or neck. The people most likely to develop skin cancer are those whose fair skin contains smaller amounts of a protective pigment called melanin.

Types of skin cancers Basal cell carcinoma, the most common type of skin cancer, is relatively harmless. Basal cells damaged by ultraviolet light form precancerous growths that can be removed easily in a doctor's office. Precancerous growths produced by sun-damaged basal cells can also become squamous cell carcinomas, the second most common type of skin cancer. Squamous cell cancers take the form of red or pink tumors that can grow rapidly and spread. Both types of skin cancer respond to treatment such as surgery, chemotherapy, and radiation therapy. By far the most lethal skin cancer is malignant melanoma. Although melanomas currently make up only three percent of all skin cancers, their incidence is growing faster than that of any other kind of skin cancer. Melanomas develop in the pigment-producing cells found deep within the skin.

Two kinds of UV rays There are two kinds of ultraviolet rays: UV-A and UV-B. Sunlight contains both kinds of rays, but it contains much more UV-A than UV-B. Both kinds can cause melanoma. Most sunscreens don't protect the skin from UV-A rays, although they do block UV-B. Thus, sunbathers might falsely think they are protected from damaging radiation.

Since suntanning salons have become increasingly popular, there has been concern over the fact that they claim to use only safe wavelengths of ultraviolet light. Many medical authorities believe this radiation is still potentially dangerous and will increase the risk for skin cancer and premature aging of the skin.

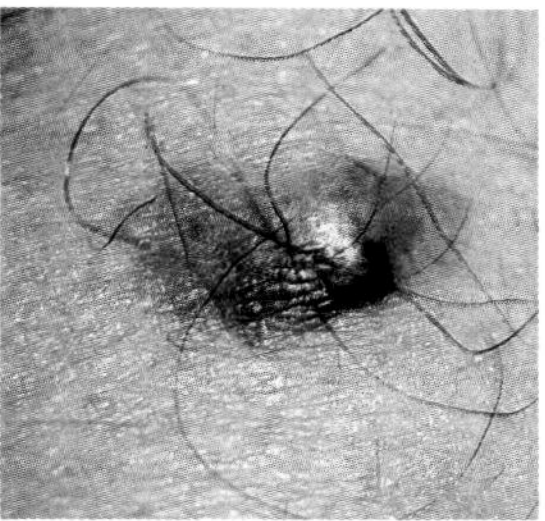

Melanoma

CONNECTION TO Biology

Many people still consider a tan both healthy and attractive. What are some ways these perceptions might be changed?

PORTFOLIO

LS **Linguistic** Have students research the types of radiation therapy or chemotherapy used to treat cancer. Ask them to write a report for their portfolios. L1 P

2 Teach

Health

Skin Cancer

Purpose

Students will examine the types and causes of skin cancer.

Teaching Strategies

- Begin a discussion on the safety of tanning salons that claim to use only "safe" wavelengths of ultraviolet light.

Possible Answer

When people learn the facts about the role of the sun in causing skin cancer, their perceptions may change.

Enrichment

Have students interview the school nurse to find out the warning signs of cancer. Ask students to present their findings to the class. The class can then discuss how these signs relate to the rapid cell division that occurs in cancer. L1

GLENCOE TECHNOLOGY

Videodisc

The Infinite Voyage: Miracles by Design
Burn Patients and Artificial Skin

Videotape

The Secret of Life
Use the videotape *On the Brink: Portraits of Modern Science* to show how uncontrolled reproduction in cells of slime molds is being used in cancer research.

3 Assess

Check for Understanding

Have students compare a normal cell cycle and a cancer cell cycle. L1

Reteach

Have students make two pie graphs, using the data presented for discussion under Focus. Ask students to compare the graphs. L1 SAE

Extension

Ask students to find out how chemotherapy drugs work. *They target areas of the body where rapid cell division occurs.* Ask them to find out why hair follicles and the lining of the digestive system are affected by these drugs, resulting in hair loss and nausea. L3

Assessment

Oral: Ask students to sequence the events that regulate the cell cycle and describe how these events change in the growth of cancer cells. L1

4 Close

Activity

Ask students to select an organ of the body and find out what types of cancer can affect the organ and what treatments are available. L1

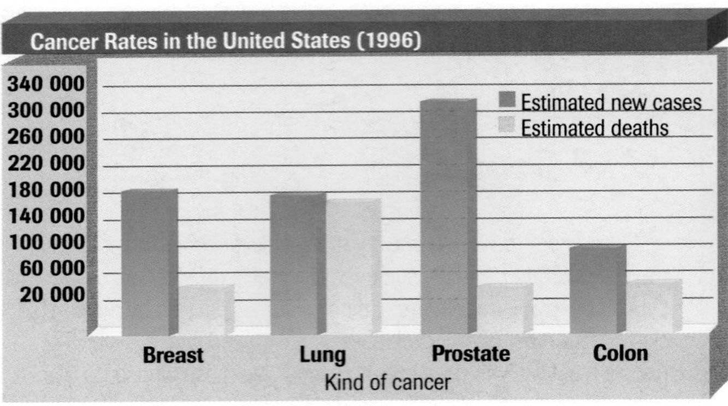

Figure 11.12
Some types of cancer are more treatable than others. Compare the number of new cases with the number of deaths for each type of cancer. Notice that prostate cancer appears to be the most treatable.

Cancer is the second leading cause of death in the United States, exceeded only by heart disease. Cancer can affect any tissues of the body. In the United States, the main cancer killers are lung, colon, breast, and prostate cancers. *Figure 11.12* shows the numbers of people in the United States that were estimated to develop each of these kinds of cancers for 1996 and the estimated numbers of people who would die from them.

The causes of cancer

The causes of cancer are difficult to pinpoint because both genetic and environmental factors are involved. The environmental influences of cancer are clear when it is noted that people in different countries get different types of cancers at different rates. For example, the rate of breast cancer is relatively high in the United States but relatively low in Japan. Similarly, stomach cancer is common in China but rare in the United States. Yet, when people move from one country to another, their cancer rates follow the pattern of the country in which they are currently living, not their country of origin. Environmental factors such as cigarette smoke, air and water pollution, and exposure to ultraviolet radiation from the sun are all known to damage the genes that control the cell cycle. Cancer may also be caused by infection with particular viruses.

Connecting Ideas

Mitosis is an orderly cell process that assures the genetic continuity of the cells and tissues of an organism. If executed without problem, mitosis results in newly formed daughter cells that are genetically identical to the parent cell. Another orderly cell division process that produces sex cells not only guarantees the continuity of the cells involved, but also promotes the genetic continuity of a species.

Section Review

Understanding Concepts

1. Describe the way genes control the cell cycle.
2. How can disruption of the cell cycle result in cancer?
3. How does cancer affect normal cell functioning?

Thinking Critically

4. How do we know that the environment influences the formation of cancer?

Skill Review

5. **Observing and Inferring** Although there are more cases of breast cancer than lung cancer per year, more deaths are caused by lung cancer than breast cancer. Using your knowledge of how cancer spreads and factors influencing cancer, provide an explanation for this difference. For more help, refer to Thinking Critically in the *Skill Handbook.*

Answers to Section Review

1. Genes control the synthesis of the enzymes that monitor a cell's progress in the cell cycle.
2. Uncontrolled cell division could cause tumor formation and cancer.
3. Cancer cells form masses of tissue that deprive normal cells of nutrients.

Thinking Critically

4. People in different countries get different types of cancers at different rates. People who have moved from one country to another have the cancer pattern of the country they are currently living in.

Chapter 11 Review

REVIEWING MAIN IDEAS

11.1 Cell Growth and Reproduction

- Cell size is limited because of physical factors, such as the rate of diffusion of materials into and out of the cell, the amount of DNA available to program the cell's metabolism, and the cell's surface area-to-volume ratio.
- The life cycle of a cell can be divided into two general periods: a period of active growth and metabolism known as interphase, and a period of cell division known as mitosis.
- Mitosis is divided into four phases: prophase, metaphase, anaphase, and telophase.

11.2 Control of the Cell Cycle

- The cell cycle is controlled by the production and secretion of key enzymes at specific times.
- Cancer is caused by an interaction between environmental factors and changes in the genes that control the cell cycle.

Key Terms

Write a sentence that shows your understanding of each of the following terms.

anaphase	interphase
cancer	metaphase
cell cycle	mitosis
centriole	prophase
centromere	sister chromatid
chromosome	spindle
gene	telophase

Understanding Concepts

1. Which of the following is NOT a factor that limits cell size?
   a. time required for diffusion
   b. elasticity of plasma membrane
   c. presence of only one nucleus
   d. surface area-to-volume ratio
2. When calculating the surface area-to-volume ratio, what does the surface area of a cell represent?
   a. cytoplasm
   b. mitochondria
   c. endoplasmic reticulum
   d. plasma membrane
3. What is the surface area of a cube-shaped cell that is 3 mm on a side?
   a. 9 mm^2 c. 54 mm^2
   b. 27 mm^2 d. 56 mm^2
4. Which is the first phase of mitosis?
   a. prophase c. telophase
   b. metaphase d. anaphase
5. The ____________ is the sequence of growth and division of a cell.
   a. centromere
   b. cell theory
   c. cell cycle
   d. surface area-to-volume ratio
6. The growth period of a cell cycle is known as ____________.
   a. interphase c. prophase
   b. mitosis d. anaphase
7. Most of a cell's life is spent carrying on the activities of ____________.
   a. the cell cycle c. mitosis
   b. interphase d. cell division
8. Chromosomes are made of ____________.
   a. cytoplasm c. RNA
   b. ATP d. DNA
9. The process of nuclear division, followed by division of the cytoplasm, is known as ____________.
   a. the cell cycle c. metaphase
   b. mitosis d. interphase

Chapter 11 Review

Reviewing Main Ideas

Summary statements can be used by students to review the major concepts of the chapter.

Key Terms

Answers should go beyond defining the terms. Accept any answer that uses the term correctly and in the proper context.

Understanding Concepts

1. b
2. d
3. c
4. a
5. c
6. a
7. b
8. d
9. b

Skill Review

5. Lung tissue has a much larger blood supply than breast tissue. Lung cancer spreads more rapidly than does breast cancer.

Chapter 11 Review

10. d
11. c
12. d
13. a
14. d
15. a
16. a
17. d
18. b
19. c
20. c

Applying Concepts

21. 78
22. Because nerve cells seldom divide, a spinal cord injury will not repair itself.
23. The cytoskeleton, which maintains the shape of the cell during interphase, is broken down to form the spindle during mitosis.
24. In animal cells, the plasma membrane pinches inward to divide the cytoplasm. In plant cells, a cell plate forms across the cell. Plant cells evolved a rigid cell wall that could not divide in the same manner as animal cells.
25. By being able to control the enzymes, scientists may be able to modify the rapid cell division in cancer cells.

Thinking Critically

26. See below.

Chapter 11 Review

10. Which of the following does NOT occur during interphase?
    a. excretion of wastes
    b. cell repair
    c. protein synthesis
    d. nuclear division
11. Of the following, which is NOT a phase of mitosis?
    a. prophase c. interphase
    b. metaphase d. anaphase
12. During metaphase, the chromosomes move to the equator of the ___________.
    a. poles c. centrioles
    b. cell plate d. spindle
13. The separation of the sister chromatids marks the beginning of ___________ in mitosis.
    a. anaphase c. prophase
    b. telophase d. metaphase
14. Structures inside cells that contain DNA and become darkly colored when stained are ___________.
    a. centrioles c. centromeres
    b. centrosomes d. chromosomes
15. During which phase of the cell cycle are the chromosomes being doubled?
    a. interphase c. metaphase
    b. prophase d. anaphase
16. Plant cell division of the cytoplasm at the end of mitosis differs from animal cell division in that plants form a _________.
    a. cell plate
    b. deep furrow
    c. centriole pair
    d. new plasma membrane
17. Which of the following is NOT a known cause of cancer?
    a. environmental influences
    b. certain viruses
    c. cigarette smoke
    d. bacterial infections
18. If a cell that has 8 chromosomes goes through mitosis, how many chromosomes will the daughter cells have?
    a. 4 c. 16
    b. 8 d. 32
19. What are the football-shaped, cagelike structures consisting of microtubules that play a vital role in separation of sister chromatids during mitosis?
    a. centrioles c. spindle fibers
    b. centrosomes d. chromosomes
20. A cell with 8 chromosomes has how many chromatids during prophase?
    a. 4 c. 16
    b. 8 d. 32

Applying Concepts

21. The body cells of a dog contain 78 chromosomes. How many chromosomes do new dog cells have following mitosis?
22. Human muscle cells and nerve cells seldom divide after they are formed. Explain how this fact might affect an individual who has received a spinal cord injury.
23. Explain why animal cells change their shape during mitosis. Refer to the normal function of the cytoskeleton and what happens to it during mitosis.
24. Compare the division of the cytoplasm in plants and animals. Why do you think different mechanisms might have evolved?
25. Suppose that all of the enzymes that control the normal cell cycle were identified. Suggest some ways that this information might be used to fight cancer.

Thinking Critically

26. ***Concept Mapping*** Make a concept map that relates the following terms and phrases. Supply the appropriate linking words for your map.
    cell cycle, interphase, prophase, metaphase, anaphase, telophase, chromosome replication, spindle, chromosome, cancer

26.

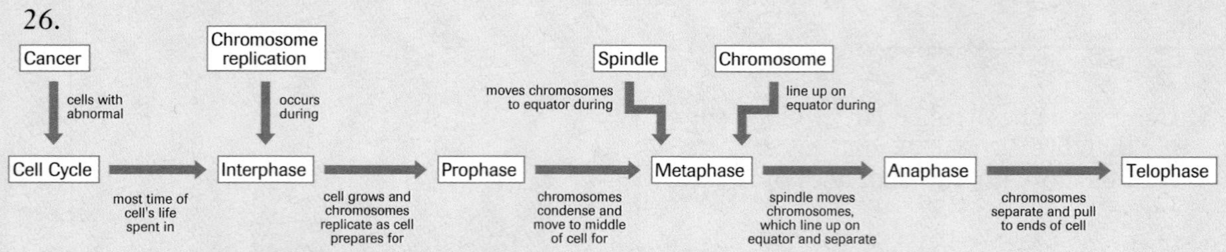

27. *Observing and Inferring* Why is cell division in adult animals needed to maintain homeostasis?
28. *Observing and Inferring* Fertilized eggs of animals have a large amount of cytoplasm, and the first few mitotic divisions occur rapidly. What do you think happens to the size of the cells during this series of divisions? Explain.
29. *Observing and Inferring* Some cells of fungi have many nuclei. Suggest a way such a cell could have evolved. Use your knowledge of mitosis to help you.
30. *Interpreting Data* The following photograph is of a blastula, magnified 320×, undergoing mitosis. Identify the phase of mitosis. Are these cells animal or plant cells? How do you know?

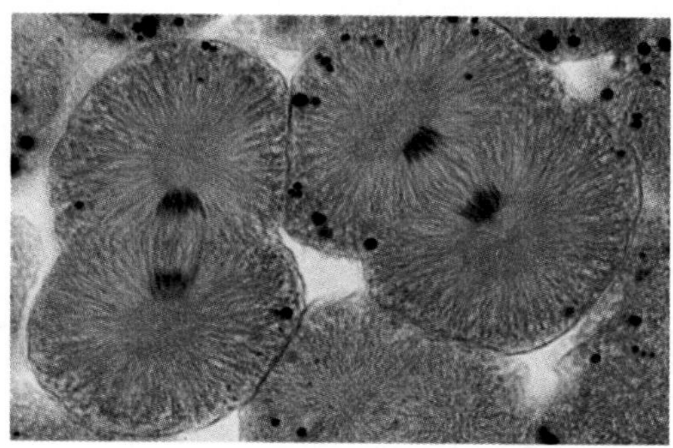

31. *Interpreting Data* An unidentified group of plant cells was found to have a cell cycle averaging 24 hours. A second set of plant cells had a cell cycle averaging 88 hours. As a scientist, you were asked to identify the tissues from which the cells came. Using your knowledge of mitosis, predict which set of cells came from the plant's root tip. From which part of the plant might the other set have come? Explain.

ASSESSING KNOWLEDGE & SKILLS

Different species of organisms vary in the number of chromosomes found in body cells.

Chromosome Comparison of Four Organisms

Organism	Number of chromosomes in body cells	Number of chromatids during metaphase of mitosis	Number of chromosomes in daughter cells
Human	46	92	46
Rye	14	A	14
Potato	48	96	48
Guinea Pig	64	128	64

Interpreting Data Examine the table, then answer the following questions.

1. During late interphase, the chromosomes double to form chromatids that are attached to each other. During which phase do the chromatids separate?
   a. prophase
   b. metaphase
   c. anaphase
   d. telophase
2. What number belongs in the space labeled *A* in the third column of the table?
   a. 14
   b. 28
   c. 7
   d. 21
3. If one pair of chromatids failed to separate during mitosis in rye cells, how many chromosomes would end up in the daughter cells?
   a. 28 and 28
   b. 14 and 14
   c. 7 and 8
   d. 15 and 13
4. *Interpreting Data* Using information presented in the table, explain how the number of chromosomes in body cells is related to the complexity of an organism.

Chapter 11 Review

27. Adult animals need cell division to replace dead cells.
28. The size of the cells is reduced because the cells do not have time to double in size before dividing.
29. Mitosis without division of the cytoplasm would produce cells with many nuclei.
30. telophase; animal cells, because there is no cell wall or cell plate forming
31. The cells with the cell cycle averaging 24 hours came from the root tip because this is an area of rapid division. The other cells may have come from leaf or stem tissue.

ASSESSING KNOWLEDGE & SKILLS

1. c
2. b
3. d
4. There is no relationship between the number of chromosomes in body cells and complexity of an organism.

Program Resources

Chapter Assessment, pp. 61-66 L1
Alternate Assessment in the Science Classroom
Computer Test Bank L1
Content Mastery, pp. 41-44 L1

UNIT

4 Genetics

Unit Overview

Unit 4 presents an overview of genetics and its role in determining the traits of organisms. Chapter 12 introduces genetics through a short historical presentation of the work of Gregor Mendel. Meiosis is then introduced and discussed.

In Chapter 13, students learn about genes and chromosomes. The concept of heredity is the main focus of Chapter 14 as students examine patterns of heredity that appear to be exceptions to Mendel's laws.

Students examine how principles of genetics apply to humans through a discussion of human heredity in Chapter 15 and DNA technology in Chapter 16.

Theme Development

The major themes of this unit are *systems and interactions* and *evolution*. These themes are intertwined throughout the unit as students explore how genetic traits from two parents combine to result in offspring with traits that are not identical to either parent. The *nature of science* also becomes apparent through the explorations of the work of Mendel, and the discovery and development of an understanding of DNA.

UNIT

4 Genetics

It's difficult to believe that these simple looking cellular structures will one day be the ultimate source for understanding human biology and inherited genetic disorders. Contained within these human chromosomes is a lengthy chemical message that makes up the spiraling DNA strands inside the nucleus of each of your cells—a message that scientists hope will one day help explain birth, development, growth, and death.

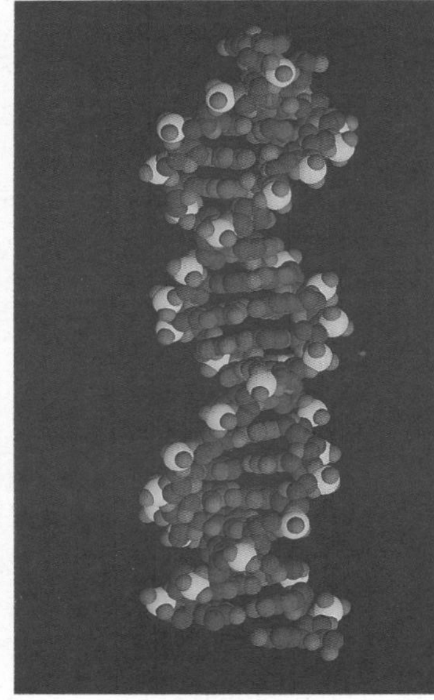

DNA, the basis of all life, resides in every cell of every organism on Earth. How is the chemical code in the double helix of DNA translated into cells, tissues, and organs?

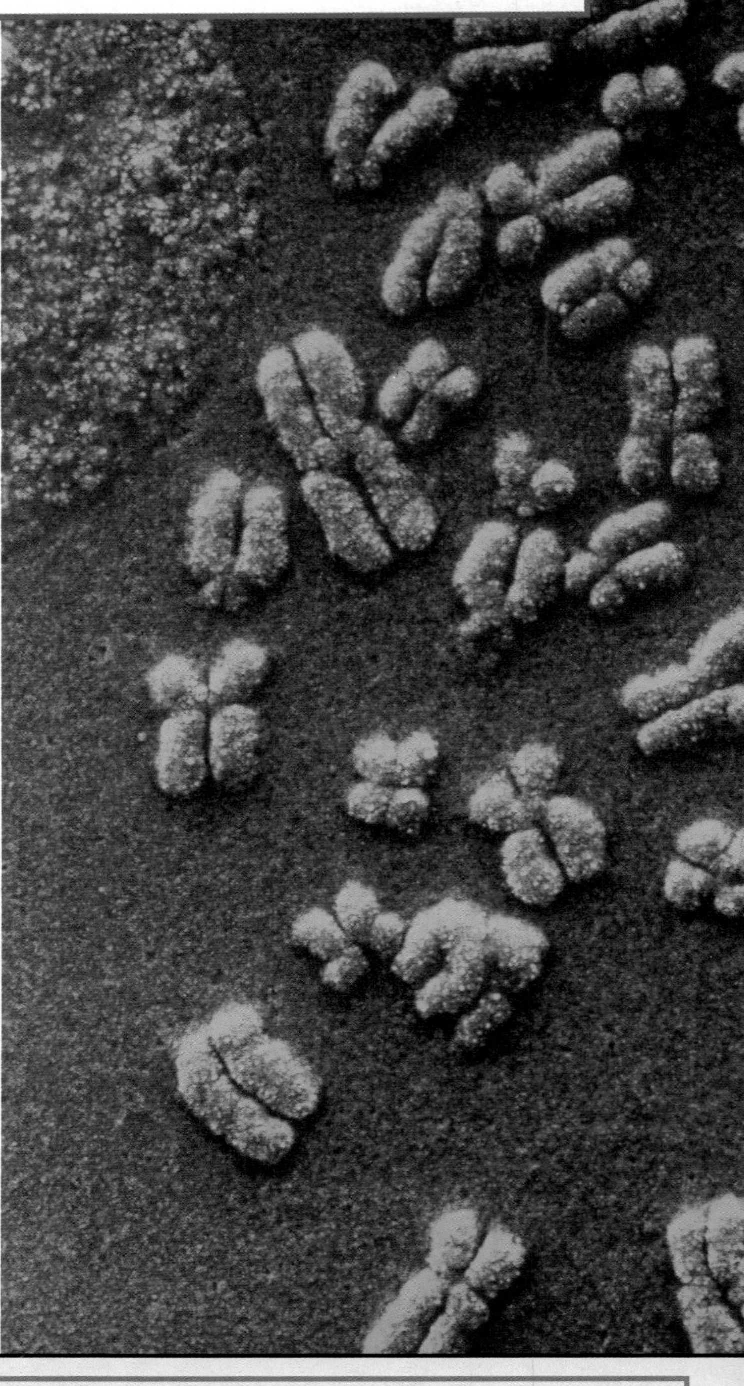

Advance Planning

Chapter 12

- Order genetic and regular tobacco seeds for the Biolab and purchase small flower pots and seedling flats.
- Obtain flowers and gather dissecting microscopes for the Minilab on page 287.

Chapter 13

- Purchase salmon testes for the Minilab on page 311.

Chapter 14

- Order *Brassica rapa* seeds for the Biolab.

Chapter 15

- Gather magnifying lenses and colored pencils for the Minilab on page 364.

Chapter 16

- Obtain plastic-wrapped wire for the Minilab on page 379.
- Gather colored pencils and scissors for the Biolab.

UNIT 4

Although individuals in a population, such as a penguin colony, may appear identical, each has almost undetectable differences. How is DNA related to the variation in a species? How is it inherited?

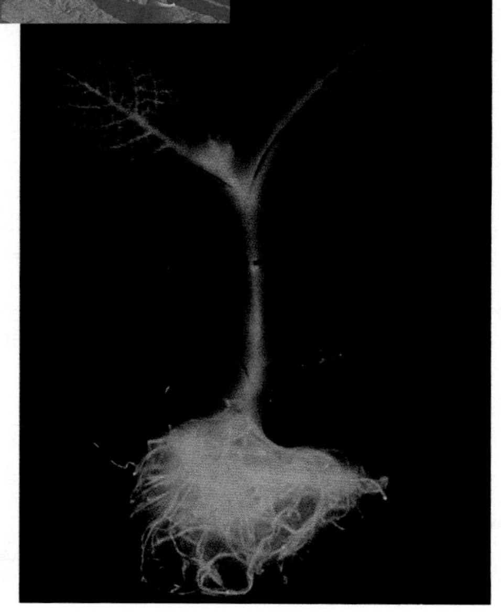

Through the study of DNA, scientists have already learned how to manipulate the fundamental building blocks of life. This tobacco plant has been given the genes of a firefly. What are some of the possible uses for genetic engineering techniques?

Unit Contents

Introducing the Unit

Elicit from students what they know about DNA and its behavior during meiosis. Explain to students that the traits of organisms present in their DNA are carried on the chromosomes.

Point out that although each penguin making up the population looks just like any other, each individual has traits that are not present in the others. Relate these individual traits to DNA. Explain that by studying or carrying out experiments with DNA, scientists have been able to learn more about the traits of organisms and how certain traits may be controlled or manipulated to change the characteristics of organisms.

Motivational Activity

Project a 35-mm slide of human chromosomes onto the screen (or use a video camera and prepared slide or projected slide with microprojector). Ask students to count the number of chromosomes present. Compare this number to those in a slide of onion cells in mitosis and/or fruit fly salivary gland chromosomes. Ask what the significance of these different chromosome numbers might be. L1

Unit Project

Class Investigation: Hair color is a rather easily observed characteristic. Hair width can be observed under the microscope and allows for measuring and accumulating of quantitative data. Hair width also correlates with other observable phenotypes such as facial features, skin color, and hair color.

Show students how to measure the diameter of their low-power field using a metric ruler. The diameter of their high-power field is equal to the low-power field diameter divided by 4. Knowing the diameter of their high-power field allows students to estimate rather closely the width of any hair strand wet mount.

Have students use the microscopic technique described to determine the average width of a strand of their hair, based on widths obtained from five samples. Ask students to compile their individual data to form a table that lists all class members. Ask students to use the class data to determine if hair width correlates to any other observable traits. L1 COOP LEARN

Chapter Organizer

SECTION	OBJECTIVES	ACTIVITIES/FEATURES
12.1 Mendel's Laws of Heredity National Science Standards: UCP.1-3; A.1, A.2; G.1-3	1. **Analyze** the results obtained by Gregor Mendel in his experiments with garden peas. 2. **Predict** the possible offspring of a cross by using a Punnett square.	**Minilab:** How do you cross plants?, p. 287 **Thinking Lab:** How did Mendel analyze his data?, p. 289 **Math Connection:** A Solution from Ratios, p. 290 **Biolab:** Design Your Own Experiment—How can phenotypes and genotypes of plants be determined?, p. 292 **Minilab:** How does sample size affect results?, p. 294 **A Broader View:** The Laws of Probability, p. 295
12.2 Meiosis National Science Standards: UCP.2, UCP.3; C.1, C.2; E.1, E.2; F.6; G.1, G.3	3. **Analyze** how meiosis maintains a constant number of chromosomes in the body cells of the members of a species. 4. **Infer** how meiosis leads to variation in a species. 5. **Relate** Mendel's laws of heredity to the events of meiosis.	

ACTIVITY MATERIALS

BIOLAB	MINILABS	ALTERNATE LAB
page 292 potting soil small flowerpots or seedling flats tobacco seeds, 2 groups hand lens light source thermometer plant watering bottle	**page 287** flower brush dissecting microscope **page 294** calculator (optional)	**page 300** 9 sheets of unlined paper or poster board (30-cm square) wool, long strip paper clips string toothpicks tape or glue scissors

Chapter 12 Mendel and Meiosis

TEACHER CLASSROOM RESOURCES

Reproducible Masters	Transparencies
Section Focus Master 25: Predicting Combinations L1 SAE (block) **Reinforcement and Study Guide,** pp. 45-46 L1 (block) **Biolab and Minilab Worksheets,** pp. 43-46 L1 (block) **Critical Thinking/Problem Solving:** Tracing a Family Tree and Calculating Probabilities, p. 12 L3 **Content Mastery,** pp. 45-48 L1	**Basic Concepts Transparency #14:** Monohybrid Cross L1 SAE (block) **Reteaching Transparency #12:** Dihybrid Cross L1 SAE (block)
Section Focus Master 26: Chromosome Numbers L1 SAE (block) **Reinforcement and Study Guide,** pp. 47-48 L1 (block) **Concept Mapping:** Mitosis/Meiosis, p. 12 L1 **Laboratory Manual:** Observation of Meiosis, pp. 73-74 L2	**Basic Concepts Transparency #15:** Meiosis L1 SAE (block) **Basic Skills Transparency #11:** Mitosis versus Meiosis L1 SAE (block)

ASSESSMENT MATERIALS

Chapter Assessment, pp. 67-72 (block)
Performance Assessment in the Biology Classroom, p. 13
MindJogger Videoquiz (block)
Alternate Assessment in the Science Classroom
Computer Test Bank

Spanish Resources SAE
English/Spanish Audiocassettes SAE
Cooperative Learning in the Science Classroom COOP LEARN
Lesson Plans (block)
Great Developments in Biology: DNA: The Code of Life L1 SAE

KEY TO TEACHING STRATEGIES

- L1 Level 1 activities should be within the ability range of all students including those with learning difficulties.
- L2 Level 2 activities are within the ability range of average to above-average students.
- L3 Level 3 activities are designed for the ability range of above-average students.
- SAE SAE activities should be within the ability range of Students Acquiring English.
- COOP LEARN Cooperative Learning activities are designed for small group work.
- P These strategies represent student products that can be placed into a best-work portfolio.
- (block) These strategies are useful in a block scheduling format.

GLENCOE TECHNOLOGY

The following multimedia resources are available from Glencoe.

Biology: The Dynamics of Life
CD-ROM SAE
Videodisc Program (block)

The Infinite Voyage Series
The Keepers of Eden

National Geographic Society Series
STV: Plants
The Cell

Science and Technology Videodisc Series (STVS)
Plants & Simple Organisms
Breeding Fruit Flies
Selective Breeding in Cows
Selective Breeding in Micro-pigs

CHAPTER

12 Mendel and Meiosis

Chapter Overview

Students begin their study of Mendelian genetics. They explore his rules of unit factors and dominance, law of segregation, and law of independent assortment. Probability is also considered. Students are then shown how to solve genetic problems using a Punnett square.

In the second section, Mendel's observations are correlated with our current understanding of meiosis. Meiosis is explained in detail with emphasis on the changes occurring during each specific phase or step. Finally, genetic recombination is described, and its importance to the survival of a species is discussed.

Key Terms

allele
crossing over
dihybrid cross
diploid
dominant
egg
fertilization
gamete
genetic recombination
genetics
genotype
haploid
heredity
heterozygous
homologous chromosome
homozygous
law of independent assortment
law of segregation
meiosis
phenotype
pollination
recessive
sexual reproduction
sperm
trait
zygote

284

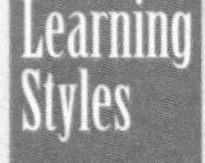

Learning Styles

Look for the following logo for strategies that emphasize different learning modalities. 

Kinesthetic	**Minilab, p. 287; Meeting Individual Needs, p. 294; Alternate Lab, p. 300**
Visual-Spatial	**Portfolio, pp. 290, 302; Reinforcement, p. 295; Meeting Individual Needs, pp. 295, 299**
Intrapersonal	**Student Journal, p. 289**
Linguistic	**Portfolio, p. 290; Student Journal, p. 303**
Logical-Mathematical	**Project, p. 288; Thinking Lab, p. 289; Minilab, p. 294; Concept Development, p. 296; Math Connection, p. 299**

LS

CHAPTER

12 Mendel and Meiosis

Do you look just like one of your parents? Have you ever heard a teacher absentmindedly call a student by the name of an older brother or sister whom he or she resembles? For thousands of years of human history, people have noticed family resemblances and wondered at them. People have even noted physical characteristics that are transferred from parent to offspring in other species, and have learned to produce unique strains of plants and breeds of animals.

Yet, it was not until the mid-1800s that a scientific study of how physical characteristics are inherited was carried out. Surprisingly, the first answers came, not from the study of family resemblances in people or even in dogs, but from the study of resemblances among garden peas. The scientist who did this pioneering work was an Austrian monk named Gregor Mendel.

Concept Check

You may wish to review the following concepts before studying this chapter.

- Chapter 8: nucleus
- Chapter 11: chromosomes, mitosis

Chapter Preview

Laboratory Activities

Biolab: Design Your Own Experiment

- How can phenotypes and genotypes of plants be determined?

Minilabs

- How do you cross plants?
- How does sample size affect results?

This Shar-Pei dog and her puppies are the result of our knowledge of inheritance. What biological mechanism lies behind the transfer of physical characteristics from parent to offspring? The answer lies at the much less obvious, microscopic level of sex cells and the bits of hereditary material within them called chromosomes.

Magnification: 3500×

Assessment Planner

Choose assessment strategies from the following pages to evaluate the progress of your students.

Assess, pp. 296, 304
Alternate Lab, pp. 300–301
Minilabs, pp. 287, 294
Portfolio, pp. 290, 302
Thinking Lab, p. 289
Biolab, pp. 292–293
Chapter Review, pp. 305–307

Introducing the Chapter

Have students examine the photograph on this page. Ask them why the puppies look so much like one another. *Some students may recognize that the puppies are all the same breed of dog and therefore all share traits characteristic of the breed.* Ask students how puppies might appear if the two parent dogs did not have many similar traits. *They might have some traits, (color, shape of the nose, length of legs, and so on) that are similar but some that are different, as well.*

Theme Development

The theme of *nature of science* is developed within the chapter as Mendel's laws of segregation and independent assortment are explained. Mendel's laws are shown to be supported by current knowledge of meiosis. The theme of *homeostasis* is illustrated by the explanation that diploid chromosome numbers are maintained when gametes join at fertilization. The gametes are formed as the result of meiosis, and their chromosome numbers are half the diploid number.

Concept Check

Ask students why the process of mitosis covered in Chapter 11 cannot produce sex cells that can fuse to form a new individual in sexual reproduction. Have them think in terms of maintaining the species chromosome number during mitosis.

SECTION

12.1

Prepare

Key Concepts

Students are led through mono- and dihybrid crosses, applying Mendel's laws of segregation and independent assortment.

Block Scheduling

Look for this symbol for strategies that are useful in a block scheduling format. For more information on block scheduling, refer to Section 12.1 in the **Lesson Plans** booklet.

Materials

- Purchase fresh or preserved flowers and gather brushes for the first Minilab.
- Purchase the following seed types from a biological supply house for the Biolab: pure breeding green tobacco seeds and green–albino tobacco seeds 3:1 ratio (heterozygous parents). Also, gather soil and pots. Locate thermometers, hand lenses, and possible light banks.

1 Focus

Bellringer

Before presenting the lesson, display **Section Focus Master 25** on the overhead projector and have students answer the accompanying questions. L1 SAE

Discussion

Ask students to name several traits that appear to be passed from parent to child. *Typically the answers will center on eye, hair, and skin color.* Ask them if the following are also traits passed from parent to child: handedness, voice, general body shape, AIDS. *All except AIDS are inherited.*

SECTION

12.1 Mendel's Laws of Heredity

Section Preview

Objectives

Analyze the results obtained by Gregor Mendel in his experiments with garden peas.

Predict the possible offspring of a cross by using a Punnett square.

Key Terms

heredity
genetics
trait
gamete
fertilization
pollination
allele
dominant
recessive
law of segregation
phenotype
genotype
homozygous
heterozygous
dihybrid cross
law of independent assortment

An Austrian monastery in the mid-19th century might seem an unusual place to begin your search for the answer to why offspring resemble their parents. Yet, during the 1800s, the Augustinian monastery in Austria was an important center of scientific study. It was this community of scholars that Gregor Mendel joined as a young man in 1843.

Why Mendel Succeeded

Gregor Mendel carried out the first important studies of **heredity,** the passing on of characteristics from parents to offspring. Although people had noticed that family resemblances were inherited from generation to generation, an explanation required the careful study of **genetics**—the branch of biology that studies heredity. Characteristics that are inherited are called **traits.** Mendel was the first person to succeed in predicting how traits are transferred from one generation to the next. How was he able to solve the problem of heredity?

Mendel chose his subject carefully

Mendel studied many plants before deciding to use the garden pea in his experiments. Garden pea plants reproduce sexually. That means they have two distinct sex cells—male and female. Sex cells are called **gametes.** In peas, both male and female gametes are in the same flower. The male gamete is in the pollen grain, whereas the female gamete is in the ovule. **Fertilization,** the uniting of male and female gametes, occurs when the male gamete in the pollen grain meets and fuses with the female gamete in the ovule. The transfer of the male pollen grains to the female organ of a flower is called **pollination.** After the ovule is fertilized, it matures into a seed.

To study inherited traits, Mendel had to transfer pollen from one plant to another plant with different traits, *Figure 12.1.* This is called making a cross. In these crosses, he had to be sure of the identity of the parents. Mendel chose the pea for the crosses because the structure of the plant's flower made this identity relatively easy.

Program Resources

Section Focus Master 25 L1 SAE
Reinforcement and Study Guide, pp. 45-46 L1
Biolab and Minilab Worksheets, pp. 43-46 L1
Critical Thinking/Problem Solving, p. 12 L3
Basic Concepts Transparency 14 and **Master** L1 SAE
Reteaching Transparency 12 and **Master** L1 SAE

Mendel was a careful researcher

Mendel carefully controlled his experiments and the peas he used. He studied only one trait at a time to control variables, and he analyzed his data mathematically. The tall pea plants he worked with were from populations of plants that had been tall for many generations and had always produced tall offspring. Such plants are said to be true breeding for tallness. Likewise, the short plants he worked with were true breeding for shortness.

Mendel's Monohybrid Crosses

What did Mendel do with the tall and short pea plants he so carefully selected? He crossed them to produce new plants. Mendel's first experiments are called monohybrid crosses because the two parent plants differed by a single trait—height.

The first generation

Mendel selected a six-foot-tall pea plant that came from a population of pea plants, all of which were over six feet tall. He cross-pollinated this tall pea plant with a short pea plant that was less than two feet tall and which came from a population of pea plants that were all short. He found that all of the offspring in the first generation were as tall as the taller parent. It was as if the shorter parent had never even existed!

MiniLab

How do you cross plants?

Mendel did his work by crossing pea plants. This procedure involves manipulating flower parts in order to fertilize the female gamete with the male gamete from the desired parent plant.

Procedure

1. Obtain a flower from your teacher. Identify the anthers, the male parts that contain pollen. Locate and identify the pistil, the female structure. The topmost part of the pistil, on which the pollen must land, is called the stigma. Locate and identify the stigma.
2. Remove the anthers from your flower.
3. Dust the pollen from an anther of another flower onto the stigma of the first flower. You may use a brush, as Mendel did.
4. If a dissecting microscope is available, examine the anthers under 10x or 20x power so that you can see the pollen grains.

Analysis

1. Why did Mendel need to remove the anthers from the flowers he was cross-pollinating?
2. After performing these steps, Mendel covered the flower with a bag. Why did he do this?

Figure 12.1

The reproductive parts of the pea flower are tightly enclosed in petals, preventing the pollen of other flowers from entering. As a result, peas normally reproduce by self-pollination. In many of Mendel's experiments, this is exactly what he wanted. When he needed to cross one plant with another, Mendel removed the male parts from an immature flower. He then dusted the female part of the flower with pollen from the plant he wished to cross it with and covered the flower with a small bag. In this way, Mendel could be sure of the parents in his cross.

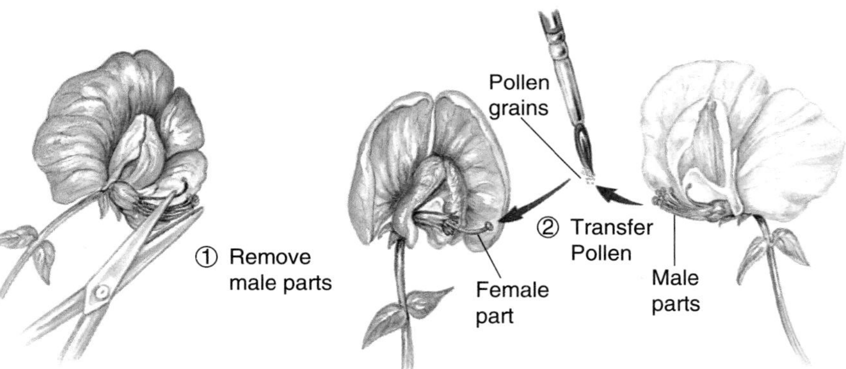

SECTION 12.1

2 Teach

MiniLab

Purpose

Kinesthetic Students will become familiar with flower anatomy and model Mendel's experiments with pea plants.

Process Skills

observe, simulate, interpret data

Teaching Strategies

- Lilies or gladioli are ideal flowers to use. The stamens and pistil must be easy to identify, so do not use any composite flower such as daisies or sunflowers.
- If fresh flowers are not available, preserved materials can be used.
- Use a chalkboard diagram to illustrate flower anatomy or refer students to Chapter 27 in this text.

Expected Results

Students should be able to understand the process Mendel used in making his crosses.

Analysis

1. He wanted to be sure that the plants did not self-pollinate.
2. The bag prevented pollen with unknown traits from pollinating the female plant.

Assessment

Portfolio: Have students diagram and label the steps that were taken in the Minilab. Have them place their diagrams in their portfolios. L1 P

Meeting Individual Needs

Visually Impaired Obtain a large flower model. Allow those students who are visually impaired to study the model. Direct their attention to the stamen and pistil. Then, have students examine an actual flower after having studied the model. L1

Students Acquiring English Point out that the reproductive process in plants and animals is very similar. Sex cells, or gametes, of both plants and animals are called egg and sperm. Sperm are present within pollen whereas eggs are present within the ovule. Ask students what structures in animals have the same functions as anthers and ovules. L1 SAE

SECTION 12.1

Reinforcement

Ask students to interpret what the generation designation would be for their grandparents, their parents, and themselves. *Grandparents, P_1; parents F_1; themselves, F_2* L1

Audiovisual

Show the filmstrip *Elementary Genetics*, Carolina.

Different Viewpoints in Biology

Mendel's Numbers

According to Robert Tamarin, in *Principles of Genetics* (1993), some researchers have doubted Mendel's honesty, despite the accuracy of his conclusions. One claim is that Mendel deliberately excluded data from traits that do not independently assort, since he had only one chance in 6000 of randomly selecting one gene on each of seven chromosomes. Another claim is that Mendel fabricated some of his data in order to make them fit expected ratios.

Countering these claims are other analyses of Mendel's data that concluded he actually had a one in three chance of choosing only traits that independently assort, and a suggestion that Mendel did not intend to present an exact experiment but something more like a demonstration. Tamarin accepts these counterclaims and concludes that Mendel did not "cheat."

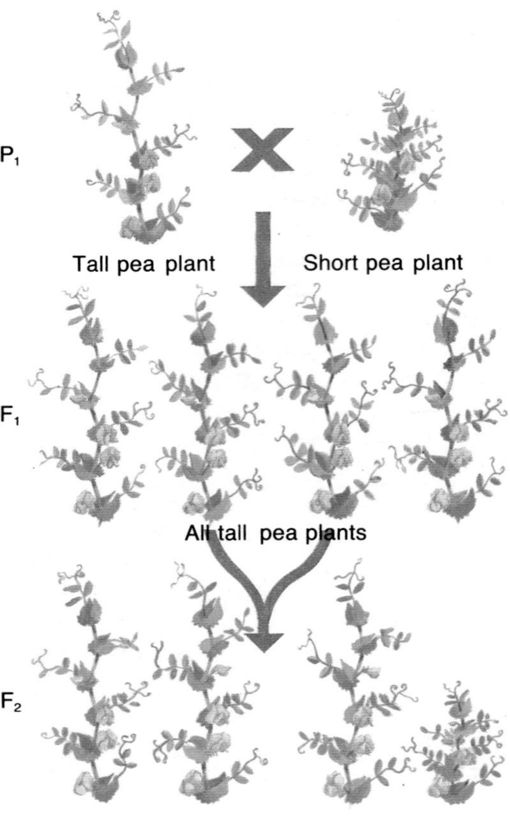

Figure 12.2

When Mendel crossed true-breeding tall pea plants with true-breeding short pea plants, all the offspring were tall. When he allowed first-generation tall plants to self-pollinate, three-fourths of the offspring were tall and one-fourth were short. Mendel counted more than 1000 tall and short plants in this second generation and found that tall and short plants occurred in a ratio of approximately three tall plants to one short plant.

The second generation

Next, Mendel crossed two tall offspring plants with each other. This was easy to do because he could simply let them self-pollinate. Mendel found that three-fourths of the offspring in the second generation were as tall as the tall plants in the parent and first generations, whereas one-fourth of the offspring were as short as the short plants in the parent generation, as shown in *Figure 12.2.* It was as if the short trait had reappeared from nowhere!

The original parents, the true-breeding tall and short plants, are known as the P_1 generation. The *P* stands for "parent." The offspring of the parent plants are known as the F_1 generation. The *F* stands for "filial"—son or daughter. When you cross two F_1 plants with each other, their offspring are called the F_2 generation—the second filial generation.

Mendel did similar monohybrid crosses with a total of seven pairs of traits, shown in *Figure 12.3.* In every case, he found that one trait seemed to disappear in the F_1 generation, only to reappear unchanged in one-fourth of the F_2 plants.

allele:
allelon (GK) of each other
Genes exist in alternative forms called alleles.

The rule of unit factors

Mendel concluded that each organism has two factors for each of its traits. We now know that these factors are genes and that they are located on chromosomes. Genes exist in alternative forms. We call these different gene forms **alleles.** For example, each of Mendel's pea plants had two alleles that determined its height. A plant could have two alleles for tallness, two alleles for shortness, or one allele for tallness and one for shortness. Alleles are located on different copies of a chromosome—one inherited from the female parent and one from the male parent.

Follow the link for this chapter on the Glencoe Homepage at **www.glencoe.com/sec/science** to find out more about Mendel and meiosis.

PROJECT

Experimental Crosses

Seeds of a plant called *Brassica rapa* are available from Carolina Biological Supply Co. under the name of Wisconsin Fast Plants. These plants, grown from seed, complete their life cycle in 30–35 days. They are ideal for use in the classroom because within this short time, the plants flower and form seeds for the next generation. Genetic studies can be carried out using different traits such as petalless flowers or hairy stems. Students can actually do the pollinating between plants. These plants are an ideal experimental organism for genetic studies. L2

The rule of dominance

Remember what happened when Mendel crossed a tall P_1 plant with a short one. The F_1 offspring were all tall. In other words, only one trait was observed. In such crosses, Mendel called the observed trait **dominant** and the trait that disappeared **recessive.** We now know that in Mendel's pea plants, the allele for tall plants is dominant to the allele for short plants. Pea plants that had two alleles for tallness were tall, and those that had two alleles for shortness were short. Plants that had one allele for tallness and one for shortness were tall because the allele for tallness is dominant to the allele for shortness. Expressed another way, the allele for short plants is recessive to the allele for tall plants. You can see in *Figure 12.4* on the following page how the rule of dominance explained the resulting F_1 generation.

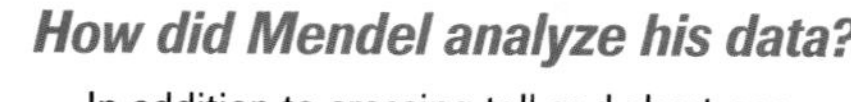

How did Mendel analyze his data?

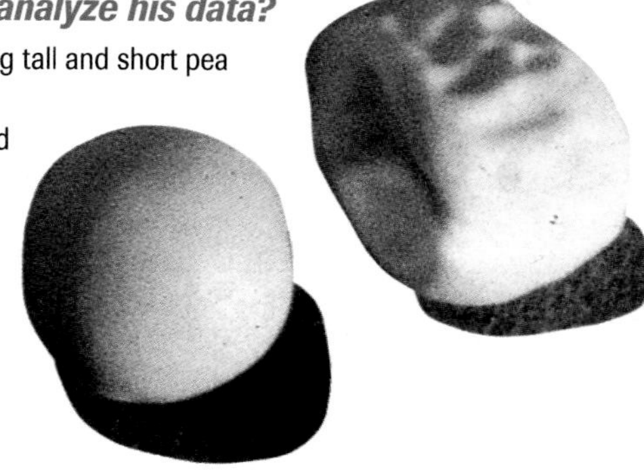

In addition to crossing tall and short pea plants, Mendel crossed plants that formed round seeds with plants that formed wrinkled seeds. He found a 3:1 ratio of round-seeded plants to wrinkled-seeded plants in the F_2 generation.

Analysis

Mendel's actual results in the F_2 generation are shown below.

Kind of Plants	Number of Plants
Round-seeded	5474
Wrinkled-seeded	1850

1. Calculate the actual ratio of round-seeded plants to wrinkled-seeded plants. To do this, divide the number of round-seeded plants by the number of wrinkled-seeded plants. (Round to the nearest hundredth.) Your answer tells you how many times more round-seeded plants resulted than wrinkled-seeded plants.
2. To express your answer as a ratio, write the number from step 1 followed by a colon and the numeral *1*.

Thinking Critically

What was the actual ratio Mendel observed for this cross? How does this compare to the expected 3:1 ratio? Why was the actual ratio different from the expected ratio?

Figure 12.3

Mendel chose seven traits of peas for his experiments. Each trait had two clearly different forms; no intermediate forms were observed.

	Seed shape	Seed color	Flower color	Flower position	Pod color	Pod shape	Plant height
Dominant trait	round	yellow	purple	axial (side)	green	inflated	tall
Recessive trait	wrinkled	green	white	terminal (tips)	yellow	constricted	short

ThinkingLab Analyze the Procedure

Purpose

LS Logical-Mathematical Students will learn how to convert and express two sets of related numbers as a ratio.

Process Skills

use and interpret data, calculate

Teaching Strategies

- Have students work in groups. Place students with poor mathematical skills with students working at grade level.
- Provide students with an example if they are having difficulty in determining a ratio.

Further Background

The round seeds contain an abundance of well-formed starch grains and have higher water retention than the wrinkled seeds. A lack of well-formed starch grains and lower water retention account for the wrinkled appearance.

Thinking Critically

The ratio was 2.96:1, slightly lower than expected. The observed and expected ratios may differ slightly due to chance.

Assessment

Skill: Provide students with other examples of traits and data. Have them calculate ratios. L1

STUDENT JOURNAL

Dominant and Recessive Traits Provide students with the following traits and corresponding letter representations. Have students record which trait is dominant and which is recessive. (a) Oriental eye shape = *O*, non-Oriental eye shape = *o;* (b) blonde hair = *b*, black hair = *B;* (c) yellow pea seed = *Y*, green pea seed = *y*. L1 LS

Math

A Solution from Ratios

Purpose

Students will gain insight into the importance of mathematics to the study of biology. They will learn how mathematics aided Mendel in understanding the laws of heredity.

Teaching Strategies

- Students may be surprised to learn that a biologist of Charles Darwin's stature missed a great opportunity in his brilliant career. Had he interpreted his data about snapdragon flower shapes correctly, the whole world might have understood the laws of heredity forty years earlier.
- Discuss how ratios helped Mendel see that definite factors were being passed on from parents to offspring. He didn't know what these factors were, nor how they operated. He knew nothing about meiosis and yet was able to show how traits were transmitted due to his mathematical analysis.

Possible Answers

- The ratio in Darwin's experiment was 2.38:1, similar to Mendel's, but it would have been closer to a 3:1 ratio if he had crossed a larger number of plants. Ratios for other flower traits are: Seed color: 3.01:1; Flower position: 3.14:1; Pod color: 2.82:1; Pod shape: 2.95:1.
- Students may say that the ratios revealed that a dominant trait showed up three times more often because it was always able to overcome the effect of a recessive trait that accompanied it.

Math

A Solution from Ratios

Gregor Mendel had three qualities that led to his discovery of the laws of heredity. First, he was curious, impelled to find out why things happened. Second, he was a keen observer. Third, he was a skilled mathematician. Mendel was the first biologist who relied heavily on statistics for solutions to how traits are inherited.

Darwin Missed His Chance About the same time that Mendel was carrying out his experiments with pea plants, Charles Darwin was gathering data on snapdragon flowers. When Darwin crossed plants that had normal-shaped flowers with plants with odd-shaped flowers, all the offspring had normal-shaped flowers. He thought the two traits had blended. When he allowed the F_1 plants to self-pollinate, his results were 88 plants with normal-shaped flowers and 37 plants with odd-shaped flowers. Darwin was puzzled by the results and did not continue his studies with these plants. Lacking Mendel's statistical skills, Darwin failed to see the significance of the ratio of normal-shaped flowers to odd-shaped flowers in the F_2 generation. What was this ratio? Was this ratio similar to Mendel's ratio of dominant to recessive traits in pea plants?

Finding the Ratios for Four Other Traits *Figure 12.3* shows seven traits that Mendel studied in pea plants. You have already looked at Mendel's data for plant height and seed shape. Now use the data for seed color, flower position, pod color, and pod shape to find the ratios of dominant to recessive for these traits in the F_2 generation.

Ratios of Inherited Traits

Seed color	Flower position	Pod color	Pod shape
Yellow 6022	Lateral 651	Green 428	Inflated 882
Green 2001	Terminal 207	Yellow 152	Constricted 299

CONNECTION TO Biology

Why were ratios so important to understanding how dominant and recessive traits are inherited?

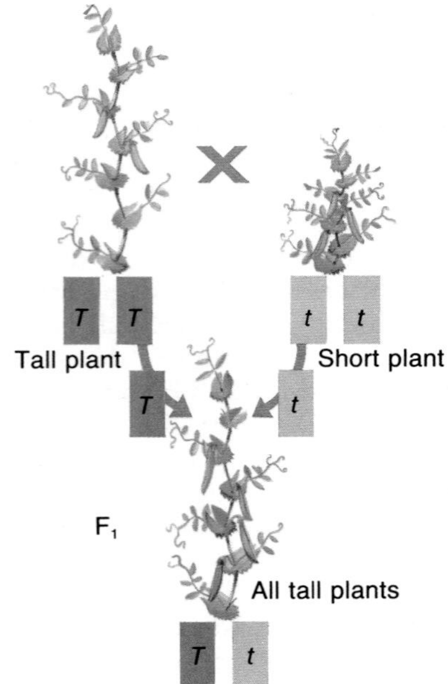

Figure 12.4

The rule of dominance explains the results of Mendel's cross between P_1 tall and short plants. When recording the results of crosses, it is customary to use the same letter for different alleles of the same gene. An uppercase letter is used for the dominant allele, and a lowercase letter for the recessive allele. The dominant allele is always written first. Therefore, the allele for tallness is represented by *T* and the allele for shortness by *t*.

The law of segregation

Now recall the results of Mendel's cross between F_1 tall plants when the trait of shortness reappeared. To explain this result, Mendel formulated the first of his two laws of heredity. He concluded that the two alleles for each trait must separate when gametes are formed. A parent, therefore, passes on at random only one allele for each trait to each offspring. This conclusion, illustrated in *Figure 12.5,* is called the **law of segregation.**

PORTFOLIO

Mendel's Laws Have students imagine they are Gregor Mendel, and they have just formulated the laws of dominance and segregation. Have them write a short article describing their findings for the only science journal in existence. Remember, the year is 1863. L3 P

Martian Traits Have the students imagine they have encountered their first Martian. Martian traits are inherited exactly the same way as Earthling traits. Have students provide examples of five Martian traits through the use of a diagram and a written statement describing the possible genotypes and phenotypes for these five traits. L2 P

Phenotypes and Genotypes

Mendel's experiments showed that tall plants are not all the same. Some tall plants, when crossed with each other, yielded only tall offspring. These were Mendel's original P_1 true-breeding tall plants. Other tall plants, when crossed with each other, yielded both tall and short offspring. These were the F_1 tall plants that came from a cross between a tall plant and a short plant.

Two organisms, therefore, can look alike but have different underlying gene combinations. The way an organism looks and behaves make up its **phenotype.** The phenotype of a tall plant is tall, regardless of the genes it contains. The gene combination an organism contains is known as its **genotype.** The genotype of a tall plant that has two alleles for tallness is *TT.* The genotype of a tall plant that has one allele for tallness and one allele for shortness is *Tt.*

You can see that you can't always know an organism's genotype simply by looking at its phenotype. An organism is **homozygous** for a trait if its two alleles for the trait are the same. The true-breeding tall plant that had two alleles for tallness (*TT*) would be homozygous for the trait of height. Because tallness is dominant, a *TT* individual is homozygous dominant for that trait. A short plant would always have two alleles for shortness (*tt*). It would, therefore, always be homozygous recessive for the trait of height.

An organism is **heterozygous** for a trait if its two alleles for the trait are different. Therefore, the tall plant that had one allele for tallness and one allele for shortness (*Tt*) is heterozygous for the trait of height.

homozygous:
homo (GK) same
zygotos (GK) joined together
Homozygous individuals have two identical alleles for a trait.

heterozygous:
hetero (GK) the other of two
zygotos (GK) joined together
Heterozygous individuals have two different alleles for a trait.

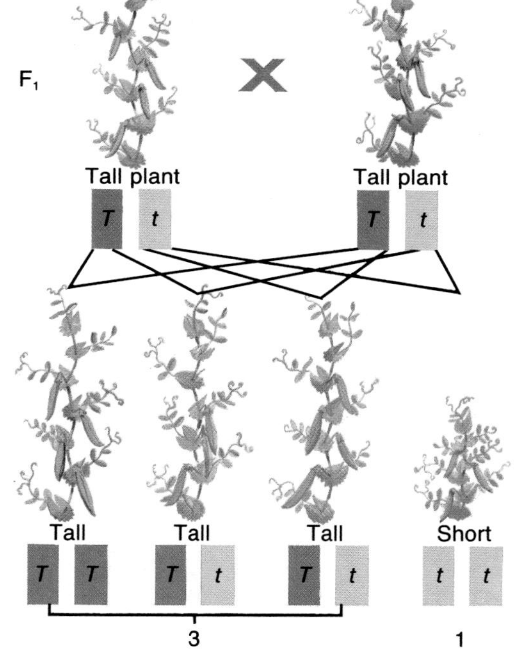

Mendel's Dihybrid Crosses

Next, Mendel performed another set of crosses in which he used peas that differed from each other in two traits rather than only one. A cross involving two different traits is called a **dihybrid cross.**

Figure 12.5

Mendel's law of segregation explains the results of his cross between F_1 tall plants. Each tall plant in the F_1 generation carried one dominant allele for tallness and one unexpressed recessive allele for shortness. It received the allele for tallness from its tall parent and the allele for shortness from its short parent in the P_1 generation. Because each F_1 plant has two different alleles, it can produce two different types of gametes—"tall" gametes and "short" gametes. During fertilization, gametes randomly pair to produce four combinations of alleles.

Using Science Words

It is sometimes easier for students to remember the meaning of the term *genotype* by reversing the word so it becomes "type of gene." When stated this way, students can associate the term *genotype* with its definition. *Pheno* means "to show." Thus, the phenotype shows the type of trait or how it appears.

Concept Development

Ask students to supply the correct term—*genotype* or *phenotype*—to the following examples: (a) *LL*, (b) blond hair, (c) dimpled chin, (d) blue eyes, (e) *Dd*, (f) *ss*, (g) white and green leaves. *a, e, and f are genotypes; b, c, d, and g are phenotypes.* L1

Have students provide the following information for this example: green pea pod = *G*, Yellow pea pod = *g*. (a) Give the phenotypes of plants with these genotypes: *Gg*, *GG*, and *gg*. (b) Use the terms *homozygous* or *heterozygous* to describe each of the three examples above. *a) green, green, yellow; b) heterozygous, homozygous, homozygous.* L1

Visual Learning

Figure 12.4 Have students write out the three important written conventions that are described with this diagram. *Use the same letter for different alleles of the same gene; use uppercase letters for dominant alleles and lowercase letters for recessive alleles; and always write the dominant allele first.* L1

CULTURAL DIVERSITY

Everett Anderson

Introduce students to the contribution of African American cell biologist, Everett Anderson, to the modern understanding of the meiotic process. Anderson (1928-), who received his Ph. D. in 1955, has been one of the leading researchers in developing electron microscopic techniques to study meiosis.

Obtain a copy of Anderson's 1972 publication, *The Meiotic Process: Pairing, Recombination, and Chromosome Movements,* and discuss with students the methodologies used in studying the process of meiosis.

BioLab Design Your Own Experiment

How can phenotypes and genotypes of plants be determined?

Time Allotment

Initial session: one class period; followup session: 5 minutes each day for watering, 20 minutes on last day for counting

Objectives

Review objectives with students before they begin the Biolab.

Process Skills

hypothesize, infer, observe, collect data

Safety Precautions

- Some seed materials are poisonous. Do not allow students to eat the seeds.

PREPARATION

Alternate Materials

- Seeds can be germinated and observed in petri dishes. This will eliminate the need for soil, flats, or pots. Place seeds on moistened paper towels in the bottom of the dish and keep the dish covered.

Possible Hypotheses

- If the parent plants were true breeding for green color, then all offspring will be green.
- If the parent plants were heterozygous for green color, then offspring will show an approximate ratio of 3 green to 1 white.

BioLab Design Your Own Experiment

How can phenotypes and genotypes of plants be determined?

It's difficult to predict the traits of plants if all that you see is their seeds. But if these seeds are planted and allowed to grow, certain traits will appear. By observing these traits, you might be able to determine the possible phenotypes and genotypes of the parent plants that produced these seeds. In this lab, you will determine the genotypes of plants that grow from two groups of tobacco seeds. Each group of seeds came from different parents. Plants will be either green or albino (white) in color. Use the following genotypes for this cross.

CC = green, *Cc* = green, and *cc* = albino

PREPARATION

Problem

Can the phenotypes and genotypes of the parent plants that produced two groups of seeds be determined from the phenotypes of the plants grown from the seeds?

Hypotheses

Have your group agree on a hypothesis to be tested that will answer the problem question. Record your hypothesis.

Objectives

In this Biolab, you will:

- **Analyze** the results of growing two groups of seeds.
- **Draw conclusions** about phenotypes and genotypes based on those results.

Possible Materials

potting soil
small flowerpots or seedling flats
two groups of tobacco seeds
hand lens
light source
thermometer
plant-watering bottle

Safety Precautions

Always wash your hands after handling plant materials.

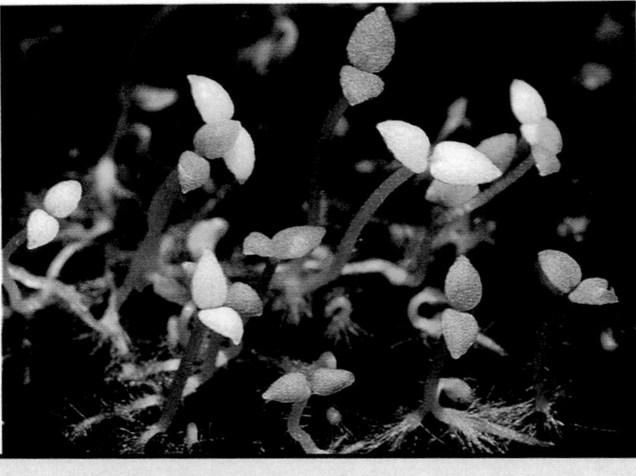

PLAN THE EXPERIMENT

Teaching Strategies

- When supplying seeds to students, make sure that the two types are kept separate from one another. Stick seeds onto tape for dispensing.
- An ideal quantity of seeds to use is 20–30 per type.
- Cotyledons (seed leaves) will appear after about 10 days.

Possible Procedures

- Keep soil moist at all times. Natural window light should be sufficient.
- Plant seeds about 1 cm below the soil. Planting is easier if the soil is moist.
- Mark the seed type planted in the flat or pot. Popsicle sticks can serve as markers.

PLAN THE EXPERIMENT

1. Examine the materials provided by your teacher. As a group, make a list of the possible ways you might test your hypothesis.
2. Agree on one way that your group could investigate your hypothesis.
3. Design an experiment that will allow you to collect quantitative data. For example, how many plants do you think you will need to examine?
4. Prepare a list of numbered directions. Include a list of materials and the quantities you will need.
5. Design and construct a data table for recording your observations.

Check the Plan

1. Carefully determine the data you are going to collect. How many seeds do you think you will need? How long will you carry out the experiment?
2. What variables, if any, will have to be controlled? (Hint: Think about the growing conditions for the plants.)
3. Does your table allow for all the various kinds of data you need?
4. ***Make sure your teacher has approved your experimental plan before you proceed further.***
5. Carry out your experiment. Make any needed observations, such as the numbers of green and albino plants in each group, and complete your data table. Design and then complete a graph or other visual representation of your results.

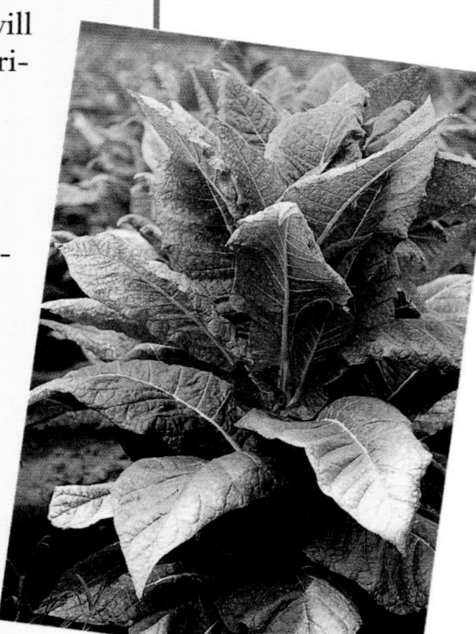

ANALYZE AND CONCLUDE

1. **Thinking Critically** Why was it necessary to grow plants from the seeds in order to determine the phenotypes of the plants that formed the seeds?
2. **Drawing Conclusions** Using the information provided in the introduction, describe how the gene for green color (*C*) is inherited.
3. **Making Inferences** For the group of seeds that yielded all green plants, are you able to determine exactly the genotypes of the parents that formed these seeds? Can you determine the genotype of each plant observed? Explain.
4. **Making Inferences** For the group of seeds that yielded some green and some albino plants, are you able to determine exactly the genotypes of the plants that formed these seeds? Can you determine the genotype of each plant observed? Explain.

Going Further

Journal Report From your results, design an experiment that would help you determine some of the genotypes you were unable to determine from the experiment you just performed. Write up your experiment in your journal.

ANALYZE AND CONCLUDE

1. Leaf color cannot be observed in the seed but appears only after the plant has emerged from the seed.
2. The gene for green color is a dominant trait.
3. No, one parent may have been true breeding for green *(CC)*, the other may have been heterozygous *(Cc)*. This would have yielded all green offspring. Offspring may be *CC* or *Cc* but still appear green.
4. Yes, both parents must be heterozygous to yield a ratio of 3 green to 1 albino. For the offspring genotypes, you can only conclude that the albino offspring are *cc*. Green are either *CC* or *Cc*.

Assessment

Portfolio: Ask students to make diagrams that show the parental and offspring genotypes and phenotypes for both groups of seeds used in this experiment. L1 P

Going Further

Ask students to determine the role that light may play in the expression of the phenotypes. Grow both seed types in the dark. Observe plants after 10 days. Place them in the light and observe 24 hours later. L2

Data and Observations

Seeds that came from true breeding plants will produce plants that are all green. Seeds from heterozygous parents will produce green and albino seedlings in the ratio of about 3 green to 1 albino. Have students recheck the Thinking Lab for help in determining the procedure for calculating the ratio of green to albino plants.

GLENCOE TECHNOLOGY

Videodisc

STVS: Plants and Simple Organisms
Disc 4, Side 1
Breeding Fruit Flies (Ch. 21)

Selective Breeding in Cows (Ch. 22)

Selective Breeding in Micro-pigs (Ch. 23)

SECTION 12.1

MiniLab

Purpose

LS Logical-Mathematical Students will observe how sample size affects agreement between expected and observed ratios.

Process Skills

collect data, record data, infer

Teaching Strategies

- If you are teaching in a setting where only males or females are in your class, either skip the first tally of classmates or flip a coin 26 times, allowing heads to represent females and tails to represent males.

Expected Results

The observed ratio of males to females will get closer to the expected 1:1 ratio as the sample size increases.

Analysis

1. The following sample data may be typical: classroom = 11 male, 15 female—ratio = 1:1.36; brothers/sisters = 35 male, 39 female—ratio = 1:1.11; cousins, aunts, uncles = 117 male, 120 female—ratio = 1:1.03.
2. The ratio approached a closer match to 1:1 as sample size increased.
3. Possibly a sample size of at least 200

Assessment

Performance: Have each student flip a coin 100 times, record the number of times heads and tails come up, and calculate the ratio of heads to tails. Gather class totals and calculate class ratios. How does the ratio of heads to tails change as the sample size increases? *It should come closer to the expected ratio of 1:1.* L1

MiniLab

How does sample size affect results?

The ratio of males to females born in a population should be 1:1. If progressively larger populations of males and females are counted, the actual numbers approach this theoretical ratio.

Procedure

1. Count the number of males and females in your class. Determine the ratio.
2. Add the brothers and sisters of your classmates to your total, and again count the number of males and females. Determine the male:female ratio.
3. Redo the count, this time adding all aunts, uncles, and cousins of the members of your class. Determine the male:female ratio.

Analysis

1. How many people were included in each of your three samples?
2. How did the male:female ratio vary as the sample size became larger?
3. Based on your data, how large a sample do you think you need in order to be fairly sure of obtaining numbers that are close to the expected ratio?

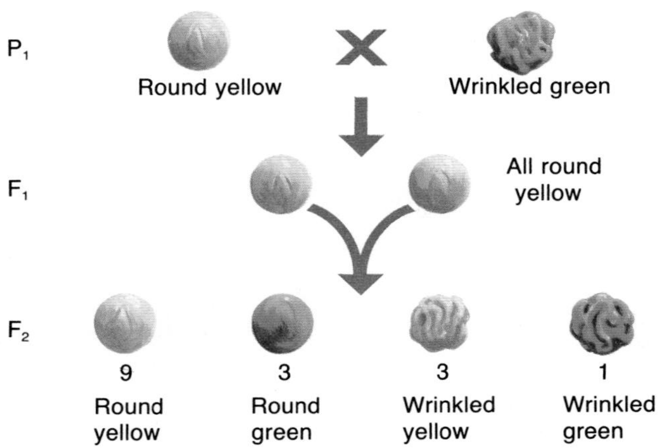

Figure 12.6

When Mendel crossed true-breeding plants with round yellow seeds and true-breeding plants with wrinkled green seeds, the seeds of all the offspring were round and yellow. When the F_1 plants were allowed to self-pollinate, they produced four different kinds of plants in the F_2 generation. If the alleles for seed shape and color were inherited together, only two kinds of pea seeds would have been produced: round yellow and wrinkled green.

The first generation

Mendel took true-breeding pea plants that had round yellow seeds (*RRYY*) and crossed them with true-breeding pea plants that had wrinkled green seeds (*rryy*). He already knew that when he crossed plants that produced round seeds with plants that produced wrinkled seeds, all the plants in the F_1 generation produced seeds that were round. In other words, round trait was dominant. Similarly, when he crossed plants that produced yellow seeds with plants that produced green seeds, all the plants in the F_1 generation produced yellow seeds—yellow was dominant. Therefore, Mendel was not surprised when he found that the F_1 plants of his dihybrid cross all had round yellow seeds, as *Figure 12.6* shows.

The second generation

Mendel then let the F_1 plants pollinate themselves. As you might expect, Mendel found some plants that produced round yellow seeds and others that produced wrinkled green seeds. But that's not all. He also found some plants with round green seeds and others with wrinkled yellow seeds. Mendel sorted and counted the plants of the F_2 generation. He found they appeared in a definite ratio—9 round yellow:3 round green:3 wrinkled yellow:1 wrinkled green. To explain the results of this dihybrid cross, Mendel formulated his second law.

The law of independent assortment

Mendel's second law states that genes for different traits—for example, seed shape and seed color—are inherited independently of each other. This conclusion is known as the **law of independent assortment.** When a pea plant with the genotype *RrYy* produces gametes, the

Meeting Individual Needs

Visually Impaired Provide students with two large and two small round pieces of candy, plus two large and two small buttons. Advise them that the large-sized candies and buttons represent dominant alleles. Ask them to prepare two sets of objects (alleles) for two traits in two parents, making both parents heterozygous for both traits. Then have them arrange their parental gene sets into independent assortments of two alleles each. Advise them that they cannot have two candies or two buttons in their final groups. Have them explain how this illustrates the law of independent assortment. L1 LS

alleles R and r will separate from each other (the law of segregation) as well as from the alleles Y and y (the law of independent assortment), and vice versa. These alleles can then recombine in four different ways. We now know that this principle is true only if genes are located on different chromosomes or are far apart on the same chromosome.

Punnett Squares

In 1905, Reginald Punnett, an English biologist, devised a shorthand way of finding the expected proportions of possible genotypes in the offspring of a cross. This method is called a Punnett square. It takes account of the fact that fertilization occurs at random, as Mendel's law of segregation states. If you know the genotypes of the parents, you can use a Punnett square to predict the possible genotypes of their offspring.

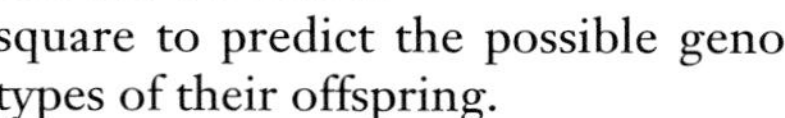

Monohybrid crosses

Consider the cross between two F_1 tall pea plants, each of which has the genotype *Tt*. Half the gametes of each parent would contain the *T* allele, and the other half would contain the *t* allele. A Punnett square for this cross is two boxes tall and two boxes wide because each parent can produce two kinds of gametes for this trait. Refer to the Punnett square on the next page and determine the possible genotypes of the offspring.

A BROADER VIEW

The Laws of Probability

Punnett squares are good for showing all the possible combinations of gametes and the likelihood that each will occur. In reality, however, you don't always get the exact ratio of results shown in the square. That's because, in some ways, genetics is like flipping a coin–it follows the rules of chance.

Rules of chance At the start of every football game, the captain of the visiting team calls either *heads* or *tails* in a coin toss. The probability of his choice coming up is expressed as a fraction–the numerator is the number of ways he could be right; the denominator is the total possible events. In this case, the probability is 1/2, or one in two.

Boys and girls Now, consider the chances of parents having a girl or a boy. Every time two parents have a child, there is one chance in two that the child will be a girl, and one chance in two that it will be a boy. Yet, you must know families that have three or even four girls and no boys, or vice versa. However, just look at a large number of people. You will find that the number of each sex is close to a 1:1 ratio. When dealing with events governed by the laws of probability, the results will approach the expected ratios when large numbers of events are counted.

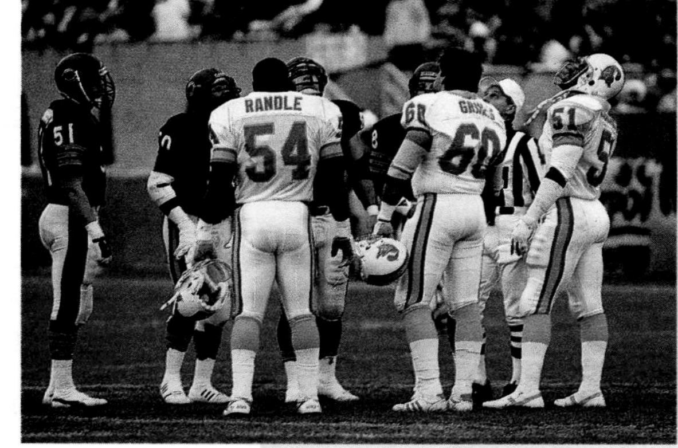

Mendel and probability When Mendel crossed his F_1 heterozygous tall pea plants, three-fourths of the offspring in the F_2 generation were tall and one-fourth were short. In other words, they were in a ratio of 3 tall:1 short. Yet, if you were to look at Mendel's actual data, you would see that he counted 787 tall plants and 277 short plants–a ratio of 2.84:1. This ratio was observed because events in genetics are governed by laws of probability. A 3:1 ratio really means that every time an F_2 plant is produced, there are three chances in four that it will be tall and one chance in four that it will be short. Each new plant is an independent event and is not affected by the traits of previous plants. When you count large numbers of plants, you will get close to the expected ratio of 3:1.

Thinking Critically

The probability of independent events happening together is the product of their individual probabilities. If you tossed three coins at the same time, what is the probability of getting all heads?

Meeting Individual Needs

Learning Disabled Provide students with copies of blank Punnett square outlines. Project a similar copy onto a screen with an overhead projector. Lead students through the steps needed for solving the problem of the *Tt* parents. Reinforce how the letters placed to the side and across the top represent all the possible gametes for each parent and how this illustrates the law of segregation. Reinforce the significance of the four squares and what the letter combinations within them represent. Provide students with a variety of problems (*TT* × *tt, Tt* × *tt, Tt* × *TT, tt* × *tt*) to solve. L1 LS

A BROADER VIEW

The Laws of Probability

Purpose

Students will learn why small data samples may not show a close match between expected and observed ratios but do show a closer match as sample size increases.

Teaching Strategies

- Review with students what the probability would be for two coins both landing heads up after two flips. The formula would be 1/2 × 1/2, because each event is independent of the other and has an even probability of landing heads or tails. The probability of these events happening together (two heads in two flips) is the product of their individual probabilities or 1/2 × 1/2 = 1/4 or 1:4.

Thinking Critically

1/2 × 1/2 × 1/2 = 1/8 or 1:8.

Reinforcement

LS **Visual-Spatial** Remind students that Mendel found that purple flower color and axial position of flowers in peas are dominant traits and white flower color and terminal flower position are recessive traits, as shown in Figure 12.3. Then have them draw a diagram similar to Figure 12.6 showing the ratio of these traits in the F_2 generation when the F_1 generation consists of true-breeding purple axial and white terminal plants. *A ratio of 9 purple axial to 3 purple terminal to 3 white axial to 1 white terminal* L1

SECTION 12.1

Concept Development

LS Logical-Mathematical Have students construct Punnett squares and solve these problems giving the genotype and phenotype ratios expected: (a) Homozygous tall plant bred to a homozygous short plant. *All offspring will be tall and heterozygous.* (b) Homozygous tall plant bred to a heterozygous tall plant. *All offspring will be tall, 1/2 being* TT *and 1/2 being* Tt. (c) Heterozygous tall plant bred to a homozygous short plant. *1/2 will be tall and 1/2 will be short; all tall offspring will be* Tt *and all short offspring will be* tt. L1 P

GLENCOE TECHNOLOGY

Videodisc

Biology: The Dynamics of Life
Disc 1, Side 1
Punnett Squares (Ch. 29)

3 Assess

Check for Understanding

Have students describe the relationship between or among the following terms:

(a) pollination—fertilization
(b) allele—dominant—recessive
(c) genotype—phenotype
(d) homozygous—heterozygous
(e) monohybrid—dihybrid L1

After the genotypes have been determined, you can determine the phenotypes. Looking at the Punnett square for this cross in *Figure 12.7,* you can see that three-fourths of the offspring are expected to be tall and one-fourth are expected to be short. Of the tall offspring, one-third will be homozygous dominant (*TT*) and two-thirds will be heterozygous (*Tt*). Note that whereas the genotype ratio is 1 *TT*:2 *Tt*:1 *tt*, the phenotype ratio is 3 tall:1 short.

Dihybrid crosses

What happens in a Punnett square when two traits are considered? Think again about Mendel's cross between peas with round yellow seeds and peas with wrinkled green seeds. All the seeds of the F_1 plants were round and yellow, and were heterozygous for each trait (*RrYy*). What kind of gametes will they form?

Mendel explained that the genes for seed shape and seed color would be inherited independently of each other. This means that each F_1 plant will produce gametes containing the following combinations of genes with equal frequency: round yellow (*RY*), round green (*Ry*), wrinkled yellow (*rY*), and wrinkled green (*ry*). A Punnett square for a dihybrid cross will then need to be four boxes on each side for a total of 16 boxes, as *Figure 12.8* shows.

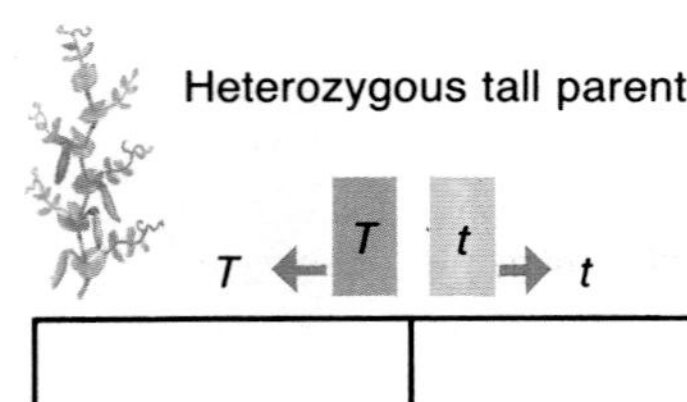

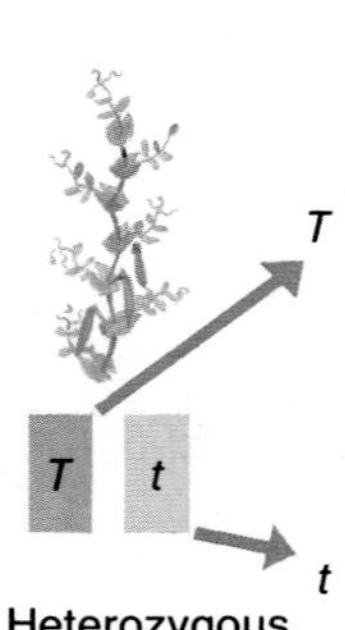

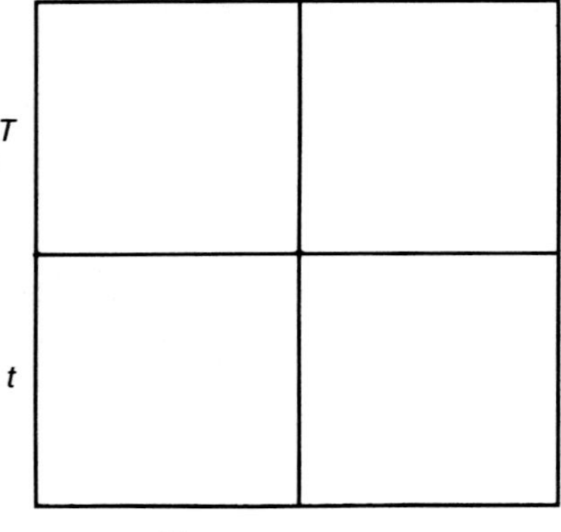

Figure 12.7

This Punnett square predicts the results of a monohybrid cross between two heterozygous tall pea plants.

A The two kinds of gametes from one parent are listed on top of the square, and the two kinds of gametes from the other parent are listed on the left side. It doesn't matter which set of gametes is on top and which is on the side, that is, which parent contributed the *T* and which contributed the *t*.

	T	*t*
T	*TT*	*Tt*
t	*Tt*	*tt*

B To find the possible genotypes of the offspring from a Punnett square, each box is filled in with the gametes above and to the left side of that box. Each box then contains two alleles—one possible genotype. You can see that there are three different possible genotypes—*TT, Tt,* and *tt*—and that *Tt* can result from two different combinations.

Meeting Individual Needs

Learning Disabled To help students predict genes within each gamete for a dihybrid cross and to illustrate the law of independent assortment, provide them with the following mnemonic: 1 and 3 and 1 and 4, 2 and 3 and 2 and 4. Have them write the numbers 1–4 across the top of a set of parental genes in a dihybrid cross. It would look like this:

1 2 3 4
R r Y y

To obtain the correct gamete combinations according to the law of independent assortment, match genes 1 and 3 (*RY*), 1 and 4 (*Ry*), then match 2 and 3 (*rY*), 2 and 4 (*ry*). L1

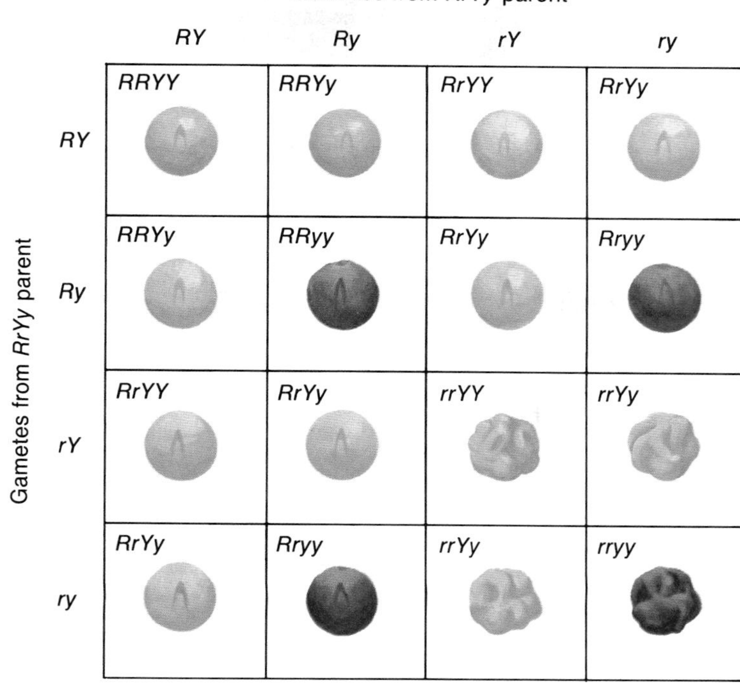

Figure 12.8

A Punnett square for a dihybrid cross between heterozygous pea plants with round yellow seeds shows clearly that the offspring fulfill Mendel's observed ratio of 9 round yellow:3 round green:3 wrinkled yellow:1 wrinkled green. How many different genotypes and phenotypes result from this cross?

Section Review

Understanding Concepts

1. Why did the flower structure of pea plants make them suitable for Mendel's genetic studies?
2. What is the genotype of a homozygous and a heterozygous tall pea plant?
3. One parent is homozygous for a certain trait and the other parent is heterozygous. Make a Punnett square to determine what fraction of their offspring is expected to be heterozygous.

Thinking Critically

4. In garden peas, the allele for yellow peas is dominant to the allele for green peas. Suppose you have a plant with yellow peas but you don't know whether it is homozygous dominant or heterozygous. What experiment could you do to find out? Draw a Punnett square to help you.

Skill Review

5. **Observing and Inferring** The offspring of a cross between a purple-flowered plant and a white-flowered plant are 23 plants with purple flowers and 26 plants with white flowers. What are the genotypes of the parent plants, and which allele is dominant? Use the letters *W* and *w* to represent the alleles. For more help, refer to Thinking Critically in the *Skill Handbook*.

SECTION 12.1

Reteach

Have students provide an example of each relationship provided in the Check for Understanding. L1

Extension

Have students list the genotypes and phenotypes resulting from (a) an *RrYy* plant cross-pollinated by an *RRyy* plant; (b) an *rrYy* plant cross-pollinated by a *RrYy* plant. (a) *genotypes:* RrYy, RRyy, RRYy, Rryy; *phenotypes: 1/2 round, yellow, 1/2 round, green* (b) *genotypes:* RrYY; RrYy, Rryy, rrYY, rrYy, rryy; *phenotypes: 3 round, yellow; 3 wrinkled, yellow; 1 round, green; 1 wrinkled, green* L1

Assessment

Knowledge: Have students explain what each of the following represents in a Punnett square: (a) the letters written at the top and side of the square; (b) the letters written within each box; (c) the four boxes. L1

4 Close

Activity

Have students illustrate Mendel's law of segregation, using as much of a Punnett square as needed. Do the same with the law of independent assortment, using as much of a Punnett square as needed. L1

Answers to Section Review

1. They are self-pollinating, and male flower parts can be easily removed to allow for cross-pollination.
2. Homozygous tall = *TT,* heterozygous tall = *Tt*.
3. One half of all offspring will be heterozygous. (Note: It will not matter if the homozygous parent is homozygous dominant or recessive.)

Thinking Critically

4. Cross-pollinate the unknown yellow plant with a recessive green parent. If the offspring are all yellow, then the unknown genotype is homozygous yellow. If half the offspring are yellow and half green, the unknown genotype is heterozygous.

Skill Review

5. Genotypes of parents are *Ww* for the purple-flowered plant and *ww* for the white-flowered plant. The purple allele is dominant.

SECTION

12.2

Prepare

Key Concepts

This section develops the concepts of diploid and haploid chromosome numbers and homologous chromosomes. An overview of meiosis is provided with a general explanation of why meiosis is necessary. Events that occur during meiosis are illustrated and the role of meiosis in genetic recombination is explained.

Block Scheduling

Look for this symbol for strategies that are useful in a block scheduling format. For more information on block scheduling, refer to Section 12.2 in the **Lesson Plans** booklet.

Materials

- Gather tape or glue, wool strands, string, paper clips, scissors, unlined paper or poster board, and toothpicks if you do the Alternate Lab.

1 Focus

Bellringer

Before presenting the lesson, display **Section Focus Master 26** on the overhead projector and have students answer the accompanying questions. L1 SAE

Discussion

Show students an egg and explain that it is actually a gamete. Based on previous information, students should be able to tell how many alleles are present for each trait. *Only one allele for each trait* Ask why an organism cannot produce gametes by mitosis. *Mitosis produces a cell with both members of each pair of chromosomes.*

SECTION

12.2 Meiosis

Section Preview

Objectives

Analyze how meiosis maintains a constant number of chromosomes in the body cells of the members of a species.

Infer how meiosis leads to variation in a species.

Relate Mendel's laws of heredity to the events of meiosis.

Key Terms

diploid
haploid
homologous chromosome
meiosis
sperm
egg
zygote
sexual reproduction
crossing over
genetic recombination

Mendel succeeded in describing how inherited traits are transmitted from one generation to the next. Yet he had no way of explaining what his factors might be, or in what physical way a factor could be passed from parent to offspring. Pieces of the heredity puzzle were still missing. As a result, Mendel's work went relatively unnoticed until almost 20 years after his death.

Today, more than 125 years later, we have a much greater understanding of what Mendel's factors are, how they determine inherited traits, and how they are transmitted between generations of organisms.

Genes, Chromosomes, and Numbers

Organisms have tens of thousands of genes that determine individual traits. Genes do not exist free in the nucleus of a cell, but are lined up in linear fashion on chromosomes. Typically, a thousand or more genes are arranged on each chromosome.

Diploid and haploid cells

If you examined the nucleus in a cell of one of Mendel's pea plants, you would find it had 14 chromosomes—seven pairs. In the body cells of animals and most plants, chromosomes occur in pairs. One chromosome in each pair came from the male parent, and the other came from the female parent. A cell with two of each kind of chromosome is called a **diploid** cell and is said to contain a diploid, or $2n$, number of chromosomes. This pairing supports Mendel's conclusion that organisms have two factors—alleles—for each trait. One allele is located on each of the paired chromosomes.

Organisms produce gametes that contain *one* of each kind of chromosome. A cell with one of each kind of chromosome is called a **haploid** cell and is said to contain a haploid, or n, number of chromosomes. This fact supports Mendel's conclusion that parent organisms give one factor, or allele, for each trait to each of their offspring.

Each species of organism contains a characteristic number of chromosomes. *Table 12.1* shows the diploid and haploid numbers of chromosomes of some species. Note the large range of chromosome numbers. Note also that the chromosome number of a species is not related to the complexity of the organism.

Program Resources

Section Focus Master 26 L1 SAE
Reinforcement and Study Guide, pp. 47-48 L1
Laboratory Manual, pp. 73-74 L2
Concept Mapping, p. 12 L1 SAE
Performance Assessment in the Biology Classroom, p. 13 L1 P
Basic Concepts Transparency 15 and **Master** L1 SAE
Basic Skills Transparency 11 and **Master** L1 SAE

Table 12.1 Chromosome Numbers of Some Common Organisms

Organism	Body Cell (2*n*)	Gamete (*n*)
Human	46	23
Garden Pea	14	7
Fruit fly	8	4
Tomato	24	12
Dog	78	39
Chimpanzee	48	24
Leopard frog	26	13
Corn	20	10
Apple	34	17
Adder's tongue fern	1260	630

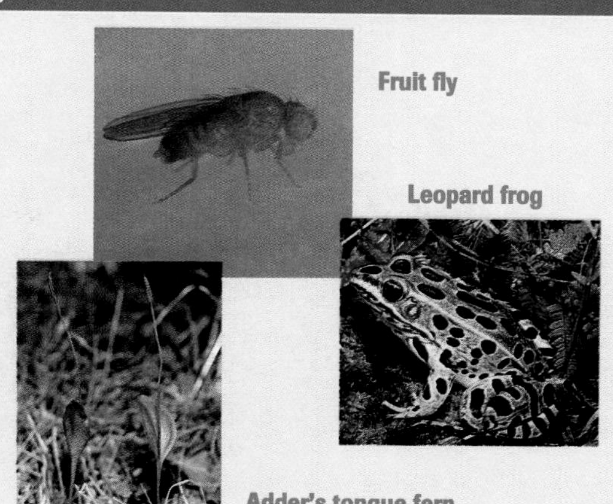

Homologous chromosomes

Together, the two chromosomes of each pair in a diploid cell help determine what the individual organism looks like. These paired chromosomes are called **homologous chromosomes.** Homologous chromosomes have genes for the same traits arranged in the same order. However, because there are different possible alleles for the same gene, the two chromosomes in a homologous pair are not identical to each other.

For example, look at one of the seven pairs of homologous chromosomes in Mendel's peas. These chromosome pairs are numbered 1 to 7. Each pair contains certain genes, located at specific places on the chromosome. Chromosome 4 contains the genes for three of the traits that Mendel studied. You can see the positions of these genes in *Figure 12.9.* Many other genes can be found on this chromosome as well.

Every pea plant has two copies of chromosome 4. It received one from each of its parents and will give one at random to each of its offspring. Remember, however, that the two copies of chromosome 4 in a pea plant may not have identical alleles. Each can have one of the different alleles possible for each gene.

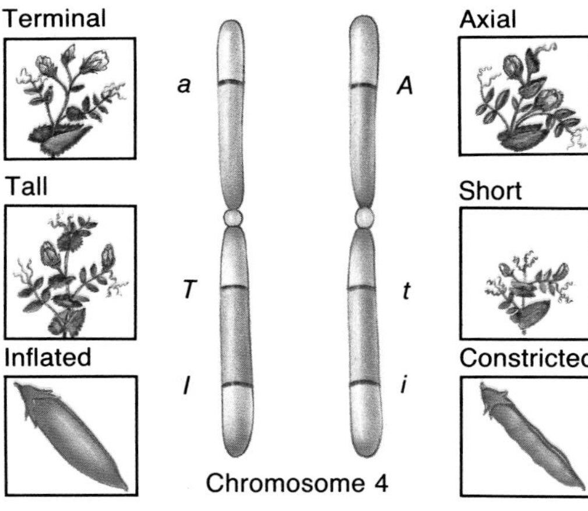

Figure 12.9

Each chromosome 4 in garden peas contains genes for flower position, height, and pod shape. The homologous chromosomes in the diagram show both alleles for each trait. Thus, the plant represented by these chromosomes is heterozygous for each of the three traits. Flower position can be either axial—flowers located along the stems, or terminal—flowers clustered at the top of the plant. As you know, plant height can be either tall or short. Pod shape can be either inflated or constricted.

2 Teach

Visual Learning

Table 12.1 Ask students if there is any evidence to support the idea that plants have fewer chromosomes than animals do. Tell them to use examples from the table to support their answers. *No; for example, apples have 34, which is more than fruit flies or frogs have.* Ask for the chromosome numbers in the skin cells of a leopard frog and a dog, *26, 78;* and in the root cells of tomatoes and garden peas, *24, 14.*

Math Connection

Calculating Genotype Ratios

LS Logical-Mathematical Algebra can be used to solve dihybrid cross problems by following these steps. Solve only for the first allele using a Punnett square of four squares. Express the genotype results as a ratio. For example: $Rr \times Rr = 1/4\ RR$, $2/4\ Rr$, $1/4\ rr$. Repeat for the second allele. For example: $Yy \times Yy = 1/4\ YY$, $2/4\ Yy$, $1/4\ yy$. Cross multiply the two totals as follows:

$$\begin{array}{l} 1/4\ RR + 2/4\ Rr + 1/4\ rr \\ \underline{\times\ 1/4\ YY + 2/4\ Yy + 1/4\ yy} \end{array}$$

The result gives you the genotype ratios for a dihybrid cross. L3

Meeting Individual Needs

Learning Disabled Prepare five strips of blue paper of varying lengths but equal widths. Prepare five matching strips of pink paper. Pair the strips according to length. Tell students the strips represent chromosomes and that the pink ones come from the female parent and the blue ones from the male parent. Ask what the diploid number is. *10* Did one parent contribute a pair? *No* Using the strips, illustrate all the possible combinations of chromosomes in gametes. What is the haploid number? *5* L1 LS

SECTION 12.2

Brainstorming

Students are to look for mitosis and meiosis occurring within a fruit fly. Have them explain where they would look for each process and why. *Mitosis occurs in all body cells; only the ovaries or testes of a fruit fly will show cells undergoing meiosis.*

Audiovisual

Show the filmstrip *Meiosis*, Biology Media.

Software

Heredity, The Life Science Programs. *Heredity Dog*, HRM Software. *Mitosis and Meiosis*, Intellimation.

Visual Learning

Figure 12.11 How are the processes different? *Events that occur in meiosis but not in mitosis include: (1) in prophase I, pairs of homologous chromosomes form tetrads and crossing over occurs; (2) in metaphase I, homologous chromosomes line up in pairs, not independently; (3) in anaphase I, homologous chromosomes separate; (4) in telophase I, each cell has only one chromosome from each pair; and (5) at the end of meiosis II, each new cell has the haploid number of chromosomes.*

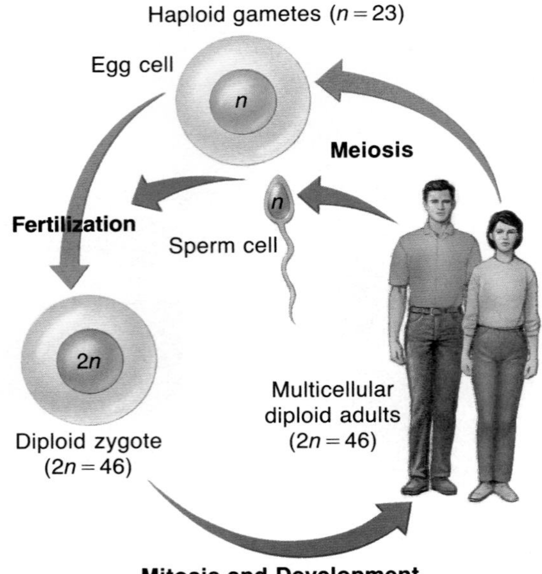

Figure 12.10

In sexual reproduction, the doubling of the chromosome number that results from fertilization is balanced by the halving of the chromosome number that results from meiosis.

zygote:
zygotos (GK) yolk
A zygote is formed by the fertilization of a gamete by another gamete.

Why meiosis?

When cells divide by mitosis, the new cells have exactly the same number and kind of chromosomes as the original cells. Imagine if mitosis were the only means of cell division. Each pea plant parent would then give each offspring a complete set of 14 chromosomes. That means each offspring would have twice the number of chromosomes as each of its parents. The F_1 pea plants would have cells with 28 chromosomes, and the F_2 plants would have cells with 56 chromosomes. They wouldn't be pea plants!

Clearly, there must be another form of cell division that allows offspring to have the same number of chromosomes as their parents. This kind of cell division, which produces gametes containing half the number of chromosomes as a parent's body cell, is called **meiosis.** Meiosis occurs in the specialized body cells that produce gametes.

Meiosis consists of two separate divisions, known as meiosis I and meiosis II. Meiosis I begins with one diploid (2*n*) cell. By the end of meiosis II, there are four haploid (*n*) cells. In animals and most plants, these haploid cells are called sex cells—gametes. Male gametes are called **sperm.** Female gametes are called **eggs.** When a sperm fertilizes an egg, the resulting cell, called a **zygote,** once again has the diploid number of chromosomes. It can then develop by mitosis into a multicellular organism. The pattern of reproduction that involves the production and subsequent fusion of haploid sex cells is called **sexual reproduction.** This pattern is illustrated in *Figure 12.10.*

The Phases of Meiosis

In meiosis, a spindle forms and cytoplasm divides in the same way as occurs in mitosis. However, what happens to the chromosomes in meiosis is very different. *Figure 12.11* illustrates interphase and the phases of meiosis.

Interphase

Recall from Chapter 11 that during interphase, the cell carries out its usual metabolic activities and replicates its chromosomes. During interphase that precedes meiosis I, the cell also replicates its chromosomes. Each chromosome then consists of two identical sister chromatids, held together by a centromere.

Prophase I

The chromosomes coil up and a spindle forms. Then, in a step unique to meiosis, each pair of homologous chromosomes comes together to form a four-part structure called a tetrad. A tetrad consists of two homologous chromosomes, each made up of two sister chromatids.

Alternate Lab: Modeling Meiosis

Purpose

Kinesthetic Students will observe the changes that occur in meiosis.

Materials

9 sheets of unlined paper or poster boards cut to 30 cm square, long length of yarn, paper clips, long length of string, toothpicks, tape or glue, scissors

Procedure

Give students the following directions.

1. Work in groups of nine students. Each student is to model one phase of meiosis. Each model is to have a phase name, labels, and an explanation of the events taking place.
2. Represent cell structures as follows: yarn strands = chromosomes, paper clips = centromeres, string = nuclear membranes, toothpicks = fibers.
3. Place models on large sheets of paper or poster board.
4. Only one pair of chromosomes is to be followed through all models. Glue or tape all parts in place.

Figure 12.11

Follow the diagrams showing the process of meiosis as you read about each phase. Compare these diagrams with those of mitosis on pages 270–271 in Chapter 11. How are the processes different?

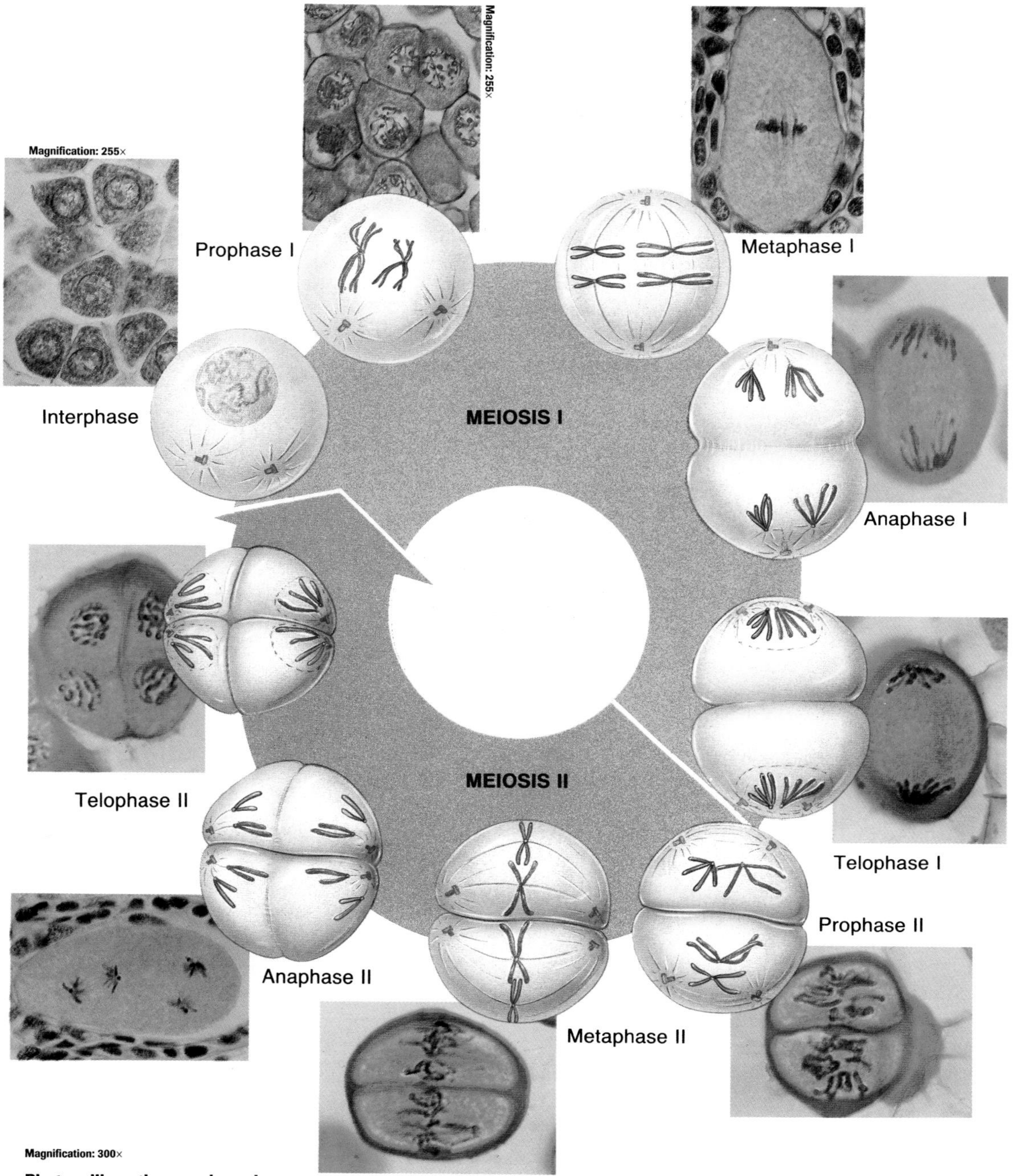

Magnification: 255×

Magnification: 255×

Interphase

Prophase I

Metaphase I

Anaphase I

Telophase I

MEIOSIS I

MEIOSIS II

Prophase II

Metaphase II

Anaphase II

Telophase II

Magnification: 300×

Photos: lily anthers and ovaries
Diagrams: animal cells

5. Model interphase, prophase I, metaphase I, anaphase I, and telophase I, prophase II, metaphase II, anaphase II, and telophase II.

Analysis

Ask students to answer the following questions.

1. What happens to chromosome number during meiosis? *reduced by 1/2*
2. How many cells are formed during meiosis? *4*
3. What is the fate of the cells formed during meiosis? *They form either egg or sperm cells.*

Assessment

Knowledge: Ask students to explain the value of making models such as the ones in this lab. L1

Using an Analogy

To reinforce the concept of homologous chromosomes, sister chromatids, tetrad formation, crossing over, and anaphase I, try the following analogy: A magic pair of shoes (left and right) is found on a shelf (homologous chromosomes in a cell). These shoes, being magic, can and do replicate (interphase replication). Each copy is tied to its original with its shoelaces (centromere; both lefts are tied together and both rights are tied together). Both rights are now called right sister shoes (sister chromatids). Both lefts are now called left sister shoes (sister chromatids).

All four shoes line up next to one another (tetrad formation). While next to one another, part of one non-sister shoe (a left shoe) exchanges its inner sole with another non-sister shoe (a right shoe) (crossing over). Right shoes move away from left shoes to different shelves (anaphase I, with homologous chromosomes separating and going to two different cells. Both rights are still tied together and both lefts are still tied together). To make the analogy work even better, attempt to locate two pairs of identical shoes to demonstrate the events as they are described.

GLENCOE TECHNOLOGY

Videodisc

Biology: The Dynamics of Life
Disc 1, Side 1
Meiosis (Ch. 30)

Using an Analogy

Continue the shoe analogy or demonstration of meiosis from the previous page to illustrate telophase I, metaphase II, and telophase II events. Start with two right shoes tied together. Both sets are separate from each other on different closet shelves (two new cells formed after telophase I). The shoes untie (centromere splits after metaphase II). They move to different shelves in the closet (two new cells formed as in telophase II). Ask students: (a) How many shoes are now on separate shelves (separate cells)? *4* (b) How many shoes were present at the start of this analogy? *2* Is the chromosome number in a cell diploid or haploid before a cell undergoes meiosis? *Diploid* (c) How many shoes are present on each shelf at the end of the process? *1* (d) How many shoes were on the original shelf? *2* (e) Is the chromosome number in each of the four new cells formed in meiosis reduced by half? *Yes*

Assessment

Performance Assessment in the Biology Classroom: p. 13, *Investigating Mitosis and Meiosis.* Have students carry out this activity to determine which slides in an unlabeled set show mitosis and which show meiosis, and also which were made from animal cells and which from plant cells. L1 P

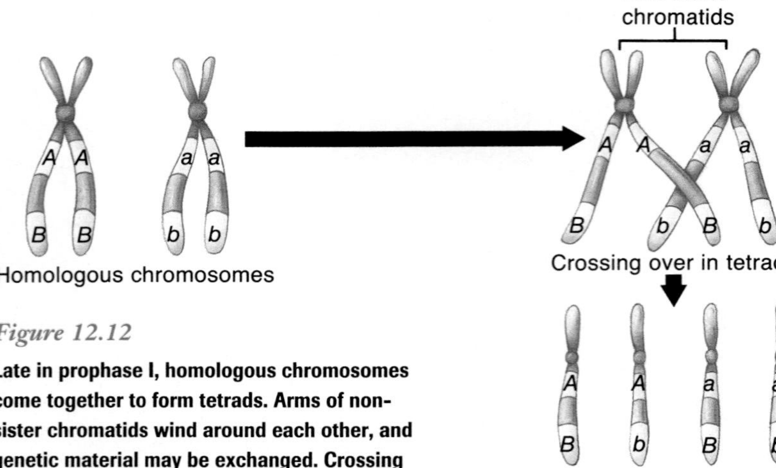

Figure 12.12

Late in prophase I, homologous chromosomes come together to form tetrads. Arms of non-sister chromatids wind around each other, and genetic material may be exchanged. Crossing over can occur at any place on the chromosome.

The chromatids in a tetrad pair tightly. In fact, they pair so tightly that non-sister chromatids from homologous chromosomes sometimes actually exchange genetic material in a process known as **crossing over.** You can see this exchange in *Figure 12.12.* Crossing over results in new combinations of alleles on a chromosome.

Metaphase I

Tetrads line up on the midline, or equator, of the spindle. This is an important step unique to meiosis. Note that homologous chromosomes are lined up together in pairs. In mitosis, on the other hand, homologous chromosomes line up on the equator independently of each other.

inter-, pro-, meta-, ana-, telo-:
inter (L) between
pro (GK) before
meta (GK) after
ana (GK) up, onward
telos (GK) end
Each division of meiosis consists of interphase, prophase, metaphase, anaphase, and telophase.

Anaphase I

Homologous chromosomes separate and move to opposite ends of the cell. This occurs because the centromeres do not split as they do during anaphase in mitosis. This critical step ensures that each new cell will receive only one chromosome from each homologous pair.

Telophase I

Events occur in the reverse order from the events of prophase I. The spindle is broken down, the chromosomes uncoil, and the cytoplasm divides to yield two new cells. Each cell has only half the genetic information of the original cell because it has only one chromosome from each homologous pair. However, another cell division is needed because each chromosome is still doubled, containing two identical sister chromatids.

The phases of meiosis II

The newly formed cells may undergo a short interphase in which the chromosomes do not replicate, or the cells may go from meiosis I directly to meiosis II. The second division in meiosis consists of prophase II, metaphase II, anaphase II, and telophase II.

During prophase II, a spindle forms in each of the two new cells. The chromosomes, still made up of sister chromatids, line up at the equator during metaphase II, just as they do in mitosis. Anaphase II begins as

PORTFOLIO

Comparing Mitosis and Meiosis Provide students with two sets of simple line drawings showing interphase and the phases of mitosis and diagrams of interphase and meiosis. Place comparable phases next to one another when possible (prophase of mitosis next to prophase I of meiosis, etc.). Ask students to describe the similarities and differences between processes. Also have them indicate the type of cell formed at the end of the process—body or gamete—and whether its chromosome number would be diploid or haploid. L2 P

the centromere of each chromosome splits, allowing the sister chromatids to separate and move to opposite poles. Finally, nuclei re-form, the spindles break down, and the cytoplasm divides during telophase II. These events are identical to those of mitosis.

At the end of meiosis II, four haploid cells have been formed from the original diploid cell. Each haploid cell contains one chromosome from each homologous pair. These haploid cells will become gametes, transmitting the genes they contain to offspring.

Meiosis Provides for Genetic Variation

Cells that are formed by mitosis are identical to each other and to the parent cell. Meiosis, however, provides a mechanism for shuffling the chromosomes and the genetic information they carry. *Figure 12.13* shows how this shuffling of chromosomes can occur.

Genetic recombination

How many different kinds of sperm can a pea plant produce? Each cell undergoing meiosis has seven pairs of chromosomes. Because each of the seven pairs of chromosomes can line up in two different ways, 128 different kinds of sperm are possible ($2^7 = 128$).

By the same reasoning, any one pea plant can form 128 different eggs. Because any egg can be fertilized by any sperm, the number of different possible offspring is 16 384 (128×128).

These numbers increase greatly as the number of chromosomes in the species increases. In humans, $n = 23$, so the number of different kinds of eggs or sperm a person can produce is more than 8 million (2^{23}). When fertilization occurs, $2^{23} \times 2^{23}$ different

Figure 12.13

If a cell has two pairs of chromosomes ($n = 2$), four kinds of gametes (2^2) are possible, depending on how the homologous chromosomes line up at the equator during meiosis I (as shown in A or B on the left). This event is a matter of chance. When zygotes are formed by the union of these gametes, 4 × 4 or 16 possible combinations may occur (right).

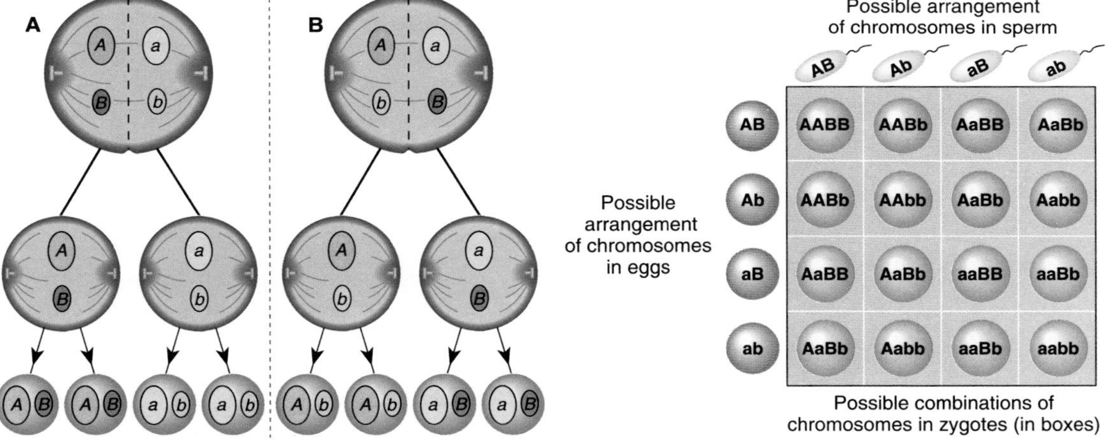

Science, Technology, and Society

Producing Desired Traits
For centuries, plants with desired traits have been obtained by selection and hybridization. This means that growers selected the traits they were looking for and then crossed plants with the desired traits. Today, through genetic engineering, it is possible to insert genes for certain desired traits into a plant. For example, scientists have developed wheat and cotton that carry a bacterial gene that makes these plants resistant to certain herbicides.

Concept Development

The chances of two humans being born exactly alike is almost an impossibility (except for identical twins). Explain why children in the same family can nevertheless resemble one another rather closely.

GLENCOE TECHNOLOGY

Videodisc

The Infinite Voyage: The Keepers of Eden
The Cheetah: Using DNA Profiles to Research Reproduction (Ch. 5)

STUDENT JOURNAL

Gamete Formation Ask students to imagine that they are pea gametes who have just met their cousin leaf cells from the same plant. You want to explain to your cousins how you are different from each other. Write a story that describes how you became a gamete and how you actually differ from your cousins. L2 LS

SECTION 12.2

3 Assess

Check for Understanding

Ask students to explain how the words in the following combinations are related:

(a) diploid—haploid
(b) homologous chromosomes—allele
(c) sperm—egg—zygote
(d) meiosis—gamete
(e) crossing over—genetic recombination L1

Reteach

Ask students to prepare a list of the important characteristics of each step in the process of meiosis. Then have them prepare a second list of all the reasons why meiosis is important to organisms that reproduce sexually. L1

Extension

Oogenesis and spermatogenesis are the terms used to describe egg and sperm cell formation in humans through meiosis. Have students research how these two processes differ and how they are alike. L2

Assessment

Performance: Provide students with simple line diagrams of the various phases of meiosis. The diagrams should not be in normal sequence. Have students place the diagrams in proper sequence, name each phase, and describe what is occurring. L1

4 Close

Discussion

Have students describe as many differences and similarities as possible between mitosis and meiosis. Verify the accuracy of their descriptions as a class. L1

Figure 12.14
Variation is of great importance to a species because some combinations of traits may be more useful to an organism's survival in a changing environment than others.

zygotes are possible, or 70 trillion! It's no wonder that each individual is unique.

In addition, crossing over can occur anywhere at random on a chromosome. Typically, two or three crossovers occur per chromosome during meiosis. This means that an almost endless number of different possible chromosomes can be produced by crossing over, providing additional variation to that already produced by the random assortment of chromosomes. This reassortment of chromosomes and the genetic information they carry, either by crossing over or by independent segregation of homologous chromosomes, is called **genetic recombination.** It is a major source of variation among organisms, *Figure 12.14.* Why is variation important to a species? Variation is the raw material that forms the basis for evolution.

Meiosis explains Mendel's results

Meiosis explains the physical basis of Mendel's results. The segregation of chromosomes in anaphase I of meiosis explains Mendel's observation that each parent gives one allele for each trait at random to each offspring, regardless of whether the allele is expressed. The random segregation of chromosomes during anaphase I explains Mendel's observation that factors, or genes, for different traits are inherited independently of each other. Today, Mendel's laws of heredity form the foundation of modern genetics.

Connecting Ideas

Genes that are transmitted from parents to offspring are responsible for the physical characteristics that the offspring inherits. You may have wondered just how genes function. How does a gene for tall plants make a plant tall? And how can a gene for wrinkled seeds cause a seed to be wrinkled? The answers to such questions lie in an understanding of the structure and processes of genes and chromosomes. They can be found in learning how to read the genetic code.

Section Review

Understanding Concepts

1. How are the cells at the end of meiosis different from the cell at the beginning of meiosis? Use the terms *chromosome number, haploid*, and *diploid* in your answer.
2. What is the role of meiosis in maintaining a constant number of chromosomes in a species?
3. Why are there so many varied phenotypes within a species such as humans?

Thinking Critically

4. How do the events of meiosis explain Mendel's law of independent assortment?

Skill Review

5. **Interpreting Scientific Illustrations** Compare *Figures 12.11* and *11.8.* Explain why crossing over between non-sister chromatids of homologous chromosomes cannot occur during mitosis. For more help, refer to Thinking Critically in the *Skill Handbook.*

Answers to Section Review

1. The chromosome number in a cell at the end is 1/2 the chromosome number in a parent cell. The original cell has a diploid number of chromosomes and each of the new cells has a haploid number.
2. The reduction of chromosome numbers by half allows for the return to the constant chromosome number when a zygote is formed at fertilization.
3. Crossing over as well as the reassortment of the 46 chromosomes both contribute to the high number of phenotypes possible.

Thinking Critically

4. After meiosis, only one member of each homologous chromosome pair can be found in a gamete. Thus, no gamete will end up with two homologues. Alleles on

Chapter 12 Review

REVIEWING MAIN IDEAS

12.1 Mendel's Laws of Heredity

- Genes are located on chromosomes and exist in alternative forms called alleles. A dominant allele can mask the expression of a recessive allele.
- When Mendel crossed pea plants differing in one trait, one form of the trait disappeared until the second generation of offspring. To explain his results, Mendel formulated the law of segregation.
- Mendel formulated the law of independent assortment to explain that two traits are inherited independently.
- Events in genetics are governed by the laws of probability.

12.2 Meiosis

- In meiosis, one diploid ($2n$) cell produces four haploid (n) cells, providing a way for offspring to have the same number of chromosomes as their parents.
- Mendel's results can be explained by the distribution of chromosomes during meiosis.
- Random assortment and crossing over during meiosis provide for genetic variation among the members of a species.

Key Terms

Write a sentence that shows your understanding of each of the following terms.

allele
crossing over
dihybrid cross
diploid
dominant
egg
fertilization
gamete
genetic recombination
genetics
genotype
haploid
heredity
heterozygous
homologous chromosome
homozygous
law of independent assortment
law of segregation
meiosis
phenotype
pollination
recessive
sexual reproduction
sperm
trait
zygote

UNDERSTANDING CONCEPTS

1. An organism that has one copy of an undesirable allele is a ____________.
   a. homozygote c. monohybrid
   b. heterozygote d. dihybrid
2. Parakeets can be many colors, but any single parakeet has only two alleles for color. Parakeets inherit color patterns through ____________.
   a. multiple alleles c. mutation
   b. monohybrids d. multiple genes
3. A pure line of pea plants is one that is ____________.
   a. recessive c. heterozygous
   b. homozygous d. dominant
4. Inherited characteristics are known as ____________.
   a. gametes c. genetics
   b. alleles d. traits
5. At the end of meiosis, how many haploid cells are formed from one cell?
   a. one c. three
   b. two d. four
6. During what phase of meiosis do sister chromatids separate?
   a. prophase I c. anaphase II
   b. telophase I d. telophase II
7. During what phase of meiosis do homologous chromosomes cross over?
   a. prophase I b. telophase I
   b. anaphase I c. telophase II

different chromosomes will sort independently from one another.

Skill Review

5. Tetrad formation does not occur during mitosis. This prevents crossing over from taking place.

Chapter 12 Review

Reviewing Main Ideas

Summary statements can be used by students to review the major concepts of the chapter.

Key Terms

Answers should go beyond defining the terms. Accept any answer that uses the term correctly and in the proper context.

Understanding Concepts

1. b
2. d
3. b
4. d
5. d
6. c
7. a

Chapter 12 Review

8. d
9. b
10. c
11. a
12. d
13. c
14. c
15. a
16. b
17. d
18. c
19. a
20. b

Applying Concepts

21. Like plants, humans reproduce sexually, have chromosomes and genes, and have traits controlled by genes.
22. The likelihood that close relatives share the same recessive genes is greater, thus raising the risk of a child being homozygous for those traits.
23. fifty percent; The probability of any one child being a certain sex is unaffected by the birth of previous children.
24. The order of lining up at the equator during metaphase I of meiosis will vary, thus providing more variation when the chromosomes separate at anaphase I.
25. Controlled experiments, such as Mendel's, require that no more than one variable be manipulated at a time. By doing many experiments, Mendel was able to determine the principles that govern genetics. These principles would not have been as easily observable without controlled experiments.

Chapter 12 Review

8. Recessive traits appear only when an organism is ___________.
a. mature
b. different from its parents
c. heterozygous
d. homozygous

9. Mendel's peas were a good choice for genetic study because ___________.
a. little was known about them
b. they were easy to grow
c. they are self-fertilizing
d. they grow slowly

10. A dihybrid cross between two heterozygous parents produces a ratio of ___________.
a. 3:1 **c.** 9:3:3:1
b. 1:2:1 **d.** 1:6:9

11. If two heterozygous organisms for a single dominant trait mate, the ratio of their young should be about ___________.
a. 3:1 **c.** 9:3:3:1
b. 1:2:1 **d.** 1:6:9

12. A trait that is hidden in the heterozygous condition is said to be a _________ trait.
a. disappearing **c.** dominant
b. carrier **d.** recessive

13. An organism that has two different alleles for a trait is called ___________.
a. dominant **c.** heterozygous
b. recessive **d.** homozygous

14. The result of ___________ is daughter cells with the same number of chromosomes as the parent cell.
a. pollination **c.** mitosis
b. meiosis **d.** fertilization

15. If a species normally has 46 chromosomes, the cells it produces by meiosis will each have ___________ chromosomes.
a. 23 **c.** 92
b. 46 **d.** 18

16. Metaphase I of meiosis occurs when ___________ appear at the equator.
a. cells
b. homologous chromosomes
c. crossovers
d. 46 chromosomes

17. Which of these is a dominant trait in garden peas?
a. wrinkled seeds **c.** yellow pods
b. green seeds **d.** purple flowers

18. In the first generation of Mendel's experiments with a single trait, the _________ trait disappeared, only to reappear in the next generation.
a. best **c.** recessive
b. dominant **d.** heterozygous

19. A cell that has successfully completed meiosis has a chromosome number called ___________.
a. haploid **c.** polyploid
b. diploid **d.** tetraploid

20. Meiosis results in the direct production of ___________.
a. zygotes
b. gametes
c. heterozygous cells
d. homozygous cells

APPLYING CONCEPTS

21. Why do you think Mendel's results are also valid for humans?

22. On the average, each human has about six recessive alleles that would be lethal, if expressed. Why do you think that human cultures have laws against marriage between close relatives?

23. Assume that a couple has four children who are all boys. What are the chances that their next child will also be a boy? Explain your answer.

24. How does separation of homologous chromosomes during anaphase I of meiosis increase variation among offspring?

25. In terms of the methods of science, why do you think it was important for Mendel to study only one trait at a time during his experiments?

THINKING CRITICALLY

26. *Concept Mapping* Make a concept map that relates the following terms and phrases. Supply the appropriate linking words for your map.
    dominant, recessive, tall pea plants, short pea plants, P_1 generation, F_1 generation, F_2 generation
27. *Recognizing Cause and Effect* Why is it sometimes impossible to determine the genotype of an organism that has a dominant phenotype?
28. *Observing and Inferring* Why is it possible to have a family of six girls and no boys, but extremely unlikely that there will be a public school with 500 girls and no boys?
29. *Interpreting Scientific Illustrations* The following drawing shows one pair of homologous chromosomes pairing tightly during prophase I of meiosis. What is the significance of the places at which the chromosomes are joined?

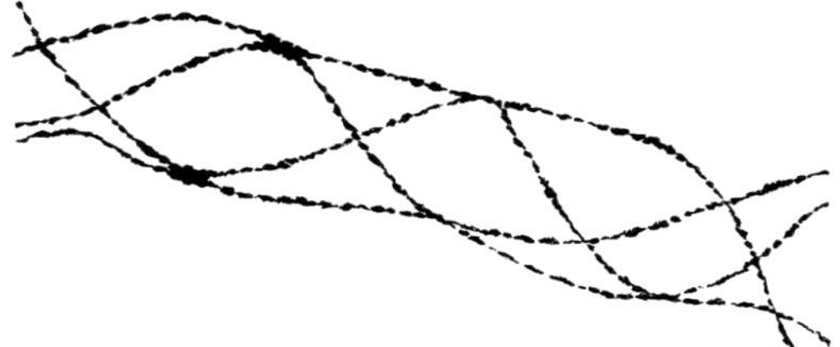

30. *Comparing and Contrasting* Compare metaphase of mitosis with metaphase I of meiosis.
31. *Recognizing Cause and Effect* Describe how each of the following events either increases or decreases variation among offspring.
    a. One allele is dominant to another allele in determining phenotype.
    b. Homologous chromosomes separate during anaphase I of meiosis.

ASSESSING KNOWLEDGE & SKILLS

In fruit flies, the allele for long wings is dominant to the allele for short wings.

	W	w
?	Ww	ww
?	Ww	ww

Interpreting Data Study the Punnett square to answer the questions that follow.

1. What term is given to the parent fly whose genotype is shown?
   a. heterozygous
   b. homozygous
   c. recessive
   d. haploid
2. What is the phenotype of each parent?
   a. both dominant
   b. both recessive
   c. one dominant and one recessive
   d. unable to tell
3. What is the genotype of each parent?
   a. WW; Ww
   b. Ww; ww
   c. Ww; Ww
   d. WW; WW
4. What are the phenotypes of the offspring?
   a. all long wings
   b. all short wings
   c. mostly long wings
   d. half short and half long
5. *Interpreting Data* Suppose the fruit fly parents in the Punnett square above were both also heterozygous for eye color trait in which X is red and x is white. What genotypes appear in the offspring? What percent of the offspring will have short wings and white eyes?

Program Resources

Chapter Assessment, pp. 67-72 L1
Alternate Assessment in the Science Classroom
Computer Test Bank L1
Content Mastery, pp. 45-48 L1

Chapter 12 Review

Thinking Critically

26. Evaluate students' concept maps. The maps should show an understanding of the relationships among the concepts listed.
27. If the dominant allele completely masks the recessive allele, you cannot tell if an organism with the dominant trait is homozygous or heterozygous for the dominant allele.
28. The chances of parents having six girls is 2^6 (about 3 percent), but the chances of a public school having only 500 girls is 2^{500} (a percentage approaching zero).
29. These are the places where crossing over actually takes place.
30. In metaphase of mitosis, all of the chromosomes have centromeres that align along the equator of the cell. In metaphase I of meiosis, pairs of homologous chromosomes align along the equator of the cell.
31. (a) Variation would decrease, a dominant allele would block expression of a recessive allele. (b) The order of lining up at the equator during metaphase I will vary, thus providing more variation when the chromosomes separate at anaphase I.

ASSESSING KNOWLEDGE & SKILLS

1. a
2. c
3. b
4. d
5. Genotypes include: WwXx, WwXX, Wwxx, wwxx, wwXX, wwXx. The genotype of a fruit fly with short wings and white eyes is wwxx. Two of the sixteen possible genotypes are wwxx, a percentage of 12.5 percent ($2/16 \times 100 = 12.5$ percent).

Chapter Organizer

SECTION	OBJECTIVES	ACTIVITIES/FEATURES
13.1 DNA: The Molecule of Heredity National Science Standards: UCP.2, UCP.3; B.2, B.3; C.2, C.5; G.3	1. **Analyze** the structure of DNA. 2. **Determine** how the structure of DNA enables it to reproduce itself accurately.	**Minilab:** What does DNA look like?, p. 311 **Social Studies Connection:** The Double Helix, p. 312 **Thinking Lab:** What evidence does chemical analysis provide about the structure of DNA?, p. 313
13.2 From DNA to Protein National Science Standards: UCP.1-3; A.1; B.2, B.3; C.1, C.2	3. **Relate** the concept of the gene to the sequences of nucleotides in DNA. 4. **Sequence** the steps involved in protein synthesis.	**Biolab:** RNA Transcription, p. 320
13.3 Genetic Changes National Science Standards: UCP.1-3; C.1, C.2; G.1, G.2	5. **Categorize** the different kinds of mutations that can occur in DNA. 6. **Compare** the effects of different kinds of mutations on cells and organisms.	**Minilab:** How do gene mutations affect proteins?, p. 326 **A Broader View:** Gene-control Operons, p. 327

ACTIVITY MATERIALS

BIOLAB	MINILABS	ALTERNATE LAB
page 320 construction paper, 5 colors scissors clear tape pencil	**page 311** 1 mL cold ethanol test tube salmon testes, prepared stirring rod **page 326** pencil and paper Table 13.1	**page 318** scissors 2 skeins of yarn box (100 cm × 50 cm)

Chapter 13 Genes and Chromosomes

TEACHER CLASSROOM RESOURCES

Reproducible Masters	Transparencies
Section Focus Master 27: DNA Structure L1 SAE **Reinforcement and Study Guide,** p. 49 L1 **Biolab and Minilab Worksheets,** p. 47 L1 **Laboratory Manual:** Chromosome Extraction and Analysis, pp. 75-78 L2	**Basic Concepts Transparency #16:** DNA Replication L1 SAE
Section Focus Master 28: Using Codes L1 SAE **Reinforcement and Study Guide,** pp. 50-51 L1 **Biolab and Minilab Worksheets,** pp. 49-52 L1 **Concept Mapping:** DNA and RNA, p. 13 L1 **Critical Thinking/Problem Solving:** Why Did Caesar Die?, p. 13 L3 **Content Mastery,** pp. 49-52 L1	**Basic Concepts Transparency #17:** DNA Transcription L1 SAE **Basic Concepts Transparency #18:** RNA Translation L1 SAE **Reteaching Transparency #13:** Translation: From RNA to Protein L1 SAE **Basic Skills Transparencies #12a, 12b:** Gene Mutations; Chromosomal Mutations L1 SAE
Section Focus Master 29: Nitrogen Base Sequence L1 SAE **Reinforcement and Study Guide,** p. 52 L1 **Biolab and Minilab Worksheets,** p. 48 L1	

ASSESSMENT MATERIALS

Chapter Assessment, pp. 73-78
Performance Assessment in the Biology Classroom, p. 17
MindJogger Videoquiz
Alternate Assessment in the Science Classroom
Computer Test Bank

Spanish Resources SAE
English/Spanish Audiocassettes SAE
Cooperative Learning in the Science Classroom COOP LEARN
Lesson Plans

KEY TO TEACHING STRATEGIES

L1 Level 1 activities should be within the ability range of all students including those with learning difficulties.

L2 Level 2 activities are within the ability range of average to above-average students.

L3 Level 3 activities are designed for the ability range of above-average students.

SAE SAE activities should be within the ability range of Students Acquiring English.

COOP LEARN Cooperative Learning activities are designed for small group work.

P These strategies represent student products that can be placed into a best-work portfolio.

These strategies are useful in a block scheduling format.

GLENCOE TECHNOLOGY

The following multimedia resources are available from Glencoe.

Biology: The Dynamics of Life
- **CD-ROM** SAE
- **Videodisc Program**

The Infinite Voyage Series
- *The Dawn of Humankind*
- *The Geometry of Life*
- *The Living Clock*

The Secret of Life Series
- *What's in Stetter's Pond: The Basics of Life*
- *Sex and the Single Gene: Cell Development*

CHAPTER

13 Genes and Chromosomes

Chapter Overview

Students begin the chapter by analyzing the structure and function of DNA. Next, they evaluate the role of DNA replication in reproduction. They will then examine the DNA replication process in which the pairing of the bases in the nucleotides follows a predictable pattern, keeping the genetic information constant during cell division.

In the next section of the chapter, the relationship of DNA to the genetic code is established. The three types of RNA and their functions in protein synthesis are described. Students then compare and contrast RNA and DNA.

The chapter concludes with a discussion of mutations resulting from changes in DNA. Students learn also about chromosomal mutations and errors in disjunction and, finally, the causes of mutations.

Key Terms

chromosomal mutation
codon
double helix
frameshift mutation
messenger RNA
monosomy
mutation
nitrogen base
nondisjunction
point mutation
replication
ribosomal RNA
transcription
transfer RNA
translation
trisomy

308

Look for the following logo for strategies that emphasize different learning modalities. LS

Kinesthetic	**Minilab, p. 311; Building a Model, p. 312; Alternate Lab, p. 318; Project, p. 326**
Visual-Spatial	**Chalkboard Example, pp. 314, 318; Display, p. 322; Demonstration, p. 325; Visual Learning, p. 327**
Linguistic	**Project, p. 312; Portfolio, pp. 313, 325**
Logical-Mathematical	**Thinking Lab, p. 313; Chalkboard Example, p. 322; Minilab, p. 326**

CHAPTER

13 Genes and Chromosomes

The forest canopy shakes violently as a pair of blue guenon monkeys reacts to the sound of a predator. Lower in this arboreal environment, one of their close relatives—a crowned guenon with its bushy, yellow hair—rests quietly after devouring a leafy dinner. Nearby on the forest floor, a young, moustached guenon squeaks and devours a fruit it has just picked.

In fact, there are at least 19 species of guenons, all remarkably diverse in color, appearance, and behavior but similar in size and body proportions. Guenons demonstrate how only a few small changes in the genetic blueprint of living organisms, DNA, can produce a variety of species, all with a common body plan. In this chapter, you'll investigate the role and structure of DNA and discover how its remarkable properties can produce the diversity of life you see on Earth.

Concept Check

You may wish to review the following concepts before studying this chapter.

- Chapter 7: protein, amino acid, DNA, RNA
- Chapter 11: chromosome, mitosis
- Chapter 12: gene

Chapter Preview

Laboratory Activities

Biolab

- RNA Transcription

Minilabs

- What does DNA look like?
- How do gene mutations affect proteins?

The many species of guenons differ in coloration, behavior, and diet, but they are remarkably similar overall. The nucleic acid DNA determines the inherited traits and behaviors of all species. How can one molecule result in the millions of species seen on Earth?

CHAPTER 13

Introducing the Chapter

Draw students' attention to the pictures of the guenon monkeys. Ask them what features these monkeys share. *They have similar body structures—heads, eyes, limbs, and so on.* What features do they have that are different? *coloration, behavior, and diet* Guide the discussion to the concept of the DNA molecule, which serves as a genetic blueprint for these similarities and differences.

Theme Development

The themes of *evolution* and *homeostasis* are developed in this chapter. Any evolutionary change is one that is inheritable. It should become obvious to students that DNA serves to both promote stability and act as an agent of change.

Concept Check

Students will build upon their knowledge gained in Chapter 7 of the monomers that make up nucleic acids, their composition, and how polymers are made. Contrast the two sugars found in DNA and RNA.

Assessment Planner

Choose assessment strategies from the following pages to evaluate the progress of your students.

Assess, pp. 315, 322, 328
Alternate Lab, pp. 318-319
Minilabs, pp. 311, 326
Portfolio, pp. 313, 325
Thinking Lab, p. 313
Biolab, pp. 320-321
Chapter Review, pp. 329-331

SECTION

13.1

Prepare

Key Concepts

The structure and composition of DNA are presented. The process of replication of DNA and its importance to organisms are emphasized.

Block Scheduling

Look for this symbol for strategies that are useful in a block scheduling format. For more information on block scheduling, refer to Section 13.1 in the **Lesson Plans** booklet.

Materials

- Order salmon testes, frog testes, or dry DNA for the Minilab. A refrigerator, ice, 95% ethanol or isopropyl alcohol, and sodium acetate are also needed.
- Collect Styrofoam and colored paper clips for Building a Model.

1 Focus

Bellringer

Before presenting the lesson, display **Section Focus Master 27** on the overhead projector and have students answer the accompanying questions. L1 SAE

Discussion

Elicit from students whether they know why each of them is a unique individual. *Students may suggest that hereditary factors determine their uniqueness.* Why has there been no other human being (except in the case of identical twins) on Earth exactly like any of them? *Each individual has different DNA and thus different traits.*

SECTION **13.1**

DNA: The Molecule of Heredity

Section Preview

Objectives

Analyze the structure of DNA.

Determine how the structure of DNA enables it to reproduce itself accurately.

Key Terms

nitrogen base
double helix
replication

Can you imagine all of the information that could be contained in 1000 books the size of this textbook? Remarkably, at least this much information is carried by the genes of an organism, even one as simple as a bacterium. Scientists have found that the substance DNA, contained in genes, is responsible for this control.

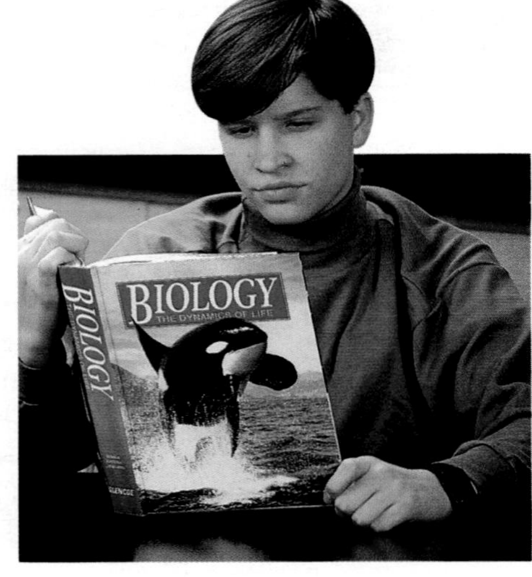

Structure and Function of DNA

In Chapter 7, you learned that DNA is an example of a complex biological polymer called a nucleic acid. You also learned that nucleic acids are made up of many smaller subunits called nucleotides. The components of a DNA nucleotide are deoxyribose (a simple sugar), a phosphate group, and a nitrogen base. A **nitrogen base** is an organic ring structure that contains one or more atoms of nitrogen. In DNA, there are four possible nitrogen bases—adenine (A), guanine (G), cytosine (C), and thymine (T). Thus, in DNA,

Figure 13.1
DNA Nucleotides

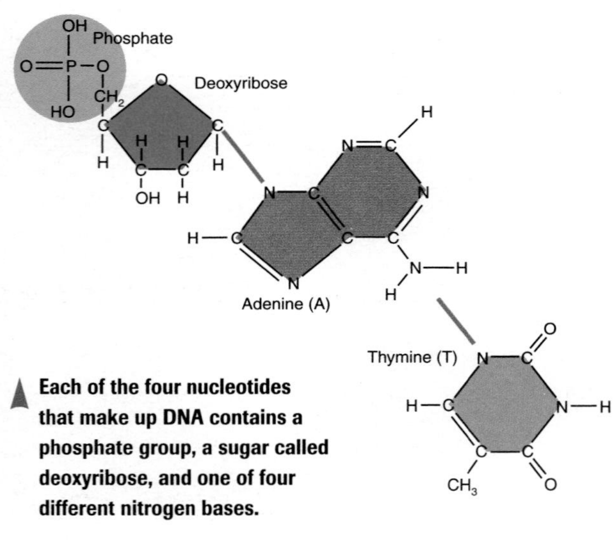

Each of the four nucleotides that make up DNA contains a phosphate group, a sugar called deoxyribose, and one of four different nitrogen bases.

Adenine and guanine (top) are double-ring bases called purines. Thymine and cytosine (bottom) are smaller, single-ring bases called pyrimidines.

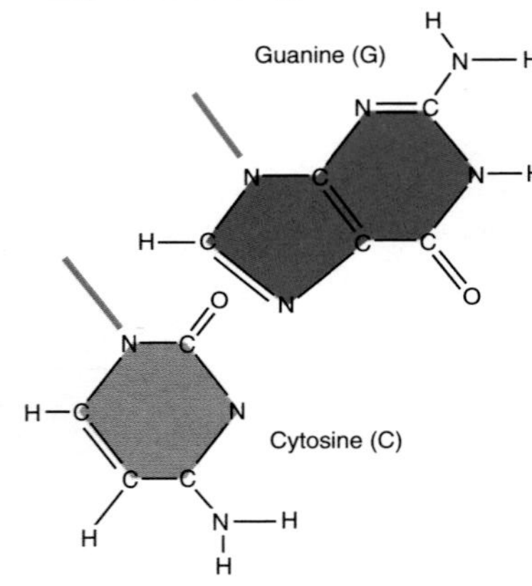

Program Resources

Section Focus Master 27 L1 SAE

Reinforcement and Study Guide, p. 49 L1

Biolab and Minilab Worksheets, p. 47 L1

Laboratory Manual, pp. 75-78 L2

Performance Assessment in the Biology Classroom, p. 17 L1 P

Basic Concepts Transparency 16 and **Master** L1 SAE

there are four possible nucleotides, each containing one of these four bases, as shown in *Figure 13.1*.

Chains of nucleotides

In nucleic acids, nucleotides do not exist as individual molecules. In DNA, nucleotides combine to form two long chains, producing one large molecule, *Figure 13.2*. Each chain of nucleotides contains single nucleotides connected to each other.

MiniLab

What does DNA look like?

DNA is pictured as a coiled helix in most diagrams. But what does the substance DNA really look like? You can extract DNA from salmon testes using the following procedure.

Procedure

1. Add 1 mL of cold ethanol to the cold test tube of salmon testes preparation given to you by your teacher. Add the ethanol by letting it run slowly down the inside of the test tube. A layer of ethanol should float on the salmon testes preparation.
2. Use a stirring rod to stir at the point where the two liquids come together. The DNA will cling to the stirring rod.
3. Withdraw the stirring rod slowly, pulling the thread of DNA with it. When the stirring rod is just above the mouth of the tube, start winding the thread of DNA onto the rod.

Analysis

1. How does the physical appearance of DNA relate to its chemical structure?
2. Why do you think that testes are used as a dependable source of DNA?

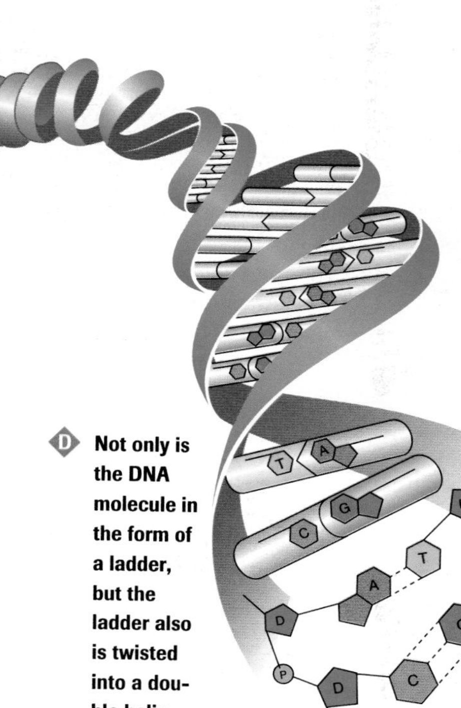

Hydrogen bonds

D Not only is the DNA molecule in the form of a ladder, but the ladder also is twisted into a double helix.

C This pairing produces a long, two-stranded molecule that is often compared to a ladder. As you can see, the sides of the ladder are formed by the sugar and phosphate units, while the rungs of the ladder are the pairs of bases.

B The two chains of nucleotides in a DNA molecule are held together by hydrogen bonds between the bases. The two bases on the same rung of the DNA are referred to as a base pair. In DNA, cytosine forms hydrogen bonds with guanine, and thymine bonds with adenine.

A In each chain of nucleotides, the sugar of one nucleotide is joined to the phosphate group of the next nucleotide by a covalent bond.

Figure 13.2
The Structure of DNA

2 Teach

MiniLab

Purpose

LS **Kinesthetic** Students will extract DNA to become acquainted with some of its physical properties.

Process Skills

observe and infer, measure in SI, interpret data, analyze

Teaching Strategies

- Salmon testes or frog testes extract, or dry DNA can be purchased from biological supply companies.
- Prepare salmon testes solution by slowly adding 0.5 mL (10 drops) sodium acetate to 0.5 mL salmon testes extract. Mix gently. Keep the solution refrigerated or on ice until used.
- Ethanol (95%) or isopropyl alcohol can be used.

Expected Results

The DNA will form a thin, white strand of a gelatinlike substance on the stirring rod.

Analysis

1. The DNA forms long strands.
2. They contain cells that are rapidly dividing and that contain a large proportion of DNA for their size and mass.

Assessment

Journal: Have students write a summary of their observations in this lab for their journals. Ask students to include their answers to the Analysis questions with their summaries. L1

GLENCOE TECHNOLOGY

Videodisc

The Infinite Voyage: The Dawn of Humankind
Bridging of Fossils and Genetic Research (Ch. 9)

History

The Double Helix

Purpose

To familiarize students with the contributions of Watson, Crick, and Franklin in determining DNA structures

Teaching Strategies

- Show students a picture of one of Franklin's X-ray diffraction images of DNA. Explain that the X-ray diffraction picture indicates that the molecule of DNA is something like a spiral staircase, an important clue in solving the mystery of the structure of DNA.

Possible Answer

Their constant exchange of ideas may have generated more ideas, the way brainstorming can.

Building a Model

LS **Kinesthetic** Help students build a model of DNA. Take a piece of Styrofoam board about 4 cm by 8 cm. You will need five each of four different colors of paper clips to represent the DNA bases. Assign letters to the colors so that you have 5A, 5T, 5C, and 5G clips. Place half of the clips in a row on the Styrofoam. On another piece of Styrofoam, line up the complementary clips in the proper order (A=T, C=G). Connect the bases with pipe cleaners or twist ties. Once the model is together it can be twisted, suggesting the double helix structure of DNA. L1

Assessment

Performance Assessment in the Biology Classroom: p. 17, *Building a Model of Replication* Have students carry out this activity after they have learned about DNA replication. L1 P

History

The Double Helix

It began as an unlikely scientific partnership. James Watson, a 24-year-old American biochemist, hated to talk to people. In fact, his ambition was to isolate himself on a wildlife refuge. His reasoning: "Birds, no people." Then, in 1951, Watson met English physicist Francis Crick, who loved to talk. Within minutes, the two were deep in conversation about DNA.

From photo to model Watson and Crick's scientific partnership developed for more than two years as they carried out an intensive study of DNA at Cambridge University in England. Words weren't the only means by which the two scientists exchanged ideas. Rosalind Franklin, a crystallographer at King's College in London, had produced a photo made by passing X rays through DNA and recording the interference pattern that resulted. By studying the photo, they deduced that DNA must have a spiral structure. They began trying to build a model that would produce the image they saw.

A new theory In February 1953, Watson and Crick's newest model resembled a rope ladder that had been given a twist. The pair of scientists was convinced that DNA had the form of a double helix with chains running in opposite directions. Two months later, the British science journal *Nature* published their letter, which contained a tentative-sounding sentence: "It has not escaped our notice that the specific pairing we have postulated immediately suggested a possible copying mechanism for the genetic material." Watson and Crick were awarded a Nobel prize in 1962 for their work. It is thought that Franklin would have shared the prize with them but for her untimely death four years earlier.

CONNECTION TO Biology

Many scientists were working on the DNA puzzle at the same time. Most of them preferred to work alone in the laboratory. How do you think Watson and Crick's working style might have given them an advantage?

The two chains of nucleotides in DNA are joined by hydrogen bonds between the bases, resulting in a structure that is like a ladder. When something is twisted like a coiled spring, the shape is called a helix. Because DNA is composed of two twisted strands, its shape is called a **double helix.**

The importance of nucleotide sequences

The genetic material of living things is made of DNA. Thus, if you compare the chromosomes of two different organisms, you will find that both contain DNA made up of nucleotides with adenine, thymine, guanine, and cytosine. How can organisms be different from each other if their genetic material is made of the same molecules? They are different because the order of nucleotides in the DNA strands of the two organisms is different. For example, a squirrel differs from a rosebush, which differs from a redwood tree, because the orders of nucleotides in their DNA are different.

The sequence of nucleotides forms the unique genetic information of an organism. The more closely related two organisms are, the more alike the order of nucleotides in their DNA will be. Scientists use this information to determine evolutionary relationships. The Human Genome Project, an international research program, is working to determine the sequences of nucleotide bases for the human species.

PROJECT DNA Pioneers

Have students carry out research on one of the following people involved in the discovery of the structure and function of DNA: Fred Griffith, O.T. Avery, Alfred Hershey, Linus Pauling, Martha Chase, Erwin Chargaff, Rosalind Franklin, or Maurice Wilkins. Have them prepare short oral reports on what these scientists contributed. L2 LS

GLENCOE TECHNOLOGY

Videodisc

The Infinite Voyage: The Geometry of Life

DNA and the Double Helix (Ch. 2)

Replication of DNA

Of course, there are mechanisms to pass exact copies of all genetic information from one cell to the next. During cell division, the mechanism is mitosis. During production of gametes for sexual reproduction, the mechanism is meiosis. Therefore, there must be a means of making duplicate copies of DNA.

Each organism on Earth today has a nucleotide sequence in its DNA that was obtained from its parents. This sequence goes back some 3.5 billion years to the first organisms on Earth.

Why must DNA replicate?

You learned in Chapter 11 that every time a cell divides, it must first make a copy of its chromosomes. In this way, each new cell can have a complete set of chromosomes. In fact, every time a cell reproduces by mitosis or a gamete is formed by meiosis, DNA is copied in a process called **replication.** Without replication, species could not survive and individuals could not successfully grow and reproduce. *Figure 13.3* shows bacterial DNA replicating.

How DNA replicates

Recall that a DNA molecule is composed of two strands, each containing a sequence of nucleotides. As you know, an adenine on one strand pairs with a thymine on the other strand. Similarly, guanine pairs with cytosine. Therefore, if you knew the order of bases on one strand, you could predict the sequence of bases on the complementary strand. In fact, part of the process of DNA replication is done in just the same way. During replication, each strand serves as a pattern to make a new DNA molecule.

ThinkingLab — Interpret the Data

What evidence does chemical analysis provide about the structure of DNA?

Much of the early research on the structure and composition of DNA was done by carrying out chemical analyses. The data from these experiments provide evidence of a relationship among the nitrogen bases of DNA.

Analysis

Examine the table below. Compare the amounts of adenine, guanine, cytosine, and thymine found in the DNA of each of the cells studied.

Percent of Each Base in DNA Samples

Source of Sample	A	G	C	T
Human liver	30.3	19.5	19.9	30.3
Human thymus	30.9	19.9	19.8	29.4
Herring sperm	27.8	22.2	22.6	27.5
Yeast	31.7	18.2	17.4	32.6

Thinking Critically

Compare the amounts of A, T, G, and C in each kind of DNA. Why do you think the relative amounts are so close in human liver and thymus cells? How do the relative amounts of bases in herring sperm compare with the relative amounts of bases in yeast? What fact can you state about the overall composition of DNA, regardless of its source?

Figure 13.3

This bacterial DNA is in the process of replication. The two loops are new copies. (The bottom loop is twisted into a figure-8 shape.) Replication is occurring at the intersections of the two loops, as indicated by the arrows.

Magnification: 200 000×

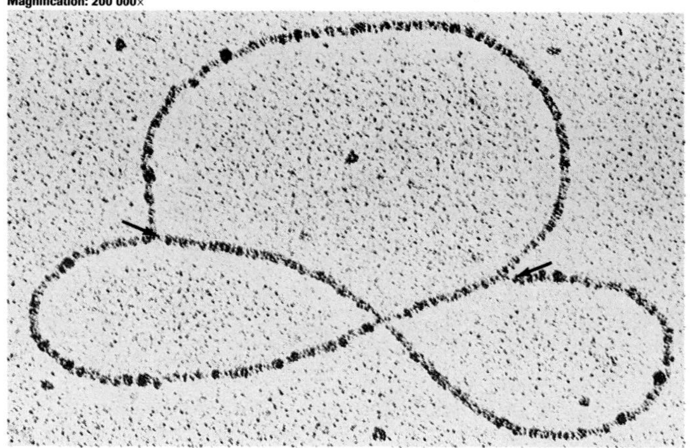

SECTION 13.1

ThinkingLab — Interpret the Data

Purpose

LS Logical-Mathematical Students will analyze a table showing the percentages of four bases in DNA samples to determine if the data show a correlation between the bases.

Process Skills

compare and contrast, observe and infer, interpret data

Teaching Strategies

- Guide students to compare the numbers within each sample before comparing different samples.

Background

In 1950, American biochemist Erwin Chargaff first showed that there is a 1:1 ratio between adenine and thymine and between guanine and cytosine.

Thinking Critically

The ratio of A:T is approximately 1:1 and the ratio of C:G is approximately 1:1. The relative amounts are so close in human liver and thymus because they are from the same species. The relative amounts of bases are different in herring and in yeast because they are different species. The ratio of A:T and G:C is 1:1 no matter what the source.

Assessment

Oral: Have students discuss why this piece of information was important in determining the structure of DNA. Ask students to summarize their analysis of the data presented. L1

GLENCOE TECHNOLOGY

Videodisc

Biology: The Dynamics of Life
Disc 1, Side 1
DNA Replication (Ch. 31)

PORTFOLIO

DNA Connection Have students read "Happy Birthday Double Helix" by Leon Jaroff, *Time,* March 15, 1993, pp. 56–59. Have them write a summary of how the discovery of DNA has been applied to other fields, such as industry and business. L1 LS P

Literature Connection

The Double Helix
James D. Watson tells the story of the discovery of DNA in his book *The Double Helix.* Watson presents a story of scientists as people with human feelings, not the stereotypical image of scientists in white lab coats who live in their laboratories. Watson also discusses the highly competitive and jealous rivalry that occurs in the scientific community.

Chalkboard Example

LS Visual-Spatial On the chalkboard, draw and label one strand of DNA. Have students copy this and draw the complementary strand. Make sure they understand that adenine pairs only with thymine and that guanine pairs only with cytosine. After the students have completed their diagrams, draw the complementary strand on the chalkboard.

Assessment

Performance Assessment in the Biology Classroom, p. 17, Building a Model of Replication. Have students carry out this activity in which they plan and build a model of the stages of DNA replication.

DNA replication begins as an enzyme breaks the hydrogen bonds between nitrogen bases that hold the two strands together, unzipping the DNA molecule. As the DNA continues to unzip, free nucleotides from the surroundings in the nucleus bond to the single strands by base pairing, as shown in *Figure 13.4.* Another enzyme bonds these new nucleotides into a chain.

This process continues until the entire molecule has been unzipped and replicated. As a result, each new strand formed is a complement of one of the original, or parent, strands. The result is the formation of two DNA molecules, each of which is identical to the original DNA molecule.

When all the DNA in all the chromosomes of the cell has been copied by replication, there are two copies of the organism's genetic information. In this way, the genetic makeup of an organism can be passed on to new cells during mitosis or to new generations through meiosis followed by sexual reproduction.

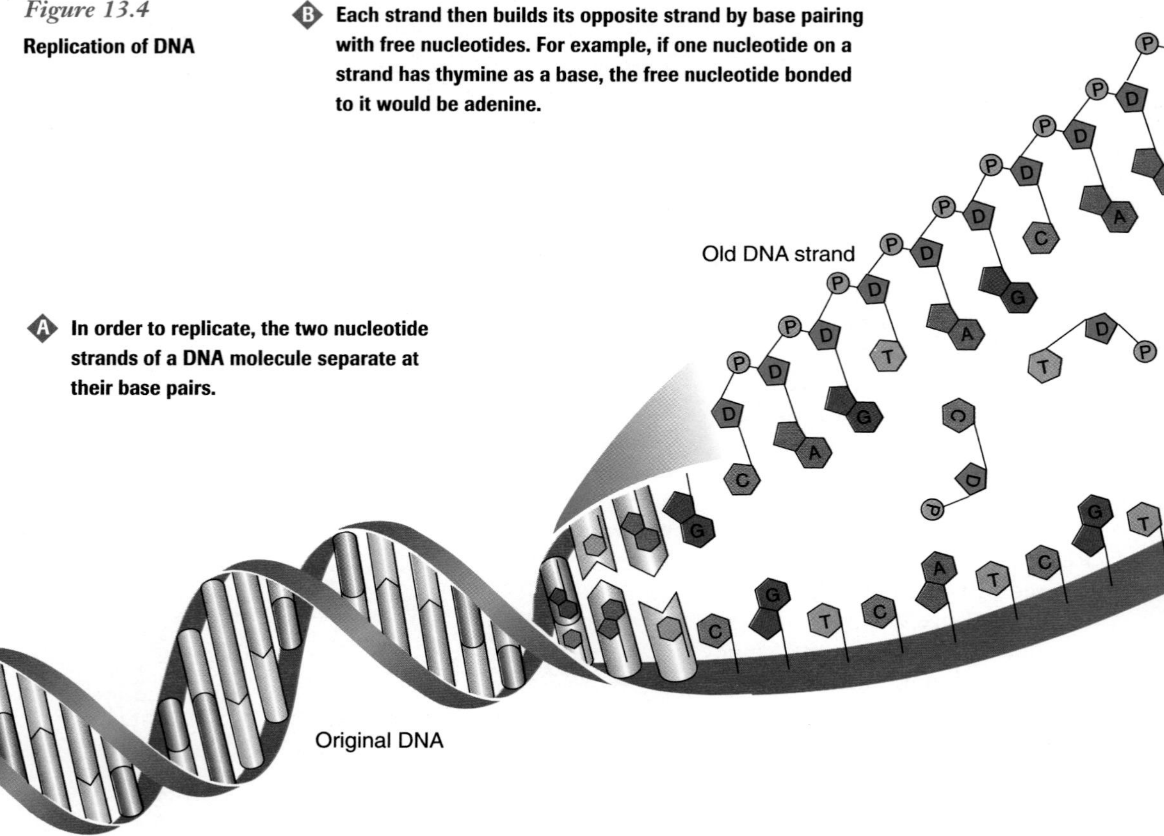

B Each strand then builds its opposite strand by base pairing with free nucleotides. For example, if one nucleotide on a strand has thymine as a base, the free nucleotide bonded to it would be adenine.
Old DNA strand
A In order to replicate, the two nucleotide strands of a DNA molecule separate at their base pairs.
Original DNA

Figure 13.4
Replication of DNA

CULTURAL DIVERSITY

Har Gobind Khorana

Teach students about some of the experimental methods scientists have used to decipher the genetic code. In particular, explain the research of Indian-American biochemist, Har Gobind Khorana.

In the 1960s, Khorana demonstrated that genes could be manufactured in the laboratory.

Khorana's artificial genes were able to code for the synthesis of proteins just as they do in living cells. This work ultimately led to the cracking of a portion of the genetic code. For this work, Khorana and colleagues received the Nobel Prize for Physiology or Medicine in 1968.

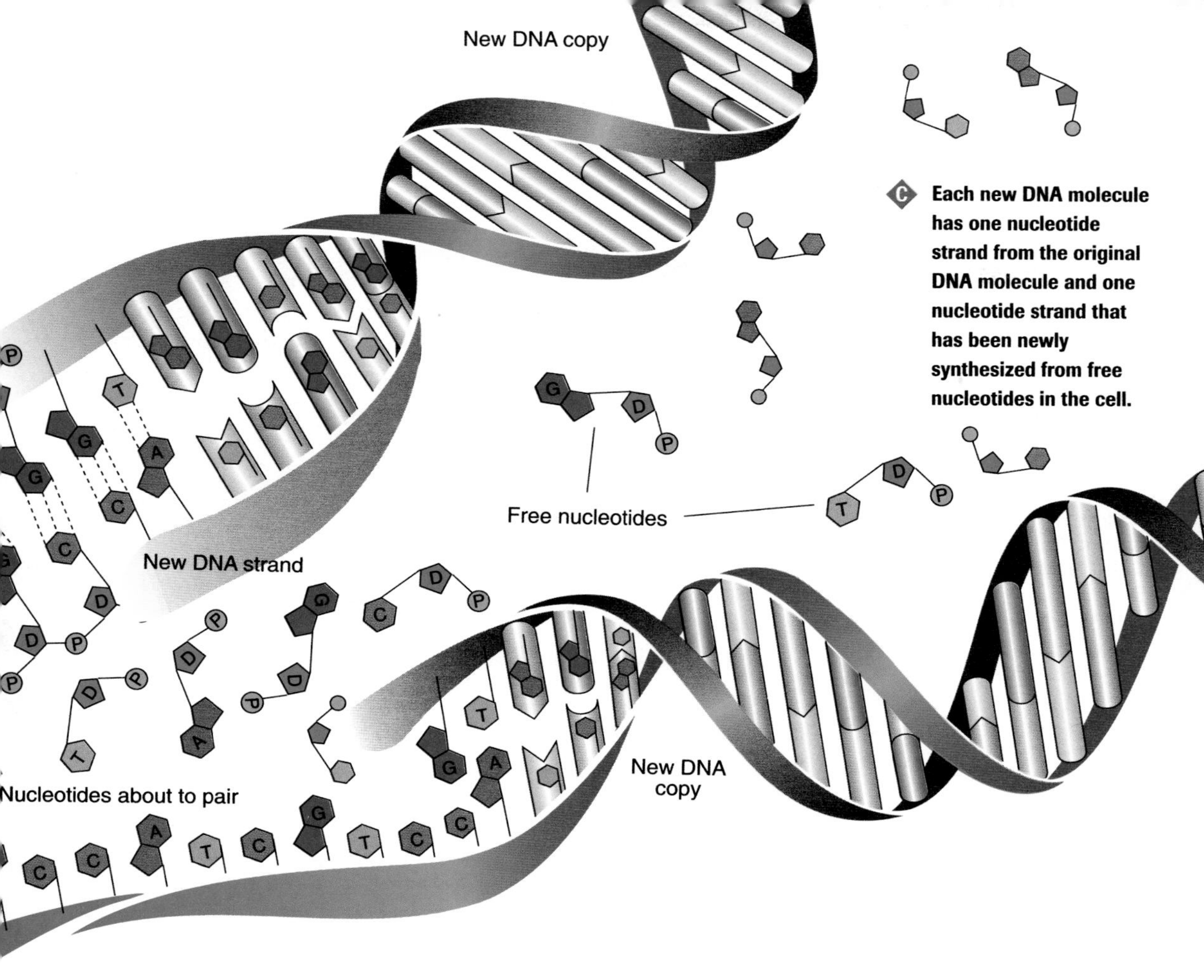

Section Review

Understanding Concepts

1. Describe the structure of a nucleotide.
2. How do the nucleotides in DNA pair?
3. Explain why the structure of a DNA molecule is often described as a ladder.

Thinking Critically

4. The sequence of bases on one strand of DNA is: GGCAGTTCATGC. What would be the sequence of bases on the complementary strand?

Skill Review

5. **Sequencing** Sequence the steps that occur during DNA replication. For more help, refer to Organizing Information in the *Skill Handbook.*

3 Assess

Check for Understanding

Have students write summaries of the process of DNA replication. Choose several students to write their summaries on the chalkboard. L1

Reteach

Count the students in class and divide them into four equal groups. Give each group a letter name—A, T, G, or C. Give each student a card with his or her letter. Write a base sequence for one half of a DNA molecule on the chalkboard. Be sure to write as many letters as there are students present. Give students two minutes to line themselves up correctly next to the appropriate complementary letter on the chalkboard. Tell them they have just replicated a DNA molecule. COOP LEARN L1

Extension

Have students research the work of F. Griffith (1928) and O. Avery and his colleagues (1944), who demonstrated that DNA is the genetic material. L3

Assessment

Knowledge: Prepare a sheet with a short section of two-stranded DNA, indicating the coding strand. Ask students to diagram the steps this short section would go through in order to replicate. L1

Answers to Section Review

1. A nucleotide consists of a sugar, a phosphate group, and a nitrogen base.
2. Cytosine forms hydrogen bonds with guanine, and thymine bonds with adenine.
3. The molecule is shaped like a twisted ladder, with the sides being formed by the sugar and phosphate molecules, while the rungs of the ladder are the pairs of bases

Thinking Critically

4. CCGTCAAGTACG

Skill Review

5. The two strands separate at the base pairs. Complementary nucleotides are attracted to those on the strands. Enzymes join the new nucleotides to form complementary chains. Two new chains separate.

4 Close

Discussion

Ask students how the DNA structure lends itself to replication. Why is accuracy so important in replication?

SECTION

13.2

Prepare

Key Concepts

Students will learn how DNA, genes, and proteins are related. The relationship between genes and the nucleotide sequences in DNA is discussed. Finally, the steps involved in the formation of mRNA and the role of tRNA in translation are discussed.

Block Scheduling

Look for this symbol for strategies that are useful in a block scheduling format. For more information on block scheduling, refer to Section 13.2 in the **Lesson Plans** booklet.

Materials

- Collect pictures of DNA replication for the Display.
- Buy yarn and bring in a box for the Alternate Lab.
- Collect cardboard for the Meeting Individual Needs.

1 Focus

Bellringer

Before presenting the lesson, display **Section Focus Master 28** on the overhead projector and have students answer the accompanying questions. L1 SAE

Discussion

Have students compare the number of letters in the alphabet to the number of amino acids available for protein formation. How many words can be made with only 26 letters? How many different proteins can be made with the amino acids available? Ask students to consider that proteins are made of many more amino acids than words are of letters.

SECTION

13.2 From DNA to Protein

Section Preview

Objectives

Relate the concept of the gene to the sequences of nucleotides in DNA.

Sequence the steps involved in protein synthesis.

Key Terms

codon
transcription
messenger RNA
ribosomal RNA
translation
transfer RNA

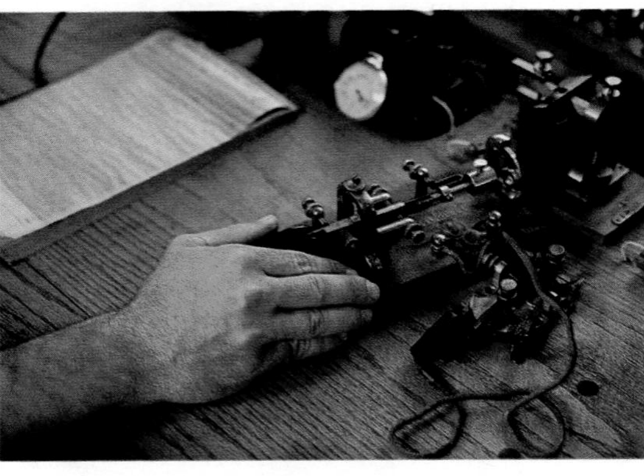

Morse code is a way of communicating that was developed in the 19th century. Even though more sophisticated means have since been developed, Morse code is sometimes used today. Morse code uses a pattern of dots and dashes to represent letters of the alphabet. In this way, long sequences of dots and dashes can produce an infinite number of different messages. Living organisms have their own code, the genetic code, in which the sequence of nucleotides in DNA represents information.

Genes and Proteins

What is the message contained within the DNA code? The answer has to do with proteins. Recall from Chapter 7 that proteins are complex polymers of amino acids. Different proteins have different functions. Some proteins, such as enzymes, control chemical reactions that perform key life functions such as building ATP or digesting food. Other proteins are produced to build and repair cell structures, such as microtubules or transport proteins in membranes. In general, proteins determine the structure and function of organisms. Perhaps you have wondered how cells are able to make these complex molecules.

Since the time of Mendel, a lot of research has gone into figuring out what genes are and how they function. A major step toward answering these questions occurred in the 1920s when it was hypothesized that each gene is responsible for the production of one protein. We also know now that genes are composed of DNA. Today, it's estimated that a human cell contains from 50 000 to 100 000 genes, each of which is made up of DNA. So, the message of the DNA code is information for building proteins.

The DNA Code

How do genes code for proteins? When scientists worked out the structure of DNA, it became clear to them that the sequence of nitrogen bases along one of the two strands was a code for the synthesis of proteins. This code is known as the genetic code.

Program Resources

Section Focus Master 28 L1 SAE
Reinforcement and Study Guide, pp. 50-51 L1
Biolab and Minilab Worksheets, pp. 49-52 L1
Concept Mapping, p. 13 L1
Critical Thinking/Problem Solving, p. 13 L3
Basic Concepts Transparencies 17, 18 and **Masters** L1 SAE
Basic Skills Transparencies 12a, 12b and **Master** L1 SAE
Reteaching Transparency 13 and **Master** L1 SAE

The genetic code

As you know, proteins are built from chains of smaller molecules called amino acids. There are 20 different amino acids, but DNA contains only four types of bases. How can these bases form a code for proteins?

Scientists working out the code realized that a single base can't represent a single amino acid because that system would code for only four different amino acids. Similarly, a sequence of two bases—such as AT, GC, or AG—yields only 16 possible combinations. Calculations will show that a sequence of three bases provides more than the 20 combinations needed to code for all amino acids. Each set of three nitrogen bases representing an amino acid is known as a **codon.** Because a sequence of three nitrogen bases forms the code for an amino acid, the DNA code is often called the triplet code. Sixty-four combinations are possible when a sequence of three bases is used; thus, 64 different codons are in the genetic code, *Table 13.1.*

The order of nitrogen bases in DNA will determine the type and order of amino acids in a protein. Of the 64 codons, 61 code for amino acids, and the remaining three are signals to stop the synthesis of a polypeptide chain. As you can see, more than one codon can code for the same amino acid. However, for any one codon, there can be only one amino acid.

Table 13.1 The DNA Code

First Base in Codon	Second Base in Codon				Third Base in Codon
	A	G	T	C	
A	phenylalanine	serine	tyrosine	cysteine	A
	phenylalanine	serine	tyrosine	cysteine	G
	leucine	serine	stop	stop	T
	leucine	serine	stop	tryptophan	C
G	leucine	proline	histidine	arginine	A
	leucine	proline	histidine	arginine	G
	leucine	proline	glutamine	arginine	T
	leucine	proline	glutamine	arginine	C
T	isoleucine	threonine	asparagine	serine	A
	isoleucine	threonine	asparagine	serine	G
	isoleucine	threonine	lysine	arginine	T
	methionine (start)	threonine	lysine	arginine	C
C	valine	alanine	aspartate	glycine	A
	valine	alanine	aspartate	glycine	G
	valine	alanine	glutamate	glycine	T
	valine	alanine	glutamate	glycine	C

SECTION 13.2

2 Teach

Tying to Previous Knowledge

Have students recall the structure of proteins, amino acids, polypeptides, and peptide bond formation from Chapter 7. Make sure students understand that most proteins require the synthesis of two or more polypeptide chains.

Audiovisual

Show the video *DNA and Genes,* Charles Clark Co., Inc.

interNET CONNECTION

Follow the link for this chapter on the Glencoe Homepage at **www.glencoe.com/sec/science** to find out more about genes and chromosomes.

Reinforcement

As your students learn to translate the bases of the DNA code to proteins via the RNA codons, explain that they can use Table 13.1 as the starting point. To convert DNA codons to RNA, they must change the base A in the table to base U, the base G to C, the base T to A, and the base C to G. Using these instructions, your students may find it helpful to make their own table of RNA bases for ready reference.

Meeting Individual Needs

Hearing Impaired Have hearing impaired students match cards showing DNA codons with cards showing the appropriate tRNA anticodons. Students will have to move through the sequence from DNA to mRNA to tRNA for each codon in order to match the cards. L1

SECTION 13.2

Chalkboard Example

Visual-Spatial On the chalkboard, show how DNA strands split and how mRNA is formed.

Science, Technology, and Society

Filming DNA

Scientists have developed a technique that uses fluorescence microscopes to film DNA. The technique involves attaching a fluorescent dye to DNA to form a complex that glows. Then, just as you can watch a firefly travel across the yard, researchers can follow the DNA and its movements. Scientists are using this method to investigate how chromosomal DNA is tightly bound to histones and other proteins and how they fold into their functional forms.

GLENCOE TECHNOLOGY

Videodisc

The Infinite Voyage: The Living Clock

Changing Circadian Rhythms Through Altering DNA (Ch. 5)

Universality of the genetic code

The genetic code was figured out by studying the DNA of the bacterium *Escherichia coli.* Yet the code is exactly the same in humans and virtually every other known organism. The code is said to be universal because the codons represent the same amino acids in all organisms. The universality of this code is powerful evidence that all organisms alive today shared a common ancestor billions of years ago.

Transcription—From DNA to RNA

As you have read in Chapter 8, proteins are made on ribosomes in the cytoplasm of the cell. Yet DNA, which contains the genetic code, is found only in the nucleus. How is information from the genetic code brought to the ribosomes for protein synthesis? Scientists began to understand how proteins are made when a second kind of nucleic acid, RNA (ribonucleic acid), was discovered.

RNA structure

Like DNA, RNA is a nucleic acid. However, RNA structure differs from DNA structure in three ways, as shown in *Figure 13.5.* First, RNA is usually composed of a single strand of nucleotides, rather than a double strand as in DNA. RNA also contains a different type of sugar molecule, ribose, instead of deoxyribose. Like DNA, RNA also contains four nitrogen bases, but rather than thymine, RNA contains a similar base called uracil (U).

Figure 13.5

The three chemical differences between DNA and RNA are shown here.

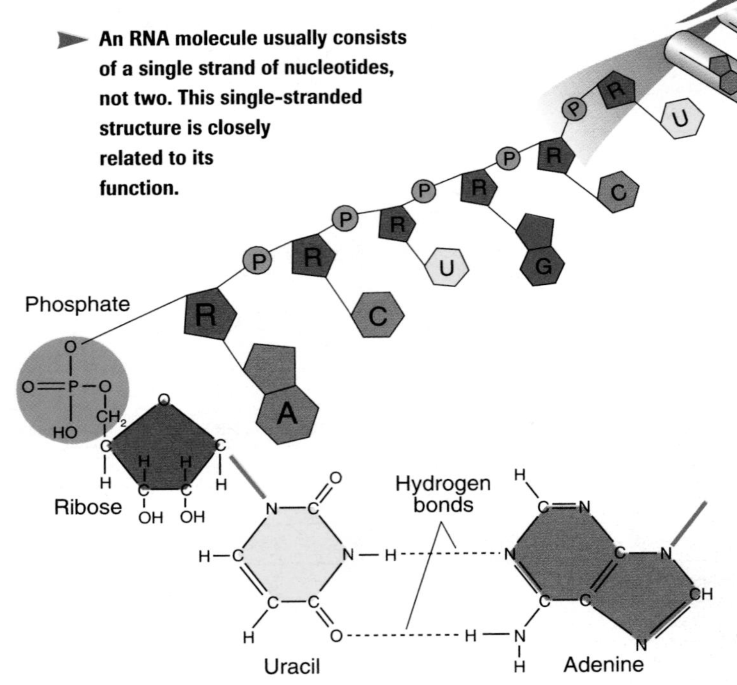

An RNA molecule usually consists of a single strand of nucleotides, not two. This single-stranded structure is closely related to its function.

The sugar in RNA is ribose sugar, rather than the deoxyribose sugar of DNA.

RNA contains the nitrogen base uracil (U) instead of thymine (T). Uracil pairs with adenine just as thymine does in DNA.

Alternate Lab: Gene and Chromosome Size

Purpose

Kinesthetic Students will use an analogy to understand gene and chromosome size.

Materials

scissors, 2 skeins of three-ply yarn for a class of 20-30, box about 100 cm × 50 cm

Background

This lab gives students an idea of the size of a gene and a bacterial chromosome by having them construct yarn models that are scaled up a million times in size. It is designed for 25 students working individually until the last part of step three, when they join individual strands to model an entire chromosome.

Procedure

Give students the following directions.

1. One nucleotide in a molecule of DNA occupies a length of 3.4×10^{-10} m. If an average gene that codes for an average-sized protein contains 1200 bases, how long is an average gene? *3.4×10^{-10} m × 1200 = 4.08×10^{-7} m* How many amino acids would these nucleotides code for? *1200 bases ÷ 3*

Making RNA by transcription

Today, we know that RNA is the form in which information moves from DNA in the nucleus to the ribosomes in the cytoplasm. Enzymes make an RNA copy of a DNA strand in a process called **transcription,** *Figure 13.6.* The process of transcription is similar to the process of DNA replication. The main difference is that the process results in the formation of one single-stranded RNA molecule, rather than one new, double-stranded molecule of DNA. This RNA copy that carries information from DNA out into the cytoplasm of the cell is called **messenger RNA (mRNA).**

Messenger RNA carries the information for making a protein chain to the ribosomes, where proteins are synthesized. Some portions of DNA code for the RNA that makes up ribosomes. This type of RNA is called **ribosomal RNA (rRNA).** Recently, scientists have shown that rRNA helps to produce enzymes needed to bond amino acids together during protein synthesis.

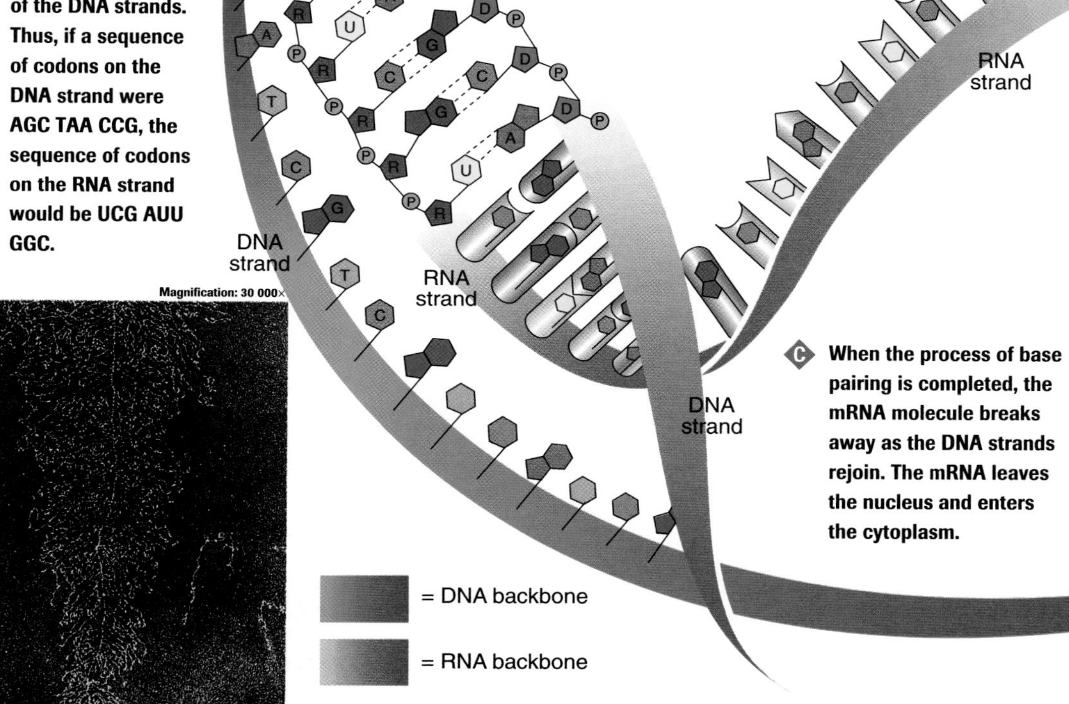
A The process of transcription begins as enzymes unzip the molecule of DNA, just as they do during DNA replication.
B As the DNA molecule unzips, free RNA nucleotides pair with complementary DNA nucleotides on one of the DNA strands. Thus, if a sequence of codons on the DNA strand were AGC TAA CCG, the sequence of codons on the RNA strand would be UCG AUU GGC.
C When the process of base pairing is completed, the mRNA molecule breaks away as the DNA strands rejoin. The mRNA leaves the nucleus and enters the cytoplasm.
DNA strand
RNA strand
RNA strand
DNA strand
Magnification: 30 000×
= DNA backbone
= RNA backbone
Multiple transcriptions from a single DNA molecule.

Figure 13.6
Transcription of DNA

SECTION 13.2

Different Viewpoints in Biology

Inheritance of Defects in Mitochondrial DNA

For many years, scientists have known that DNA in the nucleus governs inheritance. Then it was found that mitochondria possess their own DNA, quite separate from the genetic material of the nucleus. Some scientists now suggest that some defects in mitochondrial DNA may underlie several genetic disorders.

Inheritance through mitochondria is different from the usual pattern of inheritance because it cannot be inherited from the father. When the sperm meets the egg, the sperm's mitochondria do not enter the egg. Also, since each cell possesses hundreds to thousands of mitochondria, an individual could inherit various degrees of a disorder, depending on how many mitochondria are randomly distributed to each daughter cell.

GLENCOE TECHNOLOGY

Videotape

The Secret of Life

Use the videotape *What's in Stetter's Pond: The Basics of Life* to illustrate DNA's role as the controlling material of life.

bases per codon = 400 codons = 400 amino acids

2. Cut a piece of yarn 40 cm long to represent the average gene. This length is 1 000 000 times that of a gene.
3. Cut a piece of yarn that would represent 150 genes. *150 × 40 cm = 60 m* Tie the end of your 60-m piece of yarn to another in the class until all yarn in the class is connected. This length, 1500 m, represents the length of DNA in a bacterium if the cell were scaled up to the size of the box.

Analysis

Ask students the following questions.

1. Compare the length of DNA with the size of the "cell" (box). How does all of the DNA fit inside the cell? *The DNA is coiled and twisted.*
2. A human cell contains 100 000 genes. How long would the yarn be that represents all of the DNA in a human cell? *100 000 × 40 cm = 40 000 m*

Assessment

Journal: Ask students to write a summary of the lab and answers to the analysis questions and place them in their journals. L1

BioLab

RNA Transcription

Time Allotment
One class period

Objectives
Review objectives with students before they begin the Biolab.

Process Skills
sequence, observe and infer, recognize cause and effect

PREPARATION

Instead of having students copy models onto construction paper, you may wish to use the reproducible masters provided in the Biolab and Minilab Worksheets booklet. Copy these onto white paper and have students color the models with colored pencils or crayons.

GLENCOE TECHNOLOGY

Videodisc

Biology: The Dynamics of Life
Disc 1, Side 1
DNA Transcription (Ch. 32)

BioLab

RNA Transcription

DNA is the substance that makes up genes. DNA passes its information into the cytoplasm of the cell by coding for another nucleic acid, messenger RNA. This mRNA is made from the DNA pattern by base pairing in the process of transcription. In this activity, you will demonstrate the process of transcription through the use of paper DNA and mRNA models.

PREPARATION

Problem
How does the order of bases in DNA determine the order of bases in mRNA?

Objective
In this Biolab, you will:
- **Determine** how the order of bases in DNA determines the order of bases in mRNA.

Materials
construction paper, 5 colors
scissors
clear tape
pencil

Parts for DNA Nucleotides

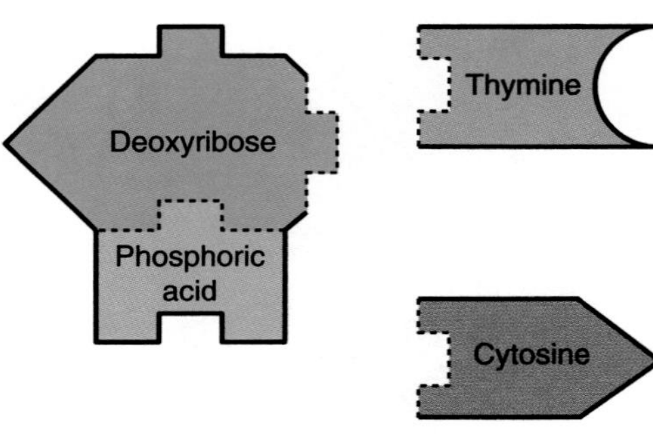

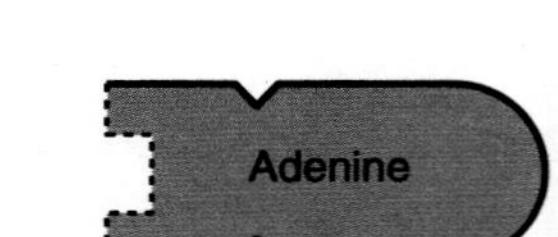

Extra Parts for RNA Nucleotides

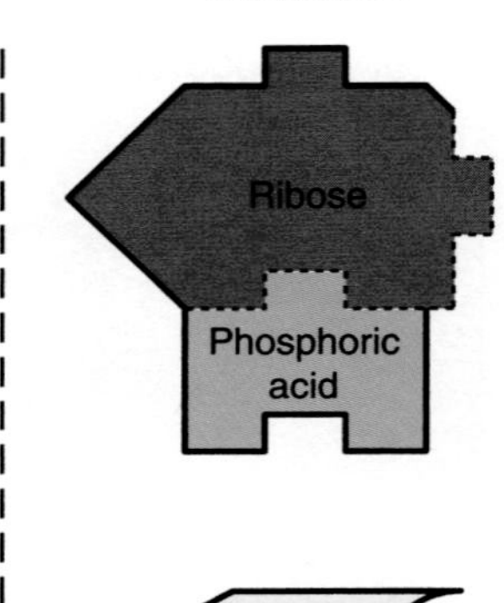

PROCEDURE

Teaching Strategies
- You may wish to give each student two envelopes to hold the nucleotide pieces. One can be used for the DNA nucleotides and one for the RNA nucleotides.
- To make larger, more varied models, the lab groups could pool their models.

Data and Observations
Students should construct a DNA molecule and a complementary mRNA molecule.

PROCEDURE

1. Copy the illustrations of the four different DNA nucleotides onto your construction paper. Make sure that each different nucleotide is on a different color of construction paper. In all, you should make ten copies of each nucleotide.
2. Using scissors, carefully cut out the shapes of each nucleotide.
3. Using any order of nucleotides that you wish, construct a double-stranded DNA molecule. If you need more nucleotides, copy them as before.
4. Fasten your molecule together using clear tape. Do not tape across base pairs.
5. As in step 1, copy the illustrations of A, G, and C nucleotides onto the same colors of construction paper. Use the fifth color of construction paper to make copies of uracil nucleotides.
6. With scissors, carefully cut out the nucleotide shapes.
7. With your DNA molecule in front of you, demonstrate the process of transcription by first pulling the DNA molecule apart between the base pairs.
8. Using only one of the strands of DNA, begin matching mRNA nucleotides with the exposed bases on the DNA model.
9. When you are finished, tape your new mRNA molecule together using cellophane tape.

ANALYZE AND CONCLUDE

1. **Observing and Inferring** Does the mRNA model more closely resemble the DNA strand from which it was transcribed or the complementary strand that wasn't used? Explain your answer.
2. **Recognizing Cause and Effect** Explain how the structure of DNA enables the molecule to be easily transcribed. Why is this important for genetic information?
3. **Relating Concepts** Why is RNA important to the cell? How does an mRNA molecule carry information from DNA?

Going Further

Journal Report
Do library research to find out more about how the bases in DNA were identified and how the base pairing pattern was determined.

ANALYZE AND CONCLUDE

1. mRNA more closely resembles the complementary DNA. They have the same base sequence except that mRNA has uracil in place of thymine.
2. Because DNA is double-stranded, sections can be unzipped to allow complimentary bases to hydrogen bond, while the remaining DNA stays zipped. Thus, only the information needed at one time is being transcribed.
3. The mRNA is formed as a complementary copy of the genetic information. The RNA copy can leave the nucleus while the "master copy" stays within the nucleus.

Assessment

Journal: Have students write a summary of the lab including answers to Analyze and Conclude. L1

Going Further

Have students use Table 13.1 to determine which amino acids they coded for in the Biolab. L1

Meeting Individual Needs

Visually Impaired Cut out very large copies of the five different nucleotides (A, T, C, G, and U) so visually impaired students can participate in the Biolab. L1

Chalkboard Example

LS Logical-Mathematical On the chalkboard, write the sequence for one strand of DNA. Have students copy the sequence and write the corresponding sequences for mRNA, the tRNA anticodons, and the coded protein. Students should follow the changes in logical steps from the original DNA through transcription and translation. Afterward, go through the correct answer on the board.

Audiovisual

Show the video *DNA: Laboratory of Life*, National Geographic.

Display

LS Visual-Spatial Use an overhead projector to carry out the following exercise. On a drawing of a tRNA molecule, show the anticodon and the area of amino acid attachment. Point out again that ATP and enzymes are required for amino acids to be attached.

GLENCOE TECHNOLOGY

Videodisc

Biology: The Dynamics of Life
Disc 1, Side 1
Translation (Ch. 33)

3 Assess

Check for Understanding

Review the processes of transcription and translation orally. Ask students to supply missing words, descriptions, and process words, including DNA, complementary codons, mRNA, tRNA, ribosomal RNA, transcription, translation, and anticodons. Discuss the definitions. L1

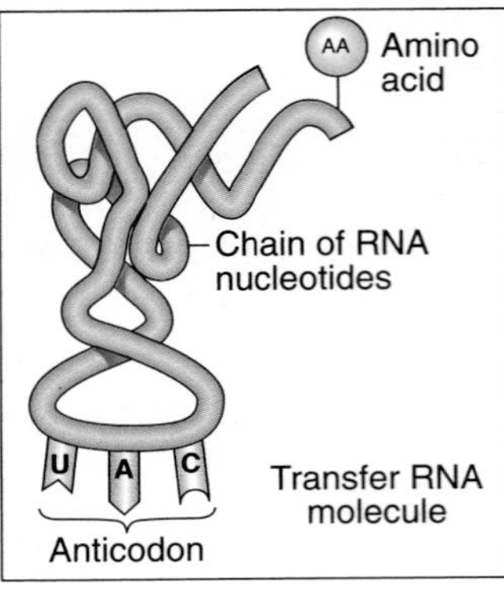

Figure 13.7
A tRNA molecule is composed of about 75 nucleotides. Each tRNA recognizes only one amino acid. The amino acid becomes bonded to the top of the tRNA molecule. Located on the bottom of the tRNA molecule are three nitrogen bases, called an anticodon, that pair up with mRNA codons during translation.

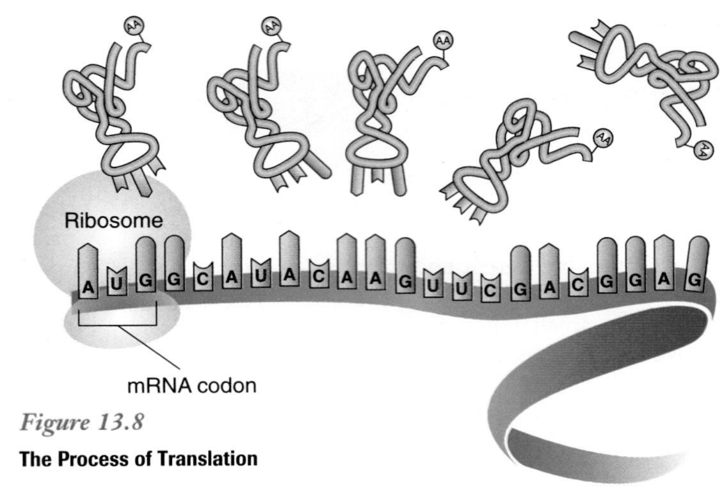

Figure 13.8
The Process of Translation

A As translation begins, the first codon of the mRNA strand attaches to a ribosome. Then, tRNA molecules, each carrying a specific amino acid, approach the ribosome. When the tRNA anticodon pairs with the mRNA codon, the two molecules join and the tRNA molecule remains there.

Translation—From RNA to Protein

When mRNA is synthesized, it carries a complementary copy of the DNA code for a protein chain. How is mRNA language used to synthesize a sequence of amino acids?

The role of transfer RNA

The process of converting the information in a sequence of nitrogen bases in mRNA into a sequence of amino acids that make up protein is known as **translation.** Translation, which occurs on ribosomes, involves a third kind of RNA. If proteins are to be built, the 20 different amino acids dissolved in the cytoplasm must be brought to the ribosomes. This is the role of transfer RNA, *Figure 13.7.* **Transfer RNA (tRNA)** brings amino acids to the ribosomes so they can be assembled into proteins. How does the cell "know" which tRNA molecules carry the proper amino acids? The answer again involves base pairing.

Translating the mRNA code

Correct translation of the code depends upon the joining of each mRNA codon with the anticodons of the proper tRNA molecules. *Figure 13.8* shows how the message in an mRNA molecule is translated into a protein. The end result of translation is the formation of the large variety of proteins that make up the structure of organisms and help them to function.

The importance of proteins

Now that you know proteins are made by the bonding together of amino acids during translation of the mRNA copy of the DNA code, can you guess how many types of proteins can be made in this way? The fact is, an almost infinite variety of proteins can be made from the same 20 amino acids. Remember, although there are only 20 amino acids, they can occur in any order and in any number. It is the sequence of amino acids in a protein that determines the characteristics of that protein.

STUDENT JOURNAL

Translation Have students write a summary of what events are being shown in Figure 13.7. Have them include what led up to translation and what will follow translation. L1

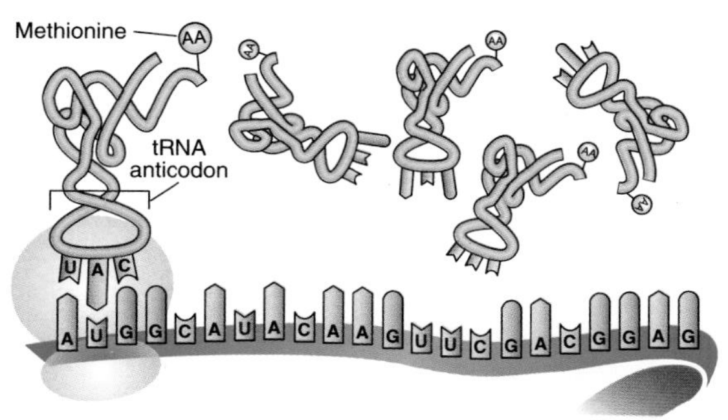

B Often, the first codon on mRNA is AUG, which codes for the amino acid methionine. AUG signals the start of protein synthesis. When this signal is given, the mRNA slides along the ribosome to the next codon.

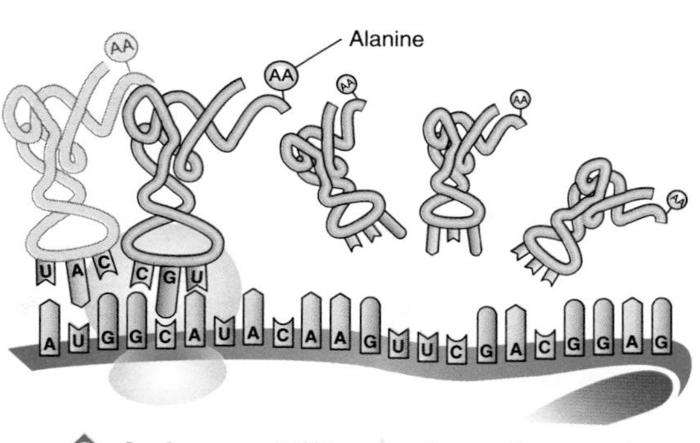

C Again, a new tRNA molecule carrying an amino acid will pair with the mRNA codon.

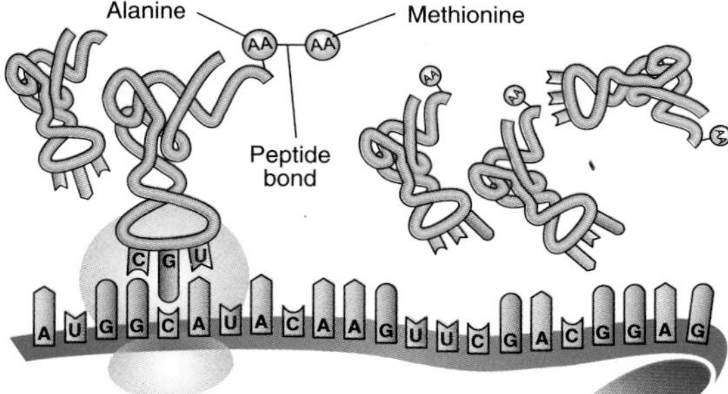

D When the first and second amino acids are in place, an enzyme joins them by forming a peptide bond between them.

E As the process continues, a chain of amino acids is formed until the ribosome reaches a stop codon on the mRNA strand.

Section Review

Understanding Concepts

1. In what ways do the chemical structures of DNA and RNA differ?
2. What is a codon, and what does it represent?
3. What is the role of transfer RNA in protein synthesis?

Thinking Critically

4. You have learned that there is a *stop* codon to signal the end of an amino acid chain. Why is it important that a signal to stop translation be part of protein synthesis?

Skill Review

5. **Sequencing** Sequence the steps involved in protein synthesis, from the production of mRNA to the final translation of the DNA code. For more help, refer to Organizing Information in the *Skill Handbook*.

SECTION 13.2

Reteach

On the chalkboard, illustrate the processes of transcription and translation of a particular portion of a strand of DNA. Have students copy the diagram and label each process. Then have them check each other's papers. L1

Extension

Have students research the structure and function of histone molecules. They could build a model of the eukaryotic chromosome. "Simulating the Eukaryotic Chromosome" by Leo E. Spencer, *Journal of College Science Teaching*, May 1985 may be helpful. L2

Assessment

Oral: Have students explain how they would be able to identify the complimentary components of a strand of DNA or mRNA if given the components of one strand. L1

4 Close

Using a Table

Ask students what would happen to a protein if one base was changed in the DNA that codes for it. Use Table 13.1 to demonstrate to students what would occur if CAT were changed to CTT. Discuss what changing one amino acid might do to protein structure.

Answers to Section Review

1. RNA contains ribose and uracil and DNA contains deoxyribose and thymine. RNA is usually single stranded.
2. A codon is a three-base unit that codes for a single amino acid.
3. Transfer RNA brings a specific amino acid to a ribosome by matching the messenger RNA strand.

Thinking Critically

4. Because a protein's 3-dimensional structure depends on its length as well as its amino acid sequence, translation must start and stop at precise positions.

Skill Review

5. The DNA strands separate and free RNA nucleotides pair with complementary DNA nucleotides on one of the DNA strands, forming mRNA. mRNA leaves the nucleus and attaches to a ribosome. tRNA molecules, carrying specific amino acids, pair with the appropriate mRNA codons as the mRNA slides along the ribosome. When two acids are in place, an enzyme joins them together.

SECTION

13.3

Prepare

Key Concepts

Point mutations and frameshift mutations and their effects on the coding of proteins are discussed. The chromosomal mutations—insertions, deletions, translocations, and inversions—are examined as well as errors in disjunction that cause monosomy and trisomy.

Block Scheduling

Look for this symbol for strategies that are useful in a block scheduling format. For more information on block scheduling, refer to Section 13.3 in the **Lesson Plans** booklet.

Materials

- Collect bar codes for the Demonstration.
- Collect materials for the Project.

1 Focus

Bellringer

Before presenting the lesson, display **Section Focus Master 29** on the overhead projector and have students answer the accompanying questions. L1 SAE

Discussion

Ask students to describe the images that the word *mutation* conjures up in their minds. They may describe fantastic beings they have encountered in movies and stories. Point out that real mutations are usually much less spectacular.

SECTION

13.3 Genetic Changes

Section Preview

Objectives

Categorize the different kinds of mutations that can occur in DNA.

Compare the effects of different kinds of mutations on cells and organisms.

Key Terms

mutation
point mutation
frameshift mutation
chromosomal mutation
nondisjunction
trisomy
monosomy

Have you ever copied a phone number incorrectly? Perhaps you've written a 2 when it should have been a 3. What are some possible consequences of changing digits in numbers? Sometimes there is little effect, but you might find that you can't call your doctor in an emergency. Mistakes in the DNA code can produce similar results. Sometimes, there is no effect on an organism, but often, mistakes in DNA can cause serious consequences for individual organisms.

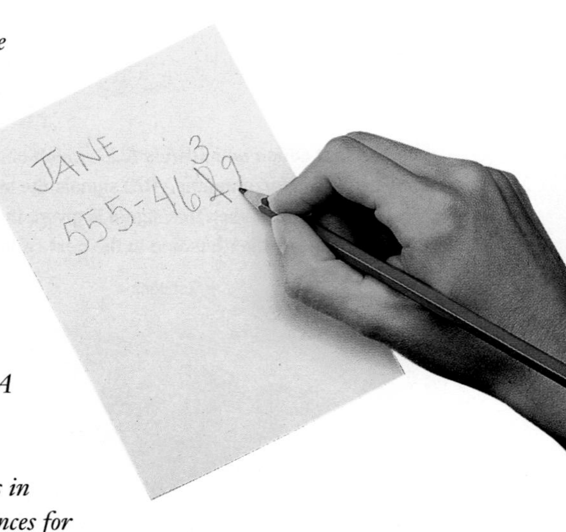

Mutation—A Change in DNA

The cell processes that copy genetic material and pass it from one generation to the next are usually accurate. Accuracy is important to ensure the genetic continuity of both new cells and offspring. However, sometimes mistakes can occur in the genetic material. Any mistake or change in the DNA sequence is called a **mutation.**

The effects of point mutations

Consider what might happen if an incorrect amino acid were inserted into a growing protein chain during translation of the DNA code. It might affect the synthesis of the entire molecule, right? Such a problem can occur if a point mutation arises. A **point mutation** is a change in a single base pair in DNA.

A simple analogy can illustrate point mutations. Read the sentences below to see what happens when a single letter in a sentence is changed.

THE DOG BIT THE CAT.
THE DOG BIT THE CAR.

As you can see, changing a single letter changes the meaning of this sentence. Similarly, a change in a single nitrogen base can change the entire structure of a protein. The top diagram in *Figure 13.9* shows what can happen with a point mutation.

Program Resources

Section Focus Master 29 L1 SAE
Reinforcement and Study Guide, p. 52 L1
Biolab and Minilab Worksheets, p. 48 L1

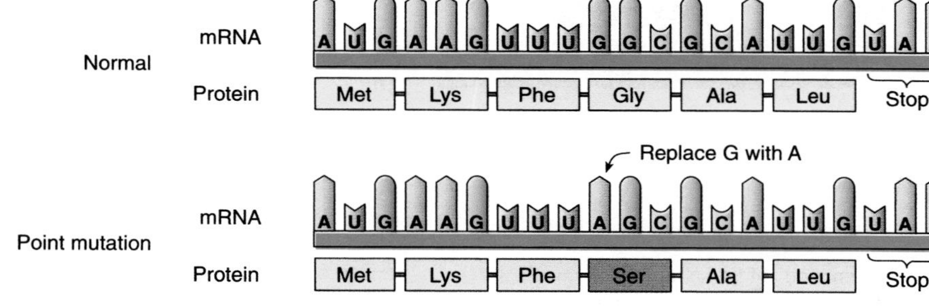

In this point mutation, the base guanine was changed to adenine. This change in the codon caused the insertion of serine, rather than glycine, into the growing amino acid chain. Sometimes the errors caused by point mutations don't interfere with protein function, but often the effect is disastrous.

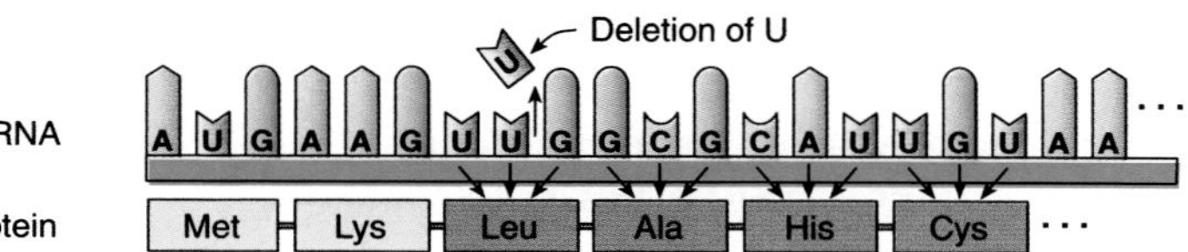

Proteins that are produced as a result of frameshift mutations seldom function properly because such mutations usually change several amino acids. Adding or deleting one base of a DNA molecule will change nearly every amino acid in the protein after the addition or deletion.

Figure 13.9
Gene Mutations

Frameshift mutations

When the mRNA strand moves across the ribosome, a new amino acid is added to the protein for every codon on the mRNA strand. What would happen if a single base were lost from a DNA strand? This new sequence with the deleted base would be transcribed into mRNA. But then, the mRNA would be out of position by one base. As a result, every codon after that base would be different, as shown at the bottom of *Figure 13.9.* This mutation would cause nearly every amino acid in the protein after the deletion to be changed. The same effect would also result from the addition of a single base. A mutation in which a single base is added or deleted from DNA is called a **frameshift mutation.**

Chromosomal Mutations

Changes may occur at the level of chromosomes as well as in genes. Mutations to chromosomes may occur in a variety of ways. For example, sometimes parts of chromosomes are broken off and lost during mitosis or meiosis. Often, chromosomes break and then rejoin incorrectly. Sometimes, the parts join backwards or even to the wrong chromosome. These changes in chromosomes are called **chromosomal mutations.**

Effects of chromosomal mutations

Chromosomal mutations occur in all living organisms, but they are especially common in plants. Such mutations affect the distribution of genes to gametes during meiosis.

2 Teach

Demonstration

LS **Visual-Spatial** Gather bar codes from various product wrappers. Use the bar codes as chromosome models by showing that the patterns of bars represent genes. Have students show how chromosome inversion occurs by cutting a piece of a bar code and rearranging it within the model chromosome. Also have them show what a linkage group is by identifying it as several bars close together. L1

Bioethics

Prenatal Diagnosis of Down Syndrome
Many pregnant women undergo a test called amniocentesis early in their pregnancy to determine whether the fetus has an extra chromosome 21. As a result of the test, a woman knows whether or not her child will have Down syndrome. Have students discuss the ethical implications of the test. Should parents have this knowledge before a child is born? What kinds of decisions might be made on the basis of the test? Have students discuss the pros and cons of prenatal testing. L1

STUDENT JOURNAL

Mutations Have students research the various mutagenic agents such as X rays, ultraviolet light, radioactive substances, and other chemicals, and report on the mechanism by which each of them causes mutations. L2

PORTFOLIO

Protein Building Have students write an essay comparing building proteins to building a house and what will occur if there are problems with the blueprint guiding the builders.
L1 LS P

SECTION 13.3

MiniLab

Purpose

Logical-Mathematical Students will examine the effect of gene mutations on proteins.

Process Skills

recognize cause and effect, analyze

Teaching Strategies

- Individual students can do this Minilab at their desks as you walk around the room and check their work.

Expected Results

Complementary strand: TTACGGTCAC-CAAGCGTG

Base sequence: mRNA: UUACGGUCAC-CAAGCGUG

Amino acids: leucine-arginine-serine-proline-serine-valine

Changing the fourth base from G to C would change the second amino acid from arginine to glycine.

Adding G in the fourth position would result in the mRNA base sequence: UUACCGGUCACCAAG-CGUG and the amino acid sequence: leucine-proline-valine-threonine-lysine-arginine

Analysis

1. Changing G to C; adding G to the chain
2. The point mutation changed only one amino acid.
3. The frameshift mutation changed every amino acid following the addition of G.

Assessment

Journal: Ask students to include a summary of the lab and answers to Analysis questions in their journals. L1

MiniLab

How do gene mutations affect proteins?

Gene mutations often have serious effects on proteins. In this activity, you will demonstrate how such mutations affect protein synthesis.

Procedure

1. Copy the following base sequence of one strand of an imaginary DNA molecule: AATGCCAGTGGTTCGCAC
2. Below the first strand, write the base sequence of the complementary DNA strand.
3. Then, write the base sequence that would appear on an mRNA strand after transcription.
4. Use *Table 13.1* on page 317 to determine the order of amino acids in the resulting protein fragment.
5. If the fourth base in the original DNA strand were changed from G to C, how would this affect the resulting protein fragment?
6. If a G were added to the original DNA strand after the third base, what would the resulting mRNA look like? How would this addition affect the protein?

Analysis

1. Which change in DNA was a point mutation? Which was a frameshift mutation?
2. In what way did the point mutation affect the protein?
3. How did the frameshift mutation affect the protein?

Gametes that should have a complete set of genes may end up with extra copies of some genes or a complete lack of certain genes.

Few chromosome mutations are passed on to the next generation because the zygote usually dies. In cases where the zygote develops, the mature organism is often sterile and thus incapable of producing offspring. The most important of these mutations—deletions, insertions, inversions, and translocations—are illustrated in *Figure 13.10*.

Errors in Disjunction

Many chromosome mutations result from the failure of chromosomes to separate properly during meiosis. Recall that during meiosis I, one chromosome from each homologous pair moves to each pole of the cell. Occasionally, both chromosomes of a homologous pair move to the same pole of the cell.

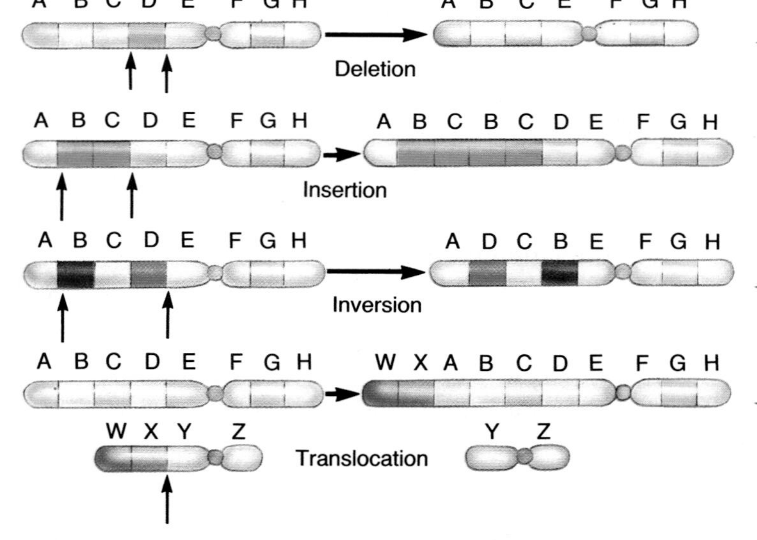

Figure 13.10
Chromosomal Mutations

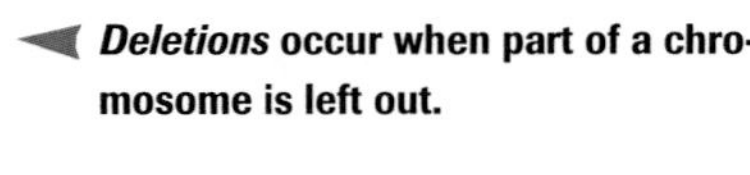

Deletions occur when part of a chromosome is left out.

Insertions occur when a part of a chromatid breaks off and attaches to its sister chromatid. The result is a duplication of genes on the same chromosome.

Inversions occur when part of a chromosome breaks out and is reinserted backwards.

Translocations occur when part of one chromosome breaks off and is added to a different chromosome.

PROJECT

Building a Model

Have students build a model demonstrating protein synthesis. They may wish to use various types of macaroni on poster board, colored pipe cleaners, beads, colored building blocks, or yarn. L1

The failure of homologous chromosomes to separate properly during meiosis is called **nondisjunction.**

Trisomy, triploidy, and monosomy

In one form of nondisjunction, two kinds of gametes result. One has an extra chromosome, and the other is missing a chromosome. The effects of nondisjunction are often seen after gametes fuse. For example, when a gamete with an extra chromosome is fertilized by a normal gamete, the zygote will have an extra chromosome. This condition is called **trisomy.** In humans, if a gamete with an extra chromosome number 21 is fertilized by a normal gamete, the resulting zygote has 47 chromosomes instead of 46. This zygote will develop into a baby with Down syndrome.

Another form of nondisjunction involves a total lack of separation of homologous chromosomes. When this happens, a gamete inherits a complete diploid set of chromosomes. When a gamete with an extra set of chromosomes is fertilized by a normal haploid gamete, offspring have three sets of chromosomes and are triploid. This condition is rare in animals but frequently occurs in plants. Often, the flowers and fruits of these plants are larger than normal. Many triploid plants, such as the sterile banana plant shown in *Figure 13.11*, are of great commercial value.

Although organisms with extra chromosomes often survive, organisms lacking one or more chromosomes usually do not. When a gamete with a missing chromosome is fertilized by a normal gamete, the resulting zygote will lack a chromosome.

A BROADER VIEW

Gene-control Operons

Is the information in genes always expressed, or can genes be turned on and off depending upon the needs of a cell? The most extensively studied system of gene regulation is the mechanism for lactose sugar metabolism in *Escherichia coli.* In order to digest lactose, *E. coli* has to produce two proteins. One of these proteins carries lactose into the cell, and the other is an enzyme that digests lactose. The genes that code for these proteins are always present in *E. coli,* but the proteins are made only when lactose is present. The expression of these genes is controlled by a genetic mechanism called the *lac* operon. How does this system work?

The* lac *operon Genes that code for proteins are called structural genes. In the diagram of the *lac* operon, z is the structural gene that codes for the enzyme that digests lactose; y is the structural gene that codes for the protein that carries lactose across the cell membrane. Next to them on the *E. coli* chromosome are two overlapping sections of DNA, the operator (o) and the promoter (p). This DNA does not code for protein. Instead, these sections determine when the structural genes, z and y, will be transcribed into mRNA. R is a regulatory gene that produces a repressor protein, which normally binds to the operator. This protein prevents the transcription enzyme from attaching to the promoter.

Switch on, switch off When lactose is present, molecules of lactose bind to the repressor protein causing it to leave the operator. Then the transcription enzyme can attach and proceed to transcribe genes z and y into mRNA. The mRNA is then translated into the proteins required to carry lactose into cells and digest it.

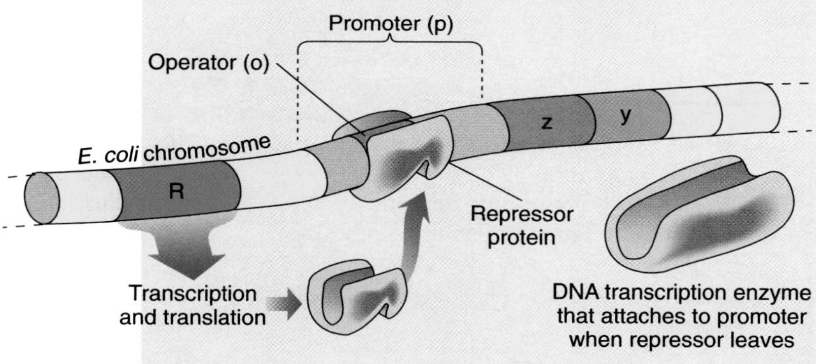

Thinking Critically
What do you think is the adaptive value of a regulatory mechanism like this one?

SECTION 13.3

A BROADER VIEW

Gene Control Operons

Purpose
Students will become acquainted with the operon concept of gene control in bacteria.

Teaching Strategies
- Ask students to consider why a skin cell might produce melanin and a pancreas cell produce insulin when they both contain the same DNA. Bring up the idea that not all genes may be turned on in all cells of the body.

Thinking Critically
It is an economy mechanism that prevents the cell from expending energy and materials making unneeded substances.

Software
Lac Operon, Queue Inc.

Visual Learning

LS **Visual-Spatial** Use an overhead projector to show Figures 13.9 and 13.10 without captions. Have students describe in their own words what is taking place in each type of mutation shown.

Meeting Individual Needs

Gifted Have gifted students research gene regulation of eukaryotic chromosomes. Have them find out about some of the work that has been done on proto-oncogenes. L3

3 Assess

Check for Understanding

Ask students to name the type of genetic or chromosomal mutation involved when (a) one kind of base in DNA takes the place of another, (b) some genes are duplicated on the same chromosome, (c) part of one chromosome breaks off and is attached to a different one, (d) a gamete receives one too few chromosomes. *(a) point mutation, (b) insertion, (c) translocation, (d) monosomy from nondisjunction* L1

Reteach

Repeat the Visual Learning activity on page 327, this time asking students to make their own labeled drawings of each type of mutation. Have them also draw and label how nondisjunction occurs and place all the drawings in their journals. L1

Extension

Ask students to research polyploidy in plants and find out what effects the extra sets of chromosomes have on plants. L2

Assessment

Oral Ask students to state and explain the effects of point, frameshift, and chromosomal mutations. L1

4 Close

Discussion

Have each student form a hypothesis about how mutations may be involved in evolution of a species. Discuss the hypotheses as a class. Correct any misconceptions students may have. L1

Figure 13.11
The banana plant is an example of a triploid organism.

This condition is called **monosomy.** Examples include human females with only a single X chromosome. Zygotes with other types of monosomy usually do not survive.

Causes of Mutations

Mutations are generally random events, and, as you will learn in Chapter 18, errors in DNA provide the variation that is fundamental to the evolution of species. Mutations that occur at random are called spontaneous mutations.

However, it is known that many environmental agents also cause mutation. Exposure to X rays, ultraviolet light, radioactive substances, or certain chemicals can cause changes in DNA. Mutations often result in sterility or the lack of normal development in an organism. If these mutations occur in human gametes, they can cause birth defects. If they occur in body cells, the mutations can lead to cancer.

Connecting Ideas

The discovery that genes are made of DNA, along with the explanation of how DNA determines inherited traits, is perhaps the greatest achievement of 20th-century biology. Unraveling the mystery of genes and the DNA within them has enabled scientists to better understand how traits are passed from one generation to the next, and has helped them explain patterns of inheritance that do not follow Mendelian principles. What are these patterns? How can they be explained by the properties of genes?

Section Review

Understanding Concepts

1. What is a mutation?
2. Describe how point mutations and frameshift mutations affect the synthesis of proteins.
3. Describe four kinds of chromosomal mutations.

Thinking Critically

4. Why do you think a low level of mutation might be advantageous to a species, while a high level of mutation might be disadvantageous?

Skill Review

5. **Interpreting Scientific Illustrations** Make a chart illustrating and summarizing the different kinds of gene and chromosomal mutations. For each kind of mutation, describe the mutation and discuss one possible effect of the mutation on an organism. For more help, refer to Thinking Critically in the *Skill Handbook.*

Answers to Section Review

1. Any mistake or change in the DNA sequence is a mutation.
2. Point mutations may change a single amino acid in a protein, whereas a frameshift mutation may alter every amino acid because the shift in bases results in different combinations all along the strand.
3. Insertions involve a part of a chromatid breaking off and attaching to its sister chromatid. Deletions occur when part of the chromosome is left out. Inversions occur when part of a chromosome breaks off and is reinserted backwards. Translocation occurs when part of one chromosome breaks off and is added to a different chromosome.

Thinking Critically

4. A change in the DNA might result in a

REVIEWING MAIN IDEAS

13.1 DNA: The Molecule of Heredity

- DNA, the genetic material of most organisms, is a large molecule composed of four different kinds of nucleotides. A DNA molecule consists of two strands of nucleotides with sugars and phosphates on the outside and bases paired by hydrogen bonding on the inside. The paired strands form a twisted-ladder shape called a double helix.
- Because adenine can pair with thymine and guanine can pair with cytosine, DNA can replicate itself with great accuracy. This process keeps the genetic information constant through cell division and during reproduction.

13.2 From DNA to Protein

- Genes are small sections of DNA code. Nearly every sequence of three bases in DNA codes for one amino acid in a protein, thus forming the genetic code.
- The order of nucleotides in DNA determines the order of nucleotides in messenger RNA in a process called transcription.
- Translation is a process through which the order of bases in messenger RNA codes for the order of amino acids in a protein.

13.3 Genetic Changes

- A mutation is a change in the DNA code. Mutations may affect only one gene, or they may affect whole chromosomes.
- Errors in disjunction result from the failure of homologous chromosomes to separate during meiosis. They result in gametes with too many or too few chromosomes.

Key Terms

Write a sentence that shows your understanding of each of the following terms.

chromosomal mutation	nitrogen base
codon	nondisjunction
double helix	point mutation
frameshift mutation	replication
messenger RNA	ribosomal RNA
monosomy	transcription
mutation	transfer RNA
	translation
	trisomy

Understanding Concepts

1. Which of the following is NOT a nitrogen base found in DNA?
   a. adenine c. cytosine
   b. guanine d. uracil
2. Nitrogen base pairs are held together by ____________ bonds.
   a. hydrogen c. ionic
   b. covalent d. nuclear
3. What nitrogen base is found in RNA but not in DNA?
   a. adenine c. cytosine
   b. guanine d. uracil
4. In each chain of nucleotides, sugars are joined to phosphate groups by __________ bonds.
   a. hydrogen c. ionic
   b. covalent d. nuclear
5. Which of these processes does NOT require DNA replication?
   a. mitosis c. meiosis
   b. cell division d. cell growth
6. In order to replicate, DNA molecules separate at their _____________.
   a. sugars c. nitrogen bases
   b. phosphate groups d. ends

better adaptation. But, a high rate of mutation could cause rapid speciation or might lead to extinction of a species. A low level of mutations provides stability to a species.

Skill Review

5. The table will summarize Figures 13.9 and 13.10.

Chapter 13 Review

Reviewing Main Ideas

Summary statements can be used by students to review the major concepts of the chapter.

Key Terms

Answers should go beyond defining the terms. Accept any answer that uses the term correctly and in the proper context.

Understanding Concepts

1. d
2. a
3. d
4. b
5. d
6. c

Chapter 13 Review

7. a
8. c
9. d
10. a
11. a
12. b
13. b
14. d
15. a
16. b
17. b
18. c
19. a
20. d

Applying Concepts

21. The deletion of one base may change all amino acids following the deletion, while a deletion of three adjoining bases might cause the deletion of one amino acid and perhaps a change in another in the protein.
22. A gene is a section of DNA that provides the information for the amino acid sequence in a protein chain.
23. A sequence of two nucleotides of four different kinds will produce 16 different combinations. The nucleotides must code for at least 20 amino acids.
24. Some point mutations result in coding for the same amino acid, such as AAA changing to AAG. Both code for phenylalanine.
25. Because of specific pairing (A to T and C to G), the DNA can separate and copy itself by bonding to complementary nucleotides.

Chapter 13 Review

7. DNA replication begins when a(n) ___________ breaks hydrogen bonds that hold the strands together.
   a. enzyme c. nucleic acid
   b. helix d. sugar
8. The order of bases in a molecule of mRNA is determined by the sequence of ___________ in DNA.
   a. sugars c. nitrogen bases
   b. phosphate groups d. covalent bonds
9. Each set of three nitrogen bases that represents an amino acid is a(n) ___________.
   a. code c. helix
   b. enzyme d. codon
10. How many strands of nucleotides does a molecule of RNA contain?
    a. one c. three
    b. two d. four
11. Of the following, which is NOT found in DNA?
    a. ribose c. deoxyribose
    b. thymine d. adenine
12. In the process of ___________, enzymes make an RNA copy of a DNA strand.
    a. translation c. transference
    b. transcription d. replication
13. The RNA copy that carries information from DNA in the nucleus into cytoplasm is ___________ RNA.
    a. ribosomal c. transfer
    b. messenger d. translation
14. DNA is copied every time a cell reproduces in the process called ___________.
    a. translation c. transference
    b. transcription d. replication
15. In the process of ___________, information on mRNA is used to make a sequence of amino acids into a protein.
    a. translation c. transference
    b. transcription d. replication
16. Which molecule brings amino acids to the ribosomes for assembling into proteins?
    a. mRNA c. rRNA
    b. tRNA d. DNA
17. The shape of DNA is described as ___________.
    a. a linear sequence of nucleotides
    b. a double helix
    c. a supercoiled peptide chain
    d. saturated or unsaturated
18. Which of the following base pairs would never be found in the cell?
    a. adenine:thymine
    b. cytosine:guanine
    c. thymine:uracil
    d. adenine:uracil
19. What does the DNA nucleotide sequence code for?
    a. proteins c. sugars
    b. phosphates d. nucleotides
20. Of the 64 triplet DNA codes, three code only for ___________.
    a. codons c. start
    b. amino acids d. stop

Applying Concepts

21. Which would cause a greater change in a protein, the deletion of one base in the DNA that coded for it, or the deletion of three sequential bases in the DNA that coded for it? Explain.
22. Describe a gene in terms of DNA.
23. Explain why the DNA code can't have sequences of two nucleotides instead of three nucleotides for each amino acid.
24. Many mutations result in changes in proteins. Give an example of a mutation in a gene that would result in no change in the protein for which the gene coded.
25. Watson and Crick, the Nobel prize winners for the structure of DNA, stated that "specific pairing . . . suggested a possible copying mechanism for the genetic material." Explain this statement.

Program Resources

Chapter Assessment, pp. 73-78 L1
Alternate Assessment in the Science Classroom
Computer Test Bank L1
Content Mastery, pp. 49-52 L1

Thinking Critically

26. ***Concept Mapping*** Make a concept map that relates the following terms and phrases. Supply the appropriate linking words for your map.
    DNA, nucleotide, mRNA, rRNA, tRNA, amino acid, protein, ribosome
27. ***Analyzing*** Identify the type of mutation illustrated in each of the following diagrams.

28. ***Making Inferences*** Suppose a DNA molecule began with the sequence ATCCTCGGA. What would happen when the resulting mRNA was translated? Why would the sequence be an unlikely one for DNA?
29. ***Interpreting Data*** The following is the sequence of bases on one strand of a DNA molecule.
    A A A T G C C A T C C G T C A
    (a) Write the sequence of bases that makes up the complementary DNA strand.
    (b) What base sequence in mRNA would the first DNA strand code for?
    (c) What sequence of amino acids would this mRNA code for?
30. ***Making Inferences*** Explain how the universality of the genetic code is evidence that all organisms alive today evolved from a common ancestor in the past.

ASSESSING KNOWLEDGE & SKILLS

The following graph records the amount of DNA in cells that have been grown in a cell culture in synchrony so that all the cells are at the same phase.

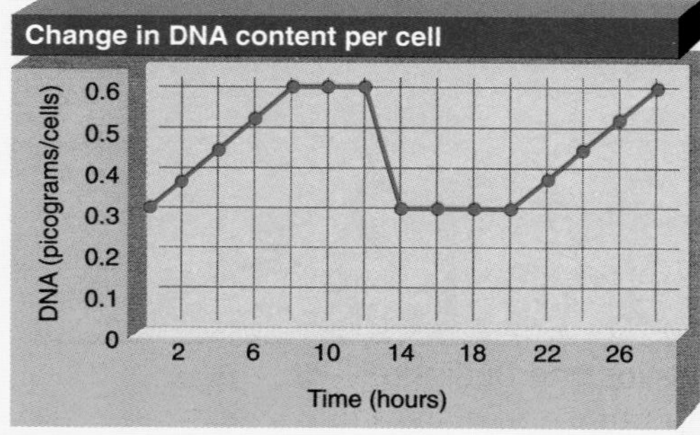

Interpreting Data Use the data in the graph to answer the following questions.

1. During the course of the experiment, these cells went through somatic cell division. What is this type of division called?
   a. cytosis **c.** mitosis
   b. cytokinesis **d.** meiosis
2. During which hours were the cells carrying out cell division?
   a. 0-8 **c.** 12-14
   b. 8-10 **d.** 14-20
3. Which phase of the cell cycle were the cells in during hours 8-10?
   a. interphase **c.** cell division
   b. DNA synthesis **d.** cytokinesis
4. If you added radioactive thymine to the culture at 0 hour, what would happen to the amount incorporated into the DNA between hours 20 and 28?
   a. stay the same **c.** double
   b. divide in half **d.** triple
5. ***Observing and Inferring*** Why was it important that the cells in this experiment be in synchrony?

Chapter 13 Review

Thinking Critically

26. See below.
27. (a) inversion; (b) deletion; (c) translocation
28. This would be an unlikely sequence because it codes for a stop codon first so translation would go no further.
29. (a) TTTACGGTAGG-CAGT
    (b) UUUACGGUAGG-CAGU
    (c) phenylalanine-threonine-valine-glycine-serine
30. All organisms' genetic codes are based on the same four nucleotides, suggesting a link to a common ancestor.

1. c
2. c
3. a
4. c
5. It is important for all of the cells to be in the same stage, whether interphase or cell division, so that the experiment can monitor what is happening in the overall culture while all cells are at the same phase of the cell cycle.

26.

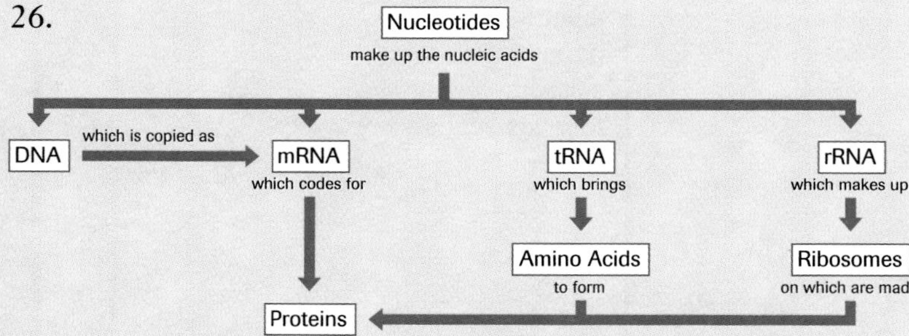

Chapter Organizer

SECTION	OBJECTIVES	ACTIVITIES/FEATURES
14.1 When Heredity Follows Different Rules National Science Standards: UCP.2; A.1, A.2; C.1, C.2; G.2, G.3	1. **Distinguish** between incompletely dominant and codominant alleles. 2. **Compare** multiple allelic inheritance and polygenic inheritance. 3. **Summarize** how internal and external environments affect gene expression.	**Thinking Lab:** How many phenotypes result from polygenic inheritance?, p. 339 **Biolab:** Design Your Own Experiment—What is the pattern of cytoplasmic inheritance?, p. 340 **Biology & Society Ethics:** Breeding Pedigreed Dogs, p. 342
14.2 Applied Genetics National Science Standards: UCP.2, UCP.3; C.2; G.1-3	4. **Interpret** testcrosses and pedigrees. 5. **Evaluate** the importance of plant and animal breeding to humans.	**Minilab:** How can you illustrate a pedigree?, p. 346 **Focus On** The Domestication of Cats, p. 348

ACTIVITY MATERIALS

BIOLAB	MINILAB	ALTERNATE LAB
page 340 *Brassica* rapa seeds, normal and variegated potting soil potting trays paintbrush forceps single-edge razor blade light source labels	**page 346** information on chosen pet animal	**page 336** petri dish label paper towels scissors tobacco seeds

Chapter 14 Patterns of Heredity

TEACHER CLASSROOM RESOURCES

Reproducible Masters	Transparencies
Section Focus Master 30: Complex Inheritance Patterns L1 SAE **Reinforcement and Study Guide,** pp. 53-54 L1 **Biolab and Minilab Worksheets,** pp. 55-56 L1 **Concept Mapping:** Patterns of Inheritance, p. 14 L1 **Critical Thinking/Problem Solving:** Mutations, p. 14 L3 **Laboratory Manual:** What Phenotype Ratio Is Seen in a Dihybrid Cross?, pp. 79-82 L2	
Section Focus Master 31: Selective Breeding L1 SAE **Reinforcement and Study Guide,** pp. 55-56 L1 **Biolab and Minilab Worksheets,** p. 54 L1 **Tech Prep Applications:** Selecting for Better Food, pp. 15-16 L2 **Content Mastery,** pp. 53-56 L1	**Basic Skills Transparency #13:** Genotypes and Phenotypes L1 SAE **Reteaching Transparency #14:** Testcross L1 SAE

ASSESSMENT MATERIALS

Chapter Assessment, pp. 79-84
Alternate Assessment in the Science Classroom
MindJogger Videoquiz
Computer Test Bank

Spanish Resources SAE
English/Spanish Audiocassettes SAE
Cooperative Learning in the Science Classroom COOP LEARN
Lesson Plans
Biology Projects: Genetics: The Secret of Life, pp. 13-16 L2

KEY TO TEACHING STRATEGIES

L1 Level 1 activities should be within the ability range of all students including those with learning difficulties.

L2 Level 2 activities are within the ability range of average to above-average students.

L3 Level 3 activities are designed for the ability range of above-average students.

SAE SAE activities should be within the ability range of Students Acquiring English.

COOP LEARN Cooperative Learning activities are designed for small group work.

P These strategies represent student products that can be placed into a best-work portfolio.

These strategies are useful in a block scheduling format.

GLENCOE TECHNOLOGY

The following multimedia resources are available from Glencoe.

Biology: The Dynamics of Life
CD-ROM SAE
Videodisc Program

The Infinite Voyage Series
The Dawn of Humankind
DNA Studies Create Controversy

Science and Technology Videodisc Series (STVS)
Plants & Simple Organisms
Disease-Resistant Tomatoes
Selective Breeding in Cows
New Grains
Salt-Resistant Crops
Breeding Fruit Flies
Microinjecting Polygenes

CHAPTER

14 Patterns of Heredity

Chapter Overview

As students begin their study of patterns of heredity that do not seem to fit Mendel's laws, they analyze incomplete dominance. Other complex hereditary patterns, such as codominance and inheritance through multiple alleles, are also explored. After the difference between autosomes and sex chromosomes is established, the inheritance of sex-linked traits is explained. Next, students learn about polygenic inheritance. The section ends with a discussion of how the external and internal environment of an organism can influence genetic traits.

In the second section, students explore applied genetics. They learn about the techniques of the testcross and the pedigree. Selective breeding programs are then discussed, with emphasis on achieving desirable traits through inbreeding. The chapter closes with examples of familiar animal breeds and plant cultivars developed through inbreeding.

Key Terms

- autosome
- carrier
- codominant alleles
- hybrid
- inbreeding
- incomplete dominance
- multiple alleles
- pedigree
- polygenic inheritance
- sex chromosome
- sex-linked trait
- testcross

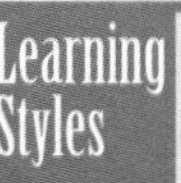

Look for the following logo for strategies that emphasize different learning modalities. LS

Kinesthetic	**Meeting Individual Needs, p. 338**
Visual-Spatial	**Visual Learning, p. 335; Chalkboard Example, p. 336; Activity, p. 344; Portfolio, p. 346**
Interpersonal	**Alternate Lab, p. 336**
Intrapersonal	**Minilab, p. 346**
Linguistic	**Student Journal, p. 347**
Logical-Mathematical	**Student Journal, pp. 335, 345; Portfolio, p. 335; Thinking Lab, p. 339; Meeting Individual Needs, p. 339; Project, p. 342**

LS

CHAPTER

14 Patterns of Heredity

Do you like to put puzzles together? Geneticists often must feel that they are solving puzzles. They need to explain how unseen factors (genes) cause visible effects (traits). After Mendel laid the groundwork explaining how traits pass from generation to generation, many of the questions that scientists had been asking could be answered. But new mysteries arose. Soon, what appeared to be exceptions to Mendel's laws were found. For instance, moss roses (*Portulaca grandiflora*) have flowers of different colors—pink, white, yellow, red, purple—all in the same species! Simple dominance and recessiveness of two alleles for flower color could not explain this trait.

Concept Check

You may wish to review the following concepts before studying this chapter.

- Chapter 12: rule of dominance, law of segregation, law of independent assortment

Chapter Preview

Laboratory Activities

Biolab: Design Your Own Experiment

- What is the pattern of cytoplasmic inheritance?

Minilabs

- How can you illustrate a pedigree?

Geneticists have extended Mendel's laws to explain unusual patterns of inheritance. Because of Mendel's work, it has been easier for others to solve new mysteries such as the inheritance of flower color in moss roses, thus fitting additional pieces into the puzzle of inheritance.

Assessment Planner

Choose assessment strategies from the following pages to evaluate the progress of your students.

Assess, pp. 342, 350
Alternate Lab, pp. 336–337
Minilab, p. 346
Portfolio, pp. 335, 346
Thinking Lab, p. 339
Biolab, pp. 340–341
Chapter Review, pp. 351–353

CHAPTER 14

Introducing the Chapter

Have students name several traits other than flower color that are apparent in moss roses. *flower petal shape or number, leaf shape, plant height, leaf color* Ask them to explain why inheritance of flower color may not be a matter of simple dominance and recessiveness. If students are having difficulties, ask them how many different phenotypes can result with a trait that is controlled by a single pair of alleles. *two* How might many colors be inherited? *More than one pair of genes, or some other mechanism, determines the colors.* L1

Theme Development

The theme of *evolution* is alluded to as students are introduced to selective breeding techniques that achieve new and different traits within offspring. The *nature of science* is illustrated with the concepts developed by Morgan as he worked with and interpreted data from sex-linked traits.

Concept Check

In Chapter 12, students learned the Mendelian rule of dominance, law of segregation, and law of independent assortment. Building on their understanding of these principles, students now study more complex patterns of inheritance. Students also learn how to use a Punnett square for the analysis of problems in applied genetics such as the interpretation of testcross and pedigree data.

SECTION

14.1

Prepare

Key Concepts

Students are shown the difference between codominance and incomplete dominance and are given examples of multiple-allelic traits, sex-linked traits, and polygenic inheritance. The section ends with a brief description of how environmental factors can affect the appearance of certain traits.

Block Scheduling

Look for this symbol for strategies that are useful in a block scheduling format. For more information on block scheduling, refer to Section 14.1 in the **Lesson Plans** booklet.

Materials

- Purchase seeds of *Brassica rapa* for the Biolab. Gather small pots, soil, and lights.
- Obtain photos of a red shorthorn bull, a white shorthorn cow, and a roan shorthorn cow for the Discussion.

1 Focus

Bellringer

Before presenting the lesson, display **Section Focus Master 30** on the overhead projector and have students answer the accompanying questions. L1 SAE

SECTION 14.1

When Heredity Follows Different Rules

Section Preview

Objectives

Distinguish between incompletely dominant and codominant alleles.

Compare multiple allelic inheritance and polygenic inheritance.

Summarize how internal and external environments affect gene expression.

Key Terms

incomplete dominance
codominant alleles
multiple alleles
autosome
sex chromosome
sex-linked trait
polygenic inheritance

Variations in the patterns of inheritance explained by Mendel became known soon after his work was rediscovered. What do geneticists do when observed patterns of inheritance do not appear to follow Mendel's laws? They often employ a strategy of piecing together bits of a puzzle until the basis for the unfamiliar inheritance pattern is understood.

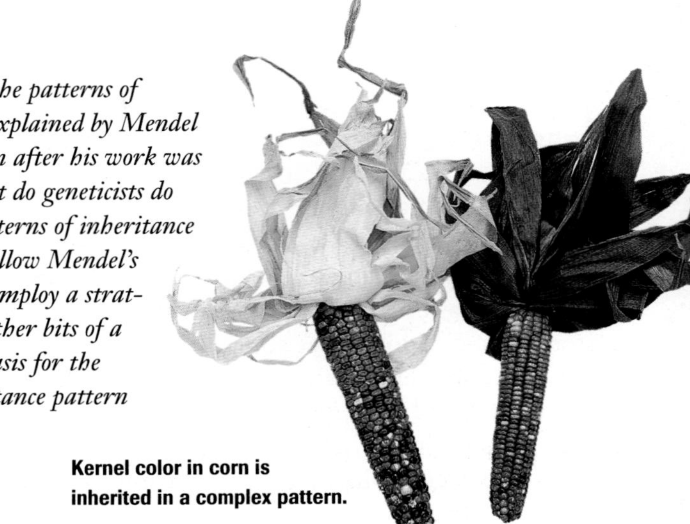

Kernel color in corn is inherited in a complex pattern.

Complex Patterns of Inheritance

Patterns of inheritance that are explained by Mendel's experiments are often referred to as simple Mendelian inheritance—the inheritance controlled by dominant and recessive paired alleles. However, many inheritance patterns are more complicated than those in garden peas studied by Mendel. As you will learn, most alleles are not simply dominant or recessive.

Incomplete dominance–Appearance of a third phenotype

When inheritance follows a pattern of complete dominance, heterozygous individuals and homozygous dominant individuals have the same phenotype. When traits are inherited in an **incomplete dominance** pattern, the phenotype of the heterozygote is intermediate between those of the two homozygotes. For example, if a homozygous red-flowered snapdragon plant is crossed with a homozygous white-flowered snapdragon plant, all of the F_1 offspring will have pink flowers, as shown in *Figure 14.1.* This intermediate form of the trait occurs because neither allele of the pair is completely dominant. Both alleles of the gene produce products, which combine to give a new trait.

Note that the segregation of alleles is the same as in simple Mendelian inheritance observed in garden peas. However, because neither allele is dominant, the plants of the F_1 generation all have pink flowers. When pink-flowered F_1 plants are crossed with each other, the offspring in the F_2 generation appear in a 1:2:1 phenotypic ratio of red to pink to white flowers. This result supports Mendel's law of independent assortment.

Program Resources

Section Focus Master 30 L1 SAE
Reinforcement and Study Guide, pp. 53-54 L1
Biolab and Minilab Worksheets, pp. 55-56 L1
Laboratory Manual, pp. 79-82 L2
Concept Mapping, p. 14 L1
Critical Thinking/Problem Solving, p. 14 L3

Meeting Individual Needs

Learning Disabled Show students why it is always possible to predict the genotype for flower color in snapdragon plants even though the parent genotypes are not known. Have them explain why this does not work when using a trait that shows complete dominance. L1

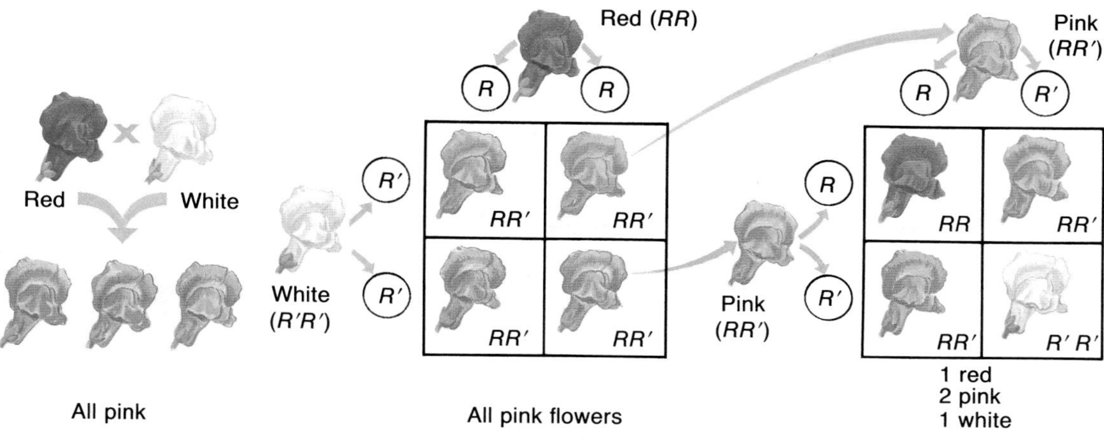

Figure 14.1

In snapdragons, the combined expression of both alleles for flower color produces a new phenotype—pink—illustrating incomplete dominance. For this reason, the letters *R* and *R′*, rather than *R* and *r*, are used here to indicate incomplete dominance. A Punnett square shows that the red snapdragon is homozygous for the allele *R*, and the white snapdragon is homozygous for the allele *R′*. All of the pink snapdragons are heterozygous, or *RR′*.

Codominance—Expression of both alleles

In chickens, black-feathered and white-feathered birds are homozygotes for the *B* and *W* alleles, respectively. Two different uppercase letters are used to represent the alleles in codominant inheritance. You might expect that heterozygous chickens, *BW*, would be black if the pattern of inheritance followed Mendel's law of dominance, or gray if the trait were incompletely dominant.

One of the resulting heterozygous offspring in a breeding experiment between a black rooster and a white hen is shown in *Figure 14.2.* Notice that the heterozygote is neither black nor gray. Instead, all of the offspring are checkered; some feathers are black and other feathers are white. In such situations, the inheritance pattern is said to be codominant. **Codominant alleles** cause the phenotypes of both homozygotes to be produced in heterozygote individuals. In codominance, both alleles are expressed equally.

Multiple phenotypes from multiple alleles

Although each trait has only two alleles in the patterns of heredity you have studied thus far, it is common for more than two alleles to control a trait in a population. This is understandable when you recall that a new

Figure 14.2

When a certain variety of black chicken is crossed with a white chicken, all of the offspring are checkered, black and white, as a result of both feather colors being produced by codominant alleles.

SECTION 14.1

Discussion

Exhibit photos of a red shorthorn bull, a white shorthorn cow, and a roan shorthorn cow. Tell students that the roan cow, offspring of the red bull and the white cow, has both red and white hairs. Ask students if Mendel's law of dominance applies to the inheritance of coat color in cattle. Why or why not? *No, because a third phenotype appears when the red and white cattle are crossed, and traits that Mendel studied had only two phenotypes.* Ask students what type of inheritance pattern explains how roan cattle express their coat color. *codominance*

2 Teach

Visual Learning

LS **Visual-Spatial** Have students examine Figures 14.1 and 14.2 and read the captions. Then have students draw Punnett squares to illustrate the inheritance of coat color in shorthorn cattle. L1

History Connection

Soviet Genetics

In the 1930s, Trofim Denisovich Lysenko became head of the Institute of Genetics of the Soviet Academy of Sciences. Lysenko rejected Mendelian genetics as a "capitalist" doctrine in favor of the idea of the inheritance of acquired characteristics. While Western agriculture was benefitting from selective breeding and hybridization, the results of the application of Lysenko's ideas were disastrous for Soviet agriculture. In 1965, after the death of Stalin, Lysenko was removed from office and Soviet biological thought returned to the international mainstream.

STUDENT JOURNAL

Codominance Provide additional practice with codominance by having students determine the phenotypes of offspring resulting from the following crosses: (a) a checkered rooster mated to a checkered hen, (b) a checkered rooster mated to a white hen, (c) a checkered rooster mated to a black hen. L1 LS

PORTFOLIO

Comparing Patterns of Inheritance Have students show through a series of Punnett squares how the offspring genotype and phenotype ratios would differ if the trait for chicken feather color was inherited through a pattern of regular dominance as compared to the actual pattern of codominance. L1 LS P

Visual Learning

Figure 14.4 Why are the X and Y chromosomes not homologous? *They do not have a matching size or shape.*

Chalkboard Example

Visual-Spatial Draw representative human chromosomes on the board. Draw them in pairs (homologues) and show differing lengths for nonhomologues. Number the pairs from 1 to 23. Place and label a bracket over the autosomes (pairs 1-22). Place a bracket and label over the sex chromosomes.

Reinforcement

Draw a Punnett square. Place sex chromosomes XX along the side and XY along the top. Ask students why there is only one sex chromosome per gamete cell along the side and top. *The chromosomes are separated during meiosis.*

GLENCOE TECHNOLOGY

Videotape

The Secret of Life
Use the videotape *Sex and the Single Gene: Cell Development* to illustrate the developmental processes that determine an individual's sex.

Figure 14.3

In rabbits, a single gene that controls coat color has multiple alleles. An enzyme that activates the production of a pigment is controlled by the *C* allele. This enzyme is lacking in *cc* rabbits.

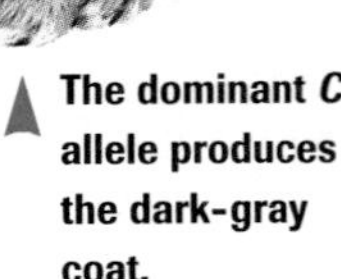

The dominant *C* allele produces the dark-gray coat.

The c^{ch} allele results in a light-gray coat called chinchilla and is dominant to c^h and *c*.

The allele c^h produces a white coat with black points, a Himalayan, and is dominant to *c*. In c^hc^h rabbits, the enzyme works only in cooler regions of the body—the ears, the feet, and the area around the nose.

The *c* allele produces a white coat.

allele could be formed any time a mutation occurs in a nitrogen base somewhere within a gene. How can there be several different types of blood among humans? How can fruit flies have so many different eye colors? These questions can be answered because more than two alleles of a gene can exist.

Traits controlled by more than two alleles are said to have **multiple alleles.** Recall that a diploid individual can possess only two alleles of each gene. Therefore, multiple alleles can be studied only in populations. The rabbits in *Figure 14.3* show the effects of multiple alleles for coat color. Four alleles of a single gene govern coat color in rabbits, although each rabbit can have only two of these alleles. The number of alleles for any particular trait is not limited to four, and there are instances in which more than 100 alleles are known to exist for a single trait!

Figure 14.4

The sex chromosomes are named for the letters they resemble. Why are the X and Y chromosomes not homologous?

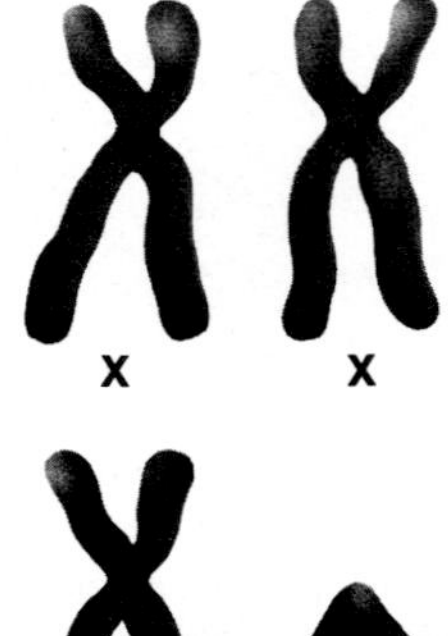

Sex determination

Recall that in humans, the diploid number of chromosomes is 46, or 23 pairs. There are 22 pairs of matching homologous chromosomes called **autosomes.** Homologous autosomes look exactly alike. The 23rd pair of chromosomes differs in males and females. These two chromosomes, which determine the sex of an individual, are called **sex chromosomes.** In humans, the chromosomes that control the inheritance of sex characteristics are indicated by the letters X and Y. If you are a human female, XX, your 23rd pair of chromosomes look alike, as shown in *Figure 14.4.* However, if you are a male, XY, your 23rd pair of chromosomes look different. Males, which have one X and one Y chromosome, produce two kinds of gametes, X and Y, by meiosis. Females have two X chromosomes and produce only X gametes. *Figure 14.5* shows that after fertiliza-

Alternate Lab: Incomplete Dominance

Purpose

Interpersonal To illustrate the phenotype ratio that appears with incomplete dominance.

Materials

petri dish, label, water, paper toweling, scissors, tobacco seeds

Procedure

Give the following directions to students.

1. Label the top of a petri dish with your name and the date. Place several layers of paper toweling inside the dish.
2. Moisten the toweling and place 20 tobacco seeds on it.
3. Cover the dish and place it where it will receive light.
4. Check the seeds for the next 10 days. Keep the toweling moist but not soaked. After 8–10 days, count the number of plants with green, yellow-green, and yellow leaves.
5. Design a data table to record your results and class totals.

Analysis

Ask students the following questions.

tion there will always be a one-to-one ratio of males to females.

Sex-linked inheritance in fruit flies

In 1910, Thomas Hunt Morgan discovered traits linked to sex chromosomes in *Drosophila* (druh SAHF uh luh), commonly known as fruit flies. Traits controlled by genes located on sex chromosomes are called **sex-linked traits.** Fruit flies usually have red eyes. Morgan noticed one day that one male had white eyes. The genotypes for eye color are shown in *Figure 14.6.* Alleles for sex-linked traits are written as superscripts of the X or Y chromosome. As you can see in the Punnett squares, the Y chromosome has no allele for the white eye color trait, so no superscript is used. Also remember that X and Y chromosomes are nonhomologous, so any allele on the X chromosome of a male will not be masked by a matching allele on the Y chromosome.

Morgan crossed the white-eyed male with a red-eyed female. All of the F_1 offspring had red eyes, indicating that the white-eyed trait is recessive. Then Morgan allowed the F_1 flies to mate among themselves. According to simple Mendelian inheritance, if the trait were recessive, the offspring in the F_2 generation would show a 3 to 1 ratio of red-eyed to white-eyed flies. As you can see in *Figure 14.6,* this is what Morgan observed. However, Morgan also noticed that the trait for white eyes was inherited only by males. Morgan concluded that because the original male was white-eyed and the trait is recessive, the red-eyed males must be heterozygous, and the dominant trait comes from the female parent. As you can see in the Punnett squares, this is the case; the dominant allele can only come in association with the X chromosome.

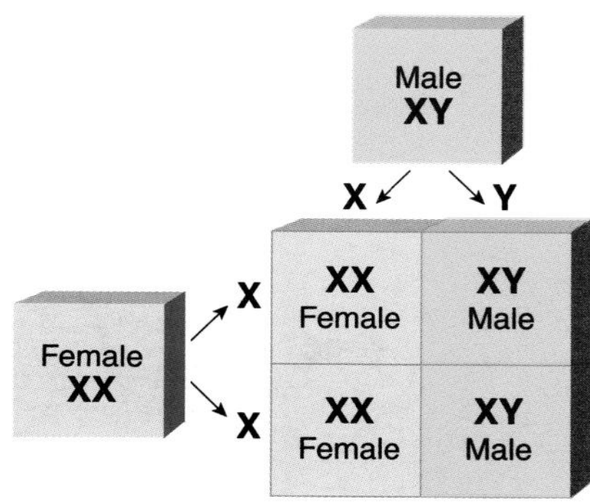

Figure 14.5

Half the offspring of any mating between humans will always have the XX genotype, which are females, and half the offspring will have the genotype XY, which are all males.

Figure 14.6

Morgan crossed a white-eyed male fruit fly with a normal red-eyed female (left). He then allowed the F_1 flies to mate (right). The superscripts R and r are the dominant and recessive alleles for eye color in fruit flies.

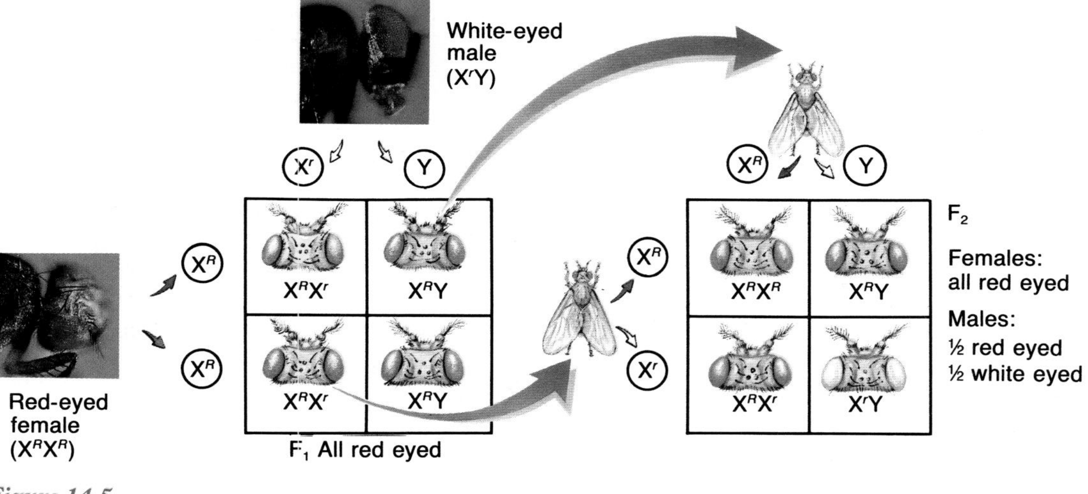

Figure 14.5

Misconception

Students may believe that all the alleles present on the X and Y chromosomes are related to maleness or femaleness. This is not the case, however. Explain that alleles for blood clotting and color vision are located on human sex chromosomes. These traits have little to do with being male or female.

Visual Learning

Figure 14.5 Make it clear to students that there is a 50 percent chance that each offspring will receive XX chromosomes, which is female, and a 50 percent chance that each offspring will receive XY chromosomes, which is male.

1. What ratio of phenotypes do your results show? Is this ratio close to the expected ratio of 1:2:1 with incomplete dominance? *ratio of phenotypes will vary; no*
2. Why might your data not be close to a 1:2:1 ratio? *small sample size*
3. What ratio of phenotypes do you get when using class totals? Why is this ratio closer to what was expected? *close to 1:2:1; large sample size*

Assessment

Knowledge: Have students assign letters to represent the genotypes of the parents and offspring. Then have them draw a Punnett square to show the expected results of a cross between plants with green leaves and plants with yellow leaves. L1 SAE

GLENCOE TECHNOLOGY

Videodisc

Biology: The Dynamics of Life
Disc 1, Side 1
Fruit Fly Genetics (Ch. 34)

Disc 1, Side 1
Sex-linked Traits (Ch. 35)

STVS: Plants & Simple Organisms
Disc 4, Side 1
Microinjecting Polygenes (Ch. 17)

Disc 4, Side 1
Breeding Fruit Flies (Ch. 21)

Visual Learning

Figures 14.6, 14.7 Use the illustrations to aid in the discussion of sex-linked traits. How did Morgan know that the gene could not be located on an autosome? *If the gene for white eye color were on an autosome, the trait would be distributed in the usual ratios, without regard to sex.*

As Morgan continued with his experiments, he eventually produced white-eyed females. When he crossed a white-eyed female with a red-eyed male, the reverse order of his original cross, all the female offspring had red eyes and all the male offspring had white eyes—quite a different inheritance pattern from his first cross, as you can see in *Figure 14.7*. Morgan concluded that the inheritance of the white-eyed trait is linked to the inheritance of the X chromosome.

In heterozygous females, the dominant allele for red eyes masks the recessive allele for white eyes. In males, however, a single recessive allele is expressed as a white-eyed phenotype. When Morgan crossed a red-eyed female with a white-eyed male, half of all the males and females inherited white eyes, as shown in *Figure 14.7*. The only explanation of these results is that alleles for eye color are carried on the X chromosome, and that the Y chromosome has no alleles for eye color. Because females have two X chromosomes, they must carry two alleles for eye color. Males carry only one allele for eye color because they have only one X chromosome, and the Y chromosome has no alleles for eye color. Traits dependent on genes that follow the inheritance pattern of the X chromosome are known as sex-linked traits. Eye color in fruit flies is an example of an X-linked trait.

Polygenic inheritance

Some traits, such as skin color and height in humans and cob length in corn, vary over a wide range. Such ranges occur because these traits are governed by many different genes. **Polygenic inheritance** is the inheritance pattern of a trait that is controlled by two or more genes. The genes may be on the same chromosome or on different chromosomes, and each gene may have two or more alleles. For simplicity, uppercase and lowercase letters are used to represent the alleles, as they are in Mendelian inheritance. Keep in mind, however, that the allele represented by an uppercase letter is not dominant. All heterozygotes are intermediate in phenotype.

In polygenic inheritance, each allele represented by an uppercase letter contributes a small, but equal, portion to the trait being expressed. The result is that the phenotypes usually show a continuous range of variability from the minimum value of the trait to the maximum value.

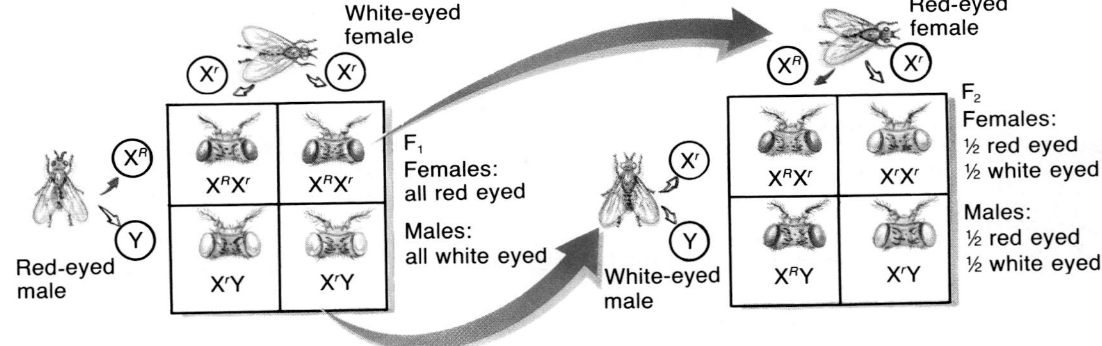

Figure 14.7
If a white-eyed female fruit fly is mated with a red-eyed male (left), the female offspring all have red eyes, and the males all have white eyes. Note the sex-linked inheritance patterns of the offspring in both the F_1 and F_2 generations.

interNET CONNECTION

Follow the link for this chapter on the Glencoe Homepage at **www.glencoe.com/sec/science** to find out more about patterns of heredity.

Meeting Individual Needs

Visually Impaired Provide large cutouts of leaves of different sizes. Have students group leaves according to size, then count the number of leaves in each pile. A partner can record the data for the visually impaired student. L1 LS

Suppose, for example, that stem length in a plant is controlled by three different genes: *A*, *B*, and *C*. Each gene is on a different chromosome and has two alleles, which are represented by an uppercase letter and a lowercase letter. Thus, each diploid plant has a total of six alleles for stem length. A plant that is homozygous for short alleles (*aabbcc*) at all three gene locations might grow to be only 4 cm tall. A plant that is homozygous for tall alleles (*AABBCC*) at all three gene locations would be 16 cm tall. The difference between the tallest possible plant and the shortest possible plant is 12 cm, or 2 cm per tall allele. You could say that each allele represented by an uppercase letter contributes 2 cm to the total height of the plant.

Suppose a 16-cm-tall plant were crossed with a 4-cm-tall plant. In the F_1 generation, all the offspring would be *AaBbCc*—intermediate (10 cm tall) in height. If they are allowed to interbreed, the F_2 offspring will show a broad range of heights. A Punnett square of this cross would show that 10-cm-tall plants are most often expected, and the tallest and shortest plants are seldom expected. Notice in *Figure 14.8* that, when these results are graphed, the shape of the graph confirms the prediction of the Punnett square.

ThinkingLab Interpret the Data

How many phenotypes result from polygenic inheritance?

Polygenic inheritance can be observed quantitatively, and it is possible to determine the number of gene pairs governing the trait if the number of genes is three or fewer. Once the number of gene pairs increases to four or more, the trait becomes continuous.

Analysis

The graphs below illustrate the number of possible phenotypes in an F_2 generation for one, two, and three pairs of genes that govern grain color in wheat. Colors range from red to white.

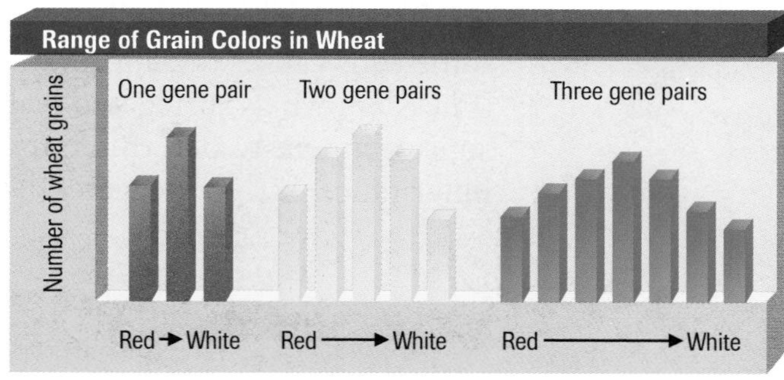

Thinking Critically

Analyze the graphs and calculate the number of phenotypes that would result from four gene pairs. Hint: It may be helpful to use letters to represent the genes and combinations.

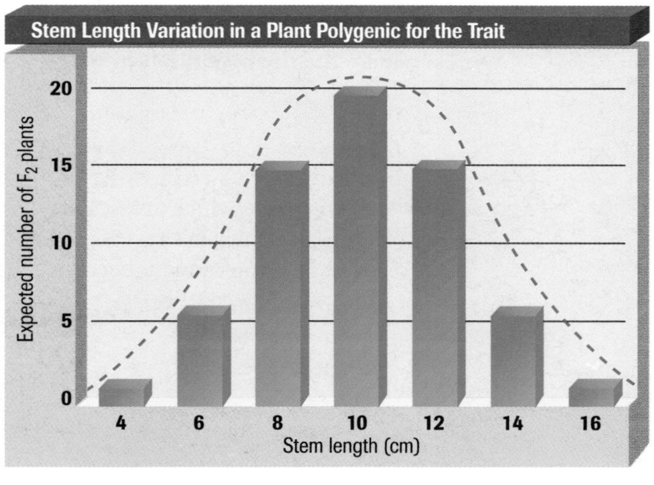

Figure 14.8

Polygenic inheritance occurs when many genes interact to produce a single trait. In this theoretical example, three genes each have two alleles that contribute to the trait. Each dominant allele contributes 2 cm to the height of the plant. When the distribution of plant heights is graphed, a bell-shaped curve is formed. Intermediate heights occur most often.

ThinkingLab Interpret the Data

Purpose

LS **Logical-Mathematical** Students will analyze graphs to calculate the number of phenotypes that would result in cases of polygenic inheritance.

Process Skills

make and use graphs, recognize cause and effect

Teaching Strategies

- Have students work in cooperative groups. Match students with excellent math skills with those working below grade level.
- Students can prepare Punnett squares for the one-gene, two-gene, and three-gene problems. Have them calculate the number of phenotypes that result to verify the graph data. L2
- Suggest to students that they may want to attempt to find a mathematical formula for determining the number of phenotypes that can result in cases of polygenic inheritance. L3

Thinking Critically

The formula $2n + 1$ will give the number of possible phenotypes. Let n = the number of gene pairs. Example: for four gene pairs, $2n + 1 = 2(4) + 1 = 9$.

Assessment

Skill: Ask students to calculate the number of phenotypes possible when five gene pairs are operating for a trait. L2

Meeting Individual Needs

Gifted Have students illustrate the cross between two plants that are each heterozygous for three genes that control a single trait, using a Punnett square that consists of 64 squares. Have students determine the gametes that appear along the sides and top of the square, using *AaBbCc* as the gene symbols for each parent. L3 LS

BioLab Design Your Own Experiment

What is the pattern of cytoplasmic inheritance?

Time Allotment

Initial: one class period for planting of P_1 generation; Daily: 5 minutes for watering; one class period for cross-pollination; 10 days later: one class period for seed collection and planting of F_1 seeds; 10 days later: one class period for examination of F_2 plants.

Objectives

Review objectives with students before they begin the Biolab.

Process Skills

form a hypothesis, observe and infer, collect and record data

Safety Precautions

Have students use caution when handling, using, and plugging in light fixtures or banks.

PREPARATION

Seeds of normal and variegated *Brassica rapa (Wisconsin Fast Plants)* can be ordered from biological supply houses. The variegated gene is carried on chloroplast DNA.

Possible Hypotheses

- If the trait is inherited through the cytoplasm, then the female parent controls and contributes the trait.
- If the trait is inherited through the cytoplasm, then the male parent controls and contributes the trait.

BioLab Design Your Own Experiment

What is the pattern of cytoplasmic inheritance?

Mitochondria and chloroplasts contain DNA. This DNA is not coiled into structures called chromosomes, but it still carries genes that control genetic traits. Because mitochondria are the site of aerobic respiration, many of the mitochondrial genes control steps in the respiration process.

The DNA in chloroplasts controls traits such as chlorophyll production. Lack of chlorophyll in some cells causes the appearance of white patches in a leaf. This trait is known as variegated leaf. In this Biolab, you will carry out an experiment to determine the pattern of this cytoplasmic inheritance of the variegated leaf trait in *Brassica rapa*.

PREPARATION

Problem

What inheritance pattern does the variegated leaf trait in *Brassica* show?

Hypotheses

Consider the possible evidence you could collect that would answer the problem question. Among the people in your group, form a hypothesis that you can test to answer the question, and write the hypothesis in your journal.

Objectives

In this Biolab, you will:

- **Determine** which crosses of *Brassica* plants will reveal the pattern of cytoplasmic inheritance.
- **Analyze** data from *Brassica* plant crosses.

Possible Materials

Brassica rapa seeds, normal and variegated
potting soil
potting trays
paintbrushes
forceps
single-edge razor blade
light source
labels

Safety Precautions

Handle the razor blade with extreme caution. Always cut away from you. Be sure to wash your hands carefully after handling the plants. Do not eat the seeds of any plants.

340 Patterns of Heredity

PLAN THE EXPERIMENT

Teaching Strategies

- Variegated plants tend to grow a little slower than the nonvariegated. Therefore, these seeds should be started about 4 days earlier.
- *Brassica* will not self-pollinate. Therefore, keep the two plant types separate from one another to avoid random cross-pollination, or have students remove those flowers that were not used in cross-pollination.
- It is critical to provide light in the form of fluorescent banks (cool-white, 40 watts/bulb) in order to achieve the complete life cycle in such a short time. Light banks should be adjustable so that they remain about 5–8 cm above the plants' growing tips at all times and are to remain on continuously for 24 hours each day.
- Students should work in cooperative groups.

PLAN THE EXPERIMENT

1. Decide which crosses will be needed to test your hypothesis.
2. Keep the available materials in mind as you plan your procedure. How many seeds will you need?
3. Record your procedure, and list the materials and quantities you will need.
4. Assign a task to each member of the group. One person should write data in a journal, another can pollinate the flowers, while a third can set up the plant trays. Determine who will set up and clean up materials.
5. Design and construct a data table for recording your observations.

Check the Plan

Discuss the following points with other group members to decide the final procedure for your experiment.

1. What data will you collect, and how will they be recorded?
2. When will you pollinate the flowers? How many flowers will you pollinate?
3. How will you transfer pollen from one flower to another?
4. How and when will you collect the seeds that result from your crosses?
5. What variables will have to be controlled? What controls will be used?
6. When will you end the experiment?
7. ***Make sure your teacher has approved your experimental plan before you proceed further.***
8. Carry out your experiment.

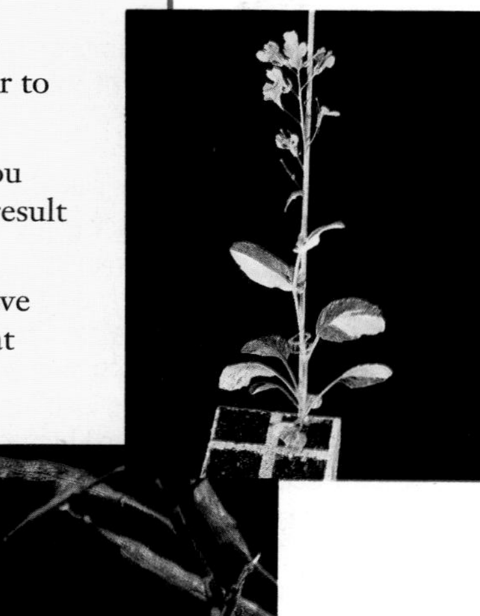

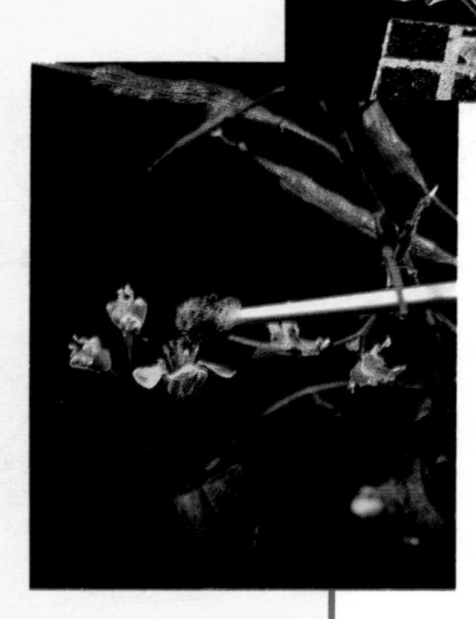

ANALYZE AND CONCLUDE

1. **Checking Your Hypothesis** Did your data support your hypothesis? Why or why not?
2. **Interpreting Observations** What is the inheritance pattern of variegated leaves in *Brassica?*
3. **Making Inferences** Explain why genes in the chloroplast are inherited in this pattern.
4. **Drawing Conclusions** Which parent is responsible for passing the variegated trait to its offspring?
5. **Making Scientific Illustrations** Draw a diagram tracing the inheritance of this trait through cell division.

Going Further

Application Make crosses between normal *Brassica* plants and genetically dwarfed, mutant *Brassica* plants to determine the inheritance pattern of the dwarf mutation.

ANALYZE AND CONCLUDE

1. Student answers will vary depending on their original hypotheses.
2. Variegation is inherited as a cytoplasmic trait from the female parent .
3. Sperm cells contain little to no cytoplasm. Cytoplasmic material containing chloroplasts is contributed by the egg cells.
4. Female
5. Diagrams should show the trait being transmitted in an egg cell of the female but not in pollen of the male.

Assessment

Skill: Have students use Punnett squares to explain what the inheritance pattern would be in offspring if this trait were inherited as a simple dominant allele. Have them show how cytoplasmic inheritance deviates from this pattern. L1

Going Further

Have students carry out crosses between nonvariegated male and female parents as well as variegated male and female parents. L1

Possible Procedures

- Both variegated and normal seed types should be planted and grown. After 13 to 15 days, the plants will flower. Students will then have to perform the following cross-pollinations depending on their hypotheses: variegated female with nonvariegated male (transfer of pollen from nonvariegated male anther via brush to pistil of variegated female) and nonvariegated female with variegated male. Pods of seeds will mature between days 28 and 30. These seeds will then be planted, and new offspring will be observed for the presence of the variegated trait.
- Soil must be kept constantly moist, especially during the seed germination period.

Data and Observations

Variegated F_1 plants will appear in only the crosses where the female parent was variegated.

Ethics
BIOLOGY & SOCIETY

Breeding Pedigreed Dogs

Purpose

This feature explains some of the reasons why people breed dogs and some of the problems that have resulted from artificial selection.

Teaching Strategies

- Review with students the meaning of terms such as *species*, *instinctive behavior*, *grandam*, and *grandsire*.
- Have students explain why any breed of dog is capable of mating and producing offspring with any other breed.

INVESTIGATING the Issue

Analyzing Consequences Do not use the same parent combination for future litters and/or do not use the offspring with genetic problems for any future breeding.

Going Further

Have students prepare a poster that traces the specific lineage of a particular dog breed, including its original function and any predisposition to genetic disease.

3 Assess

Check for Understanding

Have students explain how the following differ from one another: (a) incomplete dominance and codominance, (b) multiple alleles and polygenic inheritance, (c) autosomes and sex chromosomes, (d) sex-linked trait and autosomal trait. L1

Ethics
BIOLOGY & SOCIETY

Breeding Pedigreed Dogs

The association between dogs and humans dates back 14 000 years. Today, domestic dogs number 57 million in more than 30 million American homes. Modern dog breeds are the result of decades of selective breeding for appearance and behavior.

Selective breeding Dogs have been bred for a variety of purposes. Beagles were bred to track game because of their highly developed sense of smell. Siberian huskies, strong dogs with great stamina and tolerance for bitterly cold weather, were bred to pull sleds across the frozen Arctic. Many breeds, such as poodles, have been developed as companions to humans.

The goal of selective breeding is to firmly establish the desired genes or gene combinations in the homozygous state. Taken to its extreme form, selective breeding of closely related individuals quickly eliminates the chances of unwanted variables. Such inbreeding systems are the most reliable methods of ensuring the worth of a breeding line. The obvious drawbacks of inbreeding are the possibility of harmful, homozygous recessive gene pairs that are likely to cause genetic defects in the offspring.

Different Viewpoints

Unnatural selection Selective breeding may assist the breeder in establishing a winning pedigree line, but the results are not always in the best interest of a breed. The English bulldog has been bred for decades to an unnatural standard. This has made the dog dependent on humans for its most basic functions. With its massive skull size and low, narrow hips, it must be assisted in mating and in giving birth.

Responsible breeders try to balance the positive aspects of their trade with the negative aspects. As they develop new pure breeds that are uniquely fitted to their purpose, they try to avoid as far as possible the appearance of recessive traits.

INVESTIGATING the Issue

Analyzing Consequences How can a breeder prevent genetic problems from appearing in future generations of puppies?

Environmental Influences

Even when you have solved the puzzles of dominance and recessiveness and you understand the other patterns of heredity, the inheritance picture is not complete. The genetic makeup of an organism at fertilization determines only the organism's potential to develop and function. As the organism develops, many factors can influence how the gene is expressed, or even whether the gene is expressed at all. Two such influences are the organism's internal and external environments.

Influence of internal environment

The age or gender of an organism can affect gene function. The nature of such patterns is not well understood; however, it is known that the internal environment of an organism changes with age.

Also, the internal environments of males and females are different because of hormones and structural differences. For example, some traits of animals, such as colors of feathers in some birds, are expressed differently in the sexes. The plumage of the peacock is highly decorated and colored compared with that of the peahen. The horns of a ram are much heavier and more coiled than those of a ewe. These differences between the sexes are controlled by different hormones, which are determined by different sets of genes. In addition, each species inherits unique behavior patterns that are determined by its genes.

PROJECT
Genes and the Environment

Germinate mustard seeds (available from the condiment section in a grocery store) in petri dishes. Line the bottom of the dish with paper toweling. Soak the toweling with water and add 20 seeds. Cover the dish and place in the dark. Examine the seedlings after about 7 days. All seedlings will be white. Remove the dishes from the dark and place them in bright light for 24 hours, and then note their color. Have students write a report in which they explain how this project illustrates the role of environment in influencing gene expression. L1

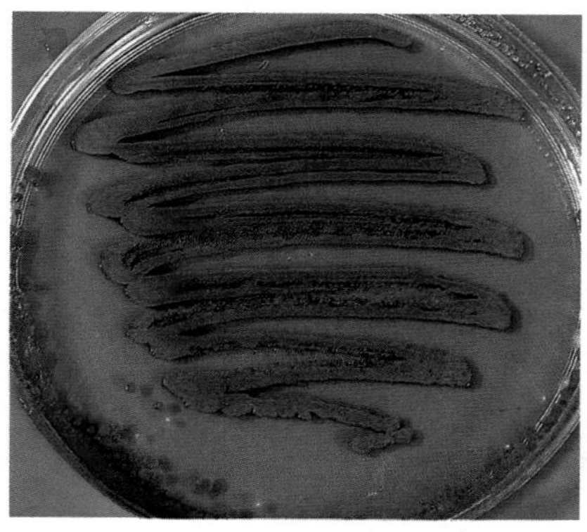

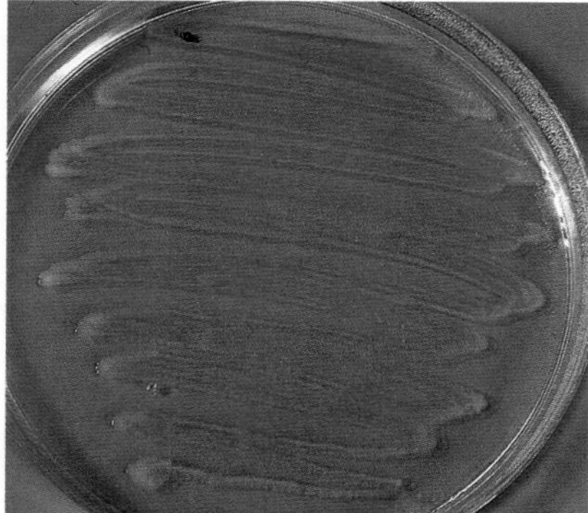

Figure 14.9

Serratia marcescens **is a bacterium that forms brick-red colonies at 25°C. However, when the same bacteria are grown at 30°C, the colonies are cream colored.**

Influence of external environment

Sometimes, individuals known to have a particular gene fail to express the phenotype specified by that gene. Temperature, nutrition, light, chemicals, and infectious agents all can influence gene expression. You have already studied the Himalayan trait of coat color in rabbits and learned that temperature has an effect on the expression of the trait. Temperature has a similar effect on the expression of color in certain bacteria, as *Figure 14.9* shows. External influences can also be seen in leaves. Leaves on a tree can have different sizes and shapes depending on the amount of light they receive.

You can now see that genes interact with each other and with the environment to form a more complete picture of inheritance. Mendel's idea that heredity is a composite of many individual traits still holds. Later researchers have filled in more details of Mendel's great contributions.

Section Review

Understanding Concepts

1. A cross between a purebred animal with red hairs and a purebred animal with white hairs produces an animal that has both red hairs and white hairs. What type of inheritance pattern is involved?
2. In a cross between individuals of a species of tropical fish, all of the male offspring have long tail fins, and none of the females possess the trait. Mating of the F_1 fish fails to produce females with the trait. Explain the inheritance pattern of the trait.
3. A red-flowered sweet pea plant is crossed with a white-flowered sweet pea plant. All of the offspring are pink. What is the inheritance pattern being expressed?

Thinking Critically

4. Armadillos always have four offspring that have identical genetic makeup. Suppose that within a litter, each young armadillo is found to have a different phenotype for a particular trait. How could you explain this phenomenon?

Skill Review

5. **Forming a Hypothesis** An ecologist observes that a population of plants in a meadow has flowers that may be red, yellow, white, pink, or purple. Hypothesize what the inheritance pattern might be. For more help, refer to Practicing Scientific Methods in the *Skill Handbook*.

SECTION 14.1

Reteach

Ask students to provide an example of each of the following patterns or types of inheritance: (a) incomplete dominance, (b) codominance, (c) multiple allelic trait, (d) polygenic trait, (e) sex-linked trait. L1

Extension

Have students describe and record some of the changes associated with human aging that alter gene expression. To get them started, provide them with a hint such as a change in hair color from black or blond to gray. L1

Assessment

Performance: Ask students to illustrate through the use of Punnett squares how sex-linked inheritance differs from autosomal inheritance. They should use the same allele letters and same parental genotypes. L1

4 Close

Activity

Ask students to prepare genotypes that illustrate the difference between pure and incomplete dominance in radish root shape. Radish roots are long, oval, or round. Have students use the symbols *L*, *l*, and *L'*. *For complete dominance, the genotypes are* LL *and* Ll *for long and* ll *for short. For incomplete dominance, the genotypes are* LL *for long,* LL' *for oval, and* L'L' *for round.* L2

Answers to Section Review

1. Genes for both red and white hairs are expressed, which is typical of a pattern of codominance.
2. The inheritance pattern of this trait is that of a sex-linked gene located on the Y chromosome, which is possessed by males only.
3. Incomplete dominance is being expressed because the heterozygous plant is an intermediate form between red and white.

Thinking Critically

4. The environment in the mother's uterus before birth may have been different for each of the offspring. The external environment can affect gene expression.

Skill Review

5. This is probably a case of polygenic inheritance.

SECTION

14.2

Prepare

Key Concepts

Students will study the role of the testcross as a tool in determining genotypes. They explore the technique for organizing and using pedigrees and investigate means of achieving desirable traits in plants and animals through the practice of breeding.

Block Scheduling

Look for this symbol for strategies that are useful in a block scheduling format. For more information on block scheduling, refer to Section 14.2 in the **Lesson Plans** booklet.

Materials

- For the Minilab, have students contact a pet shop owner or animal breeder in preparation for gathering information about a specific animal.

1 Focus

Bellringer

Before presenting the lesson, display **Section Focus Master 31** on the overhead projector and have students answer the accompanying questions. L1 SAE

Activity

LS **Visual-Spatial** Have volunteers bring in pictures of pet dogs or cats, along with pictures of their pets' offspring and/or siblings. At least some of the animals should be mixed breeds. Arrange the pictures on poster board and have students attempt to match the animals with their relatives. Discuss the characteristics used to identify the pet families. L1

SECTION

14.2 Applied Genetics

Section Preview

Objectives

Interpret testcrosses and pedigrees.

Evaluate the importance of plant and animal breeding to humans.

Key Terms

testcross
carrier
pedigree
inbreeding
hybrid

If you wanted to buy a purebred dog, how could you be certain that the animal does not have a genetic defect commonly found in its breed? Geneticists use methods developed from an understanding of Mendel's work to answer this question. Using these methods, they can find out the genotypes of individuals, anticipate the incidence of phenotypes resulting from crosses, and predict the occurrence of phenotypes and genotypes in populations.

Determining Genotypes

The genotype of an organism that is homozygous recessive for a trait is obvious to an observer because the recessive trait is expressed. However, organisms that are either homozygous dominant or heterozygous for a trait controlled by Mendelian inheritance have the same phenotype. If an organism exhibits a dominant phenotype, how can you determine whether the organism is homozygous or heterozygous for that trait?

Testcrosses can determine genotypes

One way to determine the genotype of an organism is to perform a testcross. A **testcross** is a cross of an individual of unknown genotype with an individual of known genotype in order to determine the unknown genotype. Usually, the known test organism is homozygous recessive for the trait in question.

Many traits, such as disease vulnerability in rose plants and progressive blindness in German shepherd dogs, are inherited as recessive alleles. These undesired traits are maintained in the population by carriers of the trait. A **carrier,** or heterozygous individual, appears the same phenotypically as one that is homozygous dominant. Plant and animal breeders are always cautious about introducing undesired traits into their lines of plant varieties and animal breeds. To check questionable individuals, they rely on testcrosses.

What are the possible results of a testcross? If the known test organism is homozygous recessive and the questionable organism is homozygous dominant, all of the offspring will be heterozygous for the trait and will show the dominant trait (be phenotypically dominant), as shown in *Figure 14.10.* However, if the organism being tested is heterozygous, the predicted 1:1 phenotypic ratio will be observed. If any of the offspring have the undesired trait, the parent in question must be heterozygous.

Program Resources

Section Focus Master 31 L1 SAE
Reinforcement and Study Guide, pp. 55-56 L1
Biolab and Minilab Worksheets, p. 54 L1
Tech Prep Applications, pp. 15-16 L2
Basic Skills Transparency 13 and **Master** L1 SAE
Reteaching Transparency 14 and **Master** L1 SAE

Figure 14.10

A testcross determines whether an organism is heterozygous or homozygous dominant for a trait.

A **In this testcross of Alaskan malamutes, the known test dog is homozygous recessive for a dwarf allele (*dd*), and the other dog's genotype is unknown. It can be either homozygous dominant (*DD*) or heterozygous (*Dd*) for the trait.**

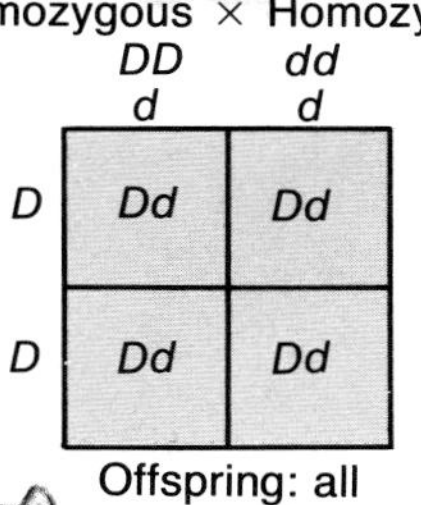

Homozygous × Homozygous
DD dd
d d
D Dd Dd
D Dd Dd
Offspring: all dominant
Dd
Dd

B **If the unknown dog's genotype is homozygous for the dominant trait, all of the offspring will be phenotypically dominant.**

C **If the unknown dog's genotype is heterozygous, half the offspring will be expected to express the recessive trait and appear dwarf. The other half will express the dominant trait and be of normal size.**

Pedigrees illustrate inheritance

Pedigree analysis is another way to solve the puzzle of whether an individual possesses a given allele, to predict the chances of an offspring receiving a trait, or to determine the inheritance pattern of a particular trait. A **pedigree** is a graphic representation of an individual's family tree, which permits patterns of inheritance to be recognized.

A pedigree is made up of a set of symbols that identify males and females, affected and unaffected individuals, matings, and other relationships. Some commonly used symbols are shown in ***Figure 14.11.*** Each horizontal row of circles and squares in a pedigree designates a generation,

Figure 14.11

Symbols are used by geneticists to make and analyze a pedigree. A circle represents a female; a square represents a male. Unshaded circles and squares designate individuals that have a normal phenotype for the trait being studied. Circles and squares with color represent affected individuals. A horizontal line connecting a circle and a square indicates a mating between those individuals. A vertical line connects a set of parents with its offspring.

2 Teach

Visual Learning

Figures 14.10, 14.11 Have students examine the illustrations. Then have them use the symbols for pedigree analysis to represent the same information that is conveyed by the Punnett square for the testcross. L1

GLENCOE TECHNOLOGY

Videodisc

STVS: Plants & Simple Organisms
Disc 4, Side 1
Disease-Resistant Tomatoes (Ch. 19)

Selective Breeding in Cows (Ch. 22)

Disc 4, Side 2
New Grains (Ch. 11)

Salt-Resistant Crops (Ch. 12)

STUDENT JOURNAL

Testcrosses Ask students to write a paragraph that tells why it would not be helpful when conducting a testcross to mate either homozygous dominant or heterozygous individuals with the animal being tested. Have students use Punnett squares to help with any technical explanations. *Mating with a homozygous dominant individual would result in all offspring showing the dominant trait, and mating with a heterozygous individual would greatly reduce the chances of an offspring that has two genes for the recessive trait.* L1 LS

MiniLab

Purpose

LS **Intrapersonal** Students will observe a specific animal trait and prepare a pedigree.

Process Skills

observe and infer

Teaching Strategies

- Students may contact local dog, cat, horse, and rabbit breeders for information.
- If contacting of breeders and/or pet shops is not practical, provide students with needed information in the form of one or two pedigrees for an animal trait.

Expected Results

Students will construct pedigrees of an animal trait. Genotype and phenotype will be marked for each animal.

Analysis

1. Answers may include: coat color, hair length, ear or tail shape. Patterns may be complete dominance, incomplete dominance, or codominance; multiple alleles; sex linked; or polygenic.
2. The number of animals may be too small to determine inheritance patterns.

Assessment

Knowledge: Provide a pedigree that consists of only a few individuals and has enough information so that students can make predictions of genotypes for the given phenotypes. L1

MiniLab

How can you illustrate a pedigree?

The pedigree method of studying family relationships uses phenotypic records extending over two or more generations. Studies of pedigrees can be used to yield a great deal of genetic information about a related group.

Procedure

1. Choose one trait in a common pet animal that interests you.
2. Ask a pet shop owner or someone who breeds the animal to provide you with information about an animal group. Perhaps you will be able to observe a litter of puppies or gerbils.

3. Collect information about your chosen animal group. Include whether each individual is male or female, does or does not have the trait, and the relationship of the individual to others.
4. Use your information to complete a pedigree for the trait.

Analysis

1. What trait did you study? From your pedigree, what is the apparent inheritance pattern of the trait?
2. How is the study of inheritance patterns limited by pedigree analysis?

with the most recent generation shown at the bottom. The generations are identified in sequence by Roman numerals, and each individual is given an Arabic numeral. Pedigrees are particularly useful if testcrosses cannot be made, if the number of offspring is small, or if the results of a testcross would take too long.

Analyzing a pedigree

Suppose that the pedigree shown in *Figure 14.12* is for a rare, recessive disorder that results in weakened bones in cattle. The owner of bull IV-1 wants to know for certain if the animal possesses the allele for the trait so that he may use the animal for future breeding.

Notice that information can be gained about the other cattle in the pedigree. You know that I-1 and I-2 are both carriers of the recessive allele for the rare trait because they have produced II-3, which shows the recessive phenotype.

Because the trait is rare, it is reasonably safe to assume that II-1 and II-6 are not carriers. For the same reason, you know that individuals II-2 and II-5 must be carriers like their parents because they each have passed on the recessive allele to the future generation IV.

You can't tell the genotype of II-4, but it has a normal phenotype. The probability that II-4 is a carrier is two out of three because its only two possible genotypes are homozygous normal and heterozygous. The homozygous recessive genotype is not a possibility because it shows an unaffected phenotype.

In generation III, individuals III-1, III-2, and III-5 are each normal and have one parent that is heterozygous. These three individuals have either a two-in-three chance or a one-in-two chance of being carriers, depending upon whether II-1 and II-6 were heterozygous or homozygous dominant.

The same situation holds for IV-1, IV-3, and IV-5 because they show a normal phenotype but have two carrier parents, III-3 and III-4; therefore, each has a two-in-three chance of being a carrier. Thus, bull IV-1 most likely has a two-in-three chance of being a carrier.

PORTFOLIO

Eye Color Provide students with a pedigree outline that will involve a sex-linked trait such as white eye color in *Drosophila.* Have students record above each symbol the sex chromosomes and genes that each fly possesses for the trait. A circle with one half darkened may be used to represent carrier females.

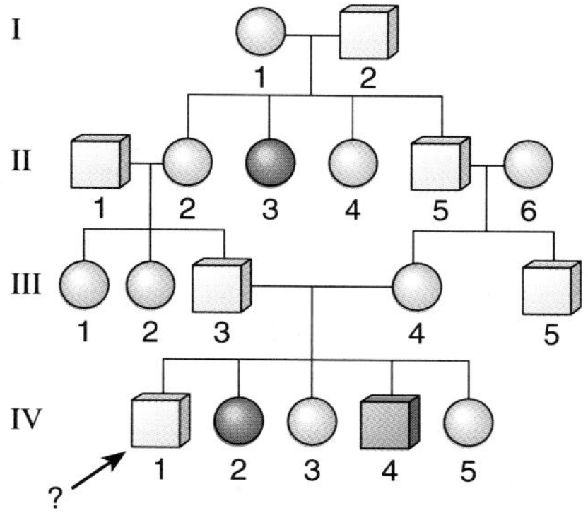

Figure 14.12
This fictional pedigree shows how a rare, recessive gene is transmitted from generation to generation. Note that the trait appears to skip generations, suggesting recessive inheritance. Also note that the trait occurs about equally in both sexes, indicating autosomal rather than sex-linked inheritance.

Selective Breeding

Heredity is particularly important in the breeding of domesticated plants and animals. Humans have been using genetics to breed and raise plants and animals for thousands of years. Most of the food you eat and the animals you take in as pets have been changed over time to suit the needs of humans.

Selective breeding produces organisms with desired traits

From ancient times, breeders have chosen the most desired plants and animals to serve as parents of the next generation. Farmers use for seed the largest heads of grain, the juiciest berries, and the most disease-resistant clover. They raise the calves of the best milk producer and save the eggs of the best-laying hen for hatching. Breeders of plants and animals want to be sure that their populations are uniform and breed predictably, that is, they have a desired trait in homozygous condition.

Horses and dogs are two examples of animals that breeders have developed as pure breeds. A breed (called a cultivar in plants) is a selected group of organisms within a species that has been bred for particular characteristics. For example, the German shepherd breed of dog has long hair, is black with a buff-colored base, has a black muzzle, and resembles a wolf.

Inbreeding develops pure lines

Breeders also want to eliminate any undesired traits from their breeding lines. One method employed by breeders in the development of pure lines is inbreeding. **Inbreeding** is mating between closely related individuals. It ensures that offspring are homozygous for most traits. However, inbreeding also brings out harmful, recessive traits because there is a greater chance of matings between carriers of rare recessive alleles than there is in random matings. Breeders are careful not to continue breeding individuals that produce offspring with undesired traits.

Hybrids are usually bigger and better

Selective breeding of plants can produce plants that have increased value as food for humans. For example, plants that are disease resistant can be crossed with others that produce fruit vigorously. The result is a plant that has greater market value. When two varieties or closely related species are crossed, their offspring are called **hybrids.** Hybrids produced by crossing two purebred plants are often larger and stronger than their parents.

STUDENT JOURNAL

Plant Breeding Have students visit a grocery store and take notes and make diagrams of squash varieties. Include any information the store might provide. Then, after the diagrams and information are recorded, advise students that summer squash (crookneck, custard marrow, pattypan, and cymling) are all the same species, *Cucurbita pepo,* and that winter squash (butternut, acorn, hubbard, butterball, spaghetti, pumpkin) are also all the same species, *Cucurbita maxima.* Ask students to write an explanation of how people have achieved such a variety of squash. L1 IN

Visual Learning

Figure 14.12 Ask students the following questions: What characteristics might be desirable in a dairy cow? *high milk production, fertility, stamina* At one time, high butterfat content was considered desirable. Why is that no longer the case? *Modern diets emphasize low fat content.*

Software

Genetics Simulation, That's Life; Explorations and Simulations in Biology. Heredity Dog, HRM Software.

Different Viewpoints in Biology

Sex Chromosomes at the Olympics

Since 1968, the Olympic Committee has tested the chromosomes in cheek cells of female athletes suspected of actually being males. In theory, an XX chromosome configuration makes an individual a female, but the chromosome makeup of some individuals with female reproductive organs is actually XY or XXY. Some geneticists and physicians object to chromosome testing on the grounds that it is the reproductive organs that determine the sex of an individual.

Reinforcement

Have students work on this problem: Albinism, or white fur, in rabbits is due to a recessive gene. Let A = normal fur pigment and a = no fur pigment. A breeder wishes to know if a male rabbit is homozygous or heterozygous for pigmented fur. Diagram the possibilities in Punnett squares and ask for an interpretation of the results. L1

Focus On

The Domestication of Cats

Purpose

Students will gain some insight into cat breeding and the influence of cats on human life throughout history.

Teaching Strategies

- Make sure that students are familiar with the meaning of the terms *domestication* and *bubonic plague*.
- Ask students to research the cause of bubonic plague, a bacterial infection caused by *Yersinia pestis*. Rodents are resistant to the disease but fleas that feed on the rodents carry the bacteria to humans. L1
- Ask students to bring in photographs of their pet cats. If any student's cat is a particular variety, have the student describe distinctive traits or qualities.
- Have students explain why it is impossible to get a Manx cat to breed true or to produce consistently offspring with no tails. The only crosses possible are *Mm* and *mm*, *Mm* and *Mm*, or *mm* and *mm*. Using Punnett squares, students should show that all crosses can produce offspring with tails. L1

Background

Evidence indicates that the dog was the first animal to be domesticated. Wildcat domestication followed about 9500 years later. An example of the results of modern breeding is the Scottish fold, a breed developed in Scotland in the 1960s. A Scottish fold cat is stocky with a large rounded head. Its most distinctive physical feature is its ears, which are folded forward.

Focus On

The Domestication of Cats

Early people had an uneasy relationship with the animals with which they shared Earth's habitats. People hunted animals for food, clothing, and whatever other necessities of life the animals could provide. Understandably, many animals came to fear these early hunters. At the same time, humans lived in fear of the animals. Quite often, the hunter became the hunted; the predator became the prey.

Egyptian mummified cat

History of cats

About 4500 years ago, people began to domesticate small wildcats for the help they could provide to humans. Cats were valued for their hunting ability. They helped control rodent and reptile populations, keeping both crops and people safe.

Cats in ancient Egypt

Ancient Egyptians were so grateful for the hunting ability of cats that they worshipped cats as sacred beings. When a pet cat died, members of its human family would shave off their eyebrows as a sign of mourning. Anyone who killed a cat in ancient Egypt was usually put to death.

CULTURAL DIVERSITY

Cultural Taboos Against Inbreeding

Although inbreeding can result in high rates of stillbirths and congenitally diseased children, scientists are not sure whether taboos against incest are culturally or naturally selected. Explain to students that taboos against inbreeding are not universal, indicating that their origins are cultural rather than genetic. On the other hand, some sociobiologists argue that individuals brought up in close proximity to each other as family members usually are less likely to develop a sexual interest in one another. This suggests incest taboos may have a biological component.

Cats in the Middle Ages

During the Middle Ages in Europe, cats were believed to be evil. Hundreds of thousands of cats were killed. During this time, the rat population exploded because its natural enemy, the cat, was being killed off. Rats transmitted bubonic plague to humans through infected rat fleas. The plague killed about one-fourth of all Europeans who lived during the 1300s.

Selective breeding

Cats have been bred mostly for their physical features and personalities. Abyssinians, one type of short-haired cat, have been bred in Ethiopia (once called Abyssinia) for thousands of years. Some experts believe that Abyssinians may be the direct descendants of the sacred cats of Egypt. Abyssinians are prized for their melodic voices and their soft, beautifully colored coats.

Big business

Today, cats are big business throughout the developed world. People spend hundreds of millions of dollars each year caring for their feline companions. Cats have their own scientifically developed diets, medical care rivaling that offered to humans, and even their own psychiatrists to help them deal with stress and depression!

EXPANDING YOUR VIEW

1. **Writing About Biology** In a short paragraph, summarize the reasons why cats are beneficial to humans.
2. **Understanding Concepts** In what ways do domestic cats reflect the wild nature of their evolutionary ancestors?
3. **Going Further** Conduct library research on Manx cats. Find out what is unique about their tails, how this trait is determined, and how the trait appeared.

Meeting Individual Needs

Gifted Have students do library research to determine the lineage of domestic cats. They can design a chart or poster that traces the taxonomy of cats and shows relatives of cats that are not domesticated. Students will require a basic understanding of levels of classification in order to do the project. L3

Focus On

Answers to Expanding Your View

1. Paragraphs may include mention of their use in rodent and reptile control, as well as serving as pets for humans.
2. They are predators by nature.
3. The tail-less Manx originated from a mutation that first appeared on the Isle of Man. A dominant gene *M* is responsible for the tail-less trait. The *MM* genotype is lethal during embryonic development, *Mm* cats have either no tail or a partial tail, and *mm* cats have a tail.

Bioethics

Undesirable Traits

What are some of the consequences of using selection to create breeds of dogs? Orthopedic problems in breeds that are extremely large or extremely small are the norm rather than the exception. One common crippler of large, fast-growing breeds is hip dysplasia, a malformation of the ball-and-socket joint. The joint loosens and deteriorates with age, usually ending in painful and debilitating arthritis. Hip dysplasia is prevalent in purebred St. Bernards, German shepherds, and Great Danes.

Eye abnormalities are common in many breeds of dogs. Pekingese, English bulldogs, pugs, and other flat-faced dogs suffer from brachycephaly, or protuberant eyes. Brachycephaly causes a tendency to "eye-out-of-socket" trauma. Abnormal development of the retina occurs often in English springer spaniels and in Labrador retrievers. Discuss with students the pros and cons of breeding that can result in undesirable traits.

SECTION 14.2

3 Assess

Check for Understanding

Have students explain: (a) the role of a testcross and a pedigree, (b) how the two differ, (c) what breeding programs attempt to accomplish. L1

Reteach

Have students prepare a concept map using the following terms: testcross, pedigree, carrier, inbreeding, determining genotypes, selective breeding, hybrid. L1

Extension

Have students explore the problems that occur when horses are bred to donkeys to produce mules and hinnies. Have them speculate as to how Mendel's laws might have differed if he had worked with these animals. L2

Assessment

Knowledge Provide students with a pedigree for a specific trait. Include individuals in which the genotypes can and cannot be predicted with 100 percent accuracy. Ask students to complete the genotypes for all individuals and to use a ? for those genotypes in which the second allele may still be in question. L2

4 Close

Discussion

Discuss with students that there is currently an interest in eggs with low cholesterol content. Then ask students how they might proceed to breed chickens that produce such low-cholesterol eggs. L1

Figure 14.13
Hybrid wheat is more nutritious and more productive than non-hybrid wheat. It was developed by selective-breeding methods.

Many crop plants such as wheat, corn, and rice, and garden flowers such as roses and dahlias have been developed by hybridization, or selective breeding. *Figure 14.13* shows one example.

Animals have been selectively bred for thousands of years. Chickens, turkeys, horses, cattle, sheep, and other livestock have been bred to increase the expression of desired traits in particular varieties. For example, at one ranch in southern Texas, where the climate is hot and semitropical, English shorthorn cattle were being raised. The cattle produced good beef but were not conditioned to the hot climate, so the rancher crossed them with Brahman cattle from India, which could tolerate hot weather. After 35 years of hybridization and selection, a pure breed was produced that combines the traits of good beef production and hot-weather tolerance—Santa Gertrudis cattle. You have benefited from selective breeding by having more eggs, milk, and meat than would have been possible otherwise.

Connecting Ideas

Geneticists discovered that Mendel had not determined all of the ways traits are inherited. The study of genetics reveals that genes interact in ways that confirm and extend Mendel's work. Animals and plants have traits that are expressed by codominant alleles, incompletely dominant alleles, and sex-linked alleles. Do these same inheritance patterns apply to humans? Can human traits be affected by internal and external environments? Answers to these questions are important as geneticists apply their understanding to traits that affect humans.

Section Review

Understanding Concepts

1. A testcross made on a cat that is suspected of being heterozygous for an undesired, recessive trait produces ten kittens, none of which has the trait. What is the presumed genotype of the cat? Explain.
2. Why is inbreeding rarely a problem among animals in the wild?
3. What effect might selective breeding of plants and animals have on the size of Earth's human population? Why?

Thinking Critically

4. Suppose you wanted to breed a variety of plants with red flowers and speckled leaves. You have two varieties, each having one of the desired traits. How would you proceed?

Skill Review

5. **Interpreting Scientific Illustrations** Examine the following pedigree and explain whether the trait is dominant or recessive. How did you arrive at your conclusion? For more help, refer to Thinking Critically in the *Skill Handbook.*

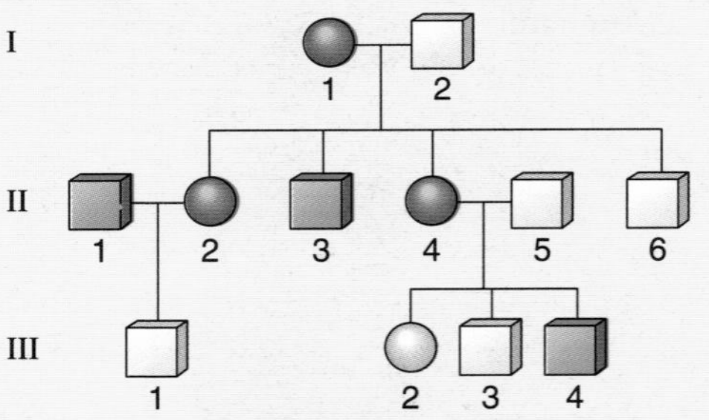

Answers to Section Review

1. The cat is probably homozygous dominant. If it were heterozygous, there would probably have been some offspring with the recessive trait.
2. In nature, mate selection is random and the chance that a mate is closely related is greatly reduced.
3. Selective breeding can increase crop size and provide for more nutritious and more disease-resistant crops, and this might increase the human population.

Thinking Critically

4. First breed the red flower plants among themselves to ensure that they are breeding true. Do the same for the speckled leaf plants. Then, cross breed the two varieties.

REVIEWING MAIN IDEAS

14.1 When Heredity Follows Different Rules

- Alleles can be completely dominant, incompletely dominant, or codominant. There may be many alleles for one trait or many genes that interact to produce a trait. Inheritance patterns of genes located on sex chromosomes are caused by differences in the number and kind of sex chromosomes in males and females.
- Interactions between genes and the environment complicate patterns of inheritance. The expression of some traits is affected by internal environments that are governed by age or sex. Expression of other traits is affected by external factors in the environment such as temperature, chemicals, and light.

14.2 Applied Genetics

- Geneticists use testcrosses and pedigrees to determine the genotypes of individuals. Both methods can also be used to predict the probability of offspring having a particular allele.
- Plant and animal breeders use genetics to selectively breed organisms with traits that are desirable to humans.

Key Terms

Write a sentence that shows your understanding of each of the following terms.

autosome	multiple alleles
carrier	pedigree
codominant alleles	polygenic inheritance
hybrid	sex chromosome
inbreeding	sex-linked trait
incomplete dominance	testcross

Understanding Concepts

1. How many alleles does a single individual carry for a trait?
   a. none c. two
   b. one d. four
2. Traits that are polygenic have ___________ form(s).
   a. one c. three
   b. two d. many
3. An organism that is heterozygous for a harmful trait is called a ___________.
   a. mutant c. lethal
   b. carrier d. hybrid
4. When two closely related species are mated, their offspring are ___________.
   a. hybrids c. mutants
   b. carriers d. inbred
5. A graphic representation of a family tree is a ___________.
   a. testcross c. pedigree
   b. Punnett square d. phenotype
6. Which of the following is the symbol for an affected female?
   a.  c. 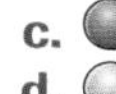
   b. d. 
7. An example of continuous variation is ___________.
   a. skin color in humans
   b. attached earlobes
   c. flower color in peas
   d. seed color in peas
8. Whose chromosomes determine the sex of offspring in humans?
   a. mother's
   b. father's
   c. equal contributions of both
   d. neither mother's nor father's
9. Coat colors in rabbits are produced by ___________ inheritance.
   a. polygenic c. environmental
   b. sex-linked d. multiple allele

Chapter 14 Review

Reviewing Main Ideas

Summary statements can be used by students to review the major concepts of the chapter.

Key Terms

Answers should go beyond defining the terms. Accept any answer that uses the term correctly and in the proper context.

Understanding Concepts

1. c
2. d
3. b
4. d
5. c
6. c
7. a
8. b
9. d

Skill Review

5. The trait is probably dominant. It was expressed in more than half the members of generation II, and in generation III after a cross with a homozygous recessive individual.

Chapter 14 Review

10. b
11. d
12. a
13. a
14. b
15. a
16. a
17. d
18. d
19. b
20. c

Applying Concepts

21. A male cannot have two alleles for fur color because it has only a single X chromosome.
22. This is a case of incomplete dominance, in which the heterozygous chicken has a feather color intermediate between black and white.
23. A carrier implies that the dog has an undesirable trait in the heterozygous condition.
24. The ratio is one white horse to two palominos to one chestnut horse.
25. *EFG, EFg, EfG, Efg, eFG, efG, efg, eFg*

Chapter 14 Review

10. Which organisms gave Morgan the first clues about sex-linked inheritance?
    a. rabbits
    b. fruit flies
    c. peas
    d. mice
11. Mating between closely related individuals is called __________.
    a. polymorphism
    b. hybridism
    c. mutation
    d. inbreeding
12. Which of the following usually increases the appearance of genetic disorders?
    a. inbreeding
    b. simple autosomal dominance
    c. incomplete dominance
    d. polygenic inheritance
13. If a trait is expressed in two ways in a single organism, the trait is inherited by __________.
    a. codominance
    b. mutation
    c. incomplete dominance
    d. sex linkage
14. What is the purpose of a testcross?
    a. produce offspring
    b. check for carriers
    c. explain recessiveness
    d. show polygenic inheritance
15. When a trait appears to blend between two types to produce an intermediate, the pattern of inheritance is __________.
    a. incomplete dominance
    b. sex linkage
    c. polygenic inheritance
    d. codominance
16. Genes that are sex linked are usually found on __________.
    a. X chromosomes
    b. any chromosome
    c. autosomes
    d. Y chromosomes
17. A chicken that is checkered black and white shows the result of __________ alleles.
    a. incomplete
    b. multiple
    c. polygenic
    d. codominant
18. Humans have 22 pairs of homologous chromosomes called __________.
    a. alleles
    b. carriers
    c. pedigrees
    d. autosomes
19. Which of these have NOT been developed by selective breeding?
    a. cattle
    b. whales
    c. roses
    d. corn
20. A __________ is a cross of an individual with an unknown genotype with an individual of known genotype.
    a. hybrid
    b. pedigree
    c. testcross
    d. carrier

Applying Concepts

21. In cats, the allele for fur color is sex linked. The allele for black is X^C, and the allele for orange is X^c. From this information, explain why calico cats, X^CX^c, are always female.
22. Suppose you mate a black rooster with a white hen. The feathers of all the offspring are "blue," a color that is intermediate between black and white. Explain the inheritance pattern in these chickens.
23. Julia purchased a puppy from a breeder. The breeder explained that the puppy should never be bred with another dog because it was a carrier for a joint defect. What did the breeder mean by this statement?
24. Inheritance of the palomino coat color in horses is a result of incomplete dominance. A white horse is *DD*, a chestnut horse is *dd*, and a palomino horse is *Dd*. What is the expected ratio of coat colors in the offspring of two palomino horses?
25. An organism has three genes—*e*, *f*, and *g*—for a trait. What are all of the possible genotypes of a random individual?

Thinking Critically

26. ***Concept Mapping*** Make a concept map that relates the following terms and phrases. Supply the appropriate linking words for your map.
multiple alleles, polygenic inheritance, pedigree, testcross, sex-linked inheritance, codominant alleles, incompletely dominant alleles

27. ***Observing and Inferring*** What part, if any, does the male parent play in cytoplasmic inheritance?

28. ***Observing and Inferring*** Why is it always better to base scientific conclusions on large amounts of data as Mendel did with his pea studies?

29. ***Recognizing Cause and Effect*** Explain why a male organism with a recessive sex-linked trait usually produces no female offspring with the trait.

30. ***Formulating Hypotheses*** After studying a particular trait in a newly discovered plant, a scientist discovered that the trait varied widely among the population so that when a graph was made, it produced a bell-shaped curve. Make a hypothesis as to the inheritance pattern for the trait.

31. ***Recognizing Cause and Effect*** In a testcross between an Alaskan malamute dog known to be homozygous recessive for a dwarf allele (*dd*) and a dog with an unknown genotype, half the offspring are of normal size and half express dwarfism. What is the genotype of the unknown parent dog? Explain how you arrived at your answer.

ASSESSING KNOWLEDGE & SKILLS

The following graph illustrates the number of flowers produced per plant by a certain plant population.

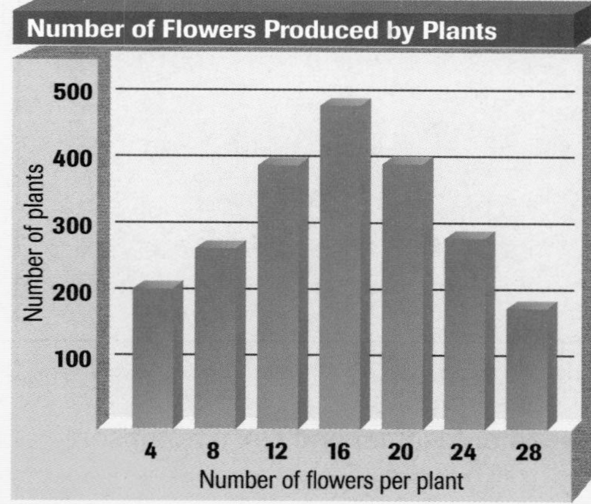

Interpreting Data Use the graph to answer the questions that follow.

1. How many flowers are produced by plants with all dominant genes for flower production?
   a. 4 c. 16
   b. 12 d. 28
2. How many flowers are produced by plants with half the possible number of dominant genes for flower production?
   a. 4 c. 16
   b. 12 d. 28
3. What pattern of inheritance is suggested by the graph?
   a. multiple alleles
   b. incomplete dominance
   c. polygenic inheritance
   d. sex linkage
4. ***Observing and Inferring*** From the above graph, estimate the number of gene pairs that control the number of flowers in these plants.

Chapter 14 Review

Thinking Critically

26. Evaluate students' concept maps. The maps should show an understanding of the relationships among the concepts listed.
27. The male plays no part in cytoplasmic inheritance.
28. Large amounts of data reduce the error due to random variations played by chance alone.
29. All X chromosomes from the male parent are contributed to female offspring. If the female parent is homozygous dominant for the trait, all female offspring will be heterozygous and will not show the trait. The female parent would have to be heterozygous for that trait in order for half her female offspring to show the trait.
30. Bell-shaped curves for traits are usually the result of polygenic inheritance.
31. The unknown dog has a genotype of *Dd;* that is, the dog is heterozygous for the dwarfism trait. You can find this out by doing a testcross using a Punnett square.

ASSESSING KNOWLEDGE & SKILLS

1. d
2. c
3. c
4. There are three gene pairs governing the number of flowers produced per plant.

Program Resources

Chapter Assessment, pp. 79-84 L1
Alternate Assessment in the Science Classroom
Computer Test Bank L1
Content Mastery, pp. 53-56 L1

Chapter Organizer

SECTION	OBJECTIVES	ACTIVITIES/FEATURES
15.1 Simple Mendelian Inheritance of Human Traits National Science Standards: UCP.2, UCP.3; C.2; F.1; G.1	1. **Predict** how a human disorder is determined by a simple dominant allele. 2. **Determine** the human genetic disorders that are caused by inheritance of a simple recessive allele.	**Math Connection:** Adaptive Genetic Variations, p. 358 **People in Biology:** Robert Murray, p. 360 **Thinking Lab:** How is Duchenne's muscular dystrophy inherited?, p. 362
15.2 Complex Inheritance of Human Traits National Science Standards: UCP.2, UCP.3; A.1, A.2; C.2; F.1; G.1, G.2	3. **Compare** multiple allelic, polygenic, and sex-linked patterns of inheritance in humans. 4. **Distinguish** between autosomal and sex chromosome aneuploidy.	**Minilab:** What colors and patterns can you detect in eyes?, p. 364 **Minilab:** How is height inherited in humans?, p. 365 **Biolab:** Constructing Pedigrees to Trace Heredity, p. 366 **Biology & Society Ethics:** Genetic Screening, p. 369

ACTIVITY MATERIALS

BIOLAB	MINILABS	ALTERNATE LAB
page 366 PTC paper	**page 364** magnifying glass colored pencils **page 365** metric tape measure or meterstick	**page 368** flat toothpick methylene blue stain microscope slide coverslip

Chapter 15 Human Heredity

TEACHER CLASSROOM RESOURCES

Reproducible Masters	Transparencies
Section Focus Master 32: Simple Dominant Human Traits L1 SAE **Reinforcement and Study Guide,** pp. 57-58 L1 **Concept Mapping:** Human Autosomal Disorders, p. 15 L1 **Tech Prep Applications:** Tracking a High-Cholesterol Gene, pp. 17-18 L2 **Laboratory Manual:** Determination of Genotypes from Phenotypes in Humans, pp. 83-86 L2	**Reteaching Transparency #15:** Analyzing Pedigrees for Huntington's Disease and Tay-Sachs L1 SAE
Section Focus Master 33: A Complex Inheritance Trait L1 SAE **Reinforcement and Study Guide,** pp. 59-60 L1 **Biolab and Minilab Worksheets,** pp. 57-60 L1 **Laboratory Manual:** How Can Karyotype Analysis Explain Genetic Disorders?, pp. 87-90 L2 **Critical Thinking/Problem Solving:** Using Genetics to Help Solve Mysteries, p. 15 L3 **Content Mastery,** pp. 57-60 L1	**Basic Skills Transparency #14:** Inheritance of Human Traits L1 SAE

ASSESSMENT MATERIALS

Chapter Assessment, pp. 85-90
Performance Assessment in the Biology Classroom, pp. 15, 21
MindJogger Videoquiz
Alternate Assessment in the Science Classroom
Computer Test Bank

Spanish Resources SAE
English/Spanish Audiocassettes SAE
Cooperative Learning in the Science Classroom COOP LEARN
Lesson Plans

KEY TO TEACHING STRATEGIES

L1 Level 1 activities should be within the ability range of all students including those with learning difficulties.
L2 Level 2 activities are within the ability range of average to above-average students.
L3 Level 3 activities are designed for the ability range of above-average students.
SAE SAE activities should be within the ability range of Students Acquiring English.
COOP LEARN Cooperative Learning activities are designed for small group work.
P These strategies represent student products that can be placed into a best-work portfolio.
These strategies are useful in a block scheduling format.

GLENCOE TECHNOLOGY

The following multimedia resources are available from Glencoe.

Biology: The Dynamics of Life
CD-ROM SAE
Videodisc Program

The Infinite Voyage Series
The Geometry of Life
A Taste of Health

The Secret of Life Series
Tinkering with Our Genes: Genetic Medicine

Science and Technology Videodisc Series (STVS)
Human Biology
Detecting Cystic Fibrosis
Obesity and Heredity

CHAPTER

15 Human Heredity

Chapter Overview

As students begin the chapter, they learn how simple Mendelian genetics applies to the inheritance of many human traits. They explore examples of disorders that are inherited by a dominant autosomal pattern, such as Huntington's disease, and examples of disorders that are transmitted by recessive autosomal heredity, such as cystic fibrosis.

In the next section, students explore the complex inheritance of human traits. They apply their knowledge of multiple alleles, polygenic inheritance, and sex-linked patterns of inheritance to human genetic disorders. They conclude their study with an examination of the mistakes that occur during meiosis. These mistakes cause aneuploidy, which results in conditions such as Down syndrome.

Key Terms

aneuploidy
fetus
karyotype

Learning Styles

Look for the following logo for strategies that emphasize different learning modalities. LS

Visual-Spatial	**Chalkboard Example, p. 357; Microscope Activity, p. 360; Minilab, p. 364; Alternate Lab, p. 368**
Intrapersonal	**Project, p. 358; Portfolio, p. 364; Meeting Individual Needs, p. 364**
Linguistic	**Student Journal, p. 365**
Logical-Mathematical	**Portfolio, p. 360; Minilab, p. 365**

LS

CHAPTER

15 Human Heredity

With one powerful swing of the bat, Ken Griffey, Jr. can send a long, hard drive over the outfield wall for a home run. So, too, could his father, Ken Griffey, Sr., when he starred as a major-league ball-player. That such athletic talent is inherited has long been believed. However, talent is not so clearly inherited as are certain physical traits, such as skin and eye color. Can you see any physical resemblance between Ken Griffey, Sr. and his son? Most people acknowledge that environment, including knowledge and training, plays a key role in developing talent. Evidence even shows that people inherit more potential for developing some traits than others. Human heredity is indeed complex.

Concept Check

You may wish to review the following concepts before studying this chapter.

- Chapter 11: chromosomes, genes
- Chapter 12: meiosis
- Chapter 14: non-Mendelian inheritance

Chapter Preview

Large photo: Ken Griffey, Jr.
Inset: Ken Griffey, Sr.

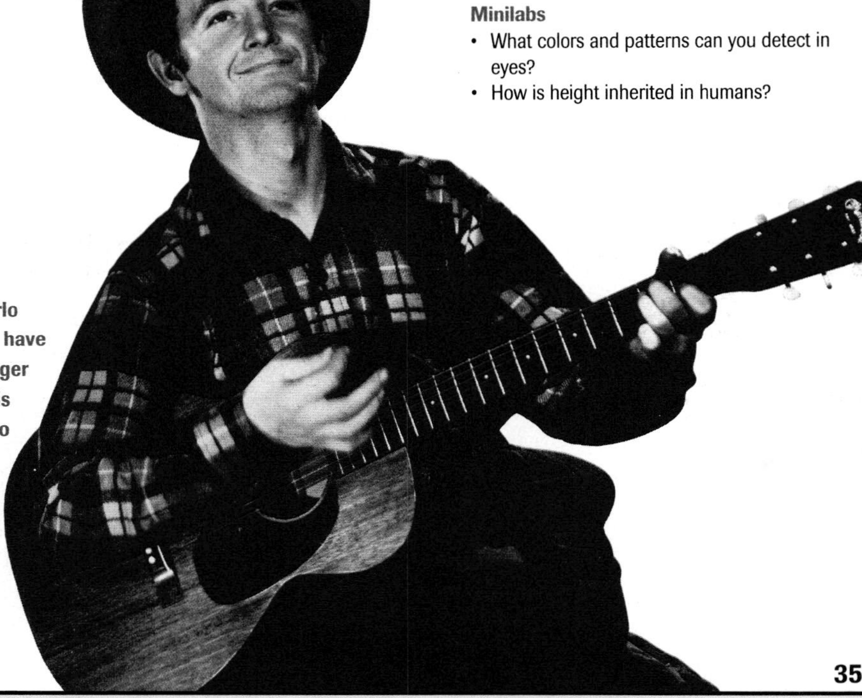

Folksinger and songwriter Arlo Guthrie's musical talent may have come from his father, folksinger Woody Guthrie, shown on this page. Unfortunately, Arlo also may have inherited a trait from his father that is better understood but highly undesirable—a fatal genetic disorder called Huntington's disease. How exactly are human traits inherited?

CHAPTER 15

Introducing the Chapter

Ask students to list traits that would be important for a good baseball player to have. Then ask them to list some traits that would not make a difference in someone who played baseball. Bring the discussion of traits to a close with a summary of what twin studies have shown about inheritance of some traits. Useful references include "All About Twins," *Newsweek*, November 23, 1987, pp. 58-69 and "The Eerie World of Reunited Twins" by Clare Mead Rosen, *Discover*, September, 1987, pp. 36-46.

Theme Development

An important theme in the discussion of genetics in humans is *systems and interactions.* For example, undesirable changes in one function of the body, due to an inherited disorder, can result in problems for the organism. A second main theme of the chapter is *homeostasis*, which is normally maintained during the transmission of genetic material but is disrupted by the inheritance of particular sets of genes that result in genetic disorders.

Concept Check

Understanding of this chapter will rely on an understanding of the process of meiosis discussed in Chapter 12 and on an understanding of the different patterns of heredity introduced in Chapter 14.

Assessment Planner

Choose assessment strategies from the following pages to evaluate the progress of your students.

Assess, pp. 361, 370
Alternate Lab, pp. 368-369
Minilabs, pp. 364, 365
Portfolio, pp. 360, 364
Thinking Lab, p. 362
Biolab, pp. 366-367
Chapter Review, pp. 371-373

SECTION

15.1

Prepare

Key Concepts

The inheritance of simple dominant autosomal traits, such as tongue curling and Huntington's disease, is discussed. Examples of autosomal recessive disorders, such as cystic fibrosis, sickle-cell anemia, Tay-Sachs, and phenylketonuria are then presented.

Block Scheduling

Look for this symbol for strategies that are useful in a block scheduling format. For more information on block scheduling, refer to Section 15.1 in the **Lesson Plans** booklet.

Materials

- Collect articles and pamphlets on genetic disorders for the Display.
- Purchase slides of sickle-cell anemia blood for the Microscope Activity.

1 Focus

Bellringer

Before presenting the lesson, display **Section Focus Master 32** on the overhead projector and have students answer the accompanying questions. L1 SAE

Discussion

Play a Woody Guthrie (1912-1967) song such as *This Land Is Your Land.* Explain to students what Huntington's disease is and tell them about Woody Guthrie. Guthrie's mother died of Huntington's disease in 1929 when he was 17 years old. It was not until 1956, when Guthrie was 44, that a neurologist diagnosed him with the same disease.

SECTION 15.1

Simple Mendelian Inheritance of Human Traits

Section Preview

Objectives

Predict how a human disorder is determined by a simple dominant allele.

Determine the human genetic disorders that are caused by inheritance of a simple recessive allele.

Key Terms

fetus

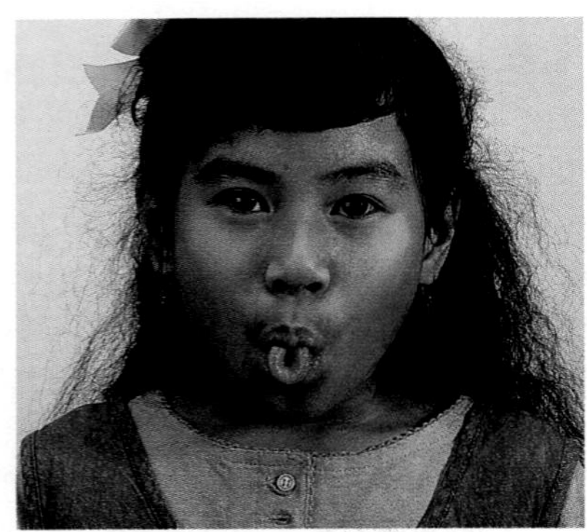

Can you curl your tongue the way this girl can? If so, you've inherited the dominant allele for tongue curling from at least one of your parents. If not, you've inherited two recessive alleles for tongue curling—one from each of your parents. Tongue curling is just one of thousands of traits humans inherit. Some traits, like tongue curling, appear to have little significance in our present environment. Others, such as hair color or eye shape, help make up an individual's personal appearance. Still others, including hereditary disorders such as cystic fibrosis, seriously affect a person's life.

Dominant Autosomal Heredity

Many traits are inherited just as the rule of dominance predicts. Remember that in Mendelian inheritance, a single dominant autosomal allele inherited from one parent is all that is needed for a person to show the dominant trait. Recessive alleles must be inherited from both parents for a person to show the recessive phenotype.

Simple dominant traits

Tongue curling is one of these simple dominant traits. Earlobe type, illustrated in *Figure 15.1,* is also determined by simple autosomal inheritance. Having earlobes that are attached to the head is a recessive trait, whereas heterozygous and homozygous dominant individuals have earlobes that hang freely.

Huntington's disease—A rare genetic disorder

Huntington's disease is a lethal genetic disorder caused by a rare autosomal dominant allele. A lethal disorder is a disease that causes death in most of the individuals who are affected. The nervous system of a person with Huntington's disease undergoes progressive degeneration, resulting in uncontrolled, jerky movements of the head and limbs and mental deterioration. No effective treatment exists.

Ordinarily, a dominant allele with such severe effects would be expected to occur only as a new mutation and not be transmitted to future generations. But because the onset of Huntington's disease usually occurs between the ages of 30 and 50, an individual may have children before knowing whether he or she carries the allele.

Program Resources

Section Focus Master 32 L1 SAE
Reinforcement and Study Guide, pp. 57-58 L1
Laboratory Manual, pp. 83-86 L2
Concept Mapping, p. 15 L1
Tech Prep Applications, pp. 17-18 L2
Performance Assessment in the Biology Classroom, p. 15 L1 P
Reteaching Transparency 15 and **Master** L1 SAE

Figure 15.1

Many traits are determined by a single dominant allele.

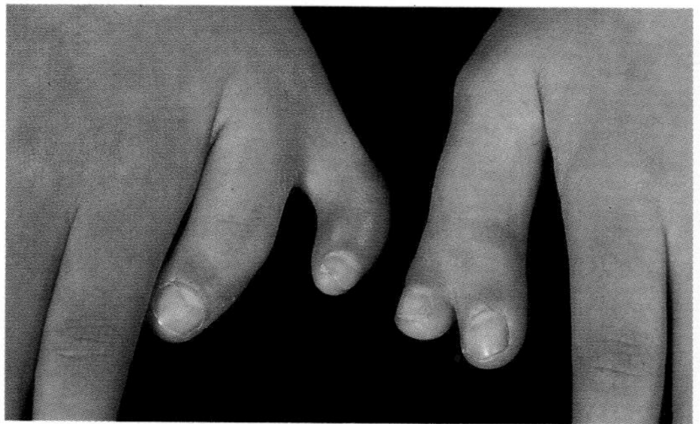

The allele for polydactyly, having more than five fingers or toes, is dominant to the allele that produces only five digits on each hand and foot.

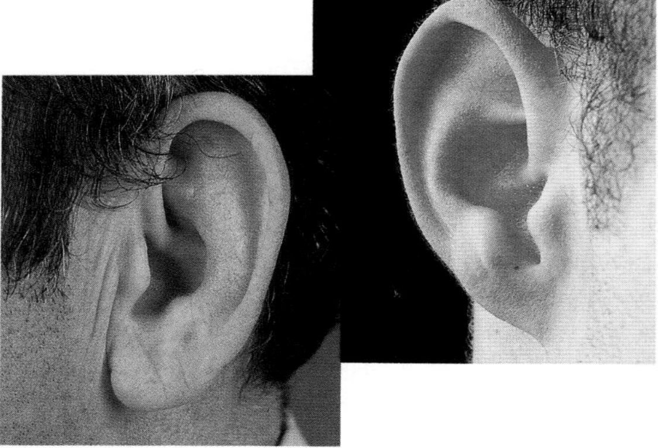

The allele for freely hanging earlobes, *F* (left), is dominant to the allele for attached earlobes, *f* (right).

A biochemical test resulting from DNA technology allows some persons who are at risk to determine whether they are indeed carriers. Persons who test positive and do not have children will eventually cause a decrease in the frequency of the allele in the population. However, not everyone who is at risk wishes to be tested because anyone who tests positive for the allele must then live with the knowledge that he or she will eventually develop the disease. It is a difficult decision. The pedigree in *Figure 15.2* shows a typical pattern of occurrence of Huntington's disease in a family.

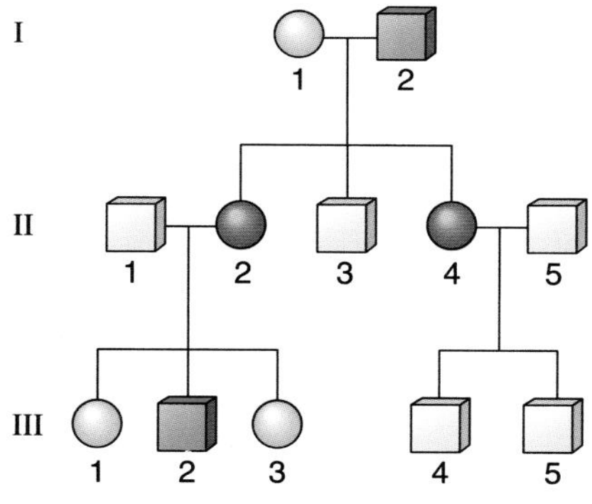

Recessive Autosomal Heredity

Unlike Huntington's disease, most genetic disorders are caused by recessive alleles. Many of these alleles are rare, but a few are common in certain ethnic groups.

Cystic fibrosis

Cystic fibrosis (CF) is the most common lethal genetic disorder among white Americans. Approximately one in 20 white Americans carries the recessive allele, and one in 2000 children born to white Americans inherits the disorder.

In a person with CF, the mucus in the lungs and digestive tract is particularly thick and viscous. Breathing is difficult because the mucus collects in the lungs, and lung infections are frequent. The thickened mucus also

Figure 15.2

A typical pedigree for Huntington's disease shows the trait in each generation and equally distributed among males and females. Every child of an affected individual has a 50 percent chance of being affected and then a 50 percent chance of passing the defective allele to his or her own child.

2 Teach

Display

Collect articles and pamphlets on various genetic disorders and post them on the bulletin board. The March of Dimes organization is a good source of materials.

Chalkboard Example

LS **Visual-Spatial** Place Punnett squares on the chalkboard to demonstrate possible inheritance patterns of each genetic disorder described in the text.

Science, Technology, and Society

Reversed Organs

Occasionally a person is born with reversed sides, or *situs inversus,* in which the heart is on the wrong side. This condition can lead to dangerous missed connections in the cardiovascular system. Scientists at the College of Medicine in Houston have uncovered a gene in mice that, when mutated, results in *situs inversus.* Scientists are now working out the structure of the gene so they can know something about the protein it codes for.

Assessment

Performance Assessment in the Biology Classroom: p. 15 *Inheritance of Human Traits.* Have students carry out this activity after they have learned about Mendelian inheritance. L1 P

GLENCOE TECHNOLOGY

Videodisc

The Infinite Voyage: The Geometry of Life

Huntington's Disease and Inheritance of the Deadly Gene (Ch. 6)

Meeting Individual Needs

Gifted Have gifted students follow up on the experimental gene therapy for cystic fibrosis as well as proposed future uses of gene therapy to treat cystic fibrosis L3

Math

Adaptive Genetic Variations

Purpose

Students will learn how the harmful mutation hemoglobin S can at times, because of interaction with the environment, produce a beneficial effect for those with the allele. They also use math to determine the percent of a population that is homozygous and heterozygous for the same trait in different environments.

Teaching Strategies

- Ask students to recall from the text how many African Americans are homozygous for the sickle-cell allele. Relate the statement "2 in 1000" to the figure 0.002 in the table. Ask students to relate the other decimals in the table to statements about the genotypes in each population.
- Review how to find percents if necessary.
- Discuss how the sickle-cell allele is an advantage in Africa, but a disadvantage in the United States.

Possible Answers

- Students will convert each of the decimals by multiplying each decimal by 100 and adding a percent sign.
  1. 8.6%; 20%
  2. 0.2%; 1%
- The fact that the harmful allele for hemoglobin S is beneficial in the heterozygous state in Central Africa shows that both genes and the environment can determine how organisms will survive. In the United States, the environment has little effect on making the harmful S allele beneficial, but other things such as amount of oxygen may affect how the harmful allele expresses itself.

Math

Adaptive Genetic Variations

Sickle-cell anemia results when a person inherits two recessive alleles for hemoglobin S, an oxygen-carrying protein in the blood. Hemoglobin S carries less oxygen than the normal allele, hemoglobin A. A person who is heterozygous for the trait has one allele for hemoglobin A and the other for hemoglobin S.

Study the table below and answer the questions.

Population	Genotype	Frequency in the Population
African American	SS	2 in 1000
African American	SA	86 in 1000
Central African	SS	1 in 100
Central African	SA	1 in 5

1. What percentage of African Americans have sickle-cell anemia? Central Africans?
2. What percentage of African Americans are heterozygous for the hemoglobin-S allele? Central Africans?

People homozygous for the sickle-cell allele almost always die before they reach puberty. Why doesn't this cause the sickle-cell gene to be eliminated? Is there some adaptive advantage to having the gene?

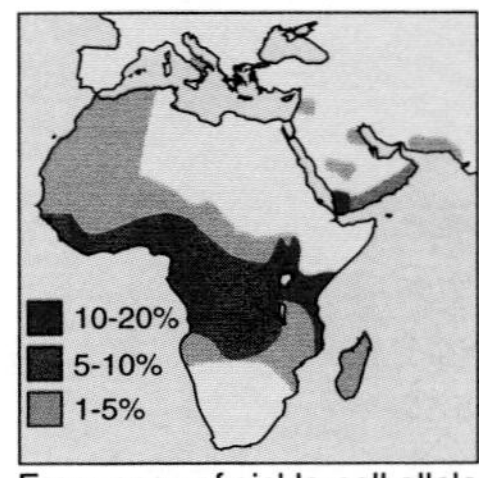

Frequency of sickle-cell allele

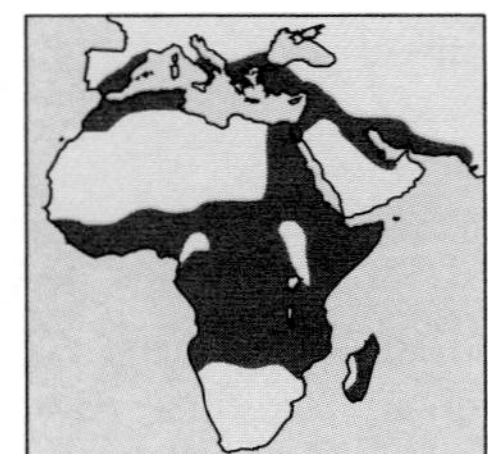
Distribution of malaria

The maps reveal that the central part of Africa is an area in which malaria is prevalent. It is also where the sickle-cell allele is prominent. People who are heterozygous for the hemoglobin-S allele are less susceptible to malaria than those who lack the S allele. The malarial parasite is unable to grow and develop within the sickled blood. Central Africans who are homozygous for the hemoglobin-S allele usually die from sickle-cell anemia, but those with just one hemoglobin-S allele have fewer symptoms associated with the disorder.

CONNECTION TO Biology

How does the history of the sickle-cell allele demonstrate that adaptations in species are determined by both their genes and environment?

slows the secretion of some digestive enzymes, so the individual cannot digest food properly. Physical therapy, special diets, and new drug therapies have raised the average life expectancy of CF patients.

Sickle-cell anemia: A blood disorder

Like cystic fibrosis, sickle-cell anemia is inherited as an autosomal recessive trait. Unlike CF, however, the disease is most common in black Americans whose families originated in Africa and in white Americans whose families originated in the countries surrounding the Mediterranean Sea. About one in 12 African Americans, a much larger proportion than in most populations, is heterozygous for the disorder.

In an individual with sickle-cell anemia, the red blood cells are shaped like a sickle, or half-moon, as shown in *Figure 15.3.* Normal red blood cells are disc shaped. Sickled cells contain hemoglobin, the oxygen-carrying protein, that differs from normal hemoglobin molecules in just one amino acid. Because sickled cells have a shorter life span than normal red blood cells, the person suffers from anemia, a low number of red blood cells. Sickled cells also clog small blood vessels, causing tissues to become damaged and deprived of oxygen and nutrients. The symptoms can include severe pain.

People with sickle-cell anemia have other serious health problems in addition to anemia because of their impaired circulation. Treatments include blood transfusions and drug therapy. Individuals who are heterozygous for the trait do not have sickle-cell anemia, but because they produce both normal and abnormal hemoglobin, they may show some symptoms if availability of oxygen is ever reduced.

PROJECT

Human Genetic Disorders

Have students make a notebook collection of articles from the newspaper, magazines, and tabloids on human genetic disorders. Ask them to write a paragraph or two about the sensitivity of the articles as they discuss human disorders. L1 LS

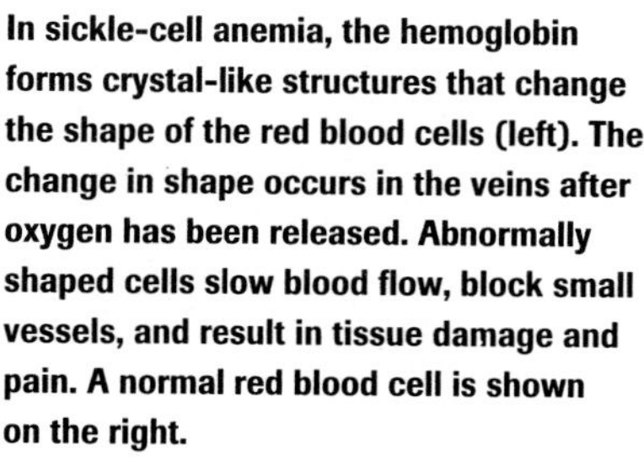

Figure 15.3

In sickle-cell anemia, the hemoglobin forms crystal-like structures that change the shape of the red blood cells (left). The change in shape occurs in the veins after oxygen has been released. Abnormally shaped cells slow blood flow, block small vessels, and result in tissue damage and pain. A normal red blood cell is shown on the right.

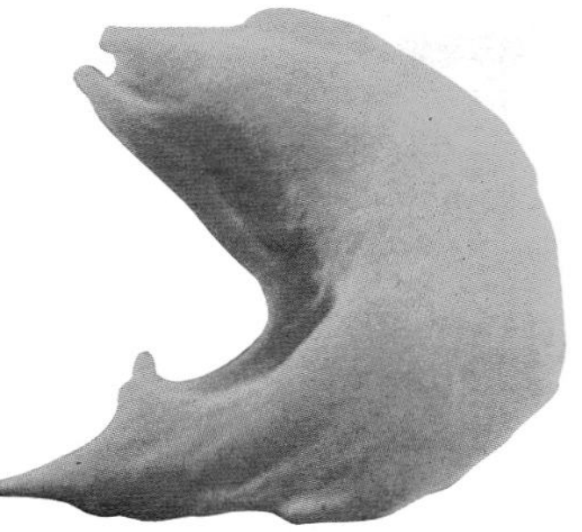

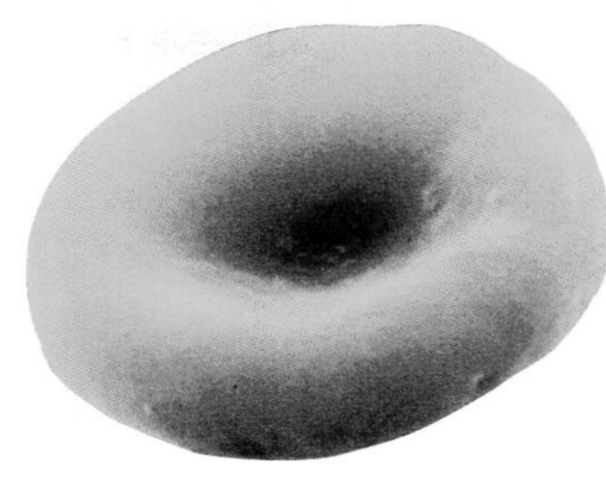

Tay-Sachs disease affects the nervous system

Tay-Sachs disease is an autosomal recessive disorder of the central nervous system. In individuals with Tay-Sachs, a recessive allele results in the absence of an enzyme that normally breaks down a lipid produced and stored in tissues of the central nervous system. Therefore, this lipid, which is found in membranes in the brain, fails to break down properly and accumulates in the cells. This accumulation results in blindness, progressive loss of movement, and mental deterioration—damage for which no treatment is available. The symptoms begin within the first year of life and result in death before the age of five. The allele for Tay-Sachs is especially common in the United States among the Pennsylvania Dutch people and among Ashkenazic Jews, whose ancestors came from eastern Europe. The Pennsylvania Dutch and Ashkenazic Jewish populations may have a Tay-Sachs allele frequency as high as one in 60. By contrast, most other populations have an allele frequency of fewer than one in 100 000. *Figure 15.4* shows a typical pedigree for Tay-Sachs disease.

Phenylketonuria

Phenylketonuria (PKU) is a recessive disorder that results from the absence of an enzyme that converts one amino acid, phenylalanine, to a different amino acid, tyrosine. Because phenylalanine cannot be broken down, it and its by-products accumulate in the body and result in severe damage to the central nervous system.

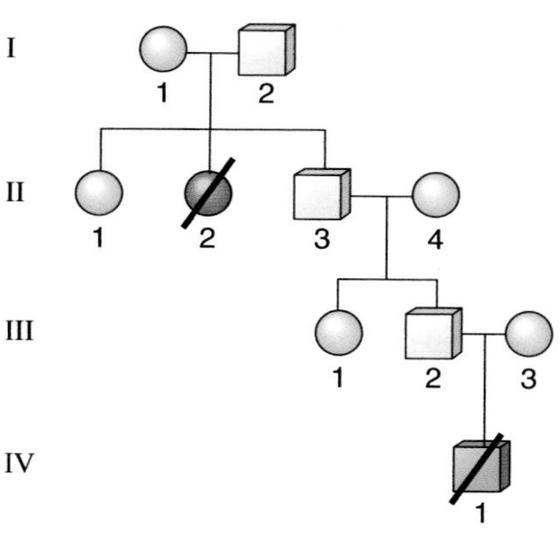

Figure 15.4

A study of families who have children with Tay-Sachs disease shows typical pedigrees for traits inherited as autosomal recessives. Note that the trait appears to skip generations. Why is this a characteristic of a recessive trait?

Visual Learning

Figure 15.4 Why is this a characteristic of a recessive trait? *An individual must have two recessive alleles to show a trait. Because these traits are rare, an affected individual most likely will mate with an individual with two dominant alleles. The offspring will be heterozygous and will not show the trait. The trait may reappear when two heterozygotes mate.*

Science, Technology, and Society

Complex Inheritance

Molecular geneticists are trying to determine the possible genetic components of conditions such as cancer, coronary heart disease, high blood pressure, and mental illnesses. Twin studies have shown that genetic susceptibility is not the only determinant of these disorders. For example, if one identical twin becomes schizophrenic, the other is affected no more than 20 to 50 percent of the time.

interNET CONNECTION

Follow the link for this chapter on the Glencoe Homepage at **www.glencoe.com/sec/science** to find out more about human heredity.

GLENCOE TECHNOLOGY

Videodisc

STVS: Human Biology
Disc 7, Side 2
Detecting Cystic Fibrosis (Ch. 17)

STUDENT JOURNAL

More Genetic Disorders Have students research the heredity disorders commonly found in the Amish population (polydactyly), or other disorders not mentioned in the chapter, such as achondroplasia, Marfan's syndrome, albinism, galactosemia, or thalassemia. L1

People in Biology

Meet Dr. Robert Murray, Geneticist

Teaching Strategies

- After students have read the interview, ask them to work with partners and role-play, taking turns as a geneticist counseling a couple. Have students assume that the man of the couple has been identified as a carrier of Tay-Sachs disease but the woman has not been tested. Have students brainstorm and make a list of questions they might ask and a summary of the information they would give this couple. L1 COOP LEARN

GLENCOE TECHNOLOGY

The Secret of Life
Use the videotape *Tinkering with Our Genes: Genetic Medicine* to show some problems of families with genetic disorders.

Microscope Activity

LS **Visual-Spatial** Have students view a prepared slide of sickled blood cells and compare it with a slide of normal blood cells. L1

people in biology

Meet Dr. Robert Murray, Geneticist

Suppose that you had a genetic defect that might one day make you very ill. Would you want to know what difficulties your future or your children's futures could hold? Or would you prefer to accept life as it comes, day by day? That's the kind of dilemma that genetic counselors help their patients face.

In the following interview, Dr. Robert Murray tells about his work as Chief of the Division of Medical Genetics in the Department of Pediatrics at Howard University in Washington, D.C.

On the Job

Q Dr. Murray, could you describe your duties as head of your division?

A Although I remain involved directing clinics that diagnose genetic conditions and provide genetic counseling, I also teach graduate courses in human genetics.

Q What are some of the genetic conditions you deal with?

A Sickle-cell anemia is a prominent one because the patient population we serve is primarily African American. We also deal with defects like cystic fibrosis and a whole host of blood disorders that relate to the red blood cells, such as hemophilia.

Q How do you and the other doctors counsel parents who may be faced with having a child with a genetic defect?

A We tell them that we have information and would be happy to share it with them and to help them with the problems that might arise as a result of having that information. Even though we might have strong feelings ourselves, it's our ethical obligation not to express those feelings in a way that would influence people's personal decisions. We're in a sensitive position, where the role of morality and ethics is very strong. Unless people are absolutely opposed, because of their religious or personal beliefs, to having the knowledge we can offer, we feel they should be counseled.

Early Influences

Q Could you tell us how you got into biology in general, and how you narrowed your specialty to genetics?

A I went into biology because of my interest in people. In high school, I completed a test that assessed students' areas of strength. My results showed strong interest in three areas: music, people, and science. For me, "Science" and "People" came together to equal "Medicine." I played piano

360

PORTFOLIO

Blue People Have students read "The Blue People of Troublesome Creek," by Cathy Trost, *Science 82,* Nov. 1982, pp. 34–39 and construct a pedigree from the article. These people have an autosomal recessive gene that causes their skin to be dark blue. Have students include the pedigree in their portfolio along with an explanation of this disorder. L2 LS P

and trumpet and sang in choruses during high school and college. As a black person, one of the reasons I decided not to go into music, even though I loved it, was because there were many minority people successful in the entertainment world but very few who worked in medicine or science, especially research. There was and still is a great need for minorities in science and medicine.

After medical school, I practiced internal medicine because I wanted to provide clinical care, but I found that many diseases were hopelessly advanced by the time they could be diagnosed. I wanted to be able to predict what might happen before the advanced stage, so that led me to genetics.

Personal Insights

Q What about your third area of strength, music?

A Music remains an important part of my life. At the Unitarian church I attend, I sing spirituals, gospel, and folk music in a group called the Jubilee Singers. The spirituals are closer to me than any other music because I grew up with those.

CAREER CONNECTION

If you'd like to help people make difficult decisions about inherited diseases, one of the following careers could interest you.

Geneticist *A bachelor's degree, four years of medical school, three years of residency (graduate medical education), master's degree in genetics*

Pharmacist *A bachelor's degree and two or more years of study at an accredited college of pharmacy*

Administrative Assistant *High-school diploma and on-the-job training*

Q Do you have any advice for students about their future careers?

A It's my personal belief that we Americans need to focus on careers that deal with helping people. In recent years, too much energy, I think, has been focused on attempts to make money and accumulate things—big houses, fancy cars, material goods. It's disappointing to have medical students ask me what specialty pays the most money and requires the least effort, although I know many of them face huge debts in paying for their education.

My advice to high-school students would be to worry less about monetary matters and more about what their lives can contribute. I think that the most satisfying choices in the long run are those that are concerned with a dedication to helping people. For myself, I could have worked in a laboratory without having to worry about dealing with people, but I feel strongly that our society suffers when we don't share what we have with others. That's why I continue to stay involved in patient care.

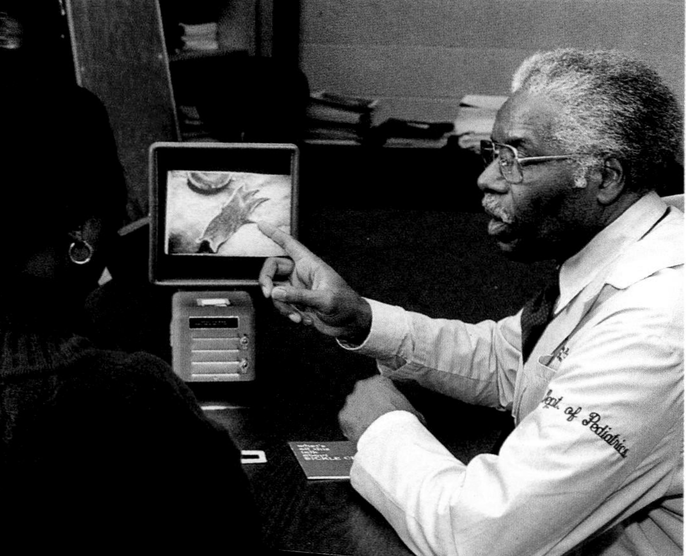

CULTURAL DIVERSITY

Mary Styles Harris

In your discussions of sickle-cell anemia, point out contributions of contemporary African-American scientists in the treatment and etiology of this disease. One of the more prominent workers in this field has been geneticist Mary Styles Harris (1949-). From 1977 to 1979, Harris was the executive director of the Sickle Cell Foundation of Georgia, and she has published many papers on this subject.

In 1980, Harris was honored as one of *Glamour* magazine's Outstanding Women Scientists. Around this time, she also wrote, narrated, and produced an educational science series for Georgia TV through a grant from the National Science Foundation.

CAREER CONNECTION

- **Career Path** Courses in High School: Earth science, biology, physics, mathematics, and chemistry.
  College: chemistry, physics, mathematics, statistics, cell biology, microbiology, genetics, psychology, sociology.
  Medical School and /or Graduate School: family medicine, pediatrics, medical genetics, medical ethics, psychology, and counseling.

For More Information

For more information on genetic counseling or on birth defects, write to the March of Dimes, 1275 Mamaroneck Avenue, White Plains, NY 10605.

3 Assess

Check for Understanding

Ask students to summarize why the study of genetics is important to couples considering having children. L1

Reteach

Have students make a table of the genetic disorders described in this section, including the type of inheritance and its effects on the person who has it. L1

Extension

Ask students to contact the local March of Dimes organization to gather information on the help it gives individuals with genetic disorders. L2

Assessment

Skill: Ask students to design a handout that tells about a genetic disorder. Students could draw Punnett squares and illustrate the chances of offspring inheriting the disorder from parents who are carriers. L1

ThinkingLab — Draw a Conclusion

Purpose

Students will determine the pattern of inheritance for Duchenne's muscular dystrophy.

Process Skills

observe and infer, recognize cause and effect

Teaching Strategies

- Ask students which individuals in the pedigree were keys to determining the type of inheritance involved.

Background

The gene that causes Duchenne's muscular dystrophy was discovered in 1986.

Thinking Critically

From the pedigree, the mode of inheritance can be inferred to be sex-linked on the X chromosome.

Assessment

Oral: Ask students what mating would have to occur to produce a female child with Duchenne's muscular dystrophy. *Mating a female carrier or an affected female with an affected male.* L1

4 Close

Discussion

Ask students to discuss whether genetic testing should be required to obtain a marriage license.

ThinkingLab — Draw a Conclusion

How is Duchenne's muscular dystrophy inherited?

Muscular dystrophy is often thought of as a single disorder, but it is actually a group of genetic disorders that produce muscular weakness, a progressive deterioration of muscular tissue, and a loss of coordination. Different forms of muscular dystrophy can be inherited as a dominant autosomal, a recessive autosomal, or a sex-linked disorder. Each pattern of inheritance appears differently when a pedigree is made.

One rare form of muscular dystrophy, called Duchenne's muscular dystrophy, affects three in 10 000 American males. People with this disorder rarely live past the age of 20.

Analysis

The pedigree shown here represents the typical inheritance pattern for Duchenne's muscular dystrophy. Refer to ***Figure 14.11*** if you need help interpreting the symbols. Analyze the pedigree to determine the pattern of inheritance exhibited by this form of the disorder.

I: 1 2
II: 1 2 3 4 5
III: 1 2 3 4
IV: 1 2 3 4 5

Thinking Critically

What can you infer from the pedigree about the way Duchenne's muscular dystrophy is inherited in families?

A homozygous PKU newborn appears healthy at first because its mother's normal enzyme level prevented phenylalanine accumulation during development. Once the infant begins drinking milk, which is rich in phenylalanine, the amino acid accumulates and damage occurs.

Mental retardation was once the usual result of PKU. Today, however, biochemical tests to detect the problem are available. Therefore, a PKU test is normally performed on all infants a few days after birth. Infants affected by PKU are given a diet that is low in phenylalanine until their brains are fully developed. With this special diet, the toxic effects of the disorder can be avoided. Ironically, the success of treating PKU infants has resulted in a new problem. If a homozygous recessive female becomes pregnant, the high phenylalanine levels in her blood can damage her **fetus**—the developing baby. This problem occurs even though the fetus is heterozygous and would be phenotypically normal. The PKU allele is most common among people whose ancestors came from Norway or Sweden.

Section Review

Understanding Concepts

1. Describe one genetic disorder that is inherited as a recessive trait.
2. How are the cause and onset of symptoms of Huntington's disease different from those of PKU and Tay-Sachs disease?
3. How does the structure of abnormal hemoglobin cause the symptoms of sickle-cell anemia?

Thinking Critically

4. The organism that causes malaria, a tropical disease, requires oxygen that is carried by the hemoglobin in red blood cells. Why might heterozygotes for sickle-cell anemia be resistant to malaria?

Skill Review

5. **Observing and Inferring** Suppose a child with free-hanging earlobes has a mother with attached earlobes. Can a man with attached earlobes be the father of the child? Explain. For more help, refer to Thinking Critically in the *Skill Handbook*.

Answers to Section Review

1. Students could describe cystic fibrosis, sickle-cell anemia, Tay-Sachs, or phenylketonuria.
2. Huntington's disease is an autosomal dominant disorder with onset between the ages of 30 and 50, whereas PKU and Tay-Sachs disease are autosomal recessive disorders with immediate onset.
3. The abnormal hemoglobin results in red blood cells that are sickle shaped. They cause anemia, clogged blood vessels, and tissue damage.

Thinking Critically

4. Because heterozygotes have both normal and abnormal hemoglobin in their blood cells, they will have less oxygen in their blood. This may make them resistant to malaria.

Skill Review

5. The man cannot be the father because the child received a gene for free-hanging earlobes from one parent; the father would have to have at least one dominant gene for this trait.

SECTION

15.2 Complex Inheritance of Human Traits

Did you ever consider whether it would be possible to predict the genetic makeup of your children? Is there some way that you can know beforehand what characteristics your children will have? Most human traits follow complex inheritance patterns, which explains why it is not an easy task to predict the genetic makeup of offspring. Also, environmental factors may influence the expression of genes. For these reasons, it is impossible to predict with any certainty the genetic makeup of a child.

Section Preview

Objectives

Compare multiple allelic, polygenic, and sex-linked patterns of inheritance in humans.

Distinguish between autosomal and sex chromosome aneuploidy.

Key Terms

aneuploidy
karyotype

Multiple Alleles

Traits that are governed by simple Mendelian heredity have only two alleles. However, more than two alleles are possible for certain traits. Blood groups are a classic example of multiple allelic inheritance.

Multiple alleles govern blood type

Human blood types are determined by the presence or absence of certain molecules on the surfaces of red blood cells. As the determinant of blood type, the gene I has three alleles: I^A, I^B, and i, often written A, B, and O. The I^A allele produces surface molecule A, and the I^B allele produces surface molecule B. The i allele produces no surface molecule. The possible combinations of blood type alleles and their phenotypes are shown in *Table 15.1*. The alleles I^A and I^B are both always expressed; that is, they are codominant. In addition, both I^A and I^B are dominant to i. So, if a person inherits an I^A allele from one parent and an I^B allele from the other parent, he or she will produce both molecules A and B and will have blood type AB. A person who is I^Ai will produce molecule A and will

Table 15.1 Human Blood Groups

Genotypes	Surface Molecules	Phenotypes
$I^A I^A$ or I^Ai	A	A
$I^B I^B$ or I^Bi	B	B
$I^A I^B$	A and B	AB
ii	none	O

Program Resources

Section Focus Master 33 L1 SAE
Reinforcement and Study Guide, pp. 59-60 L1
Biolab and Minilab Worksheets, pp. 57-60 L1
Laboratory Manual, pp. 87-90 L2
Critical Thinking/Problem Solving, p. 15 L3
Performance Assessment in the Biology Classroom, p. 21 L1 P
Basic Skills Transparency 14 and **Master** L1 SAE

SECTION 15.2

Prepare

Key Concepts

Types of complex inheritance in humans are presented. Blood types are used as an example of multiple allelic inheritance; skin color is used as an example of polygenic inheritance; color-blindness and hemophilia are used as examples of sex-linked patterns. The section concludes with a discussion of aneuploidy.

Block Scheduling

Look for this symbol for strategies that are useful in a block scheduling format. For more information on block scheduling, refer to Section 15.2 in the **Lesson Plans** booklet.

Materials

- Acquire color-blindness testing materials for the Enrichment Activity.
- Obtain flat wooden toothpicks for the Alternate Lab.

1 Focus

Bellringer

Before presenting the lesson, display **Section Focus Master 33** on the overhead projector and have students answer the accompanying questions. L1 SAE

Discussion

Ask students if they think they are genetically unique. To find out, have them calculate the chances of their parents producing another child with exactly the same genes. *The odds are greater than 1 in 2^{46}.* This number is more people than have already ever lived, making the chances of producing another identical person vanishingly small. L1

SECTION 15.2

2 Teach

MiniLab

Purpose

Visual-Spatial Students will observe the colors and patterns in the eyes of several classmates in order to hypothesize how eye color is inherited.

Process Skills

observe and infer, recognize cause and effect

Teaching Strategies

- Provide students with five cards that each have a circle with a line under it. Students should put the name of the person on the line and then draw the iris within the circle. Provide colored pencils. L1

Expected Results

Students will detect many different pigments suggesting polygenic inheritance.

Analysis

1. Students may detect brown, blue, yellow, gray, green, and black pigments.
2. Eye color is probably not inherited by simple Mendelian rules because there are so many phenotypes.
3. The pattern suggests polygenic inheritance.

Assessment

Knowledge: Ask students how they could determine the number of genes involved in the inheritance of eye color. L1

MiniLab

What colors and patterns can you detect in eyes?

Human eye color, like skin color, is determined by polygenic inheritance. You can detect several shades of eye color, especially if you look closely at the iris with a magnifying glass. Often, the pigment is deposited so that light reflects from the eye, causing the iris to appear blue, green, gray, or hazel (brown-green). In actuality, the pigment may be yellowish or brown, but not blue.

Procedure

1. Use a magnifying glass to observe the patterns and colors of pigment you detect in the eyes of five classmates.
2. Make a drawing of the iris using colored pencils.
3. Write a description of your observations in your journal.

Analysis

1. How many different pigments were you able to detect?
2. From your data, do you suspect that eye color might not be inherited by simple Mendelian rules? Explain your answer.
3. Suppose that two people have brown eyes. They have two children with brown eyes, one with blue eyes, and one with green eyes. What pattern might this suggest?

have blood type A, as will an I^AI^A individual. Only those people who are *ii* have blood type O because the *i* allele produces no surface molecules.

The importance of blood typing

Determining blood types is necessary before a person can receive a blood transfusion because incompatible blood types could clump together, causing death. Blood typing can also be helpful in cases of disputed parentage. For example, if a child has type AB blood and its mother has type A, a man with type O blood could not possibly be the father. But blood tests cannot prove that a certain man definitely is the father; they indicate only that he could be.

Polygenic Inheritance

Think of all the traits you inherited from your parents—from obvious ones such as hair and eye color, to more obscure, biochemical traits such as coding for the enzyme that converts the amino acid phenylalanine correctly. While many of your traits were inherited through simple Mendelian patterns or through multiple alleles, many other human traits are determined by polygenic inheritance. These kinds of traits are usually quantitative, representing a measurable range of variation.

Skin color is determined by polygenic inheritance

In the early 1900s, the idea that polygenic inheritance occurs in humans was first tested using data collected on skin color. Scientists found that when light-skinned people marry dark-skinned people, their offspring have intermediate skin colors. When these children marry and produce the F_2 generation, the resulting skin colors range from the light skin color to the dark skin color of the grandparents (the P_1 generation), with most children having an intermediate skin color. As shown in *Figure 15.5*, methods of measuring differences in skin colors indicate that between three and four genes are involved.

Meeting Individual Needs

Visually Impaired Many different types of color-blindness are possible. Make color-blindness testing charts available to students. Color-blindness testing kits are also available. Have pairs of students use the kits. L1 SAE

PORTFOLIO

Alzheimer's Disease Have students research the connection between chromosome 21 and Alzheimer's disease. Ask them to include a copy of their findings in their portfolio. L2 P

Sex-linked Traits

Several human traits are determined by genes that are carried on the sex chromosomes. The pattern of sex-linked inheritance is explained by the fact that males, who are XY, pass on an X chromosome to each of their daughters and a Y chromosome to each son. Females, who are XX, pass on one of their X chromosomes to each child.

Therefore, if a trait is X linked, males pass the X-linked allele to all of their daughters and to none of their sons. Heterozygous females have a 50 percent chance of passing on a recessive X-linked allele to each child. If a son receives an X chromosome with a recessive allele from his mother, he will express the recessive phenotype because he has no chance of inheriting from his father a dominant allele that would mask the expression of the recessive allele. Two traits that are known to be governed by X-linked inheritance are certain forms of color blindness and hemophilia.

Red-green color blindness

People who have red-green color blindness can't differentiate these two colors. Many people who are red-green color blind are not even aware of the fact until someone points out that they are not correctly performing a simple task such as matching the colors of their clothes.

MiniLab

How is height inherited in humans?

Because people have more than three different heights, height in humans is obviously not inherited as a simple Mendelian trait. Such environmental conditions as diet and general health also affect the expression of this trait. However, by measuring the heights of many people, you can get an indication of the inheritance pattern of the trait.

Procedure

1. Measure the heights of all your classmates to the nearest centimeter.
2. Arrange your data in 10-cm groups beginning with the shortest person and continuing to the tallest person.
3. Make a graph of your results.

Analysis

1. What was the difference in height between the tallest and the shortest person?
2. What was the average height of the group, and what group had the largest number of students?
3. From your graph, what inheritance pattern is suggested?

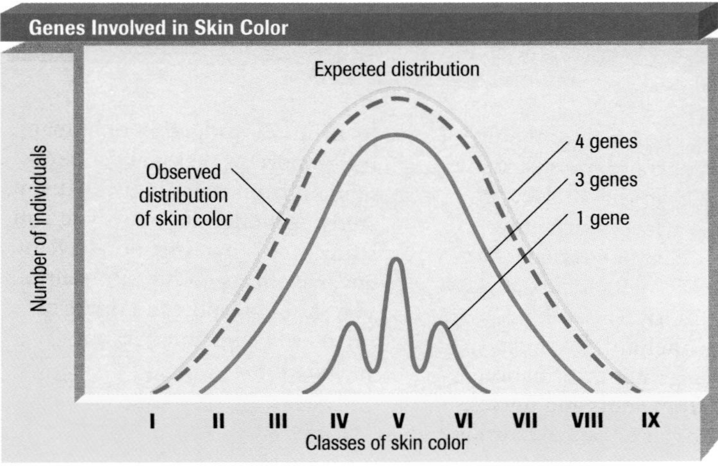

Figure 15.5

This graph shows the expected distribution of human skin color if controlled by one, three, or four genes. Which number of genes has an expected distribution of skin color that most closely matches the observed distribution?

MiniLab

Purpose

LS Logical-Mathematical Students will measure the heights of class members, graph the results, and determine the type of inheritance pattern suggested.

Process Skills

measure in SI, make a graph

Teaching Strategies

- In order to make the curve more bell-shaped, you may wish to combine data from more than one class.

Expected Results

The observed curve will be somewhat bell-shaped; the larger the number of people measured, the more bell-shaped it will be.

Analysis

1. The differences in height between the tallest and shortest person will vary, but 25 cm is not unusual.
2. The average height of the group will vary but will likely be around 165 cm.
3. The inheritance pattern suggested by the shape of the graph is polygenic inheritance.

Assessment

Portfolio: Have students write a summary and analysis of their results for their portfolios. L1 P

STUDENT JOURNAL

Improved Children Someday scientists may have the capability to change genes, making offspring smarter, physically stronger, and better looking. Ask students to write an essay on whether they would have their genes changed to "improve" their children. L1 LS

Visual Learning

Figure 15.5 Which number of genes has an expected distribution of skin color that most closely matches the observed distribution? *The actual data match the 4-gene graph best.*

SECTION 15.2

BioLab

Constructing Pedigrees to Trace Heredity

Time Allotment 
Two to three days

Objectives
Review objectives with students before they begin the Biolab.

Process Skills
observe and infer, classify

PREPARATION

- Order PTC paper from a biological supply house.
- You may wish to collect and display photos of various human genetic traits.

GLENCOE TECHNOLOGY

Videodisc

STVS: Human Biology
Disc 7, Side 2
Obesity and Heredity (Ch. 11)

The Infinite Voyage: A Taste of Health
Genetic Links to Cholesterol (Ch. 6)

The Infinite Voyage: The Geometry of Life
Hemophilia: The Effects of "Factor Eight" (Ch. 8)

BioLab

Constructing Pedigrees to Trace Heredity

Several human traits are easily observable, and constructing a pedigree is useful to determine how the traits are inherited. For instance, you may have noticed that some members of your family have attached earlobes, whereas others have earlobes that hang freely. By collecting information about family members from two or more generations, you should be able to create and analyze a pedigree that explains the inheritance pattern.

PREPARATION

Problem
How can the construction of a pedigree illustrate the heredity of human traits?

Objectives
In this Biolab, you will:
- **Construct** pedigrees for five human traits.
- **Interpret** pedigrees to determine inheritance patterns.

Materials
pencil and paper
PTC paper

Safety Precautions
PTC paper should not be ingested. After tasting, dispose of the test paper in a trash can.

PROCEDURE

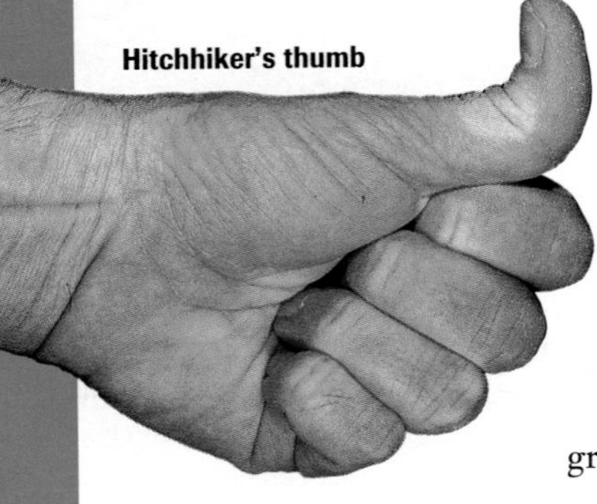
Hitchhiker's thumb

1. For each of the five traits you will study in this Biolab, collect information about as many family members as possible. It will be helpful if you can gather information from aunts and uncles, grandparents, and cousins.
2. As in any investigation of humans, it is important that you get permission from every individual you study. Although the traits you will study in this investigation have no adverse effects on human health, you should respect the rights of others to be included or not included.

PROCEDURE

Teaching Strategies
- Allow two or three days for students to collect data at home. Then they can complete their analysis and conclusions in class.
- PTC paper can be cut in half with scissors in order to double the number of people that can be tested.
- Have students review the symbols used in pedigrees and the interpretation of pedigrees. This information can be found in Figure 14.11 and on pages 345-346 of their text.

Trait Studied: ____________________

Generation number	Individual number	Relationship	Trait +/-

3. Set up a data table similar to the one above for each of the traits you study. Record all your data in the tables.
4. For each person, determine where he or she will appear in the pedigree. You need to determine each person's generation position (a Roman numeral) and individual position (an Arabic numeral).
5. Write down the relationship of each person to you.
6. Use the illustrations here and on page 357 to determine whether the person has any of the following traits: widow's peak, hitchhiker's thumb, attached earlobes, and dimple in the chin.
7. To determine whether a person can taste PTC, place the test paper flat on his or her wet tongue. After two or three seconds, the paper should be removed and discarded. PTC tasters will detect a bitter taste, and non-tasters will notice only the taste of the paper.
8. Draw a pedigree based on your data for each trait.

Chin dimple

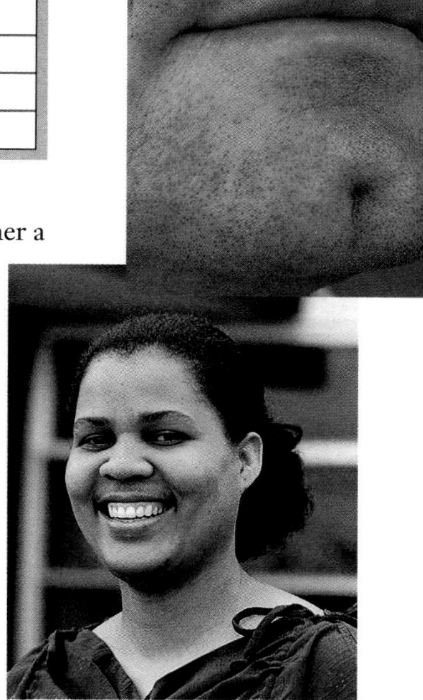

Widow's peak

ANALYZE AND CONCLUDE

1. **Interpreting Observations** Which phenotype appeared to be dominant in your investigation of the ability to taste PTC?
2. **Interpreting Observations** What appeared to be the inheritance pattern for hitchhiker's thumb?
3. **Making Inferences** If you knew that a particular trait was inherited as a recessive trait, how could a pedigree help you predict the appearance of the trait in the next generation?
4. **Analyzing Data** Were you unable to interpret the heredity pattern for any of the traits from your data? If so, explain why your results were inconclusive.

Going Further

Project Design a project to determine the inheritance pattern of a human trait that interests you. Include information from several families that express the trait. Carry out your investigation with direction from your teacher and the permission of the individuals included in the study.

Data and Observations

Student data may be similar to those shown below:

Trait Studied: PTC Tasting

Generation number	Individual number	Relationship	Trait +/-
I	1	Dad's father	+
I	2	Dad's mother	-
II	1	Dad	-
II	2	Mom	+
III	1	Self	+

ANALYZE AND CONCLUDE

1. PTC tasting appears to be the dominant trait.
2. Hitchhiker's thumb is a dominant trait.
3. A pedigree would show the genotypes of the parents.
4. Students may not be able to interpret the hereditary pattern for a trait if the sample size is too small.

Assessment

Portfolio: Students should include in their portfolios a summary of the lab, their pedigree drawings, and the answers to Analyze and Conclude. L1 P

Going Further

For background information on human traits, Victor A. McKusick's *Mendelian Inheritance of Man* is available from Johns Hopkins Press. If students want to design an investigation that uses information about another human trait, this is an excellent resource. L2

Bioethics

Preserving Confidentiality

In time, scientists will map the genes of all 46 human chromosomes. This will allow analysis of an individual's genes and prediction of genetic problems. At first it may seem that this information could only be helpful. However, some people fear that if insurance companies and employers find out about a person's genetic disorder, they could deny health insurance coverage or employment to that person. Discuss with students the pros and cons of this issue.

SECTION 15.2

Assessment

Performance Assessment in the Biology Classroom: p. 21, *Analyzing Human Pedigrees.* Have students carry out this activity after they have learned about inheritance of human traits. L1 P

History Connection

Hemophilia in High Places

Hemophilia played an important historical role in Russia during the reign of Nicholas II, the last Czar. The Czarevitch, Alexis, was hemophilic and his mother, the Czarina, was the granddaughter of Queen Victoria from whom she inherited the hemophilia allele. The Czarina was convinced that the only one who could save her son's life was the monk Rasputin, a politically unscrupulous religious fanatic. Alexis died on July 17, 1918, when the Czar and his family were murdered by order of Lenin. It is doubtful, however, that the Czarevitch would ever have succeeded to the throne. His hemophilia caused continued deterioration of his health, which neither Rasputin nor the court doctors could have prevented.

Figure 15.6

Red-green color blindness was first described in a young boy who could not be trained to harvest only the ripe, red apples from his father's orchard. Instead, he chose green apples as often as he chose red. These days, it might be a minor embarrassment if you didn't match the color of your clothes or pick only red apples, but what might happen if you were unable to match colored wires while repairing a television set? Serious consequences can result from being color blind and not realizing it.

Other more serious problems can result from this disorder, as shown in *Figure 15.6.* Color blindness is caused by the inheritance of either of two recessive alleles at two gene sites on the X chromosome that affect red and green receptors in the cells of the eyes.

Hemophilia—An X-linked disorder

Did you ever wonder at how quickly a cut stops bleeding? This human adaptation can be lifesaving. If your blood didn't have the ability to clot and you bruised yourself or scraped your knee, you would be in danger of bleeding to death, either internally from the bruise or externally from the scrape.

Hemophilia A is an X-linked disorder that causes just such a problem with blood clotting. About one male in every 10 000 has hemophilia, but only about one in 100 million females inherits the same disorder. Why? Because males have only one X chromosome. A single recessive allele for hemophilia will cause the disorder. Females would need two recessive alleles to inherit hemophilia. Males inherit the allele for hemophilia on the X chromosome from their carrier mothers. The family of Queen Victoria, who is shown in *Figure 15.7,* is the best known study of hemophilia A, also called royal hemophilia.

Hemophilia can be treated with blood transfusions and injections of Factor VIII, the blood-clotting enzyme that is absent in people affected by the condition. However, both treatments are expensive. New methods of DNA technology are being used to develop a safer and cheaper source of the clotting factor.

Figure 15.7

Queen Victoria of England was a carrier of hemophilia. Her daughter, Beatrice, transmitted the allele to the Spanish royal family, while her granddaughter, Alix, transmitted the mutant allele to the Russian royal family.

Alternate Lab: Barr Bodies

Purpose 

Visual-Spatial Students will locate Barr bodies in their epithelial cells.

Materials

flat toothpick, methylene blue stain, microscope slide, clean coverslip, microscope

Background

Females have two X chromosomes, but early in development of a female embryo, one of the X chromosomes becomes inactive in each cell. The inactive chromosome becomes condensed and can be seen as a Barr body. Male cells do not contain Barr bodies.

Procedure

Give students the following directions.

1. Place a drop of methylene blue stain on a microscope slide.
2. Gently scrape the inside of your cheek with the toothpick and stir the cells into the drop. Discard the toothpick.
3. Apply a coverslip to the drop.

Mistakes in Meiosis

You have been reading about traits that are caused by one or several genes on chromosomes. What would happen if an entire chromosome or part of a chromosome were missing from the complete set? What if there were an extra chromosome? Usually, but not always, abnormal numbers of chromosomes result from accidents of meiosis. Many phenotypic effects result from such mistakes.

Autosomal aneuploidies

You know that a human usually has 23 pairs of chromosomes, or 46 chromosomes altogether. Of these 23 pairs of chromosomes, 22 pairs are autosomes. All humans who have an unusual number of autosomal chromosomes are trisomic—that is, they have three of a particular autosome instead of just two. In other words, they have 47 chromosomes. Trisomy usually results from nondisjunction, which occurs when paired homologous chromosomes fail to separate during meiosis. Failure to separate results in **aneuploidy**—the condition of having an abnormal number of chromosomes. Trisomy is one kind of aneuploidy.

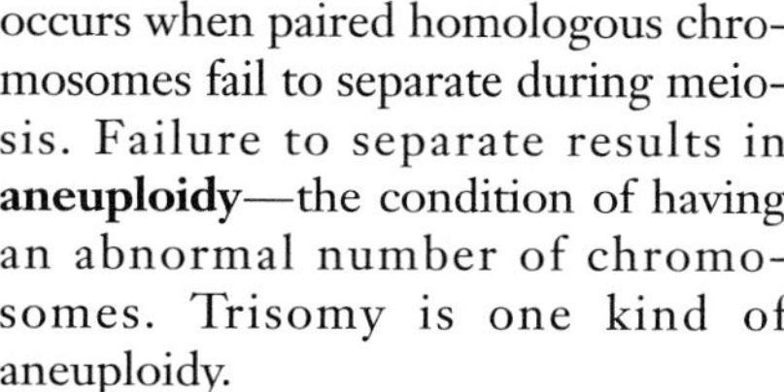

To identify an aneuploidy, a sample of cells is obtained from an individual or from a fetus. Metaphase chromosomes are photographed, and the chromosome pictures are then enlarged, cut apart, and arranged in pairs on a chart according to length and location of the centromere. This chart of chromosome pairs is called a **karyotype,** and it is valuable in pinpointing aneuploidies.

Ethics — BIOLOGY & SOCIETY

Genetic Screening

Within the nucleus of nearly all the trillions of cells in the human body is a complete copy of the human genome. That genome is all the genetic information needed as the master blueprint for building a particular person. Most people are not aware of their genetic heritage. Within the next decade, scientists are likely to change all that. The Human Genome Project, an international effort, has already located 8000 of the 100 000 genes on our chromosomes.

Genetic tests Genetic tests are becoming common. Prenatal tests examine fetal cells for genetic disorders such as Down syndrome. Prospective parents are screened for defective recessive alleles. A new test reveals a person's susceptibility to diseases such as cancer that are influenced by environmental factors.

Different Viewpoints

Genetic tests offer a person who is at risk for a disease a chance to take preventive action. In the case of heart disease, the person might exercise, quit smoking, and eat a low-fat diet to extend life expectancy. The tests might also allow for early treatment and diagnosis of diseases such as breast cancer, which can save lives. Genetic tests can also become part of our medical histories, helping in finding compatible organs for transplants.

An instrument for discrimination? Some people worry that genetic tests have the potential for being used to discriminate against those with genetic problems. Health insurers could use genetic information to refuse coverage for those who carry genes associated with diseases. Many inheritable disorders occur more often within ethnic groups. These groups may feel that screening programs are racially biased. Test accuracy in prenatal screening is often a problem, causing trauma for prospective parents.

Concerns for civil rights and the new genetics have prompted new state and national laws to protect individual privacy. Methods used in genetic testing must be accurate because of their serious consequences.

INVESTIGATING the Issue

Analyzing the Issue Brainstorm in groups why genetic testing for susceptibility to disease is a concern for employers.

Ethics — BIOLOGY & SOCIETY

Genetic Screening

Purpose

Students gain an understanding of the benefits to be derived from genetic testing. They are also presented with ethical concerns associated with the technology.

Teaching Strategies

- Have students debate whether or not insurance companies should have access to results of genetic tests. Ask some groups to pretend they own and manage a large company. Ask other groups to pretend they are officers in the employees' union. The individual's right to privacy will come up against full disclosure of genetic test results to protect the company from escalating insurance costs.

INVESTIGATING the Issue

Analyzing the Issue Genetic testing may mark some people as undesirable employees because they are at risk for developing disabling diseases.

Going Further

Have students write an essay on the following question: Would you avail yourself of the opportunity to have a genetic screening test that would reveal whether you are a carrier of a potentially fatal genetic disorder?

4. Examine cells under high power. If you are a female, a Barr body will be seen as a darkly staining mass just inside the nuclear membrane.
5. Exchange slides with a member of the opposite sex. Examine this second slide for the presence of Barr bodies.
6. Draw a cell with a Barr body and one with no Barr body.

Analysis

Ask students the following questions.

1. Do all your body cells have Barr bodies? *yes for females; no for males*
2. What percentage of the female's cells have visible Barr bodies? *all of them*
3. How many Barr bodies would you find in a cell of an XXX female? *two*

Assessment

Journal: Students should include a summary of the lab, their drawings, and the answers to the Analysis questions in their journals. L1

SECTION 15.2

3 Assess

Check for Understanding

Ask students to explain the patterns shown by multiple allelic, polygenic, and sex-linked inheritance. L1

Reteach

Have students draw the phases of meiosis to demonstrate how various aneuploidies occur. L1

Extension

Ask students to research the rare sex-linked disorder *severe combined immune deficiency (SCID)*. The "Boy in the Bubble" had this disorder. What treatments are currently being tried for this disorder? L2

Assessment

Skill: Provide students with additional genetics problems, covering both autosomal and sex-linked inheritance. L1

4 Close

Discussion

Ask students to explain why hemophilia is extremely rare in females. *Because the allele is rare in the general population, there is only a small likelihood that a male with hemophilia would marry a carrier female or a female with hemophilia.*

Figure 15.8

Down syndrome is the only autosomal trisomy in which affected individuals survive to adulthood. Individuals who have Down syndrome characteristically have a large, thick tongue and shortened stature. They have at least some degree of mental retardation. If not profoundly affected, many people with Down syndrome live happy and productive lives.

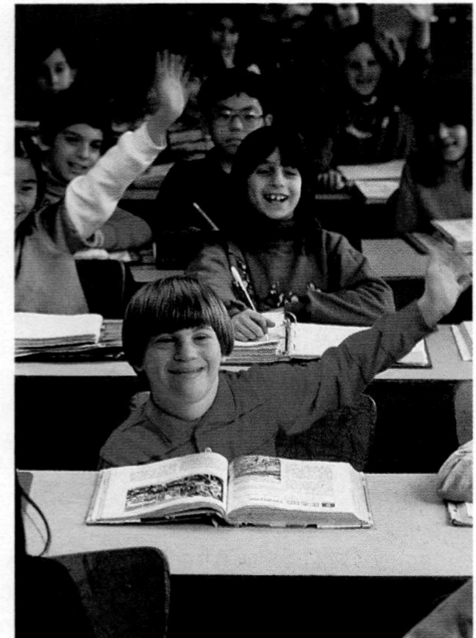

Down syndrome—Trisomy 21

Down syndrome is a group of symptoms that results from trisomy of chromosome 21. The incidence of Down syndrome births is higher in older mothers, especially those over 40. *Figure 15.8* shows a person with Down syndrome.

Other syndromes due to aneuploidy

In addition to Down syndrome, other autosomal aneuploidies can occur in humans. These syndromes result from trisomy of other chromosomes, and they cause symptoms so severe that the developing fetus dies, often even before the woman realizes she is pregnant. Fortunately, these syndromes occur only rarely.

aneuploidy:
an (GK) not
eu (GK) true
Aneuploidy is the condition of having an abnormal number of chromosomes.

Connecting Ideas

A knowledge of Mendelian inheritance and other, more complex patterns of heredity makes it easier to understand how human traits are inherited. New technology enables us to learn of the inherited traits of individual humans even before they are born. How does the application of technology affect inheritance? Can inherited characteristics be altered in individuals? How can desirable genes be used to improve other organisms? Answers to these questions depend on understanding how genes and chromosomes can be manipulated in organisms.

Section Review

Understanding Concepts

1. Why are sex-linked traits such as red-green color blindness and hemophilia more commonly found in males than in females?
2. In addition to detecting chromosome abnormalities, what information would a karyotype reveal?
3. What would the genotypes of parents have to be for them to have a color-blind daughter? Explain.

Thinking Critically

4. A man is accused of fathering two children, one with type O blood and another with type A blood. The mother of the children has type B blood. The man has type AB blood. Could he be the father of both children? Explain your answer.

Skill Review

5. **Making and Using Tables** Construct a table of the traits discussed in this section. For column heads, use Trait, Pattern of Inheritance, and Characteristics. For more help, refer to Organizing Information in the *Skill Handbook*.

Answers to Section Review

1. Males inherit only one X chromosome. If it carries a gene for a disorder, the male will show the trait.
2. Sex of the child
3. Father would have to be color-blind (have recessive allele on his X chromosome) and the mother must have at least one X chromosome with the color-blind allele since a female receives one X from each parent.

Thinking Critically

4. The mother could be I^Bi or I^BI^B and the father is I^AI^B. If a child received I^A from the father and *i* from the mother, it would be type A so the man could have fathered the type A child. This man could not father a type O child because he has no *i* allele.

Skill Review

5. Material for the table can be found on pp. 363 to 370 of the text.

Chapter 15 Review

REVIEWING MAIN IDEAS

15.1 Simple Mendelian Inheritance of Human Traits

- Many human traits are inherited according to the simple Mendelian rule of dominance.
- Most human genetic disorders are inherited as rare recessive alleles, but a few are inherited as dominant alleles.

15.2 Complex Inheritance of Human Traits

- The majority of human traits are controlled by multiple alleles or by polygenic inheritance. Because the inheritance patterns of these traits are highly variable, it is almost impossible for individuals to predict the exact phenotype or genotype of their offspring.
- Sex-linked traits are determined by inheritance of sex chromosomes. X-linked traits are usually passed from carrier females to their male offspring. Y-linked traits are passed only from male to male.
- Mistakes in meiosis result in aneuploidy, an abnormal number of chromosomes. The affected chromosomes can be autosomes or sex chromosomes. Nondisjunction during meiosis is a common cause of aneuploidy.

Key Terms

Write a sentence that shows your understanding of each of the following terms.

aneuploidy
fetus
karyotype

Understanding Concepts

1. To analyze chromosomes, geneticists construct a chart of chromosomes based on their sizes and centromere locations called a(n) ____________.
   a. pedigree
   b. karyotype
   c. inheritance pattern
   d. fetal chart
2. To determine whether an individual has a given trait, a systematic search for individuals of certain genotypes, called ____________, is performed.
   a. genetic screening
   b. building a pedigree
   c. karyotype analysis
   d. ultrasound monitoring
3. Which event in meiosis might result in trisomy 21?
   a. prophase II
   b. metaphase I
   c. prophase I
   d. nondisjunction
4. How is Huntington's disease inherited?
   a. autosomal dominant
   b. autosomal recessive
   c. sex linkage
   d. polygenic inheritance
5. How is height inherited in humans?
   a. autosomal dominant
   b. autosomal recessive
   c. sex linkage
   d. polygenic inheritance
6. How is Tay-Sachs disease inherited?
   a. autosomal dominant
   b. autosomal recessive
   c. sex linkage
   d. polygenic inheritance
7. Which of the following is inherited as an autosomal dominant?
   a. Tay-Sachs disease
   b. sickle-cell anemia
   c. cystic fibrosis
   d. polydactyly

Chapter 15 Review

Reviewing Main Ideas

Summary statements can be used by students to review the major concepts of the chapter.

Key Terms

Answers should go beyond defining the terms. Accept any answer that uses the term correctly and in the proper context.

Understanding Concepts

1. b
2. a
3. d
4. a
5. d
6. b
7. d

Chapter 15 Review

8. b
9. c
10. b
11. a
12. b
13. d
14. c
15. a
16. a
17. a
18. d
19. d
20. b

Applying Concepts

21. If the woman's father had the allele on his X chromosome, he would have had hemophilia, but he did not. Thus, the X chromosome she received from him is free of the allele. It is possible, but unlikely, that she received an X chromosome from her mother that carries the allele. In that case, she could pass it to her son.
22. A karyotype may have been prepared from fetal cells.
23. The man could be I^Bi and be the father. A blood test can show only that he could possibly be the father, but not that he is the father.

Chapter 15 Review

8. The inheritance pattern that occurs equally in males and females and skips generations is ___________.
   a. autosomal dominant
   b. autosomal recessive
   c. sex linkage
   d. polygenic inheritance
9. Which of the following is inherited by sex linkage?
   a. Huntington's disease
   b. Down syndrome
   c. hemophilia
   d. cystic fibrosis
10. To treat phenylketonuria, infants are ___________.
    a. given blood transfusions
    b. given a special diet
    c. genetically altered
    d. given phenylalanine
11. Which of these is determined by a single dominant allele?
    a. polydactyly
    b. cystis fibrosis
    c. Tay-Sachs
    d. blood type
12. Which genetic condition might be revealed by karyotyping?
    a. Huntington's disease
    b. Down syndrome
    c. polydactyly
    d. sickle-cell anemia
13. What kind of mutation results in aneuploidy?
    a. point mutation
    b. crossing over
    c. autosomal recessive
    d. nondisjunction
14. The normal karyotype of human males is ___________.
    a. XX
    b. YY
    c. XY
    d. XXY
15. Most sex-linked traits are passed from ___________ to ___________.
    a. mother; son
    b. father; daughter
    c. father; son
    d. mother; daughter
16. Normally lethal, autosomal dominant traits are eliminated from a population. This is not true of Huntington's disease because Huntington's ___________.
    a. has late onset
    b. isn't lethal
    c. isn't an autosomal trait
    d. isn't a dominant trait
17. Of these, which is NOT a result of single dominant autosomal inheritance?
    a. skin color
    b. polydactyly
    c. tongue curling
    d. free earlobes
18. A mother with blood type I^Bi and a father with blood type I^AI^B have children. Which of these is NOT a possible genotype for their children?
    a. I^AI^B
    b. I^Ai
    c. I^BI^B
    d. ii
19. If a trait is X linked, males pass the X-linked allele to ___________ of their daughters.
    a. none
    b. 25 percent
    c. 50 percent
    d. all
20. ___________ is a disorder that results from trisomy of chromosome 21.
    a. Cystic fibrosis
    b. Down syndrome
    c. Phenylketonuria
    d. Polydactyly

Applying Concepts

21. The brother of a woman's father has hemophilia. Her father was unaffected, but she worries that she may have an affected son. Should she worry? Explain.
22. A pregnant woman has learned that her fetus will have Down syndrome. How might this have been determined?
23. If a child has type O blood and its mother has type A, could a man with type B be the father? Why couldn't a blood test be used to prove whether he is the father?

Program Resources

Chapter Assessment, pp. 85-90 L1
Alternate Assessment in the Science Classroom
Computer Test Bank L1
Content Mastery, pp. 57-60 L1

Chapter 15 Review

24. How can a single mutation in a protein such as hemoglobin affect several body systems?
25. Why do certain human genetic disorders, such as sickle-cell anemia and Tay-Sachs, occur more frequently among one ethnic group than another?

Thinking Critically

26. ***Concept Mapping*** Make a concept map that relates the following terms. Supply the appropriate linking words for your map.
    gamete, meiosis, aneuploidy, homologous chromosomes, *n*, 2*n*, nondisjunction
27. ***Observing and Inferring*** Muscular dystrophy is a genetic disorder in which muscle tissue gradually breaks down. It almost always occurs in males. What does this suggest about its genetic basis? Explain.
28. ***Comparing and Contrasting*** Compare multiple allele inheritance with polygenic inheritance.
29. ***Making and Using Tables*** Make a table of all human traits studied in this chapter, including modes of inheritance. Place a check under the correct mode of inheritance for each disease listed.
30. ***Observing and Inferring*** The graph below identifies students in a class by height ranges. Assume a new student enters the class. What height range would you predict this student's height will fall into?

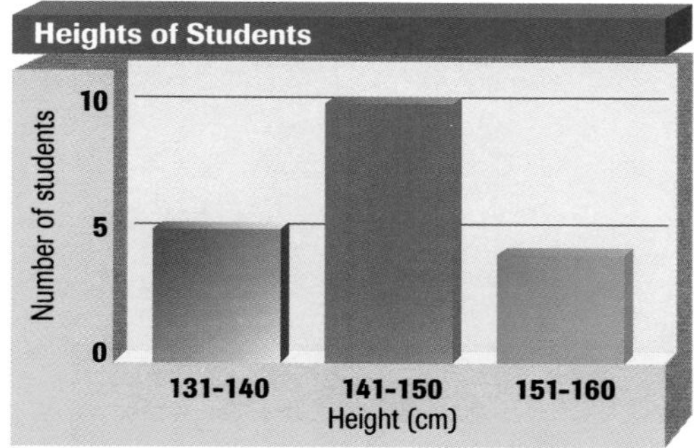

ASSESSING KNOWLEDGE & SKILLS

The following graph represents the number of children in a large family who have inherited a particular trait. Assume that no one in the history of the family has ever expressed the trait.

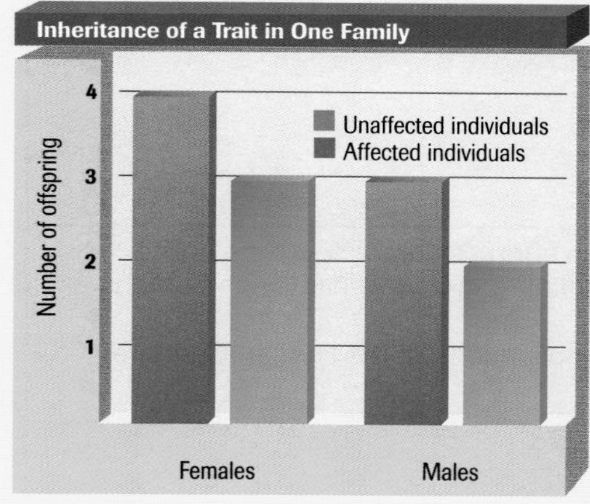

Interpreting Data Use the graph to answer the questions that follow.

1. How many total offspring were in the family?
   a. 2 c. 10
   b. 3 d. 12
2. How many total offspring were affected?
   a. 7 c. 3
   b. 4 d. 2
3. About what percentage of the offspring were unaffected?
   a. 20 percent c. 60 percent
   b. 40 percent d. 80 percent
4. ***Interpreting Scientific Illustrations*** What is the most probable inheritance pattern for the trait?

Chapter 15 Review

24. The protein may be one that is transported or used in many systems. Malfunctions of one body system usually affect others.
25. The ethnic groups tend to marry within their own ethnic group, allowing recessive genes to show up in the populations more frequently.

Thinking Critically

26. See below.
27. Because it is much more prevalent in males, sex linkage is suggested.
28. In multiple-allele inheritance, many forms of a trait may be found in a population, but they are based on only two alleles for the trait in each individual. In polygenic inheritance, the forms of the trait appear continuous, and individuals have many genes that add to the inheritance of the trait.
29. The table should list human traits in the first column and have columns for autosomal dominant, autosomal recessive, multiple alleles, polygenic, sex linked, and aneuploidy in the next columns. Checks for the modes of inheritance should be in appropriate columns for the traits discussed in the chapter.
30. between 141-150 cm in height

ASSESSING KNOWLEDGE & SKILLS

1. d
2. a
3. b
4. autosomal recessive

26.

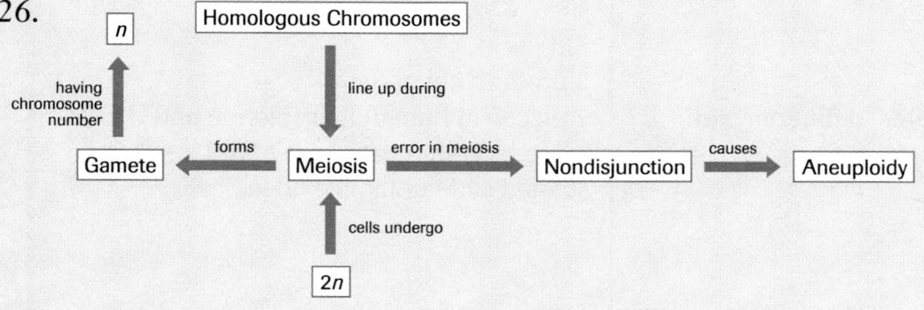

Chapter Organizer

SECTION	OBJECTIVES	ACTIVITIES/FEATURES
16.1 Recombinant DNA Technology National Science Standards: UCP.2, UCP.3; A.1, A.2; C.2; E.1, E.2; F.6; G.1, G.2	1. **Summarize** the steps used to engineer transgenic organisms. 2. **Give examples** of applications and benefits of genetic engineering.	**Minilab:** Why does recombinant DNA require palindromes?, p. 378 **Minilab:** How is splicing wires similar to splicing genes?, p. 379 **Biolab:** Modeling Recombinant DNA, p. 380 **Literature Connection:** *Jurassic Park,* p. 382
16.2 The Human Genome National Science Standards: UCP.2, UCP.3; A.1; C.2; E.1, E.2; F.1, F.6; G.1, G.2	3. **Analyze** how the effort to completely map and sequence the human genome will advance human knowledge. 4. **Predict** future applications of the Human Genome Project.	**Biology & Society Ethics:** Brave New World, p. 389 **Thinking Lab:** How do you analyze a DNA fingerprint?, p. 390

ACTIVITY MATERIALS

BIOLAB	MINILABS	ALTERNATE LAB
page 380 paper colored pencils, red and green tape scissors	**page 378** scissors paper **page 379** 2 plastic-wrapped wires, different colors wire cutters or scissors wire strippers	**page 382** microscope slide red juice 5 product bar codes 2 pieces of human hair (blonde and dark) piece of T-shirt with spot of red ink stained slide of female cheek cells

Chapter 16 DNA Technology

TEACHER CLASSROOM RESOURCES

Reproducible Masters	Transparencies
Section Focus Master 34: Making a New Food L1 SAE **Reinforcement and Study Guide,** pp. 61-63 L1 **Biolab and Minilab Worksheets,** pp. 61-64 L1 **Content Mastery,** pp. 61-64 L1	**Basic Concepts Transparency #19:** Recombinant DNA L1 SAE **Reteaching Transparency #16:** Recombinant DNA Technique L1 SAE
Section Focus Master 35: Mapping Human Genes L1 SAE **Reinforcement and Study Guide,** p. 64 L1 **Concept Mapping:** Human Genome Project, p. 16 L1 **Tech Prep Applications:** DNA Fingerprints Unite Parent and Child, pp. 19-20 L2 **Critical Thinking/Problem Solving:** The Challenges of the Human Genome Project, p. 16 L3 **Laboratory Manual:** DNA Sequencing, pp. 91-94 L2	

ASSESSMENT MATERIALS

Chapter Assessment, pp. 91-96
Performance Assessment in the Biology Classroom, p. 19
MindJogger Videoquiz
Alternate Assessment in the Science Classroom
Computer Test Bank

Spanish Resources SAE
English/Spanish Audiocassettes SAE
Cooperative Learning in the Science Classroom COOP LEARN
Lesson Plans

KEY TO TEACHING STRATEGIES

L1 Level 1 activities should be within the ability range of all students including those with learning difficulties.

L2 Level 2 activities are within the ability range of average to above-average students.

L3 Level 3 activities are designed for the ability range of above-average students.

SAE SAE activities should be within the ability range of Students Acquiring English.

COOP LEARN Cooperative Learning activities are designed for small group work.

P These strategies represent student products that can be placed into a best-work portfolio.

These strategies are useful in a block scheduling format.

GLENCOE TECHNOLOGY

The following multimedia resources are available from Glencoe.

Biology: The Dynamics of Life
CD-ROM SAE
Videodisc Program

The Infinite Voyage Series
The Geometry of Life
Selective Breeding
Manipulating Genetic Engineering
The Human Genome Project
Miracles by Design
Biodegradable Plastic: The Miracle Material?

The Secret of Life Series
In the Land of Milk and Money: Biotechnology

Science and Technology Videodisc Series (STVS)
Plants & Simple Organisms
Plant Clones
Genetic Engineering in Barley

CHAPTER

16 DNA Technology

Chapter Overview

As students begin their study of the chapter, they are introduced to recombinant DNA technology and its applications to genetic engineering. They learn about techniques scientists use to manipulate genes, using restriction enzymes and gene splicing. Students then explore the present and potential future uses of transgenic organisms in agriculture, industry, and medicine.

In the second section of the chapter, students investigate the Human Genome Project, an ambitious international effort that aims to determine the chromosome location and base sequence of every human gene. At the conclusion of the chapter, students learn about applications of chromosome maps to diagnosis of genetic disorders, gene therapy, and DNA fingerprinting.

Key Terms

- cell culture
- clone
- gene splicing
- gene therapy
- genetic engineering
- human genome
- linkage map
- plasmid
- recombinant DNA
- restriction enzyme
- transgenic organism
- vector

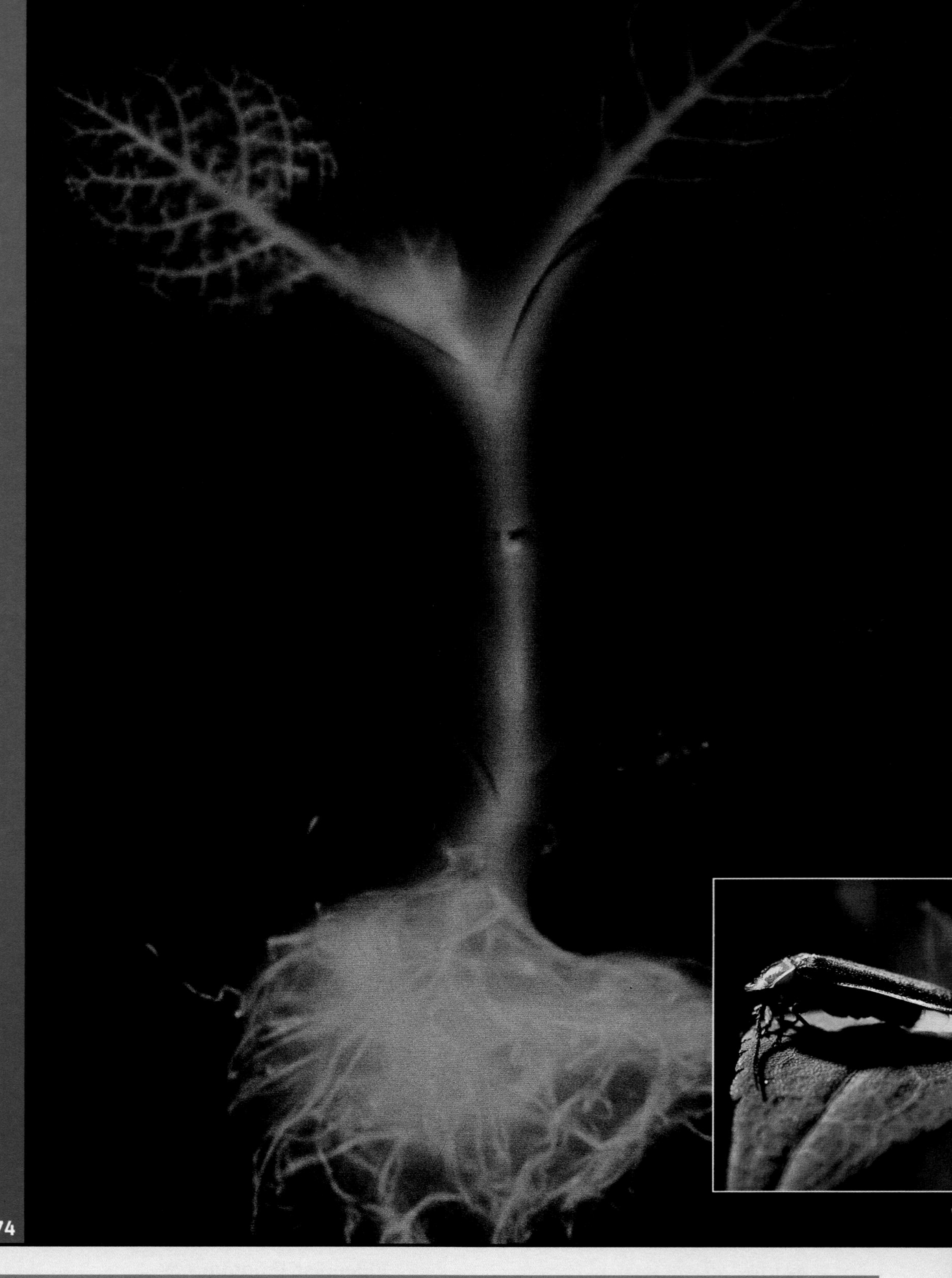

374

Learning Styles

Look for the following logo for strategies that emphasize different learning modalities. LS

Kinesthetic	**Meeting Individual Needs, p. 377; Project, p. 378; Minilab, p. 379**
Visual-Spatial	**Demonstration, p. 377; Minilab, p. 378; Thinking Lab, p. 390**
Interpersonal	**Project, p. 389**
Linguistic	**Portfolio, pp. 384, 388; Student Journal, p. 389; Meeting Individual Needs, p. 387**
Logical-Mathematical	**Alternate Lab, p. 382**

CHAPTER 16 DNA Technology

Tobacco plants that glow? How can there be such an unusual organism? You won't find glowing tobacco in nature, but it exists in a lab. Biologists now can alter the heredity of organisms by moving genes from one species to another. In a well-known experiment done in 1986, the gene for light production in fireflies was inserted into the chromosomes of a tobacco plant. When the gene was activated, the plant glowed!

On a more practical note, geneticists can produce crop plants that supply more protein. Through genetic manipulation of certain bacteria, medicines such as insulin for diabetics are being made in large quantities. Other bacteria are being custom produced with the ability to break down pollutants such as pesticides and convert them to harmless substances. Soon it may be possible to alter human heredity and perhaps "cure" such disorders as muscular dystrophy and diabetes by genetic means.

Concept Check

You may wish to review the following concepts before studying this chapter.

- Chapter 13: DNA, mutation
- Chapter 15: genetic disorders

Chapter Preview

16.1 Recombinant DNA Technology
Gene Manipulation
Applications of DNA Technology

16.2 The Human Genome
Mapping and Sequencing the Human Genome
Applications of the Human Genome Project

Laboratory Activities

Biolab

- Modeling Recombinant DNA

Minilabs

- Why does recombinant DNA require palindromes?
- How is splicing wires similar to splicing genes?

Magnification: 75×

The key that unlocks the door to genetic manipulation is DNA. The strands of synthetic DNA shown at the right were made by copying DNA extracted from animal cells. Technology that would cut, combine, and insert DNA such as this had to be developed before genetic manipulation could become a reality.

Firefly

Assessment Planner

Choose assessment strategies from the following pages to evaluate the progress of your students.

Assess, pp. 385, 389
Alternate Lab, pp. 382-383
Minilabs, pp. 378, 379
Portfolio, pp. 384, 388
Thinking Lab, p. 390
Biolab, pp. 380-381
Chapter Review, pp. 391-393

Introducing the Chapter

Explain to students that the experiment with the tobacco plants was not just an idle trick. Scientists learned a great deal from the experiment about how genes can be moved from one species to another.

The gene for luciferase (the enzyme that catalyzes the reaction of ATP, oxygen, and luciferin, producing the green light) was inserted into the DNA of a plant virus. The virus was introduced into a bacterium, which in turn infected the plant cells and carried its DNA into the host's cells. When the plants were watered with a solution containing luciferin, they would glow.

Theme Development

The theme *nature of science* is developed in this chapter as the techniques that now exist for changing the genetic makeup of organisms are discussed. Some of the techniques may be used to restore *homeostasis* to organisms afflicted with genetic diseases.

Concept Check

Review with students the double-stranded structure of DNA and how mutations occur, which they studied in Chapter 13. As students investigate the application of genetic engineering to diagnosis and treatment of genetic diseases, they will build on what they learned in Chapter 15 about how these disorders are inherited.

SECTION

16.1

Prepare

Key Concepts

Students will learn that genetic engineering involves using restriction enzymes to cleave DNA and recombining the pieces with fragments of DNA from another source. The result is recombinant DNA, which can be inserted into an organism. This organism is known as a transgenic organism. Students will explore techniques used to sequence DNA and applications of DNA technology in agriculture, industry, and medicine.

Block Scheduling

Look for this symbol for strategies that are useful in a block scheduling format. For more information on block scheduling, refer to Section 16.1 in the **Lesson Plans** booklet.

Materials

- Order Chromosome Simulation Biokit for the Demonstration.
- Collect articles on recombinant DNA technology for the Display.
- Purchase colored wire and wire strippers for the Minilab.

1 Focus

Bellringer

Before presenting the lesson, display **Section Focus Master 34** on the overhead projector and have students answer the accompanying questions. L1 SAE

SECTION 16.1

Recombinant DNA Technology

Section Preview

Objectives

Summarize the steps used to engineer transgenic organisms.

Give examples of applications and benefits of genetic engineering.

Key Terms

genetic engineering
recombinant DNA
transgenic organism
restriction enzyme
vector
plasmid
gene splicing
clone

When engineers set out to design a bridge, they consider the function of the bridge and the characteristics of the materials, such as steel and concrete, that will be used to build the bridge. Today, molecular biologists "design" organisms in much the same way, although the process is infinitely more complex than designing a bridge. First, they decide what new trait the organism should have. Then, they find the section of DNA—the gene—that will be used to modify the genetic information of an existing organism. The result of a genetic engineer's work is an organism that has been "custom designed" for a unique purpose.

Gene Manipulation

Research on moving genes in organisms requires the cutting—or cleaving—of DNA into fragments. The procedures for cleaving DNA from an organism into small fragments, and inserting the fragments into another organism of the same or a different species, are called **genetic engineering.** You may also hear genetic engineering referred to as recombinant DNA technology. **Recombinant DNA** is DNA made by connecting fragments of DNA from different sources.

Once any kind of recombinant DNA has been engineered, it can be inserted into an organism; that organism then uses the foreign DNA as if it were its own. In the example of the glowing tobacco plant, a section of DNA that codes for the light-producing enzyme in fireflies is inserted into a piece of bacterial DNA and transferred into a bacterium. When the bacterium carrying the recombinant DNA infects a tobacco plant, the recombinant DNA becomes incorporated into the tobacco plant chromosomes and causes the plant to glow when it is watered with a substrate on which the light-producing enzyme acts. The glow indicates that the cells of the plant are expressing the firefly gene by making the enzyme.

Transgenic organisms contain recombinant DNA

Organisms that contain functional recombinant DNA, such as the glowing tobacco plant, are called **transgenic organisms.** The basic techniques for producing a transgenic organism involve (1) the isolation of

Program Resources

Section Focus Master 34 L1 SAE
Reinforcement and Study Guide, pp. 61-63 L1
Biolab and Minilab Worksheets, pp. 61-64 L1
Performance Assessment in the Biology Classroom, p. 19 L1 P
Basic Concepts Transparency 19 and **Master** L1 SAE
Reteaching Transparency 16 and **Master** L1 SAE

the foreign DNA fragment to be inserted, (2) the joining of this DNA with a vehicle to transport it, and (3) the actual transfer of the recombinant DNA into a suitable host. As you read this chapter, you will learn about the many uses of transgenic organisms in agriculture, medicine, and industry.

Restriction enzymes cleave DNA

DNA can be prepared for recombination only after it has been isolated and snipped into smaller fragments. The discovery in the early 1970s of DNA-cleaving enzymes made such cutting possible. These **restriction enzymes** are bacterial proteins that have the ability to cut both strands of the DNA molecule at certain points. There are hundreds of restriction enzymes, each capable of cutting DNA at a specific point in a specific nucleotide sequence. The resulting DNA fragments are of different lengths.

You can see in *Figure 16.1* that the enzyme called *Eco*RI makes staggered cuts in the double strands of DNA. The result is a set of double-stranded DNA fragments with protruding, single-stranded ends. These ends can be joined readily to complementary single strands on DNA molecules from another organism to form recombinant DNA. Because they join so readily, the ends are described as being sticky.

Vectors transfer DNA

Once DNA has been cleaved, the fragments can be inserted into a host cell. In order to produce recombinant DNA, the cleaved fragment of DNA must recombine with something else. That something else is a vector. A **vector** is a means by which foreign DNA can be transferred into the host cell. Vectors may be classified as mechanical or biological.

vector:
vectus (L) carrier
A vector carries the foreign DNA fragment into a host cell.

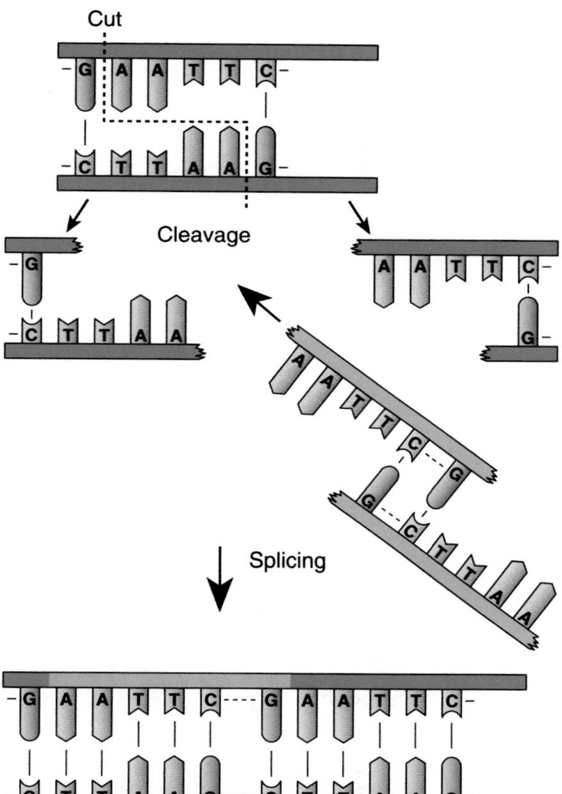

Figure 16.1

In the presence of the restriction enzyme *Eco*RI, a double strand of DNA containing the sequence–GAATTC–is cleaved between the G and the A. Notice that the same sequence of bases is found on both DNA strands, but running in opposite directions–an arrangement called a palindrome. When the DNA is cleaved, double-stranded fragments with single-stranded, sticky ends are formed. If DNA from two different sources is cleaved with the same restriction enzyme, the sticky ends of the fragments will match and pair, forming recombinant DNA.

SECTION 16.1

Discussion

Ask students what genetic changes people might want to engineer into their pets. *Students might suggest stamina, intelligence, hair retention, and disease resistance.*

2 Teach

Using Science Words

To help students remember *Eco*RI, explain that this enzyme was discovered in the bacterium *Escherichia coli. E* comes from Escherichia, *co* from coli, *R* from Restriction enzyme, and *I* indicates that this enzyme was the first restriction enzyme to be found in this bacterium.

Demonstration

Visual-Spatial The use of restriction enzymes and sticky ends can be demonstrated by using the materials in the Chromosome Simulation Biokit available from Carolina Biological Supply Co.

Audiovisual

Show *Genetic Engineering*, a 34-minute video from Charles Clark Co., Inc. 

GLENCOE TECHNOLOGY

Videodisc

The Infinite Voyage: The Geometry of Life
Selective Breeding (Ch. 7)

Manipulative Genetic Engineering (Ch. 10)

STVS: Plants & Simple Organisms
Disc 4, Side 2
Plant Clones (Ch. 13)

Meeting Individual Needs

Visually Impaired Have students who are visually impaired use large pieces of cardboard already cut into four shapes (many pieces of each shape), one to represent each of the four bases A, T, C, G. Use them to demonstrate a palindrome and to model the action of restriction enzymes. L1

MiniLab

Purpose

Visual-Spatial Students will learn why recombinant DNA engineering requires palindromes.

Process Skills

observe and infer, interpret scientific illustrations

Teaching Strategies

- Have students think of words, numbers, or phrases that are palindromes. *Possible answers include: rotator, 1991, nurses run.* L1

Expected Results

Students will examine how a specific restriction enzyme cuts a strand of DNA.

Analysis

1. Bacteria can use restriction enzymes to cut up viral DNA.
2. The ends readily join together with complementary single strands on other split DNA molecules.
3. The cut on the right does not produce sticky ends, so the DNA fragment would not combine readily with other DNA fragments.

Assessment

Student Journal: Have students write summaries explaining the use of palindromes in recombinant DNA and place them in their journals. L1

MiniLab

Why does recombinant DNA require palindromes?

A word, group of words or letters, or numerals that read the same forward and backward are called palindromes. Many restriction enzymes cut sequences of DNA that are palindromes. This property is essential for engineering recombinant DNA. As a result of cuts made by restriction enzymes, single-stranded sequences of DNA are left dangling at the ends of a fragment. These ends are available for pairing with their complementary bases in a plasmid or piece of viral DNA.

Procedure

1. Make a drawing similar to the illustration of the DNA sequence shown at the right.
2. Assume that a particular restriction enzyme cuts the chain between -G-G- so that the fragments are -CCTAG and
   G- -G
   GATCC-. Use scissors to cut your drawing in the same way that the restriction enzyme would.

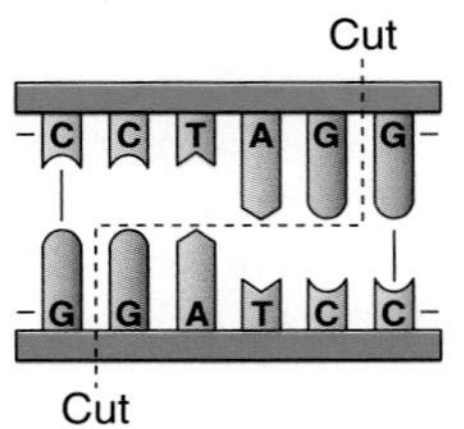

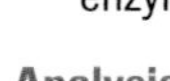

Analysis

1. Restriction enzymes were first isolated from bacteria. How might restriction enzymes be a bacterial defense against viral invaders?
2. Why are the cleaved ends of DNA called sticky ends?
3. Why would a restriction enzyme that makes a cut as shown at right be less useful than the one illustrated above?

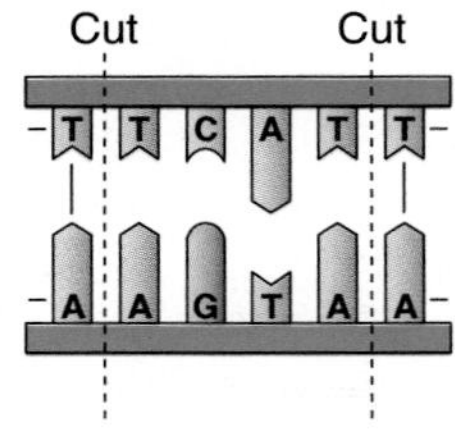

palindrome: *palindromos* (GK) running back again; A palindrome is a word, number, or phrase that reads the same backward and forward, such as *1881* or *madam.*

Two mechanical vectors are used to introduce foreign DNA directly into the nucleus of a cell. A micropipette is inserted into a cell, or tiny metal bullets coated with pieces of DNA are shot into the cell using a device called a gene gun. Biological vectors include viruses and plasmids. A **plasmid,** as shown in *Figure 16.2,* is a small ring of DNA found in a bacterial cell. It carries different genes from those of the bacterial chromosome.

Magnification: 110 000×

Figure 16.2

Plasmids are small rings of DNA found in bacterial cells. They usually carry non-essential "accessory" genes, such as those that carry resistance to antibiotics.

Gene splicing

The plasmid can be cleaved with restriction enzymes. If the plasmid and the foreign DNA have been cleaved with the same restriction enzyme, the sticky ends of each will match and they will join, reconnecting the plasmid ring. Rejoining cut DNA fragments is called **gene splicing.** The foreign DNA is recombined into a plasmid or viral DNA with the help of a second enzyme.

Gene cloning

After the foreign DNA has been spliced into the plasmid or virus vector, the recombined DNA is transferred into a host cell. When the host cell prepares to divide, it copies the recombinant DNA along with its own DNA. The process of making extra copies of recombinant DNA is a form of cloning. **Clones** are genetically identical copies. *Figure 16.3* summarizes cloning of recombinant DNA in a bacterial host cell.

Each identical recombinant DNA molecule is called a gene clone. Gene cloning is an important step in the

Display

Place articles about recombinant DNA from the newspaper and magazines on the bulletin board.

PROJECT

Enzyme Model

Build a model demonstrating the use of restriction enzymes on DNA. Students could use various materials to represent the DNA nucleotides, such as colored push pins, colored paper clips, M & M's, etc. L1

process of genetic engineering because multiple copies of the desired DNA are produced.

Why clones are possible

Cloning is possible because a foreign piece of DNA introduced into a bacterial cell has been integrated into that cell's DNA so completely that the foreign DNA is replicated as if it were bacterial DNA. Thus, when the bacteria reproduce, millions of bacterial cells are eventually formed, each of which has the desired piece of foreign DNA that can be easily purified, studied, and manipulated.

Genetic-engineering techniques provide pure DNA for use in genetic studies or are used to make a gene product in large quantities. Drug manufacturers use genetic-engineering techniques to produce certain drugs, such as human growth hormone.

MiniLab

How is splicing wires similar to splicing genes?

Electricians must often splice wires at terminal boxes. In some ways, splicing wires can be like splicing genes. However, instead of using restriction enzymes, electricians use wire cutters and strippers. Instead of using enzymes to splice the wires back together, electricians twist the wires together and tape or cover them.

Procedure

1. Obtain two plastic-wrapped wires of different colors.
2. Using wire cutters or scissors, cut one piece of wire into three sections. Strip the ends of the middle section using wire strippers, or cut around the plastic and twist it off the wire.
3. Using wire cutters or scissors, cut the other piece of wire into two sections. Strip the ends that you just cut in the same way as before.
4. Join the middle section of wire from the first piece to the two sections you cut from the second piece.
5. Twist the wires together.

Analysis

1. How do the wires represent DNA fragments?
2. Explain why the ends must be stripped.
3. How is the process of splicing wires similar to splicing genes?

Figure 16.3

In a recombinant DNA technique using a plasmid vector, foreign DNA is spliced into a bacterial plasmid. The recombined plasmid then carries the foreign DNA into a bacterial cell, where it is cloned when the cell reproduces. If the foreign DNA contains a gene for insulin production, each transgenic cell will make insulin.

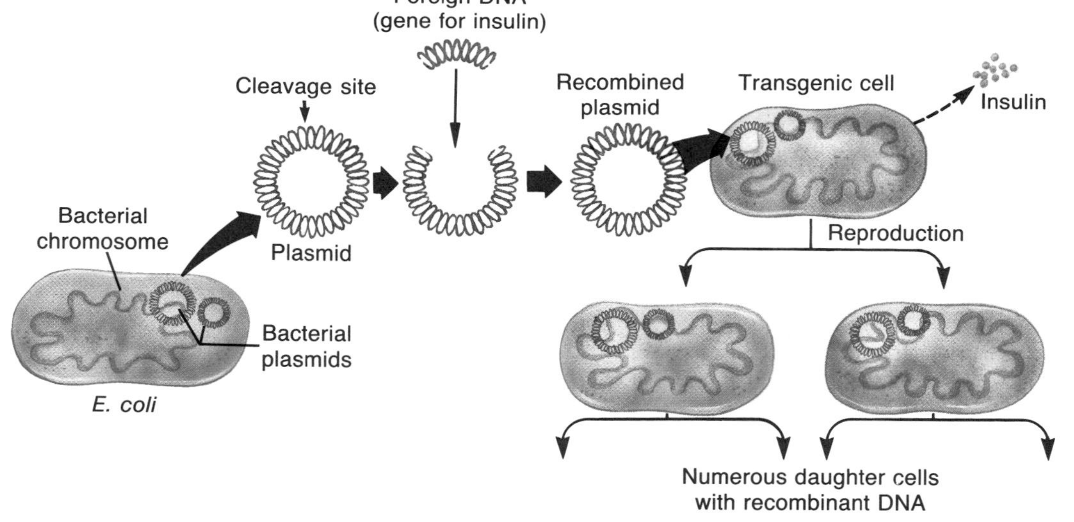

CULTURAL DIVERSITY

Flossie Wong-Staal and the AIDS Vaccine

One of the more promising applications of genetic engineering techniques is the development of vaccines, especially for AIDS. Introduce students to current research in this field, particularly the contributions of Chinese-American scientist, Flossie Wong-Staal.

Since 1990, Wong-Staal has been a leader in AIDS research at the University of California, San Diego. Currently, she is using genetic engineering techniques to manufacture mutant HIV viruses that "fool" the immune system into producing an immune response. See the December 1991 edition of *Discover* magazine for more information on Wong-Staal's research.

MiniLab

Purpose

Kinesthetic Students will compare splicing wires to splicing genes.

Process Skills

observe and infer, compare and contrast

Teaching Strategies

- This Minilab could be done as a class demonstration if there is not enough time for all students to complete it. Students can brainstorm on the analysis questions.

Expected Results

Students will splice together the two different colors of wire, representing the insertion of a gene from a different species into a strand of DNA.

Analysis

1. The wires represent DNA segments from two different species and the pieces of wire represent genes.
2. Stripping the wire is like using restriction enzymes to form sticky ends.
3. Splicing the wires together now allows the wires to function as a unit, and splicing a gene into a fragmented piece of DNA allows that gene to function within the DNA strand.

Assessment

Student Journal: Have students write summaries of the lab and the answers to the Analysis questions in their journals. L1

Audiovisual

Show *Lights Breaking: Ethical Questions About Genetic Engineering*, video from Bullfrog Films, Oley, PA.

BioLab

Modeling Recombinant DNA

Time Allotment 
One class period

Objectives
Review objectives with students before they begin the Biolab.

Process Skills
compare and contrast, observe and infer, recognize cause and effect

PREPARATION

- Review the chemical structure of DNA from Chapter 13 before students begin the Biolab.
- Additional background material on modeling recombinant DNA can be found in the article, "Recombinant Paper Plasmids" by Christie L. Jenkins, *The Science Teacher*, April 1987. You may wish to read this article before you teach the Biolab.

GLENCOE TECHNOLOGY

 Videodisc

Biology: The Dynamics of Life
Disc 1, Side 1
Recombinant DNA (Ch. 36)

BioLab

Modeling Recombinant DNA

Experimental procedures have been developed that allow recombinant DNA molecules to be engineered in a test tube. From a wide variety of restriction enzymes available, scientists choose one or two that recognize particular sequences of DNA within a longer DNA sequence of a chromosome. The enzymes are added to the DNA, which is cleaved at the recognition sites. Because the cleaved fragments have ends that are available for attachment to complementary strands, the fragments can be added to plasmids or to viral DNA that has been similarly cut. When the fragment has been incorporated into the DNA of the plasmid or virus, it is called recombinant DNA.

PREPARATION

Problem
How can you model recombinant DNA technology?

Objectives
In this Biolab, you will:
- **Model** the process of preparing recombinant DNA.
- **Analyze** a model for preparation of recombinant DNA.

Materials
paper
colored pencils, red and green
tape
scissors

Safety Precautions 
Be careful with sharp objects.

PROCEDURE

1. Cut a lengthwise strip of paper from a sheet of typing paper into a rectangle about 3 cm by 28 cm. This strip of paper represents a long sequence of DNA containing a particular gene that you wish to combine with a plasmid.
2. Cut another lengthwise strip of paper into a rectangle about 3 cm by 10 cm. When taped into a ring, this piece of paper will represent a bacterial plasmid.
3. Use your colored pencils to color the longer strip red and the shorter strip green.

PROCEDURE

Teaching Strategies
- Students can do this lab with partners or as individuals.
- If red and green colored paper is available, then colored pencils will not be necessary.
- If you make cardboard models of DNA bases for the Meeting Individual Needs on page 377, use them in this Biolab.

Data and Observations
Data for completing the table should be as follows: Gene splicing—process of taping green and red paper together; Plasmid—green strip; Restriction enzyme—scissors; Sticky ends—cut ends on paper; Recombinant DNA—red and green strips taped together.

4. Write the following DNA sequence once on the shorter strip of paper and two times about 5 cm apart on the longer strip of paper.

-G-G-A-T-C-C-
-C-C-T-A-G-G-

5. After coloring the shorter strip of paper and writing the sequence on it, tape the ends together.

6. Assume that a particular restriction enzyme is able to cleave DNA in a staggered way as illustrated here.

-G G-A-T-C-C-
-C-C-T-A-G G-

Cut the longer strand of DNA in both places as shown. You now have a cleaved foreign DNA fragment containing a gene that can be inserted into the plasmid.

7. Once the foreign gene has been cleaved, cut the plasmid in the same way.

8. Splice the foreign gene into the plasmid by taping the paper together where the sticky ends pair properly.

9. The new plasmid represents recombinant DNA.

-G-G-A-T-C-C- -G-G-A-T-C-C-
-C-C-T-A-G-G- -C-C-T-A-G-G-

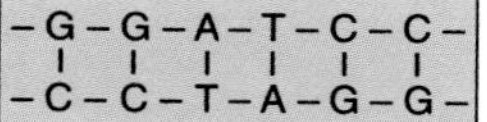

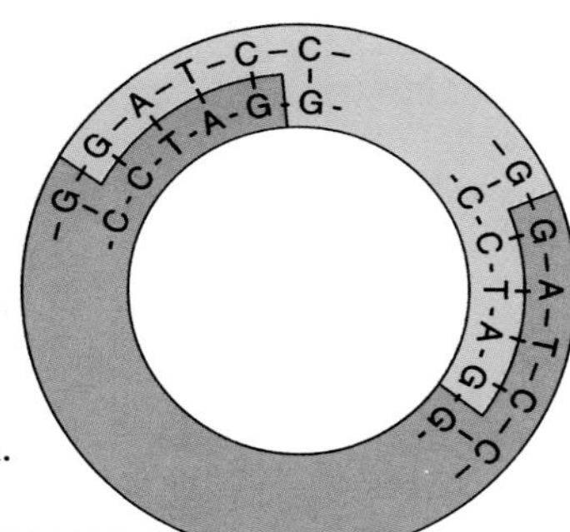

10. Copy the following table. Relate the steps of producing recombinant DNA to the activities of the modeling procedure by explaining how the terms relate to the model.

Term	Biolab Model
Gene splicing	
Plasmid	
Restriction enzyme	
Sticky ends	
Recombinant DNA	

ANALYZE AND CONCLUDE

1. **Comparing and Contrasting** How does the paper model of a plasmid resemble a bacterial plasmid?
2. **Comparing and Contrasting** How is cutting with the scissors different from cleaving with a restriction enzyme?
3. **Thinking Critically** Some restriction enzymes cut DNA at particular places but do not leave sticky ends. These enzymes cannot be used to engineer recombinant DNA. Explain why. What function might they serve in a cell?

Going Further

Project Design and construct a three-dimensional model that illustrates the process of preparing recombinant DNA. Consider using clay, plaster of paris, or other materials in your model. Label the model and explain it to your classmates.

ANALYZE AND CONCLUDE

1. It is a small circular piece of DNA.
2. The scissors can cut the DNA anywhere, but the restriction enzyme recognizes a particular sequence.
3. Without sticky ends, the pieces of DNA can't be inserted into other DNA. These enzymes may function in a cell by cutting invading viral DNA.

Assessment

Oral: Have students summarize what they have learned about the use of restriction enzymes. Discuss the table and have the students explain how the Biolab model is like the terms listed. L1

Going Further

Students could use different colors of clay, or use food coloring to dye plaster of paris, to show recombinant DNA in their models.

interNET CONNECTION

Follow the link for this chapter on the Glencoe Homepage at **www.glencoe.com/sec/science** to find out more about DNA technology.

Literature

Jurassic Park

Purpose

Students will examine the theme of the book *Jurassic Park* and consider some of the potential hazards of DNA technology.

Teaching Strategies

- Many students will be familiar with the movie made from the book *Jurassic Park.* For students who do not know the story, have others summarize what occurs. L1
- You may wish to encourage students to read the book.

Possible Answer

Insights may include clues to Earth's past environment; filling in some missing gaps in evolutionary time lines; better understanding of adaptation, mutation, and extinction; and the ability to repopulate Earth with extinct species.

Literature

Jurassic Park

by Michael Crichton

One of the most widely read adventure novels of the 1990s, Michael Crichton's *Jurassic Park* is a story of biology and technology run amok.

Ancient history brought to life Jurassic Park is an island theme park created by multi-millionaire-philanthropist John Hammond. Hammond has hired an impressive array of biologists, computer scientists, and other technological experts to create the ultimate animal preserve, one populated by previously extinct dinosaurs from the Jurassic Era!

Cloning dinosaurs Hammond's chief geneticist, Dr. Henry Wu, obtained samples of dinosaur DNA from prehistoric insects trapped in amber—insects that bit and drew blood from the dinosaurs that shared the insects' time on Earth. By cloning the dinosaur DNA and using computerized gene sequencers to structure the genetic proteins properly, Dr. Wu succeeded in re-creating long-extinct stegosaurs, triceratops, and other dinosaurs, including the fierce carnivore, *Tyrannosaurus rex.*

The best-laid plans Despite elaborately complex and seemingly fail-safe precautions and security measures, Jurassic Park suffers a terrible catastrophe. Unbeknownst to Dr. Wu, the dinosaurs, which are supposed to be sterile and all female, have begun to breed.

When a powerful tropical storm lashes the island park and electrical power is cut off, the dinosaurs are freed from their individual territories to begin a reign of terrifying death and destruction. In the end, John Hammond is literally consumed by his vision, attacked and killed by scavenger dinosaurs of his own creation.

CONNECTION TO Biology

What scientific insights might be gained from being able to re-create extinct life forms?

The reason recombinant bacteria are used is because a huge population of them can be grown. Each cell makes the desired product. For example, if a bacterial cell were engineered to produce human growth hormone, billions of cloned cells could generate the large quantities of hormone that are needed for patients who require it. It would be impossible to obtain such large amounts of the product from humans because hormones are produced in the body in minute amounts.

Sequencing DNA

Now that large numbers of the desired DNA fragment have been cloned by the procedure just described, the sequence of DNA bases in the fragment can be determined. Sequencing is of importance to biologists because it helps them understand the

Figure 16.4

Sequencing is a technique used to identify the sequence of nucleotides in DNA. The sequence of the bases can be read directly from the bands produced in the gel by electrophoresis. Determining the sequence of a fragment of DNA requires many copies of the fragment; therefore, DNA cloning is needed to produce them.

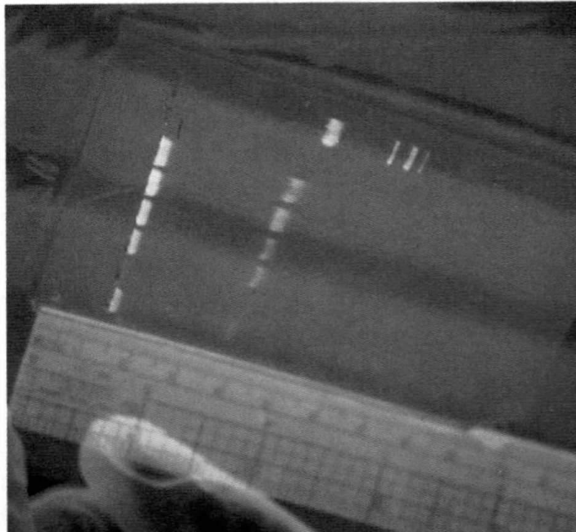

Alternate Lab: Crime Analysis

Purpose 

LS **Logical-Mathematical** Students will solve a crime using human genetics and DNA technology.

Materials

glass slide, red juice, 5 product bar codes (2 that match), 2 strands of human hair (blond and dark), T-shirt with spot of red ink, stained slide of female cheek cells, microscope

Preparation

- Table 1 has hair found at the scene.
- Table 2 has a shirt with a blood stain (red ink), and a slide with two drops of juice to represent blood testing and showing no clotting with the addition of anti-A and anti-B sera.
- Table 3 has a slide of stained cheek cells with microscope pointer on a Barr body.
- Table 4 has a bar code representing the DNA fingerprint of blood found at the crime scene (one of two that match).
- Place a bar code in each of four envelopes marked: *blond female, type O blood* (matching bar code); *blond male, type B blood*; *dark-haired male, type O blood*; and *dark-haired male, type B blood.*

fundamental basis of life—DNA. In DNA sequencing, one of the two strands of DNA is produced by cloning. Using restriction enzymes, the DNA strands are chopped up into many pieces of varying lengths. The fragments are placed in a gel that has been treated with a dye, which binds to DNA and glows pink in ultraviolet light. Then the gel is subjected to an electric field.

Each DNA piece migrates toward the electrode of opposite charge at a rate determined by its charge and length (larger pieces move more slowly than smaller pieces). The positions of the bands of dye in the gel allow a direct reading of the sequence of bases, as shown in *Figure 16.4.* The process of separating DNA fragments in an electric field is called gel electrophoresis.

Applications of DNA Technology

Once it became possible to transfer genes from one organism to another, large quantities of hormones and other products could be produced. How is this technology of use to humans?

Transgenic bacteria in agriculture

Many species of bacteria have been engineered to produce chemical compounds that are of use to humans. The three main areas proposed for transgenic bacteria are in agriculture, industry, and medicine. One species of bacteria has already been used successfully on agricultural crops. This particular bacterium normally occurs on strawberry plants and promotes frost damage of leaves and fruits because ice crystals form around a protein on the bacterium's surface. After engineering to remove the gene for this protein, frost damage is prevented.

Farmers hope that another species of bacteria that lives in soil and in the roots of plants such as peas, soybeans, and peanuts can be engineered to increase the rate of conversion of atmospheric nitrogen to nitrates, a natural fertilizer used by plants. If this can be accomplished, farmers will be able to obtain savings by cutting back on fertilizer.

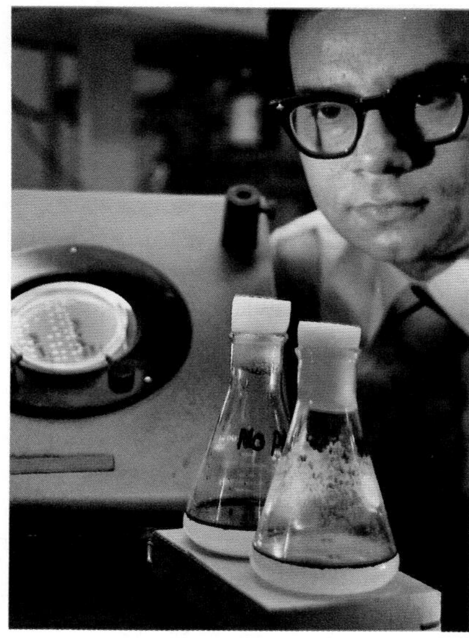

Figure 16.5
The first patented organism, a bacterium capable of breaking down oil, was engineered by Dr. Ananda Chakrabarty. A patent ensures that the originator of the organism or product owns exclusive rights to use the product. The flask in the rear contains oil and natural bacteria. The flask at the front contains the engineered bacteria and is almost free of oil.

Transgenic bacteria in industry

In industry, transgenic bacteria have been engineered to break down pollutants into harmless products. Laboratory experiments with transgenic bacteria first showed that these engineered bacteria could degrade oil more rapidly than the same species of naturally occurring bacteria, as shown in *Figure 16.5.* These transgenic bacteria were used with some success in the Gulf of Mexico to clean up an oil spill off the coast of Texas. Mining companies are interested in bioengineering bacteria to extract valuable minerals from ores.

SECTION 16.1

Chemistry Connection

Gel Electrophoresis

When the use of a gel was introduced to the technique of separating molecules by electrophoresis, the resolving power of the process was greatly improved. For a given set of conditions, each particle in electrophoresis moves at a certain velocity called the electrophoretic mobility. This quantity varies directly with the charge on the particle and inversely with the particle's size.

GLENCOE TECHNOLOGY

Videodisc

Biology: The Dynamics of Life
Disc 1, Side 1
Bioengineering (Ch. 37)

Assessment

Performance Assessment in the Biology Classroom: p. 19: Making a Model of Recombinant DNA. Have students carry out this activity after they have completed the study of recombinant DNA. L1 P

Procedure

Give the following directions to students.

1. Roger Trueblood, dark hair, type B blood was shot and killed. You must find which suspect is guilty.
2. Rotate to the tables and collect information. At the last table, decide who is guilty and open the envelope to see if the DNA fingerprint matches that at the crime scene.

Analysis

Ask students the following questions.

1. What information did you learn from the table displays? *that the clues included blond hair, type O blood, and female cheek cells (because of the presence of Barr bodies)*
2. How are the bar codes like DNA fingerprinting? *They are a series of lines of varying thickness.*

Assessment

Journal: Have students write their conclusions about the crime in their journals. *The blond female is guilty because her hair, type of blood, and cheek cells were at the scene.* L1

Science, Technology, and Society

Knockout Mice
Mice with mutant genes are particularly valuable to scientists studying human diseases. Using a grant from the National Institute of General Medical Services, scientists have developed what they call knockout mice. These mice are produced by inserting a specially altered gene into embryonic mouse cells, where it knocks out their normal counterparts. The cells with knocked-out genes are placed in mouse embryos. By selective breeding, scientists can produce a strain of mice in which nearly all cells lack the knocked-out gene, allowing the scientists to study the effect of the missing gene on conditions such as aging or cancer.

GLENCOE TECHNOLOGY

Videodisc

STVS: Plants & Simple Organisms
Disc 4, Side 1
Genetic Engineering in Barley (Ch. 20)

Bioethics

Human Genes in the Wild
Introducing human genes into plants that are grown in the field has obvious benefits in producing large amounts of proteins. Critics point out that the environmental and ethical implications of these experiments have not been explored. To invite a class discussion on this topic, ask students the following question: What might be the consequences on the environment if the new genes get into the wild gene pool?

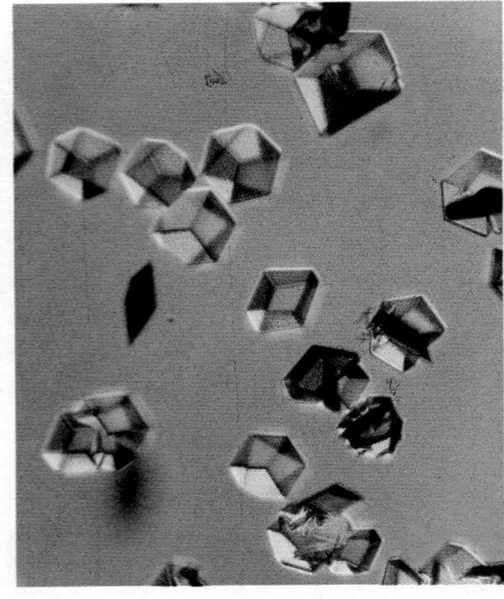

Figure 16.6
The hormone insulin, important in the treatment of diabetes, is produced commercially by bacteria. The human gene for insulin is inserted into a bacterial plasmid by genetic-engineering techniques. Transgenic bacteria produce such large quantities of insulin that they bulge. The insulin is collected as a liquid (right) and purified as crystals (left).

Transgenic bacteria in medicine

In medicine, pharmaceutical companies already are producing molecules made by transgenic bacterial species to treat human disease. Transgenic bacteria are employed in the production of growth hormone to treat dwarfism, interferon to treat cancer, and human insulin used to treat diabetes, *Figure 16.6.* Another strain of transgenic bacteria is being used to produce phenylalanine, an amino acid that is needed to make NutraSweet, an artificial sweetener found in most diet products.

Transgenic plants

Can you think of some ways that plants could be improved by genetic bioengineering? Plants are more difficult to genetically engineer than bacteria because plant cells do not have the plasmids or kinds of viruses needed for taking up foreign pieces of DNA that bacterial cells have. Because plant cells also are surrounded by thick cell walls, it is difficult to insert DNA into plant cells.

A bacterium that normally causes tumorlike galls in the tissues of certain plants, *Figure 16.7,* has been employed to carry foreign genes into plant cells. Most engineering of plants uses mechanical vectors such as the gene gun. Brief jolts of high-voltage electricity can also be used. The jolts cause temporary pores to form in the plasma membrane, through which the DNA can enter.

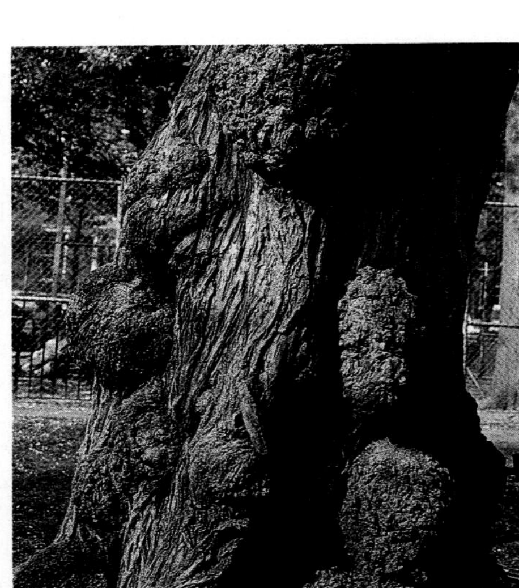

Figure 16.7
The bacterium that causes these tumorlike plant galls is the only known biological plant vector. Unfortunately, it does not work on many kinds of plants.

PORTFOLIO

Engineered Bacteria Have students research and write an essay on whether we should allow scientists to alter bacteria genetically and release them into the environment. Have students include a copy of their essays in their portfolios. L2 LS P

Plants have been genetically engineered to resist herbicides, produce internal pesticides, or increase their protein production, *Figure 16.8*. These plants have been used in field trials, and a few are now available for crop use. In the future, it is expected that plants will be engineered to be more nutritious, require less fertilizer, produce fruit that ripens later, or have the ability to grow in unfavorable conditions.

Transgenic animals

Like plant cells, most animal cells will not readily accept plasmids into their own DNA. To introduce genes into animal cells, a micropipette is sometimes used to inject DNA into the eggs of animals before they are fertilized. Animal experimentation has not progressed as far as experimentation in bacteria and plants, but some trials have succeeded in introducing genes into livestock in the hope of producing bigger varieties, more milk, and even some human proteins. Transgenic animals include goats that have been engineered to produce milk containing high levels of a human protein that dissolves blood clots.

Although these advances appear to be major breakthroughs, bioengineering is still in its infancy. It is expected that within a decade, several varieties of transgenic animals will be available for farm use.

Figure 16.8

Plants that have been engineered to produce internal pesticides will not suffer damage such as this (left). Transgenic tomato plants have foreign genes that slow the process of spoilage. This trait makes storing, shipping, and stacking tomatoes in supermarkets less of a problem (right).

Section Review

Understanding Concepts

1. How are transgenic organisms different from natural organisms of the same species?
2. How are sticky ends important in making recombinant DNA?
3. Why is it presently more difficult to engineer transgenic plants and animals than it is to engineer bacteria?

Thinking Critically

4. Many scientists consider genetic engineering to be simply an efficient method of selective breeding. What do you think? Explain.

Skill Review

5. **Sequencing** Order the steps in producing recombinant DNA in a bacterial plasmid. For more help, refer to Organizing Information in the *Skill Handbook*.

SECTION 16.1

3 Assess

Check for Understanding

Make up a sequence of DNA that includes two of the sequences GAATTC and ask students to demonstrate where *Eco*RI will cut the strand of DNA. L1

Reteach

Have students model how restriction enzymes create sticky ends, using other restriction enzymes so they can see that choice of the particular enzyme is important. For example, *Bam*HI recognizes GGATCC and cuts between the two Gs. *Hind* III recognizes AAGCTT and cuts between the two As. L1

Extension

Interested students can look into the applications of protein engineering and computer modeling to produce modified natural proteins or create entirely new proteins. L3

Assessment

Oral: Have students summarize orally the formation of recombinant DNA. L1

4 Close

Discussion

Summarize how DNA technology may provide therapy for many genetic disorders. L1

Answers to Section Review

1. Transgenic organisms contain DNA from other sources.
2. Sticky ends allow the cleaved DNA to rejoin to complementary single strands on DNA molecules from another organism.
3. Plants and animals do not accept plasmids into their DNA, and plants have cell walls that make it difficult to insert DNA.

Thinking Critically

4. Answers may include that genetic engineering involves inserting genes from very different species, whereas selective breeding is carried on only within a species.

Skill Review

5. Cleave the plasmid and foreign DNA with the same restriction enzyme, and the sticky ends will join, reconnecting the plasmid ring.

Prepare

Key Concepts

The organized effort to map and sequence the human genome using DNA technology is presented. Students will also explore applications of this project to diagnosis of genetic disorders, gene therapy, and DNA fingerprinting.

Block Scheduling

Look for this symbol for strategies that are useful in a block scheduling format. For more information on block scheduling, refer to Section 16.2 in the **Lesson Plans** booklet.

Materials

- Purchase cloning kits for the Extension.

1 Focus

Bellringer

Before presenting the lesson, display **Section Focus Master 35** on the overhead projector and have students answer the accompanying questions. L1 SAE

Brainstorming

Ask students to consider why scientists would wish to know the genetic makeup of the human genome. What might this knowledge allow scientists to do?

SECTION

16.2 The Human Genome

Section Preview

Objectives

Analyze how the effort to completely map and sequence the human genome will advance human knowledge.

Predict future applications of the Human Genome Project.

Key Terms

human genome
linkage map
cell culture
gene therapy

Scientists do not yet know on which chromosomes most human genes occur. It has been estimated that there are 3 billion base pairs in the DNA that makes up your genes. Imagine trying to identify the correct sequence! It's almost like trying to catalog all the letters in a large book. Yet, that is exactly what molecular biologists are trying to do. Current research in genetic engineering is leading to a greater understanding of similarities and differences in the human genome. Projects are underway to answer questions about the locations of all human genes, and to apply the information gained from the projects to treating, or possibly even curing, human genetic disorders.

genome:
genos (GK) offspring
A genome is the total number of genes in an individual.

Mapping and Sequencing the Human Genome

In 1990, scientists in the United States organized the Human Genome Project. It is an international effort to completely map and sequence the **human genome,** the approximately 100 000 genes on the 46 human chromosomes. The project is still underway. Maps currently are being prepared that show the locations of known genes on each chromosome. At the same time, the sequence of the 3 billion base pairs of DNA in the human genome is being analyzed and mapped. Eventually, the two maps will be synchronized.

Linkage maps

Only a few thousand of the total number of known genes have been mapped on particular chromosomes. This means that for most of these genes, scientists don't know the exact or even the approximate locations on chromosomes. The genetic map that shows the location of genes on a chromosome is called a **linkage map.** Such a map for one human chromosome is shown in *Figure 16.9.*

The historical method used to assign genes to a particular human chromosome was to study linkage data from human pedigrees. Recall from your study of meiosis that crossing-over occurs often during

Program Resources

Section Focus Master 35 L1 SAE
Reinforcement and Study Guide, p. 64 L1
Laboratory Manual, pp. 91-94 L2
Concept Mapping, p. 16 L1
Critical Thinking/Problem Solving, p. 16 L3
Tech Prep Applications, pp. 19-20 L2

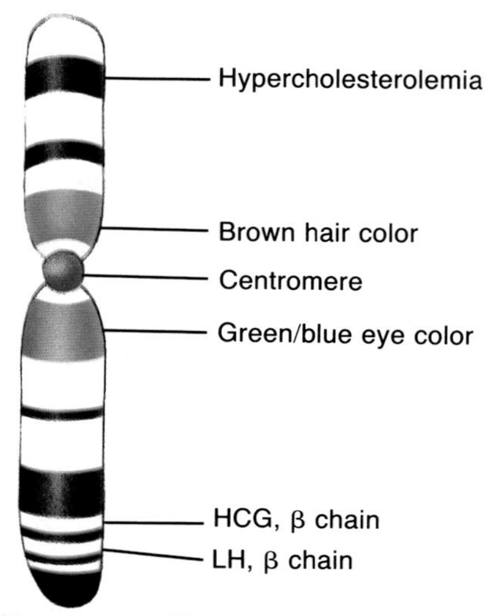

Figure 16.9

A linkage map for chromosome 19 is shown here. The DNA banding pattern is determined by staining. Identifications and locations of some of the known genes on this chromosome are indicated. The LH, ß chain is a polypeptide chain in a male sex hormone. Hypercholesterolemia is a disorder that causes high blood levels of cholesterol. HCG is a female hormone produced during pregnancy.

prophase I. As a result of crossing-over, the offspring have a combination of alleles not found in either parent. The frequency with which these alleles occur together is a measure of the distance between the genes. Genes that cross over frequently must be farther apart than genes that rarely cross over. The percentage of these crossed-over traits appearing in offspring is then used to determine the relative position of genes on the chromosome, and thus to create a linkage map.

Because humans have only a few offspring compared to other species and the generation time is so long, mapping by linkage data is extremely inefficient. Biotechnology has now provided scientists with new methods of mapping genes. By a technique called polymerase chain reaction (PCR), millions of copies of minute DNA fragments can be made in a matter of a few hours. Scientists can copy the DNA from thousands of separate sperm cells produced by one individual and analyze the results of crossing-over that occurred during the meioses that produced the sperm. Instead of examining actual offspring, they can examine sperm cells—hundreds and even thousands of potential offspring—to create linkage maps. Another method of mapping chromosomes is shown in *Figure 16.10.*

Sequencing the human genome

The difficult job of sequencing the human genome is accomplished by cleaving samples of DNA into fragments using restriction enzymes, as described earlier in this chapter. Each fragment is then individually sequenced. The short fragments are aligned in the proper order by overlapping matching sequences, thus determining the sequence of a longer fragment. Automated machines can perform this work, greatly increasing the speed of map development. It is expected that the entire human genome may be mapped within ten to 15 years.

Magnification: 4000×

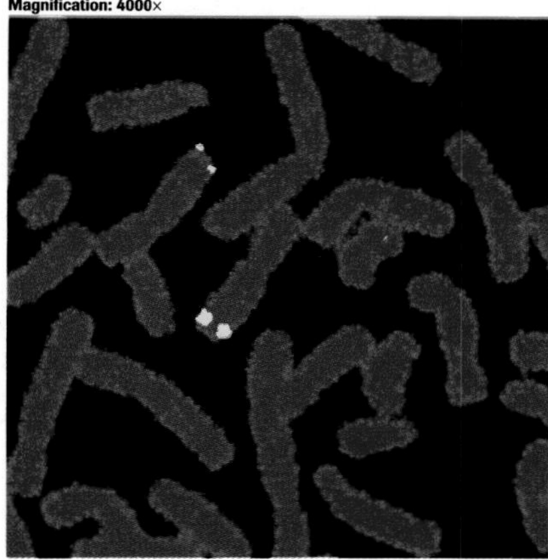

Figure 16.10

Radioactively labeled complementary DNA for the gene to be mapped is made and added to metaphase chromosomes. The labeled DNA binds to the gene on the chromosome and indicates its location as a glowing spot. Why are there four spots in this photo?

SECTION 16.2

2 Teach

Different Viewpoints in Biology

Mice For Sale

In April 1988, the U.S. Patent and Trademark Office offered the first patent ever for an animal: a mouse used in breast cancer research. DuPont now has a license to sell the mouse to others. Opponents question the right of one group to hold such a monopoly on genetically engineered animals. They feel that no one should be allowed to own life forms. For a process or product to be patentable, it must be novel, not obvious, produced by human intervention, and capable of industrial application.

Visual Learning

Figure 16.10 Why are there four spots in this photo? *Each mitotic cell has four copies of the gene, one on each chromatid of homologous chromosomes.*

GLENCOE TECHNOLOGY

Videodisc

Biology: The Dynamics of Life
Disc 1, Side 1
Gene Cloning (Ch. 38)

The Infinite Voyage: The Geometry of Life
The Human Genome Project (Ch. 11)

Meeting Individual Needs

Gifted Have gifted students look in recent issues of *Science* or *Science News* and report on the progress of the Human Genome Project and other DNA technology. L3 LS

SECTION 16.2

Visual Learning

Figure 16.11 Which cells would you expect to take up the foreign DNA? *It would be taken up by cells in the lungs.*

Science, Technology, and Society

DNA Fingerprints

DNA fingerprinting, introduced in 1985 by Alex Jeffreys, has been used in more than 2000 criminal investigations and disputed paternity cases. The DNA extracted from blood, semen, saliva, hair, urine, bone, or muscle is exposed to restriction enzymes that cut the DNA at precise places. The fragments are separated by gel electrophoresis and blotted onto a membrane. Then a radioactive probe is used to bind and target the fragments. This creates the visual pattern of bands. Fragment lengths vary from person to person, allowing the identification of the individual.

Audiovisual

Show *Genetic Fingerprinting*, a 20-minute video from EBEC.

Discussion

Some scientists and bioethicists believe DNA technology has created ethical questions too fast for society to deal with them. For example, some feel that a new emphasis on genetic factors in human performance and personality may bring back discrimination and attitudes similar to the ones that cut off immigration from parts of Europe in the 1920s. Have students discuss this problem. The discussion should make it clear that there are many factors to consider in implementing bioengineering.

Applications of the Human Genome Project

As chromosome maps are made, how can they be used? Improved techniques for prenatal diagnosis of human disorders, use of gene therapy, and development of new methods of crime detection are current areas of research.

Diagnosis of genetic disorders

Once it is clearly understood where a gene is located and the gene's DNA sequence is known, a diagnosis of a genetic disorder may be made before birth. What technique leads to making this diagnosis? The DNA of people with the trait is analyzed for common patterns that appear to be associated with the trait. A few cells are obtained from a fetus or from the fluid surrounding the fetus. To obtain a large enough sample of DNA, the cells are grown in a nutrient medium until many cells have formed, a technique known as **cell culture.** Cells in a cell culture all have the same genetic material; that is, they are clones. Thus, when fetal cells are examined and found to have DNA with the pattern associated with the disorder, there is a strong probability that the fetus will develop the trait.

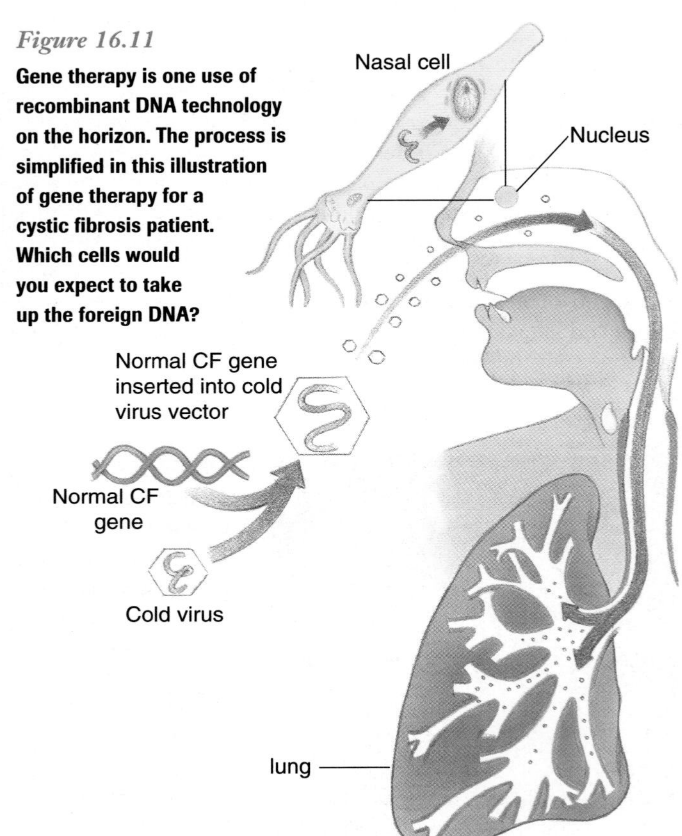

Figure 16.11

Gene therapy is one use of recombinant DNA technology on the horizon. The process is simplified in this illustration of gene therapy for a cystic fibrosis patient. Which cells would you expect to take up the foreign DNA?

Gene therapy

The next step after diagnosis is gene therapy. **Gene therapy** is the insertion of normal genes into human cells to correct genetic disorders. Gene therapy has already entered trial stages in a number of attempts to treat or cure genetic disorders; the first trials were on patients suffering from cystic fibrosis. The method used in these trials is diagrammed in *Figure 16.11.* It is hoped that copies of the normal gene introduced into the lungs by way of a nasal spray will cause lung cells to produce normal mucus. Results to date remain unclear, but it seems likely that the next decade will see the use of DNA technology to cure genetic disorders.

DNA fingerprinting

Law-enforcement workers use unique fingerprint patterns to determine whether suspects have been at a crime scene. In the past ten years, biotechnologists have developed a method that determines DNA fingerprints. DNA fingerprinting can be used to convict or acquit individuals of criminal offenses because every person is genetically unique.

PORTFOLIO

Bioethics Ask students to write an essay giving their opinions on whether parents should be able to change a child's genes before birth. Ask them to consider the following possibilities: the parents wish to correct a child's genetic disorder, they wish to make a child more attractive, they wish to increase the child's intelligence, they wish to change the child's sex. L1 P LS

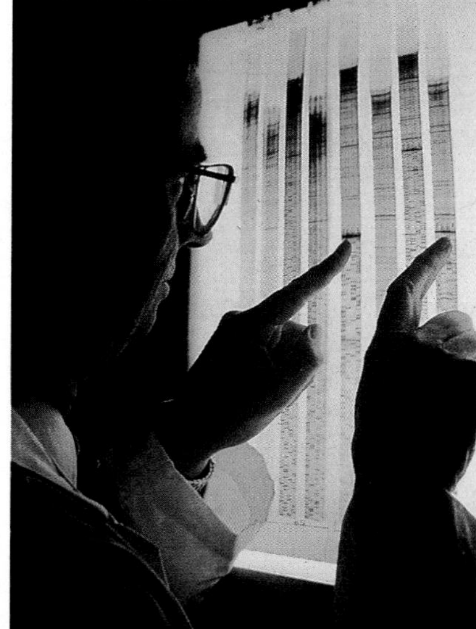

Figure 16.12

DNA fingerprinting compares the DNA fragments from a known sample to those from suspects. The bands on the fingerprints will match only if the same person donated both specimens. Can you think of a use for DNA fingerprinting other than crime analysis?

Minute DNA samples can be obtained from blood, hair, skin, or semen and copied millions of times using polymerase chain reaction techniques. When an individual's DNA is cleaved with a restriction enzyme, the resulting DNA segments are of different lengths. As shown in *Figure 16.12,* electrophoresis can be used to separate the DNA fragments so they can be compared with those obtained from a crime scene.

The genes themselves follow fairly standard patterns from person to person, but in between the genes are segments of noncoding DNA. This noncoding DNA doesn't code for proteins, but it does follow a distinct pattern in each individual—so distinct,

Ethics

BIOLOGY & SOCIETY

Brave New World

The U.S. government has pledged $5 billion toward funding the Human Genome Project. The goal of this long-term study is the mapping and sequencing of all the genes in the human body—estimated at nearly 100 000. When the Human Genome Project was announced, scientists throughout the world expressed excitement about the potential benefits of the study. Knowing the precise location and function of all the human genes would allow scientists to interpret underlying instructions that govern all aspects of human biology.

Two Treatments Unlocking the secrets of DNA offers scientists a new avenue of experimentation called germ-line therapy. Traditional gene therapy involves replacing defective genes with copies of normal genes. These implanted genes would not be passed on to future generations. Germ-line therapy involves changing the genetic instructions of the *germ cells,* or sex cells. Such changes in the eggs and sperm would be passed from one generation to the next.

Different Viewpoints

Few individuals oppose standard gene therapy to treat disorders. The benefits are obvious and the risks minimal. Gene therapy improves the quality of life for individuals and society.

Troubling aspects of DNA technology

Many people, however, want answers to questions about some aspects of gene therapy:

- Will gene mapping lead to screening of developing fetuses, prospective mates, parents, and employees?
- Will germ-line therapy allow scientists to engineer human life?
- Will gene therapy create a standard for what is considered physically normal?
- Will health insurance companies use genetic screening to deny coverage to anyone whose genes aren't in perfect working order?

INVESTIGATING the Issue

Going Further Use the points above as the basis for a class debate on the future of genetic engineering. Develop a list of ethical guidelines for the use of genetic engineering.

Ethics

BIOLOGY & SOCIETY

Brave New World

Purpose

Students will examine ethical considerations of the applications of DNA technology.

Teaching Strategies

- Provide recent articles about gene therapy for students to read. Ask them to answer these questions: How are researchers replacing defective genes? With which disorders has the most progress been made? *Answers may include that researchers are using genes spliced into viral vectors to try to replace defective genes. The most progress appears to be with cystic fibrosis.*

INVESTIGATING the Issue

Thinking Critically Students should be allowed to express their opinions freely. Guidelines developed by students may include that a particular therapy must be tested to demonstrate that it is safe before it is used.

Visual Learning

Figure 16.12 Can you think of a use for DNA fingerprinting other than crime analysis? *One possible answer is that DNA fingerprinting could be used for identification of babies in hospitals.*

3 Assess

Check for Understanding

Ask students to explain what the Human Genome Project proposes to do. *It proposes to map and sequence all the genes on human chromosomes.* L1

Reteach

Ask students to list applications of the Human Genome Project. *diagnosis of genetic disorders, gene therapy, and DNA fingerprinting* L1

PROJECT

Have groups of students design a dinosaur park on poster paper. Ask them to research what is known about the dinosaurs they want in the park and what the dinosaurs might eat. They should plan safety features for containing the dinosaurs, as well as buildings, roads, tourist activities, and tourist accommodations. L2 P LS

STUDENT JOURNAL

Gene Therapy Have students research the current progress being made on gene therapy in the treatment of cystic fibrosis or other genetic disorders. They can then write reports to include in their student journals. L2 LS

ThinkingLab Draw a Conclusion

Purpose

LS Visual-Spatial Students will compare and contrast DNA fingerprint patterns.

Process Skills

observe and infer, compare and contrast

Teaching Strategies

- Explain that all individuals, except identical twins, have unique DNA.

Thinking Critically

Number 5 matches best. Number 3 is not a perfect match. On the basis of the DNA analysis alone, there is not a perfect match and not enough evidence to convict.

Assessment

Portfolio: Have students write their personal feelings about the use of DNA fingerprinting. Ask them to support their opinions. L1 P

Extension

Cloning kits for interested students can be purchased from biological supply companies.

Assessment

Oral: Ask students to predict two or three possible future applications of the Human Genome Project. L1

4 Close

Discussion

Ask students if couples should be able to change features of their babies that are not disorders, such as eye color, hair color, or sex of the child.

ThinkingLab Draw a Conclusion

How do you analyze a DNA fingerprint?

EV 1 2 3 4 5

Assume that you have been chosen to sit on a jury in which the following DNA fingerprint evidence is presented. You must analyze the evidence and draw a conclusion about an accused individual's guilt or innocence.

Analysis

The illustration shows the results of a DNA fingerprint analysis. EV represents the electrophoresis results obtained from the evidence found at the scene of the crime. The numbers represent individuals whose DNA fingerprints are shown. Individual 3 is the accused. Analyze the DNA fingerprint patterns to determine which individual's pattern matches the pattern of the DNA obtained from the evidence.

Thinking Critically

Which individual's DNA fingerprint most closely matches the evidence? Would you convict individual 3 of the crime? Why or why not?

in fact, that DNA patterns can be used like fingerprints to identify the person (or other organism) from whom they came. Although DNA fingerprinting can give a high degree of certainty about an individual's identity, it may in certain cases fail to exclude all other non-twin individuals.

PCR techniques have been used to clone DNA from many sources. Geneticists are cloning DNA from mummies and analyzing it in order to better understand ancient life. Abraham Lincoln's DNA has been taken from the tips of a lock of his hair and studied for evidence of a possible genetic disorder. The DNA from fossils can be analyzed and used to compare extinct species with living related species, or even two extinct species with one another. The uses of DNA technology appear to be unlimited.

Connecting Ideas

We are living in a time of rapidly expanding knowledge about the code of life. The efforts of scientists involved in biotechnology and genetic engineering have resulted in a better understanding of the mechanisms of genetic disorders. As a result of their innovations, scientists may someday develop methods that eliminate inherited disorders and lead to a healthier population. Genetic changes that occur naturally also can be important to the evolution of a population.

Section Review

Understanding Concepts

1. How is a linkage map different from a sequencing map?
2. What is gene therapy?
3. Explain why DNA fingerprinting can be used as evidence in law enforcement.

Thinking Critically

4. Describe some possible benefits of the Human Genome Project.

Skill Review

5. **Observing and Inferring** Suppose a cystic fibrosis patient has been treated with gene therapy. Does this person still run the risk of passing the disorder to offspring? Explain. For more help, refer to Thinking Critically in the *Skill Handbook*.

Answers to Section Review

1. A sequencing map shows the sequence of the DNA nucleotides; a linkage map shows the position of genes on a chromosome.
2. Gene therapy is the insertion of normal genes into human cells to correct genetic disorders.
3. The DNA of every individual, except identical twins, is unique. It can be used to identify individuals connected with crime.

Thinking Critically

4. If genes are located, then gene therapy can possibly be applied to replace defective genes.

Chapter 16 Review

REVIEWING MAIN IDEAS

16.1 Recombinant DNA Technology

- Scientists have developed methods to move genes from one species into another. The process of genetic engineering uses restriction enzymes to cleave one organism's DNA into fragments. Other enzymes are used to splice the DNA segment into a plasmid or viral DNA. Transgenic organisms are able to manufacture genetic products foreign to themselves using recombinant DNA.
- Genetic engineering has already been applied to bacteria, plants, and animals. The transgenic organisms produced by biotechnology are engineered to be of use to humans in agriculture, industry, and medicine.

16.2 The Human Genome

- International efforts are presently underway to sequence the DNA of the entire human genome and to determine the chromosome location for every gene.
- Applications of the Human Genome Project include the goals of detecting, treating, and curing genetic disorders. DNA fingerprinting can be used to identify persons responsible for crimes and to provide evidence that certain persons are not responsible for crimes.

Key Terms

Write a sentence that shows your understanding of each of the following terms.

cell culture
clone
gene splicing
gene therapy
genetic engineering
human genome
linkage map
plasmid
recombinant DNA
restriction enzyme
transgenic organism
vector

Understanding Concepts

1. ___________ DNA is made by connecting fragments of DNA from other sources.
   a. Restriction
   b. Recombinant
   c. Transcription
   d. Translation
2. What are bacterial proteins that cut both strands of DNA called?
   a. restriction enzymes
   b. recombinant enzymes
   c. transgenic vectors
   d. cleaving proteins
3. A(n) ___________ is a means by which foreign DNA is transferred into a host cell.
   a. carrier
   b. hybrid
   c. enzyme
   d. vector
4. Of the following, which is NOT a vector used in genetic engineering?
   a. viruses
   b. plasmids
   c. cyanobacteria
   d. metal bullets
5. Which of the following is already produced by genetic engineering?
   a. insulin
   b. growth hormone
   c. interferon
   d. all of these
6. ___________ are genetically identical copies of DNA.
   a. Vectors
   b. Plasmids
   c. Clones
   d. Spliced genes
7. Plant cells do not have the kinds of ___________ needed for taking up foreign DNA pieces.
   a. amino acids
   b. plasmids
   c. clones
   d. bacteria
8. The process of gel electrophoresis separates ___________ fragments by using an electric field.
   a. DNA
   b. cell
   c. RNA
   d. enzyme

Chapter 16 Review

Reviewing Main Ideas

Summary statements can be used by students to review the major concepts of the chapter.

Key Terms

Answers should go beyond defining the terms. Accept any answer that uses the term correctly and in the proper context.

Understanding Concepts

1. b
2. a
3. d
4. c
5. d
6. c
7. b
8. a

Skill Review

5. If the genes were replaced in all cells of the body, including the sex cells, then he or she would not risk passing the disorder on to offspring. If only the affected body cells had the genes replaced, then the sex cells would still contain a CF allele.

Chapter 16 Review

9. c
10. a
11. d
12. b
13. d
14. c
15. c
16. b
17. c
18. d
19. b
20. b

Applying Concepts

21. Plants may be modified to become frost resistant and disease resistant.
22. treating and curing human genetic disorders
23. Bacteria are capable of making large quantities of human proteins when they carry human genes. These proteins have potential medicinal and industrial usages.
24. Probably a disorder caused by a mutant recessive allele, because it is a single gene defect that can be replaced by recombinant DNA technology. Aneuploidy involves a whole extra chromosome containing many genes.
25. This may upset the balance between the natural nitrogen-converting bacteria and the plants, possibly disrupting the soil ecosystem. Plant growth would increase because nitrogen is usually a limiting factor.

Chapter 16 Review

9. An organism that contains functional recombinant DNA is called a ____________ organism.
   a. recombinant c. transgenic
   b. cloned d. spliced
10. Restriction enzyme *Eco*RI cuts DNA strands, leaving ____________ ends.
   a. sticky c. double
   b. smooth d. slippery
11. Foreign DNA is recombined into plasmid or viral DNA by another ____________.
   a. carrier c. enzyme
   b. hybrid d. vector
12. ____________ is used to determine the order of the DNA nucleotides.
   a. DNA cloning c. DNA mapping
   b. DNA sequencing d. DNA splicing
13. A ____________ occurs when the same sequence of bases is found on both DNA strands, but running in opposite directions.
   a. plasmid c. cleavage
   b. clone d. palindrome
14. What is the primary source of the plasmids used in DNA technology today?
   a. plant cells c. bacterial cells
   b. protists d. animal cells
15. *Eco*RI restriction enzymes recognize the sequence GAATTC in the DNA. How many pieces of DNA would result if *Eco*RI were added to the following DNA?

G A A T T C C A G A A T T C A
C T T A A G G T C T T A A G T

   a. one c. three
   b. two d. four
16. ____________ is (are) used to make millions of copies of DNA pieces.
   a. Gel electrophoresis
   b. Polymerase chain reactions
   c. Restriction enzymes
   d. Gene therapy
17. DNA fingerprinting is different from DNA sequencing because it ____________.
   a. involves inserting normal genes into cells with a genetic disorder
   b. involves a vector, and DNA sequencing does not
   c. compares DNA fragments rather than the nucleotide order of DNA
   d. makes millions of copies of DNA pieces
18. Of these, which readily accepts plasmids into their own DNA?
   a. animal cells c. protists
   b. plant cells d. bacteria
19. The genetic map that shows the location of genes on a chromosome is a ____________ map.
   a. therapy c. clone
   b. linkage d. cleavage
20. Cells in a cell culture all have the same genetic material because they are ____________.
   a. linked c. transgenic
   b. clones d. recombined

Applying Concepts

21. What are some modifications that may be made to plants in the future?
22. What is the potential use of a map showing the sequence of DNA bases in a human chromosome?
23. Why would it be important to be able to have a human gene expressed in a bacterium?
24. Would it be easier to "cure" by gene therapy a person who had a chromosome aneuploidy, or a genetic disorder caused by a mutant, recessive allele? Explain.
25. Assume that transgenic organisms can be used to speed the conversion of nitrogen from the air into nitrates that plants can use as fertilizer. How might use of this organism affect an ecosystem?

392 DNA Technology

Thinking Critically

26. ***Concept Mapping*** Make a concept map that relates the following terms and phrases. Supply the appropriate linking words for your map.
genetic engineering, sticky ends, restriction enzymes, gene cleaving, vector, plasmid, viral DNA, gene splicing, recombinant DNA

27. ***Classifying*** Identify the four palindromes from the following list: dad; 11234; Madam, I'm Adam; ggcg; ladder; cagcag; atcgcta; 775535577

28. ***Sequencing*** This graphic illustrates a palindromic sequence of DNA bases. Show how the segment would be cut by restriction enzyme *Eco*RI to produce sticky ends.

G A A T T C
C T T A A G

29. ***Recognizing Cause and Effect*** How may using biotechnology to engineer many different transgenic organisms alter the course of evolution for a species?

30. ***Observing and Inferring*** A chimera is a mythical beast that is part lion, part goat, and part serpent. How do you think the word *chimera* is used in DNA technology?

31. ***Sequencing*** Once a foreign gene has been inserted into a plasmid to form a recombined plasmid, what would be the next step if you were to continue the model of recombinant DNA technology?

32. ***Interpreting Data*** If all human genes have similar patterns, how can DNA fragments from hair or skin be used to identify distinct individuals in DNA fingerprinting?

33. ***Observing and Inferring*** Explain why the use of bacterial plasmids for gene splicing does not interfere with normal cell functions such as growth and reproduction.

ASSESSING KNOWLEDGE & SKILLS

The following graph shows the results of an experiment using natural and bioengineered bacteria of the same species that can break down oil. Each culture had 40 mL of oil added on Day 1.

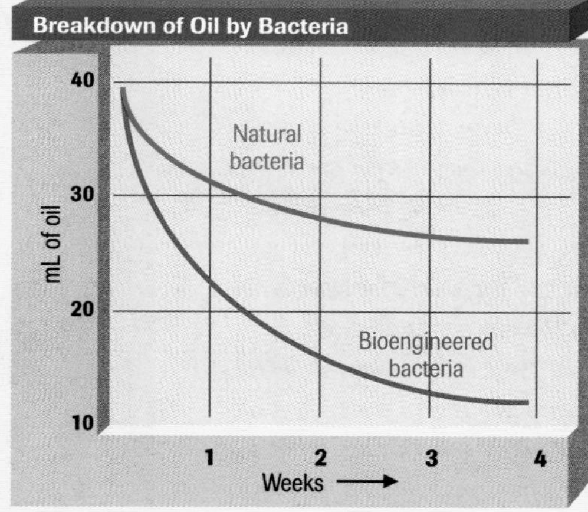

Interpreting Data Use the graph to answer the following questions.

1. Approximately how much oil had been converted into harmless products by the natural bacteria after four weeks?
a. 4 mL **b.** 14 mL **c.** 24 mL **d.** 40 mL

2. How much oil had been converted by bioengineered bacteria after four weeks?
a. 4 mL **b.** 14 mL **c.** 28 mL **d.** 40 mL

3. How much more efficient are the bioengineered bacteria than the naturally occurring species?
a. 1× **b.** 1.5× **c.** 2× **d.** 3×

4. ***Interpreting Data*** How can this technology be applied to an oil spill?

Program Resources

Chapter Assessment, pp. 91-96 L1
Alternate Assessment in the Science Classroom
Computer Test Bank L1
Content Mastery, pp. 61-64 L1

Chapter 16 Review

Thinking Critically

26. Evaluate students' concept maps. The maps should show an understanding of the relationship among the concepts listed.
27. dad; Madam, I'm Adam; atcgcta, 775535577
28. G|A A T T C
C T T A A|G
29. If many alleles were to be transferred into the normal population, their frequency would change. This would affect the genetic makeup of the population.
30. A chimera is an animal that contains genetic material from more than one animal.
31. The recommended DNA is then transferred into a host cell.
32. DNA fingerprinting uses the segments of non-coding DNA, which are in distinct, individual patterns.
33. Plasmids used for gene splicing are not part of the cells' chromosome. Thus, they can be spliced and manipulated without affecting normal cell functions.

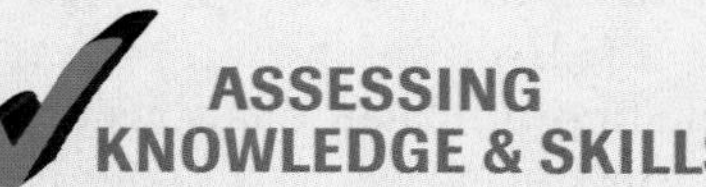

1. b
2. c
3. c
4. Because these bacteria have the ability to digest oil into a harmless product, the bacteria could be added to oil spills to clean them up quickly.

UNIT

5 Change Through Time

Unit Overview

In this unit, students will learn the concepts and principles of evolution. Chapter 17 deals with the history of life on Earth, and some of the hypotheses about how life began. Students will learn about fossils—what they are, how they are formed, and how they can be used to reconstruct the history of life on Earth. In Chapter 18, Darwin's theory of evolution by natural selection is discussed. The role of natural selection in the evolution of new species is presented. The unit concludes in Chapter 19 with a discussion of human evolution. The various lines of evidence that trace the ancestry of humans are explored.

Theme Development

Besides *evolution*, several other major themes are focused upon in this unit. The theme of *unity within diversity* is apparent as students learn how all forms of life have evolved from Earth's earliest organisms. *Systems and interactions* are central to an understanding of Chapter 18, as students discover how natural selection impacts relationships among organisms. Throughout the unit, the theme of the *nature of science* is apparent via discussions of how scientists study evolution.

UNIT

5 Change Through Time

Sharks, perhaps the most feared of all predators, are considered to be "living fossils." These fearsome fish are similar to the sharks of 350 million years ago. They've changed little during their long history. In contrast, whales were not always the graceful giants of the sea as we know them today. Long ago, their ancestors lived on land. As ancestors of whales moved into their new aquatic environment, those better adapted for living and moving in water survived. Over many generations, what was once a forelimb adapted for running became modified into an efficient flipper. This process is known as evolution—one of the most important theories of science today.

Scientists have discovered an enormous variety of fossils. What can we learn about evolution from the study of ancient life such as these fossil trilobites?

394

Advance Planning

Chapter 17

- Purchase plaster of paris for the Minilab on page 400.
- Obtain a diversity of fossils or replicas of fossils representing the major divisions of the Geologic Time Scale.

Chapter 18

- Have black paper, white paper, and hole punches available for the Minilab on page 429.
- Purchase navy beans and pinto beans for the Biolab.

Chapter 19

- Obtain casts of various fossil hominids and ape skulls for use in the Biolab.

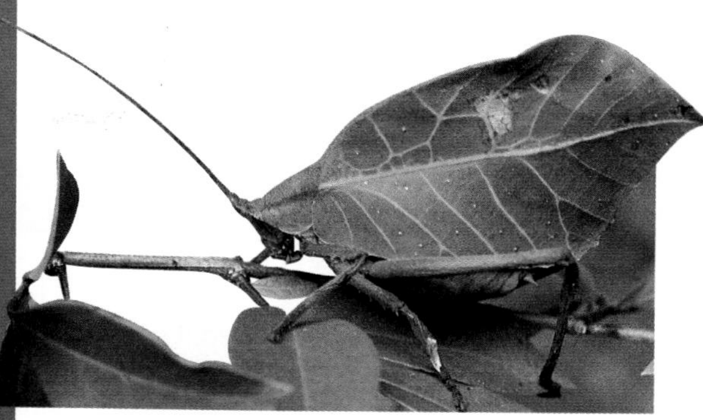

Some species, such as this katydid, mimic the appearance of other species. What advantages do organisms gain from these unusual adaptations?

The diversity among species is astonishing. What factors are responsible for the differences among these species of insects?

Unit Contents

UNIT 5

Introducing the Unit

Use the photos to introduce the concept of evolution. Lead students into a discussion about some of the anatomical and behavioral characteristics of sharks. Point out that sharks are well-designed predators that have been evolving for hundreds of millions of years. Explain that *evolution* is change in the adaptations of species over time.

After discussing sharks, ask students to think of some unusual or interesting characteristics in other organisms and have them speculate why such features exist. Use the photos of the insect collection and the mimicking of organisms to point out that all features of organisms, whether structural, behavioral, or physiological, are the result of the process of evolution.

Motivational Activity

Discussion: Ask students to imagine they could bring back an organism that is extinct. Elicit what organism different students would like to bring back and ask them to explain their choices. Discuss the types of information they might learn about an extinct organism if they could bring it back to life. L1 COOP LEARN

Unit Project

Group Project: Divide students into groups of three or four. Have the groups, for the duration of this unit, prepare a multi-media presentation about evolution. Topics you may wish to have students address include: the evolutionary history of a particular group of organisms, how scientists study fossils, the evolution of particular adaptations, or profiles of paleontologists. Emphasize cooperative learning by having each member of the group work on a different portion of the project. L1 COOP LEARN

Chapter Organizer

SECTION	OBJECTIVES	ACTIVITIES/FEATURES
17.1 The Record of Life National Science Standards: UCP.2-4; C.3; D.3; G.1-3	1. **Identify** the different types of fossils and how they are formed. 2. **Summarize** the major geological and biological events of the Geologic Time Scale.	**Minilab:** How do scientists interpret fossils?, p. 400 **People in Biology:** Robert Bakker, p. 402 **Minilab:** How can you plot the appearance of organisms on a time line?, p. 408 **Thinking Lab:** How can a clock represent Earth's history?, p. 409
17.2 The Origin of Life National Science Standards: UCP.2-4; A.1, A.2; B.2, B.3; C.1, C.3; D.2; F.3, F.4; G.1	3. **Analyze** early experiments that support the concept of biogenesis. 4. **Compare and contrast** modern theories of the origin of life. 5. **Relate** hypotheses about the origin of cells to the environmental conditions of the primitive Earth.	**A Broader View:** How Did Life Begin: Belief and Theories, p. 413 **A Broader View:** The Endosymbiont Hypothesis, p. 415 **Biolab:** Making Microspheres, p. 416

ACTIVITY MATERIALS

BIOLAB	MINILABS	ALTERNATE LAB
page 416 microscope slide coverslip glycine hot plate goggles aspartic acid 50-mL graduated cylinder 50-mL Erlenmeyer flasks (2) 400-mL beaker 1% NaCl solution watch or clock with second hand ring stand tongs clamp pipette apron oven mitts stirring rod glutamic acid	**page 400** paper milk container, pint size petroleum jelly plaster of paris **page 408** meterstick adding machine tape	**page 414** gelatin solution 3 eyedroppers gum arabic solution pH paper microscope slides coverslips hydrochloric acid test tube stirring rod

Chapter 17 The History of Life

TEACHER CLASSROOM RESOURCES

Reproducible Masters	Transparencies
Section Focus Master 36: Inferring from Fossils L1 SAE **Reinforcement and Study Guide,** pp. 65-66 L1 **Biolab and Minilab Worksheets,** pp. 65-66 L1 **Concept Mapping:** Formation of a Fossil, p. 17 L1 **Critical Thinking/Problem Solving:** Using the Law of Superposition, p. 17 L3 **Laboratory Manual:** Analyzing Fossil Molds, pp. 95-96 L2	**Reteaching Transparency #17:** Geologic Time Scale L1 SAE **Basic Skills Transparency #15:** Fossil Formation L1 SAE
Section Focus Master 37: Redi's Experiment L1 SAE **Reinforcement and Study Guide,** pp. 67-68 L1 **Biolab and Minilab Worksheets,** pp. 67-68 L1 **Content Mastery,** pp. 65-68 L1	**Basic Concepts Transparency #20:** Pasteur's Experiment L1 SAE

ASSESSMENT MATERIALS

Chapter Assessment, pp. 97-102
Alternate Assessment in the Science Classroom
MindJogger Videoquiz
Computer Test Bank

Spanish Resources SAE
English/Spanish Audiocassettes SAE
Cooperative Learning in the Science Classroom COOP LEARN
Lesson Plans
Great Developments in Biology: Evolution: An Ongoing Process L1 SAE

KEY TO TEACHING STRATEGIES

- L1 Level 1 activities should be within the ability range of all students including those with learning difficulties.
- L2 Level 2 activities are within the ability range of average to above-average students.
- L3 Level 3 activities are designed for the ability range of above-average students.
- SAE SAE activities should be within the ability range of Students Acquiring English.
- COOP LEARN Cooperative Learning activities are designed for small group work.
- P These strategies represent student products that can be placed into a best-work portfolio.
- These strategies are useful in a block scheduling format.

GLENCOE TECHNOLOGY

The following multimedia resources are available from Glencoe.

Biology: The Dynamics of Life
CD-ROM SAE
Videodisc Program

The Infinite Voyage Series

The Great Dinosaur Hunt
Where the Great Hunt Began
Alberta Badlands: A Dinosaur Graveyard
The Evolution of Extinction

Life in the Balance
Understanding Mass Extinction

The Secret of Life Series

What's in Stetter's Pond: The Basics of Life
Gone Before You Know It: The Biodiversity Crisis

Science and Technology Videodisc Series (STVS)

Chemistry
Carbon-14 Dating
Ecology
Saving the Spotted Owl

CHAPTER

17 The History of Life

Chapter Overview

Students begin their study of the history of life on Earth by learning how environmental conditions on Earth changed to become suitable for early life-forms. Next, students explore how fossils formed, what information fossils provide about organisms that once lived on Earth, and how scientists determine the ages of fossils. The section concludes with an examination of the major evolutionary events that establish the Geologic Time Scale and the ways continental movement and extinctions have shaped the diversity of life today.

In the second section of the chapter, students investigate various hypotheses concerning the origin of life on Earth. They explore the idea of spontaneous generation and the experiments that early scientists performed to test and disprove this idea. Students conclude their study with an examination of some modern hypotheses about the origin of life and the evolution of cells.

Key Terms

archaebacteria
biogenesis
fossil
plate tectonics
protocell
spontaneous generation

396

Learning Styles

Look for the following logo for strategies that emphasize different learning modalities. LS

Kinesthetic	**Minilab, p. 400**
Visual-Spatial	**Display, p. 399; Meeting Individual Needs, pp. 400, 407, 410; Visual Learning, pp. 401, 407, 411, 414; Demonstration, p. 404; Chalkboard Example, p. 405; Thinking Lab, p. 409; Activity, p. 411; Microscope Activity, p. 415**
Interpersonal	**Project, p. 408**
Intrapersonal	**Student Journal, p. 405**
Linguistic	**Student Journal, p. 406; Enrichment, p. 411; Portfolio, p. 412**
Logical-Mathematical	**Project, p. 401; Portfolio, pp. 404, 411; Minilab, p. 408; Student Journal, p. 413; Alternate Lab, p. 414**

CHAPTER 17 The History of Life

In November 1989, NASA scientists launched a satellite designed to help answer one of the most long-standing questions in science: How did the universe begin? From evidence gathered by this satellite, such as the image of the universe at the bottom of this page, scientists believe that the story began about 15 billion years ago. The main characters of the story are the galaxies, stars, and planets of the universe—among them, a small, blue-green planet known as Earth.

On Earth, there is a subplot that poses questions humans have been interested in for thousands of years: What is the history of life on Earth? How did it begin, and how has it changed? How have scientists gathered information about these questions? What are the weaknesses and strengths regarding the scientific evidences? In this chapter, you will explore some answers to these questions.

Concept Check

You may wish to review the following concepts before studying this chapter.
- Chapter 1: adaptation, evolution
- Chapter 3: autotroph, heterotroph
- Chapter 8: prokaryote, eukaryote

Chapter Preview

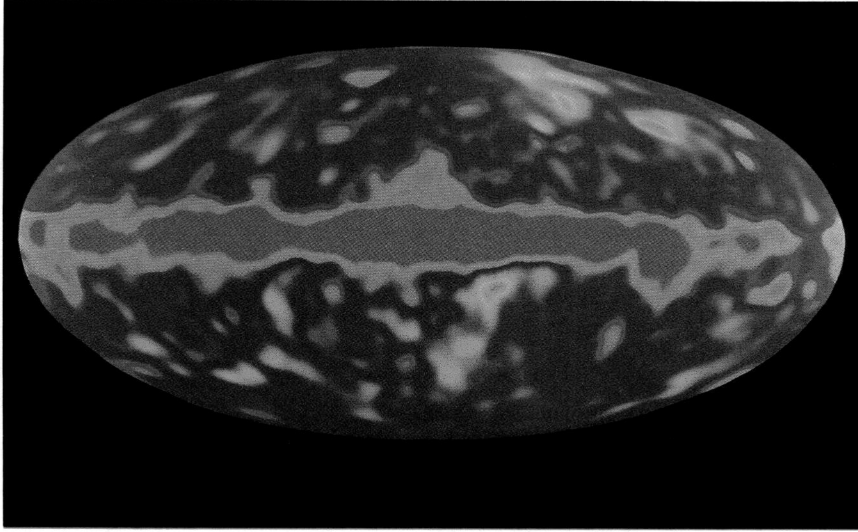

To investigate the origins of life, scientists gather fossil evidence in the field and perform tests and experiments in the laboratory. In the image of the universe (right), the pink areas represent warm temperatures where galaxies are born, and the blue areas are cooler regions.

Assessment Planner

Choose assessment strategies from the following pages to evaluate the progress of your students.
Assess, pp. 407, 418
Alternate Lab, pp. 414-415
Minilabs, pp. 400, 408
Portfolio, pp. 404, 411, 412
Thinking Lab, p. 409
Biolab, pp. 416-417
Chapter Review, pp. 419-421

Introducing the Chapter

Have students consider the importance of the computerized photo of the universe. Discuss how this photo was obtained and ask students what it might indicate about the age of the universe. *The universe must be old enough to have permitted the cooling of areas not located near the hot areas where galaxies formed.* When students begin to understand the age of the universe, they will begin to accept the fact that Earth, too, is very old, and that life on Earth has a long history.

Theme Development

In this chapter, the themes of *unity within diversity* and *evolution* are interwoven in a discussion of the amazing diversity of life today and how it is the result of evolution from single-celled organisms that lived billions of years ago.

Concept Check

In Chapter 1, students were introduced to adaptation and evolution, concepts that are further developed in this chapter. Students will also expand upon their knowledge of autotropic and heterotropic organisms as they examine how and why these types of feeding relationships originated. The differences between prokaryotic and eukaryotic cells, discussed in Chapter 8, are further explained as students examine the various hypotheses of cell evolution.

SECTION

17.1

Prepare

Key Concepts

Students will explore the different types of fossils and their scientific value. They will then consider how scientists study fossils and use the fossil record to reconstruct the history of life on Earth.

Block Scheduling

Look for this symbol for strategies that are useful in a block scheduling format. For more information on block scheduling, refer to Section 17.1 in the **Lesson Plans** booklet.

Materials

- Have students bring in small objects and pint-sized paper milk containers for the first Minilab. Purchase plaster of paris mix and petroleum jelly.
- Gather metersticks and rolls of adding machine tape for the second Minilab.

1 Focus

Bellringer

Before presenting the lesson, display **Section Focus Master 36** on the overhead projector and have students answer the accompanying questions. L1 SAE

Discussion

Elicit from students whether they have ever seen movies that use ancient Earth as their setting. Have students who respond affirmatively describe how early Earth was portrayed. *Likely responses may suggest that Earth was dry, barren, rock covered, and lifeless or jungle-like with large, ferocious animals. Explain that students will explore the question of what early Earth was like in this section.*

SECTION

17.1 The Record of Life

Section Preview

Objectives

Identify the different types of fossils and how they are formed.

Summarize the major geological and biological events of the Geologic Time Scale.

Key Terms

fossil
plate tectonics

Many popular movies, television shows, and books have explored the idea of time machines. The main characters in such stories are propelled forward and backward through time, often encountering strange people, odd creatures, and weird environments. There is no reason to expect that organisms and their environments on Earth today would be exactly like those in the past or in the future.

Early History of Earth

Step into your imaginary time machine, punch a few buttons, and get ready to explore a place to which you'll probably never want to return—primitive Earth.

Early Earth was an inhospitable place

Earth is thought to have formed about 4.6 billion years ago. It was very different from today's Earth. *Figure 17.1* illustrates what it may have looked like. Scientists theorize that Earth began as a hot ball of rock. Meteorites bombarded its surface, and volcanoes formed by the high temperatures inside Earth constantly shook the planet, shooting out gases that formed an atmosphere. Earth was much too hot for life to exist.

About 3.9 billion years ago, Earth had cooled enough for water vapor to condense, and Earth was, for the first time, experiencing violent rainstorms. Eventually, the accumulated rainfall formed Earth's oceans. It is in these oceans, about 3.5 billion years ago, that scientists hypothesize the first living organisms appeared.

Figure 17.1

The conditions on early Earth were not suitable for life. However, geological events, such as volcanic activity, set up conditions that would play a major role in the evolution of life.

Program Resources

Section Focus Master 36 L1 SAE
Reinforcement and Study Guide, pp. 65-66 L1
Biolab and Minilab Worksheets, pp. 65-66 L1
Laboratory Manual, pp. 95-96 L2
Concept Mapping, p. 17 L1
Critical Thinking/Problem Solving p. 17 L3
Reteaching Transparency 17 and **Master** L1 SAE
Basic Skills Transparency 15 and **Master** L1 SAE

A History in the Rocks

Accounts of Earth's formation and early history are based on the best scientific evidence available. However, no rocks have survived to provide direct evidence of Earth's earliest years. The oldest rocks on Earth are only about 3.9 billion years old. Though rocks tell us little about Earth's infancy, they do provide us with a historical record of life.

Fossils—Clues to the past

Anyone who has ever visited a zoo or botanical garden has seen evidence of the tremendous diversity of life. But the millions of species of organisms that live today are only a fraction of all the species that ever lived. Scientists learn about the different types of organisms that have appeared during Earth's history by studying fossils of those organisms. A **fossil** is any evidence of an organism that lived long ago.

Fossils are classified according to the way they are formed. Examples of the main types of fossils are shown in *Figure 17.2.*

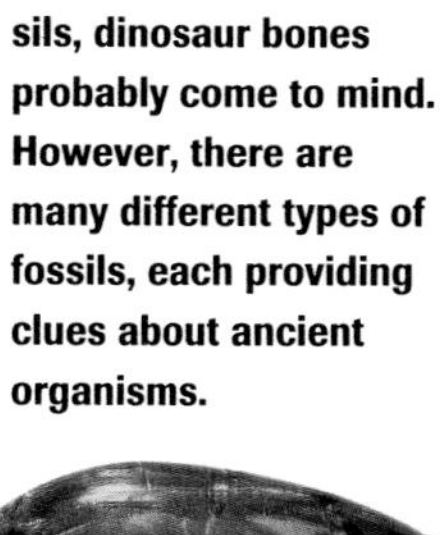

Figure 17.2

When you think of fossils, dinosaur bones probably come to mind. However, there are many different types of fossils, each providing clues about ancient organisms.

Casts **When a mold of an organism is created, it often becomes filled by minerals in the surrounding rock, producing a replica of the original organism.**

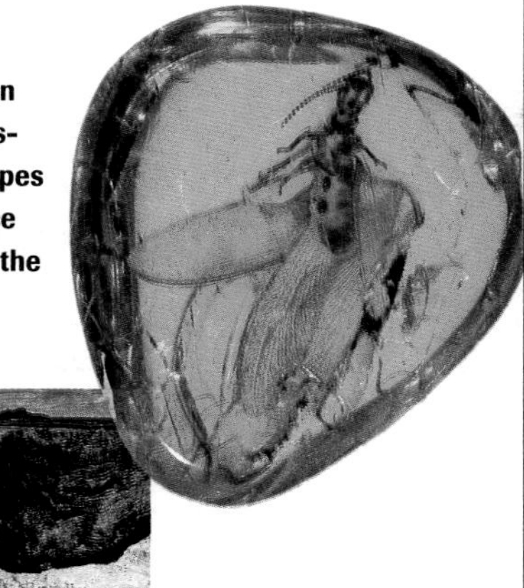

Amber-preserved and frozen fossils **Sometimes an entire, intact organism can be found frozen in ice or preserved in fossilized tree sap, such as amber. These types of fossils are rare, but valuable to science because even the most delicate parts of the organism are usually preserved.**

Trace fossils **Trace fossils are the markings or evidence of animal activities. They include footprints, trails, and burrows.**

Petrified fossils **The hard parts of organisms are sometimes penetrated and replaced by minerals, atom-for-atom. When the minerals harden, an exact stone copy of the original organism is produced.**

Imprints **Sometimes fossils form before sediments harden into rock. Thin objects, such as leaves or feathers, falling into sediments such as mud often leave imprints.**

Molds **When an organism is buried, it can decay, leaving an empty space in the rock that is the exact shape of the organism.**

2 Teach

Brainstorming

Ask students to brainstorm a list of questions that would have to be answered in order to gain a more accurate picture of ancient life.

Visual Learning

Figure 17.2 shows a variety of fossils. Ask students to list some of the information scientists might be able to obtain about each fossilized organism. *Responses may include ways in which the fossilized organism is similar to or different from an organism that is living today.*

Display

LS **Visual-Spatial** Obtain samples of the various types of fossils in Figure 17.2 to show to the class. Have students attempt to identify the types of fossils on display and explain how each may have formed.

GLENCOE TECHNOLOGY

Videodisc

The Infinite Voyage: The Great Dinosaur Hunt

Where the Great Hunt Began (Ch. 1)

Alberta Badlands: A Dinosaur Graveyard (Ch. 8)

CULTURAL DIVERSITY

Motonori Matuyama and Paleomagnetism

For reasons that are not completely understood, Earth's polarity has completely reversed on numerous occasions during its history so that the magnetic north pole becomes the south pole and vice versa. Scientists can study these polarity changes in volcanic rocks, and this is one of the methods used to establish chronology of geological strata.

Point out to students that the person credited with the discovery of Earth's magnetic reversals is Japanese geologist, Motonori Matuyama (1884-1956). Cite specific examples of how paleomagnetic studies have been important in understanding the evolution of particular organisms.

MiniLab

Purpose

IS **Kinesthetic** Students will make models to examine how two types of fossils form and how information is learned from such fossils.

Process Skills

hypothesize, observe and infer, compare and contrast, make models

Teaching Strategies

- Review how fossil molds and casts form. Have students read through the Minilab and identify which steps will produce molds and which will produce casts.

Expected Results

Students should be able to produce molds from many small objects and then use the molds to produce casts.

Analysis

1. A mold is an imprint of the outside of an object, whereas a cast forms from materials that fill a mold and harden after the object which formed the mold has been removed.
2. Some students may say that molds are easier because they provide details about the outsides of objects. Other students may say casts are easier because they are more like the real objects.
3. Students should recognize that their method is similar to those used by scientists studying real fossils.

Assessment

Knowledge: Have students hypothesize how their results may have differed if their molds and casts had been subjected to weathering factors, such as cold, heat, pressure, or erosion. L1

MiniLab

How do scientists interpret fossils?

Scientists learn about organisms that lived in the past by studying fossils. To study the details of mold fossils, casts are produced. In this activity, you will produce and analyze molds of common objects.

Procedure

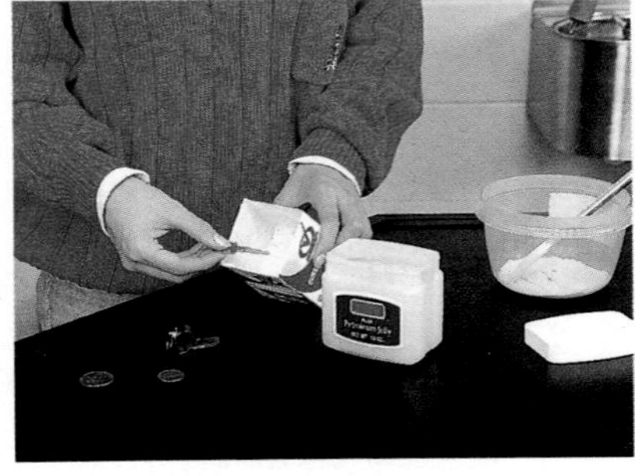

1. Open completely the top of a paper pint milk container, and grease the inside with petroleum jelly.
2. Mix plaster of paris, and pour the mixture into the milk container until it is half full.
3. When the plaster begins to thicken, grease some small objects with petroleum jelly and press them into the plaster. After the plaster has hardened, remove the objects.
4. Grease the "fossil" mold you have made. Mix and pour another layer of plaster over the hardened mold. When this layer hardens, tear away the milk container and separate the two layers. You now have casts and molds of the objects.
5. Exchange your molds and casts with those of other students. Study the molds and casts, and try to determine which objects were used.

Analysis

1. How do the molds and casts compare?
2. Which was easier to interpret, a mold or a cast? Explain.
3. Compare the way you predicted what the unknown objects were with the way a scientist predicts which organism left a fossil mold or cast.

Paleontologists—Detectives to the past

If you like to solve puzzles, you may enjoy studying fossils. Paleontologists, scientists who study ancient life, may be thought of as detectives who use fossils to understand events that happened long ago. By working with fossils, paleontologists can determine the kinds of organisms that lived in the past and sometimes draw conclusions about their behavior. For example, fossil bones and teeth can indicate how ancient animals moved around and what they ate.

The ancient climate and other environmental conditions also can be determined by studying fossils. If scientists find fossilized pollen that resembles pollen from tropical plants, they may hypothesize that an ancient environment had a tropical climate. *Figure 17.3* shows an example of fossil pollen.

By studying the condition, position, and location of fossils, paleontologists can make deductions about the geography and topography of past environments. For example, if a partial skeleton of an animal is found containing only the heaviest bones, it may mean that the lighter bones were carried away by a stream or river in the area.

Figure 17.3

This fossil pollen came from an extinct seed fern that grew in warm, moist climates about 320 million years ago. *Magnolia grandiflora* (right) probably evolved during the Cretaceous period more than 100 million years ago. It can be found growing in mild climates today.

400 The History of Life

Meeting Individual Needs

Gifted Have students obtain a small leaf and a dead insect from the local area. Have students place the leaf in a small plastic container half-filled with water and repeat this process using the insect and a second container. Have students place both containers in a freezer for twenty-four hours. Instruct students to remove their containers from the freezer and to separate the blocks of ice from their containers. Have students observe and describe the appearance of the leaf and insect specimens frozen in the ice. Ask students to allow the ice to thaw and examine the specimens again. Have students describe how fossils preserved in ice compare in appearance to those preserved as molds or casts. L3 IS

Fossils occur in sedimentary rocks

In order for fossils to form, organisms usually have to be buried in sediments—small particles of mud, sand, or clay—soon after they die. *Figure 17.4* illustrates this process of fossilization. Over time, sediments build up around and over the organism, preventing it from decaying further. Depending upon future environmental conditions and other factors, a mold, cast, or petrified fossil will then form.

Most fossils will be found in sedimentary rocks because the slow process of depositing sediments prevented damage to the organism. Fossils are unlikely to be found in other types of rock because of the way such rocks are formed. For instance, metamorphic rocks are formed when sedimentary rocks are changed by heat, pressure, and chemical reaction. Any fossils that are in such sedimentary rock are usually destroyed by the change.

Figure 17.4

Fossils are usually found in sedimentary rocks because of the way such rocks form. Follow the steps that illustrate how an organism becomes a fossil in sedimentary rocks.

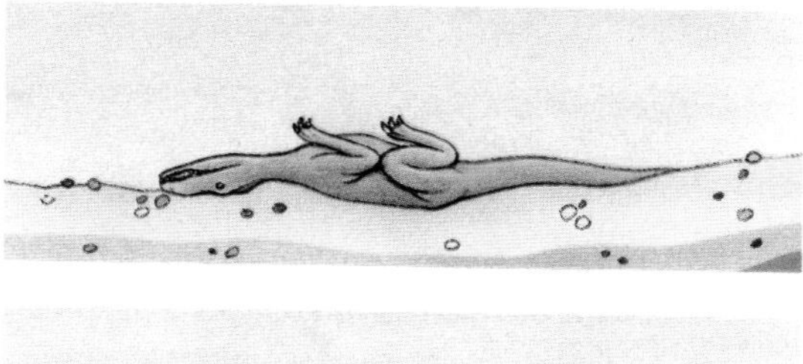

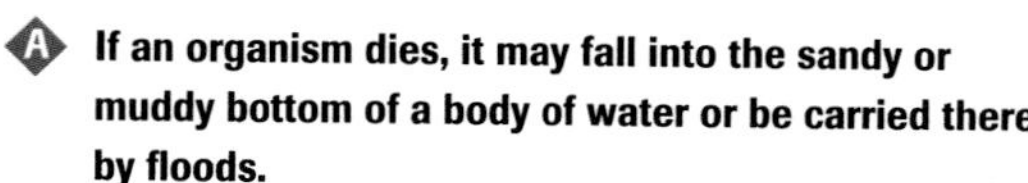

A If an organism dies, it may fall into the sandy or muddy bottom of a body of water or be carried there by floods.

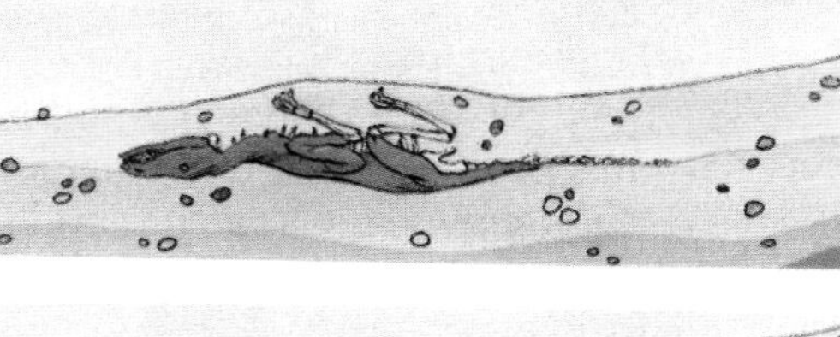

B Over time, sediments carried by the water pile on top of the carcass until it lies underground.

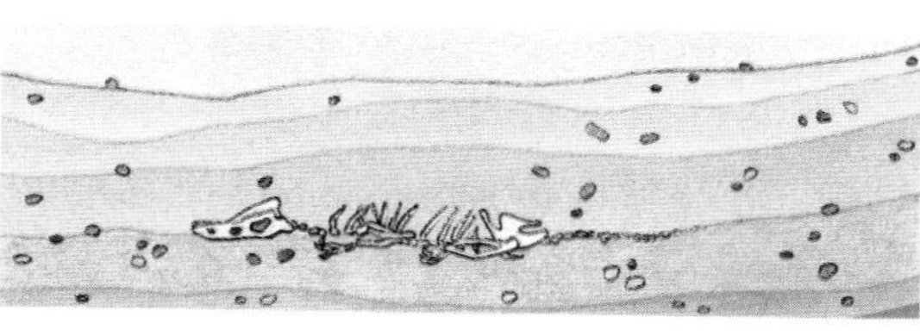

C Eventually, the muds and sands around the carcass are compressed until they form a sequence of sedimentary rocks. Geological processes have an effect on the carcass, and usually only the hardest parts survive.

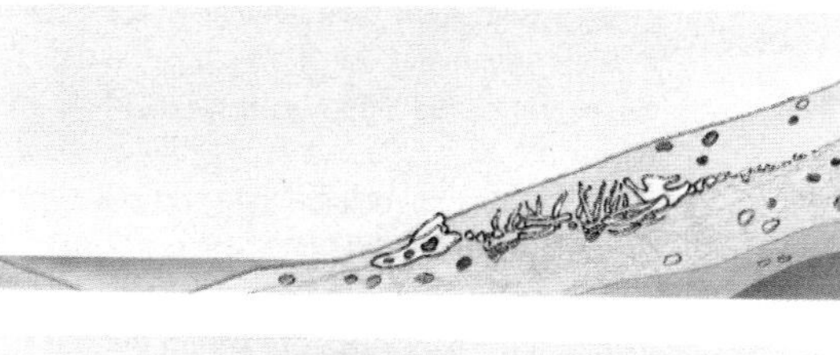

D After the fossil has become embedded in sedimentary rock, geological events, such as Earth movements and erosion, may bring the fossil to the surface.

E A scientist discovers the fossil, extracts it from the surrounding rock, and brings it to a laboratory for further study.

Concept Development

One way an organism can become fossilized in sedimentary rocks is illustrated in Figure 17.4. Point out that during fossilization an organism's remains are subjected to a variety of environmental factors, such as heat, cold, animal trampling, and others. Stress that the preservation of fossils is perhaps the greatest limiting factor in the interpretation of past life.

Visual Learning

Figure 17.4 Review with students the way sedimentation allows for the formation of fossils. You may wish to model the process shown in the diagram using objects and sediments of various sizes and weights. Place the objects and sediments in a 3-L plastic beverage container three-quarters full of water. Shake the container and have students observe how the objects become buried by the layering sediments. Explain that under pressure, such objects may be fossilized. IN

Reinforcement

Encourage students to diagram in a flowchart the sequence of events that lead to the development of fossils in sedimentary rock. L1

PROJECT

Fossil Preservation

Have students carry out the following activity to learn more about the preservation of fossils.

1. Place fresh fruit, such as strawberries, apple slices, or orange slices, in an open plastic container. Completely cover the fruit with water and place the container in the freezer.
2. Put the same amount and types of fruit in another open plastic container. Leave the container in a location where it will be undisturbed.
3. After three days, observe the fruit in each container. Prepare a report that summarizes your observations and includes a conclusion regarding any differences in the observations between the two containers. L1

SAE IN

People in Biology

Meet Dr. Robert Bakker, Paleontologist

Teaching Strategies

- Use a wall map to point out the location of the Laramie Range to students.
- Have interested students read the article "The World We Live In: 2 Billion Years of Evolution," *Life*, September 7, 1953. Encourage students to write a summary of the article in their journals and emphasize how current knowledge differs from that available in 1953.
- Remind students that they need not be field workers, like Dr. Bakker, in order to have a career involving the study of fossils.
- Have interested students conduct library research about significant fossil discoveries made in the United States. Have students identify the locations where fossils have been discovered on an outline map of the United States.
- Challenge students to find out if their state has a state fossil. Have students summarize their findings in their journals along with a description of why the fossil was chosen as the state fossil.

people in biology

Meet Dr. Robert Bakker, Paleontologist

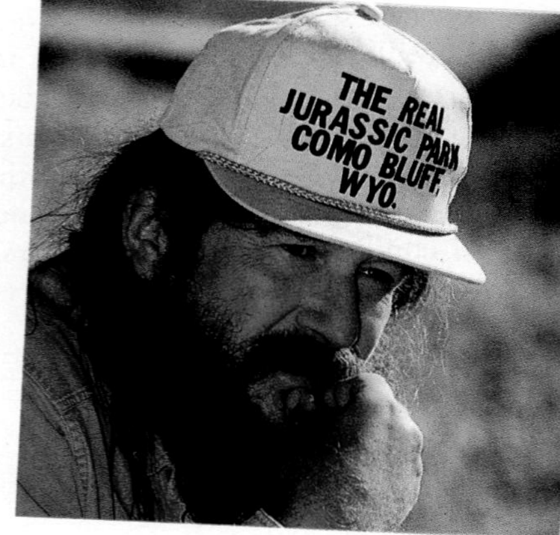

"We need an army of amateur dinosaurologists combing the hills!" exclaims paleontologist Dr. Robert Bakker. When he isn't leading a group of amateurs on a dinosaur dig, Dr. Bakker might be helping to invent a dinosaur chase video game or advising movie producers about the color and gait of an animated dinosaur.

In the following interview, Dr. Bakker tells about his life as a famous paleontologist and educator.

On the Job

Q Dr. Bakker, what did you discover on your latest dig in the Laramie Range?

A We uncovered the earliest armor-plated dinosaur skull found anywhere in the world. Right now, we're calling it Boris, the name chosen by the person who unearthed it. She's an amateur paleontologist who took my field course along with her 13-year-old son. Actually, most important finds are made by amateurs.

Q How does it work out to have young people along on your digs?

A Kids are good workers in the field. Their energy really gets focused when they're helping to uncover a bone. Fortunately, there are lots of opportunities for kids as young as ten to go on digs. Kids can call this number to find out about some family digs in the United States: 1-800-DIG-DINO. Or they can ask at libraries or museums.

Q How do you proceed when you find a dinosaur skeleton?

A It's something like investigating the scene of a crime. Here's this dead body, and we ask ourselves, How did it die? We look for broken tooth crowns from meat-eating dinosaurs, which can be direct evidence—"the smoking tooth."

Early Influences

Q How did you first get interested in paleontology?

A I have right here the magazine I first read at my grandfather's house when I was ten years old. It's a *Life* magazine dated September 7, 1953, and contains an article called *The World We Live In: 2 Billion Years of Evolution.* What struck me then and stays with me today is that dinosaurs aren't just a monster circus parade—not just funny animals with bizarre shapes. They're part of the incredible story of evolution, which stretches over 3 billion years. This great story seemed to be something I could commit my life to.

Q Did you have an opportunity to go on any digs when you were a child?

A Back then, there weren't any amateur programs, but I did spend a lot of time visiting what I thought was the greatest dinosaur show in the world at the American Museum. The digs had to wait until I was in college and went to Wyoming to dig dinos.

Q Do you remember the first fossil you found?

A It was some bits of a fossil tortoise in Nebraska. When you pick up your first fossil and hold in your hand what was once living bone, it's an electric experience. Bones are literally pulsating with life. They are quite vivid documents and very aesthetically pleasing, too.

Personal Insights

Q What advice do you give students who are interested in paleontology–dinosaurs in particular?

A The best thing they can do is to contact a local nature center, museum, or zoo and volunteer. I firmly believe in breaking down that false barrier between the zoo and the museum. Life is life, and evolution connects the living with the past. If you want to understand dinosaurs, go to the zoo and look at elephants and ostriches. If you want to understand elephants, go to the museum and look at extinct elephants.

Making your own collection of bones and videotapes is a great way to start. Here's just one idea for a project. Kids can make videos of an animal in zoos and of a brontosaurus skeleton in a museum, and then compare those videos to try to figure out how a brontosaurus would move. Whether or not it *could* move on land is a hot topic today.

CAREER CONNECTION

If you "dig" the distant past, you might want to explore these career paths.

Paleontologist *Master's degree or PhD; courses in geology and biology*

Site Surveyor *Technical institute training*

Expedition Aide *High-school degree; on-the-job training*

Q There's a lot of talk today about bringing dinosaurs back to life with genetic engineering. What do you think about that?

A Right now, I think it's impossible. However, we might see in our lifetime the woolly mammoth brought back because we have entire frozen woolly mammoths. There's a strong possibility of taking woolly mammoth DNA, inserting it into the egg of its close relative, the Indian elephant, and getting a woolly mammoth. It would be fun to comb its hair.

Q If you could go out tomorrow and discover the most exciting find of your career, what would it be?

A It would simply be the *next* thing. There isn't *one* thing I'm looking for. It may be something I've never even thought of before. For example, we found the skulls of turtles that lived with the brontosaurus. These turtles have an important story to tell because they witnessed a big extinction of dinosaurs at the end of the Jurassic period. We ask ourselves, Why would that be? and look for the answer.

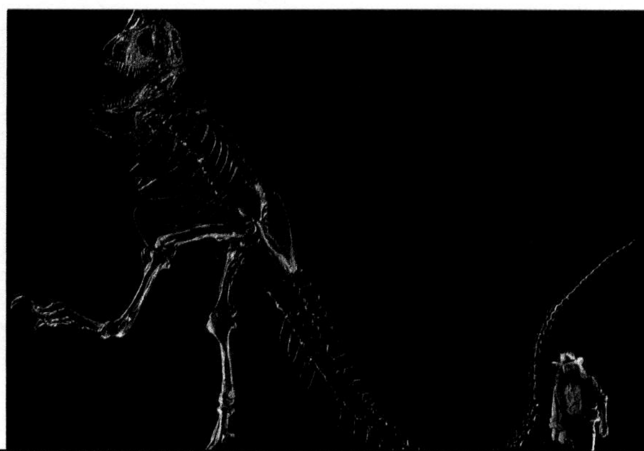

CAREER CONNECTION

- **Career Path** Students interested in a career in paleontology should have a strong background in high school and college biological science and must also have some background in chemistry, physics, and Earth science. A graduate degree in any of the biological sciences or geology would be sufficient.
- **Career Issue** Students should be aware that fossil hunting on public lands is controlled by many state and federal regulations.

For More Information

Students can contact The Paleontological Society, U.S.G.S., E/501 National Museum Building, Smithsonian Institution, Washington, DC 20006 for more information about the work of paleontologists.

Science, Technology, and Society

Cloning Dinosaurs

Ancient DNA has been extracted from a variety of amber-preserved organisms, such as bees, termites, magnolia leaves, and mollusks. Scientists have also found an amber-coated insect that may have ingested dinosaur blood. In the fictional book and movie *Jurassic Park,* blood from such an insect was used to clone dinosaurs. Have students imagine such cloning is possible and debate how bringing extinct organisms back to life may be both beneficial and harmful to society.

L1 COOP LEARN

GLENCOE TECHNOLOGY

 Videodisc

Biology: The Dynamics of Life
Disc 1, Side 2
Discovering Dinosaurs (Ch. 2)

Demonstration

Visual-Spatial Explain to students that environmental factors, such as erosion, earthquakes, and volcanic activity can fold, twist, and bend sedimentary layers, causing a disruption in the sequence of rock layers. Illustrate the difficulties in relative dating that result from such disruptions of rock layers by preparing a model of sedimentary layers using different colors of clay. Manipulate the layers of clay to simulate folded or faulted rock layers. Ask students what problems such disturbances might cause to paleontologists. *Students should indicate that such disruptions may result in older rock layers being found above younger layers. As a result, fossils contained in the layers may be incorrectly dated using relative dating methods.*

Visual Learning

Figure 17.5 Which layer is the oldest? *The layer at the bottom is oldest.* Which is the youngest? *The layer at the top is youngest.*

GLENCOE TECHNOLOGY

Videodisc

STVS: Chemistry
Disc 2, Side 2
Carbon-14 Dating (Ch. 4)

The Infinite Voyage: The Great Dinosaur Hunt
The Evolution of Extinction (Ch. 2)

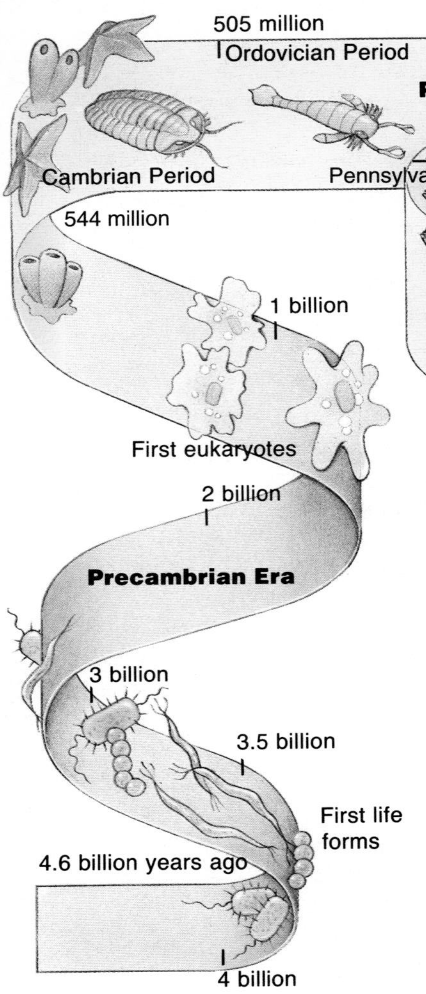

Dating methods give the approximate ages of fossils

If you think you have a problem remembering dates in history, think about a paleontologist who has just unearthed a fossil skull. Scientists use a variety of methods to date fossils in order to find out how old they are.

Perhaps the simplest dating method is a technique called relative dating, as shown in *Figure 17.5.* If you've ever stacked newspapers at home, it's easy to understand relative dating. As each day's newspaper is added to the stack, the stack grows. If it is left undisturbed, newspapers at the bottom of the stack should be older than ones toward the top.

Relative dating works in the same way. If sediments have been left undisturbed, layers closer to the surface should be younger than deeper layers. In this way, scientists can get at least an idea of the order of appearance and disappearance of organisms buried in sediments.

Relative dating techniques can determine only whether one fossil is older than another fossil. This technique cannot determine the actual age of a fossil. To find the absolute ages of fossils, scientists often use

Figure 17.5

Relative dating techniques rely on the geological law of superposition. According to this law, most sedimentary rocks are laid down in horizontal layers with the younger layers closer to the surface and the older layers buried deeper. Using this law, scientists can determine the order of appearance and disappearance of organisms. Which layer is the oldest? Which is the youngest?

PORTFOLIO

Calculating Age Using Half-Lives Have students solve the following problems by drawing a bar graph. Have them place their graphs and answers in their portfolios.

- A radioactive element has a half-life of 20 days. How much of a 16-g sample will be unchanged after 80 days? *1 g*
- A fossil contains a radioactive element that has a half-life of 10 000 years. If the ratio of radioactive element to decay element is 1:3, how old is the fossil? *20 000 years or two half-lives* L1 LS P

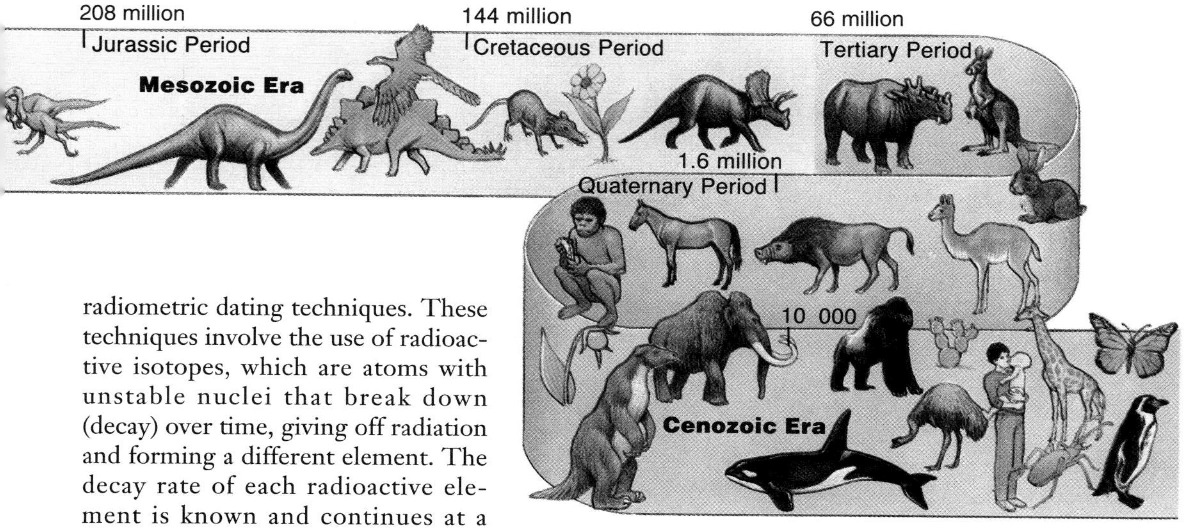

radiometric dating techniques. These techniques involve the use of radioactive isotopes, which are atoms with unstable nuclei that break down (decay) over time, giving off radiation and forming a different element. The decay rate of each radioactive element is known and continues at a steady rate. Therefore, scientists use them as a type of clock.

The principle of radiometric dating is often compared to watching the passage of time in an hourglass. You can judge how long an hourglass has been running by comparing the amounts of sand in the top and bottom. In the same way, scientists determine the ages of rocks or fossils by comparing the amount of the original radioactive element to the amount of the new element formed from decay. For example, suppose a rock contains a radioactive element that decays to half its original amount in 1 million years. If tests show that the rock contains equal amounts of the original radioactive element and the new element, then the rock is about 1 million years old.

Potassium-40, which decays to form argon-40, is an example of an isotope that is used to date the rocks in which fossils are found. It decays to half its original amount in 1.3 billion years. Fossils and archaeological artifacts that are less than 50 000 years old are dated with carbon-14. Carbon-14 breaks down to half its original amount in 5730 years.

A Trip Through Geologic Time

By dating fossils and examining layers of sediments in Earth's crust, scientists have been able to put together a time scale for the history of life on Earth. This scale, called the Geologic Time Scale, is a type of calendar that allows scientists to communicate about events that have occurred since Earth was formed.

What is the Geologic Time Scale?

The Geologic Time Scale is based on the different types of living organisms that have appeared during Earth's history. *Figure 17.6* shows that it is divided into four eras: Precambrian Era, Paleozoic Era, Mesozoic Era, and Cenozoic Era. An era is the largest division, and each era is subdivided into periods.

The Geologic Time Scale begins at the time of the formation of Earth about 4.6 billion years ago. To appreciate the immensity of this number, it is useful to scale down the history of life on Earth into a calendar year.

Figure 17.6

The Geologic Time Scale is based on the appearance of different types of life during Earth's history. The first life appeared on Earth about 3.5 billion years ago.

SECTION 17.1

Concept Development

Explain to students that radioactive dating is a precise technique for geological timekeeping. The rate of decay of a radioactive element does not change over time.

Chalkboard Example

IS **Visual-Spatial** To illustrate the concept of geological time, reproduce on the chalkboard or an overhead transparency a page from a monthly calendar. Remind students that the total area of the calendar page stands for a unit of time equal to one month.

Divide the calendar into four (or five) horizontal strips. Ask students what amount of time each strip represents. *One week* Cut one of the strips into seven pieces and ask students what each square represents. *One day* Next, ask students how minutes and seconds could be shown. Elicit from students how geologic time, like calendars, is also divided into units.

GLENCOE TECHNOLOGY

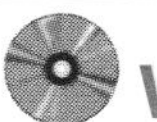

Videodisc

The Infinite Voyage: Life in the Balance
Understanding Mass Extinction (Ch. 1)

STUDENT JOURNAL

Summarizing Geologic Time Have students draw a large box on a sheet of paper in their journals. Next, have students divide the large box into 4 smaller boxes with the first box occupying about one third of the total box and the last box occupying about one-eighth of the total. Have students identify each box with the name of one of the eras of geologic time, placing the Precambrian in the largest box and the Cenozoic in the smallest. As students read about the eras of geologic time, have them write terms or phrases in each box that describe that era. Have students use their completed sheets as study tools. L1 IS SAE

Different Viewpoints in Biology

A Bird in the Hand?
Many paleontologists believe that modern birds are the living descendants of dinosaurs. However, not all scientists agree on this point. Dr. Larry Martin of the University of Kansas, for example, argues that birds evolved not from dinosaurs, but from primitive reptiles called thecodonts or from earlier reptiles.

The bird-dinosaur group, led by Dr. John Ostrom, uses the method of cladistics, which utilizes computers to map dozens of traits shared among different species to figure out how closely related organisms are. Martin and other critics claim that the features used in cladistics are actually primitive features, not uniquely shared by birds and dinosaurs. Thus, such characteristics cannot be used to establish relatedness. All of this controversy points to the difficulties inherent in the study of paleontology.

GLENCOE TECHNOLOGY

Videotape

The Secret of Life
Use the videotape *Gone Before You Know It: The Biodiversity Crisis* to illustrate the problems associated with species extinction.

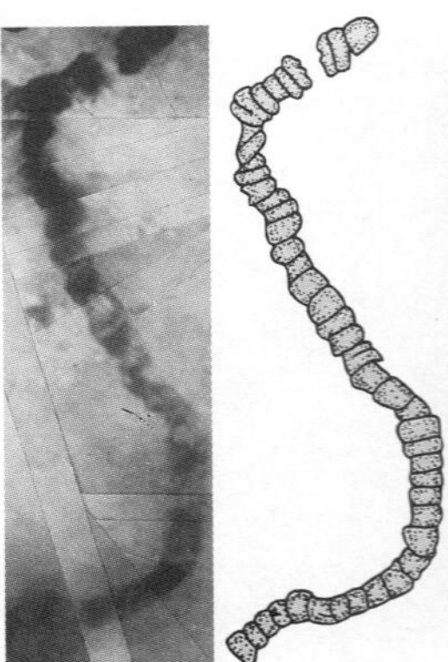

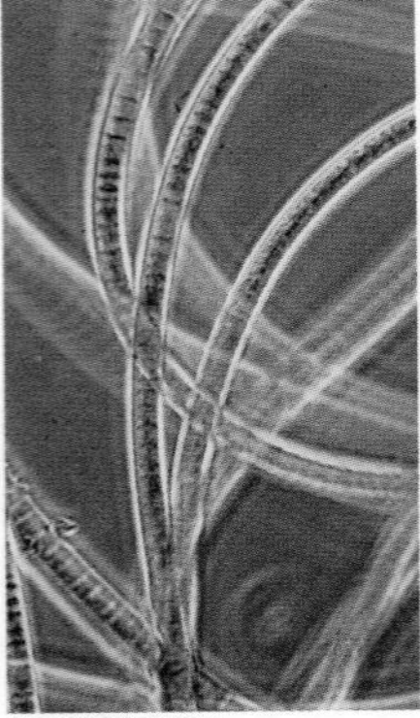

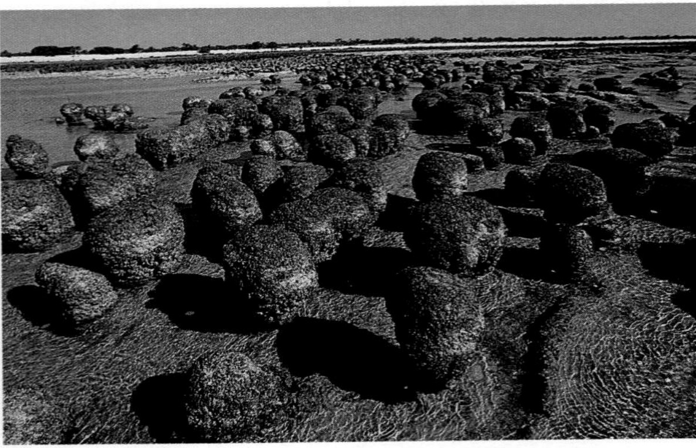

Figure 17.7
Stromatolites (right) provide evidence of photosynthetic bacteria that lived 3.5 billion years ago. Note that these filamentous fossil organisms (left) resemble modern cyanobacteria such as *Oscillatoria* (center). At least 11 different fossil species of these organisms have been identified.

Life began in the Precambrian Era

On this hypothetical calendar, Earth formed on the first of January. The oldest evidence of life on Earth is contained in Precambrian rocks dated to approximately 3.5 billion years ago, about March 20 on the hypothetical calendar. Precambrian rocks found in the hot deserts of western Australia contain fossils of an assortment of spherical and filamentous organisms that resemble present-day species of photosynthetic bacteria called cyanobacteria, as shown in *Figure 17.7.*

Also found in these rocks were dome-shaped structures called stromatolites. Stromatolites form today from thick mats of cyanobacteria; thus, their occurrence in Precambrian rocks suggests the presence of cyanobacteria or other photosynthetic bacteria 3.5 billion years ago.

The Precambrian Era accounts for about 87 percent of Earth's history—until about the middle of October in our hypothetical calendar year. Primitive prokaryotes dominated the early Precambrian, and 2 billion years passed before eukaryotic organisms, organisms with nuclei, appear in the fossil record. By the end of the Precambrian, about 544 million years ago, the oceans were filled with simple, multicellular organisms such as algae, sponges, and jellyfish.

New organisms appear in the Paleozoic Era

The Paleozoic Era, which lasted from about 544 to 245 million years ago, is characterized by the appearance of many more animal and plant phyla. Scientists often refer to a Cambrian explosion of life, which started at the beginning of this era in the Cambrian period. During this period, shallow seas were teeming with many types of invertebrates including worms, echinoderms, and primitive arthropods, such as the one shown in *Figure 17.8.*

Figure 17.8
Primitive arthropods, such as this trilobite, were among the many invertebrate species to appear during the Cambrian explosion.

406 The History of Life

STUDENT JOURNAL

Extinct Animals Have students research one of the following extinct animals: pleisiosaur, brachiosaur, pterodactyl, saber-toothed cat, woolly mammoth, or plesiadapid. Have students describe in their journals what foods, medicines, environments, special enclosures, organisms, or other factors might be required to keep the animal alive in a zoo. L1 LS

interNET CONNECTION

Follow the link for this chapter on the Glencoe Homepage at **www.glencoe.com/sec/science** to find out more about the history of life.

In the first half of the Paleozoic, fishes, the earliest vertebrates, appeared, and there is evidence of plant life such as ferns and gymnosperms on land. Around the middle of the Paleozoic, amphibians appeared, and during the second half, reptiles also appeared.

Dinosaurs, mammals, and flowering plants appear in the Mesozoic

The Mesozoic Era began about 245 million years ago, which represents about December 10 on our one-year calendar. It is a period of many changes, both in Earth's geology and its inhabitants.

The Mesozoic Era is divided into three periods; the oldest period is the Triassic. During the Triassic period, mammals made their first appearance. Early mammals were small, mouselike animals that scurried around in the shadows of huge fern forests. The first dinosaurs also appeared during this period.

The middle period of the Mesozoic is called the Jurassic. The Jurassic period began about 208 million years ago, mid-December on the hypothetical calendar, and is often referred to as the Age of the Dinosaurs. Read about how scientists have discussed how modern birds evolved toward the end of the Jurassic period in *Figure 17.9.*

The last period in the Mesozoic, the Cretaceous, which began about 144 million years ago, is characterized by the radiation of mammals and the evolution of flowering plants such as oak, fig, and elm trees.

Figure 17.9

The evolution of birds is currently under debate. *Archaeopteryx* (left) was once thought to be one of the earliest birds because it had feathers and other birdlike features. In 1992, a 135-million-year-old fossil named *Sinornis* (right) was discovered in China. Some scientists now suggest that *Sinornis* represents the earliest bird because it had more bird characteristics than did *Archaeopteryx.*

▼ **Fossil *Archaeopteryx***

▼ **The possible appearance of *Sinornis***

Meeting Individual Needs

Learning Disabled Have students compare bird and dinosaur skeletons to consider whether birds are living descendants of dinosaurs. Obtain and distribute illustrations or models of dinosaur skeletons and illustrations or actual skeletons from birds. Have students record observations of similar bones in both organisms in a data table with the headings "bird" and "dinosaur." Assist students with the examinations of the skeletons by having them look at the parts one-by-one. L1 LS SAE

Visual Learning

Figure 17.9 Have students compare the *Archaeopteryx* and *Sinornis* shown. Discuss with students what characteristics each animal shared, or did not share, with modern birds. LS

3 Assess

Check for Understanding

Prepare an in-class "game show" to check student understanding of the topics presented in this section. Prepare questions on index cards in a variety of categories, such as types of fossils, fossil formation, dating techniques, and the Geologic Time Scale. Have students working in two teams attempt to answer the questions. L1 COOP LEARN

Reteach

Have students demonstrate the concept of radiometric dating using balls of clay or other materials to represent radioactive materials. L1 SAE

Extension

Have students research the geological history of the local area to determine what organisms might have lived there in past ages. Students could exhibit some local fossils if any are available. L2

Assessment

Performance: Ask groups of students to pick one era of the Geologic Time Scale and prepare a bulletin board showing all the kinds of life that existed during that era. L1

MiniLab

Purpose

LS Logical-Mathematical Students will reinforce their knowledge about the Geologic Time Scale.

Process Skills

use a table, sequence, measure in SI

Teaching Strategies

- Review the concept of a scale of distance, and how to relate distance to a time scale.
- Review the two major divisions of the Geologic Time Scale—the era and period.

Expected Results

Students should be able to perform the calculations needed to put the adding machine tape into scale and plot major events in the history of life on their geologic time scale.

Analysis

1. The Paleozoic Era is longest, the Cenozoic Era is shortest.
2. Mesozoic Era
3. mammals

Assessment

Performance: Have students work in groups to collect pictures of 15-20 different organisms. Ask groups to illustrate the Geologic Time Scale on poster board, placing the pictures according to their time of appearance in the fossil record. L1 SAE COOP LEARN

MiniLab

How can you plot the appearance of organisms on a time line?

It may be difficult for you to visualize the 4.6 billion years that have passed during the history of Earth. In this activity, you will construct a time line for the appearance of organisms on the Geologic Time Scale.

Procedure

1. Using a meterstick, draw a continuous line down the middle of exactly 5 m of adding-machine tape.
2. For your time line, use a scale in which 1 m equals 1 billion years. Each millimeter then represents 1 million years.
3. At one end of the tape, draw a straight line and label it *The Present.* Measure to find the spot on the tape that represents 4.6 billion years ago, and label this point *Earth's Beginning.*
4. Using the table shown below, plot each event on your time line. Label the event and the number of years ago it occurred.

Event label	Estimated years ago
Earliest evidence of life	3.5 billion
Paleozoic Era begins	544 million
Mesozoic Era begins	245 million
First land plants	400 million
Triassic period begins	245 million
Jurassic period begins	208 million
First mammals and dinosaurs	225 million
First birds	150 million
Cretaceous period begins	144 million
Dinosaurs become extinct	66 million
Cenozoic Era begins	66 million
Primates appear	60 million
Humans appear	200 000

Analysis

1. Which era is the longest? The shortest?
2. In which eras did dinosaurs and mammals begin to exist?
3. Which lived on Earth the longer time, dinosaurs or mammals?

Figure 17.10

Since the start of the Mesozoic, the continents are thought to have drifted apart.

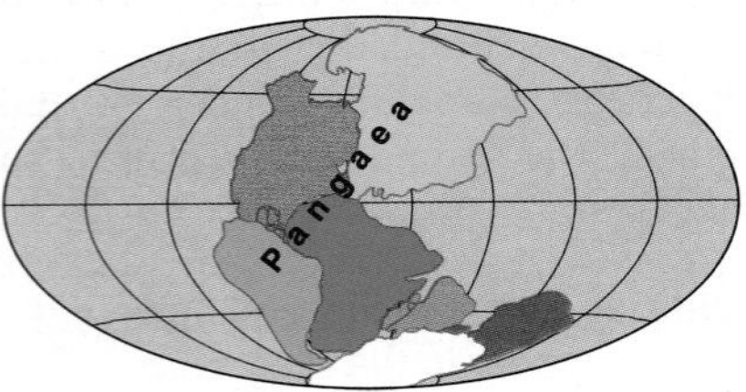

A About 245 million years ago, the continents were joined together in a single landmass known as Pangaea. Modern continents are shown in different colors.

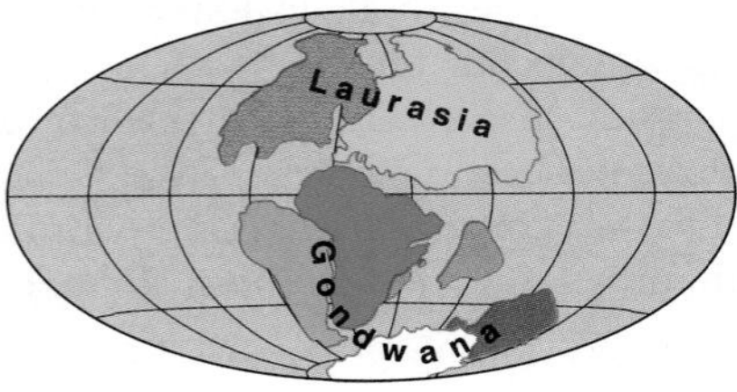

B During the Mesozoic, Pangaea began to break apart. By about 135 million years ago, the breakup of Pangaea had formed two large landmasses.

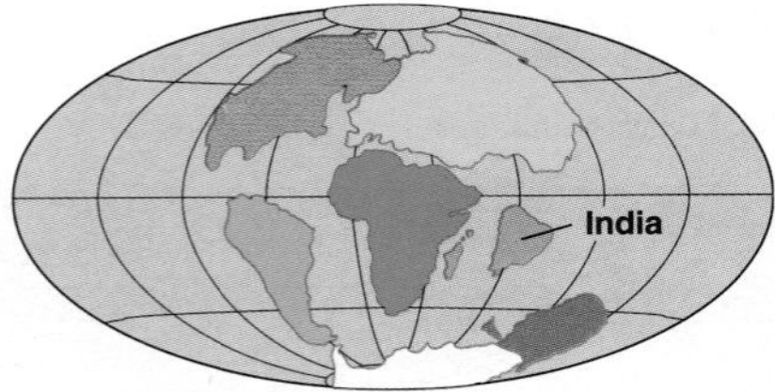

C By 65 million years ago, the end of the Mesozoic, the continents had drifted far enough apart that they were in the shapes that we recognize today.

Changes on Earth during the Mesozoic affected species

Geological events during the Mesozoic Era had a large impact on the distribution of modern species. Among the most important of these events is explained by a hypothesis of continental drift, *Figure 17.10.* The geological explanation for how the continents move is called the theory of **plate tectonics.** Earth's crust consists of several rigid plates that move through Earth's thick, plasticlike rock of the mantle. These plates are continually moving, spreading apart, and pushing against one another.

GLENCOE TECHNOLOGY

Videodisc

Biology: The Dynamics of Life
Disc 1, Side 2
Plate Tectonics (Ch. 3)

PROJECT

Evidence of Pangaea

Groups of students could prepare a report about a group of fossil animals that show a present-day distribution which supports the existence of Pangaea. Have them record the locations where their fossil animals were discovered. Encourage students to draw maps showing how plate tectonics affected their animal group. L2 COOP LEARN LS SAE

The Cenozoic Era—A world with humans

The Cenozoic Era began approximately 66 million years ago, around December 26 on our calendar of Earth history. It is the era in which you now live. Mammals began to flourish during the early part of this era. Among the mammalian groups that evolved was the order of mammals to which you belong, the primates. Primate species have radiated and diversified during this era, with the modern human species evolving perhaps as early as 200 000 years ago—the evening of December 31 on the hypothetical calendar.

Extinctions affect the diversity of species

It may seem that modern Earth has a tremendous diversity of species, but most species that have existed have also become extinct over the course of history. Billions more species than those that exist today lived in the past and have become extinct.

The fossil record also indicates several mass extinctions in the past. You've probably heard about the mass extinction of the dinosaurs at the end of the Cretaceous period, 66 million years ago. Many scientists think that this calamity was caused by the crash of a large asteroid on Earth, which sent clouds of dust into the atmosphere and caused climatic changes. Scientists estimate that nearly two-thirds of all species died during this mass extinction.

tectonics:
tekton (GK) builder
Plate tectonics can build mountains.

ThinkingLab — Interpret the Data

How can a clock represent Earth's history?

By studying fossils and analyzing geological events, scientists have constructed the Geologic Time Scale, a timetable for the appearance of organisms during the history of Earth.

Analysis

The diagram shown here compresses the history of Earth into a 12-hour clock. Note that the formation of Earth occurred at midnight on the clock, and the oceans formed at 2:00 A.M. Use this information to help you answer the following questions.

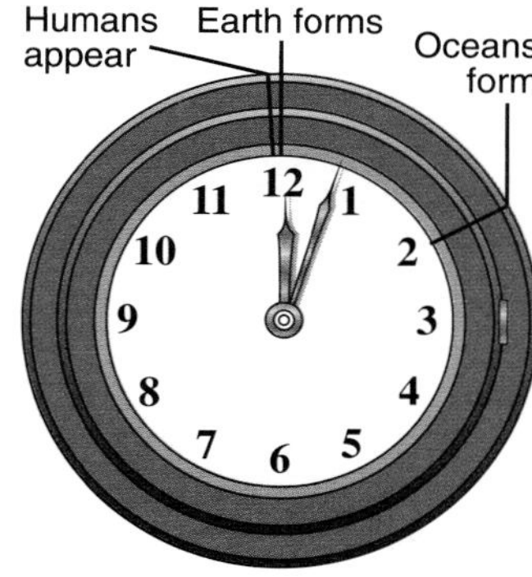

Thinking Critically

Based on fossil evidence, at approximately what time on the clock did prokaryotes evolve? What time did the first eukaryotes appear?

Section Review

Understanding Concepts

1. Describe the environmental conditions of early Earth before life arose.
2. Why are most fossils found in sedimentary rocks?
3. Suggest ways that fossils can provide evidence of the diet of animal species.

Thinking Critically

4. Suppose you are a scientist examining layers of sedimentary rock for fossils. In one layer of rock, you discover the remains of an extinct relative of the polar bear. In a layer of rock below this, you discover the fossil of an extinct alligator. What can you determine about changes over time in the climate of this area?

Skill Review

5. **Making and Using Tables** Construct a table listing the four geologic eras, their time ranges, and the major forms of life that appeared during each. For more help, refer to Organizing Information in the *Skill Handbook*.

ThinkingLab — Interpret the Data

Purpose

LS Visual-Spatial To use a model to show events on the Geologic Time Scale.

Process Skills

observe and infer, sequence, use an illustration

Teaching Strategies

- Question students about the history of life to assess their knowledge before they do this activity.

Thinking Critically

Students should estimate that prokaryotes appeared around 4:00 to 4:30 A.M. and eukaryotes around 10:00 A.M.

Assessment

Knowledge: Have students approximate the time of evolution of three other groups of organisms by finding out when they appear in the fossil record. L1

4 Close

Discussion

Ask students to summarize the general trend of development of organisms during the history of life on Earth. *Organisms became more complex, larger, and increased in diversity.* Discuss each response.

Answers to Section Review

1. Primitive Earth was hot; meteors bombarded its surface and volcanoes continually erupted.
2. The slow process of depositing sediments prevented damage to the organism.
3. Paleontologists can infer diet in fossil animals by analyzing the shapes of fossil teeth.

Thinking Critically

4. Alligators and polar bears today inhabit warm and cold environments, respectively, so one might infer that the environment of the region changed from a warm to a cold climate.

Skill Review

5. Student tables should be constructed from information in Figure 17.6 and pages 398 to 409 of the text. Check tables for accuracy.

SECTION

17.2

Prepare

Key Concepts

Students will be introduced to scientific hypotheses about the origin of life. They will explore early ideas about biogenesis and the later experiments performed to disprove them. Then, they will learn about modern experiments and hypotheses about the origin of cells.

Block Scheduling

Look for this symbol for strategies that are useful in a block scheduling format. For more information on block scheduling, refer to Section 17.2 in the **Lesson Plans** booklet.

Materials

- Obtain beef bouillon cubes for the Activity.
- Obtain the following amino acids for the Biolab: glycine, aspartic acid, glutamic acid. Gather necessary glassware, hot plates, and other equipment.

1 Focus

Bellringer

Before presenting the lesson, display **Section Focus Master 37** on the overhead projector and have students answer the accompanying questions. L1 SAE

Discussion

Discuss why people once accepted spontaneous generation. For example, why was spontaneous generation a plausible explanation for why mice were found in a bag of wheat? *People based their views on what they observed. If mice were seen emerging from a bag of wheat, people concluded that the mice developed from the wheat.*

SECTION

17.2 The Origin of Life

Section Preview

Objectives

Analyze early experiments that support the concept of biogenesis.

Compare and contrast modern theories of the origin of life.

Relate hypotheses about the origin of cells to the environmental conditions of primitive Earth.

Key Terms

spontaneous generation
biogenesis
protocell
archaebacteria

Will rotting meat give rise to flies? Can mud produce live fish? Will a bag of wheat give birth to mice? In the past, scientists asked such questions while pondering the deep mystery of science "How did life begin?" Then as today, scientists such as Louis Pasteur attempted to answer this question by relying on scientific methods of observation, hypothesis, and experimentation.

Origins: The Early Ideas

For early scientists, the ideas that mud produced fish and that rotting meat produced flies were reasonable explanations for what people observed. After all, maggots seemed to simply materialize on meat and then change into flies. These and other observations led scientists to believe in the idea of **spontaneous generation,** a process by which life was thought to be produced from nonliving matter.

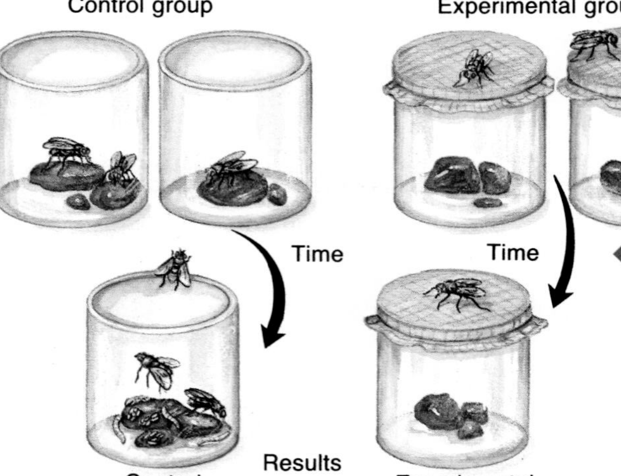

A **Rotten meat is placed in two experimental jars and two control jars. Redi used two control jars to ensure accurate results.**

B **By placing cloth over the experimental jars of rotting meat, Redi was able to test his hypothesis that only flies produce more flies.**

Figure 17.11
Francesco Redi's experiment to test the idea of spontaneous generation is a classic example of the scientific method.

C **Over time, the control jars were filled with maggots and flies. The experimental jars were free of insects, indicating that only when eggs were laid on the rotting meat were flies produced.**

Program Resources

Section Focus Master 37 L1 SAE
Reinforcement and Study Guide, pp. 67-68 L1
Biolab and Minilab Worksheets, pp. 67-68 L1
Basic Concepts Transparency 20 and **Master** L1 SAE

Meeting Individual Needs

Students Acquiring English Have students use block diagrams on paper to model the experiments of Francesco Redi or Louis Pasteur to better understand how these experiments exemplify the scientific method. Have students record their procedures and observations in writing. L1 LS SAE

Controlled experiments disprove spontaneous generation

In 1668, an Italian physician, Francesco Redi, designed a controlled experiment to disprove the idea of spontaneous generation. Redi's experiment is shown in *Figure 17.11*. The results of Redi's experiment convinced many scientists, but the spontaneous generation debate raged on.

At about the same time that Redi carried out his experiment, other scientists began using the latest tool in biology—the microscope. With the microscope, scientists discovered that the world was teeming with microorganisms. Although Redi had disproved spontaneous generation of larger organisms, microorganisms were so numerous and widespread that it was believed that they arose spontaneously from a vital force in the air.

Pasteur shuts the door on spontaneous generation

In the mid-1800s, a French scientist, Louis Pasteur, decided to test this idea of a vital force in air. To disprove spontaneous generation once and for all, Pasteur realized that he would have to set up an experiment in which only air—and no microorganisms—was allowed to come in contact with a nutrient broth. *Figure 17.12* shows how Pasteur carried out his experiment.

Pasteur showed that microorganisms do not simply arise in the broth, even in the presence of air. With his experiment, Pasteur claimed to have "driven partisans of the doctrine of spontaneous generation into the corner." **Biogenesis,** the idea that living organisms come only from other living organisms, then became a cornerstone of biology.

Figure 17.12

Pasteur's experiment to disprove spontaneous generation ended the debate. His flasks are on display today at the Pasteur Institute in Paris. They are still free from growth of microorganisms after almost 150 years.

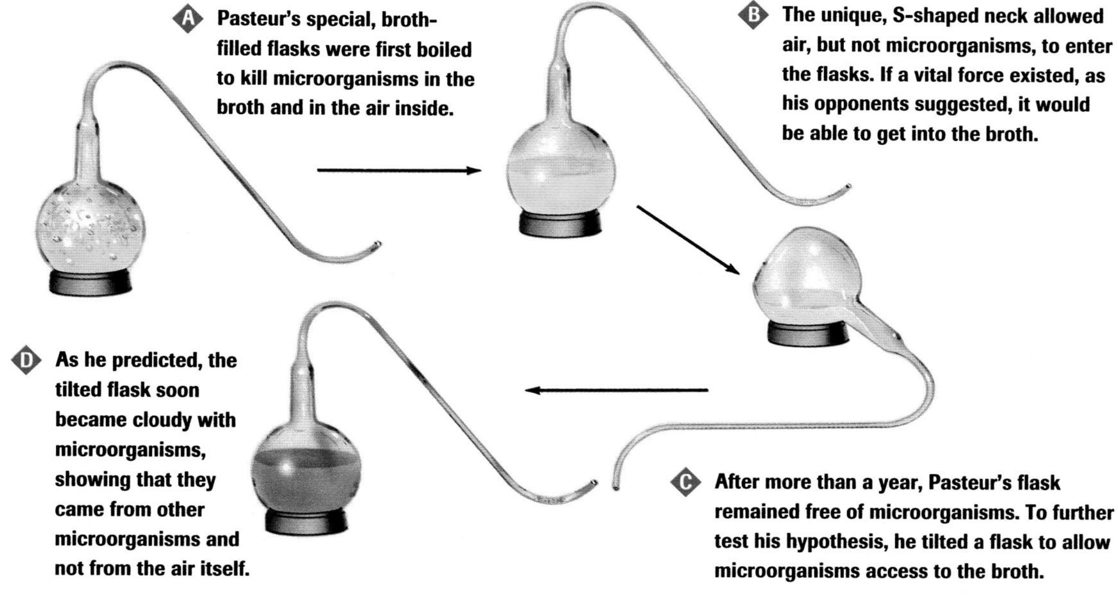

A Pasteur's special, broth-filled flasks were first boiled to kill microorganisms in the broth and in the air inside.
B The unique, S-shaped neck allowed air, but not microorganisms, to enter the flasks. If a vital force existed, as his opponents suggested, it would be able to get into the broth.
C After more than a year, Pasteur's flask remained free of microorganisms. To further test his hypothesis, he tilted a flask to allow microorganisms access to the broth.
D As he predicted, the tilted flask soon became cloudy with microorganisms, showing that they came from other microorganisms and not from the air itself.

2 Teach

Activity

LS **Visual-Spatial** Explain to students that people once believed life spontaneously arose from nonliving matter, such as broth. Dissolve two beef bouillon cubes in separate flasks of hot water. Boil the bouillon in both flasks for several minutes. After they cool, seal one flask with a rubber stopper, and label it "boiled."

Set the two flasks on a shelf in the classroom. Have students observe the flasks every day for changes and record their observations in their portfolios. SAE P

Visual Learning

Figures 17.11 and 17.12 The experiments by Redi, Pasteur, and others to disprove spontaneous generation were classic examples of the application of scientific methods. Use the illustrations to reinforce students' understanding of scientific methods. Examine and discuss each step of the procedure with students, asking them the purpose of each step. LS

Concept Development

Point out that the spontaneous generation of cells was harder to disprove than the spontaneous generation of whole organisms because microscopes were not sophisticated enough to show cells dividing until 1875.

Enrichment

LS **Linguistic** Have interested students prepare a report or poster presentation about the work of Italian scientist Lazzaro Spallanzani in the debate over spontaneous generation. L1

PORTFOLIO

Evaluating Pasteur's Experiment After setting up the Focus Activity, have students predict what they think will happen in the flask. After they observe the flask for a week, have students determine if their hypotheses were supported. Have them write their conclusions and put them in their portfolios. L1 LS P

Tying to Previous Knowledge

Review with students the basic organic chemistry concepts studied in Chapter 7, particularly the role of carbon in organic molecules. Ask students to recall the general structure of proteins and nucleic acids. Relate this information to the modern hypotheses concerning the origin of life.

Earth Connection

Earth's Atmosphere

Explain how the atmosphere has changed chemically since the formation of Earth. Have students brainstorm some possible reasons for the changes. Also discuss the possible role of meteorites and comets in the origin of life by examining evidence gathered by NASA satellites and probes.

Explain that the chemical contents of comets and meteorites have been found to contain organic molecules. Point out that some scientists, including Dr. Carl Sagan, have speculated that comets and meteorites were responsible for bringing organic molecules to Earth.

GLENCOE TECHNOLOGY

 Videotape

The Secret of Life

Use the videotape *What's in Stetter's Pond: The Basics of Life* to illustrate what early life-forms may have been like and the possible environment in which they lived.

Origins: The Modern Ideas

The concept of biogenesis has been accepted by biologists for more than 100 years. However, biogenesis does not answer one of the most basic questions: How did life begin on Earth? No one knows for sure what conditions existed on early Earth. The following hypotheses come from theoretical and experimental analysis, and not from fossils.

primordial:
primordium (L) origin
The origin of life may have been in the primordial soup.

Simple organic molecules formed from the primitive atmosphere

For life to come into being, it is generally agreed that two developments must have occurred: the formation of simple organic molecules important to life, and the organization of these molecules into complex organic molecules such as proteins.

Several billion years ago, Earth's atmosphere had no free oxygen as it does today. Instead, it was composed of water vapor, hydrogen, methane, and ammonia. How these substances could have formed simple organic compounds important to life is a challenging scientific puzzle.

In the 1930s, a Russian scientist, Alexander Oparin, proposed a widely accepted hypothesis that life began in the early oceans. He suggested that energy from the sun and from lightning triggered chemical reactions to produce simple organic compounds from the substances present. Oparin envisioned many chemical reactions occurring in the atmosphere and the products raining down into the oceans to form what is often called a primordial soup.

In 1953, two American scientists, Stanley Miller and Harold Urey, decided to test Oparin's hypothesis by simulating the conditions of early Earth in the laboratory. In an experiment similar to the one shown in *Figure 17.13*, Miller and Urey circulated water in the form of steam with ammonia, methane, and hydrogen, and subjected the mixture to electric

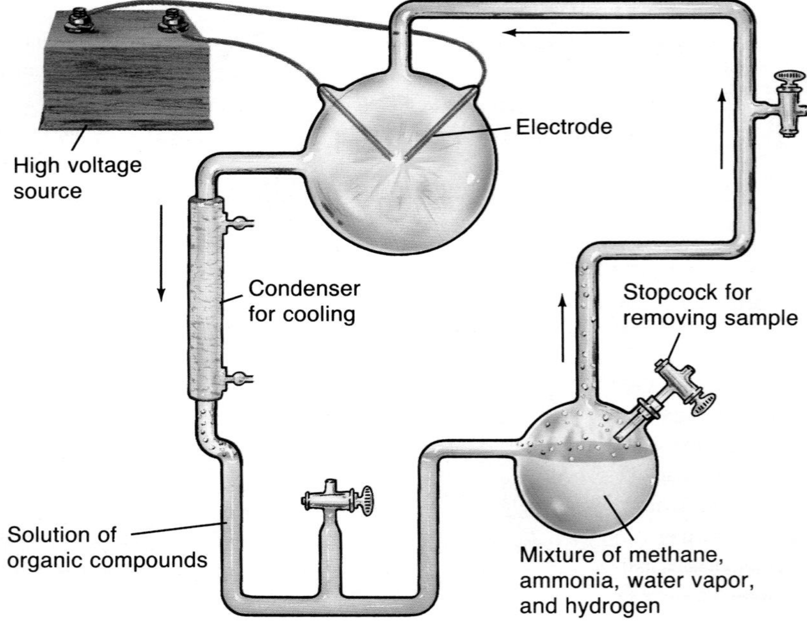

Figure 17.13
The "life-in-a-test-tube" experiment of Miller and Urey remains the cornerstone of the theories of the origin of life. At about the same time that they created amino acids in the lab, it was also learned that DNA carries the code for making proteins from amino acids. The photo (right) shows Stanley Miller with his experimental setup.

PORTFOLIO

Extraterrestrial Life Students may wish to learn more about the presence of organic molecules in space and the possibility of life existing elsewhere in the universe. Have students research and report to the class about evidence of organic molecules on planets, meteors, and comets. They may also wish to research current technology in SETI (search for extraterrestrial intelligence) projects and how such technology is used. Encourage students to prepare models, videos, or posters for a presentation to the class. L2 LS P

sparks to simulate lightning. They also repeatedly heated and cooled the mixture to simulate daily temperature fluctuations. After a week of such treatment, they analyzed the chemicals in the flask and found several kinds of amino acids, sugars, and other organic compounds, just as Oparin had predicted.

The formation of complex organic compounds and pre-cells

The next step in the origin of life, according to most scientists, was the formation of complex organic compounds and their enclosure by some type of bounding membrane. Since the 1950s, experiments have shown that amino acids will link together to form small proteins when heated in the absence of oxygen. Similar processes have also produced ATP and nucleic acids. Such experiments have led scientists to speculate that life may have originated in small pools of water, where amino acids could concentrate and be warmed.

A BROADER VIEW

How Did Life Begin: Beliefs and Theories

How life originated on Earth is one of the most fascinating and challenging questions in biology. Many ideas, hypotheses, and theories have been proposed, but the mystery has not been solved.

Divine origins Common to human cultures throughout recorded history is the belief that life on Earth did not arise spontaneously but was placed here by a creator. Most of today's major religions teach that life was created by a supreme being. Many people find it impossible to believe that life could arise without the intervention of forces beyond human understanding. Divine creation is a belief rather than a scientific theory, because it is accepted on faith.

Extraterrestrial beginnings This theory suggests that life did not begin on Earth at all, but was brought here by meteorites. Meteorites are space-borne chunks of rock that are trapped by Earth's gravitational pull, fall through the atmosphere, and then land on the surface of Earth. They are probably the remains of broken-up comets. Meteorites contain small amounts of organic matter, which could help explain how organic molecules considered necessary for the formation of living cells might have entered Earth's early oceans.

Primordial soup The most widely held view among scientists is that life arose by natural processes. This theory proposes that Earth's ancient oceans were a primordial soup full of organic molecules and that the atmosphere contained nitrogen, methane, and ammonia, but no oxygen. Energy from the sun, volcanoes, and lightning fueled chemical reactions that combined these molecules and gases into amino acids, lipids, and other complex organic molecules found in living cells. However, there are weaknesses in this theory. According to the fossil record, life may have developed more quickly than accounted for by this theory, and some scientists suspect the early atmosphere may not have contained methane or ammonia.

Bubble theory As a replacement this theory postulates that the chemical reactions of the primordial soup took place inside tiny bubbles of lipid molecules created by the action of wind, waves, and rainfall. There is evidence that methane and ammonia could have been present inside these lipid bubbles. When reactant molecules are kept close together in an enclosed space, chemical reactions take place much more quickly than when the molecules are able to move apart.

An important point that applies to the two scientific theories described above is that proteins, lipids, and other large organic molecules tend to clump together to form tiny spheres. This tendency may represent the first step in the organization of complex molecules into the structures we call cells—the building blocks of life.

Thinking Critically Use your research skills to explore strengths and weaknesses in the primordial soup theory, the bubble theory, and the theory of extraterrestrial origins. In what ways do these three ideas support each other? Do you think their weaknesses tend to support the formation of life by divine origins?

A BROADER VIEW

How Did Life Begin: Beliefs and Theories

Purpose

Students will explore a variety of ideas about the origin of life on Earth.

Teaching Strategies

- Organize students into teams for a classroom debate on the origins of life. Give each team responsibility for defending one point of view. As students do preparatory research, encourage them to explore the strengths of their viewpoint and the weaknesses of their opponent's viewpoint.
- Review with students the chemical composition of nucleic acids, amino acids, lipids, carbohydrates, and other organic molecules that make up cells.

Thinking Critically

The theory of extraterrestrial origins could explain the presence of large, complex organic molecules in Earth's oceans, even without the hundreds of millions of years of chemical reactions called for by the primordial soup theory. It may also help explain the presence of these molecules in an atmosphere lacking methane and ammonia. The bubble theory supports the primordial soup theory by supplying a protected environment for methane, ammonia, and speedier chemical processes. Weaknesses in any of these theories emphasize human uncertainty about the origins of life; uncertainty tends to support the belief that life must have been created by forces beyond human understanding.

STUDENT JOURNAL

Exploring Life's Origins Have students write about their opinions on research in the origin of life. Is the research of Oparin, Miller-Urey, Fox, and others conclusive, or should more work be done in this area? Have students support their arguments by citing evidence discussed in the text or in class. L1 LS

Meeting Individual Needs

Gifted Have students research other experiments about the origin of life. Have them write summaries in their journals of the experiments they research. Ask them to evaluate the evidence resulting from each experiment in light of other research findings. L3

Visual Learning

Figure 17.14 Have students study the illustration of Fox's experiment. Discuss how Fox's experiment relates to Oparin's hypothesis and the Miller-Urey experiment. LS

Tying to Previous Knowledge

Review the basic structure and function of cells. Emphasize the substances cells need to live and the different ways cells release energy. Relate this information to the evolution of cells and the conditions of primitive Earth.

Discussion

Have students compare the characteristics of protocells with those of actual cells. Initiate a discussion about the similarities and differences between protocells and living cells.

Discussion Question

Ask students why studying the characteristics of living archaebacteria and cyanobacteria is important for determining the origin of life. *Such bacteria are the earliest known living things and provide an indication of what conditions were required to support life.*

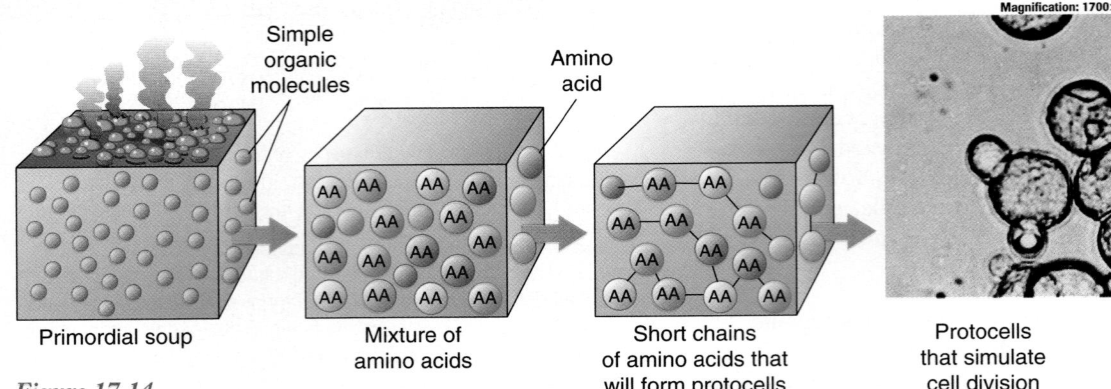

Figure 17.14

Sidney Fox took the Miller-Urey experiment one step further and showed how short chains of amino acids could cluster to form protocells. Under the microscope, protocells can look so much like living cells that scientists sometimes mistake them for new species of bacteria.

How did these chemicals combine to form the first cells? The work of American biochemist Sidney Fox showed how this may have occurred. As shown in *Figure 17.14,* Fox was able to produce protocells by heating solutions of amino acids. A **protocell** is a large, ordered structure that carries out some activities associated with life such as growth, division, or metabolism.

The Evolution of Cells

Fossils indicate that by 3.5 billion years ago, life had diversified into several different types of cells. But scientists know that these cells don't represent the earliest cells, which arose sometime between 3.9 and 3.5 billion years ago. What were the earliest cells like, and how did they evolve?

archaebacteria: *archaios* (GK) ancient Archaebacteria are similar to ancient bacteria.

Heterotrophic prokaryotes were the first true cells

Scientists speculate that the first forms of life were prokaryotes, which probably evolved from some type of protocell. Because Earth's atmosphere lacked oxygen, these organisms most likely were anaerobic, getting their energy through glycolysis and fermentation. Bathed in the warm sea, the first prokaryotes probably took in organic molecules for food. Because they obtained food from their surroundings, they were heterotrophs.

Over time, competition for nutrients by heterotrophic prokaryotes led to the evolution of Earth's first autotrophs. Organisms that could make food would have had a distinct advantage over other heterotrophic organisms. Many scientists consider that these first autotrophs were similar to present-day archaebacteria, *Figure 17.15.* **Archaebacteria** are a group of prokaryotes that live in harsh conditions with little sunlight and oxygen, such as in hot sulfur springs and deep-sea vents. The earliest autotrophs probably made glucose by chemosynthesis as archaebacteria do today rather than by photosynthesis, which requires light-trapping pigments.

Next, photosynthesizing prokaryotes evolved. The cell organization of the 3.5-billion-year-old fossils from Australia indicates that these were among the first photosynthesizing prokaryotes.

Photosynthesizing prokaryotes changed Earth's atmosphere

You will recall that the process of photosynthesis releases oxygen. With

Alternate Lab — Making Coacervates

Purpose 

LS **Logical-Mathematical** Students will investigate the conditions under which the first cells may have evolved.

Materials

gelatin solution, droppers (3), gum arabic solution, pH paper, microscope, microscope slides, coverslips, hydrochloric acid (HCl), test tube, stirring rod

Procedure

Give students the following directions.

1. Measure 5 mL of 1% gelatin solution and pour it into the test tube. Add 3 mL of 1% gum arabic solution and mix gently.
2. Measure and record the pH of the mixture.
3. Make a wet mount of the mixture. Observe under high power and record your observations.
4. Add one drop of 0.1*M* HCl solution to the mixture and stir. Measure and record the pH of the mixture.
5. Repeat step 3 for the HCl mixture.
6. Repeat steps 4 and 5 until you observe coacervates, clumps that look like cells. Record the pH at which you

the evolution of photosynthesizing prokaryotes, the concentration of oxygen in Earth's atmosphere began to increase. This increase in oxygen made possible the evolution of aerobic respiration, a more efficient method of energy conversion. The change was so drastic that scientists call it the oxygen revolution, as shown in *Figure 17.16* on page 418. An increase in the diversity of prokaryotes in the fossil record indicates that the oxygen revolution was fully underway by 2.8 billion years ago.

Figure 17.15

The earliest true cells were probably heterotrophic prokaryotes. Competition for nutrients probably led to the evolution of chemosynthetic organisms similar to archaebacteria that live in hot springs like the one shown. Ancient microfossils like the ones in the inset photo are similar to archaebacteria.

Magnification: 20 000 ×

A BROADER VIEW

The Endosymbiont Hypothesis

Many years ago, scientists noted that cyanobacteria and chloroplasts resemble each other. Likewise, they saw that mitochondria and bacteria look alike. These key observations led to the endosymbiont hypothesis. This hypothesis states that eukaryotes evolved through a symbiotic relationship between primitive prokaryotes.

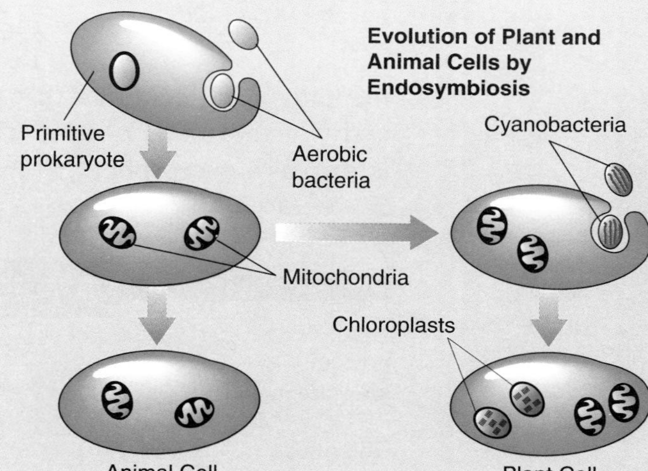

Origin of the endosymbiont hypothesis In 1962, scientists discovered DNA in chloroplasts. This finding led American biologist Lynn Margulis to develop the endosymbiont hypothesis. In the years since then, she has been compiling evidence to support it. Other observations include the fact that cholorplasts and mitochondria have their own ribosomes and grow and reproduce independently of the cell itself, and that the cell has no means to make either chloroplasts or mitochondria.

Structural evidence How cyanobacteria and bacteria evolved a symbiotic relationship and came to function as cell organelles is largely a matter of scientific speculation. However, compelling structural evidence indicates that the ancestors of mitochondria and chloroplasts were once free-living prokaryotes. For example, mitochondria and bacteria can both reproduce themselves, have similar nucleic acids, are about the same size and shape, and carry out protein synthesis on ribosomes. Chloroplasts and cyanobacteria also share important characteristics, the most important of which is the ability to carry out photosynthesis.

Thinking Critically

Some prokaryotes live in close association with eukaryotes today. Could this fact be used as evidence for the endosymbiont hypothesis? Explain your answer.

Microscope Activity

LS Visual-Spatial Have students examine slides or living cultures of bacteria and cyanobacteria and compare them to photographs of stromatolite sections and early cells, such as those discovered in western Australia. Have students note any similarities and differences and record their observations in their journals. L1

A BROADER VIEW

The Endosymbiont Hypothesis

Purpose

Students will examine the endosymbiont hypothesis of eukaryote evolution.

Teaching Strategies

- Review the differences between eukaryotes and prokaryotes and the similarities and differences between mitochondria and chloroplasts.
- Ask students to suggest alternative hypotheses for the evolution of eukaryotes.

Thinking Critically

Students may argue that a close association between prokaryotes and eukaryotes is good evidence for the endosymbiont hypothesis. Others will argue that this doesn't support the hypothesis because environmental conditions are much different today from what they were in the past.

first observe coacervates and describe their appearance.

7. Repeat steps 4 and 5 until you are unable to observe coacervates under either power of the microscope.

Analysis

Ask students the following questions.

1. What kind of living things do coacervates resemble? *amoebas*
2. At what pH were you able to observe coacervates? *around pH 5*
3. How does a mixture of gelatin (a protein) and gum arabic (a carbohydrate) simulate conditions that may have been present on early Earth? *The "primordial soup" probably contained amino acids and simple sugars as this mixture does.*

✓ Assessment

Oral: Have students present an oral report of their findings. Ask: What was the role of hydrochloric acid in this investigation? In what way does this investigation relate to hypotheses about the origin of life? L1

BioLab

Making Microspheres

Time Allotment

One full class period plus an additional half.

Objectives

Review objectives with students before they begin the Biolab.

Process Skills

observe and infer, form a hypothesis, measure in SI

Safety Precautions

Make sure students are wearing aprons, goggles, and using oven mitts when handling hot objects.

PREPARATION

Alternate Materials

- Larger Erlenmeyer flasks and beakers may be substituted.
- Note: Emphasize to students that the amino acids should not be dissolved but should be thoroughly mixed and heated as a dry powder.
- For a successful experiment, freshly purchased amino acids should be used. Amino acids decompose quickly when allowed to stand in moisture at room temperature in the light. Amino acids may be stored for long periods in a freezer. The bottles should be sealed tightly and wrapped with foil to keep out light.

BioLab

Making Microspheres

Scientists estimate that life formed sometime between 3.9 and 3.5 billion years ago from chemicals present on primitive Earth. One of the most important steps in the origin of life must have been the enclosure of organic compounds by a membrane, creating a protocell. Protocells can be produced experimentally using materials that scientists believe were present on early Earth. In this lab, you will produce a type of protocell known as a microsphere.

PREPARATION

Problem

How are microspheres produced?

Objectives

In this Biolab, you will:

- **Infer** how microspheres may have been produced on primitive Earth.
- **Compare and contrast** microspheres with living cells.

Materials

microscope
microscope slide
coverslip
50-mL graduated cylinder
50-mL Erlenmeyer flasks (2)
400-mL beaker
1% NaCl solution
aspartic acid
glutamic acid
glycine
hot plate
ring stand
ring stand and clamp
tongs
watch or clock with second hand
stirring rod
pipette
apron
oven mitts
goggles

Safety Precautions

Use care when handling hot objects. Be sure to wear oven mitts.

PROCEDURE

Teaching Strategies

- Review the modern hypotheses about the origin of life in this section.
- Have students perform this Biolab after studying protocells and Fox's experiment. Ask them whether they think the environmental conditions of primitive Earth can be adequately simulated in a laboratory. Also ask them to consider if living cells could have arisen from protocells. Why or why not?
- **Troubleshooting** If students are not able to produce microspheres, it may be due to one or more of the following reasons: inaccurate amounts of amino acids and NaCl solution were used, the amino acids

PROCEDURE

1. Pour 250 mL of water into the beaker, and place the beaker on a hot plate.
2. Clamp one of the flasks to the ring stand.
3. Add 0.5 g of each of the following amino acids to the flask: glycine, aspartic acid, and glutamic acid. Use the stirring rod to mix thoroughly.
4. When the water in the beaker begins to boil, carefully loosen the clamp and lower the flask into the water. Tighten the clamp.
5. Heat the amino acids for 20 minutes, keeping the water in the beaker at a simmer.
6. Measure and pour 5 mL of the 1% NaCl solution into the second flask, and place the flask on the hot plate.
7. When the NaCl solution begins to boil, use tongs to remove it from the hot plate. Slowly add the NaCl solution to the amino acids. Turn off the hot plate.
8. Loosen the clamp on the ring stand, and raise the flask containing the amino acid solution out of the beaker. Allow the contents of the flask to cool for ten minutes.
9. Make a copy of the data table.
10. Prepare a wet mount using one drop of the mixture, and observe under low power. Locate and diagram a microsphere.
11. Switch to high power to examine the microspheres. Draw a diagram in your data table.
12. After you have finished observing the microspheres, clean and return all equipment to its proper place.

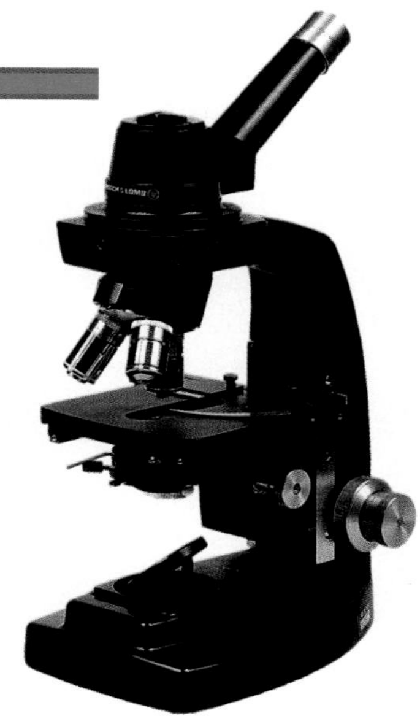

Characteristics of Microspheres

	Drawing	Description
Low power		
High power		

ANALYZE AND CONCLUDE

1. **Comparing and Contrasting** How do microspheres resemble cells? How are they different?
2. **Interpreting Observations** What features of living cells do microspheres exhibit?
3. **Thinking Critically** How does the method you used to generate microspheres compare with the conditions on primitive Earth?

Going Further

Thinking Critically Do you think your microsphere experiment would have worked if you had substituted other amino acids? How could you test this hypothesis?

were not properly mixed, the amino acids were heated too long, or the NaCl solution with the added amino acids did not cool sufficiently.

Data and Observations

Microspheres are tiny spherical objects, sometimes containing small clumps of protein. Students should be able to observe these under both low and high power. They should be able to recognize that both protocells and actual cells are discrete units with bounding membranes. Students may also observe microspheres dividing into smaller spheres.

ANALYZE AND CONCLUDE

1. Both microspheres and cells are individual units that have membranes. Microspheres contain clumps of material, but they differ from living cells in that cells have organelles, whereas microspheres do not.
2. Microspheres "reproduce" in a way that resembles division of living cells.
3. Students may point out that shallow pools may have evaporated, leaving dry mixtures of amino acids on hot rocks. Intense ultraviolet light from the sun would probably have been present on early Earth. This condition was not present in the Biolab.

Assessment

Portfolio: Have students write an evaluation of this lab to place in their portfolios. Ask students whether their initial hypotheses were supported or rejected. Ask students also to consider ways that this experiment could have better simulated the conditions of early Earth. P

Going Further

Students may wish to try this experiment using combinations of other amino acids, or manipulating the procedures to better simulate primitive Earth's conditions. Ask interested students how they might develop an experiment to test the hypothesis that prokaryotes evolved from protocells.

3 Assess

Check for Understanding

Have students list the three main conclusions of this section in their journals: life comes from existing life; life probably originated on Earth through the reaction of chemicals in Earth's atmosphere and the concentration and further reaction of those chemicals on Earth's surface; and cells probably evolved as chemicals became more organized. Ask students to list the known scientific evidence that supports each conclusion. L1

Reteach

Have students list the names of experimenters from Redi to Fox and tell what their conclusions showed about the origin of life. Have students quiz each other using this list. L1

Extension

Ask students who have mastered this section to find out how scientists think that nucleic acids may have developed from elements already present on Earth. L2

Assessment

Oral: Ask each student to present a brief oral report about how life originated. L1

4 Close

Discussion

Have students discuss whether new life could originate on Earth today. They should consider how modern Earth differs from early Earth and the problem of an oxygen-rich atmosphere.

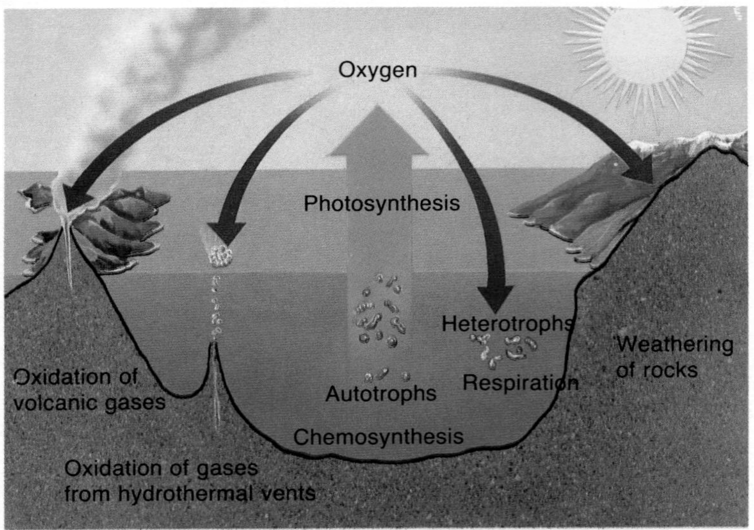

Figure 17.16
Oxygen buildup in the atmosphere resulted from photosynthesis by early prokaryotes. Oxygen was then available for aerobic respiration. By the end of the Precambrian Era 545 million years ago, oxygen content had reached approximately 10 percent of present-day levels.

The release of oxygen through photosynthesis had another effect. Lightning converted much of the atmospheric oxygen into ozone (O_3) molecules, resulting in an ozone layer 10 to 15 miles above Earth's surface. The ozone layer shielded emerging life forms from the harmful effects of ultraviolet radiation, and thus allowed for the evolution of more complex organisms. The ozone layer still exists today, protecting Earth's surface from ultraviolet radiation.

Connecting Ideas

Four-and-a-half billion years ago, a story began that is still unfolding–the story of life on Earth. Details of the story are found in the many bits and pieces of the fossil record. The fossil record of life on Earth may not be perfect, but it is a useful indicator of past events. Through the careful analysis and dating of fossils, scientists have hypothesized that life evolved from the unicellular, photosynthesizing prokaryotes of several billion years ago, to the soft-bodied animals and plants of several hundred million years ago, to the diversity of species present on Earth today.

How does evolution work, and what are the mechanisms for change? How did the unicellular organisms living in the warm oceans of 3.5 billion years ago evolve into the remarkable diversity of life forms we see today?

Section Review

Understanding Concepts

1. How did Pasteur's experiment finally disprove spontaneous generation?
2. What was Oparin's hypothesis, and how was it tested experimentally?
3. Why do scientists think the first living cells were anaerobic heterotrophs?

Thinking Critically

4. Some scientists speculate that lightning was not present on early Earth. How could you modify the Miller-Urey experiment to reflect this new idea? What energy source would you use to replace lightning?

Skill Review

5. **Sequencing** Make a time line sequencing the evolution of life from protocells to eukaryotes. For more help, refer to Organizing Information in the *Skill Handbook*.

Answers to Section Review

1. It showed that life would not arise in a sterile medium. Organisms grew only when previously existing organisms were allowed into the broth.
2. Oparin hypothesized that life began in the oceans after organic molecules created by chemical reactions between elements in the atmosphere rained down, forming an "organic soup." His hypothesis was tested by Miller-Urey.
3. Scientists believe that the first organisms were anaerobic heterotrophs because the atmosphere had no oxygen and they would have had to obtain their own food.

Chapter 17 Review

REVIEWING MAIN IDEAS

17.1 The Record of Life

- Fossils provide a record of life on Earth. Fossils come in many forms, such as the print of a leaf, the burrow of a worm, or a stone replica of a bone.
- Through the study of fossils, scientists can learn about the diversity of life. Fossils also provide clues about the behavior of organisms and the environments in which they lived.
- Scientists divide Earth's history into a series of eras and periods known as the Geologic Time Scale. This scale provides a timetable for the appearance of different groups of living organisms during the history of Earth.

17.2 The Origin of Life

- During the 17th, 18th, and 19th centuries, scientists experimented to test the idea of spontaneous generation. As a result of these experiments, scientists accept biogenesis, the theory that life comes from existing life.
- There are several competing theories about the origin of life. Many scientists agree that organic compounds were formed from elements present in Earth's primitive atmosphere and oceans.
- The earliest cells were probably anaerobic, heterotrophic prokaryotes. Over time, the ability to capture energy to make food by photosynthesis evolved, changing the atmosphere and triggering the evolution of aerobic cells and eukaryotes.

Key Terms

Write a sentence that shows your understanding of each of the following terms.

archaebacteria
biogenesis
fossil
plate tectonics
protocell
spontaneous generation

Understanding Concepts

1. Fossil markings such as footprints or animal burrows are called ________.
   a. casts c. petrified fossils
   b. trace fossils d. molds
2. About how many years ago is it theorized that Earth cooled enough for water vapor to condense?
   a. 20 million years c. 3.9 billion years
   b. 4.6 billion years d. 5.5 billion years
3. Most fossils occur in layers of ________ rocks.
   a. sedimentary c. igneous
   b. metamorphic d. volcanic
4. The mass of continents that were joined together about 245 million years ago is called ________.
   a. Laurasia c. Pangaea
   b. Gondwana d. the Mesozoic
5. The Geologic Time Scale is based on ________.
   a. different organisms that appeared during Earth's history
   b. various rock layers that occur in Europe and Canada
   c. landforms such as mountains and faults in California
   d. oceans and seas and the times of their formations
6. Shallow seas that teemed with life could be found ________.
   a. in the Precambrian Era
   b. in the Paleozoic Era
   c. in the Mesozoic Era
   d. 6.5 billion years ago
7. Dinosaurs, mammals, and flowering plants appeared during the ________ Era.
   a. Cenozoic c. Paleozoic
   b. Precambrian d. Mesozoic

Chapter 17 Review

Reviewing Main Ideas

Summary statements can be used by students to review the major concepts of the chapter.

Key Terms

Answers should go beyond defining the terms. Accept any answer that uses the term correctly and in the proper context.

Understanding Concepts

1. b
2. c
3. a
4. c
5. a
6. b
7. d

Thinking Critically

4. Possible answers include sunlight, radioactivity, or heat from volcanoes.

Skill Review

5. Protocells-anaerobic prokaryotes-chemosynthetic prokaryotes-photosynthetic prokaryotes-eukaryotes.

Chapter 17 Review

8. d
9. d
10. a
11. c
12. c
13. a
14. d
15. c
16. b
17. a
18. b
19. c
20. b

Applying Concepts

21. Because fossils are found in sedimentary rock, paleontologists need to know when and how that layer formed and what has happened to the rock since it was formed.
22. From arrangements of fossils, scientists may be able to determine whether animals lived in large groups, what they ate, and how they obtained their food, or what their home ranges were.
23. Continental movement may cause environmental changes that affect the evolution of species due to isolation and natural selection.
24. A cell with mitochondria could release more energy for a given amount of food.
25. The earliest organisms probably were heterotrophic and obtained energy through anaerobic processes. Later, organisms may have become autotrophic and stored energy by making food. The evolution of cells with mitochondria would have increased the efficiency of energy release.

Chapter 17 Review

8. The earliest evidence of life seems to have occurred in the ____________ Era.
   a. Cenozoic c. Paleozoic
   b. Mesozoic d. Precambrian
9. The geological explanation for how the continents move is ____________.
   a. biogenesis
   b. an extinction event
   c. continental genesis
   d. plate tectonics
10. The idea that life comes only from preexisting life is ____________.
    a. biogenesis
    b. spontaneous generation
    c. pangenesis
    d. primordial genesis
11. Who was the scientist who showed conclusively that life does not arrive spontaneously?
    a. Francesco Redi c. Louis Pasteur
    b. Stanley Miller d. Harold Urey
12. The group of prokaryotes that live in harsh conditions such as near-boiling water and sulfur springs is ____________.
    a. bacteria c. archaebacteria
    b. cyanobacteria d. water plants
13. Any evidence of an organism that lived long ago is a(n) ____________.
    a. fossil c. artifact
    b. stromatolite d. protocell
14. An entire, intact organism may be preserved in ____________ and ____________.
    a. casts; trace fossils
    b. molds; casts
    c. imprints; petrified fossils
    d. amber; ice
15. One scientist heated solutions of amino acids to produce ____________, which look like living cells.
    a. prokaryotes c. protocells
    b. archaebacteria d. eukaryotes
16. Beginning about 144 million years ago, the ____________ period was characterized by the appearance of flowering plants.
    a. Cambrian c. Triassic
    b. Cretaceous d. Jurassic
17. According to the Geologic Time Scale, which of these organisms evolved during the Precambrian Era?
    a. bacteria c. flowering plants
    b. mammals d. reptiles
18. The endosymbiont hypothesis suggests that ____________ may have been involved in the evolution of chloroplasts in plant cells.
    a. mitochondria c. bacteria
    b. cyanobacteria d. aerobic bacteria
19. Miller and Urey showed that ____________.
    a. life could arise spontaneously as suggested by Redi
    b. life could never arise spontaneously as suggested by Pasteur
    c. building blocks of life could be produced in an atmosphere similar to that proposed for early Earth
    d. protocells were necessary before living cells
20. Scientists theorize that oxygen buildup in the atmosphere resulted from ____________.
    a. respiration
    b. photosynthesis
    c. chemosynthesis
    d. rock weathering

Applying Concepts

21. Why is a thorough understanding of geology important to paleontologists?
22. How might paleontologists use fossils to draw conclusions about the social behavior of animals? Which social behaviors might they detect?
23. How might the movement of continents affect the future evolution of species?
24. What advantage would a primitive cell containing mitochondria have over a cell that had none?
25. How might the way organisms obtain energy have evolved throughout the history of life?

Thinking Critically

26. *Concept Mapping* Make a concept map that relates the following terms. Supply the appropriate linking words for your map.
    protocell, autotroph, heterotroph, eukaryote, prokaryote, endosymbiont hypothesis, anaerobic, aerobic
27. *Observing and Inferring* Why are amber-preserved or frozen fossils of great value to scientists?
28. *Comparing and Contrasting* What details might casts show that are not seen in molds?
29. *Formulating Hypotheses* Why do scientists hypothesize that the 3.5 billion-year-old fossils from Australia were not the first species to have evolved on Earth?
30. *Measuring in SI* Assume that a particular radioactive element has a half-life of 1 year. If 1000.0 g are present in a sample today, how much will be in the sample after 10 years?
31. *Comparing and Contrasting* The time line below was made by a group of students. Why did they make the Precambrian Era the longest era?

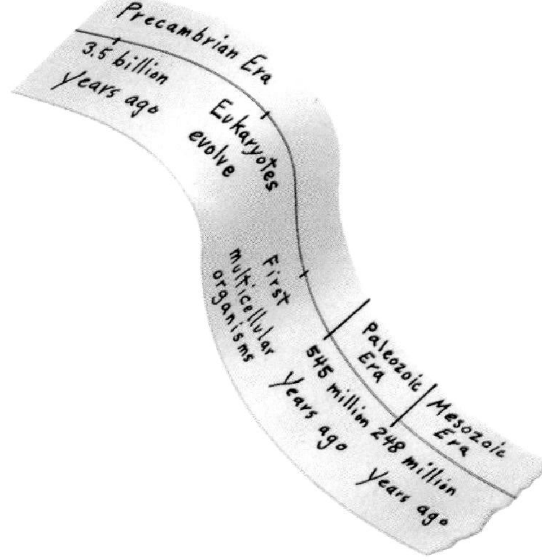

ASSESSING KNOWLEDGE & SKILLS

The following graph illustrates the decay rate of a particular radioactive element.

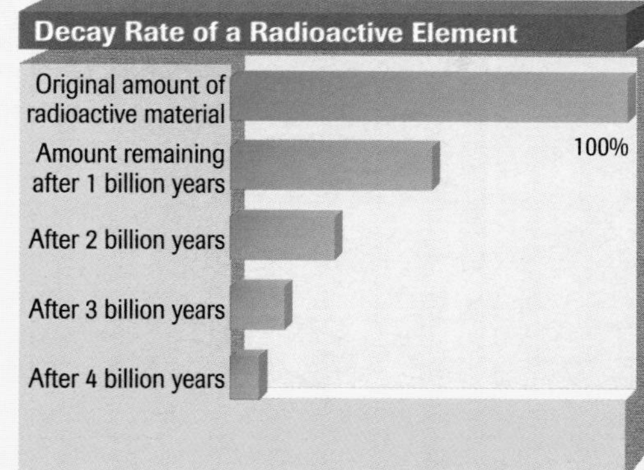

Interpreting Data Use the graph to answer the following questions.

1. How long does it take for half of the element to decay?
   a. 1 billion years c. 3 billion years
   b. 2 billion years d. 4 billion years
2. How much of the original material is left after 2 billion years?
   a. 100% c. 25%
   b. 50% d. 12.5%
3. How much of the original material is left after 4 billion years?
   a. 50% c. 12.5%
   b. 25% d. less than 10%
4. This element would best be used to date fossils that are ____________ years old.
   a. a few thousand
   b. less than a million
   c. a few million
   d. a billion
5. *Interpreting Scientific Illustrations* Using *Figure 17.6* and the above graph, identify the era in which dating fossils by the use of this radioactive element would be most useful.

Chapter 17 Review

Thinking Critically

26. Evaluate students' concept maps. The maps should show an understanding of the relationships among the concepts listed.
27. The entire organism is preserved, including soft tissues.
28. Casts are three-dimensional representations of molds and show more details, such as surface texture, than do molds.
29. The fossils from Australia are not the earliest forms of life because these organisms carried out photosynthesis. The earliest organisms were more likely to have been anaerobic heterotrophs.
30. 0.9766 grams
31. The Precambrian era is the longest because it occupies the greatest portion of geological time.

1. a
2. c
3. d
4. d
5. the Precambrian era

Program Resources

Chapter Assessment, pp. 97-102 L1
Alternate Assessment in the Science Classroom
Computer Test Bank L1
Content Mastery, pp. 65-68 L1

Chapter Organizer

SECTION	OBJECTIVES	ACTIVITIES/FEATURES
18.1 Natural Selection and the Evidence for Evolution National Science Standards: UCP.1-5; C.3, C.4, C.6; F.4; G.1, G.3	1. **Summarize** Darwin's theory of evolution by natural selection. 2. **Relate** the idea of natural selection to the origin of structural and physiological adaptations. 3. **Evaluate** the various lines of evidence for evolution.	**A Broader View:** Evolution: Interpreting the Data, p. 426 **Minilab:** How is camouflage an adaptive advantage?, p. 429 **Thinking Lab:** How can natural selection be observed?, p. 430
18.2 Mechanisms of Evolution National Science Standards: UCP.1-5; A.1, A.2; C.1-4, C.6; F.4; G.1-3	4. **Summarize** the effects of the different types of natural selection on gene pools. 5. **Relate** mechanisms of speciation to changes in genetic equilibrium. 6. **Explain** the role of natural selection in convergent and divergent evolution.	**A Broader View:** Expression of Recessive Traits, p. 437 **Math Connection:** The Mathematics of Evolution, p. 438 **Biolab:** Natural Selection and Allelic Frequency, p. 442

ACTIVITY MATERIALS

BIOLAB	MINILABS	ALTERNATE LAB
page 442 colored pencils (2) paper bag graph paper pinto beans white navy beans	**page 429** paper, white and black hole punch	**page 436** culture of *Bacillus subtilis* 3 tubes of nutrient agar tube of streptomycin agar inoculation loop 2 petri dishes Bunsen burner wax pencil test tube

Chapter 18 The Theory of Evolution

TEACHER CLASSROOM RESOURCES

Reproducible Masters	Transparencies
Section Focus Master 38: Camouflage L1 SAE **Reinforcement and Study Guide,** pp. 69-70 L1 **Biolab and Minilab Worksheets,** p. 70 L1 **Concept Mapping:** Evidence of Evolution, p. 18 L1 **Critical Thinking/Problem Solving:** Selection Pressures, p. 18 L3 **Laboratory Manual:** How Is Camouflage an Adaptive Advantage?, pp. 97-100 L2	**Reteaching Transparency #18:** Evidence for Evolution L1 SAE
Section Focus Master 39: Evolving Populations L1 SAE **Reinforcement and Study Guide,** pp. 71-72 L1 **Laboratory Manual:** Plant Survival, pp. 101-102 L2 **Biolab and Minilab Worksheets,** pp. 71-72 L1 **Content Mastery,** pp. 69-72 L1	**Basic Skills Transparency #16:** Role of Isolation in Speciation L1 SAE **Basic Concepts Transparency #21:** Genetic Equilibrium L1 SAE **Basic Concepts Transparency #22:** Variation in Populations L1 SAE

ASSESSMENT MATERIALS

Chapter Assessment, pp. 103-108
Performance Assessment in the Biology Classroom, pp. 23, 25, 35
MindJogger Videoquiz
Alternate Assessment in the Science Classroom
Computer Test Bank

Spanish Resources SAE
English/Spanish Audiocassettes SAE
Cooperative Learning in the Science Classroom COOP LEARN
Lesson Plans
Biology Projects: Learning from the Past, pp. 17-20 L2

KEY TO TEACHING STRATEGIES

- L1 Level 1 activities should be within the ability range of all students including those with learning difficulties.
- L2 Level 2 activities are within the ability range of average to above-average students.
- L3 Level 3 activities are designed for the ability range of above-average students.
- SAE SAE activities should be within the ability range of Students Acquiring English.
- COOP LEARN Cooperative Learning activities are designed for small group work.
- P These strategies represent student products that can be placed into a best-work portfolio.
- These strategies are useful in a block scheduling format.

GLENCOE TECHNOLOGY

The following multimedia resources are available from Glencoe.

Biology: The Dynamics of Life
CD-ROM SAE
Videodisc Program

The Infinite Voyage Series
To the Edge of the Earth
Exploring the Galápagos Islands
The Great Dinosaur Hunt
New Dinosaur Discoveries and Their Link with Today
Theories of Extinction
The Geometry of Life
Evolution, Molecular Genetics, and DNA Sequencing

The Secret of Life Series
It's in the Genes: Evolution

Science and Technology Videodisc Series (STVS)
Ecology
Resistance to Pesticides in Cockroaches

CHAPTER

18 The Theory of Evolution

Chapter Overview

Charles Darwin developed the idea of natural selection, his explanation of how evolution works. Students will explore the connection between natural selection and adaptations of organisms, and will examine scientific evidence that supports the theory of evolution.

In the second section, students will discover how the three general types of natural selection—stabilizing selection, directional selection, and disruptive selection—are applied to explain the evolution of unique adaptations in organisms. Students will then learn how natural selection works to form new species and about the rate of speciation.

The chapter concludes with a discussion of adaptive radiation, convergent evolution, and divergent evolution.

Key Terms

adaptive radiation
allelic frequency
analogous structure
artificial selection
camouflage
convergent evolution
directional selection
disruptive selection
divergent evolution
gene pool
genetic drift
genetic equilibrium
geographic isolation
gradualism
homologous structure
mimicry
natural selection
polyploid
punctuated equilibrium
reproductive isolation
speciation
stabilizing selection
vestigial structure

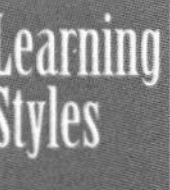

Learning Styles

Look for the following logo for strategies that emphasize different learning modalities. LS

Kinesthetic	Minilab, p. 429; Meeting Individual Needs, p. 444
Visual-Spatial	Demonstration, p. 425; Visual Learning, pp. 425, 427, 428; Display, pp. 431, 447; Chalkboard Example, p. 444
Interpersonal	Project, pp. 426, 431, 440; Alternate Lab, p. 436
Intrapersonal	Meeting Individual Needs, p. 441; Project, p. 447
Linguistic	Meeting Individual Needs, pp. 425, 427, 432, 438; Student Journal, pp. 428, 438; Portfolio, pp. 429, 430, 445, 446
Logical-Mathematical	Minilab, p. 426; Enrichment, pp. 428, 447; Thinking Lab, p. 430; Portfolio, p. 433; Visual Learning, p. 436; Reinforcement, p. 440

LS

CHAPTER

18 The Theory of Evolution

On a hot, dry day, just after the rainy season of southern Africa, the common mole-rat, *Cryptomys hottentotus*, drags across the sandy surface in search of food and a new home. Once it has selected a site, it begins the lengthy process of digging an extensive, underground burrow system. Using its large, chisel-like incisors, the common mole-rat starts a small hole as its short limbs push the dirt out of the way. When the hole is large enough, it slithers its cylindrical body in—aided by a loose layer of skin that allows it to turn and complete the burrow.

Common mole-rats don't look much like other rodents; the body, skull, and teeth of common mole-rats are adapted to life underground. But there is another variation. Common mole-rats are completely sightless. Limited or reduced vision is not uncommon in the animal kingdom.

How can underground or cave-dwelling species evolve from ancestors with normal vision? In this chapter, you'll learn how the mechanism of evolution can explain the origins of unique species such as mole-rats.

Concept Check

You may wish to review the following concepts before studying this chapter.

- Chapter 3: niche, population
- Chapter 13: mutation

Chapter Preview

Laboratory Activities

Biolab

- Natural Selection and Allelic Frequency

Minilabs

- How variable are traits?
- How is camouflage an adaptive advantage?

Members of the cave-dwelling fish genus *Amblyopsis*, the burrowing snake genus *Typhlops*, and the African mole-rat genus *Cryptomys* are examples of sightless animals whose close relatives have normal eyesight. Why are so many cave-dwelling or burrowing animals sightless, and how does evolution work to produce such species?

423

Assessment Planner

Choose assessment strategies from the following pages to evaluate the progress of your students.

Assess, pp. 433, 447
Alternate Lab, pp. 436-437
Minilabs, pp. 426, 429
Portfolio, pp. 429, 430, 433, 445, 446
Thinking Lab, p. 430
Biolab, pp. 442-443
Chapter Review, pp. 449-451

CHAPTER 18

Introducing the Chapter

Use the chapter opener photos to introduce students to the theory of evolution. Discuss general characteristics of the mole-rat, such as its classification, environment, and behavior. Then review with students the concept of adaptation and what adaptations can be observed in mole-rats. *large chisel-like incisors, shortened limbs, cylindrical body, and loose skin* Ask students why they think mole-rats have poorly developed vision and have them explain whether they think this is an adaptation to the environment. Ask similar questions about the other animals in the photos. *Students may suggest that because there is little or no light in the cave, the animals do not need to see in that environment.*

Theme Development

The theme of *unity within diversity* is apparent in this chapter as is the theme of *evolution.* The theory of evolution can explain the diversity of organisms and all their features.

Concept Check

Students will need to have a clear understanding of the concepts of adaptation, as discussed in Chapter 1, and of niche and population, as discussed in Chapter 3. Refer back to Chapter 2 and review the definition of theory with students. Students should also review the concept of mutation, from Chapter 13, and how the fossil record shows that organisms have changed over time, as discussed in Chapter 17.

SECTION

18.1

Prepare

Key Concepts

Students will be introduced to the concept of natural selection as developed by Charles Darwin. They will also learn about some of the evidence scientists use to support the theory of evolution.

Block Scheduling

Look for this symbol for strategies that are useful in a block scheduling format. For more information on block scheduling, refer to Section 18.1 in the **Lesson Plans** booklet.

Materials

- Collect photos of automobiles for the Demonstration.
- Gather black and white construction paper for the Minilab.
- Obtain models of human and ape skulls and an assortment of bird bones (for example, chicken, turkey, and quail) for the display.

1 Focus

Bellringer

Before presenting the lesson, display **Section Focus Master 38** on the overhead projector and have students answer the accompanying questions. L1 SAE

SECTION 18.1

Natural Selection and the Evidence for Evolution

Section Preview

Objectives

Summarize Darwin's theory of evolution by natural selection.

Relate the idea of natural selection to the origin of structural and physiological adaptations.

Evaluate the various lines of evidence for evolution.

Key Terms

artificial selection
natural selection
mimicry
camouflage
homologous structure
analogous structure
vestigial structure

Observations of organisms living today provoke questions about their evolutionary origins. The theory for how organisms change over time was developed close to 150 years ago, and it is still the theory most scientists base their work upon because it fits the most evidence to date.

Fossil flightless bird from the Eocene

Modern-day ostrich in Africa

Charles Darwin and Natural Selection

The modern theory of evolution is perhaps the most fundamental concept in biology. It has been said that it's impossible to understand any of the fields of biology without understanding evolution. What is the origin of the concept of evolution?

Fossils interested scientists in evolution

A rich fossil record has been important to biological science since the 18th century and formed the basis of early evolutionary concepts. Scientists wondered where these species came from, why some no longer existed, and what relationship, if any, they have to modern species.

When it became clear through geological studies that Earth was much more ancient than had been previously believed, many early 19th-century scientists were convinced that life slowly changed over time, or evolved. But how? Several ideas were proposed during the course of the century, but only one has survived to become accepted by the majority of scientists today. English scientist Charles Darwin (1809-1882) is considered to be the founder of modern evolutionary theory.

Program Resources

Section Focus Master 38 L1 SAE
Reinforcement and Study Guide, pp. 69-70 L1
Biolab and Minilab Worksheets, p. 70 L1
Laboratory Manual, pp. 97-100 L2
Concept Mapping, p. 18 L1
Critical Thinking/Problem Solving, p. 18 L3
Performance Assessment in the Biology Classroom, pp. 23, 25, 35 L1 P
Reteaching Transparency 18 and **Master** L1 SAE

Darwin studied the natural world during the voyage of the *Beagle*

In 1831, at the age of 21, Darwin was recommended for the position of naturalist on HMS *Beagle*, a ship chartered for a five-year collecting and mapping expedition to South America and the South Pacific—a voyage that would forever change the science of biology.

As the ship's naturalist, Darwin's job was to collect, study, and store biological specimens discovered along the journey. The voyage of the *Beagle* took him through a number of environments with some of the world's greatest biodiversity.

Darwin's careful study of the animal and plant life gathered along the way helped him develop his theory that would explain how evolution occurs.

Darwin's observations in the Galapagos

For the development of Darwin's hypothesis, observations made in the Galapagos Islands were among his most important. The Galapagos Islands, a group of small islands about 1000 km off the west coast of South America, support a great diversity of animal and plant life. Here Darwin studied and compared the anatomy of many species of reptiles, insects, birds, and flowering plants that are unique to the islands, yet similar to species seen in other parts of the world, *Figure 18.1*. By the end of his trip, Darwin was convinced that evolution occurs—that species change over time. However, he still wanted to test this idea before he could explain how such changes occur.

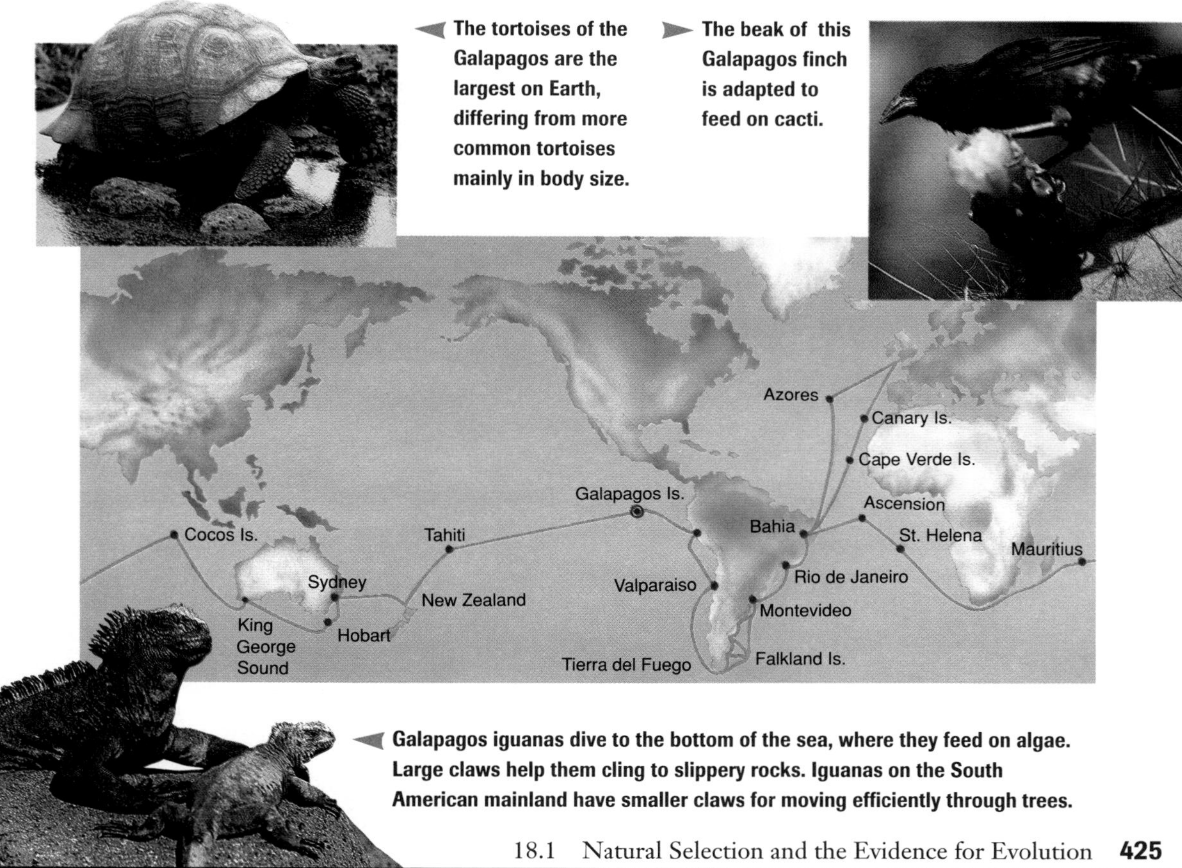

Figure 18.1
The voyage of HMS *Beagle,* which lasted for nearly five years, took Darwin from England to South America, around Cape Horn and north to the Galapagos Islands, and west through the South Pacific to Australia. Animal species in the Galapagos possess adaptations that are unique to the islands; however, they are also similar to species found in other parts of the world.

The tortoises of the Galapagos are the largest on Earth, differing from more common tortoises mainly in body size.

The beak of this Galapagos finch is adapted to feed on cacti.

Galapagos iguanas dive to the bottom of the sea, where they feed on algae. Large claws help them cling to slippery rocks. Iguanas on the South American mainland have smaller claws for moving efficiently through trees.

2 Teach

Demonstration

LS Visual-Spatial Use a series of photos from a particular model of automobile that shows how the automobile has changed over time. If you can't get a series, show a picture of one of the first automobiles and a picture of a modern automobile. Have students observe and explain ways in which automobiles are the same and ways in which they have changed over time. Point out that organisms also change over time. Elicit from them some differences in the two kinds of evolution. *Student responses should show that they recognize that the changes in models of automobiles take place much faster than evolution of organisms.*

Earth Connection

The Principles of Geology
One of the most important influences on Darwin's thinking was the book *The Principles of Geology* by geologist Charles Lyell. This book proposed that Earth is very old and that the forces that have produced changes on Earth's surface in the past are the same forces that operate in the present. You may wish to elaborate on the principle of uniformitarianism and discuss with students how other conclusions influenced Darwin.

GLENCOE TECHNOLOGY

Videodisc

Biology: The Dynamics of Life
Disc 1, Side 2
The Galapagos (Ch. 4)

Meeting Individual Needs

Students Acquiring English Have students write about how music, food, clothing, movies, and television have changed over time. Next, have students describe some changes in the natural world such as changes in weather, tides, the phases of the moon, etc. Help them relate these types of changes to changes in living organisms over time. L1 SAE LS

Visual Learning

Figure 18.1 Have students examine the photos of the Galapagos tortoise, iguanas, and finch. Discuss each organism, asking students to identify its adaptations. LS

SECTION 18.1

A BROADER VIEW

Evolution: Areas of Disagreement

Purpose

Students will compare scientific evidence for evolution and discover that areas exist within and outside the scientific community regarding the interpretation of the data. The students will then research one of these areas to better understand the reasons for the disagreements.

Teaching Strategies

- Have students form discussion groups to debate the weaknesses and strengths of the examples described in the feature.
- Ask individual students to interpret a set of class data to demonstrate how the same information can result in different interpretations.

Thinking Critically

After students have completed the research, they can report their findings to the class or they can debate the various topics.

A BROADER VIEW

Evolution: Areas of Disagreement

Most of what is known about the theory of evolution is based upon large amounts of data collected over a long period of time. However, there is often disagreement over the interpretation of these data. The following are some examples of these differences.

Gaps in the fossil record The fossil record has provided scientists with an opportunity to study the evolutionary history of various groups of organisms. However, the fossil record does not give a complete history of life on Earth. Many gaps are present and some transitional fossils are absent.

Limits of accuracy in radiometric dating Radiometric dating is another tool used to study Earth's history. Many samples have to be analyzed using as many different methods as possible because two samples may not provide consistent results. Dating methods are continually being refined, providing more accurate data.

Inconclusive origins of life Most people have wondered about how life first began. Scientists have attempted to provide answers by suggesting different possibilities. Read more about beliefs and theories on the origin of life in A Broader View, *How Did Life Begin: Beliefs and Theories,* in Chapter 17.

Similarities among embryos Biologists who have studied the development of animals have noticed that many structures are shared by different organisms during the early stages of development, but are later lost in the adult stages. For example, all vertebrates share gill slits in the embryonic stages even though the adult stage possesses lungs for breathing. Some people interpret these data to indicate that all organisms are related; others do not.

Thinking Critically Select one of the above four areas of disagreement to research. Gather information focusing on the strengths and weaknesses of each point of view.

Darwin completed his studies in England

Darwin returned from his voyage on the *Beagle* in 1836, and spent the next 22 years studying his collections and conducting experiments.

Darwin was also intrigued by an essay written at the time that suggested that the human population was growing faster than the food supply on Earth. Darwin knew that many species produce large numbers of offspring, and since Earth was not covered with such species, Darwin suspected that there must be a struggle for existence among individuals. Darwin envisioned many kinds of struggles—competition for food and space, escape from predators, and the need to find shelter. Only some of the individuals in any population survive the struggle long enough to produce offspring of their own. But which ones?

Darwin's answer to his question about the struggle for existence came from his own pigeon-breeding experiments. Darwin observed that in his populations of pigeons, individuals have different variations of traits that can be inherited. We can see this variability in all organisms, as shown, for example, in ***Figure 18.2.***

Figure 18.2

GLENCOE TECHNOLOGY

Videodisc

The Infinite Voyage: To the Edge of the Earth

Exploring the Galapagos Islands (Ch. 4)

PROJECT

Variation in Beans

Students can research the effects of variation by preparing an experimental bean garden from pinto beans. Have them measure and observe characteristics of each bean to put them in categories, such as short, long, wide, thin, etc. Ask them to hypothesize about the growth of each type of bean and then plant 3 or 4 beans from each category. Remind students to observe the plants each day and to record their observations. Have them write a brief summary after 4–5 weeks of plant growth.

L1 | SAE | COOP LEARN | IS

By breeding pigeons that had desirable variations, Darwin was able to produce offspring with these same features. Breeding experiments are an example of **artificial selection**—a technique in which a breeder selects particular traits. Darwin wondered if there were some force in nature similar to artificial selection.

Darwin's explanation for evolution

Using data he gathered from the natural world, Darwin gradually began to form his well-known idea of evolution by natural selection. **Natural selection** is a mechanism for change in populations that occurs when organisms with favorable variations for a particular environment survive, reproduce, and pass these variations on to the next generation. Organisms with less-favorable variations are less likely to survive and pass on traits to the next generation. Therefore, each new generation is made up largely of offspring from parents with the most favorable variations. The idea of natural selection is summarized in *Figure 18.3*.

Overproduction
of offspring

Variations exist
in populations

Some variations
are favorable;
some are not

Populations evolve,
or change,
over time

Figure 18.3

Darwin proposed the idea of natural selection to explain how populations of organisms evolve.

A **In nature, there is a tendency toward overproduction of offspring. Fishes, for example, lay thousands, sometimes millions, of eggs.**

B **In any population of organisms, individuals will exhibit slight variations. Fishes may differ slightly in color, fin and tail size, and speed.**

C **Individuals with variations favorable for a particular environment are more likely to survive and pass those variations on to the next generation than individuals with less-favorable variations. For example, a fast fish with a skin color that allows it to blend in better with its surroundings is more likely to survive to reproductive age than a slow fish with more obvious coloring.**

D **Gradually, offspring of survivors will make up a larger proportion of the population. Depending upon environmental factors, after many generations, a population may come to look entirely different.**

Different Viewpoints in Biology

Jean-Baptiste de Lamarck

Point out to students that before Darwin's theory of evolution by natural selection was developed, French biologist Jean-Baptiste de Lamarck (1744-1829) proposed a different mechanism for evolutionary change. Lamarck's idea rested on two assumptions: (1) the more an organism uses a particular part of its body, the stronger and better developed that part becomes, and (2) the physical characteristics that an organism develops through use and disuse can be passed on to offspring. Initiate a discussion of Lamarck's hypothesis with students, and ask them to list its inadequacies.

Visual Learning

Figure 18.3 summarizes the general principles of natural selection. Carefully review this illustration by discussing each major principle. Provide alternative examples of organisms to reinforce the concepts. IS

Assessment

Performance Assessment in the Biology Classroom: p. 23, *Investigating Variations in Populations.* Have students carry out this activity to explore what variations occur in a population. P

interNET CONNECTION

Follow the link for this chapter on the Glencoe Homepage at **www.glencoe.com/sec/science** to find out more about evolution.

Meeting Individual Needs

Students Acquiring English Review with students the meanings of the words *fit, fitter,* and *fittest.* Help them form sentences using the three words. Then have them rearrange the words *the, selects, nature,* and *fittest* to form a sentence explaining Darwin's concept of natural selection. L1 SAE IS

Visual Learning

Figure 18.4 illustrates the probable evolution of the common mole-rat from a generalized member of the rodent family *Bathyergidae*. Have students examine each step of the illustration carefully. Then ask them to list the steps that may have occurred during the evolution of the blind, cave-dwelling fish genus, *Amblyopsis*, and the sightless, burrowing snake genus, *Typhlops*. LS

Enrichment

LS **Logical-Mathematical** Using Figure 18.4 as a model, have students illustrate or describe possible evolutionary sequences of one of the following: (1) the evolution of long necks in giraffes from short-necked ancestors, (2) the evolution of whales from terrestrial carnivores, (3) the evolution of flight in birds from bipedal dinosaurs, (4) the evolution of high-speed running in cheetahs from slower cat ancestors. L2

In the 1850s, Darwin began work on a detailed, multivolume book to present his idea of natural selection to the scientific community. In 1859, Darwin published his book *On the Origin of Species by Natural Selection.* This book convinced many scientists that evolution does occur through natural selection, and today this theory is the unifying one for all biology.

Natural Selection and Adaptations

Have you ever wondered why some plants have thorns or why some animals have distinctive coloring? How do such adaptations originate? Recall that an adaptation is any trait that aids the chances of survival and reproduction of an organism. Darwin's theory of natural selection can be applied to explain the evolution of adaptations in organisms.

Structural adaptations arise over many generations

How can Darwin's theory be used to explain unique structural adaptations in species, such as those seen in common mole-rats? Common mole-rats possess a number of features that adapt them to life underground, but their ancestors lived above ground and didn't have such characteristics. How did the features of modern-day mole-rats evolve from the ancestral mole-rats? *Figure 18.4* illustrates a possible scenario for the evolution of the common mole-rat.

The adaptations of common mole-rats are examples of structural adaptations—changes in the structure of body parts. Some of the more obvious structural adaptations in organisms are used for defense against predators, such as the thorns of rosebushes or the spines of sea urchins. However, others are more subtle. For example, **mimicry** is a structural adaptation that provides protection for an organism by enabling it to copy the appearance of another species, as shown in *Figure 18.5*. Another type of defensive adaptation that involves changes to the color of organisms is camouflage. **Camouflage** is a structural adaptation that enables an organism to blend in with its surroundings. Organisms that are well camouflaged are more likely to escape predators and survive to reproduce.

Figure 18.4
The evolution of the common mole-rat can be explained by Darwin's theory of natural selection.

A **According to some scientists, the ancestors of common mole-rats resembled the African rock rat.**

B **Over time, variations arose. Large teeth and claws, for example, allowed some individuals to dig deeper holes and avoid predators.**

STUDENT JOURNAL

Evidence for Natural Selection Have students describe the main pieces of evidence Darwin used in his formulation of the concept of natural selection. Next, have them identify an organism and, in their own words, describe the four main ideas of the concept of natural selection using the organism as an example. L1 LS

GLENCOE TECHNOLOGY

Videodisc

Biology: The Dynamics of Life
Disc 1, Side 2
Adapted for Survival (Ch. 5)

Figure 18.5

Camouflage and mimicry involve changes to the external appearance of an organism.

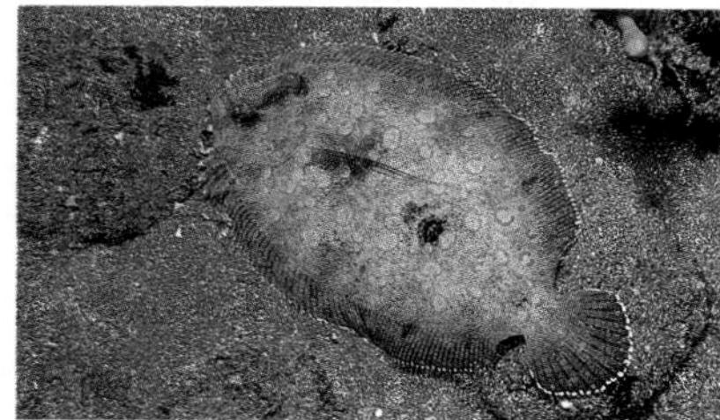

Camouflage coloration enables organisms to blend in with their surroundings. Can you see how the coloration of this flounder allows it to avoid predators?

The bee orchid is a flower that mimics the appearance of females of the bee species that pollinates it. By mimicking females, the flower attracts males, which then spread pollen to other flowers.

MiniLab

How is camouflage an adaptive advantage?

Camouflage is a structural adaptation that allows organisms to blend in with their surroundings. In this activity, you'll discover how natural selection operates to produce camouflage adaptations in organisms.

Procedure

1. Working with a partner, punch 100 dots from a sheet of white paper with a paper hole punch. Repeat with a sheet of black paper. These dots will represent black and white insects.
2. Place both white and black dots on a sheet of black paper.
3. Select one student to play the bird. That student must look away from the paper, turn back to it, and then immediately pick up the first dot he or she sees.
4. Repeat the procedure to see how many insects of each color can be collected in two minutes.

Analysis

1. What color were the dots that were most frequently collected?
2. How does color affect the survival rate of insects?
3. What might happen over time to a similar insect population in the wild?

C **If these individuals were better able to survive and reproduce, useful variations would be passed on. Over time, most individuals in the population would possess these beneficial adaptations.**

D **Over many generations, evolution may have resulted in the adaptations seen in common mole-rats of today. Blindness most likely evolved because as vision decreased in importance, sightedness no longer gave the mole-rat a selective advantage. In fact, in this environment, eyes are a handicap due to the potential for infection and injury.**

MiniLab

Purpose

LS Kinesthetic Students will investigate how camouflage adaptations enable organisms to survive.

Process Skills

observe and infer, form a hypothesis

Teaching Strategies

- Have students do this activity after studying camouflage adaptations in organisms.
- Explain to students that they will perform a simulation of how natural selection might operate on a population of insects that vary in color.

Expected Results

Most lab groups will find that they were able to pick up more white dots than black dots. Students should infer that this is because the black dots were camouflaged on the black paper.

Analysis

1. Most students will say they picked up more white dots.
2. Light-colored insects may be seen more easily than dark-colored insects. Because dark-colored insects are not preyed upon by birds as much as light-colored insects, their survival rate is higher.
3. Over time, an insect population might become all dark-colored because light-colored insects were eliminated from the population.

Assessment

Knowledge: Have students research and write a short summary about other insect adaptations that allow them to survive in particular environments. L1

PORTFOLIO

Breeds of Dogs Obtain books about breeds of dogs. Have students prepare a short report about a breed of their choice describing the characteristics of the chosen breed, reasons why it was originally bred, details about how it was bred, and the characteristics of closely related breeds. Encourage students to illustrate their breed for the report. L2 LS P

ThinkingLab Interpret the Data

Purpose

Logical-Mathematical Students will analyze data from an actual study of natural selection in peppered moths.

Process Skills

use a table, form a hypothesis

Teaching Strategies

- Remind students that these data are from an actual experiment, and have traditionally been used to support the theory of evolution by natural selection.

Background

The dark variety of peppered moth was first observed in English cities in 1848. It was hardly noticeable on the dark tree trunks near polluted areas. Over the next 100 years, scientists observed greater numbers of dark moths relative to light moths near the cities. In the 1950s, English scientist H.B. Kettlewell tested the hypothesis that the difference was due to natural selection.

Thinking Critically

Students should infer that the differences in survival rates are due to camouflage from predators. There was natural selection for the dark variation in the city, where pollution killed the lichen on trees, and natural selection for the light variation in the country, where lichens were present.

Assessment

Knowledge: Assess student knowledge by asking them to consider what might happen to the peppered moth population if the forest were invaded by a reddish-green lichen.

ThinkingLab Interpret the Data

How can natural selection be observed?

The following are data from an actual natural selection experiment by H.B. Kettlewell. Kettlewell studied camouflage adaptations in a population of light- and dark-colored moths that live on the trunks of trees. Kettlewell noticed that pollution in the area had caused lichens that usually lived on the bark of the trees to die. The tree bark had become darker because of the lack of lichens. Kettlewell was interested in discovering if this change played a role in natural selection of the moth population.

Analysis

Kettlewell captured, released, and recaptured moths from the country and from the polluted city. The number of moths recaptured indicates their survival rate in the environment. Shown here are raw data from Kettlewell's experiment.

Locations and Numbers of Moths

Location		Numbers of light moths	Numbers of dark moths
Unpolluted country	number released	469	473
	number recaptured	62	30
Polluted city	number released	137	447
	number recaptured	64	154

Thinking Critically

How can you explain the differences in survival rates of moths in the country and in the city? How is natural selection involved?

Depending on the type of adaptation and the rate of reproduction of an organism, as well as environmental factors, structural adaptations may take millions of years to develop. However, some structural adaptations can evolve relatively quickly in geologic terms. For example, a well-known study has shown that camouflage adaptations can evolve in moth species within 100 years.

Physiological adaptations can develop rapidly

Would you believe that some of the medicines developed during this century may no longer be useful for fighting disease? Only 50 years ago, the antibiotic drug penicillin was considered a wonder drug because it could kill many types of disease-causing bacteria. Today, scientists know that penicillin is not as effective as it used to be because many species of bacteria have evolved physiological adaptations that make them resistant to penicillin, *Figure 18.6.* Physiological adaptations are changes in an organism's metabolic processes.

Scientists have also observed rapid changes in pest organisms such as insects and weeds. After being exposed to particular pesticides, many species of insects and weeds have been selected for physiological resistance to the chemicals.

Figure 18.6

Examples of bacterial resistance to antibiotics, such as penicillin, provide scientists with direct evidence for evolution in progress.

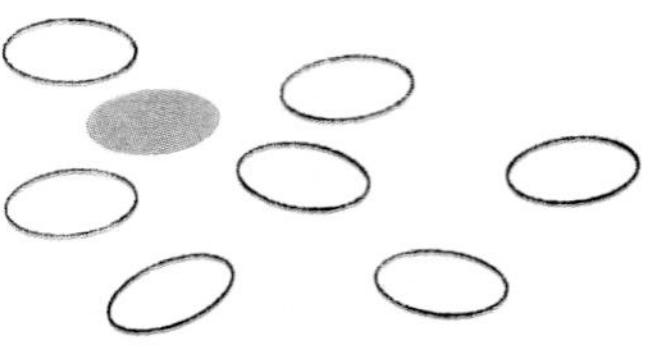

Individual bacteria in a population show variation. Some bacteria possess a gene that makes them resistant to some antibiotics.

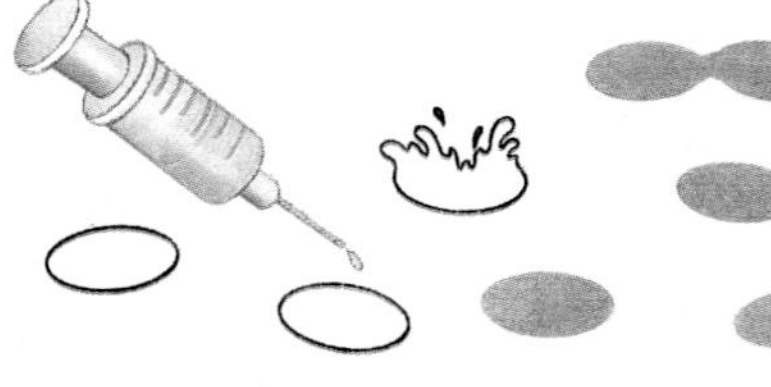

When the population is exposed to an antibiotic, some individuals die, but resistant bacteria survive.

Resistant bacteria survive and reproduce. In time, the entire population is resistant to a certain antibiotic.

430 The Theory of Evolution

PORTFOLIO

Camouflage and Mimicry Have students prepare a short report about an organism that has camouflage or mimicry adaptations. The report should include the name of the organism, details about its environment, organisms that prey on it, and a description of the camouflage or mimicry adaptations. L1 LS P

GLENCOE TECHNOLOGY

Videodisc

STVS: Ecology
Disc 6, Side 2
Resistance to Pesticides in Cockroaches (Ch. 4)

Evidence for Evolution

The evolution of camouflage in moths and the development of physiological resistance to chemicals in organisms provide examples of the process of evolution. Such examples occur rapidly enough that they can be observed in a laboratory. However, these examples are exceptions; most of the evidence for evolution comes from indirect sources.

Fossils show changes over time

Were the ancestors of whales once land-dwelling, doglike animals? Scientists have learned about the evolutionary history of whales by studying their fossils. Fossils provide a record of earlier life. Although the fossil record is incomplete, it does provide evidence that evolution has occurred.

Working with an incomplete fossil record can be compared to completing a jigsaw puzzle with missing pieces. It's great if all the pieces are there, but even if some are missing, you may still be able to understand the overall picture. It's the same with fossils. Although scientists can't trace each and every step in the evolution of some species because intermediate forms of these species cannot be found, they can still understand the general pathway of evolution. When the fossil record is sufficiently complete, scientists can show a step-by-step sequence of evolution. *Figure 18.7* shows the evolution of the camel as pieced together by fossil skulls, teeth, and limb bones. As you can see, 40 to 50 million years ago, camels were a species of small, rabbit-sized animals.

Figure 18.7

Fossils indicate that during the Eocene epoch, 37 to 54 million years ago, camels had small body size, four-toed feet, and low-crowned teeth. By Miocene times, 30 million years later, species of camels had grown larger and evolved two-toed feet and high-crowned teeth. By studying fossils, scientists have been able to trace the evolution of camels to the forms of the present-day species.

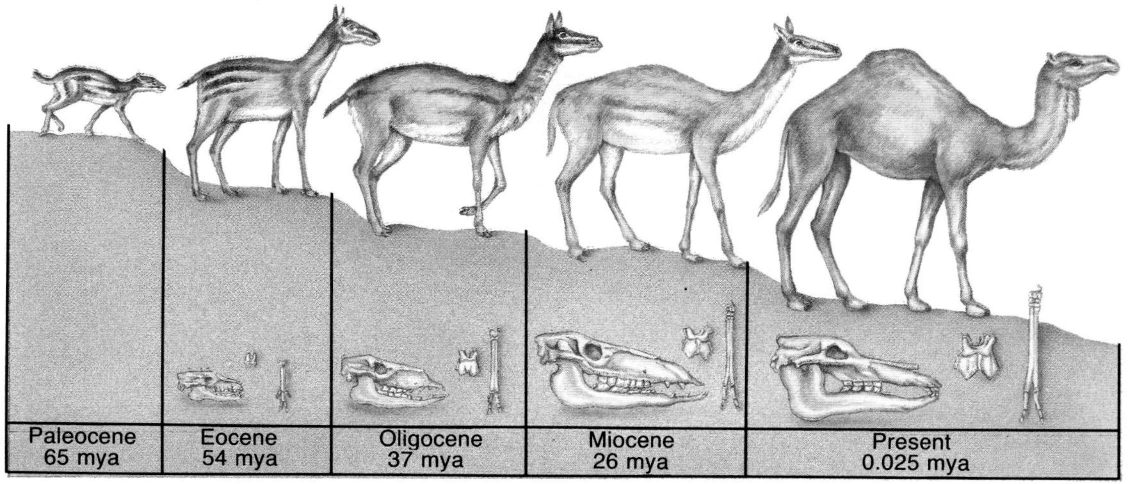

Visual Learning

Figure 18.7 illustrates that, although the fossil record can be used as evidence for evolution, a relatively complete fossil sequence for organisms, such as those that exist for camels and horses, is quite rare. Relate this to students' previous knowledge of problems in fossil preservation, dating, and interpreting the fossil record.

Display

LS **Visual-Spatial** Obtain samples of different bird wings (turkey, chicken, duck, or guinea hen) from a supermarket. Display the wings and have students identify the homologous structures. Discuss the structure and function of a bird's wing. Ask students if their observations support the idea that the organisms are closely related.

Assessment

Performance Assessment in the Biology Classroom: p. 35, *Model for Sexual Reproduction.* Have students carry out this activity to develop an awareness that natural selection will occur if the inheritance of certain traits equips some individuals better than others to cope with changes in the environment. L1 P

PROJECT Evolving Viruses

Viruses are among the fastest evolving things. Have student groups research the structural and physiological adaptations of a particular virus for class presentation. Have them describe what kind of virus it is, what organisms and cells it infects, its structural and physiological adaptations, and how it may have evolved. Encourage students to include a model of the virus made from household objects or illustrations to accompany their presentations. L2 COOP LEARN LS

SECTION 18.1

Discussion

Morphological studies of bones and teeth have traditionally been used to determine the evolutionary relationships among organisms, but now many scientists are turning to molecular studies to determine phylogeny. Discuss with students the pros and cons of each technique. What questions can they answer? What are the limitations of each technique?

Tying to Previous Knowledge

Point out that the theory of evolution predicts that organisms with similar physical characteristics will also have similarities in the structure of their DNA. Briefly review the structure and function of DNA and remind students that DNA makes up the chromosomes that determine the traits of organisms.

GLENCOE TECHNOLOGY

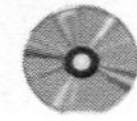

Videodisc

The Infinite Voyage: The Great Dinosaur Hunt

New Dinosaur Discoveries and Their Link with Today (Ch. 10)

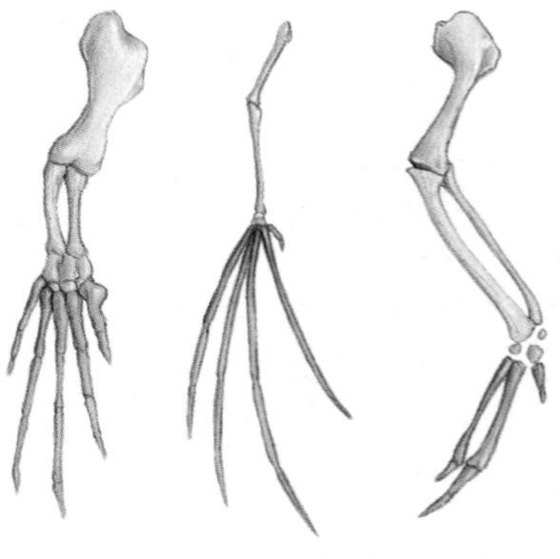

Figure 18.8

The forelimbs of crocodiles, bats, and chickens are examples of homologous structures. Can you see how the bones of each forelimb have become modified in relation to their function?

Anatomical studies indicate evolutionary relationships

Do the forelimbs of the animals shown in *Figure 18.8* look similar to you? If you said yes, then you just made a type of observation that scientists use to establish evolutionary relationships among organisms. As you can see, although the limbs of these animals would look strikingly different from the outside and vary in function, details of their skeletons are similar.

Scientists view such similarities as evidence that these organisms evolved from a common ancestor with the same basic limb structure. Over the course of evolution, vertebrates moved into different environments. Within each environment, animals faced very different needs for survival. Animals that had limbs that were most useful in their environment were probably more likely to survive. In this way, species could become adapted to different ways of life. Although different species came to use their forelimbs for different functions in different environments, these modified structures are still similar. A modified structure that is seen among different groups of descendants is called a **homologous structure.**

Homologous structures are often similar in structure, in function, or in both structure and function. However, similarity of function doesn't always mean that two organisms are closely related. Homologous structures probably had a common evolutionary origin. Compare the structures of the butterfly wing and the bird wing in *Figure 18.9.* Bird and butterfly wings are not similar in structure, but they do have the same function. The wings of birds and insects evolved independently in two distantly related groups of ancestors. Any body part that is similar in function but different in structure is an **analogous structure.**

Analogous structures can't be used to indicate evolutionary relationships, but they do provide evidence of evolution. Insect and bird wings most likely evolved when their different ancestors independently adapted to similar ways of life.

Figure 18.9

Although they have the same function, insect and bird wings are not similar in structure. Bird wings are made up of a set of bones, whereas insect wings are mainly composed of a tough substance called chitin.

Meeting Individual Needs

Learning Disabled The fossil records of some organisms, such as camels, horses, elephants, and the extinct titanotheres, are relatively complete and can show detailed evolutionary change over time. Provide students with illustrations of the fossil sequence of a particular organism, and point out the major characteristics of the species in each stage of the sequence. Have them prepare a short summary of the major changes in the species during its evolution. L1 IS SAE

Figure 18.10

Vestigial structures are considered to be evidence of evolution because they show structural change over time. Many flightless birds such as the extinct elephant bird from Madagascar (left) and the African ostrich (right) have forelimbs that appear to be extremely reduced. These birds probably attained their great sizes by foraging and nesting on the ground. Over time, they lost the ability to fly and their wings became vestigial.

Functionless structures indicate evolutionary pathways

Do you know what your appendix is used for? Your doctor may not be able to tell you what your appendix is for, but an evolutionary biologist can tell you what an appendix *was* for. That's because many organisms, including you, contain structures that have reduced functions but were once used by ancestral organisms. Any body structure that is reduced in function in a living organism but may have been used in an ancestor is known as a **vestigial structure.**

Particular structures become vestigial as species change in form and behavior. The structure, although it may serve no function, continues to be inherited as part of the body plan for that species. The eyes of sightless species, such as common mole-rats and cave fish, may be considered vestigial structures because eyes were most likely functional in their ancestors. *Figure 18.10* shows another example of vestigial structures.

Embryological development shows evolution from a common ancestor

It's easy to tell the difference between adult birds and adult mammals, but do you think you could pick them out by looking at their embryos? *Figure 18.11* illustrates an early stage of embryological development in a fish, a reptile, a bird, and a mammal. Notice the similarities in structures. In the earliest stage of development, a tail and gill slits can be seen in all species. As development continues, the embryos become more and more distinct, and in the stages before birth, they attain their distinctive forms. Similarities among vertebrate embryos suggest evolution from a common ancestor.

Figure 18.11

Comparative embryological studies of vertebrates are used to indicate their possible common ancestry. The presence of gills and tails in the early stage of all vertebrate embryos supports evidence from the fossil record that aquatic, gill-breathing vertebrates preceded air-breathing, terrestrial species.

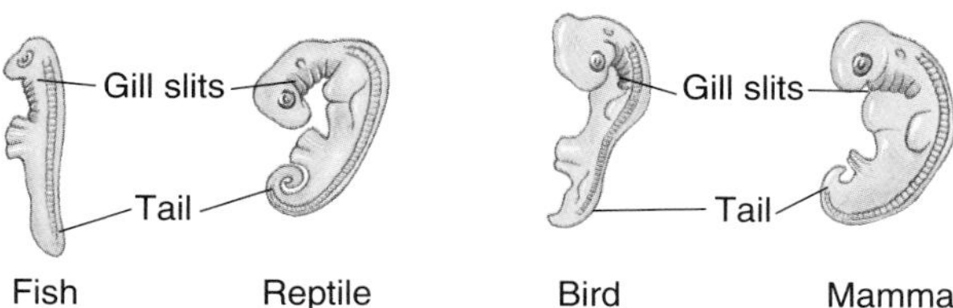

SECTION 18.1

Science, Technology, and Society

DNA Techniques

Scientists use a number of DNA techniques to determine evolutionary relationships among organisms. Amino acid sequencing compares the sequences of amino acids in individual proteins, such as myoglobin, among different organisms. In DNA hybridization, fragments of single-stranded, radioactively tagged DNA from one species are reassociated, or hybridized, to those of another species. The more difficult it is to pull the DNA apart, the more closely related two organisms are. The technique of mtDNA sequencing compares the numbers and types of base substitutions in mtDNA sequences between species.

3 Assess

Check for Understanding

Have students describe how each of the following concepts relates to natural selection: overproduction, favorable variations, population change over time. L1

Reteach

Have students identify five traits of a specific organism, such as hair color or size. Ask students to name five possible variations for each trait. Ask how each variation would make the organism better suited for its environment. L1 SAE

PORTFOLIO

Have students describe how each of the following pieces of evidence could be used to support the theory of evolution:

(1) An aphid species isn't killed by an insecticide.
(2) Geckos have specialized feet for climbing vertically on trees and rocks.
(3) Human, rabbit, chicken, and lizard embryos have gill slits during early embryological development.
(4) Human DNA is 20 percent identical to mouse DNA, and 98 percent identical to chimpanzee DNA.
(5) The fossil record shows that camels were originally rabbit-sized animals. L2 P IS

SECTION 18.1

Extension

Have students complete the following problem: South American anteaters of the rainforest eat termites. If the termites they normally feed on are replaced by termites with tougher exoskeletons, explain how the anteaters might change over time. L1

Assessment

Portfolio: Have students research a particular organism. Have them describe five adaptations of the organism. Next, have students choose a new environment for their organism. Have students predict how natural selection would affect the organism in its new environment. L1 P

4 Close

Activity

Divide the class into cooperative groups and show a slide presentation of a variety of organisms. Have groups brainstorm three different explanations for the adaptations seen in each organism. COOP LEARN

GLENCOE TECHNOLOGY

Videodisc

The Infinite Voyage: The Geometry of Life
Evolution, Molecular Genetics, and DNA Sequencing (Ch. 5)

Figure 18.12

One study of comparisons of RNA sequences led to the construction of an evolutionary tree of life that shows three major groups: the eubacteria, the archaebacteria, and eukaryotes. The lengths of the lines indicate the degrees of differences among the RNA sequences.

Archaebacteria
Halobacterium sp.
Sulfolobus sp.
Methanospirillum sp.
Thermoproteus sp.
Methanobacterium sp.
Methanococcus sp.
Human (Homo sapiens)
Frog (Xenopus laevis)
Corn (Zea mays)
Yeast (Saccharomyces sp.)
Protist (Oxytricha sp.)
Slime mold (Dictyostelium sp.)
Protist (Trypanosoma sp.)
Flavobacterium sp.
Pseudomonas sp.
Escherichia coli
Agrobacterium sp.
Mitochondrial RNA
Bacillus sp.
Anacystis sp.
Chloroplast RNA
Eukaryotes
Eubacteria

Genetic comparisons may reveal hidden relationships

Knowledge of genetics can help us understand relationships among individuals. It can also help tell us how different species are related. The most recent evidence for evolution has come from comparisons of DNA and RNA within taxonomic groups. Many scientists consider genetic studies to be more reliable than anatomical studies for sorting out evolutionary relationships. Some biologists use nucleotide sequencing studies to indicate levels of relationships among species within major taxonomic groups, as shown in *Figure 18.12.*

Since the 1970s, research has shown that by comparing nucleotide sequences in the DNA and RNA of species, it's possible to construct hypothetical evolutionary trees showing levels of relationships.

Section Review

Understanding Concepts

1. Why is the fact that snakes have skeletal remnants of legs considered evidence for evolution?
2. How does the evolution of camouflage and mimicry in animals illustrate natural selection?
3. How do homologous structures provide evidence for evolution?

Thinking Critically

4. After sequencing a section of DNA from two newly discovered species, you find that the DNA sequences differ by only two percent. Is it reasonable to assume that these two organisms are closely related? Explain.

Skill Review

5. **Sequencing** Whales are thought to have evolved from ancestors that had legs. Using your knowledge of natural selection, sequence the steps that may have occurred during the evolution of whales from their terrestrial ancestors. For more help, refer to Organizing Information in the *Skill Handbook*.

Answers to Section Review

1. They show that snakes have evolved over time and that snake ancestors may have had functional legs.
2. Camouflage and mimicry involve structural adaptations that have been naturally selected for. Ancestral populations showed variation, and over time, selection pressures produced a whole population possessing the variation.
3. Homologous structures show that organisms shared a common ancestor.

Thinking Critically

4. Yes, because the general structure of DNA is similar in all organisms.

Skill Review

5. Environmental factors, such as competition, may have forced ancestral whales to exploit another environment, such as water. Individuals that could better move and function in water survived and reproduced. Those that couldn't died and didn't pass on genes to the next generation. Eventually, variations spread throughout the population.

SECTION

18.2 Mechanisms of Evolution

If you visit the Grand Canyon National Park, you might see an interesting case of speciation. On the north side of the canyon, you will likely see a population of dark-brown squirrels with white abdomens and tails. If you visit the south side, you might see squirrels that are charcoal gray. What has happened here? The formation of the canyon itself resulted in the complete isolation, divergence, and eventual speciation of these two kinds of squirrels.

In the natural world, several mechanisms contribute to the evolution of species. Two factors that are the most consequential are geography and the effects of environmental change.

Section Preview

Objectives

Summarize the effects of the different types of natural selection on gene pools.

Relate mechanisms of speciation to changes in genetic equilibrium.

Explain the role of natural selection in convergent and divergent evolution.

Key Terms

gene pool
allelic frequency
genetic equilibrium
genetic drift
stabilizing selection
directional selection
disruptive selection
speciation
geographic isolation
reproductive isolation
polyploid
gradualism
punctuated equilibrium
adaptive radiation
divergent evolution
convergent evolution

Population Genetics and Evolution

Recall that adaptations of species, such as the shape of a tooth or the color of a flower, are determined by the genes contained in the DNA code. When Charles Darwin developed his theory of natural selection in the 1800s, he did so without including the idea of the gene. Since that time, a great deal of information about genes has been gathered and Darwin's theory has been modified. The modern understanding of evolution is based on Darwin's theory, but it now also includes principles of genetics that were not known at the time.

Populations evolve; individuals don't

Can individual organisms evolve? That is, can they acquire or lose structures or characteristics in response to natural selection? As you know, natural selection can act upon an individual's phenotype, the external expression of genes. If an individual organism possesses a phenotype that isn't adapted to the environment, it may result in the individual's inability to successfully compete. However, within the lifetime of one individual, new features cannot evolve in response to natural selection. Rather, natural selection operates only on populations over many generations.

SECTION 18.2

Prepare

Key Concepts

Students will be introduced to the concept of the gene pool and learn how natural selection affects gene pools. Students will then learn how natural selection can contribute to the speciation of populations.

The section concludes with a discussion of patterns of evolution, such as divergent and convergent evolution. Students will learn specific examples of each pattern and how they support the theory of evolution.

Block Scheduling

Look for this symbol for strategies that are useful in a block scheduling format. For more information on block scheduling, refer to Section 18.2 in the **Lesson Plans** booklet.

Materials

- Obtain bags of red pinto beans, white navy beans, and black beans for the Biolab.
- Gather slides or photographs of a variety of organisms.

1 Focus

Bellringer

Before presenting the lesson, display **Section Focus Master 39** on the overhead projector and have students answer the accompanying questions. L1 SAE

Program Resources

Section Focus Master 39 L1 SAE
Reinforcement and Study Guide, pp. 71-72 L1
Biolab and Minilab Worksheets, pp. 71-72 L1
Laboratory Manual, pp. 101-102 L2
Basic Concepts Transparencies 21, 22 and **Masters** L1 SAE
Basic Skills Transparency 16 and **Master** L1 SAE

2 Teach

Visual Learning

Figure 18.13 Have students use beans to model gene pools and allelic frequency, as shown. Pour bags of red pinto beans, black beans, and white navy beans into a large container, mix them up, and have students withdraw 20 beans at random. Tell students that they now have the gene pool of a bean population with the genotypes *BB* (black), *BB′* (white), and *B′B′* (red). Have students calculate the phenotype frequencies by dividing the total number of each phenotype by 20. Students can calculate the allelic frequencies by counting the numbers of each allele and dividing by 40. L2 IS

To understand how the genes of a population can change over time, picture the entire collection of genes among a population as its **gene pool.** Consider, for example, the population of snapdragon plants that is shown in *Figure 18.13.* Snapdragon plants show a pattern of heredity that is known as incomplete dominance, which you learned about in Chapter 14.

If you know the genotypes of all the organisms in a population, you can calculate the **allelic frequency,** the percentage of a particular allele in the gene pool. A population in which frequency of alleles does not change from generation to generation is said to be in **genetic equilibrium,** *Figure 18.13.*

Changes in genetic equilibrium lead to evolution

A population that is in genetic equilibrium is not evolving. Because allelic frequencies remain the same generation after generation, no new traits are gained or lost. Only when the genetic equilibrium of a population is disrupted can evolution occur.

What factors cause changes to genetic equilibrium? Any factor that affects genes can change allelic frequencies, and thus disrupt the genetic equilibrium of populations. You learned in Chapter 13 that one such mechanism for genetic change is mutation. Many mutations are caused by factors in the environment, such as radiation or chemicals, but others happen purely by chance.

Figure 18.13

Incomplete dominance produces three phenotypes, so in these snapdragon plants, red flowers are homozygous (*RR*) for the red trait, white flowers are homozygous (*R′R′*) for the white trait, and pink flowers are heterozygous (*RR′*) for the pink trait. Compare the parent and offspring generations of snapdragon flowers. The generations differ in phenotype frequency, but the allelic frequencies for the R and R′ alleles are the same in both generations. This population of snapdragon plants is in genetic equilibrium.

	Genotypes	Phenotype frequency	Allele frequency
First generation	*RR*, *RR*, *RR′*, *RR′*, *RR*, *RR′*, *RR*, *RR′*	White = 0 Pink = 0.5 Red = 0.5	*R* = 0.75 *R′* = 0.25
Second generation	*RR*, *RR′*, *RR*, *RR′*, *RR*, *R′R′*, *RR*, *RR*	White = 0.125 Pink = 0.25 Red = 0.625	*R* = 0.75 *R′* = 0.25

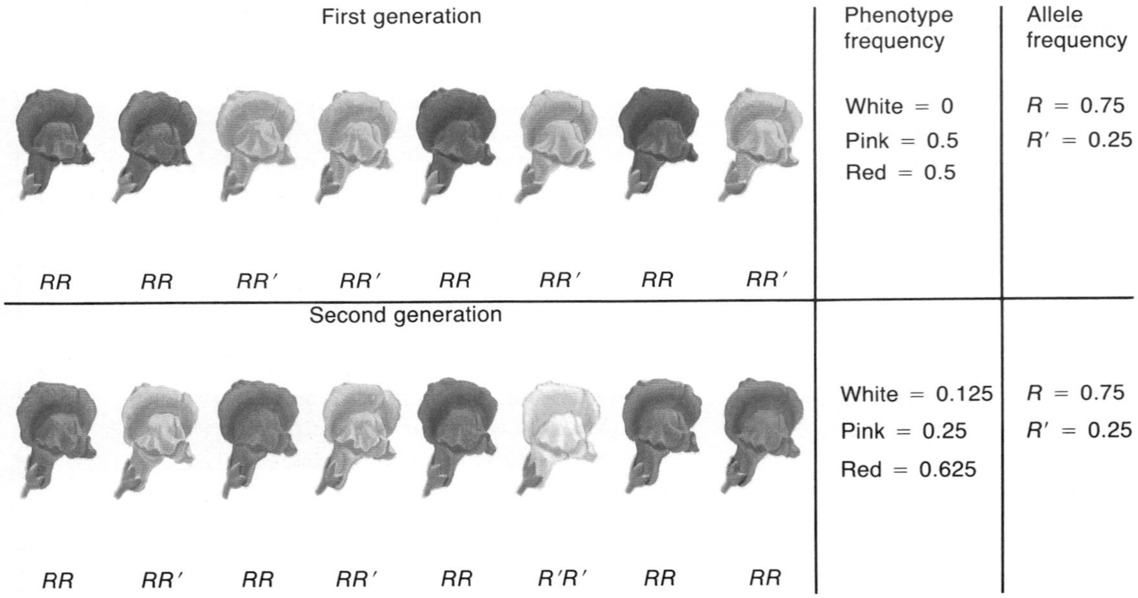

Alternate Lab — Bacterial Resistance

Purpose  

IS **Interpersonal** Students will study variation in bacterial resistance to antibiotics.

Materials

culture of *Bacillus subtilis*, 3 tubes of nutrient agar, tube of streptomycin agar, inoculation loop, 2 petri dishes, Bunsen burner, wax pencil, test tube

Procedure

Give the following directions.

1. Use a wax pencil to write the letters *A* and *B* on opposite sides of a petri dish. Repeat with the letters *C* and *D* on the other dish.
2. Quickly pour agar into each dish. Agar solidifies if it falls below 42°C and must be reheated to liquify.
3. Pour liquid streptomycin into dish A-B and cover immediately. Place a pencil under the dish to tilt the streptomycin agar to one side of the dish.
4. When the agar is solidified, remove the pencil. Pour a tube of liquid nutrient agar on top of the solidified layer. After this has solidified, cover the dish.
5. Pour two tubes of liquid nutrient agar into the C-D dish. Allow to solidify and cover.

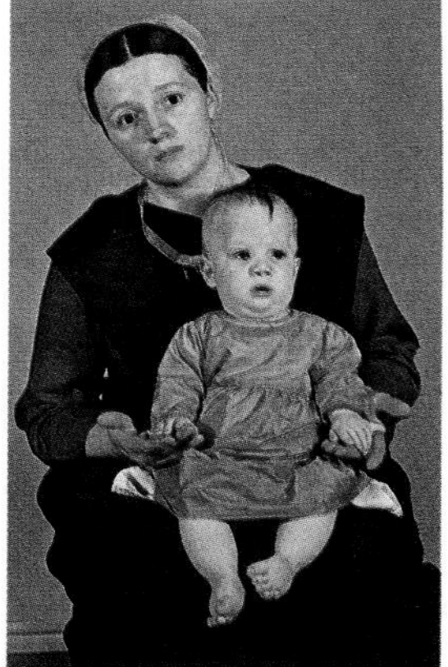

Figure 18.14

Sometimes genetic drift results in an increase, rather than a decrease, of rare alleles in a population. An example of this is seen in the Amish community of about 12 000 in Lancaster County, Pennsylvania. This community was started by about 30 individuals. Some of these people had a recessive allele that causes short arms and legs and extra fingers or toes. In the Amish community, the frequency of this allele is much higher than in other populations—one in 14 rather than one in 1000.

Mutations are important in evolution because they result in genetic changes to the gene pool. Many random mutations are harmful, and most of these are selected against and have no significance. However, once in a great while, they result in a favorable variation to an offspring, are selected for, and become part of the genetic makeup of future generations.

Another mechanism that causes changes to genetic equilibrium is genetic drift. **Genetic drift** is the alteration of allelic frequencies by chance processes. Genetic drift is more likely to occur in small populations than in large ones.

Consider, for example, a small Amish population, *Figure 18.14.* Because of chance events, such as

A BROADER VIEW

Expression of Recessive Traits

When a recessive gene is rare, its phenotype does not show up often in a population. In fact, for a rare recessive trait to appear, two people who are heterozygous for the trait must mate, and even then there is only a one-in-four chance of a homozygous recessive offspring. Thus, recessive conditions such as albinism are exceedingly rare. However, they are less rare in certain populations—usually ones that are geographically or socially isolated. For example, albinism occurs in European populations in about one in every 20 000 births. But among the San Blas Indians in Panama, a small, isolated group, the ratio is one in 132. This is because the gene pool is much smaller than that in Europe, and the number of potential mates among tribe members is more limited.

Relatives and royalty Historically, the most common examples of recessive genes being expressed were among royalty. For example, in ancient Egypt, it was the custom for the pharaoh to marry his sister. This type of practice severely limits the gene pool, can lead to genetic disorders, and was probably one of the reasons that the dynasty died out. Another famous example is the history of the descendants of the English Queen Victoria. Many of the women in this family carried a sex-linked gene that causes hemophilia, a disease in which the blood does not clot. The result was that they transmitted the disease to their sons, who were in danger of bleeding to death from minor injuries.

Relatives and religion Most religions prohibit marriage between close relatives. However, some religions are isolated by practice or lifestyle from the larger population. For example, the Amish and Mennonite populations of the United States are farmers with strong beliefs about how they should live. Because their lifestyle requires a willingness to forgo modern technology, these populations are socially isolated from the larger population. Members tend to marry within their religion, which is a relatively small gene pool. Several genetic studies have shown that these people must be careful about the genetic pedigrees of those they marry.

Thinking Critically

How do breeding-exchange programs at zoos prevent the expression of recessive traits in animal species?

SECTION 18.2

A BROADER VIEW

Expression of Recessive Traits

Purpose

Students will learn about some effects of small population size on the characteristics of populations. Examples are given of how recessive traits are expressed in small populations.

Teaching Strategies

- Point out that in small breeding populations, such as certain religious groups and royal families, changes to gene pools are accelerated because the number of potential mates is limited.
- Have students write a short summary of how genetic drift may affect small populations.

Thinking Critically

Zoos keep detailed records of the animals available for breeding to identify mating pairs that come from different gene pools.

GLENCOE TECHNOLOGY

 **Videotape**

The Secret of Life

Use the videotape *It's in the Genes: Evolution* to show how mutations in a population can benefit the population.

6. Sterilize the inoculation loop in the Bunsen burner.
7. Remove the stopper on the bacterial culture. Flame the lip of the container by gently passing it through the flame.
8. Dip the inoculation loop into the culture, remove the loop, and flame the lip of the container again. Replace the stopper.
9. Streak the agar layers of dish A-B with the loop. Do not break the surface of the agar.
10. Repeat steps 6 through 9 on the C-D dish.
11. Cover the dishes, invert them, and place them in a dark drawer or closet. Observe after 24 hours.

Analysis

Have students sketch and describe the appearance of both plates.

Assessment

Knowledge: Ask students:

1. What is the purpose of the A-B dish? The C-D dish?
2. Describe the growth of bacteria in the A-B dish and the C-D dish.
3. How do you explain the differences between the dishes? L1

Math

The Mathematics of Evolution

Purpose
Students will examine the Hardy-Weinberg principle of genetic equilibrium and learn about the five conditions necessary for genetic equilibrium.

Teaching Strategies
- Students should read this feature after learning the concept of genetic equilibrium.
- Illustrate to students how this principle is used in practice. Stress that population geneticists use these equations all the time to study the evolution in populations.

Possible Answer
Most students will say that the population of the United States is not in genetic equilibrium because new genes continually enter the population, mating is probably not random, and height is probably naturally selected for.

Tying to Previous Knowledge

Review the different types of mutations and their effects discussed in Chapter 13. Emphasize that mutations affect genetic equilibrium by producing totally new alleles for a trait and also change the frequency of alleles already in the population. For example, the mutation that causes sickle-cell trait occurs spontaneously in about five out of 100 million people.

Math

The Mathematics of Evolution

In the early 1900s, a British mathematician, G.H. Hardy, and a German doctor, W. Weinberg, discovered some basic principles about populations and how to describe them mathematically.

The Hardy-Weinberg law The Hardy-Weinberg law says that the frequencies of alleles that make up a gene pool will remain the same in a stable population. Mathematically, the law is expressed by a binomial equation. The letters p and q represent the frequency of each allele in the gene pool.

$$p + q = 1 \text{ (or 100\%)}$$

$$p^2 + 2pq + q^2 = 1$$

Providing that five conditions are met, the frequencies will stay the same in generation after generation of a sexually reproducing population. The five conditions follow.

Conditions
1. Mutations cannot occur.
2. Mating must be random.
3. Natural selection cannot occur.
4. No genes may enter or leave the population.
5. The population must be large.

These conditions are probably never met in natural populations. Mutations happen every day, mating is not random, natural selection occurs, individuals and their genes migrate in and out of populations regularly, and many breeding populations are small. Then why do we need a law that describes a situation that almost never happens?

Evolution benchmark The law is critical because it shows that populations tend to remain static without evolution. Thus, the law provides a standard by which evolution (genetic change) can be measured. In the event that gene frequency does change, it provides evidence for evolution at work.

CONNECTION TO Biology

The general population of the United States is getting taller. Assuming that height is genetically controlled, does this observation violate the Hardy-Weinberg law? Explain.

crossing-over during meiosis and the independent distribution of chromosomes to gametes, it's likely that in a small population, a larger proportion of individuals would inherit the recessive allele that causes short arms and legs and extra fingers or toes. In a large population, the frequency of recessive alleles is much lower. Suppose that these few individuals did not mate. The recessive allele would then soon be lost from the population.

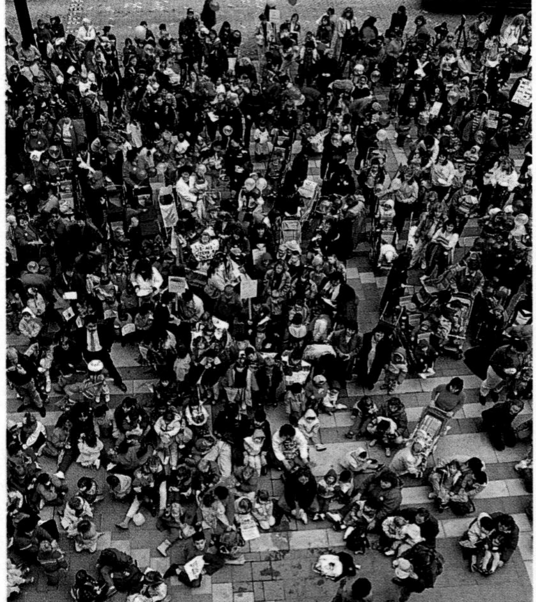

Genetic equilibrium may also be disrupted by the movement of individual organisms in and out of a population. Any time an individual leaves a population, genes are lost. When individuals enter a population, those genes are added to the gene pool.

While mutation, genetic drift, and the migration of individuals can affect allelic frequencies, they usually do not cause significant changes. The factor that causes the greatest change in gene pools is natural selection.

STUDENT JOURNAL

Populations and Natural Selection Have students write a short story about the effects of natural selection on a particular population of organisms. Have them predict what will happen to the population if it is subjected to all three types of natural selection. Encourage students to illustrate their stories . L1 LS

Meeting Individual Needs

Students Acquiring English Ask students to imagine what changes they would have to make if they were introduced to a new environment. Have them write a short story about life in the new environment. Suggest that they consider what changes they would have to make in food choice, clothing, shelter, and other factors. L1 SAE LS

Figure 18.15

These swallowtail butterflies come from different areas in North America. Despite their slight variations, they can interbreed. Therefore, all belong to the same species. Their variations and isolation from others of the same species make them subspecies.

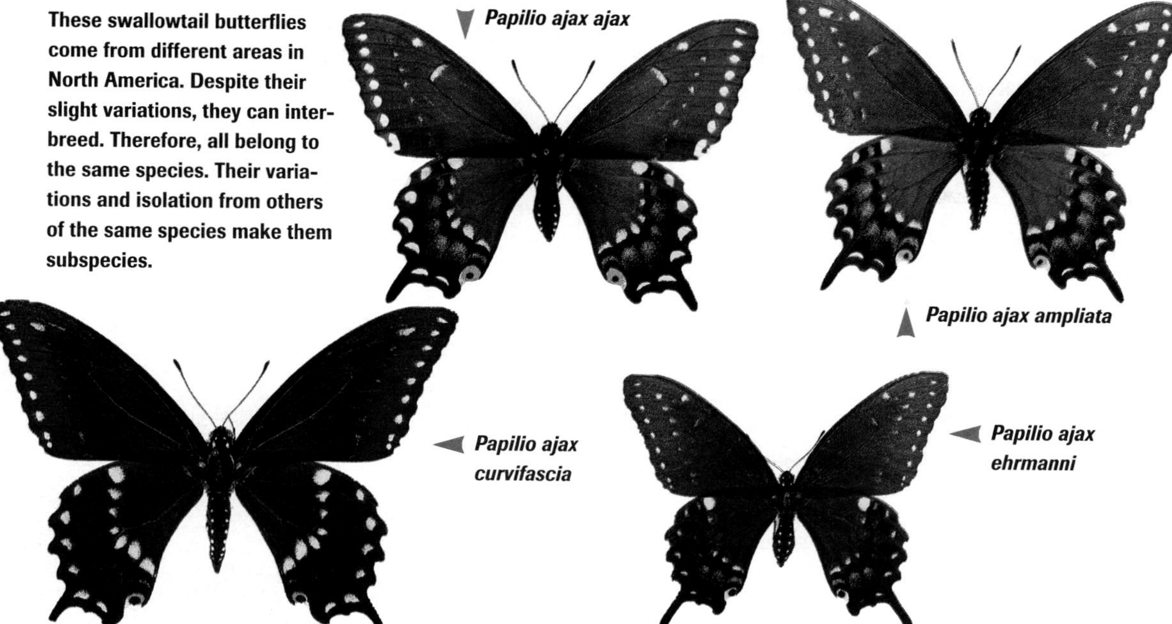

Natural selection acts upon the variation in populations

As you learned in the last section, the result of natural selection is that some members of a population are more likely to contribute their genes to the next generation than others. Thus, by the action of natural selection, allelic frequencies change from one generation to the next. Each of three types of natural selection—stabilizing, directional, and disruptive—causes changes in gene pools by acting upon the variation in populations.

As you know, all traits show variation. If you were to measure thumb length of students in your biology class, for example, you would find average, long, and short thumbs. Variation such as this can be thought of as the raw material of evolution because natural selection acts upon ranges of variation. *Figure 18.15* shows variation in a species of butterflies.

The type of natural selection that favors average individuals in a population is **stabilizing selection.** Consider a population of spiders in which average size is an advantage in terms of survival and reproduction. Spiders that are larger than average may be at a disadvantage because they can be seen and captured more easily. On the other hand, spiders that are smaller than average might not be able to catch enough prey to survive and reproduce and thus may also be at a disadvantage. In other words, average-sized spiders have a selective advantage in this particular environment.

When one of the extreme forms of a trait is favored by natural selection, it's known as **directional selection.** For example, imagine a population of woodpeckers. Woodpeckers feed by pecking holes in trees in order to get at the insects living under the bark. Suppose that one year, the trees in the woodpeckers' area are invaded by

Tying to Previous Knowledge

Review with students the process of meiosis, as discussed in Chapter 12. Point out that there are random factors involved in some of the steps of meiosis. Stress to students how such factors can contribute to genetic drift.

Using An Analogy

Use the analogy of flipping a coin to show students how small populations can be significantly affected by genetic drift. If you flip a coin 100 times, the chances of getting 100 heads and 0 tails—or even 80 heads and 20 tails—is extremely unlikely. The result will probably be close to 50-50. But if you flip the coin 10 times, the chances of getting 8 heads and 2 tails—or even 10 heads and 0 tails—are more likely. Similarly, in large populations, the likelihood of alleles becoming lost by chance processes is much lower than in small populations.

CULTURAL DIVERSITY

The Neutral Theory of Evolution

Explain to students the neutral theory of evolution developed by Japanese biologist Mootoo Kimura. This theory holds that most molecular evolution is due to random drift of selectively neutral mutations, rather than positive Darwinian selection.

The neutral theory was heavily debated upon its presentation in 1968, but today it is viewed as an improvement on classical Darwinian theory because it provides testable predictions about molecular evolution. The strongest advocate of the neutral theory is Japanese geneticist, Tomoko Ohta, head of Japan's National Institute of Genetics.

SECTION 18.2

Visual Learning

Figure 18.16 illustrates the three main types of natural selection. Refer to and offer several other examples of each type to students.

Reinforcement

LS Logical-Mathematical Have students determine which type of natural selection is operating in each of the following examples.

(1) Members of a population of tree frogs vary in leg length and hop from tree to tree to search for food in the Amazon rain forest. Environmental changes cause massive destruction in the forest, resulting in fewer trees. Over several generations, scientists discover that only long-legged tree frogs remain. *directional selection*

(2) Members of a population of grasses range in length from 8 cm to 28 cm. The 8-10 cm grass blades receive little sunlight, and the 25-28 cm grass blades are eaten quickly by grazing animals. *stabilizing selection*

(3) Members of a population of sea urchins vary in the length of their spines. Short-spined sea urchins are camouflaged easily on the seafloor. Long-spined sea urchins present a tough defense against predators. *disruptive selection*

Have students prepare illustrations for each situation. Ask students to predict what will happen to each population of organisms if natural selection continues to operate. L2

Figure 18.16

The different types of natural selection act upon the ranges of variation in organisms. The red, bell-shaped curve indicates the normal variation in a population.

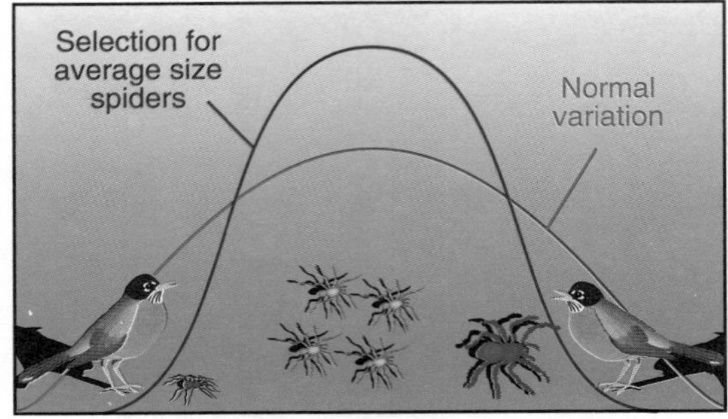

A **Stabilizing selection affects genetic equilibrium by favoring average individuals. In this way, variation in a population is reduced.**

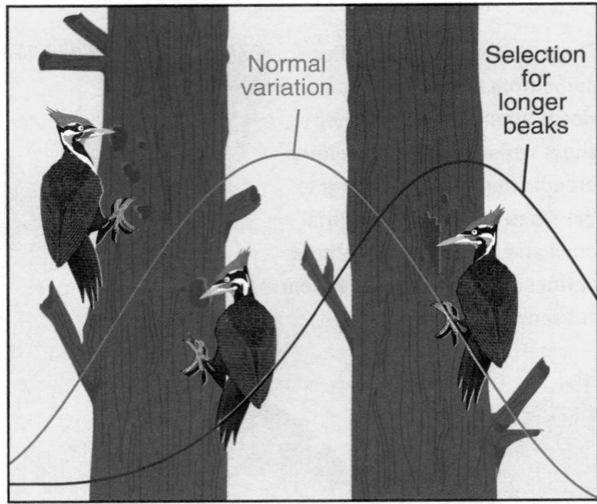

B **By favoring either of the extreme forms of a trait, directional selection can lead to the rapid evolution of a population.**

C **In disruptive selection, both extreme forms of a trait are favored. In some cases, there may be no intermediate forms, which can lead to the evolution of two new species.**

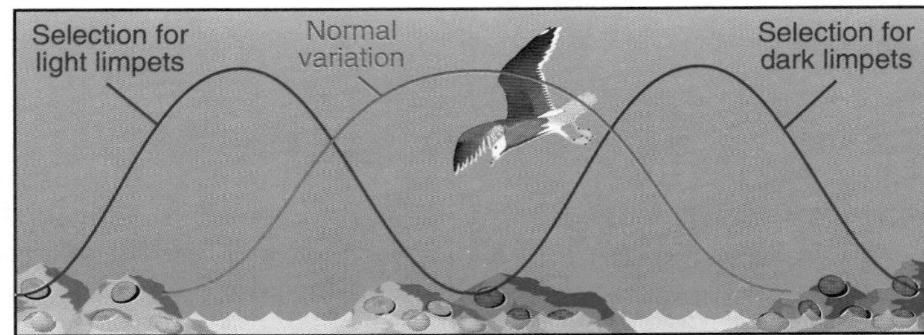

a species of insect that lives deep within trees. Only woodpeckers with long beaks would be able to reach the insects. In this scenario, shown in *Figure 18.16*, woodpeckers with long beaks have the selective advantage over those with short or average-sized beaks.

The third type of natural selection is known as disruptive selection. In **disruptive selection,** individuals with either of two extreme forms of a trait are at a selective advantage. For example, consider a population of marine organisms known as limpets. The shell color of limpets varies from white to dark brown. Limpets live their adult lives attached to rocks. On light-colored rocks, white-shelled limpets are at an advantage because birds that prey upon them have a more difficult time locating them. On dark-colored rocks, dark-colored limpets are at an advantage because they too are camouflaged. On the other hand, tan-colored limpets, the intermediate forms, are easily spotted on either the light or dark backgrounds. In other words, disruptive selection, *Figure 18.16*, eliminates the intermediate forms by favoring the extremes.

Natural selection is the most significant factor that alters genetic equilibrium and causes changes in the gene pool of a population. Significant changes in the gene pool can also lead to the formation of new species.

PROJECT

Mammalian Evolution

Have groups of students prepare a report, class presentation, or poster project about how plate tectonics has contributed to the geographic and reproductive isolation of a particular family of mammals. Projects should contain some details about the mammals, such as structure and behavior, a brief summary of the fossil evidence available for the mammals, and explanations for how they evolved. Some mammalian families to research are: Bradypodidae, Myrmecophagidae, Camelidae, Ursidae, Felidae, and Mustelidae. L2 LS COOP LEARN

The Evolution of Species

You've seen how natural processes such as mutation, random genetic drift, and natural selection can lead to changes in a population's gene pool, but how does this result in the evolution of new species? Remember that a species is defined as a group of organisms that look alike and have the ability to interbreed and produce fertile offspring in nature. The evolution of new species, **speciation,** can occur only when either interbreeding or the production of fertile offspring is somehow prevented.

Physical barriers can prevent interbreeding

Imagine a population of tree frogs living in the dense rain forests of the Amazon basin. Over time, periodic droughts begin to break up stretches of continuous forest into smaller patches, separating the original tree frog population into smaller groups. Other environmental factors cause shifts in local river courses that further isolate the patches.

In the natural world, physical barriers frequently form and break up large populations into smaller ones. Volcanic eruptions can cause lava flows, which have the potential for splitting a population. Sea-level changes along continental shelves, such as those in New Guinea, can create island environments that impose a water barrier between populations. **Geographic isolation** occurs if a physical barrier separates a population into groups.

Geographic isolation is one of the ways new species form. Think about the population of tree frogs. If small populations of tree frogs were geographically isolated from one another by areas of deforestation, they would no longer be able to interbreed, and gene exchange would cease. Over time, each small population would adapt to the local environment through the process of natural selection. Eventually, the gene pools of each group would become so different that one could be considered a new species, as shown in *Figure 18.17.*

Figure 18.17

When populations become geographically isolated, individuals from the two groups can no longer mate. Gradually, natural selection produces gene pools so distinct that individuals from the different groups can no longer produce fertile offspring, even if mating does occur.

SECTION 18.2

Discussion

Remind students that scientists used to classify organisms solely on the basis of similarities and differences in morphology. Because morphological characteristics are easy to observe, the morphological species concept is useful, but limited. For example, North American yellow flickers and red-shafted flickers can interbreed and produce fertile hybrid offspring. Using a morphological concept of a species, yellow flickers, red-shafted flickers, and the hybrid flickers could be considered three different species.

Because of such instances, scientists developed the biological species concept. According to this concept, organisms are classified by whether or not they can naturally interbreed with one another to produce fertile offspring. Elicit from students how many species of North American flickers there are under this definition. *a single species* L1

Software

Mystery Fossil (MM) Mayfield Publishing Company

GLENCOE TECHNOLOGY

Videodisc

Biology: The Dynamics of Life
Disc 1, Side 2
Geographic Isolation (Ch. 6)

Meeting Individual Needs

Gifted Have students do extended research on the topic of the rate of speciation. Students could begin by reading Gould and Eldredge's initial 1972 article about punctuated equilibrium entitled "Punctuated Equilibria: An alternative to phyletic gradualism," which can be found in Models in Paleobiology, T.J.M. Schopf (ed.), Freeman, Cooper, and Co. Have students compare punctuated equilibrium and gradualism by identifying evidence for both hypotheses. Their report should contain a conclusion about which hypothesis best supports the evidence. L3 LS P

BioLab

Natural Selection and Allelic Frequency

Time Allotment
One class period

Objectives
Review objectives with students before they begin the Biolab.

Process Skills
make and use tables, observe and infer, make and use graphs

PREPARATION

Alternate Materials
- Beads or other small objects may be substituted for beans.

BioLab

Natural Selection and Allelic Frequency

Evolution can be described as the change in allelic frequencies of a gene pool over time. Natural selection can place pressure upon specific phenotypes and cause a change in the frequency of the alleles that produce the phenotypes.

In this activity, you will simulate the effects of eagle predation on a population of rabbits, where *GG* represents the homozygous condition for gray fur, *Gg* is the heterozygous condition for gray fur, and *gg* represents the homozygous condition for white fur.

PREPARATION

Problem
How does natural selection affect allelic frequency?

Objectives
In this Biolab, you will:
- **Simulate** natural selection by using beans of two different colors.
- **Calculate** allelic frequencies over five generations.
- **Demonstrate** how natural selection can affect allelic frequencies over time.

Materials
colored pencils (2)
paper bag
graph paper
pinto beans
white navy beans

PROCEDURE

1. Make a copy of the data table shown here.
2. Place 50 pinto beans and 50 white navy beans into the paper bag.
3. Shake the bag. Reach into the bag and remove two beans. These represent an individual rabbit's genotype. Set these beans aside, and continue to remove beans until you have 50 "rabbits."

PROCEDURE

Teaching Strategies
- Tell students that they will be simulating natural selection on a population of rabbits in order to see how allelic frequency changes over time.
- You may wish to circulate throughout the room during this activity to ensure that students are following the procedure correctly.

Data and Observations
Make sure students are correctly calculating allelic frequency after each "generation" and recording these data in their data tables. Students should observe changes in the allelic frequencies of the rabbit population. Student graphs should show an increase in the frequency of the *G* allele and a decrease in the *g* allele. LS

4. Arrange the beans on a flat surface in two columns representing the two possible rabbit phenotypes, gray (genotypes *GG* or *Gg*) and albino (genotype *gg*).
5. Examine your columns, and remove 25 percent of the gray rabbits and 100 percent of the white rabbits. These numbers represent the selection pressure on your rabbit population. If the number you calculate is a fraction, remove a whole rabbit to make whole numbers.
6. Count the number of pinto and navy beans that remain, and record this number in your data table.
7. Calculate the allelic frequencies by dividing the number of beans of one type by 100. Record these numbers in your data table.
8. Begin the next generation by placing 100 beans into the bag. The proportions of pinto and navy beans should be the same as the percentages you calculated in step 7.

9. Repeat steps 3 through 8 until you have collected data for five generations.
10. Graph the frequencies of each allele over five generations. Plot the frequency of the allele on the vertical axis and the number of the generation on the horizontal axis. Use a different-colored pencil for each allele.

Allele Frequencies

	Allele G			Allele g		
Generation	Number	Percentage	Frequency	Number	Percentage	Frequency
Start	50	50	0.50	50	50	0.50
1						
2						
3						
4						
5						

ANALYZE AND CONCLUDE

1. **Calculating Results** Did allelic frequencies change over time? Why or why not?
2. **Analyzing Data** Did either of the alleles totally disappear? Why or why not?
3. **Thinking Critically** What does your graph show about allelic frequencies and natural selection?
4. **Making Inferences** What would happen to the allelic frequencies if the number of rabbit-eating eagles declined?

Going Further

Project Develop an activity using two traits instead of just one, such as fur length and color. Select a new predator and new selection pressures for the population.

ANALYZE AND CONCLUDE

1. There were changes in allelic frequency because there was heavy selection pressure against white rabbits.
2. Neither allele disappeared from the population. This is because the *g* allele is also in the heterozygous (*Gg*) rabbits.
3. The graph shows an increase in the frequency of the *G* allele and a decrease in the frequency of the *g* allele due to natural selection against white rabbits.
4. If the eagle population declined, there would be less selective pressure on white rabbits and, therefore, less decline in the frequency of the *g* allele.

Assessment

Knowledge: Ask students what would have happened to the rabbit population if only 60 percent of the white rabbits were removed from the population each generation. Would allele frequencies change as fast? Why or why not? *No, because the gene pool would contain more g alleles that could combine to make more white rabbits.* L1

Going Further

Have students repeat the activity using two traits, such as length and color of hair. In this case there would be four phenotypes. Students can also add another predator, and change the selection pressures on the population. L2

Chalkboard Example

Visual-Spatial Use hypothetical examples to illustrate the concepts of geographic and reproductive isolation on the chalkboard or overhead projector. Show a population of organisms within an environment. Then split the population with a geologic event, such as the formation of a volcano or a large canyon, that results in two new environments. Have students work in small groups to brainstorm three changes—if any—they predict will occur in each subpopulation over time.

Reinforcement

Reinforce the concept of geographic and reproductive isolation by providing students with examples of the reproductive behavior of closely related organisms. For example, wood frogs, *Rana sylvatica*, and leopard frogs, *Rana pipiens*, are species that evolved because of reproductive isolation. Wood frogs usually breed in late March or early April, and leopard frogs usually breed in mid-April.

Geographic isolation can lead to differences in mating behavior

When geographic isolation of a population occurs, gene pools are closed, and genetic material is not exchanged between the groups. Over time, as populations become more and more distinct, reproductive isolation can arise. **Reproductive isolation** occurs when formerly interbreeding organisms are prevented from producing fertile offspring.

There are many different types of reproductive isolation. One type occurs when the genetic material of the groups becomes so different that mistakes happen during development of the zygote—if a zygote forms at all. Another type of reproductive isolation is seasonal. For example, if one group of tree frogs evolves the behavior of mating in the fall, while another group mates during the summer months, these two groups are reproductively isolated from one another.

Speciation can occur when chromosome numbers change

Although geographic isolation plays an important role in most cases of speciation, it's not the only factor. Many new species of plants have evolved in the same geographic area as a result of polyploid speciation. Any species with any multiple of the normal set of chromosomes is known as a **polyploid.**

Polyploids arise when mistakes occur during meiosis. Polyploid speciation is perhaps the fastest form of speciation because reproductive isolation is instantaneous. When a polyploid individual mates with a normal diploid individual, because of the change in chromosome number, the resulting zygotes have difficulty developing normally. It's estimated that nearly half of the known flowering plant species originated in this way, as well as many important crops such as wheat, cotton, apples, and bananas. ***Figure 18.18*** illustrates how polyploid speciation occurs.

Speciation can occur quickly or slowly

You've seen how one form of speciation, polyploidy, can occur after only one generation. However, most speciation does not occur as quickly. What is the tempo of speciation?

When Darwin proposed his theory of evolution, he argued that evolution proceeds at a slow, steady rate, and that small, adaptive changes gradually accumulate over time in populations. **Gradualism** is the idea that species originate through a gradual buildup of new adaptations.

Figure 18.18

Mistakes during meiosis can result in the formation of polyploid species. If chromosomes fail to separate properly during the first meiotic division, the result can be gametes having the diploid (2*n*) condition, rather than the normal haploid (*n*) set of chromosomes. Many cultivated garden flowers, such as this chrysanthemum (below) have been developed by selecting and hybridizing polyploid individuals.

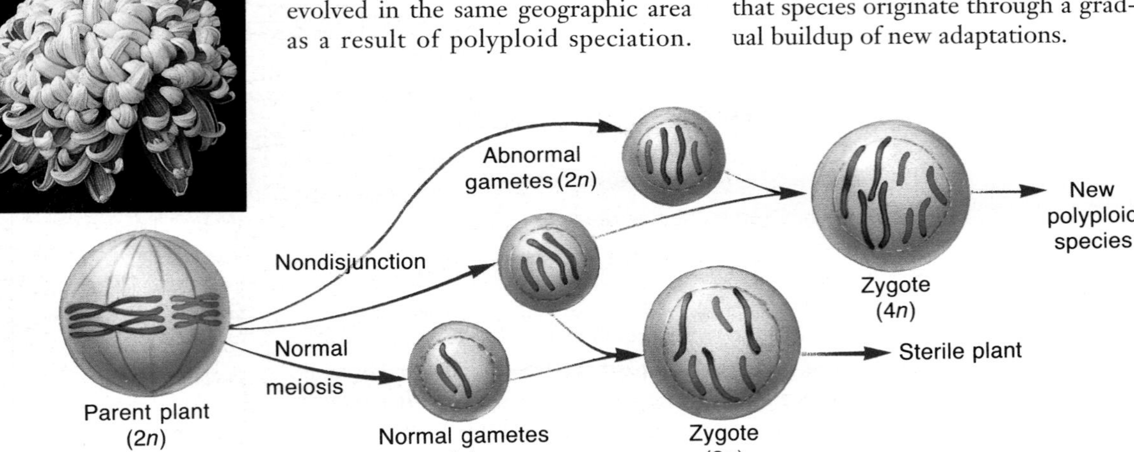

444 The Theory of Evolution

Meeting Individual Needs

Visually Impaired Have students practice calculating allelic frequency by having them work with a "population" of differently shaped objects, such as pennies, popcorn, beads, or beans. Tell students which object represents each particular genotype, and have them calculate allelic frequency by counting the numbers of each genotype. You may wish to have students practice several problems by changing the numbers of individuals in the population, as well as changing the numbers of each genotype. L2

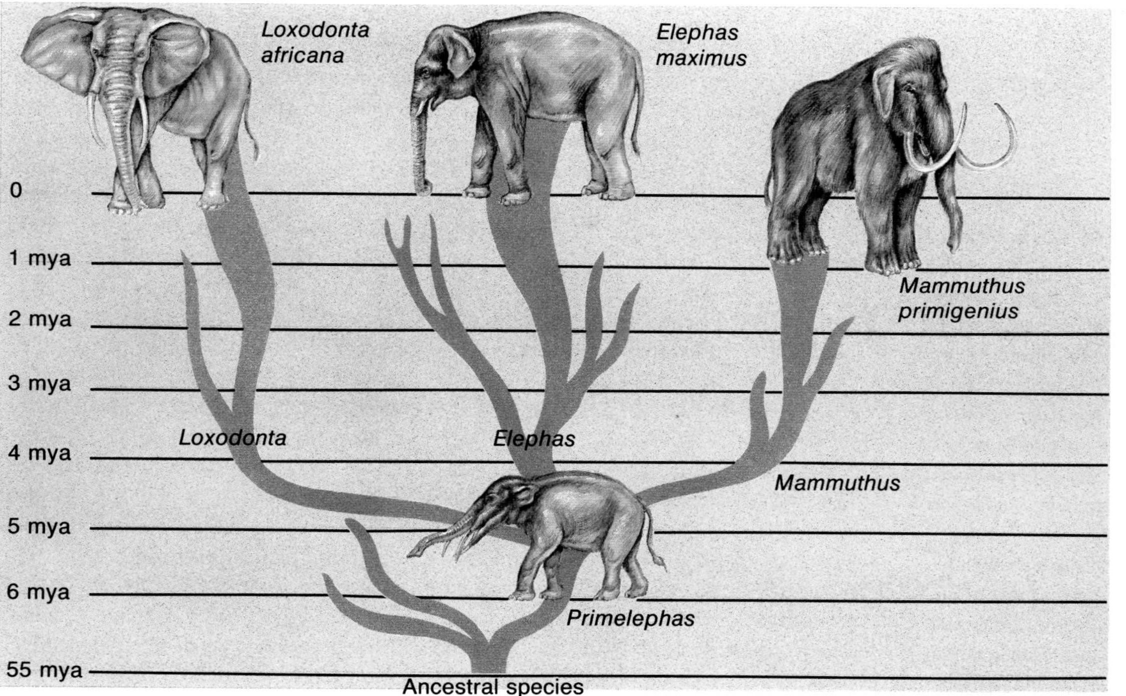

Figure 18.19

Punctuated equilibrium involves rapid evolution of a gene pool. Many supporters of the concept of punctuated equilibrium point to the fossil record of elephants to support their view of evolution. As you can see, three different types of ancient elephants evolved from an ancestral population in a short period of time.

Gradualism is supported by evidence from the fossil record. For example, the slow and steady buildup of adaptations seen in the fossils of camels and horses supports this view.

In 1971, another hypothesis about the rate of evolution was proposed. This idea, known as **punctuated equilibrium,** states that speciation occurs quickly in rapid bursts, with long periods of stability in between, *Figure 18.19.* Supporters of this hypothesis argue that populations remain at or close to genetic equilibrium for long periods of time. When environmental conditions change, such as warmer temperatures or the introduction of a new competitive species, rapid genetic change and speciation can occur, interrupting, or punctuating, genetic equilibrium. A new species can form in a span of 10 000 years or less. Like gradualism, punctuated equilibrium is supported by fossil evidence.

Whether the rate of evolution is slow or fast probably won't be resolved by fossils alone. Many scientists conclude that speciation can occur both slowly and rapidly, depending upon the circumstances. It shouldn't be surprising that alternative theories have been offered to explain observations. The nature of science is that theories are changed when new evidence becomes available.

Concept Development

Emphasize that Darwin believed species evolve slowly and steadily over long periods of time. Remind students that scientists have found little support for the idea of gradualism in the fossil record. However, point out some examples that do support it. For example, fossils show that the living members of the horseshoe crab genus *Limulus* are nearly identical to ancestors that lived hundreds of millions of years ago. If possible, show a specimen of the horseshoe crab to students and show how little it has changed over time.

Bioethics

Loss of Biodiversity
Ask students what part of the world first comes to mind when the loss of biological diversity is considered. Most will probably say "tropical rain forests." Programs to save the rain forest have had much publicity; however marine systems are in similar, if not worse, trouble. Biological diversity of the oceans is declining. Technologies such as satellite imaging and computer modeling are currently being used to study the biodiversity of the oceans.

PORTFOLIO

Abert and Kaibab Squirrels After discussion and demonstration of Abert and Kaibab squirrels, have students prepare a short summary of the discussion to put in their portfolios. Have students describe the environment that each squirrel lives in, the characteristics of each species, and some possible hypotheses for how the differences evolved. L1 LS P

Reinforcement

Reinforce the concept of the niche by asking students to write brief autobiographies in which they describe where they live, something about their school lives, and information about their activities and hobbies. SAE

Concept Development

Use the student autobiographies to develop the concept of the niche. Discuss some of the autobiographies and point out that, just as in the natural world, no two niches are alike. Stress that no two species can occupy the same place and function in a biological system and be successful. Bring up the ideas of competition and struggle for existence, and tie the concept of the niche to adaptive radiation.

GLENCOE TECHNOLOGY

Videodisc

The Infinite Voyage: The Great Dinosaur Hunt

Theories of Extinction (Ch. 11)

Patterns of Evolution

You've learned how evolution by natural selection probably results in the formation of new species. In the natural world, scientists have identified several patterns of evolution occurring throughout the world and in vastly different environments. These patterns add further evidence that natural selection is indeed the agent for evolutionary change.

Species diversify when introduced to new environments

The Hawaiian Islands are home to an extraordinary diversity of plants and animals not seen elsewhere. Among them are a family of birds commonly called Hawaiian honeycreepers, and they represent a case study in evolution. What makes this group of birds interesting is that, while similar in body size and shape, they differ sharply in plumage color, and most importantly, beak shape. Each species of honeycreeper occupies its own niche and possesses a beak adapted to the type of food it eats.

Despite the differences in beak design, however, scientists hypothesize that all honeycreepers evolved from a single ancestral species that invaded Hawaii long ago. The process of evolution of an ancestral species into an array of species that occupy different niches, ***Figure 18.20***, is called **adaptive radiation.**

Adaptive radiation is not limited to the Hawaiian Islands; examples can be seen throughout the world. The many species of finches that Darwin discovered on the Galapagos Islands during the voyage of the *Beagle* are a good example of adaptive radiation,

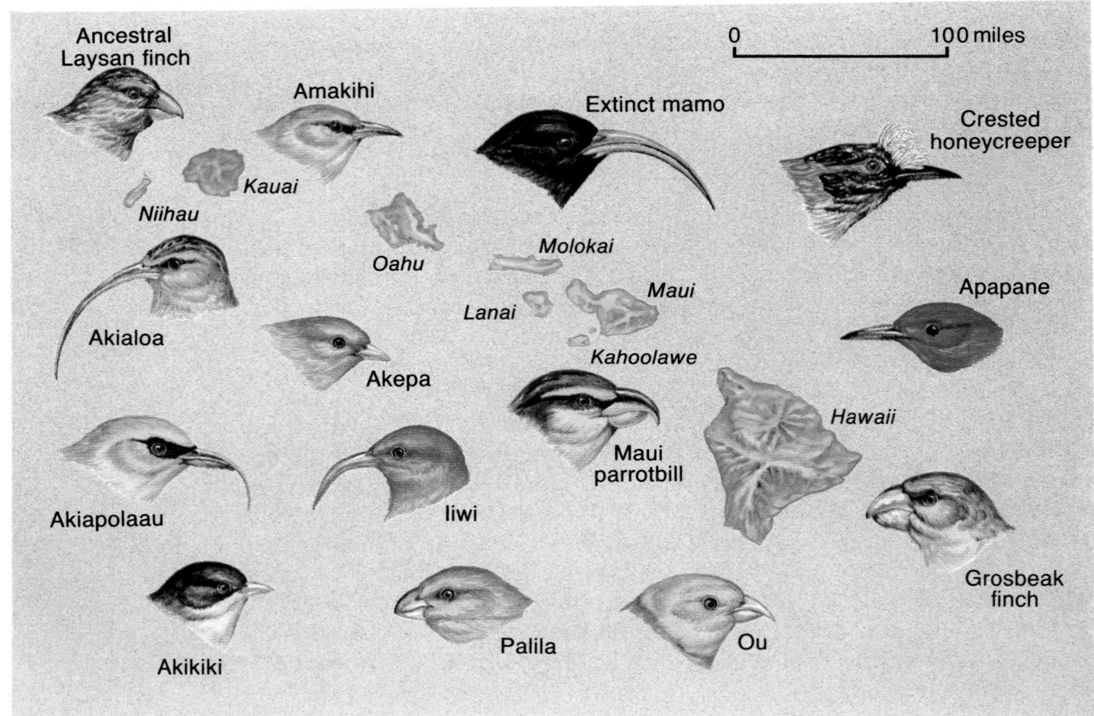

Figure 18.20

Scientists think that the ancestor of all honeycreepers somehow made it across the 2000 miles of Pacific Ocean from the American mainland about 5 million years ago. When this ancestral bird population encountered the diverse niches of Hawaii, speciation occurred, with each species adapting to a different food source.

PORTFOLIO

Adaptive Radiation Have students write answers to the following for their portfolios:
(1) Describe some possible adaptive changes that would occur in a population of gray squirrels if they were suddenly introduced to an island containing the following environments: swamps, deserts, tropical rain forest, and snow-covered mountains.
(2) How do examples of adaptive radiation support the concept of evolution by natural selection? L2 LS P

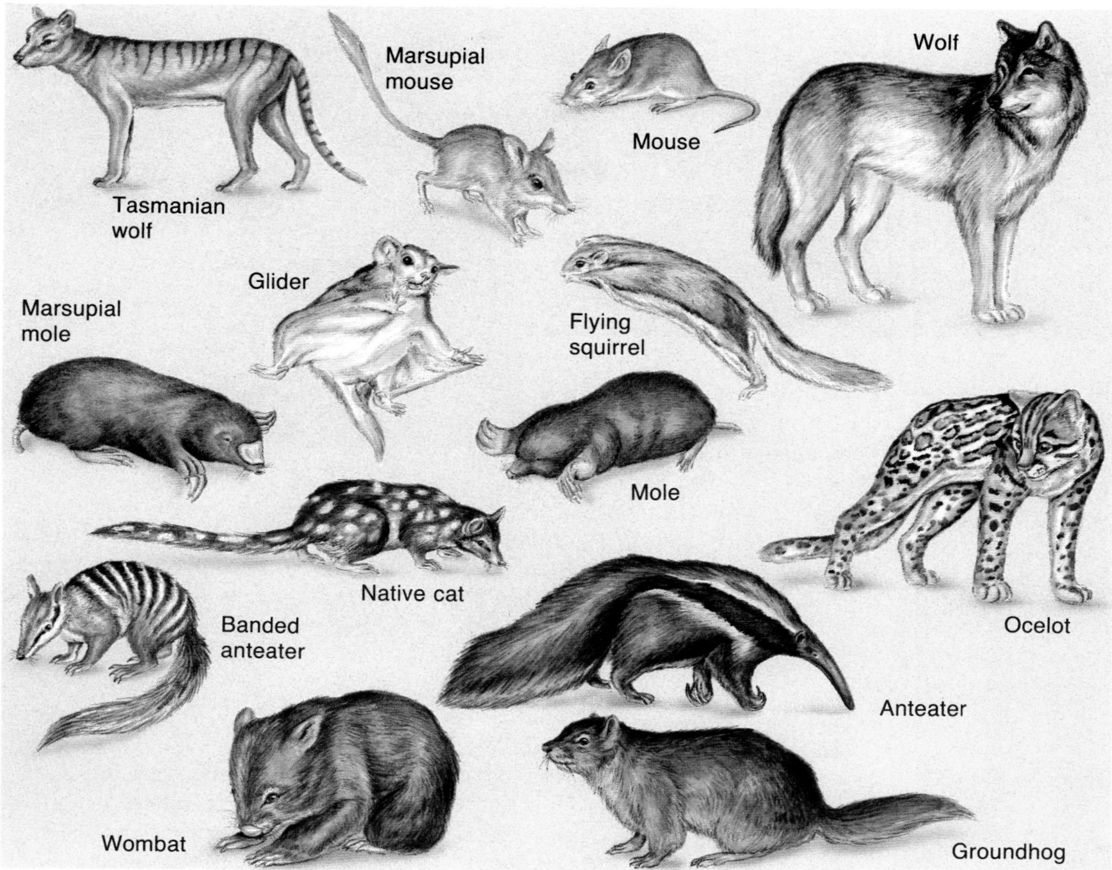

Figure 18.21

Although marsupials and placental mammals have evolved independently of one another since at least the late Cretaceous period, 100 million years ago, they are similar in habitat requirements and body form. Notice the similarities between marsupial species on the left side of the diagram, and their placental counterparts on the right.

as are the more than 300 varieties of cichlid fish in Lake Victoria, Africa.

Adaptive radiation is one example of **divergent evolution,** the pattern of evolution in which species that once were all similar to the ancestral species become more and more distinct. Divergent evolution occurs when species begin to adapt to different environmental conditions and to change, becoming less and less alike, according to the pressures of natural selection.

Distantly related species can evolve similar features

Have you ever wondered why dolphins and fishes have similar body shapes? Dolphins and fishes, although both vertebrates, are not closely related animals. The structural design of dolphins and fishes is an example of **convergent evolution,** the pattern of evolution in which distantly related organisms evolve similar traits, *Figure 18.21.*

SECTION 18.2

Enrichment

Interesting examples of adaptive radiation are the prosimians of the island of Madagascar, which supports a variety of environments such as tropical rain forests, dry woodlands, deserts, and swamps. Scientists speculate that when ancestral primates reached Madagascar they adaptively radiated into niches in the various environments.

Have students create a fictional animal with adaptations for moving, feeding, reproducing, defending against predators, and other adaptations, if they wish. Have them describe what may happen to their fictional animal upon entering Madagascar. Would there be an adaptive radiation of the species? L2 LS

Display

LS **Visual-Spatial** Display, if possible, examples of convergent structures in organisms that students can analyze during their study of this section. Obtain models or actual specimens of bird, bat, and insect wings, models or illustrations of fish and dolphin body shapes, or other examples of convergence.

3 Assess

Check for Understanding

Form five questions related to Types of Natural Selection and Patterns of Evolution. Have students work in groups to answer the questions. L1 COOP LEARN

Reteach

Have students construct a concept map to demonstrate how natural selection acts on the variation of a trait. L1

Extension

Have students write a summary of how Darwin's finches illustrate adaptive radiation. L1

PROJECT

Have students prepare a project about the evolution of flight in insects, bats, or birds. Encourage them to prepare models, posters, or videos for presentation. All projects should include: current hypotheses for the evolution of flight in the organism, fossil evidence, experimental evidence, and alternative hypotheses about how the flight mechanisms may have evolved. L2 LS

Assessment

Skill: Have students analyze the following data on rabbit population allele frequencies.
1st generation: $A = 0.5$, $a = 0.5$;
2nd generation: $A = 0.6$, $a = 0.4$;
3rd generation: $A = 0.7$, $a = 0.3$;
4th generation: $A = 0.8$, $a = 0.2$;
5th generation: $A = 0.9$, $a = 0.1$

Have students describe what is happening in the population if A represents an allele for white fur, and a represents an allele for brown fur. L1

Visual Learning

Figure 18.22 Which of these two species needs to surface to breathe air? *the dolphin*

4 Close

Discussion

Discuss with students some of the cultural adaptations of humans that "shield" us from the effects of natural selection, such as clothing, medicine, automobiles, etc.

Figure 18.22
Many species of fishes (left) and dolphins (right) share a similar streamlined body shape. Which of these two species needs to surface to breathe air?

Convergent evolution occurs when unrelated organisms occupy similar environments and face similar selection pressures. The streamlined shape of fishes and dolphins, *Figure 18.22*, is an adaptive response related to the need for moving efficiently through water. Because convergent evolution has been shown to have occurred in organisms that evolved from entirely different groups of ancestors, it is further evidence for natural selection.

Connecting Ideas

Darwin's theory of how evolution occurs explains how organisms have adapted to the land, sea, air, and every imaginable niche in between. Twelve years after *On the Origin of Species by Natural Selection,* Darwin published a book that deals with the evolution of a group of organisms that have adapted most successfully to life on this planet. This book is titled *The Descent of Man,* and it was one of the first books to discuss the evolution of the order of organisms you belong to—the Primates.

Section Review

Understanding Concepts

1. Why is the evolution of resistance to antibiotics in bacteria considered an example of directional selection?
2. How does geographic isolation result in changes to a population's gene pool?
3. Why is rapid evolutionary change more likely to occur in small populations?

Thinking Critically

4. What environmental factors may prevent some plant populations from speciating through geographic isolation mechanisms?

Skill Review

5. **Designing an Experiment** Two squirrels that live on opposite sides of the Grand Canyon are hypothesized to have evolved from a recent, common ancestor. What observations or experiments might you perform to test this hypothesis? For more help, refer to Practicing Scientific Methods in the *Skill Handbook*.

Answers to Section Review

1. Only bacteria that are totally resistant to antibiotics survive.
2. Geographic isolation separates populations into smaller groups. Eventually, each population adapts to the local environment.
3. It occurs because of genetic drift and limitations on the number of different mates.

Thinking Critically

4. Wind or organisms can spread pollen across large areas. Individual plants will interbreed and no new species will form.

Skill Review

5. Analyze the DNA, the structure, and behavior of each species, or examine possible fossils of the species.

REVIEWING MAIN IDEAS

18.1 Natural Selection and the Evidence for Evolution

- After many years of experimentation and observation of the natural world, Charles Darwin proposed the idea that species evolve through the process of natural selection—an idea that has become accepted among most scientists.
- Natural selection leads to the evolution of adaptations by favoring variations that are best suited for the environment. Individuals with favorable variations are likely to survive, reproduce, and pass these variations to future generations.
- Evolution has been observed in the lab, but much of the evidence for evolution has been gathered from studies of the fossil record, and through anatomical and genetic comparisons of organisms.

18.2 Mechanisms of Evolution

- Evolution occurs when there are changes in the genetic equilibrium of populations. Mutation, genetic drift, and migration of individuals cause minor disruptions of genetic equilibrium, but natural selection has the greatest effect.
- The separation of populations by physical barriers can prevent interbreeding and lead to speciation.
- Many patterns of evolution can be observed in the natural world. Because these patterns are observed in many different environments, they are strong evidence for evolution.

Key Terms

Write a sentence that shows your understanding of each of the following terms.

adaptive radiation
allelic frequency
analogous structure
artificial selection
camouflage
convergent evolution
directional selection
disruptive selection
divergent evolution
gene pool
genetic drift
genetic equilibrium
geographic isolation
gradualism
homologous structure
mimicry
natural selection
polyploid
punctuated equilibrium
reproductive isolation
speciation
stabilizing selection
vestigial structure

Understanding Concepts

1. Which type of evidence do fossils provide for evolution?
   a. structural c. physiological
   b. functional d. tenuous
2. Of the following, which place did Darwin NOT visit during his voyage on *H.M.S. Beagle?*
   a. China c. Mauritius
   b. Tahiti d. Australia
3. A structural adaptation that enables an organism to blend in with its surroundings is ___________.
   a. symbiosis c. mutualism
   b. camouflage d. mimicry
4. In ___________ selection, a breeder selects individuals with desirable variations and mates them.
   a. natural c. population
   b. artificial d. competitive
5. ___________ is a structural adaptation in which one organism copies the appearance of another for protection.
   a. Symbiosis c. Mutualism
   b. Camouflage d. Mimicry
6. ___________ selection is natural selection that favors average individuals in a population.
   a. Directional c. Stabilizing
   b. Disruptive d. Divergent

Chapter 18 Review

Reviewing Main Ideas

Summary statements can be used by students to review the major concepts of the chapter.

Key Terms

Answers should go beyond defining the terms. Accept any answer that uses the term correctly and in the proper context.

Understanding Concepts

1. a
2. a
3. b
4. b
5. d
6. c

Chapter 18 Review

7. b
8. b
9. c
10. b
11. b
12. d
13. d
14. d
15. c
16. a
17. b
18. d
19. b
20. c

Applying Concepts

21. In the population, there is a selective advantage for being a large male. Sexual selection is operating.

Chapter 18 Review

7. Darwin proposed the mechanism of ____________ to explain how populations of organisms evolve.
   a. artificial selection
   b. natural selection
   c. selective breeding
   d. mimicry
8. ____________ structures are similar in function but different in structure.
   a. Vestigial
   b. Analogous
   c. Embryological
   d. Homologous
9. An example of a vestigial human structure is the ____________.
   a. eye
   b. big toe
   c. appendix
   d. canine teeth
10. Bat wings and crocodile forelimbs are examples of ____________ structures.
    a. analogous
    b. homologous
    c. vestigial
    d. embryological
11. The ____________ record shows how species change over time.
    a. genetic drift
    b. fossil
    c. DNA sequencing
    d. selection pressure
12. Which of the following is NOT an example of indirect evidence for evolution?
    a. fossils
    b. embryology
    c. vestigial structures
    d. bacterial resistance to penicillin
13. The entire collection of genes among a population is its ____________.
    a. genetic drift
    b. genetic equilibrium
    c. allelic frequency
    d. gene pool
14. Which of the following is true of evolution?
    a. Individuals evolve more slowly than populations.
    b. Individuals evolve; populations don't.
    c. Individuals evolve by changing the gene pool.
    d. Populations evolve; individuals don't.
15. ____________ is the percentage of a particular allele in the gene pool.
    a. Genetic drift
    b. Genetic equilibrium
    c. Allelic frequency
    d. Speciation
16. ____________ isolation leads to speciation due to physical barriers.
    a. Geographic
    b. Polyploid
    c. Disruption
    d. Convergent
17. Speciation that occurs when chromosome numbers change is ____________.
    a. gradualism
    b. polyploidy
    c. punctuated equilibrium
    d. reproductive isolation
18. In ____________, an ancestral species evolves into an array of species that occupy different niches.
    a. convergent evolution
    b. disruptive evolution
    c. divergent evolution
    d. adaptive radiation
19. The streamlined bodies of fishes and whales are examples of ____________.
    a. adaptive radiation
    b. convergent evolution
    c. divergent evolution
    d. camouflage
20. Which kinds of evidence of evolution are considered by most scientists to be the most reliable?
    a. fossils
    b. embryology
    c. DNA sequence studies
    d. homologous structures

Applying Concepts

21. Males in a population of organisms are often larger than females. Explain how natural selection may be involved in this difference.

Program Resources

Chapter Assessment, pp. 103-108 L1
Alternate Assessment in the Science Classroom
Computer Test Bank L1
Content Mastery, pp. 69-72 L1

22. Many species, such as sharks, have changed little over the course of evolution. What evolutionary factors may be responsible for keeping sharks relatively unchanged over the generations?
23. Many poisonous plants and animals have striking color patterns. How and why might these bright colors have evolved?
24. Why is DNA ideal for determining the relatedness of organisms?
25. How is the evolution of resistance of antibiotics in bacteria similar to homeostatic mechanisms in more complex organisms?

Thinking Critically

26. ***Concept Mapping*** Make a concept map that relates the following terms. Supply the appropriate linking words for your map.
    speciation, reproductive isolation, geographic isolation, natural selection, mutation, genetic drift, migration, gene pool
27. ***Comparing and Contrasting*** Compare and contrast the wings of bats and butterflies as examples of convergent evolution.
28. ***Formulating Hypotheses*** What might happen if white insects that are easily spotted by birds evolved defense mechanisms, such as stingers, that may harm a predator?
29. ***Formulating Hypotheses*** How would a low offspring survival rate affect allele frequencies in a population of gray and white rabbits in which white rabbits were also heavily preyed upon by eagles?
30. ***Observing and Inferring*** Why is adaptive radiation considered a form of divergent evolution?
31. ***Interpreting Data*** In a population of clams, shell color is represented by two alleles. The population consists of ten *TT* clams and ten *tt* clams. What are the allelic frequencies?

ASSESSING KNOWLEDGE & SKILLS

The following graph shows leaf length in a population of maple trees.

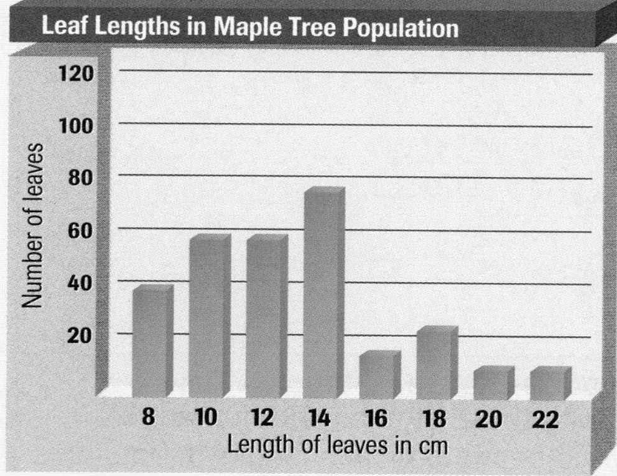

Interpreting Data Use the graph to answer the questions that follow.

1. What was the range of leaf lengths?
   a. 14 cm
   b. 8-22 cm
   c. 20-100 cm
   d. 10-14 cm
2. What was the average leaf length?
   a. 8 cm
   b. 12 cm
   c. 14 cm
   d. 6 cm
3. Which type of evolutionary pattern does the graph most closely match?
   a. artificial selection
   b. stabilizing selection
   c. disruptive evolution
   d. directional evolution
4. ***Interpreting Data*** Use the graph below to explain what might be occurring in this shark population.

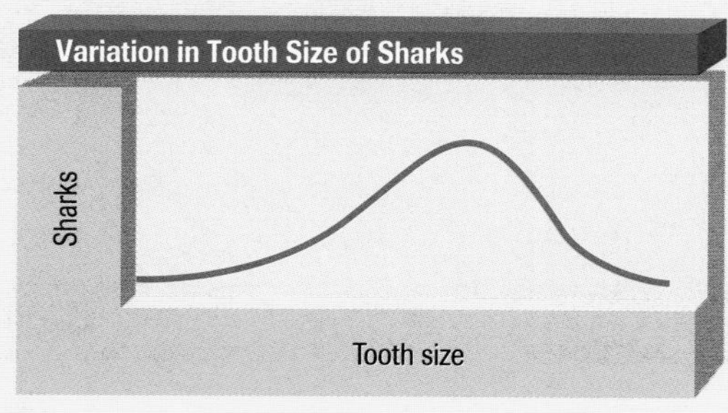

Chapter 18 Review

22. Shark populations may be close to genetic equilibrium. They have stable relationships with the environment and other organisms.
23. Many adaptations are related to escaping predators. Poisons are a natural defense. If a predator eats a brightly colored insect and becomes ill, it will avoid such an organism the next time. Bright colors advertise that the organism may be poisonous, which may deter predators.
24. It is easier to quantify differences in DNA than differences in behavior or morphology.
25. Both involve making changes in response to stimuli.

Thinking Critically

26. Evaluate students' concept maps. The maps should show an understanding of relationships among the concepts listed.
27. Bats and insects are not closely related, but they both fly by using wings. Insect wings are made of chitin and bat wings are made of bones.
28. The evolution of defense mechanisms against predators might restore genetic equilibrium to the population.
29. A low survival rate may affect the population through genetic drift, and through the diminishing numbers of alleles for the white trait in the gene pool.
30. In adaptive radiation, a generalized ancestor encounters an area of many available niches and eventually diverges into many species. This is an example of divergent evolution, in which species similar to ancestral species adapt to different environmental conditions.
31. $T = 0.5$; $t = 0.5$

ASSESSING KNOWLEDGE & SKILLS

1. b
2. b
3. d
4. Directional selection is occurring in favor of larger teeth in sharks.

Chapter Organizer

SECTION	OBJECTIVES	ACTIVITIES/FEATURES
19.1 Primate Adaptation and Evolution National Science Standards: UCP.1-5; C.3, C.4, C.6; G.2, G.3	1. **Recognize** the adaptations of primates. 2. **Compare and contrast** the diversity of living primates. 3. **Sequence** the evolutionary history of modern primates.	**Minilab:** How useful are primate adaptations?, p. 457 **Focus On** Primates, p. 460
19.2 Human Ancestry National Science Standards: UCP.2-5; A.1, A.2; C.3; E.1, E.2; G.1-3	4. **Compare and contrast** the adaptations of australophithecines with those of apes and humans. 5. **Summarize** the major anatomical changes in hominids during human evolution.	**A Broader View:** A Family of Anthropologists, p. 464 **Biolab:** Comparing Skulls of Three Primates, p. 466 **Thinking Lab:** How can tooth structure identify types of food eaten by primates?, p. 469 **Biology & Society Technology:** Communication and Computers, p. 470 **Minilab:** How do human proteins compare with those of other primates?, p. 471

ACTIVITY MATERIALS

BIOLAB	MINILABS	ALTERNATE LAB
page 466 metric ruler protractor copy of skull diagrams	**page 457** tape pencils pens keys **page 471** calculator (optional)	**page 472** metric ruler or tape measure

Chapter 19 Primate Evolution

TEACHER CLASSROOM RESOURCES

Reproducible Masters	Transparencies
Section Focus Master 40: Comparing Hands L1 SAE **Reinforcement and Study Guide,** pp. 73-74 L1 **Biolab and Minilab Worksheets,** p. 73 L1 **Concept Mapping:** Characteristics of Primates, p. 19 L1 **Critical Thinking/Problem Solving:** Interpreting the Fossil Record, p. 19 L3	
Section Focus Master 41: Skeletal Clues L1 SAE **Reinforcement and Study Guide,** pp. 75-76 L1 **Tech Prep Applications:** Analyzing Lower Back Disorders, pp. 21-22 L2 **Biolab and Minilab Worksheets,** pp. 74-78 L1 **Laboratory Manual:** Gene Frequencies and Sickle-Cell Anemia, pp. 103-106 L2 **Content Mastery,** pp. 73-76 L1	**Basic Concepts Transparency #23:** Phylogeny of Humans L1 SAE **Reteaching Transparency #19:** Human and Primate Traits L1 SAE

ASSESSMENT MATERIALS

Chapter Assessment, pp. 109-114
Alternate Assessment in the Science Classroom
MindJogger Videoquiz
Computer Test Bank

Spanish Resources SAE
English/Spanish Audiocassettes SAE
Cooperative Learning in the Science Classroom COOP LEARN
Lesson Plans

KEY TO TEACHING STRATEGIES

L1 Level 1 activities should be within the ability range of all students including those with learning difficulties.

L2 Level 2 activities are within the ability range of average to above-average students.

L3 Level 3 activities are designed for the ability range of above-average students.

SAE SAE activities should be within the ability range of Students Acquiring English.

COOP LEARN Cooperative Learning activities are designed for small group work.

P These strategies represent student products that can be placed into a best-work portfolio.

These strategies are useful in a block scheduling format.

GLENCOE TECHNOLOGY

The following multimedia resources are available from Glencoe.

Biology: The Dynamics of Life
CD-ROM SAE
Videodisc Program

The Infinite Voyage Series

Fires of the Mind
Studying the Brain Through Anthropology
Brain Development: The Range of Capability

The Dawn of Humankind
DNA Studies Create Controversy
"Out of Africa" vs. Multi-regional Debate on Origination of Modern Man
Developmemt of Modern Man
Dating Fossils: Effects of Dating Methods and Interbreeding Theories

The Search for Ancient Americans
The Ancient Native Americans

Science and Technology Videodisc Series (STVS)

Chemistry
Carbon-14 Dating

CHAPTER

19 Primate Evolution

Chapter Overview

In this chapter, students will study primate origins and evolution. They are introduced to the Order Primates and discover some of the important characteristics that unite all primates. They learn how the general primate body plan arose in response to selection pressures in different environmental conditions. Next, students investigate the evolutionary history of the major primate groups and the similarities and differences among them.

In the second section, students investigate the possible evolution of humans from primate ancestors. They learn about some of the environmental factors that may have influenced the evolution of hominids and gain a detailed appreciation of our earliest known human ancestors, the australopithecines. Next, students learn about human ancestors in the genus *Homo*. They conclude their study as they explore how some of the behaviors that make humans unique may have developed.

Key Terms

anthropoid
australopithecine
bipedal
Cro-Magnon
hominid
Neanderthal
opposable thumb
prehensile tail
primate

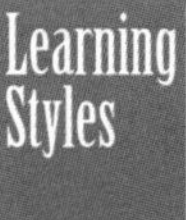

Look for the following logo for strategies that emphasize different learning modalities. LS

Kinesthetic	**Minilab, p. 457; Project, p. 468**
Visual-Spatial	**Visual Learning, pp. 455, 465; Demonstration, p. 459; Portfolio, p. 459; Meeting Individual Needs, p. 459; Time Line, p. 465**
Interpersonal	**Project, pp. 456, 471; Student Journal, p. 457**
Intrapersonal	**Student Journal, p. 455; Meeting Individual Needs, pp. 458, 469**
Linguistic	**Student Journal, p. 463**
Logical-Mathematical	**Portfolio, p. 460; Meeting Individual Needs, p. 461; Thinking Lab, p. 469; Minilab, p. 471; Alternate Lab, p. 472**

CHAPTER

19 Primate Evolution

It's mid-morning on the African savanna, and that means it's feeding time for the chimpanzees of Tanzania! In one tree, a female quietly munches on a piece of fruit, occasionally dropping a piece or two for her offspring on the ground below. In a nearby tree, young adult males gather some choice leaves. An older female snaps a dry twig from a small bush and slowly pokes it into a termite mound. Moments later, the twig, now covered with termites, is withdrawn from the nest. The female chimp is using a tool!

Despite this example, tool use is a rarity in the animal kingdom. The only animals to consistently use tools to manipulate the environment are humans. Humans also play sports, make music, and read books. How did these behaviors develop? In this chapter, you'll gain some insights into this question as you consider evidence discovered in the search for our ancestors.

Concept Check

You may wish to review the following concepts before studying this chapter.

- Chapter 3: scavenger
- Chapter 18: adaptive radiation, speciation

Chapter Preview

Laboratory Activities

Biolab

- Comparing Skulls of Three Primates

Minilabs

- How useful are primate adaptations?
- How do human proteins compare with those of other primates?

Chimpanzees, Galapagos finches, and sea otters use tools, but humans are the only animals to deliberately manufacture tools for a variety of needs. It is known that this behavior has developed during the history of humans. Did the use of tools evolve as a behavioral adaptation?

453

Assessment Planner

Choose assessment strategies from the following pages to evaluate the progress of your students.

Assess, pp. 462, 474
Alternate Lab, pp. 472–473
Minilabs, pp. 457, 471
Portfolio, pp. 459, 460
Thinking Lab, p. 469
Biolab, pp. 466–467
Chapter Review, pp. 475–477

CHAPTER 19

Introducing the Chapter

Have students look at the photos in the chapter opener. Elicit some characteristics that make humans unique among animals. *Responses may include activities such as those named in the text—making music, playing sports, reading and writing books—as well as more basic human behaviors such as making tools and using language.* List student responses on the chalkboard or on an overhead transparency.

Next, discuss tool use in sea otters, chimps, and other animals. Discuss how these examples differ from ways humans make and use tools.

Theme Development

As the chapter title suggests, *evolution* is the main theme of this chapter. Students learn how the availability of human fossils and archaeological evidence affect the development of hypotheses about human evolution. *Unity within diversity* is also a major theme that becomes apparent as the characteristics that unite primate species and make them unique among mammals are described.

Concept Check

In Chapter 18, students learned that adaptive radiation is one of the mechanisms of speciation. Now students will build on this concept as they explore evolution of primates in general and of humans in particular.

SECTION

19.1

Prepare

Key Concepts

Students discover the characteristics shared by all primates and learn how primates are distinguished from other mammals. An overview of primate evolution is presented along with some of the major characteristics of each primate group.

Block Scheduling

Look for this symbol for strategies that are useful in a block scheduling format. For more information on block scheduling, refer to Section 19.1 in the **Lesson Plans** booklet.

Materials

- Obtain pictures or specimens of primate skeletons and teeth for the Visual Learning and for the Demonstrations. Also obtain pictures or models of a variety of skeletons and teeth of nonprimate mammalian groups for the Demonstration.
- Obtain slides or videos that show examples of each primate group for Meeting Individual Needs and for Focus On Primates.
- Gather clear tape or masking tape for the Minilab.

1 Focus

Bellringer

Before presenting the lesson, display **Section Focus Master 40** on the overhead projector and have students answer the accompanying questions. L1 SAE

Discussion

Ask students why primates are such popular attractions at zoos and what makes primates unique among animals. List responses on the chalkboard.

SECTION 19.1

Primate Adaptation and Evolution

Section Preview

Objectives

Recognize the adaptations of primates.

Compare and contrast the diversity of living primates.

Sequence the evolutionary history of modern primates.

Key Terms

primate
opposable thumb
anthropoid
prehensile tail

Monkeys have always fascinated humans, most likely because of the many structural and behavioral similarities we seem to share. But only since 1871, and the publication of Charles Darwin's book The Descent of Man, *have scientists considered the possible evolutionary link among monkeys, apes, and humans. Today, scientists examine living and fossil primates for the information they can provide toward developing a theory about primate evolution.*

What Is a Primate?

If you've ever visited the monkey exhibit at a zoo or watched one of the many television programs about apes and monkeys, you have observed primates. A **primate** is a taxonomic group of mammals that includes lemurs, monkeys, apes, and humans. Primates come in a variety of shapes and sizes. Despite their diversity, all primates share some common traits.

The Inside Story on the next page shows some of the characteristics shared by all primates. Perhaps the most distinctive trait of primates is the shape of the head. Primates have a relatively rounded head and flattened face compared with other mammals. This is partly a result of the size of the brain. Relative to body size, primates have the largest brain of any terrestrial mammal. Only marine mammals are comparably brainy in terms of volume or capacity. Primate brains are also much more complex than those of other animals, and this is reflected in their diverse behaviors and social interactions.

Most primates are arboreal (meaning "tree-dwelling") animals, and they have several adaptations that have evolved for survival in trees. For example, all primates have a highly developed sense of vision. Primate eyes face forward rather than to the side as in some other mammals. This eye positioning enables primates to perceive depth and gauge distances—a type of vision known as stereoscopic vision. As you might imagine, this ability is important for any animal that moves around in a complex environment such as a forest. Primates also have color vision. How might the perception of color have an adaptive value to animals that live in trees?

Program Resources

Section Focus Master 40 L1 SAE
Reinforcement and Study Guide, pp. 73-74 L1
Biolab and Minilab Worksheets, p. 73 L1
Concept Mapping, p. 19 L1
Critical Thinking/Problem Solving, p. 19 L3

THE INSIDE STORY

A Primate

Primates are a diverse group of mammals, but they all share some common features. You can see in this diagram that primates have rounded heads and flattened faces, unlike most other mammals.

Orangutan

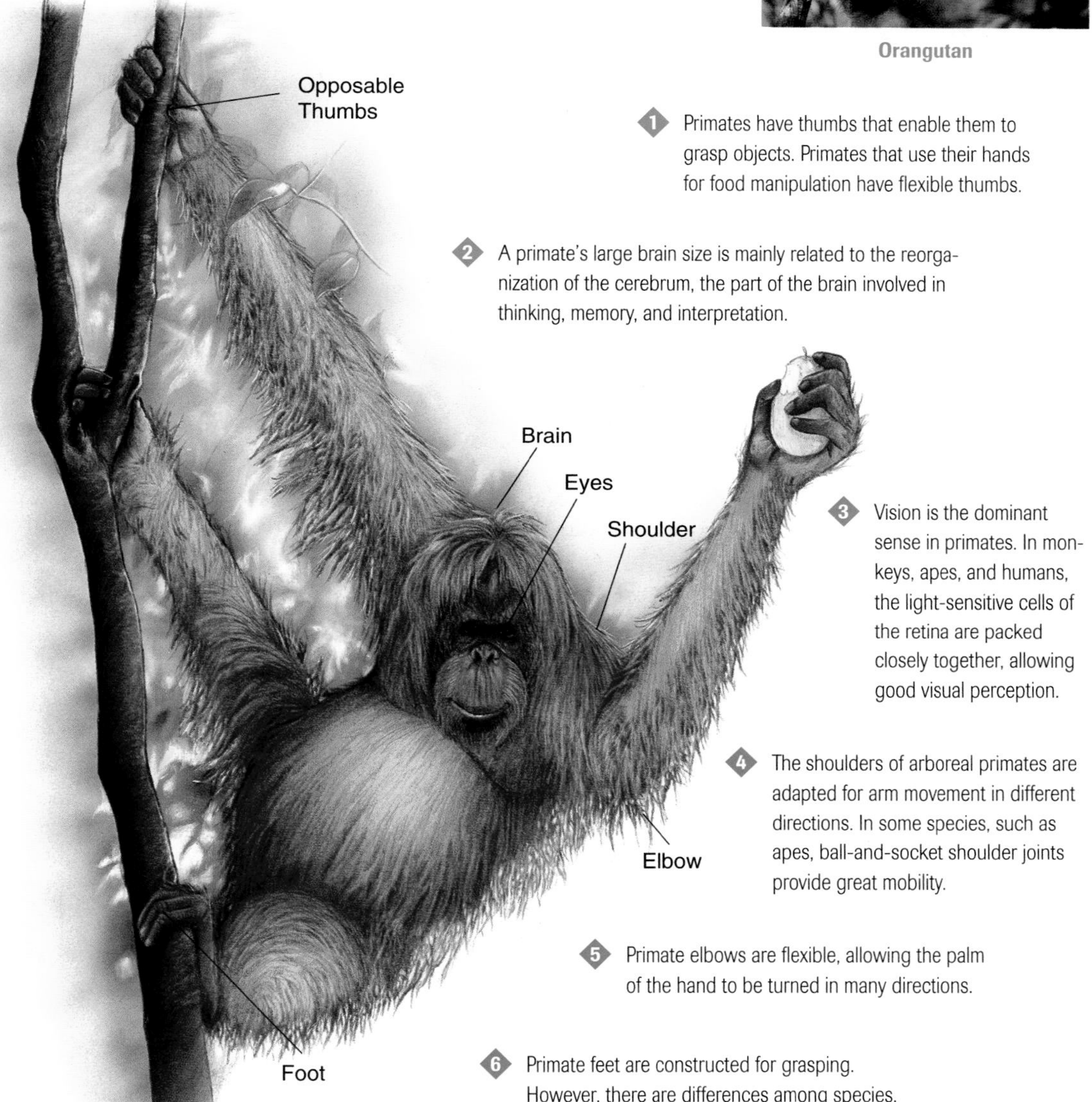

1. Primates have thumbs that enable them to grasp objects. Primates that use their hands for food manipulation have flexible thumbs.

2. A primate's large brain size is mainly related to the reorganization of the cerebrum, the part of the brain involved in thinking, memory, and interpretation.

3. Vision is the dominant sense in primates. In monkeys, apes, and humans, the light-sensitive cells of the retina are packed closely together, allowing good visual perception.

4. The shoulders of arboreal primates are adapted for arm movement in different directions. In some species, such as apes, ball-and-socket shoulder joints provide great mobility.

5. Primate elbows are flexible, allowing the palm of the hand to be turned in many directions.

6. Primate feet are constructed for grasping. However, there are differences among species, depending on their type of locomotion.

THE INSIDE STORY

A Primate

Purpose

Students will examine traits common to primates.

Visual Learning

- Remind students that humans belong to the Order Primates and that they share all the characteristics shown here. Stress that although humans are no longer arboreal, we still retain the generalized primate adaptations. Go through each major characteristic individually, and have students explain how each characteristic is an important human adaptation.
- Have students compare the teeth and skeletons of a variety of primates, including humans. Display pictures or, if possible, actual specimens of primate and nonprimate mammal skeletons to demonstrate the dental and skeletal differences among the different species.
- On the human skeleton, show students how the hands, feet, and joint structures are similar to those of other primates. LS

2 Teach

Software

Mystery Fossil, Mayfield Publishing Company

STUDENT JOURNAL

Observing a Primate Have students look up the scientific name for a primate they have seen and then write answers to the following questions: Where did you first see the primate and what was it doing? What interests or amuses you about the primate? What structures or behaviors does the primate display that you would like to know more about? L1 SAE LS

GLENCOE TECHNOLOGY

Videodisc

Biology: The Dynamics of Life
Disc 1, Side 2
Primate Characteristics (Ch. 7)

SECTION 19.1

Concept Development

Primate hands are divided into three regions: the carpus, or wrist, the metacarpus, and the phalanges. Point out each of these regions to students. As you point out each region, describe its anatomy. Explain that the wrist consists of eight or nine separate bones aligned in two rows.

Between the two rows of bones in the wrist is the mid-carpal joint, a joint that gives primate hands considerable mobility in flexion, extension, and rotation. The joints at the base of the metacarpals offer little mobility. However, the joint at the base of the thumb is extremely mobile in primates.

Remind students that, although primate hands usually have the same numbers of bones, the relative sizes of the hand elements vary with particular needs for locomotion or manipulation. The slow-climbing loris, for example, has a strong thumb and long, lateral digits for grasping branches. In contrast, gibbons and spider monkeys have long, slender fingers that enable them to hang under branches.

Tying to Previous Knowledge

Remind students of the methods used to date fossils discussed in Chapter 17.

Figure 19.1

Generally, primate hands and feet are adapted for locomotion in trees. Compare the thumbs of humans to those of other primates. Can you see that the human thumb diverges more than the thumbs of other primates?

Ring-tailed lemur

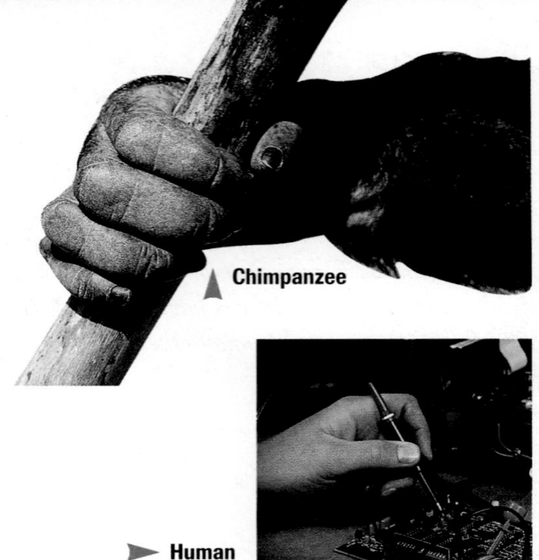

Chimpanzee

Human

The primate skeleton is well adapted for movement among trees. For example, all primates have relatively flexible shoulder and hip joints, features that are important for locomotion such as climbing, clinging, and swinging from branch to branch, as often seen in primates.

Primate hands and feet are also unique among mammals, *Figure 19.1.* Primates' hands and feet are equipped with nails, rather than the claws as seen in other mammals, and their joints are mobile. Also, most primates have an **opposable thumb;** that is, a thumb that can be brought opposite the forefinger, allowing them to grasp and cling tightly to the branches of trees. Opposable thumbs allow primates to make fine manipulations with their hands, which probably led to the ability to use tools.

In humans, the thumb is proportionately larger than in other primates, and the tip of this opposable thumb can cross over the palm to touch the other fingers. To illustrate how important this is, tape your thumb to your hand so it points in the same direction as your fingers. Now try to hold a key as if you were going to open a door. Wouldn't it be difficult to carry out your normal activities without an opposable thumb?

Primate Origins

What are the different types of primates, and how did they evolve? Primates are fairly well represented in the fossil record, and the evolutionary history of primates has been reconstructed through fossils and comparative studies of modern primates.

The earliest primates were small, squirrel-like mammals

From fossil evidence, scientists have concluded that primates evolved about 65 million years ago, but they have not yet determined which group of mammals may have preceded the primates.

The earliest identified primate from the fossil record is *Purgatorius,* a primate that lived about 60 to 65 million years ago. *Figure 19.2* illustrates how *Purgatorius* and its close relatives may have lived.

PROJECT: Interpreting Behavior in Fossil Primates

Paleoanthropologists use a variety of experimental and observational methods to interpret behavior in fossil primates. Have groups of students research and report on one of the following scientific methods: (1) Analyzing dental microwear to investigate diets of extinct primates; (2) Using cladistics to determine phylogeny; (3) Experimental analyses of muscle function in living primates.

Student reports should include one or two examples of how this research answered a question in primate evolution. L3 COOP LEARN LS

Figure 19.2

Primates such as *Purgatorius* probably evolved in response to selection pressures for life in the trees. According to one hypothesis, primates evolved keen vision and grasping hands as an adaptation for preying on tree-living insects.

MiniLab

How useful are primate adaptations?

Have you ever watched a gymnast on the rings or parallel bars? Two keys to the survival of primates in trees are opposable thumbs and stereoscopic vision. Humans don't live in trees, but these adaptations are useful to us in a variety of other ways. In this activity, you will explore the importance of these two key adaptations.

Procedure

1. Make a dot on a piece of paper. Put the paper on your desk, and move back about 0.5 m.
2. Close one eye, and quickly try to touch the point of a pencil on the dot.
3. Close the other eye and repeat step 2, and then try it with both eyes open.
4. With tape, loosely wrap your dominant hand so that your thumb points in the same direction as your fingers.
5. Try to perform a variety of activities, such as writing with a pen, holding a key, or opening a door.

Analysis

1. How is stereoscopic vision useful to humans?
2. How would the absence of an opposable thumb affect the ability of the hand to perform different tasks?

Today, there are no living species like *Purgatorius;* however, living prosimians may come close. Primates are generally divided into two subgroups: the prosimians and the anthropoids. Prosimians are a group of small-bodied primates that include the lemurs, aye-ayes, and tarsiers, *Figure 19.3.* They can be found in the tropical forests of Africa and Southeast Asia, where they leap and run through the canopy in search of insects, seeds, and small fruits. Most prosimians are nocturnal and have large eyes—important features for their nightly insect-catching activities. According to fossil evidence, prosimians evolved about 50 to 55 million years ago from ancestors similar to the earliest primates.

Figure 19.3

Prosimians are widely distributed, but most are found in areas with tropical environments.

Tarsiers, tiny nocturnal animals found in Borneo, can turn their heads nearly 180°, an advantage to an arboreal primate.

The aye-aye, a prosimian found in Madagascar, uses its long middle finger to dig grubs out of deep holes in trees.

MiniLab

Purpose

Kinesthetic Students will explore the utility of opposable thumbs and stereoscopic vision.

Process Skills

observe and infer, hypothesize

Teaching Strategies

- Have students hypothesize how the lack of opposable thumbs and stereoscopic vision might affect a particular activity.
- Make sure students tape their hands properly in step 4. They must be able to bend their hands and should not use the tips of their thumbs.

Expected Results

Students should find it is more difficult to touch the dot with the tip of a pencil with one eye closed than with both eyes open. Students should observe that it is very difficult to carry out normal activities without an opposable thumb.

Analysis

1. Stereoscopic vision is important for any activity that requires depth perception, such as driving or playing sports.
2. The absence of an opposable thumb would make it difficult for the hand to perform different tasks. For example, using keys, opening doors, and writing are examples of tasks that would be difficult without opposable thumbs.

Assessment

Knowledge: Have students write a brief essay about how their lives would be affected if their shoulder and hip joints could move only back and forth, as do those in dogs, cats, and horses. L1

STUDENT JOURNAL

Cartoon Evolution Ask students to gather cartoons that relate to human evolution. Have them work in cooperative groups to develop their own cartoons. L1 LS COOP LEARN

SECTION 19.1

Reinforcement

On an overhead transparency, develop a table that compares and contrasts anthropoids and prosimians. Be sure to include examples of animals in each group, features of the head, and features of the skeleton. Have students copy the table into their notebooks as a reference tool.

Tying to Previous Knowledge

Review the environmental characteristics and organisms present during the early Cenozoic Era, when primates evolved. Remind students that during this time many mammalian groups diversified. Discuss how characteristics of the Cenozoic may have influenced primate evolution.

Science, Technology, and Society

SEM Studies of Fossils

The scanning electron microscope (SEM) plays a major role in primate evolutionary studies. Among its more important uses is in the analysis of teeth to determine evolutionary relationships. By comparing minute differences in tooth structure, phylogenetic relationships among living and fossil primates can be determined.

Another use of the SEM is to analyze the tiny scratches on teeth that provide clues to diet in fossil animals. Scientists analyze the types and numbers of scratches on primate teeth from species with known diets. Then they use this information to infer diets eaten by fossil species.

Humanlike primates evolve

Monkeys, apes, and humans are classified in another subgroup of primates, the **anthropoids**, or humanlike primates. Anthropoids differ from prosimians in many features of the head and skeleton. In particular, anthropoids have more-complex brains than prosimians, which gives them increased intelligence and humanlike qualities.

The three major radiations of anthropoids are the New World monkeys of South and Central America; the Old World monkeys of Africa, Asia, and Europe; and the hominoids, which include Asian and African apes as well as humans, *Figure 19.4.*

New World monkeys are an entirely arboreal group and are diverse in terms of size and ecology. New World monkeys are characterized by a long tail that they use as a fifth limb. This **prehensile tail** is a muscular tail that can grasp or wrap around a branch as the monkey moves from tree to tree. Among the many New World monkeys are the tiny insect-eating marmosets and the larger, fruit-eating spider monkeys.

Old World monkeys are generally larger than New World monkeys. They include the arboreal colobus and guenon monkeys and the terrestrial baboons. Old World monkeys use their tails for balance and are adapted to a variety of environments, from the hot, dry savannahs of Africa to the cold mountain forests of Japan.

Apes include gibbons, orangutans, chimpanzees, and gorillas. Unlike prosimians and monkeys, apes lack tails and have unique adaptations for arboreal locomotion. For example, all apes have extremely flexible hips, shoulders, and elbows, as well as long, heavily muscled forelimbs for climbing and swinging from branches. The African apes, chimpanzees, and gorillas also possess skeletal adaptations for walking on their knuckles.

Brain size and complexity is also increased in apes, as evidenced by their diverse social interactions and increased parental care for offspring.

Similarities among monkeys, apes, and humans suggest that they all share a common ancestor. Fossils indicate that this anthropoid common ancestor evolved from prosimians 37 to 40 million years ago.

Golden lion tamarins are arboreal New World monkeys found in South America. They eat fruit, insects, and other small animals.

Figure 19.4

The three groups of anthropoids include New World monkeys, Old World monkeys, and apes.

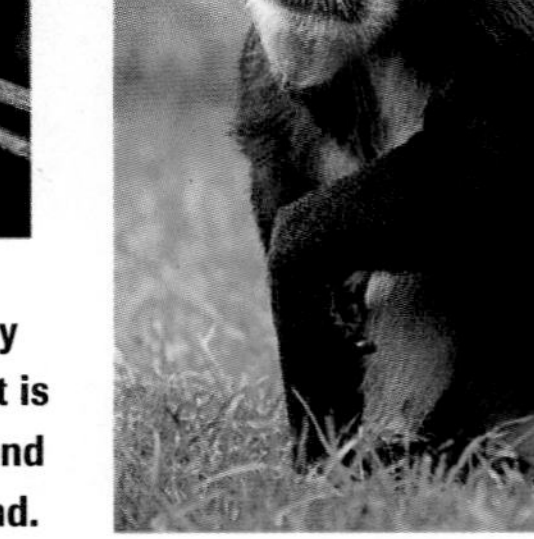

This mandrill is an Old World monkey found in the forests of West Africa. It is omnivorous, active during the day, and spends most of its time on the ground.

Meeting Individual Needs

Gifted While paleontologists know a great deal about primate evolution, many important questions about primate evolution remain unanswered. Have students choose and report on one of the following topics.
(1) What group of mammals may have given rise to the primates?
(2) What is the most likely phylogenetic place of tarsiers in primate evolution?
(3) How did New World monkeys reach South America?
(4) What is the role of *Proconsul* in primate evolution? L3 LS

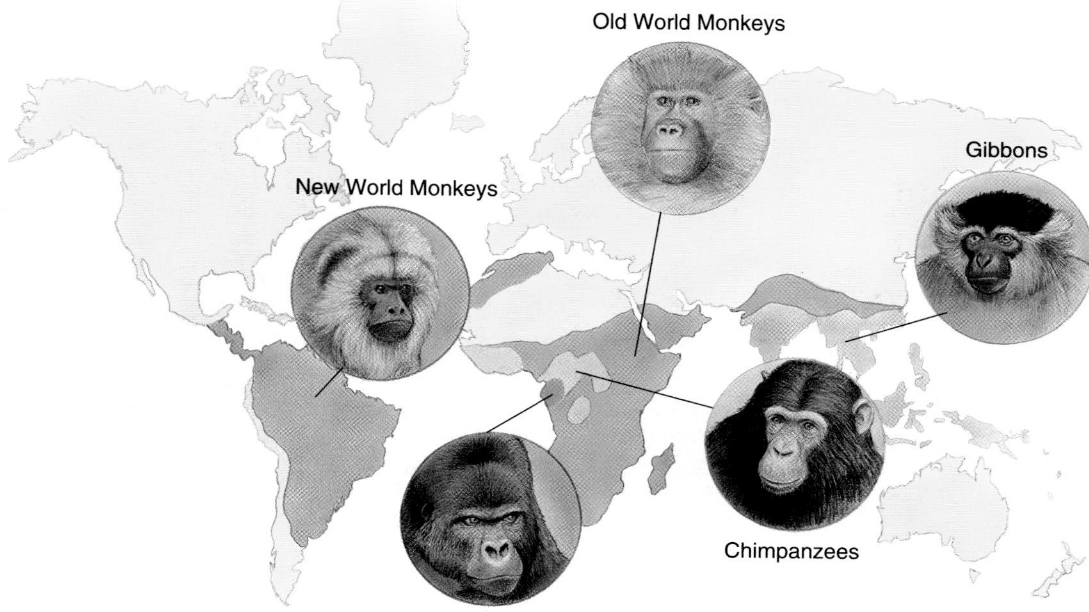

Anthropoids evolved in Africa

New World monkeys were the first modern anthropoids to evolve. According to the fossil record, New World monkeys evolved 30 to 35 million years ago when ancestral anthropoids from Africa successfully crossed the Atlantic Ocean to South America, *Figure 19.5*. Scientists speculate that they did so either by rafting on floating islands or by crossing small island bridges when sea levels were low.

Old World monkeys evolved a little later in Africa, about 20 to 22 million years ago. Fossils indicate that early Old World monkeys were arboreal. Environmental reconstructions of the Miocene period, when Old World monkeys evolved, indicate that global temperatures were dropping. These changes probably led to the shrinking of rain forests, which caused food sources to become scarce and forced some groups to exploit resources on the ground. This may have led to the eventual speciation of baboons and other ground-living apes.

Figure 19.5

Global distribution of anthropoids suggests that they all evolved from a common ancestor. The evolutionary history of primates has been reconstructed based upon the fossil record and comparison studies of modern-day primates.

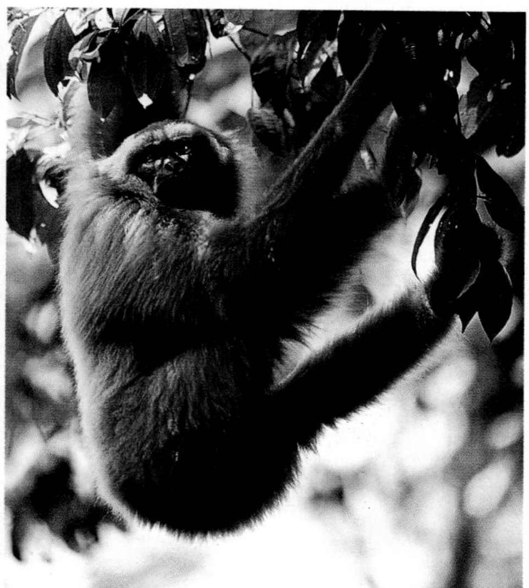

Gibbons are apes found throughout Southeast Asia to southern China and south to Indonesia. Gibbons have extremely long arms and long, curved fingers—well adapted to swinging and feeding among the branches of tall trees.

Concept Development

The division of living primates into the two suborders, Prosimians and Anthropoids, is a gradualistic classification—the two groups are viewed as stages of evolution. This classification does not indicate which group of living prosimians may be closer to the origin of anthropoids, but simply expresses that prosimians are primitive primates that lack anthropoid features.

Explain that many primatologists do not use the gradualistic classification system because of the tarsiers. These primates share a number of derived features with anthropoids, and this evidence suggests that anthropoids evolved from a tarsier-like primate. The new classification reflects a phylogenetic approach in which primates such as lemurs, indris, aye-ayes, lorises, and galagos are placed in the suborder Strepsirhini based on similarities of the nasal region. Tarsiers are grouped along with the other anthropoids in the suborder Haplorhini, based on nasal and other similarities.

Demonstration

LS **Visual-Spatial** Use photos, illustrations, models, or actual skulls to emphasize the differences between prosimians and anthropoids to students. Point out, for example, the fused frontal bone of anthropoids versus the unfused frontal bone of prosimians, as well as other skull differences.

PORTFOLIO

Ape and Baboon Evolution Have students draw the probable sequences of events in ape and baboon evolution. Their illustrations should include the period of time when apes and baboons evolved, as well as the environmental factors that may have influenced their evolution. L2 LS P

Meeting Individual Needs

Learning Disabled Provide students with photographs, illustrations, or videos of each group of primates. Have students write the names of each group in their journals, and then have them identify at least five characteristics from each group. Following this, have students review their lists to compare and contrast the characteristics. L1 LS

Focus On Primates

Purpose
Students will discover characteristics of the major groups of primates.

Background
The various species of living nonhuman primates are perhaps the most extensively studied of all mammals. By investigating the functional relationships between morphological features, such as size, tooth structure, bone shape, and behavioral habits, we can understand why primate species have evolved many of their anatomical differences.

Teaching Strategies

- Prepare video or slide presentations of one or more species of each group to provide more detail about its behavior and ecology.
- Discuss the important characteristics of each group, emphasizing how each adaptation is important for its particular environment.
- Have students use their knowledge of natural selection to hypothesize how the groups may have evolved.
- Discuss the fossil evidence for each group of primates. Show students pictures of the fossils and discuss with them how each supplies evidence in the study of primate evolution.

Focus On

Primates

Lemur, *Propithecus verreauxi*

The family Hominidae contains one species, Homo sapiens. *This species is a large, erect, omnivorous biped that lives on land. In addition, it has a large brain, vocal cords, and good manual dexterity. Biologically, these characteristics, transmitted by our DNA, are what make us human.*

Eleven other families divided into four groups make up the rest of the order Primates.

Prosimians

Small, tree-dwelling primates These primates look the least like primates. They range in size from that of a mouse to that of a large house cat. Their faces are triangular with large eyes. Because they lack the facial muscles that most primates have, they cannot make the facial changes that other primates use to communicate.

The most famous of the prosimians are the lemurs that live in Madagascar. Although lemurs do spend some time on the ground, most of their lives are spent in trees.

Mouse lemur

Spider monkey

Squirrel monkey

New World monkeys

Monkeys with an extra hand As their name implies, New World monkeys are natives of Central and South America. They all live in trees and are unique in that they have prehensile tails with which they can grasp things. These animals do not usually swing by their tails as they travel. Most often, they travel through the forests by leaping from branch to branch, using their tails as extra hands for grasping.

PORTFOLIO

Fossil Problems Have students solve the following problems.
(1) A fossil skeleton of a primate known as *Apidium* is discovered in Egypt. The fossil dates to 31 million years ago and shows similarities to New World monkeys. Could this animal be an ancestor of New World monkeys? *It is unlikely since the fossil is from the same period when the New World monkeys evolved.*
(2) A fragmentary jaw discovered in eastern Africa dates to 8 million years ago. Scientists believe that it is in the right time and place to be from the common ancestor of African apes and humans. What tests might you perform to verify if this is true? *analysis of skeletal structures and brain size* L3 LS P

Old World monkeys

Cold-weather monkeys Old World monkeys are natives of Africa and Asia. They are a varied group, with most species living in trees, some living on the ground, and still others spending time in both environments. They are the only primates, other than humans, that naturally occur outside of the tropics. The Japanese macaque and several other species live in habitats where it snows. However, most species are tropical.

Grey langur

Proboscis monkey

Mountain gorilla

Apes

Primates without tails Apes live in Africa and Southeast Asia, are tailless, and have large brains. There are five genera of apes: chimpanzees, gibbons, gorillas, orangutans, and siamangs. All apes are herbivores, although chimpanzees occasionally have been observed killing other animals for food. Apes are subject to many of the same diseases as humans, can use simple tools, and have been taught to communicate with humans in a sign language that includes a vocabulary of several hundred words.

EXPANDING YOUR VIEW

1. **Thinking Critically** Examine the photos of Old World monkeys. How is having a tail an adaptive advantage for these primates?
2. **Comparing and Contrasting** Which of the species in this feature probably live in trees? Explain your answer.

Focus On

- Discuss some of the methodologies scientists use in the analysis of primate fossils, such as biomechanical studies to infer locomotion in fossil primates and scanning-electron microscopy to infer diet in fossil species.

Answers to Expanding Your View

1. The tails aid in balance.
2. Tree-living monkeys include prosimians, New World monkeys, and some Old World monkeys. Those with long tails probably live in trees.

Going Further

Have students work in groups to prepare a presentation about a particular species or group of primates. Presentations can be models, posters, videos, skits, or other studies and must include information about behavior, anatomy, ecology, taxonomy, evolution, and importance to humans. L2 COOP LEARN

Meeting Individual Needs

Students Acquiring English Have students define the following terms: *mammal, primate, opposable thumb, stereoscopic vision, prosimian, anthropoid, New World monkey, Old World monkey,* and *ape*. Next, have students construct a concept map that reflects the relationships among these terms. Have students circle the word in each of the following groups that does not belong with the others.

(1) New World monkey, opposable thumb, Old World monkey, ape, prosimian *opposable thumb*

(2) opposable thumb, stereoscopic vision, primate, large brain *Primate* L1 SAE LS

3 Assess

Check for Understanding

Have students identify the common characteristics shared by primates. Beside each characteristic should be a description of that trait's adaptive significance. L1

Reteach

Provide students with outline maps of the world. Have students develop a key for the map to show where primates from each group discussed in this section are common. L1

Extension

Have students prepare a phylogenetic tree of primates. Ask them to identify the name of each primate group, when it evolved, and some unique characteristics of each group. L2

Assessment

Skill: Have students create in their journals a diagram that shows the possible evolutionary relationships of the primates discussed in this section. L1

4 Close

Discussion

Review with students the highlights of primate evolution. Have students identify what they think were the major evolutionary developments.

Apes retained adaptations for life in the trees

The evolutionary history of anthropoids is particularly important because of the information it can provide about human origins. Can you make a hypothesis to explain the adaptive radiation of modern apes?

According to fossils found in Africa and elsewhere, apes evolved about the same time as Old World monkeys and faced similar environmental changes, shown in *Figure 19.6*. Scientists hypothesize that instead of moving to the ground like Old World monkeys, apes remained in the trees and retained their adaptations for fruit gathering.

Fossil dating indicates that gibbons were the first apes to evolve, followed by orangutans and then African apes. This progression corresponds to evidence collected from DNA comparisons.

Figure 19.6

Apes and Old World monkeys are adapted to different environments.

Baboons are ground-dwelling Old World monkeys that live on the African savanna.

Orangutans are apes that spend all of their lives in the forests of Borneo and rarely come down to the ground.

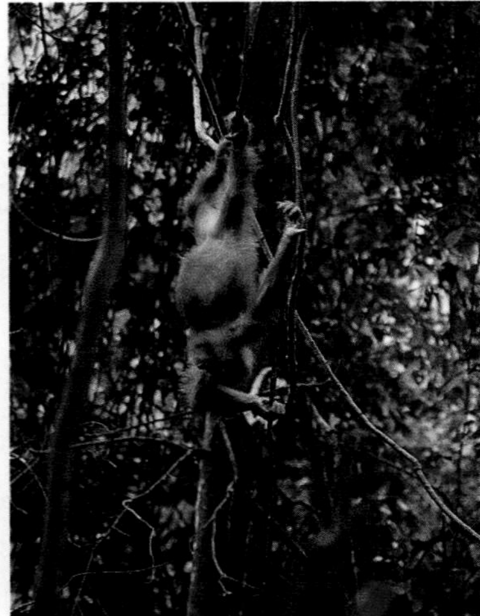

Section Review

Understanding Concepts

1. How do primates show adaptation to life in the trees?
2. What is the evolutionary significance of an opposable thumb?
3. What features distinguish anthropoids and prosimians?

Thinking Critically

4. An unidentified fossil skeleton is brought to your lab. You suspect it's a primate, but you're not sure. What observations might you make in order to identify this fossil as a primate?

Skill Review

5. **Classifying** List the different types of primates, key facts about each group, and how each group is related. For more help, refer to Organizing Information in the *Skill Handbook*.

462 Primate Evolution

Answers to Section Review

1. Primates have opposable thumbs, digits with nails rather than claws, and flexible feet, which allow them to grip branches tightly. Primates have flexible skeletons for moving their limbs in different directions, and stereoscopic vision that enables them to gauge depth and distance accurately.
2. An opposable thumb allows primates to grip branches and perform complex manipulations of objects, such as food.
3. Anthropoids have larger brains, differ in aspects of skull and skeleton structure, and are larger than prosimians.

Thinking Critically

4. Answers should include suggestions to check to see if the fossil shows an oppos-

19.2 Human Ancestry

The topic of human ancestry is fascinating but one that is still speculative and controversial. Many people have viewed early humans as dumb brutes incapable of doing anything correctly. But scientists studying the early history of human evolution have changed these views. Today, reconstructions of the anatomy and behavior of our ancestors are based on careful examination of fossils and detailed comparisons of living apes and humans.

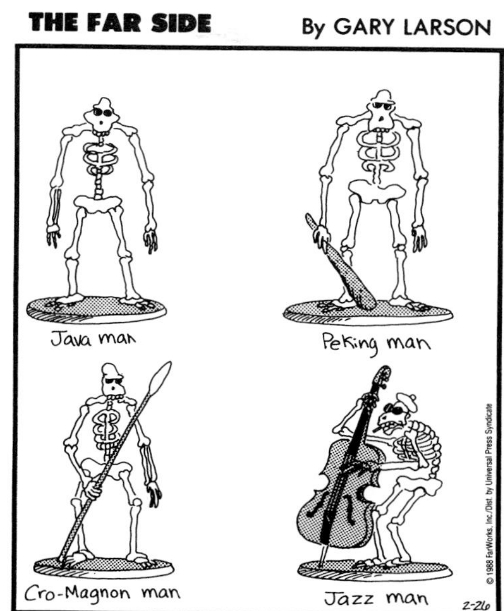

Hominid reconstructions

Section Preview

Objectives

Compare and contrast the adaptations of australopithecines with those of apes and humans.

Summarize the major anatomical changes in hominids during human evolution.

Key Terms

bipedal
hominid
australopithecine
Neanderthal
Cro-Magnon

Hominids

The story of human evolution begins between 5 and 8 million years ago in Africa. Scientists propose that during this time, a population of ancestral apes diverged into two lineages. One lineage would eventually evolve into the African apes—gorillas and chimpanzees. The other would lead to modern humans. What factors may have led to the divergence of African apes and humans?

There are few fossils that date to the time period when African apes and humans diverged. However, scientists propose that the human lineage may have begun in response to environmental changes that forced a population of apes to leave the protection of the trees to search for new food sources on the ground.

Scientists speculate that in order to move efficiently in this new environment, this early population of apes eventually became **bipedal;** that is, they evolved the ability to walk on two legs. Walking on two legs leaves the arms and hands free for other activities such as using tools. These apes were probably the earliest members of the human family of primates known as hominids. A **hominid** is a humanlike, bipedal primate. The fossil record of hominids is fairly complete, and we now have a good picture of the anatomy and behavior of early hominids.

able thumb, a large brain size, nails rather than claws, stereoscopic vision, and flexible joints.

Skill Review

5. Check student lists for accuracy and understanding of the main concepts.

STUDENT JOURNAL

Australopithecines in Primate Evolution
Have students write a short essay about the significance of australopithecines in primate evolution. Have them conclude their essays with unsolved questions about australopithecines that might be answered with the recovery of more fossils. L2 LS

SECTION 19.2

Prepare

Key Concepts

Students will study both the fossil and archaeological evidence of human evolution. They will learn how humans could have evolved from apelike ancestors and how environmental factors may have influenced this evolution. They will investigate the structural and behavioral characteristics of various human ancestors and when and how human cultural adaptations, such as tool use, fire, and language, may have originated.

Block Scheduling

Look for this symbol for strategies that are useful in a block scheduling format. For more information on block scheduling, refer to Section 19.2 in the **Lesson Plans** booklet.

Materials

- Obtain casts and/or pictures of several fossil hominid species for use in the Visual Learning.
- Gather metric rulers and protractors for the Biolab.
- Gather several round stones for the Project.

1 Focus

Bellringer

Before presenting the lesson, display **Section Focus Master 41** on the overhead projector and have students answer the accompanying questions. L1 SAE

2 Teach

A BROADER VIEW

A Family of Anthropologists

Purpose

Students will survey the contributions of the Leakey family in the search for human ancestors.

Teaching Strategies

- Point out that although Louis Leakey began searching for human fossils in the late 1930s and early 1940s, he and Mary Leakey did not actually discover a fossil of a human ancestor until 1959. Discuss how the extreme rarity of finding a fossil hominid makes paleontology difficult.
- Discuss with students how the Leakey family contributed to the growing field of paleoanthropology in other ways. Mary Leakey, for example, did a thorough analysis of the stone tool technologies at Olduvai Gorge. Louis Leakey set up the field studies of chimpanzees and gorillas carried out by Jane Goodall and Dian Fossey, respectively.

Thinking Critically

Responses may include that researchers may discover fossils that place early human origins in China and that these will provide information that can be used to test hypotheses based on current evidence that is incomplete.

A BROADER VIEW

A Family of Anthropologists

History is filled with examples of families that have pursued a common interest with extraordinary results. Science, too, has its famous families. For example, the Leakeys—husband Louis, wife Mary, and son Richard—have discovered a wealth of fossil evidence for the origin and evolution of humans.

The son of missionaries Louis Leakey was born in Kabete, Kenya in 1903. He began working as an archaeologist in eastern Africa. At Olduvai Gorge in Tanzania during the 1930s, Louis Leakey began discovering animal fossils and simple stone tools. Olduvai Gorge was to become the site of many important discoveries by the Leakey family that would change the world's view of human evolution.

From art to archaeology In 1936, Louis Leakey married Mary Douglas Nicol, an artist. In 1947, Mary found a skull of *Proconsul africanus,* structurally similar to both apes and early humans. This suggested to Mary Leakey and the scientific community that this fossil might represent the ancestor of both apes and humans. The skull is estimated to be 25 million years old. In 1978, Mary discovered fossil footprints just south of Olduvai Gorge. The footprints were made by primates that lived 3.5 million years ago and walked upright, suggesting that they were humanlike.

In his parents' footsteps Richard Erskine Frere Leakey is one of three sons born to Louis and Mary. At age 19, Richard found a fossil jaw while exploring northeastern Tanzania. Perhaps Richard's most important contribution to anthropology was his discovery in 1967 along the shore of Lake Turkana of more than 400 hominid fossils.

Thinking Critically

Over the last 70 years, China has been closed to scientific research, including archaeology. Recently, China has allowed research teams to begin digging in promising sites. How could fossils found in China change current ideas about human evolution?

Early hominids possessed ape and human characteristics

In 1924, South African scientist Raymond Dart discovered a fossil skull that had an apelike braincase and face, but it was unlike any primate Dart had ever seen, *Figure 19.7.* One feature that stood out was the position of the *foramen magnum*, the hole in the skull where the spinal column enters. In the fossil, the hole was located on the bottom of the skull, as in humans. This indicated to Dart that the organism must have walked upright. Dart classified this fossil as a new primate species, *Australopithecus africanus*, meaning "southern ape from Africa." Dating methods have shown the skull to be 2 to 3 million years old. Since Dart's discovery, scientists have recovered many more australopithecine specimens. An **australopithecine** is an early hominid that lived in Africa and had apelike and humanlike qualities.

Figure 19.7

The Taung child was the first australopithecine ever discovered. Its skull is structurally similar to both apes and humans. The braincase, face, and teeth are chimplike, but the position of the *foramen magnum* indicates that it walked upright like humans.

Program Resources

Section Focus Master 41 L1 SAE
Reinforcement and Study Guide, pp. 75-76 L1
Biolab and Minilab Worksheets, pp. 74-78 L1
Laboratory Manual, pp. 103-106 L2
Tech Prep Applications, pp. 21-22 L2
Basic Concepts Transparency 23 and **Master** L1 SAE
Reteaching Transparency 19 and **Master** L1 SAE

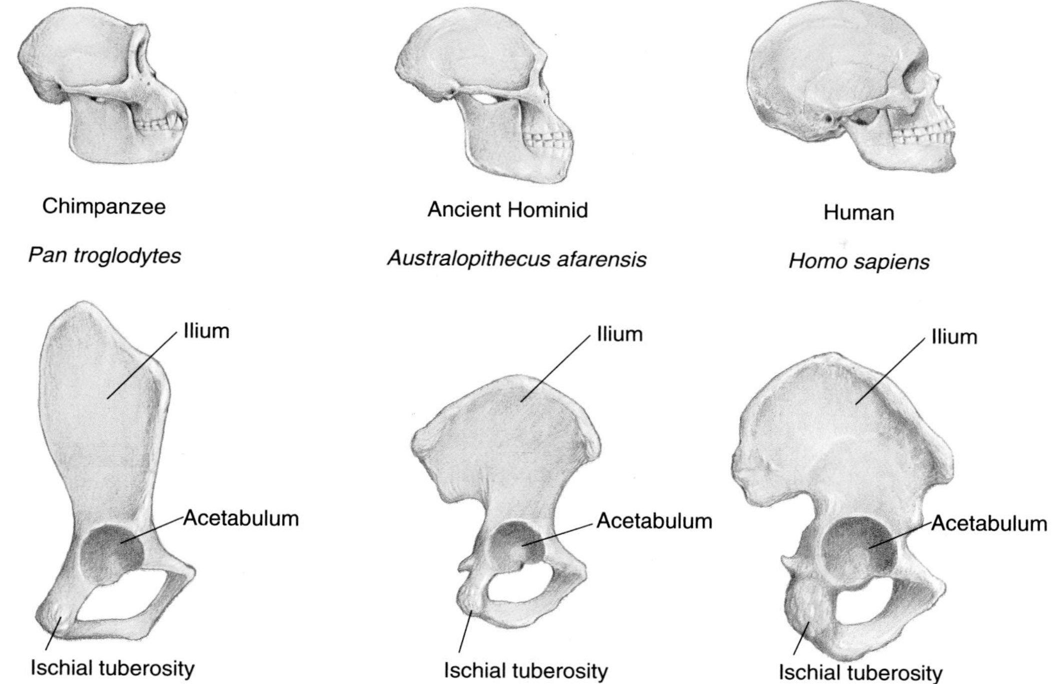

Figure 19.8

The skeleton of an australopithecine is in many ways intermediate between those of living apes and humans. Compare the skull and pelvic bone of *Australopithecus afarensis* (center) with those of the chimpanzee (left) and human (right).

Much of what scientists know about australopithecines comes from the nearly complete "Lucy" skeleton, discovered in East Africa in 1974 by American paleoanthropologist Donald Johanson. This fossil has been dated to be 3.5 million years old, and was nicknamed "Lucy" after a song by *The Beatles*, a popular rock group around the world, that was often played on a tape deck at the campsite.

The Lucy skeleton was classified as *Australopithecus afarensis*, and other fossil discoveries have shown that this species existed between 3 and 5 million years ago. *A. afarensis* is the earliest known hominid species, and, as shown in *Figure 19.8*, the structure of the pelvis, legs, and feet indicates that it was bipedal, like humans. On the other hand, the size of the braincase suggests an apelike brain. That is, walking upright occurred before the evolution of a larger brain. The shoulders and forelimbs are also apelike.

Because of this combination of features, scientists propose that *A. afarensis* and other australopithecines spent a considerable amount of time climbing in trees. When they came to the ground, they walked bipedally, but clumsily.

paleoanthropology
paleo (GK) ancient
anthropo (GK) human
logos (GK) study of
Paleoanthropology is the study of human fossils.

CULTURAL DIVERSITY

Kamoya Kimeu and the Hominid Gang

Finding human fossils is extremely difficult and requires considerable expertise. Emphasize this point to students, and introduce the contributions of Kenyan fossil expert Kamoya Kimeu. Kimeu is the leader of a team of fossil hunters known as the Hominid Gang and, since the 1960s, Kimeu's work has led to the discovery of many important hominid fossils, including "Lucy" and the twelve-year-old *Homo erectus* boy, the "Strapping Youth."

Have students read sections from *Origins Reconsidered* by Richard Leakey and Roger Lewin or other similar books to learn more about Kimeu and the techniques involved in finding human fossils.

Time Line

LS **Visual-Spatial** Encourage students to make a time line that traces the hominid to human evolution as described here. Encourage students to use ranges on their time lines, such as 3-5 million years ago for *A. afarensis.* **L1**

Visual Learning

Figure 19.8 Direct students' attention to the three sets of bones and discuss the observable differences. Elicit from students what characteristics *A. afarensis* shared with humans. *It was bipedal.* What characteristic is apelike? *the jawbone and the size of the face area* **L1** **LS**

Art Connection

Reconstructing Extinct Humans

Show students videos or have them read articles about how sculptors help paleoanthropologists learn about the past. These sculptors apply their knowledge of anatomy and anthropology to reconstruct flesh and bones of extinct humans, such as Neanderthals.

No one can say for certain just how close the reconstructions come, but the detailed work of the sculptors does help scientists make more accurate inferences about the past. Have students examine casts of fossil skulls or photographs and use clay to attempt their own reconstructions.

BioLab

Comparing Skulls of Three Primates

Time Allotment 
One class period

Objectives
Review objectives with students before they begin the Biolab.

Process Skills
observe and infer, measure in SI, hypothesize, compare and contrast, make and use tables

PREPARATION

- Give students copies of the skull diagrams from the Biolab and Minilab Worksheets book to use in this Biolab. These skulls are one half natural size.
- Alternatively, make a copy of the skull art on this page and increase it by 50 percent for students to use. The skulls on the student page are one quarter normal size.

GLENCOE TECHNOLOGY

Videodisc

The Infinite Voyage: Fires of the Mind
Studying the Brain Through Anthropology (Ch. 1)

Brain Development: The Range of Capability (Ch. 2)

BioLab

Comparing Skulls of Three Primates

Australopithecines are the earliest hominids in the fossil record, and in many ways, their anatomy is intermediate between living apes and humans. In this lab, you'll determine the apelike and humanlike characteristics of an australopithecine skull, and compare the skulls of australopithecines, gorillas, and modern humans. The diagrams of skulls shown below are one-fourth natural size. The heavy black lines indicate the angle of the jaw.

PREPARATION

Problem
How do skulls of primates provide evidence for human evolution?

Objectives
In this Biolab, you will:
- **Determine** how paleoanthropologists study early human ancestors.
- **Compare and contrast** the skulls of australopithecines, gorillas, and modern humans.

Materials
metric ruler
protractor
copy of skull diagrams

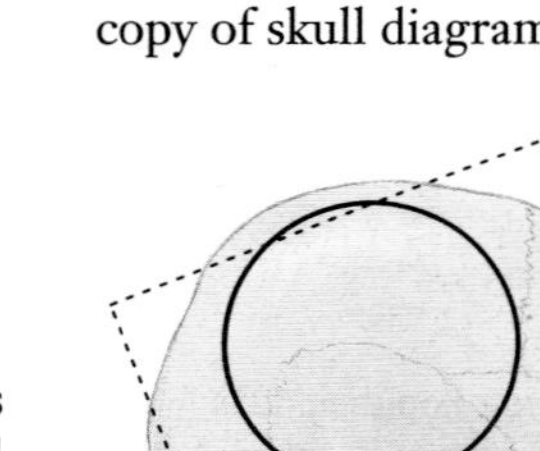

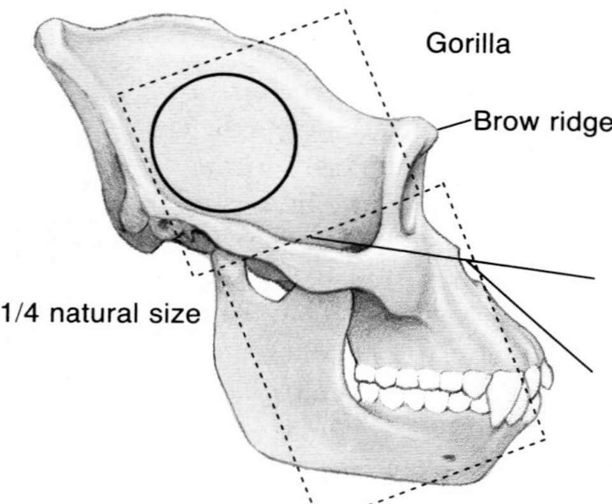

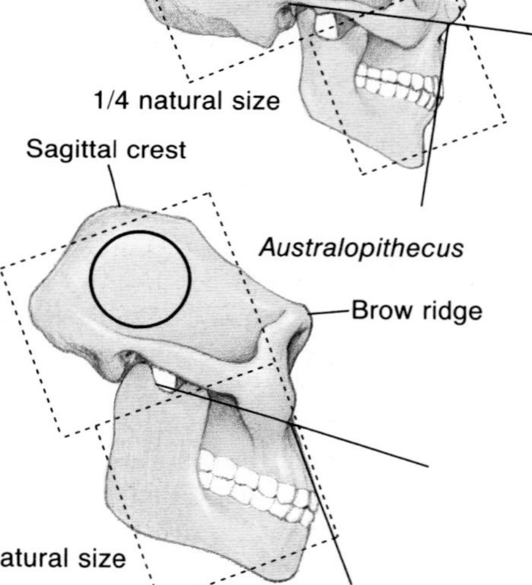

PROCEDURE

Teaching Strategies
- Remind students that australopithecines existed between 1 and 5 million years ago, close to the probable time of divergence of African apes and humans. Because of their great age, australopithecines: can be considered to be very primitive anatomically.
- To estimate cranial capacities, students should first draw circles just inside each skull on the worksheets provided or on the copied and enlarged skulls from the student page.
- Estimating cranial capacity by multiplying by the factor of 200 is suitable only for these drawings. It is a value based on the scale of the drawings at 1/2 natural size.

PROCEDURE

1. Your teacher will provide you with a copy of skull diagrams of *Australopithecus africanus, Gorilla gorilla*, and *Homo sapiens.*
2. The rectangles drawn over the skulls represent the areas of the brain (upper rectangle) and face (lower rectangle). On each skull, determine and record the area of each rectangle (length × width).
3. Measure the diameters of the circles in each skull. Multiply these numbers by 200 cm^2. The result is the cranial capacity (brain volume) in cubic centimeters.
4. The two heavy lines projected on the skulls are used to measure how far forward the jaw protrudes. Use the protractor to measure the outside angle (toward the right) formed by the two lines.
5. Complete the data table.

Comparison of Gorilla, *Australopithecus*, and Modern Human Skulls

	Gorilla	*Australopithecus*	Modern human
1. Face area in cm^2			
2. Brain area in cm^2			
3. Is brain area smaller or larger than face area?			
4. Is brain area 3 times larger than face area?			
5. Cranial capacity in cm^3			
6. Jaw angle			
7. Does lower jaw stick out in front of nose?			
8. Sagittal crest present?			
9. Browridge present?			

ANALYZE AND CONCLUDE

1. **Comparing and Contrasting** Describe the similarities and differences in face-to-brain area in the three primates.
2. **Interpreting Observations** How do the cranial capacities compare among the three skulls? How do the jaw angles compare?
3. **Drawing Conclusions** Based on your findings, what statements can you make about the place of australopithecines in human evolution?

Going Further

Application
Different parts of the australopithecine skeleton are also intermediate between apes and humans. Obtain diagrams of primate skeletons to determine the similarities and differences.

Data and Observations

Comparison of Skulls:

	Gorilla	*Australopithecus*	Modern human
1. Face area in cm^2	32 cm^2	19 cm^2	12 cm^2
2. Brain area in cm^2	23 cm^2	23 cm^2	40 cm^2
3. Is brain area smaller or larger than face area?	Smaller	Larger	Larger
4. Is brain 3 times larger than face area?	No	No	Yes
5. Cranial capacity in cm^3	600 cm^3	600 cm^3	1060 cm^3
6. Jaw angle	35°	55°	90°
7. Does lower jaw stick out in front of nose?	Yes	Yes	No
8. Sagittal crest present?	Yes	Yes	No
9. Browridge present?	Yes	Yes	No

ANALYZE AND CONCLUDE

1. Humans have a small face area compared to brain area. Apes have a large face area compared to brain area. Australopithecines were intermediate between apes and humans but closer to apes.
2. Apes have a small cranial capacity, whereas humans have a large cranial capacity, and that of australopithecines was intermediate but closer to the apes. An ape has a small jaw angle, and a human has a large jaw angle. The jaw angle of australopithecines was intermediate but closer to that of the ape.
3. Many australopithecine skull traits were intermediate between apes and humans, and some were more similar to those of apes. Australopithecines represent very early human ancestors.

Assessment

Portfolio: Have students formulate three hypotheses about what natural selection pressures they think may have been operating during australopithecine evolution. Have them place their hypotheses in their portfolios. L1 P

Going Further

Obtain illustrations or actual fossil casts of australopithecine limb bones, hands, or feet. Have students compare measurements and other observations of the postcranial skeletons of apes, humans, and australopithecines. L2

Using Science Terms

Reinforce the idea that *Homo habilis*, meaning "handy man," was so named because evidence indicates that this hominid used tools. Have students correlate this to the use of the term *handyman* today as it is applied to individuals who work with tools to make household or mechanical repairs.

Misconception

"Why, if humans evolved from apes, are there still apes alive today?" This question represents a common misconception that students have about human evolution. Explain that humans evolved from *ancestors* of apes, which were neither humans nor apes. Reiterate that the common ancestor of the two groups lived in the Miocene period.

In addition to *A. afarensis* and *A. africanus*, scientists recognize either two or three other species of australopithecines that evolved slightly later in time. These species are known from East African and South African fossil sites dated at 1 to 2.5 million years old. Overall, they were similar to the earlier species, but they were more robust and had larger teeth and jaws. They were classified in the genus *Paranthropus.*

The evolutionary relationships of australopithecines are still unclear, but fossil evidence shows that australopithecines died out by 1 million years ago, and early australopithecines were probably significant in the evolution of modern hominids.

The Emergence of Modern Humans

The discovery of australopithecines ended the long-standing debate on whether large brain size or bipedalism evolved first. When did large brains and other human characteristics, such as culture and the use of stone tools, develop?

Members of the genus *Homo* made the first stone tools

In 1964, anthropologists Louis and Mary Leakey discovered a jaw bone of another type of hominid, which created excitement in the scientific community. A skull of the same species was found in 1972 by Richard Leakey.

This skull, shown in *Figure 19.9*, was much more humanlike than australopithecine skulls. In particular, the braincase was much larger in this new skull, and the teeth and jaws were much smaller, like humans. Because of its close similarities to humans, the Leakeys classified this fossil in the genus *Homo*, the same genus in which modern humans are classified. They named the species *Homo habilis*, which means "handy human," because simple stone tools were found near the fossil.

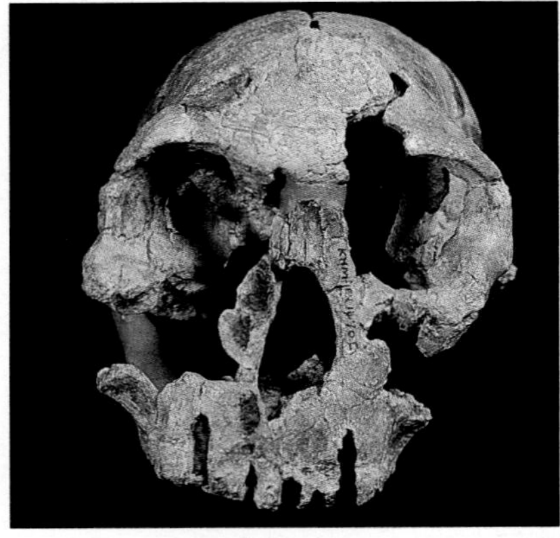

Figure 19.9

The average brain volume of *Homo habilis* has been estimated at 600 to 700 cm^3, much smaller than the 1350 cm^3 average of modern humans, but larger than the 400 to 500 cm^3 of apes and australopithecines.

PROJECT

Using Stone Tools

Provide groups of students with simple stone tools and have them evaluate the tools' effectiveness by using them for a variety of tasks. You can make the tools from quartz or flint stones. Hit one against the other to remove flakes and then use both the flakes and the cores as tools.

Have students design experiments to test the tools in a variety of tasks, such as cutting meat or branches, or breaking open fresh beef bones. Have them summarize their findings in written reports. L2 IS COOP LEARN

Homo habilis existed between 1.5 and 2 million years ago, and is considered to be the first hominid to make and use stone tools. According to scientists, *H. habilis* was probably a part-time scavenger and used stone tools to slice off bits of meat from animal carcasses found in the area.

Hunting and fire developed later

Anthropologists suggest that *Homo habilis* gave rise to another hominid species that appeared about 1.6 million years ago. This new species was *Homo erectus*, which means "upright human." This hominid had a larger brain and a more humanlike face than *H. habilis*, **Figure 19.10**, but with large browridges and a lower jaw that sloped back with no chin.

Figure 19.10

The most complete *Homo erectus* fossil ever found was discovered in East Africa in 1985. It was the skeleton of a 12-year-old boy. *H. erectus* had a brain volume of approximately 900 cm^3 and long legs like modern humans, indicating efficient bipedal locomotion.

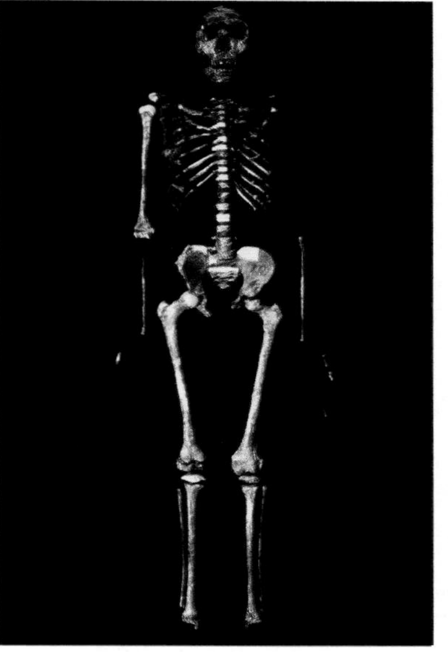

ThinkingLab — Interpret the Data

How can tooth structure identify types of food eaten by primates?

The tooth structure of living primates is often used to infer the dietary behavior of fossil primates. Primates with large molars tend to eat tough foods such as plant stems, seeds, and hard fruits.

Analysis

Examine the following data from a study comparing the molar teeth of apes with those of fossil and living hominids.

Comparing Molars in Primates

Primate	Average molar area in mm^2
Gorilla gorilla	1011
Australopithecus africanus	901
A. boisei	1010
Homo erectus	656
Homo sapiens	500

Thinking Critically

What factors may have contributed to the trend of decreasing molar area during human evolution?

ThinkingLab — Interpret the Data

Purpose

Logical-Mathematical Students will compare the molar areas of gorillas and several hominid species and hypothesize about the evolution of the human skull and jaw.

Process Skills

compare and contrast, use a table, hypothesize

Teaching Strategies

- Remind students that because teeth preserve well, making inferences about diet in fossil organisms is one of the easiest behavioral inferences scientists can make.

Thinking Critically

Answers should include the idea that *Homo* was either eating foods less tough than that which gorillas and australopithecines were eating or using fire and tools to soften food.

Assessment

Portfolio: Have students summarize their conclusions of this activity to put in their portfolios. Have them list the foods that each species eats or is thought to have eaten. Next, have students explain their hypotheses about why molar area decreased during human evolution. L1 P

interNET CONNECTION

Follow the link for this chapter on the Glencoe Homepage at **www.glencoe.com/sec/science** to find out more about human ancestry.

Meeting Individual Needs

Learning Disabled Have students construct a table that compares and contrasts *H. habilis* and *H. erectus.* Suggest that students indicate times when each group evolved as well as physical and cultural characteristics. L1 LS

GLENCOE TECHNOLOGY

Videodisc

The Infinite Voyage: The Dawn of Humankind

"Out of Africa" vs. Multiregional Debate on Origination of Modern Man (Ch. 8)

Technology
BIOLOGY & SOCIETY

Communication and Computers

Purpose
Students will see how advances in computer technology and communications promise a high-tech future for humans.

Teaching Strategies
- Ask students to discuss whether they think new technologies will advance the human species further, or if the effects of a damaged Earth will impose selective pressures that push humans toward extinction.

INVESTIGATING the Technology

Going Further Suggestions should include using the Internet for consulting up-to-date publications as well as scientists engaged in research on the topic.

Going Further
Have students use their knowledge of natural selection and human evolution to write a short essay about what they think humans will be like 1 million years from now. L1

Technology
BIOLOGY & SOCIETY

Communication and Computers

You may not realize it, but you are living in an important time in history: the early days of the Computer Age.

Personal computers It wasn't until the 1960s that electronic computers became available, but because of cost, they were purchased only by universities, governments, and big businesses. What really accelerated the pace of the computer revolution was the development of the personal computer.

Power to the people
Widespread access to personal computers has empowered ordinary people. You can now access networks that allow you to span the globe from your home or school. The largest of these networks is called the internet. The internet is a worldwide system that interconnects an estimated 10 million people. Over these connections, scientists transmit and discuss their latest findings, businesses can exchange ideas, and students can even communicate with researchers on the cutting edge of research.

From defense to news The internet was originally funded by the Department of Defense. As university researchers discovered its ease, they began to use it for other purposes. Now, anyone can use the network at very low cost, and its potential has expanded tremendously.

Applications for the Future

Recently, it has been proposed that a fiber-optic "superhighway" for information be constructed. This superhighway would carry messages on a beam of light and vastly expand the network's capacity. Already people can use computers for videoconferencing across thousands of miles.

INVESTIGATING the Technology

Going Further Discuss how you could use the internet to find out more about research in biology.

Large stone tools, called hand axes, found at some sites indicate that *H. erectus* probably hunted. In addition, charred bones and hearths found at some sites suggest that these were the first hominids to use fire, and perhaps the first to live in caves.

Unlike the earlier hominids, *H. erectus* migrated out of Africa about 1 million years ago. Fossils indicate that this hominid spread through Africa and Asia, and possibly into Europe. Scientists estimate that *Homo erectus* became extinct between 300 000 and 500 000 years ago, but first this species may have given rise to hominids that resembled modern humans.

Culture developed in modern humans

Dating the emergence of modern humans is perhaps the most controversial topic in the field of paleoanthropology today; this is because of an incomplete fossil record and controversial dating of fossils. Nevertheless, ample evidence exists to suggest a basic picture for the origin of our own species, *Homo sapiens*.

The fossil record indicates that 100 000 to 400 000 years ago, *Homo sapiens* appeared in Europe, Africa, the Middle East, and Asia. The earliest forms of our species are known as archaic *Homo sapiens* because although the skulls resembled those of *Homo erectus*, they had less-prominent browridges, more-bulging foreheads, and smaller teeth. Also, their brain size of 1000 to 1400 cm^3 was in the modern human range.

Meeting Individual Needs

Students Acquiring English Have students supply the hominid species name for each of the following descriptions: (1) earliest maker of stone tools, (2) had a chimpanzee-like brain and was bipedal, (3) earliest hominid to move out of Africa, (4) had strong bones and developed diverse tool kits. L1 SAE

Best known among the first *Homo sapiens* were the Neanderthals, illustrated in *Figure 19.11*. The **Neanderthals** were a group of early humans that lived from 35 000 to 100 000 years ago in Europe, Asia, and the Middle East. Neanderthals were a powerfully built group with thick bones and large faces with prominent noses.

Neanderthals lived in caves during the ice ages of the Pleistocene period, but the popular idea that they were brutish cavemen is incorrect. Neanderthal brains were as large as or larger than those of modern humans. Other evidence such as burial grounds suggests that they may have been the first hominids to develop religious views and communication through spoken language.

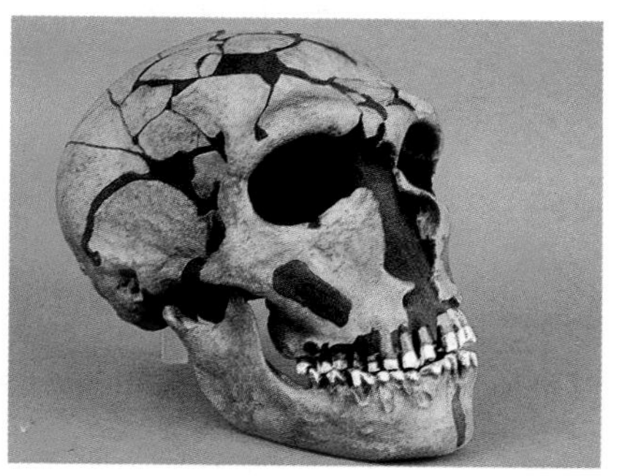

MiniLab

How do human proteins compare with those of other primates?

Scientists use the amino acid sequences in proteins to determine the evolutionary relationships of living species. In this activity, you'll compare amino acid sequences among groups of primates to determine primate evolutionary history.

Amino Acids in Primates

Amino acid sequences															
Baboon	ASN	THR	THR	GLY	ASP	GLU	VAL	ASP	ASP	SER	PRO	GLY	GLY	ASN	ASN
Chimp	SER	THR	ALA	GLY	ASP	GLU	VAL	GLU	ASP	THR	PRO	GLY	GLY	ALA	ASN
Lemur	ALA	THR	SER	GLY	GLU	LYS	VAL	GLU	ASP	SER	PRO	GLY	SER	HIS	ASN
Gorilla	SER	THR	ALA	GLY	ASP	GLU	VAL	GLU	ASP	THR	PRO	GLY	GLY	ALA	ASN
Human	SER	THR	ALA	GLY	ASP	GLU	VAL	GLU	ASP	THR	PRO	GLY	GLY	ALA	ASN

Procedure

1. Make a copy of the data table.

Comparing Primate Amino Acids

Primate	Amino acids different from humans	Percent difference
Baboon		
Chimpanzee		
Gorilla		
Lemur		

2. For each primate, count the number of amino acids that are different from the human sequence. Record these numbers in your data table.
3. Calculate the percentage differences by dividing the numbers by 15 and multiplying by 100. Record the numbers in your data table.

Analysis

1. Which primate is most closely related to humans? Least closely related?
2. Construct a phylogenetic diagram of primates that most closely fits your results.

Figure 19.11

Neanderthals were skilled hunters and were the first hominids to develop diversified tool kits that included spears and tools for scraping leather and slicing meat.

PROJECT
Social Behavior in Early Hominids

Have small groups of students develop a project about the use of nonhuman primates to infer social behavior in early hominids. Scientists assume that if modern humans and nonhuman primates share particular behaviors, then they were probably present in early hominids also. Apes and baboons, among the more social of the primates, are frequently used as models for early human behavior.

Have students observe primates at the zoo or on videotapes, list behaviors that are shared by humans and the primates, and develop models of early hominid behavior that are consistent with other information known about early hominids. L2 LS COOP LEARN

MiniLab

Purpose

LS Logical-Mathematical Students will learn how scientists determine phylogenetic relationships by comparing amino acid sequences among primates.

Process Skills

compare and contrast, hypothesize, make and use tables

Teaching Strategies

- Briefly explain some of the methodologies involved in determining phylogeny, particularly those that involve molecular and biochemical methods.

Expected Results

Chimpanzee and gorilla sequences are identical to the human sequence. Baboons show a 33 percent difference, and lemurs show a 47 percent difference from humans. Errors may be the result of inaccurate counting of the amino acid differences.

Analysis

1. Gorillas and chimpanzees are most closely related to humans. Lemurs are the least closely related.
2. The phylogenetic tree should show lemurs, and then baboons, branching off. The gorilla and chimpanzee lineages should be closer to the human lineage.

Assessment

Portfolio: Ask students to write a summary of this activity for their portfolios. Have them describe what the probable results would be if they looked at sequences from a different protein. L1 P

SECTION 19.2

Discussion Question

At a *Homo erectus* fossil site in Spain, scientists discovered what they thought were the remains of an ancient brush fire on top of a cliff. Found at the base of the cliff were the bones of a herd of elephants, and scattered among the pile of bones were stone tools.

Ask students to develop hypotheses to explain these findings. You might suggest that they list all the pertinent facts from the story and from the chapter text. Discuss each hypothesis.

Bioethics

Whose Bones and Artifacts?

Much interest in paleoanthropology is based on human curiosity about who we are and how we got this way. Knowledge about the behavior of prehistoric humans comes from the analysis of bones and archaeological artifacts. But whose bones and artifacts are these? Do scientists have the right to disturb the graves of other humans? Initiate a discussion about this issue.

What happened to Neanderthals?

The big question in anthropology today is what relationship exists between Neanderthals and present-day humans. Did Neanderthals evolve directly into modern *Homo sapiens*, or were they merely replaced by Cro-Magnons? The **Cro-Magnon** people appeared in Europe 35 000 to 40 000 years ago, and were identical to modern humans in height, skull and tooth structure, and brain size. As shown in *Figure 19.12*, Cro-Magnons were talented toolmakers and artists, and it is almost certain that they used language.

Did Neanderthals evolve directly into modern Cro-Magnon people? Recent genetic and archaeological evidence suggests that this is not likely. New fossil dates indicate that anatomically modern *Homo sapiens* were in South Africa and the Middle East approximately 100 000 years ago, before the time of Neanderthals. Genetic evidence also indicates an African origin of modern *Homo sapiens*, perhaps as early as 200 000 years ago. All of this supports the view that Neanderthals were nothing more than a side branch of *Homo sapiens*, unlikely to be ancestral to modern humans, *Figure 19.13*.

Figure 19.12

Cro-Magnon fossils have been found with cave paintings, detailed stone and bone artifacts, and human figurines. All of this evidence suggests that human culture was quite developed by 30 000 years ago.

Alternate Lab: Measuring Human Variation

Purpose

LS **Logical-Mathematical** Students will analyze variations in human height, length of left forearm, and length of left index finger.

Materials

metric ruler or tape measure

Procedure

Give students the following directions.

1. Prepare a data table with the headings, Height (cm), Length of left index finger (cm), and Length of left forearm (cm).
2. Working with a partner, measure in centimeters your height, the length of your left index finger, and length of your left forearm (from the elbow to the wrist). Record all observations in the data table.
3. List the height of each student in the class. Repeat for finger and forearm measurements.
4. Calculate a class average for each of the measurements.
5. Divide class measurements into five equal intervals and count the number of students within each. Make bar

Figure 19.13

This diagram represents the most widely accepted view among scientists of the evolution of *Homo sapiens*.

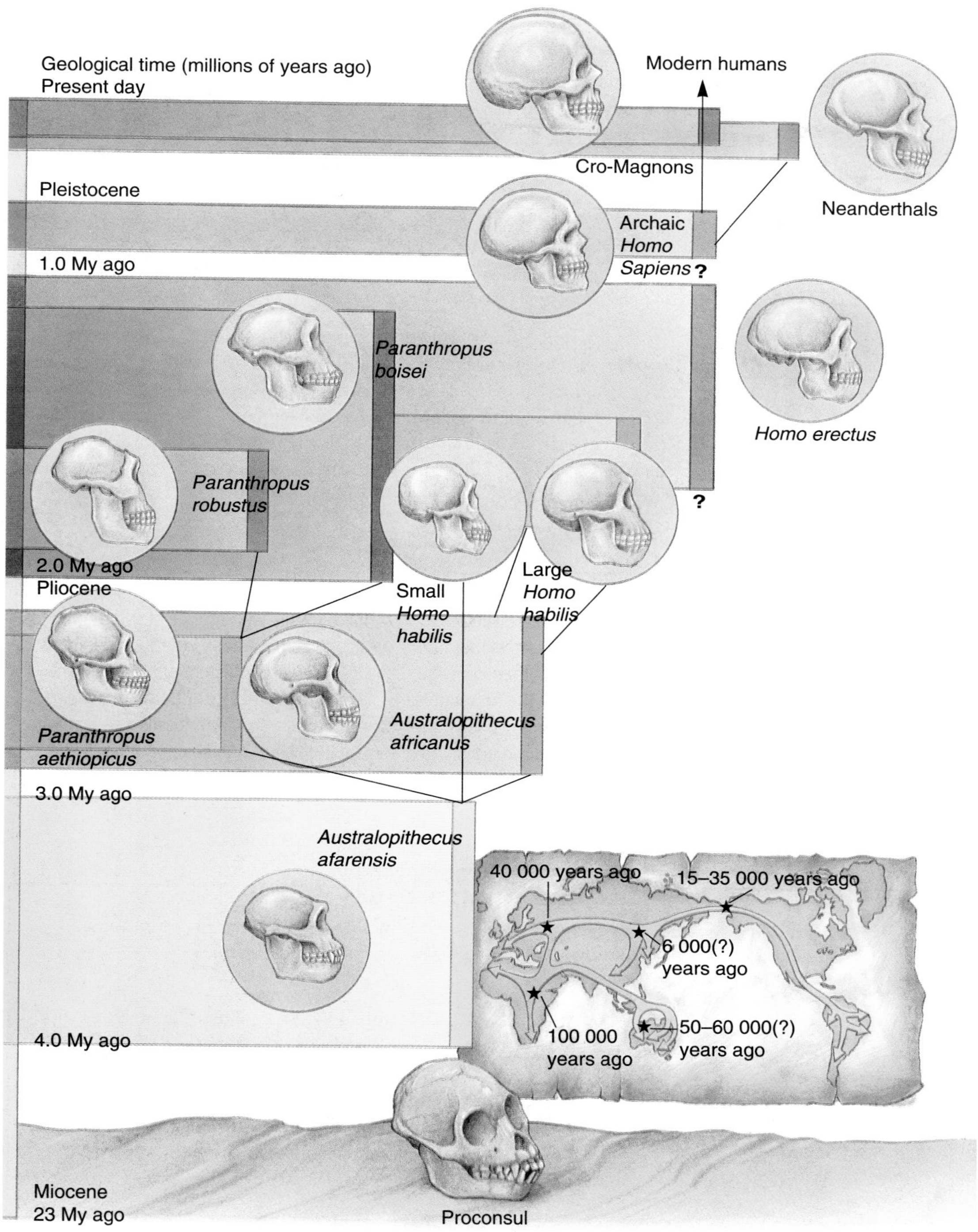

Geological time (millions of years ago)
Present day
Modern humans
Cro-Magnons
Neanderthals
Pleistocene
Archaic *Homo Sapiens* ?
1.0 My ago
Paranthropus boisei
Homo erectus
Paranthropus robustus
?
2.0 My ago
Pliocene
Small *Homo habilis*
Large *Homo habilis*
Paranthropus aethiopicus
Australopithecus africanus
3.0 My ago
Australopithecus afarensis
40 000 years ago
15–35 000 years ago
6 000(?) years ago
100 000 years ago
50–60 000(?) years ago
4.0 My ago
Miocene
23 My ago
Proconsul

Visual Learning

Figure 19.13 illustrates just one possible phylogenetic tree of human evolution based on all the available evidence. Point out to students that some paleontologists have suggested other possible pathways of human phylogeny.

Some scientists think that Neanderthals gave rise to anatomically modern humans in different regions of the world. Stress that scientists often make different hypotheses based on the same observations.

GLENCOE TECHNOLOGY

Videodisc

The Infinite Voyage: The Dawn of Humankind

Development of Modern Man (Ch. 4)

Dating Fossils: Effects of Dating Methods and Interbreeding Theories (Ch. 5)

Evolution of the Mind (Ch. 10)

graphs for each measurement.

Analysis

Ask students to answer the following questions.

1. What is the average height of your class? The average forearm length? The average finger length? *Answers will depend on actual measurements.*
2. How can you account for the variation? *There are variations in traits within humans and in the rates at which individuals attain adult sizes.*
3. How might the data change if you repeated the activity on a group of adults? *There may be fewer measurements at the low end for each trait.*

Assessment

Hypothesizing: Have students develop hypotheses about how natural selection may act on variability during future human evolution. L1

3 Assess

Check For Understanding

Have students construct a time line that shows when the following events occurred and in which hominid species they arose: use of fire, bipedalism, stone tool use, language, art, burial of the dead. L1

Reteach

Have students each make a bar graph that shows how brain size has evolved in humans. Have students summarize the graph in writing. L1

Extension

Have students illustrate a phylogenetic tree of human evolution. L1

Assessment

Oral: Have students prepare and deliver an oral presentation about a day in the life of one hominid species. Encourage them to use models, props, and illustrations for their presentations. L1

4 Close

Discussion

Discuss what paleoanthropologists of the future might determine about the behavior of people living today. Have students address what artifacts might be preserved and what information they will provide future scientists. L1

GLENCOE TECHNOLOGY

Videodisc

The Infinite Voyage: The Search for Ancient Americans

The Ancient Native Americans (Ch. 1)

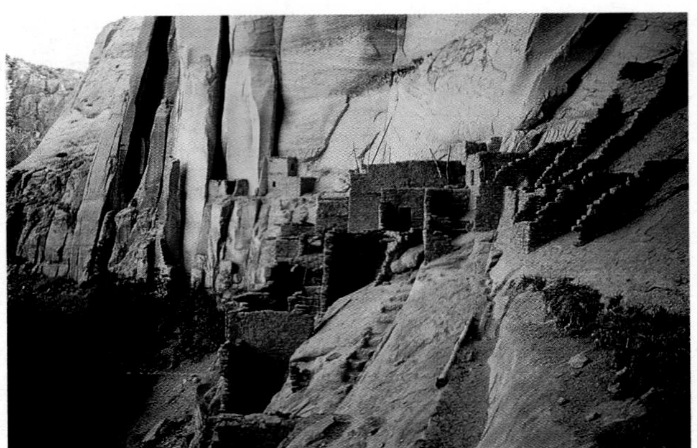

Figure 19.14

Permanent settlements in the Americas, such as this one in the Betatakin Navajo National Monument of Arizona, have been dated to 11 000 years ago.

Most anthropologists now agree that modern *Homo sapiens* evolved in Africa and spread from Africa throughout the rest of the world. Humans have remained relatively unchanged over the last 200 000 years. Humans reached North America by at least 12 000 years ago, and by 8000 to 10 000 years ago, cultural innovations such as animal domestication, agriculture, and the development of permanent settlements by Native Americans had begun, *Figure 19.14.*

Connecting Ideas

Natural selection led to a diverse group of mammals, the primates, that had characteristics that enabled them to move from life in forests to life in the grasslands. The ability to walk upright on two legs gave anthropoids the freedom to use arms and hands for other needs. By comparing primate fossils with groups of living primates, scientists have been able to understand the 65-million-year history of primates. The evolutionary history of humans is being pieced together from fossils and comparative studies of humans and other living anthropoids.

Section Review

Understanding Concepts

1. What evidence suggests that australopithecines were intermediate between apes and humans?
2. How do scientists know that Neanderthals are not ancestors of Cro-Magnon people?
3. Why are Neanderthals considered to be a side branch in human evolution?

Thinking Critically

4. What evolutionary factors may have been responsible for the increase in brain size during human evolution?

Skill Review

5. **Interpreting Scientific Illustrations** By examining figures in this section, draw a time line to show the evolution of hominids. Indicate each species of hominid that evolved, and when and where this evolution occurred. For more help, refer to Thinking Critically in the *Skill Handbook*.

Answers to Section Review

1. Australopithecines have a brain capacity slightly larger than apes and smaller than humans. Australopithecines can climb trees like apes, but also walk like humans.
2. Anatomically modern humans were in Africa long before the time of Neanderthals.
3. Neanderthal anatomy is very specialized. It is unlikely that they gave rise to modern humans.

Thinking Critically

4. Answers may include the idea that increased brain size began as a variation, and nature selected for larger brains as

Chapter 19 Review

REVIEWING MAIN IDEAS

19.1 Primate Adaptation and Evolution

- Primates are primarily an arboreal group of mammals. They possess adaptations related to survival in trees, such as stereoscopic vision, opposable thumbs, and mobile skeletal joints.
- Primates include prosimians, such as lemurs and tarsiers, and anthropoids, which include monkeys, apes, and humans.
- Fossils indicate that primates evolved approximately 65 million years ago. Major trends in primate evolution include increases in brain size and diversity in diet and methods of locomotion.

19.2 Human Ancestry

- Evidence to date indicates that the human lineage began in Africa approximately 5 to 8 million years ago. Scientists propose that australopithecines, the earliest known hominids, had the ability to walk bipedally and climb trees.
- Fossils indicate that during human evolution, brain and body size increased, bipedal locomotion became more efficient, and jaws and teeth decreased in size.
- Stone tool use coincides with the appearance of the genus *Homo* in the fossil record about 2 million years ago. Use of fire, tools, and language, as well as culture, developed later in larger-brained species of *Homo*.

Key Terms

Write a sentence that shows your understanding of each of the following terms.

anthropoid	Neanderthal
australopithecine	opposable thumb
bipedal	prehensile tail
Cro-Magnon	primate
hominid	

Understanding Concepts

1. Which of these is NOT a primate?
   a. human
   b. orangutan
   c. chimpanzee
   d. squirrel
2. Primate characteristics include shoulder joints, flat faces, stereoscopic vision, and flexible ___________.
   a. prehensile tails
   b. opposable thumbs
   c. claws
   d. small eyes
3. The science of studying fossil humans is ___________.
   a. anthropology
   b. geology
   c. paleontology
   d. paleoanthropology
4. Humanlike bipedal primates are ________.
   a. hominids
   b. apes
   c. mandrills
   d. aye-ayes
5. The first australopithecine discovered was the ___________.
   a. Taung child
   b. Neanderthal human
   c. skeleton of "Lucy"
   d. *Homo habilis*
6. "Lucy" settled which scientific debate?
   a. Both primates and hominids have color vision.
   b. Hominids are primates with opposable thumbs.
   c. Hominids had large brains before they walked upright.
   d. Hominids walked upright before they had large brains.
7. Which living primate group is similar to the earliest primates?
   a. apes
   b. monkeys
   c. prosimians
   d. hominids

early hominids competed with other organisms in the open savanna.

Skill Review

5. Evaluate student time lines for accuracy and understanding of the main concepts.

Chapter 19 Review

Reviewing Main Ideas

Summary statements can be used by students to review the major concepts of the chapter.

Key Terms

Answers should go beyond defining the terms. Accept any answer that uses the term correctly and in the proper context.

Understanding Concepts

1. d
2. b
3. d
4. a
5. a
6. d
7. c

Chapter 19 Review

8. b
9. c
10. c
11. a
12. b
13. b
14. d
15. c
16. d
17. b
18. a
19. b
20. d

Applying Concepts

21. Answers should include the idea that scientists can analyze bones to determine whether human ancestors had a particular disease. Students might also say that scientists could isolate DNA from frozen or particularly well-preserved fossils to find the origin of a disease.
22. By studying nonhuman primates, one can make inferences about extinct primates. The adaptations that evolved in primates laid the groundwork for adaptations that arose during hominid evolution.
23. Answers could include that scientists might compare DNA of modern humans and that of Neanderthals, if such DNA exists. Or scientists could compare the skulls or skeletons of Neanderthals and modern humans.

Chapter 19 Review

8. Because the eyes of primates face forward, they ____________.
   a. have color vision
   b. have stereoscopic vision
   c. can climb trees well
   d. see well to the sides
9. Fossil evidence indicates that australopithecines were ____________.
   a. vegetarians
   b. taller than modern humans
   c. tree dwellers
   d. excellent artists
10. The earliest primates were most like ____________.
    a. humans c. squirrels
    b. monkeys d. mice
11. Most primates are tree-dwelling mammals; thus, they are described as being ____________.
    a. arboreal
    b. brighter than other animals
    c. monkeys
    d. anthropoids
12. An omnivorous Old World monkey that is found in the forests of West Africa and is active on the ground during the day is a ____________.
    a. golden lion tamarin
    b. mandrill
    c. prosimian
    d. gibbon
13. Primates of Central and South America are called ____________.
    a. Old World monkeys
    b. New World monkeys
    c. apes
    d. prosimians
14. Lemurs are examples of ____________.
    a. Old World monkeys
    b. New World monkeys
    c. apes
    d. prosimians
15. The word used to describe the tails of New World monkeys is ____________.
    a. stumpy c. prehensile
    b. opposable d. bipedal
16. Hominids are ____________.
    a. similar to prosimians
    b. small, quick primates
    c. arboreal
    d. humanlike, bipedal primates
17. Evidence such as burial grounds shows that ____________ may have been the first hominids to develop religious views.
    a. *Homo erectus*
    b. Neanderthal humans
    c. Cro-Magnon humans
    d. *Homo habilis*
18. Artifacts such as stone tools and charred bones indicate that ____________ probably hunted.
    a. *Homo erectus*
    b. Neanderthal humans
    c. Cro-Magnon humans
    d. *Homo habilis*
19. ____________, the earliest hominid species known, had a pelvis, legs, and feet that show it was bipedal.
    a. *Homo erectus*
    b. *Australopithecus afarensis*
    c. *Australopithecus africanus*
    d. *Homo habilis*
20. The hole in the skull through which the spinal cord exits is the ____________.
    a. cranium
    b. browridge
    c. saggital crest
    d. foramen magnum

Applying Concepts

21. How might human fossils be important for determining the origin of particular diseases?
22. Why is it important for someone interested in human fossils to also understand nonhuman primates?
23. Some scientists suggest that Neanderthals evolved directly into the different human groups of today. How might this hypothesis be tested?

Program Resources

Chapter Assessment, pp. 109-114 L1
Alternate Assessment in the Science Classroom
Computer Test Bank L1
Content Mastery, pp. 73-76 L1

24. Why is ape evolution considered to be an example of adaptive radiation?
25. What factors may have led to the speciation of australopithecines?

Thinking Critically

26. ***Concept Mapping*** Make a concept map that relates the following terms. Supply the appropriate linking words for your map.
    australopithecine, primate, prosimian, *Homo erectus*, *Homo habilis*, *Homo sapiens*, anthropoid, Neanderthal, Cro-Magnon, ape, hominid
27. ***Observing and Inferring*** How could you tell from looking at footprints that an animal walked upright? Explain.
28. ***Formulating Hypotheses*** What tests might you perform to determine the evolutionary significance of large jaws in primates?
29. ***Formulating Hypotheses*** How would you test the idea that opposable thumbs are excellent adaptations for arboreal mammals?
30. ***Sequencing*** Make a time line that shows the proposed major events of human evolution.
31. ***Interpreting Data*** The following data are from an experiment comparing amino acid sequences in apes. What conclusions can you draw from such data?

Comparisons of Amino Acid Sequences

Primate	Percentage amino acid difference from humans
Orangutan	3.7
Chimpanzee	1.8
Gibbon	5.2
Gorilla	2.1

ASSESSING KNOWLEDGE & SKILLS

The intermembral index is the ratio of forelimb length to hind-limb length. Primates with a high intermembral index are good climbers and branch swingers. Primates with a low intermembral index tend to walk on all four legs.

Intermembral Index of Some Primates

Species	Intermembral index $\frac{\text{forelimb length}}{\text{hind-limb length}} \times 100$
Prosimians	
Indri	64
Slow loris	88
New World monkeys	
Squirrel monkey	80
Black spider monkey	105
Old World monkeys	
Pig-tailed macaque	92
Hanuman langur	83
Anthropoids	
Orangutan	139
Common chimpanzee	103

Using a Table Use the table above to answer the following questions.

1. Which group appears to have the greatest range of intermembral distance?
   a. prosimians
   b. New World monkeys
   c. Old World monkeys
   d. anthropoids
2. Which group appears to have the narrowest range of intermembral distance?
   a. prosimians
   b. New World monkeys
   c. Old World monkeys
   d. anthropoids
3. ***Interpreting Data*** You have discovered a new primate. Its intermembral index is 92. To which primate group does it belong? Explain your answer.

Chapter 19 Review

24. Apes evolved their unique traits when they began to exploit different parts of trees to get at more difficult-to-obtain food.
25. Answers may state that when forests changed into grasslands due to environmental changes, australopithecines evolved a more efficient mode of locomotion. Later, australopithecines began eating different foods because of the changing climatic conditions.

Thinking Critically

26. Evaluate students' concept maps. The maps should show an understanding of the relationships among the concepts listed.
27. Human footprints and the footprints of other animals could be compared.
28. Comparisons could be made between primates with small jaws and primates with large jaws to determine what advantage large jaws confer to primates.
29. Students should design an experiment to test and compare opposable thumbs of other arboreal mammals.
30. Students should develop a time line that shows either the dates that significant fossils were found or that begins about 25 million years ago and includes the major fossil finds in proper chronological sequence.
31. Chimpanzees are more closely related to humans than other apes. Next most closely related to humans are the gorillas, followed by the orangutans and then the gibbons.

ASSESSING KNOWLEDGE & SKILLS

1. d
2. c
3. An intermembral index of 92 could place this new primate with Old World monkeys. It could also place the primate with New World monkeys as 92 is in the range shown. More information would be needed to correctly place this new primate.

UNIT

6 The Diversity of Life

Unit Overview

Unit 6 introduces students to taxonomy and explains how and why organisms are classified. In Chapter 20, the distinguishing features of the six kingdoms of organisms—Eubacteria, Archaebacteria, Protista, Fungi, Plantae, and Animalia—are presented. The remaining chapters of the unit then introduce students to bacteria, protists, and fungi.

Chapter 21 presents viruses and the characteristics that cause biologists to question whether they are alive. Bacteria, their structure, ecology, and importance are also discussed. In the next chapter, students study the diversity and classification of protists. The unit concludes, in Chapter 23, with the study of the general characteristics and diversity of fungi.

Theme Development

The major themes of *evolution* and *unity within diversity* are woven throughout this unit. Students begin to appreciate how evolution explains the diversity of life on Earth as they begin to explore the characteristics of organisms. Students will also begin to appreciate the characteristics these diverse forms of life have in common.

UNIT

6 The Diversity of Life

How many different types of organisms could you identify in this small patch of forest? Certainly, you could pick out several species of plants. Perhaps you would also find an animal, a colony of mushrooms, or lichen growing on a rotting log. Unseen to you, however, would be the countless species of microscopic bacteria, algae, and fungi that exist in the decaying matter and soil on the forest floor.

Nearly 1.5 million species of organisms have been identified—only a fraction of the nearly 100 million species that some scientists estimate actually exist. A large portion of these uncounted species belong to kingdoms of mostly microscopic organisms that you will study in this unit.

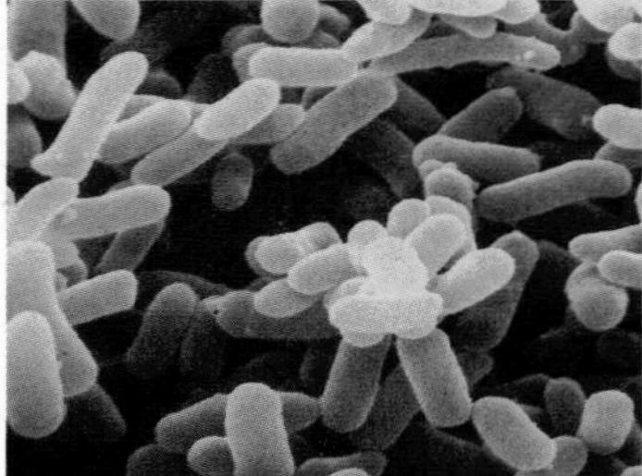

Many species of bacteria cause disease, but others play important roles in their ecosystems. Why are bacteria so important to the existence of life on Earth?

Advance Planning

Chapter 20

- Purchase ten varieties of seeds for the Minilab on page 488.
- Have dichotomous keys for classification based on leaf structure available for the Minilab on page 495. Obtain sample dichotomous keys of insect and beetle classification for the Biolab.

Chapter 21

- Purchase nuts, bolts, and wire for the Minilab on page 506.
- Order prepared slides of cocci, bacilli, and spirilla for the Minilab on page 514.
- Order sterile nutrient agar plates, bacterial cultures, and antibiotic disks for the Biolab.

Chapter 22

- Order *Paramecium* cultures for the Minilab on page 538, and for the Biolab.
- Order *Euglena* cultures and methyl cellulose for the Biolab.
- Establish a classroom aquarium with snails for the Minilab on page 549.

Chapter 23

- Obtain freshly baked bread for the Minilab on page 564.
- Purchase mushrooms for the Minilab on page 569.
- Prepare yeast mixtures for the Biolab.

Slime molds, odd cousins of the mushroom, can be found oozing over decaying branches and logs. What are the common characteristics shared by slime molds and other fungi?

An amoeba moves along the surface of a damp leaf like a living blob of jelly. How do other members of its kingdom move about in search of food?

Magnification: 70×

Unit Contents

UNIT 6

Introducing the Unit

Ask students in their groups to estimate how many different species of living things there are on Earth and ask them to speculate about why there are so many species. To help emphasize the diversity of life on Earth, direct students' attention to the large photo and ask them to list all the living things that they would expect to find in the forest. Encourage them to include both things that are visible and things they cannot see.

Call students' attention to the bacteria. Ask them to explain what a *germ* is and then decide if all bacteria would be considered germs. Use this exercise to point out that not all bacteria are harmful.

Next, ask students to look at the amoeba. Elicit in what type of water they would find this organism. Ask them why their tap water would not have amoebas.

Motivational Activity

Activity: Provide students with a fresh mushroom from the supermarket, a prepared slide of bacteria, and pond water with protozoans. Ask them to sketch each organism and explain how the organisms they have drawn are alike and how they differ. L1

Unit Project

Community Involvement: Ask students to contact their local health department to find out how restaurants, swimming pools, and other public facilities are examined to make sure they are free of harmful bacteria, fungi, and protozoans.

Organize a "day on the job" for students to accompany a health department employee who is carrying out inspections of local establishments. Ask students to come together in groups to share their experiences and to determine what other procedures the health department could carry out to better protect public health. Ask students to write a letter to the health department with their recommendations and to include this letter in their portfolios. L1 P COOP LEARN

Chapter Organizer

SECTION	OBJECTIVES	ACTIVITIES/FEATURES
20.1 Classification National Science Standards: UCP.1, UCP.2, UCP.4; C.3, C.5; G.1-3	**1. Evaluate** the history, methods, and purpose of taxonomy. **2. Demonstrate** the use of concepts in classification. **3. Explain** the purpose of a phylogenetic classification.	**Thinking Lab:** What fraction of species has been identified?, p. 485 **A Broader View:** Infinite Variety, p. 487 **Minilab:** How can you classify seeds?, p. 488
20.2 The Six Kingdoms National Science Standards: UCP.1, UCP.2, UCP.5; A.1, A.2; C.1, C.5; E.1, E.2; G.1-3	**4. Compare** the six kingdoms of organisms. **5. Distinguish** between the Kingdoms Eubacteria, Archaebacteria, Protista, Fungi, Plantae, and Animalia.	**Art Connection:** *The Terrestrial Paradise,* p. 491 **Minilab:** How is a dichotomous key used?, p. 495 **Biolab:** Making a Dichotomous Key, p. 496

ACTIVITY MATERIALS

BIOLAB	MINILABS	ALTERNATE LAB
page 496 sample keys from guidebooks metric ruler	**page 488** 10 seeds for each group of four students (use a variety of seeds) hand lens **page 495** miscellaneous leaves dichotomous key for trees paper paste field guides	**page 492** glassware: flasks, mortars, test tubes, beakers, stirring rods, pestles, thermometers, stoppers, pipettes, droppers, crucibles, funnels, petri dishes, graduated cylinders, bottles, and jars

Chapter 20 Organizing Life

TEACHER CLASSROOM RESOURCES

Reproducible Masters	Transparencies
Section Focus Master 42: Classification L1 SAE **Reinforcement and Study Guide,** pp. 77-78 L1 **Biolab and Minilab Worksheets,** p. 79 L1 **Tech Prep Applications:** A Dichotomous Key, pp. 23-24 L2 **Critical Thinking/Problem Solving:** Evaluating Methods of Classification, p. 20 L3 **Laboratory Manual:** Can a Key Be Used to Identify Organisms?, pp. 107-110 L2	
Section Focus Master 43: Classifying Organisms L1 SAE **Reinforcement and Study Guide,** pp. 79-80 L1 **Biolab and Minilab Worksheets,** pp. 80-82 L1 **Concept Mapping:** Classifying Organisms, p. 20 L1 **Content Mastery,** pp. 77-80 L1	**Basic Concepts Transparency #24:** Life's Six Kingdoms L1 SAE **Reteaching Transparency #20:** Six Kingdoms L1 SAE

ASSESSMENT MATERIALS

Chapter Assessment, pp. 115-120
Performance Assessment in the Biology Classroom, p. 27
MindJogger Videoquiz
Alternate Assessment in the Science Classroom
Computer Test Bank

Spanish Resources SAE
English/Spanish Audiocassettes SAE
Cooperative Learning in the Science Classroom COOP LEARN
Lesson Plans
Great Developments in Biology: Viruses: On the Edge of Life L1 SAE

KEY TO TEACHING STRATEGIES

- L1 Level 1 activities should be within the ability range of all students including those with learning difficulties.
- L2 Level 2 activities are within the ability range of average to above-average students.
- L3 Level 3 activities are designed for the ability range of above-average students.
- SAE SAE activities should be within the ability range of Students Acquiring English.
- COOP LEARN Cooperative Learning activities are designed for small group work.
- P These strategies represent student products that can be placed into a best-work portfolio.
- These strategies are useful in a block scheduling format.

GLENCOE TECHNOLOGY

The following multimedia resources are available from Glencoe.

Biology: The Dynamics of Life
CD-ROM SAE
Videodisc Program

The Infinite Voyage Series
Insects: The Ruling Class
Insects and Their Behavior
Life in the Balance
BioLat II: Taking Inventory of the Threatened Peruvian Rain Forest
BioLat II: Classifying the Community of Insects

Science and Technology Videodisc Series (STVS)
Animals
Naming Fish
Insect Museum

CHAPTER

20 Organizing Life

Chapter Overview

In this chapter, students first learn about the system of classification developed by Aristotle. Students then learn about the more refined system of classification developed by Linnaeus and how this system provides a more natural view of relationships among organisms. Students discover how the Linnaean system changes how organisms are classified from methods based on similar structures or behaviors to a system based on evolutionary relationships.

The use of taxonomy in various fields is discussed. Taxonomic categories are described, while the importance of developing phylogeny using embryonic development, biochemistry, and behavior is shown. The first section of the chapter ends with a discussion of how scientific names are determined.

The six-kingdom classification system is presented in the second part of the chapter. Students are introduced to the six kingdoms: Eubacteria, Archaebacteria, Protista, Fungi, Plantae, and Animalia. They then learn how organisms belonging in each kingdom are distinguished.

Key Terms

binomial nomenclature
class
classification
division
family
fungus
genus
kingdom
order
phylogeny
phylum
protist
taxonomy

480

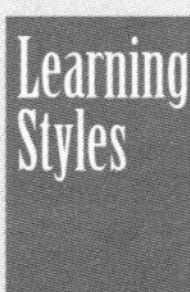

Look for the following logo for strategies that emphasize different learning modalities. LS

Kinesthetic	**Meeting Individual Needs, p. 494**
Visual-Spatial	**Portfolio, pp. 485, 486, 494; Meeting Individual Needs, p. 487; Demonstration, p. 490; Project, p. 491; Visual Learning, pp. 492, 494; Time Line, p. 493**
Interpersonal	**Minilab, p. 488**
Intrapersonal	**Student Journal, p. 485; Meeting Individual Needs, p. 488**
Linguistic	**Student Journal, pp. 483, 484, 486, 487**
Logical-Mathematical	**Thinking Lab, p. 485; Alternate Lab, p. 492; Minilab, p. 495**

CHAPTER 20 Organizing Life

Scientists estimate that between 5 and 30 million species of organisms live on our planet, although fewer than 2 million have been identified and named. Imagine trying to study life on Earth. One major problem is simply the identification of the millions of species. To that end, biologists have organized living things into related groups.

Even though the catalogs and databases of known organisms are large, there are still places on Earth where little is known about the diversity of life. Scientists estimate that only 15 percent of the species in the remaining rain forests have been identified. Perhaps one day you might make an expedition to such a place to discover more about its biodiversity. If you find a new species, you will probably use a system of classification designed by biologists to place it in relationship with other known species.

Concept Check

You may wish to review the following concepts before studying this chapter.
- Chapter 1: species
- Chapter 8: cells, prokaryotes, eukaryotes
- Chapter 18: evolution

Chapter Preview

Laboratory Activities

Biolab
- Making a Dichotomous Key

Minilabs
- How can you classify seeds?
- How is a dichotomous key used?

Think about the last time you visited a zoo. Remember the variety of organisms you saw. In one area, lions yawned in the shade of an acacia plant while vultures searched the grass for food. In a nearby area, flamingos rested by standing on one leg and hiding their heads under their wings. In the reptile exhibit, a cobra slowly lifted its head and expanded its hood as you passed by. The many animals and plants you saw represented only a tiny fraction of the variety of organisms that inhabit Earth.

Assessment Planner

Choose assessment strategies from the following pages to evaluate the progress of your students.
Assess, pp. 489, 498
Alternate Lab, pp. 492-493
Minilabs, pp. 488, 495
Portfolio, pp. 485, 486, 494
Thinking Lab, p. 485
Biolab, pp. 496-497
Chapter Review, pp. 499-501

CHAPTER 20

Introducing the Chapter

Ask students to think back to the last time they visited a zoo or a park. Have them list all the organisms they remember seeing, and ask them to place these organisms into two main groups. Many students will list only the large animals they saw and not mention plants and smaller organisms. Pose the question: If you were to group all organisms into several main groups, how many groups would you have? Ask students to brainstorm this question in their groups and prepare a hypothesis and a justification. *It is likely that students will divide organisms into some kind of plant and animal groups.*

Theme Development

The theme of *evolution* is apparent as emphasis is placed on the phylogenetic basis of classification. The theme of *unity within diversity* is emphasized in the discussion of the features that unify all organisms but distinguish organisms in each of the six kingdoms.

Concept Check

This chapter will expand students' understanding of the concepts of species and evolution by focusing on how evolutionary relationships are used in taxonomy. Students will develop a greater knowledge of the features of organisms by examining how organisms are classified based on their similarities, such as cell structures and methods of obtaining food.

SECTION

20.1

Prepare

Key Concepts

Students will examine the history, methods, and purposes of classification and taxonomy. They will compare the contributions of Aristotle and Linnaeus to systems of classification and learn about taxonomic categories.

Block Scheduling

Look for this symbol for strategies that are useful in a block scheduling format. For more information on block scheduling, refer to Section 20.1 in the **Lesson Plans** booklet.

Materials

- For the Minilab, gather ten different kinds of seeds, metric rulers, flowers from local plants, and hand lenses.

1 Focus

Bellringer

Before presenting the lesson, display **Section Focus Master 42** on the overhead projector and have students answer the accompanying questions. L1 SAE

Demonstration

Provide each group of students with two examples of living things with which they are probably unfamiliar. Examples of such organisms may include slime molds, mildew on cloth, mosses, lichens on a rock, or an unusual organism from your local area. Ask students to identify each organism as a plant or an animal and justify their reasons. Explain that although the living things familiar to most people are either plants or animals, there are actually six main groups of organisms rather than just two.

SECTION

20.1 Classification

Section Preview

Objectives

Evaluate the history, methods, and purpose of taxonomy.

Demonstrate the use of concepts in classification.

Explain the purpose of a phylogenetic classification.

Key Terms

classification
taxonomy
genus
binomial nomenclature
family
order
class
phylum
kingdom
division
phylogeny

Organization is important in everyday life. Suppose you decide to go to a movie with your friends. You don't have to look through the whole newspaper to find what's playing because the paper is organized into sections such as sports, news, entertainment, comics, and classified advertisements. The editors have grouped together articles, features, and advertisements based on similarities of their topics. Because movies are a form of entertainment, you look through the entertainment section to find out what's playing.

How Classification Began

People have always sought a better understanding of nature. One tool that early scientists used to gain that understanding is classification. **Classification** is the grouping of objects or information based on similarities. Early systems of classification had a purpose. Plants were classified as edible or toxic, based on their effects on people who first ate them. As time passed, the science of taxonomy developed. **Taxonomy** is the branch of biology concerned with the grouping and naming of organisms. Biologists who study taxonomy are called taxonomists.

taxonomy:
taxo (GK) to arrange
nomy (GK) ordered knowledge
Taxonomy is the science of classification.

Aristotle's system

The Greek philosopher Aristotle (384–322 B.C.) developed the first method of classification. He classified all living things known at that time into two major groups—plants and animals. Plants were classified as herbs, shrubs, or trees depending on their size and structure. Animals were grouped according to where they lived—on land, in the air, or in water. Later observations convinced scientists that Aristotle's system did not work. They observed that some animals, such as frogs, live both on land and in water. Scientists also realized that Aristotle's classification system did not show natural relationships among organisms. According to his system, birds, bats, and flying insects would be grouped together even though they have little in common besides the ability to fly.

Linnaeus designs a classification system

It wasn't until the late 18th century that a Swedish botanist, Carolus Linnaeus (1707–1778), developed a method of classification that is still used today. Unlike Aristotle,

Program Resources

Section Focus Master 42 L1 SAE
Reinforcement and Study Guide, pp. 77-78 L1
Biolab and Minilab Worksheets, p. 79 L1
Tech Prep Applications, pp. 23-24 L2
Critical Thinking/Problem Solving, p. 20 L3
Laboratory Manual, pp. 107-110 L2
Performance Assessment in the Biology Classroom, p. 27 L1 P

Linnaeus selected physical characteristics that led to classification based on close relationships of organisms. For example, he based his classification of flowering plants on the numbers and similarities of their reproductive structures, *Figure 20.1*. Linnaeus selected characteristics of organisms that also led eventually to classifications based on evolutionary relationships. Let's look at a bat as an example of an organism classified by this method. Whereas bats fly like birds, their origins are shared by all animals that have hair and feed milk to their young. Therefore, bats are classified with the group known as mammals, rather than with birds.

A species is known by two names

Linnaeus also invented the two-word system used to identify species. In this system, the first word identifies the genus in which the organism is classified. A **genus** consists of a group of closely related species. The genus name is immediately followed by the second word, which is often descriptive of that organism. Thus, the scientific name for each species, called a binomial, is a combination of the genus and descriptive names. The system of naming objects is known as nomenclature. The system devised by Linnaeus that gives each organism two names is called **binomial nomenclature.**

binomial:
bi (L) two
nomen (L) name
A binomial consists of two names.

Figure 20.1
Carolus Linnaeus designed a system to classify organisms based on numbers and similarities of body structures and on physical form such as size, shape, and methods of obtaining food. What features could you use to classify these flowers?

2 Teach

Different Viewpoints in Biology

Science Becomes Art
From 1833 - 1843 Louis Agassiz, a Swiss naturalist, described the world's fossil fishes in five large books of narrative and an atlas containing 391 illustrations. Agassiz did not use an evolutionary framework for his book, but defined the subgroups of fishes according to the functions they shared. As a result, many of the fish groups included members of different families and lineages.

Today's taxonomists know that separate lineages may have similar functional adaptations. Today, Agassiz's work is not given much consideration as a tool of classification. However, his atlas is viewed as a work of art showing great precision and beauty.

Visual Learning

Figure 20.1 What features could you use to classify these flowers? *Likely responses may include numbers and arrangements of petals or male and female organs (stamens and pistils).*

Meeting Individual Needs

Learning Disabled Have students examine the classified ads from the yellow pages of a phone book. Ask them to think about how difficult it would be to find the phone number of a music store for which they did not know the name if they had to look in the white pages. Ask them to think of another situation in which classification may help them. SAE

STUDENT JOURNAL

Using Binomials Have students reread the paragraph titled "A species is known by two names." Using the information in this paragraph as a guide, have students write the names of all their family members as binomials. L1 LS

SECTION 20.1

Discussion

To illustrate the changing nature of scientific discovery and classification of organisms, show students the art of *Hallucigenia* (*Natural History*, 1/92, "The Reversal of Hallucigenia," by Stephen J. Gould, p. 14). Explain that the picture shows a fossil animal discovered in 1977 that was named *Hallucigenia* because of its strange appearance. *Hallucigenia* did not seem to resemble any present-day group, and its spike-like legs were not common to any wormlike animals. Turn the picture upside-down and ask students to imagine that the spikes are appendages that may have served as protection for the animal. Explain that *Hallucigenia* was later classified as a close relative of *Peripatus*, the velvet worm.

Text Question

Based on this evidence, would you conclude that dinosaurs were more closely related to cold-blooded lizards or to warm-blooded ostriches? *Responses may suggest dinosaurs are more closely related to warm-blooded ostriches or are a link between warm-blooded and cold-blooded animals.*

GLENCOE TECHNOLOGY

Videodisc

Biology: The Dynamics of Life
Disc 1, Side 2
Museum Collections (Ch. 8)

STVS: Animals
Disc 5, Side 2
Naming Fish (Ch. 2)

Figure 20.2

The poisonous mushroom on the right, called a destroying angel, and the edible mushroom on the left, a smooth lepiota, look similar. Only a taxonomist or a person who uses taxonomy to identify the two species knows for sure which mushroom can be eaten safely.

Taxonomy—The Study of Classification

Even today, taxonomists continue to search for underlying natural relationships as a basis for classification. They compare the external and internal structures of organisms, as well as their chemical makeup and evolutionary relationships.

Why are living things organized?

Just as the editors of a newspaper organize material into categories based on similarities, taxonomists organize living things into groups by using a set of characteristics or criteria that describe how similar or different organisms are from each other. Classification provides a framework of logic and order so that relationships among living things and once-living things can be seen easily.

One question that scientists are investigating is whether dinosaurs are more closely related to warm-blooded or to cold-blooded animals. It was discovered that some dinosaur bones have large spaces inside, similar to the spaces in the bones of birds, which are warm-blooded. If you know something about classification of animals with hollow bone structure, you might conclude that some dinosaurs may have been able to control their body temperature in the same way as warm-blooded animals. Based on this conclusion, would you infer that dinosaurs were more closely related to cold-blooded lizards or to warm-blooded ostriches?

Taxonomy—A useful tool

Scientists working in agriculture, forestry, and medicine use taxonomy as a basis for their work. Assume that a child has eaten one of the mushrooms shown in *Figure 20.2*. Knowing that many poisonous mushrooms resemble harmless species, the child's parents probably would rush the child to the nearest hospital. A taxonomist working at the poison control center could identify the mushroom by examining its structures, and the doctor attending the child would then know whether treatment was necessary.

STUDENT JOURNAL

Using Scientific Names Provide students with advertisements for produce sales at a local supermarket. Ask students to rewrite the advertisement using the scientific names of the items offered. Have students exchange their ads with other students to see if they can identify the item on sale. L2 LS

Knowledge of taxonomy—Its importance to the economy

The discovery of new sources of lumber, medicines, and energy is often the result of explorations by taxonomists. Knowledge of relationships between new and existing species may lead to increased knowledge of these species. For example, we know that pine trees contain resins that can be used as disinfectants. It's possible that other evergreen trees produce useful substances. In fact, the bark of the Pacific yew, *Figure 20.3*, produces taxol, an experimental drug that shows promise in treating some forms of cancer. If we know that one group of plants has a certain kind of chemical substance, we can often assume that related plants contain similar substances.

How Living Things Are Classified

In classification systems, things are categorized for a particular purpose. Just as the classified advertising in the newspaper places all autos for sale into groups with similar characteristics such as whether they are domestic or imported, and by the model, brand, and year, biologists classify organisms using similar characteristics.

Taxonomic categories

In taxonomy, organisms are grouped into a series of categories called taxa, each one larger than the previous one. The taxa fit together like nested boxes, one inside another. You already know that organisms that look alike and successfully reproduce among themselves belong to the same species, and that a group of similar species that are alike in general features and are closely related is a genus.

ThinkingLab — Interpret the Data

What fraction of species has been identified?

The numbers of species inhabiting Earth are presently unknown. Estimates vary between 5 and 30 million or more. The number of species identified and cataloged by taxonomists has grown at different rates for different groups of organisms. Some groups of organisms have been studied extensively for a long period of time, whereas other groups have been studied only recently. Consequently, groups that have been studied for a long time are more completely known than groups that have been studied only recently.

Analysis

Examine the graphs below. Compare the percentage of bird species known by 1845 with the percentage of arthropod species (insects, spiders, crayfish) known by 1845. Note the shapes of the graphs.

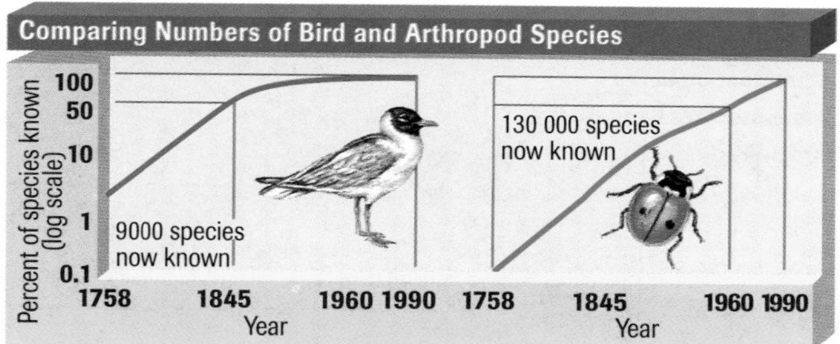

Thinking Critically

What percentage of bird species was known by 1960 compared with the percentage of arthropod species known? If the graph were extended to the year 2000, what probably would happen to the bird graph as compared with the arthropod graph, based on past trends? Explain.

Figure 20.3

The Pacific yew is a rare evergreen tree that grows only in the Pacific Northwest of the United States.

ThinkingLab — Interpret the Data

Purpose

LS **Logical-Mathematical** Students will compare graphs to identify trends and consider why some species have been studied more than others.

Process Skills

compare and contrast, interpret data

Teaching Strategy

- Before beginning the lab, ask students in their groups to list all the birds and arthropods they know. Ask them to speculate about why they know more birds than arthropods.

Thinking Critically

By 1960, nearly all bird species were named and 3/4 of all arthropods were known. The bird graph will stay about the same and the arthropod graph will continue to go up because few new birds are being discovered, whereas many arthropods are still being discovered.

Assessment

Skill: Recent studies show that, other than lichens, about 81 500 species of fungi are known, whereas there are 270 000 cataloged species of vascular plants. In areas that have been studied thoroughly, fungi outnumber vascular plants by six to one. If this relationship holds true over the entire Earth, how many species of fungi might exist? *1 620 000* L1

STUDENT JOURNAL

Classification in Daily Life Provide students with the classified ads for cars. Ask them to describe the car of their dreams and explain where it would appear in the actual classified ad pages they have. L1 LS

PORTFOLIO

Zoo and Garden Classification Have students visit a nearby zoo or botanical garden. Ask the students to present to the class a photo or video essay of classification in use at the zoo or garden. Have students place their work in their portfolios. L1 LS P

SECTION 20.1

Visual Learning

Figure 20.4 In what ways are these two animals alike? *They have whiskers, a short face, eyes on the front of their heads, short ears, tan coloration that blends with environment, four legs, similar body form, short tail.*

Different Viewpoints in Biology

Rethinking Species Designations

A common definition of a species is a population of organisms that potentially can breed with one another. Some biologists believe this definition ignores the phylogenetic relationships among species. Evidence supporting this belief is that crossbreeding between different species of plants is common, and not unheard of in animals. A phylogenetic species concept would be that a species is the smallest recognizable group of organisms that share common traits and ancestry.

Assessment

Performance Assessment in the Biology Classroom: p. 27 Designing a Classification System. Have students carry out this activity to expand their knowledge of classification. L1 P

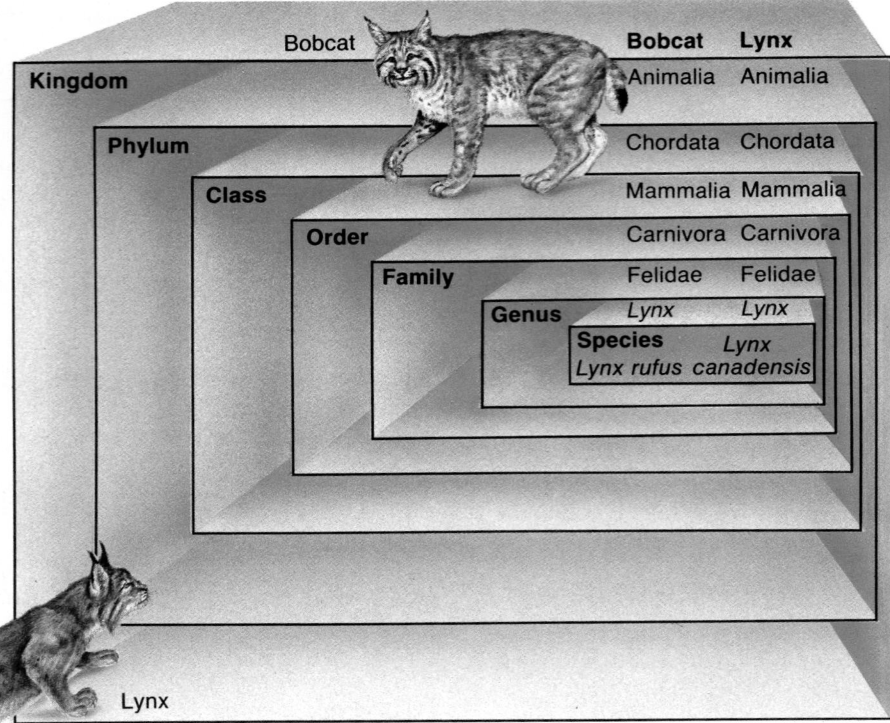

Figure 20.4
A lynx, *Lynx canadensis,* has a short tail with a black tip running all the way around the tail. It also has highly visible tufts of hair on the ears. A bobcat, *Lynx rufus,* has a short tail with black only on top of the tail's tip. It also has inconspicuous ear tufts. In what ways are these two animals alike? You know they are similar in many ways because their classification is the same in all taxa except species.

Notice by their scientific names that the lynx and bobcat shown in *Figure 20.4* belong to the same genus. The genus is indicated by the first word of the binomial. In the genus *Lynx*, the skulls of the animals have 28 teeth. The skulls of lions and mountain lions have a different number of teeth. Therefore, these animals are classified in a different genus, even though they are still obviously cats.

The next larger taxon is the family. A **family** is a group of closely related genera. The bobcat and the lynx both belong to the same family: the cat family, Felidae. Domesticated cats also belong to this family. All members of the cat family have short faces, small ears, and most have claws that can be retracted. They also have five toes on each front foot and four toes on each back foot.

The larger taxa

The remaining higher categories of classification are order, class, phylum, and kingdom. An **order** is a group of related families. A **class** is a group of related orders. A **phylum** is a group of related classes, and a **kingdom** is a group of related phyla. In this system of classification, animal groups are called phyla and plant groups are called **divisions.**

As *Figure 20.4* shows, bobcats and lynxes belong to the same order, Carnivora. The carnivores all have similar structures and arrangements of teeth. These two cats also belong to the same class, Mammalia. Mammals have hair or fur and feed milk to their young. The phylum Chordata, to which both bobcats and lynxes belong, includes animals with backbones. Kingdom Animalia includes all phyla of animals.

PORTFOLIO

Classifying Local Organisms Provide students with the taxonomic classification of a common local plant and animal. Have them make sketches similar to the nested blocks in Figure 20.4 and place the taxonomic names of each organism where they would fit in the blocks. L2 IN P

STUDENT JOURNAL

Discovering New Animals Ask students to prepare a journal report about the recent discoveries of "new" animals such as the black-faced lion tamarin, an Asian hoofed mammal, or the many new species found in an underground aquifer in Texas. ("Saving Hades' Creatures," by Elizabeth Pennisi, *Science News,* March 13, 1993.) L2 IN

How Are Relationships Determined?

Species are classified based on their similarities in structure, chemistry, and behavior. Evolutionary relationships are revealed by using all these features for classification.

Relationships can be determined by evolutionary history

You have learned that species share many characteristics, which indicates that they all evolved from a common ancestor. Because members of the cat family resemble each other more than they resemble members of other groups, taxonomists conclude that lynxes and bobcats shared a common ancestor.

When classifying species, taxonomists also compare structures of modern-day lifeforms with those found in fossils. The relationships revealed by these comparisons demonstrate the evolutionary history of organisms. The evolutionary history of a species is called **phylogeny.** Developing phylogeny can be one purpose for a system of classification. Phylogenetic classifications show the evolutionary history of the classified species.

Biologists find phylogenetic classification useful because organisms belonging to the same group can be expected to share certain characteristics. For example, if you are identifying an animal and find that it has retractile claws, you can infer that it is a member of the cat family and can assume that it has other features common to cats.

A BROADER VIEW

Infinite Variety

You might think the seven categories of kingdom, phylum or division, class, order, family, genus, and species would be sufficient to encompass the range of differences among plants and animals, but that is not the case. So seemingly infinite is the variety of characteristics among even closely related species that further classification is often considered necessary.

Subspecies Just below the level of species are the subspecies. Generally, subspecies can be defined as morphologically, physiologically, and often geographically distinguishable plant or animal groups that could interbreed if they were ever to come into contact.

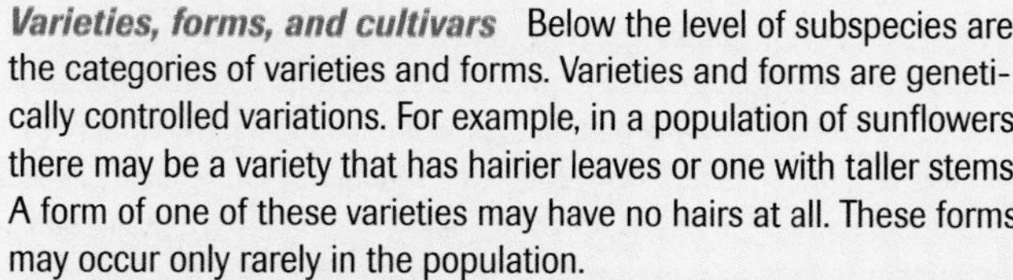

The golden whistler birds that inhabit the many islands of the Southwest Pacific are examples of subspecies. Identifiable types of golden whistlers inhabit different islands, a fact that originally prompted zoologists to classify each type of golden whistler as a separate species. When it was discovered that whistlers from different islands would mate and produce offspring when brought together, the 73 species of whistlers were reclassified as subspecies.

Varieties, forms, and cultivars Below the level of subspecies are the categories of varieties and forms. Varieties and forms are genetically controlled variations. For example, in a population of sunflowers, there may be a variety that has hairier leaves or one with taller stems. A form of one of these varieties may have no hairs at all. These forms may occur only rarely in the population.

Botanists produce cultivars by crossbreeding two varieties. Cultivated varieties are common. Suppose a botanist has two varieties of tomatoes. One variety is exceptionally sweet; another variety has an unusual creamy-orange color. By crossbreeding the two varieties, the botanist may eventually produce a cultivar that displays both the sweetness and the unusual coloration of its parents.

Thinking Critically
Some scientists estimate that Earth may hold millions of as-yet-undiscovered species. Others say billions of species are still to be named. Are these taxonomists all considering the same categories of organisms? Explain.

A BROADER VIEW

Infinite Variety

Purpose
Students will develop an understanding of taxonomic categories below the species level, such as subspecies, varieties, and forms.

Teaching Strategies

- Before students read the feature, ask them to list all the kinds of dogs they know. Ask them if all dogs belong to the same species and to justify their answers. After reading the feature, ask them the same question again. L1
- Have students research the derivation of the term *breed.* Have students explain how a breed compares to a subspecies or variety. L2

Thinking Critically
Answers should indicate that as new plants and animals are discovered, the possibility exists that some will not fit into existing taxonomic categories. New categories, including those below the species level, may have to be created for these groups.

Chemistry Connection

Using DNA for Classification

Based on similar physical features, scientists grouped red pandas in the raccoon family. Because the red panda shares some behavioral traits with the giant panda, scientists suspected that the giant panda also might be related to raccoons. However, when the DNA banding patterns of the chromosomes of these animals were compared, the red panda was found to be more closely related to raccoons and the giant panda more closely related to bears.

STUDENT JOURNAL

Discovering New Species Ask students to write a report in their journals about the procedures needed to establish that an insect found in a tropical rain forest is a "new" species that has not been described, even though it may have inhabited Earth for many thousands of years. L3 IN

Meeting Individual Needs

Hearing Impaired Ask students to prepare flash cards with the names of the taxa on one side of the card. On the other side, have students use symbols, diagrams, or analogies that serve as clues for the taxa described. Have hearing impaired students work with other students in the class to review the taxa identified on the cards. SAE IN COOP LEARN

MiniLab

Purpose

Interpersonal Students will determine what traits can be used to classify seeds.

Process Skills

observe and infer, classify, compare and contrast

Teaching Strategies

- Provide students with hand lenses and worksheets from the *Biolab and Minilab Worksheets* book.

Expected Results

Students are likely to develop classifications based on sizes, colors, surface features, and shapes.

Analysis

1. Most likely responses will include size, shape, surface features, or color.
2. Some teams used similar characteristics, while others did not. Final groups may vary.
3. Different classification systems would cause confusion and difficulty in interpreting scientific observations. One system reduces this confusion.

Assessment

Portfolio: Ask students to classify ten flowers from local plants and place the classification they develop in their portfolios. Encourage students to include drawings of the flowers they select. L1 P

Visual Learning

Figure 20.5 In what ways are a guinea pig (left) and a mouse (right) different? *Responses may include that the guinea pig is larger than the mouse, lacks a tail, and has a shorter snout.*

MiniLab

How can you classify seeds?

Seeds are plant structures that contain an embryo plant. They also contain a source of stored food and a protective seed coat. Many seeds have winglike structures or small hooks that aid in their dispersal to new areas, where they may begin to grow.

Procedure

1. Work as a team with four other students. Choose a characteristic that will allow you to classify the ten seeds your teacher gives you into two groups. Size, shape, color, or structure are some possible characteristics.
2. Record the characteristic used for your first groups in a chart provided by your teacher.
3. Within each of the two groups, form smaller categories by choosing characteristics that will allow you to separate the seeds of each group into subgroups. Record these characteristics.
4. Continue following step 3 until only one seed remains in each group.

Analysis

1. What characteristics did you use to group your seeds?
2. Did other teams use the same characteristics? How did their criteria affect their final groups compared with your final groups?
3. Why is it an advantage to scientists to use one standard classification system?

Figure 20.5

In what ways are a guinea pig (left) and a mouse (right) different? One of the most important differences cannot be seen. Of the 51 amino acids in insulin, a protein hormone, guinea pigs and mice have differences in 18. Humans and mice also differ by the same number of amino acids in their insulin molecules. So, based on amino acid comparisons, it could be concluded that guinea pigs are no more related to mice than humans are.

Relationships can be determined by development

In addition to similarities in structures between modern and fossil organisms, taxonomists have examined developmental stages of animals for similarities to determine their relationships and phylogeny. Biologists have found that even though adults of certain species may look different, the larval form of one species may show a resemblance to the adult of another species. Taxonomists may conclude, based on developmental evidence, that these two animals have a common origin.

Relationships can be determined by biochemistry

Taxonomists also use information from biochemical analyses of organisms to compare and classify species. Look at the guinea pig shown in *Figure 20.5.* Long thought to be rodents, guinea pigs were reclassified in a separate group because the amino acid sequences in some proteins of guinea pigs have been found to be significantly different from rodents. Closely related species have similar DNA, and therefore, similar proteins. In general, the more amino acid or nucleotide base sequences that are shared by two species, the more closely related they are.

Meeting Individual Needs

Gifted Have students prepare a report on the use of cytochrome C to determine the relatedness of organisms. Have the students use at least two visual aids such as transparencies or large colorful charts to present their findings to their classmates. L3 IS

Relationships can be determined by behavior

Behavioral patterns are also examined by taxonomists. Sometimes, behavioral patterns provide important clues to relationships. For example, two species of frogs, *Hyla versicolor* and *H. chrysoscelis*, live in the same area and have the same physical appearance. During the breeding season, the males of each species make distinctive calls to attract mates. Scientists used both DNA analysis and analysis of the distinctive mating calls to identify the frogs as being two separate species.

Scientific Names

The common names of organisms do not tell you how the organisms are related or classified. Common names can be misleading. A sea horse is a fish, not a horse. In addition, confusion can occur when an organism has more than one common name. The bird in *Figure 20.6* lives not only in the United States but also in several countries in Europe, and in each country it is identified by a different common name. Suppose an English scientist publishes an article on the bird's behavior and identifies it by its English common name. A Spanish scientist might not recognize the bird as being the same species that lives in Spain.

All newly discovered species are given their scientific names in Latin. Latin is the language chosen by taxonomists because it is no longer spoken and, therefore, does not change as spoken languages do. It is important that scientific names remain the same for years to come. Other naming rules include printing scientific names in italics or underlining them when they are written, and making the first letter uppercase for the genus but not for the second word of the binomial.

Figure 20.6

In the United States and England, this bird is called the house sparrow, in Spain the gorrion, in Holland the musch, and in Sweden the hussparf. Because of the potential confusion caused when organisms have more than one common name, scientists from all countries use the same language for names of organisms. The scientific name for this bird is *Passer domesticus.*

Section Review

Understanding Concepts

1. Give two reasons why binomial nomenclature is useful.
2. What kinds of data can be used to classify organisms?
3. What did Linnaeus contribute to the field of taxonomy?

Thinking Critically

4. Why is phylogenetic classification more natural than a system based on characteristics such as usefulness in medicine, or shapes, sizes, and colors of body structure?

Skill Review

5. **Classifying** Make a list of all the furniture in your classroom or your room at home. Classify it into groups based on structure and function. For more help, refer to Organizing Information in the *Skill Handbook*.

3 Assess

Check for Understanding

Have students compare and contrast classification of organisms with the classification of books in a library. *Although both systems provide a means of organizing information, the systems differ in the types of information used to create categories and in the number of categories used.* L1

Reteach

Have students list features biologists use to classify organisms. Ask them to describe each feature they list. L1

Extension

Have students find the scientific names of the wolf, fox, domestic dog, and coyote and explain which organisms are most closely related. L1

Assessment

Performance: Have students collect and classify five local weeds. Have students use field guides to identify the weeds. They should include their work in their portfolios. L1 P

4 Close

Activity

Give students working in groups of two, fruits or vegetables. Ask them to make up binomials for these "species." L1

Answers to Section Review

1. Common names do not indicate how organisms are related or classified. The genus name is the same for closely related species; the species name is descriptive of the organism.
2. Similarities in structure, chemistry, behavior, stages in development, fossil remains, and evolutionary relationships.
3. He developed the binomial system for naming organisms and the basis of classification of organisms used today.

Thinking Critically

4. The phylogenetic system is more natural because it is based on relationships and evolutionary history among organisms.

Skill Review

5. Students may classify furniture into groups based on functions such as for sitting, for writing and working, for eating, for holding books, for lighting, for entertainment and education, and for supporting objects.

SECTION

20.2

Prepare

Key Concepts

Students will examine a pictorial model of phylogenetic classification and learn how organisms in the six kingdoms are distinguished.

Block Scheduling

Look for this symbol for strategies that are useful in a block scheduling format. For more information on block scheduling, refer to Section 20.2 in the **Lesson Plans** booklet.

Materials

- For the Minilab, gather leaves and guidebooks with dichotomous keys for local trees. Your state Bureau of Forestry may be able to provide you with guidebooks.
- For the Biolab, obtain metric rulers and guidebooks with dichotomous keys of insects.

1 Focus

Bellringer

Before presenting the lesson, display **Section Focus Master 43** on the overhead projector and have students answer the accompanying questions. L1 SAE

Demonstration

LS **Visual-Spatial** Display organisms or pictures of representative organisms from each kingdom. Ask students to use their observations of the organisms to identify the traits that characterize organisms in each kingdom. L1

SECTION

20.2 The Six Kingdoms

Section Preview

Objectives

Compare the six kingdoms of organisms.

Distinguish between the Kingdoms Eubacteria, Archaebacteria, Protista, Fungi, Plantae, and Animalia.

Key Terms

protist
fungus

Do you remember looking in the newspaper to find out what movies were in town? Suppose you decide to rent a video instead of going to the movies. When you go to the video store, you find that movies are classified into categories such as comedy, drama, horror, western, and adventure. The numbers of these categories may vary from store to store. Although organisms can be classified in different ways, most taxonomists now group them into six kingdoms.

Phylogenetic Classification: A Model

phylogenetic: *phylon* (GK) related group *geny* (GK) origin A phylogenetic classification system is based on evolutionary relationships.

Aristotle's classification system and other early systems were developed before scientists had studied geologic time. Therefore, phylogenetic relationships were not originally used by taxonomists; these early systems reflected only differences in structures of organisms. As a result, the number of kingdoms in early classification schemes varied. As biologists began to unravel the trends in evolution of species, these early classification schemes were modified to reflect evolutionary relationships. Aristotle's two categories of plants and animals were eventually replaced by six kingdoms. The six-kingdom system is used by most biologists today because they think it best reflects phylogeny.

Using the fan diagram

Pictorial models like the one you'll see as you turn the page often are helpful tools for understanding scientific concepts. *Figure 20.7* is a model of the six-kingdom classification system. The model looks like a fan with all the modern-day major groups of organisms placed around the outer edges. Thus, the outermost layer of the model represents present time. Within the body of the fan, geologic time is also represented. Using similar fan diagrams throughout this book, it is possible to identify extinct groups of species.

Identify the six kingdoms in this first model. Within each kingdom, there are examples of present-day organisms. Recall from Chapter 17 that scientists have divided Earth's history into various eras and periods that make up the Geologic Time Scale. Notice that in this model, the center of the fan's base represents the origin of life. The rays of the fan show the evolution of modern-day species from a common origin. Groups joined closely together in the rays of the fan share many characteristics; groups

Program Resources

Section Focus Master 43 L1 SAE
Reinforcement and Study Guide, pp. 79-80 L1
Biolab and Minilab Worksheets, pp. 80-82 L1
Concept Mapping, p. 20 L1
Basic Concepts Transparency 24 and **Master** L1 SAE
Reteaching Transparency 20 and **Master** L1 SAE

that are far apart are very different and, therefore, probably not as closely related.

As you study the taxonomic groups in this textbook, refer to this fan diagram on the next page or to the one inside the back cover. You will learn not only about taxonomic groups, but also about their phylogenetic relationships.

How the Six Kingdoms Are Distinguished

As you have seen in the model of phylogenetic classification, the organisms comprising the six kingdoms are true bacteria, archaebacteria, protists, fungi, plants, and animals. In general, the kingdoms can be distinguished by characteristics at the cellular level and by methods of obtaining food.

Microscopic prokaryotes

Kingdoms Eubacteria and Archaebacteria contain all prokaryotes, cells without nuclei bound in membranes. Prokaryotes are microscopic, and almost all are unicellular. All are commonly known as bacteria. The first prokaryotes appeared in the fossil record about 3.5 billion years ago. More than 5000 species have been identified, and most are in Kingdom Eubacteria, the true bacteria. A few, the archaebacteria, the ancient bacteria, are found in extreme environments such as salt lakes, swamps, and deep-ocean hydrothermal vents. Many prokaryotes remain to be discovered, named, and described.

Art

The Terrestrial Paradise

by Jan Brueghel the Elder

The Terrestrial Paradise was painted by Jan Brueghel the Elder (1568–1625) in the 16th century. This pictorial inventory of exotic plants and animals was typical of the decorative style of Flemish painters of the time. Northern European artists of this period were known for their realistic style of painting, and some of the greatest paintings in this style were done by artists of Flanders, a region that lies in what is now Belgium and France.

The Terrestrial Paradise was a good representation of the diversity of life, as viewed in that period of time before people began exploring new lands and discovering the existence of vast numbers of species. Brueghel filled his canvas with rich details of the more fantastic, as well as the familiar, mammal and bird species known in Europe. As you can see, he depicts mainly the species that were the most useful or attractive to humans. Most of the animals shown are birds or mammals. The plant life is represented mainly by flowering plants. It would not be until the next century that the great natural scientist Carolus Linnaeus would develop the science of taxonomy, the systematic naming and recording of species.

CONNECTION TO Biology

How do ideas about the diversity of life in the 16th century compare with modern ideas about the diversity of life?

SECTION 20.2

2 Teach

Tying to Previous Knowledge

Review the meanings of the terms *prokaryote* and *eukaryote.* Have students speculate about what prokaryote traits may have combined to form the first eukaryotes. *The ability to photosynthesize is probably one of the most important.* L1

Art

The Terrestrial Paradise

Purpose

Students will examine how people of the 16th century viewed the diversity of organisms by examining a painting from that period.

Teaching Strategies

- List the organisms shown in *The Terrestrial Paradise* on the chalkboard. Ask students to identify the groups of organisms represented. L1
- After students have read the feature, ask them to list organisms they might include in a modern version of *The Terrestrial Paradise.* L1

Possible Answer

Today scientists think that there are between five and 30 million species. In the 16th century, knowledge of diversity was limited to species useful or attractive to humans: birds, mammals, food plants, trees, and flowering plants.

PROJECT

Classifying Organisms

Ask students to make a photo essay of the six kingdoms of life. Ask students to label each organism with its scientific name and common name. They should explain their essay to their group members who should evaluate it for accuracy, thoroughness, and interest.

interNET CONNECTION

Follow the link for this chapter on the Glencoe Homepage at **www.glencoe.com/sec/science** to find out more about organizing life.

Visual Learning

Figure 20.7 Ask students to explain how time is shown on the diagram. *Organisms that appeared on Earth earlier in its history are shown nearest the center of the fan, while those appearing later are shown nearer the rim.* LS

Bioethics

Funding Arthropod Research

Currently, few scientists have backgrounds in the systematics of medically important arthropods such as mosquitoes, flies, ticks, and chiggers. Some people believe serious problems may result in the control of these arthropods and preventive measures against diseases they carry if money is not allocated to arthropod systematics. For example, the six species of mosquitoes responsible for one million deaths each year due to malaria can be identified only by DNA probes and chromosome banding, procedures for which considerable expertise is required. Some scientists think funding for research on arthropod biosystematics needs to be increased to help ensure the future health of humans.

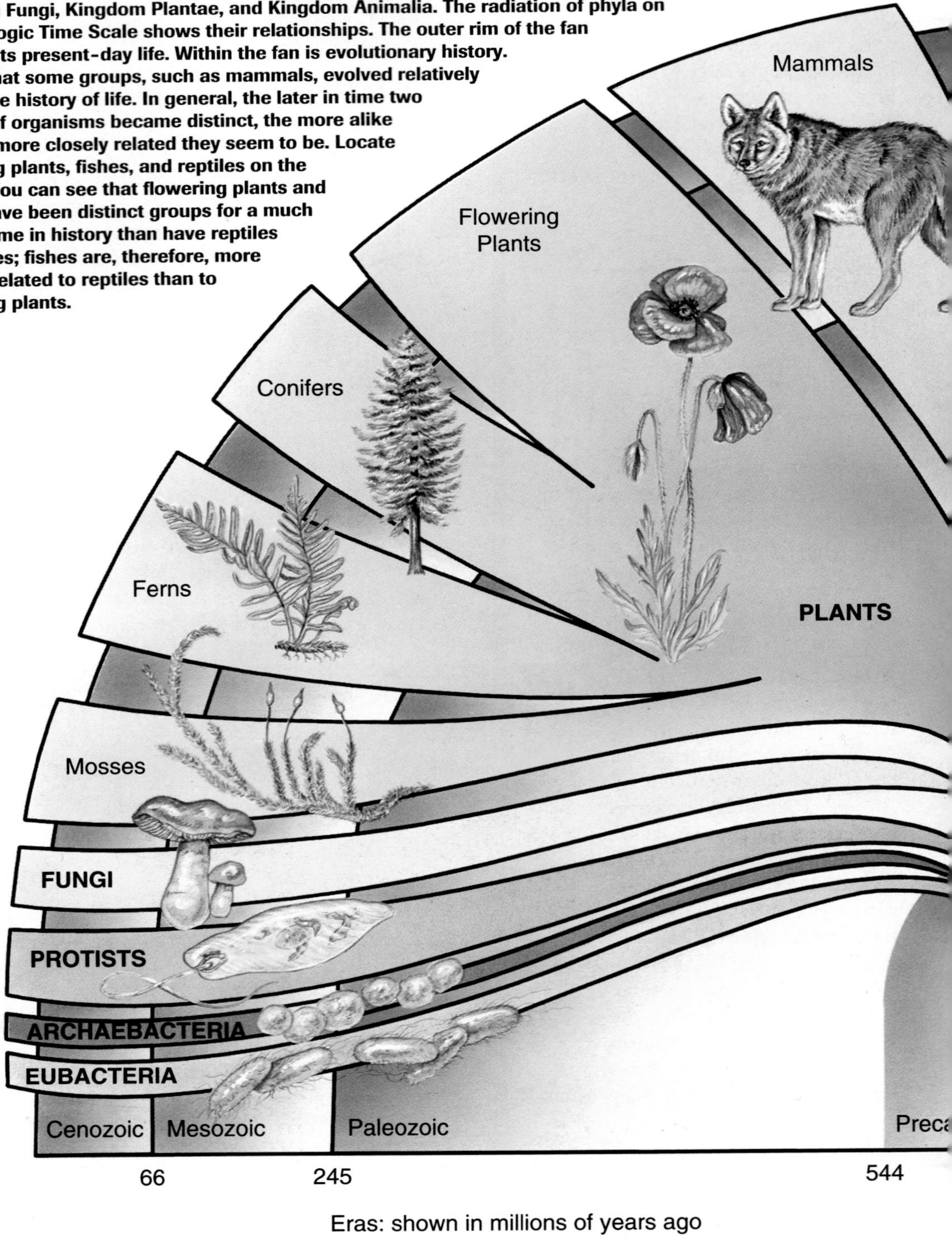

Figure 20.7

The six kingdoms of life are Kingdom Eubacteria, Kingdom Archaebacteria, Kingdom Protista, Kingdom Fungi, Kingdom Plantae, and Kingdom Animalia. The radiation of phyla on the Geologic Time Scale shows their relationships. The outer rim of the fan represents present-day life. Within the fan is evolutionary history. Notice that some groups, such as mammals, evolved relatively late in the history of life. In general, the later in time two groups of organisms became distinct, the more alike and the more closely related they seem to be. Locate flowering plants, fishes, and reptiles on the model. You can see that flowering plants and fishes have been distinct groups for a much longer time in history than have reptiles and fishes; fishes are, therefore, more closely related to reptiles than to flowering plants.

Alternate Lab: Classification of Laboratory Glassware

Purpose

LS **Logical-Mathematical** Students will classify laboratory glassware according to various characteristics.

Materials

glassware such as flasks, mortars, test tubes, beakers, stirring rods, pestles, thermometers, stoppers, pipets, droppers, crucibles, funnels, petri dishes, graduated cylinders, bottles, and jars

Procedure

Give the following directions to students.

1. Obtain a set of glassware from your teacher.
2. Classify the glassware according to structural and functional similarities.
3. Begin by classifying the glassware into two groups. Continue to classify the glassware into smaller groups and place the characteristics used and the names of the glassware into taxonomic groupings.
4. Examine the groups you created as if they were living organisms. Place the glassware in a phylogenetic model such as the fan diagram shown on this page.

Life's Six Kingdoms

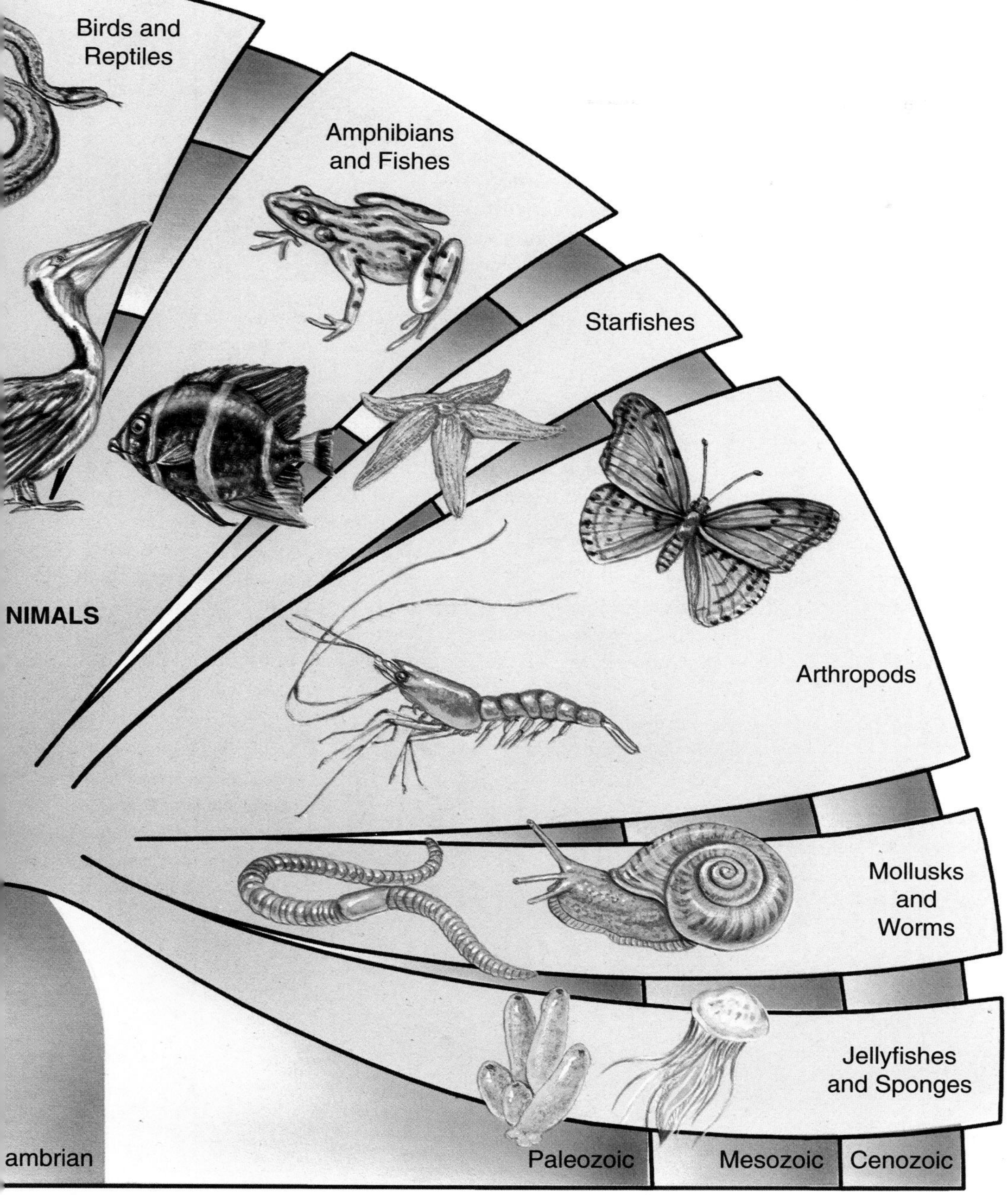

Birds and Reptiles

Amphibians and Fishes

Starfishes

NIMALS

Arthropods

Mollusks and Worms

Jellyfishes and Sponges

ambrian

Paleozoic

Mesozoic

Cenozoic

Brainstorming

Have students use the information shown in Figure 20.7 to identify other examples of organisms that belong in each group shown. Have students explain why they think the organisms belong in each group. L1

Time Line

LS **Visual-Spatial** Have students create a time line that shows the evolutionary history of life on Earth using Figure 20.7 as a model. Remind students that their time lines represent millions of years. L1

5. Make a large diagram of your phylogenetic model. Think about how the glassware may have "evolved" and why various groups may be more closely "related" than others. The groups that are "more closely related" should be placed closer together on your diagram.

Analysis

Ask students to answer the following questions.

1. What features of glassware were most useful in your classification? *Responses are likely to indicate functional traits.*
2. What were the major differences in your phylogenetic model compared with the model in your textbook? *The model created in this activity does not use actual knowledge of when materials were invented or produced. Instead, the data are likely to be completely speculative.*

Assessment

Portfolio: Have students write an essay about the "evolution" of their glassware. They should justify their statements with evidence from examination of the structures and functions of the glassware. L1 P

Science, Technology, and Society

Satellite Investigations

Marine ecosystems are among the most diverse in terms of genetics, taxonomy, and ecology. Like tropical rain forests, many marine ecosystems are in a state of crisis. Many scientists report pathogens, toxic substances, declining biodiversity, and other disruptions of marine habitats.

To determine the extent of such disruptions, scientists need to dive into the ocean using scuba gear or submarines. They also need to use state-of-the-art instruments such as satellites. Comparisons of images are obtained using satellites made to determine changes in environment. Laboratory techniques are used to assess genetic and molecular information. Computers are used for modeling ecosystems and the changes they are undergoing. Examination of such environmental changes may help preserve biodiversity for the benefit of future generations.

Visual Learning

LS Visual-Spatial Use the organisms in Figures 20.8 and 20.9 to emphasize the major differences between bacteria and protists. Be sure students understand that bacteria lack a nucleus and membrane-bound organelles, whereas protists have these structures.

Figure 20.8

Eubacteria have a variety of structures and metabolic systems.

Magnification: 123 700×

Some bacteria, such as *Nitrosomonas*, produce food by metabolizing such simple molecules as ammonia and methane.

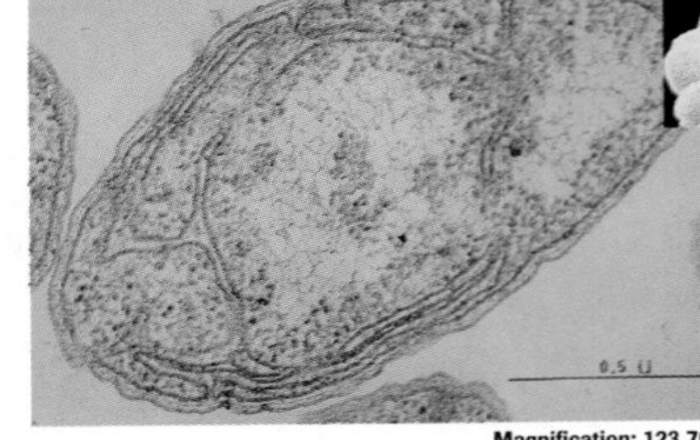

Magnification: 40 000×

The bacterium that causes strep throat, *Streptococcus*, must get its food from an outside source.

Magnification: 100×

Cyanobacteria, such as *Oscillatoria*, live in fresh water and produce food by photosynthesis.

The prokaryotes shown in *Figure 20.8* obtain their food from many diverse metabolic pathways.

Protists—A diverse group

Kingdom Protista includes unicellular and multicellular organisms with a variety of characteristics—some plantlike, some animal-like, and some funguslike. A **protist** is a eukaryotic organism that lacks complex organ systems and lives in moist environments. Protists first appeared in the fossil record about 1 billion years ago. The protists in *Figure 20.9* show the great diversity in this kingdom.

Figure 20.9

These three protists may seem to have little in common. Review the definition of a protist, and speculate about what features they share.

A slime mold can creep along the forest floor like an animal, but later produces reproductive structures, shown here, that are similar to those made by a fungus.

The paramecium is an animal-like, unicellular protist that can move rapidly as it captures smaller protists for food.

Magnification: 140×

You may have seen large kelps washed up on the shore. Although kelp may look like a plant, it does not have tissues organized into organs or organ systems.

Meeting Individual Needs

Visually Impaired Fold a sheet of construction paper into a fan shape or obtain an actual fan. Open the fan to allow visually impaired students to feel its shape. Explain that in the phylogenic model used, prokaryotes are at the place where the fan forms a point (its center). Just above this point are eukaryotes, which evolved from prokaryotes. LS

PORTFOLIO

Gathering Protists Ask students to get water samples from local ponds or streams to examine for protists. Instruct them to stir the sediments and scrape the underneath surfaces of rocks into collecting containers. Ask them to make sketches of the protists they find. They should justify why they think each organism is a protist. L2 LS P

Figure 20.10

Morels are edible fungi that are considered delicacies. These gourmet delights grow for only a few days in limited habitats.

Fungi—Earth's decomposers

Organisms from the Kingdom Fungi are consumers that do not move from place to place. A **fungus** is a unicellular or multicellular heterotrophic eukaryote that absorbs nutrients obtained by decomposing dead organisms and wastes in the environment. Fungi first appeared in the fossil record about 400 million years ago. More than 100 000 species have been named. You may be familiar with the fungus in *Figure 20.10.*

Plants—Earth's multicellular oxygen producers

The organisms in the Kingdom Plantae are stationary, multicellular eukaryotes that photosynthesize. Most have cellulose cell walls and tissues organized into organs and organ systems. Because plant materials do not fossilize as easily as the bones of animals, the oldest plant fossils that have been found are only 400 million years old. Half a million species have been described. You are probably familiar with flowering plants like the one shown in *Figure 20.11;* however, the mosses, ferns, and their relatives represent plants with some distinctly different characteristics.

MiniLab

How is a dichotomous key used?

How could you find out the name of a tree you see growing in front of your school? You might consult a local expert, or you could use a manual or field guide that contains descriptive information and keys to aid identification. A key is a set of descriptive sentences divided into steps. A dichotomous key has two descriptions at each step. The steps are followed until the key leads to the name of the tree.

Procedure

1. Using a collection of leaves from local trees and a dichotomous key for trees of your local area, determine the names of the trees from which the leaves came. To use the key, begin with a choice from the first pair of descriptions. Continue following the key until the name of each tree is found.
2. Glue each leaf on a separate sheet of paper. Record the tree name for each leaf.

Analysis

1. What is the function of a dichotomous key?
2. List three different characteristics used in your key.
3. As you used your key, did the characteristics become more general or more specific?

Figure 20.11

You can see the cell wall and chloroplasts typical of plant cells in the enlarged cell of this hibiscus plant. The structures visible in the center of the flower are male and female reproductive organs.

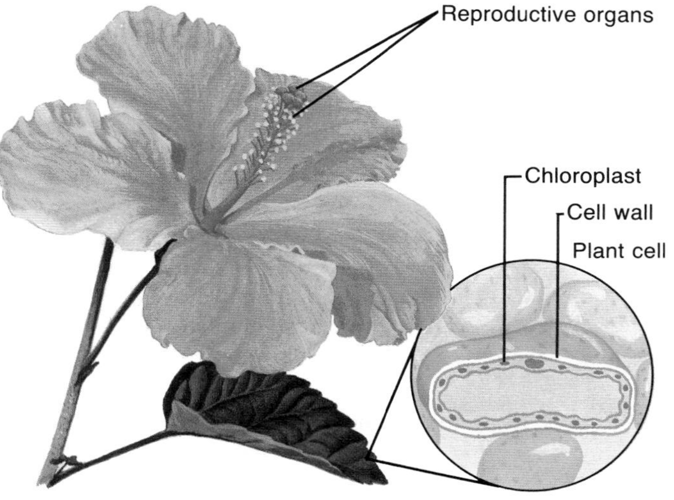

CULTURAL DIVERSITY

Adelmar Coimbra-Filho and the Classification of Lion Tamarins

Introduce students to the research of Adelmar Coimbra-Filho, Brazil's leading primatologist. Since the 1960s, Coimbra-Filho has published many articles on the systematics of the lion tamarin, *Leontopithecus,* a primate that has received much attention as a recently discovered species. Currently, Coimbra-Filho is the director of Rio de Janeiro's Primate Center. He has directed several joint efforts with the World Wildlife Fund to save lion tamarins from extinction. Elicit from students the importance of the work done by taxonomists and systematicists in discovering new species.

MiniLab

Purpose

LS Logical-Mathematical Students will learn how to use a dichotomous key to identify organisms.

Process Skills

classify, compare and contrast, observe and infer

Teaching Strategies

- Make sure to select field guides that make identification based on leaf structure rather than on bark or fruit. If you do not have trees in your area, you might try local shrubs.
- If possible, obtain leaves on the same day students will carry out the activity to avoid altering leaf appearance through wilting and changes in color.
- You may want to make a transparency of a key to show students how choices are made from each pair of descriptions.

Expected Results

Students will be able to identify local trees based on their leaves.

Analysis

1. identification of organisms
2. structure of veins or margins, shape of leaf, size of leaf, number of lobes
3. more specific

Assessment

Performance: Provide students with samples of algae and a dichotomous key for algae. Ask them to observe the algae under a microscope and diagram their observations. Have students identify several types of algae using the dichotomous key. L1 P

BioLab

Making a Dichotomous Key

Time Allotment 
One class period

Objectives
Review objectives with students before they begin the Biolab.

Process Skills
classify, observe and infer, compare and contrast, interpret data, sequence

PREPARATION

Alternate Materials
- Students can make a dichotomous key of Styrofoam packing pieces or kitchen utensils of a variety of sizes, shapes, and colors.

GLENCOE TECHNOLOGY

Videodisc

STVS: Animals
Disc 5, Side 1
Insect Museum (Ch. 22)

The Infinite Voyage: Insects: The Ruling Class
Insects and Their Behavior (Ch. 1)

BioLab

Making a Dichotomous Key

Do you remember the first time you saw a beetle? You may have asked someone nearby, "What is it?" You may still have that natural curiosity and want to know the names of insects you find in your yard. To help identify organisms, taxonomists have developed classification keys. A dichotomous key is a set of paired statements that can be used to identify organisms. When you use a dichotomous key, you choose one statement from each pair that best fits the organism. After each choice, you are directed to the next set of statements you should use. When you have made all the choices needed, you will arrive at the name of the organism or the group to which it belongs.

PREPARATION

Problem
How is a dichotomous key made?

Objectives
In this Biolab, you will:
- **Classify** organisms on the basis of structural characteristics.
- **Develop** a dichotomous key.

Materials
sample keys from guidebooks
metric ruler

PROCEDURE

1. Study the drawings of beetles.
2. Choose one characteristic and classify the beetles into two groups based on that characteristic. Take measurements if you wish.
3. Record the chosen characteristic in a diagram like the one shown. Write the numbers of the beetles in each group on your diagram.
4. Continue to form subgroups within your two groups based on different characteristics. Record the characteristics and numbers of the beetles in your diagram until you end with one beetle in each group.
5. Using the diagram you have made, make a dichotomous key for the beetles. Remember that each numbered step should contain two alternative choices for classification. Begin with 1A and 1B. For help, examine sample keys provided by your teacher.
6. Exchange dichotomous keys with another team. Use their key to classify the beetles.

PROCEDURE

Teaching Strategies
- Have students examine samples of dichotomous keys for plants and animals.
- Explain that a well-designed key has a pattern of contrasting traits for each pair of choices.
- Make sure students understand that once a beetle is identified, the key is no longer used in developing choices.
- Have groups put their keys on transparencies and explain them to the class. Ask the class to point out the good points of the keys and what needs to be corrected.

Data and Observations
Ask students to exchange their keys with other groups. If another group can use the key, the accuracy of the key can be confirmed.

1
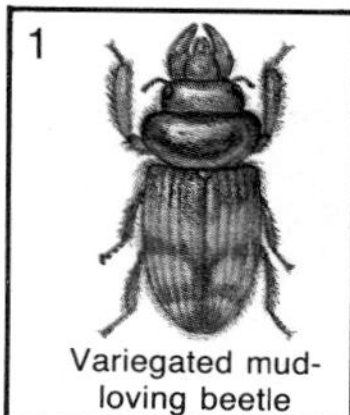
Variegated mud-loving beetle

2
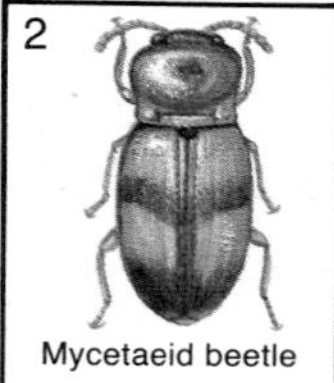
Mycetaeid beetle

3

Apricot borer

4
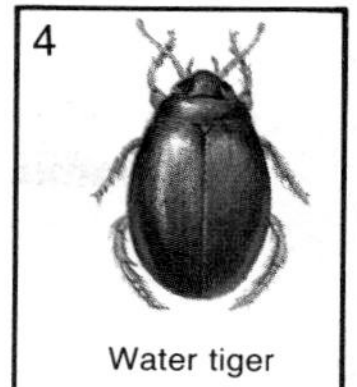
Water tiger

5

Predaceous diving beetle

6
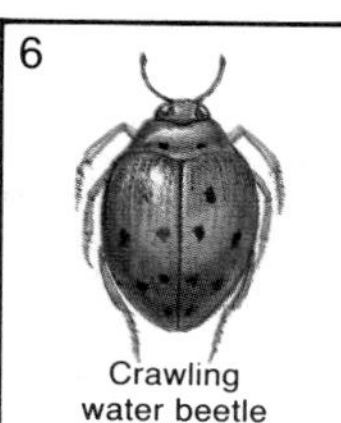
Crawling water beetle

7
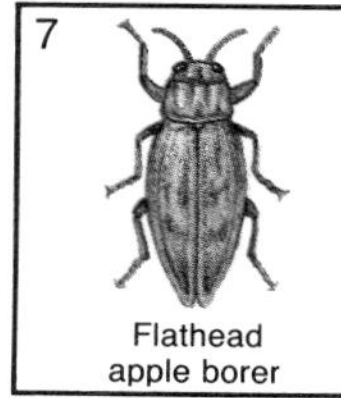
Flathead apple borer

8
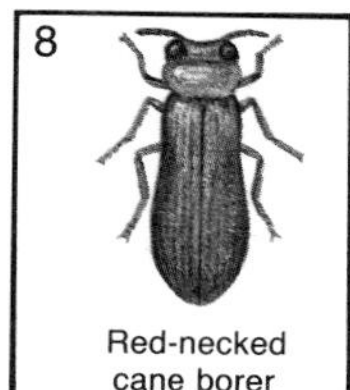
Red-necked cane borer

9
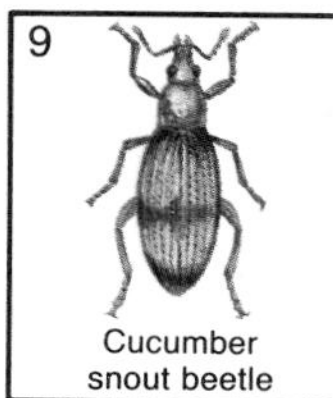
Cucumber snout beetle

10
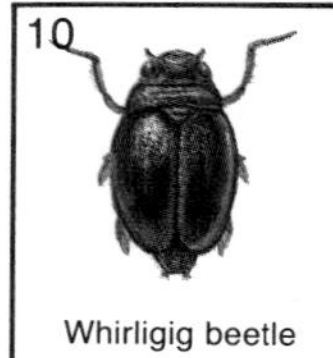
Whirligig beetle

11

Ironclad beetle

12
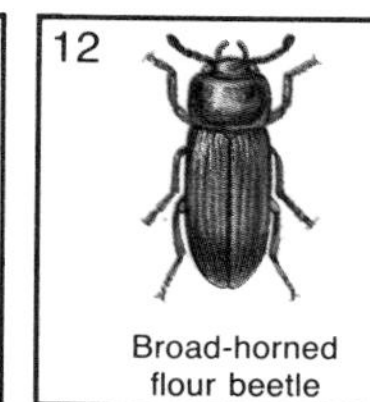
Broad-horned flour beetle

13
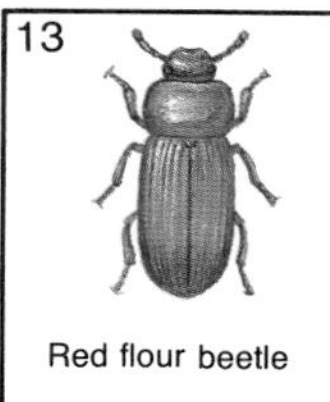
Red flour beetle

14
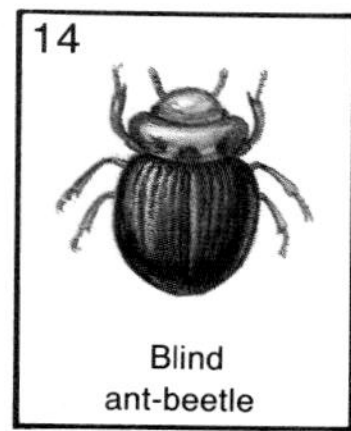
Blind ant-beetle

15
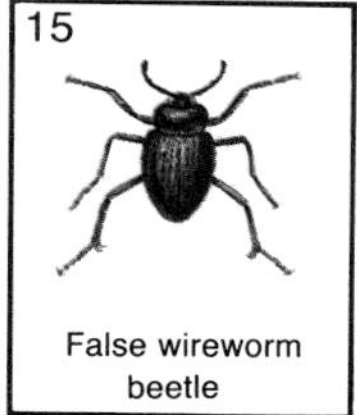
False wireworm beetle

16
White-marked spider beetle

17
Monterey cyprus beetle

18
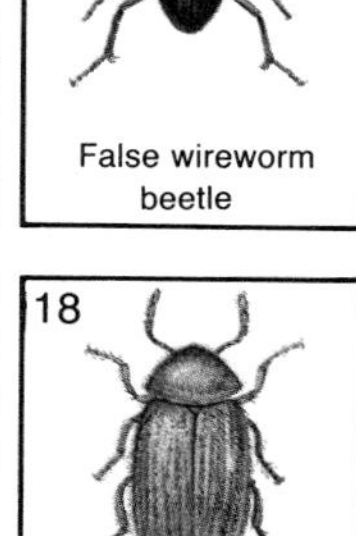
Drug store beetle

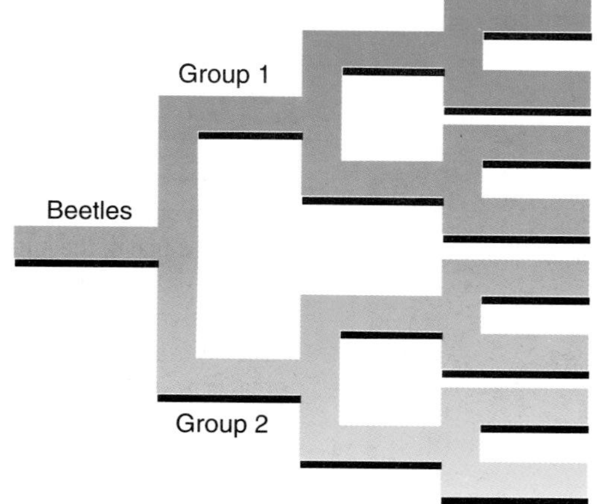

1. **Comparing and Contrasting** Was the dichotomous key you constructed exactly like those of other students? Explain.
2. **Analyzing Data** What characteristics were most useful for making a classification key for beetles? What characteristics were not useful?
3. **Thinking Critically** Why do keys typically offer two choices rather than a larger number of choices?

Going Further

Application Using the same method, make a dichotomous key to identify the students in your class.

ANALYZE AND CONCLUDE

1. The key of one group may or may not be like those of other groups. Some groups may have compared size first, or color first, and other body features later.
2. Useful: size, color, and shape of various body parts, distinctive or non-distinctive number of body sections, and features of antennae. Not useful: number of legs, number of antennae, hardness and shininess of body covering, habitat.
3. Having more than two choices would make it difficult to analyze organisms. Many keys are constructed so the choice is that the organism either has or does not have a particular trait.

Assessment

Performance: Give each group a bag of mixed beans. Ask them to make a dichotomous key for the beans. Have them exchange keys with other groups to see if their key works. Ask them to put their keys in their portfolios. L1 P

Going Further

Have students use 10 to 20 features of other students, such as hair color, or items that belong to them, such as shoes, notebooks, and backpacks. L1

GLENCOE TECHNOLOGY

Videodisc

The Infinite Voyage: Life in the Balance
BioLat II: Taking Inventory of the Threatened Peruvian Rain Forest (Ch. 4)

SECTION 20.2

Visual Learning

Figure 20.12 What senses might the cheetah use while running? *sight, touch, and smell*

3 Assess

Check for Understanding

Show students pictures of organisms from all six kingdoms. Ask them to identify the kingdom to which each belongs. L1

Reteach

Have students make flash cards with a photo of an organism on one side and the name of the kingdom and its distinguishing features on the other. Have them review the cards in groups. L1

Extension

Ask students to determine classification beyond the species level for show dogs or cats and explain why this additional classification is used. L2

Assessment

Performance: Set up fifteen stations—each displaying a numbered specimen. Use microscopes for bacteria and protists and pictures of larger organisms. Ask students to move from one station to the next and identify the kingdom to which each organism belongs. L1

4 Close

Activity

Give groups of students a large piece of butcher paper and markers. Ask them to make a quick sketch of a phylogenetic model similar to the one on pages 492 and 493. L1 SAE

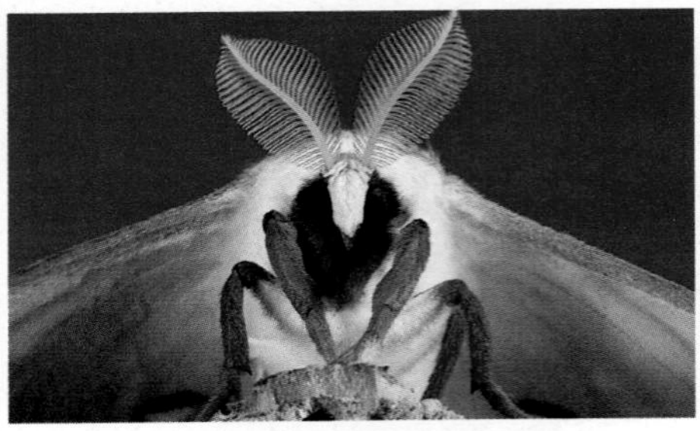

The luna moth's antennae are sense organs. With these antennae, the moth can detect minute quantities of chemical odors in the air.

The cheetah's brain, a part of its nervous system, keeps it on course as it speeds through the grasslands using its powerful muscles. What senses might the cheetah use while running?

Figure 20.12
Most animals have well-developed nervous and muscular systems.

Animals—Multicellular consumers

Animals are multicellular consumers that eat and digest other organisms for food. Animal cells have no cell walls. Nearly all animals are able to move and have tissues organized into organs and complex organ systems. Nervous systems, muscle systems, and sense organs are common to many animals, *Figure 20.12.* Animals first appeared in the fossil record 700 million years ago.

Connecting Ideas

Taxonomists have made logic and sense of much of the information gathered over the past century through the study and observation of living things. As you begin to examine phylogeny and study organisms of the six kingdoms, you will come to appreciate even more how much biology depends on the work of taxonomists. They have given us a useful tool that is continually modified as new information about organisms becomes available. The species that make up the six kingdoms of life will be presented in this text in the order of their phylogenetic relationships. These relationships represent a trend in evolution from unicellular species through the highly complex plants and animals.

Section Review

Understanding Concepts

1. Describe the six kingdoms of life based on methods that organisms use to obtain food.
2. How do prokaryotes differ from the other four kingdoms?
3. How is the fan diagram a useful model for phylogenetic classification?

Thinking Critically

4. If the present system of classification were changed to one with more kingdoms, which of the present kingdoms do you think could be further divided? Explain your answer.

Skill Review

5. **Making and Using Tables** Make a table that compares characteristics of members of each of the six kingdoms. For more help, refer to Organizing Information in the *Skill Handbook.*

Answers to Section Review

1. Prokaryotes and protists obtain food by photosynthesis or chemosynthesis. Fungi obtain food by absorption. Plants photosynthesize and animals obtain food by eating other organisms.
2. Prokaryotes lack membrane-bound organelles and have simple flagella.
3. Groups joined closer together in the rays of the fan share many characteristics. Groups that are far apart are very different and not as closely related. It also shows groups that evolved early in geological time and those that evolved later.

Thinking Critically

4. Students will probably suggest that the protist kingdom could easily be divided into more kingdoms because it has members with such diverse characteristics.

Chapter 20 Review

REVIEWING MAIN IDEAS

20.1 Classification

- Aristotle developed the first classification system based on the similarity of organisms and where they live.
- Modern classification systems began with Linnaeus's organization of living things based on their similarities in structure. This system naturally led to the phylogenetic classification most commonly used today.
- Species are named by a two-word system called binomial nomenclature.
- Classification provides a framework of logic and order so that relationships among living things can be seen easily.
- Organisms are classified by using a hierarchy of taxa. From largest to smallest, the hierarchy is kingdom, phylum or division, class, order, family, genus, and species.
- The criteria used by biologists to classify living things are similarities in phylogeny, structure, development, biochemistry, and behavior.

20.2 The Six Kingdoms

- The six kingdoms are represented by the archaebacteria and eubacteria, prokaryotic cells without nuclei bound in membranes; protists, eukaryotic organisms that lack complex organ systems and live in water or moist environments; fungi, eukaryotes that obtain food by absorption; plants, multicellular eukaryotes that produce their own food; and animals, multicellular consumers whose cells lack cell walls.

Key Terms

Write a sentence that shows your understanding of each of the following terms.

binomial nomenclature	genus
class	kingdom
classification	order
division	phylogeny
family	phylum
fungus	protist
	taxonomy

Understanding Concepts

1. Two categories in the same phylum must share what other classification category?
   a. kingdom c. genus
   b. class d. species
2. Information from which of the following would be most useful for classifying a newly discovered organism?
   a. fossil record
   b. DNA comparisons
   c. structural comparisons
   d. developmental stages
3. Linnaeus based most of his classification system on ____________.
   a. cell organelles
   b. biochemical comparisons
   c. structural comparisons
   d. embryology
4. Which of the following correctly identifies a species?
   a. *bison bison* c. *Homo sapiens*
   b. *Mimus Polyglottis* d. *quercus Alba*
5. The second part of a binomial name is called the ____________.
   a. genus c. class
   b. general d. species
6. Unlike Linnaeus, Aristotle did not base his classification system on ____________.
   a. whether an organism flies
   b. where an organism lives
   c. apparent similarities
   d. phylogenetic relationships
7. The evolutionary history of a species is its ____________.
   a. kingdom c. taxonomy
   b. ontogeny d. phylogeny

Chapter 20 Review

Reviewing Main Ideas

Summary statements can be used by students to review the major concepts of the chapter.

Key Terms

Answers should go beyond defining the terms. Accept any answer that uses the term correctly and in the proper context.

Understanding Concepts

1. a
2. b
3. c
4. c
5. d
6. d
7. d

Skill Review

5. Material for tables can be found on pages 490 to 498. Make sure students have listed all six kingdoms.

Chapter 20 Review

8. c
9. b
10. a
11. d
12. d
13. b
14. a
15. a
16. c
17. c
18. a
19. b
20. c

Applying Concepts

21. Aristotle's system was based on a few different kinds of organisms and included such identifiers as size and where they lived, characteristics which are not useful in classifying organisms or identifying phylogenetic relationships. Linnaeus grouped organisms by similarities in structure, the physical characteristics of organisms. Aristotle's system did not show natural relationships.
22. It belongs to Kingdom Fungi. It obtains food by decomposing other organisms such as the dying tree.
23. day lily and daisy: number of flower parts; lion and cheetah, presence of spots; mouse and guinea pig: DNA; mushroom and fern: method of obtaining food
24. animals—human, bear, clam, earthworm, frog, horse, jellyfish, mosquito, turtle; plants—rosebush, pine tree, tulip, carrot; fungi—mushroom, mold on fruit; protist—seaweed, pond scum
25. Behavior evolves because of natural selection, just as other animal features evolve. Classification is based on all characteristics, including behavior, that evolve. Animal structures and behaviors are a reflection of their way of life in specific environments.

Chapter 20 Review

8. Organisms are considered to be closely related when they share ____________.
   a. similar outward appearances
   b. the same kingdom
   c. more nucleotide sequences
   d. similar sizes
9. A diverse group of eukaryotic organisms lacking complex organ systems and living in moist places are in Kingdom ____________.
   a. Archaebacteria c. Fungi
   b. Protista d. Plantae
10. A group of prokaryotes found in harsh environments is the ____________.
    a. archaebacteria c. eubacteria
    b. protists d. viruses
11. Multicellular consumers with tissues and organ systems belong to Kingdom ____________.
    a. Protista c. Plantae
    b. Fungi d. Animalia
12. Animal cells have no ____________.
    a. cell membranes c. nucleus
    b. organelles d. cell walls
13. Cells of all bacteria are ____________.
    a. large
    b. prokaryotic
    c. eukaryotic
    d. disease producers
14. A multicellular heterotrophic eukaryote that absorbs food would be classified as a(n) ____________.
    a. fungus c. animal
    b. plant d. protist
15. Stationary eukaryotes that photosynthesize and have cell walls are classified as ____________.
    a. plants c. fungi
    b. animals d. protists
16. Most scientists presently agree that there are ____________ kingdoms.
    a. three c. six
    b. five d. seven
17. All organisms except for bacteria _______.
    a. are prokaryotic c. are eukaryotic
    b. contain DNA d. use oxygen
18. An example of a unicellular protist is a ____________.
    a. paramecium c. mushroom
    b. tulip d. bacterium
19. Scientific names are in ____________.
    a. Greek c. English
    b. Latin d. Italian
20. Which of the following categories contains the others?
    a. class c. phylum
    b. genus d. species

Applying Concepts

21. Explain why Linnaeus's system of classification was more useful than Aristotle's.
22. You find an unusual organism growing on the bark of a dying tree. In science class, you look at some of its cells under a microscope. It is a multicellular organism with eukaryotic cells, but no chloroplasts. To what kingdom does it belong? How does the organism obtain food?
23. Linnaeus classified organisms based on numbers and similarities of body structures, on behavior, and on physical forms such as size and shape. In the following pairs of organisms, tell what characteristics could be used to classify each organism in a pair as a different species: daylily and daisy; lion and cheetah; mouse and guinea pig; mushroom and fern.
24. Based on your own knowledge of their characteristics, classify and give your reason for grouping the following organisms by kingdom: human, mushroom, bear, rosebush, clam, seaweed, earthworm, frog, horse, pine tree, pond scum, jellyfish, mosquito, tulip, turtle, mold on fruit, carrot.
25. Explain how the behavior of organisms became a part of the classification system used today.

Thinking Critically

26. ***Concept Mapping*** Make a concept map that relates the following terms. Supply the appropriate linking words for your map.
    classification, taxonomy, Linnaeus, Aristotle, behavior, biochemistry, structural similarities, development, phylogeny
27. ***Observing and Inferring*** Explain how the work of Linnaeus illustrates the nature of science.
28. ***Classifying*** Make a list of the features you might use to classify trees in winter.
29. ***Interpreting Scientific Illustrations*** Make sketches of utensils you might find in a kitchen drawer. Classify them into groups based on their functions.
30. ***Comparing and Contrasting*** Compare the classification system used by your school library with the classification system used for living organisms.
31. ***Interpreting Data*** A researcher compared DNA nucleotide sequences of a newly discovered animal with three different animals—A, B, C. Which animal is most closely related to the newly discovered animal? Explain.

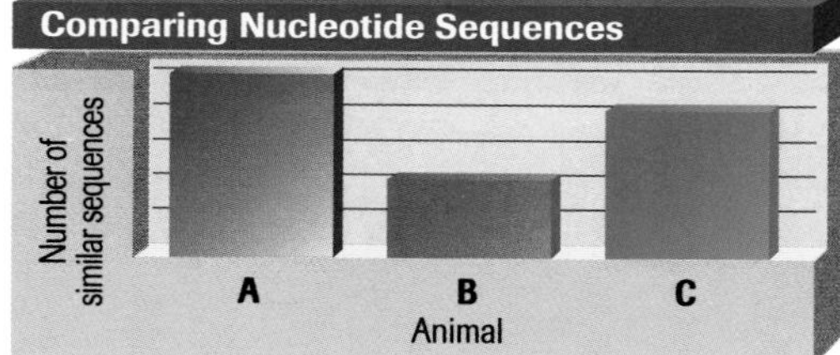

32. ***Observing and Inferring*** Suppose you find a unicellular organism swimming in a drop of water you have placed under a microscope. It is green and moves fast. You can see the nucleus clearly. In what kingdom does this organism belong? What other features would you expect this organism to have?

ASSESSING KNOWLEDGE & SKILLS

Identify the organisms in these photographs.

Key

1A Front and hind wings similar in size and shape, and folded parallel to the body when at rest..................damselflies

1B Hind wings wider than front wings near base, and extended on either side of the body when at rest........dragonflies

Classifying Use the dichotomous key above to answer the following questions.

1. ***Interpreting Data*** The insect in the photo on the right is a damselfly because it has ____________.
   a. wings that are opaque
   b. wings extended at rest
   c. smaller eyes
   d. wings similar in shape
2. The insect in the photo on the left is a dragonfly because it has ____________.
   a. wings that are opaque
   b. wings extended at rest
   c. larger eyes
   d. wings similar in shape
3. ***Classifying*** From the key and the photographs above, identify the traits that indicate dragonflies and damselflies evolved from a common ancestor.

Chapter 20 Review

Thinking Critically

26. Evaluate students' concept maps. The maps should show an understanding of the relationships among the concepts listed.
27. Linnaeus developed his classification system based on structural similarities, which are reflected in phylogeny.
28. features of the bark, twigs, buds on twigs, needle types on evergreens, height of the tree
29. utensil groups: stirring, mashing, cutting, whipping, measuring, grating, straining, basting
30. The school library groups books by similarities of their contents. The groupings do not reflect the periods in which they were written. Classification of organisms is based on similarities and evolutionary history.
31. A; animals with more nucleotide sequences that are similar are more closely related.
32. Protist; it will lack complex organ systems, live in water, and may have some plantlike, animal-like, or some funguslike characteristics.

1. d
2. b
3. two pairs of wings located on same body segment; wings transparent with clearly defined veins; long, thin body; accept any reasonable answers indicated by photographs

Program Resources

Chapter Assessment, pp. 115-120 L1
Alternate Assessment in the Science Classroom
Computer Test Bank L1
Content Mastery, pp. 77-80 L1

Chapter Organizer

SECTION	OBJECTIVES	ACTIVITIES/FEATURES
21.1 Viruses National Science Standards: UCP.1, UCP.2, UCP.5; C.5; F.1, F.6; G.2	**1. Categorize** the different kinds of viruses. **2. Compare** the different reproductive cycles of viruses.	**Minilab:** What does a bacteriophage look like?, p. 506
21.2 Bacteria National Science Standards: UCP.1, UCP.2, UCP.5; A.1, A.2; C.1, C.5; E.1, E.2; F.1, F.4-6; G.1-3	**3. Summarize** the history and adaptations of bacteria. **4. Evaluate** the economic importance of bacteria.	**Minilab:** What are the shapes of bacteria?, p. 514 **Social Studies Connection:** Miracle Cure: The Story of Penicillin, p. 517 **Biolab:** Design Your Own Experiment—How sensitive are bacteria to antibiotics?, p. 518 **A Broader View:** Living on the Edge, p. 520 **Thinking Lab:** How fast can bacteria reproduce?, p. 522 **People in Biology:** Barbara Staggers, p. 526

ACTIVITY MATERIALS

BIOLAB	MINILABS	ALTERNATE LAB
page 518 bacteria cultures sterile nutrient agar petri dishes antibiotic disks sterile disks of blank filter paper marking pen cotton swabs forceps 37°C incubator metric ruler	**page 506** bolt nuts (2 hardware) #22 gauge wire or pipe cleaners **page 514** bacteria slides microscope paper	**page 514** screw-top test tubes distilled water vinegar Shultz liquid plant food Accent seasoning baking soda 60-watt lightbulb soil samples plastic sandwich bags

Chapter 21 Viruses and Bacteria

TEACHER CLASSROOM RESOURCES

Reproducible Masters	Transparencies
Section Focus Master 44: RNA and DNA L1 SAE **Reinforcement and Study Guide,** p. 81 L1 **Biolab and Minilab Worksheets,** p. 83 L1 **Concept Mapping:** Viral Reproductive Cycles, p. 21 L1 **Laboratory Manual:** Virus Replication, pp. 111-114 L2	**Basic Skills Transparency #17:** Lytic Cycle in Viral Reproduction L1 SAE **Basic Concepts Transparency #25:** Lytic Cycle L1 SAE **Basic Concepts Transparency #26:** Lysogenic Cycle L1 SAE
Section Focus Master 45: Bacteria and the Recycling of Nutrients L1 SAE **Reinforcement and Study Guide,** pp. 82-84 L1 **Biolab and Minilab Worksheets,** pp. 84-86 L1 **Critical Thinking/Problem Solving:** Some Bacterial Diseases, p. 21 L3 **Laboratory Manual:** How Are Bacteria Affected by Heat?, pp. 115-118 L2 **Content Mastery,** pp. 81-84 L1	**Reteaching Transparency #21:** Viruses and Bacteria L1 SAE **Basic Concepts Transparency #27:** Bacterial Structure L1 SAE

ASSESSMENT MATERIALS

Chapter Assessment, pp. 121-126
Performance Assessment in the Biology Classroom, p. 29
MindJogger Videoquiz
Alternate Assessment in the Science Classroom
Computer Test Bank

Spanish Resources SAE
English/Spanish Audiocassettes SAE
Cooperative Learning in the Science Classroom COOP LEARN
Lesson Plans

KEY TO TEACHING STRATEGIES

- L1 Level 1 activities should be within the ability range of all students including those with learning difficulties.
- L2 Level 2 activities are within the ability range of average to above-average students.
- L3 Level 3 activities are designed for the ability range of above-average students.
- SAE SAE activities should be within the ability range of Students Acquiring English.
- COOP LEARN Cooperative Learning activities are designed for small group work.
- P These strategies represent student products that can be placed into a best-work portfolio.
- These strategies are useful in a block scheduling format.

GLENCOE TECHNOLOGY

The following multimedia resources are available from Glencoe.

Biology: The Dynamics of Life
CD-ROM SAE
Videodisc Program

The Secret of Life Series
Nothing to Sneeze At: Viruses

National Geographic Society Series
Newton's Apple: Life Sciences
HIV and AIDS

Science and Technology Videodisc Series (STVS)
Plants & Simple Organisms
Bacteriophage
Mapping a Virus
Viruses and Plant Disease
Bacterial Waste Treatment
Detecting Salmonella

CHAPTER

21 Viruses and Bacteria

Chapter Overview

In the first section of the chapter, students study the structures and shapes of viruses. The lytic and lysogenic reproductive cycles are explained and illustrated with emphasis on how each of these cycles results in damage to cells. Students then learn how the cycles of retroviruses and RNA lytic viruses vary from the cycles of DNA viruses. Students conclude the section by exploring the cellular origin of viruses.

In the second section of the chapter, students study bacteria, their classification, their origins and history, and their adaptations. Next, students compare the traits of archaebacteria with those of eubacteria. Students then investigate the ecological importance and diversity of bacteria, as well as their reproduction. Students conclude their study of bacteria by considering the economic importance of these organisms.

Key Terms

bacteriophage
binary fission
conjugation
endospore
host cell
lysogenic cycle
lytic cycle
nitrogen fixation
obligate aerobe
obligate anaerobe
provirus
retrovirus
reverse transcriptase
saprobe
toxin
virus

502

Look for the following logo for strategies that emphasize different learning modalities. 

Kinesthetic	**Minilab, p. 506; Meeting Individual Needs, p. 510**
Visual-Spatial	**Visual Learning, pp. 505, 507; Demonstration, pp. 508, 515; Project, pp. 512, 525; Minilab, p. 514; Microscope Activity, pp. 515, 520; The Inside Story, p. 516; Meeting Individual Needs, p. 524**
Interpersonal	**Portfolio, pp. 506, 509, 517; Meeting Individual Needs, p. 516**
Intrapersonal	**Student Journal, p. 522**
Linguistic	**Student Journal, pp. 508, 517**
Logical-Mathematical	**Math Connection, p. 505; Activity, p. 513; Alternate Lab, p. 514; Project, p. 521; Thinking Lab, p. 522**

LS

CHAPTER

21 Viruses and Bacteria

Did you ever wonder what happens to all the leaves that fall from the trees every autumn? Imagine if they just accumulated on the forest floor. In a few years, the leaves in the woods would be knee-deep, and eventually they would cover the trees themselves! This does not happen because the leaves decompose after they fall from the trees. Many bacteria, along with fungi, are involved in this decomposition. It is estimated that 10 billion bacteria exist in a spoonful of garden soil. Many of them are decomposers. Decomposers are especially important because they return nutrients that are locked up in dead organisms back to the ecosystem. These nutrients, in the form of simple, inorganic molecules, can be taken up by producers and returned to the food chains of the ecosystem.

Concept Check

You may wish to review the following concepts before studying this chapter.

- Chapter 8: cell wall, prokaryote
- Chapter 9: osmosis

Chapter Preview

Laboratory Activities

Biolab: Design Your Own Experiment
- How sensitive are bacteria to antibiotics?

Minilabs
- What does a bacteriophage look like?
- What are the shapes of bacteria?

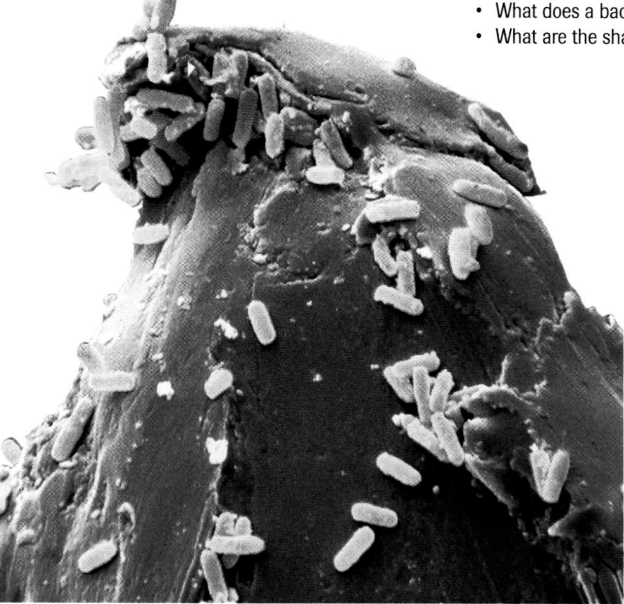

Imagine—tiny, unseen things on our skin, in our ears, even around our eyelashes. They're in the soil, in the air, in water everywhere. They even hitch rides on rocks and glacial ice. You can see at the left what they look like up close—magnified 40 000 times. The photo at the right reveals that bacteria live even on the point of a syringe needle.

Magnification: 5600×

CHAPTER 21

Introducing the Chapter

Direct students' attention to the main photograph and ask them if they observe any evidence of bacterial action. If necessary, remind students that bacteria are often involved in the process of decomposition. Reinforce students' knowledge of this process by reminding them that during decomposition, bacteria break down dead organisms and recycle organic nutrients back to the environment. Draw students' attention to the photograph of the bacteria on the point of a syringe needle. Elicit why sterilization of equipment in a doctor's office is important. *Many bacteria cause disease. Sterilization kills these bacteria.*

Theme Development

The theme of *systems and interactions* is evident through the discussions of viral replication and the recycling of nutrients by bacteria. The theme of *unity within diversity* is apparent through the presentations of the characteristics used to classify viruses and bacteria.

Concept Check

In this chapter, students will expand their knowledge of prokaryotic cells and bacteria. They will study the traits of bacteria that allow them to survive such a variety of environmental conditions.

Assessment Planner

Choose assessment strategies from the following pages to evaluate the progress of your students.

Assess, pp. 510, 528
Alternate Lab, pp. 514-515
Minilabs, pp. 506, 514
Portfolio, pp. 506, 509, 517
Thinking Lab, p. 522
Biolab, pp. 518-519
Chapter Review, pp. 529-531

SECTION

21.1

Prepare

Key Concepts

The basic structure of viruses is presented along with their reproductive mechanisms. Hypotheses regarding the origins of viruses are also presented and related to the question of whether viruses are considered to be alive.

Look for this symbol for strategies that are useful in a block scheduling format. For more information on block scheduling, refer to Section 21.1 in the **Lesson Plans** booklet.

Materials

- For the Minilab, gather #22 gauge wire, bolts (3.7 cm × 7 cm), and two nuts to fit each bolt.
- Gather a cigarette, a mortar and pestle, 100 mL of 1% glucose solution, and tomato plants for the Demonstration.
- Prepare photocopies of viral and bacterial growth curves for the Extension.

1 Focus

Bellringer

Before presenting the lesson, display **Section Focus Master 44** on the overhead projector and have students answer the accompanying questions. L1 SAE

Discussion

Elicit whether students know anyone who has ever had a cold sore. Have students describe the characteristics of a cold sore and explain whether these sores are contagious. *The cold sore appears as a red, fluid-filled blister and is contagious.*

SECTION

21.1 Viruses

Section Preview

Objectives

Categorize the different kinds of viruses.

Compare the different reproductive cycles of viruses.

Key Terms

virus
host cell
bacteriophage
lytic cycle
lysogenic cycle
provirus
retrovirus
reverse transcriptase

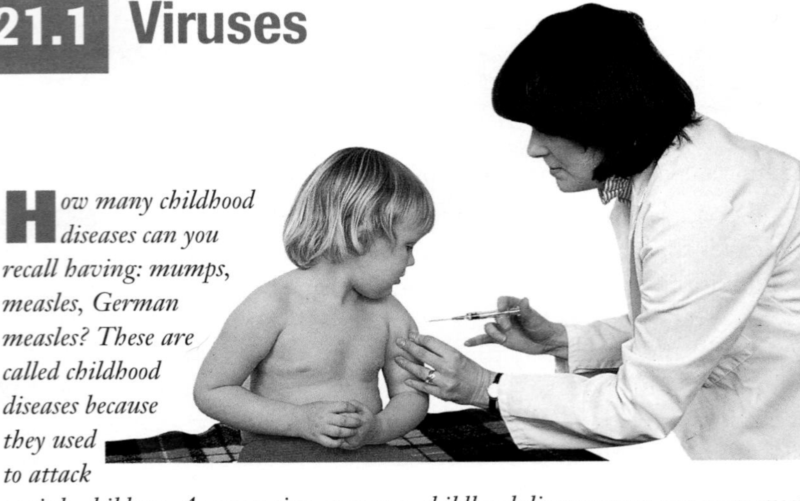

How many childhood diseases can you recall having: mumps, measles, German measles? These are called childhood diseases because they used to attack mainly children. A generation or so ago, childhood diseases were common; most children got them. Nowadays, vaccination of most infants and children has made these diseases rare. But the causes of childhood diseases have not been wiped out. They still pose a threat to those who have not been vaccinated.

Structure and Shape of Viruses

The diseases mentioned above are not caused by bacteria. They are not caused by organisms at all, but rather by tiny, nonliving particles called **viruses.** Viruses are about one-half to one-hundredth the size of the smallest bacterium. Most biologists do not consider them to be alive because they don't fulfill all the criteria for life. They do not carry out respiration or grow or move. All viruses can do is reproduce—and they can't even do that by themselves. Viruses reproduce only inside a living cell. The cell in which they reproduce is called the **host cell.**

bacteriophage:
bakterion (GK) small rod
phagein (GK) to devour
A bacteriophage is a virus that infects and destroys bacteria.

Because they are nonliving, viruses are not given Latin names, as are cellular organisms. Viruses often are named for the diseases they cause—for example, the rabies virus and polio virus. Others are named for the organ or tissue they infect—for example, adenoviruses were first detected in adenoid tissue at the back of the throat. In some cases, code numbers are used to distinguish several viruses infecting the same host. Seven viruses that infect the common intestinal bacterium, *Escherichia coli*, are named bacteriophage T1 through T7 (*T* stands for *Type*). A **bacteriophage** is a virus that infects bacteria.

Viral structure

A virus consists of an inner core of nucleic acid surrounded by one or two protein coats. Some relatively large viruses, such as the human flu virus, may have another layer, called the viral envelope, surrounding the outer coat. This layer consists primarily of phospholipids.

Program Resources

Section Focus Master 44 L1 SAE
Reinforcement and Study Guide, p. 81 L1
Biolab and Minilab Worksheets, p. 83 L1
Concept Mapping, p. 21 L1
Laboratory Manual, pp. 111-114 L2
Basic Concepts Transparencies 25, 26 and **Masters** L1 SAE
Basic Skills Transparency 17 and **Master** L1 SAE

The nucleic acid core of a virus contains the virus's genetic material. The nucleic acid in some viruses is DNA and in others it is RNA, but it is never both DNA and RNA. The nucleic acid consists of genes that contain coded instructions only for making copies of the virus. They code for nothing else.

The arrangement of the proteins in the viral coat gives different kinds of viruses different shapes, as shown in *Figure 21.1.* These shapes play a role in the infection process.

Figure 21.1

Viruses have a variety of shapes, determined by the proteins in their coats. A viral coat is called a capsid.

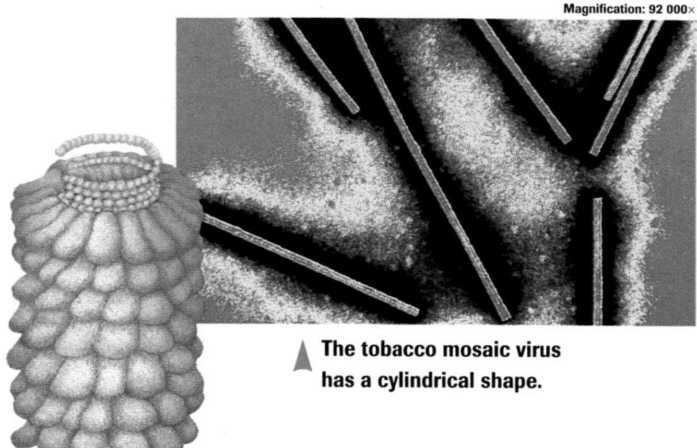

Magnification: 92 000×

The tobacco mosaic virus has a cylindrical shape.

Polyhedral viruses, such as the polio virus, resemble small crystals under the electron microscope. Many polyhedral viruses are polygons with 20 sides.

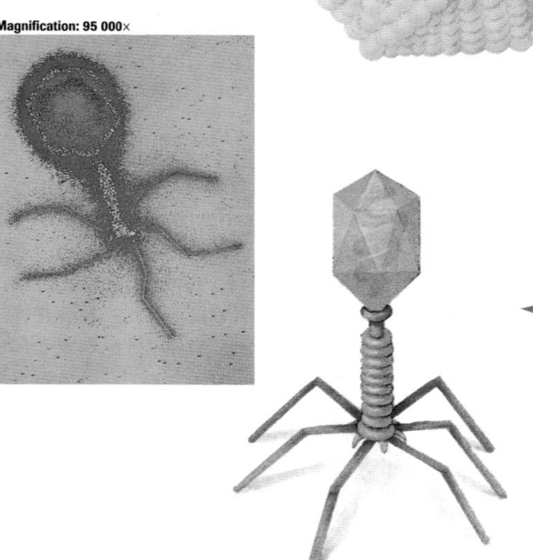

Magnification: 20 000×

Some viruses, such as influenza and HIV, the virus that causes AIDS, are covered with an envelope that is studded with projections. The diagram shows HIV; the photo shows the influenza virus.

Magnification: 95 000×

This T4 virus, which infects bacteria, resembles a minuscule lunar-landing module. It has a polyhedral head containing DNA, a protein tail, and protein tail fibers.

Magnification: 36 700×

SECTION 21.1

2 Teach

Math Connection

Calculating Viral Size

Logical-Mathematical On the chalkboard, write the following conversions:

1000 μm = 1 mm

1 000 000 nm = 1 mm

Ask students to draw a line 132-mm long. Explain that the line represents a tobacco mosaic virus that has been magnified 450 000 times. Ask students to calculate the actual length of the virus in millimeters, micrometers, and nanometers. *Its size is 0.00029 mm, 0.29 μm, and 290 nm.*

Explain that viruses are too small to be effectively studied with a light microscope.

Visual Learning

Figure 21.1 Ask students to study the viruses and describe the shape of each. *Poliovirus has a polyhedral shape; tobacco mosaic virus has a cylindrical shape; the influenza virus and HIV have a spherical shape with a studded outer covering; the T4 bacteriophage has a polyhedral head and a tail with six fibers coming from it.* L1 LS

interNET CONNECTION

Follow the link for this chapter on the Glencoe Homepage at **www.glencoe.com/sec/science** to find out more about viruses and bacteria.

STUDENT JOURNAL

Viral Time Line Have students make a time line showing the development of instrumentation used in viral studies, the discovery of components of the immune system that deal with viruses, and the discovery of viral diseases. Have them indicate when specific vaccines became available. You may want to have them make a separate time line for HIV. L2

MiniLab

Purpose

Kinesthetic Students will learn about the structure of bacteriophages and how these viruses infect only bacterial cells.

Process Skills

observe and infer, make a model, recognize and use spatial relationships

Teaching Strategies

- Pipe cleaners may be used in place of the wires.
- Wire may need to be folded and twisted to support the weight of the bolt.
- Have students work in pairs to reduce the amount of materials needed and to permit discussion of their observations.

Expected Results

Models will have a head, DNA core, and a tail with six tail fibers.

Analysis

1. The wires represent the tail fibers. The nuts and top of the bolt represent the protein head. The body of the bolt represents the tail.
2. inside the head
3. tail fibers
4. Viruses are specific in being able to infect only certain types of cells. Bacteriophages infect only bacterial cells because their tail fibers recognize receptor sites found only in certain bacterial plasma membranes.

Assessment

Performance: Provide students with references and have them build a model of another virus and explain its structure. L1

MiniLab

What does a bacteriophage look like?

Bacteriophages (*phages,* for short) are viruses that infect bacteria. A typical phage consists of a polyhedral protein head surrounding a DNA core, a tail made of several proteins, and six tail fibers.

Procedure

1. You can build a model of a bacteriophage by attaching two nuts onto the top of a bolt (3.7 cm × 0.7 cm). Screw on the nuts so that they touch the top of the bolt.
2. Take three 14-cm-long pieces of #22 gauge wire, and twist them around the bottom of the bolt. Bend down the ends of the wires so that they resemble the figure below.

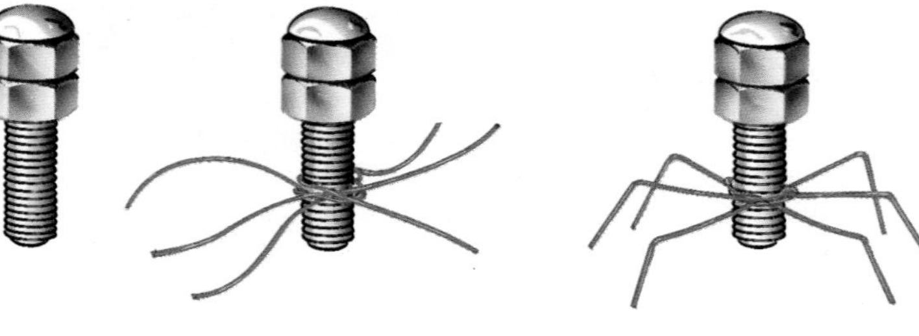

Analysis

1. What bacteriophage structure do the wires represent? What structure do the nuts and the top of the bolt represent? What structure does the body of the bolt represent?
2. Where would the nucleic acid be located in the model?
3. What structure in this model is not generally found in viruses that infect eukaryotic cells?
4. Why would a bacteriophage like the one you have modeled be unable to infect one of your own cells? Refer to a specific structure in the model.

Recognition of a host cell

Viral shape is important in viral reproduction. Before a virus can enter and reproduce in a cell, it must recognize and attach to a specific receptor site on the plasma membrane of the host cell. That's where the viral protein coat plays its role. A protein on the surface of a virus has a three-dimensional shape that matches the shape of a molecule in the plasma membrane of its host cell, like two interlocking pieces of a jigsaw puzzle. In this way, a virus recognizes its host cell. The two interlocking shapes allow the virus to land on the host and lock on, like two spaceships docking. The bacteriophage T4 uses the protein in its tail fibers to attach to the host cell it will infect. In other viruses, the attachment protein is in the envelope that surrounds the virus or in the protein coat.

Attachment is a specific process

Each attachment protein has its own particular shape due to its sequence of amino acids. Therefore, attachment is a specific process. As a result, most viruses can enter and reproduce in only a few kinds of cells. The T4 bacteriophage can infect only certain strains of *E. coli.* T4 cannot reproduce in human, animal, or plant cells, or even in other kinds of bacteria because it cannot attach to these cells. Similarly, tobacco mosaic virus can reproduce only in tobacco plant cells. Not only are viruses species specific, but some also are cell-type specific. For example, polio viruses infect only nerve cells of humans.

The fact that particular viruses can infect only a few kinds of cells has enormous significance. In the 1970s, the World Health Organization eradicated smallpox, a viral disease that has been one of the greatest killers in history. One reason why this eradication was possible was because the smallpox virus could infect only humans. It would have been much more difficult to eradicate a virus that could infect other animals as well.

PORTFOLIO

Developing Disease Awareness Have students work in groups in which they act as members of a team of doctors at the Centers for Disease Control and Prevention. Their job is to decide how to protect the public against lassa fever, which has just been identified in several people in New York City. Ask students to design a public service warning to air on the television news that describes the disease, its potential effect on the population, and what people should do to protect themselves. Provide library resources. Have students videotape their presentations and keep all their written information in their portfolios. L2 LS P COOP LEARN

Viral Reproductive Cycles

Once attached to the plasma membrane of the host cell, the virus must get inside and take over the cell's metabolism. Only then can the virus reproduce. Method of entry into a host cell depends in part on the shape of the virus. Some viruses are shaped so that they can inject their nucleic acid into the host cell like a syringe, leaving their coat still attached to the plasma membrane. Other viruses have a shape that can barge in by indenting the plasma membrane, which seals off and breaks free, enclosing the virus within a vacuole in the cell's cytoplasm. The virus then bursts out of the vacuole and releases its nucleic acid into the cytoplasm.

Lytic cycle

Once inside the cell, the virus destroys the host's DNA. It reprograms the cell's metabolic activities to copy the viral genes and make viral protein coats using the host's enzymes, raw materials, and energy. The nucleic acid and coats are assembled into new viruses, which burst from the host cell, usually killing it. The newly produced viruses are then free to infect and kill other host cells. This process is called a **lytic cycle.** *Figure 21.2* shows a typical lytic cycle for a bacteriophage.

lytic:
lyein (GK) to break down
The host cell is destroyed during a lytic cycle.

Figure 21.2
Viruses use the host cell's energy and raw materials to make new viruses. A typical lytic cycle takes about 90 minutes and produces about 200 new viruses.

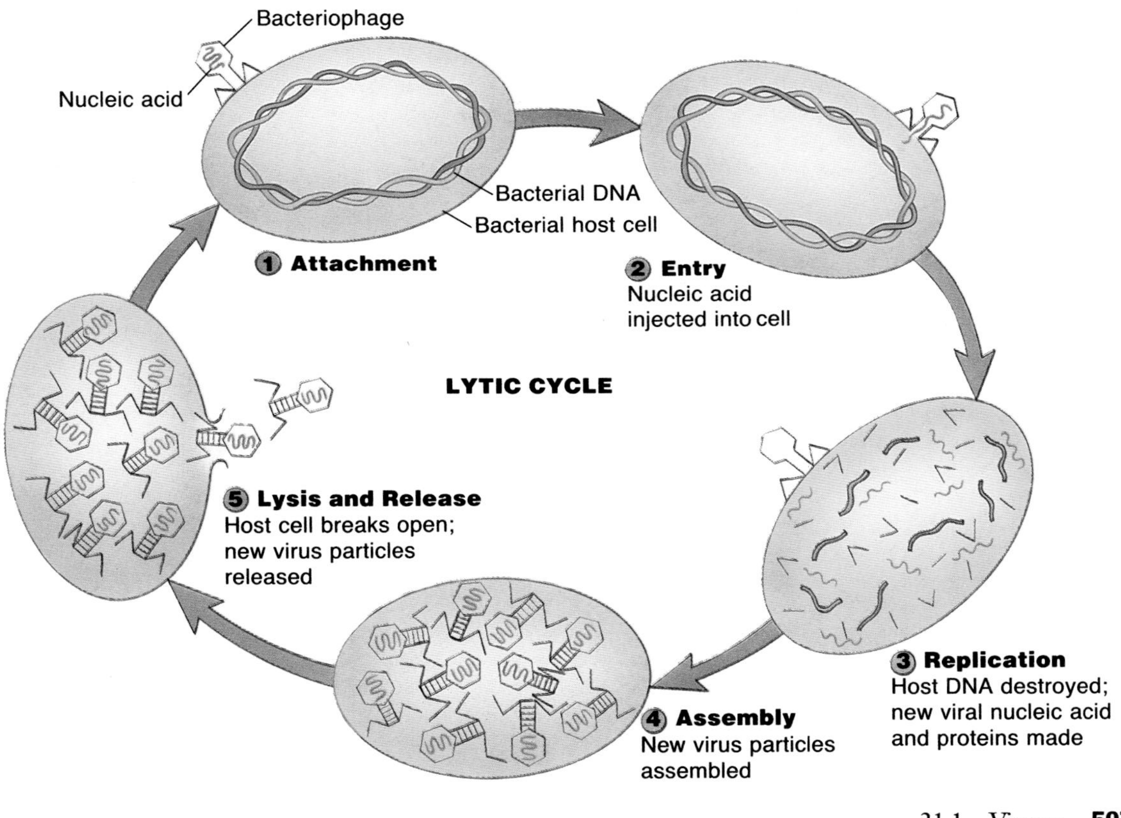

Using an Analogy

Use the following analogy to describe the actions and results of the lytic cycle. During wartime, a tank (virus) filled with enemy troops crashes through the wall (membrane) of an automobile factory (cell). The troops in the tank (nucleic acids) take over the machinery of the factory (nucleus and organelles), and force the factory to produce new tanks (viruses) instead of cars (cell parts). Explain that during the lytic cycle, a virus breaks into a cell and releases nucleic acids that direct the cell to produce new viruses instead of new cell parts.

Misconception

Some students may think that antibiotics will cure viral infections such as colds. Explain that antibiotics do work on many bacterial infections, but they do not combat viral diseases.

Visual Learning

Figures 21.2 and 21.3 On the chalkboard, draw flowcharts that show the sequence of stages in the lytic cycle and the lysogenic cycle. Encourage students to copy the flowcharts into their journals for use as a study tool. 

PROJECT Vaccine Development

Ask student groups to prepare skits on the discovery of one of the following vaccines: polio, smallpox, measles, or German measles. They should include information about the disease, the lives of the scientists involved in the vaccine's development, and the vaccine's impact on society. L2 COOP LEARN

GLENCOE TECHNOLOGY

Videodisc

Biology: The Dynamics of Life
Disc 1, Side 2
The Lytic Cycle (Ch. 9)

The Lysogenic Cycle (Ch. 10)

Bioethics

HIV and Health Care Workers

The HIV virus that causes AIDS can be transmitted by sharing needles, by sexual contact, and by coming in contact with the blood or body fluids of a person who has the virus. Dentists, physicians, and health care workers often come into contact with blood and body fluids that contain HIV. Although such workers wear gloves to protect themselves and their patients when carrying out medical procedures, some medical personnel have become infected with HIV. Initiate a class discussion or debate about whether health care workers who have tested positive for HIV should be allowed to take care of patients.

Demonstration

LS **Visual-Spatial** Explain that tobacco contains tobacco mosaic virus. Remove the tobacco from a cigarette. Use a mortar and pestle to grind the tobacco into a fine powder and then mix the powder with 100 mL of a 1 percent glucose solution. Gently scratch some leaves of a tomato plant with fine sandpaper and then drop a small amount of the solution onto the leaves. Have students observe the tomato plant each day looking for evidence of infection. In six to ten days, students should observe the mosaic pattern on the leaves of the tomato plant. L1

Lysogenic cycle

Viruses vary in the degree of damage they inflict on the host. Not all virus infections are fatal to their host cells. Some viruses can attack but not always kill a cell. These types of viruses go through a **lysogenic cycle,** a viral reproductive cycle in which the viral DNA becomes integrated into the host cell's chromosome.

A lysogenic cycle begins like a lytic cycle. As an example, let's look at a virus that contains DNA. The virus attaches to the plasma membrane of the host cell and injects its DNA into the cell. Events then become different from those in the lytic cycle. Instead of destroying the host DNA and making new viruses, the viral DNA becomes part of the host DNA.

Once the viral DNA is inserted into the host cell chromosome, it is known as a **provirus.** It does not interfere with the normal functioning of the host cell, which is able to carry out its regular metabolic activity. Every time the host cell reproduces, the provirus is replicated right along with the host cell's chromosome. This means that every descendant of the host cell will have a copy of the provirus in its own chromosome. This lysogenic phase can continue undetected for many years. However, the provirus can pop out of the host cell's chromosome at any time and enter a lytic cycle. Then the virus starts reproducing and kills the host cell. A typical lysogenic cycle is shown in *Figure 21.3.*

Figure 21.3

Some viruses do not cause the host cell to burst and be destroyed. In a lysogenic cycle, viral nucleic acid is inserted into the genetic material of the host cell and replicates with it. Eventually, the virus enters a lytic cycle and kills the cell.

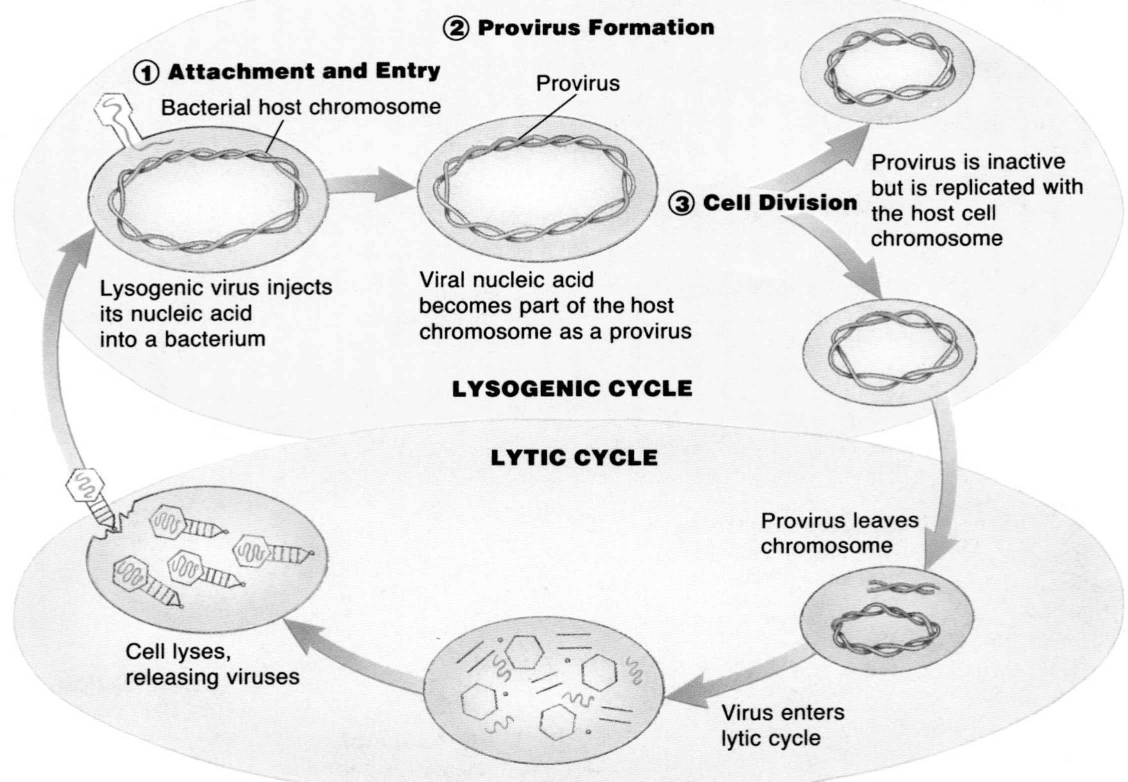
② Provirus Formation
① Attachment and Entry
Provirus
Bacterial host chromosome
Provirus is inactive but is replicated with the host cell chromosome
③ Cell Division
Lysogenic virus injects its nucleic acid into a bacterium
Viral nucleic acid becomes part of the host chromosome as a provirus
LYSOGENIC CYCLE
LYTIC CYCLE
Provirus leaves chromosome
Cell lyses, releasing viruses
Virus enters lytic cycle

STUDENT JOURNAL

Smallpox Epidemic Ask students to write a newspaper article about an epidemic of smallpox in Boston, Massachusetts, in the days of colonial America as it might have appeared in a real newspaper. L1 LS

Meeting Individual Needs

Gifted Ask your advanced students to explore the latest genetic engineering research that uses viruses as vectors to replace defective genes in people suffering from genetic disorders. Ask them to make a working model of the processes being used and to share their findings with the class. L3

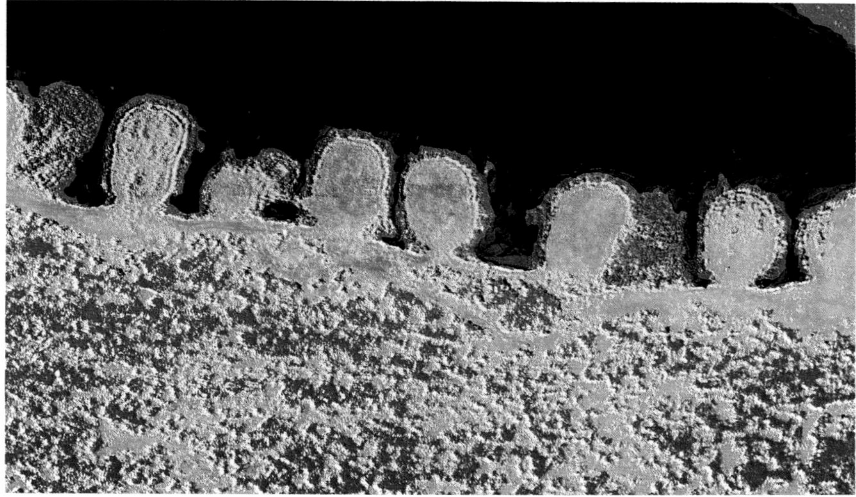

Figure 21.4

Influenza is caused by a virus that consists of an RNA core inside a protective protein coat. The virus also has an outer envelope, which it takes from the plasma membrane of the cell it infects. When the virus buds off from the host cell, it wraps itself in a piece of the host's plasma membrane as it leaves the cell.

Proviruses explain symptoms of disease

The lysogenic process explains why cold sores recur. Once you have had a cold sore, the virus causing it—herpes simplex 1—remains as a provirus in one of the chromosomes in your cells even after the cold sore has healed. When the virus pops out and enters a lytic cycle, you get another cold sore. No one knows the exact conditions that cause the provirus to pop out.

New virus particles may be released from the host cell by lysis—bursting of the cell—or they may be released without lysis by a kind of budding process, *Figure 21.4*. This process is a form of exocytosis. The plasma membrane engulfs the virus particles and then opens up to the outside, releasing them.

Sometimes, a provirus nestled in a host cell's chromosome will maintain a low level of activity by repeatedly producing small numbers of new viruses while the host cell continues to function normally. HIV can act in this manner. Small numbers of HIV viruses bud off from infected white blood cells. These viruses then can infect other white blood cells. However, at this time, the viruses do not kill the host cells.

You can see now why a person can be infected with HIV and appear to be perfectly healthy. As long as most of the infected cells are producing only a few viruses, the person will have no clinical symptoms. It also explains why most people infected with HIV will eventually get AIDS. As more and more cells become infected, and as proviruses begin to pop out, enter a lytic cycle, and kill their host cells, there is a gradual loss of white blood cells, which are important to the body's self-protection processes. This loss eventually

Science, Technology, and Society

Controlling Rabies

Humans can contract the deadly viral disease, rabies, through the bite of an infected animal such as a dog, cat, bat, raccoon, fox, or skunk. In most parts of the United States, pet owners are required by law to have their dogs and cats vaccinated against rabies. However, vaccination of pets does not protect animals in the wild from the disease. To prevent rabies in wild animals, scientists have developed a vaccine that is contained in a chicken-flavored packet. The packets are dropped from airplanes over areas with large animal populations. When wild animals find and eat the packet, they become immunized against rabies.

Concept Development

Ask students to list all the viral diseases they have had. Provide students with the resources necessary to find out how each disease is spread and what, if any, methods of prevention are available. Discuss the lists as a class to generate a classroom table of viral diseases. L2

GLENCOE TECHNOLOGY

Videotape

The Secret of Life

Use the videotape *Nothing to Sneeze At: Viruses* to show how the influenza virus was discovered.

Videodisc

STVS: Plants & Simple Organisms

Disc 4, Side 1

Bacteriophage (Ch. 2)

PORTFOLIO

Preventing the Spread of AIDS Ask students working in groups to prepare an interview for a dentist or dental assistant to find out how health care professionals protect patients and themselves from contracting HIV. Encourage students to find out about the American Dental Association's guidelines for AIDS prevention. Once questions are written, have one or two students from the group carry out the interview. These students should then report back to the group so a group report can be prepared. L2 LS P COOP LEARN

Discussion

Based on information about viral structure and the various replication methods of viruses, have students attempt to identify some approaches that might be useful in the production of vaccines against viruses.

GLENCOE TECHNOLOGY

Videodisc

STVS: Plants & Simple Organisms
Disc 4, Side 1
Viruses and Plant Disease (Ch. 18)

3 Assess

Check for Understanding

Give students out-of-sequence sketches of the steps involved in the lytic cycle and the lysogenic cycle. Have them rearrange the sketches to identify the correct sequence of events for each reproductive cycle. L1

Reteach

Ask students to create a table that lists each virus discussed in this section. For each virus, have students identify its type of nucleic acid, structure, method of reproduction, and host organism. L1

results in the breakdown of the body's disease-fighting system that is characteristic of AIDS. It also explains why you can get AIDS from a person who appears to be perfectly healthy. A person infected with HIV who has no symptoms can still transmit the virus.

Lysogenic viruses and disease

Many disease-causing viruses go through lysogenic cycles. These include the herpes viruses that cause cold sores and genital herpes, and the virus that causes hepatitis B. Another such virus is the one that causes chicken pox. An attack of chicken pox, which usually occurs before age ten, gives lifelong protection from another invasion of chicken pox viruses. However, the virus remains dormant within nerve tissues. It may flare up later in life, causing a disease called shingles—a painful infection of the nerves that transmit impulses from certain areas of the skin.

Retroviruses

Many viruses such as HIV, which causes AIDS, have RNA as their only nucleic acid. The RNA viruses with the most-complex reproductive cycles are the **retroviruses.** When retroviruses inject their nucleic acid into the host cell, they also inject a molecule of the enzyme **reverse transcriptase,** which copies viral RNA into DNA. A lysogenic cycle of a retrovirus is shown in *Figure 21.5.*

RNA lytic viruses

Not all RNA viruses are retroviruses. Some viruses contain RNA as their nucleic acid, but they do not have reverse transcriptase. Because a cell has no mechanism or enzymes for copying RNA, these viruses supply an enzyme that can make copies of their RNA molecule by base pairing. Such viruses follow a lytic cycle. They cannot make a DNA copy of their RNA to insert into the host cell's chromosome.

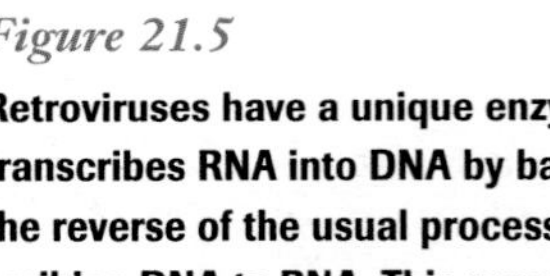

Figure 21.5

Retroviruses have a unique enzyme that transcribes RNA into DNA by base pairing, the reverse of the usual process of transcribing DNA to RNA. This process gives these viruses their name—*retro* means "backward." The viral DNA then can enter the host cell's chromosome, become a provirus, and begin a lysogenic cycle. The viral DNA lodged in the host cell's chromosome can code for small amounts of viral RNA and protein. This results in the production of small numbers of new viruses that bud off.

Meeting Individual Needs

Visually Impaired Provide visually impaired students with everyday objects that have shapes similar to the viruses studied. Have other students work with visually impaired students to associate each shape with an actual virus studied in the chapter. For the bacteriophage, students can use the model created in the Minilab. L1 LS

Plant Viruses

More than 400 viruses are known to infect plant cells, causing as many as 1000 plant diseases. In fact, the first virus to be identified was a plant virus—tobacco mosaic virus. *Figure 21.6* shows its effects.

Not all plant diseases caused by viruses are fatal or even harmful. For example, some mosaic viruses cause a striking variegation in the flowers of plants such as gladioli and pansies. The diseased flowers exhibit streaks of contrasting colors in their petals, seen in *Figure 21.6.* These viruses are easily spread by first cutting infected flower stems and then cutting healthy stems with the same tool.

Origin of Viruses

You might think that because viruses are so simple, they represent an ancient and primitive form of life. This is probably not so. Viruses require host cells for their reproduction. Therefore, scientists believe that viruses originated from their host cells. It has been suggested that viruses are nucleic acids that escaped from their host cells and developed a way to reproduce as parasites of the host cells. If this is correct, then in terms of evolution, viruses are more closely related to their host cells than they are to each other.

Figure 21.6
Tobacco mosaic virus causes yellow spotting on the leaves of tobacco plants (left) and makes them worthless as a cash crop. These Rembrandt tulips (right) get their stripes from a virus that is transmitted from plant to plant.

Section Review

Understanding Concepts

1. Why is a virus not considered to be alive?
2. What is the difference between a lytic and lysogenic cycle?
3. Why can most viruses infect only a few kinds of cells?

Thinking Critically

4. Describe the state of a herpes virus in a person who had cold sores several years ago but who does not have any now.

Skill Review

5. **Making and Using Graphs** A microbiologist added some viruses to a bacterial culture. Every hour from noon to 4 P.M., she determined the number of viruses present in a sample. The numbers of viruses in each sample beginning at noon were 3, 3, 126, 585, and 602. Graph these results. How would the graph look if there had been no living bacteria in the culture? For more help, refer to Organizing Information in the *Skill Handbook.*

Extension

Provide students with graphs showing the growth curves of bacteria and viruses. Have students compare the curves and summarize their observations. *Students should indicate that the viral curve is steplike. The numbers of viruses in the growth medium remain constant while they are being produced within the host cell. The numbers suddenly rise when the viruses are released from the host. Bacterial growth curves are exponential.* Point out that bacterial growth curves do not remain exponential indefinitely. In time, a buildup of toxic materials leads to death of the bacteria. L3

Assessment

Oral: Ask students to describe what traits viruses have in common with cells, and how viruses are different from living cells. L1

4 Close

Discussion

Ask students what is meant by the saying, "If you catch a cold, it will take seven days to get better. If you take medication, it will take a week." *Students should recognize that because a cold is a viral infection, there is no effective way to destroy the pathogens causing the illness. Thus, medications may be taken to relieve symptoms, but they will not eliminate the cause.*

Answers to Section Review

1. Viruses are not considered to be alive because they do not carry on respiration, grow, move, or carry on any life function except reproduction.
2. In both cycles, the viral nucleic acid enters the cell. In the lytic cycle, the viral nucleic acid takes over the machinery of the cell and causes it to make more viruses. These new viruses break out of the cell, destroying it. In the lysogenic cycle, the viral nucleic acid becomes part of the host cell DNA. As the host DNA replicates, so does the viral DNA.
3. Viruses must recognize a specific receptor site in their host.

Thinking Critically

4. It is a provirus incorporated in the host DNA.

Skill Review

5. Students should draw a graph showing a stepwise increase in the number of virus particles. The horizontal axis should be labeled "Time" and the vertical axis "Number of Virus Particles." If there were no living bacteria in the culture, the virus would not replicate because the virus needs a living cell in which to replicate.

SECTION

21.2

Prepare

Key Concepts

The classification, structure, and diversity of bacteria are considered along with the adaptations that allow bacteria to survive unfavorable conditions. Reproduction of bacteria is then discussed along with ways in which bacteria are helpful and harmful to people, other organisms, and the environment.

Materials

- For the Minilab on page 514, gather microscopes and prepared slides of cocci, bacilli, and spirilla.
- Purchase yogurt containing live cultures and distilled water, and gather slides for the Microscope Activity on page 515.
- Obtain two sterile nutrient agar petri dishes for the Demonstration.
- For the Biolab, gather cultures of nonpathogenic bacteria, sterile nutrient agar petri dishes, antibiotic disks, sterile disks of filter paper, marking pens, cotton swabs, forceps, a 37°C incubator, and metric rulers.
- Order prepared slides of heterotrophic and photoautotrophic bacteria, and gather microscopes and colored pencils for the Microscope Activity on page 520.
- Purchase pickles, yogurt, Swiss cheese, soybeans, peanuts, milk, sour cream, and sauerkraut or bring in pictures of these things for the Demonstration on page 524.
- Purchase prepared slides of a cyanobacterium for the Microscope Activity on page 525.

SECTION

21.2 Bacteria

Section Preview

Objectives

Summarize the history and adaptations of bacteria.

Evaluate the economic importance of bacteria.

Key Terms

saprobe
obligate aerobe
obligate anaerobe
endospore
toxin
binary fission
conjugation
nitrogen fixation

Imagine yourself going back three-and-a-half billion years. You wander around the young Earth and find yourself face-to-face with the first life on this planet. Dinosaurs? Saber-toothed tigers? No. You would be face-to-face with the most ancient and diverse form of life on Earth—bacteria.

THE FAR SIDE By GARY LARSON

Early microbiologists

Classification of Bacteria

Bacteria are classified into two very different kingdoms—Archaebacteria and Eubacteria. Many biochemical differences exist between these groups. Their cell walls and the lipids in their plasma membranes have different structures. The sequences of bases in their tRNA and rRNA are different. Also, they react differently to antibiotics.

Archaebacteria and eubacteria probably diverged from each other several billion years ago, although the exact time is unknown.

Archaebacteria—The extremists

The archaebacteria include three types of bacteria that are found mainly in extreme habitats where little else can live. One group lives in oxygen-free environments and produces methane. A second group can live only in bodies of concentrated salt water, such as the Great Salt Lake in Utah and the Dead Sea in the Middle East. A third group is found in the hot, acidic waters of sulfur springs. *Figure 21.7* introduces you to some of the unique places archaebacteria are found.

Eubacteria—The heterotrophs

Eubacteria, the second main group of bacteria, display a wide array of habitats and different types of metabolism. One group of eubacteria, the heterotrophs, are found everywhere. These bacteria need organic molecules as an energy source but are not adapted for trapping the food that contains these molecules. Thus, some bacteria live as parasites, absorbing nutrients from living organisms. Others live as **saprobes,** organisms that feed on dead organisms or organic wastes. Saprobes help recycle the nutrients contained in decomposing organisms.

PROJECT

Bacterial Microscopy

Provide students with instruction as to how to prepare and use depression slides and the hanging drop (oil suspension) method of microscopic study. Ask students to use this method of study to make microscope slides of a live culture of *Oscillatoria.* Have students record their observations both as written descriptions and in sketches. Challenge students to identify the bacterial group to which *Oscillatoria* belongs based upon its color.

L2 LS

Eubacteria—Photosynthetic autotrophs

A second group of eubacteria are photosynthetic autotrophs. They obtain their energy from light. Cyanobacteria are in this group. They can trap the sun's energy by photosynthesis because they contain a photosynthetic pigment. Most cyanobacteria are blue-green in color, which is why they are commonly called blue-green bacteria. They are not all blue-green, though; some cyanobacteria are red or yellow. Cyanobacteria are common in ponds, streams, and moist areas of land. They are composed of chains of cells—an exception to the rule that all bacteria are unicellular.

cyanobacterium:
kyanos (GK) blue
bakterion (GK) small rod
Cyanobacteria are blue-green bacteria.

Figure 21.7

All of the known archaebacteria live without oxygen. They obtain their energy from inorganic molecules or from light.

Heat- and acid-loving bacteria grow in hot sulfur springs such as this one at Yellowstone National Park. They grow best at temperatures around 60°C and at a pH of 1 to 2. This is the pH of concentrated sulfuric acid.

Salt-loving bacteria live in saturated salt solutions. They produce an unusual purple pigment that allows them to carry on photosynthesis. They can be seen in this aerial photograph as pink/purple mats. These bacteria are growing in seawater-evaporating ponds near San Francisco Bay. The ponds are used for commercial salt production.

Some methane-producing bacteria grow in the stomachs of cows, which lack the enzymes needed to break down cellulose in the grass they eat. These bacteria produce the enzymes that convert cellulose to glucose. Methane producers are also widely distributed in sewage-treatment plants, freshwater swamps, and deep-sea marine habitats. They get their energy from carbon dioxide and hydrogen gas, producing methane as a waste product.

Program Resources

Section Focus Master 45 L1 SAE
Reinforcement and Study Guide, pp. 82-84 L1
Performance Assessment in the Biology Classroom, p. 29 L1 P
Biolab and Minilab Worksheets, pp. 84-86 L1
Critical Thinking/Problem Solving, p. 21 L3
Laboratory Manual, pp. 115-118 L2
Basic Concepts Transparency 27 and **Master** L1 SAE
Reteaching Transparency 21 and **Master** L1 SAE

Block Scheduling

Look for this symbol for strategies that are useful in a block scheduling format. For more information on block scheduling, refer to Section 21.2 in the **Lesson Plans** booklet.

1 Focus

Bellringer

Before presenting the lesson, display **Section Focus Master 45** on the overhead projector and have students answer the accompanying questions. L1 SAE

Activity

LS **Logical-Mathematical** Have students cut a period from the end of a sentence that appears in a newspaper. Ask students to make a wet mount of the period. Explain that the average field of view of a microscope, using the low-power objective, is 1.44 mm. Have students use this information to estimate the size of the period. *About 0.35 mm* Then ask them to calculate how many bacteria could live on the period if the average size of a bacterium is 0.002 mm. *About 175* L1

2 Teach

Concept Development

Ask students to list the main differences between prokaryotic and eukaryotic cells. Point out that all prokaryotes are classified as bacteria. L1

MiniLab

Purpose

LS Visual-Spatial Students will examine the three typical shapes of bacteria.

Process Skills

compare and contrast, classify, observe and infer

Teaching Strategies

- You may want students to estimate the sizes of the bacteria. Provide them with the diameter of the high-power field of view of their microscopes. L2
- Encourage students to identify bacterial arrangements such as pairs, chains, and clusters. L1

Expected Results

Students will observe and draw spherical, rod-shaped, and spiral-shaped bacteria.

Analysis

1. The spherical bacteria were smallest. The spirilla were most likely largest. All bacteria were smaller than plant and animal cells.
2. Bacteria may pass information (as molecules) to each other or exchange nuclear material.

Assessment

Performance: Give students a prepared slide of bacteria they have not seen before. Ask them to identify the name for that shape. L1

MiniLab

What are the shapes of bacteria?

Bacteria come in three different shapes: spheres (coccus), rods (bacillus), and spirals (spirillum). They may appear singly or in pairs, or form chains or clusters. Each species has a typical shape and growth pattern.

Procedure

1. Obtain slides from your teacher showing the three shapes of bacteria.
2. Using low power, locate bacteria of one shape. Then switch to high power. Look for individual cells and observe their shape. Observe also the size of the cells. Then look for groups of bacterial cells to determine their growth pattern.
3. In this manner, observe all three shapes of bacteria. Compare their sizes with those of plant and animal cells on other slides.
4. Draw a diagram of each type of bacteria.

Analysis

1. How did the three bacteria compare in size? How did they compare in size with plant and animal cells?
2. What adaptive advantage might there be for bacteria to form groups of cells?

Figure 21.8

Cyanobacteria, such as *Anabaena* (left), have a blue-green color because the only photosynthetic pigment they contain is chlorophyll *a*, whereas the bacterium *Prochloron* (right) contains both chlorophylls *a* and *b* and is bright green in color. The chlorophyll in cyanobacteria is found attached to the cell membrane, not in chloroplasts.

Magnification: 160×

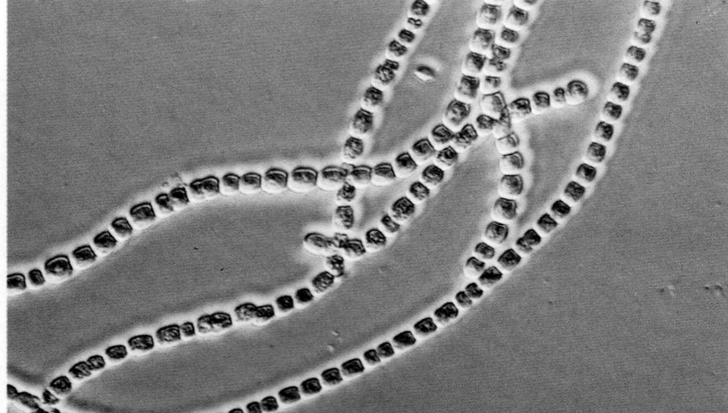

Magnification: 15 900×

For a long time, the reason why only chlorophyll *a* was present in cyanobacteria was a mystery because plants contain two forms of chlorophyll, *a* and *b*. This mystery was solved with the discovery of *Prochloron*, which contains both chlorophylls *a* and *b*, *Figure 21.8*. The existence of this bacterium indicates that chloroplasts of plants may have evolved from a similar precursor.

Eubacteria—Chemosynthetic autotrophs

A third group of eubacteria are chemosynthetic autotrophs. These bacteria obtain their energy from the chemosynthetic breakdown of inorganic substances such as sulfur and nitrogen compounds. Some of these bacteria are important in converting nitrogen in the atmosphere to forms that can be used readily by plants.

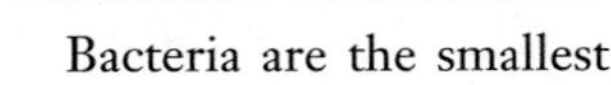

Structure of Bacteria

Bacteria are the smallest and simplest of living things. In terms of complexity, they fall between the nonliving viruses and the living eukaryotic, cellular organisms.

Alternate Lab — Photosynthetic Bacteria

Purpose

LS Logical-Mathematical Students will isolate and test photosynthetic bacteria from a variety of soil samples.

Materials

screw-top test tubes, liquid growth medium prepared as follows: [445 mL distilled water, 50 mL vinegar, 4 drops Shultz liquid plant food, 0.5 g Accent seasoning, about 4.5 g baking soda or amount that brings the pH to 7.0.], 60-watt incandescent light bulbs, soil samples, plastic sandwich bags

Procedure

Give the following directions to students.

1. Place 1 to 2 cm of soil in a screw-top test tube. Add liquid growth medium. Place top on test tube and shake well.
2. Place at room temperature about 60 cm away from a light bulb. Do not disturb.
3. Make a hypothesis about which samples will contain photosynthetic bacteria.
4. After several days, examine the color of the cultures. Red, rust, pink, or

Figure 21.9

Wash, scrub, shampoo, and gargle as you might, these actions remove only a small fraction of the bacteria that live on and in you. These and other bacteria can be classified by their cell shapes.

Magnification: 27 000×

A coccus is a round bacterium. The cocci shown here, *Streptococcus mutans,* are the bacteria that cause tooth decay by converting sugar to an acid that erodes tooth enamel.

A spirillum is a spiral-shaped bacterium. The spirillum shown here, *Treponema pallidum,* causes syphilis, a sexually transmitted disease.

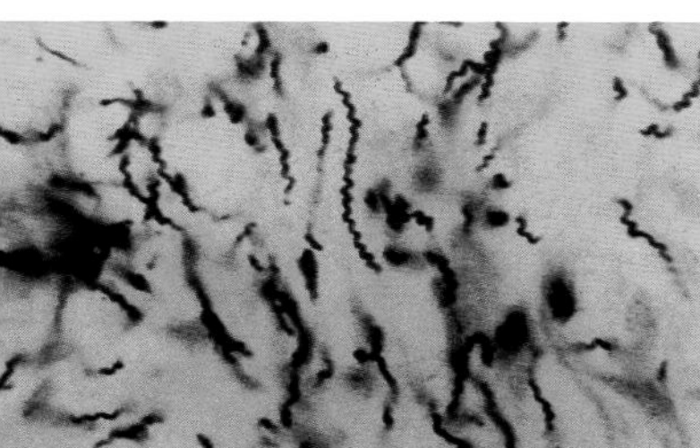

Magnification: 600×

Magnification: 11 000×

A bacillus is a rod-shaped bacterium that exists as individual cells and short chains. This bacterium, *Clostridium botulinum,* produces a potent poison that results in food poisoning when eaten.

You may recall from Chapter 8 that prokaryotic cells have no membrane-bound organelles such as a nucleus, mitochondria, or chloroplasts. Their ribosomes are smaller than those of eukaryotes. Their inherited information is held in a single circular chromosome, rather than in paired chromosomes.

Bacteria are often classified by the shapes of their cells. The three most common shapes are spheres, rods, and spirals, shown in *Figure 21.9.*

Bacterial cells are also classified by their arrangements. Although some cells live singly, others are typically grouped together, as *Figure 21.10* shows.

Figure 21.10

Bacteria are often arranged in characteristic groups.

Magnification: 52 000×

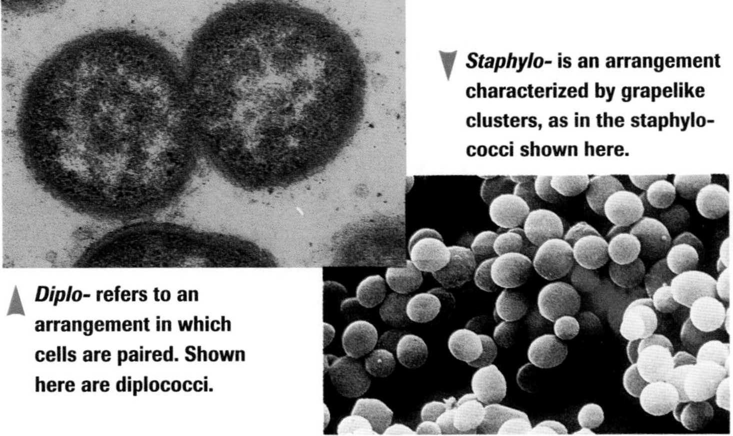

Diplo- refers to an arrangement in which cells are paired. Shown here are diplococci.

***Staphylo-* is an arrangement characterized by grapelike clusters, as in the staphylococci shown here.**

Magnification: 34 000×

Magnification: 14 000×

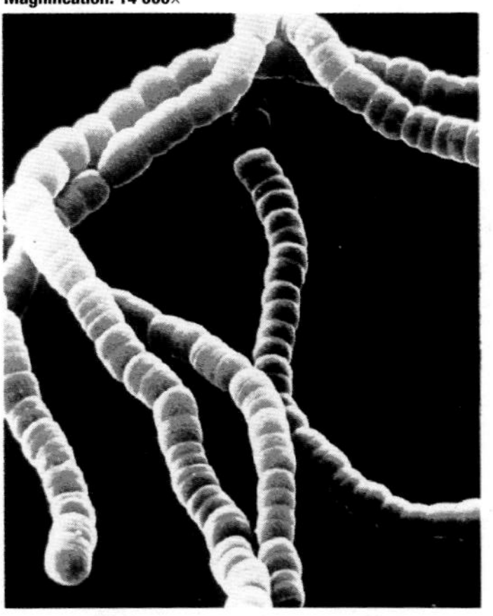

***Strepto-* is an arrangement characterized by long chains, as in the streptococci shown here.**

SECTION 21.2

Microscope Activity

Visual-Spatial Thin yogurt containing live cultures with distilled water. Have students make a wet mount of the yogurt and add a drop of methylene blue. Ask them to observe their slides and look for evidence of bacteria. *They will be able to see bacteria under high power.* L1

Demonstration

Visual-Spatial Obtain two sterile petri dishes containing nutrient agar. Have a student touch a fingertip to the agar in one plate and then close the plate. Have the student wash his or her hands thoroughly with soap and then touch the same fingertip to the agar in the second plate. Close the plate and write "Soap" on the lid.

Place the two plates upside down in a warm place where they can remain undisturbed for two to three days. Have students observe and compare the plates. Elicit a description and explanation of any differences observed. L2

orange indicates the presence of photosynthetic bacteria. You may need to wait up to two weeks for growth to occur.

Analysis

Ask students the following questions.

1. In which of your cultures did you find photosynthetic bacteria? *Perhaps in all, perhaps in none.*
2. Was your hypothesis supported by your data? Speculate why or why not. *Some students will find growth in the tubes they predicted, others will not. There are hundreds of kinds of photosynthetic bacteria, each with unique characteristics. Some may have found the growth medium adequate, and others may not.*
3. What caused the colors you saw in the test tubes? *The photosynthetic pigments*

Assessment

Performance: Have students design and carry out an experiment in which they determine the optimum temperature for growth of the photosynthetic bacteria they have isolated. They must subculture their bacteria every two weeks to keep the culture alive. L2

THE INSIDE STORY

A Typical Bacterial Cell

Purpose

Visual-Spatial Students will learn about the structure and adaptations of a representative bacterium.

Visual Learning

- Name the various bacterial structures shown in *The Inside Story.* As you name each structure, have a student use the caption to summarize the functions of the cell part. Repeat until all structures have been discussed. L1
- Challenge students to identify structures of the bacterial cell that are shared by plant and/or animal cells. Elicit from students what structure of plant and animal cells is not present in bacterial cells. *A nucleus* L1

Teaching Strategies

- Demonstrate conditions under which bacterial growth can be inhibited by pouring 100 mL of boiled beef broth into each of four sterile beakers. Label the beakers 1, 2, 3, and 4. Add 1 tsp of salt to Beaker 1; 1 tsp of sugar to Beaker 2; 1 tsp of vinegar to Beaker 3, and 1 tsp of distilled water to Beaker 4. Have students observe the clarity and color of each beaker in three days. Explain that cloudiness indicates bacterial growth. Ask students which beaker shows evidence of having the greatest amount of growth. *Beaker 2* Have students explain the effects of salt, sugar, and acids (vinegar) on bacterial growth. *Salt and acids inhibit growth. Sugar enhances growth because it is used as an energy source.* L1

THE INSIDE STORY

A Typical Bacterial Cell

Bacteria are microscopic, prokaryotic cells. The great majority of bacteria are unicellular. A typical bacterium would have some or all of the structures shown in this diagram of a cell.

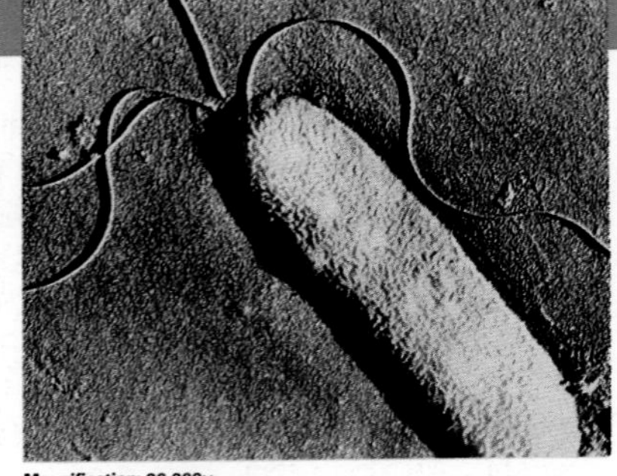

Magnification: 26 000×

This bacterium, *Pseudomonas syringae,* does not have a capsule or pili, but does have flagella.

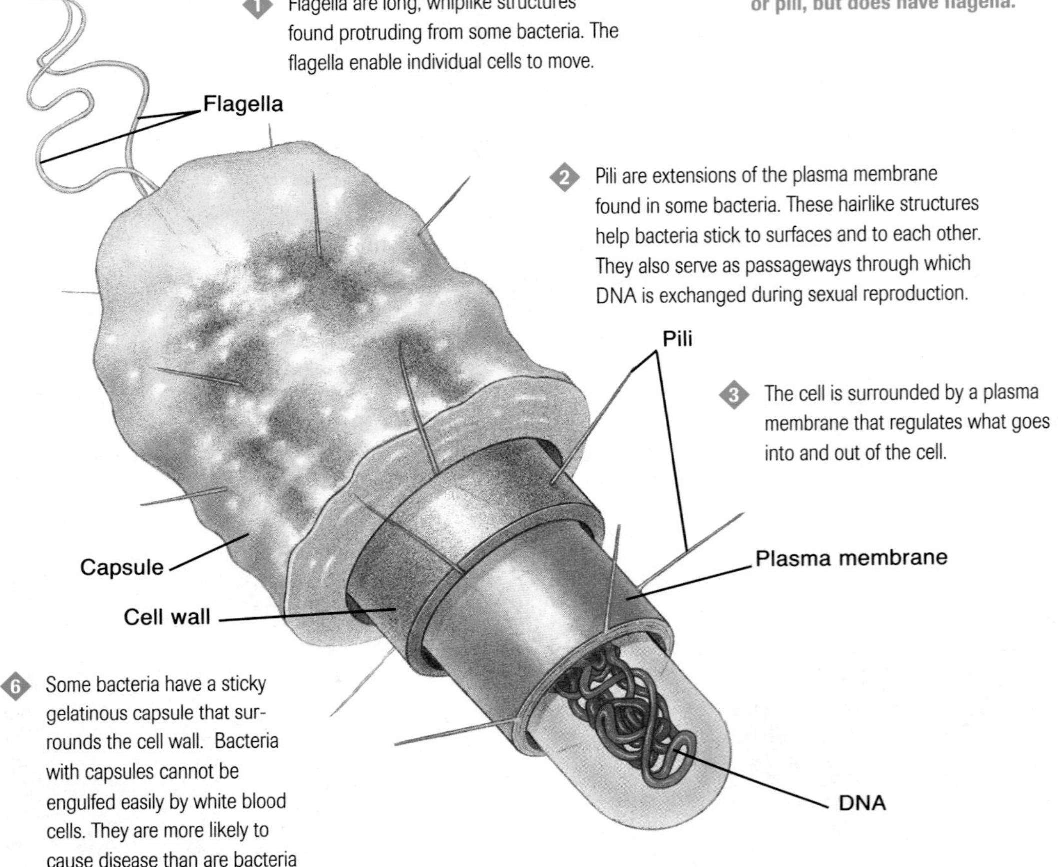

1. Flagella are long, whiplike structures found protruding from some bacteria. The flagella enable individual cells to move.

2. Pili are extensions of the plasma membrane found in some bacteria. These hairlike structures help bacteria stick to surfaces and to each other. They also serve as passageways through which DNA is exchanged during sexual reproduction.

3. The cell is surrounded by a plasma membrane that regulates what goes into and out of the cell.

4. A single DNA molecule arranged as a circular chromosome contains the genetic information of the cell. The DNA is not enclosed in a nucleus. The area of the cell containing the DNA is called the nucleoid.

5. A cell wall surrounds the plasma membrane. It is rigid, gives the cell its shape, and prevents osmotic rupture.

6. Some bacteria have a sticky gelatinous capsule that surrounds the cell wall. Bacteria with capsules cannot be engulfed easily by white blood cells. They are more likely to cause disease than are bacteria without capsules.

516 Viruses and Bacteria

Meeting Individual Needs

Learning Disabled Have students work in pairs to design flash cards that show a bacterial shape on one side of the card and the name of the bacteria associated with that shape on the other side. Students should also create cards that show bacteria of each shape arranged in clusters, chains, and pairs to gain practice in using the prefixes associated with bacterial groups. Have students use the cards to learn the names of all of the types of bacteria studied in this section. L1 LS SAE

Protection from osmotic rupture

A bacterium has a cell wall that forms a rigid outer covering for the cell. Unlike the cellulose cell walls of plants, the cell walls of bacteria are made of long chains of sugars linked by short chains of amino acids. One of the functions of the cell wall in bacterial cells is to prevent osmotic rupture. Most bacteria live in a hypotonic environment, in which there is a higher concentration of water molecules outside than inside the cell. Water is always entering a bacterial cell, but because it is surrounded by a cell wall, the cell remains intact. However, should the cell wall be damaged, the water entering the cell would cause it to rupture and die.

Penicillin—Bacterial killer?

Osmosis is also an important factor in the treatment of bacterial diseases. Penicillin kills bacteria by interfering with the enzyme that links the sugar chains in the cell wall. Bacteria growing in penicillin develop holes in their cell walls. As a result, when water enters a cell by osmosis, the bacterium ruptures and dies.

Penicillin is not effective against viruses and animal cells because neither has a cell wall. It is also nontoxic to plants because the structure of the plant cell wall is different from the structure of the bacterial cell wall.

History

Miracle Cure: The Story of Penicillin

"I need not go into details of the research on which I was engaged at the time but my culture plate was covered with staphylococcal colonies. The particular work involved opening the plate and looking at it under a dissecting microscope and then leaving it for growth to take place. That, of course, was asking for trouble by contamination, as things were always dropping from the air and sure enough, trouble came but it led to penicillin. A mould spore, coming from I don't know where, dropped on the plate. That didn't excite me. I had seen such contamination before, but what I had never seen before was staphylococci undergoing lysis around the contaminating colony. Obviously something extraordinary was happening."—Sir Alexander Fleming, "Antiseptics Old and New," 1946

Penicillium notatum The story of Fleming's accidental discovery of penicillin in 1928 and the events that followed are legendary. The bacteria-killing substance produced by his notorious airborne mold, *Penicillium notatum,* became the world's first antibiotic and prompted a golden age of antibiotics that save millions of lives from infectious disease. However, the development of this antibiotic was 12 years in coming. It was not until 1940, at the beginning of World War II, that a team of English scientists was able to purify the substance produced by Fleming's mold.

World War II production Production of the drug for human use became crucial as the war raged on in Europe. Penicillin moved from small-scale laboratory production in England to the chemical factories of America, where it helped win the war for the Allies.

Penicillin's impact Although penicillin's popularity faded somewhat when resistant bacteria appeared and allergy-induced deaths occurred, its discovery stimulated further research into lifesaving antibiotics. The post-war drug industry produced the second-most-important antibiotic—streptomycin—as well as many others.

CONNECTION TO Biology

Penicillin revolutionized medicine, providing a formidable weapon against killer diseases such as pneumonia and bacterial meningitis. Why is penicillin ineffective against cold, polio, or the HIV viruses?

History

Miracle Cure: The Story of Penicillin

Purpose

Students will learn the impact the discovery of penicillin and other antibiotics had on medicine and history.

Teaching Strategies

- Ask students to share experiences they have had with illnesses for which they received antibiotic treatment. Challenge students to speculate about how they might have been treated if antibiotics had not been available. L1
- Direct students to reread Sir Alexander Fleming's account of the discovery of penicillin. Discuss whether the account suggests that the discovery resulted from planned research or from an accidental occurrence. L1
- Have groups of students report on common antibiotics and the illnesses for which they are prescribed. L2

Possible Answer

Penicillin is not effective against viruses because they have no cell walls. Penicillin prevents cell wall formation.

STUDENT JOURNAL

Bacteria Making Headlines Ask students to bring in newspaper and magazine articles about bacterial illnesses. Ask them to summarize their articles in their journals. L1 LS

PORTFOLIO

Discovering Penicillin Ask student groups to prepare a skit about the discovery of penicillin. They should present how diseases were treated prior to penicillin, how penicillin was discovered, and how medicine changed after the discovery of penicillin. They could also include the development of penicillin-resistant bacteria. Have students place the script in their portfolios. L2 LS P

BioLab Design Your Own Experiment

How sensitive are bacteria to antibiotics?

Time Allotment
Initial session: one class period; followup session: 15 minutes, 48 hours after lab is begun.

Objectives
Review objectives with students before they begin the Biolab.

Process Skills
compare and contrast, observe and infer, recognize cause and effect, form a hypothesis, interpret data

Safety Precautions
- Students should wash their hands with soap after handling live bacterial cultures.
- When students complete the lab, they should clean their work areas with disinfectant and dispose of cultures and petri dishes as you direct.

PREPARATION
- Order *E. coli* from a biological supply house. You may want to order a culture on an agar plate.
- If students are transferring bacteria from an agar culture, they should use a sterile loop. Have them practice with the loop on plain agar before working with live cultures. A loop and solid cultures will be more difficult to work with than a cotton swab and a liquid culture, as students tend to dig up the agar with the loop.

Possible Hypotheses
- If antibiotic disks are present in a bacterial culture, no bacteria will grow.
- If no antibiotic disks are present, bacterial growth will be great.

BioLab Design Your Own Experiment

How sensitive are bacteria to antibiotics?

Doctors need to determine which antibiotic will kill disease-causing bacteria. A test similar to the one in this Biolab can be used. You will use sterile petri dishes containing nutrient agar and sterile disks that have been impregnated with antibiotics. When a disk is placed on the agar, the antibiotic diffuses into the agar. The clear ring that appears around the disk is called a zone of inhibition and represents an area where sensitive bacteria have been killed.

PREPARATION

Problem
Which antibiotic is most effective in killing a particular kind of bacteria?

Hypotheses
Have your group agree on a hypothesis to be tested. Record your hypothesis.

Objectives
In this Biolab, you will:
- **Compare** the effectiveness of different antibiotics in killing a particular strain of bacteria.
- **Determine** the most effective antibiotic for treating an infection caused by this strain of bacteria.

Possible Materials
cultures of bacteria
sterile nutrient agar petri dishes
antibiotic disks
sterile disks of blank filter paper
marking pen
cotton swabs
forceps
37°C incubator
metric ruler

Safety Precautions
Although the strains of bacterial cultures you will be working with are not pathogenic, be careful not to spill them. Wash your hands with soap immediately after handling any live bacterial culture. Be sure to clean your work area and dispose of your cultures and petri dishes as directed by your teacher.

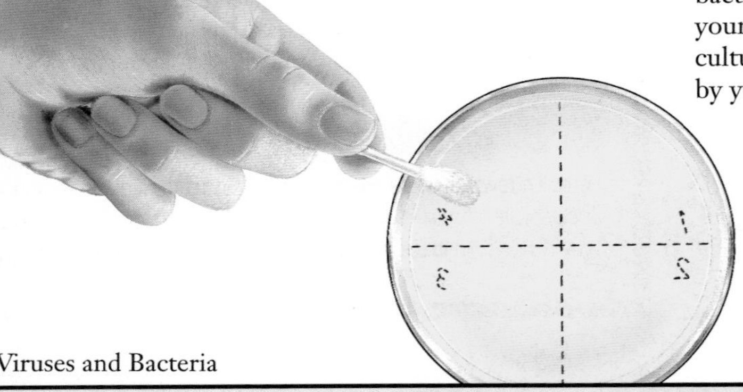

PLAN THE EXPERIMENT

Teaching Strategies
- Show students a petri dish that you inoculated two days ago and incubated.
- Teach students sterile technique and have them practice inoculating a plain agar dish with a plain cotton swab before they use live cultures.
- When students are examining their clear zones of inhibition, have them hold the dish toward the light.
- Explain that *E. coli* is a common intestinal inhabitant that ordinarily causes no harm. However, in other parts of the body, *E. coli* can cause disease.

PLAN THE EXPERIMENT

1. Examine the materials provided by your teacher, and study the illustrations on these pages. As a group, make a list of the possible ways you might test your hypothesis.
2. Agree on one way that your group could investigate your hypothesis. Design an experiment that will allow for the collection of quantitative data.
3. Prepare a list of numbered directions. Include a list of materials and the amounts you will need. Try to limit yourself to one petri dish per person.
4. Design and construct a table for recording your data. What do you think will happen around the antibiotic disks as the antibiotic diffuses into the agar? How will you measure this?

Check the Plan

Discuss the following points with other group members to decide the final procedure for your experiment.

1. Determine how you will set up your plates. How many antibiotics can you test on one plate? How will you measure the effectiveness of each antibiotic? What will be your control? Be sure you mark your petri dishes on the *bottom*. Why?
2. Do you think it would be better to add the bacteria or the antibiotic disks to the petri dish first?
3. What precautions will you take to prevent contamination of your petri dish by bacteria from around the room?
4. How often will you observe your plates?
5. ***Make sure your teacher has approved your experimental plan before you proceed further.***
6. Carry out your experiment. Make any needed observations, and complete your data table. Design and complete a graph or visual representation of your results. Do you think a bar graph or a line graph would be more appropriate?

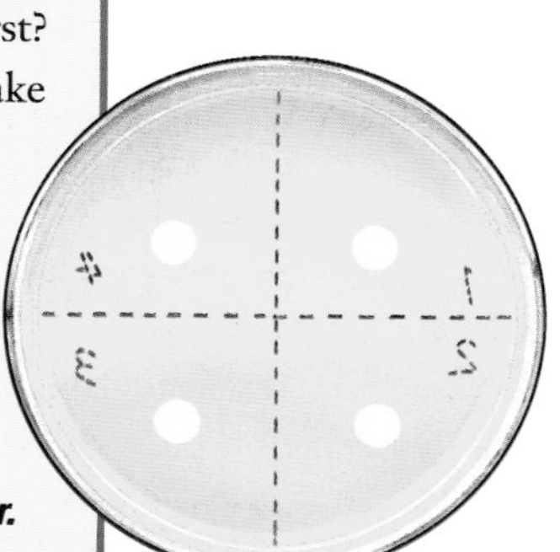

ANALYZE AND CONCLUDE

1. **Measuring in SI** How did you measure your zones of inhibition? Why did you do it this way?
2. **Comparing and Contrasting** Which antibiotic caused the largest zone of inhibition? What is the significance of this?
3. **Drawing Conclusions** If you were a physician treating a patient infected with this bacterium, which antibiotic would you use? Why?
4. **Analyzing the Procedure** Can you think of any limitations of this technique? If a real person were involved, what other tests might give you more confidence in your results?

Going Further

Application Do library research to find out how different antibiotics kill bacteria. Penicillin, for example, interferes with the production of bacterial cell walls.

ANALYZE AND CONCLUDE

1. Most students will measure the diameter of the zone of inhibition and use this diameter as a basis of comparison.
2. The largest zone of inhibition will be produced by the antibiotic that has the greatest inhibiting properties.
3. The one with the largest zone of inhibition because it has the best ability to inhibit bacterial growth
4. The antibiotic may work in an agar culture, but not in human tissue. A doctor might also do blood tests to examine the kinds of white blood cells present, or look for the bacteria themselves if they are suspected to be in blood, bladder, kidney, or intestine.

Assessment

Portfolio: Ask students to write a paragraph about the following scenario: A man arrives at the doctor's office with a bacterial infection on his hand. The doctor has ten different antibiotics she could give to the man. How can she find out which antibiotic will work best against the infection? L1 P

Performance Assessment in the Biology Classroom: p. 29, *Inhibiting Bacterial Growth.* Have students carry out this activity to determine how bacterial growth can be inhibited. L1 P

Going Further

Students may look up erythromycin, streptomycin, tetracycline, and furodantin. L2

Possible Procedures

- Using sterile technique, students may decide to make an *E. coli* culture in their petri dish.
- Students will likely divide the dish into four quadrants and place different antibiotic disks in each of three quadrants and a plain sterile disk in the fourth quadrant.
- After incubating the dish for two days, students can measure the width of the clear areas that surround the disks.

Data and Observations

All antibiotic disks should show some inhibition. Results will vary depending on the antibiotic disks used. The untreated disks should show no zone of inhibition.

A BROADER VIEW

Living on the Edge

Purpose

Students will learn about bacteria that thrive in extreme heat and extreme cold.

Teaching Strategies

- Show students a video of the geysers of Yellowstone National Park.
- If you have commercial hot springs in your area, ask a group of students to visit and find out what procedures are taken to prevent bacterial contamination.

Background

- The protein chains of bacteria that thrive in environments having extreme temperatures have more cross-links than do conventional proteins. Conventional proteins exist as spherical structures with the ends of their chains sticking out. These free ends flap about when heated and provide a spot for unwinding to begin. Heat-resistant proteins have their loose ends tied closer to the main body of the protein.

Thinking Critically

The atmosphere of early Earth did not have oxygen.

Microscope Activity

LS **Visual-Spatial** Provide students with prepared slides of heterotrophic and photoautotrophic bacteria. Ask students to observe the slides under high power and make sketches with colored pencils of their observations. Students should also estimate the size of the bacteria based on the assumption that an average high-power field of view is 0.35 mm. After students have made their observations, ask them to write a summary that compares and contrasts these two types of bacteria. L1

A BROADER VIEW

Living on the Edge

In April 1993, some residents of Milwaukee, Wisconsin, became ill after drinking contaminated tap water. Health officials advised residents to boil their water before drinking it. Why? At 100°C, the temperature at which water boils at sea level, the molecules that make up life as we know it are destroyed. Cell walls collapse, DNA unravels, and proteins unfold. Thus, any organisms in the water, including harmful bacteria, are quickly killed. Or are they?

Mighty bacteria Over the past three decades, scientists have discovered a variety of simple, unicellular organisms that actually thrive at extreme temperatures. Researchers have found colonies of cyanobacteria growing in a frozen pond just 360 km from the geographic South Pole. And on the ocean floor along the Mid-Atlantic Ridge, large colonies of bacteria thrive around volcanic vents where the water temperature soars to almost 400°C.

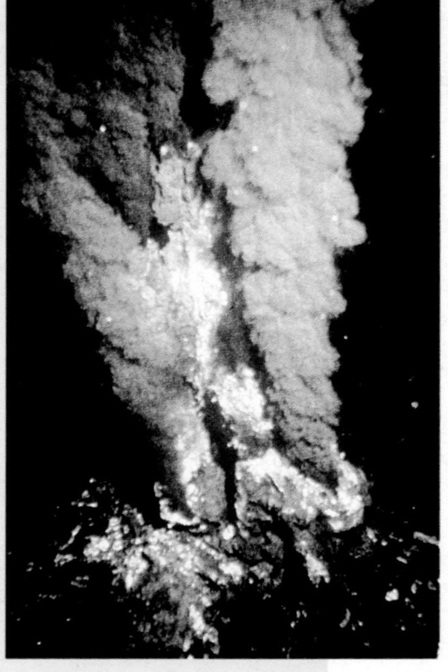

Did they adapt? Traditional science maintains that heat-loving organisms are conventional life forms that have somehow adapted to life at extreme temperatures. Recently, however, some scientists have challenged this idea. They've begun to speculate that these heat-loving organisms haven't had to adapt to anything because they are, in fact, biological relatives of some of the earliest forms of life on this planet.

Life in the primordial soup
Scientists think that life began on Earth at a time when the planet was still cooling down from its fiery birth. Extensive volcanic activity was the rule, and any organisms that were going to survive would have had to be highly resistant to heat. What's more, they would most likely have been autotrophic.

These are precisely the types of organisms scientists are finding in superhot environments on Earth today. As more is discovered about these unique bacteria, additional light will be shed on some of the mysteries surrounding the origin of life on Earth.

Thinking Critically
Why were the earliest forms of life on Earth probably obligate anaerobes as well as being heat tolerant?

Ecology and Adaptations

Some of the oldest known fossils are of bacteria. These organisms lived on Earth when it had a climate and atmosphere very different from those of today. Some of these fossils contain compounds similar to chlorophyll and so were among the first photosynthetic organisms. As Earth's atmosphere became more oxygen rich, these bacteria evolved adaptations that allowed them to survive in the changing environment.

Diversity of metabolism

You have already read that eubacteria display a broad variety of metabolic pathways. Various eubacteria obtain nutrition and energy from light, from simple inorganic molecules, and from complex organic molecules.

Most bacteria require oxygen for cellular respiration. These bacteria are called **obligate aerobes.** *Mycobacterium tuberculosis* is an obligate aerobe. Other bacteria cannot use oxygen and are killed by it. These bacteria are called **obligate anaerobes.** The bacterium that causes syphilis is an obligate anaerobe. This explains why syphilis can be transmitted only by intimate contact from one moist membrane to another. You cannot catch syphilis from a contaminated toilet seat because the bacteria would be poisoned by oxygen and die. Still other kinds of bacteria can live either with or without oxygen. They have two types of metabolism and can derive energy aerobically by cellular respiration or anaerobically by fermentation.

Meeting Individual Needs

Gifted Ask students who are gifted to do research about Gram positive and Gram negative bacteria. Ask them to examine prepared slides of these types of bacteria and present their findings to the class. L3

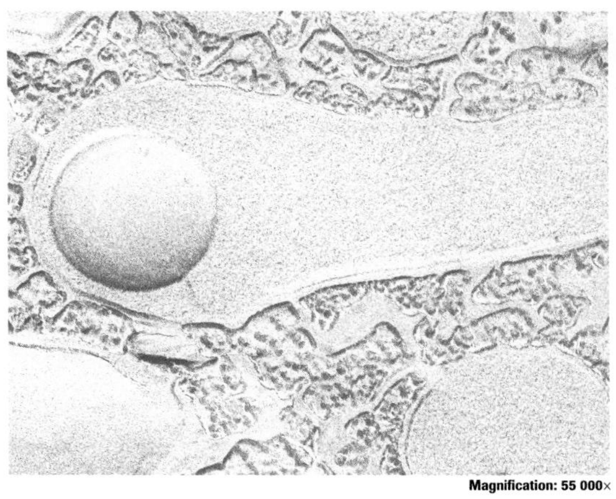

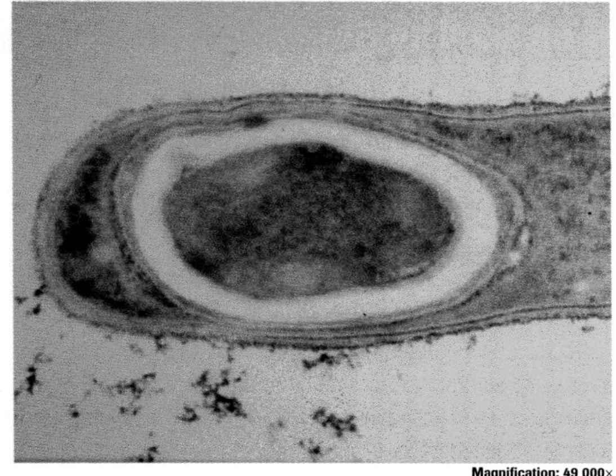

Figure 21.11

The SEM on the left shows an endospore being produced within a bacterial cell. The endospore contains a complete copy of the DNA of the bacterium. After the endospore is formed, the rest of the cell may disintegrate. The TEM on the right shows a cell containing an endospore. Notice the thickness of the wall of the endospore. Of what value to the organism is a thick wall?

Several bacteria have evolved unusual biochemical pathways. Green sulfur bacteria, for example, use hydrogen sulfide rather than water for photosynthesis. Instead of producing oxygen, they produce sulfur as a by-product. They can grow only in anaerobic environments such as lake sediments.

Adaptations for survival

Some bacteria, when faced with unfavorable environmental conditions, produce structures called **endospores,** *Figure 21.11.* Endospores have a hard outer covering and are resistant to drying out, boiling, and many chemicals. While in the endospore form, the bacterium is in a state of slow metabolism, and it does not reproduce. When it encounters favorable conditions, the endospore germinates and gives rise to a bacterial cell that resumes growth and reproduction. Some endospores have been found to germinate after thousands of years.

Endospores are of great survival value to bacteria. Because endospores can survive boiling, canned foods and medical instruments must be sterilized under high pressure. This can be done either in a pressure cooker or in an autoclave. When water is boiled under high pressure, it is hotter than the normal boiling point of water. This greater degree of heat can kill endospores.

The clostridia are a group of obligate anaerobes that can form endospores. One member of this group, *Clostridium botulinum*, produces an extremely powerful **toxin,** or poison. Instead of dying when exposed to oxygen, *Clostridium botulinum* forms endospores. These endospores can find their way into canned food. If the canned food has not been properly sterilized, the endospores will germinate and the bacteria will grow and produce their deadly toxin. Although the resulting disease, botulism, is extremely rare, it is often fatal.

Visual Learning

Figure 21.11 Of what value to the organism is a thick wall? *The thick wall provides a hard, outer covering that is resistant to drying out, boiling, and many chemicals.*

Misconception

Students may think that stepping on a rusty nail causes tetanus because of the rust on the nail. Explain that any puncture wound can harbor the anaerobic tetanus-causing bacterium. While the nail may cause the puncture wound, it is soil that is the source of the tetanus-causing bacterium.

PROJECT: Staining Techniques

Have students research the techniques involved in identifying bacteria through the use of acid-fast and Gram staining techniques. Have students use the information they find to design an experiment in which they will test five species of bacteria using the acid-fast and Gram staining techniques. After students receive approval of their experimental designs, you may wish to provide students with the materials needed to carry out their experiments. L3 LS

ThinkingLab — Draw a Conclusion

Purpose

LS **Logical-Mathematical** Students will analyze the growth pattern demonstrated by bacteria.

Process Skills

make and use a table, recognize cause and effect, interpret data, calculate, draw a conclusion

Teaching Strategies

- Exponential growth of bacteria can be modeled using a checkerboard and pennies. Bring $21.00 in pennies to class. Have students calculate how many squares you can cover on the board if you put one penny in the first square, two in the second, four in the third, and so on until you run out of pennies. *Eleven squares will be completely covered and $0.53 will remain.* Relate this demonstration to how large populations of bacteria can be produced in just a few generations. L2
- Remind students that bacteria in food may double in number every 20 minutes. Explain that food poisoning often results when foods are not properly refrigerated.

Thinking Critically

After one day there would be 2^{24} or 16 777 216 bacteria. After two days there would be 2^{48} or 2.8×10^{14}. Boils grow quickly because bacteria grow exponentially. They decrease slowly because bacteria do not die exponentially.

Assessment

Performance: How long will it take to have a bacterial population greater than 1 000 000 if the bacteria double each hour? *20 hours* L1

ThinkingLab — Draw a Conclusion

How fast can bacteria reproduce?

When they are in a favorable environment, bacteria reproduce extremely rapidly. Under ideal conditions, some bacteria can reproduce every 20 minutes. As conditions become less ideal, the rate of reproduction slows down.

Analysis

Suppose you start with a single bacterium at Hour 0, and it reproduces once an hour. How many bacteria will you have at the end of Hour 24?

To solve this problem, set up a table listing Hours 0 to 24. At Hour 0, you have one bacterium. After one hour, the bacterium will have reproduced once, so you will have two bacteria. At Hour 2, each of these bacteria will have reproduced once, giving a total of four. To find the number of bacteria after 24 hours, continue this doubling 24 times, once for each time the bacteria reproduce. If you can, calculate how many bacteria you will have after two days.

Thinking Critically

How many bacteria were there after one day? Use your knowledge of bacterial growth to explain why boils, which are caused by bacteria, swell quickly to their maximum size but decrease slowly in size.

Another member of the Clostridium group, *Clostridium tetani*, produces a powerful nerve toxin that causes the often-fatal disease, tetanus, shown in *Figure 21.12.* Because endospores of *C. tetani* are found on nearly every surface, they can easily enter a wound. A puncture wound doesn't allow air to enter, so conditions are ideal for growth of anaerobes. The endospores germinate in the wound, and the bacteria reproduce in great numbers. The bacteria produce a toxin, which enters the bloodstream and attacks the nerve cells in the spinal cord. Fortunately, there is an immunization for tetanus. You received this shot as a child. A booster shot is given as a precaution after a puncture wound. Deep wounds are hard to clean and provide ideal conditions for growth of anaerobes.

Reproduction

Bacteria cannot reproduce by mitosis or meiosis because they have no nucleus. Instead, they have evolved different methods of reproduction.

Binary fission

Bacteria reproduce asexually by a process known as **binary fission.** The bacterium first copies its single chromosome. The copies attach to the cell's plasma membrane. As the cell grows in size, the two copies of the chromosome separate. The cell then divides in two as a partition forms between the two new cells, as shown in *Figure 21.13.* Each new cell receives one copy of the chromosome. Therefore, the daughter cells are genetically identical to each other and to the parent cell.

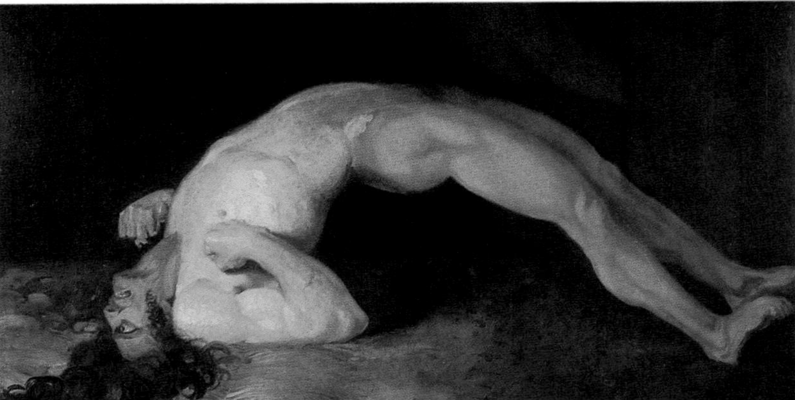

Figure 21.12
This painting by Charles Bell shows a British soldier dying of the final stages of tetanus during the Napoleonic wars in the early years of the 19th century. Note the characteristic muscle spasms.

STUDENT JOURNAL

Immunization History Have students find out about their immunization records. They should write in their journals the most recent dates of the vaccinations they have had and indicate if each disease against which they were vaccinated is bacterial or viral in origin. Encourage students to find out for how long the immunization is good. Provide library resources as necessary. If students have religious beliefs that prevent immunization, have them write about the usual times various immunizations are given during a person's life. L1 LS

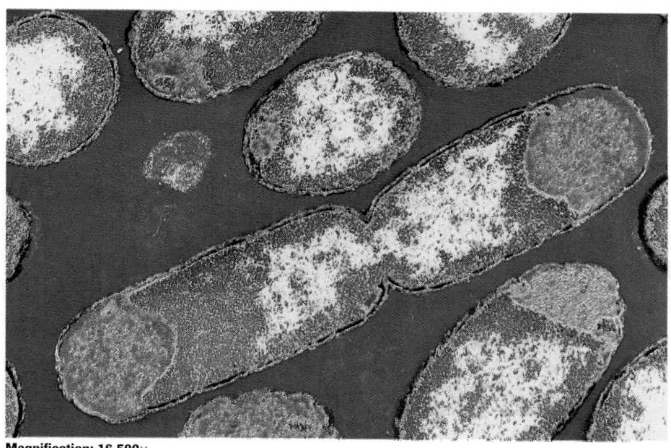
Magnification: 16 500×

Figure 21.13

This *Escherichia coli* cell is starting to divide. The newly forming partition is visible in the center of the cell. Why is cell growth needed for the process of binary fission?

Bacterial reproduction can be extremely rapid. Under ideal conditions, bacteria can reproduce every 20 minutes. Such a rate of reproduction yields enormous numbers of bacteria in a short time. If bacteria reproduced at their maximum rate, they would cover the whole planet within a few weeks. Yet, obviously, this does not happen. The reason is that, like all living things, bacteria don't always find ideal conditions. They run out of nutrients, they are eaten by predators, they dry up when no water is available, and they are poisoned by the wastes they produce.

When you have an infection, billions of bacteria grow in your body. If you are given an antibiotic for the infection, you should take the antibiotic for the full prescribed period—even though you feel better after just one or two days. Shortly after you begin to take the antibiotic, most of the bacteria are killed. However, if you stop taking the antibiotic and even a single bacterium is left, it will start reproducing. A day later, you will have millions of bacteria in your body and you will be sick again. Completing the antibiotic as prescribed ensures that all of the bacteria will be killed so you will not get sick again.

Sexual reproduction

In addition to reproduction by binary fission, some bacteria have a simple form of sexual reproduction called **conjugation.** During conjugation, one bacterium transfers all or part of its chromosome to another cell through a bridgelike structure called a pilus (plural pili) that connects the two cells. *Figure 21.14* shows this genetic transfer.

Figure 21.14

The *E. coli* bacterium on the bottom left is transferring its chromosome to the *E. coli* on the top through a long pilus. What is the advantage to the bacterium of this exchange of genetic material?

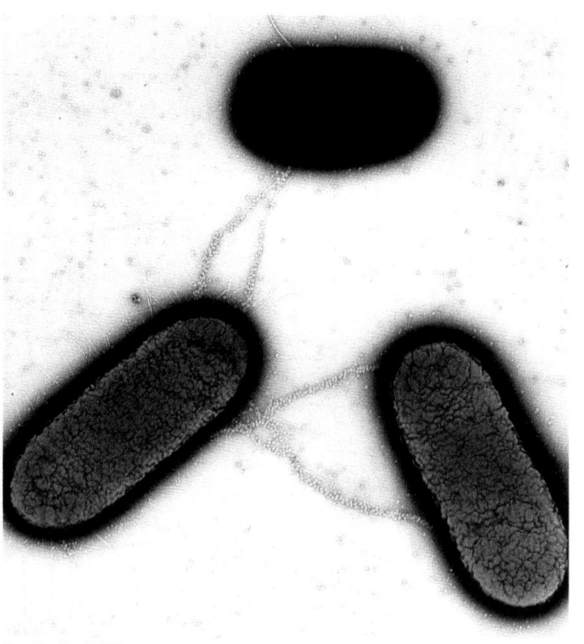
Magnification: 50 000×

Enrichment

Ask students to use library resources to research how antiseptics, disinfectants, sterilizing, canning, freezing, salting, smoking, and cooling can be used to reduce the growth of bacteria. Each student should be assigned a specific topic. Ask students to present their findings in a demonstration to the class. L2

Visual Learning

Figure 21.13 Why is cell growth needed for the process of binary fission? *As the cell grows, the two copies of the chromosomes separate.*

Figure 21.14 What is the advantage to the bacterium of this exchange of genetic material? *Exchange of genetic material allows for genetic recombination, leading to variation in the species. This variation may permit some members of the species to survive conditions that others of the species cannot.*

GLENCOE TECHNOLOGY

Videodisc

Biology: The Dynamics of Life
Disc 1, Side 2
Binary Fission (Ch. 11)

CULTURAL DIVERSITY

Medicine in Latin America

Microbiology has been perhaps the most important scientific field in Latin America. Much of this interest in microbiology is due to the founding of Brazil's Institute of Experimental Pathology (located in Rio de Janeiro) in 1902.

Discuss with students the numerous research findings from these laboratories during the 20th century, including the identification of the pathogen involved in Chagas's disease (trypanosomiasis or American sleeping sickness) by Carlos Chagas in 1909, and the solution to Carrion's disease by Alberto Barton in the same year. Also introduce students to the work of Cuban scientist Carlos Finlay, an important contributor to the etiology and pathology of yellow fever.

Demonstration

Bring to class a bag full of groceries that could not be made without bacteria. Include Swiss cheese, pickles, vinegar, sauerkraut, yogurt, peas, beans, soybeans, peanuts, milk, and sour cream. Explain the importance of bacteria to the development of each product including the role of nitrogen-fixing bacteria to the growth of certain kinds of plants.

Science, Technology, and Society

Working Bacteria

Bacteria are being used to extract lead, zinc, gold, copper, mercury, and uranium from mineral deposits. Microbial mining operations require less energy than conventional methods. In addition, the method is pollution free and mining costs are low.

High-sulfur coal is responsible for much of the acid precipitation in the world. *Thiobacillus thiooxidans* and species of *Sulfolobus* are two types of bacteria that can metabolize sulfur. When introduced into piles of high-sulfur coal, these bacteria leave behind a cleaner-burning fuel.

Other researchers are working with an organism that breaks down phenols. Phenols are by-products of wood treatment in pulp plants. By splicing the genes of this organism into *E. coli,* large amounts of phenol-destroying bacteria can be grown rapidly. The recombinant *E. coli* can change the phenol to harmless salts. If bacteria can be tailor-made to metabolize each factory's wastes, chemical companies may be able to detoxify their wastes without treatment plants.

Economic Importance

Bacteria are important to us in a number of ways. Some of our favorite picnic foods—mellow Swiss cheese, crispy pickles, tangy yogurt—would not be possible without bacteria. And once we have digested and extracted all the nutrients from those foods and eliminated them from our bodies as waste, other bacteria help break down that waste in septic tanks or at sewage-treatment plants. Still other bacteria increase crop yields when used as a biological means to control insect pests. Bacteria can increase crop yields in another way, as well.

Nitrogen fixation

Most of the nitrogen on Earth exists in the form of the nitrogen gas (N_2) that makes up 80 percent of the atmosphere. All organisms need nitrogen to produce the proteins, DNA, RNA, and ATP in their cells. Yet few organisms can use N_2 gas directly. Of all mineral elements, nitrogen is the one that most often limits the growth of plants.

Several species of bacteria have enzymes that convert N_2 gas into ammonia (NH_3) in a process known as **nitrogen fixation.** Other bacteria then convert the ammonia into nitrite (NO_2^-) and nitrate (NO_3^-), which can be used by plants. Bacteria are the only organisms that can fix nitrogen. Some nitrogen-fixing bacteria form symbiotic associations with legumes such as peas, peanuts, and soybeans, *Figure 21.15.* In nitrogen-deficient soils, legumes grow better than plants such as corn that do not have root nodules. Thus, nodules of these bacteria are a distinct advantage to a plant. The relationship is also an advantage to agriculture. When legumes are grown and then harvested, the remaining roots with nodules add considerable usable nitrogen to the soil. The following season, other crops can be grown in the newly nitrogen-rich soil. This is the basis of crop rotation.

Figure 21.15

The nodules on these soybean roots (left) contain *Rhizobium* bacteria (right) that convert nitrogen gas into ammonia. In this symbiotic association, the plant gains a source of usable nitrogen, and the bacteria can use sugars supplied by the plant for their own needs. Cyanobacteria also can fix nitrogen.

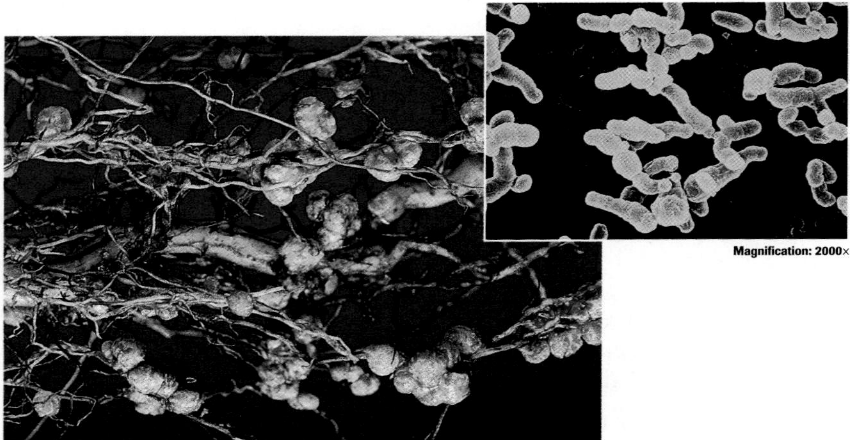

Meeting Individual Needs

Students Acquiring English Give students a diagram showing select bacteria that are representative of all three shapes, with some arranged as a single bacterium, in pairs, in clusters, and in chains. At the top of the diagram, include a key that identifies the meanings of the word parts needed to name each type of bacterium shown. Ask students to use the key to write the correct bacterial name below each drawing. Review their responses to make sure students have identified the bacteria correctly. L1 SAE

Figure 21.16

The holes in Swiss cheese are caused by bubbles of carbon dioxide produced by the bacteria that give the cheese its flavor. Cheese is made in huge vats such as the one above. The large industrial fermenter (right) holds 1500 L and is used for growing bacteria in industry.

Recycling of nutrients

You learned in Chapter 3 that life could not exist if bacteria did not break down the organic matter in dead organisms and wastes, returning the nutrients to the environment to be used by producers at the bottom of food chains. Cyanobacteria, along with plants and algae, replenish the supply of oxygen in the atmosphere. Autotrophic bacteria convert carbon dioxide in the air to the organic compounds that are passed to consumers in food chains and webs. All life depends on bacteria.

Food and medicines

Because bacteria are so metabolically diverse, different species produce a wide variety of molecules as the result of fermentation, *Figure 21.16.* Many of these molecules have distinctive flavors and aromas. As a result, the bacteria that produce them are used to make vinegar, yogurt, butter, cheese, pickles, and sauerkraut.

Strains of bacteria have evolved that produce important antibiotics, which are used as medicines to kill other bacteria. Streptomycin, erythromycin, chloromycetin, and kanmycin are some of the antibiotics that are derived from bacteria.

Bacteria cause disease

Although only a few kinds of bacteria actually cause disease, those that do have a great impact on our lives. It is estimated that about half of all human diseases are caused by bacteria.

Chemistry Connection

Methane

Explain that methane is a gas given off by some bacteria that break down organic matter. On the chalkboard, sketch the structural formula of methane and identify its molecular formula as CH_4. Point out that methane is used as a fuel. Light a Bunsen burner to illustrate the properties of methane. Gas companies add chemicals to the gas to produce an odor that can be detected by people should a leak or accident occur. Have students brainstorm a list of possible economic advantages that might be gained by utilizing bacteria that produce methane.

GLENCOE TECHNOLOGY

Videodisc

STVS: Plants & Simple Organisms

Disc 4, Side 1

Bacterial Waste Treatment (Ch. 10)

Detecting Salmonella (Ch. 5)

PROJECT

Bacteria of Decay

Have students make a mini-landfill in a jar containing moist soil. Direct students to bury a slice of carrot, paper, aluminum foil, and other items that they might expect to be put in a landfill. Make sure they do not bury any animal products.

After two weeks, have them unearth their landfill items and compare their appearance to what they looked like before burial. Ask students to record their observations and use them to discuss the importance of recycling by bacteria. L1 LS

People in Biology

Meet Dr. Barbara Staggers, Physician

Teaching Strategies

- Encourage students to talk about their current doctors and the pediatricians they may have had as children. Have them explain what they liked most and disliked most about these health care professionals. Elicit how they would determine if someone is a good doctor.
- Ask students to work in pairs to find out about various kinds of doctors such as dermatologists, endocrinologists, and cardiologists. Have students present their information to the class in the form of an interview in which one student acts as interviewer and the other assumes the role of the doctor.
- Invite a local doctor who specializes in adolescent care to speak to your class about his or her career.

Background

- Sexually transmitted diseases, teen pregnancy, and drug abuse are common youth medical problems. To combat such problems, many communities have treatment facilities that young people can go to without needing parental permission.

people in biology

Meet Dr. Barbara Staggers, Physician

"If you fall between the ages of 11 and 21, you're at risk, no matter who you are or where you live." Those words are frequently spoken by Dr. Barbara Staggers, Director of Adolescent Medicine at Children's Hospital in Oakland, California.

In the following interview, Dr. Staggers talks about the tremendous stresses on young people today and tells how she and her staff try to help teens at her clinic.

On the Job

Q Dr. Staggers, what do you do here at the teen clinic?

A We see students for everything from physicals for sports eligibility to patching them up after a fight. Kids feel comfortable here. We're known as an okay place to be. We also do a lot of outreach in the community. We employ students as peer health-care counselors to do AIDS education in local high schools and also for young people in jail. Kids would rather talk to their peers, no matter how cool I think I am. But kids seem to think I'm okay, too. Their demand for honesty is one thing I really love about adolescents.

Q Are many students interested in becoming peer counselors?

A Yes, and I think that's because it's a way for them to give back something to others. I think we underestimate how much teens need an opportunity to give. Often, because they've been there, peer counselors can be very effective in helping other kids with serious problems.

Q What are some of the major problems that you and the peer counselors see?

A The three leading causes of death in teenagers are motor vehicle injuries (of which 50 percent are alcohol and drug related), homicide, and suicide. The number-one disease in teens is mental-health illness, such as depression. And, of course, there are also the sexually transmitted diseases.

Q How do you work with kids who have such serious problems?

A I refuse to focus solely on their behavior. I try to find out where a teen is and what's happening in his or her life. I hear kids say things like, "Every boyfriend I've dated in the last five years is dead, so if I get pregnant, at least I'll have someone to be with me." Or, "I probably won't live long enough to worry about getting hooked on drugs." Then I need to deal with kids around those issues and have sort of a general life discussion. I think that having a conversation that will prevent someone from shooting himself or herself or someone else is just as valuable as taking a bullet out of a person who's been shot.

Early Influences

Q How did you decide to become a doctor?

A When I was 18 years old, I was torn between being a veterinarian and joining the Dance Theater of Harlem. Then I worked in a summer sports program at a college. I discovered that a lot of young people in the program had never been to a physician. Their access to health care was a lot different from mine because my father was a doctor. That difference made me rethink my career plans. I decided that I wanted to be in the field of medicine and work with kids of color, particularly disadvantaged, urban youth.

Q Has that decision always felt like the right one for you?

A It's clear to me, because it's so comfortable, that this is what I'm supposed to do. Not a day passes that I don't learn from one of the teens I see. They keep me honest. I can't put a price on having a teen flag me down in a mall and thank me for having given him or her some time four years ago.

Personal Insights

Q What advice do you have for teenagers who might want to pursue a medical career?

A The best advice I can give is never to give up your dream. Start by looking away from material wealth and toward human health. To be in medicine, you have to like people. You have to be able to talk to people—even sick, grouchy people—and be willing to invest in them. That part of medicine is most important. Necessary classes such as chemistry may have been hard for me, but I've always loved to work with people.

It's important to learn about yourself. You can't start too young to find out what makes you feel good about yourself and pursue those things.

CAREER CONNECTION

People in these careers work to help keep young people healthy.

Physician *A bachelor's degree, four years of medical school, three years of residency (graduate medical education)*

Social Worker *A bachelor's or master's degree in counseling, rehabilitation, or a related field*

Nurse's Aide or Orderly *High-school diploma; on-the-job training*

Q What makes you, personally, feel good about yourself?

A I've been involved in 50 million zillion things all my life—karate, ballet, jazz, poetry, photography. And I hang out with my family because that's where I get support. I know it's essential for me to have all of those things in place to make me who I am.

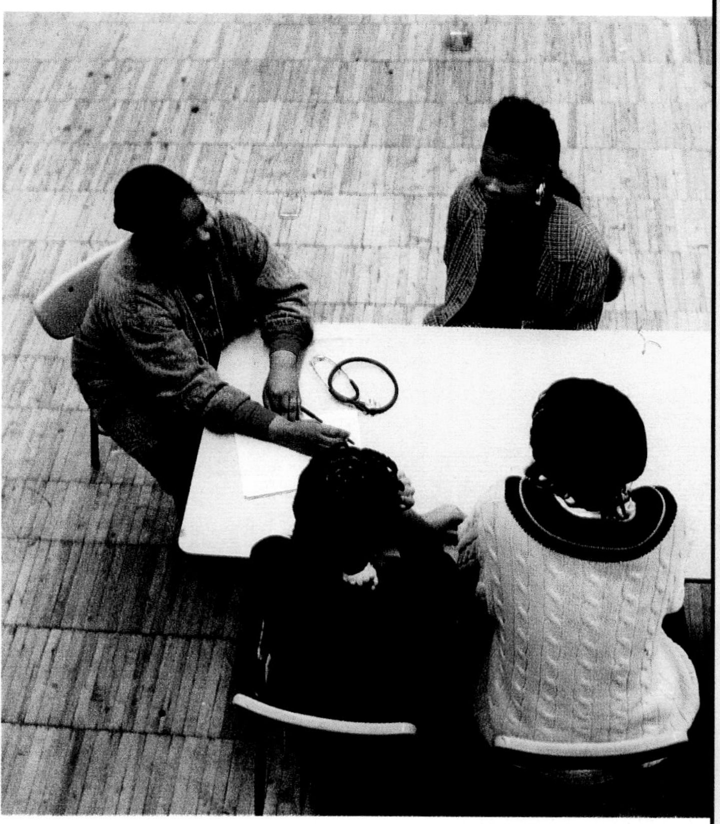

CAREER CONNECTION

- **Career Path** High School: Earth science, advanced courses in biology, chemistry, physics, and mathematics.

  College: biology, zoology, botany, anatomy and physiology, mathematics, statistics, genetics, evolution, microbiology, ecology, nutrition, population biology, biochemistry, chemistry, physics, cell biology, developmental biology, neurobiology, and immunology.

  Medical School: four years of general courses in medicine.

  Residency: three years of training in a specialized field of medicine such as pediatrics.

For More Information

American Medical Association,
515 North State St.
Chicago, Illinois 60610-4377
http://www.ama-assn.org/

American College of Physicians,
Independence Mall West,
6th and Race Sts.
Philadelphia, PA 19106.

SECTION 21.2

3 Assess

Check for Understanding

Have students list characteristics of each of the major groups of bacteria. L1

Reteach

Ask students to outline this section of the chapter using standard outline format. Beneath each of the main heads in their outlines, tell students to write a one- or two-sentence summary of the section or subsection. L1

Extension

Ask a group of students to write to the Centers for Disease Control and Prevention in Atlanta, Georgia, to get recent copies of the Morbidity and Mortality Weekly Report (MMWR). Ask students to summarize the information about diseases that are currently important in the United States and in the world. This information can also be found at http://www.cdc.gov/epo/mmwr/mmwr.html L2

Assessment

Portfolio: Have students make labeled diagrams of each type of bacteria they have studied in this chapter. Ask them to include captions that describe the adaptations of each type of bacteria. L1 P

4 Close

Brainstorming

Ask students to summarize the importance of bacteria, including their role in the environment, in causing diseases, and in industry.

In the past, bacterial illnesses took a major toll on human populations. As recently as 1900, life expectancy in the United States was only 47 years. The leading killers of the time were tuberculosis and pneumonia, both bacterial illnesses. In the intervening century, life expectancy has increased to about 75 years. This remarkable 50 percent increase means that, on average, people live nearly 25 years longer today than they did in 1900! This increase in life expectancy is due to many factors. People now have better living conditions. We have less poverty, better public health systems, improved water and sewage treatment, better nutrition, and better medical care. These improvements, combined with the presence of antibiotics, have reduced death rates from bacterial diseases to very low levels. *Table 21.1* lists some bacterial diseases.

Table 21.1

Diseases Caused by Bacteria		
Tuberculosis	Scarlet fever	Rocky Mountain spotted fever
Bacterial pneumonia	Syphilis	Tetanus
Botulism	Gonorrhea	Ear infections
Strep throat	Chlamydia	Boils
Staph infections	Diphtheria	Lyme disease

Connecting Ideas

Bacteria include all organisms with prokaryotic cells. They represent the most ancient forms of life on Earth. They were the only life forms until about 2.1 billion years ago when the first eukaryotic cells evolved. The other four kingdoms—Protista, Fungi, Plantae, and Animalia—all contain organisms with eukaryotic cells. When you compare the metabolic pathways of bacteria with those of the other four kingdoms, you may find it surprising that there is so much more metabolic diversity among bacteria than among members of the other kingdoms.

Section Review

Understanding Concepts

1. Describe six parts of a typical bacterial cell, and give their functions.
2. How do endospores and pili help bacteria survive?
3. How is a bacterial cell affected by penicillin?

Thinking Critically

4. Why is it reasonable to hypothesize that archaebacteria were the first organisms to evolve on Earth?

Skill Review

5. **Making and Using Tables** Make a table comparing archaebacteria and eubacteria. Include at least three ways they are alike and three ways they are different. For more help, refer to Organizing Information in the *Skill Handbook.*

Answers to Section Review

1. Circular chromosome contains genetic information; plasma membrane regulates what enters and exits the cell; cell wall prevents osmotic rupture and provides shape; capsule protects bacterium; flagella enable movement; pili help bacterium stick to surfaces and are involved in conjugation.
2. Endospores withstand unfavorable conditions. Pili are involved in sexual reproduction, which adds variation.
3. Penicillin kills bacteria by interfering with the enzyme that links sugar chains in the cell wall.

Thinking Critically

4. They live in conditions that may have been present on primitive Earth.

REVIEWING MAIN IDEAS

21.1 Viruses

- Viruses are made of nucleic acids and proteins. In order for a virus to reproduce in a cell, it must recognize and attach to a specific receptor molecule on the plasma membrane of the cell.
- During a lytic cycle, a virus attacks and kills the host cell. In a lysogenic cycle, a virus inserts its DNA into a chromosome of the host cell.
- Retroviruses contain RNA, which must be converted to DNA before it can be incorporated into a host cell chromosome.
- Viruses probably originated from their host cells and do not represent ancient forms of life.

21.2 Bacteria

- Bacteria can be classified either as archaebacteria or as eubacteria. Archaebacteria are more ancient and inhabit extreme environments.
- All bacteria are prokaryotes. They exhibit a tremendous diversity of metabolism and include aerobes, anaerobes, autotrophs, and heterotrophs.
- Some bacteria can form endospores that survive under unfavorable conditions.
- Bacteria reproduce asexually by binary fission. Some bacteria also have primitive forms of sexual reproduction.

Key Terms

Write a sentence that shows your understanding of each of the following terms.

bacteriophage	obligate aerobe
binary fission	obligate anaerobe
conjugation	provirus
endospore	retrovirus
host cell	reverse transcriptase
lysogenic cycle	saprobe
lytic cycle	toxin
nitrogen fixation	virus

Understanding Concepts

1. Viruses reproduce only inside a living cell known as the ____________.
   a. home cell c. vector
   b. host cell d. provirus
2. Which of these is never found as part of a virus?
   a. nucleic acid
   b. protein coats
   c. viral envelope
   d. plasma membrane
3. The nucleic acid core of a virus contains ____________.
   a. only DNA
   b. only RNA
   c. both DNA and RNA
   d. either DNA or RNA
4. Which of the following is NOT a common shape of bacteria?
   a. round c. cone
   b. rod d. spiral
5. Viruses are so species specific that the T4 bacteriophage can only infect ____________.
   a. tobacco cells c. *E. coli*
   b. nerve cells d. protists
6. In the ____________ cycle, viruses use the cell's metabolism to copy themselves, then burst from the cell.
   a. lytic c. lysogenic
   b. metabolic d. virus
7. All of the known archaebacteria live in environments without ____________.
   a. heat c. light
   b. oxygen d. carbon dioxide

Chapter 21 Review

Reviewing Main Ideas

Summary statements can be used by students to review the major concepts of the chapter.

Key Terms

Answers should go beyond defining the terms. Accept any reasonable answer that uses the term correctly and in the proper context.

Understanding Concepts

1. b
2. d
3. d
4. c
5. c
6. a
7. b

Skill Review

5. Differences: plasma membrane and cell wall structure, tRNA and rRNA base sequences, reactions to antibiotics. Similarities: prokaryotes, unicellular, single circular chromosome.

Chapter 21 Review

8. a
9. b
10. c
11. b
12. a
13. b
14. d
15. a
16. b
17. a
18. c
19. d
20. d

Applying Concepts

21. Bacteria recycle nutrients so they become available for living things.
22. Prokaryotic cells do not have membrane-bound organelles (except ribosomes) and they have one chromosome in a ring shape. Eukaryotic cells have membrane-bound organelles and linear chromosomes.
23. *Prochloron* contains both chlorophyll *a* and *b* as do chloroplasts. Cyanobacteria are probably not the precursors of chloroplasts because they contain only chlorophyll *a.*
24. Bacteria may run out of nutrients; they may be eaten by predators; they may dry up without water. All these factors limit bacterial growth.
25. If scientists cannot grow various types of bacteria in the laboratory, they will not be able to study them or conduct experiments to determine their characteristics and requirements.

Thinking Critically

26. The bacteria that cause botulism form endospores that may survive improper sterilization.

Chapter 21 Review

8. Bacteria known as ___________ are round and arranged in grapelike clusters.
   a. staphylococci c. spirillum
   b. streptococci d. diplococci
9. The cell walls of bacteria ___________.
   a. control what enters and leaves the cell
   b. prevent osmotic rupture
   c. are involved in penicillin synthesis
   d. are involved in protein synthesis
10. ___________ results in two new bacteria, each identical to the original.
    a. Meiosis c. Binary fission
    b. Lysis d. Respiration
11. What one characteristic do viruses share with living organisms?
    a. respiration c. growth
    b. reproduction d. movement
12. ___________ has been successfully eradicated as a result of the efforts of the World Health Organization.
    a. Smallpox c. Polio
    b. Tetanus d. Measles
13. During a lytic cycle, once a virus attaches and enters the cell, the virus ___________.
    a. forms a provirus
    b. replicates
    c. dies
    d. becomes inactive
14. ___________ are organisms that feed on dead organisms or organic wastes.
    a. Viruses c. Hosts
    b. Herbivores d. Saprobes
15. Prokaryotic cells have ___________.
    a. no organelles c. mitochondria
    b. a nucleus d. a cell wall
16. ___________ are hairlike extensions of bacterial plasma membranes.
    a. Capsules c. Flagella
    b. Pili d. Cilia
17. In ___________, bacteria convert gaseous nitrogen into nitrates and nitrites used by plants.
    a. nitrogen fixation c. conjugation
    b. binary fission d. attachment
18. Penicillin kills bacteria by interfering with the enzymes that link the sugar chains in the ___________.
    a. nucleus c. cell wall
    b. mitochondria d. capsule
19. Bacteria that require oxygen for cellular respiration are called ___________.
    a. obligate saprobes
    b. archaebacteria
    c. obligate anaerobes
    d. obligate aerobes
20. Some bacteria, when faced with unfavorable environmental conditions, produce structures called ___________.
    a. pili c. toxins
    b. capsules d. endospores

Applying Concepts

21. You have read that bacteria are essential to life. Why do you think this statement is true?
22. Discuss two ways that prokaryotic cells are different from eukaryotic cells.
23. Why is *Prochloron* thought to be a precursor of the chloroplasts found in eukaryotic cells? Why are cyanobacteria probably not the precursors of chloroplasts?
24. Discuss three factors that limit bacterial growth. Why do they prevent bacteria from taking over the world?
25. It has been estimated that 99 percent of all bacteria have such unusual nutritional requirements that scientists are unable to grow them in the laboratory. How do you think this inability hinders our understanding of bacteria?

Thinking Critically

26. ***Recognizing Cause and Effect*** Cases of botulism poisoning are occasionally traced to home-canned food that was not sterilized properly. Explain how improper sterilization might cause botulism.

27. ***Concept Mapping*** Make a concept map that relates the following terms and phrases. Supply the appropriate linking words for your map.
    obligate aerobe, obligate anaerobe, endospore, decomposition, nitrogen fixation, cyanobacteria, bacteria, pili, capsule, photosynthesis, *Clostridium*, toxin
28. ***Observing and Inferring*** If you were offered either a million dollars or a sum equal to a penny that doubles every day for 64 days, which choice would give you more money? Relate this to the growth rate of bacteria.
29. ***Interpreting Scientific Illustrations*** A bacterium isolated from a person with an infection was tested for its sensitivity to three different antibiotics. The results of the test are shown in the petri dish. If you were a physician treating this patient, which antibiotic would you use and why?

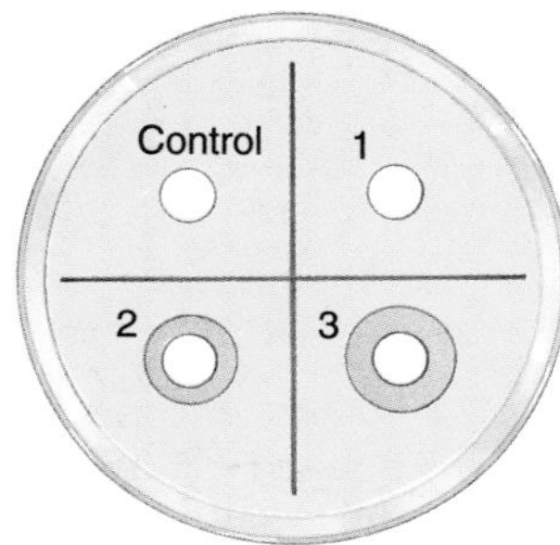

30. ***Interpreting Data*** Some bacteria produce antibiotics that kill other bacteria. Explain why antibiotic-producing bacteria would be selected for and would survive in nature.
31. ***Formulating Hypotheses*** The oldest bacteria fossils are 3.5 billion years old, whereas the oldest fossils of eukaryotic cells are 2.1 billion years old. What kind of relationship does this suggest between bacteria and eukaryotes?

ASSESSING KNOWLEDGE & SKILLS

One milliliter of *E. coli* culture was added to each of three petri plates (I, II, and III). The plates were incubated for 36 hours and the number of colonies counted.

Growth of *E. coli* under various conditions

Petri Dish Number	Medium	Colonies Per Dish
I	Agar and carbohydrates	35
II	Agar, carbohydrates, and vitamins	250
III	Agar and vitamins	0

Interpreting data Use the information in the table to answer the following questions.

1. Which of the above plates demonstrates that carbohydrates are necessary for the growth of *E. coli?*
   a. Plate I alone
   b. Plates I and II
   c. Plates I and III
   d. Plate III
2. Which of the above plates demonstrates that vitamins enhance the growth of *E. coli?*
   a. Plates I and II
   b. Plates II and III
   c. Plates I and III
   d. None of the plates
3. Which of the following is a variable in this experiment?
   a. *E. coli*
   b. agar
   c. carbohydrates
   d. number of colonies
4. ***Making a Graph*** Construct a bar graph from the data in the above table.

Chapter 21 Review

27. Evaluate students' concept maps. The maps should show an understanding of the relationships among the concepts listed.
28. Bacteria increase exponentially. The penny would yield more than $1 million in 64 days.
29. Antibiotic 3; it shows the largest zone of inhibition.
30. Antibiotic-producing bacteria would be selected for because they would be able to kill off other bacteria that were competing for a food source.
31. Bacteria were in existence prior to eukaryotes and may have been the precursors of eukaryotic cells.

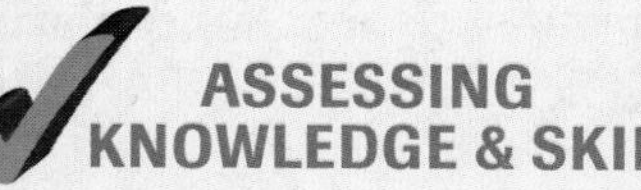

1. c
2. a
3. c
4. Student bar graphs should match the data given in the table.

Program Resources

Chapter Assessment, pp. 121-126 L1
Alternate Assessment in the Science Classroom
Computer Test Bank L1
Content Mastery, pp. 81-84 L1

Chapter Organizer

SECTION	OBJECTIVES	ACTIVITIES/FEATURES
22.1 The World of Protists National Science Standards: UCP.1, UCP.2, UCP.5; C.1, C.4, C.5; F.1, F.5, F.6	1. **Identify** the characteristics of the Kingdom Protista. 2. **Compare and contrast** the four groups of protozoans.	**A Broader View:** Beware the Protists, p. 537 **Minilab:** How does a swimming paramecium respond to obstacles in its path?, p. 538 **Thinking Lab:** How do digestive enzymes function in paramecia?, p. 540
22.2 Algae: Plantlike Protists National Science Standards: UCP.1, UCP.5; A.1, A.2; C.1, C.5; F.1, F.5	3. **Compare** the variety of plantlike protists. 4. **Analyze** the concept of alternating sporophyte and gametophyte generations in algae.	**Biolab:** Design Your Own Experiment—How do *Paramecium* and *Euglena* respond to light?, p. 544 **Minilab:** What do aquarium snails eat?, p. 549
22.3 Funguslike Protists National Science Standards: UCP.1, UCP.5; C.1, C.3-5; F.1, F.4, F.5; G.1	5. **Contrast** the two types of slime molds with respect to their cellular differences and life cycles. 6. **Discuss** the economic importance of the downy mildews and water molds.	

ACTIVITY MATERIALS

BIOLAB	MINILABS	ALTERNATE LAB
page 544 *Euglena* culture *Paramecium* culture microscope slides dropper methyl cellulose coverslips metric ruler index cards scissors toothpicks	**page 538** *Paramecium* culture wheat seed particles microscope slide coverslip microscope toothbrushes or small pieces of cotton **page 549** snails, live toothpick slide coverslip microscope aquarium	**page 552** petri dish microscope *Physarum polycephalum* agar plate oat cereal distilled water

Chapter 22 Protists

TEACHER CLASSROOM RESOURCES

Reproducible Masters	Transparencies
Section Focus Master 46 : Protozoans L1 SAE **Reinforcement and Study Guide,** p. 85 L1 **Biolab and Minilab Worksheets,** p. 87 L1 **Critical Thinking/Problem Solving:** The Effects of Chemicals on Protozoan Populations, p. 22 L3 **Laboratory Manual:** How Can Digestion Be Observed in Protozoans?, pp. 123-126 L2	**Basic Concepts Transparency #30:** Phylogeny of Protists L1 SAE **Reteaching Transparency #22:** Structure of a Paramecium L1 SAE **Basic Skills Transparency #18:** Life Cycle of Plasmodium L1 SAE
Section Focus Master 47: Giant Kelp L1 SAE **Reinforcement and Study Guide,** pp. 86-87 L1 **Biolab and Minilab Worksheets,** pp. 88-90 L1 **Tech Prep Applications:** Algae Appetizers, pp. 25-26 L2 **Laboratory Manual:** Observing Algae, pp. 119-122 L2	**Basic Concepts Transparency #28:** Life Cycle of an Alga L1 SAE **Basic Concepts Transparency #30:** Phylogeny of Protists L1 SAE
Section Focus Master 48: Slime Molds L1 SAE **Reinforcement and Study Guide,** p. 88 L1 **Concept Mapping:** Slime Molds, p. 22 L1 **Content Mastery,** pp. 85-88 L1	**Basic Concepts Transparency #29:** Life Cycle of a Slime Mold L1 SAE **Basic Concepts Transparency #30:** Phylogeny of Protists L1 SAE

ASSESSMENT MATERIALS

Chapter Assessment, pp. 127-132
MindJogger Videoquiz
Alternate Assessment in the Science Classroom
Computer Test Bank

Spanish Resources SAE
English/Spanish Audiocassettes SAE
Cooperative Learning in the Science Classroom COOP LEARN
Biology Projects: Life on a Small Scale, pp. 21-24 L2
Lesson Plans

KEY TO TEACHING STRATEGIES

- L1 Level 1 activities should be within the ability range of all students including those with learning difficulties.
- L2 Level 2 activities are within the ability range of average to above-average students.
- L3 Level 3 activities are designed for the ability range of above-average students.
- SAE SAE activities should be within the ability range of Students Acquiring English.
- COOP LEARN Cooperative Learning activities are designed for small group work.
- P These strategies represent student products that can be placed into a best-work portfolio.
- These strategies are useful in a block scheduling format.

GLENCOE TECHNOLOGY

The following multimedia resources are available from Glencoe.

Biology: The Dynamics of Life
CD-ROM SAE
Videodisc Program

The Infinite Voyage Series
Secrets from a Frozen World
The Southern Ocean—A Rich Marine Ecosystem
The Antarctic Peninsula: Pack Ice and Life Cycles
Effect of UV Radiation on Phytoplankton

The Secret of Life Series
On the Brink: Portraits of Modern Science

Science and Technology Videodisc Series (STVS)
Plants & Simple Organisms
New Uses for Algae

CHAPTER

22 Protists

Chapter Overview

In this chapter, students study the general characteristics and importance of protists. Students learn how protists contribute to oxygen production and food chains or webs. The medical importance of protists as agents of disease is also discussed. The traits of the four phyla of protozoans—sarcodines, flagellates, ciliates, and sporozoans—are explored. The section concludes with a detailed study of protozoans—the animal-like protists.

In the second section, six phyla of algae are discussed, including examples and details regarding their life cycles. The process of alternation of generations is introduced as part of the discussion of green algae.

The chapter concludes with an overview of three phyla of funguslike protists. The reproductive strategies of plasmodial molds and slime molds are included.

Key Terms

- alga
- alternation of generations
- asexual reproduction
- ciliate
- colony
- flagellate
- fragmentation
- gametophyte
- plasmodium
- protozoan
- pseudopodia
- spore
- sporophyte
- sporozoan
- thallus

532

Look for the following logo for strategies that emphasize different learning modalities. LS

Kinesthetic	**Meeting Individual Needs, pp. 535, 546; Demonstration, p. 543**
Visual-Spatial	**Activity, p. 535; Portfolio, p. 535; Microscope Activity, pp. 536, 543, 546, 553; Minilab, pp. 538, 549; Meeting Individual Needs, p. 538; The Inside Story, p. 539; Alternate Lab, p. 552**
Intrapersonal	**Meeting Individual Needs, p. 542**
Linguistic	**Student Journal, p. 539; Portfolio, p. 547; Using Science Terms, p. 547**
Logical-Mathematical	**Project, pp. 536, 543; Student Journal, p. 537; Thinking Lab, p. 540; Reinforcement, p. 543; Concept Development, p. 548; Meeting Individual Needs, p. 549; Portfolio, p. 549**

CHAPTER

22 Protists

Imagine sitting on an old log by the edge of a small pond. Except for the buzzing of a few insects, few sounds can be heard and nothing seems to be moving. Not much going on, right? Not so!

Although you don't see them, you are surrounded by a host of tiny organisms. In the soil and rotting leaves at your feet live countless numbers of protists. In the pond, there are even more—hundreds of different species. Protists are part of nearly every type of moist habitat on Earth, from high mountain peaks and rain forests to rivers, lakes, and oceans.

Of what importance are they? They form the foundation of many ecosystems. Countless other species depend either directly or indirectly on them or on what they do. In this chapter, you'll discover what vital roles protists play, how they interact with other organisms, and how they affect you.

Concept Check

You may wish to review the following concepts before studying this chapter.

- Chapter 3: autotrophs, heterotrophs, parasitism

Chapter Preview

Laboratory Activities

Biolab: Design Your Own Experiment

- How do *Paramecium* and *Euglena* respond to light?

Minilabs

- How does a swimming paramecium respond to obstacles in its path?
- What do aquarium snails eat?

Protists are all around us, but we seldom notice them. Despite their small, often microscopic size, protists are surprisingly complex and are vital components of nearly every type of ecosystem. The algae on the rocks (left) are protists as are *Volvox* and *Spirogyra* (right).

Magnification: 22×

CHAPTER 22

Introducing the Chapter

Ask students to create a simple food chain for an aquatic ecosystem that includes fishes and humans. Ask volunteers to share their food chains with their classmates by drawing them on the chalkboard or on an overhead transparency. Elicit from the examples what autotrophic organisms are in the food chains. *Many students may not have included phytoplankton or algae in their chains.* Use this activity as an opportunity to remind students that algae and phytoplankton serve as the producers for most aquatic food chains. Explain that these organisms are members of the Kingdom Protista. **L1** **COOP LEARN**

Theme Development

The theme of *unity within diversity* is illustrated throughout the chapter as traits that unite organisms in the Kingdom Protista are examined. The diversity of protists is stressed through the discussions of the traits used to classify protists into various phyla. The theme of *homeostasis* is illustrated through discussions of how the various unicellular and multicellular protists carry out their life functions.

Concept Check

Students will expand upon their knowledge of eukaryotic and prokaryotic cells. In addition, symbiotic patterns, such as parasitism, will be reviewed.

Assessment Planner

Choose assessment strategies from the following pages to evaluate the progress of your students.

Assess, pp. 540, 550, 553
Alternate Lab, pp. 552-553
Minilabs, pp. 538, 549
Portfolio, pp. 535, 547, 549
Thinking Lab, p. 540
Biolab, pp. 544-545
Chapter Review, pp. 555-557

SECTION

22.1

Prepare

Key Concepts

This section presents the general characteristics of organisms in the Kingdom Protista. The section then focuses on the four phyla of animal-like protists. Features, anatomical traits, and representative organisms of the phyla Sarcodina, Zoomastigina, Ciliophora, and Sporozoa are described.

Block Scheduling

Look for this symbol for strategies that are useful in a block scheduling format. For more information on block scheduling, refer to Section 22.1 in the **Lesson Plans** booklet.

Materials

- Order *Paramecium* from a biological supply house or culture your own *Paramecium* for the Minilab. Gather pieces of cotton or bristles from a toothbrush as well.
- Purchase prepared slides of representative protozoans for the Microscope Activity.
- Purchase prepared slides of each protozoan discussed in *A Broader View*.
- Gather a strong acid, a weak acid, test tubes, and Congo red powder for the Thinking Lab.

SECTION

22.1 The World of Protists

Section Preview

Objectives

Identify the characteristics of Kingdom Protista.

Compare and contrast the four groups of protozoans.

Key Terms

protozoan
pseudopodia
asexual reproduction
flagellate
ciliate
sporozoan
spore

In just a few drops of pond water, you can find an amazing collection of protists. Some will be moving actively, searching for food. Others, such as this diatom, will be photosynthesizing, using the sun's energy to transform carbon dioxide and water into food. Still others will be playing an active role in decomposing. The protists are a diverse group.

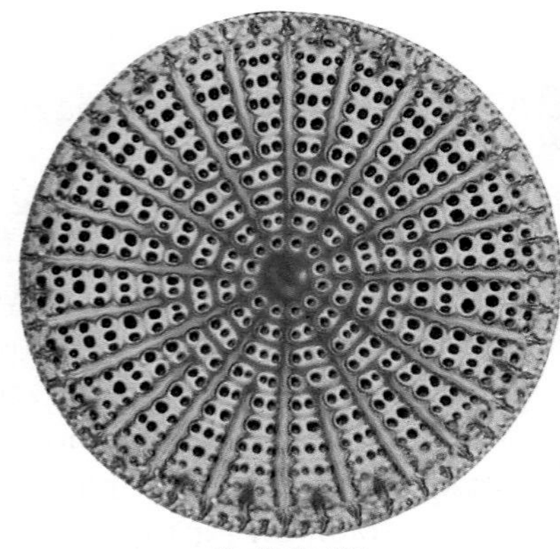

Magnification: 300×

Protist Diversity

Kingdom Protista is the most diverse of the six kingdoms. In fact, there is no such thing as a typical protist. The more than 200 000 species in Kingdom Protista come in a broad array of different shapes, sizes, and colors. *Figure 22.1* shows several members of this kingdom.

Figure 22.1

Members of Kingdom Protista can be organized into three general groups—animal-like, plantlike, and funguslike protists.

Animal-like protists are all unicellular heterotrophs. Those that move do so in a variety of ways.

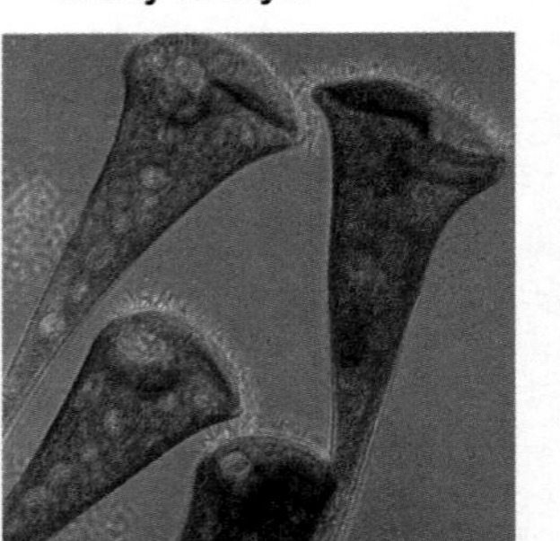

Magnification: 30×

Plantlike protists are photosynthetic autotrophs.

During part of their life cycle, funguslike protists look remarkably like some types of fungi.

Program Resources

Section Focus Master 46 L1 SAE

Reinforcement and Study Guide, p. 85 L1

Biolab and Minilab Worksheets, p. 87 L1

Laboratory Manual, pp. 123-126 L2

Critical Thinking/Problem Solving, p. 22 L3

Basic Concepts Transparency 30 and Master L1 SAE

Reteaching Transparency 22 and Master L1 SAE

Basic Skills Transparency 18 and Master L1 SAE

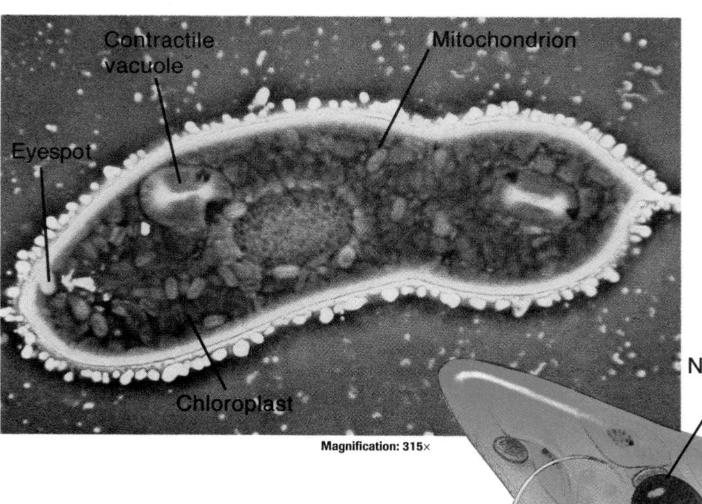

Figure 22.2

Protists are eukaryotic organisms with a wide variety of complex organelles.

Protists range from small, unicellular amoebas that prowl for food among old, wet leaves to giant, multicellular ocean seaweeds that may grow to more than 100 m long. Some protists are slow-moving, jellylike blobs that constantly change shape. Others are rigid, with hard but delicate shells. Still others undergo transformations from living as single cells to becoming part of a coordinated colony of many cells.

At first glance, most protists might look like simple organisms, but appearances can be deceiving. Protists are much more complex than prokaryotes. All protists are eukaryotes, which means that most metabolic processes in these organisms take place in membrane-bound organelles, shown in *Figure 22.2.*

Protozoans: Animal-like Protists

As you sit by the pond, you notice clumps of dead leaves at the water's edge. Under a microscope, a bit of that wet, decaying leaf litter becomes its own world, very likely inhabited by animal-like protists.

Animal-like protists are known as **protozoans.** All protozoans are unicellular heterotrophs that meet their energy requirements by feeding on other organisms or dead organic matter. Protozoans are grouped into phyla according to the way they move. Some motile protozoans propel themselves with cilia or flagella. Others get around by sending out cytoplasm-containing extensions of their plasma membranes called **pseudopodia.** These protozoans use their pseudopodia to engulf food.

Nonmotile protozoans have no way of pursuing or actively capturing food. They live as parasites. Parasitic protozoans are usually found in a part of a host that has a constant and readily available food supply, such as an animal's bloodstream or intestine.

1 Focus

Bellringer

Before presenting the lesson, display **Section Focus Master 46** on the overhead projector and have students answer the accompanying questions. L1 SAE

Activity

LS **Visual-Spatial** Have students diagram a typical bacterium based on their study of Chapter 21. Then have them diagram what they believe would be a typical protist. Ask them to make their diagrams to scale and label any cell organelles that might be present. *This task will be difficult because they have no reference yet as to what comprises a protist. It should, however, allow you to initiate a discussion about the characteristics of protists as you compare and contrast protist and bacterial traits.* L1

2 Teach

Tying to Previous Knowledge

Ask students to explain the similarities or differences between terms in each of the following pairs: *eukaryotic cells* and *prokaryotic cells, heterotrophs* and *autotrophs, single cell* and *colony, motile* and *nonmotile. Eukaryotes have membrane-bound organelles; prokaryotes lack membrane-bound organelles. Heterotrophs take in food from the environment; autotrophs produce their own food. A single cell exists on its own; a colony is made up of many cells living together. A motile organism is capable of movement; a nonmotile organism cannot move.* L1 SAE

Meeting Individual Needs

Visually Impaired Design models that will allow visually impaired students to experience the difference between cilia and flagella. The single side of Velcro will allow for a tactile impression of cilia on a ciliate. Use pieces of string taped to the side of a Ping-Pong ball to simulate flagella. L1 LS

PORTFOLIO

A Moving Analogy Have students describe or diagram typical everyday items that are analogous in structure or appearance to protist structures and their functions. Terms to include are: *cilia, flagella,* and *pseudopodia.* L1 LS P

Science, Technology, and Society

New Treatment for Malaria

Malaria is the world's most common and deadliest protozoan infection. The disease results when *Plasmodium falciparum* infects and brings about changes in the blood cells of its host. One change is the appearance of a protein in the infected red blood cells that causes them to stick to the inner linings of blood vessels. This prevents the cells and parasite from being carried to the spleen where both would be destroyed. The proteins also show a very high mutation rate. Thus, they are constantly evading the immune system. Scientists hope to use the information of how proteins in infected red blood cells react to find new approaches for both the treatment and prevention of malaria.

Microscope Activity

LS Visual-Spatial Allow students to observe prepared slides of representative protozoans such as amoebas, foraminiferans, and radiolarians. Ask students to make labeled diagrams of the organisms and to indicate the magnification at which each organism was viewed. Have them add captions to describe the function of each labeled structure. L1

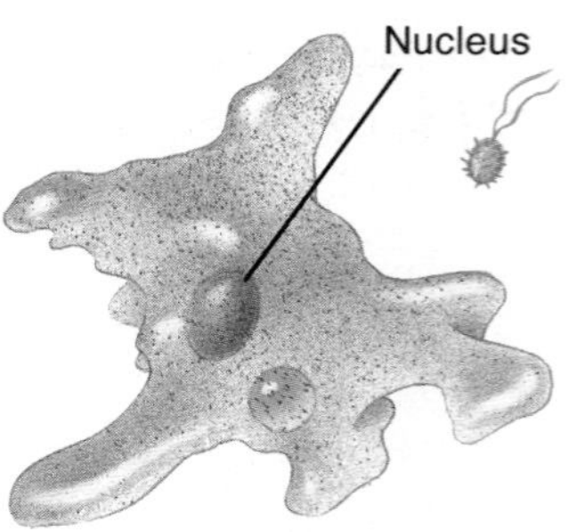

A An amoeba senses food in its immediate environment. As it approaches the food, pseudopodia begin to extend toward and around it.

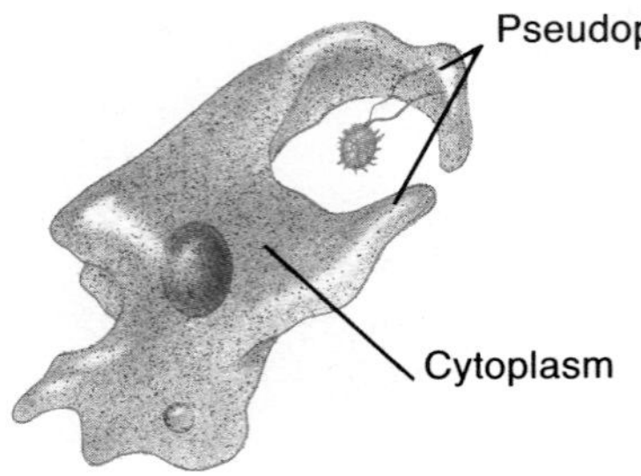

B The amoeba engulfs the food, which becomes enclosed in a food vacuole.

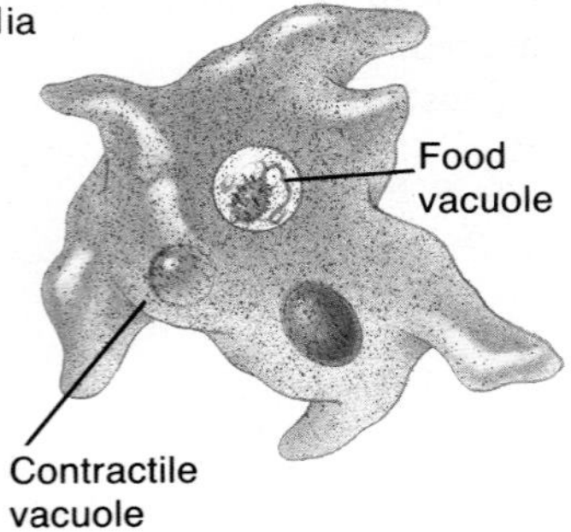

C Digestive enzymes secreted into the vacuole break down the food, and the nutrients are released into the rest of the cell.

Figure 22.3
When an amoeba happens upon a bit of food, pseudopodia stream out to engulf the morsel and take it into the cell to be digested in a food vacuole.

Amoebas: The bloblike protists

The phylum Rhizopoda includes hundreds of species of amoebas and amoeba-like organisms. Amoebas are protists that have no wall outside their cell membrane and move by forming pseudopodia. As new pseudopodia form, the shape of the cell constantly changes, so an amoeba looks different from one moment to the next. As they move, amoebas engulf bits of food by flowing around and over them, *Figure 22.3.*

Most members of this phylum are marine, but there are freshwater amoebas that live in the ooze of ponds and slow-moving streams, in wet patches of moss, and even in moist soil. Because amoebas live in water or moist places, nutrients dissolved in the water around them can diffuse directly through their cell membranes. However, freshwater amoebas face the problem of living in a hypotonic environment; they constantly take in water. To solve this problem, they have contractile vacuoles that collect and pump out excess water.

Amoebas with shells

Two forms of amoebas have shells. Found throughout the oceans, foraminiferans, like the one pictured in *Figure 22.4,* are marine amoebas that have a hard, outer shell made of calcium carbonate. Foraminiferans are extremely abundant. Much of the bottom ooze that covers the seafloor is made up of their shells. Fossil

Figure 22.4
Both foraminiferans (left) and radiolarians (right and far right) are unicellular amoebas that live inside shells. They extend pseudopodia through the tiny holes you can see in the shells. Their pseudopodia act as a sticky net in which bits of food are trapped.

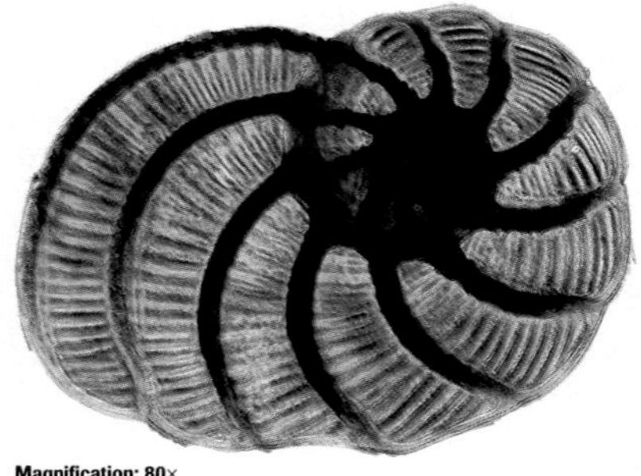

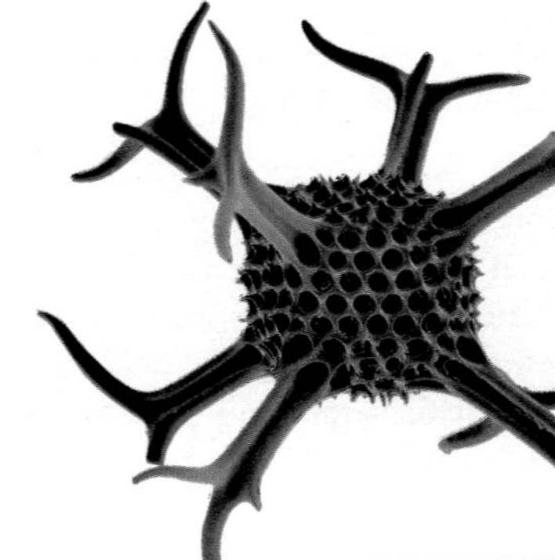

PROJECT

Comparing Speed of Locomotion in Protists and Humans

Have students calculate the speed at which different types of protists move. Have them observe a variety of protists under low-power magnification. While one student observes a protozoan move directly across the center of the field of view, his or her teammate should use a stopwatch to time the event based on oral commands given by the observer. Instruct the students to divide the number of seconds needed to travel across the diameter of the field of view by the size of the field of view (usually 1.5 mm). The result is speed in millimeters per second. Have students design a method to estimate the distance a person travels in mm/sec. L1 LS P

forms of these protists provide geologists with clues to the ages of rocks and sediments.

Most radiolarians have shells made of silica. Seen under a microscope, the great complexity of these tiny shells becomes apparent, as shown in *Figure 22.4*.

Reproductive strategies of amoebas

Most amoebas reproduce by **asexual reproduction,** in which a single parent produces one or more identical offspring. When environmental conditions become unfavorable, some types of amoeba form a cyst that can survive extreme conditions. One cyst-forming species causes dysentery, an unpleasant intestinal illness known for its sharp pains and diarrhea. Amoebic dysentery is one reason people should avoid drinking tap water in less-developed parts of the world where sanitary conditions may be poor.

Flagellates: Protozoans that move with flagella

The phylum Zoomastigina is made up of protists known as **flagellates,** so named because they have one or more flagella. Flagellated protists move by whipping their flagella from side to side.

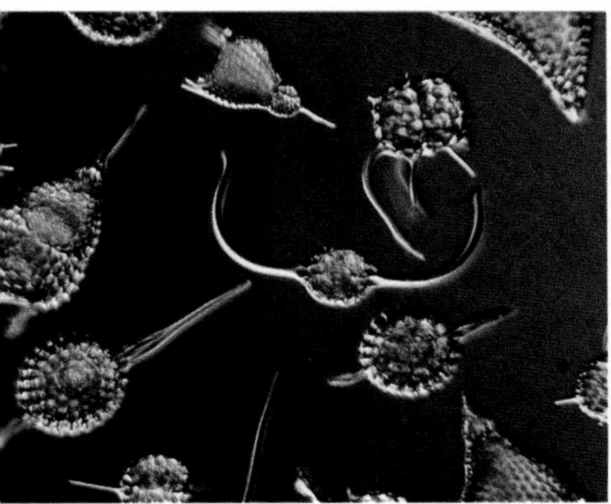

Magnification: 160×

A BROADER VIEW

Beware the Protists

When Donna Huber left Chicago for a seven-day photographic safari in Kenya, she expected to return home with hundreds of breathtaking shots of African wildlife. Donna got her pictures, but she brought home something unexpected as well.

By the time Donna's plane landed at O'Hare International airport, she was feeling listless and out of sorts. "Jet lag," she told the friend who met her at the plane. "I just need some rest."

Magnification: 9800×

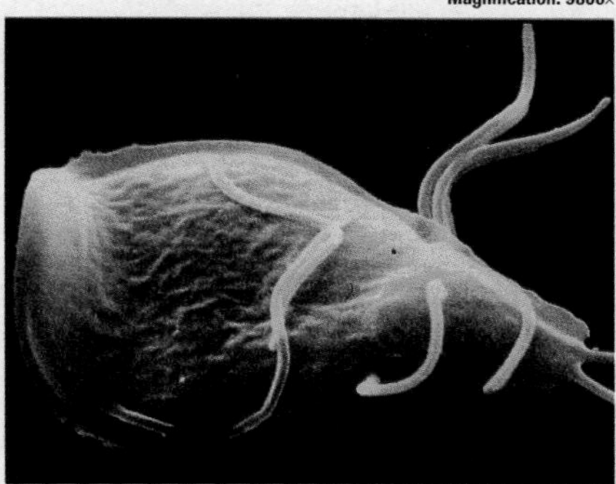

Giardia

Two days later, however, she awoke in the middle of the night with a dangerously high fever—shivering, sweating, and feeling delirious. Her physician ordered blood tests, and the diagnosis was confirmed. Donna had contracted malaria, a serious and sometimes fatal disease caused by a sporozoan protist, *Plasmodium.*

A plague of protists *Plasmodium,* transmitted by mosquitoes, is just one of many protists that can cause deadly disease in humans. *Trypanosoma* is a parasite that is transmitted from one host to another by the African tsetse (TEHT zee) fly. The result is African trypanosomiasis, or sleeping sickness. People with sleeping sickness develop high fevers and swollen lymph nodes. Eventually, the parasites make their way to the brain, where they cause uncontrollable sleepiness.

Giardiasis Muskrats and beavers are responsible for transmitting giardiasis, the symptoms of which include extreme fatigue, cramps, diarrhea, and weight loss. People contract the disease by drinking water contaminated by the protist *Giardia,* which is carried by muskrats and beavers but which harms only humans.

Dysentery Amoebas, too, can cause an unpleasant and, if left untreated, fatal digestive tract malady called amoebic dysentery. Amoebas usually live in pond and stream water; however, the dysentery amoeba prefers life in the large intestine of humans, where it feeds on the intestinal lining, causing bleeding, diarrhea, vomiting, and sometimes death. Amoebic dysentery is spread through infected water and food.

Thinking Critically
What are some ways humans can protect themselves against diseases caused by protists?

SECTION 22.1

A BROADER VIEW

Beware the Protists

Purpose
Students are introduced to a variety of pathogenic protozoa and the diseases they cause.

Teaching Strategies
- Review the concept of symbiotic relationships studied in Chapter 3. Have students identify the types of symbiotic relationships shown by each organism discussed.
- Have students observe prepared slides of each protozoan described in the feature. You may request diagrams of their observations. Students will require guidance in identifying and determining the location for malaria protists within red blood cells. L1 P

Thinking Critically
By being aware of whether an area is prone to parasitic infections caused by protozoans; by taking drugs in advance of travel into infected regions to prevent the infection; and by avoiding intermediate hosts or vectors by using netting or repellents. By not drinking untreated water from ponds, lakes, or rivers.

Software
The Microorganism Simulator, Carolina.
Protozoa, Carolina.

STUDENT JOURNAL

Charting Protist Diseases Using the information on this page as well as on other pages from Section 22.1, have students prepare a table that describes the following for each organism in the *A Broader View* feature. Have them list the following diseases along the side: Malaria, Sleeping sickness, Chagas disease, Giardiasis, and Amoebic dysentery. Instruct students to list these heads across the top: Kingdom name, Phylum name, Means of locomotion, Method by which transmitted, Parasitic or free-living, and Symptoms of disease. LS L1

GLENCOE TECHNOLOGY

Videodisc

Biology: The Dynamics of Life
Disc 1, Side 2
Protists (Ch. 12)

MiniLab

Purpose

LS **Visual-Spatial** Students will observe the response pattern of paramecia when they contact solid objects.

Process Skills

observe and infer, use the microscope

Teaching Strategies

- Do not use methyl cellulose to slow the protists. This will interfere with the normal response of *Paramecium.*
- Toothbrush bristles or small pieces of cotton may be used as objects for blocking the path of the protist.

Expected Results

Paramecium typically shows an avoidance reaction when it bumps into an object. They reverse the direction in which their cilia are beating and back off from the object. If the object is food, they may sweep small food particles into their oral grooves.

Analysis

1. It backs up, then proceeds forward in a new direction.
2. Cilia briefly reverse their direction of beating, and then move forward again.
3. The cell turns on its long axis or it may bend.

Assessment

Portfolio: Have students diagram the events that occur as a *Paramecium* encounters an object and then backs off. Have students caption their diagrams with explanations of the events that are occurring. L1

MiniLab

How does a swimming paramecium respond to obstacles in its path?

The beating cilia on the surface of a paramecium move in a coordinated way so that the cell normally swims through the water with its front (anterior) end directed forward. But when the front end bumps into an obstacle in the paramecium's path, the organism responds by changing its swimming behavior.

Procedure

1. Observe a *Paramecium* culture that has had boiled, crushed wheat seeds in it for several days.
2. Carefully place a drop of water containing wheat seed particles on a microscope slide, and gently add a coverslip.
3. Using low power (10× or 12.5×), scan across the slide until you locate a paramecium near some wheat seed particles.
4. Watch the paramecium as it swims around among the particles. Record your observations of the organism's responses each time it contacts a particle.

Analysis

1. How does a paramecium respond when it encounters an object?
2. How does it change its swimming direction in response to an obstacle? How long does this response last? What does the cell do?
3. Did you see any changes in the shape of the paramecium as it moved among the particles?

Although some flagellate species are parasites that cause diseases in animals, others are helpful. Termites like those in *Figure 22.5* can survive on a diet of wood. Without the help of a certain species of flagellate, termites could not digest the cellulose in wood. However, within the intestines of termites are flagellates that can ingest cellulose. In a mutualistic relationship, these minute protozoans convert cellulose into a carbohydrate that both they and their termite hosts can use.

Ciliates: The most complex protozoans

The roughly 8000 members of the phylum Ciliophora, known as **ciliates,** move by the synchronized beating of cilia that cover their bodies.

Ciliates are found in every kind of aquatic habitat, from ponds and streams to oceans and sulfur springs.

Paramecia usually reproduce asexually, with the cell dividing crosswise and separating into two daughter cells. But when food supplies dwindle or environmental conditions degenerate, paramecia can also undergo a form of conjugation. In this complex process, two paramecia come together and exchange genetic material through their oral grooves. After becoming genetically altered in this way, the two individuals separate, and each goes on to divide asexually.

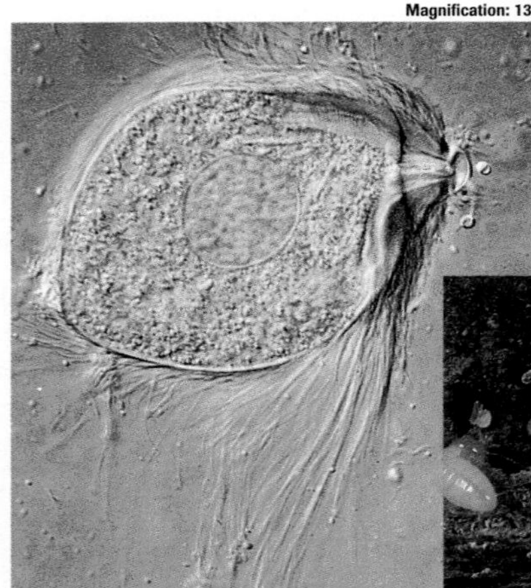

Figure 22.5

The flagellated protozoans (left) that live in the guts of termites (right) possess enzymes that digest wood, making nutrients available to their hosts.

Meeting Individual Needs

Gifted Order termites from a biological supply house and have students observe the flagellates present within the termite's intestine. Direct students to place the insect on a glass slide, grasp its head with a forceps, and gently pull to separate the head and intestines from the body. Students should next add several drops of Ringer's solution or water and a coverslip to the termite intestines. They should then press gently on the coverslip to prepare an intestine squash. Have students observe the slide under low and high power magnification. Ask students to diagram the flagellates they observe and place their observations in their portfolios. L3 P LS

THE INSIDE STORY

Inside a Paramecium

Paramecia are unicellular organisms, but that does not mean they are simple. Within the cell of a paramecium are a variety of complex organelles and structures that are each adapted to carry out a distinct function.

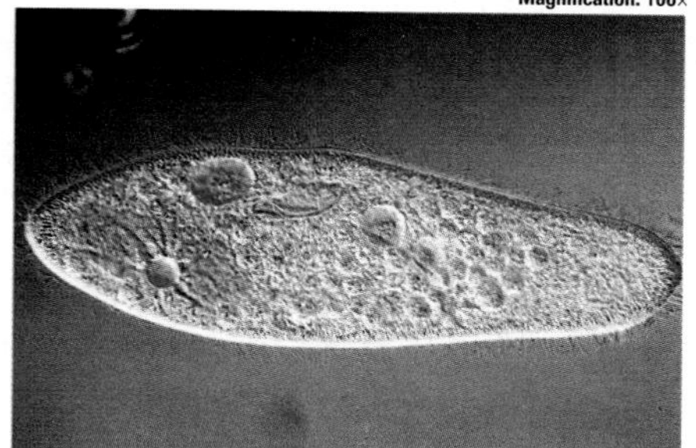

Paramecium caudatum

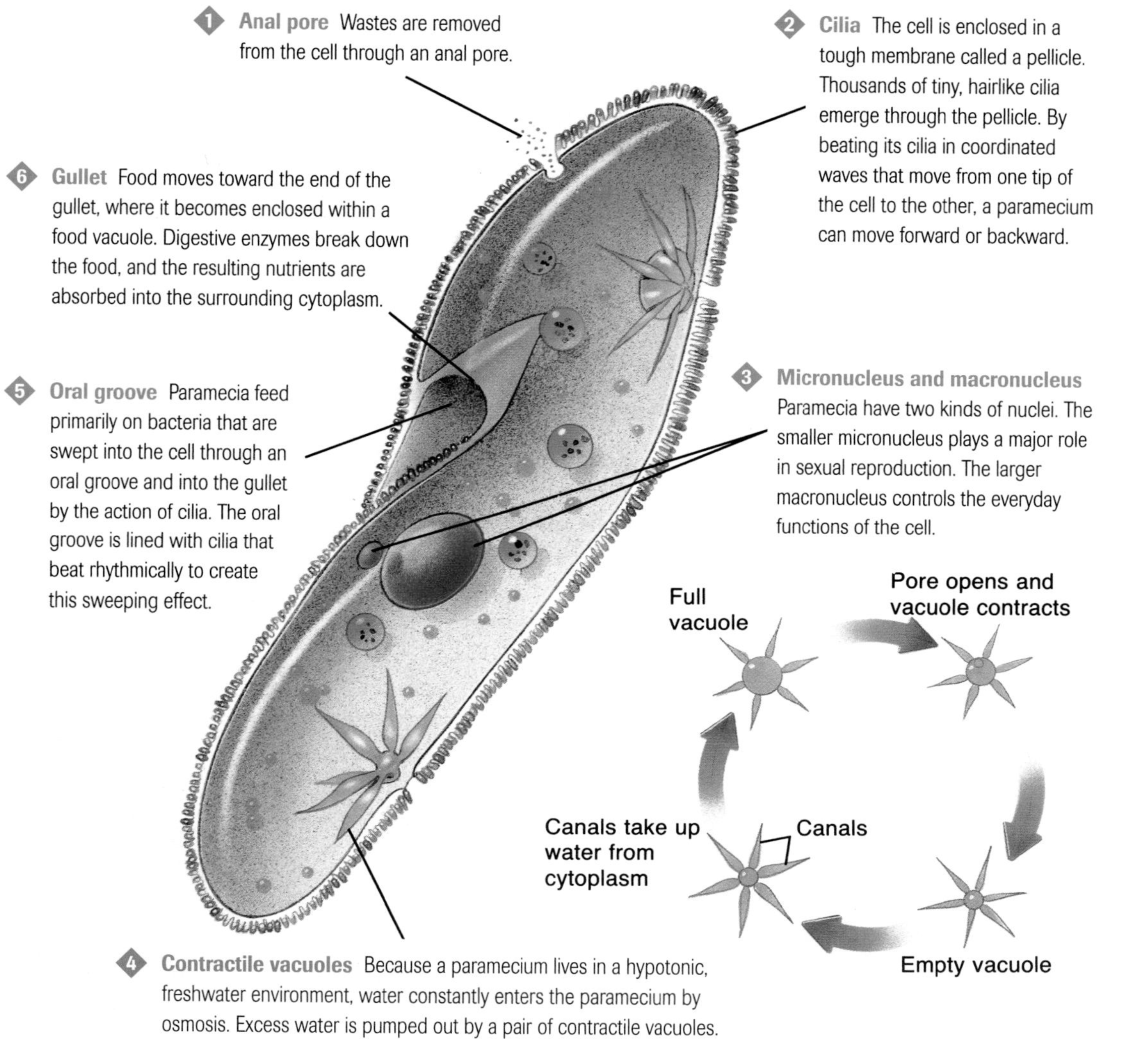

1 Anal pore Wastes are removed from the cell through an anal pore.

2 Cilia The cell is enclosed in a tough membrane called a pellicle. Thousands of tiny, hairlike cilia emerge through the pellicle. By beating its cilia in coordinated waves that move from one tip of the cell to the other, a paramecium can move forward or backward.

3 Micronucleus and macronucleus Paramecia have two kinds of nuclei. The smaller micronucleus plays a major role in sexual reproduction. The larger macronucleus controls the everyday functions of the cell.

4 Contractile vacuoles Because a paramecium lives in a hypotonic, freshwater environment, water constantly enters the paramecium by osmosis. Excess water is pumped out by a pair of contractile vacuoles.

5 Oral groove Paramecia feed primarily on bacteria that are swept into the cell through an oral groove and into the gullet by the action of cilia. The oral groove is lined with cilia that beat rhythmically to create this sweeping effect.

6 Gullet Food moves toward the end of the gullet, where it becomes enclosed within a food vacuole. Digestive enzymes break down the food, and the resulting nutrients are absorbed into the surrounding cytoplasm.

STUDENT JOURNAL

Malaria and Sickle-Cell Anemia Have students use references to prepare a report about the correlation between the genetic disease sickle-cell anemia and malaria. Have them include the evolutionary significance of such a relationship between the two diseases as well as how malaria got its name. L3 LS

THE INSIDE STORY

Inside a Paramecium

Purpose

LS **Visual-Spatial** Students study the role or function of the organelles of a *Paramecium.*

Visual Learning

- Have students describe each organelle shown in terms of its location, shape, and function. L1
- Have students create a table with the following heads: *Function* and *Organelle.* Beneath the *Function* head, have students list: *Digestion, Locomotion, Protection, Excretion, Maintaining Osmotic Balance,* and *Reproduction.* Have students use the diagram to complete their tables. L1

Bioethics

Using Biological Controls *Nosema locustae* is a spore-producing, protozoan parasite of grasshoppers and locusts. Scientists are experimenting with using *Nosema locustae* to keep grasshopper and locust populations in check. The release of the protist's spores into wheat fields is expected to kill more than 50 percent of the grasshoppers and locusts, preventing the insects from destroying valuable crops.

One problem of using *Nosema locustae* is that it may kill harmless or even helpful insects. The destruction of grasshoppers and locusts may also adversely affect food chains or webs. Many believe further research is needed before *Nosema locustae* is used to eliminate pest populations.

ThinkingLab — Draw a Conclusion

Purpose

LS **Logical-Mathematical** Analyze the change in food as it passes through digestive stages within the food vacuole of *Paramecium.*

Process Skills

observe and infer, relate cause and effect, draw a conclusion

Teaching Strategies

- Review pH with students. Advise them that there are a number of liquid chemical indicators that function in a similar manner to pH paper.

Thinking Critically

Initially, food in the vacuole has a pH near 3 as indicated by the blue color. The pH becomes less acidic (near 5) as indicated by the red color. Each particular enzyme released into the food vacuole operates best in a specific pH range.

Assessment

Performance: Provide students with numbered samples of liquids with different pH values. Have them use pH paper to determine which samples best match the contents of the food vacuole at the start and end of digestion in *Paramecium.*

3 Assess

Check for Understanding

Have students create a concept map that shows the division of the Kingdom Protista into three large subgroups. Have students continue their concept maps to show where the four phyla of protists classified as protozoans belong on their concept maps. L1

ThinkingLab — Draw a Conclusion

How do digestive enzymes function in paramecia?

Paramecia ingest food particles and enclose them within food vacuoles. Each food vacuole circulates in the cell as the food is digested by enzymes that are added to the vacuole. Nutrients made available during digestion are absorbed into the cytoplasm.

Analysis

1. Some digestive enzymes function best at higher pH levels, while others function best at lower (more acidic) pH levels.
2. Congo red is a pH indicator dye; it is red when the pH is above 5 and blue when the pH is below 3 (very acidic).
3. Yeast cells that contain Congo red can be produced by adding dye to the solution in which the cells are growing.
4. When paramecia feed on dyed yeast cells, the yeast is visible inside food vacuoles.
5. Examine the drawing below. The appearance of a yeast-filled food vacuole *over time* is indicated by the colored circles inside the paramecium. Each arrow indicates movement and the passing of time.

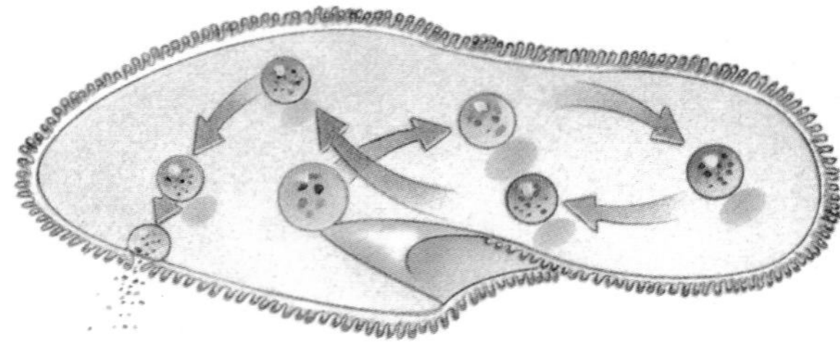

Thinking Critically

Analyze what happens to the pH in the food vacuole over time. Explain your conclusions about the sequence of different digestive enzymes that function in paramecium digestion.

Sporozoans: The parasitic protozoans

Protists that are grouped in the phylum Sporozoa are called **sporozoans.** They are all parasitic, nonmotile protozoans. Sporozoans get their name from the fact that many of them produce spores. A **spore** is a reproductive cell that can produce a new organism without fertilization.

Living as internal parasites in one or more hosts, sporozoans have complex life cycles. Probably the best-known sporozoans are members of the genus *Plasmodium.* Different species of *Plasmodium* cause the disease malaria in people, some other mammals, and birds.

Sporozoans and malaria

Malaria is a disease that is common in tropical climates. Around the world, more than 300 million people are afflicted with this debilitating disease. Malaria is caused by microscopic protozoans that are spread from person to person by mosquitoes. As you can see in *Figure 22.6,* malaria-causing *Plasmodium* species spend part of their life in humans and part in mosquitoes.

For years, medical researchers have been trying to develop a vaccine for malaria, which kills 2 to 4 million people each year. Despite many advances, however, an effective vaccine has not yet been developed.

interNET CONNECTION

Follow the link for this chapter on the Glencoe Homepage at **www.glencoe.com/sec/science** to find out more about protists.

Figure 22.6

The life cycle of *Plasmodium* begins when a hungry female *Anopheles* mosquito bites a person who already has malaria. With its blood meal, the mosquito also takes in *Plasmodium* reproductive cells.

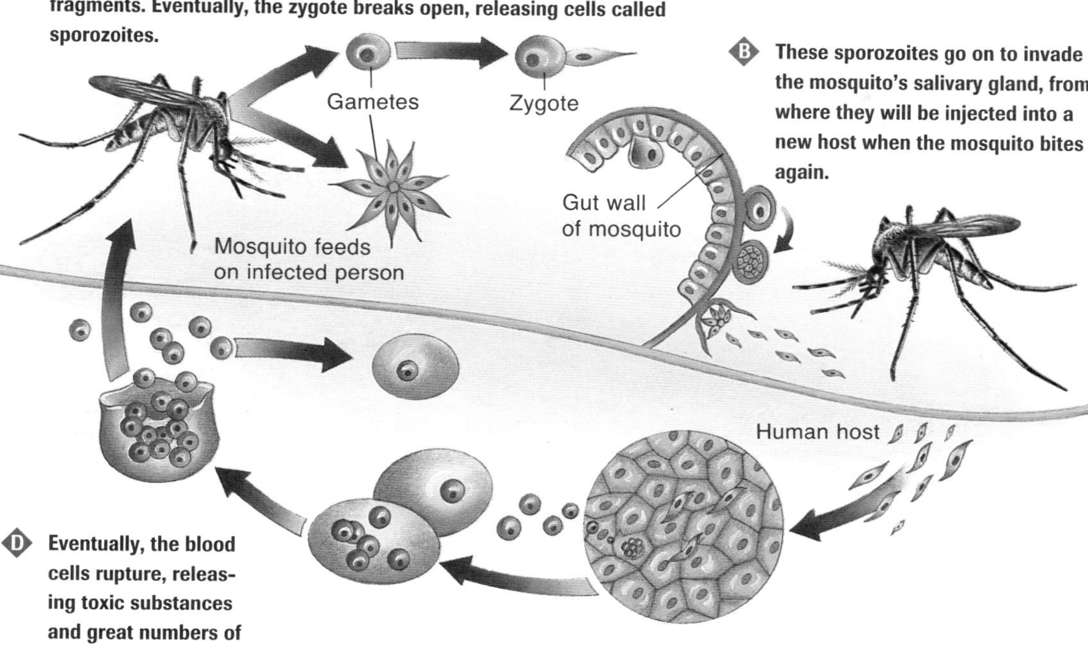

A **The reproductive cells move into the mosquito, where they fuse to form a zygote. The zygote divides many times to form numerous sporelike cell fragments. Eventually, the zygote breaks open, releasing cells called sporozoites.**

B **These sporozoites go on to invade the mosquito's salivary gland, from where they will be injected into a new host when the mosquito bites again.**

C **Once inside the host, the sporozoites reproduce asexually in the liver to form a second type of sporelike cell. From the liver, these new spores enter the bloodstream, invade red blood cells, and multiply rapidly inside them.**

D **Eventually, the blood cells rupture, releasing toxic substances and great numbers of spores. These spores go on to infect more red blood cells, and the process is repeated.**

Section Review

Understanding Concepts

1. Describe how amoebas move.
2. How do ciliates differ from flagellates?
3. What makes a sporozoan different from other protozoan groups?

Thinking Critically

4. What role do contractile vacuoles play in helping freshwater protozoans maintain homeostasis with their environment?

Skill Review

5. **Sequencing** Retrace the life cycle of *Plasmodium*, identifying the different forms of the sporozoan that are present in the host, and the role each form plays in the disease malaria. For more help, refer to Organizing Information in the *Skill Handbook*.

SECTION 22.1

Reteach

Have students list five protist traits. Beside this list, have them generate a second list that identifies traits which are shared by the organisms in the four phyla that make up the animal-like protists. L1

Extension

Have students research the phylum to which the organism *Trypanosoma cruzi* belongs. Have students create a diagram of this protist that includes labels identifying its important structures. L2 P

Assessment

Knowledge: Have students identify the word that does not belong in each of the following groups: L1 SAE

(a) amoeba, foraminifera, *Paramecium*, radiolarian.

(b) *pseudopodia*, cilia, oral groove, pellicle.

(c) dysentery, *AIDS*, sleeping sickness, Chagas disease.

4 Close

Activity

Make five sets of index cards that contain one of each of the following words on each card: *locomotion, pseudopodia, flagella, cilia, sarcodines, asexual reproduction, cyst, sexual reproduction, micronucleus, spore.* Give student groups a set of the cards. Have them arrange the cards in a logical sequence and then explain their groupings to the class. L1 COOP LEARN

Answers to Section Review

1. Cell membrane extensions, called pseudopodia, fill with cytoplasm.
2. Flagellates have one or more long, whiplike flagella. Ciliates have many short surface extensions.
3. All are parasitic, nonmotile, and produce spores.

Thinking Critically

4. Contractile vacuoles pump out excess water that diffuses into the cell, helping to keep osmotic balance.

Skill Review

5. Reproductive cells in the mammal host's blood enter a mosquito when feeding. These cells form a zygote, which divides to form spores. Spores enter the mosquito's salivary glands where they are transmitted to a mammal host. Inside the mammal, these spores move to the liver and form sporelike cells. They pass from liver to blood cells where new spores are formed. Some spores form reproductive cells that are picked up again by a new mosquito.

SECTION

22.2

Prepare

Key Concepts

This section focuses on algae, a group of autotrophic organisms belonging to the Protist Kingdom. Algae are commonly called the plantlike protists and are divided into phyla largely based on their cellular organization and the pigments they contain. Three phyla—the euglenoids, diatoms, and dinoflagellates—are unicellular. Multicellular phyla include the brown algae, the red algae, and the green algae, which contain several unicellular species. The adaptations organisms in these phyla have for survival in both freshwater and marine environments are presented along with both sexual and asexual reproductive strategies. Alternation of generations in algal species is also discussed.

Block Scheduling

Look for this symbol for strategies that are useful in a block scheduling format. For more information on block scheduling, refer to Section 22.2 in the **Lesson Plans** booklet.

Materials

- Order *Paramecium* and *Euglena* cultures for the Biolab.
- Set up an aquarium with snails and water plants in advance of the Minilab. Allow the tank to receive light to promote growth of algae.

SECTION

22.2 Algae: Plantlike Protists

Section Preview

Objectives

Compare the variety of plantlike protists.

Analyze the concept of alternating sporophyte and gametophyte generations in algae.

Key Terms

algae
thallus
colony
fragmentation
alternation of generations
gametophyte
sporophyte

As you sit by the pond, a gust of wind blows toward you bringing the scents of flowers and wet vegetation. Each time you inhale, you breathe in oxygen, some of which is being produced right in front of you by the algae in the pond. Algae hold an important position in the varied world of living things. Just about every living thing depends either directly or indirectly on these plantlike protists for oxygen. They also form, along with other eukaryotic autotrophs, the foundation of Earth's food chains.

What Are Algae?

Photosynthetic protists are known as **algae.** Even though some kinds of algae look like plants because they are big and green, they have no roots, stems, or leaves. All algae contain one or more of four kinds of chlorophyll, as well as other photosynthetic pigments. These pigments are responsible for the colors you see in different types of algae, from purplish or rusty-red to olive-brown, golden-brown, and yellow. Pigments are also important in the classification of algae. The different species of algae in *Figure 22.7* show the diversity of this group of organisms.

Photosynthesizing protists, known as phytoplankton, are so numerous that they rank as the major producers of nutrients in aquatic ecosystems and as releasers of oxygen in the world. As such, they form the critical first link in aquatic food chains. It's been estimated that algae produce more than half of the oxygen generated by all of Earth's photosynthesizing organisms.

Phyla of Algae

Algae are classified into six phyla. Three of these phyla—the euglenoids, diatoms, and dinoflagellates—are composed of only unicellular species. The other three phyla of algae include some unicellular members, but most are multicellular. These phyla are the green, red, and brown algae.

Program Resources

Section Focus Master 47 L1 SAE
Reinforcement and Study Guide, pp. 86-87 L1
Biolab and Minilab Worksheets, pp. 88-90 L1
Laboratory Manual, pp. 119-122 L2
Tech Prep Applications, pp. 25-26 L2
Basic Concepts Transparencies 28, 30 and **Masters** L1 SAE

Meeting Individual Needs

Students Acquiring English Have students prepare a table that compares protozoans and algae. Have students use the following heads: Similarities Between Animal-like Protists and Plantlike Protists and Differences Between Animal-like Protists and Plantlike Protists. L1 LS SAE

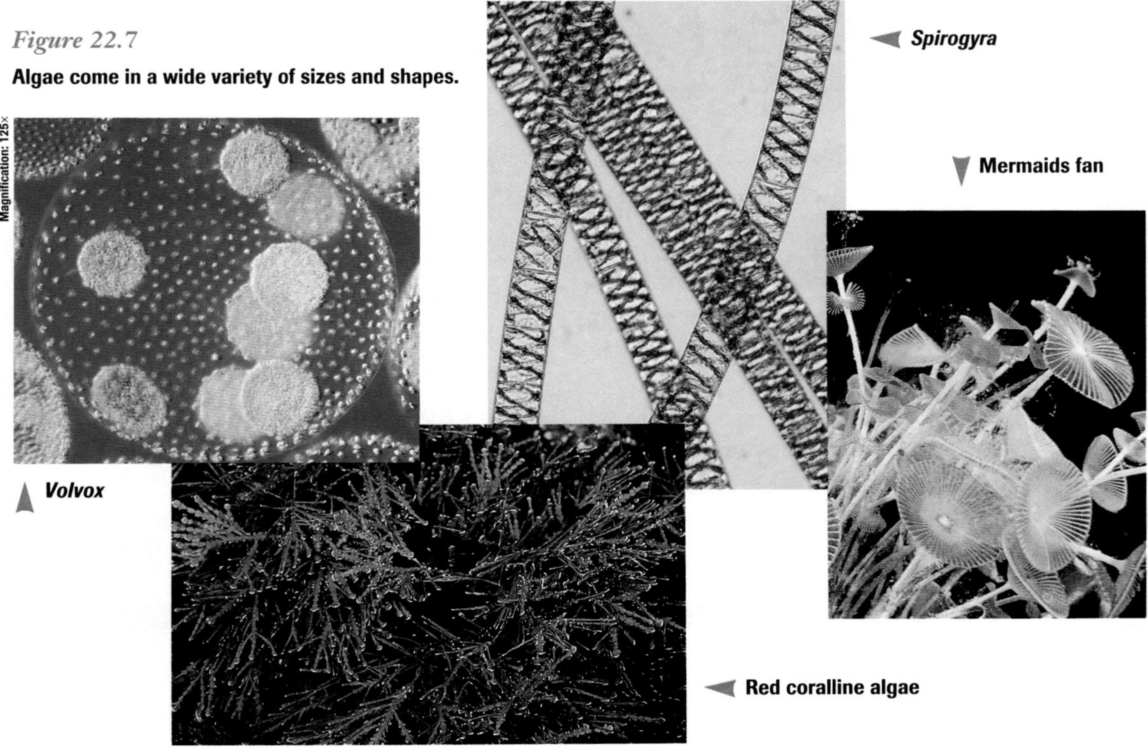

Figure 22.7

Algae come in a wide variety of sizes and shapes.

Euglenoids: Autotrophs as well as heterotrophs

Hundreds of species of euglenoids make up the phylum Euglenophyta. Euglenoids are unicellular, aquatic protists that have traits of both plants and animals. Although they lack a cellulose cell wall as in plant cells, euglenoids have a flexible pellicle made of protein strips inside the cell membrane. Euglenoids are like plants in that most contain chlorophyll and carry out photosynthesis. They are also like animals because they are responsive and move —some very actively—by using one or two flagella for locomotion. When light is not available for photosynthesis, *Euglena* can ingest food from the surrounding water the same way that many protozoans do; in short, they can be heterotrophs. *Figure 22.8* shows an example of a euglenoid.

Diatoms: The golden algae

Diatoms, members of the phylum Bacillariophyta, are unicellular organisms with shells made of silica. Diatoms are photosynthetic autotrophs, and they are abundant in both marine and freshwater ecosystems, where they make up a large component of the phytoplankton.

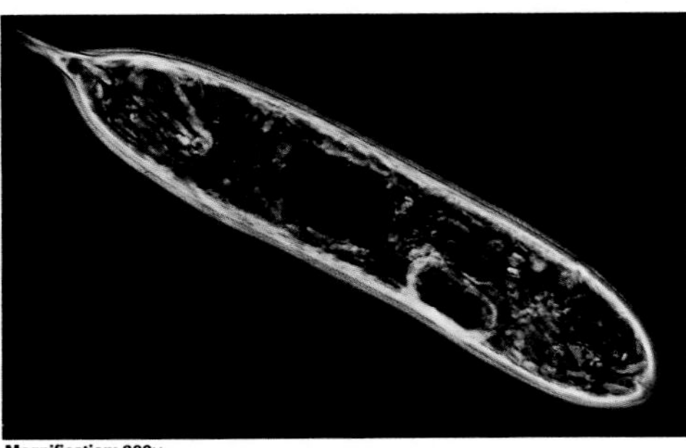

Figure 22.8

***Euglena gracilis* is one of the most famous and well-studied members of this group of protists. Notice the eyespot—a red, light-sensitive structure. This photoreceptor helps *Euglena* orient itself toward areas of bright light, where photosynthesis can occur most efficiently.**

PROJECT

Culturing *Euglena*

Allow students to culture *Euglena* using the following media and procedure. Add 20 rice grains and 5 mL dry skim milk to 1 L of spring water. Boil for 5 minutes. Add 3 L of spring water and allow to cool overnight. Add 100 mL of *Euglena* culture (purchased from a supply house) and maintain in a well-lit area, avoiding direct sunlight. After students have prepared their cultures, have them design a technique for determining either quantitatively or qualitatively how the number of *Euglena* changes over a predetermined period of time. After getting approval for their experimental designs, have students carry out the procedure. L2 LS

1 Focus

Bellringer

Before presenting the lesson, display **Section Focus Master 47** on the overhead projector and have students answer the accompanying questions. L1 SAE

Demonstration

LS **Kinesthetic** Show the class a petri dish containing agar. Explain what agar is and allow students to feel its texture. Next, show an ice cream carton label that lists alginate, agar, or carrageenan as an ingredient. Elicit the relationship between the label contents and the agar within the petri dish. *Both are products made from algae.* L1 COOP LEARN

2 Teach

Reinforcement

LS **Logical-Mathematical** Have students prepare a concept map which shows how the six algae phyla are related. Ask students to use the following terms: *plantlike protists, autotrophic, unicellular, mostly multicellular, diatoms, euglenoids, dinoflagellates, green algae, brown algae, red algae.* Have volunteers share their concept maps with the class by drawing them on the chalkboard or an overhead transparency. COOP LEARN

Audiovisual

Show the filmstrip *The Algae*, Carolina. L1

Microscope Activity

LS **Visual-Spatial** Allow students to observe prepared slides or living cultures of *Euglena gracilis*. Have them diagram their observations and estimate the size of the organism. For help, have them refer to the Skill Handbook. L2

BioLab Design Your Own Experiment

How do *Paramecium* and *Euglena* respond to light?

Time Allotment
One class period.

Objectives
Review objectives with students before they begin the Biolab.

Process Skills
observe and infer, experiment, use the microscope, form a hypothesis, use scientific methods

Safety Precautions
- Have students wash their hands after working with the protist cultures.

PREPARATION

Alternate Materials
- Analysis of light reactions may be achieved through the use of test tubes filled with cultures of *Euglena* or *Paramecium*. For *Euglena*, cover the test tube with black paper that has a thin slit cut into it, allowing only a small band of light to enter the tube. After it remains overnight in bright light, students will see the congregation of organisms (a band of green) where light entered the slit once the black paper cover is removed. For *Paramecium*, the tube can be placed on its side with half the tube covered with black paper. After it remains overnight, removal of the paper will show a concentration of organisms toward the dark side.

BioLab Design Your Own Experiment

How do *Paramecium* and *Euglena* respond to light?

Members of the genus *Paramecium* are ciliated protozoans—unicellular, heterotrophic protists that actively swim around in search of small food particles. *Euglena* are unicellular algae—autotrophic protists that usually contain numerous chloroplasts. How do these two different kinds of protists respond to light in their environment? Are they attracted to or repelled by light? What might be an explanation for their behavior?

PREPARATION

Problem
Do *Paramecium* and *Euglena* respond to light? Do *Paramecium* and *Euglena* respond in different ways?

Hypotheses
Form hypotheses about the potential responses of these two protists to light. Remember that *Paramecium* is heterotrophic. State a testable hypothesis about its possible responses to light and darkness. Remember, too, that *Euglena* is a photosynthetic autotroph. State a testable hypothesis about *Euglena's* possible responses to light and darkness.

Objectives
In this Biolab, you will:
- **Prepare** slides of *Paramecium* and *Euglena* cultures, and observe swimming patterns in the two organisms.
- **Compare** how these two different protists respond to light.

Possible Materials
Euglena culture
Paramecium culture
microscope
microscope slides
dropper
methyl cellulose
coverslips
metric ruler
index cards
scissors
toothpicks

Safety Precautions
Be careful when placing coverslips on slides. They break easily and can produce sharp fragments of glass.

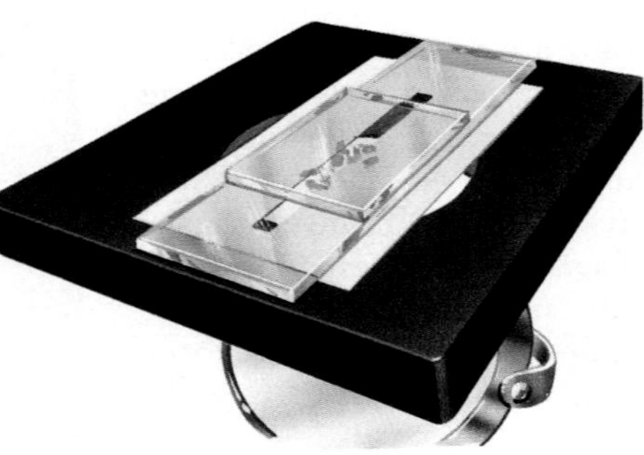

PLAN THE EXPERIMENT

Teaching Strategies
- Allow students enough time to observe both organisms under the microscope before actually starting the experiment. This will give them an opportunity to note the organisms' size, speed, and mobility.
- Low power will be the choice for observing *Paramecium*, while high power will be needed for *Euglena*. A 5× objective is preferable for observing *Paramecium*.
- Encourage students to consider several trials for each organism.
- When they are removing the paper barrier to make a quantitative count, advise students that once the index card is removed, counts must be made quickly.

PLAN THE EXPERIMENT

1. Decide on an experimental procedure that you can use in testing your hypothesis. Keep the materials list in mind as you plan your procedure.
2. Record your planned procedure, step by step, and make a list of all the materials you will be using.
3. Design and construct a data table for recording your observations. This table should include enough spaces for both organisms, and for the results of as many repetitions of the experiments as you plan to do.

Check the Plan

Discuss the following points as you work out the details of the procedure for your experiment:

1. What variables will you measure?
2. What will be your control?
3. What will be the shape of the light-controlled area(s) on your slide?
4. Decide who will prepare materials, make observations, and record data.

5. ***Make sure your teacher has approved your experimental plan before you proceed further.***
6. In order to carry out your experiments, you will need to mount drops of *Paramecium* culture and *Euglena* culture on microscope slides. This is done by using a toothpick to place a small ring of methyl cellulose on a clean microscope slide. Place a drop of *Paramecium* or *Euglena* culture within the ring of methyl cellulose. Place a coverslip over the ring and drop of culture. Methyl cellulose is a syrupy material that slows movement of organisms for easy observation.
7. Make preliminary observations of swimming *Paramecium* and *Euglena* cells. Then think again about the observation times that you have planned. Maybe you will decide to allow more or less time between your observations.
8. Carry out your experiment.

ANALYZE AND CONCLUDE

1. **Checking Your Hypothesis** Did your data support your hypothesis? Why or why not?
2. **Comparing and Contrasting** Compare and contrast the responses of *Paramecium* and *Euglena* to light and darkness. What explanations can you suggest for their behavior?
3. **Making Inferences** Can you use your results to suggest what sort of responses to light and darkness you might observe using other heterotrophic or autotrophic protists?

Going Further

Project You may want to extend this experiment by varying the shapes or relative sizes of light and dark areas, or by varying the brightness or color of the light. In each case, make hypotheses before you begin. Keep your data in a notebook, and draw up a table of your results at the end of your investigations.

Possible Procedures

- A light and dark zone will be needed once the slide is placed under the microscope. Use of index cards will provide such a zone. Index cards will have to be cut so that they fit over part of the coverslip. Students will have to focus on the edge of the card to observe the direction of each organism's movement. Coverslips that cover a greater slide area (22 × 30 mm or 22 × 40 mm) may be preferable.
- Methyl cellulose may be prepared or purchased commercially.

Data and Observations

Euglena are attracted to light; *Paramecium* are not. Data should reflect these differences in response.

Possible Hypotheses

- If *Paramecium* are attracted to light, then placing them on a glass slide that contains both light and dark zones will show an accumulation of protists toward the light side.
- If *Euglena* are attracted to light, then placing them on a glass slide that contains both light and dark zones will show an accumulation of protists toward the light side.

ANALYZE AND CONCLUDE

1. Answers may vary. Data must be used to either support or reject student hypotheses.
2. *Euglena* are attracted to light. *Paramecium* avoid bright light. Being autotrophic, *Euglena* benefit from light. Being heterotrophic, *Paramecium* may find a food supply in dim or dark conditions.
3. Most autotrophic organisms show a positive response to light whereas most heterotrophic organisms show a negative response.

Assessment

Portfolio: Have students write an evaluation of what they learned from this investigation. L1 SAE P

Going Further

Students may wish to test the response of these protists to different concentrations of salt solutions. L2

SECTION 22.2

Microscope Activity

LS **Visual-Spatial** Obtain a sample of diatomaceous earth from a biological supply house. Allow students to prepare wet mounts of this material and have them diagram each diatom shape they observe. An alternate source of living diatoms is to scrape the inside of a fish tank and observe wet mount preparations microscopically. A number of different types of living diatoms should be observed. L2 SAE

Different Viewpoints in Biology

Classification Schemes

Biologists do not always agree on classification schemes or names for taxonomic groups. For example, because the term *protist* traditionally refers to one-celled organisms, not all scientists agree with the kingdom name Protista. Some taxonomists also divide the kingdom into more than 25 phyla, twice as many as the 13 described in this chapter.

GLENCOE TECHNOLOGY

Videodisc

The Infinite Voyage: Secrets from a Frozen World

The Southern Ocean—A Rich Marine Ecosystem (Ch. 1)

The Antarctic Peninsula: Pack Ice and Life Cycles (Ch. 6)

Effect of UV Radiation on Phytoplankton (Ch. 8)

Magnification: 14×

Figure 22.9

Many diatom shells look like delicately carved, microscopic pillboxes. They come in an innumerable array of forms.

The delicate shells of many diatoms, like those in *Figure 22.9*, are like small pillboxes with lids. Each species has its own unique shape, decorated with exquisitely fine grooves and pores. Diatoms are classified into two major groups according to their basic shape. Some are radially symmetrical, while others are elongated and have distinct right and left sides.

Diatom reproduction

When diatoms reproduce asexually, the two halves of the box separate; each half then produces a new half to fit inside itself. With each new generation, half of the offspring are always smaller than the parent cells. When asexually reproducing diatoms reach the point at which they are roughly one-quarter of their original size, sexual reproduction takes place. Gametes are produced and released, then they fuse with those from another individual to produce a zygote, *Figure 22.10*. The zygote develops into a full-sized diatom, which goes on to divide asexually to start the entire process again.

As photosynthesizing autotrophs, diatoms contain chlorophyll as well as other pigments called carotenoids that give the majority of them a golden-yellow color. Carotenoids also give carrots and other yellow vegetables their distinctive colors. The food that diatoms manufacture photosynthetically is stored in the form of oils rather than starch. These oils give fishes that feed on diatoms an unpleasant, oily taste.

When diatoms die, their glasslike remains sink to the ocean floor to join huge deposits of diatom shells that have been accumulating there for many millions of years. These deposits are dredged up or mined and used as abrasives in tooth and metal polishes, or added to paint to give the sparkle that makes pavement lines more visible at night.

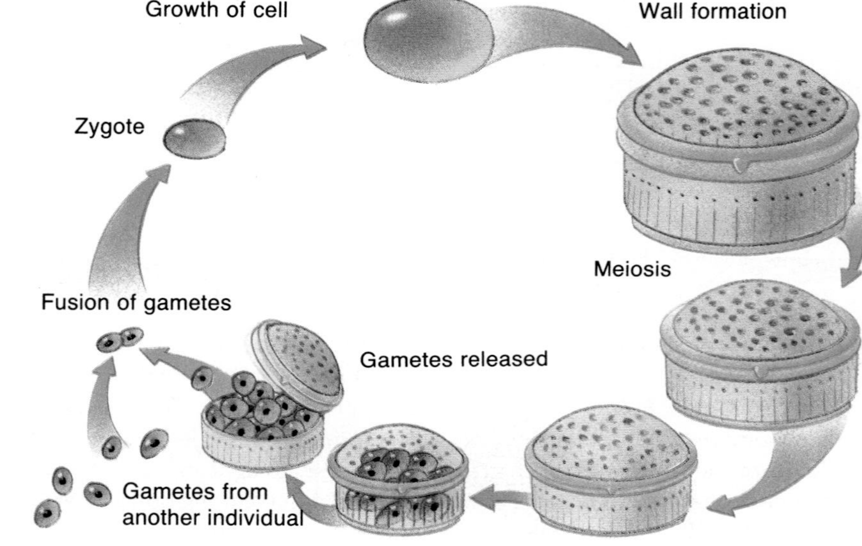

Figure 22.10

Diatoms will reproduce asexually for several generations before sexual reproduction takes place.

Meeting Individual Needs

Visually Impaired Allow students with visual difficulties to hold a petri dish. The top and bottom halves fit together in a manner that is similar to the two halves that form the cellular organization of a pillbox diatom. Extend the concept to show the decrease in diatom size during asexual reproduction by using petri dish sizes that become progressively smaller in diameter than the largest size (150-, 100-, and 60-mm diameter sizes are available). The concepts of symmetry and general shape can also be modeled using petri dishes to represent radial symmetry and a pencil or pen to represent elongated shapes. L1 SAE LS

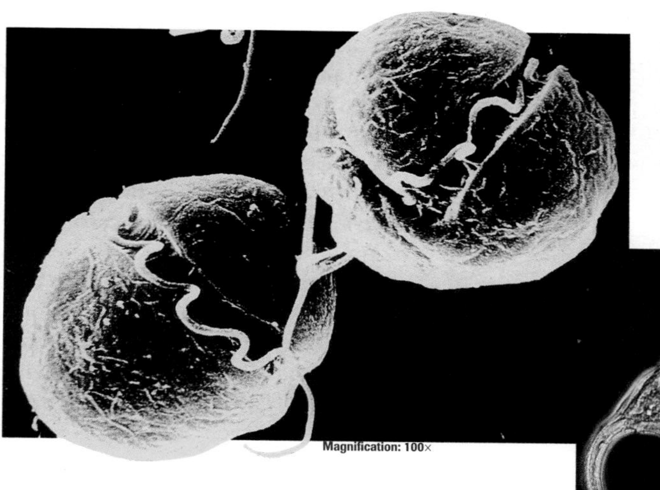

Figure 22.11

Dinoflagellates are protected by a set of armored plates. Notice that the grooves that encircle the plates are at right angles; these grooves are where the flagella are found.

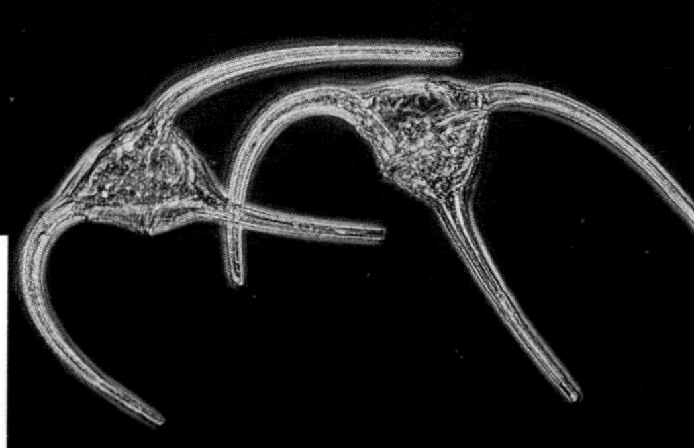

Dinoflagellates: The spinning algae

Dinoflagellates, members of the phylum Dinoflagellata, are unicellular algae that have cell walls made up of thick cellulose plates. They come in a great variety of shapes and styles; some resemble helmets, while others look like bizarre suits of armor. *Figure 22.11* shows highly magnified images of two different species of dinoflagellates.

Dinoflagellates have two flagella located in grooves at right angles to each other. When these flagella beat, the cell spins slowly. A few species of dinoflagellates are found in fresh water, but most are marine and are a major component of ocean phytoplankton.

Red tides

Dinoflagellates are autotrophic and contain chlorophyll, carotenoids, and red pigments. Many species live symbiotically with jellyfish, mollusks, and corals. Some free-living species are bioluminescent.

Several species of dinoflagellates produce poisonous toxins. One species, *Gonyaulax catanella*, produces an extremely strong nerve toxin that can be lethal. In the summer, these minute organisms may undergo tremendous population explosions. The dinoflagellates become so numerous that the ocean may actually turn red or orange-red, as you can see in *Figure 22.12*. Toxins released into the water during these red tides kill tons of fish. People who eat shellfish that have fed on these algae may risk being poisoned, too. During red tides, the harvesting of shellfish is usually banned.

Figure 22.12

This toxic red tide off the coast of Baja California, was caused by a bloom of dinoflagellates. In a red tide like this, the concentration of dinoflagellates can reach an incredible 40 to 60 million cells per liter of seawater.

SECTION 22.2

Using Science Terms

LS **Linguistic** Advise students that the prefix *dino* means "whirling" and the suffix *flagellates* means "whip." Ask students to explain why the phylum name for dinoflagellates is appropriate. *These protists spin about or whirl as their flagella propel them through water.* SAE

Brainstorming

Explain that coral is a heterotrophic organism. Have students explain the role that dinoflagellates play when living in symbiotic relationships with coral. *Food is made by the dinoflagellate through photosynthesis. Some of this food may be provided to the coral by the protist. The protist, in return, is provided with a safe home away from predators within a coral reef.* COOP LEARN

Reinforcement

Verify that students understand the meanings of the terms *bioluminescent*, *toxins*, and *shellfish*. Discuss why the term *shellfish* is a misnomer. *Shellfish are not truly fishes, but are instead either arthropods, such as shrimp and crabs, or mollusks, such as clams and squid.* SAE

GLENCOE TECHNOLOGY

Videodisc

Biology: The Dynamics of Life
Disc 1, Side 2
Kelp Forests (Ch. 13)

STVS: Plants & Simple Organisms
Disc 4, Side 1
New Uses for Algae (Ch. 12)

PORTFOLIO

Algal Names Algae are often called "seaweed, scum, and sea moss." However, none of these terms are scientifically correct. Have students find the meaning of each term and then explain why it is scientifically incorrect. Ask students to describe why each term is not helpful when describing the traits of algae as a group. L2 LS

Audiovisual

Show the filmstrip *Making Seaweed Worth Eating*, Carolina.

Demonstration

Purchase preserved specimens of a typical brown alga and a red alga to show to the class. If possible, purchase dry algae in an Asian food shop. Allow students to taste the samples.

Physics Connection

Buoyancy

Buoyancy is the ability of a fluid to exert an upward force on an object immersed in it. This force is equal to the weight of the fluid displaced by the object. Certain species of brown algae contain bladders filled with air. These bladders enable the algae to float on the ocean surface where they can capture maximum sunlight. Air bladders increase the volume of the alga, allowing it to displace more water. Thus, enough water is displaced to equal the weight of the alga, and it floats.

Concept Development

LS Logical-Mathematical Have students compare and contrast red and brown algae. *Differences: Red algae contain pigments called phycobilins, brown algae contain fucoxanthin. Similarities: Both are multicellular and mostly marine, both contain chlorophyll.*

Figure 22.13

There are some 4000 species of red algae. Some species are edible and are popular foods in Japan and other parts of the world.

Red algae

Members of the phylum Rhodophyta are the red seaweeds, all of which are multicellular marine organisms. ***Figure 22.13*** shows an example of a red alga. Red algae grow in tropical waters or along rocky coasts in colder water. They attach to rocks by structures appropriately called holdfasts.

Deep-water adaptations

In addition to chlorophyll, red algae also contain red and blue pigments called phycobilins, which are involved in photosynthesis. These pigments absorb green, violet, and blue light waves—the only part of the light spectrum that penetrates water below depths of 100 m. Red algae that live in deep water have a lot of these pigments, and so can carry on photosynthesis where light is limited.

phaeophyta:
phaios (GK) dusky
phyton (GK) plant
Phaeophyta are brown algae.

Brown algae

Brown algae make up the phylum Phaeophyta. There are approximately 1500 species of brown algae. Almost all of these species live in salt water along rocky coasts in cool areas of the world. Brown algae contain chlorophyll as well as a yellowish-brown carotenoid called fucoxanthin, which gives them their brown color. Diatoms also contain the carotenoid fucoxanthin.

The largest and most complex of all brown algae are kelp. Found along cold-water coasts, kelp anchor themselves to rocks or the sea bottom with sturdy holdfasts. The body of a multicellular alga such as a kelp is called a thallus. A **thallus** is a simple plant or algal body without roots, stems, or leaves. Some giant kelps may grow up to 60 m long. In kelp, the thallus is divided into the holdfast, stipe, and blade. Many species of brown algae have air bladders that help keep the thallus afloat near the surface, where it can get the light it needs for photosynthesis.

In some parts of the world, such as off the California coast, giant kelps form dense, underwater forests. These kelp forests are rich ecosystems, ***Figure 22.14***, and are home to a wide variety of marine organisms.

Another species of brown algae is the Sargasso seaweed, *Sargassum nitans*. This seaweed forms extensive masses that cover the Sargasso Sea in the Atlantic Ocean northeast of the Caribbean.

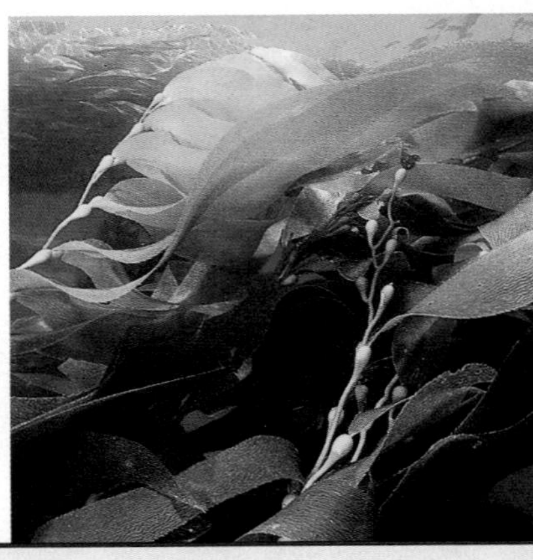

Figure 22.14

Growing close together like trees in a forest, giant kelps stretch dozens of meters from the seafloor to the ocean's surface.

CULTURAL DIVERSITY

Algae Harvesting in Japan

Discuss with students the nutritional benefits of algae and their use in cooking in many areas of the world, particularly Asia. Point out that algae contain many nutrients, including protein, and other assorted vitamins and minerals. They are eaten fresh, boiled, or fried in many Asian recipes. One of the more common types of algae used in cooking is *Porphyra*, a red alga commonly called nori. Since the 17th century, Japanese aquaculturalists have developed a huge industry around harvesting this alga from Tokyo Bay. Have students research Japanese algae harvesting techniques or bring in samples of Japanese foods that contain algae for the class to sample. L1 COOP LEARN

Green algae

Green algae make up the phylum Chlorophyta. Of all the types of algae, the green algae are the most diverse, with more than 7000 species. The major pigment in green algae is chlorophyll, but some species also produce yellow pigments that give them a yellow-green color.

Most species of green algae live in fresh water, but some live in the oceans, in moist soil, on tree trunks, in snow, and even in the fur of sloths—large, slow-moving mammals that live in the rain forest canopy.

A large variety of species

Green algae come in a wide variety of forms. *Figure 22.15* shows you two of the different types of green algae. *Chlamydomonas* is a unicellular and flagellated green alga. Other species are multicellular, forming slender filaments, flat sheets that you may have seen in many seaweeds, or colonies. A **colony** is a group of cells that lives together in close association. A *Volvox* colony, *Figure 22.15*, is a hollow ball composed of hundreds, even thousands, of flagellated cells arranged in a single layer. The cells are held together by strands of cytoplasm, and their flagella face outward. When the flagella beat—in a remarkably coordinated fashion—the colony spins through the water.

MiniLab

What do aquarium snails eat?

Many aquatic animals feed on photosynthesizing organisms that live in the water around them. Snails in an aquarium crawl around scraping off and eating the greenish film that forms on the inner surface of the glass walls, the bottom, and objects in the tank.

Procedure

1. Observe the activity of snails in your classroom or school aquarium, and record your observations.
2. Use a toothpick to scrape some green film from the glass and other surfaces. Place each sample on a slide.
3. Mix the scraped material with a drop of water. Carefully place a coverslip on each slide.

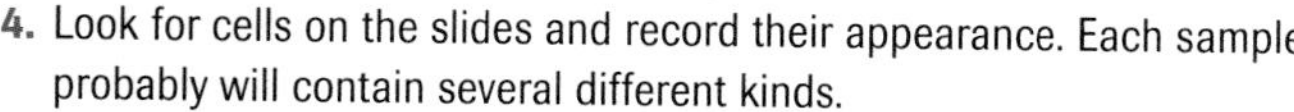

4. Look for cells on the slides and record their appearance. Each sample probably will contain several different kinds.

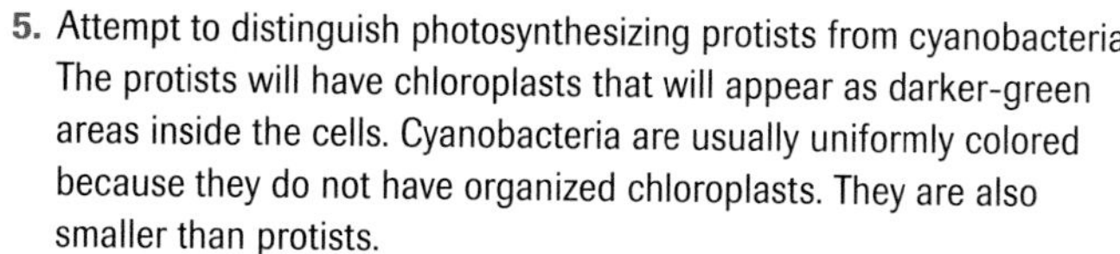

5. Attempt to distinguish photosynthesizing protists from cyanobacteria. The protists will have chloroplasts that will appear as darker-green areas inside the cells. Cyanobacteria are usually uniformly colored because they do not have organized chloroplasts. They are also smaller than protists.

Analysis

1. What evidence did you see that snails were consuming photosynthesizing organisms in the tank?
2. What did the cells that you collected look like? Did they contain chloroplasts?
3. What conclusions can you draw about the photosynthesizing organisms eaten by snails in your aquarium?

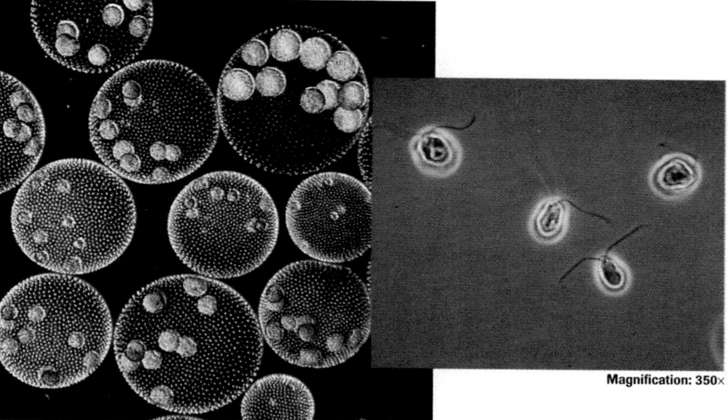

Figure 22.15

The wall of the *Volvox* sphere contains hundreds of individual cells (left). The smaller balls inside the sphere are daughter colonies. When these daughter colonies reach a certain size, the wall of the sphere breaks open. The daughter colonies are released. *Chlamydomonas* is a unicellular species of green algae (right).

MiniLab

Purpose

LS **Visual-Spatial** Students will observe food types consumed by snails and review food chains.

Process Skills

observe and infer, use the microscope, compare and contrast

Teaching Strategies

- Allow a classroom aquarium to become slightly overgrown with algae and cyanobacteria before the lab begins.
- Have students draw diagrams to scale to help distinguish protists and cyanobacteria.

Expected Results

Students will observe a variety of algae, including a number of different diatom species. In addition, evidence of cyanobacteria should be present. Students will also note that the snail leaves a clear trail on the aquarium sides as it scrapes off and consumes the underlying growth of protists and bacteria.

Analysis

1. A clear path will be noted where the snails have glided, indicating their having eaten the organisms present.
2. Cells were small and existed as individuals or colonies. Some show distinct chloroplasts, whereas others show only a general green color.
3. These organisms form the bottom autotrophic level in this aquarium food chain.

Assessment

Performance: Ask students to diagram the food chain studied within this aquarium ecosystem. L1

Meeting Individual Needs

Learning Disabled Have students prepare a pie graph that shows the number of species that comprise each phyla of the plantlike protists. Provide the following data: Euglenoids = 800, Diatoms = 10 000 Dinoflagellates = 2000, Red Algae = 4000, Brown algae = 1500, Green algae = 7000. L1 SAE P LS

PORTFOLIO

Picturing *Ulva's* Life Cycle Have students design a flowchart that demonstrates their understanding of the life cycle of *Ulva.* Have students include the following terms in their flowcharts: *alternation of generations, gametophyte, sporophyte, mitosis, meiosis, zygote,* and *gamete.* L1 LS

SECTION 22.2

3 Assess

Check for Understanding

Have students provide a brief description of three unicellular phyla of algae and three multicellular phyla of algae. L1

Reteach

As a class, prepare a structured overview or concept map for the first and second sections of this chapter. L1 SAE COOP LEARN

Extension

Ask students to conduct research to compile a list of foods that use algae in their manufacture or processing. L2 P

Assessment

Knowledge: Prepare a set of six simple line drawings showing a representative alga from each of the six phyla. Provide each student with a set of diagrams and ask them to record a minimum of three facts or ideas regarding each type of alga. L1 P

4 Close

Discussion Questions

Have each student prepare three questions based on information from this chapter. Collect the questions and present them to the class for answering. L1 COOP LEARN

Reproductive Strategies of Algae

Green algae reproduce both asexually and sexually. In *Spirogyra*, a multicellular, filamentous green alga, the filaments are haploid. They reproduce asexually by fragmentation. During **fragmentation,** an individual breaks up into pieces and each piece grows into a new individual.

The life cycles of algae such as *Ulva* show **alternation of generations.** These organisms alternate between haploid and diploid generations. The haploid form of an alga is called the **gametophyte** because it produces gametes. The diploid form of the organism—the **sporophyte**—develops from the zygote. Certain cells in the sporophyte undergo meiosis to become haploid spores. Each haploid spore can go on to develop into a gametophyte.

Look at *Figure 22.16* to see the life cycle of *Ulva*, a multicellular alga.

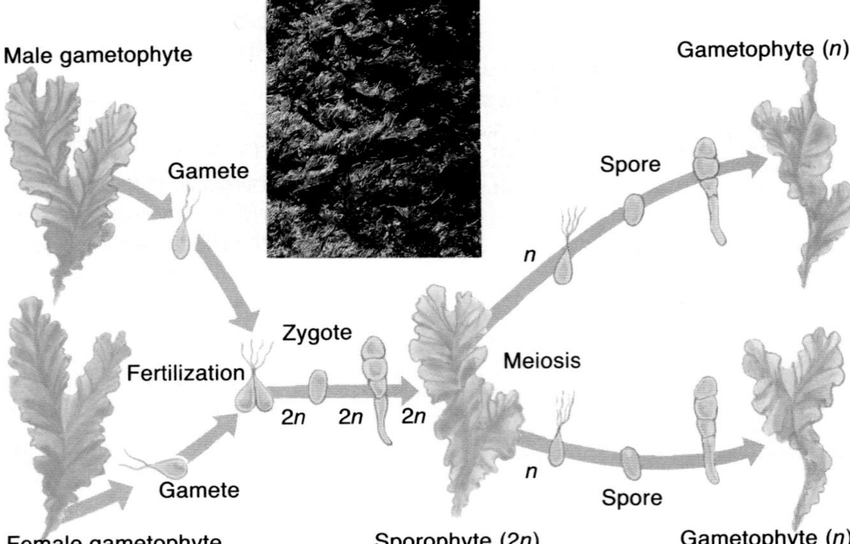

Figure 22.16

In the life cycle of *Ulva*, sea lettuce, notice how the generations alternate from haploid (gametophyte) to diploid (sporophyte). You will also see this pattern of reproduction in fungi and plants.

Section Review

Understanding Concepts

1. How are algae important to all living things?
2. Give examples that show why green algae are the most diverse of the algae.
3. How do the sporophyte and gametophyte generations of an alga such as *Ulva* differ from each other?

Thinking Critically

4. Do you think euglenoids should be classified with protozoans or algae? Explain.

Skill Review

5. **Making and Using Tables** Construct a table listing the different phyla of algae. Indicate whether they have one or more cells, their color, and give an example of each. For more help, refer to Organizing Information in the *Skill Handbook.*

Answers to Section Review

1. They are producers of oxygen and are essential autotrophs of many food chains.
2. *Chlamydomonas* is a unicellular example. *Volvox* is a colonial form. Others are filamentous like *Spirogyra*, and *Ulva* is a large alga that carries out alternation of generations.
3. Gametophytes are haploid and form gametes. Sporophytes are diploid and form spores.

Thinking Critically

4. They could be classified as either. Their chloroplasts are algae-like while their locomotion is protozoan-like.

Skill Review

5. Euglenophyta are unicellular, green, euglenas. Bacillariophyta are unicellular, golden, diatoms. Dinoflagellata are unicellular, green, yellow and red, dinoflagellates. Rhodophyta are multicellular, red, red seaweeds. Phaeophyta are multicellular, brown, brown seaweeds. Chlorophyta are unicellular or multicellular, green, green algae.

SECTION

22.3 Funguslike Protists

As you get up from the fallen log you've been sitting on by the pond's edge, a spot of color at its base catches your attention. Turning the log over, you uncover a glistening mass of yellow-orange slime that fans out over the underside of the log. What you've found is a slime mold, one of a variety of funguslike protists. Slime molds, along with water molds and downy mildews, obtain energy by decomposing organic materials, and play an important role in recycling nutrients in ecosystems.

Section Preview

Objectives

Contrast the two types of slime molds with respect to their cellular differences and life cycles.

Discuss the economic importance of the downy mildews and water molds.

Key Term

plasmodium

Different Kinds of Funguslike Protists

Fungi are classified in a distinct kingdom that you will study in Chapter 23. However, certain groups of both the slime molds and water molds show some features of fungi and protists. They form delicate, netlike structures on the surfaces of their food supplies. These organisms obtain energy by decomposing organic materials.

There are three phyla of funguslike protists, and they all obtain energy by decomposing organic materials. Slime molds make up two of these phyla. Slime molds have characteristics of both protozoans and fungi and are classified by the way they reproduce. The third phylum of funguslike protists comprises the water molds and downy mildews. Although we rarely notice in our everyday lives, some disease-causing varieties have the potential to do great damage to crops.

Slime Molds

Many slime molds are beautifully colored, ranging from brilliant yellow or orange to rich blue, violet, and jet black. They live in cool, moist, shady places where they grow on damp, organic matter such as rotting leaves or decaying tree stumps and logs.

There are two major types of slime molds: plasmodial slime molds, which belong to the phylum Myxomycota, and cellular slime molds, which make up the phylum Acrasiomycota. Slime molds are distinctly animal-like during much of their life cycle, moving about and engulfing food in the same way that amoebas do. They reproduce by making spores, a funguslike characteristic.

SECTION 22.3

Prepare

Key Concepts

This section discusses the three phyla of funguslike protists. All are composed of multicellular heterotrophs that obtain their nutrients by decomposing organic materials. Slime molds make up two phyla based on their forming either a plasmodium stage, which contains no distinct separate cells, or a cellular stage that shows a cellular organization. The third phylum consists of protists that are funguslike in that their bodies consist of a mass of threadlike fibers.

Block Scheduling

Look for this symbol for strategies that are useful in a block scheduling format. For more information on block scheduling, refer to Section 22.3 in the **Lesson Plans** booklet.

Materials

- Purchase the slime mold *Physarum polycephalum* in its plasmodium form from a biological supply house for the Alternate Lab.

Meeting Individual Needs

Students Acquiring English Stress the terms students need to remember to recall the three main subgroups of the Kingdom Protista. For animal-like protists, use *unicellular, motile,* and *heterotroph.* For plantlike protists, use *photosynthetic, autotroph, unicellular,* and *multicellular,* and for funguslike protists, *heterotroph* and *decomposer.* L1 SAE

Program Resources

Section Focus Master 48 L1 SAE
Reinforcement and Study Guide, p. 88 L1
Concept Mapping, p. 22 L1
Basic Concepts Transparencies 29, 30 and **Masters** L1 SAE

SECTION 22.3

1 Focus

Bellringer

Before presenting the lesson, display **Section Focus Master 48** on the overhead projector and have students answer the accompanying questions. L1 SAE

Audiovisual

Show the videos *Non-Cellular Slime Molds, Cellular Slime Molds,* and *Water Molds,* Carolina.

2 Teach

Reinforcement

Prepare a structural overview that illustrates the relationship of the phyla Myxomycota and Acrasiomycota to the Kingdom Protista. As an aside, review the relationship of the algae and protozoans to the kingdom.

GLENCOE TECHNOLOGY

Videodisc

Biology: The Dynamics of Life
Disc 1, Side 2
Slime mold (Ch. 14)

Figure 22.17

The moving, feeding form of a plasmodial slime mold is a multinucleate blob of cytoplasm. To reproduce, the plasmodium reorganizes itself into stalked, spore-producing structures. Minute spores are released and are carried great distances.

Plasmodial slime molds

Plasmodial slime molds get their name from the fact that they form a **plasmodium,** a mass of cytoplasm that contains many diploid nuclei but no cell walls or membranes. This slimy, multinucleate mass, like the one shown in *Figure 22.17,* is the feeding stage of the organism. It creeps by amoeboid movement, forming a netlike structure on the surfaces of decaying logs or leaves. As it moves—at the rate of about 2.5 cm per hour—it engulfs microscopic organisms and digests them in food vacuoles.

A plasmodium may reach more than a meter in diameter and contain thousands of nuclei. However, when its surroundings dry up, a plasmodium transforms itself into many separate reproductive structures. Meiosis takes place within these structures. Haploid spores form, which are dispersed by the wind. Spores germinate into either flagellated or amoeboid cells that serve as gametes. The diploid zygote grows into a new plasmodium.

Cellular slime molds

During the feeding stage, cellular slime molds exist as individual, haploid, amoeboid cells that feed, grow, and divide by cell division, *Figure 22.18.* When food becomes scarce, these independent cells come together with hundreds or thousands

Figure 22.18

Cellular slime molds spend part of their life cycle as an independent, amoeba-like cell that feeds, grows, and divides. At times, many hundreds of these amoeboid cells gather to form a multicellular clump. When it moves, this clump of cells compacts itself into a blob that looks like a small garden slug. Eventually, the slug forms a stalked reproductive structure (left). Cellular slime molds are haploid organisms during their entire life cycle.

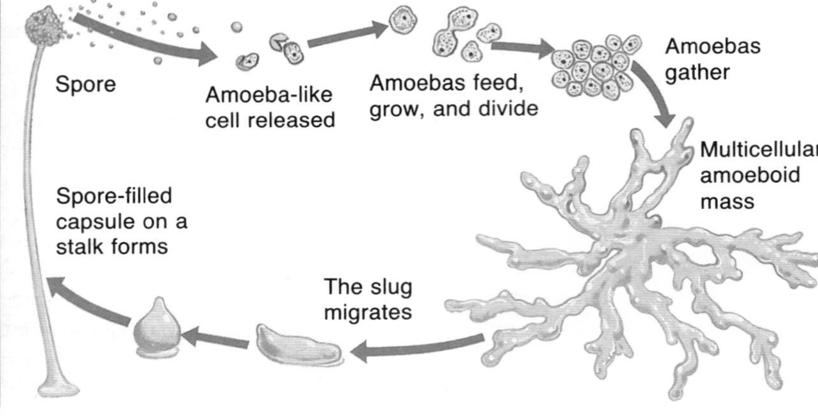

Alternate Lab — Observing Slime Mold

Purpose

LS **Visual-Spatial** Provide students with an opportunity to observe a plasmodium-type slime mold.

Materials

petri dish with plasmodium (purchase plasmodium stage of *Physarum polycephalum* on agar plate—subculture by cutting sections from the plate and putting them on sterile agar plates; add a few flakes of oat cereal and moisten with distilled water—prepare two days in advance of the day needed), microscope

Procedure

Give students the following directions.

1. Observe a petri dish containing the slime mold, *P. polycephalum.* Watch the slime mold for several minutes.
2. Design a means for judging whether the organism does or does not move.
3. Record as many macroscopic observations as you can make regarding its appearance and behavior.
4. Place the dish on your microscope stage and observe the slime mold using only low power. Continue to watch it for at least several minutes.

of their own kind to reproduce. Such an aggregation of amoeboid cells superficially resembles a plasmodium. However, this mass is multicellular; it is made up of many individual amoeboid cells, each with a distinct cell membrane.

Phylogeny of Protists

How are the many different kinds of protists related to each other and to fungi, plants, and animals? *Figure 22.19* shows the relationships of protists to each other. Although taxonomists are now comparing the RNA and DNA of these groups, there is little conclusive evidence to indicate whether ancient protists were the evolutionary ancestors of fungi, plants, and animals or whether protists emerged as evolutionary lines that were separate. Many biologists agree that ancient green algae were probably the ancestors of plants.

Figure 22.19

The radiation of the different protist phyla on the Geologic Time Scale shows their relationships.

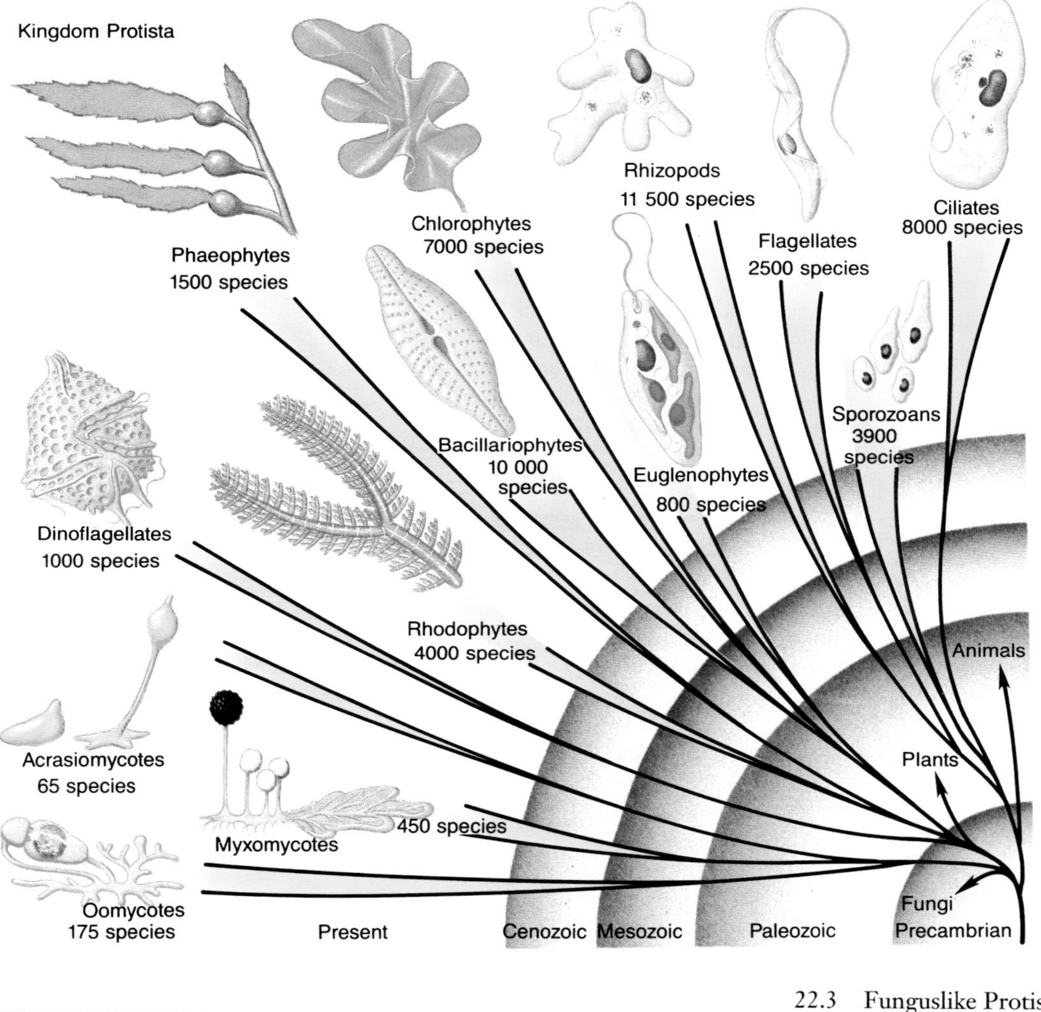

SECTION 22.3

Microscope Activity

LS Visual-Spatial Attach a piece of meat to a piece of string. Place the meat into a fish tank and allow the meat to remain there for several days. Remove the meat from the tank and have students examine it both macroscopically and microscopically. Have students describe their observations. *The meat should be covered by a white fuzz that comprises the body of a typical water mold.* L2

Visual Learning

Figure 22.19 Take a few moments to review the relationships among the different protist phyla shown.

3 Assess

Check for Understanding

Have students compare and contrast the three funguslike protist phyla. Ask them to explain why these organisms are classified as protists and not in the Kingdom Fungi even though their name suggests a relationship to the fungi. L1

5. Record as many observations as you can make regarding its microscopic appearance and behavior.

Analysis

1. Describe the general appearance and behavior of this slime mold. Is it a plasmodium or cellular slime mold? *yellow, blob-like, stringy, moves very slowly—plasmodium*
2. Describe the microscopic appearance and behavior of this slime mold. *Material could be seen flowing within slime mold, stopping, then flowing in the opposite direction.*

Assessment

Oral: Explain how you know this was the feeding rather than reproductive stage of the slime mold.

Reteach

Have students work in groups to prepare a chart that compares and contrasts the phyla Myxomycota, Acrasiomycota, and Oomycota. L1 COOP LEARN

Extension

Ask students to prepare flowcharts that trace the stages in the life cycles of organisms classified as Myxomycota and Acrasiomycota. Ask students to include the chromosome number for each particular stage and indicate if mitosis or meiosis is occurring. L2

Assessment

Oral: Ask students to list several traits shown by the funguslike protists that are typically protozoan and several traits that are typically fungal. L1

4 Close

Activity

Have students write five words related to funguslike protists on a sheet of paper. Explain that four of the words must show some relationship to one another, while the other word is unrelated. Collect the lists. Place several lists on the chalkboard. Have the class decide which word does not belong and how the remaining words are related. L1 SAE COOP LEARN

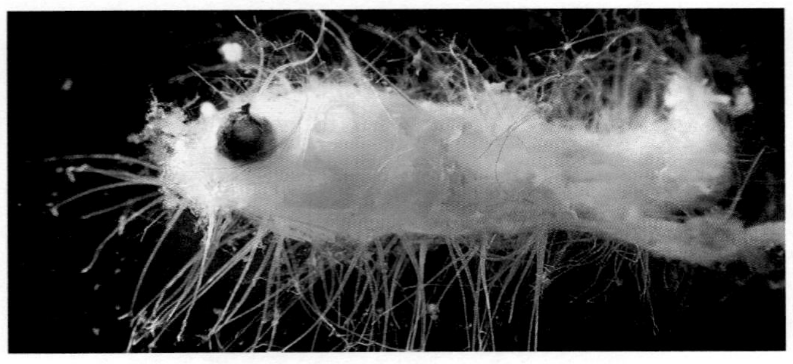

Figure 22.20
Water molds grow over the surface of dead organisms, such as this fish, gradually decomposing the tissues and absorbing the nutrients released in the process.

Water Molds and Downy Mildews

Water molds and downy mildews are members of the phylum Oomycota. Most members of this large and diverse group of funguslike protists live in water or moist places. As shown in *Figure 22.20*, some feed on dead organisms and others parasitize plants.

Most water molds appear as fuzzy, white growths on decaying matter. They resemble a fungus in that they grow as a mass of threads over a substrate, digest it, and then absorb the predigested nutrients. But at some point in their life cycle, water molds produce flagellated reproductive cells—something that true fungi never do. This is why water molds are classified as protists rather than fungi.

One economically important member of the phylum Oomycota is a downy mildew that causes a serious disease in many plants.

Connecting Ideas

Kingdom Protista is made up of a variety of species of unicellular and multicellular eukaryotic organisms. While protists may look like simple organisms, many have complex workings. Protists play some extremely important roles in the environment. Diatoms produce a large percentage of the oxygen we breathe. Unicellular algae form the first link in aquatic food chains.

Large, multicellular species can form habitats for other organisms as well as supply them with food. The funguslike protists—slime molds, water molds, and downy mildews—with their characteristics of both fungi and protists, provide a link between the two groups.

Section Review

Understanding Concepts

1. What characteristics of slime molds might cause difficulty in their classification?
2. Explain why plasmodial slime molds are sometimes referred to as acellular slime molds.
3. How would a water mold growing on a fish ultimately kill it?

Thinking Critically

4. What kinds of environments would you consider to be favorable for the growth of slime molds? Explain.

Skill Review

5. **Observing and Inferring** If you know that a plasmodium consists of many nuclei within a single cell, what can you infer about the process that formed the plasmodium? For more help, refer to Thinking Critically in the *Skill Handbook*.

Answers to Section Review

1. They exhibit protozoan characteristics during reproduction as they form spores and flagellated or amoeboid gamete cells.
2. The plasmodium is a mass of cytoplasm that contains many nuclei but no distinct cell walls or membranes separating the cells.
3. Tiny threads grow into the body of the fish and digest its tissues as food for the water mold.

Thinking Critically

4. A moist environment would be favorable. Moisture would prevent the plasmodium from drying out and would provide conditions needed for maintaining food for the slime mold.

Skill Review

5. Mitosis is cell division that usually results in two identical diploid cells. The process responsible for many nuclei within a diploid stage indicates that mitosis occurs without cell division.

Chapter 22 Review

REVIEWING MAIN IDEAS

22.1 The World of Protists

- Kingdom Protista is a diverse group that contains animal-like, plantlike, and funguslike organisms.
- Some protists are heterotrophs, some are autotrophs, and some get their nutrients by decomposing organic matter.
- Amoebas move by extending pseudopodia. Some marine amoebas have shells.
- Flagellates have one or more flagella, which they use to propel themselves. Ciliates move by the synchronized beating of their cilia. Sporozoans are nonmotile and live as parasites.

22.2 Algae: Plantlike Protists

- Algae are photosynthetic autotrophs. There are both unicellular and multicellular species of algae. Unicellular species include the euglenoids, diatoms, dinoflagellates, and some green algae. Multicellular species include red, brown, and green algae. Euglenoids have characteristics of both plants and animals.
- Green, red, and brown algae, often called seaweeds, have complex life cycles that alternate between haploid and diploid generations.

22.3 Funguslike Protists

- Slime molds, water molds, and downy mildews are funguslike protists that obtain nutrients by decomposing organic material. Both plasmodial and cellular slime molds undergo changes in appearance and behavior to produce reproductive structures.

Key Terms

Write a sentence that shows your understanding of each of the following terms.

algae
alternation of generations
asexual reproduction
ciliate
colony
flagellate
fragmentation
gametophyte
plasmodium
protozoan
pseudopodia
spore
sporophyte
sporozoan
thallus

Understanding Concepts

1. How do protists differ from bacteria?
   a. Bacteria have specialized organelles.
   b. All protists are unicellular.
   c. Protists lack cell membranes.
   d. Protists are eukaryotic.
2. Protozoans are classified based on their ____________.
   a. nutrition
   b. method of locomotion
   c. reproductive abilities
   d. size
3. Producers in aquatic food chains include ____________.
   a. algae
   b. protozoans
   c. slime molds
   d. amoebas
4. Euglenophytes are unique because of their ____________.
   a. flagella
   b. cilia
   c. flexible pellicle
   d. heterotrophic nature
5. Which of the following organisms is the cause of red tides?
   a. dinoflagellates
   b. diatoms
   c. green algae
   d. red algae
6. Which of the following organisms has a silica cell wall?
   a. dinoflagellates
   b. diatoms
   c. green algae
   d. red algae
7. Which organisms cause malaria?
   a. ciliates
   b. viruses
   c. flagellates
   d. sporozoans

Chapter 22 Review

Reviewing Main Ideas

Summary statements can be used by students to review the major concepts of the chapter.

Key Terms

Answers should go beyond defining the terms. Accept any answer that uses the term correctly and in the proper context.

Understanding Concepts

1. d
2. b
3. a
4. c
5. a
6. b
7. d

Chapter 22 Review

8. c
9. d
10. c
11. b
12. a
13. d
14. d
15. b
16. b
17. a
18. c
19. a
20. c

Applying Concepts

21. finding a suitable host; producing many spores will improve the possibility of survival because some of the spores may find a host
22. oak forest—dry conditions stimulate spore-producing structures
23. Answers will vary, but may include cilia, pseudopodia, flagella, eyespots, contractile vacuoles, or others.
24. Mosquitoes breed in water. Elimination of breeding grounds reduces the mosquito population and reduces the risk of contracting malaria.
25. Sea urchins are a food source for many organisms. If they were removed or if their numbers increased dramatically, the entire food chain would be disrupted.

Thinking Critically

26.

Chapter 22 Review

8. Which protist may have many unbound nuclei in a mass of protoplasm?
   a. flagellate c. slime mold
   b. ciliate d. brown alga
9. Which of the following contains species that are able to digest cellulose?
   a. termites c. slime molds
   b. bacteria d. flagellates
10. Which group moves by using pseudopodia?
    a. flagellates c. rhizopods
    b. ciliates d. sporozoans
11. Which of the following are not protists?
    a. algae c. amoebas
    b. mushrooms d. slime molds
12. Which group is entirely parasitic?
    a. sporozoans c. ciliates
    b. flagellates d. rhizopods
13. Which organelle in protists is able to eliminate excess water?
    a. anal pore
    b. mouth
    c. gullet
    d. contractile vacuole
14. Which of the following groups has members with calcium carbonate shells?
    a. ciliates c. sporozoans
    b. flagellates d. rhizopods
15. A long, hairlike organelle that can be moved to propel a protist through the water is a ___________.
    a. cilium c. spore
    b. flagellum d. pseudopod
16. The algae that can survive in the deepest water are the ___________.
    a. brown algae c. diatoms
    b. red algae d. fire algae
17. The largest and most complex of brown algae are the ___________.
    a. kelp c. sea lettuce
    b. *Chlamydomonas* d. *Spirogyra*
18. Which of the following are protected by a set of armored plates?
    a. kelp c. dinoflagellates
    b. fire algae d. diatoms
19. The greatest diversity is found among the ___________.
    a. diatoms c. red algae
    b. brown algae d. fire algae
20. Fuzzy, white growths on decaying matter may be ___________.
    a. slime molds c. water molds
    b. downy mildews d. sporozoans

Applying Concepts

21. What do you think a major disadvantage might be for the parasitic behavior of a sporozoan? How might a sporozoan's method of asexual reproduction offset this disadvantage?
22. In what type of ecosystem would you expect plasmodial slime molds to transform themselves into spore-producing structures more frequently: a forest in the Pacific Northwest that has heavy rainfall or a dry, oak forest in the Midwest? Explain your answer.
23. Give three examples of organelles that help protists maintain homeostasis in their environments.
24. Up to the late 1800s, malaria was common in the extreme southeastern part of the United States. In an attempt to fight the disease, ponds and wetlands were often filled in or drained. How do you suppose this action helped cut down on the number of malaria cases?
25. How may sea urchins be important to the entire food chain of a kelp forest?

Thinking Critically

26. ***Concept Mapping*** Make a concept map that relates the following terms. Supply the appropriate linking words for your map.
    protists, algae, diatoms, flagellates, protozoans, amoebas, ciliates, slime molds, *Chlorophyta*, *Rhodophyta*

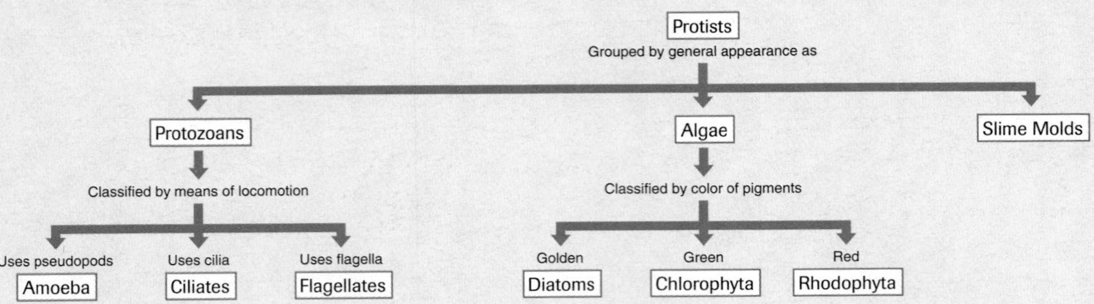

27. *Observing and Inferring* Why do you suppose many people who own aquariums add snails to them?
28. *Formulating Hypotheses* In agricultural regions where farmers apply large amounts of nitrogen fertilizers to their fields, local ponds and lakes often develop a thick, green scum of algae and cyanobacteria in late summer. Hypothesize why this happens.
29. *Sequencing* Sequence the stages of both sexual and asexual reproduction in diatoms.
30. *Observing and Inferring* Explain why it would be necessary to monitor the water for dinoflagellates in some areas where shellfish are harvested for human consumption.
31. *Interpreting Data* The following table was presented by a pair of students, showing the data they collected while observing how *Paramecium* and *Euglena* differ in their response to light. What pattern is most obvious from analysis of the table? How did the reaction of *Paramecium* and *Euglena* differ in their response to light?

Response of *Euglena* and *Paramecium* to Light

Organism	Time in minutes	% of individuals in the light
Paramecium	3	45
Paramecium	6	30
Paramecium	9	15
Euglena	3	55
Euglena	6	70
Euglena	9	90

ASSESSING KNOWLEDGE & SKILLS

During a summer ecology class, a group of high school students studied unicellular algae at a site in the middle of a pond. For three days and nights, they measured the number of cells in the water at various depths. They produced the following graph based on their data.

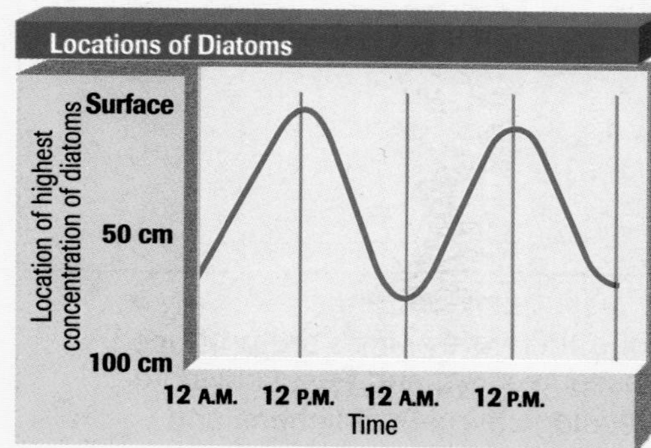

Interpreting Data Use the graph to answer the questions that follow.

1. At what time were the highest concentrations of diatoms at the surface?
   a. midnight
   b. noon
   c. 3 A.M.
   d. 6 P.M.
2. At what time were the highest concentrations of diatoms about a meter below the surface?
   a. midnight
   b. noon
   c. 3 A.M.
   d. 6 P.M.
3. Which of the following is a good description of the movement of diatoms in the water column?
   a. 6-hour cycle
   b. 12-hour cycle
   c. 24-hour cycle
   d. irregular cycling
4. *Interpreting Data* Why might the diatoms show the pattern found by the group of high school students?

Chapter 22 Review

27. Snails help keep the growth of unwanted algae in check.
28. The nitrogen fertilizers run off into water sources, where they stimulate growth and reproduction of these organisms.
29. Diatoms reproduce asexually for several generations, with the cell walls getting smaller and smaller. Eventually the cells undergo meiosis, release gametes, and grow a new cell with a cell wall, repeating the process.
30. High levels of some species of dinoflagellates can cause toxicity in the shellfish. If the shellfish are eaten by people, they may also be ingesting the toxins.
31. The longer the light remained, the fewer *Paramecium*. In contrast, the longer the light remained, the higher the concentration of *Euglena*.

ASSESSING KNOWLEDGE & SKILLS

1. b
2. a
3. c
4. They are photosynthetic and move to the surface to utilize sunlight during daylight hours.

Program Resources

Chapter Assessment, pp. 127-132 L1
Alternate Assessment in the Science Classroom
Computer Test Bank L1
Content Mastery, pp. 85-88 L1

Chapter Organizer

SECTION	OBJECTIVES	ACTIVITIES/FEATURES
23.1 The Life of Fungi National Science Standards: UCP.1, UCP.5; C.1, C.4, C.5	1. **Identify** the basic characteristics of fungi. 2. **Explain** the importance of fungi as decomposers and in the flow of energy and nutrients through food chains.	**A Broader View:** The Fungus that Ate Michigan, p. 562 **Minilab:** Are there mold spores in your classroom?, p. 564
23.2 The Diversity of Fungi National Science Standards: UCP.1, UCP.3, UCP.5; A.1, A.2; C.1, C.4, C.5; F.1, F.5; G.1-3	3. **Identify** the four major phyla of fungi. 4. **Distinguish** among the ways spores are produced in zygomycotes, ascomycotes, and basidiomycotes. 5. **Summarize** the ecological roles of lichens and mycorrhizae.	**Minilab:** How are gills arranged in a mushroom?, p. 569 **Biolab:** Design Your Own Experiment—Does temperature affect the metabolic activity of yeast?, p. 572 **Thinking Lab:** How are new antibiotics discovered?, p. 574

ACTIVITY MATERIALS

BIOLAB	MINILABS	ALTERNATE LAB
page 572 bromothymol blue solution (BTB) straw small test tubes (4) large test tubes (3) one-hole stoppers with glass tube inserts (3) yeast/molasses mixture water/molasses mixture water/yeast mixture test-tube rack 250-mL beakers (3) ice cubes Celsius thermometer hot plate 50-mL graduated cylinder glass-marking pencil 10-cm rubber tubing (3) aluminum foil	**page 564** freshly baked bread plate plastic self-sealing bags forceps slide coverslip microscope **page 569** edible mushrooms paper bag white paper	**page 562** small paper cups macaroni aluminum foil cardboard oatmeal flakes swab with mold spores

Chapter 23 Fungi

TEACHER CLASSROOM RESOURCES

Reproducible Masters	Transparencies
Section Focus Master 49: Fungi L1 SAE **Reinforcement and Study Guide,** pp. 89-90 L1 **Biolab and Minilab Worksheets,** p. 91 L1 **Concept Mapping:** Feeding Relationships of Fungi, p. 23 L1 **Tech Prep Applications:** Fungus Among Us, pp. 27-28 L2 **Laboratory Manual:** Identification of Common Molds, pp. 127-128 L2	
Section Focus Master 50: Reproduction in Fungi L1 SAE **Reinforcement and Study Guide,** pp. 91-92 L1 **Biolab and Minilab Worksheets,** pp. 92-94 L1 **Laboratory Manual:** Which Foods Can Bread Mold Use for Nutrition?, pp. 129-130; Lichens, pp. 131-132 L2 **Critical Thinking/Problem Solving:** Fields of Fungi, p. 23 L3 **Content Mastery,** pp. 89-92 L1	**Basic Concepts Transparency #31:** The Life of a Mushroom L1 SAE **Basic Concepts Transparency #32:** Phylogeny of Fungi L1 SAE **Reteaching Transparency #23:** Life Cycle of a Mushroom L1 SAE

ASSESSMENT MATERIALS

Chapter Assessment, pp. 133-138
Alternate Assessment in the Science Classroom
MindJogger Videoquiz
Computer Test Bank

Spanish Resources SAE
English/Spanish Audiocassettes SAE
Cooperative Learning in the Science Classroom COOP LEARN
Lesson Plans

KEY TO TEACHING STRATEGIES

- L1 Level 1 activities should be within the ability range of all students including those with learning difficulties.
- L2 Level 2 activities are within the ability range of average to above-average students.
- L3 Level 3 activities are designed for the ability range of above-average students.
- SAE SAE activities should be within the ability range of Students Acquiring English.
- COOP LEARN Cooperative Learning activities are designed for small group work.
- P These strategies represent student products that can be placed into a best-work portfolio.
- These strategies are useful in a block scheduling format.

GLENCOE TECHNOLOGY

The following multimedia resources are available from Glencoe.

Biology: The Dynamics of Life
CD-ROM SAE
Videodisc Program

National Geographic Society Series
Newton's Apple: Life Sciences
Mold

Science and Technology Videodisc Series (STVS)
Plants & Simple Organisms
Wood Decay
Fungal Collection for Research

CHAPTER

23 Fungi

Chapter Overview

In this chapter, students will learn about the diverse organisms that make up the Kingdom Fungi. Students will learn about the major characteristics of fungi, including their means of obtaining energy. The general anatomy of fungi is then described and the feeding relationships common to fungi—decomposition, parasitism, and mutualism—are presented. The section ends with a description of reproductive strategies of fungi, including fragmentation, budding, and spore formation.

The second section introduces the four phyla of fungi: Zygomycetes, Ascomycetes, Basidiomycetes, and Deuteromycetes. Examples of typical fungi classified in each phylum are then provided in conjunction with their reproductive structures, economic importance, and environmental importance. The section concludes with a discussion of the mutualistic relationships between lichens and autotrophs as well as those between mycorrhizae and trees.

Key Terms

ascospore	haustorium
ascus	hypha
basidiospore	lichen
basidium	mycelium
budding	mycorrhiza
chitin	rhizoid
conidiophore	sporangium
conidium	stolon
gametangium	zygospore

558

Learning Styles

Look for the following logo for strategies that emphasize different learning modalities. LS

Visual-Spatial	**Making a Model, p. 561; Microscope Activity, pp. 561, 568; Minilab, pp. 564, 569; Portfolio, p. 574; The Inside Story, p. 570; Meeting Individual Needs, p. 570**
Interpersonal	**Activity, p. 561; Alternate Lab, p. 562; Project, p. 570**
Linguistic	**Using Science Terms, pp. 563, 567**
Logical-Mathematical	**Meeting Individual Needs, pp. 561, 566, 575; Student Journal, p. 568; Thinking Lab, p. 574**

LS

CHAPTER

23 Fungi

Have you ever had a ring of mushrooms mysteriously appear in your yard overnight? You know they weren't there the day before. How did they form so quickly? And what are they doing?

Members of the Kingdom Fungi are found in nearly every type of habitat on Earth. However, what we see of them is often just the tip of the iceberg. Hidden in the soil beneath those mushrooms in your yard, for example, lies the rest of the fungus that produced them.

Fungi may not always be noticeable, but their actions are. They live by decomposing living and nonliving organic matter.

In one way or another, fungi affect nearly all other forms of life. Some cause diseases and destroy important food crops. But these negative aspects are far outweighed by the importance of fungi as decomposers, and by the crucial role they play in the biosphere as nutrient recyclers.

Concept Check

You may wish to review the following concepts before studying this chapter.

- Chapter 3: parasitism, symbiotic relationships
- Chapter 7: organic compounds

Chapter Preview

Laboratory Activities

Biolab: Design Your Own Experiment
- Does temperature affect the metabolic activity of yeast?

Minilabs
- Are there mold spores in your classroom?
- How are gills arranged in a mushroom?

Mushrooms are a visible sign of fungi at work in the soil. Mushrooms are just one of many different kinds of fungi. As decomposers, fungi form a critical link in the web of life on Earth.

CHAPTER 23

Introducing the Chapter

Have students attempt to offer an explanation for why mushrooms, such as those shown on the previous page, grow in circular patterns. Ask them to think in terms of an organism that spreads out evenly from a central area or location. Have them speculate as to: (a) the role that the visible portion of mushrooms may actually play in mushroom growth and reproduction, and (b) why rings of mushrooms appear only at certain times of the year. *Students may suggest that the visible portion of the mushroom drops spores from which new mushrooms form. The mushroom patterns appear at times of the year when conditions are conducive to growth.*

Theme Development

The theme of *unity within diversity* is emphasized in discussions of the enormous diversity of fungi. The theme of *energy* is illustrated as students learn how fungi obtain their energy needs via decomposition of organic matter.

Concept Check

Students will revisit the concept of decomposition as it relates to fungi. They will also review the symbiotic relationship of mutualism when they study lichens.

Assessment Planner

Choose assessment strategies from the following pages to evaluate the progress of your students.

Assess, pp. 565, 575
Alternate Lab, pp. 562-563
Minilabs, pp. 564, 569
Portfolio, pp. 574, 575
Thinking Lab, p. 574
Biolab, pp. 572-573
Chapter Review, pp. 577-579

SECTION

23.1

Prepare

Key Concepts

This section introduces the major characteristics of organisms classified in the Kingdom Fungi. The section first reviews the basic structure of fungi including the structure of their cell walls, which contain chitin rather than cellulose. Next, the methods by which fungi obtain nutrients are presented. The section concludes with a discussion of how fungi reproduce through strategies such as budding, fragmentation, and spore production.

Block Scheduling

Look for this symbol for strategies that are useful in a block scheduling format. For more information on block scheduling, refer to Section 23.1 in the **Lesson Plans** booklet.

Materials

- Purchase fresh mushrooms from a supermarket for the Focus Activity.
- Purchase fresh bakery bread and plastic bags for the Minilab.

1 Focus

Bellringer

Before presenting the lesson, display **Section Focus Master 49** on the overhead projector and have students answer the accompanying questions. L1 SAE

SECTION

23.1 The Life of Fungi

Section Preview

Objectives

Identify the basic characteristics of fungi.

Explain the importance of fungi as decomposers and in the flow of energy and nutrients through food chains.

Key Terms

hypha
mycelium
chitin
haustoria
budding
sporangium

You would probably recognize a mushroom growing in your yard as a fungus, but what about the yeast cells in this photograph? Yeasts are fungi too, but they don't look much like mushrooms. More than 100 000 different fungus species have been described by mycologists, scientists who study fungi. Each type of fungus has its own distinctive features. But all fungi are similar in their cellular composition.

Magnification: 70×

General Characteristics

Fungi are everywhere—in the air, in the water, on damp basement walls, in the garden, on foods, even between our toes. Some are large, bright, and colorful, while others are easily overlooked, *Figure 23.1.* Many have descriptive names such as stinkhorn, puffball, rust, or ringworm. Most species grow best at temperatures between 20° and 30°C, but many thrive at cooler temperatures. How many times have you opened the refrigerator and pulled out a piece of fruit or a chunk of cheese, only to find that it has already become a meal for a thick, furry mass of pink, green, white, or black mold?

Figure 23.1

Fungi come in many different forms, sizes, and colors.

These coral fungi resemble ocean corals and often sport vivid colors.

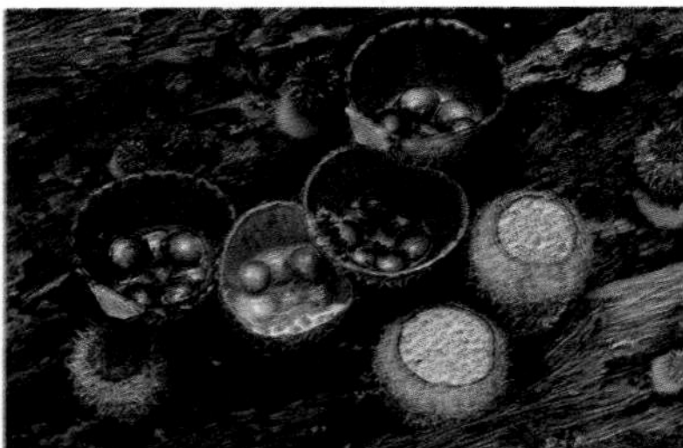

Bird's nest fungi are appropriately named. The tiny cups look like nests complete with eggs.

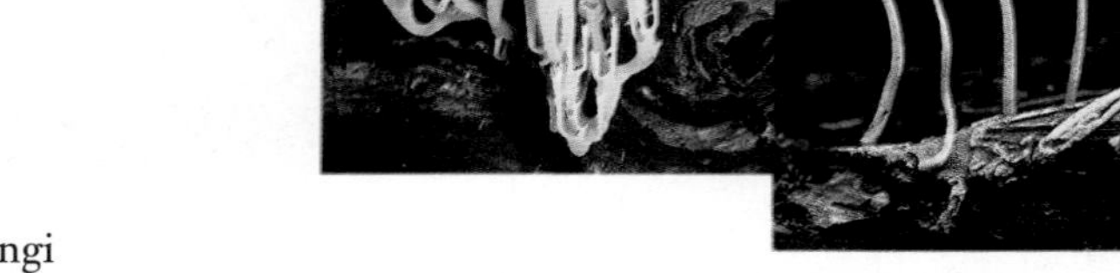

This insect has been attacked and gradually killed by the parasitic fungus that you can see emerging all over its body.

Program Resources

Section Focus Master 49 L1 SAE
Reinforcement and Study Guide, pp. 89-90 L1
Biolab and Minilab Worksheets, p. 91 L1
Laboratory Manual, pp. 127-128 L2
Concept Mapping, p. 23 L1
Tech Prep Applications, pp. 27-28 L2

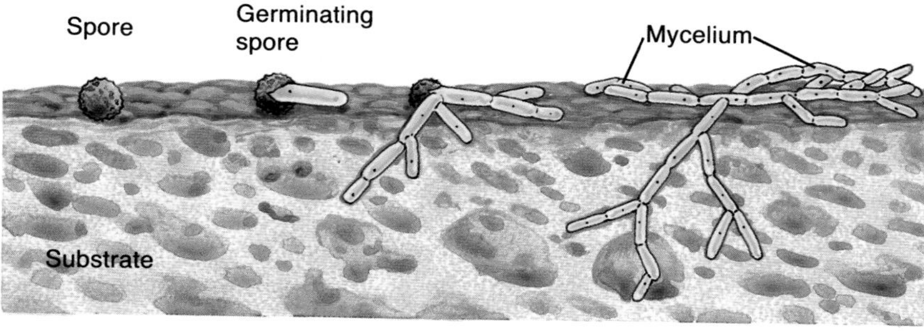

Figure 23.2
Hyphae are the basic structural units of fungi that grow to form mycelia—complex masses of filaments. Some hyphae in a mycelium are adapted to anchor, invade, or produce reproductive structures.

Fungi used to be classified as members of the plant kingdom. This early classification was based on the fact that, like plants, many fungi grow anchored in soil and have cell walls. However, as biologists came to learn more about fungal structure and how fungi live, they realized that fungi make up a distinct kingdom.

The structure of fungi

A few unicellular types of fungi exist, such as yeasts, but most fungi are multicellular. As you can see in *Figure 23.2,* the basic structural units of a multicellular fungus are threadlike filaments called **hyphae,** which develop from fungal spores. Hyphae elongate at their tips and branch extensively to form a network of filaments called a **mycelium.** With a magnifying glass, you can see individual hyphal filaments in molds that grow on bread. However, the hyphae that make up the mushrooms growing in your yard are much more difficult to see. That's because hyphae in mushrooms are tightly packed to form a dense mass.

Unlike plants, which have cell walls made of cellulose, the cell walls of most fungi contain a complex carbohydrate called **chitin.** As you will see in Chapter 31, chitin is also found in the external skeletons of some animals such as lobsters, crabs, insects, and spiders.

Inside hyphae

In many types of fungi, hyphae are divided by cross walls, called septa, into individual cells that contain one or more nuclei, *Figure 23.3.* Septa usually have holes, or pores, in them. Through these pores, cytoplasm and organelles can flow from one cell to the next. One advantage of this free-flowing cytoplasm is that it helps move nutrients from one part of a fungus to another.

Some fungi have hyphae with no septa. When you look at these hyphae under a microscope, you can see hundreds of nuclei streaming along in an undivided mass of cytoplasm.

Figure 23.3
Many types of fungi have hyphae that are divided into cells by septa (top). Hyphae without such cross walls look like giant, multinucleate cells (bottom).

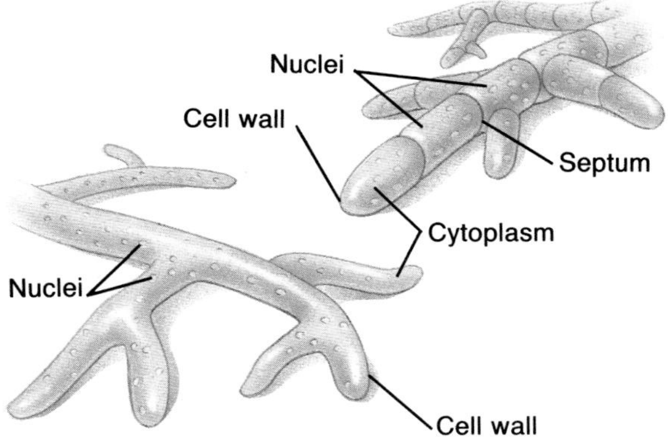

SECTION 23.1

Activity

Interpersonal Have students assemble in small cooperative groups. Provide each group with a mushroom purchased from a supermarket. Ask the groups to record as many observations as they can about this fungus. After four or five minutes of observing and recording, have student groups share their lists with their classmates. COOP LEARN

2 Teach

Discussion Question

Early classification schemes placed fungi in the Plant Kingdom. Ask students to describe a major trait or characteristic that fungi do not share with plants. *Answers should focus on fungi lacking chlorophyll. Thus, fungi are not able to make their own food through photosynthesis.*

Making a Model

Visual-Spatial Obtain several pieces of string of about the same length. Hold up one piece of string and tell students it represents a hypha. Twist the remaining pieces of string together to form one unit. Explain to students that this unit represents a mycelium, which is made of several hyphae put together.

Microscope Activity

Visual-Spatial A sample of bread mold (or any mold growing on food) can be used to illustrate the appearance of hyphae. Have students prepare wet mounts of the mold and view it under low- and then high-power magnification. L1

GLENCOE TECHNOLOGY

Videodisc

Biology: The Dynamics of Life
Disc 1, Side 2
Fungal Decay (Ch. 15)

Meeting Individual Needs

Learning Disabled Have students prepare a table that compares and contrasts fungi and plants, based on the information in the text. Have students include the following information in their tables: Cellular Organization, Method of Obtaining Food, Material of Cell Wall, Nutrient Obtaining Structures, and Habitat. L1

A BROADER VIEW

The Fungus that Ate Michigan

Purpose

Students will discover the extent to which a single fungus can grow. They will also see how DNA fingerprinting may be used to solve biological problems.

Teaching Strategies

- Make sure that students understand what is meant by testing or analyzing DNA fragments. Advise them that this is the same as performing DNA fingerprinting on a sample.
- Provide a visual comparison as to what 100 metric tons of fungus would be comparable to. For example, an elephant weighs close to 5.5 metric tons. Therefore, the fungus has a mass equal to about 18 elephants.

Thinking Critically

Answers will vary. "One individual" supporters may claim that because all cellular material is the same, it is all one individual. "Multiple individual" supporters will claim that the individuals germinated at different times and in separate places, making them multiple organisms.

Audiovisual

Show the filmstrip *The Five Kingdoms of Life—The Fungi*, Carolina.

A BROADER VIEW

The Fungus that Ate Michigan

It is huge—truly a humongous fungus. This single individual weaves through 38 acres of forest floor and is estimated to weigh at least 100 tons! Located at the western end of Michigan's Upper Peninsula, an enormous example of *Armillaria bulbosa* is estimated to be at least 1500 years old. It is surrounded by neighbors of the same species.

The fungus consists of an underground network of long, connecting bundles of mycelia that give rise to occasional mushrooms. The mushrooms produce and spread spores. Thus, the bulk of the organism is hidden in the soil and, except for the reproductive structures of the mushrooms, it is seldom noticed.

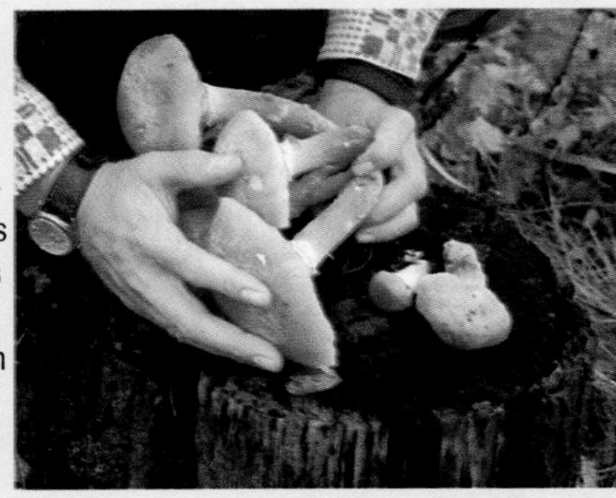

Identity check If the fungus is buried and the mycelia can't be seen, how do we know this is one individual? To answer this question, researchers at the University of Toronto and Michigan Technological University conducted several experiments. In one experiment, they gathered many specimens from a 75-acre study area, and then analyzed the DNA of 16 genes from each specimen. They found that the DNA in all the specimens from a 38 acre area was identical, but different from surrounding specimens of *A. bulbosa.* This was strong evidence that these specimens were from the same individual.

Even bigger? The Michigan fungus is certainly not the largest of its kind. Two forest pathologists from the state of Washington may have identified an even larger *Armillaria.* About 20 years ago, these scientists found one they believed traversed 1500 acres. They did not perform DNA tests to confirm that it is one individual. Instead, they depended on observations of the organism. One property of this fungus is that the hyphae of an individual will not intermingle with the hyphae of other individuals of the same species. The mycelia of this Washington fungus seemed to be well intermingled, leading the researchers to conclude that all the mycelia were from the same fungus.

Thinking Critically

How can the Michigan fungus and the Washington fungus be said to be over a thousand years old, when we know fungus cells live for only a few years at most? Defend your answer.

In thinking about fungi, it's usually the negative things that come to mind: spoiled food, plant and animal diseases, and poisonous mushrooms. But fungi have an important role in the interactions of organisms on Earth. Fungi decompose a large amount of the world's waste material.

The great decomposers

Fungi perform an indispensable function in their role as decomposers. Imagine a world without fungi. Huge mounds of wastes, dead organisms, and debris would be everywhere. Because of fungi, the leaves that carpet a forest floor each fall are mostly gone in the spring. Slowly but surely, fallen leaves, animal carcasses—whatever dies or becomes waste—is decomposed, *Figure 23.4.* Fungi, along with several species of bacteria and protists, are constantly at work transforming complex organic substances into the raw materials that other organisms can use.

Figure 23.4

These patches of lush, green grass are growing where pads of horse dung have been decomposed by fungi. By breaking down animal wastes and other organic matter, fungi make essential nutrients such as nitrogen available for plants.

Alternate Lab | **Mold Growth**

Purpose 

LS Interpersonal To demonstrate that mold will grow on any moist, organic material.

Materials

small paper cups, macaroni, aluminum foil, water, cardboard, oatmeal flakes, swab with mold spores (grow mold source one week before needed)

Procedure

Give students the following directions.

1. Number eight small paper cups 1–8. Then fill each cup about 1/8 full of the following: cup #1 chunk of cardboard, #2 well-moistened chunk of cardboard, #3 dry oatmeal flakes, #4 well-moistened oatmeal flakes, #5 macaroni, #6 well-moistened macaroni, #7 ball of aluminum foil, #8 ball of wet aluminum foil.
2. Obtain a cotton swab with mold spores. Rub the swab over the surface of the contents of all cups.
3. Cover each cup with aluminum foil and label each with your name and the date.

How Fungi Get Their Food

Fungi cannot produce their own food as do photosynthesizing protists or plants. A fungus depends on other sources for its supply of energy.

Extracellular digestion

Fungi are heterotrophs that obtain nutrients through a process called extracellular digestion; that is, food is digested *outside* a fungus's cells. The products of digestion are then absorbed. Take a fungus-covered orange as an example. As hyphae grow over and into the cells of the orange, they release powerful digestive enzymes. These enzymes break down the large organic molecules in the orange into smaller molecules. These small molecules diffuse into the fungus, where they are used to synthesize materials for growth and repair. The more extensive a mycelium becomes, the more surface area there is through which nutrients can be absorbed.

Different feeding relationships

Different kinds of fungi use different types of food sources. A fungus may be a decomposer, a parasite, or a mutualist that lives symbiotically with another organism, *Figure 23.5.* The most specialized parasitic fungi absorb nutrients from the living cells of their hosts. They do this using specialized hyphae called **haustoria,** which penetrate and grow into host cells without killing them.

Many fungi live in a symbiotic relationship with plant roots. These mutualistic fungi absorb nutrients from living hosts, but they also benefit their hosts in some way. For example, a fungus in a mutualistic relationship might help its host retain water, or obtain minerals from the soil in exchange for organic food.

haustoria:
haurire (L) to drink
Haustoria are hyphae that invade the cells of a host to absorb nutrients.

Figure 23.5
Feeding relationships

What you see along this dead tree branch are the visible parts of a turkey-tail fungus *(Trametes versicolor)*. What you don't see deep within the tissue of the branch is the extensive mycelium of this fungus.

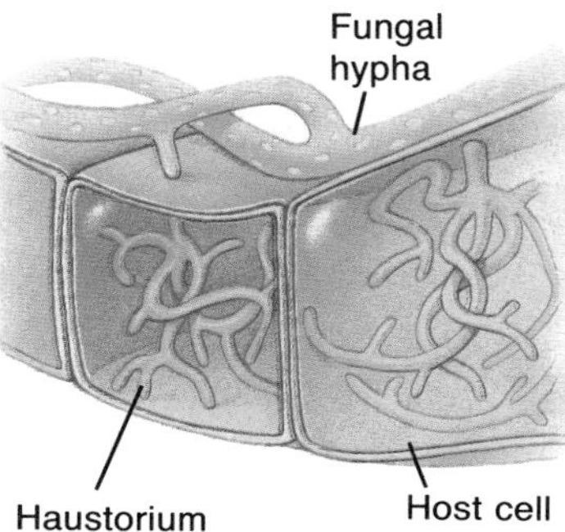

Nutrients absorbed across the walls of a haustorium are transported to the rest of the fungus.

Organisms attacked by parasitic fungi, such as the American elm tree below, are harmed by the relationship and may eventually die.

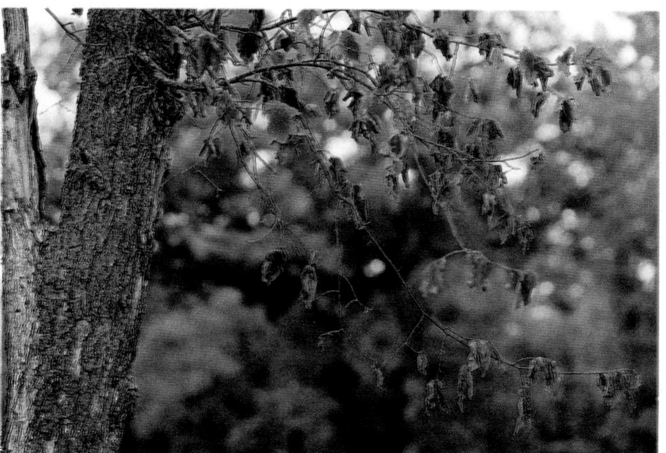

Health Connection

Aspergillus flavus

A species of fungus called *Aspergillus flavus* commonly invades stored grains of soybean, wheat, corn, rice, barley, bran, and peanuts. It produces a chemical called aflatoxin, which is a known carcinogen and can also destroy the liver if it is ingested.

What are the chances that a cereal or peanut butter is contaminated with this poison? It's not likely. The FDA and USDA monitor for this chemical. Also, grain foods and peanuts are usually roasted or heated—a process that reduces the level of aflatoxin present.

The greatest cause of aflatoxin poisoning is peanuts still in the shell. When in doubt, toss away any whose shells look moldy.

Using Science Terms

LS **Linguistic** Have students compare and contrast the terms *extracellular* and *intracellular*. Ask them to provide an example of intracellular digestion.

4. Place the cups in the designated location. Check for mold growth each day for the next week.
5. Design a data table to record your observations. Indicate the extent of mold growth in your table.

Analysis

Ask students to answer the following questions.

1. What evidence suggests that mold feeds on organic matter? *All materials that showed mold growth were organic. Aluminum foil, the only inorganic material, showed no growth.*
2. Explain the role of the cups without water. *The cups were controls. The controls show that mold needs water to grow.*

Assessment

Knowledge: Have students predict if the following will or will not support mold growth and explain why: dry cereal, penny, wet penny, wet cracker, cooked and uncooked rice. L1

MiniLab

Purpose

Visual-Spatial To show that mold spores are present in the air.

Process Skills

observe and infer, experiment

Teaching Strategies

- Fresh-baked bakery bread is not likely to have preservatives. If this type of bread is not available, commercial bread from store shelves will show mold growth, in time, in spite of preservatives having been added.
- Make photocopies of the ingredients label for any commercial bread. Point out that calcium propionate is a mold inhibitor and ask students to look for this ingredient. Have students determine if this type of bread is less likely to mold than bread without mold inhibitors.
- Mold growth can be sped by placing bags in a warm area of the classroom or in an incubator set at 30°C.

Expected Results

Only the bread slices with water added will show mold growth after 4–6 days.

Analysis

1. Only moistened bread shows mold growth.
2. Hyphae and spores may be seen.
3. The source of mold spores is the air in the classroom. Molds require water and food for growth.

Assessment

Performance: Have students design an experimental plan to show that it was the air that contained the mold spores that grew on the moistened bread. L1

MiniLab

Are there mold spores in your classroom?

Any mold spore that lands in a favorable place can germinate, produce hyphae, and begin developing into a mycelium. Can you demonstrate that there are airborne bread mold spores in your classroom by exposing bread to the air?

Procedure

1. Place two slices of freshly baked bakery bread on a plate. Sprinkle some water on one slice to make its surface moist. Leave both slices uncovered for several hours.
2. Sprinkle a little more water on the moistened slice, and place it in a plastic, self-sealing bag. Place the dry slice of bread in another self-sealing plastic bag. Capture a little air into each bag so that the plastic does not lay against the bread's surface, and then seal the bags.
3. After four or five days, take out the bags and look for mold growing on the bread slices.
4. Remove a small piece of bread mold with a forceps, place it on a slide in a drop of water, and add a coverslip. Observe the mold under a microscope, first on low power and then on high power.

Analysis

1. Did you observe mold growth on the moistened bread? On the dry bread?
2. Describe what you saw on the slide that you made.
3. How does this experiment demonstrate that there are mold spores in your classroom? What conclusions can you draw about conditions required for the growth of bread molds?

Reproduction in Fungi

Depending on the species and on environmental conditions, a fungus may reproduce asexually or sexually. Fungi reproduce asexually by fragmentation, budding, or by producing spores.

sporangium:
sporos (GK) seed
anggeion (GK) vessel
A sporangium produces spores.

Fragments and buds

In fragmentation, pieces of hyphae that are broken off or torn away from a mycelium are capable of growing into new mycelia. Suppose you are preparing your garden to plant seeds. Every time your shovel cuts down into the dirt, it slices through mycelia. Most of those hyphal fragments can grow and branch to form new mycelia.

Budding is a form of asexual reproduction in which mitosis takes place and a new individual—such as the one shown in *Figure 23.6*—grows out and eventually separates from a parent cell.

Reproducing by spores

As you learned in Chapter 22, a spore is a reproductive cell that germinates and develops into a new organism. Most fungi produce spores. When a fungal spore lands in a place that has all the conditions necessary for growth, a threadlike hypha emerges and begins to grow and branch to form a mycelium.

Eventually, some hyphae may grow upward from the substrate and produce a spore-containing structure called a sporangium. A **sporangium** is a sac or case in which spores are produced. Mushroom caps contain

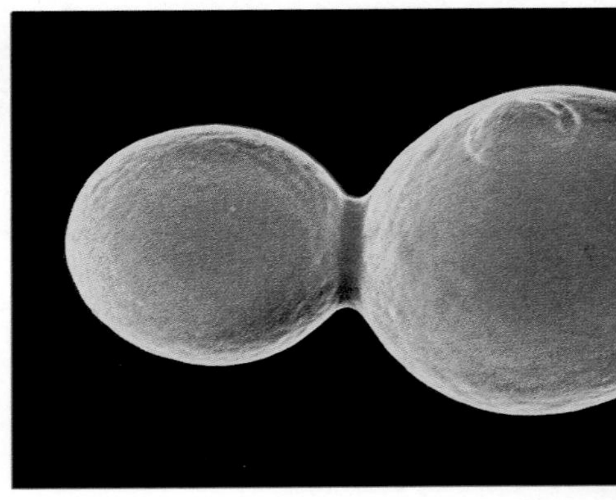

Figure 23.6

Most yeasts reproduce asexually by budding. Wherever a bud pinches off from its parent cell, a tiny bud scar is left behind. Can you spot the bud scar on the cell on the right in this photograph?

1. Materials such as food are easily and quickly transported through the entire fungus body.
2. Hyphae grow into a food source. Enzymes are released from hyphae into the food source and digestion occurs outside the fungus. Digested foods are then absorbed into the hyphae.

one type of sporangium. The tiny black spots you can see on bread mold are another type of sporangium. In fact, sporangia are often the only part of a fungus you can see, and they usually represent only a small portion of the total organism.

Fungi may produce spores by either mitosis or meiosis. The way in which spores are formed during sexual reproduction is the basis on which fungi are classified into their major groups.

Adaptations for Survival

Fungi have evolved a variety of strategies that help them adjust to changing conditions and exploit new environments and food sources. Many of these survival adaptations involve sporangia and spores.

Sporangia protect spores and, in some cases, keep them from drying out until they are ready to be released.

Spores and spore dispersal

Fungal spores are small and extremely lightweight. When released from sporangia, spores are easily swept along by the wind. Look at the puffball releasing its spores in *Figure 23.7*. The slightest breeze may carry such dust-fine spores a great distance. Spores released in one place can be blown many hundreds of kilometers.

Figure 23.7

Wind, water, and animals help disperse fungal spores. A passing animal or falling raindrops will cause some fungi, like these puffballs, to discharge a cloud of spores that will be carried away by the wind.

Most fungi produce a lot of spores—another adaptation for survival. A single puffball that measures 25 cm in circumference produces roughly 1 trillion spores. Producing such huge numbers of spores improves the chances that at least some will survive.

Fungal spores can also be dispersed by water, as well as by animals such as birds and insects. The fungus that causes Dutch elm disease, a usually fatal disease of elm trees, is carried by bark beetles. Spores of the fungus stick to the bodies of young beetles as they emerge from infected trees. When spore-covered beetles move to healthy elms to feed on the bark, they transfer spores in the process.

Section Review

Understanding Concepts

1. In what way are pores in septa an advantage for a fungus with a large mycelium?
2. Explain how a fungus obtains nutrients from the environment.
3. What role do fungi play in food chains?

Thinking Critically

4. Explain why you might expect to find several different types of fungi growing in a bird's nest.

Skill Review

5. **Measuring in SI** Outline the steps you would take to calculate the approximate number of spores contained in a puffball fungus that had a circumference of 10 cm. For more help, refer to Practicing Scientific Methods in the *Skill Handbook*.

Answers to Section Review

3. Fungi are decomposers providing nutrients to autotrophs. They break down and recycle nutrients from dead organic material.

Thinking Critically

4. Fungi can use a variety of organic materials as their food source. Some fungi may feed on bird feathers, bird excrement, or twigs of the nest.

Skill Review

5. Remove a very small section from the puffball. Calculate the number of similar sized sections that remain. Estimate the number of spores in the small section using a sampling technique. Multiply the number of spores in the section by the number of remaining sections.

3 Assess

Check for Understanding

Have students explain how the following are related.

(a) hyphae - mycelium
(b) cell wall - chitin
(c) extracellular digestion - decomposers
(d) reproduction - sporangium
(e) spores - budding L1

Reteach

Have students prepare an outline to describe fungi that includes the following: characteristics, structures, nutrition, fungi reproduction. L1

Extension

Yeast cells are aerobic at certain times and anaerobic at other times. Have students review these two processes from Chapter 10. Then have them compare and contrast both processes. L1

Assessment

Performance: Provide students with a simple diagram showing a fungus with its hyphae embedded into a slice of bread. The diagram should also show a sporangium with spores. Have students label the following: site of reproduction, site of extracellular digestion, nutrient site, site of enzyme production for digestion. L1

4 Close

Activity

Have each student list five words or phrases about fungi. Four of the words or phrases should be related while one is not. Have them exchange lists with their classmates and identify the unrelated term. Ask them to explain how the remaining four are related. SAE

SECTION

23.2

Prepare

Key Concepts

The Kingdom Fungi is divided into four phyla. The first is Zygomycota, which reproduce asexually by forming spores and sexually through zygospores. The second is the phylum Ascomycota, which reproduce asexually from spores and sexually from ascospores. The third is the phylum Basidiomycota. They carry out sexual reproduction by forming basidiospores. The last group is the phylum Deuteromycotes. These fungi reproduce only asexually. The section ends with a discussion of mutualistic associations involving lichens and mycorrhizae.

Block Scheduling

Look for this symbol for strategies that are useful in a block scheduling format. For more information on block scheduling, refer to Section 23.2 in the **Lesson Plans** booklet.

Materials

- Purchase grocery store mushrooms for the Minilab.
- Purchase yeast and prepare a BTB solution for the Biolab. Gather beakers, plastic tubing, test tubes, and stoppers. Prepare the short glass tubes needed to fit each stopper.

SECTION

23.2 The Diversity of Fungi

Section Preview

Objectives

Identify the four major phyla of fungi.

Distinguish among the ways spores are produced in zygomycotes, ascomycotes, and basidiomycotes.

Summarize the ecological roles of lichens and mycorrhizae.

Key Terms

stolon
rhizoid
zygospore
gametangium
ascus
ascospore
conidiophore
conidium
basidium
basidiospore
mycorrhiza
lichen

How are fungi classified? If you took one of the mushrooms that had so abruptly appeared in your yard to a mycologist, he or she would look for certain key features in that mushroom to help identify it. In particular, the mycologist would look at the spores it contained. Fungi are classified into phyla according to the way in which spores are produced. As you'll see, the name of each phylum is derived from the types of sexual structures that are characteristic of each group.

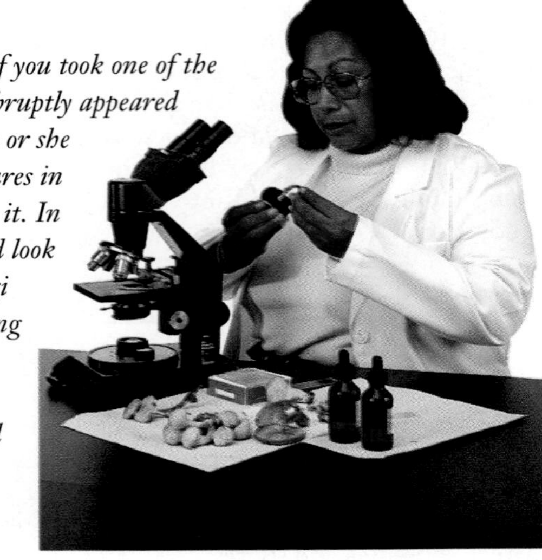

Zygospore-Forming Fungi

Have you ever pulled out the last slice of bread in a bag, only to find that it was covered with rather unappetizing black spots and a bit of fuzz? If so, then you have had a close encounter with *Rhizopus stolonifer*, the common black bread mold. *Rhizopus* is probably the most familiar member of the phylum Zygomycota. Many of the 1500 or so species that belong to this phylum are decomposers. Zygomycotes reproduce asexually by producing spores, and sexually by forming thick-walled spores. Hyphae of zygomycotes do not have septa.

Growth and asexual reproduction

When a *Rhizopus* spore settles on a piece of bread, it germinates and hyphae begin to grow. Some hyphae called **stolons** grow horizontally along the surface of a food source, in this case the bread, and rapidly produce a mycelium. Other hyphae form **rhizoids** that penetrate down into the food and anchor the mycelium to its substrate. They also are the site of most extracellular digestion and nutrient absorption.

Asexual reproduction begins when certain hyphae grow upward and develop sporangia at their tips, *Figure 23.8*. These sporangia mature to form black, rounded sporangia loaded with asexual spores. When each sporangium splits open, hundreds of spores are released into the air. Those that land on a moist food supply germinate, form new hyphae, and start the asexual cycle again.

Producing zygospores

Suppose that your piece of bread fell behind the kitchen garbage can and began to dry out. This change in environmental conditions could trig-

1 Focus

Bellringer

Before presenting the lesson, display **Section Focus Master 50** on the overhead projector and have students answer the accompanying questions. L1 SAE

Meeting Individual Needs

Students Acquiring English Have students prepare a table that lists the three different ways in which fungi reproduce asexually. Ask students to include a brief description of each process, a diagram showing what it could look like, and an example of a fungus that carries it out. L1 SAE LS

Learning Disabled Have students prepare a concept map that uses the following structures and functions: stolons, sporangia, rhizoids, absorb nutrients, form spores, produce mycelia. The concept map can begin with the statement *hyphae names and functions.* L1 LS

Figure 23.8

The life cycle of the black bread mold, *Rhizopus stolonifera* **Asexual reproduction is most common, but sexual reproduction takes place when different mating types (+ and -) come together and form zygospores.**

In this photomicrograph, you can see dark zygospores forming where gametangia have come together and fused during sexual reproduction.

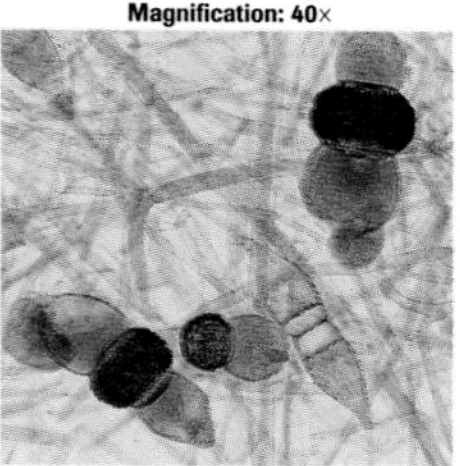

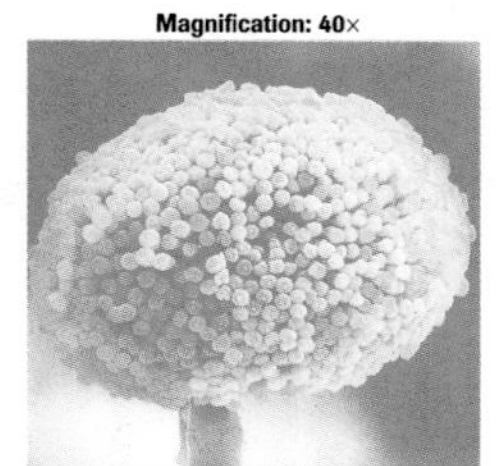

This scanning electron micrograph shows a *Rhizopus* sporangium covered with thousands of haploid spores.

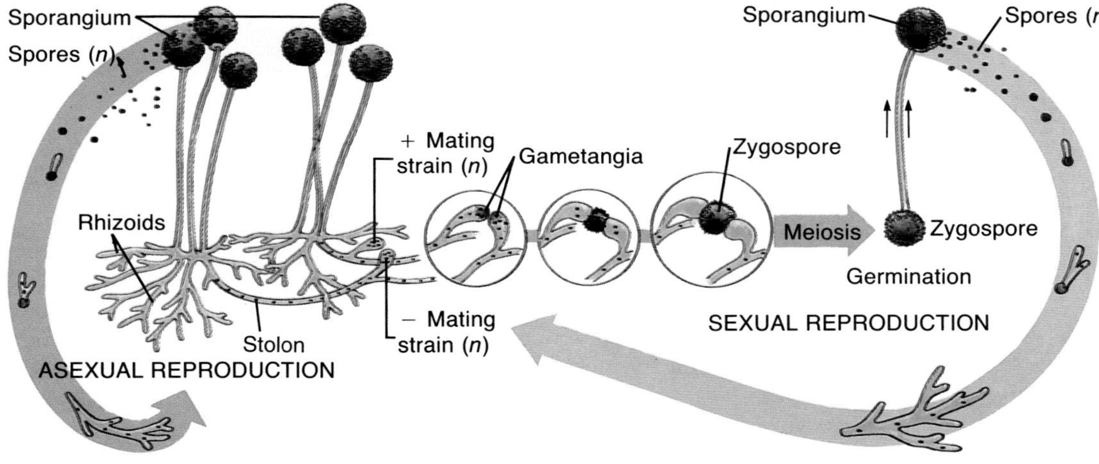

ger the fungus to reproduce sexually and produce **zygospores**—thick-walled spores that are adapted to withstand unfavorable conditions.

Look again at *Figure 23.8.* Sexual reproduction in *Rhizopus* takes place when the tips of hyphae from two different mycelia come together and fuse. Where the hyphae fuse, two gametangia form, each with a haploid nucleus inside. A **gametangium** is a structure with a haploid nucleus and in which gametes are produced. When the contents of the two gametangia fuse, a diploid zygote forms. This diploid cell then develops into a thick-walled zygospore.

A zygospore may lie dormant for many months and can survive periods of drought, cold, and heat. When conditions improve, the zygospore absorbs water, undergoes meiosis, and germinates to produce an upright hypha with a sporangium. Each haploid spore formed in this sporangium is capable of growing into a new mycelium.

Sac Fungi

The phylum Ascomycota is the largest group of fungi, with about 30 000 species. The ascomycotes are also known as sac fungi. Both names refer to the little saclike structures, each known as an **ascus,** in which the sexual spores of these fungi develop.

Discussion

Explain that mushrooms are often grown in chambers or rooms that are below ground and dark. Point out that often they are grown in manure instead of soil. Elicit why mushrooms are grown in these conditions. *Mushrooms grow best in warm, dark, moist areas. The manure provides an organic food source for the mushrooms.*

2 Teach

Reinforcement

Explain that hyphae are called by different names depending on their specialization. Examples are rhizoids, stolons, and sporangia. Have students describe the job of each of these hyphae. Have them also define the term *mycelium* as a fourth term that refers to hyphae.

Different Viewpoints in Biology

Fungal Classification

Biologists do not all agree about the classification of fungi. Some divide the kingdom into only two phyla, Mastigomycota and Amastigomycota. The Mastigomycota includes water molds that are sometimes considered Protists. All members of this phylum form swimming zoospores and are typically water dwellers. The members of the Amastigomycota have no zoospores and are further divided into four subphyla, which are the same as the four phylum groups described in this section.

interNET CONNECTION

Follow the link for this chapter on the Glencoe Homepage at **www.glencoe.com/sec/science** to find out more about fungi.

Using Science Terms

LS **Linguistic** Have students find the meanings of the terms *sporangium, gametangium, ascospores,* and *zygospore.* Ask students to use their findings to explain why the name of each structure is well suited to its function or origin. L1

Microscope Activity

LS **Visual-Spatial** Purchase preserved specimens of *Peziza* from a biological supply house. Have students remove a fingernail-sized portion of the fungus and prepare a wet mount slide. They should squash the mount with either their thumbs or pencil erasers and observe the specimen under both low- and high-power magnification. Ask students to sketch what they see. *Students should see perfect asci filled with exactly eight ascospores.* L2

Science, Technology, and Society

Using Fungi in Surgical Procedures

Heart, kidney, and liver transplant operations owe much of their success to the chemical cyclosporine, which is formed from a soil fungus called *Tolypocladium inflatum.* This chemical suppresses the immunity that causes organ rejection. Before the discovery of cyclosporine in 1979, transplant patients who survived surgery often died soon afterwards when their bodies rejected the transplanted organ. Today, more than 90 percent of transplant patients survive the rejection process thanks to this fungal product.

Figure 23.9

Many ascomycotes are cup shaped or have multiple cup-shaped indentations. Asci line the inside of these cup-shaped surfaces.

The scarlet cups of an ascomycote grow on wood.

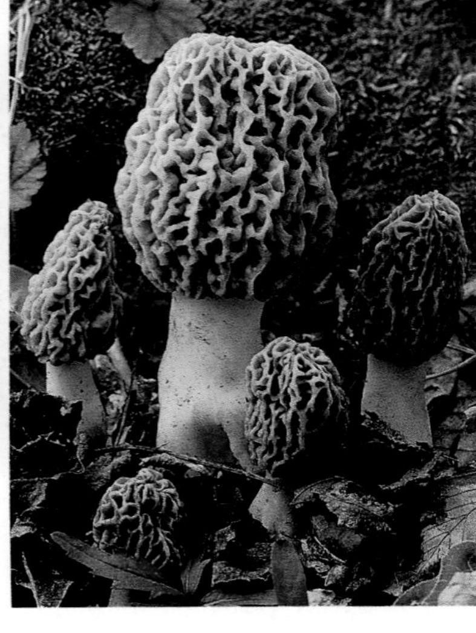

Morels are often served as a delicacy.

Because they are produced inside an ascus, the spores are called **ascospores.** Ascomycotes can take many different forms, *Figure 23.9.*

During asexual reproduction, ascomycotes produce a different kind of spore. Hyphae rise up from the mycelium and elongate to form **conidiophores,** *Figure 23.10.* Chains or clusters of asexual spores called **conidia** develop from the tips of conidiophores. Once released, these haploid spores are dispersed by wind, water, and animals.

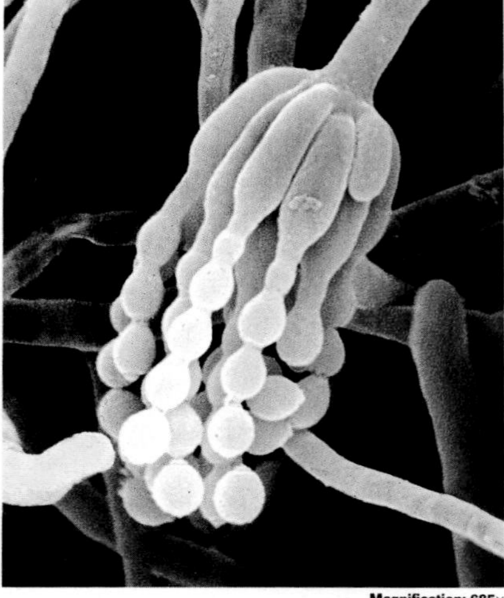

Figure 23.10

Most ascomycotes reproduce asexually by producing conidia. Conidia and conidiophores differ from species to species, making them important clues in identification.

Important ascomycotes

You've probably encountered a few types of sac fungi in your refrigerator or in the grocery store in the form of blue-green, red, and brown molds that decorate decaying foods. Other sac fungi are well-known by farmers and gardeners because they cause plant diseases such as apple scab, Dutch elm disease, and ergot on rye. Quite a few ascomycotes are also the cause of some serious fungal diseases of animals and people.

Not all sac fungi have a bad reputation. Morels, *Figure 23.9,* and truffles are edible members of this phylum. Perhaps the most economically important ascomycotes are the yeasts. Yeasts are unicellular sac fungi that usually do not produce hyphae. Most of the time, yeasts reproduce asexually by budding.

Perhaps the best-known yeasts are those that are used in brewing and baking. Grown anaerobically, these minute cells can ferment sugars to produce carbon dioxide and ethyl alcohol. Some yeasts are used to make wine and beer. Yeasts used in baking produce the carbon dioxide that causes bread dough to rise and take on a light, airy texture.

STUDENT JOURNAL

Important Ascomycotes Have students construct a table for their journals that lists fungi of the phylum Ascomycota that are important to people. Have students identify the columns of their tables as Helpful and Harmful. Encourage students to explain why each fungus listed is either helpful or harmful. L1 LS

GLENCOE TECHNOLOGY

Videodisc

STVS: Plants & Simple Organisms

Disc 4, Side 1

Fungal Collection for Research (Ch. 13)

Yeasts are also important tools for research in genetics because they have large chromosomes. A vaccine for the disease hepatitis B is produced by splicing human genes with those of yeast cells. Because yeasts can be produced quickly, they are an important source of the vaccine.

Club Fungi

Of all the different kinds of fungi, some of the 25 000 species in the phylum Basidiomycota are probably most familiar to you. Mushrooms, puffballs, stinkhorns, bird's nest fungi, and bracket fungi are all basidiomycotes. So are the rust and smut fungi, which cause billions of dollars worth of damage to grain crops every year.

Basidia and basidiospores

The spore-producing cells of basidiomycotes are club-shaped hyphae called **basidia.** It's from these basidia that the members of this division get their more common, general name—the club fungi. Basidia usually develop on a short-lived reproductive structure, which in the case of mushrooms is the mushroom cap and stalk. During sexual reproduction, basidia produce spores called **basidiospores,** *Figure 23.11.*

The reproductive cycle of a typical basidiomycote, such as a mushroom, is complex. You see evidence of this reproductive process only when mushrooms appear on lawns and forest floors. Read *The Inside Story* for an in-depth look at the structure and life of a mushroom.

Figure 23.11

Most basidiomycotes produce haploid basidiospores on club-shaped basidia. Typically, four basidiospores develop at the end of each basidium, as you can see here.

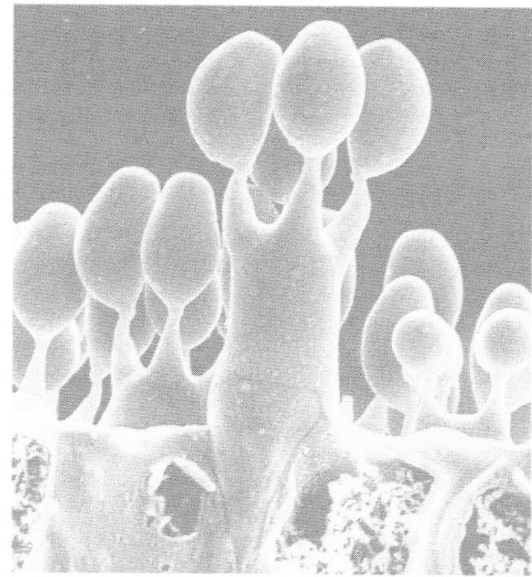

Magnification: 775×

MiniLab

How are gills arranged in a mushroom?

Even experts sometimes find it difficult to identify mushroom species. Spore prints can often help in mushroom identification by clearly revealing the organizational pattern of a mushroom's gills and the color of its spores. Can you demonstrate how gills are arranged using the spore-print technique and an ordinary grocery store mushroom?

Procedure

1. Allow several edible, grocery-store mushrooms to age for a few days, keeping them covered in a paper bag.
2. When the undersides of your mushrooms are very dark brown, carefully break off the stalks and set the caps, gill side down, on a white sheet of paper. Be sure that the gills are touching the surface of the paper.
3. Leave the caps in place overnight on a flat surface where they won't be disturbed.
4. The following day, carefully lift the caps from the paper and observe the results.

Analysis

1. What did you see on the paper? What color do the spores of these mushrooms appear to be?
2. How does the pattern of spores on the paper compare with the arrangement of gills on the underside of the mushroom cap that produced it?

MiniLab

Purpose

LS **Visual-Spatial** To have students use spore prints to infer the shape or arrangement of mushroom gills.

Process Skills

observe and infer, experiment

Teaching Strategies

- It is critical that mushrooms not be disturbed by other students. Find an area in the room where air currents also will not disturb the spore patterns.
- After students have made their spore prints, have them make wet mounts of the spores and examine them under low- and high-power magnification. Have students sketch their observations.

Expected Results

Spores will have formed a pattern on the paper that corresponds to the location of gills.

Analysis

1. Spores; brown
2. Spores are formed toward the edges of the gills. The pattern of spores follows the outline or pattern of the gills.

Assessment

Skill: Provide students with spore patterns from different mushroom types. Have students reconstruct the appearance or arrangement of gills responsible for the spore patterns. L1

Program Resources

Section Focus Master 50 L1 SAE
Reinforcement and Study Guide, pp. 91-92 L1
Biolab and Minilab Worksheets, pp. 92-94 L1
Laboratory Manual, pp. 129-130, 131-132 L2
Critical Thinking/Problem Solving, p. 23 L3
Basic Concepts Transparencies 31, 32 and **Masters** L1 SAE
Reteaching Transparency 23 and **Master** L1 SAE

THE INSIDE STORY

The Life of a Mushroom

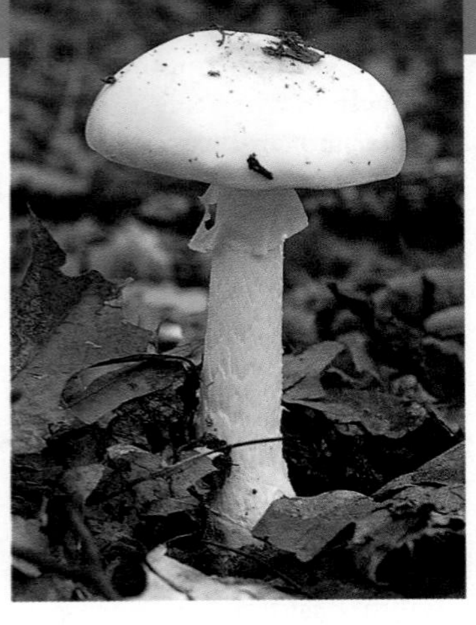

What we call a mushroom is just the above-ground sexual structure of the fungus. A single mushroom can contain hundreds of thousands of spores, which are produced as a result of sexual reproduction. Most types of mushrooms have no asexual reproductive stages in their life cycle.

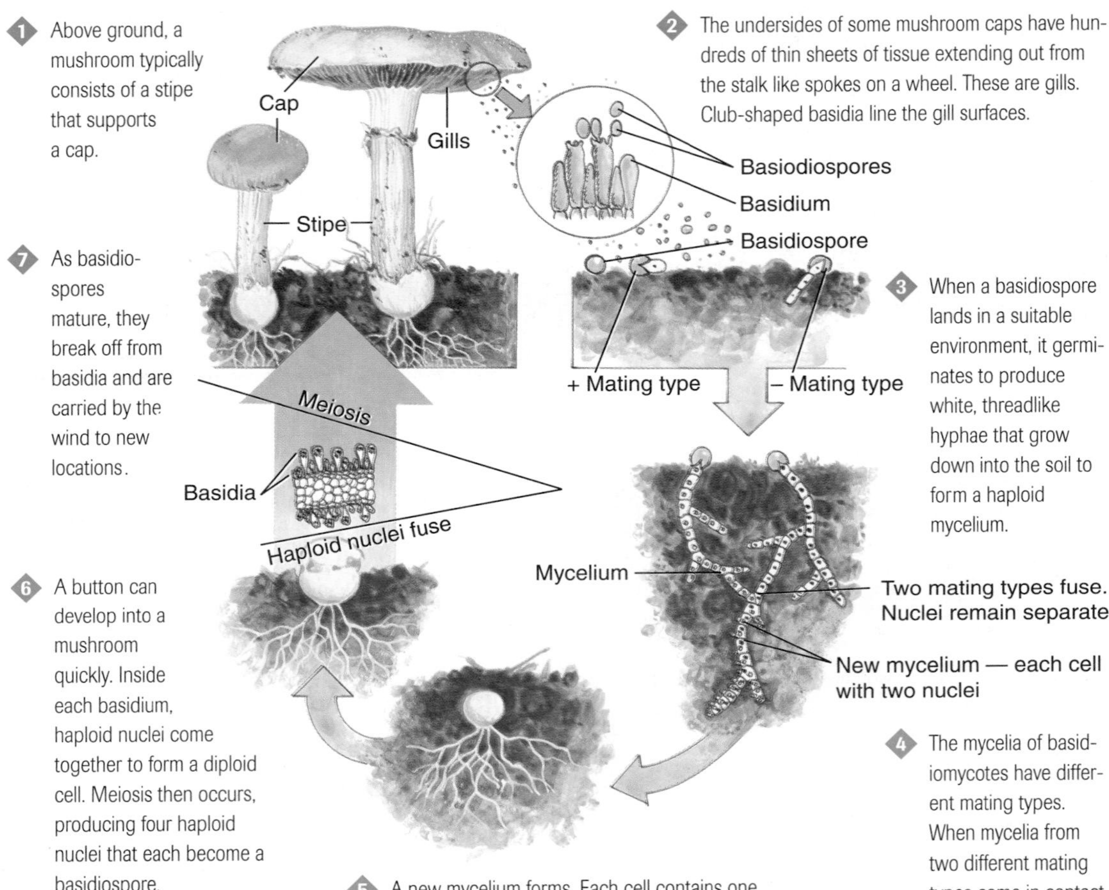

1 Above ground, a mushroom typically consists of a stipe that supports a cap.

2 The undersides of some mushroom caps have hundreds of thin sheets of tissue extending out from the stalk like spokes on a wheel. These are gills. Club-shaped basidia line the gill surfaces.

3 When a basidiospore lands in a suitable environment, it germinates to produce white, threadlike hyphae that grow down into the soil to form a haploid mycelium.

4 The mycelia of basidiomycotes have different mating types. When mycelia from two different mating types come in contact, the hyphae fuse.

5 A new mycelium forms. Each cell contains one haploid nucleus from each of the two mating strains. Eventually, compact masses of hyphae, called buttons, are formed just under the soil's surface.

6 A button can develop into a mushroom quickly. Inside each basidium, haploid nuclei come together to form a diploid cell. Meiosis then occurs, producing four haploid nuclei that each become a basidiospore.

7 As basidiospores mature, they break off from basidia and are carried by the wind to new locations.

THE INSIDE STORY

The Life of a Mushroom

Purpose

IS **Visual-Spatial** Students will follow the steps in the life cycle of a typical mushroom or fungus from the Phylum Basidiomycetes.

Teaching Strategies

- Remind students that spores are the result of sexual reproduction.

Visual Learning

- Point out that in Step 4, the nuclei from each original + and - hypha are in the same cell but have not fused. Ask if the cells are haploid, diploid, or haploid + haploid in chromosome number. *haploid + haploid*
- Ask students to use the diagram to explain when fertilization occurs. *Just prior to spore formation in the basidium*
- Have students explain the adaptive value to mushrooms of producing so many basidiospores from just a single mushroom. *The chance of a basidiospore landing on a suitable surface that will provide proper conditions for growth is slight. Thus, the large number of spores ensures the continued life cycle of this fungus.*

GLENCOE TECHNOLOGY

Videodisc

Biology: The Dynamics of Life
Disc 1, Side 2
Life Cycle of a Mushroom
(Ch. 16)

Meeting Individual Needs

Learning Disabled Have students prepare a flowchart that shows the mushroom life cycle. Have students use *The Inside Story* as a guide. L1 IS

PROJECT

Mushroom Farming

Mushroom farm kits are available from biological supply houses. Purchase such a kit and have students follow directions for growing these organisms. A log or record is to be kept, which keeps track of observed changes that occur over time. L1 IS

Figure 23.12

Many mushrooms are edible. Many more are parasites or very poisonous.

Smuts are parasitic club fungi that attack plants such as corn.

These mushrooms belong to the genus *Amanita*. Some types of *Amanita* mushrooms are extremely poisonous; eating just one can be fatal.

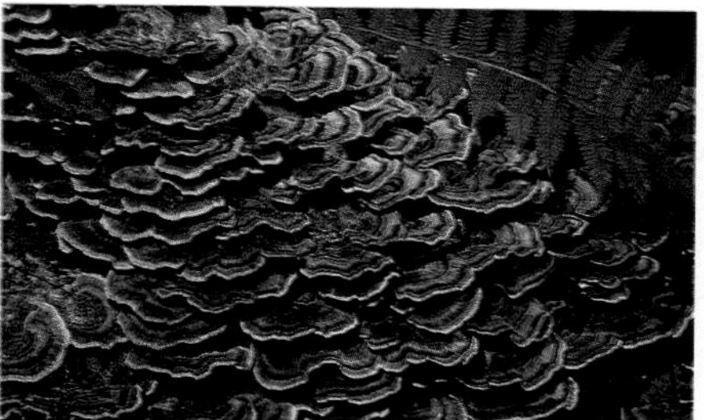

Shelf and bracket fungi are basidiomycotes often seen growing on tree branches and fallen logs.

The visible reproductive structures of members of the phylum Basidiomycota come in a variety of different shapes and sizes, as *Figure 23.12* shows.

The Deuteromycotes

Zygomycotes, ascomycotes, and basidiomycotes reproduce both asexually and sexually. About 25 000 species of fungi, the deuteromycotes, have no known sexual stage in their life cycle. These fungi reproduce only asexually or their sexual phase has not yet been observed by mycologists.

Diverse deuteromycotes

If you've ever had strep throat, pneumonia, or some other kind of infection, your doctor probably prescribed penicillin as a treatment. Penicillin is an antibiotic produced from a deuteromycote, one that is commonly seen growing on fruit, *Figure 23.13*. Other deuteromycotes are used in making soy sauce and some kinds of blue-veined cheese. Citric acid, which gives jams, jellies, soft drinks, and fruit-flavored candies their tart taste, is produced in huge quantities using deuteromycote fungi.

Figure 23.13

Many deuteromycotes are useful.

The antibiotic penicillin is derived from *Penicillium* mold, shown here on an orange.

Many people enjoy the strong, distinctive flavors of cheeses such as this Roquefort. The blue veins and splotches are patches of conidia.

Discussion

Point out to students that the best advice for avoiding poisonous mushrooms is never to eat any collected from fields. Explain that some species of mushrooms are so difficult to identify that even mycologists are, at times, unsure of the genus and species of a particular mushroom.

Enrichment

Have interested students report to the class on the folklore of mushrooms. A suggested reference is: "Who put the toad in toadstools?" A. Morgan, *New Scientist*, Volume 112, December 86/January 87. L3

Reinforcement

Review briefly the methods by which zygomycotes, ascomycotes, and basidiomycotes reproduce sexually. Emphasize to students that the main feature that distinguishes deuteromycotes from the three other phyla of fungi is that no method of sexual reproduction has been observed in these fungi.

Enrichment

Encourage students to research the discovery of penicillin. Ask students to explain how the discovery of this mold affected the field of medicine. L1

CULTURAL DIVERSITY

Which Fungus Caused Witch Behavior?

Chemicals called *ergotamine* and *ergometrine* are present in a fungus called *ergot*. This fungus grows on rye and has been known to cause severe problems if it accidentally ends up in milled flour. It causes vomiting, diarrhea, lesions on the skin, impairment of mental functions, hysteria, and hallucinations. Ergot poisoning may have been responsible for the bizarre behavior of the women accused of witchcraft in colonial Massachusetts during the Salem witch-hunt episode. Ingestion of ergot causes a condition known as Saint Anthony's fire. This condition is characterized by uncontrolled behavior and redness of the face.

BioLab Design Your Own Experiment

Does temperature affect the metabolic activity of yeast?

Time Allotment

Initial session: Twenty minutes to review procedure and set up equipment; Second session: full class period.

Objectives

Review the objectives with students before they begin the Biolab.

Process Skills

form a hypothesis, experiment, observe and infer, record data, relate cause and effect

PREPARATION

Alternate Materials

- Use a 20% sucrose solution instead of the molasses.
- Use yeast cake (available in supermarkets) rather than "dry" yeast found in packets. It will begin fermentation faster than the packet form.
- Prepare BTB solution as follows: add 0.5 g BTB powder to 500 mL distilled water to make stock solution. Dilute 10 mL of stock in 500 mL distilled water. Use diluted solution for the experiment.
- Pretest diluted BTB solution. If it fails to change from blue to either dark or light green while you are exhaling into it for about 60 seconds, then pH of stock solution may have to be adjusted. Add one or two drops of an acid (concentrated hydrochloric acid) to the stock solution. Mix and retest by exhaling through diluted solution using the straw. Add more acid to stock solution until desired change occurs within 60 seconds. If stock solution turns green, add one or two drops of base (concentrated ammonium hydroxide) and mix.

BioLab Design Your Own Experiment

Does temperature affect the metabolic activity of yeast?

Does temperature affect the rate of carbon dioxide production by yeast? Are there differences in yeast metabolism at cold and warm temperatures?

Look at the experimental setup pictured here. As yeast metabolizes in the stoppered container, the carbon dioxide that is produced is forced out through the bent tube into the open tube, which contains a solution of bromothymol blue (BTB). Carbon dioxide causes chemical reactions that result in a color change in BTB. Differences in the time required for this color change provide an indication of relative rates of carbon dioxide production.

PREPARATION

Problem

Do cold temperatures slow down yeast metabolism? Does warmth speed it up?

Hypotheses

What is your group's hypothesis? One hypothesis might concern the specific effects of cold or heat on yeast metabolism.

Objectives

In this Biolab, you will:

- **Measure** the rate of yeast metabolism using a BTB color change as a rate indicator.
- **Compare** the rates of yeast metabolism at several temperatures.

Possible Materials

bromothymol blue solution (BTB)
straw
small test tubes (4)
large test tubes (3)
one-hole stoppers for large test tubes with glass tube inserts (3)
yeast/molasses mixture
water/molasses mixture
water/yeast mixture
test-tube rack
250-mL beakers (3)
ice cubes
Celsius thermometer
hot plate
50-mL graduated cylinder
glass-marking pencil
10 cm rubber tubing (3)
aluminum foil

Safety Precautions

Be careful in attaching rubber tubing to the glass tube inserts in the stoppers. Avoid touching the top of the hot plate. Wash your hands carefully after cleaning out test tubes at the end of your experiments.

572 Fungi

PLAN THE EXPERIMENT

Teaching Strategies

- Carbon dioxide in the presence of water results in the formation of carbonic acid. BTB is an acid indicator.
- Students may wish to measure the amount of gas collected in each tube over a given period of time rather than rely on BTB color change. An inverted tube filled with water and placed in the beaker of water with the rubber tube extending into the mouth of the inverted tube will allow for the collection of carbon dioxide gas through water displacement.

Possible Procedures

- Control tubes may consist of no yeast-molasses mixture within the stoppered tube.

PLAN THE EXPERIMENT

1. Decide on ways to test your group's hypothesis.
2. Record your procedure, and list the materials and amounts of solutions that you plan to use.
3. Design and construct a data table for recording your observations.
4. Pour 5 mL of BTB solution into a test tube. Use a straw to blow gently into the tube until you see a series of color changes. Cover this tube with aluminum foil, and set it aside in a test-tube rack. Record your observations of the color changes caused by carbon dioxide in your breath.

Check the Plan

Discuss the following points with other group members to decide the final procedure for your experiment.

1. What data on color change and time will you collect? How will you record your data?
2. What variables will have to be controlled?
3. What control will you use?
4. Assign tasks for each member of your group to be carried out during this investigation.
5. ***Make sure your teacher has approved your experimental plan before you proceed further.***
6. Carry out your experiment.

ANALYZE AND CONCLUDE

1. **Checking Your Hypotheses** Explain whether your data support your hypothesis. Use specific experimental data to support or reject your hypothesis concerning temperature effects on the rate of yeast metabolism.
2. **Making Inferences** Explain what you infer about the function of molasses in this experiment.
3. **Identifying Variables** Describe some variables that had to be controlled in this experiment. How were they controlled?
4. **Drawing Conclusions** Describe the control used in your experiment and how your experimental results enabled you to make conclusions about the effect of temperature on yeast metabolism. Did your experiment clearly show that differences in rates of yeast metabolism were due to temperature differences?

Going Further

Project To carry this experiment further, you may wish to examine the effects of temperatures as warm as 45°C on the rate of yeast metabolism. Formulate a hypothesis, and suggest an appropriate control for your experiment. Use data from your experiment to make conclusions about the rate of yeast metabolism at a very warm temperature.

Data and Observations

Yeast immersed in warmer water temperatures will result in more rapid production of carbon dioxide gas. Therefore, these tubes will change color more rapidly than those at cooler temperatures.

- Prepare yeast/molasses mixture as follows: one yeast cake and 20 mL molasses to 250 mL water. Note: the more molasses and less water used, the more rapid the generation of gas and the more rapid the observed changes with BTB.

Possible Hypotheses

- If yeast respiration is independent of temperature, then BTB solution will turn green at the same time at all experimental temperatures.

ANALYZE AND CONCLUDE

1. Data must be used to support or reject the original hypothesis. Typically, the warmer the water, the more rapid the color change of BTB.
2. Molasses served as the food for the respiring yeast.
3. The volume of water/molasses and the amount of yeast used were variables. These were kept constant in all tubes.
4. Controls will vary, may be a tube with no yeast-molasses. Data showing a measurable difference in the time needed to change BTB color indicates that rates of metabolism related to water temperature.

Assessment

Performance: Have students design an experiment to measure the actual volume of gas given off by yeast-molasses at different temperatures. L1

Going Further

Have students test the effectiveness of different foods on yeast respiration. L3

ThinkingLab Analyze the Procedure

Purpose

LS Logical-Mathematical Students will analyze the procedures of a series of experiments.

Process Skills

relate cause and effect, analyze an experimental procedure

Teaching Strategies

- Diagram the appearance of a zone of inhibition for students so that they can better understand how the three procedures could be evaluated.
- Have students work in cooperative groups as they analyze the three scenarios.

Thinking Critically

Experiment #1 - shows the effectiveness of "something" on soil bacteria but not on animal disease-causing bacteria. Lack of a control makes it difficult to conclude that inhibition is caused by fungi or something in the soil.

Experiment #2 - shows the effectiveness of "something" on animal bacteria. No control makes it difficult to show that inhibition is caused by fungi or something else in soil.

Experiment #3 - Results can be directly correlated to action of a specific fungus on a specific human bacterium. Data may or may not apply to infections in animals other than humans.

Assessment

Knowledge: Provide students with diagrams showing different widths of zones of inhibition around a fungus sample. Have them describe which antibiotic is least effective and which is most effective. L1

ThinkingLab Analyze the Procedure

How are new antibiotics discovered?

In 1928, Alexander Fleming discovered that a *Penicillium* mold interfered with the growth of bacteria he was studying. This chance observation later led to the discovery of the antibiotic penicillin. Biologists continue to investigate the effects of various fungi on the growth of bacteria in hopes of finding new antibiotics that can be used in human and veterinary medicine.

Analysis

Study the experimental procedures followed in these three sets of experiments.

1. A number of small samples of soil were placed on culture dishes containing soil bacteria and nutrients needed for the bacteria to grow. The cultures were examined after a few days for inhibition zones—areas in which bacteria failed to grow around any of the soil samples.
2. A number of small samples of soil were placed on culture dishes containing bacteria that can infect humans along with nutrients needed for the bacteria to grow. The cultures were examined after a few days for inhibition zones around any of the soil samples.
3. Soil samples were placed in culture dishes that contained nutrients required for the growth of a variety of fungi. Different fungi that grew on these dishes were then isolated, and a sample of hyphae from each fungus was placed on a culture dish in which possible human-infecting bacteria were growing. The bacterial cultures were examined after a few days for inhibition zones around any of the fungus samples.

Thinking Critically

Analyze each of the experiments described above. Explain what information the results of each experiment would give you. Explain how you think each experiment would or would not be helpful in discovering new antibiotics useful in human or veterinary medicine.

Mutualism with Fungi: Mycorrhizae and Lichens

Most trees live in a mutualistic association with fungi. A **mycorrhiza** is a symbiotic relationship in which a fungus lives in close contact with the roots of a plant partner. Most of the fungi that take part in mycorrhizae are basidiomycotes, but some zygomycotes also form these important relationships.

A beneficial partnership

How does a plant benefit from a mycorrhizal relationship? The fungal hyphae increase the amount of nutrients that move into the plant by increasing the absorptive surface of the plant's roots, *Figure 23.14.* Fine, threadlike hyphae from the fungus surround and often grow into the plant's roots without harming them. Phosphorus, copper, and other minerals in the soil are absorbed by the hyphae and then released into the roots. The fungus also may help to maintain water in the soil around the plant. The fungus in a mycorrhiza benefits by receiving organic nutrients, such as sugars and amino acids, from the plant. It has been suggested that evolution of trees has been positively influenced by the presence of mycorrhizae.

mycorrhiza:
mykes (GK) fungus
rhiza (GK) root
Mycorrhizae are fungi that live in close association with the roots of a plant.

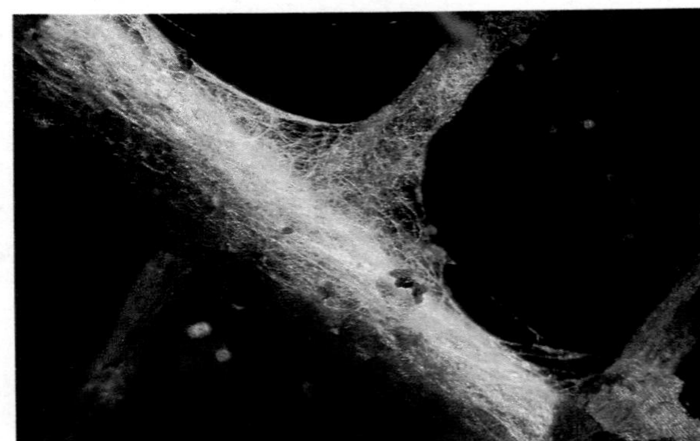

Figure 23.14
A mycorrhiza is formed with a tree root and a fungus. White strands of fungal hyphae encase the root and help supply it with mineral nutrients.

PORTFOLIO

Checking Out Moldy Cheese Supply students with samples of Roquefort and Camembert cheese. Have them prepare wet mounts for microscopic examination of the fungus that grows within each cheese. Ask them to diagram what they see and make written observations of the fungus. L2 P LS

Figure 23.15

Lichens can be found in a variety of different shapes.

Others form crustlike growths on bare rocks and stone walls.

Some lichens resemble leaves, like this one growing on a dead twig.

Each stalk of a British soldier lichen is about 3 cm tall.

In addition to trees, 80 to 90 percent of all plant species have mycorrhizae associated with their roots. Mycorrhizae are extremely important in agriculture. Having this relationship with fungi makes plants larger and more productive. Without mycorrhizae, most plants do not grow well. In fact, some species cannot survive without them. Orchid seeds, for example, will not germinate without a mycorrhizal fungus to provide them with water and nutrients.

Lichens

It's sometimes hard to believe that the orange, green, and black blotches you see on rocks, trees, and stone walls are alive, *Figure 23.15*. They may look like flakes of old paint or dried moss, but they are lichens. A **lichen** is a symbiotic association between a fungus, generally an ascomycote, and a green alga or cyanobacterium. The fungus portion of a lichen forms a dense web of tangled hyphae in which the algae or cyanobacteria grow. Together, the fungus and its microscopic, photosynthetic partners form a spongy structure that looks like a single organism.

About 20 000 species of lichens exist. They range in size from less than 1 mm to several meters in diameter. Lichens grow slowly, increasing in diameter only 0.1 to 10 mm per year. Very large lichens are thought to be thousands of years old.

Living together

Lichens need only light, air, and minerals to grow. The photosynthetic partner in the lichen provides itself and the fungus with food. The fungus, in turn, helps retain moisture, provides the alga or cyanobacterium with water and minerals absorbed from rainwater and air, and protects it from intense sunlight.

Found worldwide, lichens live in some of the harshest, most barren habitats on Earth. Lichens are true pioneer species, being among the first to colonize a barren area. You can find lichens in arid deserts, on bare rocks exposed to the blazing sun or bitter-cold winds, and just below the timberline on mountain peaks. On the arctic tundra, where large plants are scarce, lichens are the dominant form of vegetation. Caribou and musk oxen graze on lichens there, much like cattle graze on grass elsewhere.

Bioethics

Fungi Used for Worship
The fungus *Psilocybe strichper* has been used by Mexican Indians for hundreds of years in certain sacred rites. Lysergic acid diethylamide (LSD) is a chemical derived from ergot, or *Claviceps purpurea*, a fungus found on grain. Both *Psilocybe strichper* and LSD result in hallucinations when ingested.

The use of psilocybin is acceptable in Mexico because it is associated with a religious ceremony. However, the use of LSD in the United States is illegal. There have been court cases to determine whether the use of hallucinogenic fungi in the United States should be legal when it is part of religious worship.

Enrichment

LS **Visual-Spatial** Soak lichens overnight in a petri dish. Have students examine the water with a binocular microscope. Students should observe numerous water bears, also known as tardigrades. These minute animals belong to an animal phylum called the Tardigrada. Have students make a diagram of any tardigrades they observe and conduct library research to find out about the characteristics of these organisms. L2

3 Assess

Check for Understanding

Have students explain the relationships between or among the terms in the following groups: (a) stolon, rhizoid, hyphae (b) mating types, zygospores, zygomycota (c) ascomycota, asci, spores (d) basidiomycota, basidia spores (e) penicillin, deuteromycota (f) lichen, mycorrhiza, mutualism. L1

PORTFOLIO

Your Experimental Procedure Have students describe the steps they would need to take to experimentally show the effects of acid rain on the growth of mushrooms. Remind them to include the type of controls that would be needed. L1 P

Meeting Individual Needs

Students Acquiring English Have students make a table that compares mycorrhiza and lichens. The left side of the table should list the two terms. At the top of the table, they should list these phrases: Type of symbiotic relationship, Organisms forming partnership, Where found, Advantage provided to both organisms. L1 SAE LS

SECTION 23.2

Reteach

Have students working in cooperative groups explain how fungi in each phylum obtain food, reproduce asexually, reproduce sexually, and enter into symbiotic relationships. L1

Extension

Have students compile a list of different foods that are derived from fungi. L2

Assessment

Knowledge: Ask students to provide the following for each phylum of fungi: (a) an example, (b) brief description of the structure that forms spores, (c) the economic importance of the group. L1

4 Close

Discussion

Have students describe the type of buildings and special equipment needed to grow mushrooms commercially.

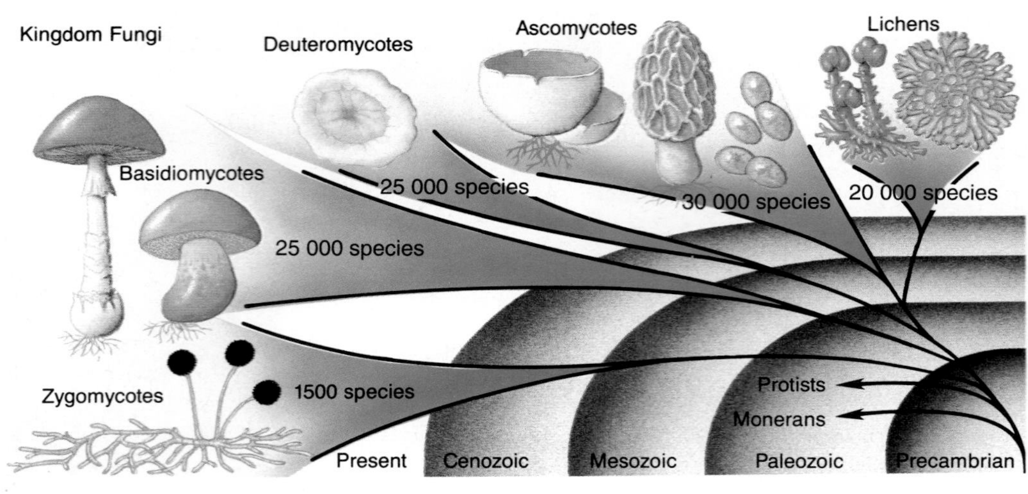

Figure 23.16

The radiation of fungus phyla on the Geologic Time Scale shows their relationships. Which group showed up latest in the fossil record?

Origins

Many researchers hypothesize that the ascomycotes and the basidiomycotes evolved from a common ancestor and that the zygomycotes evolved earlier, as you can see in ***Figure 23.16.***

Fossils can provide clues as to how organisms evolved, but fossils of fungi are rare because they are composed of soft materials. The oldest fossils that have been definitely identified as fungi are between 450 and 500 million years old.

Connecting Ideas

Fungi have evolved, along with some bacteria and protists, as the major decomposers of the biosphere. Much of what fungi decompose is plant material. As fungi break down organic matter, they release nitrogen compounds and other nutrients into the soil, where these molecules can be picked up and used again by plants. How do plants utilize the nutrients made available by fungi and other decomposers? Why have plants been so successful in adapting to their environments?

Section Review

Understanding Concepts

1. What happens between the time when a basidiospore germinates and a button forms?
2. How does a lichen obtain its nutrients?
3. Describe how each partner benefits in a mycorrhizal relationship.

Thinking Critically

4. You are working with a team of archaeologists on Easter Island in the Pacific Ocean. Huge stone statues were carved and erected on the island by a now-vanished civilization. How might you use lichens to help determine when the statues were carved?

Skill Review

5. **Observing and Inferring** Describe how lichens can be used as biological indicators of air quality. For more help, refer to Thinking Critically in the *Skill Handbook.*

Answers to Section Review

1. A hypha grows from the spore and meets with a hypha of the opposite mating type. These hyphae fuse and form mycelia with cells having a nucleus from each mating type. Eventually, compact masses of mycelia form buttons.
2. It has a partner that is capable of carrying out photosynthesis and thus receives food from it.
3. The protist receives minerals from the soil through the fungus. The fungus receives food from the protist.

Thinking Critically

4. Lichens are very slow growers. Determine the annual rate of growth for the lichen on the stones. Divide this into the size of the lichen to determine its approximate age.

Skill Review

5. Number of lichens surviving could be measured against air pollution indicators.

REVIEWING MAIN IDEAS

23.1 The Life of Fungi

- The basic structural units of a fungus are hyphae, which grow to form mycelium.
- Fungi are heterotrophs that carry out extracellular digestion. A fungus may be a saprobe, a parasite, or a mutualist in a symbiotic relationship with another organism.
- Most fungi produce asexual and sexual spores. The way in which sexual spores are produced is the basis for fungal classification.

23.2 The Diversity of Fungi

- Fungi are classified based on reproductive structures. Zygomycotes form asexual spores in a sporangium. They reproduce sexually by producing zygospores.
- Ascomycotes reproduce asexually by conidia and sexually by forming ascospores in saclike asci.
- In basidiomycotes, sexual spores are borne on club-shaped basidia.
- Deuteromycotes reproduce only asexually, usually by producing conidia.
- Fungi play an indispensable role in decomposing organic material and recycling nutrients.
- A lichen, a symbiotic association of a fungus and an alga or cyanobacterium, can survive in a variety of inhospitable habitats.
- Certain fungi associate with plant roots to form mycorrhizae, relationships in which both the plant and fungus benefit.

Key Terms

Write a sentence that shows your understanding of each of the following terms.

ascospore
ascus
basidiospore
basidium
budding
chitin
conidiophore
conidium
gametangium
haustorium
hypha
lichen
mycelium
mycorrhiza
rhizoid
sporangium
stolon
zygospore

Understanding Concepts

1. The basic structural units of multicellular fungi are ____________.
   a. psuedopods　c. spores
   b. cilia　d. hyphae
2. A ____________ is a network of filaments made up of threadlike hyphae.
   a. septum　c. thallus
   b. mycelium　d. colony
3. What complex carbohydrate is found in the cell walls of most fungi?
   a. cellulose
   b. calcium carbonate
   c. sugar
   d. chitin
4. Most fungi function as ____________ in their environments.
   a. consumers　c. decomposers
   b. producers　d. autotrophs
5. Specialized parasitic fungi use hyphae called ____________, which penetrate and grow into host cells.
   a. haustoria　c. buds
   b. septa　d. spores
6. Which of the following is a type of asexual reproduction in fungi?
   a. sporangium　c. mycelium
   b. budding　d. haustoria

Chapter 23 Review

Reviewing Main Ideas

Summary statements can be used by students to review the major concepts of the chapter.

Key Terms

Answers should go beyond defining the terms. Accept any answer that uses the term correctly and in the proper context.

Understanding Concepts

1. d
2. b
3. d
4. c
5. a
6. b

Chapter 23 Review

7. b
8. a
9. c
10. d
11. a
12. c
13. b
14. d
15. a
16. b
17. d
18. a
19. b
20. b

Applying Concepts

21. The bulk of the mushroom lies below ground and therefore cannot be destroyed this way.
22. Fewer lichens may be an indicator of an increase in the amount of pollutants.
23. Fungi return many nutrients back to the soil as they decompose once-living organic matter.
24. This may provide spores with an opportunity to land on new or different food sources.

Chapter 23 Review

7. Fungi sometimes live in a mutualistic relationship with a plant. They might help their host by ___________.
   a. using the food supplied by the host
   b. supplying water to the host
   c. using energy made by the host
   d. providing the host with spores
8. What are ideal growing temperatures for fungi?
   a. 20-30°C
   b. 30-40°C
   c. 50-60°C
   d. 10-20°C
9. Fungi used to be classified as plants because ___________.
   a. they are green
   b. they have stems
   c. they are anchored in the soil and have cell walls
   d. they take in carbon dioxide and give off oxygen
10. Fungi obtain food by ___________.
    a. intracellular digestion
    b. using the sun's energy
    c. an omnivorous feeding pattern
    d. extracellular digestion
11. By breaking down animal wastes and other organic matter, fungi provide _________, an essential nutrient for plants.
    a. nitrogen
    b. carbon dioxide
    c. oxygen
    d. sulfur dioxide
12. A reproductive cell in fungi that develops into a new organism without the fusion of gametes is called a(n) ___________.
    a. egg
    b. sperm
    c. spore
    d. sporangium
13. Fungi such as yeasts and blue molds that occur on decaying food belong to the group called ___________.
    a. club fungi
    b. sac fungi
    c. zygomycotes
    d. deuteromycotes
14. Suppose a piece of bread with mold on it began to dry out. The bread mold might produce ___________.
    a. hyphae
    b. mycelia
    c. asci
    d. zygospores
15. Mushrooms, puffballs, and bracket fungi belong to the group called ___________.
    a. club fungi
    b. sac fungi
    c. zygomycotes
    d. deuteromycotes
16. Club fungi get their name from the club-shaped ___________ they form.
    a. spores
    b. hyphae
    c. haustoria
    d. conidia
17. Soy sauce, citric acid, and penicillin all come from ___________.
    a. club fungi
    b. sac fungi
    c. zygomycotes
    d. deuteromycotes
18. Unlike other fungi, ___________ are only known to reproduce asexually.
    a. deuteromycotes
    b. zygomycotes
    c. ascomycotes
    d. basidiomycotes
19. Ascomycotes, the sac fungi, are named for the little structures called __________ in which their spores develop.
    a. stolons
    b. asci
    c. gametangia
    d. mycelia
20. Of these, which is NOT a part of the symbiotic association called a lichen?
    a. fungus
    b. plant
    c. alga
    d. cyanobacterium

Applying Concepts

21. Your neighbor is pulling up mushrooms that are growing in his lawn. He tells you that he heard mushrooms won't come back again if they are quickly removed. What would you tell him?
22. While hiking along a trail through a woods near your rapidly growing city, you notice that there are fewer lichens on the rocks and trees than there used to be. How might you interpret this change in the forest ecosystem?
23. In an effort to have healthy lawns, some home owners apply chemicals to their grass to kill fungi. How might these practices result in a nutrient-starved lawn?
24. Why is being able to produce spores that can be dispersed far from their source such an important adaptation for fungi?

25. When you transplant flowers, shrubs, or trees, why is it a good idea to leave the soil intact around a plant's roots?

Thinking Critically

26. ***Concept Mapping*** Make a concept map that relates the following terms and phrases. Supply the appropriate linking words for your map.
    reproductive structures, ascus, ascospores, basidium, basidiospores, conidia, conidiophores, phyla, Kingdom Fungi
27. ***Recognizing Cause and Effect*** When making bread, yeast is usually activated by combining it with sugar and warm water, then adding it to the rest of the ingredients. The result is a loaf of bread that has risen due to carbon dioxide released by the yeast cells. How would mixing yeast with sugar and ice water affect the way the bread turns out?
28. ***Comparing and Contrasting*** Both fungi and animals are classified as heterotrophic organisms. Contrast the interactions of fungi and plants with the interactions of animals and plants.
29. ***Interpreting Scientific Illustrations*** To what phyla could the fungus in the photomicrograph below belong? What additional information would you need before being able to place this fungus in the proper phylum?

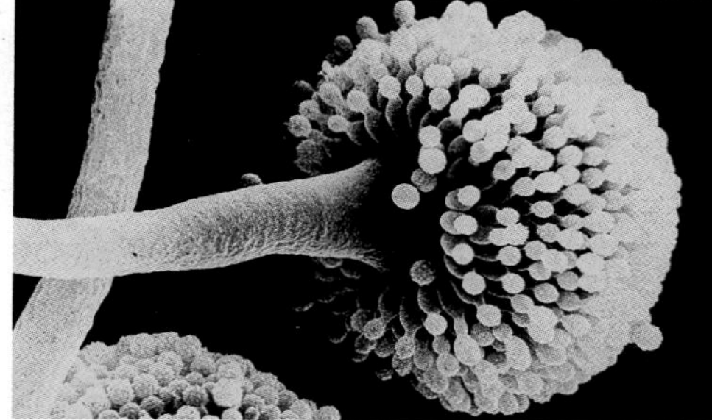

ASSESSING KNOWLEDGE & SKILLS

The metabolic activity of yeasts at various temperatures is shown in the table below. A chemical indicator added to the yeast solution changed color when yeast cells were metabolizing.

Metabolic Activities of Yeasts

Test Tube #	Temperature of yeast solution	Time elapsed until color change
1	2°C	no color change
2	25°C	44 minutes
3	37°C	22 minutes

Interpreting Data Use the table to answer the following questions.

1. At what temperature was the yeast most active?
   a. 2°C c. 37°C
   b. 25°C d. 22°C
2. No color change indicated that yeasts were ___________.
   a. metabolizing c. too hot
   b. not metabolizing d. dead
3. In which test tubes were yeast cells metabolizing?
   a. Numbers 1, 2, and 3
   b. Numbers 1 and 3
   c. Number 1
   d. Numbers 2 and 3
4. ***Observing and Inferring*** How was temperature related to the rate of yeast metabolism?

Chapter 23 Review

25. This could prevent destruction or disruption of any mycorrhiza that may be present.

Thinking Critically

26. Evaluate students' concept maps. The maps should show an understanding of the relationships among the concepts listed.
27. The bread would be flat. Ice water tends to slow yeast respiration and therefore reduces the amount of carbon dioxide gas produced.
28. A few fungal species are mutualists, but fungi are mainly decomposers or parasites. They live off dead or living organic material. Plants are autotrophs and supply the food used by most fungi. Animals depend on plants either directly or indirectly for their food.
29. The fungus could belong to the phyla Basidiomycota or Ascomycota. You would have to know if the spores are held in asci or basidia to classify it correctly.

ASSESSING KNOWLEDGE & SKILLS

1. c
2. b
3. d
4. Yeast takes longer to begin metabolizing at lower temperatures.

Program Resources

Chapter Assessment, pp. 133-138 L1
Alternate Assessment in the Science Classroom
Computer Test Bank L1
Content Mastery, pp. 89-92 L1

UNIT 7 Plants

Unit Overview

Unit 7 introduces the plant kingdom. The unit begins in Chapter 24 with a brief discussion of plant origins and the adaptations of plants to life on land.

In Chapter 25, students learn about seedless vascular plants and gymnosperms. Students then learn about angiosperms in Division Anthophyta in Chapter 26. The major focus of this chapter is on the anatomy of roots, stems, and leaves.

The unit concludes in Chapter 27 with an examination of the flower. The chapter ends with a discussion of the role of plant hormones.

Theme Development

Unity within diversity is most apparent through the discussion of how all plants are similar in their ability to photosynthesize and through the process of alternation of generations. *Evolution* is discussed as the major differences for each plant group are explained. The theme of *systems and interactions* is illustrated as the importance of reproductive structures are presented for each plant group.

UNIT 7 Plants

Orchids represent the most advanced level of evolutionary form and function in flowering plants. Orchids are the second largest and most diverse family of plants, varying widely in structure, color, size, and fragrance.

Most orchids are found in the tropics, but what makes them so successful is that they have adapted to a variety of habitats, from the sandy soils near oceans to the acidity of mountain bogs. For each locality, you will find an orchid species with a set of characteristics best suited for that environment. A study of orchids would reveal most of the structural and physiological adaptations characteristic of the plant kingdom.

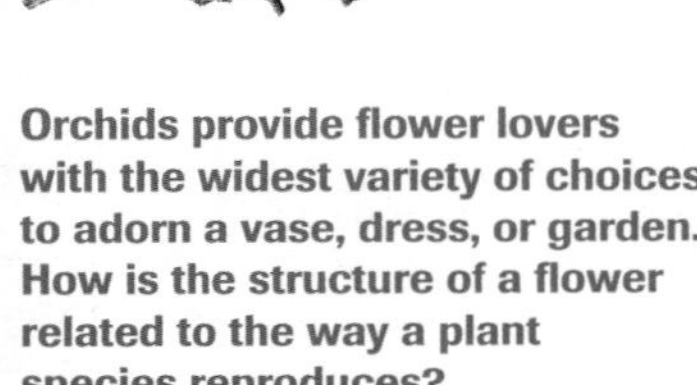

Orchids provide flower lovers with the widest variety of choices to adorn a vase, dress, or garden. How is the structure of a flower related to the way a plant species reproduces?

Advance Planning

Chapter 24

- Start growing plants for the Minilab on page 588.
- Order moss and algae specimens for the Biolab and the Minilab on page 595.

Chapter 25

- Grow or order fern prothalli for the Minilab on page 610.
- Collect or purchase male pinecones and fern fronds with mature sporangia for the Minilab on page 616. Collect conifer branches for the Biolab.

Chapter 26

- Obtain monocot and dicot leaf samples for the Minilab on page 634.

Chapter 27

- Collect or purchase apple flowers or cherry flowers for the Minilab on page 672.
- Purchase bean and corn seeds for the Minilab on page 674.

Many plants live in areas, such as deserts, that would be inhospitable to most organisms. What adaptations enable plants to inhabit such environments?

Plants make food by capturing the energy in sunlight. What structural features allow plants to make the most of this energy?

Unit Contents

UNIT 7

Introducing the Unit

Explain that orchids are plants that grow in all parts of the world. Ask students why they think orchids are so successful.

Next, ask students to observe the desert scene. Explain that, unlike orchids, some other plants are restricted to one kind of environment. Elicit from students the names of plants that are adapted to dry environments.

Motivational Activity

Activity: Ask what a plant can do that people and other animals cannot. *Use light energy to make food via photosynthesis.* Elicit how the ability to carry out photosynthesis is important. Encourage students to find photographs from newspapers and magazines to create a photo essay on why plants are important to people. L1

Unit Project

Group Project: Have students work in cooperative groups to determine whether plants need fertilizer to grow.

Purchase bean seeds at a local garden store. Have students grow plants from the seed by placing the seeds in small pots filled with sand. Have students keep the sand moist until the seeds start to germinate. Instruct students to record the date when the seedlings are first observed and to begin their experiment on this day. For the experiment, students are to water a control group using only distilled water for two weeks. The experimental group should be watered with a solution of distilled water mixed with a liquid fertilizer purchased from a local plant store. Instruct students to use the amount of fertilizer recommended on the package. Have students identify any differences they observe (height and overall appearance) in the two groups of plants after two weeks. Instruct students to be as quantitative as possible in their data. Ask students to use their data to write a report about whether the use of fertilizers has any significant effect on plant growth. L1 COOP LEARN

Chapter Organizer

SECTION	OBJECTIVES	ACTIVITIES/FEATURES
24.1 Adapting to Life on Land National Science Standards: UCP.1-5; B.2; C.5; G.1-3	1. **Relate** the adaptive value of plant characteristics to the demands of living on land. 2. **Assess** theories concerning the origin of plants.	**Chemistry Connection:** Rayon: A Natural Fiber, p. 585 **Biology & Society Issues:** Bioengineered Food, p. 586 **Minilab:** How do roots, stems, and leaves compare?, p. 588 **Thinking Lab:** What is the function of trichomes?, p. 589
24.2 Bryophytes National Science Standards: UCP.1, UCP.5; A.1; C.1, C.5; G.1, G.2	3. **Identify** the structures of a typical bryophyte. 4. **Explain** the life cycle of a moss or liverwort.	**Minilab:** How do mosses and green algae compare?, p. 595 **Biolab:** Alternation of Generations in Mosses, p. 598

ACTIVITY MATERIALS

BIOLAB	MINILABS	ALTERNATE LAB
page 598 microscope slides single-edged razor blade forceps dropper paper towels moss plants	**page 588** plants hand lens or low-power microscope **page 595** moss samples algae samples microscope slides coverslips	**page 586** lettuce leaf glass slide microscope coverslip ruler

Chapter 24 What Is a Plant?

TEACHER CLASSROOM RESOURCES

Reproducible Masters	Transparencies
Section Focus Master 51: Life on Land L1 SAE **Reinforcement and Study Guide,** pp. 93-94 L1 **Biolab and Minilab Worksheets,** p. 95 L1 **Concept Mapping:** Adaptations of Land Plants, p. 24 L1 **Tech Prep Applications:** Analyzing Dietary Fiber, pp. 29-30 L2 **Laboratory Manual:** Roots and Stems, pp. 133-138; How Do Gymnosperm Stomata Vary?, pp. 139-142 L2	**Basic Concepts Transparency #33:** Phylogeny of Simple Plants L1 SAE **Basic Skills Transparency #19:** What Is the Function of Trichomes? L1 SAE
Section Focus Master 52: Bryophytes L1 SAE **Reinforcement and Study Guide,** pp. 95-96 L1 **Critical Thinking/Problem Solving:** Solving Simple Plant Problems, p. 24 L3 **Biolab and Minilab Worksheets,** pp. 96-98 L1 **Content Mastery,** pp. 93-96 L1	**Basic Concepts Transparency #34:** The Life Cycle of a Moss L1 SAE **Reteaching Transparency #24:** Alternation of Generations L1 SAE

ASSESSMENT MATERIALS

Chapter Assessment, pp. 139-144
Alternate Assessment in the Science Classroom
MindJogger Videoquiz
Computer Test Bank

Spanish Resources SAE
English/Spanish Audiocassettes SAE
Cooperative Learning in the Science Classroom COOP LEARN
Lesson Plans
Great Developments in Biology: Discovering the Living World, L1 SAE

KEY TO TEACHING STRATEGIES

L1 Level 1 activities should be within the ability range of all students including those with learning difficulties.

L2 Level 2 activities are within the ability range of average to above-average students.

L3 Level 3 activities are designed for the ability range of above-average students.

SAE SAE activities should be within the ability range of Students Acquiring English.

COOP LEARN Cooperative Learning activities are designed for small group work.

P These strategies represent student products that can be placed into a best-work portfolio.

These strategies are useful in a block scheduling format.

GLENCOE TECHNOLOGY

The following multimedia resources are available from Glencoe.

Biology: The Dynamics of Life
CD-ROM SAE
Videodisc Program

National Geographic Society Series
STV: Plants
What Is a Leaf?

Science and Technology Videodisc Series (STVS)
Plants & Simple Organisms
What Is a Plant?
Growing Plants in Space
Sound of Thirsty Plants
Disease-Resistant Tomatoes
Genetic Engineering in Barley
New Grains

CHAPTER

24 What Is a Plant?

Chapter Overview

Students begin this chapter with an overview of what a plant is and how plants may have originated. The first section establishes the evidence supporting the idea that land plants evolved from green algae. Next, the major adaptations needed for survival on land are addressed.

In the latter part of the section, adaptations of leaves, roots, and stems are considered. Patterns of reproduction and alternation of generations are then explained. The first section ends with a review of the phylogeny of ten plant divisions.

The second section of the chapter focuses on the division Bryophyta. Physical characteristics of bryophytes as well as differences between mosses and liverworts are described. Attention is then turned to alternation of generations in mosses. The section ends with a review of why mosses and liverworts are restricted in size and distribution.

Key Terms

antheridium
archegonium
cuticle
gemmae
leaf
nonvascular plant
protonema
root
seed
stem
stomata
vascular plant
vegetative reproduction

582

Look for the following logo for strategies that emphasize different learning modalities. 

Kinesthetic	Meeting Individual Needs, p. 591
Visual-Spatial	Portfolio, pp. 585, 596, 597; Microscope Activity, p. 590; Minilab, pp. 588, 595; Thinking Lab, p. 589; Display, p. 591; Demonstration, p. 591; Activity, p. 594; The Inside Story, p. 597
Linguistic	Cultural Diversity, p. 589; Meeting Individual Needs, p. 597
Logical-Mathematical	Alternate Lab, p. 586; Project, p. 588; Student Journal, pp. 590, 592; Meeting Individual Needs, pp. 592, 595, 596

LS

CHAPTER

24 What Is a Plant?

The damp, misty bank of a rushing stream is the perfect environment for a lush growth of mosses. Mosses may be similar to some of the first plants that lived on land. Like all members of the plant kingdom, mosses probably evolved from green algae that lived in ancient swamps and oceans.

One of the challenges plants faced when they moved to land was the need for water. Plants have evolved a variety of adaptations for obtaining and conserving water. The roots of giant redwood trees obtain water from deep, underground reservoirs. The waxy covering on the leaves of a magnolia tree helps retain moisture. These are just a few of the plant adaptations that enable them to live on land.

Concept Check

You may wish to review the following concepts before studying this chapter.

- Chapter 8: cell wall, cellulose
- Chapter 10: photosynthesis
- Chapter 12: diploid, haploid, meiosis

Chapter Preview

Individual moss plants are usually no more than a few centimeters in height, and most parts of the plant are only one cell thick. Giant redwoods, on the other hand, have a circumference of up to 30 m and soar to heights of 90 m or more.

This magnolia tree absorbs water through organized tissues in its roots, which then transport the water through its trunk to its branches and leaves.

583

CHAPTER 24

Introducing the Chapter

Ask students to look at the photographs and compare and contrast the moss and redwood tree. *They most likely will comment on height and woody versus nonwoody traits.* To focus their attention on similarities, elicit if the plants share any color in common and what the significance of this color might be. *Yes, both have green coloring that is related to photosynthesis.* Ask if both plants require gases from the air and how these gases are put to use. *Yes; the plants need carbon dioxide for photosynthesis and oxygen for respiration.* Be sure to stress that both respiration and photosynthesis occur in both organisms.

Theme Development

The theme of *unity within diversity* is well illustrated in this chapter. Differences among plants in various divisions are discussed, and their similarity in reproduction via alternation of generations is stressed. *Evolution* is also a major theme in the discussion of the evolution of land plants from green algae.

Concept Check

Students will build upon their knowledge of photosynthesis as they learn about the plant structures that aid in photosynthesis. Other characteristics that unite plants in one kingdom, such as cell structures, will be reinforced.

Assessment Planner

Choose assessment strategies from the following pages to evaluate the progress of your students.

Assess, p. 592, 600
Alternate Lab, pp. 586-587
Minilabs, pp. 588, 595
Portfolio, pp. 585, 596, 597
Thinking Lab, p. 589
Biolab, pp. 598-599
Chapter Review, pp. 601-603

SECTION

24.1

Prepare

Key Concepts

Evidence for land plants evolving from algae and the adaptations needed for land survival are covered. A detailed look at leaf, stem, and root anatomy is presented in conjunction with the adaptations of these organs for land survival. Finally, an overview of the ten plant divisions is provided.

Block Scheduling

Look for this symbol for strategies that are useful in a block scheduling format. For more information on block scheduling, refer to Section 24.1 in the **Lesson Plans** booklet.

Materials

- Purchase or grow bean plants from seed for the Minilab.
- Prepare a 6% salt solution for the Microscope Activity.
- Purchase lettuce for the Alternate Lab.

1 Focus

Bellringer

Before presenting the lesson, display **Section Focus Master 51** on the overhead projector and have students answer the accompanying questions. L1 SAE

Activity

Provide students with several coniferous and deciduous leaves. For each leaf sample, ask them to describe any differences they observe. *The leaves had two basic differences in shape, and the thicknesses varied, although most were relatively thin.* L1

SECTION

24.1 Adapting to Life on Land

Section Preview

Objectives

Relate the adaptive value of plant characteristics to the demands of living on land.

Assess theories concerning the origin of plants.

Key Terms

cuticle
stomata
leaf
root
stem
vascular plant
nonvascular plant
seed

When you studied ecology in Chapter 4, you learned that plants are such an important part of life on Earth that they are used to define biomes, ecosystems, and communities. Plants are a major group of Earth's producers. They trap the energy of sunlight and store it as chemical energy in food, which supplies the fuel that makes all life possible. Multitudes of organisms, including humans, rely on plants for both food and shelter. Because plants take in carbon dioxide and release oxygen during photosynthesis, they are responsible for maintaining the supply of oxygen needed by most organisms for respiration.

Origins of Plants

What is a plant? A plant is a multicellular eukaryote, with cells surrounded by cell walls made of cellulose and with a waxy waterproof coating called a cuticle.

A billion years ago, plants had not yet begun to appear on land. No ferns, mosses, trees, grasses, or wildflowers existed. The land was barren except for some algae at the edges of inland seas and oceans. However, the shallow waters that covered much of Earth's surface at that time were teeming with bacteria, algae and other protists, and simple animals such as corals, sponges, jellyfishes, and worms. Among these organisms were those green algae that would slowly become adapted to life on land.

The first plants began to appear around 500 million years ago. These early plants may have looked something like present-day mosses. As you learned in Chapter 4, mosses are pioneer organisms that help turn bare rock into rich soil. Mosslike plants might have helped lead the way for other plants by being among the first soil builders. There is no fossil record of the first land plants, in part because their tissues were probably delicate and decayed too easily to

Figure 24.1
The oldest fossil psilophyte is probably more than 400 million years old. The plant was made up of leafless stems without roots. Underground portions of the stem bore rootlike hairs that absorbed water and nutrients from the soil.

Program Resources

Section Focus Master 51 L1 SAE
Reinforcement and Study Guide, pp. 93-94 L1
Biolab and Minilab Worksheets, p. 95 L1
Laboratory Manual, pp. 133-138; 139-142 L2
Concept Mapping, p. 24 L1
Tech Prep Applications, pp. 29-30 L2
Basic Concepts Transparency 33 and **Master** L1 SAE
Basic Skills Transparency 19 and **Master** L1 SAE

become fossilized. The earliest known plant fossils are those of plants called psilophytes, some of which still exist today, *Figure 24.1.*

All plants probably evolved from filamentous green algae that dwelt in the ancient oceans. Both green algae and plants have cell walls that contain cellulose. Both groups have the same types of chlorophyll used in photosynthesis. Both algae and plants store food in the form of starch. All other major groups of organisms store food in the form of glycogen and other complex sugars.

Adaptations of Plants

Life on land has advantages as well as challenges. A filamentous green alga floating in a pond does not need to conserve water. The alga is completely immersed in a bath of water and dissolved nutrients, which it can absorb directly into its cells. For most land plants, the only available supply of water and minerals is in the soil, and only the portion of the plant that penetrates the soil can absorb these nutrients.

Algae reproduce by releasing unprotected gametes into the water, where fertilization and development take place. The gametes of most land plants are protected from drying out by a waterproof covering of thick-walled cells. Land plants must also withstand the forces of wind and weather and be able to grow upright against the force of gravity. Over the past 500 million years or so, plants have developed a huge variety of adaptations that reflect both the challenges and advantages of living on land.

Chemistry

Rayon: A Natural Fiber

Look at the labels of your favorite clothes. Chances are, they are made from natural fabrics—cottons, linen, wool, and silk. Cotton comes from the cotton plant. Linen is made from the fibers of the flax plant. Wool is the hair of sheep. Silk is the fiber that silkworm moths make to create their cocoons. Natural fibers "breathe," absorbing moisture so they feel cooler, and they soften with each washing.

What is rayon? Rayon is a natural fabric, too. In the 1800s, a silkworm epidemic nearly ended the French silk industry. A prize was offered to anyone who could produce an artificial silk. Louis Pasteur's assistant, Chardonnet, created cellulose nitrate from soft wood. This artificial silk, called rayon, was abandoned in the late 1890s due to its flammability, but further research led to the discovery that fabrics could be made from the cellulose fibers that coat all cell walls of plants.

How rayon is made The most common form of rayon made today is xanthate rayon; it is chemically identical to cotton. It is made by dissolving cotton seed cellulose and forcing the solution through small holes in a nozzle. The extruded cellulose fibers are long, smooth filaments like silk and give rayon a similar texture and shine.

Other uses for rayon Most lightweight, sheer fabrics in use today, such as satin, are made from rayon. Rayon holds dye well and is found in carpet fibers, coverings for home furnishings, surgical gauze, and tire cords.

CONNECTION TO Biology

Use a microscope to compare textures of natural fibers from plants such as cotton and linen. How might fiber texture affect the properties of the cloth?

2 Teach

Tying to Previous Knowledge

Have students review the classification of algae. Ask them to compare the phylum characteristics of protists to the characteristics of plants. *Students will likely state that the organisms are similar in that they contain chlorophyll and carry out photosynthesis. Likely differences will relate to overall size and the fact that algae live in water, whereas most plants live on land. Students may also indicate that plants have more complex structures than algae. Plants are multicellular. Some algae are multicellular, but most are not.* L2

Chemistry

Rayon: A Natural Fiber

Purpose

To show how a plant product is used.

Teaching Strategies

- Review the structural formula for glucose. Illustrate how this single molecule—a monomer—joins to form cellulose—a polymer.
- Compare the chemical composition of rayon to nylon. Explain that nylon is formed from the non-cellulose materials hexanedoic acid and 1, 6-diaminohexane.

Possible Answers

Cotton and linen will appear fuzzy compared to rayon. Smooth fibers will likely produce a shinier, smoother fabric than rougher fibers.

GLENCOE TECHNOLOGY

Videodisc

STVS: Plants & Simple Organisms
Disc 4, Side 2
Growing Plants in Space (Ch. 8)

PORTFOLIO

Evolution of Land Plants Have students prepare an illustrated time line that indicates when plants first evolved onto land, and what early land plants looked like. Have students indicate how the appearance of plants differed 100 million years after they first became land organisms. L1 P LS

Issues

BIOLOGY & SOCIETY

Bioengineered Food

Purpose

Students will discover how scientists genetically manipulate plant materials to improve or modify existing characteristics.

Teaching Strategies

- Have students work in cooperative groups to complete the following:

  (a) list traits they would like improved in food plants they are familiar with;

  (b) use the analogy of splicing together sections from different videotapes to compare the technique used in altering plant genes.
- Ask students if they would want to know if the food they are eating had been genetically altered. Have them explain why or why not. *Students may cite fear of the unknown, concern that some foods cause allergies, and ethnic or religious beliefs that might not allow foods that have a gene from a plant or animal that may not be eaten by the group.*

INVESTIGATING the Issue

Applying Concepts Both techniques alter or change traits to improve on a certain quality or characteristic; both techniques cause changes due to the action of genes.

BIOLOGY & SOCIETY

Issues

Bioengineered Food

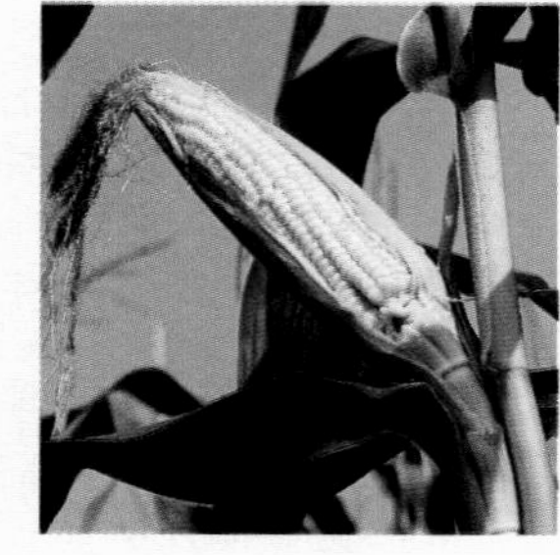

Researchers joke that farmers will soon be able to purchase a genetically engineered corn seed that plants itself, produces its own fertilizer and pesticide, requires no water, and grows tortilla chips, high-fructose sweetener, and diesel fuel. Although these corn seeds do not yet exist, you can already find high-tech tomatoes at the supermarket. They look like regular tomatoes, but inside their DNA—the genetic code that determines the way they look, feel, and taste—is an extra gene.

The added gene is to control a ripening enzyme. It is formed in reverse order so it will make backward RNA. When the backward RNA binds to the normal RNA that controls fruit softening, it makes that gene inactive and unable to form the ripening enzyme. This slows down the rotting process. The tomatoes taste better and last longer than normal tomatoes.

Different Viewpoints

Bioengineers are using gene-splicing technology as a means of solving world hunger and poverty. Nutrition-enhanced, longer-lasting produce and crops that are resistant to drought, frost, and pests can relieve famine and improve the economies of many developing countries. In addition, plant breeders who choose genetic engineering over traditional crossbreeding are able to get desired traits faster and with more consistency.

How safe are engineered foods? Consumers who have food allergies worry that genetically altered foods will cause allergic reactions. What if a gene from a wheat plant is used to provide resistance to disease in corn plants? Will someone who is allergic to wheat have a reaction after eating the genetically altered corn? Vegetarians may object to genetically engineered plants that include any animal genes.

What are the ecological consequences? What are the ecological consequences of altering genetic material? What would happen if altered plant or animal species without natural predators were introduced into a new environment?

INVESTIGATING the Issue

Applying Concepts What do traditional artificial selection and bioengineering techniques have in common?

Preventing water loss

If you run your fingers over the surface of an apple, a maple leaf, or the stem of a houseplant, you'll find that it feels smooth and slightly slippery. Most fruits, leaves, and stems are covered with a protective, waxy layer called the **cuticle.** Waxes and oils are lipids, and you have read in Chapter 7 that lipids do not dissolve in water. The waxy cuticle helps prevent the water in the plant's tissues from evaporating into the atmosphere.

Figure 24.2

The opening and closing of a stoma is regulated by guard cells that surround the pore.

Guard cells are balloonlike cells with their end walls fixed in place. During the day, water moves into the guard cells by osmosis and they swell. But at the ends, this swelling is restricted, so the stoma opens.

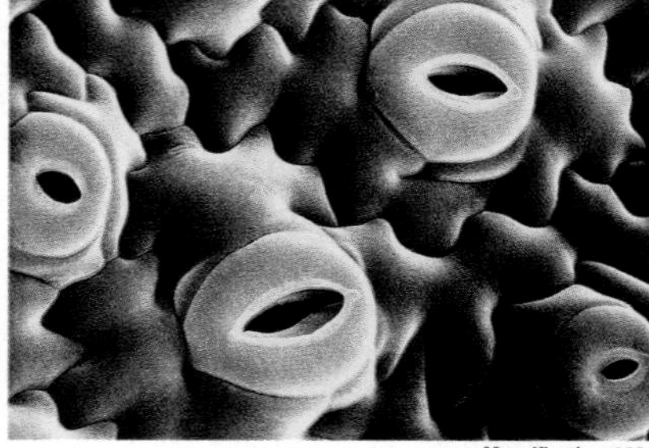

Magnification: 300×

At night, the guard cells lose water into surrounding cells and relax, partially closing the pore.

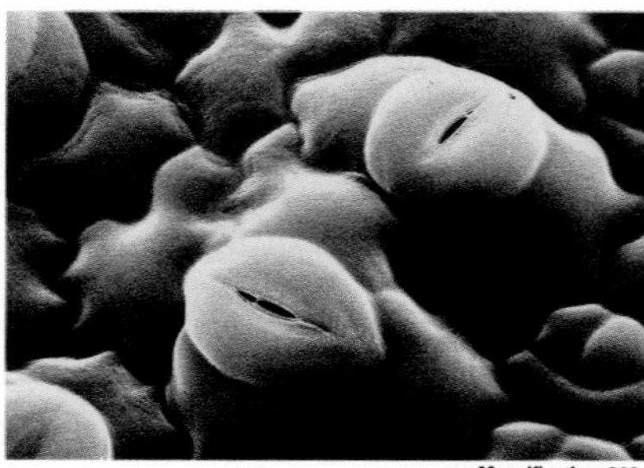

Magnification: 300×

Alternate Lab | **Observing Stomata**

Purpose 

LS Logical-Mathematical To have students estimate the number of stomata present on a leaf.

Materials

lettuce leaf, glass slide, microscope, coverslip, water, ruler

Procedure

Give students the following directions.

1. Prepare a wet mount of a small section from the thinnest area of a lettuce leaf. Save the remaining leaf.
2. Observe the wet mount under high power. Count the number of stomata seen in your field of view. Record this number.
3. Calculate the number of stomata present on the entire lettuce leaf. Measure the total area of the lettuce leaf using mm units. Record your answer. Divide the number you recorded by 0.08 mm^2 (area of a typical high power field of view). Multiply your answer by the number of stomata in one field of view from step 2.

Figure 24.3

Leaves take advantage of the plentiful supply of sunlight and carbon dioxide available to land plants.

Most leaves are thin and allow sunlight to penetrate throughout the organ's tissues.

Most plants have an enormous number of leaves, which provide a large surface area for trapping sunlight and exchanging gases during photosynthesis.

There are openings called **stomata** in the cuticle of the leaf that allow the exchange of gases. The pores of stomata open during the day while photosynthesis is taking place, *Figure 24.2.* Stomata that open during the daytime release water and oxygen and take in the carbon dioxide needed for photosynthesis. During the night, these openings partly close down to prevent too much water loss. Although the plant is coated with a waxy cuticle, plants lose up to 90 percent of the water they contain through these openings in the plant's epidermis, the outer layer of cells.

Leaves carry out photosynthesis

All the cells of a filamentous green alga carry out photosynthesis. But in most land plants, the leaves are the organs usually responsible for photosynthesis. A **leaf** is a broad, flat organ of a plant that traps light energy for photosynthesis, *Figure 24.3.* Leaves also exchange gases through their stomata. Leaves are supported by the stem and grow upward toward sunlight. They have both an upper and a lower surface.

Putting down roots

Most plants depend on the soil as their primary source for water and other nutrients. A **root** is a plant organ that absorbs water and minerals from the soil, transports those nutrients to the stem, and anchors the plant in the ground. Some roots, such as those of radishes or sweet potatoes, also accumulate starch reserves and function as organs of storage.

Mosses and their relatives have rhizoids rather than roots. Rhizoids in mosses are usually only one cell thick and do not extend far into the soil.

Bioethics

Air Purification

Plants have been called "clean air machines," but how effective are they in cleaning the air? NASA has discovered that plants do in fact remove small quantities of such chemicals as toluene, benzene, and formaldehyde from the air. However, plants perform these functions very slowly, so an enormous number of them would be needed to do an effective job of cleaning the air. For this reason, NASA recommends a properly operating ventilation system rather than plants as air purifiers.

GLENCOE TECHNOLOGY

Videodisc

STVS: Plants & Simple Organisms

Disc 4, Side 2

Sound of Thirsty Plants

(Ch. 6)

Analysis

1. Name two changes in the procedure that could improve data accuracy. *Calculate actual area of high-power field for the microscope and make several counts of stomata and use an average.*

Assessment

Skill: Have students calculate the number of stomata on an 80 × 20-mm leaf with 26 stomata observed per high-power field. *80 mm × 20 mm = 1600 mm² divided by 0.08 mm² = 20 000 × 26 = 520 000* L1

MiniLab

Purpose

Visual-Spatial To have students compare roots, stems, and leaves.

Process Skills

observe and infer, compare and contrast, record observations

Teaching Strategies

- Bean plants grown from seed are ideal organisms for this Minilab. Gently remove the plants from soil and rinse with water so students can observe the root system.
- Other ideal plants might be *Coleus* or *Tradescantia.* Do not use any monocots because students will have difficulty differentiating stems from leaves.

Expected Results

Depending on the type of plant used, students may find that both leaves and stems are green, suggesting both are photosynthetic. Students may also observe epidermal cells, and stomata with guard cells on leaves. Roots may show root hairs and an extensive root system.

Analysis

Leaves are flat and green with an extensive surface area. Stomata and guard cells are present. Stems may be green in color, containing long, tubular cells for transporting materials. Roots are not green, they have tubular cells, and root hairs are present.

Assessment

Knowledge: Have students list those adaptations that enable a leaf, stem, and root to carry out their different functions. L1

MiniLab

How do roots, stems, and leaves compare?

As land plants evolved, they developed structures that carry out different functions. In this lab, you will examine roots, stems, and leaves to identify their similarities and differences.

Procedure

1. Obtain one or more plants from your teacher.
2. Observe your plant carefully. Identify the root, stem, and leaves. Determine where the root ends and the stem begins, and where the stem ends and the leaf begins.
3. Use a hand lens or low-power microscope to observe small parts or features of each structure. Draw or take down a written description of what you see.

Analysis

What differences in shape, size, and color did you observe in the different parts of the plant? What microscopic details did you observe? What similarities and differences did you observe among structures used for photosynthesis, water absorption, and transport of food and water?

Transporting materials

Water moves from the roots of a tree to its leaves—and the sugars produced in the leaves move to the roots—through the stem. A **stem** of a plant provides structural support for upright growth and contains tissues for transporting food, water, and other materials from one part of the plant to another, *Figure 24.4.* Stems may also serve as organs for food storage. Green stems contain chlorophyll and take part in photosynthesis.

The stems of most plants contain vascular tissues made up of tubelike, elongated cells through which water, food, and other materials are transported. Plants that possess vascular tissues are known as **vascular plants.** Most of the plants you are familiar with—including pine and maple trees, ferns, rhododendrons, rye grasses, English ivy, and sunflowers—are vascular plants.

Mosses and several other small, less-familiar plants are nonvascular plants. **Nonvascular plants** are plants that do not have vascular tissues. The tissues of nonvascular plants are usually no more than a few cells thick, and water and nutrients travel from one cell to another by the relatively slow processes of osmosis and diffusion. The evolution of vascular tissues was of major importance in enabling plants to survive in the many habitats they now occupy on land. Vascular plants can live farther

Figure 24.4

Stems such as the trunk of this eucalyptus tree provide support that enables plants to grow to great heights.

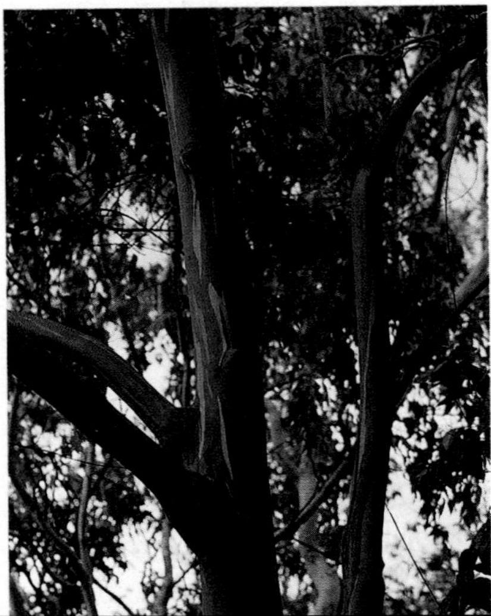

PROJECT

Comparing Plant Part Growth Patterns

Have students determine if new root growth occurs more rapidly than stem growth. Their project should involve the study of several different seed types such as grass, corn, bean, and mustard. Have students first design a chamber, possibly using a cut-away, plastic soda bottle or an empty jelly jar, that will allow for daily viewing and measuring of both stem and root growth. Instruct students to record their observations in scale diagrams and in daily measurements plotted on a graph. L1

away from water than nonvascular plants. Also, because vascular tissues include thickened cells called fibers that help support upright growth, vascular plants can grow much larger than nonvascular plants.

Reproductive strategies

Adaptations in land plants include the evolution of spores and seeds. These structures protect the zygote or embryo and keep them from drying out before they encounter conditions needed for growth and development. A **seed** contains an embryo, along with a food supply, covered by a protective coat. In contrast, as you learned in Chapter 22, a spore consists only of a single haploid cell with a hard, outer wall.

In spore-releasing plants, which include mosses and ferns, the sperm swim through a film of water to reach the egg. In seed-producing plants, which include all conifers and flowering plants, sperm are able to reach the egg without swimming through a film of water. This difference explains why non-seed plants require moister habitats than most seed producers do.

Alternation of generations

The lives of all plants consist of two alternating stages, or generations. The gametophyte generation of a plant is responsible for the development of gametes. All cells of the gametophyte, including the gametes, are haploid (n). The sporophyte generation is responsible for the production of spores. All cells of the sporophyte are diploid ($2n$). The spores are produced in the sporophyte plant body by meiosis, and therefore are haploid (n).

ThinkingLab — Make a Hypothesis

What is the function of trichomes?

Trichomes are hairlike or scaly outgrowths of the epidermis of stems and leaves of some plants. Leaves with trichomes can feel prickly, sticky, fuzzy, or woolly.

Analysis

Examine the different kinds of trichomes on the leaves shown here and think about what their function might be.

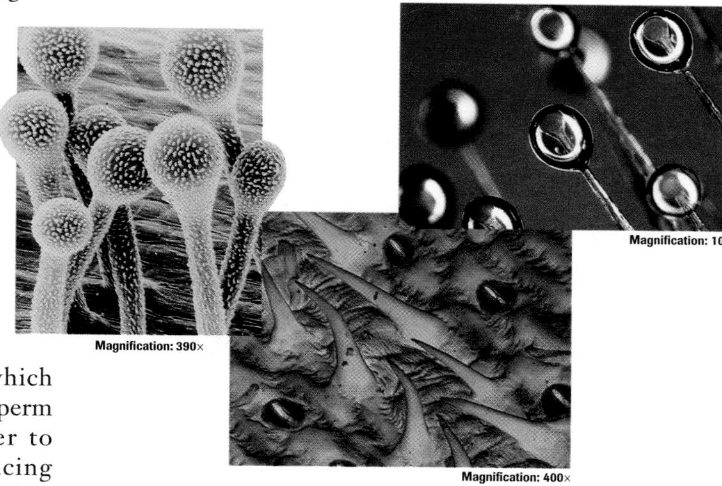
Magnification: 390×

Magnification: 10×

Magnification: 400×

Thinking Critically

Propose a hypothesis about the adaptive value of these structures to the survival of each plant.

All plant life cycles include the production of spores. In non-seed plants such as ferns, the spores have hard outer coverings and are released directly into the environment, where they grow into gametophytes. In other plants, such as conifers and wildflowers, the spores are retained by the parent plant and develop into gametophytes of only a few cells retained within the sporophyte. These plants, usually called seed-producing plants, release the new sporophytes into the environment in the form of seeds.

ThinkingLab — Make a Hypothesis

Purpose

LS **Visual-Spatial** Students will form a hypothesis about the functions of leaf structures.

Process Skills

observe and infer, form a hypothesis

Teaching Strategies

- If available, have students use a stereomicroscope or hand lens to observe trichomes on leaf specimens. Tomato, geranium, or snapdragon leaves will provide good examples of trichomes.

Background

Stinging nettle *(Urtica dioica)* trichomes contain formic acid. When touched, their tips break off and the sharp spines pierce the skin, producing a burning irritation.

Thinking Critically

Typical answers may suggest that the role of trichomes is to reduce evaporation from the leaf surface and to protect the leaves from herbivores.

Assessment

Performance: Have students design an experimental procedure to test one of their suggested hypotheses. L1

CULTURAL DIVERSITY

Ethnobotany

Introduce students to the growing field of ethnobotany, the study of how native cultures use plants. Ethnobotany has become increasingly important to the development of new drugs by pharmaceutical companies. One firm, Shaman Pharmaceuticals in California, sends researchers into South American, Asian, and African tropical forests to ask healers about the medications they derive from plants. Have students research and write reports about the medicinal uses of plants in different cultures, both in contemporary and historical perspectives. One source of such material is the December 1993 issue of *Discover* magazine. LS

SECTION 24.1

Microscope Activity

Visual-Spatial Allow students to observe moss leaves under the microscope. Have them prepare wet mounts and ask them to note the chloroplasts within cells, thickness of leaf, and any evidence of guard cells or stomata. Have students diagram what they observe. L2 P

Science, Technology, and Society

Coal Formation

Paleozoic swamps abounded in giant horsetails and tree ferns. As these large plants fell into swamps, they became waterlogged and sank into the mud. In time, this plant material became peat. As more plant material accumulated and the peat became compacted, lignite coal, a porous fossil fuel, was produced. With further compaction, bituminous and finally anthracite coal was formed.

interNET CONNECTION

Follow the link for this chapter on the Glencoe Homepage at **www.glencoe.com/sec/science** to find out more about plants.

Phylogeny of Plants

Many changes have taken place since the first plants became adapted to life on land. Landmasses have moved from place to place over Earth's surface, climates have changed, and bodies of water have formed and disappeared. Hundreds of thousands of plant species have evolved, and countless numbers of these have become extinct as conditions continually changed. Plants have adaptations that enable them to survive in almost every type of environment found on Earth. Taxonomists have classified plants into ten divisions. The five non-seed plant divisions are shown in *Figure 24.5* and *Figure 24.6.*

Bryophyta

Bryophytes are nonvascular, non-seed plants that include mosses and liverworts. These two groups of species are, for the most part, small and limited to moist habitats. Their leaves are only one to two cells thick. Spores of the bryophytes are formed in capsules.

Psilophyta

Psilophytes, also known as whisk ferns, consist of thin, green, leafless stems and are thought to represent the first land-dwelling, non-seed vascular plants. Each stem is covered with small, leaflike scales. Most of the 30 species of psilophytes are tropical or subtropical, although one genus is found in the southern United States.

Figure 24.5

The relationships of divisions of non-seed plants on the Geologic Time Scale show that the nonvascular bryophytes and vascular plants are closely related.

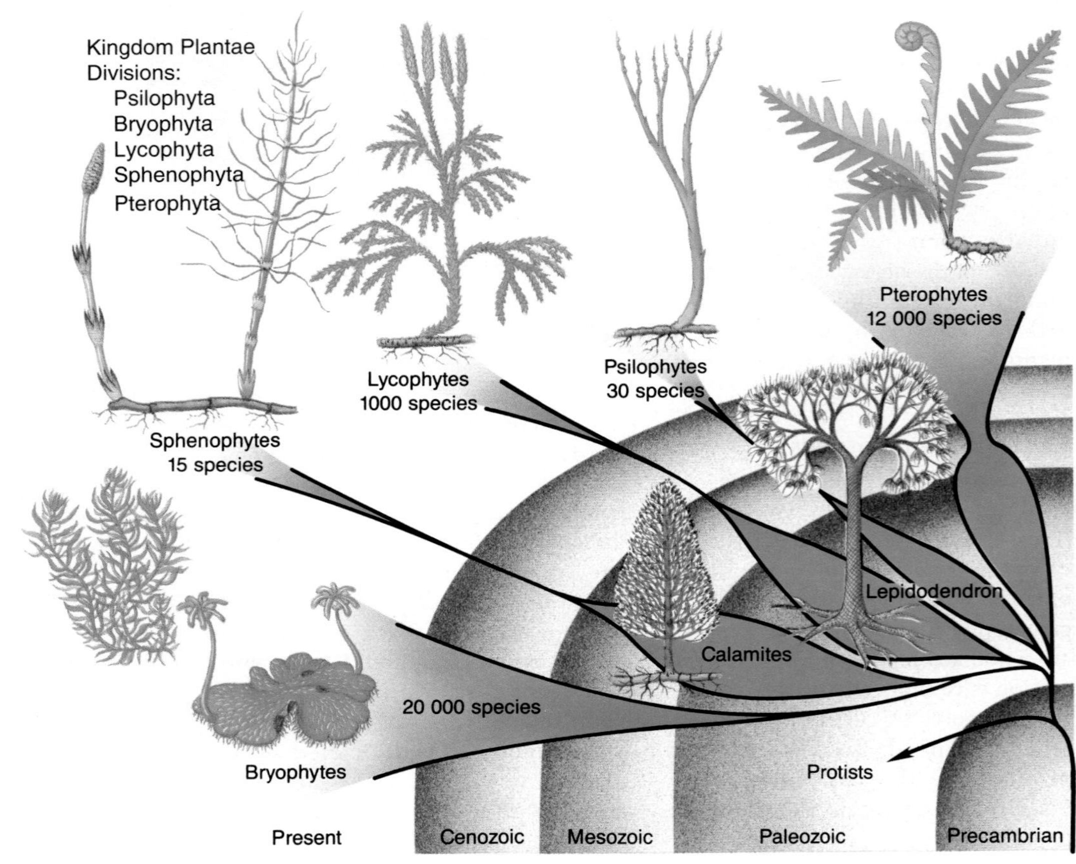

590 What Is a Plant?

STUDENT JOURNAL

Making a Classification Key Ask students to prepare a biological key that will enable others to identify members of the first three divisions of plants. As an alternate activity, students may wish to prepare a concept map that serves as a key. L1

Lycophyta

Lycopods, the club mosses, are simple, non-seed vascular plants adapted primarily to moist environments. Those species that exist today are only a few centimeters high, but their ancestors grew as tall as 30 m and formed a large part of the vegetation of Paleozoic forests. These ancient forests are now used by people in the form of coal.

Sphenophyta

Sphenophytes, the horsetails, are non-seed vascular plants. They have hollow, jointed stems surrounded by whorls of scalelike leaves. Although primarily a fossil group, about 15 species of sphenophytes exist today. All present-day horsetails are small, but their fossil relatives were the size of trees.

Pterophyta

Pterophytes, ferns, are the most well-known and diverse group of non-seed vascular plants. Ferns were abundant in Paleozoic and Mesozoic forests. Ferns have leaves that vary in length from 1 cm to 500 cm. Most ferns grow in the tropics.

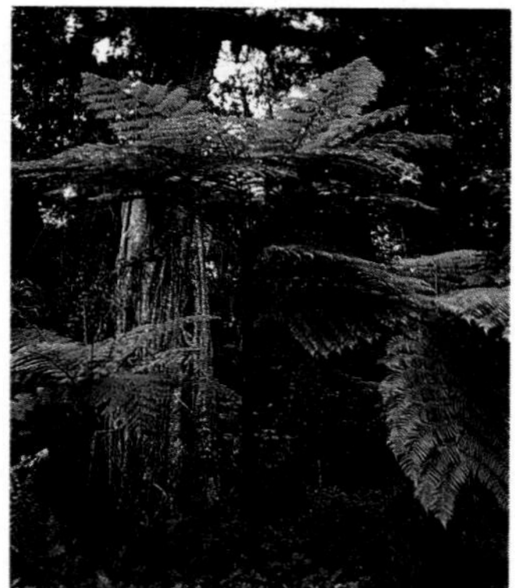

Tree ferns like this *Cyathea arborea* can still be found growing in damp, tropical forests.

Figure 24.6

The plant kingdom includes five divisions of non-seed plants. Can you name the correct division of each of the plants shown here?

***Selaginella,* a spike moss, produces two types of spores that develop into male and female gametophytes.**

***Equisetum* has true roots, stems, and leaves, but the stems are hollow and appear jointed.**

***Marchantia* is a common liverwort found on damp soil. It has air chambers on the upper surface.**

***Psilotum* sporophytes have simple stems but no leaves. They have underground rhizomes that produce rhizoids.**

Meeting Individual Needs

Visually Impaired Have students work in pairs to construct simple three-dimensional models that identify the traits of plants classified in the first three divisions. Assign students to groups consisting of those who are sighted and those who are visually impaired. Provide students with materials such as straws, coffee stirrers, thin pieces of cardboard, pieces of sponge, construction paper, string, and clay. Have students label their models and explain why they represented each structure as they did. L1 COOP LEARN LS

Visual Learning

Figure 24.6 Can you name the correct division of each of the plants shown here? *Marchantia = Bryophyta, Psilotum = Psilophyta, Selaginella = Lycophyta, Equisetum = Sphenophyta, Cyathea = Pterophyta.*

Display

LS **Visual-Spatial** Have students photograph a variety of plants or bring in pictures of various plants cut from magazines. As a class, develop a bulletin board display that places each plant pictured, along with its name, in its correct division. Use the completed display as a teaching tool throughout the unit. L1 COOP LEARN

Demonstration

LS **Visual-Spatial** Illustrate the water-holding capacity of sphagnum moss by doing the following. Add 100 mL of water to a beaker. Add a large ball of sphagnum moss (available from garden shops) that has been wrapped in cheesecloth. Allow the moss to remain in the water for a few minutes. Remove the moss from the beaker and have students note the volume of water remaining in the beaker. Ask the following questions: (a) Why is moss added to soil? *It enables soil to retain water.* (b) How can the amount of water that was absorbed by the cheesecloth be measured? *Squeeze the cheesecloth into a graduated cylinder and measure the volume of water.* (c) What hypothesis can you state about what structure actually absorbs the water? *Absorption of water occurs in the many empty cells between the leaf cells, which form a sponge-like pattern of growth.*

Health Connection

Anticancer Plants

The National Cancer Institute has reported that soybeans contain two chemical compounds that appear to have anticancer agents. These chemicals are isoflavones and phytosterols. Isoflavones and phytosteroles are present in many other plants, but are highly concentrated in soybeans.

So far, direct studies on the anticancer properties of these chemicals have not been conducted on humans, but animal studies show promise. In addition, a recent study has shown that soy products appear to protect younger women against breast cancer. This evidence is supported in Japanese women, who show a rate of breast cancer that is one-fourth that of Americans. The Japanese diet is rich in soy products.

3 Assess

Check for Understanding

Have students explain the difference between the following word groups:

(a) seed - spore
(b) vascular - nonvascular plant
(c) stomata - guard cells
(d) sporophyte - gametophyte
(e) roots - rhizoids L1

Reteach

Have students prepare an outline that addresses the following: (a) evidence that land plants evolved from green algae, (b) adaptations evolved by land plants that enabled them to survive out of water, and (c) the major functions of leaves, roots, and stems. L1

Figure 24.7

Plants classified into the five divisions of seed-producing plants produce seeds covered by tough, protective seed coats. In gymnosperms, the seeds are held on woody scales that form cones. In angiosperms, seeds are protected inside a fruit.

▲ *Welwitschia mirabilis* is found in harsh desert environments in Africa. The leaves of this plant can grow to 2 m long.

▲ Cycads are often mistaken for ferns or small palm trees.

► *Ginkgo biloba*, the maidenhair tree, is no longer found in the wild, although it continues to be cultivated in many countries, including the United States.

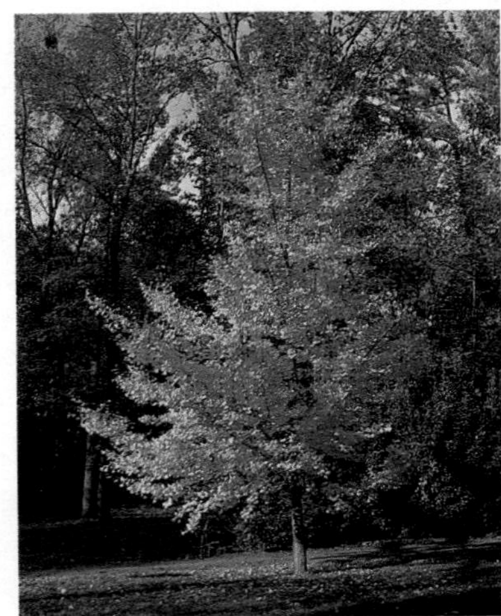

Examples from the seed-producing plant divisions are shown in *Figure 24.7.*

Cycadophyta

Cycads were abundant during the Mesozoic Era. Today, there are about 100 species of cycads, all of them short, palmlike trees with scaly trunks. Seeds are produced in cones, with male and female cones on separate trees. Cones of cycads may be as long as 1 m.

Gnetophyta

There are three genera of gnetophytes, each of which is quite distinct. *Gnetum* includes species of trees and climbing vines, *Ephedra* includes shrubby species, and *Welwitschia* is found only in South Africa. *Welwitschia* has a short stem that grows as a large, shallow cap. The leaves grow from the base of the stem. These seed plants grow in the desert and can live to be 100 years old.

Ginkgophyta

This division has only one living species, *Ginkgo biloba*, a distinctive tree with fan-shaped leaves. Like cycads, ginkgoes have separate male and female trees. The seeds have an unpleasant smell, so ginkgoes planted in city parks are usually male trees. Ginkgoes are hardy and resistant to insects and to air pollution.

Meeting Individual Needs

Learning Disabled Ask students to prepare a table that compares and contrasts the plants classified as angiosperms and gymnosperms. Encourage students to include comparisons for size, diversity, species numbers, and economic use by humans, as well as the reproductive structure and seed differences. L1 LS

STUDENT JOURNAL

Classifying Plants Have students prepare a key or concept map that can be used to identify members of the five seed-producing plant divisions. L2 LS

Wildflowers can be found in nearly every environment on Earth. There are more species of anthophytes than of any other plant division. This is a wildflower called chicory.

Pinus banksiana, the jack pine, keeps its cones closed until a fire passes over them, an example of an adaptation to extreme conditions.

Coniferophyta

These are the conifers—cone-bearing trees such as pine, fir, cypress, and redwood. Conifers are vascular seed plants that produce naked seeds in cones. Species of conifers can be identified by the characteristics of their needlelike or scaly leaves. The oldest living trees in the world are members of this plant division.

Anthophyta

Anthophytes, commonly called the flowering plants, are the largest, most diverse group of seed plants living on Earth. Fossils of the Anthophyta only date to the Cretaceous period, 130 million years ago. Anthophytes produce seeds enclosed in a fruit. This division has two classes: the monocotyledons and dicotyledons.

Section Review

Understanding Concepts

1. What is the primary difference between seeds of conifers and anthophytes?
2. Explain how development of the cuticle, stomata, and the vascular system influenced the evolution of plants on land.
3. List the sequence of events involved in the alternation of generations in land plants.

Thinking Critically

4. Explain why the alternation of generations is of adaptive value for plants living on land.

Skill Review

5. **Comparing and Contrasting** How does the life of a non-seed, nonvascular plant differ from the life cycle of a flowering plant? For more help, refer to Thinking Critically in the *Skill Handbook.*

Extension

Have students prepare a key that will enable them to distinguish among the ten plant divisions. Encourage students with well-developed keys to photocopy their keys and share them with their classmates. L2 COOP LEARN

Assessment

Knowledge: Have students explain how each of the following has enabled plants to survive on land: (a) leaves that are thin (b) stomata and guard cells (c) a waxy cuticle (d) roots or rhizoids (e) supportive tissue in stems (f) thick covering around gamete cells. L1

4 Close

Using an Analogy

Have students correlate the following items with an appropriate plant structure. Ask them to give reasons for their choice. Items: drinking straw, wax paper, skin pores, Velcro, steel beams, drawers of a cabinet, bag of sugar, roof shingles. L1

Answers to Section Review

1. Seeds of conifers are produced in cones and are naked. Anthopyhte seeds are formed within flowers and are enclosed in a fruit.
2. A cuticle reduced water loss, stomata allowed for gas exchange within the leaf, a vascular system allowed for transport of food, water, and minerals to all plant parts.
3. Sporophytes form spores via meiosis, spores become the gametophyte which forms gametes via mitosis, gametes fuse to form a new sporophyte generation.

Thinking Critically

4. Sporophyte generations enabled plants to evolve complex structures such as vascular roots, stems, and leaves, which allowed them to adapt to a land environment.

Skill Review

5. Non-seed nonvascular plants show a dominant or visible gametophyte generation with sporophyte reduced in size. Flowering plants have a dominant sporophyte and a reduced gametophyte.

SECTION

24.2

Prepare

Key Concepts

Bryophyte characteristics are examined, with emphasis on their lack of vascular tissue and the limitations this places on mosses and liverworts. Both sexual and vegetative reproduction are discussed and alternation of generations is described in detail.

Block Scheduling

Look for this symbol for strategies that are useful in a block scheduling format. For more information on block scheduling, refer to Section 24.2 in the **Lesson Plans** booklet.

Materials

- Order algae and moss specimens from a biological supply house for the Minilab.
- Order preserved or living moss specimens for the Biolab. Specimens must contain gametophyte and sporophyte generations.

1 Focus

Bellringer

Before presenting the lesson, display **Section Focus Master 52** on the overhead projector and have students answer the accompanying questions. L1 SAE

Activity

LS **Visual-Spatial** Have students prepare a simple labeled diagram of a "typical plant." *Labels will consist of root, stem, leaves, and flowers.* Ask students to prepare a second labeled diagram of a "typical moss." Use the two diagrams as a way to introduce the traits of Bryophytes. L1

SECTION

24.2 Bryophytes

Section Preview

Objectives

Identify the structures of a typical bryophyte.

Explain the life cycle of a moss or liverwort.

Key Terms

protonema
antheridium
archegonium
vegetative reproduction
gemmae

An observant hiker in this shady forest is sure to come across patches of soft, feathery mosses covering soil, stones, rotting wood, or tree bark with a velvety layer of green. The hiker might also notice shiny liverworts growing along the stony bank of a stream. Mosses and liverworts are bryophytes. Bryophytes are spore-producing plants that have no vascular tissue. They usually live in moist, cool environments.

Characteristics of Bryophytes

Nonvascular plants are not as common or as widespread as vascular plants because they must carry on photosynthesis, reproduction, and other life functions where there is a steady supply of moisture. Adequate water is not available everywhere, so most bryophytes are limited to moist habitats such as by streams and rivers or in humid tropical forests. Lack of vascular tissue also limits the size of bryophytes. They cannot compete with neighboring vascular plants, which can easily overgrow them and cut them off from sunlight and gases. But even with these limitations, bryophytes are successful in

Figure 24.8

Mosses have a central stem surrounded by small, thin leaves.

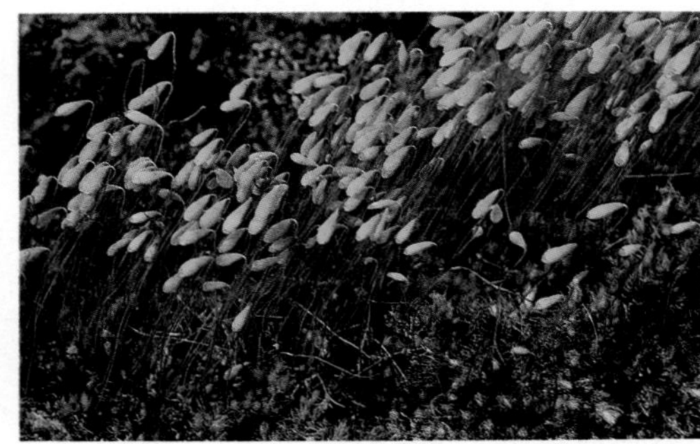

Brown stalks and spore capsules of the sporophyte generation can be seen growing from the green, leafy gametophyte of this moss.

Program Resources

Section Focus Master 52 L1 SAE
Reinforcement and Study Guide, pp. 95-96 L1
Biolab and Minilab Worksheets, pp. 96-98 L1
Critical Thinking/Problem Solving, p. 24 L3
Basic Concepts Transparency 34 and **Master** L1 SAE
Reteaching Transparency 24 and **Master** L1 SAE

habitats with adequate water. The division Bryophyta includes mosses and liverworts.

Mosses grow in sheltered places

Mosses are more familiar and more numerous than liverworts. Mosses are small plants with leafy stems and usually grow in dense carpets or tufts. Some have upright stems; others have creeping stems that lie along the ground or hang from steep banks or tree branches. *Figure 24.8* shows some typical mosses.

▼ **Mosses with creeping stems form extensive mats that retard erosion on exposed slopes.**

▼ **This tufted moss is usually found growing on the rocks of mountain slopes. Like most mosses, this plant becomes dormant during dry spells. When rainy weather returns, the plant revives and resumes photosynthesis and reproduction.**

▼ **Peat moss, *Sphagnum,* is probably the most well-known moss because of its usefulness. Compressed, dead peat moss can be dried and cut into bricks for fuel. Dried peat moss absorbs large amounts of water, so florists and gardeners use it to increase the water-holding ability of soil.**

MiniLab

How do mosses and green algae compare?

Biologists hypothesize that the first plants to migrate to land had many similarities with filamentous green algae that live in water. In this lab, you will observe similarities and differences between mosses and algae.

Procedure

1. Make a table to compare the following observations of a moss and a filamentous green alga: overall size, structures for obtaining water and nutrients, structures for support, and structures for photosynthesis.

Magnification: 50×

2. Obtain samples of both types of organisms from your teacher.
3. Make wet mounts of the structures used for photosynthesis, and observe them under low and high power. Describe similarities and differences.
4. Make wet mounts of structures used for absorbing water and nutrients, and observe them under the microscope. Describe similarities and differences.

Analysis

How do the structures of algae and mosses compare? What evolutionary advances do mosses show over algae?

2 Teach

MiniLab

Purpose

LS **Visual-Spatial** To observe and then compare the characteristics of algae and mosses.

Process Skills

compare and contrast, observe and infer, record data

Teaching Strategies

- Preserved or living moss and algae specimens may be used. It may be easier to use preserved materials.
- Suggested specimens are: algae such as *Cladophora* or *Spirogyra,* and moss such as *Mnium* with antheridia or archegonia.

Expected Results

Tables will show that algae have no specialized structures for support, obtaining water, or obtaining nutrients, and use individual cells with chloroplasts for photosynthesis. Mosses have rhizoids for obtaining water and minerals, and have leaves with chloroplasts for photosynthesis. The plant is upright and has a cuticle covering the leaves.

Analysis

Mosses and algae show similarities on a cellular level. Mosses show some specialization, with organs for obtaining water and carrying out photosynthesis.

Assessment

Journal: Have students list three observations that serve as evidence for mosses having evolved from green algae. L1

Meeting Individual Needs

Students Acquiring English Have students list as many characteristics as possible that distinguish the division Bryophyta from other plant groups. Their list should include traits of the plants and those that describe where these plants typically grow. *Nonvascular, nonseed plants, small size, live in moist, cool areas.* L1 LS SAE

Different Viewpoints in Biology

Bryophyte Classification

Some biologists have suggested that Bryophytes are vascular plants. One order of bryophytes, the Polytricales of New Zealand, show vascular tissues comparable to xylem and phloem. These mosses grow to heights of 0.5 meters making vascular tissue essential. The water-conducting cells of these mosses are called hydroids and resemble xylem tracheid cells. Food-conducting cells are called leptoids and resemble sieve phloem cells. Leptoids and hydroids have also been observed in the sporophyte stalks of most mosses.

Misconception

Students learned in Chapter 12 that the process of meiosis forms gamete cells. This is true for animals. However, in plants, spores are produced directly through meiosis and then gametes are formed through mitosis of these haploid cells.

Why the difference between plants and animals? Actually there is no difference. Plants just have an added stage or step that results from alternation of generations. In this process, meiosis still forms haploid spore cells—the gametophyte generation—that remain haploid. The gametophyte is equivalent to one large male or female gamete as is found in animals.

Figure 24.9

Liverworts have a flattened thallus or flattened leaves borne on a stem.

▲ **A leafy liverwort like this one is difficult to distinguish from a moss. Leafy liverworts have two rows of larger leaves and another row of smaller ones.**

▲ **A thallose liverwort has a distinctive appearance. Thallose liverworts called hornworts are of interest to biologists because each cell has just one large chloroplast.**

Liverworts have a flattened appearance

Like mosses, liverworts are small plants that usually grow in clumps or masses in moist habitats. However, liverworts occur in many environments, from the Arctic to the Antarctic. Some are found in water and others in deserts. They include two groups: the thallose liverworts and the leafy liverworts, *Figure 24.9.* The body of a thallose liverwort is a thallus. It is a broad, ribbonlike body that resembles a lobed leaf. Leafy liverworts are creeping plants with three rows of flat, thin leaves attached to a stem. Most liverworts have an oily or shiny surface.

Liverworts can respond to small changes in their environments, and as a result, they exhibit a wide variety of forms. The same species may be found in a compact form in its normal habitat, but have a more slender and elongated form in an environment with more moisture or diffuse light.

Reproduction in Bryophytes

As in all plants, the life cycle of bryophytes includes an alternation of generations between the diploid sporophyte and the haploid gametophyte. However, bryophytes are the only plant division in which the gametophyte generation is dominant.

Mosses produce a protonema

In mosses, the haploid spore germinates to form a structure called a protonema. The **protonema** is a small, green filament of cells that develops into either a male or female gametophyte or a gametophyte containing both kinds of reproductive structures. Liverworts have no protonema; the spore germinates and grows directly into the plant body. In both mosses and liverworts, gametophytes produce two kinds of reproductive structures, male and female. The **antheridium** is the reproductive structure in which sperm are produced. The **archegonium** is the reproductive structure in which eggs are produced.

Meeting Individual Needs

Students Acquiring English/Hearing Impaired Have students use Figure 24.9 and the text to make a table that compares liverworts and mosses. The table should have two columns labeled *Similarities* and *Differences.* Have students include both physical and reproductive differences and similarities. L1 SAE IN

PORTFOLIO

Diagramming Alternation of Generations
Have students prepare a simple flow chart diagram that depicts alternation of generations. The diagram must include these terms: *gametophyte generation, sporophyte generation, spore, gamete, diploid, haploid, mitosis,* and *meiosis.* L1 P IN

THE INSIDE STORY

Life Cycle of a Moss

Although mosses alternate between the haploid gametophyte generation and the diploid sporophyte generation, it is fairly easy to find huge carpets of mosses made up only of gametophytes.

Haircap moss

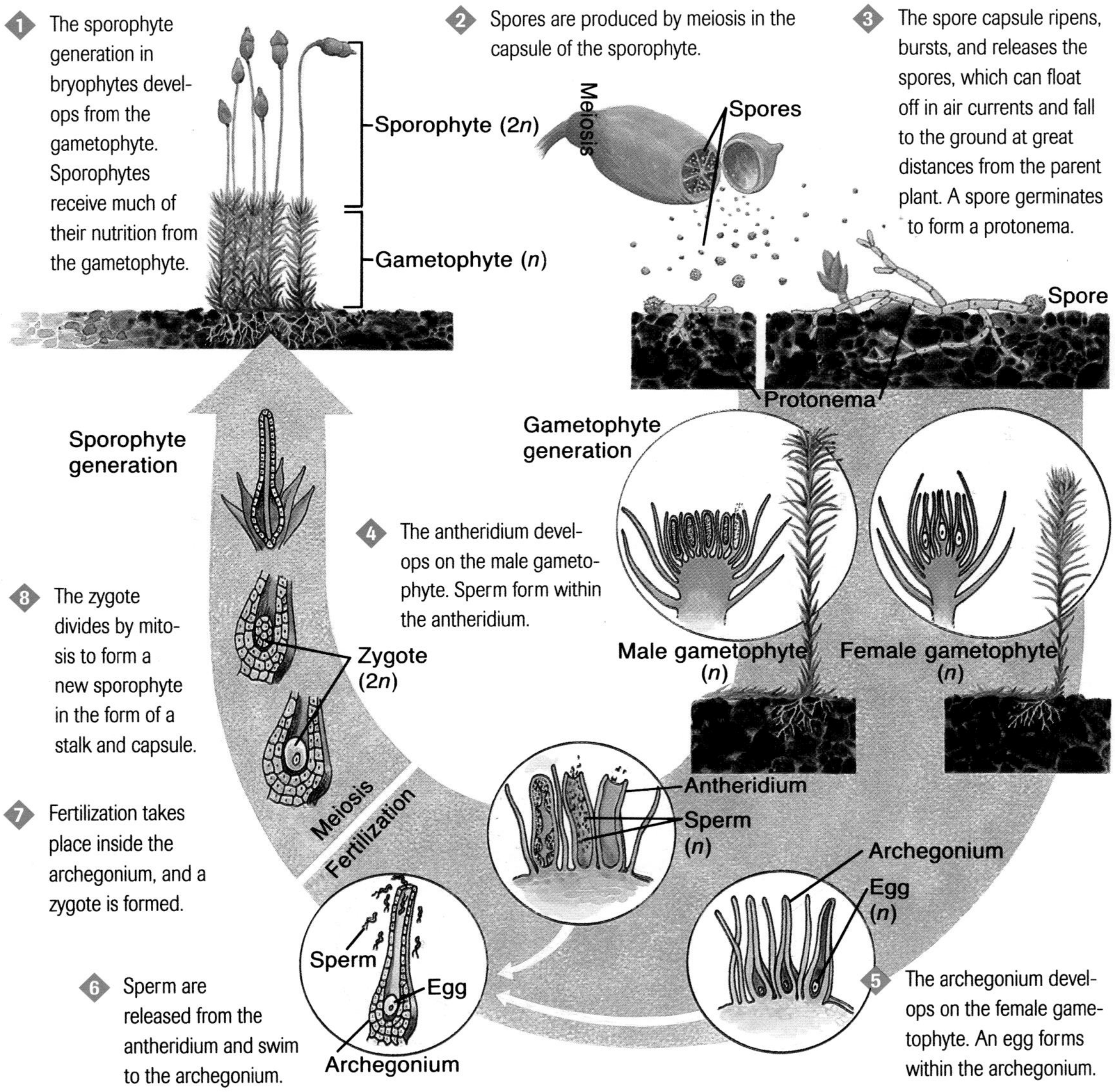

1. The sporophyte generation in bryophytes develops from the gametophyte. Sporophytes receive much of their nutrition from the gametophyte.

2. Spores are produced by meiosis in the capsule of the sporophyte.

3. The spore capsule ripens, bursts, and releases the spores, which can float off in air currents and fall to the ground at great distances from the parent plant. A spore germinates to form a protonema.

4. The antheridium develops on the male gametophyte. Sperm form within the antheridium.

5. The archegonium develops on the female gametophyte. An egg forms within the archegonium.

6. Sperm are released from the antheridium and swim to the archegonium.

7. Fertilization takes place inside the archegonium, and a zygote is formed.

8. The zygote divides by mitosis to form a new sporophyte in the form of a stalk and capsule.

Sporophyte ($2n$)
Gametophyte (n)
Meiosis
Spores
Spore
Protonema
Sporophyte generation
Gametophyte generation
Male gametophyte (n)
Female gametophyte (n)
Zygote ($2n$)
Antheridium
Sperm (n)
Archegonium
Egg (n)
Meiosis
Fertilization
Sperm
Egg
Archegonium

THE INSIDE STORY

Life Cycle of a Moss

Purpose

LS **Visual-Spatial** Students will examine the reproductive cycle, the process of alternation of generations in a moss.

Teaching Strategies

- Ask students to explain the meaning of the terms *protonema* (first thread), *archegonium* (to begin life), and *antheridium* (flower). Make sure students connect antheridium to male and archegonium to female.

Visual Learning

- Have students, working in cooperative groups, determine the chromosome condition (either diploid or haploid) for each numbered event. *1, 7, 8 are diploid, all others are haploid.*
- To reinforce the concept of diploid versus haploid, ask student groups to assign a specific chromosome number to the cells of stages 1-8. Explain that the diploid or *n* chromosome number for this species is 18. *1, 7, 8 are 18; all others are 9.*

GLENCOE TECHNOLOGY

Videodisc

Biology: The Dynamics of Life
Disc 1, Side 2
Life Cycle of a Moss (Ch. 17)

PORTFOLIO

Moss Life Cycle Have students prepare a simple life cycle diagram showing the steps that occur in the life cycle of a moss. Help them get started by showing them what the cycle may look like by providing them with this sample: sporophyte → spores by meiosis → protonema → etc. L1 P LS

Meeting Individual Needs

Students Acquiring English Have students review the key terms associated with alternation of generations by reinforcing the correlation between the name of each generation and its function. For example, sporophytes form spores, while gametophytes form gametes. L1 SAE LS

BioLab

Alternation of Generations in Mosses

Time Allotment
One class period

Objectives
Review objectives with students before they begin the Biolab.

Process Skills
observe and infer, diagram, sequence

Safety Precautions
- Slide squashes are a double thickness and are therefore too thick for high power. Only low power is to be used with each squash. If high power is desirable, remove top glass slide and replace with a coverslip after squashing.

PREPARATION

Alternate Materials
- Purchase fresh or preserved moss plants from a biological supply house. You will need three different specimens—moss antheridia, moss archegonia, and moss sporophyte. Recommended species for purchase are either *Minium* or *Polytrichum.*
- Preserved materials will be easier to maintain. Collected materials may be difficult to identify as to sex; sporophyte generations may not be available.
- Prepared slides may be substituted for all three samples, thus avoiding the need for students to prepare squashes.

BioLab

Alternation of Generations in Mosses

Most plants have life cycles involving *alternation of generations.* This term means that a gametophyte, or gamete-bearing generation, alternates with a sporophyte, or spore-bearing generation. In more complex plants, the gametophyte is greatly reduced in size and the sporophyte predominates. In mosses, both generations are visible to the naked eye, although the gametophyte is larger. A moss gametophyte is the conspicuous, leafy, green plant you are most familiar with. The sporophyte is the stalklike structure with a capsule at the tip.

PREPARATION

Problem
What are the similarities and differences of the gametophyte and sporophyte reproductive structures in mosses?

Objectives
In this Biolab, you will:
- **Distinguish** between gametophyte and sporophyte generations in a moss plant.
- **Diagram** the alternation of generations in mosses.

Materials
microscope
microscope slides (4)
single-edged razor blade
water
forceps
dropper
moss plants with male and female gametophytes and sporophytes
paper towels

Safety Precautions
Use care in working with razor blades.

PROCEDURE

Part A: Reproductive Structures of the Gametophyte

1. Obtain moss gametophyte plants with both male and female reproductive structures.
2. With forceps, remove all the leaves from the upper 1-cm portion of the stem of each plant. Be careful not to remove the reproductive structures at the tip of the stem.

598 What Is a Plant?

PROCEDURE

Teaching Strategies
- Do not advise students as to the sex of their samples. Have them determine if a specific sample is the male antheridium or female archegonium.
- Demonstrate the squash technique for students who may not know the proper procedure.
- Allow one group of students to perform the entire lab as a demonstration using a video camera and microscope.
- Have students work together in small groups to help conserve materials and time. Place physically challenged students with classmates who can help with squash preparation.

3. Place each plant onto opposite ends of a microscope slide. With a razor blade, cut each plant stem 0.5 cm from the tip end. **CAUTION:** *The razor blade is sharp. Cut away from your body.*
4. Add several drops of water to each tip end. Place a second glass slide over the first slide. Press down firmly on the top of the slide with your thumb so that each tip end is slightly squashed. To prevent thumbprints on the slide, place a piece of paper towel over the slide before pressing on it.
5. Place the stacked slides on the stage of the microscope, keeping both slides together. Use the stage clips to keep the slides from slipping.
6. Observe the moss tip ends *only under low-power magnification.*
7. Identify each of your moss tips as having antheridia or archegonia.
8. Draw and label the following structures: male gametophyte, female gametophyte, antheridium, archegonium.

Part B: Reproductive Structure of the Sporophyte

1. Obtain another specimen of a moss plant in the sporophyte stage. This plant will have a thin stalk and a capsule sticking out from the tip end of the gametophyte.
2. Remove the small capsule from the tip of the stalk, and mount it in several drops of water on a microscope slide. Add a second slide as you did for the gametophyte, and squash the capsule.
3. Observe the squashed capsule *under low-power magnification.*
4. Draw and label the following structures: sporophyte, capsule, spore.

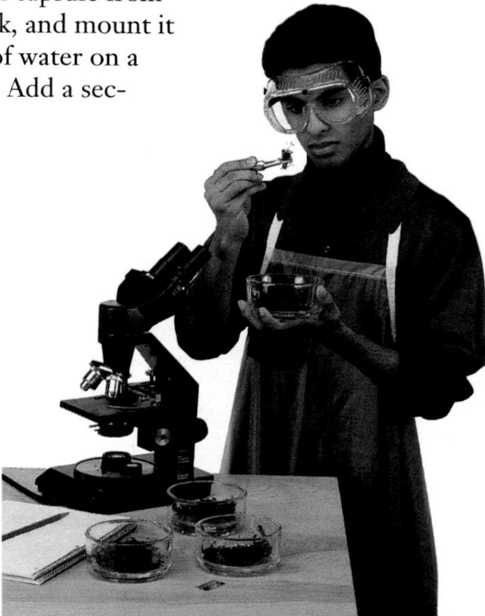

ANALYZE AND CONCLUDE

1. **Observing and Inferring** What type of reproductive cell is formed by the archegonium? By the antheridium? By the sporophyte? Which of these cells are haploid and which are diploid?
2. **Formulating Models** Using your observations of moss reproductive structures and information from the text, diagram the life cycle of a moss. Make sure your diagram includes the following: gametophyte, archegonium, antheridium, egg, sperm, zygote, sporophyte, spore capsule, and spore. Label each stage as haploid or diploid, and mark where meiosis and fertilization take place.
3. **Observing and Inferring** Which kind of gamete is produced in greater numbers, eggs or sperm? What is the adaptive advantage of this?

Going Further

Application If petri dishes and a growth medium (agar) are available, germinate moss spores and examine the developing moss gametophyte under the microscope.

Data and Observations

- Antheridia are typically round and will contain many small sperm cells (sperm may be difficult for students to see).
- Archegonia are vase shaped and will contain a single egg cell at each base (eggs may be difficult for students to see).
- Spores will be easily observed within the squashed capsule of the sporophyte.
- Check to see that student diagrams are correctly labeled.

ANALYZE AND CONCLUDE

1. Archegonia form haploid egg cells; antheridia form haploid sperm cells; sporophyte forms haploid spores.
2. Gametophyte forms antheridia or archegonia (both haploid); antheridia form sperm and archegonia form eggs (both haploid); sperm fertilizes egg forming a zygote (now diploid); zygote forms sporophyte that produces a spore capsule (all are diploid); spores are formed in capsule via meiosis; spores (haploid) leave capsule and form gametophyte (haploid).
3. Sperm are produced in greater numbers. In swimming to the eggs, many sperm will not survive. With greater numbers, the chances of successful fertilization are higher.

Assessment

Skill: Provide students with a life cycle diagram of a liverwort. Have them correctly label the structures present on the diagram. The structures are quite similar to those of mosses, and all structure names will be the same. L1 P

Going Further

Have students observe *Sphagnum* moss under the microscope. Challenge students to use their observations to explain why this moss absorbs large amounts of water. L2

SECTION 24.2

3 Assess

Check for Understanding

Ask students to explain the relationships of the following word groups:

(a) sperm - antheridium

(b) egg - archegonium

(c) protonema - spore L1

Reteach

Have students make a flowchart of the life cycle of a moss. Included in the events chain should be the ploidy condition of the chromosomes for each structure. L1

Extension

Have students use references to complete a diagram of the life cycle of *Selaginella*, a plant belonging to the Division Lycophyta. L1

Assessment

Oral: ask students to supply the next stage or structure in the moss life cycle. Sporophyte capsule, *protonema;* antheridium, *archegonium;* zygote, *gametophyte.* L1

4 Close

Activity

Have students write three questions based on the material covered in this section. Have students read their questions while their classmates provide the answers. L1

COOP LEARN

Figure 24.10

Small cups filled with tiny gemmae have formed on the thallus of this liverwort. Rainfall washes the gemmae onto the soil, where they develop rhizoids and grow into new thalluses.

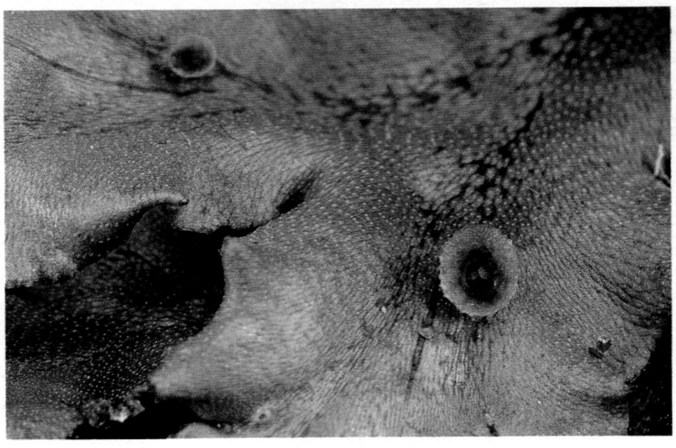

Vegetative reproduction in bryophytes

Like all plants, mosses and liverworts can also reproduce asexually, as shown in *Figure 24.10.* In asexual, or **vegetative reproduction,** a plant gives rise to new individuals without going through the alternation of generations. Some mosses can break up into pieces when the plant is dry and brittle. With the arrival of wetter conditions, these pieces each become a whole plant. Vegetative reproduction in liverworts and some mosses involves the formation of **gemmae,** which are tiny, haploid bodies that grow in cups on the surface of the gametophyte.

Connecting Ideas

Bryophytes represent the only division of plants that lack vascular tissue. This lack of a transport mechanism for water and nutrients limits the size and distribution of mosses and liverworts. Vascular plants, on the other hand, have tissues adapted for such transport. The ability to transport needed materials from the soil to plant parts above ground level enabled vascular plants to grow taller and led to further adaptations to life on land.

Section Review

Understanding Concepts

1. How can you tell a moss and a thallose liverwort apart?
2. In what way is the sporophyte generation of a moss dependent on the gametophyte generation?
3. How is the formation of spores an adaptation to life on land?

Thinking Critically

4. Explain why reproduction in a moss depends upon water.

Skill Review

5. **Sequencing** Sequence the events in the life cycle of a moss, beginning with the protonema. For more help, refer to Organizing Information in the *Skill Handbook.*

Answers to Section Review

1. Thallose liverworts have a flat, broad, lobed, ribbonlike body while mosses have small leaves on an upright stem.
2. Sporophyte generations receive their food, water, and minerals from the gametophyte.
3. Spores have a tough outer covering that allows them to resist dry conditions that may be encountered on land.

Thinking Critically

4. Sperm formed by an antheridium must swim to an archegonium in order for fertilization to occur. Sperm are therefore dependent on water for this event.

Skill Review

5. Protonema forms male or female gametophyte, sperm form in antheridium, eggs form in archegonium, fertilization of egg by sperm forms a zygote, zygote forms sporophyte generation, spores are released from sporophyte, spores form protonema.

REVIEWING MAIN IDEAS

24.1 Adapting to Life on Land

- Plants are multicellular eukaryotes with a cuticle and cells that are surrounded by cell walls. Plants have chlorophyll for photosynthesis and store food in the form of starch.
- All plants on Earth probably evolved from filamentous green algae that lived in ancient oceans. The first plants to make the move from water to land may have been leafless.
- Adaptations for life on land include a waxy cuticle that helps prevent water loss; stomata that open to allow gas exchange; development of leaves, roots, and stems; development of spores and seeds; and alternation of the gametophyte and sporophyte generations in the life cycle.
- The plant kingdom includes five divisions of spore-producing plants and five divisions of seed-producing plants.

24.2 Bryophytes

- Bryophytes are spore-producing plants that have no vascular tissue and reproduce by forming spores. They usually live in moist, cool environments, and the gametophyte generation is dominant.
- Mosses are small plants with leafy stems. Spores germinate to form a protonema, from which the gametophyte grows.
- Thallose liverworts resemble a leathery, ribbonlike or lobed leaf. Leafy liverworts have flat, narrow leaves attached to a stem.
- Liverwort spores germinate to form plant bodies; there are no protonemas as there are in mosses.

Key Terms

Write a sentence that shows your understanding of each of the following terms.

antheridium	root
archegonium	seed
cuticle	stem
gemmae	stomata
leaf	vascular plant
nonvascular plant	vegetative reproduction
protonema	

Understanding Concepts

1. Which of the following characteristics is NOT found in plants?
   a. eukaryotic cells
   b. cellulose cell walls
   c. prokaryotic cells
   d. waxy cuticle
2. Plants and green algae share all of these traits EXCEPT ____________.
   a. reproduce by fission
   b. cellulose cell walls
   c. store food as starch
   d. same kind of chlorophyll
3. The opening and closing of a stoma is regulated by ____________.
   a. the cuticle
   b. guard cells
   c. rhizomes
   d. light
4. A ____________ is the broad, flat organ of a plant that is responsible for photosynthesis.
   a. root
   b. leaf
   c. stem
   d. mitochondrion
5. The plant organ that absorbs water and minerals from the soil is the ____________.
   a. root
   b. leaf
   c. stem
   d. stoma

Chapter 24 Review

Reviewing Main Ideas

Summary statements can be used by students to review the major concepts of the chapter.

Key Terms

Answers should go beyond defining the terms. Accept any answer that uses the term correctly and in the proper context.

Understanding Concepts

1. c
2. a
3. b
4. b
5. a

Chapter 24 Review

6. a
7. b
8. b
9. a
10. a
11. c
12. c
13. d
14. c
15. c
16. b
17. b
18. a
19. b
20. d

Applying Concepts

21. Bryophytes provide nutrients to the soil when they die and decompose. They also retain water on the forest floor and prevent erosion.
22. Once moistened, certain moss species tend to swell as they retain water. This swelling enabled them to be used for filling cracks in boats and buildings.
23. If the rise in temperature is accomplished by more moisture, then mosses and liverworts will flourish. If there is only a rise in temperature, mosses and liverworts may become endangered due to water loss from higher rates of evaporation.
24. conservation of water in thick mats, ease of fertilization for sperm cells, recycling of nutrients from dead plants
25. The pioneer plants that first appear in succession are often mosses and liverworts. The adaptations that allow for survival on barren land environments may be the same adaptations that allowed for survival from water to a land environment.

Chapter 24 Review

6. Which of the following is NOT part of a seed?
   a. haploid cell   c. food supply
   b. protective coat   d. embryo
7. All plant life cycles include the production of ____________.
   a. seeds   c. roots
   b. spores   d. flowers
8. Moss gametophytes are ____________ and form gametes by ____________.
   a. diploid; meiosis
   b. haploid; mitosis
   c. diploid; mitosis
   d. haploid; meiosis
9. An important characteristic that all bryophytes share is ____________.
   a. alternating generations
   b. thick cuticles
   c. vascular tissue
   d. that they store food in stems
10. The male gametophyte of a liverwort is called a(n) ____________.
    a. antheridium   c. gemma
    b. archegonium   d. stoma
11. Which of the following is a vegetative reproductive structure of liverworts?
    a. antheridium   c. gemma
    b. archegonium   d. stoma
12. Which of the following is a non-seed plant?
    a. conifer   c. fern
    b. wildflower   d. cycad
13. Which group is theorized to be ancestral to land plants?
    a. cyanobacteria   c. archaebacteria
    b. bryophytes   d. green algae
14. Which of the following adaptations was critical for plants to adapt to life in drier areas?
    a. production of spores
    b. loss of cuticle
    c. vascular tissue
    d. alternation of generations
15. Which of the following organisms is characterized by having vascular tissue?
    a. bacteria   c. ferns
    b. green algae   d. bryophytes
16. The germinating stage of a moss spore is called a ____________.
    a. stoma   c. gemma
    b. protonema   d. cuticle
17. The waxy covering of a leaf is called a(n) ____________.
    a. stoma   c. gemma
    b. cuticle   d. epidermis
18. Liverwort spores germinate to produce the ____________.
    a. plant body   c. roots
    b. protonema   d. rhizoids
19. Reproduction without meiosis is called ____________.
    a. protonema
    b. vegetative reproduction
    c. vascular reproduction
    d. rhizoid production
20. The protonema of a moss most resembles ____________.
    a. the root of a vascular plant
    b. the leaf of a vascular plant
    c. the stem of a vascular plant
    d. a filamentous green alga

Applying Concepts

21. In what ways are bryophytes important to the ecosystem of a forest?
22. Centuries ago, dried mosses were used to fill in the cracks between timbers in log buildings and between the planks of boats. Explain why you think moss would work well for these purposes.
23. How might the distribution of mosses and liverworts on Earth be affected by a significant increase in worldwide temperatures?
24. What is the advantage to mosses of living in dense populations?
25. Explain why biologists think the first plants to adapt to life on land may have been similar to mosses and liverworts.

Thinking Critically

26. ***Concept Mapping*** Make a concept map that relates the following terms. Supply the appropriate linking words for your map.
    sporophyte, gametophyte, spore, diploid, haploid, meiosis, antheridium, archegonium, gamete, mitosis, fertilization
27. ***Comparing and Contrasting*** Compare what might happen to a moss and a liverwort after exposure to a long period of drought.
28. ***Observing and Inferring*** In parts of Australia, moss species with a normal height of 20 cm can sometimes be found growing as tall as 70 cm. In what type of environment would you search for these especially tall specimens?
29. ***Observing and Inferring*** You have collected a twig from an unknown tree. It has huge thorns sticking out from all directions around the stem. What functions might these thorns have?
30. ***Comparing and Contrasting*** Compare and contrast mosses and liverworts.
31. ***Classifying*** Name the distinguishing characteristics of the organism below that identify it as a plant.

ASSESSING KNOWLEDGE & SKILLS

Peat moss is often used to improve the water retention properties of soil. Students evaluated three different soil mixes, one of which contained peat moss. They poured 100 mL of water into a sample of each mix, measured the water that was not absorbed, and used their data to produce the graph.

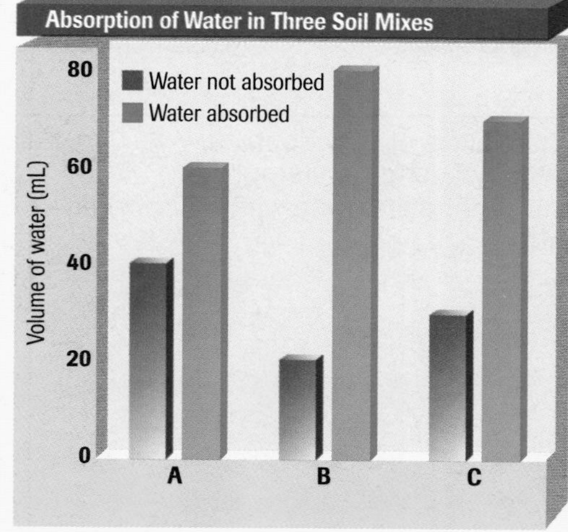

Interpreting Data Use the graph to answer the following questions.

1. What was the largest amount of water absorbed in any sample?
   a. 20 mL **c.** 60 mL
   b. 40 mL **d.** 80 mL
2. Which soil mix was the most efficient absorber of water?
   a. A
   b. B
   c. C
   d. They are equal.
3. ***Designing an Experiment*** How might you determine whether the other mixes could be improved by the addition of moss?

Program Resources

Chapter Assessment, pp. 139-144 L1
Alternate Assessment in the Science Classroom
Computer Test Bank L1
Content Mastery, pp. 93-96 L1

Chapter 24 Review

Thinking Critically

26. Evaluate students' concept maps. The maps should show an understanding of the relationships among the concepts listed.
27. Both plants may either die or become dormant. Certain species of mosses are capable of storing water between their cells and may therefore survive more easily than liverworts.
28. The environment would have to have a constant source of water, good supply of nutrients, and a reasonable amount of sunlight —possibly near the banks of a stream, lake, or brook within a forest.
29. Thorns may be protective, preventing herbivorous animals from eating the plant.
30. Mosses and liverworts are both bryophytes. They both require water for fertilization and neither grows very tall. Usually, mosses are more upright with leaflike and stemlike parts. Liverworts are usually leafy and flattened.
31. roots, stems, leaves, flowers, fruit

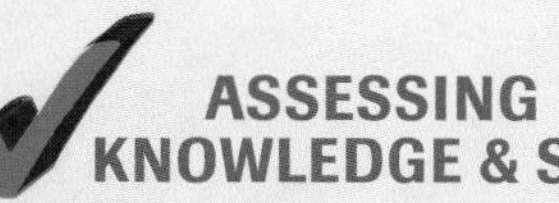

ASSESSING KNOWLEDGE & SKILLS

1. d
2. b
3. You could add additional moss and retest their absorbency.

Chapter Organizer

SECTION	OBJECTIVES	ACTIVITIES/FEATURES
25.1 Seedless Vascular Plants National Science Standards: UCP.1, UCP.3, UCP.5; C.1, C.5; G.3	**1. Explain** the importance of vascular tissues to life on land. **2. Identify** the characteristics of the Lycophyta, Spenophyta, and Pterophyta.	**Minilab:** What does a fern prothallus look like?, p. 610
25.2 Gymnosperms National Science Standards: UCP.1, UCP.3, UCP.5; A.1; C.1, C.5; F.3-6; G.1-3	**3. Identify** the characteristics of Cycadophyta, Ginkgophyta, Gnetophyta, and Coniferophyta. **4. Compare** the gymnosperm life cycle with that of seedless vascular plants.	**Earth Science Connection:** How Coal Was Formed, p. 615 **Minilab:** What are the similarities and differences of spores and pollen grains?, p. 616 **Thinking Lab:** How much acid in rain is too much?, p. 619 **Biology & Society Issues:** Should we let fires burn?, p. 620 **Biolab:** Design Your Own Experiment—How can you make a key for identifying conifers?, p. 622

ACTIVITY MATERIALS

BIOLAB	MINILABS	ALTERNATE LAB
page 622 twigs conifer branches conifer cones	**page 610** fern prothallus microscope slide coverslip pencil with eraser **page 616** pinecone, male fern frond wet mounts of pine pollen and fern spores microscope	**page 618** microscope prepared slide of gymnosperm leaf cross section

Chapter 25 Ferns and Gymnosperms

TEACHER CLASSROOM RESOURCES

Reproducible Masters	Transparencies
Section Focus Master 53: Fern Life Cycle L1 SAE **Reinforcement and Study Guide,** pp. 97-98 L1 **Biolab and Minilab Worksheets,** p. 99 L1 **Concept Mapping:** Life Cycle of a Fern, p. 25 L1 **Laboratory Manual:** How Are Ferns Affected by Lack of Water?, pp. 143-146 L2	**Basic Concepts Transparency #35:** Life Cycle of a Fern L1 SAE
Section Focus Master 54: Gymnosperm Cones L1 SAE **Reinforcement and Study Guide,** pp. 99-100 L1 **Biolab and Minilab Worksheets,** pp. 100-102 L1 **Critical Thinking/Problem Solving:** Distribution of *Ginkgo biloba,* p. 25 L3 **Content Mastery,** pp. 97-100 L1	**Basic Concepts Transparency #36:** Life Cycle of a Gymnosperm L1 SAE **Basic Concepts Transparency #37:** Phylogeny of Gymnosperms L1 SAE **Reteaching Transparency #25:** Life Cycle of a Pine L1 SAE

ASSESSMENT MATERIALS

Chapter Assessment, pp. 145-150
Alternate Assessment in the Science Classroom
MindJogger Videoquiz
Computer Test Bank

Spanish Resources SAE
English/Spanish Audiocassettes SAE
Cooperative Learning in the Science Classroom COOP LEARN
Lesson Plans

KEY TO TEACHING STRATEGIES

L1 Level 1 activities should be within the ability range of all students including those with learning difficulties.

L2 Level 2 activities are within the ability range of average to above-average students.

L3 Level 3 activities are designed for the ability range of above-average students.

SAE SAE activities should be within the ability range of Students Acquiring English.

COOP LEARN Cooperative Learning activities are designed for small group work.

P These strategies represent student products that can be placed into a best-work portfolio.

These strategies are useful in a block scheduling format.

GLENCOE TECHNOLOGY

The following multimedia resources are available from Glencoe.

Biology: The Dynamics of Life
CD-ROM SAE
Videodisc Program

National Geographic Society Series
GTV: Planetary Manager
Forest

Science and Technology Videodisc Series (STVS)
Plants & Simple Organisms
Pollen Atlas
Super Trees
Ecology
Acid Rain and Plants

CHAPTER

25 Ferns and Gymnosperms

Chapter Overview

Students begin the chapter by examining the characteristics of three divisions of seedless vascular plants: the Lycophyta, club mosses; the Sphenophyta, horsetails; and the Pterophyta, ferns. The environmental conditions required for success of seedless vascular plants are also examined.

The second section of the chapter focuses on gymnosperms. An overview of their major traits, including seeds borne on scales, formation of two different spores, the elimination of water as a needed element for fertilization, and the advantage of the structure of gymnosperm seeds is provided. In addition, the plants that are classified as gymnosperms—the cycads, Cycadophyta; the ginkgo tree, Ginkgophyta; and the Gnetophyta and Coniferophyta—are introduced. The chapter concludes with a detailed examination of conifer traits, their life cycle, adaptations, and economic uses.

Key Terms

cone
cotyledon
deciduous plant
embryo
evergreen plant
frond
gymnosperm
megaspore
microspore
ovule
phloem
pollen grain
prothallus
rhizome
sorus
strobilus
tracheid
vascular tissue
xylem

604

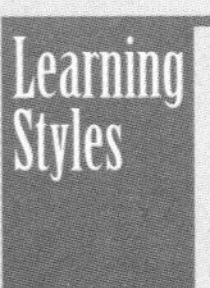

Learning Styles

Look for the following logo for strategies that emphasize different learning modalities. LS

Kinesthetic	Meeting Individual Needs, p. 608
Visual-Spatial	Demonstration, pp. 607, 609; Enrichment, p. 607; Portfolio, pp. 608, 610; Minilab, pp. 610, 616; The Inside Story, pp. 611, 621; Activity, p. 617; Alternate Lab, p. 618; Project, p. 620
Interpersonal	Meeting Individual Needs, p. 610
Intrapersonal	Student Journal, pp. 611, 614; Cultural Diversity, p. 617
Linguistic	Portfolio, p. 607; Using Science Terms, p. 614
Logical-Mathematical	Meeting Individual Needs, pp. 606, 621; Chalkboard Activity, p. 609; Time Line, p. 614; Student Journal, p. 615; Thinking Lab, p. 619

CHAPTER

25 Ferns and Gymnosperms

How is this stately sugar pine different from a tiny moss? The largest of all the pines, this sugar pine can grow to a height of 75 m, with a trunk diameter of more than 3 m. The sugar pine gets its name from resin that oozes from cracks in the bark and dries into white patches resembling sugar crystals. Huge cones hanging at the ends of branches are almost 0.5 m long and may weigh several kilograms. Each cone contains meaty seeds that will grow into new trees or perhaps provide a meal for a squirrel. Hikers beware—the sugar pine's cones make a formidable crash when they fall to the ground, and you don't want to be in the way!

The sugar pine is one example of the success with which the seed plants have adapted to life on land. In this chapter, your investigation of plant evolution continues with the study of vascular plants, including the spore-producing ferns and the seed-producing gymnosperms.

Concept Check

You may wish to review the following concepts before studying this chapter.
- Chapter 12: diploid, haploid
- Chapter 22: gametophyte, sporophyte
- Chapter 24: vascular plant

Chapter Preview

Laboratory Activities

Biolab: Design Your Own Experiment
- How can you make a key for identifying conifers?

Minilabs
- What does a fern prothallus look like?
- What are the similarities and differences of spores and pollen grains?

Ferns are vascular plants, but they share two important similarities with nonvascular bryophytes. Ferns produce spores, and they have free-swimming sperm that must travel through a surface film of water to reach the egg. Possession of vascular tissue has enabled ferns to adapt to a larger variety of habitats than their nonvascular cousins. Pines are gymnosperms. Unlike ferns, they produce seeds in cones. Pines can grow to great heights due to vascular tissue. How are ferns and gymnosperms alike?

605

Assessment Planner

Choose assessment strategies from the following pages to evaluate the progress of your students.
Assess, pp. 612, 624
Alternate Lab, pp. 618-619
Minilabs, pp. 610, 616
Portfolio, pp. 607, 608, 610
Thinking Lab, p. 619
Biolab, pp. 622-623
Chapter Review, pp. 625-627

CHAPTER 25

Introducing the Chapter

Have students examine the photo of the sugar pine and explain how its appearance differs from an oak or maple. *Likely responses will suggest that the leaves of the pine are needle-like and thin, whereas those of an oak or maple are broad and flat.*

Ask students how the appearance of the ferns differs from that of the pine, oak, or maple. *The fern is not as tall as any of the other trees and does not have a woody stem.* L1

Theme Development

The theme *unity within diversity* is carried out within this chapter as students first learn how airborne spores and vascular tissue unite ferns and gymnosperms, whereas reproductive structures—presence or absence of seeds—separate these plants. Diversity is also illustrated through the traits that divide the seedless vascular plants into three divisions. *Evolution* is apparent through the discussions of the many adaptations that all of the plants presented in the chapter have evolved for survival.

Concept Check

Students will expand upon their knowledge of haploid and diploid cells as they continue their study of alternation of generations. The structural differences that distinguish vascular plants from nonvascular plants are also reviewed and expanded upon.

SECTION

25.1

Prepare

Key Concepts

The vascular plants that comprise the divisions Lycophyta, Sphenophyta, and Pterophyta are presented along with the traits that distinguish these plants from each other. The reproductive strategies of the seedless vascular plants are presented and the stages of their life cycles emphasized through a visual learning strategy that uses the fern as a representative example.

Block Scheduling

Look for this symbol for strategies that are useful in a block scheduling format. For more information on block scheduling, refer to Section 25.1 in the **Lesson Plans** booklet.

Materials

- Obtain a thin-diameter glass tube, a petri dish, and colored water for the Focus Demonstration.
- Order living fern prothallia from a biological supply house for the Minilab.
- Order preserved specimens of horsetails for the Demonstration on page 609.

1 Focus

Bellringer

Before presenting the lesson, display **Section Focus Master 53** on the overhead projector and have students answer the accompanying questions. L1 SAE

SECTION

25.1 Seedless Vascular Plants

Section Preview

Objectives

Explain the importance of vascular tissue to life on land.

Identify the characteristics of the Lycophyta, Sphenophyta, and Pterophyta.

Key Terms

strobilus
prothallus
vascular tissue
xylem
phloem
rhizome
frond
sorus

Imagine traveling back in time to look at the vascular plants that formed Earth's forests more than 300 million years ago. The land is damp and swampy. There are huge flying insects and giant amphibians, but no birds or mammals. Even dinosaurs won't appear for another 50 million years or so. Everywhere you look, there are leafy plants and many incredibly tall, unusual-looking trees. These plants will eventually be transformed into the coal that will provide humans with fuel. How do we know what these plants looked like? Many of these ancient plants were preserved as fossils. There are plants living on Earth today that resemble these fossils of ancient vascular plants. Among them are the club mosses, horsetails, and ferns.

Lycophyta

The Lycophyta, club mosses and spike mosses, first appeared on Earth about 390 million years ago. Ancient species grew as tall as 30 m and were extremely abundant in the warm, moist forests that dominated Earth during the Carboniferous period. Most species of lycophytes died out about 280 million years ago, when Earth's climate became drier and cooler.

Lycophytes are small vascular plants

Modern lycophytes are much smaller than their ancestors. They grow close to the ground and are found mostly in damp forests, though some live in desert or mountain climates. The lycophytes are commonly called club mosses and spike mosses because their leafy stems resemble moss gametophytes, and their reproductive structures are club shaped, as shown in ***Figure 25.1***. However, these plants are not mosses, nor are they closely related to mosses. As with other vascular plants, but unlike mosses, the sporophyte generation of the lycophytes is dominant and has roots, stems, and leaves.

Leaves are adapted for reproduction

A major advance in this group of vascular plants was the adaptation of leaves into structures that protect the reproductive cells. Spore-bearing

Program Resources

Section Focus Master 53 L1 SAE
Reinforcement and Study Guide, pp. 97-98 L1
Biolab and Minilab Worksheets, p. 99 L1
Laboratory Manual, pp. 143-146 L2
Concept Mapping, p. 25 L1
Basic Concepts Transparency 35 and **Master** SAE

Meeting Individual Needs

Learning Disabled Have students prepare a list of the similarities and differences between club mosses and the mosses presented in Chapter 24. Have students use their lists to make a table with the two headings: "Bryophyte and Lycophyte Similarities" and "Bryophyte and Lycophyte Differences." L1 LS

Figure 25.1

Lycophytes are simple vascular plants with upright or creeping stems. Roots usually grow from the base of the stem. A single vein of vascular tissue runs through the center of each narrow leaf.

This club moss, *Lycopodium*, is commonly called ground pine because it is evergreen and has a treelike growth habit. The plant shown here is the sporophyte. The gametophyte is smaller and remains buried in the soil.

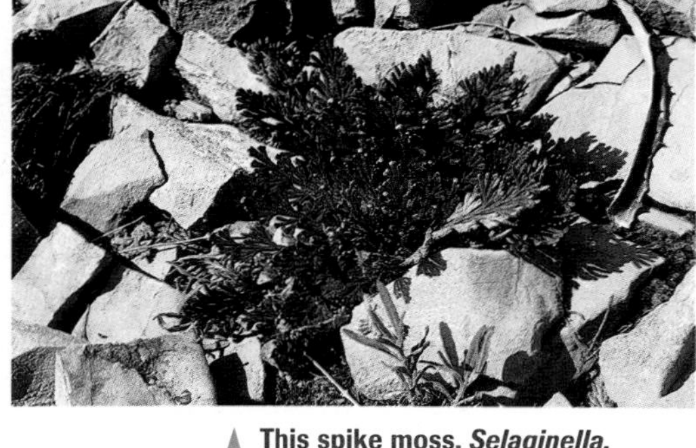

This spike moss, *Selaginella*, grows on prairie soils or dry, rocky outcrops. During dry periods, the plant appears dead, but after a rainfall the leaves turn green and the plant continues its life cycle.

leaves form a compact cluster called a **strobilus,** which is located at the end of the stem. Leaves of lycophytes occur in spirals, whorls, or pairs. For a more detailed look at the structure of a strobilus, see *Figure 25.2.* In a lycophyte life cycle, the spore germinates to form a gametophyte, which is called a **prothallus.** The prothallus is relatively small, lives in or on the soil, and produces either antheridia or archegonia. Sperm swim from an antheridium, through a film of water on the surface of the prothallus, to the egg in an archegonium. The sporophyte plant grows from the fertilized zygote and then becomes larger and dominant.

Figure 25.2

The lycophyte life cycle is similar to that of all other spore-producing plants.

Spores are produced in sporangia at the bases of small, modified leaves that form the conelike strobilus.

Some species of lycophytes have two kinds of sporangia, each of which produces a different type of spore. Small spores germinate to form a male prothallus. Large spores germinate to form a female prothallus.

A young sporophyte grows from the archegonium in a female prothallus. The prothallus with its rhizoids resembles a root with root hairs.

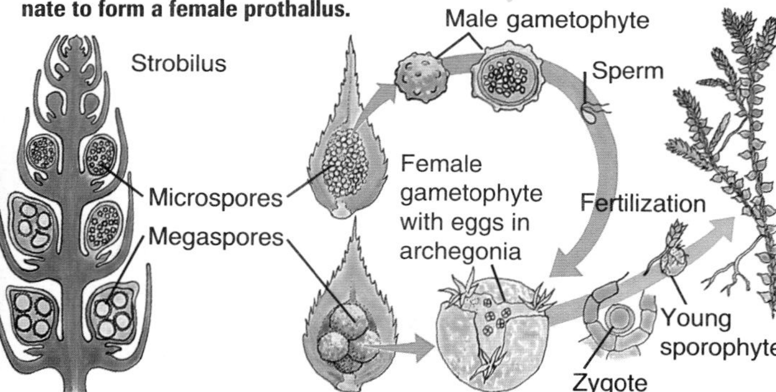

SECTION 25.1

Demonstration

Visual-Spatial Stand a glass tube with a very thin diameter in a petri dish filled with colored water. Have students note the rise of water within the tube. Explain that this movement occurs due to capillary action. Ask students to correlate the movement of water in the glass tube to the movement of water in vascular plants. Use this demonstration to explain the relationship between the evolution of vascular tissue and the increased size of plants, and the efficiency of water and mineral movement throughout the plants.

2 Teach

Audiovisual

Show the filmstrip *Plant Life Cycles: Alternation of Generations in Plants*, Carolina.

Visual Learning

Figure 25.1 Have students explain why the identification of *Lycopodium* as a "ground pine" and the identification of *Selaginella* as a "moss" is misleading. *The* Lycopodium *is not a gymnosperm as are pines. The* Selaginella *is not a bryophyte as are mosses.*

Enrichment

Visual-Spatial Purchase a resurrection plant *(Selaginella lepidophylla).* Have students examine the plant and decide if it is still alive. Soak the plant in water overnight and place it beneath a plant light. Have students observe the plant again and consider if they should revise their original conclusion about whether or not the plant is alive.

PORTFOLIO

Spike and Club Moss Questions Have students write a brief essay to explain why club mosses and spike mosses are not common in desert habitats and why most people never see the gametophyte generation of these plants. Have students explain why these plants are not related to the mosses that were discussed in Chapter 24. L2 P LS

Different Viewpoints in Biology

Classifying Club Mosses

Taxonomists do not agree on the classification of club mosses. One biologist puts them in the division Lycopodophyta, a name that means "wolf-footed plant." The name is derived from the resemblance between the shoots of the plant and the foot of a wolf.

Other biologists put them in a division called the Microphyllophyta. This division name means "small leaf plant." The disagreement regarding the naming of plant divisions is common among taxonomists who try to show relationships by placing different groups of plants or animals into species, genera, families, and divisions or phyla.

Visual Learning

Figure 25.4 Point out that the strobili at the tips of the horsetail stems are formed by spore-forming leaves. Have students use the appearance of the plant to speculate as to why it was named a horsetail. *The plant looks roughly like the tail of a horse.*

Figure 25.3

This cross section of the stem of a club moss shows tissues of xylem and phloem. Both xylem and phloem are located in the center of the stem. The central mass of phloem tissue is interrupted by irregular shaped strands of xylem tissue.

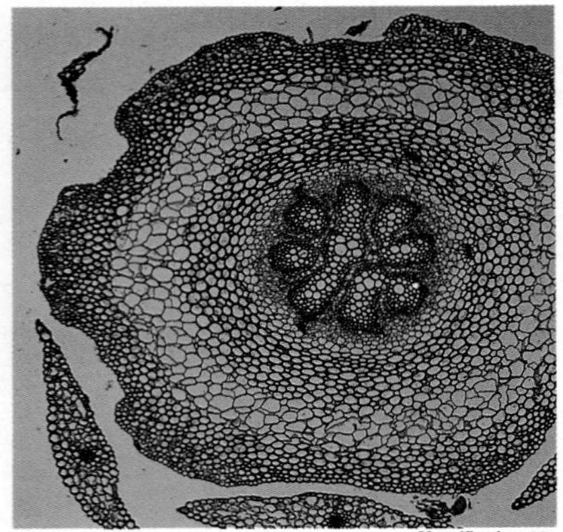

Vascular tissues transport materials

Unlike bryophytes, lycophytes contain vascular tissues. **Vascular tissues** are tissues that transport materials from one part of a plant to another. As shown in *Figure 25.3,* vascular tissues consist of xylem and phloem. **Xylem** tissue is made up of a series of dead tubular cells, joined end to end, that transport water and dissolved minerals upward from roots to leaves. **Phloem,** made up of a series of tubular cells that are still living, transports sugars from the leaves to all parts of the plant. Both xylem and phloem extend from close to the ends of the root tips, through the stem, and into the leaves of a vascular plant.

Sphenophyta

Sphenophytes, horsetails, represent a second group of ancient vascular plants. Like the lycophytes, early horsetails were tree-sized members of the forest community. There are only about 15 species in existence today, all of the genus *Equisetum.* The name *horsetail* refers to the bushy appearance of some species. These plants are also called scouring rushes because they contain silica, an abrasive substance, and were used to scour cooking utensils. If you run your finger along a horsetail stem, you can feel how rough it is.

Sphenophytes have jointed stems

Today's sphenophytes are much smaller than their ancestors, usually growing to about 1 m tall. Most horsetails, like the one shown in *Figure 25.4,* are found in marshes, shallow ponds, stream banks, and other areas with damp soil. Some species are common in the drier soil of fields and roadsides. The stem structure of horsetails is unlike most other vascular plants. The stem is ribbed and hollow, and appears jointed. At each joint, there is a whorl of tiny, scalelike leaves.

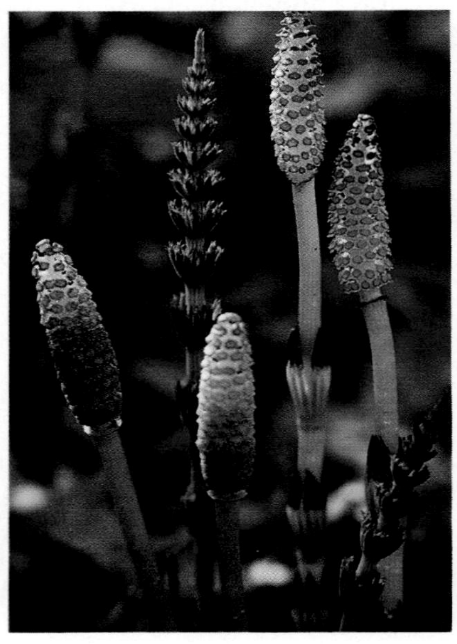

Figure 25.4

This is the sporophyte generation of a horsetail, *Equisetum.* It has thin, narrow leaves that circle each joint of the slender, hollow stem. Sporangia-bearing leaves form a strobilus at the tips of some stems.

Meeting Individual Needs

Visually Impaired Tape two straws together side by side. Have visually impaired students manipulate the model as you explain that one cell type, xylem, moves material in one direction while another cell type, phloem, moves material in an opposite direction. Point out that the tissues are located next to one another in a plant. L1 LS

PORTFOLIO

Horsetails Cross Section Have students use information they have read about horsetails to prepare what they believe would be cross-section slice diagrams of the stem of this plant. Have them add appropriate labels to structures where needed. L2 P LS

Reproduction similar to Lycophyta

Sphenophytes, like other plants, exhibit an alternation of generations. Their reproduction is similar to that of lycophytes. Sphenophytes produce spores in strobili that develop into gametophytes. A gametophyte of a sphenophyte is also called a prothallus. Eggs are produced in the archegonia, and sperm are produced in the antheridia. Sperm swim along the surface of the prothallus to the archegonia, where fertilization takes place. If more than one egg is fertilized, several sporophytes may grow from one prothallus.

Pterophyta

Ferns first appeared in the fossil record nearly 400 million years ago, at about the same time that club mosses and horsetails were the prominent members of Earth's plant population. Ferns, division Pterophyta, grew tall and treelike, forming vast fern forests. You are probably more familiar with ferns, *Figure 25.5*, than with club mosses and horsetails, primarily because ferns evolved into many more species and are more abundant. Ferns can be found in many types of environments.

Figure 25.5

There are about 12 000 species of living ferns. Many require damp, shady environments, but others live in drier, sunnier habitats.

Most modern ferns are fairly small and leafy, but many species of tall tree ferns still exist, primarily in the tropics.

This deer fern is found only in wet, shady, coastal evergreen forests.

The bracken fern thrives in the partial sun of open forests. One of the most common ferns in the world, it often takes over large areas of abandoned pasture or agricultural land. Bracken ferns usually grow about 1.5 m tall, though they can reach a height of 5 m.

PROJECT

Growing Fern Prothallia from Spores

Have students grow their own fern prothallia by following the procedure given. Purchase fern spores from a supply house. Have students gather small clay flowerpots and stuff them tightly with paper towels. Instruct students to sterilize the pot and paper toweling by immersing them in boiling water for 10 minutes. When the pot has cooled, have them place it upside-down into a saucer or shallow dish. Next, add Knop's solution to the saucer to maintain a wet pot.

Have students sprinkle fern spores on top of the inverted pot, and cover the pot with the top half of a 2-L plastic soda pop bottle that has been cut in half. They should observe prothallia in 10-20 days. L2

SECTION 25.1

Demonstration

Visual-Spatial Purchase preserved species of sphenophytes such as *Equisetum*. Have students examine the plants and note their general appearance. Ask students to sketch the plants and add labels and captions that identify their most distinguishing features.

Tying to Previous Knowledge

Have students review the function of antheridia and archegonia in mosses (Chapter 24). Elicit whether these structures carry out similar functions in bryophytes and sphenophytes. *Yes, they do.*

Chalkboard Activity

Logical-Mathematical Ask students to use information about the number of plant species to prepare a bar graph. On the chalkboard, write the following information.

Bryophyta = 20 000
Psilophyta = 30
Lycophyta = 1000
Sphenophyta = 15
Pterophyta = 12 000

Explain that psilophytes, the whisk ferns, are a small division of seedless vascular plants not covered in the text. Tell students that these numbers represent the approximate number of living species for each non-seed plant division studied so far. L1

Software

Plants, The Life Sciences Programs

MiniLab

Purpose

LS **Visual-Spatial** Students will observe the structures present on a fern gametophyte (prothallus).

Process Skills

observe and infer, care and use of the microscope

Teaching Strategies

- Prepared slides of a fern prothallus showing both antheridia and archegonia, or preserved prothallia may be used in place of living prothallia. Note that only living tissue will allow students to observe sperm cell movement.
- To aid students in locating reproductive structures, use a diagram to show where these structures are located on the prothallus.

Expected Results

Students will observe the location and structure of gametophyte antheridia (round structures) and archegonia (vaselike structures). If live materials are used and the archegonia are mature, pressing on the coverslip will release sperm cells into the water.

Analysis

1. archegonium
2. antheridium
3. haploid
4. mitosis
5. Sperm cells showed tail-like parts (flagella) that were used for a swimming-type movement.

Assessment

Knowledge: Have students explain why it is easier to grow a sporophyte fern plant from the prothallus rather than from spores. L1

MiniLab

What does a fern prothallus look like?

The gametophyte generation in spore-producing vascular plants is much smaller than the gametophytes of nonvascular bryophytes. In this lab, you will investigate the structure of the fern gametophyte.

Procedure

1. Obtain a mature fern prothallus. Make a wet mount, with coverslip, and observe the prothallus at low power under the microscope.
2. Near the notch in the heart-shaped prothallus, locate the reproductive structures that produce eggs.
3. Near the pointed base of the prothallus, locate the reproductive structures that produce sperm. Use the eraser end of a pencil to press gently on the coverslip. Watch closely to see if sperm are released. If they are, observe their movement.

Analysis

1. What is the name of the structure that produces eggs?
2. What is the name of the structure that produces sperm?
3. Is the prothallus haploid or diploid?
4. What type of cell division takes place within the prothallus?
5. If you were able to observe sperm, describe their movement.

Sporophyte is dominant in ferns

As with all vascular plants, it is the sporophyte generation of the fern that has roots, stems, and leaves, and this is the plant we commonly recognize. The gametophyte in most ferns is a thin, flat structure that is independent of the sporophyte. In most ferns, the main stem is underground, *Figure 25.6.* This thick, underground stem is called a **rhizome.** The leaves of a fern are called **fronds** and grow upward from the rhizome. The fronds of ferns are often divided into leaflets called pinnae. Ferns are the first of the vascular plants to have evolved leaves with branching veins of vascular tissue. The branched veins in ferns transport water and food to and from all the cells.

Figure 25.6

Most ferns in warm climates are perennial plants that live from year to year, gradually enlarging in size. The fronds of ferns that live in temperate climates die back during the winter.

The creeping, underground rhizome of a fern is a modified, condensed stem, with roots growing downward from each joint of stem and frond. The rhizome contains many starch-filled cells and is a storage organ for overwintering.

A fern frond has a stemlike stipe and green, often finely divided leaflets called pinnae.

Young fern fronds unfurl as they grow and are called fiddleheads because their shape is similar to the neck of a violin.

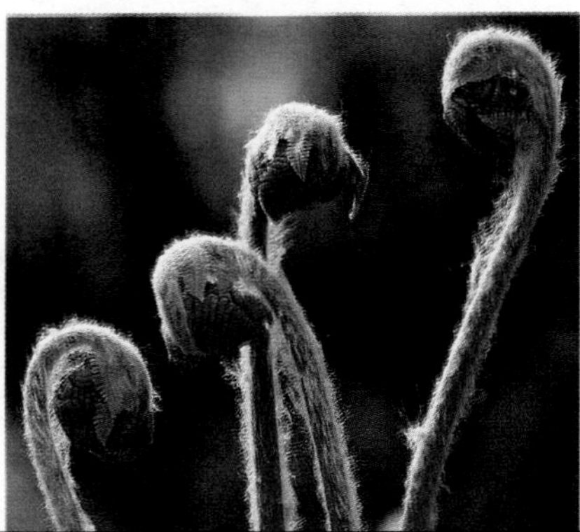

PORTFOLIO

Fern Anatomy Review Have students prepare a simple diagram of a fern. Ask them to add the following labels to their diagram: rhizome, frond, stipe, pinnae, and sori. Have them refer to Figure 25.6 if they need help. L1 P LS

Meeting Individual Needs

Students Acquiring English Pair students with limited English proficiency and students with strong English skills. Have them review together the functions and structure of fern rhizomes. Ask the students to design a concept map that shows the structure and function of this fern part. L1 SAE LS

Life Cycle of a Fern

In the life cycle of a fern, the sporophyte generation is independent of the gametophyte. As in mosses, meiosis in a fern takes place in sporangia, and sperm must swim through a surface film of water to reach the egg.

Bracken fern

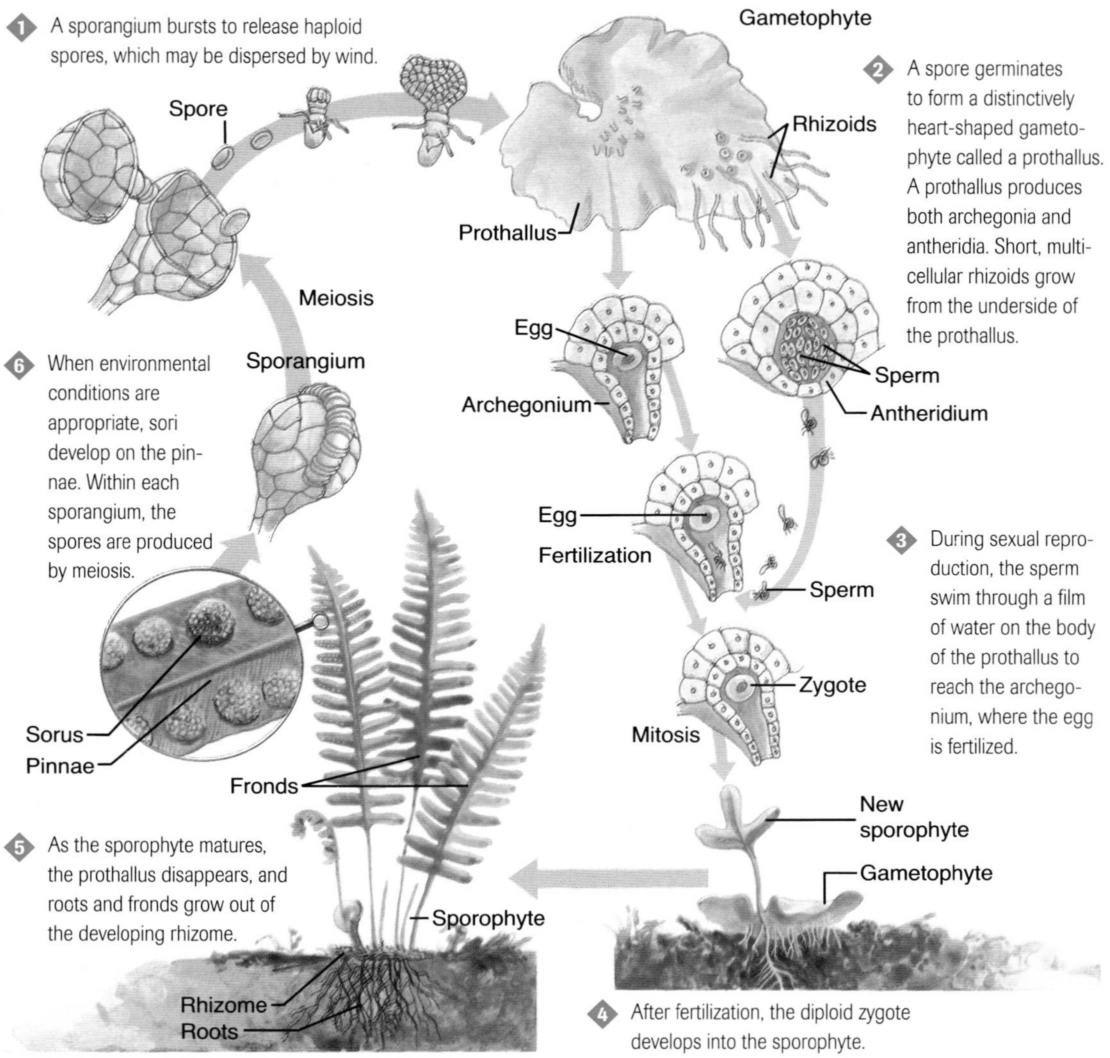

1. A sporangium bursts to release haploid spores, which may be dispersed by wind.

Spore

Gametophyte

2. A spore germinates to form a distinctively heart-shaped gametophyte called a prothallus. A prothallus produces both archegonia and antheridia. Short, multicellular rhizoids grow from the underside of the prothallus.

Rhizoids

Prothallus

Meiosis

Egg

Archegonium

Sperm

Antheridium

Sporangium

6. When environmental conditions are appropriate, sori develop on the pinnae. Within each sporangium, the spores are produced by meiosis.

Egg

Fertilization

Sperm

3. During sexual reproduction, the sperm swim through a film of water on the body of the prothallus to reach the archegonium, where the egg is fertilized.

Zygote

Mitosis

Sorus

Pinnae

Fronds

New sporophyte

Gametophyte

5. As the sporophyte matures, the prothallus disappears, and roots and fronds grow out of the developing rhizome.

Sporophyte

Rhizome

Roots

4. After fertilization, the diploid zygote develops into the sporophyte.

THE INSIDE STORY

Life Cycle of a Fern

Purpose

IS **Visual-Spatial** Students will examine the complete life cycle of a typical fern.

Teaching Strategies

- Compare the life cycle of a fern to the life cycle of a moss (Chapter 24). Challenge students to diagram both life cycles and summarize the differences in a caption. L1
- **Audiovisual** Show the filmstrip *Fern Life Cycle*, Carolina.

Misconception

Students frequently think the sori present on the underside of fern fronds are some type of insect infestation. Explain that these structures are the spore-producing parts of the plant.

Visual Learning

- Have students use the diagram as a model to draw the fern life cycle in flowchart form. L1
- Ask students to identify the part of the fern in which fertilization occurs. *Archegonia*
- Write the following terms on the chalkboard: *spore, zygote, egg, sporophyte,* and *sperm.* Have students identify whether each structure is haploid or diploid. *Haploid: spore, egg, sperm; diploid: zygote, sporophyte.*

GLENCOE TECHNOLOGY

Videodisc

Biology: The Dynamics of Life
Disc 1, Side 2
Fern Development (Ch. 18)

STUDENT JOURNAL

Have students make their own diagrams of the stages of the fern life cycle in their journals for use in review. Encourage students to use a line on their diagram to divide the diagram so that all sporophytes are on one side and all gametophytes are on the other side of the line. Have them label the diagram Sporophyte Stages, Gametophyte Stages. L1 IS

SECTION 25.1

3 Assess

Check for Understanding

Have students explain how the terms in each of the following groups are related:
(a) strobilus/spore/sorus
(b) sporophyte/frond/rhizome
(c) vascular tissue/xylem/ phloem L1

Reteach

Have students list five characteristics for these plant divisions: Lycophyta, Sphenophyta, and Pterophyta. L2

Extension

Have students make a concept map, flowchart, or other diagram to explain the changes in size that occur for the sporophyte and gametophyte generations of the Bryophyta, Lycophyta, and Pterophyta. L2

Assessment

Skill: Provide students with an unlabeled life cycle of a typical fern. Have students correctly label the diagram. L1

4 Close

Discussion

Ask students to compare the gametophyte generation of mosses and ferns. *Moss is large and green; fern is very small and green.* Have them do the same for the sporophyte generations. *Moss is small and nongreen; fern is very large and green.*

Figure 25.7
Sori found on the underside of fern fronds look like brown- or rust-colored dust.

Most sori are found as round clusters on the pinnae, although the shape of the clusters and arrangement on the pinnae vary with the species.

Some species of ferns have sori on the edges of the pinnae, such as this *Osmunda* fern.

Most ferns require moist soil

Most modern ferns are much smaller than their ancestors. You may have seen shrub-sized ferns, such as those pictured in *Figure 25.5*, on the damp forest floor or along stream banks. Some species of ferns float in water or are rooted in mud, whereas others cling to the sides of rocky cliffs or live in cold regions above the Arctic Circle. Some ferns inhabit dry areas, becoming dormant when moisture is scarce and resuming growth and reproduction when water again becomes available.

Fern spores are held in sori

A fern life cycle is similar to that of other spore-producing vascular plants. Spores are produced in structures called sporangia. Clusters of sporangia form a structure called a **sorus** (plural *sori*). The sori are usually found on the pinnae, *Figure 25.7*, but in some ferns, spores are borne on whole, modified fronds.

Section Review

Understanding Concepts

1. Why do most spore-producing vascular plants live in moist habitats?
2. What are the major differences between spore-producing vascular plants that exist today and those that lived in the Carboniferous forests?
3. Compare and contrast the structure of the sporophyte in the lycophytes and pterophytes.

Thinking Critically

4. Why do you think there are fewer spore-producing vascular plants on Earth today than there were 300 million years ago?

Skill Review

5. **Sequencing** Sequence the stages in the life cycle of a fern. For more help, refer to Organizing Information in the *Skill Handbook.*

Answers to Section Review

1. Sperm cells must be capable of swimming through water to fertilize egg cells within archegonia.
2. Ancient species were much taller and much more abundant.
3. Lycophytes are small, grow close to the ground, stems are covered by small, narrow leaves, resembling mosses. Roots, stems, and leaves are present. Pterophytes have roots, stems, and leaves. The stem is underground, supporting leaves called fronds that are usually finely divided into pinnae. Spores are found within sori on the undersides of fronds.

Thinking Critically

4. Earth's climate has changed, reducing the number of damp areas in which spore-producing plants could survive.

Skill Review

5. Sporophyte generation → spores → spores germinate to each form a prothallus or young gametophyte → gametophyte forms egg and sperm → fertilization of egg by sperm forms a young sporophyte.

SECTION

25.2 Gymnosperms

Although Earth's ancient forests were dominated by spore-producing vascular plants, early seed-bearing plants had also evolved. About 280 million years ago, when club mosses, ferns, and other spore producers had reached their greatest numbers and diversity, Earth's climate changed. Long periods of drought and freezing weather caused many spore-producing plants to become extinct, but a few of the seed-bearing plants had adaptations that enabled them to survive.

Section Preview

Objectives

Identify the characteristics of Cycadophyta, Ginkgophyta, Gnetophyta, and Coniferophyta.

Compare the gymnosperm life cycle with that of seedless vascular plants.

Key Terms

gymnosperm
cone
microspore
megaspore
embryo
cotyledon
pollen grain
ovule
evergreen plant
deciduous plant
tracheid

What Are Gymnosperms?

Among the early seed-bearing plants were the gymnosperms. **Gymnosperms** are vascular plants that produce seeds on the scales of woody strobili called **cones.** The seeds of gymnosperms are not protected by a fruit, nor do gymnosperms produce flowers. The term *gymnosperm* is used to describe four divisions of plants that bear naked seeds: Cycadophyta, Ginkgophyta, Gnetophyta, and Coniferophyta.

Gymnosperms produce spores

Gymnosperms are similar to all other plants in that spores are produced by the sporophyte generation. **Microspores** are produced in the male cone and give rise to the male gametophyte, eventually developing into pollen grains. **Megaspores** are produced in the female cone and give rise to the female gametophyte. Each female gametophyte is contained within an ovule and produces archegonia with egg cells. In conifers, such as a pine, pollen is carried by wind to the ovule and the male gametophyte contained in the pollen grain produces a pollen tube, which grows into the archegonium and provides a way for the sperm cell to reach the egg. This occurs without water.

After fertilization, the zygote develops into an embryo. An **embryo** is an organism at an early stage of growth and development. In plants, an embryo is the young, diploid sporophyte of a plant. A gymnosperm embryo can have many cotyledons. **Cotyledons** are food-storage organs of a plant embryo that become the plant's first leaves. As the embryo develops, the tissues of the ovule form the food supply and seed coat of the pine seed.

SECTION 25.2

Prepare

Key Concepts

Gymnosperm is a general term used to describe and include those plants that produce unprotected seeds in cones. Four divisions of Gymnosperms are described. The process of alternation of generations in gymnosperms is described.

Block Scheduling

Look for this symbol for strategies that are useful in a block scheduling format. For more information on block scheduling, refer to Section 25.2 in the **Lesson Plans** booklet.

Materials

- Order or gather a variety of gymnosperm cones for the Focus Demonstration.
- Gather or purchase male pinecones and fern fronds containing sori for the Minilab.
- Gather branches of a variety of conifers for the Biolab. Make sure that the branches include cones.

1 Focus

Bellringer

Before presenting the lesson, display **Section Focus Master 54** on the overhead projector and have students answer the accompanying questions. L1 SAE

Program Resources

Section Focus Master 54 L1 SAE
Reinforcement and Study Guide, pp. 99-100 L1
Biolab and Minilab Worksheets, pp. 100-102 L1
Critical Thinking/Problem Solving, p. 25 L3
Basic Concepts Transparencies 36, 37 and **Masters** L1 SAE
Reteaching Transparency 25 and **Master** L1 SAE

Demonstration

Provide students with a variety of cones. Ask them to describe what might be found within or upon each wooden scale. *Some students may say seeds.* Have them gently pry the scales open to reveal the seeds within. L1

2 Teach

Time Line

LS **Logical-Mathematical** Have students use information from Chapters 24 and 25 to make a time line that marks the era when different types of land plants evolved. Have students include a description of what climatic conditions were like during each time period. L2

Using Science Terms

LS **Linguistic** Challenge students to use their knowledge of the prefixes *micro-* and *mega-* to describe the relative sizes of microspores and megaspores. *Microspores are smaller than megaspores.* Have students speculate why it is important for microspores to be small. *They develop into pollen grains that must be light enough to be dispersed by wind.*

GLENCOE TECHNOLOGY

Videodisc

Biology: The Dynamics of Life
Disc 1, Side 2
Giant Redwoods (Ch. 19)

STVS: Plants & Simple Organisms
Disc 4, Side 2
Pollen Atlas (Ch. 5)

Figure 25.8
A pine seed shows the structures and some of the adaptive advantages of seeds.

The winged shapes of these pine seeds aid in their dispersal. The wind may carry them several feet away from the parent tree.

Both the embryo and its food supply are protected by a tough, outer seed coat. The seed coat breaks down, and growth begins only when certain conditions of temperature, moisture, and light are met.

The plant embryo is surrounded by seven to nine cotyledons, which will serve as an initial food supply when growth begins.

Advantages of seeds

A seed consists of an embryo with a food supply enclosed in a tough, protective coat, *Figure 25.8.* Plants that bear seeds have several important advantages over spore producers. The seed contains a supply of food to nourish the young plant during the early stages of growth. This food is used by the plant until its leaves are developed sufficiently to carry out photosynthesis. In gymnosperms, the food supply is stored in cotyledons. The embryo is protected during harsh conditions by a tough seed coat. The seeds of many species are also adapted for easy dispersal to new areas, so the young plant does not have to compete with its parent for sunlight, water, soil nutrients, and living space.

Fertilization without water

Seed plants also have another important advantage over spore producers: fertilization does not require water. Spore-producing vascular plants are not fully adapted to life on dry land because the sperm needs a surface film of water in which to swim to the egg during sexual reproduction. In gymnosperms, the male gametophyte develops inside a structure called a **pollen grain,** which includes sperm cells, nutrients, and a protective outer covering. The female gametophyte develops inside a structure called an ovule. An **ovule** contains a megaspore cell, one or two layers of tissue, and a protective covering. Sperm are carried in the pollen grain to the ovule by wind rather than water.

STUDENT JOURNAL

Structure and Function of Plant Parts
Have students describe the structure and function for the following: seed, cotyledons, seed coat, pollen grain, and ovule. Ask students to group the five terms into sporophyte or gametophyte generation. *First three structures belong to sporophyte generation; rest belong to gametophyte generation.* L1 LS

Cycadophyta

Cycads are seed plants that shared Earth's forests with the dinosaurs during the Triassic and Cretaceous periods. About 100 species exist today, exclusively in the tropics and subtropics. The only present-day species that grows wild in the United States is found in Florida, although you may see cycads cultivated in greenhouses or botanical gardens, as shown in *Figure 25.9.* The trunks and leaves of many species resemble palm trees, but cycads are not closely related to palms.

Figure 25.9

Cycads have a terminal rosette of leaves and bear seeds in cones. All cycads have separate male and female plants.

The male plant bears cones that produce pollen grains, which are released in great masses into the air.

The female plant bears cones that contain ovules with eggs. The moisture in the ovule is sufficient for the sperm, which are released from the pollen grains, to swim to the eggs.

Earth Science

How Coal Was Formed

More than 300 million years ago, the huge leaves of ferns swayed in the moist breezes of vast bogs and swamps. As the lush vegetation died and was replaced by succeeding generations of plants, rich layers of plant material collected on the bottom of the swamp. In some places, there was not enough oxygen or bacteria to produce decay. Over a long period of time, these swamps drained and filled again, producing many layers of plant material and sediment pressing down on one another. This combination of plant material and pressure produced a fossil fuel called coal.

Plants turned into fuel More than half of all the known reserves of coal are found in the U.S. The plant material that accumulated 300 million years ago during the Carboniferous period ultimately became coal. Pterophytes, gymnosperms, and lycopods were the dominant plants during this period.

From peat to lignite to coal As the plants died and fell into boggy waters lacking oxygen and bacteria, the layers of packed vegetation became peat. As more and more dead plant material built up on top of the peat, it became compressed and began to harden, turning into a brown material called lignite coal. Under continuous pressure, lignite coal becomes first bituminous and, finally, anthracite coal—a shiny, hard, black material that is burned in coal-fired furnaces to produce heat and electricity today.

Sand and water in coal Mud or silt often was washed or blown into the swamp during plant growth. This produced a higher ash content in today's coal beds. Vertical fractures filled with mineral deposits from groundwater also can be found in coal beds. Impurities such as these result in lower-grade coal, that is, coal that burns less efficiently.

CONNECTION TO Biology

Where does the energy contained in coal come from?

Earth Science

How Coal Was Formed

Purpose

Students will discover the role of early plants in the formation of coal.

Teaching Strategies

- Obtain "coal balls" directly from either coal mining companies or a science supply house. Use a lapidary saw to prepare thin slices of the balls to reveal the many fossilized plant parts present in coal. Have students examine the samples.
- Have students working in cooperative groups research and report on one of the following: (a) why a lack of oxygen was critical to coal formation, (b) the difference between soft and hard coals, and how each is used, (c) where peat is used as an alternative fossil fuel. L1 COOP LEARN

Possible Answer

Energy was supplied to coal in the form of sugars manufactured by the plant during photosynthesis.

Discussion Question

Elicit from students why Florida is the only state in which cycads grow naturally. *It is the only part of the continental United States with a tropical climate.*

STUDENT JOURNAL

Is It a Cycad? Have students imagine that they are traveling through Florida with a friend. Their friend identifies a plant as a cycad that is forming seeds. They search the surrounding area for more cycads and find no other cycads within a 20-kilometer radius. Have students provide as many reasons as possible in their journals to explain why they believe their friend incorrectly identified the plant as a cycad. *Cycads are either male or female, therefore the male plant must be close to the female to provide pollen; palms can self-fertilize—being both male and female. Palms can grow alone and reproduce, thus making it more likely that this was a palm.* L1 IN

MiniLab

Purpose

Visual-Spatial Students will observe and compare spores and pollen.

Process Skills

observe and infer, record observations, care and use of the microscope

Teaching Strategies

- Preserved materials will work as well as fresh plant materials.
- Use forceps to remove only one small microsporangium from each male pinecone. This is all the material each student will need for proper pollen cell observation.
- Use a single-edged razor blade to remove the fern sori from fern leaves. It may be necessary for students to squash the sorus wet mount to release spores from their sporangia.

Expected Results

Both cell types are microscopic, but pine pollen cells are larger than fern spores. Pine pollen will show small "wings" resembling Mickey Mouse hats with ears, which aid in wind dispersal. Fern spores form a macroscopic gametophyte, whereas pine pollen grains form a microscopic gametophyte generation.

Analysis

1. Both are formed by meiosis.
2. Each cell is haploid.
3. Each cell is part of the gametophyte generation.

Assessment

Knowledge: Have students trace the changes that occur to a fern spore and pine pollen as they mature. Have them explain how these changes are alike. L1

MiniLab

What are the similarities and differences of spores and pollen grains?

Reproduction in ferns involves the release of spores, and the gymnosperm life cycle includes the release of pollen. Are the roles of these reproductive cells the same or different? In this lab, you will compare the structure and function of spores and pollen.

Procedure

1. Obtain a male pine cone and a fern frond with mature sporangia. Prepare separate wet mounts of pine pollen and fern spores.
2. Observe each cell type under a microscope, and make diagrams of each.
3. Prepare a data table that compares the structure and function of pollen and spores. Use the text and your observations to include answers to the following questions in your table.

Magnification: 114×

Fern spores

Magnification: 3000×

Pine pollen

Analysis

1. By what cell division process is each cell formed?
2. Is each cell haploid or diploid?
3. Is each cell part of the sporophyte or the gametophyte generation?

Ginkgophyta

Members of the Ginkgophyta were most numerous during the Jurassic period, about 200 million years ago. They were an important part of the vegetation that lived alongside the dinosaurs but, like the dinosaurs, most members of the Ginkgophyta died out by about 65 million years ago. Today, the division is represented by only one living species, *Ginkgo biloba*, **Figure 25.10**. The ginkgo tree is considered sacred in China and Japan, and has been cultivated in temple gardens for thousands of years. These Asian temple gardens seem to have prevented the tree from becoming extinct.

Figure 25.10

The ginkgo is sometimes called the maidenhair tree because its lobed leaves resemble the fronds of a maidenhair fern.

In addition to being attractive, ginkgo trees are resistant to air pollution, so they are often planted in urban parks and gardens.

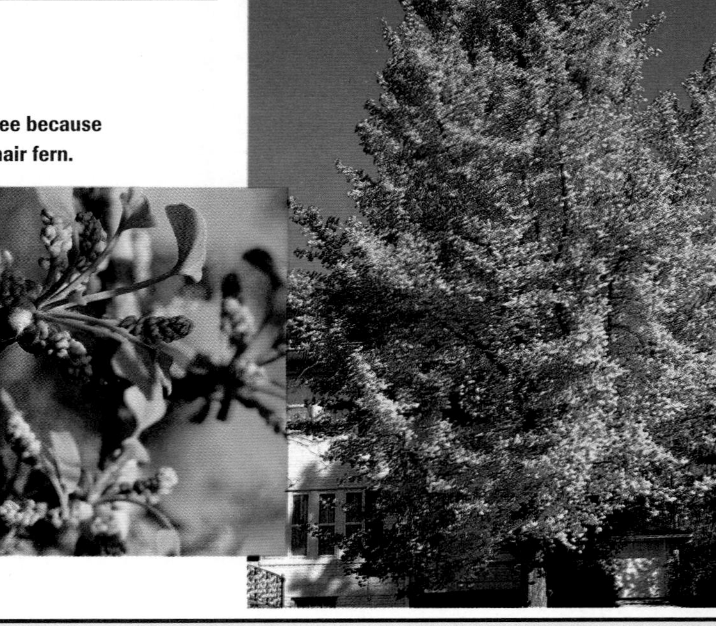

Like cycads, ginkgos bear male and female cones on separate plants. The male ginkgo produces pollen in strobilus-like cones that grow from the bases of leaf clusters.

interNET CONNECTION

Follow the link for this chapter on the Glencoe Homepage at **www.glencoe.com/sec/science** to find out more about ferns and gymnosperms.

Figure 25.11

Like the cycads and the ginkgo, most gnetophytes have separate male and female plants.

One species of *Ephedra,* also known as Mormon tea, is a shrubby, cone-bearing plant with scalelike leaves similar to those of horsetails. It is found in dry regions such as the white sands of New Mexico. Members of this genus are a source of ephedrine, a medicine used to treat asthma, emphysema, and hay fever.

The single species of *Welwitschia* is a bizarre-looking desert dweller that grows close to the ground. It has a large tuberous root, and though it may live 100 years, this plant has only two leaves, which continue to lengthen and become tattered by weathering as the plant grows older.

Gnetophyta

Fossil gnetophytes are unknown, except for pollen from Permian, upper Cretaceous, and tertiary rock formations. Most living gnetophytes can be found in the deserts or mountains of Asia, Africa, and Central or South America. The division Gnetophyta contains only three genera, which are all different in structure and adaptations. The genus *Gnetum* is composed of tropical climbing plants. The genus *Ephedra* contains shrublike plants and is the only gnetophyte genus that can be found in the United States. The third genus, *Welwitschia*, is found only in South Africa. *Ephedra* and *Welwitschia* are pictured in *Figure 25.11.*

The female *Ginkgo* bears the seeds, which develop a fleshy outer covering and resemble orange-yellow cherries as they ripen. The male trees are usually preferred by gardeners because the fleshy seed coat of the ginkgo has an unpleasant rancid smell.

SECTION 25.2

Activity

LS **Visual-Spatial** Preserved *Ephedra* leaves are available from supply houses. Order samples of the leaves and allow students to examine them.

Concept Development

Ask students to compare and contrast Gnetophyta with Ginkgophyta. As a means for doing this, have students prepare a chart with the headings "Gnetophyta and Gingkophyta Similarities" and "Gnetophyta and Gingkophyta Differences." Have students include comparisons of reproductive structures as well as general traits. Ask them to also include where each plant is found in the world, the number of species alive today, the number of genera present today, and economic uses of the plants. L1

Audiovisual

Show the filmstrip *The Gymnosperms*, Carolina.

Teacher FYI

The drug ephedrine has been used by the Chinese for more than five thousand years. It is called *ma huang*, which means "astringent and yellow." The words refer to the drug's taste and color. Ephedrine depresses smooth and cardiac muscle action, raises blood pressure, dilates the pupils, and reduces congestion.

CULTURAL DIVERSITY

Gymnosperm Importance

Many species of gymnosperms play significant roles as timber trees, food plants, and sources of essential oils and soaps in the economies of countries. Initiate a discussion about the cultural importance of the ginkgo tree in China and Japan. *Ginkgo biloba* is the only living representative of the gymnosperm family Ginkgoaceae, and does not exist in the wild. It has been cultivated for thousands of years for use as an ornamental plant in Chinese and Japanese temple gardens. Have students choose types of gymnosperms and present reports on their economic value or cultural significance in different countries. L1 LS

Chemistry Connection

Properties of Wood

Wood is a versatile material. Wood is light yet strong enough to be cut and molded. These two properties make it an ideal building material.

Density is the mass of a material per unit of volume or d = m/v. Have students use this formula to calculate the densities of several blocks of wood. Balances and a metric ruler will be needed. As a conclusion, ask students to relate density to the term *compactness.* L2

GLENCOE TECHNOLOGY

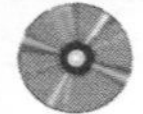

Videodisc

STVS: Ecology
Disc 6, Side 2
Acid Rain and Plants (Ch. 10)

Figure 25.12

Conifers are named for the woody cones in which the seeds of most species develop.

Yews are popular ornamental trees because they are graceful in shape and have attractive, glossy, green needles. A *Taxus* species from the Pacific Northwest is used to produce a cancer-fighting drug called taxol. Yews do not have cones. They produce single seeds with a fleshy covering.

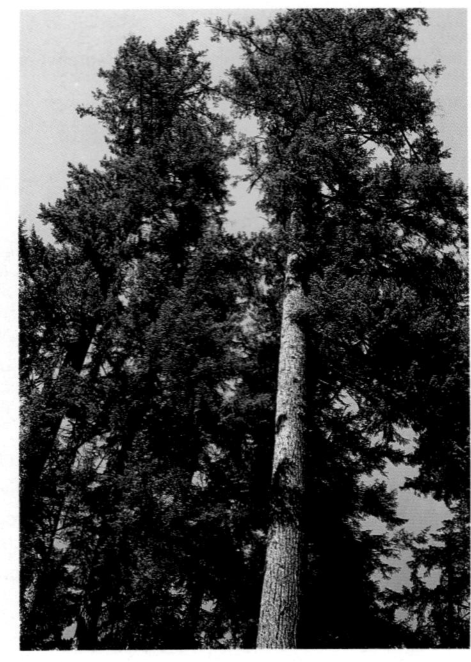

The Douglas fir, a member of the pine family, is one of the most important lumber trees in North America. It grows straight and tall, to a height of 100 m.

Coniferophyta

The sugar pine is one of many familiar-looking forest trees that belong to the division Coniferophyta. The conifers—the largest, most diverse division of the gymnosperms—are trees and shrubs with needle- or scalelike leaves. Most conifers produce seeds in woody cones. They are abundant in forests throughout the world, and include pine, fir, spruce, juniper, cedar, redwood, yew, and larch. A few representative conifers are shown in *Figure 25.12.*

The first conifers probably emerged around 280 million years ago. During the Jurassic period, 150 million years ago, conifers became prominent forest inhabitants and remain so today. Although the fossil record indicates that many species have become extinct, *Figure 25.13,* the conifers continue to evolve and flourish.

Figure 25.13

Petrified Forest National Park in Arizona contains fossilized remains of an extinct species of conifer from the Triassic period. When the trees died, they were buried in mud and volcanic ash. Instead of decaying, the organic matter was replaced with minerals. The wood has become petrified, or turned to stone.

Alternate Lab: The Gymnosperm Leaf

Purpose

LS **Visual-Spatial** Students will observe, diagram, and label the structures of a typical gymnosperm leaf.

Materials

microscope, prepared slide of gymnosperm leaf cross section

Procedure

Give students the following directions.

1. Observe the prepared slide under low power.
2. Diagram what you see. Label the following structures using these descriptions as a guide: (a) cuticle: noncellular layer, thin covering over leaf (b) epidermis: below cuticle, one cell in thickness (c) mesophyll (spongy layer): large cells, thick layer just below epidermis (d) endodermis: one-cell-layer thick, surrounds large oval central area (e) palisade layer: many cells in thickness, surrounded by endodermis (f) vascular tissue: very center of leaf, two types are present.
3. Look again at the leaf using the microscope. Check for evidence of stomata on the epidermis and the

Conifers are adapted to cold climates

Conifers form large forests in many regions of the world and have many adaptations that enable them to live in cold or dry habitats. Under dry or freezing conditions, plant roots cannot absorb water from the soil. The needlelike leaves of conifers are covered with a thick cuticle and have sunken stomata that help retain water in the tissues of the tree, *Figure 25.14.* The bark of conifers and other seed-bearing trees also helps reduce water loss by forming a protective covering over the stem.

Conifers are evergreens

Most conifers are evergreen plants. **Evergreen plants** retain their leaves all year. Although individual leaves drop off as they age or are damaged, the tree never loses all its leaves at once. Pine needles, for example, may remain on the tree for anywhere from two to 40 years, depending on the species. Trees that retain their leaves can begin photosynthesis in the early spring as soon as the temperature warms. They usually have a heavy coating of cutin, an insoluble waxy material. Evergreens are often found where the warm growing season is short, so keeping leaves year-round gives these plants a head start on growth. They are abundant where nutrients are scarce; evergreens thus eliminate the need to grow a whole new set of leaves each year. The branches and needles of conifers are flexible. They allow snow and ice to slide off the tree, so they don't build up on branches and cause them to break.

ThinkingLab — Analyze the Procedure

How much acid in rain is too much?

Scientists hypothesize that acid rain may have contributed to a decline in the number of red spruce trees in forests in the northeastern United States. When exposed to acid rain, needles of the tree turn brown and fall off, and new ones grow in slowly. Eventually, the tree may die. Scientists wanted to know exactly what concentrations of acid caused damage.

Analysis

Groups of spruce seedlings were misted with acidic water at pH levels of 2.5, 3.0, 3.5, 4.0, 4.5, 5.0, and 5.5, and effects were noted in all groups. If fewer effects were noted in plants misted with water at a higher pH, can scientists say conclusively that acidity causes problems?

Thinking Critically

What would need to be done in order to make the above experimental procedure scientifically sound?

Figure 25.14

In cross section, the needle of a conifer resembles a cylindrical leaf.

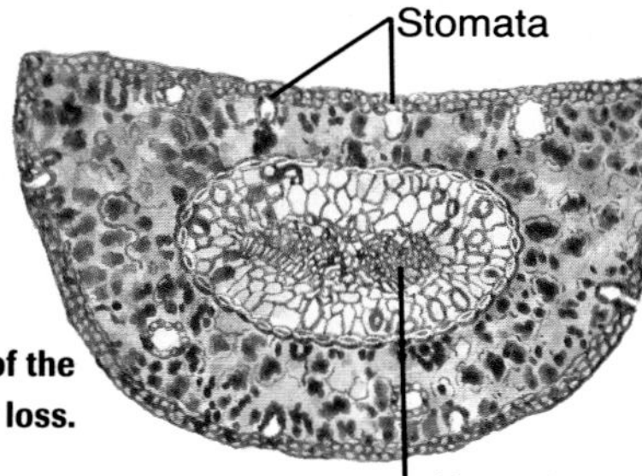

The narrow shape reduces the surface area from which water can evaporate.

Stomata

The stomata are sunken within folds of the epidermis, which helps prevent water loss.

Vascular tissues

ThinkingLab — Analyze the Procedure

Purpose

LS Logical-Mathematical Students will evaluate the procedure of an experiment.

Process Skills

using scientific methods, interpret data, observe and infer, relate cause and effect

Teaching Strategies

- Allow students to work in cooperative groups.
- Refer students to the Skills Handbook to have them review the steps of scientific methods and related science process skills.
- Remind students that as the numbers on the pH scale increase, acidity decreases. A pH of 5.5 is less acidic than a reading of 4.5.

Thinking Critically

Answers may vary. Students should suggest controlling variables, setting up controls, and expanding the procedure to see if misting affects mature plants as it did seedlings.

Assessment

Knowledge: Have students identify the following for the experiment: (a) independent variable, (b) dependent variable, (c) what variables need to be controlled, (d) what results might be expected if the experiment were carried out on plants other than red spruce. L1

presence of several ringlike parts, called resin ducts, in the mesophyll. Add these parts and labels to your diagram.

Analysis

Ask students to answer the following questions.

1. How does the cross-section view of a gymnosperm leaf differ from what you would see in a deciduous tree leaf? *Round rather than flat and long*
2. What may be the role of normally green mesophyll and palisade layers? *Responsible for food production*
3. What is the function of the cuticle, stomata, epidermis, vascular tissue? *Retain water, gas exchange, protection, transport*

Assessment

Skill: Diagram or describe how a cross-section slice of a gymnosperm leaf would be made. L1

Issues
BIOLOGY & SOCIETY

Should we let fires burn?

Purpose

Students discuss the controversial issue regarding the policy of "let-it-burn" when it comes to managing forest fires.

Background

- It is a common practice for ecologists interested in prairie restoration to prepare periodic management-set fires to help maintain the integrity of these ecosystems.

Teaching Strategies

- Use a map to show the location of Yellowstone National Park and the area affected by the 1988 fire.
- Review the meaning of the following terms or allow students to work in cooperative groups to determine meanings on their own: *crown fire, forest regeneration, naturally ignited fires,* and *management-set fires.*
- Ask students to research the 1988 Yellowstone fires to find out whether the damage would have been less severe had the fires been suppressed immediately. *Biologists tend to think that the damage was severe because the earlier policy of fighting all forest fires had allowed the buildup of flammable forest floor litter and shrubby plants.*

INVESTIGATING the Issue

Applying Concepts Students may form two groups to debate the benefits and hazards of prescribed burning.

Going Further

Contact a representative of the National Park Service in your local area. Invite someone from the service to speak to your class regarding the "let-it-burn" issue.

Issues
BIOLOGY & SOCIETY

Should We Let Fires Burn?

Devastating fires swept through Yellowstone National Park in the summer of 1988. One-third of Yellowstone's 2.2 million acres burned. The firefighters, planes, and bulldozers needed to stop the blazes cost taxpayers millions of dollars.

Policy and nature Until 1972, forest fires were put out as soon as they were detected. Since then, forestry officials and biologists have become convinced that fire is important to forest regeneration. Low-intensity ground fires that do not burn tree branches or crowns clean the forest floor of debris that could fuel larger and more damaging fires. These discoveries led to a "let-it-burn" policy, allowing some naturally ignited fires to burn unless they threatened developed areas or were out of control.

The lodgepole pines in Yellowstone Park provide an example of how fires renew life in a forest. Lodgepole pines produce two types of female cones. One type of female cone develops on the tree for two years and then opens and sheds its seeds on the ground. The second type of female cone is covered with a resin that seals the cone shut and traps the seeds inside. The cones remain sealed for years until heated by a fire. Heat dissolves the resin, letting the cone open up and release its seeds onto the forest floor. Because the fire has burned away the dense forest canopy, as well as all the vegetation on the forest floor, the pine seeds have an excellent chance to become established.

Different Viewpoints

After the 1988 fires in Yellowstone, homeowners, small business owners, and local citizens dependent on tourists protested that the "let-it-burn" policy was wrong. They argued that if the fires had been put out immediately, they would not have become so costly and would not have left the land unprotected from erosion.

Deliberately set fires Forest Service managers claim that managed fires are one way to reduce major fires. Burning small areas over many years divides the forest into growth areas of different ages. Younger areas help contain fires that begin in older, drier areas.

INVESTIGATING the Issue

Applying Concepts Debate the benefits and hazards of prescribed burning.

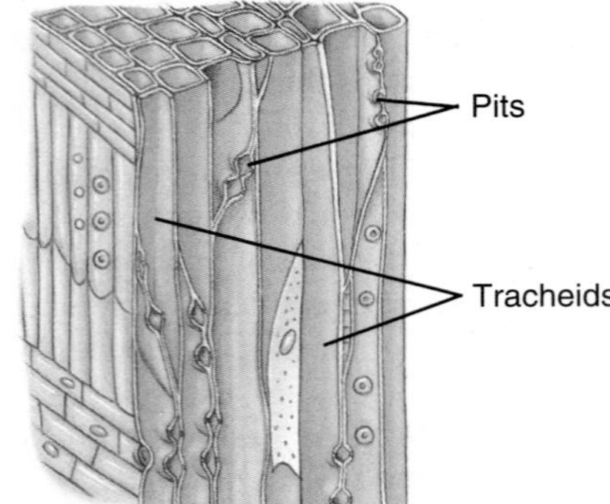

Figure 25.15

This view of tracheids shows that these cells have an elongated shape and closed, tapered ends. The average length of a tracheid cell is about 4 mm. Water moves from one tracheid to another by passing through the pits in connecting tracheid walls.

Deciduous trees lose their leaves

A few conifers, including larches and bald cypress trees, are deciduous plants. **Deciduous plants** lose all their leaves at the same time. Dropping all leaves is another adaptation for reducing water loss when water is unavailable during the winter. Plants lose most of their water through the leaves; very little is lost through bark or roots. However, a tree with no leaves cannot photosynthesize and must remain dormant during the winter months.

How do conifers grow so tall? The tissue that makes up much of the trunk of a conifer—the tissue we usually refer to as wood—is composed of thick-walled, nonliving cells called **tracheids,** *Figure 25.15.* They form the xylem of the vascular systems of ferns and gymnosperms. In addition to providing support, tracheids transport water and dissolved minerals from the roots to all other parts of the plant.

PROJECT

Pine Wood Maceration Technique

Wood cells can be separated to observe tracheids. Have students make a macerated preparation by doing the following: Obtain a 2-cm section of pine twig. Remove the bark and mince the wood into small pieces using a single-edged razor. Place the pieces into an acid solution and allow them to remain in the solution for two days. **Caution:** *Wear goggles when working with acid solution, and avoid contact with skin and clothes.* Remove the wood pieces using a medicine dropper and prepare wet mounts for microscopic examination. LS

THE INSIDE STORY

Life Cycle of a Pine

A pine, which is a common conifer found all over the northern hemisphere, can be used as a model for the seed-production process of all gymnosperms. The pine shown here is a white pine.

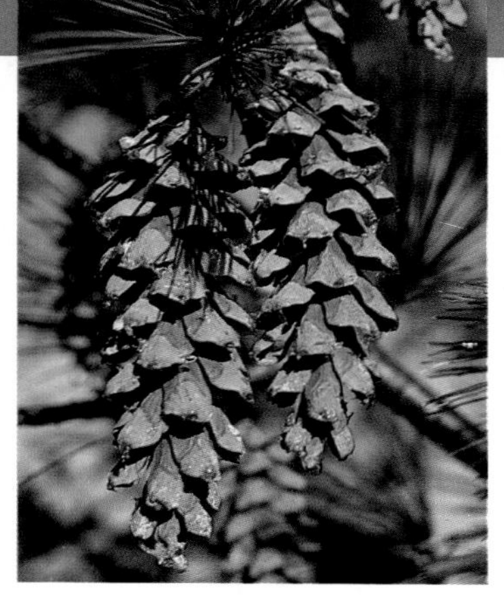

White pine

1. The adult sporophyte develops male and female cones on separate branches of the tree.

2. Female cones develop two ovules on the upper surface of each cone scale. Each ovule contains haploid megaspores and remains attached to the sporophyte.

3. Male cones produce microspores by meiosis; the microspores develop into pollen grains. Each pollen grain, with its hard, water-resistant outer covering, is a four-celled male gametophyte.

4. The female gametophyte grows, producing two or more archegonia, each of which contains an egg. Scales of the female cone are closed during this time.

5. During pollination, a wind-borne pollen grain falls near the opening in one of the ovules of the female cone. Each male gametophyte forms a pollen tube that penetrates the tissue of the female gametophyte.

6. When the pollen tube has grown into the archegonium, a sperm cell from the male gametophyte fertilizes the egg.

7. The zygote develops into an embryo with many cotyledons, the ovule generates starch tissue and a seed coat, and a mature seed is produced.

8. The female cone opens, releasing the seeds. When conditions are favorable, the seed germinates into a new, young sporophyte—a pine tree seedling.

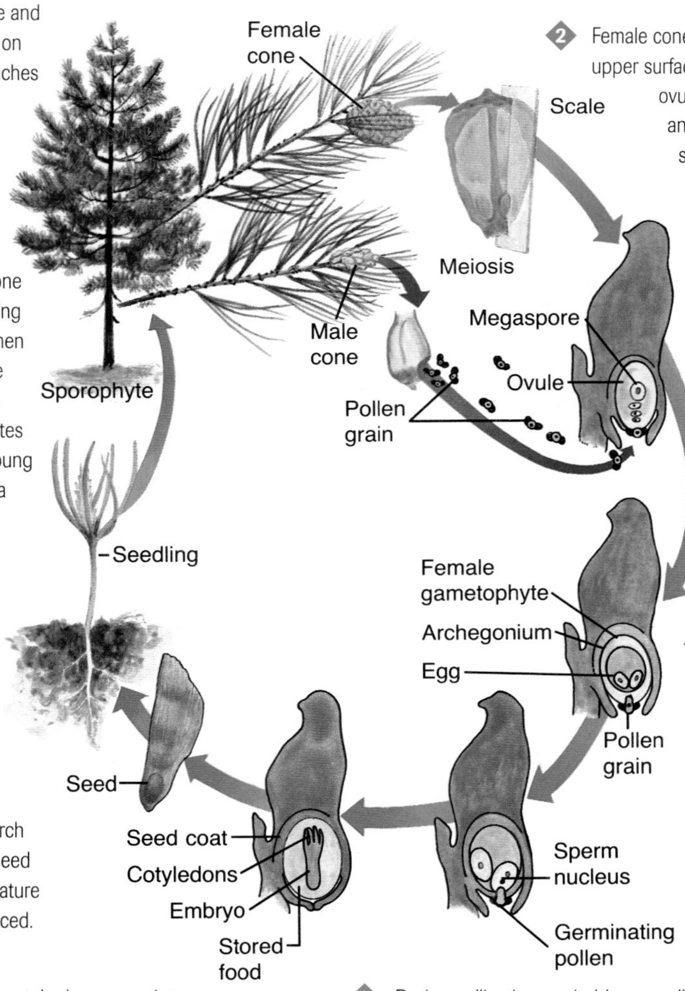

Female cone
Scale
Meiosis
Male cone
Megaspore
Ovule
Pollen grain
Sporophyte
Seedling
Female gametophyte
Archegonium
Egg
Pollen grain
Seed
Seed coat
Cotyledons
Embryo
Stored food
Sperm nucleus
Germinating pollen

Meeting Individual Needs

Learning Disabled Prepare individual cards with a drawing showing on each one a single stage of the life cycle of the pine. Provide a packet of cards to students who are learning disabled. Have students arrange the cards in proper sequence to illustrate the complete life cycle for a typical pine. L1 LS

THE INSIDE STORY

Life Cycle of a Pine

Purpose

LS **Visual-Spatial** Students examine and review the complete life cycle of a typical member of the Division Coniferophyta.

Teaching Strategies

- Before reviewing the life cycle, review the terms *alternation of generations, gametophyte*, and *sporophyte*. Have student volunteers define each term in their own words.

Visual Learning

- Have students identify the haploid structures shown in the diagram. *microspores, megaspores, pollen grains, egg*
- Have students identify the structures that make up the sporophyte and gametophyte generations. *Cones and embryo are sporophytes; all other structures are gametophytes.*

Bioethics

Tree Farms

Tree farms allow foresters to meet increased demands for wood and wood products. Logging these farms also reduces the demand for trees from natural old-growth forests.

As good as the tree farm plan sounds, some people oppose this monoculture technique. Monoculture refers to the practice of growing only one species of tree. Monocultures are very susceptible to insect pests and viral or bacterial diseases.

BioLab Design Your Own Experiment

How can you make a key for identifying conifers?

Time Allotment

One class period

Objectives

Review objectives with students before they begin the Biolab.

Process Skills

observe and infer, classify, compare and contrast, organize data

Safety Precautions

Some of the needle and cone specimens are very sharp. Warn students of the possibility of sticking themselves with the pointed ends.

PREPARATION

Alternate Materials

- Branches of conifers are ideal for this Biolab. Specimens should contain both leaves and cones. If unavailable, substitute scale diagrams of a variety of conifer leaf and cone specimens.

Possible Hypotheses

- The number of leaves per bundle, length of leaves, shape of leaves, color of leaves, and appearance of a sheath at the base of the leaves are useful in classifying conifers.
- Cone shape, length, and diameter may be used to classify conifers.

BioLab Design Your Own Experiment

How can you make a key for identifying conifers?

Most conifers have cones and needlelike or scalelike leaves. Different species have cones of different sizes, shapes, and thicknesses. The leaves of different species also have different characteristics. How would you go about identifying a conifer you are unfamiliar with? You would probably use a biological identification key. Biological keys list features of related organisms in a way that allows you to determine each organism's scientific name. Below is an example of the selections that could be found in a key that might be used to identify trees.

Needles grouped in bundles
Needlelike leaves
Needles not grouped in bundles
Leaves composed of three or more leaflets
Flat, thin leaves
Leaves not made up of leaflets

PREPARATION

Problem

What kinds of characteristics can be used to create a key for identifying different kinds of conifers?

Hypotheses

State your hypothesis in terms of the kinds of characteristics you think will best serve to distinguish among several conifer groups. Explain your reasoning.

Objectives

In this Biolab, you will:

- **Compare** structures of several different conifer specimens.
- **Identify** which characteristics can be used to distinguish one conifer from another.
- **Communicate** to others the distinguishing features of different conifers.

Possible Materials

twigs, branches, and cones from several different conifers that have been identified for you

PLAN THE EXPERIMENT

Teaching Strategies

- Label the samples with either their scientific names or generic labels such as Conifer A, Conifer B, etc. This will allow students to establish whether or not a key actually works when identification is attempted by another group.
- Illustrate on the chalkboard or an overhead what an ideal key may look like. Point out the nature of the opposite trait being used at each fork.

Possible Procedures

- Student procedures will vary. However, the general organization of the key itself will be similar from group to group. Making branches with opposite characteristics will help with the key design.

PLAN THE EXPERIMENT

1. Make a list of characteristics that could be included in your key. You might consider using shape, color, size, habitat, or other factors.
2. Determine which of those characteristics would be most helpful in classifying your conifers.
3. Determine in what order the characteristics should appear in your key.
4. Decide how to describe each characteristic.

Check the Plan

1. The traits described at each fork in a key are often pairs of contrasting characteristics. For example, the first fork in a key to conifers might compare "needles grouped in bundles" with "needles attached singly."
2. Someone who is not familiar with conifer identification should be able to use your key to correctly identify any conifer it includes.
3. ***Make sure your teacher has approved your experimental plan before you proceed further.***
4. Carry out your plan by creating your key.

ANALYZE AND CONCLUDE

1. **Checking Your Hypothesis** Have someone outside your lab group try using your key to identify your conifer specimens. If they are unable to make it work, try to determine where the problem is and make improvements.
2. **Making Inferences** Is there only one correct way to design a key for your specimens? Explain why or why not.
3. **Relating Concepts** Give one or more examples of situations in which a key would be a useful tool.

Going Further

Project Design a different key that would also work to identify your specimens. You may expand your key to include additional conifers.

Data and Observations

Have students record their keys and turn them in at the end of class. Make transparencies of sample keys and use them in class the following day as a means for illustrating correct and incorrect key design.

Meeting Individual Needs

Gifted Have students use references to determine and record the scientific names for all conifers listed in this feature. Have them describe the traits that are represented in the scientific names. L3

ANALYZE AND CONCLUDE

1. By using student key designs on the overhead, the entire class can determine whether the key does or does not work.
2. Student key designs placed on the overhead will illustrate the diversity of student trait choices used to organize and design each key.
3. It can distinguish poisonous from nonpoisonous plants, and harmful from nonharmful insects.

Assessment

Skill: Provide students with a bag of common laboratory items such as: glass slide, coverslip, paper clip, thumbtack, rubber band, staple, and dissecting needle. Ask them to prepare a key that identifies each item. L1

Going Further

Ask students to explain why a key design based on leaf odor or one based on conifer habitats may not be workable or practical. L2

GLENCOE TECHNOLOGY

Videodisc

Biology: The Dynamics of Life
Disc 1, Side 2
Life Cycle of a Pine (Ch. 20)

STVS: Plants & Simple Organisms
Disc 4, Side 2
Super Trees (Ch. 3)

SECTION 25.2

3 Assess

Check for Understanding

Have students explain the differences between: deciduous and evergreen trees, micro- and megaspores, seeds and spores. L1

Reteach

Have students prepare a concept map that uses the following terms: *gymnosperm reproduction, cones, microspore, megaspore, female gametophyte, male gametophyte, sperm, egg, fertilization, embryo, seed.* L1 SAE

Extension

Have students prepare an illustrated report that depicts uses for five types of gymnosperms. Have students identify each gymnosperm with its common and scientific name. L2

Assessment

Oral: Have students explain the role of the following as they relate to gymnosperms: (a) cones, (b) tracheids, (c) needle-shaped leaves, (d) being evergreens, (e) forming seeds, and (f) using pollen. L1

4 Close

Activity

Have students describe the adaptations for survival in a land environment that gymnosperms have made. L1

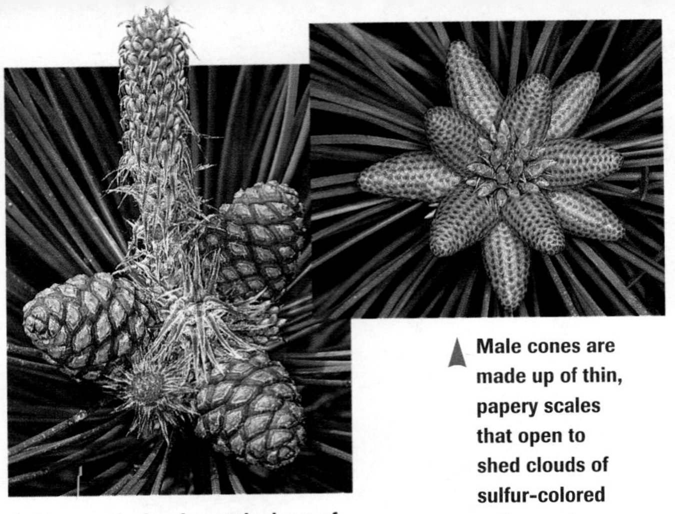

Two seeds develop at the base of each of the woody scales that make up a female cone. Cones differ in size depending on the species.

Male cones are made up of thin, papery scales that open to shed clouds of sulfur-colored pollen grains into the wind.

Figure 25.16

Each scale of a cone is a modified leaf or branch.

Conifers are named for their cones

The reproductive structures of most conifers are produced in cones. Conifers have two types of cones: male cones that produce pollen and female cones that ultimately produce seeds, as shown in *Figure 25.16*. Male cones are small and easy to overlook. They usually drop off the tree soon after the pollen they contain has been released. Female cones are much larger. They stay on the tree until the seeds have matured, which may take from several months up to two years. In the genus *Pinus*, the pollen tube grows part of the way to the egg in one year, then stops until the next spring, when it begins to grow again. Most conifers bear male and female cones on different branches of the same tree.

Connecting Ideas

With the evolution of vascular seedless plants, the prominence of the gametophyte generation was lost. The sporophyte became dominant, and roots, stems, and leaves evolved. The gametophyte generation became modified in the first seed-bearing plants to just a few cells within the pollen grain and the ovule. Fertilization without a surface film of water became possible. In the gymnosperms, plants finally became fully adapted to life on dry land. There was one further development in the evolution of land plants: development of the flowering plants–the most diverse, widespread, and numerous plants living on Earth today.

Section Review

Understanding Concepts

1. Compare the structure of the seed in a ginkgo with that of conifers.
2. Name two ways that seeds are an important adaptation to life on land.
3. How are needlelike leaves an adaptation to life in cold climates?

Thinking Critically

4. How do you think the development of the seed might have affected the lives of herbivorous animals living in Earth's ancient forests?

Skill Review

5. **Comparing and Contrasting** Compare the formation of a spore in ferns and a seed in conifers. For more help, refer to Thinking Critically in the *Skill Handbook*.

Answers to Section Review

1. Conifer seeds have a tough protective cover and contain eight cotyledons and "wings" to aid dispersal. Ginkgo seeds have a fleshy outer covering and several cotyledons.
2. Seeds contain a food supply for the developing embryo and are protected against harsh conditions by a seed coat.
3. Leaves have a thick cuticle covering that helps retain water, are flexible and are prevented from breaking if there is snow buildup.

Thinking Critically

4. Herbivores were provided with a food supply that was intended as food for the young plant embryo.

Skill Review

5. Spores are formed from meiosis. Seeds result from fertilization of gamete cells that were formed originally from mitosis.

REVIEWING MAIN IDEAS

25.1 Seedless Vascular Plants

- The spore-producing vascular plants were prominent members of Earth's ancient forests. All are represented by modern species.
- Vascular tissues, including xylem and phloem, provide the structural support that enables vascular plants to grow taller than nonvascular plants.
- Spore-producing vascular plants are classified as Lycophyta, the club mosses; Sphenophyta, the horsetails; and Pterophyta, the ferns.
- Fertilization in seedless vascular plants requires a surface film of water through which sperm swim to the egg.

25.2 Gymnosperms

- The term *gymnosperm* is used to describe the four divisions of plants that bear unprotected seeds: Cycadophyta, the cycads; Gnetophyta, the gnetophytes; Ginkgophyta, the ginkgoes; and Coniferophyta, the conifers.
- Seeds contain a supply of food to nourish the young plant, protect the embryo during harsh conditions, and provide methods of dispersal to new areas.
- Fertilization in seed-bearing plants does not require water.
- The xylem in conifers and other gymnosperms is composed of thick-walled, nonliving cells called tracheids, which transport water and dissolved nutrients and provide support.

Key Terms

Write a sentence that shows your understanding of each of the following terms.

cone	phloem
cotyledon	pollen grain
deciduous plant	prothallus
embryo	rhizome
evergreen plant	sorus
frond	strobilus
gymnosperm	tracheid
megaspore	vascular tissue
microspore	xylem
ovule	

Understanding Concepts

1. In lycophytes, spore-bearing leaves form a cluster at the end of the stem called a ___________.
   a. sorus c. strobilus
   b. frond d. prothallus
2. Leaves of lycophytes occur as all of these EXCEPT ___________.
   a. pairs c. whorls
   b. loops d. spirals
3. The ___________ is the gametophyte of a lycophyte.
   a. sorus c. strobilus
   b. frond d. prothallus
4. Lycophytes include ___________.
   a. ferns c. mosses
   b. conifers d. club mosses
5. Vascular tissue includes ___________.
   a. cotyledons
   b. rhizomes and phloem
   c. megaspores and microspores
   d. xylem and phloem
6. ___________ tissue is made up of dead, tubular cells that transport water and minerals.
   a. Cambium c. Phloem
   b. Xylem d. Nonvascular
7. About 280 million years ago, many of the spore-producing plants became extinct because of ___________.
   a. long periods of drought and freezing
   b. a change to a warm, wet climate
   c. environmental pollution by humans
   d. being eaten by dinosaurs

Chapter 25 Review

Reviewing Main Ideas

Summary statements can be used by students to review the major concepts of the chapter.

Key Terms

Answers should go beyond defining the terms. Accept any answer that uses the term correctly and in the proper context.

Understanding Concepts

1. c
2. b
3. d
4. d
5. d
6. b
7. a

Chapter 25 Review

8. c
9. c
10. d
11. b
12. c
13. c
14. b
15. c
16. d
17. b
18. a
19. b
20. a

Applying Concepts

21. Fertilization may occur at any time because it is no longer dependent on water availability.
22. Conifer leaf adaptations allow them to survive in cold, harsh climates.
23. Evergreens retain leaves all year long and can start to photosynthesize as soon as the growing season starts. Deciduous plants lose their leaves, reducing water loss during the winter.
24. The sporophyte protects the gametophyte and provides nourishment.
25. Probably not. These plant types prefer to grow in moist to wet areas, providing a clue that this land is moist and, at times, wet.

Chapter 25 Review

8. The living, tubular cells of ___________ transport sugars from the leaves to all plant parts.
   a. cambium c. phloem
   b. xylem d. algae
9. Unlike the stem of other vascular plants, the stem of the horsetail is ___________.
   a. solid c. hollow
   b. vascular d. smooth
10. What division does a tree with lobed leaves and cones belong to?
   a. Lycophyta c. Cycadophyta
   b. Sphenophyta d. Ginkgophyta
11. ___________ is a division that includes plants with a terminal rosette of leaves and cones at the center.
   a. Sphenophyta c. Coniferophyta
   b. Cycadophyta d. Gnetophyta
12. Trees that have woody cones, inhabit cold forests, and have needlelike leaves belong to the ___________.
   a. Sphenophyta c. Coniferophyta
   b. Cycadophyta d. Gnetophyta
13. The thick, underground stem of a fern is the ___________.
   a. frond c. rhizome
   b. rhizoid d. pinna
14. Vascular plants exhibit alternation of generations in which the sporophyte is ___________.
   a. the tiny prothallus
   b. the plant we commonly recognize with roots, stems, and leaves
   c. the same as the gametophyte
   d. responsible for transport of food and water
15. The wood of a conifer is made up of ___________, the cells that make up xylem tissue.
   a. sori c. tracheids
   b. cones d. megaspores
16. In ferns, clusters of sporangia form a structure called a ___________.
   a. prothallus c. cycad
   b. cone d. sorus
17. ___________ are produced in male cones of gymnosperms and become pollen grains.
   a. Megaspores c. Embryos
   b. Microspores d. Cotyledons
18. In gymnosperms, ___________ eventually develop into archegonia with egg cells.
   a. megaspores c. embryos
   b. microspores d. cotyledons
19. When a deciduous tree loses all its leaves in the autumn, ___________.
   a. it will die
   b. water loss will be reduced during the cold months that follow
   c. it will lose a great deal of sap
   d. water is lost through its bark
20. ___________ are food-storage organs of a plant embryo that become the plant's first leaves.
   a. Cotyledons c. Pollen grains
   b. Cones d. Ovules

Applying Concepts

21. What is the adaptive advantage of fertilization that no longer requires water for the sperm to reach the eggs?
22. Why are conifers more abundant in Canada, Alaska, and Siberia than are flowering trees?
23. Explain how evergreen and deciduous plants differ, and describe the adaptive value of each.
24. What might be some evolutionary advantages of having a gametophyte that is dependent on the sporophyte?
25. You are looking for land upon which to build a house. You are shown a beautiful plot of land that is lush with ferns. It also has some club mosses and horsetails. Do you think it would be wise to build a house on this land? Why or why not?

Thinking Critically

26. ***Concept Mapping*** Make a concept map that relates the following terms and phrases. Supply the appropriate linking words for your map.
vascular plants, seeds, spores, lycophytes, sphenophytes, pterophytes, cycads, ginkgoes, gnetophytes, conifers

27. ***Classifying*** Use the information in the table to develop a key to identify the five different species of pine.

Five Different Species of Pine

Pine Species	Number of Needles per Bundle	Length (cm)	Type of Spines on Cones
Red	2	10-15	none
Longleaf	3	30-45	small
Pitch	3	8-12	sharp
Ponderosa	2 or 3	15-25	sharp
White	5	8-14	none

28. ***Designing an Experiment*** You have a large collection of a variety of pinecones. You know that some pinecones must be opened by heat in order to release their seeds. Design an experiment without the use of an open flame that would help you determine the temperatures at which various cones open. Make sure you plan to use a control and gather quantitative data.

29. ***Comparing and Contrasting*** Compare the functions of strobili on a club moss with the functions of the sori on a fern frond.

30. ***Observing and Inferring*** What are the niches of pines and ferns in the forests in which they are found?

ASSESSING KNOWLEDGE & SKILLS

The germination rate of seeds is the percentage of planted seeds that eventually sprout to produce new plants. The seeds of the bristlecone pine must be exposed to cold temperatures before they will sprout.

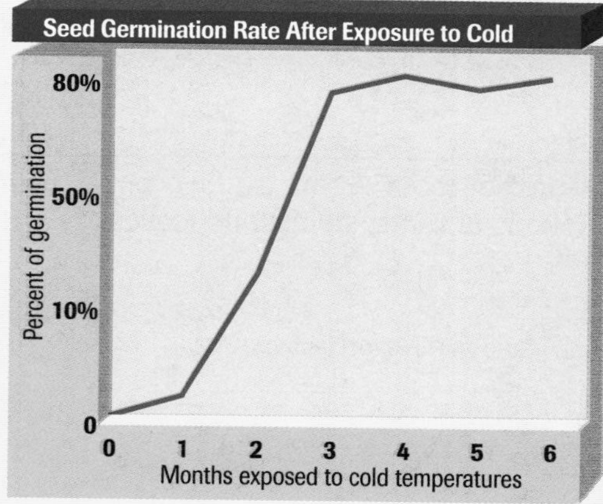

Using a Graph Answer the following questions based on the graph above.

1. What would be the minimum time you would keep bristlecone pine seeds under refrigeration before planting?
a. one month **c.** 6 months
b. 3 months **d.** 80 months

2. How long does it take to get 50 percent germination?
a. 2-1/2 months **c.** one month
b. 6 months **d.** 80 months

3. ***Interpreting Scientific Illustrations*** Examine the cross section of a conifer needle in *Figure 25.14* and answer the following question. How do the positions of stomata and vascular tissues in the leaf help to prevent water loss?

Chapter 25 Review

Thinking Critically

26. Evaluate students' concept maps. The maps should show an understanding of the relationships among the concepts listed.
27. Check students' keys for logic and accuracy. Traits to be compared will include number of needles per bundle, length of needles, and types of spines on cones.
28. Students might hang cones above a beaker of boiling water and measure the time it takes each cone to open.
29. Both structures form or contain spores produced via meiosis.
30. Both contribute organic matter to the soil as they die and decompose, supply food to herbivores in the form of leafy matter or seeds, supply oxygen to the air, and are home to many animals.

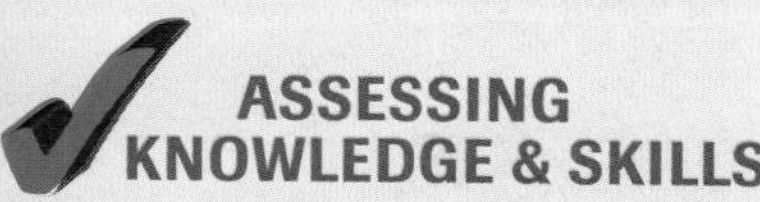

ASSESSING KNOWLEDGE & SKILLS

1. b
2. a
3. Sunken stomata and internal vascular tissue help prevent water loss.

Program Resources

Chapter Assessment, pp. 145-150 L1
Alternate Assessment in the Science Classroom
Computer Test Bank L1
Content Mastery, pp. 97-100 L1

Chapter Organizer

SECTION	OBJECTIVES	ACTIVITIES/FEATURES
26.1 What Is an Angiosperm? National Science Standards: UCP.1, UCP.5; C.5; G.1, G.2	1. **Differentiate** between gymnosperms and angiosperms. 2. **Compare and contrast** structures of monocots and dicots.	**Social Studies Connection:** Beans of the World, p. 632 **Minilab:** How can you tell monocots from dicots?, p. 634
26.2 Angiosperm Structures and Functions National Science Standards: UCP.1, UCP.2, UCP.5; A.1; C.1, C.5; F.3; G.1-3	3. **Identify** the structures of roots, stems, and leaves. 4. **Describe** the functions of roots, stems, and leaves.	**Minilab:** What do stomata look like?, p. 642 **Thinking Lab:** What determines the number of stomata on leaves?, p. 643 **Biolab:** Growth of Stems, p. 644 **A Broader View:** The Fall of the Food Factories, p. 646

ACTIVITY MATERIALS

BIOLAB	MINILABS	ALTERNATE LAB
page 644 art of one-, two- and three-year-old stems or prepared microscope slides	**page 634** plants, monocots and dicots prepared slides of monocot and dicot stems microscope labels **page 642** leaves slides coverslips microscope 5% salt solution	**page 638** petri dish black paper scissors microscope hand lens glass slide coverslip forceps mustard seeds labels

Chapter 26 Flowering Plants

TEACHER CLASSROOM RESOURCES

Reproducible Masters	Transparencies
Section Focus Master 55: Angiosperms L1 SAE **Reinforcement and Study Guide,** p. 101 L1 **Biolab and Minilab Worksheets,** p. 103 L1 **Laboratory Manual:** What Is the Effect of Light Intensity on Transpiration?, pp. 147-150 L2	**Basic Concepts Transparency #38:** Phylogeny of Flowering Plants L1 SAE
Section Focus Master 56: Angiosperm Structures L1 SAE **Reinforcement and Study Guide,** pp. 102-104 L1 **Biolab and Minilab Worksheets,** pp. 104-106 L1 **Concept Mapping:** Characteristics of Anthophyta, p. 26 L1 **Critical Thinking/Problem Solving:** Using a Key to Identify Trees, p. 26 L3 **Content Mastery,** pp. 101-104 L1	**Basic Concepts Transparency #39:** Functions of Stomata L1 SAE **Basic Concepts Transparency #40:** Leaf Structure L1 SAE **Basic Skills Transparency #20:** Internal Structure of a Leaf L1 SAE **Reteaching Transparency #26:** A Plant's Vascular System L1 SAE

ASSESSMENT MATERIALS

Chapter Assessment, pp. 151-156
Performance Assessment in the Biology Classroom, p. 9
MindJogger Videoquiz
Alternate Assessment in the Science Classroom
Computer Test Bank

Spanish Resources SAE
English/Spanish Audiocassettes SAE
Cooperative Learning in the Science Classroom COOP LEARN
Lesson Plans
Biology Projects: Grown in the U.S.A., pp. 25-28 L2

KEY TO TEACHING STRATEGIES

L1 Level 1 activities should be within the ability range of all students including those with learning difficulties.

L2 Level 2 activities are within the ability range of average to above-average students.

L3 Level 3 activities are designed for the ability range of above-average students.

SAE SAE activities should be within the ability range of Students Acquiring English.

COOP LEARN Cooperative Learning activities are designed for small group work.

P These strategies represent student products that can be placed into a best-work portfolio.

These strategies are useful in a block scheduling format.

GLENCOE TECHNOLOGY

The following multimedia resources are available from Glencoe.

Biology: The Dynamics of Life
CD-ROM SAE
Videodisc Program

The Infinite Voyage Series
A Taste of Health
Pima Indians: Old Traditions in Nutrition
Pima Indians: Coping with Disease
Life in the Balance
Discovering Patterns of Extinction in Space
Rondonia: Home of a Dying Rain Forest

National Geographic Society Series
STV: Plants
What Is a Leaf?
Newton's Apple: Life Sciences
Plant Growth
How do plants regulate water?

Science and Technology Videodisc Series (STVS)
Plants & Simple Organisms
Detecting Climate Changes in Tree Rings

CHAPTER

26 Flowering Plants

Chapter Overview

Students begin their study of flowering plants by exploring the general nature and diversity of angiosperms. The classification of angiosperms is considered with an emphasis on the structural differences between monocots and dicots. Next, the many adaptations of angiosperms are described.

The second section of the chapter focuses on angiosperm roots, stems, and leaves. Root structure and function are described with emphasis on the traits that distinguish monocots and dicots. Stem anatomy and physiology follow, including discussions of dicot woody stems and how secondary growth results in an increase in girth. Water transport in stems is also explained. The chapter concludes with a discussion of leaf anatomy and physiology and adaptations for carrying out photosynthesis and conserving water.

Key Terms

angiosperm
annual
apical meristem
bark
biennial
cambium
companion cell
cortex
dicotyledon
endodermis
epidermis
guard cell
mesophyll
monocotyledon
parenchyma
perennial
pericycle
petiole
root cap
root hair
sieve cell
sink
transpiration
vessel cell

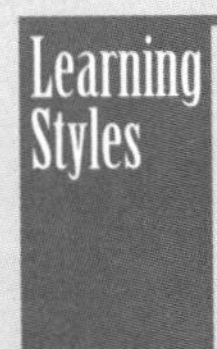

Learning Styles

Look for the following logo for strategies that emphasize different learning modalities. LS

Kinesthetic	Meeting Individual Needs, pp. 632, 642, 646; Reinforcement, p. 640
Visual-Spatial	Activity, p. 631; Visual Learning, p. 631; Demonstration, p. 633; Minilab, pp. 634, 642; Microscope Activity, p. 637; Alternate Lab, p. 638; The Inside Story, pp. 641, 647; Portfolio, p. 641
Interpersonal	Cultural Diversity, p. 631; Project, p. 633
Intrapersonal	Portfolio, p. 643; Student Journal, p. 647
Logical-Mathematical	Meeting Individual Needs, pp. 640, 646; Activity, p. 637; Project, p. 637; Thinking Lab, p. 643

CHAPTER 26 Flowering Plants

People in temperate regions of the world associate springtime with flowers. As days get longer and warmer, bright crocuses push up through the melting snow. Dry, brown grass turns green and soft again, and new leaves and young flowers unfurl on bare tree branches. The flowering plants are the most widespread, diverse, and colorful members of the plant kingdom. Flowering plants have become adapted to all corners of the world, and their species far outnumber the nonflowering plants.

Most of the plants you are familiar with are flowering plants. Fruit trees, vegetables, cereal grains, and wildflowers are all flowering plants. Why are flowering plants so successful? In this chapter, you will investigate the enormous variety of adaptations that have enabled flowering plants to fill virtually any kind of habitat, from dry salt deserts to snowy mountains and steamy tropics.

Concept Check

You may wish to review the following concepts before studying this chapter.

- Chapter 10: chlorophyll, photosynthesis, respiration
- Chapter 24: cuticle, leaf, root, seed, stem

Chapter Preview

Laboratory Activities

Biolab

- Growth of Stems

Minilabs

- How can you tell monocots from dicots?
- What do stomata look like?

How are the wildflowers, the early leaf buds, and the crocus alike? The spring crocus emerges long before the soil has become warm enough for most seeds to sprout. Wildflowers grow year after year in the same field. New leaves and flowers on a tree begin the process that eventually leads to production of seeds in a fruit. All of these organisms are flowering plants.

Assessment Planner

Choose assessment strategies from the following pages to evaluate the progress of your students.

Assess, pp. 635, 648
Alternate Lab, pp. 638-639
Minilabs, pp. 634, 642
Portfolio, pp. 641, 643
Thinking Lab, p. 643
Biolab, pp. 644-645
Chapter Review, pp. 649-651

CHAPTER 26

Introducing the Chapter

Ask students to list ten common plants. Combine the individual lists to create a class list and use this list to point out that most of the plants named are flowering plants belonging to the division Anthophyta. Elicit why the lists named mainly flowering plants. *Likely responses will suggest that flowering plants are the most common of all plants and are also sources of food.* L1

To emphasize the importance of plants as food sources, have students identify foods that come directly from plants. *Bread, cereal, and vegetables.* Next, ask where sugars and spices come from. *They come mostly from plants.* Ask where the foods eaten by animals that are consumed by people come from. *Plants.* L1

Theme Development

The theme of *systems and interactions* is apparent as the anatomy of roots, stems, and leaves is examined and the interdependence among these structures is discussed. *Evolution* is stressed through the descriptions of the various survival adaptations of angiosperms.

Concept Check

This chapter builds upon students' understanding of ecological principles, such as parasitism and epiphytic relationships discussed in Unit 2. In addition, students will add to their understanding of the process of photosynthesis.

SECTION

26.1

Prepare

Key Concepts

Angiosperms are introduced as the most well-known and diverse plant group. A general description of how angiosperms differ from gymnosperms is presented, and the classification of angiosperms into two classes—monocots and dicots—is discussed. The numerous adaptations of angiosperms are then discussed along with the roles of roots, corms, tubers, and bulbs in the different life patterns of plants.

Block Scheduling

Look for this symbol for strategies that are useful in a block scheduling format. For more information on block scheduling, refer to Section 26.1 in the **Lesson Plans** booklet.

Materials

- Obtain a gymnosperm sprig containing a cone and an angiosperm sprig containing a flower for the Focus Activity.
- Obtain iodine-potassium iodide solution and potato slices for the Demonstration.
- Obtain samples of monocot and dicot leaves from a grocery store for the Minilab.

1 Focus

Bellringer

Before presenting the lesson, display **Section Focus Master 55** on the overhead projector and have students answer the accompanying questions. L1 SAE

SECTION

26.1 What Is an Angiosperm?

Section Preview

Objectives

Differentiate between gymnosperms and angiosperms.

Compare and contrast structures of monocots and dicots.

Key Terms

angiosperm
monocotyledon
dicotyledon
annual
biennial
perennial

Did you have cereal or fruit for breakfast this morning? You've probably already eaten several flowering plants today. Check the labels on your clothing; if your clothes contain cotton or linen fibers, you are wearing flowering plants as well. A common name for all flowering plants is angiosperms.

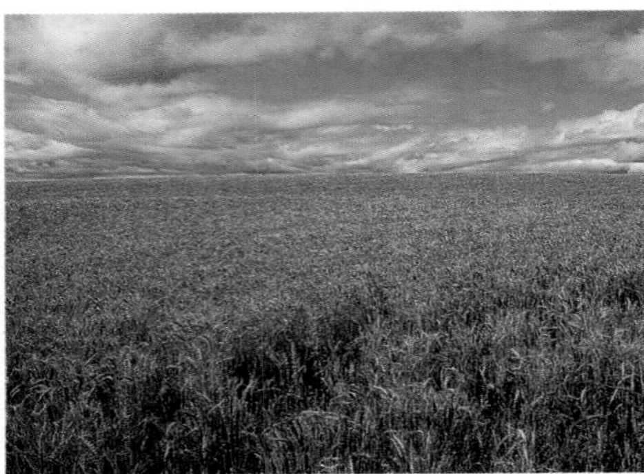

Diversity of Angiosperms

Angiosperms are the most well-known plants on Earth today. Like gymnosperms, angiosperms have seeds, roots, stems, and leaves. But unlike the other seed plants, an **angiosperm** produces flowers and develops seeds that are enclosed in a fruit, *Figure 26.1.* One of the advantages of producing fruit-enclosed seeds is the added protection the fruit provides for the newly developing plants. The flowering plants are classified in the Division Anthophyta.

Carnations belong to the family Caryophyllaceae. This family includes cultivated flowers such as pinks and baby's breath.

Peanuts are the seeds of this flowering plant. Classified in the Family Fabaceae, peanuts are related to peas, soybeans, alfalfa, and beans. They are often called legumes.

Figure 26.1
There are more than 230 000 species of Anthophyta. Most of the plants you are familiar with probably belong to this division.

Program Resources

Section Focus Master 55 L1 SAE
Reinforcement and Study Guide, p. 101 L1
Biolab and Minilab Worksheets, p. 103 L1
Performance Assessment in the Biology Classroom, p. 9 L1 P
Laboratory Manual, pp. 147-150 L2
Basic Concepts Transparency 38 and **Master** L1 SAE

Figure 26.2
Monocots have leaves with parallel veins and flower parts in multiples of three.

The lily family Liliaceae includes asparagus and onions as well as the ornamental lilies grown by many home gardeners.

The grass family Poaceae includes important cereal grains, such as rice and wheat, as well as bamboo and the sugarcane shown here.

Anthophytes are made up of two classes, the monocotyledons, or monocots, and the dicotyledons, or dicots. The two classes are named for the number of seed leaves, or cotyledons, contained within the seed. **Monocotyledons** have one seed leaf; **dicotyledons** have two seed leaves. Monocots are the smaller group, with about 60 000 species that include families such as grasses, orchids, lilies, and palms, *Figure 26.2.* Dicots make up the majority of flowering plants with about 170 000 species. They include nearly all the familiar shrubs, trees (except conifers), wildflowers, garden flowers, and herbs. Familiar dicots are shown in *Figure 26.3.*

Figure 26.3
Dicots have leaves with netted veins and flower parts in multiples of four or five.

The daisy family Asteraceae includes sunflowers, lettuce, dandelions, chrysanthemums, and goldenrod.

The mustard family Brassicaceae includes many important food plants such as cabbage, mustard, broccoli, radish, turnip, collards, and kale.

The rose family Rosaceae includes blackberries, raspberries, apples, plums, peaches, pears, and hundreds of cultivars of garden roses.

SECTION 26.1

Activity

Visual-Spatial Provide students with a gymnosperm sprig that has a cone and an angiosperm sprig that includes a flower. Ask students to examine each sprig and to describe the similarities and differences they observe. As you point to various structures on the gymnosperm sprig, challenge students to identify parts of the angiosperm sprig that have similar functions. L1

2 Teach

Visual Learning

Visual-Spatial Direct students' attention to Figure 26.2 and Figure 26.3. Elicit how the vein patterns in dicot leaves differ from those of monocots. *Dicot leaves have netted veins and monocot leaves have parallel veins.* Next, ask students to explain how the flowers of the two groups differ. *Dicot flowers have parts in multiples of four or five. Monocot flower parts exist in multiples of three.*

GLENCOE TECHNOLOGY

Videodisc

Biology: The Dynamics of Life
Disc 1, Side 2
Blooming Flowers (Ch. 21)

CULTURAL DIVERSITY

Origin and Cultural Significance of Corn

Corn was one of the earliest crops to be domesticated, and has been a major food source for Native Americans of both North and South America for about 7000 years. In addition to its importance as a food source, corn also occupies a symbolic place in the culture of many Native American tribes. Ask students working in groups to research uses for corn other than as a food source. Ask each group to prepare an illustrated essay of its findings. L1 COOP LEARN

Social Studies

Beans of the World

Purpose
Students are introduced to the variety of beans grown and used as food throughout the world.

Teaching Strategies
- Ask students to identify the structures of legumes that provide a habitat for the bacteria that absorb nitrogen from the air. *Nodules on the roots of legumes serve as the habitat for bacteria that convert gaseous nitrogen to usable forms of nitrogen.*
- Bring different beans to class. Ask students to make a poster that identifies each bean with both its common and scientific name. Scientific names will require research and references. L1

Possible Answer
Eating protein from plants is using food directly from the first trophic level, the autotrophs. Animal protein must pass through several trophic levels.

Audiovisual

Show the videos *Ancient Forests*, National Geographic, *Plants in Peril*, Films for the Humanities, or *In Celebration of Trees*, available from the Discovery Channel.

Assessment

Performance Assessment in the Biology Classroom: p. 9, Coloring Tips. Have students carry out this activity to observe the results of transport of liquids in plants. L1 P

Social Studies

Beans of the World

Beans are vegetables that are the seeds of plants with pods or legumes. The Fabaceae family includes more than 14 000 species, but only 22 are grown in quantity for human food. Even poor soils can produce a bean crop and become more fertile at the same time. This is because the root systems of beans absorb nitrogen from the air and leave it in the soil.

Beans for protein Beans are a good source of the protein that a human body needs. Protein from legumes is just as healthful as protein from animal sources. However, no one type of bean has all of the amino acids needed by humans. That is why vegetarians eat grains with beans to complement the amino acid deficiencies.

Beans around the world The following are just a few of the legumes enjoyed by people. Black-eyed peas are used in many African and Indian dishes. Cannellini beans are an important part of Italian cuisine. Channa dal are grown and eaten in India and parts of Southeast Asia. Chickpeas are India's most important legume. Lentils provide protein for the people in North Africa and Asia. Lima beans spread from Central America to North America, where Native Americans combined them with corn to make succotash. Peas originated in the Middle East around 6000 B.C. and are now eaten almost everywhere. Pinto beans are used in South American and Mexican dishes. Soybeans have been cultivated in northern China for more than 5000 years and are often made into tofu.

CONNECTION TO Biology

Why might it be more efficient to obtain our protein from beans instead of from animal products?

Angiosperm Adaptations

Flowering plants can be trees, shrubs, herbs, vines, floating plants, epiphytes, and even parasitic plants that may or may not have their own chlorophyll. They can be found in deserts, tropical forests, temperate forests, tundra, snowy mountains, lakes, ponds, and freshwater and saltwater marshes. Angiosperms are the most successful plants on Earth because they have evolved sophisticated adaptations that enable them to survive and reproduce in almost any kind of habitat.

Roots and stems have become adapted to storing food during periods of drought, cold, or limited sunlight. Roots and stems that function

Figure 26.4

Roots and stems are sometimes modified into storage organs that enable a plant to survive through a winter season underground.

Tulips, daffodils, and crocuses grow from bulbs that have stored food in the bases of last year's leaves.

632 Flowering Plants

Meeting Individual Needs

Visually Challenged Bring samples of roots, tubers, corms, and bulbs to class. Allow visually impaired students to handle and examine the variety of samples. Ask students to identify differences among the samples that might be used to distinguish them. L1 LS

interNET CONNECTION

Follow the link for this chapter on the Glencoe Homepage at **www.glencoe.com/sec/science** to find out more about flowering plants.

Figure 26.5

Angiosperms are more complex, and more adaptable, than any other plant group.

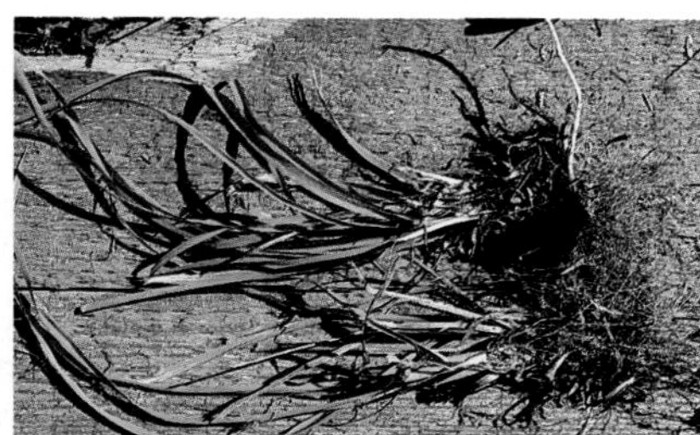

Some grasses form dense mats of roots that hold water and enable the plants to survive long periods of drought.

Mistletoe is an evergreen plant with roots that penetrate the tissues of the branches of deciduous trees. This parasite absorbs water and nutrients from the host tree.

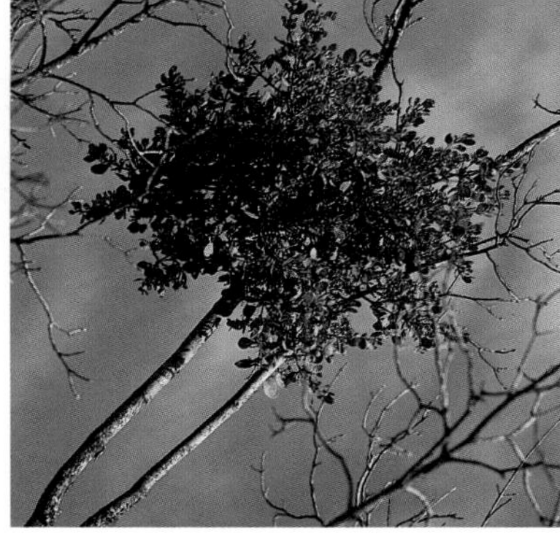

Many species of orchids are epiphytes that live on the branches of trees in a tropical forest.

as food-storage organs include bulbs, corms, and tubers. A bulb is a short stem enclosed in fleshy leaf bases. A corm is a short, thickened, underground stem that is not enclosed in leaf scales. A tuber is a swollen root or stem with buds that will sprout new plants. *Figure 26.4* illustrates some of the plants that form food-storage organs.

Leaves of different shapes and sizes have evolved in response to the amount of sunlight and moisture available. *Figure 26.5* describes a few of the adaptations that have contributed to the survival and success of flowering plants.

Growers plant cut sections of potato tubers, another kind of underground stem, that have one or more eyes, or buds, each of which will grow into a new shoot and eventually a full-grown plant.

The corm of a gladiolus is a thickened, underground stem from which leaf and flower buds arise.

The rhizome of an iris or a ginger plant is an underground stem.

Different Viewpoints in Biology

Hydroponics—Success or Failure?

Hydroponics is an agricultural process in which plants are grown in a nutrient solution rather than soil. Commercial hydroponics was begun in an attempt to reduce famine and eliminate the need for farmers to contend with poor soil conditions and constant weather changes. Unfortunately, hydroponics has not worked well on a large scale.

Hydroponic plants must be supplied with nutrient solutions that require continuous monitoring. Heating and lighting costs for maintaining the plants are high. In addition, the roots of hydroponic plants require constant aeration which also adds to the expense. While hydroponics has had some success, it is unlikely that this agricultural process will ever replace conventional farming practices.

Demonstration

LS **Visual-Spatial** Demonstrate starch storage in root tubers by adding a few drops of an iodine-potassium iodide solution to several slices of potato tuber. Explain that the blue-black color indicates the presence of starch.

Visual Learning

Figure 26.4 Have students explain whether the growth of new plants from rhizomes, corms, or tubers are examples of sexual or asexual reproduction. *Because the growth of the new plants results from cell division, each process is an example of asexual reproduction.*

PROJECT

Comparing Plant Growth

Provide student groups with a bulb (garlic or pearl onion), root section (ginger root), and tuber (sweet potato or yam). Have students gather the materials needed to grow each structure into a mature plant. Students must devise a schedule for maintaining their plants as well as a means for observing and recording changes that occur during the growth of each plant. Encourage students to make weekly diagrams or take photographs of their plants throughout the project and include these materials along with a brief report in their portfolios.

L2 P LS

MiniLab

Purpose

LS Visual-Spatial To illustrate the similarities and differences between monocot and dicot leaves.

Process Skills

observe and infer, use the microscope, compare and contrast

Teaching Strategies

- Leaf samples must be tagged with labels and names so students can complete their tables.
- Leaf samples from supermarket produce can be used (Dicots: brussels sprouts, lettuce, spinach, celery; Monocots: corn husks, green onions, chives).

Expected Results

Tables will show that dicots have a branching vein pattern, while monocots have a parallel vein pattern. The microscopic leaf cross sections will show monocot vascular bundles in a linear, regular pattern along the leaf length, while the vascular bundle pattern will be irregular in dicot leaves.

Analysis

1. Veins are parallel in monocots and netlike or branching in dicots.
2. Veins will appear in a linear, even order in monocots and irregularly in dicots.

Assessment

Skill: Have students collect and bring to class two examples each of monocot and dicot leaves. Ask students to label each leaf and explain what characteristics were used for its identification. L1

MiniLab

How can you tell monocots from dicots?

The two classes of flowering plants are monocots and dicots. An easy way to tell them apart is by looking at the pattern of veins in their leaves. The veins of monocot leaves are parallel; dicot leaves have a network of branching veins.

Procedure

1. Examine the pattern of veins in the leaves of each plant your teacher has provided.

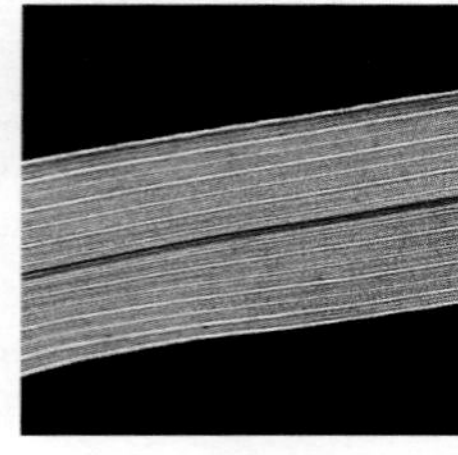

2. Determine whether each plant is a monocot or dicot. Make a table to record your answers.
3. Examine the prepared microscope slide of monocot and dicot stems. Describe what you see.

Analysis

1. What differences did you observe between the external appearance of monocot and dicot leaves?
2. What differences did you observe between the internal organization of monocot and dicot stems?

Figure 26.6

Anthophytes may be annuals, biennials, or perennials.

Vegetable gardeners grow biennial Swiss chard for its leaves, which taste a little like spinach. During cold winter months, the leaves and stems die back, but long, branching roots survive to support the growth of flowers the following spring.

Life spans of anthophytes

Why do some plants live longer than people, while others live only a few weeks? The life span of a plant reflects its strategies for surviving periods of cold, drought, or other harsh conditions.

Annual plants live for only a year or less. They sprout from seeds, grow, reproduce, and die in a single growing season. Most annuals are herbaceous, which means their stems are green and do not contain woody tissue. Many food plants such as corn, wheat, peas, beans, and squash are annuals, as are many weeds of the temperate garden. Annuals form drought-resistant seeds that survive the winter.

Biennials have a life span that lasts two years. Many biennials are plants that develop large storage roots, such as carrots, beets, and turnips. During the first year, biennials grow many leaves and develop a strong root system. Over the winter, the aboveground portion of the plant dies back, but the roots remain alive. Underground roots are able to survive conditions that leaves and stems cannot endure. During the second spring, food stored in the root is used to produce new shoots that bear flowers and seeds.

Perennials live for several years, producing flowers and seeds periodically—usually once each year. They survive harsh conditions by dropping

1. These plants have a built-in food supply and do not have to rely on photosynthesis or very warm temperatures to initiate rapid growth.
2. Leaf dropping reduces the amount of water lost through these structures. In the cold winter months, water is less available.

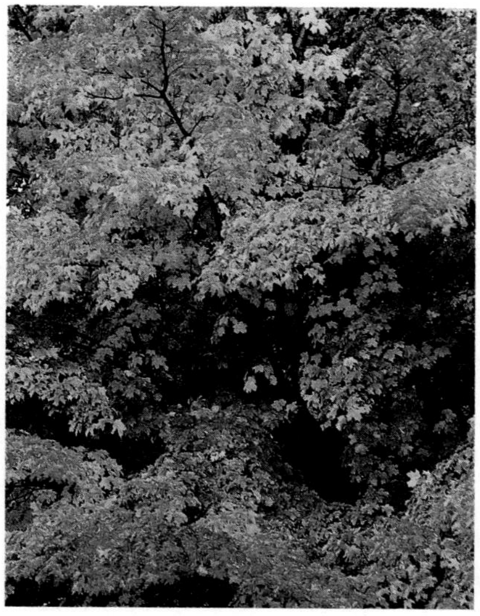

Woody perennials, like this maple, drop their leaves and become dormant during the winter. New leaves grow when the weather warms and food stored in the tree's roots and trunk moves up into the branches.

Herbaceous perennials often have underground storage organs used for overwintering.

These blue lupines and orange poppies are annual plants that live out their lives during spring and summer, leaving behind seeds for next year's growth.

their leaves or dying back to soil level, while their woody stems or underground storage organs remain intact and dormant. Perennials include the deciduous trees of the world's temperate forests. Although many perennials are large, woody trees and shrubs, there are also herbaceous perennials, including numerous species of grasses and spring wildflowers. What kinds of plants make up a grassy lawn? *Figure 26.6* gives examples of annual, biennial, and perennial anthophytes.

Section Review

Understanding Concepts

1. Why do most early spring flowers grow from bulbs, corms, or tubers?
2. Explain the adaptive value for a maple tree to lose its leaves in the fall.
3. Explain the adaptive value of annual, biennial, and perennial life spans.

Thinking Critically

4. What characteristics would you look for if you were examining an unknown plant to determine whether it is an angiosperm or a gymnosperm?

Skill Review

5. **Observing and Inferring** You are examining a plant with long, thin leaves and flowers that have six petals. Is this plant a monocot or a dicot? For more help, refer to Thinking Critically in the *Skill Handbook*.

Answers to Section Review

3. Annuals rapidly complete their life cycle and provide seeds for the following season. Biennials survive harsh winters by storing food, which becomes available for new and rapid growth the second year. Perennials may either drop their leaves and/or become dormant over winter.

Thinking Critically

4. Leaf shape and loss, reproductive structures being cones or flowers

Skill Review

5. Monocot—long thin leaves are typically associated with parallel veins as are flower parts in sixes.

3 Assess

Check for Understanding

Have students compare the following:

(a) monocots and dicots
(b) corms and bulbs
(c) epiphyte and parasite
(d) annual and biennial L1

Reteach

Have students name plants that are examples of each term listed in the word pairs of the Check for Understanding. Ask students to explain why each plant is representative of the term. L1

Extension

Have students research the diversity of the Rosaceae family. Ask them to prepare a visual essay that identifies both decorative and edible forms of the plants in this family. L2

Assessment

Performance: Have students describe as many differences and similarities as possible between monocots and dicots. *Differences should center on vein patterns and flower part numbers. Similarities are general anatomy, multicellularity, tissues and organs, and the ability to photosynthesize.* L1

4 Close

Discussion

Write the terms *annual, perennial, anthophyte, shrub,* and *herb* on the chalkboard. Ask volunteers to: (a) define each term, (b) use the term in a sentence that shows its application, (c) use the term in a way that might apply to their lives. L1 SAE COOP LEARN

SECTION

26.2

Prepare

Key Concepts

The structures and functions of roots, stems, and leaves are the focus of this section. The function of roots as water and mineral absorbers is discussed in conjunction with the tissues that permit these functions. Next, the function of stems as conduits for water and minerals is presented with a discussion of the specialized cells and tissues that carry out these jobs. The section concludes with a discussion of leaves. Both external and internal leaf structures are described and leaf function as it relates to photosynthesis is examined.

Block Scheduling

Look for this symbol for strategies that are useful in a block scheduling format. For more information on block scheduling, refer to Section 26.2 in the **Lesson Plans** booklet.

Materials

- Have thin slices of woody stems available for the Focus Activity and The Inside Story. Slices of woody stems can be obtained from landscape supply houses.
- Order prepared slides of root cross sections for the Microscope Activity.
- Gather plants or purchase spinach for the Minilab. Prepare a 5% salt solution.

1 Focus

Bellringer

Before presenting the lesson, display **Section Focus Master 56** on the overhead projector and have students answer the accompanying questions. L1 SAE

SECTION 26.2

Angiosperm Structures and Functions

Section Preview

Objectives

Identify the structures of roots, stems, and leaves.

Describe the functions of roots, stems, and leaves.

Key Terms

epidermis
root hair
cortex
parenchyma
endodermis
pericycle
cambium
apical meristem
root cap
bark
sink
vessel cell
sieve cell
companion cell
petiole
transpiration
guard cell
mesophyll

You've learned that vascular plants have roots, stems, and leaves, and that these structures have become adapted to distinct functions. Recall from Chapter 24 that roots absorb water and minerals from the soil, the stem supports the plant and transports materials, and photosynthesis takes place in the leaves.

Plants are not as active or as complex as animals, but like all multicellular organisms, plants are made up of a variety of tissues. These tissues have functions that support the growth and development of flowering plants.

Roots

Roots are the underground parts of a plant. They anchor the plant in the ground, absorb water and minerals from the soil, and transport these materials to the base of the stem. Some plants such as carrots also accumulate and store food in their roots. The total surface area of a plant's roots may be as much as 50 times greater than the surface area of its leaves. As *Figure 26.7* illustrates, roots may be short or long, thick or thin, massive or threadlike. Some roots even extend above ground.

Figure 26.7

Root systems vary according to the needs of the plant and the texture and moisture content of the soil. The two main types of root systems are taproots and fibrous roots.

Corn is a monocot with shallow, fibrous roots. As the plants grow to maturity, roots called adventitious roots grow from the stem to help keep the tall plants upright.

The fleshy taproot of the biennial beet plant serves as a food-storage organ.

Program Resources

Section Focus Master 56 L1 SAE
Reinforcement and Study Guide, pp. 102-104 L1
Biolab and Minilab Worksheets, pp. 104-106 L1
Concept Mapping, p. 26 L1
Critical Thinking/Problem Solving, p. 26 L3
Basic Concepts Transparencies 39, 40 and **Masters** L1 SAE
Basic Skills Transparency 20 and **Master** L1 SAE
Reteaching Transparency 26 and **Master** L1 SAE

Figure 26.8

The root structures of dicots and monocots differ in the arrangement of xylem and phloem.

In dicots, xylem forms a star-shaped mass at the center of the root, with phloem nestled between the rays of the star.

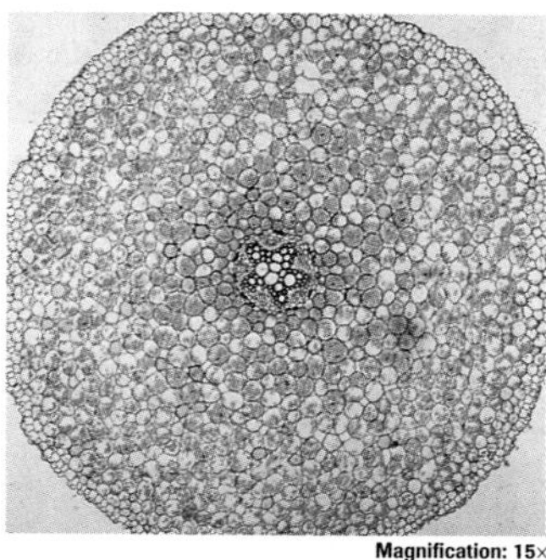
Magnification: 15×

In monocots, strands of xylem alternate with strands of phloem. Monocots usually have a central core of cells called pith. Pith is composed of parenchyma.

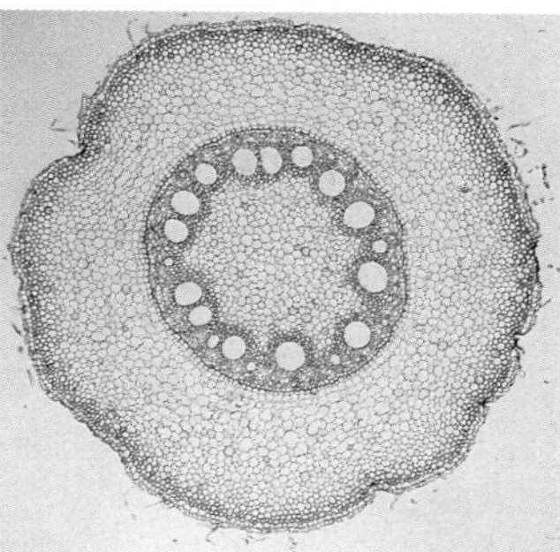
Magnification: 14×

The structure of roots

If you look at the cross section of a typical dicot root in *Figure 26.8,* you can see that the **epidermis** forms the outermost cell layer. A **root hair** is a tiny extension of a single epidermal cell that increases the surface area of the root and its contact with the soil, and absorbs water and dissolved minerals. The next layer is the **cortex,** which is involved in the transport of water and ions into the vascular core at the center of the root. The cortex is made up of packed cell layers of **parenchyma,** a tissue that sometimes acts as a storage area for food and water. Parenchyma occurs throughout most plants.

At the inner limit of the cortex lies the **endodermis,** a single layer of cells that forms a waterproof seal that surrounds the root's vascular tissue. The endodermis controls the flow of water and dissolved ions into the root. *Figure 26.9* traces the two pathways by which water and mineral ions move into the root. Just within the endodermis is the pericycle. The **pericycle** is a tissue that gives rise to lateral roots.

Figure 26.9

Water and mineral ions enter the root either by absorption into root hairs and across the cells or by flowing between the cells of the epidermis. Because of the waterproof seal between each cell of the endodermis, all water and minerals are forced to pass through the cells of the endodermis.

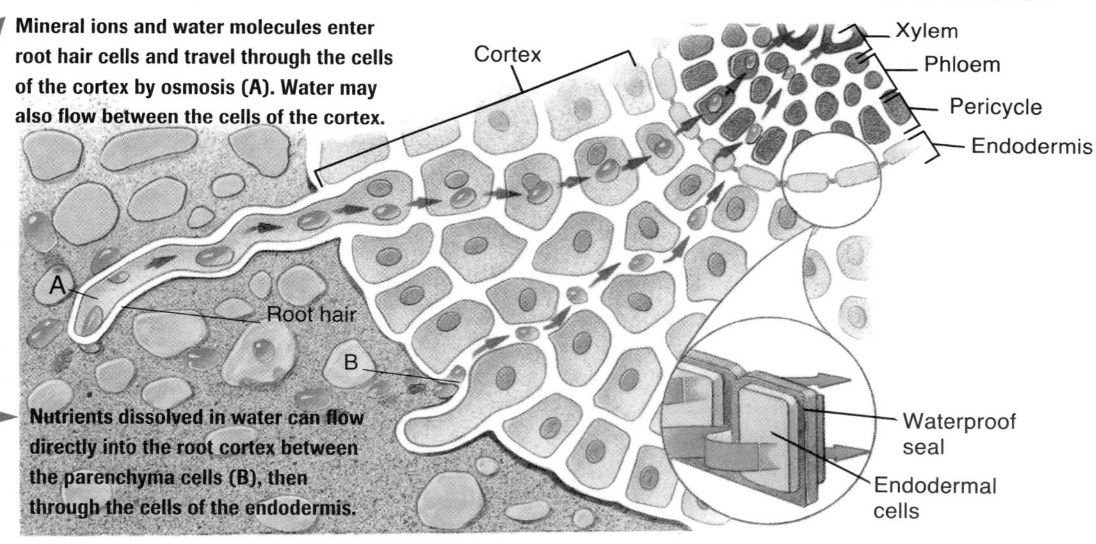
Mineral ions and water molecules enter root hair cells and travel through the cells of the cortex by osmosis (A). Water may also flow between the cells of the cortex.
Cortex
Xylem
Phloem
Pericycle
Endodermis
A
Root hair
B
Nutrients dissolved in water can flow directly into the root cortex between the parenchyma cells (B), then through the cells of the endodermis.
Waterproof seal
Endodermal cells

Activity

Logical-Mathematical Provide thin slices of tree trunk sections to small student groups. Have students count the rings to determine the age of their sample. Ask students to mark with a labeled pin the ring that represents the year when most group members were born, assuming that the tree was cut this year. L1

2 Teach

Microscope Activity

Visual-Spatial Prepared slides of root cross sections are available from supply houses. Allow students to view such slides to help them become familiar with the various tissue names and their locations. Encourage students to sketch their observations. L1

Software

Plants, The Life Science Program; *Leaf: Structure and Function,* IBM; *Photosynthesis and Transport,* Biology Program Series; *Exploring the Amazing Food Factory: The Leaf,* Carolina. L1

GLENCOE TECHNOLOGY

Videodisc

Biology: The Dynamics of Life
Disc 1, Side 2
Water Uptake in Roots (Ch. 22)

PROJECT

Do Roots Respire?

Have students carry out the following experiment to answer the question: Do roots carry out respiration and release carbon dioxide? Have students germinate bean seeds and allow them to grow into young plants (this takes about 10 days). Have students remove plants from soil and place the root portion into tubes of a bromothymol blue solution. Have students seal the roots from the air using a cotton or clay plug. Suggest that a control tube also be prepared. Advise students that BTB turns from blue to light blue or green/yellow in the presence of carbon dioxide. Have students observe the liquid 24 hours later. Ask them to write a report summarizing what conclusions can be made regarding respiration of root tissue. L2

Health Connection

Eating Potato Skins

Are potato skins safe to eat? A chemical called solanine, which is produced by potato skins, can cause headaches, fever, cramps, and diarrhea. A clue that indicates solanine production is a green color under the potato skin. The green coloring indicates that the potato was exposed to too much light after harvesting. Too much light results in solanine production. Is there any benefit to eating potato skins? *Yes. Most of the potato's iron, calcium, and fiber are contained in its skin.*

Tying to Previous Knowledge

Have students recall the differences among facilitated diffusion, active transport, diffusion, and osmosis. If necessary, briefly review these concepts.

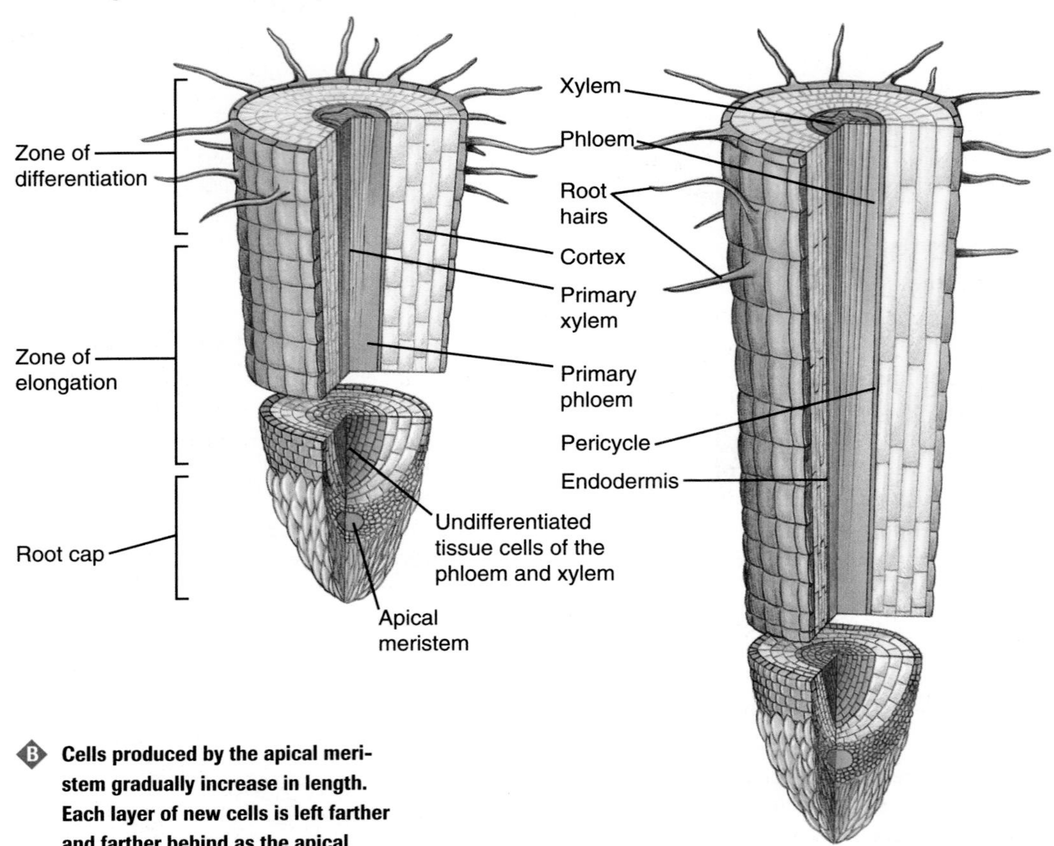

A As root cells mature, they begin to differentiate. Epidermal cells develop root hairs, and internal cells develop into cortex, endodermis, pericycle, and vascular tissues. Cells that form the cambium soon begin contributing to the root's growth by adding cells that increase its size.
C As cells of the root cap wear away, new cells are added by the tip end of the apical meristem. Cells produced further behind the apical meristem will differentiate into other types of root tissue.
Zone of differentiation
Zone of elongation
Root cap
Xylem
Phloem
Root hairs
Cortex
Primary xylem
Primary phloem
Pericycle
Endodermis
Undifferentiated tissue cells of the phloem and xylem
Apical meristem
B Cells produced by the apical meristem gradually increase in length. Each layer of new cells is left farther and farther behind as the apical meristem continues adding new cells and moving forward through the soil.
D The root cap protects the root tip as it grows down into the soil.

Figure 26.10

Roots develop by both cell division and growth. As cells build up, the root increases in size.

Xylem and phloem are located in the center of a root. Between the xylem and phloem tissues, there is a single layer of cells called the vascular cambium. **Cambium** is growth tissue that produces additional xylem and phloem cells.

Root growth

Remember from Chapter 11 that the cells responsible for root growth in an onion root lie just behind the tip. Cells that are capable of mitosis form a growth tissue that remains just behind the root tip and is called the **apical meristem.** As cells produced by the apical meristem begin to mature, they develop root hairs, as *Figure 26.10* indicates. The tip of each root in a plant is covered by a tough, protective layer of living parenchyma cells called the **root cap.**

Alternate Lab — Root Hairs

Purpose

Visual-Spatial To allow students to examine the number and nature of root hairs.

Materials

petri dish, black paper, scissors, microscope, hand lens, glass slide, coverslip, water, forceps, mustard seeds, label

Procedure

Give the following directions to students.

1. Cut out several thicknesses of black paper to fit the inside bottom of a petri dish. Moisten the paper.
2. Add 10 mustard seeds to the dish and cover. Label your dish.
3. Examine the dish each day to look for evidence of germination. Record your observations with daily diagrams.
4. Use a hand lens to check for the appearance of root hairs.
5. After several days, use forceps to remove several root hairs. Prepare a wet mount for the hairs and examine them under the microscope.

Stems

Stems are the aboveground parts of plants that support leaves and flowers. Their form ranges from the thin, herbaceous stems of daisies, which die back every year, to the massive woody trunks of trees that may live for centuries. Green, herbaceous stems are soft and flexible and usually carry out some photosynthesis. Petunias, marigolds, impatiens, and carnations are examples of plants with herbaceous stems. Trees, shrubs, some woody perennials such as oaks, maples, lilacs, and roses have woody stems. Woody stems are hard and rigid and contain a large number of strands composed of xylem.

Stems have several important functions. They provide support for all the aboveground parts of the plant. The vascular tissue that runs through the stem includes xylem and phloem and transports water, mineral ions, and sugars to and from roots and leaves. Leaves and flower buds are produced in the apical meristem of the shoot. The buds at each node located at intervals along the length of the stem produce lateral stems or flowers.

Internal structure of stems

As you learned in Chapter 25, xylem is the vascular tissue that transports water upward from the roots, and phloem is the vascular tissue that transports sugars made during photosynthesis away from the leaves. Although the stem contains many of the same tissues as a root, the tissues are arranged differently. As you can see in *Figure 26.11*, monocots and dicots differ in the arrangement of vascular tissues in their stems.

Woody stems

Most herbaceous stems live only for a single growing season. They remain soft and green because they never develop woody tissue. The stems of gymnosperms and some herbaceous and perennial dicots survive for several years. As the stems of these plants grow in height, they also grow in thickness. This added thickness, called secondary growth, results from cell division in the vascular cambium of the stem. The new vascular tissue produced by secondary growth results in annual growth rings that can be used to determine the age

Figure 26.11

One of the primary differences between roots and stems is that stems have a bundled arrangement of vascular tissues within a surrounding mass of parenchyma tissue.

In young herbaceous dicot stems and those that do not increase in thickness, xylem and phloem are arranged in vascular bundles in the cortex. In older stems, including all woody stems, the vascular tissues form a continuous cylinder between the cortex and the pith.

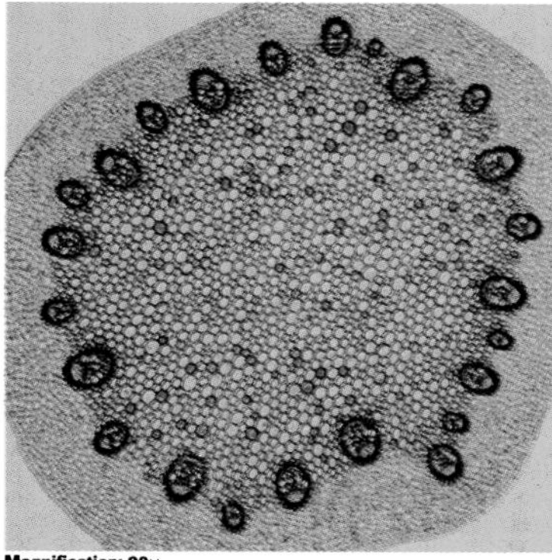

Magnification: 20×

The vascular bundles in a monocot are scattered throughout the stem.

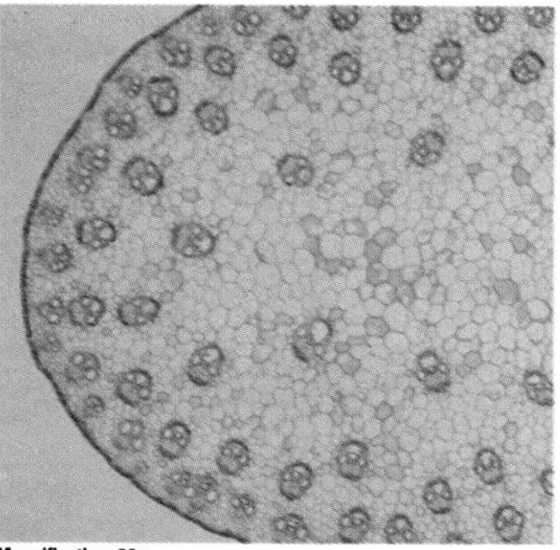

Magnification: 20×

SECTION 26.2

Using Science Terms

The term *phloem* is from the Greek meaning "bark." *Xylem* is Greek for "wood." Explain to students why these terms may have been used to describe the tissues when these names do not describe the functions of these tissues.

Science, Technology, and Society

Chemical Value of Trees

The neem tree is an angiosperm that grows in India and Thailand. The tree produces a flower with a honey scent and fruit that resembles olives. The neem tree has been called a "cornucopia tree" because of the diverse chemical products made from it. For example, the tree is used to make soap and oils for lubricants and fuel. It has also been used to make an antimalarial agent.

Plants may also act as insecticides. It is this action that has caught the interest of scientists.

After extensive testing, the tree was found to have a harmful effect on fleas, houseflies, head lice, gypsy moths, boll weevils, and cockroaches. Today, chemicals from the tree are being marketed as a home-and-garden insecticide under the name Bioneem. The plant may prove to be a valuable tool as part of the ever growing natural insecticide arsenal.

Determine if each root hair is multicellular or unicellular.

Analysis

Ask students to answer the following questions.

1. How many days were needed for root hairs to appear? *Two to three*
2. Estimate the number of root hairs that were visible. Explain why this number is adaptive. *Hundreds: increases area for water absorption*
3. Was a single root hair multicellular or unicellular? *unicellular*

Assessment

Knowledge: Have students name and describe the tissue that forms a plant's roots hairs. *Epidermis* L1

SECTION 26.2

Reinforcement

Kinesthetic Provide students with scissors, a dissecting needle, tape, and several paper straws. Ask them to use these materials to construct models of vessel cells and sieve cells with their companion cells. Have students draw the nuclei on the straws/cells that contain them. Advise students that most xylem cells are dead and, therefore, lack a nucleus. Have students add labels to the straws to identify each type. *Models should have open ends on vessels and closed ends with holes on sieve cells with a companion cell taped next to and alongside it.* L2

Bioethics

Alien Species

A potential problem exists when a plant species, either by accident or for some specific purpose, is introduced into an area where it is not native. Such was the case with *Salsola kali* or tumbleweed. This symbol of the American frontier is actually a native thistle from Russia that was introduced into the United States around 1877. The tumbleweed quickly took over the vast plains of the Midwest, proving to be a major problem for farmers, cattle, and ranchers.

How the plant got to this country remains a mystery. However, the tumbleweed illustrates how introduction of a foreign species can affect native plants, soil, and people.

Figure 26.12

Xylem carries water up from roots to leaves. Phloem transports sugars from the source in the leaves to sinks located throughout the plant.

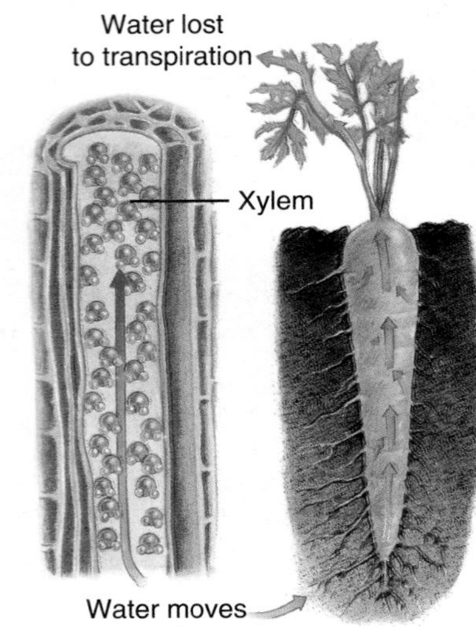

The open ends of xylem vessel cells form complete pipelike tubes. The water molecules form an unbroken column within the xylem. As transpiration from the leaves pulls water molecules from the top of the water column, the rest of the water in the xylem is pulled up through the plant.

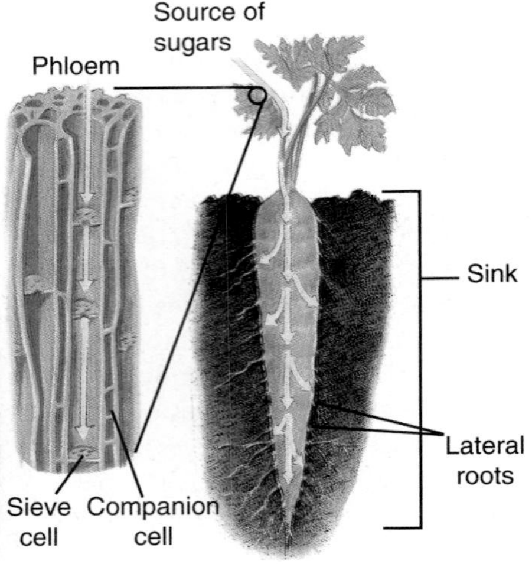

Sugars in the phloem of this carrot plant are moving to sinks. Companion cells help manage the one-way traffic that flows through the sieve cells of the phloem.

of the plant. Find out more about secondary growth in woody stems in *The Inside Story.* Some annual stems such as those of sunflowers start out herbaceous, but over the growing period also develop woody tissues. Vascular tissues contain woody fibers for support.

As secondary growth continues, the outer portion of a woody stem develops bark. **Bark** is made up of phloem tissue and a tough, corky layer that protects the phloem from damage by burrowing insects and herbivores.

Vascular tissue

Whenever a plant produces more sugars during photosynthesis than are needed for plant processes, some may be stored for future use. Any portion of the plant that uses or stores these sugars is called a **sink.** The parenchyma in the cortex of a stem or root is often a sink.

Both xylem and phloem are composed of tubular cells joined end to end. The xylem of angiosperms is made up of vessel cells and fibers. A **vessel cell** is an elongated cell with open ends. Fibers are elongated, elastic cells. Both vessel cells and fibers contain lignin, a polymer that makes cell walls more rigid. In angiosperms, vessels conduct water and fibers provide extra support.

Phloem is made up of **sieve cells,** which are thin-walled cells with sievelike plates on their end walls. Sieve cells have cytoplasm but no nuclei. Each sieve cell is next to a **companion cell** that does contain a nucleus and helps control movement through the sieve cell. Sugars move into and out of the phloem by active transport and osmosis. The structure of vascular tissue cells and the movement of water and sugars through them is shown in *Figure 26.12.*

Meeting Individual Needs

Students Acquiring English Have students prepare a table for their portfolios that compares and contrasts xylem and phloem. Have students include information about structure and function. L1 SAE P

Growth Rings of a Tree

The inner portion of the trunk of a tree is composed primarily of dead xylem cells from the growth of previous years. A woody stem also includes bark, an outer layer of cells that protects phloem cells.

Growth rings

1. After its first year of growth, a tree trunk is stiff and green but without bark. The xylem is on the inside adjacent to the pith, whereas the phloem is on the outside adjacent to the cortex.

2. During the second year of the stem's growth, the vascular cambium produces a second layer of xylem tissue. As the cambium is pushed outward by the newly formed layer of xylem tissue, the phloem is also pushed outward.

3. Another layer of cambium, called the cork cambium, develops outside the ring of phloem. The cork cambium produces cork cells, which replace the epidermis and form the tough, waterproof, outer covering of the bark.

6. Tree rings can be used to study the history of the climate of a region. Thicker rings indicate years when temperature, rainfall, and other environmental conditions were particularly favorable.

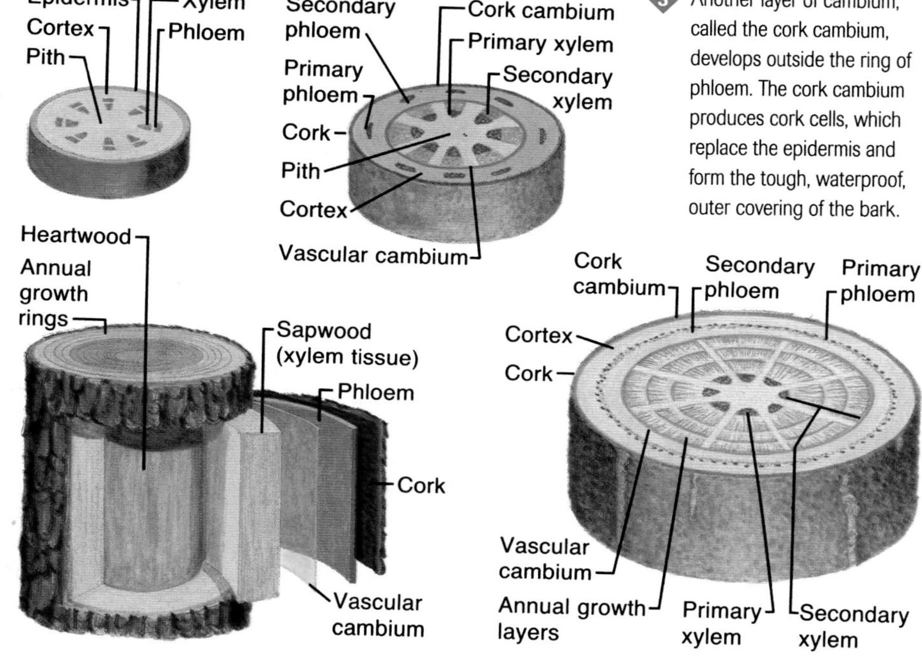

5. Over time, the xylem cells toward the center of the trunk stop functioning as vascular tissue. This tissue forms the darker heartwood of the tree. The outer layers of cells that remain actively involved in transport form the lighter-colored sapwood of the tree.

4. The vascular cambium does not divide during long periods of extreme cold or drought. Xylem cells produced in the spring tend to be larger than those produced later in the growing season because there is more water available in the spring. This alternation in the sizes of xylem cells produces a pattern of annual growth rings that in temperate regions reveals the age of the tree.

THE INSIDE STORY

Growth Rings of a Tree

Purpose

IS Visual-Spatial To illustrate and describe the location, function, and growth patterns that result in the formation of annual rings.

Teaching Strategies

- Obtain a tree stem cross section of considerable age. Show students the locations and demarcation between the heartwood and sapwood. On the stem, point out the locations of bark, cork cambium, vascular cambium, and pith.
- Provide small stem slices to each student. Have them examine the xylem using a hand lens and note the difference between the cell size of spring xylem in one ring and the cell size of xylem formed late in the season in the same ring. L1

Visual Learning

- Have students working in groups use the diagram to sequence the tissues found in a mature woody stem from outside to inside. L1 COOP LEARN
- Ask students to use the information in *The Inside Story* to list the tissues that comprise bark. Have them describe the function of each tissue layer. L1

PORTFOLIO

A Tree Has Good Years and Bad Years
Annual ring thickness is an indication of annual growing conditions. For example, good growing conditions yield a wider ring than years when conditions are poor. Have students use this information to make simple stylized diagrams of a woody stem cross section that illustrates years of both good and poor growth. Have students label their diagrams to show which ring corresponds to which type of growing condition. Encourage students to include additional labels that identify the general location of bark, phloem, cambium, and xylem. L1 P IS

GLENCOE TECHNOLOGY

Videodisc

STVS: Plants & Simple Organisms
Disc 4, Side 2
Detecting Climate Changes in Tree Rings (Ch. 22)

MiniLab

Purpose

LS **Visual-Spatial** To have students observe stomata and determine how a salt solution affects stomata.

Process Skills

recognize cause and effect, observe and infer, compare and contrast, use the microscope

Teaching Strategies

- Geranium, *Coleus*, or *Tradescantia* leaves work well. If unavailable, purchase fresh spinach.
- The epidermis will appear as a clear strip of tissue at the jagged torn edge. Have students use the lower epidermis of the leaf.

Expected Results

Chloroplasts will be seen only in guard cells. The saltwater mount will show guard cells closed. Thus, stomata will appear closed in comparison to the plain water wet mount.

Analysis

1. Cells appear to be like interlocked puzzle pieces. The cells are protective.
2. Guard cells are sausage shaped and contain chloroplasts. Epidermal cells contain no chloroplasts and are irregular in shape.
3. Stomata are closed in salt solution. Higher water concentration inside cells compared to that outside cells resulted in water moving out of guard cells causing them to collapse and close the stomata.

Assessment

Performance: Provide students with a diagram of leaf epidermis cells, guard cells, and stomata. Ask them to label each cell type and to indicate which cells are capable of carrying out photosynthesis. L1

MiniLab

What do stomata look like?

The lens-shaped openings in the epidermis of a leaf allow gas exchange and help control water loss.

Magnification: 3600×

Procedure

1. Make a wet mount by tearing a leaf at an angle to expose a thin section of epidermis. Use tap water to make wet mounts of both the upper and lower epidermis.
2. Examine each of your slide preparations under the microscope. Draw or take down a written description of what you see.
3. Make another wet mount using a five percent salt solution instead of tap water. Examine the slide under the microscope and record your observations.

Analysis

1. What do the cells of the leaf epidermis look like? What is their function?
2. How do the epidermal cells differ from guard cells? Which cells, if any, contain chloroplasts?
3. What differences in the stomata did you notice when you used a salt solution to prepare your wet mount? Can you explain what happened in terms of osmosis?

Growth of the stem

Primary growth in a stem is similar to primary growth in a root. The tissue responsible for this growth is the apical meristem, which lies at the tip of a stem. Lateral meristem, located at nodes along the stem and at the axils of leaves, gives rise to new branches, which each have their own apical meristem.

Leaves

The primary function of the leaves is to trap light energy for photosynthesis. Most leaves have a relatively large surface area so they can receive plenty of sunlight. They are also often flattened, so sunlight can penetrate to the photosynthetic tissues just beneath the surfaces of the leaves.

Types of leaves

When you think of a leaf, you probably think only of the flat, broad, green structure known as the leaf blade. Some leaves are joined directly to the stem, such as grass blades. In other leaves, there is a stalk that joins the leaf blade to the stem. An actual part of the leaf, this stalk is called the **petiole.** The petiole contains vascular tissues that extend into the leaf to form veins. If you look closely, you will notice these veins as lines or ridges running along the leaf blade. *Figure 26.13* gives one example of the variety of shapes in leaves.

Figure 26.13

A simple leaf has one entire leaf blade. A compound leaf has a divided leaf blade. The leaves of maple trees are simple leaves. The leaves of walnut trees, below, are compound leaves.

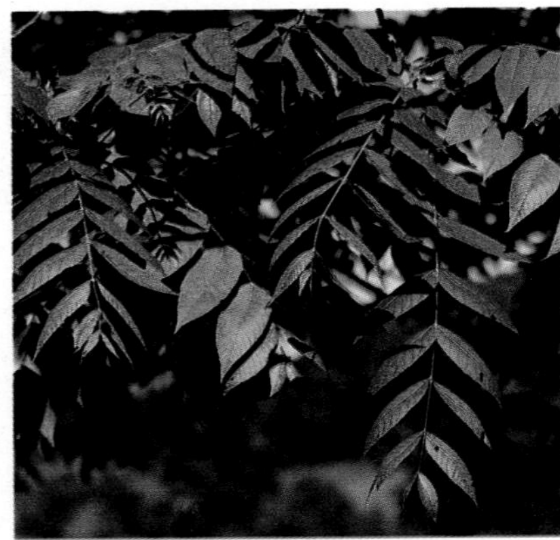

Meeting Individual Needs

Visually Impaired Construct a model leaf blade from heavy cardboard. Use straws taped to the cardboard to represent the veins of the leaf and a piece of straw extending beyond the blade to represent the petiole. Allow visually impaired students to compare the model to an actual leaf to gain understanding of leaf anatomy. L1 LS

Internal structure of leaves

You read in Chapter 24 that leaves have a waxy cuticle and stomata that help prevent water loss. The evaporation of water from these stomata in the leaves is called **transpiration.** Plants lose up to 90 percent of all the water they transport from the roots through transpiration. To help reduce water loss from transpiration, the size of the pore is reduced by guard cells. **Guard cells** are cells that surround and control the size of the opening in a stomata. The operation of guard cells is described in *Figure 26.14.*

ThinkingLab — Analyze the Procedure

What determines the number of stomata on leaves?

Stomata are openings on the surface of a leaf that enable a plant to exchange gases. Assume that you have examined a large number of leaves from individual trees of the same species in your area under the microscope. You observe that they all have about the same number of stomata in a certain area. You are curious to know whether that number is determined genetically or if it varies for trees of the same species that grow in different environments. You hypothesize that in drier climates, there would be fewer stomata on the leaves of individual trees in this species. You think that fewer stomata in a drier climate would prevent the leaves from drying out.

Analysis

You ask a relative who lives in a drier state to send you some leaves from the same species of tree. After examination, you find that there are fewer stomata on the leaves of these trees.

Thinking Critically

Can you conclude, based on your data, that dry climates cause fewer stomata to form on leaves? Why or why not? If not, what else would you need to do to determine whether the variation is related to the climate?

Figure 26.14

Guard cells prevent a plant from drying out. They regulate the size of the openings of the stomata according to the amount of water in the plant.

A **Guard cells are modified cells of the leaf epidermis that contain chloroplasts. These cells also contain microfibrils made of cellulose that are arranged in belts around the circumference of the cells.**

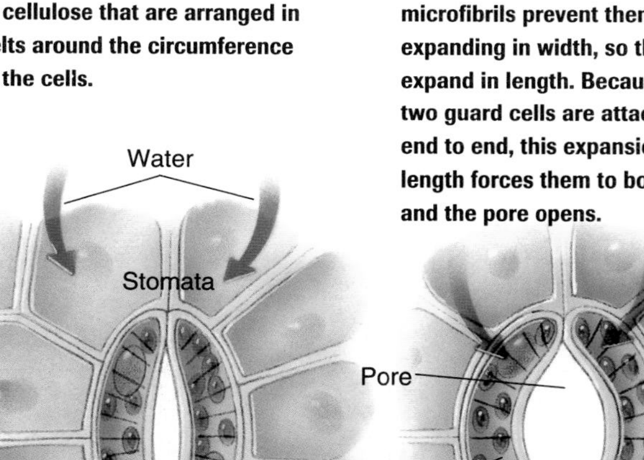

B **When there is plenty of water in surrounding cells, guard cells take in water by osmosis. When water enters the guard cells, microfibrils prevent them from expanding in width, so they expand in length. Because the two guard cells are attached end to end, this expansion in length forces them to bow out and the pore opens.**

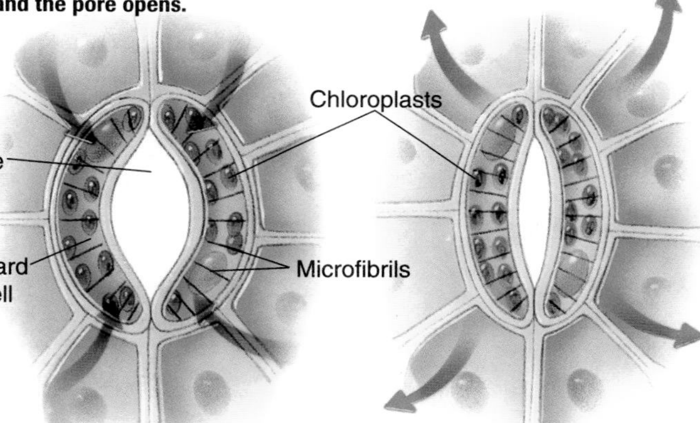

C **When the plant becomes dry, there is less water in tissues surrounding the guard cells. Water leaves the guard cells, thus lowering turgor pressure. The cells return to their previous shape, reducing the size of the pore.**

ThinkingLab — Analyze the Procedure

Purpose

LS **Logical-Mathematical** Students examine the experimental procedure used to answer a question to determine if proper scientific methods were followed.

Process Skills

form a hypothesis, use scientific methods, recognize cause and effect

Teaching Strategies

- Have students review the steps used to solve a problem using scientific methods.
- Allow students to work in small cooperative groups to complete this lab in class or have them complete the lab as a homework assignment.

Thinking Critically

Answers may suggest that responses of stomata to changes in climate in one plant species may not indicate how all plant species respond. Answers may also suggest that plants should be examined for stomata numbers first when they are grown in one climatic condition and then reexamined when they are grown under differing climatic conditions. Students should state that no control was used in the experiment, the sample size may not have been large enough, that leaf samples may not have all been from the same species, and that research should be expanded to include other species.

Assessment

Portfolio: Have students design a procedure that would allow one to determine the total number of stomata on a leaf without actually counting them. L2 P

PORTFOLIO

Tree Identification Through Leaf Shape
Easy-to-use leaf keys are available at local botanical gardens. The following booklets or guides may also be useful: *Master Tree Finder* by Nature Study Guild Publishers, *Golden Field Guide: Trees* by Golden Press, New York. Provide students with a variety of leaf specimens (both gymnosperm and angiosperm specimens). Give students biological keys and have them use the keys to identify the specimens you have given them. Ask students to tag the specimens with a letter or number identification that correlates to the biological key so you can verify if their identifications are correct. L1 P LS

BioLab

Growth of Stems

Time Allotment 
One class period

Objectives
Review objectives with students before they begin the Biolab.

Process Skills
observe and infer, recognize cause and effect

PREPARATION

Alternate Materials

- Prepared microscope slides that will allow students to examine the actual tissues shown in the diagrams are available from biological supply houses.

GLENCOE TECHNOLOGY

Videodisc

The Infinite Voyage: A Taste of Health
Pima Indians: Old Traditions in Nutrition (Ch. 1)

Pima Indians: Coping with Disease (Ch. 2)

BioLab

Growth of Stems

Plant stems contain living tissues that grow and change as the plant matures. Major stem functions include support for the plant as it grows and transport of food, water, and minerals throughout the plant. Vascular tissues of stems continue to grow as the tree matures. The woody stem of a tree provides protection for plant tissues and contains a record of the tree's age.

PREPARATION

Problem
How do the stems of one-, two-, and three-year-old trees compare?

Materials
art of one-, two-, and three-year-old stems

Objectives
In this Biolab, you will:

- **Identify** stem tissues in one-, two-, and three-year-old tree stems.
- **Relate** the functions of these stem tissues.

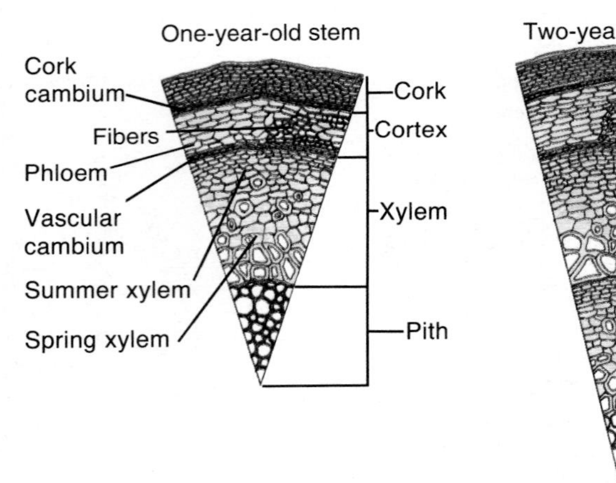

PROCEDURE

Teaching Strategies

- Provide students with a photocopy of the three diagrams. Allow them to add labels to the diagrams based on the descriptions provided in the table. You may wish to verify that all labels are correct before students proceed with the activity. L1
- As an alternative, or as a means for verifying that tissues have been correctly identified, show students 35-mm slides of tree stem cross sections. Elicit tissue names as you point to various bands or cell groups.
- You may wish to use a microprojector to review and verify that tissues have been correctly identified.

PROCEDURE

1. Examine the diagram of a cross section through a one-year-old tree stem.
2. Study the descriptions of stem tissue given in the table, and identify these tissues on the diagram.
3. Compare the first diagram with the diagram of a cross section through a two-year-old stem.
4. Note that a difference in the size of spring and summer xylem cells creates a line separating one year's xylem from the next year's xylem.
5. Compare the diagram of the three-year-old tree stem with the other two diagrams.
6. Identify the stem tissues shown on the three-year-old tree stem, including bark and wood.

Tissue	Description
Cork	Outermost layer, about eight cells thick
Cork cambium	Single layer of cells inside cork layer
Cortex	First layer inside cork cambium, about ten cells thick
Pith	Tissue at center of stem with large, thin-walled cells
Xylem	Thick layer of cells next to pith, widest layer of cells in stem
Vascular cambium	Single layer of cells at top edge of xylem
Phloem	Groups of thin-walled cells inside cortex
Fibers	Groups of thick-walled cells that surround phloem

ANALYZE AND CONCLUDE

1. **Drawing Conclusions** How can you tell the age of a tree that has grown in a temperate climate by looking at a cross section of its stem?
2. **Identifying Variables** Annual rings in a tree stem vary in thickness in response to environmental factors that influence growth. What kinds of factors during a year might influence a tree's growth?
3. **Drawing Conclusions** Give a possible reason why the cell diameter of spring xylem cells is greater than the cell diameter of summer xylem cells.

Going Further

Application
Examine a cross section of twigs from local trees with a hand lens. Determine differences in anatomy of different species.

ANALYZE AND CONCLUDE

1. The number of bands or rings of xylem correspond to its age.
2. Amount of water, minerals, and light available; temperature variations
3. The cambium produces larger cells in spring in response to better growing conditions such as more available water and minerals.

Assessment

Performance: Provide students with a drawing or prepared microscope slide of a stem cross section of a tree greater than three years old. Have students determine the age of the tree from the slide or drawing. L1

Going Further

Provide students with a piece of stained wood that has a distinct grain. Have them describe: (a) how the grain pattern is formed in the wood and (b) how the tree was sliced to achieve the grain pattern seen. L1

- Explain that the diagrams show cross-section slices. Help students visualize the stem and diagram orientation by slicing off a section of celery stem (the leaf petiole) to illustrate how a cross section differs from a longitudinal cut.

Data and Observations

Tissues that change from one- to two- to three-year-old stems include a new ring of xylem for each year of growth. All other tissues remain generally the same. Verify that students realize that the oldest xylem ring is closest to pith and the youngest ring of xylem is closest to vascular cambium.

SECTION 26.2

A BROADER VIEW

The Fall of the Food Factories

Purpose

Students will understand why leaves turn colors in the fall and then drop from the tree.

Teaching Strategies

- Have students work in small cooperative groups to review the meanings of the following terms: *dehydrating, abscission, carotenoids, ethylene, cellulose, enzyme.* L1 SAE COOP LEARN
- Have students working in cooperative groups sequence the series of changes that trigger leaf color changes and eventual leaf dropping. L1 COOP LEARN
- On the chalkboard, draw the structural formula of ethylene ($CH_2 = CH_2$). Advise students that this gas is a plant hormone.

Thinking Critically

Conduct an experiment using a series of control and experimental plants subjected to only one of the environmental factors being tested. Such experiments would provide clues as to which factor is responsible for leaves dropping.

GLENCOE TECHNOLOGY

Videodisc

Biology: The Dynamics of Life
Disc 1, Side 2
Clear Felling Forests (Ch. 23)

A BROADER VIEW

The Fall of the Food Factories

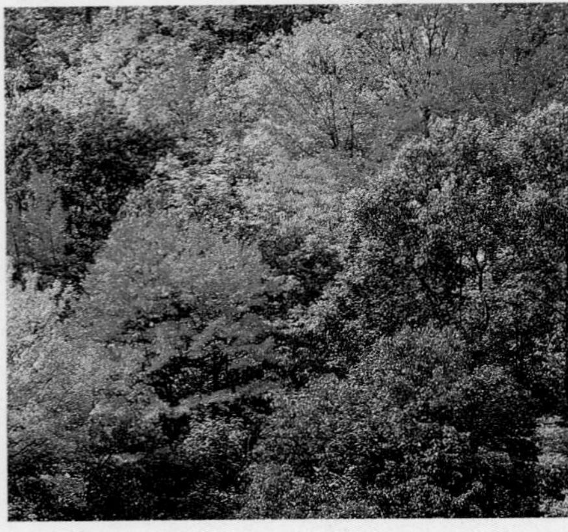

When evenings become crisp, you can feel that autumn is near. In a few weeks, trees will be surrounded by piles of dry, papery leaves in shades of scarlet to brown. This yearly event in northern temperate regions of the world is an adaptation that keeps deciduous trees from dehydrating during winter, when roots cannot absorb water from the frozen ground.

Leaf colors don't change The process of leaf drop, called abscission, begins with nitrogen and minerals moving from leaves for storage in stems and roots. Photosynthesis slows as the amount of sunlight decreases, plants stop producing chlorophyll, and water and nutrients stop moving through the plant. Fall colors result when the dominant pigment, chlorophyll, which is green, stops being produced. Other pigments that are always present in the leaf are then revealed. Carotenoid pigments present in a leaf reflect the colors red, orange, and yellow.

Hormones and enzymes After leaves change color, what causes them to fall off the tree? The answer has to do with the plant's production of a gas called ethylene, which increases in the fall. When you put unripe fruit in a paper bag, the fruit continues to produce ethylene and continues to ripen. In a tree, ethylene causes swelling on the stem side of a leaf, pushing it away from contact with the petiole. At the same time, in the abscission layer, an enzyme attacks cellulose in cell walls, making the layer loose and weak; the leaf usually falls of its own weight, and a protective layer is left to guard against disease.

Time to go Leaf abscission in deciduous trees is environmentally controlled by the shorter day length and cooler temperatures that are characteristic of autumn. City trees near streetlights may actually retain their leaves longer than usual because the lights alter the plant's response to changes in day length.

Thinking Critically

How could you determine whether it is temperature, day length, or availability of water that causes leaves to fall?

Just beneath the cuticle-covered epidermis are two layers of mesophyll. **Mesophyll** is the photosynthetic tissue of a leaf; it is made up of two types of parenchyma. The veins of vascular tissue run through the mesophyll of leaves, ***Figure 26.15***. You can read more about the internal structures of a leaf and how they function in photosynthesis in *The Inside Story*.

Figure 26.15

The mesophyll of leaves contains veins of vascular bundles. The pattern veins make in a leaf can help to identify the species.

Leaves of corn plants have parallel veins, a characteristic of monocots.

Leaves of lettuce plants have veins that form a branching network, a characteristic of dicots.

Meeting Individual Needs

Learning Disabled Ask students to review the meaning of turgor pressure. Have students sequence the steps involved in the opening and closing of stomata. L1 IN

Visually Impaired Use two long, inflated balloons as a model of guard cells for visually impaired students. Place the balloons next to one another. Typically a bow will exist in the center to simulate the stoma opening. Allow students to examine the component parts, guard cells and stoma opening, as you explain what they represent. Push the two balloons together to close the stoma and allow students to reexamine the model. L1 IN

Internal Anatomy of a Leaf

The tissues of a leaf are adapted for photosynthesis, gas exchange, limiting water loss, and transporting sugars. Most leaves have an upper and lower surface and are attached to small branches.

Pin oak

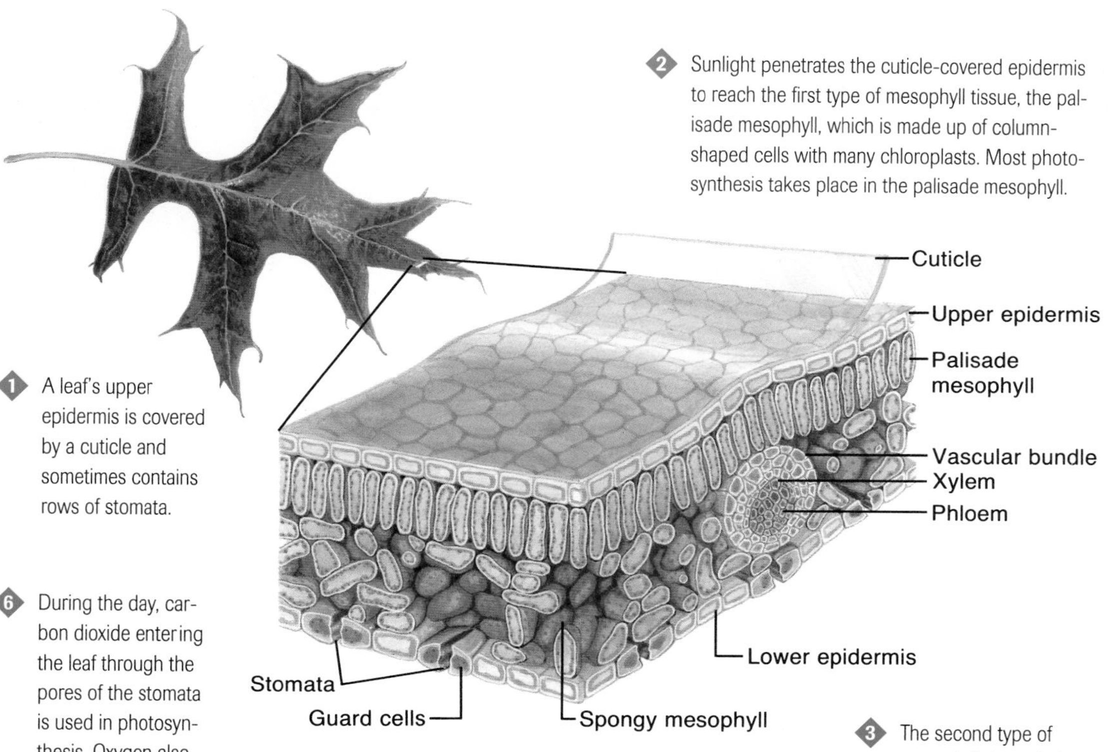

1. A leaf's upper epidermis is covered by a cuticle and sometimes contains rows of stomata.

2. Sunlight penetrates the cuticle-covered epidermis to reach the first type of mesophyll tissue, the palisade mesophyll, which is made up of column-shaped cells with many chloroplasts. Most photosynthesis takes place in the palisade mesophyll.

3. The second type of mesophyll tissue is the spongy mesophyll, which is made up of loosely packed, irregularly shaped cells surrounded by air spaces. There are fewer chloroplasts in the spongy layer because this tissue is usually on the underside of the leaf, where less light is absorbed.

4. Below the spongy mesophyll is the lower epidermis, with many stomata. Spongy mesophyll allows gases to move between palisade cells and out of the leaf.

5. Within the spongy mesophyll can be seen cross sections of bundles of vascular tissue that compose the veins of the leaf. The veins are usually surrounded by a sheath of parenchyma that helps regulate the flow of materials into and out of the vascular tissue.

6. During the day, carbon dioxide entering the leaf through the pores of the stomata is used in photosynthesis. Oxygen also enters the pores for the process of respiration. The products of photosynthesis, water vapor and oxygen, exit through the stomata.

THE INSIDE STORY

Internal Anatomy of a Leaf

Purpose

LS **Visual-Spatial** Students will study leaf anatomy and will correlate leaf structures with their functions.

Teaching Strategies

- Have students describe common everyday structures or objects that are analogous to structures found in the leaf. *Students may correlate epidermis to waxed paper, or stomata to skin pores.* L1

Visual Learning

- Photocopy the diagram of the leaf cross section without the labels. Have students add labels to the tissues described in steps 1-5. Verify that their labels are correct. L1
- Prepared slides showing leaf cross sections are available from biological supply houses. Allow students to view the slides and ask them to compare the slides to the leaf cross section shown in *The Inside Story.* L2
- Ask students to use the diagram to prepare a concept map which relates leaf structures and tissues to their locations within a leaf. L1

Audiovisual

Show the filmstrip *Adaptations of Leaves,* Carolina.

STUDENT JOURNAL

A Leaf Analogy Ask students to describe in their journals how a cross sectional view of a leaf can be represented by a sandwich. Using an analogy technique, ask students to describe the makeup of a sandwich and relate its component parts to those in the leaf. To get them started, tell them to compare the upper epidermis to the top slice of bread. L1 LS

SECTION 26.2

3 Assess

Check for Understanding

Ask students to explain the difference between the terms in each of the following word groups.

(a) epidermis - endodermis
(b) xylem - phloem
(c) cork cambium - vascular cambium
(d) vessel cell - sieve cell
(e) guard cells - stomata
(f) spongy mesophyll - palisade mesophyll
(g) sieve cell - companion cell
(h) sapwood - heartwood L1

Reteach

Have students explain how the words in each word pair listed in the *Check for Understanding* are alike or related. L1

Extension

Have students design an experiment to show that growth occurs at the tips of branches. L3

Assessment

Skill: Provide students with diagrams of root, stem, and leaf cross sections. Ask them to color code those tissues that are identical in each organ. Ask them to label their diagrams. L1

4 Close

Discussion

Bring in samples of bark from different types of trees and display them where they can be seen by students. Discuss how bark can be used to identify trees.

Figure 26.16

Modified leaves serve many functions in addition to photosynthesis.

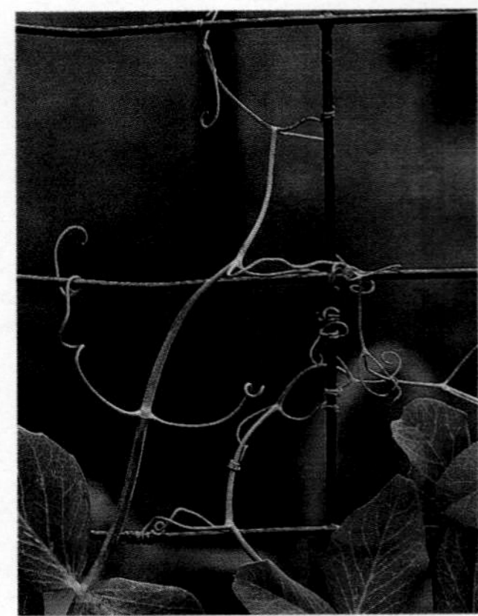

The tendrils of pea plants are leaflets that are modified for climbing.

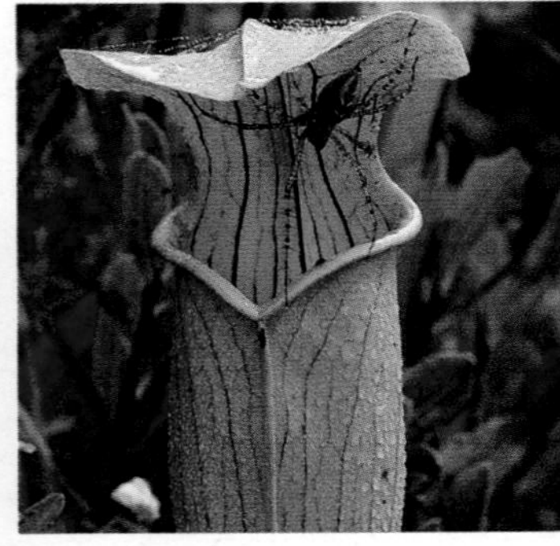

The leaves of the pitcher plant are modified for trapping insects.

The thick leaves of this *Aloe vera* plant are adapted to store water in a dry desert environment.

Modified leaves

Many plants have leaves that are modified for other functions besides photosynthesis. For example, cactus spines are leaves that protect the plant from herbivores and help reduce water loss. The fleshy leaf bases of onion and daffodil bulbs are modified for food storage. Some additional examples of modified leaves are shown in *Figure 26.16.*

Connecting Ideas

Angiosperms are the most diverse group of plants, including more than 230 000 species. These plants have moved entirely away from a dependence on water for reproduction and are adapted to life in nearly every environment found on Earth. Angiosperms have adaptations that enable them to prevent water loss from their tissues and store food under adverse environmental conditions. They also have evolved ways to attract other organisms to help in both reproduction and seed dispersal.

Section Review

Understanding Concepts

1. Compare and contrast the arrangement of xylem and phloem in dicot roots and stems.
2. In a plant with leaves that float on water, such as a water lily, where would you expect to find stomata? Explain.
3. You are given the leaf of an unknown plant and asked to determine whether it is a monocot or a dicot. Explain how you would determine this.

Thinking Critically

4. Some animals such as squirrels damage trees by stripping off portions of the bark. Of what use is bark to a squirrel?

Skill Review

5. **Observing and Inferring** You are looking at a slide of the cross section of a leaf. The label has fallen off. How can you tell which way to orient the slide on the microscope stage? Hint: Look at the numbers of chloroplasts in cells. For more help, refer to Thinking Critically in the *Skill Handbook.*

Answers to Section Review

1. In dicot roots, xylem forms a star-shaped figure in the center with phloem between the star rays. In stems, xylem and phloem are arranged in bundles that form a circle. In each bundle, the phloem is outside the xylem.
2. Stomata would be on the upper surface that is out of water. This will allow for gas exchange with air.
3. Examine the vein pattern of the leaf. If veins are parallel, the leaf is a monocot; if netlike or branching, it is a dicot.

Thinking Critically

4. Bark contains phloem, which transports food through the stem. The food typically is a sugar and serves as food for the squirrel.

Chapter 26 Review

REVIEWING MAIN IDEAS

26.1 What Is an Angiosperm?

- Angiosperms make up Division Anthophyta, which contains two classes: the monocotyledons and dicotyledons.
- Angiosperms are an extremely diverse group of plants that includes species adapted for virtually all kinds of habitats.

26.2 Angiosperm Structures and Functions

- Roots grow downward into the soil as cells elongate. Cell division in the apical meristem produces new cells that add to the length of the root. Cell division in the cambium adds xylem and phloem tissue.
- Stems also contain apical meristem, vascular cambium, and vascular tissue. The primary function of the stem is to support upright growth and transport food and water from one part of the plant to another.
- Leaves contain layers of mesophyll tissue that have chloroplasts and perform photosynthesis. Guard cells control the opening and closing of stomatal pores.

Key Terms

Write a sentence that shows your understanding of each of the following terms.

angiosperm	mesophyll
annual	monocotyledon
apical meristem	parenchyma
bark	perennial
biennial	pericycle
cambium	petiole
companion cell	root cap
cortex	root hair
dicotyledon	sieve cell
endodermis	sink
epidermis	transpiration
guard cell	vessel cell

Understanding Concepts

1. The plants most familiar to you, including fruit trees, vegetables, cereal grains, and wildflowers, are ____________.
   a. conifers c. cycads
   b. anthophytes d. lycophytes
2. Angiosperms and gymnosperms are alike in that they both have ____________.
   a. spores, rhizomes, stems, and leaves
   b. needlelike leaves, seeds, and fruits
   c. seeds, roots, stems, and leaves
   d. naked seeds, embryo plants, stems, and leaves
3. Angiosperms ____________.
   a. produce flowers and also have their seeds in fruits
   b. produce cones with naked seeds
   c. are all dicotyledons
   d. are all perennials
4. In what division are flowering plants classified?
   a. Bryophyta c. Coniferophyta
   b. Cycadophyta d. Anthophyta
5. What are the two classes of anthophytes?
   a. monocotyledons and dicotyledons
   b. biennials and perennials
   c. annuals and biennials
   d. angiosperms and gymnosperms
6. Grasses, orchids, lilies, and palms all have one seed leaf and are called ____________.
   a. dicotyledons c. annuals
   b. monocotyledons d. biennials
7. ____________ have two seed leaves and include many flowering trees, wildflowers, and garden flowers.
   a. Dicotyledons c. Annuals
   b. Monocotyledons d. Perennials

Chapter 26 Review

Reviewing Main Ideas

Summary statements can be used by students to review the major concepts of the chapter.

Key Terms

Answers should go beyond defining terms. Accept any answer that uses the term correctly and in the proper context.

Understanding Concepts

1. b
2. c
3. a
4. d
5. a
6. b
7. a

Skill Review

5. The upper epidermis usually has a thicker cuticle and fewer stomata; spongy mesophyll cells have fewer chloroplasts than the cells of the palisade layer do. The palisade layer is toward the leaf's upper or top side.

Chapter 26 Review

8. d
9. a
10. b
11. c
12. d
13. b
14. a
15. b
16. d
17. c
18. a
19. c
20. b

Applying Concepts

21. Transpiration occurs most rapidly during the day and aids in the movement of water through xylem. Thus, water movement is greater during the day than at night.
22. Palisade cells contain more chloroplasts per cell and are column shaped. Spongy mesophyll cells have fewer chloroplasts per cell and are irregular in shape.
23. Tree girth increases each year as cambium activity forms new rings of xylem.
24. Insects provide a source of minerals that are not available through the soil.
25. Sap is the result of food storage in the roots of maple trees. Sugars are formed during photosynthesis and are stored in this same chemical form.

Chapter 26 Review

8. Which plants have leaves with a branching network of veins and flower parts in multiples of four or five?
   a. monocotyledons
   b. annuals
   c. perennials
   d. dicotyledons
9. Plants with parallel veins and flower parts in multiples of three are ____________.
   a. monocotyledons
   b. annuals
   c. perennials
   d. dicotyledons
10. ____________ are the most successful plants on Earth.
    a. Gymnosperms
    b. Angiosperms
    c. Bryophytes
    d. Lycophytes
11. One advantage to plants of having fruit-enclosed seeds is that ____________.
    a. people will eat them
    b. they can store food for the winter
    c. they have added protection for the newly developing plant
    d. they will have a longer life span
12. A(n) ____________ plant is a parasite with roots that penetrate the tissues of the deciduous trees upon which it lives.
    a. orchid
    b. gladiolus
    c. potato
    d. mistletoe
13. Plants that complete their life cycles in a year or less are called ____________.
    a. biennials
    b. annuals
    c. perennials
    d. monocots
14. A cross section of a root shows a star-shaped mass of xylem when the plant is a(n) ____________.
    a. dicot
    b. monocot
    c. annual
    d. perennial
15. If you observe a cross section of a root under the microscope and see that strands of xylem alternate with strands of phloem and that there is a central core of pith, the plant is a(n) ____________.
    a. dicot
    b. monocot
    c. annual
    d. perennial
16. Water and mineral ions enter the root by absorption into the ____________.
    a. phloem
    b. cuticle
    c. stomata
    d. root hairs
17. Cambium and apical meristem are examples of ____________.
    a. photosynthetic tissue
    b. protective tissue
    c. growth tissue
    d. transport tissue
18. One of the primary structural differences between roots and stems is the ____________.
    a. arrangement of vascular tissue within the stem and root
    b. arrangement of pith within the stem and root
    c. differences in the functions of xylem and phloem
    d. differences in the numbers of stomata
19. What is the primary function of leaves?
    a. to provide protection for the plant
    b. to provide water for the plant
    c. to trap light energy for photosynthesis
    d. to enable the plant to grow taller
20. ____________ cells regulate the size of the openings of the stomata.
    a. Cuticle
    b. Guard
    c. Epidermal
    d. Mesophyll

Applying Concepts

21. Compare the expected rates of water movement in xylem tissues during the day and at night.
22. What is the difference between palisade and spongy mesophyll?
23. A tree in a park grew around a piece of barbed-wire fence until the fence was covered by tree tissues. Explain how this could happen.
24. The sundew is a carnivorous plant that captures and digests insects. The plant has chlorophyll and produces sugars by photosynthesis. Why do you think this plant captures insects?
25. Every spring, sap is collected from sugar maple trees to make maple syrup. Why is sap so full of sugars in the spring?

Program Resources

Chapter Assessment, pp. 151-156 L1
Alternate Assessment in the Science Classroom
Computer Test Bank L1
Content Mastery, pp. 101-104 L1

Thinking Critically

26. ***Concept Mapping*** Make a concept map that relates the following terms and phrases. Supply the appropriate linking words for your map.
    cortex, root hair, xylem, active transport, endodermis, phloem, mineral ions, simple diffusion, epidermis, water
27. ***Interpreting Scientific Illustrations*** This diagram was drawn based on the observation of a cross section of a plant structure through a microscope. Identify the plant structure, and identify the lettered parts.

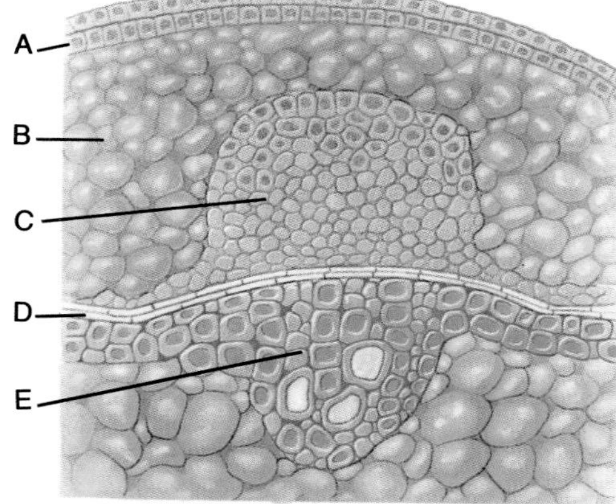

28. ***Comparing and Contrasting*** Compare and contrast the movement of water through the xylem to the movement of juice through a soda straw.
29. ***Formulating Hypotheses*** One spring, after the tulips in your garden bloomed, you cut off the leaves that were still green to tidy up the garden. The next spring, only a few of the tulips bloomed. Form a hypothesis about why this happened.
30. ***Recognizing Cause and Effect*** To allow some trees to grow straighter, foresters may remove the bark of surrounding trees in a circle around the trunk of the tree, a process called girdling. How does girdling eventually kill a tree?

ASSESSING KNOWLEDGE & SKILLS

The graph below illustrates data on daisies collected by scientists in four different states. Remember that stomata are the openings in the leaves through which water vapor escapes. Examine the graph and answer the following questions.

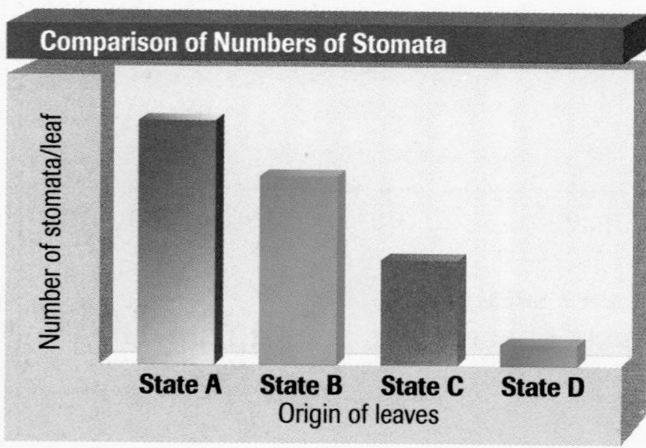

Interpreting Data

1. In which state might there be the most rainfall?
   a. state A c. state C
   b. state B d. state D
2. In which state might there be the least rainfall?
   a. state A c. state C
   b. state B d. state D
3. How is rainfall correlated to numbers of stomata on leaves?
   a. the more stomata, the less rain
   b. the fewer the stomata, the less rain
   c. the more stomata, the more rain
   d. the fewer the stomata, the more rain
4. ***Making a Graph*** Make a bar graph similar to the one on this page that shows how the thickness of the leaf cuticle of a particular plant varies from one state to another. In state A, cuticles are thickest; state B, thinner than state A; state C, thicker than state B but not as thick as state A; state D, thinner than state B.

Chapter 26 Review

Thinking Critically

26. See below.
27. This is a stem. A = epidermis; B = cortex; C = phloem; D = vascular cambium; E = xylem.
28. Water is aided by the pull of transpiration. Juice is pulled up through pressure generated in a person's mouth.
29. Tulips rely on bulb formation for the growth of next year's plant. Removal of leaves prevents photosynthesis from forming enough food to be stored in the bulbs.
30. Girdling interrupts the passage of nutrients in phloem cells from leaves to roots. Roots are unable to receive food and eventually will die, resulting in death of the tree.

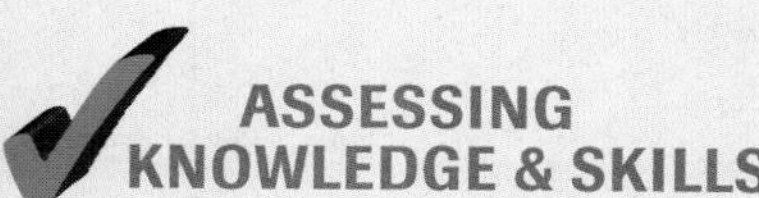

1. a
2. d
3. c
4.

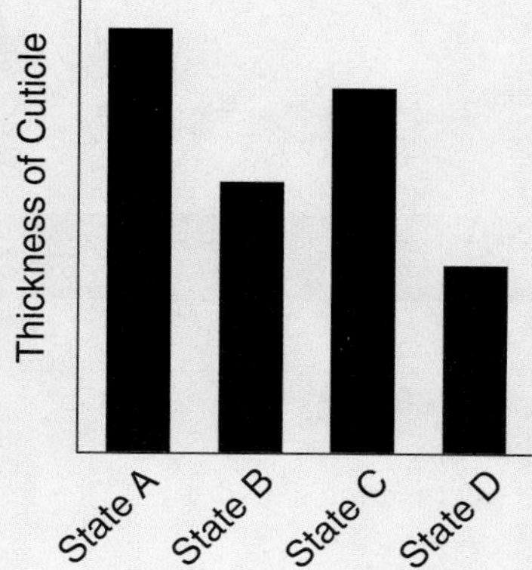

26.

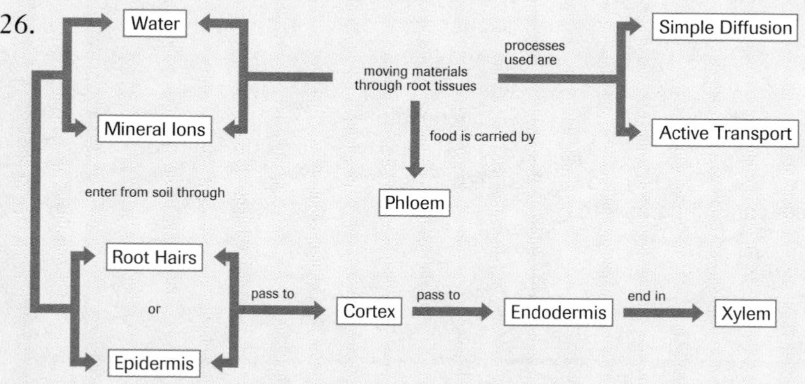

Chapter Organizer

SECTION	OBJECTIVES	ACTIVITIES/FEATURES
27.1 What Is a Flower? National Science Standards: UCP.1, UCP.2, UCP.5; A.1; C.1, C.4-6; G.1	1. **Identify** the structures of a flower. 2. **Recognize** the adaptive advantages of insect pollination.	**A Broader View:** Pollination in Orchids, p. 657 **Art Connection:** *Red Poppy,* p. 658 **Biolab:** Examining the Structure of a Flower, p. 660
27.2 Flowers and Reproduction National Science Standards: UCP.2, UCP.3, UCP.5; C.1, C.5; E.1, E.2; F.3, F.6; G.1	3. **Outline** the processes of seed and fruit formation and seed germination. 4. **Explain** strategies for seed dispersal. 5. **Describe** the effects of hormones on plant growth.	**Focus On** Plants for People, p. 668 **Minilab:** How do fruits and flowers compare?, p. 672 **Minilab:** What does a germinating seed look like?, p. 674 **Thinking Lab:** How does fruit ripen?, p. 677

ACTIVITY MATERIALS

BIOLAB	MINILABS	ALTERNATE LAB
page 660 flowers 2 microscope slides dropper single-edged razor blade 2 coverslips microscope hand lens colored pencils (red, blue, green)	**page 672** fruit flowers hand lens scalpel tweezers **page 674** corn kernels bean seeds (germinating and ungerminated) microscope single-edged razor blades paper towels plastic zipper bags	**page 672** kidney beans, canned kidney beans, dried paper cup waxed paper labels distilled water 2, 3, 5-triphenyl tetrazolium chloride dropper bottle

Chapter 27 Reproduction in Flowering Plants

TEACHER CLASSROOM RESOURCES

Reproducible Masters	Transparencies
Section Focus Master 57: Flowers L1 SAE **Reinforcement and Study Guide,** pp. 105-106 L1 **Biolab and Minilab Worksheets,** pp. 109-110 L1 **Concept Mapping:** The Structure of a Flower, p. 27 L1 **Content Mastery,** pp. 105-108 L1	**Basic Concepts Transparency #41:** Flower Structures L1 SAE
Section Focus Master 58: Seed Dispersal L1 SAE **Reinforcement and Study Guide,** pp. 107-108 L1 **Biolab and Minilab Worksheets,** pp. 107-108 L1 **Critical Thinking/Problem Solving:** Reproduction in Flowering Plants, p. 27 L3 **Laboratory Manual:** Do Dormant and Germinating Seeds Respire?, pp. 151-154; How Do Hormones Affect Plant Growth?, pp. 155-158 L2	**Basic Concepts Transparency #42:** Life Cycle of a Flowering Plant L1 SAE **Basic Concepts Transparency #43:** Germination of a Bean Seed L1 SAE **Basic Skills Transparency #21a, b:** Fruit Ripening L1 SAE **Reteaching Transparency #27:** Fruit Formation L1 SAE

ASSESSMENT MATERIALS

Chapter Assessment, pp. 157-162
Performance Assessment in the Biology Classroom, pp. 31, 47
MindJogger Videoquiz
Alternate Assessment in the Science Classroom
Computer Test Bank

Spanish Resources SAE
English/Spanish Audiocassettes SAE
Cooperative Learning in the Science Classroom COOP LEARN
Lesson Plans

KEY TO TEACHING STRATEGIES

L1 Level 1 activities should be within the ability range of all students including those with learning difficulties.
L2 Level 2 activities are within the ability range of average to above-average students.
L3 Level 3 activities are designed for the ability range of above-average students.
SAE SAE activities should be within the ability range of Students Acquiring English.
COOP LEARN Cooperative Learning activities are designed for small group work.
P These strategies represent student products that can be placed into a best-work portfolio.
These strategies are useful in a block scheduling format.

GLENCOE TECHNOLOGY

The following multimedia resources are available from Glencoe.

Biology: The Dynamics of Life
CD-ROM SAE
Videodisc Program

National Geographic Society Series
STV: Plants
What Is a Flower?
What Is a Seed?
Newton's Apple: Life Sciences
Plant Growth
Phototropism

Science and Technology Videodisc Series (STVS)
Plants & Simple Organisms
Seed Banks
Bats as Pollinators

CHAPTER

27 Reproduction in Flowering Plants

Chapter Overview

Students learn what a flower is and examine the roles of flower structures. The differences between cross-pollination and self-pollination are examined in conjunction with their genetic advantages and disadvantages. Variations in flower design are discussed. The section concludes with a discussion of photoperiodism and its influence on flowering.

The second section focuses on fertilization and the growth and development of flowering plants. Following a discussion of double fertilization, students explore the processes involved in the formation of seeds and fruits. Next, they learn about adaptations of seeds that aid in their dispersal. The chapter concludes with a discussion of seed germination and the growth and development of plants.

Key Terms

- anther
- complete flower
- day-neutral plant
- dormancy
- double fertilization
- endosperm
- fruit
- germination
- hormone
- incomplete flower
- long-day plant
- micropyle
- nastic movement
- ovary
- petal
- photoperiodism
- pistil
- sepal
- short-day plant
- stamen
- tropism

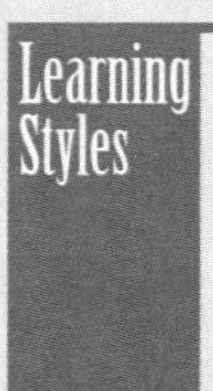

Learning Styles

Look for the following logo for strategies that emphasize different learning modalities. LS

Kinesthetic	**Meeting Individual Needs, p. 655; Building a Model, p. 662; Portfolio, pp. 670, 675**
Visual-Spatial	**The Inside Story, p. 656; Portfolio, pp. 657, 658; Microscope Activity, p. 662; Building a Model, p. 666; Enrichment, p. 667; Minilab, pp. 672, 674**
Intrapersonal	**Meeting Individual Needs, p. 659**
Logical-Mathematical	**Meeting Individual Needs, pp. 656, 657, 667; Enrichment, p. 663; Alternate Lab, p. 672; Project, p. 674; Thinking Lab, p. 677**

CHAPTER

27 Reproduction in Flowering Plants

Flowering plants are the most colorful as well as the most widespread and diverse members of the plant kingdom. This enormous variety of colors, sizes, shapes, and fragrances may be due to the close evolutionary partnership between flowers and insects. The pollen and sweet nectar produced by flowers provide food for butterflies, bees, and other insects. Brightly colored or fragrant petals help guide the insects to the nectar. While feeding, they carry pollen from flower to flower, causing fertilization and contributing to the success of seed production.

As flowers and insects evolved, they adapted to each other, as well as to other factors in their environment. In this chapter, you will study the structure of a flower, learn how flowers are pollinated, and explore the adaptations of flowers that encourage animals to help with pollination. You will also investigate the formation, dispersal, and germination of seeds.

Concept Check

You may wish to review the following concepts before studying this chapter.

- Chapter 12: meiosis
- Chapter 25: cotyledon, megaspore, microspore, ovule, pollen grain, pollen tube

Chapter Preview

Laboratory Activities

Biolab

- Examining the Structure of a Flower

Minilabs

- How do fruits and flowers compare?
- What does a germinating seed look like?

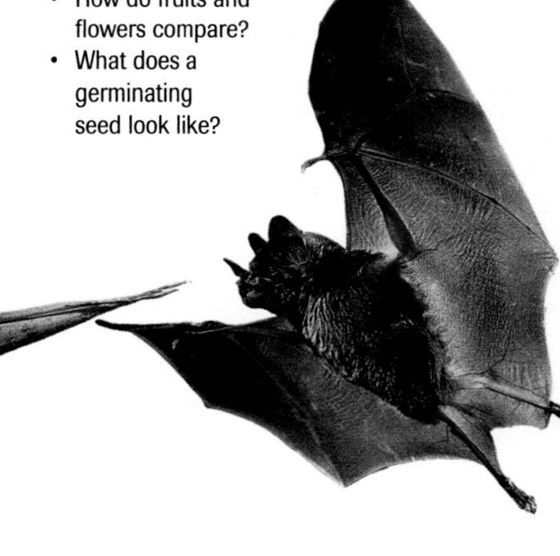

Mammals and flowering plants appeared on Earth at about the same time. Like insects, mammals also have a close evolutionary partnership with anthophytes. Fruit provides food for mammals and, by eating the fruit, mammals help disperse seeds. Why is it an adaptive advantage for a plant to produce fruit that looks and tastes good to mammals?

CHAPTER 27

Introducing the Chapter

Initiate a discussion about the role adaptations play in the survival of plants. Ask if a primate will more likely eat a fruit that is sweet or one that tastes bad. *One that is sweet.* Use responses as a springboard to a discussion about how fruit-eating animals aid in the dispersal of some seeds. Explain that seeds of a fruit eaten by an animal may pass through the digestive system and be excreted miles away from the parent plant. The seed may then grow into a new plant. Point out that this series of events will not occur if a fruit is not eaten because it has a bad taste.

Theme Development

Evolution is a theme that occurs throughout this chapter, especially as it relates to the coevolution of pollinators and flowers. *Homeostasis* is stressed through the discussion of photoperiodism and how plant hormones help maintain balance in anthophytes.

Concept Check

Remind students that each adaptation discussed in this chapter illustrates the evolutionary developments that aid in species survival.

Assessment Planner

Choose assessment strategies from the following pages to evaluate the progress of your students.

Assess, pp. 664, 678
Alternate Lab, pp. 672-673
Minilabs, pp. 672, 674
Portfolio, pp. 657, 658, 670, 675
Thinking Lab, p. 677
Biolab, pp. 660-661
Chapter Review, pp. 679-681

SECTION

27.1

Prepare

Key Concepts

Flower anatomy and the role of flowers in reproduction are the main focus of this section. Variation in flower shape is explained as resulting from adaptations that coevolved with different pollinators. Cross-pollination and self-pollination are also discussed as adaptive variations. The role of photoperiodism in anthophyte reproduction is explained.

Block Scheduling

Look for this symbol for strategies that are useful in a block scheduling format. For more information on block scheduling, refer to Section 27.1 in the **Lesson Plans** booklet.

Materials

- Have a flower available for the Demonstration.
- Gather flowers or order preserved specimens for the Biolab. Have red, blue, and green pencils available.
- Order dry bees or those in liquid preservative from a biological supply house for the Microscope Activity.

1 Focus

Bellringer

Before presenting the lesson, display **Section Focus Master 57** on the overhead projector and have students answer the accompanying questions. L1 SAE

SECTION

27.1 What Is a Flower?

Section Preview

Objectives

Identify the structures of a flower.

Recognize the adaptive advantages of insect pollination.

Key Terms

petal
sepal
stamen
anther
pistil
ovary
complete flower
incomplete flower
photoperiodism
short-day plant
long-day plant
day-neutral plant

How would you choose flowers for a bouquet or a garden? Perhaps you would start with fragrant roses, jasmine, or gardenias. You might add color with tall spikes of gladioli, cushions of marigolds, bright tulips, or daisies. Grasses would contribute a graceful shape, though their flowers may be so small we completely overlook them. All of these flowers are beautiful to look at and some have delicate scents as well. In what other ways are all of these flowers alike?

The Structure of a Flower

The process of sexual reproduction in flowering plants takes place in the flower, which is a complex structure made up of several parts. Some parts of the flower are directly involved in fertilization and seed production, whereas other parts have functions in pollination. There are probably as many different shapes, sizes, colors, and configurations of flower parts as there are species of flowering plants. In fact, features of the flower are often used in plant identification. *Figure 27.1* shows some examples of the variety in flower forms.

Even though there is an almost limitless variation in flower shapes and colors, all flowers share a simple, basic structure. A flower is made up of four kinds of organs: sepals, petals, stamens, and pistils.

The flower parts you are probably most familiar with are the petals. **Petals** are leaflike, usually colorful structures arranged in a circle called a corolla around the top of a flower stem. **Sepals** are also leaflike, usually green, and encircle the flower stem beneath the petals. Inside the circle of petals are the stamens. A **stamen** is the male reproductive structure of a flower. At the tip of the stamen is the **anther,** which produces pollen containing sperm.

At the center of the flower, attached to the tip of the flower stem, lie one or more pistils. The **pistil** is the female structure of the flower. The bottom portion of the pistil enlarges to form the **ovary,** the part of the flower in which the ovules containing eggs are formed. As you read in Chapter 25, the female gametophyte develops inside the ovule.

Program Resources

Section Focus Master 57 L1 SAE
Reinforcement and Study Guide, pp. 105-106 L1
Biolab and Minilab Worksheets, pp. 109-110 L1
Performance Assessment in the Biology Classroom, p. 31 L1 P
Concept Mapping, p. 27 L1
Basic Concepts Transparency 41 and **Master** L1 SAE

The Inside Story on page 658 shows how these structures are arranged in a typical flower.

Modifications in Flower Structure

A flower that has all four organs—sepals, petals, stamens, and pistils—is called a **complete flower.** The morning glory and tiger lily shown in *Figure 27.1* are examples of complete flowers, as is the phlox shown in *The Inside Story.* A flower that lacks one or more organs is called an **incomplete flower.** For example, squash plants have separate male and female flowers. The male flowers have stamens but no pistils; the female flowers bear pistils but no stamens. Plants such as sweet corn that are adapted for pollination by wind rather than animal pollinators have no corolla.

Adaptations for Pollination

You learned in Chapter 12 that pollination is the process of transferring pollen grains from the anther to the stigma. Plant reproduction is most successful when the rate of pollination is high, which means that the pistil of a flower receives enough pollen of its own species to fertilize the egg in each ovule.

Figure 27.1

The diversity of flower forms is evidence of the success of flowering plants.

The male flowers of the walnut tree form long catkins.

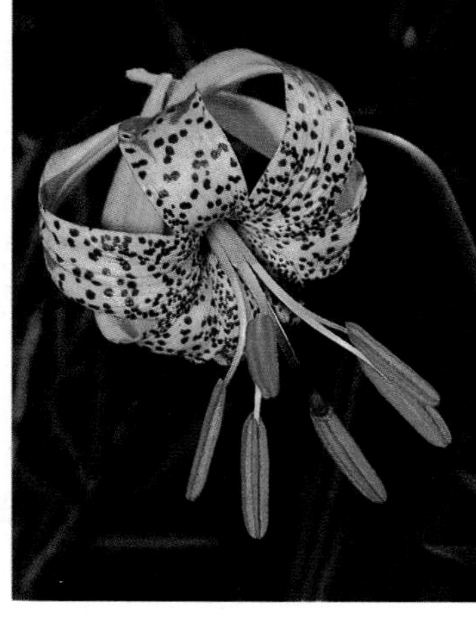

The spotted petals of the tiger lily curl away from the reproductive structures at the center of the flower.

Thistles bear clusters of tiny, tubular flowers within a mass of spiny bracts.

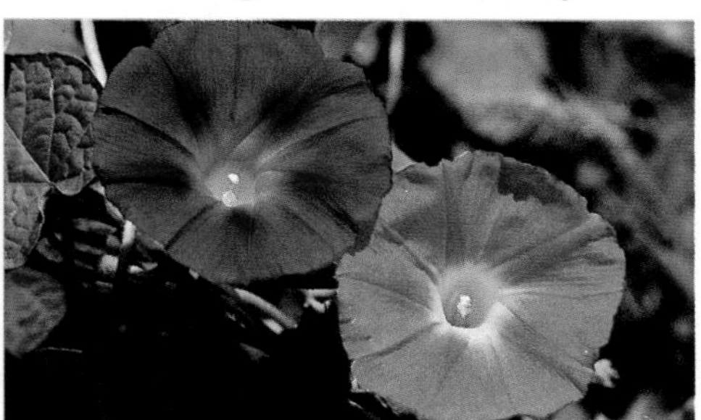

The petals of the morning glory are fused together to form a bell shape.

SECTION 27.1

Demonstration

Display a flower such as a rose or daffodil. Elicit what the function of the flower is. *It is a reproductive structure.* Ask if grass also reproduces through flowers. *Most students will answer no.* Explain that grass is a flowering plant but is often not recognized as such because its flowers do not have showy petals. L1

2 Teach

Brainstorming

Ask students to explain how a plant can reproduce sexually if its flowers are incomplete. *Pollen from the male flower can be carried to the female flower by wind, insects, or other animals.*

Tying to Previous Knowledge

Ask students to answer the following questions. (a) Why is pollen needed for fertilization? *contains sperm cells* (b) Why must the pollen that fertilizes a flower be from the same species as the female flower? *Chromosome numbers and alleles must be identical for fertilization to occur.* (c) What will happen if pollen lands on the wrong species of flower? *No pollen tube will form and fertilization will not occur.*

Meeting Individual Needs

Visually Impaired Purchase or borrow a large flower model. Allow visually impaired students to manipulate the model as you name and describe the function of each part. Have the student then tell you if the part is male, female, or neither in terms of flower function. L1 IS

interNET CONNECTION

Follow the link for this chapter on the Glencoe Homepage at **www.glencoe.com/sec/science** to find out more about flowering plants.

THE INSIDE STORY

Parts of a Flower

Purpose

LS **Visual-Spatial** To describe flower anatomy and identify the role flower parts play in reproduction.

Teaching Strategies

- Ask students to explain why stamens and pistils are described as the fertile structures in a flower. *These structures are involved in the production of egg and sperm.* Elicit what role is played by flower organs that are not important in fertilization. *These organs attract pollinators or protect young or immature fertile flower parts.* L1

Visual Learning

- Ask students to use the captions to explain how the following pairs of terms are related: (a) sepals and calyx (b) petals and corolla (c) stamen and anther (d) stigma and style (e) ovary and ovule.
- Bring in diagrams that show variation among flowers and their parts in terms of either number or anatomical differences.

THE INSIDE STORY

Parts of a Flower

Of the four major organs of a flower, only two—the stamens and pistils—are fertile structures directly involved in seed development. Sepals and petals support and protect the fertile structures and help attract pollinators. The structure of a typical flower is illustrated here by a phlox flower.

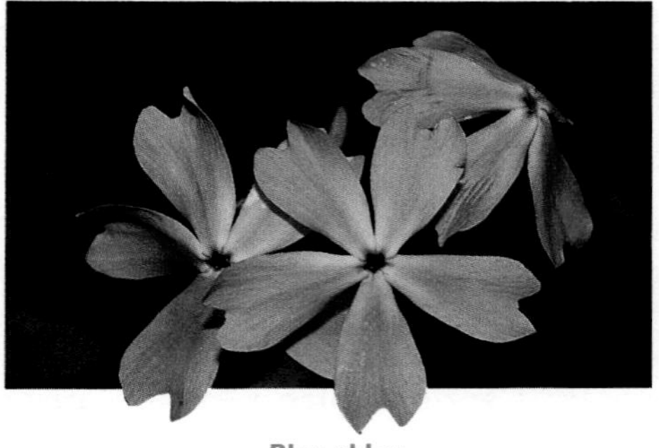

Blue phlox

1. Petals are usually brightly colored and often have perfume or nectar at their bases to attract pollinators. The corolla is made up of the flower's petals. In many flowers, the corolla also provides a surface for insect pollinators to rest on while feeding. Petals may be fused to form a tube, or shaped in ways that make the flower more attractive to pollinators.

2. The stigma, at the top of the pistil, is a sticky or feathery surface on which pollen grains land and grow. The style is the slender stalk of the pistil that connects the stigma to the ovary. The pollen tube grows down the length of the style to reach the ovary. The ovary, which will eventually become the fruit, contains the ovules. Each ovule, if fertilized, will become a seed.

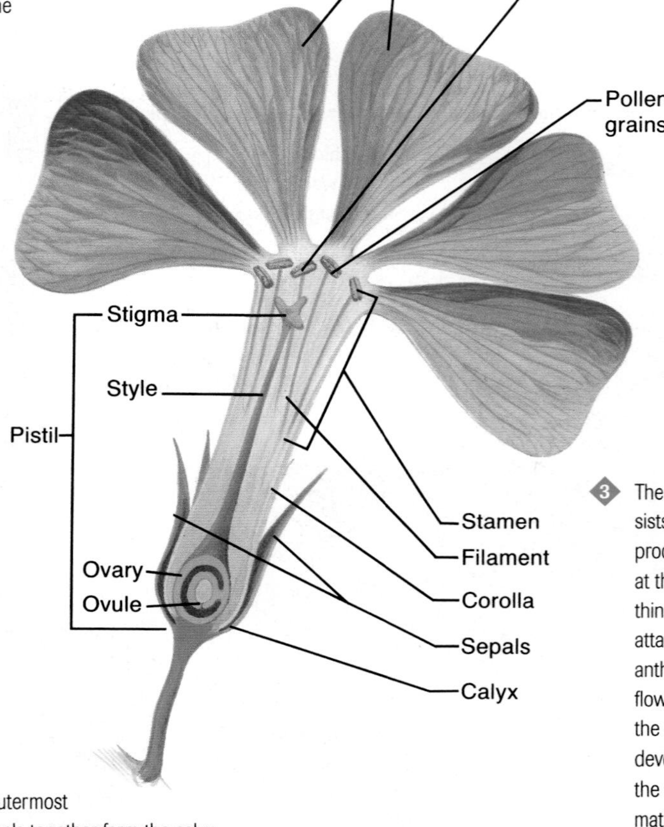

3. The stamen consists of the pollen-producing anther at the tip and a thin filament that attaches the anther to the flower stem. When the pollen grains developing inside the anther reach maturity, the anther splits open to release them.

4. The ring of sepals makes up the outermost portion of the flower. All of the sepals together form the calyx. The calyx serves as a protective covering for the flower bud, helping to protect it from insect damage and prevent it from drying out. Sepals sometimes are colored and resemble petals.

Meeting Individual Needs

Gifted Write the following phrases on the chalkboard: contains chlorophyll, contains anthocyanin, carries out meiosis, contains gametophyte, is part of the sporophyte, contains diploid cells, forms spores, forms microspores, forms megaspores. Ask students to identify the flower organ or organs described by each phrase. L3 LS

Students Acquiring English Have students create a table with the heads Male, Female, Neither male nor female. Ask students to list the following structures beneath the appropriate head: stamen, pistil, anther, calyx, corolla, ovary, stigma, petal, sepal, pollen, egg, ovule, style. L1 SAE LS

Pollination by wind is random

As you know, gymnosperms depend only on the wind for pollination. So do many flowering plants. But wind scatters pollen randomly, and pollen grains can land many places besides the stigma of the proper flower. Many anthophytes have elaborate mechanisms that help ensure that pollen grains are deposited in the right place. These pollination mechanisms involve relationships with animals, including beetles, butterflies, moths, bees, flies, hummingbirds, and even bats. These relationships have helped make flowering plants the most successful plant group on Earth.

Nectar attracts animal pollinators

Most flowers produce nectar, which serves as a valuable, highly concentrated food for animal pollinators. Nectar is a liquid, made up of proteins and sugars, that may be produced by any of the flower organs. It usually collects in the cuplike area at the base of the petals. For example, butterflies and moths unwind their long, curled proboscises to suck nectar from flowers that are often bell or tube shaped. The stamens of these flowers may extend past the petals. As the animal positions itself for a meal, it brushes against the anthers. During a visit to another flower, some of the pollen grains sticking to the pollinator's body are brushed off onto the stigma of a different flower, resulting in pollination.

A BROADER VIEW

Pollination in Orchids

Orchids make up the second largest family of flowering plants. Orchids are found in every habitat except the frozen polar regions. Some orchid flowers are many centimeters across, while some are extremely small. There are orchids that grow in dense clusters, while others are borne singly on a thick stalk. Many are exquisitely beautiful, some are green, and others can only be described as bizarre.

A unique flower Orchid flowers vary tremendously in shape, size, and color. But they all have three sepals and three petals. However, one of the petals is different from the other two. This special petal, called the lip, is usually large. It may also sport distinctive markings, be shaped into a tube or pouch, or be decorated with hairs or warts.

Orchid flowers are unique in that the male and female reproductive structures—the stamen and the pistil—are fused together into a single structure called the column. Pollen grains in orchids are not loose and powdery like they are in most other flowering plants. They are clumped together in waxy masses called pollinia.

Master of deception A group of tropical American orchids that belongs to the genus *Coryanthes* is known as bucket orchids. The lip of the flower forms a sort of bucket that is filled with liquid, and bees are attracted to the flower by a sweet-smelling oil that coats the rim of this structure. Invariably, as a bee tries to collect the tempting oil, it slips and falls into the liquid. The drowning bee eventually discovers the only escape route—a narrow tunnel that contains the pollinia. As the bee struggles to safety, the pollinia adhere to its back.

Several different kinds of orchids, including *Trichoceros antennifera* and some members of the genus *Ophrys,* have flowers that mimic the look and smell of female flies, bees, or wasps. When the appropriate male insect comes along, he vigorously tries to mate with the flower, picking up pollen in the process. When he forces his attentions on the next flower, pollination takes place.

Thinking Critically

Many orchid flowers are adapted for pollination by only a particular insect species, such as a single type of fly or wasp. What potential disadvantage can you see in having only one type of pollinator?

A BROADER VIEW

Pollination in Orchids

Purpose

To illustrate reproductive strategies in orchids.

Teaching Strategies

- Order preserved orchid specimens for classroom observation.
- Ask students to determine if orchids are complete or incomplete flowers. Have them explain their answers. *Orchids are complete flowers; they contain both male and female reproductive parts.* Elicit what the adaptive value may be for forming pollinia rather than individual pollen cells. *Pollinia are more likely to stick to an insect, helping to ensure pollination.*

Thinking Critically

Being dependent on only one species could threaten the survival of a plant if the pollinator species became extinct.

Software

Pollination and Fertilization: Seeds, Fruits, and Embryos, IBM.
Plant Reproduction, Biology Program Series.

GLENCOE TECHNOLOGY

 **Videodisc**

STVS: Plants & Simple Organisms
Disc 4, Side 2
Bats as Pollinators (Ch. 9)

Meeting Individual Needs

Students Acquiring English Have students prepare a concept map or outline that illustrates the classification scheme for roses up to the family level. For each taxon group, students should include a statement that describes a trait or characteristic specific for that group. Refer students to Appendix A for help. L1 SAE LS

PORTFOLIO

Pollination May Not Work All the Time
Have students prepare a series of diagrams to illustrate the events of a bee visiting a flower and having pollen deposited on its body. Have students make a second diagram that shows a flower design in which the bee will not be able to reach any nectar or receive any pollen. L2 P LS

Art

Red Poppy

Purpose
To illustrate the relationship between nature and art.

Teaching Strategies
- Posters of O'Keefe's works are available in art shops and museum gift shops. Bring in several examples to illustrate more of her work.
- Have students visit an art gallery to view paintings by artists who use flowers as their subject matter. As an alternative, have students examine art books to find the names of artists who use flowers. Ask students to present oral reports on their findings. L2 COOP LEARN

Possible Answer
Color attracts pollinators to the flower.

Visual Learning

- Ask students to determine if the poppy is a monocot or a dicot. Have them explain their answers.

Art

Red Poppy

by Georgia O'Keeffe (1887–1986)

"When you take a flower in your hand and really look at it," she said, cupping her hand and holding it close to her face, "it's your world for the moment. I want to give that world to someone else. Most people in the city rush around so, they have no time to look at a flower. I want them to see it whether they want to or not."

American artist Georgia O'Keeffe attracted much attention when the first of her many floral scenes was exhibited in New York in 1924. Everything about these paintings—their color, size, point of view, and style—overwhelmed the viewer's senses, just as their creator had intended.

The viewer's eye is drawn into the flower's heart In this early representation of one of her familiar poppies, O'Keeffe directed the viewer's eye down into the poppy's center, much as the flower naturally attracts an insect for reproduction purposes. By contrasting the light tints of the outer ring of petals with the darkness of the poppy's center, the viewer's eye is pulled beelike into the heart of the flower. The overwhelming size and detailed interiors of O'Keeffe's flowers give an effect similar to the photographer's close-up camera angle.

By the time of her death in New Mexico in 1986, O'Keeffe's gargantuan blossoms—including irises, lilies, and poppies—had become her signature work.

CONNECTION TO Biology

Color plays a prominent role in the artistic effect of O'Keeffe's flowers. What role does color play in the life of a real flower?

Flower colors signal appropriate pollinators

Nectar-feeding pollinators are attracted to a flower by its color or scent, or both. Flowers that attract butterflies usually have petals with bright, vivid colors, such as daisies, phlox, rhododendrons, and zinnias. Butterflies alight while feeding, so they visit flowers with a platform or cluster of petals. Because moths are active at night, flowers that attract moths stay open all night. They include tobacco, many orchids, night-blooming cereus, and honeysuckle. Moth-pollinated flowers are usually pale in color but have a strong, sweet scent that attracts the insects from some distance away. Moths hover while feeding, so they can visit blossoms that do not have a landing platform. Bees collect pollen as well as nectar. They are attracted to yellow or blue flowers with a sweet scent, such as peas, mints, primroses, irises, and lupines.

Figure 27.2
The shape, color, and size of a flower reflect its relationship with a pollinator.

The butterfly uses its long proboscis to sip nectar that the shorter tongues of bees and flies cannot reach.

PORTFOLIO

Cross- and Self-Pollination Have students prepare a series of drawings that explain the concepts of cross-pollination and self-pollination. Instruct students to include proper labels and titles for the drawings. Nonartistic students may be provided with a flower outline that they can trace for their diagrams. L1

P LS 

Scent attracts other pollinators

Flowers pollinated by beetles and flies have a strong scent but are often dull in color and may not produce nectar. Examples include magnolias, calla lilies, and wild parsley. Beetles and some fly species chew on flower parts, picking up and depositing pollen as they crawl about. The flowers aren't seriously harmed by this activity because they have such a large number of stamens and pistils that many are left undamaged. The scent of some fly-pollinated flowers such as skunk cabbage and trillium resembles rotting meat and attracts female flies searching for places to lay their eggs. *Figure 27.2* shows how flowers are adapted to attract their pollinators.

Pollination and Variation

You learned in Chapter 14 that sexual reproduction, which involves the processes of meiosis and fertilization, provides a mechanism for mixing the genetic material in a population. In anthophytes, the process of pollination contributes to the possibility of variations in genetic information. Pollination includes self-pollination and cross-pollination.

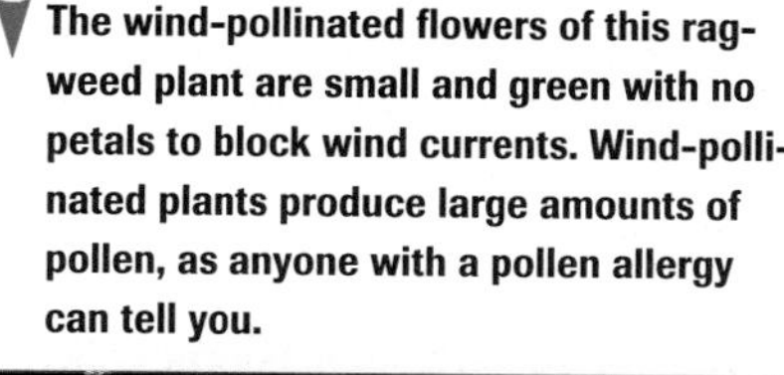

The wind-pollinated flowers of this ragweed plant are small and green with no petals to block wind currents. Wind-pollinated plants produce large amounts of pollen, as anyone with a pollen allergy can tell you.

Flowers pollinated by hummingbirds are often colored bright red or yellow but may have little scent. Birds do not have a well-developed sense of smell.

Bats sip nectar from night-blooming flowers with a strong, musty odor, such as bananas and some cacti. Notice how the anthers are positioned to rub against the bat's head as the animal reaches for nectar at the base of the flower.

Misconception

Students may believe that "hay fever" is an allergy to hay. However, hayfever is an allergic reaction to the protein present in pollen. Often, the pollen is produced by plants that bloom in the early fall or spring.

Tying to Previous Knowledge

Ask students to recall the processes of crossing over and independent assortment from their study of genetics. Explain when these events occur in plants that reproduce sexually. Then discuss why vegetative reproduction, such as spreading by rhizomes, results in offspring that are genetically identical. Self pollination is a sexual process involving gametes and although the offspring will have the same genetic makeup as the single parent, they will not be identical because of recombination of alleles.

Audiovisual

Show the videos *Sexual Encounters of the Floral Kind*, Oxford Scientific Films, and *Pollination: The Insect Connection*, Carolina.

Meeting Individual Needs

Gifted Corn has male and female flowers found on different parts of the same plant. Have students explain what a corn tassel is, where the female corn flower is located, and what corn silk is. *The corn tassel is composed of all the stamens of many male flowers. The corn silk is composed of all the styles of many female flowers. The corn kernels on the cob are the fertilized fruits of the plant.* L3 LS

BioLab

Examining the Structure of a Flower

Time Allotment
One class period

Objectives
Review objectives with students before they begin the Biolab.

Process Skills
observe and infer, use the microscope, compare and contrast, interpret data

Safety Precautions
- Some students may be allergic to the chemicals used in preserved specimens.
- Caution students against rubbing their eyes with their hands if the preservative gets onto their hands. Always have students wash their hands thoroughly after handling preserved specimens.

PREPARATION

Alternate Materials
- Do not use composite species of flowers such as sunflower, daisy, or dandelion.
- Prepared slides that show eggs within the ovule are available from supply houses.
- Preserved flowers are available at any time of the year from biological supply houses. A disadvantage to such flowers is that the preservative tends to discolor flower parts, especially petals and sepals which will appear a dull green.

BioLab

Examining the Structure of a Flower

Flowers are the reproductive structures of anthophytes. Seeds that develop within the flower are carried inside a fruit. Seeds enable the plant to reproduce. Flowers come in many colors and shapes. Often their colors or shapes are related to the manner in which pollination takes place. The major organs of a flower include the petals, sepals, stamens, and pistils. Some flowers are incomplete, which means they do not have all four kinds of organs. You will study a complete flower.

PREPARATION

Problem
What do the parts of a flower look like, and how are they arranged?

Objectives
In this Biolab, you will:
- **Observe** the structures of a flower.
- **Identify** the functions of flower parts.

Materials
flower—any complete flower that is available locally, such as phlox, lily, or tobacco flower
2 microscope slides
water
dropper
single-edged razor blade
2 coverslips
microscope
hand lens (or stereomicroscope)
colored pencils (red, blue, green)

Safety Precautions
Razor blades are sharp. Always use caution with razor blades.

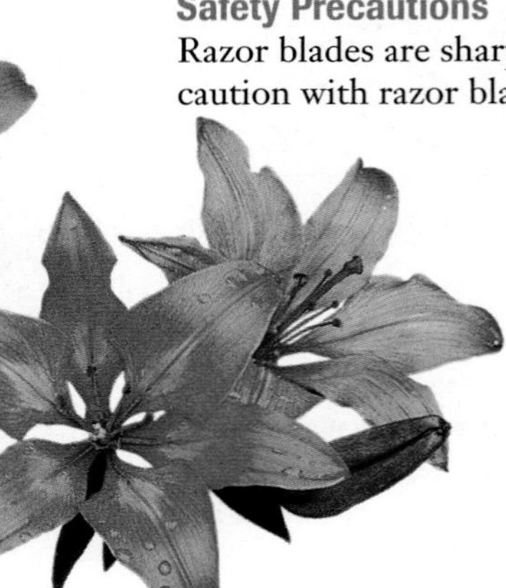

PROCEDURE

Teaching Strategies
- When preparing the anther wet mount, it may be helpful for students to squash the preparation using their thumb and a piece of lens paper over the coverslip. Advise students to press down firmly to avoid cracking the coverslip, or use plastic coverslips for this wet mount.
- Eggs will not be visible. Students should not attempt to observe ovules under the microscope in an attempt to observe egg cells.
- Students can work in groups of two to three for this Biolab.
- Show students how to remove the petals and sepals.

PROCEDURE

1. Examine your flower. Locate the sepals and petals. Note their numbers, size, color, and arrangement on the flower stem.
2. Remove the sepals and petals from your flower by gently pulling them off the stem. Locate the stamens, each of which consists of a thin filament with a pollen-filled anther on top. Note the number of stamens.
3. Locate the pistil. The stigma at the top of the pistil is sticky. The style is a long, narrow structure that leads from the stigma to the ovary.
4. Place an anther from one of the stamens onto a microscope slide and add a drop of water. Cut the anther into several pieces with the razor blade. **CAUTION:** *Always take care when using a razor blade.*
5. Examine the anther under low and high power of your microscope. The small, dotlike structures are pollen grains.
6. Slice the ovary in half lengthwise with the razor blade. Mount one half, cut side facing up, on a microscope slide.
7. Examine the ovary section with a hand lens or stereomicroscope. The many small, dotlike structures that fill the two ovary halves are ovules. Each ovule contains an egg cell that is not visible under low power. A tiny stalk, called a funiculus, connects each ovule to the ovary wall.
8. Identify the ovary, ovules, and a funiculus.
9. Make a diagram of the flower, labeling all its parts. Color the female reproductive parts red. Color the male reproductive parts green. Color the remaining parts blue.

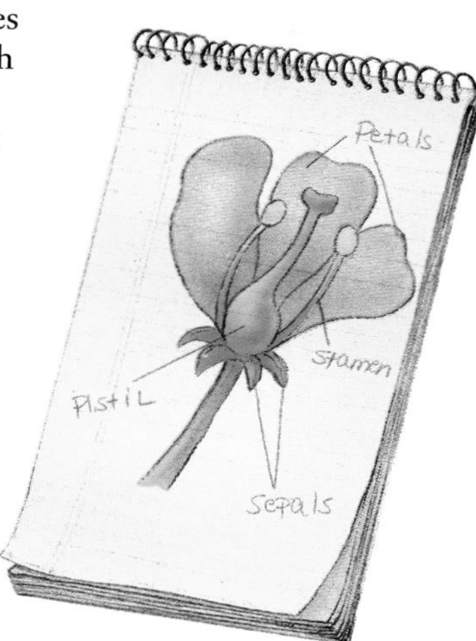

ANALYZE AND CONCLUDE

1. **Observing** How many stamens are present in your flower? How many pistils, ovaries, sepals, and petals?
2. **Comparing and Contrasting** Make a reasonable estimate of the number of pollen grains in an anther and the number of ovules in an ovary of your flower.
3. **Interpreting Data** Are there more pollen grains produced by one anther than ovules produced by one ovary? Give a possible explanation for your answer.

Going Further

Project Use a field guide to identify common wildflowers in your area. Most field identifications are made on the basis of color, shape, numbers, and arrangement of flower parts. If collecting is permitted, pick a few common flowers to press and make into a display of local flora.

ANALYZE AND CONCLUDE

1. Stamens, petals, and sepals will be in multiples of three if a monocot, multiples of four or five if a dicot. One pistil and ovary will be present.
2. thousands of pollen grains; ten to 100 ovules
3. Yes; the flower increases the probability of a pollen cell landing on the correct stigma by producing and releasing a large number of pollen grains.

Assessment

Knowledge: Provide students with diagrams of a flower other than that used in this Biolab. Have them label and indicate the general function of all important structures. L1

Going Further

Provide students with a composite flower such as a sunflower or daisy. Have them determine the locations of the individual florets and their various flower parts. L1

- You may wish to appoint certain students as lab helpers. These students can be called upon by their classmates for help and guidance.
- If necessary, review proper use of the stereomicroscope.

Data and Observations

Student diagrams, when colored properly, should show the petals and sepals as blue, pistil as red, and stamens as green.

Building a Model

LS Kinesthetic Supply students with a wad of clay, a pipe cleaner, several pieces of uncooked spaghetti, a sheet of colored paper, and scissors. Have them use these materials to model the appearance of a monocot flower that prevents self-pollination by having its pistil extend above the height of its anthers. L2

Microscope Activity

LS Visual-Spatial Have students examine bees under a stereomicroscope or hand lens. Ask students to describe the body modifications of a bee that enable it to trap pollen efficiently. L2

Assessment

Performance Assessment in the Biology Classroom: p. 31, A Living Cladogram. Have students carry out this activity to classify plants. L1 P

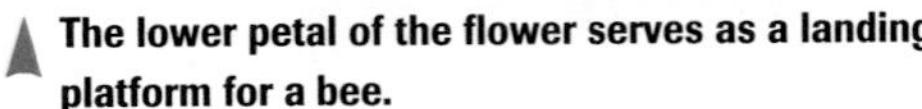

▲ **The lower petal of the flower serves as a landing platform for a bee.**

▲ **The weight of the insect crawling into the flower causes the anther to drop and deposit pollen on the bee's back.**

Figure 27.3
The anthers of this *Salvia* flower mature before the stigma. The immature stigma is the long, thin structure on the top lip of the flower in the photo.

Self-pollination is less adaptive than cross-pollination

If a stigma receives pollen from the same flower, or from another flower on the same plant, it is self-pollinated. Self-pollination results in progeny with the same genetic makeup as the parent, although the progeny will not be identical to the parent genetically due to independent assortment of chromosomes and crossing-over. If a stigma receives pollen from another flower of the same species, the flower is cross-pollinated. Cross-pollination results in an exchange of genetic material. Most flowers have adaptations that favor cross-pollination over self-pollination. For example, the male flowers of wind-pollinated sweet corn are borne at the top of the plant, while the female flowers are farther down the plant stem. This arrangement gives pollen grains a chance to be blown away from the parent plant before drifting down onto a female flower. In plants with both male and female flowers, the males on one plant often develop before the female flowers do. Self-pollination is prevented because the pollen has already been dispersed to other plants by the time the female flowers appear.

Structural adaptations favor cross-pollination

Plants with complete flowers may have several structural adaptations for cross-pollination. In some species, chemical factors may prevent successful fertilization with pollen from the same plant. Anthers may release pollen long before the stigma matures, ensuring that the stigma will receive pollen from a different plant's flower. The stigma may extend farther from the petals than the anthers, making it difficult for pollen to reach the stigma of the same flower.

662 Reproduction in Flowering Plants

STUDENT JOURNAL

Keeping a Plant Diary Have students imagine that they are a chrysanthemum plant. It is now springtime and they have just started growing. Ask students to make a spring entry that describes what they look like, explains why they are not forming flowers, and indicates the relative lengths of day and night. Have students repeat the process for each of the other seasons. L1

Insect pollination is particularly effective in favoring cross-pollination. The stamens, pistils, and nectaries of insect-pollinated flowers are usually arranged in such a way that insects pick up pollen when visiting one flower and deposit it when visiting another. *Figure 27.3* illustrates one such relationship between a flower and its insect pollinator.

Photoperiodism

The relative length of day and night has a significant effect on the rate of growth and the timing of flower production in many species of flowering plants. For example, chrysanthemums produce flowers only during the fall, when the days are getting shorter and the nights longer. A grower who wants to produce chrysanthemum flowers in the middle of summer drapes black cloth over the plants to artificially increase the length of night. The response of flowering plants to the difference in the duration of light and dark periods in a day is called **photoperiodism.**

Photoperiodism depends on length of night

Plants can be placed into three categories, depending on the day length they require for flower production, as shown in *Figure 27.4.* **Short-day plants,** such as strawberries, some

Figure 27.4

Photoperiodism refers to a plant's sensitivity to the changing length of night.

Most plants are day-neutral. Flowering in cucumbers, tomatoes, and corn, for example, is influenced more by temperature than by day length.

Spinach and lettuce are long-day plants that flower in midsummer, when days are longer than nights.

Short-day plants such as pansies and goldenrod flower in late summer and fall or early spring, when days are shorter than nights.

STUDENT JOURNAL

Identification of Flowers Provide students with the following data for different plants.

Plant A is a monocot and is day-neutral; Plant B is a dicot, is day-neutral, and has long green fruit; Plant C is a prairie flower, is a dicot, and is a short-day plant; Plant D goes well on salad, is a dicot, has large leaves curled into a bunch, and is a long-day plant, Plant E is a legume, is a dicot, is a long-day plant, and is Mendel's favorite. Have students identify a plant that fits each description. Advise students to use Figure 27.4 for more help. L2 LS

Enrichment

LS **Logical-Mathematical** Advise students that the light which triggers flower production does not have to reach the plant's flower buds, but must reach and be detected by the leaves. Have students design an experiment to prove that it is the leaf that must be stimulated by light for the plant to achieve flowering. Suggest that it is possible to place parts of the plant behind light barriers while other parts of the same plant are exposed to light. L2

Reinforcement

Explain to students that lily growers get their greenhouse plants to flower early in spring by subjecting the plants to artificial lighting that simulates longer days and shorter nights. Ask students if lilies are short-day plants or long-day plants. *Long-day*

Different Viewpoints in Biology

Carbon Dioxide Fertilization

In a technique called carbon dioxide fertilization, plants are provided with an atmosphere high in carbon dioxide to increase their rates of photosynthesis. As levels of carbon dioxide in the air rise from human activity, the uptake of the gas by plants should be beneficial. However, this may not be the case for all plants.

When grown under high carbon dioxide concentrations, some plants decline in numbers. A plant called *Abutilon* that was exposed to increased CO_2 levels showed a yellowing of its leaves. Determining whether excess carbon dioxide in the atmosphere is beneficial to plants is an area that requires further study.

SECTION 27.1

3 Assess

Check for Understanding

Have students explain the relationship among these words.

(a) stamen - anther
(b) pistil - stigma - ovary
(c) ovary - ovule
(d) pollination - pollen L1

Reteach

Obtain a flower model. As you point out structures on the model, have students name each structure and explain its function. L1

Extension

Have students determine the meaning of the terms *androecium*, *gynoecium*, and *perianth*. Have them assign the term *sterile* or *nonsterile* to each flower part. L2

Assessment

Performance: Provide students with a diagram of a typical flower. Have them label the parts of the flower and identify reproductive parts. L1

4 Close

Discussion

Show students diagrams of flowers that have different means of pollination. Have students explain their reasons for associating a specific means of pollination with a specific flower shape.

Figure 27.5

The responses of flowering plants to day length influence both their adaptations for pollination and where they are able to grow and reproduce.

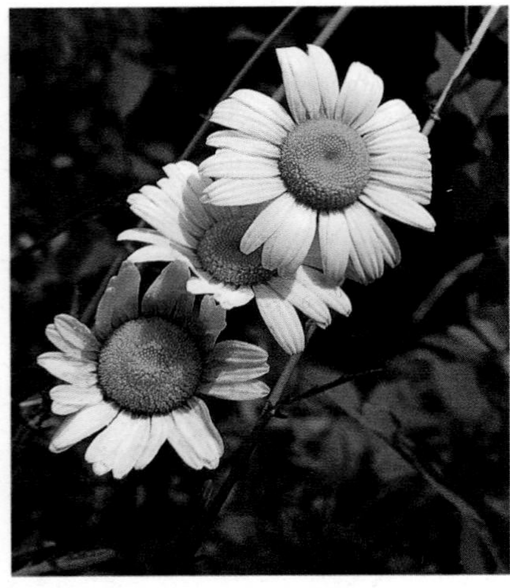

How would the seed production of these daisies be affected if they bloomed in early spring when butterflies were still just caterpillars?

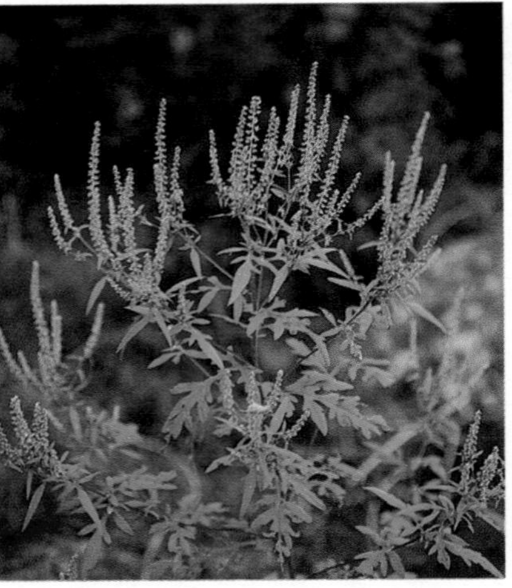

Ragweed is never found growing in northern Maine. Can you explain why in terms of photoperiodism?

chrysanthemums, cockleburs, primroses, ragweed, and poinsettias bloom when days are shorter than the nights. **Long-day plants,** such as carnations, peppermint, petunias, clover, potatoes, and garden peas, bloom as days get longer than nights. Most plant species are **day-neutral plants,** which means their blooming times are controlled by temperature, moisture, or environmental factors other than day length.

The effect of photoperiodism on pollination

Fields of wildflowers bloom during late spring and summer, when bees and butterflies are most active and numerous, *Figure 27.5*. The photoperiodism of wildflowers may ensure that a plant produces its flowers at a time when there is an abundant population of pollinators. Photoperiodism also affects the distribution of flowering plants.

Section Review

Understanding Concepts

1. Name the four organs of a flower and the functions of each.
2. How is photoperiodism of flowers related to pollination?
3. What is the adaptive value of cross-pollination for plants?

Thinking Critically

4. What benefit does an insect pollinator receive from a flower? What benefit does the flower receive from the pollinator? What type of ecological relationship is this?

Skill Review

5. **Comparing and Contrasting** Compare and contrast important features of wind-pollinated, bee-pollinated, and moth-pollinated flowers. For more help, refer to Thinking Critically in the *Skill Handbook.*

Answers to Section Review

1. Petal; attract pollinators with their coloration. Sepal; protective cover for inner flower parts. Pistil; female portion of flower, forms egg cells. Stamen; male portion of flower, forms pollen.
2. Pollen production and flowering are both correlated with photoperiodism. A plant depends on pollinators to be present to carry out their task.
3. Cross-pollination provides for greater genetic variation.

Thinking Critically

4. Pollinators receive a meal. Flowers are able to reproduce as a result of the pollinator's action. This is a mutualistic relationship.

Skill Review

5. Wind-pollinated flowers are small, lack petals, form much pollen. Bee-pollinated flowers have sweet scents, provide nectar, and their petals may serve as landing platforms. Moth-pollinated flowers are tubular in shape, have nectar, strong odors, stay open at night, are pale in color.

27.2 Flowers and Reproduction

Transferring pollen from anther to stigma is just one step in the life cycle of a flowering plant. How does pollination lead to the development of seeds enclosed in a fruit? How do sperm cells in the pollen grain reach the egg cells in the ovary? These steps in the reproductive cycle of anthophytes take place without water—an evolutionary step that enabled flowering plants to occupy nearly every environment on Earth.

Section Preview

Objectives

Outline the processes of seed and fruit formation and seed germination.

Explain strategies for seed dispersal.

Describe the effects of hormones on plant growth.

Key Terms

micropyle
double fertilization
endosperm
fruit
dormancy
germination
tropism
nastic movement
hormone

Pollen Growth and Fertilization

As you learned in Chapter 24, the life cycle of all plants, from bryophytes to anthophytes, involves the alternation of a diploid sporophyte generation with a haploid gametophyte generation. The life cycle of flowering plants resembles that of gymnosperms in that both produce seeds. In both groups, small gametophytes are retained within the body of the sporophyte.

Pollen grains grow pollen tubes

Once a pollen grain has reached the stigma, several events take place before fertilization occurs. Inside each pollen grain are two haploid sperm cells and one haploid tube cell. The tube cell elongates, forming a pollen tube that grows down the length of the style and into the ovary. The two sperm cells move through the pollen tube into the ovule through a tiny opening in the ovule called the **micropyle.**

Double fertilization occurs inside the ovule

Inside the ovule is the female gametophyte, the embryo sac, which contains several cells surrounded by cytoplasm. One of the cells is a haploid egg cell. Another is the central cell, which contains two haploid nuclei. When the tube cell reaches the female gametophyte, the nucleus of one of the sperm cells unites with the nucleus of the egg cell to form a diploid zygote. The second sperm unites with the central cell to form a cell with a triploid ($3n$) nucleus. This process, in which two sperm cell nuclei unite with two cell nuclei of the female gametophyte, is called **double fertilization.** The triploid

Prepare

Key Concepts

In this section, double fertilization and the processes and stages that lead to seed and fruit development are presented. The variety of hormones responsible for growth and development of plants is also discussed.

Block Scheduling

Look for this symbol for strategies that are useful in a block scheduling format. For more information on block scheduling, refer to Section 27.2 in the **Lesson Plans** booklet.

Materials

- Purchase or gather apple flowers and apple fruit for the Minilab.
- Soak bean and corn seeds overnight, wrap in wet paper towels, then place inside plastic sandwich bags to allow germination to take place 48 hours before needed for the Minilab. Soak another batch of seeds overnight prior to class use. Gather hand lenses or stereoscopic microscopes.

1 Focus

Bellringer

Before presenting the lesson, display **Section Focus Master 58** on the overhead projector and have students answer the accompanying questions. L1 SAE

Program Resources

Section Focus Master 58 L1 SAE
Reinforcement and Study Guide, pp. 107-108 L1
Biolab and Minilab Worksheets, pp. 107-108 L1
Laboratory Manual, pp. 151-154, 155-158 L2
Critical Thinking/Problem Solving, p. 27 L3
Basic Skills Transparencies 21a, 21b and **Master** L1 SAE
Basic Concepts Transparencies 42, 43 and **Masters** L1 SAE
Reteaching Transparency 27 and **Master** L1 SAE

Demonstration

Show students a tomato and an apple. Ask how the two plant parts are alike and how they differ. *Students may suggest that both have seeds but that one is a fruit and one is a vegetable.* Ask students to provide definitions for *fruit* and *vegetable.* Use their definitions to point out that a tomato is a fruit. If students are still not sure of the relationship, ask how the seeds got inside the two structures and what was present on the plant before the tomato or apple appeared. L1 SAE

2 Teach

Building a Model

LS **Visual-Spatial** Use a plastic sandwich bag to represent a female gametophyte. Place three small balls of clay of the same color within the bag and explain that each ball represents a haploid nucleus. (One is the egg and the other two are the central cell.) Introduce two small balls of clay of a different color to the bag. Explain that each ball represents the two sperm nuclei that enter the female gametophyte. Fuse one sperm with the egg and explain that this represents the zygote, which is now diploid. Fuse the other three nuclei. Explain that this represents the triploid nucleus that forms endosperm. Point out that together the fusings of the clay illustrate double fertilization.

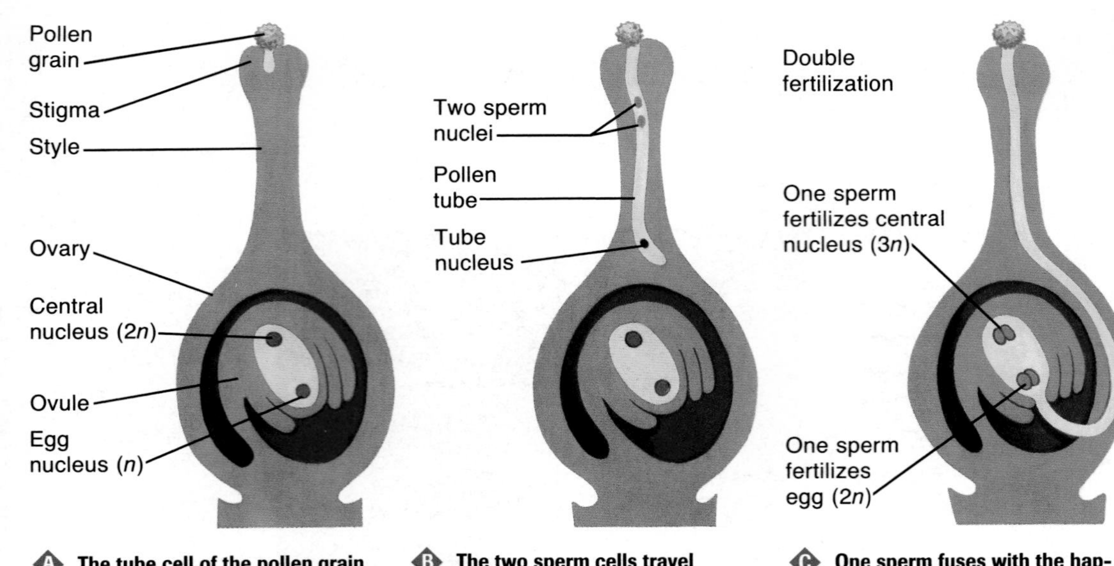

A The tube cell of the pollen grain elongates to form a pollen tube.

B The two sperm cells travel through the pollen tube to reach the female gametophyte.

C One sperm fuses with the haploid egg cell to form the zygote. The other sperm unites with the diploid central cell to form a triploid cell that will develop into endosperm.

Figure 27.6

In flowering plants, the male gametophyte grows through the pistil to reach the female gametophyte. Double fertilization involves the fusion of two sperm cell nuclei with two cell nuclei of the female gametophyte.

nucleus will divide many times, eventually forming the endosperm of the seed. The **endosperm** is food-storage tissue that supports development of the embryo.

Many flowers contain more than one ovule. Pollination of these flowers requires that at least one pollen grain land on the stigma for each ovule contained in the ovary. In a watermelon plant, for example, hundreds of pollen grains are required to pollinate a single flower. In most plants, pollen tube growth and fertilization take place fairly quickly. In barley, less than an hour elapses between pollination and fertilization. In corn, the process takes about 24 hours; in tomatoes, about 50 hours; and in cabbages, about five days. *Figure 27.6* shows the processes of pollen tube formation and double fertilization.

Seed Formation

After fertilization takes place, most of the flower parts die and the seed begins to develop. The wall of the ovule becomes the hard seed coat, which helps protect the embryo until it begins growing into a new plant. Inside the ovule, the zygote divides and grows into the plant embryo. The triploid central cell develops into the endosperm. The ovary then develops into a fruit.

PROJECT
Germinating Pollen Grains

It is possible to observe pollen grains as they germinate and form pollen tubes. The technique, however, requires some trial and error to determine the correct sugar concentration needed to bring about germination for the pollen of a particular species. Have students prepare sucrose solutions of 1, 2, 5, 10, 20, 30 and 40%. Instruct students to dust pollen onto the surfaces of these solutions. Have them remove a small sample of grains periodically, prepare wet mounts of the grains and observe the samples with a compound microscope. Have students repeat the process for several days looking for signs of pollen tube formation. Ask students to draw what they observe. Encourage students to try different solution strengths. L3 LS

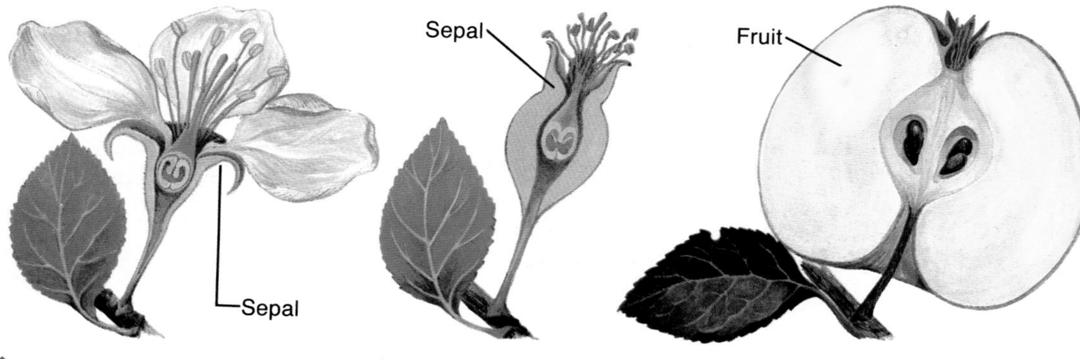

A When the eggs in the ovules of an apple flower have been fertilized, the petals, stamens, and upper portion of the pistils wither and fall away.

B The walls of the ovary in an apple harden and the supporting stem becomes fleshy and grows up and around the ovary as the seeds develop inside each ovule.

C The remains of the sepals can be seen at the end of the fruit. The fruit is attached to the plant at the opposite, or stem, end of the fruit.

Figure 27.7

A fruit consists of the seeds enclosed in the mature ovary of a flowering plant.

Fruit Formation

As the seed develops, the ovary that surrounds the seeds enlarges and becomes the fruit. A **fruit** is the structure that contains the seeds of an anthophyte. *Figure 27.7* shows how the fruit of an apple tree develops from the ovary inside the flower.

Fruits can be fleshy or dry

A fruit is as unique to a plant as its flower, and many plants can be identified by examining the structure of their fruit. You are familiar with plants that develop fleshy fruits, such as apples, grapes, melons, tomatoes, and cucumbers. Other plants develop dry fruits such as peanuts and walnuts. Some plant foods that we call vegetables or grains are actually fruits, as shown in *Figure 27.8*.

Figure 27.8

A fruit is the ripened ovary of a flower that contains the seeds of the plant. The most familiar fruits are those we consume as food.

Fleshy fruits develop a juicy fruit wall full of water and sugars. They include oranges, peaches, and watermelons, as well as foods commonly referred to as vegetables, such as tomatoes and squash.

Dry fruits have dry fruit walls. The ovary wall may start out with a fleshy appearance, as in hickory nuts or bean pods, but when the fruit is fully matured, the ovary wall is dry. Dry fruits include pecans, walnuts, and other nuts, as well as grains such as wheat, barley, and rice.

SECTION 27.2

Audiovisual

Show the filmstrip *Double Fertilization and Embryo Development in Flowering Plants*, Carolina.

Concept Development

Show students an apple. Explain that an apple is actually a "false fruit." Explain that unlike most fruits, which form from the ovary wall, the fleshy part of an apple forms from the swollen receptacle.

Enrichment

LS **Visual-Spatial** Have students research the type of fruits: cypsela, berry, pome, follicle, legume, capsule, and aggregate. Ask students to create a visual that defines each fruit type and shows examples of each. L2

GLENCOE TECHNOLOGY

Videodisc

Biology: The Dynamics of Life
Disc 1, Side 2
Double Fertilization (Ch. 24)

Disc 1, Side 2
Fruit Formation (Ch. 25)

STVS: Plants & Simple Organisms
Disc 4, Side 2
Seed Banks (Ch. 10)

Meeting Individual Needs

Learning Disabled Have students prepare a table with the heads *Haploid, Diploid,* and *Triploid* across the top. Ask students to group the following terms, phrases, or numbers beneath their correct head: sperm, 3*n*, pollen tube nucleus, egg, *n*, endosperm, central cell, fertilized egg, 2*n*. L1 LS

Focus On

Plants for People Agriculture

Purpose

To illustrate the importance of plants as an essential component of our diet and to familiarize students with agriculturally important plant species.

Background

Plant Names

Scientific names for species described in this feature are as follows: Taro, *Colocasia esculenta;* wheat, *Triticum durum var. vulgare;* corn, *Zea mays;* potato, *Solanum tuberosum;* oats, *Avena sativa;* barley, *Hordeum vulgare;* rice, *Oryza sativa;* sorghum, *Sorghum bicolor.*

Teaching Strategies

- Have students read this feature. Then have them work in cooperative groups to prepare a table of all the plants described. Along the left side have them list the crop species. Across the top, have them use the headings: *Kingdom, Division, Class, Family.* Advise students to refer to the Classification Appendix A for help in the completion of their table. **Note:** Not all plants are listed in the appendix. L1 COOP LEARN
- Have students write a short paragraph that describes the relative dependence of humans on monocots versus dicots as food sources. L2

Focus On

Plants for People

Agriculture was probably the single most significant development in human history. It is not an accident that the beginnings of civilization occurred in productive farming areas. Today, many farmers concentrate on growing just one species of food crop, often a grain, in their fields. Monocultures enable farmers to concentrate on one species of plant and allow them to use heavy equipment for planting, cultivating, and harvesting. However, monocultures also leave farming areas vulnerable to disease. A pest infestation can attack a monoculture and spread quickly from field to field, wiping out an entire season's crop. Farmers have begun to use more native species in their fields. Native species are less susceptible to pests and are more hardy than other crops used in monocultures. The preservation of native species for use as crops is gaining attention from farmers and biologists alike.

Agriculture

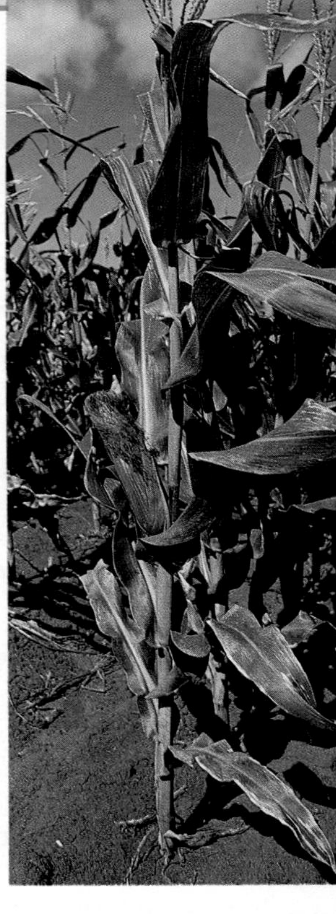

Corn More land in the United States is used to grow corn, family Poaceae, than any other crop. Most of the corn is used for livestock feed, but a significant portion is used to manufacture starch, oil, sugar, meal, breakfast cereals, and alcohol.

Rice For the majority of humanity, rice means survival. In Asia, rice, family Poaceae, is the basis of almost all diets. More than 95 percent of the world's rice crop is used to feed humans. Rice is the only grain that can grow submerged in water.

Taro The roots of this plant provide an important staple to Asia and Pacific Ocean island areas. Taro, a member of the arum family Araceae, can be made into cakes that can be baked or toasted.

Oats Oats, family Poaceae, are an important food for both man and beast. Containing from ten to 16 percent protein, oats are low in fat but high in carbohydrates, proteins, B vitamins, fiber, and minerals. They make an excellent food for growing humans and young livestock. A native of northern Europe, oats grow well in poor soils and cool, wet climates.

PORTFOLIO

Have students research the origin of a particular commercial crop and write a brief report on their findings. The report can be placed in their portfolios. L1 P

Meeting Individual Needs

Students Acquiring English Ask students to prepare a concept map that illustrates a variety of main products obtained through agriculture. Allow students to pick the word choices that are needed for the map rather than you supplying them with the key terms. L1 SAE

Wheat Nearly one-third of all the land in the world used for crop production is planted in wheat. Wheat, family Poaceae, probably originated in the Middle East and was an important food of the ancient Mesopotamian, Egyptian, and Indus civilizations.

Barley Barley, family Poaceae, is probably the oldest grain crop used by humans. The Egyptians grew barley in 6000 B.C., and it is the fourth-largest cereal crop today. Of all the grains, it grows the fastest and can stand the toughest conditions such as in Lapland and high in the Himalayas.

Potato This South American native arrived in the United States by a circuitous route–via Europe. The potato, family Solanaceae, is very nutritious, containing more of the essential amino acids than whole wheat. In addition, potatoes carry appreciable amounts of vitamins B and C, as well as minerals such as calcium and iron.

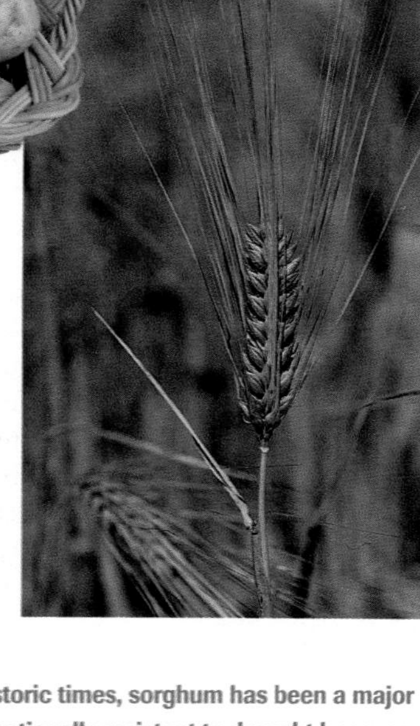

Sorghum Since prehistoric times, sorghum has been a major food crop in Africa. Exceptionally resistant to drought because of its extensive root system, sorghum, family Poaceae, provides not only food for people but also hay for cattle.

Focus On

Teaching Strategies

- Supply students with blank maps of the United States. Ask them to color in those areas that are primarily corn- and wheat-growing zones. Have them add labels to identify each crop. L2
- Supply students with a global outline map. Have them indicate on the map where the areas of early agriculture began—the Tigris and Euphrates, Nile Valley, and Indus River region. Have them label each area. L1
- Teosinte is available from biological supply houses. Purchase this type of corn and allow students to examine and compare its kernels with those of the more familiar variety.

Tying to Previous Knowledge

- Have students review the process of fermentation and allow them to infer how corn can be used in the process to produce alcohol. *Corn serves as the food source for yeast.*

Enrichment

Ask students to determine the differences between the wheat used for pasta (*T. durum*) and that used for bread (*T. vulgare*). Have them find out what some of the different physical properties are between these starches and why one is more suitable than the other for bread or pasta products. L2

GLENCOE TECHNOLOGY

Videodisc

STVS: Plants & Simple Organisms
Disc 4, Side 1
Disease-Resistant Tomatoes (Ch. 19)

Disc 4, Side 1
Genetic Engineering in Barley (Ch. 20)

Disc 4, Side 2
New Grains (Ch. 11)

Focus On

Plants for People
Silviculture

Purpose

To illustrate the importance of trees as producers of foods, drugs, lumber, fuel, and chemicals and to identify trees that supply these products.

Background

Products from Trees

Quinine, an antimalarial chemical derived from the bark of various species of *Cinchona* trees, was used by the Andean Indians before the European discovery of the New World.

Cinnamon, derived from the bark of *Cinnamomum zeylanicum*, cloves, from the flower buds of *Eugenia caryophyllata*, and nutmeg, from the seeds of *Myristica fragrans*, are examples of spices derived from trees.

Turpentine is derived from the gymnosperm *Pinus palustris*. It is an oleoresin that forms in ducts within the sapwood. Rubber comes from *Hevea brasiliensis*.

The most familiar tree fluid is probably chicle, which is the basic ingredient in chewing gum. Chicle is the dried latex of *Manilkara sapota*, a flowering tree found in Central America. A different latex from a tree indigenous to the West Indies and South America, *Mimusops balata*, supplies the basic ingredient for bubble gum.

Teaching Strategies

- Have students work in cooperative groups to compile a list of different nuts that are from trees, and draw a map that shows where these are grown around the world. *Responses may include almond, Brazil nut, cashew, hazelnut, pecan, pistachio, walnut, and chestnut.* L2 COOP LEARN

Focus On

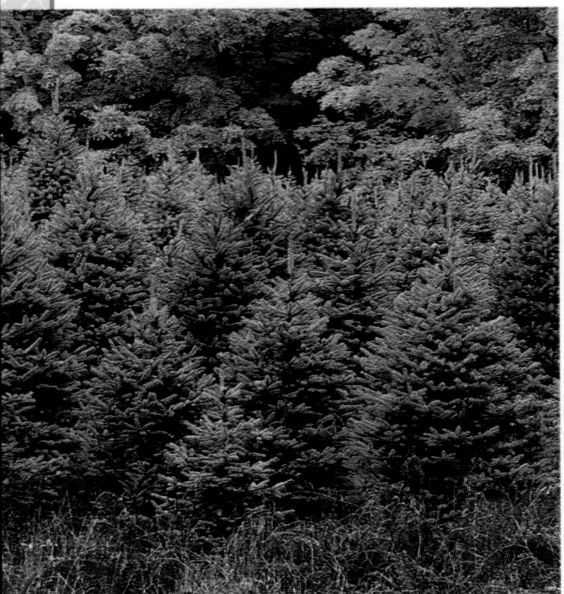

Silviculture

Silviculture, the growing of trees, is sometimes confused with forestry. Silviculture includes the growing of trees for lumber and paper, but it also refers to the growing of trees for food crops—for example, apple and pecan orchards. Oranges, nuts, olives, and cloves are some of the foods that silviculture provides. Trees also are sources of medicine, such as quinine that was used to combat malaria, or aspirin, originally derived from the willow tree. Trees even help replenish the oxygen we breathe.

Chemicals and Spices Wood is the source of many chemicals. Wood alcohol, the rosin used on the bows of violins, cellulose used to make paper, charcoal, and rayon are just a few of the chemicals derived from trees. Spices such as cinnamon and drugs such as aspirin are derived from the bark of trees. An extract from the bark of a Pacific Coast yew tree (taxol) has been used to fight cancer.

Tree fluids Fluids from trees are the source material for maple syrup, latex rubber, and turpentine used as solvent in paint. Tree fluids can be extracted year after year, providing a continuous crop with minimal injury to the tree.

Lumber and Fuel When most people think of forest products, they think of lumber. It is one of our most common building materials. Easy to work, durable, relatively abundant, and lightweight, it is close to ideal for construction of homes. Throughout much of the world, wood is the major source of fuel. It is high in heat content and easy to transport and store.

Meeting Individual Needs

Learning Disabled Ask students to prepare a concept map that illustrates the variety of main products obtained through silviculture. Have students include at least one example of each of the main products. Allow students to pick the word choices that are needed for the map rather than you supplying them with the key terms. L1

PORTFOLIO

A Graft Model Provide students with two tree twigs, paper, clay, and tape. Have them demonstrate the technique that would be used to graft a rose plant. Ask them to label the parts. *Scion is the top part; stock is the bottom rooted part.* Have them explain the advantage of grafting roses over growing them from seed. L1 P LS

Horticulture

Most of the plants we are familiar with were originally brought back from distant lands by naturalists and great explorers of the world. Using these exotic plants, horticulturists have left us a legacy of beautiful and useful varieties of flowers and food plants by their dedication to selective-breeding programs, grafting methods, and more recently, genetic engineering.

Fruits An obvious, and delicious, product of trees is fruit—both dry fruit and fleshy fruit. Dry fruits include acorns, walnuts, pecans, and other nuts used as food by wildlife and humans. Fleshy fruits include oranges, apples, pears, apricots, and cherries.

Vegetables Vegetables were some of the first plants cultivated by humans. Many of those belonging to the mustard family—such as green cabbage, watercress, and radishes—were known to the Egyptians and Romans in the Bronze Age.

Root crops such as beets, carrots, and turnips have always been especially valuable because they could be stored easily during the winter months. They also provide excellent nutrition and lots of fiber.

Legumes Some plants improve the soil. Legumes such as peas, beans, alfalfa, and clover have nitrogen-fixing bacteria in small nodules on their roots. They "fix" nitrogen in the soil. Having clover growing in your lawn benefits your grass.

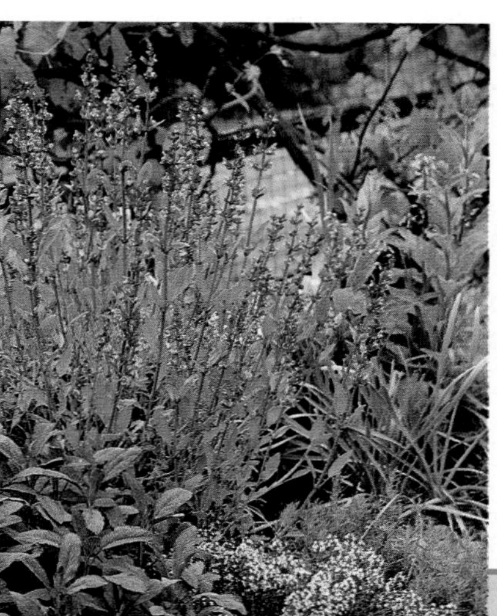

Flowers and Medicinal Plants Flowering bulbs, roses, scented herbs, and garden flowers of all kinds have been a source of joy and delight to people for generations. Plants, in eras past, supplied medicines for people who could not afford doctors.

EXPANDING YOUR VIEW

1. **Applying Concepts** Take a piece of paper and make two columns. In the left column, list all of the products essential to your life that come from plants and/or agriculture. In the right column, list all of the essential products that *do not* come from plants and/or agriculture. Now write a paragraph summarizing the role of plants in your life.

Focus On

Plants for People Horticulture

- Have students test a variety of plant roots for the presence of starch. Have them use thin slices of radish, turnip, and potato and add a few drops of iodine/potassium iodide solution to the root. A blue-black color indicates the presence of starch. As a control, have them add a drop of iodine solution to a small clump of corn starch. Ask them to write a report of their observations for their portfolios. L2 P

Answers to Expanding Your View

Applying Concepts Except for transportation, communication, certain services, and products composed of metals, almost all human basic needs are tied to plants.

CULTURAL DIVERSITY

The History of Chocolate

Theobroma cacao is a native tree from Central America. After flowering, it produces large pods that contain about 40 cacao beans. Because of its seeds, cacao was domesticated by the Mayas and Aztecs who turned the seeds into a rich brown drink called chocolate.

Researchers are suspicious that the chemicals in chocolate, theobromine and methylxanthin, may have an addictive quality. This may explain why some individuals crave this ripened anthophyte ovule.

MiniLab

Purpose

LS Visual-Spatial To have students correlate changes in flower parts to their related fruit parts.

Process Skills

observe and infer, compare and contrast

Teaching Strategies

- Preserved apple or cherry flowers from biological supply houses can be used.

Expected Results

Students will find the dried flower parts still intact on the apple when the flower scar end is viewed with a hand lens.

Analysis

1. stamens, petals
2. Diagrams should show a correlation between ovary and fruit, ovules and seeds, and flower parts such as stamens and pistil.
3. Number of seeds within the fruit will correlate to the number of ovules within the ovary.

Assessment

Performance: Show students an orange and pear. Have them indicate what plant parts the fruit and seeds inside were originally, and predict the number of ovules in each flower. L1

MiniLab

How do fruits and flowers compare?

Some parts of the flower develop into fruits and seeds, while others wither and fall away. The form of a particular flower is reflected in the structure of the fruit it produces.

Procedure

1. Carefully examine the fruits and flowers provided by your teacher.
2. Locate and identify the major structures of the flower.
3. Examine the fruit to see if any flower structures are still attached.
4. Carefully cut open both flower and fruit, and compare their internal structures.

Analysis

1. Which flower structures can be seen on the exterior of the fruit?
2. Diagram the interior of a flower and its fruit. Use a color code to label similar structures on each diagram.
3. What kind of correlation can you make between the number of seeds in the fruit and the internal structure of its flower?

Seed Dispersal

A fruit not only protects the seeds inside it, but also may aid in dispersing those seeds away from the parent plant and into new habitats. Dispersal of seeds, ***Figure 27.9,*** is important because it reduces competition for sunlight, soil, and water between the parent plant and its offspring. Animals such as raccoons, deer, bears, and birds help distribute many seeds by eating fleshy fruits. They may carry the fruit some distance away from the parent plant before consuming it and spitting out the seeds. Or they may eat the fruit, seeds and all. Seeds that are eaten pass through the digestive system unharmed and are deposited in the animal's wastes. Squirrels, birds, and

Figure 27.9

A wide variety of seed-dispersal mechanisms has evolved among flowering plants.

Natives of Russia, tumbleweeds are common plants of the American Southwest. When seeds mature, the dried leaves and stems break off, and seeds are scattered as the plant tumbles across the prairie landscape.

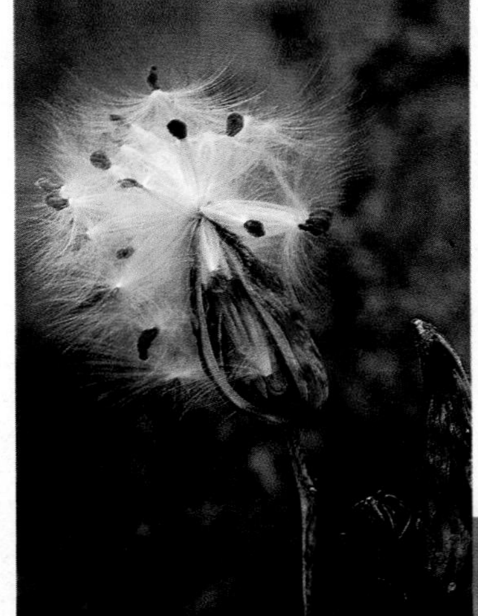

Wind-dispersed seeds have adaptations that enable them to be held aloft while they drift away from their parent plant. Milkweed seeds have hairy plumes that act as parachutes. Elm and maple trees produce winged fruit.

Clinging fruits, like those of the cocklebur and burdock, produce seeds surrounded by hooks that stick to the fur of passing animals or the clothes of passing humans.

Alternate Lab: Seed Germination

Purpose

LS Logical-Mathematical To compare the rate of respiration in germinating and nongerminating seeds.

Materials

canned kidney beans, paper cup, water, dried kidney beans, wax paper, labels, tetrazolium solution in dropper bottles (add 5 g of 2,3,5-triphenyl tetrazolium chloride to 150 mL distilled water)

Procedure

Give the following directions to students.

1. Soak ten kidney bean seeds overnight in water. The next day, split each seed in half. Split ten canned kidney bean seeds in half.
2. Place all seeds split-side up onto two sheets of wax paper. Mark each paper with a label indicating seed treatments.
3. Add a drop of tetrazolium to each cut surface. Wait 20 to 30 minutes. Record in which seeds and where in the seeds a pink color appears.

Note: Tetrazolium can be diluted and discarded down the sink.

Figure 27.10

Seeds can remain dormant for long periods of time. Lupine seeds like these may germinate after remaining dormant for thousands of years.

other nut gatherers may drop and lose some of the seeds they collect, or even bury them only to forget where.

Plants that live in or near water, such as water lilies and coconut palms, produce fruits or seeds with air pockets in the walls, which enable them to float and drift away from the parent plant. The ripened fruits of many plants split open to release seeds designed for dispersal by the wind or by clinging to animal fur. Orchid seeds are so tiny that they resemble dust grains and are easily blown away in the wind. The fruit of the poppy flower forms a seed-filled capsule that releases sprinkles of tiny seeds like a salt shaker as it bobs about in the wind. Tumbleweed seeds are scattered by the wind as the whole plant rolls along the ground.

Seed Germination

At maturity, seeds are fully formed. The seed coat dries and hardens, enabling the seed to survive conditions that are unfavorable to the parent plant. Once seeds are dispersed, they can remain in the soil until conditions are favorable for growth and development of the new plant. This period of inactivity in a mature seed is called **dormancy.**

Dormant seeds survive unfavorable conditions

Metabolic activity slows during dormancy, which may last for a few months or many years. Some seeds, such as willow, poplar, magnolia, and maple, remain dormant for only a few weeks after they mature. These seeds cannot survive harsh conditions for long periods of time. If conditions remain unfavorable for growth, the embryo within the seed may die. Most flower and vegetable seeds can survive for two or three years if kept cool and dry. Seeds of plants that live in dry or cold environments, such as desert wildflowers or conifers including spruce, fir, and pine, can survive dormant periods of 15 to 20 years. The seeds of evening primrose and curly dock, two common weeds, can survive for 100 years. The seed coat acts as a mechanical barrier in some plants. In other plants, dormancy is related to chemical inhibitors in the seed coat.

The longest-lived seeds are reported, by some scientists, to be lupine seeds, *Figure 27.10.* Found buried in the dry, frozen burrows of small arctic tundra mammals, these seeds were estimated to be 10 000 years old. This date, however, has not been confirmed by radiocarbon dating.

Chemistry Connection

Water and Seed Germination

Water must be absorbed by seeds prior to germination in a process called imbibition. The volume change that occurs during imbibition can be measured through water displacement.

Place 50 mL of water in a graduated cylinder. Add 10 bean seeds. The new level of water less the original volume equals the volume of the 10 seeds. Remove the seeds from the cylinder and soak them in water overnight. The following day, add 50 mL of water to the graduated cylinder. Add the soaked beans and calculate their new volume. Have students use the water-displacement technique to calculate the percent of volume change that occurs through imbibition. L2

GLENCOE TECHNOLOGY

Videodisc

Biology: The Dynamics of Life

Disc 1, Side 2
Seed Dispersal (Ch. 26)

Disc 1, Side 2
Germination (Ch. 27)

Analysis

1. Tetrazolium indicates cell respiration when it turns pink. Which seed treatment carried on cell respiration? Why? *Uncanned seeds are germinating and require energy; canned seeds are dead.*
2. The darker the pink color, the greater the rate of respiration. Which seed part carries on the greatest rate of respiration? Why? *Embryo; rapid growth requires more energy*

Assessment

Portfolio: Design an experiment to determine how many seeds in a seed packet are alive. L1 P

MiniLab

Purpose

LS **Visual-Spatial** To compare dormant monocot and dicot seed embryos with germinating seed embryos.

Process Skills

observe and infer, compare and contrast

Safety Precautions

- Advise students to use caution with razor blades.

Teaching Strategies

- Soak some seeds 24 hours and others 72 hours prior to use. Wrap the 72-hour seeds in moist paper toweling after soaking for 24 hours and place in sealed plastic bags.
- Advise students to lay the corn on its flat side while it is being cut open.

Expected Results

Germinating seeds will have larger embryos than those of dormant seeds.

Analysis

1. Labels should include cotyledon(s), embryo.
2. Diagrams should include cotyledons, embryo, epicotyl, radicle, plumule, and hypocotyl on dicot, but only cotyledon and embryo in monocot.
3. Monocot has one cotyledon, small embryo, is slow to germinate, and its embryo parts are hard to differentiate. Dicot seed has two cotyledons, larger embryo, and easily seen embryo parts.

Assessment

Performance: Provide students with monocot and dicot seeds. Have students classify each seed type after removing the seed coat. L1

MiniLab

What does a germinating seed look like?

Seeds are made up of a plant embryo and food-storage tissue inside a seed coat. Monocot and dicot seeds differ in their internal structures.

Procedure

1. Obtain from your teacher a soaked, ungerminated corn kernel (monocot), a bean seed (dicot), and corn and bean seeds that have begun to germinate.
2. Remove the seed coats from each of the ungerminated seeds, and examine the structures inside. Use low-power magnification. Locate and identify each structure of the embryo and any other structures you observe.
3. Examine the germinating seeds. Locate and identify the structures you observed in the dormant seeds.

Analysis

1. Diagram the dormant embryos in the soaked seeds, and label their structures.
2. Diagram the germinating seeds, and label their structures.
3. List at least three major differences you observed in the internal structures of the corn and bean seeds.

Requirements for germination

Dormancy ends when the seed is ready to germinate. **Germination** is the beginning of the development of the seed into a new plant. The absorption of enough water and the presence of oxygen and favorable temperatures usually end dormancy, but there may be other requirements. Water is important because it activates the embryo's metabolic system. Once metabolism has begun, the seed must continue to receive water or it will die. Just before the seed coat breaks open, the plant embryo begins to respire rapidly. Many seeds germinate best at temperatures between 25°C and 30°C. Arctic species germinate at lower temperatures than do tropical species. At temperatures below 0°C or above 45°C, most seeds won't germinate at all.

Some seeds have special requirements for germination, *Figure 27.11.* For example, some germinate more readily after they have passed through the acid environment of an animal's digestive system. Others require a period of freezing temperatures as do apple seeds, extensive soaking in water as do coconut seeds, or certain day lengths. You may remember reading in Chapter 25 that the seeds of some conifers will not germinate unless they have been

Figure 27.11

The seeds of the desert tree *Cercidium floridum* have hard seed coats that must be cracked open. This occurs when the seeds tumble down arroyos in sudden rainstorms.

PROJECT

Speedy Seeds

Ask students to design an experiment to compare the rates at which monocot and dicot seeds germinate. Suggest that students use the following seed types: grass and corn (monocots) bean or mustard (dicots). Have students design the procedure to be used and the type of data to be recorded. L1 LS

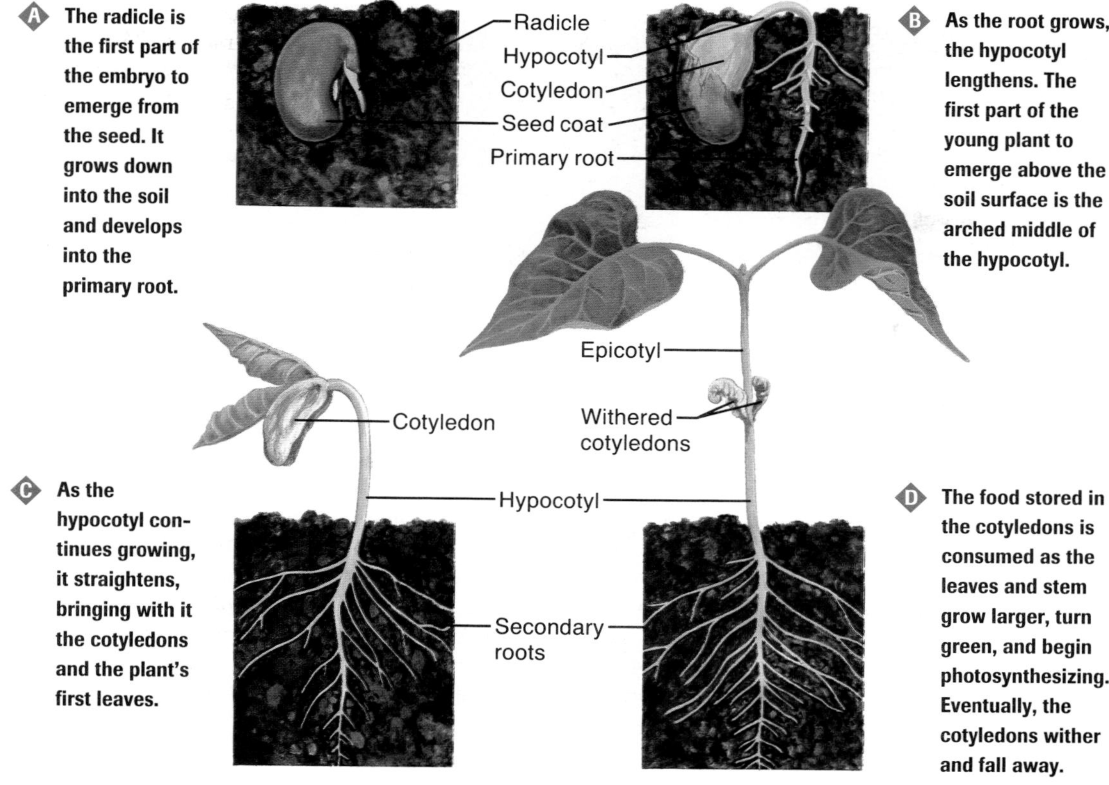
The radicle is the first part of the embryo to emerge from the seed. It grows down into the soil and develops into the primary root.
Radicle
Hypocotyl
Cotyledon
Seed coat
Primary root
As the root grows, the hypocotyl lengthens. The first part of the young plant to emerge above the soil surface is the arched middle of the hypocotyl.
Epicotyl
Cotyledon
Withered cotyledons
As the hypocotyl continues growing, it straightens, bringing with it the cotyledons and the plant's first leaves.
Hypocotyl
The food stored in the cotyledons is consumed as the leaves and stem grow larger, turn green, and begin photosynthesizing. Eventually, the cotyledons wither and fall away.
Secondary roots

exposed to fire. The same is true of certain wildflower species, including lupines and gentians. *Figure 27.12* illustrates the germination of a typical plant embryo.

Growth and Development

Why do roots grow down into the soil and stems grow up into the air? Although plants lack a nervous system and usually cannot make quick responses to stimuli, they do have mechanisms that enable them to respond to their environment. Plants grow, produce flowers and seeds, and shift the position of their roots, stems, and leaves in response to environmental conditions such as gravity, sunlight, temperature, and day length.

Responsive movement in plants

A **tropism** is a plant's response to an external stimulus that comes from a particular direction. If the tropism is positive, the plant grows toward the stimulus. If the tropism is negative, the plant grows away from the stimulus. The tendency of stems to bend toward light is called phototropism. The positive phototropism exhibited by stems is an adaptation that enables the plant to obtain a maximum amount of sunlight.

There is another tropism associated with the upward growth of stems and the downward growth of roots. The stimulus is gravity and the tropism is called gravitropism. Stems show negative gravitropism, while roots show positive gravitropism. Do leaves exhibit gravitropism?

Figure 27.12
Germination of a bean seed is stimulated by water, which softens the seed coat, and warm temperatures.

Bioethics

Lifting the Ban on Illegal Drugs
Cannabis sativa, a weedlike anthophyte, was grown as a source of hemp fiber important in the caulking of wooden ships and in the manufacture of sack cloth. Today, its use as an illegal drug is well known. Although using *Cannabis* is illegal, this plant has been shown to have medicinal value in helping patients with glaucoma, cancer, and AIDS. Patients with these diseases sometimes grow the plant to use as part of the regimen for alleviating symptoms of their disease. However, the federal government maintains that this is against the law.

Have students research the history of laws on use of *Cannabis.* L1 COOP LEARN

Assessment

Performance Assessment in the Biology Classroom: p. 47, Controlling the Rate at Which Fruit Ripens. Have students carry out this activity to observe plant hormones at work. L1 P

PORTFOLIO

Modeling Seed Dispersal Provide students with a strip of Velcro, a piece of fabric, cotton batting, tape, string, piece of clay, stiff paper, and scissors. Have students use the materials to prepare models that show different means of seed dispersal. Models may imitate attaching to animal fur, floating in air, and being transported like a helicopter. L1 P LS

Science, Technology, and Society

Can plants communicate? It appears that plants may communicate through hormones. An example of this is seen in the ability of the acacia tree of Africa to form ethylene gas. The formation of ethylene gas is not unusual; however, the acacia emits high concentrations of ethylene in response to its leaves being damaged by foraging antelope.

Ethylene released by the acacia causes nearby trees to increase their production of tannin, a chemical compound that makes acacia unpalatable to antelope. This increase takes only 30 minutes and serves as a quick protection from nearby antelopes.

Using Science Terms

Have students explain how plant responses to hormones differ from those that are nastic movements. Ask them to provide an example of both types of responses. The word *nastic* means "to press." Have students explain how the term describes the movement of *Mimosa* leaves.

Figure 27.13
Mimosa pudica **is also known as the sensitive plant. When touched, it folds its leaves in less than one-tenth of a second. This nastic response might provide some protection from insect damage or help reduce water loss in drying winds.**

Because tropisms involve growth, they are not reversible. For example, the position of a stem that has grown several inches in a particular direction cannot be changed. But if the direction of the stimulus is changed, the stem will begin growing in another direction.

A responsive movement of a plant that is not dependent on the direction of the stimulus is called a **nastic movement.** An example of a nastic movement is the sudden drooping of the leaves of a *Mimosa* when the plant is touched, as shown in *Figure 27.13.* Because nastic movements do not involve growth, they are reversible.

Plant hormones include auxins

Plants, like animals, have hormones that regulate growth. A **hormone** is a chemical that is produced in one part of an organism and transported to another part, where it causes a physiological change. Only a small amount of the hormone is needed to make this change.

Growth hormones called auxins affect many aspects of plant development. For example, auxins are responsible for regulating phototropism in plants. *Figure 27.14* describes how auxins regulate this response.

A Stems grow toward light, or show positive phototropism, because auxin collects in the cells on the shaded side of the stem.

B Auxin is a hormone that stimulates the elongation of cells, so the cells on the auxin-rich, shaded side of the stem grow longer than the cells on the other side, causing the stem to bend toward the light.

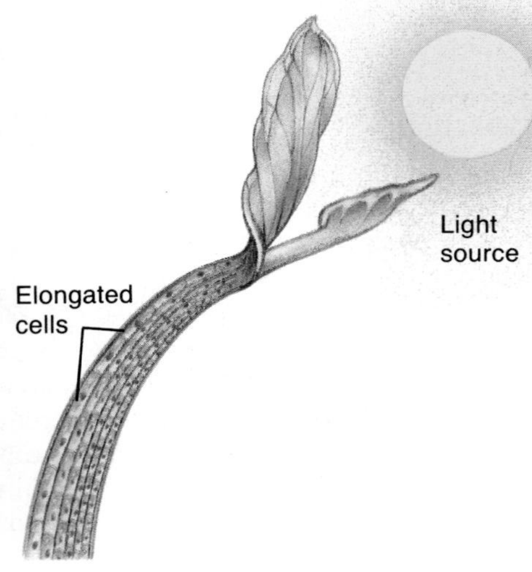

Figure 27.14
Tropisms are the result of unequal stimulation, as when light comes to a plant from only one side.

676 Reproduction in Flowering Plants

GLENCOE TECHNOLOGY

Videodisc

Biology: The Dynamics of Life
Disc 1, Side 2
Plant Tropisms (Ch. 28)

Disc 1, Side 2
Hydroponics (Ch. 29)

Auxins have a number of other effects on plant growth and development. Have you ever seen a gardener pinch the tip off a plant to encourage the growth of side branches? Auxin produced in the apical meristem inhibits the growth of side branches. Pinching the tip reduces the amount of auxin present in the stem and allows side branches to form, *Figure 27.15*. High concentrations of auxin promote development of fruit and inhibit the dropping of fruit from the plant. When auxin concentrations decrease in the autumn, ripened fruit falls to the ground and trees begin to drop their leaves.

Gibberellins promote growth

Gibberellins are growth hormones that cause plants to grow taller because, like auxins, they stimulate cell elongation. Gibberellins also increase the rate of seed germination and bud development, and may stimulate the formation of flowers and fruits in some plants.

Figure 27.15

Side branches developed on this bush when the apical meristem of the main stem was removed.

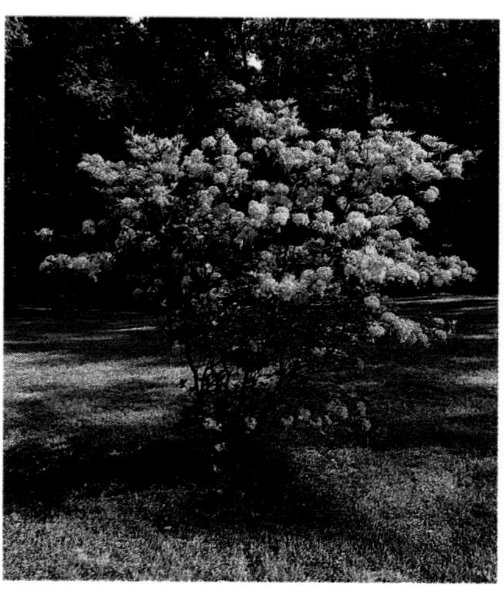

ThinkingLab — Make a Hypothesis

How does fruit ripen?

Ethylene is a gas made by plants that causes ripening of fruit. Ethylene is also given off in small amounts by the fruits themselves, causing ripening to occur even after they are picked.

Analysis

Assume that you buy peaches from the supermarket and find that they are not completely ripe. Someone tells you that if you put them in a paper bag with a ripe banana, they will ripen more quickly than if you just leave them out on your counter.

Thinking Critically

Make a hypothesis about what you think might happen to unripe peaches if you put them in a paper bag with a banana compared to what would happen if you left them on the counter. Explain your reasoning.

Some hormones inhibit growth

While auxins and gibberellins stimulate growth, other plant hormones, called inhibitors, have the opposite effect. One inhibitor is abscisic acid. Inhibitors are responsible for producing dormancy in seeds and buds during periods of unfavorable environmental conditions. Abscisic acid also plays a role in closing stomata during periods of water shortages. This hormone is often thought of as the hormone produced to protect the plant in times of stress.

ThinkingLab — Make a Hypothesis

Purpose

LS **Logical-Mathematical** Hypothesize about whether bananas can speed the ripening of other fruits.

Process Skills

form a hypothesis, recognize cause and effect

Teaching Strategies

- Allow students to work in cooperative groups or have them do this Thinking Lab as homework. COOP LEARN
- Allow students to test their hypotheses. Pears and green bananas may be placed in the bag with a yellow or ripe banana. Remind students to use a control.

Background

Bananas are picked green off trees and are shipped in this condition to whatever country is receiving them. When ready for sale, the bananas are placed into chambers and subjected to ethylene gas to start the ripening process.

Thinking Critically

If ethylene is given off by ripening bananas, then placing one in a bag with an unripe peach will speed up its ripening. The ethylene gas given off by the banana should have the same effect on other fruits as it does on itself.

Assessment

Portfolio: Have students explain how they could modify the experiment to verify that ethylene gas causes ripening of fruits. L1 P

Meeting Individual Needs

Learning Disabled Ask students to explain what is meant by the sentence that reads, "This hormone is often thought of as the hormone produced to protect the plant in times of stress." Have students provide examples of factors that are stressful to a plant and to humans. Elicit whether hormones help regulate stress levels in humans. L1

SECTION 27.2

3 Assess

Check for Understanding

Have students explain how the words in each of the following pairs are related.

(a) pollen tube, double fertilization
(b) epicotyl, plumule
(c) tropism, nastic movement
(d) dormancy, germination L1

Reteach

Have students indicate when or where in the life cycle each term in the *Check for Understanding* applies. L1

Extension

Have students research the mechanism responsible for roots growing down and stems growing up. L2

Assessment

Performance: Provide students with a sliced apple half and a sliced open flower. Have them locate and name the following structures: ovule before and after fertilization, ovary before and after fertilization, embryo. L1

4 Close

Discussion

Watermelons have numerous black seeds and some small white seeds. Ask students to speculate what the small white seeds found in a watermelon might be. *They are unfertilized ovules.*

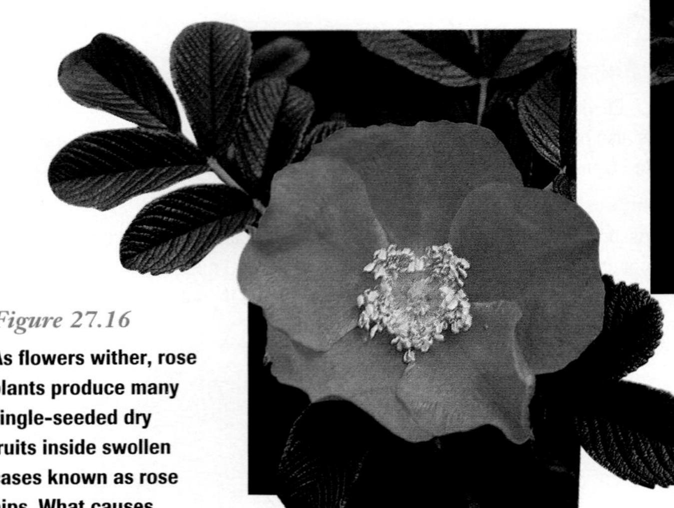

Figure 27.16

As flowers wither, rose plants produce many single-seeded dry fruits inside swollen cases known as rose hips. What causes flowers to wither?

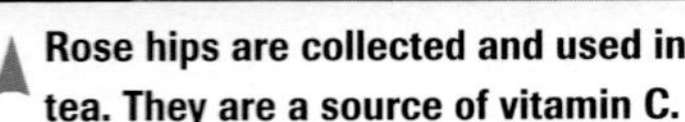

Rose hips are collected and used in tea. They are a source of vitamin C.

Roses such as this *Rugosa* belong to the family Roseaceae.

Ethylene gas promotes ripening

Ethylene gas, a simple compound of carbon and hydrogen, is a hormone that speeds the ripening of fruits. This hormone is produced on the plasma membrane. Ethylene is also responsible for the withering of flower parts after fertilization takes place, *Figure 27.16.* It also promotes the dropping of leaves in the fall.

Connecting Ideas

You've had an opportunity to take a broad look at the plant kingdom. The journey started with the simplest nonvascular plants, continued through the ferns and gymnosperms, and ended with the anthophytes—the most complex, colorful, and numerous plants on Earth today. Now it's time to take a similar phylogenetic journey through the animal kingdom, starting with the invertebrates.

Section Review

Understanding Concepts

1. How does pollination of a flower lead to development of the seed?
2. What part of the flower becomes the fruit?
3. Name two plant hormones, and describe how each one influences growth and development.

Thinking Critically

4. What does a pollen grain have in common with a spore?

Skill Review

5. **Making and Using Tables** Make a table that indicates whether each structure of a flower is involved in pollination, fruit formation, seed production, seed dispersal, or growth and development. For more help, refer to Organizing Information in the *Skill Handbook.*

Answers to Section Review

1. Pollination supplies sperm cells to the female gametophyte. Inside the ovule, double fertilization occurs to form the seed.
2. the ovary wall or receptacle
3. Gibberellin stimulates cell growth and promotes elongation of the stem, and auxin inhibits growth of side branches and allows for phototropic responses.

Thinking Critically

4. Both are microscopic; both are haploid; a spore forms a pollen grain

Skill Review

5. Pollination: anther, stamen, stigma; fruit formation: ovary and ovule; seed production: ovule, egg; seed dispersal: ovary; growth and development: ovule, ovary

Chapter 27 Review

REVIEWING MAIN IDEAS

27.1 What Is a Flower?

- Flowers are made up of four organs: sepals, petals, stamens, and pistils.
- Flowering plants have a wide variety of adaptations for pollination.
- Most flowers have adaptations that favor cross-pollination.
- The relationship between the number of hours of daylight and darkness in a day, called photoperiodism, affects the timing of flower production in many anthophytes.

27.2 Flowers and Reproduction

- A pollen tube grows downward from the stigma through the style to the ovary. Sperm move into the ovule, where fertilization takes place.
- Flowering plants undergo double fertilization. One sperm cell nucleus joins with the egg cell nucleus to form the zygote. The second sperm cell nucleus joins with the central cell nucleus in the ovule to form a triploid cell.
- The zygote develops into the plant embryo. The triploid cell develops into the endosperm. The ovary wall becomes the fruit, which may be fleshy or dry.
- When the seed matures, it separates from the ovary and may be dispersed by animals or by the wind.
- Seeds can remain dormant for weeks, months, or years. Dormancy ends and germination begins when water availability, temperature, and other factors become favorable for growth.
- Auxins and gibberellins are plant hormones that stimulate cell elongation. These and other hormones regulate flower and seed development, the ripening of fruit, and the timing of fruit and leaf drop in the autumn.

Key Terms

Write a sentence that shows your understanding of each of the following terms.

anther	micropyle
complete flower	nastic movements
day-neutral plant	ovary
dormancy	petal
double fertilization	photoperiodism
endosperm	pistil
fruit	sepal
germination	short-day plant
hormone	stamen
incomplete flower	tropism
long-day plant	

Understanding Concepts

1. Pollen and sweet nectar produced by flowers provide ___________ for butterflies and bees.
   a. protection c. shelter
   b. food d. fruit
2. ___________ are leaflike, colorful structures arranged in a circle around the top of a flower.
   a. Sepals c. Petals
   b. Stamens d. Anthers
3. By eating fruit, mammals help ___________.
   a. fertilize flowers c. photoperiodism
   b. nastic movement d. disperse seeds
4. While feeding, butterflies and bees carry pollen from flower to flower, causing ___________.
   a. pollination
   b. dormancy
   c. nastic movement
   d. photoperiodism

Chapter 27 Review

Reviewing Main Ideas

Summary statements can be used by students to review the major concepts of the chapter.

Key Terms

Answers should go beyond defining the terms. Accept any answer that uses the term correctly and in the proper context.

Understanding Concepts

1. b
2. c
3. d
4. a

Chapter 27 Review

5. a
6. b
7. c
8. d
9. c
10. a
11. a
12. d
13. c
14. a
15. b
16. c
17. d
18. a
19. b
20. a

Applying Concepts

21. Seeds store food intended for use by the embryo plant.
22. Seeds will remain alive until water becomes available for germination.
23. to ensure that each female flower receives enough pollen to bring about fertilization of all ovules (kernels)
24. Orchids and other epiphytes depend upon wind for seed dispersal. Producing many seeds guarantees that some will land in a place that is conducive to growth.

Chapter 27 Review

5. What are the leaflike, green structures encircling the flower stem beneath the petals?
   a. sepals c. petals
   b. stamens d. anthers
6. The male reproductive structure of a flower is called the ____________.
   a. sepal c. pistil
   b. stamen d. ovary
7. What is the female reproductive structure of the flower called?
   a. sepal c. pistil
   b. stamen d. anther
8. A(n) ____________ is one that has all four organs—sepals, petals, stamens, and pistils.
   a. short-day plant
   b. long-day plant
   c. incomplete flower
   d. complete flower
9. Flowers that are dull in color and have no nectar yet have a strong scent might be pollinated by ____________.
   a. bees c. beetles
   b. butterflies d. hummingbirds
10. ____________ results when a stigma receives pollen from the same flower, or from another flower on the same plant.
    a. Self-pollination
    b. Cross-pollination
    c. Germination
    d. Exhibition of a tropism
11. The response of flowering plants to the difference in the duration of light and dark periods in a day is called ____________.
    a. photoperiodism c. pollination
    b. nastic movement d. dormancy
12. After pollination, a pollen tube grows downward from the stigma through the style to the ____________.
    a. anther c. fruit
    b. pistil d. ovary
13. Flowering plants undergo ____________.
    a. spore formation
    b. asexual reproduction
    c. double fertilization
    d. single fertilization
14. What does a plant zygote develop into?
    a. a plant embryo c. a hormone
    b. an endosperm d. an ovary
15. After fertilization, the triploid cell becomes the ____________.
    a. plant embryo c. hormone
    b. endosperm d. ovary
16. A(n) ____________ consists of the seeds enclosed in the mature ovary of a flowering plant.
    a. plant embryo c. fruit
    b. endosperm d. pollen tube
17. The substances that regulate flower and seed development and the timing of fruit and leaf drop in the autumn are called ____________.
    a. enzymes c. plasma
    b. ions d. hormones
18. Auxins and gibberellins are plant hormones that stimulate ____________.
    a. growth c. fruit ripening
    b. seed germination d. tropisms
19. ____________ is a hormone that speeds the ripening of fruit.
    a. Abscisic acid c. Auxin
    b. Ethylene gas d. Gibberellin
20. The period of inactivity in a mature seed is known as ____________.
    a. dormancy c. resting
    b. germination d. maturity

Applying Concepts

21. You eat peas, beans, corn, peanuts, and cereals. Why are seeds a good source of food?
22. How does dormancy contribute to the survival of a plant species in a desert ecosystem?
23. Gardening books recommend that gardeners with small yards plant corn in blocks of at least 12 plants, spaced a foot or so apart. If fewer plants are grown, cobs do not develop full sets of kernels. What do you think is the basis for this advice?

Program Resources

Chapter Assessment, pp. 157-162 L1
Alternate Assessment in the Science Classroom
Computer Test Bank L1
Content Mastery, pp. 105-108 L1

24. Many orchids are epiphytes that live high above the ground on trees. Orchids produce millions of tiny seeds. What is the adaptive value to an epiphytic orchid of producing millions of seeds?
25. Explain why a scientist might hypothesize that the eating habits of herbivorous mammals affected the evolution of fruits in flowering plants.

Thinking Critically

26. ***Concept Mapping*** Make a concept map that relates the following terms and phrases. Supply the appropriate linking words for your map.
    pollen, ovule, anther, stigma, style, insect, ovary, sperm, pollen tube, double fertilization, embryo, triploid cell, zygote, endosperm, seed, fruit
27. ***Observing and Inferring*** One plant species produces heavy, spiky pollen grains. A second plant species produces light, smooth pollen. What conclusions can you draw about the method of pollination of each species?
28. ***Formulating Hypotheses*** Botanists determined the average number of pollen grains produced by each anther and the average number of ovules within each ovary of several different species of flowering plants. Before conducting their experiment, what hypothesis might the botanists have formed about the relationship between the numbers of pollen grains and ovules produced by any given species?
29. ***Designing an Experiment*** Explain how you would conduct an experiment to test whether or not auxin inhibits leaf abscission. Plan to collect quantitative data.
30. ***Formulating Hypotheses*** Form a hypothesis that explains why the primary root is the first part of the plant to emerge from a germinating seed.

ASSESSING KNOWLEDGE & SKILLS

The graph below provides data from an experiment that tests the effects of ionizing radiation on the germination of seeds.

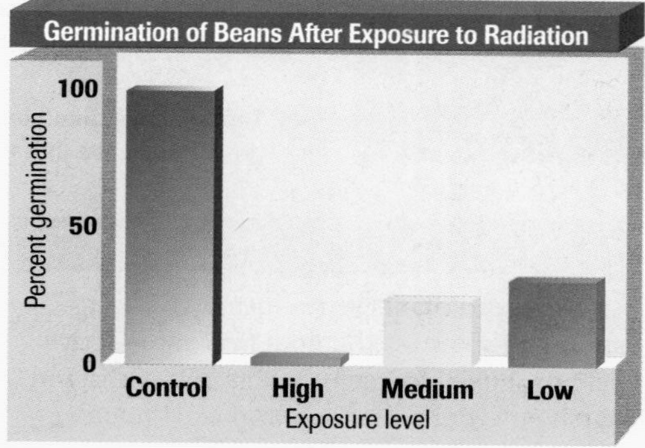

Using a Graph Use the graph to answer the following questions.

1. Which group of beans had the highest percentage of germination?
   a. control
   b. high exposure level
   c. medium exposure level
   d. low exposure level
2. As radiation dose increases, germination ____________.
   a. increases   c. stops
   b. decreases   d. is not affected
3. When beans are given a low dose of radiation, ____________ germinate.
   a. 25 percent   c. none
   b. 50 percent   d. 100 percent
4. When beans are given a medium dose of radiation, ____________ germinate.
   a. 12.5 percent   c. 37.5 percent
   b. 25 percent   d. 50 percent
5. ***Designing an Experiment*** Design an experiment on bean plants in which the following hypothesis is tested: bean plants exposed to ionizing radiation will not grow as tall as those that are not exposed.

Chapter 27 Review

25. Mammals may have selected only certain fruits for their diet and thus aided in dispersal and survival of these species.

Thinking Critically

26. See below.
27. Spiky pollen is carried by an insect or mammal pollinator, while light pollen is dispersed by the wind.
28. As the number of ovules increases in a species, so does the number of pollen grains produced.
29. Use a plant species that is ready to drop its leaves in the autumn. Apply auxin to one plant's leaves; apply water to the leaves of a control plant. Allow both plants to remain in identical environmental conditions and note and record the numbers of leaves that drop from the plants. Repeat the experiment several times.
30. The primary root anchors the seed and obtains needed water from the soil for further growth and development of the plant.

ASSESSING KNOWLEDGE & SKILLS

1. a
2. b
3. a
4. a
5. Bean plants that have just emerged from the soil are divided into four groups: one the control that is not irradiated, another given a high dose, another given a medium dose, and another a low dose of radiation. The heights of the plants are measured and recorded for the duration of the experiment.

26.

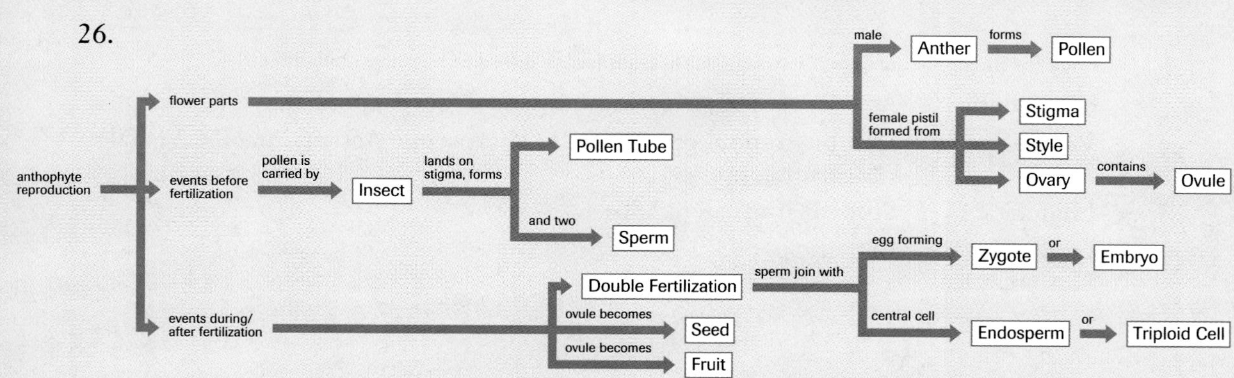

PREPARE

Purpose

This Biodigest can be used as an introduction to or as an overview of the structures and functions of plants. If time is limited, you may wish to use this unit summary to teach about plants in place of the chapters in the Plants unit.

Key Concepts

Students learn about the characteristics of the major plant divisions. They are introduced to the alternation of generations; the distinctions between non-seed and seed plants; pollination and fertilization; and the adaptive value of flowers, seeds, and fruits.

Block Scheduling

Look for this symbol for strategies that are useful in a block scheduling format. For more information on block scheduling, refer to Biodigest 1 in the **Lesson Plans** booklet.

1 Focus

Bellringer

Before presenting the lesson, display **Section Focus Master 59** on the overhead projector and have students answer the accompanying questions. L1 SAE

Demonstration

LS **Visual-Spatial** Bring a variety of live plants into the classroom, including mosses, horsetails, and club mosses, if available. Include a small potted pine or other conifer, and several potted flowering plants. Ask students to observe each plant and make a list of its characteristics in their journals. SAE

PLANTS

Earth is virtually covered with plants. Plants provide food and shelter for multitudes of organisms. They are one of Earth's most important groups of producers. Through the process of photosynthesis, they transform the radiant energy of sunlight into chemical energy in food and release oxygen to the atmosphere. All plants are multicellular eukaryotes. Plant cells are surrounded by a cell wall made of cellulose and coated with a waterproof cuticle. Chlorophyll is contained in cell organelles called chloroplasts.

Though individual moss plants are small, they usually grow in masses that form carpets of green.

NON-SEED PLANTS

Non-seed plants reproduce by forming spores. A spore is a haploid (*n*) reproductive cell, produced by meiosis, that can withstand harsh environmental conditions. When conditions become favorable, a spore can develop into the haploid, gametophyte generation of a plant. A spore will become either a female or male gametophyte.

Mosses and Liverworts

Mosses and liverworts are non-seed, nonvascular plants in the division Bryophyta that live in cool, moist habitats. Because they have no vascular tissues to move water and nutrients from one part of the plant to another, they cannot grow more than a few inches tall.

VITAL STATISTICS
NON-SEED PLANTS

Numbers of species:
Bryophyta—mosses and liverworts, 20 000 species
Lycophyta—club mosses, 1000 species
Sphenophyta—horsetails, 15 species
Pterophyta—ferns, 12 000 species

FOCUS ON ADAPTATIONS

Alternation of Generations

The life cycle of most plant species alternates between two stages, or generations. The sporophyte generation produces spores, which develop into the gametophyte generation. The gametophyte produces gametes. In nonvascular plants, the gametophyte is larger and more conspicuous than the sporophyte. In vascular plants, the sporophyte dominates. The gametophyte of a vascular plant is extremely small and may remain buried in the soil or inside the body of the sporophyte.

Gametophyte Generation A gametophyte is haploid (*n*) and produces eggs and sperm. In mosses, the gametophyte is the familiar soft, green growth that covers rotting logs or moist soil. The tiny moss gametophytes produce male and female branches. Sperm cells produced by the male branches must swim through rain or dew to reach the egg cells produced by the female branches. Fertilization takes place inside the female reproductive organ and a diploid (2*n*) zygote is produced.

682

Learning Styles

Look for the following logo for strategies that emphasize different learning modalities. LS

Kinesthetic	**Activity, p. 683; Meeting Individual Needs, p. 683**
Visual-Spatial	**Demonstration, pp. 682, 683; Microscope Activity, p. 683; Visual Learning, pp. 685, 686**
Linguistic	**Student Journal, p. 686**
Logical-Mathematical	**Activity, p. 685**

As in all vascular plants, the sporophyte of this club moss is the dominant generation. Spores develop at the base of special leaves that form cone-shaped structures called strobili.

Club Mosses

Club mosses are non-seed plants in the division Lycophyta. They possess vascular tissue and are found primarily in moist environments. Species that exist today are only a few centimeters high, but they are otherwise similar to fossil Lycophytes that grew as high as 30 m and formed a large part of the vegetation of Paleozoic forests.

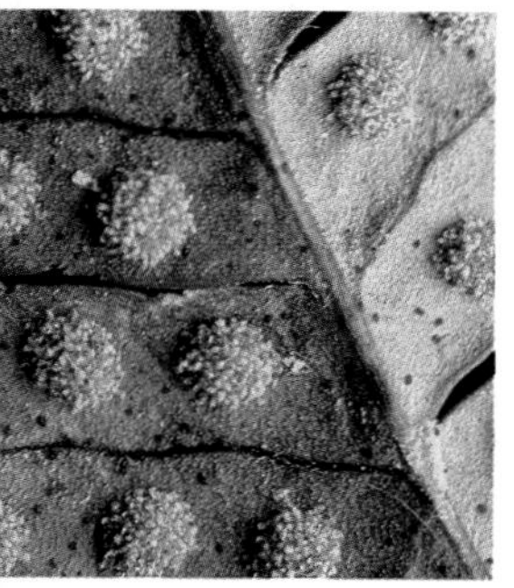

Fern spores develop in clustered structures called sori, usually found on the undersides of fronds. The gametophyte generation of the fern is small and disappears as the larger, more familiar sporophyte develops.

Horsetails

Horsetails are non-seed vascular plants in the division Sphenophyta. They are commonly found growing in areas with damp soil, such as stream banks and sometimes along roadsides. Present-day horsetails are small, but their ancestors were treelike.

The hollow, jointed stems of horsetails are surrounded by whorls of scale-like leaves.

Ferns

Ferns, division Pterophyta, are the most well-known and diverse of the non-seed vascular plants. They have leaves called fronds that grow up from an underground stem called the rhizome. Ferns are found in many different habitats, including shady forests, stream banks, roadsides, and abandoned pastures.

Sporophyte Generation The zygote develops into an embryo, which grows into the sporophyte generation of the moss. The sporophytes grow out of the tip of the female branches of the gametophyte and consist of capsule-topped stalks. Cells inside each capsule undergo meiosis to form haploid (n) spores.

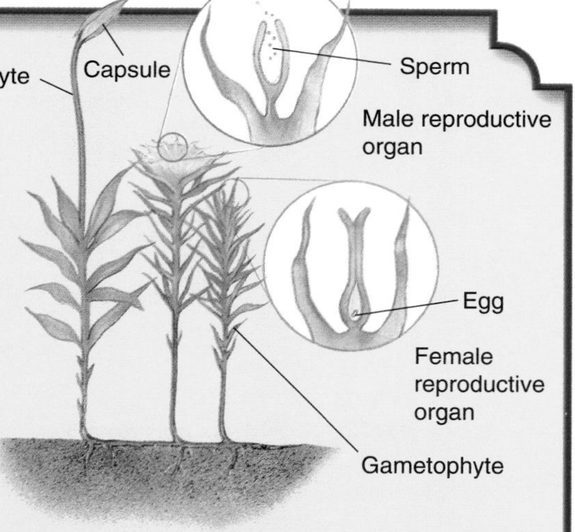

The green, leafy growth of this moss is the gametophyte. The brown stalks topped with spore-filled capsules are the sporophyte.

683

Meeting Individual Needs

Visually Impaired As part of the Focus demonstration activity, have students with visual impairments handle and smell the plants. If necessary, break off small branches or flowers for students to handle. Ask them to describe what they feel as partners write down their observations. Make sure that students cannot injure themselves by handling the plants. LS

BIODIGEST

2 Teach

Microscope Activity

LS **Visual-Spatial** Have students view moss plants with a stereomicroscope. Ask students to describe their observations orally or in writing in their journals. If possible, supply plants with male and female reproductive structures and sporophytes and ask students to compare the characteristics of these structures.

Activity

LS **Kinesthetic** If horsetails grow in your area, bring several stems into the classroom. Show students that the stems come apart fairly easily at the nodes, and that the stems are hollow. Have students examine the spore-containing strobili, if these are present. Point out that the silica in the stems gives horsetails a scouring-pad quality and is the source of the common name "scouring rush." L1 SAE

Demonstration

LS **Visual-Spatial** Show students young fiddleheads of a fern and the sori on the underside of a fern frond. In the spring, fiddleheads are sometimes available in specialty markets as a food item. If a live fern is available, show students the underground stem (rhizome) from which roots grow down and fronds grow up. SAE

Demonstration

Bring to class and allow students to handle several varieties of conifer cones. Include male cones if available. Tell students the common name and/or Latin name of the tree each cone is from. Ask students to record their observations of each species in their journals. L1 SAE

Guest Speaker

Have a pharmacist or physician visit the class to discuss plants that are important in the production of medicines.

SEED PLANTS

A seed is a reproductive structure that contains a sporophyte embryo and a food supply enclosed in a protective coating. The food supply nourishes the young plant during the first stages of growth. Like spores, seeds can survive harsh conditions. The seed develops into the sporophyte generation of the plant. Seed plants include conifers and flowering plants.

Conifers

Conifers, division Coniferophyta, produce seeds, usually in woody strobili called cones, and have needle-shaped or scale-like leaves. Conifer seeds are not enclosed in a fruit. Most conifers are evergreen plants, which means they bear leaves all year round.

Seeds of conifers develop at the base of each woody scale of female cones.

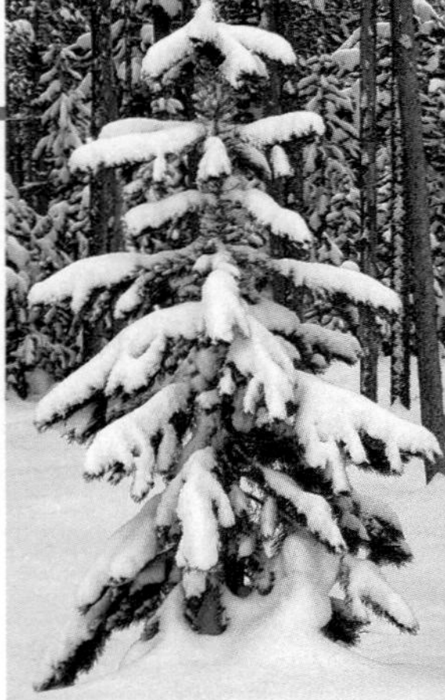

The leaves and branches of conifers are flexible. They bend under the weight of snow and ice, allowing any buildup to slide off before it becomes heavy enough to break the branch.

Adapted for Cold Climates

Conifers are a major plant feature in cold or dry habitats throughout the world. Conifer needles have a compact shape and a thick, waxy covering that helps reduce evaporation and conserve water. Conifer stems are covered with a thick layer of bark that insulates the tissues inside. These adaptations enable conifers to carry on life processes even when temperatures are below freezing.

VITAL STATISTICS CONIFERS

Examples: pine, spruce, fir, larch, yew, redwood, juniper
Numbers: 400 species
Size range: Giant sequoias of central California grow to 99 m tall, and are the most massive organisms in the world; coast redwoods of California grow to 117 m and are the tallest trees in the world.
Distribution: Worldwide; dominant plants of the taiga

FOCUS ON ADAPTATIONS

Moving from Water to Land

All plants probably evolved from filamentous green algae that lived in the nutrient-rich waters of Earth's ancient oceans. An ocean-dwelling alga can absorb water and dissolved minerals directly into its cells. As land plants evolved, new structures developed for absorbing and transporting water and minerals from the soil to all the aerial parts of the plant.

Nonvascular Plants In nonvascular plants, water and nutrients must travel from one cell to another by the relatively slow processes of osmosis and diffusion. As a result, nonvascular plants are limited to environments where plenty of water is available.

interNET CONNECTION

Follow the link for this Biodigest on the Glencoe Homepage at **www.glencoe.com/sec/science** to find out more about plants.

FLOWERING PLANTS

The flowering plants, division Anthophyta, form the largest and most diverse group of plants on Earth today. They provide much of the plant food eaten by humans. Anthophytes produce flowers and develop seeds enclosed in a fruit.

Monocots and Dicots

The Anthophytes are classified into two classes: the monocotyledons and the dicotyledons. Cotyledons, or "seed leaves," are food storage organs contained in the seed along with the plant embryo. Monocots have a single seed leaf. Dicots have two seed leaves.

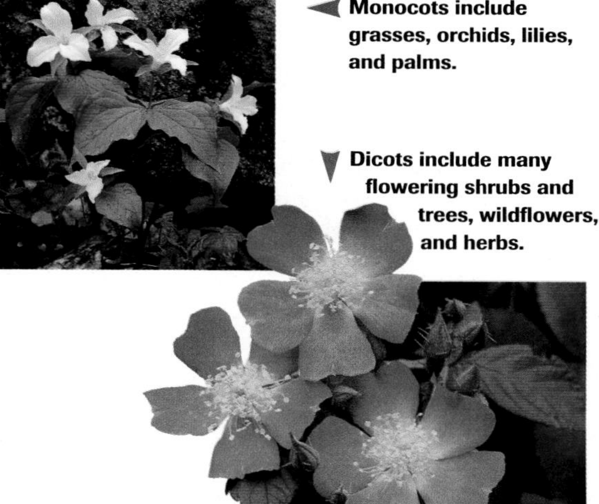

Monocots include grasses, orchids, lilies, and palms.

Dicots include many flowering shrubs and trees, wildflowers, and herbs.

Flowers

Flowers are the organs of reproduction in anthophytes. Sepals enclose the flower bud and protect it until it opens, and petals, which are often brightly colored or perfumed, attract pollinators. Inside the circle of petals are the pistil and stamens.

The pistil is the female reproductive organ. Inside the ovary at the base of the pistil are the ovules. Ovules are the female gametophyte generation of the plant. Female gametes—egg cells—form in each ovule.

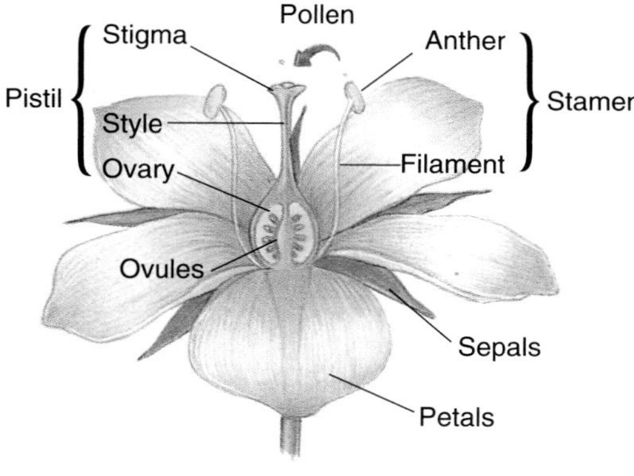

The stamen is the male reproductive organ of a flower. Pollen grains containing male gametes form inside the anther.

Vascular Plants The stems of most plants contain vascular tissues made up of tube-like, elongated cells through which water, food, and other materials move from one part of the plant to another. Because vascular tissues include thickened fibers that support upright growth, vascular plants can grow much larger than nonvascular plants.

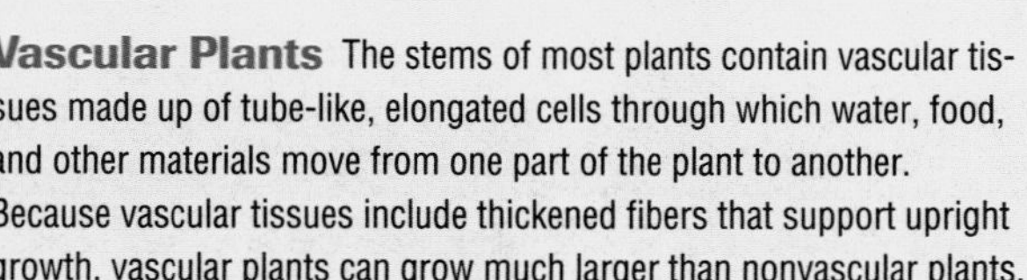

An unbroken column of water travels from the roots in xylem tissues. Sugars formed by photosynthesis travel around the plant in phloem tissues.

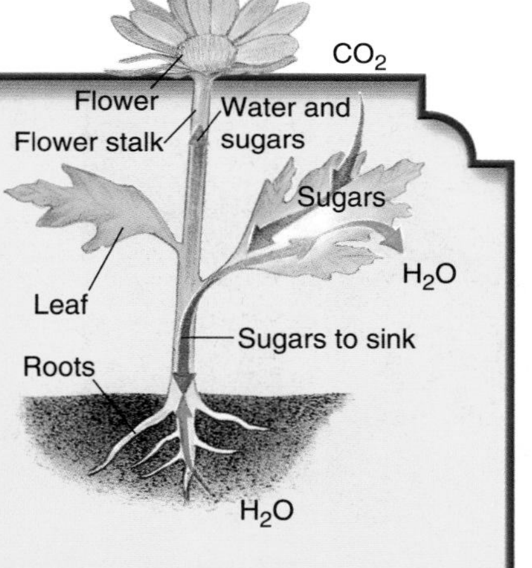

Activity

Logical-Mathematical Have students collect branches or leaves of local trees and use field guides to identify them. The lesson can be extended by trading botanical specimens with teachers in other areas. L1

Visual Learning

Visual-Spatial Explain to students that monocots have leaves with parallel veins and flower parts in multiples of three. Dicots have leaves with netted veins and flower parts in multiples of four or five. Have students use these characteristics to classify photographs or cut flowers brought into the classroom. Extend the lesson by having students discuss the sizes and shapes of the flowers and their adaptations for pollination. L1 SAE

BIODIGEST

Brainstorming

- Ask students to compare and contrast pollination and fertilization in conifers and flowering plants. *In conifers, pollen is carried by wind to the female cone. In flowers, pollen may also be transported by animals. In flowers, the pollen is carried to the stigma. In both conifers and flowers, the pollen grows a tube through which the sperm travels to reach the ovary. In both, sperm and egg unite to form a zygote that develops into a seed.*
- After students have observed flowers and fruits from the same plants, ask them to discuss which parts of the flower become which parts of the fruit, and to identify the parts of the flower that wither away and do not form part of the fruit. *The ovary swells to become the fruit. The ovule may be visible around the seeds, as in apples. The sepals may be visible on the blossom end of the fruit. Petals, pistil, and stamen wither away.*

Visual Learning

LS **Visual-Spatial** Bring into the classroom several different fruits, including both dry and fleshy varieties. If possible, bring fruits that develop from some of the flowers used in the above activity. Have students examine the exterior of each fruit, then cut it open to observe the arrangement of seeds. Ask students to describe their observations in their journals. L1 SAE

Pollen

In seed plants, the sperm are enclosed in the thick-coated pollen grains, which are the male gametophyte generation of the plant. Pollen is one of the important adaptations that has enabled seed-bearing plants to live in a wide variety of land habitats.

Pollinators

Flowers can be pollinated by wind, insects, birds, and even bats. Some flowers have colorful or perfumed petals that attract pollinators. Flowers may also contain sweet nectar, as well as pollen, which provides pollinators with food.

Plants that depend on the wind to carry pollen from anther to stigma tend to have small, inconspicuous flowers. The flowers of grasses and this alder are pollinated by the wind.

Plants that depend on insects for pollination may be brightly colored and fragrant. Pollen rubs off on a bee that visits a flower to feed on nectar. When it moves to another flower, some of the pollen may rub off onto the stigma.

Pollen tube
Ovary
Sperm nuclei
Ovule
Egg nucleus and endosperm nucleus

Pollen is carried by wind or animals to the stigma of a flower. The uneven surface of the stigma traps the pollen grain, which begins to grow a tube down the style to the ovary. Two sperm travel down the tube to fertilize the eggs in the ovule. Seeds begin to develop.

Fruit

Following fertilization, the ovary develops into a fruit with seeds inside. Some flowering plants develop fleshy fruits, such as apples, melons, tomatoes, and squash. Other flowering plants develop dry fruits, such as peanuts, almonds, or sunflowers. Fruits help protect seeds until they are mature. Fruits also help scatter seeds into new habitats.

The fruit of the coconut palm floats. It can be transported from one beach to another by ocean currents.

VITAL STATISTICS
FLOWERING PLANTS

Examples: grasses, oaks, maples, palms, irises, orchids, roses, beans
Numbers: 230 000 species (60 000 monocots; 170 000 dicots)
Size range: a few millimeters to 75 m
Distribution: worldwide; most abundant plant group

STUDENT JOURNAL

LS **Linguistic** Tell students a hypothetical story involving the development of a local open space. They are to write letters to the city council either supporting the development or calling for conservation of the area. Tell students they must include reasons for their views in their letters. L1

Maple trees produce fruits with a winglike shape that can be carried long distances by the breeze.

Some plants produce seeds with dry fruits with tiny hooks or barbs that attach to the fur of passing animals. The seeds may be transported long distances.

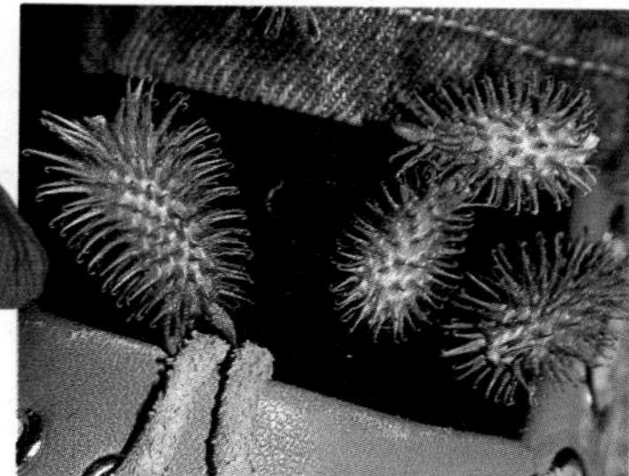

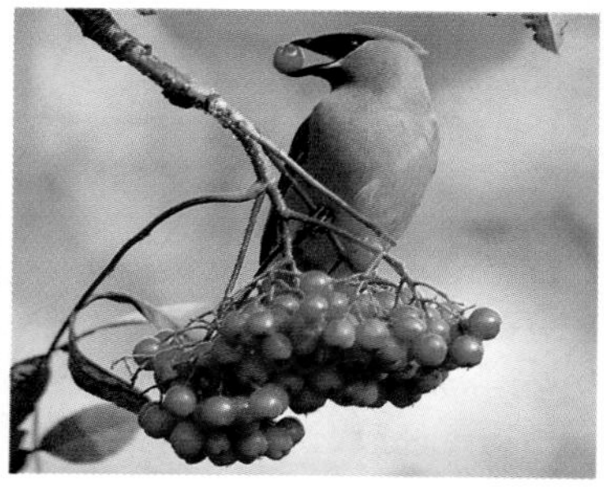

Many plants produce fruits that are eaten by animals.

BIODIGEST REVIEW

Understanding Concepts

1. Which of the following is a Bryophyte?
   a. moss
   b. horsetail
   c. club moss
   d. conifer
2. The term for a mature fern leaf is ______.
   a. leaf
   b. scale
   c. frond
   d. needle
3. Nonvascular plants would most likely be found growing _______.
   a. in sandy desert soil
   b. on an ocean beach
   c. on a snowy mountain slope
   d. on a rotting log on the forest floor
4. Which plant group has leaves adapted for life in cold environments?
   a. Anthophyta
   b. Sphenophyta
   c. Pterophyta
   d. Coniferophyta
5. Dicots have _______.
   a. naked seeds in cones
   b. flowers and seeds enclosed in a fruit
   c. needle-like leaves
   d. spores borne in cone-shaped strobili
6. Reproductive structures of conifers are _______.
   a. flowers
   b. cones
   c. fruits
   d. sori
7. Mosses, ferns, and club mosses are alike because they require ______.
   a. water for fertilization
   b. adaptations for conserving water
   c. insects for pollination
   d. warm, sunny habitats
8. Lycophytes, sphenophytes, and conifers have specialized leaves that form reproductive structures known as ______.
   a. sori
   b. flowers
   c. cones
   d. strobili
9. Vascular plants do not include the _____.
   a. Lycophytes
   b. Bryophytes
   c. Sphenophytes
   d. Pterophytes
10. Which plant group produces flowers and seeds enclosed in a fruit?
    a. Anthophyta
    b. Coniferophyta
    c. Pterophyta
    d. Lycophyta

Thinking Critically

1. Compare the spore-bearing structures of ferns with the seed-bearing structures of conifers.
2. Compare and contrast spores and seeds.
3. Describe three ways in which seeds may be dispersed.

BIODIGEST

Answers to BioDigest Review

Understanding Concepts

1. a
2. c
3. d
4. d
5. b
6. b
7. a
8. d
9. b
10. a

Thinking Critically

1. Ferns produce spores in sori on the underside or edges of fronds. Conifers develop seeds at the base of woody, scale-like leaves that form cones.
2. A spore is a haploid reproductive cell that grows into a gametophyte. A seed contains a diploid plant embryo and food supply, and grows into a sporophyte.
3. Seeds may be blown away by the wind. They may be carried by water when they fall in the ocean or in streams. They may also be transported by animals who eat the fruits and discard the seeds, or by animals who pass seeds through their digestive systems and deposit them with their droppings.

Program Resources

Section Focus Master 59 L1 SAE

Reinforcement and Study Guide, pp. 109-110 L1

Chapter Assessment, pp. 163-166 L1

Content Mastery, pp. 109-112 L1

UNIT 8 Invertebrates

Unit Overview

In this unit, students become familiar with invertebrates. Chapter 28 introduces the general characteristics of animals as well as their body plans.

In Chapter 29, students begin their examination of specific invertebrate groups through the study of the structure, adaptations, ecology, and phylogeny of sponges, cnidarians, flatworms, and roundworms. In Chapter 30, students examine the characteristics, ecology, and phylogeny of mollusks and segmented worms.

In Chapter 31, students explore the single largest group of animals—the arthropods. In this chapter, the diversity of arthropods is examined, while features that enabled many arthropods to become land dwellers are explained. The unit concludes in Chapter 32, with a presentation of echinoderms and invertebrate chordates.

Theme Development

The themes of *evolution* and *unity within diversity* provide the essential framework of this unit. Students study the adaptations of invertebrates as they discover how these organisms have changed over time.

UNIT 8 Invertebrates

Sand-colored starfish inch along on tube-shaped feet over the barnacle- and mussel-encrusted rocks. Tiny sand fleas hop among scattered patches of salt-encrusted seaweed.

A tidal pool along the edge of a beach is an excellent place for an exploration of different kinds of life. That's because tide pools and the oceans that sustain them are home to many of the world's invertebrates, the most common and diverse forms of life on Earth. They live in almost every habitat imaginable—from the bottom of the deepest oceans to the tops of the highest mountains. Thus, it's not surprising that invertebrates comprise about 95 percent of all animal species.

Mussels and many other invertebrate species remain in one place during their adult lives. How do these animals obtain food?

Advance Planning

Chapter 28

- Obtain live animals such as a mouse, earthworm, frog, hermit crab, sea urchin, cricket, goldfish, lizard, planarian, hydra, snail, and crustacean such as *Daphnia* for the Biolab. Have suitable containers and food for each animal made available.
- Order nematode cultures and earthworms for the Minilab on page 704.

Chapter 29

- Purchase hydra for the Minilab on page 718.
- Order *Planaria* cultures and pond water for the Biolab.
- Obtain garden soil for the Minilab on page 731.

Chapter 30

- Order live clams and prepare the carmine solution for the Minilab on page 742. Gather dichotomous keys for the Minilab on page 743.
- Obtain live earthworms for the Biolab.

Chapter 31

- Gather examples of local insects for the Biolab.

Chapter 32

- Order prepared slides of sea urchin development and *Amphioxus* for the Biolab.
- Order prepared slides of *Amphioxus* for the Minilab on page 796.

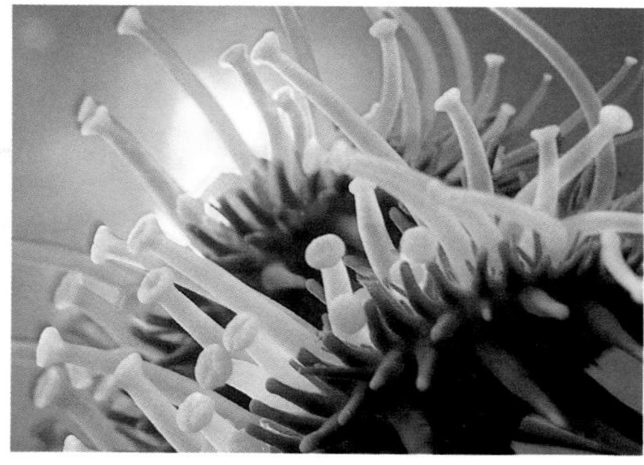

Tube feet and spiny skins are common characteristics of starfishes and many of their relatives. What other features make spiny-skinned animals unique?

The hard, outer skeleton of this fiddler crab protects it from predators. What advantages and problems do exoskeletons present for arthropods, the most diverse of the invertebrate groups?

Unit Contents

UNIT 8

Introducing the Unit

Ask students to examine the photos of the mussels, starfish, and fiddler crab. Have the class brainstorm to create lists of how these organisms are alike and how they are different. Try to lead students to include characteristics of each organism that show its adaptation to its environment.

Have students speculate as to how the presence or absence of a backbone may affect an organism. Then review the names of the animal groups that will be studied as students explore the traits and diversity of invertebrates.

Motivational Activity

Activity: Ask students to formulate a plan for setting up a marine aquarium. They should do research about marine aquariums using library resources, visiting pet stores, or viewing public aquariums. Ask them to include in their plans descriptions of the care and maintenance that will be needed for the aquarium on a daily basis. L1 COOP LEARN

Unit Project

Community Involvement: Ask students to select an area in the community where there may be a problem with either declining invertebrate populations or increasing invertebrate populations. Students may wish to consider a stretch of marsh that is scheduled for drainage, a coastline area where tourists take echinoderms from tidepools, or a lake being overpopulated by a particular invertebrate due to pollution problems.

After students choose a problem on which to focus, ask them to research the background of the problem and the animals involved. Have them brainstorm possible solutions to the problem and write to city officials to find out what is being done about the problem. L1 COOP LEARN

Chapter Organizer

SECTION	OBJECTIVES	ACTIVITIES/FEATURES
28.1 Typical Animal Characteristics National Science Standards: UCP.1, UCP.2, UCP.5; A.1, A.2; C.1, C.5; G.3	1. **Compare** the characteristics of animals. 2. **Sequence** the development of a typical animal.	**Minilab:** Where does this animal live?, p. 693 **Math Connection:** Museums: More than Stuffed Animals, p. 695 **Biolab:** Design Your Own Experiment—What is an animal?, p. 696 **Thinking Lab:** How does mesoderm influence development of nervous tissue?, p. 700
28.2 Body Plans and Adaptations National Science Standards: UCP.1, UCP.2, UCP.5; C.5; E.1, E.2; F.6; G.1-3	3. **Distinguish** among the body plans of different classes of animals. 4. **Trace** the phylogeny of animal body plans. 5. **Compare** body plans of acoelomate, pseudocoelomate, and coelomate animals.	**Minilab:** How do body plans affect locomotion?, p. 704 **A Broader View:** The First Landlubbers, p. 705

ACTIVITY MATERIALS

BIOLAB	MINILABS	ALTERNATE LAB
page 696 microscope depression slides metric ruler live animals (mouse, frog, earthworm, goldfish, etc.) containers for animals (cages, pans, etc.)	**page 693** "gooey grabbers" or "sticky whippers" meterstick beakers ice **page 704** earthworms hand lens or binocular microscope microscopic nematodes dissecting needles	**page 702** microscope prepared slides of cross sections of: planarian nematode earthworm hydra

Chapter 28 What Is an Animal?

TEACHER CLASSROOM RESOURCES

Reproducible Masters	Transparencies
Section Focus Master 60: Animal Characteristics L1 SAE **Reinforcement and Study Guide,** pp. 111-112 L1 **Biolab and Minilab Worksheets,** pp. 111, 113-114 L1 **Critical Thinking/Problem Solving:** Control of Cell Differentiation During Development, p. 28 L3 **Laboratory Manual:** Symmetry, pp. 159-162 L2	**Basic Concepts Transparency #44:** Animal Development L1 SAE **Reteaching Transparency #28:** Stages of Development in an Animal Egg L1 SAE
Section Focus Master 61: Symmetry L1 SAE **Reinforcement and Study Guide,** pp. 113-114 L1 **Biolab and Minilab Worksheets,** p. 112 L1 **Concept Mapping:** Body Structure of Animals with Bilateral Symmetry, p. 28 L1 **Content Mastery,** pp. 113-116 L1	

ASSESSMENT MATERIALS

Chapter Assessment, pp. 167-172
Alternate Assessment in the Science Classroom
MindJogger Videoquiz
Computer Test Bank

Spanish Resources SAE
English/Spanish Audiocassettes SAE
Cooperative Learning in the Science Classroom COOP LEARN
Lesson Plans
Great Developments in Biology: Discovering the Living World L1 SAE

KEY TO TEACHING STRATEGIES

- L1 Level 1 activities should be within the ability range of all students including those with learning difficulties.
- L2 Level 2 activities are within the ability range of average to above-average students.
- L3 Level 3 activities are designed for the ability range of above-average students.
- SAE SAE activities should be within the ability range of Students Acquiring English.
- COOP LEARN Cooperative Learning activities are designed for small group work.
- P These strategies represent student products that can be placed into a best-work portfolio.
- These strategies are useful in a block scheduling format.

GLENCOE TECHNOLOGY

The following multimedia resources are available from Glencoe.

Biology: The Dynamics of Life
- **CD-ROM** SAE
- **Videodisc Program**

The Infinite Voyage Series
The Keepers of Eden
- The Transformation of Zoos
- Extinction and the National Zoo's Tamarin
- Naturalist Gerald Durrell and His Special Zoo
- Preserves of Endangered Species: San Diego and Kenya

Secrets from a Frozen World
- The Southern Ocean—A Rich Marine Ecosystem

The Secret of Life Series
Sex and the Single Gene: Cell Development

Science and Technology Videodisc Series (STVS)
Animals
- Rattlesnakes

CHAPTER

28 What Is an Animal?

Chapter Overview

Chapter 28 provides students with a general overview of the characteristics shared by organisms classified in the Animal Kingdom. Students begin this chapter by studying animal cell types and ways animals obtain food, move, and carry out digestion. Next, students learn about development of animals. As part of the discussion of development, a comparison is drawn between protostomes and deuterostomes.

In the second section, students explore the body plans and symmetry that characterize animals of different classes. The phylogeny resulting in different body plans is also explored. The chapter concludes with a discussion of the nature of body cavities and animal adaptations related to protection and support.

Key Terms

acoelomate
anterior
asymmetry
bilateral symmetry
blastula
coelom
deuterostome
dorsal
ectoderm
endoderm
endoskeleton
exoskeleton
gastrula
invertebrate
mesoderm
posterior
protostome
pseudocoelom
radial symmetry
sessile
symmetry
ventral
vertebrate

690

Look for the following logo for strategies that emphasize different learning modalities. LS

Kinesthetic	**Meeting Individual Needs, pp. 693, 701; Making a Model, p. 699**
Visual-Spatial	**Activity, p. 693; Microscope Activity, p. 694; Demonstration, p. 698; Meeting Individual Needs, p. 698; Portfolio, p. 699; Alternate Lab, p. 702; Reinforcement, p. 703; Minilab, p. 704**
Interpersonal	**Project, p. 694; Reinforcement, p. 702**
Intrapersonal	**Minilab, p. 693**
Linguistic	**Student Journal, p. 704**
Logical-Mathematical	**Thinking Lab, p. 700**

CHAPTER

28 What Is an Animal?

CHAPTER

28

Dusk falls and soon shadows flit quickly by streetlights. Insects attracted to light become food for bats as they steer a zigzag course across the nighttime sky. Flying silently on leathery wings as they scoop up their insect dinners, bats avoid obstacles by using sounds that humans cannot hear.

On the ground, night spells sleep for a chipmunk. These striped rodents work hard all day, collecting seeds and berries and storing them in hidden burrows. By evening, the chipmunks settle down underground for the night, safe from predators like the owl that waits to snatch up their nocturnal mouse cousins.

Supported by warm currents, a jellyfish floats along, tentacles trailing behind. A small fish brushes past one tentacle; suddenly it is caught, immobilized, and pulled up toward the jellyfish's mouth. Even jellyfish must find and capture food in order to eat.

Concept Check

You may wish to review the following concepts before studying this chapter.

- Chapter 1: adaptation
- Chapter 8: cells
- Chapter 18: evolution

Chapter Preview

Laboratory Activities

Biolab: Design Your Own Experiment

- What is an animal?

Minilabs

- Where does this animal live?
- How do body plans affect locomotion?

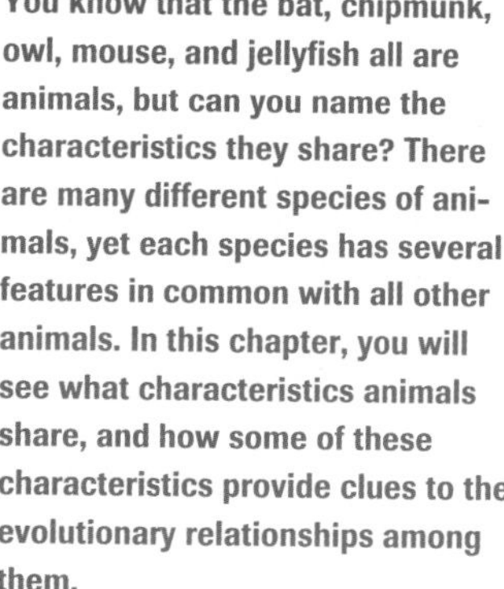

You know that the bat, chipmunk, owl, mouse, and jellyfish all are animals, but can you name the characteristics they share? There are many different species of animals, yet each species has several features in common with all other animals. In this chapter, you will see what characteristics animals share, and how some of these characteristics provide clues to the evolutionary relationships among them.

691

Introducing the Chapter

Ask students to examine the photographs and make lists of the features of the bat, chipmunk, and jellyfish. Review the lists as a class, focusing on the similarities among the organisms. As part of this review, bring out that all of the organisms are multicellular, capable of locomotion, and obtain their energy by breaking down organic molecules gained by eating other organisms.

Once the similarities of the organisms have been established, focus on the differences among them. Use this exercise as the basis of a discussion of diversity in animals. L1 COOP LEARN

Theme Development

This chapter stresses the themes of *unity within diversity* and *evolution*. Students will gain the understanding that within the tremendous variety of animals, there are common threads of shared features. They will also learn how animal characteristics and adaptations have evolved.

Concept Check

Students will bring their understanding of taxonomy, the kingdoms of living things, and evolution to the study of this chapter. In addition, students will need to recall the common features of cells and how plant and animal cells differ.

Assessment Planner

Choose assessment strategies from the following pages to evaluate the progress of your students.

Assess, pp. 699, 706
Alternate Lab, pp. 702-703
Minilabs, pp. 693, 704
Portfolio, p. 699
Thinking Lab, p. 700
Biolab, pp. 696-697
Chapter Review, pp. 707-709

SECTION

28.1

Prepare

Key Concepts

Students will study typical animal characteristics and the development of animals from eggs. They will compare and contrast animal features and the differences in types of development.

Block Scheduling

Look for this symbol for strategies that are useful in a block scheduling format. For more information on block scheduling, refer to Section 28.1 in the **Lesson Plans** booklet.

Materials

- For the Focus Activity, obtain several live animals from different phyla and clear plastic containers.
- For the Minilab, gather metersticks, "gooey-grabbers" (also called "sticky whippers") beakers or plastic cups, and ice.
- For the Biolab, gather microscopes, depression slides, metric rulers, a variety of live animals such as a mouse, earthworm, frog, hermit crab, cricket, hydra, goldfish, snail, or others, and suitable containers and food for the animals.

1 Focus

Bellringer

Before presenting the lesson, display **Section Focus Master 60** on the overhead projector and have students answer the accompanying questions. L1 SAE

SECTION

28.1 Typical Animal Characteristics

Section Preview

Objectives

Compare the characteristics of animals.

Sequence the development of a typical animal.

Key Terms

sessile
blastula
gastrula
ectoderm
endoderm
protostome
deuterostome
mesoderm

When you hear the word animal, *do you picture an organism with hair or fur and a bony skeleton? More than 1.5 million species of animals have been described, yet 95 percent of these have neither bones nor hair.*

If you saw the organism in the photograph for the first time, would you classify it as an animal? This organism is a sponge, an animal that remains attached to rocks or coral reefs in the ocean for all of its adult life. It doesn't move around in search of food. It doesn't have a bony skeleton or hair. It doesn't even have a mouth, a stomach, or intestines, yet it is still an animal.

Characteristics of Animals

All animals have several characteristics in common. Animals are multicellular organisms that feed on other organisms and have ways of moving that enable them to obtain food. Once they have consumed their food, they break it down for use as energy or as raw material for building tissue. Unlike plants, animals are composed of cells that do not have cell walls.

Methods for obtaining food vary

Examine the animals shown in *Figure 28.1.* One characteristic common to all animals is that they are heterotrophic, meaning they obtain energy and nutrients from outside sources. Animals such as lizards and birds can move from place to place in an active search for food. Other animals such as barnacles remain stationary and are adapted to draw food toward them.

Figure 28.1
The lizard, hummingbird, and barnacle each get their food in different ways.

A lizard captures flies with its long, sticky tongue.

A hummingbird flies to flowers to drink nectar.

A barnacle extends bristles from its shell to catch small organisms as they drift by in the water.

Program Resources

Section Focus Master 60 L1 SAE
Reinforcement and Study Guide, pp. 111-112 L1
Biolab and Minilab Worksheets, pp. 111, 113-114 L1
Laboratory Manual, pp. 159-162 L2
Critical Thinking/Problem Solving, p. 28 L3
Basic Concepts Transparency 44 and **Master** L1 SAE
Reteaching Transparency 28 and **Master** L1 SAE

An animal's ability to move about is directly related to its method of obtaining food. Although the heterotrophic way of life and the ability to move are characteristics shared by animals, there are many different methods of locomotion, as shown in *Figure 28.2.* The fish, osprey, and sidewinder snake are examples of vertebrates, animals with backbones, whereas the starfish is an invertebrate, an animal without a backbone. Yet all these animals have to find food to survive. As you can see, methods of locomotion have been refined and elaborated over the course of evolution.

MiniLab

Where does this animal live?

Assume you are in charge of a section of a natural history museum where people occasionally bring organisms to be identified. Someone has just brought in an unfamiliar organism. You are interested in characterizing the habitat in which it lives. So far, you have discovered that the ability to stretch and the ability to stick are important characteristics of this organism.

Procedure

1. Attach your organism to your tabletop. Place a meterstick adjacent to it, and stretch the organism along the length of the meterstick until its body detaches from the surface of the table. Record your measurement in a table.
2. Place the organism in a beaker set in ice for 5 minutes. Next follow the procedure for step 1.
3. Place the organism in a beaker, then in a warm water bath for 5 minutes, then follow the procedure for step 1.
4. Wet the organism in water at room temperature, then follow the procedure for step 1.

Analysis

1. Assume that the organism's stickiness and elasticity are important adaptations to its way of life. Under which conditions was it able to stretch most?
2. Describe a habitat in which this organism might thrive.

Figure 28.2

Animals move in a remarkable variety of ways.

A fish uses its fins and powerful tail muscles to swim after smaller fish or other prey.

The soaring osprey dives to snatch a fish from the waters of a lake or stream.

A sidewinder rattlesnake barely touches the ground as it follows the trail of a mouse.

A starfish moves slowly along the ocean floor using a unique system of tube feet that act like tiny suction cups.

SECTION 28.1

Activity

LS Visual-Spatial Ask students to observe a live earthworm, a hamster, and a cricket or grasshopper in clear plastic containers. Ask them to list ways in which these organisms are both alike and different.

2 Teach

MiniLab

Purpose

LS Intrapersonal Students will investigate the characteristics of an unknown "animal" to try to predict its habitat.

Process Skills

observe and infer, measure, compare and contrast

Teaching Strategies

- Purchase "gooey grabbers" from gumball machines or order them from A + A Co./ Parkway Machines, 1-800-638-6000. They are made of a type of plastic that is stretchy and feels sticky to the touch.

Expected Results

Students will find that the "organisms" stretch more when they are heated than when they are cool or at room temperature. They can be stretched farther when wet.

Analysis

1. It stretched more when it was warm and wet.
2. Habitat: warm and wet

Assessment

Portfolio: Ask students to summarize what they have learned in this lab about identifying unknown organisms. Have them put this information in their portfolios. L1 P

Meeting Individual Needs

Visually Impaired Provide students who are visually impaired with dried specimens of a variety of invertebrate animals such as sponges, insects, and starfish. Students, using their sense of touch, will be able to determine the shapes and textures of these animals. L1 LS

Bioethics

Preservation of Animals

Show students a piece of coral jewelry or other ornament made from coral. Explain that coral reefs are highly diverse and fragile ecosystems. Point out that when a reef is destroyed, thousands of plant and animal species lose their habitat and may become threatened. Also explain that coral reefs protect some coastlines from storms. Discuss ways in which reef mining, inexperienced divers, boating, and increased tourism threaten some reefs. Be sure to emphasize that some of those activities provide income to people living in underdeveloped regions. Point out that many scientists are in favor of restricting mining and tourism activities near coral reef ecosystems. Elicit from students how human need for money derived from the reefs can be balanced with the need to preserve reef ecosystems.

GLENCOE TECHNOLOGY

Videodisc

The Infinite Voyage: The Keepers of Eden

The Transformation of Zoos (Ch. 3)

Naturalist Gerald Durrell and His Special Zoo (Ch. 7)

Microscope Activity

LS **Visual-Spatial** Ask students to examine prepared slides of animal cells such as nerve, muscle, blood, and cells lining the stomach. Ask them to make labeled sketches of these cells in their journals. L1

Figure 28.3

Because adult sponges and corals are sessile, they rely on water currents to bring food to them.

Corals, with tentacles extended, snag food drifting by in the water.

Sponges filter food out of the water by setting up a stream of water through their bodies.

Some water-dwelling animals, such as the sponges and corals shown in *Figure 28.3,* are able to move about only during the early stages of their lives. They hatch from eggs into free-swimming larval forms, but when they become adults, they attach themselves to rocks or other objects. Organisms that don't move from place to place are known as **sessile** organisms.

Animals must digest food

Animals break down, or digest, their food once they have consumed it. In some animals—including earthworms, frogs, and monkeys—digestion takes place in an internal cavity; in other animals, such as sponges and flatworms, digestion is carried out within individual cells. Much of the food an animal consumes and digests is stored as fat or glycogen and used as an energy supply when food is not available.

Examine the structure of the digestive tracts of a flatworm and an earthworm in *Figure 28.4.* Notice that there is only one opening to the flatworm's digestive tract. An earthworm has a more complex digestive tract with two openings, one at either end.

Figure 28.4

Planarians and earthworms are animals that digest food in separate digestive tracts.

Have you ever turned over a rock in the shallow water of a lake or stream? If you looked closely, you might have seen tiny planarians attached to the underside of that rock. Planarians feed on small, live organisms or on the dead bodies of larger animals. The planarian's digestive tract has only one opening through which food enters and wastes exit.

Digestive tract

Mouth

Pharynx

694

PROJECT

What do *Daphnia* eat?

Ask students to maintain *Daphnia* cultures for two weeks. Each culture should have ten *Daphnia.* Have students choose different food materials to feed the *Daphnia.* For example, one group might use cooked egg white, and another wheat grains, and another group might use live protozoans. At the end of the two-week period, students should determine which cultures seem to be most healthy (as measured by activity). Ask them to write a report of their experimental designs, observations, and conclusions and include this report in their portfolios. L1 P LS

Animal cell adaptations

Animals are multicellular. Just as animal bodies are adapted to a variety of environments, most animal cells are adapted to carry out different functions. For example, in the human body, nerve cells conduct information, red blood cells transport oxygen, muscle cells make movement possible, and cells lining the stomach secrete digestive juices.

Earthworms ingest soil and digest the organic matter contained in it. The earthworm's digestive system has two openings and contains different structures for storing, grinding, and dissolving food. Food travels along the digestive tract in only one direction, and indigestible waste is eliminated at the second opening. Do you think a digestive tract with two openings is more efficient than one with a single opening?

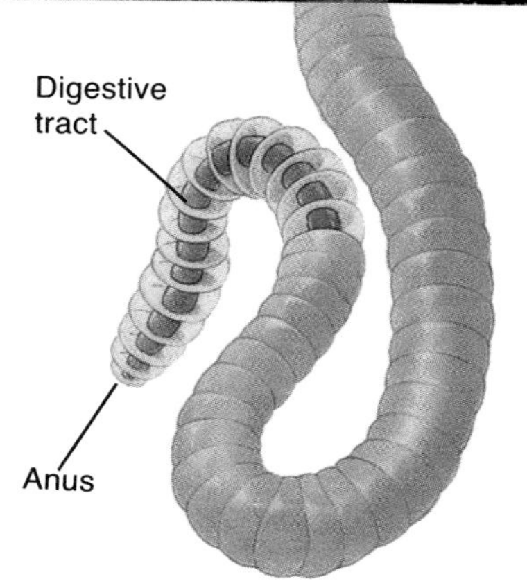

Math

Museums: More Than Stuffed Animals

Natural history museums have an important role in a community. They provide extensive and often unusual exhibits of preserved animals and fossils that help people appreciate the wildlife of the world.

Natural history museums also provide scholars access to research materials and special laboratories and libraries. For example, a graduate student studying the evolution of the shape of the thyroid gland in a family of lizards, the Lacertidae, could check out preserved specimens of the lizards from the museum just as you might take out books from a library. The student would be allowed to dissect and photograph the shape of the thyroid gland in each specimen. He or she would then sew up the skin and return the lizards to the museum. Another researcher may want to compare the DNA of lizards of several different families.

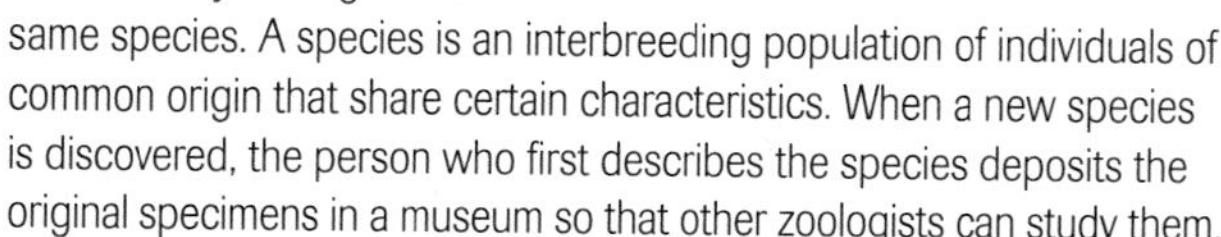

Museums as reservoirs A natural history museum maintains collections of specimens against which similar specimens can be compared to determine whether they belong to the same species. A species is an interbreeding population of individuals of common origin that share certain characteristics. When a new species is discovered, the person who first describes the species deposits the original specimens in a museum so that other zoologists can study them.

Using math at a museum Let's say you are a photographer who has just been commissioned to photograph one specimen of every animal species that has been described so far. The museum requesting the photos has at least one specimen for every one of the approximately 1 050 000 species of animals. The museum staff will take care of setting up each specimen for you. You judge that each photo will take 30 seconds to complete. If you devote 12 hours a day to the task, how many days will you schedule to complete your task?

CONNECTION TO Biology

Each species has certain traits that separate it from other species. What are some ways that a member of a species of animal can be recognized?

Math

Museums: More Than Stuffed Animals

Purpose

Students learn about museums of natural history and use math to carry out a hypothetical project in a museum.

Teaching Strategies

- Encourage discussion of the museum as a source of information for students and scientists.
- Ask students why scientists photograph their specimens.

Possible Answer

Using the number of animal species as 1 050 000, students should multiply it by 30 seconds, the time required to photograph each specimen (31 500 000 seconds). Then they should find the number of seconds in a 12-hour day (43 200 seconds). Divide 31 500 000 by 43 200 = 729 days or about two years.

CONNECTION to Biology

Animals use scent, color, displays, grooming, greetings, and calls to assure that they are dealing with a member of their own species with whom they can successfully mate.

Visual Learning

Figure 28.4 Do you think a digestive tract with two openings is more efficient than one with a single opening? *Yes*

CULTURAL DIVERSITY

Animals in a Cross-cultural Perspective

Discuss the diverse views different societies and cultures have about the place of animals in nature. Begin your discussion by asking students how animals are viewed and used in the United States. Next, point out that not all cultures view animals in this way. For example, Native American traditions often portray animals as important as or even more important than humans. Obtain books about the legends and myths of different cultures, and discuss with students the role of animals in these stories.

BioLab Design Your Own Experiment

What is an animal?

Time Allotment

One class period

Objectives

Review objectives with students before they begin the Biolab.

Process Skills

form a hypothesis, design an experiment, compare and contrast, observe and infer, make and use tables

Safety Precautions

- Do not use animals that bite or are poisonous. Be sure students treat animals in a humane fashion at all times.
- Ask students to wash their hands at the beginning and at the end of the lab.

PREPARATION

Alternate Materials

Use live animals common to your location that are easily maintained.

Possible Hypotheses

- Students may hypothesize that their organism can be identified as belonging to a specific animal group because of the way it moves, responds to a stimulus, or consumes food, or because it has certain body structures.

GLENCOE TECHNOLOGY

Videodisc

Biology: The Dynamics of Life

Disc 1, Side 2

Embryo Development (Ch. 30)

BioLab Design Your Own Experiment

What is an animal?

What do you think of when you hear the word *animal?* A lion, tiger, dinosaur, or perhaps a bird? An animal is defined as a multicellular, eukaryotic heterotroph whose cells lack cell walls. Animals can be distinguished by the way they feed, digest food, respire, respond to stimuli, move, excrete, transport materials in their bodies, and reproduce. In this Biolab, you will examine characteristics of animals that are used to place them in the animal kingdom.

PREPARATION

Problem

What characteristics enable you to identify organisms as animals?

Hypotheses

Make a hypothesis, based on the general features of animals, that best describes why a particular animal can be placed in the animal kingdom and how that animal can be specifically identified. Consider behavior, structure, and physiology.

Objectives

In this Biolab, you will:

- **Observe** behavior of animals.
- **Compare** features of animals.

Possible Materials

microscope and depression slides
metric ruler
variety of live animals such as mouse, earthworm, frog, hermit crab, sea urchin, cricket, goldfish, lizard, planaria, hydra, *Daphnia*, snail
suitable containers for each animal such as cages, pans, watch glasses, clear deli containers with lids, culture dishes

Safety Precautions

Treat animals in a humane manner at all times. Wash your hands after handling animals.

PLAN THE EXPERIMENT

Teaching Strategies

- Display a large goldfish in a bowl. Ask students how they can tell the goldfish is an animal. *Most suggestions will deal with moving and breathing.* Next, feed the goldfish. Instruct students to time how long it takes for the fish to swim to the food. As they watch the fish eat, ask how this behavior is important in identifying the goldfish as an animal. *All animals obtain food from their environments.* Ask them what the fish does with the food. *It uses the food for energy or as raw material for building tissue.*
- Elicit how the fish is different from a protist that might move around to get food. *The fish is multicellular and has tissues and organs.* Next, ask how they can tell the animal is a fish. *Most likely they will bring up*

PLAN THE EXPERIMENT

1. Choose four animals you wish to investigate.
2. In your group, brainstorm a list of possible characteristics that could be used to categorize and identify your animal.
3. Construct a data table for the features that identify your organisms as animals and that identify them specifically.

Check the Plan

1. Recall the demonstration done by your teacher at the beginning of this investigation. Have you incorporated these ideas into your table?
2. Give reasons why you made your table the way you did.
3. What quantitative data do you plan to collect?
4. ***Make sure your teacher has seen your data table before you proceed further.***
5. Carry out your experiment.
6. Record measurements and observations in your data table.

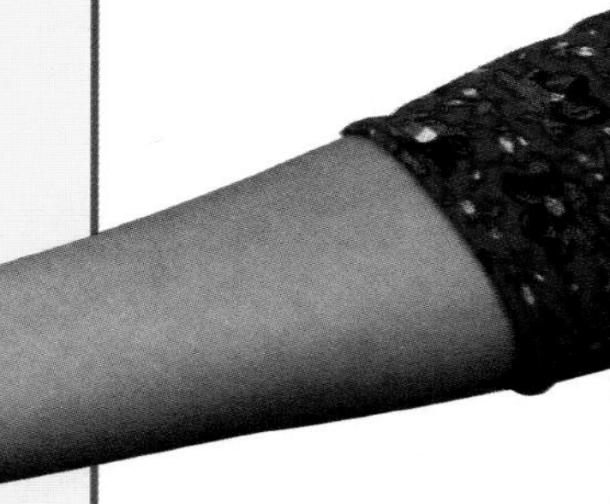

ANALYZE AND CONCLUDE

1. **Identifying Variables** On what basis did you identify your animals generally and specifically?
2. **Checking Your Hypothesis** Did your results confirm your hypothesis? Why or why not?
3. **Comparing and Contrasting** Compare your table with that of another group. How is it the same? How is it different?
4. **Drawing Conclusions** What did you learn about animals in this Biolab that you did not know before?
5. **Thinking Critically** How are animals identified?

Going Further

Project Based on this lab experience, design an experiment that would help you to answer a question about an unfamiliar animal.

ANALYZE AND CONCLUDE

1. Animals can be identified by movement, adaptations for getting food and avoiding predators, and being multicellular. They can be specifically identified by types of adaptations for locomotion, sensory structures, behavior, and physiology.
2. Students may find that they were able to characterize their animals by structure, physiology, and behavior but not by where they live, size, color, or shape.
3. The tables are different because the groups selected different animals, but are alike in the features mentioned in question number 2.
4. Students should indicate that animals can be distinguished by structural features, behavior, and physiology.
5. Animals are identified by structural characteristics, behavior, and physiology.

Assessment

Portfolio: Have students summarize what they learned about how organisms are identified as animals and how animals are classified. L1

Going Further

Ask students to design an experiment to determine what type of food an unfamiliar animal might eat. L2

ideas like swimming, having a backbone, and having scales and gills.

- Explain that they should develop ways of identifying their organisms first as animals and next as part of a particular animal group. It is not important that their answers are correct. It is important they think critically about their animals and their unique traits.

Data and Observations

Students' tables may have the following types of data: type of movement toward food, reaction to touch stimulus, structural adaptations that enable animals to get food or escape predators, time spent in lighted part of container, and description of nervous system features.

Demonstration

IS **Visual-Spatial** Project 35-mm slides of the stages of development of a variety of animals. Point out the similar stages of development of different animals. In addition, show slides of the gastrulas of protostomes such as earthworms and insects, and deuterostomes such as fish, birds, and humans.

Chemistry Connection

Cell Communication During Development

Early in the 20th century, Hans Spemann and Hilde Mangold did experiments that illustrated a mechanism called *induction.* Induction is a process in which the cells of a developing embryo bring about further differentiation. In their experiment, Spemann and Mangold removed cells from the dorsal surface of an amphibian blastula. These cells were transplanted to a location on another blastula that was originally destined to become the belly. Instead, a second notochord developed.

Later experiments showed that induction does not occur if a nonporous barrier is placed between the transplanted tissue and the tissue into which the transplant is made. Although the chemical nature of induction is not completely understood, it is believed that inducer cells make a protein that stimulates mitosis and causes changes in gene expression.

Development of Animals

Most animals develop from a single, fertilized egg cell called a zygote. But how does a zygote develop into the many different kinds of cells that make up a snail, a fish, or a human?

During development, the zygotes of different species of animals all have similar stages of development. The stages of development from zygote to fully formed embryo are shown in *Figure 28.5.* In this figure, the developing sea urchin embryo is used as an example. However, not all animals follow exactly the same patterns as the sea urchin.

Division of the egg

Initially, the unicellular zygote divides, forming two new cells in a process called cleavage. These two cells divide to form four cells and so on, until a hollow ball of cells called a blastula is formed. The **blastula** is a single layer of cells surrounding a fluid-filled space and is formed early in the development of an animal embryo. Formation of a blastula is complete about ten hours after fertilization in sea urchin development.

Forming a gastrula

The cells on one side of the blastula then fold inward to form a two-layered gastrula. The **gastrula** is a structure that is made up of two cell layers. Gastrula formation can be compared to the way a potter creates a cup or bowl from a lump of clay, *Figure 28.6.* First, the clay is formed into a solid ball. Then, the potter presses in on the top of the ball to form a cavity that will become the interior of the bowl. In the same way, the cells at one end of the blastula fold inward, forming a cavity lined with a second layer of cells. The layer of cells on the outer surface of the gastrula is called the **ectoderm.** The layer of cells lining the inner surface is called the **endoderm.**

Figure 28.5

Development of the sea urchin from a zygote is shown as an example of the sequence in which animal embryos develop.

 **Blastula formation is the earliest developmental stage of an embryo. Continuous cell divisions in sea urchin development result in a 32-cell blastula. The total number of cells present in blastula stages of different classes of animals varies. Notice that during these early developmental stages, the total amount of cytoplasm has not increased. Each cell division results in smaller cells.**

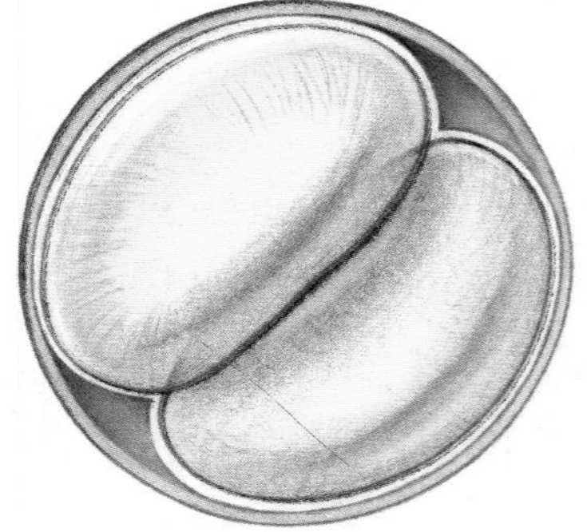

Two-celled stage

Eight-celled stage

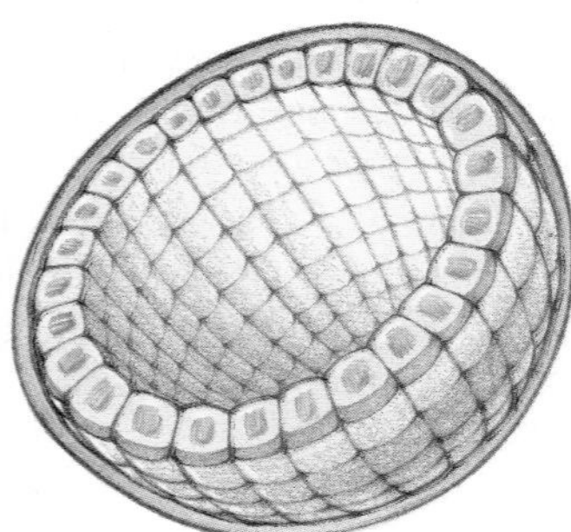

Blastula

Meeting Individual Needs

Learning Disabled/Students Acquiring English Provide students with diagrams of the human body showing bone, muscle, spinal cord, brain, kidneys, liver, lungs, and pancreas. Give them colored pencils and have them shade parts of the body that develop from ectoderm red, parts that develop from endoderm blue, and parts that develop from mesoderm green. Provide them the following information: ectoderm produces brain, spinal cord, nerves, outer skin, eye lens, nose, and ears. Mesoderm produces skeleton, muscles, excretory system, inner skin. Endoderm produces pancreas, liver, lungs, and lining of the digestive system. L1 SAE IS

GLENCOE TECHNOLOGY

 Videotape

The Secret of Life

Use the videotape *Sex and the Single Gene: Cell Development* to illustrate patterns of embryo development.

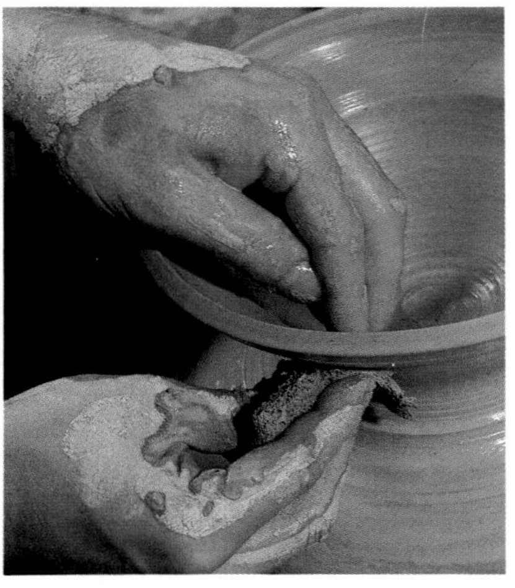

Figure 28.6

You can think of a blastula as a hollow ball of cells. By pushing in on one side, a gastrula is formed.

The ectoderm cells of the gastrula continue to grow and divide, and eventually they develop into the skin and nervous tissue of the animal. The endoderm cells develop into the lining of the animal's digestive tract.

Protostomes and deuterostomes

In some classes of animals, the opening of the indented space in the gastrula becomes the mouth. These animals, which include earthworms and insects, are called protostomes. A

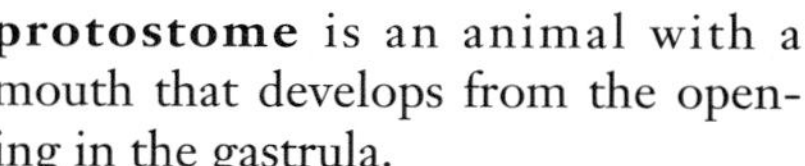

protostome is an animal with a mouth that develops from the opening in the gastrula.

In animals with more complex tissues and organ systems, such as fishes, birds, and humans, the opening of the gastrula develops into an anus, an opening through which digestive wastes are eliminated from the animal's body. A **deuterostome** is an animal in which the anus develops from the opening in the gastrula. The mouth of a deuterostome develops from cells elsewhere on the blastula.

Determining whether an animal is a protostome or deuterostome can help biologists determine the phylogeny of the animal's group. For example, both sea urchins and fishes are deuterostomes and are, therefore, more closely related than you might conclude from comparing their adult body structures.

Mesoderm develops later

The development of the gastrula progresses until a layer of cells called the mesoderm is formed. The **mesoderm,** illustrated in *Figure 28.5,* is the third cell layer formed in the developing embryo. The term *meso* means "middle," so mesoderm is found in the middle of the developing embryo; it lies between the ectoderm and endoderm.

protostome
proto (GK) before
stoma (GK) mouth

deuterostome
deutero (GK) secondary
stoma (GK) mouth
A protostome and a deuterostome differ in the location of the cells that become the organism's mouth.

B **As the embryo continues to grow, some of the cells of the blastula fold inward, forming the gastrula. All animal embryos except sponges pass through this gastrula stage.**

Mesoderm
Endoderm
Mesoderm breaks off
Ectoderm

C **The mesoderm in the gastrula develops between the endoderm and ectoderm. In deuterostomes, the mesoderm is formed from clumps of cells that break off from the endoderm. The mesoderm gives rise to muscles, reproductive organs, and circulatory vessels.**

PORTFOLIO

Spina Bifida Ask students to interview a doctor about spina bifida or any other medical condition that can be determined during embryonic development. Ask students to write a report of their findings and to prepare diagrams that show normal stages of development and stages in which the abnormal development is obvious. L2 P LS

Making a Model

LS **Kinesthetic** Give students salt dough that you have prepared (one part salt, one part flour, and one part water). Have them cover a space on their desktops with a few paper towels and dip their fingers in flour so the dough will not stick to them. Ask students to make models of a fertilized egg, a blastula, and a gastrula. They should make cross sections of the stages in sequence, with one stage turning into the next stage, rather than lining up all three stages at once. They will have to add more dough to show that the fertilized egg becomes larger; but after that time, dough should not be added since the total amount of cytoplasm does not increase as the early embryo develops. L3

3 Assess

Check for Understanding

Give students a live insect. Ask them to list the characteristics that show that the insect is an animal. L1

Reteach

Make clay models of stages of development. Ask students working in groups to place the models in sequence and describe what happens in each stage. L1 SAE

Extension

Encourage students to visit a science museum that has displays about development. Have them make a photo essay of the exhibit that can be displayed in class. Have students include captions for the photos. L2

Assessment

Portfolio: Have students summarize the development of a sea urchin egg from fertilization to the planula stage. Have them include this summary in their portfolios. L1 P

SECTION 28.1

ThinkingLab Make a Hypothesis

Purpose

LS Logical-Mathematical Students will form a hypothesis about how mesoderm influences development in the ectoderm.

Process Skills

form a hypothesis, analyze an experimental procedure, observe and infer

Teaching Strategies

- Before students do the Thinking Lab, make sure they have a thorough understanding of embryonic development. They should also have a good understanding about endoderm, mesoderm, and ectoderm.

Thinking Critically

Nervous tissue will develop in the embryo above the transplant.

Assessment

Portfolio: Ask students to write a summary of what they have learned about development in this Thinking Lab. Ask them to place their summaries in their portfolios. L1 P

4 Close

Activity

Ask students to write a "note" to a friend explaining what they understand about development and what they still do not understand. Have them exchange "notes" and provide written help to each other to explain the areas of difficulty.

ThinkingLab Make a Hypothesis

How does mesoderm influence development of nervous tissue?

Biologists wanted to determine what causes the ectoderm on the dorsal side of the embryo to develop into nervous tissue. They hypothesized that perhaps the mesoderm lying directly beneath the ectoderm is the key. To test this idea, they removed a piece of mesoderm from beneath the dorsal surface of the ectoderm of one embryo and transplanted it to another embryo. The transplant was placed in a spot on the ventral surface of the second embryo where the mesoderm had been removed. The ectoderm was sealed back over the top of the mesoderm transplant.

Analysis

Put in order the following illustrations that show the sequence of the steps the biologists took to perform the experiment.

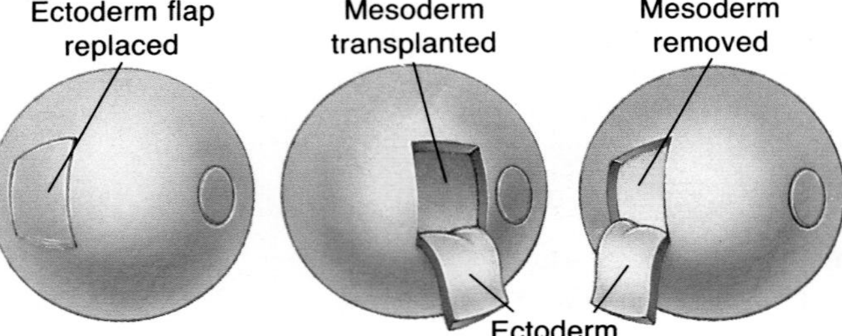

Thinking Critically

Hypothesize what might happen as the embryo develops if, as the scientists hypothesized, mesoderm influences development in the ectoderm.

Forming a larva

As cells in the developing embryo change shape and become specialized to perform different functions, the sea urchin grows into its larval form, *Figure 28.7.* Some cells begin secreting calcium, an element that is important in the formation of spines in the adult sea urchin.

Figure 28.7

A free-swimming larva (top) develops from a fertilized sea urchin egg in just 48 hours. These larvae can be found as part of the plankton floating near the surface of oceans. The larvae will develop into adult sea urchins over the next few months. An adult sea urchin (bottom) has spines that it uses for protection from predators as it moves slowly along the ocean floor.

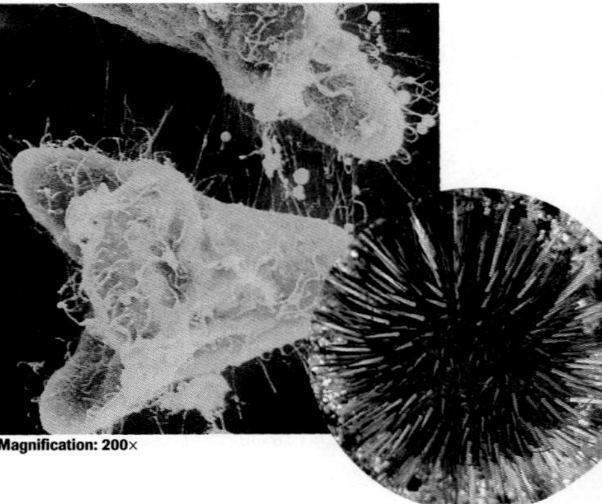

Section Review

Understanding Concepts

1. Describe the characteristics that make a mouse an animal.
2. How does an animal with only a mouth get rid of undigested waste materials?
3. Explain the difference between a protostome and a deuterostome.

Thinking Critically

4. Identical twins occur in humans when a single zygote separates into two equal halves that go on to develop into individual organisms. Explain at what point in development this separation must occur in order to result in two complete embryos.

Skill Review

5. **Sequencing** Place the following words in sequence, beginning with the earliest stage: gastrula, larva, adult, fertilized egg, blastula. For more help, refer to Organizing Information in the *Skill Handbook.*

Answers to Section Review

1. It is a multicellular heterotroph that uses food for energy or building tissue. It moves to get food and does not have cell walls.
2. It expels waste through the mouth.
3. In protostomes the mouth develops from the opening in the gastrula. In deuterostomes, the anus develops from the opening in the gastrula.

Thinking Critically

4. at the time of the first division of the fertilized egg

Skill Review

5. fertilized egg, blastula, gastrula, larva, adult

SECTION

28.2 Body Plans and Adaptations

Section Preview

Objectives

Distinguish among the body plans of different classes of animals.

Trace the phylogeny of animal body plans.

Compare body plans of acoelomate, pseudocoelomate, and coelomate animals.

Key Terms

symmetry
asymmetry
radial symmetry
bilateral symmetry
anterior
posterior
dorsal
ventral
acoelomate
pseudocoelom
coelom
exoskeleton
invertebrate
endoskeleton
vertebrate

Objects made by a potter represent many different shapes and sizes. There may be a basic plan for making each type of pottery. One plan is for a bowl, another for a vase, another for a flat plate. Each piece of pottery is suited for a particular function. Animal bodies also have basic plans or body structures that are suited to a particular way of life. In this section, you will study a variety of animal body plans and see how a specific body structure is adapted for life in a particular environment.

What Is Symmetry?

Look at the animals shown in *Figure 28.8*. You know that all animals share certain characteristics, but these animals certainly don't look like they have much in common. The sponge seems to have no particular shape, whereas the fish has a head, body, fins, and a tail. The jellyfish doesn't have a head or tail, but is circular in form. Each animal has a different shape or form because each exhibits a different kind of symmetry. **Symmetry** refers to a balance in proportions of an object or organism. All animals have some kind of symmetry. Different kinds of symmetry enable animals to move about and find food in different ways.

Figure 28.8
A sponge (left), a fish (center), and a jellyfish (right) all exhibit different kinds of symmetry, yet each animal is able to find and digest food.

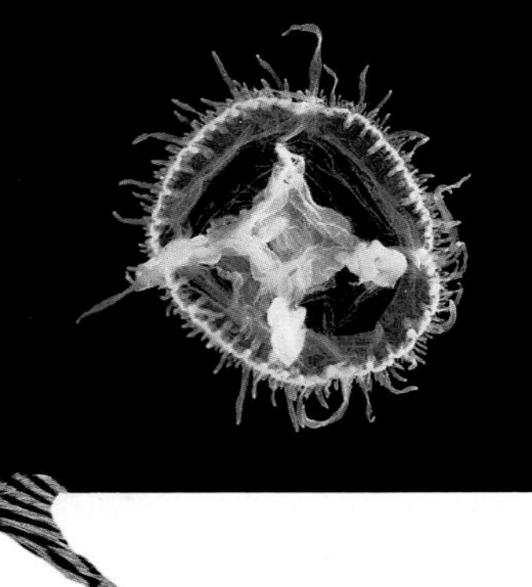

SECTION 28.2

Prepare

Key Concepts

Students compare and contrast types of symmetry and study basic body plans of animals.

Block Scheduling

Look for this symbol for strategies that are useful in a block scheduling format. For more information on block scheduling, refer to Section 28.2 in the **Lesson Plans** booklet.

Materials

- For the Minilab, gather soil nematodes or vinegar eels, earthworms, and binocular microscopes.

1 Focus

Bellringer

Before presenting the lesson, display **Section Focus Master 61** on the overhead projector and have students answer the accompanying questions. L1 SAE

Demonstration

Bring in kitchen items such as bowls, vases, and a variety of spoons. Explain that each item is suited to a particular function. For example, the bowls hold soft or liquid foods, the spoons pick up or stir soft or liquid foods, and the vases hold flowers for display. Explain that in a similar way animals have structures that are suited to specific functions such as locomotion, food-getting, and circulating materials throughout their bodies.

Meeting Individual Needs

Visually Impaired Students who are visually impaired may benefit from handling a number of objects that are asymmetric, radially symmetric, and bilaterally symmetric. At first explain to them what type of symmetry the objects have, then give them new objects and ask what kind of symmetry they have. L1 LS

Program Resources

Section Focus Master 61 L1 SAE
Reinforcement and Study Guide, pp. 113-114 L1
Biolab and Minilab Worksheets, p. 112 L1
Concept Mapping, p. 28 L1

2 Teach

Demonstration

LS **Visual-Spatial** Pass around the dried bodies of marine sponges, a live goldfish in a clear plastic container, and a large preserved jellyfish. Ask students in their groups to describe the shapes of the animals and explain if one side of the animal is equally balanced by the other. *Marine sponge, no; live goldfish, yes; preserved jellyfish, yes.* Explain to students that the quality they are observing is called symmetry.

Visual Learning

Figure 28.10 How might stinging cells be a helpful adaptation for a predator with limited ability to move? *The stinging cells can be used to paralyze prey and prevent its escape until it can be eaten.*

Reinforcement

LS **Interpersonal** Divide the class into groups. Have each group list as many items with radial symmetry as they can find in the classroom within a given time period. Repeat the activity with asymmetrical and bilaterally symmetrical objects. L1 COOP LEARN

Figure 28.9

This irregularly shaped sponge is an example of an animal with an asymmetrical body plan. A sponge's structure is adapted for filtering food from the water. The two cell layers have distinct functions. The outer layer of cells protects the sponge from predators. The cells of the inner layer keep water moving with the whiplike motion of their flagella. These inner cells also trap and digest food.

Asymmetry in a sponge

Most sponges are adapted to living on the bottom of the ocean, as seen in *Figure 28.9.* Many sponges have an irregularly shaped body. An animal that is irregular in shape has a body plan that exhibits **asymmetry.** Asymmetrical animals often are sessile organisms that do not move from place to place.

The bodies of most sponges consist of two layers of cells. Unlike most other classes of animals, the sponges' embryonic development does not include the formation of an endoderm and mesoderm, or a gastrula stage. Sponges first evolved about 600 million years ago and represent one of the oldest groups of animals on Earth—proof that their two-layer body plan makes them well adapted for life on the ocean floor.

Radial symmetry in a hydra

The hydra, a tiny predator pictured in *Figure 28.10,* feeds on small animals it snares with its tentacles. A *Hydra* has radial symmetry. Its tentacles radiate out from around its mouth. As you can see, animals with **radial symmetry** can be divided along any plane, through a central axis, into roughly equal halves. Radial symmetry is an adaptation in *Hydra* that enables the animal to detect and capture prey coming toward it from any direction.

Have you ever had your groceries double bagged at the store? The body plan of a *Hydra* can be compared to a sack within a sack. These sacks are cell layers organized into tissues with distinct functions.

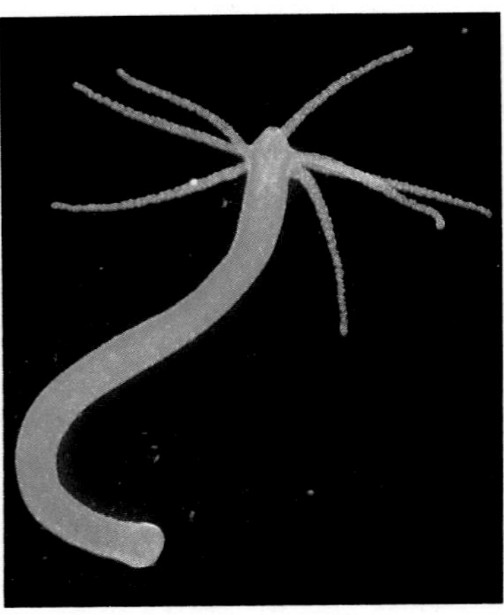

Figure 28.10

An example of an animal with radial symmetry, a hydra (left) feeds on tiny animals it immobilizes with venomous stinging cells found along its tentacles. How might stinging cells be a helpful adaptation for a predator with limited ability to move?

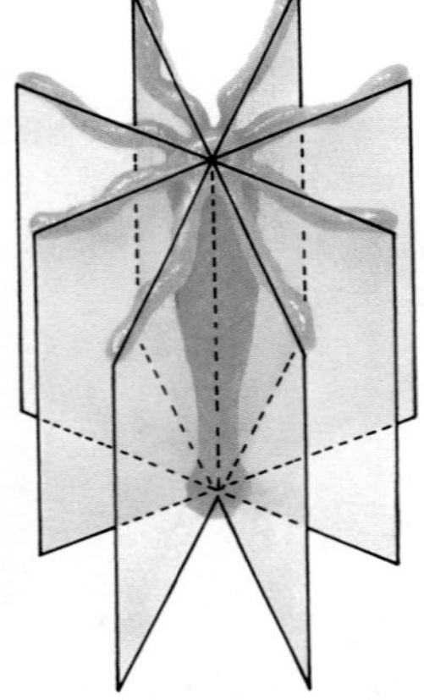

Hydra shows radial symmetry

Alternate Lab: Body Plans

Purpose 

LS **Visual-Spatial** Students will examine prepared slides of cross sections of hydra, flatworm, roundworm, and earthworm to determine if the animal is acoelomate, pseudocoelomate, or coelomate.

Materials

microscope, prepared slides of cross sections of planarian, nematode, earthworm, and hydra

Procedure

Give students the following directions.

1. Examine the cross section slides provided.
2. Sketch and label each cross section with the following labels: animal with two cell layers, pseudocoelomate animal, acoelomate animal, coelomate animal. Also label the drawings with the names of organisms.

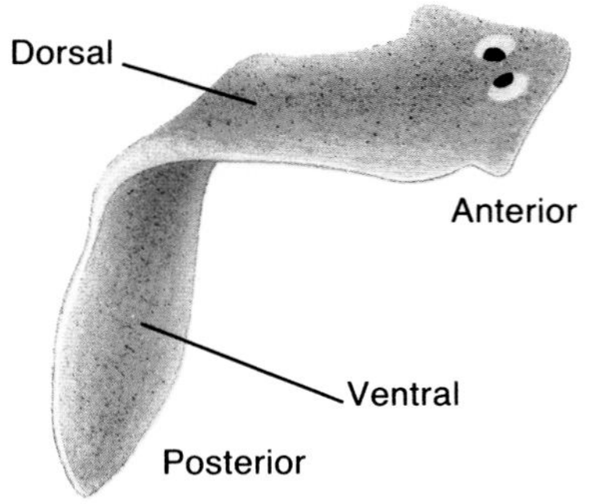

Figure 28.11

In bilaterally symmetrical animals, such as a flatworm, sensory tissue is commonly concentrated in the head, or front end. The head detects changes in light, temperature, and chemical makeup of its environment.

Bilateral symmetry

The flatworm in *Figure 28.11* has bilateral symmetry. An organism with **bilateral symmetry** can be divided down its length into similar right and left halves that form mirror images of one another. In such animals, the **anterior,** or head end, has a different appearance than the **posterior,** or tail end. The **dorsal,** or back surface, also looks different from the **ventral,** or belly surface. Animals with bilateral symmetry can find food and mates and avoid predators more efficiently than animals with radial symmetry because they have more muscular control.

Bilateral Symmetry and Body Plans

Animals that are bilaterally symmetrical also share other characteristics. What other characteristic do the animals in *Figure 28.12* have in common? All of these animals have body cavities in which internal organs are found. The development of body cavities made it possible for animals to grow larger and to move and feed more efficiently. Without a body cavity, animals depend on diffusion to take food into cells and eliminate wastes.

Acoelomate flatworms have no body cavities

The earliest group of animals in which a mesoderm evolved may have been similar to flatworms. Recall that the mesoderm of flatworms develops from cells near the opening in the gastrula. Flatworms may have been the first group of animals in which muscles and other organs evolved from the mesoderm.

Figure 28.12

Grasshoppers (left), muskrats (center), and manatees (right) are bilaterally symmetrical animals that have body cavities in which internal organs are found.

Different Viewpoints in Biology

The Phylogeny of Flatworms

Some biologists think that flatworms evolved from jellyfishes in their small, ciliated, flat, larval stage. Other biologists think that flatworms had an independent origin from the ciliate protozoans. Yet another group believes flatworms are a degenerate form related to earthworms. Similar disagreements occur regarding the phylogeny of other animal groups. New molecular techniques for determining which animals are most closely related may someday answer the question about the phylogeny of flatworms and other animals.

Reinforcement

LS **Visual-Spatial** Show students a live planarian in a deep-slide projection slide (available from biological supply companies). Fill the well with pond or stream water. The swimming planarian can then be projected by your slide projector onto a screen. Point out the features of the planarian, such as eyespots, that show it has bilateral symmetry.

Analysis

Ask students to answer the following questions:

1. Order the animals' body plans from least to most complex. *Hydra, planarian, nematode, earthworm*
2. In what way is each animal's body plan an adaptation to its environment? *Hydra and planarians are adapted to movement in water. Earthworms are adapted to life in soil.*
3. If you were to observe these animals' movements, predict what you might find. Explain in terms of their body plans. *Students' descriptions should match those given in the chapter.*

Assessment

Portfolio: Ask students to make a summary of what they have learned about body plans in this lab. Have them consider the survival value of each type of body plan. Instruct them to put their summaries in their portfolios. L1 P

MiniLab

Purpose

Visual-Spatial Students will observe how body plans affect roundworm and earthworm movement.

Process Skills

observe and infer, compare and contrast

Teaching Strategies

- Microscopic nematodes may be found in soil.
- Vinegar eels may be used in place of soil nematodes. Students may look at vinegar eel cultures in petri dishes.

Safety Precautions

- Make sure students wash their hands at the beginning and at the end of the lab.
- Remind students to treat all animals humanely.

Expected Results

An earthworm moves by stretching the front end of its body forward, then pulling the back end up. Nematodes move with a whip-like motion, lashing their bodies from side to side.

Analysis

1. whiplike motion, lashing body from side to side
2. stretches front end of body forward; pull back end up
3. Earthworms have two sets of muscles that can push against the coelom for leverage and cause more varied movements. The nematode has only one set of muscles which results in more limited movements.

Assessment

Journal: Have students place the answers to the Analysis questions in their journals.

MiniLab

How do body plans affect locomotion?

Roundworms have a pseudocoelom with one set of muscles arranged along the length of the body. Earthworms have a coelom with two sets of muscles—one arranged along the length of the body and the other arranged in a circular fashion around the body.

Procedure

1. Observe nematode movement in their culture medium through a hand lens or binocular microscope.
2. Place an earthworm on your tabletop. Observe its movement.

Analysis

1. Describe the movements of a nematode.
2. Describe the movements of an earthworm.
3. Compare the movements of a nematode and an earthworm in terms of their differences in body plans.

Flatworms have solid, compact bodies, as shown in *Figure 28.13.* Animals that have three cell layers with a digestive tract but no body cavities are called **acoelomate** animals. Flatworms are the oldest animal group that has different tissues organized into organs. The organs of flatworms are embedded in the solid tissues of their bodies.

Water and particles of digested food travel through a solid acoelomate body by the slow process of diffusion. A flattened body allows for diffusion of nutrients, water, and oxygen quickly enough to supply all body cells. Notice, too, that the digestive tract extends throughout the body to help in the process of diffusion.

Pseudocoelomates have a body cavity

A roundworm also has bilateral symmetry. However, the body plan of a roundworm is more complex than that of a flatworm. A roundworm has a body cavity, called a pseudocoelom, that develops between the endoderm and mesoderm. A **pseudocoelom** is a fluid-filled body cavity partly lined with mesoderm. Although the prefix *pseudo* means "false," a pseudocoelom is a real body cavity.

The pseudocoelom enables animals to move more efficiently. How does this work? Think about the way your muscles work. The muscles in your arm lift your hand by pulling against your arm bones. If there were no rigid bones in your arms, your muscles would not be able to do any work. Although the roundworm has no bones, it does have a rigid, fluid-filled space, the pseudocoelom, which its muscles can attach to and

Figure 28.13

Animals with acoelomate bodies (left) usually have a thin, somewhat flattened shape. The evolution of the pseudocoelom (center) made it possible for animals to become larger and thicker in body shape than their acoelomate ancestors. The coelom (right) provides a space for complex internal organs.

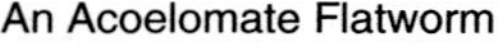

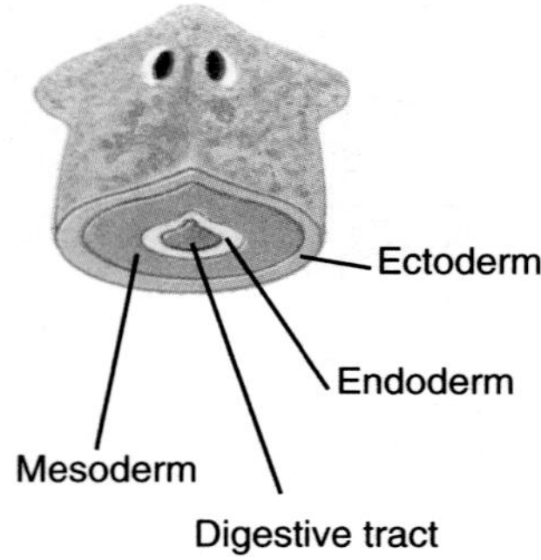

An Pseudocoelomate Roundworm

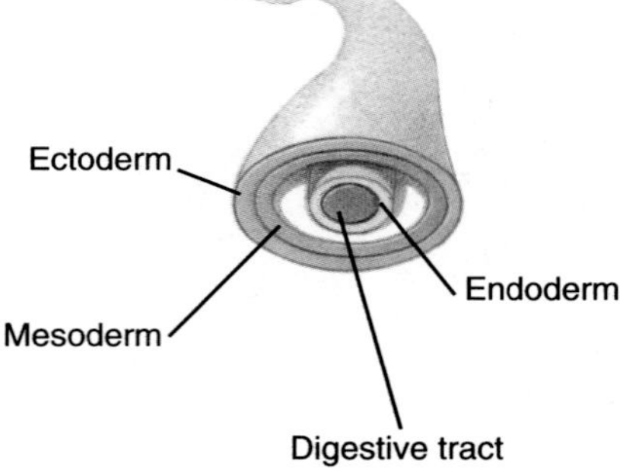

STUDENT JOURNAL

Coelom Development Have students write an account of the evolutionary development of the coelom. They should write as if they were writing for a newspaper in geologic time by first determining the time period, explaining how the coelom developed, identifying animals having coeloms, and explaining how the coelom has survival value. L3

interNET CONNECTION

Follow the link for this chapter on the Glencoe Homepage at **www.glencoe.com/sec/science** to find out more about animals.

brace against. The pseudocoelom thus makes it possible for pseudocoelomate animals to move more quickly than acoelomate animals.

A one-way digestive tract with two openings first evolved in roundworms. This innovation enabled various parts of the digestive tract to take on specific functions: the mouth takes in food, the middle region digests, and the anus expels waste.

The coelom provides space for internal organs

Earthworms have a more complex body plan than do roundworms. The body cavity of an earthworm develops from a fluid-filled space inside the mesoderm, as shown below in *Figure 28.13*. This space is called the coelom. A **coelom** is a body cavity completely surrounded by mesoderm. Humans, insects, fishes, and many other animals have a coelomate body plan.

In coelomates, the digestive tract and other internal organs are attached by double layers of mesoderm and are suspended within the fluid-filled coelom. Like the pseudocoelom, the coelom acts as a kind of watery skeleton against which muscles can work. The coelom provides space for more-complex organs.

A Coelomate Segmented Worm

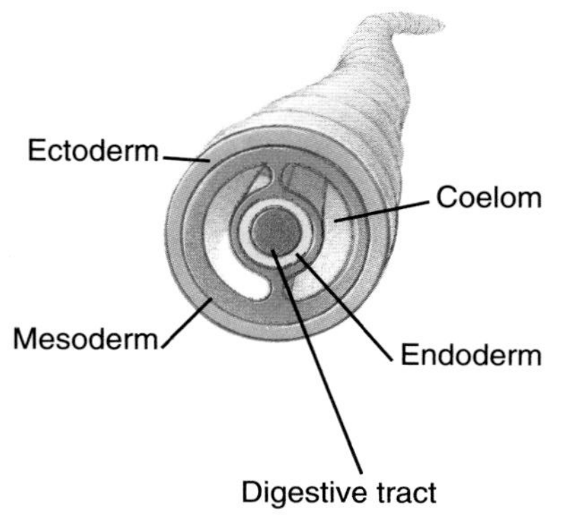

A BROADER VIEW

The First Landlubbers

If you could travel backward in time, you might be surprised to discover that Earth's oceans were the birthplace of the first organisms on this planet. By 500 million years ago, the seas were alive with red, brown, and green algae and a variety of invertebrates including sponges and crabs.

New habitats appear Life, however, did not remain in the oceans. Earth's active geology was transforming the planet, creating new continents and islands. By 300 million years ago, plants, amphibians, insects, and reptiles began populating these new landforms.

How were organisms that had evolved in water over millions of years able to survive in a completely different environment on land?

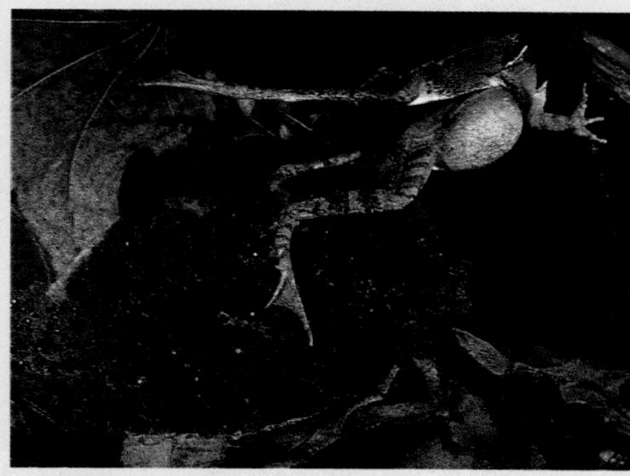

In and out of water Amphibians are animals that spend most of their time on land but must return to water to reproduce. Like fishes, amphibians have external fertilization in which eggs and sperm require water for transport. Amphibian eggs also have no protective covering; if laid on land, they would simply dry up.

A different type of egg evolved in the reptiles—an amniotic egg that surrounds the developing embryo with a protective membrane and a shell. Amniotic eggs are fertilized by internal fertilization; water is not necessary to carry sperm to the eggs.

Further adaptations Other adaptations allowed animals to survive on land. The smooth, moist skin of amphibians gave way to the horny scales or plates that form the dry skin of reptiles, to the feathers of birds, and to the hair of mammals. Reptile skin prevented water loss, whereas feathers enabled flight, and hair provided warmth and protection.

Thinking Critically

Invertebrates also moved from the sea to the land, but their adaptations varied. Mosquitoes and many other insects still lay eggs in water. What adaptations do these organisms have that allow them to survive on dry land?

A BROADER VIEW

The First Landlubbers

Purpose

Students will gain an understanding about how life evolved from water organisms to land organisms.

Teaching Strategies

- To introduce the feature, display a live frog, a snake skin, feathers, and a live mouse or mammal pelt. Ask students how these outer coverings differ and how these coverings reflect adaptations animals have to their environments. Ask them what other features of the animals displayed make them adapted to their environments. L1

Possible Answer

Mosquitoes, like other insects, have an exoskeleton that helps prevent water loss. They also have wings for moving through air.

Science, Technology, and Society

Cytochrome C

A protein called cytochrome C is present in the mitochondria of almost all organisms. Scientists study the amino acid sequence of protein samples taken from different animals to help determine how closely related the animals are. For example, when comparing cytochrome C of various animals, it has been found that there are about 30 amino acid differences between each 100 amino acids of vertebrates and insects, and about 15 differences per 100 amino acids between the cytochrome C of mammals and reptiles.

GLENCOE TECHNOLOGY

Videodisc

The Infinite Voyage: Secrets from a Frozen World
The Southern Ocean—A Rich Marine Ecosystem (Ch. 1)

3 Assess

Check for Understanding

Ask students to make a table describing the features of the three basic body plans. L1

Reteach

Give students a variety of kitchen utensils and ask them to describe what type of symmetry they have. L1

Extension

Ask a group of students to collect specimens from a freshwater pond. Have them make drawings of the specimens and demonstrate their body plans and symmetry to the class. L2

Assessment

Portfolio: Ask students to remember from their childhood their favorite stuffed animal. Have them make a sketch of it, describe its symmetry, and its basic body plan. They should include this report in their portfolios. L1 P

4 Close

Discussion

Have students examine the phylogenetic diagram in Chapter 20. Ask them to discuss the evolutionary trends of symmetry, cell layers, and patterns of development of the animal groups shown on the diagram.

Figure 28.14
The exoskeleton of this crab has been shed in order for the crab to grow. A new exoskeleton forms in a few hours to again provide protection.

Animal Protection and Support

During the course of evolution, as development of body cavities resulted in more-complex animal body plans, many animals became adapted to life on land. Some of these classes of animals developed exoskeletons. An **exoskeleton** is a hard, waxy covering on the outside of the body that provides a framework for support, ***Figure 28.14.*** Exoskeletons prevent water loss and provide protection. Exoskeletons extend into the body, provide a place for muscle attachment, and are common in invertebrates. An **invertebrate** is an animal that does not have a backbone. Some invertebrates such as crabs, spiders, grasshoppers, dragonflies, and beetles have exoskeletons, while others do not.

An endoskeleton, or internal skeleton, is a characteristic of vertebrates. An **endoskeleton** is a support framework housed within the body. A **vertebrate** is an animal with a backbone. Examples of vertebrates include fishes, birds, reptiles, amphibians, and mammals, including humans. The endoskeleton protects internal organs and provides an internal brace for muscles to pull against.

Connecting Ideas

Animal body plans increased in complexity during the course of evolution from asymmetrical sponges to the bilaterally symmetrical flatworms, roundworms, and all other classes of animals. Animals can be differentiated depending upon the presence or absence of body cavities. Some animals have exoskeletons that provide protection and support, whereas others have endoskeletons that serve the same functions. How similar, and how different, animals are will be the focus of the following chapters.

Section Review

Understanding Concepts

1. Explain the difference between radial and bilateral symmetry in animals, and give an example of each type.
2. Compare the body plans of acoelomate and coelomate animals. Give an example of an animal with each type of body plan.
3. Explain how an adaptation such as an exoskeleton enabled animals to survive on land.

Thinking Critically

4. Make a list of five household items that are asymmetrical, five that have radial symmetry, and five that have bilateral symmetry.

Skill Review

5. **Making and Using Tables** Construct a table that compares the body plans of the sponge, hydra, flatworm, roundworm, and earthworm. For more help, refer to Organizing Information in the *Skill Handbook.*

Answers to Section Review

1. Animals with radial symmetry, such as the hydra, can be divided along any plane into roughly equal halves. Animals with bilateral symmetry, such as the flatworm, can be divided down their length into similar right and left halves that are mirror images.
2. Animals with acoelomate bodies, such as the flatworm, have three cell layers with a digestive tract, but no body cavity. Animals with coelomate bodies, such as earthworms, have a coelom in which the digestive tract and other internal organs are attached by double layers of mesoderm and are suspended within the fluid-filled coelom.
3. An exoskeleton prevents water loss from body organs and provides a framework for supporting an animal's body.

Thinking Critically

4. Asymmetrical: some cookie cutters, some potholders, some keys, houseplants, some vases. Radial: lamp shade, drinking glass, pickle jar, wire whisk, plate. Bilateral: spoon, fork, spatula, broom, dry cereal box.

REVIEWING MAIN IDEAS

28.1 Typical Animal Characteristics

- Animals are heterotrophs, digest their food inside the body, typically have a type of locomotion, and are multicellular. Animal cells have no cell walls.
- Embryonic development from a fertilized egg is similar in many animal phyla. The sequence after division of the fertilized egg is as follows: the formation of a blastula, with one layer of cells; a gastrula with two cell layers, ectoderm and endoderm; and finally mesoderm, a layer of cells in between the ectoderm and endoderm.

28.2 Body Plans and Adaptations

- Animals have a variety of body plans and types of symmetry that are adaptations.
- Animal symmetry includes radial symmetry, bilateral symmetry, and asymmetry.
- Acoelomate animals such as flatworms have flattened, solid bodies with no body cavities.
- Animals with pseudocoeloms, such as roundworms, each have a body cavity that develops between the endoderm and mesoderm.
- Coelomate animals such as humans and insects have internal organs suspended in a body cavity that is completely surrounded by mesoderm.
- Exoskeletons provide a framework of support on the outside of the body, whereas endoskeletons provide internal support.

Key Terms

Write a sentence that shows your understanding of each of the following terms.

acoelomate	gastrula
anterior	invertebrate
asymmetry	mesoderm
bilateral symmetry	posterior
blastula	protostome
coelom	pseudocoelom
deuterostome	radial symmetry
dorsal	sessile
ectoderm	symmetry
endoderm	ventral
endoskeleton	vertebrate
exoskeleton	

Understanding Concepts

1. All animals are ___________ because they obtain energy and nutrients from other sources.
   a. autotrophic c. omnivores
   b. heterotrophic d. carnivores
2. Which animal obtains food by filtering it from the surrounding water?
   a. angel fish c. earthworm
   b. starfish d. barnacle
3. A(n) ___________ is an animal without a backbone.
   a. invertebrate c. fish
   b. vertebrate d. mammal
4. A(n) ___________ is an example of an animal that moves only during its larval stage.
   a. osprey c. coral
   b. flatworm d. starfish
5. Which of the following is NOT a characteristic of animals?
   a. cells with cell walls
   b. multicellular organisms
   c. are consumers
   d. break down food
6. An animal with a backbone is a(n) _____.
   a. invertebrate c. insect
   b. vertebrate d. arachnid

Chapter 28 Review

Reviewing Main Ideas

Summary statements can be used by students to review the major concepts of the chapter.

Key Terms

Answers should go beyond defining the terms. Accept any answer that uses the term correctly and in the proper context.

Understanding Concepts

1. b
2. d
3. a
4. c
5. a
6. b

Skill Review

5. Make sure the tables indicate how many cell layers there are in each animal and whether they have a coelom, pseudocoelom, or no coelom.

Chapter 28 Review

7. c
8. c
9. d
10. a
11. c
12. a
13. a
14. b
15. a
16. d
17. c
18. c
19. d
20. d

Applying Concepts

21. No, the opening in the gastrula of a bird develops into the anus.
22. Fishes have coeloms that their muscles can push against as leverage in their swimming motions. Acoelomate flatworms have thin bodies that aid diffusion through the layers of cells.
23. Asymmetrical organisms have no front or back end; thus, they have little direction to their movement and cannot pursue food. They wait for food to come to them.
24. Most animals move to get food or cause food to be drawn toward them. Most animals digest their food inside their bodies. Fungi usually live on or in their food source and do not move to find food. Fungi secrete chemicals that dissolve food outside their bodies before they take it in.
25. If it is multicellular, it is an animal rather than a protist.

Chapter 28 Review

7. ____________ are animals that can move from one location to another only in the early stages of their lives.
   a. Flatworms c. Sponges
   b. Blastulas d. Protostomes
8. Animals that have a body cavity with internal organs between the endoderm and mesoderm are ____________ animals.
   a. acoelomate
   b. pseudocoelomate
   c. coelomate
   d. sessile
9. In a process called ____________, the unicellular zygote of a sea urchin divides to form two new cells.
   a. gastrulation c. molting
   b. hatching d. cleavage
10. Which of the following animals has only one opening to its digestive tract?
    a. flatworm c. grasshopper
    b. earthworm d. frog
11. What is the earliest developmental stage of an embryo?
    a. gastrula c. blastula
    b. zygote d. endoderm
12. The ____________ cells of the gastrula eventually develop into the skin and nervous tissue of the animal.
    a. ectoderm c. mesoderm
    b. endoderm d. deuterostome
13. An animal with a mouth that develops from the opening in the gastrula is called a ____________.
    a. protostome c. zygote
    b. deuterostome d. blastula
14. The ____________ is the embryonic tissue that gives rise to muscles, reproductive organs, and circulatory vessels.
    a. endoderm c. ectoderm
    b. mesoderm d. zygote
15. ____________ are animals with flattened, solid bodies but no body cavities.
    a. Acoelomates
    b. Pseudocoelomates
    c. Coelomates
    d. Vertebrates
16. Fishes have one fin along their backs. Because fishes are bilaterally symmetrical, this fin is called the ____________ fin.
    a. radial c. ventral
    b. posterior d. dorsal
17. A ____________ is an example of an animal with radial symmetry.
    a. grasshopper c. hydra
    b. flatworm d. muskrat
18. Organisms that don't move from place to place are ____________ animals.
    a. protostome c. sessile
    b. deuterostome d. acoelomate
19. Of these, which is NOT an example of a vertebrate animal?
    a. nurse shark c. coral snake
    b. black bear d. garden spider
20. The ____________ is a framework of support on the outside of the body of an animal.
    a. endoskeleton c. ectoderm
    b. endoderm d. exoskeleton

Applying Concepts

21. During the development of an egg, a blastula forms first, followed by formation of a gastrula. The mouth of the embryo develops from the opening in the gastrula. Will the embryo develop into a bird? Explain.
22. How are body plans of animals related to the environment in which they live? Give two examples.
23. Why are asymmetrical organisms often sessile? Explain.
24. How is obtaining food different in animals and fungi?
25. You observe an unfamiliar organism from a tide pool under a microscope. When you offer live invertebrate food, the organism seems to pursue and consume this prey. How could you tell whether the organism is a protist or an animal?

Program Resources

Chapter Assessment, pp. 167-172 L1
Alternate Assessment in the Science Classroom
Computer Test Bank L1
Content Mastery, pp. 113-116 L1

Thinking Critically

26. ***Concept Mapping*** Make a concept map that relates the following terms and phrases. Supply the appropriate linking words for your map.
blastula, ectoderm, gastrula, endoderm, protostome, deuterostome, mesoderm, fertilized egg, development, larva, adult

27. ***Interpreting Scientific Illustrations*** Assume that you are a biologist observing the development of a sea urchin egg. You observe the following sequence. At which stage does the process begin to show abnormal development?

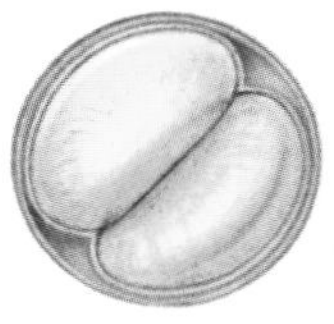
1

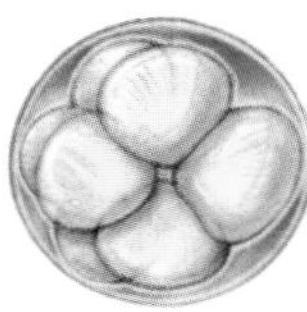
2

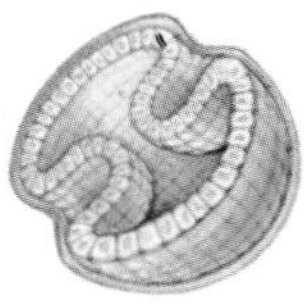
3

28. ***Observing and Inferring*** You are looking at a preserved specimen of an unidentified animal. It is easy to see that it has radial symmetry. What would you predict about its direction of movement when alive? Explain.

29. ***Designing an Experiment*** Assume that you are a zookeeper working with reptiles. You are given a group of rare lizards about which little is known. In order to provide an adequate habitat, you must find out what temperature the lizards prefer. Design an experiment safe for the lizards that would determine their preferred environmental temperature.

30. ***Observing and Inferring*** Explain why the development of a body cavity enabled animals to move and feed more efficiently.

ASSESSING KNOWLEDGE & SKILLS

The following diagrams represent three different animal body plans.

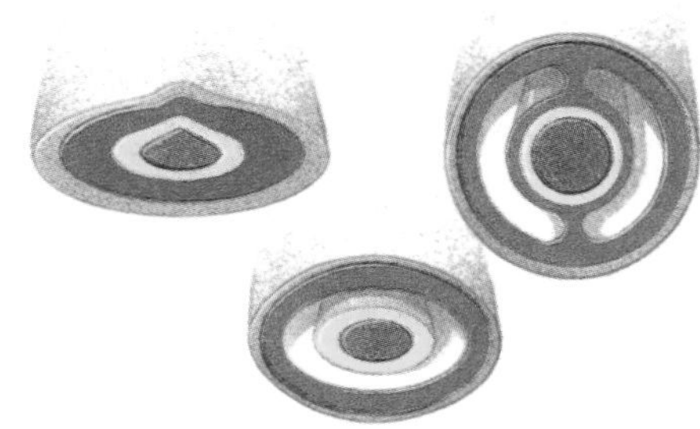

Interpreting Scientific Illustrations Use the diagram to answer the following questions.

1. Which body plan would be capable of more complex and powerful movement?
   a. 1 **b.** 2 **c.** 3 **d.** all of these
2. Which type of body plan belongs to acoelomate animals such as flatworms?
   a. 1 **b.** 2 **c.** 3 **d.** none of these
3. Which type of body plan belongs to pseudocoelomate animals such as roundworms?
   a. 1 **b.** 2 **c.** 3 **d.** none of these
4. Which type of body plan is more likely to be seen in animals that inhabit land environments?
   a. 1 **b.** 2 **c.** 3 **d.** all of these
5. ***Making a Table*** Make a table that compares the three body plans. For each body plan, indicate an example of an animal with that type, how the coelom differs in each, and probable relative speed of movement of each.

Chapter 28 Review

Thinking Critically

26. Evaluate students' concept maps. The maps should show an understanding of the relationships among the concepts listed.
27. The third stage (3) shows two areas of indentation; only one area indents in a properly developing embryo.
28. Radially symmetrical animals don't have an anterior or posterior end; they don't move forward and backward as easily as bilaterally symmetrical animals.
29. Place lizards in a long, wide pipe in which the temperature can be varied. Thermometers should be placed at close intervals. Over a 24-hour period, note the area in which most of the lizards spend most of their time.
30. Animals with no body cavities depend upon diffusion to move food into and wastes out of cells; animals with body cavities had space to develop organ systems that delivered food and took away wastes. Movement in acoelomate animals depends upon structures such as cilia, whereas coelomate animals have structures muscles can work against for locomotion.

ASSESSING KNOWLEDGE & SKILLS

1. c
2. a
3. b
4. c

5.

Comparison of Three Body Plans			
Body Plan	**Example**	**Coelom**	**Relative Speed**
Acoelomate	flatworm	none	slow
Pseudo-coelomate	roundworm	fluid-filled, partly lined with mesoderm	faster
Coelomate	fish	fluid-filled, lined with mesoderm	fastest

Chapter Organizer

SECTION	OBJECTIVES	ACTIVITIES/FEATURES
29.1 Sponges National Science Standards: UCP.1, UCP.5; C.1, C.5	**1. Relate** the sessile life of sponges to their food-gathering adaptations. **2. Relate** the structure of sponges to their reproductive adaptations.	
29.2 Cnidarians National Science Standards: UCP.1, UCP.3, UCP.5; C.1, C.3, C.5; F.3	**3. Distinguish** the different classes of cnidarians. **4. Relate** the polyp and medusa forms to the reproductive cycle of jellyfishes. **5. Analyze** the adaptations of cnidarians for obtaining food.	**Minilab:** How do nematocysts function?, p. 718 **Biology & Society Technology:** Sunscreens from Coral, p. 720 **Thinking Lab:** How does coral reef bleaching change over time?, p. 721
29.3 Flatworms National Science Standards: UCP.1, UCP.5; A.1; C.5, C.6; F.1, F.5; G.1	**6. Distinguish** the adaptive structures of parasitic flatworms and planarians. **7. Explain** how parasitic flatworms are adapted to their way of life.	**Biolab:** How Planarians Respond to Stimuli, p. 726 **Health Connection:** Pigs and Parasites, p. 728
29.4 Roundworms National Science Standards: UCP.1, UCP.5; C.5, C.6; E.1, E.2; F.1, F.5; G.1, G.2	**8. Compare** the structural adaptations of roundworms and flatworms. **9. Identify** the characteristics of four human roundworm parasites.	**Minilab:** How can you recognize a roundworm?, p. 731

ACTIVITY MATERIALS

BIOLAB	MINILABS	ALTERNATE LAB
page 726 planaria culture small pebbles eyedropper test tubes pond water ice thermometer petri dishes egg whites, cooked construction paper, black toothpick hot plate	**page 718** hydra culture slides coverslips methylene blue dropper paper towel compound microscope vinegar **page 731** garden soil petri dish hand lens toothpick	**page 712** stereomicroscope forceps balance wax marking pencil 150-mL beakers petri dish bottoms sea sponges (grass, yellow, sheep's wool, and hard head) unused synthetic sponges (four types)

Chapter 29 Sponges, Cnidarians, Flatworms, and Roundworms

TEACHER CLASSROOM RESOURCES

Reproducible Masters	Transparencies
Section Focus Master 62: Sponges L1 SAE **Reinforcement and Study Guide,** p. 115 L1 **Critical Thinking/Problem Solving:** Sponge Camouflage, p. 29 L3 **Concept Mapping:** Sponges, p. 29 L1	**Basic Concepts Transparency #45:** Body Structure of Sponges and Cnidarians L1 SAE **Basic Concepts Transparency #46:** Phylogeny of Sponges and Cnidarians L1 SAE
Section Focus Master 63: Cnidarians L1 SAE **Reinforcement and Study Guide,** p. 116 L1 **Biolab and Minilab Worksheets,** p. 115 L1 **Laboratory Manual:** Hydra Behavior, pp. 163-164 L2 **Content Mastery,** pp. 117-120 L1	**Basic Concepts Transparencies #45, 46:** L1 SAE **Basic Concepts Transparency #47:** Life Cycle of a Jellyfish L1 SAE
Section Focus Master 64: Tapeworm L1 SAE **Reinforcement and Study Guide,** p. 117 L1 **Biolab and Minilab Worksheets,** pp. 117-118 L1	**Reteaching Transparency #29:** Life Cycle of a Tapeworm L1 SAE
Section Focus Master 65: Body Plans L1 SAE **Reinforcement and Study Guide,** p. 118 L1 **Biolab and Minilab Worksheets,** p. 116 L1	

ASSESSMENT MATERIALS

Chapter Assessment, pp. 173-178
MindJogger Videoquiz
Alternate Assessment in the Science Classroom
Computer Test Bank

Spanish Resources SAE
English/Spanish Audiocassettes SAE
Cooperative Learning in the Science Classroom COOP LEARN
Lesson Plans
Biology Projects: Investigating Invertebrate Life in a Marine Aquarium, pp. 29-32 L2

KEY TO TEACHING STRATEGIES

- L1 Level 1 activities should be within the ability range of all students including those with learning difficulties.
- L2 Level 2 activities are within the ability range of average to above-average students.
- L3 Level 3 activities are designed for the ability range of above-average students.
- SAE SAE activities should be within the ability range of Students Acquiring English.
- COOP LEARN Cooperative Learning activities are designed for small group work.
- P These strategies represent student products that can be placed into a best-work portfolio.
- These strategies are useful in a block scheduling format.

GLENCOE TECHNOLOGY

The following multimedia resources are available from Glencoe.

Biology: The Dynamics of Life
CD-ROM SAE
Videodisc Program

Science and Technology Videodisc Series (STVS)
Animals
Jellyfish
Sheep Parasite
Blood Fluke Life Cycle
Nematodes
Ecology
Evaluating Artificial Reefs

CHAPTER

29 Sponges, Cnidarians, Flatworms, and Roundworms

Chapter Overview

Students begin this chapter by studying sponges and their adaptations. Next, students study cnidarians. Students explore the ecology and adaptations of these organisms as they compare and contrast the structures of hydra, sea anemones, jellyfishes, and coral. The complexity of cnidarians when compared to sponges is discussed and their evolutionary connections to other groups are considered.

In the third section of the chapter, students learn about flatworms. The free-living planarians are presented along with their parasitic relatives—the flukes and tapeworms. The adaptations of each animal to its way of life are considered and their origins and importance discussed.

The chapter concludes with roundworms. The relatively complex structures of roundworms are compared to structures in flatworms. Free-living roundworms are considered along with the parasitic *Ascaris*, hookworm, *Trichinella*, pinworm, and plant nematodes.

Key Terms

- external fertilization
- filter feeding
- gastrovascular cavity
- hermaphrodite
- internal fertilization
- medusa
- nematocyst
- nerve net
- pharynx
- polyp
- proglottid
- scolex

Look for the following logo for strategies that emphasize different learning modalities. LS

Visual-Spatial	**Activity, p. 712; The Inside Story, pp. 713, 717, 724; Demonstration, pp. 717, 719; Minilab, p. 718; Meeting Individual Needs, p. 720; Display, pp. 724, 731; Student Journal, pp. 728, 730**
Interpersonal	**Portfolio, p. 720**
Intrapersonal	**Minilab, p. 731; Portfolio, p. 731**
Linguistic	**Portfolio, p. 725**
Logical-Mathematical	**Alternate Lab, p. 712; Portfolio, pp. 714, 719; Thinking Lab, p. 721**

CHAPTER

29 Sponges, Cnidarians, Flatworms, and Roundworms

What kinds of animals would you expect to see if you could visit a 600-million-year-old zoo? Would the animals resemble those found in today's zoos? A 600-million-year-old zoo would probably be called an aquarium today. That's because the earliest animals—sponges, jellyfishes, and worms—were found in the warm, shallow seas that covered most of Earth's surface.

The ancestors of all modern invertebrates and vertebrates had simple body plans. Sponges, jellyfishes, and worms that lived in water obtained most of their food, oxygen, and other materials directly from their surroundings. But to scientists, the living descendants of these early animals are indicators of how successful these simple body plans were for food gathering, reproduction, and digestion because they still exist in animals today. By studying these surviving species, it is possible to trace the evolutionary history of body systems.

Concept Check

You may wish to review the following concepts before studying this chapter.

- Chapter 1: adaptation
- Chapter 20: phylogenetic diagram of the six kingdoms
- Chapter 28: symmetry, body plans

Chapter Preview

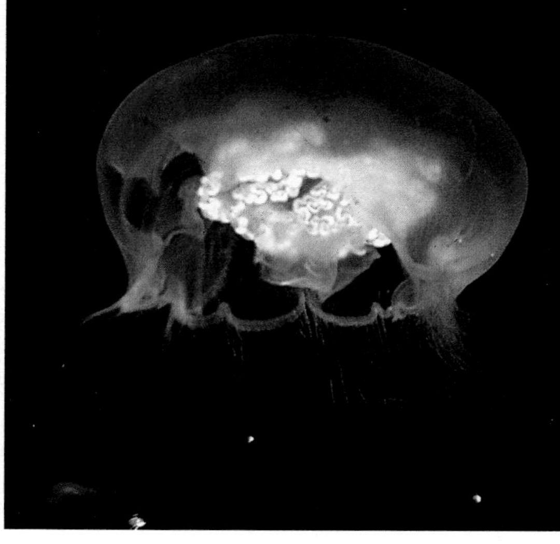

The moon jellyfish and the sponge share common traits. Both are invertebrates with simple body plans that live in water. Do all invertebrates live in water?

Assessment Planner

Choose assessment strategies from the following pages to evaluate the progress of your students.

Assess, pp. 714, 722, 728, 732
Alternate Lab, pp. 712–713
Minilabs, pp. 718, 731
Portfolio, pp. 714, 719, 720, 731
Thinking Lab, p. 721
Biolab, pp. 726-727
Chapter Review, pp. 733-735

CHAPTER 29

Introducing the Chapter

Show students slides of ocean scenes depicting coral reefs and sponges. Identify the organisms as animals and ask students to explain features they may have in common. *All are multicellular, heterotrophic, and lack cell walls in their cells.* Bring out the concepts of simple body plans and structures and the fact that these animals all have soft bodies.

Ask students to examine the photographs of the jellyfish and the sponge. Ask them to relate tales of encounters they or others they know may have had with jellyfishes.

Theme Development

The themes of *evolution* and *homeostasis* are emphasized in this chapter. Evolutionary relationships among the animal phyla are stressed, as are adaptations to the environment and the homeostatic mechanisms at work in the different animals.

Concept Check

Students will build upon their understanding of evolution and adaptations. They will also be asked to recall information learned in Chapter 28 about body plans and symmetry.

SECTION

29.1

Prepare

Key Concepts

Students will learn the main features of sponges and discuss their adaptations, origins, and ecology.

Block Scheduling

Look for this symbol for strategies that are useful in a block scheduling format. For more information on block scheduling, refer to Section 29.1 in the **Lesson Plans** booklet.

Materials

- Order live freshwater *Spongilla* and gather hand lenses for the Focus Activity.

1 Focus

Bellringer

Before presenting the lesson, display **Section Focus Master 62** on the overhead projector and have students answer the accompanying questions. L1 SAE

Activity

LS **Visual-Spatial** Have students observe the live freshwater sponge, *Spongilla*, with hand lenses. Ask students to note the asymmetrical shape of the sponge and its many pores. Explain that all sponges have a large number of pores.

SECTION

29.1 Sponges

Section Preview

Objectives

Relate the sessile life of sponges to their food-gathering adaptations.

Relate the structure of sponges to their reproductive adaptations.

Key Terms

filter feeding
hermaphrodite
external fertilization
internal fertilization

Is this red organism a plant or an animal? At first glance, it may look like a plant because it is colorful and doesn't move from place to place, but it is an animal. How do we know this organism is an animal? Like snakes, tigers, and you, this organism is multicellular and a heterotroph—two characteristics that place it in the animal kingdom. This sessile animal is a sponge.

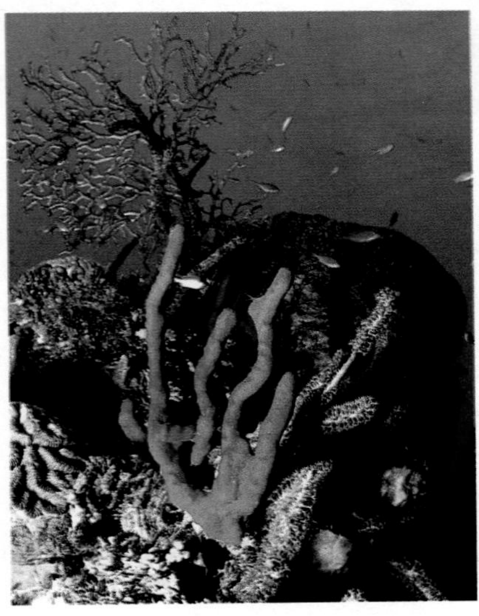

What Is a Sponge?

Sponges are asymmetrical animals that have a variety of colors, shapes, and sizes. Many are bright shades of red, orange, yellow, and green. Some sponges are ball shaped; others have many branches. Sponges can be as small as a quarter or as large as a door. Although sponges do not resemble more-familiar animals, they carry on the same life processes as all animals. *Figure 29.1* shows a natural sponge harvested from the ocean.

Figure 29.1

This heavy bath sponge is dark brown or black in its natural habitat. After harvest, it is washed and dried in the sun. When the process is complete, only a pale, lightweight skeleton made of spongin remains.

Where do sponges live?

Sponges are classified in the invertebrate phylum Porifera, which means "pore-bearer." Of the 5000 species of sponges, most live in the ocean, but a few species can be found in freshwater environments. Most occupy shallow waters, but some species have been discovered at depths up to 8500 m.

No matter where sponges live, they are sessile organisms. Remember, a sessile organism is one that remains permanently attached to a surface for all of its adult life. Because adult sponges are sessile, they can't travel in search of food. Sponges get their food by **filter feeding,** a method in which organisms feed by filtering small particles of food from water as it passes by or through some part of the organism. Filter feeding is a common feeding strategy in marine organisms.

Alternate Lab **Natural and Synthetic Sponges**

Purpose

LS **Logical-Mathematical** Students will compare the water-holding capacity and microscopic appearance of natural and synthetic sponges.

Materials

stereomicroscope, forceps, balance, wax marking pencil, 150-mL beakers, petri dish bottoms, sea sponges (one piece each of four different types: grass, yellow, sheep's wool, hard head—each piece 3 cm × 3 cm × 2 cm), unused synthetic sponges (one piece each of four different types—each piece 3 cm × 3 cm × 2 cm)

Procedure

Give students the following directions.

1. Examine each piece of sponge at its thinnest point under the microscope.
2. Draw the skeletal framework of each sponge. Label each sponge piece.
3. Predict which type of sponge will hold more water. Base your prediction on microscopic examination of the sponge structures.
4. Place each sponge piece in a petri dish and obtain its mass. Soak each

THE INSIDE STORY

Inside a Sponge

Sponges have no tissues, organs, or organ systems. The body plan of a sponge is simple, being made up of only two cell layers with no body cavity. Between these two cell layers is a jellylike substance that contains cells, as well as the components of the sponge's internal support system. Sponges have four types of cells that perform all the functions necessary to keep them alive.

Orange tube sponges

1. Water and wastes are expelled through the osculum, the large opening at the top of the sponge. A sponge no bigger than a pen can move more than 20 L of water through its body per day.

2. Surrounding each pore is a single pore cell. Pore cells bring water carrying food and oxygen into the sponge's body.

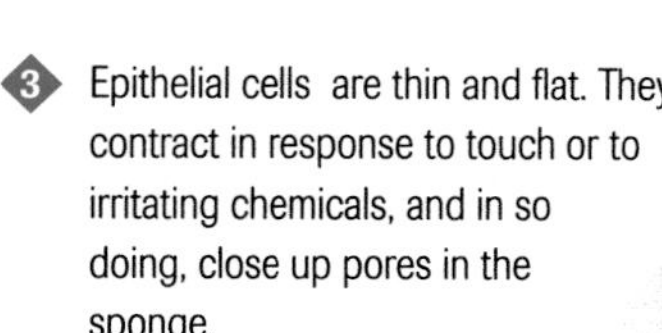

3. Epithelial cells are thin and flat. They contract in response to touch or to irritating chemicals, and in so doing, close up pores in the sponge.

Osculum
Pore cells
Collar cells
Pore cell
Direction of water flow
Amoebocytes
Epithelial cells
Spicules

4. Lining the interior of sponges are collar cells. Each collar cell has a flagellum that whips back and forth, drawing water through the pores of the sponge.

5. Between the two cell layers of a sponge's body are amoeba-shaped cells called amoebocytes. Amoebocytes carry nutrients to other cells, aid in reproduction, and also produce chemicals that help make up the spicules of sponges.

6. Spicules are structures produced by other cells that form the hard support systems of sponges. Spicules are small, needlelike structures located between the cell layers of a sponge.

THE INSIDE STORY

Inside a Sponge

Purpose

LS **Visual-Spatial** Students will observe the basic features of a sponge and examine how it accomplishes filter feeding.

Teaching Strategies

- Explain that sponges obtain food in a process called filter feeding.

Visual Learning

- Point out that food particles in water are pulled into collar cells and digested. Nutrients from food are then distributed by amoebocytes to other body cells.

2 Teach

Misconception

Students often think sponges are plants because they do not move. Review the basic characteristics of plants and animals. Explain that like all animals, sponges are heterotrophs and lack cell walls around their cells.

sponge piece in a beaker of water for 10 minutes and again obtain the mass of each sponge.

5. Calculate the mass of the water absorbed by each sponge. Compare the data for all sponges.

Analysis

Ask students the following questions.

1. Which sponges, natural or synthetic, have greater water-holding capacity? *Natural*
2. Of what adaptive value is it for a sponge to be able to take in large amounts of water? *They are filter feeders. Taking in more water increases the chances of taking in more food.*
3. Was your hypothesis supported by your data? *If students said that the natural sponge would hold more water, their data most likely supported their hypothesis.*

Assessment

Skill: Design and conduct an experiment to determine if the temperature of water affects the water-holding capacity of a sponge. **L1**

Chemistry Connection

Sponge Chemicals

Some sponges give off toxic chemicals that deter predators and the buildup of other sessile animals on their exterior surfaces. One of these chemicals, from a Caribbean sponge, is currently being used to treat cancer. Other chemicals produced by sponges are being used to fight fungus infections.

Sponges also use a Super-glue-type substance to attach themselves to surfaces. Researchers have genetically engineered this material for use in medicine for the repair of certain tissues in humans.

Visual Learning

Figure 29.2 How would these sponges compare genetically? *They would be identical genetically.*

Reinforcement

Review the end results of mitosis and meiosis. Explain that both fragmentation and regeneration involve mitosis.

Enrichment

Have your advanced students reread the description of fragmentation. Challenge these students to diagram this process and then work with less-able students to review fragmentation. L1 SAE

From one cell to cell organization

If you took a living sponge and put it through a sieve, a type of filter, you would witness a rather remarkable event. Not only would you see the sponge's many cells alive and separated out, but also, within an hour, you would be able to see these same cells coming together to form a whole sponge once again. Many biologists hypothesize that sponges evolved directly from colonial, flagellated protists, such as *Volvox*. More importantly, however, sponges demonstrate what appears to have been a major step in the evolution of animals—the change from a unicellular life to a division of labor among groups of organized cells.

Reproduction in Sponges

Sponges reproduce both sexually and asexually. Sponges reproduce asexually by fragments that break off from the parent animal and form new sponges, or by forming external buds. Buds may break off, float away, and eventually settle and become separate animals. Sometimes the buds remain attached to the parents, forming a colony of sponges. You can see a colony in *Figure 29.2.*

Most sponges reproduce sexually. Sponges are **hermaphrodites;** that is, an individual sponge can produce both eggs and sperm, though at different times. Eggs and sperm are formed from amoebocytes. During reproduction, sperm released from one sponge are carried by water currents to other sponges, where fertilization occurs.

Fertilization in sponges may be either external or internal. In **external fertilization,** the eggs and sperm are both released into the water, and fertilization occurs outside the animal's body. In **internal fertilization,** eggs remain inside the animal's body, and sperm are carried to the eggs. In sponges, the collar cells collect the sperm and transfer them to amoebocytes. The amoebocytes then transport the sperm to ripe eggs. Most sponges reproduce sexually through internal fertilization. The result is the development of free-swimming, flagellated larvae (singular, larva), *Figure 29.3.* Sponges also have the ability to regenerate lost body parts. In regeneration, lost body parts are replaced by mitosis.

Figure 29.2

Sponge colonies are the result of asexual reproduction. How would these sponges compare genetically?

3 Assess

Check for Understanding

Give students a large piece of butcher paper and colored markers. Ask them to draw a large sponge indicating epithelial cells, pore cells, collar cells, amoebocytes, spicules, and osculum. L1

Program Resources

Section Focus Master 62 L1 SAE
Reinforcement and Study Guide, p. 115 L1
Concept Mapping, p. 29 L1
Critical Thinking/Problem Solving, p. 29 L3
Basic Concepts Transparencies 45, 46 and **Masters** L1 SAE

PORTFOLIO

Sponge Symbionts Have students assume that a particular marine sponge harbors single-celled algae as symbionts in its cells. Ask students to make a sketch and describe in a paragraph what the best shape would be for this sponge to enable the algae to get maximum sunlight. L2 P LS

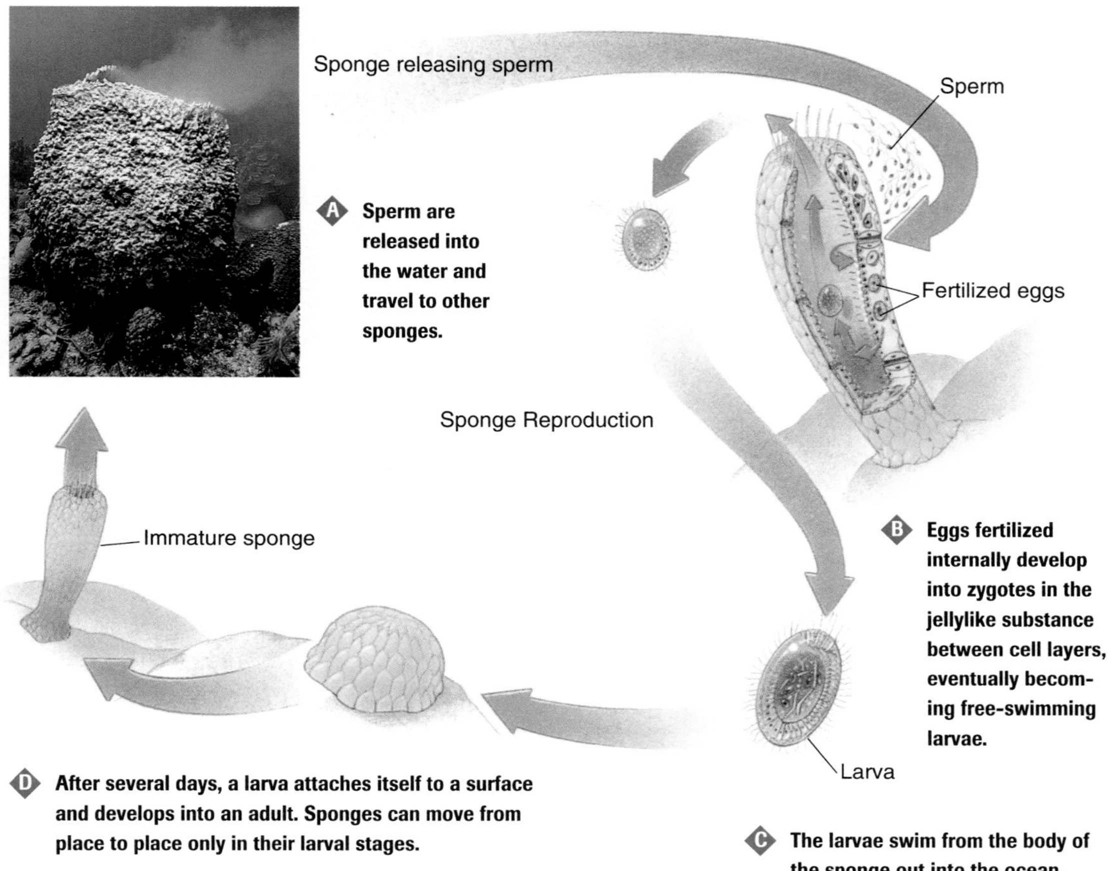
Sponge releasing sperm
Sperm
A Sperm are released into the water and travel to other sponges.
Fertilized eggs
Sponge Reproduction
Immature sponge
B Eggs fertilized internally develop into zygotes in the jellylike substance between cell layers, eventually becoming free-swimming larvae.
Larva
D After several days, a larva attaches itself to a surface and develops into an adult. Sponges can move from place to place only in their larval stages.
C The larvae swim from the body of the sponge out into the ocean.

Figure 29.3

Sponges reproduce sexually when the sperm from one sponge fertilize the eggs of another sponge.

Section Review

Understanding Concepts

1. How does a sponge obtain food?
2. Relate the functions of the four cell types in sponges.
3. Describe the steps involved in the sexual reproduction of sponges.

Thinking Critically

4. What advantages do multicellular organisms such as sponges have over unicellular organisms? Explain.

Skill Review

5. **Making and Using Tables** Make a table listing the cell types and other structures of sponges along with their functions. For more help, refer to Organizing Information in the *Skill Handbook.*

SECTION 29.1

Reteach

Have students construct an outline to summarize this section. Have them include the phylum, symmetry, habitat, food-getting process, oxygen-getting process, reproductive process, and means of protection. L1

Extension

Ask students to construct a three-dimensional, cutaway model of a sponge. Have students label the parts of their model. L2

Assessment

Performance: Have students make a flowchart that shows the events involved in sexual reproduction of sponges. L1

4 Close

Discussion

Discuss with students situations in which natural sponges are more desirable than synthetic sponges, such as for bathing. Explain that natural sponges are not used for many cleaning purposes because of their expense.

Answers to Section Review

1. Sponges take water into their bodies and filter out food. They are filter-feeders.
2. Epithelial cells perform protective functions. Pore cells bring in water. Collar cells draw water through the pores and trap food.
3. Sperm are released in water and travel to other sponges. Eggs fertilized internally develop into zygotes in the jellylike substance between cell layers, eventually becoming free–swimming larvae that later attach to a surface and develop into adult sponges.

Thinking Critically

4. Division of labor among cells enables the organism to carry out life functions more efficiently than a single cell can.

Skill Review

5. Students' tables should include the following features and their functions: epithelial cells, pore cells, collar cells, amoebocytes, spicules, and osculum.

SECTION

29.2

Prepare

Key Concepts

Students will learn about the important characteristics of cnidarians.

Block Scheduling

Look for this symbol for strategies that are useful in a block scheduling format. For more information on block scheduling, refer to Section 29.2 in the **Lesson Plans** booklet.

Materials

- Obtain a binocular microscope-projector assembly, live *Hydra*, and live *Daphnia* for the Demonstration.
- For the Minilab, gather slides, coverslips, compound microscopes, methylene blue, paper towels, vinegar, droppers, and live *Hydra*.

1 Focus

Bellringer

Before presenting the lesson, display **Section Focus Master 63** on the overhead projector and have students answer the accompanying questions. L1 SAE

Discussion

Explain to students that they are about to learn about animals called cnidarians. Point out that two characteristics shared by these organisms are soft, hollow bodies and stinging cells called cnidocytes—the structures for which the phylum gets its name. Next, explain that the phylum name for these organisms was once *Coelenterata*, which means "hollow-bodied." Explain that this name was changed because hollow bodies were not unique to the species included in this group.

SECTION

29.2 Cnidarians

Section Preview

Objectives

Distinguish the different classes of cnidarians.

Relate the polyp and medusa forms to the reproductive cycle of jellyfishes.

Analyze the adaptations of cnidarians for obtaining food.

Key Terms

polyp
medusa
nematocyst
gastrovascular cavity
nerve net

What's the largest structure ever built by living organisms? Is it the Sears Tower in Chicago? How about the Great Pyramid in Egypt? Actually, the largest structure ever built is the Great Barrier Reef, which extends for more than 2000 km off the northeastern coast of Australia, and it wasn't built by humans! This structure was built over many centuries by many different colonies of small marine invertebrate animals called corals. Corals and their relatives, jellyfishes and sea anemones, all belong to the phylum Cnidaria.

cnidarian: *knid* (GK) nettle, a plant with stinging hairs. *Cnidarians* (pronounced with a silent *c*) have stinging tentacles.

What Is a Cnidarian?

Cnidarians are a group of marine invertebrates made up of about 9000 species of jellyfishes, corals, sea anemones, and hydras. Most cnidarians are found worldwide, but some species of coral prefer the warmer oceans of the South Pacific and the Caribbean.

Though cnidarians are a diverse group, all possess the same basic body structure, which supports the theory that they had a single origin.

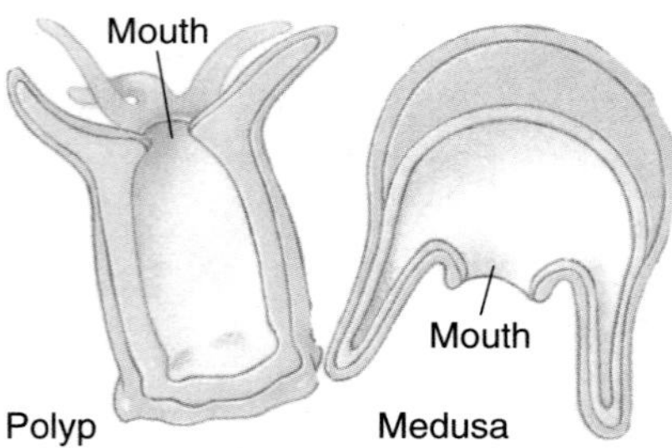

Figure 29.4
If you turn the polyp form of a cnidarian (left) upside down, you have the basic structure of a medusa (right).

All cnidarians have radial symmetry. Like sponges, a cnidarian's body has only one body opening, and the body is made up of two cell layers. Unlike sponges, however, the cell layers of cnidarians are organized into separate tissues with specific functions. Cnidarians have simple nervous systems, and both cell layers have cells that can contract like muscles. The two cell layers are derived from the ectoderm and endoderm of the embryo. The outer layer of cells serves to protect the cnidarian, whereas the inner layer is involved mostly in digestion.

Cnidarians display only two basic body forms, which occur at different stages of their life cycles. These are the polyp and medusa, *Figure 29.4*. A **polyp** is the stage with a tube-shaped body with a mouth surrounded by tentacles. A **medusa** is the stage with a body shaped like an umbrella with the tentacles hanging down.

Program Resources

Section Focus Master 63 L1 SAE
Reinforcement and Study Guide, p. 116 L1
Biolab and Minilab Worksheets, p. 115 L1
Laboratory Manual, pp. 163-164 L2
Basic Concepts Transparencies 45, 46, 47 and **Masters** L1 SAE

THE INSIDE STORY

Inside a Cnidarian

Cnidarians display a remarkable variety of shapes and sizes. Some, such as corals and hydras, can be as small as the tip of a pencil, while others, such as some jellyfishes, can be as long as half of a football field! All cnidarians, except sea anemones and corals, go through both polyp and medusa stages at some point in their life cycles. The flowerlike forms of sea anemones are often brilliant shades of red, purple, and blue, and the bodies of some jellyfishes contain glowing pigments.

Hydra

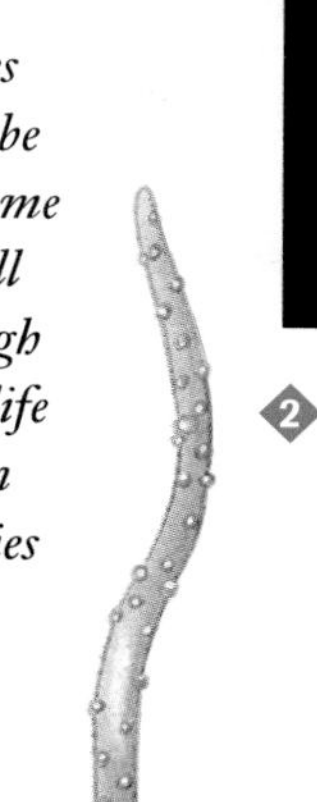

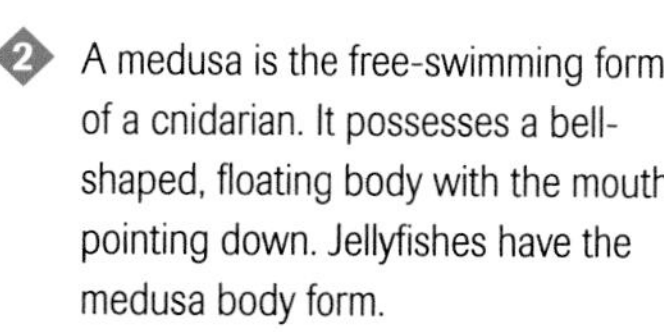

2. A medusa is the free-swimming form of a cnidarian. It possesses a bell-shaped, floating body with the mouth pointing down. Jellyfishes have the medusa body form.

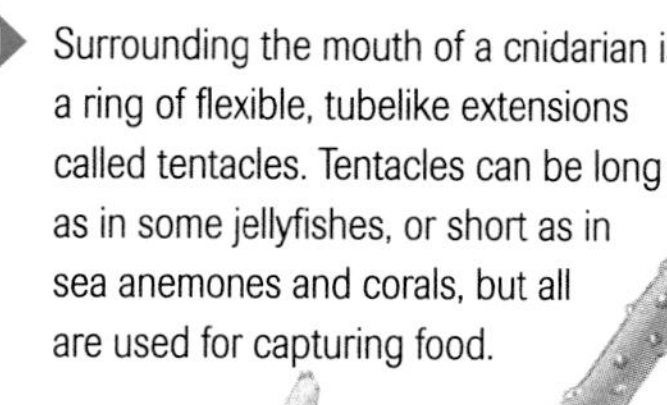

1. Surrounding the mouth of a cnidarian is a ring of flexible, tubelike extensions called tentacles. Tentacles can be long as in some jellyfishes, or short as in sea anemones and corals, but all are used for capturing food.

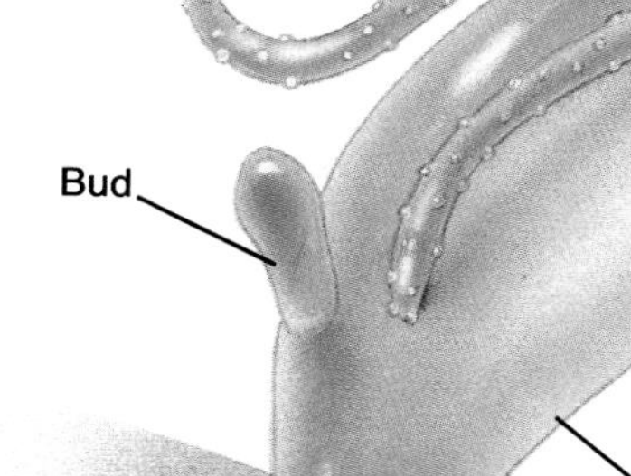

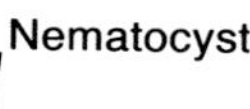

3. Located primarily at the tips of the tentacles are stinging cells that contain nematocysts. When prey touches the tentacles, the nematocyst releases a coiled tube that ejects a toxin that paralyzes the prey.

4. A polyp is the sessile form of a cnidarian. Polyps have mouths that point upward. Examples of polyps include sea anemones, corals, and hydras.

GLENCOE TECHNOLOGY

Videodisc

Biology: The Dynamics of Life
Disc 1, Side 2
Ocean Cnidarians (Ch. 31)

THE INSIDE STORY

Inside a Cnidarian

Purpose

LS Visual-Spatial Students will gain an understanding of the polyp and medusa forms of cnidarians, and learn about the tentacles and nematocysts common to these animals.

Visual Learning

- Emphasize that polyps are sessile forms of cnidarians that live attached to a surface. Be sure students recognize that the mouth of a polyp points in an upward direction so the organism can obtain food.
- Provide students with prepared slides of nematocysts on tentacles and small cnidarian polyps and medusae. Have students observe the slides and make labeled drawings of their observations.

2 Teach

Demonstration

LS Visual-Spatial Set up a binocular microscope connected to a projector. Place live *Hydra* under the microscope. Introduce *Daphnia* to the *Hydra* culture and ask students to observe the feeding behavior of *Hydra*. Discuss observations as a class.

MiniLab

Purpose

Visual-Spatial Students will observe nematocyst discharge in a hydra.

Process Skills

observe and infer, recognize cause and effect, interpret data

Teaching Strategies

- Caution students to gently cover their hydra with the coverslip.

Safety Precautions

- Explain that methylene blue will stain clothing and skin.
- Make sure students wash their hands at the end of the lab.

Expected results

Nematocysts will discharge when vinegar is added.

Analysis

1. Nematocysts paralyze or entangle potential predators. When prey touches the tentacle of a cnidarian, the nematocyst releases a coiled tube and ejects a toxin that paralyzes the prey.
2. Toxins paralyze or kill prey so the predator can consume it before it escapes.
3. Before discharge, the nematocyst would be in the epidermis. After discharge, the nematocysts are long filaments each with a pointed barb on the end.

Assessment

Portfolio: Ask students to write an evaluation of this lab for placement in their portfolios. Have them include their ideas about the survival value of nematocysts. L1 P

MiniLab

How do nematocysts function?

Some of the cells in the body surface of cnidarians contain nematocysts – tiny, harpoonlike structures with sharp barbs. These barbs are released by touch and when stimulated by certain chemicals. In this lab, you will observe nematocyst discharge in a hydra.

Procedure

1. Place a hydra on a clean, flat slide, and cover gently with a coverslip.
2. Place a drop of methylene blue at the edge of the coverslip. Draw the stain under the coverslip by placing a piece of paper towel on the opposite edge of the coverslip.
3. With a compound microscope, focus on the tentacles and nematocysts. Next, place a drop of vinegar at the edge of the coverslip, and draw it under the coverslip as you did in step 2. Vinegar will stimulate the release of nematocysts.

Analysis

1. Of what survival value are nematocysts? How do they function?
2. Why are toxins important adaptations in animals not capable of pursuing prey?
3. Draw a diagram of nematocysts before and after discharge.

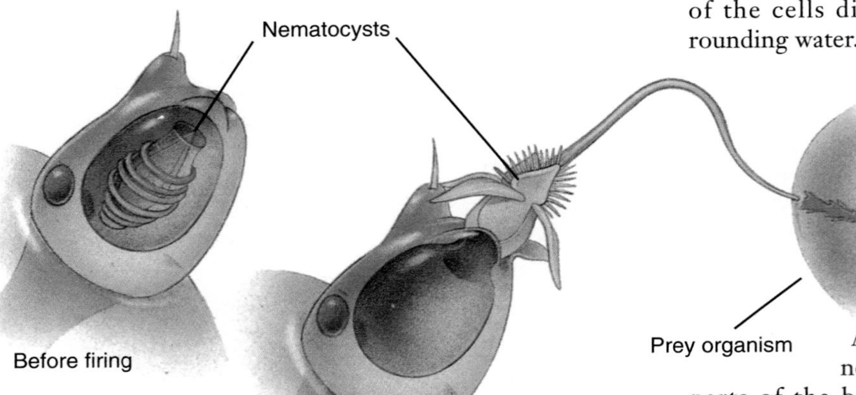

Figure 29.5

Nematocysts, located in special stinging cells in the outer cell layer of cnidarians, are discharged like toy popguns in response to touch or chemicals. Captured by nematocysts on the ends of tentacles, prey is brought to the mouth by contraction of the tentacles.

Digestion in cnidarians

Cnidarians are predators that capture or poison their prey with nematocysts, *Figure 29.5*. A **nematocyst** is a capsule that contains a coiled, threadlike tube. The tube may be sticky or barbed, or it may contain toxic substances. In cnidarians, you begin to see the origins of a digestive process similar to that of more complex animals. Cells adapted for digestion inside the bodies of cnidarians release enzymes over the newly captured prey. The inner cell layer of cnidarians surrounds a space called a **gastrovascular cavity** in which digestion takes place. Any undigested materials are ejected back out through the mouth.

Oxygen enters cells directly

Because of a cnidarian's simple, two-cell-layer body plan, no cell in its body is ever far from water. Oxygen diffuses directly into the body cells from water, and carbon dioxide and other wastes diffuse out of the cells directly into the surrounding water.

Nervous regulation in cnidarians

Cnidarians do not have a nervous system as do complex animals; rather, cnidarians possess a nerve net. A **nerve net** conducts nerve impulses from all parts of the body, but there is no control center like the brains of more complex animals. The impulses from the nerve net bring about contractions of musclelike cells in the tentacles and bodies of cnidarians. For example, when touched, a hydra will react rapidly by contracting the musclelike cells of its body.

interNET CONNECTION

Follow the link for this chapter on the Glencoe Homepage at **www.glencoe.com/sec/science** to find out more about sponges, cnidarians, flatworms, and roundworms.

Reproduction in cnidarians

Cnidarians have the ability to reproduce both sexually and asexually. Polyps, such as hydras, reproduce asexually by a process known as budding, as shown in *Figure 29.6.*

The medusa form of cnidarians, such as a jellyfish, is the sexual stage, which alternates from generation to generation with the asexual polyp stage of the life cycle, as shown in *Figure 29.7.* Medusae reproduce sexually to produce polyps, which, in turn, reproduce asexually to form new medusae.

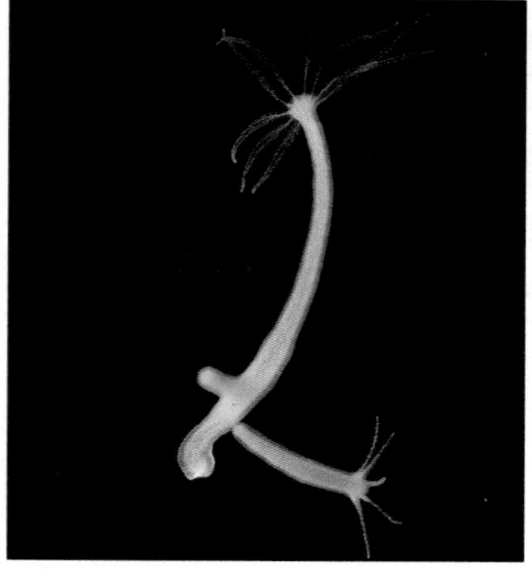

Figure 29.6

The main form of reproduction in polyps is budding. During this process, small buds grow as extensions of the body wall. As a bud grows, a mouth with tentacles develops and the new polyp breaks away from the parent.

Figure 29.7

A free-swimming larva develops into a polyp. The structure of this larva gives scientists clues about the origin of cnidarians.

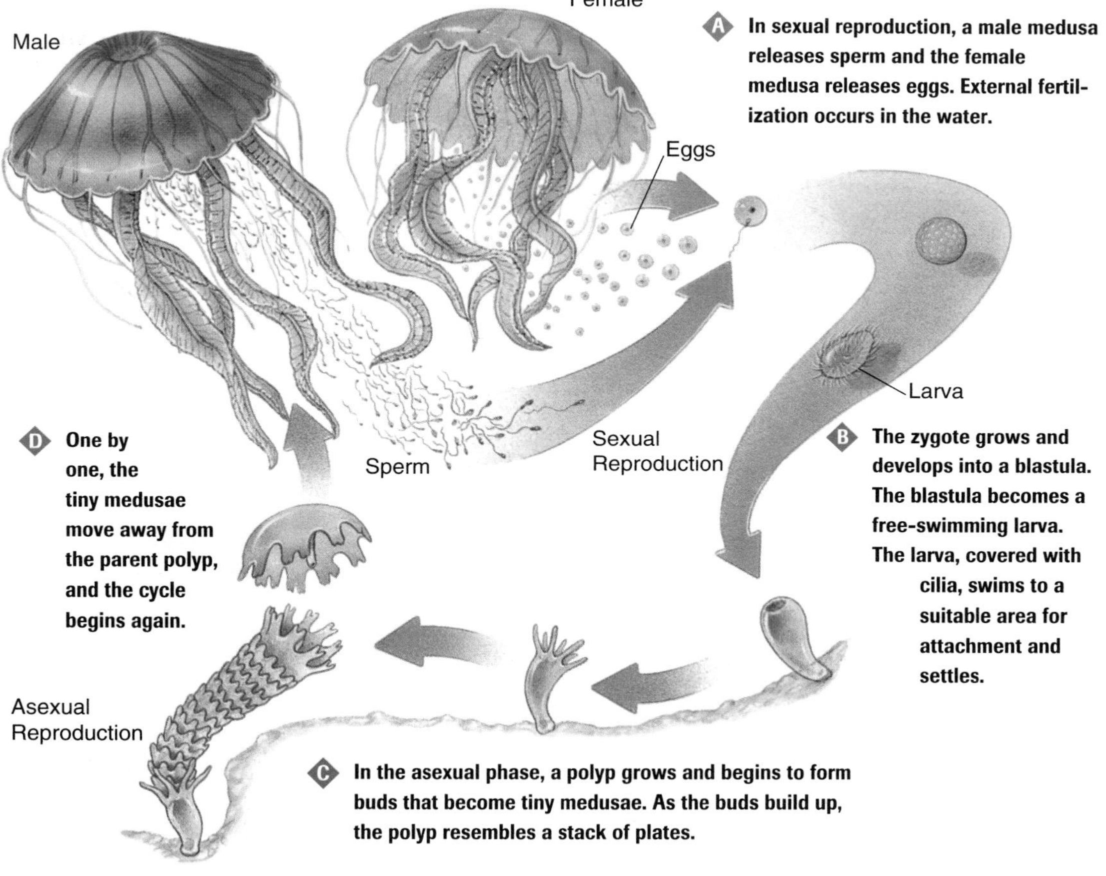

Female
Male
Eggs
Larva
Sperm
Sexual Reproduction
Asexual Reproduction
A In sexual reproduction, a male medusa releases sperm and the female medusa releases eggs. External fertilization occurs in the water.
B The zygote grows and develops into a blastula. The blastula becomes a free-swimming larva. The larva, covered with cilia, swims to a suitable area for attachment and settles.
C In the asexual phase, a polyp grows and begins to form buds that become tiny medusae. As the buds build up, the polyp resembles a stack of plates.
D One by one, the tiny medusae move away from the parent polyp, and the cycle begins again.

Science, Technology, and Society

Using Jellyfish Venom

One kind of jellyfish contains a protein that glows in the presence of calcium. Using this protein, chemists can follow the path of calcium in chemical reactions.

The venom of jellyfishes is currently under study as a potential medication for circulatory system problems. This venom has been found to cause the heart to beat with a stronger contraction, without causing it to beat faster. Such a medication will be used on patients with congestive heart failure because it will help their weak hearts beat more powerfully without speeding up.

Demonstration

LS **Visual-Spatial** Show prepared slides of budding hydras with a microprojector.

GLENCOE TECHNOLOGY

Videodisc

STVS: Animals
Disc 5, Side 1
Jellyfish (Ch. 2)

STVS: Ecology
Disc 6, Side 1
Evaluating Artificial Reefs (Ch. 11)

PORTFOLIO

Hydra Myths Ask students to look up and report about the Greek myth about the water monster, Hydra. *Students should indicate that the Hydra that Hercules slew had nine heads. As he cut each one off, a new one grew in its place.* Have students identify parts of the myth that are scientifically accurate and those that are not. L2 P LS

Technology
BIOLOGY & SOCIETY

Sunscreens from Coral

Purpose

Students will learn how corals are protected from UV radiation and how this adaptation may help humans from becoming sunburned.

Teaching Strategies

- Display several bottles of sunscreens. Review with students how to interpret SPF numbers that appear on the labels.
- Elicit from students why scientists might add UV-resistant chemicals to paints. *Such chemicals would prevent the paint from fading due to exposure to sunlight.*
- Ask students how the coral sunscreen story supports arguments for preserving habitats like coral reefs and rain forests that contain an enormous variety of as-yet-unstudied species of plants and animals. *Students should indicate that this story demonstrates that some unknown species may someday provide useful chemicals that will save lives or make our lives better.*

INVESTIGATING the Technology

Hypothesizing Probably not, because this would be an unnecessary adaptation.

Going Further

Have students do research about the relationship between UV rays and skin cancer. Ask students to write a report that includes references supporting the link between exposure to UV rays and skin cancer.

Technology
BIOLOGY & SOCIETY

Sunscreens from Coral

Sun exposure, skin cancer, and ultraviolet radiation are in the news a lot. Ultraviolet (UV) radiation is an invisible component of sunlight and a powerful form of energy that harms most organisms. When your skin is exposed to sunlight, UV rays can cause a sunburn. Prolonged and repeated exposure to UV rays can also damage your skin cells in such a way that you may eventually develop skin cancer.

Do corals get sunburned? Corals live in fairly shallow water. They are exposed to intense tropical sunshine with high levels of UV rays every day. Scientists in Australia found that corals show no signs of damage from UV rays.

Coral samples were tested A team of marine biologists collected samples of many different kinds of corals from the Great Barrier Reef. Each sample was then ground up and chemically analyzed. The researchers discovered that the samples contained chemical compounds that absorb ultraviolet radiation. These compounds act as natural sunscreens for corals by preventing UV rays from damaging cells in their delicate bodies.

Applications for the Future

An Australian pharmaceutical company is now working on developing a sunscreen lotion that incorporates these coral compounds as active ingredients. This experimental sunscreen is effective in absorbing UV rays with an SPF (sun protection factor) of at least 50. The compounds are also being tested in paints, varnishes, and plastics to determine whether they can protect these products from the damage that UV rays are known to cause.

INVESTIGATING the Technology

Hypothesizing Ultraviolet radiation can penetrate seawater to a maximum depth of 20 meters. Would you expect corals that live below 20 meters to have sunscreen compounds?

Adaptations and Ecology

Each of the 9000 known species of cnidarians belongs to one of three classes: Hydrozoa, Scyphozoa, or the class of coral reefs, Anthozoa.

Hydrozoans form colonies

It's difficult to believe that the organism shown in ***Figure 29.8*** is actually a closely associated group of individual animals. The Portuguese man-of-war is an example of a hydrozoan colony.

Each individual in the colony has a different function that helps the entire organism to survive. For example, just one individual forms a large, blue, gas-filled float. Other polyps hanging from the float have different functions, such as reproduction and feeding.

Figure 29.8

The Portuguese man-of-war, *Physalia,* is really a collection of related individuals working together to form one organism. The Portuguese man-of-war captures fish by injecting venom from nematocysts. Feeding polyps then secrete digestive enzymes over the captured prey. The venom in a large colony such as this is powerful enough to kill a human.

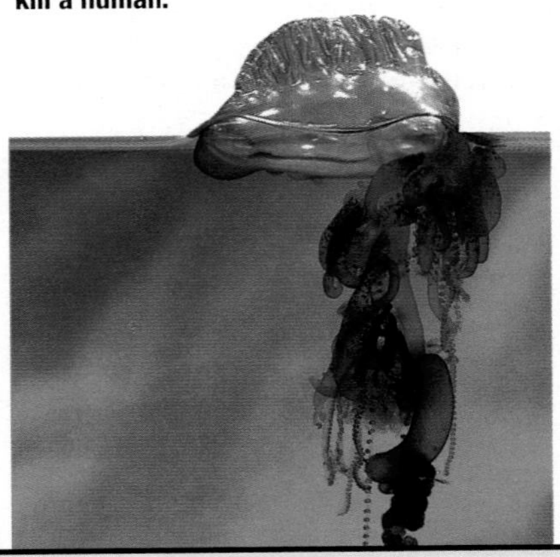

Visual Learning

Figure 29.8 Be sure students understand that the Portuguese man-of-war is a colonial hydrozoan.

PORTFOLIO

Treating Jellyfish Stings Ask a group of students to contact first aid stations on public beaches where the Portuguese man-of-war is common. Have them prepare a report on the treatment given to victims of the stings of this animal. Ask students to demonstrate the first aid procedures and explain why each procedure is performed. L3 COOP LEARN P LS

Meeting Individual Needs

Gifted Ask a group of students to visit a saltwater aquarium, either at a pet shop, restaurant, or marine park, and take photographs for a photo essay about sea anemones. You might also provide a preserved anemone for them to dissect. L3 LS

Figure 29.9

The Australian box jelly is invisible to swimmers and can kill within minutes of the sting.

Scyphozoans include jellyfishes

The fragile and sometimes luminescent bodies of jellyfishes can be beautiful, but most people know about jellyfishes more by their painful stings. Shown in *Figure 29.9* is the Australian box jelly. Many people consider this the most venomous sea animal known.

Anthozoans build coral reefs

Corals are anthozoans that live in colonies of polyps in warm ocean waters around the world. They build the beautiful coral reefs that serve as food sources and shelter for many other species of animals, including the sea anemones, *Figure 29.10*, another group of anthozoans.

Figure 29.10

Sea anemones may look like flowers, but they are animals that capture prey. Sea anemones exhibit only the polyp form.

ThinkingLab Interpret the Data

How does coral reef bleaching change over time?

Reef-building corals normally contain microscopic algae. The corals provide protection for the algae, and the corals benefit by obtaining oxygen and food produced by the algae. Sometimes, though, environmental conditions cause corals to expel their algae. The corals become pale and appear to be bleached. Corals lacking algae do not build reefs and cannot survive solely on organisms captured by nematocysts.

Analysis

Examine the map provided, and explain the trend in coral reef bleaching by answering the following question.

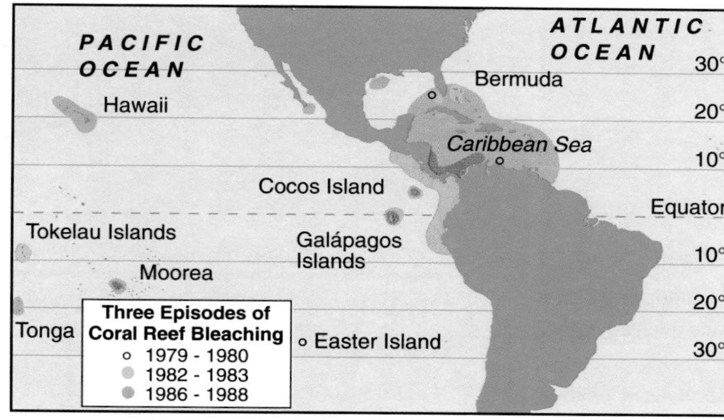

Thinking Critically

How did coral reef bleaching change over the three time periods represented on the map?

ThinkingLab Interpret the Data

Purpose

LS Logical-Mathematical Students will compare how coral reef bleaching has changed over time.

Teaching Strategies

- Show students photos from science magazines of corals before and after bleaching.

Further Background

Recent studies indicate that high seawater temperature may be the cause of coral reef bleaching. Much of the grand color of corals comes from the symbiotic algae. When the algae are expelled, the coral looks white, the color of its calcium carbonate skeleton. Although the exact mechanism by which higher water temperature damages coral is still under study, some scientists think that stressed polyps provide insufficient carbon dioxide, nitrogen, and phosphorus to the algae, causing them to leave. Another hypothesis proposes that the algae emit poisonous substances when heated and the polyp expels the damaging symbiont.

Thinking Critically

A few small places were affected in 1979-1980, more widespread bleaching occurred in 1982-1983, and in 1986-1988, larger areas in more parts of the world were affected.

Assessment

Skill: Have students examine the locations of coral reef communities on a map. Have students identify the countries and shorelines that may be affected by coral reef bleaching. L1

STUDENT JOURNAL

Coral Art Ask students to design a postage stamp that will commemorate coral animals, their importance, and their beauty. Give students colored markers and plain white paper. Remind them that very little can be written on a postage stamp. Provide library resources with photos of colorful corals. Have students place their designs in their journals. L1

SECTION 29.2

3 Assess

Check for Understanding

Have students draw a hydra cross section and add arrows to show how food reaches all cells and the exchange of oxygen and carbon dioxide. L1

Reteach

Draw a football field on the chalkboard. Divide the class into two teams. Ask questions about cnidarians. If the student answers correctly, advance the ball 10 yards toward their goalpost. If the answer is not correct, the question goes to the other team. The team that reaches its goalpost first wins. L1

Extension

Have students write about treatments for jellyfish stings and methods of preventing stings. L2

Assessment

Journal: Ask students to make a travel brochure for tourists who wish to see cnidarians. They should include all groups. L1

4 Close

Activity

Have students observe the movements of a live hydra in a deep-well 35-mm projector slide in a slide projector. Ask them to review the adaptations of the hydra to its lifestyle.

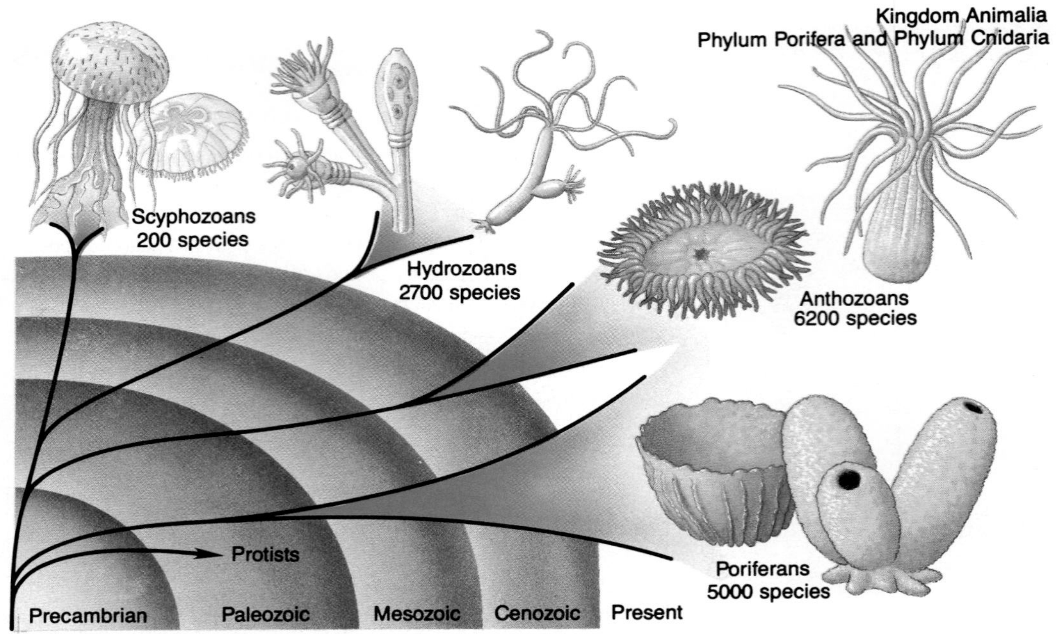

Figure 29.11

Sponges and cnidarians evolved early in geologic time. Sponges probably were the first to appear, followed by the classes of cnidarians.

Origins of Sponges and Cnidarians

As shown in *Figure 29.11*, sponges represent an old animal phylum. The earliest fossil evidence for sponges dates this group to the Paleozoic Era, about 600 million years ago. From the similarity of a group of flagellated protists that resemble the collar cells of sponges, scientists infer that sponges may have evolved directly from these protists.

The earliest known cnidarians date to the Precambrian Era, about 630 million years ago. Because cnidarians are soft-bodied animals, they do not preserve well as fossils, and their origins are not well understood. The larval form of cnidarians resembles protists, and most scientists agree that cnidarians evolved from protists.

Section Review

Understanding Concepts

1. Compare the structures of medusa and polyp forms of cnidarians.
2. Diagram the reproductive cycle of a jellyfish.
3. What are the advantages of a two-layered body in cnidarians?

Thinking Critically

4. Coral reefs are being destroyed at a rapid rate as coral is harvested. What effect would you expect the destruction of a large coral reef to have on other ocean life?

Skill Review

5. **Making and Using Tables** In a table, distinguish the three main groups of cnidarians, list their characteristics, and give examples of a member from each group. For more help, refer to Organizing Information in the *Skill Handbook*.

Answers to Section Review

1. Polyps are the sessile stage of cnidarians in which the mouth points upward. The medusa is the free-swimming stage of a cnidarian with a bell-shaped floating body with the mouth pointing down.
2. Make sure that students have the medusa stage alternating with the polyp stage and the medusa stage forming gametes that combine to form a fertilized egg that develops into a larva.
3. The cell layers of cnidarians are organized into separate tissues with specific functions. Tissues with specific functions enable an animal to be more efficient in carrying on life functions.

Thinking Critically

4. Other ocean life will be destroyed when the reef is gone because the marine life of the area is dependent on the habitat made by the coral.

Skill Review

5. Make sure students have listed the characteristics and examples of hydrozoans, scyphozoans, and anthozoans.

SECTION

29.3 Flatworms

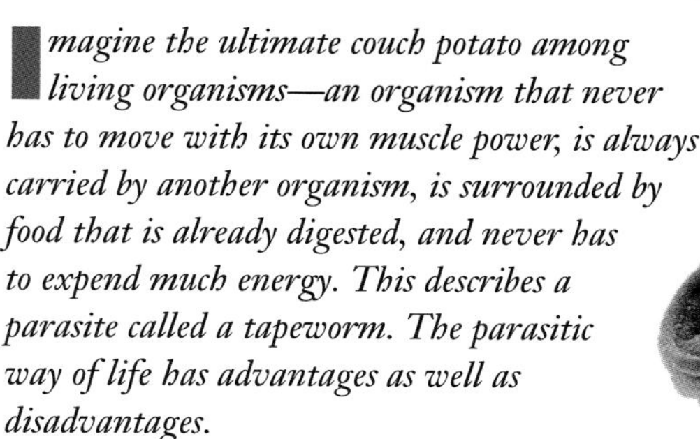

Imagine the ultimate couch potato among living organisms—an organism that never has to move with its own muscle power, is always carried by another organism, is surrounded by food that is already digested, and never has to expend much energy. This describes a parasite called a tapeworm. The parasitic way of life has advantages as well as disadvantages.

Section Preview

Objectives

Distinguish the adaptive structures of parasitic flatworms and planarians.

Explain how parasitic flatworms are adapted to their way of life.

Key Terms

pharynx
scolex
proglottid

What Is a Flatworm?

To most people, the word *worm* brings to mind a long, slimy creature without eyes and limbs. Many animals have this general structure, but now it is understood that worms can be classified into many different phyla.

Worms have more-complex structures than do sponges and cnidarians, but the least complex worms belong to the phylum Platyhelminthes, *Figure 29.12.* These flatworms are acoelomates with thin, solid bodies. The most well-known members of this phylum are the parasitic tapeworms (Class Cestoda) and flukes (Class Trematoda), which cause diseases in humans. The most commonly studied flatworms in biology classes are the free-living planarians (Class Turbellaria). Flatworms range in size from 1 mm up to several meters.

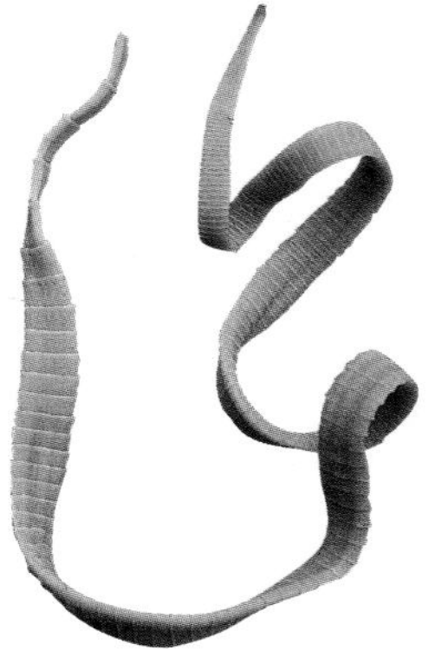

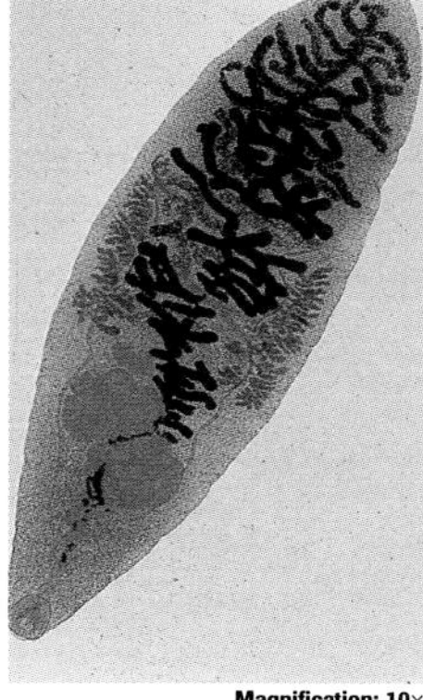

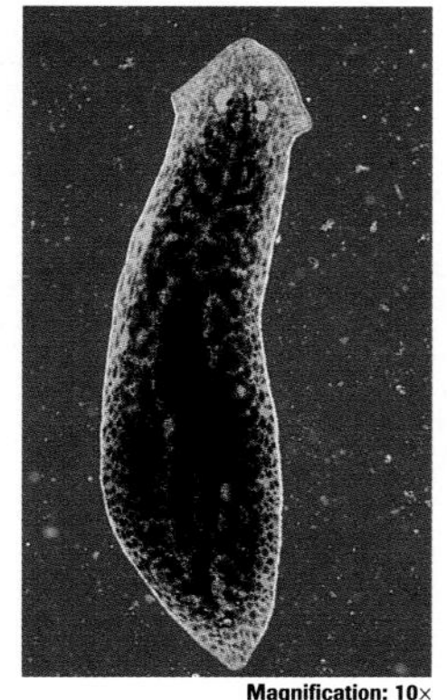

Figure 29.12

Tapeworms (left) are parasites that invade and live in host organisms. Flukes (center) usually require two hosts in a complex life cycle. Planarians (right) are not parasitic, nor do they cause diseases.

SECTION 29.3

Prepare

Key Concepts

In this section, students will study the adaptive structures of parasitic flatworms and planarians. They will learn about how these worms are adapted to their ways of life.

Block Scheduling

Look for this symbol for strategies that are useful in a block scheduling format. For more information on block scheduling, refer to Section 29.3 in the **Lesson Plans** booklet.

Materials

- For the Biolab, gather small pebbles, droppers, test tubes, pond water, ice, plain water, thermometers, petri dishes, pieces of cooked egg white, black construction paper, toothpicks, hot plates, and live planarians.

1 Focus

Bellringer

Before presenting the lesson, display **Section Focus Master 64** on the overhead projector and have students answer the accompanying questions. L1 SAE

Demonstration

Place a live planarian in water in a 35-mm deep-well slide that can be projected through a slide projector. Ask students to observe how the worm moves. Point out that the planarian has a head area. Remind them that sponges and cnidarians have no heads and ask what the survival advantage is of having a head area on the body.

Program Resources

Section Focus Master 64 L1 SAE
Reinforcement and Study Guide, p. 117 L1
Biolab and Minilab Worksheets, pp. 117-118 L1
Reteaching Transparency 29 and **Master** L1 SAE

THE INSIDE STORY

Inside a Planarian

Purpose

IS **Visual-Spatial** Students will study the structures and adaptations of planarians.

Teaching Strategies

- Allow students to examine live planarians or a prepared slide of a planarian. Have them make drawings of their observations.

Visual Learning

- Have students use the art in the feature as a model to label the planarian they draw.
- Point out various structures of the planarian. As you mention each structure, have students identify the function of the structure.

2 Teach

Display

IS **Visual-Spatial** Prepare a bulletin board display that shows the life cycles of common human flatworm parasites. College-level invertebrate zoology texts are excellent resources for such diagrams. Use the display as you discuss the life cycles.

THE INSIDE STORY

Inside a Planarian

If you've ever waded in a shallow stream and turned over some rocks, you may have found tiny, black organisms stuck to the bottom of the rocks.

These organisms were most likely planarians. Planarians have many characteristics common to all species of flatworms. The bodies of planarians are flat, with both dorsal and ventral surfaces. Unlike sponges and jellyfishes, all flatworms have bilateral symmetry.

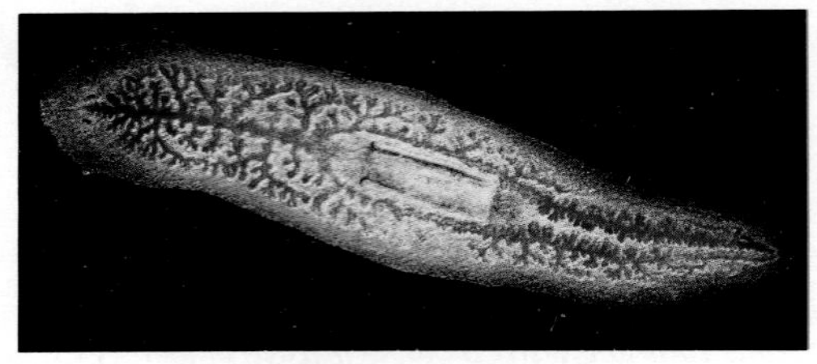

Common planarian, *Dugesia*

4 Eyespots are sensitive to light and enable the animal to respond to the amount of light present. Eyespots can't form images as do more complex eyes.

1 In contrast to sponges and cnidarians, flatworms have a clearly defined head. The head is responsible for sensing and responding to changes in the environment.

2 The pharynx is a muscular tube that can be extended outside the body. It is used to suck food into the planarian's gastrovascular cavity through the mouth.

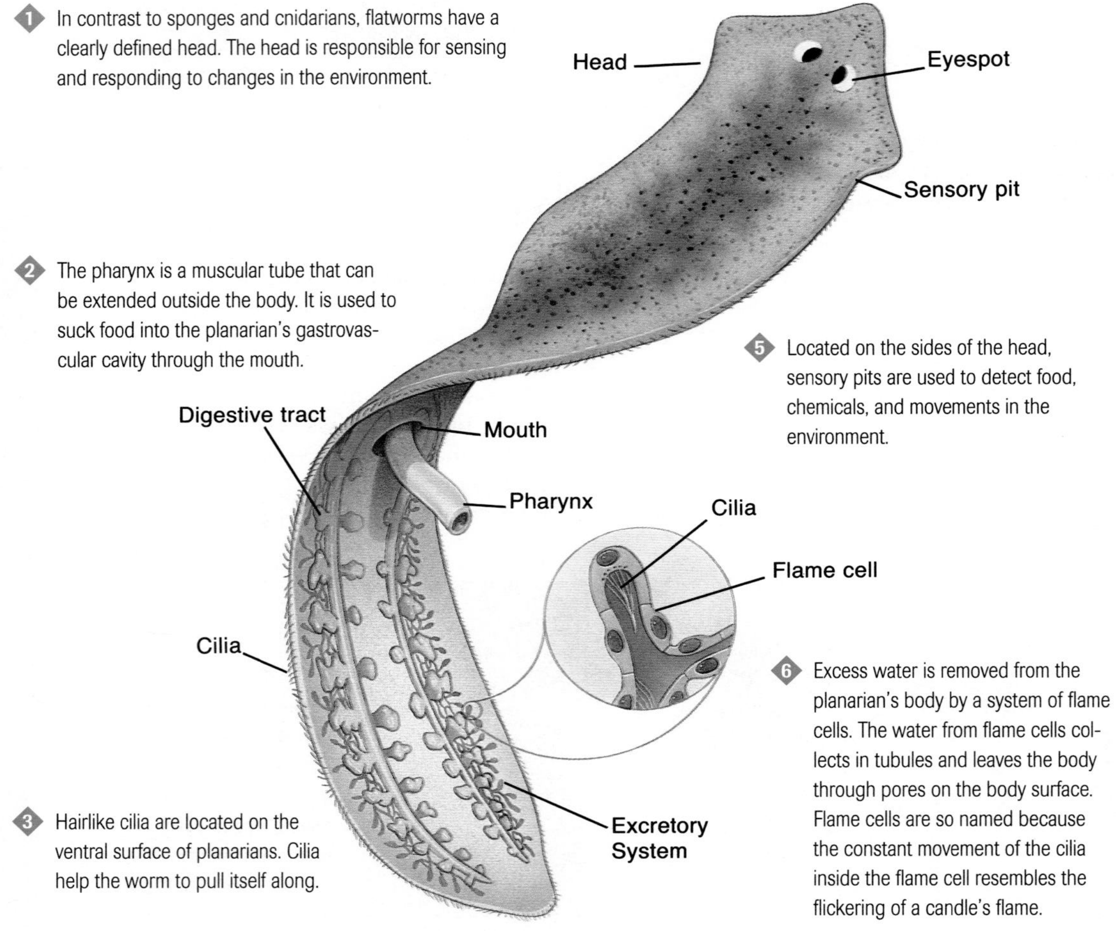

5 Located on the sides of the head, sensory pits are used to detect food, chemicals, and movements in the environment.

6 Excess water is removed from the planarian's body by a system of flame cells. The water from flame cells collects in tubules and leaves the body through pores on the body surface. Flame cells are so named because the constant movement of the cilia inside the flame cell resembles the flickering of a candle's flame.

3 Hairlike cilia are located on the ventral surface of planarians. Cilia help the worm to pull itself along.

Meeting Individual Needs

Learning Disabled Provide students with outlines of planarians. Ask students to use one outline to show the symmetry of the planarian as well as its anterior and posterior ends and dorsal and ventral sides. Have students use The Inside Story diagram as a model to show individual body systems of planarians. L1

Figure 29.13

Messages from the nerve cords trigger responses in the planarian's muscle cells. In this way, planarians are able to make adjustments to stimuli in the environment.

Feeding and digestion

A planarian feeds on dead or slow-moving organisms. It extends a tube-like, muscular organ, called the **pharynx,** out of its mouth. Enzymes released by the pharynx begin digestion of the food before it is sucked into the digestive system.

Nervous control in planarians

In addition to eyespots, planarians possess a brainlike structure called a ganglion. Located in the head, the ganglion sends messages, received from the eyespots and sensory pits, along two nerve cords that run the length of the body, *Figure 29.13.*

Reproduction in planarians

Like many of the organisms studied in this chapter, most flatworms are hermaphrodites. During sexual reproduction, planarians exchange sperm, and fertilization occurs internally. The zygotes are released in capsules into the water, where they hatch into tiny planarians.

Planarians can also reproduce asexually by fission. When a planarian is damaged, it has the ability to regenerate new body parts. If a planarian is cut horizontally, the head piece will grow a new tail, and the tail piece will grow a new head. Thus, a planarian that is damaged or cut into two pieces may grow into two new organisms—a form of asexual reproduction.

Parasitic Flatworms

Planarians are free-living organisms, but most flatworm species, such as flukes and tapeworms, are parasitic.

Although the basic structure of parasitic flatworms is similar to that of planarians, parasitic worms are adapted to obtaining their nutrients from inside the bodies of one or two kinds of hosts. Parasitic worms have mouthparts with hooks that serve to hold the worm's position inside its host. Because they are surrounded by nutrients, there is less demand for complex nervous or muscular tissue.

THE FAR SIDE
By GARY LARSON
We're still one player short... Someone's gonna have to cut themselves in half.

Planaria sports

SECTION 29.3

Teacher FYI

Studies of planarians have shown that RNA is important in recording memory. The RNA of planarians trained to respond to a light was fed to untrained planarians. These organisms had a significantly higher rate of response than a control group.

Guest Speaker

Invite a local veterinarian to be a guest speaker in your class. Ask him or her to speak about the prevention and treatment of parasitic worms in pets. Ask the speaker to bring drawings or models to show the class.

Concept Development

Display several preserved flatworms. Ask students to observe the worms and list their similarities and differences. Have them speculate which ones are parasites and ask them to explain their choices.

GLENCOE TECHNOLOGY

Videodisc

STVS: Animals
Disc 5, Side 1
Sheep Parasite (Ch. 6)

Blood Fluke Life Cycle (Ch. 4)

Nematodes (Ch. 5)

PORTFOLIO

Marine Flatworms Have students visit a marine aquarium, zoo, or pet store specializing in saltwater species to research marine flatworms. Have them write a summary of their findings to include in their portfolios. L1 LS

BioLab

How Planarians Respond to Stimuli

Time Allotment
One class period

Objectives
Review objectives with students before they begin the Biolab.

Process Skills
observe and infer, compare and contrast, recognize cause and effect, form a hypothesis, interpret data

Safety Precautions
- Students should wear goggles.
- Ask students to use caution when heating test tubes and using the hot plate.
- Make sure students treat animals in a humane fashion at all times.
- Have students wash their hands at the conclusion of the lab.

PREPARATION

Alternate Materials
- Students could work with vinegar eels instead of planarians.

Possible Hypotheses
- Students may hypothesize that planarians will exhibit movement toward food, darkness, cold, and pebbles.
- Students may hypothesize that planarians will have negative responses to light and heat.

BioLab

How Planarians Respond to Stimuli

Planarians live in temperate climates under submerged stones on the bottom of streams and ponds. In order to survive, they must detect changes in their environment such as presence of food, amount of light, and water temperature. In this lab, you will determine how planarians respond to these conditions.

PREPARATION

Problem
How do planarians respond to stimuli?

Objectives
In this Biolab, you will:
- **Determine** stimuli that cause responses in planarians.
- **Interpret** data in terms of adaptations to changes in the environment.

Materials
planaria culture
small pebbles
eyedropper
test tube with pond water
test tube with ice
test tube with heated water
thermometer
petri dishes
pieces of cooked egg white
black construction paper
toothpick
hot plate

Safety Precautions
Be careful handling the heated test tube, and be sure to wash your hands before and after handling planarians.

PROCEDURE

Teaching Strategies
- Planarians may be collected from local water sources. Try areas in streams where the water is flowing rapidly as planarians prefer high levels of oxygen. They will appear on the bottom surfaces of rocks in the water.
- Make sure planarians are kept in pond water or water that is clear of chlorine. The chlorine in tap water will kill planarians.
- Instruct students to be careful not to injure their worms when they transfer them to their dishes.

PROCEDURE

1. Make a data table with the stimuli in one column and the responses in the other column.

Stimulus	Response Positive	Response Negative	Time Spent
pebbles			
black paper			
white paper			
ice			
heat			
egg white			

2. Hypothesize whether the planarian will have a positive or negative response to each stimulus.
3. Draw a line on the bottom of the petri dish to divide it into two halves, and fill it halfway with pond water.
4. With a toothpick, transfer a planarian to the dish.
5. Place your petri dish on a white surface and add three or four pebbles to one side. Time, for two minutes, how long the worm spends on each side of the line, and record your data in your data table.
6. Remove the pebbles. Place the dish on a piece of black paper so that the dish is half on black paper and half on white. Time, for two minutes, how long the worm spends on each side of the dish.
7. Remove the black paper.
8. Place a test tube, with ice, on the inside edge of one side of the dish and a heated test tube on the opposite side of the dish. Time how long the worm spends on each side of the dish, and record this in your data table.
9. Remove the two test tubes.
10. Gently stir the water in the dish to equalize temperature. Place a small piece of egg white on one side of the dish. Record how long the worm spends on each side of the dish during a two-minute period.
11. Record your data in your table.

ANALYZE AND CONCLUDE

1. **Identifying Variables** Tell whether the planarian responded positively or negatively to each stimulus in your table. Explain.
2. **Checking Your Hypothesis** Was your hypothesis supported by your data? Why or why not?
3. **Drawing Conclusions** What might be the survival value of each of the observed responses?

Going Further

Project Design and conduct an experiment to test food preferences of planarians.

ANALYZE AND CONCLUDE

1. positive responses to dark, cold, pebbles, and egg white; negative responses to light and heat
2. Some students will find that their hypotheses were supported. Others will find that their hypotheses were not supported. If planarians are well fed prior to the lab, they may not show any response to food. If they have been in transit or maintained in a classroom for a long time prior to the experiment, they may not be healthy and may exhibit unusual behavior.
3. Planarians move toward colder areas because lower temperatures are usually associated with more oxygen. They move toward egg white because it is a food source. They stay in the dark part of the dish or under pebbles because they live under rocks in streams and ponds. In this habitat, they can find bits of decaying organic matter for food and be hidden from predators.

Going Further

Have students place other food sources on one side of a petri dish and time how long the worm spends on each side of the dish. Other food sources may include a piece of carrot, celery, or radish, or a piece of liver, chicken, or fish. The water should be changed after each food is tested. L1

- Have students record the temperature of the water at the spot where the planarians spend most of their time.
- Have each group of students test a different variable and share the class data on the chalkboard.

Data and Observations

Ask students to graph the class data.

Health

Pigs and Parasites

Purpose

Students will discover the role of pigs in transmitting the *Taenia solium* tapeworm.

Teaching Strategies

- Use a flowchart to diagram the life cycle of *Taenia solium* on the chalkboard. If necessary, obtain a college biology or zoology text to help with the identification of stages.
- Encourage students to outline the stages in the life cycle of *Taenia solium* as described in the text. Have students use the terms *Pigs* and *Humans* as their primary level heads.
- Have interested students research and diagram the life cycle of the *Trichina* worm, also associated with pigs. L3

Possible Answer

Responses may include by not eating pork, and by making sure pork is very well cooked.

3 Assess

Check for Understanding

Ask students to make and fill out a table about flatworms with the following headings: Symmetry, Habitat, Food-getting, Nervous control, Digestion, Reproduction, Free-living or Parasitic, Examples. L1

Health

Pigs and Parasites

Throughout history, people who adhere to certain religious beliefs have avoided pigs and pork products. The pig's unsavory reputation among these groups probably originated from the experiences of generations of people who learned to relate eating pork to getting sick. It is now well known that pigs are unwilling hosts to *Taenia solium,* a parasitic tapeworm that can cause serious illness and sometimes even death in humans who eat contaminated and improperly prepared pork.

From host to host The life cycle of *Taenia* is complex. Tapeworms hatch from eggs in the intestines of an intermediate host such as a pig. Worms in this larval stage are called oncospheres. The oncospheres burrow through the intestines and settle in muscle tissue. Heat and extreme cold can kill the oncospheres, so pork meat should be thoroughly cooked before it is eaten. When the intermediate host is eaten by another animal, the oncosphere either develops into a second larval stage or into a mature tapeworm.

Larvae enter bloodstream In the second larval stage, *Taenia* enters the bloodstream from the intestines of its new host and is carried to the brain. There, the larvae cause inflammation and nerve damage, often accompanied by seizures. Sometimes, the larvae even trigger a form of epilepsy.

Oncospheres also can develop into adult tapeworms that reside in the intestines. These worms can range in size from 1 mm to 15 m. They attach themselves to their host's intestines by means of suckers and hooks in their heads.

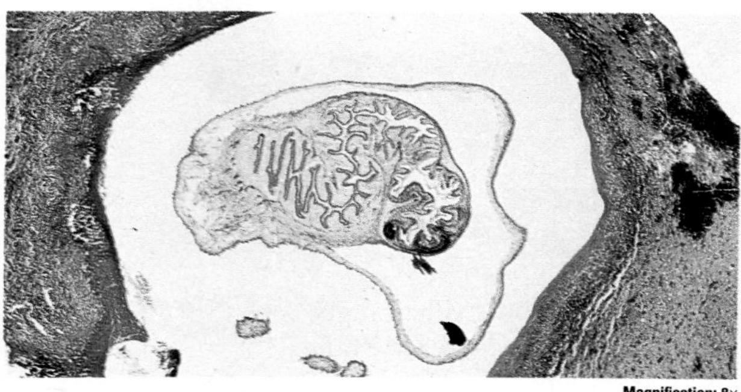

Magnification: 8×

CONNECTION TO Biology

In what ways can you avoid becoming the unwilling host of a tapeworm?

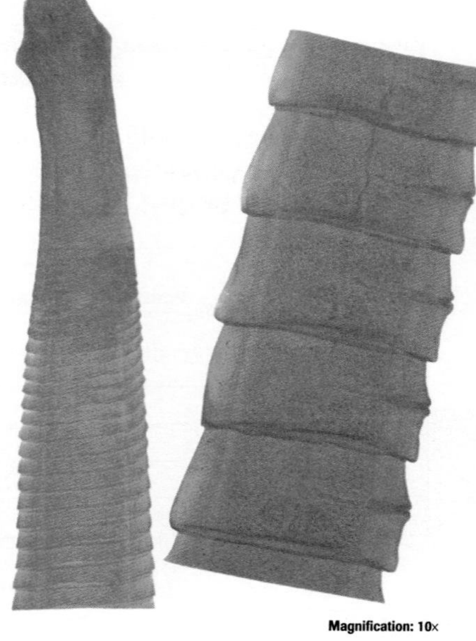

Magnification: 10×

Figure 29.14

The scolex and proglottids are adaptations to the parasitic life of a tapeworm. The scolex is covered with hooks and suckers that attach to the intestinal lining of the host.

Tapeworm adaptations

Tapeworms are parasitic flatworms of the Class Cestoda. These parasites live in the intestines of many vertebrates including dogs, cats, cattle, monkeys, and people who live in countries where sanitation is poor. Some adult tapeworms can grow to more than 10 m in length. The body of a tapeworm is made up of a head and individual repeating sections called proglottids, *Figure 29.14.* The knob-shaped head of a tapeworm is called a **scolex.** A **proglottid** is a detachable section of a tapeworm that contains muscles, nerves, flame cells, and male and female reproductive organs. Each proglottid may contain up to 100 000 eggs, and some tapeworms consist of 2000 proglottids.

STUDENT JOURNAL

Mapping Worm Parasites Give students a blank map of the world and have them color code it for where various worm parasites affect humans. Provide library resources. Ask students to put this map into their journals. L2

The life cycle of flukes

Flukes are parasitic flatworms that invade the digestive systems of vertebrates such as humans and sheep. They gain their nutrition by embedding themselves in tissues that line the intestine, where they feed on cells, blood, and other fluids of a host organism.

Blood flukes of the genus *Schistosoma*, shown in *Figure 29.15*, cause a disease in humans known as schistosomiasis. In this disease, the body is robbed of nutrition, and symptoms such as intestinal and urinary problems may occur. Blood flukes are common in parts of Africa, the Middle East, South America, and Asia—places where certain species of snails (one of the hosts) also are found.

Figure 29.15

Blood flukes have a complex life cycle that often includes two or more hosts. The *Schistosoma* fluke requires two hosts to complete its life cycle.

A **Adult flukes are about 1 cm long and live in the blood vessels of the human intestine.**

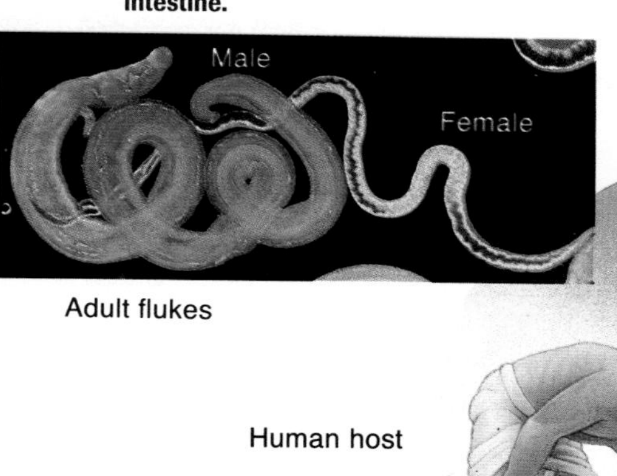

Adult flukes

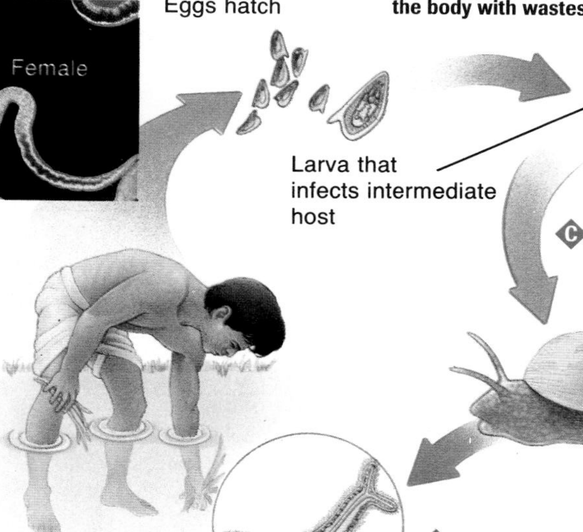

B **Fluke eggs pass out of the body with wastes.**

C **Eggs hatch into free-swimming larvae that enter their snail hosts.**

D **Larvae develop and reproduce. New larvae leave the snail and infect water.**

E **Upon contact with humans, the fluke bores through the skin, enters the bloodstream, and passes to the intestine, where the cycle begins again.**

Section Review

Understanding Concepts

1. Diagram and label the structures of a planarian.
2. Describe the adaptations of a parasitic flatworm at different stages of its life cycle.
3. What is the adaptive advantage of a nervous system to a free-living flatworm?

Thinking Critically

4. Examine the life cycle of a parasitic fluke, and suggest ways to prevent infection on a rice farm where workers must walk into water in the rice paddies during planting and harvesting.

Skill Review

5. **Observing and Inferring** What can you infer about the way of life of an organism that has no mouth or digestive system, but is equipped with a sucker? For more help, refer to Thinking Critically in the *Skill Handbook*.

SECTION 29.3

Reteach

In their groups, have some students draw and label the structures of a planarian. Have other students diagram the life cycle of *Taenia solium* or the *Schistosoma* fluke. L1

Extension

Ask a group of students to interview a veterinarian and report on the procedures for diagnosing and treating parasitic worms in pets. L2

Assessment

Knowledge: Have students assume they are preparing an exhibit for a children's museum about planarians. Ask students to draw the worm, label its parts, and explain the importance of the worm in very simple terms. L1

4 Close

Activity

Show students a selection of one-frame cartoons about animals by Gary Larson. Ask them to design a cartoon about planarians that is humorous and scientifically accurate.

Answers to Section Review

1. Make sure students have labeled the head, eyespots, sensory pits, cilia, flame cells, mouth, pharynx.
2. Parasitic flatworms produce thousands of eggs that pass out of the body of the host with its waste. The eggs may hatch and penetrate the body of an alternate host or be consumed in contaminated food depending on the type of worm. The larvae grow and bore into humans or hatch in the human depending on the type of worm. They attach themselves by means of hooks and suckers in their hosts.
3. It enables the worm to sense food and appropriate habitat as it swims.

Thinking Critically

4. Workers could wear boots and gloves. Human wastes should be kept out of water.

Skill Review

5. This organism would most likely be a parasite that attaches to its host by means of a sucker and uses food that the host has already digested.

SECTION

29.4

Prepare

Key Concepts

Students will compare and contrast the structural adaptations of roundworms and learn about the characteristics of the roundworms: *Ascaris*, hookworm, *Trichinella*, and pinworm.

Block Scheduling

Look for this symbol for strategies that are useful in a block scheduling format. For more information on block scheduling, refer to Section 29.4 in the **Lesson Plans** booklet.

Materials

- For the Minilab, gather fresh garden or meadow soil containing roots, hand lenses, and toothpicks.

1 Focus

Bellringer

Before presenting the lesson, display **Section Focus Master 65** on the overhead projector and have students answer the accompanying questions. L1 SAE

Demonstration

Show students a live culture of vinegar eels. Place several drops of culture in a deep-well 35-mm projector slide to show with a slide projector. Ask students to note how the worms move. Explain that vinegar eels are roundworms that live with the bacteria that make vinegar from fermented apple juice. They are harmless, but cannot be found in modern distilled and pasteurized vinegars.

SECTION

29.4 Roundworms

Section Preview

Objectives

Compare the structural adaptations of roundworms and flatworms.

Identify the characteristics of four human roundworm parasites.

Key Terms

none

Have you ever been to the veterinarian to have your dog tested for heartworms? Perhaps you were once warned not to eat uncooked pork products. Flatworms are not the only type of worms that can cause harm to humans and other vertebrates. It has been estimated that about one-third of the human population suffers from problems caused by roundworms.

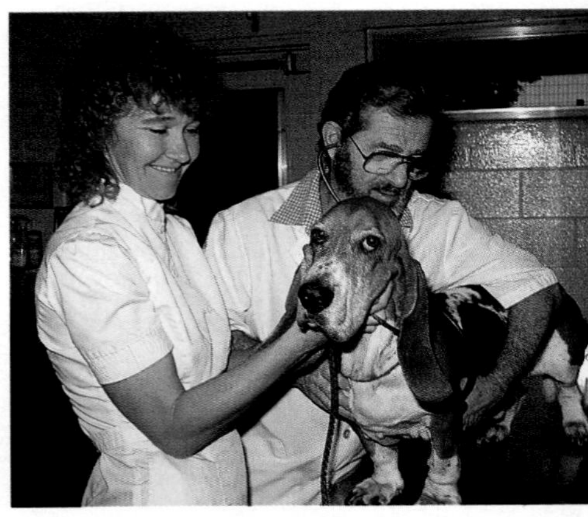

What Is a Roundworm?

Roundworms belong to the phylum Nematoda. Roundworms are widely distributed, living in soil, animals, and both freshwater and saltwater environments. Most roundworm species are free-living, but many are parasitic, *Figure 29.16.* In fact, virtually all plant and animal species are affected by parasitic roundworms.

Roundworms tend to be smaller than flatworms and are tapered at both ends. They have a thick outer

Figure 29.16

Roundworm parasites invade humans through a variety of methods. *Ascaris,* for example, is contracted by eating food grown in soil contaminated by *Ascaris* eggs. *Trichinella* is contracted by eating foods such as uncooked pork. Other parasitic roundworms, such as hookworms, can be contracted just by walking barefoot on soil containing eggs.

▲ *Ascaris* mainly infects children who swallow eggs when they put their hands into their mouths when their hands are dirty or eat vegetables that have not been washed. The eggs hatch in the intestines, move to the bloodstream, and then to the lungs, where they are coughed up and swallowed to begin the cycle again.

Magnification: 100×

Program Resources

Section Focus Master 65 L1 SAE
Reinforcement and Study Guide, p. 118 L1
Biolab and Minilab Worksheets, p. 116 L1

STUDENT JOURNAL

Roundworm Symmetry Provide students with an outline drawing of a roundworm. Have students add a line to the diagram to show the bilateral symmetry of the worm. Ask students to label the worm's anterior and posterior ends as well as its dorsal and ventral surfaces. L1

covering that protects them from being digested by their host organisms. On a flat surface, roundworms look like tiny, wriggling bits of sewing thread. Lacking circular muscles, they have pairs of lengthwise muscles. As one muscle of a pair contracts, the other muscle relaxes. This alternating contraction and relaxation of muscles causes roundworms to move in a thrashing fashion.

Roundworms have a pseudocoelom and are the simplest animals with a tubelike digestive system. Unlike flatworms, roundworms have two body openings—a mouth and an anus. The free-living species have well-developed sense organs, such as eyespots, although these are reduced in parasitic forms.

MiniLab

How can you recognize a roundworm?

Free-living roundworms feed on bacteria in all moist soils. They recycle soil nutrients and destroy some agricultural pests. One square meter of garden soil contains millions of roundworms. In this lab, you will examine garden soil with a hand lens to find roundworms.

Procedure

1. Place a small amount of garden soil in a petri dish.
2. While observing with the hand lens, tease the soil apart gently with a toothpick.
3. Roundworms will appear as tiny white, threadlike animals.

Analysis

1. Describe the movement of the roundworms you find.
2. How many roundworms did you see?
3. Explain how roundworms are adapted to their habitat.

Hookworms commonly infect humans in warm climates and in areas of poor sanitation. Hookworms cause people to feel weak and tired due to blood loss. Hookworms are contracted by walking in bare feet on contaminated soil.

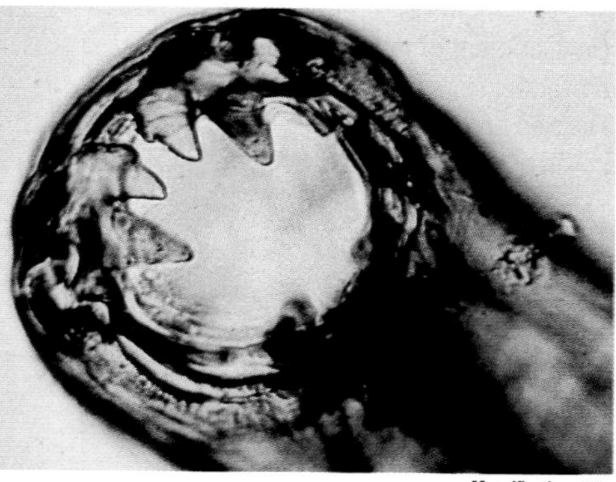

Magnification: 110×

***Trichinella* is not as common in the United States as it once was because of stricter meat inspection standards. *Trichinella* worms do still exist in areas of the world where infected, uncooked meat scraps are fed to hogs, which in turn are eaten by humans.**

Magnification: 3×

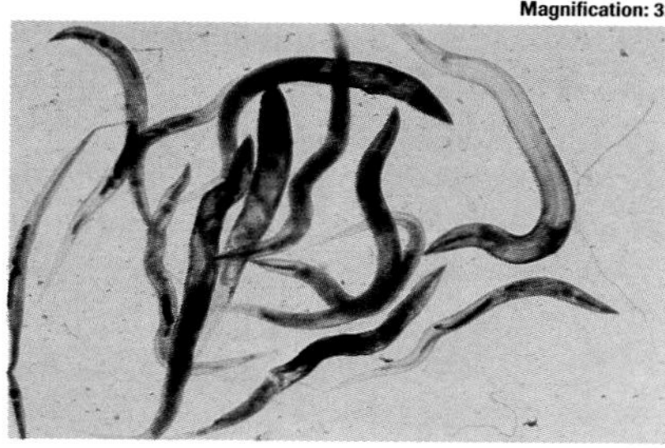

Pinworms are the most common parasites in children. Pinworms invade the intestinal tract when children eat something that has come in contact with contaminated soil. Female pinworms lay eggs near the anus, and reinfection is common because the worms cause itching.

Magnification: 100×

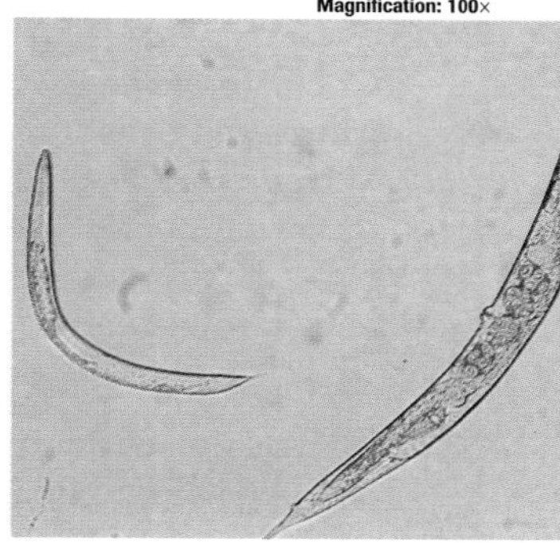

The common soil nematode, *Rhabditis,* invades roots of plants grown for food, such as tomatoes, and causes a slow decline of the plant.

PORTFOLIO

Roundworm Data Table Have students construct a table that lists the names of each roundworm discussed in this section in the left column. Have students include a sketch of the roundworm in the second column, a description of its habitat in the third column, and an explanation of how the worm affects humans in the fourth column. L1 P LS

2 Teach

Display

LS **Visual-Spatial** Have students prepare posters as part of an educational campaign to show people how to prevent the spread of *Ascaris*.

MiniLab

Purpose

LS **Intrapersonal** Students will examine soil nematodes and see how their movements differ from those of flatworms.

Process Skills

observe and infer, compare and contrast

Teaching Strategies

- Roundworms are fragile. Keep the soil warm and moist. The best kind of soil is fresh soil with roots holding it together. Do not use soil treated with pesticides.

Safety Precautions

Make sure students wash their hands after this lab.

Expected Results

Students will observe white, threadlike roundworms.

Analysis

1. Their bodies move from side to side in a whiplike motion.
2. Students may find hundreds of roundworms or only a few.
3. Roundworms move through the soil with a whiplike motion. Oxygen diffuses into their thin bodies easily.

Assessment

Portfolio: Have students write an evaluation for this lab and place it in their portfolios. Ask them to explain what they learned about nematodes in this lab. L1 P

SECTION 29.4

3 Assess

Checking for Understanding

Ask students to write one true or false question relating to the worm classes studied in the chapter. Collect the questions and read them aloud. Ask students to respond. If the answer is clear, continue. If the answer seems unclear, review and explain the correct answer. L1

Reteach

Have students draw a bingo card with 16 squares and write the name of a worm from this chapter in each square. Read a list of statements or definitions and have students cover the name of the worm to which it most relates with a small scrap of paper. The first student to cover four squares in a row wins. Check the answers by asking the winner the answer he or she had for each statement. L1

Extension

Ask students to do an in-depth report on parasitic roundworms in humans. Ask them to make a videotape presentation of the worm they researched. L3

Assessment

Portfolio: Ask students to write a paragraph to explain how the four roundworm diseases they studied might be prevented. L2

4 Close

Activity

Set up stations around the classroom with illustrations of the roundworms studied or prepared slides of these worms. Have students identify each worm.

Figure 29.17

Roundworms that are plant parasites usually enter the roots and affect the plant's ability to absorb water.

Magnification: 50×

Parasitic Roundworms

Roundworms are found as parasites in most organisms on Earth. Nematodes can infect and kill pine trees, cereal crops, and food plants such as potatoes. Nematodes are particularly attracted to plant roots. About 1200 species of nematodes cause diseases in plants, *Figure 29.17*. They also infect fungi and form symbiotic associations with bacteria.

Approximately half of the known roundworms are parasites, and about 50 species of these are human parasites. The most common human parasites are *Ascaris*, hookworm, *Trichinella*, and pinworm.

Connecting Ideas

Sponges and cnidarians are acoelomate animals that have simple body plans with two cell layers and only one body opening. Flatworms are also acoelomate animals with a mouth, but they have a simple nervous system and are able to respond to stimuli in the environment. Roundworms, however, have two body openings and a pseudocoelom; they are so successful that they are found everywhere on Earth.

Increasing complexity of body plans and structures of animals enables these organisms to survive in widely diverse environments. Development of a pseudocoelom with a digestive tract enabled roundworms to invade even inhospitable environments such as the frozen ground of the Arctic. Development of more-complex digestive, reproductive, and nervous systems, in turn, enables other animals to survive and thrive in the oceans, on land, and in the air.

Section Review

Understanding Concepts

1. Compare the body structures of roundworms and flatworms.
2. Why do parents teach children to wash their hands before eating?
3. Describe the method of infection of one human roundworm parasite.

Thinking Critically

4. An infection of pinworms is spreading to children who attend the same preschool. Make a list of precautions that could be taken to help prevent its continued spread.

Skill Review

5. **Making and Using Tables** Make a table of the characteristics of four roundworm parasites, indicating the name of the worm, how it is contracted, the action of the parasite in the body, and means of prevention. For more help, refer to Organizing Information in the *Skill Handbook*.

Answers to Section Review

1. Roundworms have a pseudocoelom while flatworms are acoelomate. Roundworms have two body openings while flatworms have one body opening.
2. It is important to wash hands before eating to prevent infection by parasitic worms and bacteria.
3. Roundworms can be contracted by eating improperly cooked pork, putting dirty hands or unwashed vegetables into the mouth, walking barefoot, and by lack of good personal hygiene.

Thinking Critically

4. good personal hygiene such as washing hands, clothing, and bedding

Chapter 29 Review

REVIEWING MAIN IDEAS

29.1 Sponges

- A sponge is an aquatic, sessile, asymmetrical, filter-feeding invertebrate.
- Sponges are made of four types of cells. Each cell type contributes to the survival of the organism.
- Sponges are hermaphroditic with free-swimming larvae.

29.2 Cnidarians

- All cnidarians are radially symmetrical, aquatic invertebrates that display two basic forms: medusa and polyp.
- Cnidarians feed by stinging or entangling their prey with cells called nematocysts, usually located at the ends of their tentacles.
- The three classes of cnidarians include the hydrozoans, hydras; scyphozoans, jellyfishes; and anthozoans, corals and anemones.

29.3 Flatworms

- Flatworms are acoelomates with thin, solid bodies belonging to the phylum Platyhelminthes. They are grouped into three classes: free-living planarians, parasitic flukes, and tapeworms.
- Planarians have well-developed nervous and muscular systems. These systems are reduced in parasitic flatworms, but flukes and tapeworms have structures adapted to their parasitic existence.

29.4 Roundworms

- Roundworms are pseudocoelomate, cylindrical worms with lengthwise muscles and relatively complex digestive systems with two body openings.
- Roundworm parasites include parasites of plants, fungi, and animals, including humans. *Ascaris*, hookworms, *Trichinella*, and pinworms are roundworm parasites of humans.

Key Terms

Write a sentence that shows your understanding of each of the following terms.

external fertilization
filter feeding
gastrovascular cavity
hermaphrodite
internal fertilization
medusa
nematocyst
nerve net
pharynx
polyp
proglottid
scolex

Understanding Concepts

1. Which of these is NOT a type of cell found in sponges?
   a. epithelial c. collar
   b. spicule d. pore
2. A ___________ is an individual animal that can produce both eggs and sperm.
   a. nematocyst c. proglottid
   b. hermaphrodite d. pharynx
3. Eggs and sperm in sponges are formed from ___________.
   a. spicules c. flagella
   b. amoebocytes d. spongin
4. In ___________ fertilization, eggs and sperm are both released into the water, where fertilization occurs.
   a. internal c. ventral
   b. external d. dorsal
5. Which of these is NOT a type of reproduction found in sponges?
   a. budding
   b. regeneration
   c. sexual reproduction
   d. asexual reproduction
6. Sponges obtain food by ___________.
   a. predation c. parasitism
   b. photosynthesis d. filter feeding

Chapter 29 Review

Reviewing Main Ideas

Summary statements can be used by students to review the major concepts of the chapter.

Key Terms

Answers should go beyond defining the terms. Accept any answer that uses the term correctly and in the proper context.

Understanding Concepts

1 b
2. b
3. b
4. b
5. b
6. d

Skill Review

5. Make sure students' tables include the worms: *Ascaris*, hookworm, *Trichinella*, and pinworms. They should indicate for each worm how it is contracted, the action of the worm in the body, and means of prevention.

Chapter 29 Review

7. c
8. d
9. a
10. a
11. b
12. d
13. a
14. b
15. b
16. c
17. c
18. a
19. b
20. c

Applying Concepts

21. The two cell layers of cnidarians are organized into tissues with specific functions. They have simple nervous systems, cells that can contract like muscles, nematocysts that are used to capture prey, and a gastrovascular cavity in which digestion occurs.
22. Sponges are filter feeders that filter food out of water. Cnidarians are predators that capture their prey with nematocysts.
23. It is a parasitic flatworm. A head with hooks is characteristic of parasitic flatworms.
24. Cook the pork thoroughly. Serve it well done.
25. Animals that are sessile are unable to move and search for mates. If they are hermaphroditic, any individual can be a mate for any other individual of the same species.
26. Predators of jellyfishes must be immune to the toxins in jellyfishes, or have a protective substance such as mucus covering the body that could absorb the nematocysts' toxins.
27. Each type of cell has a specific job; epithelial cells protect; pore cells bring in water; collar cells trap food; amoebocytes transport nutrients.

Chapter 29 Review

7. To what phylum do marine invertebrates such as jellyfishes, corals, sea anemones, and hydras belong?
a. Porifera **c.** Cnidaria
b. Platyhelminthes **d.** Cestoda

8. The two basic body forms of cnidarians are ____________.
a. nematocysts and polyps
b. nematocysts and medusae
c. internal and external fertilization
d. polyps and medusae

9. A ____________ is a free-swimming form of a cnidarian.
a. medusa **c.** nematocyst
b. polyp **d.** bud

10. Cnidarians capture prey with the help of stinging cells containing ____________.
a. nematocysts **c.** nerve nets
b. corals **d.** buds

11. The hydra digests its food in a ____________.
a. digestive tube
b. gastrovascular cavity
c. digestive cell
d. nematocyst

12. In cnidarians, medusae reproduce sexually to produce polyps, which in turn reproduce asexually to form ____________.
a. buds
b. larvae
c. hermaphrodites
d. new medusae

13. Sea anemones exhibit only the ____________ type of body form.
a. polyp **c.** bud
b. medusa **d.** colony

14. Cnidarians called ____________ live in colonies and build reefs.
a. jellyfishes **c.** sea anemones
b. corals **d.** hydras

15. Acoelomate worms called ____________ have thin, solid bodies.
a. roundworms **c.** nematodes
b. flatworms **d.** hookworms

16. Unlike sponges and cnidarians, flatworms have a clearly defined ____________.
a. stomach **c.** head
b. polyp stage **d.** larva

17. Name the planarian structures that are sensitive to light.
a. sensory pits **c.** eyespots
b. flame cells **d.** the pharynx

18. Parasitic worms have mouthparts with ____________.
a. hooks **c.** sensory pits
b. nematocysts **d.** a pharynx

19. A ____________ is a parasitic worm that uses a snail as an intermediate host and has a larval stage that can bore through the skin of humans.
a. tapeworm **c.** pinworm
b. fluke **d.** roundworm

20. Parasitic roundworms in humans include the ____________.
a. tapeworms, pinworms, and *Trichinella*
b. hookworms, pinworms, and *Planaria*
c. hookworms, pinworms, and *Ascaris*
d. tapeworms, hookworms, and *Planaria*

Applying Concepts

21. In what ways are cnidarians more complex than sponges?
22. Compare the methods for obtaining food in cnidarians and sponges.
23. You are examining a wormlike animal found in the intestines of a sheep. It has a head with tiny hooks. What kind of worm is it?
24. What could you do to ensure that people eating your pork roast dinner would NOT become hosts to pork tapeworm?
25. Of what advantage is hermaphroditism to a sessile animal?
26. Decribe the features that would be important for a predator of jellyfishes.
27. How do sponge cells illustrate the concept of division of labor?

Program Resources

Chapter Assessment, pp. 173-178 L1
Alternate Assessment in the Science Classroom
Computer Test Bank L1
Content Mastery, pp. 117-120 L1

Thinking Critically

28. ***Concept Mapping*** Make a concept map that relates the following terms and phrases. Supply the appropriate linking words for your map.

flatworm, roundworm, parasite, free-living, worm, *Planaria*, fluke, tapeworm, hookworm, *Ascaris*, pinworm

29. ***Observing and Inferring*** While examining soil from the bottom of a pond, you notice tiny red worms wriggling aimlessly in your petri dish. What kind of worms are they?

30. ***Recognizing Cause and Effect*** At what points could the life cycle of a blood fluke be interrupted so that disease would be prevented?

31. ***Observing and Inferring*** Why is the phylogeny of cnidarians so little understood?

32. ***Observing and Inferring*** Some species of sponges, as adults, are attached to the shells of crabs. Explain how this attachment might be beneficial to the sponge. Could the crab benefit as well?

33. ***Comparing and Contrasting*** Both sponges and hydras are sessile organisms that cannot pursue prey. Compare their methods of obtaining food.

34. ***Interpreting Data*** Explain the trend in coral bleaching as shown on this map.

Okinawa
Iriomoteshima
Taiwan
Andaman Islands
Maldive Islands
Equator
Mayotte Island
Réunion
INDIAN OCEAN
PACIFIC OCEAN
Enewetak
Great Barrier Reef
Fiji
New Caledonia
30° 20° 10° 10° 20° 30°

Three Episodes of Coral Reef Bleaching
○ 1979 - 1980
● 1982 - 1983
● 1986 - 1988

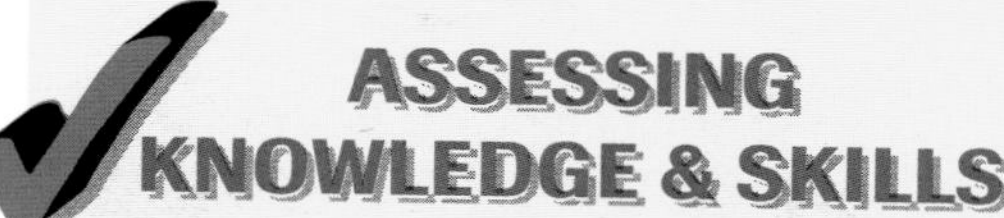

The diagram shows the life cycle for a beef tapeworm.

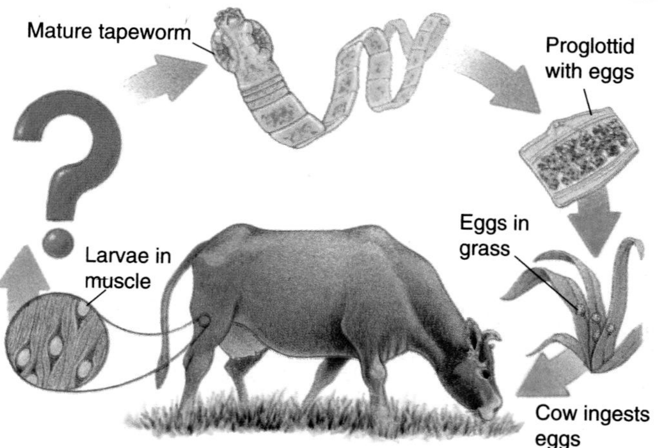

Interpreting Scientific Illustrations Use the diagram to answer the following questions.

1. Which part of the life cycle for a beef tapeworm is missing?
   a. infection of the cow
   b. infection of the grass
   c. infection of the human host
   d. infection of the tapeworm
2. How do the tapeworm eggs get into the grass?
   a. from rainwater
   b. from feces of infected cattle
   c. from snails
   d. from dead cows
3. Beef tapeworm larvae get into human hosts when humans _____________.
   a. eat beef
   b. eat pork
   c. walk barefoot
   d. go swimming
4. ***Completing a Diagram*** By making a diagram similar to the tapeworm life cycle on this page, trace the steps of a *Trichinella* infection.

Chapter 29 Review

Thinking Critically

28. Evaluate students' concept maps. The maps should show relationships among the concepts listed.
29. Roundworms. Roundworms move with a wriggling motion because they have no circular muscles as do earthworms.
30. Human wastes may be disposed of in such a way as to prevent fluke eggs from coming in contact with snails, or humans can avoid contact with water in which snails live.
31. Cnidarians are soft-bodied animals that do not preserve well as fossils. In corals, their calcium carbonate homes become fossils, showing where corals once lived, but the animals' bodies are not preserved.
32. The crab will move the sponge about, increasing its chances of finding food. The sponge may serve as a type of camouflage for the crab, allowing it to hide from prey.
33. Sponges are filter feeders. They pull water in through pores and filter out small particles of food. Hydras are cnidarians. They use nematocysts on long tentacles. When a waving tentacle touches a prey organism, the nematocyst discharges, and the tentacles bring food to the hydra's mouth for digestion.
34. Bleaching occurred in a few places in 1979-1980, became more widespread in 1982-1983, and involved even more areas between 1986 and 1988.

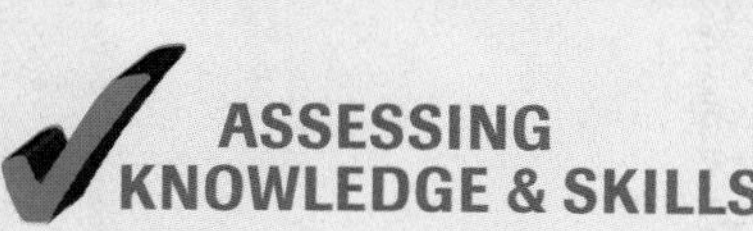

1. c
2. b
3. a
4.

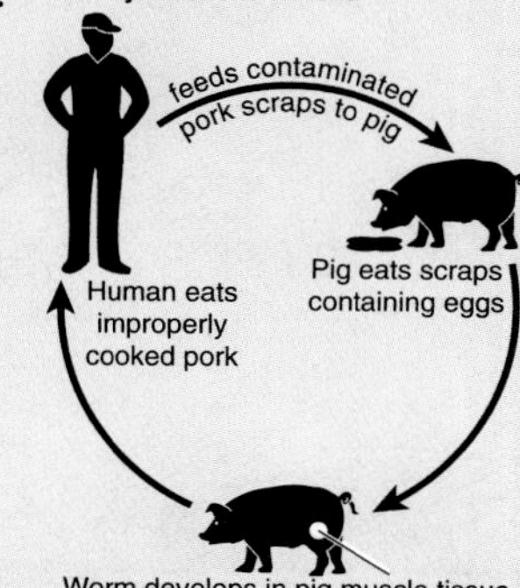

Chapter Organizer

SECTION	OBJECTIVES	ACTIVITIES/FEATURES
30.1 Mollusks National Science Standards: UCP.1-5; C.3, C.5, C.6; G.1	1. **Identify** the characteristics of mollusks. 2. **Compare** the adaptations of gastropod, bivalve, and cephalopod mollusks. 3. **Explain** the origins of mollusks.	**Thinking Lab:** What determines a snail's size before reproduction?, p. 741 **Minilab:** What path does food and water take in a clam?, p. 742 **Minilab:** How are mollusks identified?, p. 743 **Literature Connection:** *Natural Acts,* p. 744
30.2 Segmented Worms National Science Standards: UCP.1, UCP.2, UCP.4, UCP.5; A.1, A.2; C.5, C.6; E.1, E.2; F.6;	4. **Describe** the characteristics of segmented worms and their importance to the survival of these organisms. 5. **Compare and contrast** the classes of segmented worms.	**A Broader View:** Polychaetes: Feather Dusters and Christmas Trees, p. 749 **Biolab:** Design Your Own Experiment–How do earthworms respond to their environment?, p. 750

ACTIVITY MATERIALS

BIOLAB	MINILABS	ALTERNATE LAB
page 750 earthworms, live paper towels glass pan sandpaper culture dishes thermometer hand lens or stereomicroscope dropper penlight ice ruler black paper cotton swabs	**page 742** clams, live beaker carmine powder suspension dropper apron **page 743** dichotomous key transparency overhead projector shells	**page 740** land snails deli trays, clear plastic wax marking pencil lamp with 60-watt bulb crushed ice ring stand black construction paper sand paper metric ruler

Chapter 30 Mollusks and Segmented Worms

TEACHER CLASSROOM RESOURCES

Reproducible Masters	Transparencies
Section Focus Master 66: Mollusks L1 SAE **Reinforcement and Study Guide,** pp. 119-121 L1 **Biolab and Minilab Worksheets,** pp. 119-120 L1 **Laboratory Manual:** How Do Snails Respond to Stimuli?, pp. 169-172; Squid Dissection, pp. 173-176 L2	**Basic Concepts Transparency #48:** Phylogeny of Worms and Mollusks L1 SAE **Reteaching Transparency #30:** Structure of a Clam, Snail, and Squid L1 SAE
Section Focus Master 67: Segmented Worms L1 SAE **Reinforcement and Study Guide,** p. 122 L1 **Biolab and Minilab Worksheets,** pp. 121-122 L1 **Concept Mapping:** Segmented Worms, p. 30 L1 **Critical Thinking/Problem Solving:** Comparing Polychaetes, Oligochaetes, and Leeches, p. 30 L3 **Laboratory Manual:** Earthworm Dissection, pp. 165-168 L2 **Content Mastery,** pp. 121-124 L1	**Basic Skills Transparency #22:** Earthworm Responses L1 SAE

ASSESSMENT MATERIALS

Chapter Assessment, pp. 179-184
Alternate Assessment in the Science Classroom
MindJogger Videoquiz
Computer Test Bank

Spanish Resources SAE
English/Spanish Audiocassettes SAE
Cooperative Learning in the Science Classroom COOP LEARN
Lesson Plans

KEY TO TEACHING STRATEGIES

L1 Level 1 activities should be within the ability range of all students including those with learning difficulties.

L2 Level 2 activities are within the ability range of average to above-average students.

L3 Level 3 activities are designed for the ability range of above-average students.

SAE SAE activities should be within the ability range of Students Acquiring English.

COOP LEARN Cooperative Learning activities are designed for small group work.

P These strategies represent student products that can be placed into a best-work portfolio.

These strategies are useful in a block scheduling format.

GLENCOE TECHNOLOGY

The following multimedia resources are available from Glencoe.

Biology: The Dynamics of Life
CD-ROM SAE
Videodisc Program

National Geographic Society Series
GTV: Planetary Manager
Animal

Science and Technology Videodisc Series (STVS)
Animals
Conch Farming
Leeches

CHAPTER

30 Mollusks and Segmented Worms

Chapter Overview

Chapter 30 focuses on the characteristics of mollusks and segmented worms. In the first section, students learn of the traits all mollusks have. Students next compare and contrast the most common classes of mollusks—the gastropods, bivalves, and cephalopods. The discussion of mollusks concludes with an examination of mollusk origins.

In the second section of the chapter, the characteristics of segmented worms are presented. The section begins with a discussion of the traits common to all annelids and explores the traits shared by annelids and mollusks. Next, attention is focused on the earthworm, the most well-known annelid. The section concludes with a discussion of leeches.

Key Terms

- closed circulatory system
- gizzard
- mantle
- nephridia
- nocturnal
- open circulatory system
- radula

736

Learning Styles

LS

Look for the following logo for strategies that emphasize different learning modalities. LS

Kinesthetic	Meeting Individual Needs, p. 748
Visual-Spatial	Building a Model, p. 740; Minilab, p. 742; Demonstration, p. 743; Visual Learning, p. 747; Project, p. 744; The Inside Story, p. 748
Interpersonal	Student Journal, pp. 739, 747; Portfolio, p. 749
Linguistic	Student Journal, p. 744
Logical-Mathematical	Alternate Lab, p. 740; Thinking Lab, p. 741; Minilab, p. 743; Project, p. 743

CHAPTER

30 Mollusks and Segmented Worms

Animals often leave evidence showing they have been in a certain place. Broken twigs and fur on a branch indicate that a bear rested at the base of a tree. The distinctive tracks in the sand show that a sidewinder snake passed by. The fine, silvery path across a leaf is a suggestion that a slug or snail lives in your garden.

Garden slugs move slowly along a path of mucus. Both the mucous path and the thick mucus that covers the slug are produced by the slug. Mucus is an adaptation that allows the animal to adhere to and move across a surface.

The secretion of mucus is a characteristic not only of slugs, but of other animals as well. You may remember the first time you picked up an earthworm. It was soft and slippery from the mucus, without which the earthworm would not be able to burrow efficiently past rocks and through soil.

Slugs and earthworms are examples of two very different groups of animals. In this chapter, you will learn why these animals are classified in two different phyla. Snails belong to the same phylum as slugs. Snails also secrete a mucous trail over which to move. How do you suppose the snail is most like the slug, and how is it different?

Concept Check

You may wish to review the following concepts before studying this chapter.

- Chapter 18: evolution
- Chapter 28: animal body plans, development

Chapter Preview

Laboratory Activities

Biolab: Design Your Own Experiment

- How do earthworms respond to their environment?

Minilabs

- What path does food and water take in a clam?
- How are mollusks identified?

Mucus is an important adaptation for earthworms, as well as for slugs and snails. In addition to allowing earthworms to move through soil, mucus holds two earthworms together as they mate.

Assessment Planner

Choose assessment strategies from the following pages to evaluate the progress of your students.

Assess, pp. 745, 752
Alternate Lab, pp. 740-741
Minilabs, pp. 742, 743
Portfolio, p. 749
Thinking Lab, p. 741
Biolab, pp. 750-751
Chapter Review, pp. 753-755

Introducing the Chapter

Direct students' attention to the photograph of the slug. Elicit whether students have ever actually seen a slug and what impressions they recall about this animal. *Students will likely mention it was slimy.* Display several seashells and have students speculate about what kinds of animals might have lived in them. Explain that the slug and the animals that lived in the shells are classified as mollusks.

Have students look at the earthworms and ask them what these worms have in common with the slug. *Again, students may mention sliminess as a trait.* Point out that the sliminess associated with these animals is actually mucus that is secreted by the animal. Explain that this mucus is an important adaptation that aids in locomotion.

Theme Development

The theme of *unity within diversity* is evident throughout the chapter. When comparing and contrasting these animal groups, similarities are continually pointed out, while the unique characteristics of classes and species are emphasized. The theme of *evolution* is stressed through discussions of the origins of mollusks and the increasing complexity of the specialization of the body plans of mollusks and segmented worms.

Concept Check

Students will extend their knowledge of body plans, adaptations, and the use of larval forms as an indication of phylogenetic relationships.

SECTION

30.1

Prepare

Key Concepts

Students will study the general characteristics of mollusks and the traits that distinguish organisms in the three mollusk classes. They will also explore the origins of mollusks.

Block Scheduling

Look for this symbol for strategies that are useful in a block scheduling format. For more information on block scheduling, refer to Section 30.1 in the **Lesson Plans** booklet.

Materials

- Obtain live land snails, lettuce, and petri dishes for the focus activity.
- Obtain a rubber surgical glove for the Building a Model.
- For the Minilabs, gather sample dichotomous keys for mollusks that use shells as a means of classification and sample keys for another animal group, dropping pipettes, and 500-mL beakers. Order live clams from a biological supply house and prepare carmine solution.
- Obtain a whole squid from a supermarket or fish market for the Demonstration.

1 Focus

Bellringer

Before presenting the lesson, display **Section Focus Master 66** on the overhead projector and have students answer the accompanying questions. L1 SAE

SECTION

30.1 Mollusks

Section Preview

Objectives

Identify the characteristics of mollusks.

Compare the adaptations of gastropod, bivalve, and cephalopod mollusks.

Explain the origins of mollusks.

Key Terms

mantle
radula
open circulatory system
nephridia
closed circulatory system

If you are a shell collector, a walk on the beach as high tide begins to recede reveals bountiful treasures. The shell sizes, shapes, and colors are clues to the many different kinds of animals that once inhabited these structures. How could the marine animal that lived in the fan-shaped shell be related to the common garden slug?

What Is a Mollusk?

Slugs, snails, and animals that once lived in the shells on the beach are all mollusks. These organisms belong to the phylum Mollusca. Members of this phylum range from the slow-moving slug to the jet-propelled octopus. While most species live in the ocean, others live in fresh water and on land. Some have shells, while others, including slugs and squids, are adapted to life without a hard covering. All mollusks have bilateral symmetry, a coelom, two body openings, a muscular foot for movement, and a mantle. The **mantle** is a thin membrane that surrounds the internal organs of the mollusk. In shelled mollusks, the mantle secretes the shell.

Although the phylum is diverse, mollusks all share similar developmental patterns. The larval stages of all mollusks are similar, *Figure 30.1*, but they have different appearances as adults. Three classes of mollusks are shown in *Figure 30.2*.

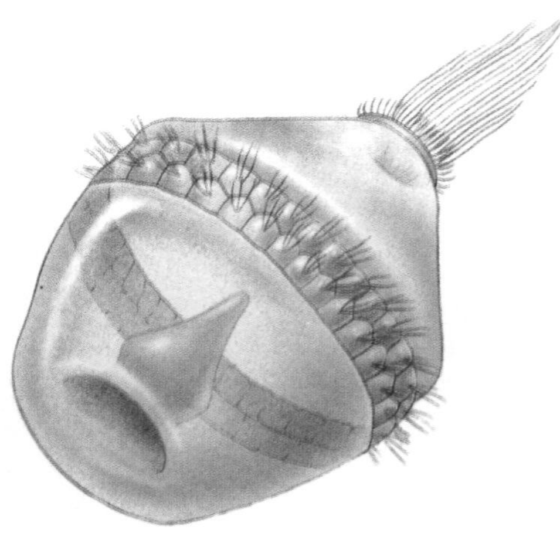

Figure 30.1

Larvae of most mollusks resemble a top with tufts of cilia. Most of these larvae are free-swimming before settling to the ocean floor for adult life. These larvae serve as food for many other organisms.

Program Resources

Section Focus Master 66 L1 SAE
Reinforcement and Study Guide, pp. 119-121 L1
Biolab and Minilab Worksheets, pp. 119-120 L1
Laboratory Manual, pp. 169-176 L2
Basic Concepts Transparency 48 and **Master** L1 SAE
Reteaching Transparency 30 and **Master** L1 SAE

Figure 30.2
With 100 000 species, phylum Mollusca is second in size only to insects and their relatives.

Bivalves: Oysters, clams, and scallops such as this one have two hinged shells and are known as bivalves. These animals strain their food from the ocean water.

Gastropods: One-shelled mollusks, called gastropods, make up the largest class of mollusks and exhibit a huge array of shell shapes and sizes. Members of this group include lung-breathing land snails and slugs, and the marine limpet shown here.

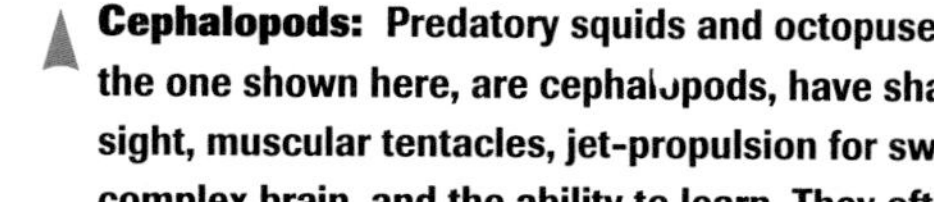

Cephalopods: Predatory squids and octopuses, such as the one shown here, are cephalopods, have sharp eyesight, muscular tentacles, jet-propulsion for swimming, a complex brain, and the ability to learn. They often apply their intelligence to capturing prey or avoiding harmful situations.

Where Do Mollusks Live?

Mollusks live in a wide variety of habitats. Most live in marine or freshwater habitats, but some live on land. A few species of mollusks can be found in cold, polar regions, and many are common in warm, tropical areas. Some aquatic mollusks, such as oysters and mussels, live firmly attached to the ocean floor or to the bases of docks or wooden boats. Others, such as the octopus, swim freely in the ocean. Land-dwelling slugs and snails can be found most frequently in moist tropical and temperate climates.

Classes of Mollusks

Within the large phylum of mollusks, there are seven classes. The three classes that include the most common and well-known species are the classes Gastropoda, Bivalvia, and Cephalopoda.

Activity

Divide the class into groups. Give each group a live land snail on one half of a petri dish. Ask students to record their observations of the snail. Instruct students to observe the snail through the underside of the dish. Have them gently touch the antenna of the snail with the eraser end of a pencil and observe and describe its reaction. Finally, have them place the snail on a piece of lettuce to see if they can observe the snail feeding. Discuss all observations as a class. COOP LEARN

2 Teach

Art Connection

Shell Shapes

An X ray technician uses a type of X ray film that shows details, without much contrast, to photograph mollusk shells. The X ray pictures are then photographed with a camera loaded with a special film. The pictures that result have become valued art. The photographs are also of interest to scientists who study how mollusks develop such intricate shells.

interNET CONNECTION

Follow the link for this chapter on the Glencoe Homepage at **www.glencoe.com/sec/science** to find out more about mollusks.

STUDENT JOURNAL

Locating Mollusks Provide students with a blank outline map of the world. Have them conduct research to find out where five species of mollusks are commonly found. For example, the Atlantic bay scallop is commonly found from North Carolina to the West Indies and Brazil. Ask students to develop a key to indicate these locations on their map. Have them locate both saltwater and freshwater species. Encourage students to combine their findings with those of two others in the class. If possible, provide students with nature and wildlife atlases to aid them in their research. L1 COOP LEARN LS

GLENCOE TECHNOLOGY

Videodisc

Biology: The Dynamics of Life
Disc 1, Side 2
Scallop Escape (Ch. 32)

Squid Locomotion (Ch. 33)

Building a Model

Visual-Spatial Fill a surgical glove with water. Squeeze the water in one of the fingers and have students observe how the water moves freely into the other parts of the glove. Explain that the glove roughly models an open circulatory system. Point out that in an open circulatory system, blood moves freely into open spaces surrounding organs, just as the water moves freely from one part of the glove to another.

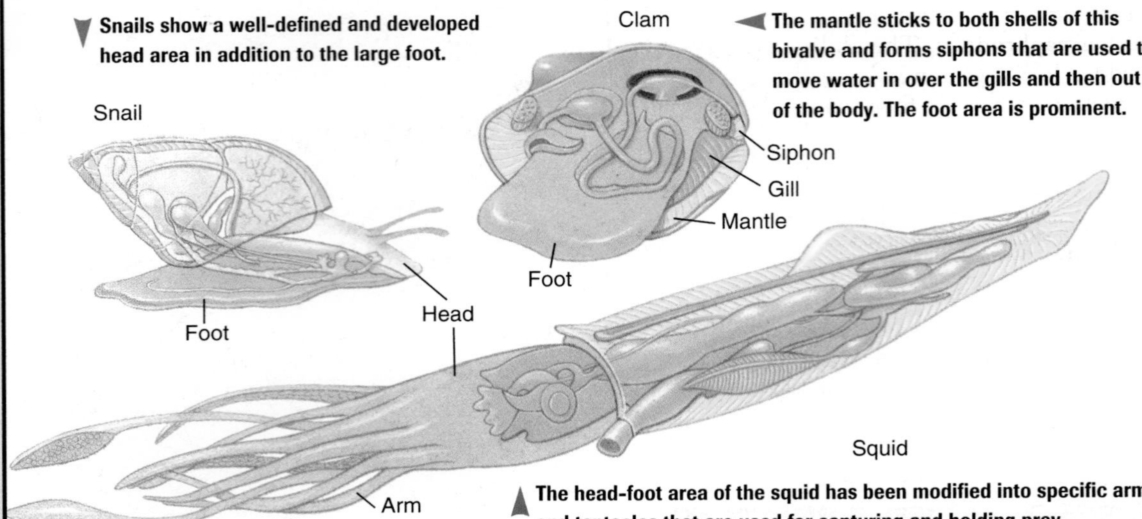
Snails show a well-defined and developed head area in addition to the large foot.
Snail
Foot
Head
Clam
The mantle sticks to both shells of this bivalve and forms siphons that are used to move water in over the gills and then out of the body. The foot area is prominent.
Siphon
Gill
Mantle
Foot
Squid
Arm
Tentacle
The head-foot area of the squid has been modified into specific arms and tentacles that are used for capturing and holding prey.
Key
Visceral mass
Mantle
Shell
Foot

Figure 30.3
Mollusks have soft bodies composed of a foot, a mantle, a shell, and a visceral mass that contains internal organs. Comparison of these three body sections shows how the bodies of the three classes of mollusks are adapted for survival.

While members of these three classes of mollusks look different from each other on the outside, they share many internal similarities. You can see the similarities and the differences in these body areas in the drawings in *Figure 30.3* as you compare the clam, snail, and squid.

Gastropods: One-shelled mollusks

The largest class of mollusks is Gastropoda, or the stomach-footed mollusks. The name comes from the way the animal's large foot is positioned under the rest of its body. Most species of gastropods have a single shell and are sometimes called univalves. Other gastropod species have no shell. Snails, slugs, and sea slugs belong to this class.

Gastropod adaptations

You may have watched a snail clean algae from the sides of an aquarium with a rasping structure called a radula. A **radula** is a tongue-like organ with rows of teeth located within the mouth of the gastropod. The radula is used to scrape, grate, or cut food.

The nervous system of gastropods is simple. Gastropods have a small brain. Associated nerves coordinate the animal's movements and behavior.

A heart is part of the gastropod's well-developed circulatory system. Blood is pumped by the heart in an open circulatory system. In an **open circulatory system,** the blood moves through vessels and into open spaces around the body organs. This adaptation exposes body organs directly to blood that contains nutrients and oxygen and removes metabolic wastes.

Mollusks are the first animals to have evolved respiratory structures. These structures are called gills. Gill tissue increases the surface area through which gases can diffuse, and it contains a rich supply of blood for the transport of gases. While aquatic gastropods have gills, the mantle cavity of land snails and slugs has evolved into a primitive lung.

Mollusks are also the first animals to have evolved excretory structures called nephridia. **Nephridia** are organs that remove metabolic wastes from an animal's body. Gastropods usually have one nephridium.

Alternate Lab **Comparing Snail Speeds**

Purpose 

Logical-Mathematical Students will compare the speed at which snails move under various environmental conditions.

Materials
land snails, clear plastic deli trays or large deli containers, wax marking pencil, lamp with 60-watt bulb, crushed ice, ring stand, black construction paper, sand paper, metric ruler

Procedure
Give students the following directions.

1. Make a table for distances traveled by the snail on a smooth surface, a rough surface, in cold conditions, and in warm conditions.
2. Make a hypothesis about the conditions under which the snail will move fastest.
3. With the wax marking pencil, mark an X in the middle of your tray. Place the snail on this X and measure how far it travels in three minutes.
4. Place a piece of black construction paper over the tray so the snail is in the dark. Measure distance traveled.

Many gastropods that live on land are hermaphrodites. The ability to produce both eggs and sperm is an adaptation commonly found in slow-moving animals. Most aquatic gastropods have either male or female reproductive organs. No mating occurs in these species. Eggs and sperm are released at the same time into the surrounding water, where external fertilization takes place.

Shelled gastropods

Snails, abalones, and conches are examples of shelled gastropods. They may live in fresh water, in salt water, or on land. Shelled gastropods may be creeping plant eaters, dangerous predators, or parasites. *Figure 30.4* shows two examples of shelled gastropods.

Gastropods without shells

Slugs are one type of gastropod without shells. Instead of being protected by a shell, the body of the slug is protected by a thick layer of mucus. The colorful nudibranchs, also called sea slugs, are protected in another way. When certain species of sea slugs feed on jellyfishes, they incorporate the poisonous nematocysts of the jellyfish into their own tissues without causing the cells to discharge. Any fish trying to eat the sea slugs are repelled when the nematocysts discharge into the unlucky predator. Other sea slugs secrete a strong, unpleasant-smelling mucus. Still others secrete a poisonous or strongly acidic mucus. The bright colors of these gastropods function to warn predators of the potential danger.

ThinkingLab Design an Experiment

What determines a snail's size before reproduction?

Biologists studying snails found that the size a particular species of snail reaches before beginning reproduction depends on the environment. If the water in which the snails live comes from an area in which snails were preyed on by crayfishes, the snails grow to be about 10 mm in length before reproducing. If the water comes from an area containing only snails, only crayfishes, or only crushed snails, the snails grow to a length of only about 4 mm before beginning reproduction.

Analysis

Biologists know that larger snails are not eaten by crayfish. Growing large quickly before reproducing has survival value for the snail, because it then will not be eaten by the crayfish. How might the young snail detect the presence of crayfish in the water?

Thinking Critically

Design an experiment that will provide information about how snails detect the presence of crayfish in their water.

Figure 30.4

Shelled gastropods vary from petite, thin-shelled species to large, thick-shelled ones.

A small, delicate gastropod species is the smooth dove shell. These organisms can be found in the Florida Keys and West Indies.

The pink conch is a large gastropod with a thick shell. Members of this species can measure up to 30 cm high. The meat of these gastropods is an important food source in the West Indian Islands. Overfishing for food and souvenirs has depleted this species in several areas.

ThinkingLab Design an Experiment

Purpose

LS Logical-Mathematical Students will analyze factors that determine a snail's size before reproduction.

Process Skills

recognize cause and effect, design an experiment

Teaching Strategies

- Ask students to sketch the sequence of testing to help them understand the problem. *They should draw an aquarium with only snails, an aquarium with snails and crayfish separated by a mesh that allows water but not animals to mix, an aquarium with snails and crushed snails, and an aquarium in which the crayfish are preying on snails.*

Thinking Critically

The snails must detect a chemical in the water that is given off by the crayfish when they eat snails. The water should be tested for chemicals that may be given off by crayfish after eating various prey species.

Assessment

Skill: Ask students to design an experiment to determine what environmental factors determine how large a freshwater mussel grows. L1

5. Cover the bottom of the tray with sandpaper that has been marked in its center with an X. Measure the distance the snail travels.
6. Place the lamp on a ring stand about 30 cm from the snail for about 3 minutes. After 3 minutes, begin timing and measuring the distance moved.
7. Place your container with the snail on another tray containing crushed ice. Wait about five minutes. Then measure the distance the snail travels in three minutes.

Analysis

Ask students to answer the following questions.

1. Was your hypothesis supported by your data? Explain. *Yes, if they hypothesized that the snail would move fastest when warm, in light, and on a smooth surface.*
2. What feature of snails aids their gliding movement? *Mucus*

Assessment

Performance: Design and conduct an experiment that would test to see if land snails prefer light or darkness. L1

SECTION 30.1

MiniLab

Purpose

LS Visual-Spatial Students will observe the process of filter feeding.

Process Skills

observe and infer, recognize cause and effect, interpret scientific illustrations

Teaching Strategies

- Instruct students not to use more than two drops of the carmine powder suspension.
- Have students wear laboratory aprons to protect their clothing from the dye.
- Students should allow the clam to open sufficiently before adding the carmine powder.
- Instruct students to release the suspension next to the siphons to prevent the suspension from floating on the water's surface.

Expected Results

The carmine powder will enter the incurrent siphon and exit through the excurrent siphon.

Analysis

1. It goes into the incurrent siphon and out of the excurrent siphon.
2. Food particles are trapped in the clam as water is filtered through the clam.
3. a sponge

Assessment

Knowledge: Ask students to diagram their clam and label its incurrent and excurrent siphons. Have them draw arrows on the diagram to show the movement of the water and add captions to their drawings to explain how a clam feeds. L1

MiniLab

What path do food and water take in a clam?

Clams are filter feeders, taking in water, small invertebrates, and other organic materials, and filtering out the food before it lets the water out. In this lab, you will observe the path water takes as it enters and exits a clam. These structures are called the incurrent and excurrent siphons.

Procedure

1. Find the incurrent and excurrent siphons of a live clam. See the photograph below.
2. Place a live clam in a beaker of water so that about 6 cm of water covers the clam. Let the clam rest undisturbed for about five minutes.
3. Place two drops of carmine powder suspension on the side of the clam that has the siphons.
4. Observe what happens to the carmine suspension.

Analysis

1. Explain what happens to the carmine suspension.
2. Explain what happens during filter feeding.
3. With what other animal could you use carmine suspension in water to observe water entering and exiting during the process of filter feeding?

Bivalves: Two-shelled mollusks

Two-shelled mollusks such as clams, oysters, and scallops belong to the class Bivalvia, *Figure 30.5*. Most bivalves are marine, but a few species live in fresh water. Bivalves show a range of sizes. Some are less than 1 mm in length and others, such as the tropical giant clam, may be 1.5 m long. Bivalves have no distinct head or radula. Most use their large, muscular foot for burrowing in the mud or sand at the bottom of the ocean or a lake. A strong ligament, like a hinge, holds their shells together; muscles allow the shell to open and close over the soft body.

One of the main differences between gastropods and bivalves is that bivalves are filter feeders. Bivalve mollusks have several adaptations for filter feeding. They have cilia that beat to draw water in through an incurrent siphon. The water moves over gills and exits through the excurrent siphon. As water moves over the gills, food and sediments become trapped in mucus. Cilia that line the gills push food particles to the stomach. Cilia also act as a sorting device. Large particles, sediment, and anything else that is rejected is transported to the mantle or to the foot, where it is eliminated.

Figure 30.5

Members of the class Bivalvia have some interesting names such as cockles, arks, angel wings, jewel boxes, and jingle shells. Whether thick or thin shelled, smooth or spiny, all these organisms have two shells held together by a hinge.

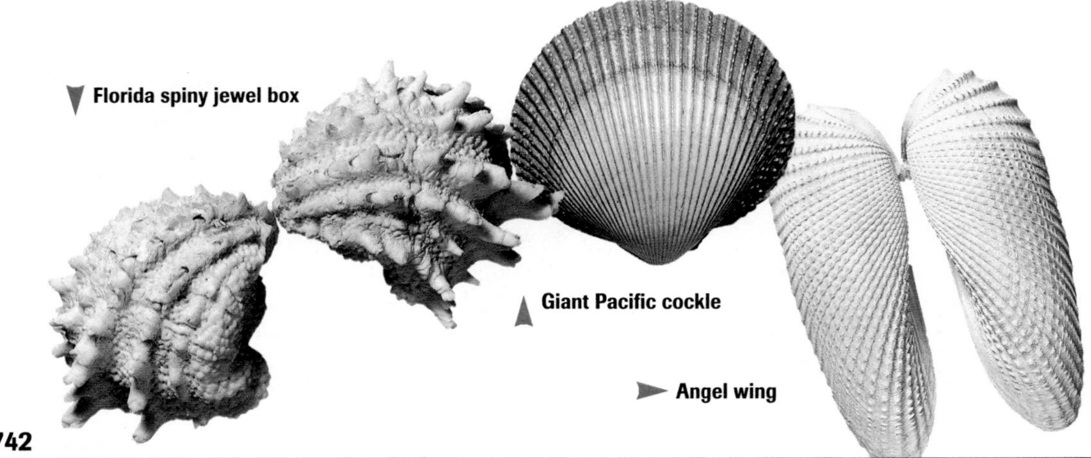

742

CULTURAL DIVERSITY

Pearl Cultivation in Japan

In your lessons on mollusk biology, discuss with students how pearls are formed and the pearl cultivation industry in Japan. Since 1893, pearl farming has been one of Japan's most famous industries.

To add further interest, introduce students to the AMA women of Japan. The AMA are a group of women who have been diving for pearls and other valuable mollusks for more than 2000 years. The divers take their name from the word *ama,* which in the ancient Japanese language meant "ocean" or "sky." The AMA of Japan have been known to dive to depths greater than 50 meters hundreds of times daily, without the use of snorkels or air tanks.

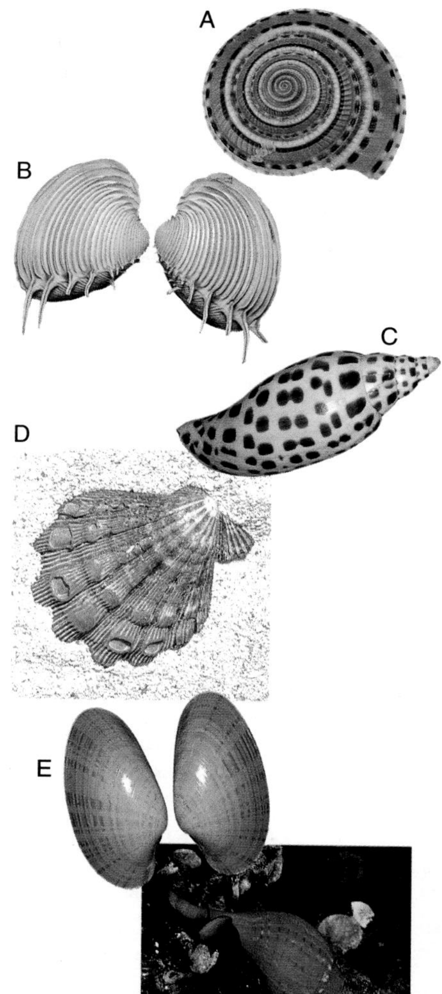

MiniLab

How are mollusks identified?

Have you ever taken a walk on the beach and filled your pockets with shells, and then as you examined them later, wondered what they all were? The shells pictured are ones that you might have picked up if you walked along the Florida coast. Use the following dichotomous key to determine the names of the shells.

Procedure

1. To use a dichotomous key, begin with a choice from the first pair of descriptions.
2. Follow the instructions for the next choice. Notice that either a scientific name can be found at the end of each description, or directions will tell you to go on to another numbered set of choices.

1A	One shelled	Gastropods see 2
1B	Two shelled	Bivalves see 3
2A	Flat coil	Sundial shell: *Architectonica nobilis*
2B	Thick coil	See 4
3A	Spines on posterior	Elegant venus shell: *Pitar dione*
3B	No spines	See 5
4A	Spotted surface	Junonia shell: *Scaphella junonia*
4B	Lined surface	Banded tulip shell: *Fasciolaria hunteria*
5A	Polished surface	Sunray shell: *Macrocallista mimbosa*
5B	Rough surface	Lion's paw shell: *Lyropecten nodosus*

Analysis

1. Why is a dichotomous key used for a variety of organisms?
2. What features were easy to pick out using the key for shells? What features were more difficult? Why?
3. What was the most general feature you used to identify shells?

Cephalopods: Head-footed mollusks

The head-footed mollusks are in the class Cephalopoda. This class includes the octopus, squid, and chambered nautilus, as shown on the following page in *Figure 30.6.* Scientists consider the cephalopods to be the most complex and most recently evolved of all mollusks. The only cephalopod with a shell is the chambered nautilus. All cephalopods are marine organisms. Instead of a single, muscular foot, cephalopods have tentacles with suckers for moving and for capturing food. The radula and sharp, beaklike jaws tear apart fish, bivalves, and crabs.

These marine predators move nutrients and oxygen through a closed circulatory system. In a **closed circulatory system,** blood moves through the body enclosed entirely in a series of blood vessels. A closed system provides a more efficient means of gas exchange within the body.

MiniLab

Purpose

Logical-Mathematical Students will use a dichotomous key to identify mollusks based on their shells.

Process Skills

observe and infer, compare and contrast, classify, use a dichotomous key

Teaching Strategies

- Make an overhead transparency of a dichotomous key. Use the transparency to demonstrate how a dichotomous key is used.
- Ask students who collect shells to bring in their shell collections and identify the shells for their classmates.

Expected Results

Students will classify the pictured shells using the dichotomous key provided.

Analysis

1. The key divides the group into smaller and smaller groups until each organism is identified.
2. easy: 1, 2, 3, 4; more difficult: 5, because it requires more interpretation and closer comparison
3. one or two shells

Assessment

Skill: Give students a simple dichotomous key for several shells. Ask them to use the key to identify the shells. L1

PROJECT

Classifying Mollusks Used as Food

Have students obtain a photocopy of a menu from a local seafood restaurant. Ask students to construct a key to identify the gastropods, bivalves, and cephalopods listed on the menu. Have students use their findings to summarize the importance of mollusks as a food source. L1

Demonstration

Visual-Spatial Obtain a whole squid from a fish market. Point out the head and tentacles of the squid. Cut the squid open to reveal its transparent cuttlebone. Explain that many scientists consider the cuttlebone to be a remnant of a shell.

Literature

Natural Acts

Purpose
Students are introduced to the writing of David Quammen.

Teaching Strategies

- Before students read this selection, ask them what they know about octopuses. *Students will likely state that the organism has a large head with eight "legs" or tentacles coming from it.*
- Explain that octopuses can be trained to distinguish between shapes such as a square and a rectangle and can remember the shapes for long periods of time. Elicit from students what this ability suggests about the brains of octopuses. *Their brains are quite well developed and appear to have some capacity for intelligence.*
- Encourage students to read other works Quammen has written about animals. Ask students to write a summary of the reading and include it in their portfolios.

Possible Answer
Questions students may indicate include: Why do some female spiders eat their mates? Why do gnats fly in swarms? Why do earthworms appear on sidewalks after rainfall? Do sharks sleep?

Literature

Natural Acts
by David Quammen

"I'm not a scientist," says naturalist writer David Quammen. "What I am is . . . a haunter of libraries and snoop." In spite of his disclaimer, Quammen, a keen observer of everything around him, has written dozens of essays on a wide variety of scientific topics. Abandoning his earlier writing style, the political thriller, Quammen seems to have settled into his favorite mode: asking an intriguing science question and then setting out to answer it in an entertaining way. In his collection of essays, called *Natural Acts: A Sidelong View of Science and Nature,* Quammen poses these kinds of questions: What are the redeeming merits, if any, of the mosquito? Are crows too intelligent for their station in life?

The eye of the octopus Another topic Quammen wondered about was the enormous eyes of the octopus, which seem to stare back at human observers in a somewhat unnerving fashion. "These animals don't just gape at you glassily, like a walleye. They make eye contact, as though they are someone you should know," he writes. The reason, he continues, is that an octopus is among the most intelligent of animals that live in the sea. In laboratory tests, an octopus combines that intellect with acute eyesight to do well in the mazes that scientists devise. Besides, adds Quammen, they have eyelids, "so they can wink at us fraternally."

More questions It seems unlikely that Quammen will run out of questions—big questions, such as "Is sex necessary?" and smaller ones, such as "Why are there so many different species of beetles?" A fellow writer calls reading about Quammen's pursuit of the answers "a wild ride on a slightly unsettled roller coaster."

CONNECTION TO Biology

What are some other questions about the natural world that you would like an entertaining writer like David Quammen to try answering?

Figure 30.6
The class Cephalopoda contains octopods, squid, and the only shelled member, the chambered nautilus.

The genus *Nautilus* is the only remaining living example of a cephalopod with a shell. All other members of this class are extinct. The shell of a *Nautilus* is divided into a series of chambers.

Octopuses, as well as all other cephalopods, have separate sexes. One tentacle of a male octopus is adapted to transferring sperm into the body of the female. Fertilization is internal, but eggs are laid outside the body.

A squid has large eyes, a well-developed nervous system, and moves by jet propulsion. With this system of movement, squids can attain speeds of 20 m per second!

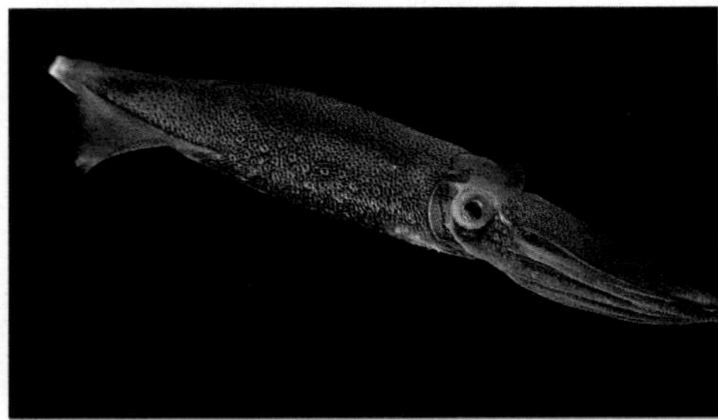

PROJECT: Dissecting a Squid

Ask students working in groups to prepare a squid dissection using clean kitchen tools, a clean aluminum tray, and a squid from a seafood market. Have them sketch the dissected squid and identify all exposed structures. Have students write a brief report that explains how squids are adapted to their environments. L2 LS

STUDENT JOURNAL

The Chambered Nautilus Ask a group of students to read the poem titled "The Chambered Nautilus" by Oliver Wendell Holmes. Ask them to explain in an essay what the poet is comparing to the nautilus, the symbolism used, and the part of the poem that is not true of a real nautilus in the sea. L3 LS

Origins of Mollusks

Fossil records show that mollusks lived in great numbers as long as 500 million years ago. Some, like the chambered nautilus, appear to have a structure similar to the related species that lived 500 million years ago. Other early species have shown radiation into a variety of forms that include land slugs and exotic marine slugs. *Figure 30.7* shows the radiation of mollusks and the closely related phylum of segmented worms.

Figure 30.7

Both mollusks and segmented worms share the same pattern of early cell division. Based on this and other evidence, biologists think that mollusks and segmented worms are closely related.

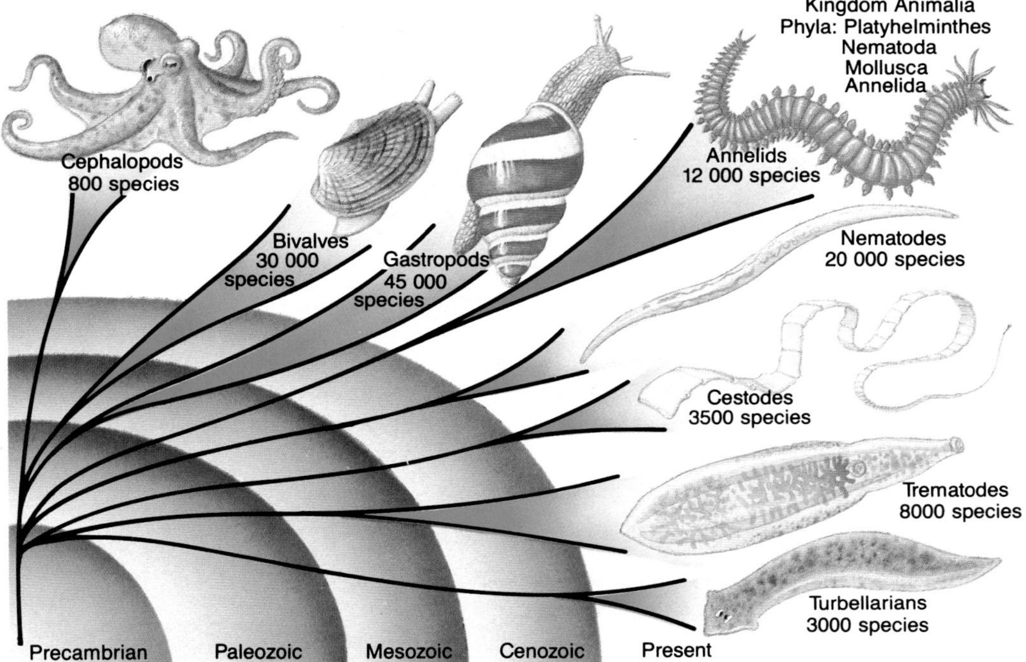

Section Review

Understanding Concepts

1. Describe how mucus is important to some mollusks.
2. What adaptations make the cephalopods effective predators?
3. Compare filter feeding with getting food by using a radula.

Thinking Critically

4. How are the methods of movement for the snail, clam, and squid related to the structure of each one's foot?

Skill Review

5. **Classifying** Construct a key to identify the three classes of mollusks discussed. For more help, refer to Organizing Information in the *Skill Handbook*.

3 Assess

Check for Understanding

Have students explain how mollusks are more advanced than the other animal groups they have studied. L1

Reteach

Have students construct a table describing mollusk characteristics. Have them use the headings: *Phylum*, *Representative Organisms*, *Symmetry*, *Habitat*, *Food Getting*, *Taking in Oxygen*, *Reproduction*, and *Protection*. L1

Extension

Ask students to prepare illustrated reports on pearl culturing. L2

Assessment

Portfolio: Ask students to write a description of a habitat that is suited to a specific mollusk from each group. Have students explain how each organism is adapted to its environment. L1 P

4 Close

Discussion

Discuss the economic importance of mollusks as a class. Include uses of mollusks as food, algae feeders in aquariums, and as souvenirs.

Answers to Section Review

1. Mucus enables mollusks to stick to surfaces and slide easily through or on materials in their habitats. Some mollusk mucus contains protective chemical poisons.
2. jet propulsion-type of swimming, tentacles with suckers, large eyes with well-developed nervous systems, and a radula for tearing apart prey
3. A filter feeder takes in water and filters out algae and other foods. The radula is a tongue-like organ that scrapes food from surfaces.

Thinking Critically

4. The muscular foot of the snail secretes mucus on which the snail glides slowly. The clam can burrow into sand with its muscular foot. The squid releases water that allows it to move as if jet propelled.

Skill Review

5. Students should construct keys that include information found in the sections titled Gastropods: One-shelled mollusks, Bivalves: Two-shelled mollusks, and Cephalopods: Head-footed mollusks.

SECTION

30.2

Prepare

Key Concepts

Students will learn of the characteristics of segmented worms that enable them to survive in their environments. The classes of segmented worms will be compared and the traits of animals that are more complex than those studied in previous chapters will be emphasized.

Block Scheduling

Look for this symbol for strategies that are useful in a block scheduling format. For more information on block scheduling, refer to Section 30.2 in the **Lesson Plans** booklet.

Materials

- Obtain a terrarium, soil, and live earthworms for the earthworm farm Demonstration.
- For the Biolab, gather live earthworms, paper towels, glass pans, sandpaper, droppers, penlights, culture dishes, ice, warm tap water, thermometers, and hand lenses or stereomicroscopes.

1 Focus

Bellringer

Before presenting the lesson, display **Section Focus Master 67** on the overhead projector and have students answer the accompanying questions. L1 SAE

Discussion

Have students name characteristics they associate with earthworms. Use the characteristics named as a springboard to introduce the characteristics of segmented worms.

SECTION

30.2 Segmented Worms

Section Preview

Objectives

Describe the characteristics of segmented worms and their importance to the survival of these organisms.

Compare and contrast the classes of segmented worms.

Key Terms

gizzard
nocturnal

Do earthworms have a front and a back end? Yes, they do. In fact, if you have ever watched one move, you know that it crawls by first stretching the front of its body forward, and then pulling the back of its body up to the front. A worm in motion looks a little like an accordion playing.

What Is a Segmented Worm?

Members of the phylum Annelida are segmented worms. They include the bristleworms, earthworms, and leeches as shown in *Figure 30.8*. The term *annelid* means "tiny rings." Segmented worms, like mollusks, are bilateral and have a coelom and two body openings. Some have a larval stage similar to the larval stages of certain mollusks.

Figure 30.8

The phylum Annelida contains about 12 000 species, which are placed in three classes. Although it has a definite anterior end, the earthworm does not have a distinct head as seen in bristleworms. Earthworms, members of the class Oligochaeta, have only a few setae on each segment.

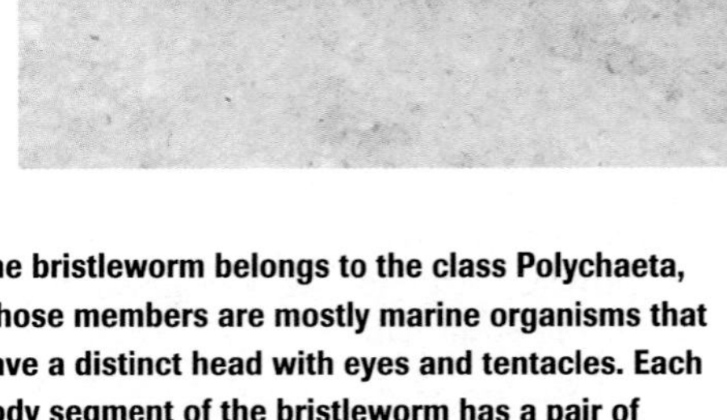

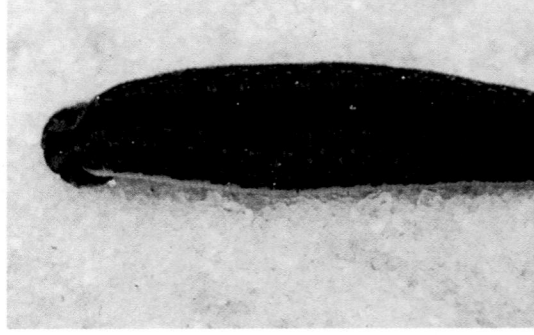

The class Hirudinea is made up of leeches that live in marine, freshwater, or land habitats. They lack parapodia and setae. All leeches have 32 segments.

The bristleworm belongs to the class Polychaeta, whose members are mostly marine organisms that have a distinct head with eyes and tentacles. Each body segment of the bristleworm has a pair of appendages called parapodia, which have many bristlelike structures called setae.

746

Program Resources

Section Focus Master 67 L1 SAE
Reinforcement and Study Guide, p. 122 L1
Biolab and Minilab Worksheets, pp. 121-122 L1
Laboratory Manual, pp. 165-168 L2
Concept Mapping, p. 30 L1
Critical Thinking/Problem Solving, p. 30 L3
Basic Skills Transparency 22 and **Master** L1 SAE

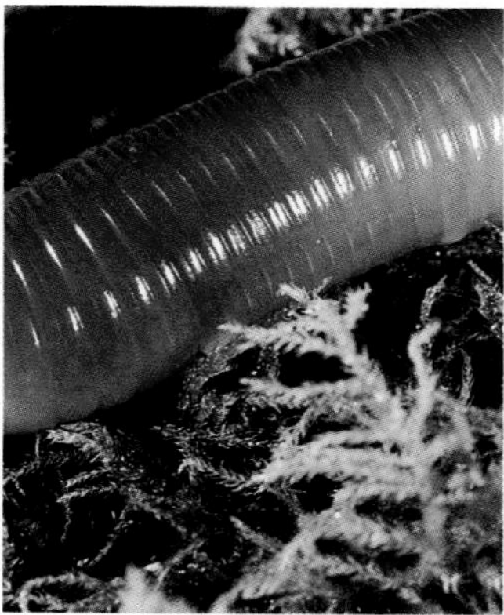

Figure 30.9

Segmentation is easily seen in the class Oligochaeta (left). Although most species in this class are 5 to 10 cm long, the giant earthworm of Australia can be more than 3 m long (right).

All annelids are made up of segments

The most distinguishing characteristic of annelids is their cylindrical bodies that appear to be divided into a series of ringed segments, as seen in the worms in *Figure 30.9.* This segmentation continues internally as each segment is separated from the others by a body partition. Segmentation is an important advantage because each segment has its own muscles, allowing shortening and lengthening of the body for movement.

The segments also allow for specialization. Groups of segments may be adapted for a particular function. If you examine each segment of most annelids, you find that most of the body is made of identical segments. Most segments contain excretory organs and nerve centers; however, a few also contain organs for digestion and reproduction.

The **gizzard** is a sac with muscular walls and hard particles that grind soil before it passes into the intestine. Some organs of the digestive system, nervous system, and circulatory system do run the length of the annelid's body. The internal structure of an earthworm can be seen in *The Inside Story* on the next page.

Where do segmented worms live?

Segmented worms live everywhere except in the frozen soil of the polar regions and the dry sand and soil of the deserts. You may be familiar with the earthworms in your garden, but these are just one of about 12 000 species of segmented worms that live in soil, fresh water, and the sea. The delicate and beautiful fan worms in *Figure 30.10* live by the shores of tropical and temperate seas. Their bodies are protected by the hard tubes they build out of sand grains. The sea mouse in *Figure 30.11* is another segmented worm that lives in the ocean.

Figure 30.10

The fan worm traps food in the mucus on its "fans." Disturbances in the water, such as a change in the direction of the current, the passage of a shadow overhead, or the passing by of an organism, cause these worms to quickly withdraw into their tubes.

2 Teach

Demonstration

Set up an earthworm farm in a terrarium. Remove one earthworm and display it so students can see it clearly. Elicit from students some uses of earthworms. *Students may cite the use of earthworms as bait.* As you place the earthworm back in the terrarium, discuss the benefits of earthworms to soil ecology. Be sure to discuss their role in aerating soil as well as their role in returning organic matter to soil.

Visual Learning

Figures 30.9 and 30.10 Direct students' attention to the giant earthworm and the fan worm shown. Have students identify the diverse environments of these two annelids. *earthworm: soil; fan worm: water (marine).* Use the photograph of the fan worm to emphasize that not all worms live in terrestrial environments. LS

Enrichment

Challenge students to research the derivation and meanings of the expressions "The early bird gets the worm" and "opening a can of worms." Ask students to share their findings with the class. L2

GLENCOE TECHNOLOGY

STVS: Animals
Disc 5, Side 1
Leeches (Ch. 7)

STUDENT JOURNAL

Earthworm Importance Ask a group of students to interview a farmer, an agriculture professor, or a representative from a local agricultural extension service about the importance of earthworms in agriculture. Have them present their findings to the class in an illustrated report. Ask the other class members to take notes on the presentation. L2 LS

THE INSIDE STORY

The Earthworm

Purpose

Visual-Spatial Students will examine the internal structures of the earthworm and their functions.

Teaching Strategies

- Ask students to write a paragraph that explains how earthworm bodies show more complexity than the bodies of flatworms and roundworms.

Visual Learning

- Make photocopies of *The Inside Story* diagram without the labels and captions. Have students label and describe the function of each structure on the photocopy as it is discussed. L1
- Obtain a plastic model of an earthworm. Point out each structure discussed in *The Inside Story* on the model as it is discussed.
- Ask students to use *The Inside Story* captions to make a table that identifies the organs that are involved in the following functions: digestion, locomotion, circulation, excretion, and sensory functions. L1
- Challenge your advanced students to make a table that compares earthworms with the free-living flatworms and roundworms studied in Chapter 29. Students should include the following functions: digestion, locomotion, circulation, excretion, and sensory functions. L2

THE INSIDE STORY

The Earthworm

As an earthworm burrows through soil, it loosens, aerates, and fertilizes the soil. Burrows provide passageways for plant roots and improve drainage.

A common earthworm, *Lumbricus terrestris*

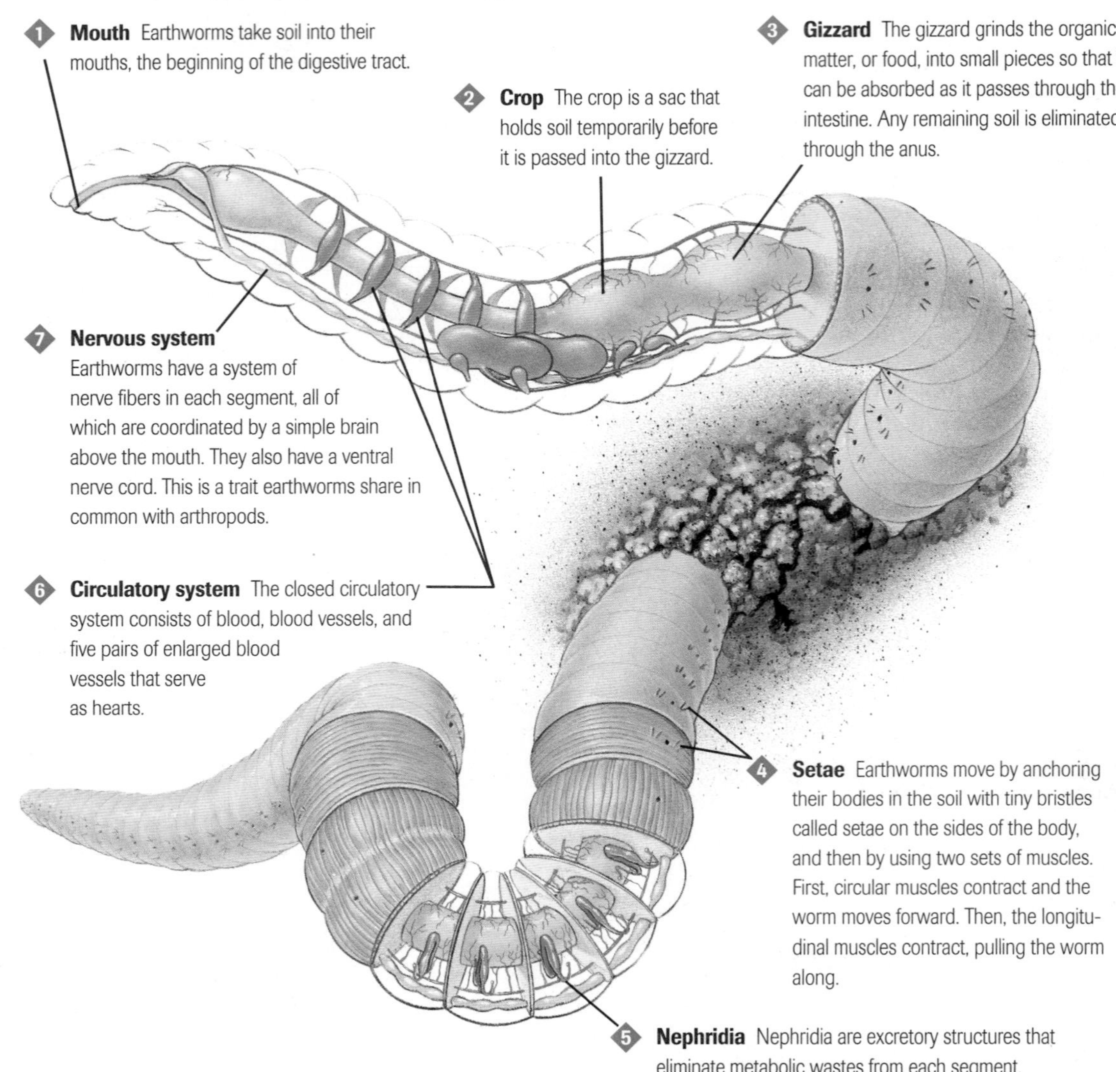

1 Mouth Earthworms take soil into their mouths, the beginning of the digestive tract.

2 Crop The crop is a sac that holds soil temporarily before it is passed into the gizzard.

3 Gizzard The gizzard grinds the organic matter, or food, into small pieces so that it can be absorbed as it passes through the intestine. Any remaining soil is eliminated through the anus.

4 Setae Earthworms move by anchoring their bodies in the soil with tiny bristles called setae on the sides of the body, and then by using two sets of muscles. First, circular muscles contract and the worm moves forward. Then, the longitudinal muscles contract, pulling the worm along.

5 Nephridia Nephridia are excretory structures that eliminate metabolic wastes from each segment.

6 Circulatory system The closed circulatory system consists of blood, blood vessels, and five pairs of enlarged blood vessels that serve as hearts.

7 Nervous system Earthworms have a system of nerve fibers in each segment, all of which are coordinated by a simple brain above the mouth. They also have a ventral nerve cord. This is a trait earthworms share in common with arthropods.

Meeting Individual Needs

Visually Impaired For students who are visually impaired, provide an earthworm for them to hold while you point out the main features of their structure and behavior. Ask the student to explain how the earthworm's shape and texture make it adapted for life in the soil. L1

Earthworms are nocturnal animals that live in burrows. A **nocturnal** animal is one that moves about primarily at night. At night, earthworms come to the surface but stay close to their burrows. The cool, moist soil provides protection for the worms during the day. Water in the soil is a source of oxygen that diffuses into an earthworm's body through the skin.

Figure 30.11

The sea mouse is a species of marine polychaete. The sea mouse is well known for its distinctive furry appearance, although it is not commonly seen.

How do annelids reproduce?

Earthworms are hermaphrodites. During mating, two worms exchange sperm. Each worm forms a capsule for the eggs and sperm. The eggs are fertilized in each worm's capsule, and the capsule is slipped off the worm into the soil. In two to three weeks, young earthworms emerge from the fertilized eggs.

Most polychaetes reproduce sexually, although mating occurs in only a few species. Eggs and sperm are released into the seawater, where fertilization occurs. All members of the class Hirudinea are hermaphrodites. Eggs and sperm are transferred from one animal to another in a way similar to that of the oligochaetes.

A BROADER VIEW

Polychaetes: Feather Dusters and Christmas Trees

Polychaetes, segmented worms that live in the sea, are like all other annelids. Each segment has a pair of swimming or crawling appendages and external gills. It's hard to imagine a worm being beautiful, but many polychaetes are spectacular because they display iridescent colors and often unusual shapes.

Worms of the shallows Many polychaetes live in shallow areas of the ocean. The name *tube worm* describes the shell-like tube that the sedentary species secrete and in which they live. Many of these worms have highly developed fans of feathery tentacles, banded in beautiful colors. This has led to descriptive names such as Magnificent feather duster, Star feather, and Christmas tree. Unlike most other annelid worms, polychaetes are not hermaphroditic and have separate sexes. Eggs and sperm are shed into the water and join to form free-swimming trochophore larvae. Many polychaetes can regenerate lost or damaged parts.

Varied feeding relationships Polychaetes are abundant in some areas and form an important link in certain marine food chains. Others have a commensal relationship with sponges, mollusks, or crustaceans—feeding on debris that the other animals leave. A few polychaetes are parasitic, and some can bite or even deliver a painful sting.

Thinking Critically
How do you think the adaptations that these species of worms have developed help them survive in their environments?

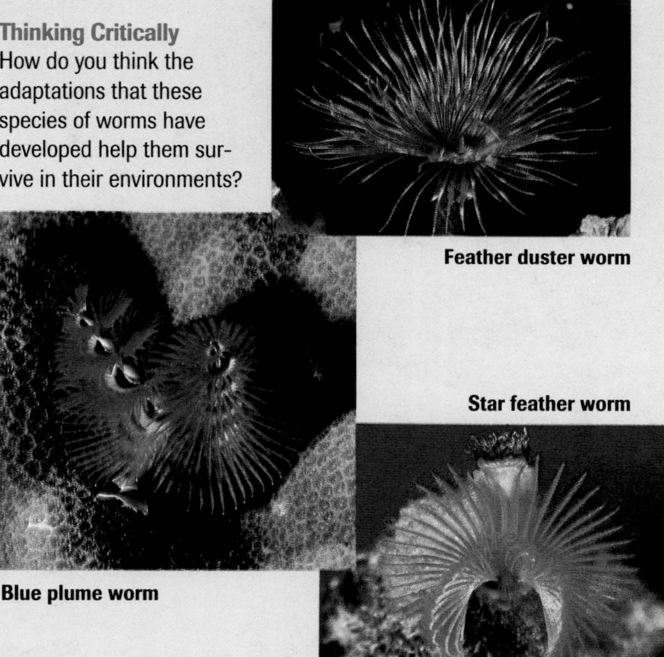

Feather duster worm

Star feather worm

Blue plume worm

A BROADER VIEW

Polychaetes: Feather Dusters and Christmas Trees

Purpose
Students are introduced to polychaetes and their various forms and ways of life.

Teaching Strategies

- Ask students to examine the colors and shapes of the worms. Have them list possible explanations for the color and shape of each worm and the survival value color and shape may have.
- Show a film clip that includes a variety of segmented worms in their natural habitats. Ask students to find features of each worm that illustrate how it is adapted to its environment. Have students describe the movements of each worm. Ask them to compare the worms and point out the similarities and differences in their movements. L1

Thinking Critically
The elaborate fans of some polychaetes help to capture food. Stinging helps protect some polychaetes from predators.

PORTFOLIO

Earthworm Terrariums Ask each group of students to prepare a large jar as a terrarium for earthworms. Have them place rocks on the bottom for drainage. They should then place alternating layers of moist sand and topsoil. Have them add about six to eight worms to the jar with dead leaves and grass on the top. Instruct students to tape black paper to the outside of the jar. After several days, have them remove the paper and observe what happened to the leaves, grass, and soil layers. Have them also describe any tunnels they observe. L1 P LS

BioLab Design Your Own Experiment

How do earthworms respond to their environment?

Time Allotment

One or two class periods

Objectives

Review objectives with students before they begin the Biolab.

Process Skills

observe and infer, compare and contrast, recognize cause and effect, form a hypothesis, interpret data, design an experiment, separate controls and variables

PREPARATION

- Keep earthworms in the refrigerator overnight, but remove them two hours prior to the lab.
- Live earthworms may be obtained from biological supply houses.

Safety Precautions

- Remind students to treat the earthworm in a humane manner at all times.
- Make sure students wash their hands when the experiment is completed.

Possible Hypotheses

- Students may hypothesize that the worm will move toward the dark, move faster on a rough surface, choose a moist surface over a dry surface, and prefer cool versus dry conditions.

BioLab Design Your Own Experiment

An earthworm spends its time eating its way through soil, digesting organic matter, and passing inorganic matter through the system and out of the body. Because earthworms are dependent on soil for food and shelter, they respond to stimuli in a way that will ensure a continuous supply of food and a safe place in which to live. These responses are genetically controlled. In this lab, you will design an experiment to determine the responses of earthworms to various stimuli.

How do earthworms respond to their environment?

PREPARATION

Problem

How do earthworms respond to light, different surfaces, moist and dry environments, and warm and cold environments?

Hypotheses

Place your worm in a tray with some moist soil. Watch your worm for about five minutes, and record what you observe. Make a hypothesis based on your observations about what the worm might do under conditions of light and dark, rough and smooth surfaces, moist and dry surfaces, and warm and cold conditions. Limit your investigation as time permits.

Objectives

In this Biolab, you will:

- **Measure** the sensitivity of earthworms to different stimuli, including light, water, and temperature.
- **Interpret** earthworm responses in terms of adaptive value to their lifestyle.

Possible Materials

live earthworm	water
paper towels	dropper
glass pan	penlight
sandpaper	ice
culture dishes	ruler
warm tap water	black paper
thermometer	cotton swabs
hand lens or stereomicroscope	

Safety Precautions

Be sure to treat the earthworm in a humane manner at all times. Wash your hands when your experiment is complete.

PLAN THE EXPERIMENT

Possible Procedures

- To test which surface enables a worm to move fastest, students may decide to measure how far the worm moves in a given period on surfaces such as sandpaper, the bottom of the dry glass pan, the bottom of a wet glass pan, and on wet and dry paper towels.
- To test the worm's reaction to light, the bottom of a pan may be covered with soil. Part of the pan may be covered with a piece of black construction paper while a penlight shines on the other side. The amount of time the worm spends in the light and dark parts of the container may then be measured.
- To determine the worm's preference for heat or cold, the glass pan could be placed

PLAN THE EXPERIMENT

1. As a group, make a list of possible ways you might test your hypothesis. Keep the available materials in mind as you plan your procedure.
2. Be sure to design an experiment that will test one variable at a time. Plan to collect quantitative data.
3. Record your procedure and list materials and amounts you will need. Design and construct a data table for recording your findings.

Check the Plan

Discuss the following points with other group members to decide the final procedure for your experiment.

1. What data will you collect, and how will they be recorded?
2. Does each test have one variable and a control? What are they?
3. Each test should include measurements of some kind. What are you measuring in each test?
4. How many trials will you run for each test?
5. Assign roles for this investigation.
6. ***Make sure your teacher has approved your experimental plan before you proceed further.***
7. Carry out your experiment.

ANALYZE AND CONCLUDE

1. **Checking Your Hypothesis** Which surface did the worm prefer? Explain.
2. **Interpreting Observations** In which temperature was the worm most active? Explain.
3. **Observing and Inferring** How did the earthworm respond to light? Of what survival value is this behavior?
4. **Observing and Inferring** How did the earthworm respond to dry and moist environments? Of what survival value is this behavior?
5. **Drawing Conclusions** Were your hypotheses supported by your data? Why or why not?

Going Further

Project Based on your experiment, design another experiment that would help to answer a question that arose from your work. You might want to try other variables similar to the ones you used, or you might choose to investigate a completely different variable.

ANALYZE AND CONCLUDE

1. a rough surface; the worm moves more easily on a rough surface
2. an intermediate temperature; an earthworm is ectothermic so its level of activity will depend on its surrounding temperature
3. moved away from light; earthworms are safer from predators in the soil, where it is dark
4. preferred a moist environment; earthworms' skin must remain moist or the animal will dry out and die
5. Students who made hypotheses that the worm would prefer moist environments, intermediate temperatures, darkness, and rough surfaces most likely would have their hypotheses supported by their data.

Assessment

Performance: Ask students to design and then carry out an experiment to determine how earthworms respond to gravity. L1

Going Further

Have students design similar experiments for other invertebrates and compare their results with those obtained for the earthworm. L2

on top of two culture dishes—one containing warm tap water and the other containing ice.

Teaching Strategies

- To save time, have groups test only one or two variables and share their class data.
- Ask students to gently rub their finger up and down the length of the ventral surface of the worm to feel the setae.
- Review the terms *anterior*, *posterior*, *dorsal*, and *ventral*. Ask students to use them when recording their observations.

Data and Observations

Most likely, earthworms will avoid light and extremes of temperature, move more quickly on a rough surface, and prefer a moist surface.

3 Assess

Check for Understanding

Show students cross-section slides of planarians, earthworms, nematodes, tapeworms, and leeches. Ask them to distinguish the segmented worms from the other worms. Have them explain their choices. L1

Reteach

Ask students to draw a large diagram that shows an earthworm's nervous, circulatory, muscular, digestive, and excretory systems. Have them label each structure and identify the system or systems to which it belongs. L1

Extension

Have students research and report on the ways leeches were used by physicians before the days of modern medicine and how they are used today. L2

Assessment

Journal: Have students create a table that compares the characteristics of the three classes of annelids. L1

4 Close

Discussion

Discuss with students the characteristics that make annelids more evolutionarily advanced than flatworms and mollusks.

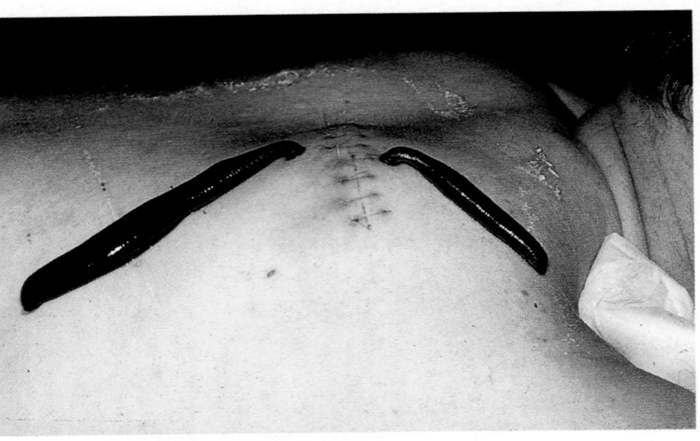

Figure 30.12

Besides anesthetic and anticlotting agents, the saliva of leeches also contains chemicals to dilate blood vessels to increase blood flow. Because of these characteristics, leeches have many medical uses.

Leeches

Leeches, as seen in *Figure 30.12,* are segmented worms with flattened bodies and no bristles. Although these animals can be found in many different habitats, most leeches live in fresh water. Most species are parasites that suck blood or other body fluids from the bodies of their hosts, which include ducks, turtles, fishes, and people. Front and rear suckers enable leeches to attach themselves to their hosts.

You may cringe at the thought of being bitten by a leech, but the bite is not painful. This is because the saliva of the leech contains chemicals that act as an anesthetic. Other chemicals prevent the blood from clotting. A leech can ingest two to five times its own weight in one meal. Once fed, it may not eat again for a year.

Connecting Ideas

In this chapter, you studied the features of two groups of invertebrate organisms, the mollusks and annelid worms. Both of these groups of protostomes show increasing specialization. All mollusks have an identifiable body plan that includes a foot, head, mantle, and gills. The major classes of mollusks include the gastropods, bivalves, and cephalopods.

The phylum Annelida includes common earthworms, marine polychaetes, and leeches. All members of this phylum show external segmentation. The annelid worms also have a closed circulatory system and a specialized digestive system. The segmentation of annelid worms was a significant step in evolution. The arthropods, descendants of annelids, became an extremely successful group due to the specialization of feeding structures and appendages that segmentation allowed.

Section Review

Understanding Concepts

1. What is the most distinguishing characteristic of bristleworms, earthworms, and leeches? Why is it important?
2. Describe how earthworms reproduce.
3. How do earthworms improve soil fertility?

Thinking Critically

4. Suppose a patient was not responding well to anticoagulant drugs administered after microsurgery. Explain why the doctor might prescribe the use of leeches.

Skill Review

5. **Interpreting Scientific Illustrations** Using *The Inside Story*, interpret how the two types of muscles in the earthworm are used to move the animal through the soil. For more help, refer to Thinking Critically in the *Skill Handbook*.

Answers to Section Review

1. segmentation; each segment has its own muscles that lengthen and shorten for efficient movement. Groups of segments may take on specific functions.
2. Even though earthworms are hermaphrodites, mating occurs. During mating, two worms exchange sperm. Eggs are fertilized in a capsule that is then shed into the soil. Young earthworms emerge in two to three weeks.
3. As the earthworm burrows through the soil, it loosens, aerates, and fertilizes the soil. Burrows provide passageways for plant roots and improve drainage.

Thinking Critically

4. Leeches applied to the surgical site would draw out clotted blood, thereby reducing swelling and allowing passage of blood through tiny vessels obstructed by clots.

Skill Review

5. Circular muscles contract to move the worm forward. Longitudinal muscles contract to pull the worm's body along.

REVIEWING MAIN IDEAS

30.1 Mollusks

- Mollusks have bilateral symmetry and a coelom. Many also have shells and similar larvae.
- Mollusks live in marine, freshwater, and land ecosystems.
- Most one-shelled mollusks have a shell, mantle, radula, open circulatory system, gills, and nephridia. Slugs are protected by a slime covering.
- Bivalve mollusks have two shells and are filter feeders.
- Cephalopods include the octopus, squid, and chambered nautilus. They have tentacles with suckers, a beaklike mouth with a radula, and a closed circulatory system.
- Fossil remains of mollusks show that they first lived 500 million years ago.

30.2 Segmented Worms

- The phylum Annelida includes the bristleworms and leeches. They are bilateral and have a coelom and two body openings; some have larvae that look like the larvae of mollusks. Their bodies are cylindrical and segmented.
- Earthworms are nocturnal and have complex digestive, excretory, muscular, and circulatory systems.
- Leeches are flattened, segmented worms. Most are aquatic parasites.

Key Terms

Write a sentence that shows your understanding of each of the following terms.

closed circulatory system
gizzard
mantle
nephridia
nocturnal
open circulatory system
radula

Understanding Concepts

1. An adaptation that allows an animal to adhere to and move across a surface is ____________.
   a. mucus c. a gizzard
   b. a radula d. nephridia
2. Slugs, snails, and animals that once lived in the shells on the beach are all ____________.
   a. gastropods c. cephalopods
   b. bivalves d. mollusks
3. The ____________ is a muscular sac that grinds soil before it passes into the intestine of an earthworm.
   a. stomach c. gizzard
   b. nephridium d. radula
4. ____________ animals are awake primarily at night.
   a. Diurnal c. Crepuscular
   b. Nocturnal d. Insomniac
5. The ____________ is a thin membrane that surrounds the internal organs of a mollusk.
   a. radula c. nephridia
   b. mantle d. gizzard
6. Oysters, clams, and scallops are ________.
   a. gastropods c. cephalopods
   b. bivalves d. nematodes
7. Snails, slugs, and limpets are __________.
   a. gastropods c. bivalves
   b. cephalopods d. cestodes
8. A ____________ is the tongue-like organ with rows of teeth in the mouth of a gastropod.
   a. foot c. mantle
   b. shell d. radula
9. Animals with bilateral symmetry, a coelom, two body openings, a muscular foot, and a mantle are ____________.
   a. segmented worms c. mollusks
   b. flatworms d. roundworms

Chapter 30 Review

Reviewing Main Ideas

Summary statements can be used by students to review the major concepts of the chapter.

Key Terms

Answers should go beyond defining the terms. Accept any answer that uses the term correctly and in the proper context.

Understanding Concepts

1. a
2. d
3. c
4. b
5. b
6. b
7. a
8. d
9. c

Chapter 30 Review

10. b
11. b
12. a
13. b
14. c
15. a
16. b
17. c
18. a
19. d
20. d

Applying Concepts

21. Cnidarian tentacles contain stinging cells that immobilize prey, and octopus tentacles have suckers for capturing prey.
22. Sponges and bivalves are filter feeders that strain food from water currents.
23. A snail helps keep the tank clean because it scrapes algae from the walls of the tank for food.
24. Gastropods can withdraw into their shells. Bivalves can close their shells. Cephalopods, such as squids, can swim very fast, and the octopus can use its tentacles for defense.
25. Nephridia remove metabolic wastes from an animal's body, thereby keeping metabolic wastes from building up in the body.

Thinking Critically

26. See below.
27. Soft-bodied worms do not fossilize as readily as mollusks, which have hard shells.
28. Bivalves filter organic matter from the water and break it down to smaller units that can be used again by marsh grasses and algae.

Chapter 30 Review

10. In a(n) ____________, the blood moves through vessels and into open spaces around the body organs.
    a. closed circulatory system
    b. open circulatory system
    c. nephridium
    d. mantle
11. Bivalves feed by ____________.
    a. scraping algae with a radula
    b. filter feeding
    c. predation
    d. extending tentacles to capture food
12. The head-footed mollusks circulate blood in a(n) ____________.
    a. closed circulatory system
    b. open circulatory system
    c. nephridium
    d. mantle
13. The only living example of a head-footed mollusk with an external shell is the ____________.
    a. scallop
    b. chambered nautilus
    c. snail
    d. octopus
14. Segmented worms and mollusks both have bilateral symmetry, a coelom, and similar ____________.
    a. gizzards c. larvae
    b. feeding methods d. setae
15. Animals distinguished by cylindrical bodies and ringed segments are called ____________.
    a. segmented worms
    b. mollusks
    c. gastropods
    d. bivalves
16. ____________ are excretory structures that remove wastes from an earthworm's body.
    a. Gizzards c. Mantles
    b. Nephridia d. Radulas
17. To what group of mollusks do the octopus, squid, and chambered nautilus belong?
    a. bivalves c. cephalopods
    b. gastropods d. annelids
18. During mating, two earthworms exchange ____________.
    a. sperm c. zygotes
    b. eggs d. capsules
19. Segmented worms with flattened bodies, suckers, and no bristles are called ________.
    a. earthworms c. fan worms
    b. bristleworms d. leeches
20. The saliva of leeches contains chemicals that act as a(n) ____________.
    a. stimulant
    b. sedative
    c. blood-clotting agent
    d. anesthetic

Applying Concepts

21. Compare how a cnidarian and an octopus use their tentacles to capture food.
22. Describe how sponges and bivalves have a similar way of obtaining food.
23. Why is it a good idea to keep a snail in an aquarium?
24. Compare the protective adaptations of gastropods, bivalves, and cephalopods.
25. Explain how nephridia enable mollusks and segmented worms to maintain homeostasis.

Thinking Critically

26. ***Concept Mapping*** Make a concept map that relates the following terms and phrases. Supply the appropriate linking words for your map.
    tentacle, filter feeder, gastropod, radula, cephalopod, mantle, slug, clam, octopus, bivalve, mollusk, snail
27. ***Observing and Inferring*** Explain why the phylogeny of worms is not as well understood as the phylogeny of mollusks.
28. ***Recognizing Cause and Effect*** Explain how bivalves in salt marshes are important for the health of the other species that live there.

754 Mollusks and Segmented Worms

26.

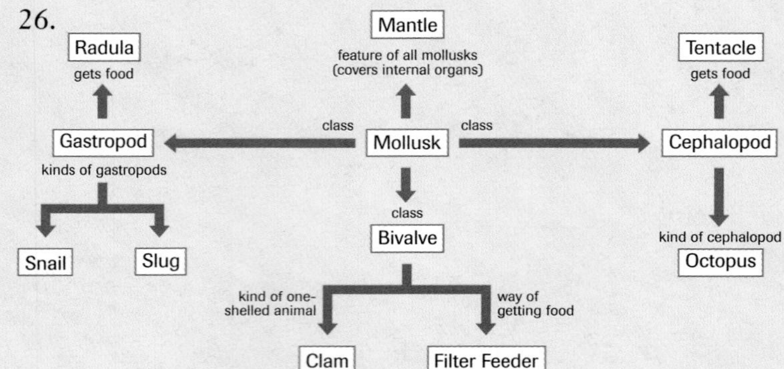

29. ***Interpreting Scientific Illustrations*** Explain what is happening in the photograph. What animal is attached to the fish? How is the body of this animal adapted to this way of life?

30. ***Interpreting Data*** The following graph illustrates data obtained on the growth rate of a particular tropical snail when kept in a tank that also contains a certain kind of shrimp. It also shows the growth rate in a tank in which there were only other snails. What is the relationship between the shrimps and the snails?

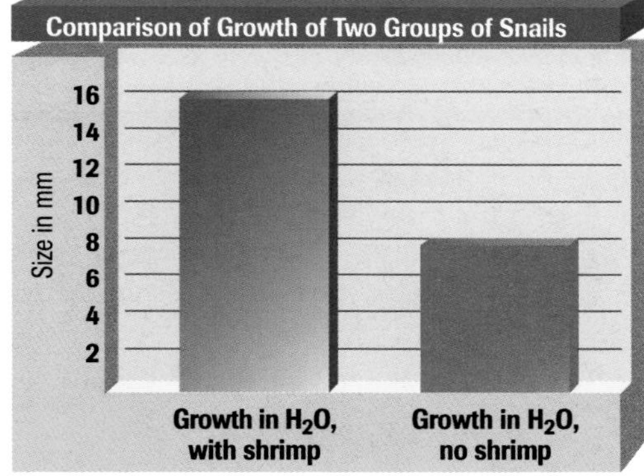

ASSESSING KNOWLEDGE & SKILLS

The graph below shows how a number of different animals respond to light.

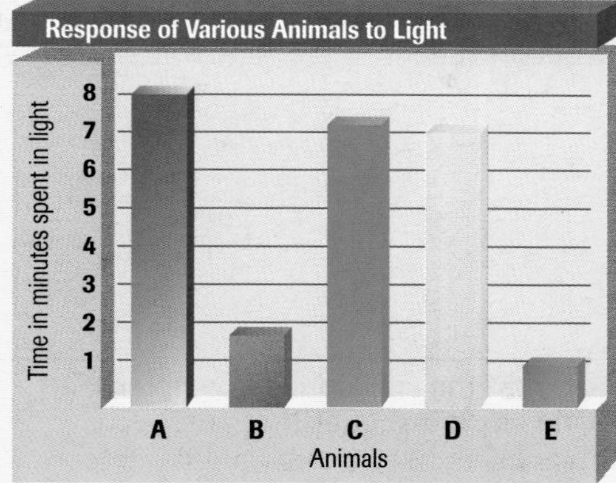

Using a Graph Use the graph to answer the following questions.

1. Which animals spend more time in the light?
   a. A, C, D **c.** A, B, C, D, E
   b. B, E **d.** C, D
2. Which animals do not spend as much time in the light?
   a. A, C, D **c.** A, B, C, D, E
   b. B, E **d.** C, D
3. Which animals might be nocturnal?
   a. A, C, D **c.** A, B, C, D, E
   b. B, E **d.** C, D
4. Which animals might live under a rock?
   a. A, C, D **c.** A, B, C, D, E
   b. B, E **d.** C, D
5. To which animal group might an earthworm belong?
   a. A **c.** C
   b. B **d.** D
6. ***Making a Graph*** Make a graph of the following data. Animal A spends 20 minutes in the dark. Animal B spends 15 minutes in the dark. Animal C spends 2 minutes in the dark. Animal D spends 7 minutes in the dark.

Chapter 30 Review

29. This is a leech feeding. The suckers on the anterior and posterior ends allow the leech to attach firmly to its food source.
30. The shrimps may release a chemical that stimulates the growth of the snails.

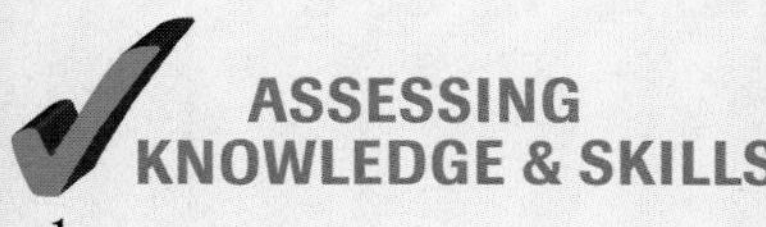

1. a
2. b
3. b
4. b
5. b
6.

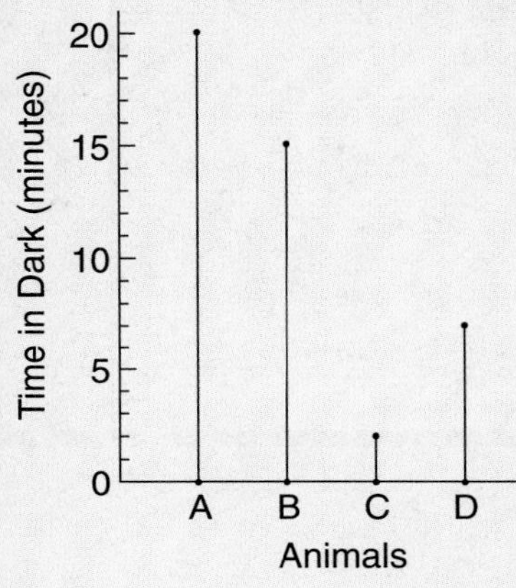

Program Resources

Chapter Assessment, pp. 179-184 L1
Alternate Assessment in the Science Classroom
Computer Test Bank L1
Content Mastery, pp. 121-124 L1

Chapter Organizer

SECTION	OBJECTIVES	ACTIVITIES/FEATURES
31.1 Characteristics of Arthropods National Science Standards: UCP.1-5; A.1, A.2; C.3, C.5, C.6	1. **Relate** the structural and behavioral adaptations of arthropods to their ability to live in different habitats. 2. **Analyze** the adaptations that make arthropods an evolutionarily successful phylum.	**Thinking Lab:** What determines the type of waxy coating produced by grasshoppers?, p. 759 **Minilab:** How strong is a spiderweb?, p. 761 **Biolab:** Design Your Own Experiment—Do flowers produce insect repellents?, p. 764
31.2 The Diversity of Arthropods National Science Standards: UCP.1, UCP.2, UCP.5; C.5, C.6; E.1, E.2; F.1, F.5, F.6; G.1	3. **Compare and contrast** the similarities and differences among the major groups of arthropods. 4. **Explain** the adaptations of insects that contribute to their success.	**Biology & Society Technology:** Milking Spiders, p. 770 **Health Connection:** Terrible Ticks, p. 771 **Minilab:** Which insects are attracted to light?, p. 773 **People in Biology:** May R. Berenbaum, p. 774

ACTIVITY MATERIALS

BIOLAB	MINILABS	ALTERNATE LAB
page 764 insects petri dishes flower extract-treated filter paper filter paper, plain forceps cotton swabs	**page 761** spiderwebs cotton swabs twigs jars human hairs beaker ruler **page 773** paper sack or box flashlight	**page 762** bess beetles cloth toweling clear tape heavy thread balance pennies plastic petri dish

Chapter 31 Arthropods

TEACHER CLASSROOM RESOURCES

Reproducible Masters	Transparencies
Section Focus Master 68: Arthropods L1 SAE **Reinforcement and Study Guide,** pp. 123-124 L1 **Biolab and Minilab Worksheets,** pp. 123, 125-126 L1 **Content Mastery,** pp. 125-128 L1	**Basic Concepts Transparency #49:** Phylogeny of Arthropods L1 SAE
Section Focus Master 69: Arthropod Diversity L1 SAE **Reinforcement and Study Guide,** pp. 125-126 L1 **Biolab and Minilab Worksheets,** p. 124 L1 **Concept Mapping:** Metamorphosis in Insects, p. 31 L1 **Tech Prep Applications:** Organic Pest Management, pp. 31-32; Insect-Plant Relationships in Your Garden, pp. 33-34 L2 **Critical Thinking/Problem Solving:** What Insect Am I?, p. 31 L3 **Laboratory Manual:** How Does Temperature Affect Mealworm Metamorphosis?, pp. 177-178; Identifying Insects, pp. 179-182; Comparing Arthropods, pp. 183-186 L2	**Basic Skills Transparency #23:** Insect Metamorphosis L1 SAE **Reteaching Transparency #31:** Structure of a Spider and a Grasshopper L1 SAE

ASSESSMENT MATERIALS

Chapter Assessment, pp. 185-190
Alternate Assessment in the Science Classroom
MindJogger Videoquiz
Computer Test Bank

Spanish Resources SAE
English/Spanish Audiocassettes SAE
Cooperative Learning in the Science Classroom COOP LEARN
Lesson Plans

KEY TO TEACHING STRATEGIES

L1 Level 1 activities should be within the ability range of all students including those with learning difficulties.

L2 Level 2 activities are within the ability range of average to above-average students.

L3 Level 3 activities are designed for the ability range of above-average students.

SAE SAE activities should be within the ability range of Students Acquiring English.

COOP LEARN Cooperative Learning activities are designed for small group work.

P These strategies represent student products that can be placed into a best-work portfolio.

These strategies are useful in a block scheduling format.

GLENCOE TECHNOLOGY

The following multimedia resources are available from Glencoe.

Biology: The Dynamics of Life
CD-ROM SAE
Videodisc Program

The Infinite Voyage Series
Secrets from a Frozen World
The Vital Link of the Food Chain
Insects: The Ruling Class
Insects and Their Behavior
Classifying Social Insects: Ants
Studying Insect Flight

National Geographic Society Series
Newton's Apple: Life Sciences
Bees

Science and Technology Videodisc Series (STVS)
Ecology
Pheromone Rope
Animals
Tick Research
Horseshoe Crab
Cockroach on a Treadmill

CHAPTER

31 Arthropods

Chapter Overview

Students begin their study of arthropods by learning about the traits they have in common: jointed appendages and exoskeletons. Their varied adaptations to life on land, water, and air are discussed. Arthropod success and diversity are discussed in terms of behavioral and structural adaptations.

In the second section of the chapter, students learn about the most common groups of arthropods. The ways of life and structural adaptations of spiders, ticks, mites, and scorpions are discussed. Next, the aquatic crustaceans including crabs, lobsters, shrimp, crayfishes, barnacles, and water fleas are presented. A comparison is made between millipedes and centipedes. Finally, the most successful arthropods in terms of diversity—the insects—are studied.

Key Terms

appendage
book lung
cephalothorax
chelicerae
compound eye
larva
Malpighian tubule
mandible
metamorphosis
molting
nymph
parthenogenesis
pedipalp
pheromone
pupa
simple eye
spinneret
spiracle
tracheal tube

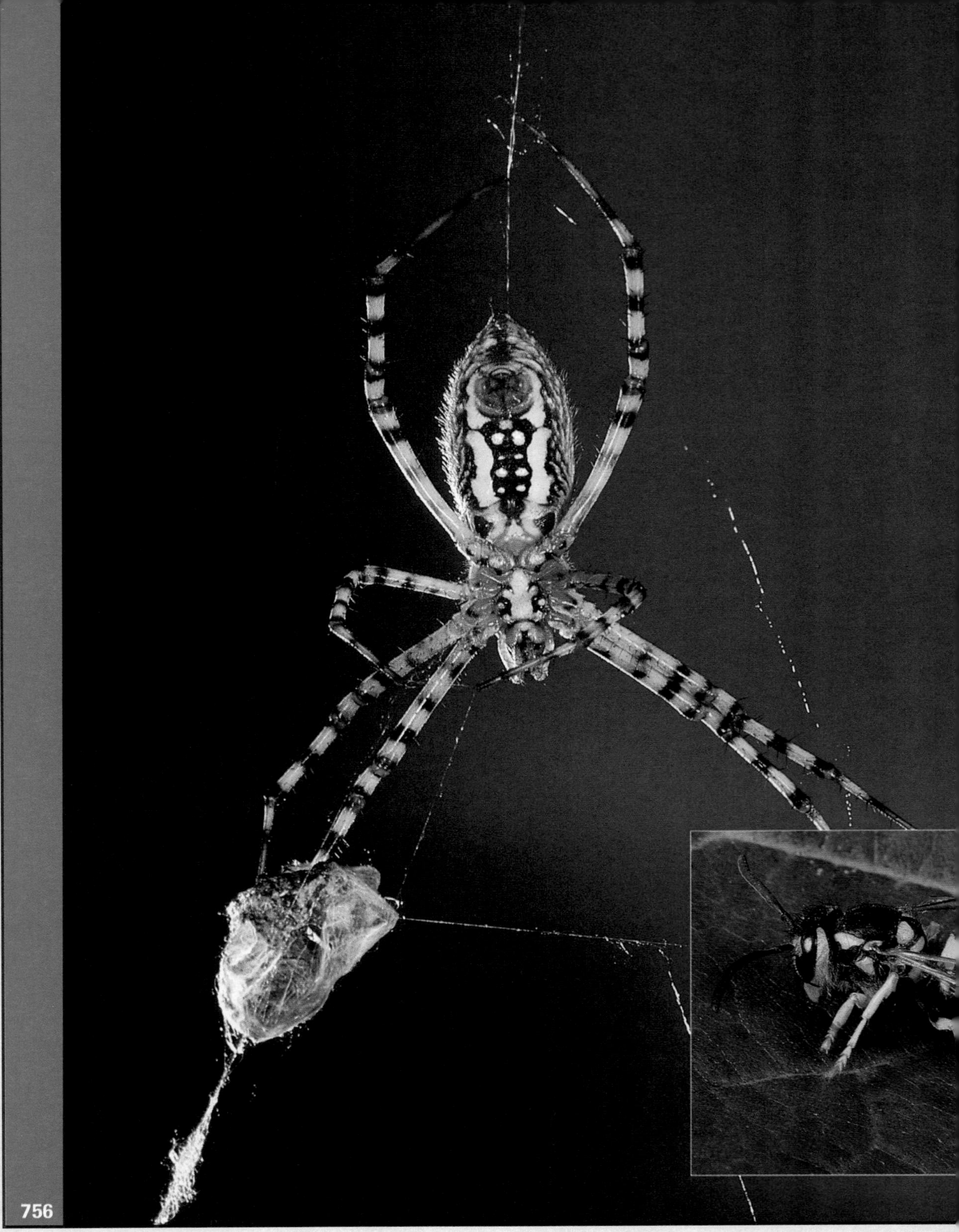

756

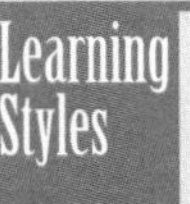

Learning Styles

Look for the following logo for strategies that emphasize different learning modalities. 

Kinesthetic	Meeting Individual Needs, p. 761
Visual-Spatial	Activity, p. 760; Visual Learning, p. 760; Project, p. 766; The Inside Story, pp. 769, 776; Demonstration, p. 777; Portfolio, p. 777
Interpersonal	Alternate Lab, p. 762; Portfolio, p. 772; Project, p. 776
Intrapersonal	Portfolio, p. 760
Linguistic	Student Journal, pp. 759, 766, 770
Logical-Mathematical	Thinking Lab, p. 759; Minilab, pp. 761, 773; Meeting Individual Needs, p. 771; Portfolio, p. 774

CHAPTER

31 Arthropods

You spread a tablecloth on the picnic table as the aroma of barbecued hamburgers attracts yellow jackets, and flies buzz around the macaroni salad. A procession of ants carries discarded potato chip pieces several times their size, while a spider descends and hovers on a glossy thread from a tree branch, waiting for its next meal. After you eat, you wade in a nearby stream and find crayfishes flipping their tails in backward lurches. Meanwhile, you hear the tiny whine of a mosquito near your ear. You knew you were going to a picnic with your friends. Did you know you would be sharing it with perhaps 1500 or more species of arthropods—the average number in a suburban garden?

Arthropods include insects, centipedes, millipedes, spiders, ticks, scorpions, mites, lobsters, shrimps, crabs, and crayfishes. They range in size from the 0.2-mm-long hairy beetle to the giant Japanese spider crab, which measures 4 m across. Phylum Arthropoda includes a diverse group of animals.

Concept Check

You may wish to review the following concepts before studying this chapter.

- Chapter 1: adaptation, evolution
- Chapter 20: classification
- Chapter 28: exoskeleton

Chapter Preview

Laboratory Activities

Biolab: Design Your Own Experiment
- Do flowers produce insect repellents?

Minilabs
- How strong is a spiderweb?
- Which insects are attracted to light?

There are about 1 million known species of arthropods, and many more remain unidentified. In fact, one scientist estimates that there may be up to 30 million species of insects in the world's tropical rain forests. How can we explain the enormous diversity of arthropods such as spiders, yellow jackets, and crayfishes? What are the adaptations that have allowed arthropods to dominate the world's habitats?

CHAPTER 31

Introducing the Chapter

After students have read the chapter introduction, ask them to work in their groups to list all the arthropods they know. Beside the name of each arthropod, ask them to identify the arthropod's habitat: land, water, or air. When they have finished, ask them to speculate about why arthropods are able to live in so many different kinds of habitats and why there are so many different kinds of arthropods. L1 COOP LEARN

Theme Development

The theme of *evolution* is stressed as the huge diversity of adaptations that arthropods have evolved is discussed. The theme of *homeostasis* is brought out through discussions of the organs that enable arthropods to maintain homeostasis with their environment.

Concept Check

Students will continue to expand on their knowledge of how structural and behavioral adaptations in organisms help them survive a variety of environmental conditions. Students will also build upon their knowledge of classification and phylogeny.

Assessment Planner

Choose assessment strategies from the following pages to evaluate the progress of your students.

Assess, pp. 766, 777
Alternate Lab, pp. 762-763
Minilabs, pp. 761, 773
Portfolio, pp. 760, 772, 774, 777
Thinking Lab, p. 759
Biolab, pp. 764-765
Chapter Review, pp. 779-781

SECTION

31.1

Prepare

Key Concepts

Characteristics common to all arthropods are presented along with their specific adaptations to land, air, and water. The ecology and origins of arthropods are also presented.

Block Scheduling

Look for this symbol for strategies that are useful in a block scheduling format. For more information on block scheduling, refer to Section 31.1 in the **Lesson Plans** booklet.

Materials

- Order live crayfish for the Focus Activity.
- For the Minilab, have students collect spiderwebs in jars of water. Provide metric rulers.
- For the Biolab, provide insects such as grasshoppers or crickets, petri dishes, plant extract-treated filter paper, plain filter paper, cotton swabs, and forceps.

1 Focus

Bellringer

Before presenting the lesson, display **Section Focus Master 68** on the overhead projector and have students answer the accompanying questions. L1 SAE

Activity

Provide students with a live crayfish in a pan of water. Ask them to describe how the crayfish moves and to describe the structures of the crayfish that make this movement possible. *Students are likely to describe the legs of the crayfish and its finlike structures that enable movement in water.* L1

SECTION 31.1

Characteristics of Arthropods

Section Preview

Objectives

Relate the structural and behavioral adaptations of arthropods to their ability to live in different habitats.

Analyze the adaptations that make arthropods an evolutionarily successful phylum.

Key Terms

appendage
molting
cephalothorax
tracheal tube
spiracle
book lung
pheromone
simple eye
compound eye
mandible
Malpighian tubule
parthenogenesis

Two out of every three animals living on Earth today are arthropods. You can find arthropods deep in the ocean and on high mountaintops. They live in polar regions and in the tropics. Arthropods are adapted to living in air, on land, and in freshwater and saltwater environments. This water flea, Daphnia, *lives in freshwater lakes and filters microscopic food from the water with its bristly legs.*

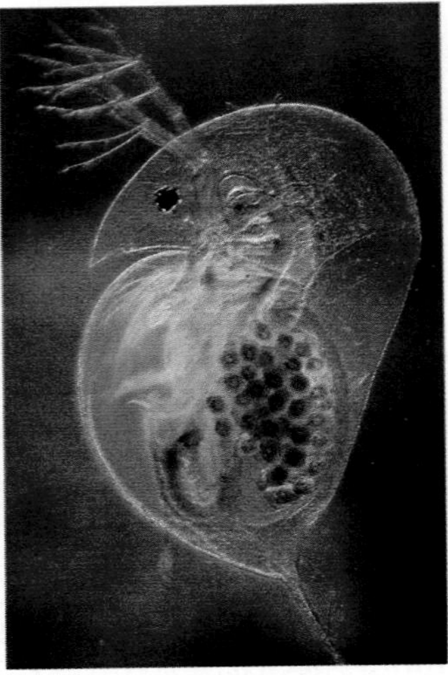

arthropod:
arthron (GK) joint
pous (GK) foot
Arthropods have jointed appendages.

What Is an Arthropod?

Despite the enormous diversity of arthropods, they all share some common characteristics. How can you recognize an arthropod?

A typical arthropod is an invertebrate animal with bilateral symmetry, a coelom, an outer covering called an exoskeleton, and jointed structures called appendages. An **appendage** is any structure, such as a leg or an antenna, that grows out of the body of an animal. In arthropods, appendages are adapted for a variety of purposes including sensing, walking, feeding, and mating. *Figure 31.1* shows some of these adaptations.

The advantage of jointed appendages

Arthropods were the first invertebrates to evolve jointed appendages. Joints are advantageous because they allow for more powerful movements during locomotion, and they enable an appendage to be used in many different ways. For example, the second pair of appendages in spiders is used for sensing and for mating. In scorpions, this pair of appendages is used for seizing prey.

Arthropod exoskeletons give protection

The success of arthropods as a group can be attributed in part to the presence of an exoskeleton. The exoskeleton is a hard, thick, outer covering made of protein and chitin. Chitin is the same substance found in fungal cell walls. In some species, the exoskeleton is a continuous covering over most of the body. In other species, the exoskeleton is made of separate plates held together by hinges. The exoskeleton protects and supports internal tissues and provides places for attachment of muscles. In many species that live on land, the exoskeleton is covered by a waxy layer that provides additional protection against water loss.

Program Resources

Section Focus Master 68 L1 SAE
Reinforcement and Study Guide, pp. 123-124 L1
Biolab and Minilab Worksheets, pp. 123, 125-126 L1
Basic Concepts Transparency 49 and **Master** L1 SAE

Why arthropods must molt

Exoskeletons are an important adaptation for arthropods, but they also have their disadvantages. First, they are relatively heavy structures. Many terrestrial and flying arthropods are adapted to their habitats by having a thinner, lighter-weight exoskeleton, which offers less protection but allows the animal more freedom to fly and jump.

More importantly though, exoskeletons cannot grow, so arthropods must shed them periodically. Shedding of the old exoskeleton is called **molting.** Before an arthropod molts, a new exoskeleton develops beneath the old one. When molting occurs, the animal contracts muscles in the rear part of its body, forcing blood forward. The forward part of the body swells, causing the old exoskeleton to

ThinkingLab Draw a Conclusion

What determines the type of waxy coating produced by grasshoppers?

Grasshopper exoskeletons are covered with a thin, waxy coating. This coating helps protect against water loss in hot, dry climates in which some grasshoppers live. Biologists raised grasshoppers at temperatures of 29°C, 32°C, and 34°C to simulate three different climates in which grasshoppers can live. After the grasshoppers molted, the scientists tested the discarded exoskeletons to find what temperature would melt the waxy layer.

Analysis

The data from the experiment were plotted on this graph.

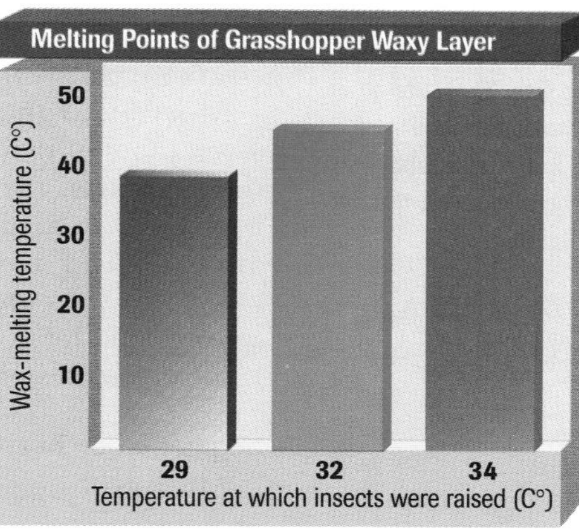

Thinking Critically

Draw a conclusion about what may determine the type of waxy layer a grasshopper exoskeleton might produce.

Figure 31.1

The development of jointed appendages was a major evolutionary step that led to the success of the arthropods.

Spiders hold their prey with jointed mouthparts while feeding.

The short, flattened appendages of the lobster, called swimmerets, are used like flippers for swimming. Large claws are modified for defense.

The powerful jointed legs of this crab are adapted for walking.

The antennae of a moth are adapted for the senses of touch and smell.

2 Teach

ThinkingLab Draw a Conclusion

Purpose

LS **Logical-Mathematical** Students will study an adaptation of grasshoppers that helps them resist drying out.

Process Skills

compare and contrast, use a graph, recognize cause and effect, draw a conclusion

Further Background

Scientists find that grasshoppers can change their lipid coat in response to changing environmental conditions. Grasshoppers must be adapted to survival in places where soil temperatures reach as high as 50°C in the summer.

Teaching Strategies

- Ask students to speculate about how a grasshopper is protected from drying out. *Students will likely cite the exoskeleton as protection from drying out.*
- Provide students with a live grasshopper. Ask them to observe the grasshopper for about 10 minutes and list any behavior or features that enable the grasshopper to survive. L1

Thinking Critically

The temperature at which the grasshopper lives determines the type of waxy layer it produces.

Assessment

Portfolio: Ask students to write a paragraph in their portfolios about how variation in the melting point of the waxy layer of grasshopper exoskeletons may help grasshoppers survive environmental changes. L1 P

GLENCOE TECHNOLOGY

Videodisc

Biology: The Dynamics of Life
Disc 1, Side 2
Molting Crab (Ch. 34)

STUDENT JOURNAL

The Importance of Chitin Ask students to read and summarize in their journals "Chitin Craze" by Elizabeth Pennisi, *Science News,* July 31, 1993, pp. 72-74. This article reviews the uses of chitin in bandages, burn dressings, food additives, cosmetics, sewage treatment, and as a compound that may help slow the spread of AIDS. L2 LS

Activity

LS **Visual-Spatial** Provide students with live or preserved specimens of insects, crayfishes, spiders, and other arthropods. Ask them to compare and contrast the exoskeletons. L1

Visual Learning

Figure 31.3 Ask students to compare the appendages of the stag beetle and the camel cleaner shrimp to speculate how their functions might be different. *The appendages of the shrimp are suited to movement in water. Those of the beetle are suited to movement on land.* LS

Activity

LS **Visual-Spatial** Ask students to examine a variety of preserved or live arthropods and draw sketches of their body segments. Provide binocular microscopes or hand lenses to aid observation. Ask students to label their diagrams with the terms: *head*, *thorax*, *cephalothorax*, and *abdomen*, as appropriate. L1

Figure 31.2

Arthropods molt several times during their development. The old exoskeleton is discarded after a new one is formed from chitin-secreting cells beneath the old exoskeleton. Before the new exoskeleton hardens, the animal swallows air or water to puff itself up in size. Thus, the new exoskeleton hardens in a larger size, allowing some room for the animal to grow.

split open, as *Figure 31.2* shows. The animal then wiggles out of its old exoskeleton.

Most arthropods molt four to seven times in their lives, and during these periods, they are particularly vulnerable to predators. Arthropods are soft and have little muscle strength after molting, so many species hide for a few hours or days while the new exoskeleton hardens.

Segmentation in arthropods

Most arthropods do not have as many segments as are found in segmented worms, but their bodies show the markings of segments. In arthropod bodies, segments have become fused into one to three body sections—head, thorax, and abdomen. In most groups of arthropods, several segments are fused, forming the head as shown in *Figure 31.3*. Other segments are fused, forming a thorax and an abdomen. In other groups of arthropods, there is an abdomen with a fused head and thorax. The fused head and thorax of an arthropod is called a **cephalothorax.**

Arthropods have efficient gas exchange

Arthropods are generally quick, active animals. They crawl, run, climb, dig, swim, and fly. In fact, some flies beat their wings 1000 times per second. As you would expect, arthropods have efficient respiratory structures that ensure rapid oxygen delivery to cells. This large oxygen demand is needed to sustain the high levels of metabolism required for rapid movements.

Figure 31.3

Fusion of the body segments is related to movement and protection. Many species such as shrimps and lobsters have a fused head and thorax, which protects the internal structures in this sensitive region but which limits movement. Other species have separate head and thorax regions that allow for more flexibility.

A stag beetle shows fusion of body segments into a distinct head, thorax, and abdomen.

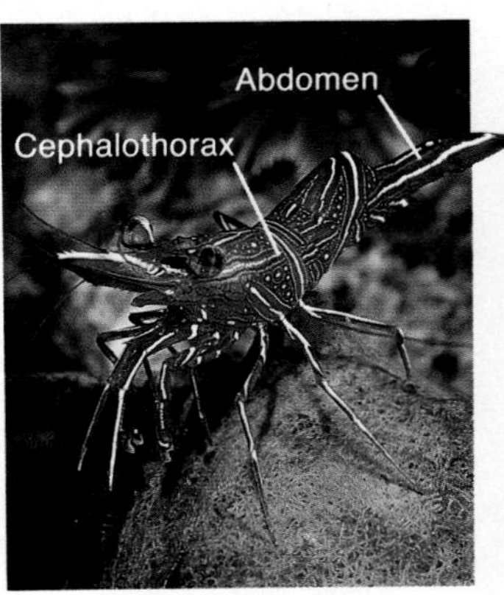

In the camel cleaner shrimp, the head and thorax are fused into a cephalothorax. The animal also has an abdomen.

PORTFOLIO

Arthropod Respiration Provide students with live or preserved arthropods that have gills, tracheal tubes, and book lungs. Ask students to explain how these arthropods get oxygen and how this method illustrates adaptations of these animals to their environments. L2 P LS

Arthropods have evolved three types of respiratory structures for taking oxygen into their bodies: gills, tracheal tubes, and book lungs. In some arthropods, air diffuses right through the body wall. Aquatic arthropods exchange gases through gills, which extract oxygen from water and release carbon dioxide into the water, *Figure 31.4.* Land arthropods have either a system of tracheal tubes or book lungs. Most insects have **tracheal tubes,** branching networks of hollow air passages that carry air throughout the body. Muscle activity helps pump the air through the tracheal tubes. Air enters and leaves the tracheal tubes through openings on the thorax and abdomen called **spiracles.**

Most spiders and their relatives have **book lungs,** air-filled chambers that contain leaflike plates. The stacked plates of a book lung are arranged like pages of a book and serve for gas exchange.

MiniLab

How strong is a spiderweb?

Imagine a fiber strong enough to withstand the impact of a bullet. Spiders produce a flexible protein thread that is five times stronger than steel. Research is underway to use spider silk in helmets, parachute cords, and other light but strong equipment, even bulletproof vests.

Procedure

1. Collect spiderwebs from the corners of your ceilings, windowsills, basement, or garden by using a cotton swab or small twig. Place the webs in a jar of water for transport. Make sure you do not disturb the webs of black widows or brown recluses.
2. Wet a few chemically untreated human hairs, at least 6 cm long, in a beaker of water.
3. Measure the lengths of the hair and a strand of the web in their relaxed positions before stretching.
4. Stretch the hair on your ruler until it breaks. Do the same with the strand of web. Note the length each stretches before breaking.

Analysis

1. Which stretched more before breaking, the spider silk or the hair?
2. Speculate about the advantage of strong, stretchable silk to a spider.
3. Why did you wet the human hair?

Figure 31.4

As you examine these arthropod respiratory structures, evaluate which are best suited for land life and which are best suited for water life.

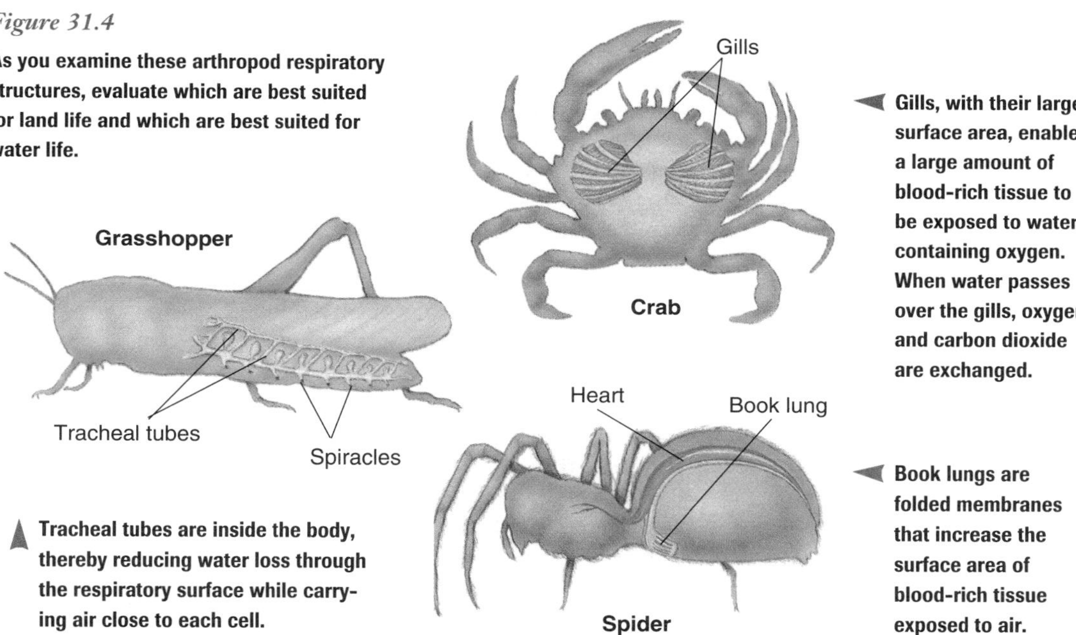

Gills, with their large surface area, enable a large amount of blood-rich tissue to be exposed to water containing oxygen. When water passes over the gills, oxygen and carbon dioxide are exchanged.

Tracheal tubes are inside the body, thereby reducing water loss through the respiratory surface while carrying air close to each cell.

Book lungs are folded membranes that increase the surface area of blood-rich tissue exposed to air.

MiniLab

Purpose

IS **Logical-Mathematical** Students will compare the strength of spiderweb silk to that of a human hair.

Process Skills

observe and infer, compare and contrast, recognize cause and effect

Teaching Strategies

- Show students photocopies of diagrams of the webs of black widows and brown recluses. Also show color photos of each spider. Explain that although most spiders are not poisonous, it is wise not to touch any spider because many common spiders can deliver a painful and perhaps somewhat dangerous bite.
- Advise students that they must place their webs in water; otherwise the web will roll up and form a ball that cannot be unwound.

Expected Results

The spider silk may stretch at least twice the length of the human hair before breaking.

Analysis

1. The spider silk stretched farther before breaking.
2. When wind blows or when other materials or animals touch the web, it will not break.
3. The hair was wetted to eliminate variables in the experiment.

Assessment

Portfolio: Ask students to write a paragraph about potential uses for a strong fiber such as spider silk. They should be both creative and scientifically accurate in their suggestions. L1 P

Meeting Individual Needs

Visually Impaired For students who are visually impaired, purchase a variety of plastic arthropods from a toy store. Have them handle each arthropod while another student names the arthropod, the group to which it belongs, and the adaptations of that group of arthropods. L1 IS

Chemistry Connection

Pheromones

Pheromones are complex chemicals made by insects. Some female insects produce pheromones that attract males when it is time to mate. Chemists have identified the molecular structures of many pheromones and synthesized them in the laboratory. These synthetic substitutes are used to bait traps to prevent insect pests from mating.

Visual Learning

Figure 31.5 Would you expect a mosquito to have more or fewer lenses than a dragonfly? *The mosquito will have fewer lenses.*

GLENCOE TECHNOLOGY

Videodisc

STVS: Ecology
Disc 6, Side 2
Pheromone Rope (Ch. 3)

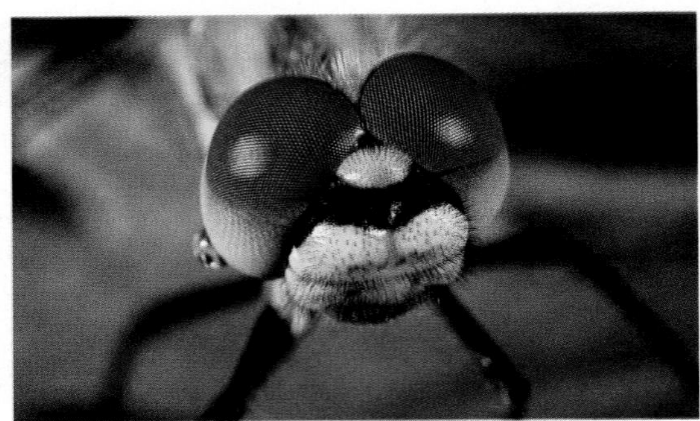

Figure 31.5

The compound eyes of this dragonfly cover most of its head and consist of about 30 000 lenses. Compound eyes form as many image parts as there are lenses. As a result, the dragonfly can see simultaneously in all directions. Multiple lenses enable a flying arthropod to analyze a fast-changing landscape during flight. Dragonflies must capture other insects in midair, a fact that may explain why their eyes have more lenses than eyes of flying insects that do not capture prey in flight. Would you expect a mosquito to have more or fewer lenses than a dragonfly?

Arthropods have acute senses

Quick movements that are the result of strong, muscular contractions enable arthropods to respond to a variety of stimuli. Movement, sound, and chemicals can be detected with great sensitivity by antennae, stalklike structures that detect changes in the environment.

Antennae are also used for communication between animals. Have you ever watched as a group of ants efficiently took apart and carried home a crumb or small piece of food? The ants were able to work together as a group because they were communicating with each other by **pheromones,** chemical odor signals given off by animals. Antennae sense the odors of pheromones, which signal animals to engage in a variety of behaviors. Some pheromones are used as scent trails, such as in the group-feeding behavior of ants, and many are important in the mating behavior of arthropods.

Accurate vision is also important to the active lives of arthropods. Most arthropods have one pair of large compound eyes, shown in *Figure 31.5,* and from three to eight simple eyes. A **simple eye** is a visual structure with only one lens. Simple eyes are used for detecting light. A **compound eye** is a visual structure with many lenses. Each lens registers light from a tiny portion of the field of view. The total image that is formed is made up of thousands of parts, somewhat like the image of dots produced on a television screen. Although the image formed by a compound eye is not as detailed as one formed by a human eye, it is better for detecting motion. Compound eyes can detect even the slightest movements of prey, mates, or predators, and can also detect colors. However, the images formed by compound eyes are fuzzy.

Arthropod nervous systems are well developed

Arthropods have a well-developed nervous system that processes information coming in from the sense organs. The nervous system consists of a double ventral nerve cord, an anterior brain, and several ganglia. Although earthworms have a ganglion for each segment, some segmental ganglia have become fused in arthropods. These ganglia act as control centers for the body section in which they are located.

Arthropods have evolved other complex body systems

Arthropod blood is pumped by one or more hearts in an open circulatory system with vessels that carry blood away from the heart. The blood

Alternate Lab: Beetle Strength

Purpose 

LS Interpersonal Students will observe and compare the pulling power of a beetle and a human.

Materials

bess beetles, cloth toweling (30-cm square), clear tape, heavy thread (30-cm long), balance, pennies, smooth tabletop, plastic petri dish

Procedure

Give students the following directions.

1. Obtain the mass of the petri dish and the mass of the penny.
2. Place a beetle on its back in the petri dish and obtain the mass of the beetle and the dish. Calculate the mass of the beetle.
3. Make a slipknot loop on one end of the thread and put the loop over the head and body of the beetle so that it acts as a harness. Tape the ends of the thread inside the rim of the petri dish. Make a hypothesis about how many pennies the beetle will be able to pull on the petri dish sled.
4. Secure the cloth to the tabletop with tape. When the beetle begins to pull or move the sled by walking, slowly

Figure 31.6

Mouthparts of arthropods exhibit tremendous variation among species. How might this variation have contributed to the success of arthropods?

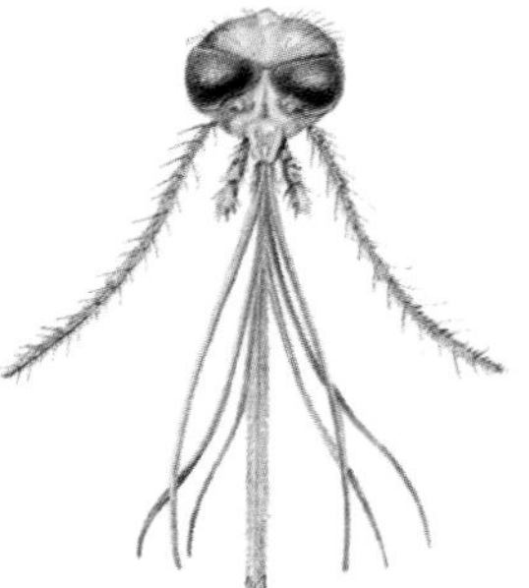

Sand flies and other insects that feed by drawing blood have piercing blades or needlelike mouthparts.

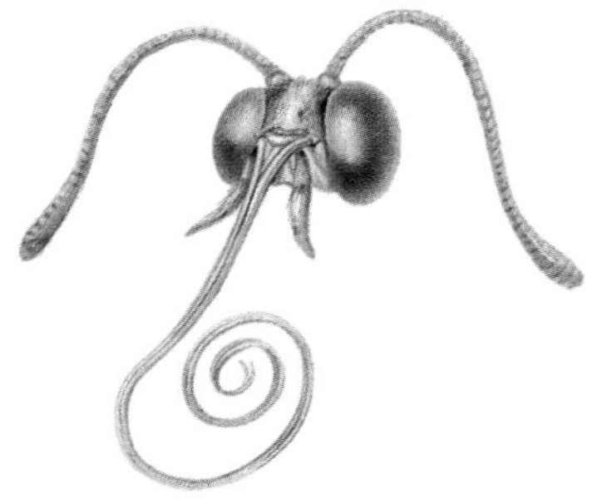

The rolled-up sucking tube of moths and butterflies can reach nectar at the bases of long, tubular flowers.

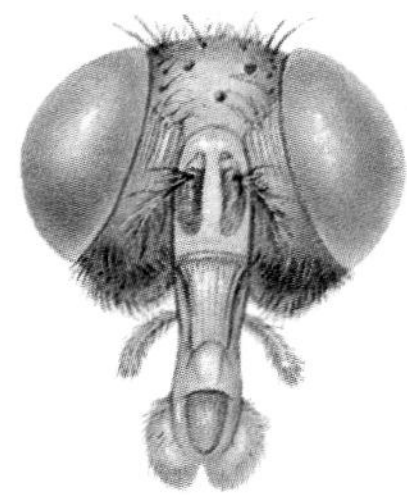

The sponging tongue of the housefly has an opening between its two lobes through which food is lapped.

flows out of the vessels, bathes the tissues of the body, and returns to the heart through open body spaces.

In addition to a complex circulatory system, arthropods have a complete digestive system with a mouth, stomach, intestine, and anus, together with various glands that produce digestive enzymes. The mouthparts of most arthropod groups include a variety of jaws—called **mandibles.** Mouthparts are adapted for holding, chewing, sucking, or biting the various foods eaten by arthropods, ***Figure 31.6.***

Most terrestrial arthropods excrete wastes through **Malpighian tubules.** In arthropods, the tubules are all located in the abdomen rather than in each segment, as they are in segmented worms. Malpighian tubules are attached to and empty into the intestine.

Another well-developed system in arthropods is the muscular system. *Figure 31.7* shows the differences in muscle attachment in vertebrate and in arthropod systems. An arthropod muscle is attached to the exoskeleton on both sides of the joint.

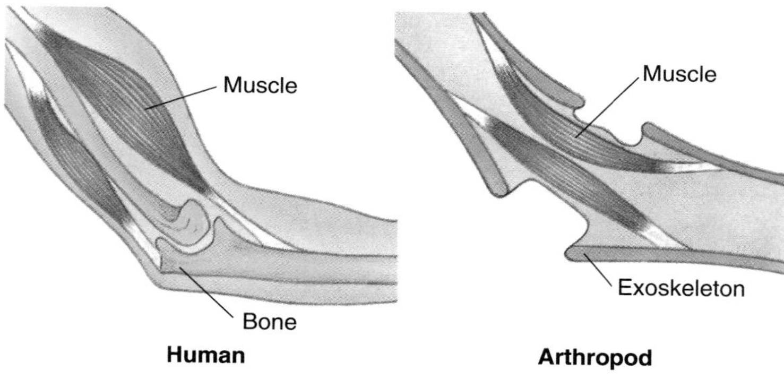

Figure 31.7

In a human limb (left), muscles are attached to the outer surfaces of internal bones. In an arthropod limb (right), the muscles are attached to the inner surface of the exoskeleton. A flea can jump a distance 100 times its own length, equivalent to a six-foot human doing a 600-foot standing high jump. Relative to body weight, a flea has stronger muscles than those of humans.

Science, Technology, and Society

Dragonfly Flight

Dragonflies can hover without effort, fly backward, make 180° turns within the length of their bodies, fly up to 35 mph, and come to a complete stop within 2 cm. How are these useful adaptations? Whereas other insects hide from birds or make poisons to deter them, dragonflies outmaneuver their enemies.

Scientists are studying dragonfly flight to develop aircraft that will be more maneuverable. The dragonflies are studied in wind tunnels containing nontoxic smoke that allows air currents to be photographed. Dragonflies twist their wings on the downstroke and create tiny whirlwinds on top of the wings. This unsteady air lowers the pressure on the wing surface, giving the dragonfly huge lift. Planes of the future may have flaps on their wings that create unsteady airflows that give them better lift.

Visual Learning

Figure 31.6 How might this variation have contributed to the success of arthropods? *Having different mouthparts helps arthropods fill a wider variety of niches and, therefore, reduces competition.*

add pennies to the petri dish, one at a time, until you find the maximum mass the beetle can pull. Do not prod the beetle.

5. Count and record the total number of pennies in the petri dish.
6. Record the relative pulling power of the insect by dividing the mass of the pennies by the mass of the beetle.

Analysis

Ask students to answer the following questions.

1. Did your data support your hypothesis? *Beetles pull about 13 pennies, more than students usually hypothesize. Some beetles pull more than 50 times their weight.*
2. Of what adaptive value is it for beetles to have such great pulling power? *They can push and pull parts of the dead logs in which they live.*

Assessment

Performance: Have students calculate how many grams they could pull if they had the same strength as their beetle. They should multiply their weight in grams by the pulling power of the beetle. L1

SECTION 31.1

BioLab Design Your Own Experiment

Do flowers produce insect repellents?

Time Allotment
One class period

Objectives
Review objectives with students before they begin the Biolab.

Process Skills
compare and contrast, recognize cause and effect, form a hypothesis, interpret data, control variables, design an experiment

Safety Precautions
Instruct students to handle the filter paper using forceps and to keep their hands away from their faces after handling the filter paper. Ask them to wash their hands at the end of the lab.

PREPARATION

Alternate Materials

- Use wild or garden flowers or flowers from a florist.
- In a chemical hood, soak petals from the flowers of three different species in separate beakers of methyl alcohol. Heat gently on a hot plate to warm the alcohol and then soak the petals overnight. Filter the petals from the extract and then soak filter paper in the extracts for 5 hours. At the same time, soak additional filter paper in plain methyl alcohol. Dry all filter paper and cut in half.
- Filter paper may be stored in labeled plastic bags.

BioLab Design Your Own Experiment

Do flowers produce insect repellents?

In many tropical countries, a mosquito bite can result in far more serious symptoms than an itching welt. Malaria, a potentially fatal disease, is caused by a protist that is carried by mosquitoes. Insecticides are used in the United States to control mosquito populations, but in many countries these control measures are too expensive. A less-expensive method might be to find plants or plant parts that produce insect-repelling substances. Keep in mind that some insecticides are made from chemicals extracted from plant parts. From plants such as these, skin creams could be produced that would repel the disease-carrying mosquitoes.

PREPARATION

Problem
Which flowers have natural insect repellents?

Hypotheses
In your group, brainstorm a list of possible hypotheses and discuss the evidence on which each hypothesis is based. Select the best hypothesis from your list.

Objectives
In this Biolab, you will:

- **Analyze** insect behavior to determine which flowers may have insect repellents.
- **Infer** why a flower might produce an insect-repelling substance.

Possible Materials
local insects such as crickets or grasshoppers
petri dishes
flower extract-treated filter paper, cut in half
plain filter paper, cut in half
cotton swabs
forceps

Safety Precautions
Use forceps to handle filter paper. Be sure to keep your hands away from your face after handling filter paper. Wash your hands at the end of the lab.

Cave cricket

764 Arthropods

PLAN THE EXPERIMENT

Teaching Strategies

- To introduce the lab, give each group a flower and ask students if they think insects would be attracted or repelled by the flower. Ask them to explain their responses.
- Elicit the role of insects in flower pollination. Ask what would happen to plants if all insects ate flowers. *The flowers would not produce seeds.* Ask what properties of flowers might make them unattractive for consumption by insects. *Students may say odor or taste.*
- Cotton swabs may be used to transfer very small insects to the petri dishes.

Possible Procedures

- Students might place the control filter paper in one half of the petri dish and the

PLAN THE EXPERIMENT

1. Make a list of the possible ways you might test your hypothesis using the materials your teacher has provided.
2. Plan to test one variable, collect quantitative data, and have a control.
3. How might you measure a particular insect behavior when the insects are exposed to the flower extract?
4. Make a table to record your data.

Check the Plan

1. Which insect behaviors will you measure and how will you measure them?
2. What controls will you use?
3. Consider the sample size you will use and how many trials you will record in each experiment.
4. ***Make sure your teacher has approved your experimental plan before you proceed further.***
5. Carry out your experiment. Make a graph of your data.

Mosquito

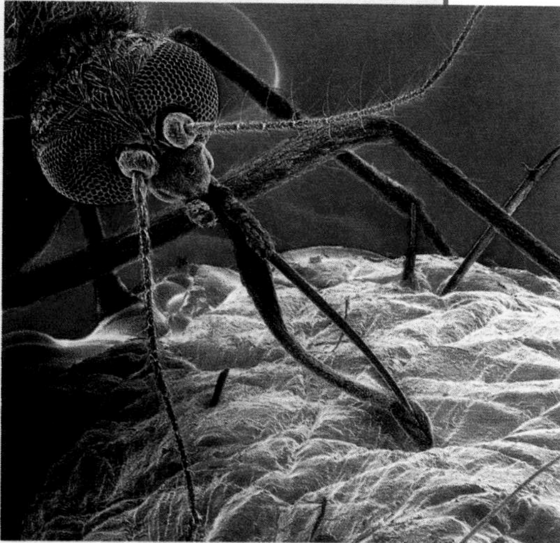

ANALYZE AND CONCLUDE

1. **Checking Your Hypothesis** Was your hypothesis supported by your data? Why or why not?
2. **Observing and Inferring** Why might a flower attract a specific insect and repel others?
3. **Drawing Conclusions** Could your experimental findings be used to help develop an insecticide? Explain.
4. **Relating Concepts** Why is repelling insects important to people?
5. **Thinking Critically** What advantage might natural repellents have over synthetic repellents?

Going Further

Application Design another experiment to determine if other parts of a plant produce insect-repelling chemicals.

Possible Hypotheses

- Insects are repelled by some, but not all flowers.
- Insects are repelled by all flowers.

ANALYZE AND CONCLUDE

1. Hypotheses may be supported if students hypothesized that some flower extracts would repel insects and others would not.
2. Flowers must be pollinated by some insects, and they must prevent other insects from eating them.
3. If the flower extract repels insects, perhaps the chemical in the flower might be a useful insecticide.
4. Insects cause disease and eat crop plants.
5. Natural insecticides might not be as expensive and would not introduce synthetic toxic chemicals into the environment.

Assessment

Skill: Give students an insect that they have not seen and have them test it to determine whether it is repelled or attracted by flower extracts. L1

Going Further

Ask students to speculate what other plant parts might have insect repellents and why this would be important.

treated paper in the other half of the dish. Then, eight insects will be added to the petri dish and the dish will be placed in a dark drawer for 5 minutes. After 5 minutes, they will count how many insects are on each side of the dish. Repeat for two or three trials.

- Students might place the filter paper as above. Insects may be added one at a time and be observed for 2 minutes while students keep track of how much time they spend on each side of the dish.

Data and Observations

Some insects will be repelled by the extracts in some filter papers. Other insects will not be repelled. For example, marigolds and chrysanthemums usually repel insects.

Different Viewpoints in Biology

Arthropod Origins

Did the various classes of arthropods evolve from a common ancestor or did they evolve independently? Some scientists believe that the arachnids, crustaceans, insects, centipedes, and millipedes all evolved from a common trilobite-like ancestor. A new hypothesis suggests that a worm-like ancestor gave rise to insects, centipedes, and millipedes, and a trilobite-like ancestor gave rise to trilobites, crustaceans, and arachnids.

Visual Learning

Figure 31.9 Which group is largest? Which are the oldest living arthropods? *Insects; merostomates*

3 Assess

Check for Understanding

Ask students to list the characteristics of arthropods. Have them describe each characteristic on their list. L1

Reteach

Ask students to make a table that summarizes the characteristics of arthropods. Students should include the following headings in their tables: Appendages, Exoskeleton, Segmentation, Gas exchange, Senses, Nervous system, Circulatory system, Digestive system, Mouthparts, Excretory system, Muscular system, Reproduction. L1

Arthropods reproduce sexually

Most arthropod species have separate males and females and reproduce sexually. Fertilization is usually internal in land species but is often external in aquatic species. A few species are hermaphrodites, animals with both male and female reproductive organs. Some species exhibit **parthenogenesis,** a form of asexual reproduction in which an organism develops from an unfertilized egg.

Ecology of Arthropods

Arthropods are found in so many habitats because of the enormous variety of adaptations that have evolved for obtaining and digesting different foods. Name any food source you can think of and there will probably be an arthropod that uses it.

Most arthropods obtain their own food, but many are parasites. As shown in *Figure 31.8,* some lay their eggs on other insects.

Arthropods and humans

Arthropods are beneficial to humans in a variety of ways. They pollinate a great many of the flowering plants and crop plants on Earth and provide food, honey, shellac, wax, and silk. Some arthropods, such as the ladybird beetle and certain spiders, provide alternatives to chemical control of insects. Research on arthropods has led to advances in the fields of genetics, evolution, and biochemistry. Research is underway to develop a chemical based on scorpion venom that will kill harmful insects but won't bother other animals. From crab shells, scientists have made artificial skin, surgical sutures, and antifungal medication.

Insects cause problems for humans by eating important crops. They also spread plant and animal diseases, including deadly human diseases such as malaria and yellow fever.

The Origins of Arthropods

Arthropods most likely evolved from the annelids. As arthropods evolved, body segments became reduced in number when they fused and became adapted for certain functions such as locomotion, feeding, and sensing the environment. Segments show more complexity of organization in arthropods than in annelids, and the head shows greater development of nerve tissue and sensory organs such as eyes.

The exoskeleton of arthropods is harder and provides more protection than the cuticle of annelids. Because arthropods have many hard parts, much is known about their evolutionary history. The trilobites shown in *Figure 31.9* were once an important group of ancient arthropods, but they have been extinct for 250 million years.

Figure 31.8

Insect parasites often lay their eggs on the larvae of other insects. You can see that as the insects hatch from the eggs, they will have an instant "fast-food" meal.

STUDENT JOURNAL

Flea Collar Safety Ask students to contact a local veterinarian, pet shop, or veterinary school and report in their journals about the safety of flea collars for pets. Students should find out about the insecticides used, how the flea collar works, why the flea collars are not dangerous to the health of a pet, and alternatives to flea collars for pets. L1 IN

PROJECT

Arthropod Origins

Ask students to create a bulletin board display of arthropod origins. Students should identify each arthropod with its common and scientific name. Instruct them to have one section of the bulletin board represent mollusk and annelid ancestors whose likenesses and differences are pointed out. L1 IN

Figure 31.9

The radiation of classes of arthropods on the Geologic Time Scale shows their relationships. Which group is largest? Which are the oldest living arthropods?

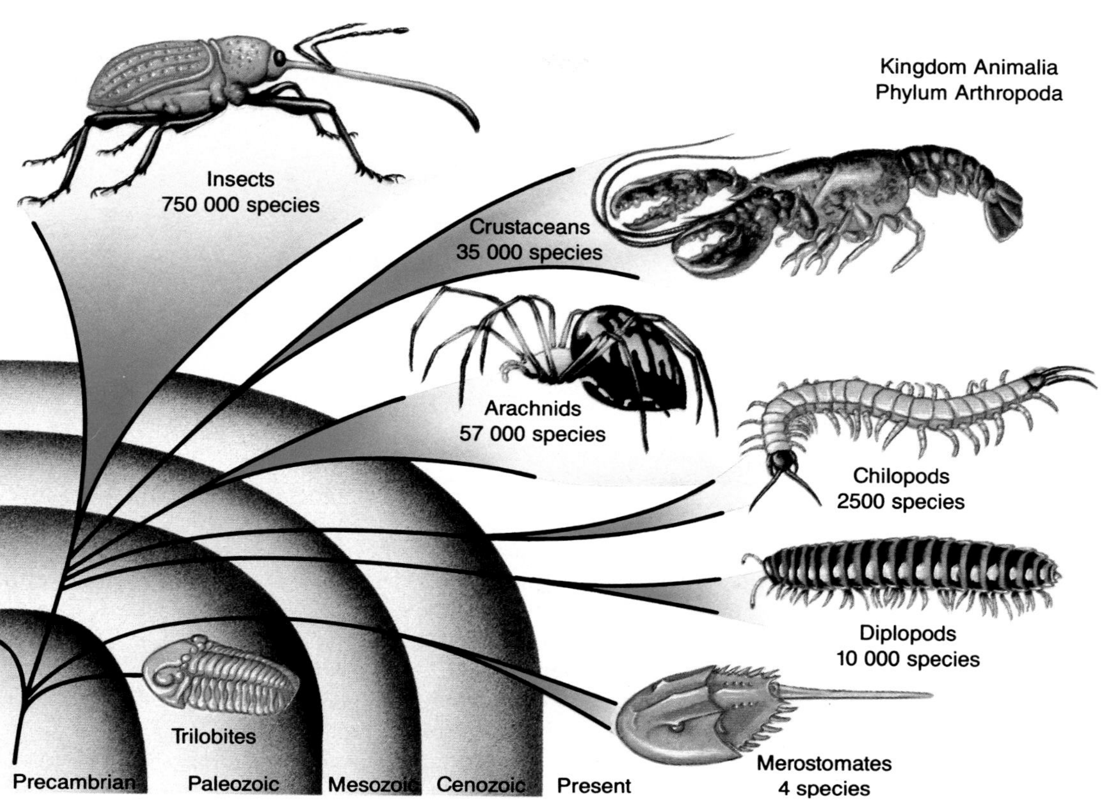

Section Review

Understanding Concepts

1. Describe the pathway taken by the blood as it circulates through an arthropod's body.
2. Describe two features that are unique to arthropods.
3. What are the advantages and disadvantages of an exoskeleton?

Thinking Critically

4. What characteristics of arthropods might explain why they are the most successful animals in terms of population sizes and numbers of species?

Skill Review

5. **Comparing and Contrasting** Compare the adaptations for gas exchange in aquatic and land arthropods. For more help, refer to Thinking Critically in the *Skill Handbook*.

SECTION 31.1

Extension

Ask students to collect and mount a variety of arthropods. Have them refer to a field guide to identify, classify, and label each specimen. L2

Assessment

Portfolio: Have students imagine that their town is scheduling mosquito spraying several times during the summer. Ask them to write a letter to the mayor, to be included in their portfolios, which states their opposition or support of the spraying. They should support their views with library research. L1 P

4 Close

Demonstration

Present a slide show of a variety of arthropods. Review the characteristics of arthropods shown by each animal. Have students speculate about the likely habitat for each arthropod. L1

GLENCOE TECHNOLOGY

Videodisc

Biology: The Dynamics of Life
Disc 1, Side 2
Arthropods (Ch. 35)

Answers to Section Review

1. Blood is pumped by one or more hearts through vessels. The blood flows out of the vessels, bathes the tissues of the body, and returns to the heart through open body spaces rather than through capillaries and veins.
2. jointed appendages, segmented bodies, exoskeletons
3. An exoskeleton is advantageous because it protects against water loss and injury. Disadvantages include its weight and inflexibility.

Thinking Critically

4. jointed appendages, exoskeletons, efficient gas exchange, wings in some, and acute senses

Skill Review

5. Aquatic arthropods exchange gases through gills, which extract O_2 from water and release CO_2 into the water. Land arthropods have either tracheal tubes or book lungs. Tracheal tubes carry air throughout the body. Book lungs are chambers with leaflike plates for gas exchange.

SECTION

31.2

Prepare

Key Concepts

Students will study the diversity and the structural and behavioral adaptations of arthropods. Spiders, ticks, mites, scorpions, crustaceans, centipedes, millipedes, and insects will all be explored.

Block Scheduling

Look for this symbol for strategies that are useful in a block scheduling format. For more information on block scheduling, refer to Section 31.2 in the **Lesson Plans** booklet.

Materials

- Order preserved animals—crayfish and lobster or spider and crab—for the Focus Demonstration.
- For the Minilab, have students gather paper bags or small boxes. Gather flashlights, hand lenses, or binocular microscopes.

1 Focus

Bellringer

Before presenting the lesson, display **Section Focus Master 69** on the overhead projector and have students answer the accompanying questions. L1 SAE

Demonstration

Display a crayfish and a lobster or a crab and a spider. Ask students to compare the animals on display. *Students should recognize differences in the structure of legs and antennae.*

SECTION

31.2 The Diversity of Arthropods

Section Preview

Objectives

Compare and contrast the similarities and differences among the major groups of arthropods.

Explain the adaptations of insects that contribute to their success.

Key Terms

chelicerae
pedipalp
spinneret
metamorphosis
larva
pupa
nymph

Female mosquitoes drink an average of 2.5 times their body weight in blood every day. Imagine yourself, weighing 120 pounds, sitting down to a steak dinner and getting up weighing 300 pounds! Other arthropods feed on nectar, dead organic matter, oil, and just about every other substance you can imagine. The varied eating habits of arthropods reflect their huge diversity.

Boll weevil

Arachnids

Do you remember the last time you saw a spider? Did you draw back with a quick, fearful breath, or did you move a little closer, curious to see what it would do next? Of the 30 000 species of spiders, only about a dozen are dangerous to humans. In North America, you need to watch out for only the two illustrated in *Figure 31.10*—the black widow and the brown recluse.

What is an arachnid?

Spiders, scorpions, mites, and ticks belong to the class Arachnida. Spiders are the largest group of arachnids. Whereas most arthropods have three body regions, the arachnids have no distinct thorax and therefore have only two body regions. Arachnids have six pairs of jointed appendages. The first pair of appendages, called **chelicerae,** is located near the mouth. Chelicerae are often modified into pincers or fangs.

Figure 31.10
The black widow spider (left) is shiny black with a red, hourglass-shaped spot on the underside of the abdomen. The brown recluse (right) is brown to yellow and has a violin-shaped mark on its body. A bite from either spider can make a person sick, but if the person gets medical treatment, the bites are rarely fatal.

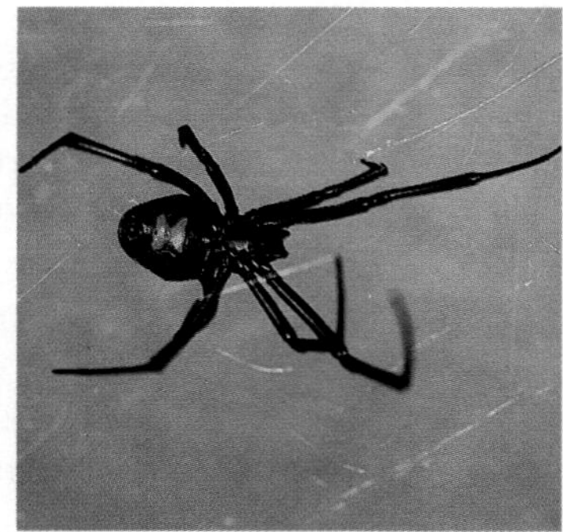

Program Resources

Section Focus Master 69 L1 SAE
Reinforcement and Study Guide, pp. 125-126 L1
Biolab and Minilab Worksheets, p. 124 L1
Laboratory Manual, pp. 177-178, 179-182, 183-186 L2
Concept Mapping, p. 31 L1
Critical Thinking/Problem Solving, p. 31 L3
Tech Prep Applications, pp. 33-34 L2
Basic Skills Transparency 23 and **Master** L1 SAE
Reteaching Transparency 31 and **Master** L1 SAE

THE INSIDE STORY

Inside a Spider

The garden spider weaves an intricate and beautiful web, dribbles sticky glue on the spiraling silk threads, and waits for insects to crash into them. Spiders are predatory animals, feeding almost exclusively on other arthropods. Each spider species builds a unique web, which is effective in trapping flying insects. Many of the structural adaptations of spiders are related to this hunting process.

Garden spider

1. The four pairs of walking legs are located on the cephalothorax of the spider.

2. Spiders have six or eight simple eyes that, in most species, detect light but do not form images. Spiders have no compound eyes.

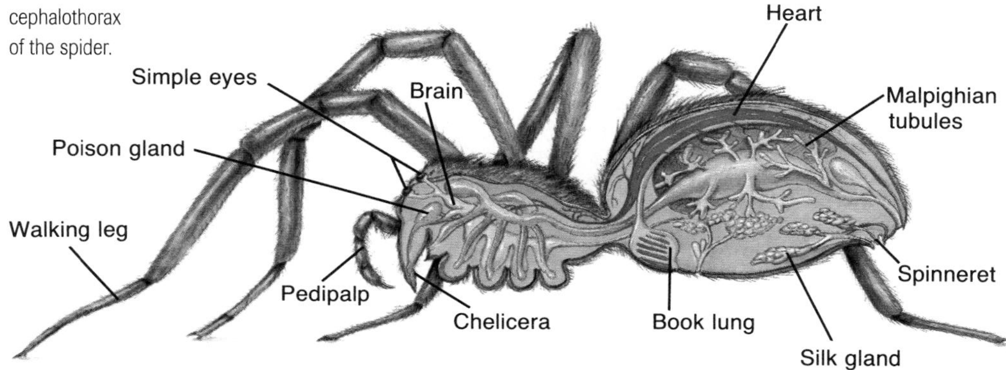

6. A pair of pedipalps is used to hold and move food and also to function as sense organs. In males, pedipalps are bulbous and are used to carry sperm.

5. Chelicerae are the two biting appendages of arachnids. In spiders, they are modified into fangs. Located near the tips of the fangs are poison glands.

3. Gas exchange in spiders takes place in book lungs.

4. Spiders have between two and six silk glands. Silk is first released from silk glands as a liquid. It then passes through as many as 100 small tubes before being spun into thread by the spinnerets.

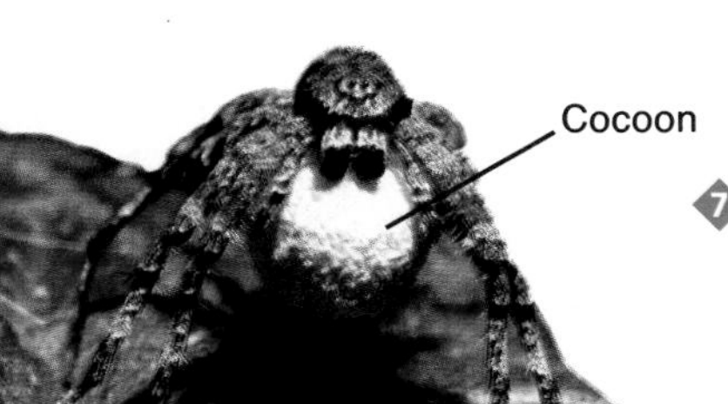

7. Female spiders wrap their eggs in a silken sac or cocoon, where the eggs remain until they hatch. Some spiders lay their eggs and never see their young. Others carry the sac around with them until the eggs hatch.

THE INSIDE STORY

Inside a Spider

Purpose

Visual-Spatial Students will learn about the structural and behavioral adaptations of spiders.

Teaching Strategies

- Explain to students that not all spiders spin webs. The tarantula is a fierce predator that stalks its prey and does not build webs. Such predators usually have thicker legs than web-building spiders.

Visual Learning

- As you point out each structure on the spider, have students use the caption to identify its function.
- Have students create a table with the column heads: *Head* and *Cephalothorax*. Beneath each head, have students list the structures of a spider that connect to that body part.

2 Teach

Discussion

Ask students to relate their experiences with spiders. Remind them that spiders are predators and are therefore able to bite.

GLENCOE TECHNOLOGY

Videodisc

Biology: The Dynamics of Life
Disc 1, Side 2
Web-Spinning Spider (Ch. 36)

Technology
BIOLOGY & SOCIETY

Milking Spiders

Purpose

Students will learn how spider venom affects nerve action and how research on spider venom may lead to treatment for victims of stroke, epilepsy, brain injuries, and Alzheimer's disease.

Teaching Strategies

- Before students read this feature, ask them to do research in groups on the following disorders and their effects: stroke, epilepsy, head injury, and Alzheimer's disease.
- Provide students with a brief overview of how nerves carry impulses.
- Ask groups of students to report about the causes, prevention, and current treatment of the following disorders: stroke, epilepsy, head injury, and Alzheimer's disease.

INVESTIGATING the Technology

Apply Students may describe centipedes or poisonous snakes common to their area. Make sure they have done the research needed to support their answers.

Going Further

Ask groups of students to report on the current medical research on venoms from other animals such as snakes, scorpions, and jellyfishes.

Literature Connection

Charlotte's Web
Many students will have read the book *Charlotte's Web* by E.B. White. Ask students to recall this story and list the scientific and factual information in this book. Have them make a second list of ideas that are not factual. Ask students to explain how this book portrays spiders. L1

Technology
BIOLOGY & SOCIETY

Milking Spiders

In an Arizona laboratory, a spider is milked for its venom. Stimulated with an electric needle, the spider secretes a tiny bit of venom, and a sample is collected with a pipette. It may take 1000 milkings to yield only 100 µL (2 drops) of venom.

Blocked channels

Some researchers use spider venom to block the action of the amino acid glutamate, which transmits nerve impulses in the brain. Normally, glutamate increases a neuron's activity in the brain by attaching to receptor sites on its plasma membrane. This action opens channels in the membrane that allow ions to flow into the cell. The cell then sends glutamate on to the next neuron. Thus, a nerve impulse passes from neuron to neuron. Spider venom has toxins that block the channels in the membrane and stop nerve transmission. It helps scientists study how glutamate transmits nerve impulses.

Applications for the Future

Glutamate is also known to play a major role in causing death after strokes. In one kind of stroke, blood and oxygen are cut off when a blood clot blocks a blood vessel in the brain. Glutamate opens the channels in the damaged nerve cells to a flood of ions that kill the cells by overworking them. Scientists hope to limit stroke damage with the glutamate-blocking toxin from spider venom.

An excess of glutamate may also play a major role in severe forms of epilepsy. Epileptic seizures are a result of hyperactive nerve cells. Experimental drugs that block glutamate receptors have been developed, but they cause side effects because they affect the entire nervous system. The specificity of glutamate blockers in spider venom gives it great potential for managing these seizures without causing side effects.

INVESTIGATING the Technology

Applying Concepts Research and describe one venomous animal other than a spider that can be found in your area. Describe the effects of this animal's toxin and any antidotes available for humans.

Pincers are used to hold food. Fangs inject prey with poison. Spiders have no mandibles for chewing. Using a process of extracellular digestion, digestive enzymes from the spider's mouth liquefy the internal organs of the captured prey. The spider then sucks up the liquefied food.

The second pair of appendages, called the **pedipalps,** are adapted for handling food and for sensing. In male spiders, pedipalps are further modified to carry sperm during reproduction. The four remaining appendages in arachnids are modified as legs for locomotion. Arachnids have no antennae.

Most people know spiders for their ability to make elaborate webs. Although all spiders spin silk, not all make webs. Spider silk is secreted by silk glands in the abdomen. As silk is secreted, it is spun into thread by structures called **spinnerets,** located at the rear of the spider.

Close relatives of spiders

Ticks and mites differ from spiders in that they have only one body section, *Figure 31.11.* The head, thorax, and abdomen are completely fused. Mites are so small that they often are not visible to the naked eye. However, you can certainly feel the bite of mites called chiggers if they get under your clothing while you are camping.

A walk in the woods may require a check for ticks. Ticks feed on blood from reptiles, birds, and mammals. They are small but capable of expanding up to 1 cm or more after a blood meal.

STUDENT JOURNAL

Spider Poetry Ask students to write a limerick about a spider. Give them examples of limericks from anthologies of poetry and explain that a limerick is a humorous poem written in a special meter with five lines. Encourage them to include both humor and scientific accuracy in the limericks. L1 LS

GLENCOE TECHNOLOGY

Videodisc

STVS: Animals
Disc 5, Side 1
Tick Research (Ch. 14)

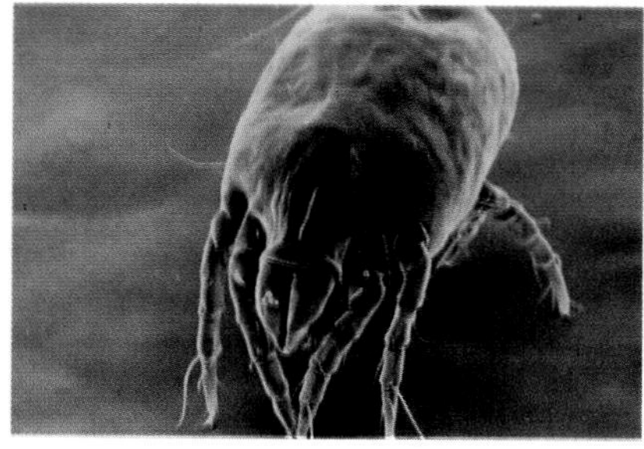

Figure 31.11

Mites, close relatives of spiders, are distributed throughout the world and in just about every habitat. House-dust mites feed on discarded skin cells that collect in dust on floors, in bedding, and on clothing. Some people are allergic to mite waste products.

Scorpions are easily recognized by their many abdominal body segments and enlarged pincers. Related to scorpions are horseshoe crabs, members of the class Merostomata. Horseshoe crabs are considered to be living fossils because they have remained relatively unchanged since the Cambrian period, about 500 million years ago. They are similar to trilobites in that they are heavily protected by an extensive exoskeleton and live a bottom-dwelling existence.

Crustaceans

Most crustaceans are aquatic and exchange gases as water flows over feathery gills. All crustaceans have mandibles for crushing food, two pairs of antennae for sensing, and two compound eyes, which are usually located on movable stalks. Unlike the up-and-down movement of your jaws, crustacean mandibles open and close from side to side. Five pairs of walking legs are used for walking, for seizing prey, and for cleaning other appendages.

Health

Terrible Ticks

Every American city, it seems, has its claim to fame. Chicago is recognized for its outstanding architecture. New York has long been thought of as the cultural center of the United States. Los Angeles is home to television production and to the nation's legendary movie industry. One American city, however, would probably just as soon forget *its* claim to fame. Lyme, Connecticut, will forever be associated with Lyme disease, a crippling bacterial malaise that was first identified in this town in 1975.

A progressive disease Lyme disease manifests itself in humans in three distinct stages. First, a circular, bull's-eye rash appears. Sometimes the rash is accompanied by chills, fever, and aching joints. This is the mildest form of the disease. If left untreated, Lyme disease progresses to a second stage. The joint pains become more severe and may be joined by neurological symptoms, such as memory disturbances and vision impairment. Stage three is the most severe form of the disease. Crippling arthritis, facial paralysis, heart abnormalities, and memory loss may result.

Tick transmission The cause of this debilitating disease is *Borrelia burgdorferi,* a corkscrew-shaped bacterium that is transmitted to humans through the bite of ticks. The bacterium infects mostly deer and white-footed mice. Ticks pick up the bacteria by sucking the blood from these animals. When the same ticks bite humans, the bacteria are passed on, and the result is Lyme disease.

Antibiotics to the rescue The only good news about Lyme disease is its response to treatment. Like most bacteria, *Borrelia burgdorferi* responds to antibiotics. Early treatment with tetracycline will usually prevent the disease from progressing to its second or third stages.

CONNECTION TO Biology

Since the turn of the century, the deer population in the United States has been increasing steadily. How might this increase affect the incidence of Lyme disease? Why?

Health

Terrible Ticks

Purpose

Students will learn about Lyme disease, the bacterium that causes it, and the ticks that transmit it.

Teaching Strategies

- Be sure students understand that ticks do not cause Lyme disease; they carry the bacterium that causes the disease from animal host to human and animal hosts.
- After students have read the feature, ask if they know how a tick should be removed. Placing several drops of vegetable oil on the tick will cause it to withdraw its head when it is no longer getting oxygen. Then, the tick can be removed.
- Ask students to report about two other diseases transmitted by ticks: Rocky Mountain spotted fever and Colorado tick fever. L2
- Provide students with a blank map of the United States. Ask them to color code the map for the areas where three common diseases are transmitted by ticks: Lyme disease, Rocky Mountain spotted fever, and Colorado tick fever. L1

Possible Answer

Because deer are primary carriers of *Borrelia burgdorferi,* an increase in the deer population probably means an increase in infected ticks. This would likely translate into an increase in the incidence of Lyme disease.

interNET CONNECTION

Follow the link for this chapter on the Glencoe Homepage at **www.glencoe.com/sec/science** to find out more about arthropods.

Meeting Individual Needs

Gifted Provide students with a crayfish. Ask them to observe the behavior of the crayfish and then design an experiment that would test some aspect of crayfish behavior such as response to stimuli, habitat, or food preferences. L3 LS

Misconception

Many students think that pill bugs, also known as wood lice, are insects. Point out that pill bugs have all the characteristics of crustaceans even though they live on land.

Bioethics

Regeneration and Toxins

Some scientists are using crabs for scientific research instead of guinea pigs, rabbits, and other mammals. Recent experiments with mud crabs show that their ability to regrow a lost leg is impaired by exposure to toxic chemicals such as pesticides. These toxins also affect development in humans. Some say that cutting down on the use of mammals as laboratory animals is important. Others say that crab biochemistry is so different from that of humans that mammal species are needed for research of this nature.

GLENCOE TECHNOLOGY

Videodisc

STVS: Animals
Disc 5, Side 1
March of the Spiny Lobster (Ch. 11)

The Infinite Voyage: Secrets from a Frozen World
Krill: The Vital Link of the Food Chain (Ch. 2)

Figure 31.12

Barnacles are distinct from other arthropods in terms of structure. Most are sessile and are covered with thick plates. A gluelike substance anchors barnacles to surfaces. Barnacles are filter feeders that trap food by extending their feathery legs out of the shell.

Crabs, lobsters, shrimps, crayfishes, barnacles, water fleas, and pill bugs are members of the class Crustacea, ***Figure 31.12***. Some crustaceans have three body sections, and others have only two.

Pill bugs, the most common land crustaceans, must live where there is moisture, which aids in gas exchange. They are frequently found in damp areas around building foundations.

Centipedes and Millipedes

Centipedes, which belong to the class Chilopoda, and millipedes, members of the class Diplopoda, are shown in ***Figure 31.13***. If you have ever turned over a rock on a damp forest floor or kicked a pile of damp, dead leaves, you may have seen the flattened bodies of centipedes wriggling along on their many tiny, jointed legs. Centipedes are carnivorous and eat soil arthropods, snails, slugs, and worms. The bite of a centipede is painful to humans. Like spiders, millipedes and centipedes have Malpighian tubules for excreting wastes. In contrast to spiders, centipedes and millipedes have tracheal tubes rather than book lungs for gas exchange.

A millipede, the so-called thousand-legger, eats mostly plants and dead material on damp forest floors. Millipedes do not bite, but they can spray obnoxious-smelling fluids from their defensive stink glands. You may have seen their cylindrical bodies walking with a slow, graceful motion rather than the wriggling, scurrying motion of a centipede.

Figure 31.13

A centipede (left) may have from 15 to 181 body segments—always an odd number. Each segment has only one pair of jointed legs. The first body segment has a pair of poison claws that secretes a toxic substance from a pair of poison glands. A millipede (right) may have more than 100 segments in its long abdomen, each with two spiracles and two pairs of legs.

PORTFOLIO

Arthropod Designs Ask each group of students to design an arthropod that would be adapted to one of the following conditions: water, soil, dry sand, snow, life in wood, life on another insect. Ask them to describe and sketch, in their portfolios, how their insect feeds, defends itself, and is adapted to its habitat. L2 LS P COOP LEARN

Insects

Have you ever launched an ambush on a fly with your rolled-up newspaper? You swat with great accuracy and speed, yet your prey is now firmly attached upside down on the kitchen ceiling. How does a fly do this?

The fly approaches the ceiling right-side up at a steep angle. Just before impact, it reaches up with its front legs. The forelegs grip the ceiling with tiny claws and sticky hairs, while the other legs swing up into position. The flight mechanism shuts off, and the fly is safely out of swatting distance. Adaptations that enable flies to land on ceilings are among the many that make insects the most successful arthropod group.

Flies, grasshoppers, lice, butterflies, bees, and beetles are just a few members of the class Insecta, by far the largest group of arthropods. *Figure 31.14* shows several insects.

MiniLab

Which insects are attracted to light?

Have you ever heard a June bug bump against your screen at night, or seen moths fluttering around your porch light? You know that some insects are nocturnal and others are active in the daytime.

Procedure

1. Construct and set up a light trap for insects by putting a paper sack with a 5-cm opening over your porch light. You may also use a flashlight in a box with a 5-cm hole.
2. Leave your light trap set up for several hours after dark.
3. When you have trapped some insects, cover the opening of the trap with tape and place the trap into a freezer for an hour. This will temporarily immobilize the insects.
4. Empty your trap, and examine the insects with a magnifying glass or dissecting microscope. Count the number of each kind of insect in your trap. Use insect identification books to try to identify them.

Analysis

1. How is nocturnal life an adaptation for some insects?
2. Why might insects be attracted to light?
3. Examine the bodies of the insects you captured. Predict the diet and habitat of each, based on its structural adaptations.

Figure 31.14

Insects generally have three pairs of legs, one pair of antennae, and three body regions.

Bees, ants, and wasps are social animals and live in large colonies.

Fireflies are grouped with more than 300 000 species of beetles in the largest insect order.

Butterflies and moths are grouped in an insect order with more than 100 000 known species.

Water striders belong to the order of true bugs.

773

SECTION 31.2

MiniLab

Purpose

Logical-Mathematical Students will study the survival value of insect behaviors.

Process Skills

recognize cause and effect, interpret data, classify, interpret scientific illustrations

Teaching Strategies

- Have students make cages from old aquariums, terrariums, large jars with small holes in lids, or milk cartons. Supply them with appropriate food and water and have students observe and record the behavior of their insects for several days.
- Students may need to make observations at home in a dark room with a flashlight covered with red plastic so only red light shines through.

Expected Results

Depending on the time of year and the part of the country you live in, your students will find a variety of insects. Moths, fruit flies, and June bugs are common night fliers.

Analysis

1. Birds do not look for insect food at night. Night flying eliminates risk from bird predators.
2. to increase the number of potential mates
3. A moth with a long proboscis may feed on nectar in flowers. Mosquitoes have piercing and sucking mouthparts.

Assessment

Portfolio: Ask students to summarize what they have learned about insects during this Minilab. L1 P

CULTURAL DIVERSITY

Charles Henry Turner

Have students report on the important contributions of African-American biologist Charles Henry Turner (1867-1923) to our modern understanding of insect behavior. Turner's research included many species of insects such as ants, bees, and cockroaches, and he often developed unique and interesting experimental techniques to study them. Turner was a very prolific scientist; between 1892 and 1923, he published 49 articles in the leading scientific journals of his time.

People in Biology

Meet Dr. May R. Berenbaum, Entomologist

Teaching Strategies

- Ask if any students ever caught and kept an insect in a jar. Ask why they did this and what they learned.
- Ask students to research the discovery of DDT, and the factors that led to the banning of its use in the United States. L3
- Have students working in groups imagine that they are a team of inventors with backgrounds in entomology. They have been given unlimited funding to design the ultimate device for insect control. Students should use their creative imaginations as well as scientific knowledge to make a model of their bug catcher and explain its function to the class. L1 COOP LEARN

people in biology

Meet Dr. May R. Berenbaum, Entomologist

In "The Deadly Mantis," a huge praying mantis gobbles up most of the population of the East Coast. The movie was a popular attraction at the Insect Fear Film Festival organized by Dr. Berenbaum at the University of Illinois. The annual attraction aims at entertaining the audience while pointing out entomological errors in popular films and interspersing some information about real-life insects.

In the following interview, Dr. Berenbaum tells about her study of insects, which, together with their close relatives, make up nearly 75 percent of the animal species alive today.

On the Job

Q Dr. Berenbaum, what is your favorite class of those you teach at the university?

A I teach a course called "Insects and People," which shows insects' influence on art, literature, and history. For example, both batik and the classic method of casting bronze would be impossible without beeswax. The product of silkworms is used in many fabrics and tapestries. Insects show up in all kinds of literature, including *Make No Bones* by Aaron Elkins, a contemporary murder-mystery writer. And disease-carrying insects have caused more deaths during war than bombs and bullets ever have.

Q What about your specialty, the interrelationships between insects and plants?

A I like to remind my students that, if it weren't for insects, the world's favorite hamburger would have no special sauce, no lettuce, no cheese, no pickles, and no onions. There would just be the bun, because wheat is wind-pollinated. The contribution of a single kind of insect, the honeybee, is worth about $19 billion to American agriculture.

My work with plants and insects began because as an undergraduate, I was strongly interested in plant chemistry. Plants contain an extraordinary diversity of chemicals including many that are toxic. I decided to combine my interests and work on insects that eat plants. Part of my work is studying how some insects switch foods. For instance, a shift in an insect's preferred food from carrots to citrus is of crucial importance to farmers.

Q Are you also involved in laboratory research?

A Yes. One project that I'm involved in uses molecular biology to study how insects cope with insecticides. We want to learn how insects have been able to develop resistance so quickly. For instance, DDT's insecticidal properties were discovered in 1939, and only six years later, there was doc-

PORTFOLIO

Insect Population Ask students to do an insect population study by sampling one species of insect in a small area and then estimating the total number over a larger area. They should measure an area 1 m^2 and collect one type of insect, such as grasshoppers. Then, knowing the size of the entire area, they can calculate how many insects of that type there are. Have students release the insects as soon as their calculations are complete. Challenge students to relate the number of insects counted to the insect's niche. Have students predict how a sudden change in insect number would impact the ecology of the area. L1 P LS

umented resistance in insect populations. We're trying to understand the molecular basis for the evolution of resistance so we can keep insects susceptible to various toxins.

Early Influences

Q How did you get involved in the study of insects?

A I actually used to be afraid of insects. When I was a freshman in college, I decided that my fear stemmed from ignorance. I thought an entomology class would at least teach me what insects I *should* be afraid of. I had a professor who showed me that insects really are absolutely amazing. Up to that point, I was truly phobic. The class altered my life.

Q Was there a particular book that helped foster your interest?

A I found Howard E. Evans's book, *Life on a Little-Known Planet,* to be a fascinating look at the diversity of the insect world. It was a great pleasure to me that Evans later wrote a blurb for the cover of my own book, *Ninety-Nine Gnats, Nits, and Nibblers,* which profiles insects that people encounter in everyday life, such as cockroaches, houseflies, and tiny insects nicknamed no-see-ums.

Personal Insights

Q What advice do you have for students about choosing a field of study?

A Having chosen a career that I never thought I would even consider, my best advice would be never to rule *anything* out! To me, insects are endlessly entertaining, even better than cable TV. Insects are everywhere, too. For example, three-fourths of the human population harbor tiny mites that live inside hair follicles. No amount of scrubbing is going to get rid of them. It's a good thing that they're harmless.

CAREER CONNECTION

Here are a few ideas for careers that relate to the insect world.

Entomologist *A doctoral degree in entomology; bachelor's degree in zoology*

Phytochemist *A bachelor's degree in chemistry; courses in botany*

Apiculturist (Beekeeper) *High-school diploma; on-the-job training*

Q Are there any careers related to insects that some people might find surprising?

A A forensic entomologist can collect insects on a corpse and calculate how long it would take the species to reach that developmental stage. Then law enforcement officers have a very good idea about when a death occurred. This university is interested in developing a program in forensic entomology.

Q Could you give us just one more amazing insect fact?

A There are almost 1 million species of known insects, and at least twice that number are as yet undiscovered. There's still plenty of work left to be done by future entomologists.

CAREER CONNECTION

- **Career Path** High School: Four years of science including biology, physics, mathematics, Earth science, and chemistry. Advanced levels of biology, chemistry, or physics would also be appropriate.

  College: chemistry, physics, botany, zoology, statistics, mathematics, entomology, ecology, geology, genetics, invertebrate zoology, cell biology, evolution, genetics, biochemistry, immunology.

  Graduate Studies: A variety of classes in advanced areas of entomology, zoology, parasitology, and original research in entomology.
- **Career Issue** The issue of which method is best used for the control of insects is one of the greatest challenges facing entomologists of the future.

For More Information

- Ask students to read and report on the following sections of Dr. Berenbaum's book, *Ninety-Nine Gnats, Nits, and Nibblers*: "Kitchen Companions," "House Guests," "Pet Peeves," "What's Eating You?" L2

GLENCOE TECHNOLOGY

Videodisc

STVS: Animals
Disc 5, Side 1
Cockroach on a Treadmill
(Ch. 18)

THE INSIDE STORY

Inside a Grasshopper

Purpose

 **Visual-Spatial** Students will learn about the structural adaptations of grasshoppers.

Teaching Strategies

- Provide students with a live grasshopper in a large, clear container. Ask students to observe and describe its behavior. Have them examine the grasshopper under a stereoscopic microscope and make a labeled sketch of its external body parts. Students should relate each structure to how the grasshopper is adapted to its environment. L1

GLENCOE TECHNOLOGY

Videodisc

The Infinite Voyage: Insects: The Ruling Class
Insects and Their Behavior (Ch. 1)

Classifying Social Insects: Ants (Ch. 3)

Studying Insect Flight (Ch. 6)

THE INSIDE STORY

Inside a Grasshopper

Grasshoppers make rasping sounds either by rubbing their wings together or by rubbing small projections on their legs across a scraper on their wings. Most calls are made by males. Some aggressive calls are made when other males are close. Other calls attract females, and still others serve as an alarm to warn nearby grasshoppers of a predator in the area.

Grasshopper structure is typical of many insects. Grasshoppers have three main body sections, a pair of antennae, two pairs of wings, and six legs.

Grasshopper

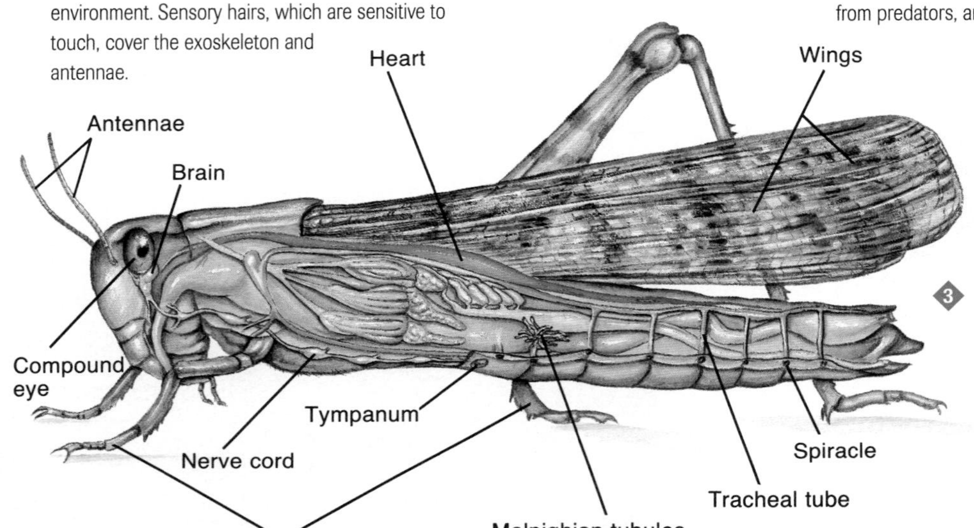

1. Insects have one pair of antennae, which is used to sense vibrations, food, and pheromones in the environment. Sensory hairs, which are sensitive to touch, cover the exoskeleton and antennae.

2. Insects are the only invertebrates that can fly. With the ability to fly, insects can move large distances, find new places to live, discover new food sources, escape quickly from predators, and find mates.

3. The spiracles in the abdomen open into a series of tracheal tubes used in gas exchange.

4. Excretion takes place by Malpighian tubules. In the grasshopper and other insects, nitrogenous wastes are in the form of dry crystals of uric acid. Producing dry waste helps insects to conserve water.

5. The structure used for hearing by an insect is a flat membrane called a tympanum.

6. Most insects have six legs. By looking at an insect's legs, you can sometimes tell how it moves about and what it eats. The grasshopper has long, thick legs for jumping large distances.

7. Grasshoppers have two compound eyes and three simple eyes. Compound eyes do not produce clear images, but they are effective for spotting movement of prey.

PROJECT
Arthropod Scavenger Hunt

Ask a group of students to prepare a class scavenger hunt for arthropods and their signs. The scavenger hunt should require that classmates observe various arthropod behaviors such as movement, sound production, features of habitat in which each arthropod lives, and physical features of the arthropod. Make sure the questions require that students actually observe the arthropod. For example: Describe two insect sounds that you hear outside. What insects are making these sounds? L1 LS

Insect reproduction

Insects mate once, or at most only a few times, during their lifetime. The eggs are fertilized internally, and shells form around them. Many female insects are equipped with an appendage that is modified for digging a hole below the surface of the ground or in wood. The female lays large numbers of fertilized eggs in the hole. Laying large numbers of eggs increases the chances that some offspring will survive long enough to reproduce.

Metamorphosis—Change in body shape and form

Growing insects undergo a series of changes in body structure as they develop. This series of changes, controlled by chemical substances in the animal, is called **metamorphosis.**

Most insects go through four stages on their way to adulthood—egg, larva, pupa, and adult. The **larva** is the free-living, wormlike stage of an insect, often called a caterpillar. As the larva eats and grows, it molts several times.

The **pupa** stage of insects is a period of reorganization in which the tissues and organs of the larva are broken down and replaced by adult tissues. After a period of time, a fully formed adult emerges from the pupa.

The series of changes that occur as an insect goes through the egg, larva, pupa, and adult stages is known as complete metamorphosis. The complete metamorphosis of a butterfly is illustrated in *Figure 31.15.* Other insects that undergo complete metamorphosis include ants, beetles, flies, and wasps.

Complete metamorphosis is an advantage for arthropods because larvae do not compete with adults for the same food. For example, caterpillars feed on leaves, but adult butterflies feed on nectar from flowers.

Figure 31.15

During complete metamorphosis, an insect undergoes a series of developmental changes from egg to adult.

A Insects begin life as a fertilized egg.
B Larvae eat huge amounts of food to supply the energy needed for tremendous growth and for the changes that take place in the pupa stage.
C The pupa stage is an outwardly inactive stage in which cells of the body are reorganized into a new body form. Most pupae are surrounded by a protective case, such as a cocoon.
D The adult insect that emerges from the pupa is sexually mature.

SECTION 31.2

Demonstration

LS **Visual-Spatial** Make a terrarium by using a 2-L soft drink bottle. Inside, place a small piece of raw meat that has been exposed to flies. Seal the terrarium. Have students observe and describe the stages of development of the flies that grow in the terrarium. L1

Enrichment

Have each student prepare a report on the life cycle of the arthropod of their choice.

Demonstration

LS **Visual-Spatial** Obtain a butterfly chrysalis from a biological supply company. Place it in a glass terrarium. Have students observe the chrysalis daily and make notes on any changes they see. L1

GLENCOE TECHNOLOGY

Videodisc

Biology: The Dynamics of Life
Disc 1, Side 2
Gradual Metamorphosis
(Ch. 37)

Complete Metamorphosis
(Ch. 38)

PORTFOLIO

Mealworm Development Ask students to observe and record the development of a colony of mealworms as they develop from larvae to adults. Direct students to focus their attention on evidence suggesting molting has occurred. L1 P LS

Mosquito Farming Have students raise mosquitoes in a bucket of pond water covered with window screening. Algae and ground-up dog biscuits will provide food for mosquito larvae. Have students observe and record in their portfolios mosquito metamorphosis by observing the eggs, larvae, pupae, and adults with hand lenses. L2 P LS

3 Assess

Check for Understanding

Give students a list of characteristics for a specific arthropod. Have them determine whether the characteristics describe an arachnid, crustacean, centipede, millipede, or insect. L1

SECTION 31.2

Reteach

Have students assemble an illustrated chart. Columns in the chart should include the classes of arthropods. You might also have students include the subphylum in their column headings. Have students include an illustration in each column that shows the number of body sections, antennae positions, and numbers of legs. The rows in the chart should be Body divisions, Number of antennae, Kinds of mouthparts, Number and location of legs, and Respiratory organs. L1

Extension

Have students use modeling clay and pipe cleaners to make small models of centipedes and millipedes. Students should use Figure 31.13 as a guide for their models. L1 COOP LEARN

Assessment

Performance: Have students develop a plan for constructing an insect exhibit for a zoo. Students should decide what insects they will display, make a diagram of their exhibit, and explain how the insects must be cared for. Have students prepare a display sign for their exhibit. L1

4 Close

Activity

Divide the class into two groups. Play the "Who am I?" game. The first member of one group will call out clues, while the first member of the second group will guess the class of the arthropod. A correct answer will earn the team one point. The team with the most points wins. L1

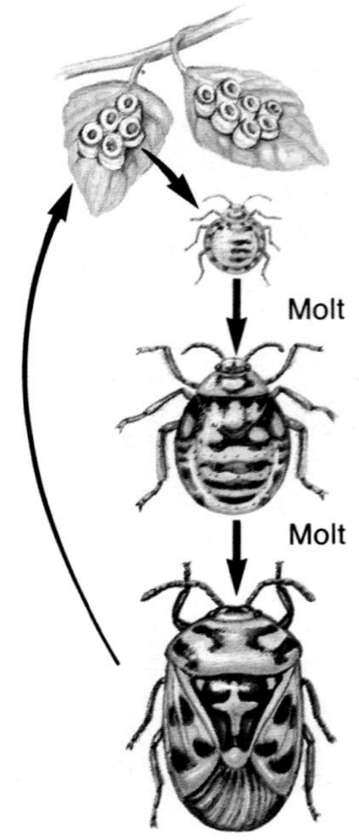

A The fertilized egg is surrounded by a protective shell that contains food.

B Because movement is limited, many nymphs have colorings that resemble their surroundings for camouflage.

C Nymphs continue to grow and molt, gradually increasing in size.

D After several molts, nymphs become adults with wings and reproductive capabilities.

Figure 31.16
Shown here is the incomplete metamorphosis of the harlequin bug.

Incomplete metamorphosis has three stages

Many insect species undergo a gradual or incomplete metamorphosis, in which the insect goes through only three stages of development—egg, nymph, and adult, as shown in *Figure 31.16.* A **nymph,** which hatches from an egg, has the same general appearance as the adult but is smaller. Nymphs lack certain appendages, such as wings, and they cannot reproduce. As the nymph eats and grows, it molts several times. With each molt, it comes to resemble the adult more and more. Wings begin to form, and an internal reproductive system develops. Gradually, the nymph becomes an adult. Grasshoppers and cockroaches are insects that undergo incomplete metamorphosis.

Connecting Ideas

Arthropods have been enormously successful in establishing themselves over the entire surface of Earth. Their ability to exploit just about every habitat is unequaled in the animal kingdom. The success of arthropods can be attributed in part to their varied life cycles, high reproductive output, and structural adaptations such as small size, a hard exoskeleton, and jointed appendages. Another group of animals, the echinoderms, followed a different evolutionary path toward success. Echinoderms are marine animals that have evolved unusual structural adaptations, which contribute to this group's success in the seas.

Section Review

Understanding Concepts

1. How are centipedes different from millipedes?
2. How are insects different from spiders?
3. Describe three sensory adaptations of insects.

Thinking Critically

4. Why might complete metamorphosis have greater adaptive value for an insect than does incomplete metamorphosis?

Skill Review

5. **Recognizing Cause and Effect** Some plants produce substances that prevent insect larvae from forming pupae. How might this chemical production be a disadvantage to the plant? For more help, refer to Thinking Critically in the *Skill Handbook*.

Answers to Section Review

1. Centipedes have only one pair of legs per body segment, eat meat, and bite. Millipedes have two pairs of legs per body segment on the thorax, eat plants and dead material, and do not bite.
2. Spiders have two body regions, six pairs of appendages, book lungs, simple eyes, and spin silk. Insects generally have three body regions, three pairs of legs, one pair of antennae, spiracles, and compound eyes.
3. Insects have compound eyes, antennae, tympanums, and sensitive hairs over parts of the body.

Thinking Critically

4. Complete metamorphosis is an advantage because the larvae do not compete with adults for food.

Chapter 31 Review

REVIEWING MAIN IDEAS

31.1 Characteristics of Arthropods

- Arthropods have jointed appendages, exoskeletons, varied life cycles, and body systems adapted to life on land, water, or air.
- Arthropods are members of the most successful animal phylum in terms of diversity. This can be attributed in part to their structural and behavioral adaptations.

31.2 The Diversity of Arthropods

- Spiders have two body regions with four pairs of walking legs. They spin silk. Ticks and mites have one body section. Scorpions have many abdominal segments, enlarged pincers, and a stinger at the end of the tail.
- Most crustaceans are aquatic and exchange gases in their gills. They include crabs, lobsters, shrimps, crayfishes, barnacles, and water fleas.
- Centipedes are carnivores with flattened, wormlike bodies. Millipedes are herbivores with cylindrical, wormlike bodies.
- Insects are the most successful arthropod class in terms of diversity. They have many structural and behavioral adaptations that allow them to exploit all habitats.

Key Terms

Write a sentence that shows your understanding of each of the following terms.

appendage	nymph
book lung	parthenogenesis
cephalothorax	pedipalp
chelicerae	pheromone
compound eye	pupa
larva	simple eye
Malpighian tubule	spinneret
mandible	spiracle
metamorphosis	tracheal tube
molting	

Understanding Concepts

1. Animals that have jointed appendages and exoskeletons are called ___________.
   a. mollusks
   b. segmented worms
   c. arthropods
   d. vertebrates
2. Arthropods are so successful because of their ___________.
   a. larvae c. diversity
   b. book lungs d. adaptations
3. Arthropods with two body regions and four pairs of walking legs are called ___________.
   a. spiders c. scorpions
   b. ticks and mites d. crustaceans
4. ___________ are arthropods with only one body section.
   a. Spiders c. Scorpions
   b. Ticks and mites d. Crustaceans
5. What arthropod has many abdominal segments, enlarged pincers, and a stinger at the end of its tail?
   a. a spider c. a scorpion
   b. a tick d. a crustacean
6. Of these, which structure is NOT used by arthropods for gas exchange?
   a. book lungs c. tracheal tubes
   b. spiracles d. skin
7. Crabs, lobsters, shrimps, crayfishes, barnacles, and water fleas are called ___________.
   a. spiders c. scorpions
   b. ticks and mites d. crustaceans
8. Name an arthropod that is a carnivore and has a flattened, wormlike body.
   a. millipedes c. crustaceans
   b. centipedes d. insects
9. Of the following, which is NOT an appendage of an arthropod?
   a. setae c. pedipalps
   b. antennae d. swimmerets

Chapter 31 Review

Reviewing Main Ideas

Summary statements can be used by students to review the major concepts of the chapter.

Key Terms

Answers should go beyond defining the terms. Accept any answer that uses the term correctly and in the proper context.

Understanding Concepts

1. c
2. d
3. a
4. b
5. c
6. d
7. d
8. b
9. a

Skill Review

5. Although the larval stage is most destructive to plants, many plants require the adult insects for pollination.

Chapter 31 Review

10. d
11. d
12. b
13. c
14. a
15. b
16. c
17. d
18. b
19. b
20. c

Applying Concepts

21. An insect larva may eat crop plants, but an adult may pollinate the flowers.
22. They cannot move around to find mates. Having both male and female sex organs in the same animal means that every individual is a potential mate.
23. Exoskeletons in arthropods that swim can be heavier as the water will help to make them buoyant and able to swim easily. Arthropods that fly have lightweight exoskeletons. Arthropods that move on land have jointed legs over which the exoskeleton extends.
24. Wings enable some arthropods to easily escape predators, find food sources inaccessible to terrestrial arthropods, and move easily to other areas to find food, mates, and nesting areas.
25. The eyes would give the rigid crustacean a greater field of view.

Thinking Critically

26. Evaluate students' maps. The maps should show an understanding of the relationships among the concepts listed.
27. Spiders in captivity may have inadequate food or water or space, or may be unable to find the materials on which they normally build webs.

Chapter 31 Review

10. The most diverse group of arthropods is the ____________ class.
a. millipede **c.** crustacean
b. centipede **d.** insect

11. Any structure, such as a leg or an antenna, that grows out of the body is called a(n) ____________.
a. spiracle **c.** thorax
b. mandible **d.** appendage

12. A(n) ____________ is the hard, thick, outer covering of an arthropod that is made of protein and chitin.
a. endoskeleton **c.** vertebrate
b. exoskeleton **d.** skeleton

13. Molting occurs when an arthropod sheds its old ____________ and grows a new one.
a. endoskeleton **c.** exoskeleton
b. skeleton **d.** appendage

14. Spiders have a fused head and thorax region called a ____________.
a. cephalothorax **c.** spiracle
b. thorax **d.** mandible

15. ____________ are the hollow passages that carry air through the body of an arthropod.
a. Book lungs **c.** Spiracles
b. Tracheal tubes **d.** Mandibles

16. When water passes over gills, ____________.
a. a response is followed by a stimulus
b. the arthropod will molt
c. oxygen and carbon dioxide are exchanged
d. arthropods filter feed

17. Of the following, which are NOT appendages used by arthropods to obtain and eat food?
a. chelicerae **c.** mandibles
b. pedipalps **d.** spiracles

18. Scientists have hypothesized that arthropods evolved from annelids because both have ____________.
a. exoskeletons
b. segmented bodies
c. setae
d. book lungs

19. Most terrestrial arthropods excrete wastes through ____________.
a. spiracles **c.** nephridia
b. Malpighian tubules **d.** chelicerae

20. Which of the following correctly lists the stages of incomplete metamorphosis?
a. egg—larva—pupa—adult
b. egg—pupa—larva—adult
c. egg—nymph—adult
d. egg—larva—nymph—adult

Applying Concepts

21. Many insects are pests to humans when they are larvae but are beneficial when they are adults. Explain.

22. Why is it an adaptive advantage for barnacles to be hermaphrodites?

23. Relate differences in exoskeleton structure to the various modes of arthropod locomotion.

24. In what ways have wings been an adaptive advantage to the success of insects?

25. Of what advantage might movable, stalked eyes be to a crustacean that has a cephalothorax?

Thinking Critically

26. ***Concept Mapping*** Make a concept map that relates the following terms and phrases. Supply the appropriate linking words for your map.
insect, appendage, leg, antennae, head, thorax, abdomen, spiracles, tracheal tubes, Malpighian tubules, compound eye

27. ***Forming a Hypothesis*** During an experiment to test spiderweb strength, you compare webs built by wild spiders with those of captive-bred members of the same species. You find that the spiders kept in captivity build webs that are not as strong and flexible. Make a list of possible hypotheses that you could test to determine why this is so.

28. ***Interpreting Scientific Illustrations*** Identify each of the arthropods below as an arachnid, crustacean, or insect. What are their distinguishing features?

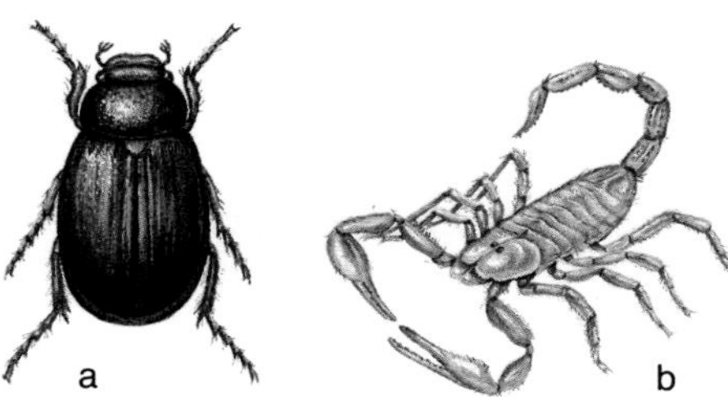

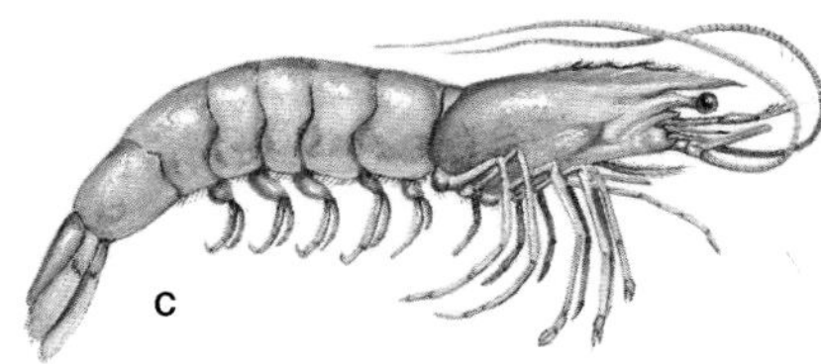

29. ***Recognizing Cause and Effect*** What is the advantage to a plant of producing a chemical that is an effective insect repellent?
30. ***Formulating Hypotheses*** You have just purchased a reptile called an anole. The anole eats crickets. When you buy a dozen crickets, the group contains large crickets that are dark brown in color and paler, smaller crickets. Are these crickets different species? Explain. (Hint: Think about what kind of metamorphosis crickets undergo.)
31. ***Recognizing Cause and Effect*** What might be the effect on plant and animal life if all insects were to die suddenly?
32. ***Observing and Inferring*** Evidence shows that deer, mice, and even household pets may harbor the bacteria that cause Lyme disease. How could pets become infected with these bacteria?

ASSESSING KNOWLEDGE & SKILLS

The melting points of the waxy layers on certain insect exoskeletons are shown in the graph below. These melting points reflect the environments in which the insects were raised. Insects raised in warmer environments have wax that melts at higher temperatures than insects raised in cooler environments.

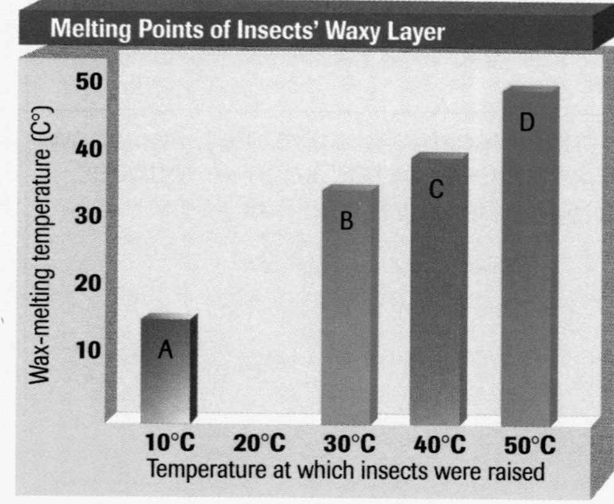

Interpreting Data Use the graph to answer the following questions.

1. What is the melting point of the wax on insects in group B?
   a. 15°C
   b. 50°C
   c. 35°C
   d. 40°C
2. What is the melting point of the wax on insects in group C?
   a. 15°C
   b. 50°C
   c. 35°C
   d. 40°C
3. Which insects were raised at the lowest temperature?
   a. A
   b. B
   c. C
   d. D
4. ***Making a Graph*** Make a graph of these data: insect exoskeletons found by a stream melt at 15°C; in a forested area at 20°C; in a grassy meadow at 40°C; and on roadside soil at 50°C.

Chapter 31 Review

28. a. insect; b. arachnid; c. crustacean. An insect has six legs and a pair of antennae. An arachnid has no antennae. A crustacean has five pairs of walking legs and stalked eyes.
29. The plant will not be eaten by insects.
30. No, the smaller crickets are probably nymphs; the crickets undergo incomplete metamorphosis.
31. Many plants that depend upon insects for pollination would be unable to reproduce, so there would be fewer plants. Plant- and insect-eating organisms would have to find new food sources or die.
32. Pets can pick up ticks when they are outdoors in an area inhabited by ticks.

ASSESSING KNOWLEDGE & SKILLS

1. c
2. d
3. a
4.

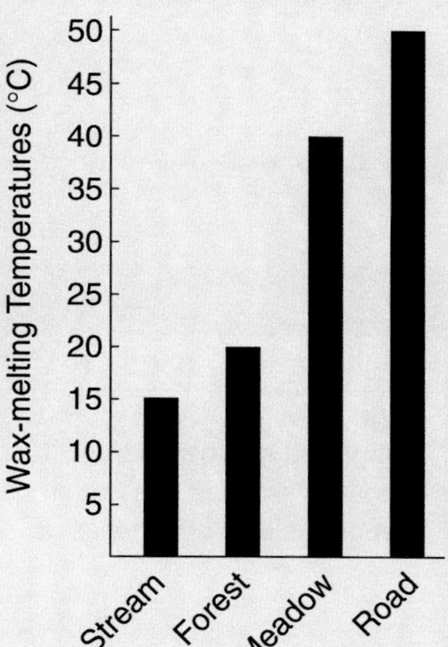

Program Resources

Chapter Assessment, pp. 185-190 L1
Alternate Assessment in the Science Classroom
Computer Test Bank L1
Content Mastery, pp. 125-128 L1

Chapter Organizer

SECTION	OBJECTIVES	ACTIVITIES/FEATURES
32.1 Echinoderms National Science Standards: UCP.1-5; C.3, C.5, C.6	1. **Compare** similarities and differences among the classes of echinoderms. 2. **Interpret** the evidence biologists have for placing echinoderms as close relatives of chordates.	**Minilab:** How do starfishes open mollusk shells?, p. 788 **Biology & Society Issues:** Too Many Starfishes?, p. 790 **Thinking Lab:** What makes sea cucumbers release gametes?, p. 791
32.2 Invertebrate Chordates National Science Standards: UCP.1, UCP.2, UCP.4, UCP.5; A.1; C.5, C.6; F.5; G.1-3	3. **Summarize** the characteristics of chordates, and show how invertebrate chordates are related to vertebrates. 4. **Distinguish** between sea squirts and lancelets.	**Biolab:** Comparing Sea Urchins and Lancelets, p. 794 **Minilab:** What does a notochord look like?, p. 796

ACTIVITY MATERIALS

BIOLAB	MINILABS	ALTERNATE LAB
page 794 compound microscope prepared slides of sea urchin development prepared slides of *Amphioxus*	**page 788** books, heavy watch or clock with second hand **page 796** prepared slide of *Amphioxus* microscope	**page 790** aquarium marine water 0.5*M* KCl solution syringes needles gravid sea urchins eyedroppers petri dish microscope depression slides coverslips

Chapter 32 Echinoderms and Invertebrate Chordates

TEACHER CLASSROOM RESOURCES

Reproducible Masters	Transparencies
Section Focus Master 70: Echinoderms L1 SAE **Reinforcement and Study Guide,** pp. 127-129 L1 **Biolab and Minilab Worksheets,** p. 127 L1 **Laboratory Manual:** Do Starfish Respond to Gravity?, pp. 187-188; Starfish Dissection, pp. 189-192; Making an Echinoderm Key, pp. 193-194 L2	**Basic Concepts Transparency #50:** Structure of a Starfish L1 SAE **Basic Concepts Transparency #51:** Phylogeny of Echinoderms L1 SAE
Section Focus Master 71: Invertebrate Chordate Structures L1 SAE **Reinforcement and Study Guide,** p. 130 L1 **Biolab and Minilab Worksheets,** pp. 128, 129-130 L1 **Concept Mapping:** Characteristics of Invertebrate Chordates, p. 32 L1 **Critical Thinking/Problem Solving:** Comparing and Contrasting Invertebrate Phyla, p. 32 L3 **Content Mastery,** pp. 129-132 L1	**Reteaching Transparency #32:** Characteristics of Chordates L1 SAE

ASSESSMENT MATERIALS

Chapter Assessment, pp. 191-196
Alternate Assessment in the Science Classroom
MindJogger Videoquiz
Computer Test Bank

Spanish Resources SAE
English/Spanish Audiocassettes SAE
Cooperative Learning in the Science Classroom COOP LEARN
Lesson Plans

KEY TO TEACHING STRATEGIES

L1 Level 1 activities should be within the ability range of all students including those with learning difficulties.

L2 Level 2 activities are within the ability range of average to above-average students.

L3 Level 3 activities are designed for the ability range of above-average students.

SAE SAE activities should be within the ability range of Students Acquiring English.

COOP LEARN Cooperative Learning activities are designed for small group work.

P These strategies represent student products that can be placed into a best-work portfolio.

These strategies are useful in a block scheduling format.

GLENCOE TECHNOLOGY

The following multimedia resources are available from Glencoe.

Biology: The Dynamics of Life
CD-ROM SAE
Videodisc Program

National Geographic Society Series
GTV: Planetary Manager
Animal

Science and Technology Videodisc Series (STVS)
Animals
Sea Urchins and Power Plants

CHAPTER

32 Echinoderms and Invertebrate Chordates

Chapter Overview

Students begin their study of echinoderms by examining the features common to all echinoderms. Next, they learn that deuterostome development and bilateral symmetry of echinoderm larvae suggest a close relationship between echinoderms and chordates. Students conclude the section by examining the diversity of echinoderms.

In the next section, students study invertebrate chordates. The features shared by invertebrate chordates as well as those they share with vertebrates are examined. Next, a comparison is made between sea squirts and lancelets.

Key Terms

ampulla
detritus
dorsal nerve cord
gill slit
madreporite
notochord
pedicellaria
ray
regeneration
tube feet
water vascular system

782

Learning Styles

Look for the following logo for strategies that emphasize different learning modalities. LS

Kinesthetic	Meeting Individual Needs, p. 784; Activity, p. 786; Minilab, p. 788
Visual-Spatial	Visual Learning, pp. 785, 786; Microscope Activity, p. 785; The Inside Story, pp. 787, 797; Portfolio, p. 789; Alternate Lab, p. 790; Minilab, p. 796
Interpersonal	Project, pp. 787, 788; Meeting Individual Needs, p. 796
Linguistic	Student Journal, p. 786
Logical-Mathematical	Thinking Lab, p. 791

CHAPTER 32 Echinoderms and Invertebrate Chordates

Beneath the cold ocean waters off the California coast, the rocky bottom is covered with brightly colored sponges, anemones, snails, and other organisms. Atop one rock, a spiny sea urchin browses on a piece of algae. A huge, purple starfish emerges from a nearby crevice and glides smoothly toward the urchin. This starfish has 24 arms—far more than the usual five—and measures almost two feet in diameter. It moves along on hundreds of tiny tube feet that line the underside of its body, coming to a stop only after it has completely covered the spiny urchin.

Now positioned for a meal, the starfish extends its stomach from its mouth and slowly engulfs the urchin. Digestion begins. Hours later, the starfish draws its stomach back in and moves away. All that's left of the urchin is the bumpy globe you see here. Even its spines are gone.

Concept Check

You may wish to review the following concepts before studying this chapter.

- Chapter 1: evolution
- Chapter 28: bilateral symmetry, endoskeleton, exoskeleton, radial symmetry

Chapter Preview

Laboratory Activities

Biolab

- Comparing Sea Urchins and Lancelets

Minilabs

- How do starfishes open mollusk shells?
- What does a notochord look like?

Feather stars

Sea urchin skeleton

Like the sea urchin and its starfish predator, the graceful feather star also is a member of the Phylum Echinodermata. Whereas sea urchins and starfishes eat algae or small animals, feather stars live on tiny particles of organic matter that drift to the ocean bottom.

Assessment Planner

Choose assessment strategies from the following pages to evaluate the progress of your students.

Assess, pp. 792, 798
Alternate Lab, pp. 790-791
Minilabs, pp. 788, 796
Portfolio, pp. 789, 797
Thinking Lab, p. 791
Biolab, pp. 794-795
Chapter Review, pp. 799-801

CHAPTER 32

Introducing the Chapter

Ask students if they have ever seen a live starfish or sea urchin. Ask those that have to describe their experiences.

Call students' attention to the opening photographs and caption. Ask them to predict what might happen to sea urchins if people collected starfishes in huge quantities along the shore. *The number of sea urchins would increase because fewer sea urchins would be eaten by starfishes.*

Next ask students to compare the starfish in the large photo with the feather star in the small photo and speculate about how their differences in structure reflect differences in their ways of life. *Responses may suggest that because it is not a predatory animal, a feather star does not need arms as thick as the arms of a starfish.*

Theme Development

Evolution is a major theme in this chapter; students study how echinoderm larvae and invertebrate chordates show similarities to certain vertebrates. The theme of *systems and interactions* is obvious as students learn how the echinoderms and invertebrate chordates are adapted to and interact with their environments.

Concept Check

Students will expand on their knowledge about how animal development is used to determine phylogeny. They will also continue to trace the evolution of invertebrate groups and the evolution of chordates.

SECTION

32.1

Prepare

Key Concepts

The characteristics common to echinoderms are presented. Deuterostome development is examined in terms of its evolutionary significance for this group. Finally, the diversity of echinoderms is considered.

Block Scheduling

Look for this symbol for strategies that are useful in a block scheduling format. For more information on block scheduling, refer to Section 32.1 in the **Lesson Plans** booklet.

Materials

- Order live or preserved starfishes or sea urchins for the Focus Demonstration and a live starfish for *The Inside Story*.
- Order sea urchin pedicellarias for the Microscope Activity.
- For the Minilab, gather heavy books.

1 Focus

Bellringer

Before presenting the lesson, display **Section Focus Master 70** on the overhead projector and have students answer the accompanying questions. L1 SAE

Demonstration

Use live or preserved starfishes or sea urchins to point out the physical characteristics of echinoderms. Elicit from students how the meaning of the term *echinoderm* relates to the features of animals in this group. *Echinoderm means "spiny skin." The animals classified in this phylum have spinelike structures covering their bodies.*

SECTION

32.1 Echinoderms

Section Preview

Objectives

Compare similarities and differences among the classes of echinoderms.

Interpret the evidence biologists have for placing echinoderms as close relatives of chordates.

Key Terms

ray
pedicellaria
tube feet
ampulla
water vascular system
madreporite
detritus
regeneration

Think about what the best defense might be for a small animal that moves slowly on the bottom of tide pools on the seashore. Did you think of armor, spines, or perhaps poison as methods of protection? Sea urchins are masters of defense—some use all three methods. The sea urchin looks very different from the starfish and feather star pictured on the previous pages, yet all three belong to the same phylum. What characteristics do they have in common? What features determine whether or not an animal is an echinoderm?

What Is an Echinoderm?

Echinoderms have a number of unusual characteristics that easily distinguish them from members of any other phylum. Nowhere else in the animal kingdom will you find creatures that move by means of hundreds of hydraulic, suction cup-tipped appendages or that have skin covered with tiny, jawlike pincers. Echinoderms live only in salt water and are found in all the oceans of the world.

echinoderm:
echinos (GK) spiny, like a hedgehog
derma (GK) skin
Echinoderms are spiny-skinned animals.

Echinoderms have an internal skeleton

If you were to examine the skin of several different echinoderms, you would find that they all have a hard, spiny, or bumpy endoskeleton covered by a thin epidermis. The long, pointed spines on a sea urchin are obvious. Some starfishes may not appear spiny at first glance, but a close look reveals that their long, tapering arms, called **rays,** are covered with short, rounded spines. The spiny skin of a sea cucumber consists of soft tissue embedded with small, platelike structures that barely resemble spines at all. The endoskeleton of all echinoderms is made primarily of calcium carbonate, the compound that makes up limestone.

Some of the spines found on starfishes and sea urchins have become modified into pincerlike appendages called **pedicellarias.** These jawlike pedicellarias are used for protection.

Echinoderms have radial symmetry

You may remember that radial symmetry is an advantage to animals that are stationary or move slowly. Radial symmetry enables these ani-

Program Resources

Section Focus Master 70 L1 SAE
Reinforcement and Study Guide, pp. 127-129 L1
Biolab and Minilab Worksheets, p. 127 L1
Laboratory Manual, pp. 187-194 L2
Basic Concepts Transparencies 50, 51 and **Masters** L1 SAE

Meeting Individual Needs

Visually Impaired Provide visually impaired students with dried specimens of various echinoderms. Allow students to handle the specimens so they can feel the shapes, sizes, and characteristic spiny skins of these animals. L1 LS

mals to sense potential food, predators, and other aspects of their environment from all directions. Observe the radial symmetry, as well as the various sizes and shapes of spines, of each echinoderm pictured in *Figure 32.1.*

Echinoderms have a water vascular system

Another characteristic unique to echinoderms is the system that enables them to move, exchange gases, capture food, and excrete wastes. Look at the close-up of the ventral side of a starfish on the next page. From the area of the starfish's mouth run grooves filled with tube feet. **Tube feet** are hollow, thin-walled tubes that each have a suction cup on the end. Tube feet look somewhat like miniature eyedroppers. The round, muscular structure called the **ampulla,** which is located on the opposite end from the suction cup, corresponds to the bulb of the eyedropper. The ampulla has muscles that contract and relax, similar to the squeezing movement of the eyedropper. Each tube foot works independently of the others, and the animal moves along slowly by alternately pushing out and pulling in its tube feet.

Tube feet function in gas exchange and excretion, as well as in locomotion. Gases are exchanged and wastes are eliminated by diffusion through the thin walls of the tube feet.

pedicellaria:
pediculus (L) little foot
Pedicellarias resemble little feet.

Figure 32.1
All echinoderms have radial symmetry and an endoskeleton composed primarily of calcium carbonate.

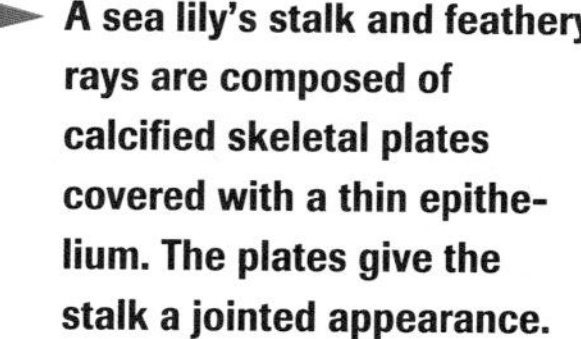

A sea lily's stalk and feathery rays are composed of calcified skeletal plates covered with a thin epithelium. The plates give the stalk a jointed appearance.

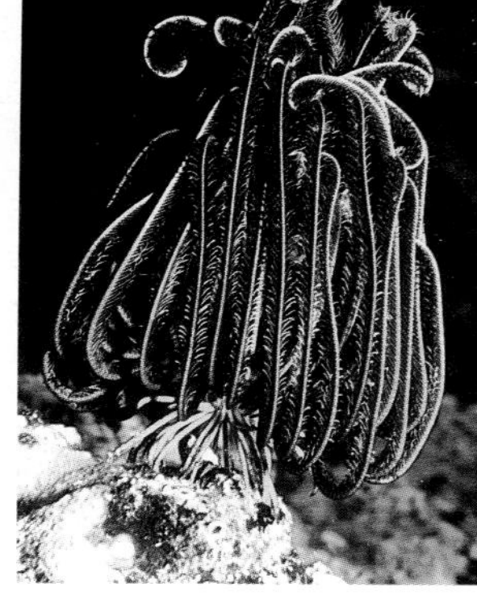

A brittle star's long, snakelike rays are composed of overlapping, calcified plates covered with a thin layer of skin cells.

A living sand dollar has a solid, immovable skeleton composed of flattened plates that are fused together.

A sea cucumber may not appear to have an endoskeleton at all. Its spines have been reduced to tiny, calcified plates embedded in its fleshy skin.

2 Teach

Visual Learning

Figure 32.1 Have students examine the animals closely. Challenge them to identify the lines of symmetry for each organism that has radial symmetry. LS

Misconception

Many scientists use the term *sea star* instead of *starfish*. Be sure students recognize that the name *starfish* is a misnomer because these animals are not closely related to fishes.

Science, Technology, and Society

Glass from Sea Urchins
Sea urchins make a hard, glasslike form of calcium carbonate. Scientists hope to determine the exact nature of this material at the molecular and genetic level. Once determined, commercial applications for this material can be explored.

Microscope Activity

LS **Visual-Spatial** Have students observe the pedicellarias of a sea urchin under a stereoscopic microscope and make sketches of their observations. Have them touch the pedicellarias with a toothpick to observe the structure's response. L1

CULTURAL DIVERSITY

Japan and the Sea Urchin Industry in Maine

Introduce students to what is rapidly becoming a $100 million international industry—the importing of Maine sea urchins to Japan. Once a remote destination in winter, coastal Maine has become a magnet for Japanese business-people because of the sea urchin, a creature once considered to be a pest to fishers.

Sea urchins are prized in Japan as *uni,* a sushi delicacy. The Japanese value the sea urchin for the quality of its eggs and ovaries, called roe. Today, more than 95 percent of all sea urchins harvested in Maine are sent to Japan. The harvesting of sea urchins is also becoming big business in California.

Visual Learning

Figure 32.2 Ask students to trace the movement of water through the starfish's water vascular system in a flowchart. L1 LS

Using Science Terms

Elicit from students why the water vascular system can be described as a *hydraulic* system. *The system works because of water pressure.* L2

Activity

LS **Kinesthetic** Provide each student with a dropper. Have students squeeze the air from the dropper and then, while still applying pressure to the rubber end of the dropper, touch the dropper tip to their finger. They should release the pressure on the dropper and observe how the dropper holds to their finger. Explain that this is similar to the suction activity of the tube feet of a starfish. L1

Bioethics

Ocean Harvests
Sea urchin roe is a popular food delicacy in Japan. In fact, roe may sell for as much as $100 a pound. As demand for the roe increases, concern has arisen about how to maintain harvesting and year-round processing without overfishing the resource. Present controls to prevent this from happening include issuing licenses to divers, closing harvest areas for one week each month, and having a minimum size requirement for sea urchins that are harvested.

madreporite:
mater (L) mother
poros (GK) channel
the main channel through which water flows into and out of a starfish

Water for the operation of an echinoderm's tube feet comes from the animal's water vascular system, *Figure 32.2*. The **water vascular system** is a hydraulic, or water pressure, system that regulates locomotion, gas exchange, food capture, and excretion in an echinoderm. Water enters and leaves the water vascular system of a starfish through the **madreporite,** a sievelike, disk-shaped opening in the echinoderm's body. You can think of this disk as being like the little strainer that fits into the drain in a sink and keeps large particles out of the pipes.

Figure 32.2
Tube feet enable starfishes and other echinoderms to creep along the ocean bottom or to pry open the shells of bivalves. The tube feet attach to the two halves of the shell by suction, and the starfish pulls open the shell just enough to insert its stomach.

The muscular ampulla contracts and relaxes with an action similar to the squeezing of an eyedropper bulb. When the ampulla contracts, it pushes water into the suction-cup portion of the tube foot, causing it to lengthen and stick tightly to the surface it is touching. When the ampulla relaxes, water flows back out of the suction cup, causing it to shorten and release its grip.

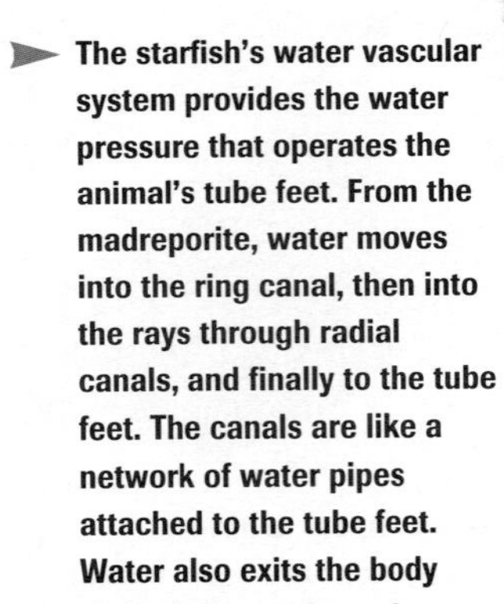

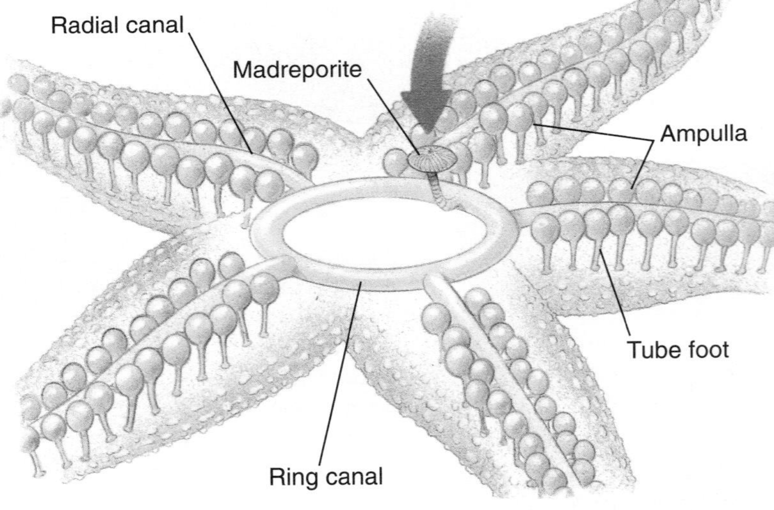

The starfish's water vascular system provides the water pressure that operates the animal's tube feet. From the madreporite, water moves into the ring canal, then into the rays through radial canals, and finally to the tube feet. The canals are like a network of water pipes attached to the tube feet. Water also exits the body through the madreporite.

786 Echinoderms and Invertebrate Chordates

STUDENT JOURNAL

Starfish Life Have students write an essay in which they describe one day in the life of a starfish. Allow a few students to read their essays. Have other students discuss which parts of the essays were scientific and which parts were creative fiction. L1 LS

interNET CONNECTION

Follow the link for this chapter on the Glencoe Homepage at **www.glencoe.com/sec/science** to find out more about echinoderms.

THE INSIDE STORY

Inside a Starfish

If you ever tried to pull a starfish from a rock where it is attached, you would be impressed by how unyielding and rigid the animal seems to be. Yet at other times, the animal shows great flexibility, such as when it rights itself after being turned upside down.

Starfish

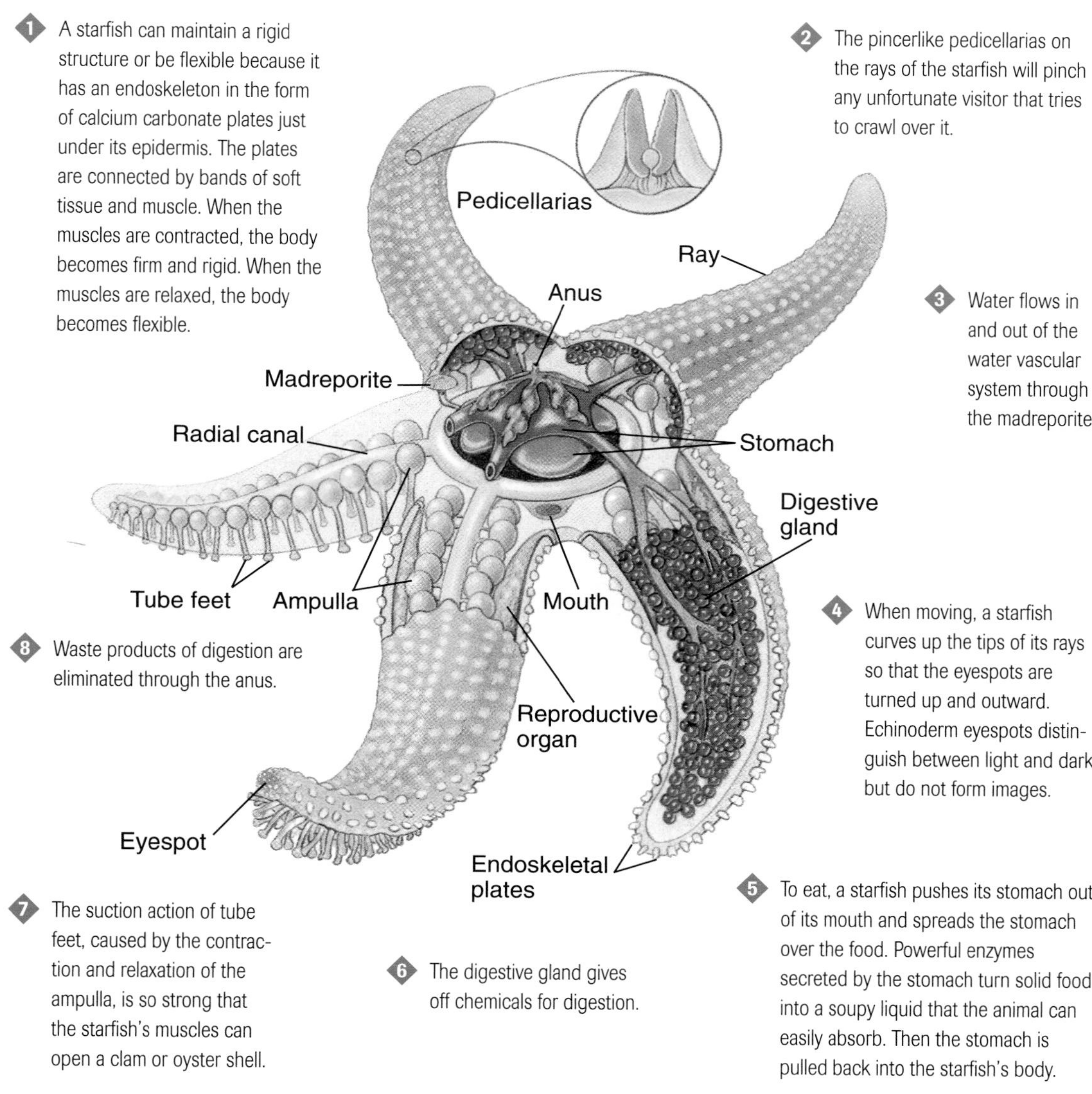

1. A starfish can maintain a rigid structure or be flexible because it has an endoskeleton in the form of calcium carbonate plates just under its epidermis. The plates are connected by bands of soft tissue and muscle. When the muscles are contracted, the body becomes firm and rigid. When the muscles are relaxed, the body becomes flexible.

2. The pincerlike pedicellarias on the rays of the starfish will pinch any unfortunate visitor that tries to crawl over it.

3. Water flows in and out of the water vascular system through the madreporite.

4. When moving, a starfish curves up the tips of its rays so that the eyespots are turned up and outward. Echinoderm eyespots distinguish between light and dark but do not form images.

5. To eat, a starfish pushes its stomach out of its mouth and spreads the stomach over the food. Powerful enzymes secreted by the stomach turn solid food into a soupy liquid that the animal can easily absorb. Then the stomach is pulled back into the starfish's body.

6. The digestive gland gives off chemicals for digestion.

7. The suction action of tube feet, caused by the contraction and relaxation of the ampulla, is so strong that the starfish's muscles can open a clam or oyster shell.

8. Waste products of digestion are eliminated through the anus.

THE INSIDE STORY

Inside a Starfish

Purpose

LS **Visual-Spatial** Students will learn about the structural and behavioral adaptations of a starfish.

Teaching Strategies

- Ask students to observe a live starfish in action. They can use a hand lens to observe the tube feet as the starfish "walks" or "climbs" a surface of a marine aquarium. L1

Visual Learning

- Make photocopies of *The Inside Story* diagram without the names of the structures. Number the parts and explain to students how each part functions and how the parts enable the starfish to survive in its environment. Have students use this information to label the diagram. L1

GLENCOE TECHNOLOGY

Videodisc

Biology: The Dynamics of Life
Disc 1, Side 2
Starfishes (Ch. 39)

PROJECT

Echinoderm Display

Ask a group of students to create a bulletin board display of echinoderms. Have them label the pictures to highlight the features of echinoderms. Ask them to find examples of everyday objects that show similar traits. Students may use sandpaper to model the texture of echinoderm skin. L1 LS

SECTION 32.1

MiniLab

Purpose

LS **Kinesthetic** Students will learn how a starfish uses force to open a clam.

Process Skills

recognize cause and effect, interpret data, compare and contrast

Teaching Strategies

- Ask students to predict how long they think they could hold up a heavy book at arm's length. Explain that this is analogous to how a starfish causes a clam's muscles to tire.

Safety Precautions

- Students who have problems with their arms, shoulders, or backs should not do this lab. These students should be the partner for someone who will hold the book.

Expected Results

Some students will not be able to hold up the book more than a few seconds. Others may hold it for several minutes.

Analysis

1. The starfish can apply steady pressure on the clam's muscles for a long time until it opens.
2. The mollusk's muscles are broken down and digested.
3. It may take longer to open a larger clam as the muscles will be larger and have more strength.

Assessment

Journal: Ask students to explain what they learned about how starfishes feed on clams. Have them include a summary of this information along with their answers to the Analysis questions in their journals. L1

MiniLab

How do starfishes open mollusk shells?

If you have ever pried open the shell of an oyster or clam, you know it's not an easy job. Starfishes feed on clams by wrapping their rays around the mollusk and using thousands of tube feet to apply suction to the shell. Then, starfish muscles exert force to pry open the tightly closed shell just far enough so that the starfish stomach can be inserted into the clam. The clamshell is held together by powerful muscles that must be overcome by the starfish. In this lab, you will see how starfish muscles overcome the force exerted by mollusk muscles.

Procedure

1. Hold your arm straight out, palm up. Your arm muscle represents the clam's muscles.
2. Place a heavy book on your hand. The book represents the force applied by the starfish.
3. Have a partner time how long you can hold your arm up with the book on it.

Analysis

1. Explain how this method of getting food works for the starfish.
2. Why do you think the mollusk shell opens wider as digestion progresses?
3. Which might take longer for a starfish to open, a small clam or a large clam? Why?

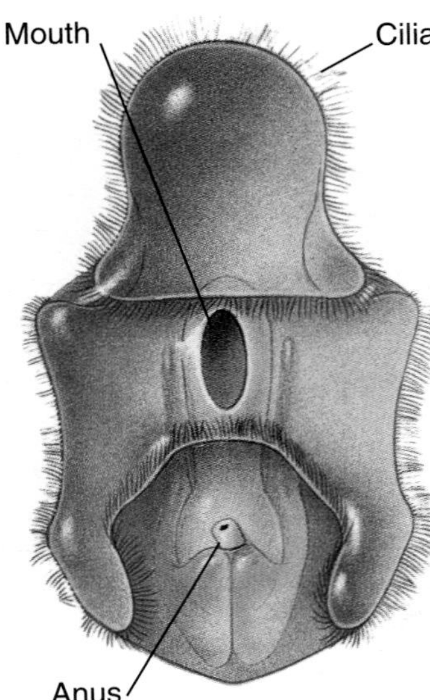

Figure 32.3
This sea-urchin larva is only 1 mm in size. The larval stage of echinoderms is bilateral, even though the adult is radial. Through metamorphosis, the free-swimming larvae make dramatic changes both in body parts and in symmetry. The bilateral symmetry of echinoderm larvae indicates that echinoderm ancestors also may have had bilateral symmetry, suggesting a close relationship to the chordates.

Echinoderm larvae have bilateral symmetry

Tube feet, rays, and pedicellarias may make echinoderms seem completely unrelated to chordates, the animal phylum you will begin to study later in this chapter. However, if you examine the larval stages of echinoderms, you will find that they, like chordates, have bilateral symmetry. The ciliated larva that develops from the fertilized egg of an echinoderm is shown in *Figure 32.3.*

Echinoderm classes have varied nutrition

All echinoderms have a mouth, stomach, and intestines, but their methods of obtaining food vary. Starfishes are carnivorous and prey on worms or on mollusks such as clams. Most sea urchins are herbivores and graze on algae. Brittle stars, sea lilies, and sea cucumbers feed on dead and decaying matter called **detritus** that drifts down to the ocean floor. Sea lilies capture detritus with their tentacle-like tube feet and move it to the mouth.

Echinoderms have a simple nervous system

Echinoderms have no head or brain, but they do have a nerve net and nerve ring. Most echinoderms have cells that detect light and touch, but they do not have sensory organs. Starfishes are an exception. A starfish's body consists of long, tapering rays that extend from the animal's central disk. At the tip of each ray, on the ventral surface, is an eyespot, a sensory organ consisting of a cluster of light-detecting cells. When walking, starfishes curve up the tips of their rays so that the eyespots are turned up and outward. This enables a starfish to detect the intensity of light coming from every direction.

PROJECT Raising Starfishes

Ask students to set up a saltwater aquarium with one or two starfishes and sea urchins. Have them observe and record starfish feeding behavior when live clams or oysters are added to the tank. Ask students to record daily observations of the tank and record all activities involved in maintaining the tank. L1 LS

Phylogeny of Echinoderms

The origin of echinoderms is subject to much speculation. Many biologists believe that the earliest echinoderms were members of an extinct class of organisms that had bilateral symmetry in their adult forms. Like modern sea lilies, members of this extinct class were sessile and lived attached to the ocean floor by stalks. Other biologists believe that modern echinoderms evolved from extinct species that were bilateral and free-swimming. Today, all adult echinoderms have radial symmetry, but nearly all their larvae have bilateral symmetry. The development of bilateral larvae is one piece of strong evidence biologists have for placing echinoderms in the evolutionary record as the closest invertebrate relatives of the chordates.

Perhaps the strongest evidence for placing the echinoderms close to the chordates is the fact that echinoderms have deuterostome development. You read in Chapter 28 that most invertebrates show protostome development, whereas deuterostome development is seen mainly in chordates. The echinoderms represent the only major group of deuterostome invertebrates. This type of development, as well as bilateral larvae and an internal skeleton, is the most important evidence for the phylogenetic relationship between echinoderms and chordates.

Echinoderms, as a group, date from the Paleozoic Era, *Figure 32.4.* Because the endoskeletons of echinoderms were easily fossilized, there is a good record of this phylum. More than 13 000 fossil species have been identified.

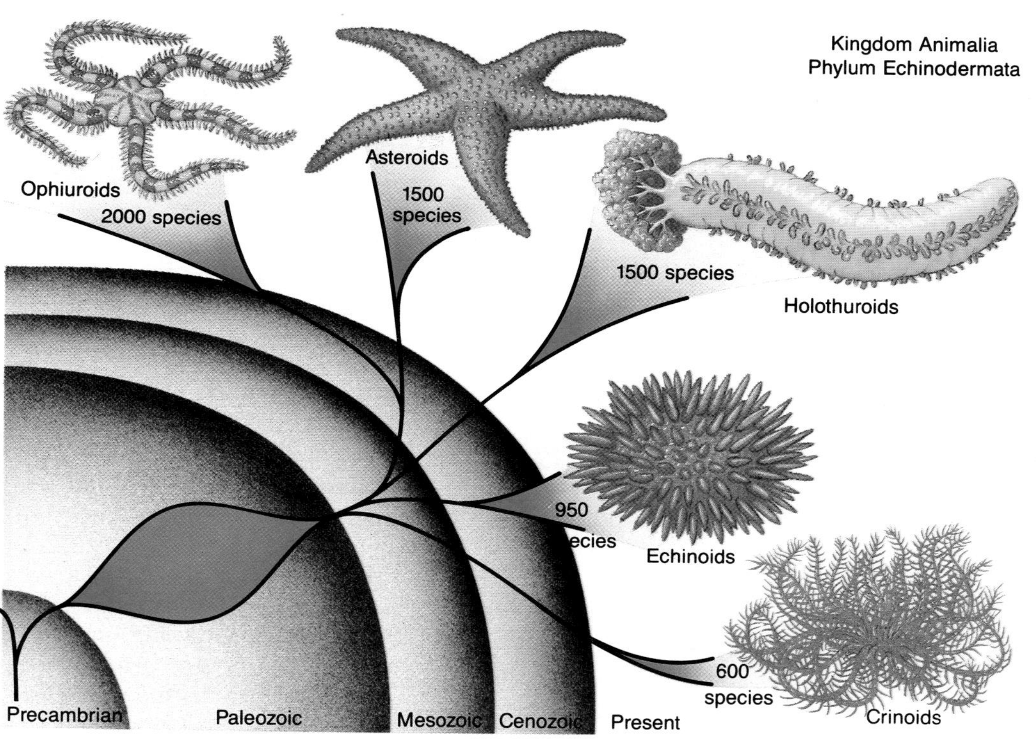

Figure 32.4

Sea cucumbers, sea urchins, and starfishes have all been found as fossils from the early Paleozoic Era. Fossils of brittle stars, on the other hand, are found beginning at a later period. From which echinoderm group are the brittle stars most likely to have evolved?

Different Viewpoints in Biology

Echinoderm Classification

Scientists have discovered about 15 extinct echinoderm classes. Crinoids were animals with five arms; some were stalked. Echinoids were shaped like flattened spheres; some had spines. Edrioasteroids were round and flat and less than 2 mm in diameter, with five bent rays. Blastoids had stalks with many small arms.

Carpoids were the strangest group. Unlike most echinoderms, many carpoids exhibited bilateral symmetry. They were also flat with large plates and a large plated appendage. Paleontologists can't agree on which is the dorsal and which is the ventral surface of the carpoids. They also do not agree about which end is the anterior and which is the posterior on bilateral specimens.

Tying to Previous Knowledge

Review protostome and deuterostome development from Chapter 28. Ask students to draw diagrams that represent each type of development. L1

Visual Learning

Figure 32.4 From which echinoderm group are the brittle stars most likely to have evolved? *from the asteroids, the starfishes*

PORTFOLIO

Echinoderm Phylogeny Provide students with echinoderm fossils. Include as many different kinds as possible and use pictures of fossils you don't have. Ask students to arrange the fossils and pictures to illustrate the phylogeny of this group. L1 IN

Issues

BIOLOGY & SOCIETY

Too Many Starfishes?

Purpose

Students will examine the relationship between the crown-of-thorns starfish and the Great Barrier Reef.

Teaching Strategies

- After students read the feature, ask them to list the possible things that could be happening to the coral reef.
- Ask students what actions should be taken, if any, if crown-of-thorns outbreaks are found to be natural events. *If outbreaks are due to natural events, nothing would need to be done. Some researchers think that long-term coral growth is enhanced if other species periodically colonize the reef during decreases in coral populations.*

INVESTIGATING the Issue

Going Further Because these starfishes have flexible bodies, they are able to wrap themselves around irregularly shaped corals to feed.

Issues

BIOLOGY & SOCIETY

Too Many Starfishes?

In the early 1960s and again in the late 1970s, huge numbers of crown-of-thorns starfishes appeared on coral reefs along part of Australia's Great Barrier Reef. Millions of these starfishes advanced across the reefs, consuming every coral polyp in their path. Eventually, the starfish armies dwindled and disappeared.

An elusive explanation Coral reefs do eventually recover from crown-of-thorns starfish outbreaks, but the cause of outbreaks remains a mystery. Are these periodic population explosions a natural event? Or are they the result of human activities near coral reefs?

Different Viewpoints

Many coral-reef biologists think that crown-of-thorns starfish outbreaks are natural events. Each female starfish produces 40 to 60 million eggs during every spawning season.

When crown-of-thorns starfishes die, tiny pieces of their skeletons become buried in reef sediments. Sediment cores that are obtained by drilling deep into reefs and then analyzed for the presence of skeletal remains suggest that crown-of-thorns starfishes may have undergone natural periodic population explosions on the Great Barrier Reef for hundreds of years.

Unnatural events Other biologists think crown-of-thorns starfish outbreaks may be the result of human activities. Outbreaks may occur because shell collectors have nearly wiped out one of the starfish's few natural enemies—the giant triton. Perhaps sewage discharged into the sea or fertilizer washed off from agricultural lands adds nutrients to the water, which can cause an increase in phytoplankton. In turn, phytoplankton provides food to support the survival of the higher-than-normal numbers of crown-of-thorns starfish larvae.

INVESTIGATING the Issue

Going Further Find out why crown-of-thorns starfishes are so much more efficient at eating corals than are other types of starfishes.

The Diversity of Echinoderms

Approximately 6000 species of echinoderms exist today. More than one-third of these species are in the class Asteroidea, to which the starfishes belong. The four other classes of living echinoderms are Ophiuroidea, the brittle stars; Echinoidea, the sea urchins and sand dollars; Holothuroidea, the sea cucumbers; and Crinoidea, the sea lilies and feather stars.

Starfishes

Most starfishes have five rays, but some species may have more than 40. The rays are tapered and come out gradually from the central disk. You have already read about the characteristics of starfishes that make them a typical example of echinoderms.

Brittle stars

As their name implies, brittle stars are extremely fragile. If you try to pick up a brittle star, parts of its rays will break off in your hand. This is an adaptation that helps the brittle star survive an attack by a predator. While the predator is busy with the broken-off ray, the brittle star can escape. A new ray will regenerate within weeks. **Regeneration** is the replacement or regrowth of missing parts; it is a common feature of echinoderms.

Brittle stars do not use their tube feet for locomotion. Instead, they use the snakelike, slithering motion of their flexible rays to propel them. The tube feet are used to pass particles of food along the rays and into the mouth in the central disk.

Alternate Lab | **Sea Urchin Fertilization**

Purpose 

LS Visual-Spatial Students will learn how sea urchin eggs are fertilized and how egg development progresses.

Materials

aquarium or other container for sea urchins, Instant Ocean (available at pet stores), 0.5*M* KCl solution, syringes and needles, gravid sea urchins, droppers, petri dish, microscope, depression slides, and coverslips

Procedure

Give students the following directions.

1. Caution students to handle sea urchins with care. Some have venom that can irritate the skin.
2. Turn the sea urchin oral side up to allow your teacher to inject 0.2 to 0.5 mL of 0.5*M* KCl solution in the soft area next to the mouth.
3. Place the urchin upside down in a petri dish. Within 5 to 10 minutes, it should shed eggs (orange color) or sperm (white).
4. If you observe sperm, dilute immediately with a few drops of seawater. If

Sea urchins and sand dollars

Sea urchins and sand dollars are globe- or disk-shaped animals covered with spines, as *Figure 32.5* shows. They do not have rays. A living sand dollar looks very different from the dead, dried-out specimens you may have seen in seashell collections or washed up on the beach. Their circular, flat skeletons look like clay sculptures with a five-petaled flower pattern on the surface. When alive, the sand dollar is covered with minute, hairlike spines that are lost when the animal dies. The living sand dollar has tube feet that protrude from the petal-like markings on its dorsal surface. These tube feet are modified to act as gills. Tube feet on the animal's ventral surface aid in bringing food particles to the mouth.

Whereas sand dollars live on the sandy ocean bottom, sea urchins primarily inhabit rocky areas. They look like living pincushions, bristling with long, usually pointed spines. They have long, slender tube feet that, along with the spines, aid the animal in locomotion.

ThinkingLab — Design an Experiment

What makes sea cucumbers release gametes?

The orange sea cucumber lives in groups of 100 or more per square meter. In the spring, these sea cucumbers produce large numbers of gametes, which they shed in the water all at the same time. The adaptive value of such behavior is that fertilization of many eggs is assured. When one male releases sperm, the other sea cucumbers in the population, both male and female, also release their gametes. Biologists do not know whether the sea cucumbers release their gametes in response to a seasonal cue, such as increasing day length or increasing water temperature, or whether they do this in response to the release of sperm by one sea cucumber.

Analysis

Design an experiment that will help to determine whether sea cucumbers release eggs and sperm in response to the release of sperm from one individual or in response to a seasonal cue.

Thinking Critically

If you find that female sea cucumbers release 200 eggs in the presence of male sperm and ten eggs in the presence of water that is warmer than the surrounding water, what would you do in your next experiment?

Figure 32.5

Echinoderms have adapted to life in a variety of habitats.

Sea urchins often burrow into rock to protect themselves from predators and rough water. Most urchins browse on algae.

Sand dollars burrow into the sandy ocean bottom. They feed on tiny organic particles found in the sand.

Basket stars, a kind of brittle star, live on the soft substrate found below deep ocean waters. They feed by using their tube feet to pass particles of detritus to the mouth.

791

ThinkingLab — Design an Experiment

Purpose

LS **Logical-Mathematical** Students will design an experiment to determine the stimulus for release of sea cucumber gametes.

Process Skills

design an experiment, separate controls and variables

Teaching Strategies

- Ask students to think of strategies animals may have to ensure aquatic fertilization. *Likely answers will include releasing large numbers of eggs and sperm, and releasing eggs and sperm at the same time.* L1

Thinking Critically

To design an experiment, students may decide to keep a group of female sea cucumbers in a laboratory setting and control all environmental variables and then release sperm into the water to see if it causes the release of eggs.

Assessment

Portfolio: Explain to students that starfishes have been found to spawn on the same day as sea cucumbers. Ask if this fact indicates that the stimulus for spawning is environmental or is in response to one male first releasing sperm. Ask students to write their responses in their portfolios. L1 P

you observe eggs, place the urchin over a beaker half-full of seawater so the eggs will fall into the beaker.

5. Place a drop of seawater with eggs into a depression slide. After focusing on eggs, add a drop of sperm solution. When the egg is fertilized, a membrane will appear around the egg.
6. Watch one of the fertilized eggs until it has divided into two cells. This may take one hour or more.
7. Watch development each day. You should have a larva after 72 hours.

Analysis

Ask students to answer the following questions:

1. How could you distinguish sperm from eggs? *Sperm are motile and smaller, with "tails." Eggs are large, with a round shape and are nonmotile.*
2. Speculate about why a fertilization membrane forms around the egg so quickly. *It prevents the egg from being fertilized by more than one sperm.*

Assessment

Journal: Explain what you have learned about sea urchin fertilization in this lab. Write a summary in your journal. L1

SECTION 32.1

3 Assess

Check For Understanding

Ask students to use Figure 32.4 to explain the phylogenetic history of echinoderms. L1

Reteach

Give students photocopies of the inner, shaded part of the phylogenetic diagram in Figure 32.4. Have them place each class in its correct position. Ask them to list also the identifying features of each class. L1

Extension

Have students research and report on the relationship between the crown-of-thorns starfish and coral reefs. L2

Assessment

Performance: Have students prepare a demonstration for third graders that explains the water vascular system of a starfish. Ask them to present this demonstration to the class using a working model. L2

4 Close

Audiovisual

Show a film about echinoderms and their interactions with other marine organisms.

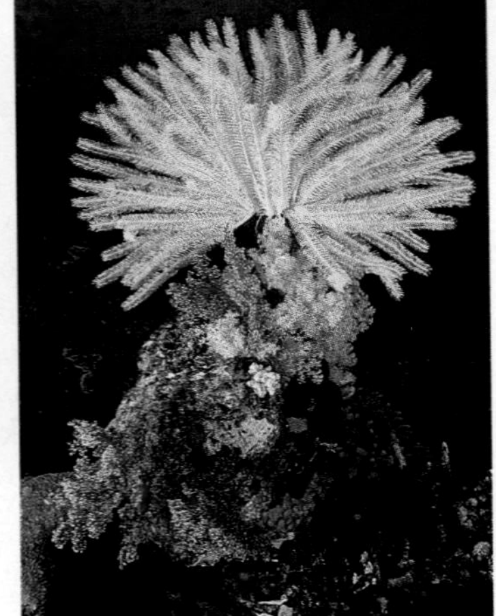

Figure 32.6

Sea cucumbers (left) trap detritus particles by sweeping their mucous-covered tentacles over the ocean bottom. Sea lilies and feather stars (right) use their feathery rays to capture downward-drifting organic particles.

The sea urchin's spines protect it from predators. In some species, sacs located near the tips of the spines contain a poisonous fluid that is injected into an attacker, further protecting the urchin. The spines also aid in locomotion and in burrowing. Burrowing species move their spines in a circular motion that grinds away the rock beneath them. This action, which is aided by a chewing action of the mouth, forms a depression in the rock that helps protect the urchin from predators and from wave action that could wash it out to sea.

Sea cucumbers

Sea cucumbers are so called because of their vegetable-like appearance, ***Figure 32.6.*** Their leathery covering allows them to be more flexible than other echinoderms; they pull themselves along the ocean floor using tentacles and tube feet. When sea cucumbers are threatened, they show a curious behavior. They may expel a tangled, sticky mass of tubes through the anus, or they may rupture, releasing some internal organs that are regenerated in a few weeks.

Sea lilies and feather stars

Sea lilies and feather stars resemble plants in some ways. Sea lilies are the only sessile echinoderms. Feather stars are sessile only in larval form. The adult feather star uses its feathery arms to swim from place to place.

Section Review

Understanding Concepts

1. How does a starfish move?
2. Describe the differences in symmetry between larval echinoderms and adult echinoderms.
3. How are sea cucumbers different from other echinoderms?

Thinking Critically

4. How do the various defense mechanisms among the echinoderm classes help them to deter predators?

Skill Review

5. **Classifying** Prepare a key that distinguishes among classes of echinoderms. Include information on presence and shape of rays, presence of spines, body shape, and other features you may find significant. For more help, refer to Organizing Information in the *Skill Handbook.*

Answers to Section Review

1. A starfish moves by regulation of its water vascular system. Tube feet attach to a surface, the starfish moves itself forward, and the suction is released.
2. Larval echinoderms are bilaterally symmetrical, whereas adult echinoderms are radially symmetrical.
3. Sea cucumbers are tubular and have a leathery outer covering instead of hard plates.

Thinking Critically

4. The rigid endoskeleton helps protect echinoderms from their enemies. Spines and poison glands also protect echinoderms. Adult echinoderms move by walking, whereas larval forms are free-swimming. If an echinoderm such as a starfish loses part of a ray, it can be regenerated. Sea cucumbers can expel their digestive tracts and grow new ones.

Skill Review

5. Student keys will vary considerably, but all should utilize the branching nature of keys described in the Skill Handbook.

SECTION

32.2 Invertebrate Chordates

The brightly colored object pictured here is a sea squirt. As one of your closest invertebrate relatives, it is placed, along with humans, in the phylum Chordata. At first glance, this sea squirt may seem to resemble a sponge more than its fellow chordates. It is sessile, and it filters food particles from water it takes in through the large opening in the top of its body. What characteristics could a human—or a fish or a lizard, for that matter—possibly share with this squishy, colorful, ocean-dwelling creature?

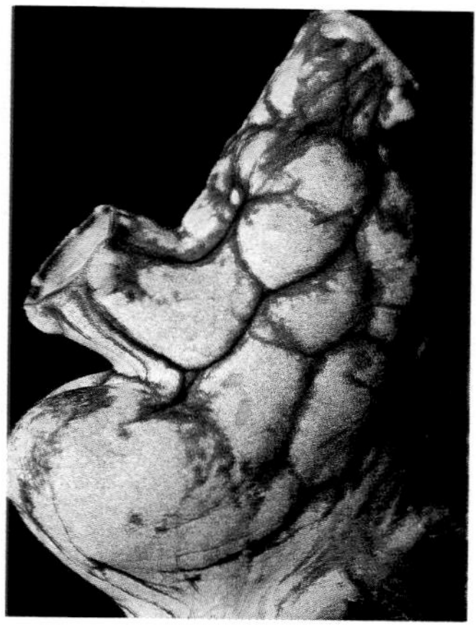

Section Preview

Objectives

Summarize the characteristics of chordates, and show how invertebrate chordates are related to vertebrates.

Distinguish between sea squirts and lancelets.

Key Terms

notochord
dorsal nerve cord
gill slit

What Are Invertebrate Chordates?

The chordates most familiar to humans are the vertebrate chordates—chordates that have backbones. However, a few invertebrate chordates have no backbones. What characteristics do these invertebrate animals share with vertebrates? Several features diagrammed in *Figure 32.7* are shared by all chordates at some time during their development.

All chordates have notochords

The term *chordata* refers to the long, semirigid, rodlike structure called a **notochord,** which is common to all members of the phylum Chordata. During the embryologic development of vertebrate chordates, this structure becomes the backbone. Invertebrate chordates have a notochord but do not develop a backbone.

The physical support of a notochord enables invertebrate chordates to make powerful side-to-side movements of the body. These movements propel the animal through the water at a much faster speed than would be possible if the body had no support.

Chordates have a dorsal nerve cord

The **dorsal nerve cord** is a bundle of nerves housed in a fluid-filled canal that lies above the notochord. In most adult chordates, the posterior portion of the dorsal nerve cord develops into the spinal cord. The anterior portion develops into the brain. A pair of nerves goes from the cord to each block of muscles.

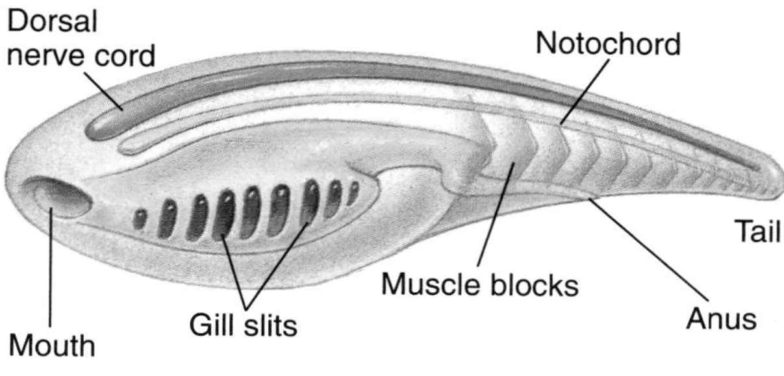

Figure 32.7
Chordate innovations—the notochord, dorsal nerve cord, gill slits, and muscle blocks—had a dramatic effect on the evolution of animals. In addition, all chordates have bilateral symmetry, a well-developed coelom, and segmentation.

SECTION 32.2

Prepare

Key Concepts

Students will learn about invertebrate chordates. They will distinguish between sea squirts and lancelets and study the relationships of these animals to the vertebrates.

Block Scheduling

Look for this symbol for strategies that are useful in a block scheduling format. For more information on block scheduling, refer to Section 32.2 in the **Lesson Plans** booklet.

Materials

- Order preserved specimens of tunicates and lancelets for the Focus Activity.
- For the Biolab, gather microscopes and prepared slides of sea urchin development and *Amphioxus.*
- For the Minilab, gather compound microscopes and cross sections of *Amphioxus.*

1 Focus

Bellringer

Before presenting the lesson, display **Section Focus Master 71** on the overhead projector and have students answer the accompanying questions. L1 SAE

Activity

Ask students to observe preserved or mounted specimens of tunicates and lancelets. Ask them if they can see any characteristics that link these animals to vertebrate chordates. Elicit whether the animals might have unseen characteristics that link them to vertebrate chordates. L1

Program Resources

Section Focus Master 71 L1 SAE
Reinforcement and Study Guide, p. 130 L1
Biolab and Minilab Worksheets, pp. 128-130 L1
Concept Mapping, p. 32 L1
Critical Thinking/Problem Solving, p. 32 L3
Reteaching Transparency 32 and **Master** L1 SAE

2 Teach

Activity

Have students run their fingers along their backs to feel their backbones. Explain that the backbone is unique to vertebrates and forms from the notochord—the structure that identifies all members of the phylum Chordata. Although all chordates have a notochord at some time during their development, not all chordates have backbones. L1

BioLab

Comparing Sea Urchins and Lancelets

Time Allotment

One class period

Objectives

Review objectives with students before they begin the Biolab.

Process Skills

compare and contrast, observe and infer, interpret data

PREPARATION

Alternate Materials

Slides of development of echinoderms other than the sea urchin may be used.

BioLab

Comparing Sea Urchins and Lancelets

Sea urchins are echinoderms that move slowly on the bottom of the sea by means of spines and tube feet. They have separate sexes, which shed their eggs and sperm into the water, where fertilization takes place. The fertilized eggs become free-swimming larvae with bands of cilia extending onto their long, graceful arms. The bilaterally symmetrical, free-swimming larvae of echinoderms are one indication of their evolutionary relationship to chordates.

Lancelets belong to the phylum Chordata. They are scaleless, fishlike animals that spend most of their time partly buried in the sand with only their mouths protruding.

PREPARATION

Problem

How do sea urchin larvae and lancelet adults compare?

Objectives

In this Biolab, you will:

- **Analyze** the stages of sea urchin development.
- **Identify** the features of chordates.
- **Compare** a sea urchin larva with an adult lancelet.

Sea urchin embryo

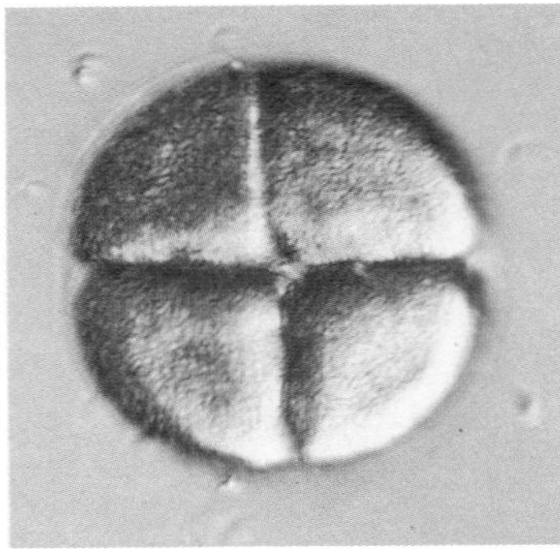

Materials

compound microscope
prepared slides of sea urchin development
prepared slide of *Amphioxus*

Safety Precautions

Use care when handling the microscope.

Sea urchin larvae

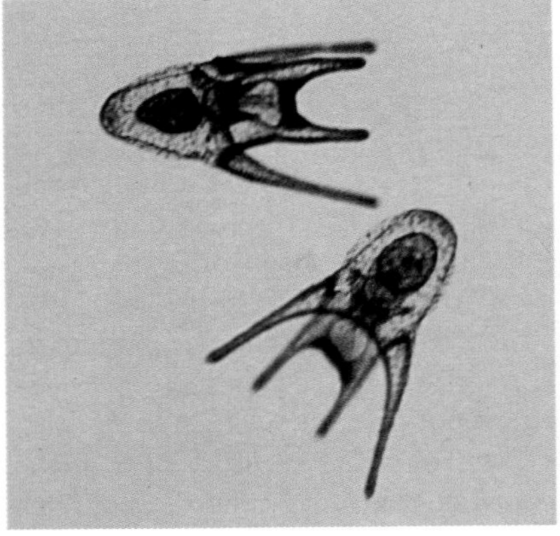

PROCEDURE

Teaching Strategies

- You may want to review development from Chapter 28.
- Emphasize to students that *Amphioxus* is not a fish.

Data and Observations

Students will make diagrams of sea urchin development. Make sure their diagram of *Amphioxus* is labeled with dorsal nerve cord, notochord, gill slits, muscle blocks, and tail.

PROCEDURE

Part A: Sea Urchin Development

1. Obtain prepared slides showing sea urchin eggs at different stages of development.
2. Prepare a data sheet consisting of eight circles, each about 5 cm in diameter.
3. Observe slides of the unfertilized egg, zygote, 2-cell stage, 4-cell stage, 8-cell stage, blastula, gastrula, and sea urchin larva.
4. Examine the slides under low power. When you have one stage in focus, switch to high power.
5. Draw a diagram of each stage. Label each diagram with the name of the stage. Examine the stages in the order listed above, and draw your diagrams in the same order.
6. Note how the size of the blastula compares with the size of the fertilized egg. Note also the size of the cells at each stage.

Part B: Lancelet Structure

1. Examine a prepared slide of a lancelet, *Amphioxus.*
2. Find the following structures: dorsal nerve cord, notochord, gill slits, muscle blocks, and tail.
3. Draw a diagram of the lancelet on a second data sheet. Label your diagram with the name of the animal and the parts listed in step 2.
4. Note the fishlike form of the lancelet.

Lancelet

ANALYZE AND CONCLUDE

1. **Comparing and Contrasting** How does the size of the sea urchin blastula compare with that of the fertilized egg? How does the size of the cells change as development proceeds?
2. **Comparing and Contrasting** At what stage do the cells look different from one another?
3. **Comparing and Contrasting** Look at a photo of an adult sea urchin in this chapter, and compare the symmetry of the larva with the symmetry of the adult. How do they compare?
4. **Making Inferences** What does the symmetry of the larva imply about the evolutionary relationship between echinoderms and chordates?
5. **Interpreting Observations** What is the function of each part you have identified in the lancelet? What is the importance of these features in placing the lancelet in the phylum Chordata?

Going Further

Application
Examine prepared slides of protostome development in *Ascaris,* and compare it with deuterostome development in sea urchins. What differences do you notice?

ANALYZE AND CONCLUDE

1. Blastula and fertilized egg are the same size. The cells become smaller as development proceeds.
2. The cells look different from one another in the late gastrula when differentiation begins.
3. The larva is bilaterally symmetrical whereas the adult is radially symmetrical.
4. Echinoderms, in terms of evolutionary origins, are closely related to chordates.
5. Dorsal nerve cord: Provides centralized coordination and control. In chordates, the posterior portion develops into the spinal cord, whereas the anterior portion develops into the brain. Notochord: Provides physical support. In chordates, the notochord becomes the backbone. Gill slits: Used to strain food from water. In chordates, gill slits develop into gills. Muscle blocks: Aid movement of the tail. Chordates, with muscle blocks are more powerful swimmers. Tail: Propels the animal through the water. In chordates, the tail enables the animal to be a powerful swimmer.

Assessment

Skill: Give students a lab practical in which you set up microscopes with various features of *Amphioxus* and sea urchin development. Ask students to identify the part or sequence of development and state the importance of the part for chordates. L1

Going Further

Have students compare slides of other deuterostomes to determine how closely development matches that of sea urchins and *Amphioxis.* L2

GLENCOE TECHNOLOGY

Videodisc

STVS: Animals
Disc 5, Side 1
Sea Urchins and Power Plants (Ch. 9)

MiniLab

Purpose

Visual-Spatial Students will study the notochord in *Amphioxus.*

Process Skills

compare and contrast, observe and infer

Teaching Strategies

- Review the differences in structure and function of the dorsal nerve cord and the notochord.
- In general, the notochord will appear solid, whereas the dorsal nerve cord will appear hollow.
- If students are looking at their cross sections with the ventral side up, the notochord will appear to be above the dorsal nerve cord.

Expected Results

Students will see the dorsal nerve cord above the notochord in their slides. The nerve cord will appear hollow or lightly stained, whereas the notochord will take on a darker stain.

Analysis

1. The notochord is stained darker and is below the dorsal nerve cord if the dorsal side of the *Amphioxus* cross section faces up.
2. The notochord enables invertebrate chordates to make powerful side-to-side movements of the body.
3. In chordates, the notochord becomes the backbone.

Assessment

Journal: Have students write a summary of what they learned about the notochord during this lab in their journals. L1

MiniLab

What does a notochord look like?

The notochord is a flexible rod that forms on the dorsal side of the early embryo of all chordates. The notochord provides support for the animal. In lancelets, the notochord is present throughout the life of the animal. In adult vertebrates, the notochord is replaced by a backbone.

Procedure

1. Examine a prepared slide of a cross section of *Amphioxus,* and observe the position of the notochord.
2. Locate the dorsal nerve cord above the notochord and compare it with the notochord.

Analysis

1. How can you tell the difference between the notochord and the dorsal nerve cord?
2. What is the function of the notochord?
3. What is the significance of the notochord in terms of evolution of chordates?

All chordates have gill slits

The **gill slits,** or gill pouches, of a chordate are paired openings located in the pharynx, behind the mouth. Many chordates have several pairs of gill slits only during embryonic development. Invertebrate chordates that have gill slits as adults use these structures to strain food from the water. In some vertebrates, especially the fishes, the gill slits develop into internal gills and become modified for gas exchange.

All chordates have muscle blocks

Muscle blocks are modified body segments that consist of stacked muscle layers. You can see these muscle blocks in the easily separated flakes of meat of a cooked fish. Muscle blocks are anchored by the notochord, which gives the muscles a firm structure to pull against. As a result, chordates tend to be more muscular than members of other phyla. The muscle blocks aid in movement of the tail. At some time during their lives, all chordates have a muscular tail. In humans, the tail appears only in the developing embryo.

Sea Squirts and Lancelets

Sea squirts, also called tunicates, are members of the subphylum Urochordata. The larval stage, as shown in *Figure 32.8,* has a tail that makes it look similar to a tadpole.

Adult sea squirts retain only gill slits as indicators of their chordate relationship. These small, tubular animals range in size from microscopic to several centimeters long. Most adult sea squirts live attached to objects on the seafloor. If you remove one from its sea home, it might squirt out a jet of water—hence the name *sea squirt.*

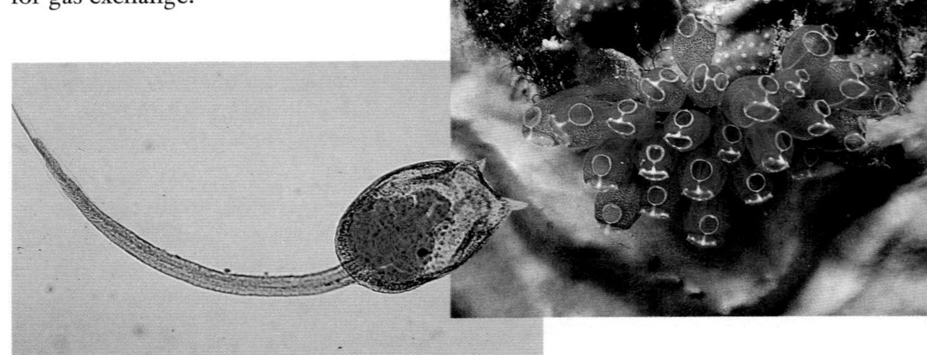

Figure 32.8
As larvae, sea squirts possess all the features common to chordates but have only gill slits and muscle blocks as adults. Sea squirt larvae (left) are about 1 cm long and swim freely through the water. As adults (right), sea squirts become stationary filter feeders enclosed in a tough, baglike layer of tissue called a tunic.

Meeting Individual Needs

Students Acquiring English Make flash cards of vocabulary words, concepts, and animals included in this section. Put a diagram on one side of the card and a word or phrase on the other side. Have students work in groups to review the cards. L1 SAE COOP LEARN 

THE INSIDE STORY

Inside a Sea Squirt

Sea squirts, or tunicates, are a group of about 1250 species that live in the ocean. They may live near the shore or at great depths. They may live individually, or several animals may share a tunic to form a colony.

Sea squirt colony

2. Water comes into the animal through the mouth, or incurrent siphon.

1. Water leaves the body of the animal through the excurrent siphon. When a sea squirt is disturbed, it may forcefully spout water from its mouth and excurrent siphon simultaneously.

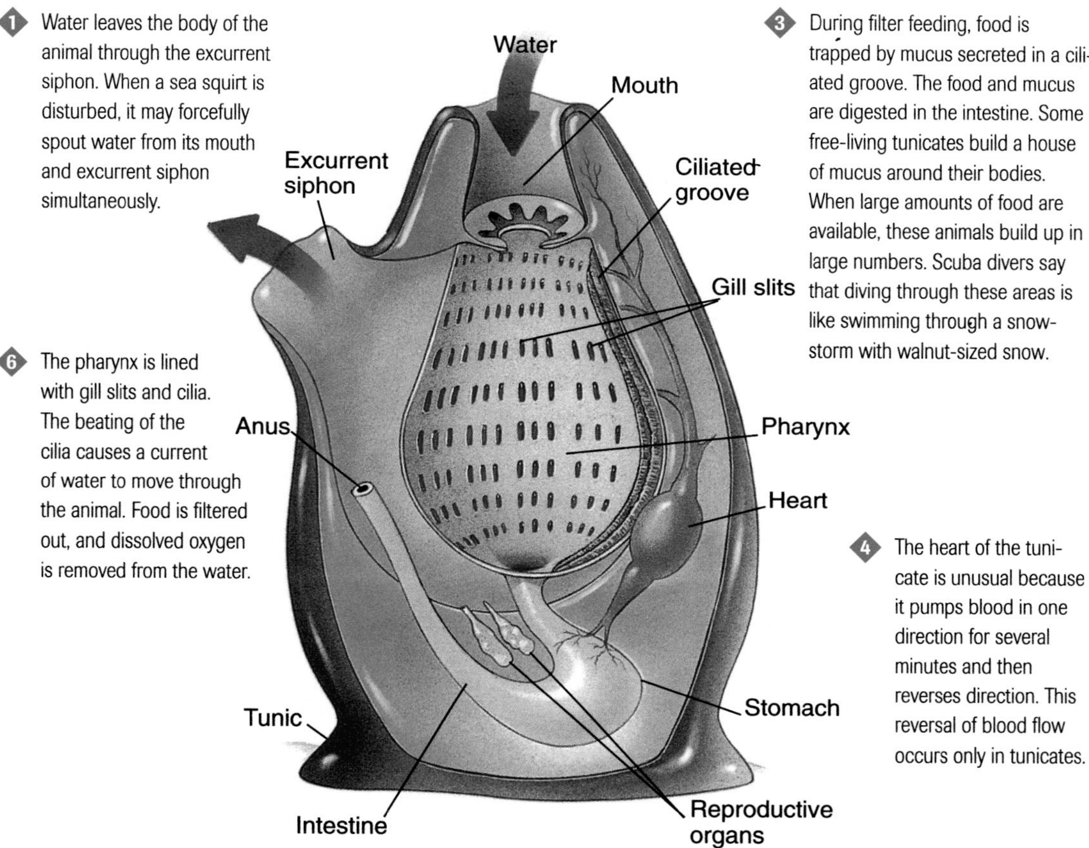

3. During filter feeding, food is trapped by mucus secreted in a ciliated groove. The food and mucus are digested in the intestine. Some free-living tunicates build a house of mucus around their bodies. When large amounts of food are available, these animals build up in large numbers. Scuba divers say that diving through these areas is like swimming through a snowstorm with walnut-sized snow.

6. The pharynx is lined with gill slits and cilia. The beating of the cilia causes a current of water to move through the animal. Food is filtered out, and dissolved oxygen is removed from the water.

4. The heart of the tunicate is unusual because it pumps blood in one direction for several minutes and then reverses direction. This reversal of blood flow occurs only in tunicates.

5. Sea squirts are covered with a layer of tissue called a tunic. Some tunics are thick and tough, and others are thin and translucent. All serve to protect the animal from predators.

THE INSIDE STORY

Inside a Sea Squirt

Purpose

LS **Visual-Spatial** Students will learn about the structural and behavioral characteristics of tunicates.

Teaching Strategies

- Ask students to recall the filter-feeding strategies of sponges and bivalve mollusks. Ask them to compare the filter feeding of tunicates to the filter feeding of the other two types of animals.

Visual Learning

- Use the diagram to compare the circulatory systems of tunicates and other invertebrates.
- Review with students the characteristics of tunicates that make them invertebrate chordates.
- Show students live specimens of tunicates and have them compare the specimens with the diagram. L1

PORTFOLIO

Water Movement in Tunicates Have students use the diagram in The Inside Story to make a flowchart that traces the path of water through a sea squirt. Have students place their completed flowcharts in their portfolios.

L1 P

SECTION 32.2

Visual Learning

Figure 32.9 Why is filter feeding such a successful adaptation for aquatic animals? *Because water naturally flows and contains numerous small organisms, animals that filter feed do not need to expend large amounts of energy to obtain food.*

3 Assess

Check for Understanding

Ask students to list invertebrate chordate characteristics. L1

Reteach

Give student groups a large piece of paper. Have them diagram an *Amphioxus* and label its typical invertebrate chordate parts and their functions. L1 COOP LEARN

Extension

Ask students to research the phyla Hemichordata and Chaetognatha. Ask them to report about the structural and behavioral adaptations of these animals and their evolutionary relationships with echinoderms and chordates. L3

Assessment

Portfolio: Have students plan an exhibit for tunicates at a public aquarium. Ask them to include in their plan suggestions about capture, transport, habitat, feeding, and general maintenance. L2 P

Figure 32.9

Lancelets, such as this *Amphioxus,* are shaped like fishes and are capable of swimming freely. But they usually spend most of their time buried in the sand with only their heads sticking out so they can filter tiny morsels of food from the water. Why is filter feeding such a successful adaptation for aquatic animals?

Lancelets belong to the subphylum Cephalochordata. They are small, streamlined, marine animals, usually about 5 cm long, as *Figure 32.9* shows. They spend most of their time buried in the sand with only their heads sticking out. Like tunicates, lancelets are filter feeders. Unlike tunicates, however, lancelets retain all their chordate features throughout life.

Because sea squirts and lancelets have no bones, shells, or other hard parts, their fossil record is incomplete. Biologists are not sure where sea squirts and lancelets fit in the phylogeny of chordates. According to one hypothesis, echinoderms, invertebrate chordates, and vertebrates all arose from an ancestral sessile animal that fed by capturing food in its tentacles. This hypothesis proposes that the ancestral organism may have evolved into a sessile filter feeder that eventually gave rise to the lancelets and tunicates. Perhaps modern vertebrates arose from the free-swimming larval stages of ancestral invertebrate chordates.

Connecting Ideas

Studying the animal phyla in order, from simplest to most complex, gives one a feeling for the progress of events during evolution. It almost seems as though evolution follows the pattern of a dramatic story line as it moves from sponges and flatworms, through mollusks and segmented worms, and on to arthropods. But evolution doesn't always follow a straight path. The echinoderms add a side plot that goes off in a different direction. You've learned that we must study the larvae of both echinoderms and invertebrate chordates to continue following evolution's story. The structural adaptations of the invertebrate chordates led to the evolution of all vertebrates.

Section Review

Understanding Concepts

1. Describe the four features common to all chordates.
2. How are invertebrate chordates different from vertebrates?
3. Compare the physical features of sea squirts and lancelets.

Thinking Critically

4. What features of invertebrate chordates suggest that you are more closely related to invertebrate chordates than to echinoderms?

Skill Review

5. **Designing an Experiment** Assume that you have found some tadpolelike animals in the water near the seashore and that you can raise them in a laboratory. Design an experiment in which you will determine whether the animals are larvae or adults. For more help, refer to Practicing Scientific Methods in the *Skill Handbook.*

4 Close

Activity

As a class, develop a table on an overhead transparency that compares the features of lancelets and tunicates. Discuss which features are common to vertebrates and which are more common to invertebrates. L1

Answers to Section Review

1. notochord, gill slits, dorsal nerve cord, muscle blocks
2. In invertebrate chordates the notochord is not replaced by a backbone.
3. Sea squirts are small, tubular, stationary filter feeders. Lancelets are shaped like fishes and can swim freely, but they spend most of their time buried in the sand.

Thinking Critically

4. notochord, gill slits, muscle blocks, and dorsal nerve cord

Skill Review

5. Watch the animals for several weeks to see if they change their body shape and symmetry. Watch for reproductive behavior or release of gametes.

REVIEWING MAIN IDEAS

32.1 Echinoderms

- Echinoderms have spines or bumps on their endoskeletons, radial symmetry, and water vascular systems. Most move by means of the suction action of tube feet.
- Deuterostome development, an internal skeleton, and bilaterally symmetrical larvae are indicators of the close phylogenetic relationship between echinoderms and chordates.
- Echinoderms include starfishes, sea urchins, sand dollars, sea cucumbers, sea lilies, and feather stars.

32.2 Invertebrate Chordates

- Chordates have a dorsal nerve cord, a notochord, gill slits, and a tail composed of muscle blocks at some stage during development.
- Sea squirts and lancelets are invertebrate chordates.

Key Terms

Write a sentence that shows your understanding of each of the following terms.

ampulla
detritus
dorsal nerve cord
gill slit
madreporite
notochord
pedicellaria
ray
regeneration
tube feet
water vascular system

Understanding Concepts

1. Starfishes, sea urchins, sand dollars, sea cucumbers, sea lilies, and feather stars are examples of ___________.
   a. invertebrate chordates
   b. chordates
   c. vertebrates
   d. echinoderms
2. Sea squirts and lancelets are examples of ___________.
   a. invertebrate chordates
   b. chordates
   c. vertebrates
   d. echinoderms
3. Animals that have spines or bumps on their endoskeletons, radial symmetry, and water vascular systems are ___________.
   a. invertebrate chordates
   b. chordates
   c. vertebrates
   d. echinoderms
4. A close phylogenetic relationship between echinoderms and chordates is indicated by the fact that both have similar ___________.
   a. habitats c. sizes
   b. larvae d. gills
5. Of the following, which is NOT a characteristic of chordates?
   a. dorsal nerve cord
   b. notochord
   c. pedicellaria
   d. tail composed of muscle blocks
6. Spines on starfishes and sea urchins have become modified into pincerlike ___________.
   a. tube feet c. gills
   b. pedicellarias d. nematocysts
7. The ___________ system operates the tube feet of starfishes and other echinoderms.
   a. water vascular c. nervous
   b. digestive d. circulatory

Chapter 32 Review

Reviewing Main Ideas

Summary statements can be used by students to review the major concepts of the chapter.

Key Terms

Answers should go beyond defining the terms. Accept any answer that uses the term correctly and in the proper context.

Understanding Concepts

1. d
2. a
3. d
4. b
5. c
6. b
7. a

Chapter 32 Review

8. c
9. d
10. b
11. a
12. b
13. b
14. a
15. b
16. d
17. d
18. b
19. a
20. c

Applying Concepts

21. Starfishes regenerate and the cut-up parts may become new starfishes.
22. Sea squirts are sessile filter feeders that can shoot jets of water to protect themselves. Lancelets swim with muscle blocks in powerful tails.

Chapter 32 Review

8. Tube feet, in addition to functioning in locomotion, also function in __________.
   a. gas exchange and digestion
   b. digestion and circulation
   c. gas exchange and excretion
   d. excretion and digestion
9. How does water enter and leave the water vascular system of a starfish?
   a. through the radial canal
   b. through the ampulla
   c. through the tube feet
   d. through the madreporite
10. Starfishes have __________ at the tip of each ray.
    a. madreporites
    b. eyespots
    c. pedicellarias
    d. mouths
11. The replacement or regrowth of missing parts is called __________.
    a. regeneration
    b. reproduction
    c. metamorphosis
    d. parthenogenesis
12. How does a sea cucumber behave when threatened?
    a. It breaks off one ray and escapes by walking away.
    b. It ruptures, releasing internal organs that are later regenerated.
    c. Sacs located near its spine tips release poisons and it injects them into attackers.
    d. It protects itself by burrowing into sand.
13. The __________ is a semirigid, rodlike structure common to all members of the phylum Chordata.
    a. spinal cord
    b. notochord
    c. vertebral column
    d. dorsal nerve cord
14. In chordates, the bundle of nerves housed in a fluid-filled canal that lies above the notochord is called the __________.
    a. dorsal nerve cord
    b. ventral nerve cord
    c. spinal cord
    d. chordate
15. Muscle blocks attached to the notochord enable chordates to be more __________.
    a. strong and muscular than invertebrates
    b. strong and flexible than invertebrates
    c. nocturnal than invertebrates
    d. intelligent than invertebrates
16. How do tunicates and lancelets get food?
    a. by predation
    b. by grazing
    c. by extending their tentacles
    d. by filter feeding
17. Which of the following animals is more likely to leave a more abundant fossil record?
    a. sea squirts
    b. lancelets
    c. sea cucumbers
    d. starfishes
18. A flat, disc-shaped echinoderm without rays, and only minute hairlike spines, is called a __________.
    a. sea urchin
    b. sand dollar
    c. starfish
    d. sea cucumber
19. Which of the following is NOT a characteristic of sea urchins?
    a. have five rays
    b. inhabit rocky areas
    c. have long, pointed spines
    d. browse on algae
20. Invertebrate chordates have no __________.
    a. digestive systems
    b. notochords
    c. backbones
    d. dorsal nerve cords

Program Resources

Chapter Assessment, pp. 191-196 L1
Alternate Assessment in the Science Classroom
Computer Test Bank L1
Content Mastery, pp. 129-132 L1

Applying Concepts

21. If you were an oyster farmer, why would you be advised not to break apart and throw back any starfishes that were destroying the oyster beds?
22. How are invertebrate chordates adapted to their way of life?
23. Relate the various functions of the water vascular system to the environment in which echinoderms live.
24. How might the ability of echinoderms to regenerate be of use to medical scientists?
25. Explain how a sea squirt maintains homeostasis.

Thinking Critically

26. ***Concept Mapping*** Make a concept map that relates the following terms and phrases. Supply the appropriate linking words for your map.
    echinoderm, radial symmetry, tube feet, water vascular system, spines, starfish, madreporite, larva, sea urchin, sand dollar
27. ***Interpreting Scientific Illustrations*** Examine *Figure 32.4* and suggest which group of echinoderms is the oldest.
28. ***Comparing and Contrasting*** Compare the pedicillarias of echinoderms with the nematocysts in cnidarians.
29. ***Observing and Inferring*** Suggest how the unusual defense mechanism of the sea cucumber may have evolved.
30. ***Observing and Inferring*** If over the course of thousands of years, starfishes evolved tube feet and muscles that could exert even more pressure on clams, what would you expect might happen to clams over the same time period?

ASSESSING KNOWLEDGE & SKILLS

The diagrams below are cross sections of larvae. The intestines are shown in red and the nerve cords are in blue.

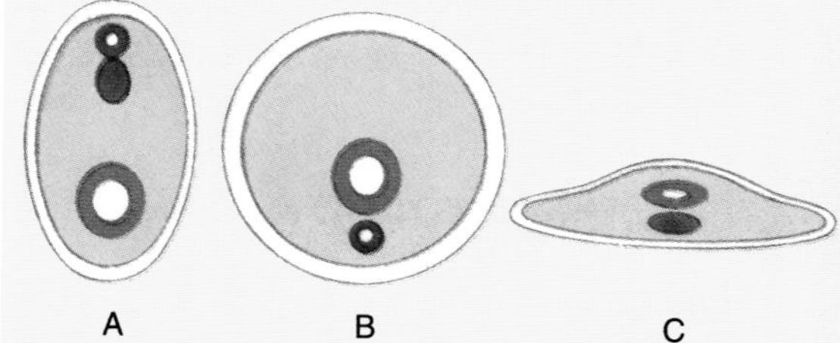

Interpreting Scientific Illustrations Study the diagram and answer the following questions.

1. Which of the diagrams shows a cross section of a lancelet?
   a. A
   b. B
   c. C
   d. none of the diagrams
2. Which of the diagrams would represent segmented worms and echinoderms?
   a. A
   b. B
   c. C
   d. none of the diagrams
3. What does the green, solid area represent?
   a. nerve cord
   b. intestines
   c. notochord
   d. spinal cord
4. What is wrong with diagram C if it represents an invertebrate chordate?
   a. notochord is ventral
   b. nerve cord is ventral and there is no notochord
   c. It is too flat.
   d. The intestine should be round.
5. ***Interpreting Scientific Illustrations*** Using the same color code and the same three organs, draw a diagram of a cross section of a larval sea squirt, starfish, and earthworm.

Chapter 32 Review

23. Tube feet enable starfishes to move and open bivalve mollusks for food. Tube feet are also used for gas exchange and excretion. This water vascular system works well in a water environment.
24. By studying regeneration in echinoderms, scientists might learn how to regenerate human cells.
25. Sea squirts filter feed and take in oxygen from the water. They maintain a balance with their watery environment by filtering out what they require and giving off waste into the water.

Thinking Critically

26. See below.
27. crinoids, the sea lilies and feather stars
28. Pedicellarias of echinoderms pinch potential predators but are not used in capturing prey and do not have immobilizing toxin as do nematocysts.
29. When internal organs exploded from a sea cucumber, it may have scared away the predator. The sea cucumber survived to reproduce more animals that would exhibit the same behavior. Perhaps at first the predator had to break the outer covering, and after many hundreds of years the response could be initiated by just a touch.
30. Clams might evolve larger and stronger muscles.

ASSESSING KNOWLEDGE & SKILLS

1. a
2. b
3. c
4. b
5. The sea squirt will look like A, the starfish like B, and the earthworm like B.

26.

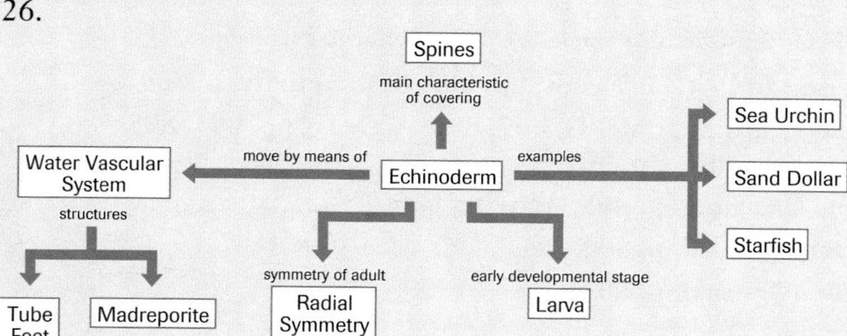

Prepare

Purpose

This Biodigest can be used as an introduction to or as an overview of the invertebrate phyla. If time is limited, you may wish to use this unit summary to teach the invertebrates in place of the chapters in the Invertebrate unit.

Key Concepts

Students learn about the phyla of invertebrates, including sponges, cnidarians, flatworms, roundworms, mollusks, segmented worms, arthropods, and echinoderms, as well as the invertebrate chordates. They learn about cell organization in animals and observe differences in physical form as organisms become more complex, with cells organized into tissues, and tissues organized into organs and organ systems.

Block Scheduling

Look for this symbol for strategies that are useful in a block scheduling format. For more information on block scheduling, refer to Biodigest 2 in the **Lesson Plans** booklet.

1 Focus

Bellringer

Before presenting the lesson, display **Section Focus Master 72** on the overhead projector and have students answer the accompanying questions. L1 SAE

INVERTEBRATES

How are jellyfishes, earthworms, starfishes, and butterflies alike? All of these animals are invertebrates—animals without backbones. The ancestors of all modern invertebrates had simple body plans. They lived in water and obtained food, oxygen, and other materials directly from their surroundings, just like present-day sponges, jellyfishes, and worms. Some invertebrates have external coverings such as shells and exoskeletons that provide protection and support.

SPONGES

Sponges, phylum Porifera, are invertebrates made up of two cell layers. Most sponges are asymmetrical. They have no tissues, organs, or organ systems. Adult sponges do not move from place to place.

VITAL STATISTICS — Cnidarians

Size ranges: Smallest: *Haliclystus salpinx*, jellyfish, diameter, 25 mm; largest: giant jellyfish medusa, diameter, 2 m; largest coral colony: Great Barrier Reef, length, 2027 km
Most poisonous: The sting of an Australian box jelly can kill a human within minutes.
Distribution: Worldwide in marine, brackish, and freshwater habitats.
Numbers of species:
Phylum Cnidaria:
- Class Hydrozoa, hydroids: 2700 species
- Class Scyphozoa, jellyfishes: 200 species
- Class Anthozoa, sea anemones and corals: 6200 species

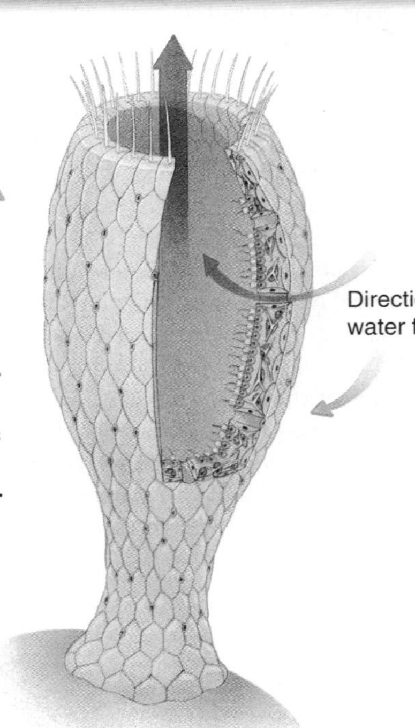

Sponges are filter feeders. A sponge takes in water through pores in the sides of its body, filters out food, and releases the water through the opening at the top.

Learning Styles

Look for the following logo for strategies that emphasize different learning modalities. LS

Kinesthetic	**Meeting Individual Needs, pp. 803, 806; Activity, p. 804**
Visual-Spatial	**Visual Learning, pp. 803, 807, 808; Activity, pp. 803, 805; Demonstration, p. 807**
Interpersonal	**Visual Learning, p. 805; Activity, p. 806**
Linguistic	**Student Journal, p. 804**
Logical-Mathematical	**Student Journal, p. 803**

interNET CONNECTION

Follow the link for this Biodigest on the Glencoe Homepage at **www.glencoe.com/sec/science** to find out more about invertebrates.

CNIDARIANS

Like sponges, cnidarians are made up of two cell layers and have only one body opening. The cell layers of a cnidarian, however, are organized into tissues with different functions. Jellyfishes, corals, sea anemones, and hydras belong to phylum Cnidaria.

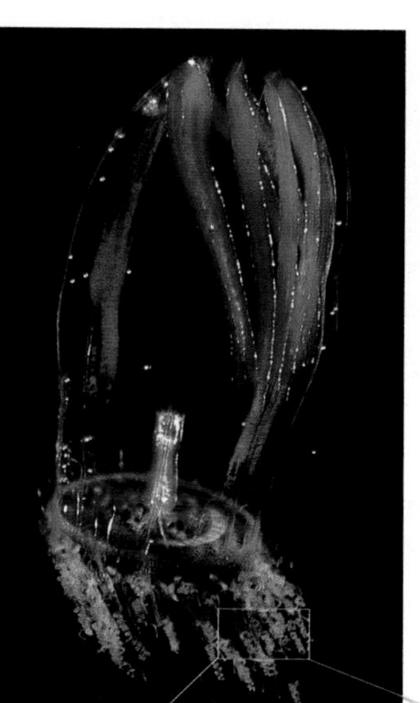

Jellyfishes and other cnidarians have nematocysts—stinging cells that are used to capture prey.

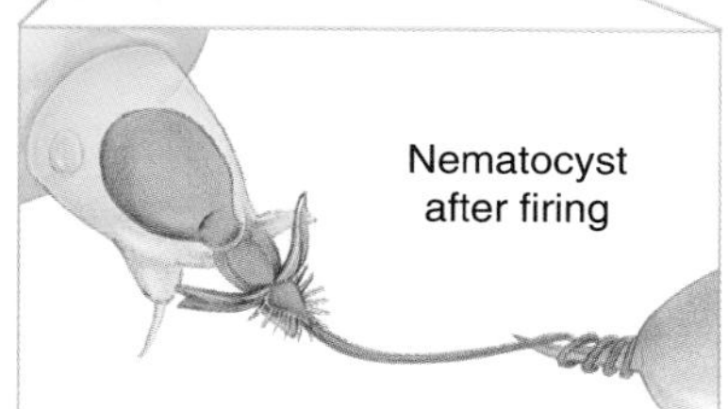

ROUNDWORMS

Roundworms, phylum Nematoda, have a pseudocoelom and a tubelike digestive system with two body openings. Most roundworms are free-living, but many plants and animals are affected by parasitic roundworms.

Parasitic roundworms such as this *Trichinella* are contracted by eating improperly cooked pork. Other roundworms can be contracted by walking barefoot on contaminated soil.

FLATWORMS

Flatworms, phylum Platyhelminthes, include free-living planarians, parasitic tapeworms, and parasitic flukes. Flatworms are bilaterally symmetrical animals with flattened solid bodies and no body cavities. Flatworms have one body opening through which food enters and wastes leave.

Free-living flatworms have a head end with organs that sense the environment. Flatworms can detect light, chemicals, food, and movements in their surroundings.

VITAL STATISTICS FLATWORMS

Size ranges: Largest: beef tapeworm, length, 30 m
Distribution: Worldwide in soil, marine, brackish, and freshwater habitats.
Numbers of species:
Phylum Platyhelminthes:
- Class Turbellaria, free-living planarians: 3000 species
- Class Cestoda, parasitic tapeworms: 3500 species
- Class Trematoda, parasitic flukes: 8000 species

2 Teach

Film Festival

Many of the invertebrate phyla are best observed in nature films. Select several nature films that highlight invertebrates. Ask students to list as many invertebrates as they see in the films, then group the animals into phyla. Ask them to identify the characteristics they used to classify the animals.

Visual Learning

LS **Visual-Spatial** Ask students in groups to make a labeled diagram of a planaria worm on the chalkboard. Ask them how the shape of a planaria worm reflects its living habits. *The planaria worm is flattened, thereby enabling it to slip easily under rocks and debris in streams.* L1 SAE COOP LEARN

Activity

LS **Visual-Spatial** Provide students with binocular microscopes, watch glasses, toothpicks, dropping pipets, and a planaria worm culture. Ask them to place a few drops of water in the watch glass, then gently move, with a toothpick, a planaria worm to the watch glass. Have students observe the worm under the microscope and describe its movement. L1 SAE

STUDENT JOURNAL

LS **Logical-Mathematical** Ask students to make a table that compares the features of sponges and the features of cnidarians. Ask them to make a labeled diagram of a sponge and a jellyfish to illustrate this page of their journal. P

Meeting Individual Needs

Visually Impaired For students who are visually impaired, provide poster board cutouts of flatworms and roundworms. Ask them to describe their tactile impressions of each worm. LS

Activity

LS **Kinesthetic** Ask students to observe the movement of a live land snail. As the snail moves, ask students to touch the antennae with a cotton swab. Ask them to describe the behavior of the snail when the antennae are touched. L1

SAE

MOLLUSKS

Slugs, snails, clams, squids, and octopuses are members of phylum Mollusca. All mollusks are bilaterally symmetrical and have a coelom, two body openings, a muscular foot for movement, and a mantle, which is a thin membrane that surrounds the internal organs. In shelled mollusks, the mantle secretes the shell.

VITAL STATISTICS
MOLLUSKS

Size ranges: Largest: tropical giant clam, length, 1.5 m; North Atlantic giant squid, length, 18 m; Pacific giant octopus, length, 10m; smallest: seed clam, length, less than 1 mm

Distribution: Worldwide in salt, fresh, and brackish water, and on land in moist temperate and tropical habitats.

Numbers of species:
Class Bivalvia, bivalves: 30 000 species
Class Gastropoda, snails and slugs: 45 000 species
Class Cephalopoda, octopuses, squids, nautiluses: 800 species

▲ Gastropods, such as snails, have a tongue-like radula used to scrape algae from surfaces.

▲ Bivalves, such as clams, strain food from water by filtering it through their gills.

Classes of Mollusks

The three major classes of mollusks are bivalves with two hinged shells; gastropods with one shell or no shell; and cephalopods. Cephalopods include octopuses, squids, and shelled nautiluses that all have muscular tentacles and are capable of swimming by jet-propulsion.

FOCUS ON ADAPTATIONS

Body Cavities

The type of body cavity an animal has determines how large it can grow, and how it takes in food and eliminates wastes. Acoelomate animals, such as planarians, have no body cavity. Water and digested food particles travel through a solid body by the slow process of diffusion.

Animals such as roundworms have a fluid-filled body cavity called a pseudocoelom that is partly lined with mesoderm. Mesoderm is a layer of cells between the ectoderm and endoderm that differentiates into muscles, circulatory vessels, and reproductive organs. The pseudocoelom

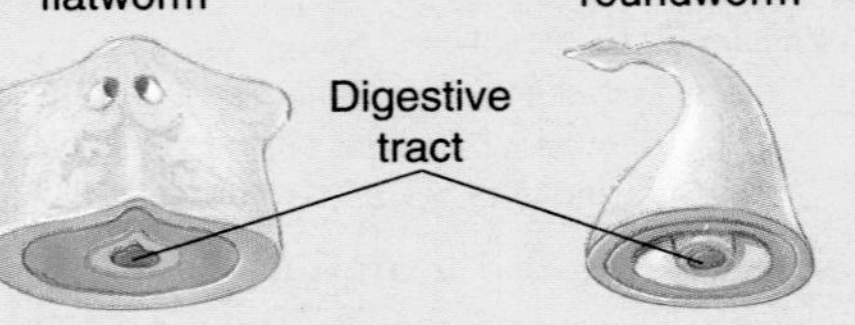

804

STUDENT JOURNAL

LS **Linguistic** Ask students to prepare an advertisement as if it were for the classified section of the newspaper for a homesite for three animals, one in each of the classes of mollusks. Provide students with a page of the real estate section of the newspaper to review how advertisements for homes are written. Students should be creative with their language but keep the precise habitat requirements of each animal in mind. L2

Cephalopods, such as octopuses, are predators. They capture prey using the suckers on their long tentacles.

SEGMENTED WORMS

Bristleworms, earthworms, and leeches are members of phylum Annelida, the segmented worms. Segmented worms are bilaterally symmetrical, coelomate animals that have segmented, cylindrical bodies with two body openings.

Most bristleworms have a distinct head and a body with bristle-like structures called setae.

Segmentation is an adaptation that provides these animals with great flexibility. Each segment has its own muscles. Groups of segments have different functions such as digestion or reproduction.

Classes of Segmented Worms

Phylum Annelida has three classes: Hirudinae, the leeches; Oligochaeta, the earthworms; and Polychaeta, the bristleworms.

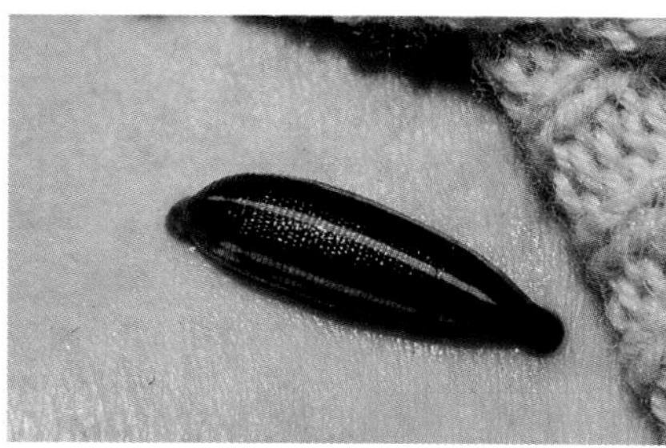

Leeches have flattened bodies with no setae. Most species are parasites that suck blood and body fluids of ducks, turtles, fishes, and humans.

provides support for the attachment of muscles, making movement more efficient. Earthworms have a coelom, a body cavity surrounded by mesoderm in which internal organs are suspended. The coelom acts as a watery skeleton against which muscles can work.

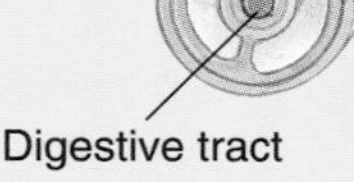

VITAL STATISTICS
SEGMENTED WORMS

Size ranges: Largest: giant tropical earthworm, length, 4 m; smallest: freshwater worm, *Aeolosoma*, length, 0.5 mm
Distribution: Terrestrial and marine, brackish, and freshwater habitats worldwide except polar regions and deserts.
Numbers of species:
Class Polychaeta, bristleworms: 8000 species
Class Oligochaeta, earthworms: 3100 species
Class Hirudinea, leeches: 500 species

Activity

LS Visual-Spatial Give each group of students a live earthworm in one dish and a bristleworm in another dish. Ask them to observe their behavior and movements. Tell students to make a table that shows the similarities and differences between earthworms and bristleworms.

Earthworms may be obtained from bait shops and bristleworms can be obtained from biological supply houses or bait shops near the sea. L1 SAE

Visual Learning

LS Interpersonal Ask students to find objects in the room that have bilateral and radial symmetry.

Ask them to explain why they have identified particular objects as having radial or bilateral symmetry. Objects such as a pencil and a test tube have radial symmetry. If an object has radial symmetry, it can be divided along any plane through a central axis, into roughly equal halves. Objects such as books and chairs have bilateral symmetry. If an object has bilateral symmetry, the right and left halves form mirror images when divided down its length. L1 COOP LEARN

Activity

Interpersonal Ask students to bring in jars of many types of arthropods. Number the jars and pass them from group to group asking students to write the name of the arthropod and obvious arthropod features beside the name. For each arthropod, ask them to infer what features helped to make that particular arthropod successful. L1

ARTHROPODS

Arthropods are bilaterally symmetrical, coelomate invertebrates with tough outer coverings called exoskeletons and jointed appendages that are used for walking, sensing, feeding, or mating. Exoskeletons protect and support their soft internal tissues. Jointed appendages allow for powerful and efficient movements.

The evolution of jointed appendages with many different functions probably led to the success of the arthropods as a group. For example, a scorpion uses its jointed stinger as a defense mechanism.

Like other members of class Arachnida, the black widow spider has chelicerae, a pair of biting appendages near the mouth.

Arthropod Diversity

Two out of three animals on Earth today are arthropods. The success of arthropods can be attributed to adaptations that provide efficient gas exchange, acute senses, and varied types of mouthparts for feeding. Arthropods include organisms such as spiders, crabs, lobsters, shrimps, crayfishes, centipedes, millipedes, and the enormously diverse group of insects.

FOCUS ON ADAPTATIONS

Insects

Insects have many adaptations that have led to their success in the air, on land, in fresh water, and in salt water. For example, insects have complex mouthparts that are well adapted for chewing, sucking, piercing, biting, or lapping. Different species have mouthparts adapted to eating a variety of foods.

If you have ever been bitten by a mosquito, you know that mosquitoes have piercing mouthparts that cut through your skin to suck up blood. In contrast, butterflies and moths have long, coiled tongues that they extend deep into tubular flowers to sip nectar. Grasshoppers and many

Meeting Individual Needs

Students Acquiring English For students who are kinesthetic learners, purchase plastic arthropods. Ask them to work in groups and point out features of the arthropods as they pass them around their group. Features that they might be able to find include head, thorax, abdomen, spiracles, claws, flippers, jointed legs, antennae, cephalothorax, mouthparts, pedipalps, chelicerae, walking legs, compound and simple eyes, wings, tympanum, and mandibles. L1 SAE

BIODIGEST

Most insects, such as this moth, have one pair of antennae for sensing their environments.

Lobsters, class Crustacea, have antennae and two compound eyes on movable stalks. Their mandibles move from side to side to seize prey.

Millipedes, class Diplopoda, are herbivores. Millipedes have up to 100 body segments, and each segment has two pairs of legs.

Arthropod Origins

Arthropods most likely evolved from segmented worms; they both show segmentation. However, an arthropod's segments are fused and have a greater complexity of structure than those of segmented worms. Because arthropods have exoskeletons, fossil arthropods are frequently found, and consequently more is known about their origins than about the phylogeny of worms.

VITAL STATISTICS ARTHROPODS

Size ranges: Largest insects: tropical stick insect, length, 33 cm; Goliath beetle, mass, 100 g; smallest insect: fairyfly wasp, length, 0.21 mm

Distribution: All habitats worldwide.

Numbers of species:
Class Arachnida, spiders and their relatives: 57 000 species
Class Crustacea, crabs, shrimps, lobsters, crayfishes: 35 000 species
Class Chilopoda, centipedes: 2500 species
Class Diplopoda, millipedes: 10 000 species
Class Insecta, insects: 750 000 species

beetles have hard, sharp mandibles used to cut off and chew leaves. But the heavy mandibles of staghorn beetles no longer function as jaws; instead, they have become defensive weapons used for competition and mating purposes.

Different Foods for Different Stages Because insects undergo metamorphosis, they often utilize different food sources at different times of their life cycles. For example, monarch butterfly larvae feed on milkweed leaves, whereas the adults feed on milkweed flower nectar. Apple blossom weevil larvae feed on the stamens and pistils of unopened flower buds, but the adult weevils eat apple leaves. Some adult insects, such as mayflies, do not eat at all! Instead, they rely on food stored in the larval stage for energy to mate and lay eggs.

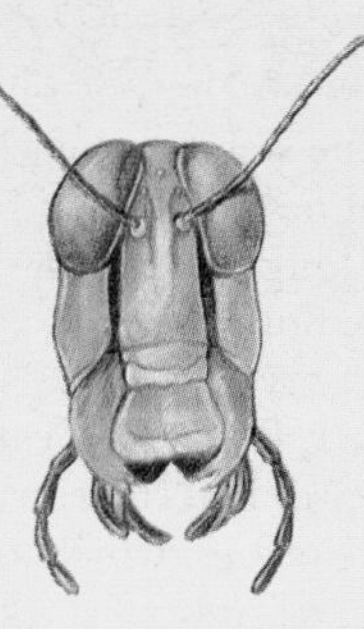

Grasshopper mouthparts

Visual Learning

Visual-Spatial Ask students to examine the photo of the moth and explain how it is adapted to its way of life. *The color provides camouflage from predators. Large wings permit long-distance flight while conserving energy during the floating parts of the flight.*

Demonstration

Visual-Spatial Borrow a collection of preserved butterflies from a nearby college or museum and call students' attention to the enormous variety in wing shapes, sizes, and colors.

Visual Learning

Visual-Spatial Ask students to make their own labeled diagrams of starfishes or other echinoderms by giving each group an acetate transparency and a transparency marker for tracing photos from the book or other materials you provide. Ask each group to do a short presentation of their echinoderm, explaining the functioning of the labeled parts. L1

SAE P

ECHINODERMS

Echinoderms, phylum Echinodermata, are radially symmetrical, coelomate animals with hard, bumpy, spiny endoskeletons covered by thin epidermis. The endoskeleton is comprised of calcium carbonate. Echinoderms move using a unique water vascular system with tiny, suction-cuplike tube feet. Some echinoderms have long spines also used in locomotion.

Echinoderm Diversity

There are five major classes of echinoderms. They include starfishes, brittle stars, sea urchins, sand dollars, sea cucumbers, sea lilies, and feather stars.

Echinoderm larvae are bilaterally symmetrical, a condition that suggests a close relationship to the chordates.

The tube feet of a starfish operate by means of a hydraulic water vascular system. Starfishes move along slowly by alternately pushing out and pulling in their tube feet.

VITAL STATISTICS
ECHINODERMS

Size ranges: Largest: sea urchin, diameter, 19 cm; longest: sea cucumber, length, 60 cm

Distribution: Marine habitats worldwide.

Numbers of species:

Phylum Echinodermata
- Class Asteroidea, starfishes: 1500 species
- Class Crinoidea, sea lilies and feather stars: 600 species
- Class Ophiuroidea, brittle stars: 2000 species
- Class Echinoidea, sea urchins and sand dollars: 950 species
- Class Holothuroidea, sea cucumbers: 1500 species

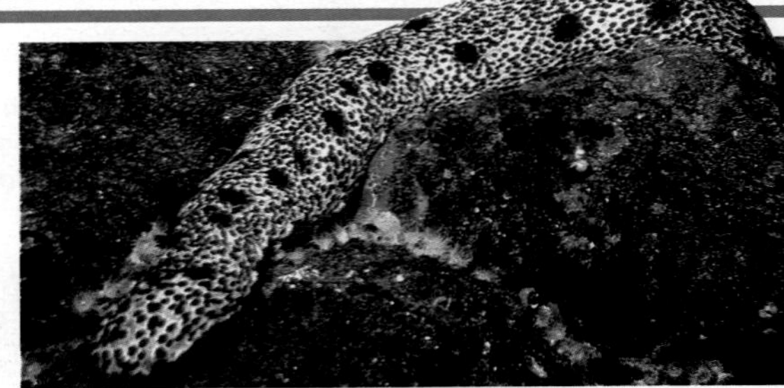

Sea cucumbers have a leathery skin and are flexible. Like most echinoderms, they move using tube feet.

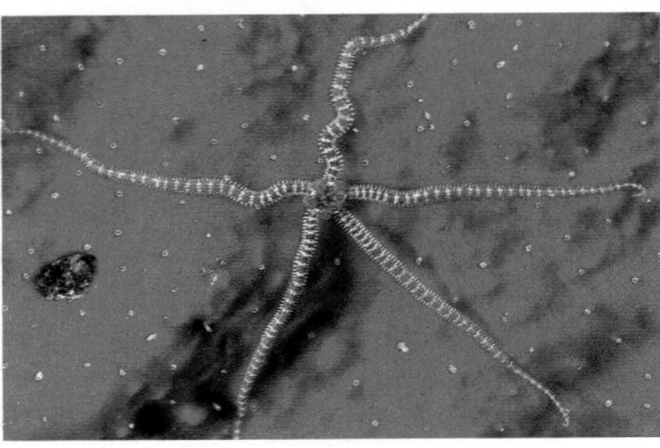

The long, thin arms of brittlestars are fragile and break easily, but they grow back. Brittlestars use their arms to walk along the ocean bottom and their tube feet for feeding.

INVERTEBRATE CHORDATES

All chordates have, at one stage of their life cycles, a notochord, dorsal nerve cord, and gill slits. A notochord is a long, semirigid, rodlike structure along the dorsal side of these animals. The dorsal nerve cord is a bundle of nerves in a hollow, fluid-filled canal that lies above the notochord. Invertebrate chordates also have muscle blocks, which are modified body segments consisting of stacked muscle layers. Muscle blocks are anchored by the notochord.

BIODIGEST

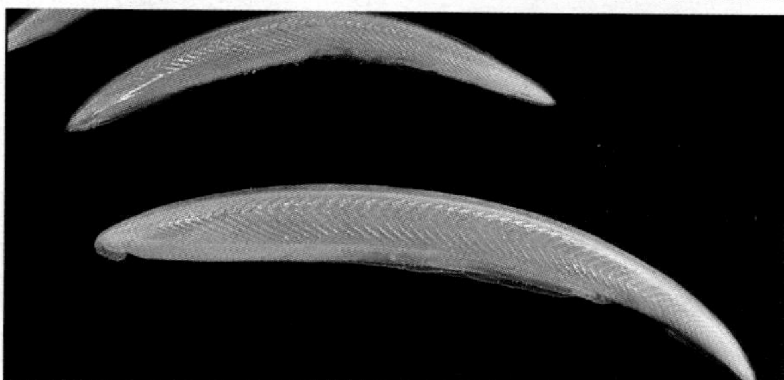

The lancelet is an example of an invertebrate chordate. Notice that the lancelet's body is shaped like that of a fish even though it is a burrowing filter feeder.

BIODIGEST REVIEW

Understanding Concepts

1. An animal that is a filter feeder, takes in water through pores in the sides of its body, and releases water from the top is a ___________.
   a. roundworm c. sponge
   b. gastropod d. lancelet
2. Nematocysts are unique to ___________.
   a. sponges c. annelids
   b. mollusks d. cnidarians
3. An example of a free-living flatworm is a ___________.
   a. planarian c. nematode
   b. tapeworm d. vinegar-eel
4. An octopus belongs to phylum ___________.
   a. Porifera c. Mollusca
   b. Cnidaria d. Arthropoda
5. Leeches feed by ___________.
   a. grazing on aquatic plants
   b. stinging prey
   c. filter feeding
   d. sucking blood
6. Which of the following characteristics is unique to arthropods?
   a. nematocysts
   b. jointed appendages
   c. filter feeding
   d. tube feet
7. Which phylum includes animals that, as adults, do not move about?
   a. Arthropoda c. Annelida
   b. Mollusca d. Porifera
8. Which of the following are invertebrate chordates?
   a. sea anemones c. bivalves
   b. lancelets d. squids
9. Parasitism is a way of life for most ___________.
   a. flukes c. cnidarians
   b. sponges d. annelids
10. Animals with no body cavity are called ___________.
    a. coelomate c. pseudocoelomate
    b. acoelomate d. asymmetric

Thinking Critically

1. A radula is to a snail as a ___________ is to an octopus. Explain your answer.
2. If you were examining a free-living animal that had a thin, solid body with two surfaces, what might it be? Explain.
3. In what two ways are spiders different from insects?
4. Why is more known about animals with hard parts than is known about animals with only soft parts?
5. In what ways are echinoderms more similar to vertebrates than to other invertebrates?

Program Resources

Section Focus Master 71 L1 SAE
Reinforcement and Study Guide, pp. 131-132 L1
Chapter Assessment, pp. 197-200 L1
Content Mastery, pp. 133-136 L1

BIODIGEST

Answers to BioDigest Review

Understanding Concepts

1. c
2. d
3. a
4. c
5. d
6. b
7. d
8. b
9. a
10. b

Thinking Critically

1. tentacle: A snail gets food by scraping surfaces with its radula. An octopus gets food by snaring it with its tentacles.
2. planarian: Planarians are free-living worms with thin solid bodies with two surfaces.
3. Spiders have four pairs of walking legs and chelicerae while insects have three pairs of legs, one pair of antennae, and three body regions.
4. Fossils are more easily formed from animals with hard parts.
5. Echinoderms have an endoskeleton covered by a thin epidermis, traits found in vertebrates but not in other invertebrates. Echinoderm larvae are bilaterally symmetrical, which also suggests a link to vertebrates.

UNIT

9 Vertebrates

Unit Overview

Students learn about the diversity of animals with backbones in Unit 9. In Chapter 33, students begin to appreciate the evolutionary movement of animals from water to land environments as they explore the structure, adaptations, ecology, and phylogeny of fishes and amphibians. The movement of animals to land is completed with the reptiles, studied in Chapter 34. In this chapter, students also examine body adaptations that enable birds to fly.

Included in Chapter 34 is Focus on Dinosaurs, a detailed examination of these remarkable, extinct animals.

In Chapter 35, students discover why mammals are able to occupy nearly all environments on Earth. The unit concludes in Chapter 36 with an exploration of animal behavior.

Theme Development

Throughout the unit, the themes of *evolution* and *unity within diversity* are apparent. Students trace the evolution of animals from ancestral water-dwelling vertebrates to the modern-day radiation into all habitat types. Within each group, students will see not only the common features of animals but also their diversity.

UNIT

9 Vertebrates

Every summer, the few visitors allowed to view the annual congregation of grizzly bears at McNeil River Falls, Alaska, are treated to a thrilling sight. At this unique sanctuary, a single sweep of the eyes may reveal up to sixty Alaskan grizzly bears playing and feeding on the river's migrating salmon. Anxious gulls wait for scraps of discarded fish.

Bears, salmon, gulls—and humans—belong to the group of animals known as vertebrates. Their ability to survive the rigors of changing environments such as the Alaskan tundra, and engage in complex behaviors such as migration, are just a few of the characteristics of vertebrates.

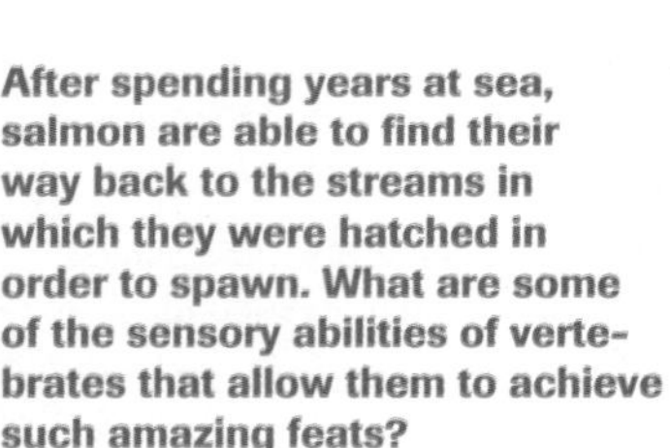

After spending years at sea, salmon are able to find their way back to the streams in which they were hatched in order to spawn. What are some of the sensory abilities of vertebrates that allow them to achieve such amazing feats?

Advance Planning

Chapter 33

- Set up a classroom aquarium containing several varieties of fishes for the Minilab on page 819.
- Have live frogs available for the Minilab on page 827.
- Order Ringer's solution, culture dishes, and *Rana pipiens* frog eggs for the Biolab. Gather binocular microscopes.

Chapter 34

- Purchase down and contour feathers for the Minilab on page 855.
- Purchase varieties of seeds for the Minilab on page 861.

Chapter 35

- Obtain various types of fur or hair for the Minilab on page 873 or have students gather samples.
- Purchase shortening for the Biolab.

Chapter 36

- Purchase tagboard for the Minilab on page 911 and the Biolab.

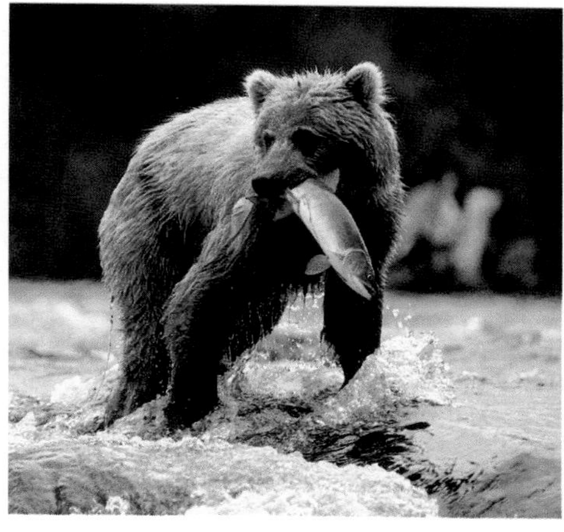

Mammals such as bears inhabit a wide range of environments. What characteristics of mammals allow them to survive in such diverse climates?

Physical obstacles such as waterfalls, rocks, and hungry bears present few problems for seagulls. What adaptations make bird flight possible?

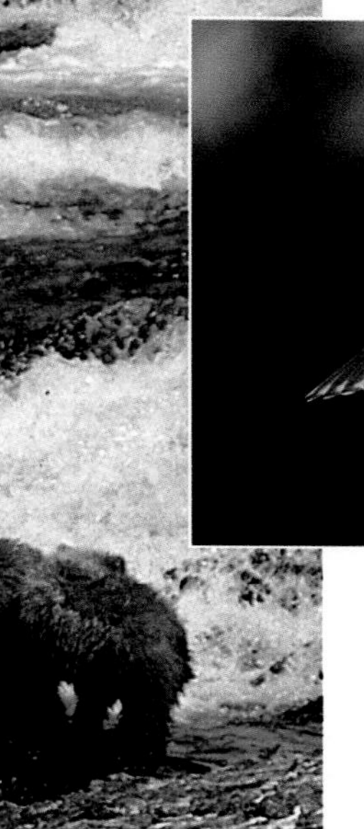

Unit Contents

UNIT 9

Introducing the Unit

Ask students to list the adaptations of a grizzly bear that enable it to survive in Alaska's cold environment.

Direct students' attention to the photo of the salmon and explain how this animal migrates to breed. Elicit names of other animals that migrate. Ask students to speculate on the adaptive value of migration.

Ask students to compare and contrast the ways of life of the three vertebrate animals shown here. Elicit how the adaptations of each animal are important to their survival in this harsh environment.

Motivational Activity

Demonstration: Have examples of live vertebrates from different classes available for observation. Examples may include: mouse, frog, lizard, and goldfish. Ask students to work in groups and make a table comparing how the animals are alike and how they are different. Vertebrate skeletons may be purchased from biological supply houses. Ask students to compare and contrast the skeletons. L1 COOP LEARN

Unit Project

Community Involvement: Ask students to select an endangered or threatened vertebrate in the local area. Have them make a proposal for action to preserve the habitat of the animal. Ask them to develop a petition for the faculty, student body, and community to sign.

Have students design and produce a colorful poster illustrating the animal and its problem. Have students organize a public meeting or attend an appropriate local government meeting to present their ideas about how the species can be preserved. L1 COOP LEARN

Chapter Organizer

SECTION	OBJECTIVES	ACTIVITIES/FEATURES
33.1 Fishes National Science Standards: UCP.1-5; C.3, C.5, C.6	1. **Relate** the structural adaptations of fishes to their environments. 2. **Compare and contrast** the adaptations of the different groups of fishes. 3. **Interpret** the phylogeny of fishes.	**Minilab:** How do fishes swim?, p. 819 **A Broader View:** Variety in the Sea, p. 821
33.2 Amphibians National Science Standards: UCP.1-5; A.1, A.2; C.3, C.5, C.6; F.3-6; G.1-3	4. **Relate** the demands of a terrestrial environment to the adaptations of amphibians. 5. **Relate** the evolution of the three-chambered heart to the amphibian lifestyle.	**Thinking Lab:** Is egg size related to survival of salamander larvae?, p. 826 **Minilab:** How does leg length affect how far a frog can jump?, p. 827 **Biolab:** Development of Frog Eggs, p. 828 **Chemistry Connection:** Killer Frogs, p. 830

ACTIVITY MATERIALS

BIOLAB	MINILABS	ALTERNATE LAB
page 828 Ringer's solution 4 culture dishes lightbulbs thermometer binocular microscope frog legs, *Rana pipiens* or *Xenopus laevis* flashlight	**page 819** fishes in an aquarium fish food **page 827** frog, live meterstick pencil with eraser aquarium or pail	**page 820** aquarium setup tropical fishes timer or stopwatch construction paper, black masking tape

Chapter 33 Fishes and Amphibians

TEACHER CLASSROOM RESOURCES

Reproducible Masters	Transparencies
Section Focus Master 73: Fish in the Sea L1 SAE **Reinforcement and Study Guide,** pp. 133-134 L1 **Biolab and Minilab Worksheets,** p. 131 L1 **Concept Mapping:** Phylum: Chordata, p. 33 L1 **Laboratory Manual:** Capillary Circulation in Fish, pp. 201-202 L2 **Tech Prep Applications:** Aquaculture, pp. 35-36 L2	**Basic Concepts Transparency #52:** Structure of a Bony Fish L1 SAE **Basic Concepts Transparency #53:** Phylogeny of Fishes and Amphibians L1 SAE
Section Focus Master 74: Living in Air and Water L1 SAE **Reinforcement and Study Guide,** pp. 135-136 L1 **Biolab and Minilab Worksheets,** pp. 132-134 L1 **Critical Thinking/Problem Solving:** Frog Development: The Early Stages, p. 33 L3 **Laboratory Manual:** Frog Dissection, pp. 195-200 L2 **Content Mastery,** pp. 137-140 L1	**Basic Concepts Transparency #53:** Phylogeny of Fishes and Amphibians L1 SAE **Reteaching Transparency #33:** Structure of a Frog L1 SAE

ASSESSMENT MATERIALS

Chapter Assessment, pp. 201-206
Performance Assessment in the Biology Classroom, p. 33
MindJogger Videoquiz
Alternate Assessment in the Science Classroom
Computer Test Bank

Spanish Resources SAE
English/Spanish Audiocassettes SAE
Cooperative Learning in the Science Classroom COOP LEARN
Lesson Plans
Great Developments in Biology: Discovering the Living World L1 SAE

KEY TO TEACHING STRATEGIES

- L1 Level 1 activities should be within the ability range of all students including those with learning difficulties.
- L2 Level 2 activities are within the ability range of average to above-average students.
- L3 Level 3 activities are designed for the ability range of above-average students.
- SAE SAE activities should be within the ability range of Students Acquiring English.
- COOP LEARN Cooperative Learning activities are designed for small group work.
- P These strategies represent student products that can be placed into a best-work portfolio.
- These strategies are useful in a block scheduling format.

GLENCOE TECHNOLOGY

The following multimedia resources are available from Glencoe.

Biology: The Dynamics of Life
CD-ROM SAE
Videodisc Program

The Infinite Voyage Series
To the Edge of the Earth
Exploring the Galapagos Islands

National Geographic Society Series
STV: Rain Forest
Forest Floor
Poison-Arrow Frog
Watery World
Upward and Onward
Garden Toads

Science and Technology Videodisc Series (STVS)
Animals
Naming Fish
Sea Skaters
Raising Super Fish

CHAPTER

33 Fishes and Amphibians

Chapter Overview

In this chapter, students study the structural adaptations shared by all fishes and compare the features that distinguish jawless fishes, cartilaginous fishes, and bony fishes. Next they examine the characteristics of the three groups of bony fishes: the lobe-finned fishes, the lungfishes, and the ray-finned fishes.

In the second part of the chapter, the move from an aquatic lifestyle to land life is presented through a discussion of the evolution of amphibians, the first land vertebrates. Students will relate the adaptations of amphibians to the demands of life on land. Finally, students study the main groups of amphibians as they compare frogs, toads, and salamanders.

Key Terms

cartilage
ectotherm
fin
lateral line system
scales
spawning
swim bladder
vocal cords

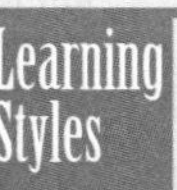

Learning Styles

Look for the following logo for strategies that emphasize different learning modalities. LS

Visual-Spatial	**Chalkboard Example, p. 815; Meeting Individual Needs, p. 817; Visual Learning, p. 817; Microscope Activity, p. 818; Portfolio, p. 818; Minilab, p. 819; The Inside Story, pp. 820, 831; Student Journal, p. 825**
Interpersonal	**Project, pp. 815, 830; Activity, p. 825**
Intrapersonal	**Student Journal, p. 830**
Linguistic	**Student Journal, pp. 816, 818; Project, p. 827**
Logical-Mathematical	**Alternate Lab, p. 820; Thinking Lab, p. 826; Portfolio, p. 826, Minilab, p. 827**

CHAPTER

33 Fishes and Amphibians

On a bright spring day in Idaho, a female chinook salmon brushes the clean river gravel into a depression nest and deposits about 6000 eggs. Soon a male salmon swims past, depositing his sperm. Eighteen months later, the young salmon begin their 4800-km migration downstream to the Pacific Ocean. As they migrate, the salmon store memories of the chemical odors in the stream water so that they may return two or three years later to the place where they hatched on that bright spring day. There they will begin the cycle again.

Like salmon, amphibians also undergo a series of structural changes during development that will enable them to survive in a different environment as adults. When leopard frog eggs hatch, they are wiggling tadpoles with bulging heads and squirming tails. A few months later, they become the four-legged, jumping, croaking, swamp-roving animals that have inspired cartoonists and writers of folktales throughout history.

Concept Check

You may wish to review the following concepts before studying this chapter.

- Chapter 1: adaptation, evolution
- Chapter 28: development, dorsal, ventral, vertebrate

Chapter Preview

Many species of fishes lay great numbers of eggs. This ocean sunfish is said to lay more than 300 million eggs. Why do fishes lay so many eggs?

Tadpoles go through many changes on the way to becoming frogs. What happens to the tadpole's tail as it becomes an air-breathing animal? The development of frogs from tadpoles provides clues about the origin of land vertebrates.

CHAPTER 33

Introducing the Chapter

Have students look at the ocean sunfish shown in the photograph. Ask them to list the adaptations they observe that this fish has to its way of life. *Fins, streamlined body, gills for obtaining oxygen dissolved in water.* Elicit what characteristics frogs share with fishes. *Both have adaptations for an aquatic lifestyle.* Next ask how frogs differ from fishes. *Frogs have adaptations, such as legs, for life on land as well as in water. Fishes have adaptations for life only in water.*

Theme Development

Evolution is a major theme of this chapter. Emphasis is placed on how animals evolved adaptations to life on land. The theme of *unity within diversity* is apparent through the discussion of how varied the features of animals within a classification group can be even though all members of the group share many characteristics.

Concept Check

Students will build upon their knowledge of vertebrates and expand their knowledge of development and adaptations.

Assessment Planner

Choose assessment strategies from the following pages to evaluate the progress of your students.

Assess, pp. 822, 832
Alternate Lab, pp. 820-821
Minilabs, pp. 819, 827
Portfolio, pp. 818, 826
Thinking Lab, p. 826
Biolab, pp. 828-829
Chapter Review, pp. 833-835

813

SECTION

33.1

Prepare

Key Concepts

Students learn about the characteristics all fishes have in common, while developing an understanding of the characteristics that distinguish the three classes of fishes: jawless fishes, cartilaginous fishes, and bony fishes.

Block Scheduling

Look for this symbol for strategies that are useful in a block scheduling format. For more information on block scheduling, refer to Section 33.1 in the **Lesson Plans** booklet.

Materials

- Obtain a large goldfish, a bowl, and fish food for the Focus Demonstration.
- Purchase mounts of ganoid, cycloid, ctenoid, and placoid scales from a biological supply house or a hatchery and gather stereomicroscopes for the Microscope Activity.
- Access to an aquarium with a variety of tropical fishes is necessary for the Minilab. A classroom aquarium may be used or students may make observations at a local pet shop.
- Purchase a fresh, uncleaned fish from a fish market and obtain a scalpel or razor for use during the *Inside Story*.

SECTION

33.1 Fishes

Section Preview

Objectives

Relate the structural adaptations of fishes to their environments.

Compare and contrast the adaptations of the different groups of fishes.

Interpret the phylogeny of fishes.

Key Terms

cartilage
fin
lateral line system
scale
swim bladder
spawning

Have you ever visited an aquarium to see the amazing diversity of fishes? As you pass tank after tank, you can see fishes of all shapes, sizes, and colors. What's interesting is that even though fishes share a common environment, they have evolved a variety of different adaptations. But although fishes may show considerable variety in structure and behavior, they all share common characteristics.

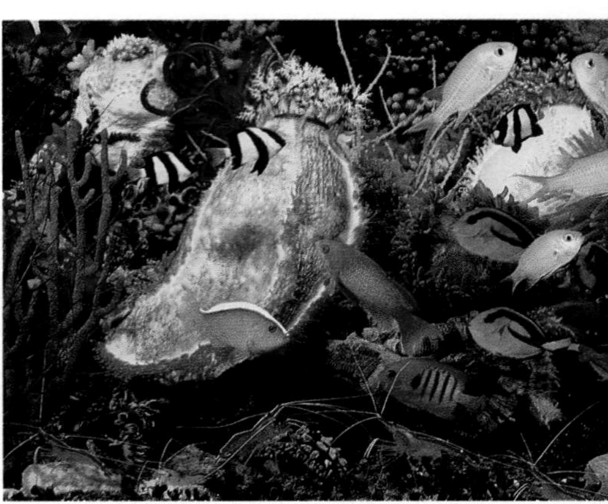

What Is a Fish?

Fishes, like all vertebrates, are classified in the phylum Chordata. This phylum includes three subphyla: Urochordata, the sea squirts; Cephalochordata, the lancelets; and Vertebrata, the vertebrates. In addition to fishes, the subphylum Vertebrata includes amphibians, reptiles, birds, and mammals. You remember from Chapter 32 that all chordates have three traits in common—a notochord, gill slits, and a dorsal nerve cord. In vertebrates, the notochord of the embryo is replaced by a backbone. All vertebrates are bilaterally symmetrical, coelomate animals that have endoskeletons, closed circulatory systems, nervous systems with complex brains and sense organs, and efficient respiratory systems.

Three classes of fishes

Fishes comprise three classes of the subphylum Vertebrata: Class Agnatha, the lampreys and hagfishes; Class Chondrichthyes, the sharks and rays; and Class Osteichthyes, the bony fishes. Far more variety can be found among the classes of fishes than in any other vertebrate group.

The diversity of fishes

Fishes inhabit nearly every type of aquatic environment on Earth. They are adapted to living in shallow, warm water and deeper cold and sunless water. They are found in fresh and salt water, and some fishes can survive heavily polluted water.

Fishes range in size from the tiny dwarf goby that is less than 1 cm long, to the huge whale shark that can reach a length of 15 m—the length of three school buses.

1 Focus

Bellringer

Before presenting the lesson, display **Section Focus Master 73** on the overhead projector and have students answer the accompanying questions. L1 SAE

Program Resources

Section Focus Master 73 L1 SAE
Reinforcement and Study Guide, pp. 133-134 L1
Performance Assessment in the Biology Classroom, p. 33 L1 P
Biolab and Minilab Worksheets, p. 131 L1
Concept Mapping, p. 33 L1
Laboratory Manual, pp. 201-202 L2
Basic Concepts Transparencies 52, 53 and **Masters** L1 SAE
Tech Prep Applications, pp. 35-36 L2

More than 30 000 species of fishes exist. In fact, there are more fish species than all other kinds of vertebrates added together!

Agnathans Are Jawless Fishes

Lampreys and hagfishes, shown in *Figure 33.1,* belong to the class Agnatha. Though they do not have jaws, they are voracious feeders. Hagfishes, for example, have a slit-like, toothed mouth and feed on dead or dying fish by drilling a hole and sucking the blood and insides from the animal. Parasitic lampreys attack other fishes and attach themselves by their suckerlike mouths. They use their sharp teeth to scrape away the flesh and then suck out the prey's blood. The skeletons of agnathans, as well as of sharks and their relatives, are made of a tough, flexible material called **cartilage.**

Agnathans breathe using gills

Agnathans, like all fishes, have gills made up of feathery gill filaments that contain tiny blood vessels. Gills are an important adaptation for fishes and other vertebrates that live in water. As a fish takes water in through its mouth, water passes over the gills and then out through slits at the side of the fish. Oxygen and carbon dioxide are exchanged through the capillaries in the gill filaments.

Agnathans reproduce sexually

Like most fishes, lampreys and hagfishes have separate sexes. Fertilization is external in most fishes, with eggs and sperm deposited in protected areas such as on floating aquatic plants or in shallow nests of gravel on stream bottoms. Although most fishes produce large numbers of eggs at one time, hagfishes produce small numbers of large eggs.

Two-chambered hearts

Agnathans have two-chambered hearts, *Figure 33.2,* like all fishes. One chamber receives deoxygenated blood from the body tissues, whereas the second chamber pumps blood directly to the capillaries of the gills, where oxygen is picked up and carbon dioxide released. Oxygenated

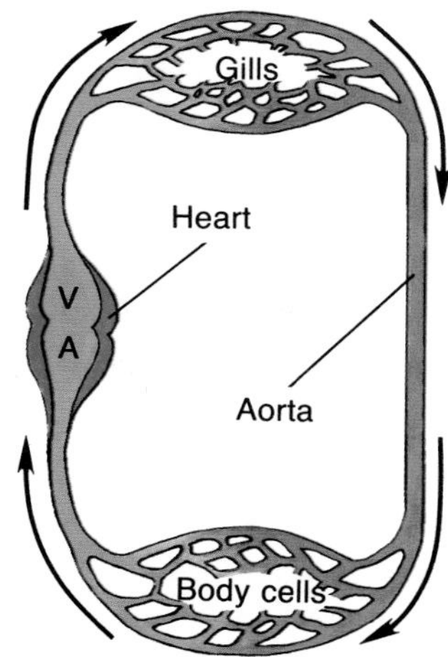

Figure 33.2

In a fish's heart, deoxygenated blood flows from the first chamber to the second chamber, then on to the gills where it picks up oxygen. Blood in a fish flows in a one-way circuit throughout the body.

Figure 33.1

Hagfishes (left) and their relatives, the lampreys (right), have long, tubular bodies without paired fins, and no scales. When touched, a hagfish's skin gives off a tremendous amount of mucus, thus allowing the fish to slither away without becoming a meal.

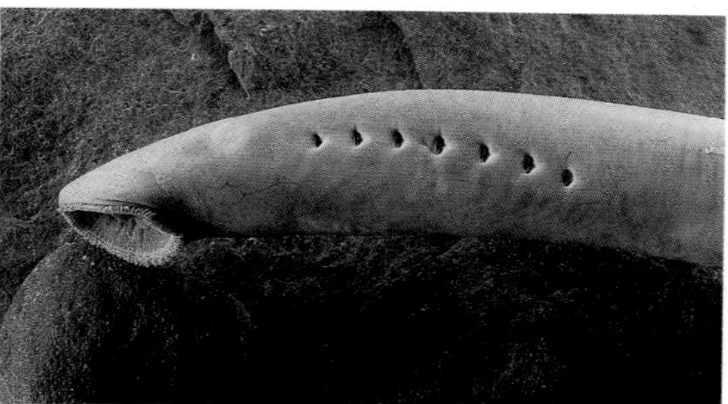

SECTION 33.1

Demonstration

Place a large goldfish in a bowl where it is observable by all students. Ask students to list the structural and behavioral adaptations of the goldfish. Feed the fish and have students observe its behavior.

2 Teach

Chalkboard Example

IS **Visual-Spatial** Draw the two-chambered heart of a fish on the chalkboard. Beside this drawing, sketch the four-chambered heart of a mammal. Briefly trace the path of blood through each heart and explain that the four-chambered heart is more efficient for an animal that must move quickly and maintain a constant body temperature. L1

Chemistry Connection

Nervous Tissue Regeneration

Unlike humans, lampreys can regenerate damaged or severed spinal nerves. To determine whether these regenerated nerves actually carry signals to coordinate swimming movements, scientists removed the regenerated spinal cord tissue from a larval lamprey and stimulated it with a neurotransmitter.

The research showed that the nerves had not only regrown, but had also established connections with other nerves. In adult lampreys, abnormalities were detected in the information sent along the regenerated nerve tissue. While the abnormalities were not significant enough to disrupt swimming in lampreys, such abnormalities could disrupt the fine motor coordination required for walking in humans.

PROJECT

The Aquarium

Ask students to prepare a written plan that identifies all materials needed to set up and maintain an aquarium, along with an explanation of why each material is needed. Allow students to set up and maintain their aquariums according to their plans. L1

COOP LEARN IS

Using an Analogy

Explain to students that fishes use their lateral line systems much like sonar is used to detect the presence and location of submerged objects. The lateral line system of fishes responds to changes in water pressure and currents as it functions to detect the sizes, shapes, and locations of objects in the water. Fishes that swim in schools use their lateral line systems to turn in unison. By acting as one body that appears to be one large fish, the individual fishes can avoid predators.

Misconception

Students often think of sharks as being blood-craving predators of people. Explain that of the 370 species of sharks, only 25 have been known to attack humans. When sharks attack people, this occurs largely because the shark's lateral line system detects movement in the water. Such movement usually signals to a shark that food in the form of a fish is near.

GLENCOE TECHNOLOGY

Videodisc

Biology: The Dynamics of Life
Disc 2, Side 1
Fish Schooling (Ch. 2)

STVS: Animals
Disc 5, Side 1
Sea Skaters (Ch. 17)

Disc 5, Side 2
Studying Sharks (Ch. 3)

Figure 33.3
The paired fins of a fish include the pectoral fins and the pelvic fins. Fins found on the dorsal and ventral surfaces include the dorsal fin and anal fin. A fish may have all of these fins or just a few. Most fishes use their body fins like the rudders and stabilizers of boats.

blood is carried from the gills to body tissues. Blood flow through the body of a fish is relatively slow because most of the heart's pumping action is used to push blood through the gills.

Sharks and Rays Are Cartilaginous Fishes

Sharks, skates, and rays, like agnathans, possess skeletons composed entirely of cartilage. Sharks, skates, and rays belong to the class Chondrichthyes. Because living sharks, skates, and rays are classified in the same genera as species that swam the seas more than 100 000 years ago, they are considered living fossils.

Sharks and rays have paired fins

Fishes in the class Chondrichthyes have paired fins. **Fins** are fan-shaped membranes, supported by stiff spines called rays, that are used for balance, swimming, and steering. Fins are attached to and supported by the endoskeleton and are important in locomotion. The paired fins of fishes, *Figure 33.3*, foreshadowed the development of limbs for movement on land and wings for flying.

Jaws evolved in fishes

Perhaps one of the most important events in vertebrate evolution was the evolution of jaws in primitive fishes. The advantage of a jaw is that it enables an animal to grasp and crush its prey with great force. Sharks are able to eat large chunks of food. This, among other factors, explains why some early fishes were able to reach such great size. *Figure 33.4* shows the evolution of jaws in fishes.

When you think of a shark, do you imagine gaping jaws and rows of razor-sharp teeth? Sharks have six to 20 rows of teeth that are continually replaced. The teeth point backwards, preventing prey from escaping once caught. Sharks are among the most streamlined of all fishes and are well adapted for life as predators.

Sharks and rays have developed sensory systems

Success as a predator is also the result of the shark's highly developed sensory system. Like salmon, sharks have an extremely sensitive sense of smell and can detect small amounts

Figure 33.4
You can see how jaws evolved from the cartilaginous gill arches of early jawless fishes in this series of illustrations. Teeth evolved from skin.

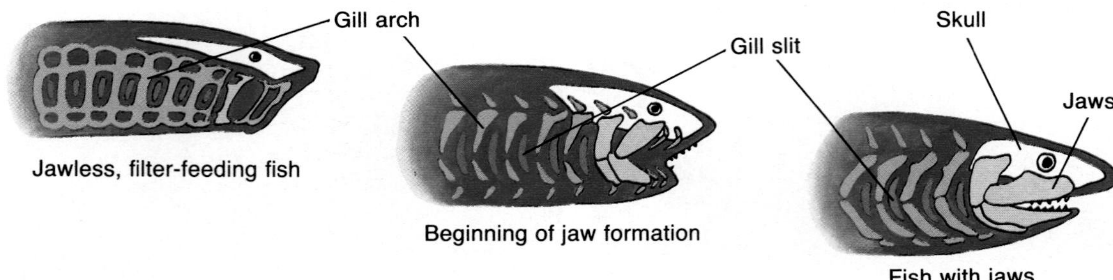

816 Fishes and Amphibians

STUDENT JOURNAL

Shark Photography Have students read the article, "On Photographing Sharks" in *Sea Frontiers,* Nov./Dec. 1992, by Marty Snyderman. Ask students to write their impressions of the article and to write their thoughts about the kind of personality a shark photographer would need. L2 LS

interNET CONNECTION

Follow the link for this chapter on the Glencoe Homepage at **www.glencoe.com/sec/science** to find out more about fishes and amphibians.

Figure 33.5

Fishes can be classified by the type of scales present in the skin.

Diamond-shaped scales are common to primitive bony fishes, such as gars, that evolved early in geologic history about 200 million years ago.

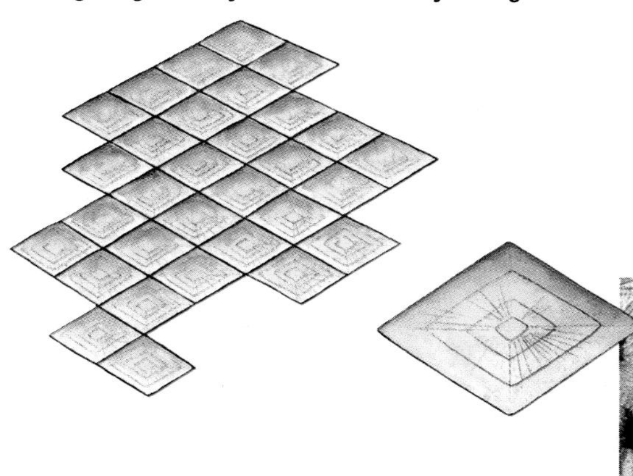

Tooth-shaped scales are characteristic of the shark group.

Bony fishes that evolved later, such as chinook salmon, have either cone-shaped or round scales.

of chemicals in the water. Sharks can follow a trail of blood through the water for several kilometers. This ability helps them locate their prey.

Another adaptation that allows sharks and other fishes to detect their prey is the lateral line system. The **lateral line system** is a line of fluid-filled canals running along the sides of a fish that detects movement and vibrations in the water.

Like sharks, most rays are predators and feed on or near the ocean floor. Rays have flat bodies and broad pectoral fins on their sides. By slowly flapping their fins up and down, rays can glide, searching for mollusks and crustaceans, along the ocean floor. Some species of rays have sharp spines with poison glands on their long tails for defense, as anyone who has stepped on one can attest. Other species have organs that generate electricity to kill both prey and predators.

Cartilaginous fishes have scales

Most fishes have skin covered by overlapping rows of scales. **Scales** are thin bony plates formed from the skin. Scales, *Figure 33.5*, can be toothlike, diamond-shaped, cone-shaped, or round. Shark scales are similar to teeth found in other vertebrates. The age of some species of fishes can be estimated by counting annual growth rings in their scales.

Sharks and rays have internal fertilization

Some fishes have internal fertilization. For example, some female sharks and rays produce as few as 20 eggs and keep them inside their bodies until they have hatched and developed to about 40 cm in length. These young, when released, behave like miniature adults, and many survive.

Bioethics

Endangered Sharks

Pandas, sea otters, and dolphins have been the traditional symbols of the movement to save endangered animals. The shark may soon follow. Some estimates indicate that more than 4 billion sharks exist. However, hunters and commercial fisheries kill more that 200 million sharks each year, which could result in sharks becoming endangered by the year 2000.

Shark populations are declining and no protection currently exists for these animals. This is especially problematic because sharks, unlike other fishes, do not produce large numbers of young, making their natural replacement a slow process. The problem is exacerbated because people who fear sharks do not often view decreasing numbers of this animal as a problem.

Visual Learning

Figure 33.5 Direct students' attention to the different types of scales. Ask students how they could determine that a scale came from a shark simply by observing the shape of the scale. *The scales of a shark have a toothlike shape. No other fish scales have this shape.* LS

Meeting Individual Needs

Learning Disabled Use visualization strategies for students who have learning disabilities. Prepare a set of large cards with visual depictions of concepts in the chapter. Ask students to look carefully at the cards for about 10 seconds. Next, they should close their eyes and imagine the cards on a bulletin board in their minds. After placing about 10 concepts on the bulletin boards in their minds, they should close their eyes and focus on their boards for about 10 seconds. The next day, you can put a list of concepts on the chalkboard and ask students to recall their imaginary bulletin boards and explain what they remember. They may not remember all the cards; you may need to show them again. L1 LS

SECTION 33.1

Microscope Activity

Visual-Spatial Have students make and examine mounts of scale types using a stereomicroscope. Briefly discuss how the age of a fish can be determined by counting the rings on its scales. Have students determine the age of the fish from which the scales they are observing came. Ask students to draw the scales they observe and include a summary in their portfolios that identifies the types of scales observed and the ages of the fishes from which the scales came. L2

Science, Technology, and Society

Gene-altered Fish

Scientists have altered the genetic structure of goldfish, carp, and channel catfish to make them grow larger and faster. The growth habits of each of these fishes was altered by inserting foreign growth hormone genes into the eggs of the fishes. The foreign gene is incorporated into the genome of the fish, causing the fish and its offspring to grow to a larger size than normal. It is hoped that the new fish strains can help feed people in nations with limited food supplies. Scientists soon hope to genetically engineer fishes that will withstand extreme temperatures, disease, and toxins that pollute water.

Bony Fishes Have Skeletons Made of Bone

The majority of the world's fishes belong to the class Osteichthyes, or bony fishes. Bony fishes are a successful and widely distributed class, differing greatly in habitat, size, feeding behavior, and shape, as *Figure 33.6* shows.

All bony fishes have a bony skeleton, gills, paired fins, and highly developed sense organs. As the name indicates, these fishes have skeletons made of bone, rather than the cartilage skeletons of the other classes of fishes. Bone is the hard, mineralized, living tissue that makes up the endoskeleton of most vertebrates. The appearance of bone was important for the evolution of fishes and vertebrates in general because it allowed fishes to adapt to a variety of aquatic environments, and finally even to land.

Vertebrae provide flexibility

The evolution of a backbone composed of separate, hard segments called vertebrae was significant in providing the major support structure of the vertebrate skeleton. Separate vertebrae provide great flexibility. This is especially important for fish locomotion, which involves continuous flexing of the backbone.

You can see how modern bony fishes propel themselves in water in *Figure 33.7.* Some fishes are effective predators, in part because of the fast speeds they can attain as a result of having a flexible skeleton.

Figure 33.6

Bony fishes vary in appearance, behavior, and way of life.

Predatory bony fishes, such as this pike, have sleek bodies with powerful muscles and tail fins for fast swimming.

Anglerfishes use a wormlike filament attached to the jaw that encourages their prey to come closer.

Eels are long and snakelike and can wriggle through mud and crevices in search of food.

Sea horses move slowly through the underwater forests of seaweed where they live. They are unusual in that the males brood their young in stomach pouches.

STUDENT JOURNAL

The Geologic History of Fishes Ask students to look up fishes in geologic history. Ask them to choose a fish that especially interests them and to write about a day in the life of their fossil fish. They should be careful to describe other aquatic organisms in the environment at the time and incorporate science facts into their creative writing. L2

PORTFOLIO

Exploring Traits of Bony Fishes Have students each cut a picture of a bony fish from a magazine. Ask students to use *The Inside Story* as a model to label and identify the functions of the structures of their fishes. L1

Development of a swim bladder

Another key to the evolutionary success of bony fishes was the evolution of the swim bladder. A **swim bladder** is a thin-walled, internal sac found just below the backbone in bony fishes; it can be filled with mostly oxygen or nitrogen that diffuse out of a fish's blood. Fish with a swim bladder control their depth by regulating the amount of gas in the bladder. The gas works like the gas in a blimp that adjusts the height of the blimp above the ground.

Fishes that live in oxygen-poor water or in ponds or rivers that dry up in the hot season often have other ways to get oxygen. The African lungfish, for example, has a structure that allows it to obtain oxygen by gulping air. This structure is a modified swim bladder. The modified swim bladder is connected to the fish's mouth by a tube.

Reproduction in bony fishes

In most bony fishes, sexes are separate and fertilization is external. Breeding in fishes and some other animals is called **spawning.** During spawning, some female bony fishes, such as cod, produce as many as 9 million eggs, of which only a small percentage will survive.

Some bony fish species are live-bearers; that is, offspring are born fully developed. In these species, such as guppies, mollies, and swordtails, fertilization is internal and young fishes develop within the mother's body. After hatching, other fishes, such as the mouth-brooding cichlids, stay with their young and scoop them into their mouths if they are threatened.

MiniLab

How do fishes swim?

As fishes have been swimming for about 500 million years, it is not surprising that they have mastered this method of movement in a watery world. Most fishes move in one of three ways. Some swim in an S-shaped pattern, alternately tightening muscles on either side of the body. These fishes do not have well-developed fins. Another swimming pattern involves keeping the front portion of the body rigid and moving the tail and back portion of the body back and forth. The third swimming pattern involves keeping the entire body rigid and moving only the tail back and forth.

Procedure

1. Observe several different fishes from the top of an aquarium for several minutes each to determine which swimming patterns they follow.
2. Add fish food to the aquarium and observe any changes in swimming patterns.

Analysis

1. Compare the swimming patterns of the fishes you observed.
2. What might be the survival value for each type of swimming pattern?
3. How might the swimming patterns of fishes reflect their feeding habits?

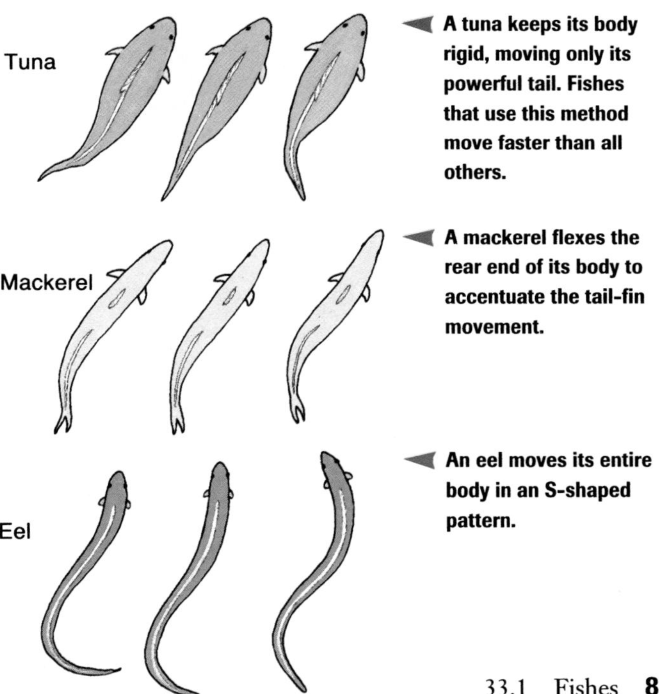

A tuna keeps its body rigid, moving only its powerful tail. Fishes that use this method move faster than all others.

A mackerel flexes the rear end of its body to accentuate the tail-fin movement.

An eel moves its entire body in an S-shaped pattern.

Figure 33.7

Most bony fishes swim in one of three possible ways.

MiniLab

Purpose

Visual-Spatial Students will determine the adaptive advantage of swimming patterns of fishes.

Process Skills

observe and infer

Teaching Strategies

- If you do not have an aquarium, students may observe fishes at a local pet shop.

Expected Results

The s-shaped pattern of movement is common in bottom-feeding fishes; moving the back and the tail is common in tetras, black ghosts, and painted glass fishes; and tail-only movement is common in black moors, goldfishes, and guppies.

Analysis

1. Patterns of movement observed include an s-shaped pattern, moving the tail and back, or moving only the tail.
2. The s-shaped pattern enables fishes to move among rocks and weedy growth. A fish that moves its tail and back has more maneuverability than a fish that moves only its tail. A fish that moves only its tail moves faster than the others.
3. S-shaped swimmers feed on the bottom. Tail- and body-movers can maneuver around objects. Tail-only movers may be predators.

Assessment

Performance: Show students a fish they have not seen before. Ask them to describe its movement. L1

CULTURAL DIVERSITY

Samuel Milton Nabrit

African American zoologist, Samuel M. Nabrit (1905-) conducted many experiments on tail regeneration in fishes. Nabrit earned a Ph.D. in Zoology in 1932 and became a researcher at Woods Hole Marine Biology Laboratory. Papers describing his work may be found in the *Journal of Experimental Zoology, Anatomical Record,* and *Biological Bulletin.* Nabrit is perhaps best known for his work toward promoting African Americans in science. He helped to organize the National Institute of Science in 1943, an organization designed to explore the teaching and research problems of African American scientists.

THE INSIDE STORY

A Bony Fish

Purpose

 **Visual-Spatial** Students will study features of bony fishes and examine the adaptive advantages of these features.

Teaching Strategies

- On the chalkboard, make a table that compares the features of bony fishes to those of jawless and cartilaginous fishes.

Visual Learning

- Ask students to study the features shown. Have students identify the external features discussed on a live goldfish.
- Purchase a fresh fish that has not been cleaned. Carefully, cut open the fish to show students its swim bladder. The swim bladder is very fragile. Care is necessary as you cut into the fish with a razor blade or scalpel.

GLENCOE TECHNOLOGY

 Videodisc

STVS: Animals
Disc 5, Side 2
Raising Super Fish (Ch. 7)

THE INSIDE STORY

A Bony Fish

The bony fishes, class Osteichthyes, include some of the world's most familiar fishes, such as the bluegill, trout, minnow, bass, swordfish, and tuna. Though diverse in general appearance and behavior, bony fishes share some common adaptations with other fish classes.

Rainbow trout

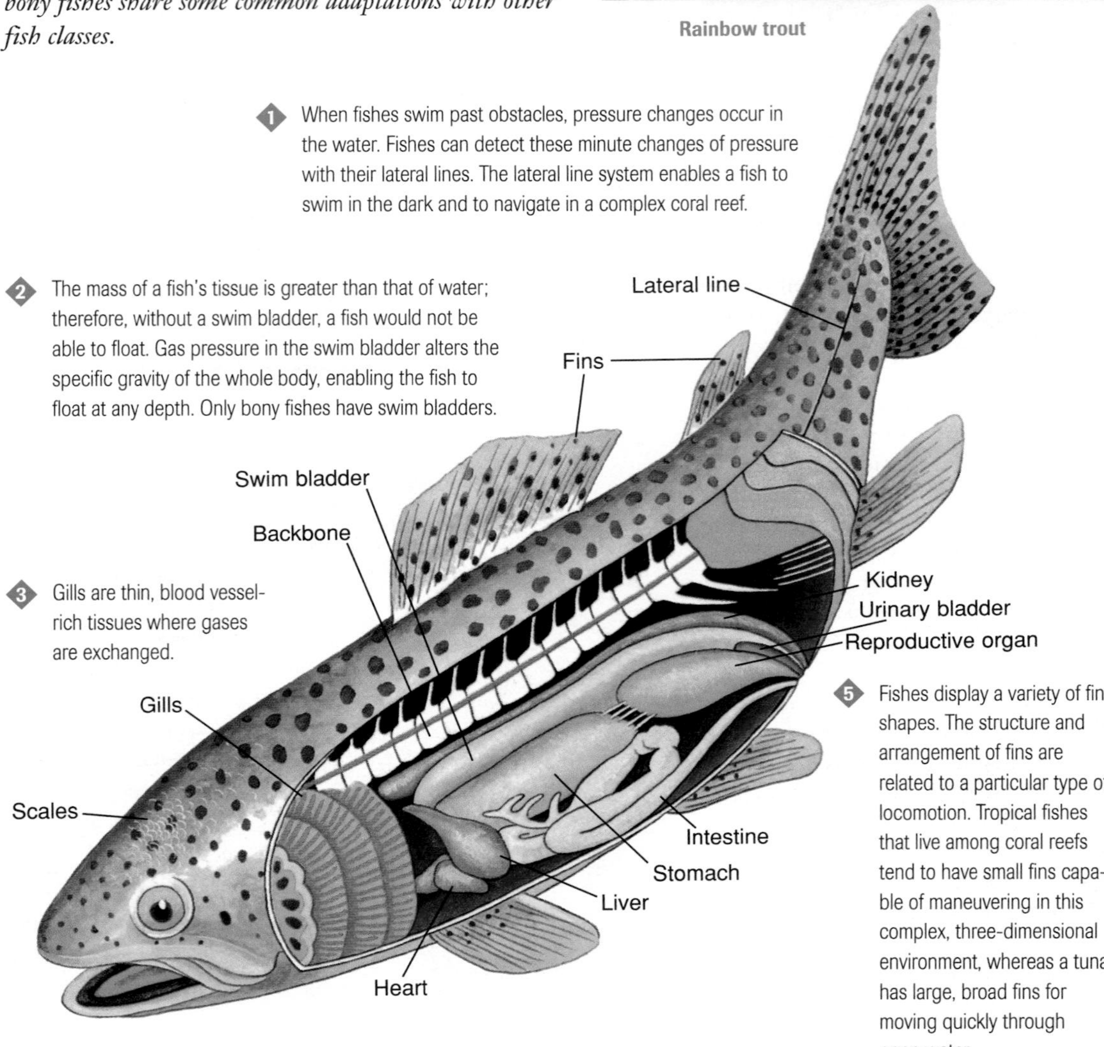

1. When fishes swim past obstacles, pressure changes occur in the water. Fishes can detect these minute changes of pressure with their lateral lines. The lateral line system enables a fish to swim in the dark and to navigate in a complex coral reef.

2. The mass of a fish's tissue is greater than that of water; therefore, without a swim bladder, a fish would not be able to float. Gas pressure in the swim bladder alters the specific gravity of the whole body, enabling the fish to float at any depth. Only bony fishes have swim bladders.

3. Gills are thin, blood vessel-rich tissues where gases are exchanged.

4. Scales are covered with slippery mucus, allowing a fish to move through water with minimal friction. Fishes in the class Agnatha do not have scales.

5. Fishes display a variety of fin shapes. The structure and arrangement of fins are related to a particular type of locomotion. Tropical fishes that live among coral reefs tend to have small fins capable of maneuvering in this complex, three-dimensional environment, whereas a tuna has large, broad fins for moving quickly through open water.

Alternate Lab: Fish Behavior

Purpose 

Logical-Mathematical Students will gain an understanding of how the behavior of fishes is adaptive.

Materials

aquarium setup, tropical fishes (zebra, catfish, gourami), timer or stopwatch, black construction paper, masking tape

Procedure

Give students the following directions.

1. Prepare a data table with columns indicating time spent over white gravel, over black gravel, at the top, bottom, or middle of the tank, and in light areas and dark areas.
2. Place a fish into an aquarium that has white gravel on one side and black gravel on the other.
3. Develop a hypothesis about in which part of the tank the fish will spend most of its time.
4. For two minutes, time how long the fish spends over each color of the gravel. Record your data.
5. Cover the sides of the half of the aquarium that contains black gravel

Three subclasses of bony fishes

Scientists recognize three subclasses of bony fishes. *Figure 33.8* shows one subclass, the lungfishes. Another subclass is the lobe-finned fishes, represented by only one living species. The third subclass, the ray-finned fishes such as catfish, perch, salmon, and cod, are probably the most familiar to you because many are fishes eaten by humans. Ray-finned fishes have fins supported by rays.

Lobe-finned fishes, or coelacanths, are an ancient group, appearing in the fossil record about 395 million years ago. They are characterized by lobelike, fleshy fins, and live at great depths, where they are difficult to find. The limblike skeletal structure of the fleshy fins of coelacanths is thought to be an ancestral condition of all tetrapods (animals with four limbs). The earliest tetrapod discovered also had gills and therefore was still aquatic.

Figure 33.8

Lungfishes represent an ancient subclass, having arisen close to 400 million years ago. Lungfishes have both gills and lungs.

A BROADER VIEW

Variety in the Sea

Four hundred million years ago, a type of jawless fish lived in the bottom of the ocean, rooting out food from the mud. One descendant of that ancient fish is the lamprey. It is also jawless, but it doesn't grub about in the mud for food. Lampreys are parasites of other fishes. Parasitism is probably the adaptation that enabled it to survive.

The alien that invaded America The sea lamprey is an unpleasant creature in both appearance and behavior. Lampreys have no jaws, but they have up to 125 sharp teeth set in a circle. Using these teeth, they rasp through the scales of their host fish. Then, with the aid of a powerful sucker, they attach themselves permanently. The lamprey sucks the blood and, eventually, the life out of the fish.

Aquatic dinosaur Another link to the past is a fish that most people thought was long extinct. No one believed the story of the strange fish brought in to East London, South Africa, in December 1938. The description of the paired, lobed fins and large, round scales sounded like an organism scientists had previously believed to be extinct. The coelacanth had died out during the Cretaceous period. Or had it? Ichthyologists who examined the fish found that this was indeed a coelacanth. No other fishes they had seen, except as fossils, had the lobed fins and the strange tuft at the end of its tail. Lobe-finned fishes were believed to represent a group of fishes that eventually evolved into animals that could live on land.

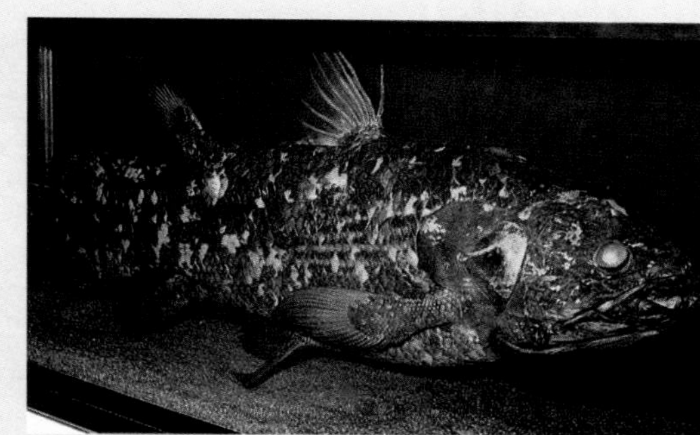

A fish that fishes But what about modern fishes? How have they survived and adapted to life in the abyss? The angler fish has perfected a unique way to catch a meal. It lives deep in the sea where there is total darkness, and uses a "rod and bait" to attract and catch other fish. The rod is actually a bony growth that hangs over the fish's mouth. The fleshy bait looks like a morsel of food dangling from the rod. Making the lure even more effective is its luminescence, which can be switched on and off.

Thinking Critically

Scientists were astonished when they discovered living coelacanths. Hypothesize how a species of fish could survive when many of the land animals that may have evolved from it are long gone.

A BROADER VIEW

Variety in the Sea

Purpose

To provide students with information about the unusual adaptations some fishes have for survival.

Teaching Strategies

- After they have read about the adaptations of each fish, ask students to speculate how each adaptation may have evolved.
- Explain that the male angler fish attaches itself to a female shortly after it hatches. Eventually, the male parasitizes the female. The union becomes permanent as the male's mouthparts fuse to the female's body. Elicit how this strategy aids in reproduction. *The fishes do not have to search for mates.*

Thinking Critically

Conditions in the sea such as temperature, salinity, and food sources are slow to change and are relatively constant, allowing marine species long periods to adjust to environmental changes.

with black construction paper.

6. Develop a hypothesis about where the fish will spend its time, in the light or the dark. Then, observe the fish for two minutes and time in which part of the tank the fish spends its time. Record your data.
7. Compare your data with those of your classmates.

Analysis

Ask students to answer the following questions.

1. Which fishes preferred dark gravel and the dark half of the tank? *Catfish*
2. Do your data for each habitat type support your hypotheses? Explain. *Some students will find that their data supported their hypotheses.*

Assessment

Performance: Give students a fish they have not seen and ask them to determine the fish's preferred habitat. Ask them to explain how this preference may have adaptive value. L1

Different Viewpoints in Biology

Ancestors to Land Animals

Scientists have suggested that lungfishes are the ancestors of ancient amphibians. However, a group of scientists who compared the blood proteins of coelacanths and frogs concluded that the coelacanth was the nearest fish relative of a frog. This research was criticized because the protein comparison involved only two species.

In later studies, however, DNA sequences of lungfishes were compared with those of frogs, and the results showed that lungfishes and frogs had three times as many similar points of comparison as coelacanths and frogs.

Text Question

To which group of fishes might amphibians be closely related? *Coelacanths are shown to be most closely related in Figure 33.9.*

3 Assess

Check for Understanding

Ask students to summarize in writing the characteristics of jawless fishes, cartilaginous fishes, and bony fishes. L1

Reteach

Have students work in groups to make a table that compares the characteristics of the three fish classes. Make sure they include features such as unique body structures, scale types, methods of getting food, and skeletal features. Have other groups construct a table that compares the three subclasses of bony fishes. Review the tables as a class. L1 COOP LEARN

Figure 33.9

The radiation of classes of fishes and orders of amphibians on the Geologic Time Scale shows their relationships.

Origins of Fishes

Scientists have identified fossils of fishes that existed during the early Devonian period, 400 million years ago. Although the fossil record for fishes is incomplete, most scientists agree that the relationships shown in *Figure 33.9* represent the best fit for the available evidence. To which group of fishes might amphibians be closely related?

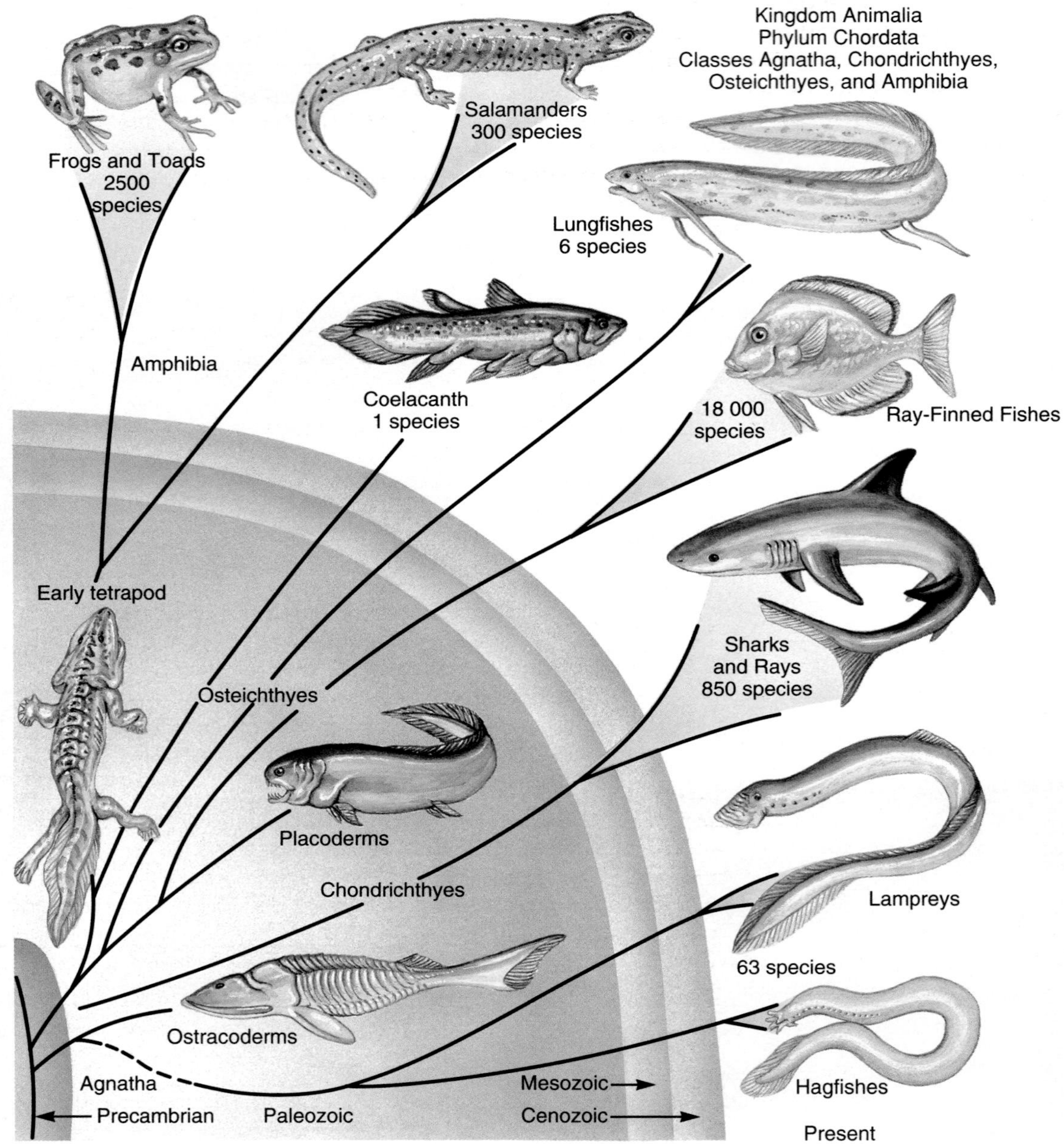

Meeting Individual Needs

Students Acquiring English To help students who speak English as a second language, have them work with native English speakers and practice their language by explaining the terms and concepts in this section to their partners. SAE

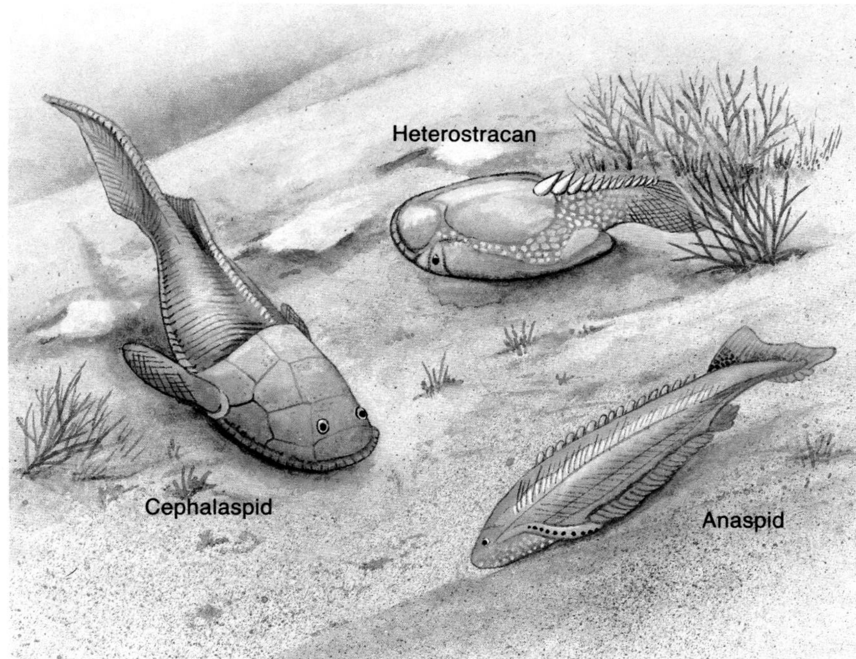

Figure 33.10
Ostracoderms, the earliest vertebrate fossils found, were characterized by bony, external plates covering the body and a jawless mouth. Lacking jaws, ostracoderms obtained food by sucking up bottom sediments and sorting out the nutrients.

Weighed down by heavy, bony external armor, ancestral fishes, shown in *Figure 33.10,* were fearsome-looking animals that swam sluggishly over the murky seafloor. The development of bone in these animals was an important evolutionary step because bone provides a place for muscle attachment, which improves locomotion. In ancestral fishes, bone that formed into plates provided protection as well.

Scientists hypothesize that the jawless ostracoderms were the common ancestors of all fishes, including the present-day jawless fishes. Modern cartilaginous and bony fishes evolved later from an ancient class of cartilaginous, jawed fishes known as placoderms.

Section Review

Understanding Concepts

1. List three characteristics of fishes.
2. Compare how jawless fishes and cartilaginous fishes feed.
3. Why was the evolution of a swim bladder important to fishes?

Thinking Critically

4. Why was the development of jaws an important step in the evolution of fishes?

Skill Review

5. **Making and Using Tables** Construct a table to compare the characteristics of the jawless, cartilaginous, and bony fishes. For more help, refer to Organizing Information in the *Skill Handbook.*

Extension

Have interested students interview a fly-fishing expert to find out about the sport of fly-fishing. Encourage students to videotape the interview or take photographs of the techniques used in the sport. Have students report their findings to the class and include a demonstration of fly tying. L1

Assessment

Knowledge: Ask students to list adaptations fishes might have for each of the following niches: to live on the muddy bottom of a pond; to feed on other fishes; to feed on clams and other mollusks; to live in crevices in a coral reef; to live in deep water where there is no light. L1

4 Close

Discussion

Ask students to speculate about why there are more fishes than any other type of vertebrate. *Responses may state that all life began in the oceans and this environment has undergone fewer changes than have terrestrial habitats, thus reducing the risk of extinction from the inability to adapt to new conditions.* Have students predict how the extinction of fishes would affect world food supplies. *Because people in many parts of the world rely on fishes as a major food source, starvation would occur in these areas.*

Answers to Section Review

1. They have gills, fins, and scales.
2. Lampreys are parasites. Hagfishes are scavengers. Cartilaginous fishes are predators.
3. It enables fishes to control their depth by regulating the amount of gas in the bladder.

Thinking Critically

4. Jaws enable fishes to grasp and crush prey with great force. With jaws they can eat large chunks of food.

Skill Review

5. Make sure that students compare in their tables structural features such as skeleton, lateral line, scales, and swim bladder, as well as behavioral features such as ways of getting food and ways of swimming.

SECTION

33.2

Prepare

Key Concepts

Students will relate the demands of a terrestrial environment to the adaptations of amphibians. They will also study the diversity of amphibians as they compare adaptations of frogs, toads, and salamanders.

Block Scheduling

Look for this symbol for strategies that are useful in a block scheduling format. For more information on block scheduling, refer to Section 33.2 in the **Lesson Plans** booklet.

Materials

- Obtain a recording of frog calls from a public library or a biological supply house for the Focus Audiovisual.
- Purchase live frogs *(Rana pipiens)* for use in the Activity and the Minilab. Obtain large plastic containers for use in the Activity.
- For the Minilab, gather the live frogs used in the activity, metric rulers, beakers, and metersticks.
- For the Biolab, gather Ringer's solution, culture dishes, light bulbs, thermometers, binocular microscopes, frog eggs *(Rana pipiens* or *Xenopus laevis)*, and flashlights.

SECTION

33.2 Amphibians

Section Preview

Objectives

Relate the demands of a terrestrial environment to the adaptations of amphibians.

Relate the evolution of the three-chambered heart to the amphibian lifestyle.

Key Terms

ectotherm
vocal cords

If an alien visitor to our planet were to watch our television programs and read our children's literature, it might return home with wondrous stories of how frogs on Earth can talk and change by the touch of a kiss into princes. Frogs and toads don't talk, but they do change—from fishlike tadpoles to four-legged animals with bulging eyes, long tongues, loud songs, and remarkable jumping ability.

The Move to Land

Imagine a time 350 million years ago when the inland, freshwater seas were filled with carnivorous fishes. One type of tetrapod had evolved that retained gills for breathing and a finned tail for swimming. In later fossils, the four limbs are placed further below the body to lift it off the ground. Any animal that could move over land from the mud of a drying stream to another water source might survive. Most likely, amphibians arose as their ability to breathe air through well-developed lungs evolved.

Challenges of life on land

Life on land held many advantages for early amphibians. There was a large food supply, shelter, and no predators. In addition, there was much more oxygen in air than in water. However, land life also held many dangers. Unlike the temperature of water, which remains fairly constant, air temperatures vary a great deal. In addition, without the support of water, the body was clumsy and heavy. Some of the efforts at moving on land by early amphibians were like movements of modern-day salamanders. The legs extended out from the body, then down, allowing for better support than limbs straight out from the sides. You can see in *Figure 33.11* why the bellies of these animals dragged on the ground.

Adaptations improved success

The success of inhabiting the land depended on adaptations that would provide support, protection for membranes involved in respiration, and efficient circulation.

1 Focus

Bellringer

Before presenting the lesson, display **Section Focus Master 74** on the overhead projector and have students answer the accompanying questions. L1 SAE

Program Resources

Section Focus Master 74 L1 SAE
Reinforcement and Study Guide, pp. 135-136 L1
Biolab and Minilab Worksheets, pp. 132-134 L1
Critical Thinking/Problem Solving, p. 33 L3
Laboratory Manual, pp. 195-200 L2
Basic Concepts Transparency 53 and **Master** L1 SAE
Reteaching Transparency 33 and **Master** L1 SAE

What Is an Amphibian?

The striking transition from a completely aquatic larva to an air-breathing, semiterrestrial adult gives the class Amphibia its name, which means "double life." The class Amphibia includes three orders: Caudata, the salamanders and newts; Anura, the frogs and toads; and Apoda, the legless caecilians, as shown in *Figure 33.12.* Although most adult amphibians are capable of a terrestrial existence, nearly all of them rely on water for breeding. Amphibian eggs lack protective membranes and shells and must be laid in water to keep them moist. Fertilization in most amphibians is external. Therefore, water is needed as a medium for transporting sperm.

Figure 33.12

Caecilians, order Apoda, are long, limbless amphibians. They look like worms but have eyes that are covered by skin.

Figure 33.11

Adaptation to life on land involves the positioning of limbs. The limbs of early fishes could have evolved into the limbs of early amphibians. The evolution of primitive tetrapods led to the diversification of land vertebrates.

In mammals, the fastest-moving land animals, the body is raised above the ground, with the legs positioned underneath.

The salamander has legs that extend straight out from its body.

Reptiles have legs on the sides of their bodies, but the limbs are flexed.

SECTION 33.2

Audiovisual

Play a recording of frog calls. Use the recording to initiate a class discussion about how frogs call using vocal cords and sacs. Ask students to speculate about why frogs call. *Responses may indicate to identify a territory, to attract mates, or to signal distress.*

2 Teach

Activity

LS **Interpersonal** Divide the class into groups. Give each group a live frog inside a large, clear container. Ask students to describe any behaviors they observe in the frog. *Students are likely to describe the frog jumping, moving about, or breathing.* Elicit from students how these activities may be suited to life on land. L1 COOP LEARN

Different Viewpoints in Biology

Decline of Amphibians
Amphibian populations are decreasing worldwide. Biologists do not agree about the causes for these declines. Some attribute the losses to environmental problems such as acid rain, increased CO_2, global warming, increased ultraviolet light, or toxic chemicals in water. Others say that much of the problem results when human activities, such as stocking lakes and streams with fishes, increases the sizes of amphibian predator populations.

Discuss hypotheses related to the decline of amphibian populations with the class. Elicit from students what they consider the major causes may be. L1 COOP LEARN

STUDENT JOURNAL

Mapping Amphibian Habitats Give students a blank world map and ask them to use an atlas of world wildlife to find where various amphibians live. Have them color code their maps for various types of amphibians and place their maps in their journals. L2 LS

ThinkingLab Draw a Conclusion

Purpose

Logical-Mathematical Analyze data to conclude whether egg size is important to the survival of salamanders.

Process Skills

recognize cause and effect, interpret data

Teaching Strategies

- Show students photographs of salamanders in their natural habitats.
- Explain that salamanders, unlike frogs, look very much like adults when they hatch. These amphibians do not undergo metamorphosis as do frogs.

Thinking Critically

Larger eggs may enhance survival in general, but do not affect survival under dry conditions.

Assessment

Oral: Have students working in groups develop three hypotheses about why larger egg size might enhance survival of salamander larvae. Ask each group to read their hypotheses aloud and discuss the plausibility of each hypothesis as a class. L1 COOP LEARN

Visual Learning

Figure 33.13 How could having two different forms in a life cycle have survival value for a species? *Responses may suggest that the two forms help the species survive during adverse or changing climate conditions on land or in water.*

ThinkingLab Draw a Conclusion

Is egg size related to survival of salamander larvae?

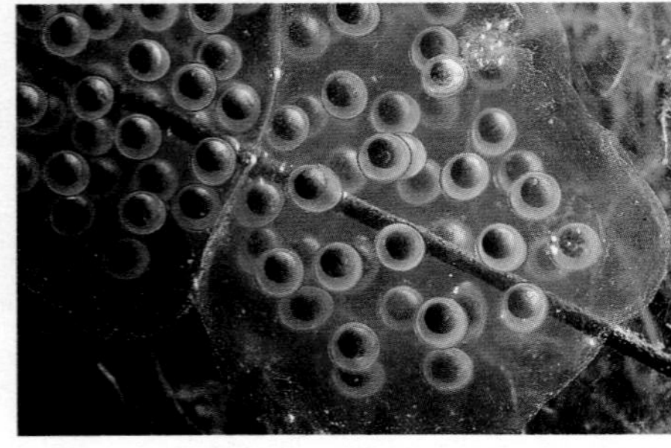

Salamanders and other amphibians are decreasing in numbers worldwide. Studies are underway to determine the cause. Biologists wanted to find out if egg size in salamanders had any effect on survival of the larvae under adverse conditions. They grew large and small salamander eggs from the same species under constant conditions, and under conditions that simulated a drying pond in the summer.

Analysis

Biologists found that the survival rates for larvae from large eggs were higher than for larvae from smaller eggs under similar conditions. No difference in survival rates was observed in the drying treatment.

Thinking Critically

From the results of this experiment, explain whether larger eggs enable larvae of salamanders to survive adverse conditions in a pond.

Amphibians undergo metamorphosis

Unlike fish, amphibians go through the process of metamorphosis. Fertilized eggs hatch into tadpoles, the aquatic stage of most amphibians. Tadpoles possess fins, gills, and a two-chambered heart as seen in fishes. As tadpoles grow into adult frogs and toads, they develop legs, lungs, and a three-chambered heart. *Figure 33.13* shows this life cycle. Young salamanders resemble adults, but they have gills and usually have a tail fin. Amphibians have thin, moist skin and no claws. Most adult salamanders lack gills and fins. Instead, they breathe through their moist skin or with lungs. Up to one-fourth of all salamanders have no lungs and breathe only through their skin. Salamanders also have four legs for moving about.

Figure 33.13

The amphibian life cycle includes an aquatic tadpole stage and a terrestrial adult stage. How could having two different forms in a life cycle have survival value for a species?

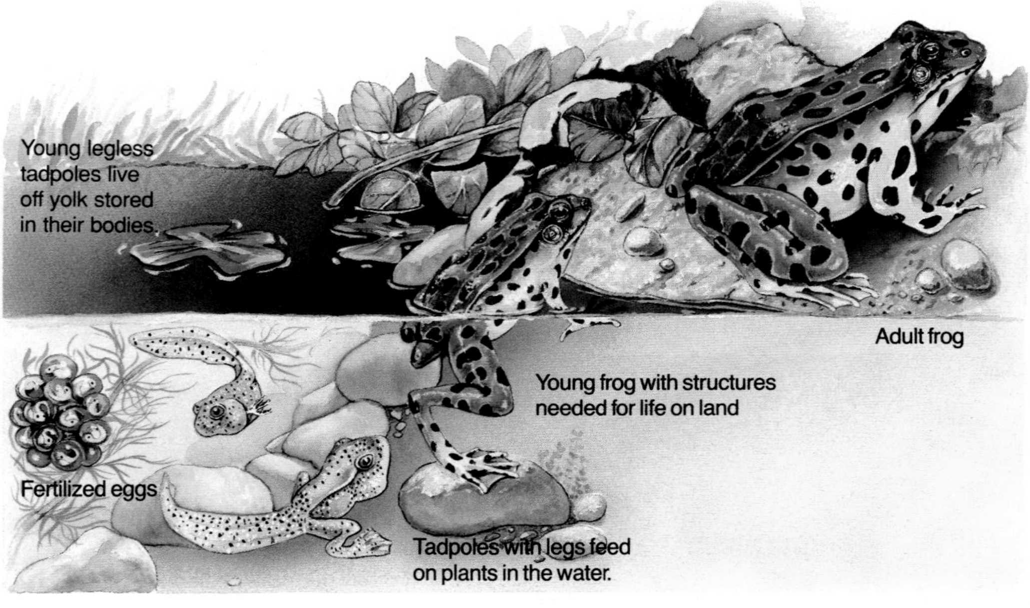
Young legless tadpoles live off yolk stored in their bodies.
Adult frog
Young frog with structures needed for life on land
Fertilized eggs
Tadpoles with legs feed on plants in the water.

826 Fishes and Amphibians

PORTFOLIO

Frog Life Cycle Have students make a flowchart that traces the stages in the frog life cycle. Ask students to describe the features of each stage of the life cycle and describe how each feature benefits the organism at that stage. L2 P LS

Amphibians are ectotherms

Amphibians are more common in regions that have warm temperatures all year because they are ectotherms. An **ectotherm** is an animal in which the body temperature changes with the temperature of its surroundings. Because many biological processes require particular temperature ranges in order to function, amphibians become dormant in regions that are too hot or cold for part of the year. During such times, many amphibians burrow into the mud and stay there until suitable conditions return.

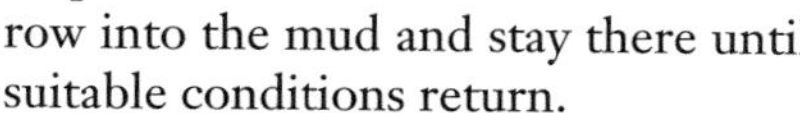

Walking requires more energy

The laborious walking of early amphibians required a great deal of energy from food and large amounts of oxygen for aerobic respiration. The evolution of the three-chambered heart in amphibians ensured that cells receive the proper amount of oxygen. This was an important evolutionary transition from the simple circulatory system of fishes.

In the three-chambered heart of amphibians, one chamber receives oxygen-rich blood from the lungs and skin, and another chamber receives oxygen-poor blood from the body tissues. Blood from both chambers then moves to the third chamber, which pumps oxygen-rich blood to body tissues and oxygen-poor blood back to the lungs and skin so it can pick up more oxygen. In amphibians, the skin is much more important than the lungs as an organ for gas exchange.

MiniLab

How does leg length affect how far a frog can jump?

Frogs have long legs relative to their body weight—a key adaptation that allows them to jump long distances and move away from predators rapidly. In this activity, you will determine how leg length is related to the jumping ability of a frog.

Procedure

1. Measure the length of one hind leg of a frog by gently extending the leg. The measurement should start on the ventral surface where the frog leg joins the body and extend to the tip of the longest toe.
2. Place a meterstick on the floor.
3. Place your frog so the anterior part of its head is at the beginning of the meterstick as it sits on the floor.
4. Make a hypothesis about how leg length affects the length of the jump.
5. Lightly tap the rear portion of the frog with the eraser end of your pencil until the frog jumps a measurable distance. The distance recorded should be marked at the position of the frog's nose. Conduct four trials with each frog.
6. Place your data in a table on the chalkboard. Make a graph of the class data. Draw conclusions.

Analysis

1. Use class data or your data from several frogs to explain how leg length is correlated to the distance a frog can jump.
2. What other factors might be important in determining how far a frog can jump?
3. Describe stimuli in the natural habitat of a frog that might cause it to jump.
4. Was your hypothesis supported by your data?

Because the skin of an amphibian must stay moist to exchange gases, most amphibians are limited to life on the water's edge. However, some newts and salamanders remain totally aquatic. Amphibians such as toads have thicker skin. Although toads live primarily on land, they still must return to water to breed.

MiniLab

Purpose

LS **Logical-Mathematical** Students will determine how leg length affects the distance a frog can jump.

Process Skills

observe and infer, hypothesize, compare and contrast

Safety Precautions

- Caution students to treat the frog in a humane manner.
- Frogs should be returned to the aquarium or pail of water periodically.

Teaching Strategies

- Use *Rana pipiens* or another frog you may already have.
- On the chalkboard, create a class data table. Have students average their trials and record the average distance jumped and the leg lengths of their frogs. Discuss the results as a class.

Expected Results

Length of leg for *Rana pipiens* vary from 8-14 cm. Distances jumped vary from 20-40 cm.

Analysis

1. Most long-legged frogs will jump farther than short-legged frogs. Some data may be inconclusive.
2. thickness of the leg muscles, temperature, light, size, distractions
3. fish swimming nearby, overhead hawk, sudden movement, loud noise
4. Some hypotheses will be supported, others will not.

Assessment

Portfolio: Ask students to design an experiment to test how food availability or temperature might affect how far a frog jumps. L1 P

PROJECT

Frog Reproduction

Ask students to read the article, "Reproductive Strategies of Frogs" by William E. Duellman in the July 1992 *Scientific American.* Ask them to prepare a report about the various strategies frogs use in reproduction. L1 LS

BioLab

Development of Frog Eggs

Time Allotment

Initial session: one class period. Followup sessions: 10 minutes daily for two weeks

Objectives

Review objectives with students before they begin the Biolab.

Process Skills

observe and infer, compare and contrast, form a hypothesis, interpret data, recognize cause and effect

Safety Precautions

- Encourage students to treat living organisms in a humane fashion at all times.
- Students should wash their hands after each observation.

PREPARATION

Alternate Materials

- Make sure to use frog eggs that have a short development time. It is best not to collect from the wild as amphibians are declining worldwide. Order from a biological supply house.
- Use sterile Ringer's solution to help prevent contamination by bacteria and fungi. Ask students to avoid touching the solution to prevent contamination.

Possible Hypotheses

- If water temperature is warm, frogs will develop from eggs more quickly.
- If water temperature is cool, few frogs will develop from eggs.

BioLab

Development of Frog Eggs

Most frogs breed in water. The male releases sperm over the female's eggs as she lays them. Some frogs lay up to a thousand eggs. A jellylike casing protects the eggs as they grow into embryos. The embryos then hatch and develop into aquatic larvae commonly called tadpoles. Tadpoles feed by scooping algae from the water or by scraping algae from water plants with small, toothlike projections in their mouths.

During metamorphosis, tadpoles lose their tails and develop legs. Many internal changes take place as well. Gills are reabsorbed by the body, and lungs form. Development and metamorphosis to the adult frog take from three weeks up to several years, depending upon the species.

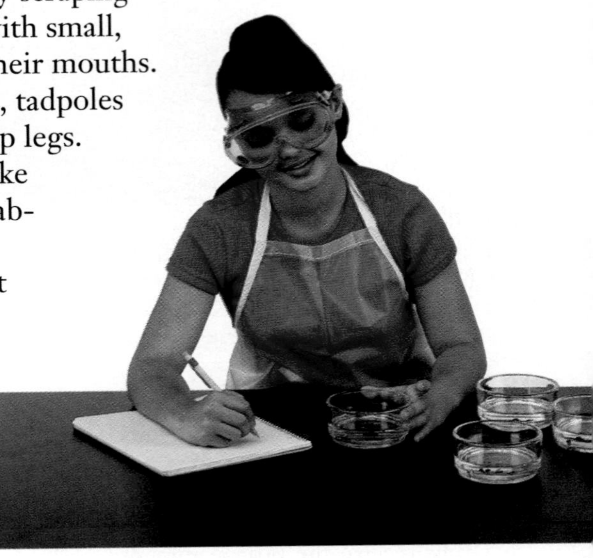

PREPARATION

Problem

How does temperature affect the development of frog eggs?

Objectives

In this Biolab, you will:

- **Compare** development of frog eggs at varying temperatures.
- **Distinguish** among various stages of development.

Materials

Ringer's solution
4 culture dishes
light bulbs with source of electricity
thermometer
binocular microscope
frog eggs, *Rana pipiens* or *Xenopus laevis*
flashlight

Safety Precautions

Wash hands after each observation.

PROCEDURE

Teaching Strategies

- Review frog development before students carry out the lab.
- Have students work in cooperative groups to reduce the materials needed. Each person in the group should have a specific job during the data collecting period.
- Explain that the white part of the egg is the yolk and the black part is the developing tadpole. The yolk is used for food by the developing tadpole. The jellylike substance surrounds and protects the egg.
- Students should be able to observe movement of the tadpole just before it hatches.

Data and Observations

Eggs develop more quickly at warmer temperatures and more slowly at lower temperatures.

PROCEDURE

1. Obtain four culture dishes of Ringer's solution and fertilized frog eggs.
2. Make a chart similar to the one shown for drawing sketches of stages of development of the eggs.
3. Observe your eggs and determine what stage of development you have. At room temperature, you should see the two-cell stage about 3.5 hours after fertilization, the eight-cell stage at 5.7 hours, the 32-cell stage at 7.5 hours, the late-gastrula stage at 42 hours, and a visible head area between 72 and 84 hours. Hatching will occur at about 140 hours (about five to six days), depending on the species used. Development time varies according to the species.
4. Set up the appropriate numbers of light bulbs of different wattages over the water to keep the temperatures at 20, 25, and 30°C.
5. Place a dish of eggs in the refrigerator. Measure the temperature of your refrigerator. Keep a flashlight on in your refrigerator at all times so that you have only one variable, the temperature.
6. Make a hypothesis about how temperature will affect development of the eggs.
7. Set up a time schedule for observations and making sketches based on what stage you are observing. Make observations until the eggs hatch. Record your observations in a journal.
8. Observe your eggs under the microscope according to the schedule you have made. Draw sketches of your eggs in the table.

Temperature	Day 1	Day 2	Day 3
30 degrees			
25 degrees			
20 degrees			
refrigerator:			
-degrees			

ANALYZE AND CONCLUDE

1. **Interpreting Observations** Which eggs develop the fastest? The slowest? Explain.
2. **Interpreting Observations** Did your data support your hypothesis?
3. **Drawing Conclusions** What advantage is it for frogs to have eggs that develop at different rates that correspond to different temperatures?
4. **Thinking Critically** What would happen if frog eggs developed when the weather was still cold in the spring?

Going Further

Developing a Hypothesis Make a hypothesis about what other environmental factors would affect the development of frog eggs. Explain your reasoning.

ANALYZE AND CONCLUDE

1. Egg development and cell division occur faster at warmer temperatures. Eggs at lower temperatures develop more slowly. Cell division is slower at lower temperatures.
2. Check students' data to see how they supported their hypotheses.
3. The tadpoles will hatch when temperature is most conducive for survival. Perhaps more foods are available at warmer temperatures.
4. The tadpoles might not survive or adequate food may not be available when the water is cold.

Assessment

Portfolio: Ask students to summarize in their portfolios what they learned in this lab.

Going Further

Students' hypotheses may include that UV light or chemical pollution of water may destroy developing egg cells; or lack of oxygen in water due to decomposition of organic matter may affect egg development.

GLENCOE TECHNOLOGY

Videodisc

Biology: The Dynamics of Life
Disc 2, Side 1
Frog Behavior (Ch. 3)

Chemistry

Killer Frogs

Purpose

Students will study the nature of poisonous frogs and how their secretions affect cell chemistry of the prey.

Teaching Strategies

- Review details of ion channels in cell membranes from Chapter 9.
- Have students diagram the two different actions frog toxins may have on cells. L3

Possible Answer

Responses will vary depending upon the human diseases researched.

Using an Analogy

A frog senses sound and other vibrations by means of tympanic membranes, circular structures located just behind the eyes of the frog. The tympanic membrane vibrates much like the skin of a drum to alert the frog to sounds. Ask students to hypothesize why the sense of hearing is important to a frog. *The sense of hearing is important to a frog for hearing the calls of mates and to avoid predators.* L1

Chemistry

Killer Frogs

The most colorful frogs in the world are found in South and Central America. Poisonous frogs, including 130 species of the Dendrobatidae family, range in size from 1 to 5 cm. Although all frogs have glands that produce secretions, these frogs secrete toxic chemicals through their skin. A predator will usually drop the foul-tasting frog when it feels the numbing or burning effects of the poison in its mouth. The frogs advertise their poisonous personalities by bright coloration; they may be red or blue, solid colored, marked with stripes or spots, or have a mottled appearance. The poison secreted by these frogs is used by native peoples to coat the tips of the darts they use in their blow guns for hunting.

Biochemistry of the toxin The secretions of these frogs are alkaloid toxins. An alkaloid toxin is a compound that includes a ring consisting of five carbon atoms and one nitrogen atom. The toxins secreted by poisonous frogs act on an ion channel between nerve and muscle cells. Normally, the channel is open to allow movement of sodium, potassium, and calcium ions. The toxins can block the flow of potassium and stop or prolong nerve impulse transmission and muscle contraction. One group of alkaloids affects the transport of calcium ions, which are responsible for muscle contraction. Current research indicates that these alkaloids may have clinical applications for muscle diseases.

CONNECTION TO Biology

Research on newly discovered organisms such as poisonous frogs may result in drugs to treat specific disorders in human patients. Find out what human diseases are caused by problems in the transmission of nerve impulses and write an essay identifying the disorders that might be treated by toxins from poisonous frogs.

Figure 33.14

Most male frogs have throat pouches that increase the volume of their calls.

The Diversity of Amphibians

The movement of early amphibians from water to a life on land has not yet been completed. Because amphibians still complete part of their life cycle in water, they are limited to the edges of ponds, lakes, streams, and rivers or to areas that remain damp during part of the year.

Frogs and toads belong to the order Anura

Frogs and toads have vocal cords that are capable of producing a wide range of sounds. **Vocal cords** are sound-producing bands of tissue in the throat. As air moves over the vocal cords, they vibrate and cause molecules in the air to vibrate. In many male frogs, air passes over the vocal cords, then passes into a pair of vocal sacs lying underneath the throat, *Figure 33.14.*

Salamanders belong to the order Caudata

Unlike a frog or toad, a salamander has a long, slender body with a neck and tail. Salamanders resemble lizards. However, salamanders have smooth, moist skin and lack claws. Some salamanders are totally aquatic, whereas others live in damp places on land.

PROJECT The Frog Bowl

LS **Interpersonal** Ask a group of students to design a review game for concepts in this section. Tell them to base their game on a TV quiz show and to prepare cards or other materials needed so that the class or groups in the class could play. L1

STUDENT JOURNAL

Researching Local Amphibians Provide students with sources about local amphibians. Ask them to draw sketches and write a paragraph about several local amphibians. Ask students to place this information in their journals. L2 LS

THE INSIDE STORY

A Frog

Many species of frogs look similar. As adults, they have short, bulbous bodies with no tails. This adaptation allows them to jump more easily.

Green frog, *Rana clamitans*

1. Vibrations from water or air are picked up by the tympanic membrane and transmitted to the inner ear and then to the brain. Each ear is connected to the mouth by a channel called the eustachian tube. Eustachian tubes help to maintain equal air pressure on both sides of the tympanic membrane.

2. Some frogs' eyes protrude from the tops of their heads—an adaptation that enables them to stay submerged in the water with only their eyes above water. Frog eyes possess one upper eyelid that does not move and a lower movable eyelid that is folded to form the nictitating membrane, which helps to keep the eye moist.

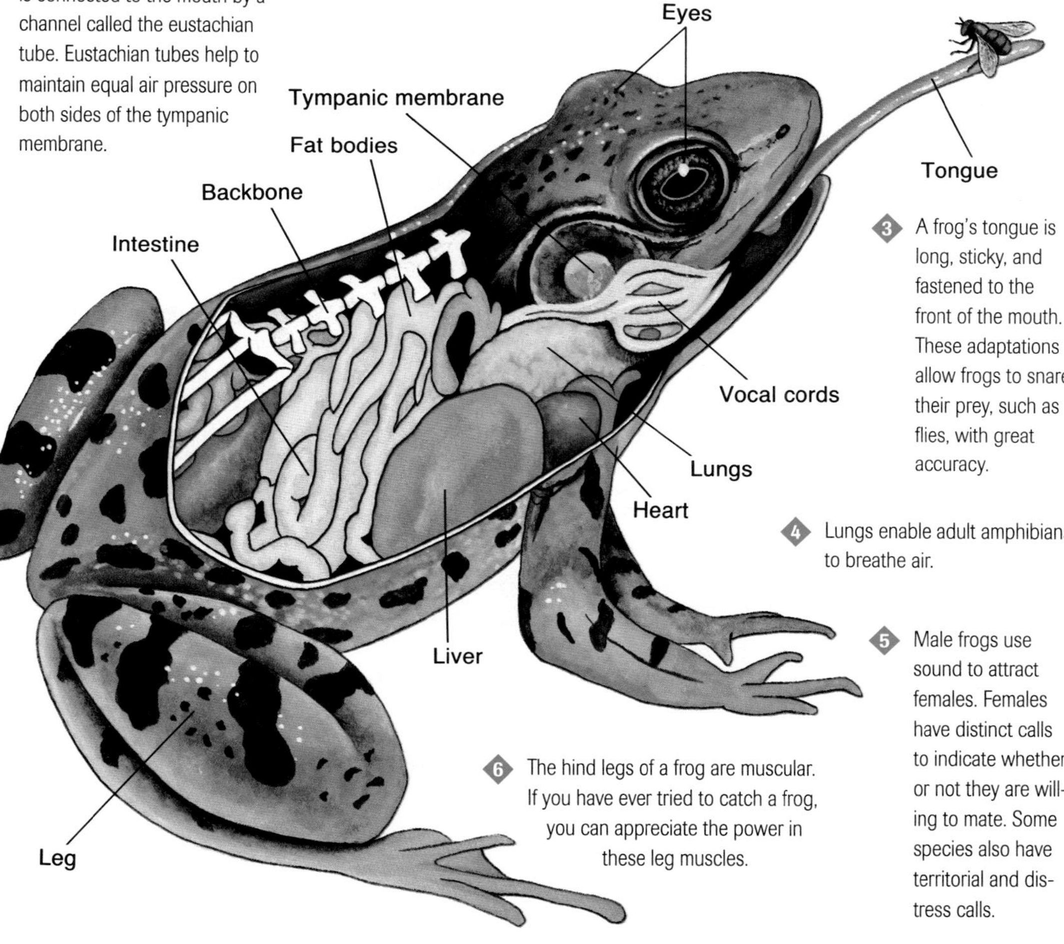

3. A frog's tongue is long, sticky, and fastened to the front of the mouth. These adaptations allow frogs to snare their prey, such as flies, with great accuracy.

4. Lungs enable adult amphibians to breathe air.

5. Male frogs use sound to attract females. Females have distinct calls to indicate whether or not they are willing to mate. Some species also have territorial and distress calls.

6. The hind legs of a frog are muscular. If you have ever tried to catch a frog, you can appreciate the power in these leg muscles.

THE INSIDE STORY

A Frog

Purpose

LS Visual-Spatial Students will learn the important features of frogs that enable them to survive in their habitats.

Teaching Strategies

- Have a live frog available as well as wall charts showing the internal and external anatomy of a frog. As each feature is discussed in the Inside Story, point out the corresponding feature on the live frog and wall charts.

Visual Learning

- Provide each student with an acetate transparency and markers. Have students lay the transparency on the frog diagram shown and trace the outline of the frog on the acetate. Encourage students to label each of the structures shown in the *Inside Story*. L1
- Ask students to use the transparencies they made to prepare a presentation about frogs for an elementary school science class. Have students outline the features to be covered in their presentations. L1
- Give students unlabeled diagrams of a frog. Ask them to use the diagram in their book to label the features of the frog on their diagrams. Beside each label, students should explain the function of the structure. L1

GLENCOE TECHNOLOGY

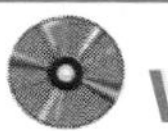

Videodisc

Biology: The Dynamics of Life
Disc 2, Side 1
Feeding Frog (Ch. 4)

SECTION 33.2

3 Assess

Check for Understanding

Have students develop a table with the headings: Trait, Anura, and Caudata. Beneath the Trait heading, have students list: Symmetry, Habitat, Food getting, Reproduction, Locomotion, Breathing, Protection, Examples. Have students review this section to complete their tables. L1

Reteach

Prepare six sets of flash cards with the vocabulary terms and names of the anatomical structures discussed in this section on one side and the meaning of the term or function of the structure on the other side. Have students working in groups use the cards to review the terms and ideas of the section. L1 SAE COOP LEARN

Extension

Ask students to prepare a cartoon that includes a frog. The cartoon should be humorous and accurately portray some aspect of the frog's environment, life history, or evolution. Have students examine some "Far Side" cartoons by Gary Larson as examples. L1

Assessment

Skill: Have students write a description for a perfectly adapted toad that lives in a dry region with puddles that form after rain and with many hawks that prey on toads. Students' descriptions should consider how the toad would protect itself, reproduce, prevent water loss from its body, and get food. L2

4 Close

Audiovisual

Show slides or a filmstrip of amphibians from all over the world. L1

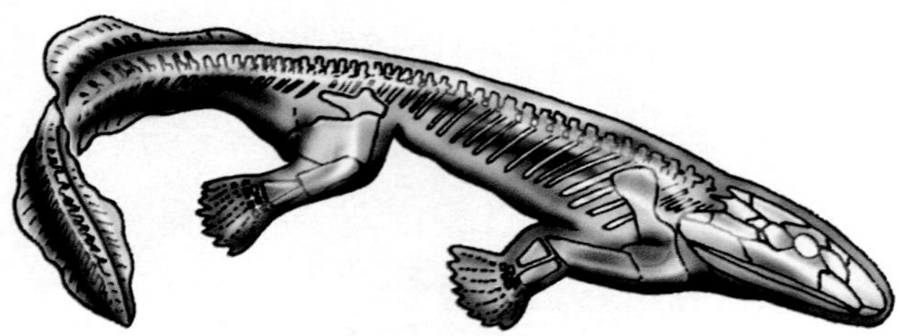

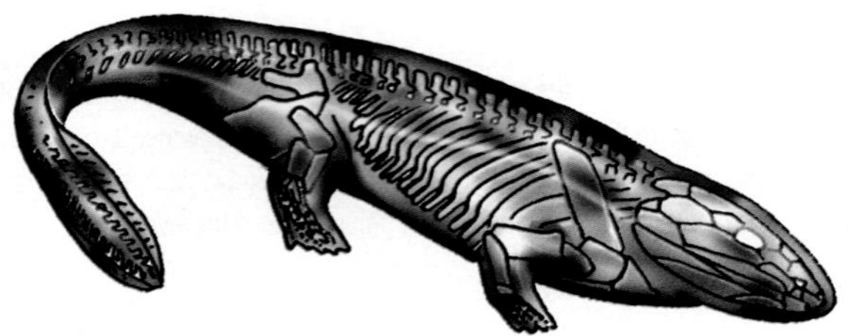

Figure 33.15

The oldest fossil with four limbs from the Devonian period shows that they were well-adapted to an aquatic existence (top). The limbs of the first land tetrapods became positioned further beneath the body to provide support on land (bottom).

Origins of Amphibians

Amphibians first appeared about 360 million years ago. It is widely accepted that amphibians evolved from an aquatic tetrapod, *Figure 33.15*, around the middle of the Paleozoic Era. At that time, the climate on Earth became warm and wet, a climate ideally suited for an adaptive radiation of amphibians. Amphibians, able to breathe through their lungs, gills, or skin, became a transitional group of vertebrates. Many new species evolved, and for a time, amphibians were the dominant vertebrates on land.

Connecting Ideas

Early fishes evolved from mud-sucking swimmers to aquatic tetrapods to air-gulping animals that crawled from one pond to another, and finally to fully developed amphibians that lived mainly on land. Amphibians faced dangers of dehydration and temperature extremes.

You've read about the adaptations that allow amphibians to survive on land. Yet amphibians have not completely made the transition from water to land. The need to protect their young and keep their eggs moist prevents amphibians from striking out to inhabit dry places on land.

Section Review

Understanding Concepts

1. Describe the events that may have led early animals to move to land.
2. List three characteristics of amphibians.
3. Name two ways that amphibians depend on water.

Thinking Critically

4. How does a three-chambered heart enable amphibians to maintain homeostasis?

Skill Review

5. **Sequencing** Trace the evolutionary development of amphibians from lobe-finned fishes. For more help, refer to Organizing Information in the *Skill Handbook.*

Answers to Section Review

1. Inland seas were full of predators, there were periodic droughts, and early tetrapods with gills were able to move briefly across land to another water source for survival.
2. three-chambered heart; eggs without shells laid in water; smooth, moist skin
3. breathing and external fertilization

Thinking Critically

4. Cells obtain oxygen quickly in a three-chambered heart system. The three-chambered heart is a more efficient pump, enabling oxygen to reach cells quickly, thereby enabling the animal to move quickly.

Chapter 33 Review

REVIEWING MAIN IDEAS

33.1 Fishes

- Fishes are vertebrates with backbones and nerve cords that have expanded into brains.
- Fishes belong to three groups: the jawless lampreys and hagfishes, the cartilaginous sharks and rays, and the bony fishes. Bony fishes are divided into three groups: the lobe-finned fishes, the lungfishes, and the ray-finned fishes.
- Early jawless fishes may be the ancestors of present-day jawless fishes. Cartilaginous and bony fishes may have evolved from ancient placoderms.

33.2 Amphibians

- Land animals face problems of dehydration, gas exchange in the air, and support for heavy bodies. Amphibians possess adaptations well suited for life on land.
- Adult amphibians have three-chambered hearts that provide oxygen to body tissues, but most gas exchange takes place through the skin.
- Amphibians may have evolved from ancient aquatic tetrapods.

Key terms

Write a sentence that shows your understanding of each of the following terms.

cartilage
ectotherm
fin
lateral line system
scale
spawning
swim bladder
vocal cords

Understanding Concepts

1. An animal with gill slits, a dorsal nerve cord, and a backbone is a(n) __________.
   a. invertebrate
   b. invertebrate chordate
   c. vertebrate
   d. echinoderm
2. In addition to fishes, the subphylum Vertebrata includes __________.
   a. amphibians, reptiles, birds, and mammals
   b. echinoderms, reptiles, birds, and mammals
   c. sea squirts, lancelets, reptiles, and birds
   d. sea squirts, reptiles, birds, and mammals
3. External fertilization in bony fishes is also known as __________.
   a. budding
   b. regeneration
   c. metamorphosis
   d. spawning
4. Of all the vertebrate groups, there is more diversity among __________ because they comprise three classes.
   a. birds
   b. mammals
   c. echinoderms
   d. fishes
5. How do jawless fishes obtain food?
   a. by injecting prey with poison from their hooklike fangs
   b. by using their round mouths like vacuum cleaners to suck up detritus
   c. by drilling a hole and sucking out blood and insides of a prey animal
   d. by using their sharp teeth to grab and swallow smaller fishes
6. Which of the following is NOT a characteristic of most fishes?
   a. have scales
   b. have a two-chambered heart
   c. breathe using gills
   d. exchange gases through thin, moist skin

Chapter 33 Review

Reviewing Main Ideas

Summary statements can be used by students to review the major concepts of the chapter.

Key Terms

Answers should go beyond defining the terms. Accept any answer that uses the term correctly and in the proper context.

Understanding Concepts

1. c
2. a
3. d
4. d
5. c
6. d

Skill Review

5. Amphibians first appeared about 350 million years ago, probably evolving from lobe-finned fishes. Over time, the lobed fins evolved into legs and feet more suited to a land environment. Animals that became amphibians developed lungs that enabled them to obtain oxygen from the air.

Chapter 33 Review

7. a
8. b
9. b
10. c
11. a
12. b
13. d
14. a
15. c
16. a
17. b
18. d
19. a
20. b

Applying Concepts

21. They occupy different niches.
22. Advantages: Egg and sperm are assured of coming together. Male and female do not have to produce as many gametes. Disadvantages: Male and female must find each other and be ready for reproduction at the same time.

Chapter 33 Review

7. The skeletons of lampreys, hagfishes, sharks, and their relatives are made of ___________.
   a. cartilage c. scales
   b. bone d. calcium
8. In all fishes, gas exchange takes place in the ___________.
   a. lungs c. fins
   b. gills d. skin
9. All fishes have a ___________ and a closed circulatory system.
   a. one-chambered heart
   b. two-chambered heart
   c. three-chambered heart
   d. four-chambered heart
10. Structures that fishes use like the rudders and stabilizers of boats are called _______.
    a. lateral line systems
    b. swim bladders
    c. fins
    d. gills
11. What structure enables a fish to detect movement and vibrations in the water?
    a. lateral line system c. fins
    b. swim bladder d. gills
12. Fishes control their depth in the water by means of their ___________.
    a. lateral line systems
    b. swim bladders
    c. fins
    d. tails
13. Scientists hypothesize that the common ancestors to all fishes are the _________.
    a. amphibians c. lancelets
    b. echinoderms d. ostracoderms
14. The class Amphibia is well-named because amphibians ___________.
    a. spend part of their lives on land and part in water
    b. lay shelled eggs on land but develop in water
    c. spend part of their lives in the air and part on land
    d. use swim bladders for breathing in water but use lungs on land
15. Frogs and toads are capable of producing sound with their ___________.
    a. gills c. vocal cords
    b. feet d. tongues
16. Reptiles have legs on the sides of their bodies, but the limbs are flexed. Salamanders have legs that ___________.
    a. extend straight out from the body and do not flex
    b. extend out from the sides of the body and flex
    c. are missing some of the key bones of reptiles
    d. evolved from shark's fins
17. What structures do adult frogs and most adult salamanders use for gas exchange?
    a. gills and lungs
    b. lungs and skin
    c. gills and lining of the mouth
    d. skin and gills
18. What is an animal called in which the body temperature changes with the temperature of its surroundings?
    a. an endotherm
    b. a mesotherm
    c. an aquatic animal
    d. an ectotherm
19. The evolution of a three-chambered heart in amphibians ensured that cells receive the proper amount of ___________.
    a. oxygen c. blood
    b. carbon dioxide d. heat
20. Frogs produce a range of sounds using their ___________.
    a. tympanic membrane
    b. vocal cords
    c. skin
    d. tongues

Applying Concepts

21. What accounts for the different body shapes of fishes?
22. What are the advantages and disadvantages of internal fertilization?

23. Describe the importance of the evolution of bone in fishes to the evolution of vertebrates.
24. Explain why biologists hypothesize that early forms of fishes evolved into amphibians.
25. A male sea horse incubates its eggs in a brood pouch. A codfish lays its eggs in the open sea. Which of these two types of fishes would need to lay more eggs? Why?

Thinking Critically

26. ***Concept Mapping*** Make a concept map that relates the following terms and phrases. Supply the appropriate linking words for your map.
    fishes, lateral line, fins, cartilage, scales, bone, rays, gills, swim bladder
27. ***Using a Graph*** The following graph shows the number of leopard frogs in a wetland on a farm. One year, there was a prolonged drought in the farmer's area. What year did the drought occur? Explain.

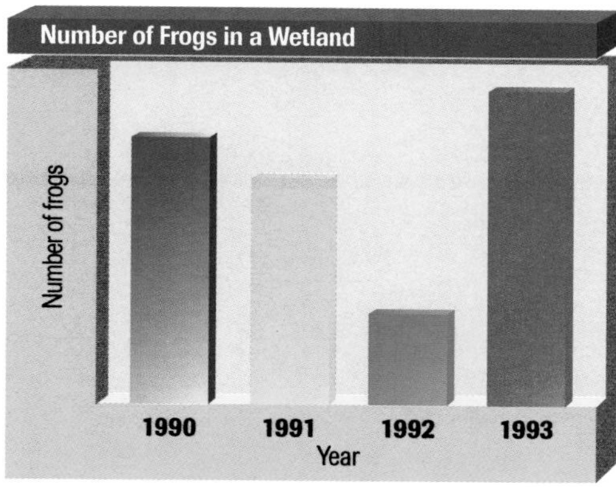

28. ***Drawing Conclusions*** In the past, some scientists hypothesized that either lungfishes or lobe-finned fishes were ancestral to amphibians. What adaptation does each group of fishes have that may have led scientists to this conclusion?

ASSESSING KNOWLEDGE & SKILLS

Amphibian populations are declining in many parts of the world. To reintroduce native amphibians into your state, you have a grant to breed frogs on a farm. To find out what temperature you need for optimum hatching of frog eggs, you test eggs at four different temperatures.

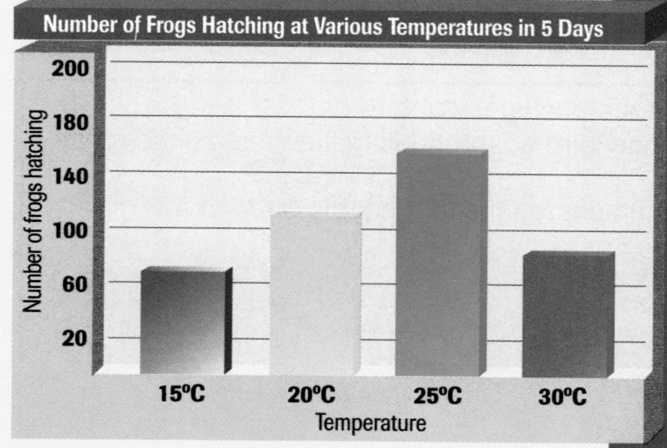

Interpreting Data Answer the following questions based on the graph.

1. How many eggs hatched at 25°C?
   a. 150 c. 110
   b. 70 d. 15
2. At which temperature did the fewest eggs hatch?
   a. 15°C c. 25°C
   b. 20°C d. 30°C
3. Based on the results of your experiment, what temperature will you use for optimum hatching of eggs?
   a. 15°C c. 25°C
   b. 20°C d. 30°C
4. ***Designing an Experiment*** You are trying to find out what the optimum pH is for hatching frog eggs. Design a controlled experiment that would give you quantitative data.

Chapter 33 Review

23. Bone provides the support land vertebrates need for locomotion on land.
24. Lungfishes and lobe-finned fishes have some characteristics of ancestral amphibians.
25. Cod are a type of fish that lays larger numbers of eggs. The eggs are unprotected in the ocean and most hatchlings die before they reach adult size.

Thinking Critically

26. Evaluate students' maps. The maps should show an understanding of the relationships among the concepts listed.
27. probably in 1992 because there was a decrease in the size of the frog population that year
28. Lungfishes have a modified swim bladder connected to the mouth by a tube, whereas lobe-finned fishes have fleshy, paired fins with bony skeletons that could support their weight on land.

ASSESSING KNOWLEDGE & SKILLS

1. a
2. a
3. c
4. Grow frog eggs in separate solutions with various pH levels. Count how many eggs hatch in each solution after a specific number of days.

Program Resources

Chapter Assessment, pp. 201-206 L1
Alternate Assessment in the Science Classroom
Computer Test Bank L1
Content Mastery, pp. 137-140 L1

Chapter Organizer

SECTION	OBJECTIVES	ACTIVITIES/FEATURES
34.1 Reptiles National Science Standards: UCP.1-5; C.3, C.5, C.6	1. **Compare** the characteristics of different groups of reptiles. 2. **Explain** how reptile adaptations make them suited to life on land.	**Thinking Lab:** Why are tortoises becoming frail?, p. 843 **Biology & Society Issues:** Zoos: Preserves or Breeding Grounds?, p. 844 **Focus On** Dinosaurs, p. 848
34.2 Birds National Science Standards: UCP1-5; A.1, A.2; C.3, C.5, C.6; D.3; G.1-3	3. **Interpret** the phylogeny of birds. 4. **Explain** how bird adaptations make them suited to life on land. 5. **Relate** the adaptations that enable birds to fly.	**Minilab:** How do down feathers and contour feathers compare?, p. 855 **Physics Connection:** Hot Birds, p. 856 **Biolab:** Design Your Own Experiment—What is the ideal length and width for a bird's tail?, p. 858 **Minilab:** What types of foods do local birds prefer?, p. 861 **Focus On** Big Birds, p. 862

ACTIVITY MATERIALS

BIOLAB	MINILABS	ALTERNATE LAB
page 858 tagboard tracings of birds' bodies and tails pencil paper fastener protractor metric ruler scissors	**page 855** contour feathers down feathers hand lens or binocular microscope **page 861** plastic bleach bottles, empty scissors wire sunflower seeds hulled oats cracked corn wheat thistle millet bird guide	**page 840** thermometer black paper white paper tape gooseneck lamp metric ruler clay frozen juice cans, empty with holes in one end

Chapter 34 Reptiles and Birds

TEACHER CLASSROOM RESOURCES

Reproducible Masters	Transparencies
Section Focus Master 75: Living on Land L1 SAE **Reinforcement and Study Guide,** pp. 137-138 L1 **Content Mastery,** pp. 141-144 L1	**Basic Concepts Transparency #54:** Amniotic Egg L1 SAE **Basic Concepts Transparency #55:** Phylogeny of Reptiles L1 SAE **Reteaching Transparency #34:** Structure of the Amniotic Egg L1 SAE
Section Focus Master 76: Into the Air L1 SAE **Reinforcement and Study Guide,** pp. 139-140 L1 **Biolab and Minilab Worksheets,** pp. 135-138 L1 **Concept Mapping:** Adaptations for Flight, p. 34 L1 **Tech Prep Applications:** Bird Wings and the Principles of Flight, pp. 37-38 L2 **Critical Thinking/Problem Solving:** Maintaining Stability in a Song Sparrow Population, p. 34 L3 **Laboratory Manual:** Examining Bird Feet, pp. 203-208; How Do Densities of Bird and Mammal Bones Compare?, pp. 209-212 L2	**Basic Concepts Transparency #56:** Adaptations for Flight L1 SAE **Basic Concepts Transparency #57:** Phylogeny of Birds L1 SAE **Basic Skills Transparency #24:** Flight L1 SAE

ASSESSMENT MATERIALS

Chapter Assessment, pp. 207-212
Performance Assessment in the Biology Classroom, p. 37
MindJogger Videoquiz
Alternate Assessment in the Science Classroom
Computer Test Bank

Spanish Resources SAE
English/Spanish Audiocassettes SAE
Cooperative Learning in the Science Classroom COOP LEARN
Lesson Plans
Biology Projects: Living with and Caring for Vetebrates, pp. 33-36 L2

KEY TO TEACHING STRATEGIES

L1 Level 1 activities should be within the ability range of all students including those with learning difficulties.
L2 Level 2 activities are within the ability range of average to above-average students.
L3 Level 3 activities are designed for the ability range of above-average students.
SAE SAE activities should be within the ability range of Students Acquiring English.
COOP LEARN Cooperative Learning activities are designed for small group work.
P These strategies represent student products that can be placed into a best-work portfolio.
These strategies are useful in a block scheduling format.

GLENCOE TECHNOLOGY

The following multimedia resources are available from Glencoe.

Biology: The Dynamics of Life
CD-ROM SAE
Videodisc Program

The Infinite Voyage Series
The Great Dinosaur Hunt

Science and Technology Videodisc Series (STVS)
Animals
Reptilian Sex Change
Sea Turtle Mystery
Rattlesnakes
Sexing Birds
Peregrine Falcon
Ecology
Bald Eagles

CHAPTER

34 Reptiles and Birds

Chapter Overview

Students begin their study of reptiles by discovering what snakes, turtles, alligators, and lizards have in common. Students next learn about reptile adaptations to land and how each reptile order differs from the others. The structure of the amniotic egg is explored visually and its importance is emphasized. Finally, students study the origins of reptiles.

The second section of the chapter focuses on birds. After learning what makes birds unique, students study the adaptations birds have that enable and aid flight. Students explore the classification of birds, and they conclude the section by learning about the origins of birds and their ancient relationships to reptiles.

Key Terms

amniotic egg
endotherm
feather
Jacobson's organ
sternum

836

Learning Styles

Look for the following logo for strategies that emphasize different learning modalities. LS

Kinesthetic	Meeting Individual Needs, p. 848; Using an Analogy, p. 860
Visual-Spatial	Demonstration, p. 839; The Inside Story, pp. 840, 857; Portfolio, pp. 844, 856, 860; Visual Learning, p. 845; Minilab, p. 855; Student Journal, p. 855; Activity, p. 865; Meeting Individual Needs, p. 865
Interpersonal	Project, p. 843; Activity, pp. 846, 864; Student Journal, p. 860; Minilab, p. 861
Linguistic	Student Journal, pp. 842, 845
Logical-Mathematical	Alternate Lab, p. 840; Thinking Lab, p. 843; Meeting Individual Needs, p. 857

CHAPTER

34 Reptiles and Birds

The setting sun turns daylight to dusk. You are walking along a shrubby knoll when you hear a sinister, rattling buzz coming from a large rock nearby. Rattlesnakes inhabit the area, and you know that they move onto rock surfaces in the evening because the rocks retain the warmth of the day as the evening air cools. When you look around carefully, you see the nested coils with raised rattle, and the menacing head with a flicking tongue. Knowing that the venom of this snake can kill a full-grown human, you slowly back away. Calming yourself, you recall that more people in the United States are hit by lightning than are bitten by poisonous snakes.

What does a legless, cold-blooded viper with poisonous fangs have in common with a bird? Scientists think that reptiles were the ancestors of birds. How do they know? The next time you see a bird up close, look at its scaly legs and the claws at the end of its toes. Think about the shelled eggs it lays. These are all characteristics of the early lizardlike animals from which birds evolved.

Concept Check

You may wish to review the following concepts before studying this chapter.

- Chapter 1: adaptation, evolution
- Chapter 28: vertebrate
- Chapter 30: gizzard
- Chapter 33: ectotherm

Chapter Preview

Laboratory Activities

Biolab: Design Your Own Experiment

- What is the ideal length and width for a bird's tail?

Minilabs

- How do down feathers and contour feathers compare?
- What types of foods do local birds prefer?

Reptiles and birds share other characteristics besides scales and claws. You can see that this is a nest. Can you tell that it was made by a crocodile? Both reptiles and birds lay shelled eggs.

837

CHAPTER 34

Introducing the Chapter

Ask students to observe the photos of the snake and bird and list the features they have in common. *Students may mention scales and egg laying.* Next, discuss features of birds and reptiles that differ. Point out the crocodile nest and explain how it differs from the nests of ground-nesting birds such as ducks. You may want to show students a photo of a duck nest. L1

Theme Development

Evolution is one theme woven throughout the chapter and is apparent in the discussions of the movement of animals to land and how dinosaurs may have evolved into birds. *Unity within diversity* is exemplified by the features reptiles have in common and a discussion of adaptations that led to the classification of reptiles into different orders. Similarly, birds have features in common but also have adaptations that make them suited for particular ways of life.

Concept Check

Students will use their knowledge of amphibians to develop an appreciation for the complete move to land by reptiles. They will use their knowledge of classification and taxonomy and expand on their knowledge of speciation, adaptive radiation, and convergence.

Assessment Planner

Choose assessment strategies from the following pages to evaluate the progress of your students.

Assess, pp. 846, 865
Alternate Lab, pp. 840-841
Minilabs, pp. 855, 861
Portfolio, pp. 844, 851, 856, 860
Thinking Lab, p. 843
Biolab, pp. 858-859
Chapter Review, pp. 867-869

SECTION

34.1

Prepare

Key Concepts

Students will study the features reptiles have in common and learn about the adaptations of crocodiles, alligators, lizards, and snakes. Origins of reptiles and their amniotic eggs are considered and discussed in terms of the movement of animals to land.

Block Scheduling

Look for this symbol for strategies that are useful in a block scheduling format. For more information on block scheduling, refer to Section 34.1 in the **Lesson Plans** booklet.

Materials

- Borrow reptilian skulls or skeletons from a college or museum for the Demonstration.
- Purchase chicken eggs for examination during *The Inside Story.*
- Purchase toy dinosaurs for use in the Activity.

1 Focus

Bellringer

Before presenting the lesson, display **Section Focus Master 75** on the overhead projector and have students answer the accompanying questions. L1 SAE

Discussion

Ask students to share their experiences and feelings about reptiles, especially snakes. If they express fear, ask why they think they have this fear. Point out that most primates show an instinctive fear of snakes.

SECTION

34.1 Reptiles

Section Preview

Objectives

Compare the characteristics of different groups of reptiles.

Explain how reptile adaptations make them suited to life on land.

Key Terms

amniotic egg
Jacobson's organ

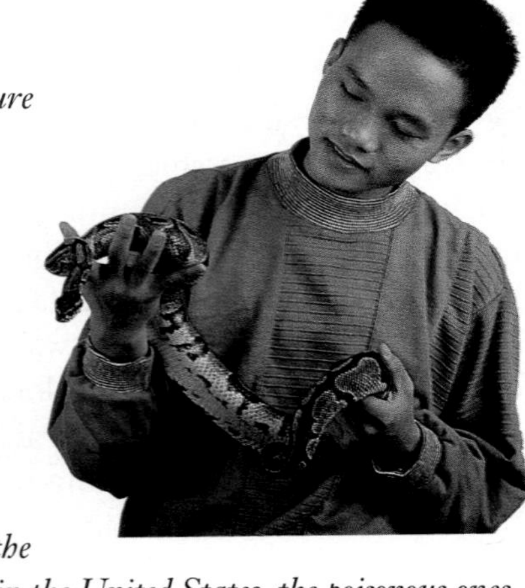

You may remember seeing an adventure movie in which a ferocious crocodile devours a villain who is trying to swim across a jungle river. Moviemakers often use crocodiles, alligators, lizards, and snakes in their films to convey a sense of fear to the audience. Fear of these "cold-blooded" reptiles may have been inspired more by legends, stories, and movies than by actual experiences. However, a few snakes are capable of killing humans. Of the approximately 120 species of snakes found in the United States, the poisonous ones include the rattlesnakes, moccasins, copperheads, and coral snakes. Even people who collect and study reptiles have a healthy respect for snakes.

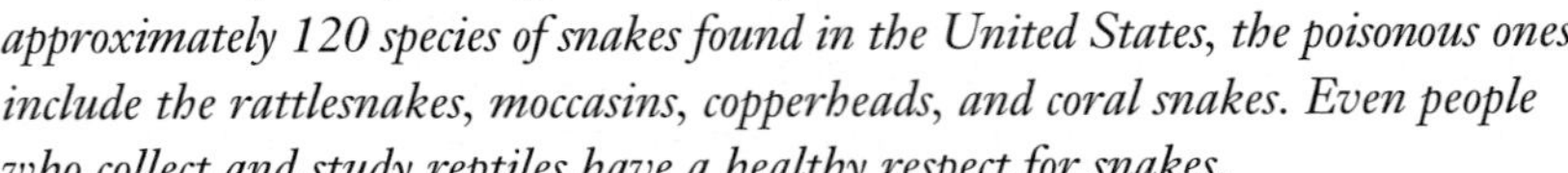

What Is a Reptile?

At first glance, it may be difficult to determine how a legless snake is related to a tortoise. Snakes, turtles, alligators, and lizards seem to be an extremely diverse group of animals, yet all share certain traits that place them in the class Reptilia.

Early reptiles, such as the cotylosaur shown in *Figure 34.1,* evolved important innovations that freed them from dependence on water. Reptiles have adaptations that enable them to complete their life cycles entirely on land. These adaptations released the cotylosaurs and other reptiles from the need to return to swamps, lakes, rivers, ponds, or oceans and allowed them to evolve into successful land animals.

Figure 34.1
Cotylosaurs, the earliest known reptiles, were probably the ancestors of the long-extinct dinosaurs as well as of today's living reptiles, birds, and mammals.

Reptiles have scaly skin

Unlike the moist, thin skin of amphibians, reptiles have a dry, thick skin covered with scales. Scaly skin, *Figure 34.2,* prevents the loss of body moisture and provides additional protection from predators. Because gas exchange cannot occur through scaly skin, reptiles are entirely dependent on lungs as their primary organ of gas exchange.

Program Resources

Section Focus Master 75 L1 SAE
Reinforcement and Study Guide, pp. 137-138 L1
Performance Assessment in the Biology Classroom, p. 37 L1 P
Basic Concepts Transparencies 54, 55 and **Masters** L1 SAE
Reteaching Transparency 34 and **Master** L1 SAE

Figure 34.2

Scales on a reptile's skin overlap like tiles on a roof.

The scales of reptiles, unlike the separate glossy scales of fish, are part of the skin itself. The scales are all connected to one another by hinges of skin.

To grow, young reptiles molt. Old scaly skin is replaced by new skin when reptiles molt.

Some reptiles have four-chambered hearts

Most reptiles, like amphibians, have three-chambered hearts. Some reptiles, notably the crocodilians, have a four-chambered heart that completely separates the supply of blood with oxygen from blood without oxygen. The separation enables more oxygen to reach body tissues. This separation is an adaptation to life on land that supports a higher level of energy use in reptiles than that found in amphibians.

Skeletal changes in reptiles

Look again at the illustration of the cotylosaur. This reptile had legs that were placed more under the body rather than out to the sides as in early amphibians. This positioning of the legs provides greater body support and makes walking and running on land easier for the reptiles. They have a better chance of catching prey or avoiding predators. Reptiles also have claws that facilitate food getting and protection. Additional changes in the structure of the jaws and teeth of reptiles allowed them to exploit other resources and niches on land.

Reptiles reproduce on land

Reptiles reproduce by laying eggs on land. Unlike amphibians, reptiles have no aquatic larval stage, and thus are not as vulnerable to water-dwelling predators as young amphibians are. Reptile hatchlings look just like adults, only smaller.

Although all of the adaptations discussed so far enabled reptiles to live successfully on land, the evolution of the amniotic egg was the adaptation that liberated reptiles from a dependence on water for reproduction. An **amniotic egg** provides nourishment to the embryo and contains membranes that protect it while it develops in a terrestrial environment. The egg functions as the embryo's total life-support system.

All reptiles have internal fertilization. The eggs are laid after fertilization, and embryos develop after eggs are laid. Most reptiles lay their eggs under rocks, bark, grasses, or other surface materials, but a few dig holes or collect materials for a nest. Most reptiles provide no care for hatchlings, but female crocodiles have been observed guarding their nests from predators.

2 Teach

Demonstration

LS **Visual-Spatial** Display reptilian skulls or skeletons. Have students compare and contrast them. Point out the locations of the appendages on the skeletons, or, if a snake is displayed, the absence of appendages. Discuss the location of the appendages in relation to life on land. L1

Different Viewpoints in Biology

Snake Locomotion

Traditionally, biologists thought snakes did not use much energy for locomotion because they do not have to hold their bodies upright or lift their weight up and down. However, studies have shown that snakes use as much energy to move as comparably sized birds, mammals, and other reptiles.

To determine how much energy an animal uses, a gas mask may be placed on the animal and oxygen consumption measured while the animal moves on a treadmill. To conduct the experiment on snakes, biologists used plastic tubing that fit over the snakes' nostrils. The tubing was connected to a gas analyzer, and the "snake treadmill" was covered with Astroturf to provide the snakes with the friction needed for movement.

GLENCOE TECHNOLOGY

Videodisc

Biology: The Dynamics of Life
Disc 2, Side 1
Sea Turtle (Ch. 5)

STVS: Animals
Disc 5, Side 2
Reptilian Sex Change (Ch. 10)

THE INSIDE STORY

The Amniotic Egg

Purpose

Visual-Spatial Students will study the amniotic egg and learn how this adaptation enabled reptiles to live out of water.

Visual Learning

- Challenge students to describe the differences between frog eggs and turtle eggs. *The frog eggs are surrounded by a jelly-like substance that requires moisture. The reptile eggs are covered by a shell.* Ask how these differences might reflect the way of life of these animals. *Frogs must live near water to permit development of their eggs. Reptile eggs can develop on land.* L1
- Ask students to carefully open a raw chicken egg and note the shell, albumen, yolk, and the tough membrane inside the shell. Ask what parts of the egg shown in *The Inside Story* are missing from their egg. L1

Assessment

Performance Assessment in the Biology Classroom: p. 37, Model of Chicken Egg Membranes. Have students carry out the activity to develop a model of amniotic egg membranes. L1 P

THE INSIDE STORY

The Amniotic Egg

The evolution of the amniotic egg was a major step in reptilian adaptations to land environments. Amniotic eggs enclose the embryo in amniotic fluid, provide a source of food in the yolk, and surround both embryo and food with membranes and a tough, leathery shell. These innovations in the egg help prevent injury and dehydration of the embryo as it develops on land.

Leatherback turtle

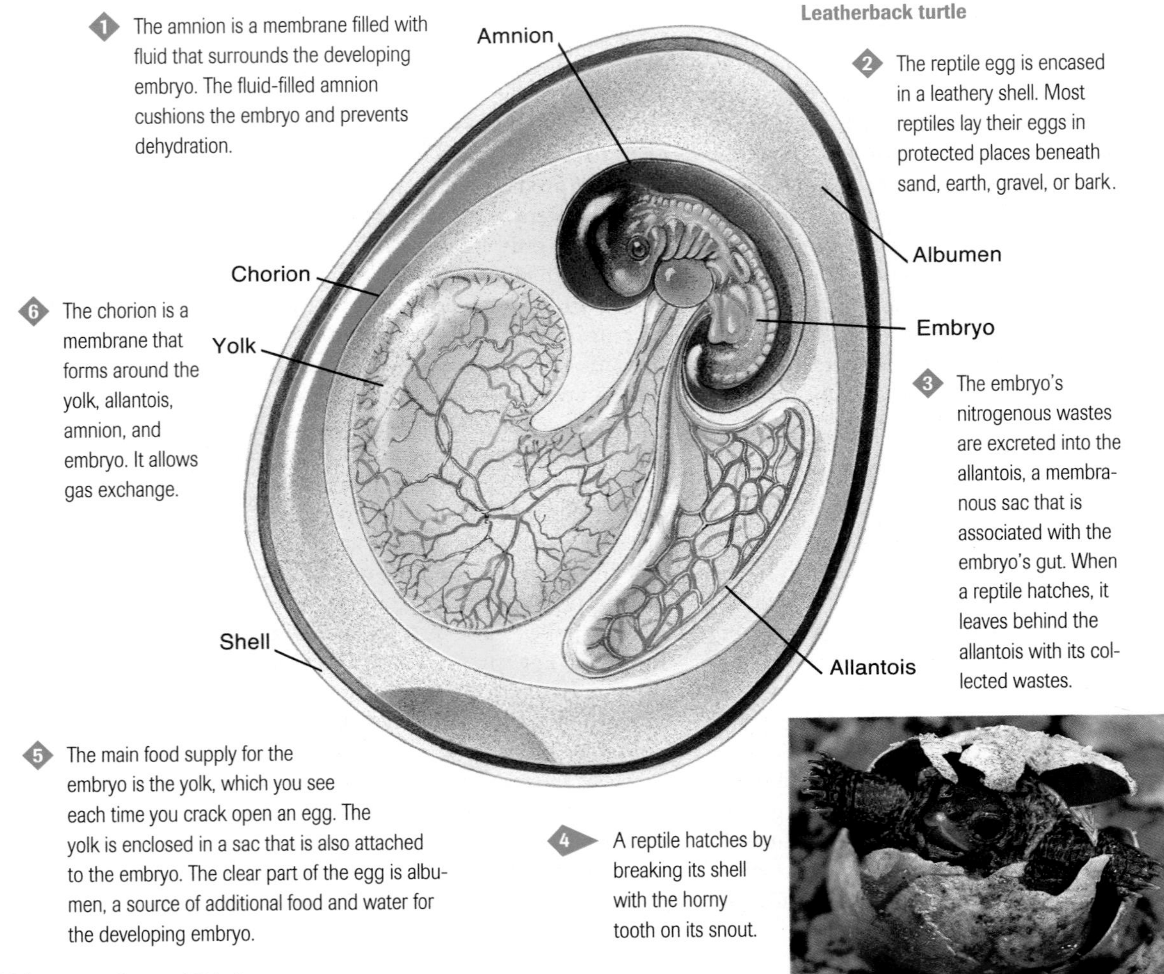

1 The amnion is a membrane filled with fluid that surrounds the developing embryo. The fluid-filled amnion cushions the embryo and prevents dehydration.

2 The reptile egg is encased in a leathery shell. Most reptiles lay their eggs in protected places beneath sand, earth, gravel, or bark.

3 The embryo's nitrogenous wastes are excreted into the allantois, a membranous sac that is associated with the embryo's gut. When a reptile hatches, it leaves behind the allantois with its collected wastes.

4 A reptile hatches by breaking its shell with the horny tooth on its snout.

5 The main food supply for the embryo is the yolk, which you see each time you crack open an egg. The yolk is enclosed in a sac that is also attached to the embryo. The clear part of the egg is albumen, a source of additional food and water for the developing embryo.

6 The chorion is a membrane that forms around the yolk, allantois, amnion, and embryo. It allows gas exchange.

Alternate Lab | **How is body color adaptive?**

Purpose

Logical-Mathematical Students observe how color affects the rate at which organisms absorb heat.

Materials

thermometer, black paper, white paper, tape, lamp, metric ruler, empty frozen juice cans with holes in one end, clay

Procedure

Give students the following directions.

1. Make a data table for starting temperature, final temperature, and total temperature change for a black can and a white can.
2. Cover one can with black paper and the other with white paper. Tape the paper in place.
3. Place a thermometer into the hole in each can top and secure it with clay. Make sure the hole is tightly sealed with clay.
4. Place both cans about 5 cm away from the bulb of a lamp. Do not turn on the lamp.
5. Record the starting temperature of both cans in your data table.

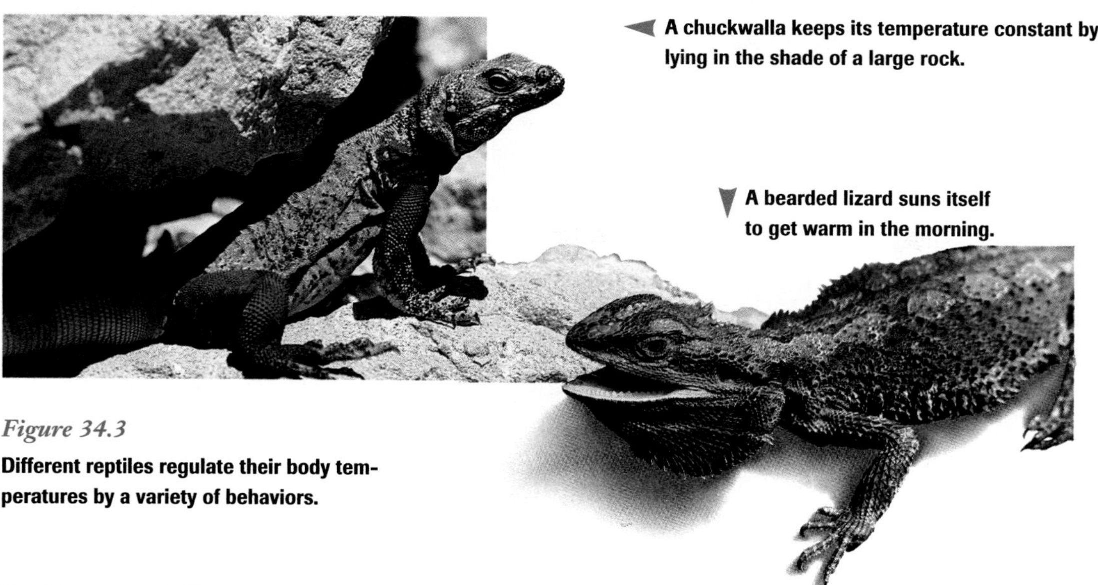

A chuckwalla keeps its temperature constant by lying in the shade of a large rock.
A bearded lizard suns itself to get warm in the morning.

Figure 34.3
Different reptiles regulate their body temperatures by a variety of behaviors.

Reptiles are ectotherms

Even though reptiles are different from amphibians in many ways, they are similar in one way. Both amphibians and reptiles are ectotherms. Their body temperatures depend on the temperatures of their environments. In the cool morning, a turtle might pull itself out of the pond or swamp and bask on a log in the sunlight until noon. Then, when the temperature gets a little too warm, the turtle may slip back into the cool water. This example shows that even though reptiles cannot control their body temperatures internally, they can use behavioral adaptations to compensate for changes in environmental temperature. *Figure 34.3* shows other examples of behavioral adjustment of body temperature.

Because reptiles are dependent on the temperatures of their surroundings, they do not inhabit extremely cold regions. Reptiles are numerous in temperate and tropical regions, where climates are warm, and in hot desert climates. Many reptiles become dormant in moderately cold environments such as in the northern United States.

Reptile Adaptations

Like other animals, reptiles have adaptations that enable them to find food and to sense the world around them.

How reptiles obtain food

Most turtles and tortoises are too slow to be effective predators, but that doesn't mean they go hungry. Most are herbivores, and those that are predators prey on worms and mollusks. Snapping turtles, however, are extremely aggressive. They attack fishes and amphibians, and also pull ducklings under water.

Lizards eat insects primarily. The marine iguana of the Galapagos Islands is one of the few herbivorous lizards, feeding on marine algae. The Komodo dragon, the largest lizard, is an efficient predator, sometimes even of humans. Although lizards such as the Komodo dragon may look slow, they are capable of bursts of speed, which they use to catch their prey.

Science, Technology, and Society

Saving Turtles

Turtles live a long time, are slow to mature, and do not reproduce quickly. Many turtles are endangered due to habitat loss, pollution, and harvesting for food and trade.

Most sea turtles are threatened with extinction because their nesting beaches have become unusable or they have been killed for their meat, shells, or skins.

A group of 150 scientists have formed a coalition to coordinate research and conservation efforts regarding turtles. At this time, 28 different projects are underway, from the study of the extremely rare angonoka of Madagascar to the small bog turtle of the United States.

Biologists study loggerhead and leatherback turtles using transmitters attached to the turtles. Tracking the turtles by satellite, biologists can determine if the turtles are following their traditional migration routes and how deep the turtles dive in the ocean.

6. Develop a hypothesis as to which can will show a higher temperature at the end of 10 minutes.
7. Turn on the lamp and allow it to shine on the cans for 10 minutes.
8. Record the final temperature of both cans. Calculate the total change in temperature for each can and record it in your data table.

Analysis

Ask students to answer the following questions.

1. Which color, black or white, had the greater temperature change after being heated for ten minutes? *Black*
2. Which color absorbs light better and warms up faster? *Black*
3. Did your data support your hypothesis? Explain. *Yes, if students said that the black can would absorb more heat.*
4. Which should absorb more light from the sun, a dark animal or a light animal? Why? *Dark; dark colors absorb more light and heat up more quickly.*

Assessment

Portfolio: The ability to change skin color is an adaptation in many reptiles. Ask students to explain how this adaptation may help reptiles survive. L1 P

Misconception

Many people believe that the age of a rattlesnake can be determined by counting the number of rattles on its tail. However, because a new rattle forms each time the snake sheds its skin, and shedding can occur many times each year, there is no correlation between the number of rattles and the age of the snake. L1

Health Connection

Treating Snake Bites

Herpetologists recommend that when people confront a poisonous snake, they should back away quietly. If bitten, experts say that the most important thing to do is to remain calm to prevent an increase in circulation rate. Medical help should be sought as soon as possible. You should not suck, cut, or apply ice to the bite, nor should a tourniquet be applied. The person with the bite should remain lying down with the bitten extremity at heart level. In addition, a constriction bandage should be applied between the bite and the heart.

Visual Learning

Figure 34.4 Can you figure out why the snapping turtle is an effective predator? *The snapping turtle has a long, flexible neck and a sharp, horny tooth on its snout.*

Figure 34.4

Many reptiles are skillful predators that obtain prey in a variety of ways.

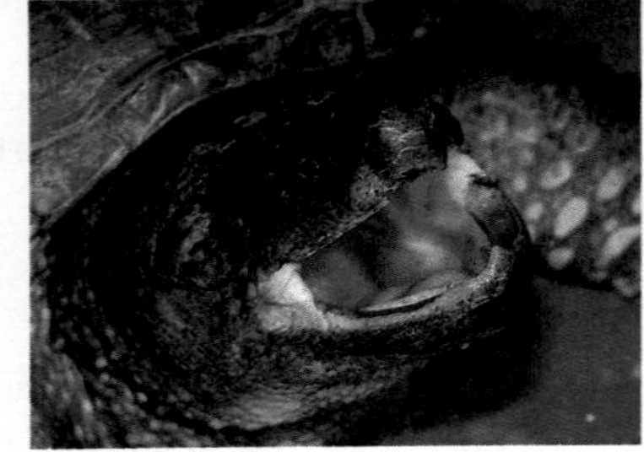

The snapping turtle is common in North America. Look at its head. With its long, flexible neck, the snapping turtle is an effective predator.

Some snakes can swallow eggs or whole animals that are larger than their heads because their lower jaws are loosely attached to their skulls.

The Komodo dragon is a predator that captures its prey in large jaws. It can kill animals as large as a deer or even a water buffalo.

Snakes are also effective predators. Some, like the rattlesnake, have poison fangs that they use to subdue or kill their prey. A constrictor wraps its body around its prey, tightening its grip each time the prey animal exhales. Several of these reptiles are shown in *Figure 34.4*.

How reptiles use their sense organs

Reptiles have a variety of sense organs that can detect danger or potential prey. How does a rattlesnake know you are nearby? The heads of some snakes, as shown in *Figure 34.5*, have heat-sensitive organs or pits that enable them to detect tiny variations in air temperature brought about by the presence of warm-blooded animals.

Snakes and lizards are equipped with a keen sense of smell. Remember the rattlesnake's flicking tongue? The tongue is picking up chemicals in the air. The snake then draws its tongue back into its mouth and inserts it into a structure called **Jacobson's organ,** *Figure 34.5*.

Figure 34.5

Snakes have sense organs that enable them to detect prey or identify chemicals in the environment.

Rattlesnakes have heat-sensitive pits below their eyes that enable them to detect prey in total darkness. The snake can tell the exact distance and direction in which to strike because the pits are paired.

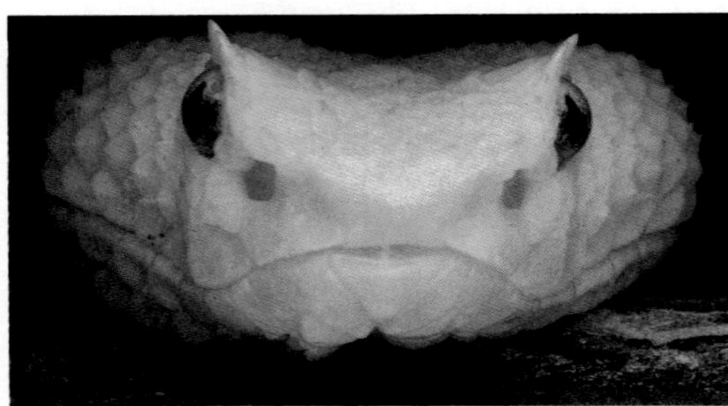

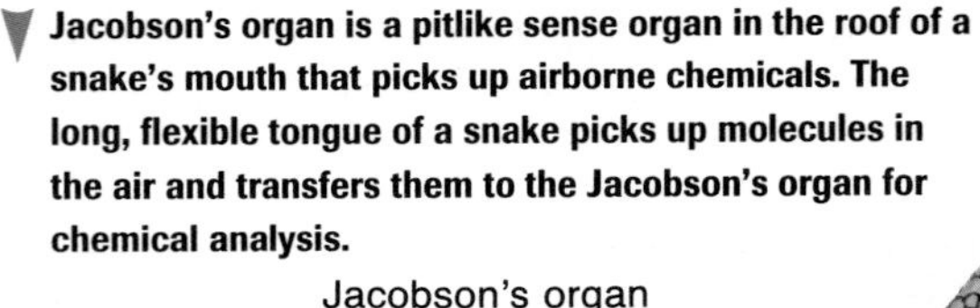

Jacobson's organ is a pitlike sense organ in the roof of a snake's mouth that picks up airborne chemicals. The long, flexible tongue of a snake picks up molecules in the air and transfers them to the Jacobson's organ for chemical analysis.

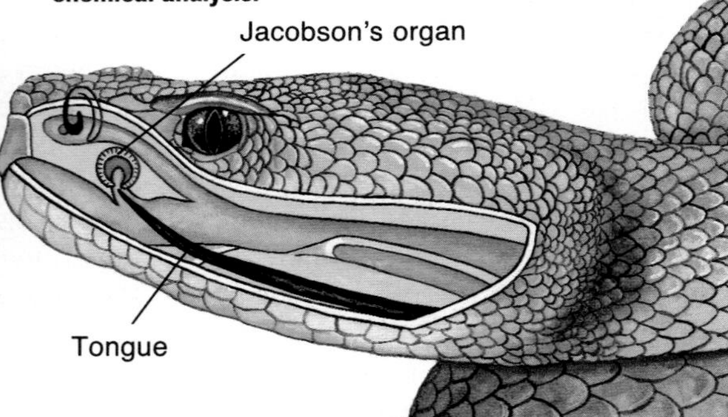

STUDENT JOURNAL

Snake Bites Ask students to read the article, "Snakebite Succor" by Rick Weiss on pages 360-362 in *Science News,* December 8, 1990. Have them summarize the article, indicating the current problems with snakebite antivenoms and the ways scientists hope to resolve these problems. L2 LS

GLENCOE TECHNOLOGY

Videodisc

STVS: Animals
Disc 5, Side 2
Sea Turtle Mystery (Ch. 8)

The Diversity of Reptiles

Gracefully gliding snakes and quickly darting lizards are grouped together in the order Squamata. Turtles, slowly plodding and carrying heavy shells, belong to the order Chelonia. Basking crocodiles and alligators, classified in the order Crocodilia, may look clumsy but are surprisingly quick hunting machines that snap up fishes and lunge out at antelopes and other large animals that come to the river to drink.

Turtles have shells

Some turtles are aquatic, and some live on land. Turtles that live on land are called tortoises. Most turtles can draw their limbs, tail, and head into their shells for protection against predators. Although turtles have no teeth, they do have powerful jaws with a beaklike structure that is used to crush food.

Figure 34.6

In the past, sailors killed Galapagos tortoises for food. As a result, their numbers declined rapidly. When just a few tortoises were left, they rarely met one another, and reproduction nearly ceased. Scientists collected the remaining tortoises and put them in an enclosure where they could easily find a mate. Soon, adult and hatchling tortoises were released back to their home areas. The tortoises are now protected by law.

ThinkingLab Analyze the Procedure

Why are tortoises becoming frail?

The number of desert tortoises of the southwestern United States has recently begun to decline. These tortoises are not growing as large as tortoises in the past, and their shells and bones are more frail and easily fractured. Scientists hypothesize that the reason for this frailty is a deficiency of calcium in the tortoises' diets. Calcium is important for the growth and development of shells and bones. Over the last century, many of the native weeds and succulent plants in the tortoises' habitat have been replaced with cultivated plants.

Analysis

In order to find out if tortoises' diets are lacking in calcium, decide which of the following procedures should be followed.

1. Measure the calcium in native and cultivated grasses.
2. Measure the calcium in the bones of tortoises.
3. Offer a group of tortoises a calcium-rich diet in one dish and a calcium-deficient diet in another dish, and measure the amount eaten from each dish.

Thinking Critically

Analyze the above possibilities for ways to determine if turtles have calcium-deficient diets. Choose the procedure that will produce data that will support the hypothesis. Explain why you made your choice.

Tortoises live on land, foraging for fruit, berries, and insects. The largest tortoises in the world, shown in *Figure 34.6*, are found on the Galapagos Islands off the coast of Ecuador.

Some adult marine turtles swim enormously long distances to lay their eggs. Like salmon, these turtles return from their feeding grounds to the place where they hatched. For example, green turtles travel from the coast of Brazil to Ascension Island in the Atlantic, a distance of more than 4000 km.

PROJECT
Protecting the Environment

Ask students to role-play a citizen group deciding how to protect a beach that is used for nesting green turtles. Have students work in groups of eight to ten and assume the role indicated on a card you have prepared. Cards should identify the following roles: teenage surfer, mayor, condominium developer, beach homeowner, turtle biologist, president of the turtle protection society, beach food-and-souvenir concession owner, state wildlife protection officer. Prior to role-playing, each student should prepare an outline of points that will be presented by the character he or she assumes. L1 COOP LEARN LS

ThinkingLab Analyze the Procedure

Purpose

LS **Logical-Mathematical** Select a procedure that will provide data to test the hypothesis that cultivated plants supply less calcium than native grasses.

Process Skills

recognize cause and effect, separate variables, analyze procedures

Teaching Strategies

- Show students slides or photos of desert tortoises.
- Before they read the Thinking Lab, ask students to explain what they know about the need for calcium in the diets of vertebrates. L2
- Ask students working in groups to provide background information about the plants that were native to the Southwest and the cultivated plants that replaced them. Have other groups report on desert tortoises. L1 COOP LEARN

Thinking Critically

Best procedure: Measure the calcium in native and cultivated grasses. If native grasses have more calcium than cultivated grasses, then a diet lacking in calcium could be a factor in frail shells and bones. This would not be absolute proof, however. Turtles might be fed the two different diets in a controlled setting and measurements taken of calcium in their bones.

Assessment

Portfolio: Have students assume that they have done the experiment described in the *Thinking Critically* and discovered there is no difference in the calcium of native and cultivated grasses. Ask them to suggest new hypotheses about what the problem might be. L2 P

Zoos: Preserves or Breeding Grounds?

Purpose

Students will explore problems faced by modern zoos and analyze what critics say about zoos.

Teaching Strategies

- Ask students to recount why they think zoos should be preserved. Have them consider a zoo's educational value and the opportunity it affords for the community.
- Explain that most zoos are financially unable to reform their programs to meet their new mission as wildlife conservationists. Ask what kind of fund-raising efforts are appropriate for zoos. They might solicit donations from people who visit the zoo and from the people in the community. They might also raise money through the sale of books, magazines, or objects created to increase awareness about animals and about zoo functions.

INVESTIGATING the Issue

Debating the Issue One side should defend the idea that the zoo is a wildlife preserve helping to preserve endangered species by captive breeding. The other side should defend the idea that captive species promote species favoritism and that the money spent on zoos should be used to preserve natural wildlife habitats.

Going Further

If you have a local zoo, ask students to find out about the zoo's captive-breeding programs or other ongoing research.

Issues
BIOLOGY & SOCIETY

Zoos: Preserves or Breeding Grounds?

Faced with rising costs and negative press from critics who claim that zoos are no better than prisons, modern zoos are struggling to survive. Many metropolitan zoos are questioning their existence and redefining their mission. They know that like the animals they exhibit, the zoos must either adapt or die. The most dedicated zoos—and the richest—have started natural habitat programs and are experimenting with captive breeding.

Zoos help educate Originally created for public entertainment, zoos began educational programs and species conservation in the 1940s. Zoo reform began in the 1980s in response to public outrage over poor conditions. Today's most modern zoos—such as the San Diego and Cincinnati Zoos—are using the latest reproductive techniques, such as artificial insemination and test-tube fertilization, to save endangered species and improve the gene pools of others. Through such innovative efforts, captive-bred female Siberian tigers have successfully been used as surrogate mothers for Bengal tigers.

Less need to capture wild animals Through captive-breeding programs, zoos are able to lessen their reliance on wild populations for their exhibits. The more sophisticated breeding laboratories are able to freeze and save genetic materials from endangered species for future breeding purposes.

Different Viewpoints

Many biologists argue that the mission that modern zoos have set for themselves is unattainable. Even though there are well over 1 million described species of animals on the planet, one-third of these may become extinct over the next century. As regrettable as that may be, the most strident zoo critics feel extinction is better than confinement in an artificial environment.

Those against zoo policy say that biodiversity is more likely to be preserved through public education than through captive-breeding programs. They accuse zoo managers of species favoritism, saying that because the managers are unable to save all endangered species, they tend to select only the most attractive or popular species.

INVESTIGATING the Issue

Debating the Issue What should be the role of modern zoos?

Crocodiles include the largest living reptiles

In contrast to marine turtles, crocodiles don't migrate. They may spend their days alternately basking in the sun on a riverbank and floating like motionless logs. Only their eyes and nostrils remain above water. Crocodiles can be identified by their long, slender snouts, whereas alligators have short, broad snouts, as shown in *Figure 34.7.* Both animals have powerful jaws with sharp teeth that can drag prey underwater and hold it there until it drowns. One species of alligator is found throughout many of the water habitats of the southeastern United States. The one species of crocodile in the United States can be found only in southern Florida. Alligators and crocodiles lay eggs in nests on the ground. Unlike other reptiles, these animals stay close to their nests and guard them from predators.

Figure 34.7

Alligators and crocodiles use their powerful tails to swim, moving so rapidly that few swimming animals can escape. On land, the tail is a formidable weapon, able to knock most opponents to the ground. Even though these animals have stubby legs, they can move quickly (up to 17 km per hour) over short distances.

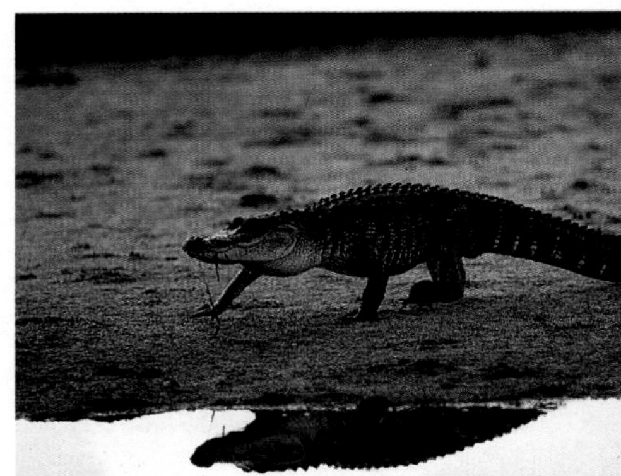

PORTFOLIO

Comparing Alligators and Crocodiles
Students often confuse alligators and crocodiles. Have students study pictures of alligators and crocodiles shown in magazines or reference books. Ask students to study the features of the two animals, paying attention to the shapes of their heads. Direct students to make a drawing of an alligator and a crocodile on the same sheet of paper. Ask students to include captions for their drawings that explain how alligators and crocodiles can be distinguished from one another based upon the shapes of their heads. Have them include the habitat ranges of the American crocodile and alligator. L2 P LS

Figure 34.8

Lizards have many adaptations that enable them to live in a variety of different habitats.

Geckos are small, nocturnal lizards that live in warm climates, such as those of the southern United States, West Africa, and Asia. The toe pads of some geckos enable them to walk across walls and ceilings.

Chameleons are tree-dwelling lizards that can change color.

Only the Gila monster of the southwestern United States and Mexico and the beaded lizard of Mexico are poisonous lizards.

Snakes and lizards are found in many environments

Lizards, shown in *Figure 34.8,* are found in many types of habitats in all but the polar regions of the world. Some live on the ground, some burrow, some live in trees, and some are aquatic. Many are adapted to hot, dry climates.

Snakes, in contrast to most vertebrates, have no limbs and lack the bones to support limbs. Exceptions are pythons and boas, which retain bones of the pelvis. The many vertebrae of snakes permit fast undulations through grass and over rough terrain. Some snakes even swim and climb trees!

Snakes usually kill their prey in one of three ways. Remember that constrictors wrap themselves around their prey. If you ever watch someone handle a constrictor, you will notice that the handler never lets the snake start to wrap around his or her body. The snake is always held carefully so that its tail does not cross over its own head to begin a coil. Common constrictors include boas, pythons, and the anacondas.

Venomous snakes use poison to kill their prey. These include rattlesnakes, cobras, and vipers, which inject poison from venom glands, shown in *Figure 34.9.* Most snakes are neither constrictors nor poisonous. They get food by grabbing it with their mouths and swallowing it whole. Snakes eat rodents, amphibians, insects, fishes, and eggs.

Figure 34.9

Many poisonous snakes have venom glands and hollow fangs for injecting venom. Venom may either paralyze the prey so it cannot run away or kill the prey immediately.

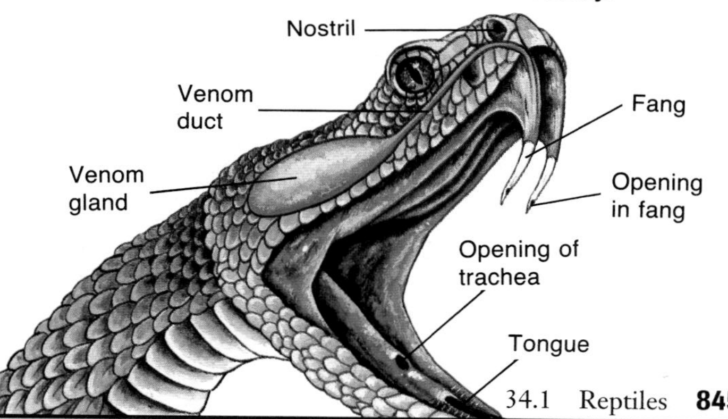

Bioethics

Saving the Green Iguana In Central America, the green iguana is considered a delicacy and has been hunted since the mid-1500s. Today, the green iguana faces extinction in many parts of Central America. Green iguanas live in the treetops of tropical rain forests that are being cut down.

Recent efforts by iguana researchers have shown that green iguanas can be raised in enclosures within rain forest trees. If people can be convinced to develop iguana ranches, the natural iguana population may be saved from extinction. In addition, parts of the rain forest will be preserved because the iguana ranchers will require natural rain forest areas to raise their animals.

Visual Learning

Figure 34.9 Ask students to study the diagram. Elicit what structures shown would be absent from or different in nonpoisonous snakes. *Nonpoisonous snakes would not have vemon glands or vemon ducts. The teeth or fangs of a nonpoisonous snake would not have an opening or be hollow.* LS

PROJECT Raising Lizards

Have students set up and maintain a terrarium for a lizard. Prior to setting up the habitat, have them determine what type of lizard will be maintained and conduct research to determine requirements for proper temperature, light, and diet. Have students present a plan that outlines the procedures and materials needed prior to setting up the terrarium. L1

STUDENT JOURNAL

Protecting Local Reptiles Ask students to contact their state or local division of wildlife to find out about programs underway to protect locally endangered reptiles. Ask them to prepare an illustrated report of their findings that identifies the endangered reptiles in your area. L1 LS

SECTION 34.1

Activity

LS Interpersonal Purchase a variety of plastic dinosaurs from a toy store. Have students form cooperative groups and give each group a different dinosaur. Ask students to select one of the features of their dinosaur and speculate about how it evolved. Ask them to present their explanations to the class.

Follow this exercise with a review of the basic ideas of evolution. Challenge students to revise their explanations by including the terms *variation* and *natural selection* in their explanations. Have students conclude their explanations with the idea that only the dinosaurs best suited to their environments survived to produce offspring like themselves. L1 COOP LEARN

Visual Learning

Figure 34.10 Which group of modern reptiles is most closely related to dinosaurs? *Crocodiles*

3 Assess

Check for Understanding

Have students brainstorm a list of concepts they learned about reptiles. Write the list on the chalkboard and ask students to classify each concept as being related to adaptations, characteristics, origins, or the importance of reptiles. L1

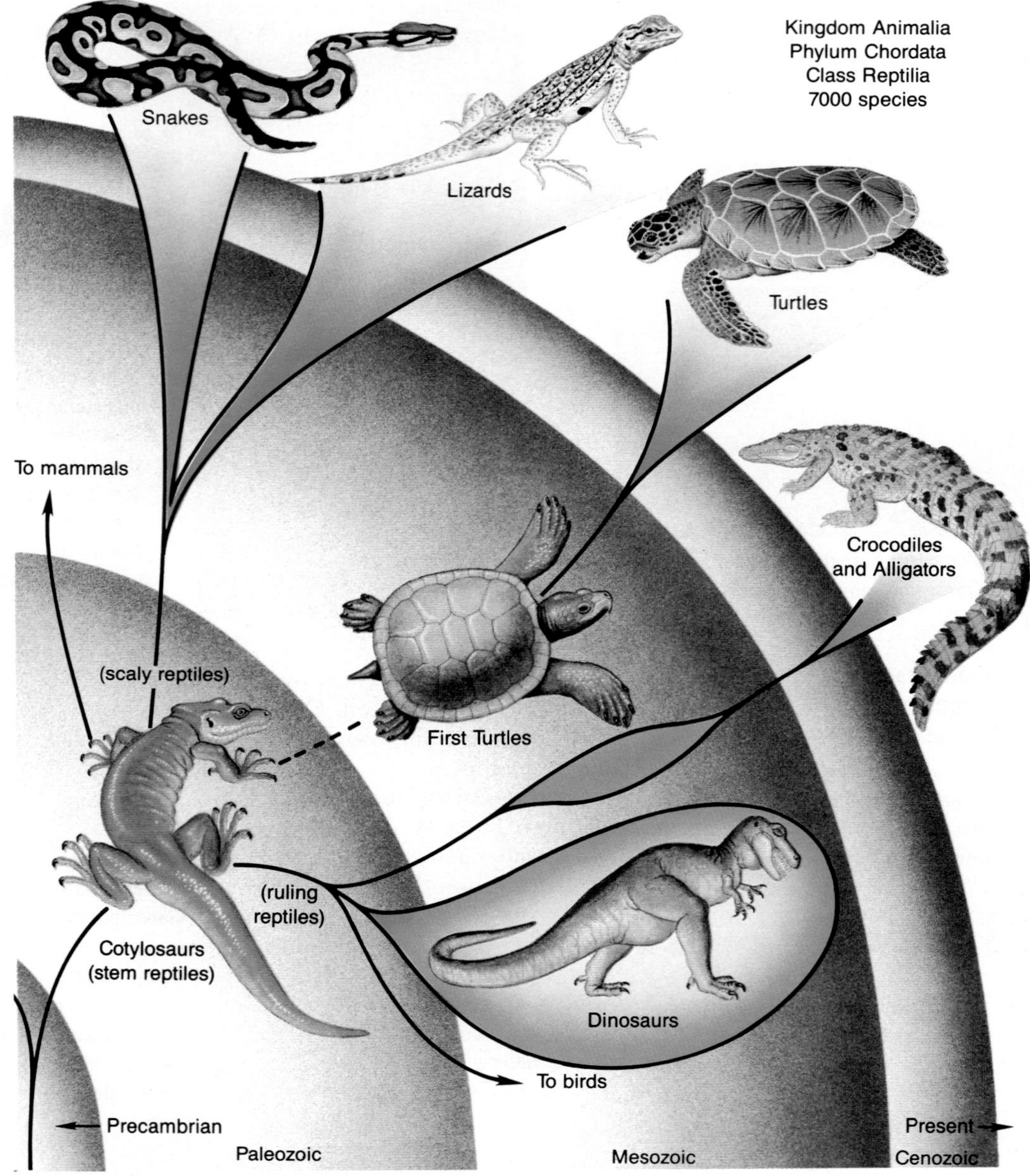

Figure 34.10

Reptiles evolved from amphibians 300 million years ago during the late Paleozoic Era. Most orders of reptiles have remained similar to their fossil ancestors. The radiation of orders of reptiles on the Geologic Time Scale shows their relationships. Which group of modern reptiles is most closely related to dinosaurs?

GLENCOE TECHNOLOGY

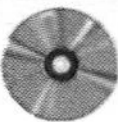

Videodisc

Biology: The Dynamics of Life
Disc 2, Side 1
Snake Feeding (Ch. 6)

STVS: Animals
Disc 5, Side 2
Rattlesnakes (Ch. 11)

Origins of Reptiles

You may have marveled at dinosaurs ever since you were very young. These animals were the most numerous land vertebrates during the Mesozoic Era. Some were the size of chickens, and others were the largest land dwellers that ever lived. What do today's reptiles have in common with these giants of the past?

The ancestors of snakes and lizards are traced to a group of early reptiles, called scaly reptiles, that branched off from the ancient cotylosaurs. The name "scaly reptiles" may be misleading because it implies that other reptiles lacked scales—which is not true. Although the evolutionary history of turtles is incomplete, scientists have suggested that they may also be descendants of cotylosaurs. Dinosaurs and crocodiles are the third group to descend from cotylosaurs, *Figure 34.10.* Although scientists used to think that birds arose as a separate group from this third branch, many now theorize that birds actually are the living descendants of the dinosaurs. The fourth order of reptiles, *Rhynchocephalia*, is represented by one living species, the tuatara, *Figure 34.11.* The tuatara is the only survivor of a primitive group of reptiles that became extinct 100 million years ago.

Figure 34.11

The tuatara, *Sphenodon punctatus*, is found only in New Zealand. It has ancestral features, including teeth fused to the edge of the jaws, and a skull structure similar to that of early Permian reptiles.

Section Review

Understanding Concepts

1. Choose one adaptation of early reptiles and explain how it enabled these animals to live on land.
2. Describe two ways in which turtles protect themselves.
3. How do snakes use the Jacobson's organ for finding food?

Thinking Critically

4. Explain how the development of a dry, thick skin was an adaptive advantage for reptiles.

Skill Review

5. **Classifying** Set up a classification key that allows you to identify a reptile as a snake, lizard, turtle, or crocodile. For more help, refer to Organizing Information in the *Skill Handbook*.

SECTION 34.1

Reteach

For each reptile group, have students develop a table with the following heads: Representative organisms, Habitat, Food acquisition, Reproduction, Locomotion, Breathing, and Protection. Have students complete their tables and then review them as a class. L1

Extension

Ask a group of students to find a children's story, folktale, or nursery rhyme that depicts a reptile in a negative way. Ask them to rewrite the story so the reptile is depicted in a positive way. Ask them to read the before and after versions of the story to the class. L2

Assessment

Portfolio: Ask students to prepare a new zoo exhibit for an unusual reptile. Explain that they should decide which reptile they wish to display, how they will obtain it, and describe the kind of exhibit space and care the reptile will need for survival. L1 P

4 Close

Activity

Ask students in their groups to develop a commercial in which a nonpoisonous reptile speaks about how it wants to be treated by people, what its habitat and food requirements are, and why it should not be feared. L1

Answers to Section Review

1. Legs are located under the body rather than out to the sides, and they have claws. Body structure and claws enable movement across land.
2. Turtles can draw limbs, heads, and tails into their hard shells. Turtles may use their powerful jaws to crush other animals.
3. The tongue picks up chemicals in the air. The snake then draws its tongue back into its mouth and inserts it into the Jacobson's organ for chemical analysis.

Thinking Critically

4. Dry, thick skin prevented reptiles from drying out.

Skill Review

5. Make sure students start with general features of reptiles such as scaly dry skin, claws, amniotic eggs, and being ectothermic. Next, they should use the main features for each group: Turtles: shells, no teeth; Crocodiles: jaws with teeth, powerful tails, four-chambered hearts, aquatic; Snakes: limbless; Lizards: long tails, legs.

Focus On
Dinosaurs

Purpose

Students will learn about dinosaurs, including their ways of life, how they are classified, and how their fossils are excavated and reconstructed. Students will also explore theories about whether dinosaurs were warm-blooded or cold-blooded and why they became extinct.

Teaching Strategies

Field Trip

Arrange a class trip to a museum that has a dinosaur exhibit. Ask students to examine the skulls and feet of the dinosaurs on exhibit and hypothesize what their diets may have been. If you do not have access to a museum, you might have students use diagrams of skulls and feet of various dinosaurs to make their hypotheses.

Focus On
Dinosaurs

The worlds of the dinosaurs

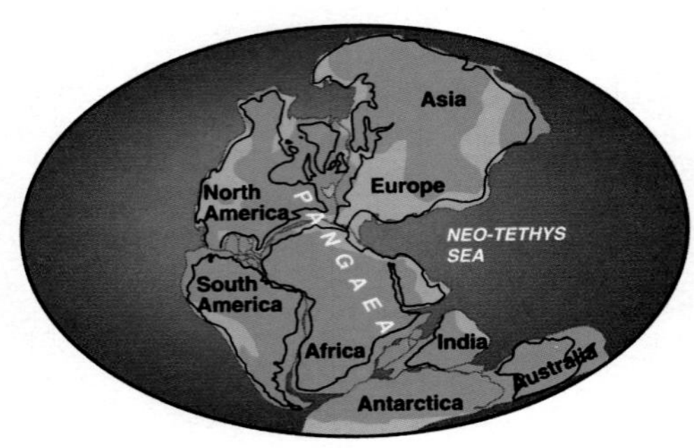

One landmass During the late Triassic period, about 210 million years ago, the continents as we know them today did not exist. The surface of Earth included vast seas and one large landmass, Pangaea. Almost imperceptibly, however, Pangaea was splitting apart, and separate continents began drifting across the globe. New species of animals began to emerge. These animals, descended from ancient reptiles and destined to rule Earth for the next 130 million years, were the dinosaurs. The large dinosaur at the far right is a *Plateosaurus.* The two smaller ones are *Coelophysis.* An early crocodile is in the water.

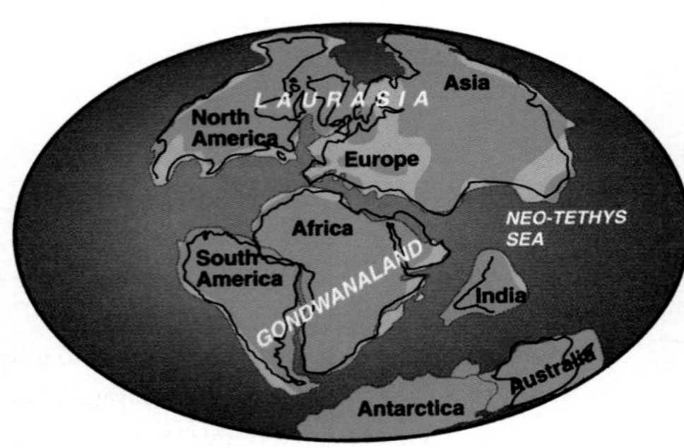

New species emerge By the late Jurassic period, the modern continents were beginning to take shape, although Africa still clung stubbornly to South America, and Australia remained attached to Antarctica. The temperature was warming up, enabling life to flourish in new areas around the world. The dinosaurs, too, were flourishing. The dominant plant life during this time included tree ferns and conifers. Examples of dinosaurs living during the Jurassic period were *Stegosaurus* (middle), *Diplodocus* (background), and *Allosaurus* (right).

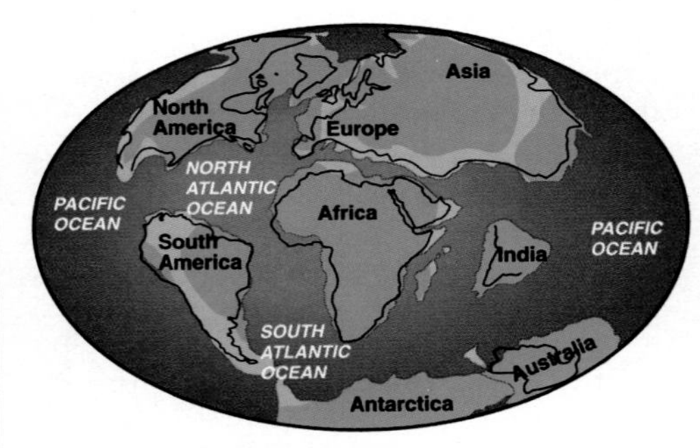

The beginning of the end By the late Cretaceous period, 100 million years ago, the continents had almost assumed the shape and position they have today. New vegetation, including flowering plants, and new species of dinosaurs, such as *Tyrannosaurus rex* (right) appeared. Other dinosaurs that lived during this period included *Pachycephalosaurus* and the carnivorous *Deinonychus* (left). *Sinornis,* a seven-inch-long link to modern birds, is also shown. About 65 million years ago, the dinosaurs died out. Exactly why is still a mystery. Today, only fossil evidence remains to tell their fascinating story.

Meeting Individual Needs

Visually Impaired Purchase plastic reptiles with scientifically accurate shapes and features from toy stores. Allow visually impaired students to handle the toys to develop an understanding of the shapes and features of these animals. Provide students with some type of scale to allow them to estimate the true size of the actual organisms. L1 LS

Plateosaurus

Coelophysis

Early crocodile

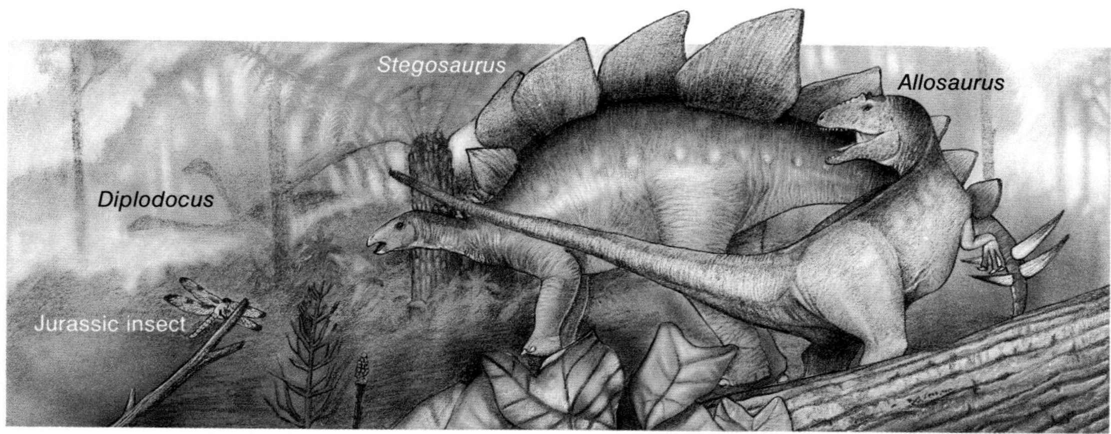

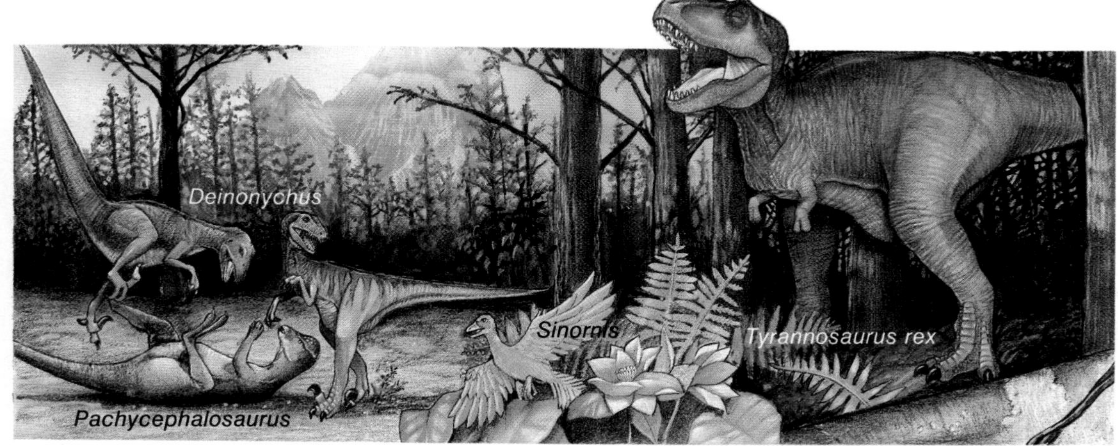

Focus On

Misconception

Explain to students that movies often lead people to believe that people and dinosaurs existed at the same time. Direct students' attention to the diagrams representing the Triassic period, Jurassic period, and Cretaceous period. Emphasize that the last of these periods, the Cretaceous, ended about 65 million years ago. Explain that humans have been on Earth for less than 2 million years. Use this fact to emphasize that humans and dinosaurs did not occupy Earth at the same time.

Enrichment

Have interested students research and report on animals living at the same time as the dinosaurs that were not dinosaurs. Suggestions for such animals may include: the flying reptiles or pterosaurs, the aquatic plesiosaurs, giant sea turtles, and sailback reptiles such as *Dimetrodon.* Have students indicate in their reports in what parts of the world fossils of the animal they have researched have been found. L1 P

Time Line

Have students develop a time line that traces the development of organisms from the earliest reptiles to the beginning of the Age of Mammals. Instruct students to include both plants and animals on their time lines. L1 P

STUDENT JOURNAL

A Historical Account Have students imagine they have gone back in time to the Cretaceous Period. Ask them to write an account of what they will see for a newspaper article. Challenge students to be creative as well as scientifically accurate. L1

Focus On

Using Science Terms

On the chalkboard, write the following prefixes and suffixes with their meanings. *Brachio-* means "arm"; *diplo-* means "two"; *-docus* means "beam"; *ankylo-* means "curved"; and *-saurus* means "lizard." Display pictures of *Ankylosaurus*, *Brachiosaurus*, and *Diplodocus*. Ask students to observe each dinosaur and use the meanings of the prefixes and suffixes to explain why it was given its name. **SAE**

Enrichment

Have students conduct library research to classify the following dinosaurs as carnivores or herbivores: *Allosaurus*, *Apatosaurus*, *Brachiosaurus*, *Diplodocus*, *Stegosaurus*, *Ankylosaurus*, *Triceratops*, and *Tyrannosaurus rex*. *Carnivores: Allosaurus, Ankylosaurus, Tyrannosaurus rex. Herbivores: Apatosaurus, Brachiosaurus, Diplodocus, Stegosaurus, and Triceratops.* Challenge students to explain how scientists studying dinosaurs know whether the dinosaurs ate plants or meat. *By examining their teeth.* **L1** **P**

GLENCOE TECHNOLOGY

Videodisc

The Infinite Voyage: The Great Dinosaur Hunt
Chapter 10
New Dinosaur Discoveries and Their Link with Today

Focus On

Classifying dinosaurs

Paleontologists have identified several hundred species of dinosaurs, some as small as the seven-inch *Sinornis* and others as huge as the *Siesmosaurus,* more than 100 feet long. Some were meat eaters, and others browsed on vegetation; some had relatively smooth skin, while others were covered with horns and bony plates.

Despite such diversity, scientists have devised a straightforward means of classifying dinosaurs. The system first classifies dinosaurs into two main groups according to the skeletal structure of their hips: the ornithischians and the saurischians.

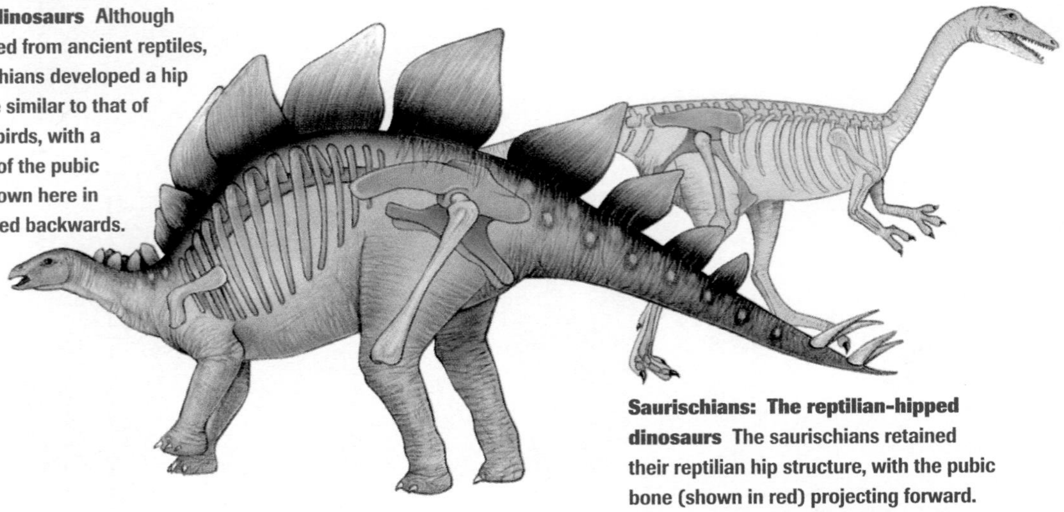

Ornithischians: The bird-hipped dinosaurs Although descended from ancient reptiles, ornithischians developed a hip structure similar to that of modern birds, with a process of the pubic bone, shown here in red, angled backwards.

Saurischians: The reptilian-hipped dinosaurs The saurischians retained their reptilian hip structure, with the pubic bone (shown in red) projecting forward. Ironically, modern birds are thought to be more closely related to reptile-hipped saurischians than to the bird-hipped ornithischians!

Typical of the ornithischians was *Iguanodon,* an early Cretaceous species.

Two hundred million years ago, the saurischian *Dilophosaurus* roamed North America.

Finding, excavating, and reconstructing dinosaurs

At the site Dinosaur fossils are found in sedimentary rock, layers of rock that have built up over millions of years. When paleontologists stumble across a major fossil find, they set up a dig site that can include everything from dynamite to sophisticated biotechnical instruments. Workers may camp at the site for many months as they painstakingly extract fossils from rock.

Tools of the trade Freeing fossils from sedimentary rock requires patience and the right equipment. Tools include hammers, saws, chisels, brushes, and, for extremely delicate work, dental picks. Workers wear goggles to shield their eyes, gloves to protect their hands, and sometimes masks to avoid breathing in too much dust or fumes from toxic chemicals used to preserve fossils.

On display This *Diplodocus* stands in the Museum of Natural History, Houston, Texas. Paleontologists often disagree about how to pose dinosaur skeletons, but as new information and new discoveries come to light, such reconstructions offer people an increasingly accurate picture of these prehistoric creatures.

Back to the lab Today, paleontologists have access to sophisticated biomedical equipment with which to study their finds. This *Nanotyrannus* skull was x-rayed using a CT scanner that produces three-dimensional images. The pictures revealed the size and location of the carnivore's braincase.

Focus On

Display

Prepare a large outline map of the world. Post the map on a classroom wall or bulletin board. Provide students with the resources necessary to identify where in the world different types of dinosaur fossils have been found. Have students work together to create a key that identifies each location according to dinosaur type on the map. COOP LEARN

Enrichment

Ask students to read one of the following articles and report back to the class about what they learned. Articles: "Jurassic Sea Monsters" by Robert Bakker, *Discover*, September 1993; "How Dinosaurs Ran" by R. McNeill Alexander, *Scientific American*, April 1991; "The Real Jurassic Park" by Jim Robbins, *Discover*, March 1991; "Dinosaur Doctors" by Karen Wright, *Discover*, November 1991. L3

PORTFOLIO

Make a Model Have students conduct research or interview an expert at a nearby museum to get information on the sizes and shapes of dinosaur bones. Ask students to work in small groups to use the information they obtain to build a life-size model of a dinosaur bone using papier-mâché and chicken wire. L3 COOP LEARN

Focus On

Discussion

Display a plastic model of *Stegosaurus.* Ask students to speculate about how the plates on the back of the *Stegosaurus* evolved. Encourage them to consider aspects of defense, mate attraction, and maintenance of body temperature. Read the description of how *Stegosaurus* used its plates for defense. Elicit from students whether they think their speculations are supported by this information.

Activity

Supply students working in groups with pictures of a variety of dinosaurs. Ask them to describe an adaptation that a dinosaur has in common with a modern animal. Have students explain how each adaptation they name enabled the dinosaur to survive. Have them repeat this process for the modern animal. As an example, point out that an *Apatosaurus* and a giraffe both have long necks. Each animal uses its long neck to reach vegetation (leaves) on trees. L1 COOP LEARN

Focus On

Dinosaur behavior

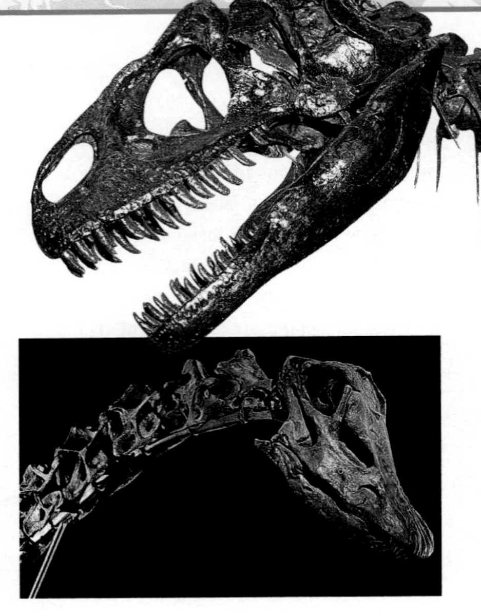

Feeding time Some dinosaurs were carnivores, others were herbivores, and some, like humans, were omnivores. Carnivores such as *Allosaurus* (top) dined on fishes, insects, birds, reptiles, and, of course, other dinosaurs. Their jaws were equipped with rows of curved, sawlike teeth.

Herbivores such as *Diplodocus* (bottom) often browsed on treetop vegetation such as evergreens. Lacking many teeth and unable to chew their food, most herbivores swallowed vegetation whole. In the stomach, the food would be ground up by gizzard stones and shed teeth the dinosaur had swallowed, and then fermented by bacteria.

Nesting Dinosaurs laid their hard-shelled eggs in hollowed-out nests. Sometimes, several dinosaurs would nest close together, indicating that some dinosaurs lived in herds.

Duck-billed dinosaurs probably covered their nests with vegetation, which would decay and thereby provide warmth to hatch the eggs.

Social or solitary? Fossil bones and footprints strongly suggest that many dinosaur species traveled in great herds. Paleontologists have discovered many bone beds of ornithischian dinosaurs. Some contain the remains of more than 10 000 individuals. Trackways that record the passage of hundreds of dinosaurs have been discovered.

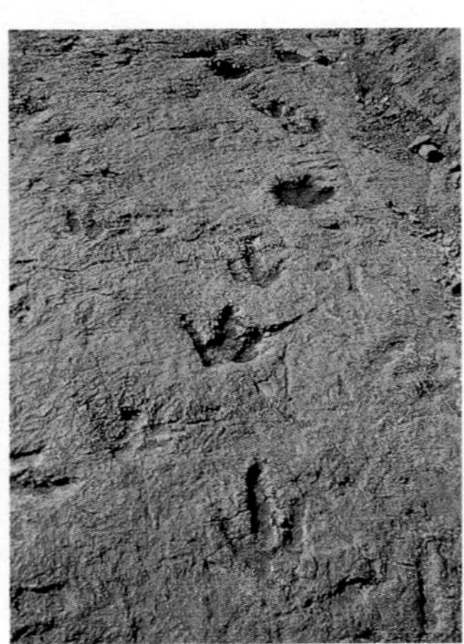

Offense and defense Dinosaurs were well equipped for attacking prey and for defending themselves against attacks from predators. *Velociraptor* used its razor-sharp, sickle-shaped claws to slash at prey. *Stegosaurus* had sharp tail spikes that could be whipped back and forth to protect it from predators.

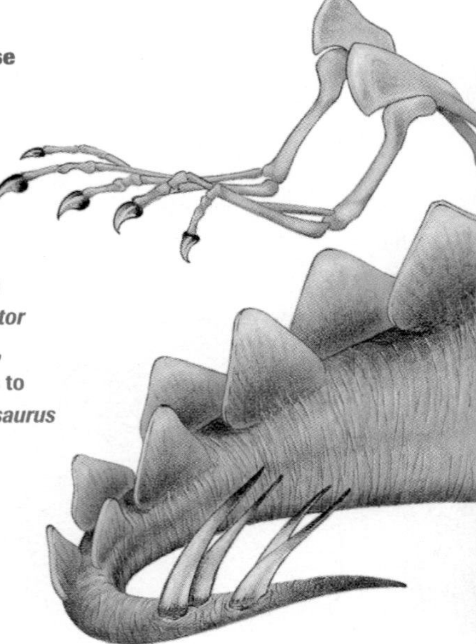

STUDENT JOURNAL

Dinosaur Stories Ask students to each write a science fiction story about dinosaurs. Give them ideas for titles such as *Why Dinosaurs Became Extinct* or *The Life of a Dinosaur.* As they write their stories, have them present evidence that supports theories about whether dinosaurs were warm-blooded or cold-blooded. L2

Theories about dinosaurs

Allosaurus

Warm or cold? Scientists long believed that dinosaurs, like lizards and other reptiles, were cold-blooded creatures dependent on the sun and shade for warming and cooling. However, noted paleontologist Robert Bakker is convinced that all dinosaurs were warm-blooded, energetic animals. Dinosaur bones, like those of birds, are laced with small canals. Reptile bones have no such features. Furthermore, dinosaur fossils have been found on land that would have once been within the Arctic circle, the wrong environment for cold-blooded land dwellers but perfectly hospitable to warm-blooded creatures.

Extinction Near the end of the Cretaceous period, about 66 million years ago, dinosaurs suddenly disappeared from Earth's landscape. Why? Some scientists are convinced that a large meteorite struck Earth, filling the atmosphere with huge clouds of dust and chemical pollutants that may have circled the planet and blocked sunlight from reaching the ground for several years. As vegetation disappeared, the herbivores would have died out. Lacking herbivores on which to feed, the carnivores would have disappeared also. Compelling evidence exists in the rocks that supports this theory.

Dinosaur descendants Many scientists think that modern birds, such as this emu, are living dinosaurs. They share similar skeletal structures. Early birds, like dinosaurs, had teeth. Many dinosaurs had feet similar to modern-day bird feet. The structures of reptile scales and bird feathers share enough similarities that some scientists think feathers may have evolved from reptilian scales.

EXPANDING YOUR VIEW

1. **Understanding Concepts** Explain some ways in which dinosaurs were adapted to their environment. What was the principal result of these advantageous adaptations?
2. **Going Further** Investigate the theory that the dinosaurs disappeared as the result of a mass extinction caused by a giant meteor colliding with Earth. What evidence exists to support this theory? Have scientists discovered a possible impact site? Share your findings with the class in an oral report. Provide graphic aids to enhance the impact of your report.

STUDENT JOURNAL

The Future of Dinosaurs Have students interested in dinosaurs read the novel *Jurassic Park* by Michael Crichton. Ask students to prepare written reports on the book that indicate what parts of the book were accurate science and what parts were creative science fiction. Ask students to explain why it would be unlikely that a dinosaur could be brought to life today using the technology described in the book. Have them speculate if such technological developments might be possible in the future. L3

Focus On

Debate

Divide the class into two groups. Ask each group to take a different point of view about why the dinosaurs became extinct. Have students include the ideas suggested in the feature as well as other ideas they discover using library resources. L1 COOP LEARN

Answers to Expanding Your View

1. Students should note some of the following: specialized teeth for eating vegetation or meat; long necks for reaching vegetation on trees; various means of defense, including sickle claws, bony plates, horns, tail clubs, and tail spikes; plates for regulating body temperature (*Stegosaurus*); the use of gizzard stones for digesting vegetation. The major result of these adaptations was the long, successful reign of the dinosaurs as the principal vertebrate life-form on Earth.
2. Reports will vary but should include information that both supports and rebuts the mass-extinction theory.

Going Further

Have students choose a single dinosaur on which to conduct additional research. Ask students to find the following information about the dinosaur: its size, what it ate, where its remains have been discovered, who named the dinosaur, and during what period it lived on Earth. Ask students to draw the dinosaur they researched or make a three-dimensional model from papier-mâché or clay. Have all students combine their information to create a classroom bulletin board display. L1 COOP LEARN

SECTION

34.2

Prepare

Key Concepts

Students will study the adaptations that make birds suited to life on land and enable them to fly. Bird classification will be presented and the origins of birds will be considered.

Block Scheduling

Look for this symbol for strategies that are useful in a block scheduling format. For more information on block scheduling, refer to Section 34.2 in the **Lesson Plans** booklet.

Materials

- Obtain recordings of bird calls for the Focus Demonstration.
- For the Minilab, obtain down and contour feathers from craft shops. Hand lenses, empty bleach bottles, bird field guides, and birdseed are also needed.
- For the Biolab, gather tagboard, pencils, paper fasteners, protractors, metric rulers, and scissors.

1 Focus

Bellringer

Before presenting the lesson, display **Section Focus Master 76** on the overhead projector and have students answer the accompanying questions. L1 SAE

Demonstration

Play selections from a recording of bird songs that includes courtship, territorial, distress calls, and calls made to the young. Ask students to list the stimuli that might cause birds to sing. L1

SECTION

34.2 Birds

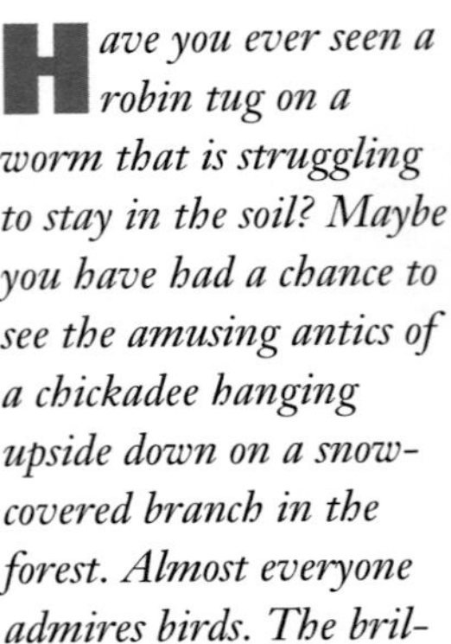

Section Preview

Objectives

Interpret the phylogeny of birds.

Explain how bird adaptations make them suited to life on land.

Relate the adaptations that enable birds to fly.

Key Terms

feather
sternum
endotherm

Have you ever seen a robin tug on a worm that is struggling to stay in the soil? Maybe you have had a chance to see the amusing antics of a chickadee hanging upside down on a snow-covered branch in the forest. Almost everyone admires birds. The brilliant flash of a bluebird's wings, the uplifting sound of a bird's song that fills the woods on a spring morning, and the effortless soaring of a redtail hawk have always fascinated and delighted people.

What Is a Bird?

After conquering the sea and land, vertebrates took to the air, where there was a huge source of insect food and a refuge from land-dwelling predators. The existence of 8600 species of modern birds, classified in the class Aves, shows that flight was a successful strategy for survival. Except for domestic animals and humans, the most common vertebrates you see in your daily life are birds. Biologists sometimes refer to birds as feathered dinosaurs. Fossil evidence seems to indicate that birds have evolved from small, two-legged, lizardlike animals, called thecodonts, *Figure 34.12.* While present-day reptiles and birds have very different physical appearances, some slight resemblances can still be seen. Like reptiles, birds have clawed toes and scales on their feet. Fertilization is internal and shelled amniotic eggs are produced in both groups.

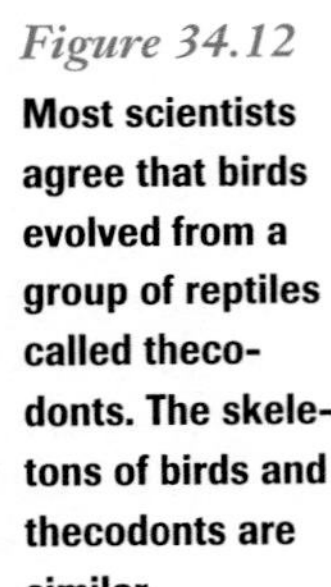

Figure 34.12
Most scientists agree that birds evolved from a group of reptiles called thecodonts. The skeletons of birds and thecodonts are similar.

Program Resources

Section Focus Master 76 L1 SAE
Reinforcement and Study Guide, pp. 139-140 L1
Biolab and Minilab Worksheets, pp. 135-138 L1
Laboratory Manual, pp. 203-212 L2
Concept Mapping, p. 34 L1
Critical Thinking/Problem Solving, p. 34 L3
Tech Prep Applications, pp. 37-38 L2
Basic Skills Transparency 24 and **Master** L1 SAE
Basic Concepts Transparencies 56, 57 and **Masters** L1 SAE

Birds have feathers

Birds can be defined simply as the only organism with feathers. A **feather** is a lightweight, modified scale that provides insulation and enables flight, *Figure 34.13*. You may have seen a bird running its bill through its feathers while sitting on a tree branch or on the shore of a pond. This process, called preening, keeps the feathers in good flying condition. The bird also uses its beak to rub oil from a gland near the tail onto the feathers. This process is especially important for water birds as a way to waterproof the feathers.

Even with good care, feathers wear out and must be replaced. The shedding of old feathers with the growth of new ones is called molting. Most birds molt in late summer. However, most do not lose their feathers all at once and are able to fly while they are molting. Wing and tail feathers are usually lost in pairs so that the bird can maintain its balance in flight.

MiniLab

How do down feathers and contour feathers compare?

Birds have two kinds of feathers. Contour feathers used for flight are found on a bird's body, wings, and tail. Down feathers lie under the contour feathers and insulate the body.

Procedure

1. Examine a contour feather with a hand lens, and make a sketch of how the feather filaments are hooked together.
2. Examine a down feather with a hand lens. Draw a diagram of the filaments of the down feather.
3. Fan your face with each feather separately. Note how much air is moved past your face by each type of feather.

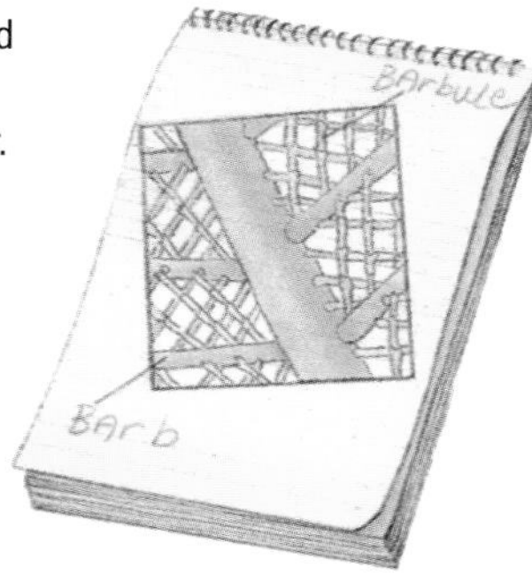

Analysis

1. How does the structure of a contour feather help a bird fly?
2. How does the structure of a down feather help keep a bird warm?
3. How can you explain the differences you felt when fanning with each feather?

Figure 34.13

Feathers streamline a bird's body, making it possible for the bird to fly.

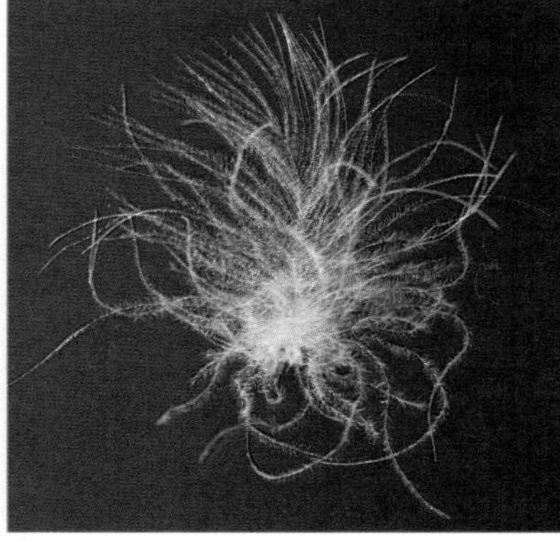

Fluffy down feathers have no hooks to hold the filaments together. Down feathers act as insulators to keep a bird warm. Have you ever worn a down coat or used a down bed cover? If you have, you know how warm down feathers can be.

A large bird can have 25 000 or more contour feathers with a million tiny hooks that interlock and make the feathers hold together. To see how feathers are held together, turn one palm up and make a hook of your fingers on that hand. Turn your other palm down and make a hook of the fingers; then hook the fingers of both hands together. All the filaments in the feather are held together by hooks, making a "fabric" suited for flight.

STUDENT JOURNAL

Observing Birds Have a live canary, parakeet, or parrot in class. Have students observe behavior such as how it uses its beak in feeding and drinking. Have them also observe perching and reactions to objects in its cage such as mirrors and bells. Ask students to write their observations in their journals. L1 LS

2 Teach

MiniLab

Purpose

LS **Visual-Spatial** Students will compare the structures of down and contour feathers and determine how each is adapted to a specific function.

Process Skills

observe and infer, compare and contrast

Teaching Strategies

- Use a binocular microscope or a hand lens.

Expected Results

- Contour feathers consist of barbules with hooks that connect the barbs, forming a streamlined feather.
- Down feathers do not have hooks on their barbules. They are soft and do not take on a specific shape.
- More air can be moved with the contour feather.

Analysis

1. Contour feathers are sleek and streamlined.
2. Down feathers are irregular in shape and when piled together form air spaces that trap body heat.
3. The contour feather barbules stay together and move air. The down feather barbules cannot move air because they do not hold together.

Assessment

Skill: Explain that some birds have feathers called powder-down feathers. The tips of these feathers disintegrate into powder that has waterproofing characteristics and luster. Have students hypothesize how this type of feather may help some birds survive. L1

Physics

Hot Birds

Purpose

Students will learn how birds keep their body temperatures within a narrow range.

Teaching Strategies

- Elicit if anyone has ever worn a down jacket or slept under a down comforter. Ask how down compares to synthetic fabric. *Down is lighter in weight, softer, and warmer than most synthetic fabrics.*
- Show students photos of penguins. Ask them to speculate about adaptations of these birds that enable them to live in very cold climates. *Body fat and minimal surface area.* Point out that penguins are short and that this shape reduces heat loss by lowering the ratio of body mass to surface area. Birds that are chunky in shape have a large mass in comparison to their surface area. The reduced surface area helps reduce heat loss.
- Review the differences between down feathers and contour feathers.
- Point out the air sacs in the diagram included in *The Inside Story*.

Possible Answer

Birds eat a lot to maintain body temperature.

Physics

Hot Birds

Birds are models of efficiency in motion. They are able to fly thousands of miles, soar high in the atmosphere, and hover for long periods in front of a flower. Of course, no single bird can do all those things, but these adaptations demonstrate the versatility exhibited by this fascinating class of animals. In order to do all these high-energy tasks, birds must respire with maximum efficiency, and this means that they must control their temperatures within narrow limits. How do birds keep their temperatures within the necessary range?

Two types of feathers help insulate Most birds live in environments where the temperature is lower than their body temperature. Heat energy tends to move from hot to cold areas. Any heat lost by a bird must be replaced by burning more Calories. Birds need to retain as much heat as possible. One of the most efficient insulators on Earth is the layer of down feathers close to a bird's skin. These are covered by a tight coat of contour feathers. Working together, these two trap a layer of air and prevent heat from moving from the skin outward.

Circulation helps regulate body temperatures When it is cold, the blood vessels in a bird's skin contract and lessen the flow of blood. When the vessels contract, less blood goes to the surface. This cuts heat loss. Conversely, if the bird is too hot, the blood vessels dilate, releasing heat to the body surface. Birds can also hold their contour feathers apart to release heat.

Air sacs increase oxygen supply Birds have an efficient respiratory system. It includes not only lungs but also a number of air sacs. Water in the air sacs evaporates to cool birds. When the air is exhaled, the heat is carried away with water vapor. This is similar to the way you perspire. The difference is that you perspire on the outside, but the bird does it internally.

CONNECTION TO Biology

People often use the expression, "You eat like a bird!" to mean that a person eats very little. From a biological viewpoint, explain why this expression doesn't make sense.

Birds' bodies are adapted for flight

A second adaptation for flight in birds is the modification of the front limbs into wings. Powerful flight muscles are attached to a large breastbone called the **sternum** and to the upper bone of each wing. The sternum looks like the keel of a sailing boat and is important because it supports the enormous thrust and power produced by the muscles as they move to generate the lift needed for flight.

Flight requires high levels of energy. Several factors are involved in maintaining these high energy levels. First, a bird's four-chambered, rapidly beating heart moves oxygenated blood quickly throughout the body. This efficient circulation supplies the cells with the oxygen needed to produce energy. Second, birds are able to maintain high energy levels because they are endotherms. An **endotherm** is an animal that maintains a constant body temperature that is not dependent on the environmental temperature.

Birds have a variety of ways to save or give off their body heat in order to maintain a constant body temperature. Feathers reduce heat loss in cold temperatures. The feathers fluff up to trap a layer of air that limits the amount of heat lost. Responses to high temperatures include flattening the feathers and holding the wings away from the body. Birds will also pant to increase respiratory heat loss.

A major advantage of being endothermic is that birds can live in all environments, from the hot tropics to the frigid Antarctic. However, birds and other endotherms must eat large amounts of food to sustain these higher levels of energy.

GLENCOE TECHNOLOGY

Videodisc

Biology: The Dynamics of Life
Disc 2, Side 1
Penguins (Ch. 7)

PORTFOLIO

Simulating Flight Ask students to make a "flip book" of a flying bird. Have them cut eight 3 × 5 index cards in half and draw a sequence of a bird in flight on the 16 cards. Have them arrange the cards in sequence, and then staple them together on one side. Have students flip through the cards quickly to observe the sequence of a bird flying. L2 P LS

Flight

Blue jay

You may have envied birds and their ability to fly. Humans have always dreamed of being able to rise free of earthbound problems. The popularity of hang gliding and parachute jumping may reflect these dreams. For birds to be able to fly, they were subjected to complex selective pressures that resulted in the evolution of many adaptations.

2 Birds' bones are thin and hollow, thereby maintaining low weight and making flight easier. The hollow bones of birds are strengthened by bony crosspieces. The sternum is the large breastbone to which powerful flight muscles are attached.

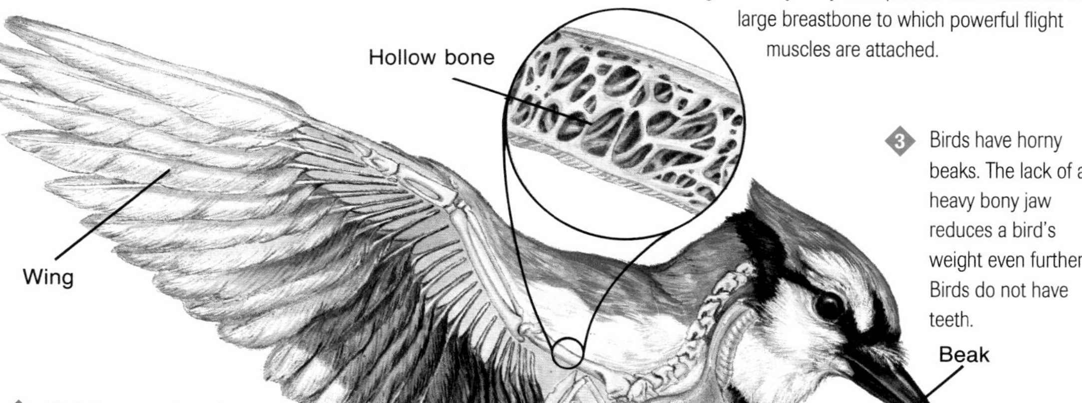

3 Birds have horny beaks. The lack of a heavy bony jaw reduces a bird's weight even further. Birds do not have teeth.

1 Birds have a variety of wing shapes and sizes. Some birds have wings adapted for soaring on updrafts, whereas others have wings adapted for quick, short flights among forest trees.

4 The digestive system of a bird is adapted for dealing with large quantities of food that must be eaten to maintain the level of energy necessary for flight. Because birds have no teeth, many swallow small stones that help to grind up food in the gizzard.

6 About 75 percent of the air inhaled by a bird passes directly into the air sacs rather than into the lungs. When a bird exhales, oxygenated air in the air sacs passes into the lungs. Birds receive oxygenated air when they breathe in and when they breathe out.

5 The legs of birds are made up of mostly skin, bone, and tendons. The feet are adapted to perching, swimming, walking, or catching prey.

THE INSIDE STORY

Flight

Purpose

LS **Visual-Spatial** Students will gain an understanding of the adaptations that enable birds to fly.

Teaching Strategies

- Ask students to bring bones from beef and chicken. Have them compare the density of the two types of bones by finding the masses of the bones and dividing the mass by the volume of each bone as determined by measuring how much water it displaces in a graduated cylinder. *Students will find that the density of chicken bones is less than the density of beef bones.* L1
- Obtain a turkey gizzard from a meat market. Cook it and demonstrate the thick muscle in the gizzard. Elicit reasons why the gizzard has such thick muscles. L1

Visual Learning

- Obtain or borrow a mounted skeleton of a bird. Point out how the bones are thin and lightweight.
- Obtain slides of birds in flight. Ask students to examine the wing shapes and determine the type of flight unique to each bird. Explain that elliptical wings are adapted for quick movements. Wings that sweep back and taper to a slender tip promote high speed. Soaring birds have broad wings.

Meeting Individual Needs

Gifted Ask students who have an interest in mathematics to make up a series of math problems dealing with wing beats for other students to solve. Provide students with the following data: Number of wing beats per 10 seconds: crow 20, robin 23, pigeon 30, starling 45, chickadee 270, hummingbird 700. An example of a problem may be: If a crow, robin, pigeon, and starling each flew in the same direction at a speed of 48 km per hour, how many times would each flap its wings if they flew 24 km? L3 LS

SECTION 34.2

BioLab Design Your Own Experiment

What is the ideal length and width for a bird's tail?

Time Allotment
One class period

Objectives
Review the objectives with students before they begin the Biolab.

Process Skills
compare and contrast, recognize cause and effect, form a hypothesis, interpret data, use a model, separate controls and variables, design an experiment

Safety Precautions
Caution students to handle scissors with care.

PREPARATION

Alternate Material
- Instead of tagboard, file folders may be used.

Possible Hypotheses
- Students may hypothesize that a bird's tail that is longer than its width would be best for balance.
- Students may state that the angle between the back of the bird and the tail must be less than 100° to balance well.

BioLab Design Your Own Experiment

What is the ideal length and width for a bird's tail?

Have you ever watched a bird balance itself on a telephone pole wire? Some birds have feet that are adapted for gripping tree branches. This adaptation prevents them from falling off a branch even when they are asleep.

The tail of a bird is an important structure for balance. When a bird first touches down on a branch, it flicks its tail up and down a few times. The bird uses its tail to find its balance point. Ducks have short tails because a duck lands in the water. Magpies have tails as long as their bodies. When a magpie lands on the ground, it must land with precision to avoid falling over on its bill. The long tail gives precise balance. In this lab, you'll design the best tail for a perching bird.

PREPARATION

Problem
For a perching bird of a particular size, what type of tail will be best for balance? Will a broad tail be better than a narrow tail? Will a long tail be better than a short tail?

Hypotheses
Make a hypothesis about the kind of tail that would best balance a medium-sized bird on a thin branch. Explain the evidence you have used to develop your hypothesis.

Objectives
In this Biolab, you will:
- **Determine** some of the factors in a bird's tail that are important for balance.
- **Compare** the importance of width, length, and angle of attachment for a bird's tail.

Possible Materials
tagboard
cutouts of various sizes of birds' bodies and tails
pencil
paper fastener
protractor
metric ruler
scissors

PLAN THE EXPERIMENT

Teaching Strategies
- You may want to take students to a place where they can watch birds perched on tree branches or telephone wires.
- Make sure students test only one variable at a time.
- Provide students with bird outlines at twice the size shown in their textbooks.
- You may want different groups to test different variables and put their data on the chalkboard to share with the class.
- Suggest to students that they experiment with different sizes of birds.

PLAN THE EXPERIMENT

1. Trace the outline of the bird's body and tail onto tagboard.
2. Punch out the holes indicated by the dotted lines, and attach the tail to the body with a paper fastener.
3. Insert a pencil through the larger hole. Make sure the bird can move easily on the pencil.
4. Bend the tail back and forth until the bird balances. Measure the angle at which the balance point is achieved.
5. Design an experiment in which you test the effect of different tail lengths or widths on the balance point.

Check the Plan

1. Make sure that you are using only one variable.
2. How many trials will you conduct?
3. How will you make your measurements consistent from one trial to the next?
4. What is your control?
5. How will you show your data graphically?
6. ***Make sure your teacher has approved your experimental plan before you proceed further.***
7. Carry out your experiment.
8. Record your data in a table you have prepared.

ANALYZE AND CONCLUDE

1. **Drawing Conclusions** In your experiment, how did length or width of the tail affect the balance point?
2. **Checking Your Hypothesis** Was your hypothesis supported by your data? Why or why not?
3. **Comparing and Contrasting** Compare your results with those of another group. How do they compare?
4. **Making Inferences** Live birds are three dimensional. Would this fact alter your results? Explain.
5. **Thinking Critically** Name functions, other than balancing, of some birds' tails.

Going Further

Changing Variables Conduct another experiment with bird tails, but this time choose another variable you may have thought about during your experiment. This might be body size, weight or length of the tail, or position of attachment of the tail.

Possible Procedures

- Students might test different tail lengths, while keeping the width of the tail and the angle of the tail constant.
- Students might test different tail widths, while keeping the tail length and the angle of the tail constant.
- Students might test different angles for the same-sized tail.

Data and Observations

At about 100°, the most stable tails will be 11 to 13 cm long for the given width. As the angle of the tail increases, the tail must be longer to maintain balance. Increased tail width enables the tail to balance with more variation in the angle of the tail. Students may also determine other relationships.

ANALYZE AND CONCLUDE

1. The bird could balance best with only a slight variation in tail length, but could balance easily with a great variation in tail width.
2. If students hypothesized that the tail should be longer than it is wide in order to balance the bird, data supported the hypothesis.
3. If another group tested the same variable, students will find that the data are similar.
4. Yes; more variables would be introduced.
5. Some birds have brightly colored tails that attract mates. The male peacock is an example. Birds' tails are important in flight as well as in balancing after landing.

Assessment

Performance: Give students a paper clip and tell them it represents a tiny transmitter placed on a bird so biologists can study their movements. Ask them to determine where the transmitter should be placed on a bird to eliminate problems caused by the weight of the transmitter interfering with perching balance. L1

Going Further

Ask students if they have ever seen owls, eagles, or hawks perching. Show pictures of these birds perched. Ask how the positions of the large birds when perching differ from the way song birds perch. Have students speculate on the reasons for the differences. L2

SECTION 34.2

Using an Analogy

Kinesthetic Bring to class a variety of tools such as a nutcracker, vise, large net, chisel, hammer, tweezers, toothpicks, key-type can opener, tongs, and a spatula. Have students imagine that each tool represents a bird's beak. Ask them to work in their groups to identify what type of wild food a bird would eat if it had a beak similar to each tool. Ask them to use field guides to identify a bird that has each type of beak. L1

Text Question

Can you guess why owls' eyes aren't on the sides of their heads? *Eyes on the front of the head make owls better able to see prey they are capturing.*

Bioethics

Saving the Spotted Owl

The endangered northern spotted owl must, by law, be protected on land owned by lumber companies. In the northwestern United States, a large lumber company has restricted logging on more than 300 000 acres of spotted owl habitat. The timber industry claims it has had to lay off many workers as a result of the restrictions. Others believe the loss of 26 000 jobs in the Northwest had nothing to do with the spotted owl. Such people claim increased efficiency within the logging industry due to mechanization resulted in a reduction in the need for workers.

Figure 34.14
Examine these birds and infer where they live and how they are adapted to their environments.

Bird Adaptations

Unlike reptiles, which take on a wide variety of forms from legless snakes to shelled turtles, birds are all very much alike in their basic form and structure. You have no difficulty recognizing a bird.

In spite of the basic uniformity of birds, they do exhibit specific adaptations, depending on the environment in which they live and the food they eat. As shown in *Figure 34.14*, ptarmigans have feathered legs that serve as snowshoes in the winter, making it easier for the birds to walk in the snow. Penguins are flightless birds with wings and feet modified for swimming and a body surrounded with a thick layer of insulating fat. Large eyes, an acute sense of hearing, and sharp claws make owls well-adapted, nocturnal predators, able to swoop with absolute precision onto their prey. Can you guess why owls' eyes aren't on the sides of their heads?

STUDENT JOURNAL

Observing Local Birds Give students working in groups field guides to birds. Ask them to list all the birds found in your area. Next, ask them to check off the names of the birds they have actually observed. Explain that this activity forms the basis of the hobby known as birding or bird-watching. L1

PORTFOLIO

Making a Flyway Map Provide students with a blank map of the world. Ask them to use a field guide to find the bird migration routes for the migrating birds of your area. Ask students to plot these data on their maps using a different color of pencil for each route. L2 P

The shape of a bird's beak gives clues to the kind of food the bird eats, *Figure 34.15.* For example, a swallow is adapted to feeding on insects but not on fishes. Other birds spear fish with long, pointed beaks. Hummingbirds have long beaks that are used for dipping into flowers to obtain nectar. Hawks have large, curved beaks that are adapted for tearing apart prey. Pelicans have huge beaks with pouches that they use as nets for capturing fish. The short, stout beak of a goldfinch is adapted to cracking seeds.

MiniLab

What types of foods do local birds prefer?

In the winter, it may be difficult for some birds to find food, especially if you live in an environment often blanketed with snow. Making a bird feeder and watching birds feed can be an enjoyable activity for you and may save birds from starvation.

Procedure

1. Obtain several large, plastic bleach bottles. Cut two holes 5 cm from the base on opposite sides of each bottle, each about 8 cm^2. These are the openings birds will use to find the food inside.
2. Place small drainage holes in the bottom of each bottle. Hang the bottles from wires strung through small holes in the neck of each one.
3. Place a different kind of seed (sunflower seeds, hulled oats, cracked corn, wheat, thistle, millet) in each bottle. Add new seed every other day.
4. Using a bird guide, make a list of numbers and kinds of birds that frequent each feeder, noting the type of food offered.

Analysis

1. What type of seed attracted the largest variety of bird types?
2. Did any birds visit more than one feeder?
3. What do you think an ideal bird food would be?

Brown pelican

Figure 34.15

The beaks of hummingbirds, hawks, pelicans, and goldfinches are adapted to different kinds of food. Based on what they eat, how do you think their feet might be modified to assist in getting food?

Calliope hummingbird

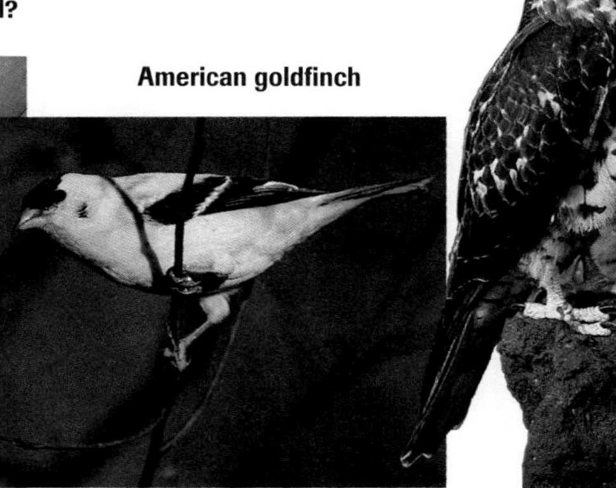

American goldfinch

Galapagos hawk

Visual Learning

Figure 34.15 Based on what they eat, how do you think their feet might be modified to assist in getting food? *The feet of birds are adapted for food getting. For example, birds of prey have long, sharp claws for grasping prey, and birds that eat insects in trees usually have feet adapted to holding onto branches.*

MiniLab

Purpose

LS **Interpersonal** Students will construct a bird feeder and learn what types of foods birds prefer.

Process Skills

compare and contrast, recognize cause and effect, interpret data, classify

Teaching Strategies

- Show photos or slides of common local birds.
- Have students use binoculars to observe the birds that visit the feeder.

Expected Results

Depending on region, students may find cardinals, jays, woodpeckers, nuthatches, juncos, and chickadees.

Analysis

1. Sunflower seeds will attract a large variety of birds.
2. It is likely that birds will visit the same feeder over and over again, unless the feeder runs out of food.
3. Ideal bird food would have a mixture of a variety of seeds that would appeal to many different birds.

GLENCOE TECHNOLOGY

Videodisc

STVS: Animals

Disc 5, Side 2

Sexing Birds (Ch. 15)

Assessment

Portfolio: Ask students to summarize data collected by watching birds. They should record not only the types of birds seen and foods preferred, but other interactions of birds such as aggressive behavior and feeding methods. L1 P

Focus On
Big Birds

Purpose

Students will learn about unusual birds that grow large and are flightless. They will learn about the adaptations of those that survived and about why others became extinct.

Background

Elephant Birds

Elephant birds were easy prey for invading sailors and islanders of Madagascar who hunted them for their meat, eggs, and feathers. New settlers to Madagascar brought livestock to the islands. The livestock competed with the elephant birds for food, speeding up the process of extinction.

Ostriches

Unlike the elephant bird, the ostrich can outrun and escape most of its predators. Ostrich eggs are the biggest single cells in the animal world today.

The Bald Eagle

The bald eagle was placed on the endangered species list in 1967. Causes for the decline of this once-populous species included overhunting, loss of habitat, and contact with chemicals. Chemical contamination of bald eagles occurred in four main ways: lead poisoning caused by the ingestion of spent pellets from shotgun shells; loss of food supplies resulting from acid rain; contamination with PCBs; and contamination with the pesticide DDT. DDT was not harmful to the eagles directly, but rather interfered with the incubation of bald eagle eggs. Because of conservation efforts, the status of the bald eagle has improved from endangered to threatened.

Focus On
Big Birds

Of the 130 species of birds known to have become extinct over the past 300 years, many were strange island birds that developed into flightless giants. Carnivores, which quickly came to dominate the animal kingdom on the continents, were not native to islands such as New Zealand and Madagascar, homes of the largest birds known. These birds, which could live without fear of predators, gained great size while foraging on the ground for food. Over time, their wings became vestigial, and they lost the ability to fly.

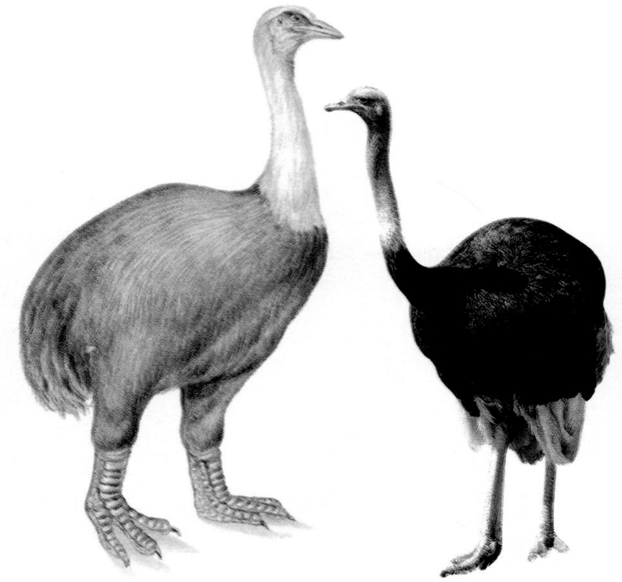

The biggest eggs

Elephant bird eggs Compared to the common hen's egg, the egg laid by the extinct elephant bird was huge (about 8 kg and able to hold up to 8 L). Fragments of these eggs and even entire eggs are still being found washed up on the beaches and lakeshores in Madagascar.

Which is the largest bird?

The African ostrich and the elephant bird The largest living bird, the African ostrich, is no match for the extinct elephant bird of Madagascar, the largest bird ever known. Three meters tall and weighing up to 456 kg, the elephant bird weighed at least three times as much as the African ostrich. Standing more than 2.5 m tall and weighing more than 150 kg, the African ostrich is the world's fastest bird, reaching speeds up to 37 km/h.

What is a dodo?

The trusting bird When 16th-century sailors landed on the beach at Mauritius, an island in the Indian Ocean, they found a curious, flightless bird they called dodo, or silly. Weighing some 22 kg, it seemed to have no fear of people. These early colonists caught dodos for food and brought predators such as dogs, which quickly decimated dodo populations. Within 200 years, all of the dodos were killed.

STUDENT JOURNAL

Endangered Birds Ask students to report on endangered birds in your area. Have them explain each bird's way of life and what has caused it to become endangered. Ask them to find out if any efforts are underway to help these birds. L2

interNET CONNECTION

Follow the link for this chapter on the Glencoe Homepage at **www.glencoe.com/sec/science** to find out more about reptiles and birds.

Largest nests built

Eagles add on yearly The endangered bald eagle is the builder of the largest tree nest in the world. Measuring up to 3 m wide and 6 m deep, an eagle's nest can be active for more than 20 years. Unfortunately, pesticide use has severely lowered the success rate of the bald eagle's reproductive efforts. About 1200 breeding pairs of bald eagles exist in the United States today.

The greatest wingspan

The wandering albatross The wandering albatross has the greatest wingspan of any living bird; it measures nearly 3.7 m.

Too fat to fly

The great auks Like the dodo, the great auk was another clumsy, flightless bird, but one with remarkable swimming skills. The great auks had plentiful supplies of fat, for which they were hunted to extinction. Standing about 0.5 m high, with great webbed feet and black-and-white markings, the great auk looked much like a penguin. The auks were unrelated to penguins and were isolated on islands in the northern hemisphere. On Eldey Island, off the coast of Iceland, the last of the great auks was killed by sailors on June 4, 1844.

EXPANDING YOUR VIEW

1. Understanding Concepts How is the evolution of flightless birds related to their environment?
2. Writing about Biology In a short paragraph, explain the major differences between the modern flightless birds such as the penguin and an extinct bird such as the dodo.

Focus On

Teaching Strategies

- Review with students the background of DDT use and its effects on bird populations.
- Direct students' attention to the picture of the ostrich. Elicit how this bird is adapted for running. *Ostriches have long legs and large feet.* L1
- Ask students who need enrichment to find out how the bald eagle came to be our national bird. L1

Answers to Expanding Your View

1. Flightless birds were common to areas in which they had no predators. Over time, these birds gained great size and lost their ability to fly.
2. Modern flightless birds such as the penguin live in regions isolated from people. These birds, unlike the extinct dodo, are small in size and are well suited to swimming as well as life on land.

Going Further

Ask students to look up information about other extinct birds and find out the details of events surrounding their extinction. L1

PROJECT

Saving Endangered Birds

Ask students to develop a campaign to save a local bird from extinction. They should prepare a list of recommendations, a name for their group, publicity in the form of radio and TV commercials, bumper stickers, and posters, and suggest an original idea to raise money to create a preserve habitat. L2 COOP LEARN

GLENCOE TECHNOLOGY

Videodisc

Biology: The Dynamics of Life
Disc 2, Side 1
Eagles (Ch. 8)

STVS: Ecology
Disc 6, Side 1
Bald Eagles (Ch. 13)

SECTION 34.2

Activity

LS Interpersonal Ask students to photograph native birds and prepare a slide presentation for the class. Challenge students to use field guides to identify each bird, its habitat, its food source, and unusual habits. L2

Teacher FYI

There are about 8600 species of modern birds. In the last 300 years, more than 87 bird species have become extinct. Causes of extinction include habitat destruction; collecting of bird parts such as feathers, beaks, and talons; illegal pet trade; lead shot poisoning; competition by introduced species; and toxic effects of pesticides, oil spills, and chemical dumping.

History Connection

Pigeons

Pigeons have had a close connection with humans for many years. More than 4000 years ago in the Middle East, people used pigeons as messengers or kept them for food. The first recorded use of pigeons as messengers was when Pharaoh Ramses III assumed the throne in 1200 B.C. Four birds were sent out in different directions to carry the message that a new Pharaoh had assumed power.

Figure 34.16

The complete evolutionary history of birds is not clear. The fossil record of birds is incomplete because bird skeletons are light and delicate and, therefore, are easily destroyed by carnivores and other natural causes before they can be preserved. The radiation of orders of birds on the Geologic Time Scale shows their relationships.

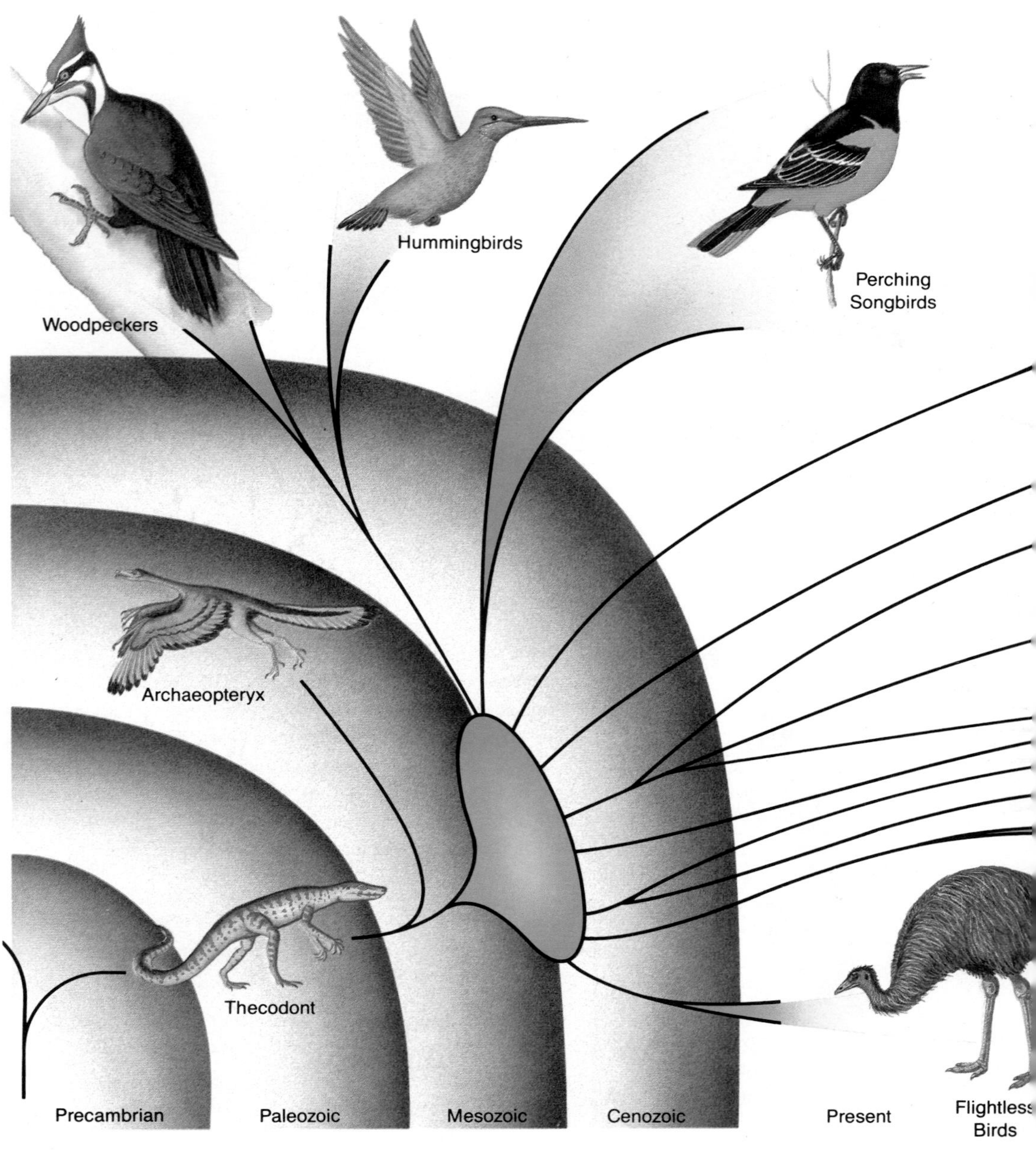

CULTURAL DIVERSITY

Sankar Chatterjee and Bird Evolution

The evolutionary history of birds is currently under debate by scientists. In your discussions of bird evolution, introduce students to the research and hypotheses of Indian-American paleontologist, Sankar Chatterjee.

Chatterjee is best known for his 1986 discovery of Protoavis, a 225 million-year-old fossil that may turn out to be the earliest known bird. Have students research the various hypotheses of bird evolution and initiate a discussion about the evidence used in each hypothesis. For more information on Chatterjee's work, see *Discover* magazine, May 1992. L2

The Orders of Birds

Kingdom Animalia
Phylum Chordata
Class Aves
8600 species

Owls

Quails, Turkeys, and Chickens

Parrots

Pigeons

Gulls, Sandpipers, and Puffins

Hawks, Eagles, and Falcons

Penguins

Swans, Geese, and Ducks

Pelicans

SECTION 34.2

Activity

LS **Visual-Spatial** Give students a blank map of the United States and field guides of birds. Ask them to look up each group of birds pictured on the phylogenetic diagram and indicate the range of a particular species in each order. They should make a color-coded key for their map. L1

Enrichment

Ask students to make a table of the unique characteristics of each of the bird orders depicted in the phylogenetic diagram. L1

3 Assess

Check for Understanding

Ask students to write a letter that begins, "This is everything I know about birds. . . ." Have them summarize what they have learned about birds. Have them write a second paragraph that begins, "What I still don't understand or haven't learned about birds is. . . ." Have students exchange notes with someone else. Each student should write a response to the other in which they try to explain what their partner needs to learn. If they both have the same areas of weaknesses, they should exchange information with someone else. L1 COOP LEARN

STUDENT JOURNAL

Raising an Issue Have students write letters in their journals to elected officials about national and local issues regarding birds and ways to protect habitat in your area. Students should address such issues as habitat destruction; lead-shot poisoning of waterfowl; illegal trade in pet birds; and poisoning through pesticides, oil spills, and chemical dumping. L2

Meeting Individual Needs

Learning Disabled Provide students with the equipment to photograph or videotape local birds in their habitats. Have students display their work and discuss in groups the habitat of each bird, the type of food it eats, and adaptations it has for its way of life. L1 LS

GLENCOE TECHNOLOGY

Videodisc

Biology: The Dynamics of Life
Disc 2, Side 1
Shorebirds (Ch. 9)

Reteach

Give students field guides to birds of your area. Ask them to observe birds for several days, identify them, and explain how each is adapted to its way of life. L1

Enrichment

Have students obtain small birds such as Cornish game hens, chickens, or quail from a meat market or grocer. The birds should be boiled and all meat removed from the bones. Have students assemble the skeleton using sturdy glue and thin wire. L3

Assessment

Performance: Have students sketch a scale map of the school grounds or a spot near the school on a sheet of butcher paper. Have students indicate how the area could be made a more suitable bird habitat. Have students working in groups add features to their maps that would make their areas more attractive to birds. Books about attracting birds in all types of environments from the inner city to the suburbs are available at libraries. L1 COOP LEARN

4 Close

Activity

Ask students to develop a hypothesis and experimental plan to determine why flamingos stand on one leg. L1

Figure 34.17
Archaeopteryx **had teeth and a long tail, as did many reptiles.** ***Archaeopteryx*****, however, had feathers. The colors shown here are imagined by the artist.** ***Protoavis*** **lived in the tropical forests of what is now West Texas about 225 million years ago. Some scientists doubt that it was a bird because no fossil feathers have been found.**

Origins of Birds

Current thoughts about bird evolution are illustrated in *Figure 34.16*. You know that some scientists think today's birds are related to an evolutionary line of dinosaurs that did not become extinct. *Figure 34.17* shows the earliest known bird in the fossil record, *Archaeopteryx*. At first, scientists thought that *Archaeopteryx* was a direct ancestor of modern birds; however, some paleontologists now think that it most likely did not give rise to any other bird groups. *Archaeopteryx* was about the size of a crow and had feathers and wings like a modern bird. But it also had teeth, a long tail, and clawed front toes, much like a reptile.

Another fossil discovered in 1984 has hollow bones and a well-developed breastbone with a keel, all characteristics of birds. Named *Protoavis*, this fossil is thought to be similar to that of a bird, yet it is 75 million years older than fossils of *Archaeopteryx*. However, *Protoavis* had hind legs, a pelvis, and a bony tail that resembled the bones of dinosaurs, *Figure 34.17*.

Connecting Ideas

You have seen how reptiles dominated the Mesozoic Era. Modern reptiles also have forms as varied as legless snakes, turtles with rocklike shells, and crocodiles with huge snapping jaws.

Birds, however, are characterized by their uniformity. Flying necessitates uniformity; feathers, lightweight bones, air sacs, beaks, and other features are modifications for flight. Flight restricts the diversity commonly seen in other vertebrates. For example, mammals include such diverse groups as whales, mice, porcupines, bats, and buffalo.

Section Review

Understanding Concepts

1. What body structure found in reptiles has replaced the function of a bird's gizzard?
2. Explain why a bird's respiratory system is extremely efficient.
3. Why does being an endotherm have adaptive value?

Thinking Critically

4. Large, flightless birds are most common in areas that do not have large, carnivorous animals. What hypothesis can you suggest for the evolution of large, flightless birds?

Skill Review

5. **Making and Using Tables** Make a table that summarizes the adaptations birds have that enable them to fly. For more help, refer to Organizing Information in the *Skill Handbook*.

Answers to Section Review

1. Teeth
2. In addition to lungs, it has air sacs that can be used for taking in and exchanging air and that can hold large amounts of air with oxygen. In addition, the four-chambered heart moves the blood with its oxygen supply efficiently.
3. Animals that are endotherms can live in a wide variety of habitats from very cold to very warm and still maintain the same body temperature.

Think Critically

4. They filled a niche not otherwise occupied (ground feeder that was large) and wings became unimportant to survival, thus becoming vestigial over time.

Skill Review

5. Heads of table columns might include internal features and external features. Make sure students' tables include all features listed on pages 855-857.

Chapter 34 Review

REVIEWING MAIN IDEAS

34.1 Reptiles

- Reptiles are ectotherms that have dry, scaly skin; legs under the body; internal fertilization; and amniotic eggs. Most reptiles have three-chambered hearts. Some reptiles have four-chambered hearts.
- Present-day reptiles belong to one of four groups. Turtles have shells and no teeth. Crocodiles and alligators have streamlined bodies and powerful, toothed jaws. Lizards have a variety of adaptations including long bodies, tails, and short limbs. Snakes have no limbs.
- The ancestors of present-day reptiles arose from ancient cotylosaurs.

34.2 Birds

- Birds have adaptations for flight including feathers; keel-shaped sternum; four-chambered heart; endothermy; thin, hollow bones; a beak; and air sacs.
- Birds may be related to a line of dinosaurs that did not become extinct. The fossils of *Protoavis* and *Archaeopteryx* reflect features of modern birds.

Key Terms

Write a sentence that shows your understanding of each of the following terms.

amniotic egg
endotherm
feather
Jacobson's organ
sternum

Understanding Concepts

1. Scientists think that ____________ were the ancestors of birds.
   a. fishes
   b. amphibians
   c. reptiles
   d. mammals
2. Of the following, which is NOT an example of a reptile?
   a. snake
   b. turtle
   c. salamander
   d. lizard
3. ____________ are the earliest known reptiles.
   a. Dinosaurs
   b. Cotylosaurs
   c. Thecodonts
   d. Therapsids
4. Which of the following is NOT a characteristic of reptiles?
   a. has three- or four-chambered heart
   b. lays amniotic eggs
   c. has legs flexed under the body
   d. has external fertilization
5. For gas exchange, reptiles are dependent on ____________.
   a. gills
   b. skin
   c. skin and lungs
   d. lungs
6. Why don't reptiles inhabit extremely cold regions on Earth?
   a. They have moist skin that would freeze in the cold.
   b. They are ectotherms.
   c. They lay eggs in water and water would freeze in the cold.
   d. They are endotherms.
7. ____________ are the only reptiles to have four-chambered hearts.
   a. Snakes
   b. Lizards
   c. Crocodilians
   d. Turtles
8. Eggs that cushion the embryo in fluid and protect it with membranes and a shell developed first in ____________.
   a. reptiles
   b. birds
   c. amphibians
   d. mammals
9. A(n) ____________ is covered by a shell and contains protective membranes, fluids, and an embryo.
   a. pupa
   b. seed
   c. jellylike egg
   d. amniotic egg

Chapter 34 Review

Reviewing Main Ideas

Summary statements can be used by students to review the major concepts of the chapter.

Key Terms

Answers should go beyond defining the terms. Accept any answer that uses the term correctly and in the proper context.

Understanding Concepts

1. c
2. c
3. b
4. d
5. d
6. b
7. c
8. a
9. d

Chapter 34 Review

10. b
11. d
12. c
13. a
14. b
15. b
16. a
17. c
18. b
19. a
20. c

Applying Concepts

21. Birds are endotherms and maintain a constant body temperature regardless of the temperature of the environment.
22. The ecological balance would be disrupted because the snakes would be predators that might decimate certain prey populations such as rare endemic birds; and the snakes will not have natural predators to hold their populations in check.
23. Sea turtles might not recognize their nesting beaches, and they may not lay eggs. Sea turtles migrate and come back to the beaches where they hatched to lay eggs. The release of chemicals into the water may alter the water in such a way that turtles will not be able to recognize the area.
24. The allantois collects nitrogenous wastes. The amniotic egg gets food from the yolk; the chorion permits gas exchange; the shell, fluids, and membranes cushion and protect the developing embryo.
25. *Protoavis* is a fossil with some birdlike characteristics, such as hollow bones and a breastbone with a keel. No feathers were found with this fossil. *Archaeopteryx* is the earliest fossil bird with feathers and wings. *Protoavis* is an older fossil than *Archaeopteryx*.

Thinking Critically

26. Evaluate students' maps. The maps should show an understanding of the relationships among the concepts listed.
27. The chorion surrounds the amnion, allantois, and yolk sac. The yolk is the main supply of food for the embryo. The allantois is a sac containing the embryo's wastes, and the amnion is a sac that surrounds the embryo.
28. They may live in marshes where their long legs keep their bodies out of the water.

Chapter 34 Review

10. When a snake flicks out its tongue, it is using its sense of ___________.
    a. vision c. hearing
    b. smell d. touch
11. The habitat of a tortoise is the ___________.
    a. water c. beach
    b. ocean d. land
12. Most snakes get food by ___________.
    a. suffocating prey by constriction
    b. injecting venom into prey
    c. grabbing it with their mouths and swallowing it whole
    d. shooting out poison from their tongues
13. Which group of reptiles includes more poisonous members?
    a. snakes c. crocodiles
    b. lizards d. turtles
14. Birds can be defined as the only organisms with ___________.
    a. scales c. shelled eggs
    b. feathers d. wings
15. From what group of reptiles do most scientists agree birds evolved?
    a. cotylosaurs c. therapsids
    b. thecodonts d. tuataras
16. The function of down feathers is ___________.
    a. insulation c. preening
    b. flight d. molting
17. Birds ___________ when they shed old feathers and grow new ones.
    a. migrate c. molt
    b. preen d. balance
18. In a bird, powerful flight muscles are attached to a large breastbone called the ___________.
    a. Jacobson's organ c. gizzard
    b. sternum d. wing bone
19. A(n) ___________ is an animal that can maintain a constant body temperature.
    a. endotherm c. mesotherm
    b. ectotherm d. endoskeleton
20. The earliest known fossil bird with feathers is ___________.
    a. *Protoavis* c. *Archaeopteryx*
    b. an ostrich d. a pigeon

Applying Concepts

21. Why are birds able to inhabit more diverse environments than reptiles?
22. No snakes live on the Hawaiian Islands. How would introducing snakes affect the environment on these islands?
23. Newly developed industries that locate on shorelines often release chemicals into the ocean. How might this affect sea turtles that use these areas to breed? Explain.
24. Explain how the amniotic egg maintains homeostasis.
25. Discuss why the fossils of *Archaeopteryx* and *Protoavis* are significant in explaining the evolutionary history of birds.

Thinking Critically

26. ***Concept Mapping*** Make a concept map that relates the following terms and phrases. Supply the appropriate linking words for your map.
    amnion, chorion, allantois, yolk, shell, embryo, amniotic egg
27. ***Interpreting Scientific Illustrations*** Examine the diagram of the amniotic egg in *The Inside Story*, and explain the relationship among the amnion, chorion, yolk sac, and allantois.
28. ***Formulating Hypotheses*** Assume that you are walking through an aviary in a zoo. You notice that the cranes have long, thin legs. Make a hypothesis about the type of habitat in which these birds might live. Justify your hypothesis.

Chapter 34 Review

29. ***Recognizing Cause and Effect*** Every year since you were a child, your family has taken a vacation near the Florida everglades. Over the years, modern hotels, homes, and roads have been built in the area. Last year, you noticed that you rarely saw the herons, ibises, and other native birds that were once common to the area. What might have caused the reduced number of sightings?
30. ***Interpreting Data*** A biologist counts the feathers on the bodies of two different species of birds. The data are represented in the graph below. What might be inferred about the type of environments in which the birds live?

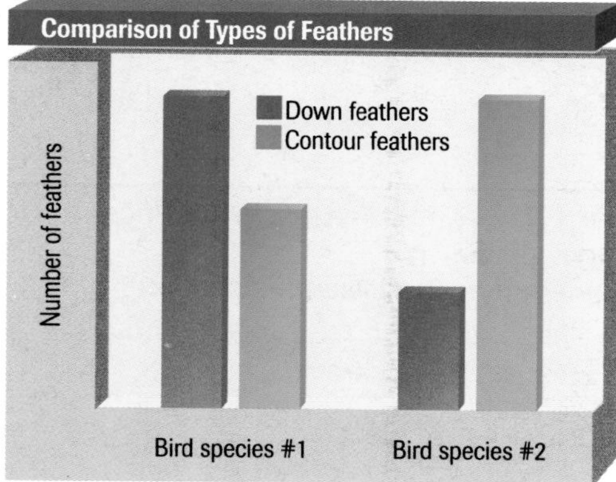

31. ***Comparing and Contrasting*** Most reptiles lay between one and 200 eggs at a time. Amphibians lay thousands of eggs at a time. Is there an adaptive advantage to laying fewer eggs on land? Explain.
32. ***Recognizing Cause and Effect*** You have just acquired a pet parakeet. Besides buying dishes with food and water, the pet store owner suggests that you also buy a container for gravel. Explain why.
33. ***Observing and Inferring*** Why don't birds freeze when they perch on tree branches in the winter?

ASSESSING KNOWLEDGE & SKILLS

A biologist is comparing yearly censuses of owls and mice found in one area. The data obtained are represented in the graph below.

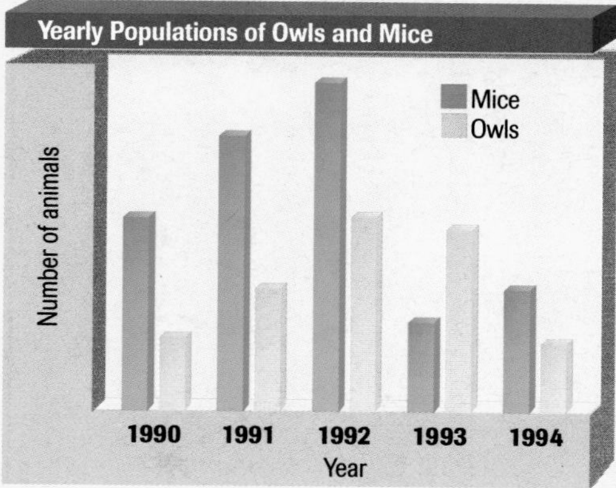

Interpreting Data Use the information in the graph to answer questions 1 through 3.

1. In most years, there are ___________.
   a. more mice than owls
   b. more owls than mice
   c. the same number of owls as mice
   d. twice as many owls as mice
2. The best explanation for the fluctuations in owl populations is that the owl population increases and decreases in response to ___________.
   a. the size of the mouse population
   b. the decrease in the mouse population
   c. the increase in the mouse population
   d. the 4-year cycle of the mouse population
3. ***Formulating Hypotheses*** Examine the graph again and assume that the key is changed to the following: green represents a certain type of eagle and yellow represents rabbits. Make a hypothesis about the relationship between eagles and rabbits.

Chapter 34 Review

29. Their feeding and nesting habitat has diminished due to increased human activity and construction.
30. Bird species 1 probably lives in a colder environment than does bird species 2.
31. The shelled egg of the reptile is more protected than the amphibian egg, so it has a greater chance of producing live young. Also, energy is conserved when fewer eggs are produced.
32. Birds have no teeth. Swallowed gravel allows them to grind food up in the muscular gizzard.
33. When it is cold, the blood vessels in a bird's skin contract and reduce the flow of blood. Less blood goes to the skin's surface, which cuts heat loss.

1. a
2. a
3. There are more eagles than rabbits in any one year except 1993. Eagles exert pressure on the rabbit population, but because eagles also eat many other animals, they do not completely decimate the rabbit population in any year. In years with fewer eagles, more rabbits survive.

Program Resources

Chapter Assessment, pp. 207-212 L1
Alternate Assessment in the Science Classroom
Computer Test Bank L1
Content Mastery, pp. 141-144 L1

Chapter Organizer

SECTION	OBJECTIVES	ACTIVITIES/FEATURES
35.1 Mammal Characteristics National Science Standards: UCP.1-5; A.1, A.2; B.2; C.3, C.5, C.6; F.1	1. **Distinguish** mammalian characteristics. 2. **Compare** the characteristics of mammals with those of reptiles and birds.	**Minilab:** How does the hair of mammals compare?, p. 873 **Thinking Lab:** How can antelopes maintain high running speeds?, p. 874 **Minilab:** What can you tell about a mammal from its tracks?, p. 875 **Biolab:** Design Your Own Experiment—Is blubber a good insulator?, p. 878 **Chemistry Connection:** Fats and Cholesterol, Good or Bad?, p. 880
35.2 The Diversity of Mammals National Science Standards: UCP.1-5; C.3, C.5, C.6; D.3; F.3-6; G.1-3	3. **Distinguish** among the three groups of living mammals. 4. **Compare** reproduction in egg-laying, pouched, and placental mammals.	**Biology & Society Issues:** What Price Beauty?, p. 887 **Focus On** Placental Mammals, p. 888

ACTIVITY MATERIALS

BIOLAB	MINILABS	ALTERNATE LAB
page 878 test tubes plastic spoons shortening thermometer plastic sandwich bag large beaker rubber bands watch with second hand	**page 873** mammal hairs slides coverslips dropper microscope glycerine **page 875** metric ruler meterstick field guides overhead transparencies	**page 880** mouse, live aquarium sawdust or sand timer objects for burying

Chapter 35 Mammals

TEACHER CLASSROOM RESOURCES

Reproducible Masters	Transparencies
Section Focus Master 77: Scales, Feathers, Hair L1 SAE **Reinforcement and Study Guide,** pp. 141-142 L1 **Biolab and Minilab Worksheets,** pp. 139-142 L1 **Concept Mapping:** Mammalian Adaptations for Feeding, p. 35 L1 **Tech Prep Applications:** Chewing Mechanics of Mammals, pp. 39-40 L2 **Critical Thinking/Problem Solving:** Solving a Mammal Mystery, p. 35 L3 **Laboratory Manual:** Mammal Teeth, pp. 213-218; How Does Insulation Affect Thermal Homeostasis?, pp. 219-222 L2	**Basic Concepts Transparency #58:** A Mammal L1 SAE **Basic Concepts Transparency #59:** Phylogeny of Mammals L1 SAE **Reteaching Transparency #35:** How Can Antelopes Maintain High Running Speeds? L1 SAE
Section Focus Master 78: Mammals: Land, Sea, and Air L1 SAE **Reinforcement and Study Guide,** pp. 143-144 L1 **Content Mastery,** pp. 145-148 L1	

ASSESSMENT MATERIALS

Chapter Assessment, pp. 213-218
Alternate Assessment in the Science Classroom
MindJogger Videoquiz
Computer Test Bank

Spanish Resources SAE
English/Spanish Audiocassettes SAE
Cooperative Learning in the Science Classroom COOP LEARN
Lesson Plans

KEY TO TEACHING STRATEGIES

L1 Level 1 activities should be within the ability range of all students including those with learning difficulties.

L2 Level 2 activities are within the ability range of average to above-average students.

L3 Level 3 activities are designed for the ability range of above-average students.

SAE SAE activities should be within the ability range of Students Acquiring English.

COOP LEARN Cooperative Learning activities are designed for small group work.

P These strategies represent student products that can be placed into a best-work portfolio.

These strategies are useful in a block scheduling format.

GLENCOE TECHNOLOGY

The following multimedia resources are available from Glencoe.

Biology: The Dynamics of Life
CD-ROM SAE
Videodisc Program

The Infinite Voyage Series
The Keepers of Eden
Cincinnati Zoo's Cat House
A Taste of Health
Consequences of a Fatty Diet
The Health Conscious Chef

Science and Technology Videodisc Series (STVS)
Animals
Temperature Regulation in Dogs
Kit Fox
How Bats Hear
Ecology
Red Wolf

CHAPTER

35 Mammals

Chapter Overview

Students begin their study of mammals by learning about general characteristics of mammals. Mammal traits are emphasized through a discussion of how these characteristics enable mammals to live in every environment on Earth. Students then explore the adaptations of mammals, including mammal intelligence and how mammals care for their young.

In the second part of the chapter, students learn to distinguish the three main groups of mammals after considering the origins of mammals. Students are introduced to the great diversity of mammals, as emphasized by the discussion of varying traits of egg-laying, pouched, and placental mammals.

Key Terms

cud chewing
diaphragm
gestation
gland
mammary gland
marsupial
monotreme
placenta
placental mammal
therapsid
uterus

interNET CONNECTION

Follow the link for this chapter on the Glencoe Homepage at **www.glencoe.com/sec/science** to find out more about mammals.

870

Learning Styles

Look for the following logo for strategies that emphasize different learning modalities. LS

Kinesthetic	**Project, p. 875**
Visual-Spatial	**Minilab, p. 873; Portfolio, p. 874; Visual Learning, pp. 876, 882; The Inside Story, p. 881; Student Journal, p. 886**
Interpersonal	**Student Journal, p. 874; Portfolio, p. 877**
Intrapersonal	**Enrichment, p. 885**
Linguistic	**Student Journal, p. 876; Meeting Individual Needs, p. 877; Portfolio, p. 884**
Logical-Mathematical	**Thinking Lab, p. 874; Minilab, p. 875; Alternate Lab, p. 880**

CHAPTER

35 Mammals

Why are gorillas so fascinating? Perhaps they remind us of ourselves. These gentle giants of the African forests have fingers so dexterous that they easily pick out the tasty parts of a fruit or tear off branches they use to make nests for sleeping. Gorilla mothers spend years protecting and nurturing their young. Perhaps you have had the opportunity to see a mother gorilla cradle her young in her arms.

Far from the natural habitat of the gorilla, manatees move slowly and gracefully through Florida rivers. Although the outward appearance of these animals in their watery habitat is different from that of gorillas, there are actually many physical and behavioral similarities between these two species. Like gorillas, manatees take good care of their young. They are also peaceful herbivores.

Gorillas, manatees, and humans are similar because they are all mammals. In this chapter, you will study the characteristics shared by mammals and learn why they are so successful.

Concept Check

You may wish to review the following concepts before studying this chapter.

- Chapter 1: adaptation, evolution
- Chapter 6: endangered species
- Chapter 34: birds, endotherms, reptiles

Chapter Preview

Laboratory Activities

Biolab: Design Your Own Experiment

- Is blubber a good insulator?

Minilabs

- How does the hair of mammals compare?
- What can you tell about a mammal from its tracks?

Communication is important between these young and their mothers. Manatee mothers chirp like birds or make high-pitched squeaks and squeals. Primates also use vocal sounds to communicate with each other.

Assessment Planner

Choose assessment strategies from the following pages to evaluate the progress of your students.

Assess, pp. 882, 892
Alternate Lab, pp. 880-881
Minilabs, pp. 873, 875
Portfolio, pp. 874, 877, 884
Thinking Lab, p. 874
Biolab, pp. 878-879
Chapter Review, pp. 893-895

CHAPTER 35

Introducing the Chapter

Have students examine the photographs of the gorillas and the manatees. Ask students to recount their experiences of seeing gorillas at a zoo. Then show parts of the film *Gorillas in the Mist* or read selections from Dian Fossey's book with the same title.

Contact a manatee protection group in Florida and ask for pamphlets about the way of life of the manatee and the problems associated with its decline. Ask students to examine the manatee photo and analyze why a sailor from long ago might have seen a manatee and imagined that it was a mermaid. Ask students to list traits that are shared by manatees, humans, and gorillas. Follow up with a discussion of traits that are unique to each of these mammals. **L1** **COOP LEARN**

Theme Development

The theme of *unity within diversity* is obvious in the discussions of the traits all mammals share in spite of their differences. The theme of *homeostasis* is woven throughout the chapter via a discussion of mammal endothermy. This mammalian trait is one of the characteristics that enables mammals to inhabit such a wide variety of habitats.

Concept Check

The concept of endothermy, first studied in connection with birds, will be related to mammals. Students will also expand upon their knowledge and understanding of evolution and adaptations.

SECTION

35.1

Prepare

Key Concepts

Mammalian characteristics such as hair, endothermy, glands, diaphragm, mammary glands, and intelligence are presented. Adaptations for obtaining and consuming food are discussed for specific mammals.

Block Scheduling

Look for this symbol for strategies that are useful in a block scheduling format. For more information on block scheduling, refer to Section 35.1 in the **Lesson Plans** booklet.

Materials

- Obtain a live mammal and a suitable cage and food for the Focus Demonstration.
- For the Minilabs, gather transparent metric rulers, mammal hair or fur, glycerin, and microscopes.
- For the Biolab, gather test tubes, plastic spoons, shortening, thermometers, plastic sandwich bags, large beakers with ice, rubber bands, and watches with a second hand.

1 Focus

Bellringer

Before presenting the lesson, display **Section Focus Master 77** on the overhead projector and have students answer the accompanying questions. L1 SAE

SECTION

35.1 Mammal Characteristics

Section Preview

Objectives

Distinguish mammalian characteristics.

Compare the characteristics of mammals with those of reptiles and birds.

Key Terms

gland
diaphragm
cud chewing
mammary gland
therapsid

Have you ever read a children's story in which the children feared a big, bad wolf? The wolf characters in stories like Little Red Riding Hood *or* The Three Little Pigs *are usually portrayed as evil, cunning, and vicious killers. But real wolves kill only prey organisms in order to eat, as all animals must. Wolves prey mostly on small animals such as rabbits, mice and other rodents, and occasionally on deer or elk. Both wolves and their prey are mammals.*

What Is a Mammal?

Mammals, like birds, are endotherms. The ability to maintain a fairly constant body temperature enables mammals to live in every possible environment on Earth. Polar bears of the Arctic, tigers in tropical jungles, and dolphins that roam the Atlantic Ocean all are able to live in these varied environments because they are endotherms. Mammals also share other important characteristics.

Mammals have hair

Have you ever heard someone complain about a pet that is shedding its hair? There's no doubt that such a pet is a mammal because only mammals have hair. You have read that birds' feathers probably evolved from scalelike structures such as those of reptiles. Mammalian hair is also thought to have evolved from scales. The structure of hair provides insulation and waterproofing and thereby conserves body heat. If you have ever worn a wool sweater made from the hair of a sheep, you know how warm wool can be on a cold day. As shown in *Figure 35.1*, hair also serves other functions.

Figure 35.1

Hair helps in maintaining a constant body temperature. It also serves a variety of other purposes.

▼ **The black stripes of a tiger's fur aid in camouflaging this beautiful cat as it hunts for prey.**

Program Resources

Section Focus Master 77 L1 SAE
Reinforcement and Study Guide, pp. 141-142 L1
Biolab and Minilab Worksheets, pp. 139-142 L1
Laboratory Manual, pp. 213-222 L2
Concept Mapping, p. 35 L1
Critical Thinking/Problem Solving, p. 35 L3
Tech Prep Applications, pp. 39-40 L2
Basic Concepts Transparencies 58, 59 and **Masters** L1 SAE
Reteaching Transparency 35 and **Master** L1 SAE

How mammals cool off

Although hair helps retain body heat, mammals also have adaptations that aid in cooling off the body when it gets too warm. Have you ever seen a dog running next to its jogging owner? When the jogger stops, the dog flops down, opens its mouth, and breathes rapidly in and out with its tongue hanging down. Why do dogs pant when they are too warm? Panting is one adaptation for cooling off the body. It releases water from the lungs, which results in a loss of body heat.

Mammals also have organs called glands that secrete various substances needed by the animal. A **gland** is a cell or group of cells that secretes fluids. Mammals have several kinds of glands, including glands that produce saliva, milk, digestive enzymes, and hormones. Some mammals have sweat glands that help regulate body temperature by secreting water onto the surface of the skin. As the water evaporates, it transfers heat from the body to the surrounding air.

MiniLab

How does the hair of mammals compare?

Many mammals have more than one type of hair. For example, animals with fur have guard hairs that make up the coarser surface layer of hair and also a layer of softer underhair. The guard hairs provide protection to the underhair and also may provide adaptive camouflage coloration. The soft underhair insulates the animal and helps to maintain body temperature.

Procedure

1. Make slides of various types of mammal hair. You could compare the hair types on one animal, or compare the hair of a variety of mammals. Prepare the slide by placing a drop of glycerine on the slide, then the animal hair, then the coverslip.
2. Find the hair under low power, and then switch to high power.
3. Draw diagrams and make brief descriptions of the differences you observe.

Analysis

1. How do the hairs from different parts of the same animal compare?
2. How do the hairs from different kinds of animals compare?
3. In what ways are the mammal hairs you have observed alike?

The sharp, barb-tipped spines of porcupines are a type of modified hair. They are an effective method of defense. If a predator attacks, the quills are easily shed for the porcupine's quick escape. They continue to move deeper into the wounded animal, causing infection or even death.

The white patch of hair on the rump of the fleeing pronghorn signals danger to other members of the herd.

Demonstration

Show students a mouse, rat, gerbil, or hamster. Offer the animal food, and ask students in groups to observe its behavior for five minutes. Ask students to list the adaptations the animal has to its environment. L1 COOP LEARN

2 Teach

MiniLab

Purpose

LS **Visual-Spatial** Students will compare the hair of various mammals.

Process Skills

observe and infer, compare and contrast

Teaching Strategies

- Have students bring in hair from farm animals and pets.

Safety Precautions

- Students with allergies to mammals should not handle the fur or hair.

Expected Results

Hair contains an inner layer that is usually darker than the outer layer. Lines, blocks, or surface scales seem to cut through the layers. Surface scales vary depending on the species and type of hair.

Analysis

1. The longitudinal layers in the hairs are of different thicknesses and have differently shaped lines.
2. The layers have different thicknesses, and differently shaped lines cutting through, or no lines or blocks at all.
3. All mammal hairs have longitudinal layers and are covered by lines, blocks, or surface scales that vary according to species.

GLENCOE TECHNOLOGY

Videodisc

STVS: Animals
Disc 5, Side 2
Temperature Regulation in Dogs (Ch. 19)

Assessment

Performance: Give students hair from a variety of animals. Ask them to make a dichotomous key based on hair characteristics. L2

SECTION 35.1

ThinkingLab Make a Hypothesis

Purpose

LS **Logical-Mathematical** Hypothesize what adaptations of an antelope's lungs and heart might enable it to run at high speeds for long periods.

Process Skills

form a hypothesis, analyze, interpret data

Teaching Strategies

- Before they read the selection, ask students to speculate about which animal would win a 33-meter dash, a pronghorn antelope or a cheetah. *The cheetah* Which animal might win a long-distance marathon? *The pronghorn antelope* L1
- After completing the lab, explain that pronghorn antelopes have lungs twice as big as other animals their size.

Thinking Critically

Students may hypothesize that the antelopes' lungs and hearts are larger than those of other animals the same size. Pronghorn antelopes are able to consume more oxygen and, therefore, run faster and longer than other species because more oxygen is being delivered to their cells.

Assessment

Portfolio: Ask students to write a paragraph explaining how the running ability of the pronghorn antelope may have evolved. Emphasize that this antelope lives on the open plains. L2 P

ThinkingLab Make a Hypothesis

How can antelopes maintain high running speeds?

When a pronghorn antelope is chased by a predator, it can average 64 km/h for more than 10 km. This speed is greater than the top speed of thoroughbred racehorses. Biologists were interested in knowing how the pronghorn antelope sustains these high speeds over such a long distance. Antelopes with masks that allowed scientists to measure the amount of oxygen inhaled were placed on treadmills. Blood samples showed the amount of oxygen delivered to cells. Scientists also examined the sizes of the antelopes' hearts and lungs.

Analysis

The graph shows data generated by the experiment. Make a hypothesis about the rate of oxygen consumption and the size of the pronghorn's heart and lungs compared to other animals the same size.

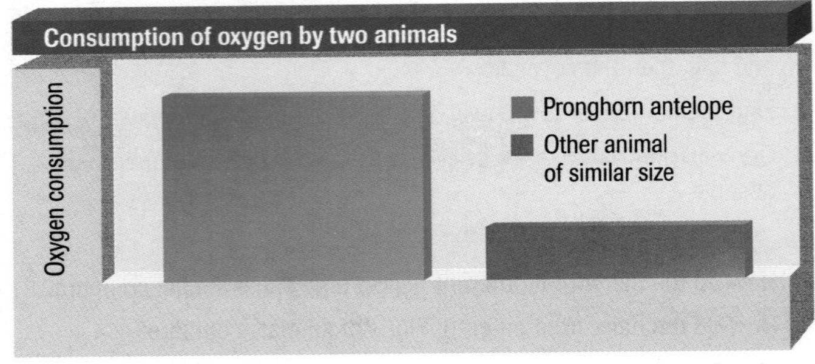

Thinking Critically

Explain the data in the graph in terms of the running ability of the pronghorn antelope compared with other species.

Mammals have a diaphragm

Mammals need a high level of energy for heating and cooling their bodies, as well as for locomotion. This high level of energy is sustained by enabling large amounts of oxygen to enter the body and reach all the cells. One way mammals accomplish this is by using a diaphragm that expands and contracts the chest cavity. A **diaphragm** is the sheet of muscle located beneath the lungs. The diaphragm separates the chest cavity from the abdominal cavity, where other organs are located.

To visualize how the diaphragm works, study *Figure 35.2* and think about how a trampoline works. When you land on the trampoline, it goes down making more room above it. When you come back up, the trampoline returns to its normal position, reducing the space above it. When a diaphragm contracts, the chest cavity above it enlarges and air moves into the lungs by suction. When the diaphragm relaxes, the chest cavity becomes smaller and air moves out of the lungs. This cycle of inhalation and exhalation is called breathing.

Figure 35.2
The diaphragm separates the chest cavity from the abdominal cavity.

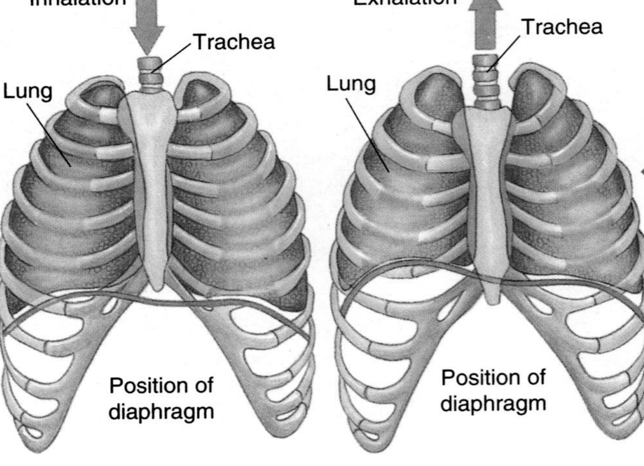

A During inhalation, the dome-shaped diaphragm contracts, enlarging the chest cavity and decreasing the pressure within the cavity. The decreased pressure allows air to move into the lungs.

B During exhalation, the diaphragm relaxes, increasing the pressure in the chest cavity and thus forcing air out of the lungs.

STUDENT JOURNAL

Products from Mammals Ask a group of students to visit a local furrier or leather shop and find out about the variety of fur and leather used in the manufacture of clothing and accessories. Ask students to find out about the processes needed to make fur and leather clothing, and have them present an illustrated report to the class. L2 LS

PORTFOLIO

Mammal Photo Essay Have students take photographs of local mammals. Have students write a caption that identifies each mammal, indicates the type of food it eats, describes its habitat, and explains how it is adapted to its habitat. Encourage students to use field guides to identify the mammal with both its common and scientific names. L2 P LS

Adaptations for obtaining and consuming food

A supply of food must be available to produce the level of energy required to maintain a steady body temperature. Mammals have several adaptations that help them meet their nutritional needs. For example, mammal limbs are adapted for a variety of methods of food gathering. Remember from Chapter 19 how primates use their opposable thumb to grasp objects including fruits and other foods they eat. *Figure 35.3* illustrates other limb modifications in mammals.

MiniLab

What can you tell about a mammal from its tracks?

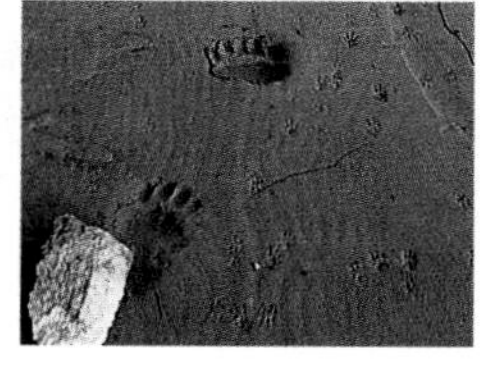

Because wild mammals are so secretive and avoid humans, it is difficult to observe them in their natural habitat. However, much can be learned about the mammals and their actions by studying their tracks. Think of reading mammal tracks as a detective game. As you read the tracks, you are gathering and analyzing clues about the mammal that made them.

Procedure

1. Find mammal tracks in the snow, wet sand, or mud by a stream or lake.
2. The distance between the steps of the same foot is the stride length. It's usually the length of the animal's body between its hip and its shoulder if the mammal is walking normally. Measure this distance to determine how large the animal is that made the tracks.
3. Determine if the mammal was moving quickly or slowly by comparing the foot length with the length of the stride. For example, small, far-apart tracks indicate that a small animal was moving quickly.

Analysis

1. What sizes were the mammals you found?
2. Were the animals moving slowly or quickly?
3. Were there other clues in the area that might give you ideas about what the mammals may have been doing? Explain.

Figure 35.3

Mammals have modified limbs that are adapted to their habitats and to their methods of getting food.

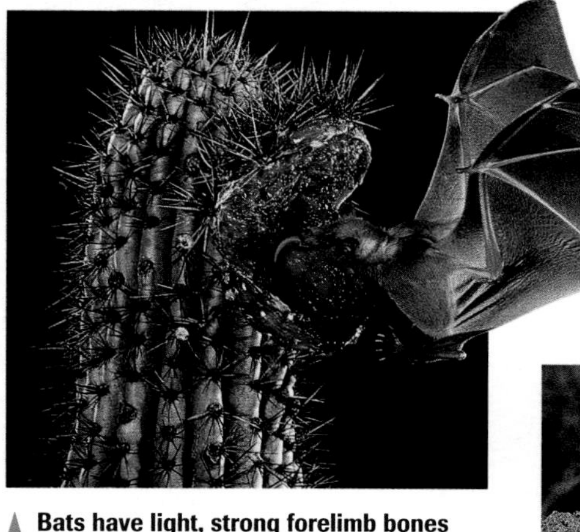

Bats have light, strong forelimb bones with greatly elongated finger bones that support the flight membranes.

In moles and other digging mammals, the front limbs are powerful and short with large claws.

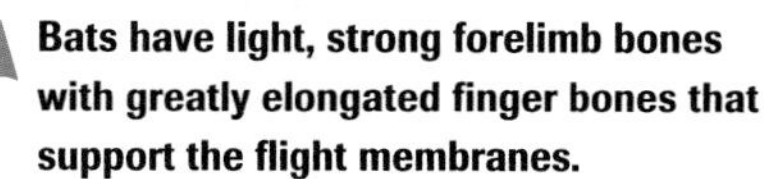

Antelope have strong, slender bones that enable them to run fast.

SECTION 35.1

MiniLab

Purpose

Logical-Mathematical Students will discover what can be learned about an animal based on its tracks.

Process Skills

compare and contrast, recognize cause and effect, form a hypothesis, interpret data

Teaching Strategies

- Make available to students field guides that interpret animal tracks.
- This is an ideal winter activity and can even be conducted in the city after a snowstorm.

Expected Results

Results will vary depending on the habitat. In the city, students will find animals the sizes of dogs, cats, mice, squirrels, and rats. In a meadow, they may find small animals such as mice, woodchucks, prairie dogs, deer, elk, or marmots. Stream tracks may be found for animals of all sizes because many animals drink from the stream. Others such as mink, muskrats, and beaver inhabit the banks.

Analysis

1. In most environments, mammals of all sizes will be found.
2. Some animals will be moving slowly, others will move quickly. Around areas of water, they may not move as quickly.
3. Tracks of two mammals may be superimposed on each other indicating a chase. Tracks may come from different directions; and where they join they may be scuffed and contain pieces of fur, indicating a fight.

PROJECT

Opposable Thumbs

Ask students to try to shell a peanut under the following conditions: (a) Tape hands with fingers and thumbs together. (b) Tape hands so that fingers move, but not thumbs. (c) Tape hands so only the top joint of the thumb can move. Have students write a report to summarize the importance of opposable thumbs. L2 LS

Assessment

Performance: Give students copies of animal tracks that are actual size and a ruler. Ask them to tell all they can about the animal. L1

Bioethics

Finding Jobs for Poachers

In past years, elephant populations in North Luangwa National Park in Zambia were being decimated by poachers. The elephants were killed for their ivory. For a long time, the illegal ivory trade sustained the economy of local villages surrounding the park.

Efforts were made to create alternative jobs for poachers. As a result, the number of elephants poached in the park has fallen from 1000 to 12 per year. With money raised by nonprofit groups, former poachers were given the means necessary to open small businesses such as butcher shops, grinding mills, fish farms, sewing stores, and carpentry shops. Now, poachers of the past are working in the park to save the elephants, which they were shooting by the hundreds only two years ago. Now, in some parts of Africa, elephants may have reached the carrying capacity of the habitat, and managers are faced with the problem of controlling habitat destruction.

Visual Learning

Figure 35.4 Have students compare the teeth of the mammals shown. Elicit how the teeth of a horse are adapted to its diet. *Horses are herbivores. Their predominant teeth are premolars and molars adapted to grinding.* How are the teeth of a wolf adapted to its diet? *The sharp canine teeth stab and pierce food. The food is then ground by the molars.* L1 LS

Figure 35.4

Mammal groups are distinguished by number and types of teeth.

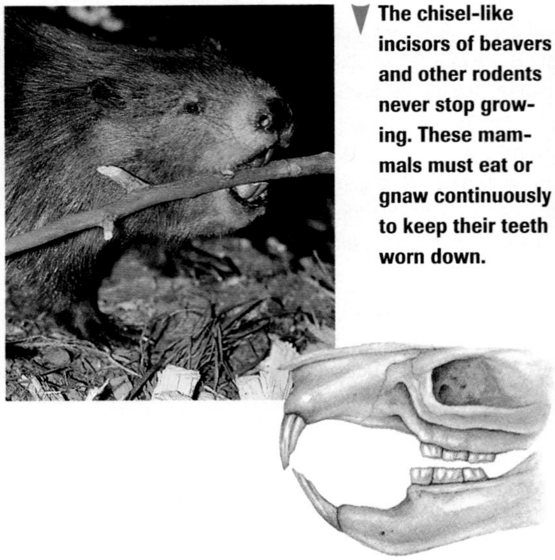

The chisel-like incisors of beavers and other rodents never stop growing. These mammals must eat or gnaw continuously to keep their teeth worn down.

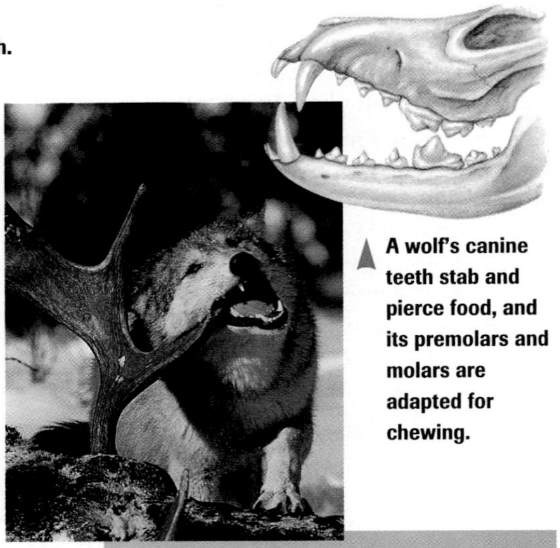

A wolf's canine teeth stab and pierce food, and its premolars and molars are adapted for chewing.

Premolars and molars are the predominant teeth in horses. These crushing and grinding teeth are covered with a hard enamel suited for heavy wear.

Adaptations for consuming food

Teeth are a distinguishing feature of most mammals. By examining the teeth of a mammal, a scientist can determine what kind of food it eats. Although fishes and reptiles have teeth, their teeth are relatively uniform and are used primarily for tearing, grasping, and holding prey.

Mammals have different kinds of teeth adapted to the type of food they eat. Think of the different tools you might use to build a piece of furniture, such as a chisel for scraping or a saw for cutting. Like a cabinetmaker's tools, teeth are shaped to match the types of jobs they do. The pointed incisors of moles grasp and hold small prey. Chisel-like incisors of beavers are modified for gnawing. A tiger's canines puncture and tear the flesh of its prey. Premolars and molars are used for slicing or shearing, crushing, and grinding. The various types of teeth in different mammals can be seen in *Figure 35.4*.

Many hoofed mammals have an adaptation called cud chewing that enables them to break down the cellulose of plant cell walls into nutrients that they can use and absorb. In **cud chewing,** plant material that has been swallowed is brought back up to the mouth and chewed again. Have

STUDENT JOURNAL

Agricultural Mammals Ask students to visit a dairy farm or sheep ranch and report in their journals about the latest techniques and strategies used in caring for and enhancing production by farm animals. If there are no farms nearby, provide library resources and information from a university cooperative extension unit. L2 LS

you ever seen cows slowly chewing and chewing while lying in a pasture? When grass is swallowed, cellulose in the cell walls is broken down by bacteria in one of several pouches in the stomach. The food, called cud, is then brought up into the mouth. After more chewing, the cud is swallowed again and passed to three other stomach areas, where digestion continues.

Mammals nurse their young

One reason mammals are successful is that they guard their young fiercely and teach them survival skills. Mammals also feed their young from mammary glands. In female mammals, the **mammary glands** secrete milk, enabling the mothers to nourish their young until they are mature enough to find food and protect themselves. *Figure 35.5* shows that the number of young each mother has and the length of time she nurses her young vary among species.

Figure 35.5

Large mammals usually have few young. Mammals that are prey for many predators tend to have larger litters.

Mice have four to nine offspring in each litter and up to 17 litters in a year. The blind, naked, and helpless young mice nurse for only a few weeks before being mature enough to go out and forage for food on their own.

The elephant normally has one calf at a time. Newborns weigh about 100 kg and drink 11 L of milk each day. Some elephant calves nurse for ten years.

Seals usually have one pup in the spring or summer. Depending on the species, seal pups nurse from a few days up to a year. The rich milk is 50 percent fat, which enables the pups to grow quickly and helps add a thick layer of blubber.

Different Viewpoints in Biology

Endangered Mammals

Seventeen years ago, the Mexican gray wolf was placed on the endangered species list. Current estimates indicate that there may be as few as 60 Mexican wolves in their original range in Arizona, New Mexico, and Texas. Most of the wolves were killed because they were thought to be a threat to livestock.

Mexican wolf populations were decimated before any observations of their behavior were made. Current studies of wolves from Canada indicate that there is only one livestock killing each year for every 100 wolves. Studies also show that wolves are likely to be nocturnal, shy, and secretive. Activities of people generally deter wolves from coming close.

The U.S. Fish and Wildlife Service proposes to reintroduce, from captive populations, the Mexican wolf into its original range. However, ranchers do not want this to happen. Breeders of registered cattle may lose up to $20 000 if just one prize bull is lost to wolves. In addition, years of scientifically controlled breeding will be lost as well. Wolf advocates say that people must learn to share Earth with wolves. They believe the wolf should be returned to its natural habitat. These people feel that the preservation of the species is more important than the loss of a cow or several sheep per year.

Meeting Individual Needs

Gifted Ask students to report in their portfolios on genetic engineering techniques that enable cows to produce more milk, and cows, sheep, and goats that have been genetically engineered to produce milk with human proteins. Ask them to discuss the importance and economic implications of these developments. L3 P LS

PORTFOLIO

Care of Young Ask students to visit or contact a zoo nursery and compare the care provided for young zoo animals. Ask students to compare these methods to the care mammals in the wild provide to their young. Students should each do an illustrated report for the class. L1 P LS

BioLab Design Your Own Experiment

Is blubber a good insulator?

Time Allotment
One class period

Objectives
Review objectives with students before they begin the Biolab.

Process Skills
compare and contrast, observe and infer, recognize cause and effect, form a hypothesis, interpret data, separate controls and variables, design an experiment

Safety Precautions
- Provide students with hand protection when they are handling heating materials or instruments.
- Direct students not to touch shortening with their fingers as the grease will spread over all their materials.

PREPARATION

Alternate Materials
- Small jars may be used in place of test tubes as long as all are of the same size.

Possible Hypotheses
- Insulated test tubes will cool more slowly than uninsulated test tubes. Students may make other hypotheses and should be allowed to test their hypotheses.

BioLab Design Your Own Experiment

Is blubber a good insulator?

A whale's skin is adapted to life in the water. It is smooth and almost hairless. Hair would slow down the streamlined movement of a whale in the water. Whales must maintain their body temperature in cold water. A layer of fat, called blubber, beneath the skin insulates the animal so that it can survive in the coldest oceans. The bowhead whale that lives in the coldest arctic region has blubber up to 25 cm thick.

PREPARATION

Problem
How well does blubber insulate an animal?

Hypotheses
Make a hypothesis about the rate at which an uninsulated test tube will cool compared to a test tube insulated with fat.

Objectives
In this Biolab, you will:
- **Measure** the rate of cooling of insulated and uninsulated test tubes.
- **Compare** the rates of cooling of insulated and uninsulated test tubes.

Possible Materials
test tubes
plastic spoons
shortening
thermometers
plastic sandwich bags
large beaker with ice water
rubber bands
watch with a second hand

Safety Precautions
Be sure to protect your hands if you use heating devices in your experiment.

PLAN THE EXPERIMENT

Teaching Strategies
- Do the following demonstration the day before students do the Biolab. Place one test tube of hot water in a wool sock and another in a test-tube rack. Take temperature readings of the two setups after 15 minutes and again after 30 minutes. Compare the results.
- Demonstrate how to put several spoonfuls of shortening into a plastic bag and then place a test tube in it. Have students spread the shortening over the test tube by moving it with their fingers on the outside of the plastic.
- When they complete their experiment, advise students to throw away their plastic bags and place their test tubes in a bucket containing soapy water you have provided.

PLAN THE EXPERIMENT

1. Design a controlled experiment that will enable you to determine the rates of cooling of insulated versus uninsulated test tubes.
2. Plan to collect quantitative data.
3. Design a data table.
4. In a numbered list, similar to a recipe, make a set of directions that anyone could follow to do your lab.

Check the Plan

1. What is your control?
2. Why did you need three test tubes?
3. What will you measure?
4. How often will you take this measurement?
5. Have you made a table for collecting data?
6. ***Make sure your teacher has approved your experimental plan before you proceed further.***

7. Carry out the experiment you have designed, and record your data in your table.
8. Make a bar graph of your data with temperature on the *y* axis and the test tube description on the *x* axis.

ANALYZE AND CONCLUDE

1. **Checking Your Hypothesis** Did your data support your hypothesis? Why or why not?
2. **Comparing and Contrasting** How did the rate of cooling of the uninsulated tube compare with that of the others? Explain.
3. **Making Inferences** Is blubber a good insulator for mammals that swim in cold seawater?

Going Further

Application What other materials insulate animals or objects? Plan an experiment with materials that might be good for insulating a water heater.

ANALYZE AND CONCLUDE

1. Hypotheses will be supported if they stated the shortening-insulated tube would lose heat more slowly.
2. The rate of cooling of the uninsulated tube was about twice that of the insulated tube because there was no insulation to prevent heat loss.
3. Blubber is similar to the shortening that covered the test tube in the experiment and it is a good insulator for mammals that swim in cold seawater.

Assessment

Performance: Ask students to design and carry out an experiment that compares the insulating properties of down feathers and wool. L1

Going Further

Students may want to compare various fabrics used in thermal underwear for their ability to retain heat, or they may obtain insulating materials from a building supply company and compare which is best for the same thickness. L2

Possible Procedures

- Students may decide to fill three test tubes with hot water. They could cover one with a 1-cm layer of shortening and a plastic bag, one with just a plastic bag, and one with neither shortening nor a plastic bag. They would use thermometers to measure the temperature every 5 minutes for 30 minutes.

Data and Observations

In general, it will take about twice as long for the shortening-insulated test tube to lose heat and take on the temperature of the ice bath. The tube with just the plastic bag will lose heat more slowly at the beginning, but catch up to the plain test tube about 10 minutes after the warm air that it may trap has cooled.

Chemistry

Fats and Cholesterol: Good or Bad?

Purpose

Students will gain an understanding of the relationship between saturated and unsaturated fat and cholesterol and the importance of each to animals.

Teaching Strategies

- Make a transparency of the molecular structure of saturated and unsaturated fats and cholesterol. Point out the differences among the structures as you display them.
- Ask students to obtain labels from margarine containers and determine the amounts of saturated and unsaturated fats and cholesterol for several brands. L1
- Ask students to research and report on the details of the importance of cholesterol to cells and the harm it can cause in the blood vessels. L1 P
- Provide students with a chart of amounts of fat in certain foods, and ask them to prepare a day's menu that has less than 10 g of fat. L1

Possible Answer

Individual responses will vary depending upon the foods eaten. Many students are likely to report that their intake of fats exceeded the recommended levels.

Chemistry

Fats and Cholesterol, Good or Bad?

You've probably heard all the bad news about fats and cholesterol. But did you know that small amounts of fat and cholesterol play an essential role in our metabolic processes? Fat cushions the internal organs, helps to insulate the body from heat loss, and is an efficient source of energy. Cholesterol helps cell membranes work properly.

Types of fats Fats are made up of long chains of carbon atoms. Fats may be saturated, monounsaturated, or polyunsaturated. Saturated fats have only single bonds between the carbon atoms, whereas unsaturated fats have one or more double bonds. Cholesterol is another type of lipid.

Saturated fats Animal fats, palm oil, coconut oil, and cocoa butter are made up primarily of saturated fats and have been associated with an increased risk of heart disease and some cancers. Cholesterol, which is produced by animals and to a much lesser extent by plants, is found in beef, poultry, pork, and dairy products.

Unsaturated fats Unlike saturated fats, mono- and polyunsaturated fats tend to be liquids at room temperature. Avocados, peanuts, cashews, and olives are sources of monounsaturated fats. Fish, corn, safflower, and soybean oils are polyunsaturated and have been shown to decrease blood levels of cholesterol.

Less fat is needed Experts recommend that 10 percent or less of a person's diet be composed of fats. It may not be necessary to eliminate saturated fats, but reducing the amount of cheese, butter, and egg yolks; trimming the visible fat from lean meats before cooking; and cutting back on junk foods containing palm and coconut oils will help bring your fat and cholesterol intake to recommended levels.

CONNECTION TO Biology

Keep a log of what you eat for two days. Use references to determine your (a) total Calorie intake per day and (b) total fat intake per day. How do your totals compare with recommended allowances?

Mammals are intelligent

Have you ever attended an aquarium show or watched a movie about performing dolphins and whales? Dolphins exhibit a wide variety of learned behavior, including the tricks performed for films or in aquarium shows. Mammals can accomplish complex behaviors, such as learning and remembering what they have learned. Primates, including humans, are perhaps the most intelligent animals. Chimpanzees, for example, can use tools, *Figure 35.6,* work machines, and use sign language to communicate with humans. Mammalian intelligence is a result of complex nervous systems and highly developed brains. A mammalian brain often is folded, forming ridges and grooves; therefore, it has a large surface area.

Figure 35.6

Scientists used to think that only humans had enough intelligence to make and use tools. A chimpanzee using a stick to get grubs out of a tree proves that other primates are also intelligent enough to make and use tools.

Alternate Lab: Burying Behavior

Purpose 

LS **Logical-Mathematical** Students will observe defensive burying behavior in mice and determine the types of objects that cause this behavior.

Materials

mouse, terrarium with 3 to 5 cm of sawdust or sand, objects for burying, timer

Procedure

Provide students with the following information and then give them the directions that follow it.

When mice are confronted by potential predators such as snakes, they sometimes spray sand or soil at the predator. This behavior is called defensive burying. Defensive burying can be elicited by objects that resemble predators, as well as by predators themselves.

1. Make a data table that will indicate a list of objects, the time it takes the mouse to start burying, and the time spent burying.
2. Develop a hypothesis about which objects a mouse will spend the most time burying.
3. Place one of the objects on the floor of the terrarium at the end farthest

THE INSIDE STORY

A Mammal

Like other mammals, a red fox must catch prey and consume it to survive. Studies show that movement attracts foxes. In the wild, most of the prey animals get away when the fox attacks a group. However, in a henhouse, the birds can't run away. The continual fluttering stimulates the fox's instinctive behavior so it kills many birds.

Red fox

1. Dense, soft underhair insulates the fox by trapping warm air next to its body. The coarse, long guard hairs protect against wear and may be colored for camouflage. The fox sheds its coat little by little during the summer.

2. The diaphragm is a muscle that helps the chest cavity expand to take in large amounts of oxygen used to maintain the high metabolism of all mammals.

3. A fox's teeth are indicative of what it eats and how it gets food. Because a fox preys on anything it can catch, it is found in unlikely environments, such as suburban areas where it can easily catch mice, shrews, and even small birds.

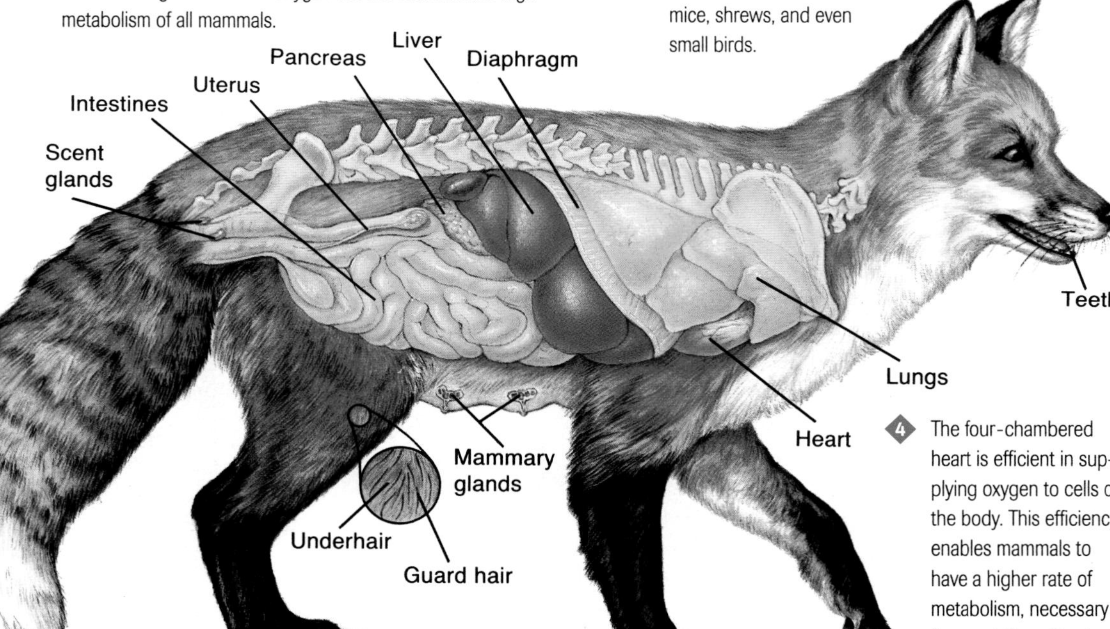

4. The four-chambered heart is efficient in supplying oxygen to cells of the body. This efficiency enables mammals to have a higher rate of metabolism, necessary for regulation of body temperature.

6. In addition to mammary glands, most mammals have sweat, oil, and scent glands. Foxes use their scent glands to mark new territories. Oil glands lubricate the hair and skin.

5. Like all female mammals, the fox nourishes her young with milk from the mammary glands.

THE INSIDE STORY

A Mammal

Purpose

Visual-Spatial Students will review the main characteristics of mammals.

Visual Learning

- Have volunteers read each caption aloud. Discuss each caption after it has been read. L1 COOP LEARN
- Obtain models that show the teeth of mammals that eat different foods. Have students examine the models. Next ask them to identify the model that most closely represents the structure of the teeth of a fox. L1

Teaching Strategies

- Ask groups of students to do detailed reports on the life history of various types of foxes in the United States.
- Ask students to speculate about the adaptive advantage of the fox's bushy tail. Ask students to explain in terms of adaptation and evolution how the fox may have come to have such a bushy tail.

from the animal. **CAUTION:** *Always use care when handling live animals.*

4. Note the time it takes for the mouse to begin burying behavior and the time it spends burying.
5. Test one object at a time, following the same procedure.
6. After filling out the data table, make a bar graph to illustrate the total amount of time the mouse spent burying each object.

Analysis

Ask students to answer the following questions.

1. Which objects did your mouse spend the most time burying? The least time? *Objects that looked more like predators; least like predators.*
2. How do your results compare with those of the class as a whole? *Answers should be similar.*
3. Do your data support your hypothesis? *Data will support hypotheses that stated "predator-like" objects would cause more burying behavior.*

Assessment

Performance: Ask students to design and conduct an experiment that will determine the types of objects that cause more defensive burying behavior. L1

Visual Learning

Figure 35.7 Elicit from students what animal is the closest relative to modern-day mammals. *Therapsids* Challenge students to explain why therapsids are a link between mammals and reptiles. *Tharapsids have characteristics that are both mammal-like (endothermic) and reptile-like (scales and claws).* L1 LS

Misconception

Some students may think that unicorns are prehistoric mammals that lived at some time during geologic history. Explain that unicorns are mythological animals. Ask a group of students to look up the role of the unicorn in myths and folklore and report their findings to the class. L1 COOP LEARN

3 Assess

Check for Understanding

Ask students to summarize in two paragraphs why mammals live everywhere. Have them include a discussion of mammals that live below the surface of Earth and in lakes, oceans, and deserts. L1

The Orders of Mammals

Figure 35.7

Most scientists agree that all mammals evolved from therapsids, mammal-like reptiles. Of the three mammalian subclasses, only the placental mammals have evolved into orders that exploit niches on land, in the sea, and in the air. This diagram represents the orders of mammals and shows their evolutionary relationships.

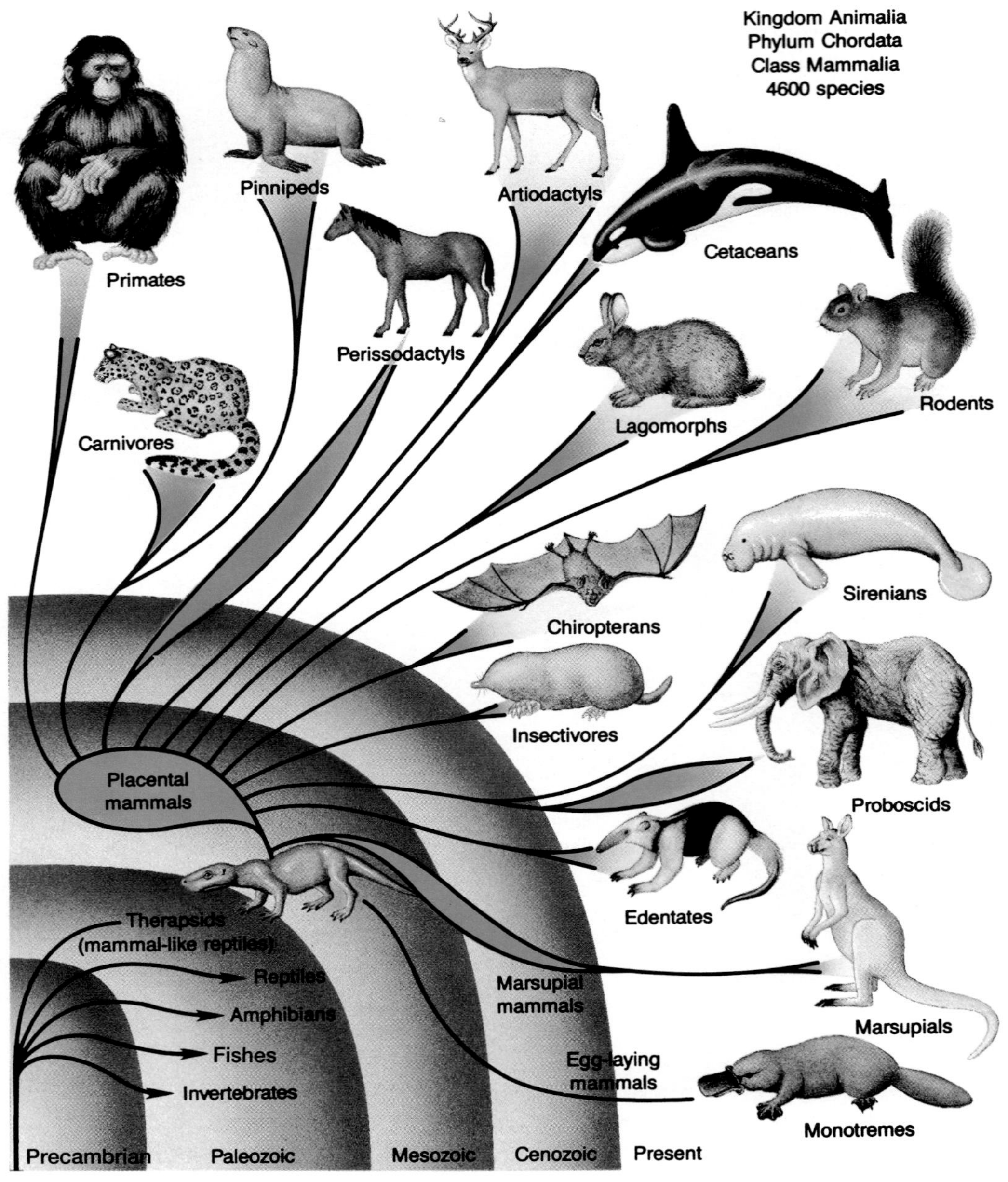

CULTURAL DIVERSITY

Protecting Mammal Wildlife

Mammals are perhaps the most exploited of all animals. African elephants, apes, pandas, manatees, dolphins, and rhinos are only some of the mammals likely to be extinct within fifty years due to poaching and the destruction of habitat.

As an example of how government intervention is essential for protecting endangered mammals, discuss with students the history of China's efforts to protect and breed pandas. Extend the discussion by asking students to bring in articles from recent newspapers and science magazines about the exploitation of mammals and conservation efforts to save them.

Origins of Mammals

The first true mammals appeared about 200 million years ago. Scientists can trace the origins of mammals back from an insect-eating animal like the one shown in *Figure 35.8,* to a group of reptilian ancestors called **therapsids.** Therapsids, represented in *Figure 35.8,* were fierce-looking, heavy-set animals that had features of both reptiles and mammals. They roamed Earth between 270 and 180 million years ago.

The mass extinction of the dinosaurs at the end of the Mesozoic Era, along with the breaking apart of Pangaea and changes in climate, opened up new niches for early mammals to fill. The appearance of flowering plants at the end of this era supplied new living areas, food, and shelter. Some mammals that moved into the drier grasslands became fast-running grazers, browsers, and predators. In fact, the Cenozoic Era is sometimes called the golden age of mammals because of the dramatic increase in their numbers and the variety of diverse forms of mammals. Present-day mammal orders are shown in *Figure 35.7.*

Figure 35.8

Scientists use fossils to trace the history of mammals back to early reptiles.

Therapsids were the reptilian ancestors of mammals. The lower jaw and middle ear bones of therapsids were like those of reptiles. However, they had straighter legs than reptiles and held them closer to the body. Straight legs enabled therapsids to run faster. Therapsids may have been endotherms.

Fossils of these insect-eating early mammals have been found throughout the world. After examining the remains of their teeth, scientists determined that these mammals ate insects. They also probably had good senses of hearing and smell.

Section Review

Understanding Concepts

1. Name four characteristics of all mammals.
2. What are therapsids, and what is their relationship to mammals?
3. Describe how endothermy has contributed to the success of mammals.

Thinking Critically

4. Suppose you are a mammal that feeds on pine seeds and lives in a forest in a cold region. Describe the adaptations that would help you survive.

Skill Review

5. **Observing and Inferring** On an archaeological dig, you find a skull about 5 cm long with two chisel-shaped front teeth and several flattened back teeth. Is this a skull from a mammal? Explain your answer. For more help, refer to Thinking Critically in the *Skill Handbook.*

SECTION 35.1

Reteach

Bring in a variety of pictures of the limbs of mammals. Ask students working in groups to examine the pictures and tell as much as they can about the life of that mammal based on the structure of its limbs. L1 COOP LEARN

Extension

Dolphins, sea lions, and beluga whales are being trained by the Navy to watch over Navy submarines while they are at base. Have students research the work of the U.S. Navy in training porpoises for various tasks, such as spying and delivering messages. L2

Assessment

Portfolio: Ask students to prepare an illustrated essay on an ideal mammal pet created by genetic engineering in the year 3000. L2 P

4 Close

Discussion

Have students propose a hypothesis to explain why deer eat more in winter than in summer. L1

Answers to Section Review

1. Hair, diaphragm, and in the female, mammary glands and uterus.
2. Therapsids are the reptilian ancestors of mammals.
3. Endothermy enables mammals to live in all environments regardless of temperature.

Thinking Critically

4. Forefeet with sharp claws that can hold a pinecone; sharp front teeth to pull apart cones to get seeds inside; thick fur; short extremities including ears, legs, and nose; a long, bushy tail that can help with balance as the mammal moves in the trees obtaining pinecones. The bushy tail might also help to protect the face from cold as the animal sleeps.

Skill Review

5. Yes. One mammal may have varied teeth types in its skull. Reptilian ancestors had skulls with teeth that were all very much alike.

SECTION

35.2

Prepare

Key Concepts

The three groups of living mammals are introduced and discussed. In addition, reproduction in egg-laying, pouched, and placental mammals is compared.

Block Scheduling

Look for this symbol for strategies that are useful in a block scheduling format. For more information on block scheduling, refer to Section 35.2 in the **Lesson Plans** booklet.

Materials

- Gather cartoons featuring different types of mammals for the Focus Activity.

Focus

Bellringer

Before presenting the lesson, display **Section Focus Master 78** on the overhead projector and have students answer the accompanying questions. L1 SAE

Activity

Provide students with cartoons that feature mammals. Ask them to compare the characteristics of cartoon mammals with the adaptations of real mammals. L1

SECTION

35.2 The Diversity of Mammals

Section Preview

Objectives

Distinguish among the three groups of living mammals.

Compare reproduction in egg-laying, pouched, and placental mammals.

Key Terms

uterus
placental mammal
placenta
gestation
marsupial
monotreme

All the animals in this photograph are mammals, yet they don't look much alike. What characteristics do zebras and wildebeests share? You know that both of these mammals have hair, mammary glands, and various kinds of teeth. Unlike other animals, the young of mammals are nourished and protected by their mothers. They learn how to survive—get food, find shelter, and evade predators—by observing their parents or other adults of their species. A major characteristic that separates mammals from all other animals is parental care and nourishment of the young. Yet even among mammals, there are differences in methods of reproduction and care of the young. These differences aid scientists in tracing the phylogeny of mammal groups.

How Mammals Are Classified

Can you make a list of all the mammals you see in a day? The majority of mammals you are familiar with probably belong to only one of the three groups, or subclasses, of the class Mammalia. Scientists place mammals into one of three subclasses based on their method of reproduction.

Placental mammals—A great success

The puppies shown in *Figure 35.9* were born after a period of development within the uterus of their mother. The **uterus** is a hollow, muscular organ in which the development of offspring takes place. Development inside the mother's body is an adaptation that played a major role in the

Figure 35.9

The length of gestation varies from species to species. These puppies were born after two months of gestation. Gestation in mice is 21 days, whereas gestation for a rhinoceros is 19 months. Gorilla gestation is about nine months, similar to the gestation of a human.

Program Resources

Section Focus Master 78 L1 SAE
Reinforcement and Study Guide, pp. 143-144 L1

PORTFOLIO

Zookeeper Have students assume they are directors of a zoo that will be adding an unusual mammal exhibit. Ask students to select a mammal; design the exhibit; and determine the animal's diet, its feeding schedule, and other special needs the mammal will have. Finally, they should prepare an exhibit sign for visitors. L1 P LS

success of mammals as they spread throughout the world. It ensures that the offspring are protected from predators and the environment during the early stages of growth.

A mammal that carries its young inside the uterus until development is almost complete is known as a **placental mammal.** Nourishment of the young inside the uterus occurs through an organ called the **placenta,** which develops during pregnancy. The placenta is also instrumental in passing oxygen to and removing wastes from the developing embryo. The time during which placental mammals develop inside the uterus is called **gestation.** About 95 percent of all mammals are placentals.

Mammals with a pouch

Marsupials make up the second subclass of mammals. A **marsupial** is a mammal in which the young have a short period of development within the mother's body, followed by a second period of development inside a pouch made of skin and hair found on the outside of the mother's body. Most marsupials are found in Australia and surrounding islands. The theory of continental drift, *Figure 35.10,* explains why most marsupials are found in Australia today. Scientists have found fossil marsupials in the continents that once made up Gondwana. These fossils support the idea that marsupial mammals originated in South America, moved across Antarctica, and populated Australia before Gondwana broke up.

Figure 35.10

Based on geological and fossil evidence, the movements of the continents through time have been reconstructed. About 200 million years ago, all the landmasses of Earth constituted one large supercontinent called Pangaea.

The formation of Pangaea changed the environment by destroying many shallow seas. This led to the evolution of species that took advantage of newly opened niches on land. Pangaea began to break up about 190 million years ago during the Mesozoic era, around the time mammals first began to appear.

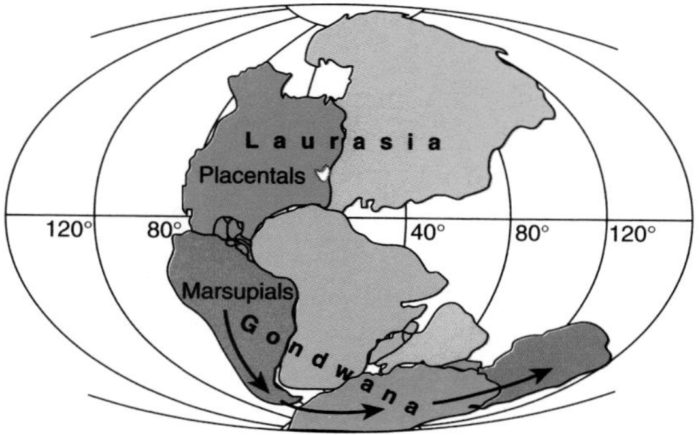

190 million years ago

65 million years ago

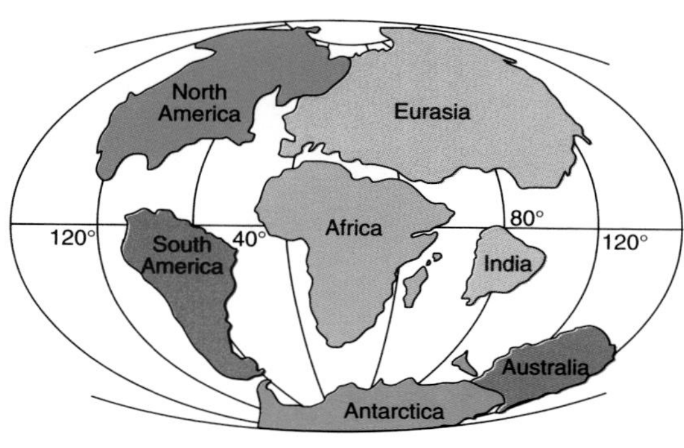

This breakup resulted in two landmasses: Laurasia, the northern group of continents, and Gondwana, the southern group of continents. By the end of the Cretaceous period 65 million years ago, Africa and South America had moved apart. By 43 million years ago, Australia had separated from Antarctica and moved to its present location.

SECTION 35.2

2 Teach

Science, Technology, and Society

Using Technology to Save Mammal Lives

In just one year, 45 bears were killed on the roads of Florida. Accidents between bears and cars are most common in October when hungry bears are foraging for food. The U.S. Fish and Wildlife Service estimates that there are only about 1000 to 1500 bears left in the state.

The Florida highway department tested the use of concrete box culverts that allow bears to travel under rather than across roads. Engineers placed video cameras inside the culverts to see whether the bears used them. Fences were installed beside the road to funnel the bears to the culvert crossings. Although the installation of the culverts was expensive, they were effective in protecting the bears.

Enrichment

IS **Intrapersonal** Have interested students conduct research to find out what young mammals of different species are called. Encourage students to prepare a photo essay of their findings. L1

GLENCOE TECHNOLOGY

STVS: Ecology
Disc 6, Side 1
Red Wolf (Ch. 12)

The Infinite Voyage: The Keepers of Eden
Cincinnati Zoo's Cat House: Innovative Breeding Techniques (Ch. 6)

Science, Technology, and Society

Fingerprinting Koalas

Australian scientists are hoping to save their wild populations of koalas by fingerprinting them. In the past, koalas were killed by the thousands for their skins. Today, loss of habitat and illegal capturing of the animals to replace those that die in zoos are the causes of much of the koala decline.

Like humans, koalas have distinct fingerprint patterns. The "thumb" on the koala's back foot provides a unique pattern that can be used for identification purposes. Scientists are fingerprinting captive animals in hopes that this will enable them to identify animals that come into zoos illegally.

Discussion Question

Ask students to explain why the introduction of grazing animals such as sheep to areas of Australia would threaten the survival of marsupials that are native to the area. *The sheep are outcompeting the marsupials for grasses and water. In addition, ranchers who raise sheep drive the native marsupials off their land to protect their investment.* L2

GLENCOE TECHNOLOGY

Videodisc

Biology: The Dynamics of Life
Disc 2, Side 1
Duck-Billed Platypus (Ch. 10)

Figure 35.11

In Australia and Tasmania, many marsupials fill niches that are occupied by placental mammals on other continents.

The Tasmanian devil, a marsupial, lives in burrows in the thick rain forest of Tasmania. It is a fearsome predator and scavenger with badgerlike teeth that are used to seize and crush its prey. African hyenas (above) are scavengers and predators that fill a niche similar to that of Tasmanian devils, but they live in grasslands.

Woodchucks (left) are common burrowing rodents found throughout North America. Wombats (right) are marsupials found in Australia. They burrow underground and eat grass and tree bark.

Ancestors of today's marsupials were able to populate the landmass that became Australia without having to share the area with the competitive placental mammals that evolved in other places. They successfully spread out and filled niches similar to those that placental mammals filled in all other parts of the world, *Figure 35.11*. In fact, because humans introduced sheep and other placental mammals to Australia, many of the native marsupial species have become threatened, endangered, or even extinct.

Egg-laying mammals

Do you think the animal shown in *Figure 35.12* is a mammal? It has hair and mammary glands, yet it lays eggs. The duck-billed platypus is a monotreme. A **monotreme** is a mammal that reproduces by laying eggs. Monotremes are found only in Australia, Tasmania, and New Guinea. Spiny anteaters, also called echidnas, belong to this subclass as well. One of the two species of spiny anteaters can be found only in New Guinea. Only three species of monotremes are alive today.

STUDENT JOURNAL

A Marsupial Adventure Have students make a travel brochure of a photo safari to Australia. Students should describe the animals they promise to show tourists and include a map of the trip. L2 LS

The platypus, a mostly aquatic animal, looks as if it were designed by a committee. It has a broad, flat tail, much like that of a beaver. Its rubbery snout resembles the bill of a duck. The platypus has webbed front feet for swimming through water, but it also has sharp claws on its front and hind feet for digging and burrowing into the soil. Much of its body is covered with thick, brown fur.

The spiny anteater is only slightly less strange in appearance. It has coarse, brown hair, and its back and sides are covered with sharp spines that it can erect for defensive purposes when threatened by enemies. From its mouth, located at the end of a long snout, the anteater extends its long, sticky tongue to catch insects.

Figure 35.12

The duck-billed platypus seems like a jumble of physical features that belong to a variety of other animals. Only one species survives today.

Issues

BIOLOGY & SOCIETY

What Price Beauty?

Before the Food, Drug, and Cosmetic Act of 1938 was passed, cosmetics commonly contained toxic chemicals such as arsenic, mercury, and quinine. Although the Food and Drug Administration (FDA) does not require cosmetic manufacturers to test their products before marketing, it strongly urges such testing. The FDA also supports the use of animals in testing the safety of beauty aids.

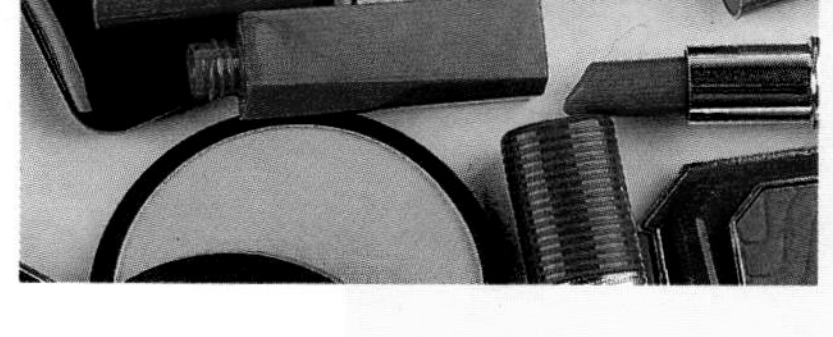

Protecting consumers

Manufacturers of cosmetics have complied with the FDA's request for testing in order to ensure public safety and avoid costly lawsuits. Consumers, manufacturers, and the government all agree that testing is needed but not on how it's done.

Guinea pigs For decades, companies specializing in health-and-beauty aids routinely tested their products on animals. During the 1980s, many people began to object to this. Some concerned groups asked the Cosmetic, Toiletry & Fragrance Association to urge its member companies to suspend all tests on animals. A few manufacturers stopped all animal testing, and some called for a temporary delay in such testing, but others still use animals in their tests.

Different Viewpoints

People who claim that using animals to test the safety of cosmetics is unnecessary and cruel have suggested new procedures, including computer modeling and the use of cultured artificial skin as possible alternatives. Some companies are even exploring the use of protists and other unicellular organisms as test subjects. Still, it is not known if results from tests on these organisms can be applied directly to human use of a product.

Where do consumers stand on the issue? A 1990 consumer survey of 1000 adults found that 60 percent opposed using animals to test the safety of cosmetics and toiletries. Furthermore, 89 percent said that they would buy products that had not been tested on animals.

INVESTIGATING the Issue

Writing About Biology Develop a public opinion questionnaire to find out people's views on the use of cosmetics. Tabulate the data by the age, gender, education, and occupation of your respondents. Prepare a written analysis of the results.

Meeting Individual Needs

Gifted Ask students who are working above grade level to take a survey of a particular mammal in your area. Ask them to look up survey methods for mammals in a wildlife management-techniques manual. Ask them to make a management plan for the area that would ensure that their mammal has adequate habitat over the next 25 years. L3

Issues

BIOLOGY & SOCIETY

What Price Beauty?

Purpose

Students will examine the issue of using animals for testing the safety of health and beauty aids.

Background

Prior to FDA regulation of cosmetic testing in 1938, some products caused injury to their users. For example, one hair dye caused inflammation of the skin, a blistered scalp, and blindness. The Draize test is the standard industry test used on animals. The test involves placing a cosmetic substance in a rabbit's eye. Then the substance is evaluated based on how much irritation it caused to the animal's eye.

Alternatives for the Draize test include using tissue cultures. However, these do not as yet reflect what would happen in the whole animal. One step in reducing harm to animals has been taken by laboratories that have developed prescreening alternative testing that has reduced the number of rabbits needed by 87 percent.

Teaching Strategies

- Because animal rights is a volatile issue, make sure that students are respectful of views different from their own.

INVESTIGATING the Issue

Writing About Biology Questionnaires should include data for all groups requested. Data will vary among individual questionnaires.

Going Further

Ask students to interview scientists who use animals in their research to find out why they feel that animal testing is necessary.

Focus On

Placental Mammals

Purpose

Students will examine the main orders of placental mammals.

Teaching Strategies

- **Activity** Ask students to solve the following problem: A little brown bat's diet consists of 20 percent mosquitoes. It eats 4 grams of food per night. How many mosquitoes does it eat in one night if a mosquito weighs 2.2 milligrams? *364 per night* How many mosquitoes will it eat in one summer during June, July, and August? *33 488 per summer* L1
- **Activity** Borrow a collection of mammal skulls from a local college or museum. Ask students to examine the teeth in each skull. In their groups, they should hypothesize about the type of food each mammal eats. L2
- **Display** Have students working in cooperative groups create a photo essay that shows examples of mammals classified in each order discussed in this *Focus On* feature. Each cooperative group should focus on a different mammal group. Have students combine their materials in a class bulletin board display. L1 COOP LEARN

Background

Lemur Classification

The Dermoptera, or gliding lemurs, have been placed in several different orders, including Insectivora, Chiroptera, and Primates. Currently, scientists classify them in a separate order. More than half the gliding lemur's body length is its tail. The gliding lemur looks like a furry kite as it spreads its limbs and leaps from tree to tree.

Focus On

Placental Mammals

Most of the more than 4300 species of mammals are placental mammals. Although all placental mammals develop within the mother's uterus, the condition of a newborn placental mammal varies. Newborn gazelles can run fast enough to keep up with the herd shortly after birth. Young kittens are blind and helpless. A human baby spends many years with its parents before it can take care of itself. There are 17 orders of placental mammals, of which 13 are shown here. The remaining four orders are pictured in Figure 35.13 following this feature.

Order Insectivora

Although they all eat insects, insectivores are a mixture of mammals that have few other common characteristics. Like other insectivores, the shrew has an extremely high metabolic rate.

Order Chiroptera

All bats are classified in this order. The bat is the only flying mammal. It has skin that stretches from the body, legs, and tail to the arms and fingers to form thin, membranous wings.

Bats continuously emit high-pitched sounds while in flight in a process called echolocation. Upon hitting an object, the sound waves bounce back to the bat. The waves are picked up by the bat's ears, allowing it to accurately locate the object. The vampire bats of South America suck blood from domestic animals such as cattle and horses. They carry and transmit diseases such as rabies to these animals.

GLENCOE TECHNOLOGY

 **Videodisc**

STVS: Animals
Disc 5, Side 2
How Bats Hear (Ch. 18)

Biology: The Dynamics of Life
Disc 2, Side 1
Shrew (Ch. 11)

Disc 2, Side 1
Bat (Ch. 12)

Gorilla (Ch. 13)

Order Primates

Pygmy marmosets live in the upper canopy of the Amazon rain forest. Except for nursing, the father takes care of the offspring. The outstanding characteristic of these mammals is their keen intelligence. Most primates have complex social lives. Members of this order can be divided into three groups: lemurlike, monkeylike, and humanlike.

Order Edentata

Giant anteaters have no teeth and feed on insects. They have strong forelegs and claws suited for digging in termite mounds. Anteaters, armadillos, and tree sloths are the only living members of this order. They are confined to Central and South America and to the southern regions of North America.

Order Lagomorpha

Pikas live in large communities in mountains and grasslands. They collect stacks of dried vegetation that they place next to their rocky homes. Rabbits, pikas, and hares are among the most numerous mammals. Lagomorphs are characterized by a fusion of bones in their hind legs where they move against the ankle bones. This fusion allows the bounding and leaping movement seen in these animals.

Order Rodentia

The golden hamster is a popular pet. Larger species have been known to destroy entire fields of wheat in Europe and Russia. The largest order of mammals, rodents live in all environments. They are identified by their continuously growing, razor-sharp teeth. They must gnaw on hard seeds, trees, bark, twigs, and roots. Rodents include beavers, porcupines, and chipmunks, as well as the more common mice and rats.

Focus On

Chemistry Connection

Lead and Development

Mammaliam development is adversely affected by high levels of lead. Mammal bones and teeth absorb lead because lead is similar to calcium. Regulation of muscle contraction and production of hemoglobin are also dependent on calcium. When calcium is replaced by lead, toxic effects result. In a study of California sea otters, it was found that their teeth contain 40 times the amount of lead that their ancestors' teeth contained. The contamination may result from a lead slag heap on Monterey Bay that has caused contamination in mussels. The otter studies could have important implications for humans, who have 750 to 1000 times more lead in their bodies than did prehistoric humans.

Background

Armadillos

Armadillos are the only members of the Order Edentata that live in the United States. Although the armadillo, like all mammals, has hair, its most distinguishing feature is its hardened skin, which is arranged in a pattern of either six or nine bands around the armadillo's body. This tough, banded skin acts like a suit of armor to protect the armadillo from predators. When frightened, the animal rolls itself into a tight ball, with its exterior armor protecting the animal's head and limbs.

GLENCOE TECHNOLOGY

Disc 2, Side 1
Three-Toed Sloth (Ch. 14)

Hare (Ch. 15)

Disc 2, Side 1
Beaver (Ch. 16)

Focus On

Background

Whale Groups

Whales are divided into two subgroups based upon how they feed. One group of whales has teeth; this group includes the killer whale shown here. The toothed whales feed on fishes, mollusks, and other aquatic mammals. The other subgroup of whales is made up of whales that have baleen rather than teeth.

Baleen acts like a giant strainer that captures tiny organisms, such as krill, present in the water. As a baleen whale swims, it gulps water containing small organisms. As the water passes back out through the baleen, the organisms become trapped, serving as the whale's next meal.

The Pinnipedia

The term *pinnipedia* means "wing-footed." The seals, sea lions, and walruses that make up this mammalian order have modified front and hind legs that take the form of flippers. The flippers permit these animals to swim through water, while allowing the animal to also walk on land. The pinnipeds are social animals that gather in areas called rookeries to breed and raise their young.

Anteaters

The scaly anteaters, or pangolins, belong to the order Pholidota, page 892. The backs and sides of these animals are covered with horny scales. Whenever danger threatens, these mammals turn on their sides and roll into tight balls. A pangolin uses its tongue, which is half the length of its body, to catch insects. Pangolins are found in Africa and Asia.

Focus On

Order Cetacea

The killer whale is the fastest swimmer and most maneuverable of all whales. They hunt for seals and porpoises in groups known as pods. Dolphins, porpoises, and whales are placed in this order. They have little or no hair, and breathe through blowholes on the tops of their heads.

Order Carnivora

Like all carnivores, a weasel has long, pointed canines and incisors, strong jaws, and long claws. A weasel can kill an animal several times its own size. Some of the best-known mammals, including dogs and cats, are placed in this group. While most of these mammals are flesh eaters, some, such as bears, do consume plant material as well.

Order Pinnipedia

The walrus is a huge mammal that lives in the Arctic. Males use their tusks to find mollusks and sea urchins, which they swallow whole. They also swallow pebbles to grind up the food in their stomachs. This order includes seals, sea lions, and walruses. They mostly live in cold-water habitats.

STUDENT JOURNAL

Endangered Mammals of the United States
Provide students with a current list of endangered mammal species in the United States. Identify the states in which each mammal is found. Have students write the names of the mammals in their correct states on an outline map. Next, explain that the state in which a particular mammal lives wants to remove the animal from the endangered species list in order to develop the area for housing, malls, and businesses. Ask students to assume they are going to testify as expert mammalogists in favor of preserving the endangered animal's habitat. Have students write their testimony in their journals. L1

Order Proboscidea

Proboscideans have flexible trunks that are used to gather vegetation for eating and to suck up water for drinking. One pair of incisors is modified into large tusks used for digging up roots and stripping bark from trees. The largest land animals now living, African elephants can be distinguished from Asian elephants by their larger ears. They spend most of their time eating.

Order Perissodactyla

The wild Przewalski's (Pruz WOL skeez) horse from Mongolia looks similar to the ancestor of all modern horses. These hairy horses have thick legs and sturdy bodies. Zoos throughout the world have breeding programs to save this species. Several other species of mammals had toenails that became modified into hooves. Most hoofed mammals are herbivores with molars used for grinding. Mammals with an odd number of toes belong to the Order Perissodactyla.

Order Sirenia

Dugongs live in the oceans. They have tails and front flippers like whales, distinct heads with a snout that points downward, and short necks. Nicknamed sea cows, dugongs and manatees are aquatic mammals that are distantly related to elephants. Only four species survive today.

Order Artiodactyla

Found along African rivers, hippos eat enormous amounts of vegetation at night. They spend the warm days laying in the water so that their skin doesn't dry out. About 200 species of hoofed, even-toed mammals are alive today. These mammals have multiple stomachs.

EXPANDING YOUR VIEW

1. **Understanding Concepts** Explain, by using examples, how mammals have become so successful.
2. **Writing about Biology** Go to a toy store and inventory the kinds of mammals in the stuffed animal section. Write a report detailing what the data might reflect about the view of society concerning mammals.

Focus On

Background

The Aardvark

The aardvark, shown on page 892, is the only animal in the order Tubulidentata. This termite-eater is a nocturnal animal with powerful limbs and claws. The aardvark is found only in Africa. Using its extremely acute sense of hearing to detect termites moving under the ground, the aardvark can dig into the hardest soil with unbelievable speed. Its thick skin is immune to the razor-sharp, cutting pincers of the soldier termites that try to protect their nests from its invasion.

GLENCOE TECHNOLOGY

Videodisc

Biology: The Dynamics of Life

Disc 2, Side 1

Dolphin (Ch. 17)

Bear (Ch. 18)

Walrus (Ch. 19)

Elephant (Ch. 20)

Zebra (Ch. 21)

Hippopotamus (Ch. 22)

Manatee (Ch. 23)

Answers to Expanding Your View

1. Responses should indicate that mammals have adapted to a large variety of environments. For example, mammals in the Order Pinnipedia have evolved modified flippers that permit life in water as well as on land. Others, such as the carnivores, have developed teeth such as canines and incisors that are well adapted to the foods they eat.
2. Likely responses will indicate that people appear to favor bears and other furry animals over other animal groups because most stuffed toys are made to resemble mammals.

SECTION 35.2

3 Assess

Check for Understanding

Ask students to list the major orders of mammals and explain how each is adapted to its niche. L1

Reteach

Have students draw a bingo card with 16 squares and the name of a mammal written in each square. Call out different mammal characteristics such as "lays eggs." As you call out each characteristic, have students cover the squares of the mammals that have that trait. The first person to cover four squares in a row wins. L1

Extension

Ask students to investigate and report on two mammals in the same order that are adapted to very different habitats, such as the jackrabbit and the Arctic hare. Have students report on the adaptations each animal has for survival in its habitat. L2

Assessment

Portfolio: Ask students to write a paragraph speculating why there are fewer big mammals than small mammals. L2 P

4 Close

Discussion

Ask students to decide which mammal they would most like to have as a pet. Have them explain their choices. L1

Figure 35.13

Four smaller orders of placental mammals are represented by only one or a very few species.

Hyraxes, order Hyracoidea, look like rabbits but have teeth like rhinoceroses and hooves on their toes.

Aardvarks, order Tubulidentata, dig up termites and eat them.

Gliding lemurs, order Dermoptera, have fangs and wide incisors.

Pangolins, order Pholidota, are anteaters that are covered with horny scales. They are powerful animals.

Connecting Ideas

Mammals are an integral part of your life. The gorilla intrigues and delights you with its humanlike antics. You laugh at the kitten playing with a ball of string. You may have ridden a horse or taken part in a whale watch. It's astonishing to think that all modern mammals evolved from the ancestors of insectivores in the time when dinosaurs moved across the land. During the evolution of mammalian structures, how do you suppose the evolution of behavior contributed to the success of mammals?

Section Review

Understanding Concepts

1. Describe the characteristics of placental mammals.
2. Compare monotremes and marsupials.
3. Name three characteristics that are used to classify mammals into one of the orders illustrated in the Focus On Placental Mammals feature.

Thinking Critically

4. You are a zoologist observing animals with a wildlife photographer in Australia. A local rancher brings you a rabbitlike animal and asks you to identify it. By looking at the skin covering the animal's abdomen, you know it is not a rabbit. How did you know? Explain.

Skill Review

5. **Observing and Inferring** You find a mammal fossil and observe the following traits: hooves, flattened teeth, skeleton the size of a large dog. What can you infer about its way of life? For more help, refer to Thinking Critically in the *Skill Handbook.*

Answers to Section Review

1. A placental mammal carries its young inside the uterus until development is nearly complete.
2. Monotremes lay eggs. Marsupials give birth to immature young that continue development in the mother's pouch.
3. Mammals are classified into orders based on similarities of body structures and functions, and on biochemical similarities.

Thinking Critically

4. The animal probably has a pouch on its abdomen as do most native Australian mammals. If it has a pouch, it is a marsupial.

Skill Review

5. It is an herbivore that can run fast. It may have multiple stomachs and chew its cud.

Chapter 35 Review

REVIEWING MAIN IDEAS

35.1 Mammal Characteristics

- The first mammal-like animals were therapsids.
- Mammals are endotherms with hair, diaphragms, modified limbs and teeth, highly developed nervous systems and senses, and mammary glands.

35.2 The Diversity of Mammals

- Mammals are classified into three subclasses—monotremes, marsupials, and placentals—based on the way they reproduce.
- Monotremes are egg-laying mammals found only in Australia, Tasmania, and New Guinea.
- Marsupials carry partially developed young in pouches on the outside of the mother's body. Most marsupials are found in Australia and surrounding islands.
- Placental mammals carry young inside the uterus until development is nearly complete. The young are nourished through an organ called the placenta.
- Most mammal species alive at present are placental mammals.
- There are 17 orders of placental mammals that include animals found on land, in the air, and in the sea.
- The orders of placental mammals are Insectivora, Chiroptera, Primates, Edentata, Lagomorpha, Rodentia, Sirenia, Cetacea, Proboscidea, Carnivora, Pinnipedia, Artiodactyla, Perissodactyla, Hyracoidea, Pholidota, Tubulidentata, and Dermoptera.

Key Terms

Write a sentence that shows your understanding of each of the following terms.

cud chewing	monotreme
diaphragm	placenta
gestation	placental mammal
gland	therapsid
mammary gland	uterus
marsupial	

Understanding Concepts

1. Which of the following is NOT a characteristic of mammals?
   a. endothermic
   b. three-chambered heart
   c. hair
   d. mammary glands
2. Which of these is NOT an endothermic animal?
   a. rattlesnake **c.** cat
   b. penguin **d.** gorilla
3. Mammalian hair probably evolved from ___________.
   a. feathers **c.** horn
   b. skin **d.** scales
4. Hair helps mammals by providing camouflage and helping them to maintain ___________.
   a. evolution
   b. running speed
   c. reproduction
   d. body temperature
5. A ___________ is a cell or group of cells that secretes fluids.
   a. diaphragm **c.** gland
   b. placenta **d.** Golgi body
6. The ___________, a sheet of muscles located beneath the lungs, enables mammals to maintain a high level of energy.
   a. diaphragm **c.** gland
   b. placenta **d.** trachea

Chapter 35 Review

Reviewing Main Ideas

Summary statements can be used by students to review the major concepts of the chapter.

Key Terms

Answers should go beyond defining the terms. Accept any answer that uses the term correctly and in the proper context.

Understanding Concepts

1. b
2. a
3. d
4. d
5. c
6. a

Chapter 35 Review

7. The kind of food a mammal eats can be determined by examining its ________.
   a. glands c. forelimbs
   b. diaphragms d. teeth
8. Cud chewing is an adaptation in many hoofed mammals that enables them to digest ________.
   a. sugar c. keratin
   b. cellulose d. chitin
9. Mammals nourish their young by producing milk in ________ glands.
   a. sweat c. digestive
   b. salivary d. mammary
10. Of the following, which type of tooth is NOT found in mammals?
   a. incisors c. fangs
   b. canines d. molars
11. The reptilian ancestors of mammals may have been ________.
   a. therapsids c. cotylosaurs
   b. thecodonts d. insectivores
12. The ________ is a hollow, muscular organ in which fetal development takes place in most mammals.
   a. gestation c. placenta
   b. mammary gland d. uterus
13. A mammal that carries its young inside a uterus until development is nearly complete is a(n) ______.
   a. amniotic mammal c. marsupial
   b. placental mammal d. monotreme
14. Most mammals nourish their young during gestation through the ________.
   a. uterus c. placenta
   b. embryo d. chorion
15. The time it takes for the mammalian embryo to develop is called ________.
   a. metamorphosis c. molting
   b. incubation d. gestation
16. Which of these are pouched mammals?
   a. placental mammals
   b. egg-laying mammals
   c. marsupials
   d. monotremes
17. Most marsupials are found in ________.
   a. South America c. North America
   b. Australia d. Africa
18. An egg-laying mammal is called a ______.
   a. placental mammal
   b. pouched mammal
   c. marsupial
   d. monotreme
19. ________ are examples of egg-laying mammals.
   a. Tasmanian devils and wombats
   b. Anteaters and shrews
   c. Platypuses and spiny anteaters
   d. Seals and whales
20. The first true mammals to evolve are thought to have been the size of a ______.
   a. rat c. horse
   b. cat d. whale

Applying Concepts

21. What is the advantage of bearing live young over laying eggs?
22. African elephants live in hot grasslands of Africa. They can often be seen using their trunks to slap watery globs of mud onto their heads and backs. Make a hypothesis that explains this behavior.
23. You find the skeleton of an animal. What features would indicate that it is a mammal rather than a reptile?
24. Explain the relationship between homeostasis and endothermy.
25. Trace the main paths of evolution from fishes to mammals to explain how mammals may have become adapted to life on land.

Thinking Critically

26. ***Concept Mapping*** Make a concept map that relates the following terms and phrases. Supply the appropriate linking words for your map.
   mammal, rodent, monotreme, marsupial, placental, kangaroo, duck-billed platypus, diaphragm, endothermy

Chapter 35 Review

7. d
8. b
9. d
10. c
11. a
12. d
13. b
14. c
15. d
16. c
17. b
18. d
19. c
20. a

Applying Concepts

21. The mother's body provides a safe environment in which offspring can develop. Eggs laid outside the body are vulnerable to predators and environmental changes.
22. They do this to cool themselves. The evaporating water cools the skin and the mud protects the skin, which has little hair, from the sun.
23. Lower jaws and middle ear bones are different from those of reptiles. Teeth are different from those of reptiles. Mammals also have straighter legs that are held nearer the body.
24. Animals that are endotherms can maintain their body temperatures at a constant level no matter what the environmental temperature is, whereas homeostasis involves the mechanisms within the body that allow for endothermy.
25. First, some early fishes probably evolved with appendages suited to propelling themselves in shallow water and muddy wetlands. Amphibians have lungs that enable them to breathe air and legs that enable them to move on land, but they are still restricted to life in the water because their eggs must be laid in water. Reptiles do not require water because their eggs are in leathery shells. Legs under the body make locomotion on land more efficient. The first four-chambered hearts, enabling more oxygen to reach cells to produce more energy and quicker movements, appear in reptiles. Mammals are endotherms and so can maintain body temperature in a wide variety of environments, especially when protected from heat loss by fur or hair. Young develop inside the body, thereby giving them protection against predation and harsh environmental conditions.

Thinking Critically

26. Evaluate students' maps. The maps should show an understanding of the relationships among the concepts listed.

27. ***Interpreting Scientific Illustrations*** Examine the teeth in the diagram of a skull and infer what the mammal might eat.

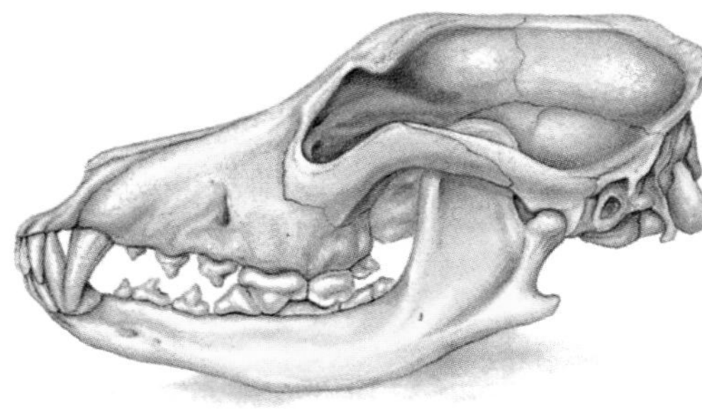

28. ***Formulating Hypotheses*** A marine biologist discovers a whale that has washed up on the shore. The fatty blubber layer is thin throughout the body of the whale. Make a hypothesis about why the whale might have died.

29. ***Interpreting Data*** A biologist graphs data on the energy needs of three mammals. A herbivore has lower energy needs than a carnivore that must hunt for and kill its food. Which of the three mammals is the hoofed herbivore?

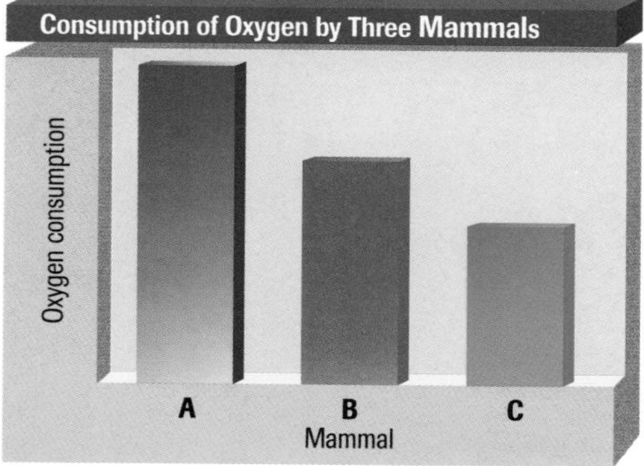

30. ***Comparing and Contrasting*** Mammals and insects are both considered to be extraordinarily successful animals. Explain the criteria used in both cases that give them this distinction.

ASSESSING KNOWLEDGE & SKILLS

Interpreting Scientific Illustrations Examine *Figure 35.7* and answer the following questions.

1. Egg-laying mammals, placental mammals, and marsupials developed from ____________.
   - **a.** therapsids
   - **b.** insectivores
   - **c.** edentates
   - **d.** Proboscids
2. During which geological time period did placental mammals evolve?
   - **a.** Precambrian
   - **b.** Paleozoic
   - **c.** Mesozoic
   - **d.** Cenozoic
3. To which group are perissodactyls more closely related?
   - **a.** chiropterans
   - **b.** edentates
   - **c.** artiodactyls
   - **d.** lagomorphs
4. Which group of animals was not common in the Paleozoic era?
   - **a.** fishes
   - **b.** amphibians
   - **c.** invertebrates
   - **d.** mammals
5. Which group of placental mammals had more diversity during an earlier time?
   - **a.** primates
   - **b.** edentates
   - **c.** perissodactyls
   - **d.** sirenians
6. ***Making a Graph*** Using the data in the following table, make a graph that shows the numbers of species in each mammal order.

Number of Mammal Species

Order	Number of Species
Insectivora	428
Chiroptera	925
Primates	233
Edentata	29
Lagomorpha	80
Rodentia	2021
Cetacea	78

Program Resources

Chapter Assessment, pp. 213-218 L1
Alternate Assessment in the Science Classroom
Computer Test Bank L1
Content Mastery, pp. 145-148 L1

Chapter 35 Review

27. The canine teeth can stab and pierce. The molars can grind and slice. The animal eats meat.
28. starvation leading to fat loss and exposure to cold, and an inability to maintain its body temperature
29. Animal C would most likely be the hoofed herbivore because it uses less oxygen, and therefore less energy than the others.
30. They both have enormous diversity of species and live in all types of environments.

1. a
2. c
3. c
4. d
5. c
6. Make sure students' graphs reflect the data and relationships shown in the table.

Chapter Organizer

SECTION	OBJECTIVES	ACTIVITIES/FEATURES
36.1 Innate Behavior National Science Standards: UCP. 2-4; C.3, C.6; G.1, G.2	1. **Distinguish** among the types of innate behavior. 2. **Demonstrate**, by example, the adaptive value of innate behavior.	**People in Biology:** Lisa Stevens, p. 900 **Thinking Lab:** Why do starlings use wild carrot leaves in their nests?, p. 903 **Earth Science Connection:** Finding Their Way Home, p. 907
36.2 Learned Behavior National Science Standards: UCP.2-4, A.1; C.6; E.1; F.6; G.1	3. **Distinguish** among types of learned behavior. 4. **Demonstrate**, by example, types of learned behavior.	**A Broader View:** Imprinting, p. 910 **Minilab:** How do you learn by trial and error?, p. 911 **Biolab:** Design Your Own Experiment—What makes a good feeding puppet for sandhill crane chicks?, p. 912 **Minilab:** What is a habit?, p. 915

ACTIVITY MATERIALS

BIOLAB	MINILABS	ALTERNATE LAB
page 912 photos of sandhill cranes fabric (black, red, and white) black plastic trash bags needle thread socks, long gray clothespins, spring type black markers stapler tape tagboard colored beads colored map tacks glue scissors	**page 911** tagboard puzzles envelopes **page 915** shoes with laces	**page 906** glass terrarium or aquarium, with cover sand 5 male field crickets 1 female cricket (same species) dry oatmeal jar lid apple slices sponge pieces 4 match boxes 5 jars, with lids nail polish, 4 different colors

Chapter 36 Animal Behavior

TEACHER CLASSROOM RESOURCES

Reproducible Masters	Transparencies
Section Focus Master 79: Behaving to Survive L1 SAE **Reinforcement and Study Guide,** pp. 145-146 L1 **Concept Mapping:** Territoriality, p. 36 L1 **Critical Thinking/Problem Solving:** Is Migration Learned or Inherited?, p. 36 L3 **Laboratory Manual:** How Is Response Related to Nervous System Complexity?, pp. 223-226 L2	
Section Focus Master 80: Learning a Lesson L1 SAE **Reinforcement and Study Guide,** pp. 147-148 L1 **Biolab and Minilab Worksheets,** pp. 143-146 L1 **Laboratory Manual:** Conditioning in Guinea Pigs, pp. 227-230 L2 **Content Mastery,** pp. 149-152 L1	**Basic Concepts Transparency #60:** Conditioning L1 SAE **Reteaching Transparency #36:** Types of Learning L1 SAE

ASSESSMENT MATERIALS

Chapter Assessment, pp. 219-224
Alternate Assessment in the Science Classroom
MindJogger Videoquiz
Computer Test Bank

Spanish Resources SAE
English/Spanish Audiocassettes SAE
Cooperative Learning in the Science Classroom COOP LEARN
Lesson Plans

KEY TO TEACHING STRATEGIES

- L1 Level 1 activities should be within the ability range of all students including those with learning difficulties.
- L2 Level 2 activities are within the ability range of average to above-average students.
- L3 Level 3 activities are designed for the ability range of above-average students.
- SAE SAE activities should be within the ability range of Students Acquiring English.
- COOP LEARN Cooperative Learning activities are designed for small group work.
- P These strategies represent student products that can be placed into a best-work portfolio.
- These strategies are useful in a block scheduling format.

GLENCOE TECHNOLOGY

The following multimedia resources are available from Glencoe.

Biology: The Dynamics of Life
CD-ROM SAE
Videodisc Program

The Infinite Voyage Series
The Keepers of Eden
The Great Dinosaur Hunt
Insects: The Ruling Class

National Geographic Society Series
Newton's Apple: Life Science
Butterfly Migration

Science and Technology Videodisc Series (STVS)
Animals
Alligator Courtship
Development of Walking in Chicks
Ecology
Pheromone Rope

CHAPTER

36 Animal Behavior

Chapter Overview

Students will first examine behavior in terms of stimulus and response. Then, students will examine inherited behavior, beginning with simple reflexes. Students will conclude the section by examining instinctive behavior patterns such as courtship, territoriality, aggression, dominance hierarchies, circadian rhythms, migration, hibernation, and estivation.

In the second part of the chapter, students will consider learned behavior. They begin their study with habituation and proceed through more complex learned behavior patterns such as imprinting, trial-and-error learning, conditioning, behavior involving insight, and finally behavior involving communication and language.

Key Terms

aggression
behavior
circadian rhythm
communication
conditioning
courtship behavior
dominance hierarchy
estivation
fight-or-flight response
habituation
hibernation
imprinting
innate behavior
insight
instinct
language
migration
motivation
territory
trial-and-error learning

896

Learning Styles

Look for the following logo for strategies that emphasize different learning modalities. LS

Kinesthetic	Demonstration, p. 902; Minilab, p. 911
Interpersonal	Portfolio, p. 904; Project, p. 910
Logical-Mathematical	Demonstration, p. 899; Activity, p. 899; Student Journal, p. 902; Thinking Lab, p. 903; Meeting Individual Needs, p. 902; Enrichment, p. 905; Project, pp. 905, 911; Alternate Lab, p. 906; Minilab, p. 915
Auditory-Musical	Student Journal, p. 910

LS

CHAPTER

36 Animal Behavior

The sun comes up over the prairie and makes the dew look like shining crystals. A herd of antelope wanders slowly over the landscape, heads down, as they begin their daily foraging. Nearby, a furry head pops up out of a hole in the ground. The prairie dog is soon joined by a host of its neighbors, all alert for possible danger before coming out of their underground burrows. Up in the brightening sky, a vulture slowly circles, searching the ground below for a dead or dying animal that could become its morning meal.

A female coyote suddenly appears from behind a pile of rocks. She stretches, yawns, and, with ears pricked, canters toward the prairie dog town. Several prairie dogs bark, and the rest dive for cover. The noise alerts an antelope; head raised, he turns and signals the herd with tail up and white rump showing. The herd springs into action, leaping quickly away. Soon, the coyote seems to be alone on the prairie. Overhead, the vulture wheels away to look for breakfast somewhere else.

Concept Check

You may wish to review the following concepts before studying this chapter.

- Chapter 1: adaptation, evolution, response, stimulus
- Chapter 2: experiment, hypothesis

Chapter Preview

Laboratory Activities

Biolab: Design Your Own Experiment

- What makes a good feeding puppet for sandhill crane chicks?

Minilabs

- How do you learn by trial and error?
- What is a habit?

Prairie dogs live in a community. Vultures can often be seen feeding together on dead or dying animals. Coyotes pursue prey. Each animal has set patterns of behavior that allow it to survive and pass on its genes to offspring. How much of this behavior is directly inherited from parents? How much of it is learned in a lifetime of trial and error?

CHAPTER 36

Introducing the Chapter

Ask students what may cause each animal shown to behave as it does. *Some students may suggest animals are born knowing how to behave in certain ways, while others will suggest behaviors are learned from parent organisms.*

Ask students to consider their pets. Have them list behaviors of their pets that they think are learned and those that are not learned.

Theme Development

Students will examine the theme of *unity within diversity* as they consider the kinds of behaviors animals have in common and behaviors that are unique to a species. The theme of *evolution* is important to the study of behavior because of the adaptive value of behavior and the fact that behavior, just like physical features of animals, evolves.

Concept Check

Students will extend their ideas of evolution to behavior. Students will use the term *instinctive behavior* and recognize that it refers to a specific type of behavior pattern.

Assessment Planner

Choose assessment strategies from the following pages to evaluate the progress of your students.

Assess, pp. 908, 916
Alternate Lab, pp. 906-907
Minilabs, pp. 911, 915
Portfolio, pp. 899, 904
Thinking Lab, p. 903
Biolab, pp. 912-913
Chapter Review, pp. 917-919

SECTION

36.1

Prepare

Key Concepts

Response to a stimulus is presented with examples of inherited behavior. Simple reflexes and fight-or-flight responses are considered along with the more complex patterns of courtship, territoriality, dominance hierarchies, and biological rhythms.

Block Scheduling

Look for this symbol for strategies that are useful in a block scheduling format. For more information on block scheduling, refer to Section 36.1 in the **Lesson Plans** booklet.

Materials

- Have a variety of small animals and a bright light for the Demonstration.
- Purchase mealworms for the Activity.

1 Focus

Bellringer

Before presenting the lesson, display **Section Focus Master 79** on the overhead projector and have students answer the accompanying questions. L1 SAE

Brainstorming

Ask students to brainstorm the following questions: How would animals be different if there were no inherited behavior? *Many animals would not survive long enough to reproduce, resulting in extinction of a species.* If humans relied on instinctive behavior, what kinds of instinctive behavior patterns might have evolved? *Those essential to survival, such as feeding and reproductive strategies* L1 COOP LEARN

SECTION

36.1 Innate Behavior

Section Preview

Objectives

Distinguish between the types of innate behavior.

Demonstrate, by example, the adaptive value of innate behavior.

Key Terms

behavior
innate behavior
fight-or-flight response
instinct
courtship behavior
territory
aggression
dominance hierarchy
circadian rhythm
migration
hibernation
estivation

Have you ever watched a bird feed its young? Baby birds greet a parent returning to the nest with cries and open beaks. Parent birds practically stuff the food down their offsprings' throats, then fly off to find ever more food. Why do baby birds open their beaks wide? Why do parent birds respond to open beaks by feeding their offspring? These actions are examples of behavior that appears in birds without being taught or learned. Animals exhibit many kinds of behavior in nature.

What Is Behavior?

A peacock displaying his colorful tail, a whale spending the winter months in the ocean off the coast of southern California, and a lizard seeking shade from the hot, desert sun are all examples of animal behavior. **Behavior** is anything an animal does in response to a stimulus in its environment. The presence of a peahen stimulates a peacock to open its tail feathers and strut. Environmental cues, such as change of temperature and length of daylight, might be the stimuli that cause the whale to leave its summertime arctic habitat. Heat stimulates the lizard to seek shade. Look at the illustrations in *Figure 36.1* that show two examples of stimuli that affect animal behavior.

Figure 36.1
Animals exhibit a variety of behavioral responses.

This butterfly exposes eyespots on its wings, and a predatory bird stops its pursuit of the insect. The eyespots look like the eyes of an owl.

The onset of short days and cold weather stimulates squirrels to collect acorns and walnuts and store them. What is the adaptive value of the squirrel's behavior?

Program Resources

Section Focus Master 79 L1 SAE
Reinforcement and Study Guide, pp. 145-146 L1
Laboratory Manual, pp. 223-226 L2
Concept Mapping, p. 36 L1
Critical Thinking/Problem Solving, p. 36 L3

Animals carry on many activities—such as getting food, avoiding predators, caring for young, finding shelter, and attracting mates—that enable them to survive. These behavior patterns, therefore, have adaptive value. For example, a parent sea gull that is not incubating eggs or caring for chicks joins a noisy flock of gulls to dive for fishes. If the parent did not catch a fish, not only would it die, but its chicks would not survive either. Therefore, this feeding behavior has adaptive value for the sea gull.

Figure 36.2

Each species has its own courtship ritual. Male and female black-headed gulls dance in unison side by side and turn their heads away from each other as if doing a tango. The female taps the male's bill, who then regurgitates a fish into her mouth. Soon the dancing and dining are over, and the pair mate.

Behavior that Is Inherited

Inheritance plays an important role in the ways animals behave. You don't expect a duck to tunnel underground or a mouse to fly. Yet, why does a mouse run away when a cat appears? Why does a mallard duck fly south for the winter? These behavior patterns are genetically programmed. An animal's genetic makeup determines how that animal reacts to certain stimuli.

Natural selection favors certain behaviors

Often, a behavior exhibited by an animal species is the result of natural selection. The variability of behavior among individuals affects their ability to survive and reproduce. Individuals with behavior that makes them more successful at surviving and reproducing will produce more offspring. These offspring will inherit the genetic basis for the successful behavior. Individuals without the behavior will die or fail to reproduce.

Inherited behavior of animals is called **innate behavior.** Consider the sea gulls in *Figure 36.2.* If the female is not able to recognize an appropriate mate of her own species, and is confused by several other species of sea gulls that also live in the same location, she will not be successful in reproducing. Courtship in the form of a ritualized dance helps the female recognize an appropriate mate.

Genes form the basis of behavior

Through experiments, scientists have found that an animal's hormonal balance and its nervous system—especially the sense organs responsible for sight, touch, sound, or odor identification—affect how sensitive the individual is to certain stimuli. Because genes control the production of an animal's hormones and development of its nervous system, it's logical to conclude that genes indirectly control behavior. Innate behavior includes both automatic responses and instinctive behaviors.

2 Teach

Visual Learning

Figure 36.1 What is the adaptive value of the squirrel's behavior? *This behavior ensures the squirrels will have food if the ground becomes covered with snow.*

Demonstration

LS Logical-Mathematical Have a variety of animals in class, such as a mouse, frog, lizard, earthworm, goldfish, snail, planarian, fruit fly, and crayfish. Expose each animal to a single stimulus, such as bright light. Have students observe their varied responses. Ask students to speculate about the survival value of each animal's response to that stimulus. *Animals that may be dried out by the sun, such as an earthworm, will avoid the light.*

Activity

LS Logical-Mathematical Have students design experiments to determine the preferences of mealworms for light or dark, damp or dry, and smooth or rough surfaces. Have students obtain approval for their designs and then conduct their experiments. You may wish to have students conduct experiments on both the larval stage and the adult beetle to determine if preferences differ at each stage. L2

PORTFOLIO

Design an Experiment Ask students to hypothesize how different animals will respond to a loud noise. Have students design experiments with several trials that would determine this. Once their designs are approved, have them carry out their experiments and write their observations in their portfolios. Ask them to speculate whether the behavior they observed is innate. L2 P

GLENCOE TECHNOLOGY

Videodisc

Biology: The Dynamics of Life
Disc 2, Side 1
Bird Courtship (Ch. 24)

People in Biology

Meet Lisa Stevens, Assistant Zoo Curator

Teaching Strategies

- Plan a trip to a nearby zoo and have students list behaviors of animals they observe. Ask students to identify each behavior as either innate or learned. If possible, arrange for the curator or a zookeeper to address the class about his or her job responsibilities.
- Advise students that most large zoos get animals on loan from other zoos. Explain that curators would have to be involved in the decision making for this process.
- Ask students to contact a nearby zoo to find out what questions the zoo poses to potential employees. Have students use this information to create a position description of a particular job at the zoo, such as keeper of the elephants, and make up a list of questions for an applicant. L1
- Ask students to write to the San Diego Wild Animal Park, the National Zoo, or some other large zoo in the country to find out what they are doing to preserve endangered animals. Have students write a report of their findings. L1

people in biology

Meet Lisa Stevens, Assistant Zoo Curator

The giant panda cannot speak for itself to TV and newspaper reporters, so Lisa Stevens becomes its voice. Acting as press contact for the black-and-white animal is just one of the duties of this assistant curator for mammals at the National Zoo, which is part of the Smithsonian Institution in Washington, D.C.

In the following interview, Ms. Stevens describes her profession and the important role of today's zoos.

On the Job

Q Ms. Stevens, could you tell us about a typical day on the job for you?

A The first thing to understand about this profession is that there *isn't* a typical day. As an assistant curator in the mammal department, I supervise the day-to-day care of three groups of animals—the primates, the camels, and the giant pandas. Here's an idea of my other duties. The first thing tomorrow morning, I'll oversee the release of a new group of monkeys to an island exhibit. Then I'm chairing a committee of our education council, which evaluates the zoo's educational programs. Later, I have a meeting about setting up a new exhibit in the primate house.

Q What will be in that new exhibit?

A It's called the "Think Tank." In this exhibit, we'll address the question of whether or not animals have a capacity to think. It will be an interactive exhibit designed to increase people's appreciation and respect for animals' cognitive abilities.

Q Does the zoo have a captive-breeding program?

A Like other zoos in North America, the National Zoo participates in over thirty Species Survival Plans. One of our captive-breeding programs involves the white-cheeked gibbon, which has a natural habitat in southern China and Vietnam. We've also had gorillas born in 1991 and 1992. They were our first gorilla births in 20 years. However, we won't, in most cases, be able to release animals into the wild because their habitats are gone. The goal of these programs is to maintain a diverse genetic population as a hedge against extinction in the wild.

Early Influences

Q How did you get interested in animals?

A As a child, I wasn't afraid of anything. I was the kind of kid who let bugs crawl up and down my arms and caught lizards and snakes. I had the good fortune to spend part of my childhood in the tropics, in Thailand and Okinawa. My exposure to all

900

GLENCOE TECHNOLOGY

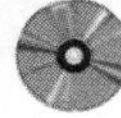

Videodisc

The Infinite Voyage: The Keepers of Eden
Extinction and the National Zoo's Tamarin Monkey Project
(Ch. 4)

interNET CONNECTION

Follow the link for this chapter on the Glencoe Homepage at **www.glencoe.com/sec/science** to find out more about animal behavior.

kinds of animals was probably quadrupled in that environment. My parents took me places where I could fulfill my interest in animals. For example, once we visited a Buddhist snake temple in Penang, Malaysia, that had all kinds of snakes draped in artificial trees with gold leaves. As a child, I also got to go on a safari in Nairobi National Park, in Kenya.

Q How did you first come to the National Zoo?

A I got a college degree in zoology and wanted to have some firsthand experience with animals before entering veterinary school. I had worked for a vet and in stables and pet stores. I thought, Hmmm, I've never worked in a zoo, so maybe that would be kind of interesting. Actually, I was going to volunteer, but before I could do that, I was hired as a keeper of lions, tigers, and bears. I first thought zookeepers were sort of like janitors. I had no idea that there was such a level of professionalism involved. Ultimately, I realized that the zoo was a whole world in itself, and I've never left that world.

Personal Insights

Q Do you have a favorite animal?

A I'm asked that a lot, but it's really important to be flexible. I'm lucky that I'm able to work with all different kinds of animal species. I truly appreciate and find delight in the diversity of species.

Q What advice do you have for students who are interested in careers related to animals?

A Often, school programs steer people with animal interests into veterinary medicine, but there are many other animal-related careers to explore. It's important to think of the complex issues that are facing wildlife, especially the issue of disappearing animal habitats. For example, there's a need for people to work in the field of human population biology and the economics of agriculture. Biologists now have to think in terms of helping the people who live in countries where many species of wildlife also live and of the interrelations of species, both animal and human.

For any kind of a career, volunteering may be an opportunity to look at a profession. Here at our zoo, we use teenage volunteers to answer visitors' questions. The young volunteers enjoy their work and learn a lot about both animals and people.

CAREER CONNECTION

If you have a love of animals and you're interested in working at a zoo, you may wish to follow one of these career paths.

Zoo Director *Master's or doctoral degree in zoology*

Animal Breeder *College-level courses in zoology; bachelor's degree*

Zookeeper *College-level courses in zoology; bachelor's degree*

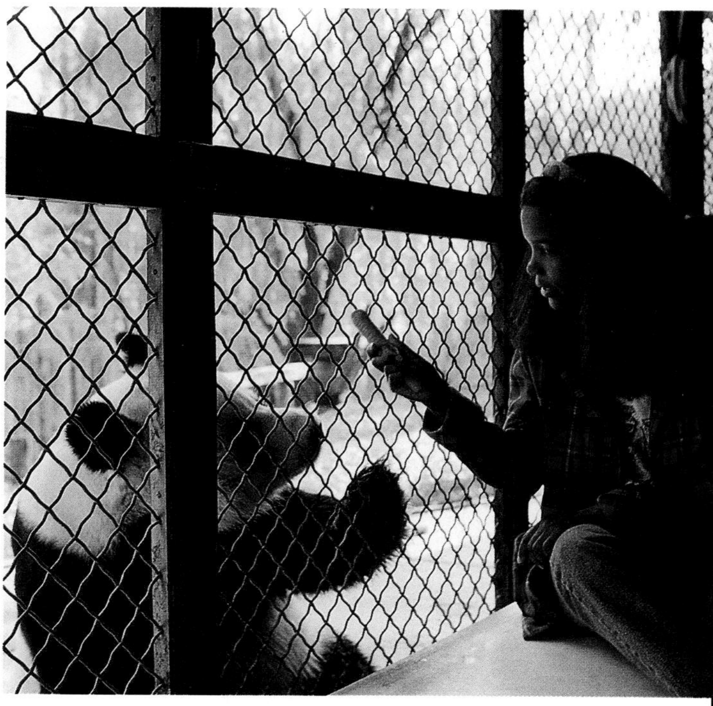

CAREER CONNECTION

- **Career Path**

  High School: Four years of science including Earth science, biology, physics, mathematics, and chemistry.

  College: chemistry, physics, botany, biology, zoology, statistics, mathematics, ecology, geology, genetics, comparative vertebrate biology, invertebrate zoology, cell biology, evolution, microbiology, and genetics.

  Graduate School: zoology, animal behavior, biogeography, wildlife biology, genetics, cell biology, evolution, ecology, herpetology, ichthyology, ornithology, mammalogy, entomology, marine biology.

- **Career Issue** Some people may think that zoos are unnecessary entertainment for people and unsuitable places for wild animals. Others think that in order to preserve endangered animals, it is important to have them in breeding programs in zoos. Ask students to write an essay defending or advocating one of these positions. L1

For More Information

For more information about requirements for specific positions at a zoo, students may want to write to the following:

Animal Caretakers Information, The Humane Society of the United States, Companion Animals Division, Suite 100, 5430 Grosvenor Lane, Bethesda, MD 20814

Demonstration

LS Kinesthetic Have students examine the blink response by asking students to hold a large piece of plexiglass in front of their faces and try to keep their eyes open when a partner gently tosses a crumpled piece of notebook paper at the glass.

Different Viewpoints in Biology

Why do animals play? Play in animals is generally exhibited in three ways: mock chasing and fighting, repeating locomotor skills, and young animals taking life-threatening risks.

Some researchers say that mammals play most actively when the brain cells and nervous system are developing most quickly. They say that play acts to "fire up the engines of survival." These researchers say that animal play is a means of rehearsing the elements of capturing prey, escaping a predator, or fighting members of the same species for mates, or dominance. Other animal behavior experts say that much behavior called "play" in animals is really immature sexual or practice predatory behavior.

To answer questions about play in animals, scientists are breeding a lazy, "couch potato" strain of mice and a "wild and crazy" strain. They will study and compare the behavior and development of both strains.

Automatic Responses to Stimuli

What happens if something quickly passes in front of your eyes or if something is thrown at your face? Your first reaction is to blink. Even if a protective clear shield is placed in front of you, you can't keep your eyes open when the object is thrown. This eye blink is the simplest form of innate behavior, called a reflex. A reflex is a simple, automatic response that involves no conscious control. *Figure 36.3* shows two examples of reflexes.

The adaptive value of another automatic response is obvious. Think about a time when you were suddenly scared. Immediately, your heart began to beat faster. Your skin got cold and clammy, your respiration increased, and maybe you trembled. You were having a fight-or-flight response. A **fight-or-flight response** mobilizes the body for greater activity. Your body is being prepared to either fight or run from the danger. A fight-or-flight response is automatic and controlled by internal chemical mechanisms.

Figure 36.3

Reflexes have survival value for animals.

When you accidentally touch a hot stove, you jerk your hand away from the hot surface. The movement saves your body from serious injury.

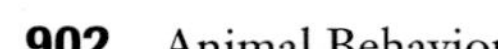

When a mollusk is touched, it withdraws into its shell. The touch stimulus may be that of a mollusk predator.

Instinctive Behavior Patterns

Compare the mating dance of the sea gull with your fight-or-flight response. The dance does not happen as quickly as a reflex does. The mating dance is an example of an instinct. An **instinct** is a complex pattern of innate behavior. A reflex can happen in less than a second, whereas instinctive behavior patterns may have several parts and may take weeks to complete. Instinctive behavior begins when the animal recognizes a stimulus and continues until all parts of the behavior have been performed.

As shown in *Figure 36.4*, greylag geese instinctively retrieve eggs that have rolled from the nest and will go through the motions of egg retrieval even when the eggs are taken away. You can see that survival of the young may be dependent on this behavior.

Courtship behavior ensures reproduction

Much of an animal's courtship behavior is instinctive. **Courtship behavior** is a behavior that males and females of a species carry out before mating. Like other instinctive behaviors, courtship has evolved through natural selection. Imagine what would happen to the survival of a species if members were unable to recognize other members of that same species. Individuals often can recognize one another by the behavior patterns each performs. In courtship, behavior ensures that members of the same species find each other and mate. Obviously, such behavior has an adaptive value for the species. Different species of fireflies, for example, can be

STUDENT JOURNAL

Observing Bird Behavior Ask students to observe local birds for one week and record their observations in their journals. Ask them to speculate about which behaviors are instinctive. Have them compare their observations with those of other students. Elicit whether students classified similar behaviors in the same ways. L1 LS

Meeting Individual Needs

Gifted Ask students to do research to determine how animal sounds are studied electronically. Ask students to present an illustrated report to the class and devise a way to study the sounds made by a local animal. L3 LS

Figure 36.4

The female greylag goose instinctively retrieves an egg that has rolled out of the nest by arching her neck around the stray egg and moving it like a hockey player advancing a puck. The female goose will retrieve many objects outside the nest, including baseballs and tin cans.

ThinkingLab Design an Experiment

Why do starlings use wild carrot leaves in their nests?

Starlings build their nests in the cavities of hollow trees. They construct a nest of dried grasses and twigs, but they also include freshly gathered wild carrot leaves. Biologists were interested in why starlings seemed to decorate their nests with this greenery. They hypothesized that the birds may have been camouflaging their nests from predators, insulating the eggs with material that would keep in warmth, or using material that would poison or repel parasites that might attack nestlings.

Analysis

Design an experiment that would test one of the hypotheses. Make sure your experiment has a control and a method to gather quantitative data.

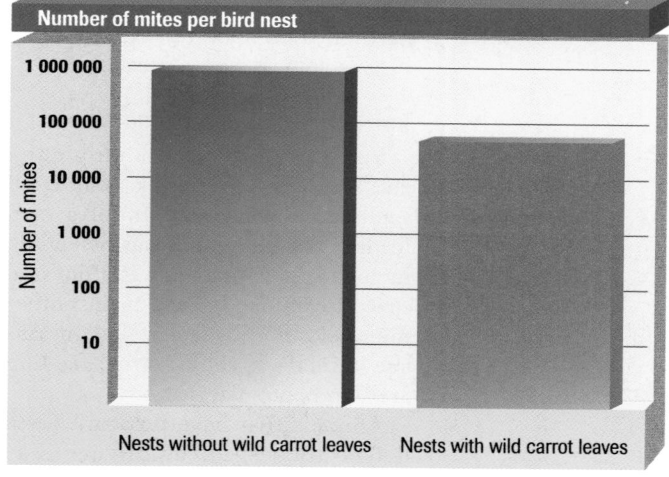

Thinking Critically

Biologists tested the hypothesis that the plant material may act as an insect repellent. They obtained the data in the graph by counting the mites in nests with wild carrot leaves and in nests without wild carrot leaves. Explain the graph in terms of how wild carrot leaves affect mites in nests.

seen at dusk flashing distinct light patterns. However, female fireflies of one species respond only to those males exhibiting the species-correct flashing pattern.

Some courtship behaviors help prevent females from killing males before they have had the opportunity to mate. For example, in some spiders, the male is smaller than the female and risks the chance of being eaten if he approaches her. Before mating, the male presents the female with a nuptial gift, an insect wrapped in a silk web. While the female is unwrapping and eating the insect, the male is able to mate with her without being attacked. After mating, however, the male may be eaten by the female anyway.

ThinkingLab Design an Experiment

Purpose

LS Logical-Mathematical Students will examine several hypotheses about why starlings use fresh, wild carrot leaves in nest construction, and they will design an experiment to test one of the hypotheses.

Process Skills

compare and contrast, recognize cause and effect, analyze hypotheses, design an experiment

Teaching Strategies

- Show students photos of a variety of birds' nests. Explain that each species of bird always builds the same type of nest.

Expected Results

If students hypothesized that the greenery repels parasites, they might have said that they could gather parasites of nestlings and place them in a long container with the wild carrot leaves at one end. They could see in which end the parasites spend the most time.

Thinking Critically

Wild carrot leaves in a nest repel mites.

Assessment

Knowledge: Show students a picture of the hanging nest of a Baltimore oriole. Ask them to make a hypothesis and design an experiment that would examine the survival value of a hanging nest for this bird. L1

GLENCOE TECHNOLOGY

Videodisc

STVS: Animals
Disc 5, Side 2
Alligator Courtship (Ch. 9)

Science, Technology, and Society

Polygamous Birds
Using DNA fingerprinting, ornithologists have found that many birds believed to be monogamous actually are not. The number of offspring resulting from birds having more than one mate may be as high as 50 percent. In many cases, a clutch of eggs from one female were fertilized by several males.

Evolutionary biologists are considering new ideas about why this happens. Perhaps a male bird that is fast, strong, and capable of having many offspring is not necessarily a good provider for his mate and offspring. Females may be selecting some males that are good providers as mates and other males that are stronger and larger.

The DNA studies have also shown that some birds actually are monogamous. These birds include masked boobies and seagoing fulmars.

Visual Learning

Figure 36.5 What adaptive value does this behavior have? *The behavior helps to ensure that all offspring can be provided for.*

GLENCOE TECHNOLOGY

Videodisc

Biology: The Dynamics of Life
Disc 2, Side 1
Territorial Behavior (Ch. 25)

Figure 36.5
Female hanging flies instinctively favor the male that supplies the largest nuptial gift—in this case, a moth. The amount of sperm the female will accept from the male is determined by the size of the gift. What adaptive value does this behavior have?

In some species, nuptial gifts play an important role in allowing the female to exercise a choice as to which male to choose for a mating partner. The hanging fly, shown in *Figure 36.5,* is such a species.

Territoriality reduces competition

You may have seen a chipmunk chase another chipmunk away from seeds on the ground under a bird feeder. The chipmunk was defending its territory. A **territory** is a physical space an animal defends against other members of its species. It contains the animal's breeding area, feeding area, and potential mates.

Animals that have territories will defend their space by driving away other individuals of the same species. For example, a male sea lion will patrol the area of beach where his harem of female sea lions rests. He will not bother a neighboring male that has a harem of his own because both have marked their territories, and each respects the common boundaries. But if an unattached, young male tries to enter the sea lion's territory, the owner of the territory will attack and drive the intruder away from his harem.

Although it may not appear so, setting up territories actually reduces conflicts, controls population growth, and provides for efficient use of environmental resources by spacing out animals so they don't compete for the same resources within a limited space. This behavior improves the chances of survival of the young produced, and, therefore, survival of the species. If the male has selected an appropriate site and the young survive, they may inherit his ability to select an appropriate territory. Therefore, territorial behavior has survival value, not only for individuals, but also for the species. The male stickleback shown in *Figure 36.6* is another animal that exhibits territoriality, especially during breeding season.

Figure 36.6
The male three-spined stickleback displays a red belly to other breeding males near his territory. The red belly is a signal or stimulus indicative of the breeding condition of the fish. The male will instinctively respond to other red-bellied males by attacking and driving them away.

PORTFOLIO

Animal Psychology Ask students to interview an animal psychologist and report about what he or she does, the kinds of phone calls received from pet owners, and the kinds of treatments provided. Have students include a complete transcript of the interview in their portfolios. L1 P LS

Figure 36.7

In many species, such as bighorn sheep, individuals have innate inhibitions that make them fight in relatively harmless ways among themselves.

Remember from Chapter 31 that pheromones are chemicals that communicate information between individuals of the same species. Many animals produce pheromones to mark territorial boundaries. For example, wolf urine contains pheromones that warn other wolves to stay away. The male pronghorn antelope uses a pheromone secreted from facial glands. One advantage of using pheromones is that they work during both the day and night.

Aggression threatens other animals

Animals occasionally engage in aggression. **Aggression** is behavior that is used to intimidate another animal. Animals fight or threaten one another in order to defend their young, their territory, or a resource such as food. Aggressive behaviors, such as bird singing, teeth baring, or growling, deliver the message to others of the same species to keep away.

When a male gorilla is threatened by another male moving into his territory, for example, he does not kill the invader. Animals of the same species rarely fight to the death. The fights are usually symbolic, as shown in *Figure 36.7*. Male gorillas do not usually even injure one another. Why does aggression rarely result in serious injury? One answer is that the defeated individual shows signs of submission to the victor. These signs inhibit further aggressive actions by the victor. Continued fighting might result in serious injury for the victor; thus, its best interests are served by stopping the fight.

Submission leads to dominance hierarchies

Do you have an older or younger sibling? Who wins when you argue? In animals, it is usually the oldest or strongest that wins the argument. But what happens when several individuals are involved in the argument? Sometimes, aggression among several individuals results in a grouping in which there are different levels of dominant and submissive animals. A **dominance hierarchy** is a form of social ranking within a group in which some individuals are more subordinate than others. Usually, one animal is the top-ranking, dominant individual. This animal often leads others to food, water, and shelter. There might be several levels in the hierarchy, with individuals in each level subordinate to the one above.

SECTION 36.1

Enrichment

LS **Logical-Mathematical** Crickets chirp as part of their courtship and territorial behaviors. The number of chirps in a specific amount of time decreases as the temperature gets colder. Have students set up a means for observing this behavior. Crickets are available in pet supply shops and from biological supply companies.

Discussion

Elicit from students whether they have ever observed aggressive behavior in dogs or cats. Allow students who have observed such behavior to recount the details of the aggression. Elicit whether the animal who lost the battle showed signs of submission at the conclusion of the struggle and what those signs were.

PROJECT

Pheromone Studies

Have students observe how pheromones influence behavior by carrying out the following activity. Place three snails of one species on one side of a pan, and three snails of the second species on the opposite side. Allow the snails to move about on their respective sides of the pan for about 10 minutes and then remove the snails from the pan. Immediately place all the snails back in the pan at the center of the pan. Observe which way each snail moves. Clean the pan with warm, soapy water. Repeat the process three times and record your observations. Write a summary of your observations that includes a conclusion about why the snails moved as they did. L1 LS

Chemistry Connection

Natural Antifreeze

When temperatures in the body drop low enough, ice forms in cells, causing them to stretch and burst. Some animals can be "supercooled," a condition in which ice crystals do not form. For example, some insects in Alaska do not freeze even when temperatures are 70°F below zero. Biologists have also discovered that box turtles and garter snakes can be frozen solid for the entire winter and then thaw out unharmed in spring. When the temperature drops and ice begins to form in the animals' skins, the animals' tissues become flooded with glucose. The high concentration of sugar seems to protect the cells from ice damage. The high level of glucose also slows cellular metabolism to near zero. The same mechanism would not work in humans because the high glucose levels would cause diabetic coma and death.

Figure 36.8

A dominance hierarchy often prevents energy from being wasted on continuous fighting because submissive birds give way peacefully in confrontations. The hierarchy also may be adaptive in providing a way for females to choose the best males.

The term *pecking order* comes from a dominance hierarchy that is formed by chickens, *Figure 36.8.* The top-ranking chicken can peck any other chicken. The chicken lowest in the hierarchy is pecked at by all other chickens in the group.

Behavior resulting from internal and external cues

Some instinctive behavior is exhibited in animals in response to internal, biological rhythms. Behavior based on a 24-hour day/night cycle is one example. Many animals, humans included, sleep at night and are awake during the day. Other animals, such as owls, reverse this pattern and are awake at night. A 24-hour cycle of behavior is called a **circadian rhythm.** Most animals come close to this 24-hour cycle of sleeping and wakefulness. Experiments have shown that in laboratory settings with no windows to show night and day, animals continue to behave in a 24-hour cycle.

Rhythms also can occur yearly. Migration is a yearly rhythm. **Migration** is the instinctive, seasonal movement of animals, *Figure 36.9.* In the United States, about two-thirds of bird species fly south in the fall to areas such as South America where food is available during the winter. The birds fly north in the

Figure 36.9

A variety of animals respond to the urge to migrate.

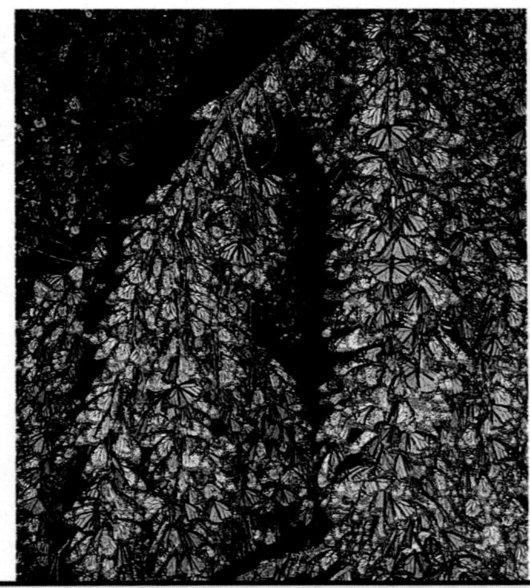

Adult monarch butterflies fly southward. In the spring, their young fly back north.

Both the freshwater eel and all species of salmon migrate to their spawning grounds.

Alternate Lab: Cricket Hierarchies

Purpose

LS **Logical-Mathematical** Students will observe crickets setting up a dominance hierarchy.

Materials

glass terrarium or aquarium with cover, sand, 1 female and 5 male field crickets, dry oatmeal in a jar lid, apple slices, clean sponge pieces soaked in water, 4 matchboxes, 5 jars with lids, 4 different colors of nail polish

Procedure

Give the following directions to students.

1. Set up a terrarium with about 2 cm of sand in the bottom and numbered matchboxes in corners.
2. Place different colored spots of nail polish on the thoraxes of four male crickets. One male will not need polish. The female can be identified by her long ovipositor at the end of her abdomen.
3. Keep the five crickets in separate jars for at least one day prior to beginning the experiment.
4. Place four males in the terrarium and

spring to areas where they breed during the summer. Whales migrate seasonally, as well. Change in day length is thought to stimulate the onset of migration in the same way that it controls the flowering of plants.

Migration calls for remarkable strength and endurance. The arctic tern migrates between the Arctic Circle and the Antarctic, a one-way flight of almost 18 000 km.

Animals navigate in a variety of ways. Some use the positions of the sun and stars to navigate. They may use geographic clues, such as mountain ranges. Some bird species seem to be guided by Earth's magnetic field. You might think of this as being guided by an internal compass.

Biological rhythms are clearly governed by a combination of internal and external cues. Animals that migrate might be responding to colder temperatures and shorter days, as well as to hormones. You can easily see why animals migrate from a cold place to a warmer place, yet most animals do not migrate. How animals cope with winter is another example of instinctive behavior.

Canadian and Alaskan caribou migrate from their winter homes in the boreal forests to the tundra for the summer.

Earth Science

Finding Their Way Home

The arctic tern, a seabird, breeds on the coasts and islands of northern North America. Yet every August, the terns and their offspring begin a monumental journey to their winter home in Antarctica, a flying distance of almost 18 000 km! Come spring, the terns begin the return flight back to the Arctic. Among birds, the arctic tern is the gold medalist of long-distance flight. Many other birds also migrate annually. How do birds navigate such great distances without losing their way?

Finding home Homing pigeons are birds trained to return to their home lofts from wherever they are released—even at distances up to 80 km. They are bred for speed; their owners often hold races to see whose pigeons can return home most quickly.

Backup systems Scientists have found that pigeons use a variety of environmental and sensory cues to guide them on their journeys. Pigeons use the sun as a compass; they allow their internal clocks to compensate for the sun's movement during the day. But how does a pigeon locate its own home so precisely? Apparently, pigeons also use Earth's magnetic field to find home on cloudy days. But pigeons raised near areas where the magnetic field is disturbed seem to rely on many other factors to find their way home. Among these are smells, winds, pressure changes, infrasonic sounds, ultraviolet light, and polarized light patterns. Perhaps all migrating birds rely on an array of navigational aids to find their way home.

CONNECTION TO Biology

Some migrating birds follow river valleys, coastlines, or mountain ranges, whereas others seem to navigate by the stars. How could scientists test which navigational cues a particular bird species uses for migration?

Earth Science

Finding Their Way Home

Purpose

Students will learn about the variety of strategies that birds use as navigational aids during migration.

Teaching Strategies

- Ask students what kinds of navigational aids birds might use for migration. Write their answers on the chalkboard. After they read the feature, ask them to add to this list.
- Have students research migration paths of local birds. On a blank map, ask students to plot the migration paths. They should make their maps using colored pencils and a color-coded key. 

Possible Answer

If birds navigate primarily by the stars, they should become disoriented or pause in their migration when it is overcast.

GLENCOE TECHNOLOGY

Videodisc

Biology: The Dynamics of Life
Disc 2, Side 1
Salmon Migration (Ch. 26)

observe and record their behavior for 15 minutes.

5. Examine the terrarium for about 10 minutes each day for 5 days. Note which cricket becomes dominant and the behavior it exhibits.

Analysis

Ask the students to answer the following questions:

1. How could you tell that your crickets set up a dominance hierarchy? *One cricket chirped more, was initially more aggressive, and later others avoided him.*
2. Describe the differences in behavior of the crickets before and after the hierarchy was set up. *Before: Much aggression and chirping. After: Other crickets avoided the dominant male.*

Assessment

Performance: Ask students to hypothesize how the crickets will respond to the female and then introduce her to the terrarium. Ask students to record the behavior and draw conclusions. *The dominant male may mate with the female and spend more time with her than the other crickets do.* L1

SECTION 36.1

3 Assess

Check for Understanding

Ask students to prepare a concept map using all the vocabulary words in this section. L1

Reteach

Have students make a table that lists types of innate behaviors down the left side. Across the top, have them write the following heads: Definition, Example, Outcome of behavior, Survival value. L1

Extension

Ask students to go to a nearby zoo. As they observe animals, have them note innate behaviors and explain their survival values. L1

Assessment

Portfolio: Have students list each major group (phylum or class) of animals they have studied. For each phylum, ask them to identify one innate behavior and explain its adaptive value. L1 P

4 Close

Discussion

Male katydids sing to attract females. In Panamanian forests where bats are common, male katydids on plants shake their bodies vigorously to attract females. The females detect the shaking of the plant and respond to the male. The bats that could detect singing of potential prey cannot detect the shaking. Ask students to explain the behaviors in terms of natural selection and evolution of behavior.

Figure 36.10

The golden-mantled ground squirrel has a body temperature of around 37°C during normal activity. When the surrounding temperature drops to 0°C, the ground squirrel's heart rate drops. Its temperature drops to 2°C, and it goes into hibernation. Hibernation conserves the animal's energy.

You know that many animals store food in burrows and nests. But other animals survive the winter by undergoing physiological changes that reduce their need for energy. Many mammals, some birds, and a few other types of animals go into hibernation during the cold winter months. **Hibernation** is a state in which the body temperature drops substantially, oxygen consumption decreases, and breathing rates decline to a few breaths per minute. Animals that hibernate typically eat vast amounts of food before entering hibernation. This extra food fuels the animal's body while it is in this state. The golden-mantled ground squirrel shown in *Figure 36.10* is an example of an animal that hibernates.

What happens to animals that live year-round in hot environments? Some of these animals respond in a way that is similar to hibernation. **Estivation** is a state of reduced metabolism that occurs in animals living in conditions of intense heat. Desert animals appear to estivate sometimes in response to lack of food or periods of drought. However, Australian long-necked turtles, *Figure 36.11*, will estivate even when they are kept in a laboratory with constant food and water. Clearly, estivation is an innate behavior that depends on both internal and external cues.

Figure 36.11

Australian long-necked turtles are among the reptiles and amphibians that respond to hot and dry summer conditions by estivating.

Section Review

Understanding Concepts

1. What is behavior?
2. How is a reflex different from instinct?
3. Explain by example two types of innate behavior.

Thinking Critically

4. How is innate behavior an advantage to a species in which the young normally hatch after the mother has left?

Skill Review

5. **Designing an Experiment** Design an experiment to test what stimulus causes an earthworm to return to its burrow. For more help, refer to Practicing Scientific Methods in the *Skill Handbook*.

Answers to Section Review

1. Behavior is anything an animal does in response to a stimulus in its environment.
2. A reflex is a simple physical response to a stimulus. An instinct is a complex pattern of innate behavior.
3. Courtship rituals in birds and salmon migration have innate behavior components.

Thinking Critically

4. The young will instinctively find food and shelter and avoid predators.

Skill Review

5. Determine which stimuli cause an earthworm to move: light, heat, cold, touch, and water, and time how long it takes the worm to move a specific distance under each condition.

36.2 Learned Behavior

You were born knowing how to cry, but were you born knowing how to tie your shoes or read? Behavior controlled by instinct, as you now know, occurs automatically. However, some behavior is the direct result of the previous experiences of an animal. A dog that has been hit by a car and survives may avoid streets. You may know how to build a model ship after being shown only once or it may take a hundred trials. These behaviors are a result of learning.

Section Preview

Objectives

Distinguish among types of learned behavior.

Demonstrate, by example, types of learned behavior.

Key Terms

habituation
imprinting
trial-and-error learning
motivation
conditioning
insight
communication
language

What Is Learned Behavior?

Learning, or learned behavior, takes place when behavior changes through practice or experience. The more complex an animal's brain, the more elaborate the patterns of its learned behavior. As you can see in *Figure 36.12,* innate types of behavior are more common in invertebrates, and learned types of behavior are more common in vertebrates. In humans, many behaviors are learned. Speaking, reading, writing, and playing a sport are all learned.

Learning has survival value for all animals in a changing environment because it permits behavior to change in response to varied conditions. Learning allows an animal to adapt to change, an ability that is especially important for animals with long life spans. The longer an animal lives, the greater the chance that its environment will change and that it will encounter unfamiliar situations.

Figure 36.12

Examine the graph and tell which groups of animals demonstrate the most learned behavior. Which groups of animals demonstrate the most innate behavior?

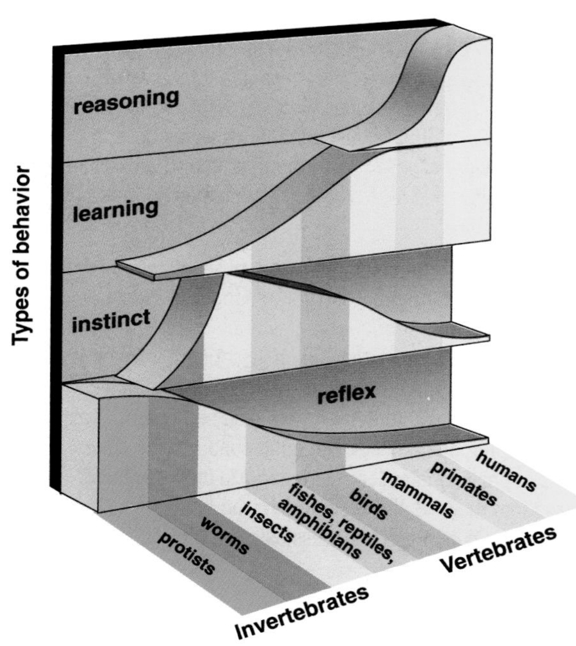

SECTION 36.2

Prepare

Key Concepts

Students will study various types of learned behavior. They will learn about the adaptive value of each, including habituation, imprinting, trial-and-error learning, conditioning, insight, and communication.

Block Scheduling

Look for this symbol for strategies that are useful in a block scheduling format. For more information on block scheduling, refer to Section 36.2 in the **Lesson Plans** booklet.

Materials

- For the Minilab on habits, ask students to wear a pair of tie shoes.
- For the Minilab on trial and error, you will need tagboard and envelopes.
- For the Biolab, gather photos of sandhill cranes, black fabric, black plastic trash bags, red fabric, white fabric, needle and thread, long gray socks, spring-type clothespins, black markers, stapler, tape, tagboard, various colored beads, various colored map tacks, glue, and scissors.

1 Focus

Bellringer

Before presenting the lesson, display **Section Focus Master 80** on the overhead projector and have students answer the accompanying questions. L1 SAE

Visual Learning

Figure 36.12 Which groups of animals demonstrate the most innate behavior? *The invertebrates*

Program Resources

Section Focus Master 80 L1 SAE
Reinforcement and Study Guide, pp. 147-148 L1
Biolab and Minilab Worksheets, pp. 143-146 L1
Laboratory Manual, pp. 227-230 L2
Basic Concepts Transparency 60 and **Master** L1 SAE
Reteaching Transparency 36 and **Master** L1 SAE

SECTION 36.2

2 Teach

A BROADER VIEW

Imprinting

Purpose

Students will learn how imprinting occurs and what the survival value of this behavior is.

Teaching Strategies

- Show photos of mother animals and their young. Ask students if they know how a mother is able to identify its own young and vice versa.
- Ask students to explain why imprinting would be important for animals that form herds.
- Explain to students that recent research has shown that young birds are not completely undiscriminating. If chicks are given a choice of objects in an imprinting experiment, they prefer some types of objects to others. The discovery of such innate preferences means that imprinting is probably a more complex process than earlier studies indicated.

Thinking Critically

If captive-raised birds see humans taking care of them, they will form attachments that will remain. When released into the wild, these animals will approach humans for food.

Software

Animal Pathfinders, Scholastic.

A BROADER VIEW

Imprinting

Have you ever seen a mother duck waddling down to a pond with a line of fluffy ducklings following close behind her? It's an endearing sight, but also an example of a learned behavior called imprinting. Many kinds of birds and mammals do not innately know how to recognize members of their own species. Instead, they learn to make this distinction early in life.

In birds such as ducks, imprinting takes place during the first day or two after hatching. A duckling rapidly learns to recognize and follow the first conspicuous moving object it sees. Normally, that object is the duckling's mother. Learning to recognize their mother and follow her ensures that food and protection will always be nearby.

Imprinting experiments In the 1930s, behavioral biologist Konrad Lorenz studied imprinting in geese. In his experiments, Lorenz showed that if newly hatched goslings were kept isolated for two or three days, they would imprint on just about any conspicuous, moving object that they were exposed to during that time. The birds followed and became attached to this object, just as they normally would to a mother goose.

Practical applications When people hand raise animals, as they do in many captive-breeding programs for endangered species, imprinting becomes an important issue. California condor chicks, for example, are isolated as soon as they hatch. Great care is taken that the young birds do not imprint on people or inappropriate objects. With their bodies and faces hidden, workers feed and care for each chick using hand puppets that look like adult condors. The chicks imprint on the puppets, and so learn to recognize their own kind. When released into the wild, these captive-raised birds are independent of humans and can interact successfully with other condors.

Thinking Critically

Why is it important that captive-raised birds destined for release into the wild do not form attachments to humans? Explain your answer.

Kinds of Learned Behavior

Just as there are several types of innate behavior, there are several types of learned behavior. Some learned behavior is simple and some is complex. Which group of animals do you think carries out the most-complex type of learned behavior?

Habituation is one of the simplest forms of learning

Horses normally shy away from an object that suddenly appears from the trees or bushes, yet after a while they disregard noisy cars that speed by the pasture honking their horns. This lack of response is called habituation. **Habituation,** *Figure 36.13,* occurs when an animal is repeatedly given a stimulus that is not associated with any punishment or reward. An animal has become habituated to a stimulus when it finally ceases to respond to the stimulus.

Figure 36.13

Habituation is a loss of sensitivity to certain stimuli. Young horses are often afraid of cars and noisy streets. Gradually horses become habituated to the city and ignore normal sights and sounds.

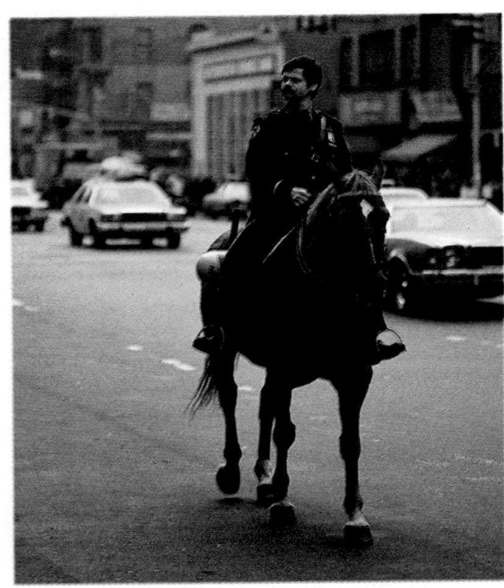

STUDENT JOURNAL

Whale Communications Certain whales are known for their melodious songs. Play a recording of whale songs to the class. Ask students to describe in their journals the songs of the whales. Provide library resources and ask them to figure out whether the sounds the whale makes are innate or learned. L1 LS

PROJECT Imprinting

LS **Interpersonal** Konrad Z. Lorenz has been called the "father of ethology." During the 1930s, he studied the behavior of birds and developed a theory of animal behavior that stressed its inherited aspects. Ask a group of students to present a skit about the work of Konrad Lorenz. L2 COOP LEARN

Imprinting involves a permanent attachment

Have you ever seen young ducklings following their mother? This behavior is the result of imprinting. **Imprinting** is a form of learning in which an animal, at a specific critical time of its life, forms a social attachment to another object. Imprinting takes place only during a specific period of time in the animal's life and is usually irreversible. For example, birds that leave the nest immediately after hatching, such as geese, imprint on their mother. They learn to recognize and follow her within a day of hatching.

Learning by trial and error

Have you ever watched a young child learning to ride a bicycle? The child tried many times before being able to successfully complete the task. Nest building, like riding a bicycle, may be a learning experience. The first time a jackdaw builds a nest, it uses grass, bits of glass, stones, empty cans, old light bulbs, and anything else it can find. With experience, the bird finds that grasses and twigs make a better nest than do light bulbs. The jackdaw has used **trial-and-error learning,** a type of learning in which an animal receives a reward for making a particular response. When an animal tries one solution and then another in the course of obtaining a reward, in this case a suitable nest, it is learning by trial and error.

Learning happens more quickly if there is a reason to learn or be successful. **Motivation,** an internal need that causes an animal to act, is necessary for learning to take place. In most animals, motivation often involves satisfying a physical need, such as hunger or thirst. If an animal isn't motivated, it won't learn. Animals that aren't hungry won't respond to a food reward. Mice living in a barn, *Figure 36.14,* discover that they can eat all the grain they like if they first chew through the container the grain is stored in.

MiniLab

How do you learn by trial and error?

Did you ever try to train a dog to sit up? If so, at first you probably rewarded any behavior that remotely resembled sitting up. The dog learned by trial and error what you expected, until it finally sat up in the appropriate way each time you gave the command. Humans also learn by trial and error, as you will see in this lab.

Procedure

1. Place the four tagboard pieces in front of you and flat on the desk.
2. Begin to put them together so that they form a solid figure with all sides matching and no pieces left out of the figure.
3. Time how long it takes you to complete the figure.
4. Time how long it takes for three additional trials.

Analysis

1. How long did it take to put the pieces together on the first trial?
2. How long did it take on the next three trials?
3. Why were the times different for each trial?

Figure 36.14

Mice soon learn where grain is stored in a barn and are motivated to chew through the storage containers.

PROJECT — Rodent Behavior

Have students construct a simple maze for small rodents such as mice and gerbils. Students should determine how many trials it takes for each type of rodent to learn the maze. Students may extend the trials to compare differences in the number of trials between young and adults and between males and females. L1 LS

MiniLab

Purpose

LS **Kinesthetic** Students will learn about trial-and-error learning.

Process Skills

compare and contrast, sequence

Teaching Strategies

- Prepare tagboard puzzles and place the pieces in envelopes prior to the lab. Cut a 10-cm piece of tagboard into four pieces.
- Ask students to make a list of activities they have learned by trial and error.

Expected Results

Students may take 5 to 10 minutes to put their puzzle together the first time, depending on the difficulty of the puzzle pieces you have created. The subsequent trials will take less time.

Analysis

1. About 5 to 10 minutes
2. Less time each time; the last trial may take only 10-15 seconds.
3. With experience, the task was learned and became easier.

Assessment

Performance: Ask students to spend a week learning to juggle three tennis balls. Provide them with a set of instructions. Encourage them to practice each day. Ask them to make an entry in their journals each day describing their experience. At the end of the week, have students show what they have learned. L1

BioLab Design Your Own Experiment

What makes a good feeding puppet for sandhill crane chicks?

Time Allotment
One class period

Objectives
Review objectives with students before they begin the Biolab.

Process Skills
compare and contrast, interpret scientific illustrations, make a model

Safety Precautions
Tell students to use care with scissors and needles.

Background
Give students this additional background: When this experiment was conducted, researchers used tape-recorded crane brooding calls to attract the attention of the chicks. For the first three months of life, the chicks lived in a pen with a stuffed surrogate mother crane equipped with speakers that made taped brooding calls. In the fall, outfitted with radio transmitters, many of the mature, hand-reared chicks left the refuge and migrated with wild cranes to Wisconsin.

PREPARATION

Alternate Materials
- If you live near a place where people hand-raise endangered birds for release, you may want to contact them and have students make feeding puppets for those birds.

BioLab Design Your Own Experiment

What makes a good feeding puppet for sandhill crane chicks?

In an innovative experiment on the Seney National Wildlife Refuge in Michigan, sandhill cranes were hatched from eggs and raised by puppet mothers, scientists with their arms in puppets that resembled the mother sandhill crane. This experiment was undertaken to develop a strategy for future use in hand raising endangered whooping cranes and releasing them into the wild. It would be important for hand-raised cranes not to become imprinted on humans who raised them. Otherwise, they would have no fear of humans. Therefore, the cranes were fed by humans who were covered with sheets the color of cranes and who used hand puppets. Biologists thought that they could perfect the technique with sandhill cranes, which are not endangered but are similar to whooping cranes.

PREPARATION

Problem
What characteristics must be present for you to make a successful feeding puppet for sandhill crane chicks?

Objectives
In this Biolab, you will:
- **Analyze** the features of sandhill crane heads and distinguish important characteristics.

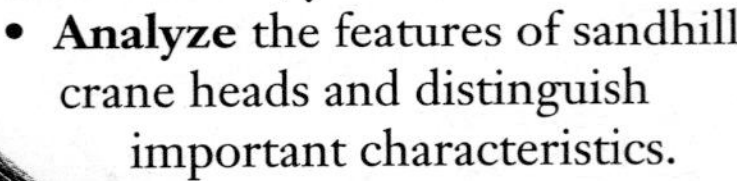

- **Design** a hand puppet suitable for feeding insects to young sandhill cranes.

Possible Materials
photos of sandhill cranes
black fabric and black plastic trash bags
red fabric
white fabric
needle and thread
long, gray socks
spring-type clothes pins
black markers
stapler
tape
tagboard
various colored beads
various colored map tacks
glue
scissors

Safety Precautions
Use care with the scissors and needles.

PLAN THE EXPERIMENT

Teaching Strategies
- Emphasize to students that they should be able to hold an insect in the bill of their puppet for a young crane to take. It will not be sufficient to make only a model of the head of a sandhill crane.
- Students who do not know how to sew may use staples.

Possible Procedures
- Students will see that the crane has a long neck and will most likely use the gray sock for the neck of their puppet. They should make the top of the head red with the red fabric, the cheeks white, and the bill black. They may want to use a spring-type clothespin or the spring mechanism alone to manipulate the beak, or make a beak

PLAN THE EXPERIMENT

1. Examine carefully the photos or diagrams of sandhill cranes.
2. Make a list of the important features of the head of the crane.
3. Brainstorm a list of ideas about how a hand puppet for feeding young cranes could be constructed.
4. Make a detailed diagram of exactly how your puppet will look. Label it with the material you will use.
5. Describe in the form of a list exactly how you will put your puppet together.

Check the Plan

1. Have you considered all the important features of the crane's head?
2. With your hand in the puppet, will you be able to feed insects to young cranes?
3. ***Make sure your teacher has approved your plan before you proceed further.***
4. Put your puppet together.

ANALYZE AND CONCLUDE

1. **Interpreting Scientific Diagrams** What features of the crane's head were important for identification of a sandhill crane?
2. **Making Predictions** How did you ensure your puppet's ability to feed insects to a young sandhill crane?
3. **Applying Concepts** Why would it be important for endangered whooping cranes being hand raised and released into the wild not to see humans?

Going Further

Project Contact an environmental group that works with endangered birds. Ask a representative of the group to visit the class or send materials that explain his or her work and some of the group's results.

ANALYZE AND CONCLUDE

1. Black bill, red on top of head, white cheeks, long gray neck, eyes
2. Tested to see if it would hold a paper clip or piece of thick string.
3. To be sure they are not imprinted on humans

Assessment

Portfolio: Ask students to explain in their portfolios what they have learned about the behavior of sandhill cranes by preparing a feeding puppet. L1 P

Going Further

Have students contact the cooperative wildlife research unit of your state university to discover more about the research being done for endangered species in your state. L1

with tagboard colored black. The eyes might be beads sewn on or map tacks glued to the head.

Data and Observations

Students' hand-feeding puppets should look like the photo of the head of the female crane and be able to hold an insect for a young crane to remove.

Bioethics

Sea Otter Behavior

In the 1800s, sea otters were abundant on the California Coast. They ate shellfish and sea urchins, which in turn fed on kelp. The kelp beds supported a rich diversity of life with huge numbers of fishes, crabs, shrimps, snails, and a variety of waterbirds that fed on these animals.

By 1910, the sea otter was nearly exterminated as a result of hunting. Kelp beds disappeared as sea urchins grew to large numbers and consumed the kelp. The California Coast became "urchin barrens." Two laws were enacted to protect the remaining otters, and a comeback began.

Today, the sea otter thrives, as do the kelp beds. However, people who harvest shellfish are not pleased with the comeback because the sea otter is destroying commercial and sports fishery.

GLENCOE TECHNOLOGY

Videodisc

The Infinite Voyage: Insects: The Ruling Class
Classifying Social Insects: Ants (Ch. 3)

Biology: The Dynamics of Life
Disc 2, Side 1
Bees (Ch. 27)

Ants (Ch. 28)

Figure 36.15
In 1900, Ivan Pavlov, a Russian biologist, first demonstrated conditioning in dogs.

A He noted that dogs salivate when they smell food. Responding to the smell of food is a reflex, an example of innate behavior.

B By ringing a bell each time he presented food to a dog, Pavlov established an association between the food and the ringing bell. The dog smelled food while a bell was rung. The dog salivated.

C Eventually, the dog salivated at the sound of the bell alone. The dog had been conditioned to respond to a stimulus that it did not normally associate with food.

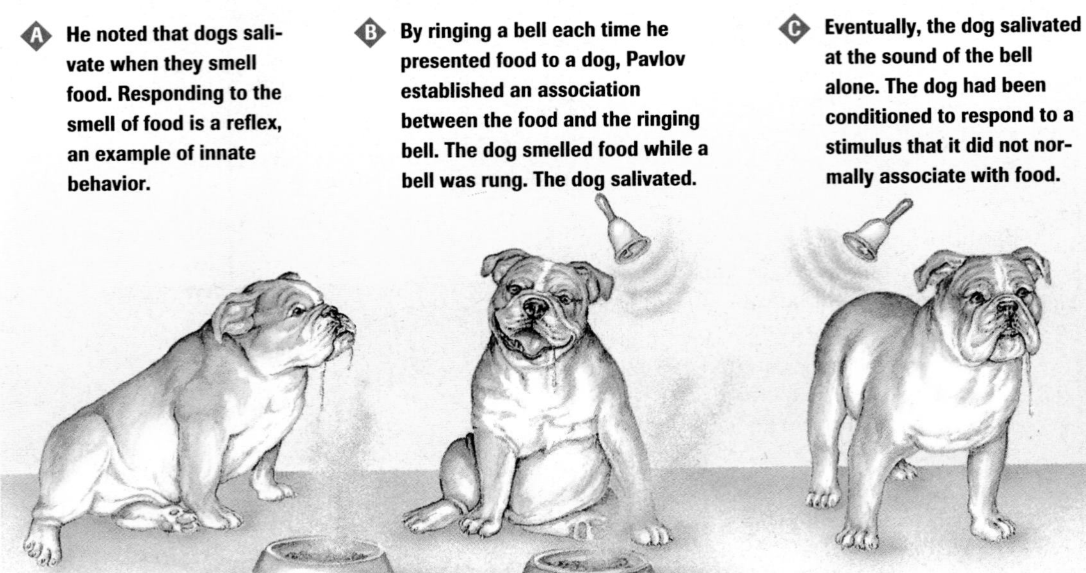

Conditioning is learning by association

When you first got a new kitten, it would meow and rub against your ankles as soon as it smelled the aroma of cat food in the can you were opening. After a few weeks, the sound of the can opener alone attracted your kitten, causing it to meow and rub against your ankles. Your kitten had become conditioned to respond to a stimulus other than the smell of food. **Conditioning** is learning by association. A well-known example of an early experiment in conditioning is illustrated in *Figure 36.15.*

Insight is the most complex type of learning

In a classic study of animal behavior, a chimpanzee was given two bamboo poles, neither of which was long enough to reach some fruit placed outside its cage. By connecting the two short pieces to make one longer pole, the chimpanzee learned to solve the problem of how to reach the fruit. This type of learning is called insight. **Insight** is learning in which an animal uses previous experience to respond to a new situation.

Much of human learning is based on insight. When you were a baby, you learned a great deal by trial and error. As you grew older, you relied more on insight. Solving math problems is a daily instance of using insight. Probably your first experience with mathematics was when you learned to count. Based on your concept of numbers, you then learned to add, subtract, multiply, and divide. Years later, you continue to solve problems in mathematics based on your past experiences. When you encounter a problem you have never experienced before, you solve the problem through insight.

Meeting Individual Needs

Gifted Ask students to conduct an experiment to see whether ants learn. They should set a food source farther and farther from an ants' nest at regular intervals. Ask students to determine whether the ants will walk to the next spot in anticipation of the food. Have students record their observations and conclusions.
L3

The Role of Communication

When you think about the interactions that happen between animals as a result of their behavior, you realize that some sort of communication has taken place. **Communication** is an exchange of information that results in a change of behavior. Black-headed sea gulls visually communicate their availability for mating with instinctive courtship behavior. The pat on the head from a dog's owner after the dog retrieves a stick signals a job well done.

Most animals communicate

Animals have several channels of communication open to them. You remember from Chapter 2 that elephants use sounds to communicate over long distances. Other animals also signal each other by sounds, sights, touches, or smells. Sounds radiate out in all directions and can be heard a long way off. The sounds of the humpback whale can be heard 1200 km away. Sounds such as songs, roars, and calls are best for communicating a lot of information quickly. For example, the song of a male cricket tells his sex, his location, his social status, and, because communication by sound is usually species specific, his species.

MiniLab

What is a habit?

How many things do you do automatically without even thinking about how to do them? Do you have to think about how to brush your teeth or button a shirt? Once you learn how to carry out a procedure that is repeated often, it becomes a habit. Doing something by habit has value because you don't waste time relearning the procedure each time you perform it. The same is true for an animal in the wild.

Procedure

1. Without actually tying your shoes or even looking at the laces, write a set of directions that explain step by step how to tie a shoe.
2. Have your partner read your directions and then try to follow exactly your procedure to tie one shoe.

Analysis

1. What is the advantage of forming a habit?
2. Why was it difficult to describe how to tie a shoe?
3. How is forming a habit like habituation?

Signals that involve odors may be broadcast widely and carry a general message. Ants, *Figure 36.16,* leave odor trails that are followed by other members of their nest. Some odors may also be species specific. As you know, pheromones, such as those of moths, may be used to attract mates. Because only small amounts are needed, other animals, especially predators, don't detect the odor.

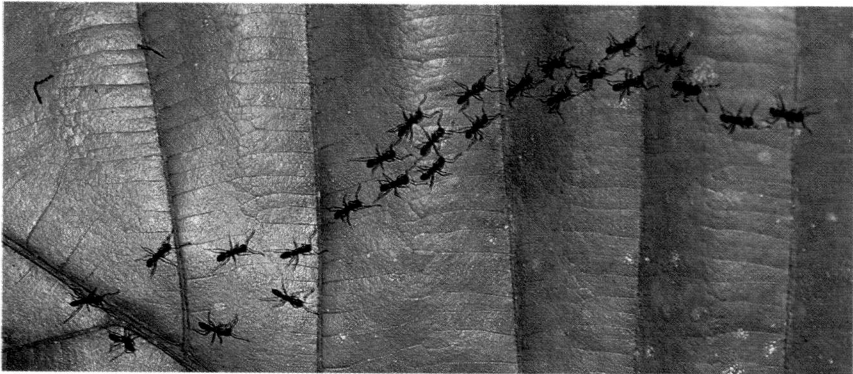

Figure 36.16

Did you ever watch as a group of ants moved in single file to a piece of potato chip left on a picnic table? Ants follow chemical trails left by other ants to find food sources. In what way does this behavior have adaptive value?

Visual Learning

Figure 36.16 In what way does this behavior have adaptive value? *This behavior helps ensure that all ants can find food.*

MiniLab

Purpose

LS Logical-Mathematical Students will learn the adaptive advantage of a habit.

Process Skills

recognize cause and effect, sequence, observe and infer

Teaching Strategies

- When the partners are trying to follow the directions, instruct them to do only what the directions indicate. They should not "read between the lines" and add steps that are not written.

Expected Results

Steps for tying will have been omitted from the written instructions and students may not be able to tie the laces.

Analysis

1. You don't have to relearn a process each time you do it.
2. It is a complex procedure with many little detailed steps that must be done in sequence. Some people may have developed variations that are different from the way others tie shoes.
3. Habituation is loss of sensitivity to a repeated stimulus that is not associated with punishment or reward. A habit is formed when a behavior is repeated over and over until it can be done without much thought.

CULTURAL DIVERSITY

Bertram Fraser-Reid and the Production of Artificial Pheromones

Discuss with students the important contributions of African-American organic chemist, Bertram O. Fraser-Reid. His most important work involved synthesizing artificial insect pheromones as substitutes for dangerous insecticides. In Canada, the western pine beetle causes billions of dollars of damage to trees each year. Fraser-Reid reasoned that if he released artificial pheromones of female pine beetles in a part of the forest that contained no females, it might attract male pine beetles to the spot, thus preventing them from mating. Fraser-Reid's initial research laid the groundwork for future studies.

3 Assess

Check for Understanding

Write a list of behaviors on the chalkboard. Ask students to identify each as innate or learned. L1

Reteach

Have students play animal behavior charades. Write a number of behaviors on slips of paper. Have students draw a slip and act out the behavior. Class members will try to guess the name of the behavior and whether it is innate or learned. L1 COOP LEARN

Extension

Ask students to accompany a wildlife specialist, a fishery biologist, or a veterinarian on the job for a day. Have them report about how the knowledge of animal behavior is important to being able to carry out that particular job. L1

Assessment

Portfolio: Ask students to select a local wild animal species to observe for a period of time each day for a week. Students should look for and record examples of feeding behavior, movement, communication, and social interactions. They should try to categorize the behaviors as innate, learned, or a combination of both. L1

4 Close

Audiovisual

Show a film about animal behavior. Interesting films about dolphins, killer whales, baboons, cats, honeybees, and penguins are available from National Geographic.

Figure 36.17
English and other languages are made up of words that have specific meanings. People who use the language put the words together to convey messages and ideas. An unlimited number of meanings can be communicated using the same words.

Using both innate and learned behavior

Some communication is a combination of both innate and learned behavior. Male songbirds automatically sing when they reach sexual maturity. Their songs are specific to their species, and singing is innate behavior. Yet members of the same species that live in different regions learn different variations of the song. They learn to sing with a regional dialect.

Some animals use language

Language, the use of symbols to represent ideas, is present primarily in animals with complex nervous systems, memory, and insight. Humans, with the help of spoken and written language, can benefit by what other people and cultures have learned and don't have to experience everything for themselves. They can use accumulated knowledge as a basis on which to build new knowledge, *Figure 36.17*.

Connecting Ideas

The environment selects for genes that direct the development of animals' nervous systems in ways that promote certain behavior patterns. In all animals, some behaviors are innate, and others are learned. Whether the behavior is innate or learned, it increases the animal's ability to adapt and pass on its genes to its own offspring. Humans, like other animals, have complex nervous systems and other body systems that work together to increase the individual's chances of survival.

Section Review

Understanding Concepts

1. What is the difference between imprinting and other types of learned behavior?
2. How does learning have survival value in a changing environment?
3. Explain by example the difference between trial-and-error learning and conditioning.

Thinking Critically

4. How would a cat respond if the mice in a barn no longer entered at the usual places?

Skill Review

5. **Observing and Inferring** Two dog trainers teach dogs to do tricks. One trainer gives her dog a bone-shaped treat whenever the dog correctly performs the trick. The other trainer does not. Which trainer will be more successful at dog training? Why? For more help, refer to Thinking Critically in the *Skill Handbook*.

Answers to Section Review

1. Imprinting can take place only during a specific time during an animal's life. Other learned behaviors may develop throughout life.
2. The behavior can be modified to meet an animal's needs in a changing environment.
3. You learn to ride a bicycle by trial-and-error learning, while an animal may be conditioned to the sound of a can opener.

Thinking Critically

4. The cat might wait at the usual places for a while; after not having any success finding mice, the cat would look for new places mice might enter.

Skill Review

5. The dog will be more easily trained if it is rewarded when it responds correctly. Food is a motivator for most animals.

REVIEWING MAIN IDEAS

36.1 Innate Behavior

- Behavior is anything an animal does in response to a stimulus. Behaviors have adaptive value and are shaped by natural selection.
- Innate behavior is inherited. Innate behaviors include automatic responses and instincts. Automatic responses include reflexes and fight-or-flight responses.
- An instinct is a complex pattern of innate behavior.
- Behaviors such as courtship rituals, displays of aggression, territoriality, dominance hierarchies, hibernation, and migration are all forms of instinctive behavior.

36.2 Learned Behavior

- Learning takes place when behavior changes through practice or experience. Learned behavior has adaptive value.
- Learning includes habituation, imprinting, trial and error, and conditioning. The most complex type of learning is learning by insight.
- Some animals use language, whereas most communicate by either visual, auditory, or chemical signals.

Key Terms

Write a sentence that shows your understanding of each of the following terms.

aggression
behavior
circadian rhythm
communication
conditioning
courtship behavior
dominance hierarchy
estivation
fight-or-flight response
habituation
hibernation
imprinting
innate behavior
insight
instinct
language
migration
motivation
territory
trial-and-error learning

Understanding Concepts

1. Anything an animal does in response to a stimulus in its environment is called ____________.
   a. an instinct
   b. aggression
   c. behavior
   d. conditioning
2. A peacock spreads his tail feathers and displays them to a peahen. What kind of behavior is this?
   a. courtship
   b. dominance hierarchy
   c. aggression
   d. trial-and-error learning
3. Animals with behavior that makes them more successful at surviving and reproducing will produce more ____________.
   a. offspring
   b. aggression
   c. territory
   d. eggs
4. A(n) ____________ is the simplest form of innate behavior.
   a. dominance hierarchy
   b. circadian rhythm
   c. instinct
   d. reflex
5. A female goose retrieves an egg that has rolled out of the nest. This behavior is an example of ____________.
   a. learned behavior
   b. imprinting
   c. a reflex
   d. an instinct
6. All inherited behavior of animals is called ____________ behavior.
   a. instinct
   b. innate
   c. conditioning
   d. insight

Chapter 36 Review

Reviewing Main Ideas

Summary statements can be used by students to review the major concepts of the chapter.

Key Terms

Answers should go beyond defining the terms. Accept any answer that uses the term correctly and in the proper context.

Understanding Concepts

1. c
2. a
3. a
4. d
5. d
6. b

Chapter 36 Review

7. d
8. b
9. a
10. d
11. c
12. c
13. a
14. c
15. b
16. a
17. b
18. b
19. d
20. d

Applying Concepts

21. A dominance hierarchy would reduce aggression at common feeding sites.
22. No, most likely they would have already imprinted to their own mother by the time they are five days old.
23. Squirrels that store extra food when the days get shorter and the weather gets colder will be more likely to survive than squirrels that do not store food. The squirrels that store food will survive and produce offspring that have the same genetic makeup for behavior that includes storing winter food.
24. Pavlov observed dogs' natural behavior when food was presented. He hypothesized that the dogs could be conditioned to respond to the sound of a bell as if it were food. He designed an experiment to test his hypothesis. He rang a bell each time he fed the dog, then again when the dog smelled food. Finally, when he rang the bell with no food stimulus, the dog salivated. He concluded that the dog had been conditioned to respond to a stimulus that it did not normally associate with food.
25. learned behavior

Chapter 36 Review

7. A physical space that contains the breeding area, feeding area, shelter, or potential mates of an animal is called its ____________.
   a. community
   b. habitat
   c. ecosystem
   d. territory
8. Your adult dog is chewing on a bone when a puppy approaches. Your dog growls at the puppy. What behavior is your dog exhibiting?
   a. fighting
   b. aggression
   c. habituation
   d. conditioning
9. A dominance hierarchy may prevent energy from being wasted on continuous ____________.
   a. fighting
   b. courtship
   c. feeding
   d. learning
10. A(n) ____________ is a 24-hour cycle of behavior.
    a. instinct
    b. fight-or-flight response
    c. reflex
    d. circadian rhythm
11. Caribou are ____________ when they move from their winter homes in the forests to the tundra for the summer.
    a. hibernating
    b. imprinting
    c. migrating
    d. learning
12. The golden-mantled ground squirrel reduces its need for energy during the winter by going into ____________.
    a. estivation
    b. deep sleep
    c. hibernation
    d. motivation
13. Learning has survival value for animals in a changing environment because it permits behavior to ____________.
    a. change
    b. remain the same
    c. become imprinted
    d. become instinctive
14. The use of symbols to represent ideas is called ____________.
    a. conditioning
    b. learning
    c. language
    d. motivation
15. One of the simplest forms of learned behavior is ____________.
    a. conditioning
    b. habituation
    c. trial and error
    d. insight
16. Learning to ride a bicycle is an example of what kind of learning?
    a. trial and error
    b. insight
    c. conditioning
    d. habituation
17. ____________ is an internal need that causes an animal to act and is necessary for learning.
    a. A habit
    b. Motivation
    c. Conditioning
    d. Instinct
18. Your cat runs for its food dish when it hears the sound of the can opener. What kind of behavior is your cat exhibiting?
    a. insight
    b. conditioning
    c. habituation
    d. imprinting
19. When you learned to do long division, you were using ____________.
    a. conditioning
    b. instinct
    c. habituation
    d. insight
20. Moths use pheromones to ____________.
    a. estivate
    b. migrate
    c. imprint
    d. signal

Applying Concepts

21. What would be the advantage of a dominance hierarchy in members of a species that are not defending a territory?
22. If you found a nest of five-day-old goslings, would they imprint on you and follow you home? Explain.
23. Explain by example how natural selection is important in determining behavior of animals.
24. Explain how Ivan Pavlov used the methods of science to study conditioning behavior.
25. By accident, a gull drops a snail on the road. The snail's shell breaks, and the gull eats the snail. The gull continues to drop mollusks on the road. What type of behavior is this?

Thinking Critically

26. ***Concept Mapping*** Make a concept map that relates the following terms and phrases. Supply the appropriate linking words for your map.
innate behavior, learned behavior, imprinting, reflex, habituation, courtship behavior, insight, migration, conditioning, trial-and-error learning

27. ***Formulating Hypotheses*** You have prepared a hand puppet for hand feeding young sandhill cranes. You have made it look similar to the female crane's head. You try to feed beetles to the young chicks, and they refuse the food. Make a hypothesis about why the chicks refuse the food.

28. ***Comparing and Contrasting*** Ducklings display an alarm reaction when a model of a hawk is flown over their heads, and no alarm reaction when a model of a goose is flown over their heads. After several days, neither model causes any reaction. Compare the effects of the two models during the first two days with the effects of the same models two weeks later.

29. ***Recognizing Cause and Effect*** When Charles Darwin visited the Galapagos Islands in 1835, he was amazed that the animals would allow him to touch them. Why were they not afraid?

30. ***Observing and Inferring*** You notice that birds sitting on a telephone wire always maintain an equal distance from one another. If a new bird approaches and perches on the wire, all the birds move slightly so that this distance is again established. What kind of instinctive behavior are they exhibiting? Explain.

ASSESSING KNOWLEDGE & SKILLS

Chickens that show submission find themselves at lower levels in the barnyard pecking order.

Interpreting Scientific Illustrations Examine the dominance hierarchy diagram above and answer the following questions.

1. Which chicken is dominant?
- **a.** 1
- **b.** 2
- **c.** 3
- **d.** 4

2. Which chicken or chickens would chicken number 4 peck?
- **a.** only 3
- **b.** only 2 and 3
- **c.** 1, 2, and 3
- **d.** none

3. Which chickens could be described as submissive?
- **a.** only 2, 3, and 4
- **b.** only 3 and 4
- **c.** only 4
- **d.** only 1

4. What type of behavior is illustrated in the diagram?
- **a.** learned behavior
- **b.** courtship behavior
- **c.** instinctive behavior
- **d.** reflex behavior

5. ***Creating a Model*** Under crowded conditions, mice form dominance hierarchies. Create a model showing six mice that have set up a dominance hierarchy in a small cage. Write a caption that explains what is happening.

Chapter 36 Review

Thinking Critically

26. Evaluate students' concept maps. The maps should show an understanding of the relationships among the concepts listed.
27. Possible hypotheses might include: beetles may not be the usual food of the chicks, or the hand puppet does not look enough like the mother.
28. They show alarm with the hawk model, but not with the goose because a hawk is a potential predator of the duckling, and the goose is not a predator. This is instinctive behavior. When there are no dangerous results, the ducklings become habituated to the hawk model.
29. They had learned not to fear large animals because there were no large predators on the islands.
30. Individual distance is a form of territoriality—an innate behavior.

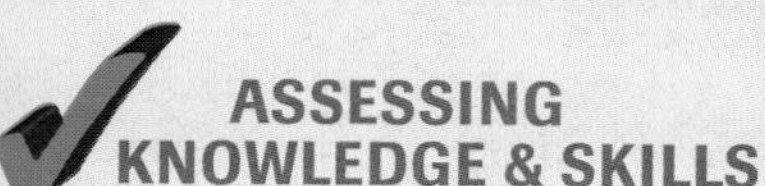

ASSESSING KNOWLEDGE & SKILLS

1. a
2. d
3. a
4. c
5. Make sure that the mice are exhibiting threatening or aggressive posture from one to another and that the order of the dominance hierarchy is clear.

Program Resources

Chapter Assessment, pp. 219-224 L1
Alternate Assessment in the Science Classroom
Computer Test Bank L1
Content Mastery, pp. 149-152 L1

Prepare

Purpose

This Biodigest can be used as an introduction to or as an overview of the vertebrate classes. If time is limited, you may wish to use this unit summary to teach the vertebrates in place of the chapters in the Vertebrate unit.

Key Concepts

Students will study vertebrates by comparing and contrasting the features of fishes, amphibians, reptiles, birds, and mammals. Students learn that there are three classes of fishes as compared with one class each of amphibians, reptiles, birds, and mammals. Students will trace the evolution of vertebrates and identify the similar traits that are found within all seven vertebrate classes. Students will see differences in form that correspond to adaptations to different habitats as vertebrate ancestors moved from life in the water to life on land.

Block Scheduling

Look for this symbol for strategies that are useful in a block scheduling format. For more information on block scheduling, refer to Biodigest 3 in the **Lesson Plans** booklet.

1 Focus

Bellringer

Before presenting the lesson, display **Section Focus Master 81** on the overhead projector and have students answer the accompanying questions. L1 SAE

BIODIGEST VERTEBRATES

How are fishes, amphibians, reptiles, birds, and mammals alike? All of these animals are vertebrates. Like all chordates, vertebrates have a notochord, gill slits, and a dorsal hollow nerve cord. However, in vertebrates the notochord is replaced during development by a backbone. All vertebrates are bilaterally symmetrical, coelomate animals that have an endoskeleton, a closed circulatory system, an efficient respiratory system, and a complex brain and nervous system.

FISHES

All fishes are ectotherms, animals with body temperatures dependent upon the temperature of their surroundings. Fishes are grouped into three different classes.

Lacking jaws, this sea lamprey obtains food by clamping its round mouth onto the side of a fish, using its rasping tongue to make a wound, then sucking out the prey's blood.

VITAL STATISTICS FISHES

Size ranges: Whale shark, length, 15 m; dwarf goby, length, 1 cm
Unusual adaptations: Electric eels can deliver an electrical charge of 650 volts, which stuns or kills their prey. Some deep-sea fishes have their own bioluminescent lures to catch prey.
Longest-lived: Lake sturgeon, 80 years
Distribution: Fresh, salt, and brackish water worldwide.
Numbers of species:
Class Agnatha, jawless fishes: 63 species
Class Chondrichthyes, cartilaginous fishes: 850 species
Class Osteichthyes, bony fishes: 18 000 species

Jawless Fishes

Lampreys and hagfishes are jawless fishes. Like all fishes, jawless fishes have a two-chambered heart and breathe with gills.

Cartilaginous Fishes

Sharks, skates, and rays are cartilaginous fishes. Fossil evidence shows that jaws first evolved in these fishes. Cartilaginous fishes have paired fins and a lateral line system that enables them to detect movement and vibrations in water.

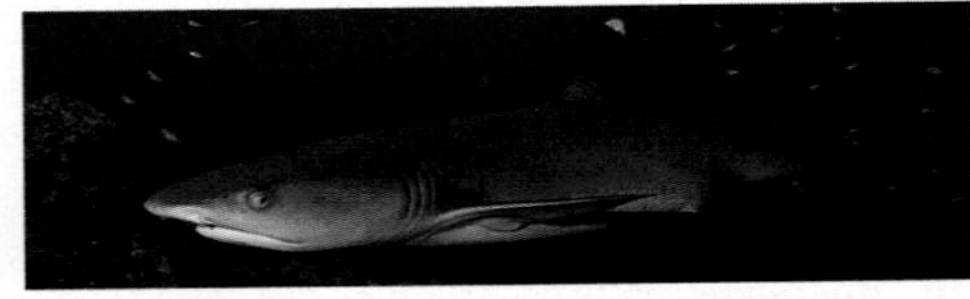

Cartilaginous fishes such as this whitetip reef shark are more dense than water. This shark will sink if it stops swimming.

Bony Fishes

Most fish species belong to the bony fishes. All bony fishes have a bony skeleton, gills, paired fins, flattened bony scales, and a lateral line system. Bony fishes breathe by drawing water into their mouths, then passing it over gills where gas exchange occurs. They adjust their depth in the water by regulating the amount of gas that diffuses out of their blood into a swim bladder.

Learning Styles

Look for the following logo for strategies that emphasize different learning modalities. LS

Kinesthetic	Meeting Individual Needs, p. 922; Student Journal, p. 925
Visual-Spatial	Student Journal, p. 921; Visual Learning, pp. 921, 922, 927, 928; Meeting Individual Needs, p. 921; Demonstration, p. 922; Activity, pp. 927
Interpersonal	Activity, p. 923; Project, pp. 926, 928
Intrapersonal	Student Journal, p. 928
Logical-Mathematical	Visual Learning, p. 925; Student Journal, p. 926
Auditory-Musical	Meeting Individual Needs, p. 924

BIODIGEST

Most fishes fertilize their eggs externally and leave their survival to chance. Sea horses are unusual compared with most other fishes in that the females deposit their eggs directly into brood pouches found underneath the tails of the males. The eggs develop inside the partially sealed brood pouch. After hatching, the tiny sea horses emerge from the pouch.

AMPHIBIANS

Amphibians are ectothermic vertebrates with three-chambered hearts; lungs; and thin, moist skin. Although they have lungs, most gas exchange in amphibians is carried out through the skin. As adults, amphibians live on land but rely on water for breeding and egg development. Almost all amphibians go through metamorphosis, in which the young hatch into tadpoles, gradually losing their tails and gills as they develop legs, lungs, and other adult structures.

Amphibian Classification

Amphibians are classified into three orders: Caudata, the salamanders and newts; Anura, the frogs and toads; and Apoda, the legless caecilians. Frogs and toads have vocal cords that can produce a wide range of sounds. Frogs have thin, smooth, moist skin, and toads have thick, bumpy skin with poison glands. Salamanders have long, slender bodies with a neck and tail. Caecilians are amphibians with long, wormlike bodies and no legs.

Vibrations from air or water are picked up by the frog's tympanic membrane and transmitted to the inner ear. The tympanic membrane is located behind and below the frog's eye.

VITAL STATISTICS AMPHIBIANS

Size ranges: Largest: Goliath frog, length, 30 cm; Chinese giant salamander, length, 1.8 m; Smallest frog: *Psyllophryne didactyla*, length, 9.8 mm

Distribution: Tropical and temperate regions worldwide; generally moist, but sometimes dry or aquatic environments.

Numbers of species:

Order Anura, frogs and toads: 2500 species

Order Caudata, salamanders and newts: 300 species

Order Apoda, caecilians: 168 species

Although salamanders resemble lizards, they have smooth, moist skin and lack claws on their toes, which are features used to classify salamanders as amphibians.

BIODIGEST

Field Trip

Arrange a class trip to a local zoological park, nature center, or other park or natural area where students will be able to see representatives of the classes of vertebrates discussed in this unit. Ask students to list as many vertebrates as they see, then divide the animals into classes. Ask them to identify the characteristics they used to classify the animals.

2 Teach

Visual Learning

LS **Visual-Spatial** Direct students' attention to the legs of the frog and legs of the salamander. Ask students to describe how each type of leg is an adaptation to the way of life of the animal. *The frog has powerful hind legs that enable it to leap quickly away from predators. The salamander has small legs on the sides of its body that assist with crawling movements into small spaces or under debris.*

STUDENT JOURNAL

LS **Visual-Spatial** Provide each group of students with a goldfish in a clear container with water. Ask them to observe its features and movements and write their observations in their journals. Also, ask them to make a diagram of the goldfish and label key features. SAE

Meeting Individual Needs

Students Acquiring English Ask students to prepare a bulletin board display of bony fishes by cutting out pictures of bony fishes from outdoor sports magazines. Fishes should be identified and each fish should have one part labeled. Have students use the following labels: scales, lateral line, gills, swim bladder, and fins. L1 SAE LS

Visual Learning

Visual-Spatial Direct students' attention to the photos of the snake and the crocodile. Ask them to explain how the snake and the crocodile are alike and how they are different. *They are both ectotherms with skin that is dry, thick, and covered with scales. They both have lungs and lay eggs with leathery shells. The snake does not have legs while the crocodile has four legs. The snake has a three-chambered heart and the crocodile has a four-chambered heart.* SAE

Demonstration

Visual-Spatial Show students a live snake, lizard, or turtle from the local environment or borrowed from a pet shop. Have students point out the reptile features of the animal. SAE

REPTILES

Reptiles are ectotherms with dry, scaly skin and clawed toes. They include snakes, lizards, turtles, crocodiles, and alligators. With the exception of snakes, all reptiles have four legs that are positioned somewhat underneath their bodies.

Snakes capture their prey using a variety of methods. Constrictors such as this emerald tree boa hold prey with their mouths, wrap coils around the prey's body, and then squeeze until it suffocates.

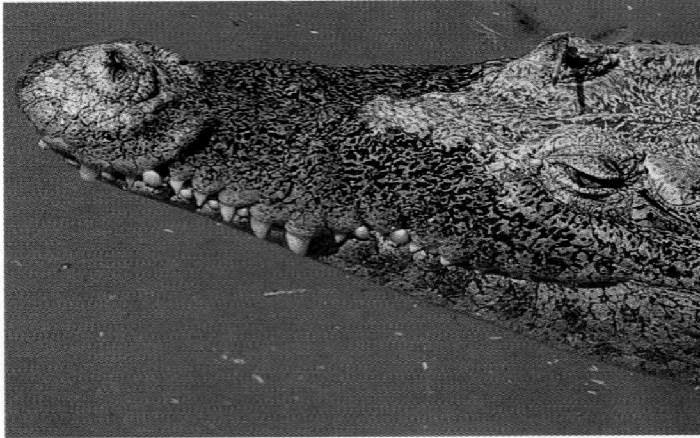

All crocodilians have nostrils and eyes that extend above the rest of their faces. This enables the animal to breathe and see while most of its body is under water.

Most reptiles have a three-chambered heart, but crocodilians have a four-chambered heart in which oxygenated blood is kept entirely separate from blood without oxygen. The scaly skin of reptiles reduces the loss of body moisture on land, but scales also prevent the skin from absorbing or releasing gases to the air. Reptiles are entirely dependent upon lungs for this essential gas exchange.

FOCUS ON ADAPTATIONS

The Amniotic Egg

Reptiles were the first group of vertebrates to live entirely on land. They evolved a thick, scaly skin that prevented water loss from body tissues. They evolved strong skeletons, with limbs positioned somewhat underneath their bodies. These limbs enabled them to move quickly on land, avoiding or seeking the sun as their body temperatures demanded. But perhaps their most important adaptation to life on land was the development of the amniotic egg.

Protecting the Embryo An amniotic egg encloses the embryo in amniotic fluid; provides the yolk, a source of food for the embryo; and surrounds both

Meeting Individual Needs

Students Acquiring English Give each group of students several scientifically accurate reptiles that are available in toy stores. As they handle the models, ask them to make a list of the reptilian features of each model. L1 SAE

BIODIGEST

Internal Fertilization

All reptiles have internal fertilization and lay eggs. The development of the amniotic egg enabled reptiles to move away from a dependence upon water for reproduction. The amniotic egg provides nourishment to the embryo and protects it from drying out as it develops.

The shell of a turtle is unique among vertebrates. Made out of bony plates covered with horny shields, the turtle's shell provides protection from most predators, but prevents turtles from making rapid movements.

Marine turtles, such as this olive ridley, come ashore only to lay eggs in nests they dig on sandy beaches. Once the eggs are laid and covered by sand, mother turtles head back to sea.

VITAL STATISTICS — REPTILES

Size ranges: Largest reptiles: anaconda snake, length, 9 m; leatherback turtle, weight, 680 kg; Smallest: thread snake, length, 1.3 cm
Reptile causing most human deaths: King cobra, 7500 per year
Distribution: Temperate and tropical forests, deserts, and grasslands, and fresh, salt, and brackish water.
Numbers of species:
Order Squamata, snakes and lizards: 6800
Order Chelonia, turtles: 250 species
Order Crocodilia, crocodilians: 25 species
Order Rhynchocephalia, tuataras: 1 species

embryo and food with membranes and a tough, leathery shell. These structures in the egg help prevent dehydration and injury to the embryo as it develops on land. Most reptiles lay their eggs in protected places beneath sand, earth, gravel, or bark.

Membranes Inside the Egg Membranes found inside the amniotic egg include the amnion, the chorion, and the allantois. The amnion is a membrane filled with fluid that surrounds the developing embryo. The embryo's nitrogenous wastes are excreted into a membranous sac called the allantois. The chorion surrounds the yolk, allantois, amnion, and embryo. With this egg, reptiles no longer needed water for reproduction, and the move from water to land was complete.

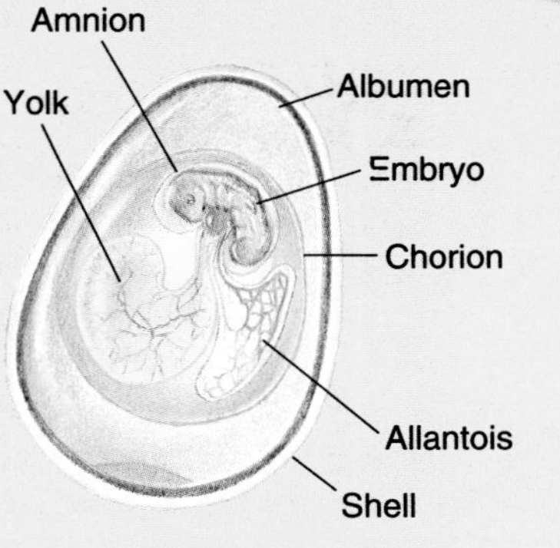

923

Activity

Interpersonal Ask students in their groups to make a model of an amniotic egg using the following materials: a small lump of clay; two small, zipper-type, clear plastic bags; two pill vials or plastic film containers; and a larger, lunch-size paper bag.

Do not tell them how to make their model, but give hints if they have trouble. Ask each group to explain their model. A model might be constructed as follows. The lump of clay represents the embryo. The pill vials stick into the lump of clay, one representing the yolk, the other the allantois. The clay, with attached vials, is placed in one plastic bag that represents the amnion. The amniotic sac and contents are placed in the other plastic bag, which represents the chorion. The chorion and contents are placed into the paper bag, which is folded closed to represent the shell. L1 SAE P COOP LEARN

interNET CONNECTION

Follow the link for this Biodigest on the Glencoe Homepage at **www.glencoe.com/sec/science** to find out more about vertebrates.

BIODIGEST

BIRDS

Birds are the only class of organisms with feathers. Feathers, which are lightweight, modified scales, help insulate the bird and enable it to fly. Birds have forelimbs that are modified into wings. Birds, like reptiles, have scales on their feet and clawed toes. Unlike reptiles, birds are endotherms, which are animals that maintain a constant body temperature. Endotherms must eat frequently to provide energy needed for producing body heat.

Feathers keep birds warm and streamline them for flight. Feather colors are often important in courtship or camouflage. The peacock attracts the peahen with its display of tail feathers.

Penguins are flightless birds with wings and feet modified for swimming and a body surrounded by a thick layer of insulating fat. This baby emperor penguin may reach a height of 1 m and weigh nearly 34 kg.

Bird Flight

Besides feathers and wings, bird flight requires several other adaptations. Birds have thin, hollow bones with cross braces that provide support for strong flight muscles while reducing the bird's body weight. Birds also have a four-chambered heart and an efficient respiratory system.

FOCUS ON ADAPTATIONS

Bird Flight

What selection pressure may have resulted in bird flight? Maybe an early bird's need to escape from a predator caused it to run so fast its feet left the ground. Whatever caused birds to evolve an ability to fly, there must first have been adaptations that made flight possible. What are some of these adaptations? A bird's body is lighter in weight than any other animal's of the same size because it has hollow bones and air sacs throughout its body. It also has a beak instead of a heavy jaw with teeth, and its legs are made mainly of only skin, bone, and tendons.

Meeting Individual Needs

Students Acquiring English Play a recording of various types of bird songs and calls. Ask students to distinguish between distress calls, calls made to the young, courtship songs, and territorial calls. L1 SAE LS

▲ **Hummingbirds are not only quick as they fly forward, but agile as they fly up, down, sideways, and backwards. Their wings beat 20 to 50 times per second, enabling them to hover while feeding on nectar from flowers.**

VITAL STATISTICS BIRDS

Size ranges: Largest: ostrich, height, 2.4 m, mass, 156 kg; smallest: bee hummingbird, length, 57 mm, mass, 1.5 g
Longest yearly migration: Flown by the Arctic tern, 40 000 km
Widest wing span: 3.7 m in the wandering albatross
Fastest flyer: White-throated spinetail swift, 171 kph
Largest egg: Ostrich egg, length, 13.5 cm, mass, 1.5 kg
Distribution: Worldwide in all habitats.
Numbers of species:
Class Aves: 8600 species in 27 present-day orders
- Order Passeriformes, perching song birds: 5400 species
- Order Ciconiiformes, herons, bitterns, ibises: 127 species
- Order Anseriformes, swans, ducks, geese: 161 species
- Order Falconiformes, eagles, hawks, falcons: 298 species

Nest Builders

Like reptiles, birds lay amniotic eggs. Unlike reptiles, birds incubate their eggs in nests, keeping eggs warm until the baby birds hatch.

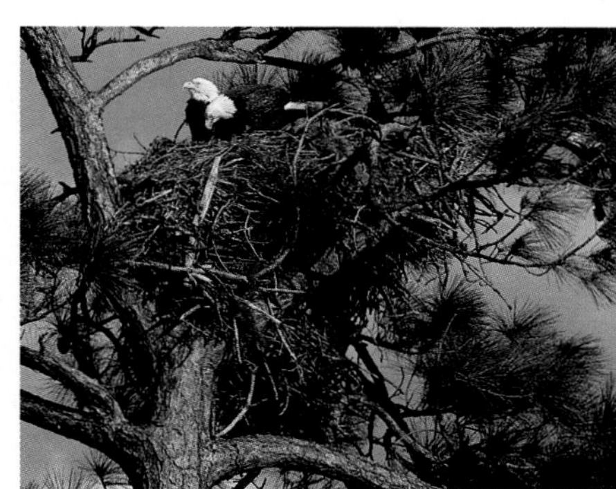

The largest bird's nest is built by the bald eagle. Every year eagles add another layer of sticks to the nest until some nests are 2 m across and 2 tons in mass. ▶

Efficient Respiration Birds receive oxygenated air when they breathe in as well as when they breathe out. Air sacs enable birds to get more oxygen because 75 percent of the air inhaled by a bird passes directly into the air sacs rather than into the lungs. When a bird exhales, oxygenated air in the air sacs passes into the lungs. The ability to get large amounts of oxygen helps to generate the energy needed to sustain flight.

Wings Adapted for Flight Flight is also supported by feathers that streamline a bird's body and shape the wings. Wing shape and size determine the type of flight a bird is capable of. Birds that fly through the branches of trees in a forest have elliptically shaped wings adapted to quick changes of direction. Wings of swallows and terns have shapes that sweep back and taper to a slender tip, promoting high speed. The broad wings of hawks, eagles, and owls provide strong lift and low speeds. These birds are predators that carry prey while in flight.

BIODIGEST

Visual Learning

LS Logical-Mathematical Show students slides or pictures of birds in flight and ask them to hypothesize about what types of flight each bird could undertake: quick movements, high speed, or soaring. *Elliptical wings are adapted to quick movements. Wings that sweep back and taper to a slender tip promote high speed. Soaring birds have broad wings.* L1

STUDENT JOURNAL

LS Kinesthetic Provide students with contour and down feathers and a magnifying glass. Feathers may be purchased from craft shops. Ask students to examine both feathers with a magnifying glass and draw diagrams of them in their journals. Ask them to hypothesize how the function of each feather differs. *Down feathers keep birds warm by trapping warm air near their bodies. Contour feathers are used for flight.* L1 SAE

Teacher FYI

Classification of mammals changes as new information is added to the database about specific organisms. Although all mammals can be divided into three groups based upon their method of reproduction, taxonomists differ as to classification within orders. For example, the order Pinnipedia—consisting of seals, sea lions, and walruses—is sometimes combined with the order Carnivora because of similarities in teeth, and in how and what the animals eat. Recent classification schemes divide marsupials into seven orders.

Occasionally, a group is removed from one order and placed into its own new order. Elephant shrews now are classified into their own order—Scandentia.

Students who are interested in taxonomy may be referred to *Mammal Species of the World*, Second Edition, Wilson, Don E. and Dee Ann M. Reeder, eds. Smithsonian Institution Press, Washington and London, 1993.

MAMMALS

Mammals are endotherms. Mammals are named for their mammary glands, which produce milk. Most mammals have hair that helps insulate their bodies and sweat glands that help them cool off. Mammals need a high level of energy for maintaining body temperature and high speeds of locomotion. An efficient four-chambered heart and the muscular diaphragm beneath the lungs help to deliver the necessary oxygen for these activities.

Mammal Diversity

Mammals are classified into three groups by their method of reproduction. Monotremes are mammals that lay eggs. Marsupials are mammals in which the young complete a second stage of development after birth on the outside of the mother's body in a pouch made of skin and hair. Placental mammals carry their young inside the uterus until development is nearly complete.

Female mammals feed their young milk secreted from mammary glands. Often, the young are cared for until they become adults.

FOCUS ON ADAPTATIONS

Endothermy

Both birds and mammals are endotherms. Endotherms have internal processes that maintain a constant body temperature. Just as a thermostat controls the temperature of your home, internal processes cool endotherms if they are too warm, and warm them if they are too cool, thus maintaining homeostasis.

A variety of adaptations enables mammals to maintain body temperature. Hair helps many mammals conserve heat. The thick coat of a polar bear is an adaptation to living in a cold climate. Small ears and an accumulation of body fat under the skin also prevent heat loss. Small ears have less surface area than large ears from which body heat can escape.

PROJECT

LS **Interpersonal** Ask students to develop a campaign to save a local mammal from extinction. They should prepare a list of recommendations, a name for their group, publicity in the form of radio and TV commercials, bumper stickers, and a poster, and suggest an original idea to raise money to create a sanctuary. L2 P COOP LEARN

STUDENT JOURNAL

LS **Logical-Mathematical** Give students a blank map of the United States and field guides. Ask them to make a list of local mammals of the area. Ask them to look up each mammal on their list and indicate the range of each mammal on their maps. They should make a color-coded key for their map. L1 P

The duck-billed platypus is a monotreme with webbed front feet adapted to swimming, and sharp claws for digging and burrowing into the soil.

Most mammals are placental mammals. They have extraordinary ranges in sizes and body structures. Many hoofed mammals, such as this moose, have an adaptation known as cud chewing in which swallowed food is brought up later and chewed again. In this way, the cellulose of plants is broken down and nutrients are made available to the animal.

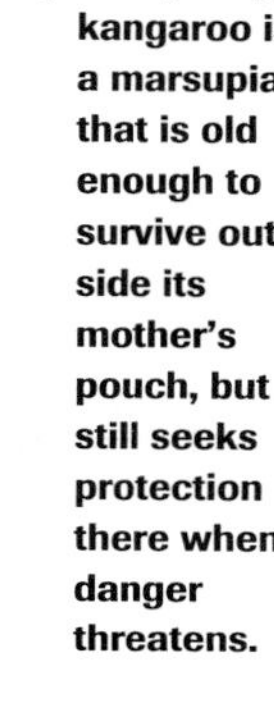

This young kangaroo is a marsupial that is old enough to survive outside its mother's pouch, but it still seeks protection there when danger threatens.

Hibernation Many rodents hibernate during periods of extreme cold. During hibernation, the body temperature lowers. For example, when the surrounding temperature drops to about 0°C, a ground squirrel's temperature drops to 2°C, and it goes into hibernation, which conserves the animal's energy.

Estivation In hot desert environments, where water is limited, some small rodents survive without drinking. They obtain enough water from the foods they eat. Other desert mammals, such as the fennec fox, have large ears that aid in heat loss. During periods of intense heat, some desert mammals go into a state of reduced metabolism called estivation. As a result, the animal's body temperature lowers and energy is conserved.

Activity

LS **Visual-Spatial** Have students sketch, on a sheet of butcher paper, a scale map of the school grounds or a spot near the school. Have students indicate how the area could be made more suitable for local wildlife. Have students, working in groups, add features to their maps that would make their area more attractive to wildlife. Books about wildlife habitat restoration are available in libraries. L1 P

COOP LEARN 

Visual Learning

LS **Visual-Spatial** Show students slides of local mammals and ask them to point out how they are adapted to their habitats. *Make sure they point out features that help them move, feed, and give them protection from predators and environmental elements.* SAE

Meeting Individual Needs

Gifted Ask students who are working above grade level to develop a wildlife-management plan for a particular mammal in your area. They should develop and carry out a census. Then, based on the results of their census, they should develop a management plan that will protect and maintain appropriate habitat for 25 years. L3 P

Visual Learning

Visual-Spatial Show students a variety of mammal skulls or transparencies of mammal skulls you have made from field guides to mammals. Direct their attention to how teeth reflect what the animal eats. Ask them to make comparisons of teeth in carnivores and herbivores. *Carnivores have sharp, pointed canines that enable them to tear apart prey. Herbivores have premolars and molars adapted to grinding the plant materials they eat.* L1

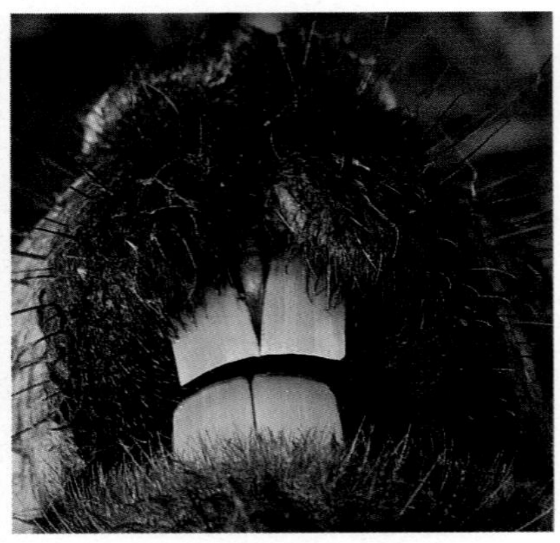

The chisel-like incisors of beavers and other rodents never stop growing.

Grazing animals, such as this donkey, rely on incisors to cut grasses and molars to grind and crush their food. You can see that donkeys, like many other herbivores, lack canine teeth.

Carnivores such as this coyote have canine teeth that stab and pierce food.

Mammal Teeth

Mammals can be classified by their number and types of teeth. All mammals have diversified teeth used for different purposes. Incisors are used to cut food. Canines—long, pointed teeth—are used to stab or hold food. Molars and premolars have flat surfaces with ridges and are used to grind and chew food.

VITAL STATISTICS MAMMALS

Size ranges: Largest: blue whale, length, 30 m, mass, 190 metric tons; smallest: Etruscan shrew, length, 6 cm, mass, 1.5 g
Fastest mammal: Cheetah, 110 kph
Longest-lived: Asiatic elephant, 80 years
Distribution: Worldwide in all habitat types.
Numbers of species:
Class Mammalia: 4563 species
Order Monotremata, egg-laying mammals: 3 species
Order Marsupialia, pouched mammals: 260 species
Orders of Placental Mammals: 4300 species

PROJECT

Interpersonal Ask a group of students to go to a nearby zoo, take photos, and present their trip in a slide show to the rest of the class. Ask them to describe any research going on at the zoo, and discuss educational programs and breeding programs for endangered animals. L2 COOP LEARN

STUDENT JOURNAL

Intrapersonal Ask students to use their creativity and knowledge of mammals to create the ideal mammal pet. They should write a description and draw a diagram of this pet in their journals. Have them label their diagrams with all features unique to mammals.

BIODIGEST REVIEW

Understanding Concepts

1. Which of the following animals are ectotherms?
   a. fishes, amphibians, reptiles, and birds
   b. birds and mammals
   c. fishes, amphibians, and reptiles
   d. fishes, amphibians, reptiles, birds, and mammals
2. Which of the following fishes have jaws?
   a. lampreys, cartilaginous fishes, bony fishes
   b. lampreys only
   c. cartilaginous fishes only
   d. bony and cartilaginous fishes only
3. Which of the following animals have eggs without shells?
   a. lizards, snakes, turtles
   b. lizards, frogs, toads
   c. frogs, toads, salamanders
   d. frogs, snakes, alligators
4. Which amphibian has thick, bumpy skin with poison glands?
   a. toads c. lizards
   b. frogs d. salamanders
5. The first animals to lay amniotic eggs were ____________.
   a. fishes c. reptiles
   b. amphibians d. mammals
6. Which reptile has a four-chambered heart?
   a. duck-billed platypus
   b. lizard
   c. snake
   d. crocodile
7. The air sacs of birds enable them to ____________.
   a. eat more food
   b. receive more oxygen
   c. hide from predators
   d. incubate eggs
8. Both birds and reptiles lay shelled, amniotic eggs; however, only birds ____________ their eggs, keeping them warm until they hatch.
   a. guard c. protect
   b. incubate d. nurse
9. The hearts of all mammals have ____________ chambers.
   a. two c. four
   b. three d. five
10. Mammals are classified into subclasses based on their method of ____________.
   a. locomotion c. feeding
   b. reproduction d. breathing

Thinking Critically

1. Why are endothermic animals able to live in areas of extreme temperatures such as the Arctic and the tropics? Explain.
2. Explain how the development of the amniotic egg was important for the transition of animals from life in the water to life on land.
3. How are reptiles and amphibians alike? How are they different?
4. If you found a mammal skull with chisel-like incisors, what type of feeding habits might this animal have?
5. Describe three structures fishes have that mollusks do not have.

Program Resources

Section Focus Master 81 L1 SAE
Reinforcement and Study Guide, pp. 149-150 L1
Chapter Assessment, pp. 225-228 L1
Content Mastery, pp. 153-156 L1

BIODIGEST

Answers to BioDigest Review

Understanding Concepts

1. c
2. d
3. c
4. a
5. c
6. d
7. b
8. b
9. c
10. b

Thinking Critically

1. When an animal can control its body temperature, it can live in habitats with temperature extremes without upsetting homeostasis of its body.
2. A food source, protective membranes and fluid, and a tough outer shell on the egg help prevent injury and dehydration of the embryo as it develops on land.
3. They are all ectotherms. All amphibians have a three-chambered heart, as do reptiles with the exception of crocodilians. Reptiles have claws on their toes while amphibians do not. Reptiles have thick, dry skin covered with scales, while amphibians have smooth, moist skin.
4. It might be an animal that gnaws on woody branches and bark.
5. Fishes have internal skeletons, fins, and lateral line systems.

UNIT 10 Human Biology

Unit Overview

Unit 10 describes the organs and systems of the human body and how they interact with one another. Chapter 37 describes the skin, skeletal, and muscular systems. Digestion and nutrition, respiration, circulation and excretion follow in Chapters 38 and 39.

Chapter 40 looks at the nervous control of the body and the effects of drugs on body systems. Included is a Focus on the Brain, which describes the brain in more detail.

Chapter 41 provides a discussion of human reproduction and development. Chapter 42 outlines the function of the immune system and ends with a discussion of how AIDS affects the immune system.

Theme Development

The major themes of this unit are *homeostasis* and *systems and interactions.* The themes are developed as students gain an understanding of how all the body systems function together.

UNIT 10 Human Biology

The crowd falls silent as the starter raises the pistol. With a bang, the runners are off—their leg muscles flexing powerfully as they race down the track. The crowd cheers. As the leader crosses the finish line, her heart is pounding hard. For a few moments, she finds it difficult to catch her breath, but when she recovers, she's exalted and ready to run another race.

Winning a race may be the result of individual effort, but any successful dash to the finish line requires the interaction of all of the body's systems—systems that don't work independently, but are guided by hundreds of complex interactions.

Magnification: 5000×

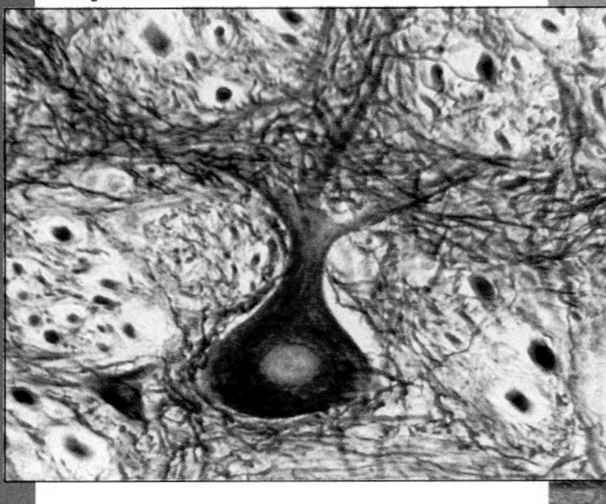

Nerve cells are part of the body's control center. How might nerve cells function during a complex activity such as running?

930

Advance Planning

Chapter 37

- Collect pictures of skin diseases and disorders for the Display Activity.

Chapter 38

- Gather food samples and prepare iodine and biuret solutions for the Biolab.

Chapter 39

- Order *Daphnia* for the Chapter 39 Microscope Activity and for the Chapter 40 Biolab.
- Purchase balloons for the Biolab.

Chapter 40

- Prepare solutions for the Biolab.

Chapter 41

- Order sea urchins for the Chapter 41 Minilab on page 1051.

Chapter 42

- Gather apples for the Minilab on page 1074.
- Order prepared blood cells for the Minilab on page 1080.
- Gather colored pencils for the Biolab.

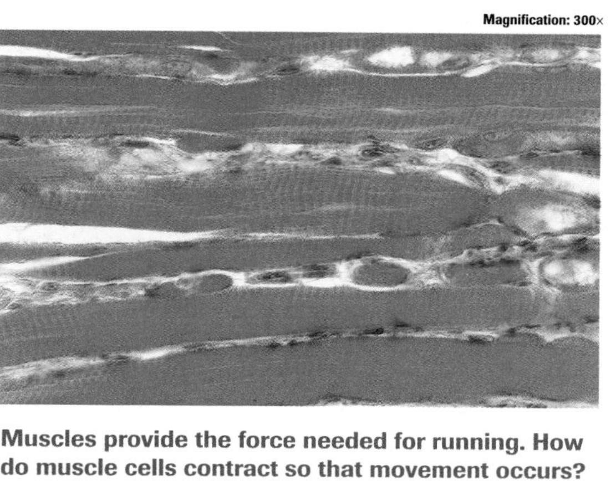
Magnification: 300×

Muscles provide the force needed for running. How do muscle cells contract so that movement occurs?

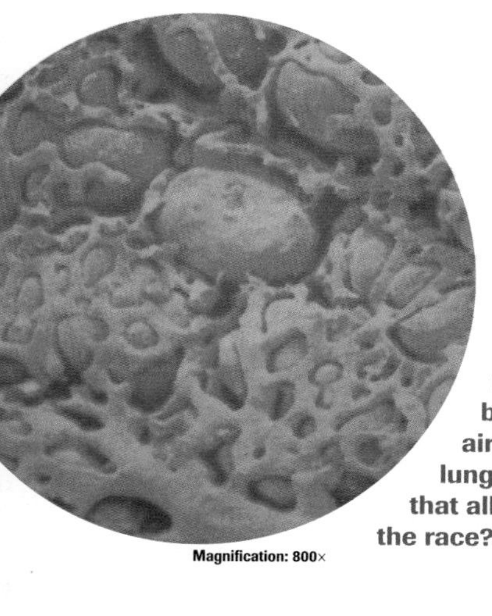
Magnification: 800×

What happens between these microscopic blood vessels and air sacs in the lungs of the runner that allows her to finish the race?

Unit Contents

UNIT 10

Introducing the Unit

Ask students what systems are important for a runner to succeed at a track meet. Then discuss the following questions with the class. What systems are involved in the excitement and tension the runner feels before the race? What systems are involved when the runner is racing?

Direct students' attention to the other photographs of nerve, muscle, and blood vessels. As a class, discuss the questions associated with each photo.

Motivational Activity

Discussion: Put up pictures of several different stages of skyscraper construction. Point out various things such as the foundation, girders, wiring, windows, exterior or interior walls, and elevators. Explain that the structure of the building depends on each of these elements and that if these elements are not working properly, the building may not be sound.

Draw an analogy between the building and the human body. Explain that, like the building, the body organs depend on other tissues and organs to keep the body in homeostasis and functioning properly. Have students brainstorm and provide analogies between parts of the building and parts of the body that have similar functions. L1 COOP LEARN

Unit Project

Class Activity: Divide the class into 11 groups. Assign each group one of the body systems (nervous, endocrine, integumentary, skeletal, muscular, digestive, respiratory, circulatory, urinary, lymphatic, and reproductive). Have groups make a large poster summarizing the role of this system in the body and demonstrating some of the major anatomical features of the system. L1 COOP LEARN

Chapter Organizer

SECTION	OBJECTIVES	ACTIVITIES/FEATURES
37.1 Skin: The Body's Protection National Science Standards: UCP.1, UCP.2, UCP.5; C.5; F.1, F.5; G.1	1. **Summarize** the importance of the skin in maintaining homeostasis in the body. 2. **Outline** the healing process that takes place when the skin is injured. 3. **Summarize** the effects that environmental factors and aging have on the skin.	**Minilab:** Does each of your fingers have a unique fingerprint?, p. 937 **A Broader View:** Acne, p. 938
37.2 Bones: The Body's Support National Science Standards: UCP.1, UCP.2, UCP.5; B.2, B.6; C.5; D.1; E.1; F.1, F.6	4. **Summarize** the structure and functions of the skeleton. 5. **Compare** the types of movable joints. 6. **Explain** how the skeleton forms.	**Physics Connection:** X rays—The Painless Probe, p. 942 **Minilab:** What structures can be seen in compact bone?, p. 944
37.3 Muscles for Locomotion National Science Standards: UCP.1-3, UCP.5; A.1, A.2; C.5; E.1, E.2; F.1, F.6; G.1	7. **Distinguish** among the three types of muscles. 8. **Explain** the structure of a myofibril and summarize the sliding filament theory.	**Thinking Lab:** How is rigor mortis used to estimate time of death?, p. 947 **Biolab:** Design Your Own Experiment—Does fatigue affect the ability to do work?, p. 948

ACTIVITY MATERIALS

BIOLAB	MINILABS	ALTERNATE LAB
page 948 watch with second hand graph paper small weights	**page 937** ink pad index cards magnifying glass nail polish remover or acetone **page 944** slides of compact bone microscope	**page 936** microscope slide of human skin

Chapter 37 Protection, Support, and Locomotion

TEACHER CLASSROOM RESOURCES

Reproducible Masters	Transparencies
Section Focus Master 82: To Cover, Contain, and Protect L1 SAE (block) **Reinforcement and Study Guide,** p. 151 L1 (block) **Biolab and Minilab Worksheets,** p. 147 L1 (block) **Content Mastery,** pp. 157-160 L1	**Basic Concepts Transparency #61:** Skin Deep L1 SAE (block) **Basic Skills Transparency #25:** Repairing Damages L1 SAE (block)
Section Focus Master 83: Flexible Support L1 SAE (block) **Reinforcement and Study Guide,** pp. 152-153 L1 (block) **Biolab and Minilab Worksheets,** p. 148 L1 (block) **Concept Mapping:** Joints in the Human Body, p. 37 L1 **Critical Thinking/Problem Solving:** Joints in Action, p. 37 L3 **Laboratory Manual:** The Skeletal System, pp. 231-234 L2	**Basic Concepts Transparency #62:** Skeleton and Joints L1 SAE (block) **Basic Concepts Transparency #63:** Structure of Bone L1 SAE (block)
Section Focus Master 84: Built to Move L1 SAE (block) **Reinforcement and Study Guide,** p. 154 L1 (block) **Biolab and Minilab Worksheets,** pp. 149-150 L1 (block) **Tech Prep Applications:** Biceps Biomechanics, pp. 41-42 L2	**Reteaching Transparency #37:** The Muscular System L1 SAE (block) **Basic Concepts Transparency #64:** Muscle Contraction L1 SAE (block)

ASSESSMENT MATERIALS

Chapter Assessment, pp. 229-234 (block)
Performance Assessment in the Biology Classroom, p. 51
MindJogger Videoquiz (block)
Alternate Assessment in the Science Classroom
Computer Test Bank

Spanish Resources SAE
English/Spanish Audiocassettes SAE
Cooperative Learning in the Science Classroom COOP LEARN
Lesson Plans (block)

KEY TO TEACHING STRATEGIES

L1 Level 1 activities should be within the ability range of all students including those with learning difficulties.

L2 Level 2 activities are within the ability range of average to above-average students.

L3 Level 3 activities are designed for the ability range of above-average students.

SAE SAE activities should be within the ability range of Students Acquiring English.

COOP LEARN Cooperative Learning activities are designed for small group work.

P These strategies represent student products that can be placed into a best-work portfolio.

(block) These strategies are useful in a block scheduling format.

GLENCOE TECHNOLOGY

The following multimedia resources are available from Glencoe.

Biology: The Dynamics of Life
CD-ROM SAE
Videodisc Program (block)

The Infinite Voyage Series
Unseen Worlds
Miracles by Design
The Champion Within

National Geographic Society Series
STV: Human Biology Vol. 2
Muscular and Skeletal Systems
Newton's Apple: Life Sciences
Hip Replacement

Science and Technology Videodisc Series (STVS)
Human Biology
Orthopedic Implants
Arthritis Research

CHAPTER

37 Protection, Support, and Locomotion

Chapter Overview

Students begin their study of human body systems by exploring the structure and functions of the skin. They examine what occurs when the skin is injured and learn how the body restores this organ to health.

The second section of the chapter acquaints students with the structure and functions of the skeletal system. Students are introduced to joints and their role in movement. They also examine the homeostatic maintenance and aging of the skeletal system.

The last section of the chapter presents the three types of muscles and their characteristics. The current sliding filament theory of skeletal muscle contraction is explained. Students also learn the importance of exercise in maintaining a healthy muscular system.

Key Terms

actin
appendicular skeleton
axial skeleton
bursa
cardiac muscle
compact bone
dermis
epidermis
hair follicle
involuntary muscle
joint
keratin
ligament
marrow
melanin
myofibril
myosin
osteoblast
sarcomere
skeletal muscle
sliding filament theory
smooth muscle
spongy bone
tendon
voluntary muscle

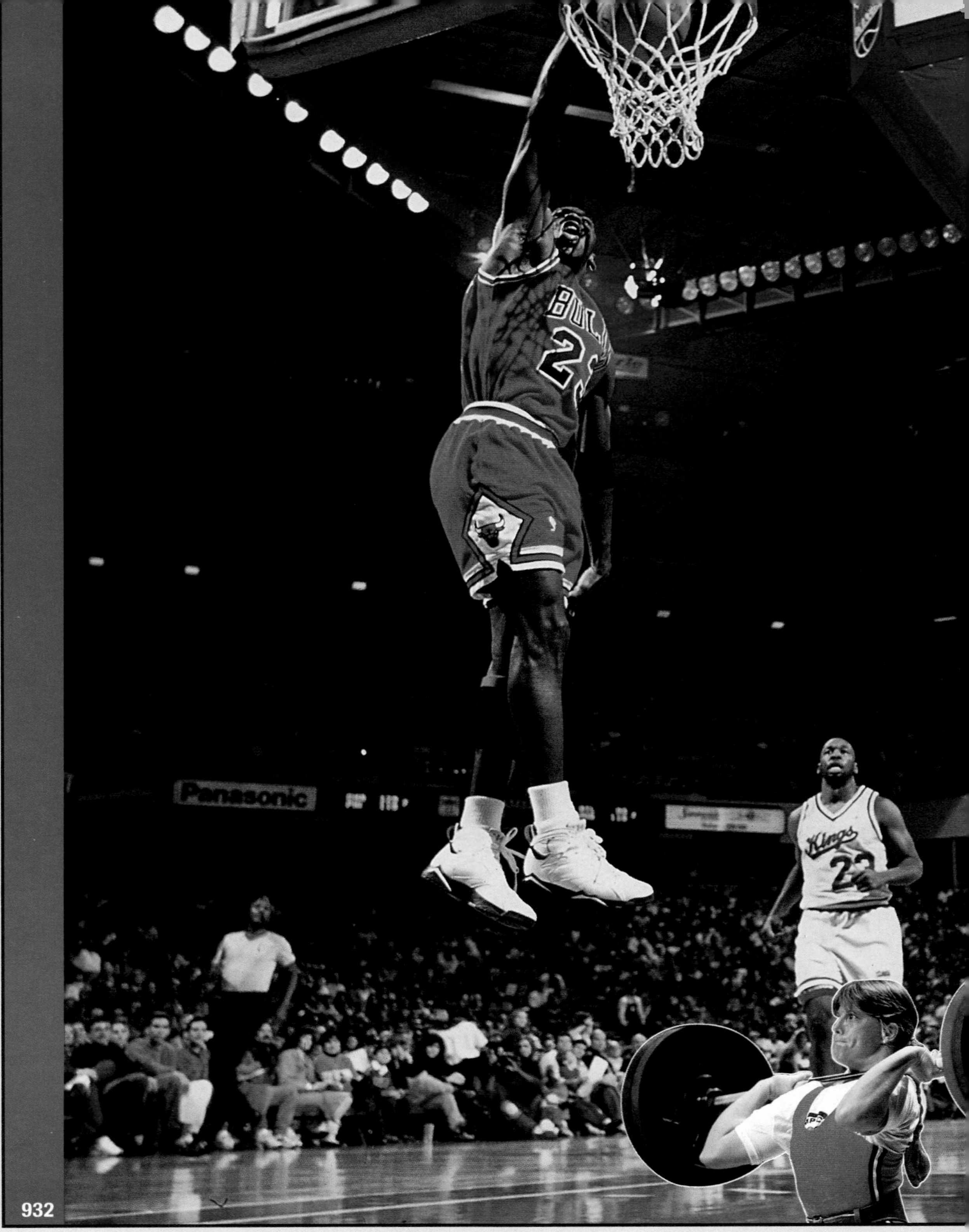

932

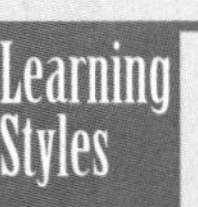

Learning Styles

Look for the following logo for strategies that emphasize different learning modalities. LS

Kinesthetic	Portfolio, p. 935
Visual-Spatial	Display, p. 935; The Inside Story, p. 936; Alternate Lab, p. 936; Minilab, pp. 937, 944; Using an Analogy, p. 941; Demonstration, pp. 941, 943, 947; Portfolio, p. 944; Visual Learning, pp. 950, 951
Interpersonal	Project, p. 950
Linguistic	Using Science Terms, p. 944
Logical-Mathematical	Thinking Lab, p. 947

CHAPTER 37 Protection, Support, and Locomotion

Some people believe Michael Jordan can defy the laws of gravity. With his body drenched in sweat, the team feeds him the ball; he dribbles, lunges, and hangs in the air, seemingly forever. Down comes his hand. SLAMMMM! Another basket and the fans are on their feet.

How do you suppose Jordan is able to perform such feats? His conditioning probably involves running to increase muscle tone and strengthen parts of his skeletal system, as his body's natural cooling system, skin, works overtime. Perhaps he lifts weights to increase the mass of the muscles that move his skeleton into the incredible positions that you see. Jordan may be an athlete who has reached peak physical condition, but even if you are not an athlete, your bones, muscles, and skin work in the same way. In this chapter, you'll investigate the structure of these important parts and learn how they function to protect, support, and move your body.

Concept Check

You may wish to review the following concepts before studying this chapter.

- Chapter 8: tissue, organ, organ system
- Chapter 10: aerobic respiration, lactic acid fermentation

Chapter Preview

Laboratory Activities

Biolab: Design Your Own Experiment

- Does fatigue affect the ability to do work?

Minilabs

- Does each of your fingers have a unique fingerprint?
- What structures can be seen in compact bone?

Animals differ from plants and fungi in that they are able to move from place to place. In vertebrates, the muscular and skeletal systems are mainly responsible for this function of locomotion, and skin protects muscles and bones and helps maintain homeostasis. How are bones, muscles, and skin adapted for these important processes?

Introducing the Chapter

Direct students' attention to the photograph of Michael Jordan. Elicit whether Michael Jordan can truly hang in the air longer than anyone else. *No* Explain that the amount of time any hurled object, remains in the air is directly related to the height of the object's trajectory (curved path). The higher the trajectory, the longer the object stays airborne. Much of Michael's power for his high jumps come from his strong thigh muscles.

Theme Development

The major themes of the chapter are *homeostasis* and *systems and interactions*. The restoration of the skin after injury and the control of the calcium level in the bones and blood illustrate homeostasis. Systems and interactions is emphasized as each system is discussed in terms of interactions that occur within the system and with other systems.

Concept Check

This chapter expands upon the students' understanding of tissues, organs, and organ systems. Students also expand upon their knowledge of aerobic respiration and lactic acid fermentation.

Assessment Planner

Choose assessment strategies from the following pages to evaluate the progress of your students.

Assess, pp. 939, 945, 952
Alternate Lab, pp. 936-937
Minilabs, pp. 937, 944
Portfolio, pp. 935, 944, 947
Thinking Lab, p. 947
Biolab, pp. 948-949
Chapter Review, pp. 953-955

SECTION

37.1

Prepare

Key Concepts

Students will examine how skin structures perform their functions and how the skin helps maintain homeostasis. Students will also consider the effects injuries and factors such as sunburn and aging have on the skin.

Block Scheduling

Look for this symbol for strategies that are useful in a block scheduling format. For more information on block scheduling, refer to Section 37.1 in the **Lesson Plans** booklet.

Materials

- Collect pictures of skin diseases and disorders for the Display. Materials for this display may be available through the American Cancer Society or from a local dermatologist.
- Gather ink pads, index cards, and hand lenses for the Minilab.

1 Focus

Bellringer

Before presenting the lesson, display **Section Focus Master 82** on the overhead projector and have students answer the accompanying questions. L1 SAE

Discussion

Ask students to compare the skin that covers various parts of their bodies and describe whether the skin is the same in all places. *Likely responses will state that in different parts of the body, skin has different thicknesses, varies in moisture, is tight or loose, and is either covered with hair or hairless.* L1

SECTION

37.1 Skin: The Body's Protection

Section Preview

Objectives

Summarize the importance of the skin in maintaining homeostasis in the body.

Outline the healing process that takes place when the skin is injured.

Summarize the effects that environmental factors and aging have on the skin.

Key Terms

epidermis
keratin
melanin
dermis
hair follicle

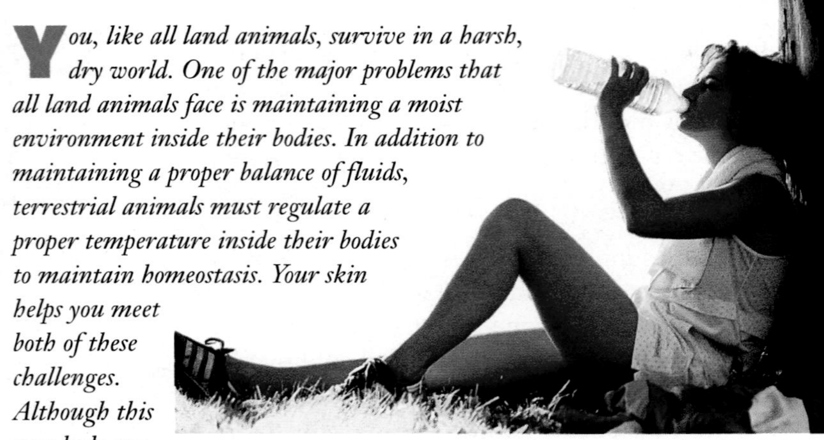

You, like all land animals, survive in a harsh, dry world. One of the major problems that all land animals face is maintaining a moist environment inside their bodies. In addition to maintaining a proper balance of fluids, terrestrial animals must regulate a proper temperature inside their bodies to maintain homeostasis. Your skin helps you meet both of these challenges. Although this vast body covering may seem like just a wrapping on the surface of the body, you'll see that it is a complex organ that performs a variety of functions.

Structure and Function of the Skin

Skin is composed of many layers of cells of each of the four types of body tissues: epithelium, connective, muscle, and nerve. Recall from Chapter 28 the origins of different tissue types during embryo development. Epithelial cells are derived from the ectoderm layer of the embryo and function to cover surfaces of the body. Connective tissue is a fibrous tissue usually made of collagen and elastin protein fibers, and it connects cell layers to each other. Nerve tissue helps to detect stimuli from the external environment. Muscles in skin are associated with hairs and respond to some stimuli such as cold and fright. As a result of these four types of tissues, skin is an elastic, flexible, and responsive organ.

epidermis:
epi (GK) on
derma (GK) skin
The epidermis covers other layers of skin.

Skin is composed of two principal layers—the epidermis and dermis. Each layer performs a different function in the body.

Epidermis—The outer layer of skin

The layer of skin that you see covering your body is called the epidermis. The **epidermis** is the outer, thinner portion of the skin. It's composed of layers of both dead and living cells. The top layer consists of 25 to 30 layers of flattened, dead cells that are continually shed. Although dead, these cells still serve an important function. Dead epidermal cells contain a protein called **keratin** that helps waterproof and protect the living cell layers underneath.

Program Resources

Section Focus Master 82 L1 SAE
Reinforcement and Study Guide, p. 151 L1
Biolab and Minilab Worksheets, p. 147 L1
Basic Concepts Transparency 61 and **Master** L1 SAE
Basic Skills Transparency 25 and **Master** L1 SAE

interNET CONNECTION

Follow the link for this chapter on the Glencoe Homepage at **www.glencoe.com/sec/science** to find out more about skin, bones, and muscles.

The inner layer of the epidermis contains living cells that continually divide by mitosis to replace the dead cells. These cells contain **melanin,** a cell pigment that colors the skin and protects the cells from damage by solar radiation. As the newly formed cells are pushed up toward the surface, the nuclei degenerate and the cells die. Eventually, as they reach the top, they are shed.

Look at your fingertips. The epidermis on the fingers and palms of your hands and on the toes and soles of your feet contains ridges and grooves formed before birth, *Figure 37.1*. The epidermal ridges are important for gripping because they increase friction. Prints of these patterns are used to identify individuals.

Dermis—The inner layer of skin

The second major component of the skin is the dermis. The **dermis** is the inner, thicker portion of the skin. The thickness of the dermis varies in different parts of the body, depending on the function of that part. For example, dermis is 3 to 4 mm thick on the palms of the hands and soles of the feet, providing padding and protection. The 0.5-mm thickness of skin on the surface of the eye allows it to be transparent.

The dermis is adapted to a broader range of functions than those of the epidermis. These adaptations include structures such as blood vessels, nerves, nerve endings, sweat glands, and oil glands. Have you ever wondered why some areas of your skin are looser and more flexible than others? It's because fat deposits lie underneath the dermis in the subcutaneous layer. The amount of fat deposited in this layer varies in different areas of the body and among individuals. These deposits function to cushion the body, to insulate and help the body retain heat, and to store food for long periods of time.

Although skin is a well-adapted organ of protection, you know that skin is not a continuous, unbroken layer. Hair, for example, grows out of narrow cavities in the dermis called **hair follicles.** As hair follicles develop, they are supplied with blood vessels and nerves and become attached to muscle tissue.

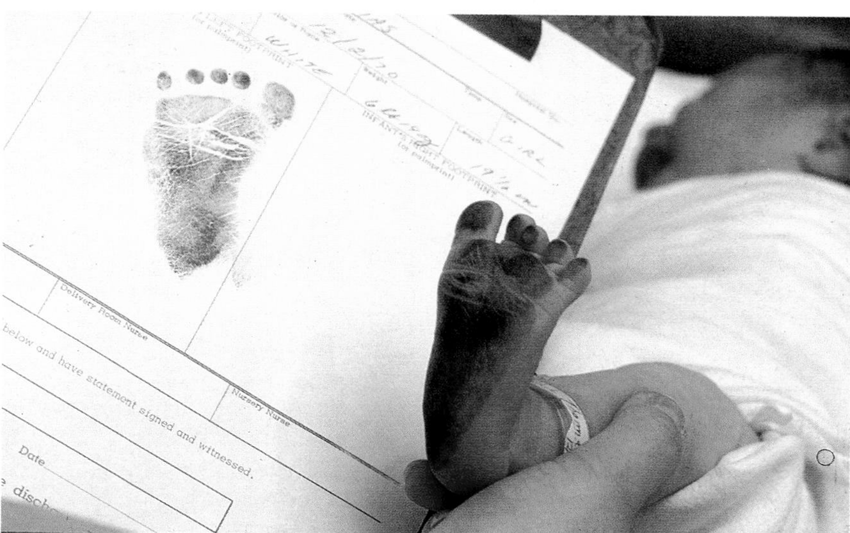

Figure 37.1

Babies have their footprints recorded at birth to establish an identification record. The patterns of fingerprints, footprints, and even ear prints are genetically determined and are often used for identification.

SECTION 37.1

2 Teach

Display

IS **Visual-Spatial** Set up a bulletin board display that shows the various kinds and stages of skin cancer and other skin diseases. If available, include a poster that lists the warning signs of skin cancer. Materials about skin cancer can be obtained free of charge from the American Cancer Society.

Enrichment

Have students conduct research on artificial skin. Students should report on how artificial skin is made and how its structure compares with that of natural skin. L2 P

Audiovisual

Show *The Skin: Its Structure and Function*, EBEC. L1

Science, Technology, and Society

Brushing Up on Hair

Hair growth remains a part of biology that is not well understood. An area of current research focuses on the cycle of hair growth. Individual human hairs grow for periods of two to five years. The hairs then rest for four to six months. Scientists do not understand what makes hair grow or stop growing. One hypothesis as to why hair stops growing suggests that immunologic factors may be involved. During a hair's resting phase, hair follicle cells gain cell surface markers that can trigger an attack by the immune system. Scientists hope that by studying the hair cycle, they will learn to control hair growth and regeneration. Someday immune-altering drugs may help prevent baldness or enhance hair growth.

PORTFOLIO

Modeling Skin Structure Have students use a variety of materials to create three-dimensional models of the structure of the skin. Encourage students to use the illustration of skin structure in the *Inside Story* as a guide. Students should include labels that identify each structure in their models as well as its function. L1 P IS

THE INSIDE STORY

Skin Deep

Purpose

LS **Visual-Spatial** Students will gain a further understanding of how structures of the skin perform their functions.

Visual Learning

- Direct students' attention to the diagram of the skin. Discuss the functions of the skin, emphasizing that skin is much more than a body wrapping. As students examine the structures in the skin, ask them to relate a function to each structure. L1
- If you set up the bulletin board display showing skin problems, challenge students to identify which skin structure is involved in each problem shown in the display. L1

Audiovisual

Show the video *The Integument*, Charles Clark Co., Inc.

THE INSIDE STORY

Skin Deep

The skin is an organ because it consists of tissues joined together to perform specific activities. It is the largest organ of the body; the average adult's skin occupies one to two square meters.

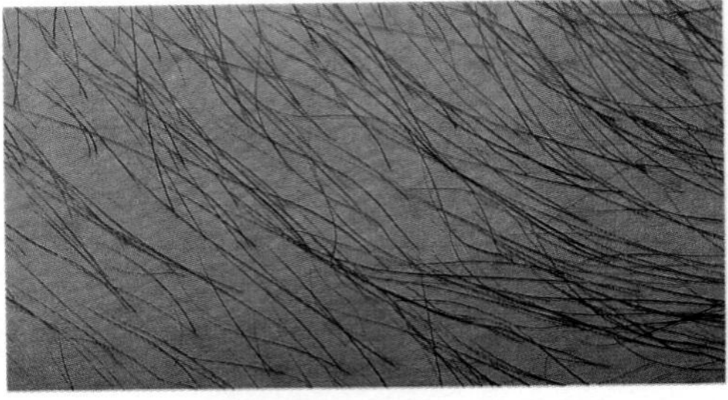

Surface view of skin

1. All people have the same number of melanin-filled cells. Differences in skin color are due to the amount of pigment produced by the cells. Exposure to sunlight causes an increase in melanin production, and the skin becomes darker.

2. Most oil glands are connected to hair follicles. Skin oil is a mixture of fats, cholesterol, proteins, and inorganic salts. Oil keeps hair from drying out and keeps the skin soft and pliable. Oil also inhibits the growth of certain bacteria.

4. Hair's primary function is protection of the skin from injury and sun. When hairs stand up, as occurs when you get goose bumps, they provide an insulating layer of air on the surface of the skin, thus reducing heat loss.

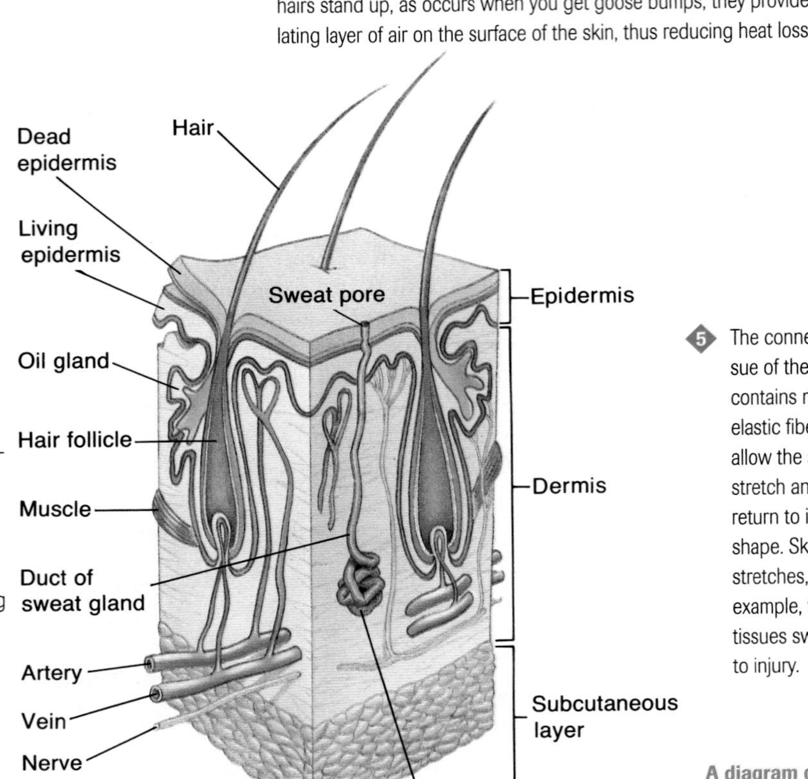

5. The connective tissue of the dermis contains many elastic fibers that allow the skin to stretch and then return to its original shape. Skin stretches, for example, when tissues swell due to injury.

A diagram of human skin

3. Sweat glands are located in the dermis and open up through pores onto the surface of the skin. Sweat is a mixture of water, salt (mostly NaCl), sugar, lactic acid, ascorbic acid (vitamin C), and small amounts of organic waste products. A person usually loses about 900 mL of sweat each day, depending on the type of activity and the temperature and humidity of the environment.

Alternate Lab: Structure of the Skin

Purpose

LS **Visual-Spatial** Students will observe the structure of human skin.

Materials

microscope slide of human skin, microscope

Procedures

Give students the following directions.

1. View the human skin slide on low power. Focus first on the epidermis and switch to high power.
2. Switch back to low power and locate the dermis of the skin. Switch to high power.
3. Find the following structures: hair follicle, hair shaft, sweat gland, blood vessels, fat deposits, oil gland, and elastic fibers.
4. Notice that the outer portion of the dermis forms numerous projections, similar to hills and valleys. These projections, or papillae, form no pattern except on the fingertips, palms, and soles of the feet, where they form parallel ridges that improve frictional characteristics in these areas.

The skin's vital functions

Skin, shown in *Figure 37.2*, has several vital roles in maintaining homeostasis. Think about how your body warms up as you exercise. A major function of skin is to regulate your body temperature. When your body heats up, the many capillaries in the dermis dilate, blood flow increases, and body heat is lost by radiation. This mechanism also works in reverse. When you are cold, the blood vessels in the skin constrict and heat is conserved.

Another noticeable thing that happens to your skin as your body heats up is that it becomes wet. Glands in the dermis produce sweat in response to increases in body temperature. As sweat evaporates, the body cools. This is because when water changes state from liquid to vapor, heat measured in Calories is lost.

Another function of skin is to serve as a protective layer to underlying tissues. Skin protects the body from physical and chemical damage and from the invasion of microbes.

Of course, anyone who has ever stepped on a sharp object or been burned by the sun knows that skin also functions as a sense organ. Nerve cells in the dermis receive stimuli from the environment and relay information about pressure, pain, and temperature.

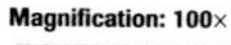
Magnification: 100×

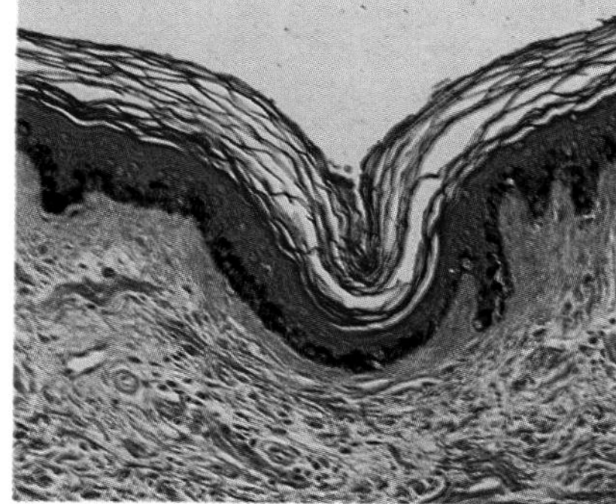

MiniLab

Does each of your fingers have a unique fingerprint?

Fingerprints play a major role in any police or detective story investigation. Because a fingerprint is an individual characteristic, extensive FBI fingerprint files are used for identification in criminal cases.

Procedure

1. Press the tip of your thumb onto the surface of an ink pad.
2. Roll your thumb from left to right across the corner of an index card, then immediately lift your thumb straight up from the paper.
3. Repeat the steps above for your other four fingers, placing them in order on different spots on the card.
4. Examine your fingers with a magnifying glass, and identify the patterns of each fingerprint by comparing them with the diagrams.
5. Compare your fingerprints with those of all others in the class.

Arch

Whorl

Loop

Combination

Analysis

1. Are the fingerprint patterns the same on your five fingers?
2. Do any of your fingerprints have the same patterns as those of a classmate?
3. Why is a fingerprint a good way to identify a person?

Skin also functions to maintain the balance of chemicals in the body. When exposed to ultraviolet light, dermis cells produce vitamin D, a nutrient that aids the absorption of calcium into the bloodstream. Because an individual's exposure to sunlight varies, a daily intake of vitamin D may be needed.

Figure 37.2

Photomicrograph of a cross section of human skin

MiniLab

Purpose

LS Visual-Spatial Students will observe, describe, and compare similarities and differences among fingerprint patterns to determine the uniqueness of such patterns.

Process Skills

observe and infer

Teaching Strategies

- Students may need to use acetone to clean their fingers. Fingernail polish remover may also be used.

Safety Precautions

- Be sure students do not breathe the fumes of acetone or fingernail polish remover. When these substances are in use, the room should be well ventilated. Keep chemicals away from open flames.

Expected Results

Each finger has a unique fingerprint and each individual has a unique set of fingerprints.

Analysis

1. Although each finger has a unique fingerprint, the general patterns may be the same on different fingers.
2. It is possible that some students may share the same pattern, but not identical fingerprints.
3. It is unique for each person.

Assessment

Performance: Have students design an experiment to determine if the print patterns of the toes are unique to each individual. Have students work in cooperative groups to carry out their experiments. L1 COOP LEARN

Analysis

Ask students to answer the following questions.

1. What structures form the fingerprints on the fingers? *The outer portion of the dermis forms projections, or papillae, that form fingerprints.*
2. What is the function of the oil glands of the skin? *Oil secreted by the glands keeps hair from drying out and keeps the skin soft and pliable.*
3. How does sunlight affect the elastic fibers of the skin? *Sunlight causes the elastic fibers to lose their elasticity.*

Assessment

Journal: Have students draw, label, and color the human skin as seen under the microscope. Students should place their drawings in their journals. L1

SECTION 37.1

Bioethics

A Healthy Tan?
Many people claim that the coloring provided through tanning gives them a "healthy" look. However, scientists have found that "looks" may be deceiving. Tanning and burning are actually harmful to the skin. For example, the UV rays given off by the sun have been linked to skin cancer.

Have students discuss the relative pros and cons of various tanning lotions and sun blocks. Have them address the problems that might be encountered by using tanning salons year-round. Refer students to the Health Connection in Chapter 11 on skin cancer. L1 COOP LEARN

A BROADER VIEW

Acne

Purpose
Students investigate the causes and possible treatments for acne.

Teaching Strategies
- After students read *A Broader View,* ask them to describe how each of the four factors (hormones, oil, bacteria, and abnormal development within hair follicles) contributes to the development of acne. Then ask them to answer the *Thinking Critically* question. L1

Thinking Critically
Antibiotics will kill the bacteria that contribute to acne. Synthetic hormones can decrease oil production. Using medicines that remove skin cells helps to remove excess cells around hair follicles. Cleaning reduces bacteria and removes dead skin cells.

A BROADER VIEW

Acne

Have you ever had a pimple or blackhead? Acne is a disorder of the hair follicles and oil glands that affects almost everyone at some time or another. The exact cause of acne is unknown, but it is believed that four factors are involved: hormones, oil, bacteria, and abnormal development within hair follicles.

How acne starts An acne lesion begins when the epidermis produces too many cells surrounding a hair follicle. These cells then stick to each other, forming a mass that mixes with oil and blocks the hair follicle. A blackhead results if the opening of the hair follicle is blocked by the accumulating cells and oil. A pimple develops if the wall of the hair follicle ruptures, stimulating an immune response that results in a red pimple filled with pus. Acne is more of a problem during the teen years because the increase in sex hormones increases the secretions of oil into the hair follicles.

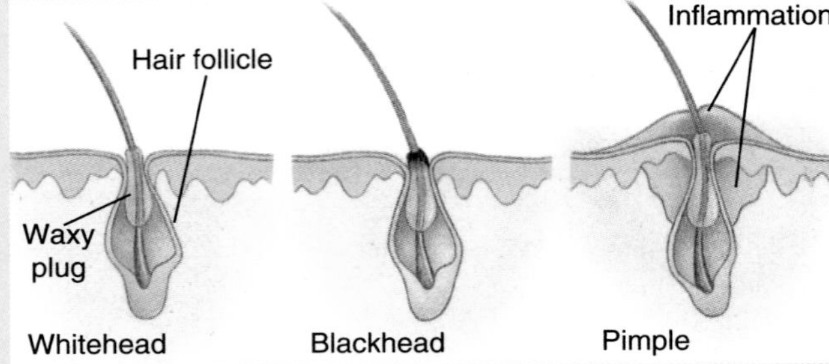

Keeping acne at bay Acne can be treated by regular cleansing of the skin. By keeping skin clean, acne-causing bacteria living on the skin can be destroyed, and dead cells and oil can be reduced.

Sometimes physicians prescribe drugs to control acne. Antibiotics to manage bacteria and synthetic hormones to decrease the production of oil are often used. Over-the-counter medicines that produce peeling of the skin are also available. These include salicylic acid, benzoyl peroxide, and synthetic vitamin A.

Thinking Critically
Consider each form of treatment. How does each control acne?

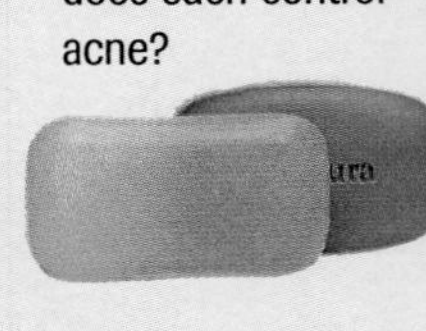

Figure 37.3
When skin is injured into the dermis, the first reaction of the body is to restore the continuity of the skin. This prevents the invasion of harmful bacteria that live on the skin.

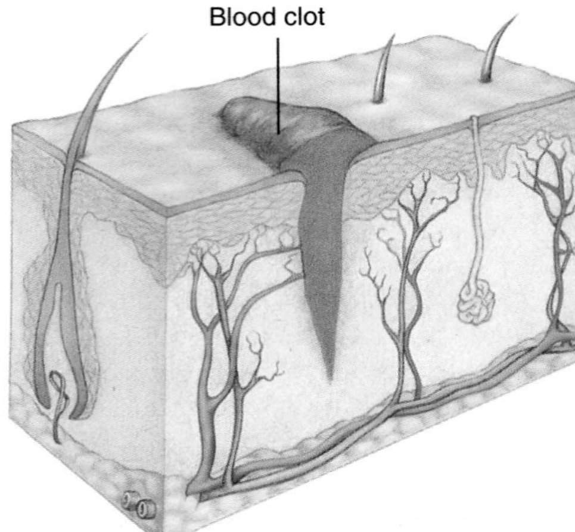

A Blood flows out of the wound until a clot forms over it.

Skin Injury and Restoration of Homeostasis

If you've ever had a mild scrape on your arms or legs, you know that it doesn't take long for the scrape to heal. When small injuries occur to the epidermis, such as a scrape, epidermal cells divide by mitosis and fill in the gap left by the abrasion. However, when the skin is injured more deeply into the dermis, bleeding usually occurs. The skin goes through a series of stages to heal itself and maintain the body's homeostasis. *Figure 37.3* shows the stages involved in skin repair.

Have you ever suffered a painful burn? Burns can result from exposure to the sun or contact with chemicals or hot objects. Burns are rated according to their severity. Most people have received a first-degree burn at one time or another.

GLENCOE TECHNOLOGY

Videodisc

The Infinite Voyage: Miracles by Design
Burn Patients and Artificial Skin (Ch. 5)

B The wound is closed by the formation of a scab. The scab unites the wound edges to prevent bacteria from entering. Blood vessels dilate, and infection-fighting cells speed to the wound site.

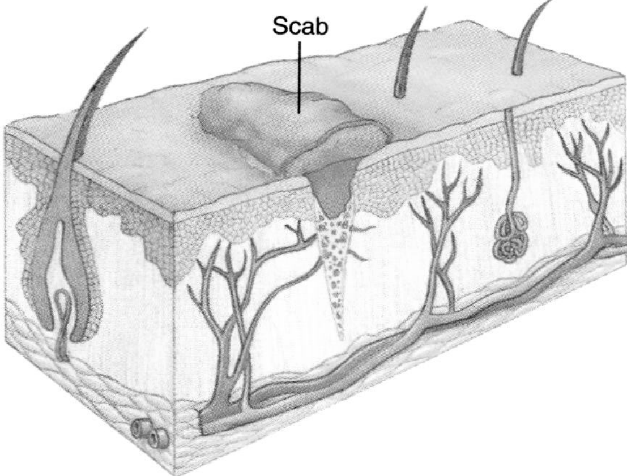

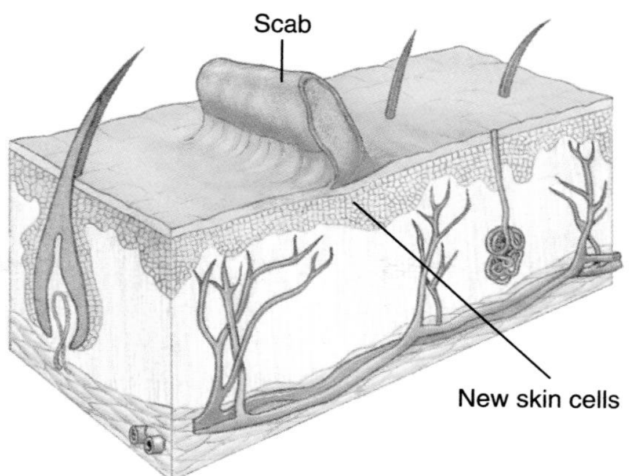

C Skin cells beneath the scab begin to multiply and fill in the gap. Eventually, the scab falls off to expose new skin. If a wound is large, a scar may result from the formation of large amounts of dense connective tissue fibers.

First-degree burns are characterized by redness and mild pain and involve the death of epidermal cells. Second-degree burns involve damage to skin cells of the dermis and can result in blistering and scarring. The most severe burns are third-degree burns, which destroy the epidermis and dermis. With this type of burn, skin function is lost, and regrowth of the skin is slow with much scarring. Skin grafts may be required to replace lost skin. In most cases, skin can be removed from another area of the patient's body and transplanted to a burned area.

As people get older, the appearance and function of the skin changes. Aging of the skin is evidenced by an increase in wrinkles and sagging. Wrinkles appear because the skin becomes less elastic with age. As aging progresses, the oil glands produce less oil and the skin becomes drier. These changes are natural, but prolonged exposure to ultraviolet rays from the sun can damage skin cells and accelerate the aging process.

Section Review

Understanding Concepts

1. Compare the structures of the epidermis and dermis.
2. How does skin control body temperature?
3. How does skin protect the underlying tissue layers from environmental factors?

Thinking Critically

4. Why is it uncomfortable to wear wool and nylon in hot climates and cotton in cold climates?

Skill Review

5. **Sequencing** Outline the steps that occur when a cut in the skin heals. For more help, refer to Organizing Information in the *Skill Handbook*.

SECTION 37.1

3 Assess

Check for Understanding

On the chalkboard, list the structures of the skin and ask students to copy the list. Have students identify the function of each structure. L1

Reteach

Have students organize the following information in a table. The thickness of the epidermis on the back is 0.25 mm, on the face and scalp 0.12 mm, on the palm 0.5 mm. The thickness of the dermis on the back is 3.75 mm, on the face and scalp 1.6 mm, and on the palm 1.0 mm. Discuss why skin thickness varies in different parts of the body by relating this phenomenon to the function of the skin in each area. L1

Assessment

Journal: Ask students to write a paragraph listing the functions of the skin as they relate to skin structures. L1

4 Close

Discussion

Discuss the relationship between the structures of the skin and the skin's functions. Relate the presence of sweat glands, hair, oil glands, nerves, and blood vessels to the functions of skin. Ask students what problems a burn victim faces. *infection, fluid loss* L1

Answers to Section Review

1. The epidermis is made of outer layers of dead epithelial cells and inner layers of live epithelial cells. The dermis contains live epithelial cells surrounding blood vessels, nerves, nerve endings, sweat glands, oil glands, and hair follicles.
2. by producing sweat and by constriction and dilation of skin capillaries
3. The skin layers provide protection against water loss, physical and chemical damage, and microbe invasion.

Thinking Critically

4. Wool and nylon do not easily permit air to reach the skin to cool the body as one perspires. Cotton easily allows air to pass through it. Thus, air reaches the skin to cool the body.

Skill Review

5. (1) formation of blood clot, (2) blood vessels dilate and infection-fighting blood cells rush to the wound, (3) epidermal cells divide to fill in the wound

SECTION

37.2

Prepare

Key Concepts

Students will become acquainted with the structure of bones and their functions. They will be introduced to joints and will compare the types of movable joints. Students also discover how the skeleton forms.

Block Scheduling

Look for this symbol for strategies that are useful in a block scheduling format. For more information on block scheduling, refer to Section 37.2 in the **Lesson Plans** booklet.

Materials

- Gather slides of compact bone for the Minilab.
- Obtain X rays showing different bones, joints, and fractures or knee and hip replacements from hospitals or doctors for the Focus Demonstrations.
- Borrow a human skeleton from the health department of a local college for the Demonstration on page 941 and Meeting Individual Needs.
- Save two toilet paper tubes for Using an Analogy.
- Soak a chicken bone to demineralize it for the first Demonstration on page 943.
- Have beef bones cut by a butcher for the second Demonstration on page 943.

SECTION

37.2 Bones: The Body's Support

Section Preview

Objectives

Summarize the structure and functions of the skeleton.

Compare the types of movable joints.

Explain how the skeleton forms.

Key Terms

axial skeleton
appendicular skeleton
joint
ligament
bursa
tendon
osteoblast
compact bone
spongy bone
marrow

Because bones are hard, many people believe that they are not living tissue. Would you believe that you had more bones when you were born than you have now? That's because bones are living tissue and have grown together since you were born. Your head, for example, had soft spots when you were an infant. Your head feels solid now because the bones of the skull have fused together. Remodeling of the skeleton occurs throughout life. In fact, your skeleton hasn't completely fused yet. You will not have a solid, fused skeleton until about age 25.

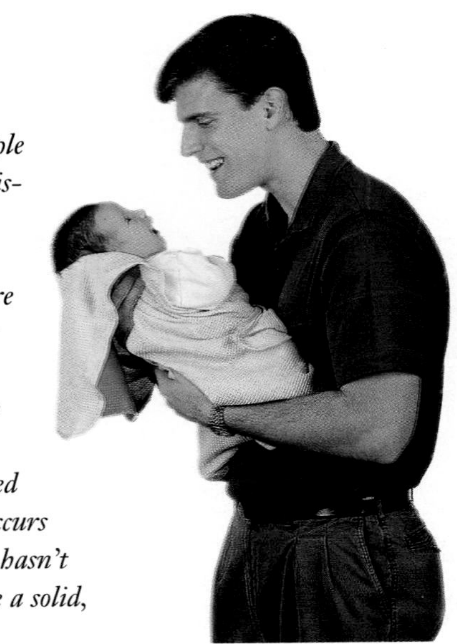

Structure and Function of the Skeletal System

The adult human skeleton has 206 named bones and is composed of two main parts, as shown in *Figure 37.4*. The **axial skeleton** includes the skull and the bones that support it, such as the vertebral column, ribs, and sternum. The **appendicular skeleton** includes the bones of the arms and legs and structures associated with them, such as the shoulders and pelvic girdle.

Joints—Where bones meet

Every time you open a door, you're using a joint. A door is connected to a door frame by a hinge joint. In vertebrates, a **joint** is where two or more bones meet. Most joints allow movement, but some, such as the joints of the skull, are fixed. Fixed joints are joints that do not move. *Figure 37.4* shows the types of movable joints in the skeleton and the different actions that they allow.

Joints are supported and surrounded by structures that aid in their movement. The joints are held together and often enclosed by ligaments. A **ligament** is a tough band of connective tissue that connects bones to bones. Joints with large ranges of motion, such as the knee, typically have more ligaments surrounding them. In addition, the ends of bones are covered with a layer of cartilage, which allows for smooth movement between them. In movable joints such as the shoulder and knee joints, there is also a fluid-filled sac called a **bursa** located between the bones. The bursa acts as a cushion to absorb shock and keep bones from rubbing against each other. **Tendons,** which are thick bands of connective tissue, attach muscles to bones.

Program Resources

Section Focus Master 83 L1 SAE
Reinforcement and Study Guide, pp. 152-153 L1
Biolab and Minilab Worksheets, p. 148 L1
Laboratory Manual, pp. 231-234 L2
Concept Mapping, p. 37 L1
Critical Thinking/Problem Solving, p. 37 L3
Basic Concepts Transparencies 62, 63 and **Masters** L1 SAE

Skull
Clavicle
Scapula
Sternum
Ribs
Humerus
Ulna
Radius
Carpals
Metacarpals
Phalanges
Vertebrae
Pelvis
Femur
Patella
Tibia
Fibula
Tarsals
Metatarsals
Phalanges

Axial Skeleton
Appendicular Skeleton

Figure 37.4

The axial skeleton includes the bones of the head, back, and chest. The bones of the skull and face contain the sinuses, which are cavities lined with cells that are continuous with the nasal cavity. Bones in the appendicular skeleton are related to movement of the limbs. To which skeletal group do the phalanges belong?

◆ **Body movements are made possible by joints that move in several different ways.**

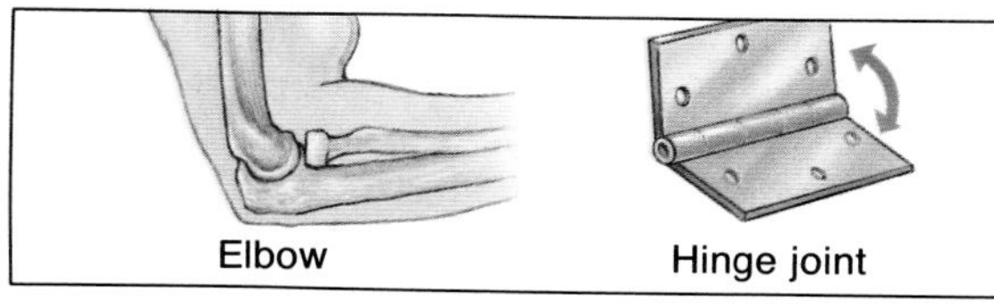

Hinge joints are present in the elbows, knees, fingers, and toes. They allow back-and-forth movement like that of a door hinge.

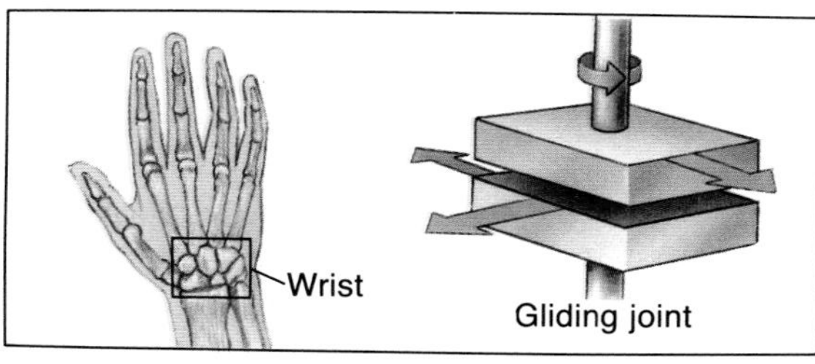

Gliding joints occur in the wrists and ankles and allow bones to slide against each other.

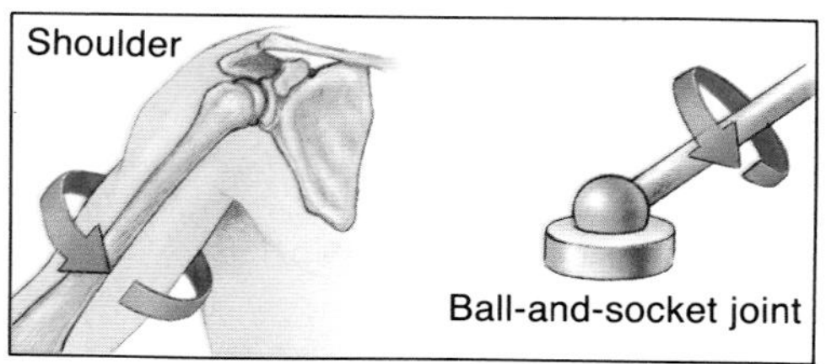

Ball-and-socket joints allow rotational movements. The joints of the hips and shoulders are ball-and-socket joints.

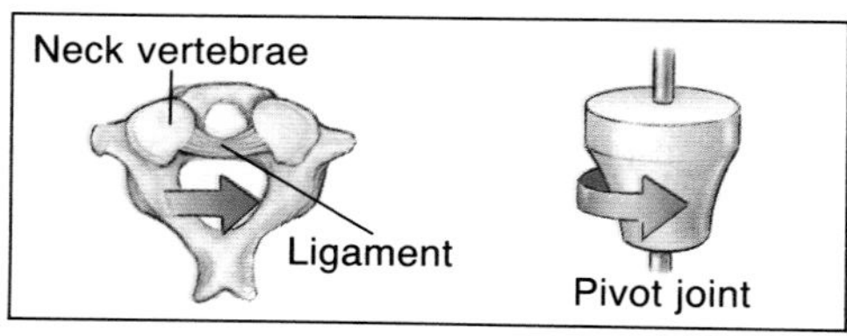

Pivot joints allow bones to twist against each other. The joint between the first two vertebrae of the neck is a pivot joint.

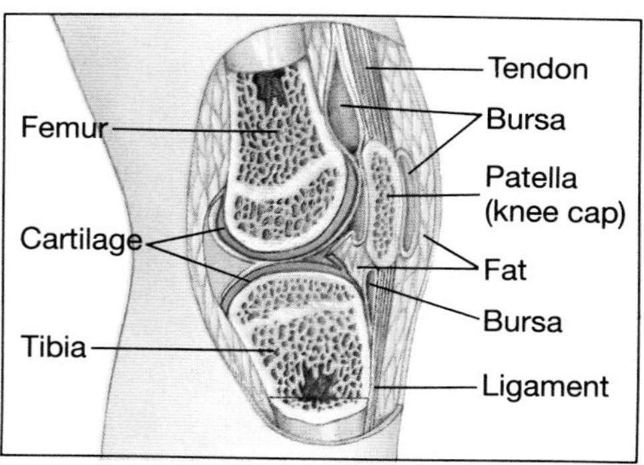

Structures of the knee aid in its movement and reduce wear and tear.

SECTION 37.2

1 Focus

Bellringer

Before presenting the lesson, display **Section Focus Master 83** on the overhead projector and have students answer the accompanying questions. L1 SAE

Demonstration

Project X rays or hang them in windows on the day that you begin discussing the skeleton. Identify the bones shown in each X ray and the types of information X rays can reveal about bones.

2 Teach

Visual Learning

Figure 37.4 To which skeletal group do the phalanges belong? *appendicular*

Using an Analogy

LS **Visual-Spatial** To demonstrate the strength of hollow bones, balance a heavy book on the open ends of two toilet paper tubes placed side by side. L1

Misconception

Many students believe that males have fewer ribs than females. Explain to students that males and females have an equal number of ribs. L1

Demonstration

LS **Visual-Spatial** Point out the differences between a male and female pelvis by using a skeleton.

Tying to Previous Knowledge

Compare the anatomy of the front limbs of a frog, bird, cat, whale, and bat with that of the human arm. Compare the number, shapes, and arrangements of bones. L1

CULTURAL DIVERSITY

Human Skeletal Variation

Variation exists in human body form, especially within the skeletal system. Differences in skeletal morphology are related to the geographical origins of populations. For example, a leaner body form is often observed in people who live in arid regions where greater skin-surface area in proportion to body weight facilitates heat loss. A stockier build is more adaptive for inhabitants of cold areas. Differences in body form and skin color have been used as a justification for racism. Have students write short essays about how their attitudes of racism have been affected by knowledge of the influences of human variation. L1 P

Audiovisual

Show the video *Locomotion and Skeletons*, Charles Clark Co., Inc. L1

Physics

X Rays—The Painless Probe

Purpose

Students will learn about some uses of X rays in medicine.

Teaching Strategies

- Ask students to share experiences about when they have had X rays taken. Ask students to explain why the X rays were needed. L1

Possible Answer

An X ray will show the image of a broken bone because X rays don't pass through dense bone. It won't show if a ligament is sprained because the X rays pass through soft tissue and can't show a sprain.

Different Viewpoints in Biology

Osteoporosis

Osteoporosis is a condition in which the bones become porous and thin due to reduced bone mass. Factors that increase the risk of osteoporosis include being a woman, being Caucasian, chronic low calcium intake, lack of exercise, being underweight, and smoking.

One treatment for osteoporosis is estrogen replacement therapy, which reduces further bone loss. However, this therapy increases the risk of endometrial cancer, abnormal bleeding, and other medical complications. Some scientists suggest that any activity involving the muscles working against gravity may increase bone strength in postmenopausal women and reduce bone loss.

Physics

X rays—The Painless Probe

X rays are a form of light emitted by X-ray tubes and by some astronomical objects such as stars. The short wavelength and high energy of X rays make it possible to discern a broad range of objects from the tiniest atoms to the largest galaxies. X-ray machines are so common that you have probably had contact with one recently. Dentists use them to examine teeth, doctors to inspect bones and organs, and airports to look inside your carry-on baggage.

Non-invasive diagnosis In medicine, X rays are passed through soft tissues to photographic film placed behind the area; bones and other dense objects show up as white areas on the film. The position or nature of a break is clearly visible. The contours of organs such as the stomach are seen when a patient ingests a high-density liquid; other organs can be marked with special dyes. CAT scans are 3-D images made by X-raying an area many times at different angles. They are useful for viewing the brain and observing tissues shadowed by overlying organs.

Radiation treatments As X rays bombard atoms of tissues, electrons are knocked from their orbits, resulting in damage to the exposed tissue cells. To protect healthy tissues, absorptive metals are used as shields. You've probably had a dental X ray where the dental assistant spread a heavy lead apron across your chest. The destructive nature of X rays has proven useful in the treatment of cancers, where cancerous cells are targeted and destroyed.

CONNECTION TO Biology

Why would an X ray be useful to diagnose a fractured bone but not a sprained ligament?

Forcible twisting of a joint, called a sprain, can result in injury to the bursa, ligaments, and tendons. A sprain most often occurs at joints with large ranges of motion such as the wrist, ankle, and knee.

Besides injury, joints can also be subjected to disease. One common joint disease is arthritis, an inflammation of the joints. It can be caused by infections, aging, or injury. One kind of arthritis results in large bone spurs or bumps of bone inside the joints. Such arthritis is painful, and joint movement may become limited. The causes of all types of arthritis are not completely known.

The formation of bone from cartilage

The skeleton of a vertebrate embryo is made of cartilage. By the ninth week of human development, bone begins to replace cartilage. The cartilage framework formed during embryonic life is covered by a membrane. Blood vessels penetrate the membrane and stimulate cells in the cartilage to enlarge and become potential bone cells called **osteoblasts.** These bone cells secrete a protein called collagen in which minerals are deposited. The deposition of calcium salts and other ions hardens in the cartilage and transforms it to bone. The adult skeleton is almost all bone, with cartilage found in regions such as the nose tip, external ears, discs between vertebrae, and lining the movable joints.

Bones grow in both length and diameter. Growth in length occurs at the ends of bones in cartilage plates. Growth in diameter occurs on the outer surface of the bone. The increased sex hormones produced during the teen years cause the

Meeting Individual Needs

Visually/Hearing Impaired Have visually impaired and hearing impaired students work with other students to review the names and functions of various bones of the body. For visually impaired students, have an articulated human skeleton available for this activity. L1 COOP LEARN

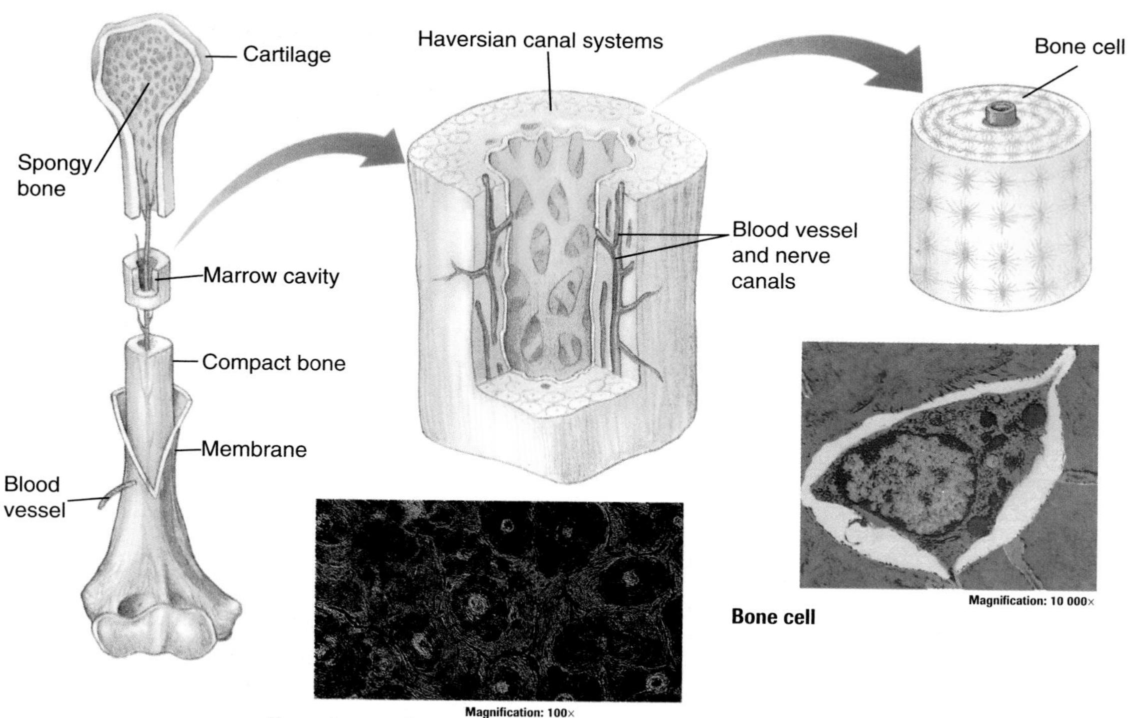

Figure 37.5

A bone has several components, including compact bone, spongy bone, and a surrounding membrane. Notice in the cross section (center) that the compact bone looks like a series of rings. The rings, along with the canals they surround, make up what are called Haversian canal systems. Bone cells (right) receive oxygen and nutrients from small blood vessels running within the Haversian canals. Nerves in the canals conduct impulses to and from each bone cell.

osteoblasts to divide more rapidly, resulting in a growth spurt. By age 20, 98 percent of the growth of the skeleton is completed. However, these same hormones also cause the growth centers at the ends of bones to degenerate. As these cells die, growth slows. After growth in length stops, bone-forming cells are involved mainly in repair and remodeling of bone.

The compact and spongy structure of bone

Although bones may appear uniform, they are actually composed of two different types of bone tissue, as shown in *Figure 37.5*. Surrounding every bone is a layer of hard bone, or **compact bone.** Compact bone is covered by a nerve and blood vessel-filled membrane that supplies nutrients and oxygen to bone cells. Compact bone surrounds less-dense bone known as **spongy bone,** so called because it is filled with many holes and spaces, like those seen in a sponge. The center cavity of a bone is filled with a soft tissue called **marrow.** Marrow fills the cavities of the ribs, sternum, vertebrae, skull, and the long bones of the arms and legs.

SECTION 37.2

Demonstration

Demineralize a chicken bone by soaking it in vinegar or 10 percent HCl for several days. Have students compare a regular bone to the demineralized bone. L1

Enrichment

Have students use a medical dictionary to find what is meant by the term *slipped disk* or *herniated disk*. Have students find out the causes of these problems and report their findings to the class. L1

Demonstration

LS **Visual-Spatial** Have a beef bone cut in half longitudinally and another one cut crosswise. Point out the compact bone, spongy bone, and marrow. If the bone is fresh, you can show the periosteum by pulling it away from the bone. L1

GLENCOE TECHNOLOGY

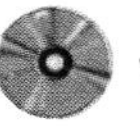

Videodisc

STVS: Human Biology
Disc 7, Side 2
Arthritis Research (Ch. 19)

The Infinite Voyage: Unseen Worlds
Digital X Rays, 3-D X Rays: Detection Made Easy (Ch. 8)

STUDENT JOURNAL

Marfan's Syndrome Have advanced students report on Marfan's syndrome, which affects connective tissue, bones, muscles, and ligaments. It is possible that Abraham Lincoln was mildly afflicted by this syndrome. Have students relate this syndrome to how it affects normal functioning of the musculoskeletal system. L3

MiniLab

Purpose

Visual-Spatial Students will observe a Haversian canal system and identify the parts of the system.

Process Skills

observe and infer, recognize and use spatial relationships

Teaching Strategies

- Have students work in pairs, with each student observing the slide.
- Have students use only the low-power objective. Bone slides often are too thick to fit under a high-power objective.

Expected Results

Students will locate a Haversian canal system similar to that in the illustration.

Analysis

1. location of blood vessels and nerves that service bone cells; carry fluids between blood vessels and bone cell
2. This system provides pathways for blood vessels in compact bone.
3. in the compact bone

Assessment

Journal: Have students compare their drawings with that shown in the text. Ask students to summarize the similarities and differences between the drawings. L1

MiniLab

What structures can be seen in compact bone?

The skeletal system of vertebrates is composed of bones like those in your skeleton. Bone-forming cells originate in cartilage and secrete bone tissue. Pathways for blood vessels and nerves form in the bone.

Procedure

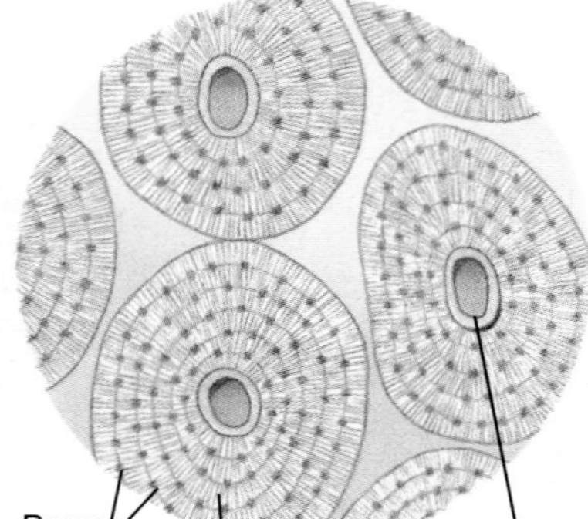

1. Obtain a slide of compact bone, and use the low-power objective to focus on the tissue.
2. Compact bone is made up of many Haversian canal systems. Look at a Haversian canal system shown in the illustration, and locate a similar structure on your slide.
3. A Haversian canal is located in the center of each system. Locate these structures on your slide. Blood vessels and nerves are found in each Haversian canal.
4. Bone cells are embedded in mineral salts in layers around Haversian canals. Find these layers on your slide. Within each layer are small canals called canaliculi that carry fluids between the blood vessels and bone cells.

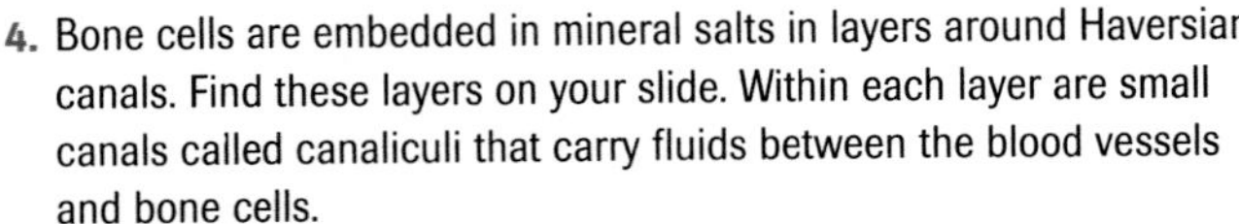

5. Draw the Haversian canal system as observed on your slide. Label a Haversian canal, a bone cell, and canaliculi.

Analysis

1. What is the function of Haversian canals? The canaliculi?
2. How does the Haversian canal system allow for efficient delivery of oxygen and nutrients to bone cells?
3. Where do you find the majority of calcium salts in the Haversian canal system?

Figure 37.6
Dairy products and leafy vegetables are good sources of calcium.

Skeletal system functions

The primary function of your skeleton is to provide a framework for the tissues of your body, much like a building's inner, steel framework. The skeleton is also designed to protect the internal organs including the heart, lungs, and brain.

Besides support and protection, the skeleton is designed for efficient movement. Joints allow for this movement, and muscles that move the skeleton need firm points of attachment to pull against so they can work efficiently.

Bones are also responsible for producing blood cells. Red marrow—found in the humerus, femur, sternum, ribs, vertebrae, and pelvis—produces red blood cells, some white blood cells, and cell fragments that are involved in blood clotting. Yellow marrow, found in many other bones, stores fats and aids in producing red blood cells when there is a massive blood loss due to severe injury.

Finally, the bones of the skeleton serve as storehouses for minerals, including calcium and phosphorus. Calcium is a critical part of the diet for healthy, strong bones. Sources of calcium are shown in *Figure 37.6*. Calcium is also important for transmission of nerve impulses and muscle contraction. However, most of these minerals are used during growth and formation of the skeleton.

Homeostasis, Aging, and the Skeletal System

Bones will remodel themselves throughout life. While osteoblasts maintain their action of depositing calcium, another group of bone cells acts to remove calcium. The calcium removed from bone is needed by many tissues of the body, such as

Using Science Terms

Linguistic Discuss the derivation of the word *osteoporosis* with students. Ask students to find other terms in the chapter that make use of the prefix *osteo-*. Ask them to explain how the prefix relates to the meaning of each term. L1 SAE

PORTFOLIO

Joint Movement Have students gather examples of everyday objects that model the movement of each type of joint. Have students describe on index cards the type of movement each object allows and to identify the type of joint that allows this movement in the body. Encourage students to include a sketch of the modeled joint on their index cards. L1 P LS

1. The axial skeleton includes the skull and the bones that support it and is centrally located in the body. The appendicular skeleton includes the bones associated with the appendages.
2. ball-and-socket, shoulder or hip; pivot joint, between the first two vertebrae; hinge joint, elbow, knees, fingers and toes; fixed, skull bones; gliding, wrists and ankles

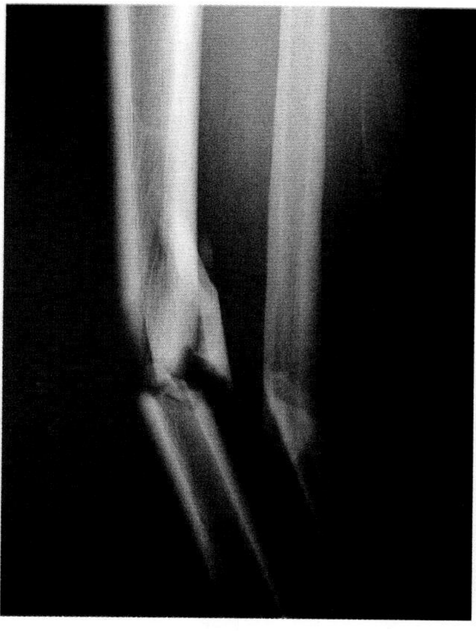

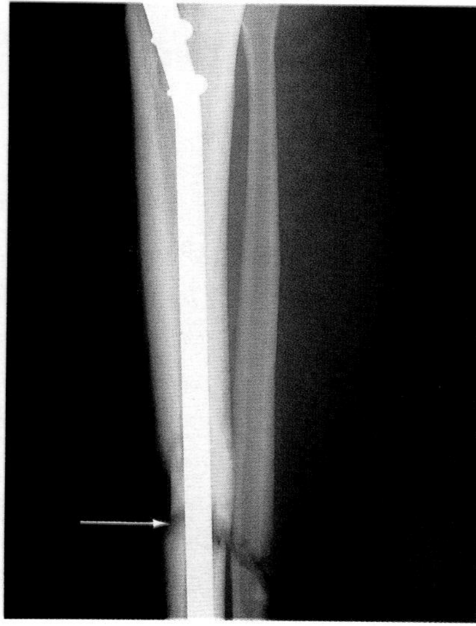

Figure 37.7

The X ray on the left shows a leg bone that has completely fractured. The X ray on the right shows the bone with a supporting rod and after the bone has healed. The arrow indicates the area where the fracture healed.

nerves and muscles, to perform their functions. The removal of calcium also prevents bone from becoming thick and heavy. This remodeling of the skeleton occurs as you age, gain or lose weight, or change your level of activity.

Bone cells are also involved in the process of repair when a bone is broken. Perhaps you have injured a bone and had the injured area set. When bones are broken, *Figure 37.7*, a doctor moves them back into place, and they are kept immobile until the bone tissue regrows.

The composition of bone changes as a person ages. Minerals are continually deposited in the bones. These minerals, such as calcium and phosphorus, make the bones hard. A child's bones have more collagen protein and fewer minerals than the bones of adults and, as a result, are less brittle. Bones tend to become more brittle as their composition changes with age. For example, a disease called osteoporosis is a condition in which there is loss of bone mass. The reduced bone mass results in the bones becoming more porous and brittle. Osteoporosis is most common in older women because they produce lesser amounts of a hormone that aids in bone formation.

Section Review

Understanding Concepts

1. Distinguish between the appendicular skeleton and the axial skeleton.
2. List the five main kinds of joints and provide an example of each.
3. In what way do bones help regulate mineral levels in the body?

Thinking Critically

4. Why would it be impossible for bones to grow from within?

Skill Review

5. **Sequencing** Outline the steps involved in bone formation from cartilage to bone. For more help, refer to Organizing Information in the *Skill Handbook*.

3 Assess

Check for Understanding

Have students fill in the names of the bones and label an example of each of the major types of joints on a diagram of the skeleton. L1

Reteach

Have students make flash cards containing the names of bones. On the opposite side of each card, have students describe where each bone is located and identify its common name. Have students work in pairs to review the information on the flash cards. L1 SAE

Extension

Have students conduct research on the various types of arthritis and their causes and treatments. Have students construct a table of their findings. L1

Assessment

Portfolio: Provide students with an outline showing the bones of the human skeleton. Have students create a color key to show which bones make up the appendicular skeleton and which make up the axial skeleton. P

4 Close

Activity

Play bone and bone anatomy Password. Divide the class into teams and have four students play at a time. After they guess the word, switch players. L1 SAE COOP LEARN

Answers to Section Review

3. Minerals are stored in bones and removed when needed by other tissues in the body.

Thinking Critically

4. The inflexible structure of compact bone will not allow growth from the inside.

Skill Review

5. (1) embryo skeleton is cartilage (2) 9th week bone begins to replace cartilage (3) within the cartilage, osteoblasts secrete a material in which calcium salts and other ions are deposited and harden to form bone (4) 20 years of age—all cartilage that will be replaced has been replaced.

SECTION

37.3

Prepare

Key Concepts

Students will learn to distinguish among the three types of muscles. Emphasis is placed on the structure and function of skeletal muscle and the current sliding filament theory of muscle contraction.

Block Scheduling

Look for this symbol for strategies that are useful in a block scheduling format. For more information on block scheduling, refer to Section 37.3 in the **Lesson Plans** booklet.

Materials

- Obtain chicken feet and wings for the Demonstration.

1 Focus

Bellringer

Before presenting the lesson, display **Section Focus Master 84** on the overhead projector and have students answer the accompanying questions. L1 SAE

Discussion

Ask students which skeletal muscle they think is the strongest and why they think it is the strongest. Ask them to explain their answers by giving examples. *They will probably choose the thigh or biceps muscles due to their size.* L1

2 Teach

Audiovisual

Color 35-mm slides can be purchased to demonstrate the cellular features of the three basic muscle types.

SECTION

37.3 Muscles for Locomotion

Section Preview

Objectives

Distinguish among the three types of muscles.

Explain the structure of a myofibril and summarize the sliding filament theory.

Key Terms

smooth muscle
involuntary muscle
cardiac muscle
skeletal muscle
voluntary muscle
myofibril
myosin
actin
sarcomere
sliding filament theory

Perhaps you have seen the Olympic games on television. Think about the different athletes involved in the games. You may be able to tell what sports some of them participate in just by looking at their body shapes. For example, swimmers and ice skaters have different shapes because ice skaters develop strong leg muscles over many months of training, whereas swimmers develop larger shoulder muscles.

Three Types of Muscles

A muscle consists of a mass of protein fibers grouped together. Almost all of the muscle fibers you will ever have were present at birth. Nearly half of your body mass is muscles. The muscles in the body are of three main kinds, *Figure 37.8.* One type of muscle, **smooth muscle,** is found in internal organs and blood vessels. Smooth muscle is made up of sheets of cells that are ideally shaped to form a lining for organs such as the digestive tract and reproductive tract. The most common function of smooth muscle is to squeeze, exerting pressure on the space inside the tube or organ it surrounds. Contractions of smooth muscle are slow and prolonged compared with the contractions of the other two kinds of muscles. These contractions are not under conscious control, so smooth muscle is an example of an **involuntary muscle.**

Figure 37.8

Muscles are under either voluntary or involuntary control and differ in their structure and appearance.

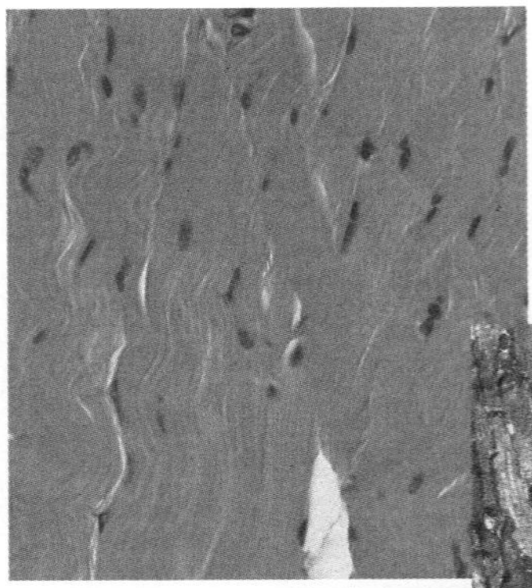

Smooth muscle cells that make up involuntary muscle are spindle shaped and have a single nucleus.

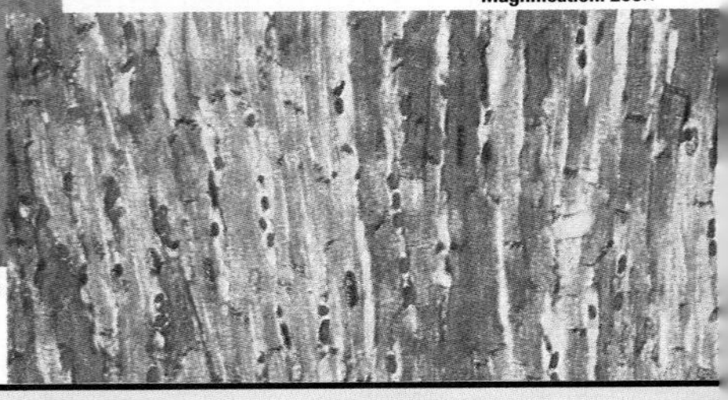

Magnification: 200×

Cardiac muscle cells, which also form involuntary muscle, appear striated or striped when magnified.

Magnification: 200×

Program Resources

Section Focus Master 84 L1 SAE
Reinforcement and Study Guide, p. 154 L1
Biolab and Minilab Worksheets, pp. 149-150 L1
Performance Assessment in the Biology Classroom, p. 51 L1 P
Tech Prep Applications, pp. 41-42 L2
Reteaching Transparency 37 and **Master** L1 SAE
Basic Concepts Transparency 64 and **Master** L1 SAE

Another type of involuntary muscle is the muscle that makes up the heart, or **cardiac muscle.** Cardiac muscle cells are interconnected and form a network that helps the heart muscle contract efficiently. Cardiac muscle is found only in the heart, and is adapted to conduct the electrical impulses necessary for rhythmic contraction. The third type of muscle makes up the largest mass of muscle in the body, skeletal muscle. **Skeletal muscle** is the type of muscle that is attached to bones and moves the skeleton. Skeletal muscle cell contractions are short and strong, providing the force needed for movement. The majority of the muscles in your body are skeletal muscles, and, as you know, you can control their contractions. A muscle that contracts under conscious control is called **voluntary muscle.**

ThinkingLab Draw a Conclusion

How is rigor mortis used to estimate time of death?

When a person dies, the respiratory and circulatory systems that obtain and deliver oxygen to the muscles stop functioning. Rigor mortis is the stiffening of both voluntary and involuntary muscles due to lack of ATP after death. This condition generally develops about an hour or two after death, and full rigor mortis takes 10 to 12 hours. Stiffening usually proceeds from the upper body to the lower body, and it is believed to be related to fiber length. Rigor mortis affects smaller fibers in the face area first; medium-length fibers in the neck, chest, and arms next; and longer fibers in the legs last. As time proceeds, rigor mortis gradually subsides in the same order it develops. It disappears completely after 24 to 36 hours.

Analysis

Coroners determine the time of death, based on the degree of rigor mortis, by examining the body and its tissues under a microscope.

Thinking Critically

A murder victim is found with relaxed face muscles, but a stiff upper and lower body. How long might it have been since the crime?

Skeletal Muscle Contraction

Whether you are playing tennis, pushing a lawn mower, or writing, your muscles are contracting as they do work. When the muscle contracts, the bones are pulled by tendons. *Figure 37.9* shows the movement of the lower arm.

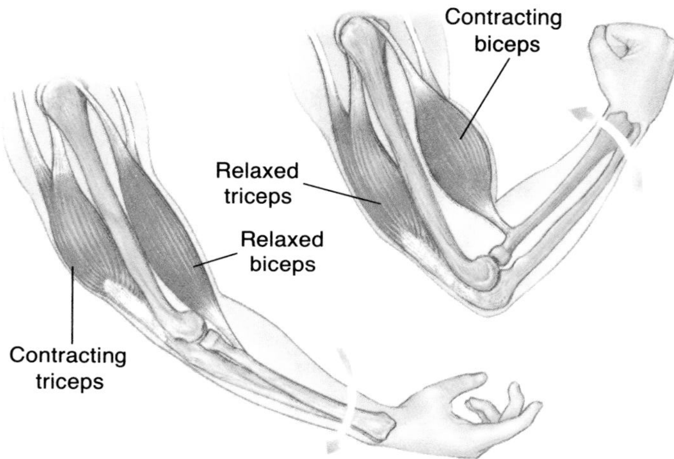

Skeletal muscle is also striated but makes up voluntary muscle. Each skeletal muscle cell has more than one nucleus and may be up to 30 cm long in tall humans.

Figure 37.9

When the biceps muscle contracts (top), the lower arm is moved upward. When the triceps muscle on the back of the upper arm contracts (bottom), the lower arm moves downward. Muscles work in opposing pairs, like the biceps-triceps pair, to move bones.

ThinkingLab Draw a Conclusion

Purpose

LS **Logical-Mathematical** Conclude how rigor mortis is used to estimate time of death.

Process Skills

analyze, draw a conclusion

Teaching Strategies

- Draw an outline of a human body on the chalkboard. On the diagram, trace the path of the stiffening of muscles in the body that indicates the progression of rigor mortis.

Thinking Critically

Accept answers in the range of 12-16 hours, reflecting the initial subsiding of rigor mortis from the head down.

Assessment

Knowledge: Have students list the order in which rigor mortis subsides in the body and indicate the approximate number of hours represented by this relaxing of muscles. L1

Demonstration

LS **Visual-Spatial** Obtain some chicken feet from a butcher. Cut the upper end so that the tendons are free. Pull on each tendon to demonstrate the movement of the toes. Use a chicken wing to demonstrate muscle pairing in the same manner.

PORTFOLIO

Sports Medicine Ask students to prepare and conduct an interview with a person who works in the field of sports medicine or in physical education. Likely candidates for interviews include sports coaches or assistant coaches, physical therapists, trainers, or dance instructors. Students should be encouraged to record their interviews on audiotape or videotape and should include questions on the type of training the specialists have and how their programs relate to muscle fitness. Have students find out what types of sports injuries may require the use of replacement structures such as artificial joints. L1 P

SECTION 37.3

BioLab Design Your Own Experiment

Does fatigue affect the ability to do work?

Time Allotment 

One class period

Objectives

Review objectives with students before they begin the Biolab.

Process Skills

observe and infer, communicate, measure in SI, predict, form a hypothesis, design an experiment

PREPARATION

Alternate Materials

Stopwatches are not necessary if a clock with a second hand is visible to all students.

Possible Hypotheses

- If a muscle becomes fatigued, then the muscle will not be able to do as much work.
- If a muscle becomes fatigued, its capacity to do work will not be diminished.

GLENCOE TECHNOLOGY

Videodisc

The Infinite Voyage: The Champion Within
Exercise Programs: The Good, The Bad (Ch. 3)

Biology: The Dynamics of Life
Disc 2, Side 1
Paired Skeletal Muscles
(Ch. 29)

BioLab Design Your Own Experiment

Does fatigue affect the ability to do work?

The movement of body parts results from the contraction and relaxation of muscles. In this process, muscles use energy from aerobic respiration and lactic acid fermentation. When exercise is continued for a long period of time, the waste products of fermentation accumulate and muscle fibers are stressed, causing fatigue. Fatigue affects the various muscles differently. It also affects individuals differently, even when they are performing the same tasks. Muscular strength, muscular endurance, and the amount of effort required to perform a task are variables to consider.

PREPARATION

Problem

How does fatigue affect the number of repetitions of an exercise you can accomplish? How do different amounts of resistance affect rate of fatigue?

Hypotheses

Hypothesize whether muscle fatigue affects the amount of work muscles can accomplish. Consider whether fatigue can occur within minutes or hours.

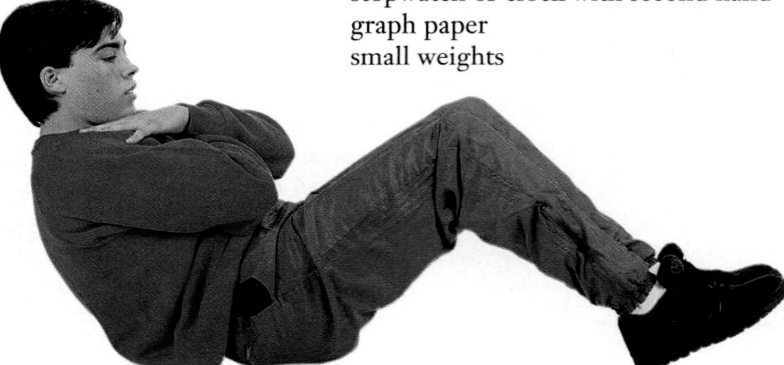

Objectives

In this Biolab, you will:

- **Hypothesize** whether muscle fatigue affects the amount of work muscles can accomplish.
- **Measure** the amount of work done by a group of muscles.
- **Prepare** a graph to show the amount of work done by a group of muscles.

Possible Materials

stopwatch or clock with second hand
graph paper
small weights

PLAN THE EXPERIMENT

Teaching Strategies

- One student should act as timekeeper. The exerciser should count the number of times he or she can carry out the exercise in a given time period, such as three minutes.
- Trials should be performed as closely together as possible, so that little muscle rest occurs between trials.
- Place a sample graph on the chalkboard to help students prepare their graphs.

Possible Procedures

- Students should choose an exercise such as how many times they can lift a weight in 3 minutes. They should repeat the experiment four or five times so that the muscles become fatigued.

PLAN THE EXPERIMENT

1. Design an exercise for a group of muscles that can be counted over time.
2. Work in pairs, with one member of the team being a timekeeper and the other member performing the exercise.
3. Consider setting up your experiment so that the amount of resistance is the independent variable. Compare your results with those of other groups.

Check the Plan

1. Be sure that the exercises are ones that can be done rapidly and cause a minimum of disruption to other groups in the classroom.
2. Consider how long you will do the activity and how often you will record measurements.
3. ***Make sure your teacher has approved your experimental plan before you proceed further.***
4. Make a table that records the number of exercise repetitions versus time intervals.
5. On the graph paper, plot the number of repetitions on the vertical axis and the time on the horizontal axis.
6. Carry out the experiment.

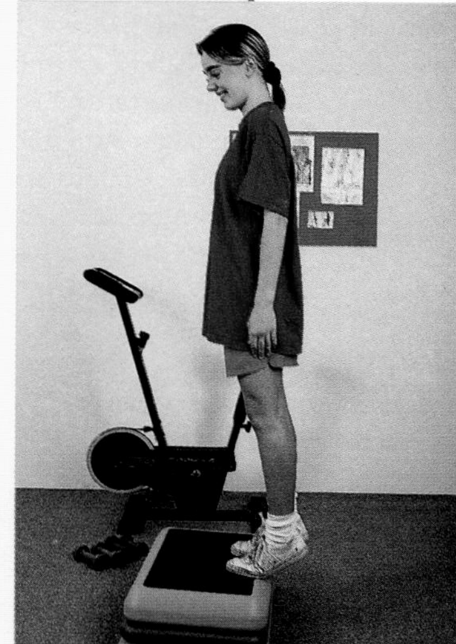

ANALYZE AND CONCLUDE

1. **Making Inferences** What effect did repeating the exercise over time have on the muscle group?
2. **Comparing and Contrasting** As you repeated the exercise over time, how did your muscles feel?
3. **Recognizing Cause and Effect** What physiological factors are responsible for fatigue?
4. **Thinking Critically** How well do you think your fatigued muscles would work after 30 minutes of rest? Explain your answer.

Going Further

Project Design an experiment that will enable you to measure the strength of muscle contractions.

ANALYZE AND CONCLUDE

1. The muscle groups become fatigued and the number of repetitions of the activity went down as more trials were run.
2. The muscles felt tired, and may have been slightly painful toward the end. It became harder to do the activity and the strength of contraction was reduced.
3. Cells are running out of oxygen and accumulating toxic products such as lactic acid and carbon dioxide, as the muscles change to anaerobic processes.
4. Fatigued muscles should work as well after a rest of 30 minutes as they did before fatigue became a factor. The accumulated lactic acid has been broken down as the oxygen supply in the muscle cells has been replenished during the rest.

Assessment

Student Journal: Have students place their laboratory reports, including their tables and graphs, in their journals. L1

Going Further

Ask students how athletes such as marathon runners continue to exercise at a particular level for 2 hours. Why do their muscles not fatigue after a few minutes? L2

Data and Observations

- Student graphs should show that the number of exercise repetitions of the activity goes down over time as the muscles become tired.

Meeting Individual Needs

Learning Disabled/Students Acquiring English Have students who are having difficulty with the anatomical terms in the chapter make flash cards that show the anatomical term on one side of the card and its meaning and location in the body on the other side. Have students work in pairs to review the terms and meanings on the cards. L1 SAE

Software

Mechanical Properties of Active Muscle, a simulation, Queue, IBM.

Audiovisual

Show *Muscular and Skeletal Systems*, video, National Geographic.

Physics Connection

Levers in the Body
To produce body movement, the muscles use bones as levers. A lever is a rigid structure that moves about a fixed point called a fulcrum. The joints act as the fulcrum. A lever is acted upon by two forces: resistance and effort. The resistance is the force (or load) that must be overcome for movement to occur. The resistance may be as little as the weight of the part of the body to be moved. The effort is the amount of force required to overcome the resistance. The elbow joint and the biceps are good examples of how muscles and bones act as levers.

Visual Learning

Figure 37.10 Ask students what makes up a single muscle fiber. *many myofibril units* Elicit what chemical stimulates the formation of attachments between myosin and actin filaments. *calcium* LS

Concept Development

Outline the steps involved in the sliding filament process of muscle contraction as a flowchart on the chalkboard or on an overhead transparency. Spend a few minutes reviewing and answering questions about the process. Encourage students to copy the diagram for use as a study tool. L1

myofibril:
mys (GK) muscle
fibrilla (L) small fiber
A myofibril is a small muscle fiber.

How muscles contract is related to their structure. Skeletal muscle tissue is made up of muscle cells and fibers. Each cylindrical muscle fiber is made up of smaller fibers called **myofibrils.** Myofibrils are composed of even smaller protein filaments. Filaments can be either thick or thin. The thick filaments are made of the protein **myosin,** and the thin filaments are made of the protein **actin.** The arrangement of myosin and actin gives skeletal muscle its striated appearance. Notice in *Figure 37.10* that each myofibril appears to be divided into sections. Each section of a myofibril is called a **sarcomere** and is the functional unit of muscle.

The sliding filament theory is currently the best explanation for how muscle contraction occurs. The **sliding filament theory** states that the actin filaments within the sarcomere slide toward one another during contraction. The myosin filaments do not move.

Figure 37.10
Locomotion is an essential body function that results from the contraction and the relaxation of muscles. The structure of muscle helps explain the sliding filament theory of how muscles contract.

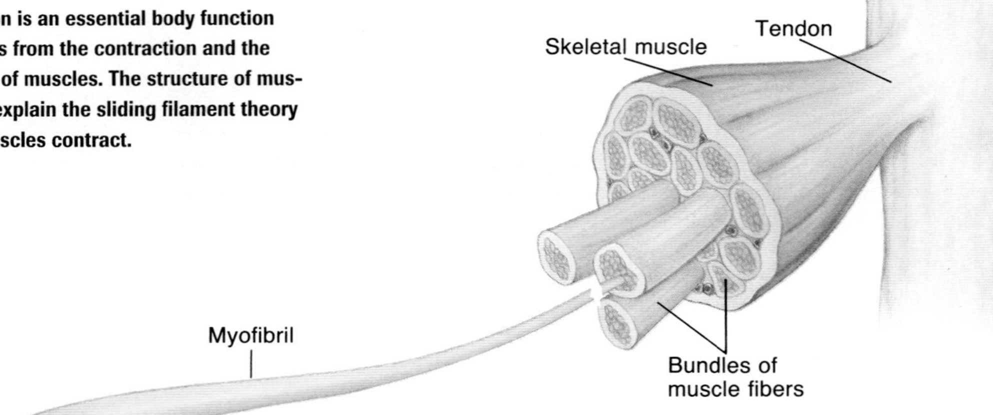

When you tease apart a typical skeletal muscle and view it under a microscope, you can see that it consists of bundles of fibers. A single muscle fiber is made up of many myofibril units called sarcomeres.

When a nerve signals a skeletal muscle to contract, calcium is released inside the muscle fibers. The presence of calcium causes attachments to form between the myosin and actin filaments. The actin filaments are pulled inward toward the center of each sarcomere, shortening the sarcomere. As the attachment is formed between myosin and actin, ATP is broken down to provide the energy for the muscle contraction. When the muscle relaxes, the filaments slide back into their original positions.

Meeting Individual Needs

Gifted Have students interview a medical doctor who specializes in muscle diseases. Ask them to find out about such diseases as muscular dystrophy, Lou Gehrig's disease, and myasthenia gravis. Have them research the depolarization of the membrane of the muscle fibers that occurs when a muscle contracts. L3

PROJECT

Modeling Body Movement

Have students working in cooperative groups construct models to show how bones and muscles work together to move appendages. More advanced students may wish to work together to prepare a model that demonstrates the sliding filament theory of muscle contraction. L1 COOP LEARN LS

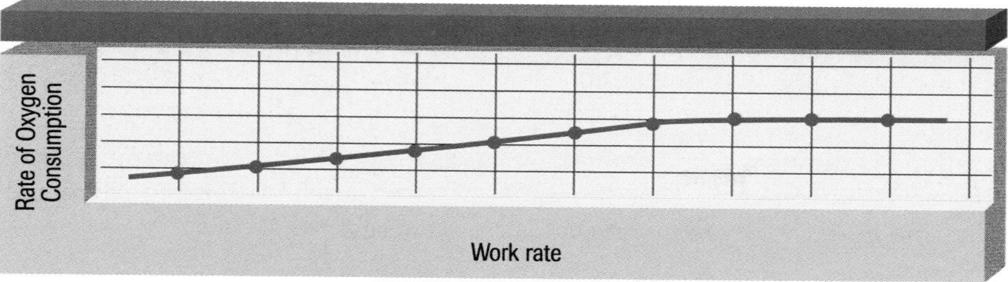

As an individual increases the intensity of the exercise activity, the need for oxygen goes up in almost exact, predictable increments.

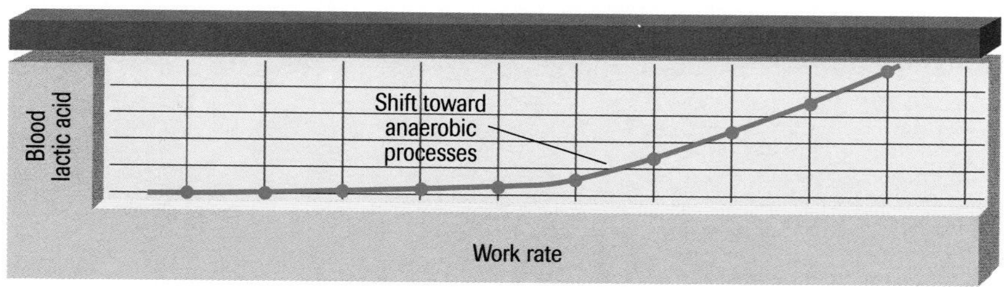

At a certain intensity, which differs among individuals, the body shifts from aerobic respiration to the anaerobic process of lactic acid fermentation for the energy needed to accomplish the activity. Because lactic acid is produced, an upswing in its presence in the bloodstream can be used to indicate the point at which lactic acid fermentation dominates.

Figure 37.11

Athletic trainers use information about muscle functioning during exercise to establish appropriate levels of intensity for training. These data allow athletes to get the most out of their workouts without becoming overly tired.

Muscle Strength and Exercise

How can you increase the strength of your muscles? Muscle strength does not depend on the number of fibers in a muscle. It has been shown that this number is basically fixed before you are born. Rather, muscle strength depends on the thickness of fibers and how many of them contract at one time. Thicker fibers are stronger and contribute to muscle mass. Regular exercise stresses muscle fibers slightly, and to compensate for this added workload, the fibers increase in size.

Recall from Chapter 10 that ATP is produced during cellular respiration. Muscle cells are continually supplied with ATP from both aerobic and anaerobic energy systems. However, the aerobic energy system for producing ATP dominates when adequate oxygen is delivered into a muscle cell to meet its energy production needs, as when a muscle is at rest or during slow or moderate activity. When an inadequate supply of oxygen is available to meet the muscle cell's energy needs, the anaerobic process of lactic acid fermentation, *Figure 37.11*, becomes the primary source of ATP during vigorous activity.

Think about what happens when you are running in gym class or around the track at school. When your muscles are working hard, they are not able to get oxygen fast enough to sustain aerobic respiration to supply adequate ATP. Thus, the amount of ATP available becomes limited. For your muscle cells to get the energy they need, they must rely on lactic acid fermentation as well.

Meeting Individual Needs

Physically Challenged Find a physically challenged spokesperson who has a musculoskeletal disease who is willing to speak to the class about the disorder. The focus of the discussion should be on the nature of the disease, what restrictions the disease places on the person, and what the treatments are for the disease. L1

Visual Learning

Figure 37.11 Direct students' attention to the bottom graph. Elicit whether an increase in the amount of lactic acid in the blood indicates that aerobic respiration or fermentation is occurring. *fermentation* Ask whether lactic acid fermentation occurs during light, moderate, or strenuous exercise. *strenuous* L1 LS

Discussion

As a class, debate the use of jogging in a fitness program. Encourage students to identify the benefits and limitations of this form of exercise. L1 COOP LEARN

Tying to Previous Knowledge

Point out that myofibrils of the muscles are composed of microfilaments like those discussed in Chapter 8. This is a good time to review ATP, along with aerobic and anaerobic metabolism from Chapter 10.

Text Question

How can you increase the strength of your muscles? *by participating in a regular exercise program*

GLENCOE TECHNOLOGY

Videodisc

The Infinite Voyage: The Champion Within
Physical Training and Senior Citizens (Ch. 8)

Biology: The Dynamics of Life
Disc 2, Side 1
Sliding Filament Theory (Ch. 30)

3 Assess

Check for Understanding

Ask students to diagram, and explain in their own words, the sliding filament theory of muscle contraction. L1

Reteach

Review the three types of muscles and where in the body each is found. L1

Extension

Have interested students research diseases and disorders that affect the muscular system. Have students prepare a report on one of the disorders for their portfolios. L1 P

Assessment

Performance: Provide students with a diagram of muscles of the human body. Have students make a key to identify each type of muscle. L1

4 Close

Discussion

Discuss the function of each type of muscle. Help students recognize the relationship between the skeleton and skeletal muscles. Ask students to list the general features of muscles. L1

During exercise, lactic acid builds up in muscle cells. As the excess lactic acid is passed into the bloodstream, the blood becomes more acidic, rapid breathing is stimulated, and cramping can occur. As you catch your breath following exercise, oxygen is supplied adequately to your muscles and lactic acid is broken down. Regular exercise can result in improved functioning of muscles. *Figure 37.12* shows aerobic training, which, for example, can affect a muscle's ability to share and use energy.

Figure 37.12
Muscles that are exercised regularly become stronger and able to store more fuel for sustained activity.

Connecting Ideas

The skin of animals plays an important role in maintaining homeostasis of the body. Because skin is in direct contact with the external environment, it often is the first organ to respond to any changes in the environment. However, the majority of responses are made by the muscular and skeletal systems to produce movement from place to place. The size and arrangement of muscles, and the structure of bones and joints, work to produce an efficient mode of locomotion for animals. Animals move for a variety of reasons. Perhaps most important of these is to support their heterotrophic way of life. The energy you require for all the muscle activity in your daily life is delivered from the food you eat.

Section Review

Understanding Concepts

1. Compare the structures and functions of the three main types of muscles.
2. Summarize the sliding filament theory of muscle contraction.
3. How does exercise change muscle strength?

Thinking Critically

4. Why would a disease that causes paralysis of smooth muscles be life threatening?

Skill Review

5. **Interpreting Scientific Illustrations** Outline the composition of muscle fibers as shown in *Figure 37.10*. For more help, refer to Thinking Critically in the *Skill Handbook*.

Answers to Section Review

1. Smooth muscle cells are spindle-shaped cells that form the linings of organs. They apply pressure and squeeze. Cardiac muscle cells are striated, interconnected, and form a contraction network in the heart. Skeletal muscle cells are striated, multinucleated, and may be very long. They move the body.
2. Calcium is released into the muscle fiber, actin and myosin filaments form attachments, and the actin filaments are pulled inward toward the center of each sarcomere.
3. Exercise stresses muscles and causes muscle fibers to increase in size.

Thinking Critically

4. Internal organs utilize smooth muscle for movement. Internal movement, such as digestion, would stop.

Skill Review

5. Muscle fibers are made of tiny cylinders called myofibrils. Each myofibril is made of myosin and actin filaments arranged in a regular pattern within the sarcomere.

REVIEWING MAIN IDEAS

37.1 Skin: The Body's Protection

- Skin plays a major role in maintaining homeostasis in the body. It regulates body temperature, protects the body, functions as a sense organ, and produces vitamin D.
- Skin responds to injury by producing new cells by mitosis and signaling a response to fight infection.

37.2 Bones: The Body's Support

- The skeleton is made up of the axial and appendicular skeletons. The skeleton supports the body, provides a place for muscle attachment, protects vital organs, manufactures blood cells, and serves as a storehouse for calcium and phosphorus.
- Bones remodel themselves throughout life.

37.3 Muscles for Locomotion

- There are three types of muscles: smooth, cardiac, and skeletal.
- Skeletal muscles contract as the filaments within the myofibrils slide toward one another.
- Muscle strength is a result of muscle fiber thickness and the number of fibers contracting.

Key Terms

Write a sentence that shows your understanding of each of the following terms.

actin
appendicular skeleton
axial skeleton
bursa
cardiac muscle
compact bone
dermis
epidermis
hair follicle
involuntary muscle
joint
keratin
ligament
marrow
melanin
myofibril
myosin
osteoblast
sarcomere
skeletal muscle
sliding filament theory
smooth muscle
spongy bone
tendon
voluntary muscle

Understanding Concepts

1. Which of these is NOT a type of tissue found in the skin?
   a. connective　c. muscle
   b. epithelial　d. brain
2. ___________ tissue is fibrous tissue usually made of collagen and elastin protein fibers.
   a. Connective　c. Muscle
   b. Epithelial　d. Brain
3. The ___________ is the outer, thinner portion of the skin.
   a. keratin　c. epidermis
   b. dermis　d. melanin
4. ___________ is a skin pigment that protects cells from solar radiation damage.
   a. Keratin　c. Epidermis
   b. Melanin　d. Dermis
5. Which of the following is NOT found in the dermis?
   a. blood vessels　c. keratin
   b. nerves　d. oil glands
6. What protein helps waterproof skin and protect the living cell layers underneath?
   a. epidermis　c. collagen
   b. keratin　d. insulin
7. When your body heats up, it becomes wet. This is a function of the ___________.
   a. skin cells
   b. oil glands
   c. sweat glands
   d. subcutaneous layer
8. Epithelial cells are derived from the ___________ of the embryo.
   a. ectoderm　c. mesoderm
   b. endoderm　d. metaderm

Chapter 37 Review

Reviewing Main Ideas

Summary statements can be used by students to review the major concepts of the chapter.

Key Terms

Answers should go beyond defining the terms. Accept any answer that uses the term correctly and in the proper context.

Understanding Concepts

1. d
2. a
3. c
4. b
5. c
6. b
7. c
8. a

Chapter 37 Review

9. c
10. b
11. a
12. b
13. d
14. a
15. c
16. c
17. a
18. d
19. d
20. b

Applying Concepts

21. Third-degree burns destroy the epidermis and dermis, while first-degree burns involve only epidermal cells.
22. The membrane surrounding the bone as well as the Haversian canal systems can be bruised because they contain blood vessels.
23. If red bone marrow were destroyed, the production of red blood cells would stop, white cell production would be impaired, and clotting would be impaired.
24. Degree of bone fusion can be used to identify the age of a skeleton. The more fusion, the older the skeleton. The amount of cartilage present and brittleness of bone also may determine age.
25. Sweat evaporates from the body and cools the body. The blood vessels in the skin can constrict to conserve heat or dilate to increase heat loss.

Thinking Critically

26. Evaluate students' concept maps. The maps should show an understanding of the relationships among the concepts listed.
27. Answers might include addition of human hormones to bone cells or using a different material (other than ceramic).

Chapter 37 Review

9. _______ burns are the most severe, resulting in destroyed epidermis and dermis.
   a. First-degree c. Third-degree
   b. Second-degree d. Fourth-degree
10. Which of the following is NOT a function of the skeletal system?
   a. provide a framework for body tissues
   b. produce vitamin D
   c. produce blood cells
   d. act as a storehouse for minerals
11. A _________ is where two or more movable bones meet.
   a. joint c. bursa
   b. suture d. ligament
12. Which type of joint allows back-and-forth movement and is found in the elbows?
   a. gliding c. pivot
   b. hinge d. ball-and-socket
13. The fetal skeleton is mostly made of ___________.
   a. collagen c. bone
   b. insulin d. cartilage
14. The _________ skeleton includes the skull, ribs, vertebral column, and sternum.
   a. axial c. bursa
   b. appendicular d. dermal
15. _________ fills the cavities of the ribs, sternum, vertebrae, skull, and the long bones of the arms and legs.
   a. Compact bone c. Marrow
   b. Spongy bone d. Melanin
16. In movable joints, a fluid-filled sac called a _________ prevents bones from rubbing against each other.
   a. ligament c. bursa
   b. tendon d. marrow
17. Which of the following is NOT a type of muscle?
   a. filamentous c. smooth
   b. cardiac d. skeletal
18. Muscles are attached to bones by ___________.
   a. ligaments c. filaments
   b. joints d. tendons
19. A muscle that contracts under conscious control is a(n) ___________ muscle.
   a. smooth c. impulse
   b. cardiac d. voluntary
20. Muscle filaments ___________ when the muscle relaxes.
   a. slide in toward each other
   b. slide back into their original positions
   c. are digested into their component proteins
   d. lengthen and become invisible to the microscope

Applying Concepts

21. Why is recovery from third-degree burns slower than from first-degree burns?
22. A bruise is due to a broken blood vessel. What part of a bone can be bruised?
23. How would the destruction of red marrow affect other systems of the body?
24. How could you use the skeleton to determine the age of a person?
25. The skin is an important body organ in maintaining homeostasis. Outline how the skin helps the body maintain body temperature.

Thinking Critically

26. ***Concept Mapping*** Make a concept map that relates the following terms and phrases. Supply the appropriate linking words for your map.
   actin, cardiac muscle, involuntary muscle, myosin, myofibrils, sarcomere, skeletal muscle, smooth muscle, voluntary muscle
27. ***Designing an Experiment*** A new study with rat bone cells has shown that bone-building cells will mature and grow masses of tendrils when placed in small, ceramic cubes that are permeated with tunnels. The filled ceramic cubes can then be inserted into a bone injury where the cells will continue to build new bone.

Chapter 37 Review

Suppose you discover that human bone cells grow slower than rat bone cells. Design an experiment to test why the results in the two species differ.

28. ***Interpreting Scientific Illustrations*** A muscle physiologist studied the effect of load or stress on the shortening of a muscle. The more a muscle shortens, the more work a muscle can do. Based on the graph, describe the relationship between muscle shortening and its load.

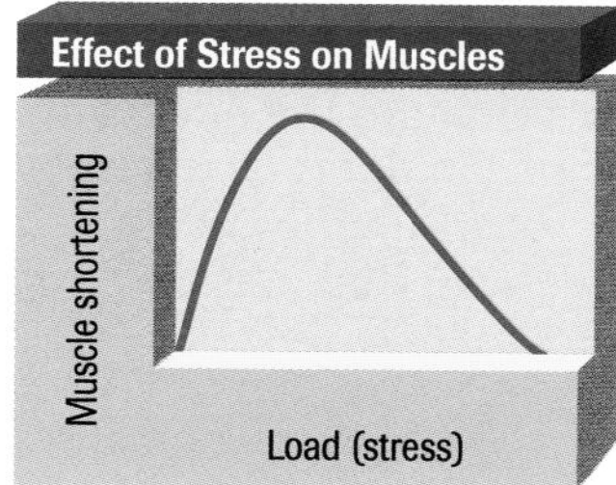

29. ***Interpreting Data*** The graph below shows the time relationship between muscle force and calcium level inside the muscle cell. Relate the cause-and-effect relationship between development of muscle force and calcium level with increasing load.

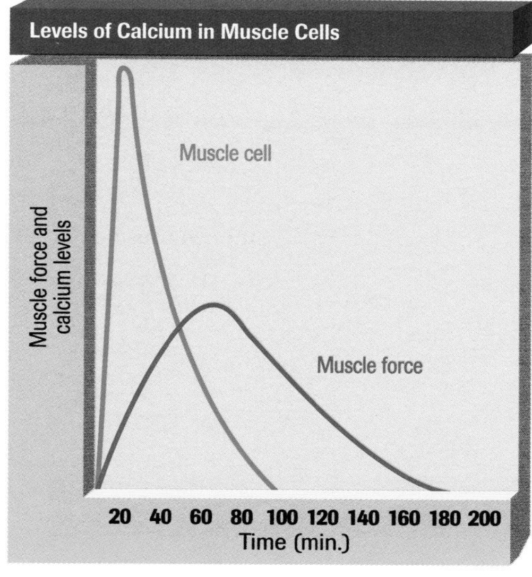

ASSESSING KNOWLEDGE & SKILLS

Bone is living tissue that includes different kinds of cells and blood vessels.

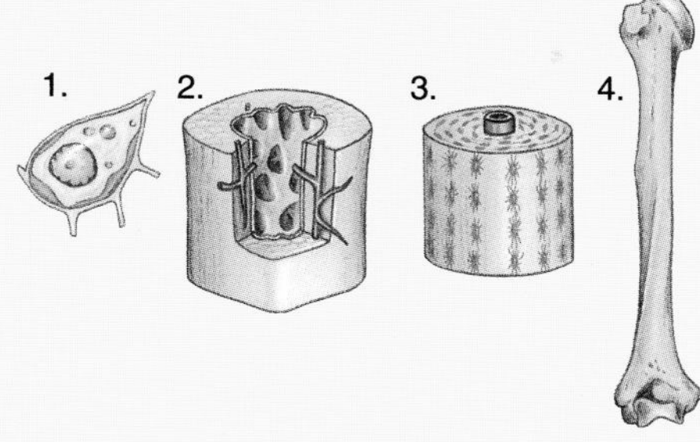

Interpreting Scientific Illustrations Study the diagram above, then answer these questions.

1. Place the structures of the human body shown in the diagram in order from least to most complex.
   a. 1, 2, 3, 4 **c.** 4, 3, 2, 1
   b. 1, 3, 2, 4 **d.** 4, 2, 3, 1
2. Which diagram represents one Haversian Canal System?
   a. 1 **c.** 3
   b. 2 **d.** 4
3. Which of the following demonstrates that bones are alive?
   a. Bones grow in both length and diameter.
   b. Bones are able to repair themselves.
   c. Bones contain living cells called osteocytes.
   d. all of the above
4. ***Interpreting Data*** The center of most long bones is hollow. What advantage does having bones with hollow centers confer to humans and other mammals?

Chapter 37 Review

28. Muscles shorten more as load increases up to a point and then muscles shorten less and less as the load increases.
29. A large increase in muscle cell calcium precedes (and initiates) muscle contraction.

ASSESSING KNOWLEDGE & SKILLS

1. b
2. c
3. d
4. Hollow bones allow humans and other mammals to have strong bones with reduced weight.

Program Resources

Chapter Assessment, pp. 229-234 L1
Alternate Assessment in the Science Classroom
Computer Test Bank L1
Content Mastery, pp. 157-160  L1

Chapter Organizer

SECTION	OBJECTIVES	ACTIVITIES/FEATURES
38.1 Following Digestion of a Meal National Science Standards: UCP.1-3, UCP.5; B.3; C.5; F.1; G.1	1. **Summarize** the digestive functions of the organs of the digestive system. 2. **Outline** the pathway food follows through the digestive tract.	**Biology & Society Issues:** Megavitamins, p. 963
38.2 The Control of Digestion and Homeostasis National Science Standards: UCP.1-3; B.3; C.5; F.1	3. **Explain** how the nervous system and hormones control the digestive process. 4. **Identify** the functions of the glands that affect digestion.	**Thinking Lab:** What are the effects of glucagon and insulin during exercise?, p. 966 **Health Connection:** Diabetes—A Sugar Problem, p. 967
38.3 Nutrition National Science Standards: UCP.2, UCP.3; A.1, A.2; B.2, B.6; C.5; F.1; G.1, G.2	5. **Summarize** the role of the six classes of nutrients in body nutrition. 6. **Relate** Calories and metabolism.	**Minilab:** Where does the food in a spaghetti dinner fit in the food pyramid?, p. 970 **Minilab:** Does a bowl of soup provide a complete meal?, p. 973 **Biolab:** Testing for Nutrients in Foods, p. 974

ACTIVITY MATERIALS

BIOLAB	MINILABS	ALTERNATE LAB
page 974 food samples mortar and pestle test tubes droppers soapy water test-tube rack test-tube holder brown paper iodine solution biuret solution apron goggles	**page 970** plain paper colored pencils (optional) **page 973** calculator (optional)	**page 962** Lactaid (liquid lactose digestive aid) glucose test paper glucose solution milk graduated cylinders eyedroppers test tubes toothpicks

Chapter 38 Digestion and Nutrition

TEACHER CLASSROOM RESOURCES

Reproducible Masters	Transparencies
Section Focus Master 85: Food Processors L1 SAE **Reinforcement and Study Guide,** pp. 155-156 L1 **Tech Prep Applications:** An Inside Look at the Heimlich Maneuver, pp. 43-44 L2 **Content Mastery,** pp. 161-164 L1	**Reteaching Transparency #38:** Function of the Small Intestine L1 SAE
Section Focus Master 86: Managing the Food Supply L1 SAE **Reinforcement and Study Guide,** p. 157 L1 **Laboratory Manual:** Endocrine Gland Studies, pp. 235-238 L2 **Critical Thinking/Problem Solving:** Interpreting a Blood Analysis Printout, p. 38 L3	**Basic Concepts Transparency #65:** Regulation of Blood Sugar Concentration L1 SAE
Section Focus Master 87: All the Things You Eat L1 SAE **Reinforcement and Study Guide,** p. 158 L1 **Biolab and Minilab Worksheets,** pp. 151-154 L1 **Concept Mapping:** Carbohydrates, Fats, and Proteins in Nutrition, p. 38 L1 **Laboratory Manual:** Caloric Content of a Meal, pp. 239-240; How Much Vitamin C Are You Getting?, pp. 241-244 L2	**Basic Skills Transparency #26:** Information on a Food Label L1 SAE

ASSESSMENT MATERIALS

Chapter Assessment, pp. 235-240
Performance Assessment in the Biology Classroom, p. 39
MindJogger Videoquiz
Alternate Assessment in the Science Classroom
Computer Test Bank

Spanish Resources SAE
English/Spanish Audiocassettes SAE
Cooperative Learning in the Science Classroom COOP LEARN
Lesson Plans
Biology Projects: You Are What You Eat, pp. 37-40 L2

KEY TO TEACHING STRATEGIES

- L1 Level 1 activities should be within the ability range of all students including those with learning difficulties.
- L2 Level 2 activities are within the ability range of average to above-average students.
- L3 Level 3 activities are designed for the ability range of above-average students.
- SAE SAE activities should be within the ability range of Students Acquiring English.
- COOP LEARN Cooperative Learning activities are designed for small group work.
- P These strategies represent student products that can be placed into a best-work portfolio.
- These strategies are useful in a block scheduling format.

GLENCOE TECHNOLOGY

The following multimedia resources are available from Glencoe.

Biology: The Dynamics of Life
CD-ROM SAE
Videodisc Program

The Infinite Voyage Series
To the Edge of the Earth
A Taste of Health

National Geographic Society Series
STV: Human Biology Vol. 1
Digestive System

Science and Technology Videodisc Series (STVS)
Human Biology
Insulin Pills
Measuring Calcium Deficiency
Measuring Body Fat

CHAPTER

38 Digestion and Nutrition

Chapter Overview

Students begin their study of digestion by tracing the path of a meal through the digestive system while examining the structure and function of each organ of the system. As part of this survey, both mechanical digestion and chemical digestion are discussed.

In Section 38.2, nervous and endocrine control of the digestive process is explained. The discussion focuses on how the digestive system maintains homeostasis and the relationship between homeostasis and metabolism.

As the chapter concludes, students examine nutrition. They learn about the vital nutrients and how nutrients are involved in the metabolic processes of the body.

Key Terms

amylase
bile
Calorie
endocrine gland
epiglottis
esophagus
exocrine gland
gallbladder
large intestine
liver
mineral
pancreas
parathyroid gland
pepsin
peristalsis
rectum
small intestine
stomach
target tissue
thyroid gland
villus
vitamin

Look for the following logo for strategies that emphasize different learning modalities. LS

Visual-Spatial	**Demonstration, pp. 959, 963; Meeting Individual Needs, pp. 959, 971; The Inside Story, p. 960; Portfolio, p. 960; Minilab, p. 970**
Interpersonal	**Student Journal, p. 972**
Logical-Mathematical	**Project, pp. 961, 973; Alternate Lab, p. 962; Thinking Lab, p. 966; Portfolio, pp. 967, 973; Student Journal, p. 971; Minilab, p. 973**

CHAPTER

38 Digestion and Nutrition

Have you ever been sitting in a quiet classroom full of students, all silently working, when your stomach started to rumble and gurgle? As the noise got louder and louder, you may have wished you could crawl under your desk. As common and natural as this event is, it still can be quite embarrassing. What causes an empty stomach to be so noisy?

Maybe you've also experienced butterflies in your stomach—a tense feeling accompanied by mild nausea. Perhaps the butterflies occurred before you had to take a big exam, deliver a speech, or play in an important game. What makes your stomach feel this way?

Your stomach is just one part of your digestive system—the group of organs that is responsible for bringing food into your body and breaking that food into microscopic, usable molecules. How they accomplish that change is the story of digestion.

Concept Check

You may wish to review the following concepts before studying this chapter.

- Chapter 7: enzymes, carbohydrates, fats, proteins

Chapter Preview

Laboratory Activities

Biolab

- Testing for Nutrients in Foods

Minilabs

- Where does the food in a spaghetti dinner fit in the food pyramid?
- Does a bowl of soup provide a complete meal?

You probably don't pay much attention to your stomach—unless it begins to rumble like an earthquake or to feel like a nest of butterflies. When this embarrassed student has an opportunity to eat a good, nutritious meal, his stomach (inset) will stop gurgling and start digesting.

CHAPTER 38

Introducing the Chapter

On the chalkboard, list different "feelings" associated with your stomach such as: having butterflies, emptiness, hunger, nausea, and fullness. Initiate a discussion about the role of food in maintaining a feeling of well-being. Lead the discussion to the importance of the digestive tract in taking food into the body.

Theme Development

The theme of *homeostasis* is emphasized through development of the role of the digestive system in regulating body functions. *Systems and interactions* is focused on through a discussion of the diffusion of molecules obtained from food into the bloodstream, and the nervous and endocrine system's control of digestion.

Concept Check

An understanding of the concepts of diffusion, osmosis, and active transport are essential in this chapter, especially as they relate to absorption.

Assessment Planner

Choose assessment strategies from the following pages to evaluate the progress of your students.

Assess, pp. 964, 968, 976
Alternate Lab, pp. 962-963
Minilabs, pp. 970, 973
Portfolio, pp. 960, 967, 973
Thinking Lab, p. 966
Biolab, pp. 974-975
Chapter Review, pp. 977-979

SECTION

38.1

Prepare

Key Concepts

The structures and functions of the organs of the digestive tract are presented as students follow a meal through the digestive tract.

Block Scheduling

Look for this symbol for strategies that are useful in a block scheduling format. For more information on block scheduling, refer to Section 38.1 in the **Lesson Plans** booklet.

Materials

- Acquire old X rays for the Demonstration on page 959.
- Prepare materials for the second Demonstration on page 959.
- Acquire a pancreas from a meat packing plant and a hard-boiled egg for the Demonstration on page 963.
- Purchase milk and Lactaid for the Alternate Lab.

1 Focus

Bellringer

Before presenting the lesson, display **Section Focus Master 85** on the overhead projector and have students answer the accompanying questions. L1 SAE

Activity

Have two students hold a 9-m long piece of string. Explain that the string represents the length of the human digestive tract. Briefly discuss how food moves along the length of the digestive tract and the changes that occur within the digestive tract. Ask students to speculate on why plant eaters must have longer digestive tracts than meat eaters. L1

SECTION

38.1 Following Digestion of a Meal

Section Preview

Objectives

Summarize the digestive functions of the organs of the digestive system.

Outline the pathway food follows through the digestive tract.

Key Terms

amylase
esophagus
peristalsis
epiglottis
stomach
pepsin
small intestine
pancreas
liver
bile
gallbladder
villus
large intestine
rectum

Sometimes it may seem like your stomach has a mind of its own. But it's really one part of a system that is, ultimately, controlled by your brain and by hormones that are secreted along the digestive pathway. As in many animals you have studied, the human digestive system is a tube that develops in the embryo. Different parts of the tube have evolved into different digestive organs, each with a special function to perform as food passes along.

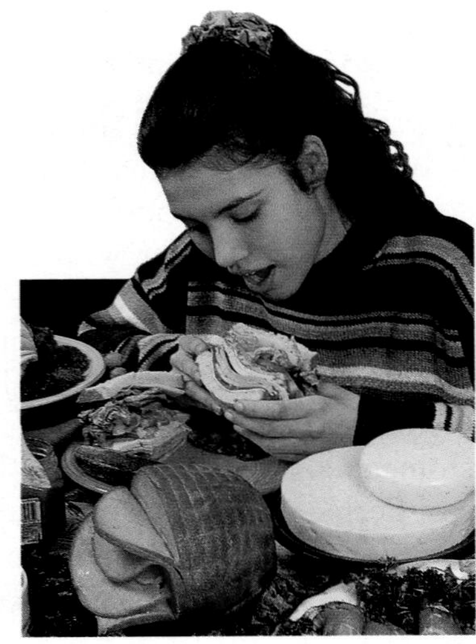

Functions of the Digestive System

The digestive system performs several functions in the preparation of food for cellular utilization. First, the system ingests food. The next two functions take place at the same time; the system digests food while it is being forced along the digestive tract. Fourth, the system absorbs the digested food into the cells of your body, and fifth, it eliminates undigested materials from your body.

Refer to *Figure 38.1* as you read about the digestive system and follow a meal through the organs of digestion.

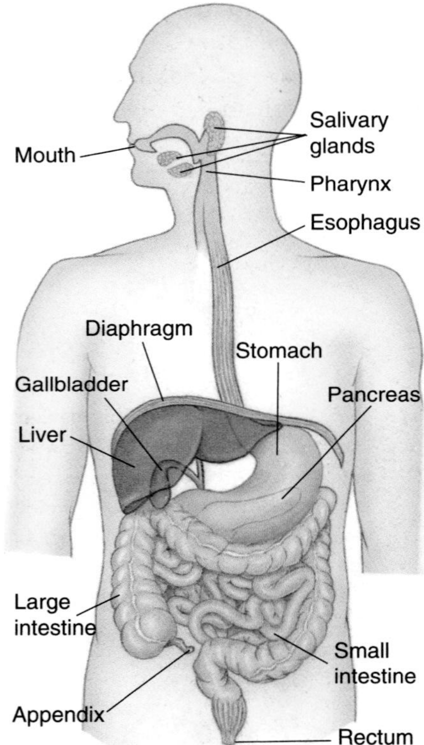

Figure 38.1
All the digestive organs work together to break down food into simpler compounds that can be absorbed. As you read about each digestive organ, use this diagram to locate its position within the system.

Program Resources

Section Focus Master 85 L1 SAE
Reinforcement and Study Guide, pp. 155-156 L1
Performance Assessment in the Biology Classroom, p. 39 L1 P
Tech Prep Applications, pp. 43-44 L2
Reteaching Transparency 38 and **Master** L1 SAE

Remember from Chapter 8 how a cell is like a place of business in which all essential parts are assembled to form an efficient working system? The digestive system works in the opposite way. Food is disassembled into its component molecules so that the body can have a supply of resources for building new cells.

The Mouth

The first stop in the digestive disassembly line is the mouth, *Figure 38.1.* Suppose it's lunchtime, and you just prepared a bacon, lettuce, and tomato sandwich. The first thing you do is bite off a piece and chew it for ten to 20 seconds.

What happens as you chew?

In your mouth, your tongue moves the food around and helps position it between your teeth as you chew. Chewing is a part of mechanical digestion, the physical process of breaking food into smaller pieces. Mechanical digestion increases the surface area of food particles and prepares them for chemical digestion. Chemical digestion is the process of structurally changing food molecules through the action of enzymes. Digestive enzymes, listed in *Table 38.1,* act to break the molecules apart.

Chemical digestion begins in the mouth

Some of the nutrients in your sandwich are starches, which are large polysaccharides. As you chew your bite of sandwich, salivary glands around your mouth secrete saliva. Saliva is a slightly alkaline solution that adds water and a digestive enzyme to the food. This enzyme, called **amylase,** breaks down starch into smaller sugar molecules called disaccharides. Even though food is swallowed too quickly for all the starches to be reduced in your mouth, amylase in the swallowed food continues to digest the starches in the stomach for about 30 minutes.

Table 38.1 Digestive Enzymes

Organ	Enzyme	Molecules Digested	Product
Salivary glands	Salivary amylase	Starch	Disaccharide
Stomach	Pepsin	Proteins	Peptides
Pancreas	Pancreatic amylase	Starch	Disaccharide
	Trypsin	Protein	Peptides
	Pancreatic lipase	Fats	Fatty acids and glycerol
	Nucleases	Nucleic acids	Sugar and nitrogen bases
Small intestine	Maltase	Disaccharide	Monosaccharide
	Sucrase	Disaccharide	Monosaccharide
	Lactase	Disaccharide	Monosaccharide
	Peptidase	Peptides	Amino acids
	Nuclease	Nucleic acids	Sugar and nitrogen bases

SECTION 38.1

2 Teach

Demonstration

LS **Visual-Spatial** Show students old barium X rays of the digestive tract (often available at hospitals). Use the X rays to discuss and identify parts of the digestive tract. X rays of jawbones may also be obtained and used for discussions on teeth.

Visual Learning

Using the Table Explain that the pancreas is 15 cm long and produces 1 L of pancreatic juice each day. Have students use Table 38.1 to list the chemicals that compose pancreatic juice. *Pancreatic amylase, trypsin, pancreatic lipase, nucleases*

Demonstration

LS **Visual-Spatial** To demonstrate both digestion and diffusion, fill a dialysis bag with 1-5% starch solution. In a second dialysis bag, add 2 mL of pancreatic enzymes and fill the remainder of the bag with starch solution. Allow each bag to sit in a beaker of distilled water for one day. Then, run a Benedict's test on the fluid in each beaker. Explain that the enzymes will digest the starch and change it to glucose, which diffuses out of the dialysis bag. The glucose will produce a positive Benedict's test.

Meeting Individual Needs

Learning Disabled Use two pieces of string to demonstrate how surface area is increased by folding. Cut a 15-cm length of string and lay it atop a desk fully stretched out. Take a second 15-cm piece of string and form a zigzag pattern with it. Have students observe the total horizontal distance between the two ends of each piece of string. Use the activity to point out that more string is needed to cover a given distance if it has folds. Compare this to the increased surface area created by villi. L1 LS

THE INSIDE STORY

Inside Your Mouth

Purpose

[LS] **Visual-Spatial** Students will examine the structure of teeth and other structures of the mouth.

Teaching Strategies

- Bring in models of animal skulls, including a human skull, to compare teeth. Ask students whether other mammals have baby or milk teeth. *Yes. Some students may recall their puppies or kittens teething and losing the baby teeth.*

Misconception

Students may believe that cavities are the greatest threat to teeth. Explain that while cavities do destroy teeth, periodontal disease of the gums is much more serious and far more prevalent today.

Visual Learning

- Point out each structure of the mouth shown in the diagram. As each structure is pointed out, have a volunteer read the description of the structure.
- Ask students to name the three types of teeth shown. *incisors, cuspids, and molars*
- Elicit from students the function of the tonsils and their location. *The tonsils are located in the back of the mouth and remove bacteria that enter the mouth or nose.*

Audiovisual

Show the video *Digestive System*, National Geographic.

THE INSIDE STORY

Inside Your Mouth

Your mouth houses many structures that are involved in other functions besides digestion. For example, these structures protect against foreign materials invading your body and help you taste the food you eat.

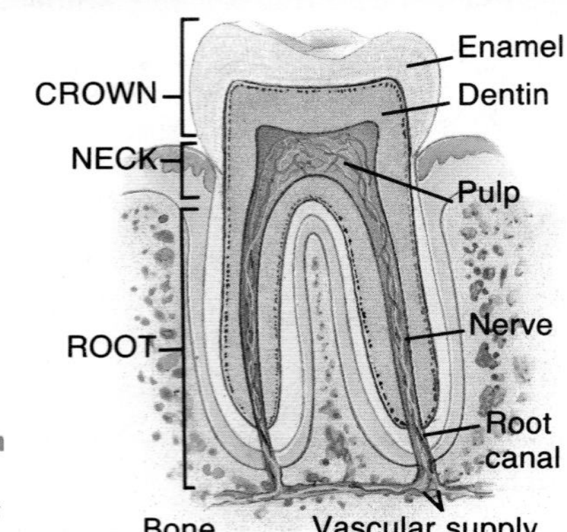

Structure of a tooth

Teeth are made mainly of dentin, a bonelike substance that gives a tooth its shape and strength. The dentin encloses a space filled with pulp, a tissue that contains blood vessels and nerves. The dentin of the crown is covered with an enamel that consists mostly of calcium salts. Tooth enamel is the hardest substance in the body.

1. The incisors are adapted for cutting food. The cuspids, or canines, tear or shred food. The three sets of molars can crush and grind food. Often, there is not enough room for the third set of molars, called wisdom teeth, which then have to be removed.

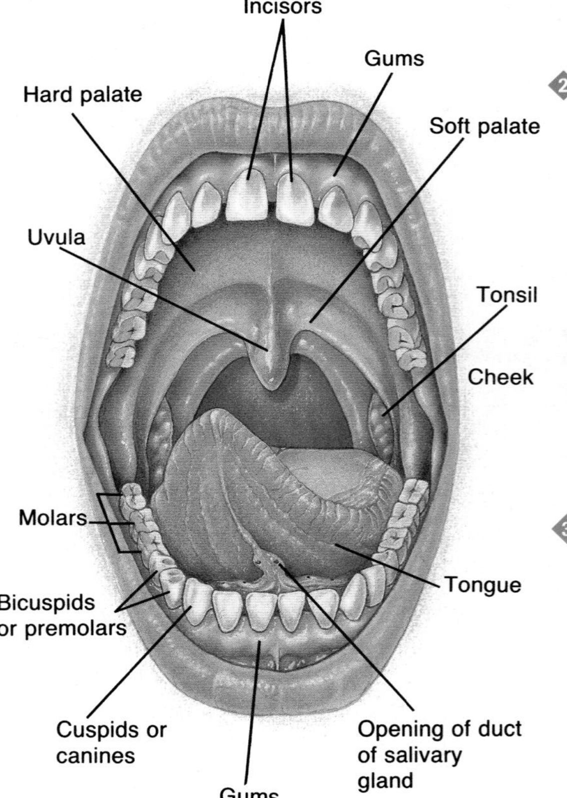

Structures of the oral cavity

2. A pair of tonsils are located at the back of the mouth. They help remove bacteria that have entered the mouth and nose.

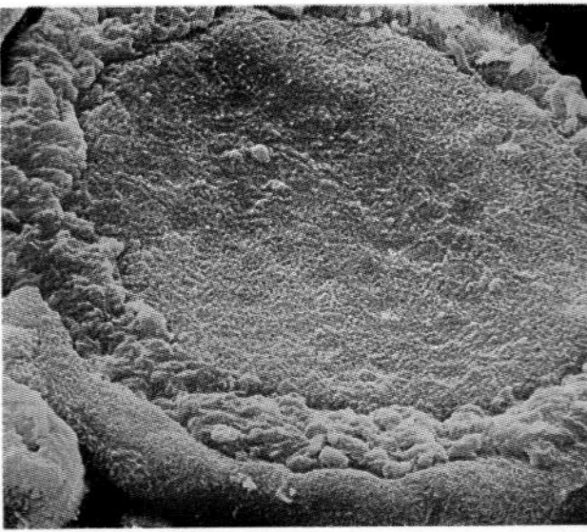

3. Attached to the floor of the mouth is the tongue. It is made of skeletal muscle covered with a mucous membrane.

   The upper surface of the tongue has many projections that form the taste buds, shown in the micrograph above, which contain numerous taste receptor cells.

PORTFOLIO

Primary Teeth Have students investigate how the number and location of primary teeth in humans differ from the secondary teeth. Ask students to diagram a mouth that shows only primary teeth. Have students identify the types of teeth shown in the diagram. L1 P LS

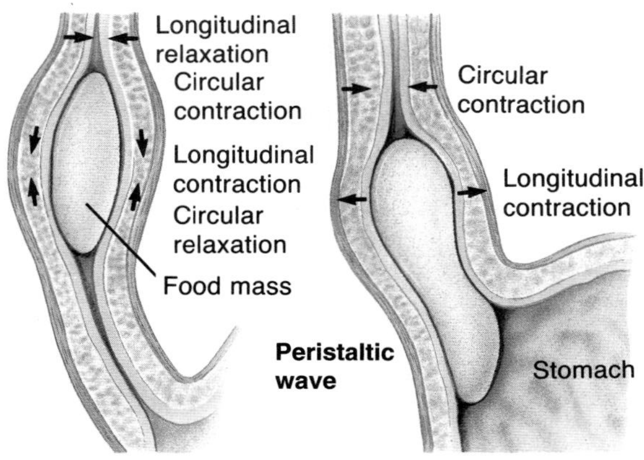

Figure 38.2

Smooth muscle contractions are responsible for moving food through the esophagus. The contractions occur in waves; first, circular muscles relax and longitudinal muscles contract, then circular muscles contract and longitudinal muscles relax.

Swallowing your food

Once you've thoroughly chewed your bite of sandwich, your tongue shapes it into a ball and moves it to the back of your mouth to be swallowed. Swallowing forces food from your mouth to the esophagus. The **esophagus** is the muscular tube that connects the mouth to the stomach. Like a busy expressway, the esophagus moves the food into the stomach in five to eight seconds. *Figure 38.2* shows how the esophagus moves food along by **peristalsis,** a series of involuntary muscle contractions along the walls of the digestive tract.

Have you ever had food go down the wrong way? When you swallow, the food is diverted by the larynx and passes over the windpipe, the tube that leads to the lungs. Usually, when you swallow, a flap of skin called the **epiglottis** covers the opening to the windpipe, and breathing is temporarily interrupted. After the food passes into your esophagus, the epiglottis opens again. But if you talk or laugh as you swallow, the epiglottis opens and the food may enter the windpipe. Your response, a reflex, is to choke and cough, forcing the food out of the windpipe.

The Stomach

When the chewed food reaches the bottom of your esophagus, a valve lets the food enter the stomach. The **stomach** is a muscular, pouchlike enlargement of the digestive tract.

peristalsis:
peri (GK) around
stellein (GK) to draw in
Peristalsis propels food in one direction.

Muscular churning

Three layers of muscles, lying across one another, make up the wall of the stomach. When these muscles contract, as in *Figure 38.3*, they mix the food in the stomach—another step in mechanical digestion.

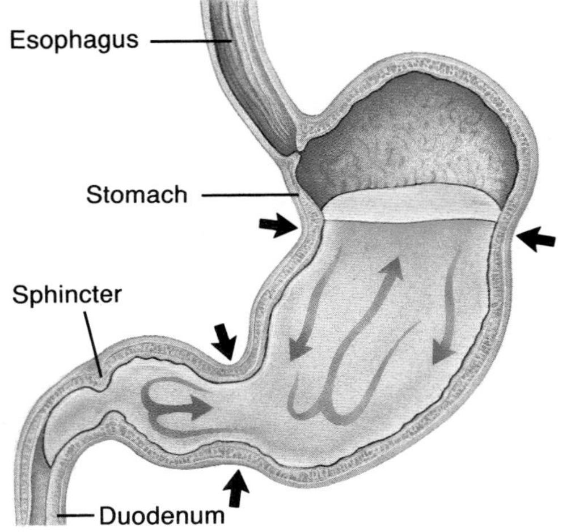

Figure 38.3

When food enters the stomach, peristaltic movements pass through the stomach walls every 15 to 25 seconds. These waves soften the food, mix it with digestive juices, and reduce it to a thin liquid. The waves slowly become more vigorous and force a small amount of liquid out of the muscular opening at the lower end of the stomach. What type of muscle makes up the walls of the stomach?

Health Connection

Impacted Wisdom Teeth

Impacted (blocked) wisdom teeth can lead to abscesses, jaw problems, and can cause intense pain. Surgery to solve these problems is common. The patient is given a local anesthetic. Then the dentist cuts the gums and extracts the teeth. If this is done early, it is usually routine. However, if one waits too long, the wisdom teeth may push other teeth out of alignment or grow down into the jawbone.

Visual Learning

Figure 38.3 What type of muscle makes up the walls of the stomach? *smooth muscle*

GLENCOE TECHNOLOGY

Videodisc

Biology: The Dynamics of Life
Disc 2, Side 1
X Ray of Swallowing (Ch. 31)

PROJECT

Meat Tenderizers

Have students conduct library research to find out how meat tenderizer works. Have them use the information they gather to design a demonstration that can be used to explain the process to others. Encourage students to write out the procedural steps for their demonstration and explain the purpose for each step. Finally, have students complete their demonstration with information that relates the function of meat tenderizer to its complementary organ of the digestive system. L2 LS 

Different Viewpoints in Biology

Use of Laparoscopy

Using a tiny camera inserted into the abdomen, doctors have revolutionized gallbladder surgery. Today, approximately 90 percent of gallbladder operations use laparoscopy. Some doctors now use this technique to remove appendixes and repair hernias, the two most common general surgeries in the United States.

Using laparoscopy in gallbladder surgery reduces pain and requires shorter hospital stays. However, many doctors question its use in other surgeries. Laparoscopy is more expensive than conventional surgery. Yet, there is no significant difference in pain or length of stay between laparoscopy and conventional surgery in appendix and hernia surgery.

Chemical digestion in the stomach

The lining of the stomach contains millions of glands that secrete a mixture of chemicals called gastric juice. Gastric juice contains hydrochloric acid and pepsin. **Pepsin** is a digestive enzyme that begins the chemical digestion of proteins in food. The enzyme pepsin works only in an acidic environment. This necessary environment is provided by the hydrochloric acid. The hydrochloric acid reduces the pH of the stomach contents to 2.

Knowing that the stomach secretes acids and enzymes, you may be wondering why the stomach doesn't digest itself. It does to an extent; however, the stomach lining secretes a mucus that forms a protective layer that limits damage, and cells lining the stomach that are damaged are constantly replaced.

Food remains in your stomach for two to four hours. Serving as a holding chamber, the stomach slowly releases food into the small intestine. By the time the food is ready to leave the stomach, it is in the form of a thin liquid about the consistency of tomato soup.

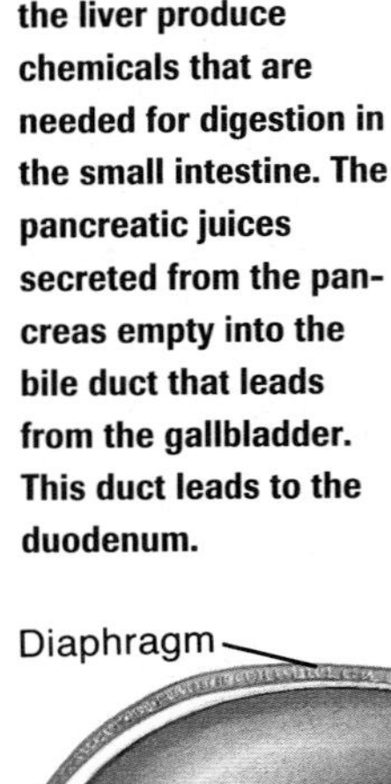

Figure 38.4

Both the pancreas and the liver produce chemicals that are needed for digestion in the small intestine. The pancreatic juices secreted from the pancreas empty into the bile duct that leads from the gallbladder. This duct leads to the duodenum.

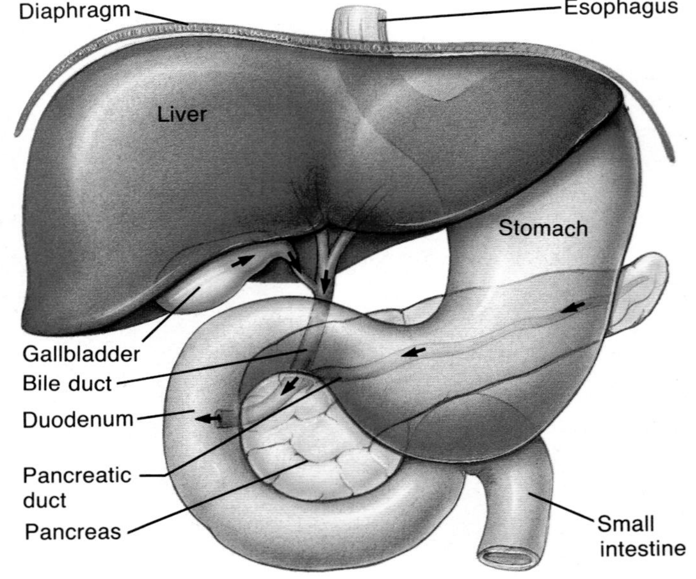

The Small Intestine

From your stomach, the partially liquid food moves into your **small intestine,** a muscular tube about 6 m long. This section of the intestine is called *small,* not because of its length, but because it has a narrow diameter of only 2.5 cm. Digestion of your meal is completed within the small intestine. Muscle contractions contribute to further mechanical breakdown, and at the same time, carbohydrates and proteins undergo further chemical digestion.

Chemical action

The first 25 cm of the small intestine is called the duodenum. Although the inner walls of the duodenum secrete enzymes, most of the enzymes and chemicals that function in the duodenum enter it through a duct from the pancreas and the liver. These organs, shown in *Figure 38.4,* have important roles in digestion, even though food does not pass through them.

Secretions of the pancreas

The **pancreas** is a soft, flattened gland that secretes both digestive enzymes and hormones. The mixture of enzymes it secretes breaks down carbohydrates, proteins, and fats. The pancreas also secretes sodium hydrogen carbonate, which makes the pancreatic juice alkaline. Recall that the liquid leaving the stomach—a mixture of partially digested food and gastric juice—is acidic. The alkalinity of pancreatic juice neutralizes this acidity and stops any further action of pepsin.

Alternate Lab: Digesting Lactose

Purpose

LS **Logical-Mathematical** Students will test whether Lactaid or other lactose digestive aid will digest lactose.

Preparation

Do not use the tablet form of Lactaid; it tests positive for glucose. If you use another liquid lactose digestive aid, test it for glucose first.

Materials

Lactaid or other liquid lactose digestive aid, glucose test paper, glucose solution, milk, graduated cylinders, eyedroppers, test tubes, toothpicks

Procedure

Give students the following directions.

1. Label four test tubes #1, 2, 3, and 4.
2. Place 2 mL of glucose in test tube #1, 2 mL of milk in test tube #2, 2 mL of milk in test tube #3, and 2 mL of Lactaid in test tube #4.
3. Place 5 drops of Lactaid into test tube #3.
4. Allow all tubes to sit for 5 minutes.

Secretions of the liver

The **liver** is a complex gland that, among its many important functions, produces bile. **Bile** is a chemical that breaks fats into small droplets and is also involved in neutralizing stomach acids. Once made in the liver, bile is stored in a small organ called the **gallbladder,** from which bile passes into the duodenum. Although bile is a chemical, this step is part of mechanical digestion. Large drops of fat are broken apart into smaller droplets, but the fat molecules are not changed chemically by the bile.

Absorption of food

Liquid food stays in your small intestine for three to five hours and is slowly moved along by peristalsis. As food moves through the small intestine, it passes over thousands of fingerlike structures called villi. A **villus** is a projection on the lining of the small intestine that functions in the absorption of digested food. The structure and number of the villi greatly increase the surface area of the small intestine.

Notice in *Figure 38.5*, on the next page, that a network of blood vessels and a lymph duct go into and out of each villus. The villi, then, are the link between the digestive system and the circulatory system (described in Chapter 39), and between the digestive system and the lymphatic system (described in Chapter 42).

As your lunch comes to the end of its passage through the small intestine, only the materials from your sandwich that could not be digested remain in the digestive tract.

Issues BIOLOGY & SOCIETY

Megavitamins

Almost half of the American population takes vitamins. Multivitamins are the most popular, followed by vitamins C, E, and B complexes. This extraordinary and widespread use of vitamins is not for fear of diet-deficiency diseases such as rickets; vitamin D has been added to milk and margarine for many years to prevent rickets. Most vitamin takers are seeking greater energy levels, better complexions, stronger hair and nails, and a longer life.

RDAs Traditionally, doctors recommended vitamins only if a person's diet was deficient in specific vitamins. Their patients took only the Recommended Daily Allowances (RDAs), the nutritional levels recommended to prevent deficiencies. Today, the Food and Drug Administration (FDA) has estimated that almost 10 percent of those who use vitamins take more than the RDA. Currently, the benefits and risks of taking large doses of vitamins, called megavitamins, are the subject of debate among scientists.

Different Viewpoints

Pauling and vitamin C Two-time Nobel prize-winning chemist Linus Pauling was the first well-known scientist to favor the megavitamin-as-preventive-medicine theory. He claimed that large doses of vitamin C could ward off the common cold and help prevent heart disease and cancer. His theories have been supported by studies in which large daily doses of vitamin C cut heart disease deaths by nearly half in men and by one-fourth in women.

Antioxidants Megavitamins may also protect people from chronic diseases and boost their immune systems. High doses of antioxidants—vitamins such as C and E and beta-carotene—destroy free radicals, substances that can damage cells and lead to cancer.

How much is too much? Some doctors caution that megavitamins should be treated like all medicines. Some vitamins, when taken in large doses and/or over a long period of time, can be toxic. Cholesterol patients who were not monitored by their doctors suffered severe liver damage from extended use of niacin (vitamin B_3).

INVESTIGATING the Issue

Writing About Biology Research the use and abuse of amino acids as dietary supplements. Write a report in your journal.

5. Using the glucose test paper, test each tube for glucose.

Analysis

Ask students the following questions.

1. What did Lactaid do to the lactose in milk? *It broke down the lactose.*
2. What is the function of test tubes #1, 2, and 4? *They are controls.*
3. How will Lactaid help people who are lactose intolerant? *Lactaid will break down the lactose present in dairy products.*

Assessment

Journal: Have students write a summary of the lab and the answers to the Analysis questions in their journals. L1

Issues BIOLOGY & SOCIETY

Megavitamins

Purpose

Students will become familiar with the debate over megadosing vitamins.

Teaching Strategies

- Ask students to make a vitamin table listing for each vitamin the adult RDA, food sources, benefits, and possible risks of overdosing.

INVESTIGATING the Issue

Writing About Biology Studies in lab animals have shown that some amino acid supplements cause decreased growth, changes in brain chemistry, and result in offspring with abnormally small brains. Methionine in high doses has been found to damage the spleen, pancreas, and kidneys. Other amino acid supplements, such as lysine, were found to be fairly harmless. Concerns are for pregnant women, diabetics, people with chronic liver disease, and for those with a genetic inability to metabolize particular amino acids.

Going Further

Ask students to visit a health store. Have them speak with a salesperson to get his or her opinion on this issue. Ask students to report to the class.

Demonstration

LS **Visual-Spatial** Acquire a sheep or pig pancreas. Blend the pancreas with 150 mL of 30 percent ethyl alcohol. Allow the solution to stand for 14 hours, shaking occasionally. Strain the solution through cheesecloth and then filter. Neutralize with KOH until you get near the end point, then use 0.5% sodium carbonate. Use the resulting solution on a chopped, hard-boiled egg to demonstrate the action of pancreatic enzymes.

3 Assess

Check for Understanding

Play the "Digestive System Game." Have a student name an organ of the digestive system. The next student must name a type of digestion that occurs in the organ (chemical or mechanical). The next student must further define this action by listing the enzymes or juices involved or defining the mechanical action (chewing or stomach churning). A fourth student should tell how the digestive action is controlled. Repeat the process until all organs have been discussed. L1 COOP LEARN

Reteach

Have students prepare a chart of the digestive organs and their types of digestion. L1 SAE

Extension

Have interested students research and report on cancers of the digestive tract and their possible causes and treatments. L2

Assessment

Knowledge: Have students label a diagram of the digestive system and identify the functions of each part. L1

4 Close

Using an Analogy

Ask students how the digestive system is like the plumbing system of a house. Have them explain where the faucets of the digestive system are located. L1

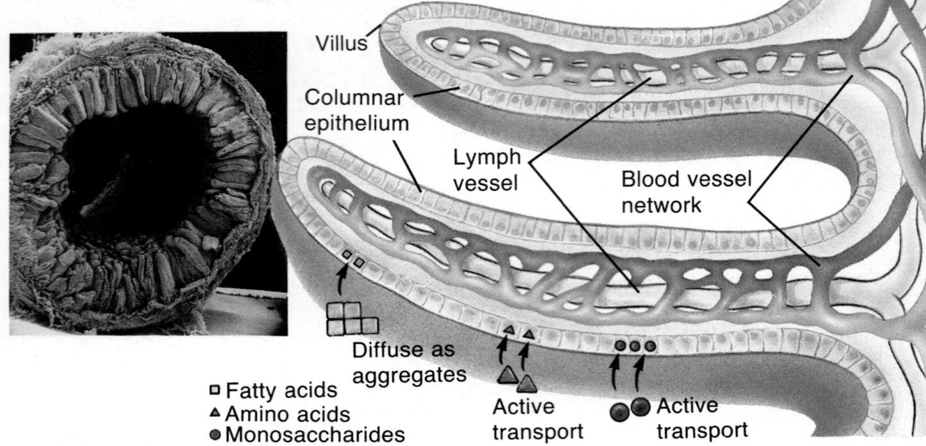

Figure 38.5
Once food has been fully digested in the small intestine (cross section shown on left), it is composed of molecules small enough to enter the body's bloodstream. Most monosaccharides and amino acids are actively transported into the epithelial cells of the villi. The molecules diffuse into the blood vessels that supply each villus (right). Within the cells of each villus, the fat molecules diffuse into the lymphatic system, which provides the body with tissue fluids.

appendix:
ad (L) to
pendere (L) to hang
The appendix hangs from the intestine.

The Large Intestine

The indigestible material from your meal passes into your **large intestine,** a muscular tube that is also called the colon. Even though the large intestine is about 1.5 m long, and therefore much shorter than the small intestine, it is much wider—about 6.5 cm. The appendix, an extension at the junction of the small and large intestines, is thought to be an evolutionary remnant from our herbivorous ancestors. No function in digestion has been discovered.

Vitamin synthesis

Water is absorbed from the indigestible mixture through the walls of the large intestine, leaving behind a more solid material. Anaerobic bacteria residing in the large intestine digest more of this material and also synthesize some B vitamins and vitamin K, which are absorbed as needed by the body.

Elimination of wastes

After 18 to 24 hours in the large intestine, the remaining indigestible material, now called feces, reaches the rectum. The **rectum** is the last section of the digestive system. Feces are eliminated from the rectum through the anus. The journey from the beginning of the digestive system to the end has taken your meal from 24 to 33 hours.

Section Review

Understanding Concepts

1. Sequence the organs of your body through which food passes.
2. In which two parts of the digestive system are starches digested?
3. How do villi of the small intestine enhance the rate of nutrient absorption?

Thinking Critically

4. What could happen to a person if the number of bacteria in the intestine were lowered because of chronic diarrhea?

Skill Review

5. **Making and Using Graphs** Prepare a pie graph representing the time food remains in each part of the digestive tract. For more help, refer to Organizing Information in the *Skill Handbook.*

Answers to Section Review

1. mouth, esophagus, stomach, small intestine, large intestine, rectum
2. mouth, small intestine
3. By increasing the surface area, villi provide more surface on which absorption can occur.

Thinking Critically

4. The B vitamins and vitamin K that they synthesize will not be available.

Skill Review

5. Students' graphs should show approximately these percentages:
large intestine 75%
small intestine 13%
stomach 11%
mouth and esophagus 1%

SECTION 38.2 The Control of Digestion and Homeostasis

Remember the predicament of the student embarrassed in class by his rumbling stomach? The rumbling can be explained when we examine how digestion is controlled by the interaction of the nervous system, the endocrine system, and the digestive system.

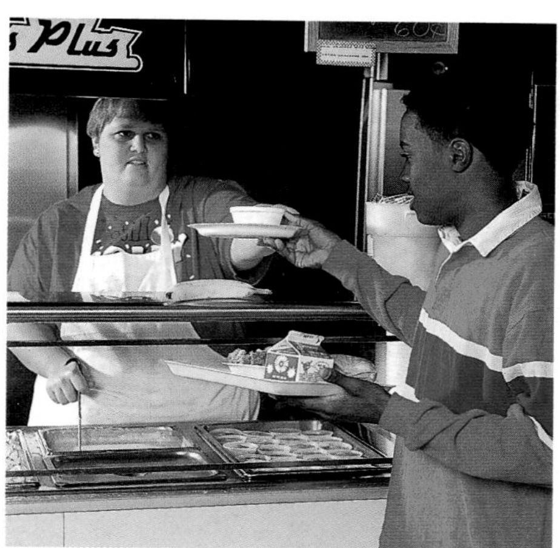

Section Preview

Objectives

Explain how the nervous system and hormones control the digestive process.

Identify the functions of the glands that affect digestion.

Key Terms

exocrine gland
endocrine gland
target tissue
thyroid gland
parathyroid gland

Control of Digestion

When you smell food, look at a picture of food when you're hungry, or just think about your favorite food, you may notice that your mouth waters. The smell, sight, or thought of food increases the secretions of saliva in preparation for eating. This response involves a part of the brain and the memories that are associated with food.

The secretion of gastric juice in the stomach is under the control of both the nervous and endocrine systems. The sight or smell of food stimulates the secretion of gastric juices. This response is controlled by the nervous system. Once food is in the stomach, nerve receptors respond to the stretching of the stomach and signal the medulla of the brain. The medulla then stimulates the stomach glands to continue the secretion of gastric juice. Recall that gastric juice is responsible for beginning the digestion of proteins. Protein in the stomach stimulates the stomach lining to secrete gastrin. Gastrin is a hormone that is absorbed into the blood and further stimulates the glands in the stomach to secrete gastric juice. As food passes into the intestine, hormones are secreted by the intestine. These hormones inhibit the secretion of gastric juice in the stomach, thus slowing its action, and stimulate pancreatic secretions. They also stimulate the production and release of bile.

Endocrine Control of Homeostasis and Metabolism

Just as hormones of the digestive tract are directly involved in digestion, hormones also play key roles in maintaining the necessary levels of nutrients in the blood and controlling the energy exchange between cells in the body.

SECTION 38.2

Prepare

Key Concepts

Students will study how the nervous system and hormones control the digestive and metabolic processes.

Block Scheduling

Look for this symbol for strategies that are useful in a block scheduling format. For more information on block scheduling, refer to Section 38.2 in the **Lesson Plans** booklet.

1 Focus

Bellringer

Before presenting the lesson, display **Section Focus Master 86** on the overhead projector and have students answer the accompanying questions. L1 SAE

Discussion

Have students identify causes of stress in their lives and elicit what bodily responses they notice as a result. *increase in breathing rate, heart rate, alertness, a churning feeling in the stomach, or an increase in blood pressure*

List the stresses and responses on the chalkboard. Point out to students that such responses are under the control of the nervous system and hormones.

Program Resources

Section Focus Master 86 L1 SAE
Reinforcement and Study Guide, p. 157 L1
Laboratory Manual, pp. 235-238 L2
Critical Thinking/Problem Solving, p. 38 L3
Basic Concepts Transparency 65 and **Master** L1 SAE

2 Teach

Using Science Terms

Explain the meaning of the following terms: *Endo* is Greek for "within" and *crine* (krinein) is Greek for "to separate." Insulin gets its name from the Latin *insula*, meaning "island." Insulin is made in the small islands or islets of beta cells in the pancreas. *Hypo* is Greek for "under" and *thalamus* is Greek for "the inner room." The *thalamus* was an inner room in a Greek ship.

ThinkingLab Interpret the Data

Purpose

LS **Logical-Mathematical** Interpret the changes in plasma insulin and glucagon during prolonged exercise and relate these actions to getting glucose to the body cells.

Process Skills

recognize cause and effect, interpret data, analyze

Teaching Strategies

- Ask students to list on the chalkboard the changes that occur in the body during exercise. Identify the changes that require increased glucose inside cells. L1

Thinking Critically

Glucagon causes blood glucose to increase by increasing the conversion of glycogen into glucose. Insulin decreases because it acts to lower blood glucose level and accelerate the conversion of glucose to glycogen.

Assessment

Journal: Ask students to write in their journals a summary of the effects of prolonged exercise on plasma insulin and glucagon. Have them explain how the actions of exercise help get glucose to body cells. L1

ThinkingLab Interpret the Data

What are the effects of glucagon and insulin during exercise?

Exercise represents a special example of rapid fuel mobilization in the body. The body must gear up to supply great amounts of glucose and oxygen for muscle metabolism, while also providing a steady delivery of glucose to the central nervous system. The glucose use in a resting muscle is generally low but changes dramatically with exercise. Within ten minutes after exercise has started, glucose uptake from the blood may increase up to fifteenfold, and by 60 minutes up to thirtyfold.

Insulin
Glucagon
Insulin (μU/mL): 5, 10
Glucagon (pg/mL): 150, 300, 450
Hours: 0, 1, 2, 3, 4

Analysis

The graph here shows the effects of prolonged exercise on plasma insulin and glucagon in humans.

Thinking Critically

Explain why glucagon concentration goes up and insulin concentration goes down during exercise, and how these actions help get glucose to the body cells.

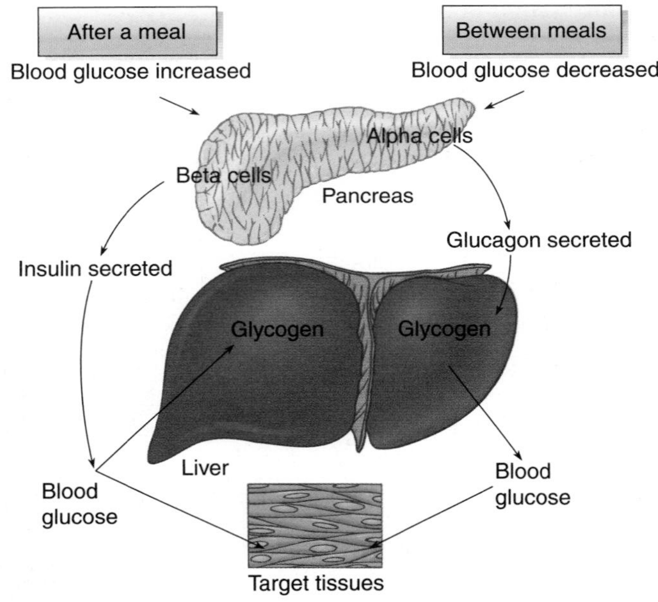

Glucose regulation: The pancreas

The maintenance of blood glucose is accomplished by two means—by your eating meals at regular intervals during the day and by hormones making finely tuned adjustments in the blood. Digested sugars from a meal just eaten may pass directly to the liver, where they are absorbed and converted into glycogen. Glycogen is a polysaccharide and is the form in which sugars are stored in the liver as well as in skeletal muscle. When your body needs energy, glycogen is broken down into glucose, which is transported by blood to fuel cell processes throughout the body. The release of sugar into the bloodstream is controlled by the portion of the pancreas that produces the hormones glucagon and insulin.

Recall that the pancreas is involved in the digestive process by secreting enzymes into the small intestine. In this role, it functions as an **exocrine gland,** secreting its enzymes through ducts. The portion of the pancreas that secretes hormones functions as an endocrine gland. An **endocrine gland** is a ductless organ that releases hormones directly into the bloodstream. The hormones released by endocrine glands convey information to specific cells in the body called **target tissue.** *Figure 38.6* illustrates how the hormones of the pancreas regulate the body's blood sugar level before and after a meal. Physical activity also will cause secretions of these hormones to vary.

Figure 38.6

The endocrine portion of the pancreas is made of about 1 million clusters of cells called islets. The islets contain alpha and beta cells. The alpha cells secrete glucagon, which raises blood sugar level. The beta cells secrete insulin, which lowers blood sugar level.

CULTURAL DIVERSITY

Food Preferences Across Cultures

Food preferences develop as practical solutions to the problem of providing essential nutrients to populations that live under stable natural conditions. Have students find out how different cultures choose combinations of foods and flavors. Have them bring in samples of foods from various cultures. L1

After a meal, blood sugar levels rise. Insulin secreted by the pancreas enables glucose in the blood to enter the body's cells, especially muscle fiber cells. The level of blood sugar is then lowered. Insulin accelerates the conversion of excess glucose into glycogen, which can then be stored in the liver.

When the blood sugar level becomes low between meals, the alpha cells of the pancreas secrete glucagon. Glucagon accelerates the conversion of stored glycogen in the liver to glucose. The liver then releases the glucose into the blood and the blood sugar level rises. The opposite action of glucagon and insulin on the blood sugar level is an example of a feedback mechanism of the body that helps to maintain its homeostasis.

Metabolic control: The thyroid gland

The **thyroid gland** regulates metabolism and energy balance, growth and development, and the general activity of the nervous system. The main metabolic and growth hormone of the thyroid is thyroxine. This hormone affects the rate at which the body uses energy and determines your food intake requirements.

The thyroid also secretes the hormone calcitonin. Calcitonin is one of two hormones that regulate calcium and phosphate levels in the blood. The other hormone is secreted by the parathyroid glands.

Health

Diabetes—A Sugar Problem

To understand diabetes, you need to understand how sugar and other carbohydrates are metabolized.

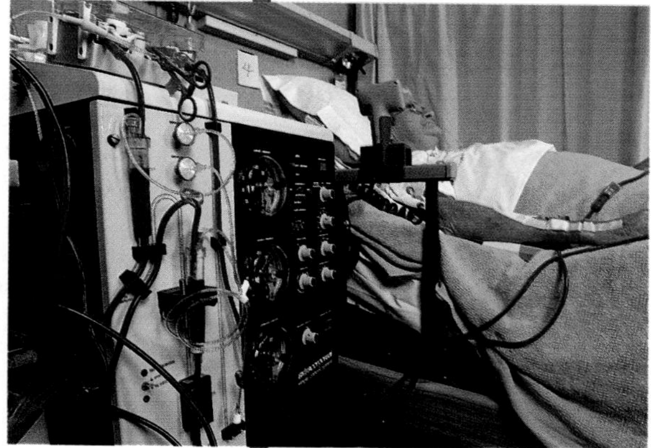

Insulin is the key During digestion, carbohydrates and sugar are changed to glucose and travel through the bloodstream nourishing the cells of the body. But glucose can't do its job without insulin. It's an important hormone because it enables glucose to enter the cells. Sometimes the pancreas doesn't produce enough insulin, or the cells of the body become resistant to insulin. When either happens, the glucose level in the blood rises, and large amounts of sugar are excreted in the urine.

The hypothalamus is also affected by the lack of insulin. The hypothalamus is a gland located at the base of the brain that regulates the appetite. A lack of insulin prevents glucose in the blood from entering the cells of the hypothalamus, so the person with diabetes gets hungry and eats constantly, but he or she is actually starving because glucose can't enter the cells and nourish the body. Rapid weight loss, along with great thirst and hunger, are the first signs of this disorder.

Causes of diabetes Diabetes can occur in young children. When it does, it is probably due to genetics, a virus infection, or an autoimmune reaction. An autoimmune reaction is one in which the body's immune system attacks its own tissue, the pancreas in this case, as if it were a foreign substance.

Diabetes can also affect adults in their middle years. Unlike the children with diabetes, they have higher-than-normal amounts of insulin as well as sugar in their blood. How does this happen? The pancreas is producing insulin normally, but the body cells are resistant to insulin.

CONNECTION TO Biology

Hypothesize how genetic engineering could be used to attack childhood diabetes.

Health

Diabetes — A Sugar Problem

Purpose

Students will become acquainted with the causes, effects, and treatments of diabetes.

Teaching Strategies

- Review the endocrine functions of the pancreas.
- Ask students to predict what would happen if the pancreas did not produce insulin. *The glucose level of the body would rise.* L1

Possible Answer

Answers may include the idea of inserting the human insulin gene into the diabetic's pancreas.

Teacher FYI

Insulin-dependent diabetes is also called Type I, or juvenile, diabetes. One major complication of Type I diabetes is loss of vision due to cataracts. The excessive blood glucose chemically attaches to the lens proteins, clouding the lens. Type I diabetes often causes kidney damage also. Non-insulin-dependent diabetes is called Type II diabetes. Because this type of diabetes is most common in elderly people, it is sometimes called late-onset diabetes.

GLENCOE TECHNOLOGY

Videodisc

STVS: Human Biology
Disc 7, Side 2
Insulin Pills (Ch. 9)

Meeting Individual Needs

Learning Disabled Have students who are having difficulty keeping track of hormones and glands, prepare a table with the columns Gland, Hormone, Action, and Target Tissue. Beneath the Gland head, have students list the names of the glands presented in this section and then complete the table for each gland they list. L1

PORTFOLIO

Thyroid Hormone Have students sequence the pathway of thyroid hormone from the thyroid gland to its target tissues. Have students caption their flowcharts with a summary of the effects the thyroid hormone has on its target tissues. L1 P LS

SECTION 38.2

3 Assess

Check for Understanding

Have students make a diagram that summarizes the control of calcium level in the body. L1

Reteach

Show the video *Blood Sugar Regulation & Diabetes*, EBEC. L1

Extension

- Have students look up information on scientists who have discovered or synthesized endocrine hormones. F.G. Banting and C.H. Best discovered insulin, E.C. Kendall isolated thyroxine and cortisone, and P.S. Hench discovered that cortisone had a beneficial effect on inflamed tissues. L2

Assessment

Journal: Ask students to summarize the control of blood sugar level. L1

4 Close

Discussion

Discuss with students what will happen if the thyroid gland is overactive or underactive. L1

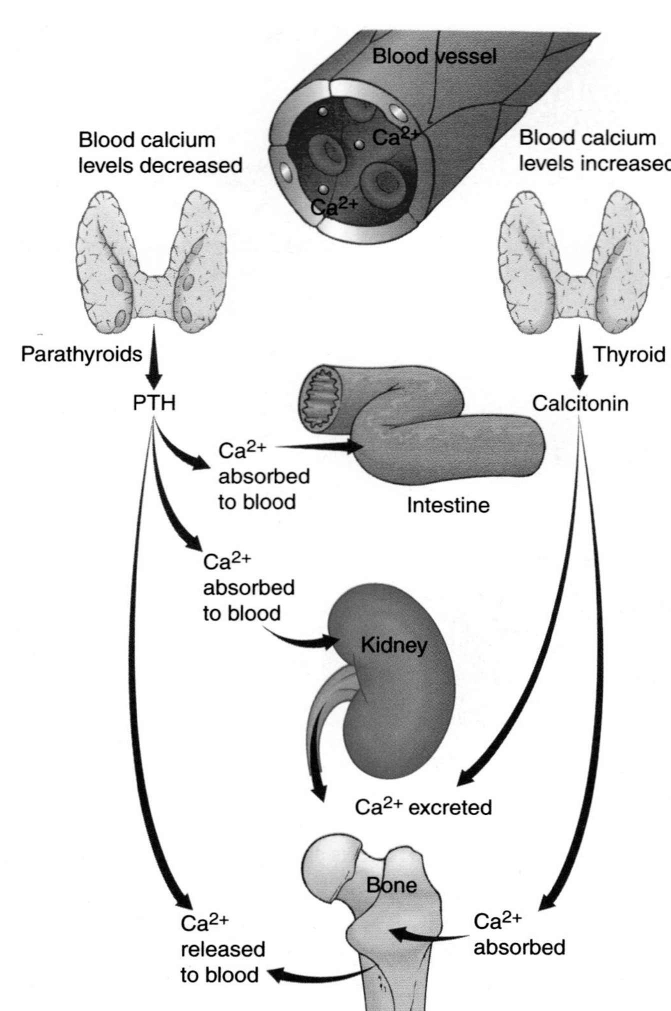

Calcium regulation: The parathyroid glands

Recall that calcium is necessary for blood clotting, formation of bones and teeth, and normal muscle function. It is also needed for normal nerve activity. Parathyroid hormone (PTH) from the **parathyroid glands** increases the rate of calcium, phosphate, and magnesium absorption in the intestines and causes the release of calcium and phosphate from bone tissue. It also increases the rate at which the kidneys remove calcium and magnesium from urine and returns them to the blood. The overall results of secretion of parathyroid hormone are the decrease of the blood phosphate level and the increase in calcium and magnesium levels. *Figure 38.7* shows the effects of the interaction of parathyroid hormone and calcitonin from the thyroid on the calcium level of blood.

Parathyriod hormone also increases the amount of vitamin D that the body makes. Vitamin D is required for the intestines to absorb calcium ions.

Figure 38.7

Parathyroid hormone and calcitonin have opposite actions in regulating calcium levels in the bloodstream.

Section Review

Understanding Concepts

1. Outline the process by which digestion is controlled.
2. Explain how the pancreas is both an exocrine and an endocrine gland.
3. What is the effect of parathyroid hormone on bone tissue?

Thinking Critically

4. Hormones continually make adjustments in blood glucose. Why must the blood glucose level be kept fairly constant?

Skill Review

5. **Comparing and Contrasting** What effects do calcitonin and PTH have on blood calcium levels? For more information, refer to Thinking Critically in the *Skill Handbook*.

Answers to Section Review

1. Digestion begins with the production of gastric juice that is stimulated by the smell, sight, or thought of food. Production continues to be stimulated by the stretching of the stomach and the presence of protein in the stomach. Food entering the small intestine causes a different hormone to inhibit the secretion of gastric juices and stimulate pancreatic secretions and the production and release of bile.
2. The pancreas secretes both digestive enzymes into a duct (exocrine function) and hormones into the blood (endocrine function).
3. Parathyroid hormone causes the release of calcium and phosphorus from bone tissue.

Thinking Critically

4. Glucose is the fuel for body cells and a constant level needs to be maintained for normal body functions.

Skill Review

5. Parathyroid hormone ultimately raises blood calcium levels and calcitonin lowers blood calcium levels.

SECTION

38.3 Nutrition

ZIGGY®
PROCESSED CHEESE FOOD?...GIMME A BREAK!

What's your favorite food? Does it taste salty, sour, sweet, bitter, or a combination of these? Does it have a wonderful aroma? Of what nutritional value is it? To be considered a food, a substance must provide energy or building materials, or it must assist in some body process. In other words, it must contain at least one of six essential nutrients.

Section Preview

Objectives

Summarize the role of the six classes of nutrients in body nutrition.

Relate Calories and metabolism.

Key Terms

mineral
vitamin
Calorie

The Vital Nutrients

Six kinds of nutrients can be found in foods: carbohydrates, fats, proteins, minerals, vitamins, and water. These substances are essential to chemical reactions in the body. You supply your body with these nutrients when you eat foods from the five main food groups, shown in *Figure 38.8.*

Carbohydrates: The body's preferred energy source

Perhaps your favorite food is pasta, fresh-baked bread, or corn on the cob. If so, your favorite food contains carbohydrates. Recall from Chapter 7 that carbohydrates are starches and sugars. Starches are complex carbohydrates found in bread, cereal, potatoes, rice, corn, beans, and pasta. Sugars are simple carbohydrates found mainly in fruits such as plums, strawberries, and oranges, as well as in syrups and jellies.

Figure 38.8

Select foods from the five food groups and you'll have a healthful diet that supplies the six essential nutrients your body needs.

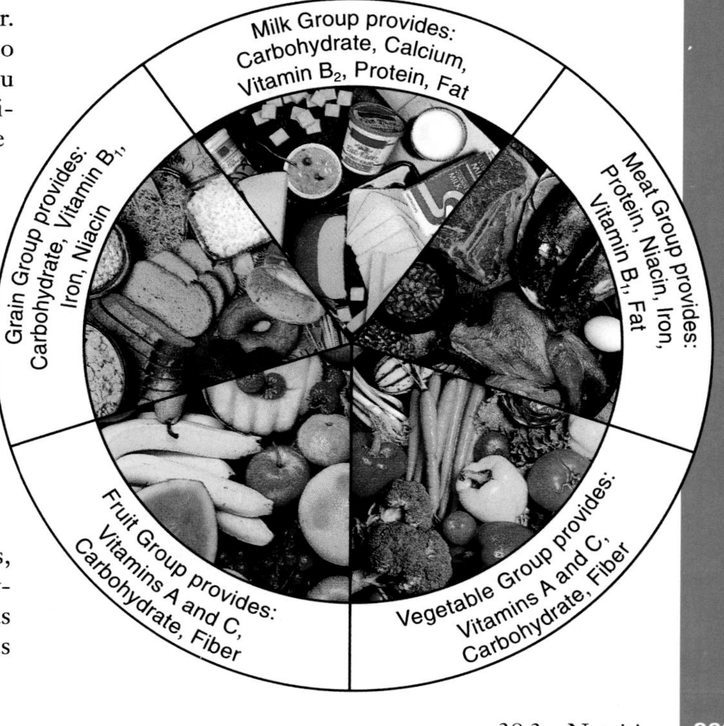

Milk Group provides: Carbohydrate, Calcium, Vitamin B_2, Protein, Fat
Meat Group provides: Protein, Niacin, Iron, Vitamin B_1, Fat
Vegetable Group provides: Vitamins A and C, Carbohydrate, Fiber
Fruit Group provides: Vitamins A and C, Carbohydrate, Fiber
Grain Group provides: Carbohydrate, Vitamin B_1, Iron, Niacin

Program Resources

Section Focus Master 87 L1 SAE
Reinforcement and Study Guide, p. 158 L1
Biolab and Minilab Worksheets, pp. 151-154 L1
Laboratory Manual, pp. 239-244 L2
Concept Mapping, p. 38 L1
Basic Skills Transparency 26 and **Master** L1 SAE

SECTION 38.3

Prepare

Key Concepts

Students will become familiar with the role of the six classes of nutrients in body nutrition. They will relate these nutrients to Calories and metabolism.

Block Scheduling

Look for this symbol for strategies that are useful in a block scheduling format. For more information on block scheduling, refer to Section 38.3 in the **Lesson Plans** booklet.

Materials

- Save food labels and ingredient lists for the Display.
- Purchase foods for the Biolab.
- Purchase a school lunch for Meeting Individual Needs.

1 Focus

Bellringer

Before presenting the lesson, display **Section Focus Master 87** on the overhead projector and have students answer the accompanying questions. L1 SAE

Using an Analogy

Ask students how the body is like an engine. What fuel is used to run this engine? What are the waste products from the engine? *Like an engine, the body needs fuel to meet its energy needs. The fuel of the body is glucose. The waste products of the body's engine are carbon dioxide and water.* L1

2 Teach

Display

Prepare a bulletin board display of food labels. Refer to the display when discussing foods rich in various nutrients.

MiniLab

Purpose

Visual-Spatial Students will evaluate a meal using a food pyramid.

Process Skills

compare and contrast, analyze, interpret a scientific illustration

Teaching Strategies

- Explain that a food pyramid demonstrates the relative amounts of each food group in a balanced diet.

Expected Results

Students should recognize that the fruit and vegetable groups are missing.

Analysis

1. Fruit and vegetable groups
2. An apple and a green salad
3. The meal was nutritious except for the missing fruit and vegetables.

Assessment

Skill: Have students use the food pyramid to list the foods that make up a well-balanced dinner. L1 P

MiniLab

Where does the food in a spaghetti dinner fit in the food pyramid?

A food pyramid describes pictorially the number of servings a person should have from each food group daily. The small top of the pyramid contains fat, oils, and sweets because these foods should be eaten most sparingly. Most Americans' diets are too high in fat. Note that some fat and sugar symbols are shown in all the food groups. This is because even foods that are naturally low in fat are often prepared in ways that add fat or sugar.

Procedure

1. Construct a food pyramid showing the foods that make up the following dinner:
   spaghetti and meatballs, garlic bread, glass of milk
2. Compare your pyramid with the pyramid shown.

Analysis

1. What food groups are missing in the dinner?
2. What could be added to this dinner to make it fit the food pyramid better?
3. Overall, do you think the spaghetti was a nutritious dinner? Explain.

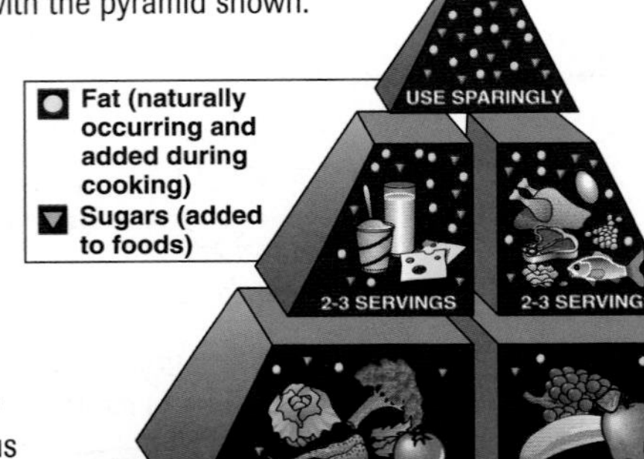

Table sugar is derived from sugar cane or sugar beet, and is generally not considered an essential nutrient. Carbohydrates are important sources of energy for your body cells.

During digestion, complex carbohydrates are broken down into simple sugars such as glucose, fructose, and galactose. Absorbed into the bloodstream through the villi of the small intestine, these sugar molecules circulate to fuel body functions. Remember that some sugar is carried to the liver where it is converted to and stored as glycogen.

Cellulose, another complex carbohydrate, is found in all plant cell walls and is not digestible by humans. However, cellulose is still an important item to include in your diet. It provides roughage, often called fiber. By eating such cellulose-containing foods as oranges, celery, and spinach, you'll provide bulk in your diet that stimulates your digestive tract and helps in the elimination of wastes.

Fats: Energy sources and building materials

To many people, eating fat means getting fat. Yet fats are an essential nutrient. Fats provide energy for your body and are also used as building materials. Recall from Chapter 7 that fats are essential building blocks of the cell membrane. Fats also are important for the synthesis of hormones, for protecting body organs against injury, and for insulating the body from cold. Sources of fat in the diet include meats, nuts, and dairy products, as well as cooking oils. In the digestive system, fats are broken down into fatty acids and glycerol and absorbed by the villi of the small intestine. Eventually, some of these fatty acids end up in the liver. The liver converts them to glycogen or stores them as fat. Unused fats are stored as fat deposits throughout your body.

Proteins: Building materials

Your body has many uses for proteins. Enzymes, antibodies, many hormones, and other substances that help the blood to clot are proteins. Proteins are part of muscles and many cell structures, including the cell membrane.

GLENCOE TECHNOLOGY

Videodisc

The Infinite Voyage: To the Edge of the Earth
Exploring Tibet: The Gateway of Exchange (Ch. 2)

The Infinite Voyage: A Taste of Health
Consequences of a Fatty Diet (Ch. 4)

During digestion, proteins are broken down into amino acids. After the amino acids have been absorbed by the small intestine, they enter the bloodstream and are carried to the liver. The liver can convert amino acids to fats or glucose, both of which can be used for energy. However, your body uses amino acids for energy only if other energy sources are depleted. Most amino acids are absorbed by cells and used for protein synthesis. The human body needs 20 different amino acids to carry out protein synthesis, but it can make only 12 of them. The rest must be consumed in the diet and are called essential amino acids. Sources of essential amino acids include meats, dried beans, whole grains, eggs, and dairy products.

Minerals and vitamins

When you think of minerals, you may picture substances that people mine, or extract from Earth. The same minerals can also be extracted from foods and put to use by your body.

A **mineral** is an inorganic substance that serves as a building material or takes part in chemical reactions in the body. Minerals make up about four percent of your total body weight. Most of the mineral content of your body is in your skeleton. Calcium and phosphorus form much of the structure of bone. *Table 38.2* lists the functions of some of the minerals that are needed by humans and the foods that provide them. Notice that the body does not use minerals as energy sources.

Table 38.2 Vital Minerals

Mineral	Function	Source
Calcium	formation of bones and teeth, blood clotting, normal muscle and nerve activity	milk, cheese, nuts, whole grains
Phosphorus	formation of bones and teeth, regulation of blood pH, muscle contraction and nerve activity, component of enzymes, DNA, RNA, and ATP	milk, whole-grain cereals, meats, vegetables
Iron	component of hemoglobin (carries oxygen to body cells) and cytochromes (ATP formation)	liver, egg yolk, peas, enriched cereals, whole grains, meat, raisins, leafy vegetables
Iodine	part of thyroid hormone, required by thyroid gland	seafood, eggs, milk, iodized table salt
Sodium	regulation of body fluid pH, transmission of nerve impulses	bacon, butter, table salt, vegetables
Potassium	transmission of nerve impulses, muscle contraction	vegetables, bananas, ketchup
Magnesium	muscle and nerve function, bone formation, enzyme function	potatoes, fruits, whole-grain cereals, vegetables
Fluorine	dental cavity reduction	fluoridated water
Manganese	enzyme activator for carbohydrate, protein, and fat metabolism; important in growth of cartilage and bone tissue	wheat germ, nuts, bran, leafy green vegetables
Copper	ingredient in several respiratory enzymes, needed for development of red blood cells	kidney, liver, beans, whole-meal flour, lentils
Sulfur	component of insulin; builds hair, nails, skin	nuts, dried fruits, barley, oatmeal, eggs, beans, cheese

Science, Technology, and Society

Can food affect your behavior?

Certain foods are known to alter the synthesis of the neurotransmitters serotonin, dopamine, norepinephrine, and acetylcholine. All these neurotransmitters are precursor dependent. For example, serotonin is dependent on the concentration of the amino acid tryptophan.

When protein is digested, tryptophan enters the blood and is transported by a carrier to the brain. The more tryptophan that enters the brain, the more serotonin that is made. One would expect high-protein meals to increase brain serotonin, but this is not the case. However, a meal high in carbohydrates results in an insulin release that facilitates the uptake of all amino acids except tryptophan.

Serotonin neurons participate in a wide range of behaviors, including sleep, feeding, locomotor activity, aggression, and pain sensitivity. Dietary manipulations that alter brain tryptophan concentrations have been found to affect many of these behaviors in animals.

Visual Learning

Refer to the periodic table on page A10 as you discuss Table 38.2. Discuss the sources of the minerals listed in Table 38.2.

Meeting Individual Needs

Students Acquiring English/Hearing Impaired Bring in a sample school lunch. Ask students to list, on a sheet of paper, the nutrients found in each food in the lunch. Next, have them point out on a diagram of the digestive system where each type of nutrient is digested. L1 SAE LS

STUDENT JOURNAL

Nutrient Analysis Have students keep track of everything they eat for one day. Have students use the data to analyze their nutrient intake. L1 LS

interNET CONNECTION

Follow the link for this chapter on the Glencoe Homepage at **www.glencoe.com/sec/science** to find out more about digestion and nutrition.

Bioethics

Dietary Supplements
Since December 31, 1993, the Food and Drug Administration has restricted access to certain dietary supplements that were previously available in health food stores. These products include such things as antioxidant nutrients, digestive enzymes, essential fatty acids, chromium, selenium, protein powders, amino acids, high potency vitamins, herbal products, and weight gain and loss products. The FDA considers that such products need to be controlled because of questionable safety. This concern arose as a result of deaths from tryptophan overdoses.

Many users of these substances believe they have a right to purchase these supplements. These people also believe they have a right to what they feel are preventative and beneficial health-care methods.

Enrichment

Have students work in groups to investigate and prepare a report on one vitamin deficiency, such as scurvy, beriberi, rickets, night blindness, polyneuritis, or pellagra. L1 COOP LEARN

Visual Learning

Using the Table Have students use Table 38.3 to identify seven vitamins contained in liver. *vitamins A, D, B_6, B_{12}, niacin, folic acid, and biotin* Have students identify the vitamins that are formed in the body. *vitamins D and K* L1

Table 38.3 Vitamins

Vitamin	Function	Source
Fat-soluble		
A	maintain health of epithelial cells; formation of light-absorbing pigment; growth of bones and teeth	liver, broccoli, green and yellow vegetables, tomatoes, butter, egg yolk
D	absorption of calcium and phosphorus in digestive tract	egg yolk, shrimp, yeast, liver, fortified milk; produced in the skin upon exposure to ultraviolet rays in sunlight
E	formation of DNA, RNA, and red blood cells	leafy vegetables, milk, butter
K	blood clotting	green vegetables, tomatoes; produced by intestinal bacteria
Water-soluble		
B_1	sugar metabolism; synthesis of neurotransmitters	ham, eggs, green vegetables, chicken, raisins, seafood, soybeans, milk
B_2 (riboflavin)	sugar and protein metabolism in cells of eyes, skin, intestines, blood	green vegetables, meats, yeast, eggs
Niacin	energy-releasing reactions; fat metabolism	yeast, meats, liver, fish, whole-grain cereals, nuts
B_6	fat metabolism	salmon, yeast, tomatoes, corn, spinach, liver, yogurt, wheat bran, whole-grain cereals and breads
B_{12}	red blood cell formation; metabolism of amino acids	liver, milk, cheese, eggs, meats
Pantothenic acid	aerobic respiration; synthesis of hormones	milk, liver, yeast, green vegetables, whole-grain cereals and breads
Folic acid	synthesis of DNA and RNA; production of red and white blood cells	liver, leafy green vegetables, nuts, orange juice
Biotin	aerobic respiration; fat metabolism	yeast, liver, egg yolk
C	protein metabolism; wound healing	citrus fruits, tomatoes, leafy green vegetables, broccoli, potatoes, peppers

Unlike minerals, a **vitamin** is an organic nutrient that is required in small amounts to maintain growth and metabolism. The two main groups of vitamins are fat-soluble vitamins and water-soluble vitamins, *Table 38.3.* Fat-soluble vitamins must be dissolved in fat before they can be absorbed. Fat-soluble vitamins can be stored in the liver, where excess amounts can become toxic. Water-soluble vitamins dissolve readily in water and cannot be stored in the body.

Vitamin D, a fat-soluble vitamin, is synthesized in your skin. Vitamin K and some B vitamins are made by bacteria in your intestine. The rest of the vitamins must be consumed in your diet.

STUDENT JOURNAL

Nutritional Issues Divide the class into groups. Have each group prepare a presentation on a nutrition issue such as cholesterol, HDLs versus LDLs, anorexia, bulimia, weight gain, junk foods, vegetarian diets, or food additives. Have students record their notes for the group report in their journals. L1 COOP LEARN

Cancer and Nutrients Have students research how antioxidants destroy free radicals that can damage cells and lead to cancer. Ask them to find out how vitamins C and E and beta-carotene destroy these substances. L3

Water: Abundant and essential

Water is the most abundant substance in your body. It makes up 60 percent of red blood cells and 75 percent of muscle cells. Water plays a role in many chemical reactions in the body and is necessary for the breakdown of foods in digestion. Water is an excellent solvent, and oxygen and nutrients from food could not enter your cells without water.

Water absorbs and releases heat slowly. It is this characteristic that helps water regulate your body's temperature. A large amount of heat is needed to raise the temperature of water. Because the body contains so much water, this slow-to-heat property of water helps keep body temperature nearly constant. Your body loses about 2.5 L of water per day through exhalation during breathing, and through sweat and urine. As a result, water must be replaced constantly.

MiniLab

Does a bowl of soup provide a complete meal?

As a consumer, you will be bombarded by advertising that promotes the nutritional benefits of specific food products. Choosing a food to eat on the basis of such ads may not make nutritional sense. By examining the ingredients of processed foods, you can learn important things about their nutritional content.

Percentage of Daily Value (DV)

Carbohydrates	60%
Fat	30%
Saturated Fats	10%
Cholesterol	1.5%
Protein	10%
Total Calories	2000

NUTRITION FACTS
Serving Size: 2 cups (452g)
Serving Per Container: 1

Amount Per Serving

Calories 140	Calories from Fat 54
	% Daily Value*
Total Fat 8g	12%
Saturated Fat 6g	30%
Cholesterol 20mg	7%
Sodium 1640 mg	68%
Total Carbohydrate 22g	7%
Dietary Fiber 5g	20%
Sugars 5g	
Protein 6g	
Vitamin A 50% • Vitamin C	4%
Calcium 2% • Iron	2%

*Percent Daily Values are based on a 2,000 calorie diet. Your daily values may be higher or lower depending on your calorie needs:

		Calories 2,000	2,500
Total Fat	Less than	65g	80g
Sat Fat	Less than	20g	25g
Cholesterol	Less than	300mg	300mg
Sodium	Less than	2,400mg	2,400mg
Total Carbohydrate		300g	375g
Fiber		25g	30g

Calories per gram:
Fat 9 • Carbohydrates 4 • Protein 4

Procedure

1. Examine the information in the above table listing the daily value (DV) of various nutrients. DV expresses what percent of Calories should come from certain nutrients. For instance, in the proposed diet of 2000 Calories, 60 percent of the Calories should come from carbohydrates.
2. Examine the nutritional information on the soup can label and compare it with the DV table.

Analysis

1. Does your bowl of soup provide more than 30 percent of any of the daily nutrients? Which ones?
2. Is soup a nutritious meal? Explain your answer.
3. What could be eaten along with the soup to provide more carbohydrates? Protein?

Metabolism and Calories

Think of all the chemical reactions you've learned about that are involved in your body's metabolism. Some of these chemical reactions break down materials such as the bacon in your sandwich into fatty acids, glycerol, and amino acids. Other chemical reactions involve synthesizing. For example, once in your cells, the amino acids from your digested sandwich are put together to form the proteins that make up your body.

The energy content of food is measured in units of heat called **Calories,** each of which represents a kilocalorie, or 1000 calories (written with a small *c*). A calorie is the amount of heat required to raise the temperature of 1 mL of water by 1°C. Some foods, especially those with fats, contain many more Calories than others. In general, 1 g of fat contains nine Calories, while 1 g of carbohydrate or protein contains four Calories.

Software

Snackmaster, Learning Seed, *The Diet Balancer,* Nutridata Software Corp., or *MacDiet,* Intellimation.

MiniLab

Purpose

LS **Logical-Mathematical** Students will evaluate the nutritional value of a food based on its label.

Process Skills

observe and infer, analyze, interpret data

Teaching Strategies

- Explain the concept of daily value to students.
- You may wish to have students work in pairs.

Expected Results

Students should discover that the soup represented is basically a healthy meal.

Analysis

1. yes; sodium
2. Basically yes, although the sodium content is high.
3. Enriched crackers or bread; milk

Assessment

Performance: Have students repeat the Minilab for three foods found in their homes. L1 P

PORTFOLIO

Analyzing Food Labels Have students list the ingredients of their favorite pizza or frozen food. Have students analyze each ingredient and determine in which part or parts of the digestive system it would be digested. Have students present their findings visually or in a table. L1 P LS

PROJECT

Calories

Ask students to list all the foods they eat during a 24-hour period. Distribute Calorie charts. Have students use the charts to figure out how many Calories they ate, as well as calculate the grams of carbohydrates, fats, and proteins they consumed. Have students use their data to evaluate the nutritional value of their meals. L2 LS

BioLab

Testing for Nutrients in Foods

Time Allotment
One class period

Objectives
Review objectives with students before they begin the Biolab.

Process Skills
observe and infer, measure in SI, form a hypothesis, collect and organize data, interpret data, experiment, analyze

Safety Precautions
- Be sure students understand that iodine solution and biuret solution can be toxic.

PREPARATION

- Prepare sufficient samples of several foods so each group may choose three different foods.

Possible Hypotheses
- Hypotheses will vary with particular foods used, but students should hypothesize that at least one nutrient can be found in each food.

BioLab

Testing for Nutrients in Foods

As heterotrophs, humans need to take in vital nutrients. Carbohydrates, such as sugars and starches, are used by the body for energy. Lipids, or fats, may be used directly as an energy source, or they may be stored as a future source of energy. Proteins are used in metabolic reactions and as building blocks to make new cells or repair old ones, or they are converted to fat or glucose to be used as a source of energy. Simple chemical or physical tests can be performed on foods to see which of these nutrients they contain.

PREPARATION

Problem
What nutrients are found in common foods?

Objectives
In this Biolab, you will:
- **Hypothesize** which nutrients are present in each food tested.
- **Demonstrate** by means of chemical tests the presence of carbohydrates and proteins in foods.
- **Determine** by a physical test whether foods contain lipids.

Materials
food samples (3)
mortar and pestle
test tubes (8)
droppers (4)
water
soapy water
test-tube rack
test-tube holder
brown paper (3 pieces)
iodine solution in dropper bottle
biuret solution in dropper bottle
laboratory apron
safety goggles

Safety Precautions
If solutions are spilled, rinse with water and call your teacher immediately.

PROCEDURE

Teaching Strategies
- Have students work in groups of four. Students should add the drops of the testing solutions carefully to count the correct number of drops.

Data and Observations
Tables will vary depending on food samples chosen. Food samples containing starch will be blue or blue-black when tested for starch. Foods not containing starch will be yellow or tan. Foods containing lipids will leave a translucent spot when tested for lipid, while foods with no lipids should not leave a spot. Foods containing proteins will be purple or lavender when tested for protein; foods with no protein will be blue.

PROCEDURE

Part A: Preparation of Samples

1. Prepare data tables like the one shown below for Parts B and C.
2. Choose three food samples. Grind each solid with about 5 mL of water. Return the ground food to its container.
3. Label your food samples 1, 2, and 3.
4. Make a hypothesis to predict which nutrients are present in each food tested.

Part B: Test for Starch

1. Label four test tubes 1, 2, 3, and 4. Using a different dropper for each food sample, add ten drops of food sample 1 to test tube 1, ten drops of food sample 2 to test tube 2, and ten drops of food sample 3 to test tube 3. Add ten drops of water to test tube 4.
2. Record the color of each test tube's contents in your Data Table.
3. Add three drops of iodine solution to each test tube.
4. Record the color of each tube's contents. If the color of the solution is deep blue, starch is present.

Part C: Biuret Test for Proteins

1. Label four test tubes, 1, 2, 3, and 4, and add the food samples and water as in step 1 of Part B.
2. Record the color of each test tube's contents.
3. Add ten drops of biuret reagent to each test tube. **CAUTION:** *Biuret reagent is extremely caustic to the skin and clothing.*
4. Record the color of each tube's contents in your data table. If the color of the solution is a shade of purple, protein is present.

Part D: Brown Paper Test for Lipids

1. Label four pieces of brown paper 1, 2, 3, and 4.
2. Place one drop of each food on each piece of paper.
3. Wait for the papers to dry.
4. Hold the pieces of paper up to a light, and observe each to see if light passes through it. If light passes through it, the spot is translucent and lipids are present. Record this for each food.

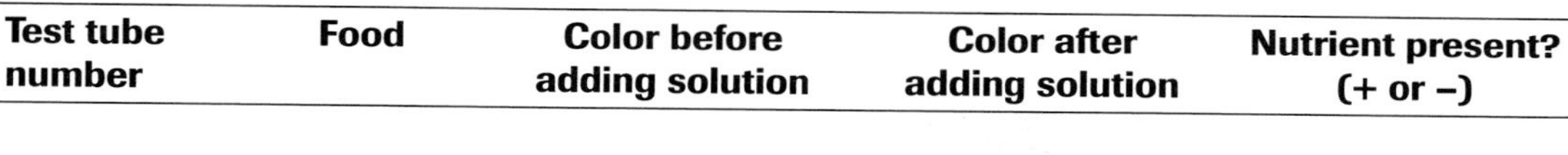

Test tube number	Food	Color before adding solution	Color after adding solution	Nutrient present? (+ or –)

ANALYZE AND CONCLUDE

1. **Analyzing Data** Which of the food samples gave a positive test for starches? For proteins? For lipids?
2. **Analyzing Data** Which foods contain more than one nutrient?
3. **Checking Your Hypothesis** Do your data support your hypothesis? Why or why not?

Going Further

Making Predictions
Predict which nutrients are present in one of your school lunches, and then test the food.

ANALYZE AND CONCLUDE

1. Answers will vary with foods. Breads, desserts, and pasta will give a positive test for carbohydrates. Butter, oils, and potato chips will give a positive test for lipids. Meats, milk, and cheeses will give a positive test for protein.
2. Many foods will contain more than one nutrient.
3. Students who hypothesize correctly will have test results to support their hypotheses.

Assessment

Oral: Ask students to summarize the results of their food tests. Once a nutrient has been named in a food, ask the students what the body uses that nutrient for. L1

Going Further

Ask students why a nutritionist working for a food company might need to use tests such as the ones in this lab. Ask why nutritionists might need to also quantify the amounts of each nutrient.

GLENCOE TECHNOLOGY

Videodisc

STVS: Human Biology
Disc 7, Side 2
Measuring Body Fat (Ch. 12)

Disc 7, Side 2
Measuring Calcium Deficiency (Ch. 14)

3 Assess

Check for Understanding

Give the students a sample lunch that has nutritional deficiencies. Ask them to evaluate the lunch. Ask what nutrients are in excess and what nutrients should be added. Elicit what foods could provide these nutrients. L1

Reteach

Have students prepare a chart of the six nutrients. Have them identify the functions of these nutrients and identify foods that contain each. L1 P

Extension

Have students interview a nutritionist to find out what criteria the nutritionist uses when planning meals to meet the needs of the people who eat them. L2

Assessment

Portfolio: Ask students to write a summary of what substances need to be included in a daily balanced diet and what each nutrient does. L1

4 Close

Discussion

Ask students to evaluate an average American's diet compared to that of someone living in a developing nation. Have students consider differences in protein, mineral, and vitamin content. L2

Figure 38.9

When the energy taken in (Calories consumed) is greater than the energy expended (Calories burned), the result will be that the person gains weight.

The actual number of Calories needed each day varies from person to person depending on the person's body mass, age, sex, and level of physical activity. In general, males need more Calories per day than females do, teenagers need more Calories than adults need, and active people need more Calories than inactive people need.

What happens if you eat more Calories than your body can metabolize? As *Figure 38.9* shows, you'll store the extra energy as body fat and gain weight. If you eat fewer Calories than your body can metabolize, you will use some of the energy stored in body fat and lose weight.

Connecting Ideas

Your body needs a constant supply of energy, yet you don't need to eat constantly. As a meal is digested and the nutrients are absorbed, your blood has more than enough nutrients in it. When blood passes from the intestine to the liver, the liver absorbs extra nutrients from the blood. Hormones released directly into blood regulate the blood's levels of nutrients and keep it fairly constant. What other roles does blood play in the body? How does the body get the oxygen it needs along with food to produce energy? And how else does it rid itself of wastes it cannot use? Answers to these questions lie in an understanding of the circulatory, respiratory, and excretory systems of the human body.

Section Review

Understanding Concepts

1. In what ways are proteins used in the body?
2. Why can an excess of fat-soluble vitamins be toxic?
3. A person can live several weeks without food, but can live only days without water. Why is the constant intake of water necessary for the body?

Thinking Critically

4. How does the liver help maintain homeostasis?

Skill Review

5. **Classifying** Prepare a chart of food groups high in each of the six nutrients. For more help, refer to Organizing Information in the *Skill Handbook*.

Answers to Section Review

1. form enzymes, antibodies, hormones, clotting chemicals, and cell structures
2. Fat-soluble vitamins accumulate in the body and excess vitamins can be toxic to certain organs.
3. Water is needed for oxygen and nutrients to enter cells and to maintain body temperature.

Thinking Critically

4. The liver converts sugar to glycogen, converts fatty acids to glycogen and stores them, and converts amino acids to glucose or fats.

Skill Review

5. fruit-vegetable: carbohydrates, vitamins, minerals, protein (beans); grains: protein, carbohydrates, minerals, vitamins; dairy: protein, fat, vitamins, minerals, carbohydrates; meat: protein, fat, vitamins, minerals

Chapter 38 Review

REVIEWING MAIN IDEAS

38.1 Following Digestion of a Meal

- Digestion begins in the mouth with both mechanical and chemical action on food. The esophagus transports food from the mouth to the stomach.
- Chemical and mechanical digestion continue in the acidic environment of the stomach.
- In the small intestine, digestion is completed and food is absorbed. The liver and pancreas play key roles in digestion.
- The large intestine absorbs water before indigestible materials are eliminated.

38.2 The Control of Digestion and Homeostasis

- Digestion is controlled by both the nervous system and the endocrine system.
- Hormones regulate the blood levels of products of digestion.

38.3 Nutrition

- Carbohydrates are the main source of energy for the body. Fats are used as a source of energy and as building blocks. Proteins are used as building materials, enzymes, hormones, antibodies, and a source of energy. Minerals serve as structural materials or take part in chemical reactions. Vitamins are needed for growth and metabolism. Water serves many vital functions in the body.
- Metabolic rate determines how quickly energy is burned. Together with Calorie intake, it helps determine a person's weight.

Key Terms

Write a sentence that shows your understanding of each of the following terms.

amylase	pancreas
bile	parathyroid gland
Calorie	pepsin
endocrine gland	peristalsis
epiglottis	rectum
esophagus	small intestine
exocrine gland	stomach
gallbladder	target tissue
large intestine	thyroid gland
liver	villus
mineral	vitamin

Understanding Concepts

1. Which of these is NOT a function of the digestive system?
   a. eliminating wastes
   b. absorbing nutrients
   c. digesting food
   d. regulating metabolism
2. Salivary glands in your mouth produce ____________, an enzyme that breaks down starches.
   a. saliva c. pepsin
   b. amylase d. bile
3. Which of these is an example of mechanical digestion?
   a. peristalsis c. chewing
   b. coughing d. epiglottis
4. The ____________ is a muscular tube that connects the mouth to the stomach.
   a. esophagus c. palate
   b. epiglottis d. tongue
5. The ____________ prevents swallowed food from entering the windpipe.
   a. esophagus c. palate
   b. epiglottis d. tongue
6. Even if you were standing on your head, this process would move food along your digestive tract.
   a. peristalsis c. secretion
   b. swallowing d. absorption
7. What is the pH of your stomach during digestion?
   a. 2 c. 6
   b. 4 d. 8

Chapter 38 Review

Reviewing Main Ideas

Summary statements can be used by students to review the major concepts of the chapter.

Key Terms

Answers should go beyond defining the terms. Accept any answer that uses the term correctly and in the proper context.

Understanding Concepts

1. d
2. b
3. c
4. a
5. b
6. a
7. a

Chapter 38 Review

8. b
9. c
10. c
11. a
12. d
13. b
14. b
15. d
16. c
17. a
18. c
19. b
20. c

Applying Concepts

21. Proteins would not be digested in the stomach.
22. diet pop: no nutrients, mostly excreted, sodium absorbed; regular pop: contains water and carbohydrates that are absorbed into the bloodstream and converted to glycogen in the liver; diet pop appears safe but is not a food
23. Because the large intestine removes water, the fecal materials would have a high liquid content.
24. The parathyroid glands secrete PTH, which increases the calcium level in the blood. Calcium is required for muscle contraction. Removal of the gland could greatly reduce blood calcium concentration and interfere with muscle contraction.
25. Responses should include a nutritious meal that is low in sugar and does not include too many starches.

Thinking Critically

26.

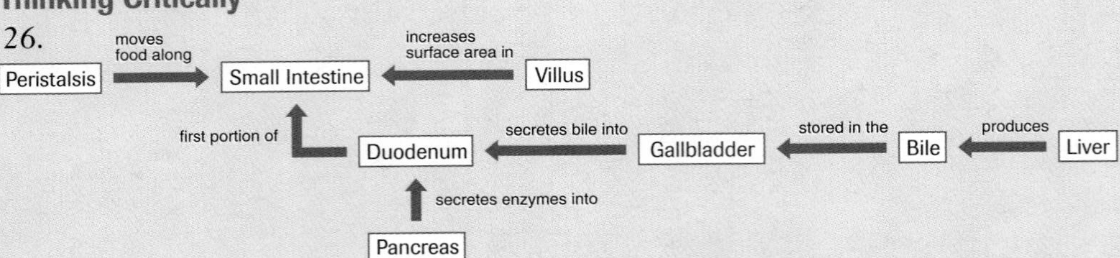

Chapter 38 Review

8. The first 25 cm of the small intestine is known as the ____________.
   a. pancreas
   b. duodenum
   c. gallbladder
   d. rectum
9. Digested food is absorbed in the small intestine by what structures?
   a. the epiglottis
   b. tonsils
   c. villi
   d. sphincters
10. Which of these enzymes functions best in the acid pH of the stomach?
    a. lipase
    b. lactase
    c. pepsin
    d. amylase
11. Of these, which organ is NOT directly involved in digestion?
    a. lung
    b. pancreas
    c. liver
    d. duodenum
12. What is the primary function of the large intestine?
    a. nutrient absorption
    b. food digestion
    c. vitamin synthesis
    d. water absorption
13. When your body needs energy, it breaks down ____________ and releases sugar into the bloodstream.
    a. calcium
    b. glycogen
    c. glucagon
    d. insulin
14. Which of the following contains taste receptors?
    a. the cheek
    b. the taste buds
    c. the taste palate
    d. the tonsils
15. The pancreas functions as a(n) ____________ gland.
    a. exocrine
    b. endocrine
    c. digestive
    d. all of these
16. Hormones released by endocrine glands affect specific body cells known as ____________.
    a. body fluids
    b. lymph glands
    c. target tissues
    d. epithelium
17. Which of these is NOT a function of the thyroid gland?
    a. regulates blood glucose levels
    b. regulates development
    c. regulates metabolism
    d. regulates nervous system
18. What is used to measure the energy content of food?
    a. temperature
    b. grams
    c. Calorie
    d. mass
19. A(n) ____________ is an inorganic substance that serves as a building material or takes part in chemical reactions in the body.
    a. enzyme
    b. mineral
    c. vitamin
    d. Calorie
20. What is the most abundant molecule in the human body?
    a. carbohydrates
    b. vitamins
    c. water
    d. proteins

Applying Concepts

21. Achlorhydria is a condition in which the stomach fails to secrete hydrochloric acid. How would this condition affect digestion?
22. Compare the digestion of regular soda pop with sugar-free soda pop. Is sugar-free soda pop a food?
23. Various medical conditions can make it necessary to remove portions of the large intestine. What consequences would this have on the digestive process?
24. How could removal of the parathyroid glands affect muscle contraction?
25. Provide a sample menu for one day (breakfast, lunch, dinner) that would be healthful for an individual with diabetes.

Thinking Critically

26. ***Concept Mapping*** Make a concept map that relates the following terms and phrases. Supply the appropriate linking words for your map.
    liver, gallbladder, bile, duodenum, small intestine, villus, peristalsis, pancreas

27. ***Observing and Inferring*** Why was the development of a tubular digestive system with a mouth and anus important to the evolutionary success of animals?

28. ***Recognizing Cause and Effect*** How is the liver's role in glucose regulation important for homeostasis?

29. ***Interpreting Data*** The relationship between parathyroid hormone (PTH) secretion and plasma calcium is shown in the graph below. What level does the plasma calcium of the blood have to fall to in order to get maximum parathyroid hormone secretion?

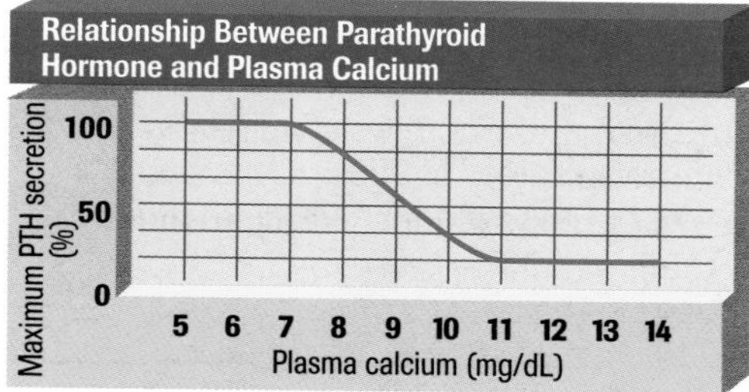

30. ***Comparing and Contrasting*** Evaluate the content of the following bowl of cereal with two percent milk, using the Daily Value table and the food pyramid in the MiniLabs in Section 38.3.

1 cup of bran with raisins and 1/2 cup two percent milk
Calories 125
Protein 10%
Carbohydrate 9%
Fat 2%

31. ***Comparing and Contrasting*** Write down the contents of your meals and snacks for a day. Compare them with the food pyramid in the MiniLab in Section 38.3. Is there a pattern to your meals? Does the pattern fit the recommended food pyramid?

ASSESSING KNOWLEDGE & SKILLS

The following table contains information concerning a meal of macaroni and cheese.

Nutrient Content of Macaroni and Cheese
Serving Size – 8 ounces
Calories per serving – 280

Nutrient	Grams per Serving	Calories per gram
Protein	7	4
Carbohydrate	35	4
Fat	12	9
Sodium	1.540	0

Interpreting Data Use the data in the table to answer these questions.

1. How many Calories are there in one serving of macaroni and cheese?
   a. 7 c. 280
   b. 35 d. 1540
2. George eats five servings each day, being fond of macaroni and cheese. How many Calories does George eat each day?
   a. 8 c. 300
   b. 35 d. 1400
3. What percent of George's Calories are derived from fat?
   a. less than 1 percent
   b. approximately 10 percent
   c. approximately 38 percent
   d. more than 50 percent
4. George should eat only 1800 Calories a day. What proportion of his daily diet is derived from his five servings of macaroni and cheese?
   a. 10 percent c. 50 percent
   b. 38 percent d. 77 percent
5. ***Interpreting Data*** The recommended daily allowance of sodium per day is approximately 2.4 g. Make a statement that describes George's sodium intake.

Chapter 38 Review

27. The development of a mouth and anus allowed specialization of digestion along the tube.
28. The liver acts as a reservoir for glycogen. Glycogen, under stimulation of glucagon, is broken down to glucose, which is absorbed into the bloodstream. Glucose, under the stimulation of insulin, leaves the blood.
29. Plasma calcium level must drop below 8 mg/dL to get maximum parathyroid hormone secretion.
30. Bran cereal with raisins is low in fat and a good source of protein.
31. Students need to evaluate their meals and snacks in comparison with the relative amounts of foods shown in a food pyramid. Individual responses will be dependent upon the foods students list.

ASSESSING KNOWLEDGE & SKILLS

1. c
2. d
3. c
4. d
5. George's diet contains too much sodium.

Program Resources

Chapter Assessment, pp. 235-240 L1
Alternate Assessment in the Science Classroom
Computer Test Bank L1
Content Mastery, pp. 161-164 L1

Chapter Organizer

SECTION	OBJECTIVES	ACTIVITIES/FEATURES
39.1 The Respiratory System National Science Standards: UCP.1, UCP.2, UCP.5; A.1; C.5; F.4, F.5	1. **List** the structures involved in external respiration. 2. **Explain** the mechanics of breathing. 3. **Contrast** external and cellular respiration.	**Biolab:** Measuring Respiration, p. 984
39.2 The Circulatory System National Science Standards: UCP.1, UCP.2, UCP.5; B.3; C.1, C.5; F.1, F.5; G.1-3	4. **Distinguish** among the various components of blood and among blood types. 5. **Trace** the route blood takes through the body and heart. 6. **Explain** how heart rate is controlled.	**Literature Connection:** *Julius Caesar,* p. 989 **A Broader View:** Maintaining Homeostasis of Body Temperature, p. 992 **Minilab:** What effect does swallowing have on pulse rate?, p. 993 **Thinking Lab:** How does exercise affect heart rate?, p. 995
39.3 The Urinary System National Science Standards: UCP. !, UCP.2, UCP.5; C.5; F.1, F.5; G.1-3	7. **Describe** the structures and functions of the urinary system. 8. **Explain** the kidneys' role in maintaining homeostasis.	**Minilab:** What structures are visible in a sheep's kidney?, p. 999

ACTIVITY MATERIALS

BIOLAB	MINILABS
page 984 round balloon string metric ruler watch with second hand straws	**page 993** paper cups watch with second hand **page 999** sheep's kidney knife

Chapter 39 Respiration, Circulation, and Excretion

TEACHER CLASSROOM RESOURCES

Reproducible Masters	Transparencies
Section Focus Master 88: Take a Breath L1 SAE **Reinforcement and Study Guide,** p. 159 L1 **Biolab and Minilab Worksheets,** pp. 157-158 L1 **Tech Prep Applications:** The Biology of a Hiccup, pp. 45-46 L2	
Section Focus Master 89: The Blood Goes Around L1 SAE **Reinforcement and Study Guide,** pp. 160-161 L1 **Biolab and Minilab Worksheets,** p. 155 L1 **Concept Mapping:** Circulation in Humans, p. 39 L1 **Content Mastery,** pp. 165-168 L1	**Basic Concepts Transparency #66:** Blood Types L1 SAE **Basic Concepts Transparency #67:** Your Blood Vessels L1 SAE **Basic Concepts Transparency #68:** Your Heart L1 SAE **Reteaching Transparency #39:** Circulatory Path Through the Heart L1 SAE
Section Focus Master 90: Saving and Discarding L1 SAE **Reinforcement and Study Guide,** p. 162 L1 **Biolab and Minilab Worksheets,** p. 156 L1 **Critical Thinking/Problem Solving:** Solving Respiratory, Circulatory, and Excretory Problems, p. 39 L3 **Laboratory Manual:** How Does Exercise Affect Heart Rate?, pp. 245-248 L2	**Basic Concepts Transparency #69:** The Urinary System L1 SAE

ASSESSMENT MATERIALS

Chapter Assessment, pp. 241-246
Performance Assessment in the Biology Classroom, pp. 41, 43
MindJogger Videoquiz
Alternate Assessment in the Science Classroom
Computer Test Bank

Spanish Resources SAE
English/Spanish Audiocassettes SAE
Cooperative Learning in the Science Classroom COOP LEARN
Lesson Plans

KEY TO TEACHING STRATEGIES

- L1 Level 1 activities should be within the ability range of all students including those with learning difficulties.
- L2 Level 2 activities are within the ability range of average to above-average students.
- L3 Level 3 activities are designed for the ability range of above-average students.
- SAE SAE activities should be within the ability range of Students Acquiring English.
- COOP LEARN Cooperative Learning activities are designed for small group work.
- P These strategies represent student products that can be placed into a best-work portfolio.
- These strategies are useful in a block scheduling format.

GLENCOE TECHNOLOGY

The following multimedia resources are available from Glencoe.

Biology: The Dynamics of Life
CD-ROM SAE
Videodisc Program

The Infinite Voyage Series
A Taste of Health
To the Edge of the Earth

National Geographic Society Series
STV: Human Body Vol. 1
Circulatory and Respiratory Systems

Science and Technology Videodisc Series (STVS)
Human Biology
Heart Rehabilitation
Measuring Blood Pressure
Nicotine and the Lungs
Modeling Blood Flow
Hypothermia Research

CHAPTER

39 Respiration, Circulation, and Excretion

Chapter Overview

In this chapter, students study the processes of respiration, circulation, and excretion. The mechanics of breathing and the exchange of gases in the lungs and tissues are presented along with the control of respiration.

In the second part of the chapter, students learn that the circulatory system is divided into three sections: blood, blood vessels, and the heart. The anatomy and physiology of each part of the circulatory system is examined.

The chapter concludes with an exploration of the anatomy and function of the urinary system. After examining how the kidneys clean wastes from the body, the role of the urinary system in maintaining homeostasis is presented.

Key Terms

alveoli
antibody
antigen
aorta
artery
atrium
blood pressure
capillary
hemoglobin
hypothalamus
kidney
nephron
plasma
platelet
pulse
red blood cell
trachea
ureter
urethra
urinary bladder
urine
vein
vena cava
ventricle
white blood cell

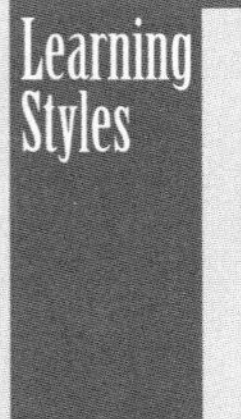

Look for the following logo for strategies that emphasize different learning modalities. LS

Kinesthetic	**Meeting Individual Needs, p. 983**
Visual-Spatial	**Display, p. 983; Meeting Individual Needs, p. 988; The Inside Story, p. 994, Portfolio, pp. 994, 999; Microscope Activity, p. 995; Minilab, p. 999**
Interpersonal	**Student Journal, p. 994; Meeting Individual Needs, p. 998**
Linguistic	**Student Journal, pp. 982, 991, 999**
Logical-Mathematical	**Portfolio, pp. 992, 993; Minilab, p. 993; Thinking Lab, p. 995**

CHAPTER 39 Respiration, Circulation, and Excretion

These mountain climbers have to be in good physical condition. But hiking up the mountain has begun to have its effect on their bodies. Their breathing has become deeper as their lungs strive to take in more vital oxygen. Their hearts now beat a little faster to help transport oxygen and nutrients to cells of their bodies. This enables them to continue to produce the energy needed to complete their adventure. As the breathing rates and heart rates of the climbers increase, so do wastes within their body cells, the end products of energy production. A slow, methodical pace, with rest stops to catch one's breath and enjoy the view, is the safest way to hike the mountains.

As the hikers climb higher, the demands on their bodies will increase. How does the human body adapt to these changing conditions? The answers lie in the workings of the human respiratory, circulatory, and urinary systems.

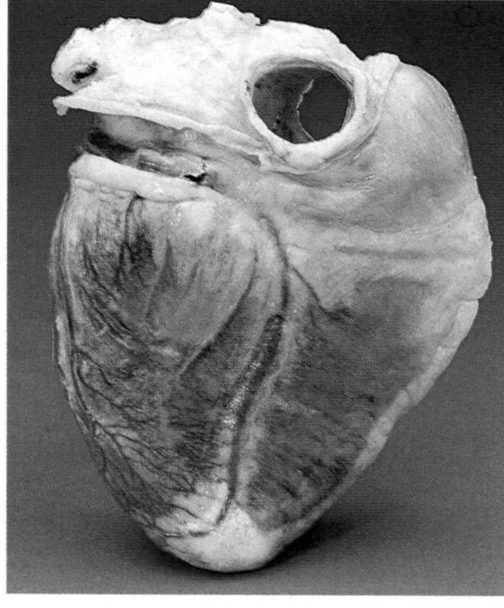

The view at the top will be the hikers' reward for all the hard work their bodies have done. Of course, their feet and legs have brought them this far. But without the work of their lungs, heart, and other organs of their respiratory and circulatory systems, their feet and legs wouldn't have the energy to take them any further.

Concept Check

You may wish to review the following concepts before studying this chapter.

- Chapter 9: diffusion
- Chapter 10: respiration
- Chapter 15: multiple alleles, blood groups

Chapter Preview

Laboratory Activities

Biolab
- Measuring Respiration

Minilabs
- What effect does swallowing have on pulse rate?
- What structures are visible in a sheep's kidney?

Assessment Planner

Choose assessment strategies from the following pages to evaluate the progress of your students.

Assess, pp. 986, 996, 1000
Minilabs, pp. 993, 999
Portfolio, pp. 985, 992-994, 999
Thinking Lab, p. 995
Biolab, pp. 984-985
Chapter Review, pp. 1001-1003

CHAPTER 39

Introducing the Chapter

Ask students to imagine how the hikers felt as they climbed. Elicit what changes occurred in their bodies. *Breathing was deeper; heart rate was faster.* Ask them to relate these changes to what they learned in Chapter 10 about respiration. *Respiration provides the cells of the body with energy. As cells require more energy, the body responds by taking in additional oxygen through faster and heavier breathing and the heart pumps faster to deliver this oxygen to the cells.* L1

Theme Development

Homeostasis is the major theme of the chapter. Emphasis is placed on how the respiratory, circulatory, and urinary systems maintain balance in the body. *Systems and Interactions* is shown by the interactions among the three systems discussed and how these systems work in conjunction with the nervous system.

Concept Check

Students will need to recall the biochemistry of cellular respiration from Chapter 10 and relate these principles to respiration in the human body. Students will expand upon their knowledge of diffusion and its role in maintaining homeostasis in the body.

SECTION

39.1

Prepare

Key Concepts

Students will learn about the mechanics of breathing, and the cellular exchange of gases in the lungs and between the blood and body cells. The control of the respiratory system by the nervous system is also discussed.

Block Scheduling

Look for this symbol for strategies that are useful in a block scheduling format. For more information on block scheduling, refer to Section 39.1 in the **Lesson Plans** booklet.

Materials

- Obtain a model of the respiratory system for the Display and the Meeting Individual Needs.
- Buy balloons, straws, and string for the Biolab.

1 Focus

Bellringer

Before presenting the lesson, display **Section Focus Master 88** on the overhead projector and have students answer the accompanying questions. L1 SAE

Discussion Question

Explain to students that oxygen concentration doesn't decrease with an increase in altitude. Pressure, however, does decrease with altitude, resulting in the lowering of the density of the air. Have students try to explain why it is harder to breathe at higher altitudes such as mountaintops. Use this question to begin a discussion of the diffusion of oxygen in the alveoli. L2

SECTION

39.1 The Respiratory System

Section Preview

Objectives

List the structures involved in external respiration.

Explain the mechanics of breathing.

Contrast external and cellular respiration.

Key Terms

trachea
alveoli

Breathing underwater is easy—if you're a fish. With their gills, fishes are adapted to extract oxygen from water for their life processes. Humans, on the other hand, are not. But we are equipped to extract oxygen from air using our own adapted respiratory system.

Passageways and Lungs

Your respiratory system is made up of a pair of lungs, a series of passageways into your body, and a thin sheet of smooth muscle called the diaphragm, all shown in *Figure 39.1*. When you hear the word *respiration*, you probably think of breathing. But breathing is just part of the process of respiration that an oxygen-dependent organism carries out. Respiration includes all the mechanisms involved in getting oxygen to the cells of your body and getting rid of carbon dioxide. You learned in Chapter 10 that respiration also involves the formation of ATP within cells.

The path air takes

The first step in the process of respiration involves taking air into your body through either your nose or your mouth. It passes to the pharynx, moves past the epiglottis, and passes through the larynx. Air then travels down the **trachea**, the passageway that leads to the lungs. Recall that at the time of swallowing food, the epiglottis covers the trachea. It prevents food and other large materials from getting into the air passages.

Cleaning dirty air

The air you breathe is far from clean. It is estimated that an individual living in an urban area breathes in 20 million particles of foreign matter each day. To prevent most of this material from reaching your lungs, the trachea and bronchi are lined with cilia and cells that secrete mucus. The cilia constantly beat upward in the direction of your throat, where foreign material can be expelled or swallowed.

Alveoli: The place of gas exchange

The trachea divides into two narrower tubes called bronchi. Each bronchus branches into bronchioles, which in turn branch into numerous microscopic tubules that eventually

Program Resources

Section Focus Master 88 L1 SAE
Reinforcement and Study Guide, p. 159 L1
Biolab and Minilab Worksheets, pp. 157-158 L1
Performance Assessment in the Biology Classroom, p. 43 L1 P
Tech Prep Applications, pp. 45-46 L2

STUDENT JOURNAL

Requesting Information Have students compose letters to the American Lung Association and American Cancer Society asking for posters and information on the respiratory system and diseases of the respiratory system. Have students include the letters in their journals. L1 LS

expand into thousands of thin-walled sacs called alveoli. **Alveoli** are the sacs of the lungs where oxygen and carbon dioxide are exchanged by diffusion between the air and blood. Diffusion takes place easily because the wall of each alveolus is only one cell thick. External respiration, as shown in *Figure 39.1,* is the exchange of oxygen or carbon dioxide between the air that is in the alveoli and the blood that supplies these air sacs.

Figure 39.1

As air passes through the respiratory system, it travels through narrower and narrower passageways until it reaches the alveoli. The clusters of alveoli are surrounded by networks of tiny blood vessels. Blood in these vessels has come from the cells of the body and contains wastes from cellular respiration. What are the products of cellular respiration?

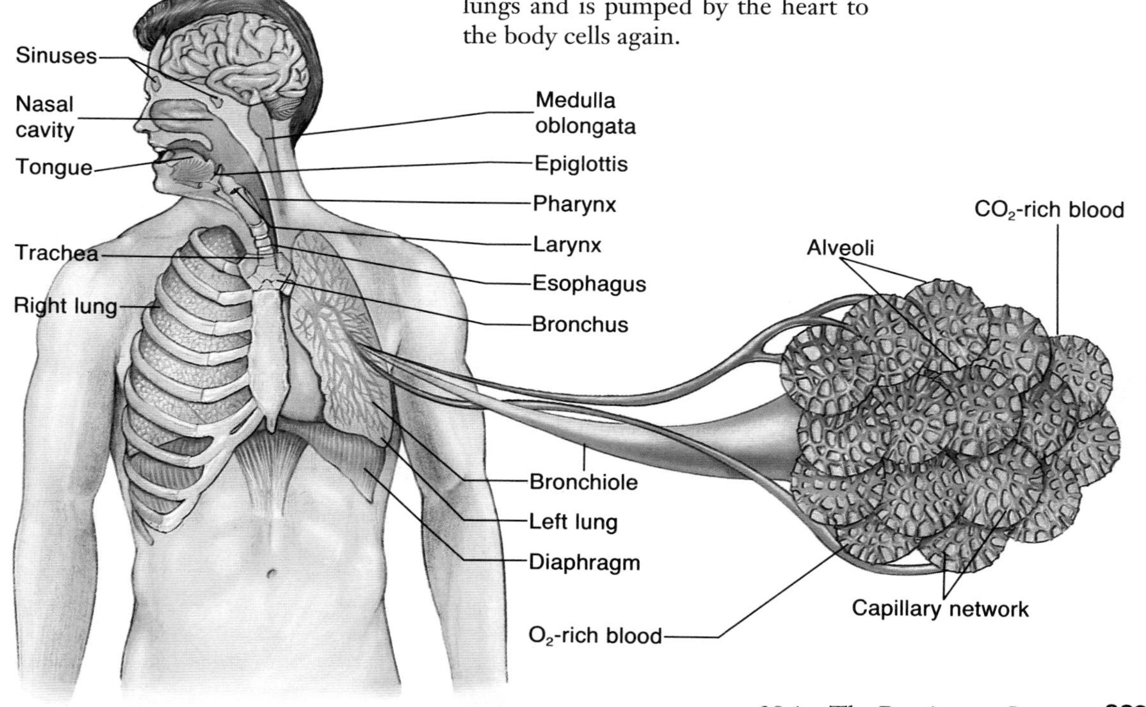

Blood transport of gases

Once oxygen from the air diffuses into the blood vessels surrounding the alveoli, it is transported to the cells of your body, where it is used for cellular respiration. In Chapter 10, you learned that cellular respiration uses oxygen during the process by which glucose is broken down and results in energy in the form of ATP. Carbon dioxide and water are waste products of this process. The water can stay in the cell or diffuse into the blood. The carbon dioxide diffuses into the blood, which carries it back to the lungs.

As a result, the blood that comes to the alveoli from the body's cells is high in carbon dioxide and low in oxygen. As carbon dioxide from the body diffuses from the blood into the air in the alveoli to be carried out of your body, oxygen diffuses from the air in the alveoli into the blood, making the blood rich in oxygen. This oxygen-rich blood then leaves the lungs and is pumped by the heart to the body cells again.

2 Teach

Misconception

Students may believe that the diaphragm is a smooth muscle because of its role in breathing. Point out that the diaphragm is striated skeletal muscle. L1

Display

LS **Visual-Spatial** Obtain an anatomical model of the respiratory organs to display in the classroom. Use the model to point out the parts of the respiratory system as they are mentioned in the chapter.

Audiovisual

Show the video *Respiration*, Charles Clark Co., Inc.

Health Connection

Asthma

Asthma is a condition in which the bronchi suddenly constrict due to spasms of the smooth muscles that line the walls of the smaller bronchi. The patient has trouble exhaling. Many asthma attacks are caused by allergies to edible or airborne substances as common as dust or pollen.

Visual Learning

Figure 39.1 What are the products of cellular respiration? *carbon dioxide, water, and energy in the form of ATP*

Meeting Individual Needs

Visually/Hearing Impaired Pair visually impaired or hearing impaired students with other students. Have them use a model of the respiratory system to explain the mechanics of breathing. Have students point to the various anatomical parts as they discuss them. L1 LS

Tying to Previous Knowledge

Ask students to explain why exhaled breath contains water vapor. *Water is an end product of respiration. Also, the alveoli are wet in order to allow oxygen to diffuse into the cells. Both of these contribute to the loss of water via exhaled breath.*

SECTION 39.1

BioLab

Measuring Respiration

Time Allotment 
One class period

Objectives
Review objectives with students before they begin the Biolab.

Process Skills
communicate, measure in SI, use numbers, interpret data, experiment, formulate a model

PREPARATION

Safety Procedures
Provide plenty of balloons; students should not share balloons.

GLENCOE TECHNOLOGY

Videodisc

The Infinite Voyage: To the Edge of the Earth
Exploring Tibet: The Gateway of Exchange (Ch. 2)

STVS: Human Biology
Disc 7, Side 2
Nicotine and the Lungs (Ch. 15)

BioLab

Measuring Respiration

The exchange of oxygen and carbon dioxide among the atmosphere, the blood, and the body cells is commonly called respiration. This external respiration should not be confused with cellular respiration, the chemical reactions that take place within cells to provide energy. In external respiration, the lungs serve as the area of gas exchange between the atmosphere and the blood. The amount of air exchanged during breathing can be measured using a clinical machine called a spirometer. It also can be measured, although less accurately, using a balloon.

PREPARATION

Problem
How can you measure volume of respiration?

Objectives
In this Biolab, you will:
- **Measure** the resting breathing rate.
- **Measure** tidal volume by exhaling into a balloon.
- **Calculate** the amount of air inhaled per minute.

Materials
round balloon
string (1 m)
metric ruler
clock or watch with second hand

PROCEDURE

Part A: Breathing Rate at Rest
1. Copy Table 1.
2. Have your partner count the number of times you inhale in 30 s.
3. Repeat step 2 two more times.
4. Calculate the average number of breaths.
5. Multiply the average number of breaths by two to get the average resting breathing rate per minute.

Table 1 Resting Breathing Rate

Trial	Inhalations in 30 s
1	
2	
3	
Average	
Inhalations per minute	

PROCEDURE

Teaching Strategies
- Have students work in pairs. If time allows, have students reverse roles.
- To decrease the number of balloons needed, cut a straw in half and use a rubber band to attach the balloon to the straw. Blow up the balloon through the straw, then replace only the straw for the next student.
- Ask students to bring calculators to lab for use in making calculations.
- The balloons should be large-capacity balloons.

Part B: Tidal Volume

1. Copy Table 2.
2. Take a regular breath and exhale normally into the balloon. Pinch the balloon closed.
3. Have a partner fit the string around the balloon at the widest part.
4. Measure the length of the string, in centimeters, around the circumference of the balloon. Record this measurement.
5. Repeat steps 2–4 four more times.
6. Calculate the average circumference of the five measurements.
7. Calculate the average radius of the balloon by dividing the average circumference by 6.28 ($\approx 2\pi$).
8. Tidal volume is the amount of air expelled during a normal breath. Tidal volume can be determined using the balloon radius and the formula for determining the volume of a sphere

$$\text{Volume} = \frac{4\pi r^3}{3}$$

where r = radius and $\pi = 3.14$. Calculate the average tidal volume using the average balloon radius.

9. Your calculated volume will be in cubic centimeters; 1 cm^3 = 1 mL.

Table 2 Tidal Volume

Trial	String measurement
1	
2	
3	
4	
5	
Average circumference	
Average radius	
Average tidal volume	

Part C: Amount of Air Inhaled

1. Copy Table 3.
2. Multiply the average tidal volume by the average number of breaths per minute to calculate the amount of air you inhale per minute.
3. Divide the number of milliliters of air by 1000 to get the number of liters of air you inhale per minute.

Table 3 Amount of Air Inhaled

mL/min	
L/min	

ANALYZE AND CONCLUDE

1. **Making Comparisons** Compare your average number of breaths per minute and tidal volume per minute with those of other students.
2. **Thinking Critically** An average amount of air inhaled per minute for an adult is 6000 mL. Compare your average volume of air with this figure. What factors could be responsible for any differences?
3. **Making Predictions** What would you predict would happen to your resting breathing rate right after you exercise?

Going Further

Applying Concepts The largest amount of air that can be inhaled is called vital capacity. Determine your average vital capacity by following a procedure similar to the one you used to determine average tidal volume.

ANALYZE AND CONCLUDE

1. Average breaths per minute and tidal volume per minute will probably differ among students.
2. Answers may vary from the average due to type of balloon, age, sex, size, and athletic condition.
3. After exercising, the breathing rate will be higher.

Assessment

Journal: Have students write a summary of the lab, including the three tables and the answers to the Analyze and Conclude. L1

Going Further

Students can calculate their tidal volume after exercising and compare this with their tidal volume during rest. L1

Enrichment

There are about 300 million alveoli in a human lung with a surface area of more than 50 square meters. Have students measure the classroom, calculate the number of square meters, and compare this figure with that of a human lung. L1

Assessment

Performance Assessment in the Biology Classroom: p. 43, Making a Model of Inhalation and Exhalation. Have students carry out this activity to model the workings of the respiratory system. L1 P

Data and Observations

Answers among students could vary greatly. Average breathing rate is 11 to 12 breaths per minute. Tidal volume should be approximately 280 mL. The amount of air inhaled should be 3 to 4 L per minute.

PORTFOLIO

Tracing the Path of Air Ask the students to make a flowchart demonstrating the pathway of air as it moves into and out of the respiratory system. L1 COOP LEARN P

When you inhale, the muscles between your ribs contract and your rib cage rises. Your diaphragm contracts, becomes flattened, and moves lower in the chest cavity. As a result, the space in the chest cavity becomes larger and a slight vacuum forms. Air now rushes into your lungs because the air outside your body is under more pressure than the air inside your lungs.

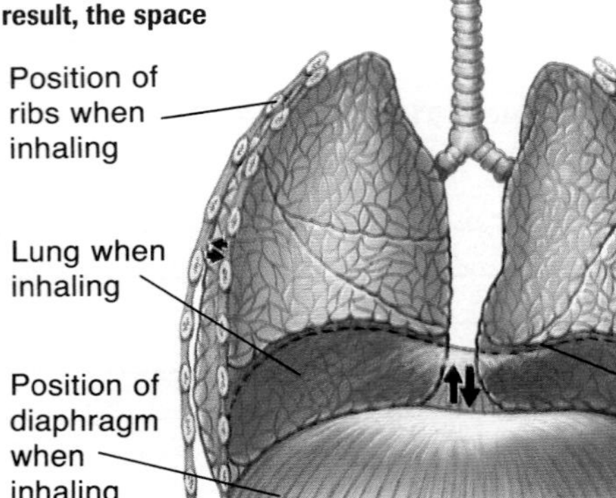

When you exhale, the muscles over the ribs relax, and your ribs drop down in the chest cavity. Your diaphragm relaxes and returns to its resting, dome-shaped position (dashed line). The relaxation of these muscles decreases the volume of the chest cavity, putting pressure on the air and forcing most of it out of the alveoli.

Figure 39.2

The pressure in the lungs is varied by changes in the volume of the chest cavity. When relaxed, your diaphragm lies in a dome shape under your lungs.

The Mechanics of Breathing

The action of your diaphragm and the muscles between your ribs enable you to breathe in and breathe out. *Figure 39.2* shows how air is drawn in or is forced out of the lungs as a result of the diaphragm's position.

The alveoli in healthy lungs are elastic, like balloons. They stretch as you inhale and return to their original size as you exhale. Like a balloon that has had the air let out of it and does not go totally flat, the alveoli still contain a small amount of air when you exhale.

Control of Respiration

Breathing is usually an involuntary process. It is controlled by the chemistry of your blood as it interacts with a part of your brain called the medulla oblongata. The medulla oblongata helps maintain homeostasis. It responds to higher levels of carbon dioxide in your blood by sending nerve signals to the rib muscles and diaphragm. As a result, these muscles contract and you inhale. As breathing becomes more rapid, as during exercise, a more rapid exchange between air and blood occurs.

Section Review

Understanding Concepts

1. Outline the path of an oxygen molecule from your nose to a cell.
2. Compare and contrast external respiration and cellular respiration.
3. Explain the process by which gases are exchanged in the lungs.

Thinking Critically

4. During a temper tantrum, four-year-old Josh tries to hold his breath. His parents are afraid that he will be harmed by this behavior. How will Josh be affected by holding his breath?

Skill Review

5. **Sequencing** What is the sequence of muscle actions during inhaling and exhaling? For more help, refer to Organizing Information in the *Skill Handbook*.

SECTION 39.1

3 Assess

Check for Understanding

Have students trace the pathway of a carbon dioxide molecule from a body cell to the nose. L1

Reteach

Have students label a diagram of the respiratory system and draw arrows along the pathway of air molecules. L1

Extension

Have interested students find out what causes nitrogen narcosis. L2

Assessment

Portfolio: Ask students to make a circular flowchart describing the pathway and processes involved in respiration. L1 P

4 Close

Discussion Question

Discuss why a gas mixture of 95 percent oxygen and 5 percent carbon dioxide is used to revive someone who has fainted. L2

GLENCOE TECHNOLOGY

Videodisc

Biology: The Dynamics of Life
Disc 2, Side 1
Mechanics of Breathing (Ch. 32)

Answers to Section Review

1. nose, pharynx, epiglottis, larynx, trachea, bronchi, alveoli, blood, cells
2. External respiration is the exchange of oxygen and carbon dioxide in the alveoli. The process that uses oxygen to break down glucose within the cells is cellular respiration.
3. In the alveoli, oxygen from the air diffuses into the blood and carbon dioxide diffuses out of the blood and into the air of the alveoli.

Thinking Critically

4. As carbon dioxide builds up in the blood, the child's medulla will stimulate his muscles to contract so he inhales.

Skill Review

5. During inhalation, muscles between the ribs and the diaphragm contract. During exhalation, muscles between the ribs and the diaphragm relax.

SECTION 39.2 The Circulatory System

Blood flowed freely from this injury until direct pressure was applied. This stopgap measure is needed only until the blood's adaptive ability to clot takes over. Your blood has other life-supporting qualities. As it travels throughout your body, it carries oxygen from your lungs and nutrients from your digestive system to the cells of your body, then hauls away cell wastes. Together, your blood, your heart, and a mazelike network of passageways make up the components of your circulatory system.

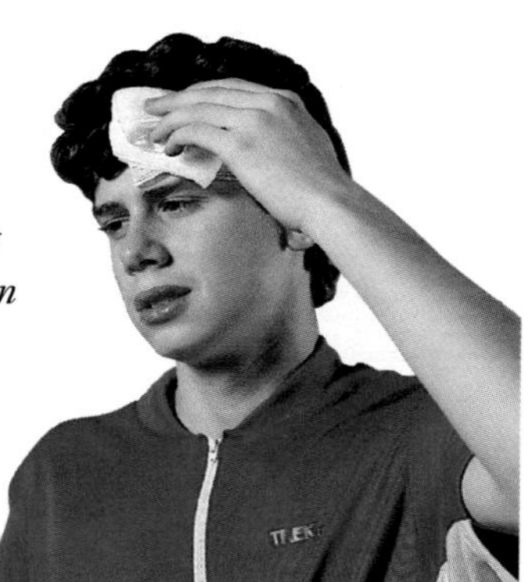

Section Preview

Objectives

Distinguish among the various components of blood and among blood types.

Trace the route blood takes through the body and heart.

Explain how heart rate is controlled.

Key Terms

plasma
red blood cell
hemoglobin
white blood cell
platelet
antigen
antibody
artery
capillary
vein
atrium
ventricle
vena cava
aorta
pulse
blood pressure

Your Blood: Fluid Transport

Your blood is a tissue composed of fluid, cells, and fragments of cells. *Table 39.1* summarizes information about these human blood components. The fluid portion of blood in which blood cells move is called **plasma.** Plasma is straw colored and makes up about 55 percent of the total volume of blood. Blood cells—both red and white—and cell fragments are suspended in plasma.

Red blood cells: Oxygen carriers

The round, disk-shaped cells in blood are **red blood cells.** Red blood cells carry oxygen to body cells. They make up 44 percent of your blood.

Red blood cells are produced in the red bone marrow of your ribs, humerus, femur, sternum, and other long bones. At the same time, old red blood cells are being destroyed in your liver and in your spleen, an organ of the lymphatic system.

Table 39.1 Blood Components

Components	Characteristics
Red blood cells	Transport oxygen and some carbon dioxide; lack a nucleus; contain hemoglobin
White blood cells	Large; several different types; all contain nuclei; defend the body against disease
Platelets	Cell fragments needed for blood clotting
Plasma	Liquid; contains proteins; carries red and white blood cells, platelets, nutrients, enzymes, hormones, gases, and inorganic salts

SECTION 39.2

Prepare

Key Concepts

Students will examine the three major components of the circulatory system: the blood, the vessels, and the heart. They will examine the composition and functions of blood and learn the importance of blood groups. Students will trace the path of blood through the body and heart. They will examine how heart rate is controlled.

Block Scheduling

Look for this symbol for strategies that are useful in a block scheduling format. For more information on block scheduling, refer to Section 39.2 in the **Lesson Plans** booklet.

Materials

- Borrow a stethoscope and a heart model for *The Inside Story.*
- Order *Daphnia* for the Microscope Activity.
- Borrow a sphygmomanometer for the Focus On Feature.
- Purchase straws for the Alternate Lab.
- Order the National Cardiovascular quiz from the American Heart Association for the Reteach.

Program Resources

Section Focus Master 89 L1 SAE

Reinforcement and Study Guide, pp. 160-161 L1

Biolab and Minilab Worksheets, p. 155 L1

Performance Assessment in the Biology Classroom, p. 41 L1 P

Concept Mapping, p. 39 L1

Reteaching Transparency 39 and **Master** L1 SAE

Basic Concepts Transparencies 66, 67, 68 and **Masters** L1 SAE

1 Focus

Bellringer

Before presenting the lesson, display **Section Focus Master 89** on the overhead projector and have students answer the accompanying questions. L1 SAE

Demonstration

Discuss the role of the heart and arteries in creating a pulse. Have students take their pulse rates for one minute. Record the data needed to calculate the class average. Use the class average to introduce and discuss the meanings of average, high, and low pulse rates. L1 COOP LEARN

Different Viewpoints in Biology

High-Altitude Life

Why do people living at high elevations seem largely unaffected by the oxygen restrictions posed by this harsh environment? One possibility is that people who spend their lives in the mountains become acclimated to their environment. Recent research indicates that these people may have inherited fundamental changes in their genes that help them survive such conditions. One study of six men who work and live at 13 000 feet found that their brains use an average of 14 percent less glucose and therefore less oxygen. Another study showed that, during exercise, these high-altitude dwellers produce less lactic acid than would be expected. Somehow, the enzymes in their mitochondria have developed a way to use oxygen more efficiently. Most interesting is that these traits remained stable when these people stayed in North America lowlands for five weeks.

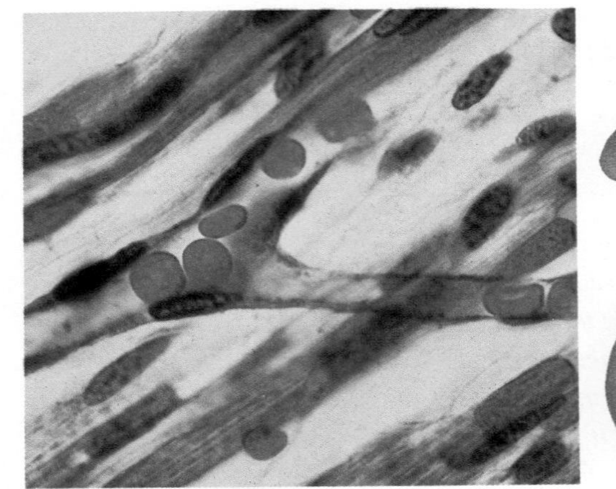

Side view
Red blood cell
2.0 micrometers
Top view
7.5 micrometers

Figure 39.3
Red blood cells in humans have nuclei only at an early stage in development. The nucleus is lost as the cell enters the bloodstream. As a result, red blood cells have a limited life span. They remain active in the bloodstream for only about 120 days before they break down and are removed as wastes.

Oxygen in the blood

How is oxygen carried by the blood? Red blood cells, *Figure 39.3*, are equipped with an iron-containing protein molecule called **hemoglobin**. Hemoglobin picks up oxygen after it enters the lungs. There, oxygen becomes loosely attached to hemoglobin and is carried to the body's cells. As blood passes tissues where oxygen concentration is low, the oxygen attached to the hemoglobin in the red blood cells is released and it diffuses into the tissues.

Carbon dioxide in the blood

Hemoglobin carries some carbon dioxide as well as oxygen. You know that once biological work has been done in a cell, wastes in the form of carbon dioxide diffuse into the blood and are carried in the bloodstream to the lungs. About 70 percent of this carbon dioxide combines with water and sodium in the blood plasma to form sodium hydrogen carbonate. The remaining 30 percent travels back to the lungs dissolved in the plasma or attached to hemoglobin.

White blood cells: Infection fighters

Shown in *Figure 39.4*, **white blood cells** make up only one percent of your blood, yet they play a major role in protecting your body from foreign substances and from organisms that cause disease. The role of white blood cells and their function in immunity will be discussed in detail in Chapter 42.

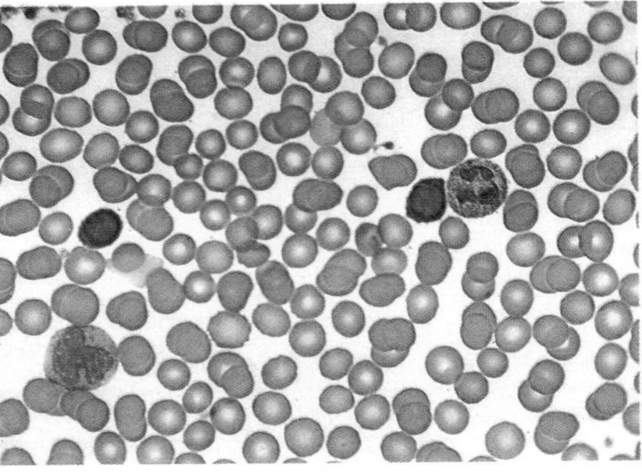

Figure 39.4
Compared with red blood cells, white blood cells have a nucleus, are larger, and are far fewer in number.

Magnification: 500×

Meeting Individual Needs

Gifted Have students conduct research to find information about the different types of white blood cells. Ask students to create a visual that includes sketches of each of the different types of cells with their names and functions. L3 P IS

Blood clotting

Think about what happens when you cut yourself. If the cut is slight, you usually bleed for a short while and then stop bleeding. That's because in addition to red and white blood cells, your blood also contains some cell fragments called **platelets,** which help blood clot after an injury, *Figure 39.5.* Platelets are produced from cells in bone marrow. They have a short life span, remaining in the blood for only about one week.

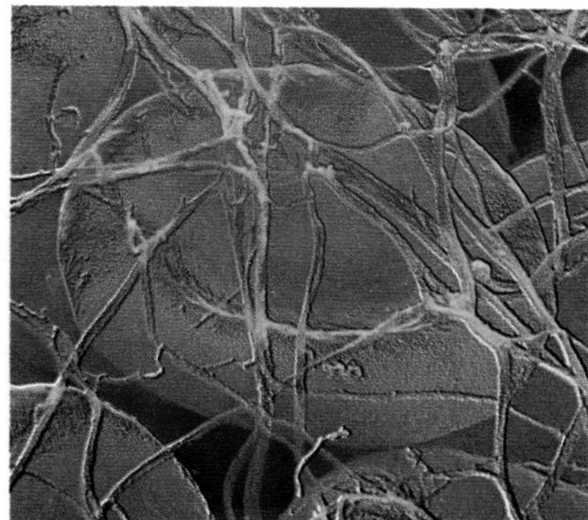

Figure 39.5

When you cut yourself, platelets in your blood help clot the blood, preventing you from bleeding to death. A sticky network of protein fibers called fibrin forms a web over the wound, trapping escaping blood components. Eventually, a dry, leathery scab forms.

ABO Blood Types

If a person is injured so severely that a massive amount of blood is lost, a transfusion of blood from a second person may be necessary. Whenever blood is transfused from one person to another, knowing blood types becomes important. In Chapter 15, you learned of the four human blood types—A, B, AB, and O. You've inherited one of these blood types from your parents.

Literature

Julius Caesar

by William Shakespeare

Amidst all the talk of treason and intrigue in his play *Julius Caesar,* William Shakespeare interjected a tender scene between Brutus and his wife, Portia. Brutus says to Portia, "You are my true and honorable wife; As dear to me as are the ruddy drops that visit my sad heart." These words must have puzzled the people in the audience watching this play in 1599. Perhaps they thought, "Ruddy drops of blood going to the heart? Nonsense! Doesn't Shakespeare know about Galen?"

Galen's theory Galen, a Greek physician who died more than 1000 years before Shakespeare began writing, had originated the accepted theory for the circulation of blood. In his view, the liver changed food into blood. Then the blood flowed through the veins to other parts of the body, where it was used up.

Galen's theory stood until 1628, when English physician William Harvey published a study about blood circulation. Unlike Galen, Harvey made firsthand observations of the circulatory system and dissected dozens of animals and humans. Harvey's experiments demonstrated that the heart acts as a pump and that blood circulates endlessly through a system of arteries and veins.

A meeting of the minds? How did Shakespeare, writing 29 years before Harvey, come to make an accurate comment about the flow of blood? Perhaps the quote was just poetic, or maybe it was a guess about the workings of the body. There's another possibility, too. Shakespeare and Harvey *may* have been in London at the same time, probably between 1592 and 1593, when Harvey was a student at nearby Cambridge and Shakespeare was a London actor and playwright. London's taverns have always been places where ideas are exchanged and theories are argued. Could Shakespeare have overheard Harvey's theory of blood circulation being discussed? Or might the two great men even have discussed these ideas? Of course, we'll probably never know, but it's intriguing to speculate.

CONNECTION TO Biology

Read further about the works of William Harvey. Why do you think Harvey's theory on blood circulation withstood attacks by followers of the ancient Greek physician Galen?

2 Teach

Tying to Previous Knowledge

Review diffusion from Chapter 9. Relate this process to how oxygen and carbon dioxide get into the blood.

Literature

Julius Caesar

Purpose

Students will be introduced to the work of Galen and Harvey and speculate about how they may have affected Shakespeare's writing.

Teaching Strategies

- Write the following on the chalkboard:

  Galen (A.D. 130-200)

  Harvey (1578-1657)

  1593–1599 Studied at Cambridge

  1602 - Began medical practice in London

  1628 - Wrote "An Anatomical Study of the Motion of the Heart and of the Blood in Animals"

  Shakespeare (1564-1616)

  1589 - Moved to London

  1599 - Wrote *Julius Caesar*

  1610 - Retired to Stratford

  Have students use this information to make a time line before reading the Literature Connection. L1

Possible Answer

Possibly, it withstood attacks because it was firsthand experimental evidence published on the circulation of blood. He showed that the heart acts as a pump and the blood circulates endlessly within the body. He also demonstrated the valves in the heart and veins, showing that blood can flow in only one direction. He also showed that blood pressure comes from the left ventricle of the heart.

Meeting Individual Needs

Learning Disabled For students having difficulty, use the review tutorial called *Circulation and Respiration* from Queue, Apple II, MS-DOS and Mac. L1

History Connection

Karl Landsteiner

Karl Landsteiner (1868-1943), Austrian-born pathologist of the United States, was awarded a Nobel prize in 1930 for discovering the major blood groups and developing the ABO system of blood typing. He was the first to reveal specific differences between the cells of one individual and another.

Misconception

A common misconception is that arteries carry only oxygenated blood and veins carry only unoxygenated blood. Point out that in the case of the pulmonary artery and pulmonary vein, this generalization does not hold true.

Enrichment

Have a representative of the American Red Cross speak to students on blood donation, blood transfusions, blood tests, and AIDS.

Audiovisual

Show the video *The Heart of Circulation,* Charles Clark Co., Inc.

Bioethics

Animal Heart Transplants

At Loma Linda University Medical School in 1984, Dr. Leonard Bailey transplanted a baboon heart into a 12-day-old girl. The child survived only 20 days. This surgery stirred an ethical debate that still continues today. Elicit from students their views on the use of organs and tissues from animal donors. L1 COOP LEARN

Figure 39.6

Blood contains both antigens and antibodies. The plasma in your blood contains specific antibodies that will *not* react to antigens on your own red blood cells.

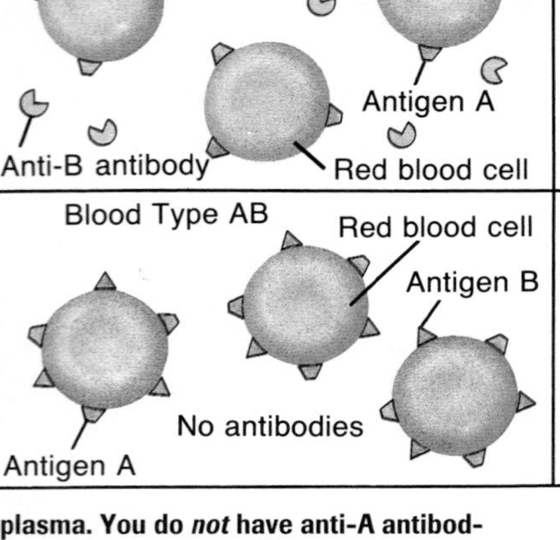

For example, if you have type A blood, you have the A antigen on your red blood cells and the anti-B antibody in your plasma. You do *not* have anti-A antibodies because they would react with your red blood cells and your blood would clump together. Antigen-antibody reactions are used to type blood.

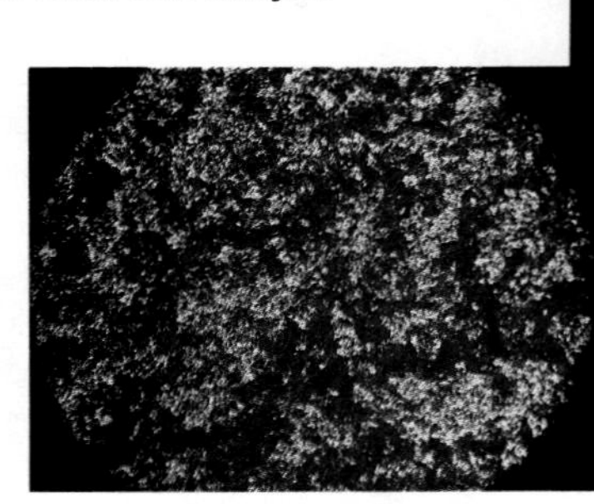

If you have type A blood and anti-A is added to it by way of transfusion, the blood will form clumps as shown on the left. If you add anti-B to type A blood, no clumps will form (right).

Antigens determine blood type

Differences in blood type are due to the presence or absence of proteins, called antigens, on the membranes of red blood cells. **Antigens** are normally foreign substances that stimulate an immune response in the blood. The letters *A* and *B* stand for the types of antigens found in a person's blood.

Blood plasma contains proteins with shapes that correspond to the different antigens, ***Figure 39.6.*** These proteins are called **antibodies.** One type of antibody will react with its matching antigen on a red blood cell if they are ever brought into contact with one another. However, each type of blood contains antibodies for the antigens found only on the red blood cells of the other blood types—*not* of its own red blood cells.

Rh factor in the blood

Another blood group can cause complications in some pregnancies. Illustrated in ***Figure 39.7,*** this problem involves the antigen called Rh, or Rhesus factor, also an inherited characteristic. People are Rh positive (Rh^+) if they have the Rh antigen factor on their red blood cells. They are Rh negative (Rh^-) if they don't.

Treatment for this problem is now available. At 28 weeks and shortly after each Rh^+ baby is born, the Rh^- mother is given a substance that prevents the production of antibodies to the Rh factor. As a result, the next fetus will not be in danger.

CULTURAL DIVERSITY

Carlos Monge and High-Altitude Physiology

Much of what is known about high-altitude or environmental physiology is the result of research by Peruvian scientist Carlos Monge (1884-1970), the first to describe high-altitude sickness, or hypoxia, and the founder of the Institute of Andean Biology and Pathology. Introduce students to Monge's research and major findings. Discuss with students the physiological adaptations of people who live at high altitudes, such as the Aymara and Quechua Indians of South America. Have students relate these adaptations to the proper functioning of the respiratory system. L1

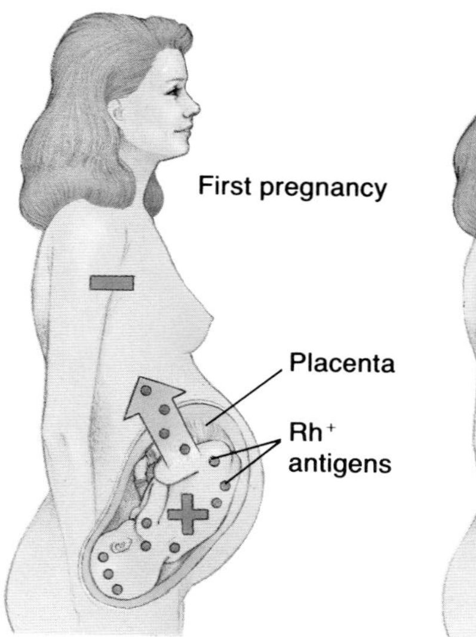

▲ Rh$^+$ baby's blood mixes with Rh$^-$ blood of mother at birth.

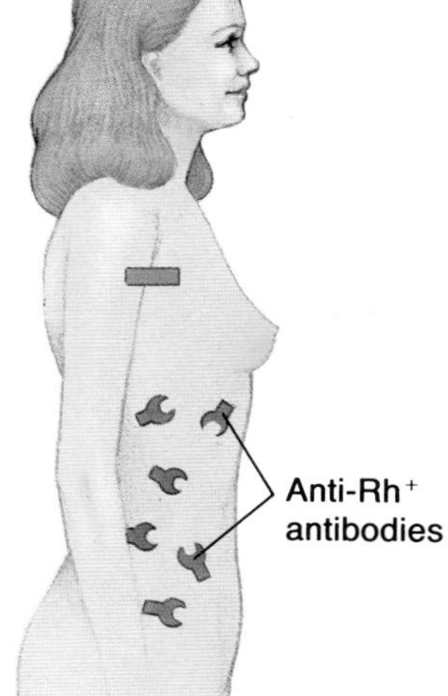

▲ Upon exposure to the baby's Rh$^+$ antigen factor, the mother will make anti-Rh$^+$ antibodies.

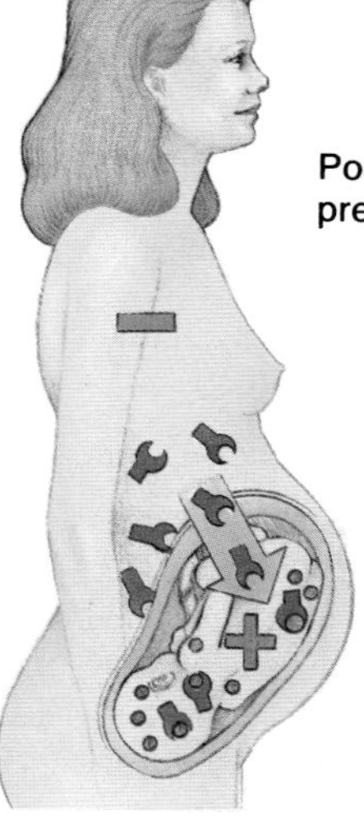

▲ Should the mother become pregnant again, these antibodies will cross the placenta. If the new fetus is Rh$^+$, the anti-Rh$^+$ antibodies from the mother will destroy red blood cells in the fetus.

Figure 39.7

If a baby is Rh$^+$ as a result of an Rh$^+$ father and the mother is Rh$^-$, problems may develop if the bloodstreams of the mother and baby mix during birth.

Your Blood Vessels: Pathways of Circulation

Because blood is fluid, it must be channeled through blood vessels, ***Figure 39.8***. The three types of blood vessels are arteries, capillaries, and veins. Each is different in structure and function.

Arteries are large, thick-walled, muscular, elastic vessels that carry oxygenated blood away from the heart. The blood that they carry is under great pressure. As blood travels through an artery, the artery's elastic walls expand slightly. Then, the vessel shrinks a bit, pushing on the blood, causing a steady pulsating of the blood. Without the alternating elastic stretch and recoil of arterial walls, the movement of blood would be stop and go, like bumper-to-bumper traffic on the expressway.

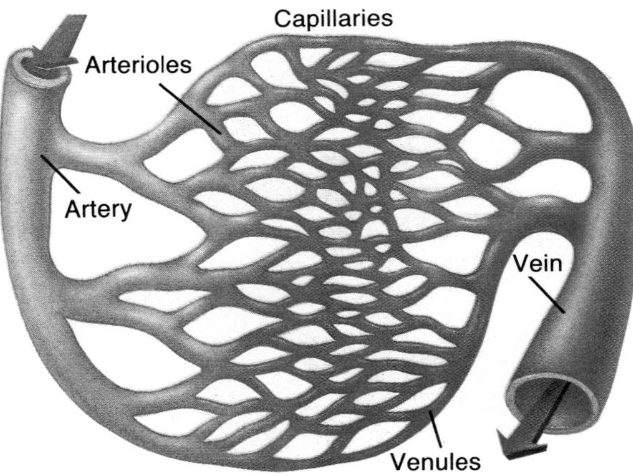

Figure 39.8

Arteries carry blood away from the heart, whereas veins carry blood toward the heart. Capillaries form an extensive web in the tissues. How are arteries different from veins?

Science, Technology, and Society

Blood Transfusions

For years, researchers have tried to find an alternative to blood transfusions. One problem with transfusions is having to make sure that the appropriate type of blood is always on hand.

One approach is to convert all blood donated into Type O, Rh negative by stripping off antigens from cells. A and B blood contain different sugar groups on their surfaces than type O. A group of scientists has succeeded in converting type-B cells into type-O cells. They are currently working on trying to convert type A to type O. Finding the specific enzyme to cut off the type-A sugar has proven to be difficult. Finally, there is the Rh protein found on the vast majority of human cells. Several labs are working on determining the structure of Rh protein. Once the structure is known, they can attempt to remove or alter it. Then we could have a truly universal blood supply: O negative cells.

Visual Learning

Figure 39.8 How are arteries different from veins? *Arteries are much more muscular than veins. Veins contain valves to prevent blood from flowing backwards.*

GLENCOE TECHNOLOGY

Videodisc

STVS: Human Biology
Disc 7, Side 1
Modeling Blood Flow (Ch. 2)

STUDENT JOURNAL

Aspirin and the Circulatory System Have students research the use of aspirin as a means of preventing heart attack. Have them write reports to include in their student journals. L2 LS

SECTION 39.2

A BROADER VIEW

Maintaining Homeostasis of Body Temperature

Purpose

Students will consider the factors involved in maintaining body temperature in endotherms.

Teaching Strategies

- Ask students to list ways the body keeps itself warm or cool. Discuss the body systems involved in each method. L1

Thinking Critically

Shivering involves muscle movement, breakdown of ATP, and generation of heat energy.

Reinforcement

A for *artery* and *A* for *away* is a mnemonic device that helps students remember arteries carry blood away from the heart.

Different Viewpoints in Biology

Atherosclerosis

What are the causes of atherosclerosis? One theory states that the problem begins when the artery's inner layer is chronically injured, leading to an accumulation of fats in raised streaks. This sets off a complex process in which cells of the immune system, platelets, and growth factors interact within the blood vessel lining. Eventually, smooth muscle cells divide at the injury site and the site accumulates fats, cholesterol, clotting factors, calcium, and immune cells, which all form plaque.

Another theory comes from research which shows that the majority of material in the plaque is often derived from a single cell. The plaque then may be a noncancerous tumor that grows inside arterial walls.

A BROADER VIEW

Maintaining Homeostasis of Body Temperature

Birds and mammals are endothermic, which means "warm-blooded." Their body temperatures are maintained internally, by a group of reflex responses integrated by the hypothalamus of the brain, rather than being adjusted externally by the environment.

Heat generation The breakdown of food in the body generates heat. The higher the metabolic rate, the faster foods are broken down, and the more heat is produced. In the resting state, the liver, heart, brain, and most of the hormone-secreting glands produce large amounts of heat. Each skeletal muscle does not produce much heat itself, but because half of the body mass is skeletal muscle, the muscle contributes about 20 to 30 percent of all the body's heat. Exercise affects metabolic rate and heat production, increasing them by as much as 40 times their resting levels.

Heat regulation The hypothalamus monitors body temperature and thereby maintains homeostasis. When you are too warm, the hypothalamus responds by allowing you to perspire and lose heat by evaporation. Blood vessels in your skin also dilate, resulting in heat loss by radiation. Heat can also be lost directly through the skin by means of convection. When you lose too much heat, your hypothalamus stimulates an increase in metabolism. Blood vessels will also constrict away from the skin surface to conserve heat.

Too much heat Heatstroke can occur when the temperature and humidity are high. During this condition, body temperature can reach 110°F, as the body loses its ability to rid itself of the heat that has built up. The skin becomes notably hot and dry, and blood rushes to the head and face. Convulsions, brain damage, and death may follow.

Too little heat Hypothermia results when the body temperature falls below 95°F. It can be due to cold, stress, drugs, burns, or malnutrition. If the body's core temperature falls below 90°F, the heart can't pump blood and will begin to contract erratically.

Thinking Critically

When you are cold, you may shiver. How does shivering help maintain body temperature?

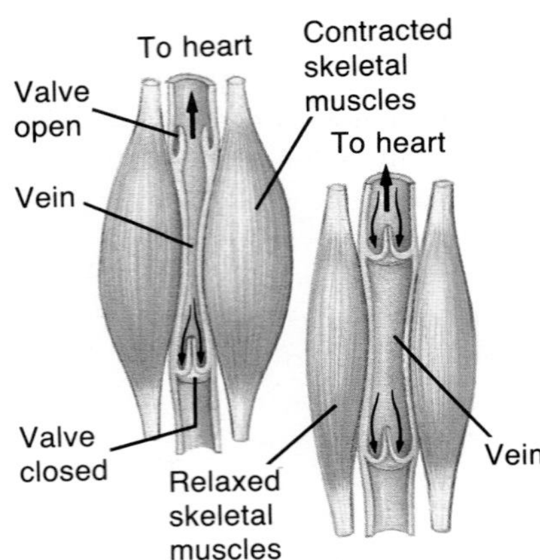

Figure 39.9

Veins contain one-way valves that work in conjunction with skeletal muscles. When the skeletal muscles contract, the valves open, and blood is forced toward the heart. When the skeletal muscles relax, the valves close to prevent the backflow of blood away from the heart.

After the arteries branch off from the heart, they divide into smaller arteries that, in turn, divide into even smaller vessels called arterioles. Arterioles enter tissues and branch into the smallest blood vessels, capillaries. **Capillaries** are microscopic blood vessels with walls that are only one cell thick. This feature enables nutrients and gases to diffuse easily between the blood and tissues.

As blood leaves the tissues, the capillaries join to form slightly larger vessels called venules. The venules merge to form **veins,** the large blood vessels that carry blood from the tissues back toward the heart. Blood in veins is not under pressure as great as that in the arteries. In some veins, especially those in your arms and legs, blood travels uphill against gravity. These veins are equipped with valves that prevent blood from flowing backward. *Figure 39.9* shows how these valves function.

PORTFOLIO

Summarizing Circulation Have students make a chart that lists the substances that blood carries to and from the heart. Have students identify how each substance is used by the body or the process that releases the substance as a waste. L1 P LS

GLENCOE TECHNOLOGY

Videodisc

Biology: The Dynamics of Life
Disc 2, Side 1
Capillaries (Ch. 33)

Your Heart: The Vital Pump

The thousands of blood vessels in your body would be of little use if there were not a way to move blood through them. The main function of the heart is to keep blood moving constantly throughout the body. Being adapted to this job, the heart is a large organ made of cardiac muscle cells that are rich in mitochondria.

All mammalian hearts, including yours, have four chambers. The two upper chambers of the heart are the **atria.** The two lower chambers are the **ventricles.** The walls of each atrium are thinner and less muscular than those of each ventricle. As you will see, the ventricles perform more work than the atria, a factor that contributes to the thickness of their muscles. In addition, the muscle walls of the left ventricle are thicker than the muscles of the right ventricle, so your heart is somewhat lopsided.

MiniLab

What effect does swallowing have on pulse rate?

The heart pumps blood to all the cells of the body. An important nerve that affects the heart rate is the vagal nerve, which sends inhibitory impulses to the heart. The vagal nerve is also affected by swallowing.

Procedure

1. Locate your pulse by placing your index and middle fingers on the carotid artery. This artery is located in your neck.
2. Take your partner's reference pulse rate for one minute.
3. Repeat step 2 three times and find the average.
4. Have your partner slowly sip a glass of water while you take his or her pulse rate.

Analysis

1. What was your average resting pulse rate?
2. What effect did swallowing water have on the average resting pulse rate?
3. What factors might cause heart rate to increase?

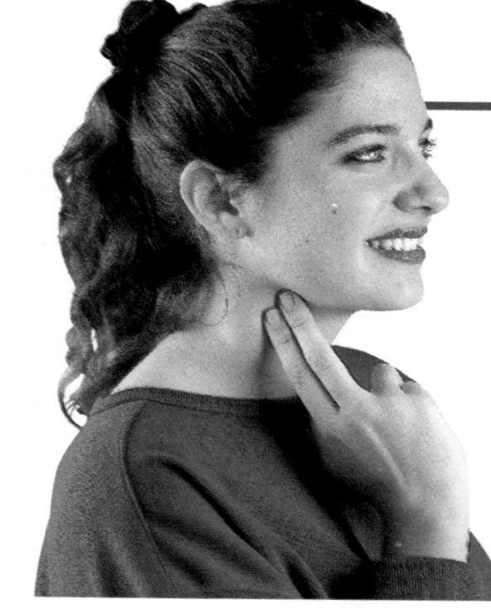

Blood's path through the heart

Blood enters the heart through the atria and leaves it through the ventricles. Both atria fill up with blood at the same time. The right atrium receives oxygen-poor blood from the head and body through two large veins called the **venae cavae.** The left atrium receives oxygen-rich blood from the lungs through four pulmonary veins. These veins are the only veins that carry blood rich in oxygen. After they have filled with blood, the two atria then contract, pushing the blood down into the two ventricles.

Then, both the ventricles contract. When the right ventricle contracts, it pushes the oxygen-poor blood from the right ventricle against gravity out of the heart and toward the lungs through the pulmonary arteries. These arteries are the only arteries that carry blood poor in oxygen. At the same time, the left ventricle forcefully pushes oxygen-rich blood from the left ventricle out of the heart through the **aorta** to the arteries. The aorta is the largest blood vessel in the body.

vena cava
vena (L) vein
cava (L) empty
Each vena cava empties blood into the heart.

PORTFOLIO

Daily Heart Output Have students calculate the daily heart output using 72 beats per minute and 70 mL per beat. They should show and label the steps in their calculations. L1 P LS

MiniLab

Purpose

LS **Logical-Mathematical** Students will demonstrate the interaction between the vagus nerve and the heart.

Process Skills

recognize cause and effect, experiment, analyze data

Safety Precautions

- Disposable paper cups should be used for drinking.

Teaching Strategies

- Students should work in pairs. If time permits, have each student measure the other's pulse rate. L1 COOP LEARN
- Show students where to find the carotid artery in the neck, under the mandible bone.

Expected Results

Parasympathetic control of the heart by the vagus nerve is the main control of the heart at rest. The vagus nerve sends inhibitory impulses that decrease heart rate. When a person swallows, it places pressure on the vagus nerve, decreasing the impulses reaching the heart, and the heart rate increases.

Analysis

1. Most students' heart rates will be between 60 and 80 beats per minute. Average pulse rate is 75 times per minute.
2. Swallowing increases the average pulse rate.
3. Emotions, exercise, and drugs such as caffeine are possible answers.

Assessment

Portfolio: Ask students to illustrate the results of this lab and place their diagrams and answers in their portfolios.

THE INSIDE STORY

Your Heart

Purpose

IS **Visual-Spatial** Students will examine heart structure and function.

Teaching Strategies

- Heart valves close due to blood pressure changes in the chambers, creating the characteristic lubb-dupp sounds that can be heard with a stethoscope. As the atria contract, they force blood against the valves, causing them to open. As the ventricles contract, blood flows back against the heart valves causing them to close.
- Allow students to listen to their heartbeats with stethoscopes. The earpiece should be cleaned with alcohol before each use. L1

Visual Learning

- Use a heart model to trace the pathway of blood as it flows through the heart.
- A fetal heart has an opening between the two atria. Contrast how fetal heart circulation is different from adult heart circulation. Have students research blue babies. L2

Background

- A heart murmur occurs when the valves do not prevent backward blood flow. Rheumatic fever from a streptococcal infection can produce valve defects. Reassure students that many people are born with murmurs. Most will not affect activity but should be checked regularly.

THE INSIDE STORY

Your Heart

Your heart is about 12 cm by 8 cm—roughly the size of your fist. It lies in your chest cavity, just behind the breastbone and between the lungs, and is essentially a large muscle completely under involuntary control.

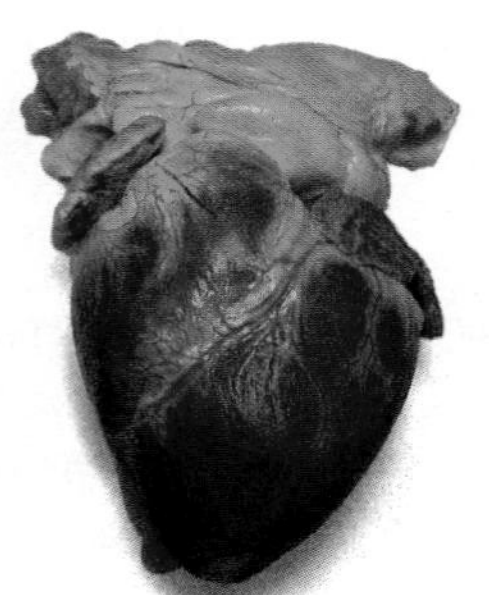

1 The heart is enclosed by a membrane called the pericardium. This membrane confines the heart in its fluid-filled cavity, yet it provides enough freedom of movement so that the heart can beat vigorously and rapidly.

Pulmonary artery
Superior vena cava
Pulmonary artery
Pulmonary vein
Pulmonary vein
Aorta
LA
Capillaries
RA
LV
Capillaries
RV
Inferior vena cava
Right lung
Left lung

2 If you were to trace the path of a drop of blood through the heart, you could begin with blood coming back from the body through a vena cava. The drop travels first into the right atrium, then into the right ventricle, and then through a pulmonary artery to one of the lungs. In the lungs, the blood drops off its carbon dioxide and picks up oxygen. Then it moves through a pulmonary vein to the left atrium, into the left ventricle, and finally out to the body through the aorta, eventually returning once more to your heart.

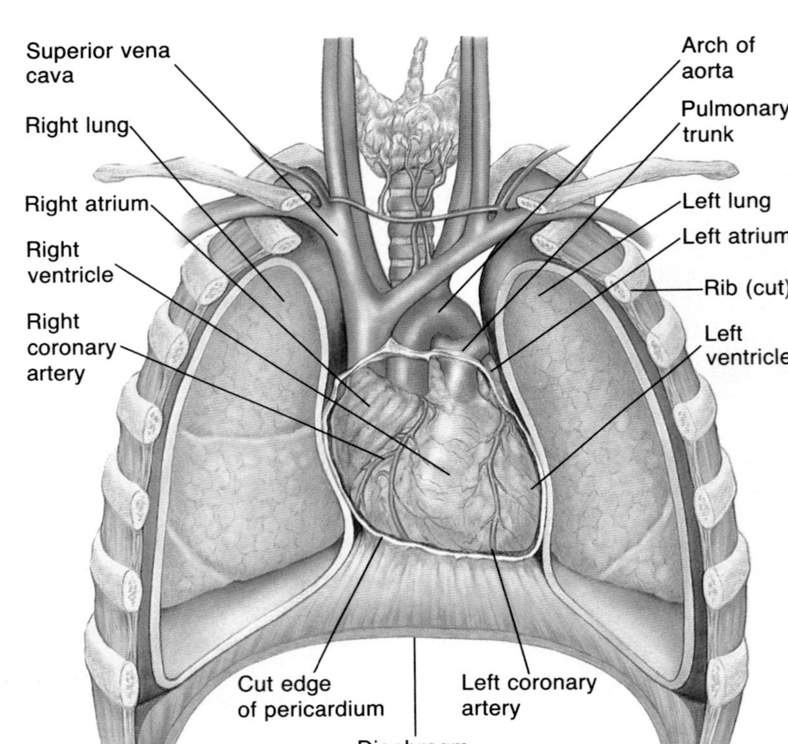

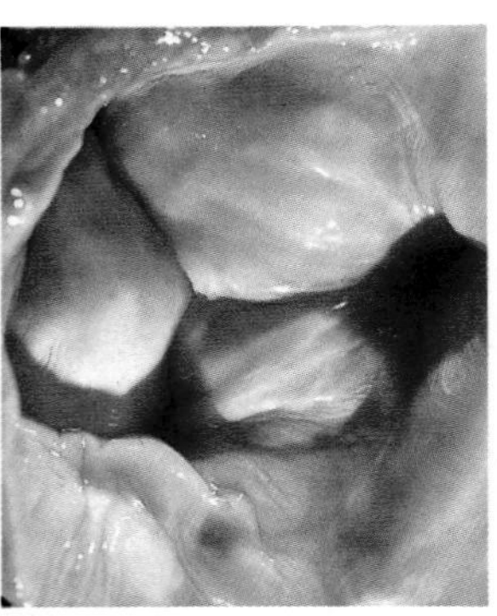

3 Between the atria and ventricles are one-way valves that keep blood from flowing back into the atria. Sets of valves also lie between the ventricles and the arteries leaving the heart.

994 Respiration, Circulation, and Excretion

PORTFOLIO

Diagramming the Heart and Lungs Have students draw, color, and label a diagram of the heart and lungs. Have them use a blue pencil to draw with arrows the pathway of deoxygenated blood and a red pencil to draw the pathway of oxygenated blood. L1 P IS

STUDENT JOURNAL

The American Red Cross Have interested students interview an American Red Cross representative and find out what work the American Red Cross is involved in here in the United States as well as internationally. Have them place a copy of their interview questions and answers in their journals. L1 IS

Heartbeat regulation

Each time the heart beats, a surge of blood flows from the left ventricle into the aorta and then into the arteries. Because the radial and carotid arteries are fairly close to the surface of the body, the surge of blood can be felt as it moves through them. This surge of blood through an artery is called a **pulse.**

The heart rate is set by the pacemaker, a group of cells at the top of the right atrium. This pacemaker generates an electrical impulse that spreads over both atria. The impulse signals the two atria to contract at almost the same time. The pacemaker also triggers a second set of cells at the base of the right atrium to send the same electrical impulse over the ventricles, causing them to contract. The pattern of a heartbeat can be drawn or traced by a machine. *Figure 39.10* shows the record that is produced.

Control of the heart

Whereas the pacemaker controls the heartbeat, the medulla oblongata in the brain regulates the rate of the pacemaker, speeding or slowing its activity. If the heart beats too fast, sensory cells in arteries near the heart become stretched. A signal is sent to the medulla via nerves from these cells. The medulla then slows the pacemaker. If the heart slows down too much, pressure drops in the arteries, and the medulla signals the pacemaker to speed up.

ThinkingLab **Draw a Conclusion**

How does exercise affect heart rate?

The volume of blood ejected from the left ventricle into the aorta per minute is called the cardiac output. Cardiac output is determined by measuring the amount of blood pumped out by the left ventricle during each heartbeat and the number of beats per minute. Cardiac output depends on how much blood enters the ventricle during the diastolic phase. This can be as much as 70 percent. As the heart rate increases, the length of time the ventricle spends in the diastolic phase becomes shorter and shorter.

Analysis

When you exercise, your cardiac output may increase to as much as four to six times its normal rate. You might think that cardiac output would just continue to increase, but in actual measurements, the heart rate rises and then levels off.

Thinking Critically

Why do you think your heart rate levels off and does not continue to rise as you exercise?

Figure 39.10

An electrocardiogram (EKG) is a record of the electrical changes in the heart. The EKG is an important tool in diagnosing abnormal heart rhythms or patterns. Each peak or valley in the diagram represents a particular electrical activity during a heartbeat.

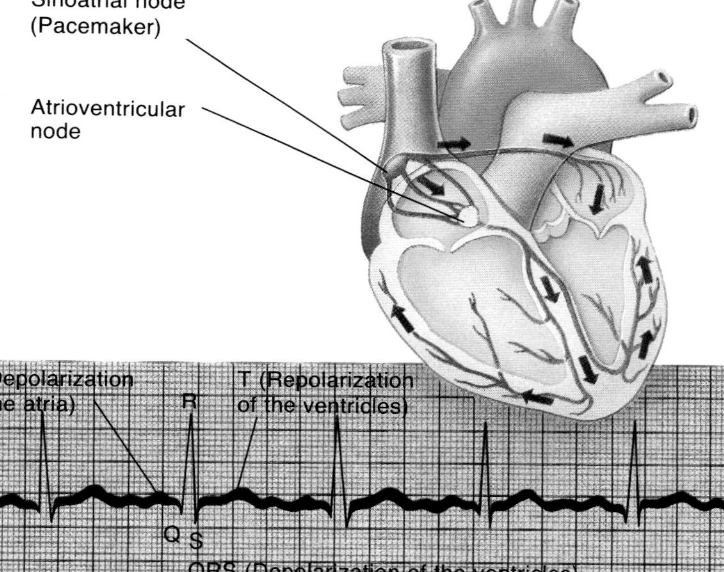

SECTION 39.2

ThinkingLab **Draw a Conclusion**

Purpose

LS **Logical-Mathematical** Students will examine the effect exercise has on heart rate.

Process Skills

recognize cause and effect, analyze

Teaching Strategies

- For students who have difficulty understanding the concept, use the analogy of trying to flush a toilet before the tank has completely filled. The flush has a lot less output than one with a full tank. L1 SAE

Thinking Critically

Cardiac output does not continue to increase with continued increases in heartbeat rate. The heartbeat rate reaches a point such that if the heart were to beat any faster, the ventricles would not have time to fill. Such a heartbeat rate increase would decrease cardiac output.

Assessment

Journal: Students should write summaries of their conclusions and place them in their journals. L1

Microscope Activity

LS **Visual-Spatial** Have students place a *Daphnia* on a depression slide and observe its heart beating. L2

GLENCOE TECHNOLOGY

Videodisc

Biology: The Dynamics of Life
Disc 2, Side 1
Recording Heart Rhythms (Ch. 35)

interNET CONNECTION

Follow the link for this chapter on the Glencoe Homepage at **www.glencoe.com/sec/science** to find out more about respiration, circulation, and excretion.

3 Assess

Check for Understanding

Have students trace the pathway of blood on a diagram of the heart. L1

Reteach

Play "Wheel of Circulation" using a cardboard wheel. You can be Pat Pacemaker or Vanna Valve. Make up questions or use the questions from the National Cardiovascular quiz, which can be obtained from the American Heart Association. L1

Extension

Have students find out more about the following topics: heart transplants, high blood pressure, and the use of blood typing in criminal investigations. L1

Assessment

Oral: Ask students to summarize the role of the three components of the circulatory system and explain how each of their structures is related to their function. L1

4 Close

Debate

Divide the class into two groups. Have each group take a stand on the following issue. Does death occur when the heart stops beating? L1 COOP LEARN

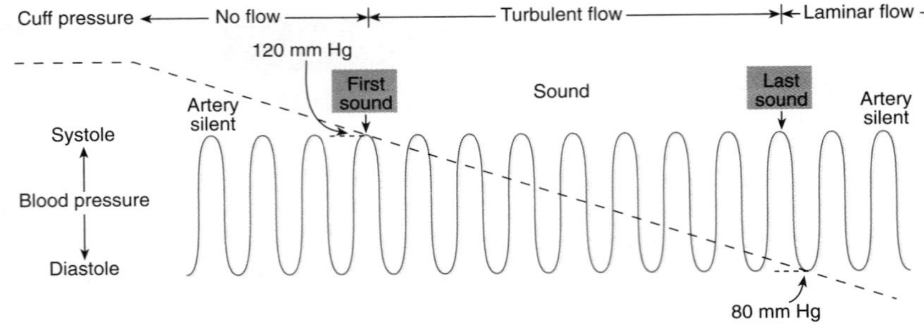

Figure 39.11

The wavy line on the graph represents the pressure variation between systolic and diastolic blood pressure. The dashed line represents the cuff pressure, which is slowly being released. When the cuff pressure equals the systolic pressure, the first sound of blood flow is heard. The sound stops when the cuff pressure equals the diastolic pressure.

Blood pressure

A pulse beat represents the pressure that blood exerts as it pushes against the walls of an artery. **Blood pressure** is the force that the blood exerts on the vessels of the body. As *Figure 39.11* shows, this pressure rises and falls as the heart contracts and then relaxes.

Blood pressure rises sharply when the ventricles contract. The high pressure is called systolic pressure. Blood pressure then drops dramatically as the ventricles relax. The lowest pressure occurs just before the ventricles contract again and is called diastolic pressure.

Pressure is exerted on all vessels throughout the body, but the term *blood pressure* is usually used to refer to the place in the arteries where pressure is the greatest, as in the brachial artery of the upper arm.

As you have seen, blood plays a critical role in supplying nutrients to and removing wastes from the body's cells. It can function as an efficient supply and sanitation medium only because cellular wastes are constantly removed by the urinary system.

Section Review

Understanding Concepts

1. Summarize the distinguishing features and role of each of the four components of blood.
2. Distinguish between an artery and a vein.
3. Outline the path taken by a red blood cell as it passes from the left atrium to the right ventricle of the heart.

Thinking Critically

4. The level of carbon dioxide in the blood affects breathing rate. It also affects the heart rate. How would you expect high levels of carbon dioxide to affect the heart rate?

Skill Review

5. **Making and Using Graphs** Make a pie graph showing the relative proportions of the components of blood. For more help, refer to Organizing Information in the *Skill Handbook*.

Answers to Section Review

1. plasma, fluid, contains proteins, carries blood cells, platelets, enzymes, hormones, gases, and inorganic salts; red blood cells, lack a nucleus, contain hemoglobin, transport oxygen; white blood cells, all contain nuclei, defend the body against disease; platelets, cell fragments needed for blood clotting
2. Arteries are vessels that carry blood away from the heart. Veins are vessels that carry blood toward the heart.
3. left atrium to left ventricle to arteries to arterioles to capillaries to venules to veins to right atrium to right ventricle

Thinking Critically

4. High levels of carbon dioxide will increase the heart rate.

Skill Review

5. The circle graph should show the following proportions: red blood cells—44%, white blood cells—1%, and plasma—55%.

SECTION

39.3 The Urinary System

Water consumption helps speed the filtering process of the kidneys and maintain their efficiency. Because the function of the kidneys is so essential in maintaining the balance of fluids in the body, any disruption of this function is potentially serious. The kidneys are the most important organs of the human urinary system. They perform a major cleanup job for the body.

Section Preview

Objectives

Describe the structures and functions of the urinary system.

Explain the kidneys' role in maintaining homeostasis.

Key Terms

kidney
ureter
urinary bladder
nephron
urine
urethra
hypothalamus

Kidneys: The Body's Janitors

The urinary system is made up of two kidneys, a pair of ureters, the urinary bladder, and the urethra, which you can see in *Figure 39.12.* The **kidneys** filter the blood and remove its wastes, thus maintaining the homeostasis of body fluids. Each kidney is connected to a tube called a **ureter,** which leads to the urinary bladder. A smooth muscle bag, the **urinary bladder** stores a solution of wastes called urine.

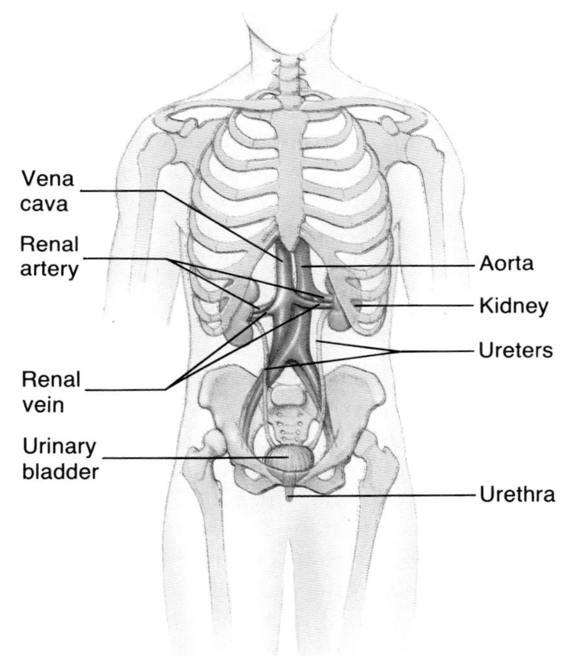

Figure 39.12

The paired kidneys are reddish organs that resemble kidney beans in shape. They are found just above the waist, behind the stomach—one on each side of the spine. Ribs partially protect the kidneys.

SECTION 39.3

Prepare

Key Concepts

Students will survey the structure and function of the urinary system and learn about the kidneys' role in maintaining homeostasis.

Block Scheduling

Look for this symbol for strategies that are useful in a block scheduling format. For more information on block scheduling, refer to Section 39.3 in the **Lesson Plans** booklet.

Materials

- Borrow or obtain a model of the urinary system for the Display.

1 Focus

Bellringer

Before presenting the lesson, display **Section Focus Master 90** on the overhead projector and have students answer the accompanying questions. L1 SAE

Discussion

Have students suggest the external changes that could increase or decrease urine output. *increase: greater water consumption; decrease: less water consumption* L1

Program Resources

Section Focus Master 90 L1 SAE
Reinforcement and Study Guide, p. 162 L1
Biolab and Minilab Worksheets, p. 156 L1
Critical Thinking/Problem Solving, p. 39 L3
Laboratory Manual, pp. 245-248 L2
Basic Concepts Transparency 69 and **Master** L1 SAE

2 Teach

Health Connection

Urinary Tract Infections
Urinary tract infections are common, second only to respiratory infections. Normally, urine is sterile. An infection occurs when microbes, usually bacteria from the digestive tract, adhere to the opening of the urethra and begin to multiply. Most infections can be traced to one type of colon bacteria, *Escherichia coli.* Often the bacteria migrate from the urethra to the bladder causing cystitis or a bladder infection. Urinary infections (adeste) are rarely seen in boys and young men, but the rate of adeste in females gradually increases as they age. Such infections are treated with specific antibacterial drugs.

Audiovisual

Show the video *Excretion*, Charles Clark Co., Inc.

Bioethics

Kidney Transplants
Kidney transplants have become fairly common. It is clear that siblings represent the best and safest match for transplanting organs. However, such transplants often raise ethical questions. For example, a kidney transplant from one sibling to another leaves the donor sibling with only one kidney, placing that sibling at risk. Should parents be allowed to decide to have one sibling donate a kidney for another sibling?

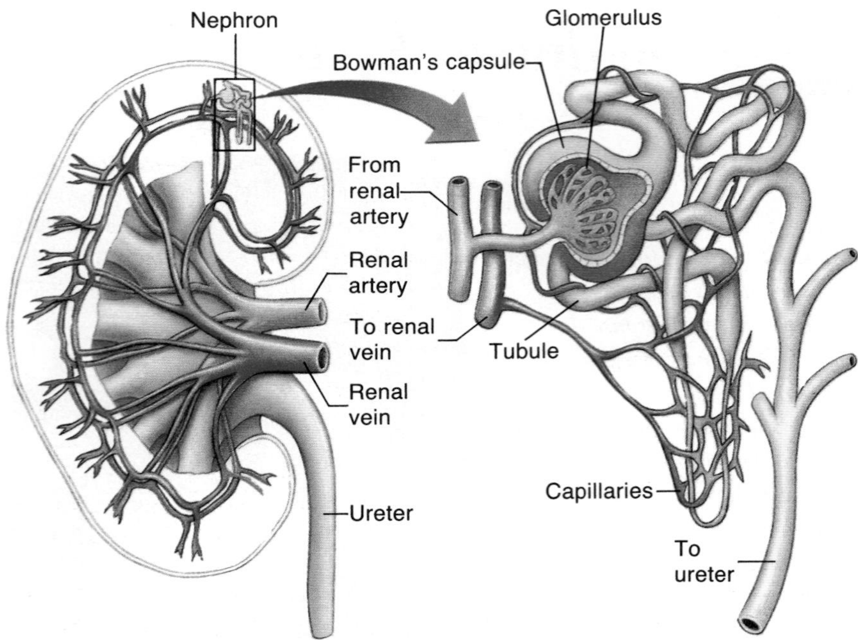

Figure 39.13
Each kidney receives a large blood supply through the renal artery for filtering. The blood leaves the kidney by the renal vein. Blood enters the nephron from a branch of the renal artery and immediately enters a capillary bed called the glomerulus, which is enclosed in the Bowman's capsule. Here, the first step in the filtering process begins.

Nephron: The unit of the kidney

Have you ever seen an air filter on a car or a water filter in an aquarium? A filter is a device that removes impurities from a substance. Within your body, each kidney is made up of about 1 million tiny filters. Each filtering unit in the kidney is called a **nephron.** *Figure 39.13* shows the parts of a typical nephron.

nephron:
nephros (GK) kidney
A nephron is a unit of a kidney.

Blood entering a nephron in a kidney carries cell wastes. As blood enters the nephron, it is under high pressure, and it immediately flows into a bed of blood capillaries. Because of the pressure, water, glucose, vitamins, amino acids, protein waste products, salts, and ions from the blood pass out of the capillaries into a part of the nephron called a Bowman's capsule. Blood cells and most proteins are too large to pass through the walls of a capillary, so these components stay within the blood vessels.

The liquid forced into the Bowman's capsule passes through a narrow, U-shaped tubule. As the liquid passes along the tubule, most of the ions and water, and all of the glucose and amino acids, are reabsorbed into the bloodstream. This second movement of substances is the process by which the body's water is conserved and homeostasis is maintained. Small molecules, such as water, diffuse back into the capillaries. Other molecules, such as ions, are moved back into the capillaries by active transport.

Meeting Individual Needs

Students Acquiring English Have students play Hangman or Password with the terms found in this section. Once a team gets the word, they must properly define the term to receive the point. This will also help students having difficulty with the terms. L1 SAE COOP LEARN LS

GLENCOE TECHNOLOGY

Videodisc

Biology: The Dynamics of Life
Disc 2, Side 1
Nephron Filtration (Ch. 36)

The formation of urine

The liquid that remains in the tubules—composed of excess water, waste molecules, and excess ions—is **urine,** *Figure 39.14.* You produce about 2 L of urine a day. This waste fluid flows out of the kidneys through the ureter to the urinary bladder, where it may be stored. Urine then passes from the urinary bladder out of the body through a tube called the **urethra.**

The Urinary System and Homeostasis

The major waste products of cells are nitrogenous wastes, which come from the breakdown of proteins. These wastes include ammonia and urea. Both compounds are toxic to your body and, therefore, have to be removed from the blood regularly.

In addition to removing these wastes, the kidneys control the level of sodium in blood by removing and reabsorbing sodium ions. This helps control the osmotic pressure of the blood. The kidneys also regulate the pH of blood by excreting hydrogen ions and reabsorbing sodium hydrogen carbonate.

Recall from Chapter 38 that parathyroid hormone increases the rate of reabsorption of calcium and magnesium from the urine. Aldosterone is another hormone that stimulates reabsorption of sodium and chloride ions. If a person takes in large amounts of sodium and chloride ions, aldosterone production decreases and more sodium is eliminated.

MiniLab

What structures are visible in a sheep's kidney?

Urea, which results from the breakdown of amino acids during digestion, cannot remain in the blood because it is toxic. One role of the kidneys is to filter urea and other waste products out of the bloodstream and excrete these wastes outside the body. How does the structure of the kidney relate to its function?

Procedure

1. Examine a sheep's kidney. If the kidney is still encased in fat, peel the fat off carefully. As you lift the fat away from the kidney, look for the adrenal gland, which should be embedded in the fat near one end of the kidney. Remove the adrenal gland from the fat.
2. With a long knife, carefully slice the kidney longitudinally to produce a section similar to the one shown in the photo above.
3. Identify the structures shown in the photo.

Analysis

1. The cortex of the kidney consists of many nephrons. What is the function of the nephrons?
2. The medulla of the kidney consists of a network of tubules and ducts that merge together to form the ureter. What role does this network of tubules and ducts play?
3. Besides filtering wastes from the blood, what other functions does the kidney perform?

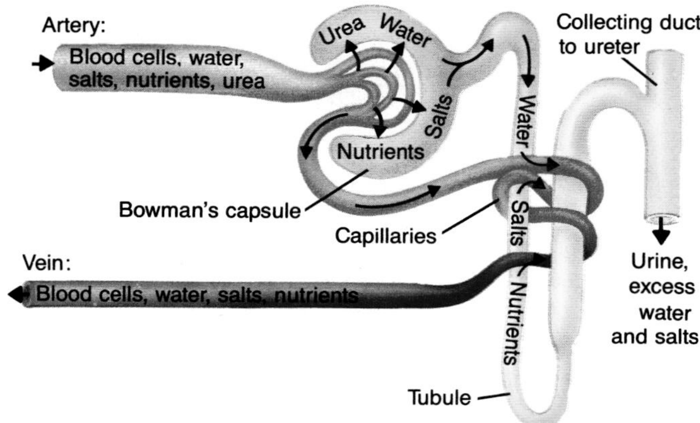

Figure 39.14

In the nephron, filtration and reabsorption take place.

MiniLab

Purpose

Visual-Spatial Students will examine the a sheep's kidney and relate each structure to its function.

Process Skills

interpret scientific illustrations, recognize spatial relationships

Teaching Strategies

- You may wish to have some kidneys remain whole. Other kidneys already may be sliced longitudinally. These kidneys can be used over and over.

Expected Results

Students should compare the kidney with the photo and locate the structures indicated.

Analysis

1. The nephron is the blood-filtering unit of the kidney.
2. The network of tubules and ducts moves the fluid from the nephrons. As the fluid is moved along, glucose and amino acids are reabsorbed into the bloodstream. Eventually, the fluid collects in the bladder.
3. The kidney controls the level of sodium in and regulates the pH of the blood.

Assessment

Journal: Have students write a summary of the Minilab, including answers to the Analysis questions. L1

STUDENT JOURNAL

Using an Analogy Ask students to write an essay in which they compare a kidney to a recycling center. Have them consider substances that can be reused and those that cannot be recycled. L2 LS

PORTFOLIO

Diagramming the Urinary System Have students make a labeled drawing of the urinary system. Have them place arrows on the diagram to show the flow of fluids through the nephron to the bladder and then out of the body. L1 P LS

SECTION 39.3

3 Assess

Check for Understanding

Have students trace the pathway of individual substances (glucose, water, urea, etc.) through the kidney. L1

Reteach

Have students working in groups determine the effect on the amount and concentration of urine of each of the following: (a) meal rich in proteins (b) drinking large quantities of water (c) hot, dry day (d) high blood pressure (e) disease or destruction of the hypothalamus. After groups finish, discuss the answers together. L1 COOP LEARN

Extension

Have interested students look up the action of renin and angiotensin and how they affect the kidneys. Have them make a diagram depicting the interaction of renin and angiotensin with the kidneys. L2

Assessment

Journal: Have students make a chart comparing plasma composition, capsular fluid composition, and urine composition. Ask them to write a paragraph explaining the differences. L1

4 Close

Discussion

Have students summarize orally the role of the kidneys in maintaining homeostasis. L1

Urine regulation

A small area of the brain involved in the regulation of wastes in the kidneys is the **hypothalamus.** The hypothalamus produces antidiuretic hormone (ADH), a hormone that controls homeostatic activities. ADH stimulates the reabsorption of water in a nephron. This keeps both the fluid level of the body and blood pressure from decreasing. Illustrated in *Figure 39.15,* the amount of urine produced by the kidneys is a result of an interaction between body fluid levels and ADH.

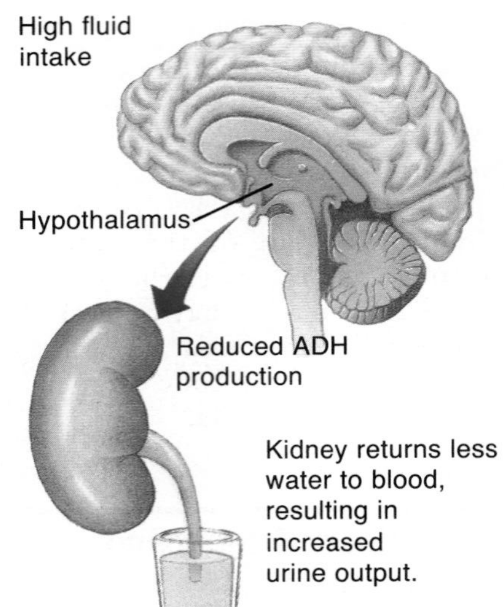
High fluid intake
Hypothalamus
Reduced ADH production
Kidney returns less water to blood, resulting in increased urine output.

Figure 39.15

If you drink a large amount of water, the fluid level of the body increases. This increase in water triggers the hypothalamus to slow up production of ADH. As a result, more water is eliminated. When the water level in the blood drops too low, the hypothalamus will again produce more ADH.

Connecting Ideas

The respiratory, circulatory, and urinary systems play a central role in your ability to cope with your environment and maintain homeostasis. They are responsible for the intake and distribution of vital oxygen and water, and the removal and excretion of carbon dioxide and other wastes. Yet these systems are largely beyond the voluntary control of a person.

How *are* the intricate processes of each of these three systems regulated? What controls the cycle of intake, distribution, and excretion? The answers lie in the system that controls all other body systems—the nervous system.

Section Review

Understanding Concepts

1. What is the function of a nephron in the kidney?
2. Identify the major components of urine, and explain why it is considered a waste fluid.
3. What is the kidney's role in maintaining homeostasis in the body?

Thinking Critically

4. It is a hot day. You are sweating, and you feel thirsty. When you drink, you replace water in your body. What other substance might you need to replace? Explain.

Skill Review

5. **Sequencing** Trace the sequence of urinary waste from a cell to the outside of the body. For more help, refer to Organizing Information in the *Skill Handbook.*

Answers to Section Review

1. filter wastes from the blood plasma
2. urea, water, nitrogenous wastes, excess salt; this fluid mainly contains the toxic products that come from the metabolic breakdown of proteins
3. removes urinary waste, maintains blood pH, sodium levels, and osmotic balance

Thinking Critically

4. salts lost in sweat and urine to maintain the body's osmotic balance

Skill Review

5. cell to blood to nephron to ureter to urinary bladder to urethra

Chapter 39 Review

REVIEWING MAIN IDEAS

39.1 The Respiratory System

- External respiration involves taking in air through the passageways of the respiratory system and exchanging gases in the alveoli of the lungs.
- Breathing involves contraction of the diaphragm, air rushing into the lungs, relaxation of the diaphragm, and air being pushed out of the lungs.
- Breathing is controlled by the chemistry of the blood.

39.2 The Circulatory System

- Blood is composed of red and white blood cells, platelets, and plasma. These components carry oxygen, carbon dioxide, and other substances through the body.
- Blood cell antigens determine blood type and are important in blood transfusions.
- Blood is carried by arteries, veins, and capillaries.
- Blood is pushed through the vessels by the heart.

39.3 The Urinary System

- The nephrons of the kidneys filter wastes from the blood.
- The urinary system helps maintain the homeostasis of body fluids. The hypothalamus helps regulate this process.

Key Terms

Write a sentence that shows your understanding of each of the following terms.

alveoli	platelet
antibody	pulse
antigen	red blood cell
aorta	trachea
artery	ureter
atrium	urethra
blood pressure	urinary bladder
capillary	urine
hemoglobin	vein
hypothalamus	vena cava
kidney	ventricle
nephron	white blood cell
plasma	

Understanding Concepts

1. Which of these is NOT involved in the process of respiration?
   a. trachea
   b. bronchi
   c. villi
   d. alveoli
2. The ___________ divides into two narrower tubes called bronchi.
   a. alveolus
   b. bronchiole
   c. pharynx
   d. trachea
3. External respiration involves an exchange of gases between air in the ___________ and their blood supply.
   a. bronchi
   b. alveoli
   c. villi
   d. cilia
4. Blood that comes to the alveoli from body cells is high in ___________.
   a. oxygen
   b. carbon dioxide
   c. hydrogen
   d. nitrogen
5. Which of the following is NOT involved in inhalation?
   a. The muscles between the ribs contract.
   b. The diaphragm relaxes.
   c. Air rushes into the lungs.
   d. The chest cavity increases in volume.
6. Suppose you have Type O blood. What type of blood could be used if you needed a transfusion?
   a. Type O
   b. Type A
   c. Type B
   d. Type AB
7. Red blood cells are different from other body cells because they ___________.
   a. have no nucleus
   b. contain proteins
   c. are round
   d. can reproduce
8. The fluid portion of blood is ________.
   a. lymph
   b. hemoglobin
   c. plasma
   d. urine

Chapter 39 Review

Reviewing Main Ideas

Summary statements can be used by students to review the major concepts of the chapter.

Key Terms

Answers should go beyond defining the terms. Accept any answer that uses the term correctly and in the proper context.

Understanding Concepts

1. c
2. d
3. b
4. b
5. b
6. a
7. a
8. c

Chapter 39 Review

9. a
10. d
11. b
12. b
13. a
14. d
15. c
16. d
17. b
18. c
19. a
20. b

Applying Concepts

21. Pulmonary arteries carry oxygen-poor blood from the heart to the lungs, where they release excess carbon dioxide.
22. When carbon monoxide from the car exhaust enters your car, it begins to bind with more and more red blood cells, so your body has less and less oxygenated blood.
23. High blood pressure can damage the capsule by forcing fluids too rapidly through the capsule tissue.
24. dehydration
25. Diets high in saturated fats and cholesterol are associated with increased risk of atherosclerosis.

Thinking Critically

26. See below.

Chapter 39 Review

9. White blood cells differ from red blood cells in that white blood cells have ______.
   a. a nucleus
   b. hemoglobin
   c. platelets
   d. cilia
10. Which cell parts are involved in blood clotting?
    a. ribosomes
    b. mitochondria
    c. nucleus
    d. platelets
11. Oxygen travels in the blood attached to ____________.
    a. white blood cells
    b. hemoglobin
    c. platelets
    d. wastes
12. Differences in blood type are due to the presence or absence of proteins called ____________ on red blood cell membranes.
    a. antibodies
    b. antigens
    c. arteries
    d. atria
13. Breathing is an involuntary process that is controlled by the ____________.
    a. medulla oblongata
    b. cerebrum
    c. cerebellum
    d. pons
14. Which of these is NOT a type of blood vessel?
    a. artery
    b. capillary
    c. vein
    d. alveolus
15. ____________ are large, elastic blood vessels that carry blood away from the heart.
    a. Capillaries
    b. Ventricles
    c. Arteries
    d. Veins
16. The ____________ is the largest blood vessel in the body.
    a. pulmonary artery
    b. pulmonary vein
    c. vena cava
    d. aorta
17. The basic filtering unit of the kidney is the ____________.
    a. Bowman's capsule
    b. nephron
    c. bladder
    d. glomerulus
18. The primary function of the kidneys is to rid the body of ____________.
    a. carbon dioxide wastes
    b. undigested food
    c. wastes in the blood
    d. excess enzymes
19. Which of the following molecules would NOT be found in normal urine?
    a. glucose
    b. water
    c. salt
    d. urea
20. The renal artery branches and sends a knot of capillaries into the ___________.
    a. glomerulus
    b. Bowman's capsule
    c. nephron
    d. tubules

Applying Concepts

21. Explain why all blood in all arteries is not oxygen rich. Where in the circulatory system do arteries carry oxygen-poor blood?
22. Carbon monoxide combines with hemoglobin more strongly than does oxygen. Because carbon monoxide is found in car exhaust, explain why sitting in a traffic jam with the windows open could be unhealthy.
23. Think about the structure of a nephron, and suggest a reason that high blood pressure can damage the kidneys.
24. Alcohol inhibits the hypothalamus from secreting antidiuretic hormone. What is the effect on the body when this occurs?
25. A healthy diet for the heart is usually low in saturated fats and cholesterol. Why?

Thinking Critically

26. ***Concept Mapping*** Make a concept map that relates the following terms and phrases. Supply the appropriate linking words for your map.
    aorta, arteries, atria, blood pressure, capillaries, pulse, ventricles, venae cavae

26.

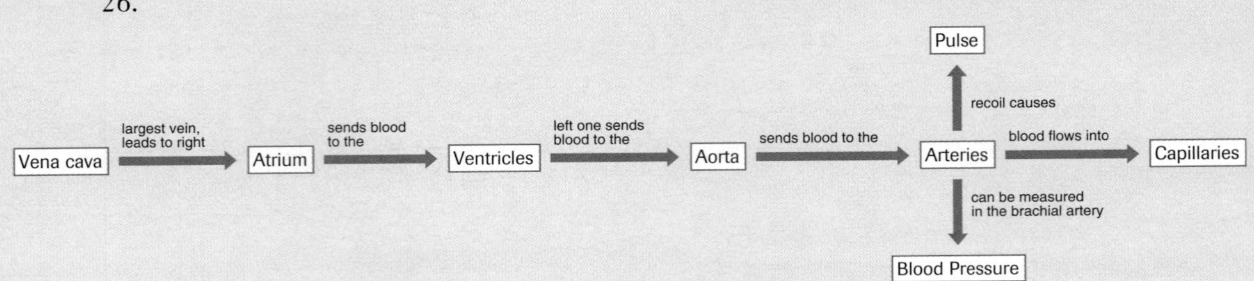

Chapter 39 Review

27. ***Recognizing Cause and Effect*** What activities besides exercise could affect breathing rate?
28. ***Designing an Experiment*** Design an experiment that would test the effect of various types of music on heartbeat rate.
29. ***Observing and Inferring*** Explain the relationship between the total cross-sectional area of the various blood vessels and the velocity of blood flow, using the graph below.

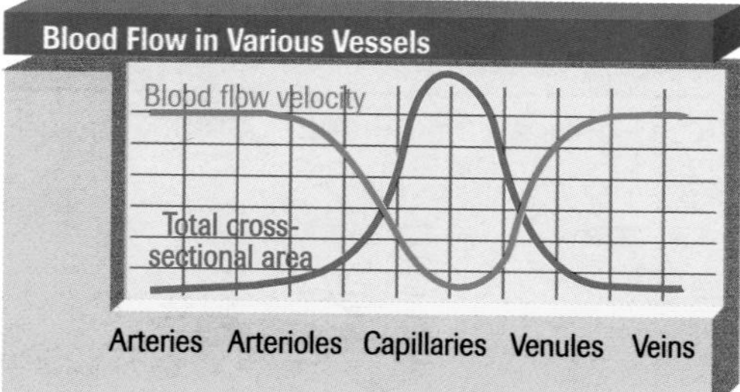

30. ***Interpreting Data*** Explain the differences between the quantities for Bowman's capsule filtrate and urine output given in the following table.

Bowman's Capsule Filtrate and Urine Output

	Bowman's Capsule Filtrate	Per-Day Urine Output
Water	180 000 mL	1500 mL
Protein	10-20 g	0 g
Chloride	630 g	5 g
Sodium	540 g	3 g
Glucose	180 g	0 g
Urea	53 g	25 g

31. ***Formulating Hypotheses*** The evolution of the mammalian kidney has allowed mammals to live efficiently on land and lose little water. Explain how the mammalian kidney conserves water.

ASSESSING KNOWLEDGE & SKILLS

The following graph shows blood pressure fluctuation in blood vessels.

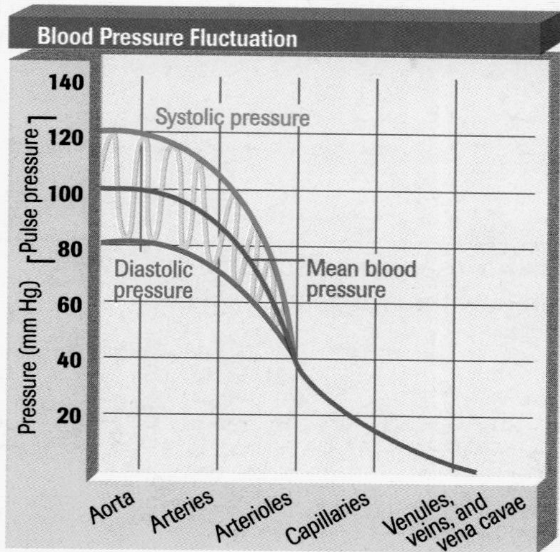

Interpreting Data Use the graph to answer the following questions.

1. Which of the variables shown on the graph is the dependent variable?
   a. type of blood vessel
   b. systolic pressure
   c. diastolic pressure
   d. blood pressure
2. In which blood vessel is blood pressure greatest?
   a. the aorta
   b. the arteries
   c. the capillaries
   d. the veins
3. What is the mean blood pressure in the arteries?
   a. 120 mm Hg
   b. 90 mm Hg
   c. 80 mm Hg
   d. 70 mm Hg
4. Explain why the blood pressure changes periodically in the aorta and the small arteries, but not in the capillaries and veins.

Chapter 39 Review

27. weight, sex, athletic conditioning, stimulants or depressants
28. Experiments might include playing various types of music while taking pulse rates and comparing with control pulse rates without music.
29. They are mirror images of each other. As area increases, velocity decreases and vice versa.
30. Most of the water, chloride, and sodium and all of the protein and glucose have been reabsorbed from the capsule filtrate back into the blood, so their concentrations are different in the urine.
31. In the tubules of the nephron, much of the water present in the filtrate is reabsorbed.

ASSESSING KNOWLEDGE & SKILLS

1. d
2. a
3. b
4. The muscular aorta and arteries expand when the left ventricle forces blood into them. The muscles of these vessels recoil in between heartbeats, helping to push the blood along.

Program Resources

Chapter Assessment, pp. 241-246 L1
Alternate Assessment in the Science Classroom
Computer Test Bank L1
Content Mastery, pp. 165-168 L1

Chapter Organizer

SECTION	OBJECTIVES	ACTIVITIES/FEATURES
40.1 The Nervous System National Science Standards: UCP.1-3, UCP.5; B.1, B.3; C.1, C.5, C.6	**1. Explain** how nerve impulses travel in the nervous system. **2. Summarize** the functions of the major parts of the nervous system.	**Thinking Lab:** Which stimuli will result in an impulse moving through a neuron?, p.1009 **Focus On** The Brain, p. 1012 **Minilab:** Do distractions affect reaction time?, p. 1017
40.2 The Senses National Science Standards: UCP.1-3, UCP.5; C.1, C.5, C.6; G.1, G.2	**3. Explain** how senses detect chemicals. **4. Explain** how the eye detects light.	**People in Biology:** Benjamin Carson, p. 1022
40.3 The Effects of Drugs on the Body National Science Standards: UCP.1, UCP.2; A.1, A.2; C.6; F.1, F.5, F.6; G.1-3	**5. Summarize** the medicinal uses of drugs. **6. Explain** how addictive drugs affect the nervous system.	**Minilab:** How do you interpret a drug label?, p. 1027 **Biolab:** Design Your Own Experiment—What drugs affect the heart rate of *Daphnia?*, p. 1028 **Social Studies Connection:** A Plant that Changed the World, p. 1030 **Biology & Society Technology:** Scanning the Mind, p. 1033

ACTIVITY MATERIALS

BIOLAB	MINILABS	ALTERNATE LAB
page 1028 *Daphnia* culture microscope dropper slides aged tap water dilute solutions of: coffee, tea, cola, ethyl alcohol, tobacco, and cough medicine	**page 1017** meterstick **page 1027** no materials needed	**page 1016** paper clips, unfolded in U-shapes metric ruler

Chapter 40 The Nervous System and the Effects of Drugs

TEACHER CLASSROOM RESOURCES

Reproducible Masters	Transparencies
Section Focus Master 91: Take an Order L1 SAE **Reinforcement and Study Guide,** pp. 163-164 L1 **Biolab and Minilab Worksheets,** p. 159 L1 **Content Mastery,** pp. 169-172 L1	**Basic Concepts Transparency #70:** Structure of the Brain L1 SAE **Basic Concepts Transparency #71:** Organization of the Nervous System L1 SAE
Section Focus Master 92: Common Senses L1 SAE **Reinforcement and Study Guide,** p. 165 L1 **Concept Mapping:** The Sense of Touch, p. 40 L1 **Tech Prep Applications:** Looking Far and Near, pp. 47-48 L2 **Critical Thinking/Problem Solving:** Analyzing Sensory-Somatic Responses, p. 40 L3 **Laboratory Manual:** What Are the Locations of Taste and Smell Receptors?, pp. 249-252 L2	**Basic Concepts Transparency #72:** Structure of the Eye L1 SAE **Basic Concepts Transparency #73:** Process of Hearing L1 SAE **Basic Skills Transparency #27:** Structure of the Skin L1 SAE **Reteaching Transparency #40:** Structure of the Ear L1 SAE
Section Focus Master 93: A Question of Poppies L1 SAE **Reinforcement and Study Guide,** p. 166 L1 **Biolab and Minilab Worksheets,** pp. 160-162 L1 **Tech Prep Applications:** Relief from a Patch, pp. 49-50 L2	

ASSESSMENT MATERIALS

Chapter Assessment, pp. 247-252
Performance Assessment in the Biology Classroom, p. 49
MindJogger Videoquiz
Alternate Assessment in the Science Classroom
Computer Test Bank

Spanish Resources SAE
English/Spanish Audiocassettes SAE
Cooperative Learning in the Science Classroom COOP LEARN
Lesson Plans

KEY TO TEACHING STRATEGIES

L1 Level 1 activities should be within the ability range of all students including those with learning difficulties.

L2 Level 2 activities are within the ability range of average to above-average students.

L3 Level 3 activities are designed for the ability range of above-average students.

SAE SAE activities should be within the ability range of Students Acquiring English.

COOP LEARN Cooperative Learning activities are designed for small group work.

P These strategies represent student products that can be placed into a best-work portfolio.

These strategies are useful in a block scheduling format.

GLENCOE TECHNOLOGY

The following multimedia resources are available from Glencoe.

Biology: The Dynamics of Life
- **CD-ROM** SAE
- **Videodisc Program**

The Infinite Voyage Series
Prisoners of the Brain
Fires of the Mind

National Geographic Society Series
STV: Human Body Vol. 2
- Nervous System

Newton's Apple: Life Science
- Novocain

Science and Technology Videodisc Series (STVS)
Human Biology
- Vision Diagnosis
- Brain Development
- Ear Implants
- Hearing by Touch

CHAPTER

40 The Nervous System and The Effects of Drugs

Chapter Overview

Students begin their study of the nervous system by examining neurons. They also examine the organization of the nervous system into the central and peripheral nervous systems.

The second part of the chapter focuses on the senses. The structures and functions of the senses that detect chemicals, light, and mechanical stimulation are presented.

In the last part of the chapter, both the medicinal use of drugs and the misuse and abuse of drugs are explored. The various classes of abused drugs are also presented.

Key Terms

addiction
autonomic nervous system
axon
central nervous system
cerebellum
cerebrum
cochlea
cones
dendrite
depressant
drug
hallucinogen
medulla oblongata
narcotic
neuron
neurotransmitter
parasympathetic nervous system
peripheral nervous system
reflex
retina
rods
semicircular canal
somatic nervous system
stimulant
sympathetic nervous system
synapse
taste bud
tolerance
withdrawal

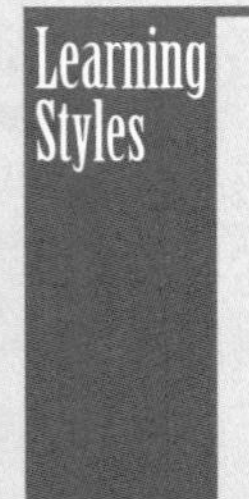

Look for the following logo for strategies that emphasize different learning modalities. LS

Kinesthetic	Demonstration, p. 1016; Project, p. 1020
Visual-Spatial	Using an Analogy, p. 1007; Visual Learning, pp. 1007, 1010; Reinforcement, p. 1008; Demonstration, p. 1020; Display, p. 1021; Portfolio, p. 1021
Interpersonal	Alternate Lab, p. 1016; Meeting Individual Needs, p. 1030; Student Journal, p. 1032
Intrapersonal	Student Journal, p. 1011; Minilab, p. 1027
Linguistic	Student Journal, pp. 1007, 1021, 1027, 1031; Portfolio, pp. 1009, 1032; Project, p. 1010
Logical-Mathematical	Thinking Lab, p. 1009; Minilab, p. 1017
Auditory-Musical	Demonstration, pp. 1014, 1021

CHAPTER 40 The Nervous System and the Effects of Drugs

Zap! Pow! Slam! With lightning-quick speed, or so you hope, you push the buttons in response to the moving lights and electronic sounds of your video game. After all, if you don't respond quickly enough, you might lose this one fight or even the entire game.

Your success at a video game depends on your response time to the stimuli in the game. In the time you take to respond, electrical impulses have bolted from your eyes and ears to your brain and down through your arms to your fingertips. All the while, similar electrical impulses have been coursing through your body, controlling critical body functions. Unlike your moves in the game, however, body actions such as your heartbeat and breathing are beyond your control.

How does your body manage all this activity? All complex machines require a system of control. In your body, that system is the nervous system.

Concept Check

You may wish to review the following concepts before studying this chapter.

- Chapter 8: cell structure
- Chapter 9: ion channels, structure of cell membrane, exocytosis

Chapter Preview

Laboratory Activities

Biolab: Design Your Own Experiment
- What drugs affect the heart rate of *Daphnia?*

Minilabs
- Do distractions affect reaction time?
- What information can be found on the label of an over-the-counter drug?

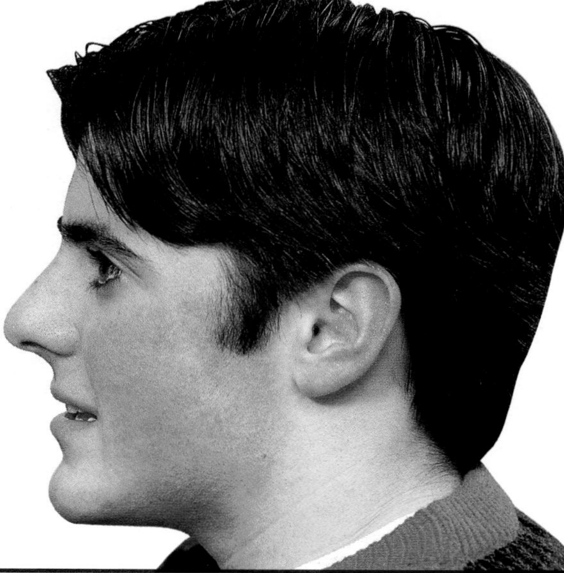

Playing a video game requires not only fast-moving fingers, but also the use of millions of microscopic nerve cells and sensory organs such as your eyes and ears. How do all these parts of your body work together to help you defeat your video foe?

Magnification: 640×

CHAPTER 40

Introducing the Chapter

Most students have played video games in an arcade. Elicit what information is needed to play successfully. *good hand-eye coordination, quick thinking and reflexes* Discuss how the nervous system and senses work together to enable a person to play successfully. L1

Theme Development

The major themes in this chapter are *homeostasis* and *systems and interactions.* Stress that the nervous system receives information from all body systems that are in constant flux. The nervous system responds in order to maintain homeostasis in the body.

Concept Check

Review the differences in complexity of the nervous systems in the various phyla of the animal kingdom. Have students recall the nerve net of a cnidarian and the simple brain of a planarian. You may wish to have students look at the evolutionary comparison in the *Focus On the Brain* feature that appears in this chapter.

Assessment Planner

Choose assessment strategies from the following pages to evaluate the progress of your students.

Assess, pp. 1018, 1024, 1034
Alternate Lab, pp. 1016-1017
Minilabs, pp. 1017, 1027
Portfolio, pp. 1009, 1021, 1023, 1032
Thinking Lab, p. 1009
Biolab, pp. 1028-1029
Chapter Review, pp. 1035-1037

SECTION

40.1

Prepare

Key Concepts

The method by which nerve impulses travel in the nervous system, including the electrical transmission of the impulse along the neuron and the chemical transmission at the synapse, are covered in this section. The organization of the nervous system and the functions of the major parts of the nervous system are discussed.

Block Scheduling

Look for this symbol for strategies that are useful in a block scheduling format. For more information on block scheduling, refer to Section 40.1 in the **Lesson Plans** booklet.

Materials

- Obtain a piece of cable for Using an Analogy.
- Acquire a rubber hammer for the Demonstration on page 1008.
- Borrow an oscilloscope for the Thinking Lab.
- Obtain a piece of plexiglass for the Demonstration on page 1009.

1 Focus

Bellringer

Before presenting the lesson, display **Section Focus Master 91** on the overhead projector and have students answer the accompanying questions. L1 SAE

Discussion

Have students name some of the reflexes an infant has from birth. List these reflexes on the chalkboard. Ask students how these behaviors are different from conscious behaviors such as walking and talking. *Reflexes are unlearned behaviors.* L1

SECTION

40.1 The Nervous System

Section Preview

Objectives

Explain how nerve impulses travel in the nervous system.

Summarize the functions of the major parts of the nervous system.

Key Terms

neuron
dendrite
axon
synapse
neurotransmitter
central nervous system
peripheral nervous system
cerebrum
cerebellum
medulla oblongata
somatic nervous system
reflex
autonomic nervous system
sympathetic nervous system
parasympathetic nervous system

What do you use the telephone for? To talk, or more precisely, to communicate. You probably know that your voice is transmitted as an electrical charge across telephone wires. Would it surprise you to know that a similar electrical charge travels through your body, helping different parts of your nervous system communicate and control other body parts?

Neurons: Basic Units of the Nervous System

The basic unit of structure and function in the nervous system is the neuron. As shown in *Figure 40.1*, **neurons** are cells that conduct impulses throughout the nervous system. You can see that a neuron is made up of dendrites, a cell body, and an axon.

Dendrites are branchlike extensions of the neuron that receive impulses and carry them toward the cell body. The **axon** is a single extension of the neuron that carries impulses away from the cell body. The axon of one neuron may branch extensively, sending its signal to many other neurons, muscles, or glands. In turn, the multiple dendrites of one neuron can receive signals from the axons of many different neurons. Although receptor sites on

Magnification: 2500×

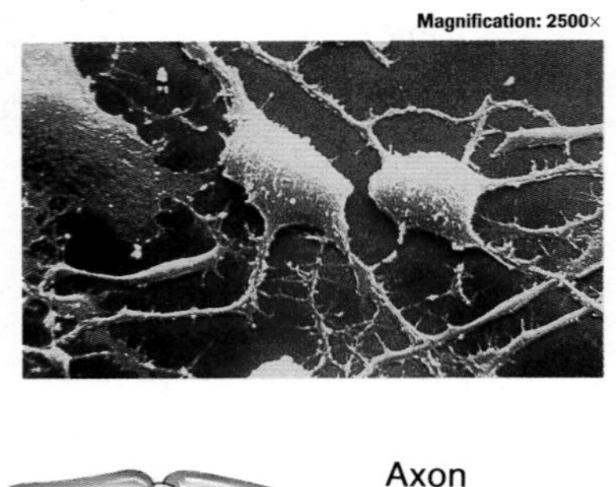

Dendrites
Nucleus
Cell body
Motor neuron
Axon
Schwann cell
Axon endings

Figure 40.1
The cell body of a neuron contains a nucleus, cytoplasm, and organelles such as mitochondria and Golgi apparatus. Dendrites and axons are extensions of the cytoplasm that branch out from the cell body.

Program Resources

Section Focus Master 91 L1 SAE
Reinforcement and Study Guide, pp. 163-164 L1
Performance Assessment in the Biology Classroom, p. 49 L1 P
Biolab and Minilab Worksheets, p. 159 L1
Basic Concepts Transparencies 70, 71 and **Masters** L1 SAE

the ends of dendrites usually are identified as the point of origin for the signals, the signals may originate at various points on a neuron.

Neurons are classified as sensory neurons, motor neurons, or interneurons. Sensory neurons carry impulses from outside and inside the body to the brain and spinal cord. Interneurons are found within the brain and spinal cord. They process these incoming impulses and may pass response impulses to motor neurons. Motor neurons carry the response impulses away from the brain and spinal cord. How do neurons carry these messages?

How neurons work

Suppose you're in a crowded, noisy store, and you feel a tap on the shoulder. Turning your head, you see the smiling face of a good friend. How did the tap on the shoulder work to get your attention? The touch stimulated sensory receptors in your shoulder, starting an impulse in these neurons. The sensory impulse was carried to the spinal cord and then up to your brain. From your brain, an impulse was sent out, causing motor neurons to transmit an impulse to muscles in the neck, and you turned to look. *Figure 40.2* shows how stimuli, such as the tap on your shoulder, are transmitted through your nervous system.

How does impulse transmission take place? Start with a resting neuron—one that is not transmitting an impulse. You have learned that the plasma membrane controls the concentration of ions inside a cell. Although sodium ions (Na^+) and potassium ions (K^+) are on both sides of the membrane, sodium ions exist in a larger concentration outside the cell, and potassium ions are in a higher concentration inside the cell.

Figure 40.2

The nervous system receives information, transmits the information, sorts and interprets the incoming information, determines a response, transmits the response, and activates the response.

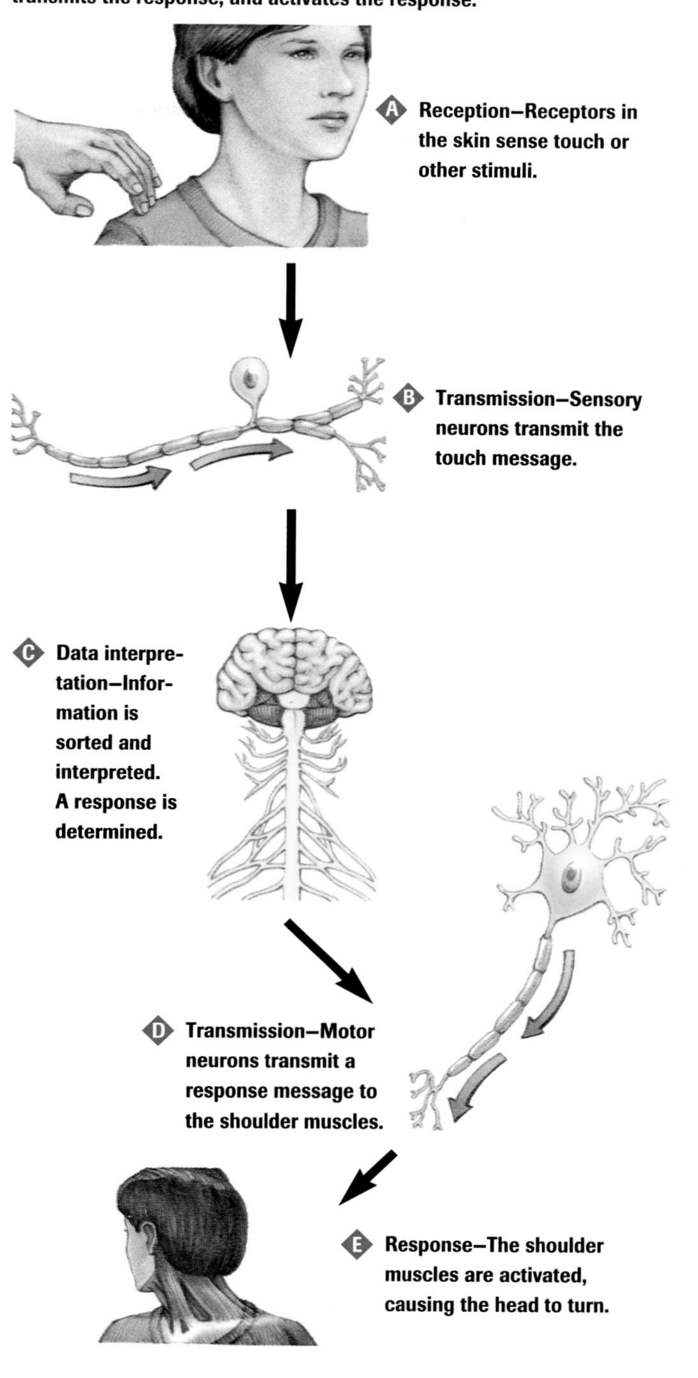

A Reception—Receptors in the skin sense touch or other stimuli.
B Transmission—Sensory neurons transmit the touch message.
C Data interpretation—Information is sorted and interpreted. A response is determined.
D Transmission—Motor neurons transmit a response message to the shoulder muscles.
E Response—The shoulder muscles are activated, causing the head to turn.

SECTION 40.1

2 Teach

Using an Analogy

LS **Visual-Spatial** To demonstrate what a nerve is like, show the students the end of a cable, such as a telephone cable. Compare each wire to an axon and the individual wire wrappings to myelin sheaths. Explain that the whole cable represents a nerve.

Audiovisual

Show the video *The Nervous System*, *The Brain*, or *The Autonomic Nervous System*, Charles Clark Co., Inc.

Software

The Body Electric, Brain Waves. HRM Software, MS-DOS or Apple.

Text Question

How do neurons carry these messages? *The messages are transmitted along a neuron much like messages are transmitted along a wire.*

Visual Learning

Figure 40.2 Spend a few minutes reviewing the sequence of events involved in the nervous system's response to a stimulus. Ask students what path the nervous system would take to turn the head in response to a sound such as a honking car horn. *The path would be the same except for the initial receptors, which would be located in the ear rather than in the skin.* LS

STUDENT JOURNAL

Nerve Analogy Have students write a paragraph that explains how a nerve is similar to a wire going from a controlling switch (stimulus) to a lightbulb (effector). L1 LS

SECTION 40.1

Science, Technology, and Society

Using Cultured Neurons

In 1992, doctors at Johns Hopkins University School of Medicine removed a portion of the brain of an 18-month-old girl suffering from megalencephaly, a disorder in which one side of the brain grows substantially larger than the other. The condition results when neurons undergo too many cell divisions before they mature.

The removed neurons were placed in a blood-rich culture medium. After 3 weeks, nearly all the cells had died, but a few continued growing well and dividing. For the first time, neuroscientists maintained mature, human brain neurons in long-term cultures. The cultured neurons will be used to study brain cell development and differentiation and should help pharmacologists design drugs for various neurological diseases.

Visual Learning

Figure 40.3 What effect does this action have on the charge of the normal resting neuron? *It maintains a positive charge on the outside of the membrane and a negative charge within the membrane of a resting neuron.*

Reinforcement

LS Visual-Spatial Have students draw neurons similar to Figures 40.3 and 40.4, label the parts, and use arrows to show the pathway of the nerve impulse. L2

These differences occur because the membrane is more permeable to potassium than it is to sodium.

The membrane also contains an active transport set of proteins called the Na^+/K^+ATPase pump. As you can see in *Figure 40.3*, the action of the pump increases the concentration of positive charges on the outside of the membrane. In addition, the presence of many negatively charged proteins and organic phosphates means that the inside of the membrane is more negative than the outside.

Under these conditions, the membrane is said to be polarized, or at rest. A polarized membrane has the potential to do work. In this case, work is the transmission of an impulse.

How an impulse is transmitted

When a stimulus excites a neuron, sodium channels in the membrane open up, and sodium ions rush into the cell. As the positive sodium ions build up inside the membrane, the inside of the cell becomes more positively charged than the outside. This change in charge, called depolarization, sets up a wave similar to an ocean wave. A wave of changing charges moves down the length of the axon, *Figure 40.4*. Because the impulse is generated anew at each stage along the axon, it does not diminish in strength during transmission. As the wave passes, the membrane immediately behind it returns to its resting state, with the inside of the cell negatively charged and the outside positively charged.

This wave of depolarization is actually the transmission of an impulse along the complete length of the axon. An impulse can move down the axon only when the stimulation of the neuron is strong enough to reach a certain threshold level. If the threshold level is not reached, the impulse quickly dies out. This is referred to as the *all-or-none principle*.

For a short time immediately after depolarization occurs, the neuron cannot be restimulated. This happens because the gated sodium channels are closed and the gated potassium channels are open.

Figure 40.3

The membrane of an axon contains open channels that allow movement of sodium (Na^+) and potassium (K^+) ions into and out of the cell, as well as gated sodium and potassium channels. In addition, a Na^+/K^+ pump uses ATP to pump three sodium ions out for every two potassium ions it pumps in. What effect does this action have on the charge of the normal resting neuron?

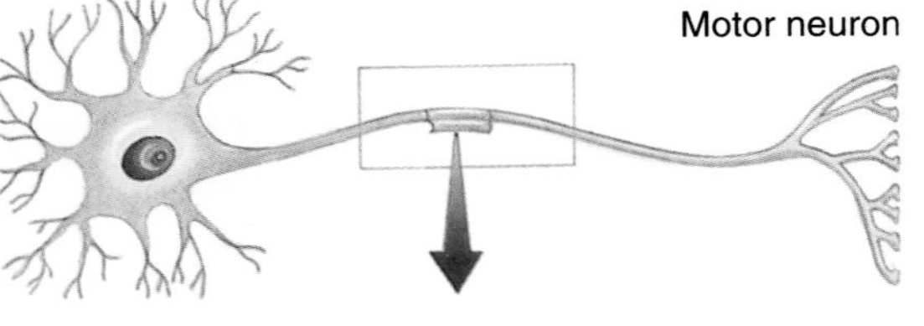

Figure 40.4

A wave of depolarization moves down the axon of a neuron.

A Gated sodium channels open, allowing sodium ions to enter and make the inside positively charged and the outside negatively charged.

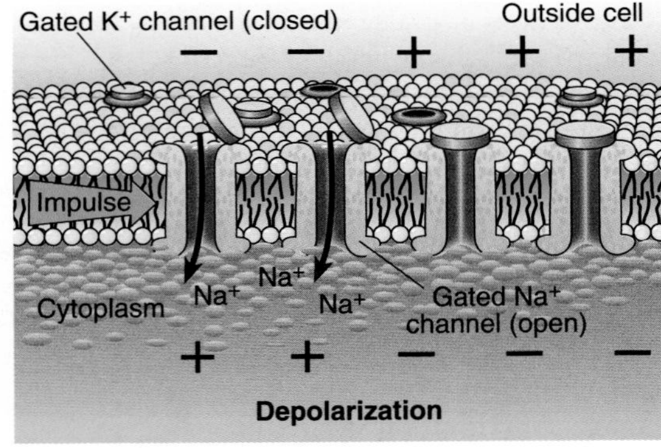

Meeting Individual Needs

Learning Disabled/Hearing Impaired For learning disabled and hearing impaired students, the review tutorial *Nervous and Hormonal Systems,* from Queue (Apple II, MS-DOS, MAC) may be helpful for reviewing major concepts. L1

GLENCOE TECHNOLOGY

Videodisc

Biology: The Dynamics of Life
Disc 2, Side 1
Impulse Transmission: Neuron (Ch. 37)

White matter and gray matter

Most axons are surrounded by a white covering called the myelin sheath. The sheath is formed by Schwann cells that grow and wrap around the axon like a jellyroll. Like the plastic coating on an electric wire, the Schwann cells tightly insulate the axon, hindering the movement of ions across the plasma membrane. The ions move quickly down the axon until reaching a gap in the sheath that occurs between each of the individual Schwann cells. At this point, the ions pass through the membrane and depolarization occurs. As a result, the impulse does not travel continuously down the length of the axon but jumps from gap to gap, greatly increasing the speed at which the impulse travels. The myelin sheath gives neurons a white appearance. In the brain and spinal cord, masses of myelinated neurons make up what is called white matter. The absence of myelin in masses of neurons accounts for the gray matter of the brain.

ThinkingLab — Interpret the Data

Which stimuli will result in an impulse moving through a neuron?

Scientists are able to determine which stimuli are strong enough to send a wave of depolarization through the axons of the squid *Loligo.* This is accomplished by inserting an electrode inside one of the squid's axons. A second electrode is placed outside the axon. The electrodes measure any changes in the charge of the neuron. They are also attached to an oscilloscope, which shows the strength of the charge changes.

Analysis

Study the deflections of the beam of light that occur on the oscilloscope screen when an axon is at rest and when it is stimulated.

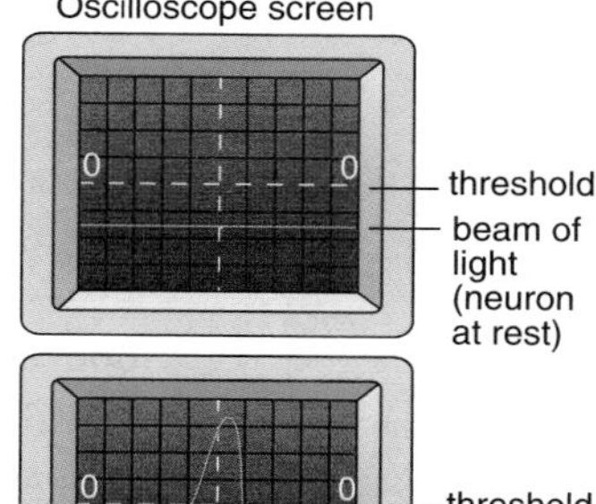

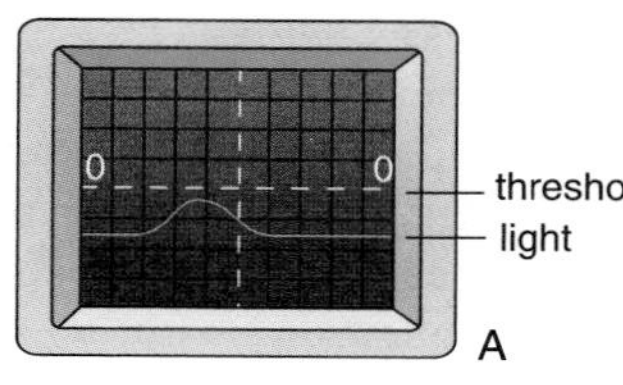

0 0 threshold light B

Thinking Critically

Analyze the charge changes shown here. Determine which will send an impulse through the neuron and which is a localized signal. Explain how you determined your answers.

B As soon as the wave passes, that part of the axon returns to a resting state. This occurs because the gated sodium channels close, stopping the influx of sodium ions, and gated potassium channels open, allowing potassium to flow out of the neuron.

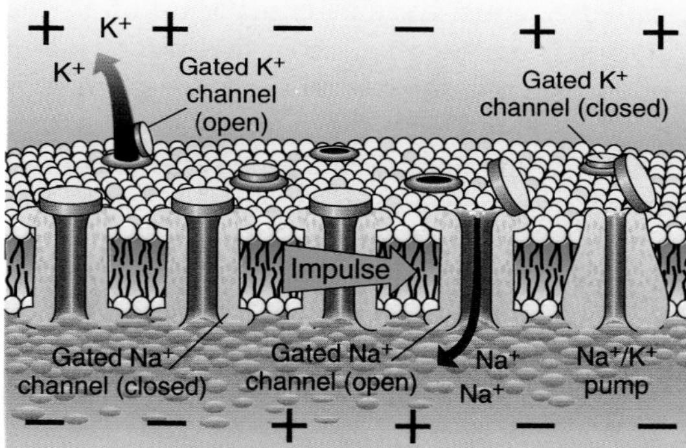

C Then, the Na^+/K^+ pump takes over again, moving sodium ions back out through the membrane and potassium ions in. The charges return to the same condition as in a resting neuron.

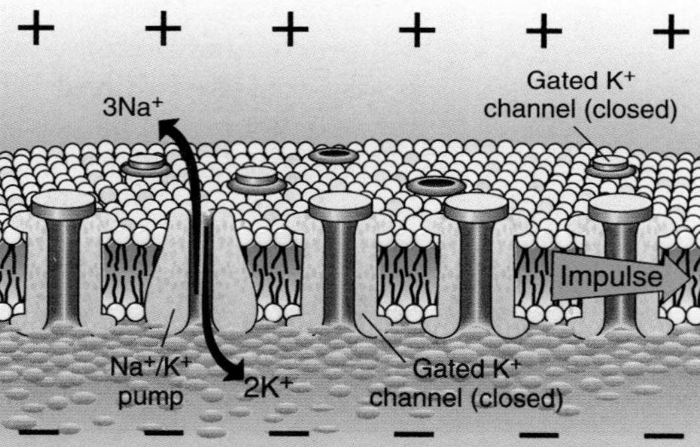

PORTFOLIO

Using a Model Ask students to obtain a small cable of wires. Have them mount the cable on a sheet of paper and draw a nerve beside it. Ask the students to write a paragraph that compares and contrasts the nerve and the cable of wires. Similarities include: Both carry messages, connect two things, contain a bundle of "wires," and have a covering. Differences include: The nerve is alive, made of cells, and can repair itself to a degree. L1 P LS

ThinkingLab — Interpret the Data

Purpose

LS **Logical-Mathematical** Students will examine the changes in charge of a stimulated neuron.

Process Skills

observe and infer, compare and contrast, interpret data, analyze

Teaching Strategies

- Borrow an oscilloscope from the physics department and show students how it works.

Background

Neurophysiologists use oscilloscopes hooked to amplifiers to measure the currents in individual neural ion channels using a technique called patch clamp. The scientist uses a tiny glass pipet. The pipet tip is manipulated onto the surface of the neuronal plasma membrane. Currents are then measured, amplified, and viewed on oscilloscopes.

Thinking Critically

The charge change in A is a localized signal and is not strong enough to send an impulse. The charge change in B will send an impulse because it crosses the threshold level.

Assessment

Portfolio: Have students draw the oscilloscope screen pattern for an axon at rest and for one that is stimulated. Ask students to write a paragraph about what is occurring in the neural plasma membrane when the neuron is stimulated. L1

Visual Learning

Figure 40.5 Draw attention to the illustration as you discuss the transmission of an impulse across a synapse. LS

Audiovisual

Show the videos *Nervous System*, National Geographic and *The Human Brain* or *The Nervous System*, from EBEC.

Different Viewpoints in Biology

Acupuncture

The Chinese have used acupuncture as a complete system of medicine for many years. Needles as long as 23 cm are inserted into the body and turned to elicit results that range from cancer cures to pain relief. A modern use of acupuncture in China is in providing anesthesia for surgery – although only three percent of operations use acupuncture.

Acupuncture has been slow to gain acceptance in Western medical practices. However, most scientists agree that acupuncture does something. Lorenz Ng, chief of the Pain Studies Program at the National Institute on Drug Abuse, found that rats had elevated levels of endorphins in the cerebrospinal fluid when subjected to electroacupuncture.

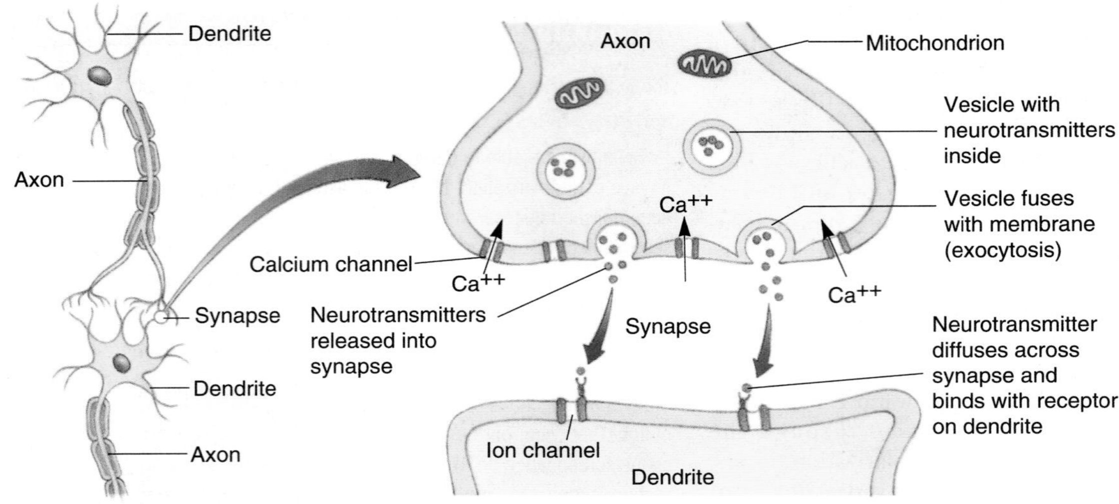

Figure 40.5
Neurotransmitters are released into a synapse by the process of exocytosis. The dendrite receiving the impulse contains receptor sites that are linked to ion channels. When the neurotransmitters diffuse across the synapse and bind to the receptors, the ion channels open, changing the polarity of the receiving dendrite. In this way, nerve impulses move from neuron to neuron.

Connections between neurons

Although the neurons of the nervous system lie end to end in bundles, axons to dendrites, their ends don't touch. A tiny space called a **synapse** lies between one neuron's axon and another neuron's dendrites. Impulses moving to and from the brain must move across these synapses. How does an impulse make this jump?

As an impulse reaches the end of an axon, the wave of changing charges opens calcium channels, allowing calcium to enter the end of the axon. As shown in *Figure 40.5*, the calcium causes small packages of chemicals first to fuse with the membrane and then to burst open into the synapse, releasing their chemicals. These chemicals, called **neurotransmitters,** diffuse across the synapse to the dendrites of the next neuron. There they stimulate polarity changes in the neuron.

A synapse is a one-way passage, allowing a nerve impulse to pass from axons to dendrites. The continuous relay of an impulse is prevented because neurotransmitters typically are broken down quickly by enzymes in the synapse.

The Central Nervous System

When you make a call to a friend, your call travels over wires to a control center, where it is switched over to wires that connect with your friend's telephone. In the same manner, an impulse traveling through neurons in your body usually reaches the control center of the nervous system—your brain. The brain and the spinal cord together make up the **central nervous system,** which acts as your body's control center and coordinates your body's activities.

Two systems work together

Another division of your nervous system, called the **peripheral nervous system,** is made up of all the nerves that carry messages to and from the central nervous system. It is similar to the telephone wires that run between a phone system's control center and the phones in the homes of the community. Together, the central nervous system (CNS) and the peripheral nervous system (PNS), shown in *Figure 40.6*, make rapid changes in your body in response to stimuli in your environment.

PROJECT: Neuroscience Research

Have students research and prepare a report, a visual device, or an audiovisual presentation on one of the following topics: sleep, split brain experiments, dementia, Alzheimer's disease, retrograde amnesia, or Penfield maps of the brain. Have students present their findings to the class. L1 LS

GLENCOE TECHNOLOGY

Videodisc

Biology: The Dynamics of Life
Disc 2, Side 1
Impulse Transmission: Synapse (Ch. 38)

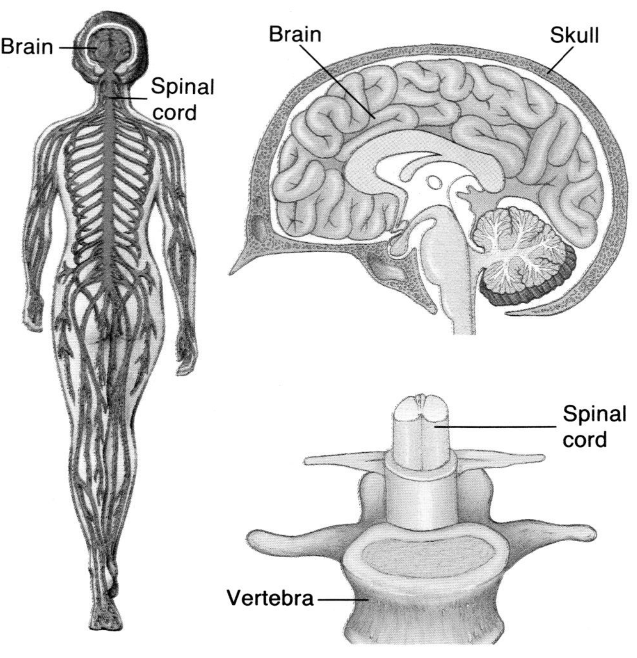

Figure 40.6

The human nervous system (left) is divided into the CNS, in red, and the PNS, in blue. The brain receives and transmits impulses by way of the spinal cord, which connects the brain to the peripheral nervous system. The brain and spinal cord are protected by your skull and the vertebrae (right).

Anatomy of the brain

The brain is the control center of the entire nervous system. For descriptive purposes, it is useful to divide the brain into three main sections: the cerebrum, the cerebellum, and the brain stem, *Figure 40.7.*

The **cerebrum** is divided into two halves, called hemispheres, that are connected by nerve tracts. Your conscious activities, intelligence, memory, language, skeletal muscle movements, and senses are all controlled by the cerebrum. The cerebrum is wrinkled with countless folds and grooves and is covered with an outer layer of gray matter called the cerebral cortex. Has anyone ever told you to "use your gray matter"? While the saying means that you should use your powers of reasoning, it refers to neurons that are not covered by myelin sheaths.

The **cerebellum,** located at the back of your brain, controls your balance, posture, and coordination. If the cerebellum is injured, your movements become jerky.

The brain stem is made up of the medulla oblongata, the pons, and the midbrain. The **medulla oblongata** is the part of the brain that controls involuntary activities such as breathing and heart rate. The pons and midbrain act as pathways connecting various parts of the brain with each other. Find out more about the brain's structure and its fascinating functions in Focus On the Brain.

Figure 40.7

The large surface area created by the many folds in the cerebrum results in more neurons being located in the cerebral cortex. This increase was important in the evolution of human intelligence. The portion of the brain that receives and sends sensory signals to the cerebrum is the thalamus. It receives all sensory information except smell.

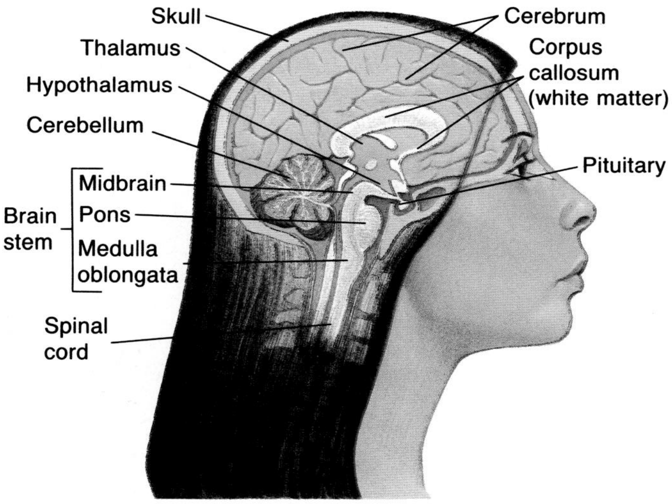

SECTION 40.1

Enrichment

Have interested students read "Wide Hats and Narrow Minds," from Stephen Jay Gould's *The Panda's Thumb,* on the topic of brain size versus intelligence. L2

Activity

Have students take a right/left hemispheric mode indicator test to access their personal approach to learning. Indicator tests are available from EXCEL INC, 200 W. Station St., Barrington, IL 60010. L2

Science, Technology, and Society

Treating Spinal Injuries

An estimated 10 000 Americans suffer spinal cord injuries each year.

Scientists have found in most cases paralysis does not result from the actual injury itself but from inflammation and swelling following the injury. A drug called methylprednisolone, a synthetic steroid, prevents some paralysis associated with spinal injuries. By giving the drug within eight hours after injury, most of the swelling is prevented. In addition, part of the damage is the release of destructive free-radical molecules as cells die. This steroid mops up free radicals, reducing the cell destruction that can be caused from their release.

interNET CONNECTION

Follow the link for this chapter on the Glencoe Homepage at **www.glencoe.com/sec/science** to find out more about the nervous system.

STUDENT JOURNAL

Involuntary Actions Ask students to list activities their bodies do without conscious thought. To get them started thinking about this, have them consider what their bodies do while they are asleep. L1 LS

Assessment

Performance Assessment in the Biology Classroom: p. 49, Preparing and Teaching a Lesson About the Nervous System. Have students carry out this activity to demonstrate their knowledge of the structure and function of the nervous system. L1 P

Focus On
The Brain

Purpose

Students will explore topics concerning the brain, its evolution, anatomy, and function.

Background

When viewing the brain embryologically, it is divided into three major lobes: prosencephalon or forebrain, mesencephalon or midbrain, and rhombencephalon, or hindbrain.

Functionally, the brain is divided into the cerebrum, hypothalamus, thalamus, midbrain, cerebellum, pons, and medulla oblongata.

Tying to Previous Knowledge

Review the evolutionary sequence of the various phyla, paying special attention to the evolutionary tree of the vertebrates.

Display

Display models or posters of various vertebrate brains. Point out the differences among the brains.

Focus On
The Brain

As animals have changed and evolved over hundreds of millions of years, what has really set humans apart from other animals is the evolution of the brain. From a bundle of nerve cells to a staggeringly complex cerebral cortex, this evolution has progressed from organisms that react instantly to changes in the environment or immediate danger to humans capable of thinking and reasoning.

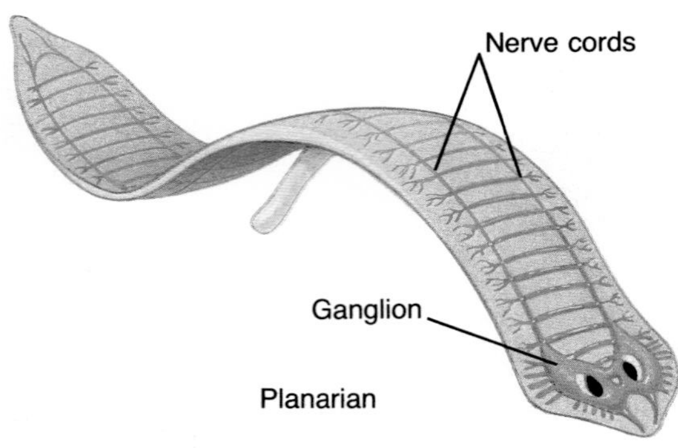

Evolution of the brain

The simplest brain Flatworms are the simplest animals that have an identifiable brain. A planarian, for example, has a mass of nerve tissue called a ganglion that lies beneath the eyespots.

Millions of years later Jumping ahead by millions of years to when the vertebrates emerged, the five brains shown here illustrate how evolution has transformed a simple ganglion to a complex brain. As the brain evolved, areas that control senses, instinct, and coordination became predominant.

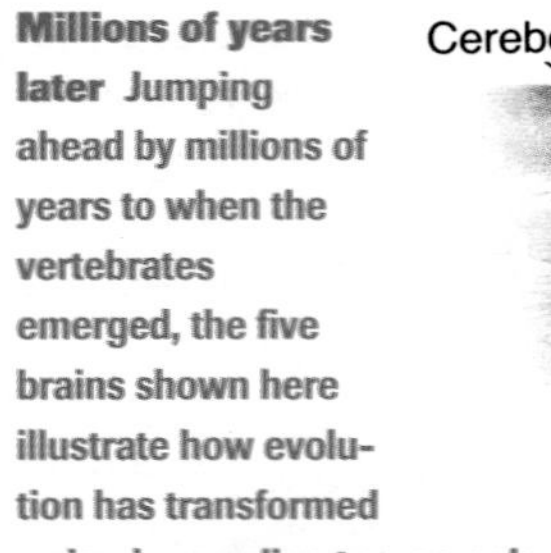

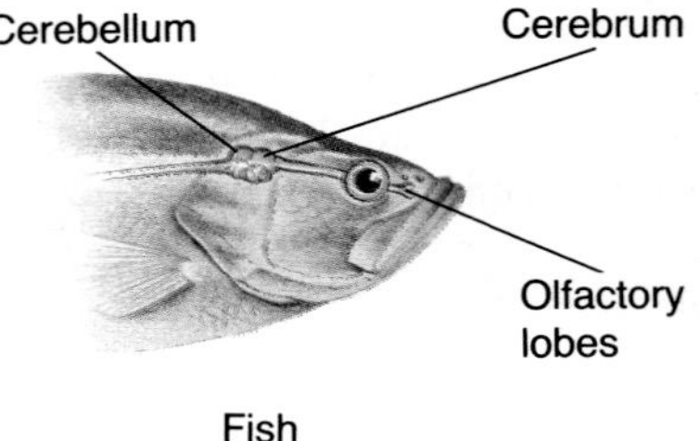

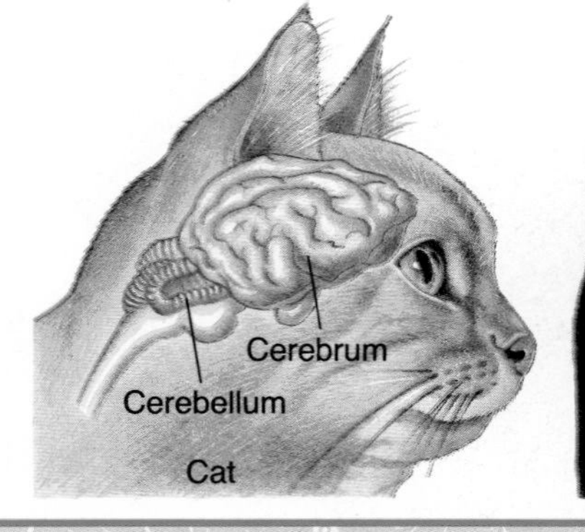

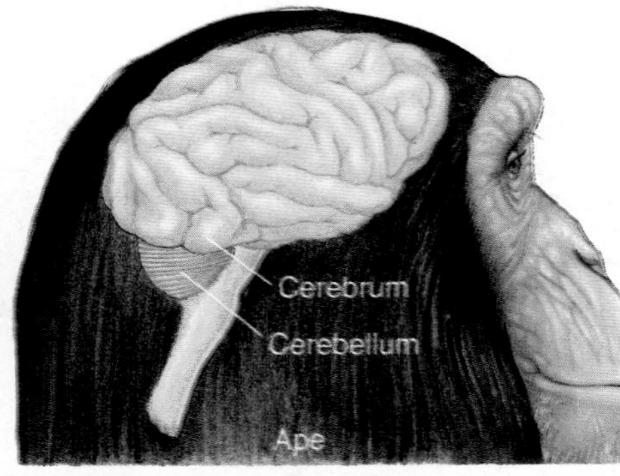

Notice that in humans (right), the brain is much larger, and the area dedicated to thinking (cerebrum) covers up and dominates everything else.

Human brain development

When a human embryo is just 21 days old and nothing but a hollow ball of cells, a groove begins to form along the outer surface. This groove is the start of the nervous system. Folds of cells come together across the groove, forming a neural tube. The top portion of the tube swells, becoming the brain, and the lower portion becomes the spinal cord. During this stage of development, the brain generates approximately 360 million new cells a day.

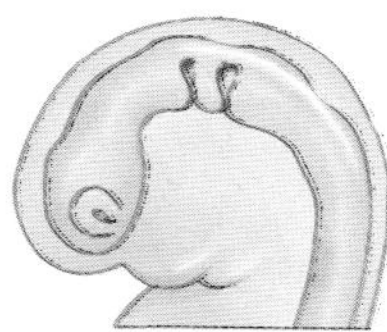

21 days

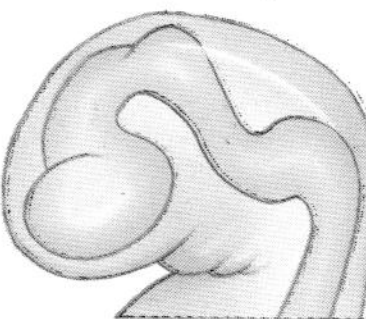

40 days

The cerebellum, medulla, and the frontal lobes of the brain begin to develop.

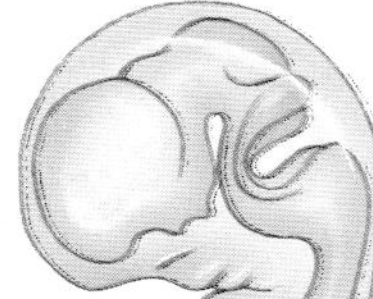

60 days

The head is almost half the size of the entire embryo.

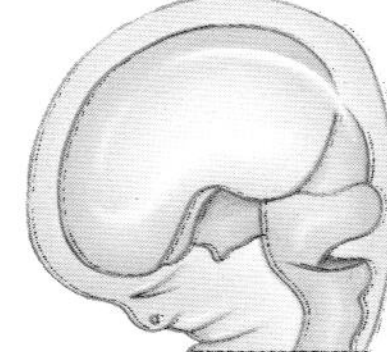

4 months

The cerebral cortex of the fetus begins to develop the ridges and grooves of the mature brain pattern.

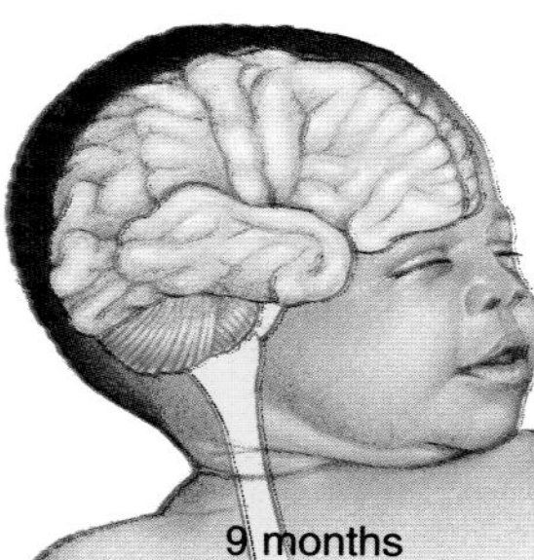

9 months

The brain can now carry on one of its most important functions—maintaining homeostasis.

The adult brain

The largest part of the human brain is the cerebrum. The cerebrum is divided into two hemispheres, left and right, connected by a band of millions of nerve cells called the corpus callosum. But the feature that makes the cerebrum unique is an outer, folded layer less than 5 mm thick—the cerebral cortex. Because of the cortex, you can remember, reason, organize, communicate, understand, and create.

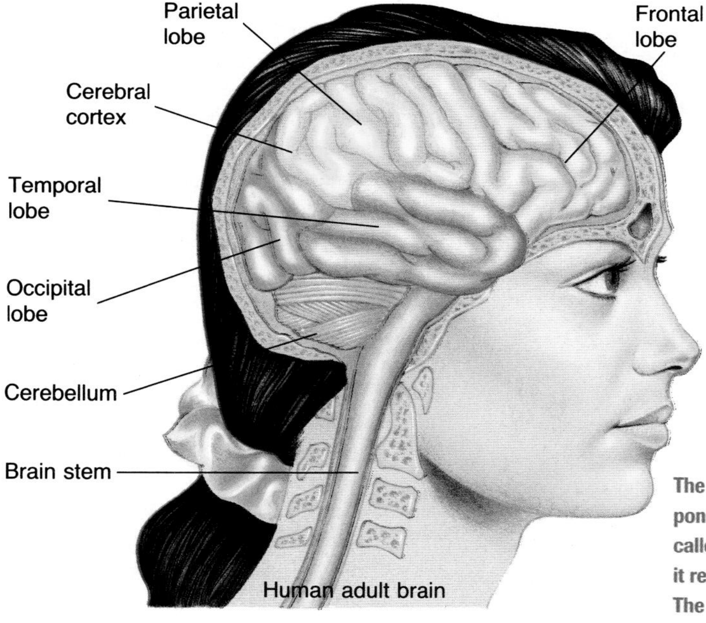

Human adult brain

When you see a gold-medal performance by an Olympic gymnast or figure skater, you are watching the work of a well-trained cerebellum. It is here that muscles are coordinated and the memories of physical skills are stored. The cerebellum has tripled in size over the last million years.

The brain stem consists of the medulla, pons, and midbrain, and is sometimes called the reptilian brain. This is because it resembles the entire brain of a reptile. The brain stem controls homeostasis.

Focus On

Demonstration

Slides of chick embryos or preserved embryos can be used to demonstrate the development of the brain over the 21-day period of chick development.

Reinforcement

Have students in groups draw a large outline of the brain and spinal cord on a piece of poster paper. Have them cut out magazine pictures that relate to the specializations of the brain and glue them on their outline. For instance, they could cut out eyes and place them in the occipital lobe to indicate that is where vision is interpreted. L1

GLENCOE TECHNOLOGY

Videodisc

STVS: Human Biology
Disc 7, Side 1
Brain Development (Ch. 17)

The Infinite Voyage: Fires of the Mind
Brain Development: The Range of Capability (Ch. 2)

STUDENT JOURNAL

Spinal Cord Injuries Have students research the latest treatments for spinal cord injuries and the progress on stimulating nerve cells to repair themselves. They should include a copy of their reports in their journals. L2

Focus On

Demonstration

Using a microprojector, project a slide of a neuron. Point out the dendrites, cell body, and axon.

Software

The Human Brain: Neurons, Projected Learning Programs, Inc.

Microscope Activity

Students could look at prepared slides of various areas of the brain to locate neurons, neuroglia, pyramidal neurons, and Purkinje's cells. A histology textbook can be used for reference. L3

Audiovisual

Show the eight videocassettes available from the PBS series "The Brain." The videocassettes are available from WNET Education, 356 West 58th Street, NY, NY 10019.

Demonstration

LS **Auditory-Musical** Create a loud noise by slamming a book on your desk. Use student reaction to begin a discussion of reflexes.

Focus On

The cerebral cortex

Left and right brain Certain higher functions are centered in either one or the other hemisphere. For example, in most right-handed people, language and speech are controlled by specific areas in the left cerebral hemisphere, while the ability to recognize and duplicate certain types of spatial relationships is located in the right hemisphere.

Speech and spatial perception are not solely determined by one or the other hemisphere. The corpus callosum allows you to perceive things such as shapes and then describe them in words.

Different thinking skills are also located in separate hemispheres. For example, logic, mathematical skills, and reasoning are the domain of the left brain. Insight (that flash of understanding), imagination, and awareness of the beauty of art and music are centered in the right brain. However, this does not make one hemisphere more important than the other. It is the blending of these abilities that makes each of us unique.

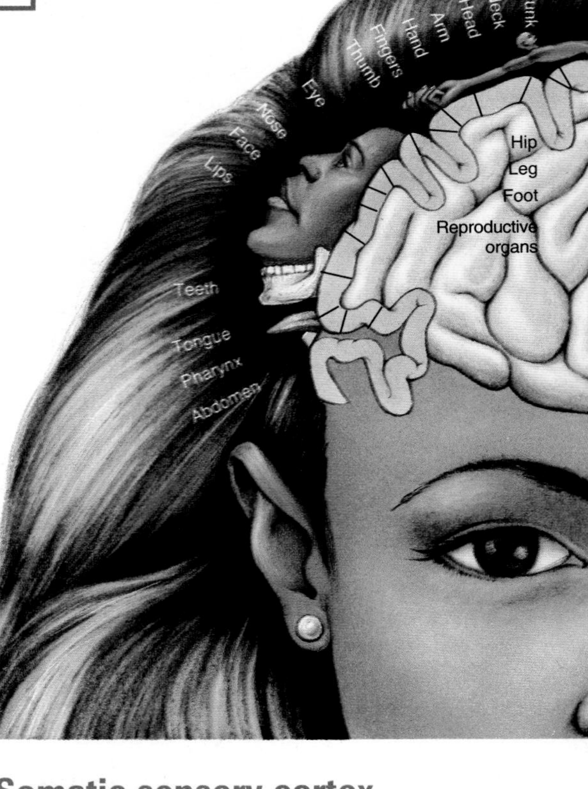

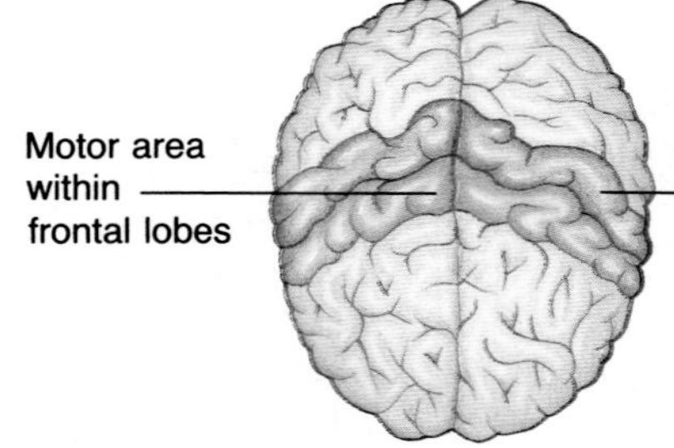

Somatic sensory cortex

Almost every part of the body sends messages to its own part of the cerebral cortex. The amount of cortex dedicated to the face and limbs is large because control of these areas requires great precision. The areas of the sensory cortex labeled here on the right hemisphere are also present in the left hemisphere.

Using the brain to study the brain—A chronology

circa 275 B.C.
Erasistratus Greek physician and anatomist—made detailed studies of the human brain comparing the convolutions to those of animals. Erasistratus related brain complexity to intelligence.

circa 1600
Descartes French philosopher and mathematician—defined thinking as a range of intellectual thoughts, feelings, sensations, and will. He believed the mental process went on even during sleep.

1871
Cajal Spanish physician and histologist—refined a method of staining nerve cells so they could be studied. He was first to establish that neurons are the basic structure of the nervous system.

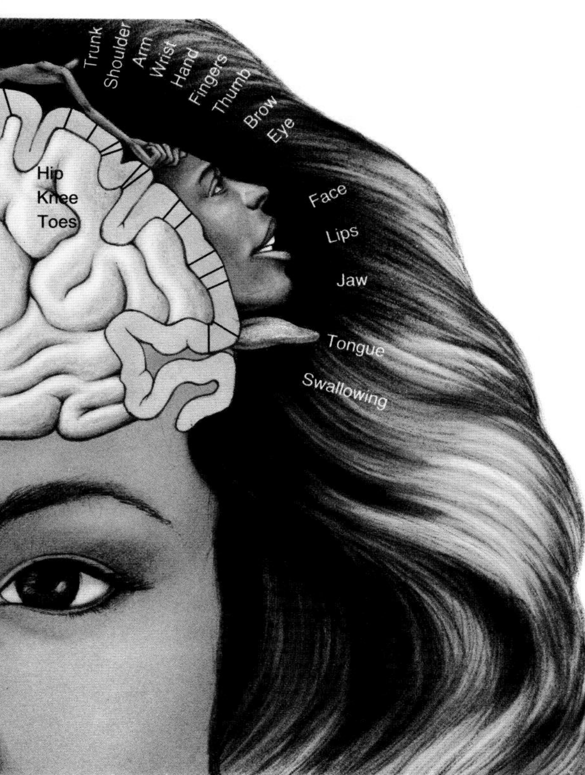

Motor cortex

When you walk, write, type, or talk, you must control millions of muscle fibers very precisely. This is where the motor control areas of your cerebral cortex take over. The areas labeled here on the left hemisphere are also present in the right hemisphere.

1992
Ronnett, *et al.* Cultured human brain cells. Team of researchers at Johns Hopkins University succeeded in culturing human brain cells for more than 19 months.

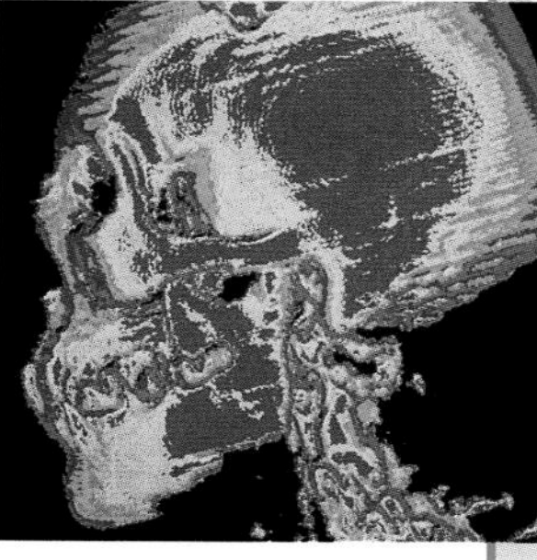

1961
Hounsfield and Cormack Developed the computerized axial tomograph (CAT scanner) that revolutionized the diagnosis of brain lesions, blood clots, tumors, and cerebral damage.

1950s
Gazzaniga and Sperry Demonstrated the role of the corpus callosum in the exchange of information between brain hemispheres. Sperry won the Nobel prize for split-brain research.

1930
Luria Soviet psychologist–studied the relationship between injured areas of the brain and aphasia (loss of the power to use or understand words).

1921
Loewi German pharmacologist–isolated acetylcholine, a chemical from the brain that transmits nerve impulses.

1929
Berger German neurologist–first to use technology to record a paper record of human brain waves on an electroencephalograph.

EXPANDING YOUR VIEW

1. **Thinking Critically** Research current ideas about the differences between the functions of the left and and right sides of the brain.
2. **Journal Writing** Write in your journal your prediction as to how a person's movements may be affected after an injury to the cerebellum.

Focus On

Activity

Have students test their cerebellar function by:

1. Walking heel-to-toe for 20 feet without losing their balance while looking straight ahead.
2. Standing with their feet together and eyes closed without losing their balance.
3. Looking straight ahead, moving the heel of one foot down the shin of the other leg.
4. With the eyes closed, touch the nose with the index finger of each hand. L1

Enrichment

Have students research the patterns and functions of the various brain waves (alpha, beta, theta, and delta). L2

Answers to Expanding Your View

1. A variety of materials is available in human anatomy and physiology textbooks and journals regarding the left and right brain functions.
2. An injury to the cerebellum affects coordination of movement, especially physical skills such as walking. Damage to this area of the brain would cause jerky, uncoordinated movements.

Going Further

Students interested in the evolution and function of the brain can read *Dragons of Eden*, by Carl Sagan.

Demonstration

LS **Kinesthetic** Demonstrate the knee-jerk reflex using a rubber hammer.

Different Viewpoints in Biology

Memory Molecules

For years, scientists maintained RNA was the "memory molecule." Although RNA has not been disproven as the memory molecule, scientists are looking at other molecules.

One theory suggests the NMDA receptor (N-methyl, D-aspartate) is a possible key component in setting up the brain's memory circuits. When activated, this receptor allows calcium to flow into brain cells, possibly strengthening the neuronal connections believed to form memory.

Another theory suggests that protein kinase C (PKC) may underlie learning and memory. Scientists at Princeton University have found that the appearance, disappearance, and reappearance of PKC coincides with learning, forgetting, and remembering, respectively. Others have found that most of the PKC in the hippocampal neurons of newborn rabbits resides in two tracks flanking the cell body. But, in older rabbits, there is a single, diffuse track of PKC situated around the dendrites, perhaps programming memories.

The Peripheral Nervous System

Remember that the peripheral nervous system carries impulses between the body and the central nervous system. For example, when a stimulus, such as cold air, is picked up by receptors in your skin, it initiates an impulse in the sensory neurons. The impulse is carried to the CNS. There, the impulse transfers to motor neurons that carry the impulse away from the CNS to a muscle and causes the muscle to contract. The peripheral nervous system consists of two separate divisions—the somatic nervous system and the autonomic nervous system.

The somatic nervous system

The **somatic nervous system** is made up of 12 pairs of cranial nerves from the brain, 31 pairs of spinal nerves from the spinal cord, and all of their branches. These nerves are actually bundles of neurons bound together by connective tissue, similar to a telephone cable. Some nerves contain only sensory neurons, and some contain only motor neurons, but most nerves contain both types.

The nerves of the somatic system relay information mainly between your skin, the CNS, and skeletal muscles. This pathway is voluntary and under conscious control, meaning that you can decide to move or not to move body parts under the control of this system.

Reflex arcs in the somatic system

Sometimes a stimulus will result in an automatic, unconscious response within the somatic system. When you touch something hot, you automatically jerk your hand away. Such an action is a **reflex,** an automatic response to a stimulus. Reflexes are the result of the shortest nerve pathways in the body called reflex arcs, as shown in *Figure 40.8.*

Figure 40.8

Reflex responses are carried out at the level of the spinal cord without assistance from the brain. The impulse travels directly to the spinal cord from the affected body part, crosses over to a small interneuron, and then moves to a motor neuron that transmits it to a muscle. The muscle contracts. The brain becomes aware of the reflex only after it occurs.

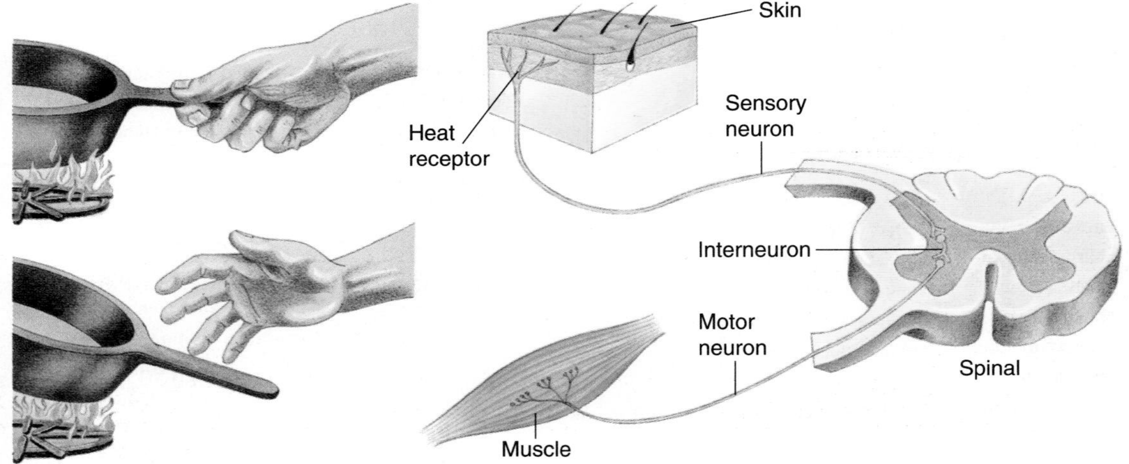

1016 The Nervous System and the Effects of Drugs

Alternate Lab: Skin Sensitivity

Purpose

LS **Interpersonal** Design and carry out an experiment to determine the sensitivity of skin.

Materials

paper clips unfolded in U-shapes, metric ruler

Background

Certain areas of the skin contain sensory neurons that are packed together, whereas other areas have sensory neurons scattered sometimes centimeters apart. When two stimuli depolarize parts of the same neuron, the brain will interpret them as if they were one stimulation.

Procedure

Give students the following directions.

1. Working with a partner, plan ten areas of the skin to test for the distance between sensory neurons.
2. Test your partner for sensitivity with his or her eyes closed.
3. Test an area of the skin by continually spreading the two ends of the paper clip farther apart until the student can feel two points rather than one point.
4. Measure the distance between sensory neurons in millimeters by measuring

The autonomic nervous system

Imagine that you are spending the night alone in a creepy, old, deserted house. Suddenly, a creak comes from the attic. You think you hear footsteps. Your heart begins to pound. Your breathing becomes rapid. Your thoughts race wildly as you try to figure out what to do—stay and confront the unknown, or run fast to get out! The reactions that make up your response in this scary situation are being controlled by your autonomic nervous system.

The **autonomic nervous system** carries impulses from the CNS to internal organs. These impulses produce responses that are involuntary and not under conscious control. For example, glands in your stomach pour out hydrochloric acid without any voluntary decision on your part.

There are two divisions of the autonomic nervous system—the sympathetic nervous system and the parasympathetic nervous system. The **sympathetic nervous system** controls many internal functions during times of stress. It is responsible for the fight-or-flight response shown in *Figure 40.9*. Without a conscious decision on your part, the body sends glucose to muscles and

MiniLab

Do distractions affect reaction time?

Have you ever tried to read while someone is talking to you? What effect does such a distracting stimulus have on reaction time?

Procedure

1. Work with a partner, and sit facing him or her. Your partner should stand.
2. Have your partner hold the top of the meterstick above your hand. Hold your thumb and index finger about 2.5 cm away from either side of the lower end of the meterstick without touching it.
3. Tell your partner to drop the meterstick straight down between your fingers.
4. Catch the meterstick between your thumb and finger as soon as it begins to fall. Measure how far it falls before you catch it. Practice several times.
5. Run ten trials, recording the number of centimeters the meterstick drops each time. Average the results.
6. Repeat the experiment using the distraction of counting backwards from 100 by fives (100, 95, 90, . . .).

Analysis

1. Did your reaction time improve with practice? Explain.
2. How was your reaction time affected by the distraction?
3. What factors, besides distractions, would increase reaction time?

Figure 40.9

During a fight-or-flight response, the sympathetic nervous system causes the hormone epinephrine, commonly called adrenalin, to be released. This hormone causes an increase in the heart and breathing rates of an individual.

MiniLab

Purpose

LS **Logical-Mathematical** Students will investigate how a distracting stimulus affects reaction time.

Process Skills

observe and infer, compare and contrast, recognize cause and effect, collect and organize data, interpret data, experiment, and analyze

Teaching Strategies

- If students find counting backwards awkward, have them whistle or sing a song while they work.

Expected Results

A distraction should increase reaction time.

Analysis

1. Students should learn to anticipate the drop, and reaction time should improve.
2. A distraction should increase reaction time.
3. Answers may include being tired or sleepy.

the distance between the two ends of the paper clips.

5. Make a data table and record the location and the distance between sensory neurons.

Analysis

Ask students to answer the following questions.

1. Which areas of the skin did you find most sensitive? Least sensitive? *Responses will vary depending upon the areas tested. Regions of the back and inner arms will be less sensitive than areas of the palms or fingers.*
2. When two stimuli are felt as one, explain why they are felt as two points when the stimuli are moved farther apart. *The signals are detected by different sensory neurons.*

Assessment

Journal: Have students include a summary of the lab, the data table, and answers to Analysis questions in their student journals. Have students also write a paragraph explaining the advantages of some areas of the skin being more sensitive than others. L1

3 Assess

Check for Understanding

Make sure students understand that some parts of the nervous system are not under conscious control. Test their understanding by asking students to list body functions that they can and cannot control. Have them identify the part of the brain that controls each function. L1

Reteach

Have students make a table that lists the three major parts of the brain and the functions of each. L1

Extension

Have interested students find out about the function of various neurotransmitters such as dopamine, serotonin, norepinephrine, adenosine, or GABA. L2

Assessment

Skill: Prepare a handout showing a resting neuron, a stimulated neuron, and a synapse. Have students label the steps of neurotransmission. L1

4 Close

Discussion

Ask students what would happen if their cerebellum were damaged. For example, could they play a video game? *probably not, because the cerebellum controls coordination*

Figure 40.10

Understanding the organization of the human nervous system can be made easier by studying the relationships of the different divisions in a chart, like the one shown here.

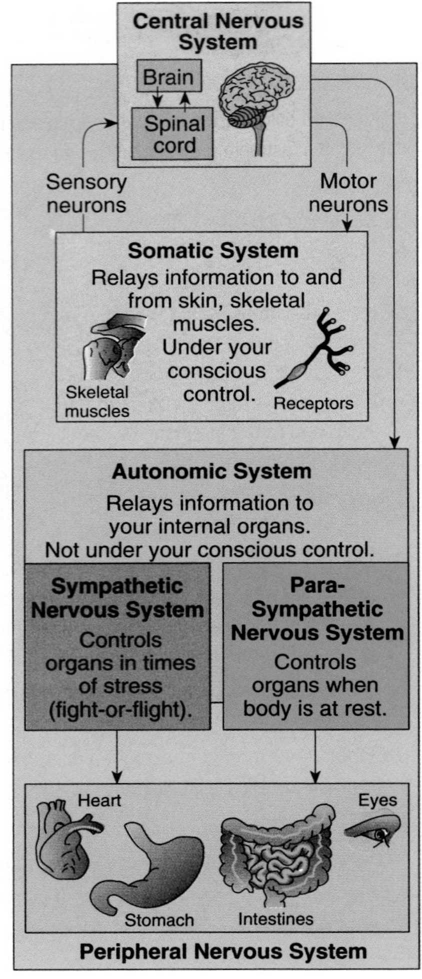

nerve tissue where extra energy is required, your heart rate increases, and blood flow is directed to your arms and legs, away from the digestive system. Can you see that your sympathetic nervous system would be at work in the scary house or when you're suddenly faced with a life-threatening emergency?

The **parasympathetic nervous system** controls many internal functions of the body at rest. The neurons in this system produce effects opposite to those of the sympathetic nervous system. Your parasympathetic nervous system is in control when you are relaxing on a warm summer day after a picnic or listening to the soothing melody of a lullaby.

Both the sympathetic and parasympathetic systems send signals to the same internal organs. The resulting activity of the organ depends on the intensities of the opposing signals. Thus, the sympathetic system increases your heart rate while you're in the scary house, and the parasympathetic slows it down when you exit.

The different divisions or subsystems of your nervous system are summarized in *Figure 40.10.* Each system plays a key role in communication and control within your body.

Section Review

Understanding Concepts

1. Summarize the charge distribution that exists inside and outside a resting neuron.
2. Outline the functions of the three major parts of the brain.
3. Contrast the functions of the two divisions of the autonomic nervous system.

Thinking Critically

4. Why is it nearly impossible to stop a reflex from taking place?

Skill Review

5. **Sequencing** Sequence the events as a nerve impulse moves from one neuron to another. For more help, refer to Organizing Information in the *Skill Handbook.*

Answers to Section Review

1. In a resting neuron, the inside of the neuron is more negatively charged than the outside.
2. The cerebrum controls conscious activities, intelligence, memory, language, skeletal muscle movements, and the senses. The cerebellum controls balance, posture, and coordination. The medulla oblongata mainly controls involuntary activities.
3. The sympathetic nervous system controls internal functions in times of stress, while the parasympathetic controls internal functions at rest.

Thinking Critically

4. A reflex is an involuntary action that does not involve conscious control by the brain.

Skill Review

5. An impulse reaches the end of an axon, opening calcium channels and allowing calcium to enter the end of the axon, which causes vesicles to fuse with the membrane and release neurotransmitters into the synapse. The chemicals diffuse across the synapse to the dendrite of the next neuron.

SECTION

40.2 The Senses

Section Preview

Objectives

Explain how senses detect chemicals.

Explain how the eye detects light.

Key Terms

taste bud
retina
rods
cones
cochlea
semicircular canal

Picture yourself in a park on a beautiful summer day. Stretching out on the grass, you look up and see white clouds float by against a background of blue. You feel grass tickling your toes and hear a breeze rustling through the trees. Sipping on a glass of lemonade, you enjoy its lemony smell and sweet, tangy taste. All these sensations are made possible by your senses: sight, touch, hearing, smell, and taste. Senses enable you to detect and respond to your environment.

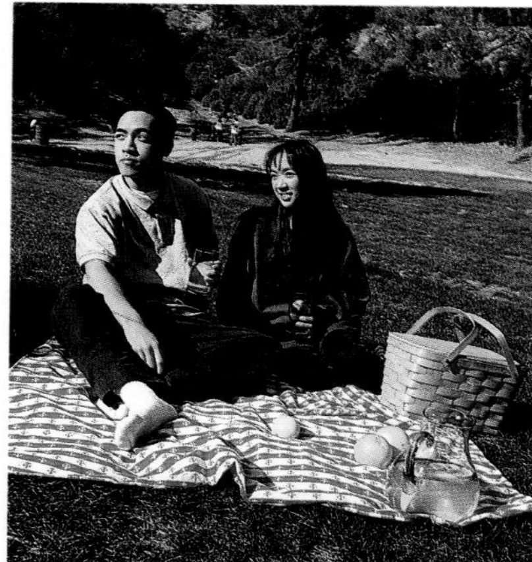

Senses that Detect Chemicals

How are you able to smell and taste the lemonade? The senses of smell and taste depend on receptors in your body that respond to chemical molecules. The receptors for smell are hairlike nerve endings located in the upper portion of the nose that respond to molecules in the air, as shown in *Figure 40.11.*

The senses of taste and smell are closely linked. Think about what your sense of taste is like when you have a cold and your nose is stuffed up. You can smell little, if anything. Because much of what you taste depends on your sense of smell, it too may be dulled. The sensation of taste occurs when chemicals dissolved in saliva make contact with sensory receptors on your tongue called **taste buds.**

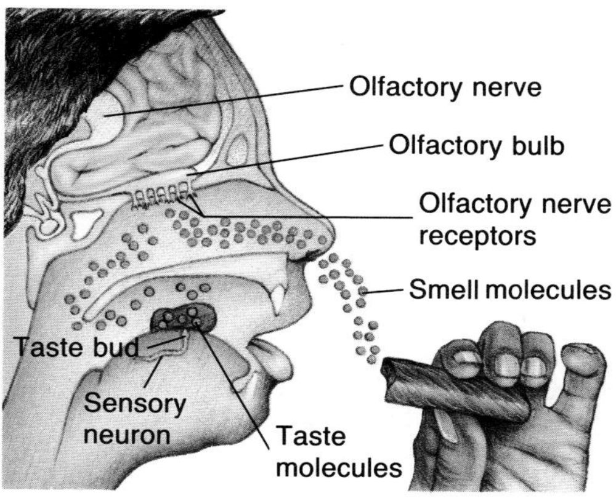

Figure 40.11

Chemicals acting on the hairlike nerve endings of your nose initiate impulses in the olfactory nerve. In the cerebrum, the signal is interpreted, and you notice a particular odor. Signals from your taste buds travel along a nerve to the medulla, and finally to the cerebrum. There, the signal is interpreted, and you experience a particular taste.

SECTION 40.2

Prepare

Key Concepts

The anatomy and physiology of the major senses are presented. The senses that detect chemicals (taste, smell), the sense that detects light (vision), and the senses that detect mechanical stimulation (hearing, touch, and balance) are examined.

Block Scheduling

Look for this symbol for strategies that are useful in a block scheduling format. For more information on block scheduling, refer to Section 40.2 in the **Lesson Plans** booklet.

Materials

- Order cow or sheep eyes for the Demonstration on page 1012.
- Acquire models of ears, eyes, skin, and nose for the Display.
- Make or purchase a tape of common sounds for the Demonstration on page 1013.

1 Focus

Bellringer

Before presenting the lesson, display **Section Focus Master 92** on the overhead projector and have students answer the accompanying questions. L1 SAE

Program Resources

Section Focus Master 92 L1 SAE
Reinforcement and Study Guide, p. 165 L1
Laboratory Manual, pp. 249-252 L2
Concept Mapping, p. 40 L1
Critical Thinking/Problem Solving, p. 40 L3
Tech Prep Applications, pp. 47-48 L2
Basic Skills Transparency 27 and **Master** L1 SAE
Basic Concepts Transparencies 72, 73 and **Masters** L1 SAE
Reteaching Transparency 40 and **Master** L1 SAE

Discussion

Ask students what kind of information the sense organs keep the body informed of. *They inform the body of changes that occur in the surroundings.* What is the reason for keeping the body informed? *The body is able to respond to changes in the environment.*

2 Teach

Demonstration

LS Visual-Spatial Dissect a cow or sheep eye. Cow and sheep eyes have a layer on the inner choroid coat that is not present in humans. Explain that this iridescent layer causes the eyes of animals to reflect light at night and enhances night vision by reflecting some light back into the retina. L2

Health Connection

Vision Problems

Nearly half of the population has one of four vision problems: farsightedness (hyperopia), when the eyeball is too short; nearsightedness (myopia), when the eyeball is too long; astigmatism, when the cornea is misshapen; and presbyopia, when the lens loses elasticity.

GLENCOE TECHNOLOGY

Videodisc

Biology: The Dynamics of Life
Disc 2, Side 1
Sense of Sight (Ch. 39)

Sense of Hearing (Ch. 40)

Tastes that you experience can be divided into four basic sensations: sour, salty, bitter, and sweet. Certain regions of your tongue react more strongly to a particular taste. Bitter is most likely to be sensed at the back of the tongue, sour on the sides, and sweet and salty on the tip.

A Sense that Detects Light

How are you able to see? The sense of sight depends on receptors in your eyes that respond to light energy. The **retina,** found at the back of the eye, is a layer of nerve tissue made up of sensory neurons. Follow the pathway of light to the retina in *Figure 40.12.*

The retina contains two types of cells—rods and cones. **Rods** are light receptors adapted for vision in dim light. They help you detect shape and movement. **Cones** are light receptors adapted for sharp vision in bright light. They also help you detect color. At the back of the eye, retinal tissue comes together to form the optic nerve, which leads to the thalamus and then to the cerebrum.

Close one eye. Everything that you can see with one eye open is the visual field of that eye. The visual field of each eye can be divided into two parts: a lateral and a medial part. As shown in *Figure 40.12,* the lateral half of the visual field projects onto the medial part of the retina, and the medial half of the visual field projects onto the lateral portion of the retina.

The projections and nerve pathways are arranged so that images entering the eye from the right half of each visual field project to the left half of the brain, and vice versa. Each eye sees about two-thirds of the total visual field, so the visual fields of the eyes partially overlap. This overlapping allows your brain to judge the depth of your visual field.

Figure 40.12

A cross section through the human eye shows where light goes as it enters the eye.

Light enters the cornea, passes through the aqueous humor, and then passes through the pupil. The colored iris muscle regulates the amount of light entering the eye. Light then passes through the lens and the jellylike vitreous humor. The lens focuses the light on the back of the eye, where it strikes the retina.

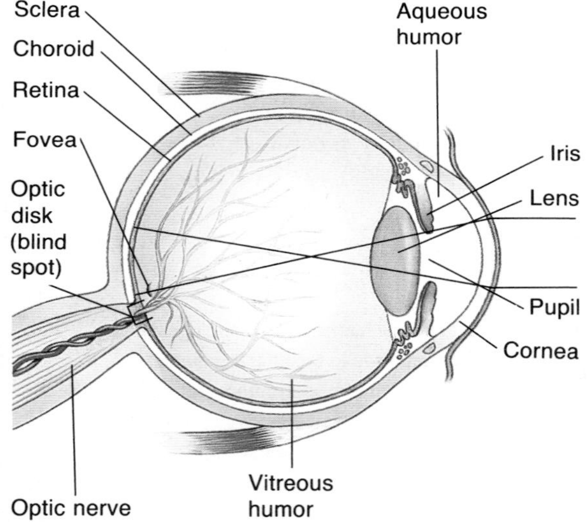

The left half of the retina in each eye is connected to the left side of the visual cortex in the cerebrum. The right halves of the eyes send signals to the right side of the brain. Images are interpreted in the cerebrum.

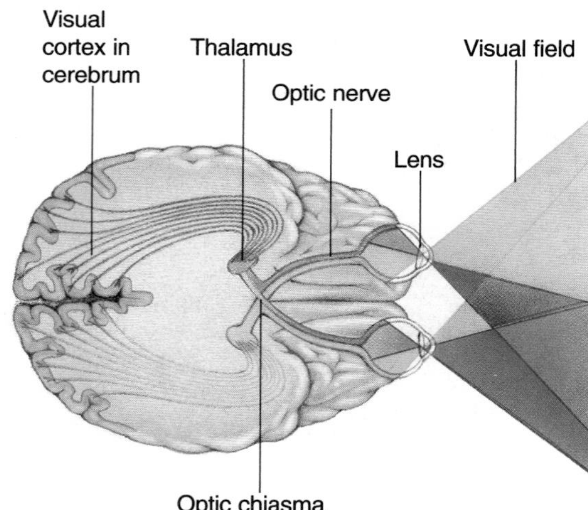

PROJECT

Modeling the Senses

Have students select and make a model of one of the senses. The model may demonstrate the anatomy of the sense or the function of the sense. Have students explain and demonstrate their models to the class. L1

LS

GLENCOE TECHNOLOGY

Videodisc

STVS: Human Biology
Disc 7, Side 1
Vision Diagnosis (Ch. 22)

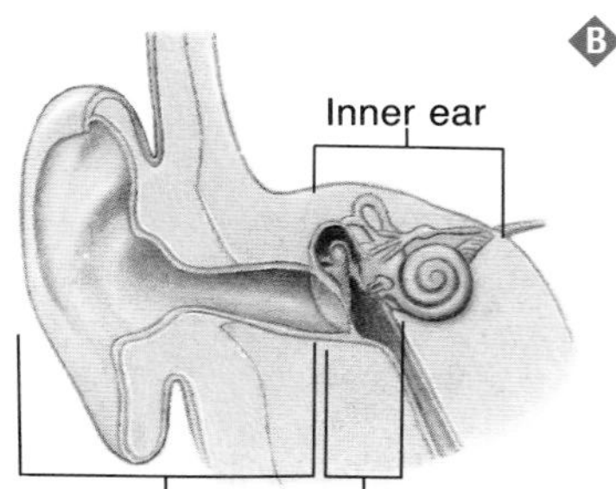

Ⓐ **Sound waves enter the outer ear and travel down to the end of the ear canal, where they strike a membrane called the eardrum and cause it to vibrate.**

Ⓑ **The vibrations then pass to three bones in the middle ear—the malleus, the incus, and the stapes.**

Ⓒ **As the stapes vibrates, it causes the membranous oval window in the inner ear to vibrate. Fluid in the cochlea picks up the vibrations and moves like a wave against the hair cells, causing them to bend.**

Ⓓ **The movement of the hairs initiates the movement of impulses along the auditory nerve. Impulses travel along this nerve to the sides of the cerebrum, and you hear the sound.**

Figure 40.13

The internal structure of the human ear is divided into three areas: the outer ear, middle ear, and inner ear. Follow the pathway sound waves take as they move through your ears.

Senses that Detect Mechanical Stimulation

How are you able to hear the leaves rustle and feel the grass as you relax in the park? These senses depend on receptors that respond to mechanical stimulation.

Your sense of hearing

Every sound causes the air around the source of the sound to vibrate. These vibrations are known as sound waves. When the sound waves enter your ear, they set up a response that stimulates a nerve impulse. This impulse then moves along the auditory nerve to the brain. It is in the **cochlea,** a fluid-filled, snail-shaped structure in the inner ear, that the mechanical stimulation of sound is converted into a nerve impulse. Trace the pathway of sound waves to the cochlea in *Figure 40.13.*

Your sense of balance

The ear also converts the physical signal of the position of your head into the nerve impulses, which travel to your brain, informing it about your body's equilibrium.

Maintaining balance is the function of the **semicircular canals.** The semicircular canals also are filled with a thick fluid and lined with hair cells. When you tilt your head, the fluid and hairs move. The mechanical movements of the hairs stimulate the neurons to carry an impulse to the cerebellum and the cerebrum. There, impulses from motor neurons stimulate muscles in the head and neck to readjust the position of your head.

Enrichment

Invite an audiologist to speak to the class about sound levels, how they are measured, and damage that may result from high levels of sound. Have the audiologist demonstrate how hearing is tested.

Display

LS Visual-Spatial Obtain display models of the ear, eye, skin, or nose.

Demonstration

LS Auditory-Musical To show how important each sense is, play a tape of various common sounds that students can identify. Discuss the dependence of humans on senses and the adjustments that are made if one sense is lost.

Audiovisual

Show the video *The Senses (smell, taste)* or *The Eye and Ear,* Charles Clark Co., Inc.

Physics Connection

Sound Travel

Sound travels through the air as waves—the higher the frequency, the closer the waves are to each other. When the waves reach the pinna, they pile up on each other, interacting in ways that diminish some of the frequencies while amplifying others. The contours of the outer ear determine which sounds are amplified and diminished. The contours of the ears encountered depend on the angle at which the waves hit the ear, so noises from one place will be modified differently from those from another.

STUDENT JOURNAL

Comparing Cameras to Eyes Ask students to write an essay comparing the eye to a camera. Students should include references to the lens, the aperture, the body, and the shutter of the camera. Review the function of these camera parts with students. Students should include a copy of the essay in their journals. L1 LS

PORTFOLIO

Creating a Flowchart Have students use Figures 40.12 and 40.13 to create flowcharts that show the series of events involved in seeing an object or hearing a sound. Encourage students to include the names of the parts of the eye, ear, and brain involved in these processes. L1 P LS

People in Biology

Meet Dr. Benjamin Carson, Neurosurgeon

Teaching Strategies

- Use this feature when discussing the brain and spinal cord.

Background

Dr. Benjamin Carson gained national recognition in 1987 when he successfully separated a pair of conjoined twins born in what was then West Germany. The twins were joined by a common blood vessel in the back of their heads. Currently, Dr. Carson serves as director of pediatric surgery at Johns Hopkins Hospital in Baltimore, Maryland.

people in biology

Meet Dr. Benjamin Carson, Neurosurgeon

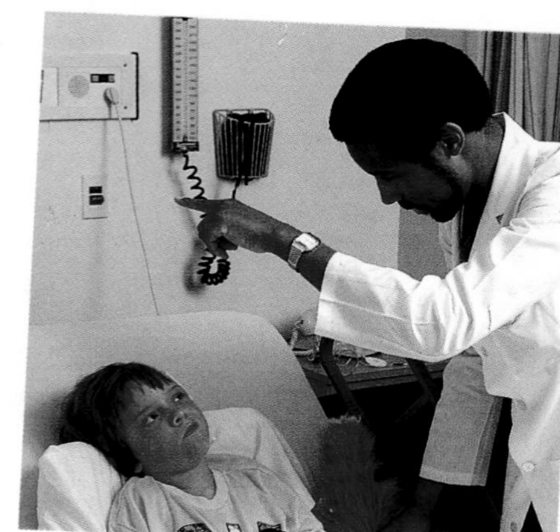

"Sit back and say, What have I always been good at? Ask other people what they see that you're good at. This is thinking big," says Dr. Benjamin Carson, chief of pediatric neurosurgery at Johns Hopkins Hospital in Baltimore, Maryland. He looked at his list of things he was good at and decided he could be a neurosurgeon.

In the interview that follows, Dr. Carson talks about both his work as a brain surgeon and his efforts to provide a much-needed role model for young people.

On the Job

Q Dr. Carson, what do you consider the most difficult brain surgery you've ever done?

A Believe it or not, even though news magazines in 1987 called it the most complex operation in the history of the world, it wasn't the separating of the Siamese twins who were joined at the head. I find that operating on a patient's brain stem is much more taxing—especially when you do it for 16 hours at a stretch. It's like trying to defuse a time bomb.

Q Those news magazines you mentioned also report that in spite of your extremely full schedule, you make time to talk to young people. Why is that so important to you?

A I believe kids need appropriate role models. Young people often look up to sports stars, but the chances of succeeding in that field are incredibly tiny. I talk not so much about kids becoming brain surgeons as about utilizing their intellectual ability and the things in their lives that they can control. For instance, I tell kids that if they put in the appropriate amount of work to be an engineer, that's what they'll be. The correlations are completely different with sports stars because only one basketball player in a million can be an NBA star.

Q What is one of the most frequent questions kids ask you about your profession?

A They often want to know if I'm frightened when operating on the brain. I tell them it's just as if you were a tightrope walker. If you're experienced, walking the tightrope isn't all that frightening, but if you haven't done it before, it will scare you to death. Brain surgeons must have a very healthy respect for the brain because they are dealing with the substance that makes people who they are. Brains don't tolerate mistakes of the knife very well! I need to be extremely aware of what I'm doing and of the consequences.

1022

Meeting Individual Needs

Hearing Impaired/Visually Impaired Students who are hearing or visually impaired may be sensitive during discussions of the senses. They also may be curious about the possible causes and treatments of their particular impairments. Be prepared to answer questions about the causes of hearing impairment or visual impairment. Some students may feel comfortable sharing with fellow students information about the treatments they have received. L1

Early Influences

Q How did you get interested in the medical field?

A From the time I was eight years old, I wanted to be a missionary doctor. The stories I heard in church made missionary doctors seem like the noblest people in the world. At great personal sacrifice, they brought not only physical but spiritual healing to people. But then, having grown up in poverty, I decided at 13 that I'd rather be rich and changed my goal to becoming a psychiatrist. I didn't know any, but the ones I saw on TV had mansions and sports cars. It looked as though all they had to do was sit around and talk to people. I majored in psychology at college. When I got to medical school, I had to ask myself, Well, what do I want to do now? I made a list of the things I had going for me: good eye-hand coordination, aptitude for thinking in three dimensions, ability to be very careful. I put all those together and said, What would that make me? A neurosurgeon, of course!

Q Were there certain people who had a great influence on you?

A My mother had only an elementary education, but she always told me "Mr. I-Can't is dead." She would never accept an excuse for not being able to do something. She worked several jobs at a time—scrubbing floors, washing windows, whatever she could find. When I complained about not having expensive clothes like my friends, she invited me to take the money she had earned, pay the bills, and use what was left to buy all the fancy clothes I wanted. I tried and soon discovered that my mother was a financial genius in keeping us housed, fed, and clothed.

CAREER CONNECTION

There are many ways to be part of the teamwork in a hospital. Here are just a few.

Neurosurgeon *Medical school plus several years of advanced training*

Surgical Technologist *High-school diploma and two-year associate degree in college*

Nursing Aide *High-school diploma and on-the-job training*

Personal Insights

Q Most people have problems in high school. Did you?

A Of course. My grades plummeted for a while because I gave in to peer pressure. Then I began to think that the here and now was not as important as the future. I decided that I could make my future better and also make other people's futures better. As a young person, I knew that time was on my side.

Q What advice do you have for students about careers in the medical field?

A I tell them that medicine is teamwork and that people in careers like nursing, medical technology, and physical therapy are crucial members of surgical teams. I also tell them that science and math, which you use all the time in medicine, are logical, not difficult. The major key to science and math is taking things one step at a time.

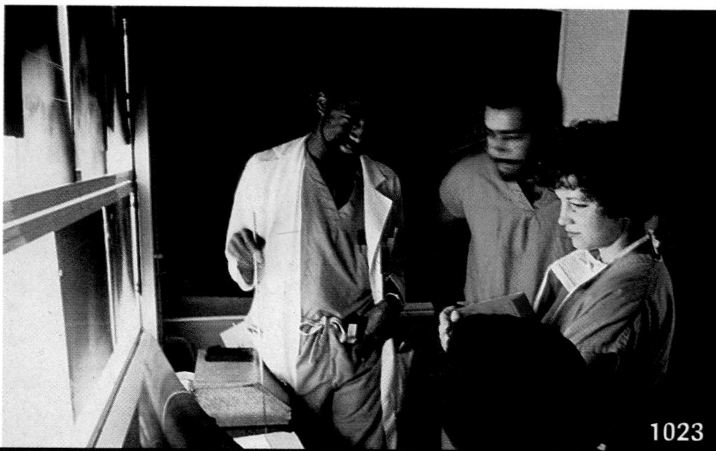

1023

CAREER CONNECTION

- **Career Path** Review the educational steps a person must take to become a physician. Advanced science classes and advanced math classes are recommended for students interested in a career in medicine. Most medical schools require a four-year college degree with emphasis on science- and math-related courses. Classes in chemistry, biology, physics, and calculus are often required to apply to medical school. Most medical schools also require hospital-related service before students apply.
- **Career Issue** The general public often feels that medical doctors make too much money. Physicians, on the other hand, point to high medical school costs, the cost of setting up a medical practice, and the high cost of medical malpractice insurance to justify their salaries. Discuss as a class whether costs of medical procedures should be determined by the government.

For More Information

For information on requirements for entrance to medical school, write to specific medical schools. For more information about the medical profession, contact a local doctor.

PORTFOLIO

Brain Dissection Have gifted students prepare a labeled dissection of the brain to be viewed by other students. They should prepare whole brains and half brains, labeling areas discussed in Chapter 40. Brains for this activity can be ordered from biological supply houses. L3 P

STUDENT JOURNAL

A Medical Career Have students who are interested in a career in medicine prepare interview questions about medical school and the different careers in medicine. Have students work in groups to arrange an interview with a local physician. Have them include a copy of the questions and the physician's answers in their journals. L1

SECTION 40.2

Text Question

Can you infer why it would be important for your body to perceive cold skin temperature? *Extremes in temperature can be harmful to the body. Also, exposure of the skin to the cold can lead to frostbite, a condition that is dangerous.*

3 Assess

Check for Understanding

Have students label diagrams of the eye, ear, and nose. L1

Reteach

Have students make a table of the functions of the various sense organs. L1

Extension

Have students research and report on the causes and treatments of cataracts, glaucoma, or vertigo. L2

Assessment

Knowledge: Provide students with a list of sensations such as pain, pressure, cold, odor, etc., that can be detected by the body. Have students match each sensation with the sense organ that detects it. L1

4 Close

Discussion

Ask students to identify which of the following professions could be undertaken by someone who has lost the sense of sight: architect, violinist, mathematician, public speaker, professional athlete, accountant, or physicist. L1

GLENCOE TECHNOLOGY

Videodisc

STVS: Human Biology
Disc 7, Side 2
Hearing by Touch (Ch. 6)

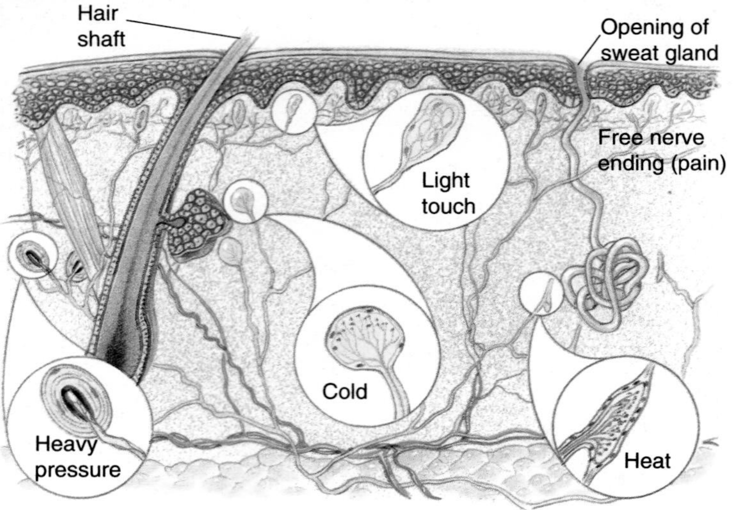

Figure 40.14

Many kinds of receptors are located throughout the dermis of your skin. When stimulated, some receptors detect gentle touches; others respond to heavy pressure. The sensations of pain, heat, and cold are sensed by other kinds of receptors.

Have you ever felt dizzy after a carnival ride that spun you around and around? When you stopped, your brain was receiving conflicting signals. The fluid in your semicircular canals was still moving, sending impulses to your brain indicating that you were still spinning.

Your sense of touch

Like the ear, your skin also responds to mechanical stimulation with receptors that convert the stimulation into a nerve impulse. Receptors in the dermis of the skin respond to changes in temperature, touch, pressure, and pain. It is with the help of these receptors, shown in *Figure 40.14,* that your body meets and responds to its environment.

Although some receptors are found all over your body, those responsible for responding to particular stimuli are usually concentrated within certain areas of your body. For example, touch receptors are numerous on your fingertips, eyelids, lips, the tip of your tongue, and the palms of your hands. When these receptors are stimulated, you perceive sensations of fine or light touch.

Pressure receptors are found inside your joints, in muscle tissue, and in certain organs. They are also abundant on the skin of your palms and fingers and on the soles of your feet. When these receptors are stimulated, you perceive heavy pressure.

Free nerve endings extend into the lower layers of the epidermis. These are most important as pain receptors. Two receptors respond to temperature. Heat receptors are found deep in the dermis, while cold receptors are found closer to the surface of your skin. Can you infer why it would be important for your body to perceive cold skin temperature?

Section Review

Understanding Concepts

1. Summarize the types of messages that the senses receive.
2. When you have a cold, why is it hard to taste food?
3. Explain how your eyes detect light.

Thinking Critically

4. Why might an ear infection lead to problems with balance?

Skill Review

5. **Sequencing** List the sequence of structures through which sound passes to reach the auditory nerve. For more help, refer to Organizing Information in the *Skill Handbook.*

Answers to Section Review

1. Sight is a response to light energy. Hearing depends on sound waves, while balance depends on movement of fluids in the semicircular canals of the ear. Touch responds to mechanical energy. Taste and smell respond to chemicals.
2. The taste of food involves both the sense of smell and the sense of taste.
3. Light stimulates the rod or cone cells in the retina, which transmit a signal to the brain by way of the optic nerve.

Thinking Critically

4. An ear infection with swelling could cause fluid in the ear to put pressure on the semicircular canals and cause the hairs in the canals to signal a false sense of balance in the brain.

SECTION 40.3 The Effects of Drugs on the Body

Nerves frazzled? Feeling tense? For a person trying to quit a smoking habit, the desire for a cigarette is almost always overpowering. Without realizing that it happened, the lighting of an occasional cigarette has grown into a pack-a-day habit. The body now cries out for a nicotine fix. Nicotine affects the body in dozens of ways—most of them harmful. In fact, nicotine is such a dangerous substance that a mere thimbleful could kill an adult if taken all at once. Why, then, do people knowingly take nicotine into their bodies?*

Section Preview

Objectives

Summarize the medicinal uses of drugs.

Explain how addictive drugs affect the nervous system.

Key Terms

drug
narcotic
stimulant
depressant
addiction
tolerance
withdrawal
hallucinogen

Drugs Act on the Body

You probably hear the word *drug* used often, maybe even every day. A **drug** is a chemical that is capable of reacting with the body's functions. Most drugs combine with some molecular structure on the surface of or within a cell. This molecular structure is a receptor site. In fact, the receptors with which drugs combine are assumed to be the same as the receptors for neurotransmitters of the nervous system or hormones of the endocrine system. How some drugs act to modify the transmission of neurotransmitters at a synapse is illustrated in *Figure 40.15*.

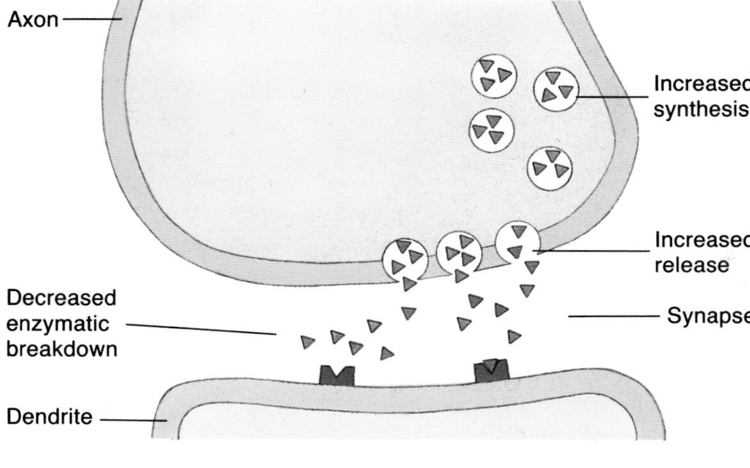

Figure 40.15
Some drugs can increase the amount of neurotransmitters at the synapse by increasing the rate at which the transmitters are synthesized and released, or by slowing the breakdown of transmitters by enzymes.

SECTION 40.3

Prepare

Key Concept

This section summarizes the medicinal uses of drugs and explains how addictive drugs affect the body. The major classes of misused and abused drugs are discussed.

Block Scheduling

Look for this symbol for strategies that are useful in a block scheduling format. For more information on block scheduling, refer to Section 40.3 in the **Lesson Plans** booklet.

Materials

- Order *Daphnia* and purchase coffee, tea, cola, unfiltered cigarettes, and cough syrup with dextromethorphen hydrobromide for the Biolab.
- Have students bring in magazines for the Student Journal activity on page 1024.

1 Focus

Bellringer

Before presenting the lesson, display **Section Focus Master 93** on the overhead projector and have students answer the accompanying questions. L1 SAE

Text Question

Why, then, do people knowingly take nicotine into their bodies? *Responses may reflect the idea that nicotine is an addictive substance.*

Skill Review

5. outer ear, eardrum, malleus, incus, and stapes, window on cochlea, fluid of cochlea, hairs of cochlea, auditory nerve to the brain

Program Resources

Section Focus Master 93 L1 SAE
Reinforcement and Study Guide, p. 166 L1
Biolab and Minilab Worksheets, pp. 160-162 L1
Tech Prep Applications, pp. 49-50 L2

2 Teach

Health Connection

Medicinal Plants Used by Animals

In 1992, The American Association for the Advancement of Science held its first symposium in zoopharmacognosy—the study of medicinal plants used by wild animals. This growing area of research involves observing animals such as chimpanzees, bears, and other wild animals eating medicinal plants. Researchers speculate that animals have learned to medicate themselves using plants. In 1989, a primatologist from Kyoto University observed a chimpanzee that was lethargic and had diarrhea and discolored urine. The chimpanzee chewed on the bark and leaves of a plant called *Vernonia amygdalina.* The leaves were saturated with an antibiotic and antiparasitic compound. The day after the chimp ate the plant, she had recovered fully.

Enrichment

Invite a speaker from Alcoholics Anonymous, Al-Anon, or Al-A-Teen to talk with students about alcohol abuse. Literature regarding this illness is available from Alcoholics Anonymous.

The Medicinal Uses of Drugs

A medicine is a drug that, when taken into the body, helps prevent, cure, or relieve a medical problem. Medicines that are used to prevent infections are called vaccines, and those that are used to cure infections may be antibiotics or antiviral drugs. You'll study these medicines in Chapter 42. Some of the many kinds of medicines that are used to relieve medical problems are discussed below.

cardiovascular:
kardia (GK) the heart
vasculum (L) small vessel
Cardiovascular drugs treat problems associated with blood vessels of the heart.

Relieving pain

Headache, muscle ache, cramps—these are some pain sensations that a person sometimes feels. You just studied how pain receptors in your body send signals of pain to your brain. Medicines that relieve pain affect these receptors. Pain relievers that do not cause a loss of consciousness are called analgesics. If you've ever taken aspirin for a headache, you've used such an analgesic. Medicines that relieve pain but also cause sleep because of their effect on the central nervous system are called **narcotics.** Many narcotics are made from the opium poppy flower, which is shown in *Figure 40.16.*

Figure 40.16

The sticky sap of the fruit of an opium poppy is used to make drugs called opiates. Opiates can be useful in medical therapy because only these narcotics are able to relieve severe pain.

Treating circulatory problems

Many drugs have been developed to treat heart and circulatory problems such as high blood pressure. These medicines are called cardiovascular drugs. In addition to treating high blood pressure, cardiovascular drugs may be used to normalize an irregular heartbeat, increase the heart's pumping capacity, or enlarge small blood vessels.

Treating nervous disorders

Several kinds of medicines are used to help relieve symptoms of problems with a person's nervous system. Among these medicines are stimulants and depressants.

Drugs that increase the activity of the central and sympathetic nervous systems are **stimulants.** Amphetamines are synthetic stimulants that increase the output of CNS neurotransmitters. Amphetamines are seldom prescribed because they can lead to dependence, which you'll read about in detail later in this chapter. However, because they increase wakefulness and alertness, amphetamines are sometimes used to treat patients with sleep disorders.

Drugs that lower, or depress, the activity of the nervous system are **depressants.** The primary medicinal uses of depressants are to encourage calmness and produce sleep. These drugs are commonly called sedatives. For some people, the symptoms of anxiety become so extreme that they interfere with the person's ability to function effectively. By slowing down the activities of the CNS, a depressant can temporarily relieve some of this anxiety.

GLENCOE TECHNOLOGY

Videodisc

The Infinite Voyage: Prisoners of the Brain
Pharmaceutical Companies Research the Effects of Drugs on the Brain (Ch. 10)

The Misuse and Abuse of Drugs

With all drugs, there are responsibilities as well as risks. Drug misuse is using a medicine in a way for which it was not intended. For example, giving your prescription to someone else, not following the prescribed dosage by taking too much or too little, and mixing medicines are all instances of drug misuse.

On the other hand, drug abuse is the inappropriate self-administration of a drug for nonmedical purposes. Drug abuse may involve use of an illegal drug, such as cocaine; use of an illegally obtained medicine, such as a depressant; or excessive use of a legal drug, such as alcohol or nicotine. Abused drugs have powerful effects on the nervous system as well as on other systems of the body, *Figure 40.17.* In addition, the abuse of drugs may lead to tolerance and addiction.

Addiction to drugs

When a person believes he or she needs a drug in order to feel good or function normally, that person is psychologically dependent on the drug. When a person's body actually develops a chemical need for the drug in order to function normally, the person is physiologically dependent. Psychological and physiological dependence are the same as **addiction.**

MiniLab

How do you interpret a drug label?

One common misuse of drugs is not following the instructions that accompany a medicine. Over-the-counter medicines have the potential of being harmful—even fatal—if they are not used as directed. The information on the label is required by the Food and Drug Administration to help the consumer use the medicine properly and safely.

Procedure

1. The photograph below shows a label from an over-the-counter drug. Read it carefully.
2. Make a data table like the one shown, and complete the table using the information on the label.

Information from a Drug Label

People with these conditions should avoid this drug	Possible side effects	This drug should not be taken with these medicines	Symptoms this drug will relieve	Correct dosage

INDICATIONS: The **decongestant** temporarily relieves nasal congestion due to the common cold, hay fever or other upper respiratory allergies, and associated with sinusitis. Helps decongest sinus openings and sinus passages. Reduces swelling of nasal passages; shrinks swollen membranes; and temporarily restores freer breathing through the nose. The **antihistamine** alleviates runny nose, sneezing, itching of the nose or throat, and itchy and watery eyes as may occur in allergic rhinitis (such as hay fever).

DIRECTIONS: ADULTS AND CHILDREN 12 YEARS AND OVER—one tablet every 12 hours. Do not exceed two tablets in 24 hours. **CHILDREN UNDER 12 YEARS OF AGE:** Consult a doctor.

Store between 2° and 25°C (36° and 77°F).

Protect from excessive moisture.

WARNINGS: Do not exceed recommended dosage. If nervousness, dizziness, or sleeplessness occur, discontinue use and consult a doctor. If symptoms do not improve within 7 days or are accompanied by fever, consult a doctor. May cause excitability especially in children. Do not take this product if you have a breathing problem such as emphysema or chronic bronchitis, or if you have glaucoma, heart disease, high blood pressure, thyroid disease, diabetes, or difficulty in urination due to enlargement of the prostate gland, or give this product to children under 12 years of age, unless directed by a doctor. May cause drowsiness; alcohol, sedatives, and tranquilizers may increase the drowsiness effect. Avoid alcoholic beverages while taking this product. Do not take this product if you are taking sedatives or tranquilizers, without first consulting your doctor. Use caution when driving a motor vehicle or operating machinery. As with any drug, if you are pregnant or nursing a baby, seek the advice of a health professional before using this product. Keep this and all drugs out of the reach of children. In case of accidental overdose, seek professional assistance or contact a Poison Control Center immediately.

DRUG INTERACTION PRECAUTION: Do not use this product if you are now taking a prescription monoamine oxidase inhibitor (MAOI) (certain drugs for depression, psychiatric or emotional conditions, or Parkinson's disease), or for 2 weeks after stopping the MAOI drug. If you are uncertain whether your prescription drug contains an MAOI, consult a health professional before taking this product.

Analysis

1. What is a side effect? What side effects are caused by this drug?
2. Why should a person never take more than the recommended dosage?
3. How are over-the-counter drugs different from prescription drugs?

Figure 40.17

The use of anabolic steroids without the careful monitoring by and a prescription from a physician is illegal. Some dangerous side effects include cardiovascular disease, kidney damage, and cancer.

MiniLab

Purpose

IS **Intrapersonal** Students will study information found on the labels of over-the-counter drugs.

Process Skills

observe and infer, analyze

Teaching Strategies

- Have students look at the packages of other over-the-counter drugs.

Expected Results

Student tables should show the following information: aspirin-sensitive people, children under 12 years, and pregnant women should avoid this drug; mild heartburn, upset stomach, and stomach pain are possible side effects; should not be taken with other non-prescription pain relievers, or drugs for depression or high blood pressure; will relieve sinusitus and symptoms of the common cold or flu; and the correct dosage is one caplet every 4-6 hours.

Analysis

1. A side effect is an effect of the drug other than the one that it is taken for. Side effects of this drug are excitability, especially in children, and drowsiness.
2. The correct dosage is the one that has been tested as safe. Over-the-counter medicines are potentially harmful if not used as directed.
3. Over-the-counter drugs are available without a doctor's prescription. They are not as strong as prescription medicines.

Assessment

Portfolio: Ask students to repeat the Minilab for other drug labels and include their tables in their portfolios. L1 P

STUDENT JOURNAL

Animal Testing of Medicines Animal testing is used to test certain drugs. Some people believe that using animals is unnecessary. Ask students to take a stand and write an editorial for the student newspaper or local newspaper and include a copy in their journals. L1 IS

BioLab Design Your Own Experiment

What drugs affect the heart rate of *Daphnia*?

Time Allotment

One class period

Objectives

Review objectives with students before they begin the Biolab.

Process Skills

observe and infer, form a hypothesis, communicate, predict, interpret data, experiment, and analyze

Safety Precautions

Caution students not to drink any of the solutions and advise them to wash their hands at the conclusion of the lab.

PREPARATION

Preparation of solutions:

ethyl alcohol—add 2 mL of ethyl alcohol to 98 mL of distilled water; nicotine—soak an unfiltered cigarette in 100 mL of warm distilled water for one hour, filter the solution; prepare weak coffee and weak tea; dilute cola 1 part water to 1 part cola; cough medicine—add 2 mL of cough medicine to 98 mL of distilled water.

Possible Hypotheses

- Students' hypotheses should categorize each drug according to whether it will increase, decrease, or not affect heart rate. Coffee, tea, cola, tobacco, and possibly cough medicine are stimulants and will increase heart rate. Ethyl alcohol and possibly cough medicine (with dextromethorphen hydrobromide) are depressants and will decrease heart rate.

BioLab Design Your Own Experiment

What drugs affect the heart rate of *Daphnia*?

Depending on their chemical composition, drugs affect different parts of your body. Stimulants and depressants are drugs that affect the central nervous system and the autonomic nervous system. Stimulants increase the activity of the sympathetic nervous system, which is responsible for the fight-or-flight response. They cause an increase in your breathing rate and in your heart rate. Depressants, on the other hand, increase the activity of the parasympathetic nervous system and decrease your breathing and heart rates.

PREPARATION

Problem

What legally available drugs are stimulants to the heart? What legal drugs are depressants? Because these drugs are legally available, are they less dangerous?

Hypotheses

From your knowledge and reading of the chapter, which drugs are stimulants? Which drugs are depressants? How will they affect the heart rate in *Daphnia?* Make a hypothesis concerning how each of the drugs listed will affect heart rate.

Objectives

In this Biolab, you will:

- **Measure** the resting heart rate in *Daphnia*.
- **Compare** the resting heart rate with the heart rate when a drug is applied.

Possible Materials

dilute solutions of coffee, tea, cola, ethyl alcohol, tobacco, and cough medicine (destromethorphen hydrobromide)
Daphnia culture
microscope
dropper
microscope slide
aged tap water

Safety Precautions

Do not drink any of the beverages used during this lab.

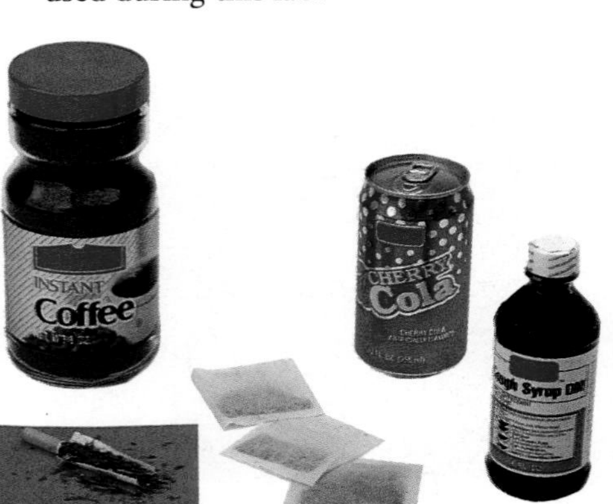

PLAN THE EXPERIMENT

Teaching Strategies

- Age tap water by leaving water in a beaker overnight.
- *Daphnia* that are placed in the aged tap water after being tested with a drug can be used later.

Possible Procedures

- Students should measure the resting heart rate of each *Daphnia* and then add a dropper of one of the drugs and measure the heart rate.

PLAN THE EXPERIMENT

1. Using a dropper, place a single *Daphnia* crustacean on a slide.
2. Observe on low power and find the heart.
3. Design an experiment to measure the effect on heart rate of four of the drug-containing substances listed in the Possible Materials list.
4. Design and construct a data table for recording your data.

Check the Plan

1. Be sure to consider what you will use as a control measurement.
2. Plan to add two drops of a drug-containing substance directly to the slide.
3. When you are finished testing one drug, you will need to flush the used *Daphnia* with the solution into a beaker of aged tap water provided by your teacher. Plan to use a new *Daphnia* for each drug-containing substance.

4. ***Make sure your teacher has approved your experimental plan before you proceed further.***
5. Carry out your experiment.

Magnification: 40×

Daphnia

ANALYZE AND CONCLUDE

1. **Making Inferences** Which drugs are stimulants? Depressants?
2. **Checking Your Hypothesis** Compare your predicted results with the experimental data. Explain whether your data support your hypotheses.
3. **Drawing Conclusions** How do the drugs affect the rate of the heartbeat of an animal?

Going Further

Changing Variables Many other over-the-counter drugs are available. You may wish to test their effect on the heart rate of a *Daphnia.*

ANALYZE AND CONCLUDE

1. Stimulants are coffee, tea, cola, and tobacco. Cough medicine may also be listed. Depressants are ethyl alcohol and cough medicine if it contains dextromethorphen hydrobromide.
2. Some students' hypotheses will be confirmed by their data; others will be rejected.
3. Stimulants speed the heart rate of an animal. Depressants slow the heart rate.

Assessment

Journal: Have students each prepare a laboratory report, including the experimental plan, the data table, and the answers to Analyze and Conclude to be placed in the journal. L1

Going Further

Have students compare the effects of four over-the-counter cough medicines on the heart rate of *Daphnia*. They should carefully make dilutions of the active drug after determining its concentration from the package. Tablet cough medications list the milligrams of medication in each tablet. L2

Data and Observations

Drug	Heart rate/min
No drug	240
Coffee	270
Cola	270
Tea	260
Ethyl alcohol	215
Tobacco	300
Cough medicine	heart rate varies with brand

History

A Plant That Changed the World

Purpose

Students will read about the history of tobacco as a world-wide crop.

Teaching Strategies

- Discuss the problems that would be created if farmers were forced to stop growing tobacco. Discuss some of the advantages to society if tobacco production were prohibited. Allow students to present their viewpoints on these issues. L1
- Provide students with outline maps of the United States. Have students use library resources to indicate areas of the country where tobacco is grown in large quantities. Have students suggest alternative crops that could be grown in each region if tobacco growing were prohibited by law. L2

Possible Answer

Answers may include laws and regulations, market forces such as reduction in demand for tobacco or higher prices for limited-nutrition food, and opening up new land to agriculture through irrigation and/or development of new varieties of food crops.

History

A Plant that Changed the World

Agriculture is the basis of our civilization. Until humans learned to farm, we could not progress past the point of hunting and gathering. Thus, it is ironic that one agricultural product that changed our world is neither an effective source of nutrition nor is it good for us.

The noxious weede This is how King James I described the tobacco brought back from his American colony of Virginia in 1604. A militant nonsmoker, he tried to pass laws prohibiting its use. However, sailors and merchants saw great profit in this native American import. Descendants of King James soon accepted its use because tobacco taxes became a major source of government revenue. First promoted as a medicine, tobacco rapidly gained wide use because of the addictive properties of nicotine. As more people around the world learned of the profit to be made from tobacco, it became widely cultivated.

The tobacco trade The economy of colonial Virginia was built on the tobacco trade. Thomas Jefferson and George Washington made their fortunes as tobacco planters. After harvesting, the tobacco was dried and then packed in barrels for transport to England. There the tobacco was processed and packaged for sale throughout the world. Thus, this native American plant became the basis of some of the world's great fortunes, triggered the building of great merchant fleets, and opened trade with the far corners of the world.

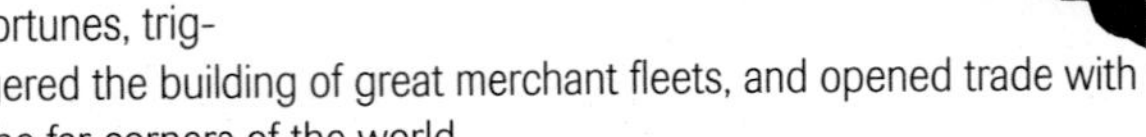

CONNECTION TO Biology

All over the world, land is used to grow crops of high commercial, but limited nutritional, value. With the malnutrition and starvation in the world, discuss how this imbalance could be corrected.

Tolerance and withdrawal

Addiction becomes evident when a drug user experiences tolerance or withdrawal. **Tolerance** occurs when a person needs larger or more frequent doses of a drug to achieve the same effect. The drug increases are needed because the body becomes more efficient at eliminating the drug from its systems. Nerve cells may also become less responsive to the dose. **Withdrawal** occurs when the person stops taking the drug and actually becomes ill, both psychologically and physiologically.

Classes of Drugs

Each class of drug produces a special effect on the body by working on body systems in its own way. Each class of drug produces its own symptoms of withdrawal, as well.

Stimulants: Cocaine, amphetamines, caffeine, and nicotine

You already know that stimulants increase the activity of the central nervous system and the sympathetic nervous system. The CNS stimulation can be seen in behaviors that range from the mild elevation of alertness to increased nervousness, anxiety, and even convulsions.

Cocaine stimulates the CNS by working on the part of the inner brain that governs emotions and basic drives such as hunger and thirst. It affects this reward center by interfering with the neurotransmitters, such as dopamine and norepinephrine. Cocaine causes levels of the neurotransmitters in the brain to increase, which results in a false

GLENCOE TECHNOLOGY

Videodisc

The Infinite Voyage: Prisoners of the Brain

Understanding Addiction (Ch. 8)

Dopamine and the Craving of Narcotics (Ch. 9)

Meeting Individual Needs

Students Acquiring English Have students with limited English proficiency make a poster that summarizes the effects of one of the drugs that is misused or abused. Have students combine their charts to create a bulletin board display. L1 SAE LS

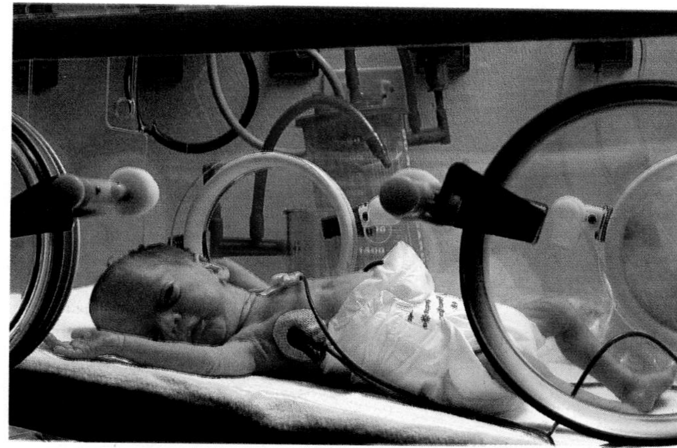

Figure 40.18

Crack is a cheap form of the drug cocaine. Pregnant women who are addicted to the drug give birth to crack babies who are also addicted to the drug. These newborns are usually low in birth weight, continually irritable, and may shake constantly.

message being sent to the reward center indicating that a basic drive has been satisfied. The user quickly feels a euphoric high called a rush. This sense of intense pleasure and satisfaction cannot be maintained, and the effects of the drug change. Physical hyperactivity follows, and the user is unable to sit still. Anxiety and depression often set in.

Cocaine also causes the sympathetic system to disrupt the body's circulatory system. It initially causes a slowing of the heart rate followed by a great increase in heart rate and a narrowing of blood vessels, called vasoconstriction. The result is high blood pressure. Heavy use of this drug causes weight loss, a compromise of the immune system, and eventually heart abnormalities. Eventually, death may also occur. Cocaine affects more than just the people who use it. As *Figure 40.18* shows, babies are sometimes born already addicted to this drug.

As you've already learned, amphetamines are stimulants that increase CNS neurotransmitters. Like cocaine, amphetamines also cause vasoconstriction, a racing heart, and increased blood pressure. Other adverse side effects of amphetamine abuse include irregular heartbeat, chest pain, paranoia, hallucinations, convulsions, coma, and sometimes death.

Not all stimulants are illegal or even difficult to get. One stimulant in particular is as close as the nearest coffee maker or soft-drink machine. Caffeine—a substance found in coffee, cola-flavored drinks, cocoa, and tea—is a CNS stimulant. Its effects include increased alertness and some mood elevation. Caffeine also causes an increase in heart rate and urine production, *Figure 40.19.*

Figure 40.19

The condition called tachycardia, when the heart beats more than 100 times per minute, can be triggered by caffeine.

Bioethics

Getting FDA Approval

Part of the procedure to get a drug approved by the Food and Drug Administration involves human drug tests. The deaths of five of 15 participants in an NIH drug trial in the summer of 1993 has raised questions as to the safety and efficacy of new therapies. Even though participants sign a consent form, desperately ill patients may grasp at any straw. This causes researchers to question whether these patients can be truly "informed" about the consequences of giving their consent.

Ask students to conduct research to find out what the procedures for consent to drug testing by the FDA involve. Ask students to also find out about how many lives have been saved by such research. Have students present their findings in a written report. L3

Enrichment

Have students find out how computers are allowing pharmacologists to design new drugs. A place to start is "Drugs by Design" by Lori Oliwenstein, *Discover,* Nov. 1991, pp. 74-77. L2

STUDENT JOURNAL

Cocaine Abuse Have students research the history of a person who has suffered as a result of cocaine abuse (Len Bias, Richard Pryor, etc). Each student should write a short essay that includes his or her personal feelings about cocaine, its effects on the body, and how this drug affected the person. L1 LS

Bioethics

Birth-Control Implants

In 1991, a U.S. congressional representative introduced a bill that would require convicted female addicts to accept Norplant birth-control inserts, which prevent pregnancies for up to five years, in order to avoid jail. Under the proposed law, the state would pay for the $500 procedure and also for its removal if the woman remained drug-free for a year. This is one example of how the tragedies of children born to drug-addicted mothers has caused state and local authorities to reevaluate the way they deal with drug addicts.

Judges and lawmakers are coming to view mothers of drug-exposed infants as criminals and are charging them with criminal child abuse or assault with a deadly weapon. Civil liberties advocates argue that criminal prosecution of mothers of drug-exposed babies is both racist and sexist. Most violators are minorities and all are women. Others feel that the get-tough approach makes it more likely that the mother will stay clean.

Enrichment

Have interested students research how the following toxins affect the nervous system and what they are used for: saxitoxin (from red tide), physostigmine, alpha-bungarotoxin, tetrodotoxin, and diisopropyl fluorophosphate. L2

Figure 40.20

Fruits such as grapes, berries, and apples may be covered with a coating of wild yeast. The yeast is an important component of the alcoholic fermentation process that is used to produce wine.

Although the moderate use of caffeine does not appear to cause problems in most people, some people may find it habit forming. Chronic coffee drinkers who suddenly abstain often experience the withdrawal symptoms of headaches and a general feeling of fatigue.

Nicotine, a substance found in tobacco, also is a stimulant. By increasing the release of the hormone epinephrine, nicotine increases heart rate, blood pressure, breathing rate, and stomach acid secretion. While nicotine is the substance in tobacco that leads to addiction, it is only one of about 3000 known chemicals found in cigarettes, many of which are also harmful. Smoking cigarettes is legal for adults. But our ever-increasing knowledge of its effects on the bodies of nonsmokers as well as smokers has made tobacco use a much less acceptable drug habit than it was in the past.

Depressants: Alcohol and barbiturates

As you already know, depressants slow down the activities of the CNS. All CNS depressants relieve anxiety, but most also produce noticeable sedation.

One of the most widely abused drugs in the world today is alcohol. Easily produced from various grains and fruits, such as the grapes shown in *Figure 40.20,* this depressant is distributed throughout the body by the blood. In organs with high blood flow, such as the brain, liver, lungs, and kidneys, it is rapidly distributed.

Unlike other drugs, alcohol probably acts on the brain by dissolving through the membranes of neurons rather than acting on specific receptors. Once inside a neuron, alcohol disrupts important cellular functions. It appears to block the movement of sodium and calcium ions that are responsible for transmitting impulses and releasing neurotransmitters.

The degree to which alcohol disrupts the normal body functions is directly proportional to the quantity consumed. Small amounts can result in disorientation, loss of muscle coordination, and diminished judgment.

Tolerance to the effects of alcohol develops as a result of heavy consumption. A number of organs are adversely affected by chronic alcohol use. For example, cirrhosis, a hardening of the tissues of the liver, is common. Addiction to alcohol—alcoholism—can also occur. Addiction can result in the destruction of nerve cells and brain damage. Research is ongoing to understand the

STUDENT JOURNAL

Evaluating Advertising Ask students to cut out cigarette ads from magazines. Post the ads on a bulletin board so all students can view them. In groups, have students discuss what groups the ads are likely to influence (the targeted audience). Ask students to choose an ad and discuss its effectiveness. L1 LS

PORTFOLIO

Alcohol in the Body Ask students to write a skit about the travel route of alcohol in the body, from the mouth, into the bloodstream, and to the brain and liver. Have them place the written lines and directions for their skit in their portfolios. L1 P LS

genetics of alcoholism. Most researchers agree that addiction has some hereditary basis, but it has not yet been linked to a specific gene.

Barbiturates are sedatives and antianxiety drugs. A barbiturate abuser generally is sluggish and has difficulty thinking, as well as paying attention. Chronic use results in both tolerance and addiction. When barbiturates are used in excess, the user's respiratory and circulatory systems are also depressed.

Narcotics: Opiates

Narcotics are opiates that are used to relieve severe pain. They act directly on the brain. In fact, narcotic receptors have been identified in the brain. The most abused narcotic in the United States is heroin. It depresses the CNS, slows breathing, and lowers the heart rate. Tolerance develops quickly, and withdrawal from heroin is painful.

Hallucinogens: Natural and synthetic

Natural hallucinogens have been known and used since ancient times, but the abuse of hallucinogenic drugs did not become widespread in the United States until the 1960s, when new synthetic types became widely available.

Hallucinogens stimulate the CNS, altering moods, thoughts, and sensory perceptions. Quite simply, the user sees, hears, feels, tastes, or smells something that is not there.

Technology BIOLOGY & SOCIETY

Scanning the Mind

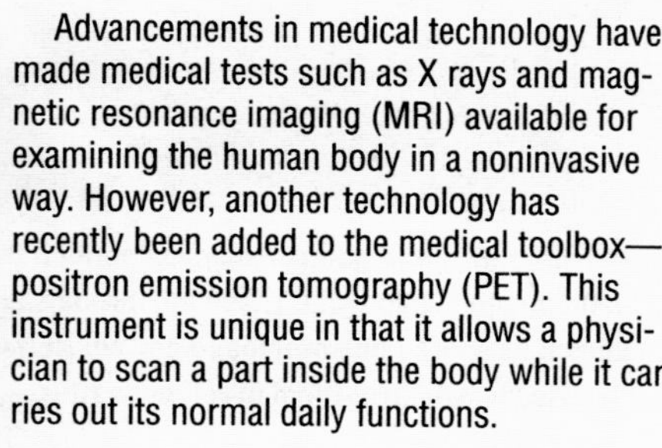

Advancements in medical technology have made medical tests such as X rays and magnetic resonance imaging (MRI) available for examining the human body in a noninvasive way. However, another technology has recently been added to the medical toolbox—positron emission tomography (PET). This instrument is unique in that it allows a physician to scan a part inside the body while it carries out its normal daily functions.

How PET scans are made PET scanners are excellent tools for studying the human brain. Monitoring either the blood flow to an area or the amount of glucose being metabolized pinpoints the active sections of the brain. The patient receives a compound containing radioactive isotopes by injection. Because these isotopes emit detectable radiation, they can be tracked by the sensitive PET scanner. Computers create a picture of brain activity by converting the energy emitted from the radioisotopes into a colorful map.

PET scanners in the United States are rare because they can cost more than $6 million. Some insurance companies still question the need for PET technology.

Applications for the Future

PET scans the regions of the brain that are taking part in a particular activity, such as listening to music or learning a new task. These scans reveal changes in the areas of the brain involved while a task is performed. The computer then prepares a visual record of the blood flow in areas of the brain during the activity. Current research with PET scans reveals how learning a subject at an early age may be critical to a person's future learning.

PET scans are also proving useful in the study of drug and alcohol addiction. A person is given the addictive drug and then is asked questions about his or her physical and emotional status while the scanner records metabolic activity in the brain. Researchers hope that information gained from the study of drug addiction will provide help in diagnosing and treating manic-depressive psychosis and schizophrenia.

INVESTIGATING the Technology

Thinking Critically Debate the question, "Under what circumstances should an insurance company pay for a PET scan?"

Technology BIOLOGY & SOCIETY

Scanning the Mind

Purpose

Students learn how radioactive isotopes are used in positron emission tomography (PET).

Background

- To make use of PET scanners, patients are injected with short-lived radioisotopes such as C-11, N-13, or O-15. As the radioisotope circulates through the body, it emits positively charged particles called positrons. The positrons collide with electrons in body tissues, causing the release of gamma rays that are detected by PET receptors.

Teaching Strategies

- Pet scanners can be used to measure activities that involve circulation of chemicals or chemical reactions in the body. Ask students to list some body activities that doctors might be able to monitor with PET scanners.

INVESTIGATING the Technology

Thinking Critically A reason for not paying for a PET scan might include that insurance companies are not involved in researching the uptake of drugs in chemically dependent people. A response in support of insurance companies paying for PET scans may include that customers who have brain tumors, epilepsy, or brain injuries may benefit from this technology.

Going Further

Students can find out how PET is used to distinguish between the two types of breast tumor: those that have estrogen receptors and those that do not have estrogen receptors.

CULTURAL DIVERSITY

Solomon Carter Fuller

Students may obtain a more thorough understanding of brain function by learning about common neuropathologies, such as Alzheimer's disease. During your discussions, emphasize the work of African-American psychiatrist and researcher, Solomon Carter Fuller (1872-1953). Fuller taught pathology, neurology, and psychiatry at Boston University for nearly 40 years. He was best known for expanding our medical knowledge in the fields of neuropathology and psychiatry, and his research on degenerative diseases of the brain, such as Alzheimer's disease, was pioneering. In 1913, Fuller became the editor of the *Westborough State Hospital Papers*, an influential publication specializing in mental diseases.

SECTION 40.3

3 Assess

Check for Understanding

Have students list stimulants and depressants and write a paragraph about how these substances affect the body. L1

Reteach

Have students make a table listing misused and abused drugs and their effects on the body. L1

Extension

Have students research information about the genetic susceptibility to alcoholism. L2

Assessment

Oral: Have one student name a drug from the chapter and a second student categorize the drug as a stimulant, depressant, narcotic, or hallucinogen. Have a third student list a side effect of the drug. Continue around the room until each student has been involved. L1

4 Close

Discussion

Ask students to discuss the consequences of introducing any type of drug into the body. Bring out that all drugs affect body function. *Some drugs help the body establish and maintain homeostasis while other drugs, those most frequently abused, alter homeostasis.*

Mushrooms of the genus *Psilocybe* contain the CNS hallucinogen psilocybin. These mushrooms are known as sacred mushrooms to certain Native Americans, who use them in religious rites.

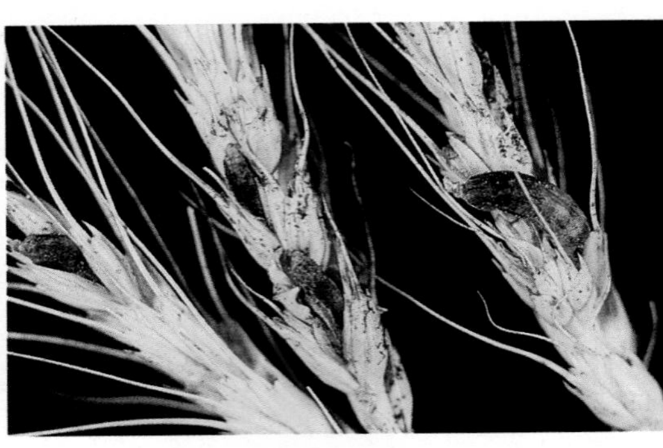

Ergot, a mold disease of cereal grains that contains a hallucinogen, killed thousands of people in the Middle Ages when they ate flour made from ergotized rye.

Figure 40.21
Some hallucinogens are found in nature.

Hallucinogens also increase heart rate, blood pressure, respiratory rate, and body temperature, and sometimes cause sweating, salivation, nausea, and vomiting. After large enough doses, convulsions may occur.

Unlike the hallucinogens shown in *Figure 40.21*, LSD is a synthetic drug. The mechanism by which LSD produces hallucinations is still debated, but it may be due to the blocking of a CNS neurotransmitter.

Connecting Ideas

The nervous system is an electrical communication and control system that senses the environment, integrates the information, and issues a command for action. Motor neurons help carry out the commands of the central nervous system by carrying an impulse to a muscle or gland. Drugs may interfere with the natural sensation-integration-command process.

How does such a complex system develop? What effect does it have on the development of a new human? Answers can be found in an understanding of human reproduction and growth.

Section Review

Understanding Concepts

1. What are the ways in which a drug can be used to treat a cardiovascular problem?
2. What is the difference between an analgesic and a narcotic?
3. How does nicotine affect the CNS?

Thinking Critically

4. Suggest why a physician but not a pharmacist is legally permitted to write a prescription.

Skill Review

5. **Comparing and Contrasting** Distinguish between a stimulant and a depressant, and compare their effects on the body. For more help, refer to Thinking Critically in the *Skill Handbook*.

Answers to Section Review

1. Drugs are used to normalize an irregular heartbeat, increase the heart's pumping capacity, or enlarge small blood vessels.
2. Analgesics relieve pain but do not cause a loss of consciousness. Narcotics relieve pain but cause sleep.
3. Nicotine will stimulate the central nervous system, causing an increase in heart rate, blood pressure, and breathing rate.

Thinking Critically

4. A physician knows the medical history of the patient and can treat problems that might occur as side effects of the drug.

Skill Review

5. Stimulants increase the activity of the central and sympathetic nervous systems. Depressants decrease the activity of the central nervous system and increase activity of the parasympathetic nervous system. Stimulants can increase alertness, nervousness, anxiety, heart rate, and breathing rate. Depressants do the opposite.

Chapter 40 Review

REVIEWING MAIN IDEAS

40.1 The Nervous System

- The neuron is the basic structural unit in the nervous system. Impulses move along a neuron in a wave of changing charges.
- The central nervous system consists of the brain and spinal cord.
- The peripheral nervous system brings messages to and from the central nervous system. It consists of the somatic and autonomic nervous systems.

40.2 The Senses

- Taste and smell are senses that respond to chemical stimulation.
- Sight is a sense that responds to light stimulation.
- Hearing and touch are senses that respond to mechanical stimulation. Receptors in the semicircular canals of the ears produce a sense of balance.

40.3 The Effects of Drugs on the Body

- Drugs act on receptor sites on neurons of the body.
- Some medicinal uses of drugs include relieving pain and treating cardiovascular problems and nervous disorders.
- The misuse of drugs involves using a medicine in a way for which it was not intended. Drug abuse involves using a drug for a nonmedical purpose.
- Harmful, commonly abused drugs include stimulants, depressants, narcotics, and hallucinogens.

Key Terms

Write a sentence that shows your understanding of each of the following terms.

addiction
autonomic nervous system
axon
central nervous system
cerebellum
cerebrum
cochlea
cones
dendrite
depressant
drug
hallucinogen
medulla oblongata
narcotic
neuron
neurotransmitter
parasympathetic nervous system
peripheral nervous system
reflex
retina
rods
semicircular canals
somatic nervous system
stimulant
sympathetic nervous system
synapse
taste bud
tolerance
withdrawal

Understanding Concepts

1. Which of the following is NOT part of the brain?
   a. cerebrum c. cerebellum
   b. cochlea d. pons
2. The basic unit of structure and function in the nervous system is the __________.
   a. Schwann cell c. axon
   b. neuron d. dendrite
3. Most axons are surrounded by a white covering called the __________ sheath.
   a. synapse c. myelin
   b. neuron d. dendrite
4. When a stimulus excites a neuron, __________ ions rush into the cell.
   a. sodium c. calcium
   b. potassium d. hydrogen
5. The __________ nervous system is made up of the spinal cord and brain.
   a. sympathetic c. central
   b. parasympathetic d. peripheral
6. A(n) __________ is a single extension of a neuron that carries messages away from the cell body.
   a. dendrite c. nucleus
   b. axon d. Schwann cell

Chapter 40 Review

Reviewing Main Ideas

Summary statements can be used by students to review the major concepts of the chapter.

Key Terms

Answers should go beyond defining the terms. Accept any answer that uses the term correctly and in the proper context.

Understanding Concepts

1. b
2. b
3. c
4. a
5. c
6. b

Chapter 40 Review

7. d
8. a
9. d
10. c
11. b
12. c
13. d
14. d
15. a
16. b
17. c
18. b
19. c
20. a

Applying Concepts

21. Your eyes see that the horizon is stationary, yet the inner ear tells you that you are moving. The conflicting messages to the brain result in seasickness.
22. Impulses continue to be transmitted.
23. sensory neuron to spinal cord, then to the brain, down the spinal cord to a motor neuron, to the muscles
24. The prey is paralyzed by the chemical.
25. Alcohol diffuses through the membranes of neurons and blocks the movement of sodium and calcium ions that are responsible for transmitting impulses and releasing neurotransmitters.

Chapter 40 Review

7. A __________ lies between one neuron's axon and another neuron's dendrites.
   a. rod c. cone
   b. reflex d. synapse
8. Chemicals called __________ diffuse across synapses and stimulate neurons.
   a. neurotransmitters c. depolarizers
   b. sodium channels d. white matter
9. Which of the following is NOT a type of neuron?
   a. interneuron
   b. sensory neuron
   c. motor neuron
   d. stimulus neuron
10. Which of the following is NOT a taste that is sensed by taste buds?
    a. sweet c. sweaty
    b. salty d. bitter
11. Which portion of the brain controls balance, posture, and coordination?
    a. cerebrum
    b. cerebellum
    c. medulla oblongata
    d. midbrain
12. Which vision cells allow humans to see color?
    a. thalamic cells c. cone cells
    b. rod cells d. cortex cells
13. Which part of the ear is involved in maintaining balance?
    a. cochlea
    b. oval window
    c. stapes
    d. semicircular canals
14. What type of medicine relieves pain but also causes sleep?
    a. stimulants c. analgesics
    b. depressants d. narcotics
15. What type of drug is nicotine?
    a. stimulant c. analgesic
    b. depressant d. narcotic
16. A(n) __________ is a psychological or physiological dependence on a drug.
    a. tolerance c. reflex
    b. addiction d. withdrawal
17. Of the following, which are NOT depressants?
    a. antianxiety drugs c. caffeine
    b. sedatives d. alcohol
18. Which of these is NOT a type of receptor found in the dermis of the skin?
    a. pain c. pressure
    b. light d. temperature
19. Which of the following drugs depresses the activities of the CNS?
    a. cocaine c. alcohol
    b. aspirin d. opiate
20. Which type of neuron carries impulses toward the brain?
    a. sensory
    b. motor
    c. association
    d. none of the above

Applying Concepts

21. You are making a ferry crossing during rough weather and the horizon seems to be moving up and down while you attempt to hold on to the railing. You begin to feel seasick. Explain the effects on the nervous system that might be the cause of this sensation.
22. A chemical stops the breakdown of enzymes that remove neurotransmitters at a synapse. How would this affect a person's body?
23. Describe the route an impulse travels from a stimulus to a skeletal muscle contraction.
24. Tetrodotoxin, a chemical produced by the puffer fish, blocks sodium channels. How does this toxin help this fish capture its prey?
25. Explain how alcohol disturbs the homeostasis inside a neuron.

Program Resources

Chapter Assessment, pp. 247-252 L1
Alternate Assessment in the Science Classroom
Computer Test Bank 
Content Mastery, pp. 169-172 L1

Thinking Critically

26. ***Concept Mapping*** Make a concept map that relates the following terms and phrases. Supply the appropriate linking words for your map.
    neuron, neurotransmitter, axon, dendrite, synapse, depolarization, central nervous system, peripheral nervous system
27. ***Interpreting Data*** Some pain impulses travel along myelinated neurons. The speed of these impulses is higher than other pain impulses, as shown in the table below. Explain why it is important for a person's survival that these particular pain sensations have a faster impulse speed.

Speed of Pain Impulses

Sensation	Speed of Impulse	Myelinated
Aching pain	1 m/s	no
Pricking pain	18 m/s	yes
Deep pressure	30 m/s	yes

28. ***Formulating Hypotheses*** The small pond-water crustacean *Daphnia* can be used to test how various drugs affect heartbeat rate. What would the overall effect on heartbeat rate be if you applied both ethyl alcohol and coffee to a *Daphnia* heart?
29. ***Observing and Inferring*** A medicine has this precaution: "Avoid driving a motor vehicle while taking this medicine. This medication may cause drowsiness." What type of drug does this medicine contain?
30. ***Comparing and Contrasting*** Explain why the nervous system is a high energy user compared with other systems in the body.

ASSESSING KNOWLEDGE & SKILLS

Paul, a construction worker, is planning on lighting a fuse hooked to some TNT to blast out a portion of rock as a part of his job in building a new highway.

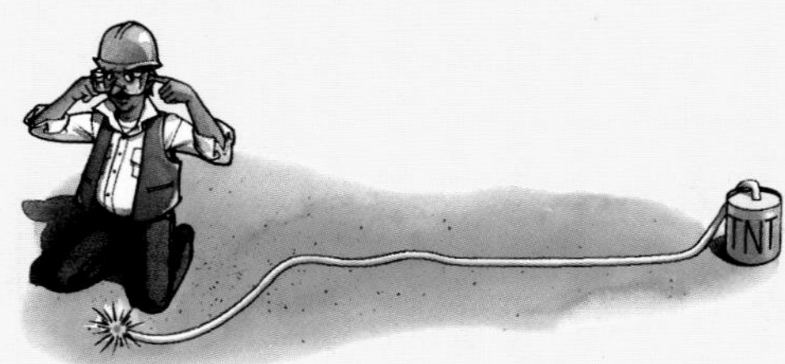

Use the illustration above to answer the following questions.

1. How does the steady burning of the fuse resemble the action potential of a nerve fiber?
   a. Both involve sodium channels.
   b. Both involve protein ion channels.
   c. Both are self-propagating.
   d. Both can repair themselves.
2. How do a nerve fiber and the fuse compare in terms of repeated use?
   a. Neither the fuse nor the nerve fiber can be used repeatedly.
   b. The fuse can be used over and over, but the nerve must regrow before being reused.
   c. The nerve fiber can be used after recovery, but the fuse is consumed and cannot be reused.
   d. Both the nerve fiber and the fuse can be used repeatedly.
3. ***Observing and Inferring*** In what ways is the action potential of a nerve similar to an electric current in a wire?

Chapter 40 Review

Thinking Critically

26. See below.
27. Pain is an indication of possible tissue damage. Pain signals the person to withdraw from the cause of the pain.
28. The addition of a stimulant (coffee) and a depressant (ethyl alcohol) together should cause little change in the heartbeat rate.
29. The medicine may contain a depressant or mild narcotic.
30. Energy is needed continually to move the sodium and potassium ions back and forth across a membrane as impulses travel through the body.

ASSESSING KNOWLEDGE & SKILLS

1. c
2. c
3. Both are electrical phenomena due to the flow of charged particles. The electrical current is due to the flow of electrons, whereas the action potential is due to the flow of charged ions.

26.

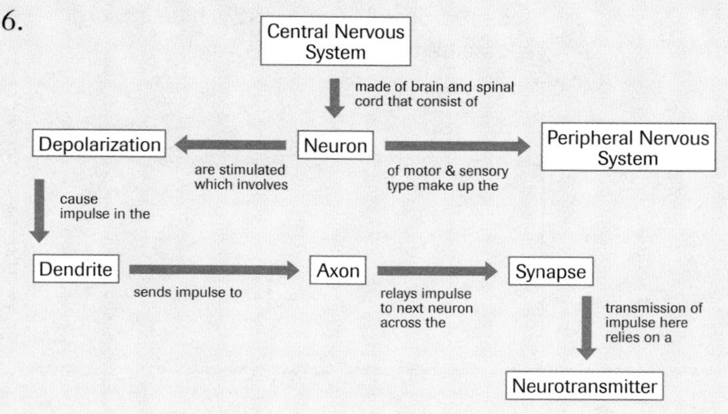

Chapter Organizer

SECTION	OBJECTIVES	ACTIVITIES/FEATURES
41.1 Human Reproductive Systems National Science Standards: UCP.1-3, UCP.5; C.1, C.5, C.6; G.1	1. **Identify** the parts of the male and female reproductive systems. 2. **Summarize** the negative feedback control of hormones. 3. **Sequence** the stages of the menstrual cycle.	**Thinking Lab:** Why is the level of testosterone an advantage for the dominant male baboon?, p. 1043
41.2 Development Before Birth National Science Standards: UCP.1-3, UCP.5; C.1, C.5, C.6; E.1, E.2; F.1; G.1	4. **Summarize** the events during each trimester of pregnancy.	**Minilab:** Are sperm attracted to eggs?, p. 1051 **Minilab:** How long is an embryo?, p. 1054 **A Broader View:** Prenatal Care, p. 1056
41.3 Birth, Growth, and Aging National Science Standards: UCP.1, UCP.3; A.1; E.1, E.2; F.1, F.6; G.1, G.2	5. **Describe** the three stages of birth. 6. **Summarize** the developmental stages of humans after they are born.	**Biolab:** Average Growth Rate in Humans, p. 1060 **Art Connection:** *Emmie and Her Child,* p. 1063

ACTIVITY MATERIALS

BIOLAB	MINILABS	ALTERNATE LAB
page 1060 graph paper pencils, red and blue	**page 1051** sea urchin eggs dropper microscope slides sea urchin sperm **page 1054** graph paper	**page 1046** graph paper colored pencils (red, yellow, blue, green) data sheet

Chapter 41 Reproduction and Development

TEACHER CLASSROOM RESOURCES

Reproducible Masters	Transparencies
Section Focus Master 94: Beginning Again L1 SAE **Reinforcement and Study Guide,** pp. 167-168 L1 **Critical Thinking/Problem Solving:** The Menstrual Cycle, p. 41 L3 **Content Mastery,** pp. 173-176 L1	**Basic Concepts Transparency #74:** Female Reproductive System L1 SAE **Basic Concepts Transparency #75:** The Menstrual Cycle L1 SAE **Basic Skills Transparency #28:** Negative Feedback Systems L1 SAE
Section Focus Master 95: From a Cell to a Human L1 SAE **Reinforcement and Study Guide,** p. 169 L1 **Biolab and Minilab Worksheets,** pp. 163-164 L1 **Laboratory Manual:** When Does a Chicken Embryo Grow the Fastest?, pp. 253-256; Human Fetal Growth, pp. 257-260 L2	**Reteaching Transparencies #41a, 41b:** Fertilization; Fetal Development L1 SAE
Section Focus Master 96: Birth, Growth, and Age L1 SAE **Reinforcement and Study Guide,** p. 170 L1 **Biolab and Minilab Worksheets,** pp. 165-166 L1 **Concept Mapping:** Human Growth, p. 41 L1 **Tech Prep Applications:** Tales from the Past, pp. 51-52 L2	

ASSESSMENT MATERIALS

Chapter Assessment, pp. 253-258
Alternate Assessment in the Science Classroom
MindJogger Videoquiz
Computer Test Bank

Spanish Resources SAE
English/Spanish Audiocassettes SAE
Cooperative Learning in the Science Classroom COOP LEARN
Lesson Plans

KEY TO TEACHING STRATEGIES

- L1 Level 1 activities should be within the ability range of all students including those with learning difficulties.
- L2 Level 2 activities are within the ability range of average to above-average students.
- L3 Level 3 activities are designed for the ability range of above-average students.
- SAE SAE activities should be within the ability range of Students Acquiring English.
- COOP LEARN Cooperative Learning activities are designed for small group work.
- P These strategies represent student products that can be placed into a best-work portfolio.
- These strategies are useful in a block scheduling format.

GLENCOE TECHNOLOGY

The following multimedia resources are available from Glencoe.

Biology: The Dynamics of Life
CD-ROM SAE
Videodisc Program

The Infinite Voyage Series
The Geometry of Life
Development of the Egg After Fertilization
The Keepers of Eden
Cincinnati Zoo's Cat House
Fires of the Mind
Learning from Birth
Positron Emission Tomography

The Secret of Life Series
Sex and the Single Gene: Cell Development
Tinkering with Our Genes: Genetic Medicine

National Geographic Society Series
STV: Human Body Vol. 3
Reproductive System

CHAPTER

41 Reproduction and Development

Chapter Overview

Students begin their study of human reproduction and development by examining the anatomy and hormonal control of the male and female reproduction systems. The changes that characterize puberty are presented, and sperm production and the menstrual cycle are discussed.

Human development before birth is the focus of the second section. The section includes a discussion of embryonic membranes, the placenta, and fetal development. Students also examine the role of genetic counseling during pregnancy.

The chapter concludes with a presentation of human development from birth through old age. Students survey the processes that occur during birth and examine the changes that characterize different parts of the growth and aging processes.

Key Terms

amniocentesis
bulbourethral gland
cervix
corpus luteum
epididymis
follicle
genetic counseling
implantation
labor
menstrual cycle
negative-feedback system
oviduct
ovulation
pituitary
prostate gland
puberty
scrotum
semen
seminal vesicle
umbilical cord
vagina
vas deferens

1038

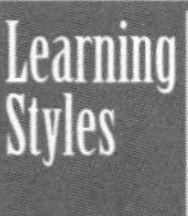

Look for the following logo for strategies that emphasize different learning modalities. LS

Kinesthetic	Project, p. 1055
Visual-Spatial	Visual Learning, pp. 1041, 1044, 1045; Portfolio, pp. 1044, 1045, 1053; Chalkboard Example, p. 1047; Student Journal, p. 1048; Minilab, p. 1051; Display, p. 1052; Meeting Individual Needs, pp. 1041, 1052
Interpersonal	Meeting Individual Needs, p. 1062; Portfolio, p. 1063
Linguistic	Meeting Individual Needs, p. 1042; Student Journal, pp. 1043, 1051, 1059; Portfolio, p. 1056
Logical-Mathematical	Meeting Individual Needs, p. 1042, 1045; Thinking Lab, p. 1043; Alternate Lab, p. 1046; Minilab, p. 1054; Time Line, p. 1055; Reinforcement, pp. 1062, 1063

LS

CHAPTER

41 Reproduction and Development

Think for a moment what it would be like to ride in the space shuttle. You float slowly, turning and moving effortlessly through space. The shuttle monitors your oxygen and temperature, providing a comfortable environment. Your food has been prepared and packaged for you. What a life!

Like an astronaut in the shuttle, at one time we all dwelt in a capsule, floating and turning in an inner, protected space. Inside your mother's uterus, you did not worry about food, water, or warmth. Your needs were provided for by your mother's body as your own body developed and grew. Of course, once outside your mother's body, you continued and are continuing to grow and develop.

How does the microscopic union of two sex cells eventually develop into a fully mature human being—one who is himself or herself capable of reproducing?

Concept Check

Review the following concepts before studying this chapter.

- Chapter 12: meiosis
- Chapter 27: hormone
- Chapter 38: endocrine gland, target tissue

Chapter Preview

It's not difficult to see what a floating astronaut and a human fetus have in common. Not only are they both fully protected in a controlled environment, but they also both developed from a single, fertilized egg. The story of how each developed is the story of human reproduction, growth, and development.

Magnification: 1500×

CHAPTER 41

Introducing the Chapter

Ask students to name the life-support systems in the tether that connects space-walking astronauts to their vehicle. *Likely responses will indicate a mechanism for gas exchange between the astronaut and the space vehicle.* Have students compare this structure with the structure that attaches a fetus to its mother. *The tether is similar to the umbilical cord that attaches a fetus to its mother. Both provide a mechanism for exchange of materials needed for life.*

Theme Development

The themes of *homeostasis* and *systems and interactions* are evident in the study of the hormone regulation of the male and female reproductive systems and in the examination of embryonic membranes, fetal development, growth, and aging.

Concept Check

Students will need to recall and review the production of sex cells by meiosis from Chapter 12. The relationship of cell structure and function from Chapter 8 will also be expanded upon.

Assessment Planner

Choose assessment strategies from the following pages to evaluate the progress of your students.

Assess, pp. 1048, 1057, 1064
Alternate Lab, pp. 1046-1047
Minilabs, pp. 1051, 1054
Portfolio, pp. 1044, 1045, 1053, 1056, 1063
Thinking Lab, p. 1043
Biolab, pp. 1060-1061
Chapter Review, pp. 1065-1067

SECTION

41.1

Prepare

Key Concepts

The anatomy and physiology of the male and female reproductive systems are presented. The negative-feedback control of the hormones involved in puberty of males and females is considered. In addition, hormonal control of the menstrual cycle is presented.

Block Scheduling

Look for this symbol for strategies that are useful in a block scheduling format. For more information on block scheduling, refer to Section 41.1 in the **Lesson Plans** booklet.

1 Focus

Bellringer

Before presenting the lesson, display **Section Focus Master 94** on the overhead projector and have students answer the accompanying questions. L1 SAE

Discussion

Explain that a human female is born with about 400 000 potential eggs in her ovaries. Beginning at puberty, one egg matures each month until menopause occurs, accounting for approximately 400 eggs. Ask students to think about the probability of any one egg being released. *1:400 000* Point out that this rate of release makes the chances that any one sperm will fertilize any given egg very small.

SECTION 41.1

Human Reproductive Systems

Section Preview

Objectives

Identify the parts of the male and female reproductive systems.

Summarize the negative feedback control of hormones.

Sequence the stages of the menstrual cycle.

Key Terms

scrotum
epididymis
vas deferens
seminal vesicle
prostate gland
bulbourethral gland
semen
puberty
pituitary
negative-feedback system
oviduct
cervix
vagina
follicle
ovulation
corpus luteum
menstrual cycle

Garfield appears to be up against stiff competition as he tries to woo the female cat away from his macho-looking, muscle-bound rival. The cartoonist was able to illustrate the essence of maleness in the rival cat by exaggerating one of his secondary sex characteristics, specifically, muscle development. In real life, how are secondary sex characteristics related to the reproductive process, and how are they controlled?

ARE YOU TRYING TO MAKE ME JEALOUS?

Human Male Anatomy

The ultimate goal of the reproductive process is the formation and union of egg and sperm, development of the fetus, and birth of the infant. The organs, glands, and hormones of the male reproductive system are instrumental in meeting this goal. Their main functions are the production of sperm—the male sex cells—and their delivery to the female. *Figure 41.1* shows the organs and glands that make up the male reproductive system.

Where sperm form

Sperm production takes place in the testes, which are located in the scrotum. The **scrotum** is a sac containing the testes and is suspended directly behind the base of the penis. Because sperm develops only in an environment with a temperature about 3°C lower than normal body temperature, the exterior position of the scrotum provides an ideal location.

Located within each testis, a fine network of highly coiled tubes are the production facilities for sperm. Sperm are produced through the meiosis of cells lining these tubes. Recall from Chapter 12 that meiosis produces haploid cells. One cell divides by meiosis and produces four cells that mature into sperm. In human males, the production of mature sperm takes about 74 days. A sexually mature male can produce about 300 million sperm per day, each day of his life.

As you can see in *Figure 41.2*, a sperm is highly adapted for reaching and penetrating the female egg. It can live for about 48 hours inside the female reproductive tract.

Program Resources

Section Focus Master 94 L1 SAE
Reinforcement and Study Guide, pp. 167-168 L1
Critical Thinking/Problem Solving, p. 41 L3
Basic Skills Transparency 28 and **Master** L1 SAE
Basic Concepts Transparencies 74, 75 and **Masters** L1 SAE

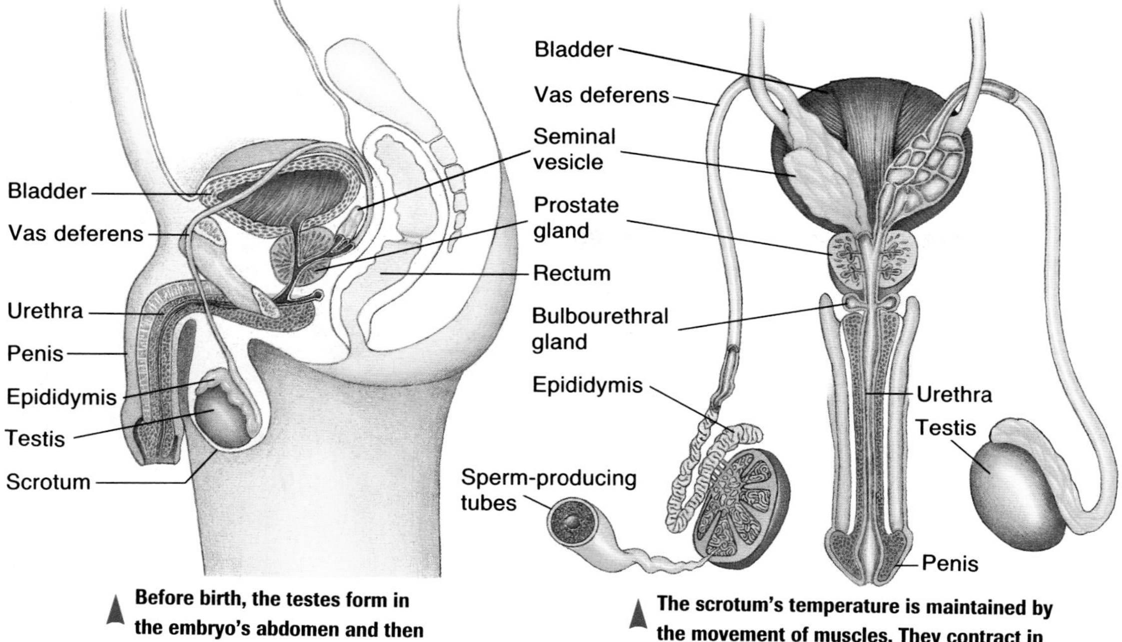

Before birth, the testes form in the embryo's abdomen and then descend into the scrotum.

The scrotum's temperature is maintained by the movement of muscles. They contract in cold weather, pulling the scrotum closer to the body to maintain warmth. In warm weather, the muscles relax to lower the scrotum, thus allowing air to circulate and cool the sperm.

Figure 41.1

The organs and glands of the male reproductive system are shown in a front and side view.

How sperm leave the testes

Before the sperm mature, they move out of the testes through a series of coiled ducts that empty into a single tube called the epididymis. The **epididymis** is a coiled tube within the scrotum in which the sperm complete their maturation. Mature sperm remain in the epididymis until they are released from the body.

When sperm are released from the epididymis, they enter the vas deferens. The **vas deferens** is a duct that transports sperm from the epididymis toward the ejaculatory ducts and the urethra. Peristaltic contractions of the vas deferens force the sperm along. The urethra is the tube that transports sperm out of the male's body. Notice in *Figure 41.1* that the urethra also transports urine from the urinary bladder. A muscle located at the base of the bladder prevents urine and sperm from mixing.

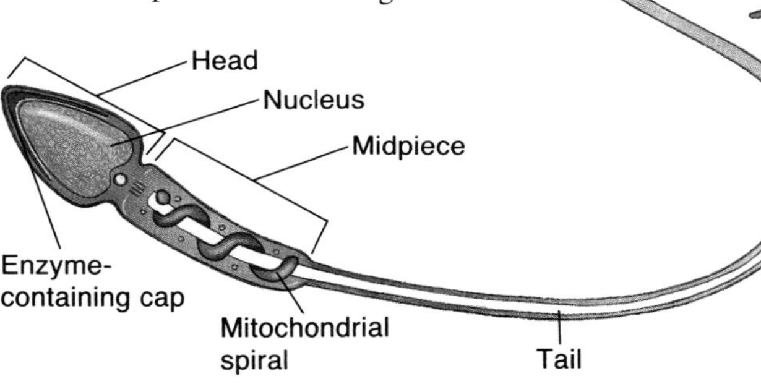

Figure 41.2

A sperm is composed of a head, a midpiece, and a tail. The head contains the nuclear material and is covered by a cap containing enzymes that facilitate penetration of the egg. A number of mitochondria are found in the midpiece; they provide energy for locomotion. The tail is a typical flagellum that propels the sperm along its way.

SECTION 41.1

2 Teach

Using Science Terms

The Latin word *testis* means "witness," while *testicle* means "little witness." Explain that the terms *testis* and *testicle* may have developed as a result of the Romans allowing only men to testify in court.

Audiovisual

Show the video, *The Male,* Charles Clark Co., Inc.

Visual Learning

Figure 41.1 Have students use the illustration of the male reproductive system to trace the path of sperm from the testes to the outside of the body. *testes, epididymus, vas deferens, urethra* LS

Text Question

In real life, how are secondary sex characteristics related to the reproductive process, and how are they controlled? *Responses may indicate that many secondary sex characteristics help the body prepare itself for parenthood. The characteristics are controlled by hormones.*

Meeting Individual Needs

Learning Disabled Provide students with tracing paper. Have them use the paper to trace the structures of the male reproductive system shown in Figure 41.1. Have students label each structure and write a description of its function beside each label. Encourage students to repeat this task for the female reproductive system shown in Figure 41.6. Instruct students to include the diagrams they have created in their journals and to use these diagrams as study tools. L1 SAE LS

Science, Technology, and Society

Tracking Sperm

The University of California at Davis has developed a computer-aided sperm analysis system. The digital image processor shows sperm as dots on a monitor. The dots are observed to trace their speed and trajectory as the sperm move.

Scientists have observed that sperm have different types of movements. For example, when they are near the cervix, sperm lash their tails rapidly and dart forward. Many seem to get nowhere; however, a few struggle through the cervical mucus using a second swimming stroke in which the tail spirals like a propeller. The sperm then move slowly, swimming to crypts on the walls of the cervix.

In time, a few sperm finally enter the oviduct. Once they are in the oviduct, calcium ions hyperactivate these sperm and they begin to thrash and flop violently. Scientists think this movement raises the odds of a sperm meeting an egg.

Discussion

Discuss some causes of infertility in men and women. In males, a low sperm count decreases the chances of a sperm reaching the egg. In females, the inability to release eggs, as a result of blocked oviducts or low levels of sex hormones, can result in infertility.

Figure 41.3

Puberty results in many physical and emotional changes. Generally, males undergo puberty sometime between the ages of 13 to 16.

Fluids that help transport sperm

As sperm travel from the testes, they mix with the fluids secreted by several different glands. The **seminal vesicles** are a pair of glands located at the base of the urinary bladder. They secrete a mucouslike fluid into the vas deferens. The fluid is rich in the sugar fructose, which provides energy for the sperm cells.

Another structure, the **prostate gland,** is a single, doughnut-shaped gland that surrounds the top portion of the urethra. The prostate secretes a thinner, alkaline fluid that helps sperm move and survive. Two tiny **bulbourethral glands** are located beneath the prostate. These glands secrete a clear, sticky, alkaline fluid that protects sperm by neutralizing the acidic environment of the vagina. The combination of sperm and all of these fluids is called **semen.**

Hormonal Control

Recall that in Chapter 38 you learned how hormones, which play a key role in the regulation of digestion, metabolism, and homeostasis, are released by glands of the endocrine system. Hormones also are the control mechanism for the male's reproductive system.

Hormones and male puberty

As you watch young children, it's obvious from their physical appearance that they are not sexually mature. However, in the early teen years, changes begin to occur to their bodies. Puberty, *Figure 41.3,* begins. **Puberty** refers to the time when secondary sex characteristics begin to develop so that sexual maturity—the potential for sexual reproduction—is reached. The changes associated with puberty are controlled by the sex hormones that are secreted by the endocrine system.

Hormones and the male reproductive system

In males, the onset of puberty causes the hypothalamus to produce several kinds of hormones that interact with the pituitary gland. The **pituitary** is a gland, located at the base of the hypothalamus, that secretes hormones used to influence many different physiological processes of the body. As shown in *Figure 41.4,* the hypothalamus secretes a hormone

Meeting Individual Needs

Gifted Have students conduct research on prostate cancer. Ask students to find out what factors place a man at risk for this type of cancer, what the possible treatments are, and what factors determine the likelihood of survival with treatment. Have students prepare written reports of their findings for inclusion in their portfolios. L3 LS

Students Acquiring English Have students make a table with the following heads: Organ, Reproductive Function. In the first column, have students list the organs of the male reproductive system. In the second column, they should describe each organ's function. L1 SAE LS

that causes the anterior lobe of the pituitary to release two other hormones: follicle-stimulating hormone (FSH) and luteinizing hormone (LH). When released into the bloodstream, FSH and LH are transported to the testes. In the testes, FSH causes the production of sperm cells. LH causes endocrine cells in the testes to produce the male hormone—testosterone.

Testosterone is the steroid hormone responsible for the growth and development of secondary sex characteristics in a male. These characteristics include the growth and maintenance of male sex organs; the production of sperm; an increase in body hair, especially on the face, under the arms, and in the pubic area; an increase in muscle mass and in the growth rates of the long bones of the arms and legs; and the deepening of the voice. Evidence suggests that testosterone may also be responsible for an increase in aggressive behavior.

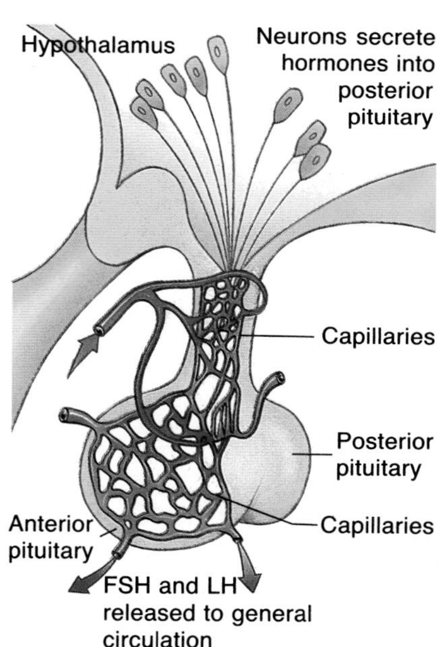

ThinkingLab Draw a Conclusion

Why is the level of testosterone an advantage for the dominant male baboon?

In baboon tribes, a social structure of dominant and subordinate males exists. The dominant males have better access to food, to the best resting spots, and to the female baboons. In contrast, the subordinate male baboons must laboriously search for food, often only to have it stolen by the dominant male.

Analysis

In males, the hormone testosterone regulates sexual behavior and aggression as well as increases the rate at which glucose reaches the muscles. The graph shows testosterone levels of dominant and subordinate male baboons. When the male baboons are at rest, the testosterone levels are essentially equal. After being exposed to the same stress, the reactions of the dominant and subordinate males differ sharply for the first few hours.

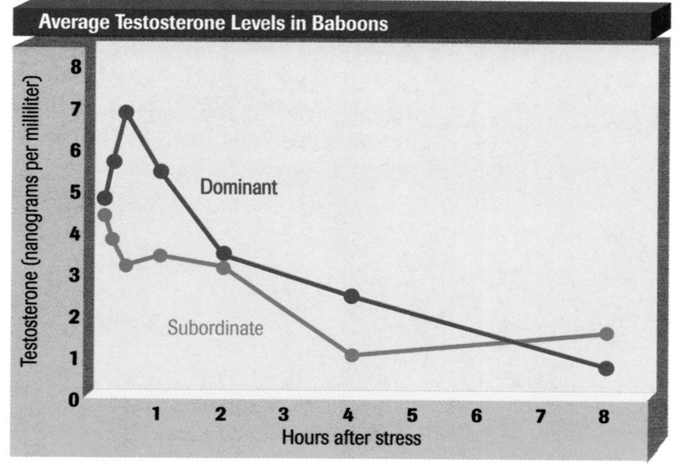

Thinking Critically

Explain the adaptive advantage of higher levels of testosterone in the dominant male during times of stress.

Figure 41.4

Because the body is controlled by two systems, the nervous system and the endocrine system, there must be coordination between the two. This coordination is centered in the hypothalamus and pituitary, which are located close together in the brain. In the hypothalamus, neurons secrete releasing and inhibiting hormones. These hormones travel in the bloodstream to the pituitary gland, where they speed up or slow down the secretion of FSH and LH.

ThinkingLab Draw a Conclusion

Purpose

LS Logical-Mathematical Analyze the adaptive advantage of high levels of testosterone in baboons.

Process Skills

observe and infer, compare and contrast, analyze

Teaching Strategies

- Explain that testosterone levels seem to be strongly linked to competitiveness and dominance. A link between triumph and testosterone levels was found in tennis players studied at the University of Nebraska (published in *Hormone and Behavior* in 1989). Studies also show that men who most try to dominate in social situations are likely to have higher testosterone levels than their peers.

Thinking Critically

Dominant males will be more aggressive and be able to fight longer and harder, thus maintaining their dominant position.

Assessment

Oral: Discuss the conclusion with students and elicit how it might apply to humans. Have students consider professions in which being more aggressive may be important. L1

STUDENT JOURNAL

Steroids Have students conduct library research to find out what steroids are and what function they perform in the body. Ask students to research the steroids that are sometimes taken by athletes to build strength and muscle and what the risks and drawbacks of using such steroids are. Have students prepare a table or a written report of their findings. L1 LS

Using an Analogy

Have students use a sharp pencil to make a small dot on a piece of paper. Explain that the dot is approximately the same size as a human egg.

Software

Human Sexuality, ABC Interactive.

Visual Learning

Figure 41.5 Have students use Figure 41.5 and the text description of negative feedback systems to create a flowchart that shows how the hypothalamus causes the pituitary to release FSH and LH in males. L1 LS

Audiovisual

Show the video *The Female*, Charles Clark Co., Inc.

Maintaining the correct level of hormones

In your home, the thermostat regulates the room temperature. When the temperature falls below a certain level, the lack of heat energy is sensed by the thermostat. The thermostat's control unit signals the heating unit to begin the production of heat. When enough heat is produced and the temperature increases to a certain point, the thermostat signals the heating unit to shut off.

In the body, the endocrine system is self-regulated by its own **negative-feedback system**, which is similar to the thermostat. In the body's system, the level of the hormone or its effect is signaled back to the hypothalamus. *Figure 41.5* shows how the feedback circuits between the hypothalamus, pituitary gland, and the testes function in human males.

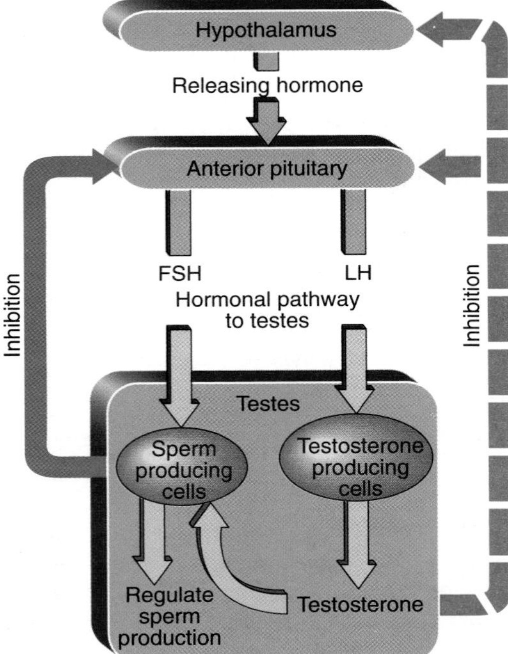

Figure 41.5

In human males, signaling molecules from the hypothalamus cause the pituitary gland to release FSH and LH, which in turn stimulate the production of sperm and testosterone. The solid blue line shows that, within the testes, cells that help in the formation of sperm then send signals back to the pituitary to inhibit or slow down the production of FSH. The dashed red line shows that an increased level of testosterone sends a signal back to inhibit or slow the production of both FSH and LH. By constantly monitoring the signals, the body maintains the correct levels of hormones.

Human Female Anatomy

The main functions of the female reproductive system are to produce eggs, the female sex cells, and to provide an environment for a fertilized egg to develop. Egg production takes place in the two ovaries. Each ovary is about the size and shape of an almond. One ovary is located on each side of the lower part of the abdomen.

As you can see in *Figure 41.6*, the open end of an oviduct is located close to each ovary. The **oviduct** is a tube that transports eggs from the ovary to the uterus. Peristaltic contractions of the muscles in the wall of the oviduct combine with beating cilia to move the egg through the tube.

You learned earlier that female mammals have a uterus in which the fetus develops during pregnancy. The uterus is situated between the urinary

PORTFOLIO

Negative-Feedback Systems Have students create diagrams that explain how the thermostat in a home works as a negative-feedback system. Beside this drawing, have students make a second diagram that explains how the endocrine system of the body works in a way comparable to the thermostat in a home. L2 P LS

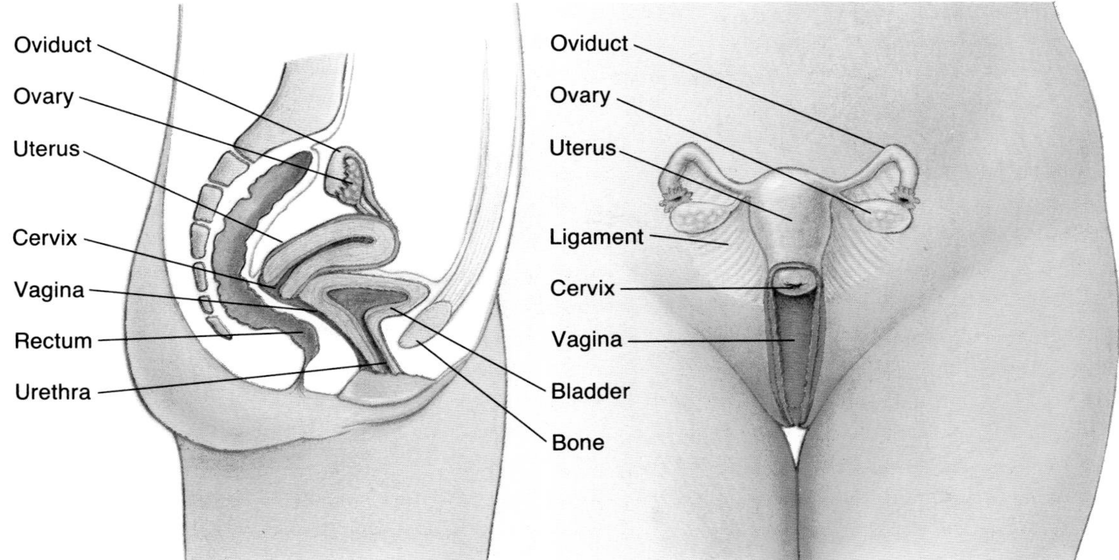

Figure 41.6

The female reproductive system includes two ovaries, two oviducts—sometimes called fallopian tubes—the uterus, and the vagina. The ovaries produce eggs and female sex hormones. The uterus is composed of two layers: a thick, muscular layer and a thin lining called the endometrium.

bladder and the rectum and is shaped like an inverted pear. Its lower end, called the **cervix,** tapers to a narrow opening into the vagina. The **vagina** is a passageway between the uterus and the outside of the female's body.

Puberty in Females

As in males, puberty in females begins when the hypothalamus signals the anterior lobe of the pituitary to produce and release the hormones FSH and LH. These are the same hormones that are produced in males; however, in females, FSH stimulates the development of a follicle in the ovary. A **follicle** is a group of epithelial cells that surround an undeveloped egg cell. FSH also causes the release of the hormone estrogen from the ovary. Estrogen is the steroid hormone responsible for the secondary sex characteristics of females. These characteristics include the growth and maintenance of female sex organs; an increase in body hair, especially under the arms and in the pubic area; an increase in the growth rates of the long bones of the arms and legs; a broadening of the hips; an increase in fat deposits in the breasts, buttocks, and thighs; and the onset of the menstrual cycle.

Production of eggs

Recall that sperm production does not begin in males until they reach puberty, after which time it continues for the rest of their lives. Egg production is different. Even before a female is born, her body begins to

Discussion

Elicit from students why drug use (such as marijuana) by the mother rather than the father creates a greater chance of producing genetic defects. *Females do not replace eggs, but are born with all their potential eggs.*

Visual Learning

Figure 41.6 Have students use the illustration of the female reproductive system to trace the path of an egg from the ovaries to the uterus. *ovary, oviduct (fallopian tube), uterus* LS

Making a Model

LS **Logical-Mathematical** Have students use the text description to develop a flowchart that shows the sequence of events involved in the production and release of the hormones FSH and LH. L2

Meeting Individual Needs

Hearing Impaired and Learning Disabled Have students prepare a table contrasting anatomy, sex cells, and puberty in males and females. Have students consult Making and Using Tables in the *Skill Handbook* if they need help. L1 LS

PORTFOLIO

Tracing the Path of Sex Cells Have students develop flowcharts or concept maps that trace the path of sperm from the testes through the reproductive system to outside the body. Have them make a similar diagram showing the path of an egg through the female reproductive system. L1 P LS

Health Connection

Disruption of the Menstrual Cycle

The normal menstrual cycle is affected by physical fitness. During intense exercise (as with Olympic athletes), the body produces endorphins, which reduce the production of sex hormones. This can stop the menstrual cycle. Smoking also has an anti-estrogenic effect.

Another factor that can disrupt the normal menstrual cycle is the fat percentage in the body. Scientists have found that a critical weight and fat percentage act like a trip wire that triggers monthly ovulation. Thus, a too-low or too-high fat percentage can interfere with normal cycling.

Misconception

Many people believe that the egg is a passive player in the fusion of egg and sperm. Explain that a team of researchers at Johns Hopkins University have determined that sperm that reach an egg are held to the egg by molecules on the egg's surface The molecules of the egg hook together with counterparts on the surface of the sperm. These receptor molecules hold a sperm tight until it can be absorbed by the egg.

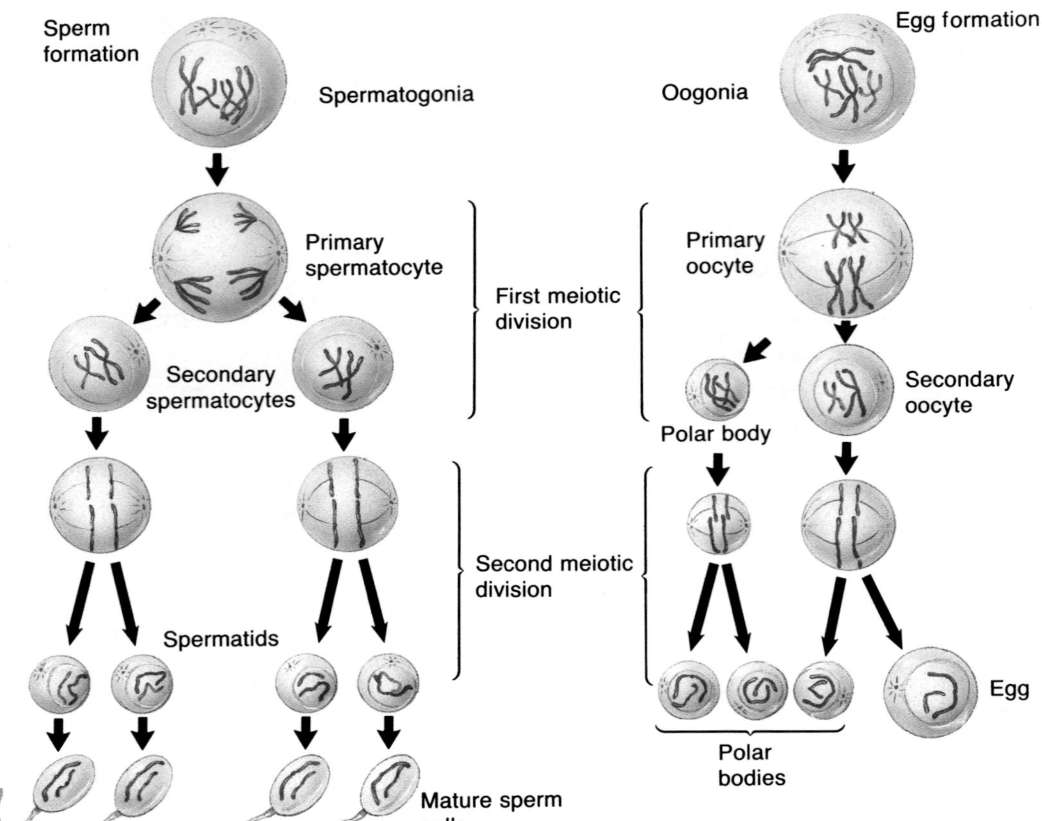

Figure 41.7

During sperm formation, the potential sperm cell divides by meiosis, resulting in the formation of four mature sperm. During egg formation, the potential egg divides by meiosis, resulting in the formation of one egg and three smaller bodies that eventually degenerate.

develop eggs. During this prenatal period, cells in her ovaries divide until the first stage of meiosis, prophase I, is reached. At this point, the cells go into a resting stage. *Figure 41.7* compares the process that produces sperm with the process that produces eggs. At birth, a female has all the potential eggs she will ever have.

How eggs are released

About once a month in a sexually mature female, the process of meiosis starts up again in one of the prophase I cells. The result is the production of an egg that ruptures from the ovary and passes into the oviduct. This process of the egg rupturing through the ovary wall and moving into the oviduct is called **ovulation.** Fertilization occurs in the oviduct if the egg and sperm unite. *Figure 41.8* shows the process leading to ovulation.

The Menstrual Cycle

All the activities of the human female reproductive system are part of a cycle. You learned that in a sexually mature female, ovulation occurs about once a month. Once the egg has been released, the remaining part of the follicle develops into a structure called the **corpus luteum.** The corpus luteum secretes the steroid hormone progesterone. Progesterone causes changes to occur in the lining of the uterus

Alternate Lab: Tracking Hormone Levels

Purpose

Logical-Mathematical Students will graph and analyze the patterns of the hormones of the menstrual cycle.

Materials

graph paper, colored pencils (red, yellow, blue, green), data sheet

Preparation

Prepare data sheets for students that include the following information.

Data

Day	2	4	6	8	10	12	14	16	18	20	22	24	26	28
LH	17	17	17	17	17	46	35	20	19	18	17	16	14	13
FSH	14	14	14	13	10	8	15	8	7	7	6	6	6	7
Estrogen	4	4	5	6	10	13	13	10	9	10	11	11	11	8
Progesterone	1	1	1	1	1	1	2	4	7	12	14	14	9	3

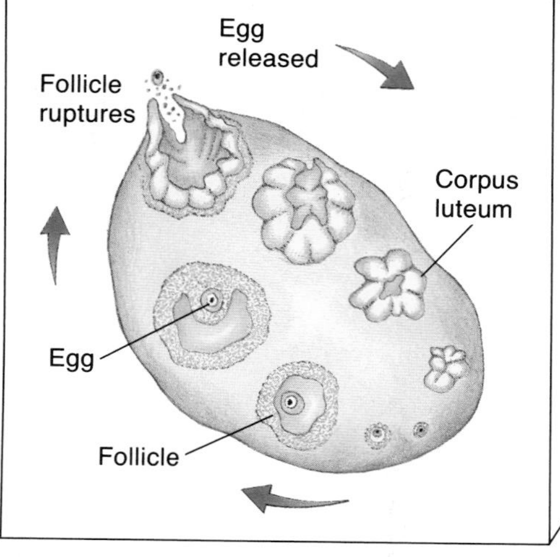

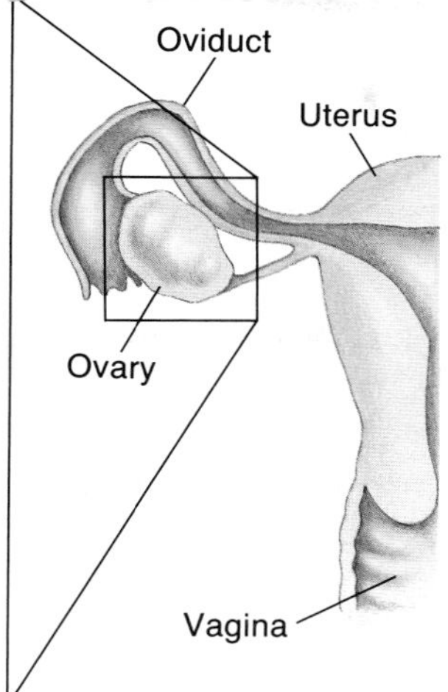

Figure 41.8

Once a female reaches puberty, follicles within her ovaries begin to mature. Usually, one follicle matures each month. As it matures, the follicle secretes the female hormone estrogen. During ovulation, the follicle ruptures, releasing the egg from the ovary. The remains of the follicle become the corpus luteum, which secretes some estrogen and the hormone progesterone.

that prepare it for receiving a fertilized egg. The series of changes in the female reproductive system that include producing an egg and preparing the uterus for receiving it is known as the **menstrual cycle.** The menstrual cycle begins during puberty and continues for 30 to 40 years until menopause. At that time, the female stops releasing eggs and hormone secretion slows.

The length of the menstrual cycle varies from female to female, with the average length being 28 days. If the egg is not fertilized, the lining of the uterus will be shed, causing some bleeding for a few days. The entire menstrual cycle can be divided into three phases: the flow phase, the follicular phase, and the luteal phase. The phases of the menstrual cycle are illustrated in *Figure 41.9.*

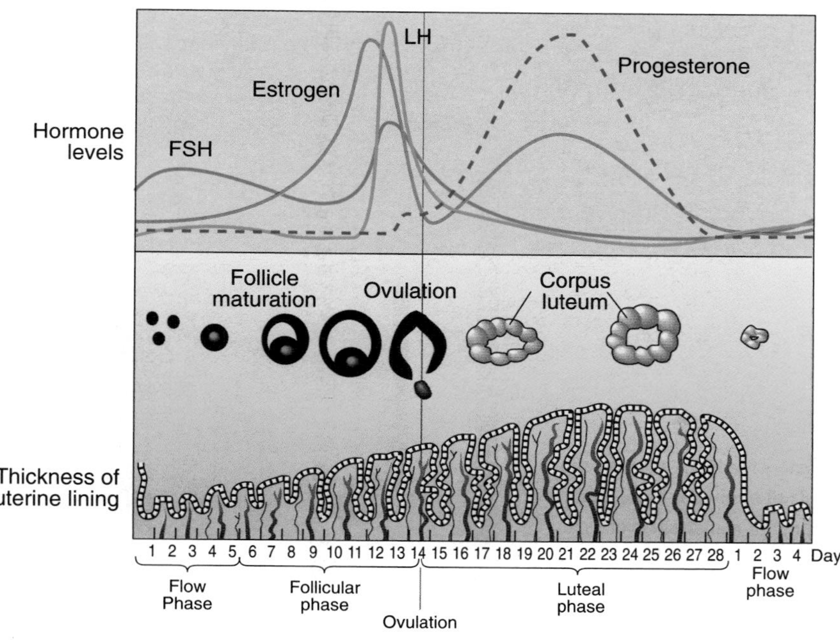

Figure 41.9

The three phases of the menstrual cycle correlate with hormone output from the anterior pituitary. Notice that estrogen levels increase during the follicular phase, and progesterone levels increase during the luteal phase. LH levels increase sharply at ovulation. Because the average cycle is 28 days, the graph of phases is based on 28 days, with the first day beginning the flow phase.

SECTION 41.1

Enrichment

Have students use medical dictionaries to find out the difference between estrous and menstrual cycles. Ask students to create a visual essay that explains the differences in these processes. L3

Chalkboard Example

LS **Visual-Spatial** Using two different colors of chalk, prepare a graph that shows the actions of the hormones LH and FSH during the menstrual cycle. Explain each phase of the cycle as it is graphed. Label the hormones represented by each colored line.

GLENCOE TECHNOLOGY

Videotape

The Secret of Life

Use the videotape *Sex and the Single Gene: Cell Development* to illustrate the developmental processes that determine an individual's sex.

Audiovisual

Show the video, *Reproductive Systems,* National Geographic.

Procedure

Give students the following directions.

1. Using the data provided, make a graph showing changes in the amounts of LH and FSH throughout the menstrual cycle. Use a yellow pencil for LH and a blue pencil for FSH.
2. Make another graph showing the levels of estrogen (in red) and progesterone (in green).

Analysis

Ask students the following questions.

1. Describe the pattern of each hormone. *LH peaks around day 12, estrogen peaks around day 12, and progesterone around day 21. FSH levels rise on days 2 through 8 and then drop off until spiking around day 14.*
2. Indicate with a dashed green line how the progesterone level would look if pregnancy occurred. *Check students' graphs for maintenance of high level.*

Assessment

Journal: Ask students to include a summary of the lab, their graphs, and answers to Analysis questions in their journals. L1

Reinforcement

List on the chalkboard various events that occur during the menstrual cycle. Such events may include rising and falling levels of LH and FSH, increased body temperature, menstrual flow beginning, and ovulation occurring. Ask volunteers to come to the chalkboard and label during which of the three phases each event occurs.

3 Assess

Check for Understanding

Display a diagram of the female reproductive system. Ask students to identify where an egg would be located within the system at various stages of the menstrual cycle. L1

Reteach

Provide students with diagrams of the male and female reproductive systems. Have students draw a line to show how sperm and an unfertilized egg travel through their respective reproductive systems. L1

Flow phase

Day 1 of the menstrual cycle is the day menstrual flow begins. Menstrual flow is the shedding of the blood, tissue fluid, mucus, and epithelial cells that made up the lining of the uterus. This flow passes from the uterus through the cervix and the vagina to the exterior. Generally, menstrual flow ends by day 5 of the cycle. During the flow phase, FSH is beginning to rise, and a follicle in one of the ovaries is continuing its development as meiosis of the prophase I cell proceeds.

Follicular phase

The second phase of the menstrual cycle varies in length more than the other phases. It lasts from day 6 to about day 14 in a 28-day cycle. FSH and LH released from the pituitary stimulate the ovary to produce more estrogen, and this increase in estrogen stimulates the repair of the lining of the uterus. The lining undergoes mitosis and begins to thicken. The increase in estrogen feeds back to the hypothalamus, which decreases its stimulation of the pituitary. The follicle continues to develop and mature. The presence of estrogen also stimulates a sharp increase in the release of LH. About day 14, this increase causes the follicle to rupture, releasing the egg into the oviduct. At this time, a detectable change in the female's body temperature occurs—about +0.5°F. In addition, the cervical cells produce large amounts of mucus in response to the increasing levels of estrogen. Some females also experience discomfort in the area of one or both of the ovaries around the time of ovulation.

Luteal phase

The last stage of the menstrual cycle, from days 15 to 28, is named for the corpus luteum. LH stimulates

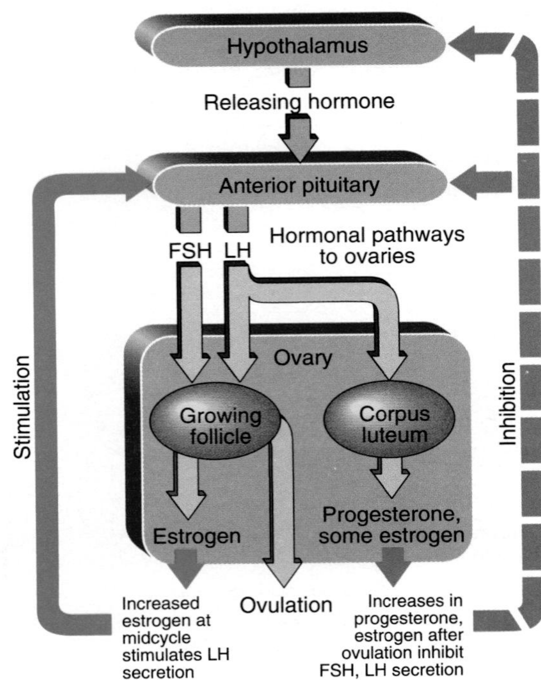

Figure 41.10

Negative feedback controls the levels of hormones in the female during her menstrual cycle. If fertilization occurs, this cycle is broken and the levels of hormones change.

STUDENT JOURNAL

Diagramming the Menstrual Cycle Have students create flowcharts or concept maps showing the sequence of events that occur during each phase of the menstrual cycle. Instruct students to include information about all hormones involved in the regulation of the cycle. L1 LS

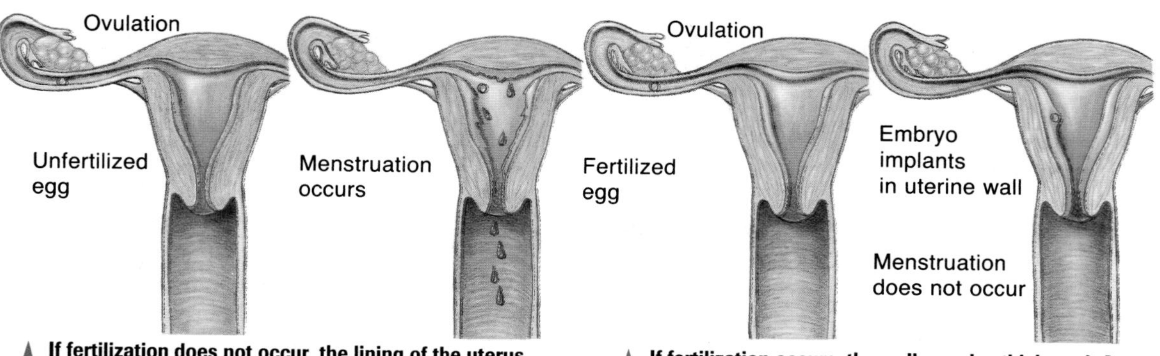

If fertilization does not occur, the lining of the uterus deteriorates because blood vessels in the uterine wall constrict, reducing the blood supply to the cells.

If fertilization occurs, the wall remains thickened. A fluid rich in nutrients for the embryo is secreted.

Figure 41.11

Events in the uterine wall after ovulation depend on whether fertilization has or has not occurred.

the corpus luteum to develop from the ruptured follicle. The corpus luteum produces both progesterone and estrogen, but progesterone is the dominant hormone during the luteal phase. This hormone causes the uterine lining to thicken, increases its blood supply, and accumulates fat and tissue fluid. These changes correspond to the anticipated arrival of a fertilized egg. Through negative feedback, progesterone prevents the production of LH.

As shown in *Figure 41.10,* if the egg is not fertilized, the rising levels of progesterone and estrogen from the corpus luteum inhibit the release of FSH and LH by inhibiting the hypothalamus. This negative feedback causes the corpus luteum to degenerate, and the progesterone levels drop. The thick lining of the uterus then begins to be shed. *Figure 41.11* shows what happens to the lining of the uterus if fertilization occurs.

Section Review

Understanding Concepts

1. Describe the pathway of an unfertilized egg as it travels through the female reproductive system.
2. Summarize the negative-feedback system in a male, including the roles of the hypothalamus and pituitary gland.
3. What is the function of the menstrual cycle?

Thinking Critically

4. What might happen to sperm production if a male has a high fever?

Skill Review

5. **Interpreting Scientific Illustrations** Study *Figure 41.1.* Using the terms *dorsal, ventral, anterior, posterior, superior,* and *inferior,* describe where the epididymis is in relation to the vas deferens. Describe where the prostate is in relation to the testes. For more help, refer to Thinking Critically in the *Skill Handbook.*

SECTION 41.1

Extension

Have students research the development and specialization of sperm cells within the testis and epididymis. Ask students to prepare a report of their findings. L2

Assessment

Performance: Ask students to label diagrams of the female and male reproductive systems. Ask students to indicate, using colored pencils and arrows, the pathways of sperm and egg through these systems. L1

4 Close

Discussion

Students may have misconceptions or questions about the reproductive process. Have each student anonymously submit three questions written on index cards. Read the questions aloud and have volunteers suggest possible answers. Elaborate on or clear up any misconceptions students may have. As an alternative, you may wish to invite the school nurse or a physician to come to class to address the questions.

Answers to Section Review

1. The egg ruptures from the ovary and is swept into the oviduct. Cilia beat, moving the egg along toward the uterus. The unfertilized egg passes through the vagina to the outside of the body.
2. Testosterone levels and cells in the testes signal the hypothalamus to increase or decrease the release of LH and FSH, which in turn control testosterone and the cells in the testes.
3. to produce an egg and prepare for possible pregnancy

Thinking Critically

4. Sperm may be killed by the high fever.

Skill Review

5. The epididymis is inferior to the vas deferens. Students may also say that it is ventral. The prostate is superior and dorsal to the testes.

SECTION

41.2

Prepare

Key Concepts

This section summarizes fertilization and implantation of the egg, and development of the human fetus. The last part of the section discusses the importance of genetic counseling.

Block Scheduling

Look for this symbol for strategies that are useful in a block scheduling format. For more information on block scheduling, refer to Section 41.2 in the **Lesson Plans** booklet.

Materials

- Save articles on unusual births for the Focus Discussion.
- Order and prepare sea urchin sperm and eggs for the Minilab.
- Gather photos for the Display.

1 Focus

Bellringer

Before presenting the lesson, display **Section Focus Master 95** on the overhead projector and have students answer the accompanying questions. L1 SAE

Discussion

Using recent newspaper and magazine articles, initiate a class discussion of how a woman's body changes during pregnancy and the importance of prenatal care. Elicit what factors might prevent women from receiving proper prenatal care and what the results of such neglect might be.

SECTION

41.2 Development Before Birth

Section Preview

Objective

Summarize the events during each trimester of pregnancy.

Key Terms

implantation
umbilical cord
genetic counseling
amniocentesis

What do you have in common with a period at the end of a sentence? You were once about the same size. You started out life as a single, microscopic fertilized egg. That one cell went through numerous mitotic divisions to produce the trillions of cells that make up your body today. It all began when an egg from your mother was fertilized by a sperm from your father.

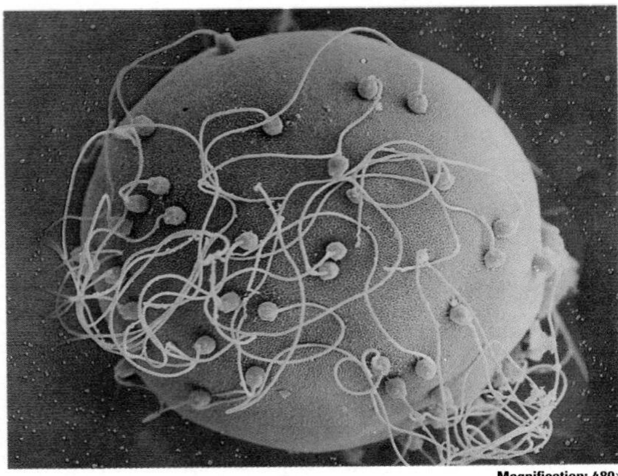

Fertilization and Implantation

After an egg ruptures from a follicle, it is able to stay alive for about 24 hours. For fertilization to occur, sperm must be present in the oviduct at some point during those first hours after ovulation. Sperm enter the female's reproductive system when strong, muscular contractions ejaculate semen from the male's body. As many as 350 million sperm are forced out of the male's penis and into the female's vagina during intercourse. Since sperm can live for 48 hours after ejaculation, fertilization can occur if intercourse occurs anywhere from a few days before to a day after ovulation.

One sperm plus one egg

How is it possible that of the millions of sperm released into the vagina during ejaculation, only one fertilizes the mature egg? One reason is that the fluids secreted by the vagina are acidic and destroy most of the delicate sperm. Yet, some sperm survive because of the buffering effect of semen. The surviving sperm swim up the vagina into the uterus. Of the sperm that reach the uterus, only a few hundred pass into the two oviducts. The egg is present in one of them.

Recall that the head of the sperm contains enzymes that help the sperm penetrate the egg. As the sperm penetrates the egg, it loses its midpart and tail. Once one sperm penetrates the egg, the egg's membrane changes its electrical charge, thus preventing other sperm from entering. After the one sperm penetrates the egg, its nucleus combines with the egg's nucleus to form a zygote.

Program Resources

Section Focus Master 95 L1 SAE
Reinforcement and Study Guide, p. 169 L1
Biolab and Minilab Worksheets, pp. 163-164 L1
Laboratory Manual, pp. 253-256; 257-260 L2
Reteaching Transparencies 41a, 41b and **Masters** L1 SAE

The fertilized egg travels to the uterus

As the zygote passes down the oviduct, it begins to divide by mitosis. During its journey, the zygote obtains nutrients from fluids secreted by the mother. By the sixth day, the zygote passes into the uterus. Continuous cell divisions cause a hollow ball of cells called a blastocyst to form, *Figure 41.12*. *Blastocyst* is the term used when discussing human embryonic development. Recall from Chapter 28 that the term *blastula* is used for the embryonic development of other animals.

The blastocyst attaches to the uterine lining seven to eight days after fertilization. This attachment of the blastocyst to the lining of the uterus is called **implantation.** A small, inner mass of cells within the blastocyst will soon become a human embryo.

MiniLab

Are sperm attracted to eggs?

Most animals that reproduce by external fertilization live in water. These animals deposit their sperm and eggs into the water surrounding them. The sperm must swim to the egg, and these tiny cells must somehow meet. One adaptive advantage to ensure that the egg and sperm meet is that these animals give off thousands of sex cells at one time. A large number of sex cells increases the odds that one of the eggs will be fertilized. Another possible adaptation would be attraction. Certainly, if the eggs attracted the sperm, it would help the sperm "find" the egg.

Procedure

1. Place a dropperful of sea urchin eggs on a microscope slide.
2. While observing the eggs under the microscope, add a drop of sea urchin sperm to the eggs.

Analysis

1. Describe the motion of the individual sperm.
2. What cell structures are involved in providing energy for the sperm motion?
3. Are the sperm attracted to the eggs? How do you know?

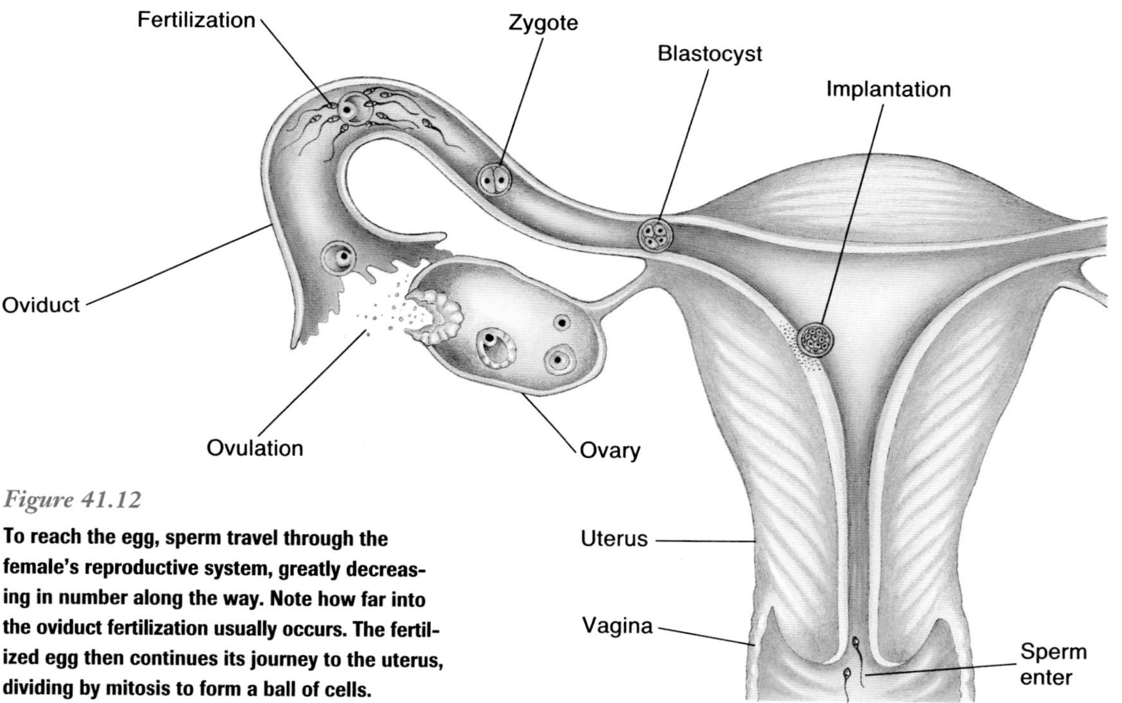

Figure 41.12

To reach the egg, sperm travel through the female's reproductive system, greatly decreasing in number along the way. Note how far into the oviduct fertilization usually occurs. The fertilized egg then continues its journey to the uterus, dividing by mitosis to form a ball of cells.

2 Teach

MiniLab

Purpose

Visual-Spatial Students will determine whether sperm are attracted to eggs.

Process Skills

observe and infer, recognize cause and effect, experiment, analyze

Teaching Strategies

- Fertilize some of the eggs a few hours before class so students can observe various stages of cleavage. If the eggs are kept at room temperature, the first cleavage takes about 50 to 60 minutes; second cleavage, one and one-half hours; third cleavage, one and three-quarter hours; and the blastula forms after about six hours.

Expected Results

Sperm will collect around the egg, indicating that they are attracted to the egg.

Analysis

1. The sperm swim like tadpoles with tails whipping back and forth.
2. mitochondria
3. Yes, they swim toward and gather around the egg.

Assessment

Journal: Have students write a summary of the Minilab and place it along with the answers to the Analysis questions in their journals. L1

STUDENT JOURNAL

Life Before Birth Ask students to write an imaginary story about what they think it is like to be an embryo or fetus in the womb. The article "Sensing in the Womb," by Jacqueline S. Palmer, *The American Biology Teacher*, Vol. 49, No. 7, October 1987, may be a helpful resource for students. L1 LS

GLENCOE TECHNOLOGY

Videodisc

Biology: The Dynamics of Life
Disc 2, Side 1
Human Fertilization (Ch. 41)

SECTION 41.2

Display

LS **Visual-Spatial** Prepare a bulletin board display that shows the development of the human fetus. Use illustrations or photographs from medical journals or from biological supply houses. Refer to the display as various stages of development are discussed.

Bioethics

Embryo Ownership
Questions have arisen regarding to whom fertilized human embryos belong. A couple who underwent in vitro fertilization had embryos frozen and stored. Later, they were divorced. During the divorce proceedings, the question was raised as to whom, if anyone, the embryos should belong. Initiate a class debate regarding this issue. Ask students what should happen to embryos when a couple divorces. If the embryos are implanted and a baby results, should one parent be permitted to sue the other for child support?

GLENCOE TECHNOLOGY

Videodisc

The Infinite Voyage: The Geometry of Life
Development of the Egg After Fertilization (Ch. 3)

The Infinite Voyage: The Keepers of Eden
Cincinnati Zoo's Cat House: Innovative Breeding Techniques (Ch. 6)

Embryonic Membranes and the Placenta

You learned in Chapter 34 about the importance of the amniotic egg to the evolutionary advancement of animals. Membranes that are similar to those of the amniotic egg form around the human embryo, protecting and nourishing it. The amnion is the thin, inner membrane filled with a clear, watery fluid called amniotic fluid. Amniotic fluid serves as a shock absorber and helps regulate the body temperature of the developing embryo.

The allantois membrane is an outgrowth of the digestive tract of the embryo. Blood vessels of the allantois form the **umbilical cord,** a ropelike structure that attaches the embryo to the wall of the uterus. The chorion is the outer membrane that surrounds the amniotic sac and the embryo within it. Eventually, part of the chorion forms the placenta, as shown in *Figure 41.13.*

Exchange between embryo and mother

To survive and grow, the embryo must obtain the proper nutrients and eliminate the wastes its cells produce. The placenta delivers these nutrients to the embryo and carries its wastes away.

In the placenta, blood vessels from the mother's uterine wall lie close to the blood vessels of the embryo's chorionic villi. They are close, but they do not connect to one another. Instead, oxygen and nutrients transported by the mother's blood diffuse into the blood vessels of the chorionic villi in the placenta. These vital substances are then carried by the blood in the umbilical cord to the embryo. In turn, waste products from the embryo travel in the umbilical blood vessels to the placenta. Here they diffuse out of the vessels in the chorionic villi into the blood of the mother. These waste products are then removed by the mother's excretory system.

Figure 41.13
About 14 days after fertilization, fingerlike projections of the chorion, called chorionic villi, begin to grow into the uterine wall (left). Together, the chorionic villi containing the embryo's blood vessels and a portion of the lining of the uterus form the placenta, the point of exchange between the mother and the embryo. This embryo is shown at approximately seven weeks of development (right).

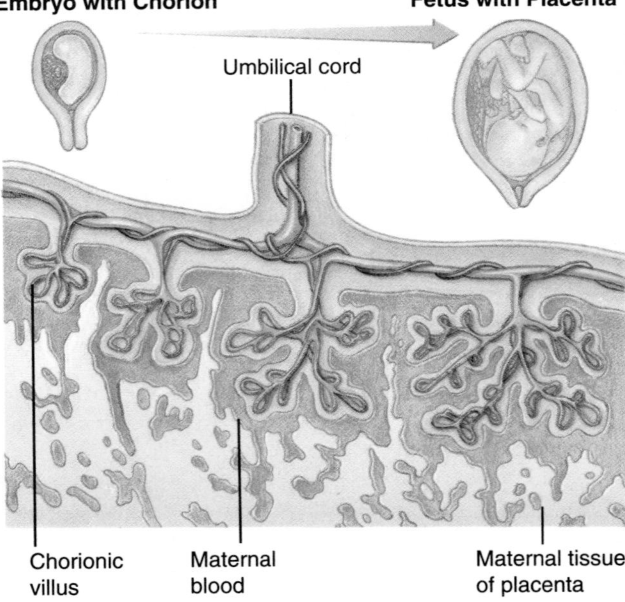

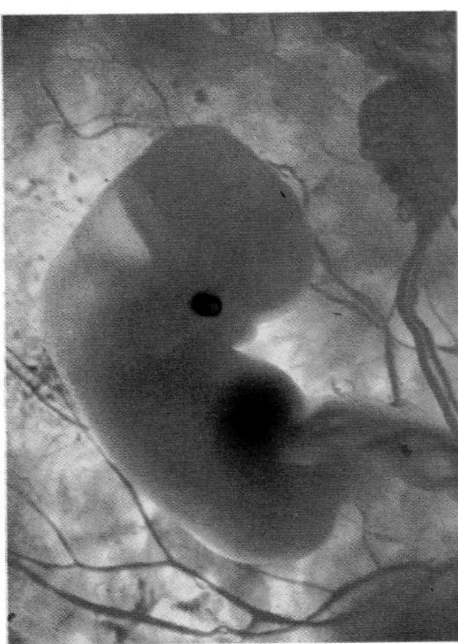

Meeting Individual Needs

Learning Disabled and Hearing Impaired
The review tutorial *Reproduction, Growth, and Development* from Queue may be helpful for hearing impaired students (Apple II, MS-DOS, and Mac). A program called *Understanding Systems of the Human Body,* also from Queue and designed for less-able students, may be helpful. (Apple II, MS-DOS) **L1** **LS**

Follow the link for this chapter on the Glencoe Homepage at **www.glencoe.com/sec/science** to find out more about reproduction and development.

Hormonal maintenance of pregnancy

Remember that estrogen, and especially progesterone, cause the uterine lining to thicken in preparation for implantation. Once the blastocyst implants, the chorion membrane of the embryo starts to secrete the hormone chorionic gonadotropin. This hormone keeps the corpus luteum alive so that it continues to secrete progesterone. By the third month, the placenta has taken over for the corpus luteum, secreting enough estrogen and progesterone to maintain the pregnancy.

Fetal Development

When you think of an embryo developing within the mother's body, you probably don't realize that the development involves three different processes: growth, development, and cellular differentiation. Growth refers to the actual increase in the number of cells. But the cells must also move within the embryo's body and arrange themselves into specific organs. In addition, the cells become specialized to perform specific tasks and functions. All three processes begin with fertilization.

Pregnancy in humans usually lasts about 280 days, calculated from the first day of the mother's last menstrual period. The baby actually develops for about 266 days, calculated from the time of fertilization to birth. This time span is divided into three trimesters, each equal to three months. Each trimester brings significant advancement in the development of the embryo.

First trimester—Body systems form

During the first trimester, all the body systems of the embryo begin to form. A five-week embryo is shown in *Figure 41.14.* During this time of development, the woman may not even realize she is pregnant. Yet, the first seven weeks after fertilization are critical because during this time, the embryo is more sensitive to outside influences, such as alcohol, smoking, and other drugs that cause malformations, than at any other time.

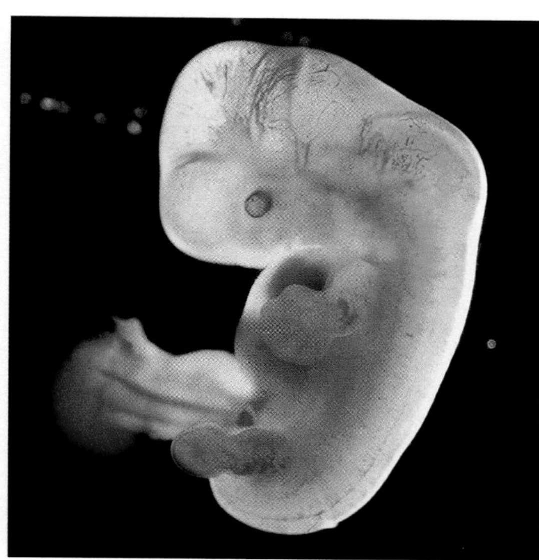

7 mm

Figure 41.14

A five-week-old embryo is about 7 mm long. The heart—the large, red, circular structure protruding out of the embryo—begins as two muscular tubes. It starts to beat around the 25th day of development. You can see the arms and legs beginning to bud, as well as the head. Note that the tissue that will form the eyes is beginning to darken.

Science, Technology, and Society

SEM Microscopy of the Oviduct

A series of pictures has shown how the oviduct permits transport of sperm and egg. The photos indicate that for three to four days following ovulation, mucus forms a block that holds the egg in the top portion of the oviduct. The mucus serves as a medium for sperm transport and protects the sperm from the cilia that will later sweep the egg toward the uterus. The process appears to be controlled by estrogen and perhaps progesterone.

History Connection

In Vitro Fertilization

On July 25, 1978, the first authenticated success of in vitro fertilization occurred with the birth of Louise Brown. This method was carried out because Mrs. Brown had blocked oviducts.

To carry out the process, an egg was removed from Mrs. Brown's ovary at the time of ovulation, fertilized by her husband's sperm, allowed to divide to the 16-cell stage, and then implanted in her uterus. In 1991, 4000 in vitro births occurred in the United States, and 16 000 occurred worldwide.

Software

Reproduction, Development, and Genetics, Intellectual Software.

GLENCOE TECHNOLOGY

Videodisc

Biology: The Dynamics of Life
Disc 2, Side 1
Fetal Development (Ch. 43)

PORTFOLIO

Body System Formation Assign each student a body system. Have students research the fetal development of the assigned body system and prepare a visual display that traces the development of the system. Have students include labels and captioned summaries of the changes that occur at different stages of development. L3 P IS

MiniLab

Purpose

LS Logical-Mathematical Students will graph and evaluate the growth of a human embryo.

Process Skills

make and use graphs, interpret data, analyze

Teaching Strategies

- Be sure students recognize the change in units between the 6th and 7th week of development.

Expected Results

Students' graphs will show that the embryo grows the fastest during the early periods of development.

Analysis

1. The embryo doubles in size during two 1-week periods—3 to 4 weeks and 7 to 8 weeks.
2. All body systems are beginning to form.
3. 5th month

Assessment

Journal: Have students write a summary of the Minilab and place it with the graph and the answers to the Analysis questions in their journals. L1

Audiovisual

Show the video, *The Miracle of Life*, TimeLife.

MiniLab

How long is an embryo?

You started out as a single cell. That cell divided by the process of mitosis to produce body systems that were able to maintain an independent existence outside your mother's uterus. During the time you were in your mother's uterus, great changes occurred. One of these changes involved your growth in length.

Procedure

1. Prepare a graph that plots time on the horizontal axis and length in centimeters on the vertical axis. Equally divide the horizontal axis first into nine months. Then equally divide each of the first three months into four weeks.
2. Plot the data in the table on your graph.

Growth in Length of a Fetus

	Time after fertilization	Size
First Trimester	3 weeks	3 mm
	4 weeks	6 mm
	6 weeks	12 mm
	7 weeks	2 cm
	8 weeks	4 cm
	9 weeks	5 cm
	3 months	7.5 cm
Second Trimester	4 months	15 cm
	5 months	25 cm
	6 months	30 cm
Third Trimester	7 months	35 cm
	8 months	40 cm
	9 months	51 cm

Analysis

1. When is the fastest period of growth?
2. What structures are developing during this period of growth?
3. At what point does growth begin to slow down?

By the eighth week, all the body systems are present, and the embryo is now referred to as a fetus. You can see this stage of fetal development in *Figure 41.15.* At the end of the first trimester, the fetus weighs about 28 g and is about 7.5 cm long from the top of its head to its buttocks. The sex of the fetus can be determined by the appearance of the external sex organs.

Second trimester—A time of growth

For the most part, fetal development during the next three months is limited to body growth. Growth is rapid at the beginning of the second trimester, but then slows by the beginning of the fifth month. At this point, the fetus can survive outside the uterus with a great amount of medical assistance, but the mortality rate is high. The fetus's body metabolism cannot yet maintain a constant temperature, and its lungs have not matured enough to provide a regular respiratory rate. *Figure 41.16* shows a fetus during the second trimester.

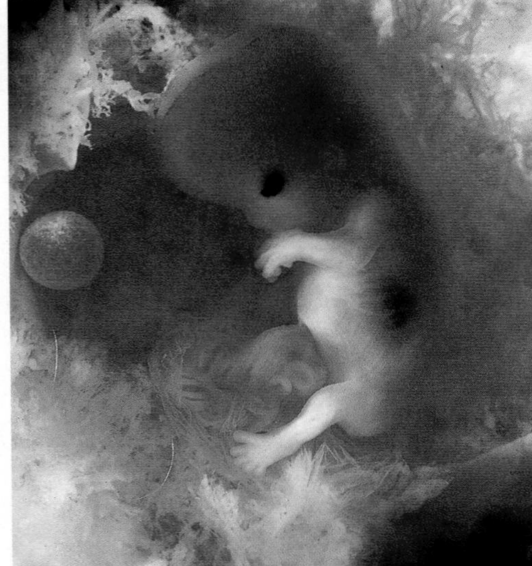

Figure 41.15
In a two-month-old fetus, the two muscular tubes have fused together to form a four-chambered heart. The limbs are beginning to elongate, and the fingers and toes appear. Notice how the eyelids have formed, and the face looks distinctly human. Bones are beginning to harden, and nearly all muscles have appeared. As a result, the fetus can move spontaneously.

CULTURAL DIVERSITY

Rites of Passage

In many cultures and religions, traditional celebrations mark the transition from childhood to adulthood. For example, in Mexican tradition, a girl celebrates her transition from childhood to adulthood on her fifteenth birthday in a celebration known as *quinceanera.* Discuss with students examples of initiation rites in various cultures. You may wish to have students research topics by asking them to identify a culture to determine whether that culture has a traditional rite of passage. L1

Figure 41.16

The facial features of a second-trimester fetus are well formed. Its skin is covered by a white, fatty substance that protects it against the amniotic fluid. Kicks and movements are commonly felt by the mother during this time as the fetus exercises its muscles. It can also suck its thumb, preparing the muscles that will allow the baby to feed later. By the end of the second trimester, the fetus weighs about 650 g and is about 34 cm long.

Third trimester—Continued growth

During the last trimester, the mass of the fetus more than triples. By the beginning of the seventh month, the fetus kicks, stretches, and moves freely within the amniotic cavity, somewhat like the astronaut in the space shuttle at the beginning of the chapter. During the eighth month, its eyes open.

During the last weeks of pregnancy, the fetus has grown large enough to fill the space within the embryonic membranes. Sometime in the ninth month, the fetus rotates its position so that its head is down, partly as a result of the shape of the uterus, but also because the head is the heaviest part of the body. By the end of the third trimester, the fetus weighs about 3300 g and is about 51 cm long. All of its body systems have developed, and it can now survive independently outside the uterus.

Genetic Counseling

Most expectant parents desperately want just one thing—a healthy, normal baby. With our increasing knowledge of human heredity and advancing technology, determining that a newborn will be healthy and normal is much more possible today than it was in the past.

Genetic disorders can be predicted

Generally, people in the industrialized nations are aware of the possible genetic disorders that can affect a child. For many, this awareness has made them eager to know whether a potential child will be healthy. As advances have been made in the detection and treatment of genetic disorders, including those you studied in Chapter 15, the demand for genetic services, especially prenatal testing, has increased.

Time Line

LS **Logical-Mathematical** Ask students to prepare a time line that plots the changes that take place in a developing embryo/fetus. Encourage students to use different-colored pencils to indicate clearly where each trimester of pregnancy begins and ends. L1

Science, Technology, and Society

Survival of Premature Infants

Great progress has been made in helping severely premature infants survive. At present, babies born before 23 or 24 weeks of pregnancy are the most unlikely to survive. Prior to this stage of development, babies are unable to breathe, even on a respirator. Another problem inhibiting the survival of babies born before 23 or 24 weeks involves the kidneys. Such babies have immature kidneys that cannot properly regulate urine.

Researchers have suggested that an artificial placenta may someday take the place of a baby's immature lungs and kidneys. The artificial placenta would function like a heart-lung machine, making it possible to save fetuses born before the 24th week.

GLENCOE TECHNOLOGY

The Secret of Life
Use the videotape *Tinkering with Our Genes: Genetic Medicine* to illustrate ethical considerations of genetic testing.

PROJECT: Modeling Fetal Development

Have students use modeling clay to show the changes a human egg undergoes from the period of fertilization to implantation. Students should label the zygote and embryo stages, and summarize on index cards the changes and approximate time period involved between changes. Have students work in groups to carry out the project by having gifted students work with less-able students. If necessary, have students refer to the sections on embryonic development in Chapter 28. L1 COOP LEARN LS

SECTION 41.2

A BROADER VIEW

Prenatal Care

Purpose
Students will evaluate the importance of prenatal care.

Teaching Strategies

- Explain that the United States has a fairly high rate of infant mortality, especially when compared to other developed nations. Initiate a class discussion about why this is true.
- Ask students to use information from this feature to write a summary about why early prenatal care is important.
- Invite a physician to speak to the class about the importance of prenatal care.

Thinking Critically
At-risk mothers are usually without health insurance. Programs that expand public health, especially in impoverished areas where the infant mortality rate is highest, should help.

Different Viewpoints in Biology

Doctors have different viewpoints about which prenatal tests should be offered and how often they should be used. For example, obstetricians estimate that more than half of all pregnant women in the United States receive a sonogram. Yet, medical questions concerning the safety of this procedure exist.

The FDA advises that sonograms be done only when there is a history of bleeding or birth defects, or some other medical reason for the procedure. However, this advice is not consistent with the current claim that sonograms provide valuable information even in low-risk pregnancies.

A BROADER VIEW

Prenatal Care

Medical experts agree that early prenatal care can save lives. By identifying and monitoring possible complications—such as poor nutrition, multiple fetuses, and maternal high blood pressure—healthier babies are born and survive their first year of life. In addition, expectant mothers can be treated for medical problems that may result from pregnancy.

The U.S. infant mortality rate Low birth weight (less than 5.5 pounds) is the primary contributor to the nation's infant mortality rate. At 9.1 deaths per 1000 births, America currently ranks 19 among all other nations in infant deaths. The infant mortality rate is also influenced largely by the number of premature births. Babies born before the 37th week of gestation are considered to be premature. These babies, if they survive, also have an increased risk for health problems later in life.

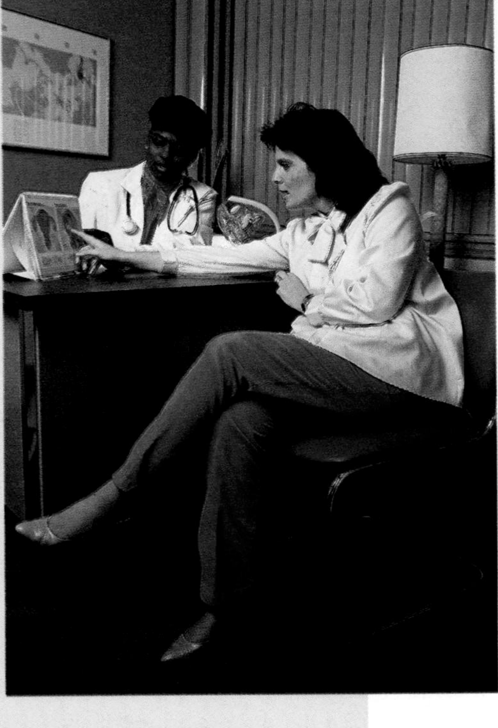

Prenatal checkups Experts believe low birth weight or prematurity can be prevented through early and regular prenatal care during pregnancy. Expectant mothers may need to make lifestyle changes, such as improving their nutrition and avoiding smoking, drugs, and alcohol—all of which contribute to low birth weight, premature birth, and birth defects. Blood and urine tests taken during regular prenatal checkups monitor both baby and mother for common problems such as iron-deficiency, known as anemia, and Rh incompatibility. Both conditions are treatable.

Thinking Critically
Unfortunately, those who need prenatal care the most—poor women, teenagers, and substance abusers—are the ones least likely to get it. What might be some of the factors that contribute to the problems these women have in getting proper prenatal care, and what can be done to improve the care they get?

Most people do not even think about genetics until they are considering having children or are already expecting one. If there is no history of genetic disorders in either family of the prospective parents, there may be no need for genetic services. However, if one or both prospective parents have a family history of some genetic disorder, both will likely want to get additional information before proceeding with a pregnancy.

The job of a genetic counselor

Couples who seek information from trained professionals about the probabilities of hereditary disorders and what can be done if they occur are receiving **genetic counseling.** Genetic counselors have a medical background with additional training in genetics. Sometimes, a team of professionals works with prospective parents. The team may include geneticists, clinical psychologists, social workers, and other consultants.

How do genetic counselors go about their work? First, they develop medical histories of both families. These histories may include pedigrees, biochemical analyses of blood, and karyotypes. Once the counselor has collected and analyzed all the available information, he or she explains the risk factors of having offspring with genetic disorders. If the probabilities of having a severely affected child are high, a couple must decide whether or not to have children of their own.

PORTFOLIO

The Story of Life Have students write a story or poem using any topic in this section. Encourage students to include some of the biology of this section in their writings. L2 P LS

Prenatal testing

How can prospective parents learn whether a child they are expecting is developing normally? A number of prenatal tests now exist that give prospective parents and their doctors valuable information about the fetus while it is in the mother's uterus. In fact, prenatal testing can detect more than 100 genetic disorders.

Amniocentesis provides cells

A common prenatal procedure that allows genetic analysis of fetal cells is **amniocentesis.** In amniocentesis, a long needle is passed through the abdominal wall of the pregnant woman to withdraw a small amount of the fluid that surrounds the 14- to 16-week-old fetus. The fluid contains cells that have sloughed off from the fetus. Since only a few fetal cells are in the fluid, the results of the tests are delayed for several weeks while the cells grow and multiply in a cell culture. Once enough cells for testing purposes are grown, a karyotype, biochemical tests, or both procedures are carried out. In *Figure 41.17,* you can see how amniocentesis is done. This procedure is often used to determine whether a child will have a genetic disorder such as Down syndrome.

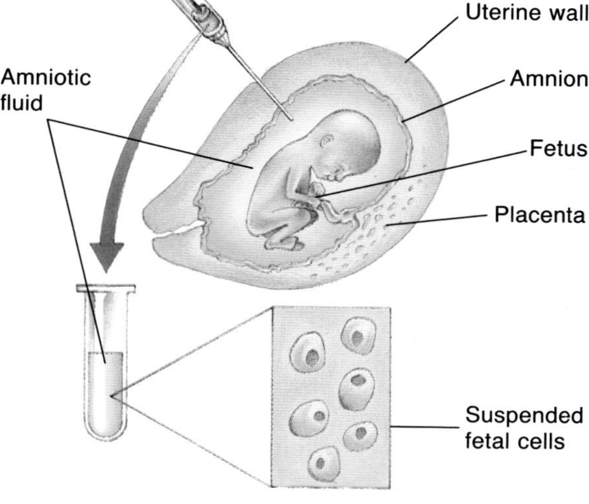

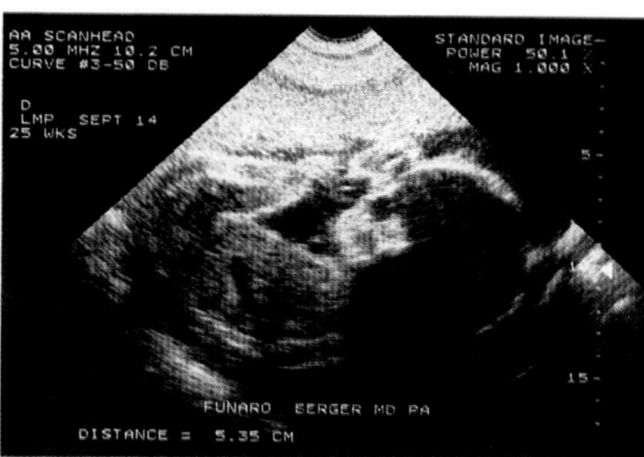

Figure 41.17
The process of amniocentesis can be used to diagnose chromosomal and genetic problems; however, the test is not risk free (left). Infection or injury to the fetus may occur. Ultrasonic techniques help locate the fetus during testing so that injury can be avoided (right).

Section Review

Understanding Concepts

1. What changes occur in the zygote as it passes along the oviduct and into the uterus?
2. What is the function of the placenta?
3. Why is an embryo most vulnerable to drugs and other harmful substances taken by its mother when it is between two and seven weeks old?

Thinking Critically

4. Compare the functions of human embryonic membranes with those inside a bird's egg.

Skill Review

5. **Sequencing** Prepare a table listing the events in the three trimesters of pregnancy. For more help, refer to Organizing Information in the *Skill Handbook.*

3 Assess

Check for Understanding

Ask students to describe orally how a fertilized egg changes after it is implanted in the uterus. L1

Reteach

Photocopy and distribute sketches showing fetal development in the later stages of pregnancy. Have students label the embryonic membranes and the placenta as you discuss them. L1

Extension

Ask students to interview their mothers and discuss the circumstances of their births. Have students determine what types of prenatal care and tests their mothers received during their pregnancies. Ask students to record the information they obtain in their journals. L1

Assessment

Knowledge: Ask students to write a summary of the role of the embryonic membranes and hormones during pregnancy. L1

4 Close

Discussion

Have students summarize the highly ordered development of the human fetus. Discuss how changes in this orderly development may result in birth defects. L1

Answers to Section Review

1. The zygote divides as it passes down the oviduct. The cells organize themselves into a small hollow ball.
2. The placenta is an area of exchange between the mother's and embryo's blood. Nutrients and wastes are exchanged.
3. This is when all body systems are forming.

Thinking Critically

4. Both types of membranes have the same functions. They help protect the developing embryo and allow for the exchange of materials.

Skill Review

5. First Trimester: All the body systems start forming. Heart begins beating around day 25. The hands, fingers, and toes form. The eyelids form and close. The fetus begins to move. Second Trimester: Growth and maturation of fetal tissues occurs. The skin is covered by a white, fatty substance. The mother becomes aware of fetal movements. Third Trimester: Growth continues. The fetus moves freely. The eyes open.

SECTION

41.3

Prepare

Key Concepts

The three stages of birth—dilation, expulsion, and the placental stage—are described. The developmental stages of humans after birth are also summarized.

Block Scheduling

Look for this symbol for strategies that are useful in a block scheduling format. For more information on block scheduling, refer to Section 41.3 in the **Lesson Plans** booklet.

Materials

- Gather additional pictures to add to the Display begun in Section 41.2.

1 Focus

Bellringer

Before presenting the lesson, display **Section Focus Master 96** on the overhead projector and have students answer the accompanying questions. L1 SAE

Discussion

Ask students to share their views on aging. Elicit what changes people can expect to undergo as they age and why students believe such changes occur. Have students explain whether aging always involves reduced mental ability or if other factors are involved in this occurrence.

SECTION

41.3 Birth, Growth, and Aging

Section Preview

Objectives

Describe the three stages of birth.

Summarize the developmental stages of humans after they are born.

Key Term

labor

How often have you heard the comment, "My, you've grown since I last saw you"? It may seem that you have grown a lot in the last few years. Yet the most rapid stage of growth in the life cycle of a human takes place within the uterus. From fertilization to birth, mass increases about 3000 times. Even so, although growth slows after birth, changes certainly do not stop. The human body changes throughout life.

Birth

Birth is the process by which a fetus is pushed out of the uterus and the mother's body and into the outside world. What triggers the onset of birth is not fully understood. However, it occurs in three recognizable stages: dilation, expulsion, and the placental stage.

Dilation of the cervix

The physiological and physical changes a female goes through to give birth are called **labor.** Labor begins with a series of mild contractions of the uterine muscles. These contractions are stimulated by oxytocin, a peptide hormone released by the posterior lobe of the pituitary. The contractions open, or dilate, the cervix to allow for passage of the baby, as shown in *Figure 41.18.* As labor progresses, the contractions begin to occur at regular intervals and intensify as the time between them shortens. This first stage of labor is usually the longest, sometimes lasting up to 24 hours.

Expulsion of the baby

Expulsion occurs when the involuntary uterine contractions become so forceful that they push the baby through the cervix into the birth canal. The mother assists with expelling the baby by contracting her abdominal muscles in time with the uterine contractions. As shown in *Figure 41.18,* the baby moves from the uterus, through the birth canal, and out of the mother's body. The expulsion stage usually lasts from 20 minutes to an hour.

Program Resources

Section Focus Master 96 L1 SAE
Reinforcement and Study Guide, p. 170 L1
Biolab and Minilab Worksheets, pp. 165-166 L1
Concept Mapping, p. 41 L1
Tech Prep Applications, pp. 51-52 L2

A When the opening of the cervix is about 10 cm, it is fully dilated. Usually, the amniotic sac ruptures and releases the amniotic fluid through the vagina, which is referred to as the birth canal.

B The birth canal is the passage through which the baby will be expelled from the mother's body, or born. The baby's head rotates as it moves through the birth canal, making it easier for its body to be expelled.

C During the placental stage, the placenta and umbilical cord are expelled from the mother's body.

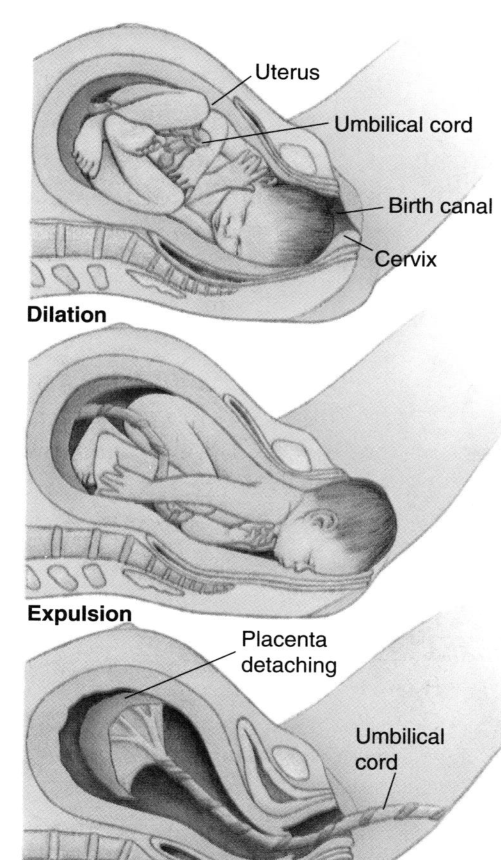

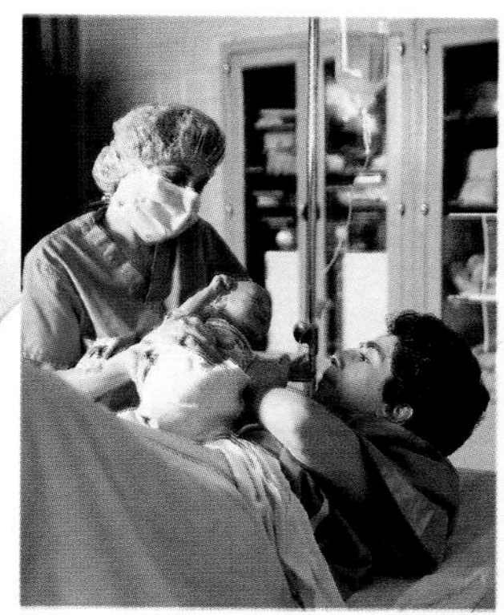

Figure 41.18

The stages of birth are dilation, expulsion, and the placental stage.

Placental stage

As shown in *Figure 41.18,* within ten to 15 minutes after the birth of the baby, the placenta separates from the uterine wall and is expelled with the remains of the embryonic membranes. Collectively, these materials are known as the afterbirth. The uterine muscles continue to contract forcefully, constricting uterine blood vessels to prevent the mother from hemorrhaging. After the baby is born, the umbilical cord is clamped and cut near the baby's abdomen. The bit of cord that is left eventually dries up and falls off, leaving an abdominal scar called the navel.

Growth and Aging

Once a baby is born, growth continues and learning begins. Human growth varies with age and is somewhat sex dependent.

A hormone controls growth

Human growth is regulated by human growth hormone (hGH), a protein secreted by the anterior pituitary. Although hGH causes all body cells to grow, it acts principally on the skeleton and skeletal muscles. The hormone works by increasing the rate of protein synthesis and the metabolism of fat molecules.

2 Teach

Science, Technology, and Society

Using Umbilical-Cord Tissues

In the late 1980s, scientists found that umbilical-cord blood is a rich source of blood stem cells (cells that have not yet differentiated). These same blood-forming cells are found in bone marrow. Recent studies have shown that cord blood can be frozen for about ten years and used for transplants in the place of the bone marrow, which is used to treat certain genetic diseases and leukemia.

Unlike bone marrow transplants, which cost from $15 000 to $20 000, umbilical-cord transplants cost only a few hundred dollars. Another advantage of umbilical-cord blood transplants is the possibility of using them for nonrelated tissue donations.

Display

LS **Visual-Spatial** Cut out pictures from magazines of humans at various life stages. Be sure to include pictures of active senior citizens. Add these pictures to the development display that was started earlier in the chapter. Refer to the display to illustrate changes in people at various stages of development.

STUDENT JOURNAL

Aging Assessment Have students write an essay expressing their personal views on aging. Ask them to explain why they feel as they do. L1 LS

BioLab

Average Growth Rate in Humans

Time Allotment

One class period

Objectives

Review objectives with students before they begin the Biolab.

Process Skills

make and use graphs, interpret data, analyze

BioLab

Average Growth Rate in Humans

Human growth is the result of more than one hormone. Human growth hormone, thyroid hormones, and the reproductive hormones that are produced during puberty are all important in human growth at various ages. Together, these hormones stimulate the growth of bone and cartilage, protein synthesis, and the addition of muscle mass. Because the reproductive hormones are involved in human growth, perhaps there is a difference in the growth rate between males and females.

PREPARATION

Problem

Is average growth rate the same in males and females?

Objectives

In this Biolab, you will:

- **Graph** the average growth rates in males and females.
- **Determine** if there are differences in the average growth rates of males and females.

Materials

graph paper
red and blue pencils

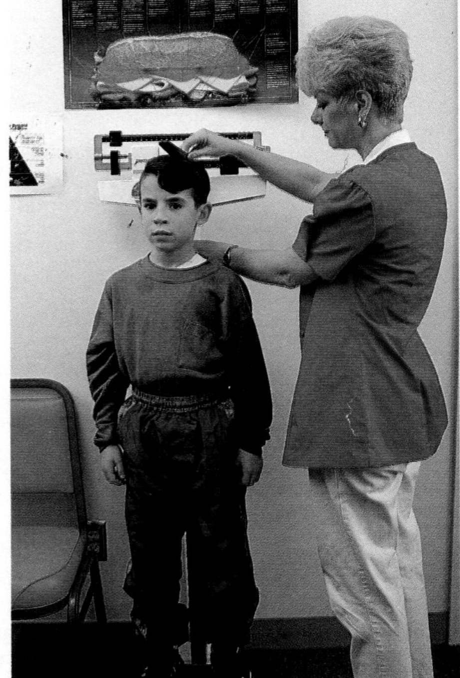

1060 Reproduction and Development

PROCEDURE

Teaching Strategies

- Have students refer to the Making and Using Graphs section of the *Skill Handbook* for help.

Data and Observations

Students will observe that average growth is about the same until puberty. At puberty, females have an earlier growth spurt than males, but on the average, males grow taller and heavier than females.

PROCEDURE

1. Construct a graph that plots mass on the vertical axis and age on the horizontal axis.
2. Plot the data in the table for the average female growth in mass from ages eight to 18. Connect the points with a red line.
3. On the same graph, plot the data for the average male growth in mass from ages eight to 18. Connect the points with a blue line.
4. Construct a separate graph that plots height on the vertical axis and age on the horizontal axis.
5. Plot the data for the average female growth in height from ages eight to 18. Connect the points with a red line. Plot the data for the average male growth in height from ages eight to 18. Connect the points with a blue line.

Averages for Growth in Humans

	Mass (kg)		Height (cm)	
Age	Female	Male	Female	Male
8	25	25	123	124
9	28	28	129	130
10	31	31	135	135
11	35	37	140	140
12	40	38	147	145
13	47	43	155	152
14	50	50	159	161
15	54	57	160	167
16	57	62	163	172
17	58	65	163	174
18	58	68	163	178

ANALYZE AND CONCLUDE

1. **Analyzing Data** During what ages do females and males increase the most in mass? In height?
2. **Thinking Critically** How can you explain the differences in growth between males and females?
3. **Analyzing Data** Interpret the data to find if the average growth rate is the same in males and females.

Going Further

Applying Concepts Correlate the range of heights in your biology class to the statistical average.

ANALYZE AND CONCLUDE

1. Mass: females, ages 11-13; males, ages 12-15. Height: females, ages 9-14; males ages 12-15
2. Times of puberty and the production of sex hormones differ between males and females. For example, testosterone increases growth in muscle and bone mass in males.
3. Average growth is the same until puberty. At puberty, females have an earlier growth spurt than males, but on the average, males grow taller and have greater mass than females.

Assessment

Journal: Have students write summaries of the Biolab and include them, their graphs, and their answers to Analyze and Conclude in their journals. L1

Going Further

Ask students to research what factors are known to affect growth rate in humans, including hormones, genetics, and diet. L2

Bioethics

Cloning

In October 1993, scientists at George Washington University Medical Center announced they had successfully cloned human eggs. The cloning of human eggs raised many ethical questions. For example, could cloned eggs be reared for spare parts, be used to allow for convenient family spacing, or be used to allow choosing and retaining the most desirable human traits? For some people, such developments are desirable. Others see the potential results of cloning as interfering with natural processes.

Audiovisual

Show the video, *Development and Aging,* Charles Clark Co., Inc.

Reinforcement

LS Logical-Mathematical An aging person's resting heartbeat stays about the same throughout life. However, the beats get weaker as the heart muscle ages. Have students graph the following data regarding the amount of blood pumped by the resting hearts of people at various ages. Ask them to summarize the data shown in their graph. L1

Age	Qts/Min
30	3.6
40	3.4
50	3.2
60	2.9
70	2.6

GLENCOE TECHNOLOGY

Videodisc

The Infinite Voyage: Fires of the Mind
Learning from Birth (Ch. 3)

Positron Emission Tomography (PET) (Ch. 4)

Like the sex hormones, testosterone and estrogen, the levels of human growth hormone within your body are controlled by a negative-feedback system. *Figure 41.19* shows what occurs if levels of hGH are abnormally high or low.

The first stage of growth—Infancy

During the first two years of life, or infancy, a child shows tremendous growth as well as increased physical coordination and mental development. Generally, an infant will double its birth weight by five months and triple its weight in a year. By two years of age, most infants weigh approximately four times their birth weight. During this time, the infant learns to control its limbs, roll over, sit, crawl, and walk. By the end of infancy, the child also utters his or her first words.

From child to adult

Childhood is the period of growth and development that extends from infancy to adolescence, when puberty begins. Physically, the childhood years are a period of relatively steady growth. Mentally, a child develops the ability to reason and to solve problems.

Figure 41.19

The amount of hGH secreted during childhood determines the height of an individual.

If a person's pituitary fails to produce enough hGH during childhood, a condition called pituitary dwarfism results.

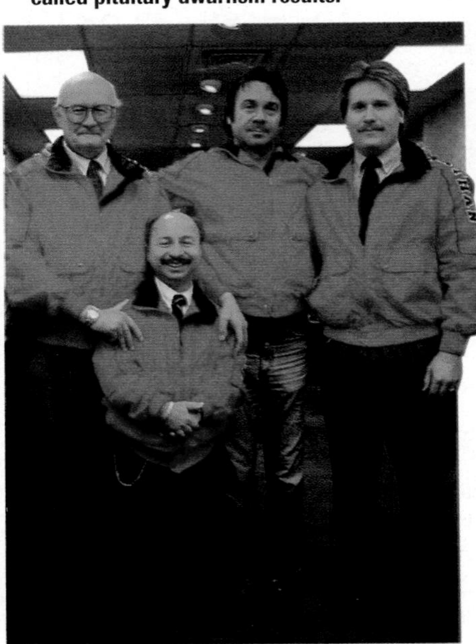

Large amounts of hGH secreted during childhood result in very tall individuals.

Abnormally large amounts of hGH result in a condition called gigantism.

Meeting Individual Needs

Students Acquiring English Have students prepare flash cards of the vocabulary terms from this chapter. Students should write the terms on one side of the card and their definitions on the other side. Have students work in groups to review the terms and their meanings. L1 SAE COOP LEARN LS

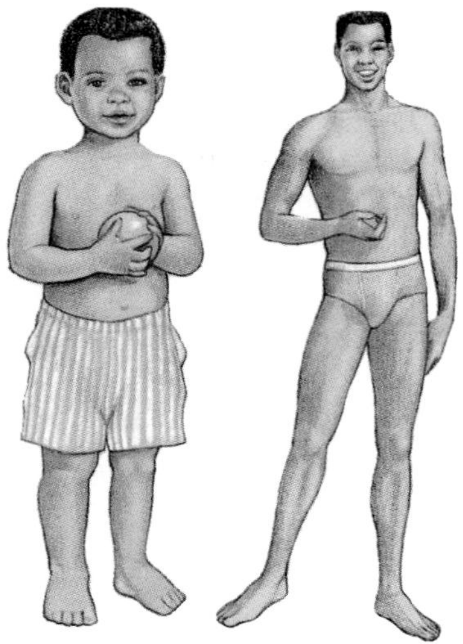

Figure 41.20

Notice the relative sizes of the head, arms, and legs as a person matures. In infancy, the head accounts for almost 25 percent of the body's length.

Adolescence follows childhood. At puberty, the onset of adolescence, there is often a growth spurt, sometimes quite a dramatic one! Increases of 5 to 8 cm of height in one year are not uncommon in teenage boys. During the teen years, adolescents reach their maximum physical stature, which is determined by heredity, nutrition, and their environment. By the time a young person reaches adulthood, his or her organs have reached their maximum mass, and physical growth is complete. You can see in *Figure 41.20* how the physical appearance of a person changes from birth to adulthood.

Art

Emmie and Her Child

by Mary Stevenson Cassatt (1855–1926)

Mary Cassatt's sensitive portrayals of the intimate relationship between mothers and their children brought her world fame. As a young woman, she chose to leave her close-knit family in Philadelphia and settle permanently in Paris. There she met Edgar Degas, and at his invitation, she became the only American to exhibit with the French Impressionists. Her achievement is all the more impressive because she was an unmarried, professional woman living in the male-dominated world of the late 19th century.

Her own individualized art Unable to attend the better French art schools, which accepted only men, Cassatt's training took place as she imitated the works of the great masters that hung in the Louvre, the famous art museum in Paris. After learning and perfecting their techniques, Cassatt began experimenting with different materials, such as pastels. She was able to combine her training with modern materials and attitudes to create her own individualized art—exemplified in her mother-child series. Never before had this age-old motif been so realistically rendered by an artist.

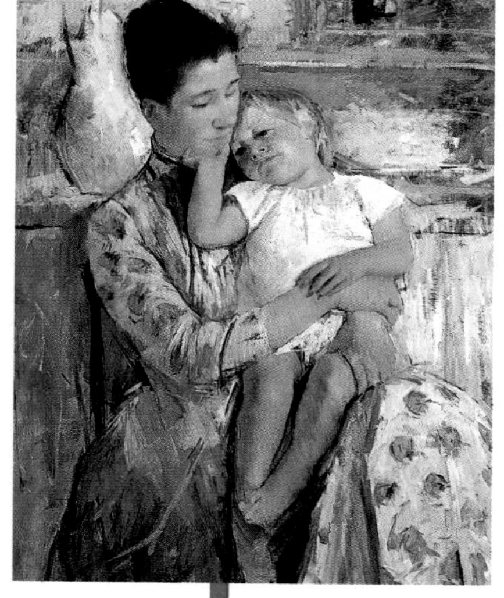

A simple embrace Mary Cassatt's favorite subjects were ordinary people in everyday settings doing everyday activities. She especially enjoyed painting mothers and children performing their daily rituals. In the beautiful painting entitled "Emmie and Her Child," Cassatt sensitively portrays a simple embrace between a mother and her child. The head and back positions of the loving pair communicate not only the tenderness of their embrace, but also feelings of complete trust and support, deep warmth, unquestioned security, and unconditional love.

CONNECTION TO Biology

Cassatt communicates the psychological benefits of the maternal instinct very well in her paintings. How might the maternal instinct serve a biological function in human reproduction?

Art

Emmie and Her Child

Purpose

Students will examine what information art communicates about the maternal instinct.

Teaching Strategies

- Ask students what the maternal instinct may involve.
- Encourage students to read and summarize "Biology Is One Key to the Bonding of Mothers and Babies," by Hara Estroff Marano, *Smithsonian,* Feb. 1981. L3

Possible Answer

The maternal instinct to protect and care for the infant increases the likelihood of survival of the infant.

Reinforcement

LS **Logical-Mathematical** Have students prepare a table that compares and contrasts the characteristics of infancy, childhood, and adolescence. Review the tables as a class. As a homework assignment, have students use the information on page 1062 to add to their tables those traits that characterize an adult. L1

PORTFOLIO

Senior Survey Have students prepare interview questions to ask senior citizens. Instruct them that their questions should focus on the biology of aging as well as how the senior citizens view aging. Collect questions from all students and photocopy them. Distribute copies of all the questions to students in their groups and ask each group to prepare a final survey containing ten questions. Have each group member use the survey to interview someone older than 65. Students should then tally the results obtained by the group. L1 P COOP LEARN LS

SECTION 41.3

3 Assess

Check for Understanding

Ask students to describe the aging process. L1

Reteach

Have students choose a system of the body and describe how it changes from birth through old age. L1

Extension

Ask students to research and report on current theories on infant learning. Students should be encouraged to include illustrations with their reports. L3

Assessment

Knowledge: Ask students to choose a hormone discussed in this chapter and diagram a negative-feedback loop for the hormone. L1

4 Close

Discussion

Ask students to state several misconceptions about aging, and list their statements on the chalkboard. As a class, discuss evidence that supports why each statement is a misconception. L1

Figure 41.21

Age has not deterred this 90-year-old woman from trying a new experience—skydiving.

An adult ages

As an adult ages, his or her body undergoes many distinct changes. Metabolism and digestion become slower. The skin loses some of its elasticity, while less pigment is produced in the hair follicles; that is, the hair turns white. Bones become thinner and more brittle, resulting in an increased risk of fracture. Stature may shorten because the disks between the vertebrae become compressed. Vision and hearing might diminish, but, as ***Figure 41.21*** shows, many people continue to be both intellectually and physically active.

Connecting Ideas

The human reproductive system is unique among body systems in that there are two—male and female. Together, the reproductive system and the endocrine system work to produce sex cells and the fully grown humans that develop from them. In fact, in order for your body to maintain homeostasis, all systems must function together. In order for your body to be healthy, whether you are an adolescent or an aged adult, your body systems must be free of infections and serious disorders.

Section Review

Understanding Concepts

1. What events occur during dilation?
2. How does the human growth hormone produce growth?
3. How does the human body change during childhood?

Thinking Critically

4. Compare the birth of a human baby with that of a marsupial mammal.

Skill Review

5. **Recognizing Cause and Effect** Someone tells you that as people age, their personalities normally change. Do you think this statement is valid? Why or why not? For more help, refer to Thinking Critically in the *Skill Handbook*.

Answers to Section Review

1. Muscle contractions expand the cervix to about 10 cm.
2. It increases the rate of protein synthesis inside cells and increases the metabolism of fat molecules.
3. Muscle coordination, reasoning, and problem-solving abilities increase; the body grows; the proportion of head size to the rest of the body changes.

Thinking Critically

4. Marsupials are born earlier in development when compared with humans. Further development of marsupials occurs within a pouch. Both humans and marsupials are dependent upon the mother at birth.

Skill Review

5. Most people function effectively throughout life without experiencing changes in personality.

REVIEWING MAIN IDEAS

41.1 Human Reproductive Systems

- The male reproductive system produces sperm and the female reproductive system produces eggs.
- Through the control of the hypothalamus and pituitary, hormones act on the reproductive system as well as on other body systems. Their levels are regulated by negative feedback.
- Changes in males and females at puberty are the result of the production of FSH, LH, and other sex hormones.
- During the menstrual cycle, hormones control the production of a mature egg and prepare the uterus for implantation of a fertilized egg.

41.2 Development Before Birth

- Fertilization occurs in the oviduct. The ball of cells that develops from the fertilized egg implants in the uterine wall.
- The embryo changes from a small ball of cells to a well-developed fetus over the course of nine months.
- Genetic counseling offers people information about their chances of having a child with a genetic disorder.

41.3 Birth, Growth, and Aging

- Birth involves dilation of the cervix, expulsion of the baby, and expulsion of the placenta.
- Infancy, childhood, adolescence, and adulthood are the stages of human development. Human growth hormone (hGH) produces growth in all body cells, especially in cells of the skeleton and muscles.

Key Terms

Write a sentence that shows your understanding of each of the following terms.

amniocentesis
bulbourethral gland
cervix
corpus luteum
epididymis
follicle
genetic counseling
implantation
labor
menstrual cycle
negative-feedback system
oviduct
ovulation
pituitary
prostate gland
puberty
scrotum
semen
seminal vesicle
umbilical cord
vagina
vas deferens

Understanding Concepts

1. Which of the following is NOT a part of the male reproductive system?
   a. scrotum
   b. vas deferens
   c. testis
   d. cervix
2. Sperm production takes place in the ___________.
   a. urethra
   b. bladder
   c. testes
   d. penis
3. The ___________ is a coiled tube inside the scrotum in which sperm become mature.
   a. epididymis
   b. vas deferens
   c. follicle
   d. prostate
4. Which of these is NOT found in a sperm?
   a. head
   b. chloroplast
   c. tail
   d. nucleus
5. What organ transports sperm from the testes to the urethra?
   a. epididymis
   b. vas deferens
   c. scrotum
   d. follicle
6. Which of these does NOT produce a fluid that surrounds sperm as they travel from the testes?
   a. seminal vesicles
   b. bulbourethral glands
   c. prostate gland
   d. pituitary gland

Chapter 41 Review

Reviewing Main Ideas

Summary statements can be used by students to review the major concepts of the chapter.

Key Terms

Answers should go beyond defining the terms. Accept any answer that uses the term correctly and in the proper context.

Understanding Concepts

1. d
2. c
3. a
4. b
5. b
6. d

Chapter 41 Review

7. a
8. b
9. d
10. c
11. d
12. a
13. c
14. b
15. c
16. b
17. b
18. a
19. a
20. c

Applying Concepts

21. The signal from the hypothalamus to the pituitary would increase due to the lack of negative feedback. The levels of FSH and LH would increase.
22. about 36 days
23. This hormone is secreted by the embryo/fetus, so it indicates pregnancy.
24. The combination of genetic material from the egg and sperm allows for a new combination resulting in a unique individual.
25. The placenta is the place of exchange between the mother and fetus, allowing the fetus to obtain nutrients and oxygen and to get rid of waste products.

Chapter 41 Review

7. Secondary sex characteristics begin to develop during ____________.
   a. puberty c. childhood
   b. adolescence d. adulthood
8. When luteinizing hormone is released by the pituitary, it causes cells in the testes to produce ____________.
   a. FSH c. progesterone
   b. testosterone d. estrogen
9. What tubule transports both urine and semen?
   a. testes c. epididymis
   b. vas deferens d. urethra
10. In females, eggs are produced in the ____.
   a. testis c. ovary
   b. cervix d. bladder
11. An undeveloped egg is surrounded by a group of epithelial cells known as a(n) ____________.
   a. oviduct c. vesicle
   b. vagina d. follicle
12. The __________ is a tube that transports the egg from the ovary to the uterus.
   a. oviduct c. cervix
   b. vagina d. urethra
13. When there is a surge of LH during the menstrual cycle, what event occurs?
   a. luteinization c. ovulation
   b. fertilization d. menstruation
14. What happens to a follicle after it releases an egg?
   a. It degenerates.
   b. It changes into the corpus luteum.
   c. It is released into the oviduct.
   d. It turns into the placenta after fertilization of the egg.
15. ___________ occurs when the blastocyst attaches to the lining of the uterus.
   a. Fertilization c. Implantation
   b. Luteinization d. Menopause
16. ___________ occurs when an egg ruptures through the ovary wall and moves into the oviduct.
   a. Fertilization c. Implantation
   b. Ovulation d. Menstruation
17. Once one sperm enters an egg, no other sperm can enter because the egg's membrane ____________.
   a. changes its shape
   b. changes its electrical charge
   c. hardens
   d. closes off its pores
18. Human development occurs most rapidly during the ____________ stage.
   a. embryo c. adolescence
   b. child d. adult
19. During which trimester does the embryo's heart start beating?
   a. first c. third
   b. second d. fourth
20. Which type of prenatal testing examines fetal cells for genetic disorders?
   a. genetic counseling
   b. ultrasound
   c. amniocentesis
   d. X ray

Applying Concepts

21. If the testes were removed from an adult male, what would happen to the signal from the hypothalamus to the pituitary? To the levels of FSH and LH?
22. A pregnant woman tells her physician that she is 50 days past the first day of her last menstrual period. How many days has the embryo been developing?
23. Why do pregnancy tests check for chorionic gonadotropin?
24. What is the adaptive advantage for an embryo to be formed from an egg and a sperm?
25. How is the placenta important in maintaining homeostasis in the developing fetus?

1066 Reproduction and Development

Thinking Critically

26. ***Concept Mapping*** Make a concept map that relates the following terms and phrases. Supply the appropriate linking words for your map.
    negative-feedback system, FSH, LH, testosterone, testes, semen, vas deferens, epididymis, puberty
27. ***Recognizing Cause and Effect*** In gigantism, the pituitary gland produces too much hGH during childhood, resulting in excessively long bones. Where does the negative-feedback system appear to fail in the case of gigantism?
28. ***Observing and Inferring*** Explain what role the circulatory system has regarding the sex hormones produced by the endocrine system.
29. ***Interpreting Data*** This graph indicates changes in cardiac output and heartbeat rate in a woman over the course of her pregnancy. Explain the changes.

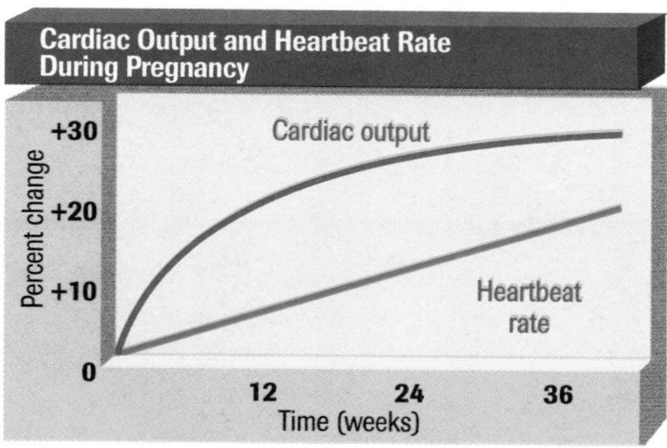

30. ***Recognizing Cause and Effect*** During embryonic development, the testes develop near the kidneys. Usually a month or two before birth, the testes descend into the scrotum. What would be the consequences if the testes remain undescended? Why?
31. ***Comparing and Contrasting*** Compare the events of the first trimester with those of the third trimester of pregnancy.

ASSESSING KNOWLEDGE & SKILLS

The following graph represents the average blood concentration of four circulating hormones collected from 50 healthy adult women who were not pregnant.

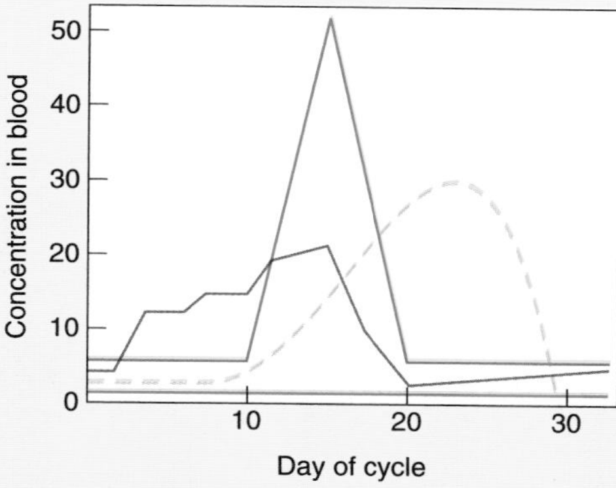

Interpreting Data Use the graph to answer questions 1-4.

1. Which line represents luteinizing hormone?
   a. red line c. yellow line
   b. blue line d. green line
2. Which hormone increases during the last half of the menstrual cycle?
   a. estrogen c. LH
   b. progesterone d. FSH
3. Which hormone is responsible for stimulating the egg development each month?
   a. estrogen c. LH
   b. progesterone d. FSH
4. ***Interpreting Scientific Illustrations*** The green line represents the hormone human chorionic gonadotropin. Explain why this hormone remained at an extremely low level during the women's menstrual cycles.

Chapter 41 Review

Thinking Critically

26. Evaluate students' concept maps. The maps should show an understanding of the relationships among the concepts listed.
27. The sensing portion of the hypothalamus does not work.
28. The circulatory system circulates sex hormones throughout the body.
29. During pregnancy, the heartbeat rate and cardiac output increase in order to compensate for blood delivery to the developing fetus.
30. If the testes remained undescended, the production of sperm would be affected. Sperm mature at a cooler temperature than that of normal body temperatures. Thus, the male would be sterile.
31. During the first trimester, all body systems are forming. During the third trimester, all body systems are developed sufficiently to allow the fetus to survive outside the mother's body.

ASSESSING KNOWLEDGE & SKILLS

1. a
2. b
3. d
4. This hormone is secreted only when women are pregnant.

Program Resources

Chapter Assessment, pp. 253-258 L1
Alternate Assessment in the Science Classroom
Computer Test Bank L1
Content Mastery, pp. 173-176 L1

Chapter Organizer

SECTION	OBJECTIVES	ACTIVITIES/FEATURES
42.1 The Nature of Disease National Science Standards: UCP.1, UCP.2, UCP.5; C.1, C.4, C.6; F.1, F.5; G.1-3	1. **Outline** the steps of Koch's postulates. 2. **Describe** how infections are transmitted. 3. **Explain** what causes the symptoms of a disease.	**Minilab:** How are diseases spread?, p. 1074 **Thinking Lab:** How does the herpes simplex virus spread?, p. 1076
42.2 Defense Against Infectious Diseases National Science Standards: UCP.1-3, UCP.5; A.1; C.1, C.4, C.5; F.1, F.5, F.6; G.1-3	4. **Explain** the various types of nonspecific defense mechanisms. 5. **Compare** antibody and cellular immunity.	**Minilab:** What are the percentages of different kinds of white blood cells?, p. 1080 **Biology & Society Ethics:** Human Guinea Pigs, p. 1083 **Health Connection:** Protecting the Future, p. 1087 **Biolab:** AIDS and Its Effect on the Immune Response, p. 1088

ACTIVITY MATERIALS

BIOLAB	MINILABS	ALTERNATE LAB
page 1088 graph paper colored pencils, red and blue	**page 1074** plastic bags apple, fresh apple, rotting cotton balls alcohol **page 1080** prepared slides of blood cells microscope	**page 1080** 3 types of mouthwash forceps toothpicks aluminum foil glass hockey sticks transparent tape alcohol, small bottles nutrient agar plates glass marker

Chapter 42 Immunity from Diseases

TEACHER CLASSROOM RESOURCES

Reproducible Masters	Transparencies
Section Focus Master 97: Don't Spread it Around L1 SAE **Reinforcement and Study Guide,** p. 171 L1 **Biolab and Minilab Worksheets,** p. 167 L1 **Content Mastery,** pp. 177-180 L1	**Basic Concepts Transparency #76:** Identifying a Pathogen L1 SAE
Section Focus Master 98: Becoming Immune L1 SAE **Reinforcement and Study Guide,** pp. 172-174 L1 **Biolab and Minilab Worksheets,** pp. 168-170 L1 **Concept Mapping:** The Lymphatic Cycle, p. 42 L1 **Critical Thinking/Problem Solving:** Diagnosing Allergies, p. 42 L3 **Laboratory Manual:** How Do Bactericides Affect the Growth of Bacteria?, pp. 261-264 L2	**Basic Concepts Transparency #77:** Response to Injury L1 SAE **Basic Concepts Transparencies #78a, 78b:** Antibody Immunity; Cellular Immunity L1 SAE **Basic Concepts Transparency #79:** The AIDS Epidemic L1 SAE **Basic Skills Transparency #29:** Cells of the Immune Response L1 SAE **Reteaching Transparency #42:** Human Body Systems L1 SAE

ASSESSMENT MATERIALS

Chapter Assessment, pp. 259-264
Performance Assessment in the Biology Classroom, p. 45
MindJogger Videoquiz
Alternate Assessment in the Science Classroom
Computer Test Bank

Spanish Resources SAE
English/Spanish Audiocassettes SAE
Cooperative Learning in the Science Classroom COOP LEARN
Lesson Plans

KEY TO TEACHING STRATEGIES

L1 Level 1 activities should be within the ability range of all students including those with learning difficulties.

L2 Level 2 activities are within the ability range of average to above-average students.

L3 Level 3 activities are designed for the ability range of above-average students.

SAE SAE activities should be within the ability range of Students Acquiring English.

COOP LEARN Cooperative Learning activities are designed for small group work.

P These strategies represent student products that can be placed into a best-work portfolio.

These strategies are useful in a block scheduling format.

GLENCOE TECHNOLOGY

The following multimedia resources are available from Glencoe.

Biology: The Dynamics of Life
CD-ROM SAE
Videodisc Program

The Infinite Voyage Series
Insects: The Ruling Class
The Rothschild Legacy
Prospecting for Healing Medicine from Insects

The Secret of Life Series
Nothing to Sneeze At: Viruses

National Geographic Society Series
STV: Human Body Vol. 3
Immune System
Newton's Apple: Life Sciences
HIV and AIDS

Science and Technology Videodisc Series (STVS)
Human Biology
White Cells in Action
Combined Immune Deficiency
Plants & Simple Organisms
Anticavity Vaccine

CHAPTER

42 Immunity from Diseases

Chapter Overview

Students will examine what infectious diseases are, what causes diseases, and how diseases are spread. They will also examine the causes of the symptoms of disease, the patterns of disease prevalence and transmission, and how diseases are treated.

In the second section of the chapter, students examine the defenses the body uses to fight diseases. The anatomy of the lymphatic system, nonspecific defense mechanisms, and specific defense mechanisms are introduced. The chapter concludes with a discussion of AIDS and its effect on the immune system.

Key Terms

antibiotic
B cell
complement
endemic disease
epidemic
immunity
infectious disease
Koch's postulates
lymph
lymph node
lymphocyte
macrophage
pathogen
phagocyte
pus
T cell
tissue fluid
vaccine

1068

Learning Styles

Look for the following logo for strategies that emphasize different learning modalities. LS

Visual-Spatial	**Display, p. 1071; Demonstration, pp. 1071, 1072, 1079; Visual Learning, p. 1072; Meeting Individual Needs, pp. 1072, 1082; Student Journal, p. 1079; Reinforcement, p. 1086**
Interpersonal	**Discussion, p. 1079; Portfolio, p. 1086**
Intrapersonal	**Student Journal, pp. 1071, 1074, 1075**
Linguistic	**Portfolio, p. 1073; Student Journal, pp. 1085, 1087, 1091**
Logical-Mathematical	**Minilab, pp. 1074, 1080; Thinking Lab, p. 1076; Visual Learning, p. 1077; Alternate Lab, p. 1080; Meeting Individual Needs, p. 1086; Portfolio, p. 1090; Time Line, p. 1091**

CHAPTER

42 Immunity from Diseases

Schooled in the art of self-defense, a tae kwon do expert can fend off almost any attacker—even more than one attacker at a time. Through skillful kicks and strikes with her hands, feet, elbows, and knees, she can protect herself from harm.

Your body, too, has a system of self-defense. The attackers it must defend against are disease-producing organisms, which can strike anywhere and at any time. Your body's chief experts in self-defense are its white blood cells, which can do everything from engulfing these foreign invaders to producing antibodies that destroy them.

How do these attacking organisms invade your body? How does your body work to defend itself against them? Your immune system holds the answers to these questions.

Concept Check

You may wish to review the following concepts before studying this chapter.

- Chapter 9: phagocytosis
- Chapter 21: viral replication, retroviruses

Chapter Preview

Laboratory Activities

Biolab

- AIDS and Its Effect on the Immune Response

Minilabs

- How are diseases spread?
- What are the percentages of different kinds of white blood cells?

Magnification: 6000×

This martial arts expert actually has two systems of self-defense: one is the art of tae kwon do and the other is her immune system. However, if an attacking disease organism should happen to temporarily elude her white blood cells and other lines of defense, she may occasionally succumb to a cold or other more serious infection.

CHAPTER 42

Introducing the Chapter

Ask students to describe what mechanisms of the body help fight disease. *Students may recall that the cilia and mucus of the respiratory system, the skin, and some blood cells carry out disease-fighting functions.* Ask them what types of foreign invaders their body must defend against and where these unseen enemies may be found. *Students may mention bacteria, viruses, protists, fungi, and chemical substances present in the environment.*

Theme Development

The theme of *systems and interactions* is emphasized throughout the chapter as the mechanisms by which parts of various body systems work together to fight disease are discussed. *Homeostasis* is emphasized as the immune system's response to disease in an attempt to maintain homeostasis in the body is discussed.

Concept Check

Students will review and build upon their knowledge of disease-causing organisms including bacteria, protists, and fungi, and the disease-causing nature of viruses and parasites. The structure of the skin will be recalled along with its anatomical ability to prevent the entry of pathogens into the body.

Assessment Planner

Choose assessment strategies from the following pages to evaluate the progress of your students.

Assess, pp. 1077, 1092
Alternate Lab, pp. 1080-1081
Minilabs, pp. 1074, 1080
Portfolio, pp. 1073, 1086, 1090
Thinking Lab, p. 1076
Biolab, pp. 1088-1089
Chapter Review, pp. 1093-1095

SECTION

42.1

Prepare

Key Concepts

The steps of Koch's postulates are presented. The spread of disease is covered with emphasis on how infections are transmitted.

Block Scheduling

Look for this symbol for strategies that are useful in a block scheduling format. For more information on block scheduling, refer to Section 42.1 in the **Lesson Plans** booklet.

Materials

- Obtain materials from the health department, local doctors' offices, hospitals, or pharmaceutical companies for the Display.
- Buy a chicken TV dinner for the Demonstration on page 1071.
- Prepare nutrient agar plates for both Demonstrations and order the needed bacterial cultures, antibiotic disks, and antiseptics.
- Purchase apples for the Minilab.

1 Focus

Bellringer

Before presenting the lesson, display **Section Focus Master 97** on the overhead projector and have students answer the accompanying questions. L1 SAE

SECTION

42.1 The Nature of Disease

Section Preview

Objectives

Outline the steps of Koch's postulates.

Describe how infections are transmitted.

Explain what causes the symptoms of a disease.

Key Terms

pathogen
infectious disease
Koch's postulates
endemic disease
epidemic
antibiotic

Occasionally everyone gets a cold. Cold viruses enter your body by way of your nose and are swept to the back of your throat by hairlike cilia. Some are washed down your esophagus and destroyed by your digestive system. Others lodge against the lining of your nasal passage, binding tightly to cell receptors. These viruses enter your nasal cells and unleash their genes, taking over your cells' reproductive machinery. Soon you have a sore throat, a stuffed and runny nose, a headache, and a mild fever. How did the cold virus get in your nose in the first place? How did the infection produce these symptoms?

What Is an Infectious Disease?

The cold virus is an example of a microbe that causes a disease—a change that disrupts the homeostasis in the body. Disease-producing agents such as bacteria, protozoans, fungi, viruses, and other parasites are called **pathogens.** The main sources of pathogens are soil, contaminated water, and infected people or animals. Any disease caused by the presence of pathogens in the body is called an **infectious disease.** Some of the infectious diseases that occur in humans are shown in *Table 42.1.*

Not all microorganisms are pathogenic. In fact, the presence of some microorganisms in your body is beneficial. You are microbe-free before birth. At birth, microbes establish themselves on your skin and in your upper respiratory system, lower urinary tract, reproductive tract, and intestinal tract. ***Figure 42.1*** shows some common microorganisms that live on your skin.

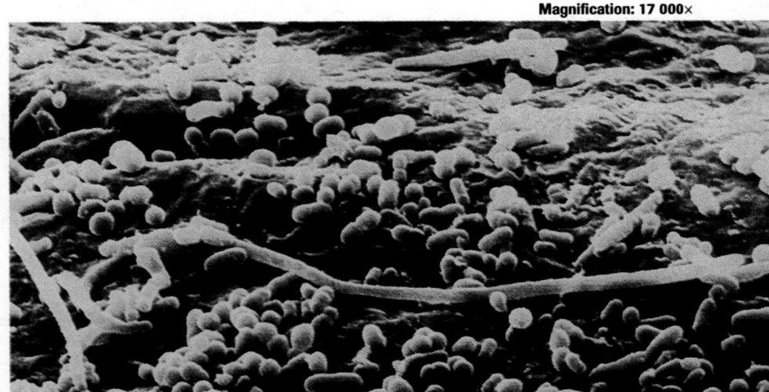

Figure 42.1

These microorganisms establish a more-or-less permanent residence in or on your skin, but do not cause disease under normal conditions. They have a symbiotic relationship with your body. However, if you become weakened or injured, these same organisms are potential pathogens.

Program Resources

Section Focus Master 97 L1 SAE
Reinforcement and Study Guide, p. 171 L1
Biolab and Minilab Worksheets, p. 167 L1
Basic Concepts Transparency 76 and **Master** L1 SAE

Table 42.1 Human Infectious Diseases

Disease	Cause	Affected Body System	Transmission
Smallpox	Virus	Skin	Droplet
Chicken Pox	Virus	Skin	Droplet
Cold Sores	Virus	Skin	Direct Contact
Rabies	Virus	Nervous System	Animal Bite
Poliomyelitis	Virus	Nervous System	Contaminated Water
Infectious Mononucleosis	Virus	Salivary Glands	Direct Contact
Colds	Viruses	Respiratory System	Direct Contact
Influenza	Viruses	Respiratory System	Droplet
HIV/AIDS	Virus	Immune System	Exchange of body fluids
Hepatitis B	Virus	Liver	Direct Contact
Measles	Virus	Skin	Droplet
Mumps	Virus	Salivary Glands	Droplet
Tetanus	Bacteria	Nervous System	Deep wound
Food Poisoning	Bacteria	Digestive System	Contaminated Food
Tuberculosis	Bacteria	Respiratory System	Droplet
Whooping Cough	Bacteria	Respiratory System	Droplet
Spinal meningitis	Bacteria	Nervous System	Droplet
Impetigo	Bacteria	Skin	Direct Contact

Determining What Causes a Disease

One of the first problems scientists face when studying a disease is finding out what causes the disease. Not all diseases are caused by pathogens. Disorders such as hemophilia, caused by a recessive allele on the X chromosome and discussed in Chapter 15, and rheumatoid arthritis are inherited. Others, such as osteoarthritis, may be caused by wear and tear on the body as it ages. Pathogens cause infectious diseases and some cancers. In fact, about half of all human diseases are infectious. In order to determine which pathogen causes a specific disease, scientists follow a standard set of procedures.

First pathogen identified

The first proof that a microbe actually caused a disease came from the work of Robert Koch in 1876. Koch, a German physician, was working on the cause of anthrax, a deadly disease that mainly affects cattle and sheep but can also occur in humans. Koch discovered a rod-shaped bacterium in the blood of cattle that had died of anthrax. He cultured the bacteria on nutrients and then injected samples of the culture into healthy animals. When these animals became sick and died, Koch isolated the bacteria in their blood and compared them with the bacteria originally isolated. He found that the two sets of blood cultures contained the same bacteria.

Discussion

Explain the term *plague* and ask why some people survived plagues, while many others perished. *Students may suggest that people in better health may have been more resistant to infection or possessed some unusual trait that prevented them from succumbing to infection. Accept all logical responses.* Initiate a discussion about the body's defenses to disease.

Text Questions

How did the cold virus get in your nose in the first place? *Likely responses will suggest that the virus was inhaled along with air.* How did the infection produce these symptoms? *Symptoms are caused as the virus brings about a response from mucous membranes and other disease-fighting structures.*

2 Teach

Display

LS Visual-Spatial Prepare a bulletin board display with information about specific diseases. Brochures on diseases may be available from pharmaceutical companies, doctors' offices, or hospitals. Refer to the display as you discuss pathogens and diseases.

Demonstration

LS Visual-Spatial Thaw a deboned chicken TV dinner and mix it with 100 mL of sterile water. Prepare a dilution series (1/10, 1/100 to 1/1 000 000) of the mixture and spread one drop of each dilution on nutrient agar. Incubate and then count the number of colonies for each dilution. Use these numbers to calculate how many bacteria were present in the TV dinner. Use the demonstration to point out that pathogens may be present in the foods people eat.

STUDENT JOURNAL

Disease Transmission Ask students to list the ways in which diseases are transmitted and give an example of a disease transmitted in each way. Students may need help giving the examples. L1 LS

SECTION 42.1

Visual Learning

Figure 42.2 Have students describe each of the four steps of Koch's postulates shown in the diagram. LS

Discussion Question

Elicit from students why Koch's postulates are not useful for identifying viral pathogens. *Viruses multiply only within living cells.*

Demonstration

LS **Visual-Spatial** Spread pure bacterial culture over several nutrient agar plates. Place different types of antibiotic disks on the agar in all plates except one. Incubate all plates for 24-48 hours. Display the plates, pointing out the zones of inhibition that develop around each antibiotic disk. Repeat the exercise using circles of filter paper that have been dipped into different antiseptics to see if a zone of inhibition will develop. Use the demonstration to discuss the effectiveness of antibiotics and antiseptics in combating or preventing bacterial infections.

Figure 42.2

Koch's postulates are steps used to identify an infectious pathogen. Step 3, injecting the pathogen into a healthy host, is not morally acceptable using human subjects, so other susceptible hosts are used.

A procedure to establish the cause of a disease

Koch established experimental steps, shown in *Figure 42.2,* for directly relating a specific microbe to a specific disease. These steps today are known as **Koch's postulates:**

1. The pathogen must be found in the host in every case of the disease.
2. The pathogen must be isolated from the host and grown in a pure culture.
3. When the pathogen from the pure culture is placed in a healthy host, it must cause the disease.
4. The pathogen must be isolated from the new host and be shown to be the original pathogen.

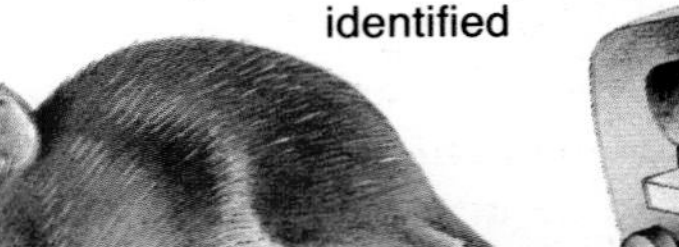

Exceptions to Koch's postulates

Although Koch's postulates are useful in determining the cause of most diseases, some exceptions exist. Some organisms, such as the pathogenic bacterium that causes the sexually transmitted disease syphilis, have never been grown on an artificial medium. Viral pathogens also cannot be cultured this way because they multiply only within cells. Finding living tissue other than human tissue for a culture medium can be difficult. For example, the bacterial pathogen that causes leprosy, a potentially fatal disease that afflicts a person's skin and nerves, was first isolated in 1870. Because the pathogen will not grow on an artificial medium, there was a problem carrying out Koch's postulates until 1969. At that time, scientists discovered that the pathogen could be grown successfully in armadillos.

The Spread of Infectious Diseases

For a disease to continue and spread, there must be a continual source of the disease organisms. This source can be either a living organism or an inanimate object on which the pathogen can survive.

Meeting Individual Needs

Hearing Impaired Ask students needing extra help in this section to use the review tutorial software *Disease, Immunity, and Health,* from Queue, (Apple II, MS-DOS, Mac) for review. L1 LS

Reservoirs of pathogens

The main source of human disease pathogens is the human body itself. In fact, the body can be a reservoir of disease-causing organisms. Many people harbor pathogens and transmit them directly or indirectly to other people. Sometimes, people can harbor pathogens without exhibiting any signs of the illness and unknowingly transmit the pathogens to others. These people are called *carriers* and are a significant reservoir of infectious diseases.

Other people may unknowingly pass on a disease during its first stage, before they begin to experience symptoms. This symptom-free period, while the microbes are multiplying within the body, is called an incubation period. Humans can unknowingly spread colds, streptococcal throat infections, and sexually transmitted diseases (STDs) such as gonorrhea and HIV/AIDS during the incubation periods of these diseases.

Animals are other living reservoirs of microorganisms that cause disease in humans. For example, some types of influenza, commonly known as *the flu*, and rabies are often transmitted to humans from animals.

The major nonliving reservoirs of infectious diseases are soil and water. Soil harbors pathogens such as fungi and the bacterium that causes botulism, a type of food poisoning. Water contaminated by feces of humans and other animals is a reservoir for several pathogens, especially those responsible for intestinal diseases.

Transmission of disease

How are pathogens transmitted from a reservoir to a human host? Pathogens can be transmitted from reservoirs in four main ways: by direct contact, by an object, through the air, or by an intermediate organism called a vector. *Figure 42.3* illustrates each of these.

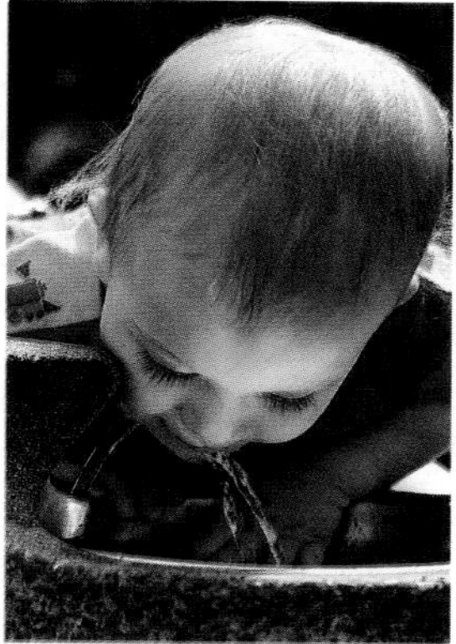

Figure 42.3

Diseases can be transmitted to humans from reservoirs in various ways.

The most common ways for a disease to be transmitted by direct contact involve touching, kissing, and/or sexual contact.

Insects and arthropods are the most common vectors.

Common inanimate objects such as food, water, drugs, toys, dishes, and intravenous needles can harbor and transmit pathogens.

Airborne transmission by droplets of water or dust spreads pathogens from the reservoir to the host.

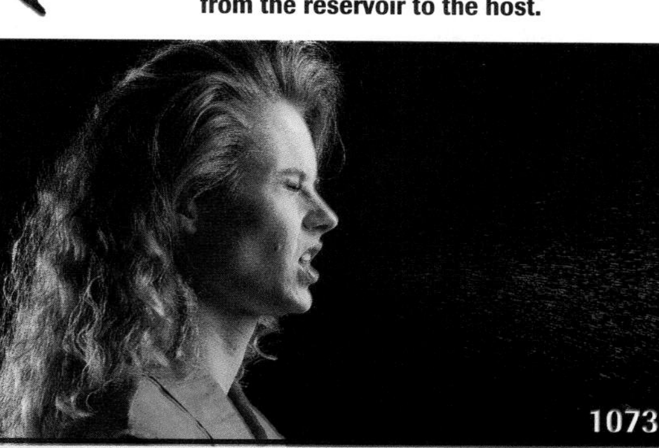

SECTION 42.1

Health Connection

The Spread of TB

Since 1986, tuberculosis (TB) infections have increased 20 percent in the United States, creating an urban epidemic. Tuberculosis is caused by an airborne bacteria. Infection can be latent or active. Once active, the lung infection, as well as the effects on other vital organs, can cause death if the disease is not cured.

Among all the world's infectious diseases, TB remains the number-one cause of death. The majority of new cases in the United States have occurred among foreign-born groups, especially Asian-born refugees. The disease has also become common in the homeless populations of many urban centers.

GLENCOE TECHNOLOGY

Videodisc

The Infinite Voyage: Insects: The Ruling Class

The Rothschild Legacy: Study of Fleas and the Bubonic Plague (Ch. 2)

PORTFOLIO

Role-Playing Have students write a story concerning the ways diseases might spread in a restaurant, from the viewpoint of a health inspector, a restaurant owner, a restaurant cook, or a disease organism. Methods of prevention should also be discussed. L1 P LS

MiniLab

Purpose

Logical-Mathematical Students will examine how microbes spread infection in apples.

Process Skills

use controls, collect and organize data, interpret data, experiment, analyze

Teaching Strategies

- Point out that the brown area under the apple skin may be caused by bacteria.

Safety Precautions

- Have students wash their hands after handling the rotting apples.

Expected Results

Apples 2 and 3 will develop brown spots and decay. Apple 1 will show no change. Apple 4 may show no change or a small change because the alcohol may have inhibited, but not completely stopped, the decay.

Analysis

1. Number 1 is a control for comparison.
2. Apple 2 decayed the most, Apple 3 had brown spots, and Apple 4 may have showed no change or only a small change.
3. Cleaning a wound with alcohol may remove and kill pathogens to prevent infection.

Assessment

Journal: Ask students to write a summary of the lab, including their results. Have them place their summaries along with their answers to the Analysis questions in their journals.

MiniLab

How are diseases spread?

Among microorganisms, the ability to move from place to place is generally limited. Microorganisms cannot travel over long distances by themselves or fly or climb on their own. Therefore, unless they are transferred from one animal or plant to another, infections will not spread. One method of transference is by direct contact with an infected animal or plant.

Procedure

1. Label plastic bags 1 to 4.
2. Put a fresh apple in bag 1 and seal the bag.
3. Rub a rotting apple over the entire surface of the remaining three apples. The rotting apple is your source of microorganisms.
4. Put one of the apples in bag 2.

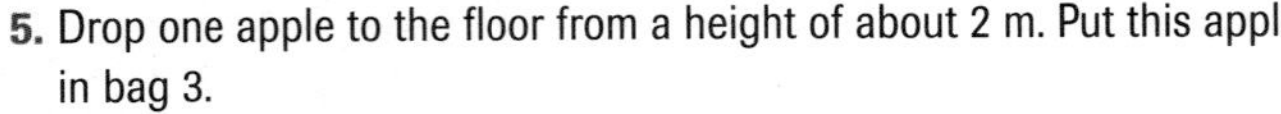

5. Drop one apple to the floor from a height of about 2 m. Put this apple in bag 3.
6. Use a cotton ball to spread alcohol over the last apple. Let the apple air-dry for a short time, and then place it in bag 4.
7. Store all of the bags in a dark place for one week. **CAUTION:** *Make sure to wash your hands after handling microorganisms.*
8. At the end of the week, compare all the apples. Record your observations. **CAUTION:** *Give all apples to your teacher for proper disposal.*

Analysis

1. What was the purpose of the fresh apple in bag 1?
2. Explain what happened to the rest of the apples.
3. Why is it important to clean a wound with a chemical such as alcohol?

The common cold, influenza, and STDs are spread by direct contact. The main way that STDs such as genital herpes and the HIV virus are transmitted is by the exchange of body fluids and the direct contact of sexual intercourse.

Food poisoning is a common example of a disease transmitted by an object. This disease is often transmitted by contamination of the food by the food handler. In order to help prevent transmission of these types of diseases, restaurants have equipment that cleans and disinfects their dishes and utensils. Today, laws require food handlers to wash their hands thoroughly before preparing food, and frequent inspections of restaurants help prevent unsanitary conditions.

Diseases transmitted by vectors are most commonly spread by insects and arthropods. In Chapter 22, you learned about malaria, which is transmitted by mosquitoes, and in Chapter 31, you read about Lyme disease, transmitted by ticks. The bubonic plague—a disease that swept through Europe in the 1600s, killing up to one-third of the population—was transmitted from infected rats to humans by fleas. Flies also are significant vectors of disease, as shown in *Figure 42.4.*

Figure 42.4

Houseflies transmit disease when they land on infected materials, such as animal wastes, and then land on fresh food that is eaten by humans.

STUDENT JOURNAL

Diseases and Their Causes Have students make a table similar to that on page 1071, to list all the infectious diseases in this section. For each disease, students should indicate the cause, its vector (if applicable), and a description of the body system the disease infects. Have students retain their tables for use as study tools. L1 LS

What Causes the Symptoms of a Disease?

When a pathogen invades your body, it initially encounters your immune system. If the pathogen overcomes your defense system, it can cause damage directly in the tissues it has invaded. As the pathogens metabolize and multiply, they can kill host cells.

Damage to the host by viruses and bacteria

As you learned in Chapter 21, viruses invade cells and take over a host cell's genetic and metabolic machinery. Many viruses also cause the eventual death of the cells they invade.

Most of the damage done to host cells by bacteria is inflicted by toxins. Toxins are poisonous substances that are sometimes produced by microorganisms. These poisons are transported by the blood and can cause serious and sometimes fatal effects. Some toxins produce fever and cardiovascular disturbances. Toxins can also inhibit protein synthesis, destroy blood cells and blood vessels, and disrupt the nervous system, causing spasms.

For example, the toxin produced by tetanus bacteria affects nerve cells and produces uncontrollable muscle contractions, *Figure 42.5.* If the condition is left untreated, paralysis and death occur. Tetanus bacteria are normally present in soil. If dirt with the bacteria is transferred into a deep wound on your body, the bacteria begin to produce the toxin in the wounded area. A small amount of this toxin, about the same amount as the ink used to make a period on this page, would kill 30 people. That is why you should be vaccinated for tetanus.

Figure 42.5

Conditions of a battlefield are ideal for tetanus bacteria. Before the days of modern medicine, wounded soldiers faced the additional, deadly danger of becoming infected with these bacteria.

Discussion Question

Elicit from students why addicts who share needles when taking drugs often spread diseases to each other. *The needles are not sterile. Pathogens present on the needles themselves or present in fluids (including blood) contained in the needles are easily passed from one person to the next.* Explain to students that this process is an example of transmission by an object.

Enrichment

Have students research a disease in which they are interested. Ask them to find the name of the pathogen, current treatments for the disease, and symptoms the disease causes. L2

Discussion

Discuss with students how the development of a vaccine for tetanus has reduced the occurrence of this disease in developed nations. Explain the need for periodic boosters to maintain effectiveness of the vaccine.

GLENCOE TECHNOLOGY

Videodisc

The Infinite Voyage: Insects: The Ruling Class
Prospecting for Healing Medicine from Insects (Ch. 8)

STUDENT JOURNAL

New Medicines Have students conduct research to find examples of medicines developed during wartime that have become important in treating civilians. Have students identify the reason the medicine was developed and explain what it is used to treat. L1 LS

ThinkingLab Design an Experiment

Purpose

LS Logical-Mathematical Students will design an experiment to determine how a virus enters a cell.

Process Skills

observe and infer, formulate models

Teaching Strategies

- Ask students to include a control setup in their procedures.

Thinking Critically

Students' experiments should include a control. One experiment might be to remove the growth factor from the cell to see if the virus can still get in. Another possibility would be to remove the receptor sites from the cell or to remove the glycoprotein spikes from the virus and see if the virus can still get in.

Assessment

Portfolio: Ask students to write a summary of the Thinking Lab. Have them include the summaries of their experimental designs in their Portfolios. L1 P

Brainstorming

Ask students to speculate about why antibiotics are often given during a viral infection if they are ineffective against viruses. *to prevent secondary bacterial infections* L1

ThinkingLab Design an Experiment

How does the herpes simplex virus spread?

Herpes simplex virus, which causes cold sores, infects a person for life, occasionally reproducing and then spreading to other cells in the body of its host. Scientists have been interested for a long time in how the herpes virus actually enters a cell.

Analysis

Scientists have found that the herpes virus infects a cell in one of two possible ways. It may latch onto a cell receptor with its own glycoprotein spike, or it may use this spike to grab a growth factor molecule that latches onto the receptor, as shown in the diagram.

Thinking Critically

Design an experiment to determine which method the herpes virus uses to get into a cell.

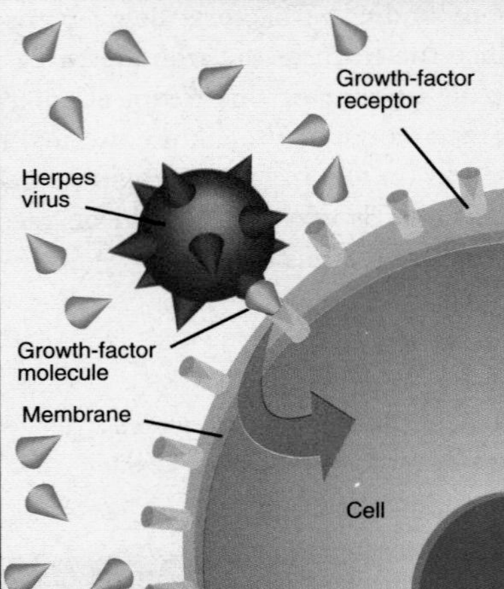

Patterns of Diseases

In today's highly mobile world, diseases can spread rapidly. Contaminated water, for example, can affect many thousands of people quickly. Therefore, identifying a pathogen, its method of transmission, and the geographic distribution of a disease is a major concern of government health departments. The Centers for Disease Control (CDC), the central source of disease information in the United States, publishes a weekly report about the incidence of specific diseases.

Some diseases, such as typhoid fever, occur only occasionally in the United States. An outbreak occurs periodically and is often due to someone traveling in a foreign country and bringing the disease back to the States. On the other hand, many diseases are constantly present in the population. Such a disease is called an **endemic disease.** The common cold is an endemic disease.

Sometimes, an epidemic breaks out. An **epidemic** occurs when many people in a given area are afflicted with the same disease in a relatively short period of time, *Figure 42.6.* Influenza is a disease that often achieves epidemic status, sometimes spreading to many parts of the world.

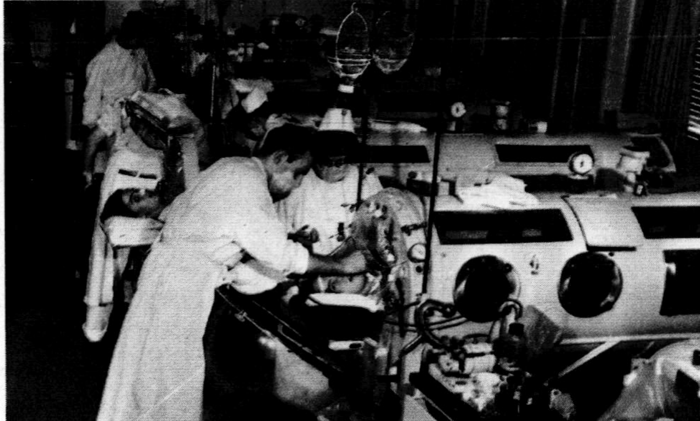

Figure 42.6

A polio epidemic spread across the United States in the 1950s. Victims of the disease were paralyzed or died when the polio virus attacked the nerves of the brain and spinal cord. Many survived only after being placed in an iron lung—a machine that allowed the patient to continue to breathe.

PROJECT

Locating Agents of Disease

Ask students to survey their school for areas that have large numbers of bacteria present. Have students swab various surfaces with a moistened cotton swab and then streak the swab across the surface of an agar plate. Have students incubate the agar plates and observe and record their results. L1

Treating Diseases

When a person becomes sick, the disease often can be treated with medicinal drugs, such as antibiotics. Any substance produced by a microorganism that, in small amounts, will kill or inhibit the growth and reproduction of other microorganisms is an **antibiotic**. Antibiotics are produced naturally by various species of bacteria and fungi. Although they work to cure some bacterial infections, antibiotics do not destroy viruses.

A problem that sometimes occurs with the continued use of antibiotics is that the bacteria become resistant to the drugs. That means the drugs become ineffective. Penicillin, an antibiotic produced by a fungus, was used for the first time in the 1940s and is still one of the most effective antibiotics used today. However, over the past 50 years that penicillin has been used, more and more types of bacteria have evolved that are resistant to it. One example of resistance is graphed in *Figure 42.7.*

The use of antibiotics is only one way to fight infections. In addition to medicinal drugs, your body has its own built-in defense system called the immune system that continually works to keep you healthy.

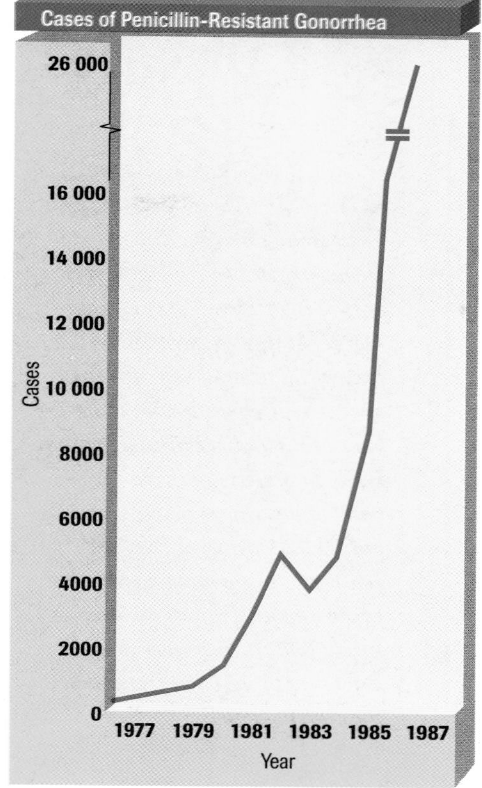

Figure 42.7

Bacteria that are resistant to penicillin produce an enzyme that breaks down this antibiotic. In certain infections, such as gonorrhea, an STD, this resistance is a problem because so far, penicillin has been the most successful drug in treating the infection. Notice the increase in the number of reported cases of gonorrhea in the United States.

Section Review

Understanding Concepts

1. What are the major reservoirs of pathogens?
2. In what way can a family member who is in the incubation period for strep throat be a threat to your good health?
3. When does a disease become an epidemic?

Thinking Critically

4. Many patients enter the hospital with one medical problem but contract an infection while in the hospital. What are possible ways in which a disease might be transmitted to a hospital patient?

Skill Review

5. **Designing an Experiment** Design an experiment to determine whether a recently identified bacterium causes a type of pneumonia. For more help, refer to Practicing Scientific Methods in the *Skill Handbook.*

SECTION 42.1

Visual Learning

Have students explain how the graph in Figure 42.7 is an example of natural selection. LS

3 Assess

Check for Understanding

Have students work in groups to draw an infection cycle that involves a vector. Ask them to describe ways of disrupting and preventing this type of infection cycle. L1 COOP LEARN

Reteach

Have students illustrate Koch's postulates for a specific disease. L1

Extension

Ask students to research the history of the development of antibiotics. L1

Assessment

Oral: Provide students with a diagram of a disease infection cycle. Ask them to provide feasible ways of disrupting the cycle. L1

4 Close

Discussion

Discuss the kinds of precautions school cafeterias must take to avoid disease transmission.

Answers to Section Review

1. infected animals, soil, and contaminated water
2. They can spread the disease to you.
3. When many people in a given area are infected with the same disease in a relatively short period of time.

Thinking Critically

4. from direct contact with employees, through contaminated objects, or from airborne transmission

Skill Review

5. You could take the suspected organisms from an infected animal and grow them in a pure culture. Then inject these organisms into another animal. If that animal develops the same disease symptoms, then you know the organisms probably caused the disease. To be sure, isolate the organisms from the animal and compare with the original organisms.

SECTION 42.2

Defense Against Infectious Diseases

Section Preview

Objectives

Explain the various types of nonspecific defense mechanisms.

Compare antibody and cellular immunity.

Key Terms

tissue fluid
lymph
lymph node
lymphocyte
phagocyte
macrophage
pus
complement
immunity
T cell
B cell
vaccine

You can't see it, but a war is going on around these teenagers. In fact, the same sort of war is occurring around you. Hordes of unseen enemies are present all around you—in the air, on your chair, and even on your books and pencils. Defenders ready to protect you from the onset of attack are inside your body. How does your body save you from the formidable foes that try to establish infectious diseases inside you? Which microscopic defenders protect you from these unseen enemies?

The Lymphatic System

Your lymphatic system is one line of defense your body has against outside pathogens. In fact, this system performs three basic functions. Within your body, the lymphatic system maintains homeostasis by keeping body fluids at a constant level. This system also absorbs fats from the digestive tract, as described in Chapter 38, and it helps the body defend itself against disease. *Figure 42.8* shows the major organs and vessels that make up the lymphatic system.

Pathways through the lymphatic system

The cells of your body are constantly bathed with fluid. This **tissue fluid** is the fluid that forms when water and dissolved substances diffuse from the bloodstream into intercellular spaces and the surrounding tissues. The tissue fluid then collects in open-ended lymph capillaries. Once the tissue fluid enters these lymphatic vessels, it is called **lymph.**

Lymphatic capillaries meet to form larger vessels called lymph veins. The flow of lymph is only toward the heart, so there are no lymph arteries. The lymph veins converge to form two lymph ducts. These ducts return the lymph to the bloodstream in the shoulder area. However, before it is returned, the lymph has been filtered through various lymph organs.

Organs of the lymphatic system

At locations along the lymphatic pathways, the lymph vessels pass

SECTION 42.2

Prepare

Key Concepts

The anatomy and physiology of the lymphatic system are discussed in this section. Various types of nonspecific defense mechanisms are described along with the specific defenses of antibody and cellular immunity. The last part of the section discusses AIDS and its effect on the immune system.

Block Scheduling

Look for this symbol for strategies that are useful in a block scheduling format. For more information on block scheduling, refer to Section 42.2 in the **Lesson Plans** booklet.

Materials

- Collect articles on AIDS for the Display and the Biology and Society Teaching Strategies.
- Buy mouthwashes for the Alternate Lab.
- Sterilize toothpicks and prepare nutrient agar plates for the Alternate Lab.

1 Focus

Bellringer

Before presenting the lesson, display **Section Focus Master 98** on the overhead projector and have students answer the accompanying questions. L1 SAE

Discussion

Ask students why their physicians feel under their chins and behind their ears and look down their throats when they complain of sore throats. *The doctors are looking for evidence of swelling in the glands and for redness and blotchy white spots on the throat that indicate bacterial infections.*

Program Resources

Section Focus Master 98 L1 SAE
Reinforcement and Study Guide, pp. 172-174 L1
Biolab and Minilab Worksheets, pp. 168-170 L1
Performance Assessment in the Biology Classroom, p. 45 L1 P
Laboratory Manual, pp. 261-264 L2
Concept Mapping, p. 42 L1
Critical Thinking/Problem Solving, p. 42 L3
Basic Skills Transparency 29 and **Master** L1 SAE
Basic Concepts Transparencies 77, 78a, 78b, 79 and **Masters** L1 SAE
Reteaching Transparency 42 and **Master** L1 SAE

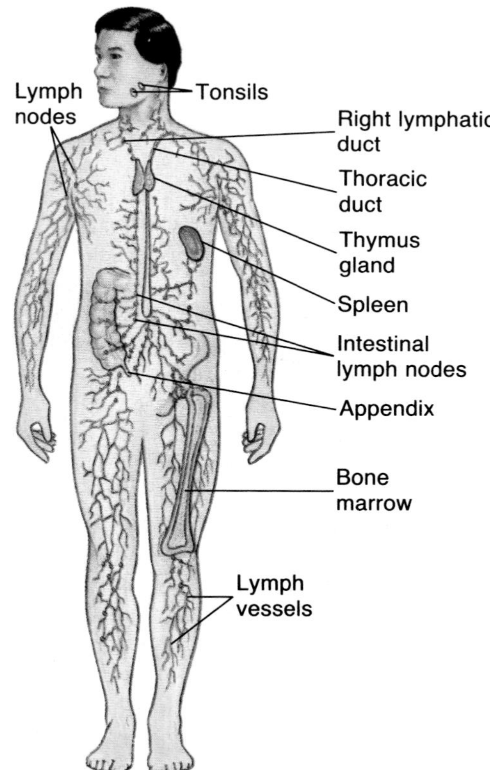

Figure 42.8

The lymphatic system consists of organs and vessels throughout the body. Each plays a role in filtering out harmful or waste particles. As you read about the various vessels, tissues, and organs of the lymphatic system, locate them on this diagram.

through lymph nodes. A **lymph node** is a small mass of tissue that filters lymph, as shown in *Figure 42.9.* Lymph nodes are made of an interlaced fiber network that holds lymphocytes. A **lymphocyte** is a type of white blood cell that defends the body against foreign substances.

Have you ever had a sore throat caused by infected tonsils? The tonsils are unusually large groups of lymph nodes located at the back of the mouth cavity and at the back of the throat. They form a protective ring around the openings between the nasal and oral cavities. Your tonsils provide protection against bacteria and other harmful material that enters your nose and mouth.

The spleen detects and responds to foreign substances in the blood. It also filters out and destroys bacteria and worn-out red blood cells and acts as a blood reservoir. Unlike the lymph nodes, the spleen does not filter lymph.

Another important organ in the lymphatic system is the thymus gland, which is located above the heart. The size of the thymus gland differs markedly depending on your age. In newborns and young children, it is quite prominent, and it continues to grow until puberty, although not as rapidly as other body structures. After puberty, it gradually decreases in size. The thymus gland processes some of the lymphocytes that are involved in the body's defense system. It is here that the lymphocytes mature and develop into cells that fight specific pathogens.

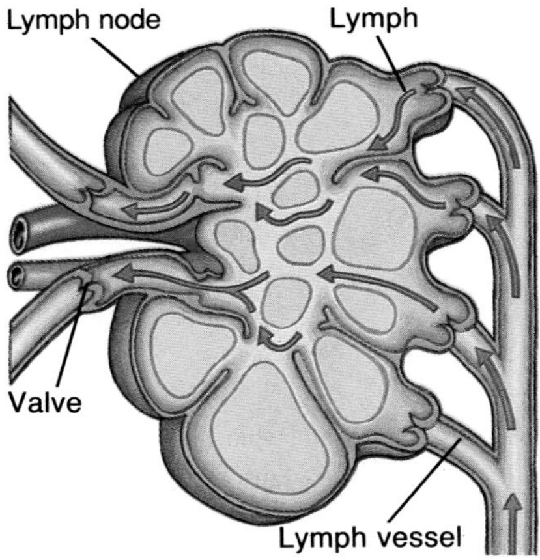

Figure 42.9

When lymph filters through a lymph node, the fiber network of the node traps microorganisms. The presence of the microorganisms stimulates the lymphocytes to divide and increase in numbers.

SECTION 42.2

2 Teach

Display

Before beginning this chapter, assemble a bulletin board display containing newspaper and magazine articles on AIDS. Refer to the articles when discussing AIDS and other viral infections.

Demonstration

LS **Visual-Spatial** Using an overhead microscope, demonstrate the differences between white blood cells (nucleated) and red blood cells. You can also use 35-mm slides or have students examine prepared slides to observe these features.

Discussion

LS **Interpersonal** Ask students to compare and contrast the functions of the circulatory system and the lymphatic system. Have students as a class discuss the two systems. L1 COOP LEARN

Software

Pathology: Disease and Defenses, IBM Biology Series.

Audiovisual

Show the video *Our Immune System*, National Geographic.

GLENCOE TECHNOLOGY

Videodisc

Biology: The Dynamics of Life
Disc 2, Side 1
Lymphocytes (Ch. 47)

STUDENT JOURNAL

Outlining the Lymphatic System Ask students to draw an outline of a human figure. Ask them to add and label all of the organs of the lymphatic system. L1 LS

MiniLab

Purpose

LS Logical-Mathematical Students will determine the percentages of different kinds of white blood cells.

Process Skills

classify, observe and infer, compare and contrast, collect and organize data

Teaching Strategies

- It would be helpful for student identification to set up a slide showing both an eosinophil and a basophil.

Expected Results

The common percentages of white blood cells are: neutrophils, 60-70%; lymphocytes, 20-25%; monocytes, 3-8%; eosinophils, 2-4%; and basophils, 0.5-1%.

Analysis

1. neutrophils, lymphocytes
2. White blood cells have a nucleus; red blood cells don't.
3. White blood cells increase in number and move to the infection site to destroy invading pathogens.

Assessment

Journal: Have students include summaries of the lab, their drawings, their data tables, and their answers to the Analysis questions in their journals. L1

MiniLab

What are the percentages of different kinds of white blood cells?

There are five types of white blood cells that defend your body: neutrophils, lymphocytes, monocytes, eosinophils, and basophils.

Procedure

1. Make a data table to record numbers of cells.
2. Mount a slide of blood cells on the microscope, and focus on low power. Turn to high power and examine the white blood cells.
3. Find a neutrophil. Its nucleus has several lobes, usually three. Neutrophils are phagocytic cells that arrive first at a wound site.
4. Find a lymphocyte. Lymphocytes have nuclei that nearly fill the cells.
5. Find a monocyte. This phagocyte is two to three times larger than the other white blood cells. Its nucleus fills about half of the cell.
6. Find a basophil. These histamine-releasing cells are covered with granules that are stained purple. Eosinophils are phagocytic cells covered with granules that are stained pink.
7. Count a total of 50 white blood cells, and record how many of each type you see.
8. Calculate the percentage by multiplying the number of each cell type by two. Record the percentages. Diagram each cell type.

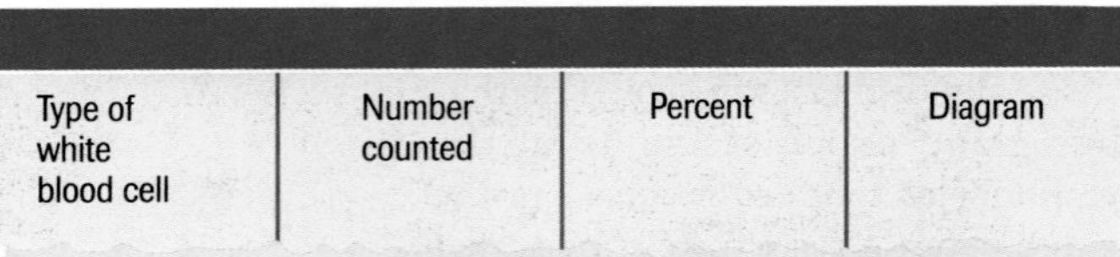

Type of white blood cell	Number counted	Percent	Diagram

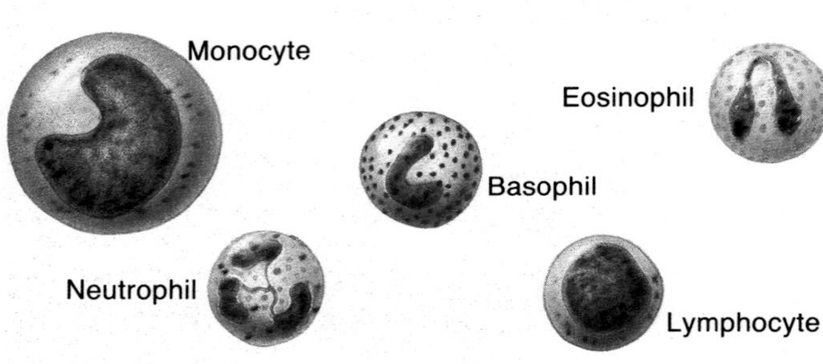

Analysis

1. Which type of white blood cell was most common? Second most common?
2. What is the major difference between red and white blood cells?
3. Why would you expect the white blood cell count to go up during an infection?

Innate Immune System

You know that pathogens are constantly bombarding your body. Yet your body is also constantly working to fight them off. Some of the defense mechanisms your body uses are effective against a wide variety of pathogens. In other words, these defense mechanisms are nonspecific and make up the innate immune system. Nonspecific defense mechanisms are always present and always fight off infection the same way.

Skin—The first line of defense

When a potential pathogen contacts your body, the first defense mechanism that it confronts is often your skin. Like the walls of a castle, intact skin is a formidable physical barrier to the entrance of microorganisms. Recall that the outer skin consists of many layers of keratinized cells that are closely packed together. Skin is also normally populated by millions of microorganisms that inhibit the multiplication of pathogens that land on the skin.

Secretions that destroy microbes

Besides the skin, pathogens encounter your body's secretions of mucus, sweat, tears, and saliva. The main function of mucus is to prevent various areas of the body from drying out. Because mucus is slightly viscous (thick), it also traps many microbes and other foreign substances that enter the respiratory and digestive tracts. Mucus is continually swallowed and passed to the stomach, where acidic gastric juice (made of hydrochloric acid and other fluids) destroys most bacteria and their toxins. Sweat, tears, and saliva all contain the enzyme lysozyme, which is capable of breaking down the cell walls of some bacteria.

Alternate Lab: Mouthwash and Bacteria

Purpose

LS Logical-Mathematical Students will test which mouthwash is the most effective in inhibiting bacterial growth.

Materials

3 types of mouthwash disks, forceps, sterile toothpicks wrapped in aluminum foil, glass hockey sticks, transparent tape, small bottles of alcohol, plates of nutrient agar, glass marker

Procedure

Give the following directions to students.

1. Using a sterile toothpick, gently scrape around your teeth.
2. Transfer the scrapings to the surface of an agar plate using sterile technique.
3. Repeat steps 1 and 2 several times.
4. Dip the glass hockey stick in alcohol and allow it to air dry. Do not allow it to touch any surface.
5. Spread the scraping material evenly over the surface of the agar plate using the hockey stick.
6. With a marker, divide the bottom of the plate into four sections.

Phagocytosis of microbes

If a pathogen manages to get past the skin and body secretions, your body has several nonspecific defense mechanisms remaining that can destroy the pathogen and restore homeostasis. Some foreign cells encounter body cells that carry on phagocytosis. Recall from Chapter 9 that phagocytosis occurs when a cell engulfs a particle. A **phagocyte** is a type of white blood cell that ingests and destroys pathogens by surrounding and engulfing them. When a pathogen is present, phagocytic cells migrate out of your blood capillaries to the infected areas. One type of phagocyte, called a **macrophage,** that combats invading pathogens is shown in *Figure 42.10.* Known as the giant scavengers or big eaters, macrophages develop from maturing monocytes.

After macrophages engulf large numbers of microbes and damaged tissue, they eventually die. After a few days, the infected area harbors a collection of dead white blood cells and various body fluids called **pus.** Pus formation usually continues until the infection subsides.

Inflammation of body tissues

When microbes damage body tissues, inflammation may result. Inflammation is a reaction to any type of injury, not just to infection. Physical force, chemical substances, extreme temperatures, and radiation can also inflame body tissues. Inflammation is characterized by four

Figure 42.10

Macrophages migrate out of capillaries by squeezing between the cells of the capillary wall. Macrophages will attack anything they recognize as foreign, including microbes and dust particles that are breathed into the lungs. Once the macrophage has ingested the foreign material, lysosomal enzymes that digest the foreign matter are released.

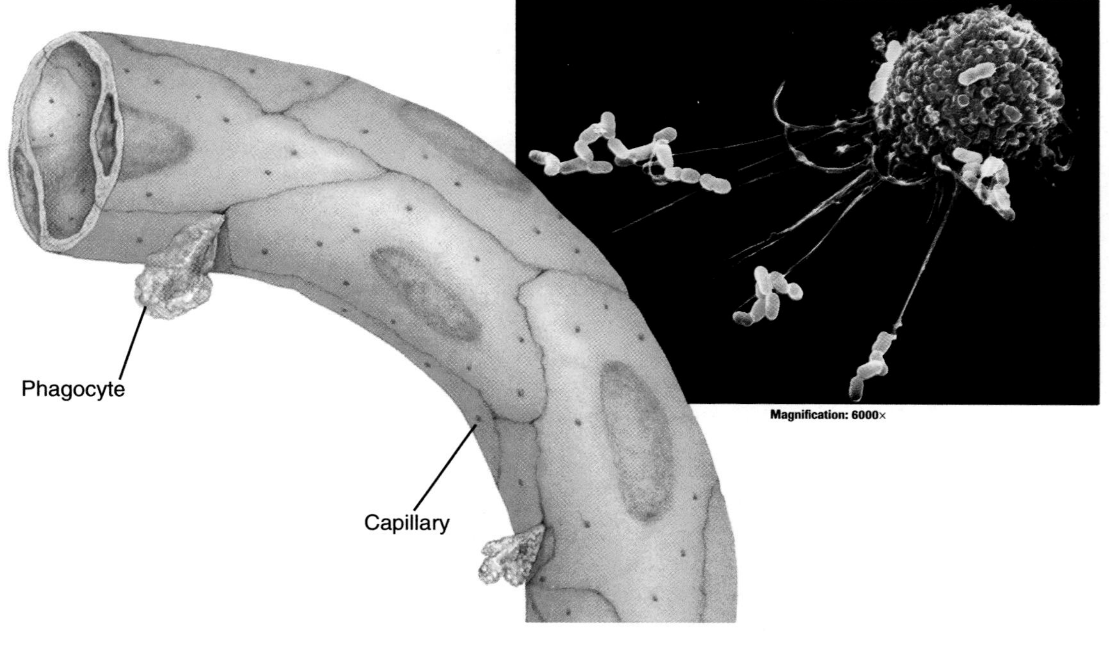

SECTION 42.2

Audiovisual

Show the video *Immunity,* Charles Clark Co., Inc.

Science, Technology, and Society

Defensins

The first cell to approach a microbe invader is a neutrophil. The neutrophil uses its own potent set of antibiotics, called defensins, to destroy invaders. These native molecules hold great promise for a new generation of antibiotics.

In the test tube, small amounts of defensins destroy pathogens almost instantly by making the bacteria leaky. Because defensins are native components of cells, they may be safer to take than many types of antibiotics. The trick now is to find out how to produce large amounts of defensins. The standard genetic engineering techniques using bacteria or yeast to make large amounts of desired proteins does not work because defensins kill these microbes.

7. Dip the tweezers into alcohol and allow it to air dry without touching any surface.
8. Label the bottom with the three types of mouthwash. Transfer two of each type of mouthwash disk into each section of the plate.
9. Place untreated disks in the fourth section.
10. Tape the plates closed and incubate for 48 hours at 37°C.
11. View and look for regions of inhibition near the disks.

Analysis

1. Why was one section left without treated disks? *This section was the control.*
2. Did any of the mouthwashes inhibit the growth of the bacteria? *Answers will vary.*

Assessment

Journal: Have students include a summary of the lab, and the answers to the Analysis questions in their Student Journal. L1

SECTION 42.2

Science, Technology, and Society

Anti-Adhesive Therapy

In order for white blood cells to leave the bloodstream to find infected tissue, they are believed to attach to cell adhesion molecules that protrude from the walls of blood vessels. In certain cases, such as the inflammatory diseases rheumatoid arthritis and multiple sclerosis, white blood cells accumulate in tissues where they should not be.

During an infection, infected tissues may produce substances such as interleukin-1 and tumor necrosis factor that arouse endothelial cells lining nearby capillaries. The activated endothelial cells then stick out their cell adhesion molecules. Researchers believe that anti-adhesive therapy might be applicable to diseases like arthritis. They hope they can develop drugs with structures to interfere with adhesions, selectively blocking the bloodstream exits for white blood cells unless they are actually needed.

Using an Analogy

Have students orally make an analogy between the reaction of the immune system and a battle scene. Ask them who or what might represent the skin, macrophage, B cells, types of T cells, lymph, and antibodies. L1

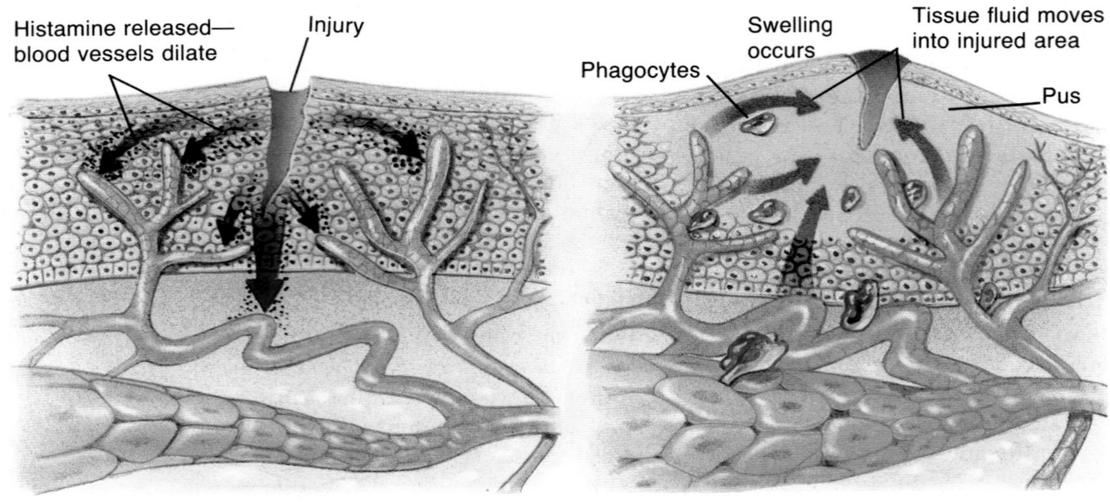

Figure 42.11

Histamine is produced when tissue damage occurs. Histamine causes blood vessels in the injured area to dilate (left). Therefore, they become more permeable to tissue fluid. This increase in tissue fluid in the injured area helps the body destroy toxic agents and restore homeostasis. The dilated blood vessels cause the redness of an inflamed area (right). The increase in tissue fluid causes the swelling.

symptoms—redness, swelling, pain, and heat. It begins when damaged cells release histamine, *Figure 42.11.*

As inflammation proceeds, phagocytes migrate into the injured area and begin to ingest pathogens. They also release a chemical that causes the hypothalamus to reset the body's temperature, causing a fever. Up to a certain point, fever is actually a defense against disease. A higher body temperature interferes with the metabolism of some microorganisms and speeds up the body's reactions, possibly helping body tissues repair themselves more quickly.

Protective Proteins

Another nonspecific defense is called complement. **Complement** is a group of proteins found in the blood that attach themselves to the surfaces of pathogens. When these protein molecules attach to pathogens, they help your body destroy the pathogens by damaging their plasma membranes and attracting an increased number of phagocytes.

When an infection is caused by a virus, your body faces a problem. Phagocytes cannot destroy the virus. Recall that a virus multiplies within a host cell. If phagocytes engulf the virus, they are destroyed after the virus multiplies within their cells. One way your body does counter viral infections is with interferons. Interferons are proteins that are host-cell specific; that is, human interferons will protect human cells from viruses but will do little to protect cells of other species.

Interferon is produced by a body cell that has been infected by the virus. The interferon diffuses into uninfected neighboring cells, which then produce antiviral proteins that often disrupt viral multiplication.

Meeting Individual Needs

Learning Disabled/Students Acquiring English Have students having difficulty with the science words in this unit color and label an illustration of the flow of tissue fluid back to the bloodstream. Have them label the parts in the diagram. You may wish to have students work in groups. L1 SAE IN

Adaptive Immune System

When your body is invaded by a pathogen, its nonspecific defenses begin. The nonspecific defenses are an immediate general defense while specific forces are being mobilized—a process that takes several days. Defense against a specific pathogen by building up a resistance to it is called **immunity.** When the body recognizes a specific foreign substance, it develops an immune response to inactivate or destroy it. Specific defense is the job of the lymphatic system, and it includes the production of two kinds of immunity: antibody immunity and cellular immunity.

Initiating antibody immunity

Normally, the immune system recognizes components of the body as self and foreign matter as nonself. Immunity occurs when the system recognizes a foreign substance and responds to it by producing specialized lymphocytes, which then produce antibodies specific to that foreign substance. Recall from Chapter 39 that organisms or substances that provoke such a response are called antigens. Antigens are usually proteins and are present on the surfaces of whole organisms, such as bacteria, or on parts of organisms, such as the pollen grains of plants. When a foreign antigen is introduced into your body, it causes the production of antibodies.

After this initial infection and recovery, if you are reinfected, you will not get sick because all the defense mechanisms are in place. You now have immunity.

Ethics

BIOLOGY & SOCIETY

Human Guinea Pigs

From the time it was first identified in 1981, AIDS quickly became the medical scourge of the late 20th century. Shortly after HIV was identified as the cause of AIDS, scientists began searching for vaccines that would prevent HIV infection or render the virus harmless in those already infected. A French scientist hoped to be the first to demonstrate the effectiveness of such a vaccine.

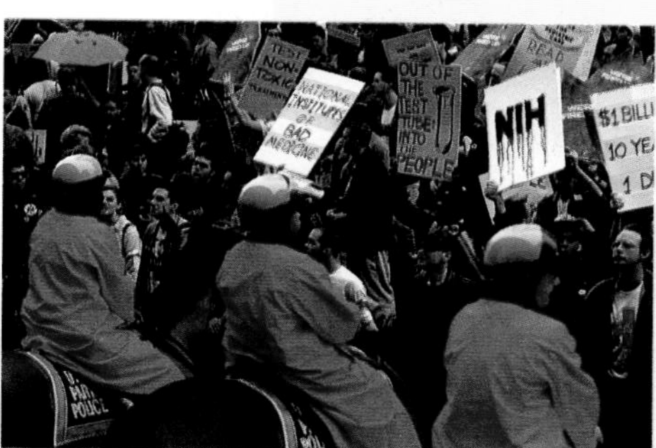

Unsuccessful trials The researcher first tested a crude version of his vaccine in Africa. The vaccine had been prepared for use in animals; however, he tested it on humans instead. His subjects were children and too young to give their informed consent to the procedure. The vaccine proved to be unsuccessful. The National Institutes of Health have imposed serious penalties on the French researcher after the death of the participants in his trials. They also criticized him for using children as experimental subjects.

Different Viewpoints

Pressure for a medical breakthrough Concerned citizens in many countries are lobbying for a quick response to the AIDS epidemic, but development of a vaccine to prevent or treat AIDS may take many years. In the meantime, society as a whole must decide at what point in a vaccine's development it is appropriate to test it on human subjects.

Testing on nonhuman primates only? Some researchers contend that any experimental AIDS drugs or vaccines should first be tested on nonhuman primates. Only if the drug or vaccine helps these primates should human trials proceed. However, some people with AIDS favor early human testing. With so many people dying, these advocates say that time should not be wasted on animal studies that might not apply to humans. When asked whether trying untested AIDS drugs on humans is ethical, the AIDS patients point out that they have little to lose.

INVESTIGATING the Issue

Writing About Biology Research and present a report on how AIDS activists lobby the Food and Drug Administration to speed up the time it takes to approve the use of experimental drugs.

Ethics

BIOLOGY & SOCIETY

Human Guinea Pigs

Purpose

Students will examine some of the problems and dilemmas concerning development of an AIDS vaccine.

Teaching Strategies

- Collect recent newspaper articles concerning AIDS research. Use these articles to discuss the development of an AIDS vaccine.
- Ask students to state their views on using humans to test unproven drugs and other therapies. Have them consider whether or not they would agree to participate in such tests if they were suffering from a fatal disease.

INVESTIGATING the Issue

Writing About Biology Students will most likely report that many groups say not enough money is being spent on AIDS research and the development of treatments.

Going Further

Have students research the controversies over orphan drugs. Tell them orphan drugs are used to treat only a small part of the population and are expensive for drug companies to develop and produce.

Reinforcement

Compare the antigen-antibody reaction to the lock-and-key fit of enzymes and substrates. L1

Assessment

Performance Assessment in the Biology Classroom: p. 45, Vaccine Models and the Common Cold. Have students carry out this activity to find out why a single vaccine will not combat the common cold.

CULTURAL DIVERSITY

Susumu Tonagawa and Immunity from Diseases

Introduce students to the important contributions of Japanese immunologist Susumu Tonagawa toward our understanding of the immune system.

Tonagawa's major contribution ended the debate on whether individuals inherited separate genes for each of the millions of antibody molecules, or whether individuals inherit a small number of genes that diversify in specialized somatic cells. Tonagawa was able to show that somatic cells use a multi-step process to rearrange fragments from different genes, in different ways, to produce many different antibodies. For this work, Tonagawa received a Nobel prize in 1987 .

Visual Learning

Figure 42.12 Ask students to write a summary of what is happening in this figure.

History Connection

The Boy in the Plastic Bubble

David, the "bubble boy" of Houston, Texas, lived in a sterile plastic bubble for all but 15 days of his life. He suffered from severe combined immune deficiency (SCID), a rare genetic disorder related to a disfunction of the lymphocyte-producing cells of the bone marrow.

At birth, David was placed in a sterile plastic bubble. As he grew, he was placed in increasingly larger bubbles. At the age of 12, David left his bubble and underwent a bone marrow transplant using tissue from his sister. David died four months later, in February, 1984, of blood cancer, which was unrelated to his transplant.

Since David's death, no other child has lived in a bubble as David did. Bone marrow transplants from family members have been successfully used to treat SCID. The treatment involves taking about a liter of bone marrow from the donor, and eliminating all the mature T cells so they don't attack the recipient's cells. The remaining immature cells are then injected into a vein of the recipient.

Figure 42.12

ANTIBODY IMMUNITY
Antibody immunity utilizes B cells and antibodies in defending your body against invading pathogens.

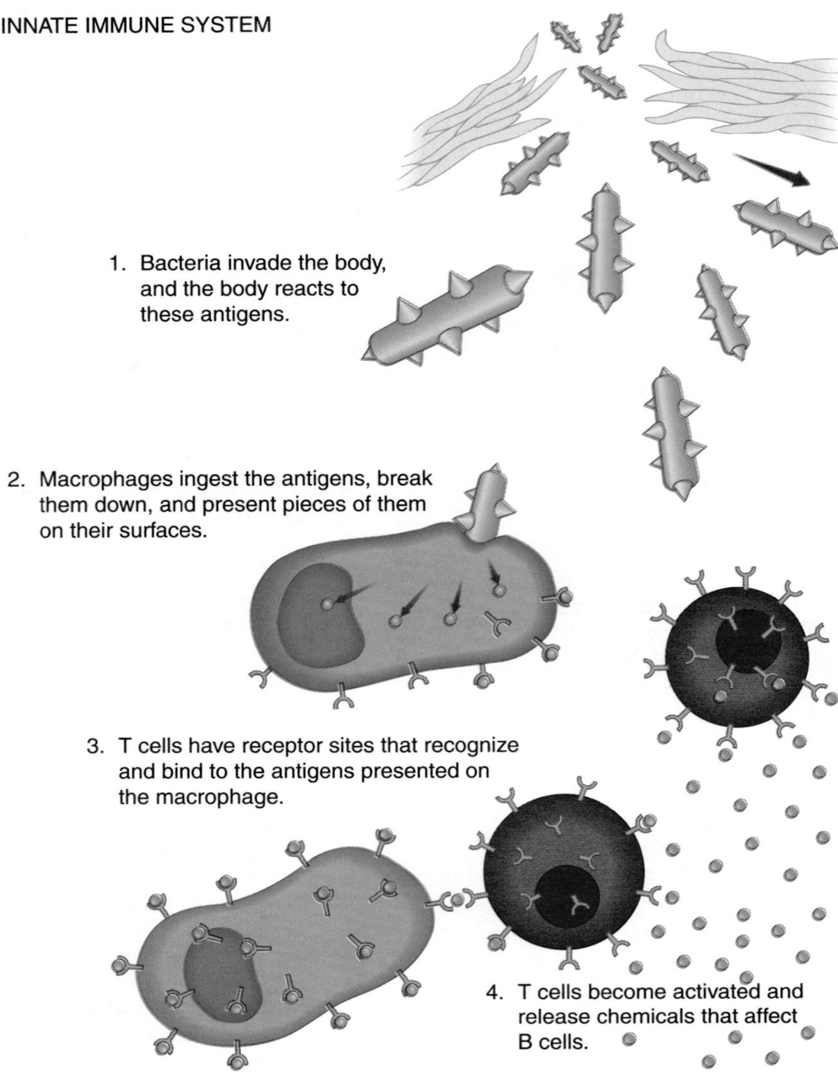

INNATE IMMUNE SYSTEM
1. Bacteria invade the body, and the body reacts to these antigens.
2. Macrophages ingest the antigens, break them down, and present pieces of them on their surfaces.
3. T cells have receptor sites that recognize and bind to the antigens presented on the macrophage.
4. T cells become activated and release chemicals that affect B cells.

Antibody immunity is a type of chemical warfare within your body. Several types of cells are involved. Follow the steps of antibody immunity illustrated in *Figure 42.12.*

When a pathogen invades your body, it is engulfed by a macrophage. Portions of the antigen are taken up by the macrophage and positioned on its plasma membrane. At this time, a type of lymphocyte called a T cell becomes involved. A **T cell** is a lymphocyte that is produced in bone marrow and processed in the thymus gland. Different kinds of T cells play different roles in immunity.

One kind of T cell, called a helper T cell, interacts with B cells. A **B cell** is a lymphocyte that produces antibodies when activated by one of the

Meeting Individual Needs

Physically Challenged Ask students who are physically challenged by allergies or asthma to speak to the class about how the disorder affects their lives. L1

Visually Impaired Photocopy and enlarge Figures 42.12 and 42.13. Use a felt-tip pen to outline the edges of the cells and the arrows in the illustrations for use by visually impaired students. L1

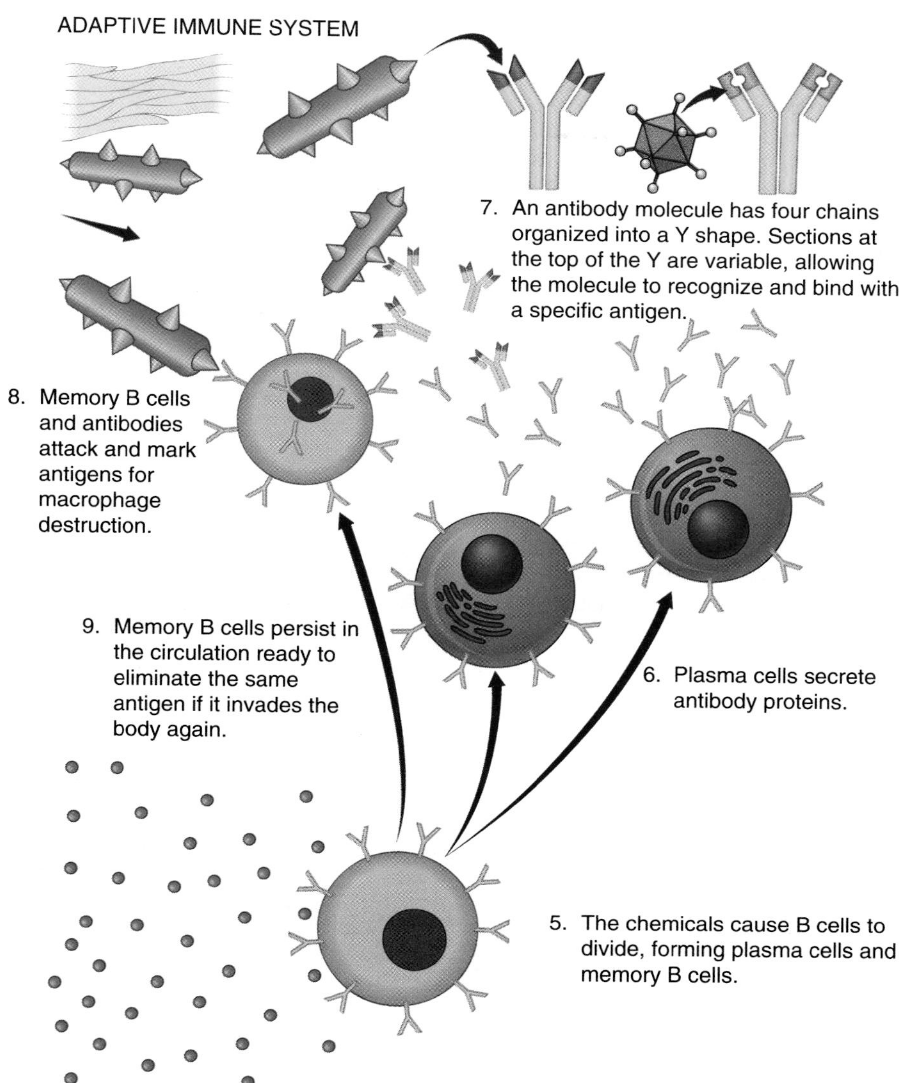

ADAPTIVE IMMUNE SYSTEM
7. An antibody molecule has four chains organized into a Y shape. Sections at the top of the Y are variable, allowing the molecule to recognize and bind with a specific antigen.
8. Memory B cells and antibodies attack and mark antigens for macrophage destruction.
9. Memory B cells persist in the circulation ready to eliminate the same antigen if it invades the body again.
6. Plasma cells secrete antibody proteins.
5. The chemicals cause B cells to divide, forming plasma cells and memory B cells.

T cells. B cells are produced from stem cells in bone marrow and released to lymphatic organs.

When an antibody is produced by a B cell, it is released into the bloodstream. Antibodies recognize those antigens to which they can fit and bind. This binding results in an antigen-antibody complex. The antigen-antibody complex protects your body in several ways. For example, a complex can neutralize bacterial toxins by blocking their active sites. It can deactivate viruses by attaching to them and preventing them from attaching to host cells. Or it can fix complement, which leads to the lysis of invading cells. Ultimately, the antigen-antibody complex is removed from your body when phagocytes destroy it.

Concept Development

Normally, the body's immune response helps the body maintain homeostasis. Sometimes, the body loses its ability to discriminate between self and nonself, which leads to autoimmunity. Autoimmunity is a response by antibodies or sensitized T cells against a person's own tissue antigens.

Autoimmunity is involved in multiple sclerosis, Graves' disease of the thyroid, and rheumatic fever, where antibodies are formed against the heart, and juvenile diabetes where antibodies are formed against the pancreas.

GLENCOE TECHNOLOGY

Videodisc

Biology: The Dynamics of Life
Disc 2, Side 1
Antibody Immunity (Ch. 45)

STVS: Human Biology
Disc 7, Side 1
White Cells in Action (Ch. 14)

interNET CONNECTION

Follow the link for this chapter on the Glencoe Homepage at **www.glencoe.com/sec/science** to find out more about the immune system.

STUDENT JOURNAL

Immune System Roles Have students write a play about the various immune system "characters" involved in the immune response to AIDS. L1 LS

Different Viewpoints in Biology

Explaining Self-Tolerance

Three different theories are used to try to explain self-tolerance. One theory presumes that there are some type of cell suppressors capable of suppressing the body from attacking itself. Two alternative ideas include clonal deletion and clonal anergy.

The clonal deletion and clonal anergy theories propose that the body becomes tolerant because mechanisms eliminate self-reactive cells (deletion) or turn them off (anergy). The idea that deletion occurs is supported by data from experiments that have demonstrated that certain self-reactive clones of maturing T cells were deleted from the thymus and never reached the rest of the body. Scientists are now trying to solve the mystery of exactly what kills the T cells. Deletion cannot explain all types of tolerance because some antigens may not be able to get to the thymus. Thus, another mechanism must deal with them.

Reinforcement

IS **Visual-Spatial** Have students make a flowchart to show the sequence of events in cellular immunity as shown in Figure 42.13. L1

Initiating cellular immunity

Like antibody immunity, cellular immunity also involves T cells and macrophages with antigens on their surfaces, but cellular immunity involves direct contact between the two. T cells, after being processed by the thymus gland, enter the lymphatic system and are stored in the lymph nodes and tonsils. Some of the T cells also circulate in the blood. However, unlike B cells, T cells do not secrete antibodies. Instead, they have antibody-like molecules called antigen receptors attached to their surfaces. These antigen receptors allow T cells to recognize and react to many different kinds of antigens.

Your body produces three main types of T cells: the helper T cells that are involved in antibody immunity, cytotoxic—or killer—T cells, and suppressor T cells. Like B cells, individual T cells seem to be specific for a single antigen, but they do not respond to antigens in the same way that B cells do. Follow *Figure 42.13* to see the steps of T cell response.

Figure 42.13
Cellular immunity

In cellular immunity, a macrophage engulfs an antigen, breaks it down, and places part of the foreign substance on its own cell surface. The macrophage then binds to the antigen receptor on helper T cells, activating the cytotoxic T cells to differentiate and produce identical clones. Some T cells remain behind in the lymph organs as memory T cells. Like memory B cells, these T cells are able to respond rapidly to a second attack. Other T cells travel out to the infected tissue to destroy the pathogen. At the infection site, cytotoxic T cells may destroy the pathogen directly or release chemicals that attract other macrophages to the site.

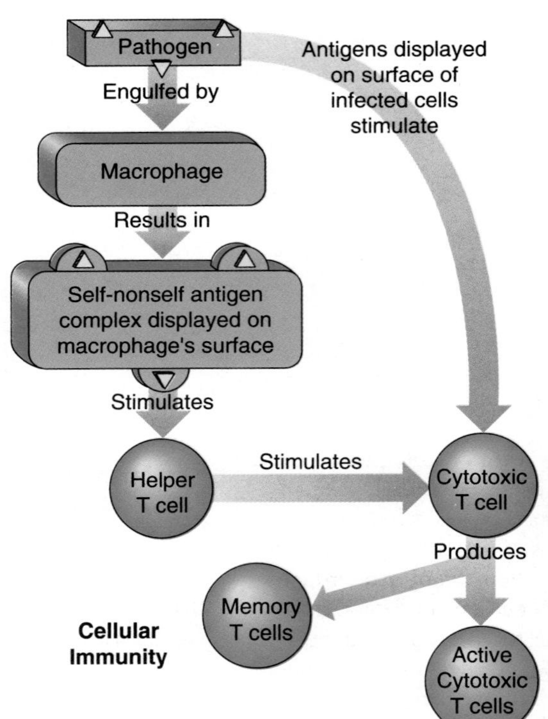

A cytotoxic T cell binds to a body cell that has been invaded by a virus and produces proteins that punch holes in the infected cell's wall.

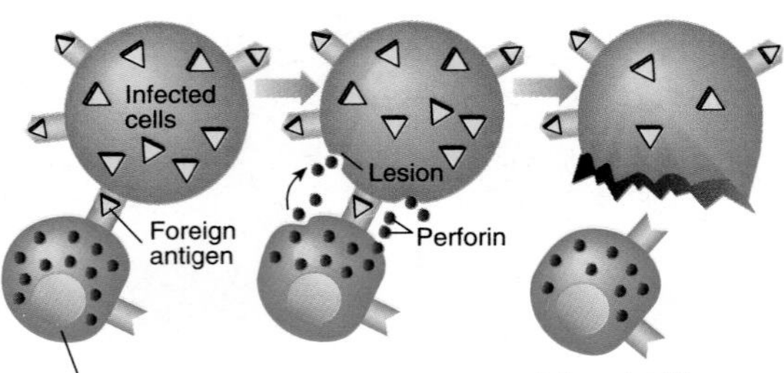

Magnification: 5000×

GLENCOE TECHNOLOGY

Videodisc

Biology: The Dynamics of Life
Disc 2, Side 1
Cellular Immunity (Ch. 46)

PORTFOLIO

Demonstrating Immune Response Divide students into groups and have them write and perform a short play demonstrating aspects of the specific immune response. They should each include a copy of the play in their portfolios. L2 COOP LEARN P IS

Meeting Individual Needs

Students Acquiring English Have students prepare a chart of the functions of T and B cells. Use the chart to check their understanding of these concepts. L1 SAE IS

Immunity from infectious diseases

Perhaps you had chicken pox as a child, *Figure 42.14*. Most children have had chicken pox by the time they enter school. Why don't you have the disease over again, as you do a cold? The answer is that you have become immune to this virus. Immunity from a disease may be either passive or active. Passive immunity develops as a result of passively acquiring antibodies. Active immunity develops as a result of exposure to antigens, which results in the production of antibodies and memory immune cells.

A glance at the immunization list shows that a chicken pox vaccine is missing. While one exists and has been approved by the FDA, a controversy exists about its use. The vaccine doesn't provide permanent immunity, and some scientists are concerned that the disease then will strike adults.

Figure 42.14

The virus that causes chicken pox can remain latent in the body for many years. Years later, it may be reactivated and cause the painful skin disease known as shingles.

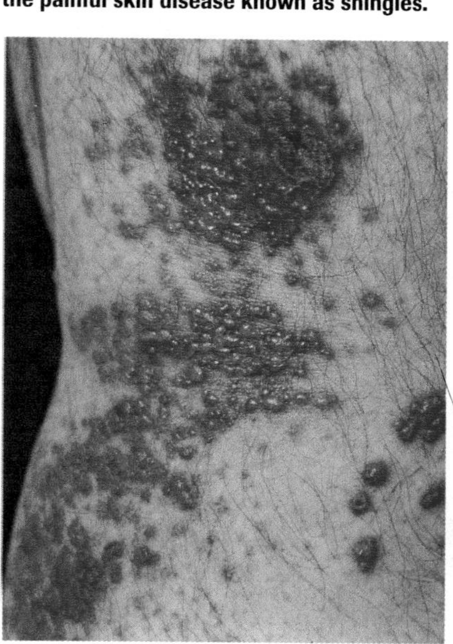

Health

Protecting the Future

In some countries, as many as half of the babies die before they can grow up. Immunity for several childhood diseases is now possible when a child is properly vaccinated. The table below shows the diseases that can be prevented by childhood immunization. Some of the information may be new to you because the number of recommended immunizations has been increased over the last few years. In some cases, immunizations have different names because they have been improved or combined to reduce the number of shots.

Recommended Childhood Immunizations

Immunization	Agent	Protection Against
Acellular DPT or Tetrammune	bacteria	diptheria, pertussis (whooping cough), tetanus (lockjaw)
MMR	virus	measles, mumps, rubella
OPV	virus	poliomyelitis (polio)
HBV	virus	hepatitis B
HIB or Tetrammune	bacteria	*Haemophilis influenzae B* (spinal meningitis)

The old, the new, and the improved Older vaccines that are still used are oral polio vaccine (OPV), which prevents the crippling disease poliomyelitis, and MMR, which protects against measles, mumps, and German measles (rubella). The newest recommended vaccine, HBV, is for hepatitis B, a serious disease that affects the liver.

Improved vaccines Acellular DPT is an enhanced version of the classic DPT shot that protects against diphtheria, whooping cough, and tetanus. Produced by improved methods, this version reduces the side effects, such as pain, fever, and fussiness, caused by the immunization. Another new twist on the classic DPT shot is called Tetrammune. It combines DPT with immunization for *Haemophilis influenzae B* and eliminates the need for a separate injection.

CONNECTION TO Biology

Vaccines are often made from dead or weakened pathogens. Why might a genetically engineered virus be safer to use for immunization than a weakened one?

Health

Protecting the Future

Purpose

Students will examine the recommended current childhood immunizations and become acquainted with new and improved vaccines.

Teaching Strategies

- Before having students read this Health Connection, ask them to get a list of the vaccinations that they have had. Have them compare their record with the recommended list. L1
- Other vaccinations are available for persons traveling outside the United States. Discuss why a person might need different vaccinations when traveling.

Possible Answer

A genetically engineered virus could have the genes removed that would allow it to replicate and cause an illness.

GLENCOE TECHNOLOGY

Videotape

The Secret of Life

Use the videotape *Nothing to Sneeze At: Viruses* to illustrate how the immune system responds to viral infection, and to show how scientific methods are used to deal with viruses.

STUDENT JOURNAL

The Nature of Scientific Discovery Ask students to prepare a report on Louis Pasteur's work with rabies or Jonas Salk's and Albert Sabin's work with polio vaccines. L1 LS

BioLab

AIDS and Its Effect on the Immune Response

Time Allotment 
One class period

Objectives
Review objectives with students before they begin the Biolab.

Process Skills
make and use graphs, formulate models, analyze

PREPARATION

Review briefly the way the human body's immune system functions before beginning this Biolab.

GLENCOE TECHNOLOGY

Videodisc
STVS: Human Biology
Disc 7, Side 1
Combined Immune Deficiency (Ch. 15)

BioLab

AIDS and Its Effect on the Immune Response

The HIV that leads to AIDS does not itself cause the life-threatening symptoms associated with the disease. Instead, the virus weakens a person's immune response to other pathogens that invade the body. This happens because the virus destroys the T cells that help in the production of antibodies. Thus, the immune system's ability to fight disease organisms is severely impaired. The person usually dies from the continually reoccurring secondary infections and cancers.

PREPARATION

Problem
How does AIDS affect the normal immune response?

Materials
colored pencils (red and blue)
graph paper (2 pieces)

Objectives
In this Biolab, you will:

- **Plot** graphs that demonstrate a healthy body's normal immune response and an AIDS-infected body's immune response.
- **Compare and interpret** these two different immune responses.

Magnification: 16 000×

PROCEDURE

Teaching Strategies

- The immune response data are actually a hypothetical index that summarizes the activity of the immune system, such as antibody production and the direct attack on the invaders. The numbers of microbes and viruses in Tables 1 and 2 are also hypothetical.

Data and Observations

See graphs on the next page. Dotted lines and straight lines have been used to distinguish between the two parts of each graph.

PROCEDURE

Part A: The Normal Immune Response

1. Make a graph of the data in Table 1. Number the vertical axis from 0 to 10 000 in multiples of 500. Label the horizontal axis with the number of days from 0 to 10. In red, plot the number of microbes on the vertical axis against the number of days. In blue, plot the immune response on the vertical axis against the number of days.
2. Label your graph *Normal Immune Response.*

Part B: The Immune Response in a Person with AIDS

1. On a second piece of graph paper, construct a graph of the data in Table 2. Number the axes as in your first graph, but label the horizontal axis with the number of years instead of days. In red, plot the number of viruses against the number of years. In blue, plot the immune response against the number of years.
2. Label your graph *Immune Response in Person with AIDS.*

Table 1 Normal Immune Response

Days	Number of viruses	Immune response (units)
0	1	0
1	1000	0
2	10 000	1
3	1000	100
4	10	1000
5	1	1000
6	0	1000
7	0	100
8	0	10
9	0	1
10	0	1

Table 2 Immune Response in Person with AIDS

Years	Number of viruses	Immune response (units)
0	1	0
1	100	1
2	10 000	50
3	10	1000
4	1	10 000
5	1	10 000
6	1	1000
7	10	500
8	100	100
9	1000	1
10	10 000	1

ANALYZE AND CONCLUDE

1. **Analyzing Data** Summarize what happens during the normal immune response and during the immune response of a person with AIDS.
2. **Comparing and Contrasting** How is the AIDS graph similar to that of the normal immune response? How is it different?
3. **Thinking Critically** Why does the number of AIDS viruses increase during years 7 through 10?

Going Further

Applying Concepts How could the HIV antibody be used for blood screening?

Normal Immune Response

10 000, 8000, 6000, 4000, 2000, 0

number of microbes

immune response

1 2 3 4 5 6 7 8 9 10

Days

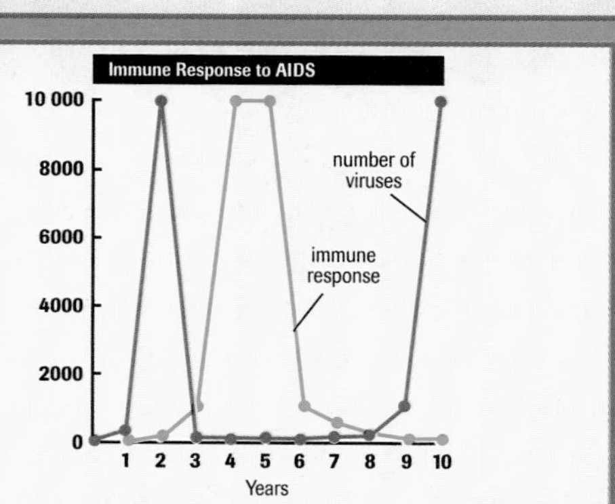

ANALYZE AND CONCLUDE

1. The normal immune response begins a day or two after infection and takes a few days to clear the body of microbes. The immune response in AIDS at first increases greatly, then decreases greatly even though the number of viruses continues to increase.
2. The immune response to the AIDS virus during the first four years is similar to the immune response to other viruses and bacteria. The immune response to AIDS, though, declines before the body is free of the AIDS virus.
3. The immune system is breaking down and is unable to control the AIDS virus.

Assessment

Portfolio: Have students write summaries of the lab and include them along with their graphs and answers to Analyze and Conclude questions in their Portfolios. L1

Going Further

Have students do research to find out about the diseases that AIDS victims frequently suffer from, such as *Pneumocystis carinii*, Kaposi's sarcoma, candidiasis, cryptococcosis, cytomegalovirus, cryptosporidiosis, Herpes simplex, and toxoplasmosis. Ask students to write a report on one of these illnesses, its cause, and the current treatment. L1 P

Teacher FYI

When tissues or organs are transplanted to replace injured or diseased ones, the immune system of the person receiving the transplant often causes the transplanted material to be rejected. The mechanism by which the transplant is destroyed may involve both cellular and antibody immunity.

In order to make transplants successful, physicians try to closely match the donor and recipient's tissue proteins. The most successful transplants occur when a person's own tissues are used in another part of the body. For example, in treating burn patients, skin grafts are frequently transplanted from one area of the patient's body to the burned area.

History Connection

Eliminating Smallpox

Smallpox was the first disease for which a vaccine was made and is the first disease to be eradicated due to a concerted vaccination effort throughout the world. The last known case of smallpox occurred in Birmingham, England, in July 1978. The infection from smallpox, spread by droplets, causes lesions in layers of the skin. The lesions leave disfiguring scars following recovery from the disease.

Approximately 86 percent of the people who contracted smallpox survived. During the Middle Ages, 80 percent of Europeans could expect to contract smallpox during their lifetimes. This disease devastated Native Americans when European settlers brought it to America. Because Native Americans had never been exposed to smallpox, many died from the disease.

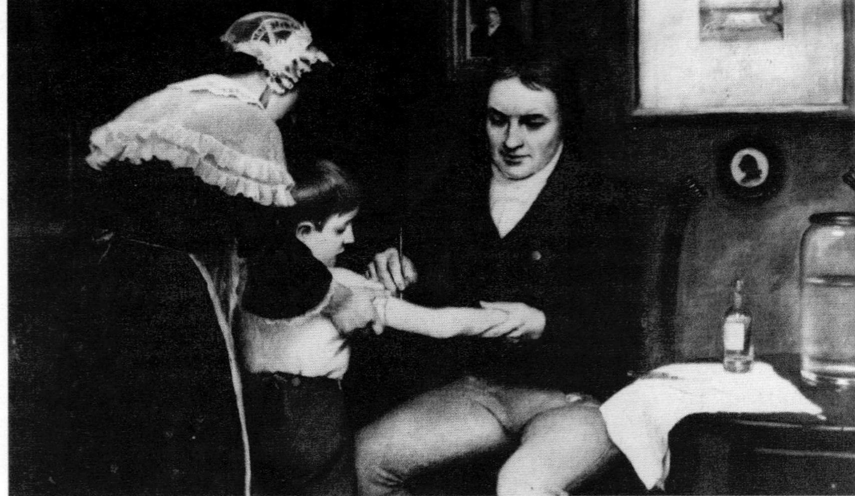

Figure 42.15

Smallpox killed thousands of people up until the middle of the 20th century. A worldwide attack on the disease through vaccinations brought an end to it. Because of the efforts of the World Health Organization, smallpox has been eliminated.

Passive immunity may develop in two ways. Natural passive immunity develops when antibodies are transferred from a mother to her unborn baby through the placenta or through the mother's milk while she is nursing the infant. Artificial passive immunity involves injecting antibodies into the body. These antibodies come from an animal or a human who is already immune to the disease. For example, a person who is bitten by a snake might be injected with antibodies from a horse that is immune to the snake venom.

Active immunity is obtained when a person is exposed to antigens. As you've just learned, the body produces antibodies as well as memory B cells and memory T cells in response to an antigen. Once the person recovers from the infection, he or she will be immune in the event of exposure to the pathogen again.

In order to produce active immunity, vaccines have been developed. A **vaccine** is a substance consisting of weakened, dead, or parts of pathogens or antigens that, when injected into the body, cause immunity. This occurs because the body reacts as if it were naturally infected.

In 1798, Edward Jenner, an English country doctor, demonstrated the first safe vaccination procedure. Jenner knew that dairy workers who acquired cowpox from infected cows were resistant to catching smallpox during epidemics. Cowpox is a disease similar to but milder than smallpox. To test whether immunity to cowpox also caused immunity to smallpox, Jenner infected a young boy with cowpox. The boy developed a mild cowpox infection. Six weeks later, Jenner scratched the skin of the boy with viruses from a smallpox victim, depicted in ***Figure 42.15***. The viruses for cowpox and smallpox are so similar that the immune system cannot tell them apart. The boy, therefore, did not get sick because he had developed antibodies and memory cells.

PORTFOLIO

Types of Immunity Have students make concept maps that identify the different types of immunity and describe how they develop. Have students begin their concept maps with the terms *Passive Immunity* and *Active Immunity.*  L1 P LS

GLENCOE TECHNOLOGY

Videodisc

STVS: Plants & Simple Organisms

Disc 4, Side 1

Anticavity Vaccine (Ch. 6)

Figure 42.16

Kaposi's sarcoma is characterized by blue-violet or brown spots on the patient's body.

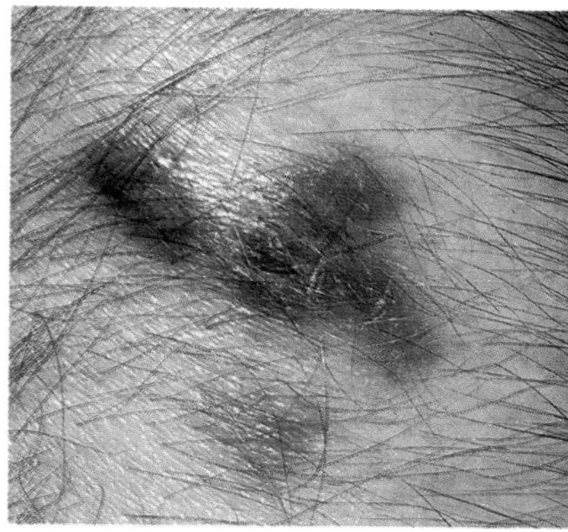

AIDS and the Immune System

In 1981, an unusual cluster of cases of a rare pneumonia caused by a protozoan appeared in the San Francisco area. Medical investigators soon related the appearance of this disease with the incidence of a rare form of skin cancer called Kaposi's sarcoma, *Figure 42.16.* Both diseases seemed associated with a general lack of function of the body's immune system. By 1983, the pathogen causing this immune system disease had been identified as a retrovirus, now known as Human Immunodeficiency Virus, or HIV. HIV kills helper T cells and leads to the disorder known as Acquired Immune Deficiency Syndrome, or AIDS.

HIV is transmitted when body fluid of an infected person is passed to an uninfected person through direct contact or by contaminated objects in contact with a wound. Intimate sexual contact, contaminated intravenous needles, and blood-to-blood contact such as in transfusions of contaminated blood are methods of transmission. Since 1985, careful screening measures have been instituted by blood banks in the United States to help keep HIV-infected blood from being given to those people who need transfusions. A pregnant woman infected with the virus can also transmit it to her fetus. Abstinence from intimate sexual contact provides protection from AIDS and other sexually transmitted diseases, in addition to preventing pregnancy.

The HIV virus is basically a set of genes wrapped in proteins, then further wrapped in a lipid coat, *Figure 42.17.* The knoblike outer proteins of the virus attach to a receptor on a helper T cell. Then the virus can penetrate the cell, where it may remain for months. The enzyme reverse transcriptase inside the virus allows HIV to synthesize viral DNA in the host cell.

Figure 42.17

HIV is a retrovirus with an outer envelope covered with knoblike proteins. Researchers are studying these proteins with the view to finding possible ways to stop the spread of this virus in humans.

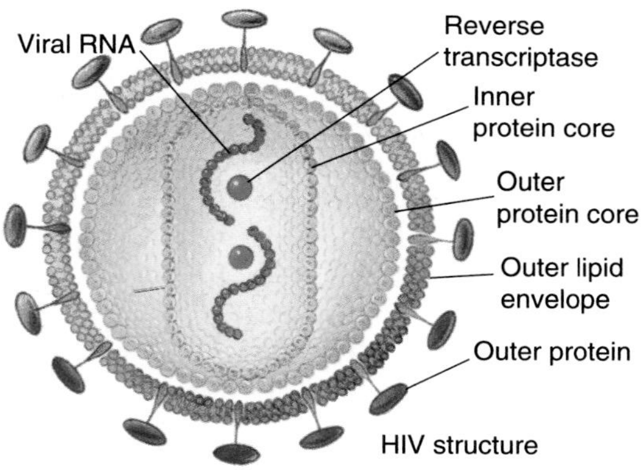

SECTION 42.2

Time Line

Logical-Mathematical Have students use the text to develop a time line that traces the history of what is known about the AIDS virus from its discovery to the present. Students also should include information about the development of drugs used to treat AIDS. L2

Software

AIDS Educational Program, National Association of Biology Teachers (MS-DOS and Mac).

Audiovisual

Show one of the three videos on AIDS available from Media Basics Video: *The AIDS Epidemic: Is Anyone Safe?*, *AIDS Update-Profile of an Epidemic*, or *AIDS: Can I Get It?*

Concept Development

Review the life cycle of a retrovirus. After discussing the AIDS epidemic, have students find out what their school policy is on AIDS.

STUDENT JOURNAL

Spreading the Word Ask students to prepare a newspaper article announcing an AIDS vaccine that will prevent people from getting AIDS. Students should include a summary of what is known about AIDS. L1

SECTION 42.2

3 Assess

Check for Understanding

Ask students to create a flowchart of B and T cell immune reactions. L1

Reteach

Have students prepare a table of the components of the lymphatic system and their functions. L1

Extension

Ask students to find out what the current progress is on an AIDS vaccine. Have them look in the latest issues of medical and scientific journals. L2

Assessment

Oral: Have students explain how the AIDS virus overcomes both nonspecific and specific immunity. L1

4 Close

Discussion

Ask students to summarize the role of B and T cells in the immune reactions. L1

The incubation period of the AIDS virus may last a long time. The first symptoms of AIDS may not appear until as long as 11 years after the initial infection of the HIV. After the initial stages, many infected persons exhibit symptoms of AIDS-Related Complex (ARC), *Figure 42.18.* Among the symptoms of ARC are swollen lymph nodes, a loss of appetite and weight, fever, rashes, night sweats, and fatigue.

It is not known what percentage of persons infected with HIV will develop AIDS, but present indications are that the majority will. Almost all who develop the AIDS syndrome die, usually because of infections that take advantage of the body's weakened immune system, such as the rare protozoan pneumonia and Kaposi's sarcoma.

Figure 42.18

ARC (AIDS-Related Complex) may develop after the number of helper T cells in the body is depleted.

Connecting Ideas

The immune system is not a system that can keep the body healthy by itself. It relies on the circulation of blood and lymph as well as communication by the nervous and endocrine systems. Like the other body systems you have studied, the immune system works to help maintain homeostasis in your body.

Section Review

Understanding Concepts

1. What role do phagocytes play in defending the body against disease?
2. What role does a lymph node play in defending your body against microorganisms?
3. What is the difference between artificial passive immunity and artificial active immunity?

Thinking Critically

4. Why is it adaptive for memory cells to remain in the immune system after an invasion by pathogens?

Skill Review

5. **Sequencing** Sequence the events that occur in the formation of antibody immunity. For more help, refer to Organizing Information in the *Skill Handbook.*

Answers to Section Review

1. Phagocytes are white blood cells that ingest and destroy pathogens by surrounding and engulfing them.
2. A lymph node is a small mass of tissue that filters lymph and traps microorganisms.
3. In artificial passive immunity, antibodies produced by animals or humans that have the disease are given to someone by injection. In artificial active immunity, the body makes its own antibodies when exposed to weakened or dead pathogens.

Thinking Critically

4. Memory cells remain in case the body encounters the same antigen again. The second response will be rapid, before the antigen can cause a disease.

Skill Review

5. Helper T cells interact with macrophages that have "self" and pathogen antigens on their surfaces. The helper T cells activate B cells to reproduce and produce antibodies, as well as produce a clone of memory B cells.

Chapter 42 Review

REVIEWING MAIN IDEAS

42.1 The Nature of Disease

- Infectious diseases are caused by the presence of pathogens in the body.
- The cause of an infection can be established by following Koch's postulates.
- Animals, including humans, and nonliving objects can serve as reservoirs of pathogens. Pathogens can be transmitted by direct contact, by an object, through the air, or by a vector.
- Symptoms of a disease are caused by direct damage or toxins from the pathogen.
- Some diseases are periodic, while others are endemic. Occasionally, a disease reaches epidemic proportions.
- Some infectious diseases can be treated with antibiotics, but pathogens may become resistant to drugs.

42.2 Defense Against Infectious Diseases

- The lymphatic system consists of the lymphatic pathways and the lymph nodes, tonsils, spleen, and thymus.
- Nonspecific defense mechanisms provide general protection against various pathogens.
- Specific defense mechanisms provide a way of fighting particular pathogens by recognizing invaders as nonself. Specific immunity includes the production of antibodies and cellular immunity.
- Caused by HIV, which damages the immune system, AIDS is a disorder in which other infections invade the body, leading to death.

Key Terms

Write a sentence that shows your understanding of each of the following terms.

antibiotic	lymph node
B cell	lymphocyte
complement	macrophage
endemic disease	pathogen
epidemic	phagocyte
immunity	pus
infectious disease	T cell
Koch's postulates	tissue fluid
lymph	vaccine

Understanding Concepts

1. Bacteria, viruses, and other disease-producing agents are called __________.
   a. parasites c. antibodies
   b. pathogens d. lymph
2. Any disease caused by disease-producing agents in the body is known as a(n) ___________ disease.
   a. infectious c. epidemic
   b. endemic d. nonspecific
3. Koch's postulates are a series of steps a scientist takes to relate a specific ___________ to a specific disease.
   a. host c. epidemic
   b. medium d. pathogen
4. Which of these diseases is caused by a pathogen?
   a. osteoarthritis c. smallpox
   b. hemophilia d. cystic fibrosis
5. Of the following, which is NOT a reservoir for pathogens?
   a. human body c. water
   b. animals d. lava
6. Which of these diseases is spread only through sexual intercourse?
   a. food poisoning c. tetanus
   b. genital herpes d. mumps
7. Which of these diseases is NOT caused by an arthropod vector?
   a. bubonic plague c. influenza
   b. Lyme disease d. malaria

Chapter 42 Review

Reviewing Main Ideas

Summary statements can be used by students to review the major concepts of the chapter.

Key Terms

Answers should go beyond defining the terms. Accept any answer that uses the term correctly and in the proper context.

Understanding Concepts

1. b
2. a
3. d
4. c
5. d
6. b
7. c

8. Viruses and bacteria damage host cells by producing ____________.
a. toxins c. hormones
b. antibodies d. tRNA

9. Diseases that are constantly present in the population are ___________ diseases.
a. infectious c. endemic
b. chronic d. epidemic

10. A(n) _____________ is a substance produced by a microorganism that inhibits the growth of other microorganisms.
a. epidemic c. antigen
b. endemic d. antibiotic

11. When tissue fluid collects in open-ended vessels, it is called _____________.
a. lymph c. pus
b. complement d. blood

12. A(n) ___________ is a type of white blood cell that defends the body against foreign substances.
a. osteoblast c. platelet
b. lymphocyte d. amoebocyte

13. Of these, which is NOT a type of a nonspecific defense mechanism in your body?
a. phagocytosis c. skin
b. immunity d. mucus

14. ____________, an enzyme produced in sweat, tears, and saliva, can break down the cell walls of some bacteria.
a. Gastric juice c. Lysozyme
b. Amylase d. Mucus

15. ___________ is a body response to an injury characterized by redness, swelling, pain, and heat.
a. Phagocytosis c. Complement
b. Pus d. Inflammation

16. Building up a resistance to a specific pathogen is called ____________.
a. immunity c. transmission
b. infection d. complement

17. A _______________ combats invading pathogens by scavenging and engulfing large numbers of microbes.
a. leukocyte c. B cell
b. macrophage d. T cell

18. Which of these is produced by a body cell that has been infected with a virus?
a. interferon c. lysozyme
b. complement d. histamine

19. What scientist demonstrated the first safe vaccination procedure?
a. Koch c. Mendel
b. Pasteur d. Jenner

20. Which type of lymphocyte is destroyed by HIV?
a. memory B cells c. plasma cells
b. helper T cells d. phagocyte

Applying Concepts

21. If the bacteria that cause tetanus are easily killed by penicillin, why doesn't penicillin cure the disease tetanus?

22. Why must severe burn victims be kept in pathogen-free isolation?

23. Cholera, a waterborne disease, often reaches epidemic proportions after a flood. Explain why this is so.

24. While working on building a tree house, you get a tiny splinter in your finger. Two days later, the area is swollen and pus leaks out when you press it. Why is there pus around the splinter? Explain.

25. A month after buying a new pet parakeet, Susan experienced pains in her legs, followed by chills, fever, diarrhea, and a headache. She recovered after two weeks of antibiotics. When she next visited the pet store, many of the parakeets were ill. How could researchers find out if Susan had the same disease as the birds?

Thinking Critically

26. ***Concept Mapping*** Make a concept map that relates the following terms and phrases. Supply the appropriate linking words for your map.
infectious disease, pathogen, incubation period, AIDS, antibody, antigen, vaccine

Chapter 42 Review

8. a
9. c
10. d
11. a
12. b
13. b
14. c
15. d
16. a
17. b
18. a
19. d
20. b

Applying Concepts

21. The disease tetanus is due to a toxin produced by the bacteria. Killing the bacteria does not affect the toxin that has already been produced.

22. A burn patient loses protective layers of skin, exposing the body to the possibility of massive infection.

23. During floods, rainwater washes sewage into water supplies, which then become contaminated by the pathogen. When people drink this untreated water, they ingest pathogens, including the ones that cause cholera.

24. Macrophages that came to the injured area engulfed microbes and damaged tissue, then died. Dead macrophages and body fluids collectively are called pus.

25. Researchers would have to isolate the organism from the parakeets and run Koch's postulates on them. The organism can also be exposed to human cell cultures to determine susceptibility.

Thinking Critically

26.

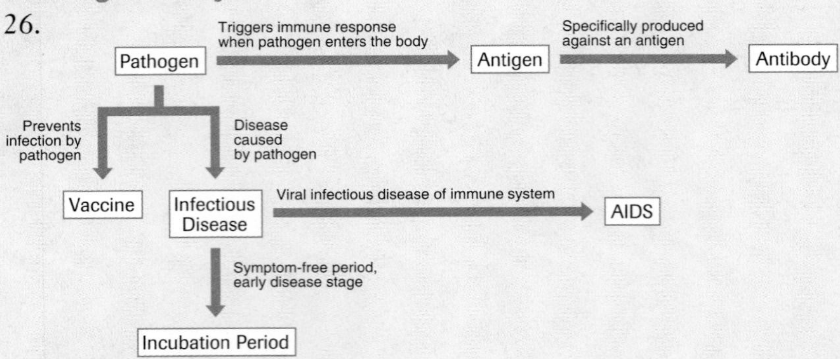

27. ***Interpreting Data*** When you are vaccinated against a disease, an immune response occurs, creating memory cells. Using the data presented, explain why the second exposure to the pathogen does not cause a disease.

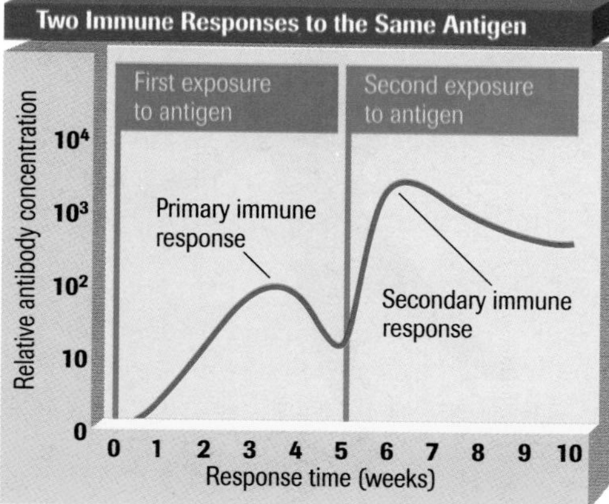

28. ***Observing and Inferring*** Why don't you develop an immunity to colds as you do to mumps?
29. ***Formulating Hypotheses*** If an infant were found to be lacking a thymus because of a developmental disorder, what would you predict about the infant's resistance to infections? Why?
30. ***Observing and Inferring*** How does AIDS upset homeostasis in the body?
31. ***Observing and Inferring*** In health class, your teacher asks everyone to list the childhood diseases they have had. Everyone lists chicken pox, but you aren't sure if you contracted this disease or not. When you ask your mother, she says that your older brother was covered with chicken pox sores when you were a baby, but she is not sure whether you had chicken pox at that time because you only had about ten spots on your body. Are you immune to the virus that causes chicken pox? Explain.

ASSESSING KNOWLEDGE & SKILLS

The graph below shows the progress of a typical HIV infection with signs of the various stages of the AIDS disease.

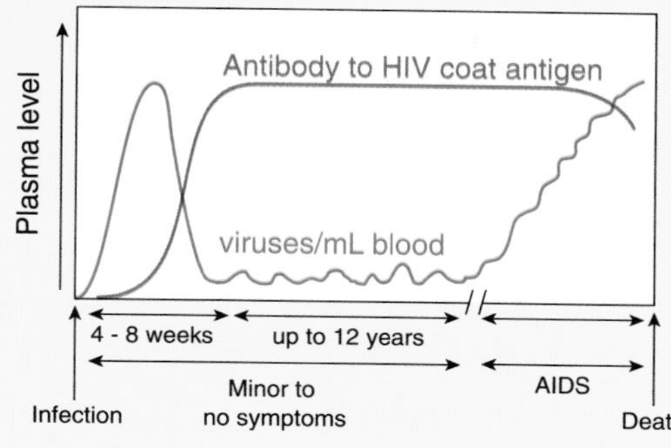

Interpreting Data Use the graph to answer these questions.

1. Which variable is the dependent variable?
   - **a.** time
   - **b.** symptoms
   - **c.** level in the blood
   - **d.** antibody
2. When does the antibody level begin to rise?
   - **a.** at about 4 weeks
   - **b.** at about 8 weeks
   - **c.** during the AIDS symptom stage
   - **d.** at death
3. The HIV virus attacks ____________.
   - **a.** red blood cells
   - **b.** B cells
   - **c.** T cells
   - **d.** epithelial cells
4. What type of molecule are antibodies?
   - **a.** carbohydrates
   - **b.** fats
   - **c.** proteins
   - **d.** nucleic acids
5. ***Observing and Inferring*** Explain why AIDS is considered a syndrome rather than one single disease.

Chapter 42 Review

27. The immune system remembers the pathogen and immediately makes antibodies and sends T cells to rid the body of the pathogen before a person gets ill.
28. Colds are caused by more than one type of virus, whereas mumps is caused by one type of virus.
29. The infant would have poor resistance to infections because it could not produce T cells.
30. AIDS destroys the immune system, which constantly monitors the body for invasion by pathogens. Without this system functioning, the body is extremely susceptible to disease.
31. Yes, you are immune because the virus that causes chicken pox was present in your body, so your immune system has retained memory cells to fight that particular virus.

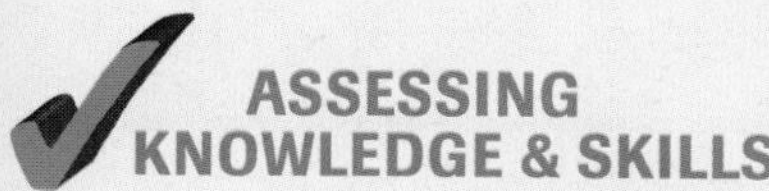

1. c
2. a
3. c
4. c
5. The variety of symptoms that characterize AIDS is caused by a syndrome of infections.

Program Resources

Chapter Assessment, pp. 259-264 **L1**
Alternate Assessment in the Science Classroom
Computer Text Bank **L1**
Content Mastery, pp. 177-180 **L1**

Prepare

Purpose

This Biodigest can be used as an introduction to or as an overview of the structures and functions of the human body systems. If time is limited, you may wish to use this unit summary to teach human biology in place of the chapters in the Human Biology unit.

Key Concepts

Students learn about the level of organization in body systems. They are then introduced to the structures and functions of 11 major body systems. Various vital statistics provide students with interesting facts about their body systems.

Block Scheduling

Look for this symbol for strategies that are useful in a block scheduling format. For more information on block scheduling, refer to Biodigest 4 in the **Lesson Plans** booklet.

1 Focus

Bellringer

Before presenting the lesson, display **Section Focus Master 99** on the overhead projector and have students answer the accompanying questions. L1 SAE

Visual Learning

LS **Visual-Spatial** Students' attention should be drawn to the levels of organization diagram showing how cells are organized into tissues, organs, and organ systems. All systems together make up an organism.

BIODIGEST HUMAN BIOLOGY

How do the human body systems function together? When an Olympic ice-skater performs on the ice, the cells of the body, the tissues, the organs, and the organ systems function together to help the athlete perform at his or her best and win a gold medal. All body systems must work together to make an award-winning performance possible.

LEVELS OF ORGANIZATION

All organisms are made of cells. In complex organisms, such as humans, most cells are organized into functional units called tissues. The four basic tissues of the human body are epithelial tissue, muscle tissue, connective tissue, and nervous tissue. Epithelial tissue covers the body and lines organs, vessels, and body cavities. Muscle tissue is contractile and is found attached to bones and in the walls of organs, such as the heart. Connective tissue is widely distributed throughout the body. It produces blood and provides support, binding, and storage. Nervous tissue transmits impulses that coordinate, regulate, and integrate body systems.

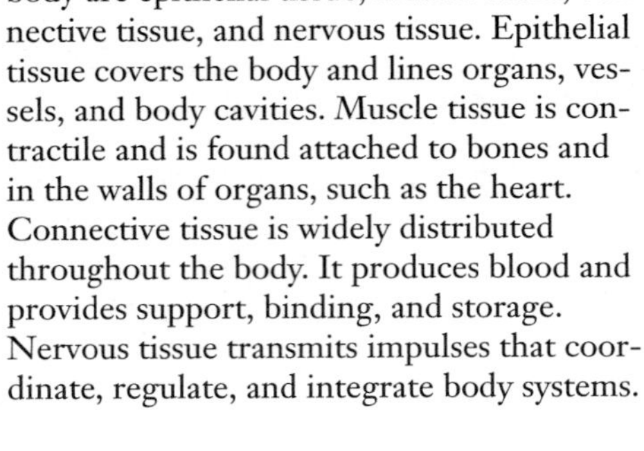

Tissues to Systems

Groups of tissues that perform specialized functions are called organs. Your stomach and eyes are both examples of organs. Both organs contain all four basic tissue types. Organs are a part of an organ system. Organ systems contain a group of organs that work together to carry out a major life function. The eleven major organ systems are described in this BioDigest.

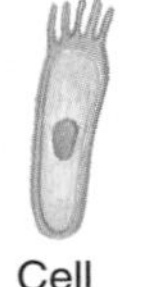
Cell

Levels of organization: cell, tissue, organ, organ system, organism

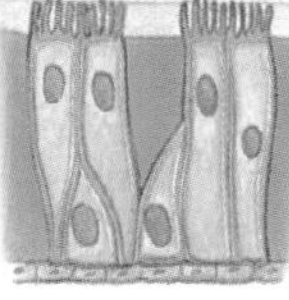
Tissue

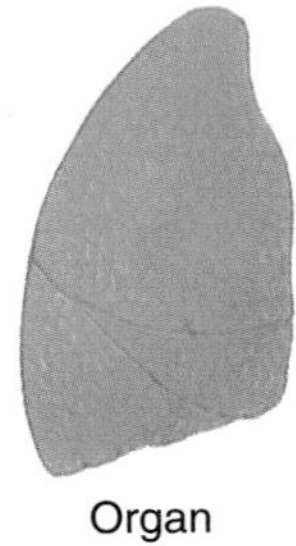
Organ

Organ system

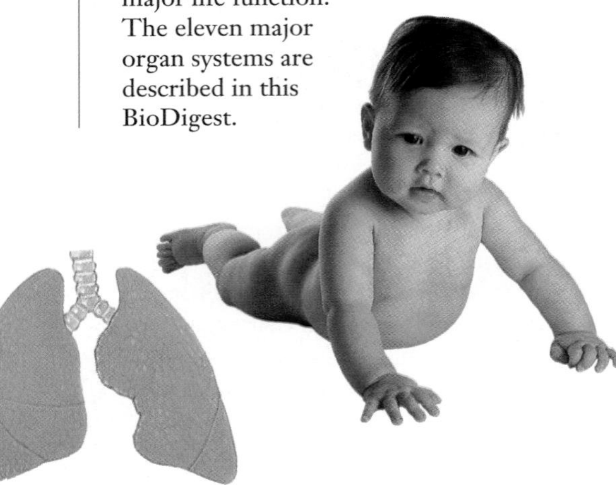

Learning Styles LS

Look for the following logo for strategies that emphasize different learning modalities. LS

Visual-Spatial	**Project, p. 1097; Display, p. 1098; Visual Learning, pp. 1096, 1097, 1100, 1102; Demonstration, p. 1097; Activity, p. 1101**
Interpersonal	**Enrichment, p. 1103; Meeting Individual Needs, p. 1098; Activity, p. 1099; Portfolio, p. 1101**
Linguistic	**Student Journal, pp. 1100, 1102, 1104**
Logical-Mathematical	**Project, p. 1099**

SKIN

The skin and its associated structures, including hair, nails, sweat glands, and oil glands, are important in maintaining homeostasis in the body. The skin protects tissues, helps regulate body temperature, produces vitamin D, and contains sensory receptors.

Skin structure varies somewhat from place to place in the body.

SKELETAL SYSTEM

The skeletal system consists of the axial skeleton and appendicular skeleton. The axial skeleton supports the head and includes the skull and the bones of the back and chest. The appendicular skeleton contains the bones associated with the limbs. The skeleton, which is made up of 206 bones, provides support for the softer, underlying tissues; provides a place for muscle attachment; protects vital organs; manufactures blood cells; and serves as a storehouse for calcium and phosphorus.

Joints: Where Bones Meet

The place where two bones meet is called a joint. Joints can be immovable such as the joints in your skull, or movable such as your shoulder joint. The shoulder joint is called a ball-and-socket joint. Your elbow joint is called a hinge joint. Your wrists have gliding joints, and your neck has pivot joints.

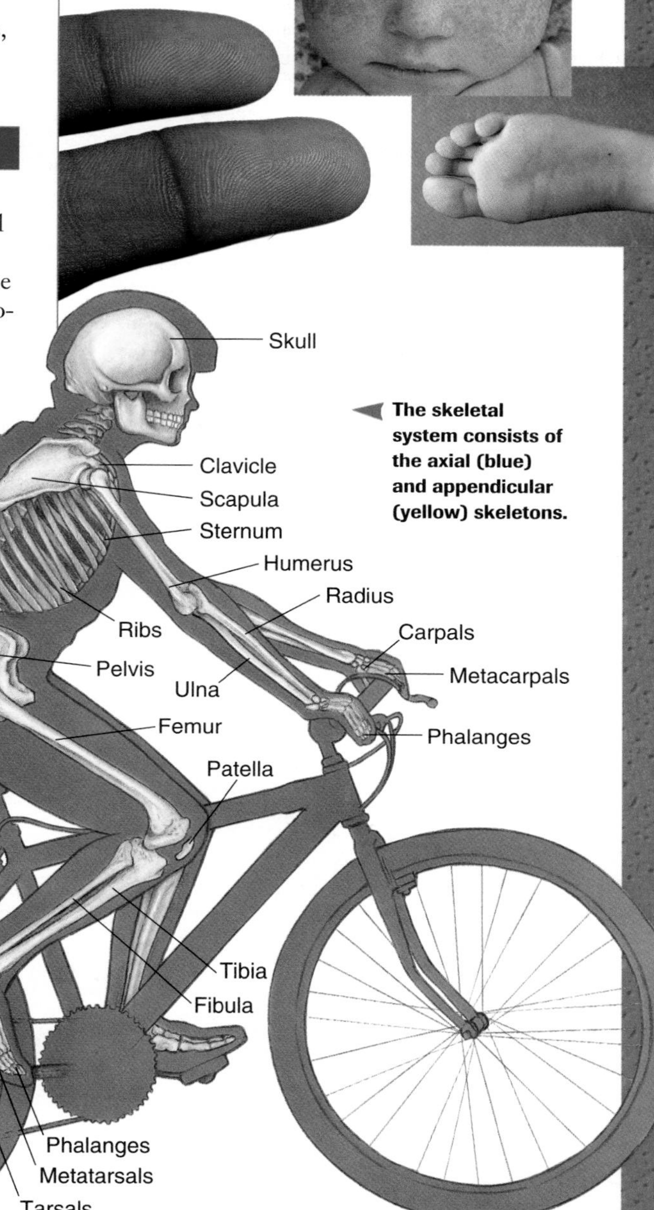

The skeletal system consists of the axial (blue) and appendicular (yellow) skeletons.

2 Teach

Audiovisual

The video *The Incredible Human Machine* summarizes the body systems.

Demonstration

Visual-Spatial A model human skeleton can be used to talk about the skeletal system and how skeletal muscles function. A human torso can be used to demonstrate the other body systems. These can often be borrowed from a nearby college. SAE

Visual Learning

Visual-Spatial Have students examine areas of their skin. The palms of the hands and soles of the feet contain no hairs. The fingertips contain large numbers of sensory nerve endings and ridges to help grip objects. Have them also note the different patterns and types of skin coloration.

Demonstration

Visual-Spatial X rays of the skeletal system and barium X rays of the digestive system can be used by hanging them on classroom windows or on the overhead projector. Use X rays to discuss the parts and functions of each of these systems. SAE

PROJECT

Visual-Spatial Have students become familiar with the different body systems by asking groups of two or three students to prepare a photo essay representing one body system using pictures from magazines. They should present their essays to the class with an oral explanation. L1 SAE P COOP LEARN

Meeting Individual Needs

Hearing Impaired Have hearing impaired students make flash cards with the system on one side and the functions on the other side of the cards. LS

Display

Visual-Spatial Students could prepare a bulletin board or posters using pictures cut out of magazines to illustrate the body systems. L1 SAE

Display

Prepare a bulletin board of magazine and newspaper articles of various disorders, sports injuries, and other problems relating to the human body systems.

MUSCULAR SYSTEM

The muscular system includes three types of muscles: smooth, cardiac, and skeletal. Smooth muscles are found in the walls of hollow internal organs, such as inside the stomach or blood vessels. These muscles are not under conscious control and are called involuntary muscles. Smooth muscle cells are spindle shaped and contain a single nucleus.

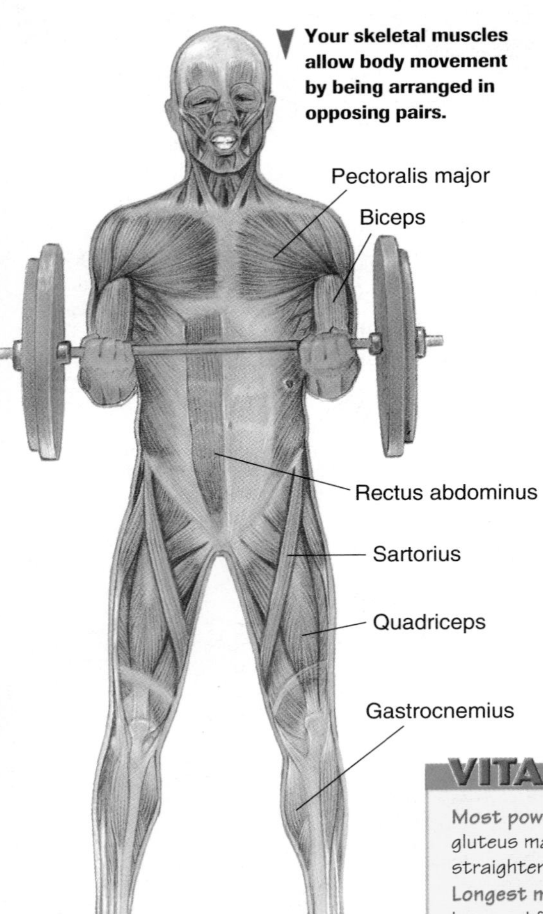

Your skeletal muscles allow body movement by being arranged in opposing pairs.

During physical activity practically every muscle can be involved, either voluntarily or by reflex actions.

Skeletal Muscle

Skeletal muscle tissue is found in muscles that are usually attached to bones. They can be controlled by conscious effort so they are called voluntary muscles. The long, threadlike cells called myofibrils have alternating dark and light striations, and each cell has many nuclei.

Heart Muscle

Cardiac muscle tissue is found only in the heart. The cells contain a single nucleus and striations due to organized protein filaments that are involved in contraction of the muscle. Like smooth muscle, cardiac muscle is involuntary muscle. Cardiac muscle has the unique ability to contract without first being stimulated by nervous tissue.

VITAL STATISTICS MUSCLES

Most powerful skeletal muscle: The muscle you sit on is the gluteus maximus; it moves the thighbone away from the body and straightens the hip joint.
Longest muscle: The sartorius muscle runs from the waist to the knee and flexes the hip and knee.
A broad smile: When you smile, you use 17 muscles in your face.

Meeting Individual Needs

Visually Impaired Pair a visually impaired student with a nonvisually impaired student to go over the human body systems using a human torso. Ask the students to consider which organs are in which system and their functions.

interNET CONNECTION

Follow the link for this Biodigest on the Glencoe Homepage at **www.glencoe.com/sec/science** to find out more about human biology.

DIGESTIVE SYSTEM

The digestive system receives, breaks down, and absorbs food and nutrients into the body and eliminates materials that are not absorbed or digested. Foods are broken down into simpler molecules so they can be transported through cell membranes into the bloodstream or into the lymphatic vessels. The digestive system includes the mouth, tongue, teeth, salivary glands, pharynx, esophagus, stomach, liver, gallbladder, pancreas, and small and large intestines.

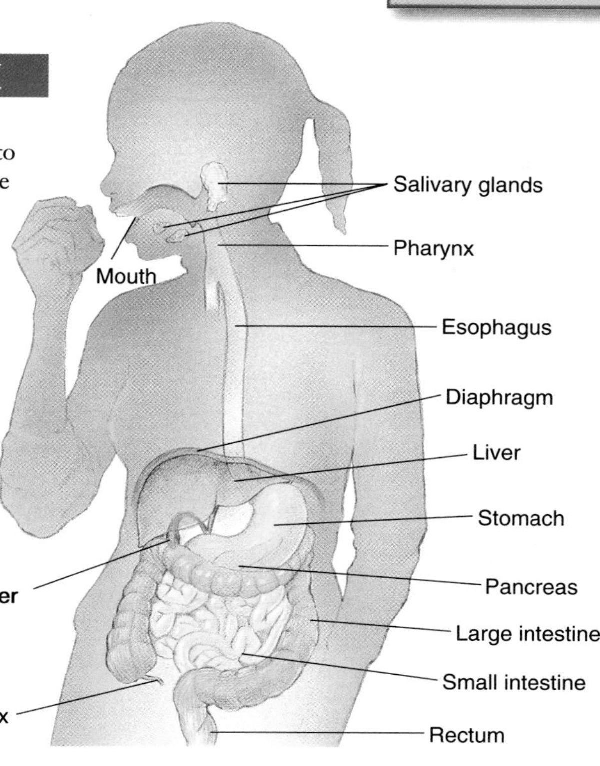

The digestive system breaks down food particles so that they can be absorbed by your body.

FOCUS ON HEALTH

Blood Glucose Levels

The levels of blood glucose in your body are maintained all day long by hormones secreted by the pancreas. After you eat a meal, the sugar in your meal is transported into your blood and either used immediately for activity or stored in the liver for later use. The pancreas secretes insulin, which helps the body's cells take up the sugar or convert it to glycogen in the liver for storage.

Between meals, when blood glucose levels go down, the pancreas secretes glucagon. Glucagon causes the glycogen in the liver to be broken down into glucose, which is then released into the bloodstream and made available to the body's cells. The control of blood sugar levels in the body is an example of a feedback mechanism that is vital to a person's homeostasis.

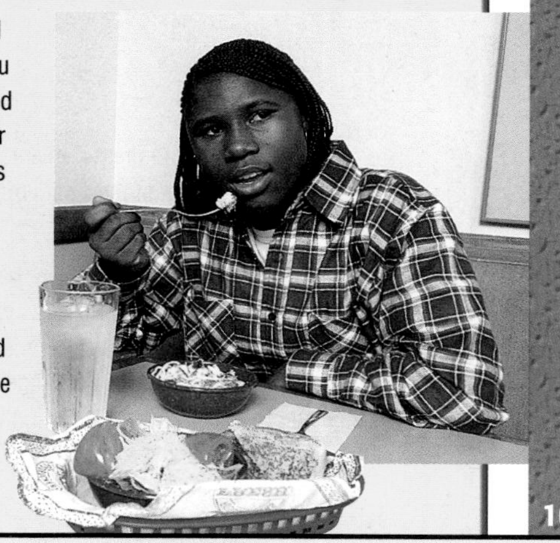

1099

BIODIGEST

Activity

LS **Interpersonal** Play systems password game. Students are divided into teams of two. Each team plays against one other team. One person on each team gives one-word clues to his or her partner in an attempt to get him or her to guess the password. Team scores are tallied at the end of the game. L1

Enrichment

Have a doctor from a sports medicine clinic speak to the class concerning fitness and the effects of sports injuries on various systems.

PROJECT

LS **Logical-Mathematical** Have students keep track of what they eat and of all of their activities (including resting, sleeping, watching TV, etc.) for 48 hours. Students can use a Calorie guide to estimate their daily Calorie intake and an activity guide to calculate their calorie usage. L1

Visual Learning

Visual-Spatial Review the process of diffusion, using the diffusion of oxygen out of the alveoli into the blood and carbon dioxide out of the blood into the alveoli.

BIODIGEST

RESPIRATORY SYSTEM

The organs of the respiratory system exchange gases between blood and the air. When you inhale, oxygen in the air passes into the blood from small air sacs called alveoli in the lungs. Oxygen is needed by your body cells in order to break down glucose to make ATP for the cell's metabolism.

Carbon dioxide (CO_2) is produced as an end product of this glucose metabolism and is transported to the lungs by the blood. It diffuses out of the blood and into the alveoli. Carbon dioxide is then forced out of the lungs when you exhale. The major organs of the respiratory system are the nasal cavity in your nose, the pharynx, larynx, trachea, bronchi, and lungs.

The respiratory system filters the air as it passes into the nose, down the air passages, and into the lungs.

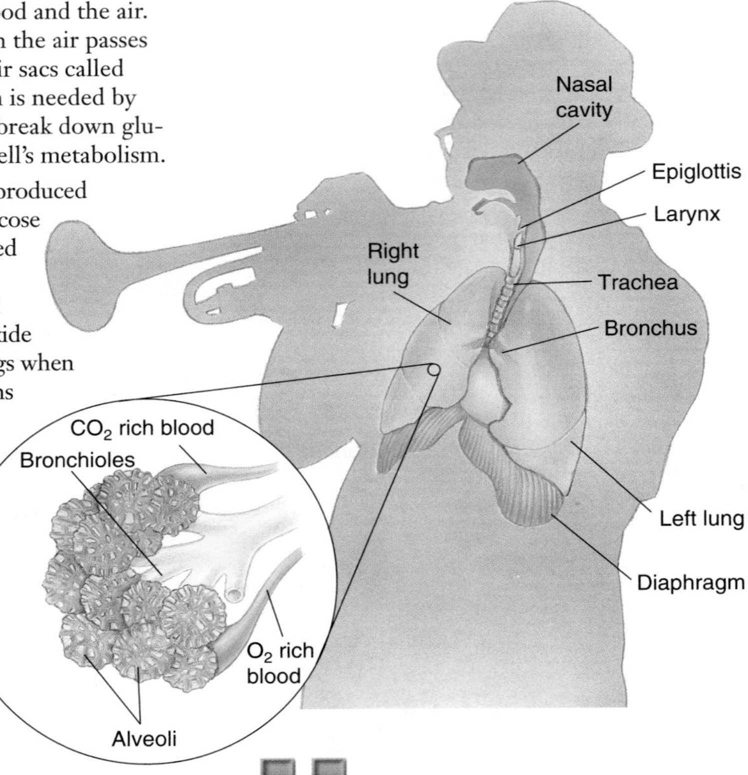

The lungs contain many small sacs called alveoli, where gas exchange with the blood occurs.

FOCUS ON HEALTH

Your Blood Pressure

Blood pressure measurements give you an indication of the health of your heart and your blood vessels.

Systolic Pressure When the cuff of the blood pressure machine squeezes your arm, it blocks the blood flow in an artery. As the pressure in the cuff is released, a gauge attached to the cuff measures the pressure in the artery as blood flows back into the artery. This is the

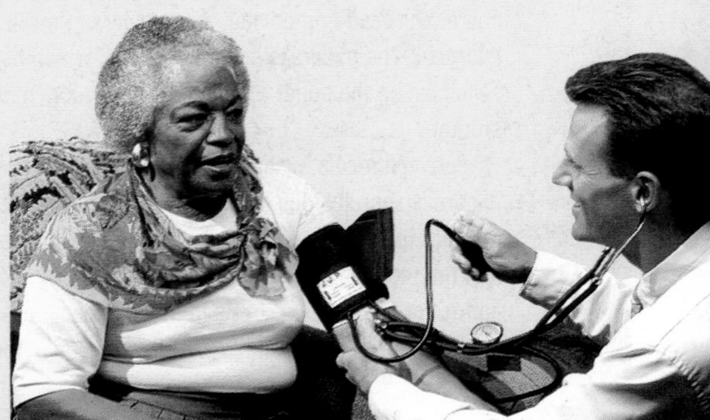

1100

STUDENT JOURNAL

Linguistic Most representatives of the tobacco industry still claim there is no proof that tobacco causes cancer. After researching the issue, have students write an essay in their journal on how they would respond to the tobacco industry's assertion that tobacco usage does not cause cancer. L2

VITAL STATISTICS
RESPIRATION

Breathing: At rest, you inhale and exhale about 12 to 20 times per minute, moving about 15 L of air per minute. Each day you inhale 21.6 cubic meters of air.
Lungs: Lungs weigh about 2.2 kg each. The right lung has three lobes and the left lung has two lobes. There are 300 million alveoli in your lungs. If you flattened them all out, they would cover 360 square meters.
Sneezes: When you sneeze, you eject particles at 165.76 km/hr.

CIRCULATORY SYSTEM

The circulatory system includes the heart, blood vessels (arteries, veins, and capillaries), and blood. The muscular heart pumps blood through the blood vessels. The blood carries oxygen from the lungs and nutrients from the digestive tract to all the body cells. Blood also carries hormones to their target tissues, waste carbon dioxide back to the lungs, and other waste products to the excretory system.

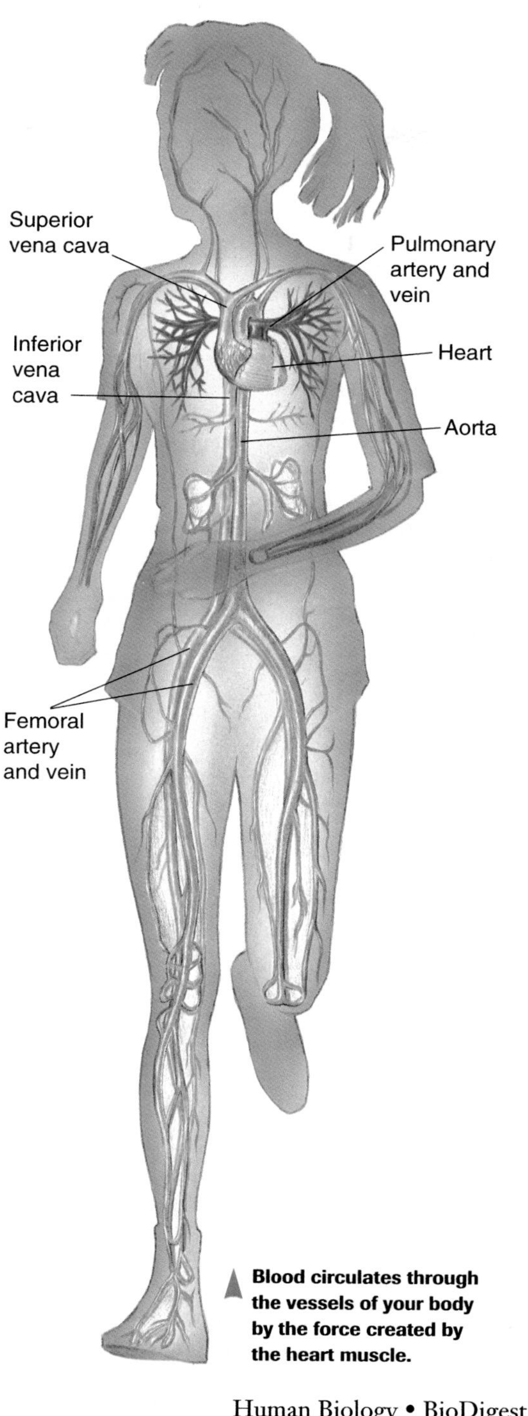

Blood circulates through the vessels of your body by the force created by the heart muscle.

systolic pressure, which is a measure of the pressure when your right and left ventricles contract.

Diastolic Pressure When the first rush of blood through the arteries slows, the gauge measures a pressure called the diastolic pressure. This is the lowest pressure in the vessels, just before the two ventricles contract again. Blood pressure readings give both the systolic and the diastolic pressure in your arteries. Blood pressure is thus an indication of the condition of your arteries.

Demonstration

Ask the school nurse to visit and demonstrate the measuring of blood pressure of a few students in the class.

Activity

LS **Visual-Spatial** Provide groups of students with diagrams of the different human body systems. Have them use colored pencils to shade various organs of particular systems, such as coloring the heart and vessels red for the circulatory system. SAE COOP LEARN

PORTFOLIO

LS **Interpersonal** Have students prepare an interview of a health professional such as a nurse, doctor, emergency room personnel, ambulance driver, physician's assistant, and phlebotomist. They should prepare questions about what the person does and what education the person needs to acquire the job. These can be shared with the whole class. L1

Visual Learning

Visual-Spatial Draw a neuron on the chalkboard and label the parts. Using arrows, indicate the direction an impulse travels along a neuron. As you discuss transmission across the synapse, bring in the action of various drugs on the synapse. You can find more information on this in the chapter on the nervous system.

NERVOUS SYSTEM

The nervous system consists of the brain, spinal cord, nerves, and sense organs. Nerve cells or neurons within these organs conduct impulses that communicate with one another and with your muscles and glands. Nerve cells conduct an impulse by changing the charges on the surfaces of their cell membranes, conducting a wave of charges.

Between two neurons, there is a small gap called a synapse. At the end of one neuron, chemicals called neurotransmitters are released into the synapse, and this stimulates a wave of charges in the next neuron cell.

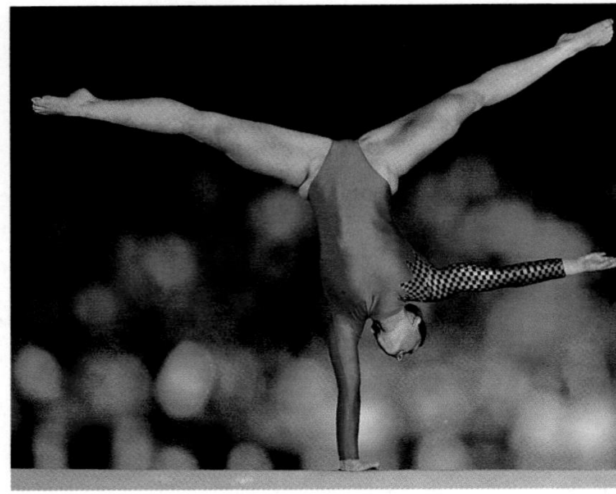

Interpreting and acting on information sent to the central nervous system (brain and spinal cord) is the major job of the nervous system.

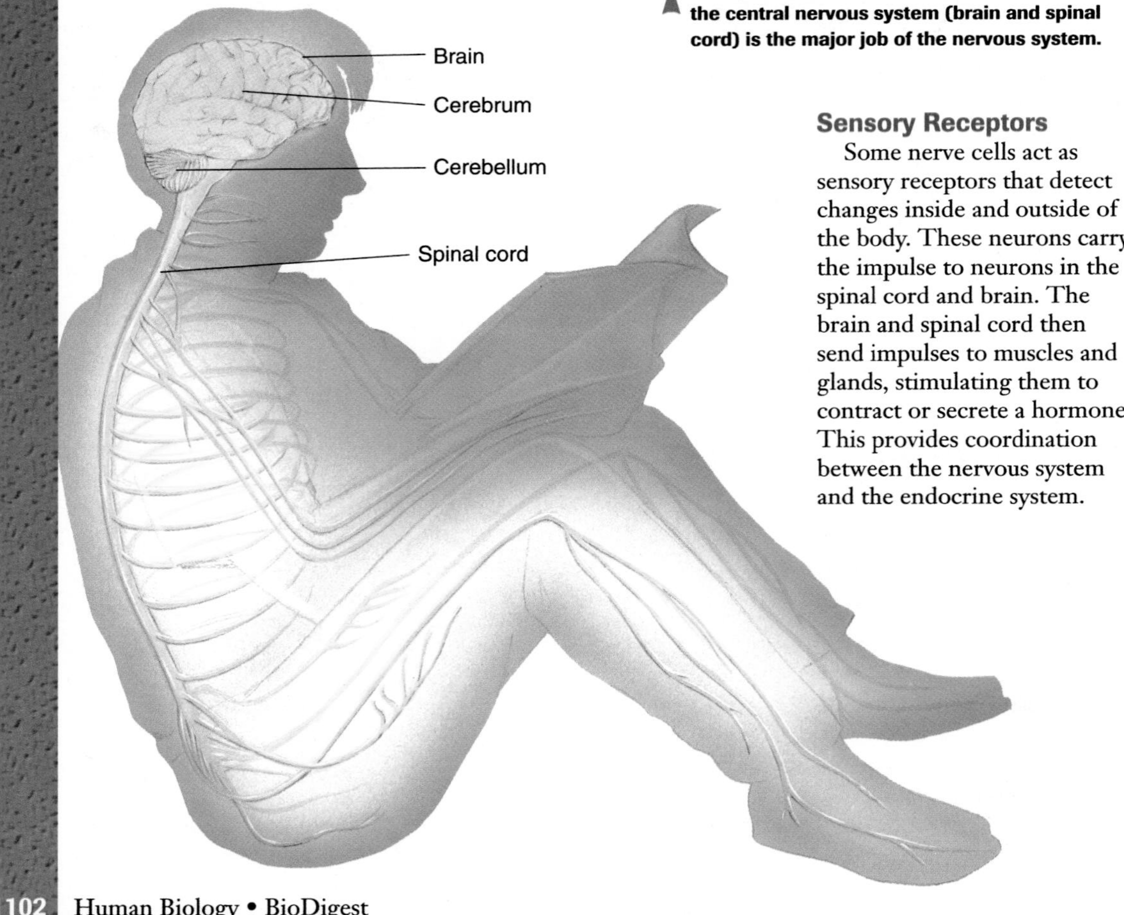

Sensory Receptors

Some nerve cells act as sensory receptors that detect changes inside and outside of the body. These neurons carry the impulse to neurons in the spinal cord and brain. The brain and spinal cord then send impulses to muscles and glands, stimulating them to contract or secrete a hormone. This provides coordination between the nervous system and the endocrine system.

STUDENT JOURNAL

Linguistic Have students write a skit in their journal with a major player being the brain and how it interacts and controls the other systems. L2

REPRODUCTIVE SYSTEM

The reproductive system is involved in the production of gametes. The male reproductive system includes the scrotum, testes, epididymis, seminal vesicles, prostate gland, bulbourethral gland, urethra, and penis. These structures are involved in producing and maintaining sperm cells and transferring sperm cells into the female reproductive tract.

The major function of the male reproductive system is to produce sperm.

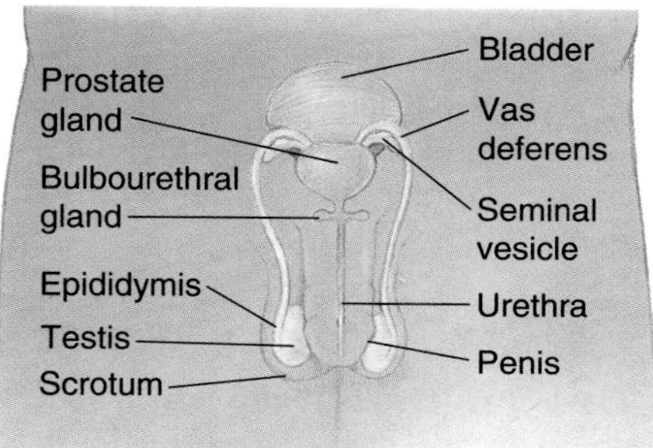

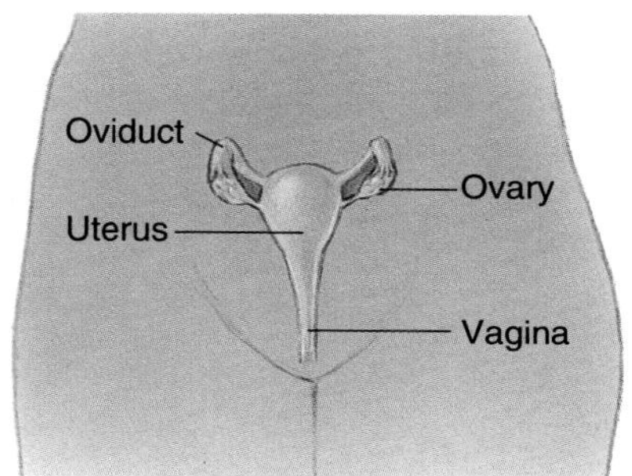

The major function of the female reproductive system is to produce an egg and support a developing embryo.

The female reproductive system includes the ovaries, oviduct, uterus, and vagina. These structures produce and maintain egg cells, receive and transport sperm cells, and support the development of the fetus.

ENDOCRINE SYSTEM

The endocrine system controls all of the metabolic activities of body structures. This system includes all of the glands in the body that secrete chemical messengers called hormones. Hormones travel in the bloodstream to target tissue where they alter the metabolism of the target tissue. Some of the major glands include the pituitary gland, thyroid gland, parathyroid glands, adrenal glands, pancreas, ovaries, and testes.

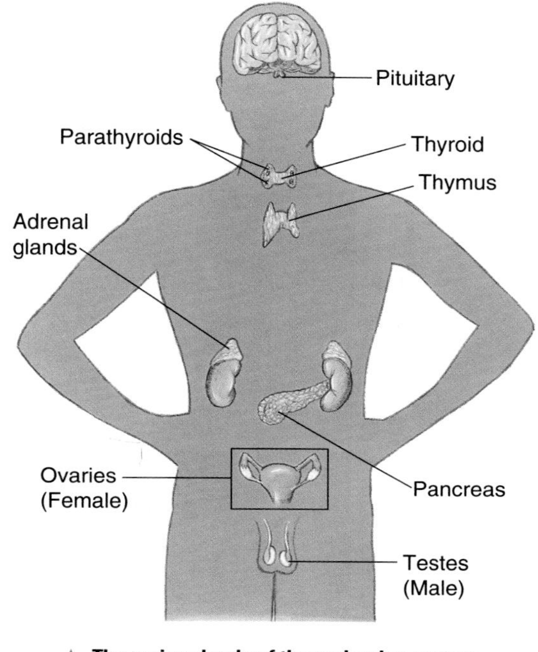

The major glands of the endocrine system secrete hormones to regulate body functions.

VITAL STATISTICS REPRODUCTION

Testes: The testes contain 244 m of tubules in which sperm cells are continually produced by meiosis. A sperm cell can swim up to 12.7 cm/hr.

Ovaries: At birth, a female has about 2 million eggs. About 300 000 survive to puberty, but only 450 or so mature and are expelled from the ovary.

Enrichment

Interpersonal Students can be broken into groups that are assigned to summarize the materials in each of the body systems and present the summaries to the class. L1 COOP LEARN

Meeting Individual Needs

Gifted Have gifted students research premature infant care costs. They can write a position paper on whether Americans should spend more or less on dramatic lifesaving measures and why. L3

Reteach

Have students make a table based on this Biodigest with three columns: System, Major Parts, and Function.

LYMPHATIC SYSTEM

Fluids leak out of your capillaries and bathe your body tissues. The lymphatic system, also known as the immune system, is involved in transporting this tissue fluid back into the bloodstream. As tissue fluids pass into lymphatic vessels and lymph nodes, any antigens, disease-causing microorganisms, or other foreign substances are filtered out and destroyed.

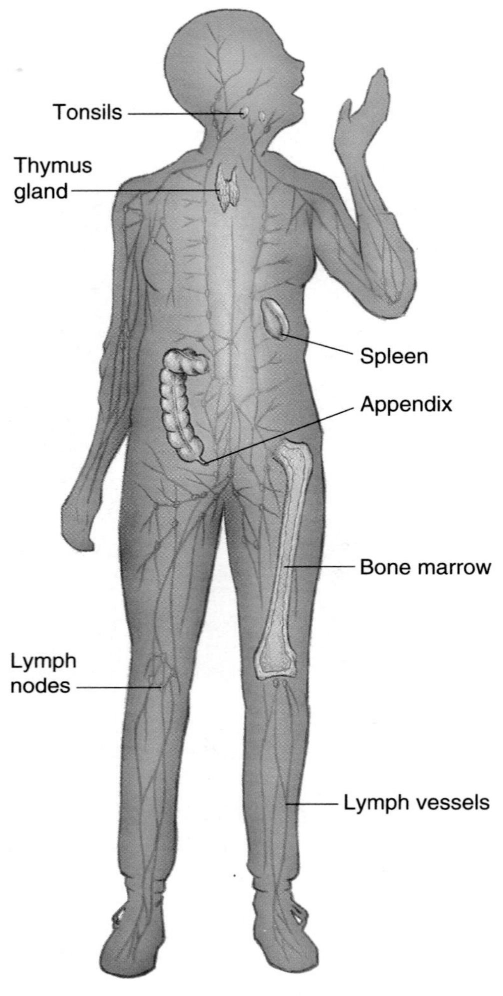

Your lymphatic system protects you from infections such as colds and flu.

White blood cells called macrophages, B lympyhocytes, and T lymphocytes constantly monitor lymph fluid for antigens, such as bacteria and viruses. Macrophages engulf foreign substances that enter the body. T cells then recognize the affected macrophages, bind to the antigens presented on the macrophage surfaces, and then pass on this chemical information to the B lymphocytes.

B cells manufacture specific antibodies to destroy the invaders. Some B cells remain in the body as memory B cells that will recognize the antigens if they ever invade the body again. This natural process provides the body with a natural immunity to diseases.

Organs of the immune system include the thymus gland and spleen. T cells mature in the thymus. The spleen has both T and B cells and is active in filtering antigens from the blood.

URINARY SYSTEM

The urinary system removes metabolic waste products of amino acid breakdown from the blood, maintains the water and salt balance of the blood, and stores and transports urine out of the body. The urinary system includes the kidneys, ureters, bladder, and urethra.

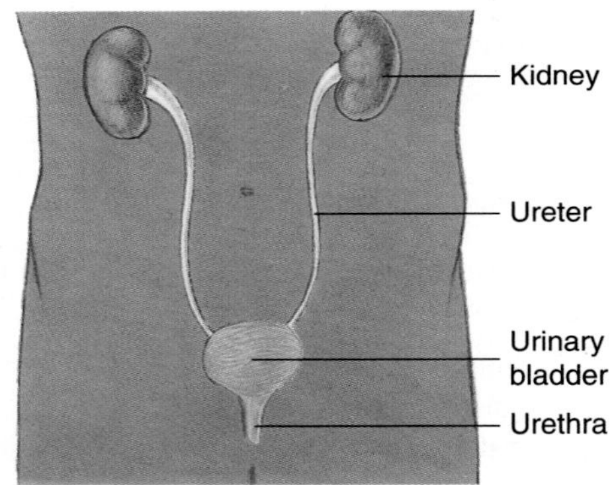

The urinary system filters the blood, collects urine, and excretes urine from the body.

PORTFOLIO

Have students look up early health myths from the 1700-1800s. History of medicine, surgery, or disease books are good sources of this information. These can be shared with the classes to discuss more modern treatments and how our understanding of the human body systems has changed medical treatments. L2 P

STUDENT JOURNAL

LS **Linguistic** Ask students to write a paragraph summarizing the function of each of the body systems. L1

BIODIGEST

BIODIGEST REVIEW

Understanding Concepts

1. Which of the following is NOT one of the levels of organization of cells in the human body?
   a. tissue
   b. organ
   c. organ system
   d. receptor
2. Which of the following systems manufactures blood cells?
   a. skin
   b. skeletal system
   c. circulatory system
   d. respiratory system
3. Which type of muscle lines hollow internal organs?
   a. smooth
   b. skeletal
   c. cardiac
   d. voluntary
4. Which of the following organs is NOT a part of the digestive system?
   a. tongue
   b. saliva glands
   c. spleen
   d. pancreas
5. Oxygen is needed by your body cells to ____________.
   a. produce carbon dioxide in the cells
   b. break down glucose to make ATP
   c. exchange gas in the alveoli of the lungs
   d. provide muscles with energy to contract
6. What type of event occurs at the synapse between two neurons?
   a. Calcium passes from one cell to another cell.
   b. A neurotransmitter passes from one neuron to the next neuron.
   c. A wave of charges passes from one cell to the next cell.
   d. Sensory receptors detect changes inside the body.
7. Which system secretes hormones to control the metabolic activities of the body structures?
   a. endocrine system
   b. nervous system
   c. circulatory system
   d. excretory system
8. Which type of immune cell secretes antibodies against foreign invaders?
   a. red blood cells
   b. T lymphocyte
   c. spleen cells
   d. B lymphocytes
9. Urine contains the metabolic waste products of the digestion of ____________.
   a. glucose
   b. fats
   c. amino acids
   d. water
10. The highest blood pressure, systolic pressure, is the force created by __________.
    a. the lungs
    b. the two atria
    c. the two ventricles
    d. the arteries

Thinking Critically

1. Describe how both the nervous system and endocrine system are involved in controlling all other body systems.
2. AIDS is a viral disease that attacks and kills T lymphocytes. Why does a person with AIDS usually die of an infection?
3. Which systems are involved in excretion of waste materials?
4. How does the respiratory system work with the circulatory system?
5. How might an injury to the skeletal system affect the circulatory system?

BIODIGEST

Answers to Biodigest Review

Understanding Concepts

1. d
2. b
3. a
4. c
5. b
6. b
7. a
8. d
9. c
10. c

Thinking Critically

1. The nervous system receives information from the inside and outside of the body, interprets it, and acts on the information by stimulating muscles or glands. The endocrine system secretes hormones that regulate metabolic activities of body structures.
2. By killing T cells, the disease AIDS removes the cells that kill and remove the agents that cause infections.
3. digestive, respiratory, and urinary
4. The respiratory system delivers oxygen, which is transported by the circulatory system to the body cells. The circulatory system delivers carbon dioxide from the body cells to the lungs for elimination from the body.
5. Breaking a bone will disrupt capillaries, causing bleeding. Blood clots must be formed until the capillaries can be healed. Breaking a bone could also disrupt blood cell production from that area until healing restores the bone.

Program Resources

Section Focus Master 99 L1 SAE
Reinforcement and Study Guide, pp. 175-176 L1
Chapter Assessment, pp. 265-268 L1
Content Mastery, pp. 181-184 L1

Chapter Organizer

SECTION	OBJECTIVES	ACTIVITIES/FEATURES
Technology and Society: Keeping the Balance National Science Standards: UCP.2, UCP.3; A.1, A.2; E.1, E.2; F.4-6; G.1	1. **Demonstrate**, by example, how different sciences work together to solve critical problems for society. 2. **Identify** the benefits, risks, and social concerns of new technology.	**Focus On** Biologists' Views of the Future, p. 1116

GLENCOE TECHNOLOGY

The following multimedia resources are available from Glencoe.

Biology: The Dynamics of Life
CD-ROM SAE
Videodisc Program

The Infinite Voyage Series
Life in the Balance
The Geometry of Life

The Secret of Life Series
Gone Before You Know It: The Biodiversity Crisis

National Geographic Society Series
STV: Biodiversity
Destroying Diversity
Preserving Diversity
GTV: Planetary Manager
Vanishing Act

Science and Technology Videodisc Series (STVS)
Plants & Simple Organisms
Selective Breeding in Cows
Managing a Forest

Chapter 43 Biology and the Future

TEACHER CLASSROOM RESOURCES

Reproducible Masters

Section Focus Master 100: Staying Informed L1 SAE
Reinforcement and Study Guide, pp. 177-178 L1
Concept Mapping: Technology and Society, p. 43 L1
Tech Prep Applications: A Future Zoo, pp. 53-54 L2
Critical Thinking/Problem Solving: Science and Language Arts: A Wildlife Preserve, p. 43 L3
Laboratory Manual: Testing Water Quality, pp. 265-268 L2
Content Mastery, pp. 185-188 L1

Transparencies

Reteaching Transparency #43: Analyzing the Value of Biodiversity L1 SAE

ASSESSMENT MATERIALS

Chapter Assessment, pp. 269-272
Alternate Assessment in the Science Classroom
MindJogger Videoquiz
Computer Test Bank

Spanish Resources SAE
English/Spanish Audiocassettes SAE
Cooperative Learning in the Science Classroom COOP LEARN
Lesson Plans

KEY TO TEACHING STRATEGIES

L1 Level 1 activities should be within the ability range of all students including those with learning difficulties.

L2 Level 2 activities are within the ability range of average to above-average students.

L3 Level 3 activities are designed for the ability range of above-average students.

SAE SAE activities should be within the ability range of Students Acquiring English.

COOP LEARN Cooperative Learning activities are designed for small group work.

P These strategies represent student products that can be placed into a best-work portfolio.

These strategies are useful in a block scheduling format.

CHAPTER 43 Biology and the Future

Chapter Overview

In this chapter, some unsolved problems related to biology are presented. Such problems include studying and preserving Earth's biodiversity, understanding the complexities of ecosystems, and learning new techniques to manipulate DNA.

The importance of basic research is stressed as a way to solve biological problems. In addition, the integration of biology with other sciences is emphasized as a necessary means of solving future problems. Students are encouraged to become informed and involved citizens in areas of decision making that involve biological concepts.

CHAPTER 43 Biology and the Future

As you've discovered throughout this course, biology is very much a part of your life. In fact, you are living at a time when biology is beginning to affect human lives in ways not even imagined just a decade or two ago. Advances in the fields of medicine, genetics, and agriculture promise a future of exciting new developments.

Solutions to the future's challenges will come about only as biologists continue to learn about the natural world and develop ways to apply this knowledge. Whether or not you pursue a career in science, you will most certainly play a part in shaping the future.

Doctors will soon have the technology to cure diseases by replacing defective genes with normal ones. What biological studies have led up to this technology?

Learning Styles

Look for the following logo for strategies that emphasize different learning modalities. LS

Kinesthetic	Activity, p. 1108
Visual-Spatial	Visual Learning, p. 1112
Interpersonal	Display, p. 1109; Portfolio, p. 1109; Meeting Individual Needs, p. 1110
Intrapersonal	Student Journal, p. 1111
Linguistic	Enrichment, p. 1110, 1111; Portfolio, p. 1112
Logical-Mathematical	Project, p. 1111; Activity, p. 1110

CHAPTER

43

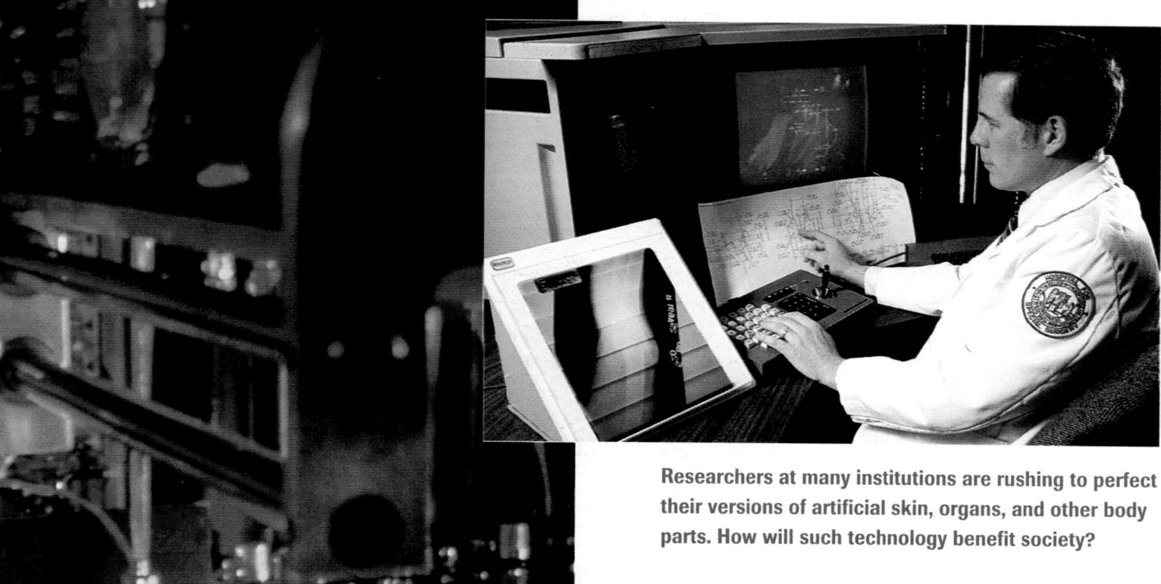

Researchers at many institutions are rushing to perfect their versions of artificial skin, organs, and other body parts. How will such technology benefit society?

People use oil for energy and a variety of important products, but oil spills can wreak havoc on ecosystems. Can we strike a balance between protecting people's livelihoods and protecting the environment?

Chapter Preview

Technology and Society: Keeping the Balance

Introducing the Chapter

As a class, discuss how the content of each photograph relates to the study of biology. Challenge students to explain the relationship between biology and technology shown in each photograph.

Ask students to answer the questions posed in each caption. What biological studies have led up to this technology? *Likely responses may include genetic engineering techniques and studies such as the Human Genome Project.* How will such technology benefit society? *People who are currently unable to perform certain tasks due to loss of a limb may regain use of the limb.* Can we strike a balance between protecting people's livelihoods and protecting the environment? *Likely responses may suggest that these tasks can be accomplished through population control and increased environmental awareness.*

Theme Development

The *nature of science* is emphasized as students explore how biology is being used to help address some of the challenges people will face in the future.

Concept Check

Students will build upon their knowledge of the themes of science as well as the nature of biology. Students will also expand upon their understanding of the role of technology in addressing the concerns of society, particularly those related to the environment, medicine, and improving quality of life.

Assessment Planner

Choose assessment strategies from the following pages to evaluate the progress of your students.

Assess, p. 1114
Portfolio, pp. 1109, 1112, 1116
Chapter Review, pp. 1118-1119

CHAPTER 43

Prepare

Key Concepts

Students learn how physics, chemistry, biology, and Earth science can be coordinated to solve such problems as oil spills, creating new artificial organs, and improving agricultural practices using genetically engineered organisms.

Block Scheduling

Look for this symbol for strategies that are useful in a block scheduling format. For more information on block scheduling, refer to Section 43 in the **Lesson Plans** booklet.

1 Focus

Bellringer

Before presenting the lesson, display **Section Focus Master 100** on the overhead projector and have students answer the accompanying questions. L1 SAE

Activity

LS **Kinesthetic** Provide students with pieces of Velcro and cockelburs or other seeds that stick to clothing. Ask them to compare the two objects carefully using hand lenses. Ask them to make a diagram of a magnified view of the seeds and Velcro. Then ask them to explain what relationship exists between the seeds and the Velcro. *Both objects have a similar structure when viewed with the hand lens. This structure is responsible for the sticking capabilities of each object.*

Technology and Society: Keeping the Balance

Section Preview

Objectives

Demonstrate, by example, how different sciences work together to solve critical problems for society.

Identify the benefits, risks, and social concerns of new technology.

Key Terms

none

Just about everyone has seen or used Velcro, but do you know the origin of this useful product? Actually, Velcro has an innocent, almost amusing origin. One day, while walking through the woods, a Swiss engineer noticed several cockleburs stuck to his clothing. Being interested in this unusual property of cockleburs, he examined them under a microscope and found hundreds of tiny hooks. As a result of this biological observation, Velcro was born!

Velcro

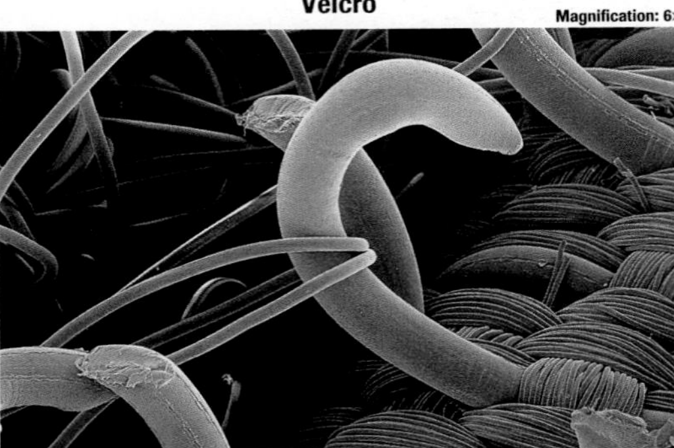

Integration of the Sciences

The development of Velcro can't be attributed only to biological science; it also required contributions from chemistry and physics. Velcro represents a perfect example of how different sciences can merge to solve problems or produce useful products.

Discovering different or better ways to fasten clothing isn't a challenge that humans will face in the future. However, as you've learned, society must address a number of critical problems in the future. These larger problems such as supplying food to the growing human population, protecting the environment, finding cures for human diseases, conserving critical resources, and improving the quality of people's lives aren't simple biological problems. The sciences of biology, chemistry, physics, and mathematics all must come into play to solve various problems that arise in our world. Just as in the development of Velcro, finding solutions to complex biological problems requires the expertise of more than one science.

Program Resources

Section Focus Master 100 L1 SAE
Reinforcement and Study Guide, pp. 177-178 L1
Concept Mapping, p. 43 L1
Critical Thinking/Problem Solving, p. 43 L3
Tech Prep Applications, pp. 53-54 L2
Reteaching Transparency 43 and **Master** L1 SAE
Laboratory Manual, pp. 265-268

Figure 43.1

The *Exxon Valdez* oil spill was the biggest in U.S. history, but oil spills are not uncommon. In the United States alone, about 10 000 oil spills of various types and sizes occur each year. Oil spills happen—the problem is how to clean them up.

Scientists work together to protect the environment

When the oil tanker *Exxon Valdez* spilled more than 11 million gallons of crude oil into the pristine waters of Alaska's Prince William Sound on the frigid morning of March 24, 1989, it resulted in the biggest oil spill in U.S. history, *Figure 43.1.*

As you may well imagine, oil spills have devastating effects on the environment and the organisms that live there. The immediate effects are on the animals, plants, and microorganisms that live near the spill. As shown in *Figure 43.2,* birds, mammals, and fish swimming in the water become covered with thick layers of oil, affecting their ability to move and maintain homeostasis. For the many species of plants, bacteria, fungi, and algae that live near or surround the area of the spill, the poisoning effects of oil are even more devastating because they absorb toxins more readily than the animals do. The toxins can also persist for years in the soil, affecting future generations.

Figure 43.2

As you've learned, the feathers of birds and the fur of mammals are adaptations that enable them to regulate their internal environments. Oil disrupts the proper functioning of fur and feathers, and animals contaminated with oil slowly freeze to death.

2 Teach

Concept Development

Although much oil enters the ocean through spills, more than half of the oil reaching the ocean comes from the land. This oil is usually dumped on land as waste oil. The oil then enters streams and is carried to the ocean. A quart of oil left after changing the oil in a car can contaminate 250 000 gallons of water.

Have students suggest ways of making the public aware of this problem. *Responses may include print, radio, and television ads.* Also ask students how average citizens might be encouraged not to dump oil onto the ground. *Responses may include providing education about proper disposal, encouraging recycling of oil products, and the imposition of heavy fines on people who are caught dumping oil.*

Display

LS **Interpersonal** Have students obtain pictures and headlines from newspapers and magazines that show biologists at work or announce advances in biology. Have students work cooperatively to create a bulletin board display titled Biology in Action. Ask students to include captions explaining the importance of what is shown by each photograph or headline. L1 COOP LEARN

GLENCOE TECHNOLOGY

Videotape

The Secret of Life
Use the videotape *Gone Before You Know It: The Biodiversity Crisis* to illustrate species diversity and ways to prevent its loss.

PORTFOLIO

Thinking Technologically To illustrate how creative thinking is important to developing new technology, provide student groups with several round sheets of paper about 30 cm in diameter, paper clips, and scissors. Explain to the group that they must create a water conservation device of the future with only those materials. Give them 20 minutes and as many pieces of paper and clips as necessary, but no other materials. Ask them to keep their devices in their portfolios. *Students may use the materials to create some type of water-collecting device that directs rainwater toward a plant or a storage container for use at a later time.* L1 P COOP LEARN LS

CHAPTER 43

Activity

LS Logical-Mathematical Give students feathers from a craft shop. Have them determine the masses of the feathers. Then have them coat the feathers with vegetable oil and again determine their masses.

Have students predict the best way to clean the feathers and then use the materials they stated to carry out this task. Once cleaned, have them again determine the masses of the feathers. They might choose paper towels, sand, sawdust, or other materials. Have them compare the masses of the three groups of feathers and explain their observations. L1

Enrichment

LS Linguistic Have students identify what they consider the most interesting concept they have studied in this course. Then have them identify a question they still have about biology. Have students conduct research to try to answer the question. Ask them to write a report of their findings. L2

interNET CONNECTION

Follow the link for this chapter on the Glencoe Homepage at **www.glencoe.com/sec/science** to find out more about biotechnology.

Perhaps worse than the immediate effects are the long-term effects of oil spills. First, the oil itself is a problem. As you know, oil doesn't mix with water, it doesn't disappear, and as long as it remains in the environment, it continues its destructive effects on organisms.

Oil spills can have disastrous consequences in the living world. Although oil spills are essentially biological problems, cleaning up oil and restoring the ecosystems requires technologies from many different scientific disciplines, as ***Figure 43.3*** indicates.

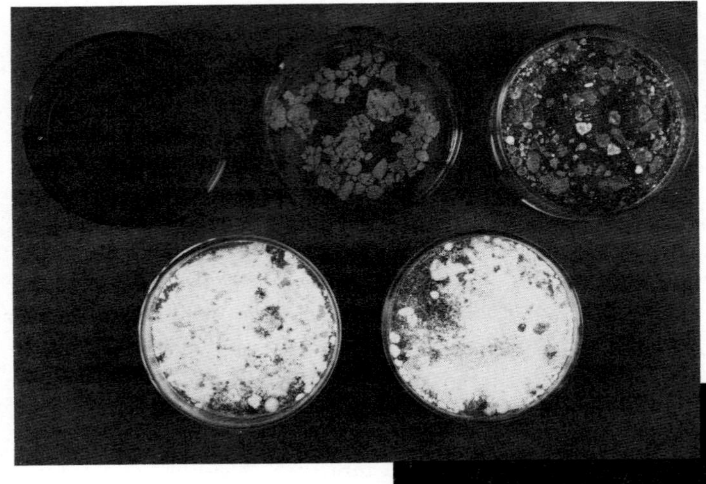

Chemistry Chemists develop methods for cleaning up oil. One important technology that is being developed by chemical engineers is the use of tiny glass beads coated with titanium dioxide (bottom row). These coated beads induce oxygen to attach to the oil, speeding its disintegration. Another type of bead actually takes up oil and forms a floating mass that can be suctioned away (top row).

Physics Physicists are involved chiefly in the containment of oil spills. Booms—a type of floating, flexible fence—are placed around a spill to prevent it from spreading.

Figure 43.3

To clean up after large-scale oil spills like the *Exxon Valdez* disaster, a number of techniques have been developed within a variety of scientific disciplines.

Meeting Individual Needs

Gifted Ask students to organize and run a school-wide competition titled "The ––– ––– High School Oil Spill." Students in groups should design or create materials or build apparatus to clean up a simulated oil spill in an aquarium. Have students plan to award prizes. L3 COOP LEARN LS

Oil spills are not the only ecological disasters scientists must deal with. Similar ecological problems occur in areas decimated by fires, floods, storms, toxic waste and acid rain contamination, and a host of other human-influenced or natural disasters. These problems require diverse solutions too, and as scientists perfect their techniques for protecting the environment, it's possible that the ravaging effects of environmental disasters such as oil spills will become a thing of the past.

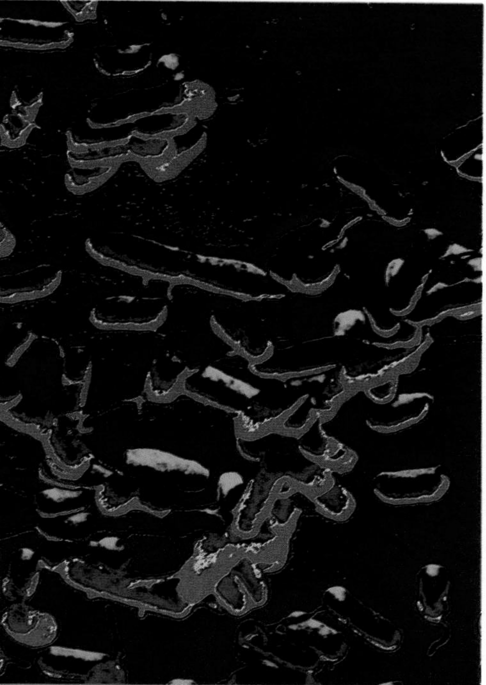

Biology Perhaps the biggest contribution of biologists is to study the ecological effects of small-scale spills and then apply this knowledge to larger ones. Biology has also been important in developing methods to use large populations of naturally occurring bacteria that consume oil such as *Pseudomonas putida,* shown here, to help clean up oil spills.

Earth science Geologists deal with the effects of oil on the physical environment, such as how oil affects the water supply. Geological studies are also important in the discovery of safer routes for transport.

CHAPTER 43

Reinforcement

Emphasize how different fields of science are involved in the cleaning of oil spills. Elicit how the same fields of science could be involved in attempts to reduce the amount of air pollution resulting from the burning of fossil fuels. *Because oil is a fossil fuel, which contributes to air pollution when burned, the process of attacking the air pollution problem involves chemical substances that are the same or similar to those involved in oil spills. Chemists are needed to analyze what chemicals are involved in the air pollution problem. Biologists examine how the problem affects organisms. Earth scientists must evaluate how air pollution alters the atmosphere and how it may affect Earth's surface. Accept all logical responses.*

Enrichment

LS **Linguistic** On individual slips of paper, list products that originated in tropical rain forests. Have each student draw one of the product slips. For the product, have them research the following questions. From what organism does your product come? How did native people use your product? How and when was your product discovered by nonnative people? What, if any, substitutes exist for your product? Have students present their findings to the class. L2

PROJECT Effects of Crowding

LS **Logical-Mathematical** Have students design and conduct an experiment to determine the effect of crowding on a population. Suggest that they use plants grown from seed. *Seeds of the same type should be planted at varying distances apart to determine the effect of crowding on growing plants.* L2

STUDENT JOURNAL

Make a Table Ask students to conduct research about specific oil spills and their effects on the environment. Have students make a table of their findings for inclusion in their journals. L1 LS

Visual Learning

Figure 43.4 Have students examine the photographs. Elicit from students what areas of science, in addition to biology, are involved in the development of the structures shown. *Chemists are likely to be involved in the development of both artificial skin and nerve chips. Computer specialists and possibly physicists are likely to be involved in the development of the nerve chips.* LS

Misconception

Many students believe that genetically engineered fruits and vegetables may be harmful. Fruits such as tomatoes have been genetically engineered so that they can be ripened on the vine and then transported. Genetic engineering has not altered the nutritional quality or the chemistry of the tomato in any way that would be harmful to people.

GLENCOE TECHNOLOGY

Videodisc

The Infinite Voyage: The Geometry of Life
The Human Genome Project
(Ch. 11)

Integrating disciplines in medicine

Environmental protection is critical for maintaining the long-term health of the human species, but for help in dealing with current health issues, people turn to medicine.

In the field of medicine, biology has always had a long-standing relationship with other sciences. The development of drugs, surgical techniques, X rays, and a variety of other medical instruments and products are just some of the examples of technologies that have been created through interdisciplinary research. However, the face of medicine is changing rapidly. Today, scientists from a variety of new fields are developing products and tools that promise a high-tech future for medicine.

One of the newest and perhaps the most exciting medical technologies to come out of this scientific teamwork is in the field of artificial organ development, or bionics. Using technology from such diverse fields as computer science, electronics, chemistry, physics, and biology, scientists have already developed a large assortment of artificial body parts, including artificial tendons, synthetic skin, replacement elbow and hip joints, and implants for the ears and eyes.

But are scientists close to producing the bionic men and women of movie and television fame? This may be several years away, but scientists predict that within the not-too-distant future, as you can see by examples in *Figure 43.4*, an artificial component will be able to take the place of almost any portion of human anatomy, with the major exception of the brain.

Figure 43.4
The high-tech body parts of the present and future will combine the technologies of many scientific fields.

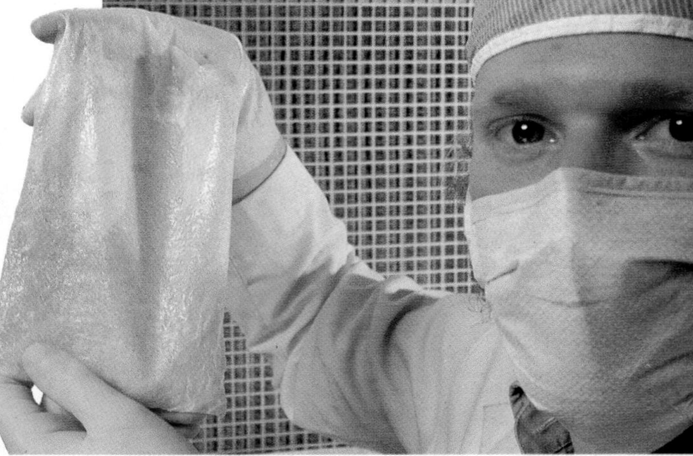

Skin Biologists hope to produce sheets of real skin cultured from human skin cells, rather than relying on this artificial skin made of silicone.

Nerve chips Computer microchips similar to this one will be able to pick up impulses from nerve fibers in the stump of a limb and transmit messages to computer-driven artificial limbs.

PORTFOLIO

Genetics in Medicine Ask students to research and report in their portfolios about the latest applications of genetic engineering to the field of human medicine. Ask them to explain how the Human Genome Project has enhanced the applications of genetic engineering. L2 P LS

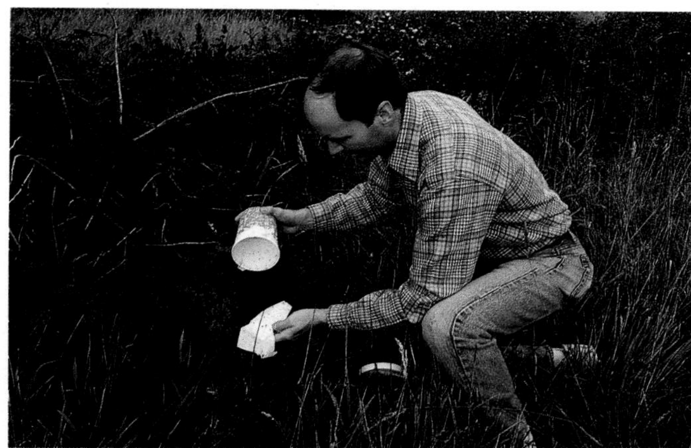

Figure 43.5

Associated with the benefits of new technologies are risks. Since the application of genetic-engineering techniques to agriculture is so new, can scientists be sure that food products are not harmful to humans, or that the release of genetically engineered pests poses no threat to the delicate balances of ecosystems?

Technology and Society

Technological advances in the fields of medicine, genetics, and ecology offer great hope for the future of humans and all life on Earth; however, new technologies sometimes raise serious social and ethical concerns for humans, as well as risks.

New technologies raise questions about benefits and risks

Genetically engineered plants and pests will transform the agriculture industry, but does this technology have limits? By genetically engineering sterile pests, spraying insecticides on crops will become a thing of the past. This will decrease pollution, but what might happen if the newly engineered traits were accidentally transferred to the wild relatives of donor species, *Figure 43.5*? How can scientists be sure that such technology will not be harmful?

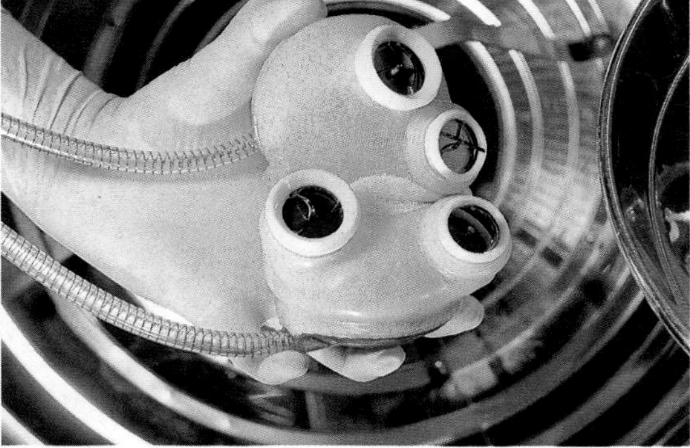

Heart **The first artificial heart, the Jarvik-7 shown here, has been replaced by more sophisticated models.**

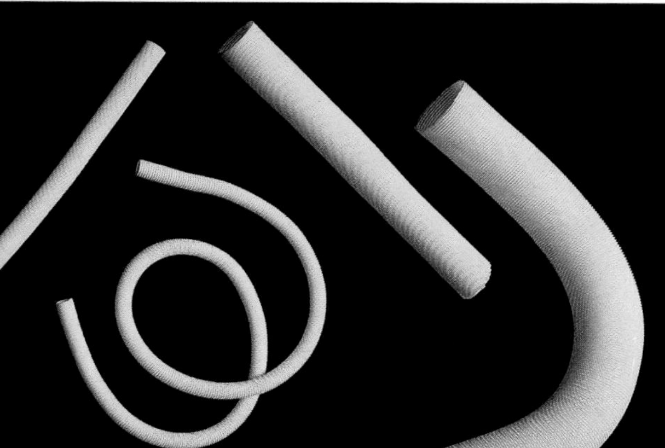

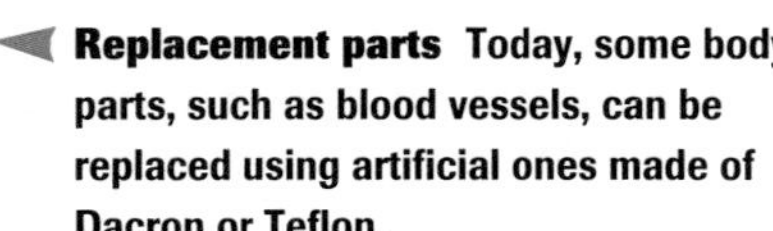

Replacement parts **Today, some body parts, such as blood vessels, can be replaced using artificial ones made of Dacron or Teflon.**

Science, Technology, and Society

Disabling Single Genes

Fifteen years ago, geneticists dreamed about disabling a single gene in a mouse to see what would happen. Today, "knockout" mice are a reality. More than 100 different strains of laboratory mice are missing a gene, and geneticists are determining what each gene does by observing what the mouse can no longer do.

"Knockout" mice will have important implications for people. Because mice and humans share 90% of the same genes, the "knockout" mice can serve as models for human diseases. If the same gene that causes disease in humans is knocked out in a mouse, the mechanism of the disease and potential effectiveness of drugs and gene therapy can be tested in the mouse model.

Visual Learning

Figure 43.5 Since the application of genetic-engineering techniques to agriculture is so new, can scientists be sure that food products are not harmful to humans, or that the release of genetically engineered pests poses no threat to the delicate balances of ecosystems? *Students may respond that scientists cannot predict the effects of genetically altered foods or pests until more experimental test results have been examined.*

Bioethics

Fighting the Tsetse Fly

The tsetse fly spreads the protozoan parasite that causes African sleeping sickness. Before insecticides were available to control this insect, tsetse fly-infested areas had to be avoided. Because large areas can now be sprayed by plane, these areas are being settled and used for agriculture so rapidly that wildlife habitat is being lost at an ever-increasing rate.

The elimination of the tsetse fly that caused untold human suffering and death is now accompanied by loss of habitat and extinction of its wild inhabitants.

Visual Learning

Figure 43.6 Will this technology ever be applied to humans? *Some students may suggest that it will be used to replace lost organs with healthy ones.* What may some of the ramifications be? *Some students will suggest improved quality of life for people. Others may suggest that it will make people too similar.*

3 Assess

Check for Understanding

Ask students to read through the local newspaper and cut out all the articles that pertain to biology. Ask them to group the articles into those that raise ethical issues, those that deal with new technology, and those that show how biology is connected to other areas of science. L1

Ethical concerns of technology

Besides risks, new technologies sometimes also raise important social questions. For instance, biologists have been using genetic-engineering techniques to clone organisms such as plants, mice, frogs, and cattle for years, *Figure 43.6,* but now scientists report that it may be possible to clone human embryos as well. Has this technology gone too far? Science has also developed many different techniques that make it possible for infertile couples to conceive. Some of these involve freezing and storing human embryos. Many controversies have already arisen as a result of this technology.

Protecting the rain forests and other delicate ecosystems is critical for preserving Earth's biodiversity, but even this raises controversy. Rain forests, for example, are diminishing at astonishing rates mainly because the cleared land is being used for agriculture. Some people say that it is not fair for highly industrialized nations such as the United States to ask developing nations, many of which are in areas with the greatest biodiversity, to suppress their own poor economies in order to save animals and plants.

Finding the proper balance between preserving the environment and protecting people's livelihoods is a societal problem that has existed for centuries. We encounter this problem in our homes by deciding which products to buy and which cars to drive. And if you choose to pursue a career in science, you'll be making critical decisions about this matter on a day-to-day basis.

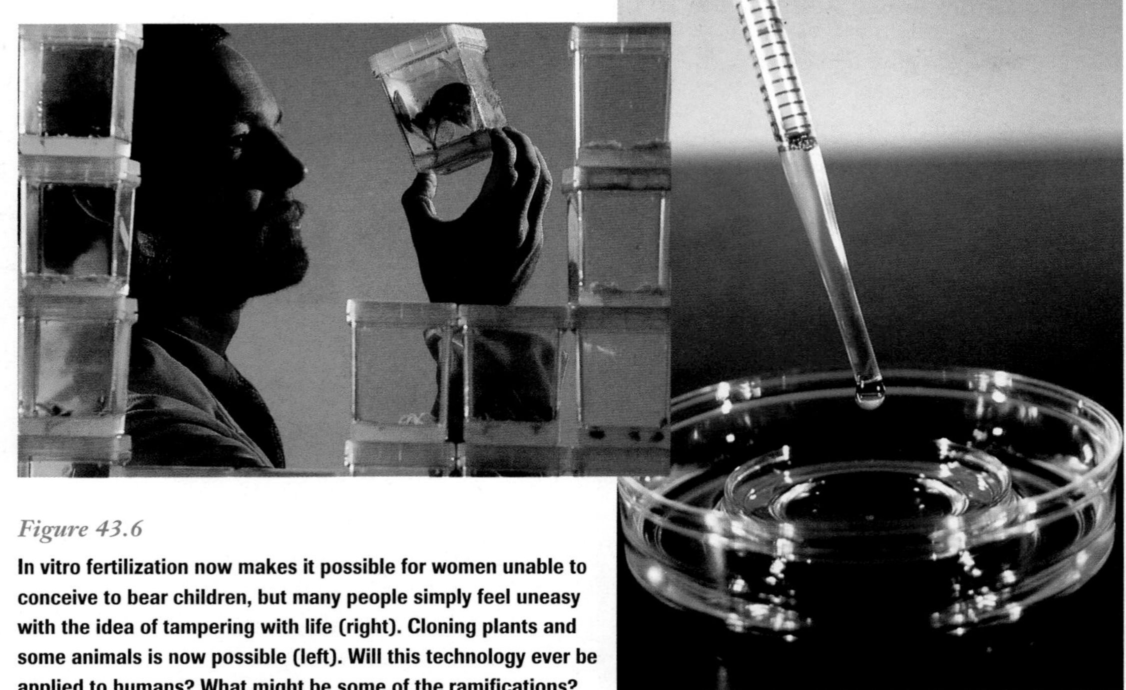

Figure 43.6

In vitro fertilization now makes it possible for women unable to conceive to bear children, but many people simply feel uneasy with the idea of tampering with life (right). Cloning plants and some animals is now possible (left). Will this technology ever be applied to humans? What might be some of the ramifications?

CULTURAL DIVERSITY

Solving China's Food Crisis

More than 1 billion people, roughly one-fourth of the world's population, currently live in the People's Republic of China. Each year, 12 million people are added to the population.

Discuss how the enormous population of China creates the problem of trying to feed all the people. Explain that scientists from many different disciplines are working on solutions to this problem. For example, engineers are developing ways to make China's waterways more navigable for the transportation of agricultural products; chemists are creating better, more effective fertilizers; and agricultural scientists are using methods to try to improve crop yields.

L1

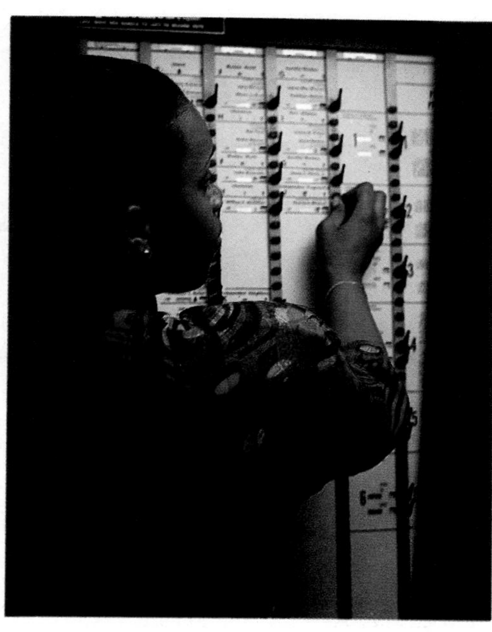

Figure 43.7

Only by educating yourself and staying informed on issues can you develop your own viewpoint. You can have an impact on the future by staying interested and involved.

Stay informed . . . Be involved

The ethical and social questions of biotechnology are outside the realm of science; however, you can ensure that science plays a positive role in your life by staying informed and involved.

School is a great place to start becoming an informed citizen. Each of the scientific disciplines offers a wealth of knowledge about you and your world. Reading newspapers and magazines and watching science programs on television are other ways you can stay informed. The more informed you are, the less complex and mysterious the world will seem.

Finally, as a citizen of the future, it's critical that you remain involved, *Figure 43.7*. You will soon be able to vote, and your vote can truly make a difference. Search out candidates who reflect your views on important scientific issues. Above all, remember that the benefits of science can be directed by the continuing efforts of concerned citizens such as you if you stay informed and involved.

Section Review

1. Write an essay about your feelings on genetic engineering. How will genetic engineering improve people's lives? What concerns, if any, do you have about this technology? What are scientists doing to lessen societal apprehension about genetic engineering?
2. Consult newspapers, science magazines, or television programs to prepare a report about a new development in technology that involved the expertise of more than one scientific discipline. What are the benefits of this technology? What sciences were involved in this technology, and what were their contributions?

Reteach

Ask students to make a table that lists and describes the focus of other sciences. Ask them to indicate examples of how each science is helping to solve biological problems. L1

Extension

Ask a group of students to visit a site where prosthetic devices are made or installed, or a physical therapy office where patients are taught how to use their prostheses. Ask students to take photographs of their experience and report back to class. L1

Assessment

Performance: Ask students to create a board game that deals with bioethical questions brought about by new technology. Have them write out instructions, cards, or other materials necessary for the game. Then have groups trade and play the games. Games can be evaluated based on the ability of another group to play the game and the biological issues brought up by the game. L1

4 Close

Activity

Ask students to plan to bury a time capsule that will be opened in the year 3000. Ask each student to bring in something that would symbolize the natural world as it exists today. Have students explain why they selected the items they did for the capsule.

Answers to Section Review

1. Students may suggest that genetic engineering could enable people to have genetic disorders corrected. Some may suggest that human organs may be grown in animals and used for transplants in people. Some students may suggest that crop plants may have better resistance to insects and adverse environmental conditions.
2. Students might report about genetic engineering, bionic limbs, electronic senses, oil spill cleanup methods, humans and other organisms in space, new technology in agriculture, or other technology.

Focus On

Biologists' Views of the Future

Purpose

Students will learn about the visions biologists have for the future.

Teaching Strategies

- Ask students to create a table with the following heads across the top: Scientist, Job or Area of Science, Job Description. Have students complete the table with information about each of the scientists included in this feature.
- Emphasize to students the importance of education in each of these careers. As a class, review the career paths for each of these scientists. Elicit from students what educational requirements the careers have in common and how requirements differ.

Visual Learning

Figure Captions Have students read the captions that accompany each photograph. As a class, discuss each scientist's view of the future and elicit whether students agree or disagree with these views of how biology will affect the future. Have students explain their responses.

GLENCOE TECHNOLOGY

Videodisc

STVS: Plants & Simple Organisms
Disc 4, Side 1
Selective Breeding in Cows
(Ch. 22)

Focus On

Biologists' Views of the Future

Most biologists would probably agree with this scientist's comment about the future: "I love my life, and I'm so glad I live when I do. My only regret is that I won't see all the exciting changes in science that will happen in the next 100 years." On these pages, a dozen biologists look ahead to the 21st century and speculate on the changes that they hope will develop in their specific fields.

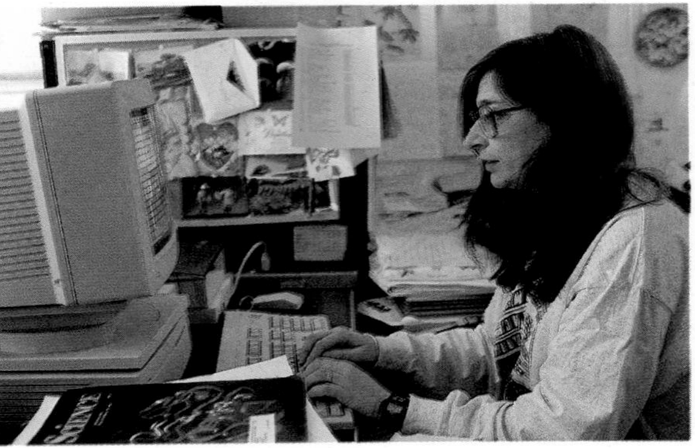

Dr. May Berenbaum, Entomologist Worldwide, the leading cause of death of children under five years of age is from insect-borne diseases. I think a definite possibility in the near future is using genetic engineering to make mosquitoes unable to carry malaria. Other radical, new forms of environmentally compatible control of insects will also improve the quality of life.

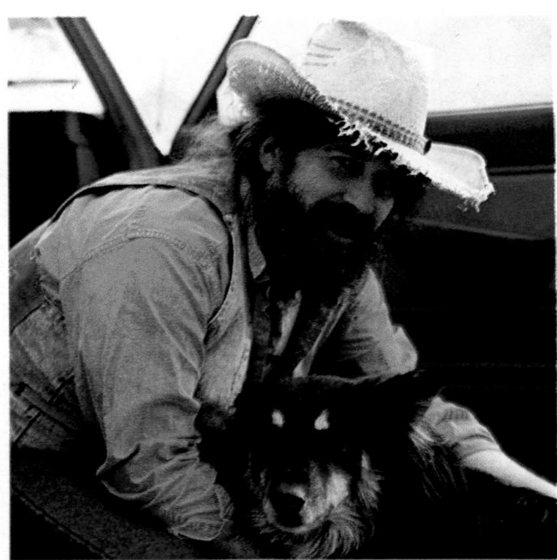

Dr. Robert Bakker, Paleontologist Many kids ask me, "Will there be anything left to discover when I'm grown up?" I always tell them that the more we dig, the more we realize we have just scratched the surface of Earth, quite literally. I think that 99 percent of the dinosaurs are still waiting to be discovered.

Dr. Benjamin Carson, Neurosurgeon I think the marriage of biomedical technology and neurosurgery will be wonderful. As computers become smaller and more biocompatible, it will probably be possible to transplant microcameras instead of artificial eyes to give some blind persons sight. I think microchips will also be inserted in the spinal cord so that paralytic nerves can work again.

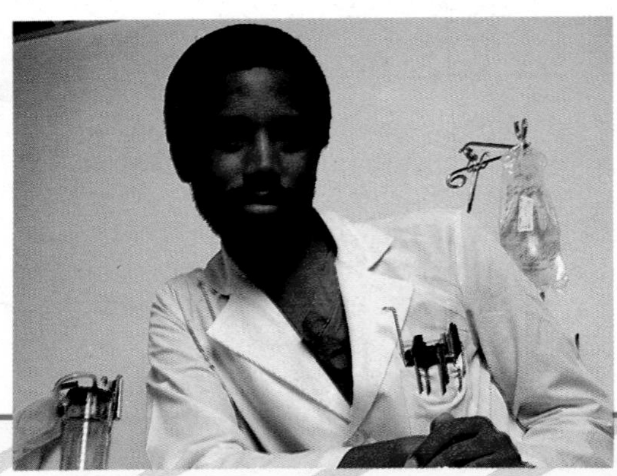

1116 Biology and the Future

PORTFOLIO

Interviewing Health Professionals Ask students to interview public health officials, nurses, or physicians to find out how these professionals envision health care in the year 2020. Ask students to include their interviews in their portfolios. L2 P

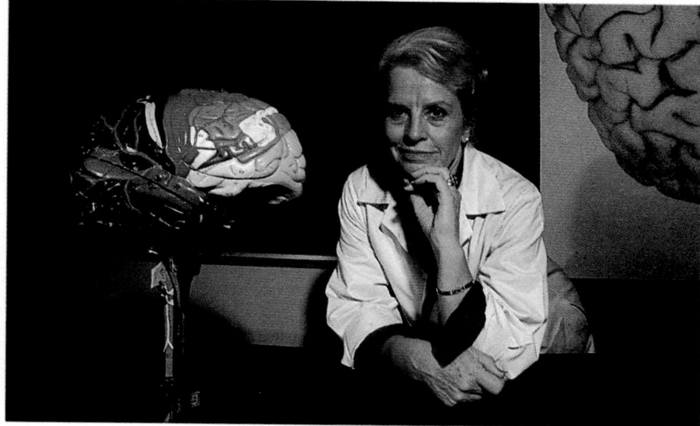

Dr. Marian Diamond, Professor of Anatomy and Museum Director I think the whole field of biology will change tremendously as we learn to understand the human brain as well as how we interact with other forms of life on Earth. After all, there's a reason it's called the universe; we're all in this together.

Dr. Robert Murray, Geneticist I see great possibilities in gene therapy; the replacing of patients' missing or defective genes with normal genetic components. Genetics holds the secret of life. Some people think that someday we'll be able to synthesize genes and use them to produce new life-forms. It's an area of enormous uncertainty because we still don't know a lot about life-forms that already exist in nature.

Ms. Lisa Stevens, Assistant Zoo Curator More and more, zoos will be a kind of repository, a genetic bank. If people can correct some of the environmental errors of the past, perhaps zoo specimens could be used to replenish habitats.

EXPANDING YOUR VIEW

1. **Applying Concepts** Choose one field of biology, such as biochemistry, genetics, or ecology, and suggest how you as a biologist might have an influence on its future.
2. **Writing about Biology** Describe what new developments in biology you would like to see during your lifetime.

Focus On

Teaching Strategies

- Invite a city planner to speak to the class about how your town or city considers conservation when planning development projects. Ask the speaker to answer questions students may have about his or her job. Also ask the speaker to discuss how biology and science affect his or her job and how he or she uses this science knowledge.

Answers to Expanding Your View

1. **Applying Concepts** Individual responses will vary depending upon the area chosen. Likely responses will focus on curing disease, improving health care, and improving environmental quality or preserving biodiversity.
2. **Writing About Biology** Students are likely to focus on improvements in health care, environmental quality, technology, and disease control. Accept all responses.

Assessment

Ask students to make a list of five important biological problems they think need to be solved in the future. Ask students to prioritize their lists and explain their reasons. L1

Meeting Individual Needs

Gifted Have students research other careers that may involve science. Examples of such careers may include environmental law, computer analysis of scientific data, biological illustration, and sports medicine. Encourage students to work together and prepare a bulletin board display that identifies each career and its requirements.

Chapter 43 Review

Reviewing Main Ideas

Summary statements can be used by students to review the major concepts of the chapter.

Key Terms

Answers should go beyond defining the terms. Accept any answer that uses the term correctly and in the proper context.

Understanding Concepts

1. c
2. b
3. a
4. d
5. c
6. a
7. d
8. c
9. d
10. c

Applying Concepts

11. Examples may include that a chemist develops toothpaste based on knowledge about the human mouth. A physicist may help design athletic shoes that will properly cushion the foot, based on the anatomy of the foot. Accept any reasonable answer.
12. Island organisms may be burned by lava as it flows to the sea, or they may be poisoned by fumes. Humans will suffer the economic consequences of volcanic eruptions and loss of plants, fisheries, and wildlife.
13. Aphids may not be able to pierce the thicker stems and may die as a result of not being able to feed.
14. No, because the doctor's role is a biological one, not a societal role. Once the embryos are available, the parents have to decide how many embryos to implant.
15. We will be able to help preserve ecosystems for our own benefit if we understand more about their complex interactions.

Chapter 43 Review

REVIEWING MAIN IDEAS

Technology and Society: Keeping the Balance

- The integration of biology and other sciences has been important for past technologies, such as those in medicine and agriculture, and will be critical for solving the challenging problems of the future.
- New technologies often introduce complex questions to society. By staying informed about the changing face of biology, you will be able to make the critical decisions that will affect your life in the future.

Understanding Concepts

1. Biology, chemistry, physics, and ____________ are sciences used to solve problems in the natural world.
a. astrology **c.** mathematics
b. marketing **d.** mythology

2. Which of the following is NOT a human-influenced disaster?
a. acid rain **c.** fire
b. a tornado **d.** toxic waste

3. The field of ______________ provides new information for dealing with current human health issues.
a. medicine **c.** mythology
b. astrology **d.** geology

4. Oil spills are more harmful to algae, fungi, and plants than to animals because they ___________.
a. are killed instantly
b. need sunlight to survive
c. cannot absorb toxins
d. absorb toxins more readily

5. By studying Earth's ______________, scientists hope to find new medicines, foods, and raw materials.
a. geology **c.** biodiversity
b. atmosphere **d.** zoos

6. Toxins from an oil spill can persist for years in the ______________ and affect later generations of organisms.
a. soil **c.** water
b. air **d.** plants

7. A floating ___________ can be placed around an oil spill to contain it and prevent it from spreading.
a. balloon **c.** bead
b. crane **d.** boom

8. The field of artificial organ development is called ______________.
a. electronics **c.** bionics
b. pathology **d.** cloning

9. Which of the following body parts has NOT been produced in an artificial form?
a. skin **c.** heart
b. joints **d.** brain

10. Which of these activities is NOT a way to keep informed about scientific issues?
a. reading newspapers
b. watching television news
c. reading comic books
d. listening to political candidates who share your views

Applying Concepts

11. Provide several examples of how scientists from other disciplines work with biologists.

12. Describe some possible consequences of volcanic eruptions on island organisms.

13. Scientists have engineered plants that are shorter and have thicker stems so that they can withstand hurricanes without uprooting. How might these altered plants affect aphids that normally feed on plant juices?

Program Resources

Chapter Assessment, pp. 269-272 L1
Alternate Assessment in the Science Classroom
Computer Test Bank L1
Content Mastery, pp. 185-188 L1

14. An infertile couple uses in vitro fertilization, which results in five embryos. They ask the doctor to say how many embryos they should implant. Is this a decision the doctor should make? Explain.
15. How will studying the complex interactions among organisms in ecosystems be important for the future?

Thinking Critically

16. ***Recognizing Cause and Effect*** Why is staying informed of new technologies important for any citizen?
17. ***Formulating Hypotheses*** At one time, most of the United States east of the Mississippi River was covered with forest. These forests were cleared for farming, an activity that is happening now in tropical rain forests around the world. Why is the clearing of land for farming in tropical areas of more concern than clearing land in more temperate environments?
18. ***Making a Table*** Make a table titled Methods of Oil Spill Cleanup. In your table, include the technologies contributed by the following sciences: physics, chemistry, biology, and Earth science.
19. ***Comparing and Contrasting*** Why is putting a genetically altered pest organism back into its environment less harmful than introducing a new pest species into the same environment?

ASSESSING KNOWLEDGE & SKILLS

Oil spills of varying sizes occur all over the world.

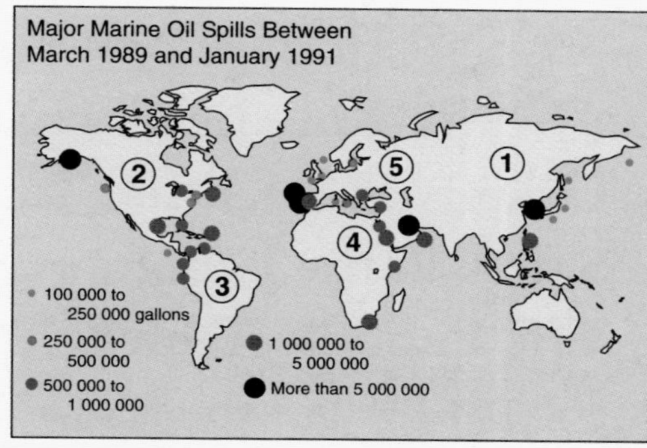

Interpreting Scientific Illustrations Examine the map and answer the following questions.

1. On which continents of the world have major oil spills (more than 5 million gallons) occurred?
   a. 1, 2, 5
   b. 1, 2, 3, 4
   c. 2, 3, 4, 5
   d. 1, 3, 4, 5
2. In which part of the world have there been the most major oil spills?
   a. 1
   b. 2, 3
   c. 4, 5
   d. 3
3. In which part of the world has there been the least amount of oil spilled?
   a. 2
   b. 3
   c. 4
   d. 5
4. On which coastlines would you expect to find oil contamination problems in wildlife?
   a. 1, 2, 3, 4, 5
   b. 1, 3, 5
   c. 1, 2, 4, 5
   d. 3
5. ***Making a Graph*** Make a bar graph showing the number of oil spills on each continent of more than 1 million gallons.

Chapter 43 Review

Thinking Critically

16. Citizens must be able to vote on issues that reflect their views and candidates who support them.
17. The soil in tropical rain forests is thin because most of the biomass is in the organisms. Cleared rain forest land only supports farming for a short time compared to cleared temperate forest soils that can be farmed continuously for many years.
18. See below.
19. Introduced species often have unexpected behavior, causing damage to native species. A genetically altered native species will fit back into the environment and be less disruptive because only one variable has been changed.

ASSESSING KNOWLEDGE & SKILLS

1. a
2. c
3. b
4. a
5.

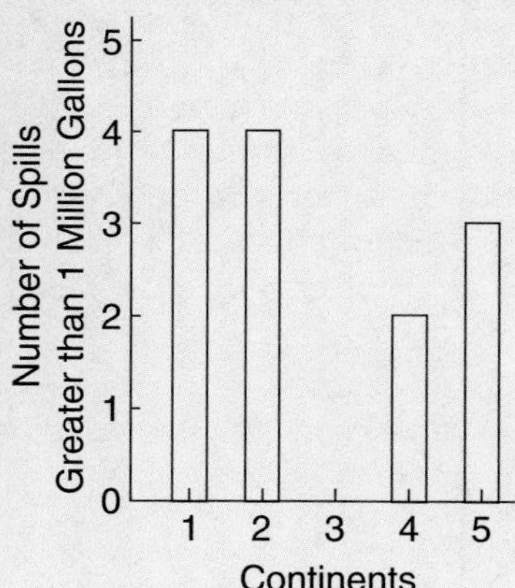

18.

Methods of Oil Spill Cleanup	
Science	**Contribution**
Physics	containment of oil in booms
Chemistry	use of beads to remove oil, or break down oil
Biology	use of oil-consuming bacteria, study of small-scale spills
Earth Science	effects of oil on physical environment, safer transport routes

Appendices *Contents*

Appendix A The Six-Kingdom Classification

The classification used in this text is one that combines information gathered from the systems of many different fields of biology. For example, phycologists, biologists who study algae, have developed their own system of classification, as have mycologists, biologists who study fungi. The naming of animals and plants is controlled by two completely different sets of rules. The six-kingdom system, although not yet ideal in reflecting the phylogeny of all life, is now currently accepted. Taxonomy is an area of biology that evolves just like the species it studies. In this Appendix, only the major phyla are listed, and at least one genus is named as an example. For more information about each phylum, refer to the chapter in the text in which the group is described.

Kingdom Eubacteria

(True Bacteria)

Phylum Actinobacteria
Example: *Mycobacterium*

Phylum Omnibacteria
Example: *Salmonella*

Phylum Spirochaetae (Spirochaetes)
Example: *Treponema*

Phylum Chloroxybacteria (Grass-green Bacteria)
Example: *Prochloron*

Phylum Cyanobacteria (Blue-green Algae)
Example: *Nostoc*

Kingdom Archaebacteria

(Ancient Bacteria)

Phylum Aphragmabacteria (Thermoacidophiles)
Example: *Mycoplasma*

Phylum Halobacteria (Halophiles)
Example: *Halobacterium*

Phylum Methanocreatrices (Methanogens)
Example: *Methanobacillus*

Kingdom Protista

Animal-like Protists

Phylum Rhizopoda (Amoebas)
Example: *Amoeba*

Phylum Ciliophora (Ciliates)
Example: *Paramecium*

Phylum Sporozoa (Sporozoans)
Example: *Plasmodium*

Phylum Zoomastigina (Flagellates)
Example: *Trypanosoma*

Plantlike Protists

Phylum Euglenophyta (Euglenoids)
Example: *Euglena*

Phylum Bacillariophyta (Diatoms)
Example: *Navicula*

Phylum Dinoflagellata (Dinoflagellates)
Example: *Gonyaulax*

Phylum Rhodophyta (Red Algae)
Example: *Chondrus*

Phylum Phaeophyta (Brown Algae)
Example: *Laminaria*

Phylum Chlorophyta (Green Algae)
Example: *Ulva*

Funguslike Protists

Phylum Acrasiomycota (Cellular Slime Molds)
Example: *Dictyostelium*

Phylum Myxomycota (Plasmodial Slime Molds)
Example: *Physarum*

Phylum Oomycota (Water Molds, Mildews, Rusts)
Example: *Phytophthora*

Kingdom Fungi

Phylum Zygomycota (Sporangium Fungi)
Example: *Rhizopus*

Phylum Ascomycota (Cup Fungi and Yeasts)
Example: *Saccharomyces*

Phylum Basidiomycota (Club Fungi)
Example: *Amanita*

Phylum Deuteromycota (Imperfect Fungi)
Example: *Penicillium*

Phylum Mycophycota (Lichens)
Example: *Cladonia*

Kingdom Plantae

Spore Plants

Division Bryophyta (Mosses and Liverworts)

Class Mucopsida (Mosses)

Example: *Polytrichum*

Appendix A The Six-Kingdom Classification

Class Hepaticopsida (Liverworts)
Example: *Marchantia*

Division Psilophyta (Whisk Ferns)
Example: *Psilotum*

Division Lycophyta (Club Mosses)
Example: *Lycopodium*

Division Sphenophyta (Horsetails)
Example: *Equisetum*

Division Pterophyta (Ferns)
Example: *Polypodium*

Seed Plants

Division Ginkgophyta (Ginkgoes)
Example: *Ginkgo*

Division Cycadophyta (Cycads)
Example: *Cycas*

Division Coniferophyta (Conifers)
Example: *Pinus*

Division Gnetophyta
Example: *Welwitschia*

Division Anthophyta (Flowering Plants)

Class Dicotyledones (Dicots)

Family Magnoliaceae (Magnolias)
Example: *Magnolia*

Family Fagaceae (Beeches)
Example: *Quercus*

Family Cactaceae (Cacti)
Example: *Opuntia*

Family Malvaceae (Mallows)
Example: *Gossypium*

Family Brassicaceae (Mustards)
Example: *Brassica*

Family Rosaceae (Roses)
Example: *Rosa*

Family Fabaceae (Peas)
Example: *Arachis*

Family Aceracea (Maples)
Example: *Acer*

Family Lamiaceae (Mints)
Example: *Thymus*

Family Asteraceae (Daisies)
Example: *Helianthus*

Class Monocotyledones (Monocots)

Family Poaceae (Grasses)
Example: *Triticum*

Family Palmae (Palms)
Example: *Phoenix*

Family Liliaceae (Lilies)
Example: *Asparagus*

Family Orchidaceae (Orchids)
Example: *Cypripedium*

Kingdom Animalia

Invertebrates

Phylum Porifera (Sponges)
Example: *Spongilla*

Phylum Cnidaria (Corals, Jellyfishes, Hydras)

Class Hydrozoa (Hydroids)
Example: *Hydra*

Class Scyphozoa (Jellyfishes)
Example: *Aurelia*

Class Anthozoa (Sea Anemones, Corals)
Example: *Corallium*

Phylum Platyhelminthes (Flatworms)

Class Turbellaria (Free-living Flatworms)
Example: *Dugesia*

Class Trematoda (Flukes)
Example: *Fasciola*

Class Cestoda (Tapeworms)
Example: *Taenia*

Phylum Nematoda (Roundworms)
Example: *Trichinella*

Phylum Mollusca (Mollusks)

Class Gastropoda (Snails and Slugs)
Example: *Helix*

Class Bivalvia (Bivalves)
Example: *Arca*

Class Cephalopoda (Octopuses, Squid)
Example: *Nautilus*

Phylum Annelida (Annelids)

Class Polychaeta (Polychaetes)
Example: *Nereis*

Class Oligochaete (Earthworms)
Example: *Lumbricus*

Class Hirudinea (Leeches)
Example: *Hirudo*

Phylum Arthropoda (Arthropods)

Class Arachnida (Spiders, Mites, Scorpions)
Example: *Latrodectus*

Class Merostomata (Horseshoe Crabs)
Example: *Limulus*

Class Crustacea (Lobsters, Crayfishes, Crabs)
Example: *Homarus*

Class Chilopoda (Centipedes)
Example: *Scutigerella*

Class Diplopoda (Millipedes)
Example: *Julus*

Class Insecta (Insects)
Example: *Bombus*

Phylum Echinodermata (Echinoderms)

Class Crinoidea (Sea Lilies, Feather Stars)
Example: *Ptilocrinus*

Class Asteroidea (Starfishes)
Example: *Asterias*

Class Ophiuroidea (Brittle Stars)
Example: *Ophiura*

Class Echinoidea (Sea Urchins and Sand Dollars)
Example: *Arbacia*

Class Holothuroidea (Sea Cucumbers)
Example: *Cucumaria*

Vertebrates

Phylum Chordata (Chordates)

Subphylum Urochordata (Tunicates)
Example: *Polycarpa*

Subphylum Cephalochordata (Lancelets)
Example: *Branchiostoma*

Subphylum Vertebrata (Vertebrates)

Class Agnatha (Lampreys and Hagfishes)
Example: *Petromyzon*

Class Chondrichthyes (Sharks, Rays)
Example: *Squalus*

Class Osteichthyes (Bony Fishes)
Example: *Hippocampas*

Subclass Crossopterygii (Lobe-finned Fishes)
Example: *Latimeria*

Subclass Dipneusti (Lungfishes)
Example: *Neoceratodus*

Subclass Actinopterygii (Ray-finned Fishes)
Example: *Acipenser*

Class Amphibia (Newts, Frogs, Toads)
Example: *Rana*

Class Reptilia (Turtles, Snakes, Lizards, Crocodiles, Alligators)
Example: *Anolis*

Class Aves (Birds)

Order Anseriformes (Ducks, Geese, Swans)
Example: *Olor*

Order Falconiformes (Hawks, Eagles)
Example: *Falco*

Order Galliformes (Ground Birds)
Example: *Perdix*

Order Passeriformes (Perching Birds)
Example: *Spizella*

Class Mammalia (Mammals)

Order Monotremata (Monotremes)
Example: *Ornithorhynchus*

Order Marsupialia (Marsupials)
Example: *Didelphis*

Order Insectivora (Insect Eaters)
Example: *Scapanus*

Order Chiroptera (Bats)
Example: *Desmodus*

Order Carnivora (Carnivores)
Example: *Ursus*

Order Rodentia (Rodents)
Example: *Castor*

Order Cetacea (Whales, Dolphins)
Example: *Delphinus*

Order Primates (Primates)
Example: *Gorilla*

Appendix B Origins of Scientific Terms

This list of Greek and Latin roots will help you interpret the meaning of biological terms. The column headed *Root* gives many of the actual Greek (*GK*) or Latin (*L*) root words used in science. If more than one word is given, the first is the full word in Greek or Latin. The letter groups that follow are forms in which the root word is most often found combined in science words. In the second column is the meaning of the root as it is used in science. The third column shows a typical science word containing the root from the first column. Most of these words can be found in your textbook.

Root	Meaning	Example
A		
a, an *(GK)*	not, without	anaerobic
abilis *(L)*	able to	biodegradable
ad *(L)*	to, attached to	appendix
aequus *(L)*	equal	equilibrium
aeros *(GK)*	air	anaerobic
agon *(GK)*	assembly	glucagon
aktis *(GK)*	ray	actin
allas *(GK)*	sausage	allantois
allelon *(GK)*	of each other	allele
allucinari *(L)*	to dream	hallucinate
alveolus *(L)*	small pit	alveolus
amnos *(GK)*	lamb	amnion
amoibe *(GK)*	change	amoebocyte
amphi *(GK)*	both, about, around	amphibian
amylum *(L)*	starch	amylase
ana *(L)*	away, onward	anaphase
andro *(GK)*	male	androgens
anggeion, angio *(GK)*	vessel, container	angiosperm
anthos *(GK)*	flower	anthophyte
anti *(GK)*	against, away, opposite	antibody
aqua *(L)*	water	aquatic
archaios, archeo *(GK)*	ancient, primitive	archaebacteria
arthron *(GK)*	joint, jointed	arthropod
artios *(GK)*	even	artiodactyl
askos *(GK)*	bag	ascospore
aster *(GK)*	star	Asteroidea
autos *(GK)*	self	autoimmune
B		
bakterion *(GK)*	small rod	bacterium
bi, bis *(L)*	two, twice	bipedal
binarius *(L)*	pair	binary fission
bios *(GK)*	life	biology
blastos *(GK)*	bud	blastula
bryon *(GK)*	moss	bryophyte
bursa *(L)*	purse, bag	bursa

Root	Meaning	Example
C		
caedere, cide *(L)*	kill	insecticide
capillus *(L)*	hair	capillary
carn *(L)*	flesh	carnivore
carno *(L)*	flesh	carnivore
cella, cellula *(L)*	small room	protocells
cervix *(L)*	neck	cervix
cetus *(L)*	whale	cetacean
chaite, chaet *(GK)*	bristle	oligochaeta
cheir *(GK)*	hand	chiropteran
chele *(GK)*	claw	chelicerae
chloros *(GK)*	pale green	chlorophyll
chondros *(GK)*	cartilage	Chondrichthyes
chondros *(GK)*	grain	mitochondrion
chorda *(L)*	cord	urochordata
chorion *(GK)*	skin	chorion
chroma, chrom *(GK)*	colored	chromosome
chronos *(GK)*	time	chronometer
circa *(L)*	about	circadian
cirrus *(L)*	curl	cirri
codex *(L)*	tablet for writing	codon
corpus *(L)*	body	corpus luteum
cum, col, com, con *(L)*	with, together	convergent
cuticula *(L)*	thin skin	cuticle
D		
daktylos *(GK)*	finger	perissodactyl
de *(L)*	away, from	decompose
decidere *(L)*	to fall down	deciduous
degradare *(L)*	to reduce in rank	biodegradable
dendron *(GK)*	tree	dendrite
dens *(L)*	tooth	edentate
derma *(GK)*	skin	epidermis
deterere *(L)*	loose material	detritus
dia, di *(GK)*	through, apart	diastolic
dies *(L)*	day	circadian
diploos *(GK)*	twofold, double	diploid
dis, di *(GK)*	twice, two	disaccharide
dis, di *(L)*	apart, away	disruptive

Appendix B Origins of Scientific Terms

Root	Meaning	Example
dormire *(L)*	to sleep	dormancy
drom *(GK)*	running, racing	dromedary
ducere *(L)*	to lead	oviduct
E		
echinos *(GK)*	spine	echinoderm
eidos, old *(GK)*	form, appearance	rhizoid
ella *(GK)*	small	organelle
endon, en, endo *(GK)*	within	endosperm
engchyma *(GK)*	infusion	parenchyma
enteron *(GK)*	intestine, gut	enterocolitis
entomon *(GK)*	insect	entomology
epi *(GK)*	upon, above	epidermis
equus *(L)*	horse	Equisetum
erythros *(GK)*	red	erythrocyte
eu *(GK)*	well, true, good	eukaryote
evolutus *(L)*	rolled out	evolution
ex, e *(L)*	out	extinction
exo *(GK)*	out, outside	exoskeleton
extra *(L)*	outside, beyond	extracellular
F		
ferre *(L)*	to bear	porifera
fibrilla *(L)*	small fiber	myofibril
fissus *(L)*	a split	binary fission
flagellum *(L)*	whip	flagellum
follis *(L)*	bag	follicle
fossilis *(L)*	dug up	microfossils
fungus *(L)*	mushroom	fungus
G		
gamo, gam *(GK)*	marriage	gamete
gaster *(GK)*	stomach	gastropoda
ge, geo *(GK)*	the Earth	geology
gemmula *(L)*	little bud	gemmule
genesis *(L)*	origin, birth	parthenogenesis
genos, gen, geny *(GK)*	race	genotype
gestare *(L)*	to bear	progesterone
glene *(GK)*	eyeball	euglenoid
globus *(L)*	sphere	hemoglobin
glotta *(GK)*	tongue	epiglottis
glykys, glu *(GK)*	sweet	glycolysis
gnathos *(GK)*	jaw	Agnatha
gonos, gon *(GK)*	reproductive, sexual	gonorrhea
gradus *(L)*	a step	gradualism
graphos *(GK)*	written	chromatograph
gravis *(L)*	heavy	gravitropism

Root	Meaning	Example
gymnos *(GK)*	naked, bare	gymnosperm
gyne *(GK)*	female, woman	gynoecium
H		
haima, emia *(GK)*	blood	hemoglobin
halo *(GK)*	salt	halophile
haploos *(GK)*	simple	haploid
haurire *(L)*	to drink	haustorium
helix *(L)*	spiral	helix
hemi *(GK)*	half	hemisphere
herba *(L)*	grass	herbivore
hermaphroditos *(GK)*	combining both sexes	hermaphrodite
heteros *(GK)*	other	heterotrophic
hierarches *(GK)*	rank	hierarchy
hippos *(GK)*	horse	hippopotamus
histos *(GK)*	tissue	histology
holos *(GK)*	whole	Holothuroidea
homo *(L)*	man	hominid
homos *(GK)*	same, alike	homologous
hormaein *(GK)*	to excite	hormone
hydor, hydro *(GK)*	water	hydrolysis
hyper *(GK)*	over, above	hyperventilation
hyphe *(GK)*	web	hypha
hypo *(GK)*	under, below	hypotonic
I		
ichthys *(GK)*	fish	Osteichthyes
instinctus *(L)*	impulse	instinct
insula *(L)*	island	insulin
inter *(L)*	between	internode
intra *(L)*	within, inside	intracellular
isos *(GK)*	equal	isotonic
itis *(GK)*	inflammation, disease	arthritis
J		
jugare *(L)*	join together	conjugate
K		
kardia, cardia *(GK)*	heart	cardiac
karyon *(GK)*	nut	prokaryote
kata, cata *(GK)*	break down	catabolism
kephale, ceph *(GK)*	head	cephalopoda
keras *(GK)*	horn	chelicerae
kinein *(GK)*	to move	kinetic
koilos, coel *(GK)*	hollow, cavity, belly	coelom
kokkus *(GK)*	berry	streptococcus
kolla *(GK)*	glue	colloid

Appendix B Origins of Scientific Terms

Root	Meaning	Example
kotyl, cotyl *(GK)*	cup	cotylosaur
kreas *(GK)*	flesh	pancreas
krinoeides *(GK)*	lilylike	Crinoidea
kyanos, cyano *(GK)*	blue	cyanobacterium
kystis, cyst *(GK)*	bladder, sac	cystitis
kytos, cyt *(GK)*	hollow, cell	lymphocyte
L		
lagos *(GK)*	hare	lagomorph
leukos *(GK)*	white	leukocyte
libra *(L)*	balance	equilibrium
logos, logy *(GK)*	study, word	biology
luminescere *(L)*	to grow light	bioluminescence
luteus *(L)*	orange-yellow	corpus luteum
lyein, lysis *(GK)*	to split, loosen	lysosome
lympha *(L)*	water	lymphocyte
M		
makros *(GK)*	large	macrophage
marsupium *(L)*	pouch	marsupial
meare *(L)*	to glide	permeable
megas *(GK)*	large	megaspore
melas *(GK)*	black, dark	melanin
meristos *(GK)*	divided	meristem
meros *(GK)*	part	polymer
mesos *(GK)*	middle	mesophyll
meta *(GK)*	after, following	metaphase
metabole *(GK)*	change	metabolism
meter *(GK)*	a measurement	diameter
mikros, micro *(GK)*	small	microscope
mimos *(GK)*	a mime	mimicry
mitos *(GK)*	thread	mitochondrion
molluscus *(L)*	soft	mollusk
monos *(GK)*	single	monotreme
morphe *(GK)*	form	lagomorph
mors, mort *(L)*	death	mortality
mucus *(L)*	mucus, slime	mucosa
multus *(L)*	many	multicellular
mutare *(L)*	to change	mutation
mykes, myc *(GK)*	fungus	mycorrhiza
mys *(GK)*	muscle	myosin
N		
nema *(GK)*	thread	nematology
nemato *(GK)*	thread, threadlike	nematode
neos *(GK)*	new	Neolithic
nephros *(GK)*	kidney	nephron
neuro *(GK)*	nerve	neurology
nodus *(L)*	knot, knob	internode
nomos, nomy *(GK)*	ordered knowledge	taxonomy
noton *(GK)*	back	notochord
O		
oikos, eco *(GK)*	household	ecosystem
oisein, eso *(GK)*	to carry	esophagus
oligos *(GK)*	few, little	oligochaeta
omnis *(L)*	all	omnivore
ophis *(GK)*	serpent	Ophiuroidea
ophthalmos *(GK)*	referring to the eye	ophthalmologist
organon *(GK)*	tool, implement	organelle
ornis *(GK)*	bird	ornithology
orthos *(GK)*	straight	orthodontist
osculum *(L)*	small mouth	osculum
osteon *(GK)*	bone	osteocyte
ostrakon *(GK)*	shell	ostracoderm
oura, ura *(GK)*	tail	anura
ous, oto *(GK)*	ear	otology
ovum *(L)*	egg	oviduct
P		
palaios, paleo *(GK)*	ancient	paleontology
pan *(GK)*	all	pancreas
para *(GK)*	beside	parenchyma
parthenos *(GK)*	virgin	parthenogenesis
pathos *(GK)*	disease, suffering	pathogenic
pausere *(L)*	to rest	decompose
pendere *(L)*	to hang	appendix
per *(L)*	through	permeable
peri *(GK)*	around	peristalsis
periodos *(GK)*	a cycle	photoperiodism
pes, pedis *(L)*	foot	bipedal
phagein *(GK)*	to eat	phagocyte
phainein *(GK)*	to show	phenotype
phaios *(GK)*	dusky	phaeophyta
phase *(GK)*	stage, appearance	metaphase
pherein, phor *(GK)*	to carry	pheromone
phloios *(GK)*	inner bark	phloem
phos, photos *(GK)*	light	phototropism
phyllon *(GK)*	leaf	chlorophyll
phylon *(GK)*	related group	phylogeny
phyton *(GK)*	plant	epiphyte
pinax *(GK)*	tablet	pinacocytes
pinein *(GK)*	to drink	pinocytosis

Appendix B Origins of Scientific Terms

Root	Meaning	Example
pinna *(L)*	feather	pinniped
plasma *(GK)*	mold, form	plasmodium
plastos *(GK)*	formed object	chloroplast
platys *(GK)*	flat	platyhelminthes
plax *(GK)*	plate	placoderm
pleuron *(GK)*	side	dipleurula
plicare *(L)*	to fold	replication
polys, poly *(GK)*	many	polymer
poros *(GK)*	channel	porifera
post *(L)*	after	posterior
pous, pod *(GK)*	foot	gastropoda
prae, pre *(L)*	before	Precambrian
primus *(L)*	first	primary
pro *(GK* and *L)*	before, for	prokaryote
proboskis *(GK)*	trunk	proboscidean
producere *(L)*	to bring forth	reproduction
protos *(GK)*	first	protocells
pseudes *(GK)*	false	pseudopod
pteron *(GK)*	wing	chiropteran
punctus *(L)*	a point	punctuated
pupa *(L)*	doll	pupa
R		
radius *(L)*	ray	radial
re *(L)*	again	reproduction
reflectere *(L)*	to turn back	reflex
rhiza *(GK)*	root	mycorrhiza
rhodon *(GK)*	rose	rhodophyte
rota *(L)*	wheel	rotifer
rumpere *(L)*	to break	disruptive
S		
saeta *(L)*	bristle	Equisetum
sapros *(GK)*	rotten	saprobe
sarx *(GK)*	flesh	sarcomere
sauros *(GK)*	lizard	cotylosaur
scire *(L)*	to know	science
scribere, script *(L)*	to write	transcription
sedere, ses *(L)*	to sit	sessile
semi *(L)*	half	semicircle
skopein, scop *(GK)*	to look	microscope
soma *(GK)*	body	lysosome
sperma *(GK)*	seed	angiosperm
spirare *(L)*	to breathe	spiracle
sporos *(GK)*	seed	microspore
staphylo *(GK)*	bunch of grapes	staphylococcus
stasis *(GK)*	standing, staying	homeostasis

Root	Meaning	Example
stellein, stol *(GK)*	to draw in	peristalsis
sternon *(GK)*	chest	sternum
stinguere *(L)*	to quench	extinction
stolo *(L)*	shoot	stolon
stoma *(GK)*	mouth	stoma
streptos *(GK)*	twisted chain	streptococcus
syn *(GK)*	together	systolic
synapsis *(GK)*	union	synapse
systema *(GK)*	composite whole	ecosystem
T		
taxis, taxo *(GK)*	to arrange	taxonomy
telos *(GK)*	end	telophase
terra *(L)*	land, Earth	terrestrial
thele *(GK)*	cover a surface	epithelium
therme *(GK)*	heat	endotherm
thrix, trich *(GK)*	hair	trichocyst
tome *(GK)*	cutting	anatomy
trachia *(GK)*	windpipe	tracheid
trans *(L)*	across	transpiration
trematodes *(GK)*	having holes	monotreme
trope *(GK)*	turn	gravitropism
trophe *(GK)*	nourishment	heterotrophic
turbo *(L)*	whirl	turbellaria
tympanon *(GK)*	drum	tympanum
typos *(GK)*	model	genotype
U		
uni *(L)*	one	unicellular
uterus *(L)*	womb	uterus
V		
vacca *(L)*	cow	vaccine
vagina *(L)*	sheath	vagina
valvae *(L)*	folding doors	bivalvia
vasculum *(L)*	small vessel	vascular
venter *(L)*	belly	ventricle
ventus *(L)*	a wind	hyperventilation
vergere *(L)*	to slant, incline	convergent
villus *(L)*	shaggy hair	villus
virus *(L)*	poisonous liquid	virus
vorare *(L)*	to devour	carnivore
X		
xeros *(GK)*	dry	xerophyte
xylon *(GK)*	wood	xylem
Z		
zoon, zo *(GK)*	animal	zoology
zygotos *(GK)*	joined together	zygote

Appendix C Safety in the Laboratory

The biology laboratory is a safe place to work if you are aware of important safety rules and if you are careful. You must be responsible for your own safety and for the safety of others. The safety rules given here will protect you and others from harm in the lab. While carrying out procedures in any of the **Biolabs,** notice the safety symbols and caution statements. The safety symbols are explained in the chart on the next page.

1. Always obtain your teacher's permission to begin a lab.
2. Study the procedure. If you have questions, ask your teacher. Be sure you understand all safety symbols shown.
3. Use the safety equipment provided for you. Goggles and a safety apron should be worn when any lab calls for using chemicals.
4. When you are heating a test tube, always slant it so the mouth points away from you and others.
5. Never eat or drink in the lab. Never inhale chemicals. Do not taste any substance or draw any material into your mouth.
6. If you spill any chemical, wash it off immediately with water. Report the spill immediately to your teacher.
7. Know the location and proper use of the fire extinguisher, safety shower, fire blanket, first aid kit, and fire alarm.
8. Keep all materials away from open flames. Tie back long hair.
9. If a fire should break out in the lab, or if your clothing should catch fire, smother it with the fire blanket or a coat, or get under a safety shower. **NEVER RUN.**
10. Report any accident or injury, no matter how small, to your teacher.

Follow these procedures as you clean up your work area.

1. Turn off the water and gas. Disconnect electrical devices.
2. Return materials to their places.
3. Dispose of chemicals and other materials as directed by your teacher. Place broken glass and solid substances in the proper containers. Never discard materials in the sink.
4. Clean your work area.
5. Wash your hands thoroughly after working in the laboratory.

First Aid in the Laboratory

Injury	Safe Response
Burns	**Apply cold water. Call your teacher immediately.**
Cuts and bruises	**Stop any bleeding by applying direct pressure. Cover cuts with a clean dressing. Apply cold compresses to bruises. Call your teacher immediately.**
Fainting	**Leave the person lying down. Loosen any tight clothing and keep crowds away. Call your teacher immediately.**
Foreign matter in eye	**Flush with plenty of water. Use an eyewash bottle or fountain.**
Poisoning	**Note the suspected poisoning agent and call your teacher immediately.**
Any spills on skin	**Flush with large amounts of water or use safety shower. Call your teacher immediately.**

Appendix C Safety in the Laboratory

Safety Symbols

These safety symbols are used to indicate possible hazards in the activities. Each activity has appropriate hazard indicators.

	DISPOSAL ALERT This symbol appears when care must be taken to dispose of materials properly.		**ANIMAL SAFETY** This symbol appears whenever live animals are studied and the safety of the animals and the students must be ensured.
	**BIOLOGICAL HAZARD** This symbol appears when there is danger involving bacteria, fungi, or protists.		**RADIOACTIVE SAFETY** This symbol appears when radioactive materials are used.
	**OPEN FLAME ALERT** This symbol appears when use of an open flame could cause a fire or an explosion.		**CLOTHING PROTECTION SAFETY** This symbol appears when substances used could stain or burn clothing.
	**THERMAL SAFETY** This symbol appears as a reminder to use caution when handling hot objects.		**FIRE SAFETY** This symbol appears when care should be taken around open flames.
	**SHARP OBJECT SAFETY** This symbol appears when a danger of cuts or punctures caused by the use of sharp objects exists.		**EXPLOSION SAFETY** This symbol appears when the misuse of chemicals could cause an explosion.
	**FUME SAFETY** This symbol appears when chemicals or chemical reactions could cause dangerous fumes.		**EYE SAFETY** This symbol appears when a danger to the eyes exists. Safety goggles should be worn when this symbol appears.
	**ELECTRICAL SAFETY** This symbol appears when care should be taken when using electrical equipment.		**POISON SAFETY** This symbol appears when poisonous substances are used.
	**PLANT SAFETY** This symbol appears when poisonous plants or plants with thorns are handled.		**CHEMICAL SAFETY** This symbol appears when chemicals used can cause burns or are poisonous if absorbed through the skin.

Appendix D The Periodic Table

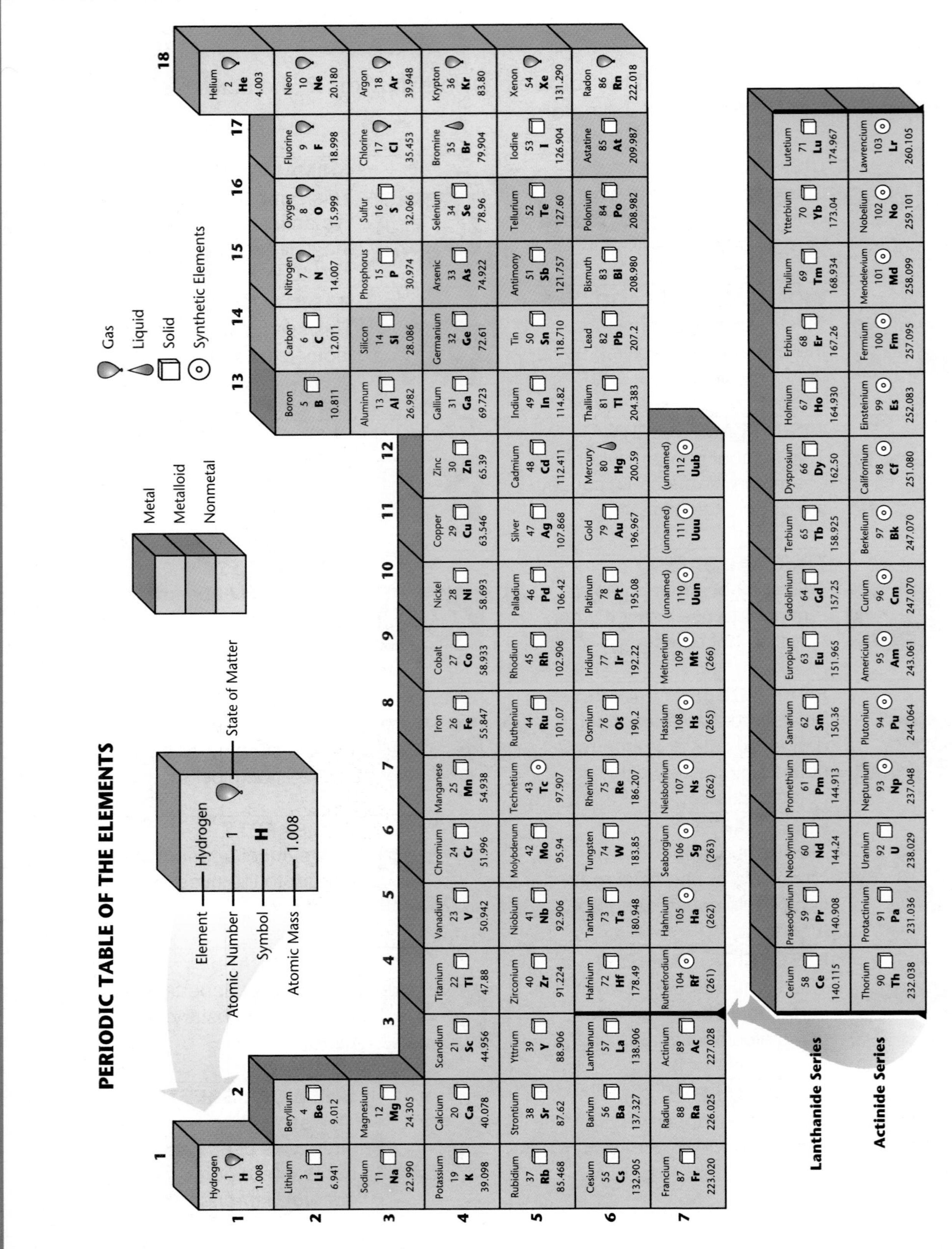

	1	2	3	4	5	6	7	8	9	10	11	12	13	14	15	16	17	18
1	Hydrogen 1 H 1.008																	Helium 2 He 4.003
2	Lithium 3 Li 6.941	Beryllium 4 Be 9.012											Boron 5 B 10.811	Carbon 6 C 12.011	Nitrogen 7 N 14.007	Oxygen 8 O 15.999	Fluorine 9 F 18.998	Neon 10 Ne 20.180
3	Sodium 11 Na 22.990	Magnesium 12 Mg 24.305											Aluminum 13 Al 26.982	Silicon 14 Si 28.086	Phosphorus 15 P 30.974	Sulfur 16 S 32.066	Chlorine 17 Cl 35.453	Argon 18 Ar 39.948
4	Potassium 19 K 39.098	Calcium 20 Ca 40.078	Scandium 21 Sc 44.956	Titanium 22 Ti 47.88	Vanadium 23 V 50.942	Chromium 24 Cr 51.996	Manganese 25 Mn 54.938	Iron 26 Fe 55.847	Cobalt 27 Co 58.933	Nickel 28 Ni 58.693	Copper 29 Cu 63.546	Zinc 30 Zn 65.39	Gallium 31 Ga 69.723	Germanium 32 Ge 72.61	Arsenic 33 As 74.922	Selenium 34 Se 78.96	Bromine 35 Br 79.904	Krypton 36 Kr 83.80
5	Rubidium 37 Rb 85.468	Strontium 38 Sr 87.62	Yttrium 39 Y 88.906	Zirconium 40 Zr 91.224	Niobium 41 Nb 92.906	Molybdenum 42 Mo 95.94	Technetium 43 Tc 97.907	Ruthenium 44 Ru 101.07	Rhodium 45 Rh 102.906	Palladium 46 Pd 106.42	Silver 47 Ag 107.868	Cadmium 48 Cd 112.411	Indium 49 In 114.82	Tin 50 Sn 118.710	Antimony 51 Sb 121.757	Tellurium 52 Te 127.60	Iodine 53 I 126.904	Xenon 54 Xe 131.290
6	Cesium 55 Cs 132.905	Barium 56 Ba 137.327	Lanthanum 57 La 138.906	Hafnium 72 Hf 178.49	Tantalum 73 Ta 180.948	Tungsten 74 W 183.85	Rhenium 75 Re 186.207	Osmium 76 Os 190.2	Iridium 77 Ir 192.22	Platinum 78 Pt 195.08	Gold 79 Au 196.967	Mercury 80 Hg 200.59	Thallium 81 Tl 204.383	Lead 82 Pb 207.2	Bismuth 83 Bi 208.980	Polonium 84 Po 208.982	Astatine 85 At 209.987	Radon 86 Rn 222.018
7	Francium 87 Fr 223.020	Radium 88 Ra 226.025	Actinium 89 Ac 227.028	Rutherfordium 104 Rf (261)	Hahnium 105 Ha (262)	Seaborgium 106 Sg (263)	Nielsbohrium 107 Ns (262)	Hassium 108 Hs (265)	Meitnerium 109 Mt (266)	(unnamed) 110 Uun	(unnamed) 111 Uuu	(unnamed) 112 Uub						

Lanthanide Series	Cerium 58 Ce 140.115	Praseodymium 59 Pr 140.908	Neodymium 60 Nd 144.24	Promethium 61 Pm 144.913	Samarium 62 Sm 150.36	Europium 63 Eu 151.965	Gadolinium 64 Gd 157.25	Terbium 65 Tb 158.925	Dysprosium 66 Dy 162.50	Holmium 67 Ho 164.930	Erbium 68 Er 167.26	Thulium 69 Tm 168.934	Ytterbium 70 Yb 173.04	Lutetium 71 Lu 174.967
Actinide Series	Thorium 90 Th 232.038	Protactinium 91 Pa 231.036	Uranium 92 U 238.029	Neptunium 93 Np 237.048	Plutonium 94 Pu 244.064	Americium 95 Am 243.061	Curium 96 Cm 247.070	Berkelium 97 Bk 247.070	Californium 98 Cf 251.080	Einsteinium 99 Es 252.083	Fermium 100 Fm 257.095	Mendelevium 101 Md 258.099	Nobelium 102 No 259.101	Lawrencium 103 Lr 260.105

Contents Skill Handbook

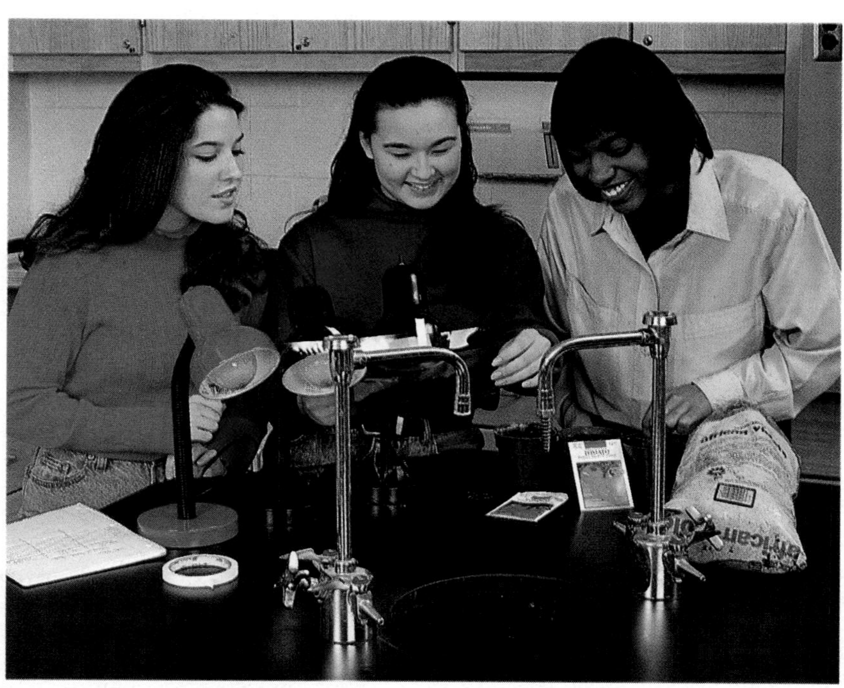

Observing and Inferring

Scientists try to make careful and accurate observations. When possible, they use instruments, such as microscopes and tape recorders, to extend their senses. Other instruments, such as a thermometer or a pan balance, are used to measure observations. Measurements provide numerical data, a concrete means of comparing collected data that can be checked and repeated.

When you make observations in science, you may find it helpful first to examine the entire object or situation. Then, using your senses of sight, touch, and hearing, examine the object in detail. Write down everything you observe.

Scientists often use their observations to make inferences. An inference is an attempt to explain or interpret observations or to determine what caused what you observed. For example, if you observed a CLOSED sign in a store window around noon, you might infer that the owner is taking a lunch break. But, perhaps the owner has a doctor's appointment or has a business meeting in another town. The only way to be sure your inference is correct is to investigate further.

When making an inference, be certain to make accurate observations and to record them carefully. Collect all the information you can. Then, based on everything you know, try to explain or interpret what you observed. If possible, investigate further to determine whether your inference is correct. What can you infer from observing the behavior of a cat that is crouched and ready to pounce?

Comparing and Contrasting

Observations can be analyzed and then organized by noting the similarities and differences between two or more objects or situations. When you examine objects or situations to determine similarities, you are comparing. Contrasting is looking at similar objects or situations for differences.

Suppose you were asked to compare and contrast a grasshopper and a dragonfly. You would start by making your observations. You then divide a piece of paper into two columns. List ways the insects are similar in one column and ways they are different in the other. After completing your lists, you report your findings in a table or in a graph.

Similarities you might point out are that both have three body parts, two pairs of wings, and chewing mouthparts. Differences might include large hind legs on the grasshopper, small legs on the dragonfly; wings held close to the body in the grasshopper, wings held outspread in the dragonfly.

Recognizing Cause and Effect

Have you ever observed something happen and then tried to figure out why or how it came about? If so, you have observed an event and inferred a reason for the event. The event or result of an action is an effect, and the reason for the event is the cause.

Suppose that every time your teacher fed fish in a classroom aquarium, she tapped the food container on the edge. Then, one day she tapped the edge of the aquarium to make a point about an ecology lesson. You observe the fish swim to the surface of the aquarium to feed.

What is the effect and what would you infer was the cause? The effect is the fish swimming to the surface of the aquarium. You might infer the cause to be the teacher tapping on the edge of the aquarium. In determining cause and effect, you have made a logical inference based on careful observations.

Perhaps the fish swam to the surface because they reacted to the teacher's waving hand or for some other reason. When scientists are unsure of the cause for a certain event, they often design controlled experiments to determine what caused their observations. Although you have made a sound judgment, you would have to perform an experiment to be certain that it was the tapping that caused the effect you observed.

Interpreting Scientific Illustrations

Illustrations are included in your textbook to help you understand, interpret, and remember what you read. Whenever you encounter an illustration, examine it carefully and read the caption. The caption explains or identifies the illustration.

Some illustrations are designed to show you how the internal parts of a structure are arranged. Look at the illustrations below of a squash. The squash has been cut lengthwise so that it shows a section that runs along the length of the squash. This type of illustration is called a longitudinal section. Cutting the squash crosswise at right angles to the length produces a cross section.

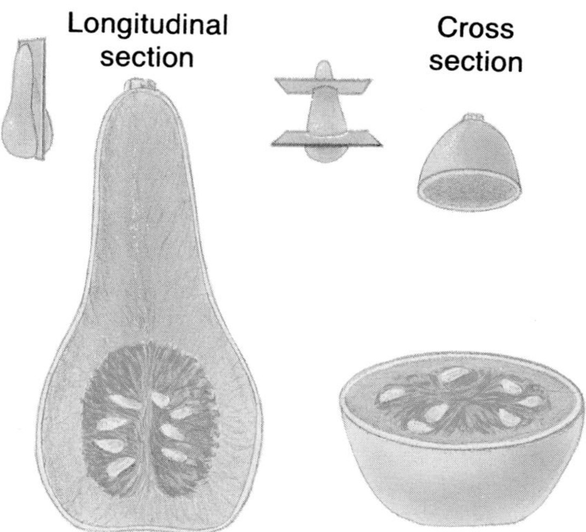

In your reading and examination of the illustrations, you will sometimes see terms that refer to the orientation of an organism. The word *dorsal* refers to the upper side or back of an animal. *Ventral* refers to the lower side or belly of the animal. The illustration of the shark shows that it has both dorsal and ventral sides.

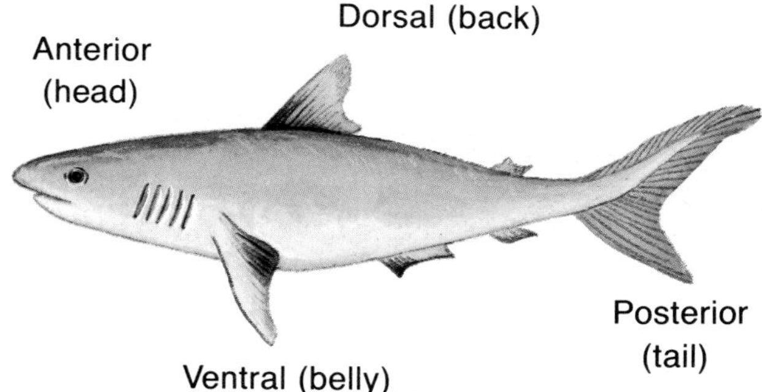

Symmetry refers to a similarity or likeness of parts. Many organisms and objects have symmetry. When something can be divided into two similar parts lengthwise, it has bilateral symmetry. Look at the illustration of the butterfly on the next page. The right side of the butterfly looks very similar to the left side. It has bilateral symmetry.

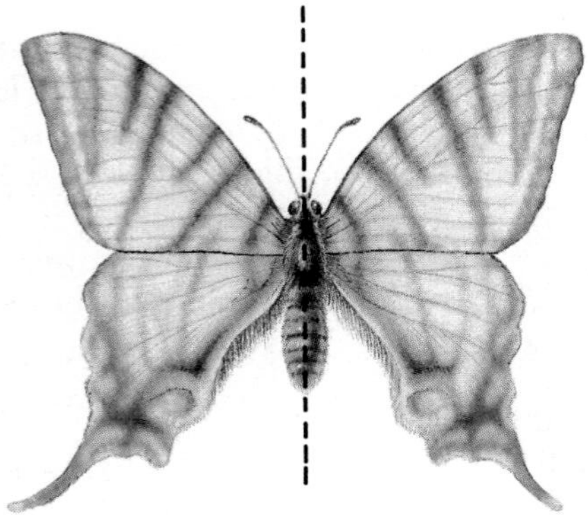
Bilateral symmetry

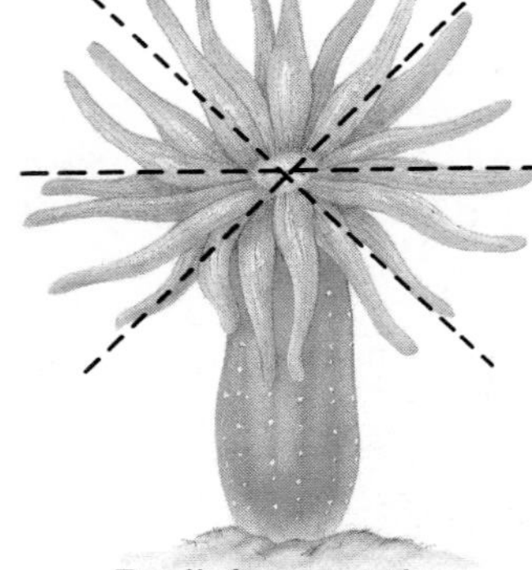
Radial symmetry

Asymmetry

Other organisms and objects have radial symmetry. Radial symmetry is the arrangement of similar parts around a central point. The anemone in the figure has radial symmetry. It can be divided anywhere through the center into similar parts.

Some organisms and objects cannot be divided into two similar parts. If an organism or object cannot be divided, it is asymmetrical. Study the sponge. Regardless of how you try to divide a sponge, you cannot divide it into two parts that look alike.

Calculating Magnification

Objects viewed under the microscope appear larger than normal because they are magnified. Total magnification describes how much larger an object appears when viewed through the microscope.

Look for a number marked with an × on the eyepiece, the low-power objective, and the high-power objective. The × stands for how many times the lens of each microscope part magnifies an object.

To calculate *total* magnification, multiply the number on the eyepiece by the number on the objective. For example, if the eyepiece magnification is 4×, the low-power objective magnification is 10×, and the high-power objective magnification is 40×:

(a) then total magnification under low power is 4× for the eyepiece times 10× for the low-power objective = 40 (4 × 10 = 40).

(b) then total magnification under high power is 4 × 40 = 160.

To measure the field of view of a microscope, you must use a unit called a micrometer. A micrometer equals 0.001 mm; in other words, there are 1000 micrometers in a millimeter. Place a millimeter section of a plastic ruler over the central opening of your microscope stage. Using low power, locate the measured lines of the ruler in the center of the field of view. Move the ruler so that one of the lines representing a millimeter is visible at one edge of the field of view.

Remember that the distance between two lines is one millimeter, and estimate the diameter in millimeters of the field of view on low power. Calculate the diameter in micrometers. For example, if the distance is 1.5 mm, then the diameter of the field of view at low power is 1500 µm. [1.5 × 1000]

To calculate the diameter of the high-power field, divide the magnification of your high power (40×) by the magnification of the low power (10×); 40 ÷ 10 = 4. Then, divide the diameter of the low-power field in micrometers (1500 µm) by this quotient (4). The answer is the diameter of the high-power field in micrometers. In this example, the diameter of the high-power field is 1500 ÷ 4 = 375 µm.

You can calculate the diameters of microscopic specimens, such as pollen grains or amoebas, viewed under low and high power by estimating how many of them could fit end to end across the field of view. Divide the diameter of the field of view by the number of specimens.

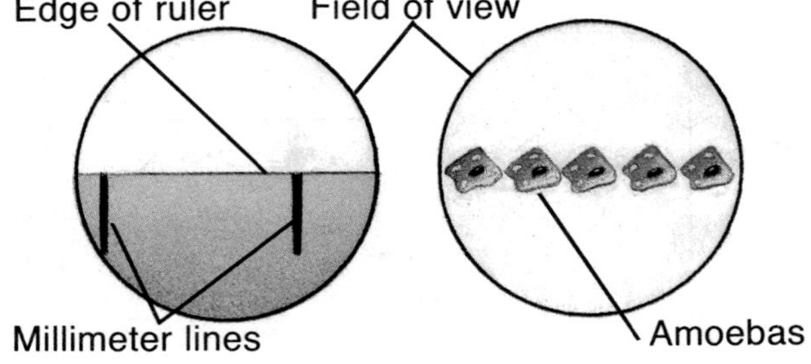

Care and Use of the Microscope

1. Always carry the microscope holding the arm with one hand and supporting the base with the other hand.
2. Place the microscope on a flat surface that is clear of objects. The arm should be toward you.
3. Look through the eyepiece. Adjust the diaphragm so that light comes through the opening in the stage.
4. Place a slide on the stage so that the specimen is in the field of view. Hold it firmly in place by using the stage clips.
5. Always focus first with the coarse adjustment and the low-power objective lens. Once the object is in focus on low power, turn the nosepiece until the high-power objective is in place. Use ONLY the fine adjustment to focus with this lens.
6. Store the microscope covered.

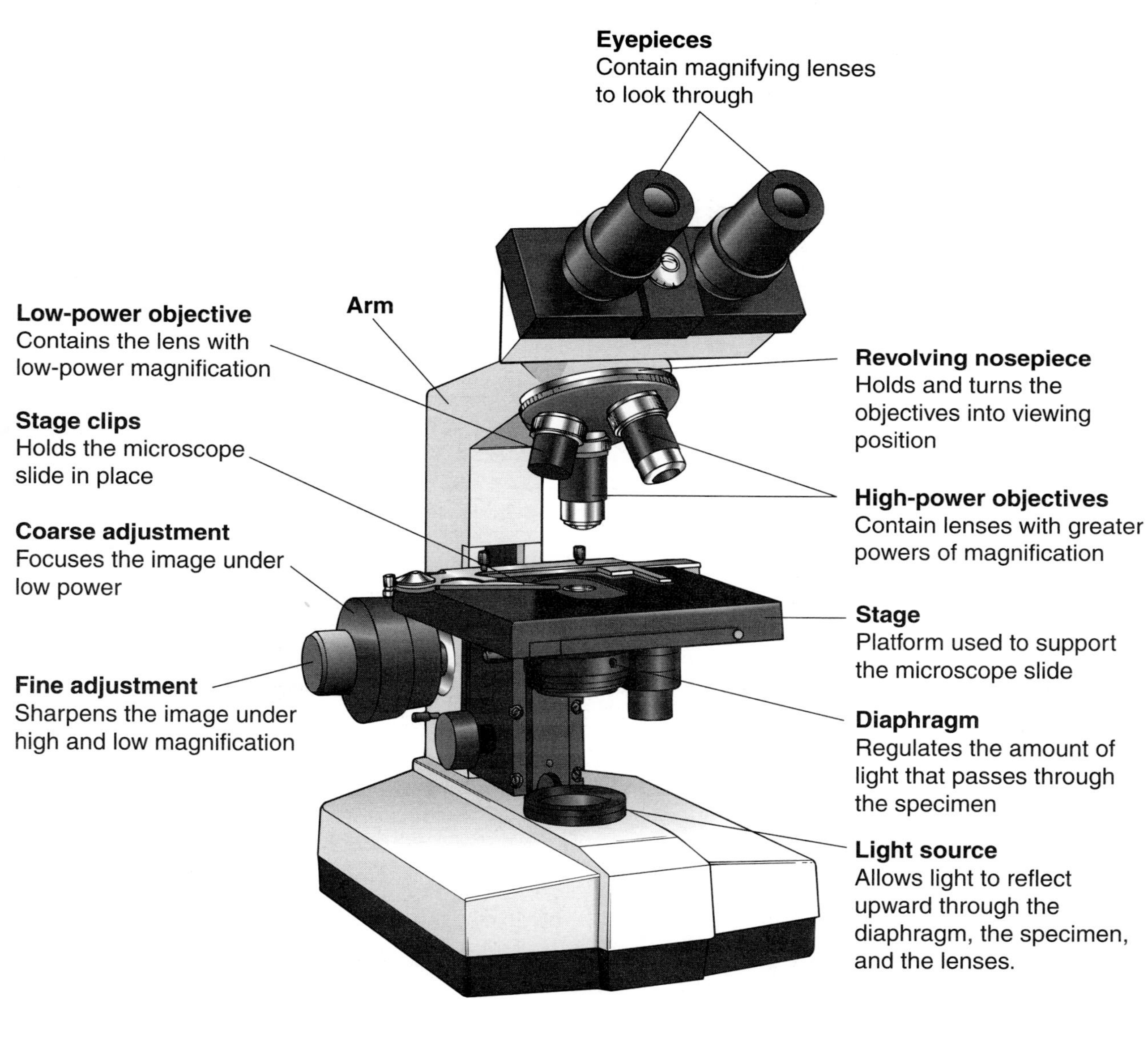
Eyepieces
Contain magnifying lenses to look through
Arm
Low-power objective
Contains the lens with low-power magnification
Stage clips
Holds the microscope slide in place
Coarse adjustment
Focuses the image under low power
Fine adjustment
Sharpens the image under high and low magnification
Revolving nosepiece
Holds and turns the objectives into viewing position
High-power objectives
Contain lenses with greater powers of magnification
Stage
Platform used to support the microscope slide
Diaphragm
Regulates the amount of light that passes through the specimen
Light source
Allows light to reflect upward through the diaphragm, the specimen, and the lenses.

SI Measurement

The International System (SI) of Measurement is accepted as the standard for measurement throughout most of the world. Four of the base units in SI are the meter, liter, kilogram, and second. The size of a unit can be determined from the prefix used with the base unit name. For example: *kilo* means one thousand; *milli* means one-thousandth; *micro* means one-millionth; and *centi* means one-hundredth. The tables below give the standard symbols for these SI units and some of their equivalents.

Larger and smaller units of measurement in SI are obtained by multiplying or dividing the base unit by some multiple of ten. Multiply to change from larger units to smaller units. Divide to change from smaller units to larger units. For example, to change 1 km to meters, you would multiply 1 km by 1000 to obtain 1000 m. To change 10 g to kilograms, you would divide 10 g by 1000 to obtain 0.01 kg.

Common SI Units

Measurement	Unit	Symbol	Equivalents
Length	**1 millimeter**	**mm**	**1000 micrometers (μm)**
	1 centimeter	**cm**	**10 millimeters (mm)**
	1 meter	**m**	**100 centimeters (cm)**
	1 kilometer	**km**	**1000 meters (m)**
Volume	**1 milliliter**	**mL**	**1 cubic centimeter (cm^3 or cc)**
	1 liter	**L**	**1000 milliliters (mL)**
Mass	**1 gram**	**g**	**1000 milligrams (mg)**
	1 kilogram	**kg**	**1000 grams (g)**
	1 metric ton	**t**	**1000 kilograms (kg)**
Time	**1 second**	**s**	
Area	**1 square meter**	**m^2**	**10 000 square centimeters (cm^2)**
	1 square kilometer	**km^2**	**1 000 000 square meters (m^2)**
	1 hectare	**ha**	**10 000 square meters (m^2)**
Temperature	**1 Kelvin**	**K**	**1 degree Celsius (°C)**

The top of the thermometer is marked off in degrees Fahrenheit (°F). To read the corresponding temperature in degrees Celsius (°C), look at the bottom side of the thermometer. For example, 50°F is the same temperature as 10°C. You may also use the formulas shown here for conversions.

Conversion of Fahrenheit to Celsius

$$°C = \frac{5}{9}(°F - 32)$$

Conversion of Celsius to Fahrenheit

$$°F = (\frac{9}{5}°C) + 32$$

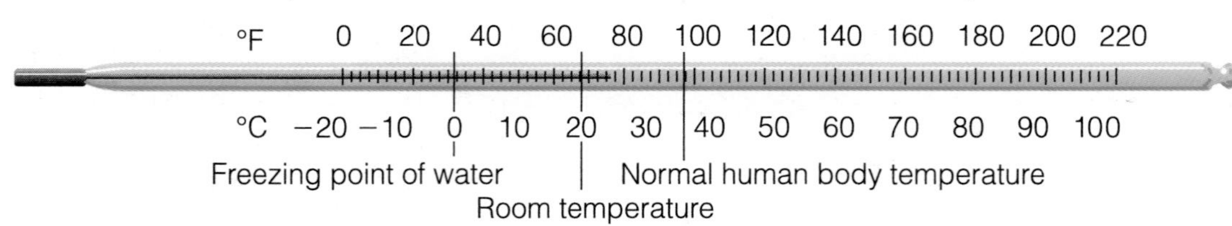

Forming a Hypothesis

Suppose you wanted to earn a perfect score on a spelling test. You think of several ways to accomplish a perfect score. You base these possibilities on past experiences and observations of your friends' results. All of the following are hypotheses you might consider that could explain how it would be possible to score 100 percent on your test:

If the test is easy, then I will get a good grade.

If I am intelligent, then I will get a good grade.

If I study hard, then I will get a good grade.

Scientists use hypotheses that they can test to explain the observations they have made. Perhaps a scientist has observed that fish activity increases in the summer and decreases in the winter. A scientist may form a hypothesis that says: If fishes are exposed to warmer water, their activity will increase.

Designing an Experiment to Test a Hypothesis

Once you have stated a hypothesis, you probably want to find out whether or not it explains an event or an observation. This requires a test. To be valid, a hypothesis must be testable by experimentation. Let's figure out how you would conduct an experiment to test the hypothesis about the effects of water temperature on fishes.

First, obtain several identical, clear glass containers, and fill them with the same amount of tap water. Leave the containers for a day to allow the water to come to room temperature. On the day of your experiment, you fill another container with an amount of aquarium water equal to that in the test containers. After measuring and recording the aquarium water temperature, you heat and cool the other containers, adjusting the water temperatures in the test containers so that two have higher temperatures and two have lower temperatures than the aquarium water temperature.

You place a guppy in each container. You count the number of horizontal and vertical movements each guppy makes during five minutes and record your data in a table. Your data table might look like this:

Number of Guppy Movements

Container	Temperature (°C)	Number of movements
Aquarium water	38	56
A	40	61
B	42	70
C	36	46
D	34	42

From the data you recorded, you will draw a conclusion and make a statement about your results. If your conclusion supports your hypothesis, then you can say that your hypothesis is reliable. If it did not support your hypothesis, then you would have to make new observations and state a new hypothesis, one that you could also test. Do the data above support the hypothesis that warmer water increases fish activity?

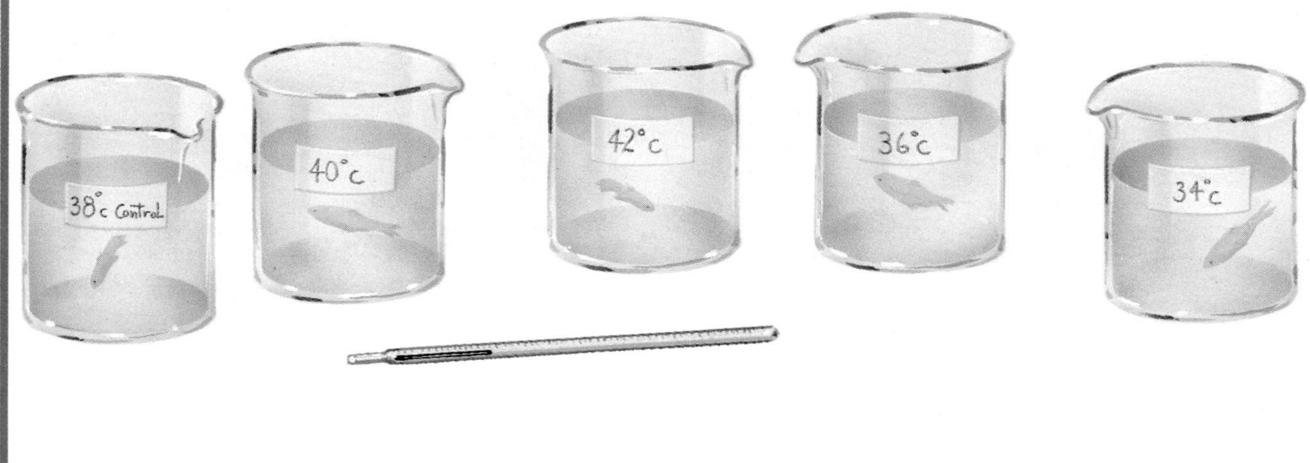

Separating and Controlling Variables

When scientists perform experiments, they must be careful to manipulate or change only one condition and keep all other conditions in the experiment the same. The condition that is manipulated is called the independent variable. The conditions that are kept the same during an experiment are called constants. The dependent variable is any change that results from manipulating the independent variable.

Scientists can only know that the independent variable caused the change in the dependent variable if they keep all other factors the same in an experiment. Scientists also use controls to be certain that the observed changes were a result of manipulation of the independent variable. A control is a sample that is treated exactly like the experimental group except that the independent variable is not applied to the control. After the experiment, the change in the dependent variable of the control sample is compared with any change in the experimental group. This allows scientists to see the effect of the independent variable.

What are the independent and dependent variables in the guppy experiment? Because you are changing the temperature of the water, the independent variable is the water temperature. Because the dependent variable is any change that results from the independent variable, the dependent variable is the number of movements the guppy makes during five minutes.

What factors are constants in the experiment? The constants are using the same size and shape containers, filling them with equal amounts of water, and counting the number of movements during the same amount of time. What was the purpose of counting the number of movements of a guppy in an identical container filled with aquarium water? The container of aquarium water is the control. The number of movements of the guppy in the aquarium water will be used to compare the movements of the guppies in water of different temperatures.

Why is it important to know the best temperature for a fish to survive? If you have an aquarium at home, you have probably learned how different fishes need different conditions to survive. They probably came from many different parts of the world and are adapted to living in very different environments.

Classifying

You may not realize it, but you impose order on the world around you. If your shirts hang in the closet together, your socks take up a corner of a dresser drawer, or your favorite CDs are stacked in groups according to recording artist, you have used the skill of classifying.

Classifying is grouping objects or events based on common features. When classifying, you first make careful observations of the group of items to be classified. Select one feature that is shared by some items in the group but not others. Place the items that share this feature in a subgroup. Place the remaining items in a second subgroup. Ideally, the items in the second subgroup will have some feature in common with one another. After you decide on the first feature that separates the items into subgroups, examine the items for other features and form further subgroups until the items can no longer be distinguished enough to identify them as distinct.

How would you classify a collection of CDs? Classify the CDs based on observable features. You might classify CDs you like to dance to in one subgroup and CDs you like to listen to in another. The CDs you like to dance to could be subdivided into a rap subgroup and a rock subgroup. Note that for each feature selected, each CD fits only one subgroup. For example, you wouldn't place a CD into both rap and rock categories. Keep selecting features until all the CDs are classified.

Remember, when you classify, you are grouping objects or events for a purpose. The purpose could be general such as for ease of finding an item. The classification of books in a library is a general-purpose classification. The classification may have a special purpose. For example, plants may be classified as poisonous or harmless to humans.

Sequencing

A sequence is an arrangement of things or events in a particular order. A common sequence with which you may be familiar is the order of steps you must follow to make an omelette. Certain steps of preparation have to be followed in order for the omelette to taste good.

When you are asked to sequence things or events, you must identify what comes first. You then decide what should come second. Continue to choose things or events until they are all in order. Then, go back over the sequence to make sure each thing or event logically leads to the next.

Suppose you wanted to watch a movie that just came out on videotape. What sequence of events would you have to follow to watch the movie? You would first turn the television set to Channel 3 or 4. You would then turn the videotape player on and insert the tape. Once the tape has started playing, you would adjust the sound and picture. Then, when the movie is over, you would rewind the tape and return it to the store. What would happen if you did things out of sequence, such as adjusting the sound before putting in the tape?

Concept Mapping

If you were taking an automobile trip, you would probably take along a road map. The road map shows your location, your destination, and other places along the way. By examining the map, you can understand where you are in relation to other locations on the map.

A concept map is similar to a road map. But, a concept map shows the relationship among ideas (or concepts) rather than places. A concept map is a diagram that visually shows how concepts are related. Because the concept map shows the relationships among ideas, it can clarify the meaning of ideas and terms and help you to understand better what you are studying.

A Network Tree

Notice how some words in the concept map below called a **network tree** are circled. The circled words are science concepts. The lines in the map show related concepts, and the words written on the lines describe relationships between the concepts.

When you are asked to construct a network tree, state the topic and select the major concepts. Find related concepts and put them in order from general to specific. Branch the related concepts from the major concept, and describe the relationship on the lines. Continue to write the more specific concepts. Write the relationships between the concepts on the lines until all concepts are mapped. Examine the concept map for relationships that cross branches, and add them to the concept map.

An Events Chain

An **events chain map** is used to describe ideas in order. In science, an events chain map can be used to describe a sequence of events, the steps in a procedure, or the stages of a process.

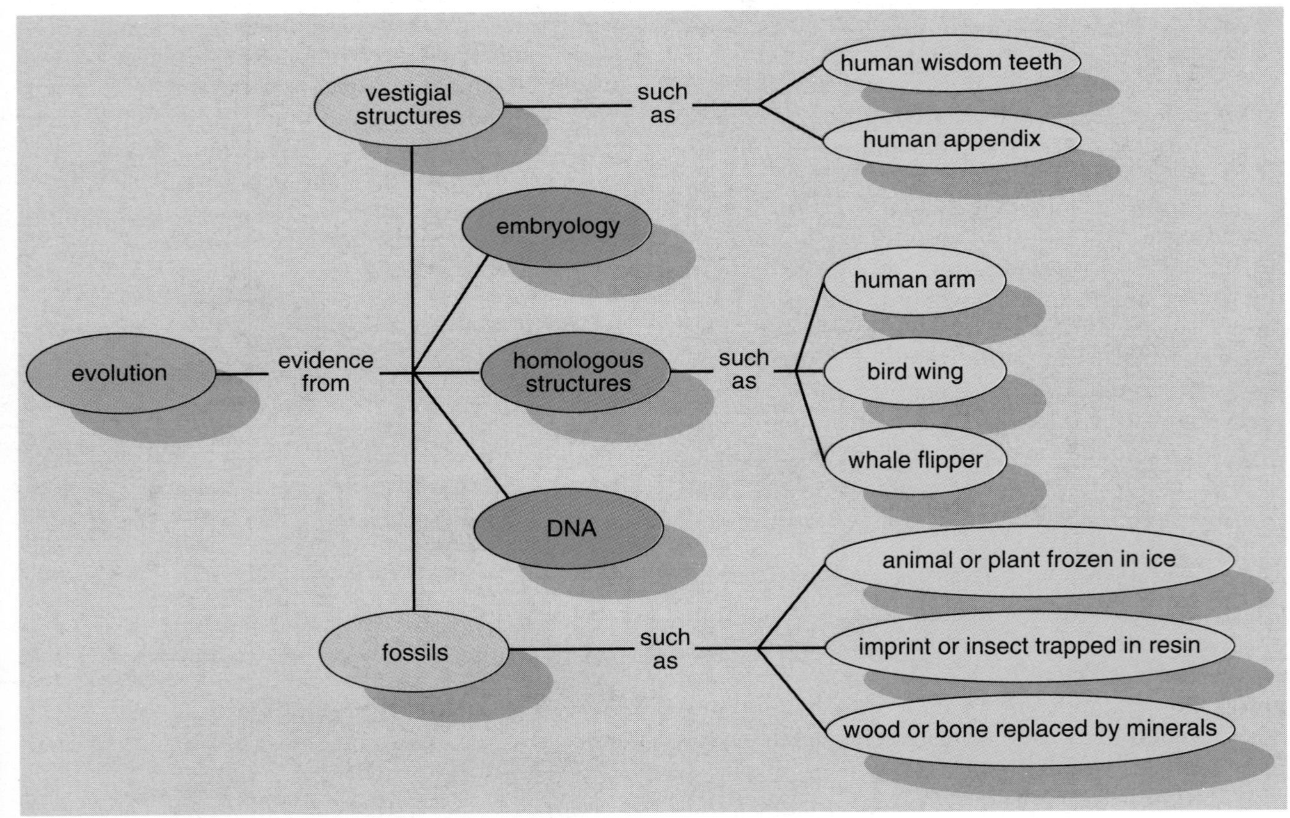

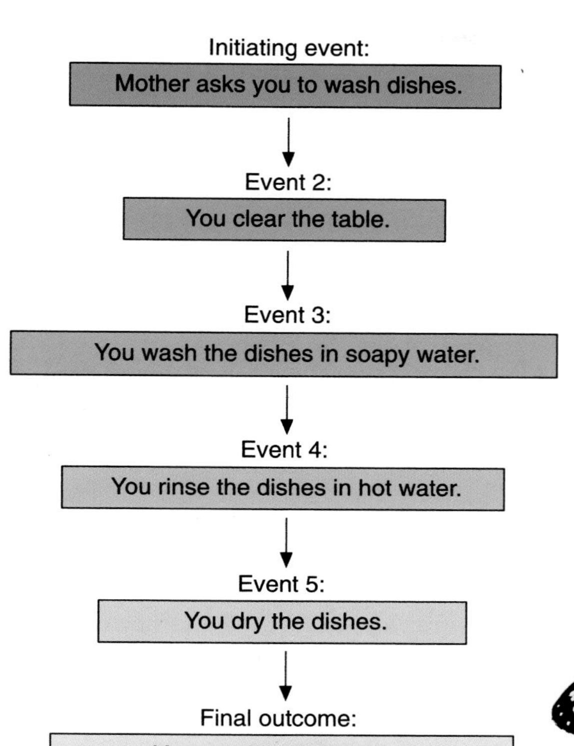
Initiating event:
Mother asks you to wash dishes.
Event 2:
You clear the table.
Event 3:
You wash the dishes in soapy water.
Event 4:
You rinse the dishes in hot water.
Event 5:
You dry the dishes.
Final outcome:
You put the dishes away.

When making an events chain map, you first must find the one event that starts the chain. This event is called the initiating event. You then find the next event in the chain and continue until you reach an outcome. Suppose your mother asked you to wash the dinner dishes. An events chain map might look like the one shown here. Notice that connecting words may not be necessary.

Cycle Concept Map

A **cycle concept map** is a special type of events chain map. In a cycle concept map, the series of events do not produce a final outcome. The last event in the chain relates back to the initiating event. Since there is no outcome and the last event relates back to the initiating event, the cycle repeats itself. Follow the stages shown in the cycle map of insect metamorphosis.

There is usually not one correct way to create a concept map. As you are constructing a map, you may discover other ways to construct the map that show the relationships among concepts better. If you do discover what you think is a better way to create a concept map, do not hesitate to change it.

Concept maps are useful in understanding the ideas you have read about. As you construct a map, you are organizing knowledge. Once concept maps are constructed, you can use them again to review and study and to test your knowledge. The construction of concept maps is a learning tool.

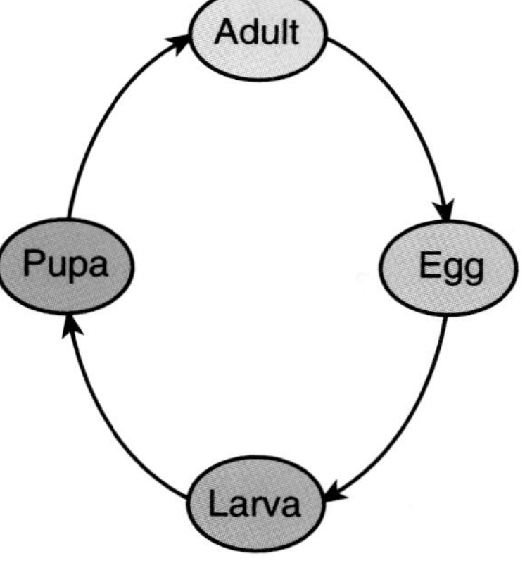
Adult
Egg
Larva
Pupa

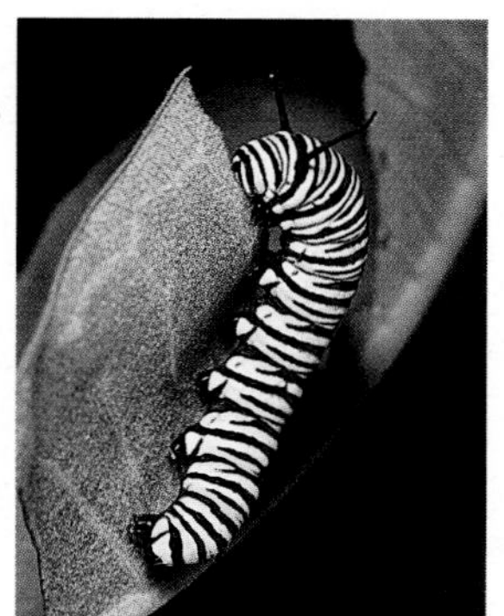

Making and Using Tables

Browse through your textbook, and you will notice many tables both in the text and in the labs. The tables in the text arrange information in such a way that it is easier for you to understand. Also, many labs in your text have tables to complete as you do the lab. Lab tables will help you organize the data you collect during the lab so that it can be interpreted more easily.

Most tables have a title telling you what is being presented. The table itself is divided into columns and rows. The column titles list items to be compared. The row headings list the specific characteristics being compared among those items. Within the grid of the table, the collected data are recorded. Look at the following table, and then study the questions that follow it.

Effect of Exercise on Heart Rate

Pulse Taken	Heart Rate	
	Individual	Class average
At rest	73	72
After exercise	110	112
1 minute after exercise	94	90
5 minutes after exercise	76	75

What is the title of this table? The title is "Effect of Exercise on Heart Rate." What items are being compared? The heart rate for an individual and the class average are being compared at rest and for several durations after exercise.

What is the average heart rate of the class one minute after exercise? To find the answer, you must locate the column labeled "class average" and the row "1 minute after exercise." The data contained in the box where the column and row intersect are the answer. Whose heart rate was 110 after exercise? If you answered "the individual's," you have an understanding of how to use a table.

Making and Using Graphs

After scientists organize data in tables, they often manipulate and organize and then display the data in graphs. A graph is a diagram that shows a comparison between variables. Since graphs show a picture of collected data, they make interpretation and analysis of the data easier. The three basic types of graphs used in science are the line graph, bar graph, and pie graph.

A **line graph** is used to show the relationship between two variables. The variables being compared go on two axes of the graph. The independent variable always goes on the horizontal axis, called the *x*-axis. The independent variable such as temperature is the condition that is manipulated. The dependent variable always goes on the vertical axis, the *y*-axis. The dependent variable such as growth is any change that results from manipulating the independent variable.

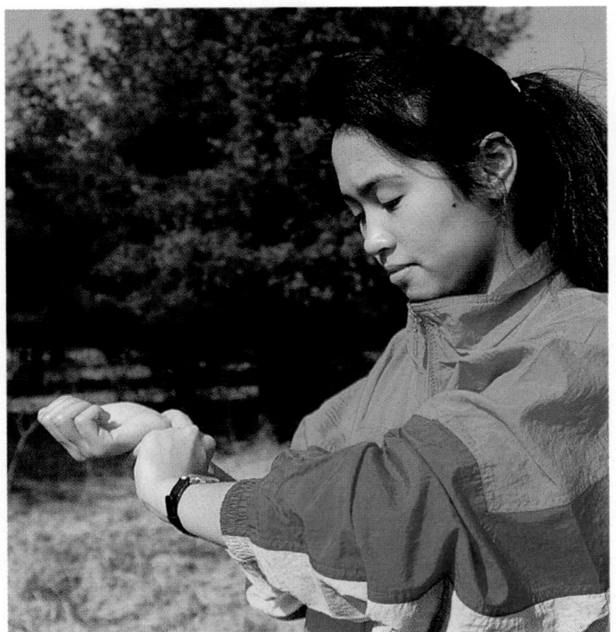

Suppose a school started a peer-study program with a class of students to see how it affected their science grades.

Average Grades of Students in Study Program

Grading Period	Average Science Grade
First	81
Second	85
Third	86
Fourth	89

You could make a graph of the grades of students in the program over a period of time. The grading period is the independent variable and should be placed on the *x*-axis of your graph. Instead of four grading periods, we could look at average grades for the week or month or year. In this way, we would be manipulating the independent variable. The average grade of the students in the program is the dependent variable and would go on the *y*-axis.

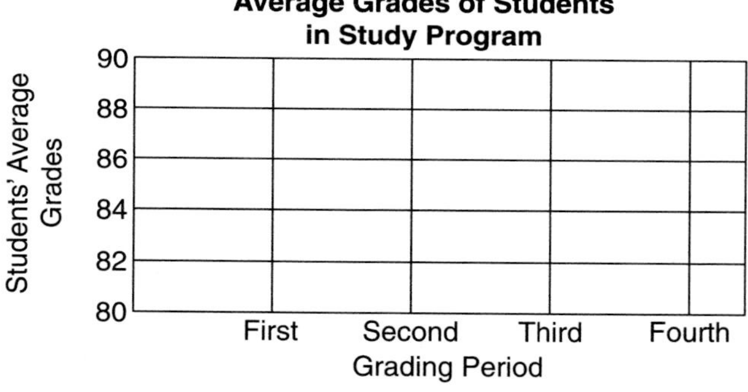

Plain or graph paper can be used to construct graphs. After drawing your axes, you would label each axis with a scale. The *x*-axis simply lists the grading periods. To make a scale of grades on the *y*-axis, you must look at the data values provided in the data table above. Since the lowest grade was 81 and the highest was 89, you know that you will have to start numbering at least at 81 and go through 89. You decide to start numbering at 80 and number by twos spaced at equal distances through 90.

You next must plot the data points. The first pair of data you want to plot is the first grading period and 81. Locate "First" on the *x*-axis and 81 on the *y*-axis. Where an imaginary vertical line from the *x*-axis and an imaginary horizontal line from the *y*-axis would meet, place the first data point. Place the other data points the same way. After all the points are plotted, connect them with a smooth line.

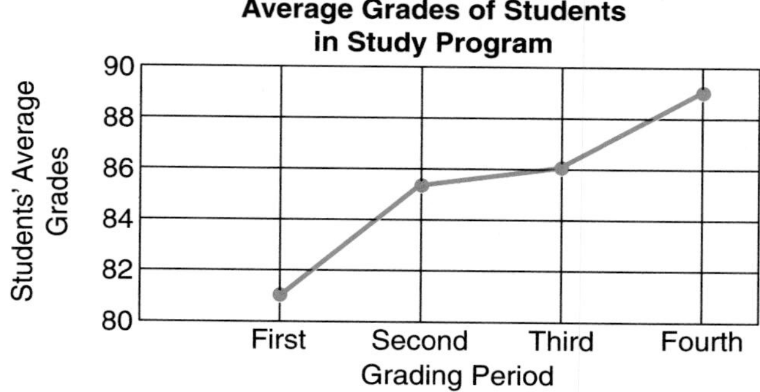

What if you wanted to compare the average grades of the class in the study group with the grades of another class? The data of the other class can be plotted on the same graph to make the comparison. You must include a key with two different lines, each indicating a different set of data.

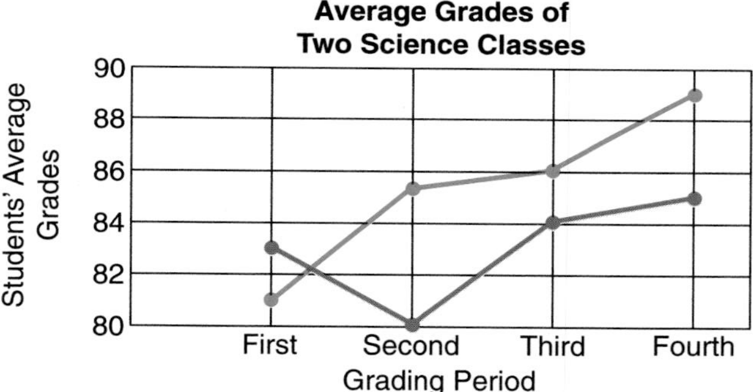

KEY: Class of study students —— Regular class ——

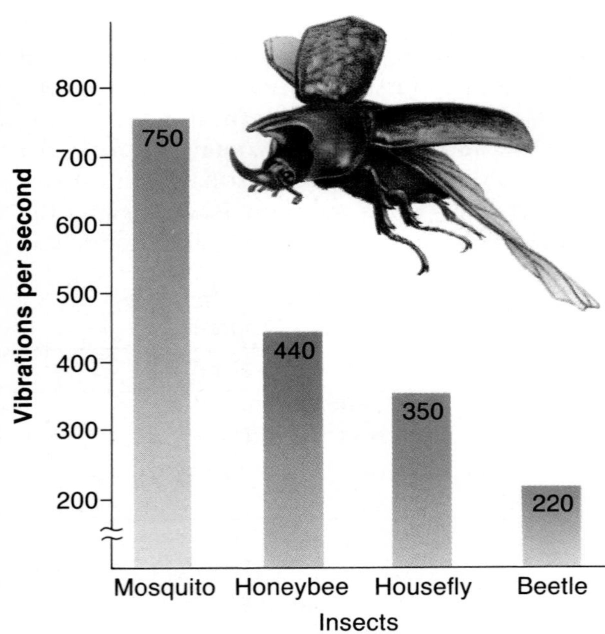

Bar graphs are similar to line graphs, except they are used to show comparisons among data or to display data that does not continuously change. In a bar graph, thick bars rather than data points show the relationships among data.

To make a **bar graph**, set up the x-axis and y-axis as you did for the line graph. The data are plotted by drawing thick bars from the x-axis up to an imaginary point where the y-axis would intersect the bar if it were extended.

Look at the bar graph above comparing the wing vibration rates for different insects. The independent variable is the type of insect, and the dependent variable is the number of wing vibrations per second. The number of wing vibrations for different insects is being compared.

A **pie graph** uses a circle divided into sections to display data. Each section represents a part of the whole. When all the sections are placed together, they equal 100 percent of the whole.

Suppose you wanted to make a pie graph to show the number of seeds that germinate in a package. You would have to determine the total number of seeds and the number of seeds that germinate out of the total. You count the seeds and find that the package contains 143 seeds. Therefore, the whole pie will represent this amount.

You plant the seeds and determine that 129 seeds germinate. The group of seeds that germinated will make up one section of the pie graph, and the group of seeds that did not germinate will make up another section.

To find out how much of the pie each section should take, you must divide the number of seeds in each section by the total number of seeds. You then multiply your answer by 360, the number of degrees in a circle. Round your answer to the nearest whole number. The number of seeds that germinated would be determined as follows:

$\frac{143}{129} \times 360 = 324.75$ or $325°$

To plot these data on the pie graph, you need a compass and a protractor. Use the compass to draw a circle. Then, draw a straight line from the center to the edge of the circle. Place your protractor on this line, and use it to mark a point on the edge of the circle at 325°. Connect this point with a straight line to the center of the circle. This is the section for the group of seeds that germinated. The other section represents the group of seeds that did not germinate. Complete the graph by labeling the sections of your graph and giving the graph a title.

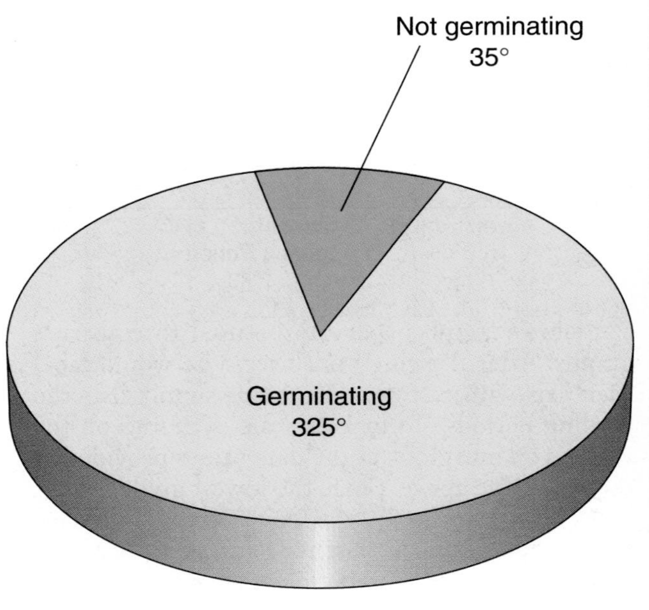

Glossary

Guide to Using the Glossary

This glossary defines each key term that appears in **bold type** in the text. It also shows the page number where you can find the word used.

The pronunciation key below can be used to sound out words in the glossary. Use this key to pronounce the key terms that have been sounded out for you in parentheses.

a	back (bak)	**ew**	food (fewd)
ay	day (day)	**yoo**	pure (pyoor)
ah	father (fahth ur)	**yew**	few (fyew)
ow	flower (flow ur)	**uh**	comma (cahm uh)
ar	car (car)	**u** (+con)	flower (flow ur)
e	less (les)	**sh**	shelf (shelf)
ee	leaf (leef)	**ch**	nature (nay chur)
ih	trip (trihp)	**g**	gift (gihft)
i (i+con+e)	idea, life (i dee uh, life)	**j**	gem (jem)
oh	go (goh)	**ing**	sing (sing)
aw	soft (sawft)	**zh**	vision (vihzh un)
or	orbit (or but)	**k**	cake (kayk)
oy	coin (coyn)	**s**	seed, cent (seed, sent)
oo	foot (foot)	**z**	zone, raise (zohn, rayz)

abiotic factors: (ay bi YAH tihk) nonliving parts of the environment such as air currents, temperature, soil, light, and moisture. (Chap. 3, p. 57)

acid: any substance that forms hydrogen ions in water; acidic solutions have a pH below 7. (Chap. 7, p. 172)

acid precipitation: rain or snow more acid than unpolluted rainwater; leaches valuable nutrients from soil, causing tree death and plant tissue injury; damages stone buildings and statues. (Chap. 6, p. 142)

acoelomate: (uh SEE luh mayt) animal without a body cavity; water and digested food pass into its solid tissues by diffusion; flatworms are acoelomates. (Chap. 28, p. 704)

actin: structural protein in muscle cells; together with the protein myosin, functions in muscle contraction. (Chap. 37, p. 950)

active transport: process requiring energy by which cells move materials against a concentration gradient. (Chap. 9, p. 230)

adaptation: (ad ap TAY shun) evolution of structural, internal, or behavioral features that help an organism better survive in an environment; the large eyes of nocturnal animals are an adaptation. (Chap. 1, p. 16)

adaptative radiation: type of divergent evolution in which an ancestral species can develop into an array of species, each specialized to fit into different niches. (Chap. 18, p. 446)

addiction: psychological or physiological drug dependence. (Chap. 40, p. 1027)

ADP: adenosine diphosphate; molecule occurring from the breaking off of a phosphate group from ATP, resulting in the release of energy to do biological work. (Chap. 10, p. 239)

aerobic process: (uh ROH bihk) oxygen-requiring biological process. (Chap. 10, p. 251)

age structure: proportions of any given population that are either in their pre-reproductive, reproductive, or post-reproductive years. (Chap. 5, p. 126)

aggression: threatening but rarely lethal behavior used to intimidate animals of the same species. (Chap. 36, p. 905)

alcoholic fermentation: anaerobic process in which pyruvic acid is changed to ethyl alcohol and CO_2; carried out by many bacteria and fungi such as yeasts. (Chap. 10, p. 254)

algae: (AL jee) photosynthetic, plantlike protists important in oxygen production; along with other eukaryotic autotrophs, they form the foundation of Earth's food chain. (Chap. 22, p. 542)

Glossary

allele: (uh LEEL) gene form for each variation of a trait of an organism; for example, Mendel's pea plants had two alleles for height— an allele for tallness and an allele for shortness. (Chap. 12, p. 288)

allelic frequency: percentage of a specific allele in the gene pool. (Chap. 18, p. 436)

alternation of generations: in plants and some protists, the reproductive life cycle in which a gametophyte (haploid, or n) generation alternates with a sporophyte (diploid, or $2n$) generation. (Chap. 22, p. 550)

alveoli: (al VEE uh lee) in the lungs, tiny, thin-walled sacs surrounded by capillaries; CO_2 diffuses from the blood into the alveoli, and oxygen diffuses from the alveoli into the blood. (Chap. 39, p. 983)

amino acids: basic building blocks of protein molecules. (Chap. 7, p. 180)

amniocentesis: (am nee oh sen TEE sus) process in which a small amount of amniotic fluid is withdrawn from around the fetus to diagnose chemical, chromosomal, or genetic abnormalities. (Chap. 41, p. 1057)

amniotic egg: (am nee YAH tihk) in reptiles, birds, and mammals, a major adaptation to land environments; encloses the embryo in amniotic fluid and membranes for protection. (Chap. 34, p. 839)

ampulla: (am POOL uh) in echinoderms, the bulblike muscular structure that contracts to push water into the tube foot and relaxes to let water flow out. (Chap. 32, p. 785)

amylase: (AM uh lays) digestive enzyme secreted by the salivary glands and pancreas; breaks starches into smaller sugars. (Chap. 38, p. 959)

anaerobic process: (an uh ROH bihk) simple biological process that does not require oxygen. (Chap. 10, p. 251)

analogous structures: structures without a common evolutionary origin that are similar in function but not in structure; wings of birds and insects are examples. (Chap. 18, p. 432)

anaphase: (AN uh fayz) stage of mitosis during which the centromeres split and sister chromatids are pulled apart to opposite poles of the cell. (Chap. 11, p. 272)

aneuploidy: (AN yoo ploy dee) abnormal number of chromosomes, usually as a result of accidents during meiosis; can result in Down syndrome and Turner syndrome. (Chap. 15, p. 369)

angiosperms: (AN jee uh spurmz) Anthophyta, flowering plants, extremely diverse plants with seeds enclosed in a fruit; most complex and highly adapted plants on Earth. (Chap. 26, p. 630)

annual: anthophyte that lives for a year or less; examples are corn and blue lupines. (Chap. 26, p. 634)

anterior: the head end in bilaterally symmetrical animals, where sensory tissue is commonly located. (Chap. 28, p. 703)

anther: in flowers, the male reproductive structure in which sperm-containing pollen grains develop and from which they burst when mature. (Chap. 27, p. 654)

antheridium: in bryophytes and pteridophytes, the male reproductive structure where sperm develops in the male gametophyte. (Chap. 24, p. 596)

anthropoids: humanlike primates; three major groups include New World monkeys, Old World monkeys, and hominids. (Chap. 19, p. 458)

antibiotic: microbial or fungal product that kills or inhibits the growth of other microorganisms. (Chap. 42, p. 1077)

antibodies: proteins in the plasma produced by plasma cells and memory B cells in reaction to foreign substances, or antigens. (Chap. 39, p. 990)

antigens: foreign substances that stimulate the production of antibodies in the blood. (Chap. 39, p. 990)

aorta: largest blood vessel in the human body; receives oxygen-rich blood from the left ventricle and sends it out to the body. (Chap. 39, p. 993)

aphotic zone: marine biome that never receives sunlight because of water depth; has intense water pressure; fish populating an aphotic zone have adapted to scarce food and life in darkness. (Chap. 4, p. 91)

apical meristem: (AY pih kul • MUR uh stem) terminal growth tissue of roots and stems that adds new cells for growth in length. (Chap. 26, p. 638)

appendage: structure growing out of an animal's body, such as a leg; arthropods were the first invertebrates to evolve jointed appendages. (Chap. 31, p. 758)

appendicular skeleton: one of the two main parts of the human skeleton; includes bones related to limb movement. (Chap. 37, p. 940)

archaebacteria: (ar kee bak TEER ee yuh) group of prokaryotes, similar to ancient fossil bacteria that produce glucose by chemosynthesis rather than by photosynthesis. (Chap. 17, p. 414)

archegonium: (ar kih GOH nee um) in bryophytes and pteridophytes, the female reproductive structure that develops eggs in the female gametophyte. (Chap. 24, p. 596)

artery: elastic, thick-walled vessel that transports blood away from the heart with a steady, pulsating rhythm. (Chap. 39, p. 991)

artificial selection: breeding of organisms selected for specific traits in order to produce offspring with those desired characteristics. (Chap. 18, p. 427)

ascospore: sexual spore of ascomycotes. (Chap. 23, p. 568)

ascus: saclike structure of ascomycotes in which the sexual spores develop. (Chap. 23, p. 567)

asexual reproduction: any reproductive process, such as budding, that does not involve the fusion of gametes. (Chap. 22, p. 537)

asymmetry: body plan exhibited by animals, such as sponges, that are irregular in shape. (Chap. 28, p. 702)

atoms: basic building blocks of all matter; smallest particles of an element with all the characteristics of that element. (Chap. 7, p. 161)

Glossary

ATP: adenosine triphosphate; energy-storing molecule that serves as the cell's "energy currency"; stored energy of glucose is used to attach phosphate groups to ADP to form ATP. (Chap. 10, p. 239)

atria: dual, thin-walled upper chambers of mammalian heart through which blood enters. (Chap. 39, p. 993)

australopithecine: (ah stray loh PIH thuh sine) early African hominid; genus *Australopithecus;* the first discovered example is the Taung child, which has a chimplike braincase, face, and teeth but walked upright like humans. (Chap. 19, p. 464)

autonomic nervous system: (ANS) portion of the peripheral nervous system that transmits messages to the internal organs from the central nervous system. (Chap. 40, p. 1017)

autosomes: in humans, the 22 pairs of matching homologous chromosomes. (Chap. 14, p. 336)

autotrophs: organisms that are able to synthesize food using sun energy or energy stored in chemical compounds; plants are the most common autotrophs; also called producers. (Chap. 3, p. 64)

axial skeleton: one of the two main parts of the human skeleton; includes bones of the head, chest, and spine. (Chap. 37, p. 940)

axon: cytoplasmic extension of a neuron; transmits impulses away from the cell body. (Chap. 40, p. 1006)

B

B cells: specialized lymphocytes produced from stem cells in the bone marrow; secrete antibodies that bind with antigens to form a pathogen-fighting antigen-antibody complex. (Chap. 42, p. 1084)

bacteriophage: (bak TEER ee yuh fayj) also called phages; viruses that infect bacteria. (Chap. 21, p. 504)

bark: outer layer of a woody stem that includes phloem tissue. (Chap. 26, p. 640)

base: substance that forms hydroxide ions in water; solutions with a pH above 7 are basic. (Chap. 7, p. 173)

basidia: (buh SIH dee uh) club-shaped hyphae of basidiomycotes; produce spores during sexual reproduction. (Chap. 23, p. 569)

basidiospore: sexual spore of basidiomycotes. (Chap. 23, p. 569)

behavior: response of an animal to an environmental stimulus; innate, or inherited, behavior is more common among invertebrates, and learned behavior is more common among vertebrates. (Chap. 36, p. 898)

biennial: anthophyte that lives for two years; examples include carrots and beets. (Chap. 26, p. 634)

bilateral symmetry: body plan in which the right and left halves form mirror images when divided down the animal's length; flatworms are bilaterally symmetrical. (Chap. 28, p. 703)

bile: chemical produced by the liver and stored in the gallbladder; breaks fat into small droplets by mechanical digestion. (Chap. 38, p. 963)

binary fission: asexual reproduction in which the cell divides into two equal parts. (Chap. 21, p. 522)

binomial nomenclature: two-word classification system originated by Linnaeus for naming species; the first word is the genus name and the second word is the descriptive name or specific epithet. (Chap. 20, p. 483)

biodegradable: solid wastes that can be broken down by natural processes; examples include wood products and food. (Chap. 6, p. 148)

biogenesis: (bi yoh JEN uh sus) idea that living organisms arise only from other living organisms. (Chap. 17, p. 411)

biology: science of life that seeks to provide an understanding of the natural world. (Chap. 1, p. 6)

biome: (BI ohm) large areas with the same type of climax community; examples include deserts and tropical rain forests. (Chap. 4, p. 91)

biosphere: (BI oh sfeer) life-supporting portions of Earth composed of air, land, fresh water, and salt water. (Chap. 3, p. 56)

biotic factors: (bi YAH tihk) all the living organisms inhabiting any of Earth's many different environments. (Chap. 3, p. 56)

bipedal: (bi PEE dul) having the ability to walk on two legs; in humans, bipedalism may have evolved in response to environmental changes that forced a population of apes to leave the trees and obtain food on the ground. (Chap. 19, p. 463)

blastula: (BLAS chuh luh) hollow ball of cells in a single layer enclosing a fluid-filled space; an animal embryo after cleavage, before the gastrula is formed. (Chap. 28, p. 698)

blood pressure: the force blood exerts on the vessels; rises when ventricles contract and falls when ventricles relax. (Chap. 39, p. 996)

book lungs: in spiders, air-filled spaces with leaflike plates that function in gas exchange and that resemble book pages. (Chap. 31, p. 761)

budding: asexual reproduction involving mitosis in which a group of cells pinches off from a parent cell and matures into a new individual; yeasts and hydras reproduce asexually by budding. (Chap. 23, p. 564)

bulbourethral glands: (bul boh yew REE thrul) in human males, the tiny glands beneath the prostate gland; secrete an alkaline, sperm-protecting fluid into the urethra. (Chap. 41, p. 1042)

bursa: (BUR suh) cushioning, fluid-filled sac found between the bones of movable joints. (Chap. 37, p. 940)

Glossary

Calorie: unit of heat used in measuring the energy content of food. (Chap. 38, p. 973)

Calvin cycle: set of reactions during photosynthesis in which simple sugars are formed from CO_2 using ATP and hydrogen from the light reactions. (Chap. 10, p. 243)

cambium: (KAM bee um) mytotic growth layer in vascular plants; vascular cambium produces xylem and phloem cells, and can result in secondary growth; cork cambium produces cork cells, which form the waterproof covering of bark. (Chap. 26, p. 638)

camouflage: (KAM uh flahzh) structural adaptation that involves an individual's external appearance, allowing it to blend in with its surroundings and avoid predators. (Chap. 18, p. 428)

cancer: uncontrolled cell division and death caused by the interaction between environmental factors and changes in the production of enzymes involved in the cell cycle. (Chap. 11, p. 276)

capillaries: (KAP uh layr eez) smallest blood vessels with walls of only one cell in thickness, through which nutrients and gases diffuse between blood and tissues. (Chap. 39, p. 992)

carbohydrate: (car boh HI drayt) organic compound used by cells to store and release energy; composed of carbon, hydrogen, and oxygen. (Chap. 7, p. 178)

cardiac muscle: involuntary muscle found only in the heart; contains interconnected cells that appear striated as a result of proteins in the cells; conducts electrical impulses that produce rhythmic contractions. (Chap. 37, p. 947)

carrier: heterozygous individual that appears phenotypically the same as a homozygous dominant individual, but which has a recessive allele for an undesirable trait. (Chap. 14, p. 344)

carrying capacity: maximum, stable population size an environment can support over time. (Chap. 5, p. 118)

cartilage: in animals such as agnathans, sharks, and skates, the tough flexible material forming the entire skeleton. (Chap. 33, p. 811); tissue from which bone develops in vertebrates. (Chap. 37, p. 940)

cell: building block of both unicellular and multicellular organisms; all living things are made of cells. (Chap. 8, p. 190)

cell culture: growth of cells in a nutrient medium; can be used to diagnose genetic disorders prenatally. (Chap. 16, p. 388)

cell cycle: continuous sequence of growth (interphase) and division (mitosis) of a cell, which is controlled by key enzymes. (Chap. 11, p. 267)

cell theory: the theory that (1) all living things are composed of one or more cells, (2) the cell is the basic unit of organization of organisms, and (3) all cells come from preexisting cells. (Chap. 8, p. 191)

cell wall: firm, fairly inflexible structure outside the plasma membrane of plants, fungi, most bacteria, and some protists; provides support and protection. (Chap. 8, p. 196)

cellulose: (SEL yew lohs) polysaccharide made of glucose units hooked together; forms plant cell walls and provides structural support for plants. (Chap. 7, p. 178)

central nervous system: (CNS) in humans, the body's control center; composed of the brain and spinal cord. (Chap. 40, p. 1010)

centrioles: (SEN tree ohlz) pair of cylindrical structures composed of microtubules that duplicate during interphase and move to opposite ends of the cell during prophase; found in animal cells but seldom in plant cells. (Chap. 11, p. 269)

centromere: (SEN truh meer) cell structure that joins two sister chromatids of a chromosome. (Chap. 11, p. 268)

cephalothorax: (sef uh luh THOR aks) structure formed from the fused head and thorax; found in some arthropods, such as shrimps. (Chap. 31, p. 760)

cerebellum: portion of the human brain that maintains balance and muscle coordination. (Chap. 40, p. 1011)

cerebrum: largest portion of the human brain; divided into two hemispheres; controls such functions as conscious activities and memory; covered with the highly folded cerebral cortex or gray matter. (Chap. 40, p. 1011)

cervix: in females, the lower end of the uterus that tapers to an opening into the vagina. (Chap. 41, p. 1045)

chelicerae: (kuh LIH sur ee) biting appendages of arachnids. (Chap. 31, p. 768)

chemosynthesis: process to obtain energy and produce food from inorganic compounds used by some organisms, such as methane-producing bacteria. (Chap. 10, p. 249)

chitin: (KI tun) complex carbohydrate found in the cell walls of most fungi and in the exoskeleton of some animals such as insects, lobsters, and crabs. (Chap. 23, p. 561)

chlorophyll: (KLOR uh fihl) embedded in the inner membrane of chloroplasts, this green pigment traps light energy from the sun and gives some protists and the leaves and stems of plants their green color. (Chap. 8, p. 204)

chloroplasts: (KLOR uh plasts) chlorophyll-containing organelle found in green plants and some protists; site where light energy is converted into chemical energy, which is stored in food molecules. (Chap. 8, p. 204)

chromatin: (KROH muh tun) long, tangled strands of DNA found in the interphase nucleus of eukaryotic cells. (Chap. 8, p. 198)

chromosomal mutation: mutation affecting gene distribution to gametes during meiosis, most commonly by deletions, insertions, inversions, or translocations. (Chap. 13, p. 325)

chromosome: (KROH muh sohm) cell structure that carries the genetic material. (Chap. 11, p. 266)

Glossary

cilia: (SIH lee uh) short, numerous, hairlike projections on a cell's surface that are composed of microtubules; their "beating" activity propels unicellular organisms and moves fluids over the cell surface in multicellular organisms. (Chap. 8, p. 206)

ciliates: (SIH lee uts) protozoans that move through their aquatic habitats by the beating of cilia in coordinated waves; paramecia are ciliates. (Chap. 22, p. 538)

circadian rhythm: (sur KAY dee un) daily, 24-hour cycle of behavior in response to internal biological cues. (Chap. 36, p. 906)

citric acid cycle: cyclic series of chemical reactions that take place during aerobic respiration; produces ATP and releases electrons. (Chap. 10, p. 252)

class: taxonomic grouping of related orders. (Chap. 20, p. 486)

classification: grouping together of objects or information based on similarities. (Chap. 20, p. 482)

climax community: stable, mature community that undergoes little ecological succession. (Chap. 4, p. 87)

clones: genetically identical DNA, cells, or organisms. (Chap. 16, p. 378)

closed circulatory system: system within an organism in which blood moves through the body enclosed entirely within blood vessels; found, for example, in cephalopods. (Chap. 30, p. 743)

cochlea: (KOH klee uh) fluid-filled structure in the ear in which sound vibrations are converted into nerve impulses. (Chap. 40, p. 1021)

codominant alleles: (koh DAH muh nunt • uh LEELZ) equal expression of both alleles; cause the phenotypes of both homozygote parent organisms to be produced in the heterozygote offspring. (Chap. 14, p. 335)

codon: (KOH dahn) in the genetic code, the set of three nitrogen bases representing a specific amino acid. (Chap. 13, p. 317)

coelom: (SEE lum) body cavity totally surrounded by mesoderm in which the digestive tract and other internal organs are suspended; humans, fish, and insects are coelomates. (Chap. 28, p. 705)

colony: (KAH luh nee) group of unicellular or multicellular organisms living together in close association. (Chap. 22, p. 549)

commensalism: (kuh MEN suh liz um) symbiotic relationship in which one species benefits and the other species is neither harmed nor helped; an orchid growing on the branch of a larger plant is an example. (Chap. 3, p. 68)

communication: information exchange among animals by signals such as sounds, sights, touches, or smells, resulting in a change in behavior. (Chap. 36, p. 915)

community: several interacting populations that inhabit a common environment and are able to function because each organism within the ecosystem depends on other organisms. (Chap. 3, p. 60)

compact bone: hard, dense bone tissue made up of Haversian canal systems containing blood vessels and nerves; protects spongy bone. (Chap. 37, p. 943)

companion cell: in vascular plants, the nucleated cell that helps manage the transport of food through the sieve cells of the phloem. (Chap. 26, p. 640)

complement: group of proteins that attaches to the surfaces of pathogens, damages their plasma membranes, and attracts phagocytes. (Chap. 42, p. 1082)

complete flower: flower with sepals, petals, stamens, and pistils; examples include phlox and delphinium. (Chap. 27, p. 655)

compound: composed of atoms of different substances that are chemically combined; for example, one atom of sodium (Na) combines with one atom of chlorine (Cl) to form table salt (NaCl), or sodium chloride. (Chap. 7, p. 164)

compound eye: in arthropods, a complex eye composed of multiple lenses that effectively spot moving prey but produce fuzzy images. (Chap. 31, p. 762)

compound light microscope: an instrument that magnifies living cells, small organisms, and preserved cells by passing visible light through the object, then through two or more glass lenses; can magnify up to about 1500×. (Chap. 8, p. 190)

conditioning: learning a type of behavior by association, for example, feeding associated with the ringing of a bell. (Chap. 36, p. 914)

cone: type of light receptor in the retina; responsible for color detection and vision under bright illumination. (Chap. 40, p. 1012); woody strobili of gymnosperms; scales support male or female reproductive structures and are the site of seed production. (Chap. 25, p. 613)

conidiophores: (kuh NIH dee uh forz) in ascomycotes, specialized hyphal branches that develop asexual spores in their tips. (Chap. 23, p. 568)

conidium: (kuh NIH dee um) asexual spore of ascomycotes. (Chap. 23, p. 568)

conjugation: in some bacteria and protists, a simple type of sexual reproduction in which genetic material is transferred from one organism to another through a connecting tube or strand of cytoplasm. (Chap. 21, p. 523)

conservation: planned management of wildlife habitats and other natural areas to prevent exploitation or destruction. (Chap. 6, p. 149)

contractile vacuole: (kun TRAK tile • VAK yew ohl) organelle of some protists such as *Paramecium* that collects excess water from the cell, then contracts and expels the water through a plasma membrane pore. (Chap. 9, p. 228)

control: the part of the experiment in which all conditions are kept constant. (Chap. 2, p. 29)

convergent evolution: evolution of similar traits in distantly related organisms resulting from adaptations to similar environments. (Chap. 18, p. 447)

corpus luteum: (KOR pus • LEW tee um) in human females, the structure stimulated by luteinizing hormone (LH) to develop from the ruptured follicle after the egg has been released; produces progesterone and estrogen. (Chap. 41, p. 1046)

Glossary

cortex: in plant stems and roots, the tissue between epidermis and vascular core involved in transport and storage of water, sugar, and mineral ions. (Chap. 26, p. 637)

cotyledon: (kah tuh LEE dun) structure of a seed plant embryo that functions as a food-storage organ; sometimes becomes the first leaves of the plant. (Chap. 25, p. 613)

courtship behavior: instinctive behavior patterns of male and female members of a species that occur before mating. (Chap. 36, p. 902)

covalent bond: (koh VAY lunt) chemical bond formed when two atoms combine by sharing electrons. (Chap. 7, p. 165)

Cro-Magnons: early *Homo sapiens;* people identical to modern humans in skull and teeth structure and brain size; appeared in Europe about 35 000 years ago and left behind cave paintings and detailed stone artifacts. (Chap. 19, p. 472)

crossing over: exchange of genetic material by non-sister chromatids during late prophase I of meiosis, resulting in new combinations of alleles. (Chap. 12, p. 302)

cud chewing: adaptation of many hoofed mammals enabling swallowed food matter to be brought up and rechewed. (Chap. 35, p. 876)

cuticle: (KYEW tih kul) waxy, protective coating on the outer surface of the epidermis of most fruits, stems, and leaves; adaptation that helps prevent water loss. (Chap. 24, p. 586)

cytoplasm: clear fluid in eukaryotic cells surrounding the nucleus and organelles; site of many important chemical reactions. (Chap. 8, p. 199)

cytoskeleton: in eukaryotic cells, a network in the cytoplasm; usually composed of microtubules and microfilaments; provides support for organelles and is important in cell locomotion. (Chap. 8, p. 205)

data: information obtained by observation, particularly during experiments; data are considered valid only if repeating the experiment several times yields similar results. (Chap. 2, p. 34)

day-neutral plant: plant in which flowering is influenced more by environmental conditions than by day length; most plants are day-neutral. (Chap. 27, p. 664)

deciduous plant: (dih SIH juh wus) plant that loses all of its leaves at the same time and can no longer perform photosynthesis and must remain dormant during winter. (Chap. 25, p. 620)

decomposers: organisms, such as many bacteria and most fungi, that play beneficial roles in all ecosystems by breaking down and absorbing nutrients from dead and decaying organic matter. (Chap. 3, p. 65)

deductive reasoning: "if...then" reasoning used after forming a hypothesis; suggests that something may be true about a specific case based on a known rule. (Chap. 2, p. 28)

demography: study of populations by mathematicians called demographers, who tally characteristics such as birthrates and death rates, fertility rates, age structure, and geographic distribution. (Chap. 5, p. 124)

dendrite: (DEN drite) cytoplasmic extension of a neuron; transmits impulses to the cell body. (Chap. 40, p. 1006)

density-dependent factors: factors that limit population density; examples include predation, disease, and competition for food, water, and territory. (Chap. 5, p. 120)

density-independent factors: most often weather-related occurrences such as storms and floods that affect populations, regardless of their density. (Chap. 5, p. 121)

dependent variable: in a controlled experiment, the measurable condition that results from changing the independent variable. (Chap. 2, p. 30)

depressant: drug that produces sedation by slowing down the activities of the central nervous system and increasing the activities of the parasympathetic nervous system; alcohol is a depressant. (Chap. 40, p. 1026)

dermis: (DUR mus) inner portion of skin; contains structures such as sweat glands and capillaries and produces vitamin D when exposed to ultraviolet light. (Chap. 37, p. 935)

desert: driest biome south of the taiga; generally receives less than 25 cm of annual rainfall, producing a range of vegetation from shrub communities to drifting dunes; animals include lizards, owls, and coyotes. (Chap. 4, p. 102)

detritus: (dih TRITE us) dead and decaying organic matter that provides food for small organisms. (Chap. 32, p. 788)

deuterostome: (DEW tuh roh stohm) animal whose anus develops from the opening in the gastrula; examples include humans, fish, and birds. (Chap. 28, p. 699)

development: rapidly or slowly occurring changes due to growth in an organism. (Chap. 1, p. 14)

diaphragm: (DI uh fram) in mammals, the sheet of muscle in the chest cavity that contracts during inhalation and relaxes during exhalation. (Chap. 35, p. 874)

dicotyledon: (di kah tuh LEE dun) plant with flower parts in multiples of four or five, seeds with two cotyledons, and a branching network of veins in the leaves; garden flowers and herbs are examples. (Chap. 26, p. 631)

diffusion: net, random movement of particles from an area of higher concentration to an area of lower concentration, eventually resulting in even distribution. (Chap. 9, p. 224)

dihybrid cross: (di HI brud) fertilization between two organisms to study the inheritance of two different traits. (Chap. 12, p. 291)

diploid: (DIH ployd) a cell with two copies of each type of chromosome is considered to be a diploid ($2n$) cell. (Chap. 12, p. 298)

directional selection: selection favoring individuals with extreme forms of a trait; can lead to rapid evolution of a population. (Chap. 18, p. 439)

Glossary

disaccharide: (di SAK uh ride) two-sugar carbohydrate formed by linking two monosaccharide molecules; sucrose is a disaccharide. (Chap. 7, p. 178)

disruptive selection: selection favoring individuals at both ends of extreme forms of a trait; can lead to the evolution of two new species. (Chap. 18, p. 440)

divergent evolution: evolution in which highly distinct species were once both similar to an ancestral species. (Chap. 18, p. 447)

division: taxonomic grouping of related classes for plants; the equivalent category of phylum in animal taxonomy. (Chap. 20, p. 486)

DNA: deoxyribonucleic acid; complex biological polymer; master copy of an organism's information code, which is passed on each time a cell divides and also from one generation to the next. (Chap. 7, p. 184)

dominance hierarchy: "pecking order" among individuals of a group in which there are several levels of dominance and submission; often prevents continuous fighting and may provide females a way to select the best mate. (Chap. 36, p. 905)

dominant: visible, observable trait of an organism that masks a recessive form of the trait. (Chap. 12, p. 289)

dormancy: in seeds, the period of inactivity allowing them to survive conditions unfavorable for growth. (Chap. 27, p. 673)

dorsal: back surface of bilaterally symmetrical animals. (Chap. 28, p. 703)

dorsal nerve cord: nerve bundle that in most chordates develops into the spinal cord and brain. (Chap. 32, p. 793)

double fertilization: in anthophytes, sexual reproduction resulting in the formation of a diploid ($2n$) zygote and a cell with a triploid ($3n$) nucleus that divides to eventually form the endosperm, or food-storage tissue for the developing embryo. (Chap. 27, p. 665)

double helix: in DNA, the two twisted, ladder-shaped nucleotide strands held together by hydrogen bonds between the bases. (Chap. 13, p. 312)

drug: chemical substance that can react with and alter body functions; examples include aspirin and caffeine. (Chap. 40, p. 1025)

dynamic equilibrium: condition of continuous movement but no overall change in concentration; movement of materials into and out of the cell at equal rates maintains its dynamic equilibrium with its environment. (Chap. 9, p. 224)

ecology: (ih KAH luh jee) scientific study of interactions between organisms and their environments; for example, ecologists study how day length influences migrating bird behavior. (Chap. 3, p. 55)

ecosystem: (EE koh sihs tum) populations in a community and abiotic factors with which they interact; examples are terrestrial and marine ecosystems. (Chap. 3, p. 61)

ectoderm: embryonic outer layer of cells in a gastrula; develops into skin and nervous tissue. (Chap. 28, p. 698)

ectotherm: animal with a body temperature regulated by its environment; amphibians and reptiles are ectotherms. (Chap. 33, p. 827)

egg: female sex cell or gamete. (Chap. 12, p. 300)

electron microscope: an instrument that allows scientists to magnify, view, and photograph dead cells or organisms by using a beam of electrons instead of light; its resolving power may be 1000 times greater than that of a light microscope. (Chap. 8, p. 194)

electron transport chain: series of molecules along which electrons are transferred; as the electrons travel, they release energy that is stored in the bonds of ATP. (Chap. 10, p. 244)

embryo: the early stage of growth and development of a plant or animal that followed fertilization of the egg by the sperm cell and the formation of a zygote. (Chap. 25, p. 613)

emigration: movement of individuals out of a population. (Chap. 5, p. 128)

endangered species: species with numbers of individuals so low that it is in danger of extinction; examples include manatees, the Florida panther, and loggerhead turtles. (Chap. 6, p. 138)

endemic disease: disease continually present in a population; the common cold is an example. (Chap. 42, p. 1076)

endocrine gland: (EN duh crun) ductless gland that releases hormones directly into the bloodstream; the thyroid functions as an endocrine gland in metabolism. (Chap. 38, p. 966)

endocytosis: (en duh si TOH sus) active transport process by which large particles enter a cell. (Chap. 9, p. 231)

endoderm: embryonic inner layer of cells in a gastrula that gives rise to the lining of the digestive tract. (Chap. 28, p. 698)

endodermis: in plant roots, the waterproof, single layer of cells that controls the flow of solutes into the root. (Chap. 26, p. 637)

endoplasmic reticulum: (en duh PLAZ mihk • rih TIHK yuh lum) folded, complex system of membranes forming a type of transport system in the cytoplasm of eukaryotic cells; can be either rough (with ribosomes) or smooth (without ribosomes). (Chap. 8, p. 199)

endoskeleton: internal skeleton of vertebrates; provides a support framework and protects internal organs. (Chap. 28, p. 708)

endosperm: in anthophytes, the triploid ($3n$) food-storage tissue used by the developing embryo. (Chap. 27, p. 666)

Glossary

endospore: (EN doh spor) in bacteria, the structure with a hard, protective outer covering formed when conditions are unfavorable; under favorable conditions, the endospore germinates and the bacterium resumes growth and reproduction. (Chap. 21, p. 521)

endotherm: animal able to maintain a constant body temperature based on metabolic processes rather than being regulated by environmental conditions; for example, birds are endotherms. (Chap. 34, p. 856)

energy: ability to do work or move things; powers life processes. (Chap. 1, p. 17)

environment: external conditions, such as weather, or internal conditions, such as the presence of infection, to which an organism must continually adjust. (Chap. 1, p. 15)

enzymes: proteins that accelerate chemical reactions but do not change themselves in the reaction; enzymes enable molecules to undergo chemical change, forming new substances called products. (Chap. 7, p. 181)

epidemic: disease pattern in which large numbers of people in a given area are sickened by the same disease over a short period of time; influenza often produces epidemics. (Chap. 42, p. 1076)

epidermis: in plants, the outermost layer of cells of the plant body; also forms hairs on the plant's surfaces. (Chap. 26, p. 639); in humans and other animals, the protective outer portion of skin composed of two cell layers; the top layer continually sheds dead cells and the inner layer forms new living cells. (Chap. 37, p. 934)

epididymus: (ep uh DIHD uh mus) in human males, the single, coiled tube in which sperm complete maturation; also provides storage for sperm. (Chap. 41, p. 1041)

epiglottis: (ep uh GLAH tus) flap of skin that covers the opening of the windpipe during swallowing. (Chap. 38, p. 961)

esophagus: (ih SAW fuh gus) muscular tube that moves food from the mouth to the stomach by smooth muscle contractions. (Chap. 38, p. 961)

estivation: (es tuh VAY shun) state similar to hibernation; affects many desert animals in response to heat or drought; an innate behavior governed by both internal and external cues. (Chap. 36, p. 908)

estuary: (ES chuh wayr ee) coastal body of water in which both freshwater and saltwater mix; provides excellent food supply and shelter for young fishes. (Chap. 4, p. 92)

ethics: study of right and wrong; in the scientific arena, ethical issues must be decided by all of society. (Chap. 2, p. 45)

eubacteria: (yew bak TEER ee uh) group of prokaryotes with a wide variety of structures and types of metabolism; examples include *Nitrosomonas* and *Oscillatoria*. (Chap. 21, p. 512)

eukaryote: (yew KAYR ee oht) cell having a true nucleus and membrane-bound internal organelles; the majority of cells are eukaryotic. (Chap. 8, p. 195)

evergreen plant: plant that retains its leaves year-round; most conifers are evergreens; often found in areas that have a warm, short growing season and scarce nutrients. (Chap. 25, p. 619)

evolution: change in the gene pool of a population in response to various stimuli exhibited by a species over time. (Chap. 1, p. 20)

exocrine gland: (EK suh crun) gland that releases its secretions through ducts; the pancreas functions as an exocrine gland during digestion. (Chap. 38, p. 966)

exocytosis: (eks oh si TOH sus) active transport process by which materials are expelled or secreted from a cell. (Chap. 9, p. 232)

exoskeleton: hard outer covering common in some invertebrates such as crabs and spiders; provides support, protection, and a place for muscle attachment. (Chap. 28, p. 708)

experiment: procedure by which scientists determine the validity of a hypothesis by collecting information under controlled conditions. (Chap. 2, p. 29)

exponential growth: explosive population growth in which the number of reproducing individuals increases by an ever-increasing rate. (Chap. 5, p. 115)

external fertilization: fertilization that occurs in water into which both eggs and sperm have been released, as in sponges, frogs, and fishes. (Chap. 29, p. 714)

extinction: occurs when the last members of a species die; may be a natural process or the result of human activity such as hunting, urbanization, and the destruction of natural habitats. (Chap. 6, p. 136)

F

facilitated diffusion: passive transport of materials, such as sugars and amino acids, across the plasma membrane by way of transport proteins. (Chap. 9, p. 230)

family: taxonomic grouping of closely related genera. (Chap. 20, p. 486)

feather: lightweight, external covering of birds; down feathers insulate and contour feathers are used for flight. (Chap. 34, p. 855)

fertilization: fusion of male and female gametes; in Mendel's pea plants, occurred when the male gamete in the pollen grain fused with the female gamete in the ovule. (Chap. 12, p. 286)

fetus: the developing mammal from nine weeks until birth. (Chap. 15, p. 362)

fight-or-flight response: automatic, chemically controlled response to a stimulus that mobilizes the individual for greater activity, such as running or fighting—for example, by increasing the heart and breathing rates. (Chap. 36, p. 902)

filter feeding: process in which food particles are filtered from water as it travels through or around the organism; for example, sponges and bivalve mollusks are filter feeders. (Chap. 29, p. 712)

fins: in fishes, the external fan-shaped membranes attached to the endoskeleton and supported by stiff rays; fins function in locomotion. (Chap. 33, p. 816)

flagella: (fluh JEL uh) long, threadlike structures composed of microtubules; project from within the plasma membrane and propel cells and organisms by a whiplike motion. (Chap. 8, p. 206)

flagellates: (FLAJ uh luts) protozoans that move by means of one or more flagella, which they whip from side to side. (Chap. 22, p. 537)

fluid mosaic model: property of the plasma membrane wherein many similar molecules are free to move sideways within their lipid bilayer. (Chap. 9, p. 220)

follicle: (FAH lih kul) in females, the epithelial cells that surround an undeveloped egg cell in the ovary; when the egg is released, the remaining part of the follicle develops into the corpus luteum. (Chap. 41, p. 1045)

food chain: a possible route for the transfer of matter and energy through an ecosystem from autotrophs through heterotrophs and decomposers. (Chap. 3, p. 70)

food web: shows all the possible feeding relationships in a community at each trophic level; represents a network of interconnected food chains. (Chap. 3, p. 72)

fossil: any evidence of organisms that lived in the past; usually found in sedimentary rock; scientists study fossils to look for evolutionary clues. (Chap. 17, p. 399)

fossil fuels: coal, oil, and natural gas formed from the buried remains of organisms. (Chap. 6, p. 136)

fragmentation: asexual reproduction in which an organism breaks up into two or more parts, each of which forms a new organism; seen, for example, in *Spirogyra.* (Chap. 22, p. 550)

frameshift mutation: error in the DNA sequence that adds or deletes a single base, causing nearly all amino acids following the mutation to be changed. (Chap. 13, p. 325)

frond: (FRAWND) the leaflike organ of a fern that grows upward from the rhizome; fern fronds have a stemlike stipe and green pinnae. (Chap. 25, p. 610)

fruit: seed-containing, ripened ovary of an anthophyte; examples are dry fruits such as pecans or fleshy fruits such as peaches. (Chap. 27, p. 667)

fungus: heterotrophic, eukaryotic consumer that absorbs nutrients from decomposing wastes and dead organisms. (Chap. 20, p. 495)

gallbladder: small, bile-storing organ of the digestive system. (Chap. 38, p. 963)

gametangium: (gam uh TAN jee um) in zygospore-producing fungi, the structure with a haploid nucleus in which gametes are produced. (Chap. 23, p. 567)

gametes: (GAM eets) male and female sex cells; sperm and eggs. (Chap. 12, p. 286)

gametophyte: (guh MEE tuh fite) haploid (n) form of a plant that produces gametes. (Chap. 22, p. 550)

gastrovascular cavity: (gas troh VAS kyuh lur) cavity in which digestion occurs; in cnidarians, which have a simple, two-cell-layer body plan, this space is surrounded by the inner cell layer. (Chap. 29, p. 718)

gastrula: (GAS truh luh) developmental stage of animal embryos in which the blastula folds inward forming a cavity with two cell layers, ectoderm and endoderm; a middle layer of cells, the mesoderm, is formed later. (Chap. 28, p. 698)

gemmae: (JEM ee) tiny haploid (n) structures of liverworts and some mosses; grow in cups on the surface of gametophytes during vegetative reproduction. (Chap. 24, p. 600)

gene: a segment of DNA located on the chromosome; directs the protein production that controls the cell cycle. (Chap. 11, p. 276)

gene pool: sum of all genes among a population. (Chap. 18, p. 436)

gene splicing: in recombinant DNA technology, the rejoining of cut DNA fragments. (Chap. 16, p. 378)

gene therapy: a goal of recombinant DNA technology, presently in trial stages; inserts normal genes into human cells to correct genetic disorders. (Chap. 16, p. 388)

genetic counseling: provided by trained professionals in various disciplines to couples seeking information about the probability of having hereditary disorders occur in their offspring. (Chap. 41, p. 1056)

genetic drift: changes in allelic frequency by chance events; results in alteration of genetic equilibrium. (Chap. 18, p. 437)

genetic engineering: process using restriction enzymes to cleave an organism's DNA into smaller fragments in order to move genes from one organism to another of the same or different species; has produced transgenic plants, bacteria, and animals. (Chap. 16, p. 376)

genetic equilibrium: condition in which allelic frequency remains constant over generations; populations in genetic equilibrium are not evolving. (Chap. 18, p. 436)

genetic recombination: major source of genetic variation resulting from crossing over or random assortment. (Chap. 12, p. 304)

genetics: branch of biology that studies heredity. (Chap. 12, p. 286)

genotype: (JEE nuh tipe) an organism's gene combination. (Chap. 12, p. 291)

genus: (JEE nus) first word of the two-part scientific name used to classify a group of closely related species. (Chap. 20, p. 483)

geographic isolation: occurs when populations become isolated by a factor such as deforestation, which results in individuals no longer being able to mate; can lead to the formation of new species. (Chap. 18, p. 441)

germination: process by which a seed begins to develop into a new organism. (Chap. 27, p. 674)

Glossary

gestation: (jes TAY shun) full term of pregnancy during which the young of placental mammals develop in the uterus; length varies from species to species. (Chap. 35, p. 885)

gill slits: in chordates, the paired openings found in the pharynx, behind the mouth; in some vertebrate chordates, the gill slits develop into internal gills used for gas exchange. (Chap. 32, p. 796)

gizzard: muscular sac containing hard particles for grinding food-containing material into small pieces so it can be absorbed by the intestine; found, for example, in annelids. (Chap. 30, p. 747)

gland: fluid-secreting cell or group of cells; in mammals, various glands produce fluids such as enzymes, hormones, and milk. (Chap. 35, p. 873)

glycogen: (GLI koh jun) polysaccharide with very highly branched chains of glucose units; animals store food as glycogen. (Chap. 7, p. 178)

glycolysis: (gli KAH luh sus) anaerobic process that splits glucose, forming two molecules of pyruvic acid; also produces hydrogen ions and electrons. (Chap. 10, p. 251)

Golgi apparatus: (GAWL jee) membrane sacs that receive, chemically modify, and repackage proteins into forms the cell can use, expel, or keep stored. (Chap. 8, p. 200)

gradualism: (GRA juh wuh lih zum) idea that species originate gradually over time through the accumulation of small, adaptive changes. (Chap. 18, p. 444)

grasslands: biome composed of large communities of grasses and other small plants; characterized by hot summers, cold winters, and uncertain rainfall; high humus content of soil is ideal for growing oats, rye, and wheat; home to buffalo, prairie dogs, and many bird and insect species. (Chap. 4, p. 103)

greenhouse effect: a natural phenomenon by which carbon dioxide and other atmospheric gases prevent heat from escaping into space; without the greenhouse effect, Earth would be too cold for life to exist. (Chap. 6, p. 144)

groundwater: fresh water from rain and surface streams that accumulates in underground reservoirs. (Chap. 6, p. 146)

growth: process that results in structural changes and increased living material in an organism. (Chap. 1, p. 14)

guard cells: pairs of modified cells of the leaf epidermis that control the opening and closing of stomatal pores. (Chap. 26, p. 643)

gymnosperms: (JIHM nuh spurmz) nonflowering vascular plants with seeds produced on the scales of cones; the four divisions are Cycadophyta, Ginkgophyta, Gnetophyta, and Coniferophyta. (Chap. 25, p. 613)

habitat: collection of niches in which an organism lives its life. (Chap. 3, p. 62)

habituation: (huh bih chuh WAY shun) simple form of learned behavior. (Chap. 36, p. 910)

hair follicles: groups of cells in the dermal skin layer from which hair grows. (Chap. 37, p. 935)

hallucinogen: (huh LEW suh nuh jun) mood-altering substance; examples are LSD and psilocybin. (Chap. 40, p. 1033)

haploid: (HAP loyd) a cell of an organism that has half the number (n) of chromosomes; one of each type of chromosome that makes up the genotype. (Chap. 12, p. 298)

haustoria: (haw STOR ee uh) hyphae of parasitic fungi adapted to penetrate the cells of hosts and absorb nutrients. (Chap. 23, p. 563)

hemoglobin: (HEE muh gloh bun) iron-containing molecule of red blood cells that transports oxygen and some CO_2. (Chap. 39, p. 988)

heredity: passing on of characteristics from parents to offspring; first major studies of heredity were done by Gregor Mendel in the 1800s. (Chap. 12, p. 286)

hermaphrodite: (hur MAF ruh dite) organism that can produce both eggs and sperm but is not necessarily self-fertilizing. (Chap. 29, p. 714)

heterotrophs: (HET uh ruh trohfs) organisms unable to make their own food, they rely on autotrophs as their nutrient and energy source; examples are rabbits and cows; also called consumers. (Chap. 3, p. 64)

heterozygous: (het uh roh ZI gus) having nonidentical alleles for a particular trait. (Chap. 12, p. 291)

hibernation: state in which metabolic needs are greatly reduced; exhibited by many mammals, some birds, and other types of animals during cold winter months. (Chap. 36, p. 908)

homeostasis: (hoh mee oh STAY sus) equilibrium of an organism's internal environment that maintains conditions suitable for life; an example is human sweating, which helps the body maintain its proper temperature. (Chap. 1, p. 16); the balance of nature in ecosystems. (Chap. 3, p. 23)

hominid: (HAW muh nud) bipedal, humanlike primate; *Australopithecus afarensis* is the earliest known example of a hominid and was bipedal with an apelike braincase. (Chap. 19, p. 463)

homologous chromosomes: (hoh MAW luh gus • KROH muh sohmz) paired chromosomes with genes for the same traits arranged in the same order. (Chap. 12, p. 299)

homologous structures: (hoh MAW luh gus • STRUK churz) structures having a common evolutionary origin; examples are the forelimbs of bats, crocodiles, and humans. (Chap. 18, p. 432)

homozygous: (hoh muh ZI gus) having identical alleles for a particular trait. (Chap. 12, p. 291)

hormone: chemical secreted by one part of a plant or animal that affects another part of the organism; in plants, hormones such as auxins and gibberellins promote growth. (Chap. 27, p. 676); in humans, hormones play a key role in regulating metabolism of digestion, growth, and reproduction. (Chap. 41, p. 1042)

Glossary

host cell: cell in which a virus reproduces. (Chap. 21, p. 504)

human genome: approximately 100 000 genes, on 46 human chromosomes, made of about 3 billion DNA base pairs, which when mapped and sequenced may enable the treatment or cure of genetic disorders. (Chap. 16, p. 386)

hybrid: offspring produced when two varieties of plants or animals, or closely related species, are mated; hybrids often exhibit greater vigor and size than their parents. (Chap. 14, p. 347)

hydrogen bond: weak bond formed by attraction of opposite charges between hydrogen and other atoms; helps to hold large molecules, such as proteins, together. (Chap. 7, p. 168)

hypertonic solution: (hi pur TAH nihk) in cells, solution in which the concentration of dissolved materials is higher in the solution surrounding the cell than the concentration inside the cell, which will shrink as water leaves the cell by osmosis. (Chap. 9, p. 229)

hyphae: (HI fee) threadlike filaments that form the basic structural units of multicellular fungi. (Chap. 23, p. 561)

hypothalamus: (hi poh THAL uh mus) area of the brain directly above the pituitary that produces antidiuretic hormone (ADH) and releasing factors that cause the pituitary to secrete hormones, which control reproduction. (Chap. 39, p. 1000)(Chap. 41, p. 1042)

hypothesis: testable explanation of a question or problem; a hypothesis may be formed by extensive reading, observation, reasoning, and knowledge of earlier experiments. (Chap. 2, p. 27)

hypotonic solution: in cells, solution in which the concentration of dissolved materials is lower in the solution surrounding the cell than the concentration inside the cell, which will swell and possibly burst as water enters the cell by osmosis. (Chap. 9, p. 227)

immigration: movement of individuals into a population; in the last 165 years, more people immigrated to the United States than to any other country. (Chap. 5, p. 127)

immunity: (ih MYEW nuh tee) overall ability of an organism to resist a specific pathogen. (Chap. 42, p. 1083)

implantation: in females, the attachment of the fertilized egg, or blastocyst, to the uterine lining. (Chap. 41, p. 1051)

imprinting: species-specific learned behavior that occurs only at a certain, critical time in an animal's life and forms a permanent social attachment. (Chap. 36, p. 911)

inbreeding: mating between closely related individuals to produce pure lines; however, undesirable recessive traits can appear more often than in random matings. (Chap. 14, p. 347)

incomplete dominance: inheritance pattern in which the phenotype of the heterozygote is intermediate between those of the two homozygotes; neither allele of the pair is completely dominant but they combine to give a new trait. (Chap. 14, p. 334)

incomplete flower: flower lacking one or more organs; examples include sweet corn and squash. (Chap. 27, p. 655)

independent variable: in a controlled experiment, the one condition that is changed. (Chap. 2, p. 30)

inductive reasoning: most common type of scientific reasoning used in developing a hypothesis; produces a general rule based on a set of observations. (Chap. 2, p. 27)

infectious disease: disease caused by pathogens in the body; chicken pox and tuberculosis are examples. (Chap. 42, p. 1070)

innate behavior: genetically programmed behavior pattern of an animal; includes both automatic responses and instinctive behaviors. (Chap. 36, p. 899)

insight: most complex type of learning resulting in the formation of a concept or idea that can be applied to new situations. (Chap. 36, p. 914)

instinct: pattern of innate, or inherited, animal behavior, such as courtship rituals. (Chap. 36, p. 902)

internal fertilization: fertilization of eggs by sperm inside an animal's body; in snakes, birds, and mammals. (Chap. 29, p. 714)

interphase: growth period of a cell during which chromosomes are duplicated. (Chap. 11, p. 267)

intertidal zone: part of the shoreline between the high and low tide lines; productivity of intertidal ecosystems is limited because of wave action even though levels of nutrients, sunlight, and oxygen are high. (Chap. 4, p. 92)

invertebrate: animal lacking a backbone; examples include crabs and grasshoppers. (Chap. 28, p. 708)

involuntary muscle: muscle whose contractions are not under conscious control; smooth muscle is involuntary muscle. (Chap. 37, p. 946)

ion: (I un) an atom or molecule that gains electrons and carries a negative electrical charge or loses electrons and has a positive electrical charge. (Chap. 7, p. 165)

ionic bond: bond formed by the mutual attraction of two ions of opposite charge. (Chap. 7, p. 166)

isomers: (I suh murz) compounds with the same formula but a different three-dimensional arrangement of the atoms, resulting in the molecules having different chemical properties. (Chap. 7, p. 176)

isotonic solution: (i suh TAH nihk) solution in cells in which dissolved materials and water occur in the same concentration as inside the cell. (Chap. 9, p. 226)

isotopes: (I suh tohps) atoms of the same element differing in the numbers of neutrons in the nucleus; examples are carbon-12 and carbon-14; some radioactive isotopes are used in medicine. (Chap. 7, p. 163)

Glossary

Jacobson's organ: pitlike olfactory organ of most reptiles that is used to detect airborne chemicals. (Chap. 34, p. 842)

joint: point where two or more bones meet; can be fixed or allow movement; examples include ball-and-socket joints and hinge joints. (Chap. 37, p. 940)

karyotype: (KAYR ee uh tipe) charted arrangement of chromosomes possessed by an individual; helpful in locating aneuploidies in humans such as Down syndrome. (Chap. 15, p. 369)

keratin: (KAYR uh tun) protein formed by the outer epidermal skin layer; helps protect underlying skin cells. (Chap. 37, p. 934)

kidneys: organs of the vertebrate urinary system; remove nitrogenous wastes, control the sodium level of blood, and regulate the pH of blood. (Chap. 39, p. 997)

kingdom: taxonomic grouping of related phyla. (Chap. 20, p. 486)

Koch's postulates: (KOHKS • PAHS tyuh luts) a set of steps for proving that a specific pathogen causes a specific disease. (Chap. 42, p. 1072)

labor: sum of physical and physiological changes that occur to the mother during the birth process. (Chap. 41, p. 1058)

lactic acid fermentation: anaerobic process in which pyruvic acid changes to lactic acid; occurs in some bacteria, plants, and most animals; can create an oxygen debt in exercised muscles because of a buildup of lactic acid, resulting in fatigue. (Chap. 10, p. 254)

language: form of communication using symbols to represent ideas. (Chap. 36, p. 916)

large intestine: muscular tube into which indigestible material is passed to the rectum for elimination; site of water absorption and synthesis of B and K vitamins. (Chap. 38, p. 964)

larva: in insects, the stage of metamorphosis in which the organism is wormlike, free-living, and molts several times. (Chap. 31, p. 777)

lateral line system: fluid-filled canals along the sides of a fish that serve as sensory receptors, enabling detection of prey and navigation in the dark. (Chap. 33, p. 817)

law of independent assortment: Mendelian principle explaining that different traits are inherited independently if on different chromosomes. (Chap. 12, p. 294)

law of segregation: Mendelian principle explaining the disappearance of a specific trait in the F_1 generation and its reappearance in the F_2 generation. (Chap. 12, p. 290)

leaf: plant organ that provides a surface area for trapping sunlight and exchanging gases through stomata during photosynthesis. (Chap. 24, p. 587)

lichen: (LI kun) organism comprising a symbiotic association between a fungus and a photosynthetic partner. (Chap. 23, p. 575)

ligament: strong band of connective tissue that holds joints together and connects bones to other bones. (Chap. 37, p. 940)

light reactions: highly complex reactions by which light energy is converted to chemical energy during photosynthesis; results in the splitting of water and release of oxygen. (Chap. 10, p. 243)

limiting factor: any factor limiting the survival and productivity of organisms; for example, the lack of water could limit grass growth in a grassland. (Chap. 4, p. 84)

linkage map: genetic map showing gene location on chromosomes. (Chap. 16, p. 386)

lipids: organic compounds commonly called fats and oils; insoluble in water, lipids are the major components of membranes surrounding all living cells. (Chap. 7, p. 179)

liver: gland that produces many chemicals, including several needed for digestion, that are delivered to the small intestine. (Chap. 38, p. 963)

long-day plant: plant that flowers in midsummer, when days are longer than nights; examples are carnations and spinach. (Chap. 27, p. 664)

lymph: (LIHMF) tissue fluid collected in lymphatic vessels to be returned to the blood. (Chap. 42, p. 1078)

lymph node: small tissue mass with a fiber network that holds lymphocytes. (Chap. 42, p. 1079)

lymphocytes: (LIHMF uh sites) white blood cells that mature and differentiate into cells that destroy specific pathogens. (Chap. 42, p. 1079)

lysogenic cycle: (li suh JEN ihk) viral reproductive cycle; the host cell's chromosome becomes integrated with viral DNA; the provirus formed is replicated each time the host reproduces; the host cell is not killed until the lytic cycle is entered. (Chap. 21, p. 508)

lysosomes: (LI suh sohmz) membrane-bound organelles containing enzymes that digest food particles, viruses and bacteria, worn-out cell parts, and sometimes the cell itself. (Chap. 8, p. 201)

lytic cycle: (LI tihk) viral reproductive cycle; host cell DNA is destroyed by the virus, which then forms new viruses that burst from the host cell, killing it. (Chap. 21, p. 507)

macrophage: (MAK ruh fayj) type of phagocyte that attacks anything recognized as foreign, including microbes and dust in the lungs. (Chap. 42, p. 1081)

Glossary

madreporite: (MAD ruh por ite) in echinoderms, the disc-shaped, sievelike opening through which water flows in and out of the water vascular system. (Chap. 32, p. 786)

Malpighian tubules: (mal PIH jee un • TEW byew ulz) waste-excreting, abdominal structures of most terrestrial arthropods. (Chap. 31, p. 763)

mammary glands: milk-secreting glands of female mammals enabling them to feed their young until they are mature enough to forage for food. (Chap. 35, p. 877)

mandible: (MAN duh bul) in most arthropods, mouthpart adapted for piercing, sucking, lapping, and chewing of food. (Chap. 31, p. 763)

mantle: thin, outer membrane of mollusks; in mollusks with shells, the mantle secretes the shell. (Chap. 30, p. 738)

marrow: soft tissue filling the cavities of most bones; functions include red blood cell production and fat storage. (Chap. 37, p. 943)

marsupial: mammal in which the young develop inside the body, followed by longer development outside the body in a pouch of skin and hair. (Chap. 35, p. 885)

medulla oblongata: (muh DUHL uh • ah blong GAH tuh) portion of the human brain that controls involuntary activities such as breathing. (Chap. 40, p. 1011)

medusa: in the life cycle of cnidarians, the free-swimming, bell-shaped stage. (Chap. 29, p. 716)

megaspores: female spores in plants that develop eventually into archegonia with egg cells. (Chap. 25, p. 613)

meiosis: (mi OH sus) cell division in which one diploid ($2n$) cell produces four haploid (n) cells called sex cells or gametes, which have half the number of chromosomes as a body cell of the parent. (Chap. 12, p. 300)

melanin: (MEL uh nun) cell pigment formed by the inner epidermal skin layer; colors the skin and helps protect against ultraviolet radiation. (Chap. 37, p. 935)

menstrual cycle: (MEN strul) in human females, the monthly cycle in which the lining of the uterus is prepared to receive an egg and is shed if the egg is not fertilized. (Chap. 41, p. 1047)

mesoderm: middle layer of embryonic cells in a gastrula, between the ectoderm and endoderm; develops into reproductive organs, muscles, and circulatory vessels. (Chap. 28, p. 699)

mesophyll: (MEZ uh fihl) photosynthetic tissue of a leaf, either palisade or spongy. (Chap. 26, p. 646)

messenger RNA (mRNA): carries protein synthesis information from DNA to the ribosomes. (Chap. 13, p. 319)

metabolism: (muh TAB uh lihz um) total of all chemical reactions that occur within a living organism. (Chap. 7, p. 169)

metamorphosis: (met uh MOR fuh sus) chemically controlled series of changes in body structure development in insects; can be complete or incomplete. (Chap. 31, p. 777)

metaphase: second stage of mitosis in which chromosomes move to the equator of the spindle and chromatids are each attached by centromeres to a separate spindle fiber. (Chap. 11, p. 272)

microfilaments: thin, solid protein fibers present in the cytoskeleton of eukaryotic cells; play a role in cell structure and motion. (Chap. 8, p. 205)

micropyle: (MI kruh pile) in an anthophyte, the tiny opening in the ovule through which two sperm cells enter. (Chap. 27, p. 665)

microspores: male spores in plants that develop eventually into pollen grains. (Chap. 25, p. 613)

microtubules: hollow, thin, protein cylinders found in the cytoskeleton of eukaryotic cells; important in cell structure and locomotion. (Chap. 8, p. 205)

migration: yearly rhythm cycle of behavior resulting in the seasonal movement of animals; governed by internal and external biological cues. (Chap. 36, p. 906)

mimicry: structural adaptation evolved in some organisms resulting in the mimicing appearance to other organisms for protection or other advantages. (Chap. 18, p. 428)

mineral: inorganic substance essential to chemical reactions or as building materials in the body; examples include calcium and potassium. (Chap. 38, p. 971)

mitochondrion: (mi tuh KAHN dree un) eukaryotic membrane-bound organelle in which food molecules are broken down to produce ATPs; contains highly folded inner membrane that produces energy-storing molecules. (Chap. 8, p. 201)

mitosis: (mi TOH sus) cell division during which chromosomes are equally distributed to the two identical daughter cells that are formed; results in growth. (Chap. 11, p. 267)

mixture: combination of substances that do not combine chemically but retain their individual properties. (Chap. 7, p. 166)

molecule: group of atoms held together by covalent bonds. (Chap. 7, p. 165)

molting: in arthropods, periodic shedding of the protective exoskeleton, allowing for a size increase. (Chap. 31, p. 759)

monocotyledon: (mah nuh kah tuh LEE dun) plant with flower parts in multiples of three seeds with one cotyledon, and parallel leaf veins; grasses and orchids are examples. (Chap. 26, p. 631)

monosaccharide: (mah nuh SAK uh ride) a simple sugar such as glucose or fructose. (Chap. 7, p. 178)

monosomy: (MAH nuh soh mee) absence of a chromosome; most monosomic organisms do not survive. (Chap. 13, p. 328)

monotreme: (MAH nuh treem) egg-laying mammal; only three species survive today. (Chap. 35, p. 886)

motivation: need such as hunger or thirst that causes an animal to act. (Chap. 36, p. 911)

multicellular: organisms made up of many cells that are highly specialized to perform specific functions of metabolism. (Chap. 8, p. 207)

multiple alleles: presence of more than two alleles for a given genetic trait in a species. (Chap. 14, p. 336)

Glossary

mutation: random error or change in the DNA sequence that may affect whole chromosomes or just one gene. (Chap. 13, p. 324)

mutualism: (MYEW chuh lih zum) symbiotic relationship beneficial to both species; acacia trees and ants have a mutualistic relationship. (Chap. 3, p. 69)

mycelium: (mi SEE lee um) in fungi, complex mass of branching hyphae, some of which anchor, invade, or produce reproductive structures. (Chap. 23, p. 561)

mycorrhiza: (mi kuh RI zuh) symbiotic association between fungi and roots of vascular plants; hyphae supply roots with water and nutrients and fungi receive organic nutrients from the plant. (Chap. 23, p. 574)

myofibril: (mi yuh FI brul) small contractile muscle part composed of thick myosin protein filaments and thin actin protein filaments. (Chap. 37, p. 950)

myosin: (MI uh sun) a structural protein in muscle cells; together with the protein actin, functions in muscle contraction. (Chap. 37, p. 950)

N

narcotic: pain reliever that also causes sleep; narcotics are often derived from the opium poppy. (Chap. 40, p. 1026)

nastic movement: reversible movement of a plant in response to a stimulus; the folding of a *Mimosa's* leaves in response to being touched is an example. (Chap. 27, p. 676)

natural resources: renewable or nonrenewable parts of the natural environment such as soil, crops, and water that are used by humans. (Chap. 6, p. 134)

natural selection: a mechanism that explains how populations evolve; changes in populations occur when organisms with favorable variations for a particular environment survive, reproduce, and pass these variations on to the next generation; can be stabilizing, directional, or disruptive. (Chap. 18, p. 427)

Neanderthals: (nee AN dur tawlz) early *Homo sapiens;* powerfully built hominids; were skilled hunters who may have been the first to develop religious views and to have spoken language. (Chap. 19, p. 471)

negative-feedback system: in the endocrine system, the means of self-regulation by which the body maintains correct hormone levels. (Chap. 41, p. 1044)

nematocysts: (nuh MAT uh sihsts) in cnidarians, the tiny, harpoonlike structures, primarily found at the tips of tentacles, that are used to poison, entangle, or stab prey. (Chap. 29, p. 718)

nephridia: (ne FRIH dee uh) excretory structures, first evolved in mollusks, that remove metabolic wastes from an animal's body. (Chap. 30, p. 740)

nephron: (NE frawn) a filtering unit in the kidney; a human kidney has about 1 million nephrons. (Chap. 39, p. 998)

nerve net: conducts nerve impulses, resulting in contraction of musclelike cells; lacks a control center or brain; found, for example, in cnidarians. (Chap. 29, p. 718)

neuron: (NEW rahn) basic structural and functional unit in the nervous system; composed of dendrites, a cell body, and an axon. (Chap. 40, p. 1006)

neurotransmitters: chemicals that diffuse across the synapses between neurons, changing the polarity of the receiving dendrite, resulting in movement of nerve impulses from neuron to neuron. (Chap. 40, p. 1010)

niche: (NIHCH) role of a particular species in a community regarding food, space, reproduction, and how it interacts with abiotic factors. (Chap. 3, p. 62)

nitrogen base: component of DNA or RNA along with a sugar and a phosphate group; can be adenine, guanine, cytosine, thymine, or uracil. (Chap. 13, p. 310)

nitrogen fixation: metabolic process in which bacteria use enzymes to convert atmospheric nitrogen gas into ammonia. (Chap. 21, p. 524)

nocturnal: describes animals that are active primarily at night; for example, earthworms and owls. (Chap. 30, p. 749)

nonbiodegradable: types of wastes that are not easily broken down and can exist in the environment for many years; examples include radioactive residues and plastics. (Chap. 6, p. 148)

nondisjunction: failure of homologous chromosomes to separate during meiosis, resulting in gametes with too few or too many chromosomes. (Chap. 13, p. 327)

nonrenewable resources: resources available in limited amounts that cannot be replaced and cannot be recycled quickly by natural means. (Chap. 6, p. 135)

nonvascular plants: plants that lack vascular tissue to efficiently transport water and nutrients, limiting their size and distribution; bryophytes are nonvascular plants. (Chap. 24, p. 588)

notochord: (NOH tuh kord) rodlike structure from which the backbone in vertebrate chordates develops; in invertebrate chordates, the notochord enables side-to-side propulsion through water. (Chap. 32, p. 793)

nucleic acid: complex macromolecule, such as DNA and RNA, that stores information in cells in coded form. (Chap. 7, p. 181)

nucleolus: (new KLEE uh lus) region within the nucleus of eukaryotic cells that produces ribosomes. (Chap. 8, p. 198)

nucleotides: subunits of nucleic acid formed from a simple sugar, a nitrogen base, and a phosphate group. (Chap. 7, p. 181)

nucleus: positively charged center of an atom; contains neutrons, positively charged protons, and is surrounded by a negatively charged electron cloud. (Chap. 7, p. 161); in eukaryotes, the largest membrane-bound organelle; contains the cell's DNA and manages cell functions. (Chap. 8, p. 195)

nymph: (NIHMF) immature stage of many insect species; the hatchling looks like the adult form but smaller and goes through successive molts, eventually becoming a sexually mature adult with wings. (Chap. 31, p. 778)

Glossary

obligate aerobes: bacteria that cannot survive without oxygen for cellular respiration. (Chap. 21, p. 520)

obligate anaerobes: bacteria that cannot use oxygen and are killed by it. (Chap. 21, p. 520)

open circulatory system: system within an organism in which blood is not confined solely to vessels but also circulates in spaces around organs; found, for example, in gastropods. (Chap. 30, p. 740)

opposable thumb: in most primates, a thumb that can be brought opposite the forefinger, allowing the hand to be used for grasping, climbing, and using tools. (Chap. 19, p. 456)

order: taxonomic grouping of related families. (Chap. 20, p. 486)

organ: group of two or more tissues that perform an activity together; examples include plant leaf and mammalian heart. (Chap. 8, p. 210)

organ system: group of organs that work together to perform a major life function; examples include vascular system in plants and circulatory system in vertebrates. (Chap. 8, p. 210)

organelles: internal membrane-bound structures in a cell. (Chap. 8, p. 195)

organism: any unicellular or multicellular form exhibiting all the characteristics of life. (Chap. 1, p. 12)

organization: orderly structure shown by living things. (Chap. 1, p. 13)

osmosis: diffusion of water molecules through a selectively permeable membrane depending on the concentration of solutes on either side of the membrane. (Chap. 9, p. 226)

osteoblast: (AH stee uh blast) potential bone-forming cell; originates in the cartilage. (Chap. 37, p. 942)

ovary: the female reproductive organ; in flowers, the structure formed at the lower end of the pistil; eggs are formed in the ovary. (Chap. 27, p. 654)

oviduct: in females, the tube that transports eggs from ovary to uterus by means of peristaltic contractions and beating cilia; sometimes called a fallopian tube. (Chap. 41, p. 1044)

ovulation: in females, the process in which the follicle ruptures, releasing the egg from the ovary. (Chap. 41, p. 1046)

ovule: in seed plants, the structure in which the female gametophyte develops and contains a megaspore cell; forms a seed after fertilization. (Chap. 25, p. 614)

ozone layer: protective layer of ozone at the top of the stratosphere; absorbs most of the sun's harmful radiation; chlorofluorocarbons (CFCs) have caused the ozone layer to thin. (Chap. 6, p. 143)

pancreas: (PANG kree us) gland that produces both hormones and digestive enzymes; pancreatic juice is alkaline, which stops further digestive action of pepsin. (Chap. 38, p. 962)

parasitism: (PAYR uh sih tih zum) symbiotic relationship in which one species benefits at the expense of the other species; examples are ticks and tapeworms. (Chap. 3, p. 69)

parasympathetic nervous system: (payr uh sihm puh THET ihk) (PNS) portion of the autonomic nervous system that acts mainly during times of relaxation. (Chap. 40, p. 1018)

parathyroid glands: (payr uh THI royd) produce parathyroid hormone, which acts to decrease blood phosphate levels and increase calcium and magnesium levels. (Chap. 38, p. 968)

parenchyma: (puh RENG kuh muh) plant tissue found throughout most plants; can store food and water. (Chap. 26, p. 637)

parthenogenesis: (par thuh noh JEN uh sus) in some arthropods, a type of asexual reproduction in which a new individual is produced from an unfertilized egg; in plants, results in seedless fruit. (Chap. 31, p. 766)

particulates: solid particles of soot contained in smoke released by burning fossil fuels. (Chap. 6, p. 141)

passive transport: in cells, movement of particles across cell membranes requiring no expenditure of energy by the cell; examples are diffusion and osmosis. (Chap. 9, p. 230)

pathogens: (PATH uh junz) disease-causing agents such as bacteria, fungi, or viruses; can be transmitted by direct contact, by an object, through the air, or by a vector. (Chap. 42, p. 1070)

pedicellarias: (ped uh suh LAYR ee uz) modified spines of echinoderms that are adapted into jawlike appendages used for protection. (Chap. 32, p. 784)

pedigree: graphic representation showing patterns of inheritance in a family or breeding group; yields genetic information about a related group. (Chap. 14, p. 345)

pedipalps: (PED uh palps) in arachnids, appendages used to hold and transport food and also to serve as sense organs; in male spiders, pedipalps are adapted to carry sperm. (Chap. 31, p. 770)

pepsin: digestive enzyme secreted by the stomach that acts on proteins to produce peptides. (Chap. 38, p. 962)

peptide bond: covalent bond linking amino acids. (Chap. 7, p. 180)

perennial: anthophyte that lives for several years; examples include oaks, chrysanthemums, and numerous grass species. (Chap. 26, p. 634)

pericycle: in roots, the plant tissue from which lateral roots arise. (Chap. 26, p. 637)

Glossary

peripheral nervous system: (PNS) transmits impulses to and from the body and the central nervous system; composed of the somatic and autonomic nervous systems. (Chap. 40, p. 1010)

peristalsis: (payr uh STAWL sus) waves of smooth muscle contraction in the digestive system that move food along the digestive tract. (Chap. 38, p. 961)

permafrost: layer of permanently frozen ground found under the topmost layer of soil in the tundra. (Chap. 4, p. 100)

petals: leaflike flower organs; usually brightly colored; can attract pollinators with perfume, nectar, or color patterns. (Chap. 27, p. 654)

petiole: (PET ee ohl) leaf part that joins the leaf blade to the stem; contains vascular tissue. (Chap. 26, p. 642)

pH: symbol to describe how acidic or basic a solution is; the pH scale ranges from 0 to 14; a solution with a pH of 7 is neutral, above 7 is basic, and below 7 is acidic. (Chap. 7, p. 172)

phagocyte: (FAJ uh site) white blood cell that migrates from the capillaries to the pathogen, then engulfs and destroys it. (Chap. 42, p. 1081)

pharynx: (FAYR ingks) in planarians, the tubelike, enzyme-releasing organ that can extend out of the mouth and suck food into the gastrovascular cavity; in humans, the area at the back of the mouth. (Chap. 29, p. 725)

phenotype: (FEE nuh tipe) outward appearance of an organism, regardless of its genes. (Chap. 12, p. 291)

pheromone: (FAYR uh mohn) chemical odor given off by animals that cues a specific behavior. (Chap. 31, p. 762)

phloem: (FLOH em) in vascular plants, the tissue composed of living tubular cells joined end to end in a series; conducts sugars from the leaves to all plant parts. (Chap. 25, p. 608)

phospholipid: (faws foh LIH pud) membrane lipid having an organic section attached to a phosphate group; plasma membranes are formed of a bilayer of phospholipids with embedded proteins. (Chap. 9, p. 220)

photic zone: portion of the marine biome shallow enough for the sun to penetrate; has abundant life and high productivity; includes coastal areas formed by rocky shores, sandy beaches, and mud flats. (Chap. 4, p. 91)

photolysis: (foh TAW luh sus) splitting of water molecules during photosynthesis. (Chap. 10, p. 245)

photoperiodism: a plant's response to the difference in day and night length. (Chap. 27, p. 663)

photosynthesis: (foh toh SIHN thuh sus) process by which autotrophs produce simple sugars from water and carbon dioxide using energy absorbed from sunlight by chlorophyll; oxygen is one of the end products. (Chap. 10, p. 242)

phylogeny: (fi LAW juh nee) evolutionary history of a species based on comparative relationships of structures, and comparisons of modern life forms with fossils. (Chap. 20, p. 487)

phylum: (FI lum) taxonomic grouping of related classes. (Chap. 20, p. 486)

pistil: in flowers, the female reproductive structure attached to the top of the flower stem; its lower portion forms the ovary. (Chap. 27, p. 654)

pituitary: (puh TEW uh tayr ee) gland situated beneath and stimulated by the hypothalamus; releases FSH and LH, which serve such functions as stimulating production of sperm and testosterone in males, and estrogen release and regular development of follicles in the ovary in females. (Chap. 41, p. 1042)

placenta: organ developed during pregnancy through which the young of placental mammals are supplied with nourishment; also involved in passing oxygen and removing wastes. (Chap. 35, p. 885)

placental mammal: mammal that carries its young inside the uterus until ready for birth. (Chap. 35, p. 885)

plankton: microscopic organisms found floating in the photic zone; form the base of all marine food chains. (Chap. 4, p. 93)

plasma: straw-colored, fluid portion of blood containing, for example, blood cells, platelets, hormones, and nutrients. (Chap. 39, p. 987)

plasma membrane: serves as the boundary between the cell and its external environment and allows materials such as oxygen and nutrients to enter and waste products to leave. (Chap. 8, p. 196)

plasmid: (PLAZ mud) in bacterial cells, the small ring of DNA that can serve as a vector; usually carries nonessential, "accessory" genes. (Chap. 16, p. 378)

plasmodium: (plaz MOH dee um) multinucleate cytoplasmic mass lacking cell walls or membranes that is the moving and feeding stage of a plasmodial slime mold. (Chap. 22, p. 552)

plasmolysis: (plaz MAH luh sus) process resulting from a drop in turgor pressure as a result of water loss in a cell, causing the plasma membrane to shrink away from the cell wall. (Chap. 9, p. 229)

plastids: (PLAS tudz) plant organelles, some of which contain pigment molecules giving fruits and flowers their color; other plastids store starch and lipids. (Chap. 8, p. 204)

plate tectonics: geological explanation for the movement of continents over Earth's thick, liquid interior. (Chap. 17, p. 408)

platelet: short-lived cell fragment contained in plasma; needed for blood clotting. (Chap. 39, p. 989)

point mutation: error in the DNA sequence that affects only a single base pair; can interfere with protein function. (Chap. 13, p. 324)

polar molecule: a molecule with a positive end and a negative end, resulting in an unequal distribution of charge. (Chap. 7, p. 168)

pollen grain: in seed plants, the structure in which the male gametophyte develops and contains nutrients for the sperm cells. (Chap. 25, p. 614)

pollination: in a flower, the process of transfer of pollen grains from the anther to the stigma. (Chap. 12, p. 286)

Glossary

pollution: contamination of air, water, or land by wastes produced in such excess that they cannot be recycled by natural processes. (Chap. 6, p. 140)

polygenic inheritance: (pah lee JEN ihk) determination of a given trait, such as skin color or height, produced by the interaction of many genes. (Chap. 14, p. 338)

polymer: large molecule formed by bonding many smaller molecules together, most often in long chains; spider silk is a biological polymer. (Chap. 7, p. 177)

polyp: (PAH lup) in the life cycle of cnidarians, the sessile, tube-shaped stage. (Chap. 29, p. 716)

polyploid: (PAH lee ployd) form of a species that can be formed from mistakes during meiosis, resulting in diploid ($2n$) gametes rather than haploid (n) gametes; many flowering plants and important crops originated by polyploidy. (Chap. 18, p. 444)

polysaccharides: (pah lee SAK uh ridez) largest carbohydrate molecules, these polymers are composed of numbers of monosaccharide subunits; examples are cellulose and glycogen. (Chap. 7, p. 178)

population: interbreeding individuals of one species that compete with one another for food, water, and mates and live in the same place at the same time. (Chap. 3, p. 58)

posterior: the tail end of bilaterally symmetrical animals. (Chap. 28, p. 703)

prehensile tail: (pree HEN sul) muscular tail that functions as a fifth limb; can be used to grasp or wrap around an object; New World monkeys have a prehensile tail. (Chap. 19, p. 458)

preservation: keeping an organism or an area from harm or destruction by the establishment of parks, wildlife habitats, and other refuges. (Chap. 6, p. 149)

primary succession: development of living communities from bare rock, where pioneer organisms such as lichens are often followed by small plants, shrubs, and then trees. (Chap. 4, p. 87)

primate: group of mammals including monkeys and humans; evolved from a common ancestor; share such characteristics as fingernails, flexible shoulder joints, flattened face, an opposable thumb or big toe, and large, complex brain. (Chap. 19, p. 454)

proglottid: (proh GLAH tud) parasitically adapted, detachable section of a tapeworm; contains male and female reproductive organs, flame cells, muscles, and nerves. (Chap. 29, p. 728)

prokaryote: (proh KAYR ee oht) cell lacking a true nucleus or membrane-bound internal organelles. (Chap. 8, p. 195)

prophase: first phase of mitosis during which chromatin coils to form visible chromosomes. (Chap. 11, p. 268)

prostate gland: doughnut-shaped gland of males that surrounds the top of the urethra and secretes an alkaline fluid that transports sperm. (Chap. 41, p. 1042)

protein: large, complex polymer essential to all life composed of amino acids made of carbon, hydrogen, oxygen, nitrogen, and sometimes sulfur; important in muscle contraction, transporting oxygen in the bloodstream, and providing immunity. (Chap. 7, p. 180)

prothallus: (proh THAL us) the gametophyte generation in the life cycle of spore-producing, vascular plants; produces either antheridia or archegonia. (Chap. 25, p. 607)

protist: eukaryotic plantlike, animal-like, or funguslike organism lacking complex organ systems and living in a moist environment; examples include protozoans, slime molds, and seaweeds. (Chap. 20, p. 494)

protocells: ordered structures formed by heating amino acid solutions, producing clusters resembling living cells that carry out some, but not all, of the functions of life. (Chap. 17, p. 414)

protonema: (proh tuh NEE muh) in mosses, filamentous structure formed by germination of the haploid spore; develops into either a male or female gametophyte or into a gametophyte containing both male and female structures. (Chap. 24, p. 596)

protostome: (PROH tuh stohm) animal in which the mouth develops from the opening in the gastrula; animals such as insects and earthworms show protostome development. (Chap. 28, p. 699)

protozoans: (proh tuh ZOH unz) animal-like protists with a variety of complex organelles; grouped into phyla according to their method of locomotion. (Chap. 22, p. 535)

provirus: virus integrated into a host cell's chromosome; can remain dormant or become activated at any time and enter a lytic cycle. (Chap. 21, p. 508)

pseudocoelom: (sew duh SEE lum) body cavity partly lined with mesoderm and filled with fluid; enables animals such as roundworms to move more efficiently. (Chap. 28, p. 704)

pseudopodia: (sew duh POH dee uh) temporary cytoplasmic extensions used by amoebas and amoebalike organisms for movement and for engulfment of food. (Chap. 22, p. 535)

puberty: in both males and females, the period characterized by development of secondary sex characteristics as a result of production of FSH, LH, and other sex hormones. (Chap. 41, p. 1042)

pulse: rhythmic surge of blood through an artery. (Chap. 39, p. 995)

punctuated equilibrium: idea that periods of speciation occur rapidly with long periods of no speciation in between. (Chap. 18, p. 445)

pupa: in insects, the stage of metamorphosis from which the larva emerges as a fully mature adult. (Chap. 31, p. 777)

pus: thick, fluid substance formed from dead white blood cells, dead microorganisms, and body fluids in an infected area. (Chap. 42, p. 1081)

Glossary

R

radial symmetry: body plan exhibited by animals, such as adult starfish, that can be divided along any plane, through a central axis, into roughly equal halves; enables slow-moving or stationary animals to sense potential food and predators from all directions. (Chap. 28, p. 702)

radula: (RAJ uh luh) in animals such as gastropods, the rasping, tonguelike organ used for cutting and grating food. (Chap. 30, p. 740)

rays: long, tapering arms of some echinoderms such as starfishes and the sea lily; rays are covered with spines or plates composed primarily of calcium carbonate protected with a thin epidermal layer. (Chap. 32, p. 784)

recessive: hidden trait of an organism that is masked by a dominant trait. (Chap. 12, p. 289)

recombinant DNA: produced when a cleaved DNA fragment is incorporated into the DNA of a plasmid or virus. (Chap. 16, p. 376)

rectum: final segment of the digestive system; passes feces out of the body through the anus. (Chap. 38, p. 964)

red blood cell: hemoglobin-containing cell in humans that transports oxygen and some CO_2; loses its nuclei as it enters the bloodstream, resulting in a limited life span. (Chap. 39, p. 987)

reflex: rapid, automatic, unconscious response to a stimulus. (Chap. 40, p. 1016)

regeneration: (rih jen uh RAY shun) ability to replace or regrow a missing body part; for example, sea cucumbers can eviscerate and later replace the lost vital organs. (Chap. 32, p. 790)

renewable resources: resources replaced or recycled by natural processes; for example, oxygen is replenished during photosynthesis. (Chap. 6, p. 134)

replication: process in which the two strands of the double helix separate and bases pair with free nucleotides to form two molecules of DNA, each identical to the original molecule. (Chap. 13, p. 313)

reproduction: production of offspring. (Chap. 1, p. 13)

reproductive isolation: occurs when organisms that formerly interbred are prevented from producing offspring—for example, by developing different mating times. (Chap. 18, p. 444)

respiration: (res puh RAY shun) process in which cells break down molecules of food to release energy; cellular respiration can be either aerobic or anaerobic. (Chap. 10, p. 251)

response: reaction to an internal or external stimulus. (Chap. 1, p. 15)

restriction enzymes: bacterial proteins that cleave DNA at specific points in the nucleotide sequence. (Chap. 16, p. 377)

retina: layer of the eye containing rods and cones; light entering the cornea is focused by the lens on the back of the eye, where it hits the retina. (Chap. 40, p. 1020)

retroviruses: viruses containing a unique enzyme, reverse transcriptase, which transcribes viral RNA into DNA, enabling the viral DNA to enter the host cell's chromosome. (Chap. 21, p. 510)

reverse transcriptase: (rih VURS • tran SKRIHP tays) enzyme that transcribes viral RNA into viral DNA. (Chap. 21, p. 510)

rhizoids: (RI zoydz) fungal structures formed by hyphae that anchor the mycelium to its food source or substrate; site of most extracellular digestion and absorption of nutrients. (Chap. 23, p. 566); hairlike extensions of a moss or liverwort gametophyte that anchor the plant to its substrate. (Chap. 24, p. 587)

rhizome: (RI zohm) underground, horizontal stem of vascular plants such as ferns; may function as a food-storage organ. (Chap. 25, p. 610)

ribosomal RNA (rRNA): (ri buh SOH mul) the RNA that composes ribosomes. (Chap. 13, p. 319)

ribosomes: eukaryotic organelles involved in protein synthesis. (Chap. 8, p. 198)

RNA: ribonucleic acid; forms a copy of DNA for use in protein synthesis. (Chap. 7, p. 184)

rods: type of light receptor in the retina; responsible for vision in low illumination. (Chap. 40, p. 1020)

root: plant organ normally found below ground that anchors the plant, absorbs water and minerals, and transports water and nutrients to the stem. (Chap. 24, p. 587)

root cap: layer of tough parenchyma cells that protects the root tip as its grows down into soil. (Chap. 26, p. 638)

root hair: in plant roots, a single-celled extension of the epidermis that absorbs water and dissolved minerals. (Chap. 26, p. 637)

S

safety symbol: warns against specific hazards such as radiation and high-voltage electricity (see Appendix C for safety symbols used in this textbook). (Chap. 2, p. 34)

saprobes: (SAP rohbz) heterotrophic eubacteria that feed on nonliving organic matter or wastes and help recycle nutrients. (Chap. 21, p. 512)

sarcomere: (SAR koh meer) each section of a myofibril in a striated muscle. (Chap. 37, p. 950)

scales: skin covering of reptiles and fishes; can be diamond-shaped, cone-shaped, tooth-shaped, or round in fishes; agnathans lack scales. (Chap. 33, p. 817)

scavenger: animal such as a vulture that plays a positive role in the ecosystem by consuming dead organisms and their refuse. (Chap. 3, p. 65)

scientific methods: common procedures used by biologists and other scientists to gather information used in problem solving and experimentation. (Chap. 2, p. 26)

scolex: (SKOH leks) parasitically adapted, knob-shaped head of a tapeworm; covered with suckers and hooks that embed in the host's intestinal lining. (Chap. 29, p. 728)

Glossary

scrotum: testes-containing sac of males; located externally; maintains sperm at a lower temperature than body temperature by means of muscle contraction and relaxation. (Chap. 41, p. 1040)

seed: adaptive reproductive structure of land plants; a protective coat prevents drying out of the embryo; also contains a food supply. (Chap. 24, p. 589)

selective permeability: property of a plasma membrane that maintains the cell's homeostasis by the taking in of needed substances, the elimination of wastes, and the prevention of harmful substances from entering. (Chap. 9, p. 217)

semen: combination of sperm and sperm-carrying fluids of the male reproductive system. (Chap. 41, p. 1042)

semicircular canals: fluid-filled structures in the ear involved in maintaining the body's balance. (Chap. 40, p. 1021)

seminal vesicles: in males, the paired glands at the base of the urinary bladder; produce a fructose-containing fluid that nourishes sperm. (Chap. 41, p. 1042)

sepals: leaflike structures at the base of a flower; protect the flower while in bud. (Chap. 27, p. 654)

sessile: (SES sile) organism that stays attached permanently to a surface during its adult life; sponges and corals are examples. (Chap. 28, p. 694)

sex chromosomes: in humans, the 23rd pair of chromosomes, which controls the inheritance of sex characteristics and differs in males and females. (Chap. 14, p. 336)

sex-linked trait: inherited trait, such as color blindness, controlled by genes located on the sex chromosomes. (Chap. 14, p. 337)

sexual reproduction: reproductive pattern in which haploid gametes fuse to produce a diploid zygote, which then develops by mitosis into a new organism. (Chap. 12, p. 300)

short-day plant: flowers in late summer, when days are becoming shorter; examples are primroses and chrysanthemums. (Chap. 27, p. 663)

sieve cell: (SIHV • sel) in vascular plants, the cytoplasmic, nonnucleated cell of the phloem involved in transporting sugars throughout the plant. (Chap. 26, p. 640)

simple eyes: in arthropods, the single-lens structures used to focus images. (Chap. 31, p. 762)

sink: in vascular plants, storage area for excess sugars produced by a plant during photosynthesis. (Chap. 26, p. 640)

sister chromatids: identical halves of the duplicated parent chromosome formed before the onset of cell division; these exact copies are joined at a centromere. (Chap. 11, p. 268)

skeletal muscle: muscle attached to bones; functions under voluntary control to move the skeleton; composed of striated, multinucleated cells. (Chap. 37, p. 947)

sliding filament theory: theory that actin filaments slide toward each other during muscle contraction, whereas the myosin filaments remain still. (Chap. 37, p. 950)

small intestine: narrow, muscular tube in which digestion is completed; connects the stomach to the large intestine. (Chap. 38, p. 962)

smog: type of urban air pollution resulting from a combination of chemical pollutants, particulate matter, and sulfur dioxide. (Chap. 6, p. 141)

smooth muscle: muscle composed of sheets of spindle-shaped cells; lines internal organs and blood vessels; produces involuntarily controlled contractions. (Chap. 37, p. 946)

solution: mixture in which a substance (solute) is dissolved easily in another substance (solvent). (Chap. 7, p. 167)

somatic nervous system: (soh MAT ihk) portion of the peripheral nervous system that transmits messages between the skin, the central nervous system, and skeletal muscles. (Chap. 40, p. 1016)

sorus: (SOR us) in ferns, the structure formed on the surface of fronds by clusters of sporangia. (Chap. 25, p. 612)

spawning: act of breeding in fishes and some other animals; results in release of large numbers of eggs to the environment. (Chap. 33, p. 819)

speciation: process by which a new species is formed; occurs when individuals of a population are unable to interbreed or produce fertile offspring. (Chap. 18, p. 441)

species: population of interbreeding organisms capable of producing fertile offspring. (Chap. 1, p. 14)

sperm: male sex cell or gamete. (Chap. 12, p. 300)

spindle: thin fiber structure that forms between the two poles or centrioles during prophase and shortens during anaphase, pulling the sister chromatids apart. (Chap. 11, p. 269)

spinnerets: (spih nuh RETS) in spiders, structures that spin silk into thread. (Chap. 31, p. 770)

spiracles: (SPEER uh kulz) openings on the abdomen and thorax of most insects through which air enters and exits the tracheal tubes. (Chap. 31, p. 761)

spongy bone: bone tissue containing numerous holes and spaces; is less dense than the compact bone surrounding it. (Chap. 37, p. 943)

spontaneous generation: idea that living organisms can arise from nonliving matter. (Chap. 17, p. 410)

sporangium: (spuh RAN jee um) in zygomycotes, the saclike structure in which asexual spores are formed at the tips of some hyphae. (Chap. 23, p. 564)

spore: in sporozoans, the haploid (*n*) reproductive cell with a hard outer wall that develops into a new organism without the fusion of gametes. (Chap. 22, p. 540)

sporophyte: (SPOR uh fite) in algae and plants, the diploid stage in an alternation of generations; develops from the zygote. (Chap. 22, p. 550)

sporozoans: (spor uh ZOH unz) parasitic, nonmotile protozoans, many of which reproduce by the production of spores. (Chap. 22, p. 540)

stabilizing selection: selection favoring average individuals, resulting in the decline of variations in a population. (Chap. 18, p. 439)

Glossary

stamen: in flowers, the male reproductive structure consisting of an anther and attaching filament. (Chap. 27, p. 654)

starch: the polysaccharide consisting of highly branched chains of glucose units used as food storage in plants. (Chap. 7, p. 178)

stem: plant structure that provides support for leaves and reproductive structures; vascular plant stems contain tissues for transport of nutrients and water. (Chap. 24, p. 588)

sternum: in birds, the large breastbone to which flight muscles are attached; supports the muscular thrust of wings as they produce the power for flight. (Chap. 34, p. 856)

stimulant: drug that increases the activity of the central nervous system and sympathetic nervous system; nicotine is a stimulant. (Chap. 40, p. 1026)

stimulus: any adjustment-requiring condition of an organism's environment. (Chap. 1, p. 15)

stolons: (STOH lunz) hyphae that grow horizontally across the surface of a food source, such as bread, and produce rhizoids that grow down and reproductive hyphae that grow up. (Chap. 23, p. 566)

stomach: pouchlike, muscular digestive organ that secretes acids and enzymes; food leaves the stomach in a thin, acidic liquid. (Chap. 38, p. 961)

stomata: (stoh MAT uh) openings in a leaf's epidermis that release water and oxygen to the air and take in oxygen and carbon dioxide for respiration and photosynthesis. (Chap. 24, p. 587)

strobilus: (STROH bih lus) in lycophytes, the conelike, spore-producing leaf cluster at the tip of the stem. (Chap. 25, p. 607)

succession: natural, orderly process in the community of an ecosystem characterized by population growth or reduction. (Chap. 4, p. 86)

swim bladder: gas-filled, internal sac of bony fishes that regulates buoyancy. (Chap. 33, p. 819)

symbiosis: (sihm bee OH sus) permanent, close association between two or more organisms of different species. (Chap. 3, p. 68)

symmetry: (SIH muh tree) balance in body proportions of animals. (Chap. 28, p. 701)

sympathetic nervous system: (sihm puh THET ihk) portion of the autonomic nervous system that acts mainly during times of stress. (Chap. 40, p. 1017)

synapse: (SIH naps) space between neurons across which impulses are chemically transmitted from axons to dendrites. (Chap. 40, p. 1010)

T cells: specialized lymphocytes produced in the bone marrow and processed in the thymus gland; have antibody-like antigen receptors on their surfaces that allow them to react to many types of antigens. (Chap. 42, p. 1084)

taiga: (TI guh) biome located just south of the tundra; characterized by acidic, mineral-poor topsoil, long, harsh winters, and short, mild summers; abundant fir and spruce trees provide food and shelter for such animals as moose and lynx. (Chap. 4, p. 101)

target tissue: cells specifically affected by endocrine hormones, which act to convey information. (Chap. 38, p. 966)

taste bud: sensory receptor of the tongue that sends messages to the cerebrum, resulting in experiencing a specific taste. (Chap. 40, p. 1019)

taxonomy: branch of biology dealing with grouping and naming organisms based on their similarities, chemical makeup, and evolutionary relationships. (Chap. 20, p. 482)

technology: scientific research to solve society's needs and problems; can have both beneficial effects and problematic side effects. (Chap. 2, p. 45)

telophase: (TEL uh fayz) final stage of mitosis during which the two daughter cells become separated. (Chap. 11, p. 272)

temperate forests: biome in which an even amount of precipitation falls in all four seasons, averaging from 70 to 150 cm annually; temperate forest type is determined by the dominant tree species, which typically includes oak, beech, maple, birch, and hickory; animal life includes squirrels, rabbits, and bears. (Chap. 4, p. 104)

tendon: cord of connective tissue that attaches muscles to bones. (Chap. 37, p. 940)

territory: physical area containing an animal's breeding and feeding areas that is actively defended against members of the same species; territoriality has survival value for the individual and the species. (Chap. 36, p. 904)

testcross: breeding technique used to determine whether an individual is homozygous dominant or heterozygous for a particular trait; often employed by plant and animal breeders to test for such traits as blindness and disease vulnerability. (Chap. 14, p. 344)

thallus: (THAL us) undifferentiated plant or algal body that has no roots, stems, or leaves. (Chap. 22, p. 548)

theory: results when a hypothesis is repeatedly verified over time and through many separate experiments; valid theories enable scientists to predict new facts and relationships of natural phenomena. (Chap. 2, p. 40)

therapsids: (thuh RAP sudz) mammal-like reptilian ancestors of all mammals. (Chap. 35, p. 883)

threatened species: species that have rapidly decreasing numbers of individuals; examples include African elephants and grizzly bears. (Chap. 6, p. 137)

thyroid gland: produces thyroxine, a hormone that regulates metabolism and growth, and calcitonin, a hormone that helps regulate calcium and phosphate levels. (Chap. 38, p. 967)

tissue: group of cells that function together to carry out an activity; examples include leaf and nerve tissue. (Chap. 8, p. 207)

tissue fluid: fluid that bathes the cells of the body. (Chap. 42, p. 1078)

tolerance: state resulting when a drug increasingly loses its effect, necessitating larger doses for the same effect. (Chap. 40, p. 1030)

toxin: a poison; for example, the obligate anaerobe *Clostridium botulinum* forms endospores when exposed to oxygen; when the endospores germinate, the resultant bacteria produce the powerful toxin that is the causative agent of botulism. (Chap. 21, p. 521)

trachea: (TRAY kee uh) the windpipe; lined with constantly beating cilia that prevent foreign particles from reaching the lungs. (Chap. 39, p. 982)

tracheal tubes: (TRAY kee ul) in most insects, the internal, air-carrying passages. (Chap. 31, p. 761)

tracheid: (TRAY kee ud) water transport cell in plants, forms xylem tissue in ferns and gymnosperms. (Chap. 25, p. 620)

trait: inherited characteristic; in simple Mendelian inheritance, can be either dominant or recessive. (Chap. 12, p. 286)

transcription: the process by which enzymes make an RNA copy of a DNA strand. The process is similar to DNA replication, but a single-stranded molecule of transfer RNA is produced. (Chap. 13, p. 319)

transfer RNA (tRNA): delivers amino acids to the ribosome for protein synthesis. (Chap. 13, p. 322)

transgenic organism: genetically engineered organism containing recombinant DNA that gives the organism the ability to create products foreign to itself; for example, some transgenic plants can produce internal insecticides. (Chap. 16, p. 376)

translation: process in which the order of bases in mRNA codes for the order of amino acids in a protein. (Chap. 13, p. 322)

transpiration: in plants, the evaporation of water from the stomata of leaves. (Chap. 26, p. 643)

transport proteins: channel proteins and carrier proteins embedded in the lipid bilayer of the plasma membrane; move ions and molecules across the membrane. (Chap. 9, p. 230)

trial-and-error learning: occurs when an animal makes various attempts before being able to successfully complete a task. (Chap. 36, p. 911)

trisomy: (TRI soh mee) presence of an extra chromosome; trisomic organisms often survive into maturity. (Chap. 13, p. 327)

trophic level: (TROH fihk) link represented by each organism in a food chain; represents a feeding step in the transfer of energy and matter in an ecosystem. (Chap. 3, p. 71)

tropical rain forest: most biologically diverse terrestrial biome; receives from 200 to 400 cm of annual rainfall and has year-round warm temperatures; plants include ferns, orchids, and trees; animal life includes parrots, monkeys, snakes, and tarantulas. (Chap. 4, p. 105)

tropism: (TROH pih zum) irreversible, responsive movement in plants toward or away from such external stimuli as light (phototropism) or gravity (gravitropism). (Chap. 27, p. 675)

tube feet: in echinoderms, hydraulic, suction cup-tipped appendages that function in locomotion, gas exchange, excretion, and capture of food. (Chap. 32, p. 785)

tundra: (TUN druh) treeless biome south of the ice cap of the north pole; has a short growing season; temperatures never rise above freezing, and its permafrost soil layer never thaws; supports grasses, small annuals, and reindeer moss; animal life includes lemmings, arctic foxes, mosquitoes, and reindeer. (Chap. 4, p. 100)

turgor pressure: (TUR gur) internal pressure of a cell due to water held there by osmotic pressure. (Chap. 9, p. 227)

U

umbilical cord: in placental mammals, the cordlike structure that attaches the developing embryo to the uterine wall; the embryo receives nutrients and oxygen and eliminates wastes through the blood of the umbilical cord. (Chap. 41, p. 1052)

unicellular: (yew nuh SEL yuh lur) organism that carries out all its life processes within its single cell. (Chap. 8, p. 207)

ureter: (YUR uh tur) tube that transports urine from each kidney to the urinary bladder. (Chap. 39, p. 997)

urethra: (yew REE thruh) tube through which urine is eliminated from the body. (Chap. 39, p. 999)

urinary bladder: smooth muscle bag in which urine is stored until it leaves the body through the urethra. (Chap. 39, p. 997)

urine: (YUR un) liquid composed of excess H_2O, ions, and waste molecules that is filtered from the blood by the kidneys, stored in the urinary bladder, and eliminated through the urethra. (Chap. 39, p. 999)

uterus: (YEW tuh rus) in female placental mammals, the muscular, hollow organ in which young are developed and protected. (Chap. 35, p. 884)

V

vaccine: (vak SEEN) immunity-producing substance formed from weakened, dead, or parts of pathogens or antigens. (Chap. 42, p. 1090)

vacuole: (VAK yew ohl) membrane-bound, fluid-filled space within the cytoplasm; temporarily stores food, enzymes, and wastes. (Chap. 8, p. 200)

vagina: (vuh JI nuh) in females of placental mammals, the passageway that leads from the uterus to the outside of the body. (Chap. 41, p. 1045)

Glossary

vas deferens: (vas • DEF uh runtz) in males, the duct through which sperm move, by peristaltic contractions, from the epididymis toward the urethra, which releases the sperm from the body. (Chap. 41, p. 1041)

vascular plants: plants that contain vascular tissues specially adapted to transport water and dissolved materials, enabling taller growth and survival in land habitats. (Chap. 24, p. 588)

vascular tissues: in vascular plants, the tissues made up of tubular cells that transport water and dissolved nutrients from one part of a plant to another; the two types are xylem and phloem. (Chap. 25, p. 608)

vector: can be biological or mechanical; viruses and plasmids are biological vectors with which DNA fragments are recombined before introduction into the host cell. (Chap. 16, p. 377)

vegetative reproduction: asexual reproduction in plants; examples are bulbs, tubers, gemmae, and rhizomes. (Chap. 24, p. 600)

veins: large blood vessels that return blood from the tissues back to the heart; contain one-way valves to prevent backflow of blood. (Chap. 39, p. 992)

vena cava: (VEE nuh • KAY vuh) [cavae-plural] two large veins that empty oxygen-poor blood from the body and head into the right atrium of the mammalian heart. (Chap. 39, p. 993)

ventral: belly surface of bilaterally symmetrical animals. (Chap. 28, p. 703)

ventricles: (VEN trih kulz) thick-walled, muscular lower chambers of the mammalian heart that receive blood from the atria and send it toward the lungs and arteries. (Chap. 39, p. 993)

vertebrate: (VUR tuh brut) animal with a backbone; examples include mammals, birds, and reptiles. (Chap. 28, p. 708)

vessel cell: open-ended, tubular cell that makes up vessels; in angiosperms, vessels conduct water, which diffuses into and out of the xylem. (Chap. 26, p. 640)

vestigial structure: (ve STIH jee ul) body structure with reduced function that may have been useful to an earlier stage of the species; provides evidence of evolution. (Chap. 18, p. 433)

villus: (VIH lus) fingerlike projection on the lining of the small intestine that increases the surface area for absorption of digested food. (Chap. 38, p. 963)

virus: (VI rus) disease-causing, nonliving particle composed of an inner core of nucleic acid enclosed by one or two protein coats; reproduces only in living cells. (Chap. 21, p. 504)

vitamin: (VI tuh mun) organic substance that regulates processes in the body; examples include vitamin E and pantothenic acid. (Chap. 38, p. 972)

vocal cords: in frogs and mammals, the bands of tissue in the throat that vibrate to produce a wide variety of sounds such as mating calls. (Chap. 33, p. 830)

voluntary muscle: muscle whose contractions are under conscious control; skeletal muscle is voluntary muscle. (Chap. 37, p. 947)

water vascular system: in echinoderms, provides water pressure to operate the tube feet; regulates locomotion, excretion, gas exchange, and capture of food. (Chap. 32, p. 786)

white blood cell: large, nucleated, infection-fighting cell in mammalian blood. (Chap. 39, p. 988)

withdrawal: psychological or physiological illness resulting from cessation of drug use. (Chap. 40, p. 1030)

xylem: (ZI lum) in vascular plants, the tissue composed of dead tubular vessels or tracheids laid end to end; conducts water and dissolved minerals upward from the plant roots to the leaves. (Chap. 25, p. 608)

zygospore: (ZI goh spor) in zygomycotes, the thick-walled, sexually produced, resting spore adapted to withstand poor environmental conditions. (Chap. 23, p. 567)

zygote: (ZI goht) fertilized egg; has a diploid ($2n$) number of chromosomes; develops into a multicellular organism by mitosis. (Chap. 12, p. 300)

Index

Index

Index

Index

Index

Index

Index

Index

Index

Index

Index

Index

Index

Index

Index

S

Index

Index

Art & Photo

Art Credits

David M. Dennis, 553, 576, 590, 722, 745, 767, 789, 846, 864-865, 882
John Edwards, 34(t), 59, 161, 832, 976, 1037
Gary Ferster, 115, 118, 181, 440, 979, 1076
Howard Fullmer/Cary & Company, 41(t)
Tom Gagliano, 85, 427, 815, 816, 817, 819, 820, 823, 831, 842, 845, SH3(rb)
Glencoe Publishing Co., 34(b), A9
Nancy Heim/158 Street Design Group, 197, 198(b), 199, 200, 201, 207, 208, 213(l), 227, 228, 229, 248, 256, 270-271, 299, 301, 302, 326, 331(l), 336, 379, 387, 444, 510, 961, 962, 964, 988, 990, 991(b), 992, 999, 1046, 1080
Tom Kennedy/Romark Illustrations, 81(r), 97, 98-99, 111(r), 125, 127, 145, 162, 165, 166, 167, 168, 169, 176, 177, 178, 179, 180, 184, 187(r), 220, 222, 230, 231, 235(r), 239, 240, 244(b), 245, 250, 252, 253, 254, 255(b), 259(r), 264(t), 267, 268, 303, 307(r), 310, 311, 314-315, 318, 319, 320, 322, 323, 325, 327, 331(r), 337(t), 345(b), 347, 350, 351, 357, 358, 359, 362, 377, 378, 381, 390, 392, 393(l), 408, 409, 414, 415, 434, 497(b), 721, 735(b), 848, 885, 909, 969, 970, 973, 996, 1008(b), 1009, 1018, 1044, 1047(b), 1048, 1053, 1067(r), 1086, 1095(r), 1119, SH10, SH11, SH13, SH14(r)
Greg LaFever/Scott Hull Associates, 172-173, 226, 243, 255(t), 277, 518, 519
Michael Maydak/Morgan-Cain & Associates, 39, 72, 73, 74, 75, 76, 77(b), 107, 146, 418
MicroPublishing, 22, 40, 41(b), 49, 81(l), 111(l), 131, 155, 187(l), 192-193(t), 235(l), 244(t), 259(l), 278, 339, 353, 365, 373, 393(r), 420(r), 451, 485, 501, 557, 603, 627, 651, 681, 755, 759, 781(r), 835, 869, 874(t), 895(b), 903(r), 951, 955, 966(t), 979, 1003, 1043(r), 1067(l), 1077, 1095(l)
Susan Moore/Lisa Freeman Inc., 198(c), 288, 289, 290, 291, 294, 296, 297, 335, 426, 436, 495, 550, 552, 561(t), 567, 570, 597, 607, 611, 619, 621, 641, 644, 647, 656, 666, 667, 675
Laurie O'Keefe, 20, 28-29, 287, 337(b), 338, 345(t), 410, 428, 429, 431, 441, 445, 446, 447, 485, 486, 492-493, 497(t), 505, 661, 702, 761, 763(t), 769, 776, 777, 778, 781(l), 822, 855, 857, 862, 863, 876, 881, 895(t), 903(l), 906, 914, 919, SH4(t)
Felipe Passalacqua, 70-71, 217, 232, 241, 263, 516, 535, 536, 539, 540, 546, 561(b), 563, 637, 643, 651, 676, 694, 695, 703, 704-705, 709, 713, 715, 716, 717, 718, 719, 724, 725, 729, 735(t), 738, 740, 748, 786, 787, 788, 793, 797, 801, 802, 825, 840, 960, 1012(t), 1024, 1082, SH3(rt), SH4(b)
David Reed, 425, A10
Rolin Graphics Inc., 9, 36, 37, 86, 122, 143, 144, 193(l), 210, 213(r), 269, 300, 307(l), 388, 398, 401, 404-405, 407, 430, 432, 433, 455, 457, 459, 465, 466, 471, 473, 507, 508, 531, 541, 606, 620, 638, 640, 683, 685, 686, 698, 699, 700, 709, 763(b), 803, 804-805, 806-807, 838, 849, 850, 852, 853, 854, 866, 874(b), 883, 923, 936, 938, 939, 941, 943, 944, 947, 950, 955, 958, 966(b), 968, 983, 986, 991(t), 994, 995, 997, 998, 1000, 1006, 1007, 1008(t), 1010, 1011, 1012(cl, cr, bl, br), 1013, 1014-1015, 1016, 1019, 1020, 1021, 1025, 1041, 1043(l), 1045, 1047(t), 1049, 1051, 1052, 1057, 1059, 1063, 1072, 1079, 1084-1085, 1096, 1097, 1098, 1099, 1100, 1101, 1102, 1103, 1104, SH5, SH14(l)
James Shough & Associates, 142, 224, 264(b), 332, 333, 411, 412, 421(l), 506, 544, 545, 573, 858, SH6, SH7
Jackie Urbanovic, 45, 238, 386, 1070
Unknown, 584, 1081

Photo Credits

List of Abbreviations:
AA=Animals Animals; AS=Allstock; BC=Bruce Coleman, Inc.; BCI=Bat Conservation International; BPS=Biological Photo Service; DRK=DRK Photo; CP=Color-Pic; EN=Ellis Nature Photography; ES=Earth Scenes; FPG=FPG International; GH=Grant Heilman Photography; MP=Minden Pictures; NHPA=Nature History Photo Agency; PA=Peter Arnold, Inc.; PN=PHOTO/NATS; PR=Photo Researchers; PT=Phototake NYC; SB=Stock Boston; SPL=Science Photo Library; TSM=The Stock Market; TSA=Tom Stack & Associates; TSI=Tony Stone Images; VU=Visuals Unlimited.

Cover, Richard Johnson/Alaska Stock; iv-v AA/James D. Watt; vi Barbara Gerlach/DRK; vi-vii Barbara Gerlach/DRK; viii (t)Johnny Johnson, (b)Doug Martin; ix (t)Tom Bean/DRK, (b)Secchi-Lecaque/Roussel-UCLAF/CNRI/SPL/PR; x (t)VU/Joe McDonald, (b)David M. Dennis; xi Stephen J. Krasemann/AS; xii Mike Severns/TSA; xiii Gail Shumway/FPG; xiv courtesy Dupont; xv David Stoecklein/TSM; xvii VU/D. Cavagnaro; xix Lynn Stone; xxi VU/Tom J. Ulrich; xxiii ©1994 The Georgia O'Keeffe Foundation/Artists Rights Society (ARS), New York; xxiv (l)Robert Young, (c)Peter Serling, (r)Helene Layne; xxv Gerard Lacz/PA; xxvi (l)Norbert Wu, (r)Carr Clifton/MP; xxvii (l)Andrew Martinez/PR, (r)Marty Cordano/DRK; xxviii (t,br)Ron Kimball, (bl)file photo; xxix (l)John Watney/PR, (r)Clinique ste Catherine/CNRI/SPL/PR; xxx-1 VU/David B. Fleetham; 2 Tom & Pat Leeson; 2-3 Tom & Pat Leeson/DRK; 3 (t)Frans Lanting/MP, (b)Tom & Pat Leeson; 4 (t)Stephen J. Krasemann/DRK, (b)David Hall/PR; 5 AA/Henry Ausloos; 6 (t)Roger K. Burnard, (b)Fletcher W.K./PR; 7 (tl)Tom McHugh/PR, (tc)John Cancalosi/DRK, (tr)Kim Heacox/DRK, (bl)AA/Raymond A. Mendez, (br)Dr. E.R. Degginger/CP; 8 (t)Robert Bateman/Mill Pond, Inc., (b)Matt Meadows; 9 VU/R. Calentine; 10 (t)Bill Groethe, (b)David Vaughan/SPL/PR; 11 (t)Stephen J. Krasemann/DRK, (c)Mark Burnett, (b)Doug Perrine/DRK; 12 (t)Alex Kerstitch, (b)Kjell B. Sandved/PR; 13 (t)Alex Kerstitch, (b)Alan D. Carey/PR; 14 (t)Wayne Lankinen/DRK, (b)Richard Harrington/Globe Photos, 15 (t)Tom & Pat Leeson/DRK, 15 (b)John Harquail/Masterfile; 16 (l)VU/T. E. Adams, (c)AA/Michael Dick, (r)David Muench/AS; 17 Brock May/PR; 18 19 Matt Meadows; 20 (t)Rod Planck/PR, (bl)Michael Fogden/DRK, (br)AA/Mickey Gibson; 21 BIOS/PA; 24 (t)AA/Anup & Manjo Shan, (b)AA/Doug Wechsler; 25 KS Studios/Bob Mullenix; 26 AA/M. Austerman; 27 (t)Dr. E.R. Degginger/CP, (b)AA/Anthony Bannister; 28 Arthur C. Smith/GH, 30 (t)KS Studios/Bob Mullenix, (b)OSU/OARDC, 31 (t)Phil Degginger/CP, (bl)Tom Pantages, (br)Philippe Plailly/SPL/PR; 32 33 Matt Meadows; 35 (l)Holly Payne, (r)Cornell University; 36 Louis Psihoyos/Matrix; 37 (l)S. Nielson/DRK, (c)Wardene Weisser/BC, (r)Allen B. Smith/TSA; 38 (t)AA/Patti Murray, (b)Aaron Haupt; 41 M. Abbey/PR; 42 (t)Luiz C. Marigo/PA, (b)Fritz Polking/PA; 43 Peter Drowne/CP; 44 Australian Institute of Marine Science; 45 Science VU/VU; 46 Dr. E.R. Degginger/CP; 47 (l)Matt Meadows, (r)VU/S. Maslowski; 50 ES/Oxford Scientific Films/George Bernard; 50-51 Frans Lanting/MP; 51 (t)David Dennis/TSA, (b)Frans Lanting/MP; 52 (t)Cameron Davidson/Comstock, Inc., (b)Hans Pfletschinger/PA; 53 (l)Dr. E.R. Degginger/CP, (r)Anthony Mercieca/PR; 54 (t)Life Images Studio, (b)Elaine Shay; 55 Wayne Lankinen/DRK, 56 (tl)Norbert Wu/PA, (tr)Dr. E.R. Degginger/CP, (b)Phil Degginger/CP; 57 Jim Stienberg/PR; 58 VU/William Grenfell; 59 Thomas Kitchin/TSA; 60 (tl)AA/Breck P. Kent, (tr)Runk/Schoenberger/GH, (bl)Grant Heilman/GH, (br)Phil Degginger/CP; 61 (tl)Hans Pfletschinger/PA, (tr)Pat O'Hara/DRK, (b)Michael P. Gadomski/PR; 62 VU/John D. Cunningham; 62-63 Matt Meadows, 63 AA/Zig Leszczynski; 64 Warner Brothers/Shooting Star; 65 (tl)Life Images, (tr)Matt Meadows, (b)Ingeborg Lippman/PA; 66 (l)VU/M. Abbey, (r)Dr. E.R. Degginger/CP; 67 Biophoto Associates/PR; 68 (tl)AA/Margot Conte, (tr)Dr. E.R. Degginger/CP, (b)Runk/Schoenberger/GH; 69 (t)VU/John Gerlach, (bl)ES/Patti Murray, (br)Gregory G. Dimijian/PR; 74 Dr. E.R. Degginger/CP; 78 Ron Sanford/AS; 82 VU/Bruce Berg; 82-83 Greg Vaughn/TSA; 83 Dr. E.R. Degginger/CP; 84 Steve Maslowski/PR; 85 (l)VU/Mack Henley, (r)Dan Guravich/PR; 86 Matt Meadows; 87 Stella Snead/BC; 88 Scott Cunningham Photography; 89 Rick Weber; 90 Jeff Foott/TSA; 91 ES/John Gerlach; 92 (t)VU/Roger Cole, (bl)Kevin Magee/TSA, (br)Norbert Wu/PA; 93 (t)VU/William C. Jorgensen, (b)ES/Robert P. Comport, (inset)AA/Dr. E.R. Degginger; 94 95 Helene Layne; 96 Doug Sokell/TSA; 100 (t)Johnny Johnson, (bl)Tom McHugh/PR, (br)AA/Dr. E.R. Degginger; 101 (t)VU/William J. Weber, (bl)Alan D. Carey/PR, (bc)Stephen J. Krasemann/PR, (br)Arthur C. Smith III/GH; 102 (t)Matt Meadows/PA, (bl)Tim Davis/PR, (br)John Cancalosi/PA; 103 (l)Graham Robertson/Auscape International, (r)Spencer Swanger/TSA; 104 (t)Ed Shay, (b)VU/John Gerlach; 105 (tl)VU/Tom J. Ulrich, (tc)Suzanne & Joseph Collins/PR, (tr)Jeff Lepore/PR, (b)Gregory G. Dimijian/PR; 106 (tr)Dr. E.R. Degginger/CP, (bl)Gunter Ziesler/PA, (br)Dr. E.R. Degginger/CP; 107 NASA; 108 111 Dr. E.R. Degginger/CP; 112 Robert Goldwitz/PR; 112-113 Will & Deni McIntyre/AS; 113 Scott Cunningham Photography, 114 (t)ES/Dr. E.R. Degginger, (b)Matt Meadows, courtesy Saint Ann's Hospital, Westerville, OH.; 116 Matt Meadows; 117 M. Abbey/PR; 118 Elaine Shay; 119 (l)Blair Seitz/PR, (r)Thomas Kitchin/TSA; 120 (t)David M. Dennis/TSA, (b)David Cavagnaro/DRK; 121, (t)Stephen P. Parker/PR, (c)ES/Ana Laura Gonzalez, (b)Johnny Johnson; 123 Matt Meadows; 124 NASA; 126 Michael P. Gadomski/PR; 128 Jan Halaska/PR; 129 Tim Davis/PR; 132 ES/G.I. Bernard; 132-133 ES/Leonard Lee Rue III; 133 Matt Meadows; 134 Johnny Johnson; 135 (tl)Dr. E.R. Degginger/CP, (tr)Larry Ulrich/DRK, (bl)Glencoe Photo, (br)Chris Lobban/BPS; 136 (t)John A. Lynch/PN, (b)Michael Mathers/PA; 137 (tl)Johnny Johnson, (tr)Doug Perrine/DRK, (bl)Johnny Johnson; (br)John Caucolosi/TSA; 138 (tl)Jeff Foott/TSA, (tr)AA/Mickey Gibson, (bl)Tom & Pat Leeson/PR, (br)Tom McHugh/PR; 139 (l)J. Pavlovsky/Sygma, (r)Bruce Brander/PR; 140 (t)Bob Pool/TSA, (c)Matt Meadows, (b)Doug Sokell/TSA; 141 Susan McCartney/PR; 143 John W. Bova/PR; 145 (l)James Westwater, (r)Charlie Ott/PR; 146 Gary Milburn/TSA; 147 (t)Jim Robinson/*The Chicago Tribune*, (b)Crown Studios; 148 (l)Louis Psihoyos/Matrix, (r)James Sugar/Black Star; 149 (t)Crown Studios, (b)Larry Ulrich/DRK; 150 (t)Crown Studios, (b)KS Studios/Bob Mullenix; 151 Matt Meadows; 152 AA/Zig Leszczynski; 153 Runk/Schoenberger/GH; 156 Biophoto Associates/PR; 156-157 ES/Dr. E.R. Degginger; 157 (t)Dr. E.R. Degginger/CP, (b)CNRI/SPL/PR; 158 Mark E. Gibson; 158-159 David Young-Wolff/PhotoEdit; 159 John Greim/The Stock Shop; 160 Vladimir Pcholkin/FPG; 161 Tom McHugh/PR; 162 Matt Meadows; 163 (inset)Southern Illinois University/PR, (l)Yoav Levy/PT, (r)Matt Meadows; 164 (t)Matt Meadows, (b)Chip Clark; 165 Stephen J. Krasemann/PA; 166 Doug Martin; 167 168 Matt Meadows; 170 (t)Hermann Eisenbeiss/PR, (c)Clyde H. Smith/The Stock Shop, (b)Matt Meadows; 171 (t)Matt Meadows, (c)Alan Pitcairn/GH, (b)KS Studios/Bob Mullenix; 172 W. Kleck/Terraphotographics/BPS; 174 175 Richard B. Young; 176 G. Ziesler/PA; 177 AA/J. C. Stevenson; 178 (t)Elaine Shay, (b)U.S. Geological Survey, Flagstaff, AZ; 179 (t)C. Allan Morgan/PA, (c)Elaine Shay, (b)Elaine Shay; 180 (l)Will & Deni McIntyre/PR, (r)Francois Gohier/PR; 182 183 Matt Meadows; 188 Dr. Brian Eyden/SPL/PR; 188-189 VU/Triarch; 189 VU/R. Kessel-G. Shih; 190 (l)Eli Lilly and Company, (r)Jim Brandenburg/MP, 191 VU/Cabisco; 192 (tl)The Science Museum, London, (tr)Biophoto Associates/PR, (bl)VU/Southern Illinois University, (br)VU/David M. Phillips, 193 (tl)Gary Retherford/Science Source/PR, (tr)Harry Przekop/Medichrome Div./The Stock Shop Inc., (bl)VU/Karl Aufderheide, (br)IBM Research/PA; 194 (t)Phil Degginger/CP, (b)A.B. Dowsett/SPL/PR; 195 VU/K. G. Murti; 196 Scott Cunningham Photography; 197 (t)W. Rosenberg, Iona College/BPS, (c)VU/D. E. Akin, (b)VU/J. Litray; 198 VU/Don W. Fawcett; 199 VU/R. Bolender-D. Fawcett; 200 (t)VU/Don Fawcett, (bl)Biophoto Associates/PR, (br)VU/David M. Phillips; 201 Keith R. Porter/PR, 202 203 Ed Kashi, 204 Dr. Jermey Burgess/SPL/PR; 205 (l)VU/Schliwa, (r)Beth Brookhart; 206 (t)VU/David M. Phillips, (bl)VU/David M. Phillips, (bc)Professor P. Motta/Dept. of Anatomy/University La Sapienza, Rome/SPL/PR, (br)VU/David M. Phillips; 208 Brent Turner; 209 (l)VU/Stanley Fleger, (r)Biophoto Associates/PR; 214 Jeffrey L. Rotman/PA; 214-215 ES/James D. Watt; 215 Manfred Kage/PA; 216 Rick Weber; 217 (t)Scott Cunningham, (b)Don Fawcett/PR; 218 219 Matt Meadows; 221 (l)VU/Steve McCutchen, (inset) Dwight Kuhn, (r)VU/NO, 223 (t)PN/Liz Ball, (b)Francois Gohier/PR, 224 Matt Meadows; 225 John Watney/PR; 227 (t)Matt Meadows, (b)Joseph Kurantsin-Mills, M.Sc. Ph.D; 228 (t)Joseph Kurantsin-Mills, M.Sc. Ph.D, (b)VU/Cabisco; 229 (t)Matt Meadows, (b)Joseph Kurantsin-Mills, M.Sc. Ph.D; 236 (t)Greg Vaughn/TSA, (b)David R. Frazier/PR; 237 Matt Meadows; 238 David M. Dennis/TSA; 239 KS Studios/Bob Mullenix; 240 ES/Zig Leszczynski; 241 (t)Dan Guravich/PR, (bl)Michael Doolittle from Rainbow, (br)A.B. Dowsett/SPL/PR; 242 Matt Meadows; 243 John Shaw/TSA; 245 (l)KS Studios/Bob Mullenix, (c)Matt Meadows, (r)Crown Studios; 246 247 Matt Meadows; 248 H. P. Merten/TSM; 249 VU/D. Foster, Woods Hole Oceanographic Institution; 250 Matt Meadows; 251 Bob Daemmrich; 253 Matt Meadows; 254 Greg Vaughn/TSA; 255 Matt Meadows; 260 J.M. Murray, Dept. of Anatomy, University of PA; 260-261 VU/K. G. Murti; 261 (l)Tom Bean/DRK, (r)Ed Reschke/PA; 262 AA/Anup & Manjo Shah; 263 VU/Michael Abbey; 265 (l)Dr. E.R. Degginger/CP, (tr)Patrick J. Lynch/PR, (c)Dr. E.R. Degginger/CP, (br)Biophoto Associates/PR; 266 Dr. Gopal Murti/SPL/PR; 269 (t)Science VU/VU, (b)Barry King, U. of California School of Medicine/BPS; 270 271 Runk/Schoenberger/GH; 272 Southern Illinois University/PR; 273 (t)VU/David M. Phillips, (b)B.A. Palevitz & E.H. Newcomb, U. of Wisconsin/BPS; 274 (t)Dr. E.R. Degginger/CP, (b)Matt Meadows; 275 Matt Meadows; 276 AA/Oxford Scientific Films;

Art & Photo

277 VU/Kenneth Greer; 280 Dr. E.R. Degginger/CP; 282 Ken Eward/PR; 282-283 Lennart Nilsson, The Body Victorious, Bonnier-Alba; 283 (t)L. Kooyman/AA, (b)courtesy of Keith Woods; 284 Larry Lefever/GH; 284-285 Omikron/PR; 285 Pete Saloutos/TSI; 286 The Bettmann Archive; 287 (t)Alvin E. Staffan; 287 (b)Thomas R. Taylor/PR; 289 VU/R. W. Van Norman; 292 Matt Meadows; 293 Ray Pfortner/PA; 295 Robert Tringali/Sportschrome East/West; 298 ES/Donald Specker; 299 (t)Grant Heilman/GH, (l)VU/John D. Cunningham, (r)David M. Dennis; 301 (t, tl)VU/John D. Cunningham, VU/Cabisco; 304 Mike Yamashita/The Stock Shop; 305 Dr. E.R. Degginger/CP; 308 AA/Patti Murray; 308-309 AA/Gerard Lacz; 309 Tom McHugh/PR; 310 Matt Meadows; 312 Will and Deni McIntyre/PR Inc.; 313 National Institute of Health/Kakefuda/PR; 316 L. Smith/H. Armstrong Roberts; 319 Professor Oscar Miller/SPL/PR; 321 324 Matt Meadows; 328 Thomas Dimock/TSM; 332 David M. Dennis; 332-333 David M. Stone/PN; 334 KS Studios; 335 Larry Lefever/GH; 336 Mak-1 Photodesign; 337 Dr. E.R. Degginger/CP; 340 Crown Studios; 341 Carolina Biological Supply/PT; 342 E.M. Vogeler/OKAPIA/PR; 343 Dr. E.R. Degginger/CP; 344 346 Matt Meadows; 347 VU/Walt Anderson; 348 (l)Michael Holford, (r)H. Schwind/OKAPIA/PR; 349 (tl)AA/Robert Maier; (tr)AA/Gerard Lacz, (bl)AA/Robert Pearcy, (br)K.S. Studio/Bob Mullinex; 350 Grant Heilman/GH; 353 Hans Reinhard/OKAPIA/PR; 354 Duomo/Al Tielemans; 354-355 Duomo/Mitchell Layton; 355 The Bettmann Archive; 356 Matt Meadows; 357 (l)Biophoto Associates/PR, (c)Runk/Schoenberger/GH, (r)David M. Dennis; 359 VU/Stanley Flegler, 360 361 Jeff Fearing, Howard University; 363 Matt Meadows Photography; 364 Martin Dohrn/SPL/PR; 365 Phil Degginger/CP; 366 VU/Dan Kline; 367 (t)Runk/Schoenberger/GH, (b)VU/Bruce Berg; 368 (tl)Elaine Shay, (tr)Matt Meadows, (b)The Bettmann Archive; 369 Biophoto Associates/PR; 370 Richard Hutchings/Science Source/PR; 371 Southern Illinois University/PR; 374 courtesy of Keith Woods; 374-375 Dr. E.R. Degginger/CP; 375 Phillip A. Harrington/PA; 376 VU/Mack Henley; 378 VU/K. G. Murti; 380 Crown Studios; 382 (l)Archive Photos, (r)VU/Michael Gabridge; 383 GE Research and Development Center; 384 (tl)Eli Lilly and Company, (tr)Eli Lilly and Company, (b)Leonard Lessin/PA; 385 (l)AA/Breck P. Kent, (r)KS Studios; 387 Peter Menzel/Yale Univ. Med. School/Dr. Peter Lichter; 389 (l)VU/Southern Illinois University, (r)Petit Format/Nestle/PR; 394 James L. Amos/PR; 394-395 Doug Perrine/DRK; 395 (t)Stephen J. Krasemann/AS, (b)VU/John D. Cunningham; 396 Phil Schofield/AS; 396-397 Richard T. Nowitz/PR; 397 NASA GSFC/SPL/PR; 399 (tl)ES/Breck P. Kent, (tr)Dr. E.R. Degginger/CP, (cl)Tom Bean/DRK, (cr)Francois Gohier/PR, (bl)ES/Breck P. Kent, (br)ES/William D. Griffin; 400 (t)Matt Meadows, (bl)Thomas N. Taylor, The Ohio State University, (br)ES/Perry D. Slocum; 402 Ken Abbott, University of Colorado; 403 Louis Psihoyos/Matrix; 406 (tl)J. William Schopf/UCLA, (tcl)J. William Schopf/UCLA, (tcr)Sinclair Stammers/SPL/PR, (tr)Fred Bavendam/PA, (b)David M. Dennis/TSA; 407 James L. Amos/PR; 410 The Granger Collection; 413 (l)Roger Ressmeyer-Starlight, (r)AA/Stouffer Prod. Ltd.; 414 VU/Sidney Fox; 415 (t)T. J. Beveridge & C. Forsberg, U. of Guelph/BPS, (b)Comstock Inc./Georg Gerster; 416 Matt Meadows; 417 David M. Dennis; 422 Tom McHugh/PR; 422-423 Michael Fogden/DRK; 423 Rob Simpson/TSA; 424 (l)Kevin Schafer/PA, (r)Barbara von Hoffmann/TSA; 425 (tl)Kevin Schafer & Martha Hill/TSA, (tr)Frans Lanting/PR, (b)David C. Fritts/AS; 426 Hans Reinhard/TIERBILD OKAPIA/PR; 429 (t)Norbert Wu/PA, (b)Hans Dieter Brandl/Okapia/PR; 432 (l)Anna E. Zuckerman/TSA, (r)AA/Stephen Dalton; 433 AA/Michael Dick; 435 (t)Tom & Pat Leeson/PR, (b)Tom & Pat Leeson/PR; 437 (t)Victor A. McKusick, (b)John G. Ross/Art Resource; 438 Aaron Haupt; 439 Matt Meadows; 442 443 Dr. E.R. Degginger/CP; 444 Leonard Wolfe/PR; 448 (l)VU/Marty Snyderman, (r)Stephen J. Krasemann/DRK; 449 Don & Pat Valenti/DRK; 452 Gunter Ziesler/PA; 452-453 Alan Root/PR; 453 Pat & Tom Leeson/PR; 454 Steve Kaufman/DRK; 455 AA/Daniel J. Cox; 456 (l)VU/Walt Anderson, (tr)Tom McHugh/PR, (br)Holt Confer/DRK; 457 (l)AA/Michael Dick, (r)Daniel Cande/PR; 458 (l)AA/Michael Dick, (r)Roy Toft/TSA; 459 Tim Laman/The Wildlife Collection; 460 (t)M. Loup/Jacana/PR, (c)AA/Michael Dick, (bl)Tom McHugh/PR, (br)Dr. E.R. Degginger/CP; 461 (tl)Dr. E.R. Degginger/CP, (tr)Stanley Breeden/DRK, (b)George Holton/PR; 462 (l)George Holton/PR, (r)AA/Mickey Gibson; 463 THE FAR SIDE ©1988 FARWORKS, INC. Dist. by UNIVERSAL PRESS SYNDICATE; 464 (t)UPI/Bettmann, (b)John Reader/PR, 468 (l)Frank T. Aubrey/VU, (r)Ronald Bowen/Robert Harding Picture Library; 469 (l)J. Beckett/D. Finnin/American Museum of Natural History, (r)VU/National Museum of Kenya; 470 Dan Bosler/TSI; 471 VU/Cabisco; 472 (t)Art Resource, (bl)VU/Science VU, (br)Ronald Bowen/Robert Harding Picture Library; 474 David M. Dennis; 478 VU/David M. Phillips; 478-479 Hans Reinhard/OKAPIA/PR; 479 (t)David M. Dennis/TSA; 479 (b)M.I. Walker/PR; 480 (t)Frans Lanting/MP, (b)Alan D. Carey/PR; 481 AA/Mella Panzella; 482 Life Images Studio; 483 (tl)Peter G. Altken/PR, (r)ES/John Lemker, (c)Andrew Henderson/PN, (bl)Dr. E.R. Degginger/CP; 484 (l)Dr. E.R. Degginger/CP, (r)Matt Meadows; 485 Gerald & Buff Corsi/TSA; 487 Academy of Natural Sciences Philadelphia/VIREO; 488 (t)Mak-1 Photodesign, (bl)Elizabeth Lemoine/Jacana/PR, (br)Rob Simpson/TSA; 489 Rudolf Hofels/TIERBILD OKAPIA/PR; 490 Life Images; 491 Scalla/Art Resource; 494 (tl)Science VU/S.W. Watson, 494 (tc)VU/David M. Phillips, (tr)VU/T. E. Adams, (bl)Michael Abby/PR, (bc)David M. Dennis/TSA, (br)Dr. E.R. Degginger/CP; 495 Ken Highfill/PR; 496 Dr. E.R. Degginger/CP; 497 Life Images; 498 (l)Runk/Schoenberger/GH, (r)David Weintraub/PR; 501 (l)VU/John Gerlach, (r)Runk/Schoenberger/GH; 502 Larry Lefever/GH; 502-503 VU/Sylvia E. Coleman; 503 Dr. Tony Brain/SPL/PR; 504 Jim Selby/SPL/PR; 505 (tl)OMIKRON/PR, (tr)Scott Camazine & Sharon Bilotta-Best/PR, (bl)Lee D. Simon/PR, (br)CNRI/SPL/PR; 509 CNRI/SPL/PR; 511 (l)Norm Thomas/PR; 511 (r)VU/Wayside; 512 FAR SIDE copyright FARWORKS, INC./Dist. by UNIVERSAL PRESS SYNDICATE. Reprinted with permission. All rights reserved.; 513 (t)ES/Breck P. Kent, (c)Helen E. Carr/BPS, (b)Grant Heilman/GH; 514 (l)Biology Media/PR, (r)T.D. Pugh & E.H. Newcomb, U. of Wisconsin/BPS; 515 (tl)VU/Sylvia E. Coleman, (tc)VU/John D. Cunningham, (tr)VU/David M. Phillips, (bl)VU/G. Musil, (bc)VU/David M. Phillips, (br)VU/David M. Phillips; 516 East Malling Research Station/SPL/Science Source/PR; 517 Andrew McClenahan/SPL/PR; 520 VU/J. Edmond, Woods Hole Oceanographic Institute; 521 (l)T.J. Beveridge, U. of Guelph/BPS, (r)VU/George B. Chapman; 522 Royal College of Surgeons, Edinburgh; 523 (t)A.B. Dowsett/SPL/PR, (b)Dr. L. Caro/SPL/PR; 524 (l)David M. Dennis/TSA, (r)Dr. Jeremy Burgess/SPL/PR; 525 (tl)Kunio Owaki/TSM, (tr)John Colwell/GH, (br)James Holmes, Cell Tech Ltd/SPL/PR; 526 527 Andree Abecassis Photography; 529 Marty Cordano/DRK; 532 Stephen G. Maka/PN; 532-533 Michael P. Gadomski/PR; 533 Dwight Kuhn; 534 (t)Manfred Kage/PA, (c)Milton Rand/TSA, (bl)Eric Grave/PR, (br)Dr. E.R. Degginger/CP; 535 Biophoto Associates/PR; 536 (l)Eric V. Grave/PR, (r)Manfred Kage/PA; 537 (t)VU/J. Paulin, (b)Eric Grave/PR; 538 (l)VU/M. Abbey, (r)Dr. E.R. Degginger/CP; 539 VU/M. Abbey; 540 CALVIN AND HOBBES copyright Watterson./Dist. by UNIVERSAL PRESS SYNDICATE. Reprinted with permission. All rights reserved.; 542 Michael P. Gadomski/PR; 543 (tl)VU/Cabisco, (tr)Runk/Schoenberger/GH, (cl)Tammy Peluso/TSA, (cr)VU/L. Sims, (b)Harris Biological Supplies LTD; 546 M.I. Walker/PR; 547 (tl)VU/David M. Phillips, (tr)Eric Grave/PR, (b)Kevin Schafer/PA; 548 (t)J.R. Waaland, U. of Washington/BPS, (b)Tammy Peluso/TSA; 549 (t)Runk/Schoenberger/GH, (bl)M. I. Walker/PR, (br)J.R. Waaland, U. of Washington/BPS; 550 Brian Parker/TSA; 551 Matt Meadows/PA; 552 (t)Patrick W. Grace/PR, (b)VU/Cabisco; 554 Noble Proctor/PR; 555 Stephen J. Krasemann/DRK; 558 Jeff LePore/PR; 558-559 Michael Fogden/DRK; 559 Stephan J. Krasemann/DRK; 560 (t)Dr. Jeremy Burgess/SPL/PR, (bl)ES/Patti Murray, (bc)Michael Fogden/DRK, (br)Michael Fogden/DRK; 562 (t)Johann Bruhn, University of Missouri, (b)Ron Watts/West Light; 563 (l)David M. Dennis/TSA, (r)D.A. Glawe, U. of Illinois/BPS, 564 J. Forsdyke & Gene Cox/SPL/PR; 565 ES/Michael Fogden; 566 Matt Meadows; 567 (tl)W.E. Schadel, Small World Enterprise/BPS, (tr)Ralph C. Eague/PR; 568 (l)David Dennis/TSA, (r)Matt Meadows; 568 (b)E. Gueho-CNRI/SPL/PR; 569 Biophoto Associates/PR; 570 Stephen J. Krasemann/DRK; 571 (tl)David M. Dennis, (tc)Scott Camazine/PR, (tr)Dr. E.R. Degginger/CP, (bl)Runk/Schoenberger/GH, (br)Dwight R. Kuhn; 572 Matt Meadows; 574 Gary Braasch; 575 (l)John Shaw/TSA, (c)Robert Lee/PR, (r)L. West/PR; 577 Dr. E.R. Degginger/CP; 578 VU/David M. Phillips; 580 Dr. Paul A. Zahl/PR; 580-581 Scott Camazin/PR; 581 (t)Phil Schermeister/AS, (b)Nicole Richards/AS; 582 ES/Ralph A. Reinhold; 582-583 Larry Ulrich/DRK; 583 Lefever/GrushowGH; 584 Bonnie Rauch/PR; 585 Runk Schoenberger/GH; 586 (t)ES/Nigel Cattlin, Holt Studio Int., (c, b)Dr. Jeremy Burgess/SPL/PR; 587 (l)Bill Brooks/Masterfile, (r)Clyde H. Smith/PA; 588 (t)Matt Meadows, (b)ES/Raymond A. Mendez; 589 (l)VU/G. Shih-R. Kessel, (c)Runk/Schoenberger/GH, (r)ES/Peter Parks/Oxford Scientific Films; 591 (tr)Len Rue, Jr./PR, (cl)ES/Patti Murray, (cr)Kevin Schafer/TSA, (bl)ES/Doug Wechsler, (br)Gregory K. Scott/PR; 592 (tr)Gunter Ziesler/PA, (l)W.H. Hodge/PA, (br)Jim Strawser/GH; 593 (l)Alvin E. Staffan/National Audubon Society/PR, (r)Angelina Lax/PR; 594 (t)Pat O'Hara/DRK, (b)Dr. E.R. Degginger/CP; 595 (tl, tr, br)Dr. E.R. Degginger/CP, (c)VU/Bayard H. Brattstrom, (bl)VU/Walt Anderson; 596 (l)Rod Planck/TSA, (r)Les Saucier/PN; 597 Wayne Lankinen/DRK; 598 Les Saucier/PN; 599 Matt Meadows; 600 Runk/Schoenberger/GH; 602 ES/George Bernard; 604 ES/Patti Murray; 604-605 VU/John D. Cunningham; 605 Runk/Schoenberger/GH; 607 (tl)Runk/Schoenberger/GH, (tr)VU/E.F. Anderson, (b)VU/James W. Richardson; 608 (t)VU/John D. Cunningham, (b)Walter H. Hodge/PA; 609 (t)ES/Doug Wechsler, (c)ES/Patti Murray, (b)VU/Doug Sokell; 610 (t)Matt Meadows, (c)VU/Walt Anderson, (b)Farrell Grehan/PR; 611 VU/Doug Sokell; 612 (l)Glenn Foss/PR, 612 (r)David M. Dennis; 613 Runk/Schoenberger/GH; 614 (l)ES/Patti Murray, (r, b)ES/Jack Wilburn; 615 (tl)VU/E.F. Anderson, (tr)VU/McCutcheon, (b)Andrew Henderson/PN; 616 (tl)R.E. Litchfield/SPL/PR, (tr)Dr. Jeremy Burgess/SPL/PR, (bl)Andrew Henderson/PN, (br)Gay Bumgarner/PN; 617 (tl)VU/Doug Sokell, (tr)ES/Michael Fogden, (b)G.R. Roberts; 618 (tl)Alfred Owczarzak/BPS, (tr)Dr. E.R. Degginger/CP, (b)ES/Ken Cole; 619 David M. Dennis; 620 Gregory G Dimijian/PR; 621 Noble Proctor/PR; 622 623 Matt Meadows; 624 (l)VU/B. Ormerod, (r)VU/B. Ormerod; 627 Doug Sokell/TSA; 628 Tom Bean/DRK; 628-629 ES/Donald Specker; 629 Sharon Gerig/TSA; 630 (t)Greg Vaughn/TSA, (bl)Jane Grushow/GH, (br)VU/John D. Cunningham; 631 (tl)ES/Richard Shiell, (tr)Joyce Photographics/PR, (bl)Dr. E.R. Degginger/CP, (bc)ES/Robert Maier, (br)Michael P. Gadomski/PR; 632 (t)Elaine Shay, (b)Doug Martin; 633 (tl)Marilyn Wood/PN, (tc)Jim Strawser/GH, (tr)Belinda Wright/DRK, (bl)J.R. Waaland, U. of Washington/BPS, (bc)Matt Meadows, (br)Dwight Kuhn; 634 (l,c)Dwight Kuhn, (r)G. Buttner/Naturbild/OKAPIA/PR; 635 (l)John Gerlach/TSA, (c)Paul Schrock, (r)Larry Ulrich; 636 (t)ES/Richard Shiell, (bl)Jane Grushow/GH, (br)Grant Heilman/GH; 637 (l)VU/S. Elems, (r)VU/John D. Cunningham; 639 (l)Visuals Unlimited/Cabisco, (r)Carolina Biological Supply; 641 Runk/Schoenberger/GH; 642 (t)Dwight Kuhn, (b)ES/Ken Lewis; 646 (t)Don Johnston/PN, (c)G.R. Roberts, (b)Dr. Wm. M. Harlow/PR; 647 ES/Carson Baldwin, Jr.; 648 (l,r)Dwight Kuhn, (c)Gwen Fidler; 649 Chip & Jill Isenhart/TSA; 652 Christine M. Douglas/PR; 652-653 Brian Parker/TSA; 653 AA/Stephen Dalton; 654 Elaine Shay; 655 (tl)Hans Reinhard/TIERBILD OKAPIA/PR, (tr)R. J. Erwin/PR, (bl)Stephen P. Parker/PR, (br)Alvin E. Staffan; 656 ES/Ken Cole; 657 Scott Camazine/PR; 658 (l)© 1994 The Georgia O'Keeffe Foundation/Artists Rights Society (ARS), New York, (r)AA/Raymond Mendez, (tr)Anthony Mercieca/PR; 659 (l)Martha Cooper/PA, (br)Merlin D. Tuttle/BCI; 660 Dwight Kuhn; 662 Hermann Eisenbeiss/PR; 663 (tl)Dwight Kuhn, (tr)Ann Reilly/PN, (bl)ES/Joyce Wilson; 664 (l)Larry Hamill, (r)John Kaprielian/PR; 665 Hans Pfletschinger/PA; 667 Elaine Shay; 668 (t)VU/S. McCutcheon, (cl)Steve King/PA, (cr)ES/K.G. Preston-Mafhan, (b)ES/Nigel Cattlin; 669 (t)Thomas Kitchin/TSA, (c)Elaine Shay, (bl)David L. Perry, (br)ES/Donald Specker; 670 (tl)Larry West/FPG, (tr)Arthur Tilley/FPG, (bl)ES/Breck P. Kent, (br)ES/Holt Studios LTD; 671 (tl, tr)Elaine Shay, (c)Dwight Kuhn, (b)Gary Braasch; 672 (tr)Alvin E. Staffan, (l)Tom McHugh/PR, (br)David M. Dennis; 673 Matt Meadows; 674 (t)David M. Dennis, (b)Doug Sokell/TSA; 676 (tl)John Kaprielian/PR, (tr)John Kaprielian/PR, (b)Dr. E.R. Degginger/CP; 677 (t)Elaine Shay, (b)Aaron Haupt; 678 (l)ES/Bertram G. Murray, Jr., (r)John W. Bova/PR; 682 ES/C.W. Perkins; 683 (t)ES/Doug Wechsler, (bl)David M. Dennis, (br)N.H. (Dan)Cheatham/DRK; 684 (l)Pat & Tom Leeson/PR, (r)Darrell Gulin/DRK; 685 (l)Darrell Gulin/DRK, (r)Alvin E. Staffan; 686 (l)Runk/Schoenberger/GH, (c)Jeremy Burgess/PR, (r)Inga Spence/TSA; 687 (l)Stephen G. Marka/DRK, (c)Ken W. Davis/TSA, (r)Gregory K. Scott/PR; 688 VU/Mack Henley, 688-689 AA/Doug Wechsler; 689 (t)VU/Dave B. Fleetham, (b)Gay Bumgarner/AS; 690 Merlin D. Tuttle/BCI/PR; 690-691 Jeff Lepore/PR; 691 VU/Bill Kamin; 692 (t)Brian Parker/TSA, (bl)Stephen Dalton/NHPA, (bc)G. C. Kelley/PR, (br)VU/Alex Kerstitch; 693 (t)Crown Studios, (cl)Craig Kramer, (cr)VU/Bayard H. Brattstrom,

Art & Photo

(bl)Alan and Sandy Carey, (br)AA/Herb Segars; 694 (tl)VU/Hal Beral, (tr)Brian Parker/TSA, (b)VU/John D. Cunningham; 695 (l)Dr. E.R. Degginger/CP, (r)Aaron Haupt; 696 Mak-1 Photodesign; 697 699 Scott Cunningham; 700 (l)VU/David M. Phillips, (r)Brian Parker/TSA; 701 (t)Scott Cunningham Photography, (bl)Charles V. Angelo/PR, (bc)AA/Max Gibbs, (br)Dr. E.R. Degginger/CP; 702 (t)Sharon M. Kurgis, (b)Runk/Schoenberger/GH; 703 (bl)Alvin E. Staffan, (bc)Wendy Shatil & Bob Ruzinski/TSA, (br)D. Holden Bailey/TSA; 705 AA/Zig Leszczynski; 706 Runk/Schoenberger/GH; 710 Tom Bean/AS; 710-711 Andrew J. Martinez/PR; 711 David Hall/PR; 712 (t)Nancy Sefton/PR, (b)Crown Studios; 713 Nancy Sefton/PR; 714 VU/Hal Beral; 715 VU/Marty Snyderman; 716 Fred Bavendam/PA; 717 AA/George Bernard; 719 Richard Campbell/BPS; 720 (t)VU/Hal Beral, (b)VU/Dave B. Fleetham; 721 (t)Ben Cropp/Auscape International, (b)VU/Richard Thom; 723 (t)Cath Wadforth/SPL/PR, (bl)Matt Meadows, (bc)Manfred Kage/PA, (br)M.I. Walker/PR; 724 Michael Abbey/PR; 725 THE FAR SIDE copyright 1987 FARWORKS, INC./Dist. by UNIVERSAL PRESS SYNDICATE. Reprinted with permission. All rights reserved.; 726 Runk/Schoenberger/GH; 727 Matt Meadows; 728 (t)Roger K. Burnard, (b)M.I. Walker/PR; 729 Sinclair Stammers/SPL/PR; 730 (t)AA/Norvia Behling, (c)Sinclair Stammers/SPL/PR, (b)VU/R. Calentine; 731 (t)VU/Arthur M. Siegelman; 731 (l)VU/Fred Marsik, (b)Runk/Schoenberger/GH; 732 Runk/Schoenberger/GH; 736 Dwight Kuhn; 736-737 David M. Dennis/TSA; 737 Dr. E.R. Degginger/CP; 738 Stephen J. Krasemann/PA; 739 (t)AA/Scott Johnson, (bl)Robert A. Ross/CP, (br)AA/Zig Leszczynski; 741 Wayne & Karen Brown/Brown & Company Photography; 742 (t)Harry Rogers/PR, (bl, bc)Wayne & Karen Brown/Brown & Company Photography, (br)VU/Don W. Fawcett; 743 (tl)Matt Meadows, (tr,cl,bl)Wayne & Karen Brown/Brown & Company Photography, (cr)AA/Dr. E.R. Degginger, (br)AA/Zig Leszczynski; 744 (t)VU/Alex Kerstitch, (c)Dave B. Fleetham/TSA, (b)Tom McHugh/PR; 746 (t)David M. Dennis/TSA, (bl)AA/W. Gregory Brown, (br)VU/Cabisco; 747 (tl)AA/Oxford Scientific Films, (tr)Ralph Buchsbaum, The Boxwood Press, (b)Richard Chesher/PR; 748 VU/David S. Addison; 749 (tl)VU/Daniel W. Gotshall, (tr)Dr. E.R. Degginger/CP, (c)Richard Chesher/PR, (br)VU/Kjell B. Sandved; 750 751 Matt Meadows; 752 St. Bartholomew's Hospital/SPL/PR; 755 J.A.L. Cooke/Globe Photos; 756 Robert J. Erwin/DRK; 756-757 Dr. E.R. Degginger/CP; 757 Tom McHugh/PR; 758 AA/Peter Parks; 759 (tl)Michael Fogden/DRK, (tr)Stephen J. Krasemann/DRK, (bl)Dwight Kuhn, (br)AA/Robert Maier; 760 (t)Robert C. Simpson/TSA, (bl)David Dennis, (br)Kerry T. Givens/TSA; 762 M. Wendler/OKAPIA/PR; 764 Alvin E. Staffan; 765 Lennart Nilsson/BonnierAlba; 766 AA/George Bernard; 768 (t)Norm Thomas/PR, (bl)Scott Camazine/PR, (br)Larry Miller/PR; 769 (t)Runk/Schoenberger/GH, (b)Kjell B. Sandved, S. I. Washington, D.C./PR; 770 J.R. Rost/©1991 The Walt Disney Co. Reprinted with permission of *Discover Magazine;* 771 (t)AA/Oxford Scientific Films, (b)Dr. E.R. Degginger/CP; 772 (t)Jett Britnell/DRK, (bl)David M. Dennis/TSA, (br)VU/Thomas Gula; 773 (tr)Carolina Biological Supply, (l)Robert J. Erwin/DRK, (c)AA/Raymond Blythe, (br)Sharon M. Kurgis; 774 775 David Riecks; 776 Dwight Kuhn; 782 F. Stuart Westmorland/AS; 782-783 Ed Reschke/PA; 783 Fred Bavendam/PA; 784 Brian Parker/TSA; 785 (tl)Jett Britnell/DRK, (tr)VU/Daniel Gotshall, (bl)AA/Scott Johnson, (br)VU/David B. Fleetham; 786 (l)Tom Wiewandt/DRK, (r)Dave B. Fleetham/TSA; 787 David Hall/PR; 790 Andrew J. Martinez/PR; 791 (t)Milton Rand/TSA, (bl)Gerald & Buff Corsi/TSA, (br)Andrew J. Martinez/PR; 792 (l)VU/Alex Kerstitch, (r)Fred Bavendam/PA; 793 Jack Stein Grove/TSA; 794 VU/Cabisco; 795 VU/William C. Jorgensen; 796 (l)VU/Stan Elems, (r)Doug Perrine/DRK; 797 David Hall/PR; 798 AA/George Bernard; 802 Larry Lipsky/DRK; 803 (tl)Norbert Wu/DRK, (tr)M. I. Walker/PR, (b)Marjorie Bank/Mo Yung Productions; 804 (t)Hans Pfletschinger/PA, (b)Sharon M. Kurgis; 805 (t)Fred Bavendam/PA, (c)Kathie Atkinson/PR, (b)Larry Lipsky/TSA; 806 (l)Marty Cordano/DRK, (r)Dan Suzio/PR; 807 (tl)David M. Dennis, (tc)Dave Fleetham/TSA, (tr)David M. Dennis/TSA, (b)Hans Pfletschinger/PA; 808 (l)Norbert Wu, (r)Mike Severns/TSA; 809 Norbert Wu; 810 VU/Nada Pecnik; 810-811 Jeff Foott/TSA; 811 (t)Johnny Johnson, (b)Gerry Ellis/EN; 812 Jeff Foott/TSA; 812-813 Richard Herrmann/The Wildlife Collection; 813 AA/Zig Leszczynski; 814 AA/Max Gibbs; 815 (l)Tom McHugh/PR, (r)Runk/Schoenberger/GH; 816 Joey Jacques; 817 Herbert Schwind/Okapia/PR; 818 (tl)Herve' Berthoule/JACANA/PR, (bl)AA/Mickey Gibson, (br)Dwight Kuhn; 820 Seattle Aquarium/PR, 821 (t)AA/L. L. T. Rhodes, (b)AA/Zig Leszczynski; 822 Norbert Wu; 824 Stephen Dalton/PR; 825 (t)AA/Michael Fogden, (cl)AA/Zig Leszczynski, (cr)AA/Betty Press, (b)Philip Kahl, Jr./DRK; 826 David M. Dennis/TSA; 827 AA/Stephen Dalton; 828 Matt Meadows; 829 T. Wiewandt/DRK; 830 (t)Mike Hopiak for the Cornell Laboratory, (b)David M. Dennis; 831 Dr. E.R. Degginger/CP; 836 (t)David M. Dennis, (b)AA/Zig Leszczynski; 836-837 Tom McHugh/PR; 837 VU/Kirley-Perkins; 838 (t)Frans Lanting/PR, (b)Matt Meadows; 839 (l)Mary Clay/PN, (r)AA/J. H. Robinson; 841 (l)AA/Charles Black, (r)John Cancaslosi/TSA; 842 (t)AA/Donald Specker, (cl)Tom McHugh/PR, (cr)Dr. E.R. Degginger/CP, (b)Gregory G. Dimijian/PR; 843 (t)Dr. E.R. Degginger/CP, (b)AA/J. & C. Kroeger; 844 (t)VU/John D. Cunningham, (b)Millard H. Sharp/PR; 845 (t)Stephen J. Krasemann/DRK, (cl)VU/Tom J. Ulrich, (cr)David M. Dennis; 847 John Cancalosi/Auscape International, LTD; 851 (t)Phil Schofield/AS, (c)Matt Meadows, (br)Andrew Leitch/©1992 The Walt Disney Co. Reprinted with permission of *Discover Magazine*, (bl)Sam C. Pierson/PR; 852 (t)VU, (cl) Louis Psihoyos/Matrix, (cr)Chip Clark, (b)Francois Gohier/PR; 853 AA/Joe McDonald; 854 Barbara Gerlach/DRK; 855 Crown Studios; 856 David Cavagnaro/DRK; 857 Russell C. Hansen/PA; 859 Crown Studios; 860 (l)Stephen J. Krasemann/DRK, (r)AA/Fred Whitehead, (b)Dave Watts/TSA; 861 (t)Tom & Pat Leeson/DRK, (bl)AA/Alan G. Nelson, (bc)AA/Bates Littlehale; 861 (br)AA/Ralph A. Reinhold; 862 (t)AA/Michael Dick, (b)AA/Stephen Dalton; 863 Thomas Kitchin/TSA; 868 Tim Davis/PR; 870-871 Doug Faulkner/AS; 871 Jon Feingersh/TSA; 872 (t)The Bettmann Archive, (b)Belinda Wright/DRK; 873 (t)Thomas Kitchin/TSA, (bl)Rod Planck/TSA, (br)Johnny Johnson/DRK; 875 (t)Stephen J. Krasemann/DRK, (cl)Merlin D. Tuttle/BCI/PR, (cr)AA/Robert Maier, (b)C. Wisiniewski/AS; 876 (tl)VU/John D. Cunningham, (tr)Chase Swift/TSA, (br)AA/Renee' Stockdale; 877 (t)AA/Oxford Scientific Films/George Bernard, (c)Fred Bruemmer/PA, (b)Stephen J. Krasemann/DRK; 878 Flip Nicklin/MP; 879 DR. Carleton Ray/PR; 880 (l)KS Studios/Bob Mullenix, (r)Warren & Jenny Jarst/TSA; 881 Tom & Pat Leeson/DRK; 884 (t)Art Wolfe/AS, (b)VU/Ron Spomer; 886 (tl)Mark Newman/TSA, (tr)Stephen J. Krasemann/DRK, (bl)VU/Joe McDonald, (br)John Cancalosi/DRK; 887 (l)Dave Watts/TSA, (r)Matt Meadows; 888 (t)John Serrao/PR, (b)Merlin D. Tuttle/Bat Conservation/PR; 889 (tl)Schafer & Hill/PA, (tr)Gunter Ziesler/PA, (bl)AA/Robert Maier, (br)Thomas Kitchin/TSA; 890 (t)Fred Felleman/AS, (c)AA/Robert Maier, (b)Johnny Johnson; 891 (t)Gerald & Buff Corsi/TSA, (cl)AA/Robert Maier, (cr)Ben Cropp/Auscape International, (b)Renee Lynn/AS; 892 (tl)Art Wolfe/AS, (tr)Peter Ward/BC, (bl)AA/Patti Murray, (br)E.P.I. Nancy Adams/TSA; 896 John Gerlach/DRK; 896-897 Jeff Lepore/PR; 897 Stephen J. Krasemann/DRK; 898 (t)Dr. E.R. Degginger/CP, (bl)John Serrao/PR, (br)Jeff Lepore/PR; 899 AA/David Thompson; 900 901 Jessie Cohen/National Zoological Park, Smithsonian Institution; 902 (l)Matt Meadows, (r)Fred Bavendam/PA; 904 (t)VU/J. Alcock, (b)Dwight Kuhn/DRK; 905 Arthur C. Smith III/GH; 906 (l)William H. Calvert/CP, (r)John W. Warden/AlaskaStock; 907 (t)Wayne Lankinen/DRK, (b)Larry B. Jennings/PR; 908 (t)AA/VJ Anderson, (b)Jean-Paul Ferrero/Auscape International; 909 Matt Meadows; 910 (t)Priscilla Connell/PN, (b)Joseph Schuyler/SB; 911 VU/William J. Weber; 912 Gerald C. Kelly/PR; 913 VU/Milton H. Tierney, Jr.; 915 Gerry Ellis/EN; 916 Matt Meadows; 920 (l)Tom Stack/TSA, (r)Ed Robinson/TSA; 921 (t)Dr. Paul A. Zahl/PR, (c)David M. Dennis, (b)David M. Dennis/TSA; 922 (tl)Gerry Ellis/ EN, (tr)Tom McHugh/PR, (b)AA/Zig Leszczynski; 923 (l)Alvin E. Staffan, (r)Fred Bruemmer/DRK; 924 (t)Wolfgang Kaehler, (c)Barbara Von Hoffmann/TSA, (b)S. Nielson/DRK; 925 (t)Anthony Mercieca/PR, (c)David Hosking/PR, (b)Bill Wilson/PR; 926 (t)Tom Brakefield/DRK, (b)Johnny Johnson; 927 (tl)Johnny Johnson, (tr)Dave Watts/TSA, (cl)Warren Garst/TSA, (cr)Mitch Reardon/PR, (b)Anthony Mercieca/PR; 928 (tl)Stephen J. Krasemann/DRK, (tr)J.C. Carton/BC, (b)Jeff Lepore/PR; 929 Stephen Dalton/PR; 930 J. & L. Weber/PA; 930-931 David Madison; 931 (t)VU/David M. Phillips, (b)VU/R. Kessel; 932 Brian Drake/Sportschrome; 932-933 Duomo/Steven E. Sutton; 933 Duomo; 934 J. Villegier/PR; 935 Guy Gillette/PR; 936 Matt Meadows; 937 file photos, (b)Runk/Schoenberger/GH; 938 Aaron Haupt; 940 Matt Meadows; 942 Tim Beddow/SPL/PR; 943 (l)Biophoto Associates/PR, (r)AA/Breck P. Kent; 944 Elaine Shay; 945 Matt Meadows; 946 (t)David Madison/DUOMO, (bl)Dr. E.R. Degginger/CP, (br)Biophoto Associates/PR; 947 VU/Fred Hossler; 948 949 Matt Meadows; 952 956 Doug Martin; 956-957 Biophoto Associates/PR; 957 958 Elaine Shay; 960 Omikron/PR; 963 Matt Meadows; 964 VU/R. Kessel-C. Shih; 965 Matt Meadows; 967 Larry Mulvehill/PR; 969 Elaine Shay, (t)ZIGGY copyright 1993 ZIGGY AND FRIENDS, INC. Dist. by UNIVERSAL PRESS SYNDICATE. Reprinted with permission. All rights reserved.; 974 975 Matt Meadows; 980 Phil Degginger/CP; 980-981 Thomas Porett/Science Source/PR; 981 Carolina Biological Supply/PT; 982 984 Matt Meadows; 987 Doug Martin; 988 (t)Lennart Nilsson/BonnierAlba, (b)Philip Harris Biological; 989 (l)NIBSC/SPL/PR, (r)Archiv/PR; 990 Runk/Schoenberger/GH, (b)Elaine Shay; 992 993 Matt Meadows; 994 (l)A. & F. Michler/PA, (r)SPL/PR; 997 David Young-Wolff/Photoedit; 999 Matt Meadows; 1004 Matt Meadows; 1004-1005 CNRI/SPL/PR; 1005 Matt Meadows; 1006 (tl,tr)Matt Meadows, (b)VU/David M. Phillips; 1015 (t)Clinique Ste Catherine/CNRI/SPL/PR, (b)James Holmes/SPL/Custom Medical Stock Photo; 1017 (t)Matt Meadows, (b)Norbert Wu; 1019 Michael Newman/PhotoEdit; 1022 Peter Serling; 1023 courtesy Johns Hopkins Children's Center; 1025 Matt Meadows; 1026 Dr. E.R. Degginger/CP; 1027 1028 Matt Meadows; 1029 Sinclair Stammers/PR; 1030 The Granger Collection, New York; 1031 (t)Henley & Savage/TSM, (b)Matt Meadows; 1032 Michael Fogdon/DRK; 1033 Tim Beddow/SPL/PR; 1034 (l)VU/F.W. Buchanan, (r)Robert Knauft/PR; 1038 NASA/PT; 1038-1039 Petit Format/Nestle/PR; 1039 David Phillips/PR; 1040 GARFIELD reprinted by permission of UFS, Inc.; 1042 Bob Daemmrich/TSI; 1050 David M. Phillips/PR; 1052 Petit Format/Nestle/PR; 1053 1054 1055 Lennart Nilsson/BonnierAlba; 1056 Blair Seitz/PR; 1057 Dr. E.R. Degginger/CP; 1058 Jon Feingersh/TSM; 1059 O'Brient & Mayor Photography/FPG; 1060 Matt Meadows; 1061 Rick Weber; 1062 (tr)Richard Mackson/FPG, (l)A. Tsiaras/PR, (br)Bettina Cirone/PR; 1063 Mary Cassatt, *Mother and Child*, The Roland P. Murdock Collection , Wichita Art Museum, Wichita Kansas; 1064 Mark Tuschman; 1068 Matt Meadows; 1068-1069 Biology Media/PR; 1069 Matt Meadows; 1070 VU/David M. Phillips; 1073 (t)J. Wengle/DRK, (c)AA/Joe McDonald, (bl)Mark E. Gibson/VU, (br)Matt Meadows; 1074 (t)Matt Meadows, (b)Elaine Shay; 1075 The Bettmann Archive; 1076 UPI/BETTMANN; 1078 Michal Heron/TSM; 1081 GMBH/Lennart Nilsson/BonnierAlba; 1083 Paul Gero/SYGMA; 1086 ©Boehringer Ingelheim International Gmbh, Lennart Nilsson, The Body Victorious, BonnierAlba; 1087 VU/Southern Illinois University; 1088 NIBSC/SPL/PR; 1090 The Bettmann Archive; 1091 Dept. of Medical Photography, St. Steven's Hospital, London/ SPL/PR; 1092 Bruce Ayres/TSI; 1096 (t)Brad Mangin/Duomo, (b)David Pollack/TSM; 1097 (l)Doug Martin, (c)Wernher Krutein/Gamma-Liaison, (r)Aaron Haupt; 1098 Lori Adamski/TSI; 1099 Aaron Haupt; 1100 Don Mason/TSM; 1102 Cyberimage/TSI; 1104 NIBSC/SPL/PR; 1106 Peter Menzel; 1106-1107 Peter Menzel; 1107 (t)James Prince/PR, (b)Vanessa Vick/PR; 1108 (t)SPL/PR, (b)John N.A. Lott, McMaster Univ./BPS; 1109 (t)Vanessa Vick/PR, (b)Kathy Bushue/AS; 1110 Adam Heller, University of Texas; 1110-1111 Ken Graham/AS; 1111 (t)Manfred Kage/PA, (b)VU/Steve McCutcheon; 1112 (l)Dan McCoy from Rainbow, (r)Doug Armand/TSI; 1113 (t)AA/Doug Wechsler, (bl)Dan McCoy from Rainbow, (br)Dan McCoy from Rainbow; 1114 (l)Richard Nowitz/PT, (r)Hank Morgan/Science Picture Library/PR; 1115 (l)Matt Meadows, (r)Superstock; 1116 (t)David Riecks, (c), University of Colorado Office of Public Relations, Photograph by Ken Abbott, (b)Peter Serling; 1117 (t)Ed Kashi, (c)Jeff Fearing, Howard University, (b)Jessie Cohen/National Zoological Park/Smithsonian Institution; A0 Francois Gohier/PR; SH1 Matt Meadows, SH2 (t)Roger K. Burnard, (bl)Claudia Parks/TSM, (br)Lynn Stone; SH3 Matt Meadows; SH8 AA/Robert Maier; SH9 Matt Meadows; SH11 (t)John R. Hicks/DRK, (cl)David Cavagnaro/DRK, (cr)AA/Breck P. Kent, (b)AA/F.E. Unverhau; SH12 Rod Joslin.

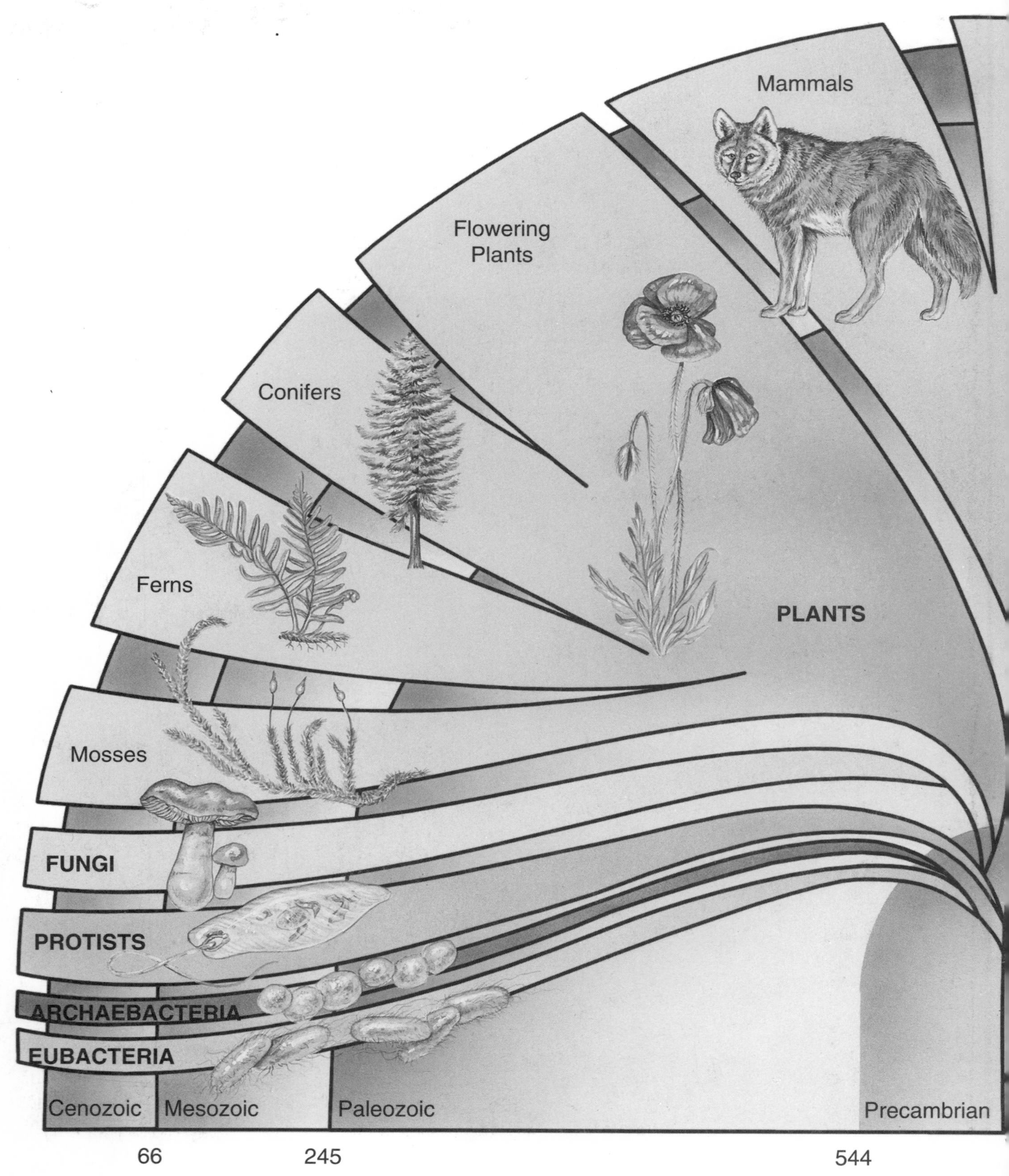
Mammals
Flowering
Plants
Conifers
Ferns
PLANTS
Mosses
FUNGI
PROTISTS
ARCHAEBACTERIA
EUBACTERIA
Cenozoic
Mesozoic
Paleozoic
Precambrian
66
245
544
Eras: shown in millions of years ago